AF687778

FLORA *of* INDIANA

BY

CHARLES C. DEAM, M.A., D.Sc., LL.D.

Research State Forester

THE BLACKBURN PRESS

2020 Reprint of the First Edition, Published 1940

Flora of Indiana

ISBN-10: 1-930665-59-8
ISBN-13: 978-1930665-59-0

Library of Congress Card Number: 2002106383

THE BLACKBURN PRESS
P. O. Box 287
Caldwell, New Jersey 07006
U.S.A.
973-228-7077
www.BlackburnPress.com

STATE OF INDIANA
DEPARTMENT OF CONSERVATION

JUNE, 1940

Department of Conservation, Division of Forestry
Indianapolis, Indiana

The Deam Oak (Quercus Deamii Trelease)

This oak is a cross between the white and chinquapin oaks (× Quercus alba × Muhlenbergii). It was discovered Oct. 9, 1904, by Lent A. Williamson and his son E. Bruce Williamson on the border of a woods along State Road 116 about 3 miles northwest of Bluffton, Wells County, Indiana. In 1904-5 the author made a collecting trip to Guatemala and at the request of William Trelease of the Missouri Botanical Garden I collected Agaves and Furcraeas for him. In recognition of this favor and without my knowledge he named this tree in my honor, although I feel the honor should have gone to the Williamsons. In 1915 the tree was blazed for cutting and in order to save it Mr. and Mrs. Chas. C. Deam bought about a half acre on which the tree stands and deeded it to the state. This area is now known as "Deam Oak Monument." In 1933 three seedlings from this tree were planted about it. The tree produced viable acorns in 1918, 1927, and 1930. In March,1939, the tree measured 90 inches in circumference at breast height.

FOREWORD to the 2002 Reprinting

"I'll leave my obituary in the books I create." Charles C. Deam (1865-1953)

Prophetic to his word, the books of Charles Clemon Deam—or simply "Charlie" as he was known to his friends—do indeed reveal much about the man. A druggist, forester, and botanist from Bluffton, Indiana, Charlie Deam was meticulous, opinionated, studious, disciplined, driven, and even, shall we say, eccentric. Simply put, he was a character. But that character produced a collection of some of the most thorough botanical works ever published. To be certain, his standards were high, as perhaps is best revealed in his magnum opus, the *Flora of Indiana* (hereafter, the *Flora*). Published in 1940, with reprintings in 1970 and 1984, the *Flora* has served as the standard by which other state floras must be compared. Now over 60 years old, it has clearly withstood the test of time, and continues to be a primary source of information for any serious student of field botany. Thus, how fortunate we are that The Blackburn Press has selected the *Flora* for reprinting, making it available again for purchase after a long hiatus.

Deam insisted upon the utmost standards for his work, and strove to make the Flora as accurate as possible. That was clearly the policy when considering a species for inclusion in the book; it was his rule that every species included must be vouched for by at least one collected specimen. He examined over 84,000 specimens in preparation for the book, and from these he prepared keys, species accounts, and range maps showing species' occurrence by county. Although these maps reflect the knowledge only as it existed in 1940, they continue to be useful today in determining a species' general range in the state. This is especially helpful for the beginner, or one not familiar with Indiana's flora, as it can reduce the field of options when trying to determine an unknown plant's identity.

"Where's Deam?" is not an uncommon cry heard about the office in the Indiana DNR Division of Nature Preserves. There's no expectation of finding a person with such a question, but rather a misplaced copy of a book that we simply can't do without. Early on we declared that the *Flora* would be our starting point in creating the first state list of rare, threatened, and endangered plant species. It continues to be one of the first choices to consult when gathering information on wild plants. However, the utility of the *Flora* extends beyond species identification, range, and habitat. In our efforts to protect Indiana's natural landscape, we often find ourselves retracing the footsteps of Charlie. In one example, information found in the *Flora*'s account of bog bluegrass (*Poa paludigina*) revealed the existence of an unusual habitat for southern Indiana, resulting in the discovery not only of the rare bluegrass, but also a site that would become a state-dedicated nature preserve.

Information in the *Flora* has also been very useful in the restoration of landscapes. Because Deam collected plants in every township of the state, we have an excellent record of what historically occurred in an area. This has been especially helpful when attempting to restore areas that no longer possess their native vegetation. A prime example involves an area in Daviess County, Indiana, where a major restoration project conducted by the Division of Nature Preserves has relied heavily on the *Flora* and Deam's plant collection for guidance. Landscape restorationists throughout the state would do well by utilizing the *Flora* in similar fashion.

The *Flora* is clearly more than a list of plants. Many treasures are found within its pages, ranging from topics on early 20th century agrarian culture to herbal cures. There are frequent references to discussions with "old timers", including some of whom were the first European settlers in the state. Also mentioned are accounts of Deam's own early activities on the family farm, such as when he used scissors to cut cockle and rye in a wheat field, or pulled common purslane by the bushel and fed it to the hogs. One of the most exciting bits of information, at least for me, was Deam's statement that he had in his possession two books owned by Dr. Asa Clapp, a pioneer botanist who lived in New Albany in the early 1800's. Dr. Clapp was Indiana's first resident botanist, and his records of the state's early flora are extremely important. Deam commonly referenced Dr. Clapp's records in the *Flora*, but their source was never clear to me. Then, during a cozy, armchair reading of the *Flora* (yes, I enjoy reading the *Flora* like some read novels), I learned that Deam had owned Dr. Clapp's books, and that Clapp's records were written in them! That revelation, found specifically in Deam's accounting of *Trautvetteria*, ultimately led to the current location of the books.

I never had the pleasure of meeting Charlie Deam—I was born the year that he died—but I often sense his presence whenever I see one of his plant specimens, retrace his footsteps into a natural area, or better yet, open a copy of the *Flora*. It should come as no surprise to you that I have a special reverence for the man and his work. Thus it brings me great pleasure that the *Flora* is available again, allowing others to share in the pursuit of understanding, and ultimately the appreciation and protection of our native plants and natural areas. So whether you're a professional botanist, ecologist, teacher, wildflower enthusiast, naturalist, forester, wildlife biologist, soil scientist, landscape architect, horticulturist, or just someone wanting to know what that plant is in your backyard, this book is for you.

Michael A. Homoya, Botanist/Plant Ecologist
Indiana Department of Natural Resources, Division of Nature Preserves

FOREWORD

It is difficult to write a suitable foreword to such a notable book.

In his *"Flora of Indiana"* Dr. Deam has set new standards of excellence in many lines.

The most casual examination shows that it was based upon painstaking field studies—field studies covering years of time and involving thousands of miles of travel. While local lists were carefully studied no plant was admitted to the Flora upon their authority, it was admitted only as these field studies proved its presence in the state, or it could be verified by actual specimens in accessible herbaria. I know of no other State Flora based upon long continued field studies and in which every plant admitted is based upon an actual and accessible specimen.

The work is notable because of its accuracy. Dr. Deam, not content to rest upon his own taxonomic acumen, has referred every critical genus and species to specialists for their confirmation or correction. Scores of shipments of such specimens to these specialists were made up to the very date of publication. It is safe to say that in no other regional Flora has such meticulous care been taken to secure absolute accuracy in determination, as well as the very latest word in these special studies. The *Flora of Indiana* is accurate and up to date in an unusual degree.

The clearness of the floral picture is increased by a series of unique distribution maps showing not only location but the time of the occurrence of various seasonal phases.

Perhaps as illuminating as any single feature of the Flora are the incidental ecological notes that appear on almost every page. From the unity of treatment that characterizes the text, plant associations stand out with amazing distinctness. It adds greatly to the value of the book that while no attempt is made to emphasize these features, they take their place in the picture of the flora of the state and aid in its interpretation, as into this book has entered the experience of former taxonomic work by the author. His *Trees of Indiana, Shrubs of Indiana,* and *Grasses of Indiana* are models of what such reports should be as to completeness, accuracy, and widespread utility.

The canvas is of course larger in *Flora of Indiana* but there has been no sacrifice of accuracy, no lessening of the purpose lying back of all these books—that they should be useful to citizens of Indiana.

The *Flora of Indiana* will be a treasure trove to education from the secondary schools to the university. It will be a stimulus and guide to nature lovers; it will be of immense practical value to every agriculturist and horticulturist. It will have its place in libraries, and it is a great book by an author whom I have been proud to claim as a personal friend for nearly half a century.

STANLEY COULTER,
Dean (Emeritus) School of Science,
Purdue University.

Table of Contents

Note for the 2020 Printing

The following sections are not printed here. They are available on BlackburnPress.com under Flora of Indiana

- **Lists of new forms and new combinations**
- **Names of collecting places not in current use**
- **List of Indiana collectors whose specimens have been seen or referred to in the Flora**
- **Bibliography**

INTRODUCTION

The first flora of Indiana was a "Catalogue of the phaenogamous and vascular cryptogamous plants of Indiana" by the Editors[1] of the Botanical Gazette and Prof. Charles R. Barnes, published in 1881. To this was added a supplement in April, 1882. These listed 1,194 species native to the state and 140 species that had been introduced.

Stanley Coulter in 1897 compiled a list of Indiana plants by families (Proc. Indiana Acad. Sci. 1897: 158-165. 1898). This list contains 124 families, 534 genera, and 1,369 species, an increase of only 35 species. The names of the species are not given and the totals include both native and introduced species.

The second flora was "A Catalogue of the flowering plants, ferns, and fern allies indigenous to Indiana" by Stanley Coulter, published in 1900. He lists 1,765 species but this number includes both native and introduced species and some erroneous reports. I have studied this catalogue and as I interpret the species, the list should read 1,400 native species, 177 established exotics, 34 not yet established, and 154 species to be excluded for various reasons. It should be borne in mind that when this catalogue was published the author was not able to verify reports as critically as has been done in the present flora. At that time reports by recognized botanists were accepted. It must be remembered that our early botanists did not have access to large herbaria and had few books or perhaps only one book to guide them in naming plants.

Since the publication of these floras much work has been done in the state by various botanists. Among the principal collectors the following persons may be mentioned: Edna Banta, A. R. Bechtel, Chas. M. Ek, Ray C. Friesner, Ralph M. Kriebel, Marcus Lyon, Jr., Scott McCoy, Madge McKee, J. A. Nieuwland, J. E. Potzger, Paul Weatherwax, Winona Welch, and T. G. Yuncker.

Improved highways and the automobile have greatly facilitated collecting. I have been collecting for 40 years. Since 1914 I have used an automobile, traveled over 125,000 miles, and collected in each of the 1,016 townships in Indiana. My accession numbers are now over 59,000.

The plan of this flora is to include all the species native to Indiana, although a few are now known only from herbarium specimens, and introduced plants that are known to be established. Introduced plants that have been reported as escapes without data concerning their establishment are carried in an excluded list with all the data which I can assemble. If one of the excluded species is later found to be established, the data here recorded may be of service. In the excluded list are included also species that are no longer regarded as segregates, species which have been erroneously reported for the state, and those which do not have sufficient data to warrant their inclusion.

[1] J. M. Coulter and Stanley Coulter.

The present flora is an attempt to bring up to date our knowledge of the ferns, fern allies, and flowering plants of Indiana. It became necessary to adopt a rule or standard by which a species could be admitted or excluded from the flora, and it was decided to admit only those species which have one or more herbarium specimens to verify their occurrence. An exception has been made in the case of *Adlumia fungosa* which I saw in a woods in La Porte County. Doubtless a few species have been excluded that do occur in the state. I refer specifically to *Podostemum ceratophyllum* and *Elatine minima* which have been reported and have Indiana within their general range. I have, however, made strenuous but unsuccessful efforts to find specimens of both these species. I have admitted a few species where I have seen no specimen but the evidence for their existence in Indiana is convincing.

My study has been made primarily from specimens in my own herbarium which numbers more than 65,000 sheets, more than 47,000 of which are from Indiana. In addition I have examined all the Indiana specimens in all of the other Indiana herbaria which total 36,936 sheets but these were studied only sufficiently to check the identification. The keys and measurements have been made from my own specimens. The ecological notes have been taken also from my specimens.

No effort or expense has been spared to have my specimens named correctly. In order that specimens belonging to critical genera be authentically named, I have sent them to specialists to be determined or to have my identifications verified. I wish here to express my sincere appreciation to the following persons who have examined my specimens in the groups upon which they are authorities: L. H. Bailey for *Rubus* and *Vitis;* C. R. Ball for *Salix;* J. H. Barnhart for *Utriculariaceae;* Ezra Brainerd (deceased) for *Viola;* Agnes Chase and A. S. Hitchcock (deceased) for *Gramineae;* H. S. Conard for *Nymphaeaceae;* Carl Epling for *Labiatae* in part; M. L. Fernald for *Potamogeton* and various species; Ray C. Friesner for *Solidago;* Frederick J. Hermann for *Carex* and *Juncaceae;* Lawrence E. Hicks for *Lemnaceae;* Milton S. Hopkins for *Arabis* in part; Theodor Just for *Chenopodiaceae;* Rogers McVaugh for *Lobelia;* P. A. Munz for *Onograceae* in part; E. J. Palmer for *Crataegus* and miscellaneous species; Francis W. Pennell for *Scrophulariaceae;* Rosendahl, Butters, and Lakela for *Heuchera* and *Sullivantia;* Paul Standley for *Houstonia* in part; E. E. Watson (deceased) for *Helianthus;* C. A. Weatherby for assistance for many years on ferns; Louis C. Wheeler for *Euphorbia;* Edgar T. Wherry for *Polemoniaceae;* K. M. Wiegand for *Amelanchier* and *Oxalidaceae;* and T. G. Yuncker for *Cuscuta.* I wish here to thank all others who named or checked over small groups or who loaned me Indiana specimens for study.

Distribution of Indiana Plants.—The general distribution of a species is given in a closing paragraph after the discussion of the species. The state distribution is shown by a map. Published records that do not cite specimens are omitted but sometimes one or more may be discussed. Some more or less complete county floras have been published without verifying specimens; no reference is made to these except that when a species is reported which does not occur in Indiana, it is discussed and placed in the excluded list where it belongs.

Those plants whose mass distribution is to the south or southwest of Indiana and always found in cultivated grounds, are probably introduced. These are discussed in the text.

The date of flowering of a species is given in the vertical column at the left of the map. No effort has been made to collect plants at their very earliest or latest flowering dates, and dates and the number of specimens have been taken from my collection only.

The distribution on the map is by counties and is indicated by letters which are symbols for the herbaria in which specimens are deposited. I have seen all the Indiana specimens in both public and private herbaria in Indiana and many specimens cited outside of Indiana. Those which I have not seen are ones cited by recent authors. Hermann has seen all of the *Carex* and *Juncaceae* cited.

It was impracticable to go through all the herbaria of the United States. The principal collectors of Indiana plants are known and I have seen their plants except those of E. J. Hill which are deposited in the herbarium of the University of Illinois, Urbana, Illinois; those of H. Walton Clark and B. W. Evermann from Marshall County which are deposited in the Field Museum, Chicago, Illinois, and the National Herbarium, Washington, D. C.; and those collected by L. M. Umbach which are in the herbarium of the University of Wisconsin, Madison, Wisconsin. Since Hill and Umbach did most of their collecting in the counties along Lake Michigan whose flora is well represented later by my own work, and by that of Marcus Lyon, Jr., J. A. Nieuwland, and others it is doubtful if these former authors found anything not later collected and reported. They reported all the rare things they collected and I have examined all of these rarities.

When the area of the county is too small to hold all the reports, those of private herbaria have been omitted.

The herbaria indicated by symbols and their location are as follows:

AA.......Arnold Arboretum, Jamaica Plain, Massachusetts.
B.........Butler University, Indianapolis, Indiana.
Ba........Private herbarium of Edna Banta, Bloomington, Indiana.
C.........University of California, Berkeley, California.
Cm.......Carnegie Museum, Pittsburgh, Pennsylvania.
Cu........Cornell University, Ithaca, New York.
D.........Deam Herbarium, Bluffton, Indiana. (Later to be located at Indiana University, Bloomington, Indiana.)
Dk.......South Dakota Agricultural College, Brookings, South Dakota.
DP.......DePauw University, Greencastle, Indiana.
F.........Field Museum of Natural History, Chicago, Illinois.
Fr........Franklin College, Franklin, Indiana.
G.........Gray Herbarium, Cambridge, Massachusetts.
H.........Private herbarium of Frederick J. Hermann, Ann Arbor, Michigan.
Hi........Private herbarium of Lawrence E. Hicks, Columbus, Ohio.
I.........University of Illinois, Urbana, Illinois.
IU........Indiana University, Bloomington, Indiana.
K.........Private herbarium of Ralph M. Kriebel, Bedford, Indiana.

L.........Private herbarium of Marcus Lyon, Jr., South Bend, Indiana.
M.........University of Minnesota, Minneapolis, Minnesota.
MC.......Private herbarium of Scott McCoy, Indianapolis, Indiana.
Mi........University of Michigan, Ann Arbor, Michigan.
MK.......Private herbarium of Madge McKee, Goodland, Indiana.
Mo.......Missouri Botanical Garden, St. Louis, Missouri.
Mw.......Milwaukee Public Museum, Milwaukee, Wisconsin.
N.........National Herbarium, Washington, D. C.
ND.......University of Notre Dame, Notre Dame, Indiana.
NW.......Northwestern University, Evanston, Illinois.
NY.......New York Botanical Garden, Bronx Park, New York.
O.........Oberlin College, Oberlin, Ohio.
P.........Purdue University, West Lafayette, Indiana.
Pa........University of Pennsylvania, Philadelphia, Pennsylvania.
Ph........Philadelphia Academy of Sciences, Philadelphia, Pennsylvania.
Po........Pomona College, Claremont, California.
S.........Private herbarium of A. S. Slavin, Rochester, New York.
Sw........State College of Washington, Pullman, Washington.
St........Stanford University, Stanford University, California.
T.........Private herbarium of R. M. Tryon, Jr., Chicago, Illinois.
W........Wabash College, Crawfordsville, Indiana.
We.......Private herbarium of Paul Weatherwax, Bloomington, Indiana.
Wi........University of Wisconsin, Madison, Wisconsin.

Botanical Descriptions.—The botanical descriptions have been drawn almost exclusively from specimens I have collected because they have been at hand. Technical terms have been avoided whenever possible and the few found necessary to use are defined in a glossary. The measurements in the keys have been taken from herbarium specimens and are given in the metric system and those in the descriptive text are in English terms. The frequent use of "more or less, usually, and generally" is objectionable to some people but to me these expressions are the shortest, the most definite, and most comprehensive way of expressing the wide limits of a qualitative or quantitative character. The ampersand (&) is used between joint authors and joint collectors.

Botanical names of native plants are printed in **bold face** type and are in accordance with the International Rules of Botanical Nomenclature. When the names given in Gray's Manual, edition 7 and Britton and Brown's, Illustrated Flora edition 2 differ from those in the bold face type for the same plant they are regarded as synonyms and are printed in italics. Botanical names in the text are printed in *italics*. Botanical names of introduced plants and common names are printed in SMALL CAPITALS.

The accented pronunciation of the botanical names is indicated as follows: the grave (`) accent indicates the long English sound of the vowel, and the acute (´) accent indicates the short or otherwise modified sound.

The use of the term "variety typica" to designate the typical form of a species is limited to those species where I have found it used as such.

The common names are those given as such in "Standardized Plant Names," with few exceptions. In many instances I do not agree with this authority but I believe it is in the best interest of uniformity for me to accept the names given in the aforementioned work. In rare instances I have given two common names and the reason for so doing. Many of our plants do not have accepted common names and I have left these without them.

In the writing of the manuscript an effort has been made to conform to some supreme rule. In spelling and in the use of the hyphen Webster's New International Dictionary, latest edition has been followed with few exceptions. Since there is no universally accepted standard of colors, although Ridgway's "Color standards and color nomenclature" is used by mammalogists, ornithologists, and some botanists, and since color terms have been loosely used by authors to convey color concepts, I believe it is in the interest of uniformity to delete all hyphens between color terms because they add nothing to clarify the concept, except where used by Ridgway when they represent a definite color. The "Style Manual of the United States Government Printing Office," 1935, edition has been followed with few exceptions. The outstanding innovation is the omission of the period after abbreviations used in the metric system. The exception is that while this authority does not begin proper names of specific and subspecific names with a capital letter, I am following the International Botanical Rules and I am using capital letters. I wish to go on record as vigorously opposing the practice of decapitalizing specific or subspecific names derived from proper nouns. Biological Abstracts has been followed in the matter of abbreviating and listing bibliographic data.

The keys and how to use them.—The key to the families has been copied with a few changes from Robinson & Fernald's Gray's Manual, edition 7, published in 1908 and adapted to the species which occur in Indiana. The reason that I have adopted this key is that I have used it since its publication and I have found it satisfactory. Other botanists with whom I have conferred upon this subject all agree that the key is all that is to be desired. I wish to express my thanks for the privilege of using it. Keys to genera and species, except those of the parts contributed by others, I have written myself and they are all artificial.

A general key is given to assist the student in learning to which family an unknown plant belongs. It is arranged in pairs of leads. The second lead of a pair repeats the data given in the first lead but in a negative form. Each succeeding set of leads is placed 2 spaces to the right and some of the sets are preceded by a pair of letters to make them more easily located, especially when one of the pair is very far from the other with many intervening leads.

To name a plant, read the first lead. If it fits your plant, proceed to the next set of leads. If it fits the first lead of this set, proceed to succeeding leads until it leads to a family or genus. If it does not fit a lead, try the opposing lead. If it fits, proceed to the first part of the next set of leads. Accept or reject leads until the key leads to a family or genus. The task is not as easy as it may seem. After you have followed the key to a family you may find the plant does not fit the family. Then you must retrace the steps taken and be more careful to be sure the terms are understood. Errors are usually the result of haste, misunderstanding of terms used, or of poor or inadequate material for naming. The key may call for a character your specimen does not have. Then outside aid must be sought. One who is interested in naming the flora of a region should have one or more manuals of botany that go into more detail than can be given in a flora of this kind. An illustrated manual will be of great assistance.

After you have reached the family name, turn to the page in the book where the family is found and proceed through the family key to the species.

Sequence of families and genera.—The sequence of families and genera and their interpretation is that of the "Genera Siphonogamarum" by C. G. de dalla Torre and Dr. H. Harms. This sequence is in accord with the "Engler and Prantl" system of classification which is in current use by most authors. I am aware that several newer systems of classification have been offered but students are not unanimous in accepting them. An exception has been made in the *Graminae* in which the sequence is that of Hitchcock's Manual of Grasses which is used by most students of grasses.

It is to be noted that the numbers that precede family and generic names in our manuals and floras differ. This disagreement follows because each author treats a different area and he numbers only the families and genera that are found within the area he considers. The innovation in this flora is that the numbers of families and genera refer to the families and genera of the whole plant kingdom and are the numbers assigned to them by dalla Torre and Harms. This system places no limit upon expansion if one wishes to build up an herbarium and makes it easy to incorporate it into a large herbarium. Plants in an herbarium should not be arranged alphabetically but according to their relationship.

Indiana, its location, drainage, and climate.

Indiana is one of the north-central states. It is about 153 miles wide and 275 miles long between the most distant points. The southern boundary is low water line of the north side of the Ohio River and the northern boundary is Lake Michigan and the state of Michigan. The most southern point is in 37°40′ north latitude and the most northern point is in 41°50′ north latitude. In longitude it lies between 84°49′ on the east and 88°2′ on the west.

The land area occupies 36,045 square miles besides 280 square miles of rivers and interior lakes and 230 square miles of Lake Michigan.

The whole of the state has been glaciated except the south-central and southwestern parts (see map on page 1164). The highest point in the state is in Randolph County, 1,285 feet above sea level, and the lowest is at the mouth of the Wabash River, 313 feet. The average elevation is about 700 feet.

About nine-tenths of the state drains westward and southwestward into the Mississippi Basin and about a tenth, located in the northern part, drains into the St. Lawrence Basin.

The average annual precipitation is about 39 inches. The average annual temperature is about 52 degrees Fahrenheit. The average growing season is about 158 days in the northern part of the state and 188 days in the southern part. (See plates on pages 1162 and 1163.)

Floral Areas of Indiana (See map on page 1164.)

To assist in understanding the distribution of a species in the state and at the same time give some idea of its habitat, I have divided the state into seven areas. These are not all strictly floral areas but for convenience they may be so considered. The limits of the ranges of certain species within the area determine one boundary of that area.

Dune area

The dune area is bounded on the north by the waters of Lake Michigan and on the south for the most part by the Michigan Central Railroad. It is about four miles wide at the west end and half a mile wide at the east end. In Lake County this area consisted of low dunes, for the most part from 5 to 15 feet high, alternating with sloughs and interdunal flats. In the extreme northwest part of it were Wolf Lake, Berry Lake (now extinct), and Lake George. The greatest variety of plants of this area were found in this county. In the east part of Lake County the dunes begin to rapidly increase in height and high dunes continue to Michigan City. The highest dune is Mount Tom in Dunes Park, Porter County and is 192 feet high. The dunes proper are almost pure sand but were formerly well wooded. The sloughs and interdunal flats are more or less mucky.

The following list is of plants known in Indiana only from this small area and all are of northern range. Those preceded by "?" are probably extinct and those preceded by "o" are now known from one colony only.

Ammophila breviligulata
? Botrychium simplex
Cakile edentula var. lacustris
Carex folliculata
o Carex Richardsonii
? Ceanothus ovatus
Cirsium Pitcheri
o Clintonia borealis
? Corallorrhiza trifida
Cornus canadensis
Cyperus Houghtonii
o Equisetum variegatum
Euphorbia polygonifolia
Hudsonia tomentosa var. intermedia
o Myosotis laxa
Oryzopsis asperifolia
? Panicum lucidum
? Panicum scoparioides
? Panicum subvillosum
Pinus Banksiana
o Polygala paucifolia
Potentilla Anserina
o Potamogeton pusillus
? Psilocarya nitens
? Pyrola secunda
Ptelea trifoliata var. Deamiana
? Rhynchospora cymosa
Salix adenophylla
o Scirpus subterminalis
Shepherdia canadensis
o Solidago Deamii
Solidago Gillmani
o Thuja occidentalis

Lake area

The lake area occupies the northern part of the state, southward to the Tipton Till Plain but is not sharply separated from it. For practical purposes the south line of this area may be considered to coincide with the north line of the Tipton Till Plain which may be given roughly as a line extending westward from Fort Wayne to Huntington, Logansport, and Monticello to the state line. South of this line are a few, nearly extinct small lakes. There is one in each of the following counties: Wells, Blackford, Grant, and Warren. Deep peat deposits in Hamilton and Madison Counties indicate extinct lakes.

The area has a great variety of habitats ranging from lakes and rivers, bogs and marshes, dry sand and gravelly places, prairies, and remnants of prairies (oak openings) to the mesophytic forest. Within this area about 300 species of a northern range find their southern limit. Within this area a small number of plants have been found also that have their mass distribution on the Coastal Plain and the Lower Mississippi Valley. Among these are *Panicum albemarlense, Panicum spretum, Panicum verrucosum, Cyperus dentatus, Eleocharis melanocarpa, Eleocharis Torreyana, Fimbristylis puberula, Scleria pauciflora* var. *caroliniana, Scleria reticularis, Scleria setacea,* and *Hypericum adpressum.* These are found in a few marshes and on their borders between low dunes in section 2 a mile east and a mile and a half south of Tefft, Jasper County, or about 4 miles south of the Kankakee River. A few of these species occur also in the dune area and in a few adjacent counties in like habitats. I have not botanized the marshes in adjacent sections to ascertain how widely these species are spread or whether additional species may be found. The whole area for a width of about 5 miles from Bass Lake in Starke County westward to the Illinois line, a distance of about 50 miles is, for the most part, a series of low dunes and interdunal marshes. I am of the opinion that these Coastal Plain plants have migrated into Indiana through the Mississippi Valley rather than through the Mohawk Valley and the Great Lakes area as Peattie and Svenson suggest. To this list of plants should be added *Styrax americana* which is found along the Kankakee River and is not found again until the Patoka River Basin is reached in Dubois County. *Mikania scandens* is found along the Kankakee River just east of Baum Bridge, Porter County. I have not found it elsewhere in Indiana although it has been reported. This very disjunct distribution suggests migration from the Mississippi Valley by streams through Illinois. Recently several Coastal Plain plants have been found in Minnesota which adds weight to the theory that our Coastal Plain plants came into Indiana through the Mississippi Valley.

Tipton Till Plain

This area is not strictly a botanical one but is given as such for the convenience of discussing distribution and habitat. Excepting the prairie area it nearly coincides with the physiographic area given it by Malott (Handbook of Geology). It is bounded on the north by the "lake area" and on the south by the southern boundary of the Wisconsin drift. The surface of this area is comparatively level although marked by many terminal moraines. The soil is mostly neutral or only slightly acid. The soil acidity factor may be the one which prevents plants from migrating into it from the Illinoian drift area where the soil is much more acid. Within this area some plants from all directions reach their limits of distribution in Indiana. This area contains the best agricultural land of the state and in the brief period of a hundred years almost all of the woodland has disappeared and the whole is now under cultivation. As a consequence it is now impossible to learn just how far plants invaded this area and what stopped them. Too, our distribution maps show few records because the plants in this area are rare or have been exterminated by cultivation. The area, however, contains some extinct lake areas and springy places which accounts for the many lake area plants in it.

Illinoian Drift Area

This area lies south of the Tipton Till Plain, north of the glacial boundary, and east of the Lower Wabash Valley area. It is divided into an eastern and a western lobe. The topography varies from level areas to deeply cut ravines. The flora of the two parts has several species not in common. The Appalachian flora has entered in a small degree the eastern part while the southwestern flora has entered the western part. In Clark, Jefferson, Jennings, and Ripley Counties are level, poorly drained areas with an acid soil that are locally known as "flats." These may be divided into high and low "flats." The principal tree species of the "high flats" are beech, sweet gum, tulip, and black gum. Often a depression a foot in depth will result in a "low flat" wooded with swamp chestnut oak, swamp white oak, pin oak, southern red oak, and red maple. Sometimes the lowest places will consist of a pure stand of pin oak. All of the species named will not be found in the same "flat" but usually two or three of them will be the dominant species. The western part has some low areas but these are usually wooded with pin oak and shingle oak, associated with hickory. In the western lobe are sand dunes that have a peculiar flora. Such a sand area forms the terrace of the Wabash River from north of Terre Haute southward to Posey County. In Knox County in places its width increases to more than a mile. On this sandy terrace are found plants not found elsewhere in Indiana which have their mass distribution in the Lower Mississippi Valley. East of the North fork of White River in the northwestern part of Daviess County are many low dunes upon which, and in the low places between them, occur several Coastal Plain plants. Among those that are restricted to this area are *Gymnopogon ambiguus* and *Gaura filipes.*

Prairie Area

This area is small and the boundary very irregular. The many small prairies and "oak openings" that occur throughout the lake and Tipton Till Plain areas are not included in this area. Our distribution maps may show a prairie species fairly well distributed over the whole of northern Indiana which does not mean that the whole area is an uninterrupted prairie. There was probably not a county in the lake and Tipton Till Plain areas that did not have one or more areas of an acre or more in prairie. The tension zone between the prairie and the forest is one of the most interesting studies in plant geography. The whole area is now devoted to agriculture and since no one made a record of its plant life before cultivation, our knowledge of it must now be gleaned from the few plants that have survived along railroads and roadsides and in cemeteries and waste places. Every year our roadsides are mowed and the rights of way of railroads are mowed and usually burned, so that the extermination of our native prairie plants will soon be complete.

Lower Wabash Valley

This is a narrow strip of alluvial land on the east side of the Wabash River from Parke County southward to the Ohio River and thence up the Ohio River to Little Pigeon Creek in Warrick County. To it belong also the short alluvial extensions of the White and Patoka Rivers. The whole area is usually inundated each year at flood stage. Among the trees restricted to these lowlands are *Acer rubrum* var. *Drummondii, Carya Pecan* (with few exceptions), *Celtis laevigata* (with few exceptions), *Forestiera acuminata* (with one exception), *Gleditsia aquatica, Gleditsia texana, Taxodium distichum,* and *Quercus lyrata* (one exception). Other plants are *Aristolochia tomentosa, Echinodorus radicans, Hottonia inflata, Leptochloa panicoides, Ludwigia glandulosa, Spigelia marilandica, Trachelospermum difforme,* and *Vitis palmata.* All these species belong to the flora of the Mississippi Valley and find their northeastern limit in this area.

Unglaciated area

This area may be divided into eastern and western parts. The western part is included by Malott in the Wabash Lowland and is bounded on the east by Anderson Creek to St. Meinrad and then extends northwestward to the glacial boundary. The eastern half of this part is hilly and wooded mostly with oaks. The western part has gently sloping or low hills and is wooded on the high ground with beech, tulip, and sugar maple and in the lowland with oak, hickory, elm, and sweet gum. I do not regard this as a botanical area but only a part of a region where some southern plants reach the northern limit of their distribution. In it, however, we have *Dicliptera brachiata* and *Crotonopsis elliptica* that have not been found outside of it.

The eastern part of the unglaciated area is mostly hilly and broken, being divided by the broad valley of White River. I think a good common name for it would be the "Chestnut Oak Upland" area, because this species of oak crowns the crests of all of the high ridges of the area and these ridges are popularly known as "chestnut oak ridges" or "knobs." Malott divides the area into three parts. The most eastern he calls the Norman Uplift, the middle the Mitchell Plain, and the western the Crawford Upland. With the exception of one small restricted area I think these uplands can be considered as one botanical unit. *Pinus virginiana,* Virginia pine, crowns the crests of the highest ridges in Floyd County, the western part of Clark County, a fragment of the southwestern part of Scott County, and a few places on the southeast boundary of Washington County. The total area of pine is quite small and might well be considered a separate botanical area if there were one more species peculiar to it.

Within the chestnut oak area many plants reach their northern limit. Some, such as *Bumelia lycioides, Oxydendrum arboreum, Ligusticum canadense, Eragrostis capillaris,* and *Aconitum uncinatum,* have merely crossed the Ohio River. Others such as *Smilax Bona-nox, Gentiana villosa, Melothria pendula, Kalmia latifolia, Galactia volubilis,* and *Cirsium virginianum* have penetrated 5 to 25 miles. Others such as *Quercus montana* and *Cunila origanoides* have covered the whole area but not beyond it except on a small knob in Jefferson County, one in Spencer County, and one in Warrick County. *Gaultheria procumbens* and *Tsuga canadensis* are evidently relicts on this old rock area. There also remains *Carex picta* which offers a problem in disjunct distribution. This *Carex* is frequent in Brown County in certain places near the glacial boundary and is found sparingly in Monroe, Jackson, Lawrence, Morgan, and Owen Counties. I have watched carefully for this species elsewhere in Indiana but have failed to discover it. It is known only in the area mentioned in Indiana, in Tennessee, Alabama, and in one place in Louisiana. Another interesting relict of this area is *Betula lutea* which has a few specimens struggling for existence on the walls of the gorges about a mile southeast of Taswell, Crawford County. It is associated here with *Tsuga canadensis.*

State Flower

The Indiana flora is rich in the number of native species that are attractive and beautiful. Out of our abundance of native flowers we should be able to select one for our state flower. I take this opportunity which may be my last to voice my protest against designating as a state flower one that is not a well known native of the state nor even a native of the United States. Our first state flower was the carnation of Europe. I assisted in having this changed in 1923 to the flower of the tulip tree which is found in every county of Indiana except in the prairies. It is recognized as one of the most stately trees of the United States. In 1931 the legislature named the blatant zinnia the state flower, *Zinnia elegans* (a native of Mexico). Why advertise some foreign country and our ignorance of our native plants? I appeal to readers to take a pride in our state and in our native plants. I hope that our next legislature will not consider the state flower only as a buttonhole bouquet and will name one of our many native flowers to represent us and cease paying homage to any other country.

Acknowledgments

I have received help and suggestions from many persons to whom I wish to make grateful acknowledgment. First to the persons previously mentioned who have examined my specimens in difficult genera, I tender my sincere thanks.

I wish especially to thank those who have contributed difficult parts of the text: Frederick J. Hermann of the University of Michigan for the text of *Carex, Juncus,* and *Luzula;* Theodor Just of the University of Notre Dame for the text of *Chenopodiaceae;* and Ernest J. Palmer of the Arnold Arboretum for the text of *Crataegus.* These authors have with few exceptions followed the phraseology of the flora.

I owe much to Stanley Coulter, until recently Dean of the School of Science, Purdue University, who encouraged me to write a flora of Indiana and who enlisted the aid of the Department of Conservation. He has also read most of the manuscript and has been helpful in many ways.

C. A. Weatherby of the Gray Herbarium, Cambridge, Massachusetts, has promptly answered my many letters relative to botanical nomenclature. I wish to express my appreciation for this special service and reading proof.

Paul Weatherwax of Indiana University has read the manuscript and given me helpful suggestions.

Frederick J. Hermann of the University of Michigan has read both the manuscript and the proof and has been exceedingly helpful in many ways.

Mrs. Leland Winch, of West Lafayette, Indiana, née Harriet M. Gragg, has typed the manuscript. She has been most helpful in the English composition and has been an accurate, earnest, and conscientious assistant.

I wish to express my sincere thanks to E. P. Wilson for his interest and efforts in having the Flora published in the best manner possible; also for the making of the county and botanical area maps.

Our thanks are also due to J. H. Armington of the U.S. Weather Bureau for the two full page maps, showing the rainfall and temperature of Indiana.

I wish to acknowledge the great assistance of my wife, Stella M. Deam, who has, during the past forty years, helped to collect and prepare specimens, has read copy and proof, and has shared the financial burden the work has entailed.

Lastly, I wish to thank the Department of Conservation for the opportunity of doing this work and publishing the results.

Conclusion

Active work of writing the flora was begun about seven years ago. Much data on the distribution of rare species yet remain to be collected but since I have just passed my seventy-third birthday it seems wise to conclude the work.

CHAS. C. DEAM.

Bluffton, Indiana, Sept. 28, 1938.

P.S. In order to keep the nomenclature up to date while the flora was going through the press it was necessary to make the changes in footnotes and omit some of the synonyms.

Feb. 15, 1940. CHAS. C. DEAM.

ABBREVIATIONS OF THE NAMES OF AUTHORS

Adans.—**Adanson,** Michel.
A. DC.—**De Candolle,** Alphonse.
Ait.—**Aiton,** William.
Ait. f.—**Aiton,** William Townsend.
All.—**Allioni,** Carlo.
Anders.—**Andersson,** Nils Johan.
Andrz.—**Andrzejowski,** Anton Lukianowicz.
Arn.—**Arnott,** George A. Walker.
Arrh.—**Arrhenius,** Johan Pehr.
Asch.—**Ascherson,** Paul.
B. & H.—**Bentham,** George, and **Hooker,** Joseph Dalton.
Bab.—**Babington,** Charles Cardale.
Baill.—**Baillon,** Henri Ernest.
Baldw.—**Baldwin,** William.
Barnh.—**Barnhart,** John Hendley.
Bart.—**Barton,** William P.C.
Bartr.—**Bartram,** William.
Beauv.—**Beauvois,** A.M.F.J. Palisot de.
Benn.—**Bennett,** Arthur.
Benth.—**Bentham,** George.
Bernh.—**Bernhardi,** Johann Jacob.
Bess.—**Besser,** Wilhelm S.J.G. von.
Bickn.—**Bicknell,** Eugene P.
Bigel.—**Bigelow,** Jacob.
Biv.—**Bivona-Bernardi,** Antonio.
Bjornstr.—**Bjornström,** Friedrich Johann.
Boeckl.—**Boeckeler,** Otto.
Boenn.—**Boenninghausen,** C.M.F. von.
Boerh.—**Boerhaave,** Hermann.
Boiss.—**Boissier,** Edmond.
Borkh.—**Borkhausen,** M.B.
Br., A.Br.—**Braun,** Alexander.
Br., P.Br.—**Browne,** Patrick.
Br., R.Br.—**Brown,** Robert.
Briq.—**Briquet,** John.
Britt.—**Britton,** Nathaniel Lord.
BSP.—**Britton,** Nathaniel Lord, **Sterns,** E. E., and **Poggenberg,** Justus F.
Buch.—**Buchenau,** Franz.
Burm. f.—**Burman,** Nikolaus Laurens.
C. & S.—**Chamisso,** Adalbert von, and **Schlechtendal,** D.F.L. von.
Carr.—**Carrière,** Elie Abel.
Casp.—**Caspary,** Robert.
Cass.—**Cassini,** Henri.
Cav.—**Cavanilles,** Antonio José.
Celak.—**Celakovsky,** Ladislav.
Chapm.—**Chapman,** Alvan Wentworth.
Chr., C.Chr.—**Christensen,** Carl.
Clairv.—**Clairville,** Joseph Phillipe de.
Clayt.—**Clayton,** John.
Coss.—**Cosson,** Ernest.
Coult.—**Coulter,** John Merle.
Cov.—**Coville,** Frederick V.
Cyrill.—**Cirillo,** Domenico.
Darl.—**Darlington,** William.
Davenp.—**Davenport,** George Edward.
DC.—**De Candolle,** Augustin Pyramus.
Dcne.—**Decaisne,** Joseph.
Desf.—**Desfontaines,** Réné Louiche.
Desr.—**Desrousseaux,** Louis Auguste Joseph.
Desv.—**Desvaux,** Augustin Nicaise.
Dietr.—**Dietrich,** Albert.
Dill.—**Dillenius,** Johann Jacob.
Dougl.—**Douglas,** David.
Dufr.—**Dufresne,** Pierre.
Duham.—**Du Hamel du Monceau,** H.L.
Dumont.—**Du Mont de Courset,** G.L.M.
Dumort.—**Dumortier,** Barthélemy C.
Eat.—**Eaton,** Amos.
Eggl.—**Eggleston,** Willard Webster.
Ehrh.—**Ehrhart,** Friedrich.
Ell.—**Elliott,** Stephen.
Endl.—**Endlicher,** Stephan Ladislaus.
Engelm.—**Engelmann,** George.
Farw.—**Farwell,** Oliver A.
Fern.—**Fernald,** Merritt Lyndon.
Fisch.—**Fischer,** F.E. Ludwig von.
Forst.—**Forster,** J.R. and George.
Fourn.—**Fournier,** Eugène.
Fresn.—**Fresenius,** J.B.G.W.
Froel.—**Froelich,** Joseph Aloys.
Gaertn.—**Gaertner,** Joseph.
Gatt.—**Gattinger,** Augustin.
Gaud.—**Gaudichaud-Beaupré,** Charles.
Germ.—**Germain,** Ernest.
Gilib.—**Gilibert,** Jean Emmanuel.
Gmel.—**Gmelin,** Samuel Gottlieb.
Gmel., J.F.—**Gmelin,** Johann Friedrich.
Gmel., J.G.—**Gmelin,** Johann Georg.
Godr.—**Godron,** Dominique Alexandre.
Grab.—**Grabowski,** Heinrich Emanuel.
Graebn.—**Graebner,** Paul.
Gren. & Godr.—**Grenier,** Charles, and **Godron,** D.A.
Grev.—**Greville,** Robert Kaye.
Griseb.—**Grisebach,** Heinrich R.A.
Gronov.—**Gronovius,** Jan Fredrik.
Guss.—**Gussoni,** Giovanni.
H. & A.—**Hooker,** William Jackson, and **Arnott,** G.A. Walker.
Hack.—**Hackel,** Eduard.
Hartm.—**Hartman,** Carl Johan.
Hassk.—**Hasskarl,** Justus Carl.
Haussk.—**Haussknecht,** Carl.
HBK.—**Humboldt,** F. Alexander von, **Bonpland,** Aimé, and **Kunth,** C.S.
Heist.—**Heister,** Lorentz.
Herb.—**Herbert,** William.

Hitchc.—**Hitchcock,** Albert Spear.
Hochst.—**Hochstetter,** Christian Frederich.
Hoffm.—**Hoffmann,** George Franz.
Hook.—**Hooker,** William Jackson.
Hornem.—**Hornemann,** Jens Wilken.
Houtt.—**Houttuyn,** M.
Hubb.—**Hubbard,** F. Tracy.
Huds.—**Hudson,** William.
Jacq.—**Jacquin,** Nicolaus Joseph.
Jord.—**Jordan,** Alexis.
Juss.—**Jussieu,** Antoine Laurent de.
Juss., B.—**Jussieu,** Bernard de.
Karst.—**Karsten,** Hermann.
Koel.—**Koeler,** George Ludwig.
Krock.—**Krocker,** Anton Johann.
Ktze.—**Kuntze,** Otto.
L.—**Linnaeus,** Carolus, or **Linné,** Carl von.
L.f.—**Linné,** Carl von (the son).
Laestad.—**Laestadius,** Lars Levi.
Lag.—**Lagasca,** Mariano.
Lall.—**Ave-Lallemant,** J.L.E.
Lam.—**Lamarck,** J.B.A.P. Monnet.
Lamb.—**Lambert,** Aylmer Bourke.
Laxm.—**Laxmann,** Eric.
Leavenw.—**Leavenworth,** Melines C.
Ledeb.—**Ledebour,** Carl F. von.
Lehm.—**Lehmann,** J.G.C.
Lesp. & Thev.—**Lespinasse,** Gustave, and **Théveneau,** A.
Less.—**Lessing,** Christian Friedrich.
Leyss.—**Leysser,** Frederich Wilhelm.
L'Her.—**L'Hériter, de Brutelle,** C.L.
Lightf.—**Lightfoot,** John.
Lindl.—**Lindley,** John.
Lodd.—**Loddiges,** Conrad.
Loisel.—**Loiseleur-Deslongchamps,** J.L.A.
Loud.—**Loudon,** John Claudius.
Lour.—**Loureiro,** Juan.
Macb.—**Macbride,** J. Francis.
Mack.—**Mackenzie,** Kenneth Kent.
MacM.—**MacMillan,** Conway.
Marsh.—**Marshall,** Humphrey.
Maxim.—**Maximowicz,** Carl Johann.
Medic.—**Medicus,** Friedrich Casimir.
Meisn.—**Meisner,** Carl Friedrich.
Merr.—**Merrill,** Elmer D.
Mert. & Koch.—**Mertens,** Franz Karl, and **Koch,** Wilhelm Daniel Heinrich.
Mett.—**Mettenius,** Georg Heinrich.
Mey.—**Meyer,** Ernest Heinrich F.
Mey., C.A.—**Meyer,** Carl Anton.
Mey., G.F.W.—**Meyer,** Georg Friedrich Wilhelm.
Mich.—**Micheli,** Pier' Antonio.
Michx.—**Michaux,** André.
Michx.f.—**Michaux,** Francois André.
Mill.—**Miller,** Philip.
Moq.—**Moquin-Tandon,** Alfred.
Muell. Arg.—**Mueller,** Jean (of Aargau).
Muench.—**Muenchhausen,** Otto Freiherr von.
Muhl.—**Muhlenberg,** H.E.
Murr.—**Murray,** Johann Andreas.
Neck.—**Necker,** Noel Joseph de.
Nees—**Nees von Esenbeck,** Christian Gottfried.
Nees & Eberm.—**Nees von Esenbeck,** T.F. L., and **Ebermaier,** K.H.
Newm.—**Newman,** Edward.
Nieuwl.—**Nieuwland,** Julius Arthur.
Nutt.—**Nuttall,** Thomas.
Pall.—**Pallas,** Peter Simon.
Parl.—**Parlatore,** Filippo.
Pers.—**Persoon,** Christian Hendrik.
Peterm.—**Petermann,** Wilhelm Ludwig.
Planch.—**Planchon,** Jules Emile.
Plum.—**Plumier,** Charles.
Poir.—**Poiret,** Jean Louis Marie.
Poll.—**Pollich,** Johann Adam.
R. & P.—**Ruiz,** Lopez Hipolito, and **Pavon,** Josef.
R. & S.—**Roemer,** J.J., and **Schultes,** August.
Raf.—**Rafinesque-Schmaltz,** C.S.
Rehd.—**Rehder,** Alfred.
Reichenb.—**Reichenbach,** H.G.L.
Richards.—**Richardson,** John.
Rivin.—**Rivinius,** August Quirinus.
Rodr.—**Rodriguez,** José Demetrio.
Roem.—**Roemer,** M.J.
Rostk.—**Rostkovius,** F.W.G.
Rottb.—**Rottboell,** Christen Fries.
Rupp.—**Ruppius,** Heinrich Bernhard.
Rupr.—**Ruprecht,** Franz J.
Rydb.—**Rydberg,** Per Axel.
Salisb.—**Salisbury,** Richard Anthony.
Sarg.—**Sargent,** Charles Sprague.
Schk.—**Schkuhr,** Christian.
Schleich.—**Schleicher,** J.C.
Schleid.—**Schleiden,** Matthias Jacob.
Schneid.—**Schneider,** Camillo.
Schrad.—**Schrader,** Heinrich Adolph.
Schreb.—**Schreber,** Johann D.C. von.
Schwein.—**Schweinitz,** Lewis David de.
Scop.—**Scopoli,** Johann Anton.
Scribn.—**Lamson-Scribner,** Frank.
Ser.—**Seringe,** Nicolas Charles.
Shuttlw.—**Shuttleworth,** Robert.
Sibth.—**Sibthorp,** John.
Sieb. & Zucc.—**Siebold,** P.F. von, and **Zuccarini,** J.G.
Sm.—**Smith,** James Edward.
Sm., J.—**Smith,** John.
Sm., J.D.—**Smith,** John Donnell.

Sm., J.G.—**Smith,** Jared Gage.
Soland.—**Solander,** Daniel.
Spreng.—**Sprengel,** **Kurt.**
Sternb.—**Sternberg,** Caspar.
Steud.—**Steudel,** Ernst Gottlieb.
St. Hil.—**St. Hilaire,** Auguste de.
Sudw.—**Sudworth,** George B.
Sulliv.—**Sullivant,** William Starling.
Sw.—**Swartz,** Olaf.
T. & G.—**Torrey,** John, and **Gray,** Asa.
Thunb.—**Thunberg,** Carl Pehr.
Tidestr.—**Tidestrom,** Ivar.
Torr.—**Torrey,** John.
Tourn.—**Tournefort,** Joseph Pitton de.
Traut.—**Trautvetter,** Ernest Rudolph.
Trel.—**Trelease,** William.
Trev.—**Treviranus,** Christian Ludolf.
Trin.—**Trinius,** Karl Bernhard.
Tuckerm.—**Tuckerman,** Edward.
Turcz.—**Turczaninow,** Nicolaus.
Underw.—**Underwood,** Lucien Marcus.
Vaill.—**Vaillant,** Sébastien.
Vent.—**Ventenat,** Etienne Pierre.
Vict.—**Marie-Victorin.**
Vill.—**Villars,** Dominique.
Wahlb.—**Wahlberg,** Pehr Frederik.
Wahl.—**Wahlenberg,** Georg.
Waldst. & Kit.—**Waldstein,** F.A. von, and **Kitaibel,** P.
Wallr.—**Wallroth,** K.F.W.
Walp.—**Walpers,** Wilhelm Gerhard.
Walt.—**Walter,** Thomas.
Wang.—**Wangenheim,** F.A.J. von.
Wats.—**Watson,** Sereno.
Wats. E. E.—**Watson,** Elba Emanuel.
Wendl.—**Wendland,** Johann Christoph.
Wettst.—**Wettstein,** Richard von.
Wieg.—**Wiegand,** Karl M.
Willd.—**Willdenow,** Carl Ludwig.
Wimm.—**Wimmer,** Friedrich.
With.—**Withering,** William.
Wormsk.—**Wormskiold,** M. von.
Wulf.—**Wulfen,** Franz Xavier.

Key to the Families[1]

(Carried out, in some cases, to genera. The numbers preceding the family and generic names refer to their sequence in the class to which they belong.)

PTERIDÓPHYTA

Plants without true flowers, reproducing by spores (without embryos); fernlike, mosslike, rushlike, or aquatic plants.

- A. Plants floating, with small, 2-ranked leaves; sporocarps borne on the under side of the stem.. SALVINIACEAE, p. 59.
- A. Plants terrestrial or submerged, not floating B.
  - B. Stems conspicuously grooved and jointed, their nodes covered by toothed sheaths; sporangia borne on the scales of terminal, dry, conelike spikes...... .... EQUISETACEAE, p. 59.
  - B. Stems not conspicuously grooved, without sheathing joints C.
    - C. Leaves closely imbricated, short or long-linear (from a cormlike base); sporangia sessile, axillary.
      - Stem short, cormlike; leaves linear, in a rosette; sporangia borne in a cavity on the inner side of the leaf-base.................. ISOETACEAE, p. 66.
      - Stem elongate, creeping or branching; leaves very short, crowded or imbricated.
        - Plants small and mosslike; spores of two sizes.... SELAGINELLACEAE, p. 65.
        - Plants not resembling mosses; spores all of one size...................... .... LYCOPODIACEAE, p. 63.
    - C. Leaves (fronds) not closely imbricated D.
      - D. Leaves (fronds) 4-foliolate, cloverlike; aquatic...... MARSILEACEAE, p. 102.
      - D. Leaves (fronds) not 4-foliolate, broad, flat, fernlike, more or less pinnately or ternately divided or entire; terrestrial E.
        - E. Sterile and fertile fronds flat, entire; the fertile ones ending in long-stalked, simple spikes...................... OPHIOGLOSSACEAE, p. 37.
        - E. Sterile and fertile fronds not entire F.
          - F. Fertile fronds or fertile portions of the fronds conspicuously unlike the sterile; sporangia not on the lower surface of green leaves G.
            - G. Rootstock almost none; the solitary (rarely 2) fronds appearing to rise from a cluster of fleshy roots; lower segments sterile, the upper ones fertile and bearing 2-ranked, globular sporangia. .... BOTRYCHIUM, p. 38.
            - G. Rootstock well developed, elongate or stout, the roots fibrous; fronds numerous H.
              - H. Sporangia globose, thin-walled, 2-valved, densely crowded, not 2-ranked .... OSMUNDACEAE, p. 40.
              - H. Sporangia within firm, 2-ranked, globose and distinct or connected in beadlike segments.......................... ONOCLEA, p. 45.
          - F. Fertile fronds or segments essentially like the sterile; sporangia borne on the lower surface or on the margins of green segments. .... POLYPODIACEAE, p. 42.

SPERMATÓPHYTA

Plants with true flowers containing stamens or pistils or both, reproducing by seed (containing an embryo).

Ovules not in a closed ovary; trees and shrubs with needlelike or scalelike, mostly evergreen leaves; flowers monoecious or dioecious (Gymnosperms).

[1] See Introduction, p. 13.

GYMNOSPÉRMAE

Flowers solitary, axillary; seed solitary, enveloped in a pulpy disk (berry-like) ....5. TAXACEAE, p. 66.

Flowers borne in catkins; fruit a cone or a several-seeded berry..6. PINACEAE, p. 66.

Ovules borne in a closed ovary which, at maturity, becomes the fruit; herbs or woody plants, with broad or narrow, evergreen or deciduous leaves (Angiosperms) J.

ANGIOSPÉRMAE

J. Embryo with a single cotyledon; early leaves always alternate (leaves sometimes whorled), mostly parallel-veined (net-veined in *Araceae and Dioscoreaceae*; parts of the flower in threes or sixes, never in fives; stems without a central pith or ringlike layers, but with woody fibers distributed through them; our species, except in the genus *Smilax,* herbaceous (Monocotyledons) K.

MONOCOTYLEDÒNEAE

K. Plant scarcely differentiated into stem and leaf, small, usually lens-shaped, ellipsoid or oblong; free-swimming aquatics without true leaves..24. LEMNACEAE, p. 279.

K. Plant with stem and leaves L.

L. Perianth free from the ovary or none M.

M. Perianth lacking, or of scalelike or bristle-form divisions N.

N. Flowers enclosed or subtended by scales (glumes); plants grasslike, with jointed stems, sheathing leaves, and 1-seeded fruit.

Stems hollow, round or flattened; leaf sheaths split; anthers attached at the middle..........................19. GRAMINEAE, p. 93.

Stems solid, usually more or less triangular; leaf sheaths not split; anthers attached at the base............20. CYPERACEAE, p. 181.

N. Flowers not enclosed in scales (though sometimes in involucrate heads) O.

O. Plants immersed aquatics, branching and leafy, the upper leaves often floating.

Leaves opposite or ternate; pistils solitary, naked...12. NAJADACEAE, p. 84.

Leaves alternate or 2-ranked; pistils aggregated into heads or clusters.

Fruit in heads, the nutlets composing it tightly compact, with prominent, conical style bases mostly 2-4 mm long....................................10. SPARGANIACEAE, p. 72.

Fruit in clusters; nutlets not tightly compact, the style bases usually short or very slender.......................................11. POTAMOGETONACEAE, p. 75.

O. Plants terrestrial or of a marsh habitat.

Leaves petiolate, the blades net-veined.........23. ARACEAE, p. 277.

Leaves not petiolate, linear or sword-shaped, parallel-veined P.

P. Flowers monoecious or dioecious.

Flowers and fruit in a cylindrical spike...8. TYPHACEAE, p. 71.

Flowers and fruit in heads.

Heads spheroidal, pubescent, involucrate..30. ERIOCAULACEAE, p. 283.

Heads globose, glabrous, not involucrate..10. SPARGANIACEAE, p. 72.

P. Flowers perfect.

Plants with flowers in a dense spike (4-7 cm long), borne on the margin of a long, 2-edged scape; rhizome aromatic...694. ACORUS, p. 277.

Plants not as above, the flowers not in spikes; rhizomes not aromatic.

Carpels 3-6, more or less united, separating at least when ripe........................14. JUNCAGINACEAE, p. 85.

Carpels 3, completely united, not separating at maturity. ....36. JUNCACEAE, p. 290.

M. Perianth always present, herbaceous or colored, neither scalelike nor bristle-form Q.

Q. Pistils numerous, in a head or ring...........15. ALISMACEAE, p. 86.

Q. Pistil one, compound (cells or placentae mostly 3) R.

R. Stamens 3.

Flowers racemose or spicate............14. JUNCAGINACEAE, p. 85.

Flowers in dense, scaly heads..............29. XYRIDACEAE, p. 282.

Flowers cymose............................36. JUNCACEAE, p. 290.

R. Stamens 4 ....1119. MAIANTHEMUM, p. 318.

R. Stamens 6 S.

S. Stamens all alike and fertile.

Ovary of 3-6 carpels, separating at maturity.................... ....14. JUNCAGINACEAE, p. 85.

Ovary not deeply cleft (often angled or lobed).

Divisions of the perianth alike or nearly so.

Plants rushlike; perianth small, greenish or purplish brown. ....36. JUNCACEAE, p. 290.

Plants not rushlike...................38. LILIACEAE, p. 303.

Divisions of the pedianth unlike; the 3 sepals green and 2 or more of the petals colored.

Stem leaves ovate or oblong, in a whorl of 3; flowers solitary, terminal........................1138. TRILLIUM, p. 321.

Stem leave of a linear type, not in whorls; flowers in umbels.. ....33. COMMELINACEAE, p. 283.

S. Stamens dissimilar, or only 3 with fertile anthers.

Perianth of 6 yellow, petaloid segments........................ ....*Erythronium americanum*, p. 314.

Perianth of 3 herbaceous sepals and 2 or 3 colored ephemeral petals (petals rarely white)...........33. COMMELINACEAE, p. 283.

Perianth tubular, 6-lobed, mostly colored........................ ....34. PONTEDERIACEAE, p. 287.

L. Perianth present, adnate to the ovary.

Stamens 1 or 2; flowers irregular; seeds many....50. ORCHIDACEAE, p. 335.

Stamens 3 or more; flowers mostly regular or nearly so.

Plants immersed aquatics...............17. HYDROCHARITACEAE, p. 91.

Plants terrestrial.

Flowers dioecious; plants twining; leaves net-veined................. ....43. DIOSCOREACEAE, p. 330.

Flowers perfect; leaves parallel-veined.

Stamens 6..........................40. AMARYLLIDACEAE, p. 328.

Stamens 3; leaves 2-ranked....................44. IRIDACEAE, 332.

J. Embryo with a pair of opposite cotyledons; leaves net-veined (except in *Eryngium*); parts of the flower mostly in fours and fives; stems formed of bark, wood, and pith, increasing in size by the annual addition of a new layer (rarely two) to the outside, next to the bark (Dicotyledons.) T.

DICOTYLEDONEAE

T. Corolla none; calyx present or lacking U.

U. Flowers monoecious or dioecious (rarely polygamous), one or both sorts in catkins or dense heads V.

V. Staminate or pistillate (not both) flowers in catkins or catkinlike heads.

Pistillate flowers in a short catkin or catkinlike head....64. MORACEAE, p. 394.

Pistillate flowers single or clustered; the staminate in slender catkins (except in *Fagus*).

Leaves pinnate; pistillate flowers and fruit naked..60. JUGLANDACEAE, p. 365.

Leaves simple; pistillate flowers 1-3 in a cup or involucre...62. FAGACEAE, p. 378.

V. Staminate and pistillate (both) flowers in catkins or catkinlike heads W.

W. Ovary many-ovuled; fruit many-seeded.

Ovary and pod 2-celled; seed not tufted.........3298. LIQUIDAMBAR, p. 523.

Ovary and pod 1-celled; seeds hairy-tufted..........56. SALICACEAE, p. 352.

W. Ovary 1- or 2-celled; cells 1-ovuled; fruit 1-seeded.

Parasitic on trees; fruit a berry................67. LORANTHACEAE, p. 402.

Trees and shrubs, not parasitic.

Calyx regular in fertile flower, succulent in fruit....64. MORACEAE, p. 394.

Calyx none or rudimentary and scalelike.

Style and stigma simple; leaves palmately angled or lobed...124. PLATANACEAE, p. 523.

Styles or long stigmas 2.

Pistillate flowers 2 or 3 at each scale of the catkin...61. BETULACEAE, p. 373.

Pistillate flowers single under each scale; nutlets naked, drupelike.......................................57. MYRICACEAE, p. 365.

U. Flowers not in catkins X.

X. Ovary or its cells containing only 1 or 2 (rarely 3 or 4) ovules Y.

Y. Pistil composed of more than one carpel; carpels distinct or nearly so.

Stamens insterted on the calyx; leaves with stipules..126. ROSACEAE, p. 524.

Stamens inserted on the receptacle.

Leaves punctate with transparent glands.....3990. ZANTHOXYLUM, p. 632.

Leaves not punctate with glands.

Calyx present, usually colored or petal-like...91. RANUNCULACEAE, p. 454.

Calyx none; flowers in a spike...............52. SAURURACEAE, p. 352.

Y. Pistil simple or compound but without distinct carpels Z.

Z. Ovary free from the calyx, which is sometimes lacking a.

a. Stipules (ocreae) sheathing the stem at the nodes.

Calyx none; trees........................124. PLATANACEAE, p. 523.

Calyx present, commonly petal-like; herbs..77. POLYGONACEAE, p. 405.

a. Stipules not sheathing the stem or lacking b.

b. Herbs c.

c. Plants aquatic, submerged or nearly so.

Leaves whorled, dissected; style 1...89. CERATOPHYLLACEAE, p. 454.

Leaves opposite, entire; styles 2; ovary 4-celled.............................148. CALLITRICHACEAE, p. 646.

c. Plants not aquatic d.

d. Styles 10; ovary and berry 10-celled..83. PHYTOLACCACEAE, p. 433.

d. Style, if any, and stigma 1.

Flowers unisexual; ovary of the fertile flowers 1-celled...65. URTICACEAE, p. 397.

Flowers perfect; pods 2-celled, 2-seeded.................................2883. LEPIDUM, p. 487.

d. Styles 2 or 3 or branched; ovary 1-4 celled e.
e. Leaves palmately lobed or divided, the terminal ones sometimes simple..............................64. MORACEAE, p. 394.
e. Leaves not palmately lobed or divided f.
f. Ovary and capsule 3-celled; juice usually milky.......................................147. EUPHORBIACEAE, p. 636.
f. Ovary 1-celled; juice not milky g.
g. Leaves stellate-pubescent beneath...4350. CROTONOPSIS, p. 638.
g. Leaves not stellate-pubescent beneath.
Stipules scarious..............2475. PARONYCHIA, p. 442.
Stipules none.
Leaves opposite.
Flowers in heads or spikes, these often panicled; anthers 1-celled.....79. AMARANTHACEAE, p. 427.
Flowers sessile in the forks of a branching inflorescence................2483. SCLERANTHUS, p. 444.
Leaves alternate.
Flowers and bracts scarious...79. AMARANTHACEAE, p. 427.
Flowers small, chiefly greenish, no scarious bracts........................78. CHENOPODIACEAE, p. 418.
b. Trees or shrubs.
Leaves opposite.
Fruit 1-celled, a single samara..............243. OLEACEAE, p. 751.
Fruit 2-celled, a double samara.............163. ACERACEAE, p. 654.
Fruit 3-celled, not winged...............169. RHAMNACEAE, p. 659.
Leaves alternate.
Ovary 3-celled.........................169. RHAMNACEAE, p. 659.
Ovary 1- or 2-celled.
Styles and stigmas 2.....................63. ULMACEAE, p. 390.
Style and stigma 1.
Anthers opening lengthwise......214. THYMELAEACEAE, p. 694.
Anthers opening by uplifted lids.......102. LAURACEAE, p. 480.
Z. Ovary inferior or so closely and permanently invested by the calyx as to appear so.
Plants parasitic on the branches of trees........67. LORANTHACEAE, p. 402.
Plants not parasitic on trees.
Plants aquatic..........................225. HALORAGIDACEAE, p. 710.
Plants not aquatic.
Herbs with calyx colored like a corolla.
Leaves opposite, simple.................80. NYCTAGINACEAE, p. 432.
Leaves alternate.
Leaves simple.........................2112. COMANDRA, p. 402.
Leaves compound.....................3381. SANGUISORBA, p. 573.
Trees or shrubs.
Leaves scurfy.......................215. ELAEAGNACEAE, p. 695.
Leaves not scurfy.
Style 1; flowers solitary, in pairs or in umbel-like clusters...6151. NYSSA, p. 728.
Styles 2.........................123. HAMAMELIDACEAE, p. 533.
X. Ovary or its cells containing many ovules h.
h. Calyx none; ovary and fruit naked.
Aquatic herbs...............................113. PODOSTEMACEAE, p. 512.
Shrubs or trees.............................123A. ALTINGIACEAE, p. 533.
h. Calyx present j.
j. Ovary superior k.
k. Ovaries 2 or more, separate...............91. RANUNCULACEAE, p. 454.

k. Ovary single m.
m. Ovary 5-celled, 5-beaked; leaves scattered...3173. PENTHORUM, p. 514.
m. Ovary 3-5-celled; leaves opposite or whorled...84. AIZOACEAE, p. 434.
m. Ovary 1- or 2-celled.
Leaves compound........................91. RANUNCULACEAE, p. 454.
Leaves simple.
Style 1.................................216. LYTHRACEAE, p. 695.
Styles 2-5..........................87. CARYOPHYLLACEAE, p. 435.
j. Ovary and pod inferior.
Ovary 1-celled; stamens 8-10...............3199. CHRYSOSPLENIUM, p. 519.
Ovary 4-celled; stamens 4.......................5793. LUDWIGIA, p. 700.
Ovary 6-celled; stamens 6-12................74. ARISTOLOCHIACEAE, p. 403.

T. Corolla and calyx both present n.

n. Corolla of separate petals o.

o. Stamens numerous, at least more than 10 (rarely 9 or 10 in *Polanisia*), and more than twice as many as the sepals or calyx lobes p.
p. Calyx entirely free and separate from the pistil or pistils q.
q. Pistils several or many, wholly distinct or united at the base into a strongly lobed or several-beaked ovary r.
r. Aquatic plants with peltate leaves...........88. NYMPHAEACEAE, p. 450.
r. Terrestrial plants.
Plants climbing.
Leaves alternate.........................94. MENISPERMACEAE, p. 477.
Leaves opposite............................2542. CLEMATIS, p. 463.
Plants not climbing.
Filaments of stamens united into a tube......175. MALVACEAE, p. 666.
Filaments not united.
Stamens on the calyx.......................126. ROSACEAE, p. 524.
Stamens on the receptacle or disk.
Trees or shrubs.
Sepals and petals imbricated; fruit aggregate...95. MAGNOLIACEAE, p. 478.
Sepals and petals valvate; fruit not aggregate...98. ANNONACEAE, p. 479.
Herbs; inflorescence simple; pistils several, simple...91. RANUNCULACEAE, p. 454.
q. Pistils strictly one as to ovary; the styles or stigmas may be several s.
s. Leaves punctate with translucent dots.......187. HYPERICACEAE, p. 671.
s. Leaves not punctate t.
t. Ovary simple, 1-celled.
Ovules 2.................................126. ROSACEAE, p. 524.
Ovules many.
Leaves 2- or 3-ternately compound or dissected..91. RANUNCULACEAE, p. 454.
Leaves peltate, lobed.................2558. PODOPHYLLUM, p. 475.
t. Ovary compound.
Ovary 1-celled.
Sepals 2 (rarely 3), caducous; sap milky or colored; placentae parietal..........................104. PAPAVERACEAE, p. 481.
Sepals 2; sap watery; placentae central...85. PORTULACACEAE, p. 434.
Sepals 4; sap watery; placentae parietal..107. CAPPARIDACEAE, p. 510.
Sepals 3 or 5, persistent; sap watery; placentae parietal...193. CISTACEAE, p. 677.

Ovary several-celled.
Calyx valvate in the bud.
Herbs or rarely shrubs; stamens united; anthers 1-celled.........
.....................................175. MALVACEAE, p. 666.
Trees; anthers 2-celled...................174. TILIACEAE, p. 665.
Calyx imbricate in the bud.
Leaves tubular with a flange at the top, radical.................
.............................110. SARRACENIACEAE, p. 511.
Leaves petiolate, mostly peltate or flattish; plants aquatic........
..............................88. NYMPHAEACEAE, p. 450.
p. Calyx more or less adherent to a compound ovary.
Ovary 7-30-celled.
Cells many-ovuled; aquatic herbs.............88. NYMPHAEACEAE, p. 450.
Cells 10, each 1-ovuled; shrubs or trees........3343. AMELANCHIER, p. 531.
Ovary 6-celled...................................2170. ASARUM, p. 403.
Ovary 1-5-celled.
Plants without leaves (in the popular sense), more or less spiny; petals many, yellow................................210. CACTACEAE, p. 694.
Plants with leaves.
Sepals or calyx lobes 2; ovules arising from the base of a 1-celled ovary..
....................................85. PORTULACACEAE, p. 434.
Sepals or calyx lobes more than 2.
Leaves opposite; stipules none............117. SAXIFRAGACEAE, p. 514.
Leaves alternate.
Stipules present...........................126. ROSACEAE, p. 524.
Stipules none; shrubs...................241. STYRACACEAE, p. 751.
o. Stamens not more than twice as many as the petals u.
u. Stamens of the same number as the petals and opposite them.
Ovaries 3-6, separate; herbaceous vines (rarely woody in Indiana)..........
....94. MENISPERMACEAE, p. 477.
Ovary only one.
Ovary 2-4-celled.
Calyx lobes minute or obsolete; petals valvate.....170. VITACEAE, p. 661.
Calyx 4- or 5-cleft; petals involute...........169. RHAMNACEAE, p. 659.
Ovary 1-celled.
Anthers opening by uplifted lids................93. BERBERIDACEAE, p. 475.
Anthers not opening by uplifted lids.
Style 1, unbranched; stigma 1...............237. PRIMULACEAE, p. 744.
Styles, style branches or stigmas more than 1.
Sepals or calyx lobes 2...................85. PORTULACACEAE, p. 434.
Sepals or calyx lobes 3-5..................4350. CROTONOPSIS, p. 638.
u. Stamens not of the same number as the petals or if of the same number alternate with them v.
v. Calyx free from the ovary, i. e. ovary wholly superior w.
w. Ovaries 2 or more, wholly separate or somewhat united x.
x. Stamens united with each other and with a large thick stigma common to the 2 ovaries..................248. ASCLEPIADACEAE, p. 764.
x. Stamens free from each other and from the pistils y.
y. Stamens on the receptacle, free from the calyx.
Leaves punctate with translucent dots........137. RUTACEAE, p. 632.
Leaves without translucent dots.
Trees...............................4124. AILANTHUS, p. 632.
Herbs.
Ovaries or lobes of the ovary 2-5, with a common style.
Ovary 2- or 3-lobed...........152. LIMNANTHACEAE, p. 647.
Ovary 5-lobed...................129. GERANIACEAE, p. 623.
Ovaries with separate styles or sessile stigmas..............
............................91. RANUNCULACEAE, p. 454.

y. Stamens inserted on the calyx.
Plant fleshy; stamens not twice as many as the pistils............ ....115. CRASSULACEAE, p. 513.
Plant not fleshy; stamens not twice as many as the pistils.
Stipules present........................126. ROSACEAE, p. 524.
Stipules none.......................117. SAXIFRAGACEAE, p. 514.

w. Ovary 1 z.
z. Ovary simple with 1 parietal placenta......128. LEGUMINOSAE, p. 582.
z. Ovary compound, as shown by the number of its cells, placentae, styles, or stigmas A.
A. Ovary 1-celled.
Corolla irregular.
Petals 4; stamens 6.................104A. FUMARIACEAE, p. 482.
Petals and stamens 5.....................198. VIOLACEAE, p. 681.
Corolla regular or nearly so.
Ovule solitary.
Trees or shrubs..................153. ANACARDIACEAE, p. 648.
Herbs..............................105. CRUCIFERAE, p. 484.
Ovules more than one.
Ovules at the center or bottom of the cell.
Petals not inserted on the calyx..87. CARYOPHYLLACEAE, p. 435.
Petals inserted on the throat of a bell-shaped or tubular calyx. ....216. LYTHRACEAE, p. 695.
Ovules on 2 or more parietal placentae.
Leaves punctate with transparent dots.................... ....187. HYPERICACEAE, p. 671.
Leaves with gland-tipped bristles...112. DROSERACEAE, p. 512.
Leaves neither punctate nor bristly-glandular.
Petals 4.
Stamens essentially equal; pod usually stipitate......... ....107. CAPPARIDACEAE, p. 510.
Stamens unequal, 2 being shorter than the other 4; pod sessile....................105. CRUCIFERAE, p. 484.
Petals 3 or 5.
Ovary stipitate.............203. PASSIFLORACEAE, p. 693.
Ovary sessile.
Calyx 5-lobed or of 5 equal sepals.................... ....117. SAXIFRAGACEAE, p. 514.
Calyx of 3 equal or 5 very unequal sepals............ ....193. CISTACEAE, p. 677.

A. Ovary 2-several-celled B.
B. Flowers irregular C.
C. Anthers opening at the top.........145. POLYGALACEAE, p. 633.
C. Anthers opening lengthwise.
Stamens 12 and petals 6 on the throat of the gibbous calyx..... ....5478. CUPHEA p. 698.
Stamens 5-10 and petals hypogynous or nearly so.
Ovary 3-celled; trees or shrubs......4721. AESCULUS, p. 658.
Ovary 5-celled; herbs.........168. BALSAMINACEAE, p. 659.
B. Flowers regular or nearly so D.
D. Stamens neither just as many nor twice as many as the petals.
Trees or shrubs.
Stamens fewer than the 4 petals.......243. OLEACEAE, p. 751.
Stamens more numerous than the petals.................... ....163. ACERACEAE, p. 654.

Herbs.
Petals 5........................187. HYPERICACEAE, p. 671.
Petals 4.........................105. CRUCIFERAE, p. 484.
D. Stamens just as many as or twice as many as the petals E.
E. Ovules and seeds only 1 or 2 in each cell.
Herbs.
Flowers monoecious or dioecious.........................
........................147. EUPHORBIACEAE, p. 636.
Flowers perfect and symmetrical.
Cells of the ovary as many as the sepals.
Ovary 2- or 3-celled......152. LIMNANTHACEAE, p. 647.
Ovary 5-celled..............129. GERANIACEAE, p. 623.
Cells of the ovary twice as many as the sepals.
Leaves abruptly pinnate............................
...................135. ZYGOPHYLLACEAE, p. 631.
Leaves simple..................132. LINACEAE, p. 629.
Trees or shrubs.
Leaves compound.
Leaves 3-foliolate, punctate........4069. PTELEA, p. 632.
Leaves pinnate, not punctate...165. SAPINDACEAE, p. 658.
Leaves simple.
Blades palmately veined.........163. ACERACEAE, p. 654.
Blades pinnately veined.
Leaves alternate.
Shrubs, climbing...........4625. CELASTRUS, p. 653.
Shrubs, erect............157. AQUIFOLIACEAE, p. 651.
Leaves opposite............158. CELASTRACEAE, p. 653.
E. Ovules, and usually seed, several or many in each cell F.
F. Leaves compound.
Trees or shrubs..............161. STAPHYLEACEAE, p. 654.
Herbs; leaves alternate or all radical...................
........................130. OXALIDACEAE, p. 626.
F. Leaves simple.
Stipules present between opposite leaves................
........................189. ELATINACEAE, p. 677.
Stipules none when the leaves are opposite.
Style 1.
Stamens free from the calyx....233. ERICACEAE, p. 733.
Stamens inserted on the calyx..216. LYTHRACEAE, p. 695.
Styles 2-5 or splitting into 2 in fruit.
Stamens free from the calyx; leaves opposite.........
..................87. CARYOPHYLLACEAE, p. 435.
Stamens inserted on the calyx..233. ERICACEAE, p. 733.
v. Calyx tube adherent to the ovary, at least to its lower half G.
G. Tendril-bearing and often succulent herbs..275. CUCURBITACEAE, p. 892.
G. Tendrils lacking H.
H. Ovules and seed only 1 in each cell.
Stamens 5 or 10.
Trees or shrubs.
Leaves simple, not prickly...............3345. CRATAEGUS, p. 533.
Leaves compound or prickly.............227. ARALIACEAE, p. 712.
Herbs.
Fruit dry, splitting at maturity; styles 2......................
..............................228. UMBELLIFERAE, p. 714.
Fruit berrylike; styles 2-5, separate or united..................
................................227. ARALIACEAE, p. 712.
Stamens 2, 4 or 8.
Style and stigma 1; fruit a drupe.........229. CORNACEAE, p. 728.

Styles or stigmatic branches or sessile stigmas usually more than 1; fruit not a drupe.
Shrubs or trees....................123. **HAMAMELIDACEAE**, p. 523.
Herbs.
Style 1; stigma 2-4 lobed.............224. **ONAGRACEAE**, p. 699.
Styles or sessile stigmas 4........225. **HALORAGIDACEAE**, p. 710.
H. Ovules and seed more than 1 in each cell.
Ovary 1-celled.
Sepals or calyx lobes 2; ovules borne at the base of the ovary...... ....85. **PORTULACACEAE**, p. 434.
Sepals or calyx lobes 4 or 5; placentae 2 or 3, parietal............ ....117. **SAXIFRAGACEAE**, p. 514.
Ovary 2-many-celled.
Anthers opening by pores at the apex...223. **MELASTOMACEAE**, p. 698.
Anthers not opening by pores.
Stamens inserted on or about a flat disk which covers the ovary. ....158. **CELASTRACEAE**, p. 653.
Stamens inserted on the calyx.
Style 1; stamens 4 or 8 (rarely 5)....224. **ONAGRACEAE**, p. 699.
Styles 2 or 3, distinct; stamens 5 or 10........................ ....117. **SAXIFRAGACEAE**, p. 514.
n. **Petals more or less united I.**
I. Stamens more numerous than the lobes of the corolla J.
J. Ovary 1-celled.
Placenta 1, parietal..............................128. **LEGUMINOSAE**, p. 582.
Placentae 2, parietal..........................104A. **FUMARIACEAE**, p. 482.
Placenta at the center or base of the ovary........241. **STYRACACEAE**, p. 751.
J. Ovary 2-celled; cells 1-ovuled...................145. **POLYGALACEAE**, p. 633.
J. Ovary 3-many-celled K.
K. Stamens free from the corolla.
Style 1; leaves simple..........................233. **ERICACEAE**, p. 733.
Styles 5; leaves 3-foliolate.....................130. **OXALIDACEAE**, p. 626.
K. Stamens attached to the base or tube of the corolla.
Saprophytic herbs without green foliage.........6169. **MONOTROPA**, p. 737.
Not saprophytic; foliage green.
Trees or shrubs; anthers mostly 2-celled.
Filaments united at the base, forming a tube.....6411. **STYRAX**, p. 751.
Filaments free from each other.
Style 1................................233. **ERICACEAE**, p. 733.
Styles 4...............................240. **EBENACEAE**, p. 751.
Herbs; anthers 1-celled.........................175. **MALVACEAE**, p. 666.
I. Stamens not more numerous than the corolla lobes L.
L. Stamens of the same number as the corolla lobes and opposite them.
Corolla appendaged with scales inside; ovary 5-celled; trees or shrubs......... ....239. **SAPOTACEAE**, p. 750.
Corolla not appendaged with scales inside; ovary 1-celled; herbs............ ....237. **PRIMULACEAE**, p. 744.
L. Stamens alternate with the corolla lobes or fewer M.
M. Ovary free from the calyx tube (superior) N.
N. Corolla regular O.
O. Stamens as many as the corolla lobes P.
P. Ovaries more than 1, or if 1, deeply lobed Q.
Q. Ovaries 2, or if 1, 2-horned.
Stamens united......................248. **ASCLEPIADACEAE**, p. 764.
Stamens distinct.
Stipules or stipular membrane or line between opposite leaves; ovary 2-horned...................245. **LOGANIACEAE**, p. 754.
Stipules none; ovaries 2...............247. **APOCYNACEAE**, p. 760.

Q. Ovary deeply 4-lobed.
Leaves alternate........................252. **BORAGINACEAE**, p. 787.
Leaves opposite...........................254. **LABIATAE**, p. 798.
P. Ovary 1, not deeply lobed R.
R. Ovary 1-celled.
Seed 1; corolla scarious..............269. **PLANTAGINACEAE**, p. 867.
Seed several-many.
Leaves entire, opposite...............246. **GENTIANACEAE**, p. 754.
Leaves toothed, lobed, or compound.
Whole upper surface of the corolla white-bearded; leaflets 3, entire........................6543. **MENYANTHES**, p. 760.
Corolla not conspicuously bearded; leaves, if compound, with toothed leaflets.............251. **HYDROPHYLLACEAE**, p. 784.
R. Ovary 2-10-celled.
Leafless twining parasites..................6968. **CUSCUTA**, p. 770.
Leaves opposite, their bases connected by a stipular line.......................................245. **LOGANIACEAE**, p. 754.
Leaves alternate or, if opposite, with no trace of stipules.
Stamens free from the corolla or nearly so.
Style 1.............................233. **ERICACEAE**, p. 733.
Style none; stamens attached to the base of the corolla....................................157. **AQUIFOLIACEAE**, p. 651.
Stamens on the tube of the corolla.
Stamens 4.
Leafy-stemmed; leaves opposite; corolla petaloid..253. **VERBENACEAE**, p. 795.
Acaulescent; corolla scarious...269. **PLANTAGINACEAE**, p. 867.
Stamens 5 or rarely more.
Fruit of 2 or 4 seedlike nutlets....252. **BORAGINACEAE**, p. 787.
Fruit a few-many-seeded pod or berry.
Styles 2.
Pod few, mostly 4-seeded...249. **CONVOLVULACEAE**, p. 770.
Pod many-seeded.........251. **HYDROPHYLLACEAE**, p. 784.
Style 1, often branched.
Branches of the style (or at least the lobes of the stigma) 3.
Plants twining................7003. **IPOMOEA**, p. 776.
Plants not twining........250. **POLEMONIACEAE**, p. 778.
Branches of the style or lobes of the stigma 2 or rarely 4, or 1 (in Solanaceae).
Seed few, mostly 4......249. **CONVOLVULACEAE**, p. 770.
Seed many..................256. **SOLANACEAE**, p. 826.
O. Stamens fewer than the corolla lobes.
Stamens with anthers 4, in pairs.
Ovary 2-celled; cells several-seeded........266. **ACANTHACEAE**, p. 864.
Ovary 2-4-celled; cells 1-seeded; ovary not lobed; style apical...253. **VERBENACEAE**, p. 795.
Ovary 4-celled, 4-lobed; style basal.............254. **LABIATAE**, p. 798.
Stamens with anthers only 2 or rarely 3.
Ovary 4-lobed..............................7326. **LYCOPUS**, p. 821.
Ovary 2-celled, not 4-lobed.
Herbs.
Acaulescent; corolla scarious.......269. **PLANTAGINACEAE**, p. 867.
Leafy-stemmed; corolla not scarious......7579. **VERONICA**, p. 845.
Trees or shrubs...........................243. **OLEACEAE**, p. 751.
N. Corolla irregular S.
S. Stamens with anthers 5.

Ovary deeply 4-lobed around the style............7118. ECHIUM, p. 794.
Ovary not deeply lobed, many-ovuled.
Filaments or some of them woolly...........7460. VERBASCUM, p. 834.
Filaments not woolly....................7396. HYOSCYAMUS, p. 1087.
S. Stamens with anthers 2 or 4.
Ovules solitary in the 1-4 cells.
Ovary 4-lobed; style arising from between the lobes.................
....254. LABIATAE, p. 798.
Ovary not lobed; style from the apex.
Ovary 1-celled; fruit pointing backwards..268. PHRYMACEAE, p. 866.
Ovary 2-4-celled; fruit not pointing backwards....................
.................................253. VERBENACEAE, p. 795.
Ovules 2-many in each cell.
Ovary imperfectly 4- or 5-celled...........260. MARTYNIACEAE, p. 860.
Ovary 1- or 2-celled.
Ovary 1-celled.
Parasites without green foliage, terrestrial; stamens 4..........
............................261. OROBANCHACEAE, p. 860.
Not parasitic, chiefly aquatic or mud plants; stamens 2...........
..........................264. LENTIBULARIACEAE, p. 862.
Ovary 2-celled.
Trees or woody climbers; placentae parietal..BIGNONIACEAE, p. 858.
Herbs, rarely trees; placentae in the axis.
Seed (mostly numerous) not borne on hooks..................
.........................257. SCROPHULARIACEAE, p. 882.
Seed (2-12) borne on hooklike processes of the placentae......
............................266. ACANTHACEAE, p. 864.
M. Ovary adherent to the calyx tube (inferior) T.
T. Tendril-bearing herbs; anthers often united...275. CUCURBITACEAE, p. 892.
T. Tendrils none U.
U. Stamens separate V.
V. Stamens free from the corolla or nearly so, as many as its lobes; stipules none; sap milky...............276. CAMPANULACEAE, p. 893.
V. Stamens inserted on the corolla.
Stamens 1-3, always fewer than the corolla lobes..................
..................................273. VALERIANACEAE, p. 890.
Stamens 4 or 5; leaves opposite or whorled.
Ovary 2-5-celled.
Leaves opposite or perfoliate but never whorled, rarely provided with true stipules...............271. CAPRIFOLIACEAE, p. 878.
Leaves either opposite and stipulate, or whorled and destitute of stipules...........................270. RUBIACEAE, p. 870.
Ovary 1-celled; flowers in dense involucrate heads.................
.................................274. DIPSACACEAE, p. 892.
U. Stamens united by their anthers, these joined in a ring or tube.
Flowers separate, not involucrate; corolla irregular....................
..................................276A. LOBELIACEAE, p. 895.
Flowers in an involucrate head.................280. COMPOSITAE, p. 899.

PTERIDÓPHYTA. FERNS and FERN ALLIES

Note: Ferns and their allies have always been an attractive subject of study and many persons have made intensive studies of them and have designated many of the minute differences by special names. No attempt has been made here to evaluate the status of these variations and the common interpretation of them has been accepted.

In this treatment the term frond is used to mean the expanded portion of the leaf of a fern.

[Students who wish to use the stipe to assist in the determination of the ferns are referred to "An analytical key for the ferns of the Northeastern States, based on the stipes," by C. E. Waters, published in 1903 and republished as a supplement to the American Fern Journal, vol. **18:** no. 2. 1928.]

1. OPHIOGLOSSÀCEAE Presl ADDER'S TONGUE FAMILY*

Sporangia cohering in a simple spike; fronds (leaves) one, rarely 2 or 3, entire; veins reticulate..1. OPHIOGLOSSUM, p. 37.
Sporangia in pinnate or compound spikes, rarely in a simple spike but not cohering; fronds (leaves) not simple; veins free....................2. BOTRYCHIUM, p. 38.

1. OPHIOGLÓSSUM [Tourn.] L. ADDER'S TONGUE

Fronds mostly rounded or obtuse at the apex, rarely acute but never apiculate...1. *O. vulgatum.*
Fronds more or less acute at the apex and apiculate...............2. *O. Engelmanni.*

1. **Ophioglossum vulgàtum** L.** COMMON ADDER'S TONGUE. Map 1. Local in various habitats in the southern half of the state. It is always found in dense shade and most commonly associated with beech, especially in low beech and sweet gum woods. Ordinarily it seems to prefer a slightly acid soil. It has been found in Lake County by several collectors, where it is evidently rather frequent. I have a specimen collected by Edwin D. Hull near Liverpool, Lake County, which was growing under some shrubs in almost pure sand with cranberry. Mr. Hull found more than 30 fruiting specimens at this time at the place mentioned above. Besides the counties shown on the map it has been reported from Crawford, Harrison, and Wayne Counties.

Markle (Proc. Indiana Acad. Sci. **1915:** 357. 1916) in 1914 found near Gary, Lake County, many plants with more than one leaf. He reports "of a total of two hundred plants, selected at random, ninety-one had one leaf above ground, one hundred and five had two leaves, and four had three leaves".

1a. **Ophioglossum vulgatum** f. **pseudópodum** Blake. (Rhodora **15:** 87. 1913.) This is a form in which the sterile blade is narrowed below into a stalklike base a fourth to two thirds as long as the expanded portion. This form has been found in St. Joseph County by R. M. Tryon, Jr.

P. E. I., Ont. to Alaska, southw. to Fla. and Mex.; also in Eurasia.

---

* R. T. Clausen checked the determination of all my specimens and rendered valuable help.

** For a discussion of this species and varieties see Rhodora 41:494-499. 1939.

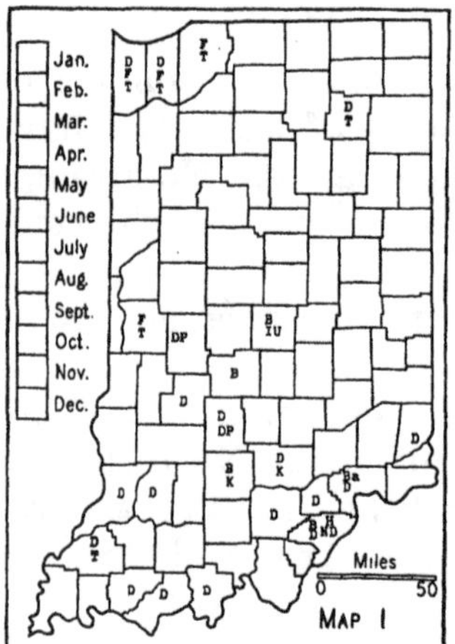

MAP 1
Ophioglossum vulgatum L.

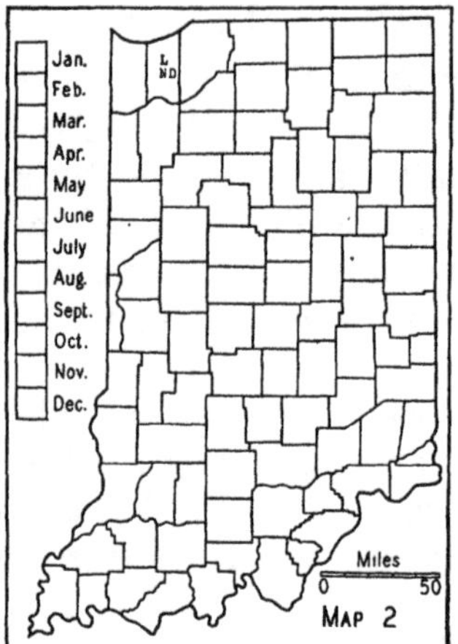

MAP 2
Botrychium simplex E. Hitchcock

MAP 3
Botrychium multifidum var. silaifolium (Presl) Brown

2. **Ophioglossum Engelmánni** Prantl. There is a fragmentary specimen in the herbarium of the New York Botanical Garden which R. T. Clausen has seen and reported in the Mem. Torrey Club **19**: no. 2:140. 1938 as belonging to this species. Clausen in a letter to me dated June 1, 1938, confirms his examination of the specimen and determination. The specimen was collected by L. M. Underwood in June, 1893, on the campus of Indiana University.

Nw. Va., s. Ohio and Ill. to Mo., southw. to cent. Fla., La., Tex., and Ariz.; cent. and s. Mex.

2. BOTRÝCHIUM Sw. GRAPEFERN

Fronds small, mostly 1-3 cm long, simple and roundish or pinnately 3-7-lobed....... ....1. *B. simplex.*
Fronds larger, more than 3 cm long, ternate.
  Fronds on long petioles (arising from near the base of the stem), bipinnate-pinnatifid.
    Sterile frond with all the segments of about the same size and shape; segments ovate or obovate, the terminal ones not elongate.......... ....2. *B. multifidum* var. *silaifolium.*
    Sterile fronds with segments of different size and shape.
      Ultimate divisions of the frond cut into linear segments; segments more or less notched at the apex..........3. *B. dissectum.*
      Ultimate divisions of the frond not dissected but variously and unevenly cut.
        Divisions of the pinnae oblong-ovate to oblong-lanceolate, more or less acute.
          Segments of frond many more than 9......3a. *B. dissectum* var. *obliquum.*
          Segments usually about 9................3b. *B. dissectum* var. *tenuifolium.*
        Divisions of the pinnae broadly ovate and obtuse.......... ....3c. *B. dissectum* var. *oneidense.*
  Fronds sessile (arising from near or above the middle of the stem), the short-stalked primary divisions once or twice pinnate and these in turn once or twice pinnatifid..........4. *B. virginianum.*

1. **Botrychium símplex** E. Hitchcock. HITCHCOCK GRAPEFERN. Map 2. I have seen specimens from three collections. The first was collected in 1910 by W. N. Clute along the Michigan Central Railroad near Glen Park, Lake County. A second specimen was collected in 1929 by Marcus W. Lyon, Jr., on the wooded border of an interdunal flat in Porter County. R. T.

Clausen has seen this specimen and confirms the identification. The third was collected by J. A. Nieuwland at Dune Park, Porter County.

P. E. I. to Pa., westw. to Oreg. and Calif.

2. **Botrychium multífidum** (Gmel.) Rupr. var. **silaifòlium** (Presl) Broun. (*Botrychium ternatum* var. *intermedium* D. C. Eaton.) Map 3. This report is based upon specimens collected by Marcus Lyon, Jr., and R. M. Tryon, Jr., in the Dunes State Park, Porter County. Tryon reported his specimens as *Botrychium dissectum* f. *elongatum*. R. T. Clausen and E. T. Wherry have seen these specimens and refer them to this species.

Maine to Que., and B. C., southw. to N. J. and Oreg.

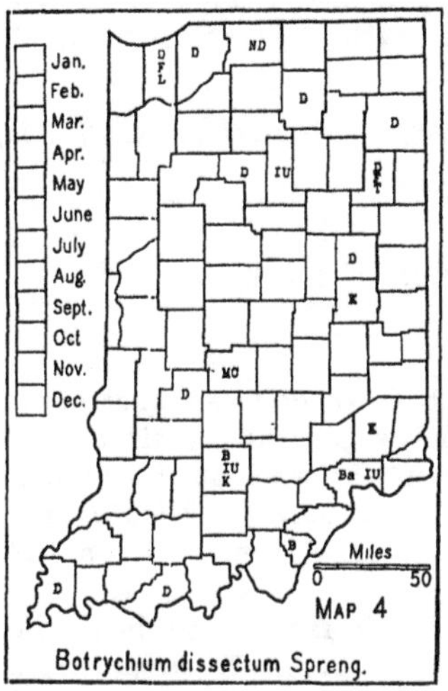

MAP 4

Botrychium dissectum Spreng.

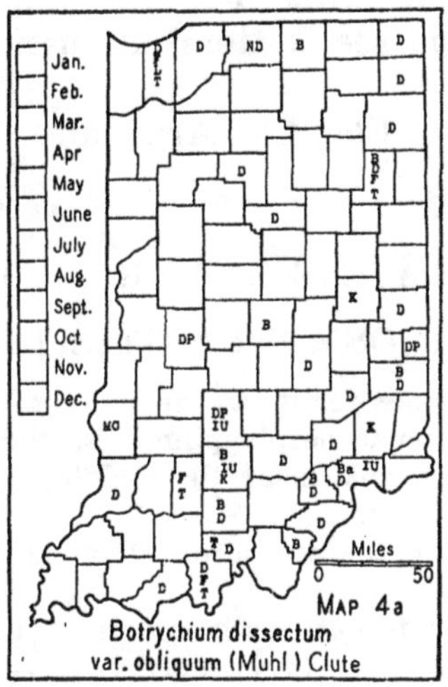

MAP 4a

Botrychium dissectum var. obliquum (Muhl.) Clute

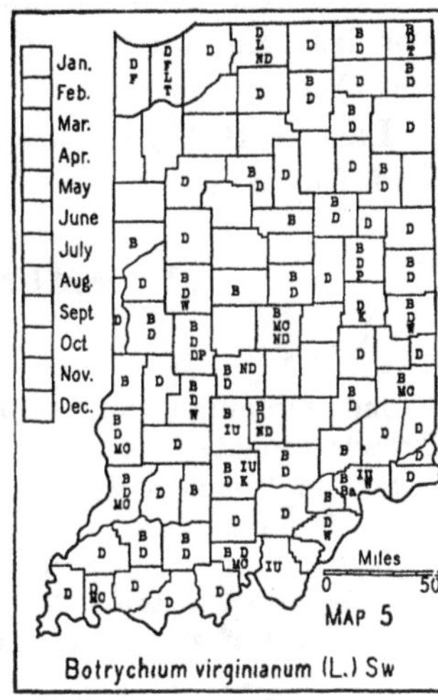

MAP 5

Botrychium virginianum (L.) Sw

3. **Botrychium disséctum** Spreng. (*Botrychium obliquum* var. *dissectum* (Spreng.) Clute.) CUTLEAF GRAPEFERN. Map 4. Local throughout the state in either dry or moist soils. All of my specimens are from woodland; some are from white oak woods, some are from beech and sugar maple woods, and one specimen was found associated with sweet gum and white elm.

N. B. and N. S. to Minn., southw. to Fla., Mo., Ark., and Mex.

3a. **Botrychium dissectum** var. **oblìquum** (Muhl.) Clute. (*Botrychium obliquum* Muhl.) OBLIQUE GRAPEFERN. Map 4a. Infrequent throughout the state in wet or dry woodland. Most of my specimens were found in low, flat woods associated with sweet gum and beech, and a few were found in dry woodland with beech and sugar maple.

A form with less divided and oblong pinnae has been described by E. W. Graves (Amer. Fern Jour. 22: 50-52. 1932) as *Botrychium obliquum* var. *oblongifolium*. Graves named one of my specimens from Marion County and one from Crawford County as belonging to this variety. Since fern students are not agreed upon the status of this fern, I record the data without comment.

N. B. to Minn., southw. to Fla., Mo., and Tex.

3b. **Botrychium dissectum** var. **tenuifòlium** (Underw.) Farw. I have a specimen of this variety collected in a low woods about 3 miles northwest of Leavenworth, Crawford County, which is referred to this variety by both R. T. Clausen and E. T. Wherry. This variety is found chiefly in the southern states.

3c. **Botrychium dissectum** var. **oneidénse** (Gilbert) Farw. According to Clausen's determination this variety occurs in De Kalb, Howard, Porter, and Steuben Counties.

4. **Botrychium virginiànum** (L.) Sw. RATTLESNAKE FERN. Map 5. This is strictly a woodland species and is found in moist, rich woods of many kinds throughout the state. For a treatment of the varieties of this species and a key to them see Butters' discussion (Rhodora **19:** 207-215. 1917).

Lab. to B. C., southw. to Fla., La., Ariz., and Wash.; also in Mex., W. I., and Eurasia.

2. OSMUNDÀCEAE R. Br. ROYAL FERN FAMILY

1. OSMÚNDA [Tourn.] L.

Fronds bipinnate, the fertile ones fertile at the summit...1. *O. regalis* var. *spectabilis.*
Fronds pinnate, the sterile pinnae deeply pinnatifid, the lobes generally entire.
 Fertile fronds with fertile pinnae near the middle; no tuft of wool at the base of the pinnae ....2. *O. Claytoniana.*
 Fertile fronds separate from the sterile ones; pinnae of sterile fronds with a tuft of wool in the axils.
  Pinnae of sterile fronds with entire segments and the fertile frond entirely fertile. ....3. *O. cinnamomea.*
  Pinnae with the basal segments on the lower side (or rarely on both sides) much elongated and deeply and sharply toothed, other segments normal or nearly so ....3a. *O. cinnamomea* f. *auriculata.*
  Pinnae (at least some of them) of fertile fronds more or less sterile (usually the lower ones) ....3b. *O. cinnamomea* f. *frondosa.*
  Pinnae (at least some of them) of sterile fronds with lobes more or less cut or pinnatifid ....3c. *O. cinnamomea* f. *incisa.*

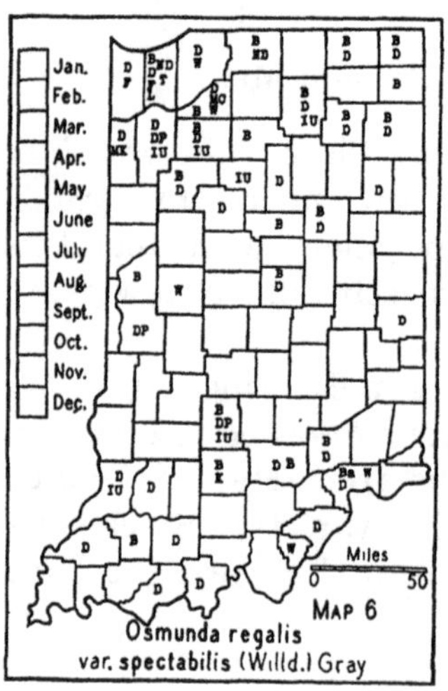

MAP 6
Osmunda regalis var. spectabilis (Willd.) Gray

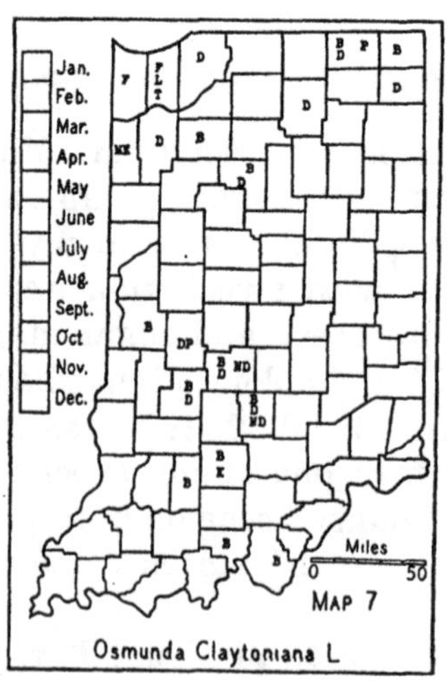

MAP 7
Osmunda Claytoniana L

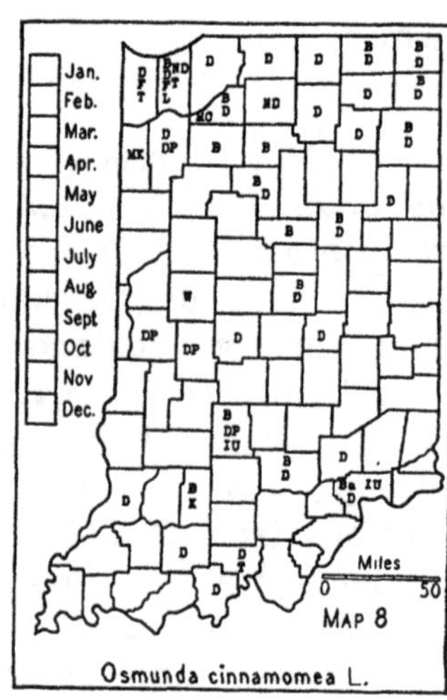

MAP 8
Osmunda cinnamomea L.

SUPPLEMENTARY KEY FOR THE SEPARATION OF STERILE FRONDS OF SOME SPECIES THAT SUPERFICIALLY LOOK MUCH ALIKE.

In my early study of ferns I was not aware that sterile fronds could be identified. Sterile specimens of *Osmunda* and *Pteretis* much resemble each other and I had never been able to find the last named genus until I was able to identify the sterile specimen. Since that time I have found several colonies and I think if all of our fern students knew how to separate these genera that many more colonies of *Pteretis* would be found. Likewise there is a possibility that sterile specimens of *Woodwardia virginica* and certain species of *Athyrium* and *Dryopteris* might be confused with *Osmunda*. Hence this key.

Veins simple, not forked; pinnules entire; vascular bundles in stipe 7. . . . .*Pteretis*.
Veins not simple, more or less forked.
  Veins usually forked once; vascular bundle in the stipe 1; stipe stramineous.
    Sterile fronds with tufts of wool at the base of the pinnae. .*Osmunda cinnamomea*.
    Sterile fronds without tufts of wool at the base of the pinnae. . . . . *Osmunda Claytoniana*.
  Veins with areolae on both sides of the midrib with which simple or rarely forked veins connect the margin; vascular bundles more than 5; stipe dark brown. . . . . *Woodwardia*.

1. **Osmunda regàlis** L. var. **spectábilis** (Willd.) Gray. (Rhodora **21**: 179. 1919.) (*Osmunda regalis* of Gray, Man., ed. 7 and Britton and Brown, Illus. Flora, ed. 2, not L.) ROYAL FERN. Map 6. Frequent in the lake area and infrequent to local south of it. It is not especially particular as to its habitat except that it must be a moist or wet one. It is found mostly in low woods, about ponds and lakes, and less frequently in the open in wet prairies.

Newf. to Sask., southw. to Fla. and Miss.

2. **Osmunda Claytoniàna** L. INTERRUPTED FERN. Map 7. Infrequent to local throughout the greater part of the state. Besides the counties shown on the map, there are reports from thirteen additional counties. It seems to prefer the moist bases of black and white oak slopes. In the southern part of the state it is found on the slopes of deep, wooded ravines.

Newf. to Minn., southw. to N. C., Ky., and Mo.; a variety also in s. Asia.

3. **Osmunda cinnamòmea** L. CINNAMON FERN. Map 8. Frequent in the lake area, becoming infrequent to local south of it. In the lake area it is usually common in tamarack bogs and swamps about lakes, and in the southern part of the state it grows in low, flat woods, associated with sweet gum and red maple. Throughout its range it is found only in wet soil in bogs or about ponds and marshes and rarely on shaded slopes.

Newf. to Minn., southw. to Fla., La., and N. Mex.; also in Mex., S. A., W. I., and Eurasia.

3a. **Osmunda cinnamomea** f. **auriculàta** (Hopkins) Kittredge. (Bull. Conn. State Geol. and Nat. Hist. Surv. **48**: 12. 1931.) This form has the basal segments much elongated and deeply and sharply toothed on the lower side or rarely on both sides. The other segments are normal or nearly so. It has been found in Porter County by R. M. Tryon, Jr.

3b. **Osmunda cinnamomea** f. **frondòsa** (T. & G.) Britt. (Cat. Plants of New Jersey, p. 312. 1890.) This form has the fertile frond partly leafy, the fertile and sterile pinnae variously intermixed. I found this unusual form in Lagrange County and Nieuwland found it in St. Joseph County.

3c. **Osmunda cinnamomea** f. **incìsa** (Huntington) Gilbert. (List North American Pteridophytes, pp. 13, 28. 1901.) This form usually has acutely toothed or lobed segments. I found it in De Kalb County.

3. POLYPODIÀCEAE R. Br. FERN FAMILY

Fronds conspicuously dimorphic, the fertile ones with divisions greatly contracted or berrylike, brown when fully mature.
  Sterile fronds pinnatifid, the veins netted; fertile fronds bipinnate, the divisions berrylike........4. ONOCLEA, p. 45.
  Sterile fronds bipinnatifid, the veins free; fertile fronds pinnate, the divisions linear, strongly ascending........3. PTERETIS, p. 44.
Fronds not conspicuously dimorphic, all green.
  Sori marginal, the indusium appearing to consist of the reflexed margin of the segments of the frond or of a marginal cup.
    Fronds with sporangia borne in minute cuplike indusia near the notches of the segments, the sori separate; fronds bipinnate, the lower surface well covered with short, erect, glandular hairs........7. DENNSTAEDTIA, p. 50.
    Fronds not as above.
      Stipes stout (2-4 mm in diameter), commonly solitary, green (stramineous or pale brown in dried specimens)........15. PTERIDUM, p. 57.
      Stipes less than 2 mm in diameter, commonly clustered (brown to blackish).
        Pinnules pubescent above and below with long, white hairs, densely so along the margins below........13. CHEILANTHES, p. 56.
        Pinnules glabrous or with a few scattered hairs.
          Indusia of pinnules continuous; fronds coriaceous, pinnate or bipinnate. ....12. PELLAEA, p. 55.
          Indusia definitely interrupted on the fanlike margin of the pinnule; fronds delicate, branched at the summit, the branches definitely pinnate. ....14. ADIANTUM, p. 57.
  Sori dorsal, not marginal (except in *Dryopteris marginalis*).
    Sori and indusia (when present) more or less circular, or reniform.
      Fronds pinnate, pinnules narrowly oblong-lanceolate with an auricle at the base of the upper margin, the stipe and rachis thickly covered with scales; pinnules of fertile fronds contracted; sori confluent..6. POLYSTICHUM, p. 50.
      Fronds not as above.
        Stipe, rachis, and lower surface of the pinnae more or less glandular-puberulent; stipe and rachis deciduously chaffy........1. WOODSIA, p. 43.
        Stipe, rachis, and lower surface of pinnae not, or not all, more or less glandular-puberulent.
          Fronds deeply pinnatifid, the divisions confluent at the base; sori naked; blades of fronds coriaceous........16. POLYPODIUM, p. 57.
          Fronds not as above.
            Indusia attached in the center or lacking, if lacking then the rachis pubescent and chaffy........5. DRYOPTERIS, p. 45.
            Indusia attached by a broad base on the side toward the midrib and partly under the sori, opening on the opposite side...2. CYSTOPTERIS, p. 43.
    Sori elongated, oblong to linear, often curved.
      Sori in rows parallel to the midribs of the pinnae and along the midveins of the segments........11. WOODWARDIA, p. 55.
      Sori not disposed as above.

Blades of fronds simple, long-attenuate at the apex, cordate at the base, entire or undulate..............................9. CAMPTOSORUS, p. 53.

Blades once to several times divided.

Sori straight or slightly curved; fronds mostly 10-40 cm long...10. ASPLENIUM, p. 53.

Sori often curved over the ends of the veins; fronds mostly 35-90 cm long...8. ATHYRIUM, p. 51.

1. WOÓDSIA R. Br. WOODSIA

1. **Woodsia obtùsa** (Spreng.) Torr. COMMON WOODSIA. Map 9. Infrequent to rare in the southern part of the state and very local northward to the counties shown on the map. Probably not found in Indiana north of the counties shown on the map. It no doubt occurs also in Wabash County but I have not been able to find it. It is usually found in shallow soil on rocky slopes. It prefers sandstone but is also found on limestone.

Cent. Maine to Wis., B. C., and Alaska, southw. to Ga., Ala., Tex. and Ariz.

2. CYSTÓPTERIS Bernh.

Fronds lanceolate, attenuate, often bulblet-bearing on the lower surface of the upper part; segments and teeth crowded; rachis not winged; pinnules mostly oblong, very obtuse; indusium truncate on the free side, minutely glandular...1. *C. bulbifera.*

Fronds ovate or oblong-lanceolate, acute, not bulblet-bearing; segments and teeth more distant, decurrent on the slightly margined rachis; pinnules mostly oval, more pointed; indusium acute or acuminate, and often lacerate on the free side, not glandular ....2. *C. fragilis.*

1. **Cystopteris bulbífera** (L.) Bernh. (*Filix bulbifera* (L.) Underw.) BERRY BLADDER FERN. Map 10. Infrequent in the southern part of the state, becoming very local to absent in the northern part. This species grows only in wet places or places that are usually constantly kept moist in shady, rocky ravines and in pockets or crevices of shaded cliffs. It is usually found along the outlets of springs in southern Indiana. My Steuben County specimen was found in an old tamarack bog.

Newf. to Man., southw. to Ga., Ala., Ark., and Iowa.

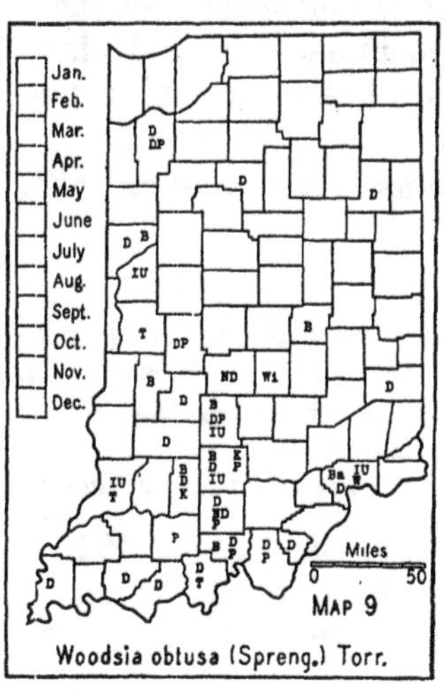

Woodsia obtusa (Spreng.) Torr.

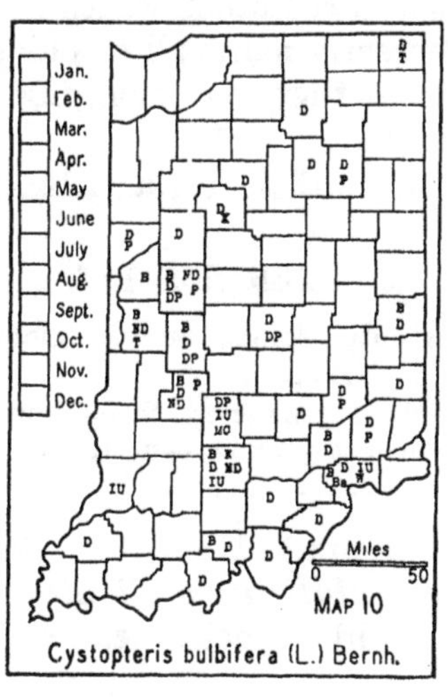

Cystopteris bulbifera (L.) Bernh.

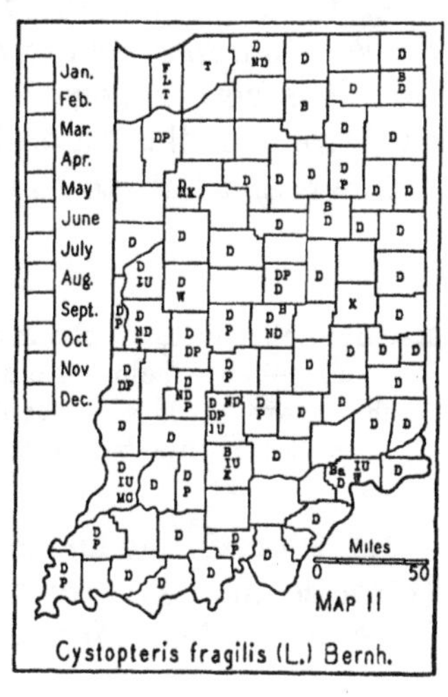

Cystopteris fragilis (L.) Bernh.

2. **Cystopteris frágilis** (L.) Bernh. (*Filix fragilis* (L.) Underw.) BRITTLE FERN. Map 11. This species prefers the deep, rich leaf mold of beech and sugar maple and white oak woods and is frequent to common throughout the state south of the Wabash River where woods of this kind are found. North of the Wabash River it is infrequent to very rare. It is absent in the southern part of the state in the areas where low, flat woods occur. It is found in exposed places on sandstone ridges and bluffs.

Students sometimes find difficulty in distinguishing this species from *Woodsia obtusa.* The stipe of the last named species is covered more or less densely with short, stipitate glands while the stipe of *Cystopteris* is entirely glabrous or with only a few glands near the summit.

The *Cystopteris fragilis* species complex has been restudied by C. A. Weatherby. He has recently described a new variety to which, in my opinion, all or most all of our specimens belong. It is described as follows:

"**Cystopteris fragilis** (L.) Bernh. var. **protrùsa** Weatherby. (Rhodora 37: 373-375. 1935.) Rootstock creeping, only sparsely beset with bases of old fronds, the growing point hardly paleaceous, produced 2-4 cm beyond the fronds of the season; well-developed blades nearly bipinnate-pinnatifid, 11-22 cm long, 5-11.5 cm wide, pinnae ovate to ovate-lanceolate, acute, pinnules toward the base of the pinnae deltoid-ovate to ovate-lanceolate, subacute, usually shortly but distinctly petiolulate, deeply pinnatifid into oblong, obtuse lobes; in juvenile or depauperate blades less lobed and more obtuse; indusium about 0.5 mm long, shallowly lobed or nearly entire at apex. Southern New York, south in the piedmont and the mountains to Alabama, west to Minnesota and Missouri."

My specimens have been examined by two fern specialists and they agree that most of them belong to this variety and some can not be determined with certainty.

A form of this species with large, abundant sori has been named f. **magnasora** Clute (Fern Bull. 9: 65. 1901).

The true species has a range to the north of Indiana. In order to refer specimens to their correct variety and form it is usually necessary for them to have the indusium and rootstock which most of our specimens lack. Since it is impossible to correctly name all of our specimens I have decided that it is best to regard all of them as belonging to a species complex and they are so indicated on the map.

Newf. and Lab. to Alaska, southw. to Ga., Ala., Kans., Ariz., and s. Calif.

3. PTERÈTIS Raf.

Rachis glabrous throughout its entire length or only glabrate above the lowest pinnae and polished below them..1. *P. nodulosa.*
Rachis more or less puberulent to pubescent throughout, at least above the lowest pinnae; rachis below the lowest pinnae usually not polished.................. ....1a. *P. nodulosa* f. *pubescens.*

1. **Pteretis nodulòsa** (Michx.) Nieuwl. (Rhodora 21: 178. 1919.) (*Onoclea Struthiopteris* and *Matteuccia Struthiopteris* of most authors.) OSTRICH FERN. Map 12. This species is, no doubt, very local in the

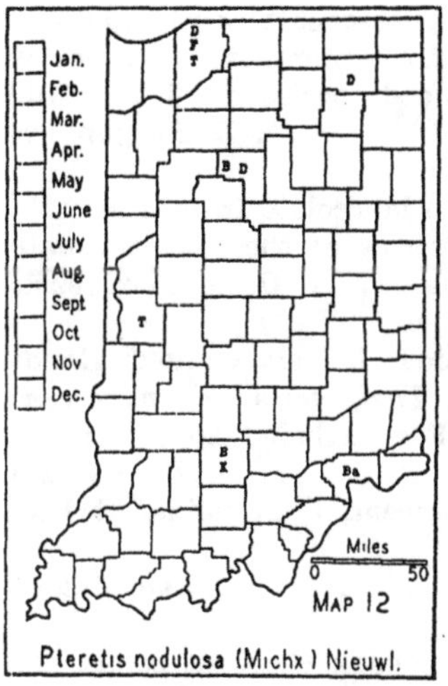

MAP 12
Pteretis nodulosa (Michx.) Nieuwl.

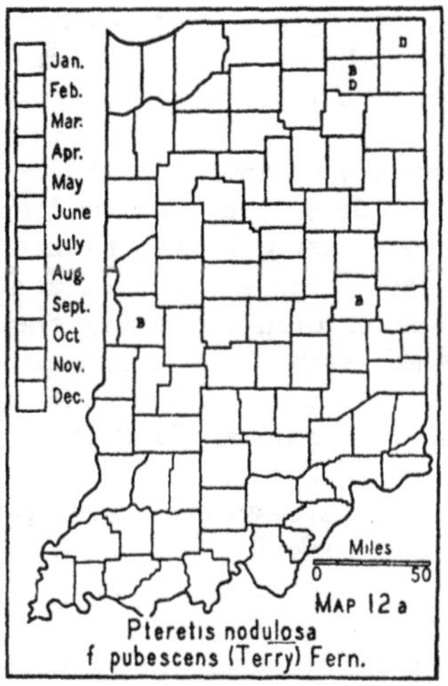

MAP 12a
Pteretis nodulosa f. pubescens (Terry) Fern.

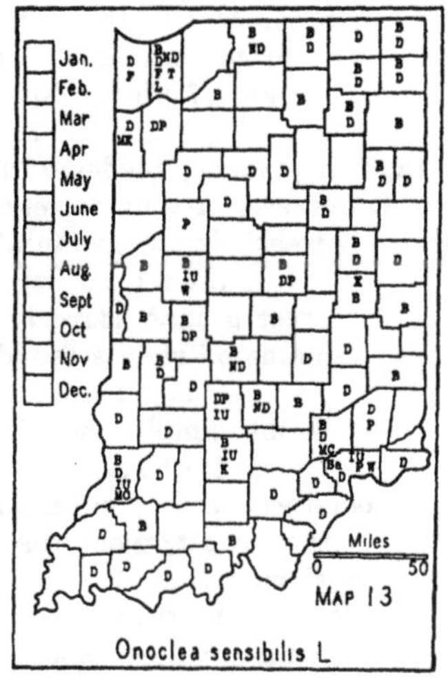

MAP 13
Onoclea sensibilis L.

state although it may have been overlooked because of its close resemblance to *Osmunda cinnamomea.* My specimens are mostly from alluvial flood plains of small streams.

Newf. to B. C., southw. to Va. and Iowa.

1a. **Pteretis nodulosa** f. **pubéscens** (Terry) Fern. (Rhodora 37: 219. 1935.) Map 12a. This form is not well marked in Indiana.

4. ONOCLÈA L.

1. **Onoclea sensíbilis** L. SENSITIVE FERN. Map 13. Frequent throughout the state in low places in woodland, about lakes, and along roadsides.

Forma **obtusilobàta** (Schkuhr) Gilbert is a form with fronds intermediate between the normal fertile and normal sterile phases, bipinnate or nearly so, the pinnules flat and nearly free-veined, rarely partly fertile. This form has been found in Porter County by R. M. Tryon, Jr., who says it is not infrequent in meadows that have been mowed in the early part of the year. There is a specimen from Porter County in the herbarium of the University of Notre Dame.

A form with the frond fertile, or somewhat so, on one side and sterile on the other is forma **hemiphyllòdes** (Kiss & Kümmerle) Weatherby (Amer. Fern Jour. **26**: 16. 1936). This form was found on the right of way of the Chicago, South Shore, and South Bend Railroad near Tremont, Porter County. The right of way was mowed earlier in the year.

Newf. to Sask., southw. to Fla. and Okla.

5. DRYÓPTERIS Adans.

Indusia lacking; blades of fronds triangular or ternate.
- Blades ternate with the divisions nearly equal and petiolate, glabrous; rachis wingless. (See excluded species no. 2, p. 1019.) .... *D. Linnaeana.*
- Blades bipinnatifid; pinnae sessile and more or less decurrent on the rachis.
  - Fronds as wide as or wider than long, usually light green, finely puberulent or glandular beneath; veins of the pinnules on the lowest pair of pinnae several times forked .... 1. *D. hexagonoptera.*

Fronds longer than wide, dark green, more coarsely pubescent beneath and with prominent brown scales along the rachis; veins of the pinnules on the lowest pair of pinnae simple or once forked. (See excluded species no. 3, p. 1019.) .... *D. Phegopteris.*

Indusia present; blades of fronds not triangular or ternate.

Rootstocks creeping; veins simple or once forked; fronds lanceolate in outline.

Lowest pinnae gradually decreasing in size toward the base; the lowest usually less than 1 cm long; veins simple; indusia glandular.....2. *D. noveboracensis.*

Lowest pinnae scarcely smaller than the middle ones.

Veins of sterile fronds generally forked; sori crowded; indusia without glands. ....3. *D. Thelypteris* var. *pubescens.*

Veins simple; sori distant; indusia glandular. (See excluded species no. 4, p. 1020.) ....*D. simulata.*

Rootstocks short, suberect; fronds cespitose, never pubescent, their veins, at least the lowest, more than once forked.

Sori marginal ....4. *D. marginalis.*

Sori not marginal.

Pinnae widest above the base; basal scales of stipe dark chestnut color; sori mostly 3-7 pairs; the largest fern of the genus (in Indiana)..5. *D. Goldiana.*

Pinnae widest at the base; basal scales of stipe not so dark colored as the preceding.

Surface of indusium glabrous.

Fronds bipinnatifid or pinnate.

Basal scales of stipe lance-linear, caudate-attenuate; segments with parallel sides, serrate at the rounded apex and obscurely so, if at all, on the sides, the teeth rarely somewhat spinulose; sori usually on the lower half of the segment. (See excluded species no. 5, p. 1020.) ....*D. Filix-mas.*

Basal scales of stipe wider; teeth of segments more or less spinulose; sori not restricted to the lower half of the segment.

Fronds linear-oblong or lanceolate in outline; pinnae 5-8 cm long, triangular-oblong or the lowest pair somewhat triangular-ovate, usually the lower half of the frond conspicuously decreasing in size toward the base.............................6. *D. cristata.*

Fronds wider; pinnae 8-15 cm long, oblong-lanceolate, the lower half of the frond not decreasing in size toward the base............. ....6a. *D. cristata* var. *Clintoniana.*

Fronds bipinnate, tripinnate, or tripinnatifid, segments with spinulose teeth.

Basal inferior and superior pinnules of the lowermost pinnae subopposite, rarely more than 4 mm apart; the inferior 1-6 cm long, if more, then twice as long as the superior; pinnules of the middle pinnae often only toothed; pinnules pinnatifid or pinnate.........7. *D. spinulosa.*

Basal inferior and superior pinnules of the lowest pinnae remote, 0.5-2 cm wider apart; the inferior 3-10 cm long, 2-4 times as long as the superior; pinnules pinnatifid or pinnate. (See excluded species no. 7, p. 1020.) ....*D. spinulosa* var. *americana.*

Surface of indusium glandular.

Frond commonly minutely glandular especially on the rachis and rachillae, tripinnatifid or sometimes tripinnate; pinnae slightly ascending to divergent, the basal inferior pinnule shorter than to rarely exceeding the second inferior one; scales of stipe usually dark brown at base.

Mature indusium 0.8-1.4 mm wide; pinnae gradually tapering to apex. ....7a. *D. spinulosa* var. *fructuosa.*

Mature indusium 0.5-0.8 mm wide; pinnae usually narrowed rather abruptly to prolonged lance-linear tips........................ ....7b. *D. spinulosa* var. *intermedia.*

Frond not minutely glandular but more or less chaffy, bipinnate or tripinnatifid ....8. *D. Boottii.*

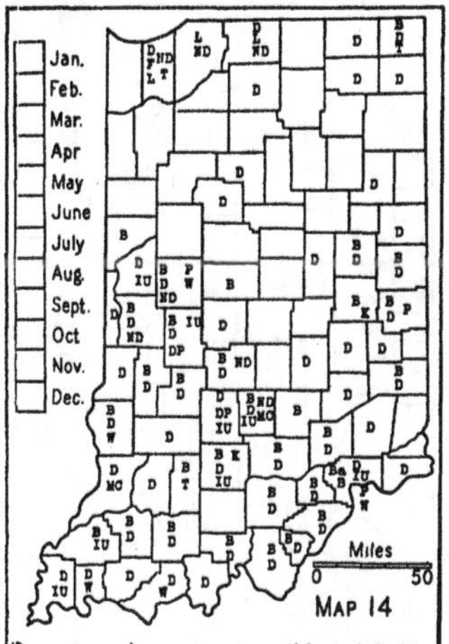

Dryopteris hexagonoptera (Michx.) C. Chr.

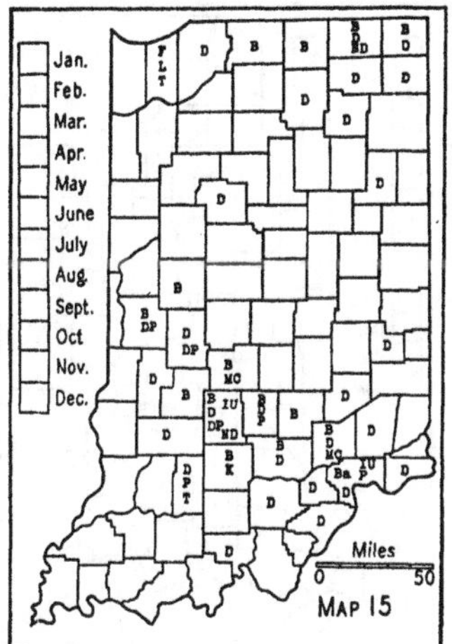

Dryopteris noveboracensis (L.) A. Gray

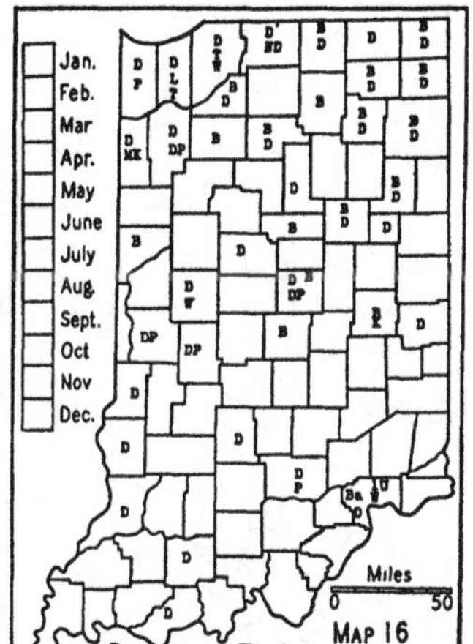

Dryopteris Thelypteris var pubescens (Lawson) A. R. Prince

1. **Dryopteris hexagonóptera** (Michx.) C. Chr. (*Phegopteris hexagonoptera* (Michx.) Fée.) WINGED WOODFERN. BROAD BEECHFERN. Map 14. Frequent in the southern half of the state, becoming less frequent northward and even rare in some of our northern counties. This is a woodland species found in all kinds of dry soils. It is found more frequently associated with black and white oak and only occasionally with beech and sugar maple.

Cent. Maine to w. Que. and Minn., southw. to Fla., La., Iowa, and Okla.

2. **Dryopteris noveboracénsis** (L.) A. Gray. (*Aspidium noveboracense* (L.) Sw.) NEW YORK FERN. Map 15. This species is found only in slightly acid soil, hence its zonal distribution. It is infrequent to local in the northern part of the state where it usually occurs in black and white oak woods. It is rare or absent in the Tipton Till Plain, becoming infrequent to frequent southward in the hard, white clay soil of beech and sweet gum woodland. In the southern part of the state is is usually closely associated with beech.

Newf. to Ont. and Minn., southw. to Ga., Ala., and Ark.

3. **Dryopteris Thelýpteris** (L.) A. Gray var. **pubéscens** (Lawson) A. R. Prince. (*Aspidium Thelypteris* of Gray, Man., ed. 7, not Sw.; *Dryopteris Thelypteris* of Britton and Brown, Illus. Flora, ed. 2, not A. Gray; and *Thelypteris palustris* of authors, not Schott.) MARSHFERN. Map 16. Common in the lake area and infrequent south of it. In the lake area it is common in tamarack bogs, sedge marshes, and on the low borders of lakes. South of this place it is found in springy and marshy areas.

Se. Newf., Que. to Man., southw. to Ga., Tenn., and Okla.

4. **Dryopteris marginàlis** (L.) A. Gray. (*Aspidium marginale* (L.) Sw.) LEATHER WOODFERN. Map 17. This species is, for the most part, restricted to the outcrops of sandstone or nearby residual soils which are the product of sandstone in the southern part of the state. Most of my

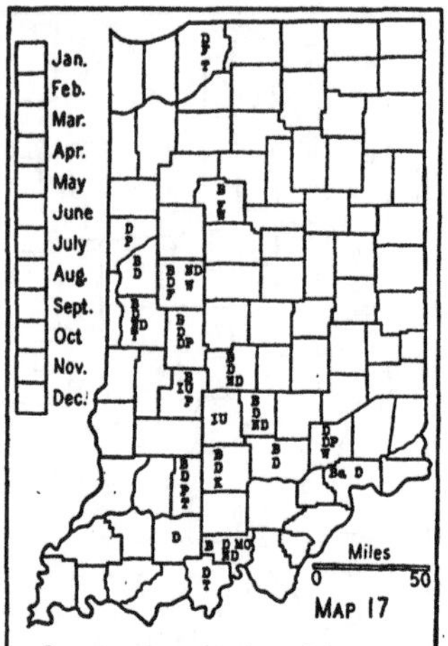

Dryopteris marginalis (L.) A. Gray.

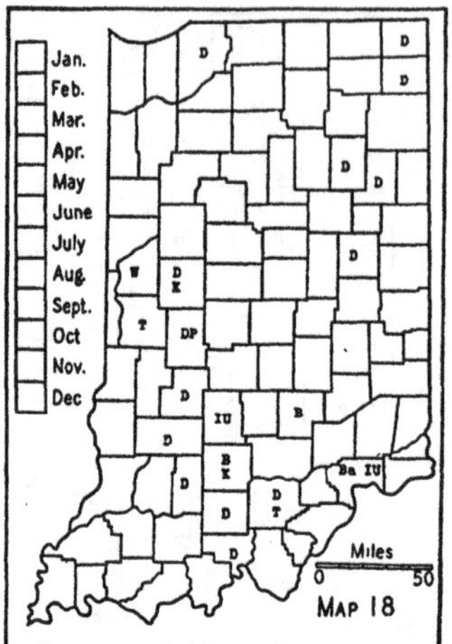

Dryopteris Goldiana (Hook.) A. Gray.

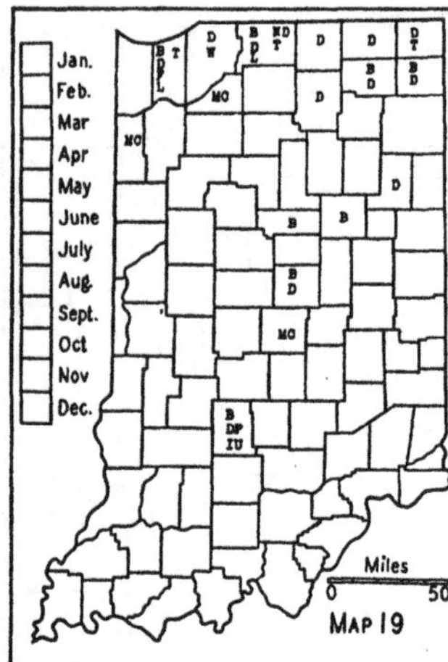

Dryopteris cristata (L.) A. Gray.

specimens are from wooded bluffs and slopes along streams. In addition to my collections, it has been reported from Clark, Floyd, Monroe, and Vigo Counties. It has been reported also from the dune area, and on May 30, 1935, R. M. Tryon, Jr. showed me large colonies of it on a north, wooded slope in Memorial Park about a mile east of Michigan City. It is, without question, a native here.

A form in which the pinnae are toothed or lobed has been named and has been reported from Indiana. I have a few specimens with some of the pinnae toothed but I do not think it is worth while to name such minor fluctuations.

N. S. to B. C., southw. to Ga., Ala., Ark., Kans., and Okla.

5. **Dryopteris Goldiàna** (Hook.) A. Gray. (*Aspidium Goldianum* Hook.) GOLDIE FERN. Map 18. Infrequent to rare throughout the state in deep humus, usually on the slopes of wooded ravines.

Cent. Maine to Minn., southw. to N. C., Tenn., and Iowa.

×**Dryopteris Goldiana × marginalis** Dowell. This hybrid was found in Martin County by R. M. Tryon, Jr. (Amer. Fern Jour. 28: 74. 1938.)

5. **Dryopteris Goldiàna** (Hook.) A. Gray. (*Aspidium Goldianum* CRESTED WOODFERN. Map 19. This species is restricted nearly to the lake area where it is frequent in tamarack bogs and in low woods, usually in masses of decaying organic matter. There are, however, reports of it from Grant, Howard, and Monroe Counties.

Newf. to Sask., southw. to N. C.

6a. **Dryopteris cristata** var. **Clintoniàna** (D. C. Eaton) Underw. (*Aspidium cristatum* var. *Clintonianum* D. C. Eaton and *Dryopteris Clintoniana* (D. C. Eaton) Dowell.) CLINTON WOODFERN. My only specimens of this fern are my no. 47776 from La Porte County and one collected by Tryon in Porter County.

N. H. to Wis., southw. to N. C.

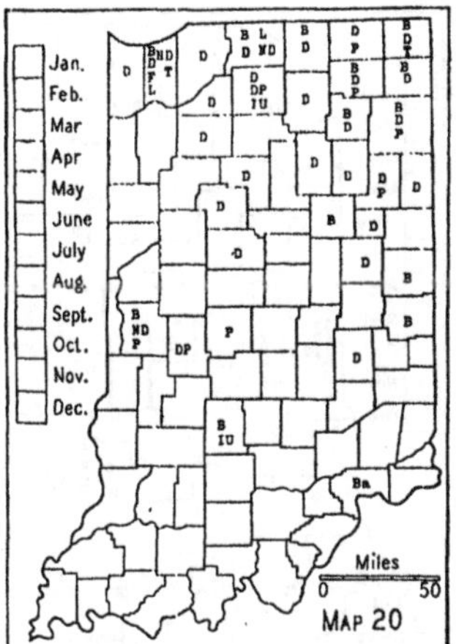

MAP 20
Dryopteris spinulosa (O.F.Muell.) Watt

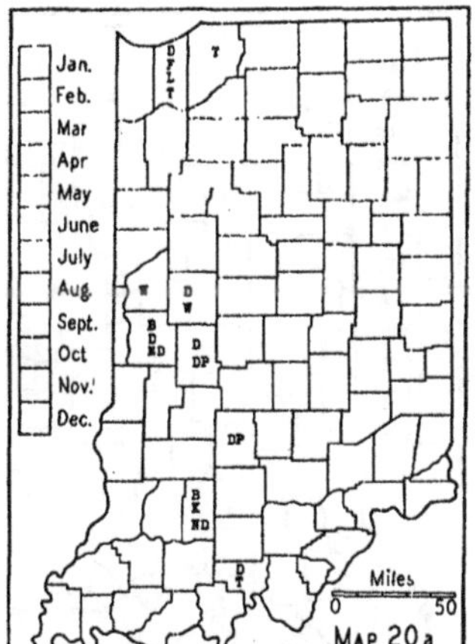

MAP 20a
Dryopteris spinulosa var. intermedia (Muhl.) Underw.

MAP 21
Dryopteris Boottii (Tuckerm.) Underw.

×**Dryopteris cristata** × **spinulòsa** C. Chr. is a closely allied form which is represented in my collection by a single specimen. It is my no. 54091 from Lagrange County, which was determined by C. A. Weatherby.

7. **Dryopteris spinulòsa** (O. F. Muell.) Watt. (*Aspidium spinulosum* (O. F. Muell.) Sw.) (Amer. Fern Jour. **26**: 65-69. 1936.) TOOTHED WOODFERN. Map 20. The greater number of specimens are from the lake area where it is usually frequent in wet woods, especially about ponds, in tamarack bogs, and on the wet, wooded borders of lakes. Sometimes it is found in dry woods after the water level has been lowered. This is one of our commonest and most attractive ferns. It usually grows in clusters of from 5 to 10 fronds.

Lab. to the Selkirks and Idaho, southw. to Va. and Ky.

7a. **Dryopteris spinulosa** var. **fructuòsa** (Gilbert) Trudell. (Rhodora 28: 146. 1926.) My specimens are from tamarack bogs and very low woods. I have no data concerning its general distribution.

7b. **Dryopteris spinulosa** var. **intermèdia** (Muhl.) Underw. (Rhodora 21: 178. 1919 and Rhodora 22: 196. 1920.) (*Aspidium spinulosum* var. *intermedium* (Muhl.) D. C. Eaton and *Dryopteris intermedia* (Muhl.) Gray.) COMMON WOODFERN. Map 20a. I have only a few specimens of this fern although it has been reported from 10 counties not shown on the map. It has a wide distribution in the state and seems to favor wooded ravines.

Newf. to Wis., southw. to N. C. and Mo.

8. ×**Dryopteris Boòttii** (Tuckerm.) Underw. (*Aspidium Boottii* Tuckerm.) BOOTT WOODFERN. Map 21. I reported this fern from Noble and Wells Counties but I now refer my specimens to other species. R. M. Tryon, Jr. has found it in La Porte and Porter Counties. His determinations have been checked by fern specialists. This species is regarded by some fern

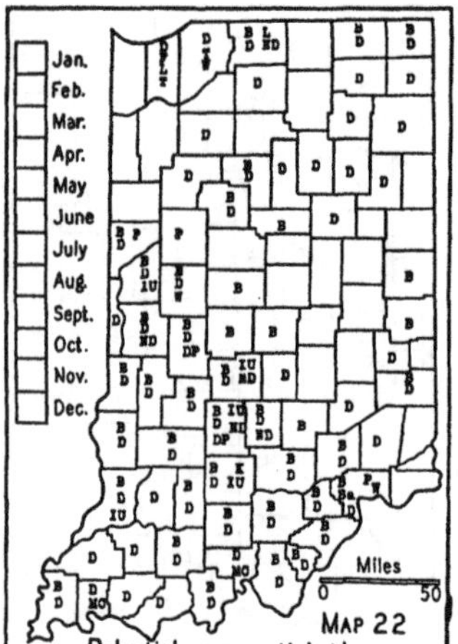

MAP 22
Polystichum acrostichoides (Michx.) Schott

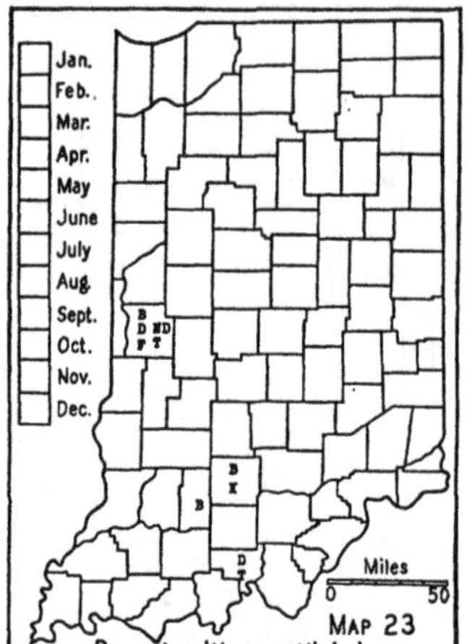

MAP 23
Dennstaedtia punctilobula (Michx.) Moore

MAP 24
Athyrium pycnocarpon (Spreng) Tidestr.

students as a hybrid between *Dryopteris cristata* and *Dryopteris spinulosa* var. *intermedia.*

N. S. to Minn., southw. to Va.

6. POLÝSTICHUM Roth

1. **Polystichum acrostichoides** (Michx.) Schott. CHRISTMAS FERN. Map 22. This is a woodland species preferring the lower part of the slopes of deep wooded ravines. It is infrequent to rare in some of the northern counties, becoming frequent to common in the southern half of the state, especially among the hills. In protected places in the southern part of the state it is evergreen.

N. S. to Ont. and Wis.; southw. to Ga. and Tex.

1a. **Polystichum acrostichoides f. incisum** (Gray) Gilbert. (*Polystichum acrostichoides* var. *Schweinitzii* (Beck) Small.) I have a specimen of this form from Daviess County. It has, however, been reported from several other counties throughout the state.

1b. **Polystichum acrostichoides f. críspum** Clute. This is a form with the margins of the pinnae crisped and ruffled. It has been found by R. M. Tryon, Jr. in Porter County.

7. DENNSTAÈDTIA Bernh.

1. **Dennstaedtia punctilóbula** (Michx.) Moore. (*Dicksonia punctilobula* (Michx.) Gray.) HAY-SCENTED FERN. Map 23. This fern seems to be rare in the state. It prefers the sandstone and shaly rocks of deep, wooded ravines. Williamson, in "Ferns of Kentucky," says it was found along Silver Creek north of Louisville, Kentucky. It was rather common in a rocky ravine in Turkey Run State Park. Outside of Indiana in suitable habitats it often becomes an annoying weed in pastures.

N. S. to Minn., southw. to Ga. and Mo.

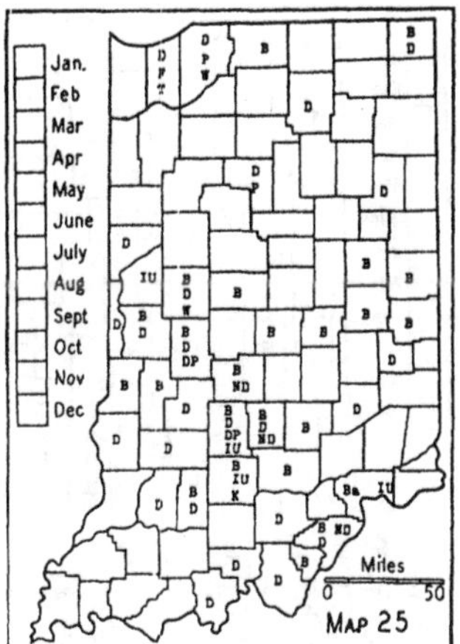

Athyrium thelypteroides (Michx.) Desv.

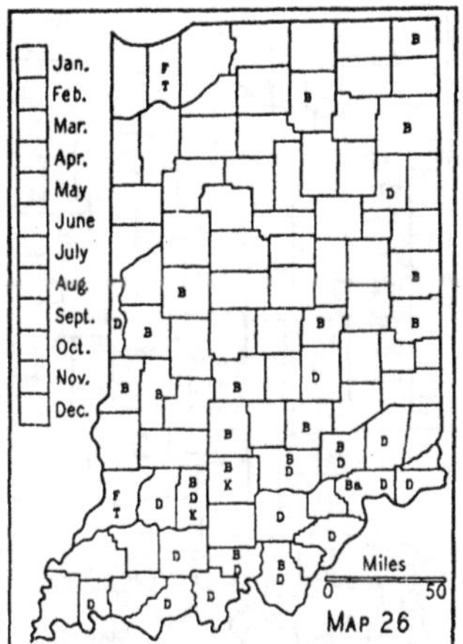

Athyrium asplenioides (Michx.) Desv.

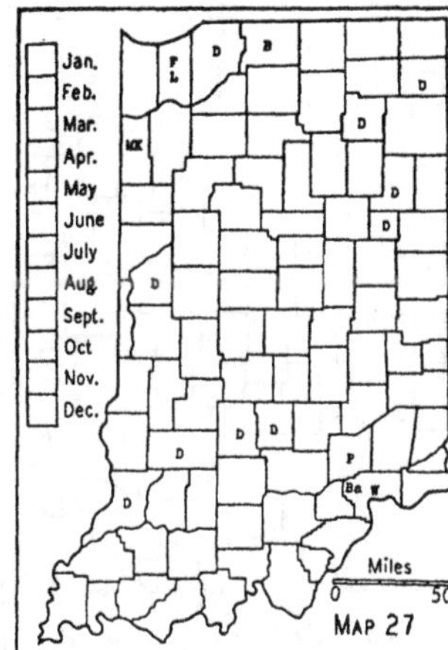

Athyrium angustum (Willd.) Presl

8. **ATHÝRIUM** Roth

[Butters. The genus Athyrium and the North American ferns allied to Athyrium Filix-femina. Rhodora **19**: 170-197. 1917. Pinkerton. Ferns of Missouri. Ann. Missouri Bot. Gard. **20**: 54-57. 1933.]

Fronds pinnate.
  Pinnae entire..1. *A. pycnocarpon.*
  Pinnae deeply pinnatifid..............................2. *A. acrostichoides.*
Fronds bipinnate.
  Rhizomes creeping, not densely covered with persistent bases of old fronds; stipe usually about as long as the blade; scales of stipes very few, rarely persistent, yellowish brown or tawny; blades widest near the base; young indusia with glandular cilia; spores somewhat nigrescent, wrinkled........3. *A. asplenioides.*
  Rhizomes horizontal, completely concealed by thick, fleshy bases of old fronds; stipe about half as long as the blade; scales of stipes varying from Mars Brown (Ridgway Standard) to nearly black; blades widest near the middle, the lower pinnae shorter and often deflexed; indusia toothed or short-ciliate, never glandular; spores yellow, smooth or slightly papillate.
    Sori confluent at maturity and usually covering the lower side of the fertile pinnules; fertile frond contracted.
      Longest pinnae of fertile frond 5-12 cm long; pinnules 4-12 mm long; pinnules of sterile fronds oblong, obtuse, slightly toothed or lobed...4. *A. angustum.*
      Longest pinnae of fertile frond 1-2 dm long; pinnules 12-25 mm long, pinnatifid; sori several on each of the lower segments, often horseshoe-shaped; pinnules of sterile fronds oblong-lanceolate, strongly toothed or pinnatifid, somewhat acute..............................4a. *A. angustum* var. *elatius.*
    Sori usually separate and distinct at maturity; fertile fronds not contracted; pinnules lanceolate, subacute, strongly toothed or pinnatifid, the segments toothed..................................4b. *A. angustum* var. *rubellum.*

1. **Athyrium pycnocárpon** (Spreng.) Tidestr. (*Asplenium angustifolium* Michx. and *Asplenium pycnocarpon* Spreng.) NARROWLEAF SPLEENWORT. Map 24. Infrequent to frequent in southern Indiana, becoming less frequent to rare northward. It prefers deep humus and is most commonly found on the slopes of ravines in beech woods.

W. Que to Wis., southw. to Ga., Ala., Mo., and Kans.

MAP 27a
Athyrium angustum var. elatius (Link) Butters

MAP 27b
Athyrium angustum var. rubellum (Gilbert) Butters

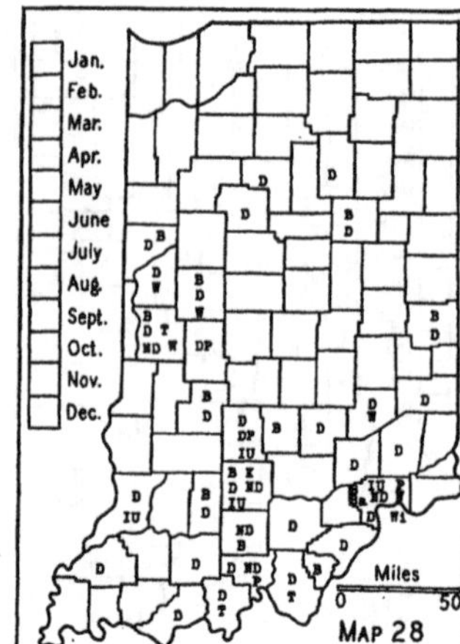

MAP 28
Camptosorus rhizophyllus (L.) Link.

2. **Athyrium thelypteroides** (Michx.) Desv. (*Asplenium acrostichoides* Sw. and *Athyrium acrostichoides* (Sw.) Diels.) SILVERY SPLEENWORT. Map 25. Infrequent in southern Indiana, becoming rare in the northern part. It prefers a moist, deep humus soil in ravines and protected places in beech and sugar maple or white oak woods.

N. S. to Minn., southw. to Ga., Ala., and Mo.; also in Asia.

3. **Athyrium asplenioides** (Michx.) Desv. Map 26. This species and the next species and its varieties are the results of dividing an aggregate that formerly had been designated as *A. Filix-femina.* For a detailed study of this group see Butters' "Synoptical treatment of the Lady Ferns of Eastern North America" (Rhodora **19**: 188-197. 1917). Butters has gone into great detail in his study of the species and discusses "sun" and "shade" forms. Some recent authors are disposed to regard some of the forms as merely ecological variations. See Wiegand's comment on varieties of the next species in "The Flora of the Cayuga Lake Basin," page 32. 1926. Pinkerton in "Ferns of Missouri" (Ann. Missouri Bot. Gard. **20**: 55. 1933) says: "This species and *A. angustum* are very difficult to distinguish. It is often necessary to have the whole plant, fruiting and not too mature, to be absolutely certain. I have taken the character of the spore as my ultimate criterion."

I can not satisfactorily separate the species and their varieties and would not publish on them were it not that C. A. Weatherby has named nearly every one of my specimens. I hereby wish to express my appreciation of the difficult task of naming so many of my specimens of this complex.

Infrequent in the southern counties but frequent in its habitat. It prefers a hard, white, moist, clay soil and is usually found in low, flat woods associated with beech and sweet gum or sweet gum and pin oak. It is also found in residual soil at the base of sandstone cliffs and in sandstone soil on wooded slopes.

Mass., Ohio to Mo., southw. to Fla. and Tex.

4. **Athyrium angústum** (Willd.) Presl. (Rhodora **19**: 190-197. 1917.) (*Asplenium Filix-femina* of most authors.) Map 27. Infrequent in moist, rich woods throughout the state.

Lab. to Man., southw. to s. N. E., the mts. of Pa., and Mo.

4a. **Athyrium angustum** var. **elàtius** (Link) Butters. Map 27a. This variety is infrequent throughout the state and found in rich beech and sugar maple and white and black oak woods.

Maine to Minn., southw. to R. I., N. Y., and Mo.

4b. **Athyrium angustum** var. **rubéllum** (Gilbert) Butters. Map 27b. This variety is infrequent throughout the state. The habitats of my specimens are notable because of lack of uniformity. I have one specimen from a tamarack bog and others from low, flat woods in hard, white clay soil, dry black and white oak woods, bluffs of the Ohio River, and rich, moist woods.

Newf. to Que., Ont., Minn., southw. to N. Y., Pa., Ohio, and Mo.

9. CAMPTOSÒRUS Link

1. **Camptosorus rhizophýllus** (L.) Link. WALKING FERN. Map 28. Infrequent in the southern part of the state, becoming rare to absent in the northern part. It grows in the shade in shallow soil on calcareous rocks on rocky ledges, usually along streams. It is not usually abundant unless it is found in deep shade and on rocks with considerable moisture.

Cent. Maine to Ont. and Minn., southw. to Ga. and Kans.

1a. **Camptosorus rhizophyllus** f. **auriculàtus** Clute. (Amer. Bot. 35: 102. 1929.) This is a named form infrequently found with the species in this state. It has the basal lobes of the leaves prolonged into slender tips.

10. ASPLÈNIUM L. SPLEENWORT

Frond long-attenuate at the apex.
- Stipe greenish............1. *A. pinnatifidum.*
- Stipe black and polished............3. *A. ebenoides.*

Frond not long-attenuate at the apex.
- Frond pinnate; stipe and rachis polished, dark reddish brown.
  - Pinnules sessile, oblong or oblong-linear, mostly 10-30 mm long, and distinctly auricled on the upper margin at the base............2. *A. platyneuron.*
  - Pinnules subsessile, roundish-oblong or oval, 3-7 mm long, not auricled on the upper margin............4. *A. Trichomanes.*
- Frond laxly 2-3-pinnate, ultimate segments long-cuneate at the base and finely toothed at the apex; stipe and rachis green............5. *A. cryptolepis.*

1. **Asplenium pinnatífidum** Nutt. PINNATIFID SPLEENWORT. Map 29. Rare in pockets of dry soil on cliffs in the area of sandstone outcrops. Usually closely associated with *Asplenium Trichomanes* but less frequent.

Se. Pa., Ohio, and Ind. to Mo., southw. to Ga.

2. **Asplenium platyneùron** (L.) Oakes. EBONY SPLEENWORT. Map 30. Infrequent in the southern half of the state where it is restricted to the unglaciated and sandstone areas. It probably reaches its greatest size on shady slopes of some of the loess banks of the southwestern counties. In

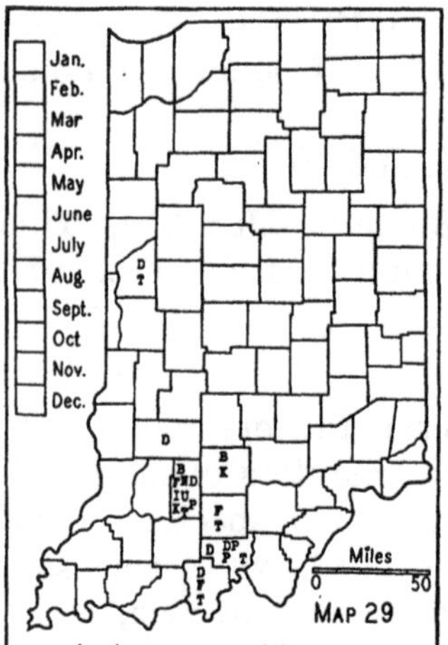

Asplenium pinnatifidum Nutt.

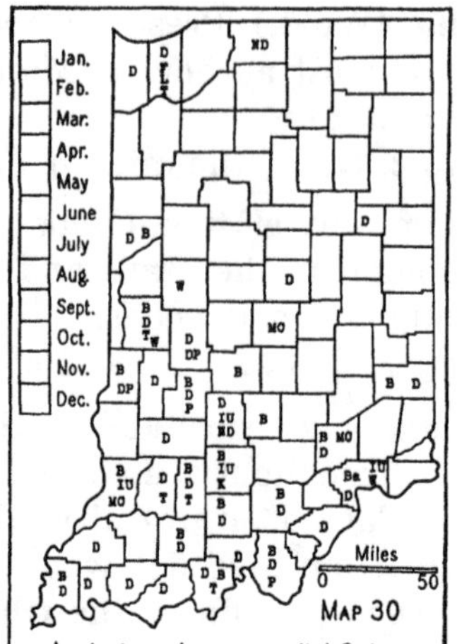

Asplenium platyneuron (L.) Oakes.

Asplenosorus ebenoides (Scott) Wherry

the northern half of the state it is either absent or restricted again to the soils of sandstone outcrops and to the sand areas about Lake Michigan where it is only local. I have never seen it except in slightly acid soil, and when transplanted into an alkaline environment, even with great care and in a half bushel of the soil in which it grew, it gradually disappeared in a few years.

S. Maine to Ont., and Colo., southw. to the Gulf States and Tex.

2a. **Asplenium platyneuron** f. **serràtum** (E. S. Miller) Hoffm. This is a form with some of the pinnae more or less deeply and irregularly serrate. I think this is merely a nutritional form. A fine example of this form was found in Perry County by R. M. Tryon, Jr.

3. × **Asplenosòrus*** **ebenoìdes** (Scott) Wherry. (*Asplenium ebenoides* R. R. Scott.) SCOTT SPLEENWORT. Map 31. This fern is a hybrid between *Asplenium platyneuron* and *Camptosorus rhizophyllus*. (Slosson. Bull. Torrey Bot. Club. **29**: 487-495. 1902.) Three colonies of this hybrid were discovered in Lawrence County by Ralph M. Kriebel who fully described them in Amer. Fern Jour. **23**: 52-59. 1933. Mr. Kriebel is one of the best amateur botanists Indiana has ever had, and it is to his discriminating collecting that we owe not only an authentic Indiana record of this hybrid fern but also the records of three hybrid oaks and many other rare plants of Lawrence County.

Vt. to Mo. and southw.

4. **Asplenium Trichómanes** L. MAIDENHAIR SPLEENWORT. Map 32. Infrequent to rare in pockets of soil on cliffs in the area of sandstone outcrops of the state.

Nearly throughout N. A. except in the extreme northern part and in Mex.; also in Eurasia.

5. **Asplenium cryptólepis** Fern. (Rhodora **30**: 41-43. 1928.) (*Asplenium Ruta-muraria* of Gray, Man., ed. 7 and Britton and Brown, Illus.

* Amer. Fern Jour. **27**: 56. 1937.

MAP 32
Asplenium Trichomanes L.

MAP 33
Asplenium cryptolepis Fern.

MAP 34
Woodwardia virginica (L.) Sm.

Flora, ed. 2, not L.) AMERICAN WALL-RUE SPLEENWORT. Map 33. My only specimens are from the rocks of the bluff of the Ohio River near Madison and in Clifty Falls State Park, Jefferson County. It was reported also from Clark and Floyd Counties by the editors of the Botanical Gazette in their list of the plants of Indiana, published in 1881. In 1939 R. M. Kriebel found it in the eastern part of Clark County.

Vt. to n. Mich., southw. to N. C., Ala., and Mo.

11. WOODWÂRDIA J. E. Smith

1. **Woodwardia virgínica** (L.) Sm. (*Anchistea virginica* (L.) Presl) VIRGINIA CHAINFERN. Map 34. This fern is infrequent to very local in the area shown on the map. Usually where it is found it is common. It grows in bogs and marshes. Its preferred habitat is old tamarack bogs and its most common associate is *Chamaedaphne.*

The sterile fronds of this species resemble those of *Osmunda, Dryopteris,* and *Athyrium,* but the fronds of *Woodwardia* may be distinguished by the areolae in the venation along the midrib.

N. S. to Fla., La., and Ark., chiefly along the coast; also inland in the Great Lake Region.

12. PELLAÈA Link CLIFFBRAKE

Stipe, rachis, and rachilla pubescent with long multicellular hairs, usually more or less densely so, especially on the rachilla, scabrous to the touch. . 1. *P. atropurpurea.*

Stipe, rachis, and rachilla glabrous or with a few scattered hairs, smooth to the touch. .... 2. *P. glabella.*

1. **Pellaea atropurpùrea** (L.) Link. PURPLE CLIFFBRAKE. Map 35. Infrequent to very local in shallow soil on calcareous rocks. These rocks usually are the perpendicular cliffs and ledges along streams but are often small or large detached fragments at the base of cliffs. Sometimes it is found in the seams of stratified rock outcrops only a few feet high. It grows in both shade and sun, preferring shade of medium density. My

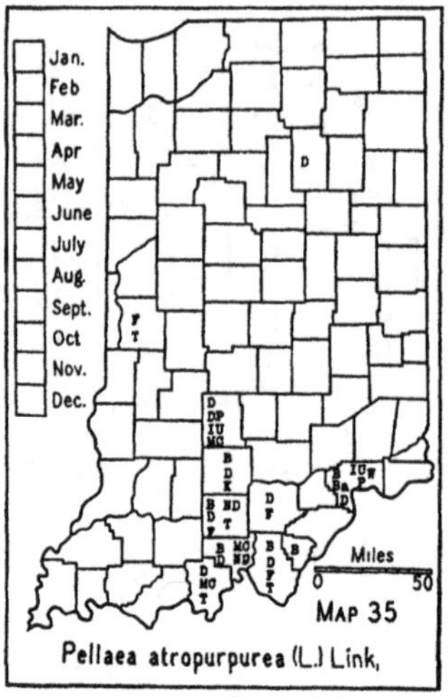

Map 35
Pellaea atropurpurea (L.) Link,

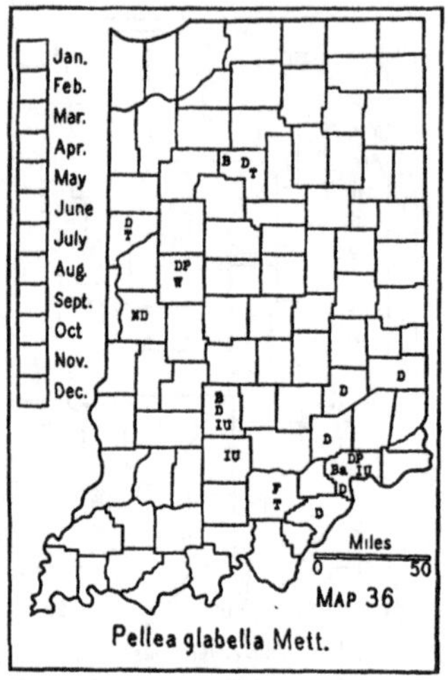

Map 36
Pellea glabella Mett.

Map 37
Cheilanthes lanosa (Michx.) Watt

Wabash County specimen was found about a mile southeast of Lagro on Hanging Rock, which is 84 feet high. It is probably extinct there now since that place has become a picnic ground.

Vt., N. Y. and n. Mich. to S. Dak., southw. to Fla., Tex., and Ariz.

2. **Pellaea glabélla** Mett. SMOOTH PURPLE CLIFFBRAKE. Map 36. This species was not separated from the preceding species even in Britton and Brown, Illustrated Flora, edition 2, published in 1913. Pickett (Amer. Fern Jour. 4: 97-101. 1914) wrote an article entitled "A peculiar form of *Pellaea atropurpurea* Link" and set forth the differences at length, but he did not give it a name until in a later article (Amer. Fern Jour. 7: 3-5. 1917.) Butters (Amer. Fern Jour. 7: 77-87. 1917) took up the subject and listed the specimens at the Gray Herbarium to show the range of the two species.

This species has the habitat of the preceding but it is less frequent. *Pellaea atropurpurea* is regarded as the southern representative of the genus in our area and has a mass distribution to the south of a line connecting Kansas and Connecticut. *Pellaea glabella* is regarded as the northern representative of the genus in our area and has its mass distribution north of that of *Pellaea atropurpurea*.

Vt., Ont. to Minn., southw. to Pa., Ohio, Ind., Mo., and Okla.

13. CHEILÁNTHES Sw. LIPFERN

Fronds bipinnate, hirsute, especially beneath; hairs straightish, jointed, and often of a rusty color, especially on the stipe............................1. *C. lanosa*.
Fronds tripinnate, tomentose with white hairs. (See excluded species no. 10, p. 1020.) .... *C. tomentosa*.

1. **Cheilanthes lanòsa** (Michx.) Watt. HAIRY LIPFERN. Map 37. I have found this species on the exposed cliffs along White River at the McBride Bluffs about 5 miles north of Shoals in Martin County. I have also found it in three places in Perry County. It is infrequent on the stones capping the high cliffs along the Ohio River about 5 miles east of

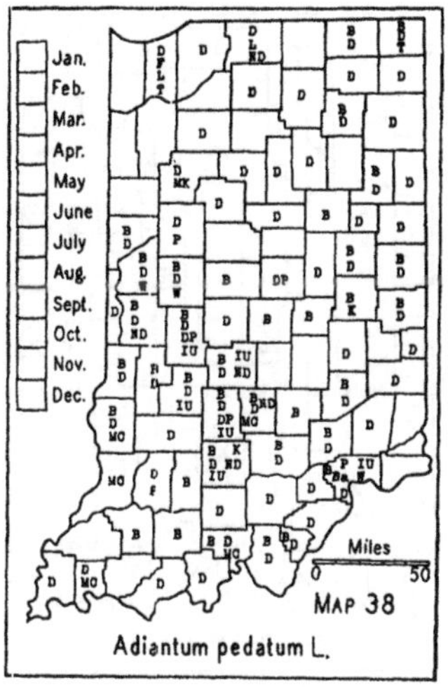

MAP 38
Adiantum pedatum L.

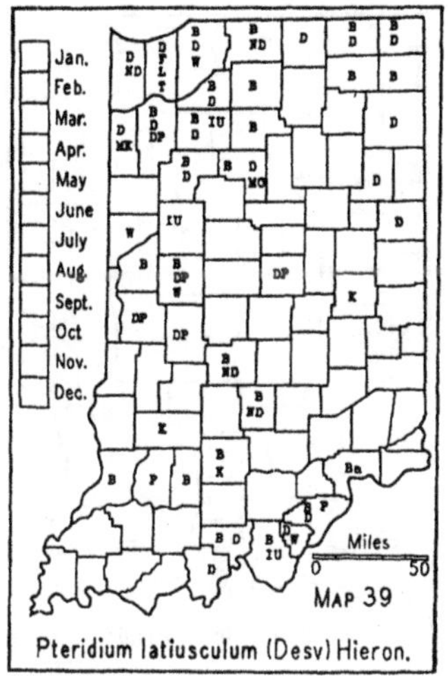

MAP 39
Pteridium latiusculum (Desv) Hieron.

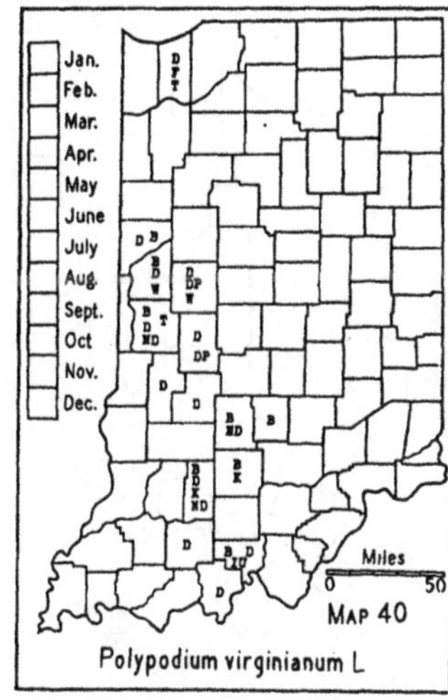

MAP 40
Polypodium virginianum L.

Cannelton, on the top of low, rocky ledges about 8 miles east of Cannelton, and in the shade on a low cliff in the woods of Wm. Stahl about 3 miles south of Mt. Pleasant. The plants were numerous here but were small (mostly less than 2 dm high) because they grew in the shade.

Conn. to Kans., southw. to Ga. and Tex.

14. ADIÁNTUM [Tourn.] L.

1. **Adiantum pedàtum** L. MAIDENHAIR FERN. Map 38. Infrequent to frequent throughout the state in deep humus in many kinds of soils and with many kinds of associates. It prefers shade and shelter from wind, hence it is most often found in protected places.

Newf. to Alaska, southw. to Ga., La., and Kans., and locally westward to Utah and Calif.; also in Asia.

15. PTERÍDIUM Scop.

1. **Pteridium latiúsculum*** (Desv.) Hieronymus. (*Pteris aquilina* of Gray, Man., ed. 7, not L. and *Pteridium aquilinum* of Britton and Brown, Illus. Flora, ed. 2, not Kuhn.) BRACKEN. Map 39. Infrequent but locally common throughout the lake area in dry, sandy soil or in dry prairie habitats. It is found also locally in a few of the southern counties on wooded sandstone ridges.

Newf. to Wis., and Wyo., southw. to D. C., W. Va., Ill., and Ariz.

16. POLYPÒDIUM [Tourn.] L.

Blades of fronds glabrous, green..................................1. *P. virginianum.*
Blades of fronds densely scaly beneath, grayish. 2. *P. polypodioides* var. *Michauxianum.*

1. **Polypoium virginiànum** L. (Rhodora 24: 125. 1922.) (*Polypodium vulgare* of American authors, not L.) COMMON POLYPODY. Map 40. Local on the ledges of rocks in the area of the state where outcrops of

* Variety **pseudocaudatum** (Clute) **Maxon** is now known from Crawford and Knox Counties.

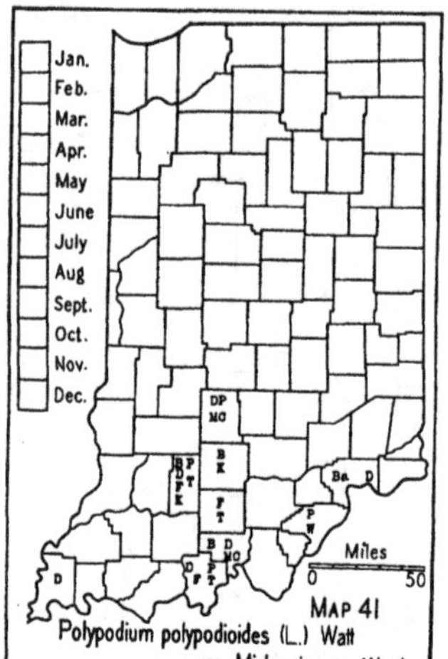

MAP 41
Polypodium polypodioides (L.) Watt var. Michauxianum Weath.

MAP 42
Azolla caroliniana Willd.

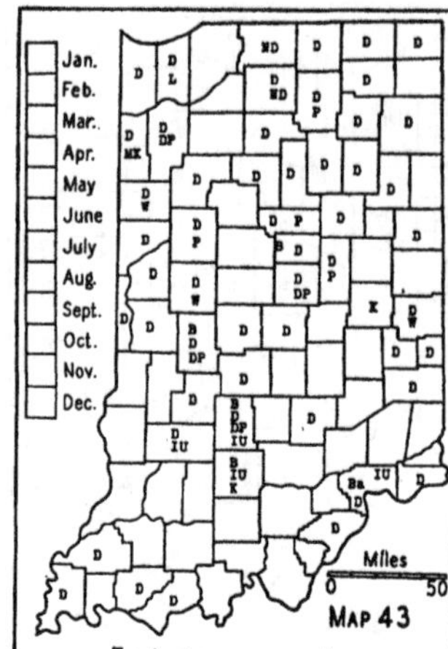

MAP 43
Equisetum arvense L.

sandstone and knobstone occur. There are, however, a few reports for it outside of this area. For example, Phinney reported it from the area composed of Delaware, Jay, Randolph, and Wayne Counties, saying: "Common. Moist woods". Van Gorder reported it from Noble County, saying: "A common plant of moist woods". Neither of these authors report the Christmas fern which occurs in their area, and, without doubt, their reports for this *Polypodium* should be referred to *Polystichum*. Bradner reported *Polypodium* from Steuben County but he also reported *Polystichum*. In this instance I think he may have had a sterile specimen of *Polystichum* and thought it was a *Polypodium*. This species was reported from the vicinity of Lake Michigan by three authors. I have always questioned these reports because my idea of the habitat of this species is that of outcrops of sandstone rocks. Doubtless Buhl had the same idea when he said (Amer. Midland Nat. **16**: 250. 1935) that this report should be deleted for lack of confirming specimens. To my great satisfaction (because I always prefer to confirm rather than to deny a report) on May 30, 1935 through the courtesy of R. M. Tryon, Jr. I was shown a colony of this species on a wooded dune in the Dunes State Park. Mr. Tryon has had this colony under observation for several years and reports that it is gradually diminishing. The plant is growing in dense shade on the north side of a high dune which is well protected from the wind. Doubtless this species was infrequent to frequent in the dunes before it had to compete with fire and civilization.

Lab., Newf. to Man., southw. to Ga., Ala., and Mo.

2. **Polypòdium polypodioìdes** (L.) Watt var. **Michauxiànum** Weath. (Contrib. Gray Herb. **124**: 31. 1939.) (*Polypodium polypodioides* (L.) Watt of recent authors.) RESURRECTION FERN. Map 41. Very local in a few counties in the southern part of the state. It is usually found in large mats, clinging to almost perpendicular cliffs or on large detached fragments of rock below the cliff. I found it once in Posey County in the crotch of a large bur oak tree which grew on the border of one of the numerous sloughs in the bottoms. It grew at a height of about 10 feet above the ground but

I did not take a specimen because I was not prepared to care for it. This is the only specimen I have ever seen on a tree in Indiana although it is common in this habitat in the South.

Md., Ill., and Mo., southw. to Fla. and Tex.; Guatemala.

4. SALVINIÀCEAE Reich. SALVINIA FAMILY

1. AZÓLLA Lam.

1. **Azolla caroliniàna** Willd. WATER FERN. Map 42. This species is found in stagnant water along streams, about lakes, and in dredged ditches. It is doubtless much more frequent than our map indicates. I did not know the species until recent years and I suspect that many collectors are not acquainted with it. It is usually found associated with duckweeds. This species was first reported from Indiana by Prince Maximilian in 1839. It has been reported so far from Starke and St. Joseph Counties.

Mass., Ont. to B. C., southw. to Fla., Ariz., and Mex.; also in tropical Amer.

5. EQUISETÀCEAE Michx. HORSETAIL FAMILY

1. EQUISÈTUM [Tourn.] L.

[Schaffner. How to distinguish the North American species of Equisetum. Amer. Fern Jour. 13: 33-40; 67-72. 1923. Diagnostic key to the species of Equisetum. Amer. Fern Jour. 22: 69-75; 122-128. 1932.]

J. H. Schaffner, our foremost authority on the genus *Equisetum,* has seen and named all of my specimens. The following key has, for the most part, been adapted from Schaffner's keys.

Stems without or with little chlorophyll, unbranched at first or permanently so, always terminating in a blunt cone.
  Sheaths bright reddish brown and translucent, their teeth comparatively long, cohering in 3 or 4 broad lobes; fertile stems finally developing whorls of compound green branches; internodal ridges sometimes with rows of siliceous spinules. (See excluded species no. 13, p. 1021.) .... *E. sylvaticum.*
  Sheaths not reddish brown and translucent, their teeth not cohering in 3 or 4 broad lobes.
    Teeth of the sheaths light brown, membranous, usually soon becoming green; stems soon developing whorls of 3-angled, green branches, with deltoid, membranous teeth; internodal ridges sometimes with rows of spinules. (See excluded species no. 14, p. 1021.) .... *E. pratense.*
    Teeth of the sheaths dark brown, rigid, only slightly membranous at the margins; stems withering promptly after the spores are shed; sheaths rarely slightly green; internodal ridges without spinules................ 1. *E. arvense.*
Stems green or with green branches, with or without cones.
  Teeth of the lower sheaths of the main stem cohering in 3 or 4 broad lobes, comparatively long, bright reddish brown, and translucent, not deciduous; branches of the whorls prominently compound, horizontal or often curving downward, especially on the fertile shoots; stomata in bands; internodal ridges with or without 2 rows of siliceous spinules; cones not apiculate. (See excluded species no. 13, p. 1021.) .... *E. sylvaticum.*
  Teeth of the sheaths of the main stem neither united in 3 or 4 broad lobes nor bright reddish brown, deciduous or persistent.

Stems usually not branched above the ground unless the plants are injured, or the branches few, irregular and sporadic; stomata in regular rows; teeth of the sheaths or their bristle-tips usually soon deciduous, but several species with persistent teeth or the teeth forming pagodalike caps; cones with or without a point.

Teeth of the sheath persistent or only their bristle-tips deciduous, white-margined, not sharply differentiated from the sheath; sheath segments and lower part of teeth distinctly quadricarinate; stems 5-10-grooved, erect in tufts, evergreen; cones apiculate.

Ridges of internodes prominently biangulate (2 ridges to a sheath tooth), with a double row of rounded tubercles.

Sheaths cylindric, tight, often crusty, partly or completely black; stems rather large to medium, sometimes rather slender....2. *E. trachyodon.*

Sheaths campanulate, usually discoloring tardily; stems mostly very slender and small....................................3. *E. variegatum.*

Ridges of internodes not biangulate, with a single row of tubercles or cross bands of silica...4. *E. Nelsoni.*

Teeth of the sheath soon deciduous, sharply differentiated from the sheath; main stem usually tall, 10-many-grooved, with a large central cavity.

Sheaths cylindrical, short, appressed, or only slightly dilated when young, at first green, but soon turning black or gray, commonly gray with black bands above and below, often split in age; stems usually very rough, evergreen; sheath segments of the main stem tricarinate; ridges of the internodes with one row of tubercles; cones apiculate.........5. *E. prealtum.*

Sheaths more or less funnel-shaped, elongate, green, the limb normally with a narrow black band, sometimes the lower sheaths with bands of gray or black below; stems evergreen or annual; cones with or without a point.

Cones tipped with a rigid point........................6. *E. laevigatum.*

Cones rounded or the tip merely acute; limb of the long green sheath dilated upwards ....7. *E. kansanum.*

Stems usually much branched with several to many whorls of branches, rarely with only few sporadic branches; stomata in broad bands or scattered in the grooves of the internodes or only on the sheaths; teeth of the sheaths persistent; cones not apiculate.

Branches hollow, usually simple, terete, both fertile and sterile stems green; plants of wet soil or growing in water, sometimes without or with only sporadic branches; sheaths of the main stem usually appressed, 15-20-toothed; stems usually many-grooved, with a very large central cavity and thin wall...8. *E. fluviatile.*

Branches solid, simple or compound, mostly sharply 3- or 4-angled; fertile stems brown and at first without branches, soon withering or developing green branches when mature; usually in moderately moist or dry situations.

Teeth of the branches with subulate tips; branches usually 4-angled (sometimes 3-angled); fertile stems withering after the spores are shed....... ....1. *E. arvense.*

Teeth of the branches not subulate-tipped, deltoid, merely acute or long-acute, usually white-membranous; branches generally 3-angled, very slender, fertile stems developing green branches after the spores are shed. (See excluded species no. 14, p. 1021.)........................*E. pratense.*

1. **Equisetum arvénse** L. FIELD HORSETAIL. Map 43. Infrequent to frequent throughout the state. Where it is found it usually forms large colonies, especially in its preferred habitat along railroad embankments. It prefers a moist, sandy soil, usually lean in organic matter, but it is also found in moist places on the borders of bogs and along streams. It grows in both shade and sun and its appearance is so erratic and it is so widespread that I am not able to tell what controls its distribution. Once I saw

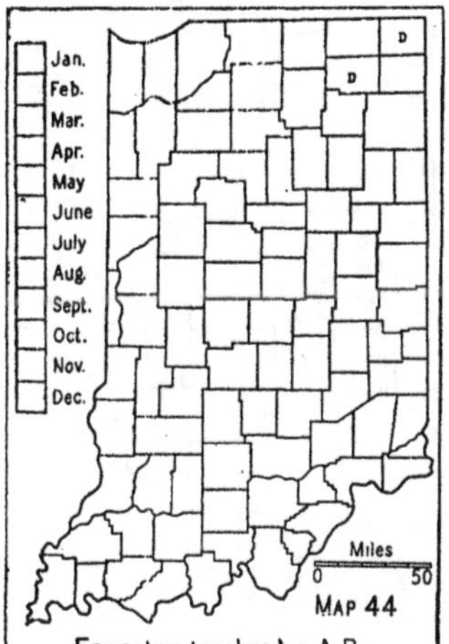

Equisetum trachyodon A Br.

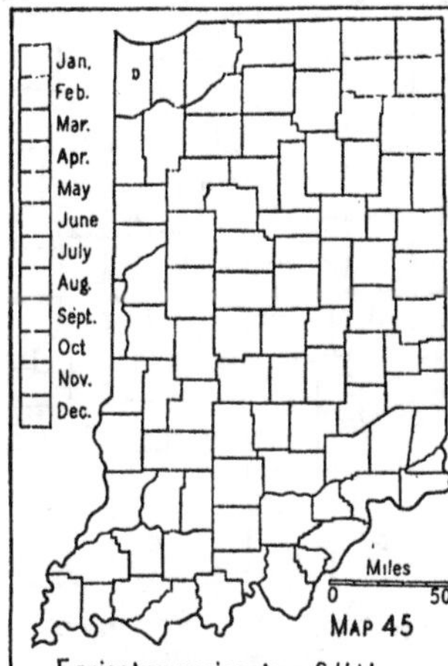

Equisetum variegatum Schleich

Equisetum Nelsoni (Eaton) Schaffner

where it had almost covered a sandy fallow field in the valley of Pigeon River. The plant is extremely variable and many varieties have been named, several of which have been reported from Indiana. According to Schaffner these variations are all ecological and not worth recognition.

Newf. to Alaska, southw. to N. C. and Calif.; also in Eurasia.

2. **Equisetum trachýodon** A. Br. (*Equisetum variegatum* var. *Jesupi* A. A. Eaton). ROUGH-TOOTHED SCOURING-RUSH. Map 44. My only specimens are from Pokagon State Park from the wet, sandy shore of the east side of Lake James and from the east side of Crooked Lake, Noble County.

Que. and Ont., southw. to Conn. and Ill.; also in Eurasia.

3. **Equisetum variegàtum** Schleich. VARIEGATED SCOURING-RUSH. Map 45. My only specimen is from the grassy border of a dried-up slough in the dunes about a quarter of a mile south of Pine, Lake County (now along Clark Street in Gary about a quarter of a mile south of Lake Michigan). It has been reported also from Porter and La Porte Counties. This

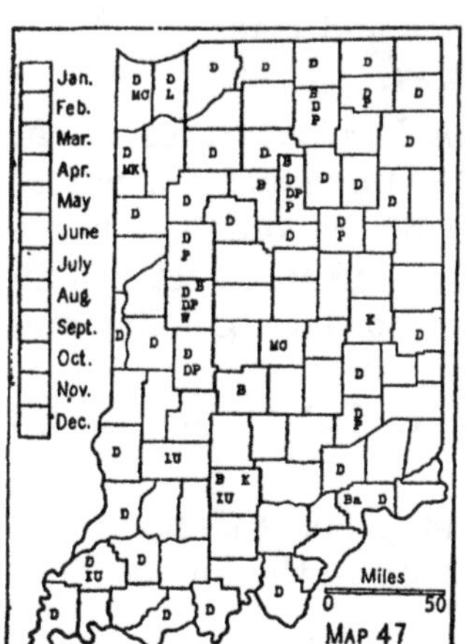

Equisetum prealtum Raf.

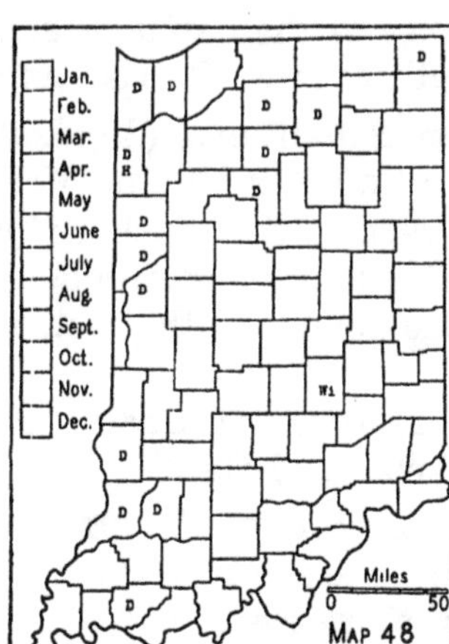

Equisetum laevigatum A.Br.

Equisetum kansanum Schaffner

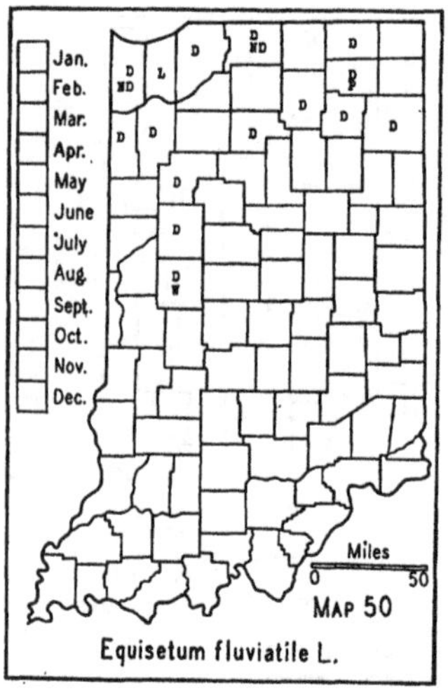

MAP 50
Equisetum fluviatile L.

MAP 51
Lycopodium Selago var. patens (Beauv.) Desv.

MAP 52
Lycopodium lucidulum Michx.

species much resembles the next one and is closely associated with it.

Lab. to Alaska, southw. to Maine, N. Y., and Wyo.; also in Eurasia.

4. **Equisetum Nélsoni** (A. A. Eaton) Schaffner. (*Equisetum variegatum* var. *Nelsoni* A. A. Eaton.) NELSON SCOURING-RUSH. Map 46. Wet, moist, or dry, sandy borders of lakes and sloughs.

N. Y. to Mich., Ind., and Ill.

5. **Equisetum preáltum** Raf. (*Equisetum hyemale* var. *affine* (Engelm.) A. A. Eaton). TALL SCOURING-RUSH. Map 47. Infrequent throughout the state. It is usually found in rather moist, sandy soil and on the slopes of the banks of streams; sometimes the habitat may even be springy. It grows in colonies, and these sometimes may extend along the bank for several rods. It rarely occurs in other habitats but is found along railroad embankments and in wet prairie habitats.

Canada to Mex.

6. **Equisetum laevigàtum** A. Br. SMOOTH SCOURING-RUSH. Map 48. Infrequent in sandy to very sandy soil in the greater part of the state. It is most commonly found on railroad embankments and less frequently in moist, sandy soil of the slopes of the banks of streams and lakes.

Conn., N. J. to B. C., southw. to N. C., La., and Mex.

7. **Equisetum kansànum** Schaffner. (*Equisetum laevigatum* of A. A. Eaton, not A. Br.) KANSAS SCOURING-RUSH. Map 49. Infrequent in northern Indiana and probably rare in the southern part of the state. It has a very wide range of habitat but is most frequently found in moist soil in prairies; it is, however, also found on the wet, marl borders of lakes and other moist habitats.

Mainly in the western Mississippi Basin, Ohio to Mont. and B. C., southw. to Mo., N. Mex., Ariz., and Calif.

8. **Equisetum fluviátile** L. WATER HORSETAIL. Map 50. Infrequent in northern Indiana in marshes and bogs, in the dune area on the low borders of sloughs, and rarely in wet prairies.

Newf. to Alaska, southw. to Va., Nebr., and Oreg.; also in Eurasia.

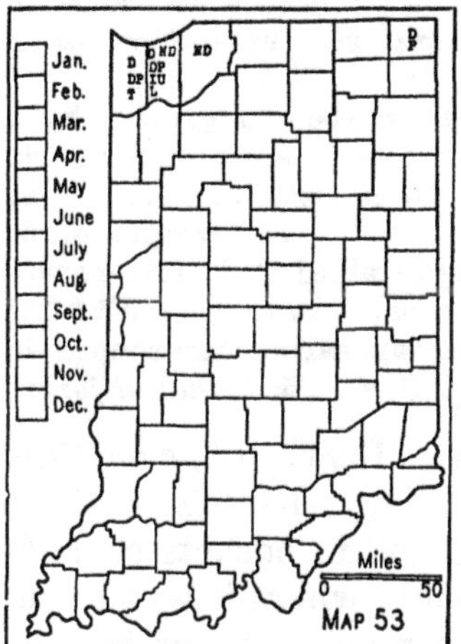

Lycopodium inundatum L.

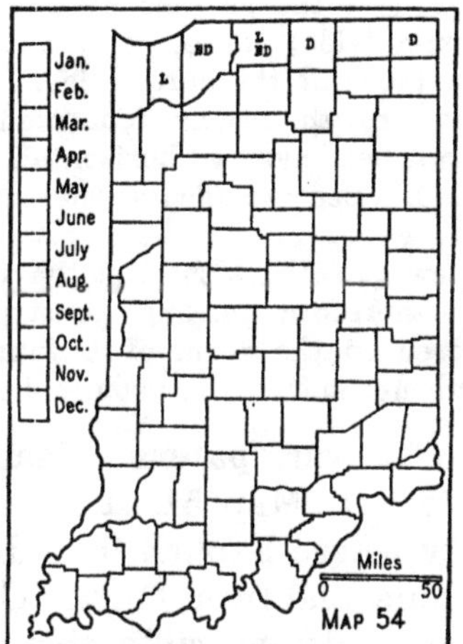

Lycopodium obscurum L.

Lycopodium flabelliforme (Fern.) Blanchard

6. LYCOPODIÀCEAE Michx. CLUBMOSS FAMILY

1. LYCOPÒDIUM L. CLUBMOSS

[Wilson. The identity of Lycopodium porophilum. Rhodora 34: 169-172. 1932. The spores of the genus Lycopodium in the United States and Canada. Rhodora 36: 13-19. 1934.]

Sporangia in the axils of normal leaves, not forming a well marked terminal spike.
  Leaves linear-attenuate to lanceolate, entire (sometimes with a few minute serrations toward the apex), usually widest below the middle; plants yellowish green, tufted, erect or slightly decumbent at the base...1. *L. Selago* var. *patens.*
  Leaves oblanceolate, widest near or above the middle, serrate or entire, arranged in alternate zones of shorter and longer leaves, the shorter ones more frequently bearing sporangia in their axils; stems bright or dark green, in loose clusters, decumbent.
    Blades of leaves serrate....................................2. *L. lucidulum.*
    Blades of leaves entire or slightly serrate, often some of them of a linear type. (See excluded species no. 17, p. 1022.).........*L. lucidulum* var. *occidentale.*
Sporangia borne only in the axils of the upper (bracteal) leaves, forming a spike.
  Bracteal leaves linear-attenuate from a distinctly broadened ovate base...3. *L. inundatum.*
  Bracteal leaves scalelike, yellowish, very different from those of sterile part of the stem.
    Ultimate sterile branches with their leaves mostly 5-10 mm wide, free portion of leaves more than 3 mm long.
      Stems creeping on the surface of the ground with short, leafy branches, the leaves linear, bristle-tipped at apex; fertile branches terminating in a slender peduncle (1-1.5 dm long), bearing 2-4 slender cylindrical spikes. (See excluded species no. 15, p. 1021.).......................*L. clavatum.*
      Stems subterranean, bearing scattered upright branches resembling miniature coniferous seedlings; leaves merely acute at the apex; spikes 1-3, essentially sessile...4. *L. obscurum.*
    Ultimate sterile branches with their leaves less than 5 mm wide; free portion of leaves less than 3 mm long.
      Horizontal stems rather deeply buried in the ground; branchlets bluish green, 1-1.75 (2) mm wide; leaves on ventral and dorsal sides of the branchlets about equal. (See excluded species no. 18, p. 1022.)........*L. tristachyum.*

Horizontal stems on or near the surface of the ground; branchlets yellowish green, (1.5) 2-3 mm wide; leaves on the ventral side of the branchlet much shorter than those of the dorsal side.

Branchlets with new growth clearly separated from the old growth by a constriction; branches mostly horizontal, or some erect, irregularly divided; spikes 1-3. (See excluded species no. 16, p. 1021.) .... *L. complanatum.*

Branchlets lacking new growth at the tips, having attained their full growth the first year, therefore lacking constrictions; branches erect, the branchlets disposed in the form of a funnel, appearing fan-shaped in herbarium specimens; spikes 1-6, usually 4............5. *L. flabelliforme.*

1. **Lycopodium Selàgo** L. var. **pàtens** (Beauv.) Desv. (*Lycopodium porophilum* Lloyd & Underw.) Map 51. I have this variety from three places in Crawford County where I found it in dry soil in pockets of cliffs of the knobstone or sandstone, and from Martin County where it was found in dry soil pockets of the sandstone cliffs about a mile north of Shoals.

Que. to Wis., southw. to n. Vt. and Ky.

2. **Lycopodium lucídulum** Michx. SHINING CLUBMOSS. Map 52. Very local. It grows in deep humus, sometimes forming large colonies. In the lake area it is generally found in decadent tamarack bogs and southward in moist, shaded woodland, although my Clay County specimen was found in the open among rocks along Croy Creek.

Newf. to B. C., southw. to N. E., N. Y., Ind., Iowa, and Wash., and in the mts. to S. C.

3. **Lycopodium inundàtum** L. Map 53. Very local. It grows in wet, somewhat acid sandy soil, usually on the borders of lakes and in the dunes. It has also been reported from Marshall County. I have twice found it associated with cranberry and hair-cap moss. In 1937 I revisited the Steuben County station and found that it has been exterminated there.

Newf. to Alaska, southw. to N. J., Pa., Ill., Idaho, and Wash.; also in Eurasia.

4. **Lycopodium obscùrum** L. GROUNDPINE. Map 54. Very local. In addition to my specimens it has been reported from Lake, Montgomery, Porter, and St. Joseph Counties. One of my specimens is from a small colony at the base of a north beech slope, bordering a soft maple swamp, and the other is also from a swamp bordering a lake.

My specimens are not typical and seem to be intermediate between the species and the var. *dendroideum* (Michx.) D. C. Eaton.

Newf. to Alaska, southw. to N. C. and Ind.

5. **Lycopodium flabellifórme** (Fern.) Blanchard. (Rhodora 13: 168-171. 1911.) (*Lycopodium complanatum* var. *flabelliforme* Fern.) Map 55. Extremely local. Found on moist, rocky slopes.

This species is regarded by many authors as a variety of *Lycopodium complanatum.* Blanchard (Rhodora 13: 168-171. 1911) made a special study of this species and *L. complanatum* in the field, and after nearly ten years' observation, concluded that the two were distinct species. Victorin (Contrib. Lab. Bot. Univ. Montreal. no. 3: 62-63. 1925) confirms Blanch-

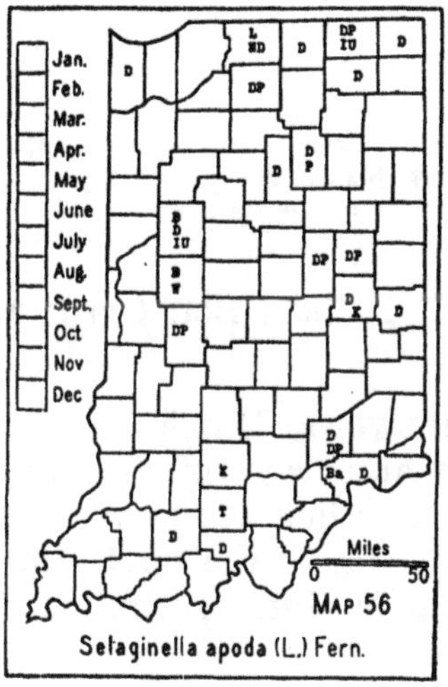

MAP 56
Selaginella apoda (L.) Fern.

MAP 57
Selaginella rupestris (L.) Spring

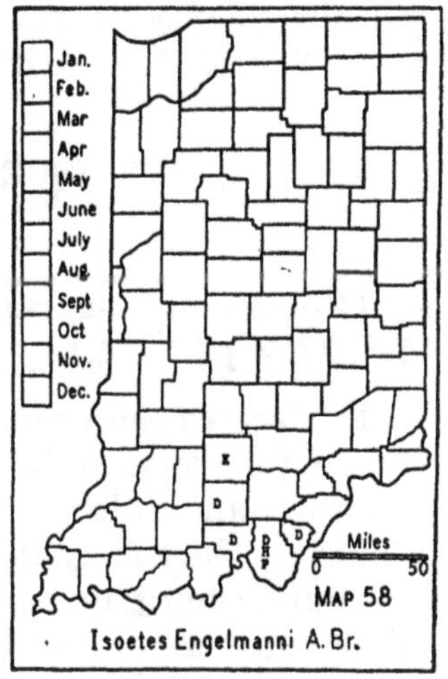

MAP 58
Isoetes Engelmanni A. Br.

ard's observation of characters which seem to me also to be sufficient to regard this form as a species rather than a variety. These two species have definite geographical ranges which add to this opinion. The range of *L. complanatum* in North America extends from Newfoundland through the greater part of Canada to Alaska and southward to northern Michigan, northern Wisconsin (not reaching New England), and Washington. *L. flabelliforme* is much more southern, occurring from Newfoundland, Nova Scotia, and the lower valley of the St. Lawrence River westward to Minnesota, southward to North Carolina and Kentucky.

Newf. to Minn., southw. to N. C. and Ky.

7. SELAGINELLÀCEAE Underw.

1. SELAGINÉLLA Beauv. SELAGINELLA

Leaves comparatively few, of 2 sizes, 4-ranked, spreading in 2 planes, ovate, acute or cuspidate; plants usually a light green, of a wet or moist habitat.....1. *S. apoda.*
Leaves very numerous, alike, appressed, widely overlapping, many-ranked, linear-lanceolate, grooved on the back, ending in a slender, whitish awn; plants grayish green, of a very dry habitat.................................2. *S. rupestris.*

1. **Selaginella ápoda** (L.) Fern. (Rhodora **17**: 68. 1915.) (*Selaginella apus* Spring.) BASKET SELAGINELLA. Map 56. Occasionally throughout the lake area, becoming infrequent to local in the southern part of the state. It is, no doubt, more frequent than our map indicates. It prefers moist, grassy places and in the lake area it is usually in calcareous, sandy soil. In Dubois County I found it in a low woods in a hard, white clay soil with sweet gum.

Maine and Ont. to the Rocky Mts., southw. to Fla. and Tex.

2. **Selaginella rupéstris** (L.) Spring. ROCK SELAGINELLA. Map 57. Local. Found only on dry, exposed sandstone rocks and in dry sand in the dune area. It has also been reported from Montgomery County. Underwood (Proc. Indiana Acad. Sci. **1893**: 257. 1894) says the report from

Gibson County in the State Catalogue was an error.
N. S. and Ont., southw. to Ga. and Mo.

8. ISOËTÀCEAE Underw. QUILLWORT FAMILY

1. ISÒËTES L. QUILLWORT

[Pfeiffer. Monograph of the Isoëtaceae. Ann. Missouri Bot. Gard. **9**: 79-232. 8 pl. 1922.]

1. **Isoëtes Engelmánni** A. Br. ENGELMANN QUILLWORT. Map 58. I have found this species in artificial ponds in hard, white clay soil in three counties, and in low woods in similar soil but richer in humus in Harrison County. The colony in Floyd County is on the east side of the road south of Martinsburg in an old mill pond on the Philip McGuirk farm. It is abundant here and of large size.
Southern N. H. and Vt. to Ga., westw. to Mo.

SPERMATÓPHYTA. SEED PLANTS OR FLOWERING PLANTS

5.[1] TAXÀCEAE Lindl. YEW FAMILY

18.[1] TÁXUS [Tourn.] L. YEW

1. **Taxus canadénsis** Marsh. CANADA YEW. Map 59. This species is local and is restricted to the sides of the steep slopes and cliffs along Sugar Creek in Turkey Run State Park, Parke County, to like habitats along Sugar Creek in the "Shades" in Montgomery County, and along Big Walnut Creek about 3 miles northeast of Bainbridge, Putnam County. It is usually found under hemlock.
Newf. to Man., southw. to Va. and Iowa.

6. PINÀCEAE Lindl. PINE FAMILY

Leaves linear, in bundles of 2, 3, 5 or more than 5.
  Leaves in bundles of 2, 3 or 5.................................22. PINUS, p. 67.
  Leaves in bundles of more than 5 (single on new shoots)..........24. LARIX, p. 68.
Leaves linear and solitary, or scalelike.
  Leaves all linear.
    Blades obtuse...27. TSUGA, p. 68.
    Blades sharp-pointed.
      Leaves green on both sides, alternate....................35. TAXODIUM, p. 69.
      Leaves glaucous beneath, opposite, or whorled............45. JUNIPERUS, p. 70.
  Leaves all scalelike, or scalelike on fruiting branchlets and linear and sharp-pointed on sterile branchlets or juvenile plants, usually green on both sides.
    Spray of branchlets flat; leaves all scalelike, the dorsal and ventral ones differing from the lateral ones; fruit a cone of 8-12 imbricated but opposite scales..42. THUJA, p. 69.
    Spray of branchlets not flat; leaves all scalelike or on most specimens some branchlets with linear and sharp-pointed leaves; fruit berrylike, bluish black, glaucous ....45. JUNIPERUS, p. 70.

[1] See paragraph 2 on page 14 of introduction.

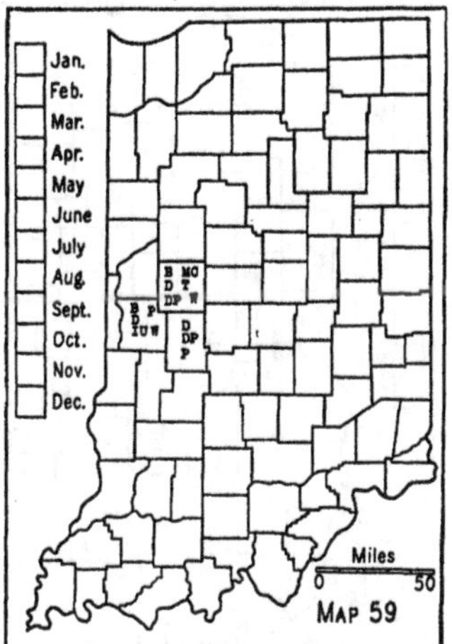

Taxus canadensis Marsh.

Pinus Strobus L.

Pinus Banksiana Lamb.

22. PÌNUS [Tourn.] L. PINE

Leaves 5 (rarely more) in a bundle, 6-12 cm long..................1. *P. Strobus.*
Leaves 2 or 3 in a bundle.
  Scales of cones unarmed; leaves in 2's, 2-5 cm long.............2. *P. Banksiana.*
  Scales of cones tipped with a short spine; leaves 2 or 3 in a bundle.
    Spine of cone-scale 2-3 mm long, curved; leaves twisted, 4-8 cm long...........
    ....3. *P. virginiana.*
    Spine of cone-scale about 1 mm long; leaves straight, 7-13 cm long. (See excluded species no. 20, p. 1022.)............................*P. echinata.*

1. **Pinus Stròbus** L. NORTHERN WHITE PINE. Map 60. This species is local and is usually found in limited numbers, except along Bear Creek, Fountain County and Big Pine Creek in Warren County where there were formerly many acres of it. In the dune area it was scattered in its distribution with a large colony here and there. There formerly were several acres of it in a bog east of Merrillville, Lake County, but it has now nearly disappeared.

In our area its favored habitat was wet woods or boggy places, on the dunes along Lake Michigan, on cliffs and high banks along Bear Creek, Fountain County, and in a like habitat including adjacent lowland in Warren County along Big Pine and Kickapoo Creeks.

Newf. to Man., southw. in the mts. to n. Ga., Tenn., and Iowa.

2. **Pinus Banksiàna** Lamb. JACK PINE. Map 61. This pine is found only on the dunes near Lake Michigan. I can recall when it was common on the low dunes in Lake County but it has now nearly disappeared on account of advancing civilization.

N. S. to n. N. Y., n. Ind. to Minn., northw.

3. **Pinus virginiàna** Mill. VIRGINIA PINE. Map 62. This species is restricted to the crests of some of the ridges of knobstone in three counties. On some of the ridges it formed dense stands, but, on the whole, the species is not a strong competitor of the other species. It, however, promptly invades abandoned fields within and adjacent to the area of its natural

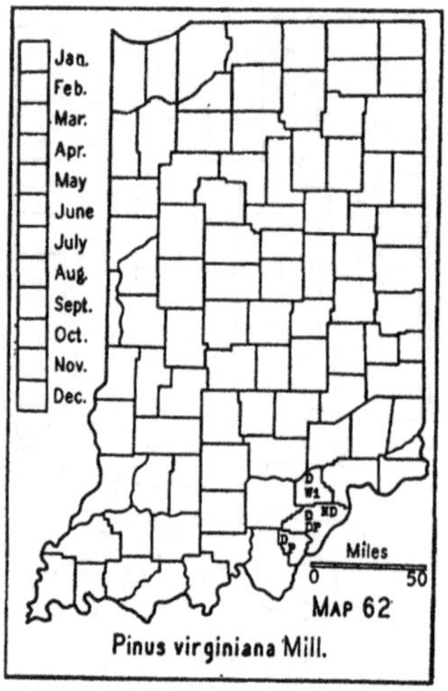

Pinus virginiana Mill.

Larix laricina (Du Roi) Koch

Tsuga canadensis (L.) Carr.

distribution. Also when planted in a favorable habitat, it freely escapes. Notable examples of its escape are on the knobstone in northern Washington County, on the bluffs along Raccoon Creek in Owen County, and in Monroe County in a grove about 4 miles northwest of Ellettsville and about Weimer's Lake 2½ miles west of Bloomington where it has been established for more than 50 years. A colony of about 3 acres in Orange County about 8 miles southeast of Paoli and just north of Danner's Chapel originated from a tree planted in the church yard. Some of the trees have already been cut for saw logs. R. M. Kriebel reports several large colonies in Lawrence County. He has traced the origin of each colony to a planted tree. In the knobstone area this species is truly "an old field" species. Within a 25-year observation I have seen it cover abandoned fields although it is a species difficult to transplant.

Long Island, N. Y., to Ind., southw. to S. C. and Ala.

24. LÀRIX [Tourn.] Mill. LARCH

1. **Larix laricìna** (DuRoi) Koch. TAMARACK. Map 63. Infrequent to frequent in bogs and on the low borders of lakes and streams throughout the lake area. It was formerly more or less common in many places that have been drained and are now farmed. It has suffered much during the past few years due to drought and is becoming scarce because of drainage and cutting.

Lab. and Newf., N. W. Territory, southw. to N. J., n. Pa., n. Ill., and cent. Minn.

27. TSÙGA [Endl.] Carr. HEMLOCK

1. **Tsuga canadénsis** (L.) Carr. EASTERN HEMLOCK. Map 64. Local in the state and usually restricted to a fringe of trees on the tops and slopes of high sandstone bluffs along streams. Rapidly disappearing in some of its stations.

N. S., N. B. to Minn., southw. to Del., s. Ind., Wis., and in the mts. to Ga. and Ala.

35. TAXÒDIUM Richard

1. **Taxodium dístichum** (L.) L. C. Richard. SOUTHERN CYPRESS. Map 65. The cypress is restricted to five counties in the southwestern part of the state. Collett (Rept. Ind. Geol. Surv. 5: 338. 1874) estimated that 20,000 acres of the southwestern part of Knox County were "covered with a fine forest of cypress." In this whole area there are now only a few straggling specimens left. In Little Cypress Swamp in the extreme southwestern corner of Knox County the species still persists and is reproducing in small numbers. There were a few cypress sloughs in Posey County but the trees have been slaughtered in most of them. There are no objections to judicious cutting but an attempt to annihilate a species without sufficient cause seems a tragedy. I found a few trees along Cypress Creek in Warrick County about 20 years ago but I was not able to find them recently. It has also nearly disappeared in Vanderburgh County. Baird & Taylor reported it from Clark County but I am excluding this report for lack of confirming specimens or convincing proof that it really did exist in this county. There is, however, some evidence to support this report. Audubon is quoted as having taken Rafinesque into extensive canebrakes in Indiana north of Louisville, and Victor Lyon, former surveyor of Clark County, also told me that he had seen large native pecan trees in the Silver Creek bottoms. I have not been able to study this area sufficiently to find other associate species of the cypress, and I leave this report to be confirmed.

I have never seen this species growing in Gibson County, but late in 1935 I met Smith White, who was 71 years old and who had always lived in the Gibson County Bottoms, and he told me, in the presence of three other persons, that it had never occurred in that area except for a single tree in a slough in a woods on the farm of C. B. Balse, about 3 miles south of East Mt. Carmel. These other three men had also seen the tree to which he referred.

Atlantic coast from Del. to Fla., westw. along the Gulf to Tex. and northw. in the Mississippi Valley to Ind.

42. THÙJA L.

1. **Thuja occidentàlis** L. NORTHERN WHITE CEDAR. Map 66. There are three old reports for this species from Lake County and I have an Umbach specimen collected near Pine. I collected it about 2 miles east of Indiana Harbor in 1906 but I have not seen it since in this county. No doubt later reports are based upon the early reports. Several authors report it from Mineral Springs bog, Porter County and Lyon reports a few trees near Tamarack. I have seen it in only two places in Porter County and, doubtless, there are only two colonies of it in the county. In the Mineral Springs bog there are quite a number of trees 4-6 inches in diameter but their number is rapidly decreasing. Buried remains of this species have been found as far south as Henry County.

E. Que. to Man., southw. to Pa., Tenn., Ill., and Minn. and in the mts. to N. C.

Taxodium distichum (L.) L.C. Richard

Thuja occidentalis L.

Juniperus communis var. depressa Pursh

45. JUNÍPERUS [Tourn.] L. JUNIPER

Leaves mostly in whorls of 3, glaucous beneath, all linear and sharp-pointed, mostly 7-15 mm long; stem divided at the surface of the ground, the several subdivisions or branches decumbent and growing to great lengths, rarely one branch becoming a leader.......................................1. *J. communis* var. *depressa.*

Leaves not in whorls, scalelike on fertile branchlets and linear on sterile branchlets, generally green on both sides, the scalelike ones 1-2 mm long and the linear ones mostly less than 10 mm long; stems erect with lateral branches like those of other trees..2. *J. virginiana* var. *crebra.*

1. **Juniperus commùnis** L. var. **deprèssa** Pursh. (*Juniperus sibirica* of Britton and Brown, Illus. Flora, ed. 2.) PROSTRATE JUNIPER. Map 67. This species has an erratic distribution and grows in widely different habitats. It is frequent in the dunes near Lake Michigan where a single plant will form a large clump. I found specimens in Steuben County in a decadent tamarack bog, one of which had a spread of about 25 feet. The branches were in a whorl and the plant was circular in shape with the tips of the decumbent branches usually 4-7 feet high. In Elkhart County I found a specimen in hard, clay soil 3 miles northwest of Goshen. This specimen maintained an erect branch with a very strong taper. It had just been cut and the upright branch was made into a small fence post. At the base where the tree was cut off it was a foot in diameter and it had many radiating branches that were several inches in diameter. I saw this variety growing in both Jefferson and Wayne Counties in shallow soil on rocky slopes. In 1923 I transplanted a seedling about 6 inches high from the dunes into a black loam soil and it grew erect until it reached a height of about 3 feet when the leader began to become decumbent and three branches at the surface began to elongate. After 12 years all the branches, numbering about 50, are decumbent and radiate in all directions, forming a circular clump 15 feet across, the branches being 4-6 feet high. This variety also occurs in Montgomery County.

Lab. to B. C., southw. to Conn., N. Y., and in the Rocky Mts. to Colo. and Utah.

Juniperus virginiana L. var. crebra Fernald & Griscom

Typha latifolia L.

Typha angustifolia L.

2. **Juniperus virginiàna** L. var. **crèbra** Fernald & Griscom. (Rhodora 37: 131-133. 1935.) (*Juniperus virginiana* f. *Bremerae* Standley & Macbride.) EASTERN RED CEDAR. Map 68. In a recent study of the species Fernald & Griscom found that our spirelike trees of the north and interior are not like the ovoid type of tree of the south. The leaves of adult branchlets of the northern form are narrower and attenuate at the apex while those of the southern form are rather broadly deltoid and obtuse or merely subacute. The mature fruit of the north has sweet flesh and the seed shallow pits at the base while those of the south have flesh with a pitchy taste and deep pits at the base. Caution must be used in separating the two forms by the character of the leaves because of transitional forms. All the specimens I have examined belong to the northern variety.

The eastern red cedar is found throughout the state although there are no records from the southwestern part. It is rare to infrequent in the northern part except along the St. Joseph River where it is frequent on its banks or close to them, becoming rare in the central part of the state, and frequent to common in the southern part in the unglaciated area and east of it. This tree seems to prefer calcareous soils, and in the unglaciated area some eroded and abandoned fields have grown up thickly with it. It has a wide range of habitats for I have seen it even in the "flats" in Clark County. I am of the opinion that in the primitive forest this species was restricted to high bluffs and banks of streams and eroded slopes where it could compete with other species. While it is tolerant of shade it is rarely found in the dense forest.

N. S. to w. Ont. and S. Dak., southw. at least to Mo. and Va.

8. TYPHÀCEAE J. St. Hil. CATTAIL FAMILY

49. TŸPHA [Tourn.] L. CATTAIL

Staminate and pistillate parts of spike contiguous; stems stout, usually 1-2 m high; leaves flat, the lower ones 12-23 mm wide; sterile flowers shorter than the hairs; pollen grains in 4's; stigmas fan-shaped; mature pistillate spikes about 2.5 cm in diameter..1. *T. latifolia.*

Staminate and pistillate parts of spike usually separated by an interval of 0.5-6 cm; stems slender, usually 8-12 dm high; leaves more or less dorsally convex, the lower ones mostly 4-7 mm wide; sterile flowers scarcely shorter than the hairs; pollen grains single; stigmas linear; mature pistillate spikes 10-18 mm in diameter..2. *T. angustifolia.*

1. **Typha latifòlia** L. COMMON CATTAIL. Map. 69. Found in ditches, ponds, marshes, gravel pits, and marshy places about lakes and along streams. It is frequent in the lake area, becoming infrequent to local in the southern part of the state where its habitat is rarely found.

Throughout temperate N. A.; cosmopolitan.

2. **Typha angustifòlia** L. NARROWLEAF CATTAIL. Map 70. This species is usually found on the borders of larger bodies of water than the preceding species, but it seems to adapt itself to nearly the same habitats. Near my home is a small gravel pit that has not been in use for about 10 years, and it is now filled with both species of cattails, this species occupying about a fourth of the space. It is to be noted that the pistillate part of the spike sometimes divides. I have one specimen with a 5-parted spike. I also have a specimen of the preceding species that has a 3-parted spike. This species, as well as the preceding one, is variable, and several varieties have been named. A giant form of this species is found on the east side of Tippecanoe Lake in the southern part of Noble County. Peattie's var. *calumetensis* seems to me to be an ecological form. Its diminutive size I attribute to the pollution of the Grand Calumet River near where it is found. In the summer when the soil along the bank is exposed it is slimy and reddish.

N. S. to Fla., mainly along the coast, and inland mostly about the Great Lakes; almost cosmopolitan.

10. SPARGANIÀCEAE Agardh BUR-REED FAMILY

54. SPARGÀNIUM [Tourn.] L. BUR-REED

[Fernald. Notes on Sparganium. Rhodora 24: 26-34. 1922.]

The following key has been adapted from this paper:

Achenes broadly obpyramidal, sessile, truncate or retuse at the summit, 4-8 mm in diameter; stigmas 2; anthers 1.5-2 mm long; sepals nearly equaling the achenes. ....1. *S. eurycarpum.*

Achenes fusiform, short-pedicelled, beaked, 1.2-3 mm in diameter; anthers 0.5-1.6 mm long; sepals from much shorter than to two thirds as long as the achenes.

Staminate heads 2-20 (rarely only 1); fruiting heads 1.2-3.5 cm in diameter; mature achenes strongly fusiform, 5.5-14 mm long, the stipe 1-4 mm long, the slender beak 1.5-6 mm long; plants erect.

Pistillate heads or branches strictly axillary; achenes with the beak abruptly contracted above the dilated base; leaves 6-12 mm wide, without a scarious margin.

Leaves stiffish, at least the middle keeled; inflorescence branched, some branches all staminate, or some both staminate and pistillate, with 1-4 pistillate heads and up to 8 staminate heads; stigmas 2-4 mm long; fruiting heads usually 3-7, 2.5-3.5 mm in diameter; achenes lustrous, the body 5-7 mm long and 2.5-3 mm thick, the beak 4.5-6 mm long; anthers 1-1.6 mm long. ....2. *S. androcladum.*

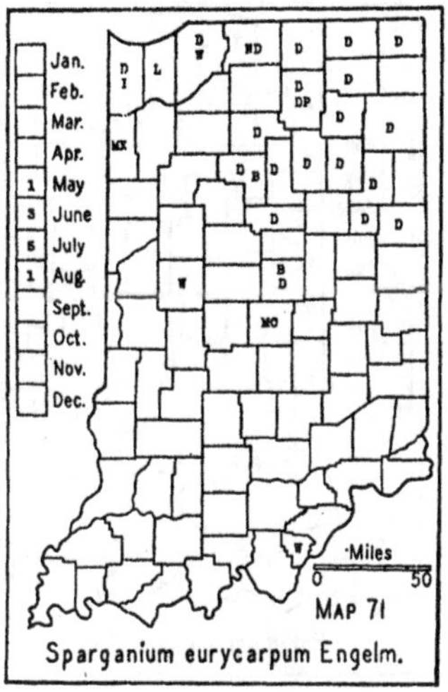

Sparganium eurycarpum Engelm.

Sparganium androcladum (Engelm.) Morong

Sparganium americanum Nutt.

Leaves soft and mostly translucent, flat or obscurely keeled; inflorescence simple or, if branched, the branches strict with 1-3 pistillate and 1-6 staminate heads; stigmas 1-2 mm long; fruiting heads 1.5-2.5 cm in diameter; achenes slightly lustrous, the body 4.5-5.5 mm long, about 2 mm thick; anthers 0.8-1.2 mm long....................................3. *S. americanum.*

Pistillate heads usually supra-axillary; achenes shining, the beak more gradually narrowed upward; leaves 3-9 mm wide with a scarious margin near the base.

Plants commonly erect and emersed; leaves flat or slightly keeled, little, if at all, dilated at the base (except for the scarious margin); staminate half of the inflorescence 2-10 cm long, of 4-9 scattered heads (if shorter and with fewer leads, the plant very low and with ribbonlike, translucent, erect, lower bracts); beak of achenes 2-4.3 mm long; sepals appressed, cuneate-spatulate, scarcely narrowed to a claw.

Pistillate heads (1) 2-4, remote or subremote, at maturity 1.5-2.7 cm in diameter, the lowest borne 1-6.5 dm above the base of the plant; staminate half of the inflorescence 2-10 cm long, of 4-9 heads.....4. *S. chlorocarpum.*

Pistillate heads 1-3, at least the upper usually approximate, at maturity 1.2-2.2 cm in diameter, the lowest borne 0.1-1.8 dm above the base of the plant; staminate half of the inflorescence 1-4 (5) cm long, of 2-5 heads. ....4a. *S. chlorocarpum* var. *acaule.*

Plants commonly submerged or floating, sometimes emersed; leaves rounded on the back, the middle and upper with dilated and subinflated sheathing bases; staminate half of the inflorescence 1-3 cm long, of 1-4 (rarely 6) crowded heads; beak of achene about 2 mm long; sepals loosely ascending, with slender claw and dilated tip. (See excluded species no. 26, p. 1023.) ....*S. angustifolium.*

Staminate head 1; fruiting heads 5-12 mm in diameter; achenes ellipsoid or slenderly obovoid-fusiform, 3.5-5 mm long; stipe obsolete or up to 1 mm long, beak obsolete or up to 1.5 mm long; inflorescence simple, the heads all axillary; sepals elliptic to cuneate-spatulate, a half to two thirds as long as the achene; plants usually floating. (See excluded species no. 27, p. 1023.)....*S. minimum.*

1. **Sparganium eurycárpum** Engelm. GIANT BUR-REED. Map 71. Infrequent to frequent in the lake area and rare or possibly absent from the southern part of the state. There are only three reports for it south of Hamilton County, and it is barely possible that these should be referred to the next species. It is found in wet places, mostly in ditches. It also

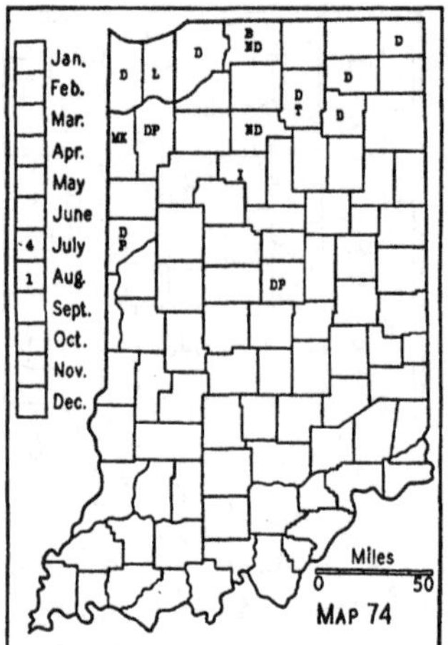

MAP 74
Sparganium chlorocarpum Rydb.

MAP 75
Sparganium chlorocarpum var. acaule (Beeby) Fern.

MAP 76
Potamogeton natans L.

occurs on the low borders of lakes, streams, and sloughs and in ponds and springy places.

N. S., Maine, Que. to B. C., southw. to Fla., Mo., Utah, and Calif.

2. **Sparganium andrócladum** (Engelm.) Morong. (*Sparganium lucidum* Fern. & Eames.) Map 72. My only specimen is from a slough about 4 miles northwest of Grayville, Sullivan County. A specimen reported from St. Joseph County should now be referred to *Sparganium chlorocarpum.* The species of this genus are not well known, hence their distribution is not, as yet, understood.

Newf. to Man., southw. to Fla. and westw.

3. **Sparganium americànum** Nutt. (Including var. *androcladum* Fern. & Eames of Gray, Man., ed. 7.) Map 73. Infrequent in the lake area and probably very local south of it. All of my specimens are from ditches, sloughs, and outlets of lakes.

Newf. to Minn., southw. to Fla. and Mo.

4. **Sparganium chlorocárpum** Rydb. (*Sparganium diversifolium* of authors.) Map 74. All of the specimens, with one exception, and reports are restricted to the lake area where it is infrequent. It is found in habitats similar to those of the preceding species.

Newf. to Iowa, southw. to N. J., N. Y., and Ind.

4a. **Sparganium chlorocarpum** var. **acaúle** (Beeby) Fern. (*Sparganium diversifolium* var. *acaule* (Beeby) Fern. & Eames and *Sparganium acaule* (Beeby) Rydb.) Map 75. All of my specimens are from the lake area except one which was found in Hancock County in a springy place along a creek. It is infrequent but probably more common in the state than the species. The habitat is that of the other species of the genus.

Newf. to N. Dak., southw. to V.a. and W. Va.

11. POTAMOGETONÀCAE Engl. PONDWEED FAMILY

Flowers perfect, borne in spikes; anthers 4; leaves alternate, or the upper ones sometimes opposite....................................58. POTAMOGETON, p. 75.
Flowers unisexual, axillary; stamens 1 (2); leaves mostly opposite, filiform, 1-nerved, entire..62. ZANNICHELLIA, p. 84.

58. POTAMOGÈTON [Tourn.] L. PONDWEED

[Morong. The Naiadaceae of North America. Mem. Torrey Bot. Club 3: 11-65. 36 pl. 1893; Fryer and Bennett. Potamogetons of the British Isles. 1-94. 60 col. pl. 1915; Hagström. Critical Researches on the Potamogetons. 1-281. 119 fig. Stockholm 1916; Fernald. The Linear-leaved North American Species of Potamogeton, Section Axillares. Mem. Gray Herb. 3: 1-183. 40 pl. 1932.]

Note: M. L. Fernald named and cited many of my specimens during the writing of his monograph, and these specimens have greatly aided me in the study of this difficult genus. I have also made free use of his monograph, and I wish to acknowledge this assistance.

I have never made a special effort to collect *Potamogetons,* and some species have probably been overlooked; some which once occurred in the state have doubtless been exterminated. Drainage has destroyed the plants in many places. Cottages now surround most of our lakes, and the dredging of all kinds of aquatic vegetation to improve bathing beaches will doubtless lead to extermination of some species. Many specimens are covered more or less with a deposit of lime which may obscure such characters as veins in the leaves. This can be removed at least in part by gently brushing with a round bristle brush (about size no. 4), or in case of badly incrusted linear-leaved specimens, it can be removed by immersing them in dilute hydrochloric acid. After such treament the specimens should be washed and dried between blotters.

A. Plants with both floating and submerged leaves; floating leaves more or less coriaceous, usually on petioles half as long to much longer than the length of the blades; submerged leaves thin, ranging from linear to ovate, or sometimes reduced to a mere petiole.
B. Submerged leaves bladeless; floating leaves large, 17-29-nerved.
Floating leaves mostly broadly elliptic, subcordate at the base (rarely rounded), 21-29-nerved; fruit mostly 4-5 mm long, stramineous or greenish..............................1. *P. natans.*
Floating leaves mostly narrowly elliptic, narrowed at the base, usually 2-5 times as long as wide, 17-23-nerved; fruit usually reddish (phase of this species, having the blades of submerged leaves rotted off)..............................2. *P. americanus.*
B. Submerged leaves lanceolate to ovate or linear.
Blades of floating leaves 31-55-nerved, rounded at the base; blades of submerged leaves mostly 25-39-nerved; fruit 4-5.5 mm long, usually reddish...............................3. *P. amplifolius.*
Blades of floating leaves with fewer than 31 nerves.
C. Submerged leaves linear, 0.2-13 mm wide.
Peduncles of spikes from the axils of submerged leaves, mostly less than 1.5 cm long; fruit up to 1.5 mm long.
Blades of submerged leaves with bristle tips, 0.2-0.6 mm wide.

Leaves thin, distinctly several-nerved; fruit with a sharp ridge on the back, the sides concave.........................4. *P. capillaceus.*

Leaves 1-nerved (under high magnification 3-nerved); fruit rounded on the back, the sides convex. (See excluded species no. 32, p. 1024.) ....*P. Vaseyi.*

Blades of submerged leaves rounded, subobtuse or acute at the tips, 0.5-2 mm wide.

Submerged leaves obtuse, usually rounded at the tip; the connate leaf sheath much longer than the free stipular tip; the space between the midrib and the faint lateral nerves usually filled with lacunae; fruit 1.3-2.2 mm long, mostly about 2 mm long. (Should be sought in northern Indiana.)................................*P. Spirillus.*

Submerged leaves subobtuse to acute; the connate leaf sheath about half the length of the free stipular tip; the midrib of leaves rarely with lacunae; fruit 1-1.5 mm long..................5. *P. diversifolius.*

Peduncles from the axils of submerged leaves, more than 1.5 cm long.

Floating leaves obtuse at the apex; submerged leaves 6-14 cm long and up to 10 mm wide, ribbonlike, the sides nearly parallel...14. *P. epihydrus.*

Floating leaves acute at the apex; submerged leaves 1-3 cm long, 2-13 mm wide, apiculate, broadest about the middle...6. *P. gramineus* var. *graminifolius.*

C. Submerged leaves lanceolate to ovate; floating leaves large.

Floating leaves broad and distinctly cordate at the base, 25-37-nerved; submerged leaves 9-19-nerved; fruit 3-4 mm long............7. *P. pulcher.*

Floating leaves mostly broadly or narrowly elliptic, rounded or narrowed at the base, generally 17-27-nerved; submerged leaves 7-29-nerved; fruit 3-4 mm long.

Blades of floating leaves usually narrowly elliptic, narrowed at the base; submerged leaves not recurved.

Submerged leaves usually strongly mucronate; mature fruiting spikes mostly 7-8.5 mm wide; fruit green...............8. *P. angustifolius.*

Submerged leaves acute or acuminate, mostly 17-23-nerved; mature fruiting spikes generally 9-11 mm wide; fruit usually tinged red...2. *P. americanus.*

Blades of floating leaves rounded at the base, 17-27-nerved; submerged leaves long, usually recurved, 13-29-nerved............9. *P. illinoensis.*

A. Plants with all the leaves submerged.

D. Blades of leaves lanceolate, oblong or broader, not linear.

Leaves sessile or short-petiolate, not clasping.

Margins of blades finely and sharply serrulate.................10. *P. crispus.*

Margins of blades entire or some with a few minute teeth.

Submerged leaves all mucronate, or long-acuminate.

Fruit 2-2.5 mm long; submerged leaves 2.5-8 cm long...6. *P. gramineus* var. *graminifolius.*

Fruit 3-4 mm long; submerged leaves 4-20 cm long.

Fruit distinctly 3-keeled..........................8. *P. angustifolius.*

Fruit with rounded, scarcely keeled sides................11. *P. lucens.*

Submerged leaves all large, not mucronate................3. *P. amplifolius.*

Leaves with blades clasping the stem for half or more of its diameter.

Blades slightly clasping, lanceolate, rounded and cucullate at the apex (in dried specimens often bifid), mostly 10-30 cm long; fruit 4-5 mm long, the middle dorsal rib prominent and sharply keeled; stipules large, usually not shredded...12. *P. praelongus.*

Blades strongly clasping, lanceolate to ovate-orbicular, 1-8 (11) cm long, obtuse or acute; fruit 2.5-4 mm long, the dorsal ribs inconspicuous and rounded; stipules short and mostly shredded...................13. *P. Richardsonii.*

D. Blades of leaves linear.

Leaves ribbonlike, 2 mm or more wide, with a broad, coarsely cellular-reticulate space on each side of the midrib, 5-7-nerved; stipules very obtuse.. 14. *P. epihydrus.*

Leaves narrower, if 2 mm wide, without broad cellular-reticulate spaces along the midrib.

E. Blades free from the stipules.

Leaves more than 7-nerved, 2-5 mm wide; peduncles stout, 1.5-5.5 cm long, 1-1.6 mm thick; fruit 3.5-5 mm long, with one strong, usually crested, keel on the back....................................15. *P. zosteriformis.*

Leaves 1-7-nerved; fruit not more than 3 mm long.

Blades 5-7-nerved, usually with a pair of glands at the base, 1.5-3.5 mm wide, usually 2-2.5 mm wide, rounded or short-mucronate at the apex; stipules 7-11 mm long; fruit 2-3 mm long, rounded on the back. ....16. *P. Friesii.*

Blades 1-3-nerved (if some leaves 5-nerved, plant not agreeing with the other characters of the preceding species).

Leaves 1-nerved (under high magnification 3-nerved); fruit strongly compressed with the sides almost flat, 1.6-2.2 mm long. (See excluded species no. 32, p. 1024.)..................................*P. Vaseyi.*

Leaves 3-nerved, rarely some of them 5-nerved.

Blades usually without basal glands; peduncles 0.4-3 cm long, clavate; spikes subcapitate, 2-6-flowered, in fruit 2-8 mm long; sepaloid connectives 0.4-1 mm long; fruit compressed, 1.8-2.5 mm long, with a thin or acute, undulate or coarsely dentate dorsal keel.

Primary leaves 4-10 cm long, 1.4-2.7 mm wide, 3-5-nerved, midnerve with 1-3 rows of lacunae on each side at the base; stipules 0.7-1.8 cm long; fruit 2-2.5 mm long, beak broad at the base, 0.2-0.4 mm long; winter buds sessile in the axils or on short (rarely elongate) branches..................17. *P. foliosus* var. *genuinus.*

Primary leaves 1-7 cm long, 0.3-1.5 mm wide, 1-3-nerved; midnerve without marginal lacunae or with a single row on each side below the middle; stipules 3-11 mm long; fruit green, 1.8-2.3 mm long, beak slender, 0.3-0.8 mm long; winter buds terminating the mostly elongate branches...............17a. *P. foliosus* var. *macellus.*

Blades usually with a pair of basal glands; peduncles 1-9 cm long; spikes interruptedly cylindric, of 2-5 remote whorls of flowers or subglobose, in fruit 0.6-1.5 cm long; sepaloid connectives 1-2.5 mm long; fruit plump, 1.9-3 mm long, rounded on the back, dorsal keel obscure.

Spikes subglobose, continuous or slightly interrupted, 2-8 mm long in fruit; leaves 3-7 cm long, rounded or acute at the apex.................................18. *P. pusillus* var. *mucronatus.*

Spikes cylindric, of 2-5 remote whorls of flowers, in fruit 0.6-1.5 cm long.

Stipules strongly fibrous, becoming whitish.

Leaves mostly rigid, obtuse or abruptly contracted to mucronate tips; stipules strongly fibrous...19. *P. strictifolius* var. *typicus.*

Leaves firm, scarcely rigid, very gradually tapering to a slender tip; stipules less strongly fibrous..19a. *P. strictifolius* var. *rutiloides.*

Stipules scarious-membranaceous or subherbaceous, greenish or brownish.

Primary leaves 1-3 mm wide....20. *P. panormitanus* var. *major.*

Primary leaves only 0.3-1 mm wide...20a. *P. panormitanus* var. *minor.*

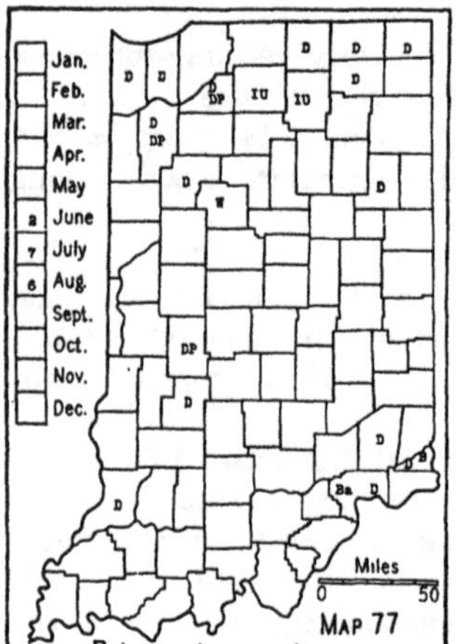

Potamogeton americanus Cham. & Schlecht.

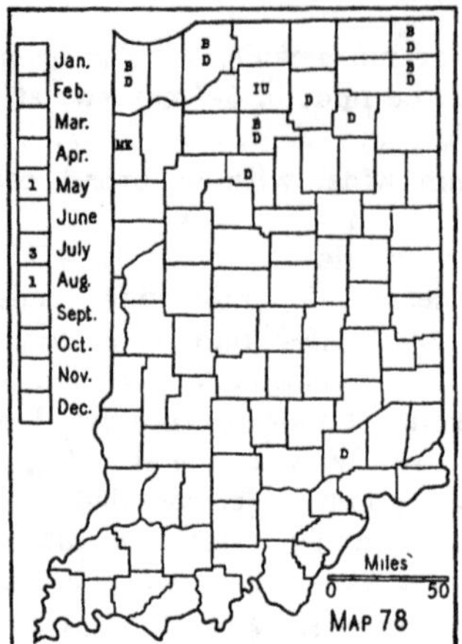

Potamogeton amplifolius Tuckerm.

Potamogeton capillaceus Poir.

E. Blades with the stipules more or less adnate to the base.
- Leaves 4-8 mm wide, auricled at the base, stiffly 2-ranked, with a cartilaginous, finely and sharply serrate margin or the margin entire.
  - Margins of blades finely and sharply serrate..............21. *P. Robbinsii.*
  - Margins entire..........................21a. *P. Robbinsii* f. *cultellatus.*
- Leaves less than 4 mm wide, not auricled at the base, their margins not finely serrate.
  - Spikes from the axils of submerged leaves subglobose, sessile or on peduncles only a few mm long; fruit compressed, with concave sides.
    - Submerged leaves obtuse, usually rounded at the tip; the connate leaf sheath much longer than the free stipular tip; the space between the midrib and faint lateral nerves usually filled with lacunae; fruit 1.3-2.2 mm long. (Should be sought in northern Indiana.).......... ....*P. Spirillus.*
    - Submerged leaves subobtuse to acute at the tip, the connate sheath about half the length of the free stipular tip; midrib of leaves rarely with lacunae; fruit 1-1.5 mm long....................5. *P. diversifolius.*
  - Spikes from the axils of submerged leaves elongate, with separated whorls of flowers, usually 1 to several cm long; fruit 3.5-4.5 mm long, usually with a beak about 0.5 mm long....................22. *P. pectinatus.*

1. **Potamogeton nàtans** L. Map 76. All of my specimens are from lakes in the northern part of the state. Usually found in all of our lakes.

Newf. to B. C., southw. to n. N. J., Pa., Nebr., and Calif.; throughout the world in temperate climates.

2. **Potamogeton americànus** Cham. & Schlecht. Map 77. Frequent in the lake area and rather local south of it. It is found mostly in streams, and less often in lakes, dredged ditches, old canals, ponds, gravel pits, and old stone quarries.

N. B. to B. C., southw. to Fla., Tex., Calif., Mex., and W. I.; also in the Old World.

3. **Potamogeton amplifòlius** Tuckerm. Map 78. Frequent in our lakes and very local elsewhere. I have it, however, from a dredged ditch in Jennings County.

N. S. to B. C., southw. to n. N. J., Ky., Mo., Kans., and Calif.

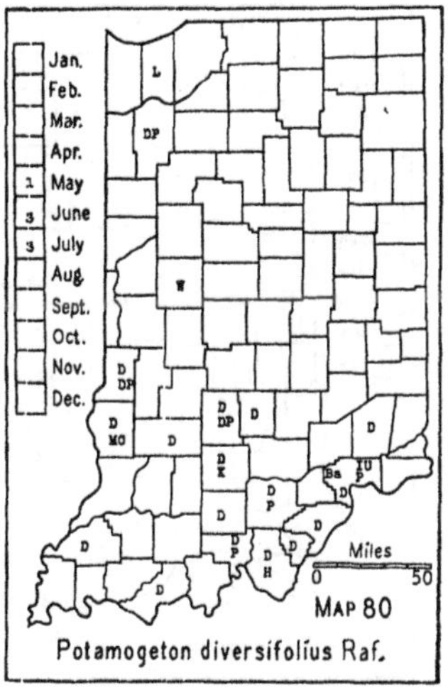

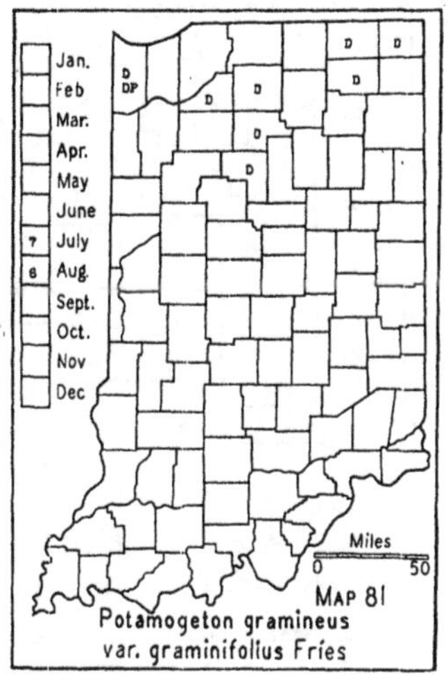

4. **Potamogeton capillàceus** Poir. Map 79. This species was reported by Fernald (Mem. Gray Herb. 3: 111. 1932) as having been found by Hill and by Chase in Goose Pond, near Dune Park, Porter County. This pond is located mostly in section 28 about 4 miles northwest of Porter.

Coastal Plain from Maine to Fla. and Tex., and in Ind. and Wis.; also in Cuba and Isle of Pines.

5. **Potamogeton diversifòlius** Raf. (*Potamogeton hybridus* Michx. of Gray, Man., ed. 7.) Map 80. All of my specimens are from the southern half of the state, although it has been reported repeatedly from the dune area. The reports from the northern part of the state should probably be referred to some other species.

L. I., Pa., s. Ind., Wis., Minn., Mont., s. Oreg., southw. to Ga., Tex., Calif., and n. Mex.

6. **Potamogeton gramíneus** L. var. **graminifòlius** Fries. (*Potamogeton heterophyllus* of recent authors.) Map 81. Rather frequent in shallow water in our lake area.

Throughout the greater part of N. A.

7. **Potamogeton púlcher** Tuckerm. Map 82. My only specimen is from a pond in Sullivan County. It has been reported from the dune area.

Maine to Fla. and westw. to Mo.

8. **Potamogeton angustifòlius** Berchtold & Presl. Map 83. Rather frequent in the lakes that I have studied and probably well distributed in the lake area. It has been reported from the dune area.

Mass., Que., Wyo. to Calif., southw. to Flà. and Tex.; also in W. I., Eurasia, and Africa.

9. **Potamogeton illinoénsis** Morong. Map 84. Infrequent in the lakes throughout the lake area.

Ind. to Minn., southw. to Mo.

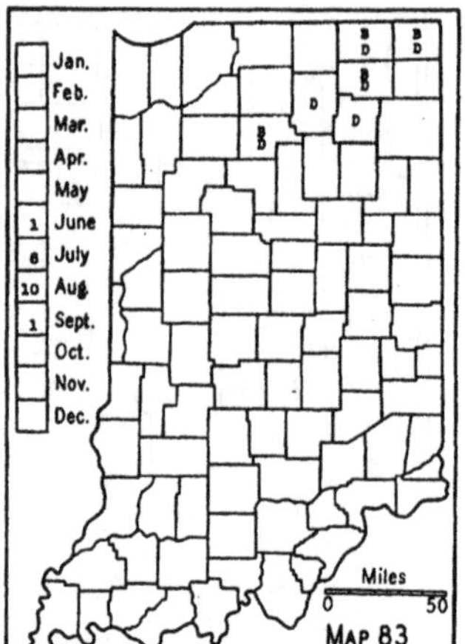

Potamogeton angustifolius Berch. & Presl

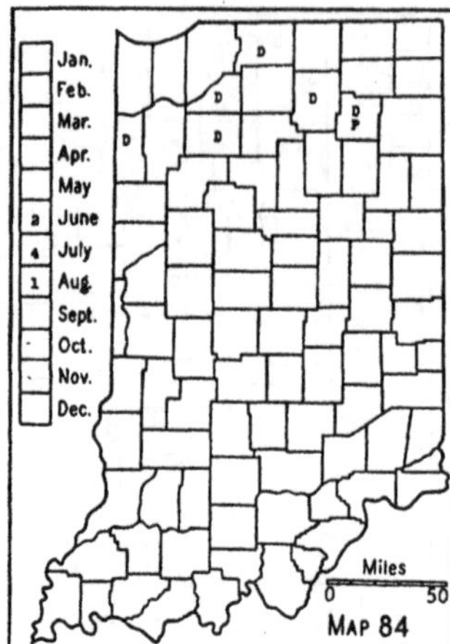

Potamogeton illinoensis Morong

Potamogeton crispus L.

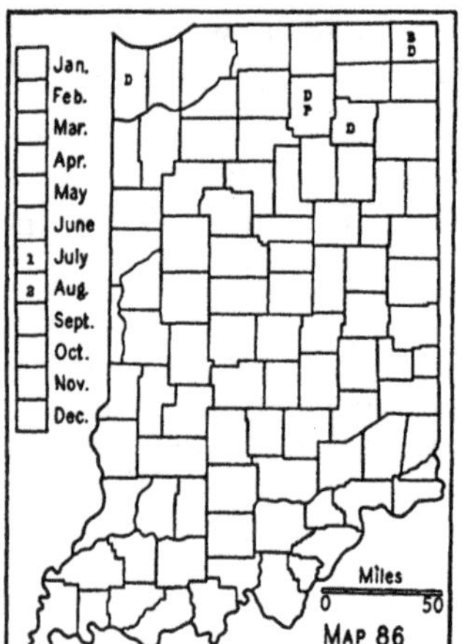

Potamogeton lucens L.

Potamogeton praelongus Wulfen

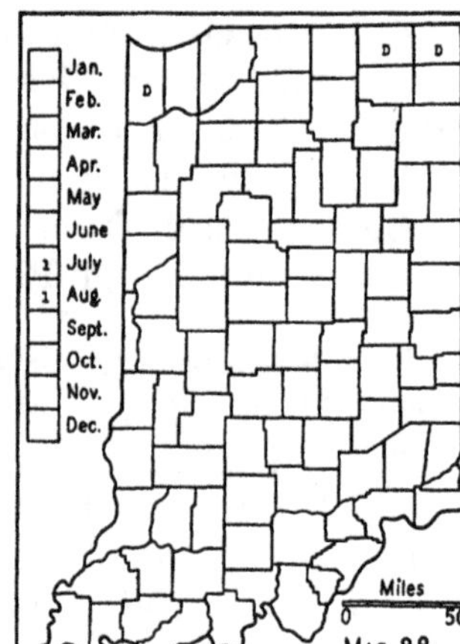

Potamogeton Richardsonii (Bennett) Rydb.

10. POTAMOGETON CRÍSPUS L. Map 85. I have found this species in both Cedar Lake and Wolf Lake in Lake County. It was reported from Wolf Lake as early as 1913. In 1937 I found a few plants in shallow water on the south side of Lake Cicott, Cass County. Doubtless it is not common in this lake because a few years ago I spent a half day in a boat in search for pondweeds in this small lake and I did not find it.

Nat. of Eu.; Mass. to Ont., southw. to Va. and Mo.

11. **Potamogeton lùcens** L. Map 86. Infrequent in our lakes. It is difficult to distinguish this species from *Potamogeton angustifolius* if floating leaves and fruits are not present.

N. S. to Calif., southw. to Fla. and Mex.; also found in W. I., Eurasia, and Africa.

12. **Potamogeton praelóngus** Wulfen. Map 87. Infrequent in the lakes of the lake area.

Newf. to B. C., southw. to Conn., N. J., Ind., Iowa, Mont., and Calif.

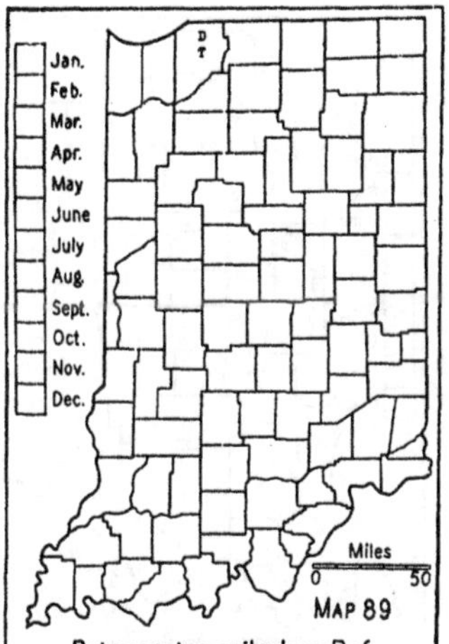

Potamogeton epihydrus Raf.

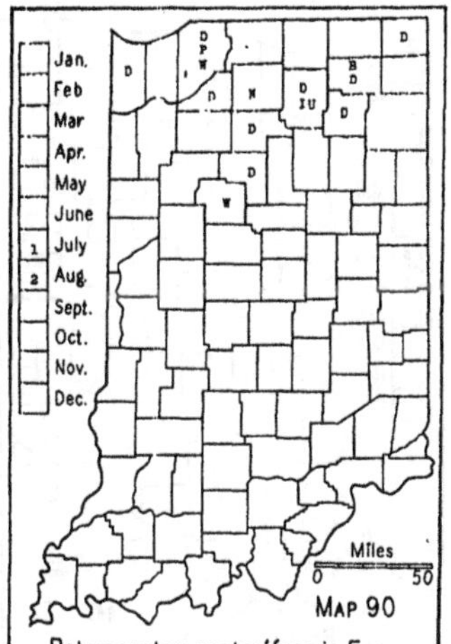

Potamogeton zosteriformis Fern.

Potamogeton Friesii Rupr.

13. **Potamogeton Richardsònii** (Bennett) Rydb. Map 88. In a few of our northern lakes.

Que. to Mack. and B. C., southw. to N. E., N. Y., Ind., and Nebr.

14. **Potamogeton epihỳdrus** Raf. Map 89. Our only specimen is one collected in 1936 by R. M. Tryon, Jr. It was found in State Line Creek in La Porte County. It has been reported but, no doubt, all other reports should be referred to other species.

N. B., Que. to Minn., southw. to N. J., W. Va., Ill., and Iowa; also on the Pacific coast from Wash. to Calif.

15. **Potamogeton zosterifórmis** Fern. (Mem. Gray Herb. 3: 36-40. 1932.) (*Potamogeton zosterifolius* of American authors.) Map 90. Frequent in the lakes of the lake area from which there are many reports. Sterile specimens of this species closely resemble sterile specimens of *Heteranthera dubia* from which they can be separated by the abruptly acute leaf tips. Those of *Heteranthera dubia* have the blades gradually tapering at the apex into a blunt tip.

Que., n. Alberta to s. B. C., southw. to Va., Ohio, n. Ind., n. Ill., n. Iowa, Nebr., nw. Mont., and n. Calif.

16. **Potamogeton Frièsii** Rupr. Map 91. In a few of our northern lakes.

Southern Lab. to B. C., southw. to N. S., Conn., N. Y., Mich., Iowa, and Wash.; also in Eu.

17. **Potamogeton foliòsus** Raf. var. **genuìnus** Fern. Map 92. Infrequent throughout the state in creeks, small rivers, ditches, and gravel pits but rarely in lakes.

Western N. Y., Ont., Mich., Wis., s. Man. to Wash., southw. through the U. S. to W. I. and Cent. Amer.

17a. **Potamogeton foliosus** var. **macéllus** Fern. (Mem. Gray Herb. 3: 46-51. 1932.) Map 92a. The distribution of the variety is indicated on

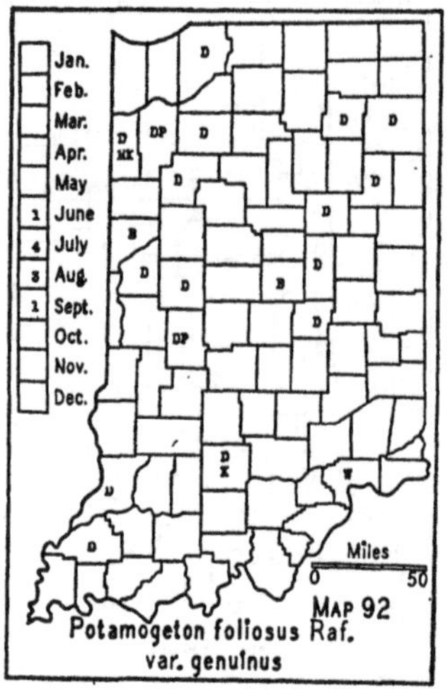

MAP 92 Potamogeton foliosus Raf. var. genuinus

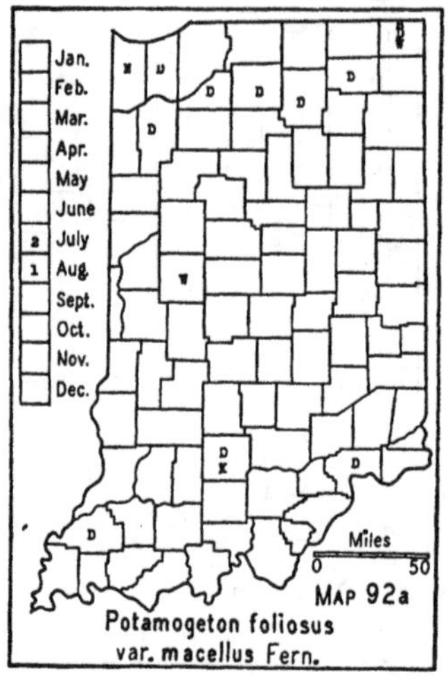

MAP 92a Potamogeton foliosus var. macellus Fern.

MAP 93 Potamogeton pusillus var. mucronatus (Fieber) Graebn.

the map. The habitat is that of the species, although I have more specimens from lakes.

Cape Breton Island, N. S., Que. to Mack., southw. to Fla., Mo., Kans., Nev., and Calif.; also in Hawaii.

18. **Potamogeton pusíllus** L. var. **mucronàtus** (Fieber) Graebn. Map 93. Our only report is that of Fernald. The specimen was collected by E. B. Williamson in Crooked Lake, Steuben County, June 17, 1900, and is deposited in the herbarium of the Missouri Botanical Garden.

Sw. Greenland, Newf. to Alaska, southw. to N. S., s. N. E., L. I., Del., s. Minn., Mont., and Vancouver Island; Eurasia.

19. **Potamogeton strictifòlius** Bennett var. **týpicus** Fern. (Mem. Gray Herb. 3: 56-57. 1932.) Map 94. There are specimens from only a few of our northwestern lakes.

Vt. to Sask., southw. to Mass., cent. N. Y., s. Ont., n. Ohio, n. Ind., n. Wis., s. Minn., n. Nebr., and Utah.

19a. **Potamogeton strictifolius** var. **rutiloìdes** Fern. (Mem. Gray Herb. 3: 57-60. 1932.) Our only Indiana specimen was collected by Scovell & Clark in Lake Maxinkuckee, Marshall County, and is deposited in the herbarium of the Field Museum.

Sw. Que. to Mack., southw. to Vt., nw. N. Y., s. Mich., nw. Ind., s. Minn., n. Nebr., and Utah.

20. **Potamogeton panormitànus** Biv. var. **màjor** G. Fischer. Map 95. My only specimens are from a dredged ditch in Jasper County and from a small lake in De Kalb County.

Magdalen Islands and Gaspé Co., Que. to n. Alberta and s. B. C., southw. to Va., Ark. to s. Calif., and south-central Mex.; Cuba, Azores, and Eurasia.

Map 94
Potamogeton strictifolius Bennett var. typicus

Map 95
Potamogeton panormitanus var. major G. Fischer

Map 95a
Potamogeton panormitanus var. minor Biv.

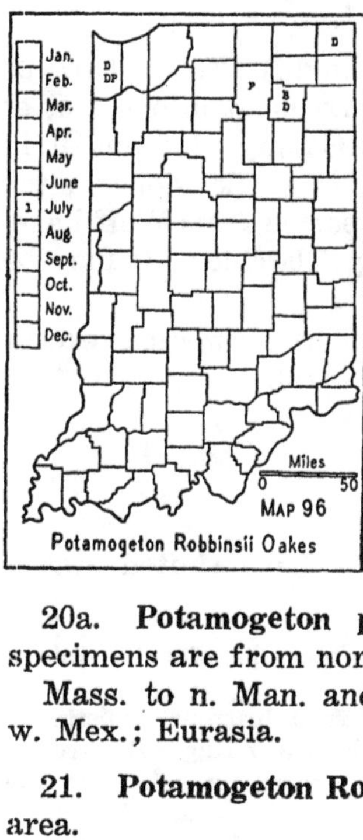

Map 96
Potamogeton Robbinsii Oakes

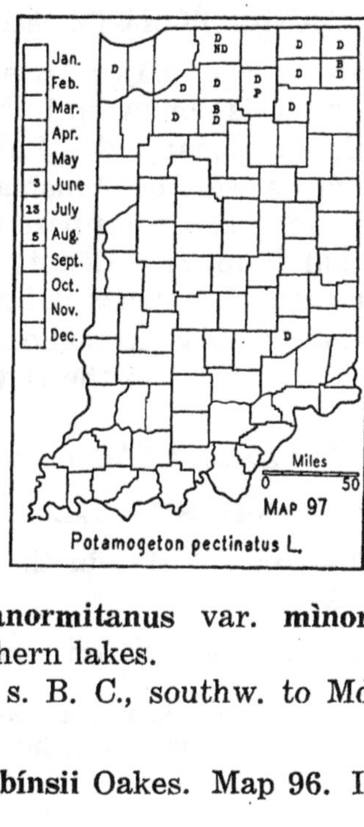

Map 97
Potamogeton pectinatus L.

Map 98
Zannichellia palustris L. var. major (Boenningh.) Koch

20a. **Potamogeton panormitanus** var. **mìnor** Biv. Map 95a. Our specimens are from northern lakes.

Mass. to n. Man. and s. B. C., southw. to Md., s. Ala., La., Tex., and w. Mex.; Eurasia.

21. **Potamogeton Robbínsii** Oakes. Map 96. In a few lakes of the lake area.

N. B. to n. Ont., southw. to Del., Pa., n. Ind.; also Wyo. and s. B. C. to Nev.

21a. **Potamogeton Robbinsii** f. **cultellàtus** Fassett. (Rhodora 35: 389. 1933.) Fassett cites a specimen of this form which was collected by J. T. Scovell in Lake Maxinkuckee and which is now in the Gray Herbarium.

Conn., Ont., Mich., Ind., and Wis.

22. **Potamogeton pectinàtus** L. Map 97. This in frequent to common in all of our lakes in the lake area.

Newf. to B. C., southw. to Fla., Tex., and Calif.; also in Eu.

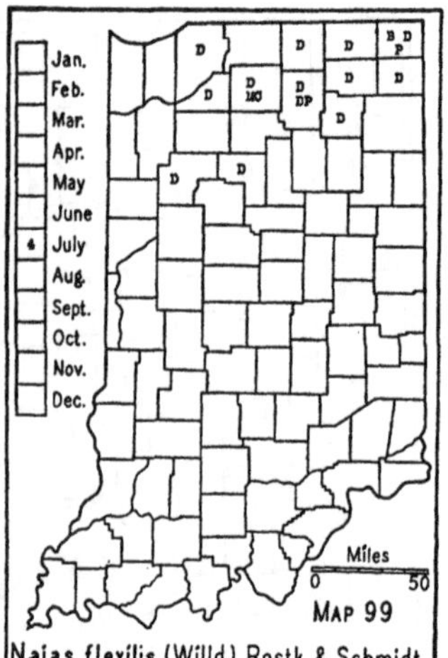

Najas flexilis (Willd.) Rostk. & Schmidt

Najas guadalupensis (Spreng.) Morong

Najas gracillima (A.Br.) Morong

62. ZANNICHÉLLIA [Micheli] L.

1. **Zannichellia palústris** L. var. **màjor** (Boenningh.) Koch. HORNED PONDWEED. Map 98. I found this pondweed to be frequent in one foot of water on the southwest side of Cedar Lake, Lake County. I found it in Pulaski County about 13 miles west of Winamac, in Little Monon ditch where it is crossed by State Road 14. It has been reported from Wolf Lake, Lake County, by Peattie and from Vigo County by Blatchley. It may be more frequent in the state than our reports indicate.

In fresh or brackish water nearly throughout North America, except the extreme north; widely distributed in the Old World.

12. NAJADÀCEAE Lindl.

64. NÀJAS L. NAIAD

[Clausen. Studies in the genus Najas in the northern United States. Rhodora 38: 333-345. 1936.]

Leaves mostly (0.4) 0.5-1 mm wide and 1-1.5 cm long, gradually widening into a clasping base.

Styles (including the stigmas) filiform, 0.8-2 mm long; fruit lustrous, indistinctly marked with 30-50 longitudinal lines, enclosing obscure hexagonal areolae; leaves gradually tapering from the base into a long drawn out point, the fine teeth numerous....................1. *N. flexilis.*

Styles (including the stigmas) stouter, 0.1-0.6 mm long; fruit dull, more distinctly marked with about 10-20 longitudinal lines which enclose rectangular areolae; leaves linear, with a rounded or merely acute apex, the teeth not so numerous as in the preceding species but more conspicuous..........2. *N. guadalupensis.*

Leaves mostly 0.25 mm wide, ranging from 0.2-0.3 mm wide and 1.5-2.5 cm long; fruit somewhat curved, dull, the surface longitudinally marked with short, oblong reticulations....................3. *N. gracillima.*

1. **Najas fléxilis** (Willd.) Rostk. & Schmidt. Map 99. So far as known, this species is restricted to the lake area of the state. It is found principally in lakes and in a few rivers. A variety *robusta* Morong is a stouter

Triglochin maritima L.

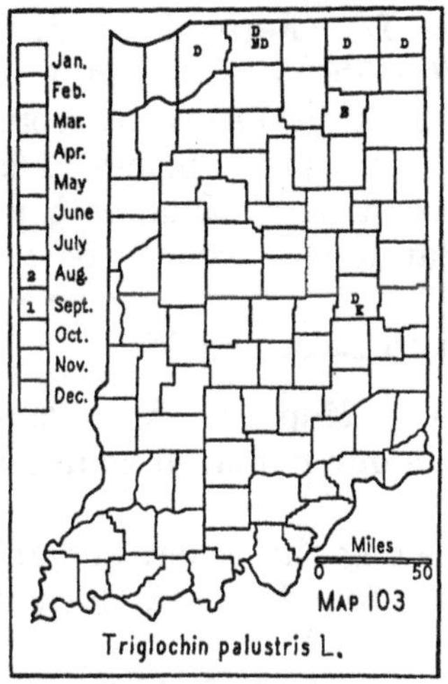

Triglochin palustris L.

Scheuchzeria palustris var. americana Fern.

form that rarely fruits, and, according to Clark, (Lake Maxinkuckee 2: 173. 1920), grows on muddy bottoms in deeper water than the species.

Md., Ohio, Ind., Ill., Iowa, Idaho to Oreg., and northw. into Canada.

2. **Najas guadalupénsis** (Spreng.) Morong. Map 100. This species is found in lakes and is restricted to our lake area. I have never taken notes concerning the habitats of this or the preceding species, but all that I have collected were found on sandy or marly bottoms in less than 4 feet of water.

Basin of the St. Lawrence River to Minn., and Oreg., southw. to Fla. and Mex., W. I., and S. A.

3. **Najas gracíllima** (A. Br.) Morong. (*Najas gracillima* (A. Br.) Magnus of Gray, Man., ed. 7 and Britton and Brown, Illus. Flora, ed. 2.) Map 101. This species was reported in 1876 by Schneck as found in the "deeper ponds" of the Lower Wabash Valley. Our only specimens were collected in 1935 by Kriebel in Lawrence County.

Maine, N. Y., Wis., and Minn., southw. to Mass., Conn., N. Y., Ind., and Mo.

14. JUNCAGINÀCEAE Lindl. ARROW-GRASS FAMILY

Stem scapose; leaves all radical; flowers bractless, many, in a spikelike raceme; ovaries 3-6, united until maturity........................66. TRIGLOCHIN, p. 85.

Stem leafy; flowers bracteate, few, in a loose raceme; ovaries 3, nearly distinct, divaricate..67. SCHEUCHZERIA, p. 86.

66. TRIGLÒCHIN [Riv.] L. ARROW-GRASS

Fruit oblong or ovoid, mostly 3-6 mm long and 2-3.5 mm wide, rounded at the base; carpels 6 (rarely 3 but none seen in Indiana), not beginning to separate first at the base...1. *T. maritima.*

Fruit linear or clavate, mostly 7-8 mm long, about 1 mm wide, tapering to a narrow base; carpels 3, separating first at the base....................2. *T. palustris.*

1. **Triglochin marítima** L. (Fernald. Some variations of Triglochin maritima. Rhodora 5: 174-175. 1903.) Map 102. Infrequent in a few counties in the lake area. It prefers calcareous soil and grows on the marly borders of lakes and in springy places. I have seen it growing with the next species in marl so strongly alkaline that only a few plants could survive. In such a habitat it will usually be associated with *Eleocharis pauciflora.*

Lab. to Alaska, southw. to N. J. and Mex.

2. **Triglochin palústris** L. Map 103. Very local in marly springy areas on marly shores of lakes in our northern counties and in a marly springy place in Henry County.

Greenland to s. Maine along the coast, and inland to the Great Lakes, westw. to Colo. and Alaska; found also in Eurasia.

67. SCHEUCHZÈRIA L.

1. **Scheuchzeria palústris** L. var. **americàna** Fern. (Rhodora 25: 177-. 179. 1923.) Map 104. Very local in some of the counties of the lake area. I have it only from two counties but it has been reported also from Cass, Fulton, Lake, Marshall, Porter, and St. Joseph Counties. It is usually found in sphagnum with pitcherplant and cranberry.

Newf. to Hudson Bay and Alaska, southw. to N. J., Pa., Wis., and Calif.

15. ALISMÀCEAE DC. WATER-PLANTAIN FAMILY

Flowers in a panicle, the branches bearing whorls of flowers in verticils of 3-10 flowers each; flowers perfect; carpels in a single series, forming a ring on a small receptacle........70. ALISMA, p. 86.

Flowers in verticils; carpels in several series on a convex receptacle.

Flowers in verticils of 3-9 or more, in plants of average vigor with some of the verticils with more than 3 flowers; leaf blades large, cordate or subcordate at the base, usually with 5-7 primary veins; flowers all perfect........75. ECHINODORUS, p. 87.

Flowers mostly in verticils of 3, or 1 or 2 at a node; leaf blades sagittate or lanceolate, usually with more than 5-7 veins.

Fruiting pedicels very thick, usually 2-5 cm long, at least the lowermost widely spreading or recurved; sepals mostly suborbicular, large, surrounding the mature fruit; lower verticils of flowers pistillate, the upper ones staminate; stamens 9-15........76. LOPHOTOCARPUS, p. 88.

Fruiting pedicels not conspicuously thick, ascending; sepals not broad and surrounding the fruit at maturity, usually reflexed; staminate flowers on separate scapes or at the top of the scape above the pistillate ones; stamens numerous........78. SAGITTARIA, p. 88.

70. ALÍSMA L. WATER-PLANTAIN

Petals 1-2 mm long; anthers subspherical, 0.3-0.5 mm long; styles 0.2-0.3 mm long, curved at the apex; achenes 1.5-2 (2.5) mm long........1. *A. subcordatum.*

Petals 3.5-6 mm long; anthers oblong, 0.6-0.9 mm long; styles 0.4-0.7 mm long, slightly curved; achenes 2.5-3 mm long........2. *A. Plantago-aquatica* var. *brevipes.*

1. **Alisma subcordàtum** Raf. (*Alisma Plantago-aquatica* of Gray, Man., ed. 7 and of Indiana authors, in part, not of L.) Map 105. Infrequent to

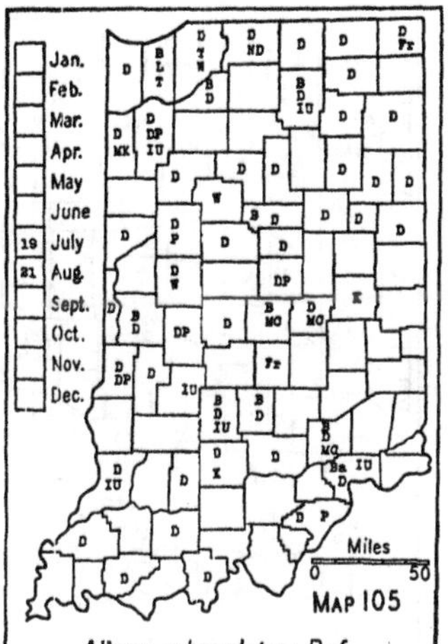

Alisma subcordatum Raf.

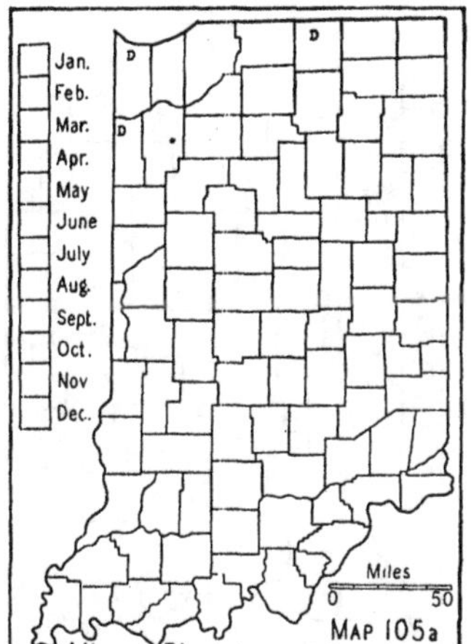

Alisma Plantago-aquatica L. var. brevipes (Greene) Samuelsson

Echinodorus radicans (Nutt.) Engelm.

frequent throughout the state, being more common in the lake area where dredged ditches are more frequent. It is found in muddy or mucky soil in ditches, ponds, and sloughs and about lakes.

N. S. to Minn., southw. to Fla. and Tex.

2. **Alisma Plantàgo-aquática** L. var. **brévipes** (Greene) Samuelsson. Found only in our northern counties with the habitat of the preceding species. This species was unknown to me until I studied my specimens. It is probable that now since I know it, I could find it in more of our northern counties. I have specimens from Lake, Elkhart, and Newton Counties. This is the boreal representative of the genus.

N. S., Maine, Col. to Wash.

75. ECHINÓDORUS Richard

Scapes (stems) reclining or prostrate, 7-15 dm long, usually rooting at the nodes; leaves cordate, blades 4-15 cm long; flowers in verticils at the nodes, their pedicels 2-5 cm long in fruit; beak of achene a fourth as long as the body...1. *E. radicans.*

Scapes erect, 10-30 cm high; leaves cordate, 2-11 cm long; pedicels stiff, 12-15 mm long in fruit; beak of achene half as long as the body...........2. *E. cordifolius.*

1. **Echinodorus radicans** (Nutt.) Engelm. Map 106. This species is restricted to the Lower Wabash Valley where it is found on the muddy borders of old river channels. Very local.

D. C. to Kans., southw. to Fla. and Tex.

2. **Echinodorus cordifòlius** (L.) Griseb. Map 107. The only specimen of this species known to have been collected in Indiana is one in the herbarium of DePauw University. It was collected by Blatchley on the south side of Conover's Pond, now drained, which was located in the southeast corner of sec. 9, now within the city limits of Terre Haute, Vigo County. This species was reported from Tippecanoe County by Wilson, but his specimen can not be located.

Ind., Ill. to Mo., southw. to Fla. and Tex.

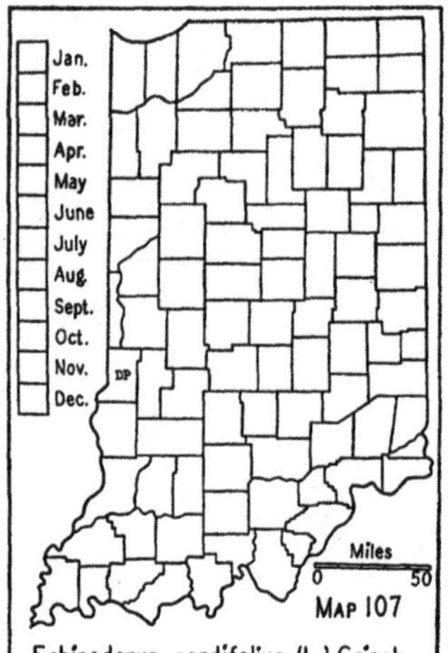

Echinodorus cordifolius (L.) Griseb.

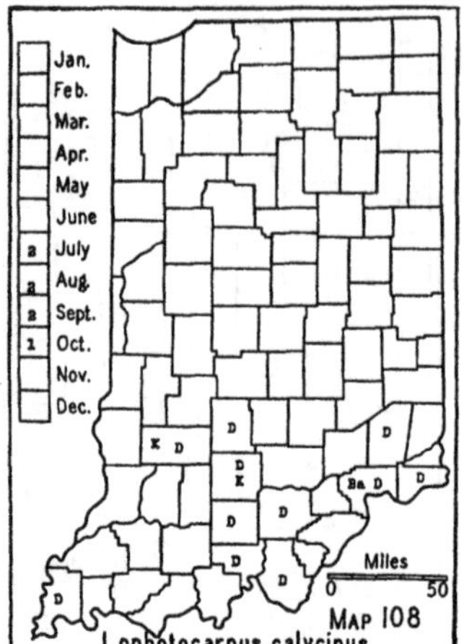

Lophotocarpus calycinus (Engelm.) J. G. Smith

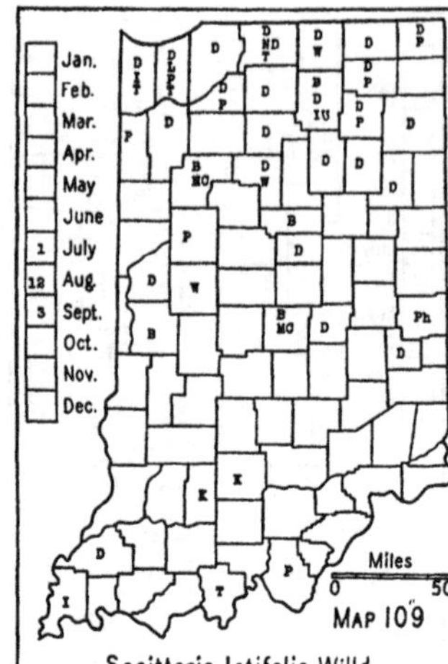

Sagittaria latifolia Willd.

76. LOPHOTOCÁRPUS Th. Durand

1. **Lophotocarpus calycinus** (Engelm.) J. G. Smith. Map 108. Restricted to the southern part of the state where it is found in artificial ponds and in sinkholes. All of my specimens were found in such habitats except one, which was from a muddy slough along White River in Greene County. When once established in a pond or sinkhole, it soon becomes the dominant plant, usually almost crowding out all other species. It is fast migrating northward, and I now find it in places where it was absent 20 years ago. None of our early authors reported it. Probably introduced.
Del. to S. Dak., southw. to Ala. and N. Mex.

1a. **Lophotocarpus calycinus** f. **máximus** (Engelm.) Fern. (Rhodora 38: 73. 1936.) This is a very wideleaf form with blades up to 3 dm wide and with 18-21 nerves. Miss Edna Banta found it in an artificial pond in Jefferson County.
Ohio and southw.

1b. **Lophotocarpus calycinus** f. **depauperàtus** (Engelm.) Fern. (Rhodora 38: 73. 1936.) I collected a specimen of this form in an artificial pond on the August Bocard farm on the road between Corydon and Milltown, about a mile south of DePauw, Harrison County.

78. SAGITTÀRIA L. ARROWHEAD

Leaves all sagittate, rarely somewhat hastate, or some without lobes, the basal lobes as long as, shorter, or longer than the terminal one; pistillate heads never sessile; filaments of stamens glabrous.

Bracts ovate, obtuse or rarely merely acute, usually 4-8 (10) mm long; achenes mostly 2-3 mm long; beaks of achenes, 0.5-2 mm long, horizontal, arising from the inner margin and pointing inward; leaf blades usually about 1.5 dm long (sometimes up to 4.5 dm long or as short as 3 cm long).

Bracts and pedicels pubescent. (See excluded species no. 35, p. 1024.) .... *S. pubescens.*

Bracts and pedicels glabrous; scape not ribbed or rarely so; faces of achenes not keeled or crested.

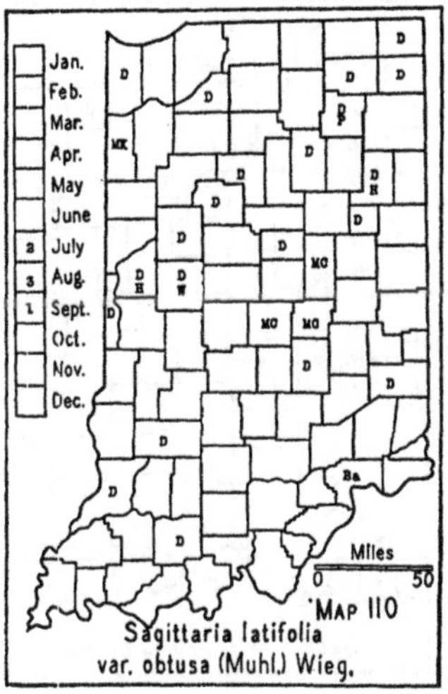

Sagittaria latifolia var. obtusa (Muhl.) Wieg.

Sagittaria cuneata Sheldon

Sagittaria brevirostra Mack. & Bush

Terminal lobes of leaves of an ovate type, wider than linear.
Apex of terminal lobe of leaves acute; plants usually monoecious; scapes usually terete..1. *S. latifolia.*
Apex of terminal lobe of leaves obtuse or rounded; plants usually dioecious. ....1a. *S. latifolia* var. *obtusa.*
Terminal lobes of leaves linear.....................1b. *S. latifolia* f. *gracilis.*
Bracts lanceolate or long-cuneate, usually long-acuminate at the apex, sometimes merely acute in *Sagittaria cuneata;* beaks of achenes erect or if curved, bent outward from an erect base.
Beaks of achenes very short, less than 0.5 mm long, usually about 0.3 mm long; achenes 2-2.5 (3) mm long, strongly obovoid, their faces not keeled or crested; leaves mostly 4-15 cm long, sagittate or sagittate-hastate; bracts mostly 6-12 mm long...2. *S. cuneata.*
Beaks of achenes 0.5-2 mm long, usually about 1.5 mm long, arising from the inner edge of the achene and forming with the top of the achene a minute sinus, sometimes some of the beaks diverging; achenes keeled on each face and their margins more or less notched.
Each face of the achene with a single keel; sinus at the top of achene (between the wing and beak) wide and rounded; scapes strongly ribbed; bracts acuminate, mostly 8-25 mm long.......................3. *S. brevirostra.*
Each face of the achene with two or more crested keels and sometimes with one or two short, intermediate ones; sinus at the top of the achene deeper and narrower, sometimes almost closed by the outwardly curved beak......... ....4. *S. australis.*
Leaves all entire or with a few leaves lobed; blades linear, lanceolate, or elliptic; filaments of stamens more or less glandular-pubescent; plants growing in shallow water or in very wet places.
Pistillate heads sessile; beak of achene about 1.5 mm long; body of achene about 3 mm long...5. *S. rigida.*
Pistillate heads pedicellate; beak of achene very short, mostly 0.3-0.75 mm long, lateral; body of achene about 2 mm long.....................6. *S. graminea.*

1. **Sagittaria latifòlia** Willd. COMMON ARROWHEAD. Map 109. The extreme variability of the leaves of this species has led authors to describe several forms, one of which has been reported from Indiana. I think that much of the variations in leaf pattern is due to habitat. This species is

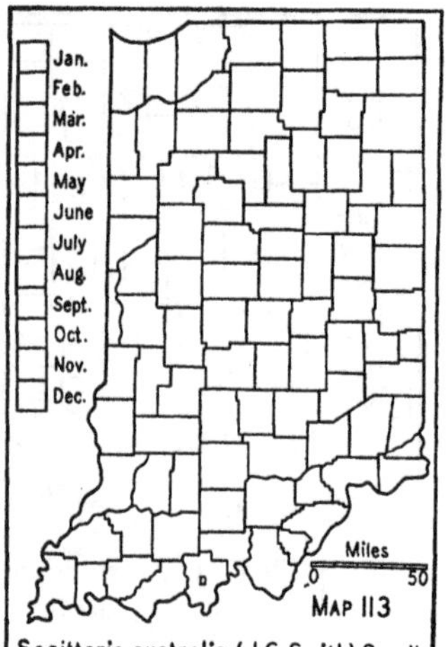

Sagittaria australis (J.G. Smith) Small

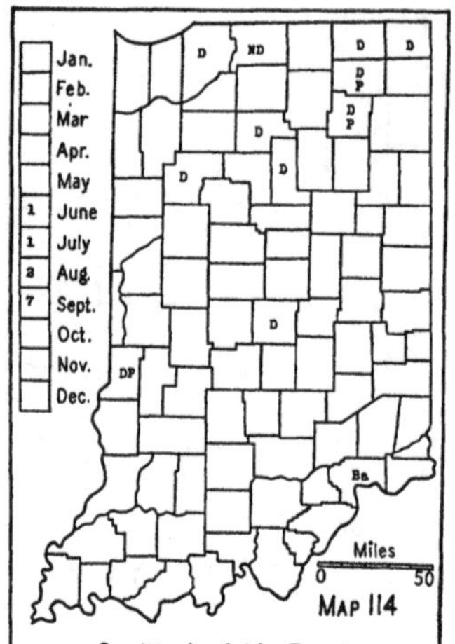

Sagittaria rigida Pursh

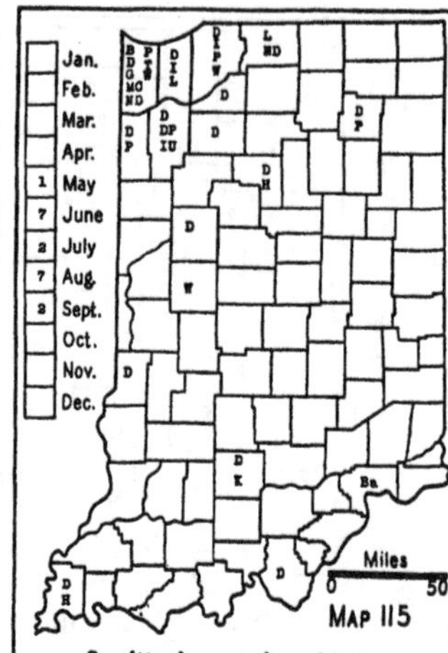

Sagittaria graminea Michx.

restricted mostly to the lake area with a few outlying stations. It has been reported in various parts of the state because, no doubt, it has not been separated from *Sagittaria brevirostra*. It is found on the muddy borders of streams, ponds, and lakes and in ditches. It is rather frequent in its habitat but its habitat is more or less local. Since there has been no recent revision of the genus, the general distribution is not definitely known and the best that can be done is to accept that of our most recent authors.

N. B. to B. C., southw. to Fla. and Calif.

1a. **Sagittaria latifolia** var. **obtùsa** (Muhl.) Wieg. (Rhodora 27: 186. 1925.) (*Sagittaria latifolia* f. *obtusa* (Muhl.) Rob.) Map 110. This form is probably local or infrequent throughout the state. The habitat is that of the species. The general distribution is not known.

1b. **Sagittaria latifolia** f. **grácilis** (Pursh) Rob. This is a rare form in our area. In 1936 I studied some large colonies on the marl border of the northwest part of Crooked Lake, Steuben County. On the shore and as far out as I could wade with boots, the typical form of the species occurred. Beyond this, which I examined with a boat, the roots of the marsh plants formed a floating mass among which the linear-lobed form was frequent. Among them could be found plants with all the leaves with two lobes. Others could be found where a single plant would have leaves with two lobes, one lobe, and others without lobes (mere phyllodia).

2. **Sagittaria cuneàta** Sheldon. (*Sagittaria arifolia* Nutt.) Map 111. All of our specimens are from the lake area where it is local, although there are no reports from the dune area. Found on the muddy or wet, sandy borders of streams, lakes, and ponds and in ditches.

N. S., Que. to B. C., southw. to Conn., Kans., N. Mex., and Calif.

3. **Sagittaria brevirόstra** Mack. & Bush. SHORTBEAK ARROWHEAD. Map 112. This plant is found probably throughout the state and is probably

our most common species. This arrowhead is more robust than *Sagittaria latifolia* with which it is sometimes associated. It is found on the muddy shores of streams, ponds, and sloughs and in ditches. I have seen specimens from Iowa, Illinois, Wisconsin, Missouri, and Tennessee.

4. **Sagittaria austràlis** (J. G. Smith) Small. Map 113. This is a southern species which is known only from Perry County. It is found on muddy shores.

Pa., Va., and Ind. to Ala.

5. **Sagittaria rígida** Pursh. (*Sagittaria heterophylla* Pursh.) STIFF ARROWHEAD. Map 114. This species is essentially northern in its distribution and is practically restricted to our lake area with a few locations south of it. It is infrequent and found on muddy borders and in ditches. The leaves are extremely variable, ranging from linear to rather broadly elliptic. Three forms have been named, but I believe these ecological fluctuations do not merit names.

Que. to Minn., southw. to N. J., Tenn., and Kans.

6. **Sagittaria gramínea** Michx. Map 115. Infrequent in a part of the lake area and local southward. This species is usually found in shallow water or in very wet places about lakes, ponds, and artificial ponds and in ditches.

Newf. to Sask., southw. to Fla. and Tex.

17. HYDROCHARITÀCEAE Asch. FROGBIT FAMILY

Leaves less than 2 cm wide.
Plants with long, leafy submerged stems; spathes very small, sessile...87A. ANACHARIS, p. 91
Plants stemless, submerged, with long narrow leaves; spathes peduncled...89. VALLISNERIA, p. 92.
Leaves more than 2 cm wide.......................................97. LIMNOBIUM, p. 92.

87A. ANÁCHARIS Bab. & Planch. WATERWEED

[Victorin. L' Anacharis canadensis. Contrib. Lab. Bot. Univ. Montreal 18: 1-43. figs. 7. 1931.]

K. M. Wiegand has made an extensive study of the species of this genus in the Cayuga Lake Basin and has published his findings in the "Flora of the Cayuga Lake Basin," by Wiegand & Eames. I have taken the following key from this work, and I here make acknowledgment for its use.

Leaves 1.2-4 mm wide (averaging 2.13 mm); spathe of the staminate flower oblong-linear, 11-13 mm long, constricted at base into a stipelike part, the orifice gaping, 2-lobed; staminate flower remaining attached by means of a long filiform peduncle; sepals or mature bud of the staminate flowers 3.8-5 mm long; anthers 2.2-2.5 mm long; sepals or mature bud of the pistillate flowers 2.3-2.7 mm long...1. *A. canadensis.*
Leaves 0.7-1.8 mm wide (averaging 1.3 mm); spathe of the staminate flower globose, apiculate, the body about 2 mm long; staminate flower sessile, breaking out of the spathe and rising free to the surface before anthesis; sepals or mature bud of the staminate flowers 2-2.5 mm long; anthers 0.8-1.1 mm long; sepals or mature bud of the pistillate flowers 1.2-1.8 mm long.................2. *A. occidentalis.*

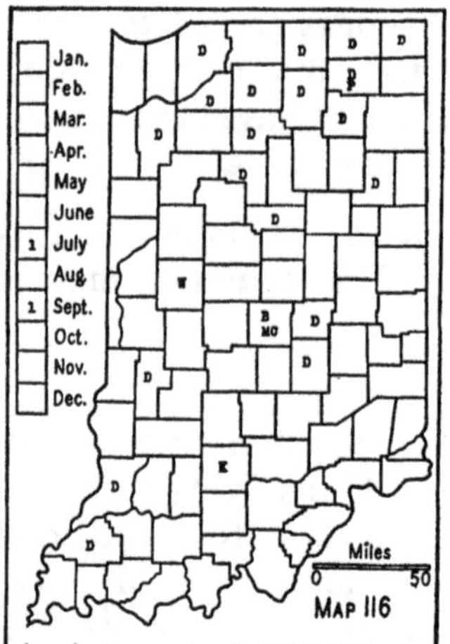

Anacharis canadensis (Michx.) Planch.

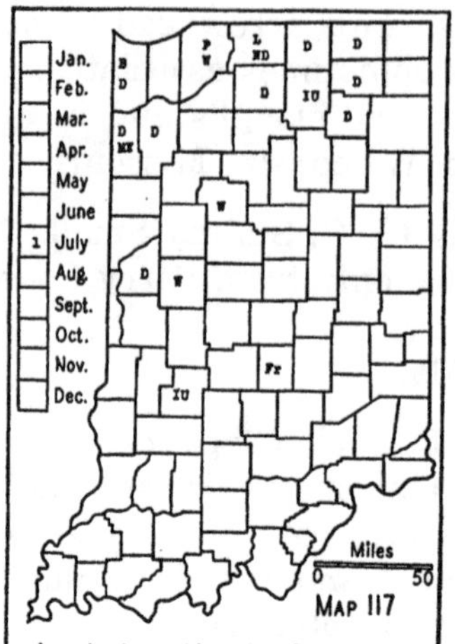

Anacharis occidentalis (Pursh) Vict.

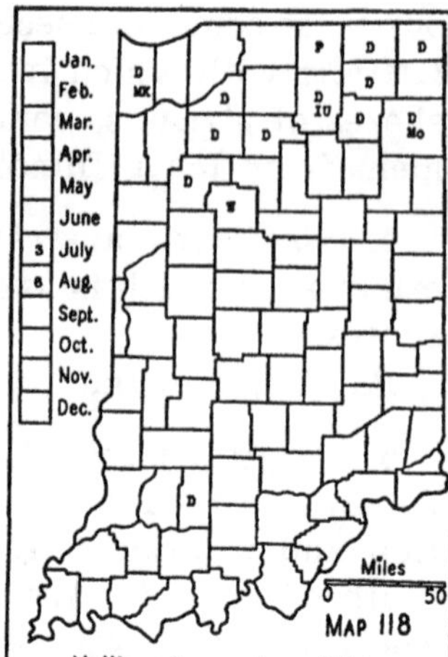

Vallisneria americana Michx.

1. **Anacharis canadénsis** (Michx.) Planch. (*Elodea* of Gray, Man., ed. 7 and *Philotria* of Britton and Brown, Illus. Flora, ed. 2.) CANADA WATERWEED. Map 116. Frequent to common in most of our lakes, ponds, slow flowing streams, and ditches of the lake area, becoming rare southward because its habitat is not found. It prefers clear and calcareous waters. Wiegand, in his study of the species, concludes that *Anacharis canadensis* is dioecious and that *Anacharis Planchonii* is the pistillate form of the species.

Que., N. E. to Sask. and Wyo., southw. to N. Y., Ky., and Ill.

2. **Anacharis occidentàlis** (Pursh) Vict. (Contrib. Lab. Bot. Univ. Montreal 18: 50: 1931.) (*Philotria angustifolia* of Britton and Brown, Illus. Flora, ed. 2 and *Elodea Nuttallii* (Planch.) St. John.) WESTERN WATERWEED. Map 117. This species has the habitat of the preceding species but is less frequent. Most of our specimens are from the lake area.

Southern Maine to Wis. and Oreg., southw. to D. C., Mo., and Nebr.

89. VALLISNÈRIA [Micheli] L.

1. **Vallisneria americàna** Michx. (Rhodora 20: 108. 1918.) (*Vallisneria spiralis* of Gray, Man., ed. 7 and Britton and Brown, Illus. Flora, ed. 2.) WILD CELERY. Map 118. Infrequent to frequent or even common in the lakes of the lake area and rare in our streams, except those of the lake area where it may be common. The sepals of my specimens and those which I have measured in the field are rounded at the apex and 3-3.5 mm wide and 3-5 mm long, usually slightly less than 4 mm long. The peduncles of the staminate inflorescences are mostly about 1 cm long and the leaves are 6-8 mm wide. The widest leaf I have been able to find was 9 mm wide.

Cent. Maine to S. Dak., southw. to Fla. and Tex.

97. LIMNÒBIUM Richard

See excluded species no. 38, p. 1024.

19. GRAMÍNEAE Juss. GRASS FAMILY

[Hitchcock. Manual of the Grasses of the United States. 1040p. 1096 fig. 1935. Deam. Grasses of Indiana. 356p. 81 pl. 1929.]

The sequence of genera, nomenclature, and concept of species are those of Hitchcock, "Manual of the Grasses of the United States." In a few instances, however, where a named form of a species is distinct in Indiana and is not given in Hitchcock's Manual, it is added here in the belief that it will be appreciated by students who are making an intensive study of the grasses.

It is to be noted that the numbers of the genera are not consecutive; this is because they are the ones used in Hitchcock's Manual. For the benefit of students who prefer to follow the sequence of genera as they occur in Dalla Torre and Harms' General Siphonogamarum, those numbers are also added, following the number used in Hitchcock's Manual.

KEY TO THE TRIBES

Plants woody, culms perennial..............................1. BAMBUSEAE, p. 94.
Plants herbaceous, culms annual.
Spikelets 1-many-flowered,[1] terete or laterally compressed; sterile lemmas or incompletely developed florets above the fertile ones, except in *Uniola* and the *Phalarideae*, in each of which the spikelet has at least 3 florets, the lower 2 sterile or rudimentary, and in *Arrhenatherum*, which has 2 florets, the upper perfect, the lower staminate.
Glumes present, rarely one of them obsolete.
Spikelets 3-flowered in plan, the uppermost floret perfect, the lower 2 staminate or represented by sterile lemmas, which may be reduced to minute scales. ....7. PHALARIDEAE, p. 144.
Spikelets 1-many-flowered, no incomplete florets below the perfect ones, except in *Uniola*, *Phragmites*, and *Arrhenatherum*, none of which has spikelets 3-flowered in plan.
Inflorescence of spikes or racemes, either solitary, digitate, racemose, or the spikelets never long pedicellate.
Spikelets solitary or in clusters of 2-6, alternate on opposite sides of the axis; spike solitary, terminal....................3. HORDEAE, p. 113.
Spikelets in 1-sided spikes or racemes, the spikes or racemes solitary or several...................................6. CHLORIDEAE, p. 141.
Inflorescence a panicle, open or contracted, sometimes spikelike.
Spikelets 1-flowered..........................5. AGROSTIDEAE, p. 125.
Spikelets 2-many-flowered.
Glumes shorter than the lowermost floret (see also *Sphenopholis*); lemmas usually awnless, if awned, the awn terminal or from a minutely bifid apex.................................2. FESTUCEAE, p. 95.
Glumes at least as long as the lowermost floret (shorter in *Sphenopholis*); lemmas awnless or with the awn attached to the back or from a bifid apex.......................................4. AVENEAE, p. 121.
Glumes obsolete.
Flowers perfect, each having a pistil and at least 1 stamen..8. ORYZEAE, p. 145.
Flowers imperfect, staminate and pistillate flowers in different spikelets....... ....9. ZIZANIEAE, p. 146.

[1] Spikelets of *Panicum* are apparently 1-flowered but examination shows them to be structurally 2-flowered. The upper flower is fertile and the lower one is represented usually only by a lemma which is the outer or loose one of the spikelet.

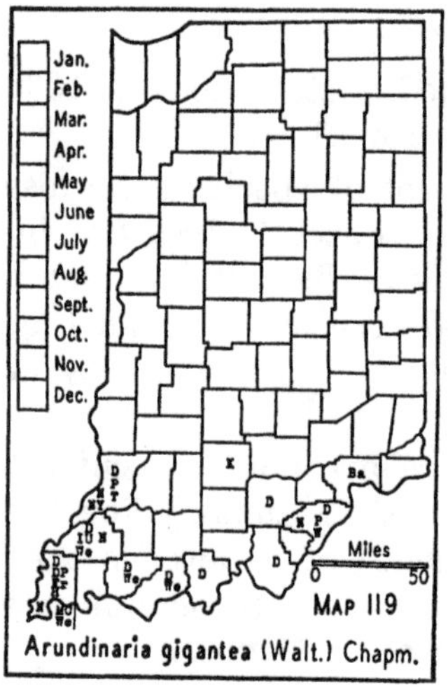

Arundinaria gigantea (Walt.) Chapm.

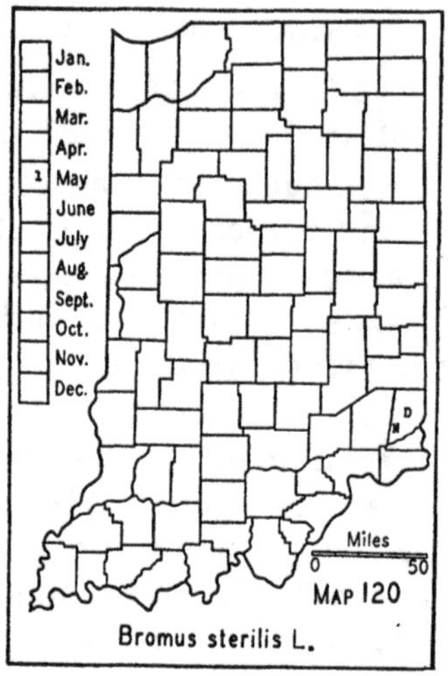

Bromus sterilis L.

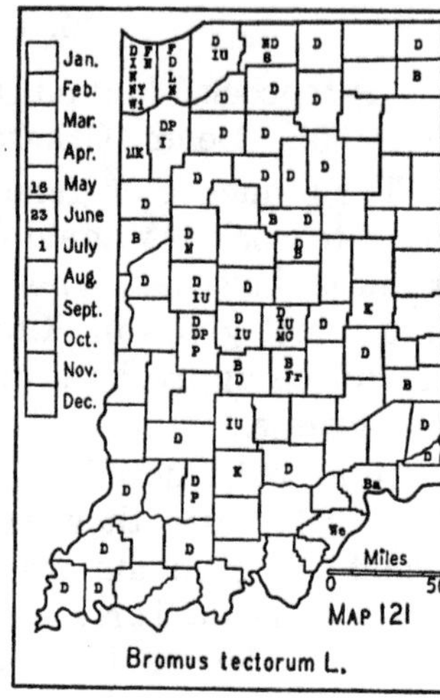

Bromus tectorum L.

Spikelets essentially 2-flowered in structural plan, the lower floret represented by a sterile lemma, the first glume sometimes lacking; various types of imperfect flowers common; spikelets never strongly compressed laterally.

Spikelets usually not in pairs; fertile lemmas thicker or firmer than the glumes and sterile lemmas.................................10. PANICEAE, p. 147.

Spikelets in pairs (sometimes in threes), one member sessile, the other (or others) pedicellate (occasionally both sessile or pedicellate), the pedicelled member often variously reduced in structure, represented by only a pedicel or a microscopic rudiment in extreme cases; fertile lemmas thin and papery; glumes firmer.

Spikelets in pairs, one sessile and perfect, the other pedicellate and usually staminate or neuter (the pedicellate one sometimes obsolete), rarely both pedicellate; lemmas hyaline................11. ANDROPOGONEAE, p. 177.

Spikelets unisexual, the pistillate below, the staminate above, in the same inflorescence or in separate inflorescences..........12. TRIPSACEAE, p. 181.

1. *BAMBUSEAE* Nees. BAMBOO TRIBE

1[1]-414[2]. ARUNDINÀRIA Michx. CANE

[Galloway. Bamboos: their culture and uses in the United States. U. S. Dept. Agric. Bull. 1329: 1-44. illus. 1925.]

Panicles on leafy branches; culms as much as 10 m high..............1. *A. gigantea*.
Panicles on leafless shoots from creeping rhizomes. (See excluded species no. 39, p. 1025.)..*A. tecta*.

1. **Arundinaria gigántea** (Walt.) Chapm. (*Arundinaria macrosperma* Michx. of Gray, Man., ed. 7, of Britton and Brown, Illus. Flora, ed. 2, and of Deam, Grasses of Ind.) SOUTHERN CANE. Map 119. This species is restricted to southern Indiana. I have found it only in the counties bordering the Ohio and Wabash Rivers. Kriebel, however, found it along Beaver Creek near Huron, Lawrence County, and there is a place named "cane marsh" in Greene County which indicates that it, at one time, did occur in that county. This species is usually found in lowlands that are periodi-

[1] The first number refers to the numbers used in Hitchcock's Manual of Grasses of the United States.
[2] The second number refers to the numbers used in Dalla Torre and Harms' Genera Siphonogamarum.

cally inundated. I have seen it on rocky wooded slopes, however, and on the top of the bluff of the Ohio River, where it was 200 feet above the water. In my opinion fire and grazing have been instrumental in limiting its distribution on the uplands.

Se. U. S. from Va. to Mo. and Okla., southw. to Fla. and Tex.

2. *FESTÙCEAE* Nees. FESCUE TRIBE

Plants stout, usually 1.5-2.5 m high; inflorescence large, plumelike; rachilla plumose. ....26. PHRAGMITES, p. 111.

Plants much shorter, rarely as high as 1.5 m; inflorescence not plumelike; rachilla not plumose.

- Lemmas prominently 3-nerved, without a cobwebby base.
  - Lemmas more or less villous on the nerves.
    - Nodes of stem glabrous; plants mostly 75-125 cm high, basal parts smooth to the touch....31. TRIODIA, p. 113.
    - Nodes of stem pubescent; plants mostly 25-60 cm high, basal parts rough to the touch....32. TRIPLASIS, p. 113.
  - Lemmas not villous on the nerves, glabrous or scabrous.
    - Lemmas less than 5 mm long; fruit less than 5 mm long..12. ERAGROSTIS, p. 108.
    - Lemmas about 8 mm long; fruit about 5 mm long......15. DIARRHENA, p. 110.
- Lemmas 5-many-nerved (the intermediate pair in some species of *Poa* obscure).
  - Spikelets with 2 or 3 empty lemmas above the 2 or 3 fertile florets, or with 1-4 sterile lemmas below the 6 or 7 fertile florets.
    - Sterile lemmas above the fertile florets....28. MELICA, p. 111.
    - Sterile lemmas below the fertile florets....20. UNIOLA, p. 110.
  - Spikelets without sterile lemmas (terminal florets often not developed).
    - Lemmas awned.
      - Lemmas awned or awn-tipped from a minutely bifid apex.
        - Grain pubescent at the summit; callus of florets not bearded....2. BROMUS, p. 95.
        - Grain not pubescent at the summit; callus of florets bearded....29. SCHIZACHNE, p. 112.
      - Lemmas awned from the tip, rounded on the back; grain not pubescent at the summit....3. FESTUCA, p. 99.
    - Lemmas awnless.
      - Spikelets strongly flattened, subsessile in 1-sided clusters at the ends of long naked branches, these spreading in anthesis, erect in fruit....21. DACTYLIS, p. 111.
      - Spikelets neither strongly flattened nor in clusters.
        - Florets cobwebby at the base....10. POA, p. 104.
        - Florets not cobwebby at the base.
          - Lemmas plainly 7-nerved, scarious at the apex......6. GLYCERIA, p. 102.
          - Lemmas 5-nerved, sometimes 2 of the nerves obscure.
            - Lemmas 8-11 mm long....2. BROMUS, p. 95.
            - Lemmas mostly less than 8 mm long.
              - Lemmas keeled on the back....10. POA, p. 104.
              - Lemmas rounded on the back....3. FESTUCA, p. 99.

2-389. BRÒMUS L. BROMEGRASS

[Shear. A revision of the North American species of Bromus occurring north of Mexico. U. S. Dept. Agric. Agrost. Bull. 23: 1-66. 1920. Wiegand. Notes on some East-American species of Bromus. Rhodora 24: 89-92. 1922.]

[Note: Measurements of spikelets, glumes, and lemmas do not include awns.]

First glume 1-nerved (rarely 3-nerved in *Bromus latiglumis*, the leaves of which have prominent flanges at the base).

Awns 12-25 mm long, straight.

Spikelets glabrous or more or less scabrous; awns about 25 mm long............ ....1. *B. sterilis.*

Spikelets pubescent; awns mostly 12-17 mm long................2. *B. tectorum.*

Awns less than 12 mm long or sometimes lacking.

Branches of panicle compact, erect or slightly spreading at maturity; glumes and lemmas glabrous or more or less scabrous but not pubescent; sheaths usually glabrous.

Creeping rhizomes present; sheaths glabrous (sometimes late shoots pubescent); lemmas awnless or with awns up to 3 mm long..............3. *B. inermis.*

Creeping rhizomes lacking; sheaths glabrous or somewhat pilose; lemmas with awns 5-6 mm long. (See excluded species no. 42, p. 1025.)........*B. erectus.*

Branches of panicle loose, drooping; glumes and lemmas more or less pubescent; sheaths usually pubescent.

Glumes glabrous except the scabrous midnerve or sometimes the whole surface more or less scabrous.

Nodes usually 4-6; lemmas strongly pubescent near the margin on the lower half to three-fourths, their backs glabrous or scaberulous; plants of a marsh or prairie habitat, flowering in July..............4. *B. ciliatus.*

Nodes 10-20; lemmas more or less pubescent, especially on the back; plants of dry woods, ravines, and dry banks of streams, flowering from July to September......................................5. *B. latiglumis.*

Glumes more or less pubescent all over; lemmas more or less pubescent, especially on the back; plants of dry woods, ravines, and dry banks; plants flowering from May to July.

Nodes 4-6; sheaths shorter than the internodes or the lower ones longer, not flaring at the summit.

Sheaths and blades more or less villous....................6. *B. purgans.*

Sheaths and blades (except the lower ones) glabrous.................. ....6a. *B. purgans* f. *laevivaginatus.*

Nodes 10-20; sheaths longer than the internodes, at least the 4 lower ones longer; plants flowering from July to September.........5. *B. latiglumis.*

First glume 3- or 5-nerved.

Sheaths glabrous..7. *B. secalinus.*

Sheaths pubescent.

Lemmas awnless or with awns less than 5 mm long.

Glumes and lemmas glabrous or scabrous on the nerves; awnless or with short awns..8. *B. brizaeformis.*

Glumes and lemmas silky-pubescent all over; awns mostly 2-3 mm long......... ....9. *B. Kalmii.*

Lemmas with awns more than 5 mm long.

Glumes and lemmas more or less silky-pubescent.................10. *B. mollis.*

Glumes and lemmas glabrous or somewhat scabrous.

Branches of the panicle rather stiffly spreading or drooping, not flexuous; awns straight.....................................11. *B. commutatus.*

Branches of the panicle slender, lax or flexuous............12. *B. japonicus.*

1. BROMUS STÉRILIS L. Map 120. Our only report of this species is of a colony which I found along Tanners Creek about a half mile southeast of Guilford, Dearborn County. It was well established here along the roadside.

Nat. of Eu.; N. E. to Ill., southw. to Va. and Ala., and in the west from B. C. to Calif. and Colo.

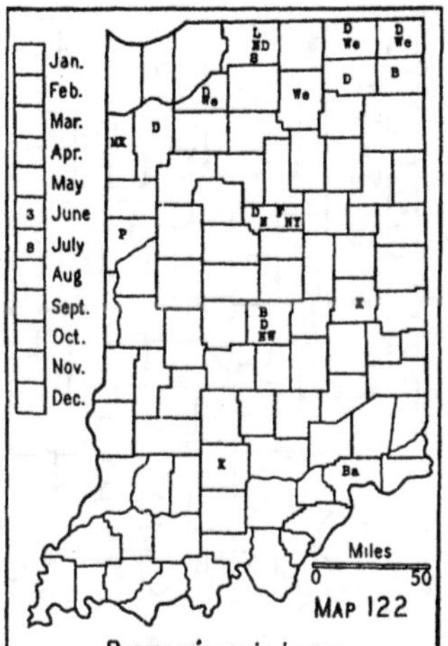

Bromus inermis Leyss.

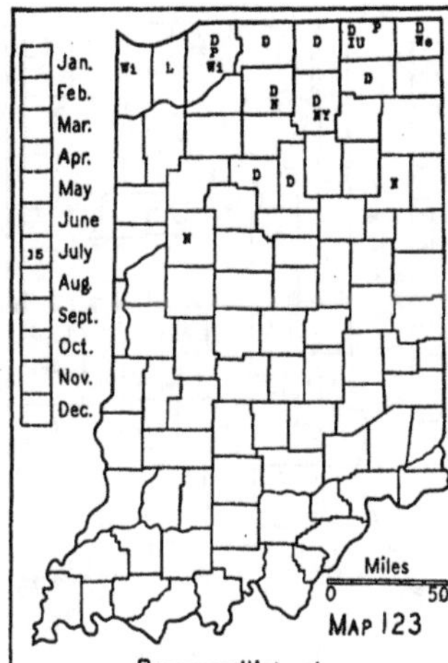

Bromus ciliatus L.

Bromus latiglumis (Shear) Hitchc.

2. BROMUS TECTÒRUM L. DOWNY CHESS. Map 121. This species is now found throughout the state and has become a pernicious weed in all the northern counties where a sandy soil is found. It is found along roadsides and in waste places, hayfields, pastures, and fallow fields.

Nat. of Eu.; throughout the U. S. as far south as Va. and Miss. Common on the Pacific coast.

3. BROMUS INÉRMIS Leyss. SMOOTH BROME. Map 122. I do not know that this species has been intentionally sown to any extent in Indiana but it is now found frequently along railroads and roadsides in sandy soil in the northern half of the state. I found one farmer in Lagrange County who had sown it with success in a field of blow-sand soil.

Native from central Europe to China; used in the western states as a hay and pasture grass and now found as an escape in the northern half of the United States.

4. **Bromus ciliàtus** L. FRINGED BROME. Map 123. Infrequent in marshes and springy areas of the lake region. I found a specimen in Steuben County with all the sheaths glabrous except the lowest one. This is *Bromus ciliatus* f. *denudatus* Wiegand (Rhodora 24: 91. 1922) which Fernald now regards as the typical form of the species. (Rhodora 32: 70. 1930.)

Newf. to Wash., southw. to N. J., Tenn., Iowa, w. Tex., and s. Calif.

5. **Bromus latiglùmis** (Shear) Hitchc. (*Bromus altissimus* Pursh, *Bromus purgans* of Britton and Brown, Illus. Flora, ed. 2, and including *Bromus incanus* (Shear) Hitchc.) Map 124. Infrequent throughout the state. This species seems to prefer dense shade and is found most often on wooded slopes along streams and in ravines, in fact, it is rarely found far distant from a stream. This species was separated from the form with densely pubescent sheaths by most authors but Hitchcock has united the two forms under this name.

Maine to e. Mont., southw. to N. C., Tenn., Tex., and N. Mex.

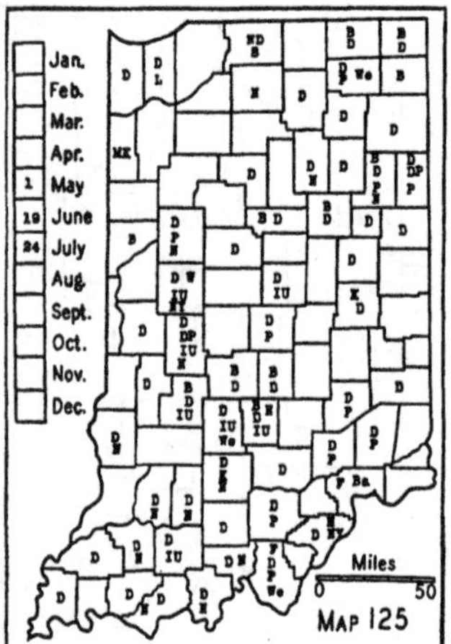

Bromus purgans L.

Bromus secalinus L.

Bromus Kalmii Gray

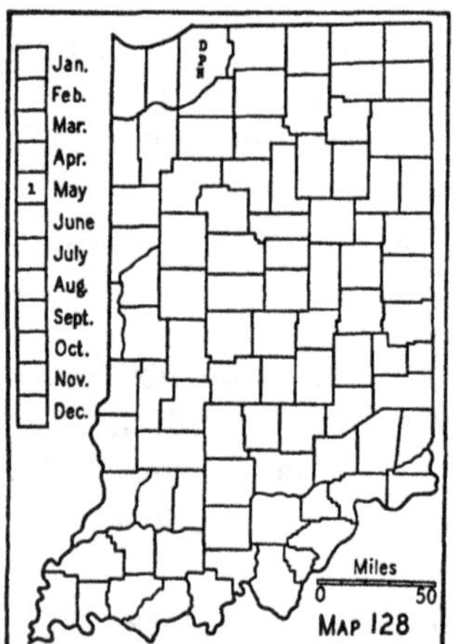

Bromus mollis L.

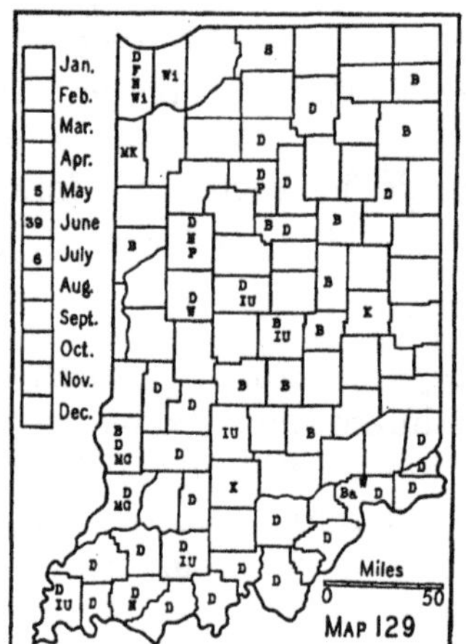

Bromus commutatus Schrad.

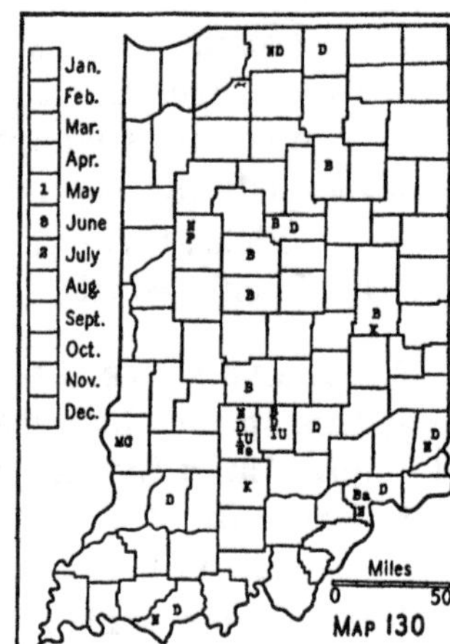

Bromus japonicus Thunb.

6. **Bromus púrgans** L. CANADA BROME. Map 125. Infrequent to frequent throughout the state in dry places, rarely in wet places, in black and white oak woods and less frequent in beech and sugar maple woods.

Mass. to Alberta, southw. to Fla. and Ariz.

6a. **Bromus purgans** f. **laevivaginàtus** Wieg. (Rhodora 24: 92. 1922.) This is a form of the species that has all the sheaths glabrous except sometimes the lowest one.

7. BROMUS SECALÌNUS L. CHESS. Map 126. Frequent to common in all parts of the state. It is found almost everywhere in cleared grounds except in pastures. It is most abundant in wheatfields and waste grounds. In Indiana it is called cheat.

Nat. of Eu.; now found throughout the U. S.

8. BROMUS BRIZAEFÓRMIS Fisch. & Mey. RATTLESNAKE CHESS. My only specimen is from a waste place near the water works, Michigan City, in La Porte County. Sometimes cultivated as an ornamental grass.

Nat. of Eu.; rare in e. U. S. from Mass. to Del. and occasional in the Pacific Coast States.

9. **Bromus Kálmii** Gray. KALM CHESS. Map 127. Infrequent on low, open dunes and in marshy and springy places in the lake region.

Maine to Minn. and S. Dak., southw. to Md. and Iowa.

10. BROMUS MÓLLIS L. (*Bromus hordeaceus* of recent authors.) SOFT CHESS. Map 128. In 1913 I found this species to be frequent along the roadside near the water works in Michigan City, La Porte County.

Nat. of Eu.; in e. U. S. from N. S. to N. C., and abundant on the Pacific coast.

11. BROMUS COMMUTÀTUS. Schrad. HAIRY CHESS. Map 129. This species is now frequent to common throughout the state and is our most common chess. It is found almost everywhere in cultivated and waste grounds and along roadsides and railroads.

Nat. of Eu.; now well established in most parts of the U. S. and abundant in the Pacific Coast States.

12. BROMUS JAPONICUS Thunb. (*Bromus patulus* Mertens & Koch of Britton and Brown, Illus. Flora, ed. 2.) JAPANESE CHESS. Map 130. This species is now found throughout the state in habitats similar to those of *Bromus commutatus*.

Native of the Old World; now found throughout the United States except the Gulf States.

3-385. FESTÙCÀ L. FESCUE GRASS

[Piper. North American species of Festuca. Contr. U. S. Nation. Herb. **10**: 1-42. 1906.]

Leaves involute, setaceous or capillary, less than 1.5 mm wide; internodes of rachilla more or less scabrous.
- Annual; some of the sheaths partly or entirely retrorsely pubescent, rarely all of them glabrous; spikelets mostly 5-13-flowered; lemmas more or less scabrous all over; stamens 1, generally included at anthesis.
  - Lower glumes 3.5-4.5 mm long; awns of lemmas 3.5-7 mm long....1. *F. octoflora.*
  - Lower glumes 2.3-4 mm long; awns of lemmas 1-3 mm long..1a. *F. octoflora* var. *tenella.*
- Perennial; sheaths glabrous; spikelets 3-8-flowered; lemmas scabrous only toward the apex; stamens 3, generally protruding at anthesis.
  - Culms in loose tufts, decumbent at the usually red, fibrillose base; awn of lemma shorter than the body; blades smooth.........................2. *F. rubra.*
  - Culms erect.
    - Lemmas 3-3.8 mm long, awnless; spikelets 5-8 mm long; leaves capillary...3. *F. capillata.*
    - Lemmas 4-5 mm long, short-awned; spikelets 7-10 mm long; leaves narrow but not capillary...4. *F. ovina.*

Leaves flat, more than 1.5 mm wide; internodes of rachilla glabrous.
- Lemmas 5-7 mm long; spikelets 9-25 mm long; panicles nearly erect or slightly curved, branches short.......................................5. *F. elatior.*
- Lemmas 4-4.5 mm long; spikelets mostly 5-7 mm long; panicles usually open and nodding at maturity if of normal size, branches long.

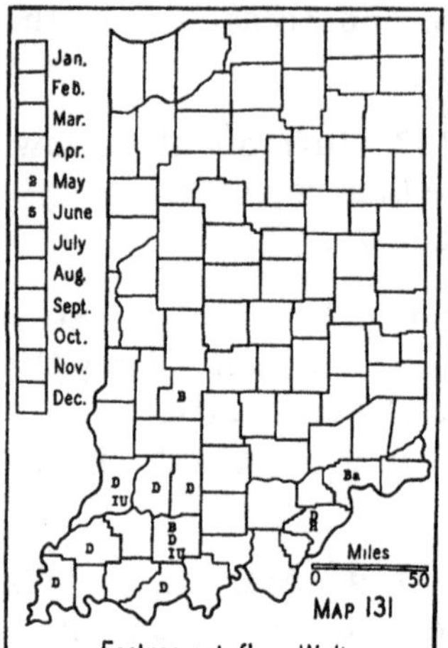

Festuca octoflora Walt.

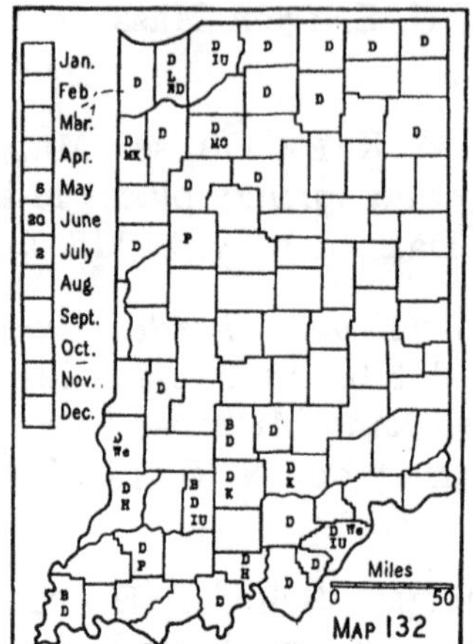

Festuca octoflora var. tenella (Willd.) Fern.

Festuca rubra L.

Lemmas mostly subacute; mature panicles strongly curved; spikelets somewhat scattered at the ends of the panicle-branches.................6. *F. obtusa.*

Lemmas more acute; mature panicles drooping; spikelets more clustered at the ends of the panicle-branches; florets more uniformly 3 or 4 to a spikelet...7. *F. paradoxa.*

1. **Festuca octoflòra** Walt. Map 131. This species, as now known, is restricted to the southern part of the state. It, and also the variety, are more or less local because their habitat is local. On the whole, it is more or less frequent and is found in bare, sandy, sometimes very sandy soil.

N. J. to Okla., southw. to Fla. and Tex.

1a. **Festuca octoflora** var. **tenélla** (Willd.) Fern. (Rhodora 34: 209-211. 1932.) Map 132. The variety is doubtless found in every county of the state where its peculiar habitat is found. The slightly acid property of the sandy soil in which it is found doubtless restricts its appearance in the Tipton Till Plain.

Maine, Que., B. C., southw. to Ga., Ark., Tex., and Calif.

2. FESTUCA RÙBRA L. Map 133. This grass was found by Madge McKee in a vacant lot in Goodland, Newton County. It was well established here. It was found in 1935 by R. C. Friesner at 3711 N. Gladstone Ave. in Indianapolis, where it had taken possession of the lawn. It is probably established in many other places throughout the state where it has been introduced in lawns in grass seed, but it has not been detected because of its close resemblance to *Festuca ovina* and *Poa pratensis.* In 1937 I found it along a roadside near a house in Noble County.

Lab. to Alaska, southw. in the mts. in the west to Ariz., in the Allegheny Mts. to Ga., and along the Coastal Plain; probably mostly introduced in the Eastern States.

3. **Festuca capillàta** Lam. (Rhodora **18**: 235. 1916.) (*Festuca ovina* var. *capillata* (Lam.) Hack.) HAIR FESCUE. Map 134. I have a letter from

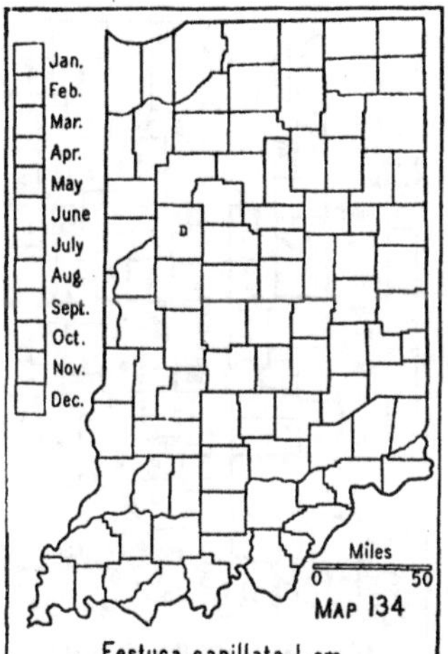

Festuca capillata Lam.

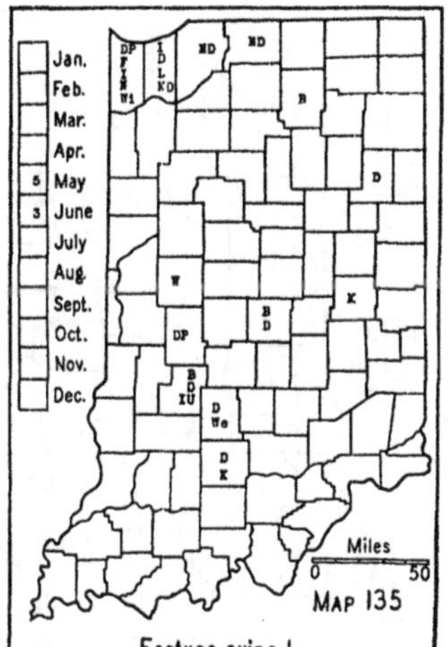

Festuca ovina L.

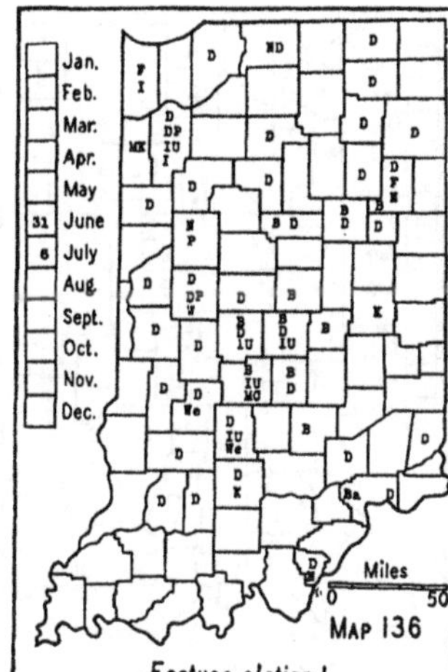

Festuca elatior L.

A. A. Hansen who says this species is established in the vicinity of Lafayette, Tippecanoe County.

Newf. to Mich., southw. to N. C. and Ill., and in Oreg.

4. FESTUCA OVÌNA L. (Fernald. The allies of Festuca ovina in eastern America. Rhodora 37: 250-252. 1935.) SHEEP FESCUE. Map 135. This species prefers sandy soil and has been found in several places in open woodland and waste places.

Nat. of Eu.; Maine, Mich. to N. Dak. and southw. to S. C. and Ill. and N. Mex.; also on the west coast from Alaska to Wash.

5. FESTUCA ELÀTIOR L. MEADOW FESCUE. ENGLISH BLUEGRASS. Map 136. Infrequent to frequent throughout the state. It is most frequent along roadsides and in waste places and has sparingly escaped to open woodland. Introduced as a forage plant. The Indiana farmers whom I have interrogated call it English bluegrass.

Nat. of Eurasia; throughout the cooler parts of N. A.

6. **Festuca obtùsa** Spreng. (*Festuca nutans* Spreng.) NODDING FESCUE. Map 137. Infrequent to frequent throughout the state in woodland of many kinds.

N. S., Que. to Man, southw. to Fla. and e. Tex.

7. **Festuca paradóxa** Desv. (Opusc. 105. 1831.) (See Amer. Jour. Bot. 24: 33. 1937.) (*Festuca Shortii* Kunth.) SHORT'S FESCUE. Map 138. This species is easily recognized in the field but herbarium material is difficult to determine. I have seen it growing in Posey County. I have herbarium material from Decatur County which I believe belongs here. In Posey County it grows in hard, white clay soil in low, open woodland with pin oak.

Pa. to Iowa, southw. to S. C. and e. Tex.

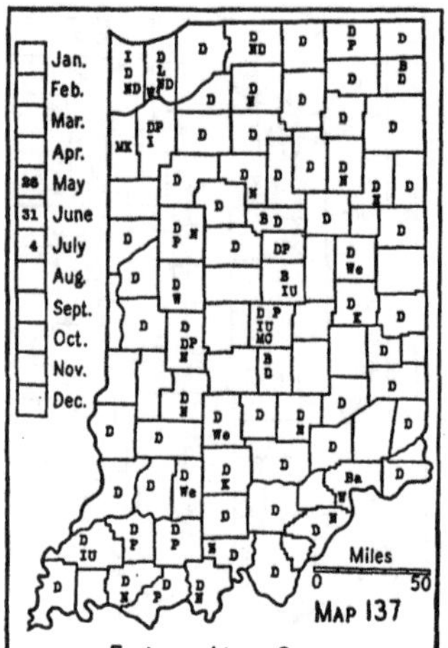

Festuca obtusa Spreng.

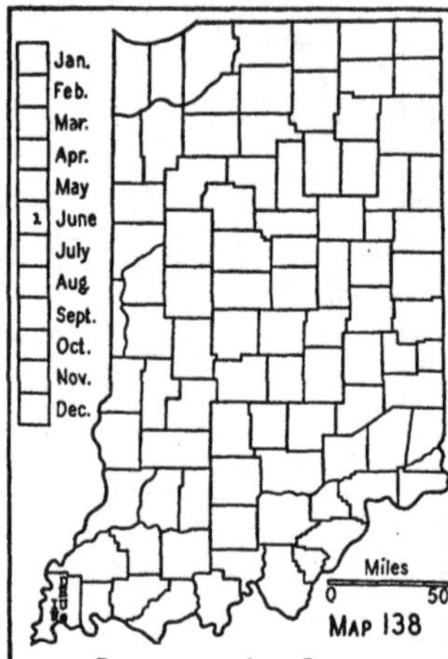

Festuca paradoxa Desv.

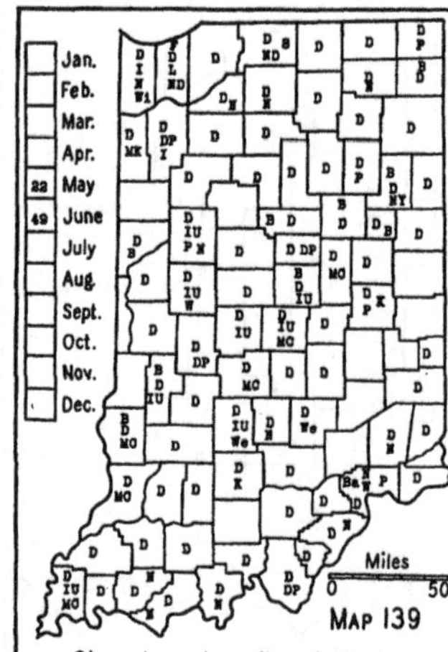

Glyceria striata (Lam.) Hitchc.

6-383. GLYCÈRIA R. Br. MANNAGRASS

Spikelets 2-8 mm long.
  Second glume about 1 mm long....................................1. *G. striata.*
  Second glume about 2 mm long.
    Lemmas 1.4-2.5 mm long.
      Leaves 2-4 mm wide; panicles contracted (less than 5 cm wide); spikelets 3- or 4-flowered. (See excluded species no. 44, p. 1025.).............*G. melicaria.*
      Leaves mostly 5-15 mm wide; panicles open (more than 5 cm wide); spikelets 4-7 flowered....................................2. *G. grandis.*
    Lemmas 3-3.5 mm long.
      Second glume 1-nerved; florets smooth and glossy; lemmas abruptly acute; anthers about 0.5 mm long..........................3. *G. canadensis.*
      Second glume 3- or 5-nerved; florets not smooth and glossy; lemmas obtuse; anthers 1-1.5 mm long....................................4. *G. pallida.*
Spikelets 10-40 mm long.
  Lemmas obtuse, about equaling the palea.
    Spikelets 10-15 mm long; lemmas hispid only on the nerves; anthers about 1 mm long; grain 1.5 mm long..................................5. *G. borealis.*
    Spikelets 15-30 mm long; lemmas hispid on the nerves and on the spaces between them; anthers 1-1.5 mm long; grain 2-2.5 mm long......6. *G. septentrionalis.*
  Lemmas acute; palea about 1.5 mm longer than the lemma.........7. *G. acutiflora.*

1. **Glyceria strìata** (Lam.) Hitchc. (Proc. Biol. Soc. of Washington **41**: 157. 1928.) (*Glyceria nervata* (Willd.) Trin. and *Panicularia nervata* (Willd.) Ktze.) FOWL MANNAGRASS. Map 139. Frequent throughout the state in wet soil in ditches, marshes, and wet woods, along streams, and about ponds and swampy places.

Newf. to B. C., southw to Fla., Tex., and n. Calif.

2. **Glyceria grándis** Wats. (*Panicularia grandis* (Wats.) Nash.) AMERICAN MANNAGRASS. Map 140. This species grows in very wet places or in shallow water in ponds or in ditches. I have found only a few plants in three counties.

P. E. I. to Alaska, southw. to Ohio, Tenn., Iowa, Nebr., N. Mex., and e. Oreg.

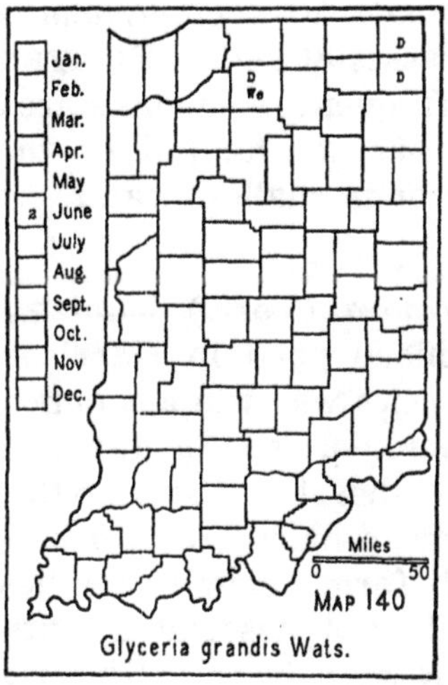

Glyceria grandis Wats.

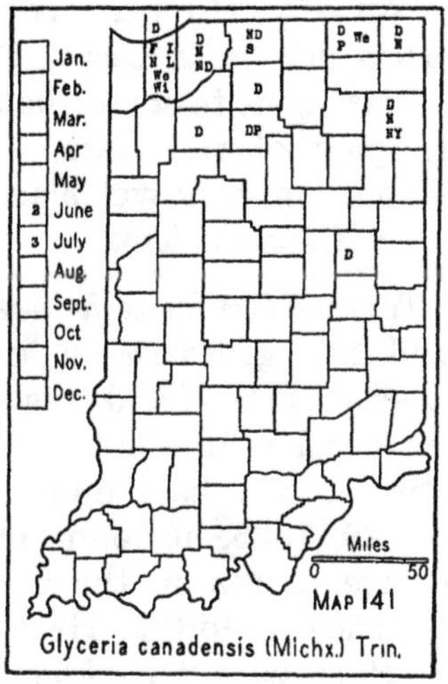

Glyceria canadensis (Michx.) Trin.

Glyceria pallida (Torr.) Trin.

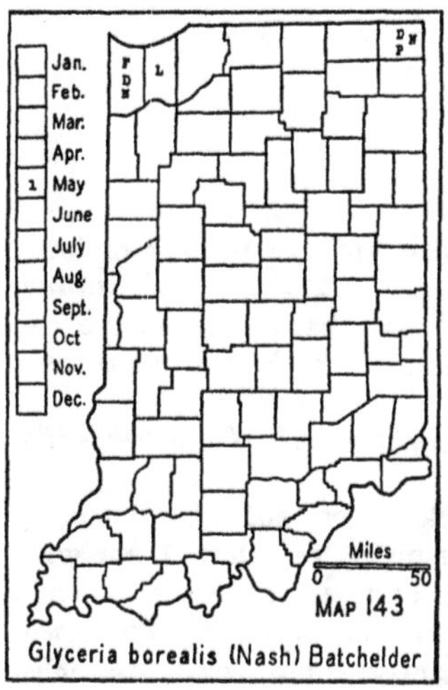

Glyceria borealis (Nash) Batchelder

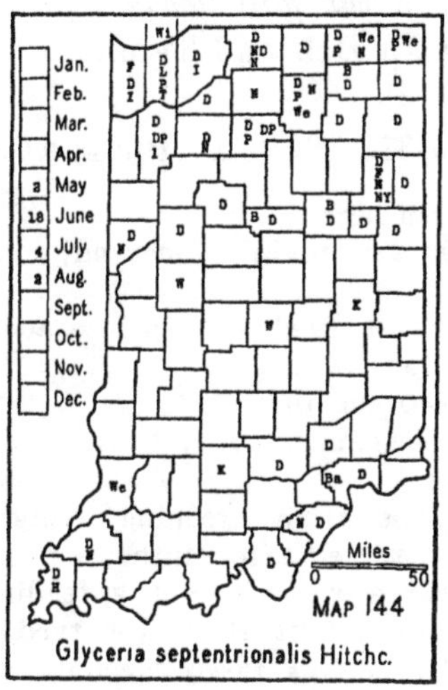

Glyceria septentrionalis Hitchc.

Glyceria acutiflora Torr.

3. **Glyceria canadénsis** (Michx.) Trin. (*Panicularia canadensis* (Michx.) Ktze.) CANADA MANNAGRASS. Map 141. Infrequent in the lake area where it is found in wet habitats in marshes, ditches, and springy places and about ponds.

Newf. to Minn., southw. to Md. and Ill.

4. **Glyceria pállida** (Torr.) Trin. (*Panicularia pallida* (Torr.) Ktze.) PALE MANNAGRASS. Map 142. This grass has been found infrequently in a few of our northern counties. It grows in a very wet habitat, usually in shallow water or in ponds that dry up in midsummer.

Maine to Wis., southw. to N. C. and Mo.

5. **Glyceria boreàlis** (Nash) Batchelder. (*Panicularia borealis* Nash.) NORTHERN MANNAGRASS. Map 143. This plant is doubtless very rare in Indiana. The habitat is the same as that of the preceding species.

Newf. to Alaska, southw. to Conn., Ind., Iowa, S. Dak., and in the mts. to N. Mex. and Calif.

6. **Glyceria septentrionàlis** Hitchc. (*Panicularia septentrionalis* (Hitchc.) Bickn. and *Glyceria plicata* of Deam, Grasses of Ind.) EASTERN MANNAGRASS. Map 144. Infrequent to somewhat frequent in the lake area and local in the southern part of the state. It has the habitat of the preceding species, growing only in very wet places or in shallow water.

Que. to Minn., southw. to S. C. and e. Tex.

7. **Glyceria acutiflòra** Torr. (*Panicularia acutiflora* (Torr.) Kuntze.) Map 145. In 1919 I found this grass in an artificial pond in Harrison County. The pond was revisited in 1935 and this species was still a common plant in it. R. M. Kriebel found it in 1934 in a sinkhole on the farm of Julius Blackwell, about two and a half miles northeast of Springville, Lawrence County. On July 29, 1935, he found about a half acre in a buttonbush swamp of about three acres on the Cobb farm about two miles northeast of Avoca, Lawrence County. Here it was associated with *Cephalanthus occidentalis, Populus heterophylla, Rosa palustris, Glyceria septentrionalis*, and *Ranunculus flabellaris*.

N. H. to Mich., southw. to Del. and Tenn.

10-378. PÒA L. BLUEGRASS

Annual, usually less than 40 cm high.
  Lemmas not cottony at the base, plainly 5-nerved; mature anthers 0.7-1 mm long. ....1. *P. annua*.
  Lemmas cottony at the base, 3-nerved or with two additional obscure ones; mature anthers about 0.2 mm long........2. *P. Chapmaniana*.
Perennial, usually more than 40 cm high.
  Lemmas not cottony at the base........3. *P. autumnalis*.
  Lemmas cottony at the base.
    Plants bluish green; culms from creeping rootstocks, not tufted, distinctly flattened; panicles contracted after anthesis and usually less than 1 cm wide (shade forms sometimes slender and spreading and as wide as 2 cm), branches of panicle erect; first glume 3-nerved........4. *P. compressa*.
    Plants green (not bluish); culms terete or only slightly compressed; panicles more or less expanded after anthesis, at least 2 cm wide, branches ascending or spreading; first glume 1-nerved except in *P. Wolfii*.
      Lemmas glabrous........5. *P. languida*.
      Lemmas pubescent or scabrous, at least on the keel.
        Marginal nerves of lemmas glabrous.
          Sheaths smooth; intermediate nerves of the flowering glume obscure; spikelets 4-6 mm long; anthers 0.4-0.7 mm long, pink; ligule 1 mm long or less; inflorescence silvery green, without spikelets close to the rachis........6. *P. alsodes*.
          Sheaths scabrous; intermediate nerves prominent; spikelets 3.2-3.6 mm long; anthers 1.6 mm long, pale; ligule 5-6 mm long; inflorescence yellowish green or purplish, with normal rays and, in addition, many spikelets on short branches closer to the rachis........7. *P. trivialis*.
        Marginal nerves of lemmas pubescent.
          Intermediate nerves of lemmas obscure.
            Plants slender, lax; ligules less than 2 mm long; anthers often purple, 0.5-0.8 (1) mm long; branches of panicles in 2's (rarely in 3's). ....8. *P. paludigena*.
            Plants robust; ligules more than 2 mm long; anthers 0.8-1.4 mm long; branches of panicles in 3's or more........9. *P. palustris*.

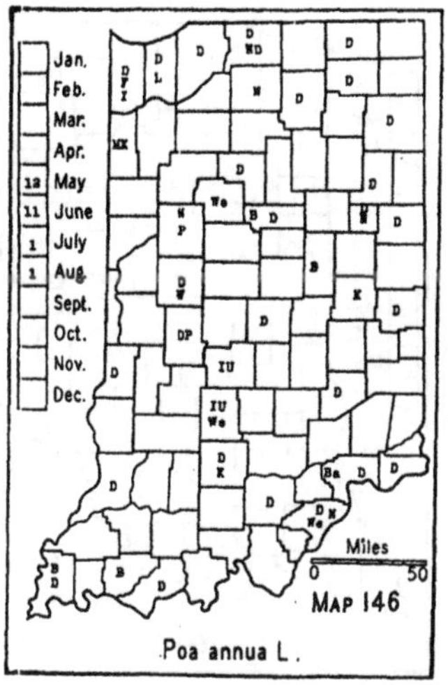

Poa annua L.

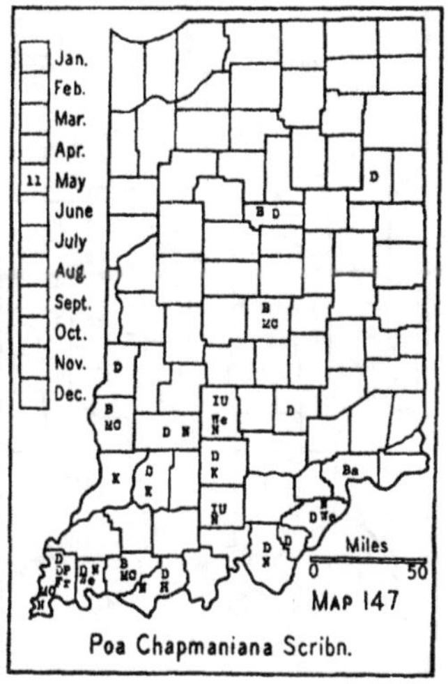

Poa Chapmaniana Scribn.

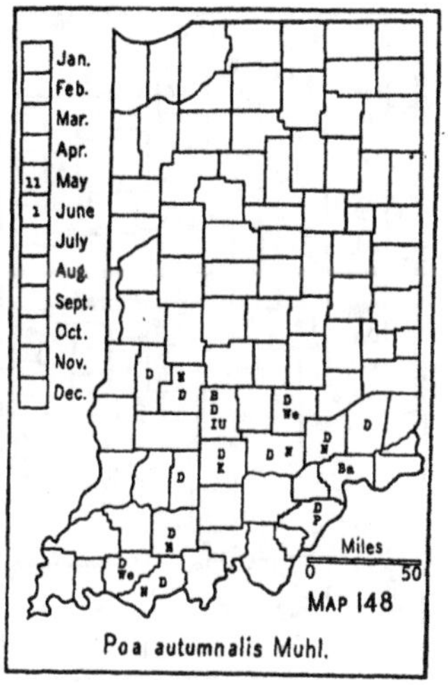

Poa autumnalis Muhl.

Intermediate nerves of lemmas prominent.
- Lemmas 2.5-3 mm long.
  - Midnerve of lemmas pubescent only on the basal half; lemmas acute or subacute; plants with creeping rootstocks; anthers purple, mostly 1.4-1.5 mm long; .... 10. *P. pratensis.*
  - Midnerve of lemmas pubescent the entire length; lemmas obtuse; plants without creeping rootstocks; anthers purple, 1.6-1.8 mm long .... 11. *P. sylvestris.*
- Lemmas 4-4.5 mm long.
  - Anthers 1-1.5 mm long; lemmas acute.................. 12. *P. Wolfii.*
  - Anthers 2-2.5 mm long; lemmas obtuse............... 13. *P. cuspidata.*

1. POA ÁNNUA L. ANNUAL BLUEGRASS. Map 146. This grass is found throughout the state in almost all sorts of habitats except in very wet places. It is most often found in lawns, gardens, orchards, and waste places about dwellings. It is, however, found in logging roads in dense woodland, in pastures, and along roadsides.

Nat. of Eu.; Newf. and Lab. to Alaska, southw. to Fla. and Calif.; also in tropical America at high altitudes.

2. **Poa Chapmanìana** Scribn. CHAPMAN BLUEGRASS. Map 147. This species is restricted mostly to southern Indiana where it is usually found in hard, white, slightly acid, clay soil in fallow fields where it is often abundant and usually associated with *Alopecurus carolinianus, Myosotis virginica,* and *Arabis virginica.* Since all of my specimens are from fallow and cultivated fields, it seems that one would be justified in assuming that it is being introduced from the area to the south of us. In 1937 it was an abundant weed in an Iris farm near Bluffton, Wells County.

Del. to Iowa, southw. to Ga. and Tex.

3. **Poa autumnàlis** Muhl. Map 148. This species, as I know it, is a deep woodland grass found in slightly acid soil in low beech and sweet gum, pin oak, and red maple woods. All of our specimens are from southern Indiana, although it is reported to occur in Michigan.

N. J. to Mich. and Ill., southw. to Fla. and Tex.

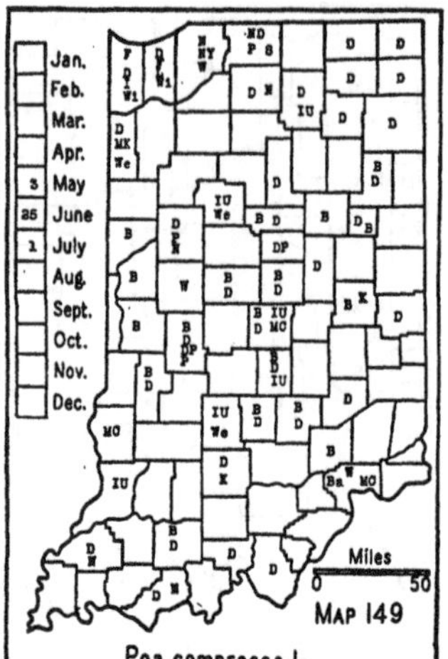

Poa compressa L.

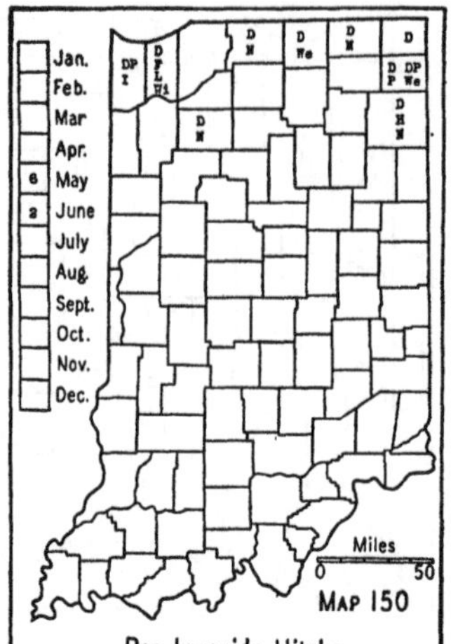

Poa languida Hitchc.

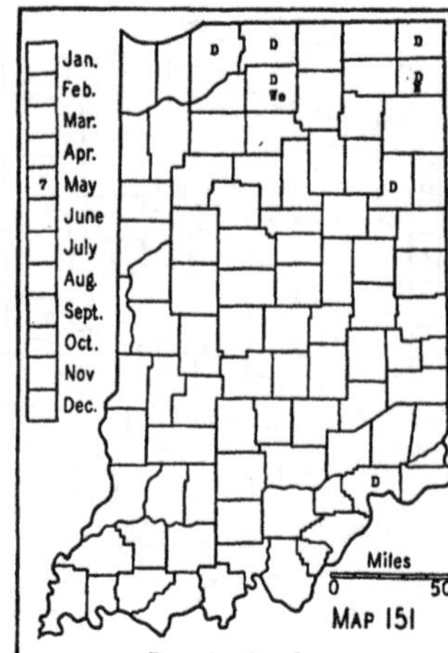

Poa alsodes Gray

4. POA COMPRÉSSA L. CANADA BLUEGRASS. Map 149. Found throughout the state almost everywhere except in very wet places and in dense woodland. It often forms a good part of permanent pastures but is inferior to Kentucky bluegrass. It is sometimes confused with the last named species from which it is easily separated by its flat stem. Roll the stem between the fingers to ascertain if flat or round.

Nat. of Eu.; Newf. to Alaska, southw. to Ga., Ala., Okla., N. Mex., and Calif.

5. **Poa lánguida** Hitchc. (Proc. Biol. Soc. of Washington **41**: 158. 1928.) (*Poa debilis* Torr. of Gray, Man., ed. 7, of Britton and Brown, Illus. Flora, ed. 2, and of Deam, Grasses of Ind.) Map 150. This is an infrequent grass in our northern counties. It is strictly a dense woodland species, and is usually found on black and white oak ridges, sometimes in moist locations.

Newf., Que. to Wis., southw. to Pa., Ky., and Iowa.

6. **Poa alsòdes** Gray. Map 151. This is a rare woodland species found in a few of our northern counties and in one southern county. It is usually found in dry soil in beech and sugar maple woods although I have one specimen that was found in a low woods associated with white elm and soft maple.

Maine to Minn., southw. to Del., and in the mts. to N. C. and Tenn.

7. POA TRIVIÀLIS L. ROUGH BLUEGRASS. Map 152. Although I have found this species only once in the state, it has been reported from five counties. In 1936 it was found in Grant County by J. E. Potzger. It is often used in mixtures of lawn grass seed, and I was told by the superintendent of parks at La Porte that it was the grass he had found to thrive in shade. It is remarkable that it has not been found more often.

Nat. of Eu.; Newf., Ont. to S. Dak., southw. to Va. and W. Va., and on the Pacific coast from s. Alaska to n. Calif.

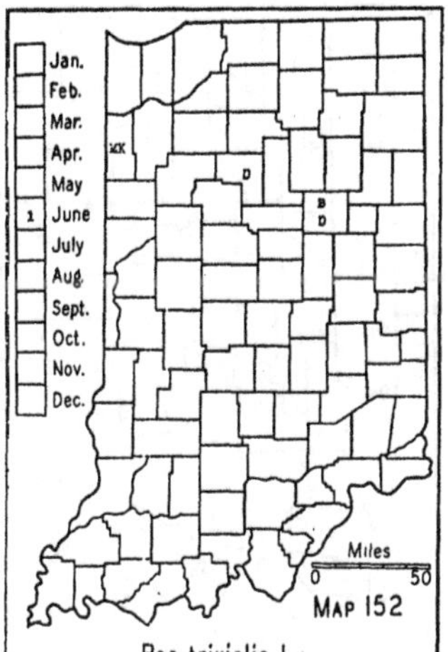

Poa trivialis L.

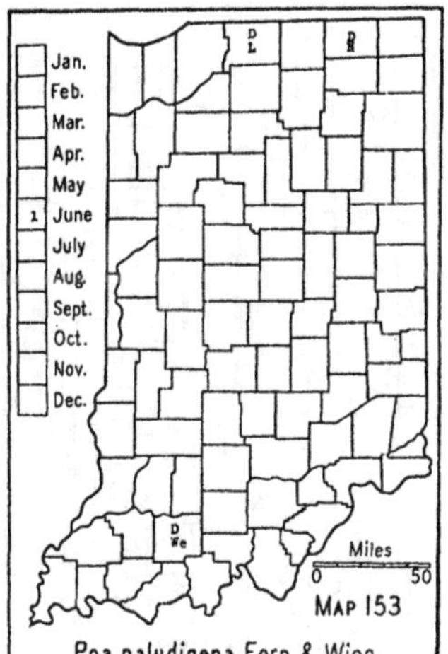

Poa paludigena Fern & Wieg.

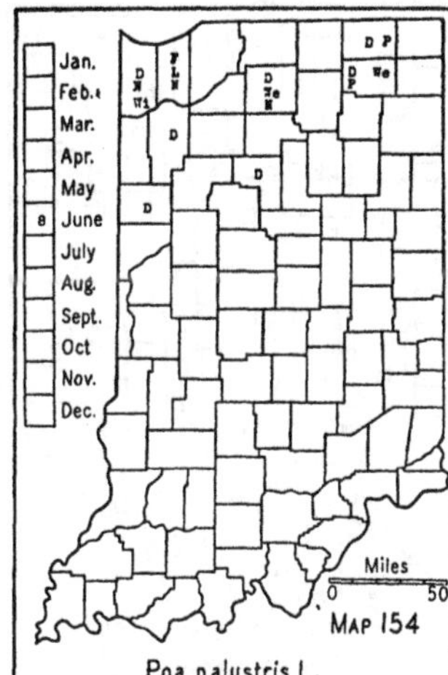

Poa palustris L.

8. **Poa paludígena** Fern. & Wieg. (Rhodora 20: 126. 1918.) (*Poa leptocoma* Trin. of Deam, Grasses of Ind.) Map 153. Only a few specimens of this rare grass have been found, and in widely separated counties. In Lagrange County it grew in tussocks of sphagnum about tamarack and in Dubois County it grew in a swamp in sphagnum about *Alnus rugosa.*

N. Y., Mich., and Wis., southw. to Pa., Ind. and Ill.

9. **Poa palústris L.** (*Poa triflora* Gilib. of Gray, Man., ed. 7 and Britton and Brown, Illus. Flora, ed. 2.) FOWL BLUEGRASS. Map 154. An infrequent grass in the lake area in marshes and in wet prairies.

Newf. and Que., southw. to Va., Ind., Mo., N. Mex., and Calif.; Eurasia.

10. **Poa praténsis** L. KENTUCKY BLUEGRASS. Map 155. Frequent to common in all of the limestone areas of the state and rare or absent from the areas of acid soil. It is our principal pasture grass and is found almost everywhere, often as a weed in gardens. This species is here regarded as a native and by others as introduced into Indiana. See the discussion in Deam's Grasses of Indiana.

Native in northern N. A. and introduced from Eu.; throughout the U. S. except in the arid regions.

11. **Poa sylvéstris** Gray. Map 156. This is strictly a woodland species and is infrequent to frequent throughout the state. It is found in moist soil and prefers beech and sugar maple woods, but it is found also in other types of woodland.

N. Y. to Wis., southw. to Fla. and Tex.

12. **Poa Wólfii** Scribn. WOLF'S POA. Map 157. I found this species in Jay County and Miss Madge McKee found it in a mesophytic forest along the Iroquois River in Newton County. In 1937 it was found by J. E. Potzger in Grant County.

Ohio to Minn. and Mo.

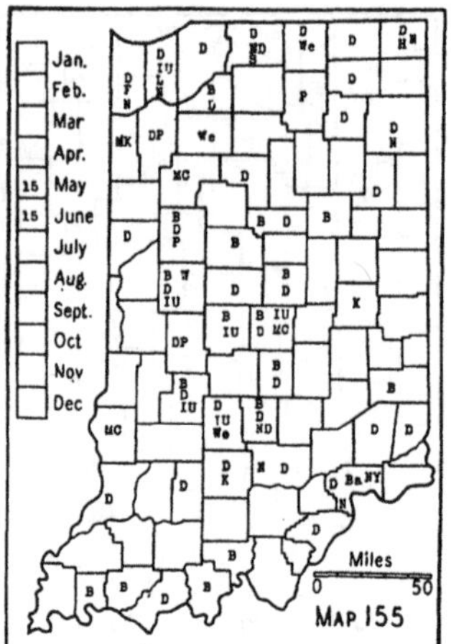

Poa pratensis L.

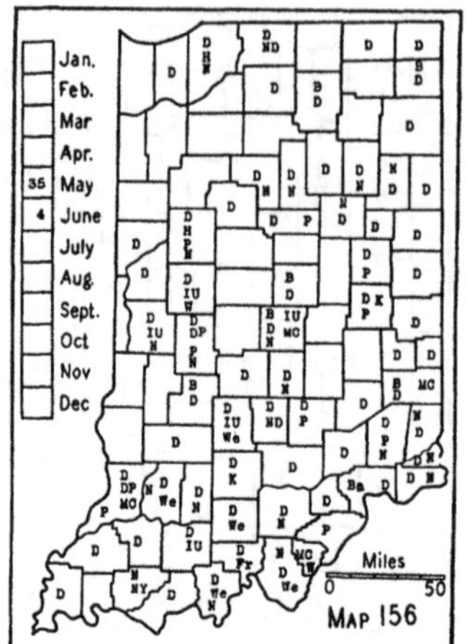

Poa sylvestris Gray

Poa Wolfii Scribn.

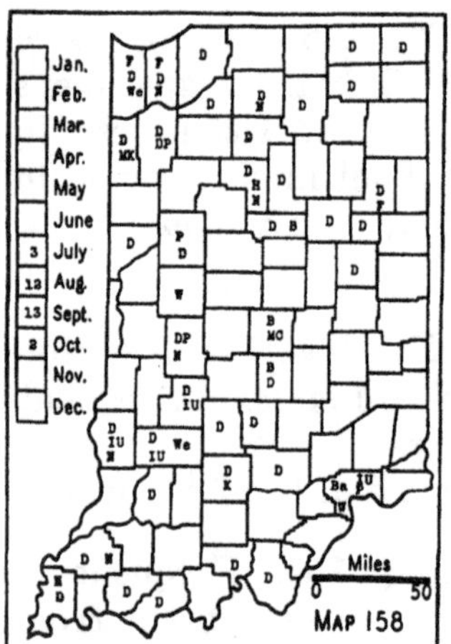

Eragrostis hypnoides (Lam.) B.S.P.

Eragrostis spectabilis (Pursh) Steud.

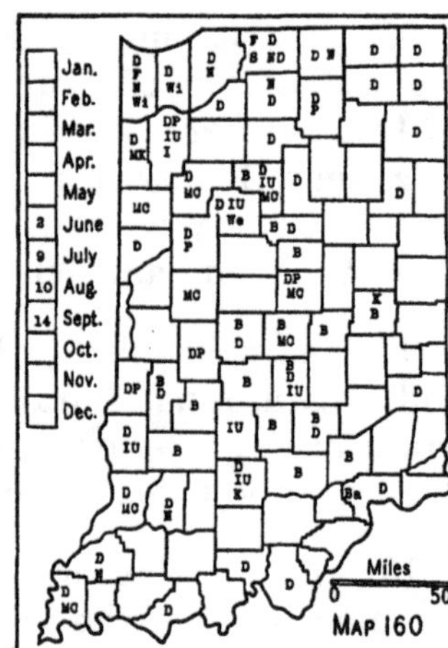

Eragrostis cilianensis (All.) Link.

13. **Poa cuspidàta** Nutt. (*Poa brachyphylla* Schultes.) Known from Indiana only by a specimen collected in 1837 near New Albany by Dr. A. Clapp, which is now in the herbarium of Wabash College. I found it in southern Ohio the last of March in a habitat that convinces me that it can still be found in Indiana if search is made in early spring in the knobs on the ridges of Virginia pine and chestnut oak.

Pa., Ohio, Ind., southw. to Ga. and e. Tenn.

12-341. **ERAGRÓSTIS** Host LOVEGRASS

Culms creeping and rooting at the nodes............................1. *E. hypnoides.*
Culms not creeping and rooting at the nodes.
  Perennials.
    Sheaths villous along the margins, rarely smooth; panicles mostly purplish, broadly spreading, more than half as wide as long; glumes about 1 mm long; lemmas 1.8-2.5 mm long, obtuse..........................2. *E. spectabilis.*
    Sheaths glabrous along the margins; panicles mostly yellow, narrow and elongate, not half as wide as long; glumes mostly 2-3 mm long; lemmas mostly 2.5-3 mm long, acute. (See excluded species, no. 50, p. 1026.)........*E. trichodes.*

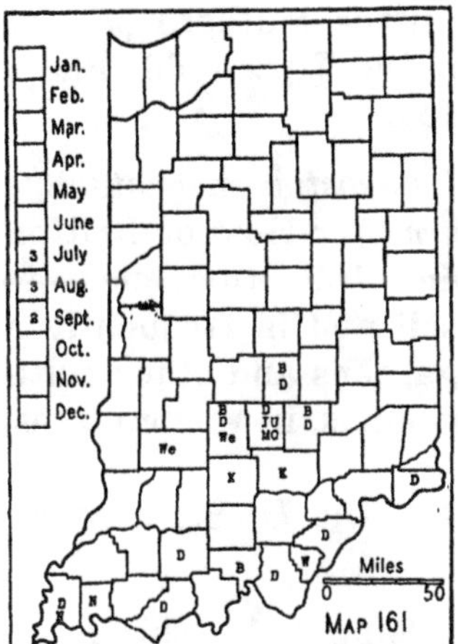

MAP 161
Eragrostis capillaris (L.) Nees

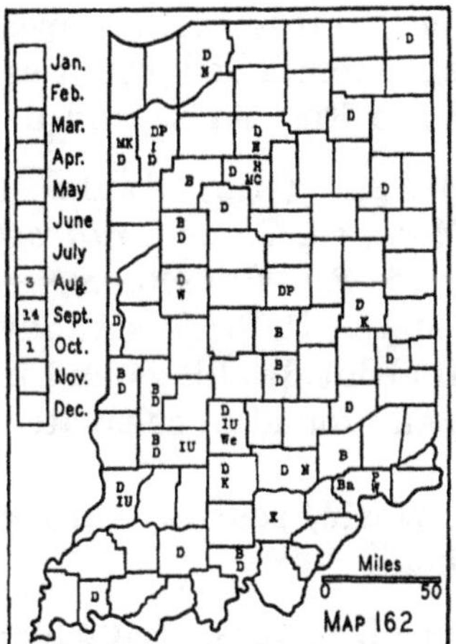

MAP 162
Eragrostis Frankii C.A.Meyer

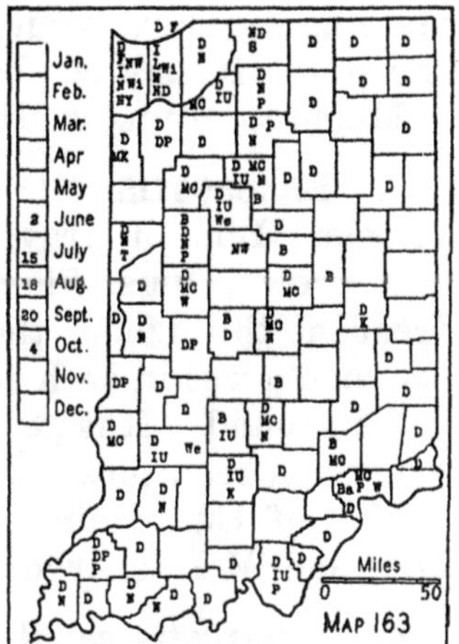

MAP 163
Eragrostis pectinacea (Michx) Nees

Annuals.
Keels of glumes and lemmas more or less glandular.
Spikelets 2.5-4 mm wide; anthers 0.5 mm long................3. *E. cilianensis.*
Spikelets about 2 mm wide; anthers 0.2 mm long. (See excluded species no. 49, p. 1026.)..*E. poaeoides.*
Keels of glumes and lemmas not glandular.
Sheaths generally longer than the internodes; spikelets of lateral branchlets spreading; spikelets of terminal panicles 2-5-flowered.
Culms branching only at the base; pedicels of lateral spikelets mostly 5-10 mm long or longer; grain with a longitudinal groove.....4. *E. capillaris.*
Culms branching at the base and at each node or nearly so; pedicels of lateral spikelets mostly 1-3 mm long; grain without a longitudinal groove..5. *E. Frankii.*
Sheaths shorter than the internodes; spikelets of lateral branchlets appressed or only slightly spreading; spikelets of terminal panicles usually 5-16-flowered (shade forms often 2-5-flowered).
Lateral nerves of the lemmas plainly visible, at least at the base...6. *E. pectinacea.*
Lateral nerves of the lemmas not plainly visible.
Lemmas obtuse, their sides glabrous. (See excluded species no. 48, p. 1026.)..*E. pilosa.*
Lemmas subacute, their sides more or less scabrous. (See excluded species no. 47, p. 1026.)......................................*E. mexicana.*

1. **Eragrostis hypnoìdes** (Lam.) BSP. CREEPING ERAGROSTIS. Map 158. Infrequent throughout the state but more frequent in the southwestern part where its habitat is more frequent. It is found on sandy or gravelly bars in ditches, creeks, and rivers and on the sandy shores of lakes. It is also found in muddy habitats along streams and in dried-up ponds and sloughs. In the latter habitats it often forms large mats.

Que. to Wash., southw. through Mex. and W. I. to Argentina; not found in the Rocky Mts.

2. **Eragrostis spectábilis** (Pursh) Steud. (*Eragrostis pectinacea* of Gray, Man., ed. 7, Britton and Brown, Illus. Flora, ed. 2, and *Eragrostis spectabilis* var. *sparsihirsuta* Farw.) PURPLE LOVEGRASS. Map 159. This species is frequent to rare in sandy to very sandy soils throughout the

state and in hard, white clay soil in certain areas in the southern part of the state. It may be absent from a few counties of the Tipton Till Plain.

Maine to Minn., southw. to Fla., Ariz., and n. Mex.

3. ERAGROSTIS CILIANÉNSIS (All.) Link. (*Eragrostis megastachya* (Koeler) Link of Gray, Man., ed. 7 and *Eragrostis major* Host of Britton and Brown, Illus. Flora, ed. 2.) STINKGRASS. Map 160. Infrequent to frequent throughout the state. It prefers sandy soil and is frequently a common grass in such soil about dwellings and in gardens and other cultivated grounds. It is generally found in cultivated grounds, in waste places, and along roadsides.

Nat. of Eu.; Maine to Wash., southw. throughout the U. S.; through Mex. and W. I. to Argentina.

4. **Eragrostis capillàris** (L.) Nees. LACEGRASS. Map 161. This is an infrequent grass of southern Indiana which is found in poor soil, mostly on the open crests and slopes of black oak and black oak-white oak ridges.

Maine to Wis., southw. to Ga. and e. Tex.

5. **Eragrostis Fránkii** C. A. Meyer. FRANK'S LOVEGRASS. Map 162. Infrequent to rare in all parts of the state. It prefers sandy soil and is most often found on sandy bars of streams, along roadsides, and in pastures and barnlots.

N. H. to Minn., southw. to Fla. and Kans.

6. **Eragrostis pectinàcea** (Michx.) Nees. (*Eragrostis Purshii* Schrad. and *Eragrostis caroliniana* (Spreng.) Scribn.) Map 163. This is our most common species of the genus and is frequent throughout the state. It prefers the open in sandy or muddy soils, and is found mostly along roadsides and railroads and in waste places and fallow fields. It is less frequent on sandy bars and muddy borders of streams and ditches.

Maine to N. Dak., southw. to Fla. and e. Tex.

15-356. DIARRHÈNA Beauv.

1. **Diarrhena americàna** Beauv. (*Diarrhena diandra* (Michx.) Wood and *Korycarpus arundinaceus* Zea.) Map 164. This is a woodland grass usually found with oak, beech, and sugar maple. It is local to infrequent and is often found on rocky wooded slopes as where it occurs in Clifty Falls State Park.

W. Va. to Mich. and S. Dak., southw. to Tenn., Ark., Okla., and e. Tex.

20-365. UNÌOLA L.

1. **Uniola latifòlia** Michx. BROADLEAF UNIOLA. Map 165. This is an open woodland species and is found mostly in our southern counties although Miss Madge McKee found it along the Iroquois River in Newton County. It is found in greatest abundance in slightly acid, hard clay soils of the bottomlands. It occurs, however, in upland woods and even on the rocky cliffs along the Ohio River.

Pa., N. J. to Ill. and Kans., southw. to Fla. and Tex.

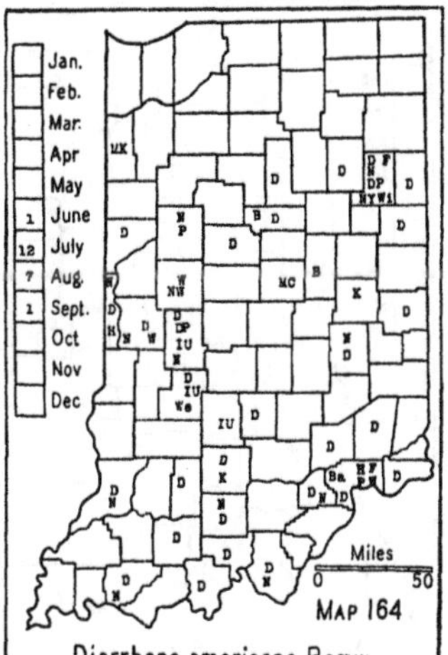

Diarrhena americana Beauv.

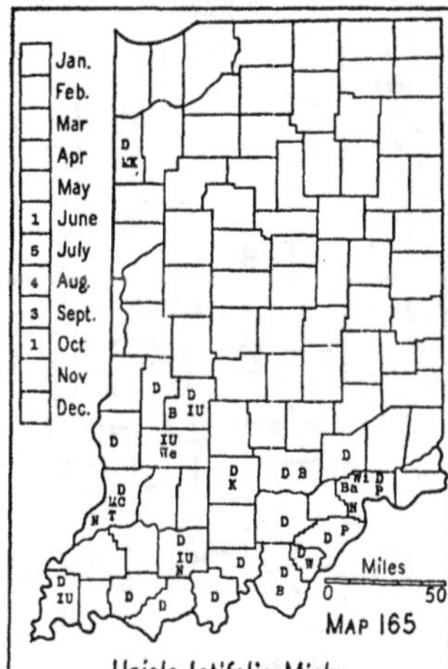

Uniola latifolia Michx.

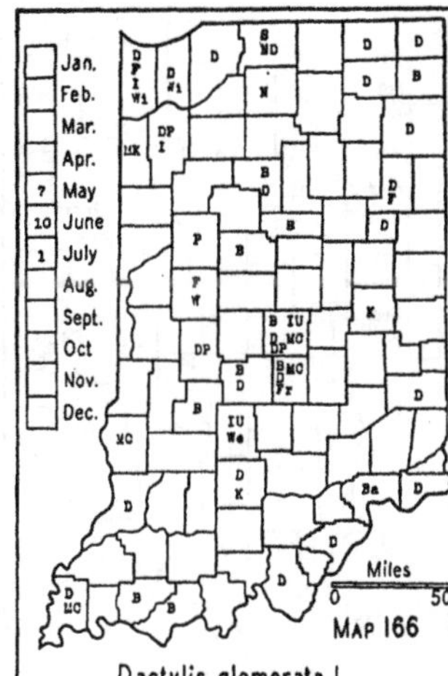

Dactylis glomerata L.

21-372. DÁCTYLIS L.

1. DACTYLIS GLOMERÀTA L. ORCHARD GRASS. Map 166. This species has now escaped in all parts of the state, commonly so in limestone areas. It has been sown for both hay and pasture. It affords early pasture and is drought resistant. I think its use is now on the decline.

Nat. of Eurasia; Newf. to se. Alaska, southw. to Fla. and cent. Calif.

26-333. PHRAGMÌTES Trin.

1. **Phragmites commùnis** Trin. COMMON REED. Map 167. This grass is found in wet marshes, on mucky borders of lakes and streams, and in springy places in general, hence it is found mostly in our lake area. Here it was once frequent, but it is now rather local on account of drainage.

N. S. to B. C., southw. to Fla. and Calif.; also in Mex., W. I. to Chile and Argentina. It is also found in Eurasia, Africa, and Australia.

28-355. MÉLICA L.

Upper surface of leaves generally glabrous and the lower surface generally pubescent; spikelets with 2 fertile florets; lateral nerves and midrib of the lemmas fading out before reaching the hyaline apex; panicles simple or nearly so........1. *M. mutica.*

Upper surface of leaves generally pubescent and the lower surface scabrous or smooth; spikelets with 2 or 3 fertile florets; lateral nerves and midrib of lemmas usually reaching the apex; panicles compound.............................2. *M. nitens.*

1. **Melica mùtica** Walt.* TWO-FLOWER MELIC. Map 168. This is a local grass in a few of the southern counties, where it is found on the rocky crests or slopes of black oak ridges, and is rarely associated with beech and sugar maple. I have seen this species a good many times but have found only a few tufts here and there and only a few culms to a tuft.

Md. to Iowa, southw. to Fla. and Tex.

2. **Melica nitens** (Scribn.) Nutt. THREE-FLOWER MELIC. Map 169. This species is very local but usually abundant where found. Its habitat is so varied that it seems worth while to give the habitat in which specimens

* Plants with spreading pubescent sheaths are *Melica mutica* f. *diffusa* (Pursh) Fern. (Rhodora 41: 501. 1939.) I have it from Crawford and Perry Counties.

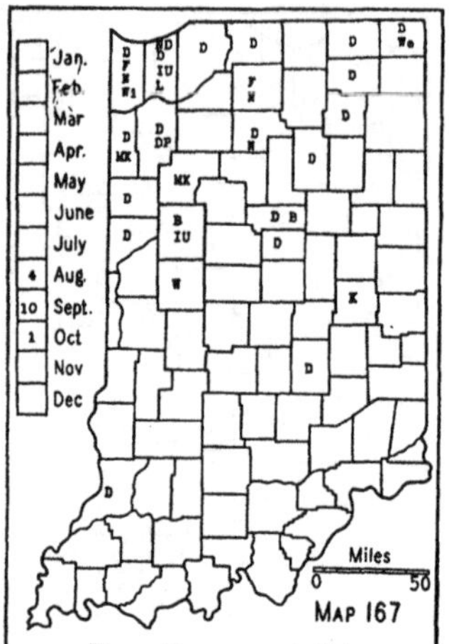

Phragmites communis Trin.

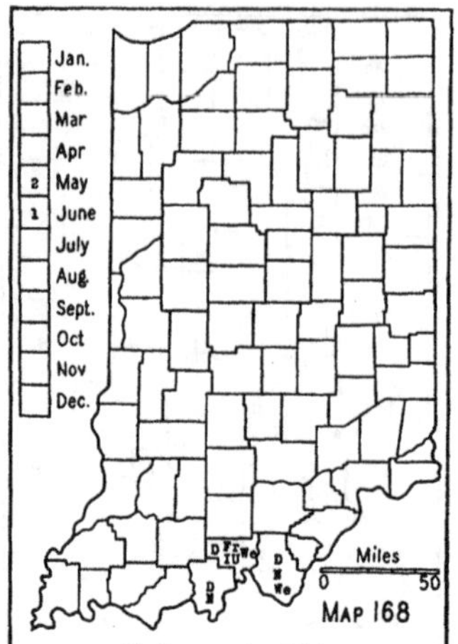

Melica mutica Walt.

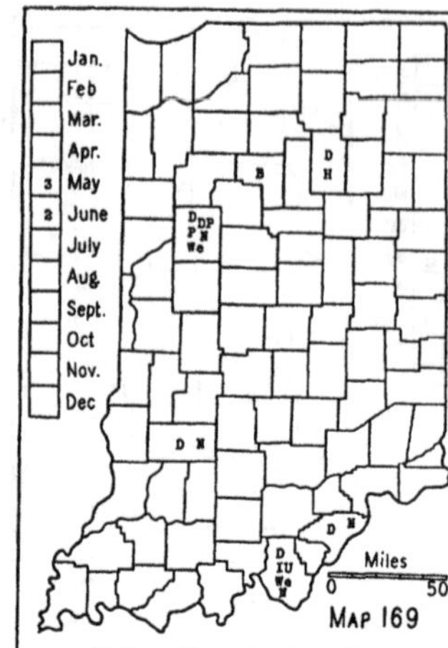

Melica nitens (Scribn.) Nutt.

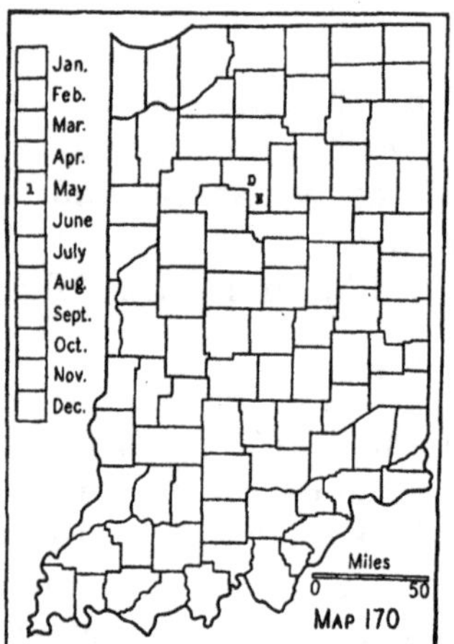

Schizachne purpurascens (Torr.) Swallen

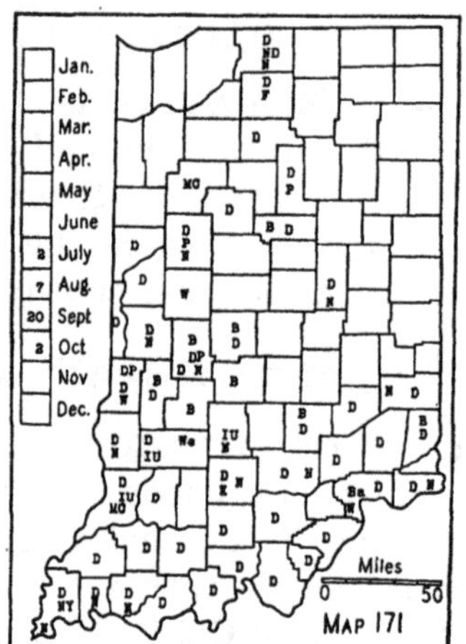

Triodia flava (L.) Smyth

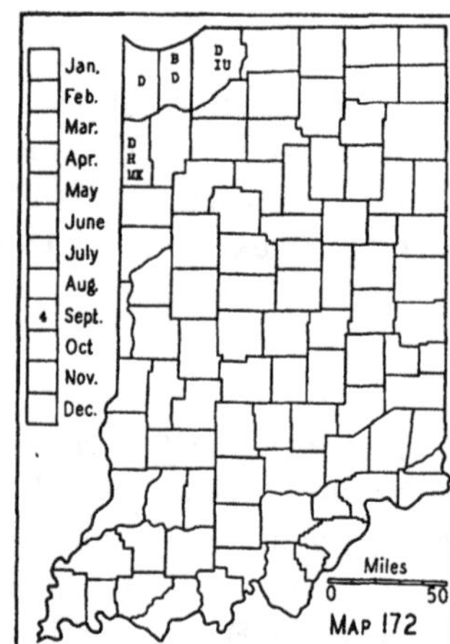

Triplasis purpurea (Walt.) Chapm.

have been found. In Harrison and Clark Counties it occurs on top of bluffs between 200 and 300 feet high along the Ohio River and at the very edge of the bluff. I found a few specimens in an alluvial flat along a small stream in Harrison County. In Greene County I found it along a railroad and I assume that this single specimen was a waif. In Tippecanoe County it occurs as a common plant near the top of the very high gravelly bank of Big Wea Creek southwest of Lafayette. In Wabash County I found a few plants on "hanging rock." This is a large rock isolated by erosion, standing 84 feet high on the low bank of the Wabash River near Lagro.

Pa. to Iowa and Kans., southw. to Ky., Ark., Tex., and Ariz.

29-355A. SCHIZÁCHNE Hackel

1. **Schizachne purpuráscens** (Torr.) Swallen. (*Melica striata* (Michx.) Hitchc. of Gray, Man., ed. 7 and *Avena Torreyi* Nash of Britton and Brown, Illus. Flora, ed. 2.) Map 170. My only specimen was found along the

Wabash River on the top of the first rocky bluff east of Georgetown or about 6 miles west of Logansport.

Newf. to s. Alaska, southw. to Pa., Ky., S. Dak., and Mont. and in the mts. from B. C. to N. Mex.; also in Siberia and Japan.

31-335. TRIÒDIA R. Br.

1. **Triodia flàva** (L.) Smyth. (*Tridens flavus* (L.) Hitchc. of Gray, Man., ed. 7 and *Tridens flava* (L.) Hitchc. of Britton and Brown, Illus. Flora, ed. 2.) PURPLETOP. Map 171. Infrequent to frequent or even locally common. Possibly absent in a few counties where the soil is neutral and there are no sandy areas. It prefers open, sandy soil; and it is usually most abundant in prairie habitats.

N. H. to Nebr., southw. to Fla. and Tex.

32-335A. TRÍPLASIS Beauv.

1. **Triplasis purpùrea** (Walt.) Chapm. Map 172. This species is local in the dry sand of the dunes about Lake Michigan and common in a similar habitat in Newton County about three miles northwest of Morocco where it occurs in open sandy woods and fallow fields over an area at least 4 miles long and a mile wide (1938).

N. H. to Minn., and Nebr., southw. to Fla. and Tex.

3. *HÓRDEAE* Lindl. BARLEY TRIBE

Spikelets solitary at each node of the rachis (rarely 2 in species of *Agropyron*, but never throughout).
  First glume (except in the terminal spikelet) lacking; spikelets placed edgewise to the rachis....47. LOLIUM, p. 120.
  First glume present; spikelets placed flatwise to the rachis.
    Glumes 1-nerved; spikelets with 2 perfect florets....42. SECALE, p. 115.
    Glumes 3-many-nerved.
      Glumes lanceolate or linear; spikelets 3-many-flowered...39. AGROPYRON, p. 113.
      Glumes ovate; spikelets 2-6-flowered....40. TRITICUM, p. 115.
Spikelets 2-6 at each node of the rachis.
  Spikelets all alike, 2-6-flowered.
    Glumes well developed, about as long as the florets, nerved; spikes densely flowered, the spikelets mostly imbricated....43. ELYMUS, p. 115.
    Glumes obsolete or bristlelike, nerveless; spikes loosely flowered, the spikelets widely spreading....45. HYSTRIX, p. 118.
  Spikelets not all alike, (rarely 2- or 3-) 1-flowered, in 3's at each joint, the lateral pair pedicellate, usually aborted....46. HORDEUM, p. 119.

39-405. AGROPÝRON Gaertn. WHEATGRASS

Creeping rootstocks present; anthers about 4 mm long.
  Spikelets mostly 4-6-flowered, 9-17 mm long....1. *A. repens.*
  Spikelets mostly 7-12-flowered, 15-28 mm long....2. *A. Smithii.*
Creeping rootstocks lacking; anthers about 1.5 mm long.
  Spikelets awnless or rarely a few with awns, the awns rarely up to 5 mm long........3. *A. pauciflorum.*
  Spikelets all awned; awns usually all 6 mm long or longer....4. *A. subsecundum.*

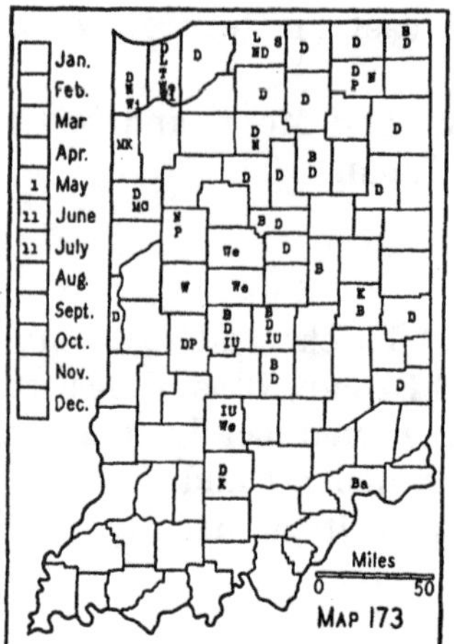

Agropyron repens (L.) Beauv.

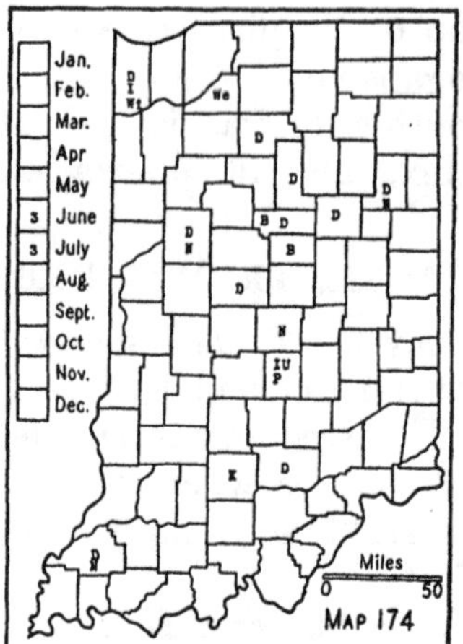

Agropyron Smithii Rydb.

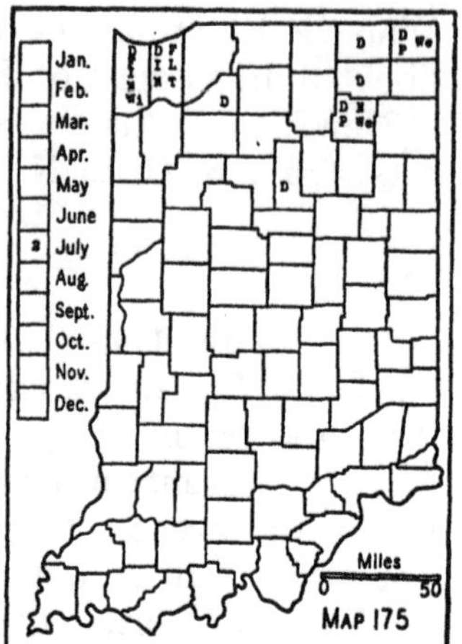

Agropyron pauciflorum (Schwein.) Hitchc.

1. AGROPYRON RÈPENS (L.) Beauv. QUACKGRASS. Map 173. This species has become well established in the northern two thirds of the state, especially along roadsides and railroads where there is no effort to exterminate it. It is most abundant in the lake area where it sometimes covers acres of cultivated fields and pastures. Most of the landowners have despaired of exterminating it and merely use control measures. It is now known that it can be eradicated by the use of chemicals, and every landowner should proceed without delay to exterminate it.

The extreme variability of this species has caused some confusion in its recognition. It has been decided to treat the varied forms as a species complex. Those who wish to divide the forms should see Fernald on the American variations of Agropyron repens in Rhodora 35: 182-185. 1933.

Nat. of Eurasia; Newf. to Alaska, southw. to N. C., Ark., and Calif.

2. AGROPYRON SMÍTHII Rydb. BLUESTEM WHEATGRASS. Map 174. All of my specimens were found along railroads, where the colonies will doubtless persist and spread. Apparently it does not propagate as vigorously as the preceding species, but, when discovered, it should be eradicated. This is a western species that has been introduced eastward of Iowa and Kansas.

N. Y., Mich. to Alberta and Wash., southw. to Ohio, Kans., Tex., Ariz., and Calif.

3. **Agropyron paucifiòrum** (Schwein.) Hitchc. (*Agropyron tenerum* Vasey of Gray, Man., ed. 7, Britton and Brown, Illus. Flora, ed. 2, and *Agropyron caninum* var. *tenerum* (Vasey) Pease & Moore of Deam, Grasses of Ind.) SLENDER WHEATGRASS. Map. 175. Very local. Found in both dry and moist habitats in a few of our northern counties.

Lab. to Alaska, southw. to the mts. of W. Va., Mo., N. Mex., Calif., and nw. Mex.

4. **Agropyron subsecúndum** (Link) Hitchc. (*Agropyron caninum* f. *pubescens* (Scribn. & Smith) Pease & Moore and *Agropyron trachycaulum* (Link) Malte.) BEARDED WHEATGRASS. Map 176. This species is local in

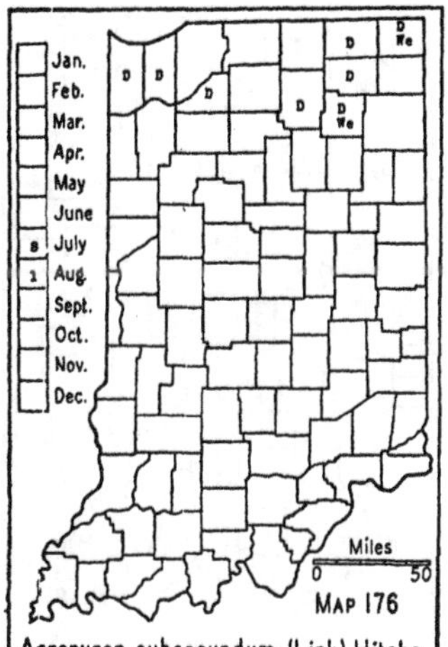

Agropyron subsecundum (Link) Hitchc.

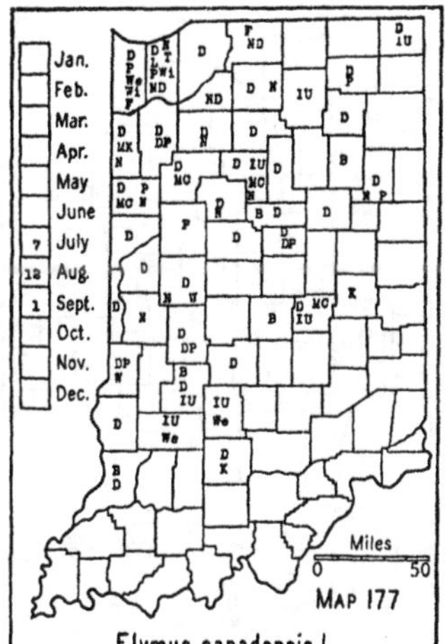

Elymus canadensis L.

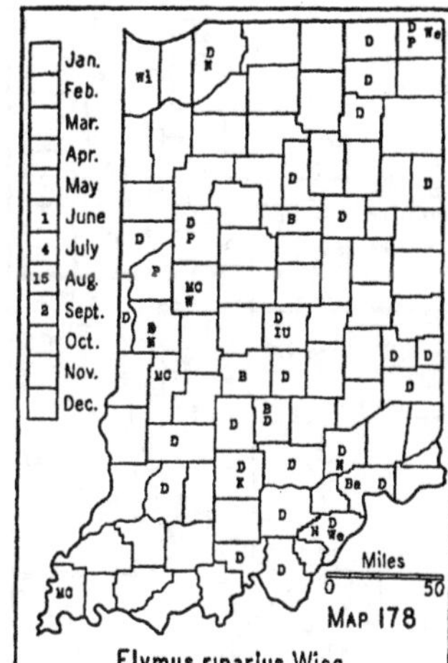

Elymus riparius Wieg.

a few of our northern counties, where it is found in dry, sandy or clayey soil on the crests of low dunes, on wooded banks about lakes, and in springy places and marshes.

Newf. to Alaska, southw. to the mts. of Md., Ind., Nebr., N. Mex., Ariz., and Calif.

40-408. **TRÍTICUM** L. WHEAT

Wheat is a winter annual and it often grows where it finds lodgment along roads, paths, fields, and waste places, but it does not persist. It has been reported from Porter County by Lyon under the name of *Triticum aestivum* and from Jasper County by Welch as *Triticum sativum.*

Wheat properly belongs with the excluded species because it fails to perpetuate itself.

42-407. **SECÀLE** L. RYE

Rye is a winter annual which springs up where it may be scattered along roads, in fields, and in waste places, but it will not persist. It has been reported from Jasper County by Welch.

Rye properly belongs with the excluded species because it fails to perpetuate itself.

43-411. **ÉLYMUS** L. WILD-RYE

[Note: Measurements of glumes and lemmas include their awns, and measurements of paleas are those of the first floret of a spikelet taken from the middle of the spike.]

Awns long and, at maturity, curved outward; paleas mostly 10-13 mm long...1. *E. canadensis.*

Awns straight; paleas mostly 6-9 mm long.

Glumes 0.5-1 mm wide (rarely up to 1.3 mm wide), straight or only slightly bowed out at the base, mostly 3-nerved above the middle; spikes long-exserted.

Blades glabrous above and beneath; paleas 7-8 mm long; grain 5-6.5 mm long. ....2. *E. riparius.*

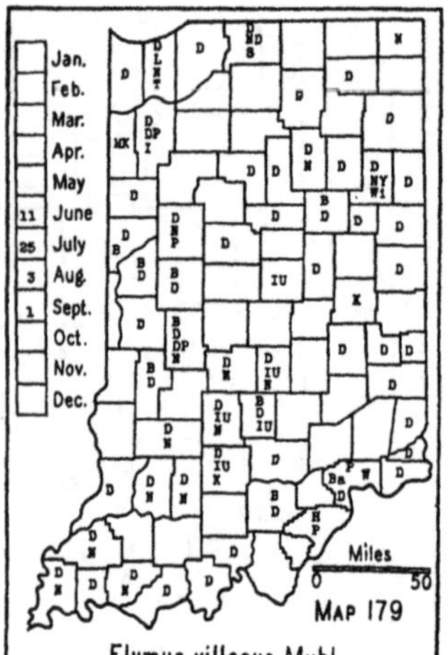

Elymus villosus Muhl.

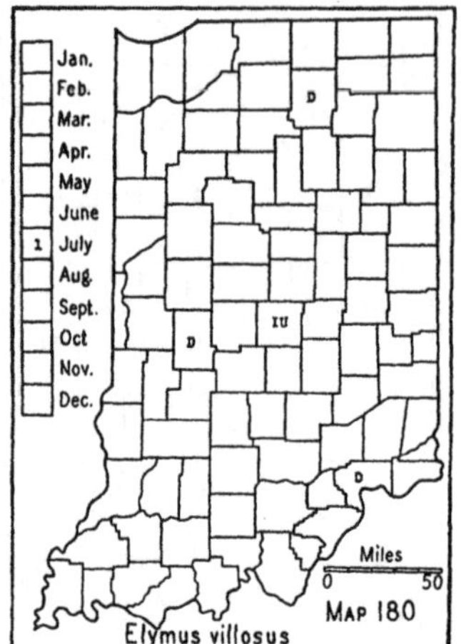

Elymus villosus f. arkansanus (Scribn. & Ball) Fern.

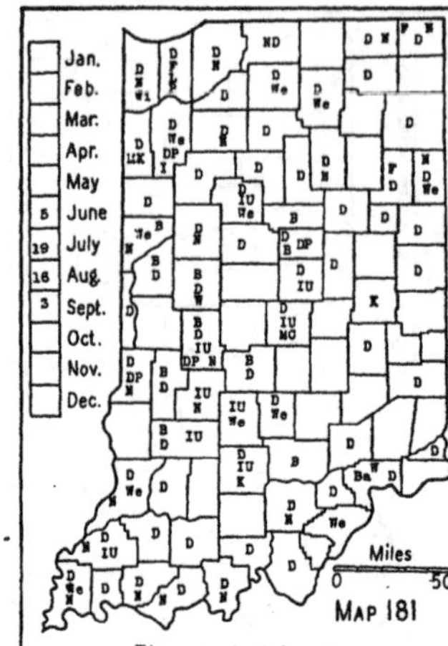

Elymus virginicus L.

Blades villous above, smooth or scabrous beneath; paleas 5.5-7 mm long; grain 4-4.5 mm long.
Lemmas and glumes hirsute.................................3. *E. villosus.*
Lemmas and glumes glabrous or sparingly strigose-hispid..3a. *E. villosus* f. *arkansanus.*
Glumes mostly 1.3-2.5 mm wide (rarely as narrow as 1 mm), generally conspicuously curved outward and indurated at the base, usually more or less dilated above and twisted, generally plainly 5-nerved on the upper half (sometimes 3- or 4-nerved), the basal part generally rounded and nerveless.
Spikes included at the base or barely exserted.
Spikelets with awns more than 3 mm long.
Glumes and lemmas glabrous or scabrous on the margins only; upper surface of leaves usually scabrous (rarely pubescent)............4. *E. virginicus.*
Glumes and lemmas hirsute; upper surface of leaves glabrous....................................4a. *E. virginicus* var. *intermedius.*
Spikelets awnless or some with awns up to 3 mm long..4b. *E. virginicus* var. *submuticus.*
Spikes generally long-exserted.
Glumes and lemmas hirsute; upper surface of leaves villous...4c. *E. virginicus* var. *australis.*
Glumes and lemmas glabrous or strigose-scabrous.
Blades glabrous above; awns generally 5-15 mm long...4d. *E. virginicus* var. *jejunus.*
Blades hirsute above; awns generally 15-30 mm long...4e. *E. virginicus* var. *glabriflorus.*

1. **Elymus canadénsis** L. CANADA WILD-RYE. Map 177. Infrequent to locally common in sandy soil along roadsides and railroads, in open woodland, on open dunes, and in prairie habitats. It becomes very local south of the lake area and probably is absent from some of the southern counties. It is extremely variable in the size of the spikes and in the density of the pubescence of the spikelets.

Que. to s. Alaska, southw. to Ky., Mo., Tex., and Ariz.

2. **Elymus ripàrius** Wieg. (Rhodora 20: 84-86. 1918.) Map 178. This is a recently described species and is infrequent probably throughout the

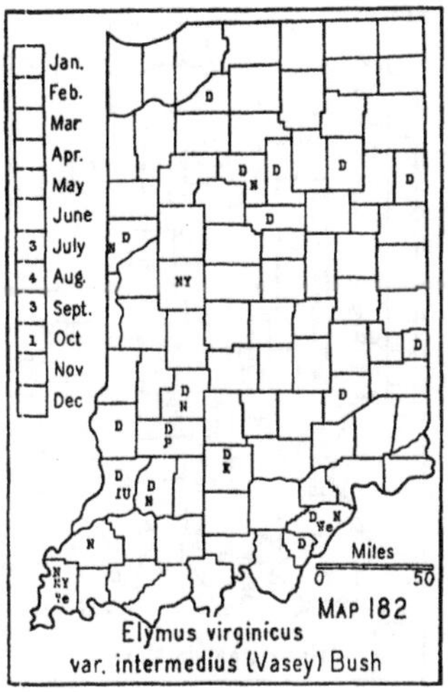

Elymus virginicus var. intermedius (Vasey) Bush

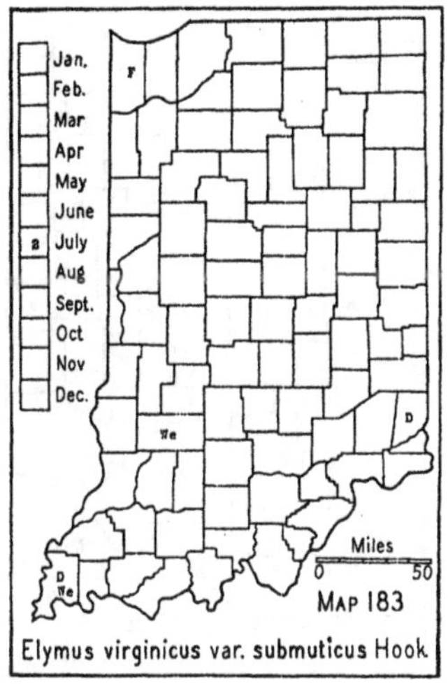

Elymus virginicus var. submuticus Hook

Elymus virginicus var. australis (Scribn. & Ball) Hitchc.

state. It is a low ground grass which is usually found on wooded, alluvial areas and along streams.

Maine, Que., and Mich., southw. to N. C., Ohio, Ind., and Mo.

3. **Elymus villòsus** Muhl. (*Elymus striatus* of recent authors, not Willd.) Map 179. Infrequent to frequent throughout the state. This species prefers a dry and rather sandy soil, although it is sometimes found in moist situations. It is found mostly on wooded slopes, crests or ridges, on alluvial banks, and rarely in the open along roadsides.

Vt. to Wyo., southw. to N. C., Ala., and Tex.

3a. **Elymus villosus** f. **arkansànus** (Scribn. & Ball) Fern. (Rhodora 35: 195. 1933.) (*Elymus striatus* var. *arkansanus* (Scribn. & Ball) Hitchc. and *Elymus arkansanus* Scribn. & Ball.) Map 180. This form has been found in only a few counties. It grows in habitats similar to those of the species.

Mass. to Ind. and Iowa, southw. to Md., Mo., and Okla.

4. **Elymus virgínicus** L. VIRGINIA WILD-RYE. Map 181. Frequent to common throughout the state. It grows in wet or moist soil and is found mostly in alluvial areas along streams and ditches, in low places in woodland, and along roadsides.

The upper surface of the leaves is usually glabrous or somewhat scabrous or rarely with a few hairs on the veins. I have, however, a few specimens with the upper surface of the blades softly pubescent. I think these plants should have a distinguishing name. They are from Fayette, Marion, Starke, and Warrick Counties.

Newf. to Alberta, southw. to Fla. and Ariz.

4a. **Elymus virginicus** var. **intermèdius** (Vasey) Bush. (Amer. Midland Nat. **10**: 60. 1926.) (*Elymus virginicus* var. *hirsutiglumis* (Scribn.) Hitchc. and *Elymus hirsutiglumis* Scribn.) Map 182. Infrequent throughout the state in habitats similar to those of the species.

Maine to Iowa, southw. to Fla. and Tex.

MAP 185
Elymus virginicus var. jejunus (Ramaley) Bush

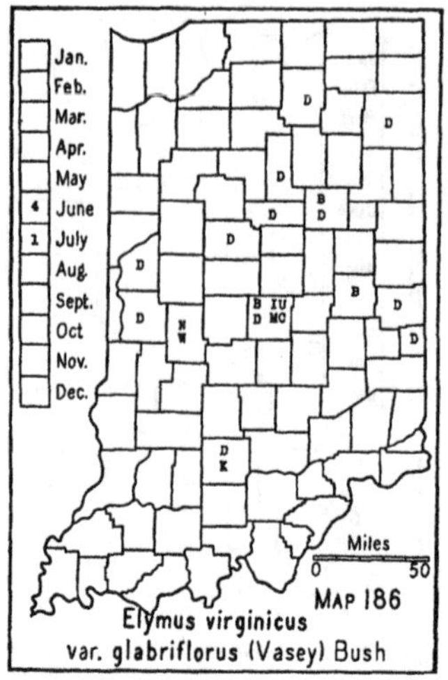

MAP 186
Elymus virginicus var. glabriflorus (Vasey) Bush

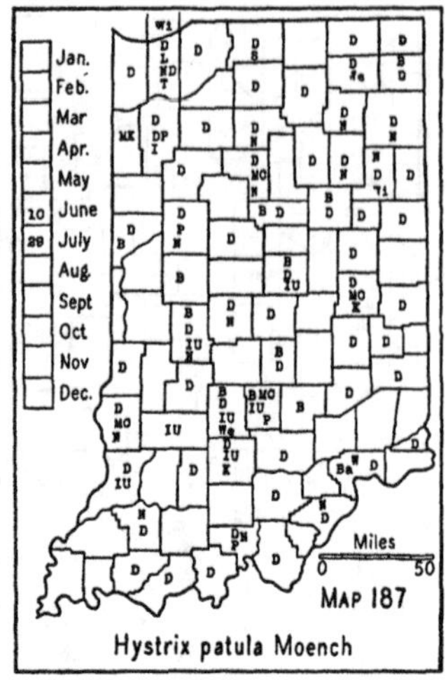

MAP 187
Hystrix patula Moench

4b. **Elymus virginicus** var. **submùticus** Hook. (*Elymus curvatus* Piper.) Map 183. A rare form with the habitat of the species.

Que. to Wash., southw. to R. I., Ohio, Ky., Okla., and Mont.

4c. **Elymus virginicus** var. **austràlis** (Scribn. & Ball) Hitchc. (*Elymus australis* Scribn. & Ball.) Map 184. This form has been found in a few places in the southern half of the state on wooded ridges and on post oak flats.

Vt. to Iowa, southw. to Fla. and Tex.

4d. **Elymus virginicus** var. **jejùnus** (Ramaley) Bush. Map 185. This rare form is known only from Umbach's specimen collected "on the sands at Pine," Lake County, on June 29, 1898. Hitchcock does not recognize this variety in his recent manual.

4e. **Elymus virginicus** var. **glabriflòrus** (Vasey) Bush. (*Elymus glabriflorus* Scribn.) Map 186. This variety has been found in several counties, and doubtless it will be found to be well distributed in the state when intensive work is done. It is a woodland grass found in both moist and dry situations.

Maine to Kans., southw. to Fla. and N. Mex.

45-412. **HÝSTRIX** Moench

Spikelets glabrous..1. *H. patula.*
Spikelets pubescent..............................1a. *H. patula* var. *Bigeloviana.*

1. **Hystrix pátula** Moench. (*Hystrix Hystrix* (L.) Millsp.) BOTTLE-BRUSH. Map 187. This is an infrequent to frequent grass throughout the state. It is a woodland species, but is often found growing in open places along fences and roadsides. It prefers dry soil and occurs in almost all types of woodland.

Maine, Ont., Mich., and Minn., southw. to Ga., Ala., and Ark.

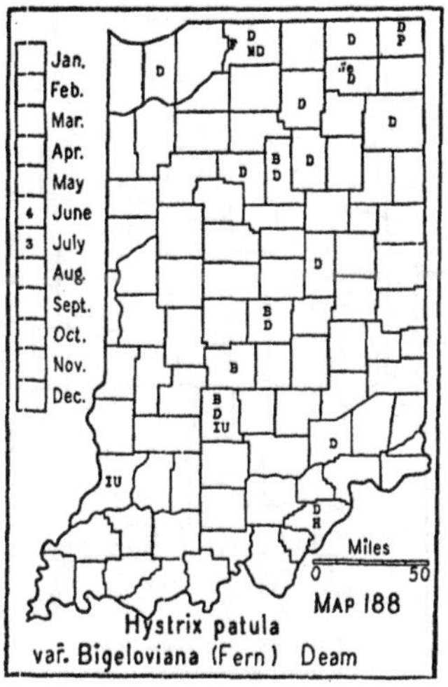

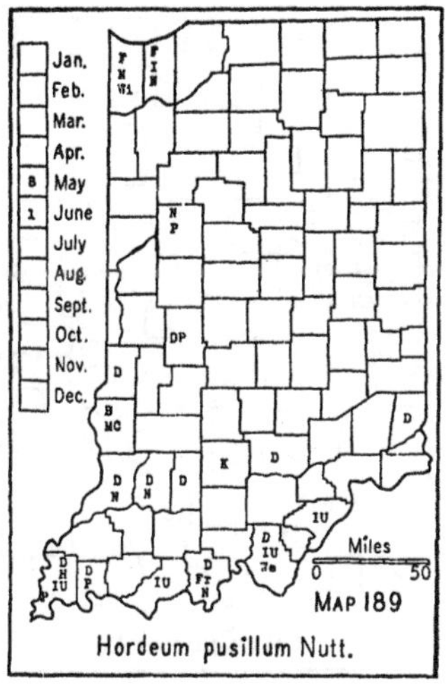

1a. **Hystrix patula** var. **Bigeloviàna** (Fern.) Deam. Map 188. The habitat of the variety is the same as that of the species. It is sparingly found in northern Indiana and is rare in the southern part of the state.

N. S. to N. Dak., southw. to Conn., Ohio, Ind., and Mo.

46-410. HÓRDEUM [Tourn.] L. BARLEY

[Wiggans. Classification of the cultivated varieties of barley. Cornell Agric. Exper. Sta. Mem. 46: 365-456. 1921.]

Rachis of spikes disarticulating.
  Awns of the glumes less than 20 mm long.
    Glumes of fertile spikelets dilated above the base..................1. *H. pusillum.*
    Glumes of fertile spikelets not dilated above the base..............2. *H. nodosum.*
  Awns of the glumes more than 20 mm long..........................3. *H. jubatum.*
Rachis of spikes not disarticulating. (See excluded species no. 55, p. 1027.)..*H. vulgare.*

1. HORDEUM PUSÍLLUM Nutt. LITTLE BARLEY. Map 189. This species is found usually in slightly acid soils in waste places and fallow fields and along railroads and roadsides. It is infrequent to local in the southern counties and has been found in four of our northern counties in waste places and along railroads. I think that this species has been introduced into the state and I am so recording it. Spillman found it in Knox County in 1890, and, although Schneck reported a wild barley earlier, there is no specimen to verify the report. Our early authors, Baird & Taylor, Barnes, Clapp, J. M. Coulter, and Young, who collected intensively in some of the Ohio River counties, did not report a wild barley of any kind. This evidence, in addition to its habitats, convinces me that it has invaded the state since that time.

Del. to Wash., southw. to Fla., s. Calif., and Mex.

2. HORDEUM NODÒSUM L. MEADOW BARLEY. Map 190. Hansen (Proc. Indiana Acad. Sci. 37: 320. 1928) reported this species from Vanderburgh County. He sent me a specimen to have his determination verified. Hansen

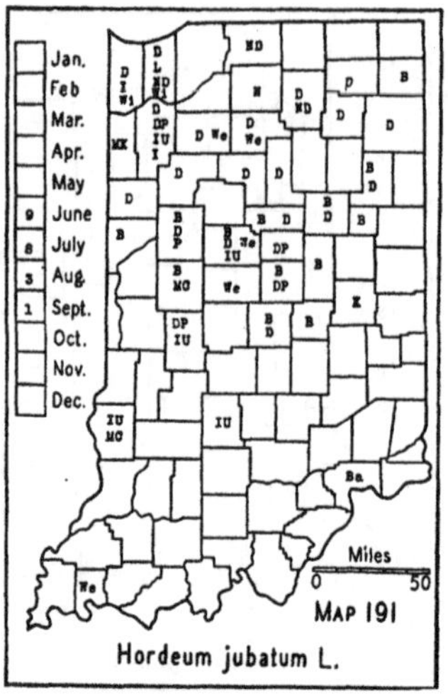

Hordeum jubatum L.

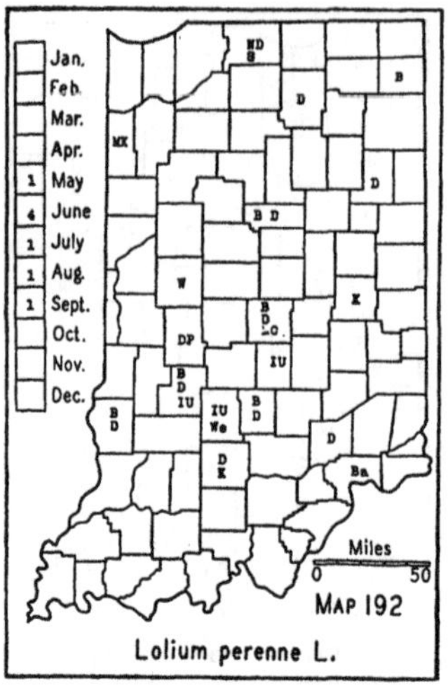

Lolium perenne L.

Lolium multiflorum Lam.

says: "Found growing abundantly in Vanderburgh County during July." I do not have any data except the county locality, so I have not been able to visit the place to ascertain whether it persists or is spreading. I am including this species in our flora upon his authority. Since this is a western species, it has been introduced here and should be so regarded.

Mont. to Alaska, southw. to N. Mex., Calif., and in S. A.; introduced in some of the eastern states.

3. **Hordeum jubàtum** L. FOXTAIL BARLEY. Map 191. This species has become well established in the northern half of the state, especially in the lake area where it has already become a veritable pest. It is found mostly along roadsides and railroads and in waste places, fallow fields, and pastures. It is extremely doubtful that this species is a native of Indiana.

Newf. and Lab. to Alaska, southw. to Md., Ill., Mo., Tex., Calif., and Mex.; introduced in the Eastern States.

47-395. LÒLIUM L. RYEGRASS

Glumes shorter than the spikelets.
  Lemmas mostly 5-6 mm long, awnless (sometimes short-awned); spikelets mostly 5-10 (12) -flowered..1. *L. perenne.*
  Lemmas mostly 7-8 mm long, awned; spikelets mostly 10-20-flowered..2. *L. multiflorum.*
Glumes as long as or longer than the spikelets. (See excluded species no. 56, p. 1027.) ....*L. temulentum.*

1. LOLIUM PERÉNNE L. PERENNIAL RYEGRASS. Map 192. This species is doubtless sparingly found throughout the state. It has been sown intentionally as an adulterant of grass seed in meadows and lawns. It is now found as an escape along roadsides, on the unkept borders of lawns, and in waste places. Besides one small colony which I have seen, I have no evidence to indicate that it is more than an occasional escape. Usually known in commerce as English Ryegrass.

Nat. of Eu.; Newf. to Alaska, southw. to Va. and Calif.

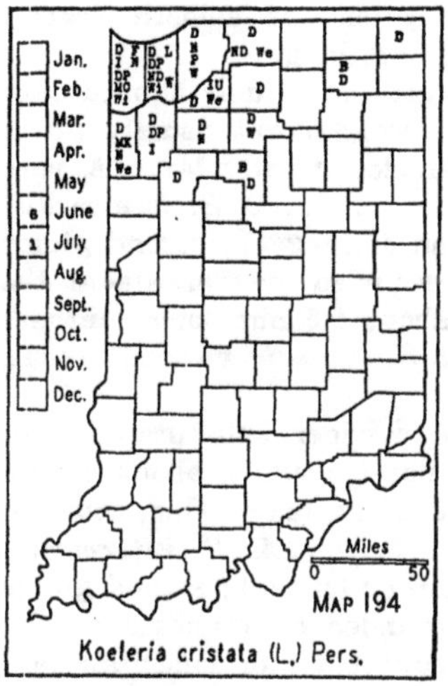

MAP 194
Koeleria cristata (L.) Pers.

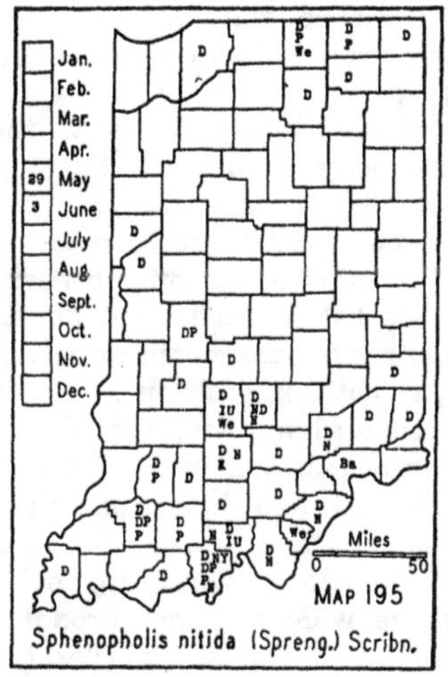

MAP 195
Sphenopholis nitida (Spreng.) Scribn.

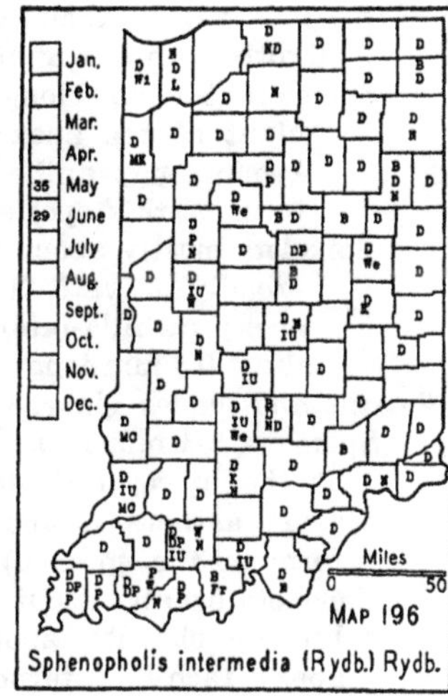

MAP 196
Sphenopholis intermedia (Rydb.) Rydb.

2. LOLIUM MULTIFLÓRUM Lam. ITALIAN RYEGRASS. Map 193. This ryegrass has been found in several counties in the state in lawns, parks, and golf grounds and may be considered established.

Nat. of Eu.; common on the Pacific coast, infrequent eastward.

4. *AVÈNEAE* Nees. OAT TRIBE

Spikelets not over 5 mm long.
  Spikelets disarticulating above the glumes.
    Inflorescence spikelike; plants of a dry, sandy habitat........52. KOELERIA, p. 121.
    Inflorescence a widely spreading panicle; plants of a springy habitat..55. DESCHAMPSIA, p. 123.
  Spikelets disarticulating below the glumes.
    Florets all perfect, awnless..........................53. SPHENOPHOLIS, p. 121.
    Florets unlike, the lower perfect, awnless, the upper staminate and bearing a hooked awn...59. HOLCUS, p. 124.
Spikelets more than 5 mm long.
  Lemmas awned from the back.
    Annual; spikelets more than 10 mm long.....................57. AVENA, p. 123.
    Perennial; spikelets less than 10 mm long............58. ARRHENATHERUM, p. 123.
  Lemmas awned from between the two apical teeth............60. DANTHONIA, p. 124.

52-346. KOELÈRIA Pers.

1. **Koeleria cristàta** (L.) Pers. JUNEGRASS. Map 194. Infrequent to local in the northwestern counties where it grows in dry sand on dunes and sand hills, rarely on gravelly hills. The species is variable. The inflorescence expands in anthesis, and becomes spikelike afterward.

Ont. to B. C., southw. to Del., Mo., La., Calif., and Mex.

53-344. SPHENÓPHOLIS Scribn. WEDGEGRASS

Sheaths and blades softly pubescent, sometimes only the sheaths pubescent.
  First glume fully a third as wide as the second; glumes subequal; lemmas more or less scabrous all over (at least the exposed apical end scabrous); anthers mostly 0.8-1.2 mm long; spikelets 3-4 mm long.........................1. *S. nitida.*

First glume less than a third as wide as the second; lemmas smooth (rarely slightly scabrous at the apex); anthers mostly 0.5-0.8 mm long.

Spikelets 3-4 mm long; second glume narrowly obovate or wider, subacute or blunt at the apex; rachilla-internode below the second floret usually about 1 mm long; anthers mostly 0.5-0.6 mm long; panicles usually lax. (A rare form of this species.)..................................2. *S. intermedia.*

Spikelets mostly about 2.5 mm long (rarely up to 3 mm or longer); second glume broadly obovate, about as wide as long, broadly rounded or truncate at the apex; rachilla-internode below the second floret about 0.5 mm long; anthers about 0.8 mm long; panicles usually contracted..3a. *S. obtusata* var. *pubescens.*

Sheaths and blades glabrous, smooth or scabrous.

Spikelets 3-4 mm long; second glume narrowly obovate or wider, subacute or blunt at the apex; rachilla-internode below the second floret usually about 1 mm long; anthers mostly 0.5-0.6 mm long; panicles usually lax. (Our common form of the species.)....................................2. *S. intermedia.*

Spikelets usually about 2.5 mm long (rarely up to 3 mm or longer); second glume broadly obovate, about as wide as long, broadly rounded or truncate at the apex; rachilla-internode below the second floret about 0.5 mm long; anthers about 0.8 mm long; panicles usually contracted..................3. *S. obtusata.*

1. **Sphenopholis nítida** (Spreng.) Scribn. Map 195. Rather frequent in the unglaciated area of southern Indiana and rare in the northern part of the state. It is generally found on black and white oak ridges and rarely with beech. It prefers a rich soil of weathered sandstone and it may be entirely absent in neutral or alkaline soils.

A glabrous form of this species has been described but it may not occur in Indiana as all of my 41 specimens are copiously pubescent.

Mass. to N. Dak., southw. to Fla. and Tex.

2. **Sphenopholis intermèdia** (Rydb.) Rydb. (*Sphenopholis pallens* of recent authors.) SLENDER WEDGEGRASS. Map 196. Infrequent to frequent throughout the state. It prefers a dry soil and is found in many habitats. Usually frequent in beech and sugar maple woods, white oak woods, and white oak and black oak woods; less frequent in moist or wet woodland, bogs, and fallow fields and along railroads. I have a specimen with pubescent sheaths and leaves, which was found growing in sphagnum in a decadent tamarack bog just east of Pokagon State Park, Steuben County. This is the only pubescent specimen I have out of 79 Indiana specimens.

Newf. to B. C., southw. to Fla. and Ariz.

3. **Sphenopholis obtusàta** (Michx.) Scribn. PRAIRIE WEDGEGRASS. Map 197. Infrequent to local throughout the state. Its habitat varies from the crests of ridges in the "knobs" to low sand ridges and old lake and river bottoms.

Maine to B. C., southw. to Fla., Ariz., Calif., and Mex.

3a. **Sphenopholis obtusata** var. **pubéscens** (Scribn. & Merr.) Scribn. This is a form with the sheaths and upper and lower surface of the leaves pubescent. I have it from only the southern part of the state where it occurs in Crawford, Perry, and Posey Counties. I segregate this form from the species for the benefit of other workers who may be interested in the geographical distribution of the form.

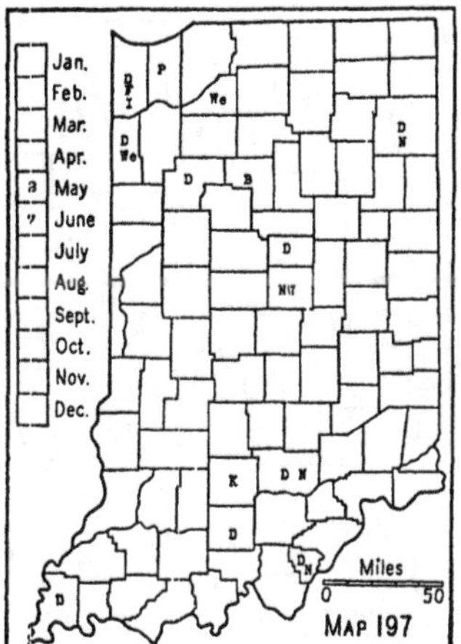

MAP 197
Sphenopholis obtusata (Michx.) Scribn.

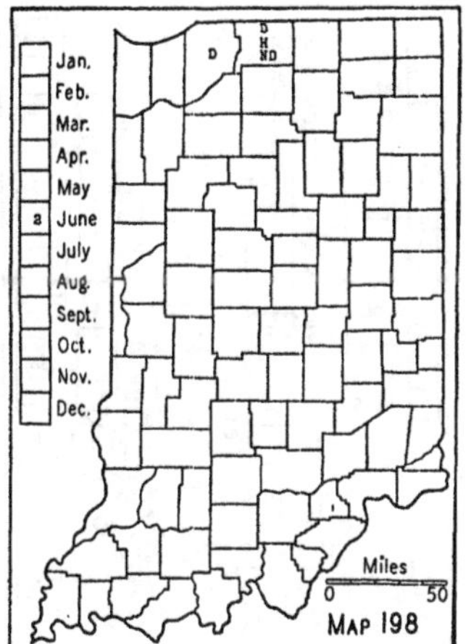

MAP 198
Deschampsia caespitosa (L.) Beauv.

MAP 199
Arrhenatherum elatius (L.) Mert. & Koch

55-270. DESCHÂMPSIA Beauv. HAIRGRASS

Blades flat or folded, stiff; awn included or slightly exserted, straight...1. *D. caespitosa.*
Blades filiform, flexuous; awn exserted, geniculate, twisted. (See excluded species no. 58, p. 1027.)..........*D. flexuosa.*

1. **Deschampsia caespitòsa** (L.) Beauv. TUFTED HAIRGRASS. Map 198. I found this species to be a frequent grass in very marly soil in the outlet of a marly, springy place about 6 miles southwest of South Bend, St. Joseph County, and in a cold, marly, springy place on the border of Mill Creek about a mile north of Mill Creek, La Porte County. Only a few plants were seen at the latter station. Bradner reported this species from Steuben County and his determination was, no doubt, correct, but no specimen has been seen.

Greenland to Alaska, southw. to N. J., W. Va., Ind., Ill., N. Dak., N. Mex., and Calif.

57-273. AVÈNA L. OAT

Lemmas pubescent with long, brown hairs. (See excluded species no. 60, p. 1027.)..........*A. fatua.*
Lemmas glabrous or nearly so. (See excluded species no. 61, p. 1027.)..........*A. sativa.*

58-275. ARRHENÁTHERUM Beauv.

1. ARRHENATHERUM ELÀTIUS (L.) Mert. & Koch. TALL OATGRASS. Map 199. This is an infrequent escape throughout the state. All of my specimens and those that I have seen are from roadsides. Usually not common where it is found although, in a few instances, it was found for a mile or more along roadsides.

Nat. of Eu.; Newf. to B. C., southw. to Ga., Tenn., Iowa, Idaho, and Calif.

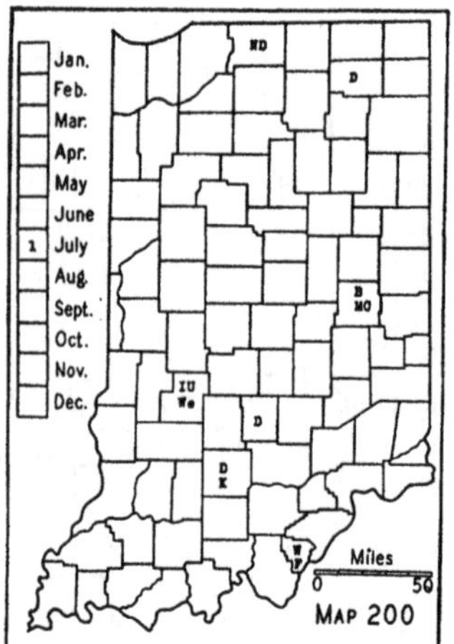

Holcus lanatus L.

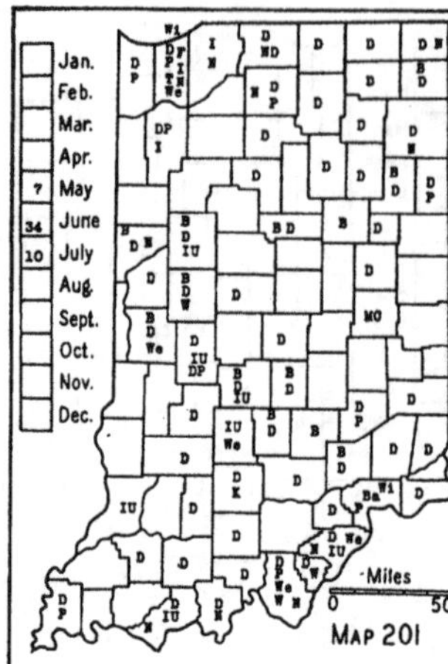

Danthonia spicata (L.) Beauv.

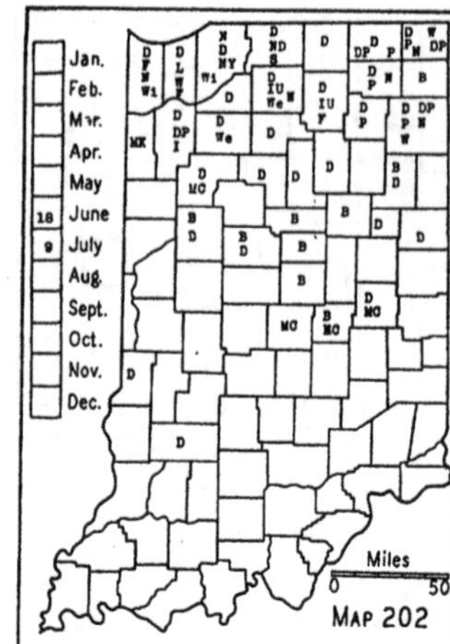

Calamagrostis canadensis (Michx.) Beauv.

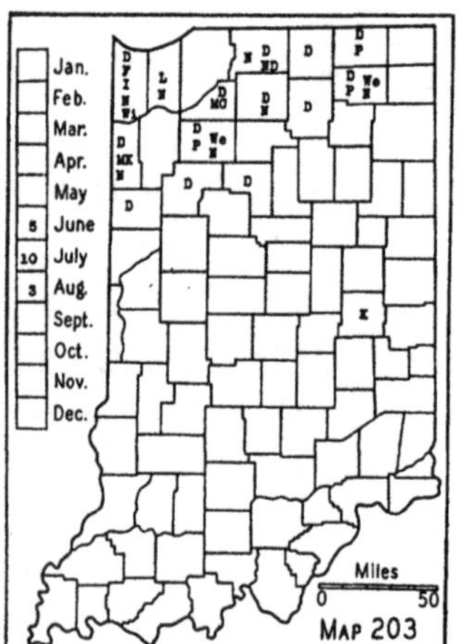

Calamagrostis inexpansa Gray

Ammophila breviligulata Fern.

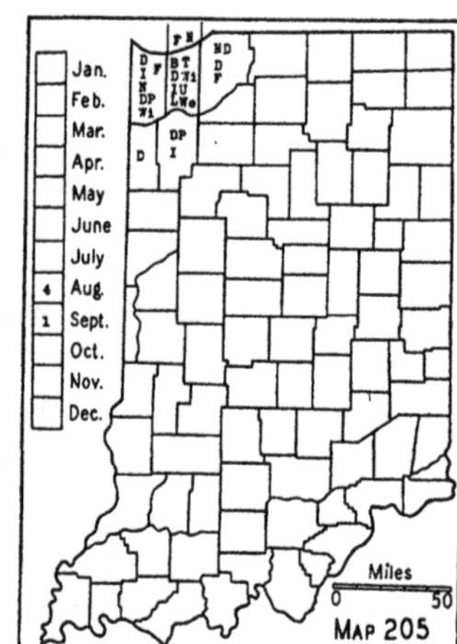

Calamovilfa longifolia (Hook.) Scribn.

59-257. HÓLCUS L.

1. HOLCUS LANÀTUS L. (*Ginnania lanata* (L.) Hub., Rhodora 18:234. 1916.) **VELVET GRASS.** Map 200. As yet, this species is a rare escape in Indiana. Weatherwax found it in a hayfield in Owen County in 1918. In 1933 he found it along a roadside in Brown County about 15 miles east of Bloomington, and in 1934, about one and a fourth miles west of Merriam, Noble County. Kriebel, in 1935, found about a dozen plants under a beech tree in an open woods in sec. 18, Pleasant Run Township, Lawrence County.

Nat. of Eu.; Maine to Iowa, southw. to Ga. and La.; common on the Pacific coast, and in B. C., Idaho, and Ariz.

60-280. DANTHÒNIA Lam. & DC.

Ligule a band of short hairs usually less than 0.5 mm long. . . . . 1. *D. spicata.*
Ligule a band of hairs usually 2-4 mm long. (See excluded species no. 62, p. 1027.) . . . . *D. compressa.*

1. **Danthonia spicàta** (L.) Beauv. POVERTY OATGRASS. Map 201. Infrequent to common in all parts of the state. It is common in poor soil in open woods on the crests of ridges in southern Indiana, becoming less frequent to rare in the rich, neutral soils of the central part of the state, and again appearing as frequent on black and white oak ridges of the northern counties. It is found also in post oak flats.

Newf. to B. C., southw. to Fla., e. Tex., e. Kans., and in the mts. of N. Mex. and Oreg.

5. *AGROSTÍDEAE* KUNTH. TIMOTHY TRIBE

Lemmas 1-nerved.
  Callus pilose..63. CALAMOVILFA, p. 126.
  Callus glabrous.
    Keels of glumes glabrous or more or less scabrous........76. SPOROBULUS, p. 135.
    Keels of glumes softly ciliate..........................79. HELEOCHLOA, p. 1027.
Lemmas more than 1-nerved.
  Spikelets articulated below the glumes.
    Inflorescence a loose panicle.................................67. CINNA, p. 129.
    Inflorescence a dense spikelike panicle...................69. ALOPECURUS, p. 129.
  Spikelets articulated above the glumes.
    First glume with 3 or 5 nerves.
      Inflorescence spikelike; lemmas about 2 mm long............72. PHLEUM, p. 130.
      Inflorescence paniculate; lemmas more than 2 mm long......85. ARISTIDA, p. 138.
    First glume 1-nerved or nerveless.
      Lemmas indurate, much firmer than the glumes.
        Lemmas awnless, glabrous..............................81. MILIUM, p. 137.
        Lemmas awned, pubescent at least at the base.
          Awns readily falling; callus blunt..................82. ORYZOPSIS, p. 137.
          Awns persistent; callus sharp-pointed, pubescent.
            Lemmas 1-awned...84. STIPA, p. 138.
            Lemmas 3-awned (sometimes the lateral pair short)..85. ARISTIDA, p. 138.
      Lemmas not indurate, thinner than the glumes.
        Spikelets (exclusive of awns) 9 mm or more long.
          Lemmas with an inconspicuous awn; glumes as long as the body of the lemma....................................62. AMMOPHILA, p. 126.
          Lemmas long-awned; glumes minute or lacking..80. BRACHYELYTRUM, p. 136.
        Spikelets not over 5 mm long, usually less.
          Second glume 3-nerved........................61. CALAMAGROSTIS, p. 125.
          Second glume 1-nerved.
            Glumes (at least the first one) slightly longer than the lemma; first glume slightly longer than the second or glumes equal in length, awnless; lemmas thin; palea obsolete or lacking in our native species..64. AGROSTIS, p. 126.
            Glumes generally shorter than the lemma, the first one obsolete, or shorter than the second; if the first glume is as long as or longer than the lemma, the glume with an awn 1-2 mm long; lemmas rather firm; paleas present in normal lengths.........75. MUHLENBERGIA, p. 131.

61-248. CALAMAGRÓSTIS Adans. REEDGRASS

[Stebbins. A Revision of some North American species of Calamagrostis. Rhodora 32: 35-57. 1930.]

Blades usually flat or sometimes involute toward the tips, mostly 4-8 mm wide; panicle usually expanded or loose; spikelets usually 3-3.5 mm long, rarely only 2.5 mm long; lemma thin, glabrous or more or less sparsely scabrous; callus hairs three fourths to as long as the lemma................................1. *C. canadensis.*

Blades involute, except sometimes near the base, usually less than 4 mm wide; panicle narrow, contracted; spikelets usually 3.5-4.2 mm long; lemmas firmer, scabrous all over; callus hairs usually two thirds to three fourths as long as the lemma...... ....2. *C. inexpansa.*

1. **Calamagrostis canadénsis** (Michx.) Beauv. (Inman. Calamagrostis canadensis and some related species. Rhodora 24: 142-144. 1922). BLUE-JOINT. Map 202. Frequent in marshes, wet prairies, and mucky places in general in the lake area, but local southward because its habitat is lacking. Where found, it often covers large areas and was formerly the source of "wild hay" in the state and known as little bluestem grass. Since most of the areas of its habitat have been drained and farmed, it has now become infrequent.

Greenland to Alaska, southw. to Md., N. C. (Roan Mt.), Mo., Kans., and Calif.

2. **Calamagrostis inexpánsa** Gray. NORTHERN REEDGRASS. Map 203. This is an infrequent species in the lake area, where it prefers marly marshes and springy places, although it is sometimes found in habitats associated with pin oak and chokeberry. It is also found in prairie habitats. Stebbins divided the species into varieties and, according to him, our Indiana specimens belong to var. *brevior* (Vasey) Stebbins. Hitchcock, in his Manual of Grasses, does not divide the species. According to Stebbins, the distribution of the variety is as follows:

Newf., Que. to B. C., southw. to N. Y., Ind., Minn., Colo., Ariz., and Calif.

62-249. AMMÓPHILA Host

1. **Ammophila breviligulàta** Fern. (Rhodora 22: 70-71. 1920.) (*Ammophila arenaria* of American authors, not Link.) BEACHGRASS. Map 204. Infrequent on the dunes bordering Lake Michigan. This species is used in this country as a soil binder.

On dunes from Newf. to N. C., and on the shores of the Great Lakes from Lake Ontario to Lake Superior and Lake Michigan.

63-250 CALAMOVÍLFA Hack.

1. **Calamovilfa longifòlia** (Hook.) Scribn. LONGLEAF REEDGRASS. Map 205. This species is found in dry, shifting sands on the dunes about Lake Michigan and on a few shifting dunes in Jasper and Newton Counties.

Mich. to Alberta, southw. to Ind., Colo., and Idaho.

64-242 AGRÓSTIS L. BENTGRASS

[Hitchcock. North American species of Agrostis. U. S. Dept. Agric. Bur. Plant Ind. Bull. **68**: 1-68. 1905. Piper. The agricultural species of bent grasses. U. S. Dept. of Agric. Bull. **692**: 1-26. 1918. Malte. Commercial bent grasses (Agrostis) in Canada. Reprinted from Annual Report for **1926**, National Museum of Canada, 105-126. 1928.]

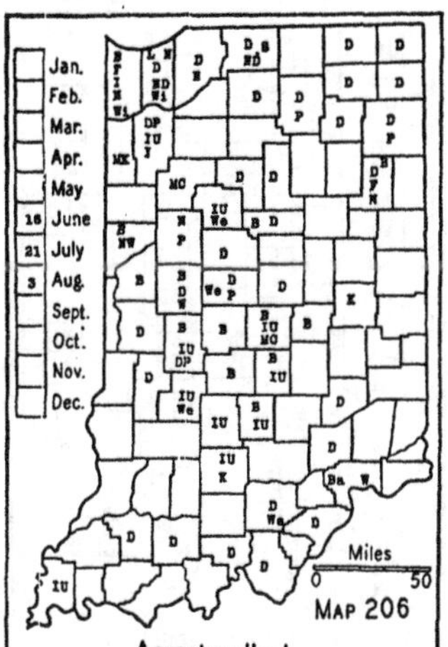

Agrostis alba L.

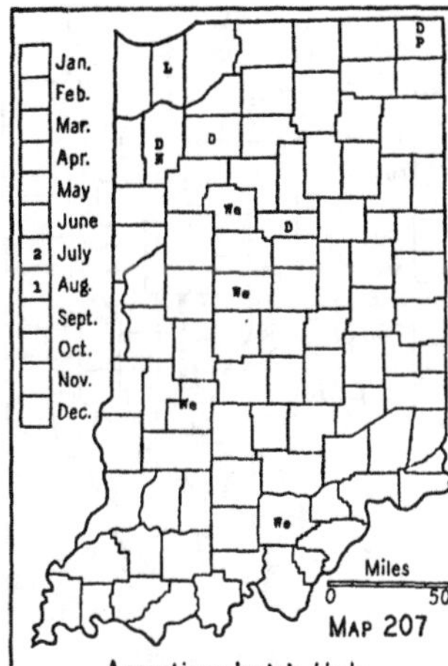

Agrostis palustris Huds.

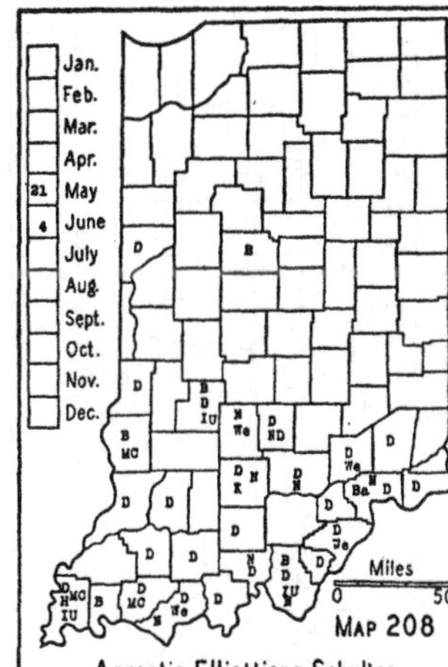

Agrostis Elliottiana Schultes

Palea about half as long as the lemma.
  Plant perennial from strong creeping rhizomes, without creeping and rooting stolons; culms erect or only slightly decumbent at the base, not rooting at the lower nodes; panicle open or spreading............................1. *A. alba.*
  Plant perennial without rhizomes, with creeping and more or less rooting stolons; culms usually decumbent at the base and rooting at the lower nodes; panicle usually contracted, sometimes open............................2. *A. palustris.*
Palea minute or lacking.
  Lemmas awned............................3. *A. Elliottiana.*
  Lemmas awnless.
    Plants generally found growing in the open, usually flowering and maturing before August 1; basal leaves narrow, stiff, mostly involute; panicles diffuse, generally purplish at maturity, the branches beginning to divide beyond the middle.
      Spikelets mostly 2-2.5 mm long; glumes connivent in fruit, covering the grain; anthers mostly 0.5 mm long; flowering in northern Indiana from about June 3 to July 10............................4. *A. scabra.*
      Spikelets mostly 1.4-1.9 mm long; glumes not connivent in fruit, exposing the grain; anthers mostly about 0.2 mm long; beginning to flower in northern Indiana the last of May and maturing the fruit usually by the middle of June............................5. *A. hyemalis.*
    Plants generally found growing in woods, usually flowering after August 1; basal leaves flat, wider than in the two preceding species, generally lax; panicles open or spreading, green or nearly so at maturity, the branches beginning to divide mostly at or below the middle............................6. *A. perennans.*

1. AGROSTIS ÁLBA L. (*Agrostis stolonifera* var. *major* (Gaud.) Farw. and *Agrostis palustris* of recent American authors, not Huds.) REDTOP. Map 206. This species has been commonly sown as a pasture and hay grass in all parts of the state, especially in the southern part. It has abundantly escaped everywhere and is found along roadsides and railroads and in fallow fields, pastures, and waste places.

Besides the commercial redtop, seed of other species of the bentgrasses have been imported and sown in lawns and on golf courses. Several strains of each species have been developed and some European authors credit one species with 15 varieties and subvarieties. The species are separated with

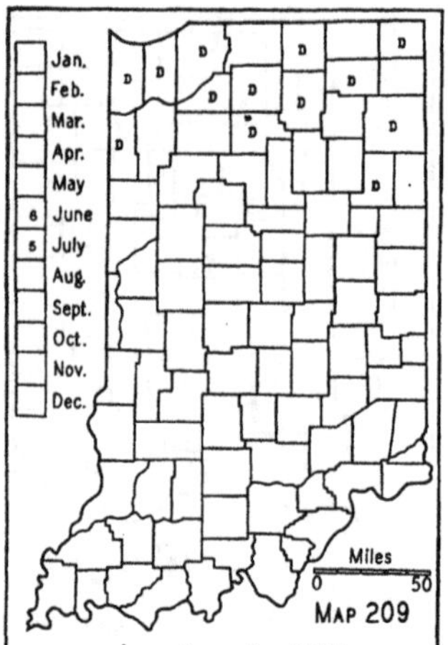

Agrostis scabra Willd.

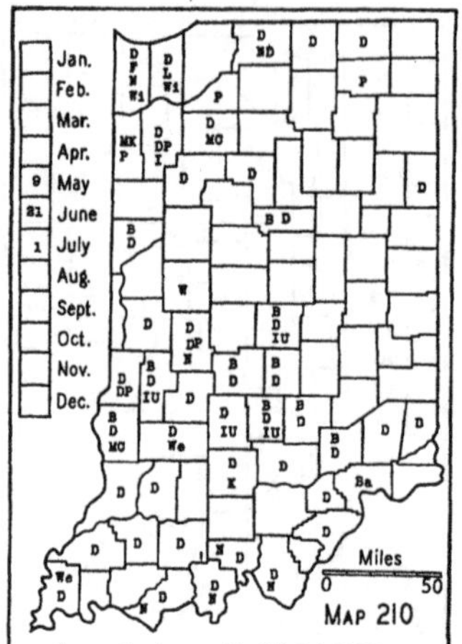

Agrostis hyemalis (Walt.) BSP.

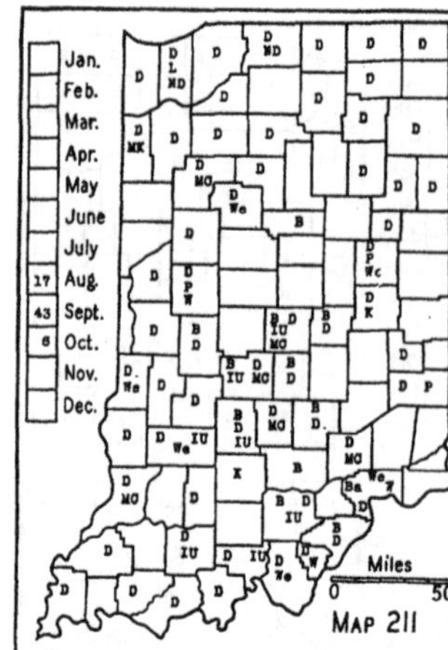

Agrostis perennans (Walt.) Tuckerm.

difficulty and the task is complicated by the addition of the many cultivated forms.

Nat. of Eurasia; in all the cooler parts of the U. S.

2. AGROSTIS PALÚSTRIS Huds. (*Agrostis alba* var. *maritima* (Lam.) G. F. W. Mey., *Agrostis maritima* Lam., and *Agrostis stolonifera* var. *compacta* Hartman of Deam, Grasses of Ind.) CREEPING BENT. Map 207. The few specimens of this species I have seen were found on the low borders of streams, usually with a part of the colony in the running water.

Nat. of Eurasia; introduced in the northern part of the U. S., and occasionally as far south as Tex. and N. Mex.

3. **Agrostis Elliottiàna** Schultes. ELLIOTT BENTGRASS. Map 208. Frequent to common throughout the area shown on the map in hard, white clay soils with a pH value ranging from 6-6.6. The mass distribution occurs in moist fallow fields and pastures. It is also found on washed slopes and on crests of ridges in open woodland. The species is usually associated with *Agrostis hyemalis* from which it is easily separated by its scabrous feel, smaller size, and awned lemmas.

Md. to Ill., Mo., and Kans., southw. to Ga., Ala., and e. Tex.; Yucatan.

4. **Agrostis scàbra** Willd. (*Agrostis hyemalis* of recent authors, in part.) (Rhodora 35: 207-209. 1933.) NORTHERN TICKLEGRASS. Map 209. In low sandy and mucky soils in the northern counties. This species very much resembles the next one but it is separated from it by its larger size, its broader and flat cauline leaves, larger panicle, longer-pedicelled spikelets, longer spikelets, longer anthers, its later flowering season, and its northern range. This species flowers, on the whole, at least a half month later than the next one.

Lab. and Newf. to Alaska, southw. to Pa., Ind., Iowa, Nebr., N. Mex., Ariz., and Calif.

5. **Agrostis hyemàlis** (Walt.) BSP. (*Agrostis antecedens* Bickn. and *Agrostis hyemalis* of recent authors, in part.) TICKLEGRASS. Map 210. This species is infrequent to common in all parts of the state. It prefers a slightly acid soil, hence it is infrequent to absent in the neutral soils of the central counties. In the southern counties it occurs in hard, white clay soil and is usually common in fallow fields, on washed slopes, along clayey roadsides, and in moist, sandy and mucky places in our northern counties.

Mass. to Iowa and Kans., southw. to Fla. and Tex.

6. **Agrostis perénnans** (Walt.) Tuckerm. AUTUMN BENT. Map 211. Infrequent to frequent in all parts of the state except in the prairie areas. This is a woodland species which seems to prefer a slightly acid soil and is found in black and white oak woods, pin oak woods, aspen thickets, at the bases of sandstone ledges, and rarely in prairie habitats or fallow fields. This species shows great variation which I assume to be the result of varying amounts of light, soil acidity, and nutriment.

Que. to Minn., southw. to Fla. and e. Tex.

67-241. CÍNNA L.

Spikelets 5 mm long; panicle rather dense, the branches ascending...1. *C. arundinacea.*
Spikelets 3.5-4 mm long; panicle loose, the branches spreading or drooping. (See excluded species no. 68, p. 1028.)....................................*C. latifolia.*

1. **Cinna arundinàcea** L. WOODREED. Map 212. Frequent to rather common in all parts of the state. It grows in wet soils in almost all kinds of habitats except in pure sand. This is a woodland species but is sometimes found in wet clearings if shaded by rank vegetation.

Maine to S. Dak., southw. to Ga. and e. Tex.

69-225. ALOPECÙRUS L. FOXTAIL

Spikelets about 5 mm long..1. *A. pratensis.*
Spikelets less than 3 mm long.
Lemmas awned on the back about midway between the base and apex, the awn usually included, sometimes exserted but not for more than 1 mm.......... ....2. *A. aequalis.*
Lemmas awned on the back at about a fourth the length of the lemma above the base, the awn exserted about 2-3 mm.......................3. *A. carolinianus.*

1. ALOPECURUS PRATÉNSIS L. MEADOW FOXTAIL. Map 213. Specimens of this species have been collected in Tippecanoe County, and I have it from Wells County, where it was well established when collected in 1932.

Nat. of Eurasia; introduced from Newf. and Lab. to Alaska, southw. to Del., Iowa, Idaho, and Oreg.

2. **Alopecurus aequàlis** Sobol. (*Alopecurus geniculatus* var. *aristulatus* Torr. of Gray, Man., ed. 7 and *Alopecurus geniculatus* Michx. of Britton and Brown, Illus. Flora, ed. 2.) SHORT-AWN FOXTAIL. Map 214. This grass is infrequent in the lake area and local south of it. It grows in shallow

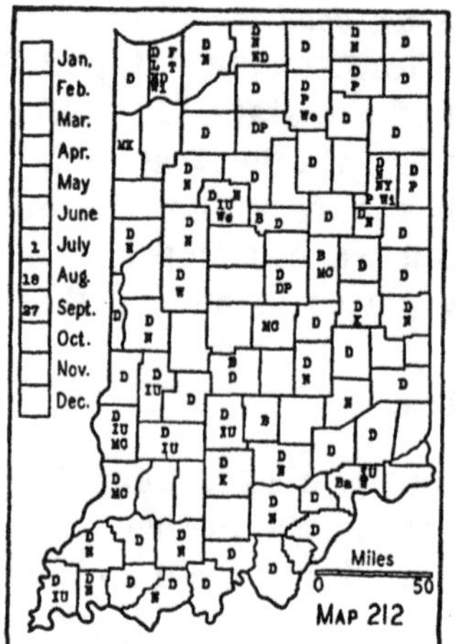

Cinna arundinacea L.

Alopecurus pratensis L.

Alopecurus aequalis Sobol.

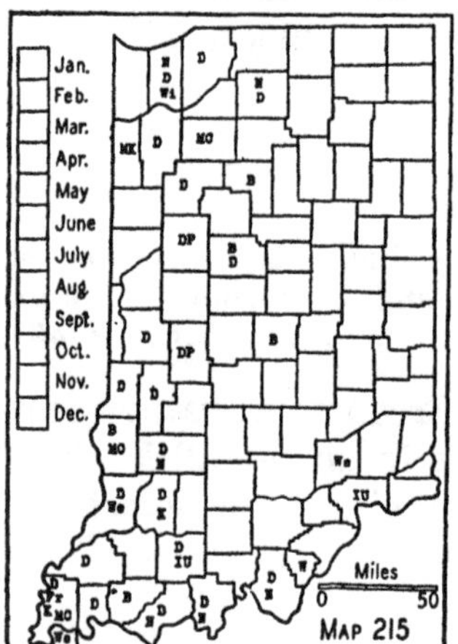

Alopecurus carolinianus Walt.

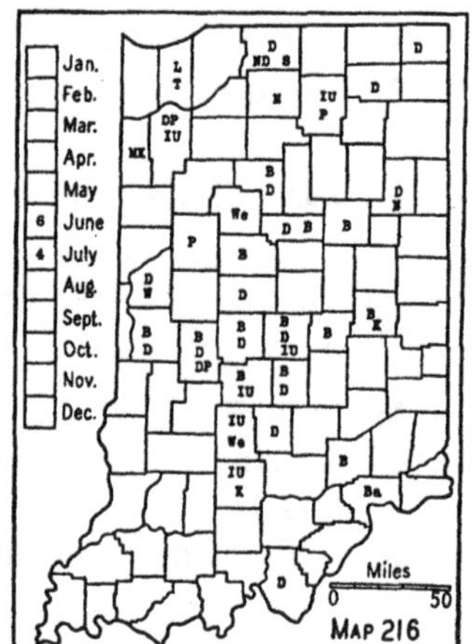

Phleum pratense L.

Muhlenbergia capillaris (Lam.) Trin.

water and on the muddy borders of ponds and swamps that usually become dry in midsummer.

Greenland to Alaska, southw. to Pa.

3. **Alopecurus caroliniànus** Walt. (*Alopecurus ramosus* Poir. of Deam, Grasses of Ind.) Map 215. Infrequent to local in the greater part of the state. In the northern part it is found in mucky soil about ponds and in ditches, and in the southern part it is usually found in slightly acid, white clay soil in fallow fields, and usually associated with one or more of the following plants: *Poa Chapmaniana, Agrostis hyemalis, Myosotis virginica,* and *Arabis virginica.*

N. J. to B. C., southw. to Fla., Tex., Ariz., and Calif.

72-223. PHLÈUM L.

1. PHLEUM PRATÉNSE L. TIMOTHY. Map 216. This species has abundantly escaped in all parts of the state. It is usually found in either dry

or moist soil along roadsides and railroads and in fallow fields, pastures, and waste places.

Nat. of Eurasia; throughout the U. S.

75-215. MUHLENBÉRGIA Schreb. MUHLY

[Note: In this genus the measurements of the spikelets, glumes, and lemmas do not include the awns, unless so stated. In observing nodes and internodes, both the lower and the upper ones of the plant should be examined.]

Panicles diffuse, more than 2.5 cm wide, usually 10-20 cm wide.......1. *M. capillaris.*
Panicles not diffuse, less than 2.5 cm wide.
  First glume obsolete; second glume not over 0.6 mm long, very obtuse............. ....2. *M. Schreberi.*
  First glume not obsolete; second glume more than 0.6 mm long, not obtuse.
    Spikelets 1.5-2.2 mm long; glumes shorter than the lemmas (if as long, see opposing lead).
      Lemmas acute to acuminate, not awned......................3. *M. sobolifera.*
      Lemmas awned.................................3a. *M. sobolifera* f. *setigera.*
    Spikelets more than 2.2 mm long.
      Lemmas not pilose at the base (on the callus).
        Culms without creeping rootstocks; anthers about 1-1.5 mm long........... ....4. *M. cuspidata.*
        Culms with creeping scaly rootstocks; anthers about 0.5 mm long........... ....5. *M. glabriflora.*
      Lemmas short-pilose at the base (on the callus).
        Nodes and infranodes glabrous.
          Panicles included at the base, rarely short-exserted; anthers about 0.5 mm long.
            Lemmas without awns, or some with short awns up to 2 mm long........ ....6. *M. mexicana.*
            Lemmas awned; awns usually 5-10 mm long........................ ....6a. *M. mexicana* f. *commutata.*
          Panicles usually very long-exserted; anthers about 0.8 mm long............ ....7. *M. brachyphylla.*
        Nodes and infranodes not glabrous.
          Nodes and infranodes puberulent; anthers about 0.8 mm long.
            Glumes longer than the lemma; panicles more than 5 mm wide.......... ....8. *M. racemosa.*
            Glumes usually two thirds to three fourths as long as the lemma; panicles generally less than 5 mm wide......................9. *M. tenuiflora.*
          Nodes glabrous; infranodes puberulent, rarely nearly all glabrous, but not polished below the node; anthers about 0.5 mm long.
            Culms usually puberulent below the panicles; spikelets crowded on the branches, glumes about as long as the lemmas.
              Lemmas awnless.....................................10. *M. foliosa.*
              Lemmas awned, awns 4-10 mm long.........10a. *M. foliosa* f. *ambigua.*
            Culms generally glabrous below the panicles; spikelets not at all crowded on the branches; glumes about two thirds as long as the lemmas.
              Lemmas awned.......................................11. *M. sylvatica.*
              Lemmas awnless.......................11a. *M. sylvatica* f. *attenuata.*

1. **Muhlenbergia capillàris** (Lam.) Trin. Map 217. My only specimen was collected October 7, 1921, about 3 miles east of Elizabeth, on an open wooded, rocky hillside, bordering the roadside of the Elizabeth Road to Stewart's Landing, Harrison County. It was still persisting here in 1938.

Mass., Ind., and Kans., southw. to Fla. and Tex.; W. I. and e. Mex.

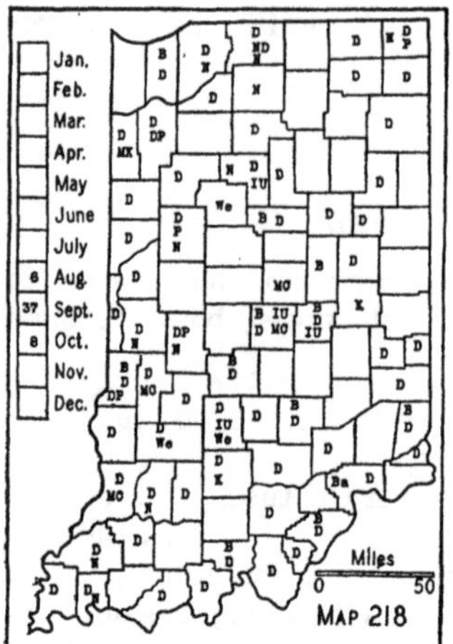

Muhlenbergia Schreberi J. F. Gmel.

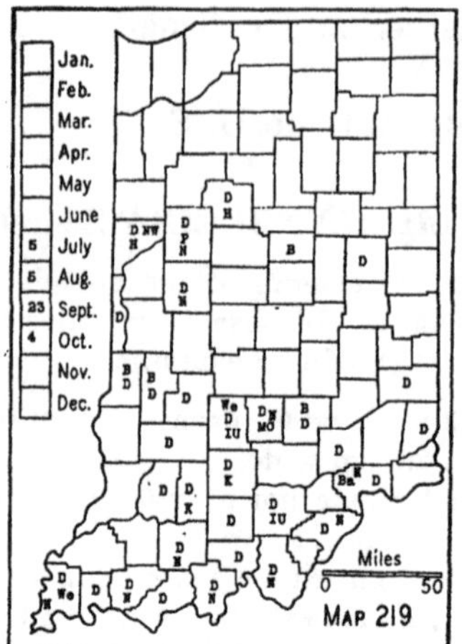

Muhlenbergia sobolifera (Muhl.) Trin.

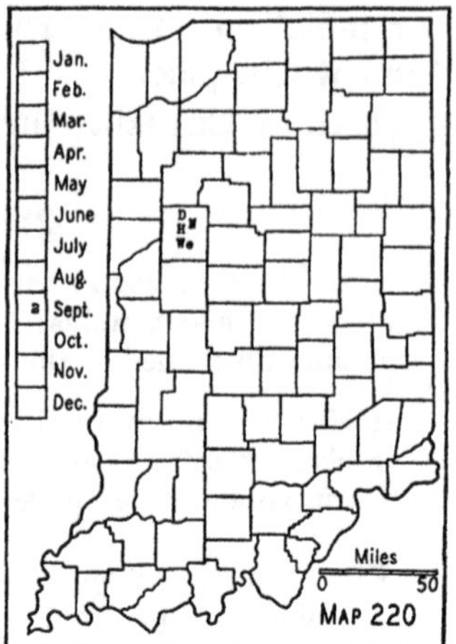

Muhlenbergia cuspidata (Nutt.) Rydb.

2. **Muhlenbergia Schrèberi** J. F. Gmel. NIMBLEWILL. Map 218. Infrequent to frequent throughout the state. It is found usually in dry soils and less frequently in moist soils in open woodland, clearings, woods pastures, and pasture fields. It is usually conspicuous in pasture fields because stock graze around it, preferring other herbage. It is also found about dwellings and in lawns and is an obnoxious weed in flower gardens.

N. H. to Wis., e. Nebr., southw. to Fla., Tex., and e. Mex.

3. **Muhlenbergia sobolífera** (Muhl.) Trin. Map 219. This species is found principally in the southern half of the state. It is strictly a woodland species and occurs on wooded slopes, preferring those along streams. It is found in both beech and sugar maple, and black and white oak woodland.

N. H. to Iowa, southw. to Va., Tenn., and Tex.

3a. **Muhlenbergia sobolifera** f. **setígera** (Scribn.) Deam. This is a form with awned lemmas. I am referring my no. 32921 from Sullivan County to this form.

Ind. to Ark. and Tex.

4. **Muhlenbergia cuspidàta** (Nutt.) Rydb. PLAINS MUHLY. Map 220. This species is infrequent on the high, gravelly bank of the north side of Big Wea Creek where the Shadeland Road crosses the creek about 4 miles southwest of Lafayette. Its associates make it certain that it is a native here.

Mich., Wis. to Alberta, southw. to Ohio and N. Mex.

5. **Muhlenbergia glabriflòra** Scribn. (Rhodora **9**: 22. 1907.) Map 221. This species, as now known, is restricted to seven of our southwestern counties. It is found in hard, white clay soil in moist or wet places, usually in pin oak woods or in the pin oak and post oak flats of the southwestern part of Posey County. Probably locally frequent and possibly well distributed in the southwestern counties where its habitat is found.

Md., Ind., Ill., Mo., and Tex.

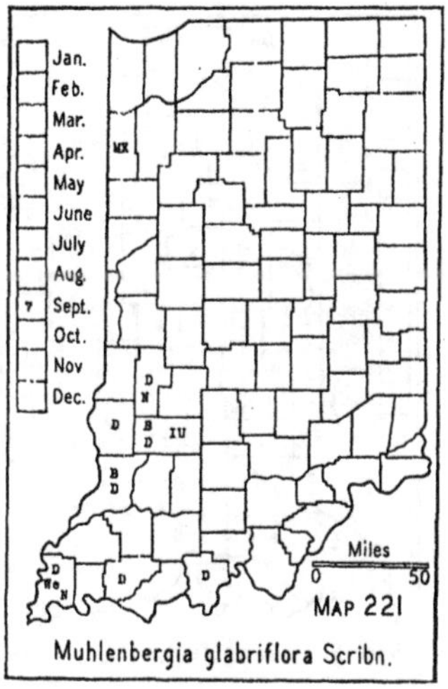

MAP 221
Muhlenbergia glabriflora Scribn.

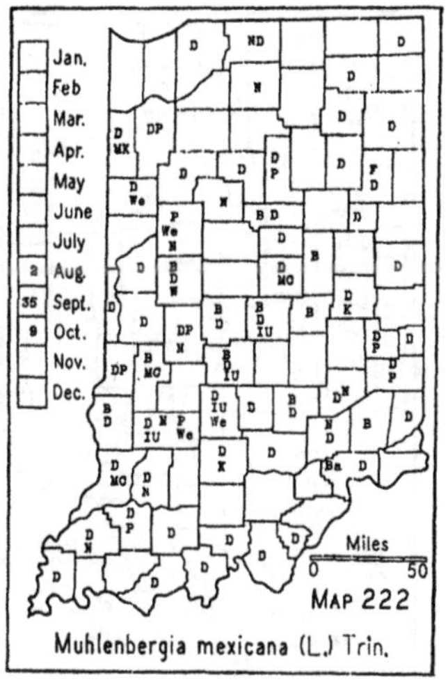

MAP 222
Muhlenbergia mexicana (L.) Trin.

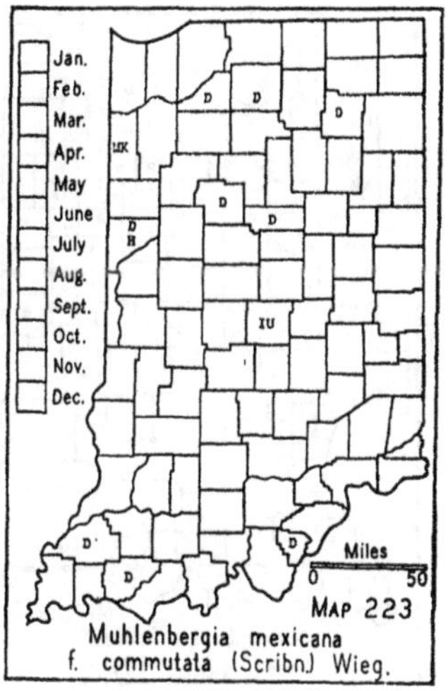

MAP 223
Muhlenbergia mexicana f. commutata (Scribn.) Wieg.

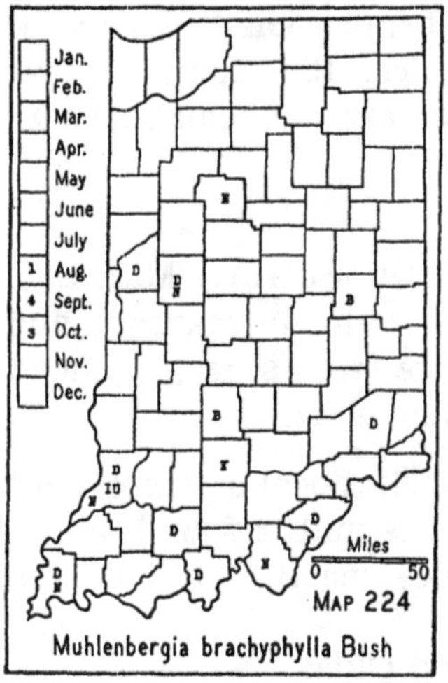

MAP 224
Muhlenbergia brachyphylla Bush

MAP 225
Muhlenbergia racemosa (Michx.) BSP.

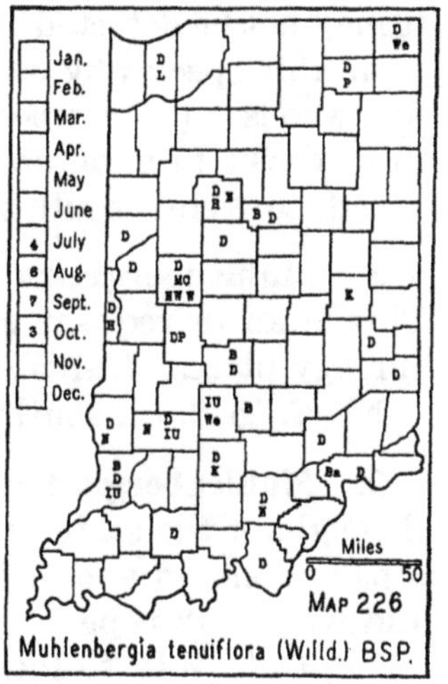

MAP 226
Muhlenbergia tenuiflora (Willd.) BSP.

6. **Muhlenbergia mexicàna** (L.) Trin. WIRESTEM MUHLY. Map 222. This species is frequent to common in all parts of the state except in our northern counties, where it becomes rare to infrequent. It prefers a moist soil but will grow and thrive in almost all kinds of soils. It prefers open, alluvial soil along streams where it often forms exclusive stands. It is an obnoxious weed when it invades cultivated fields because it propagates from underground stems.

N. B. to N. D., southw. to the mts. of Ga. and Tex.

6a. **Muhlenbergia mexicana** f. **commutàta** (Scribn.) Wieg. (Rhodora **26**: 1. 1924.) Map 223. I have specimens of this long-awned form from the counties shown on the map.

Maine, Que., and S. Dak., southw. to Va. and Mo.

7. **Muhlenbergia brachyphýlla** Bush. (Amer. Midland Nat. **6**: 41-42. 1919.) Map 224. Probably infrequent to rare in the southern part of the state. At a distance it so closely resembles *Muhlenbergia tenuiflora* that

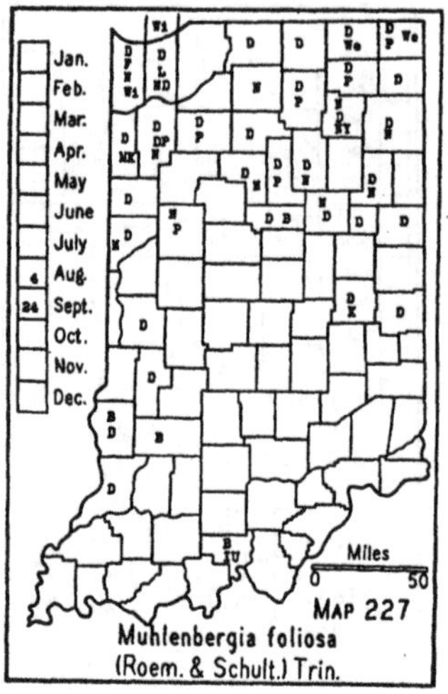

MAP 227
Muhlenbergia foliosa (Roem. & Schult.) Trin.

MAP 228
Muhlenbergia sylvatica Torr.

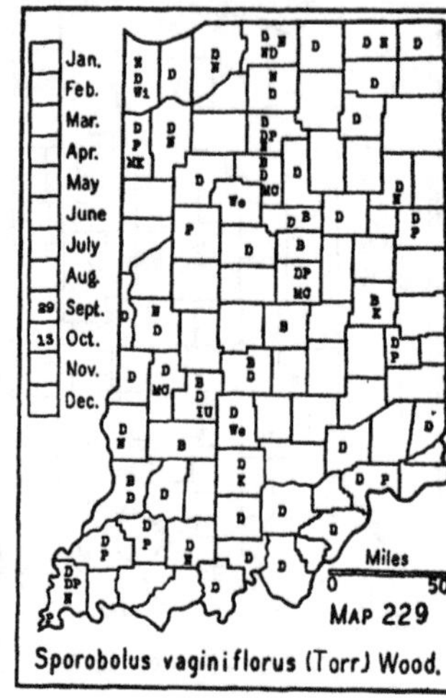

MAP 229
Sporobolus vaginiflorus (Torr.) Wood.

it may not be detected. On close observation, however, it is easily separated from this species by its glabrous nodes and infranodes. It is found in low, flat woods and on wooded slopes. I am not well enough acquainted with this species to understand its habitat.

Ind. to Nebr., southw. to Tex.

8. **Muhlenbergia racemòsa** (Michx.) BSP. MARSH MUHLY. Map 225. This is an infrequent grass of the lake area. It is found in marshes and springy places. Our plants south of the lake area are from springy places.

Newf. to B. C., southw. to Va., Md., Ky., Okla., and Ariz.

9. **Muhlenbergia tenuiflòra** (Willd.) BSP. Map 226. Local or infrequent throughout the state. It is strictly a woodland species and is found on the tops and slopes and along the bases of wooded slopes, usually of the black and white oak type.

Vt., Ont., Wis. to Iowa, southw. to Va., Tenn., and Okla.

10. **Muhlenbergia foliòsa** (Roem. & Schult.) Trin. Map 227. Infrequent in the lake area and local south of it. It is generally found in marshes and springy places, usually about lakes and in ditches.

Maine to Que. and Mont., southw. to N. C., Ind., N. Mex., and Ariz.

10a. **Muhlenbergia foliosa** f. **ambígua** (Torr.) Wieg. (*Muhlenbergia ambigua* Torr.) This form has the habitat of the species. I have it from Kosciusko, Lagrange, Marshall, Starke, Steuben, Warren, and Whitley Counties.

11. **Muhlenbergia sylvática** Torr. (*Muhlenbergia umbrosa* Scribn.) Map 228. Infrequent throughout the state. It is usually a low ground, woodland species found on the borders of streams, ponds, and swamps, and rarely on dry, wooded slopes.

Maine to S. Dak., southw. to Ala., Tex., and Ariz.

11a. **Muhlenbergia sylvaticà** f. **attenuàta** (Scribn.) Palmer & Steyermark. I have this form from only Carroll, Clark, and Posey Counties.

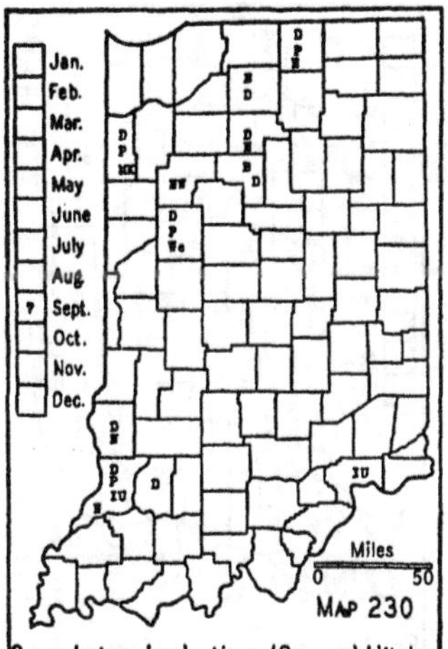

Sporobolus clandestinus (Spreng.) Hitchc.

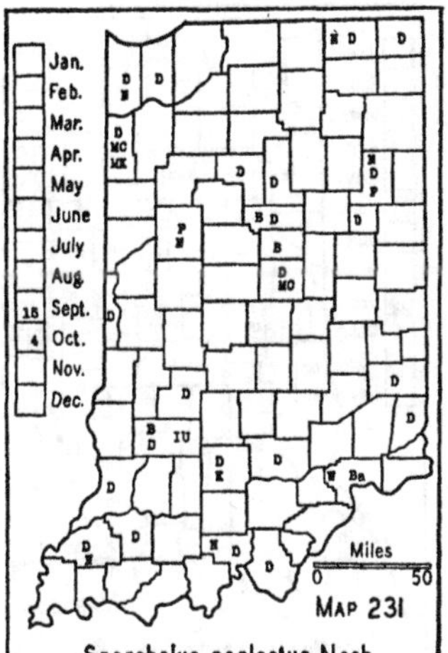

Sporobolus neglectus Nash

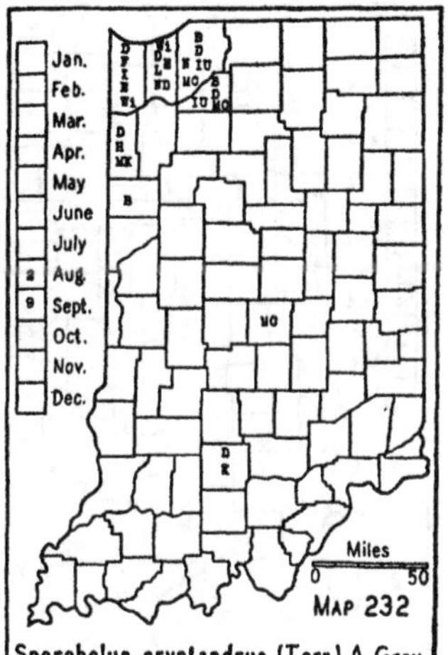

Sporobolus cryptandrus (Torr.) A. Gray

76-230. SPORÓBOLUS R. Br. DROPSEED

Lemma appressed-pubescent on the sides, at least near the base.
Plant annual, more or less decumbent at the base; terminal panicles included (very rarely one free), lateral panicles common; spikelets 3.5-6 mm long............ ....1. *S. vaginiflorus.*
Plant perennial, erect, not decumbent at the base; terminal panicles free (rarely one partly included), lateral panicles absent (rarely one or more present); spikelets about 5-6 mm long...................................2. *S. clandestinus.*
Lemma glabrous on the sides, the keel usually somewhat scabrous.
Spikelets of the terminal panicles 2-3 mm long.
Plant annual; sheaths not bearded at the throat (sometimes a few long hairs on the inside); terminal panicle less than 7 cm long, usually included, contracted and spikelike; grain about 1-1.5 mm long.....................3. *S. neglectus.*
Plant perennial; sheaths conspicuously bearded at the throat; terminal panicle more than 7 cm long, usually almost free and widely spreading at maturity (sometimes included and spikelike); grain about 1 mm long.............. ....4. *S. cryptandrus.*
Spikelets of the terminal panicles 3.5-5 mm long.
Glumes acuminate or aristate, the second one about 1-2 mm longer than the first; panicles free at maturity and widely spreading; grain orbicular, smooth. ....5. *S. heterolepis.*
Glumes obtuse, the second one about 1 mm longer than the first; panicles usually included, contracted; grain flattened, reticulate..................6. *S. asper.*

1. **Sporobolus vaginiflòrus** (Torr.) Wood. Map 229. Infrequent to common in all parts of the state. It prefers poor, dry soils and is rarely absent from the borders of limestone highways. It seems to be spreading for I have seen it spread over a fallow field and over vacant lots in a few years. It can now be found almost everywhere along roads made of crushed rock, in waste places, on washed slopes, and in poor soil in pastures.

Maine, Ont. to Minn., southw. to Ga., Tex., and Ariz.

2. **Sporobolus clandestìnus** (Spreng.) Hitchc. (Including *Sporobolus canovirens* Nash.) Map 230. Infrequent in very sandy soil in prairie habitats in the counties shown on the map. The range will doubtless be extended to include the counties about Lake Michigan.

Conn. to Ill. and Kans., southw. to Fla. and Tex.

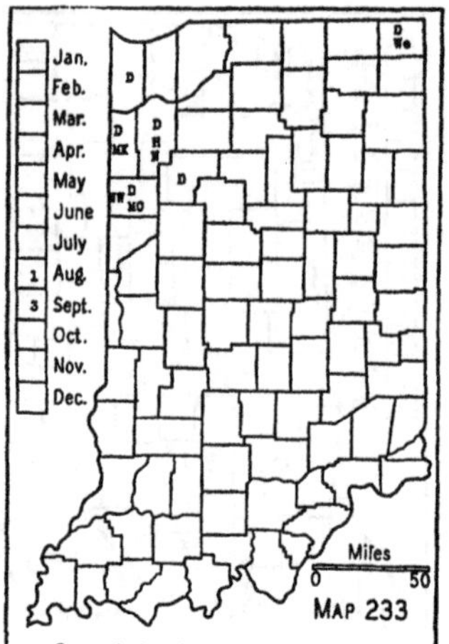

Sporobolus heterolepis A.Gray

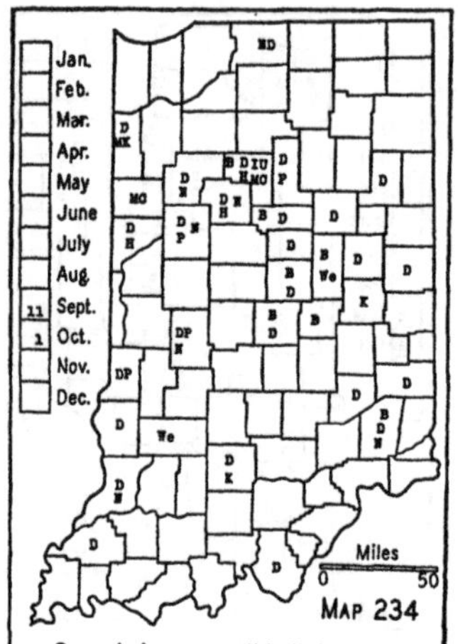

Sporobolus asper (Michx.) Kunth

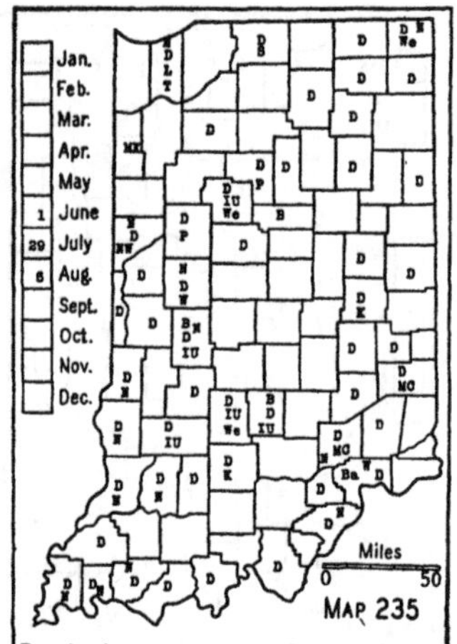

Brachyelytrum erectum (Schreb.) Beauv.

3. **Sporobolus negléctus** Nash. Map 231. Infrequent throughout the state. It seems to have much the same habitat as *Sporobolus vaginiflorus* and is often found with it.

Maine, Que. to N. Dak. and Wash., southw. to Md., Tenn., Tex., and Ariz.

4. **Sporobolus cryptándrus** (Torr.) Gray. SAND DROPSEED. Map 232. Infrequent in dry, sandy soil in the area shown on the map. The specimen from Lawrence County was found in an old stone quarry and the Marion County specimen was found in a waste place on North Meridian Street in Indianapolis. I regard these plants as waifs. The fact that the panicle sometimes remains in the sheath and does not expand has caused a form to be named. I am following Hitchcock, considering the form to be without taxonomic significance.

Maine, Ont. to Alberta and Wash., southw. to N. C., Ind., La., and Ariz.

5. **Sporobolus heterólepis** Gray. PRAIRIE DROPSEED. Map 233. This species is infrequent to very local in a few of our northern counties. It is found in dry or moist prairie habitats.

Que. to Sask. and Wyo., southw. to Conn., Ill., Ark., and e. Tex.

6. **Sporobolus ásper** (Michx.) Kunth. Map 234. This species is infrequent throughout the state. It is doubtful that this species is a native. I have noted its advent into the state during the past few years. It now often forms complete stands for rods along railroads, highways, and adjacent fields. It will no doubt, in time, become a weed.

Vt., Mich. to N. Dak. and Utah, southw. to La. and N. Mex.

80-216. BRACHYÉLYTRUM Beauv.

1. **Brachyelytrum eréctum** (Schreb.) Beauv. Map 235. Infrequent to frequent in all parts of the state where beech and sugar maple woods or black and white oak woods are found. It prefers dry slopes and, for this reason, it is often very local in some counties. I have botanized Wells

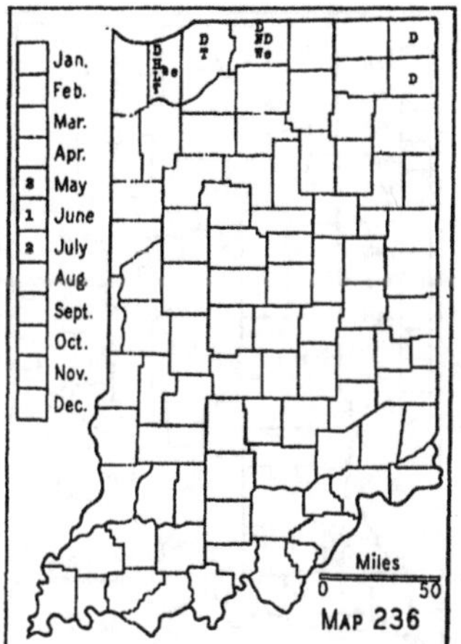

Milium effusum L.

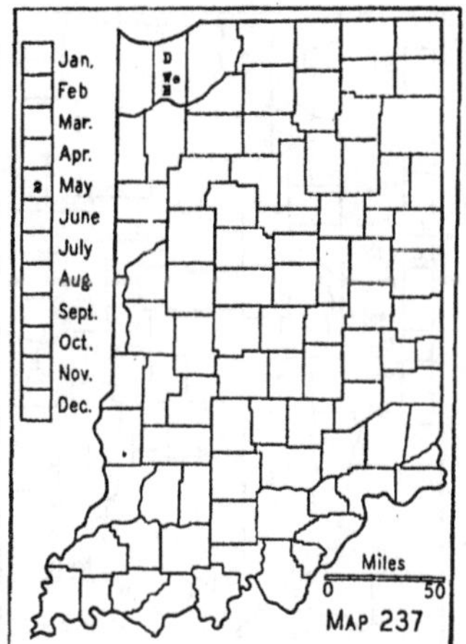

Oryzopsis pungens (Torr.) Hitchc.

Oryzopsis asperifolia Michx.

County for 40 years and I have not found it, possibly because woods in which it grew are now cultivated fields.

Newf. to Minn., southw. to Ga. and Okla.

81-213. MÍLIUM L.

1. **Milium effùsum** L. Map 236. This species is very local and is found in peaty woods with soft maple or in mucky or springy places with skunk cabbage.

N. C., Que. to Minn., southw. to Md. and Ill.; also in Eurasia.

82-210. ORYZÓPSIS Michx. RICEGRASS

Blades narrow, involute; spikelets (exclusive of awns) less than 5 mm long; awns not more than 2 mm long .... 1. *O. pungens.*
Blades broad, flat; spikelets (exclusive of awns) more than 5 mm long.
  Leaves mostly basal; blades of culm generally less than 2 cm long, scabrous above. .... 2. *O. asperifolia.*
  Leaves scattered along the culm; blades of culm more than 2 cm long, pubescent above. .... 3. *O. racemosa.*

1. **Oryzopsis púngens** (Torr.) Hitchc. Map 237. A few tufts of this species have been found in Porter County over a limited area about a quarter mile east of Waverly Beach in the Dunes State Park. This is the only locality now known in Indiana.

Lab. to B. C., southw. to Conn., Ind., S. Dak., and N. Mex.

2. **Oryzopsis asperifòlia** Michx. Map 238. This species is known only from La Porte and Porter Counties where it is found on open wooded dunes.

Newf., Man., B. C., southw. to Conn., Ind., S. Dak., and N. Mex.

3. **Oryzopsis racemòsa** (J. E. Smith) Ricker. Map 239. The specimens found in the southern part of the state are from rocky woods and those from the northern part are from moist or dry, sandy woods. It is very local and I cannot account for its widely different habitats and limited distribution.

Que. to Minn., and S. Dak., southw. to Del., Ky., and Iowa.

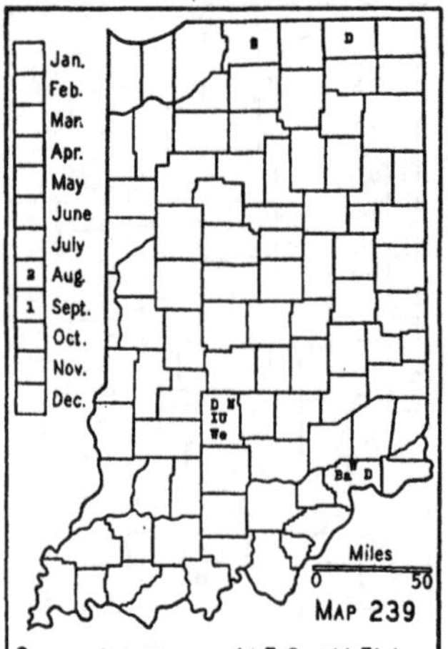

Oryzopsis racemosa (J.E.Smith) Ricker

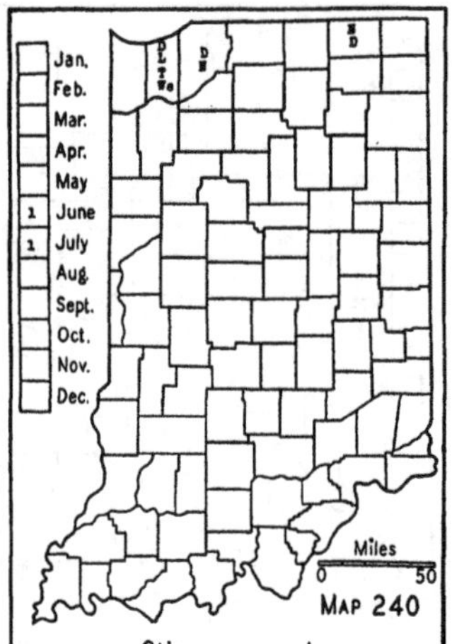

Stipa avenacea L.

Stipa comata Trin.& Rupr.

84-209. STÌPA L. NEEDLEGRASS

[Hitchcock. The North American species of Stipa. Contr. U. S. Nation. Herb. 24: 215-289. 1925.]

Glumes about 10 mm long.. 1. *S. avenacea.*
Glumes about 15-40 mm long.
  Lemmas 8-12 mm long.. 2. *S. comata.*
  Lemmas 15-22 mm long.. 3. *S. spartea.*

1. **Stipa avenàcea** L. BLACKSEED NEEDLEGRASS. Map 240. Local in dry, sandy soil in a few of our northern counties.

Mass. to Mich., southw. to Fla. and Tex., mostly on the Coastal Plain.

2. **Stipa comàta** Trin. & Rupr. NEEDLE-AND-THREAD. Map 241. This species is known only from a high gravelly hill on the northeast side of Diamond Lake, Noble County.

Ind. and Mich. to Yukon Territory, southw. to Tex. and Calif.

3. **Stipa spártea** Trin. PORCUPINE GRASS. Map 242. Local to infrequent or frequent on open sand knolls, sand ridges, and dunes, or rarely on open gravelly places in the northwestern part of the state.

Ont., to B. C., southw. to Pa., Ind., Kans., and N. Mex.

85-208. ARÍSTIDA L. THREE-AWN GRASS

[Hitchcock. North American species of Aristida. Contr. U. S. Nation. Herb. 22: 517-586. 1924. Henrard. A critical revision of the genus Aristida. vii+701p. 1928. Supplement: 702-747. 1933. Rijks Herbarium. Leiden.]

Awns of lemma united into a column, 10-15 mm long, articulated with the lemma. .... 1. *A. tuberculosa.*
Awns of lemma not united into a column and not articulated with the lemma.
  Lemmas (exclusive of awns) less than 12 mm long.
    Central awn of lemma coiled at the base at maturity; lateral awns rarely more than 1.5 mm long.. 2. *A. dichotoma.*
    Central awn of lemma not coiled at the base, but abruptly bent outward, usually to a 45-90 degree angle, sometimes with a slight twist at the base; lateral awns usually more than 1.5 mm long.

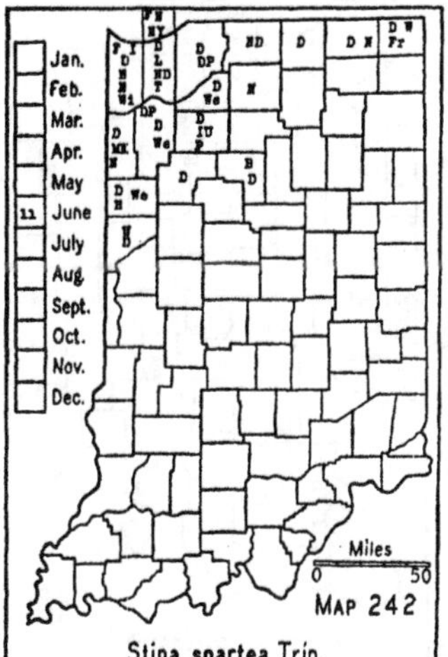

Stipa spartea Trin.

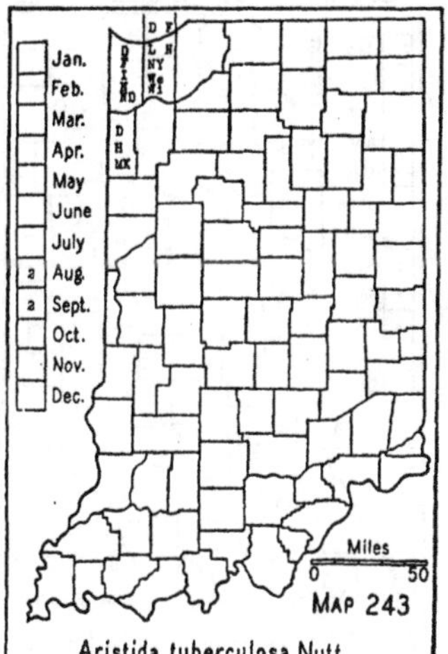

Aristida tuberculosa Nutt.

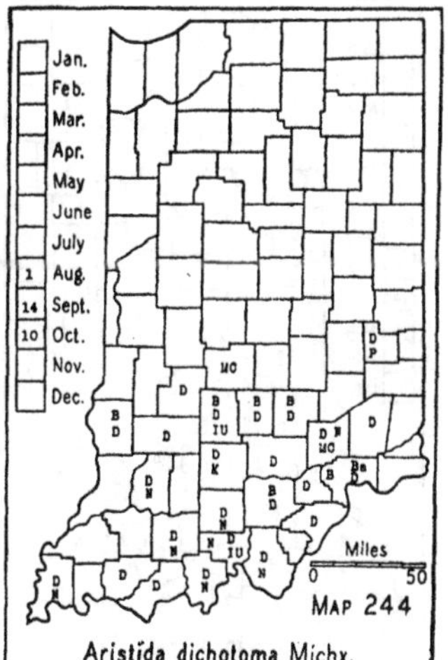

Arist͏ída dichotoma Michx.

Glumes mostly 4-5 mm long; lemmas (exclusive of awns) generally 4-5 mm long, their central awns usually 5-15 mm long............3. *A. longespica.*

Glumes mostly 6-11 mm long; lemmas (exclusive of awns) generally 6-8 mm long, their central awns usually 15-25 mm long.

Plant annual, 20-40 cm high; first glume shorter than or equaling the second. ....4. *A. intermedia.*

Plant perennial, 40-70 cm high; first glume generally longer than the second. .... 5. *A. purpurascens.*

Lemmas (exclusive of awns) more than 12 mm long.

Central awn of lemma about 20 mm long, lateral awns much shorter, usually 1.5 mm long...6. *A. ramosissima.*

Central awn of lemma about 25-70 mm long, lateral awns usually about 5-10 mm shorter than the central one.............................7. *A. oligantha.*

1. **Aristida tuberculòsa** Nutt. Map 243. Local in the northwestern part of the state where it grows in almost pure sand on old beaches and low dunes.

Mass. to Ga. and Miss. near the coast; around the southern end of Lake Michigan, and locally in Wis., Ill., Iowa, and Minn.

2. **Aristida dichótoma** Michx. Map 244. Infrequent to frequent in the southern half of the state. It is usually a common plant where it is found. It prefers hard, white clay soil in abandoned and fallow fields, on washed slopes, and along clayey roadsides.

All the species of this genus, when found in habitats similar to those just mentioned, are known in Indiana as poverty grasses.

This species is often confused with *Aristida longespica* from which it may be separated by its dichotomously branched culms, its shorter terminal panicles, its tighter second glume, the coiled central awn, and its straight lateral awns being 1 mm long. In the other species the glumes are usually much looser and the lateral awns of the lemma are much longer, diverging, or widely spreading.

This species was reported from Marshall County by Clark but there is no verifying specimen.

Maine to Mich. (*Hanes*) and e. Kans., southw. to s. Fla. and Tex.

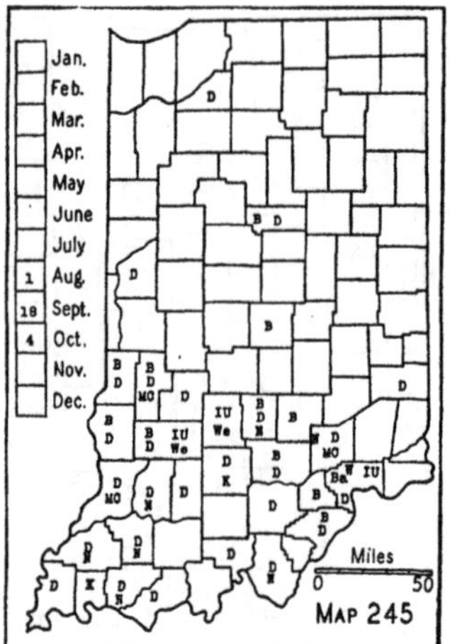

Aristida longespica Poir.

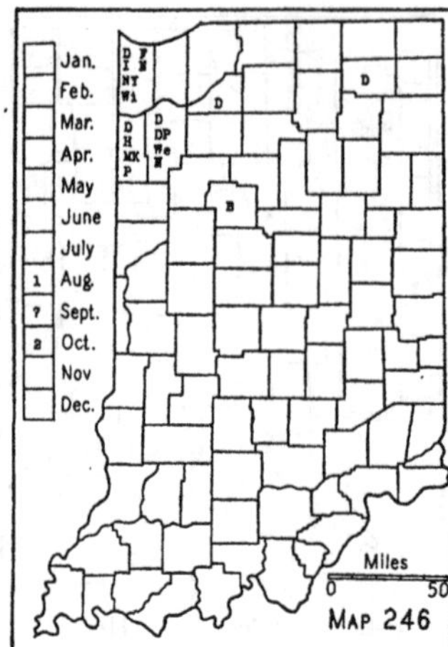

Aristida intermedia Scribn. & Ball

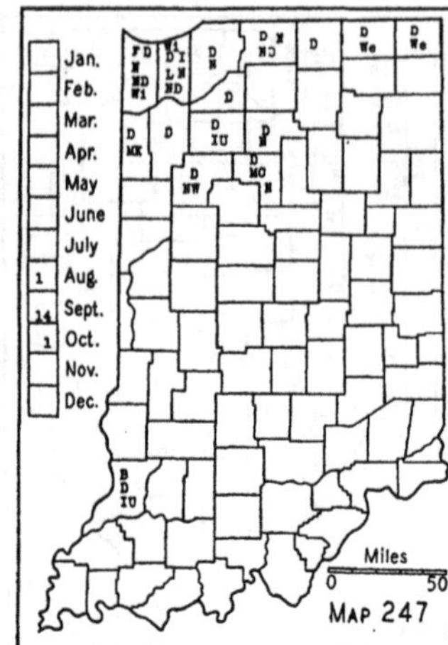

Aristida purpurascens Poir.

3. **Aristida longespìca** Poir. (*Aristida gracilis* Ell.) Map 245. Infrequent to frequent, but plentiful where found, in the southern part of the state. Usually abundant in hard, white clay soil in low, flat, fallow fields and in habitats similar to those of the preceding species. Probably also infrequent in the sandy areas of the northwestern part of the state, although there are specimens only from Starke County. Our specimens vary somewhat in the length of their awns, but I do not think the variation has any taxonomic value.

N. H. to Mich., southw. to Fla. and Tex., especially on the Coastal Plain.

4. **Aristida intermèdia** Scribn. & Ball. Map 246. This species seems to be local but abundant where it is found. I have seen acres of it in Newton County in the old lake bed, and in Noble County it forms large colonies on the former bottom of Tippecanoe Lake. Local in moist, sandy soil on interdunal flats about Lake Michigan, in moist sandy, prairie habitats, and on moist sandy borders of lakes.

Ind. to Nebr., southw. to Miss. and Tex.

5. **Aristida purpuráscens** Poir. Map 247. Infrequent in very dry sand in the northwestern part of the state and in a similar habitat in Knox County. Its habitat is found in contiguous counties, and doubtless its range will be extended.

Mass. to Kans., southw. to Fla. and Tex.

6. **Aristida ramosíssima** Engelm. Map 248. This is an infrequent grass of the southwestern counties in hard, white clay soil in abandoned and fallow fields, on washed slopes, along clayey roadsides, and infrequently in yellow clay soil.

Ind. to Iowa, southw. to Tenn., La., and Tex.

7. **Aristida oligántha** Michx. PRAIRIE THREE-AWN GRASS. Map 249. Like the other species of the genus, this species is partial to a slightly acid soil and is infrequent to frequent in the southern half of the state where

Aristida ramosissima Engelm.

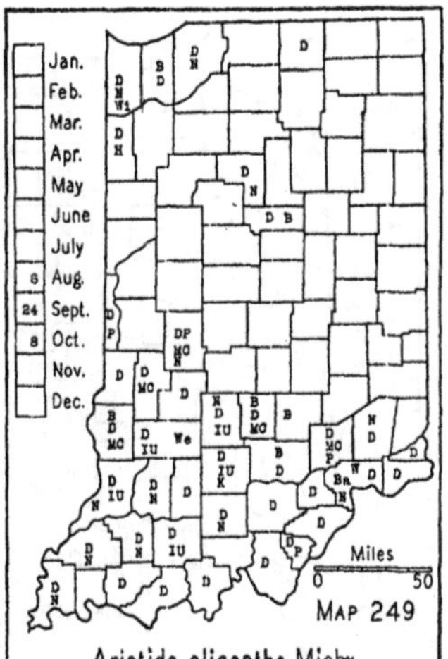

Aristida oligantha Michx.

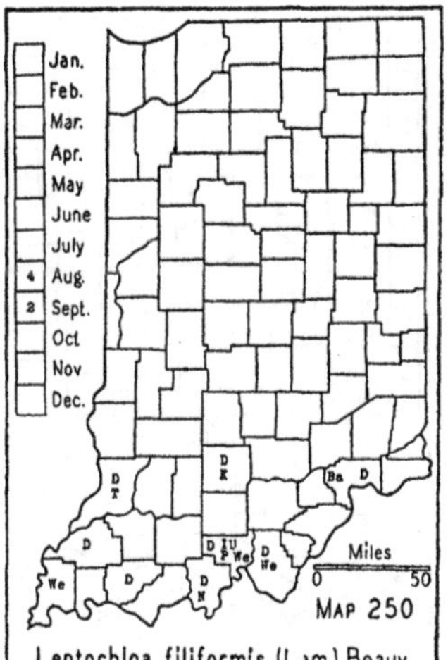

Leptochloa filiformis (Lam.) Beauv.

its habitat is found. It is usually found in hard, white clay soil in abandoned and fallow fields, on washed slopes, along clayey roadsides, and locally in sandy soil in the northern counties.

Mass. to S. Dak., southw. to Fla. and Tex.

6. *CHLORÍDEAE* Kunth. GRAMA TRIBE

Spikelets disarticulating below the glumes; large coarse grasses, usually more than a meter high..99. SPARTINA, p. 143.
Spikelets disarticulating above the glumes; grasses shorter than the preceding.
  Spikes digitate or, in *Eleusine*, one or rarely 2 spikes remote (rarely as distant as 2.5 cm).
    Spikelets 1-flowered.
      Spikelets awnless ....95. CYNODON, p. 143.
      Spikelets awned ....102. CHLORIS, p. 144.
    Spikelets more than 1-flowered.
      Rachis extending beyond the florets into a naked sharp point; second glume and at least the lowest lemma cuspidate..........94. DACTYLOCTENIUM, p. 142.
      Rachis not extending beyond the florets and not ending in a sharp point; glumes and lemmas not cuspidate..........................93. ELEUSINE, p. 142.
  Spikes racemose, on an axis more than 5 cm long.
    Spikes ascending or widely spreading, slender, elongate.
      Lemmas with an awn 4-6 mm long.................101. GYMNOPOGON, p. 143.
      Lemmas awnless ....90. LEPTOCHLOA, p. 141.
    Spikes drooping, dense, short.........................104. BOUTELOUA, p. 144.

90-307. LEPTÓCHLOA Beauv. SPRANGLETOP GRASS

[Hitchcock. North American species of Leptochloa. U. S. Dept. Agric. Bur. Pl. Ind. Bull. **33**: 1-21. 1903.]

Sheaths papillose-pilose; second glume acute; grain 3-angled, grooved on the side toward the palea...1. *L. filiformis.*
Sheaths smooth; second glume obtuse; grain compressed, not grooved..2. *L. panicoides.*

1. **Leptochloa filifórmis** (Lam.) Beauv. RED SPRANGLETOP. Map 250. Infrequent in the counties along the Ohio River. It grows in sandy soil on the slope of the bank of the Ohio River where it is washed at flood

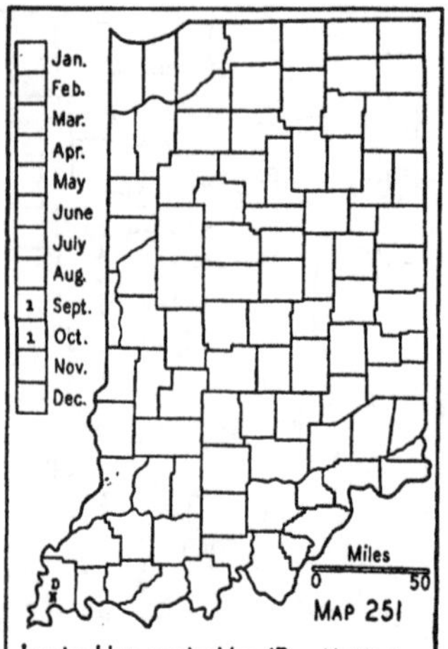

Leptochloa panicoides (Presl) Hitchc.

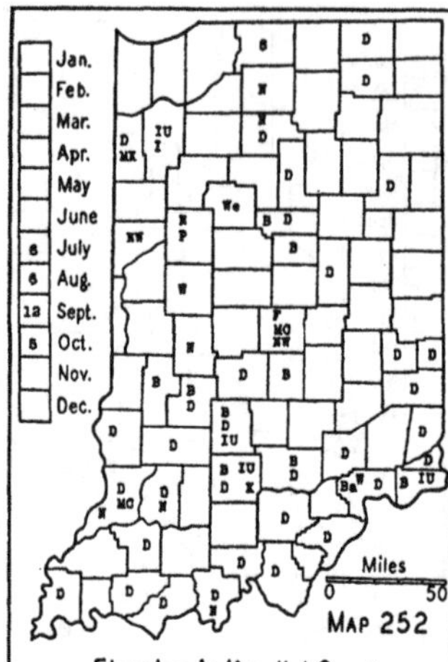

Eleusine indica (L.) Gaertn.

Cynodon Dactylon (L.) Pers.

stages. Also found in sandy, alluvial fields along the Ohio River, and rarely in a similar habitat away from the river. Usually rather plentiful where it occurs.

Va. to s. Ind. and e. Kans., southw. to Fla. and Tex., s. Calif., and throughout tropical America.

2. **Leptochloa panicoìdes** (Presl) Hitchc. (*Leptochloa floribunda* Doell of Deam, Grasses of Ind.) Map 251. In 1916 I found a few specimens of this species in a large, miry, muddy flat in what is locally known as Pitcher's Lake, about 5 miles west of Mt. Vernon, Posey County. Pitcher's Lake is in reality a shallow lagoon about 2 miles long and a half mile wide. It is filled with water during the winter months and is usually nearly or entirely dry in autumn. This grass was found with *Lindernia, Cyperus, Acnida,* and *Leersia oryzoides.* I revisited the place in 1920 and found a few more specimens. The Indiana specimens are the only ones known north of Mississippi. The species is rare, having been found only in Indiana, Mississippi, Louisiana, and Texas, and southward to Brazil.

93-304. ELEUSÌNE Gaertn.

1. ELEUSINE ÌNDICA (L.) Gaertn. GOOSEGRASS. Map 252. This species is doubtless found in every county of the state although our records are less frequent in the northern counties. It prefers a moist, sandy habitat and is found about dwellings, along roadsides and footpaths and in waste places, pastures, and cultivated fields.

Nat. of the Old World; Mass. to S. Dak., southw. to Fla. and Tex.; occasional in Oreg. and Calif.

94-305. DACTYLOCTÈNIUM Willd.

See excluded species no. 71, p. 1028.

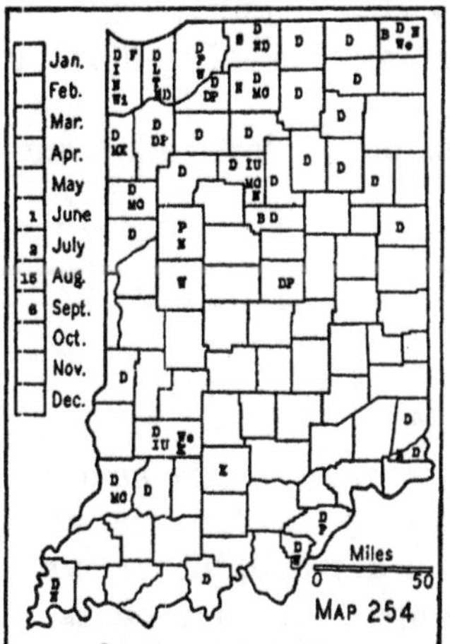

MAP 254
Spartina pectinata Link

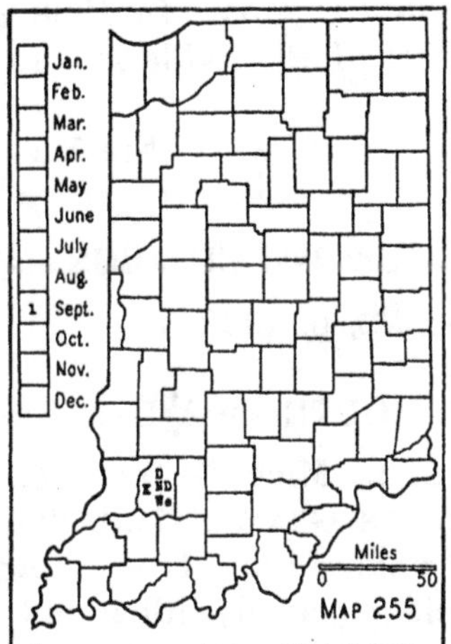

MAP 255
Gymnopogon ambiguus (Michx.) BSP.

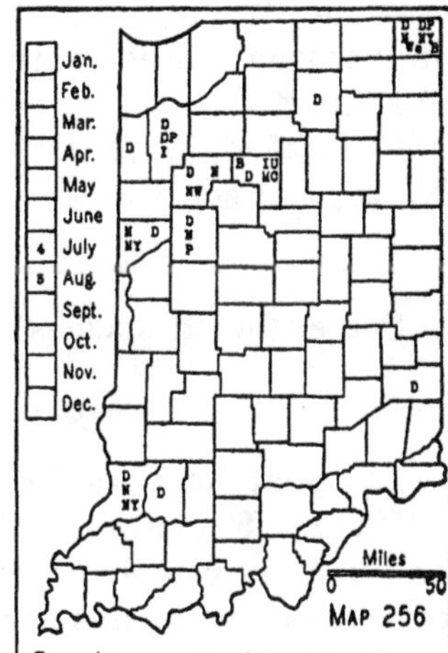

MAP 256
Bouteloua curtipendula (Michx.) Torr.

95-282. CÝNODON Richard

1. CYNODON DÁCTYLON (L.) Pers. (*Capriola Dactylon* (L.) Ktze. BERMUDA GRASS. Map 253. This grass has become sparingly established in the state and I predict that in time it will become a grass used frequently for lawns and pasturage. It thrives well in sandy soil where bluegrass will not.

In a waste, vacant lot in Bluffton, Wells County, two large colonies have been established for several years and these were not injured by a temperature of twenty-one degrees below zero of the winter of 1935-1936.

Introduced in America, and found in the warm regions of both hemispheres. Md. to Okla., southw. to Fla., Tex., and Calif.; occasionally northw. from N. H. to Mich. and Oreg.

99-283. SPARTÌNA Schreber

[Merrill. The North American species of Spartina. U. S. Dept. Agric. Bur. Pl. Ind. Bull. 9: 1-16. 1902. Saint-Yves, Alf. Monographia Spartinarum. Candollea 5: 19-100. Dec. 1932.]

1. **Spartina pectinàta** Link. (*Spartina Michauxiana* Hitchc.) PRAIRIE CORDGRASS. Map 254. This species is infrequent or rarely frequent and seems to be restricted to the lake and prairie areas and to the slope of the bank of the Ohio River, usually in crevices of shale. In the lake area, it is found on the low borders of lakes and streams and in marshy places. In the prairie area, it is found in wet places, usually closely associated with *Calamagrostis canadensis*.

Newf., Que. to e. Wash. and Oreg., southw. to N. C., Ky., Ill., Ark., Tex., and N. Mex.

101-290. GYMNOPÒGON Beauv.

1. **Gymnopogon ambíguus** (Michx.) BSP. Map 255. On September 19, 1934, I found a large colony of this species in very sandy soil on the crest of a sand ridge in an open place in a woods in sec. 35 about 5 miles north-

west of Washington, Daviess County. In 1938 Kriebel found it here and in a woods a mile southwest of Plainville and in a woods 4 miles north of Washington.

Coastal Plain, N. J., Fla., and Tex.; in the Mississippi Valley, Ind., Tenn., Kans., and southw.

102-288. CHLÒRIS Sw.

See excluded species no. 72, p. 1029.

104-195. BOUTELOÙA Lag. GRAMA GRASS

1. **Bouteloua curtipéndula** (Michx.) Torr. (*Atheropogon curtipendulus* (Michx.) Fourn.) SIDE-OATS GRAMA. Map 256. Very local in the state and usually restricted to small areas. It is found in dry soil, either sandy or clayey, on sandy knolls, gravelly hills and slopes, and on bluffs of streams.

Maine, Ont. to Mont., southw. to Md., W. Va., Ala., Tex., Ariz., and s. Calif.; introduced in S. C.

7. *PHALARÍDEAE* Link. CANARY GRASS TRIBE

Glumes 1-nerved; sterile lemmas awned................109. ANTHOXANTHUM, p. 144.
Glumes 3-nerved (sometimes the lateral pair of the first glume faint); lemmas not awned.
Glumes very thin, not keeled..........................108. HIEROCHLOË, p. 144.
Glumes firm, strongly keeled.............................110. PHALARIS, p. 144.

108-206. HIERÓCHLOË R. Br.

1. **Hierochloë odoràta** (L.) Beauv. (*Hierochloë odorata* (L.) Wahl., *Savastana odorata* (L.) Scribn., and *Hierochloë odorata* var. *fragrans* (Willd.) Richt.) SWEETGRASS. Map 257. Infrequent in some of our northern counties where it is usually found in open marshes. I have one specimen from mucky soil of a fallow cornfield.

Lab. to Alaska, southw. to N. J., Ind., Iowa, Oreg., and in the mts. to N. Mex. and Ariz.

109-205. ANTHOXÁNTHUM L.

1. ANTHOXANTHUM ODORÀTUM L. SWEET VERNALGRASS. Map 258. While this grass has been reported from all parts of the eastern United States, in Indiana it has been reported from only 2 counties in addition to those shown on the map. Found along railroads and in pastures, waste places, and meadows.

Nat. of Eurasia; Greenland and Newf. to La. and Mich., and on the Pacific coast from B. C. to n. Calif.

110-204. PHÁLARIS L. CANARY GRASS

Inflorescence 6-16 cm long; glumes not dilated above the middle; fertile floret about 3.5 mm long..1. *P. arundinacea.*
Inflorescence 2-4 cm long; glumes dilated above the middle; fertile floret about 5 mm long. (See excluded species no. 73, p. 1029.).....................*P. canariensis.*

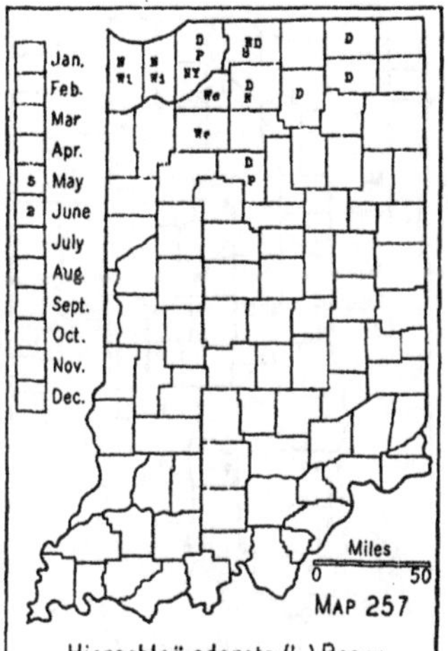

Hierochloë odorata (L.) Beauv.

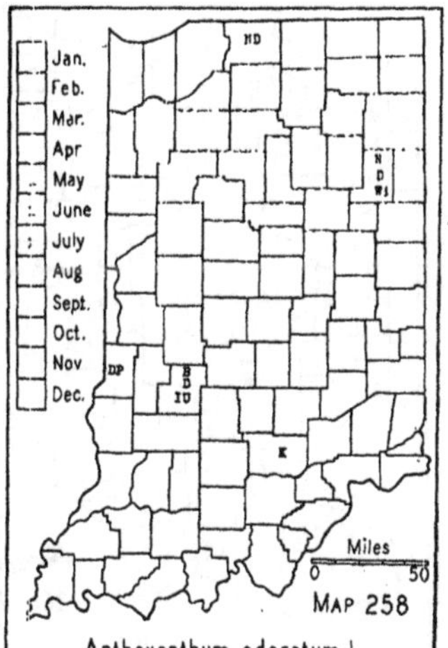

Anthoxanthum odoratum L.

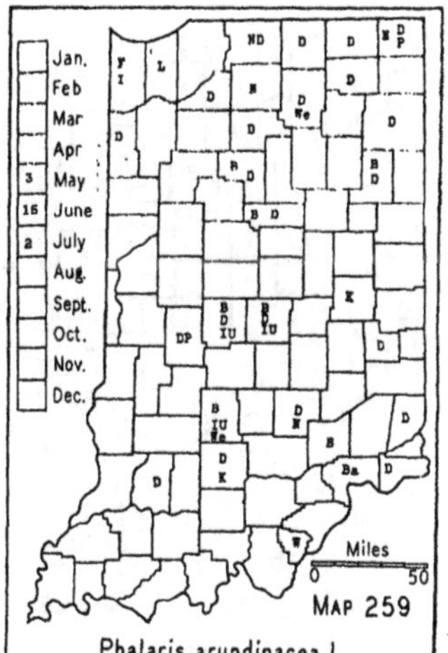

Phalaris arundinacea L.

1. **Phalaris arundinàcea** L. REED CANARY GRASS. Map 259. This species is infrequent in the lake area and local south of it. In most places in the lake area it seems to be a native while southward it is doubtless an escape. It is usually found in marshes but will thrive in almost any habitat. It is recommended as a fodder plant for low grounds, especially in the northwest. My observation and personal experience with it is that it is wise not to plant it if one wishes ever to get rid of it. I have found it as difficult to exterminate as most pernicious weeds.

Nat. of Eurasia; N. B. to se. Alaska, southw. to N. C., Ky., Okla., N. Mex., Ariz., and ne. Calif.

1a. PHALARIS ARUNDINACEA var. PÍCTA L. This is a variety with the leaves striped with white. It is often used in cultivation and found as an escape in colonies along roadsides and in waste places.

8. *ORỲZEAE* Kunth RICE TRIBE

112-194. LEÉRSIA Sw.

Culms compressed; foliage more or less scabrous or scabrous-pubescent; spikelets 3-3.5 mm long (rarely one 4 mm long), 1-1.3 mm wide; stamens 1 or 2; grain about 2.5 mm long, 1 mm wide..1. *L. virginica.*

Culms terete; foliage more or less hispid; spikelets 4.1-5 mm long, 1.5-4 mm wide; stamens 2 or 3; grain 3-3.5 mm long, 1.5-1.8 mm wide.

Spikelets oblong, 1.5-1.8 mm wide; stamens 3; grain about 3 mm long, 1.5 mm wide. ....2. *L. oryzoides.*

Spikelets broadly oval to nearly orbicular, 3-4 mm wide; stamens 2; grain about 3.5 mm long, 1.8 mm wide..3. *L. lenticularis.*

1. **Leersia virgínica** Willd. (*Homalocenchrus virginicus* (Willd.) Britt.) WHITEGRASS. Map 260. An infrequent to frequent grass in low woodland of all kinds in all parts of the state. It is usually found where the mineral soil is exposed, hence it is most frequent on old logging roads.

Que. to S. Dak., southw. to Fla. and Tex.

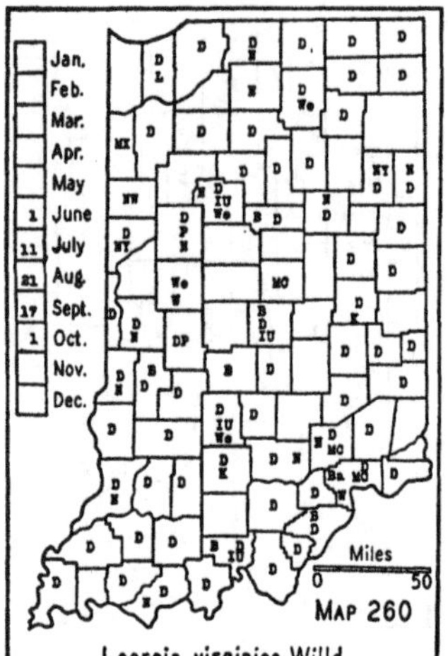

MAP 260

Leersia virginica Willd.

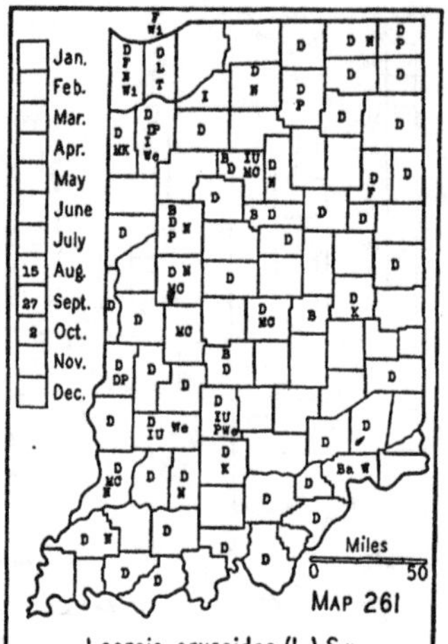

MAP 261

Leersia oryzoides (L.) Sw.

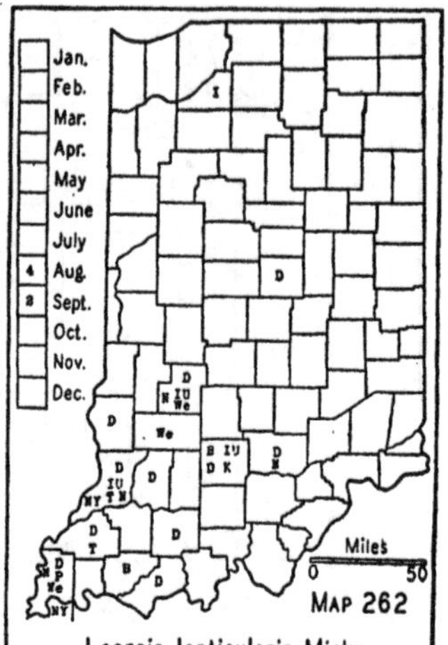

MAP 262

Leersia lenticularis Michx.

2. **Leersia oryzoìdes** (L.) Sw. (*Homalocenchrus oryzoides* (L.) Poll.) RICE CUTGRASS. Map 261. This species is found throughout the state but on account of its habitat and light requirements it is infrequent. It prefers full sunlight and is found in low, wet places, especially along ditches, old river channels, outlets of springs, in springy places, and about lakes. The species varies considerably. In late flowering forms, sometimes the panicles do not expand and are included. Rarely forms with smooth sheaths are found.

Maine, Que. to e. Wash., southw to Fla., Tex., Colo., Ariz., and s. Calif.; also in Eu.

3. **Leersia lenticulàris** Michx. (*Homalocenchrus lenticularis* (Michx.) Scribn.) CATCHFLY GRASS. Map 262. This grass seems to be restricted to the stream courses of the southwestern part of the state and the valley of the Kankakee River. It is usually found in low ground in woods, on the borders of ponds, about sloughs, and in ditches. It is infrequent but where found often plentiful.

Ind. to Minn., southw. to S. C., Fla., and Tex.

9. *ZIZANIÈAE* Hitchc. INDIAN RICE TRIBE

113-190. ZIZÀNIA L. WILDRICE

[Fassett. A study of the genus Zizania. Rhodora **26**: 153-160. 1924.]

Pistillate lemmas thin and papery, dull, finely striate, scabrous over the whole surface; aborted spikelets slender and shriveled, less than 1 mm thick.......1. *Z. aquatica.*

Pistillate lemmas firm and tough, with a lustrous and coarsely corrugate surface, scabrous on the margins, at the summit, along the awn, and sometimes along the nerves, otherwise glabrous; aborted spikelets with a distinct body 1.5-2 mm thick.

Culms 60-120 cm high; blades generally less than 10 mm wide; lower pistillate branches with 2-6 spikelets; lower or middle staminate branches with 5-15 spikelets ....1a. *Z. aquatica* var. *angustifolia.*

Culms mostly 125-275 cm high; blades generally 10-50 mm wide; lower pistillate branches with 11-29 spikelets; lower or middle staminate branches with (20) 30-60 spikelets...............................1b. *Z. aquatica* var. *interior.*

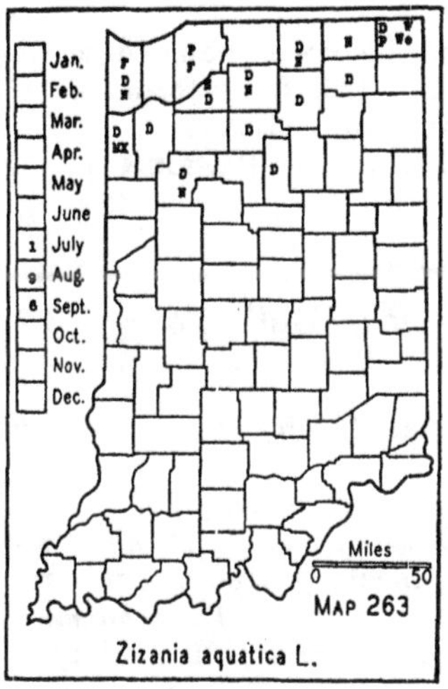

Zizania aquatica L.

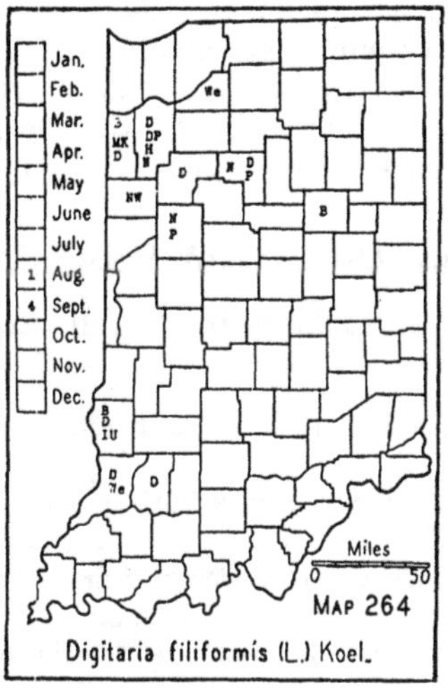

Digitaria filiformis (L.) Koel.

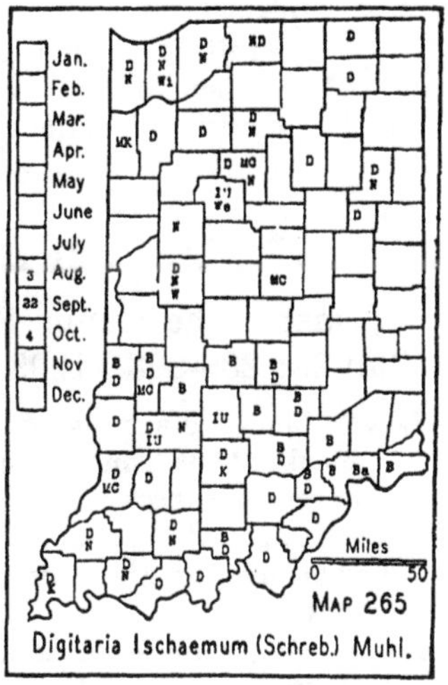

Digitaria Ischaemum (Schreb.) Muhl.

1. **Zizania aquática** L. (*Zizania palustris* of recent authors, not L.) ANNUAL WILDRICE. Map 263. Infrequent to local in the lake area in dredged ditches, sloughs, and swamps.

This grass affords protection and food for water birds, especially wild ducks, and it is now often planted for these purposes.

The following are popular publications on this subject: Wild rice; its uses and propagation, by Brown and Scofield. U. S. Dept. Agric. Bur. Pl. Ind. Bull. 50: 1-23. 7 pl. 1903; Wild rice, by Fyles. Dept. Agric. Dominion of Canada, Bull. 42, n.s. 1-20. 1920; Propagation of wild-duck foods, by McAtee. U. S. Dept. Agric. Bull. 465. 1917.

Que. to N. Dak. and Idaho, southw. to Fla., and La.

1a. **Zizania aquatica** var. **angustifòlia** Hitchc. (*Zizania palustris* L.) NORTHERN WILDRICE. This variety has the habitat of the species.

N. B., Que., and N. Dak., southw. to N. Y. and Nebr.

1b. **Zizania aquatica** var. **intèrior** Fassett. This is a form of the species which Fassett recognizes but Hitchcock does not. It seems to be distinct in our area. Its habitat is that of the species.

10. *PANÍCEAE* R. Br. MILLET TRIBE

Spikelets enclosed in globose spiny burs......................137. CENCHRUS, p. 177.
Spikelets not as above.
  Spikelets subtended by 1-several scabrous bristles; inflorescence a spikelike panicle. ....135. SETARIA, p. 176.
  Spikelets not subtended by bristles.
    Spikelets awned or with an acuminate and spiny-hispid second glume and sterile lemma..133. ECHINOCHLOA, p. 174.
    Spikelets not awned nor spiny-hispid.
      Spikelets in slender one-sided racemes, subsessile and in two rows; first glume obsolete.
        Spikelets obovate to suborbicular or, if narrowly elliptic, the rachis broadly winged...128. PASPALUM, p. 149.
        Spikelets lanceolate or elliptic, on a narrow rachis....121. DIGITARIA, p. 148.

Spikelets in an open or contracted panicle.
Fruiting lemma firm, with flat, hyaline margins; pedicels 3-angled at the summit..122. LEPTOLOMA, p. 148.
Fruiting lemma chartaceous-indurate, the margins inrolled...129. PANICUM, p. 150.

121-166A. DIGITÀRIA Heist. CRABGRASS

[Nash. The Genus Syntherisma in North America. Bull. Torrey Bot. Club 25: 289-303. 1898.]

Lower blades glabrous or nearly so; mature fertile lemmas (fruit) dark brown or black, about 2 mm long.
Lower sheaths papillose-hirsute; rachis wingless, about 0.3 mm wide...1. *D. filiformis.*
Lower sheaths glabrous or with a few straggling hairs; rachis winged, about 1 mm wide ....2. *D. Ischaemum.*
Lower blades more or less pubescent; mature fertile lemmas (fruit) light gray to light drab, about 3 mm long...................................3. *D. sanguinalis.*

1. **Digitaria filifórmis** (L.) Koel. (*Syntherisma filiforme* (L.) Nash.) Map 264. This species is known from only eleven counties and reported from Marshall County. It is very local but common enough where found. My specimens are from very sandy soil in shallow depressions on low, sandy ridges in open woodland, in a moist prairie habitat, and in dry, sandy soil in pastures.

N. H. to Iowa and Kans., southw. to Fla., Tex., and Mex.

2. DIGITARIA ISCHAÈMUM (Schreb.) Muhl. (*Digitaria humifusa* Pers. and *Syntherisma Ischaemum* (Schreb.) Nash.) SMOOTH CRABGRASS. Map 265. Infrequent in the northern part of the state and frequent to common in moist, clayey flats in the southwestern counties. Like the next species, it is found almost everywhere except in dense woodland and very wet soil. It prefers a moist, sandy soil and is found generally in cultivated fields, pastures, meadows, and waste places and along roadsides. In the southwestern counties in the moist, clayey, fallow fields, it forms dense mats over large areas.

Nat. of Eurasia; Que. to N. Dak., southw. to S. C., Tenn., and Ark.

3. DIGITARIA SANGUINÀLIS (L.) Scop. (*Syntherisma sanguinalis* (L.) Dulac.) CRABGRASS. Map 266. This species is a common weed throughout the state, especially in truck gardens, lawns, gardens, and cultivated grounds of all kinds.

Nat. of Eu.; throughout the U. S., more common in the East and South.

122-166C. LEPTOLÒMA Chase

1. **Leptoloma cognàtum** (Schult.) Chase. Map 267. This grass is found in very sandy soil on sand ridges and sandy knolls, usually in fallow fields, along roadsides, and in open woodland.

N. H. to Minn., southw. to Fla. and Tex., and westw. to Ariz.

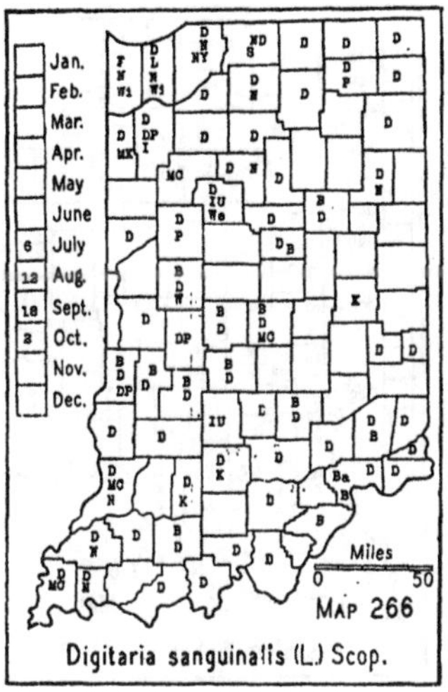

Digitaria sanguinalis (L.) Scop.

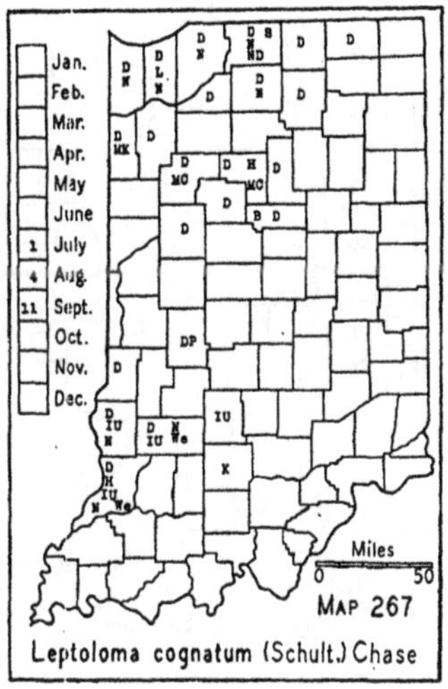

Leptoloma cognatum (Schult.) Chase

Paspalum fluitans (Ell.) Kunth

128-161. PÁSPALUM L.

[Chase. The North American species of Paspalum. Contr. U. S. Nation. Herb. 28: 1-310. 1929.]

Racemes more than 10 to a panicle....................................1. *P. fluitans.*
Racemes fewer than 10 to a panicle.
  Spikelets 2.5-3.2 mm long.
    Spikelets solitary ....2. *P. circulare.*
    Spikelets in pairs....................................3. *P. pubiflorum* var. *glabrum.*
  Spikelets 1.5-2.4 mm long.
    Blades from sparsely to rather densely pilose, rather thin.
      Spikelets 2 mm long....................................4. *P. pubescens.*
      Spikelets about 1.5 mm long. (See excluded species no. 77, p. 1029.)..*P. setaceum.*
    Blades puberulent on both surfaces, with long hairs intermixed, or the lower surface nearly or quite glabrous except a few long hairs along the midrib and margin, usually rather firm....................................5. *P. stramineum.*

1. **Paspalum flùitans** (Ell.) Kunth. (*Paspalum mucronatum* Muhl. and *Paspalum repens* Berg.) Map 268. Infrequent to local in the state and restricted to the muddy banks of ponds, sloughs, and streams. The oldest specimen seen was one collected in 1836 near New Albany by Dr. Clapp.

S. C. to Ind., Kans., and Tex., southw. to Argentina.

2. **Paspalum circulàre** Nash. Map 269. Infrequent to frequent in the southern part of the state where it is usually found in hard, white clay soil in roadside ditches, low places in woodland, and fallow fields. The specimen from Marion County was found along the Monon Railroad and doubtless was introduced.

Conn. to N. C. and Miss., northw. to Ind., Kans., and westw. to Tex.

3. **Paspalum pubiflòrum** Rupr. var. **glàbrum** Vasey. Map 270. Infrequent in a few of our southern counties. Usually found in moist, sandy soil in ditches and in low ground. It is a common plant in the street gutters in the southeastern part of Mt. Vernon.

N. C. to Ind. and Kans., southw. to Fla. and Tex.

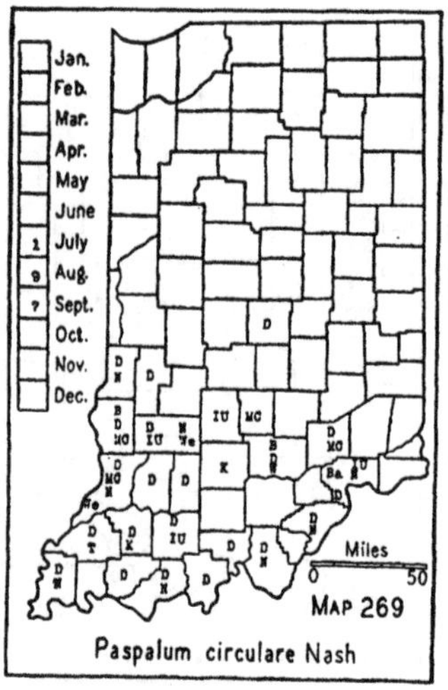

Paspalum circulare Nash

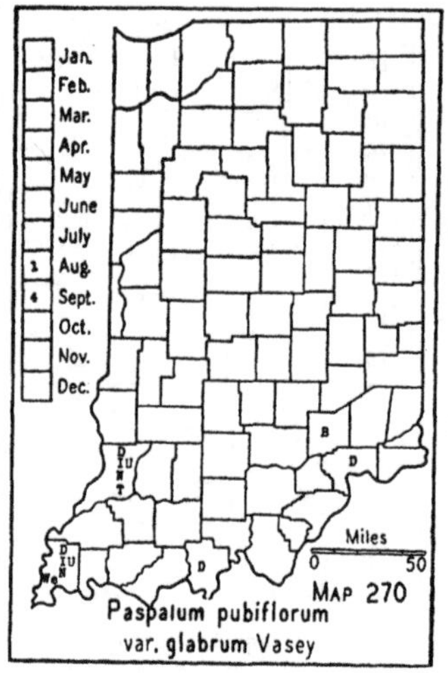

Paspalum pubiflorum var. glabrum Vasey

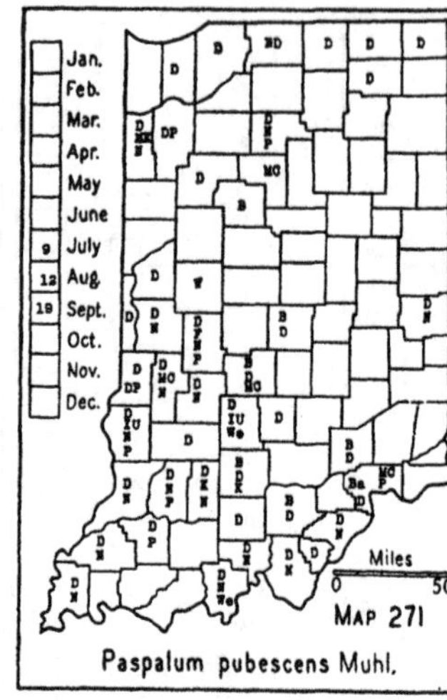

Paspalum pubescens Muhl.

Paspalum stramineum Nash

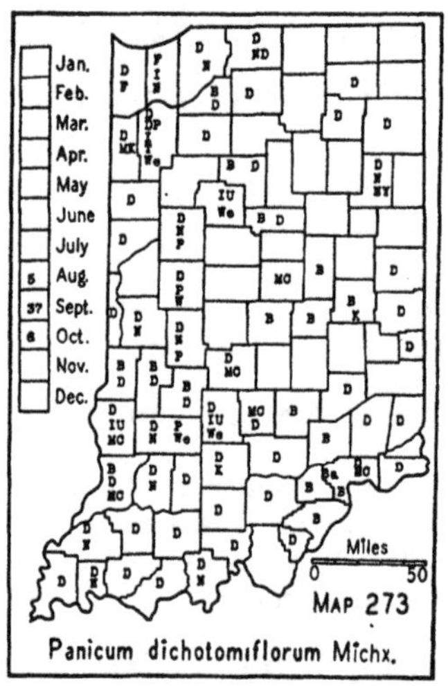

Panicum dichotomiflorum Michx.

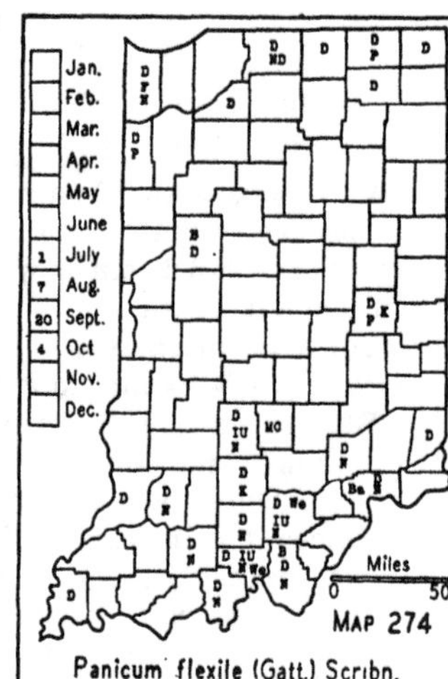

Panicum flexile (Gatt.) Scribn.

4. **Paspalum pubéscens** Muhl. (Including *Paspalum Muhlenbergii* Nash.) Map 271. This species is found sparingly in the northern two thirds of the state and is infrequent to frequent in the southern part. It prefers moist, sandy soil but adapts itself to many habitats. It is usually found in pastured fields and woodlots.

Vt. to Mich., southw. to Fla. and Tex.

5. **Paspalum stramíneum** Nash. Map 272. A few specimens of this species have been found in a few of the northwestern counties in very dry, sandy soil along roadsides and in waste places.

Ind. to Minn., southw. to Tex., Ariz., and nw. Mex.

129-166. PÁNICUM L. PANICUM

[Hitchcock and Chase. The North American species of Panicum. Contr. U. S. Nation. Herb. **15**: 1-396. 1910. Fernald. Realignments in the genus Panicum. Rhodora **36**: 61-87. 1934.]

Annual or perennial grasses of various habit, foliage, and inflorescence; spikelets disarticulating below the glumes, arranged in open or compact panicles, rarely racemose, 1- or 2-flowered, the lower flower usually represented by a sterile lemma and palea, the palea usually not developed or rarely lacking, when 2-flowered the lower staminate only; glumes 2, usually very unequal, the first smaller and often minute, the second typically equaling the sterile lemma, the latter of the same texture and simulating a third glume; stamens 3; fertile lemma chartaceous-indurate, the nerves obsolete, the margins inrolled and enclosing a palea of the same texture.

KEY TO SECTIONS OF INDIANA PANICUM.

Basal leaves similar to those of the culm; plants not forming winter rosettes.
First glume truncate or triangular-tipped, usually about a fourth (rarely longer) as long as the acute or acuminate glabrous spikelet; annual...1. DICHOTOMIFLORA, p. 156.
First glume not truncate, more than a fourth as long as the spikelet, usually a third to nearly half as long; annual or perennial.
Spikelets 2-5 mm long, smooth, or the keels more or less scabrous, but the spikelet not warty.
Annual..2. CAPILLARIA, p. 156.
Perennial.
Spikelets on long pedicels in large, open panicles; plants with creeping rootstocks..3. VIRGATA, p. 158.
Spikelets on short pedicels, arranged close together in one-sided branches in large panicles; plants without creeping rootstocks..4. AGROSTOIDEA, p. 158.
Spikelets 1.8-2 mm long, warty............................5. VERRUCOSA, p. 159.
Basal leaves not similar to those of the culm; plants forming winter rosettes.
Culm leaves elongated, not over 5 mm (rarely 6 mm) wide, more than 20 times as long as wide; spikelets 2.2-4 mm long, beaked in *P. depauperatum* and its variety; autumnal phase branching at the base.........6. DEPAUPERATA, p. 160.
Culm leaves not elongated (if elongated, glabrous on both surfaces with spikelets 2.2-3 mm long, or the blades softly pubescent on both surfaces and the spikelets 1.8-2 mm long); autumnal phase branching above the base or remaining simple.
Plants with elongate foliage aggregated at the base, light green, softly pubescent, the basal leaves not in distinct rosettes in autumn; ligules nearly obsolete; spikelets 1.8-2 mm long; autumnal phase branching near the base, forming close, flat tufts, with reduced panicles.................7. LAXIFLORA, p. 161.
Plants not as above.
Uppermost leaves elongate, generally longest, stiff, widely spreading, 3-8 (10) mm wide and up to 22 cm long, glabrous on both surfaces; sheaths glabrous or only the margins pubescent; spikelets 2.3-3 mm long...8. BICKNELLIANA, p. 162.
Uppermost leaves and spikelets not as above.
Culms glabrous or only the nodes pubescent; spikelets not over 3 mm long.
Ligules less than 1.2 mm long, usually nearly obsolete.
Culms bearded at the nodes, at least the lower ones (rarely only the upper ones puberulent in *P. mattamuskeetense*).....9. DICHOTOMA, p, 162.
Culms not bearded at the nodes.
Spikelets more than 1.8 mm long................9. DICHOTOMA, p. 162.
Spikelets less than 1.8 mm long.............13. SPHAEROCARPA, p. 170.
Ligules 2-5 mm long..............................10. SPRETA, p. 164.
Culms and sheaths more or less strongly pubescent; if glabrous except the nodes, the spikelets more than 3 mm long.
Ligules 2-5 mm long............................11. LANUGINOSA, p. 164.

Ligules not more than 2 mm long.
Spikelets nearly spherical at maturity, less than 1.8 mm long; blades glabrous, firm, cordate................13. SPHAEROCARPA, p. 170.
Spikelets elliptic or obovate, more than 1.7 mm long (except in *P. columbianum*).
Blades not cordate at the base.
Spikelets less than 3 mm long.............12. COLUMBIANA, p. 169.
Spikelets more than 3 mm long..........14. OLIGOSANTHIA, p. 171.
Blades cordate at the base.
Spikelets 2.5-2.9 mm long..................15. COMMUTATA, p. 172.
Spikelets 3-5 mm long.......................16. LATIFOLIA, p. 173.

KEY TO THE INDIANA SPECIES OF PANICUM

A. Spikelets glabrous.
Spikelets 3 mm long or longer.
Annual.
Spikelets 4-5 mm long and more than 1.8 mm wide. (See excluded species no. 82, p. 1030.)...*P. miliaceum.*
Spikelets 3-4 mm long, less than 1.8 mm wide...............2. *P. flexile,* p. 157.
Perennial.
Panicles 20-40 cm long; spikelets gaping or curved at the apex.
Ligules 2-4 mm long; first glume two thirds to three fourths as long as the spikelet.....................................6. *P. virgatum,* p. 158.
Ligules less than 1 mm long; first glume a third to half as long as the spikelet.....................................7. *P. anceps,* p. 159.
Panicles 3-8 cm long; spikelets not gaping or curved at the apex.
Blades elongated, not over 5 mm wide and more than 20 times as long as wide; spikelets beaked or pointed at the apex..11. *P. depauperatum,* p. 160.
Blades not elongated, less than 20 times as long as wide; spikelets blunt at the apex.............................43. *P. Scribnerianum,* p. 171.
Spikelets less than 3 mm long.
Spikelets warty.................................10. *P. verrucosum,* p. 159.
Spikelets not warty.
Annual; basal leaves similar to those of the culm; plants not forming winter rosettes; panicles more than 12 cm long (except in depauperate plants).
Sheaths glabrous.
Spikelets 2-3.5 mm long, usually about 2.9 mm long (rarely a few as short as 2 mm), acute; plants usually large and spreading, 50-100 cm long. ....1. *P. dichotomiflorum,* p. 156.
Spikelets 1.8-2.2 mm long, usually about 2 mm long, blunt; plants shorter and more slender than the preceding.................................. ....1a. *P. dichotomiflorum* var. *puritanorum,* p. 156.
Sheaths pubescent.
Pulvini of the panicle hispid.
Panicles included at the base, usually large, about as wide as long, generally about half as long as the whole plant; blades thickly papillose-hispid above and beneath...................3. *P. capillare,* p. 157.
Panicles exserted, ovoid, usually not as large as the preceding, about a third as long as the whole plant; blades sparsely hirsute above and beneath...........................4. *P. philadelphicum,* p. 157.
Pulvini of panicle glabrous......................5. *P. Gattingeri,* p. 157.
Perennial; basal leaves not like those of the culm; plants forming winter rosettes; panicles not over 12 cm long, except those of *Panicum agrostoides* which are much longer.
Pedicels mostly about half as long as the spikelets; spikelets subsecund on the lower side of the branchlets of the inflorescence.................. ....9. *P. agrostoides,* p. 159.

Pedicels mostly longer than the spikelets; spikelets not subsecund on the lower side of the branchlets of the inflorescence.
Spikelets not more than 1.8 mm long.
Nodes bearded; ligule less than 1 mm long; sheaths usually covered more or less with white spots................16. *P. microcarpon*, p. 162.
Nodes not bearded; ligule more than 1 mm long; sheaths without white spots..................................22. *P. spretum*, p. 164.
Spikelets 1.9-2.8 mm long.
Sheaths or some of them usually marked more or less with white spots, the overlapping margin usually glabrous; spikelets more than 2.2 mm long (mostly 2.3-2.5 mm long).......21. *P. yadkinense*, p. 164.
Sheaths not marked with white spots, the overlapping margin pubescent; spikelets 2-2.8 mm long.
Spikelets 2.3-2.8 mm long...................15. *P. Bicknellii*, p. 162.
Spikelets 2-2.2 mm long.
Plants of dry ground; culms erect, rarely autumnal plants reclining..............................17. *P. dichotomum*, p. 163.
Plants of bogs and swamps; culms weak, soon becoming decumbent and trailing..........................20. *P. lucidum*, p. 163.

A. Spikelets pubescent.
Blades mostly more than 15 mm wide.
Sheaths, at least the lower ones and those of the branches, papillose-hispid; spikelets 2.7-3 mm long (rarely longer)...........46. *P. clandestinum*, p. 173.
Sheaths not papillose-hispid.
Nodes retrorsely bearded; spikelets 4-4.5 mm long.
Blades glabrous or nearly so on both surfaces...........48. *P. Boscii*, p. 174.
Blades velvety to the touch beneath.........48a. *P. Boscii* var. *molle*, p. 174.
Nodes not retrorsely bearded, glabrous or minutely appressed-pubescent.
Spikelets 3.2-3.7 mm long..........................47. *P. latifolium*, p. 174.
Spikelets 2.5-3 mm long........................45. *P. commutatum*, p. 173.
Spikelets 1.4-1.6 mm long.........................39. *P. polyanthes*, p. 170.
Blades mostly less than 15 mm wide.
Blades elongated, not over 5 mm wide and more than 20 times as long as wide.
Spikelets beaked, mostly 3.2-3.8 mm long (rarely as short as 3 mm).
Sheaths pilose..................................11. *P. depauperatum*, p. 160.
Sheaths glabrous or nearly so..11a. *P. depauperatum* var. *psilophyllum*, p. 160.
Spikelets not beaked, 3 mm or less in length (rarely 3.2 mm long).
Spikelets 2.7-3.2 mm long; panicles narrow, usually less than a third as wide as long; ligules mostly about 1 mm long........12. *P. perlongum*, p. 161.
Spikelets 2.2-2.7 mm long; panicles usually more than a third as wide as long; ligules mostly less than 1 mm long.
Sheaths pilose.............................13. *P. linearifolium*, p. 161.
Sheaths glabrous or nearly so..13a. *P. linearifolium* var. *Werneri*, p. 161.
Blades not elongated or, if elongated, more than 5 mm wide.
Spikelets 3 mm or more long.
Spikelets beaked, somewhat curved, smooth except the scabrous keels........ ....7. *P. anceps*, p. 159.
Spikelets obovate, not curved, more or less pubescent with spreading hairs.
Ligule less than 0.5 mm long; blades papillose-hispid above and beneath; spikelets papillose-hispid.....................41. *P. Leibergii*, p. 171.
Ligule more than 0.5 mm long; blades not papillose-hispid; spikelets not papillose.
Culms and at least the lower sheaths with an appressed pubescence; ligules mostly 1.5 mm long with longer hairs intermixed; spikelets oblong-obovate, mostly 3.5-4 mm long and 1.7-1.9 mm wide......... ....42. *P. oligosanthes*, p. 171.

Culms and sheaths with a spreading pubescence; ligules about 1 mm long; spikelets bluntly obovate, mostly 3-3.5 mm long and 2 mm wide................................43. *P. Scribnerianum*, p. 171.

Spikelets less than 3 mm long.

Sheaths retrorsely pilose..........................14. *P. xalapense*, p. 161.

Sheaths not retrorsely pilose.

Basal leaves like those of the culm; plants not forming winter rosettes.

Spikelets 1.8-2.3 mm long; fruit not stalked.....9. *P. agrostoides*, p. 159.

Spikelets 2.4-2.8 mm long; fruit with a basal stalk 0.2-0.4 mm long. ....8. *P. stipitatum*, p. 159.

Basal leaves not like those of the culm; plants forming winter rosettes B. B (to left to save space).

B. Culms glabrous or only the nodes pubescent.

Ligule more than 1.5 mm long; spikelets 1.3-1.6 mm long.

Panicles narrow, a fourth to a third as wide as long (somewhat wider in anthesis); spikelets elliptic...................................22. *P. spretum*, p. 164.

Panicles open, two thirds as wide as long or longer; spikelets obovate........... ....23. *P. Lindheimeri*, p. 164.

Ligule less than 1.5 mm long; spikelets 1.4-2.9 mm long.

Spikelets 1.4-1.7 mm long.

Nodes of culms usually copiously barbed with long, lax, retrorse hairs; at least the lower sheaths more or less marked with white spots between the nerves; leaves usually glabrous, 6-14 mm wide, spreading or the upper reflexed...................................16. *P. microcarpon*, p. 162.

Nodes of culms minutely appressed-pubescent.

Upper three blades usually 10-20 cm long and 25 mm wide, the upper blade usually not much smaller than the other two, the blades below the upper three usually much smaller; anthers mostly 0.4-0.5 mm long.......... ....39. *P. polyanthes*, p. 170.

Upper three blades usually 5-10 cm long and 7-14 mm wide, the upper one usually much reduced, the blades below the upper three usually not reduced; anthers mostly 0.6-0.8 mm long....40. *P. sphaerocarpon*, p. 170.

Spikelets 1.8-2.9 mm long.

Spikelets 1.8-2.2 mm long.

Culms soon decumbent and trailing, the nodes usually glabrous or the lowest with a few soft spreading hairs; vernal blades spreading, mostly 4-6 mm wide; plants of a wet habitat......................20. *P. lucidum*, p. 163.

Culms erect, never trailing; vernal blades erect or spreading, mostly 4-14 mm wide.

Vernal blades mostly 4-8 mm wide, rarely some of them wider; lower part of culms usually more or less geniculate; lowest nodes of culms usually more or less barbed with soft hairs; plants usually of a dry habitat, often reclining in the autumnal phase and the nodes glabrous......... ....17. *P. dichotomum*, p. 163.

Vernal blades mostly 6-14 mm wide, more erect; culms not geniculate and the nodes usually all glabrous or with only a few soft hairs on the lowest; plants of a wet habitat..................19. *P. boreale*, p. 163.

Spikelets 2.3-2.9 mm long.

Blades mostly less than 8 mm wide, glabrous on both surfaces, not cordate at the base; spikelets oblong-elliptic, 2.3-2.9 mm long................. ....15. *P. Bicknellii*, p. 162.

Blades mostly 8-12 mm wide, cordate at the base, usually pubescent or the upper surface glabrous; spikelets elliptic, about 2.5 mm long.......... ....18. *P. mattamuskeetense*, p. 163.

B. Culms and sheaths more or less puberulent to strongly pubescent.

C. Ligule 2 mm or more long.

Plants grayish velvety-pubescent; spikelets 1.3-1.4 mm long.................... ....24. *P. auburne*, p. 166.

Plants pubescent, often villous but not velvety.
Culms conspicuously pilose with long horizontal hairs, branching before the expansion of the primary panicles; spikelets mostly 1.8-1.9 mm long....... ....25. *P. praecocius*, p. 166.
Culms variously pubescent, if pilose the hairs appressed or widely spreading; culm not branching before the expansion of the primary panicles.
Spikelets less than 2 mm long.
Vernal blades glabrous or nearly so above, 6-10 cm long and 5-10 mm wide. ....26. *P. tennesseense*, p. 166.
Vernal blades pubescent above or, if glabrous, smaller than the preceding, sometimes pilose above near the base and margins only.
Spikelets 1.3-1.5 mm long.
Upper surface of blades puberulent as well as long-villous........... ....27. *P. albemarlense*, p. 166.
Upper surface of blades villous but lacking the short, appressed puberulence.
Axis of panicle pilose, the lowest branches widely spreading; spikelets 1.5 mm long.................28. *P. implicatum*, p. 166.
Axis of panicle puberulent only, the lowest branches ascending; spikelets 1.3-1.4 mm long.............29. *P. meridionale*, p. 167.
Spikelets 1.6-1.9 mm long.
Pubescence on upper surface of vernal blades short-pilose, appressed at least on the apical half; first glume about a third the length of the spikelet, blunt or acute.
Blades stiff, erect........................30. *P. huachucae*, p. 167.
Blades lax, spreading...30a. *P. huachucae* var. *fasciculatum*, p. 168.
Pubescence on upper surface of vernal blades long-pilose, ascending; first glume about half as long as the spikelet, acuminate......... ....31. *P. subvillosum*, p. 168.
Spikelets 2-2.4 mm long.
Upper internodes shortened; leaves approximate, the blades often equaling the panicles; pubescence sparse and stiff....32. *P. scoparioides*, p. 168.
Upper internodes not shortened, the pubescence usually copious and rather silky.
Culms, sheaths, and lower surface of blades pilose but lacking the short pubescence; center of blades not glabrous; spikelets about 2 mm long.................................33. *P. villosissimum*, p. 168.
Culms, sheaths, and lower surface of blades puberulent as well as pilose; center of blades glabrous; spikelets 2.1-2.4 mm long.............. ....34. *P. pseudopubescens*, p. 168.
C. Ligules not over 2 mm long.
Spikelets nearly spheric at maturity, less than 1.8 mm long.
Upper three blades usually 10-20 cm long and 25 mm wide, the upper blade usually not much smaller than the other two, the blades below the upper three usually much smaller; anthers mostly 0.4-0.5 mm long............ ....39. *P. polyanthes*, p. 170.
Upper three blades usually 5-10 cm long and 7-14 mm wide, the upper one usually much reduced, the blades below the upper three usually not reduced; anthers mostly 0.6-0.8 mm long.....40. *P. sphaerocarpon*, p. 170.
Spikelets elliptic or obovoid.
Blades not cordate at the base, spikelets more than 1.7 mm long except in *P. columbianum*.
Spikelets mostly 2.8-2.9 mm long.....................35. *P. Deamii*, p. 169.
Spikelets 2-2.2 mm long..........................36. *P. Addisonii*, p. 169.
Spikelets mostly 1.8-1.9 mm long.................37. *P. tsugetorum*, p. 169.
Spikelets mostly 1.5-1.7 mm long................38. *P. columbianum*, p. 169.
Blades cordate at the base.

Culms and sheaths usually densely crisp-puberulent (sometimes sparsely so); blades generally less than 12 mm wide; spikelets 2.2-2.5 mm long..44. *P. Ashei*, p. 172.
Culms and sheaths generally nearly glabrous or only sparsely puberulent (not crisp-puberulent); blades or some of them usually more than 12 mm wide; spikelets 2.5-3 mm long, generally about 2.7 mm long..45. *P. commutatum*, p. 173.

1. DICHOTOMIFLÒRA

Annual plants with smooth culms; ligule membranous below, densely ciliate above; spikelets glabrous; fruit smooth and shining.

Spikelets 2-3.5 mm long, usually about 2.9 mm long (rarely a few as short as 2 mm), acute; plants usually large and spreading, 50-100 cm long...1. *P. dichotomiflorum.*
Spikelets 1.8-2.2 mm long, usually about 2 mm long, blunt; plants shorter and more slender than the preceding.............1a. *P. dichotomiflorum* var. *puritanorum.*

1. **Panicum dichotomiflòrum** Michx. FALL PANICUM. Map 273. This is an infrequent to frequent grass in all parts of the state, being much more frequent in the southern part. It prefers a wet or moist soil, and is found on the muddy shores of streams; in moist, open places in woodland, especially in old logging roads; and in moist places in stubblefields, cornfields, waste places, and roadside ditches.

Maine to Nebr., southw. to Fla. and Tex.

1a. **Panicum dichotomiflorum** var. **puritanòrum** Svenson. (Rhodora 22: 154-155. 1920.) My only specimen of this variety is from a dried-up pond about 3 miles southwest of Tefft, Jasper County, where it was closely associated with *Panicum spretum.* The specimen I reported from Kosciusko County I am now referring to *Panicum Gattingeri* Nash.

Mass., Conn., L. I., and Ind.

2. CAPILLÀRIA

Annuals; sheaths papillose-hispid; ligules membranous, 1-3 mm long, ciliate; panicles many-flowered, mostly diffuse; spikelets glabrous, pointed; first glume large, clasping; fruit smooth and shining.

Panicles drooping; spikelets 4.5-5 mm long. (See excluded species no. 80, p. 1030.)..*P. miliaceum.*
Panicles erect; spikelets not more than 4 mm long.
  Spikelets mostly 3-3.5 mm long; second glume and sterile lemma 7-9-nerved; pulvini glabrous..1. *P. flexile.*
  Spikelets mostly 1.8-2.9 mm long; second glume and sterile lemma 5-nerved.
    Pulvini of panicle hispid.
      Terminal panicles generally about half as long as the length of the whole plant (except when crowded by other vegetation, when the terminal panicle may be much shorter), usually large, about as wide as long, generally included at the base; blades thickly papillose-hispid above and beneath, 5-15 mm wide; spikelets mostly 2-2.5 mm long..........................2. *P. capillare.*
      Terminal panicles about a third the length of the entire plant, generally about half as wide as long, usually long-exserted; blades sparsely papillose-pubescent on both surfaces, 2-6 (8) mm wide; spikelets 1.7-2 (2.2) mm long..3. *P. philadelphicum.*

Panicum capillare L.

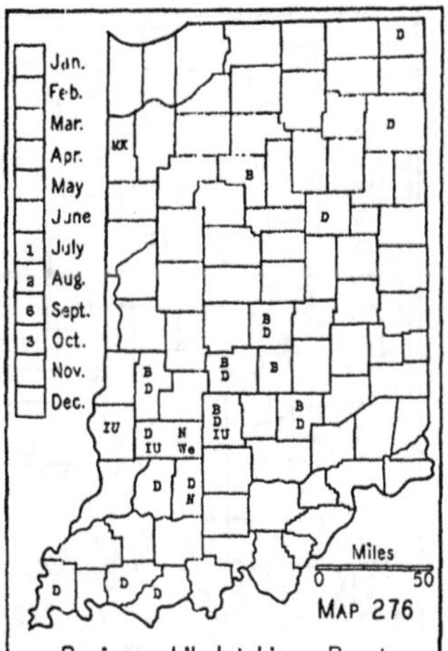

Panicum philadelphicum Bernh

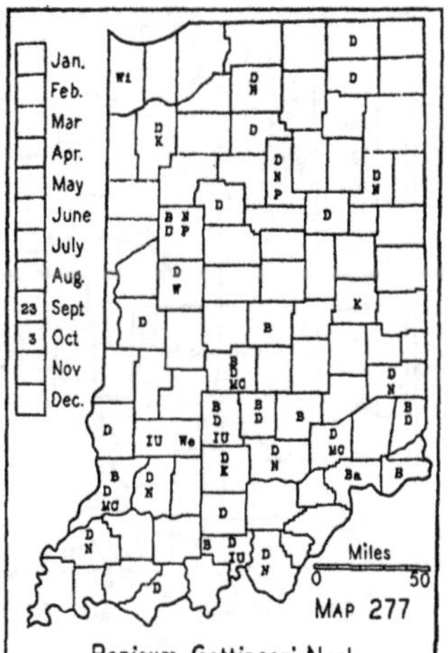

Panicum Gattingeri Nash

Pulvini of panicle glabrous (sometimes the lower ones pubescent).
Leaf blades mostly 6-10 mm wide; spikelets 2 (2.2) mm long; plants yellowish green, freely branching at the nodes.....................4. *P. Gattingeri.*
Leaf blades 2-6 mm wide (according to Hitchcock), 1-10 mm wide (according to Fernald). (Rhodora **21**: 112-114. 1919.) (See excluded species no. 82, p. 1030.)..*P. Tuckermani.*

1. **Panicum fléxile** (Gatt.) Scribn. Map 274. Infrequent in the northern and southern counties. In the north it is found in dry or moist, sandy soil, usually on the marly borders of lakes, and on interdunal flats. In the southern counties it is found in poor, dry soil in open places on the crests of ridges, on washed or rocky slopes, and in dry pastures.

N. Y., Que. to S. Dak., southw. to Fla. and Tex.

2. **Panicum capillàre** L. WITCHGRASS. Map 275. A pernicious weed in all parts of the state in all kinds of soils and in all kinds of habitats except in dense woodland. It shows great variation in size and form, depending upon how much it is crowded in growing. In dried-up ponds where it germinates late, mature plants may be only a few inches high.

Maine to Mont., southw. to Fla. and Tex.

3. **Panicum philadélphicum** Bernh. Map 276. A local to infrequent or frequent species found mostly in the southern half of the state. It is found in poor soil, probably slightly acid, generally in fallow fields and on washed slopes.

Conn. to Wis., southw. to Ga. and Tex.

4. **Panicum Gáttingeri** Nash. Map 277. Infrequent throughout the state. It is usually found in moist, sandy soil along streams, about ponds, in old logging roads, and along moist roadsides.

Panicum Tuckermani Fern. is a closely allied species which I am not able to separate from *Panicum Gattingeri*. Some of my specimens have been named for me as *Panicum Tuckermani*, but I am referring them

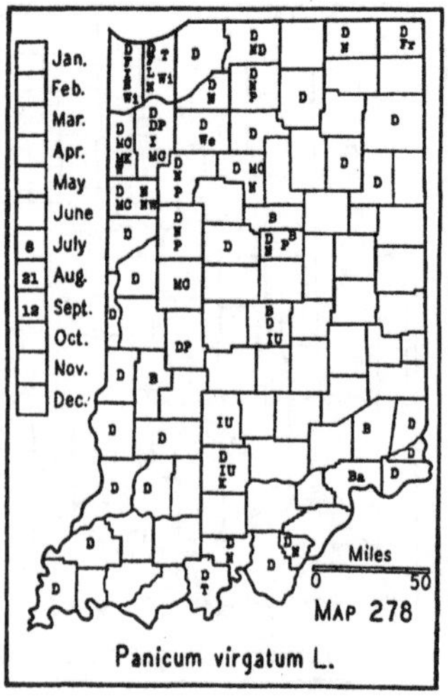

Panicum virgatum L.

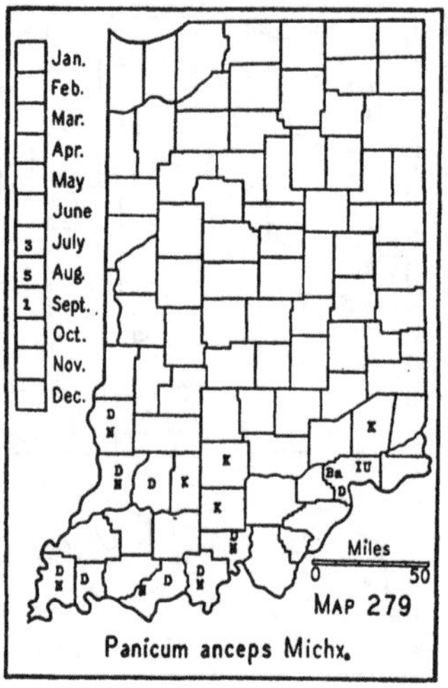

Panicum anceps Michx.

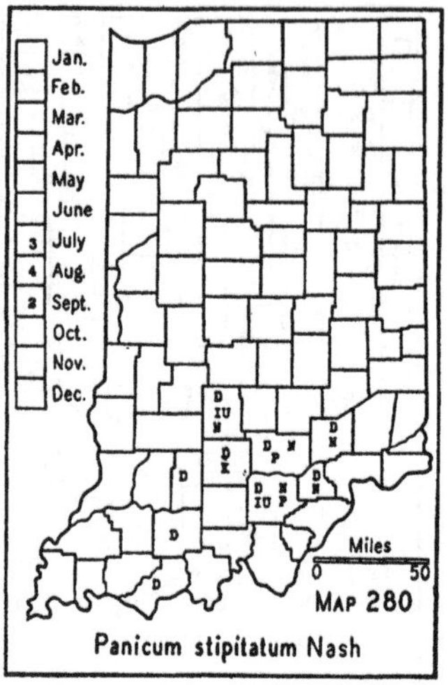

Panicum stipitatum Nash

to *Panicum Gattingeri* until satisfactory characters are found to separate them.

N. Y., Ont. to Minn., southw. to N. C. and Tenn.

3. VIRGÀTA

Perennials from stout rootstocks; spikelets gaping at the apex, owing to the well developed staminate floret and its palea in addition to the perfect one; species mostly maritime, only one in Indiana.

6. **Panicum virgàtum** L. (Linder. Some varieties of Panicum virgatum. Rhodora 24: 11-16. 1922.) SWITCHGRASS. Map 278. This species is found as a native in almost all the counties in the state and is now introduced in sand ballast along railroads in many counties. It is not a native of Wells County but I have found it along railroads in three widely separated places in the county. It prefers the open and a sandy soil. Where it is found, it is generally common over the extent of its habitat. It is found in sandy prairies, "oak openings," on gravelly banks of lakes and streams, and along the Ohio River it often grows among the cobblestones of boat landings and in the seams of outcrops of shale.

Maine, Que. to Mont., southw. to Fla., Nev., and Ariz.; Mex. and Cent. Amer.

4. AGROSTOÌDIA

Tufted perennials; culms erect, compressed; sheaths keeled; ligules membranous, 0.5-1 mm long; spikelets short-pediceled, lanceolate, pointed, glabrous, 5-7-nerved; fruit smooth and shining, with a minute tuft of stout hairs at the apex.

Rootstocks present; blades pilose above toward the base; spikelets 3-3.8 mm long..7. *P. anceps.*

Rootstocks lacking; blades not pilose above toward the base; spikelets less than 3 mm long.

Spikelets 2.4-2.8 mm long, conspicuously secund; fruit with a basal stalk 0.2-0.4 mm long...8. *P. stipitatum.*

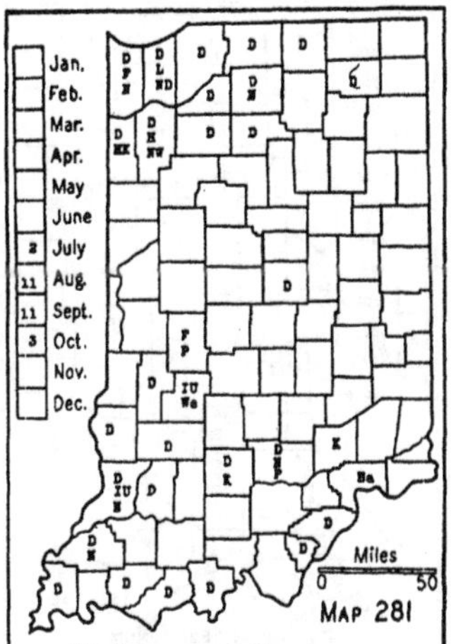

Pánicum agrostoides Spreng.

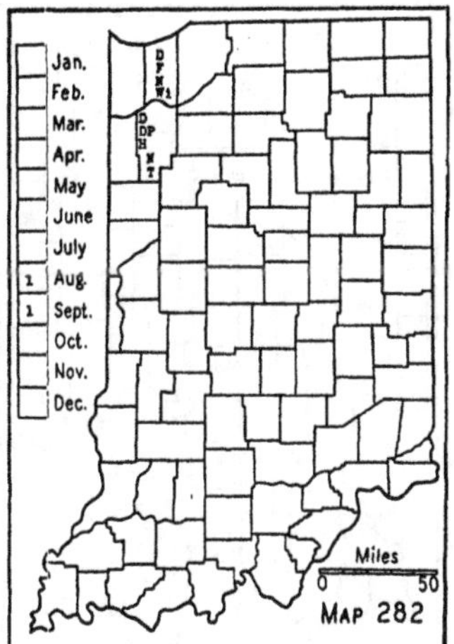

Panicum verrucosum Muhl.

Panicum depauperatum Muhl.

Spikelets 1.8-2.3 mm long, not conspicuously secund; fruit without a stalk at the base (if stalked, the stalk less than 0.2 mm long)..................9. *P. agrostoides.*

7. **Panicum ánceps** Michx. Map 279. This species is restricted to the southern part of the state where it is infrequent and found in woodland in open, wet places about ponds, swamps, and sloughs and in roadside ditches.

N. J. to Kans., southw. to Fla. and Tex.

8. **Panicum stipitàtum** Nash. Map 280. An infrequent grass in a few counties of southern Indiana. It is usually found in hard, white clay soil in wet places in swamps, clearings, fallow fields, and ditches. It is frequently associated with *Panicum agrostoides* with which it is often confused.

Conn. to Mo., southw. to Ga. and Tex.

9. **Panicum agrostòides** Spreng. Map 281. Infrequent to frequent in the sandy areas of the northwestern part of the state; more frequent in the southwestern part, where it usually grows in large clumps in hard, white clay soil in dried-up swamps and on the borders of streams, lakes, ditches, sloughs, and old canals. In our northern counties it grows in wet, sandy, or muddy soil.

Maine to Kans., southw. to Fla. and Tex.; Vancouver Island and Calif.

5. VERRUCÒSA

Annuals, glabrous; culms weak, divaricately branching, decumbent at the base; ligule ciliate; panicles divaricate, the branches capillary, spikelet-bearing toward the ends.

10. **Panicum verrucòsum** Muhl. Map 282. This species is very local and is found in wet or moist, sandy soil about sloughs near Lake Michigan and in marshes and roadside ditches in sec. 12 of Jasper County about 3 miles southeast of Tefft. We have specimens from only Jasper and

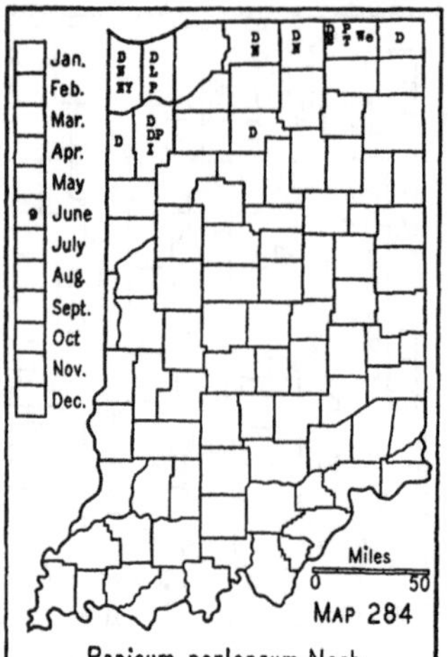

Panicum perlongum Nash

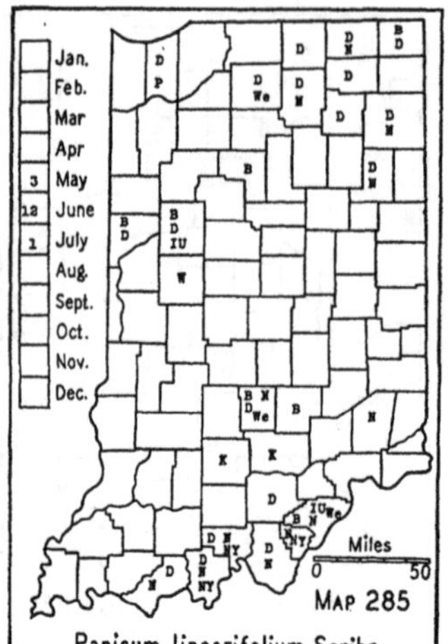

Panicum linearifolium Scribn.

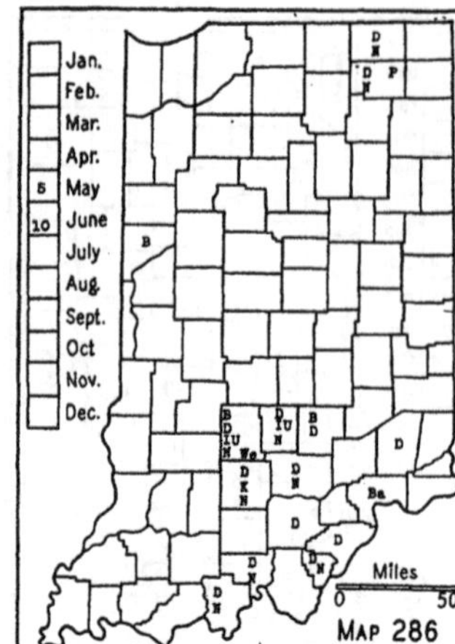

Panicum linearifolium var. Werneri (Scribn.) Fern.

Porter Counties although it has been reported from Lake County where it probably occurs or was once found. The mass distribution of this species is along the Coastal Plain.

Mass. to Fla., westw. to Mich., Tenn., and Tex.

6. DEPAUPERÀTA

Culms simple, the vernal ones generally 15-35 cm high, the nodes ascending-pilose; ligule a band of hairs of irregular length up to 1 mm long; blades long-linear; spikelets 2.2-4 mm long, somewhat shrunken at the base; palea of sterile floret usually half to two thirds as long as the fruit; fruit smooth, glossy, the lemma strongly indurated. The panicles of the autumnal phase are borne on short branches from the lower nodes.

Spikelets beaked, mostly 3.2-3.8 mm long (rarely as short as 3 mm or as long as 4 mm).
- Sheaths pilose..11 *P. depauperatum.*
- Sheaths glabrous or nearly so..............11a. *P. depauperatum* var. *psilophyllum.*

Spikelets not beaked, 3 mm long or less (rarely 3.2 mm long).
- Spikelets 2.7-3.2 mm long; panicles narrow, usually less than a third as wide as long; ligules mostly about 1 mm long......................12. *P. perlongum.*
- Spikelets 2.2-2.7 mm long; panicles usually more than a third as wide as long; ligules mostly less than 1 mm long.
  - Sheaths pilose..13. *P. linearifolium.*
  - Sheaths glabrous or nearly so..............13a. *P. linearifolium* var. *Werneri.*

11. **Panicum depauperàtum** Muhl. Map 283. Infrequent in southern Indiana in open woodland on the crests of black oak, black and white oak, and chestnut oak ridges. In the northern part of the state it is local except in the dune area, where it is frequent in very sandy soil on open, wooded dunes or on sandy knolls and ridges.

N. S., Que. to Minn., southw. to Ga. and Tex.

11a. **Panicum depauperatum** var. **psilophýllum** Fern. (Rhodora 23: 193-194. 1921.) This northern variety has the habitat of the species and is found only in sandy areas of the northern part of the state.

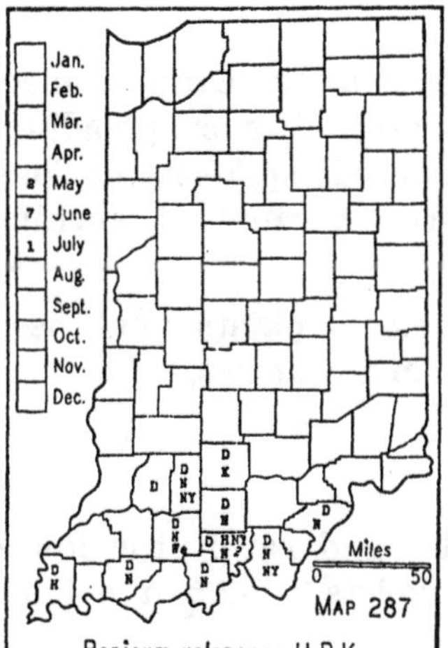

Panicum xalapense H.B.K.

Panicum Bicknellii Nash

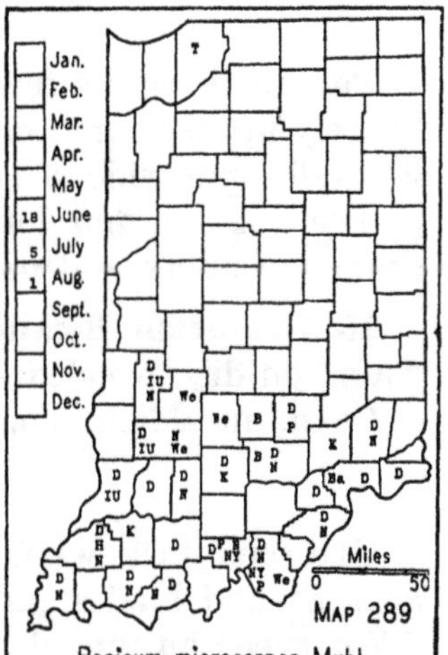

Panicum microcarpon Muhl.

12. **Panicum perlóngum** Nash. Map 284. This is an infrequent species in the sand areas of the northern part of the state. It is found in very dry soil on the crests of open dunes and on sandy knolls and ridges, sometimes in dry, sandy prairies.

Ind. to Man. and N. Dak., southw. to Colo. and Tex.

13. **Panicum linearifòlium** Scribn. Map 285. Infrequent in the unglaciated area of the southern part of the state and in sandy habitats of the lake area. In the south it is found in open woodland on the crests of ridges, and in the lake area it is found in dry, sandy soil on open dunes, sandy knolls, and sandy ridges.

Que., Maine, and Mich., southw. to Ga. and Tex.

13a. **Panicum linearifolium** var. **Wérneri** (Scribn.) Fern. (Rhodora 23: 194. 1921.) (*Panicum Werneri* Scribn.) Map 286. This variety is found with the species but is less frequent, especially in the northern part of the state.

Que., Maine to Minn., southw. to Va., Ky., and Tex.

7. LAXIFLÒRA

Vernal culms 15-50 cm high, tufted, erect to spreading; foliage aggregated toward the base, not in distinct rosettes in autumn; blades pilose on one or both surfaces or nearly glabrous, usually short-ciliate; ligules nearly obsolete; panicles sometimes reduced and exceeded by the leaves; spikelets pilose, 1.8-2 mm long.

14. **Panicum xalapénse** HBK. (*Panicum laxiflorum* of Britton and Brown, Illus. Flora, ed. 2, not Lam.) Map 287. An infrequent species in the area shown on the map. It is usually found on wooded slopes, most often at their bases.

Md. to Ill. and Mo., southw. to Fla., Tex., Mex., and Guatemala; also in Santo Domingo.

8. BICKNELLIÀNA

Perennial; culms few to several in a tuft; ligules usually nearly obsolete (rarely up to 1 mm long); blades elongated, stiffly ascending or spreading; 3-8 (10) mm wide, 7-15 cm long; panicles few-flowered; spikelets on long pedicels, 2.3-3 mm long, 7-nerved; autumnal form sparingly branching from the upper and middle nodes.

15. **Panicum Bicknéllii** Nash. Map 288. Occasional plants have been found on dry, wooded slopes in a few of the southern counties.

Conn. and Mich., southw. to Ga. and Mo.

9. DICHÓTOMA

Glabrous as a whole or nearly so, or the nodes and rarely the lower sheaths and blades pubescent; ligule minute; spikelets 1.5-2.5 mm long, 5-7-nerved; autumnal phase freely branching.

Nodes bearded (at least the lower ones).
  Spikelets 1.5-1.6 mm long....................................16. *P. microcarpon.*
  Spikelets more than 1.6 mm long.
    Spikelets 2 (2.2) mm long; blades rarely more than 8 mm wide.................. ....17. *P. dichotomum.*
    Spikelets 2.3-2.7 mm long; blades 8-12 mm wide............18. *P. mattamuskeetense.*
Nodes not bearded (glabrous or puberulent, rarely with a few long hairs).
  Spikelets pubescent.
    Culms erect, never trailing.
      Nodes glabrous (rarely a few with hairs); margins of upper sheaths glabrous; blades mostly 6-14 mm wide; spikelets 2-2.2 mm long.........19. *P. boreale.*
      Nodes puberulent or somewhat bearded; margins of upper sheaths pubescent (rarely entirely glabrous).
        Blades 3-8 mm wide; spikelets 2 (2.2) mm long..........17. *P. dichotomum.*
        Blades mostly 8-12 mm wide; spikelets 2.3-2.7 mm long.................... ....18. *P. mattamuskeetense.*
    Culms weak, soon becoming decumbent and trailing..............20. *P. lucidum.*
  Spikelets glabrous.
    Sheaths, or some of them, usually marked more or less with white spots, the margins glabrous; spikelets more than 2.2 mm long (mostly 2.3-2.5 mm long)..21. *P. yadkinense.*
    Sheaths not marked with white spots, the margins pubescent; spikelets mostly 2-2.1 mm long.
      Plants of dry ground; culms erect (rarely autumnal plants reclining)........ ....17. *P. dichotomum.*
      Plants of bogs and swamps; culms weak, soon becoming decumbent and trailing. ....20. *P. lucidum.*

16. **Panicum microcárpon** Muhl. Map 289. Rather frequent in the southern third of the state. It seems to prefer a slightly acid soil and is usually found in low, flat woods with sweet gum, pin oak, and beech, although it is sometimes found in drier situations with different associates. The Tryon specimen from La Porte County lacks the white spots on the sheaths.

The report of this species from Marshall County is evidently an error in determination; its habitat is not in that area, and the detailed description given by the collector does not apply to this species.

Mass. to Ill., southw. to Fla. and e. Tex.

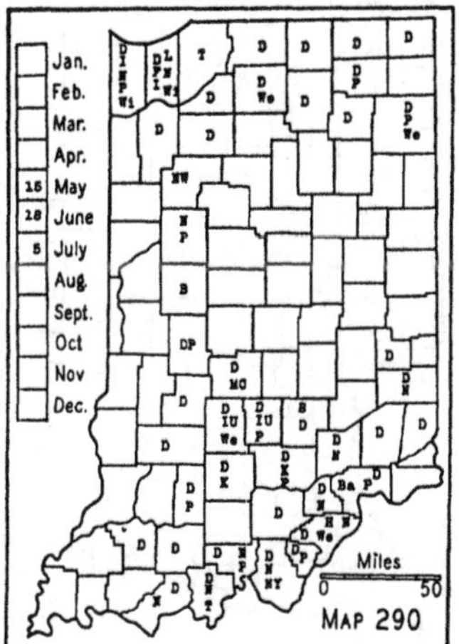

Panicum dichotomum L.

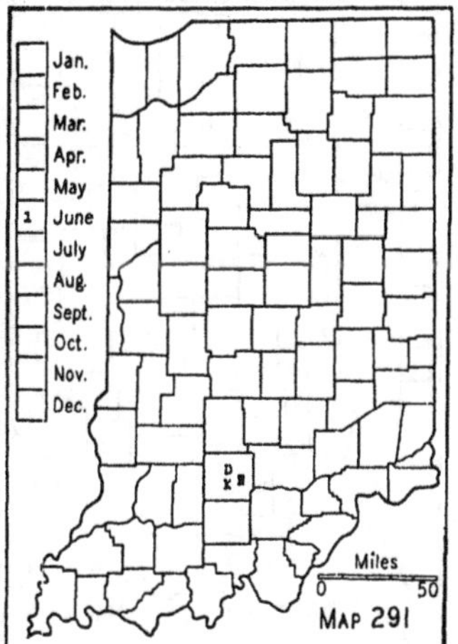

Panicum mattamuskeetense Ashe

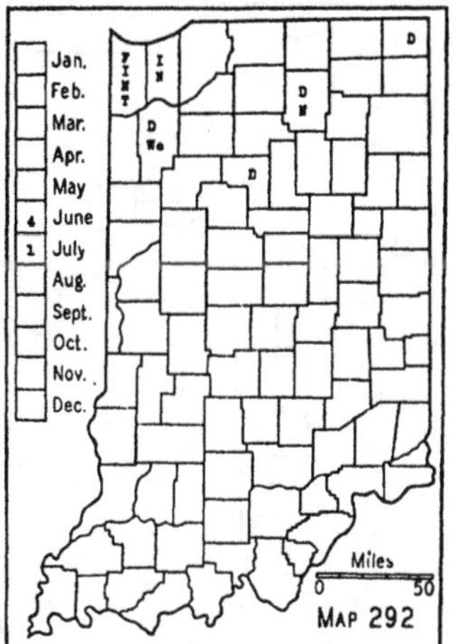

Panicum boreale Nash

17. **Panicum dichótomum** L. (Including *Panicum barbulatum* Michx.) Map 290. Frequent in the northern and southern counties. It is usually found in open places on the crests and slopes of black and white oak woods and less frequently in beech and sugar maple woods. It prefers a poor soil and is sometimes found in the dunes growing in almost pure sand.

Some authors separate from this species, under the name of *Panicum barbulatum* Michx., plants with broad leaves and pubescent nodes. In Indiana the two forms intergrade so that I cannot make a satisfactory division of them.

N. B. to Ill., southw. to Fla. and e. Tex.

18. **Panicum mattamuskeeténse** Ashe. Map 291. Our only specimens were found in 1935 by Ralph M. Kriebel in the northeastern corner of section 16 of Pleasant Run Township, Lawrence County. They were found in a shallow drainage ditch near Little Salt Creek bridge between Heltonville and Bartlettsville where they were associated with *Panicum clandestinum.* The determination was made by Agnes Chase. Since this was written Kriebel found another colony near Huron, about 20 miles distant.

N. Y., along the coast to N. C., and in Ind.

19. **Panicum boreàle** Nash. Map 292. A rare grass of marshes in the lake area. It is also occasionally found in the mucky borders of ponds and lakes.

Newf. to Minn., southw. to N. J. and Ind.

20. **Panicum lùcidum** Ashe. Map 293. Our Indiana record is based upon Umbach's specimen no. 4962 collected at Dune Park, Porter County, which is deposited in the U. S. National Herbarium. Pepoon reports it also from the same area. It is an inhabitant of wet woods and sphagnum marshes.

Coastal Plain, Mass. to Fla., Ark., and Tex.; also Ind. and Mich.

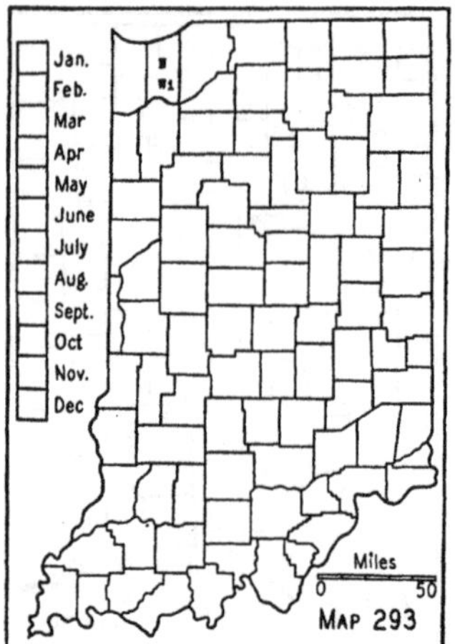

Panicum lucidum Ashe

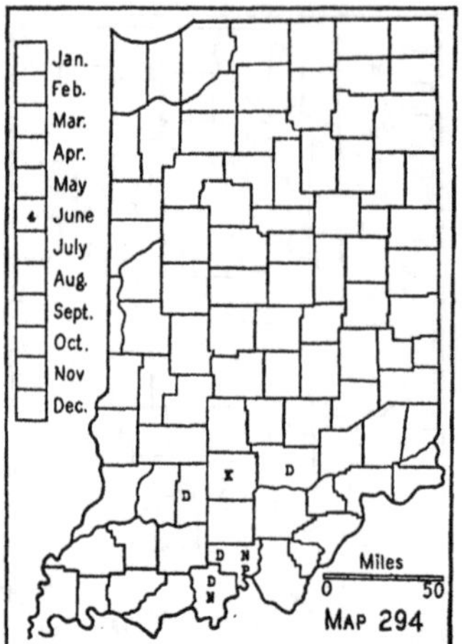

Panicum yadkinense Ashe

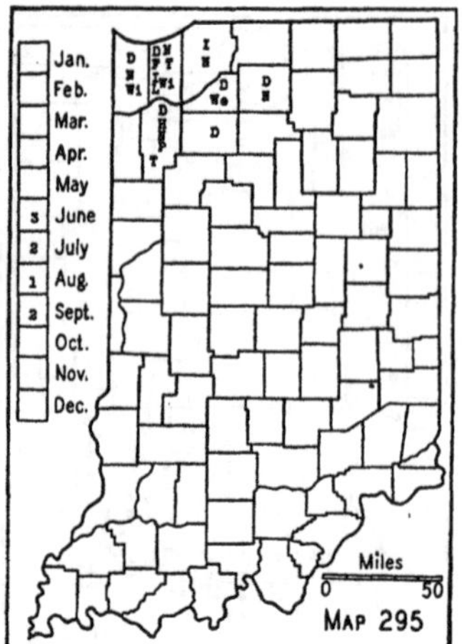

Panicum spretum Schultes

21. **Panicum yadkinénse** Ashe. Map 294. Infrequent in a few southern counties on the slopes and bases of wooded, usually high hills.

Pa. to Ill., southw. to Ga. and La.

10. SPRÈTA

Culms tufted, rather stiff, glabrous or rarely the lower internodes and sheaths ascending-pubescent; ligules mostly 2-5 mm long; blades not over 8 mm wide; spikelets pubescent, rarely glabrous; second glume and sterile lemma 5-7-nerved; autumnal form with more or less tufted branchlets, reduced blades and panicles.

Panicles narrow, a fourth to a third as wide as long (somewhat wider in anthesis); spikelets elliptic.......................................22. *P. spretum.*
Panicles open, at least two thirds as wide as long; spikelets obovate.......................................23. *P. Lindheimeri.*

22. **Panicum sprètum** Schultes. Map 295. In moist, sandy soil in open places and on the borders of marshes that do not yet have a sod of other grasses. Local but usually frequent where found.

Coastal Plain, N. S. to Tex.; Ind.

23. **Panicum Lindheìmeri** Nash. (*Panicum lanuginosum* var. *Lindheimeri* (Nash) Fern.) Map 296. This species is probably somewhat restricted to the lake area and to the hilly areas of the southern part of the state. It is usually found in dry, sandy soil in open woodland and open, dry places, or in moister situations at the bases of sandy slopes, and rarely in dry, sandy, clay soil.

Que., Maine to Minn., southw. to Fla. and N. Mex.; Calif.

11. LANUGINÒSA

Plants more or less pubescent throughout; ligules densely hairy, 2-5 mm long; blades not over 10 mm wide; spikelets 1.3-2.4 mm long, spreading-

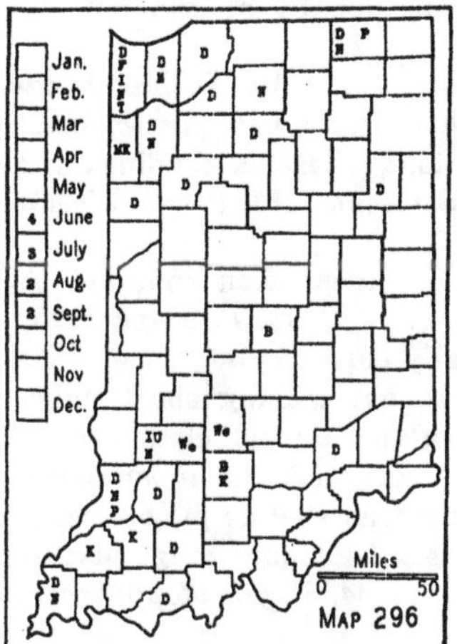

Panicum Lindheimeri Nash

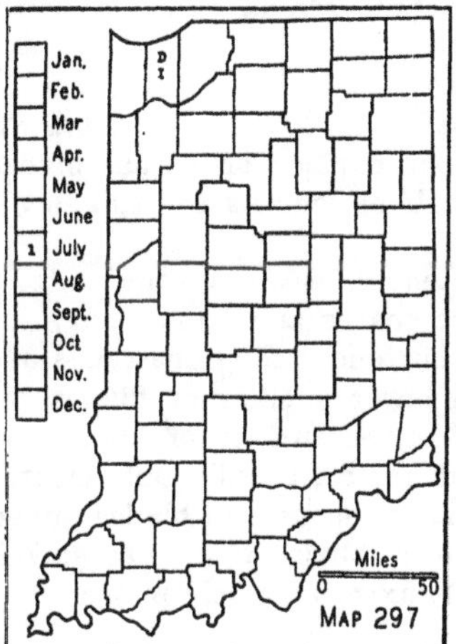

Panicum auburne Ashe

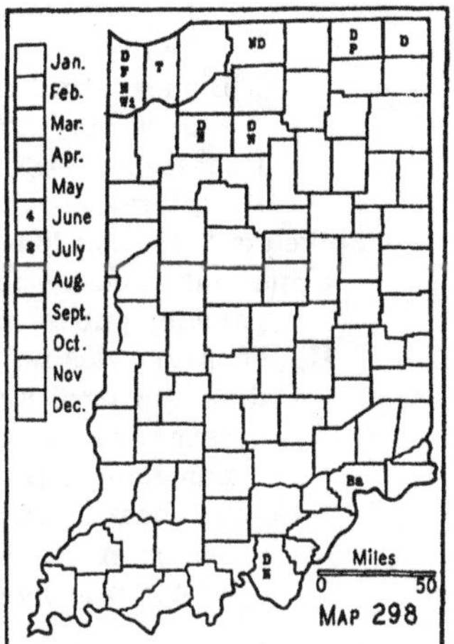

Panicum praecocius Hitchc. & Chase

pubescent; second glume and sterile lemma 5-7-nerved or 7-9-nerved on large spikelets.

Plants grayish, velvety-pubescent; spikelets 1.3-1.4 mm long..........24. *P. auburne.*
Plants pubescent, often villous but not velvety.
 Culms conspicuously pilose with long, horizontal hairs 4-5 mm long, and branching before the expansion of the primary panicles; spikelets mostly 1.8-1.9 mm long. ....25. *P. praecocius.*
 Culms variously pubescent, if pilose the hairs appressed or some widely spreading, less than 4 mm long; culms not branching before the expansion of the primary panicles.
  Spikelets less than 2 mm long.
   Vernal blades glabrous or nearly so above (6-10 cm long and 5-10 mm wide); spikelets 1.5-1.7 mm long............................26. *P. tennesseense.*
   Vernal blades pubescent above or if glabrous smaller than the preceding, sometimes pilose near the base and margins only.
    Spikelets 1.3-1.5 mm long.
     Upper surface of blades puberulent as well as long-villous.............. ....27. *P. albemarlense.*
     Upper surface of blades villous but lacking the short, appressed pubescence; vernal plants usually purplish with erect leaves, autumnal plants usually greenish; nodes with short hairs, if bearded.
      Sheaths papillose-pilose with no short, appressed pubescence in the spaces between the nerves; lower surface of blades with a subappressed, papillose pubescence and lacking a short, appressed pubescence; axis of panicle generally pilose, the lowest panicle-branches spreading and tangled..................28. *P. implicatum.*
      Sheaths more or less softly papillose-pilose, some or all of them with a short, appressed pubescence on the spaces between the nerves; blades erect; lower surface of the blades more or less short appressed-puberulent, in addition to a longer pubescence; axis of panicle generally puberulent, the lowest panicle-branches ascending and not tangled..............................29. *P. meridionale.*
    Spikelets 1.6-1.9 mm long; plants green, rarely purplish; nodes mostly bearded, usually with long, spreading hairs.
     Pubescence on upper surface of vernal blades short-pilose, rarely long-pilose, appressed at least on the apical half (rarely not appressed); nodes usually densely pilose with spreading hairs; spikelets rarely less than

1.6 mm long; first glume about a third the length of the spikelet, blunt, subacute.
Blades stiff, erect....................................30. *P. huachucae.*
Blades lax, spreading...............30a. *P. huachucae* var. *fasciculatum.*
Pubescence on upper surface of vernal blades long-pilose, ascending; first glume about half as long as the spikelet, acuminate..31. *P. subvillosum.*
Spikelets 2-2.5 mm long.
Upper internodes shortened; leaves approximate, the blades often equaling the panicle; pubescence sparse and stiff....................32. *P. scoparioides.*
Upper internodes not shortened; the pubescence usually copious and rather silky.
Culms, sheaths, and lower surface of blades pilose but lacking short pubescence; center of upper surface of blades not glabrous; spikelets about 2 mm long; axis of panicle usually pubescent..........33. *P. villosissimum.*
Culms, sheaths, and lower surface of blades puberulent as well as pilose; center of upper surface of blades glabrous; spikelets 2.1-2.5 mm long (usually 2.2-2.4 mm long); axis of panicle pilose...........34. *P. pseudopubescens.*

24. **Panicum aubúrne** Ashe. Map 297. Our only specimen is one collected by Hill, July 8, 1913, in dry sand by a woods road at Dune Park, Porter County. It is Hill's no. 7 and is deposited in the herbarium of the University of Illinois. I have a duplicate of this number.

Coastal Plain, Mass. to n. Fla. and La.; Ark. and Ind.

25. **Panicum praecòcius** Hitchc. & Chase. Map 298. This species is rare in the sands of the northern counties. I have, also, a specimen which was found in Harrison County, about 3 miles east of Elizabeth on a rocky wooded slope along the road leading from Elizabeth to Stuart's Landing on the Ohio River. This rocky slope is rich in rare Indiana plants such as *Eragrostis capillaris.*

Mich. to Minn., southw. to Mo. and e. Tex.

26. **Panicum tennesseénse** Ashe. (*Panicum languinosum* var. *septentrionale* Fern.) Map 299. This is an infrequent grass found throughout the state in various habitats. My specimens are from dry sands, moist sand on the marly shore of a lake, wooded slopes, and hard, white clay soil in a fallow field in the Wabash Bottoms.

Maine, Que. to Minn., southw. to Ga. and Tex.; westw. to Utah and Calif.

27. **Panicum albemarlénse** Ashe. (*Panicum meridionale* var. *albemarlense* (Ashe) Fern.) Map 300. I have only one specimen of this grass from Indiana and it is in the autumnal phase. I am not able to make a satisfactory study of this species from the few specimens at hand. Some authors refer it to a form of *Panicum meridionale,* to which it may belong. It is found in sandy soils.

Coastal Plain, Mass. to N. C.; n. Mich., Wis., Ind. to Tenn.

28. **Panicum implicàtum** Scribn. (*Panicum lanuginosum* var. *implicatum* (Scribn.) Fern.) Map 301. Local to infrequent but common in its habitat. It is generally found in moist, sandy soil on the marly borders of lakes, in interdunal flats, and rarely in dry, sandy soil.

I think this grass is restricted to the lake area and that all reports of it from south of this area should be referred to some other species, most

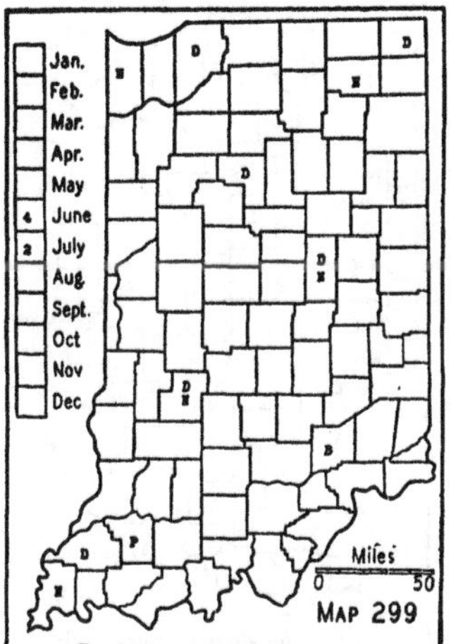

Panicum tennesseense Ashe

Panicum albemarlense Ashe

Panicum implicatum Scribn

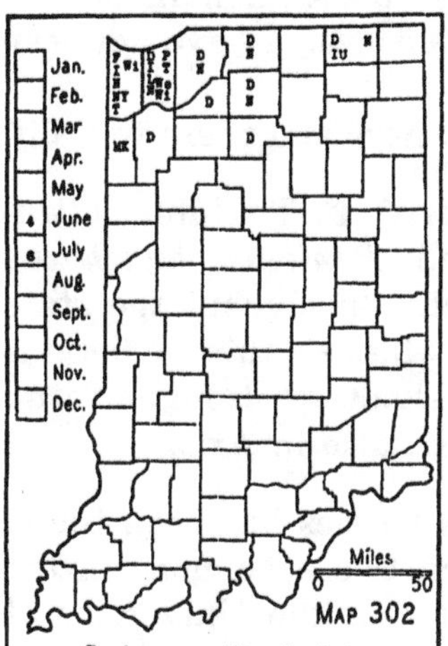

Panicum meridionale Ashe

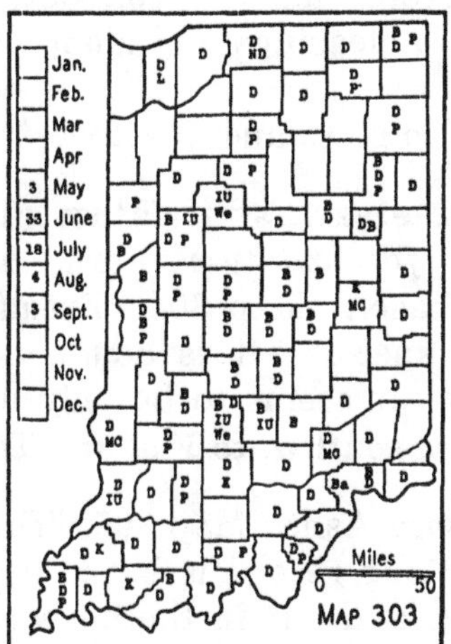

Panicum huachucae Ashe

Panicum huachucae var. fasciculatum (Torr.) F. T. Hubb.

probably to *Panicum huachucae*. This *Panicum* is difficult to separate from *Panicum huachucae*, but usually the length of the spikelet and the color of the whole plant are sufficient to distinguish them.

Newf. to Wis., southw. to Del. and Mo.

29. **Panicum meridionàle** Ashe. Map 302. Infrequent in the lake area, probably rather local. It is found in moist soil on the borders of marshes, in interdunal flats, and on the bases of wooded slopes where there are open spaces not sodded over with grasses and sedges. This plant usually can be distinguished easily from the preceding and the following species by the puberulence in the channels between the nerves of the sheaths and sometimes of the culms, and the puberulent panicle.

N. S. to Wis., southw. to Ala.

30. **Panicum huachùcae** Ashe. Map 303. This is a frequent to common species of dry ground throughout the state. It is found in open places in

MAP 305
Panicum subvillosum Ashe

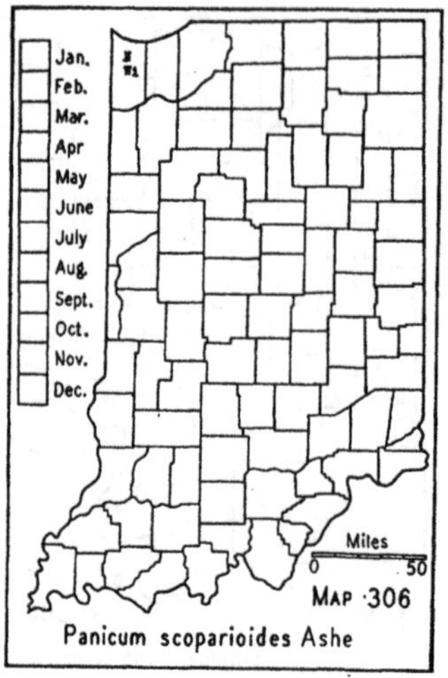

MAP 306
Panicum scoparioides Ashe

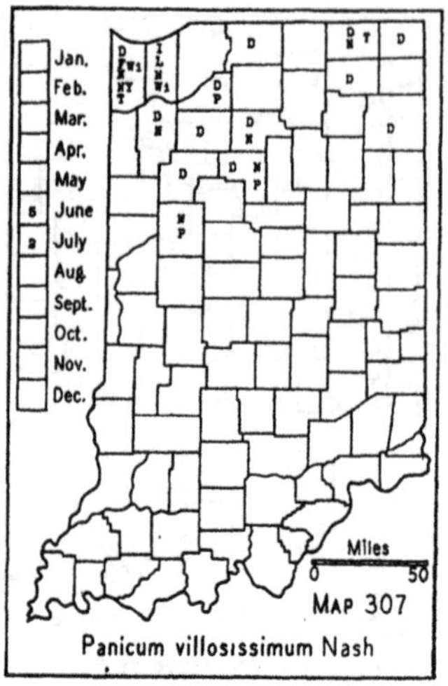

MAP 307
Panicum villosissimum Nash

all kinds of woodland, preferring dry soil but often common in bottom lands along streams and in clearings and along roadsides. I have not seen it in wet places.

N. S. to Mont., southw. to N. C. and Tex.; westw. here and there to Calif.

30a. **Panicum huachucae** var. **fasciculàtum** (Torr.) F. T. Hubb. (*Panicum lanuginosum* var. *fasciculatum* Fern. and *Panicum huachucae* var. *silvicola* Hitch. & Chase.) Map 304. Frequent throughout the state and associated with the species. It is doubtful whether this variety is distinct from the species. It seems to be only a shade or drought form.

Que. to Minn. and Nebr., southw. to Fla. and Tex.; also in Ariz.

31. **Panicum subvillòsum** Ashe. Map 305. This species has been found only in Lake County and our record is based upon two specimens in the U. S. National Herbarium and one in the herbarium of the University of Wisconsin.

N. S. to Minn., southw. to N. Y., Ind., and Mo.

32. **Panicum scoparioìdes** Ashe. (*Panicum villosissimum* var. *scoparioides* (Ashe) Fern.) Map 306. Known only from Lake County. Our record in based upon a specimen in the U. S. National Herbarium, collected by Umbach near Gary, June 29, 1909. A duplicate specimen is in the herbarium of the University of Wisconsin.

Vt. to Del.; Mich. and Ind. to Minn. and Iowa.

33. **Panicum villosíssimum** Nash. Map 307. Local probably throughout the lake area. It is found in open places in dry, sandy or gravelly soil, usually on black and white oak ridges and in the dunes.

Mass. to Minn., southw. to Fla. and Tex.; also in Guatemala.

34. **Panicum pseudopubéscens** Nash. (*Panicum villosissimum* var. *pseudopubescens* (Nash) Fern.) Map 308. As now known, this species is restricted to the northwestern counties. Further study will doubtless

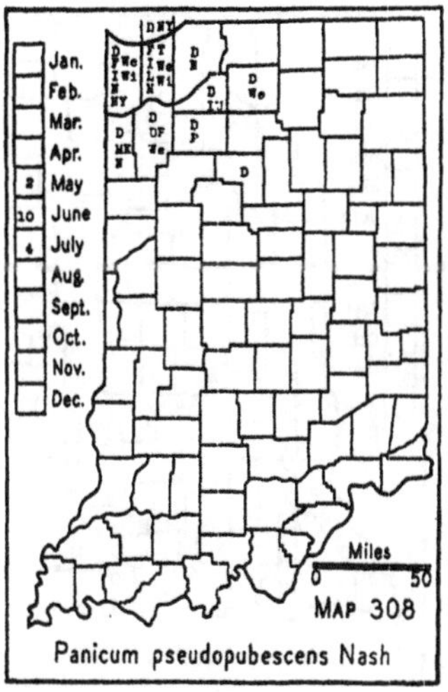

Panicum pseudopubescens Nash

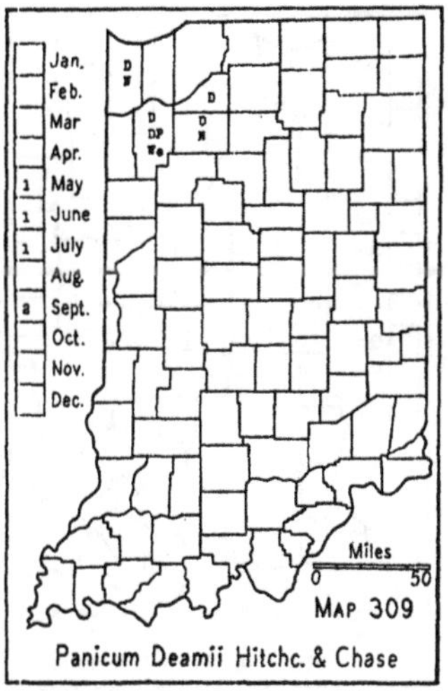

Panicum Deamii Hitchc. & Chase

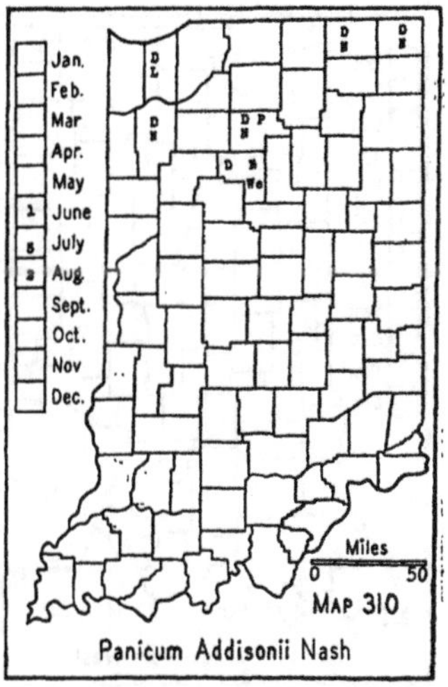

Panicum Addisonii Nash

extend its range to a few adjoining counties. It grows in very dry, sandy soil in the open on knolls, dunes, and ridges, where it is usually associated with black and white oak.

Conn. to Wis., southw. to Fla., Miss., Mo., and Kans.

12. COLUMBIÀNA

Culms tufted, stiff, crisp-puberulent to appressed-pubescent; ligules usually less than 1 mm long, rarely longer; blades firm; spikelets pubescent; branches and blades of the autumnal phase appressed or ascending.

Spikelets 2-2.9 mm long; sheaths usually copiously pilose, short hairs few or lacking.
Spikelets mostly 2.8-2.9 mm long; vernal blades 7-15 cm long.........35. *P. Deamii.*
Spikelets mostly 2-2.2 mm long; vernal blades usually all less than 8 cm long...36. *P. Addisonii.*
Spikelets 1.5-1.9 mm long; sheaths sparingly pilose but densely pubescent with short, appressed hairs.
Spikelets 1.8-1.9 mm long.................................37. *P. tsugetorum.*
Spikelets 1.5-1.7 mm long.................................38. *P. columbianum.*

35. **Panicum Dèamii** Hitchc. & Chase. Map 309. Local in a few of the northwestern counties, where it is found on open, wooded dunes and sandy knolls.

Ind. and Iowa.

36. **Panicum Addisònii** Nash. Map 310. Local in our northern counties, where it is found in dry sand on open, wooded dunes and sandy knolls.

Coastal Plain, Mass. to S. C.; Ind.

37. **Panicum tsugetòrum** Nash. Map 311. This is another *Panicum* which is restricted to the northern part of the state and is found in dry, sandy or gravelly soils on wooded slopes and dunes. It is included by some authors with *Panicum columbianum* Scribn.

Maine to Wis., southw. to Ga. and Tenn.

38. **Panicum columbiànum** Scribn. Map 312. My only specimens are from the H. H. Peele woods about a mile and a half southwest of Knox,

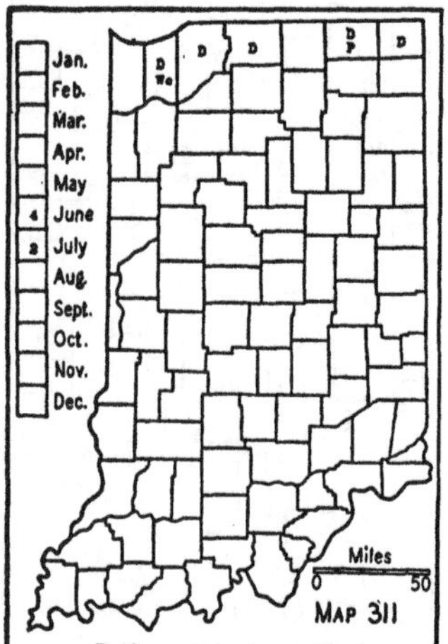

MAP 311
Panicum tsugetorum Nash

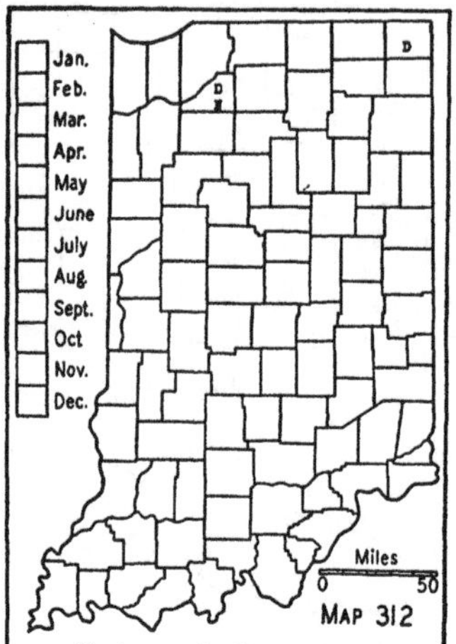

MAP 312
Panicum columbianum Scribn.

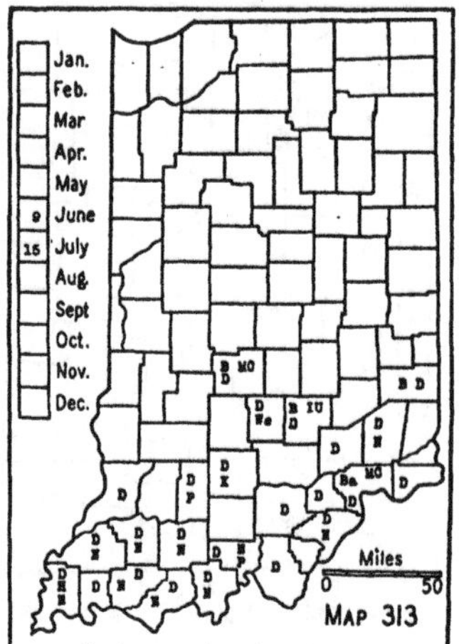

MAP 313
Panicum polyanthes Schultes

Starke County. They were found in dry, sandy soil in a flat, black and white oak woods where they were closely associated with *Panicum Deamii*.

In 1938 I found it in Steuben County.

Maine to N. C.; Ind.

13. SPHAEROCÁRPA

Culms glabrous; ligule obsolete or nearly so; blades cordate and ciliate at the base; spikelets obovoid-spherical at maturity; second glume and sterile lemma 5-7-nerved; autumnal form remaining simple or but sparingly branching; the thick, white-margined blades of the winter rosette conspicuous.

Upper three blades usually 10-20 cm long and 10-25 mm wide, the upper blade usually not much smaller than the other two, the blades below the three usually much smaller than the upper three; anthers mostly 0.4-0.5 mm long...39. *P. polyanthes.*

Upper three blades usually 5-10 cm long and 7-14 mm wide, the upper one usually much reduced, the blades below the three upper ones usually not reduced; anthers mostly 0.6-0.8 mm long....................................40. *P. sphaerocarpon.*

39. **Panicum polyánthes** Schultes. Map 313. This species is restricted to the southern half of the state and is rather frequent in the counties along the Ohio River. It prefers a slightly acid soil and is found in dry soil associated with black oak, and in moist soil associated with sweet gum. It is also found sparingly in fallow fields.

Conn., Ind. to Okla., southw. to Ga. and Tex.

40. **Panicum sphaerocárpon** Ell. Map 314. This species is infrequent in the lake area and reappears in the unglaciated area where it is rather local. In the lake area it is found in very dry, sandy or gravelly places and in the southern part of the state on black oak and black and white oak ridges.

This species much resembles the preceding from which it may easily be separated by its larger anthers and usually much reduced upper leaf. It also much resembles *Panicum microcarpon* which has the nodes of the

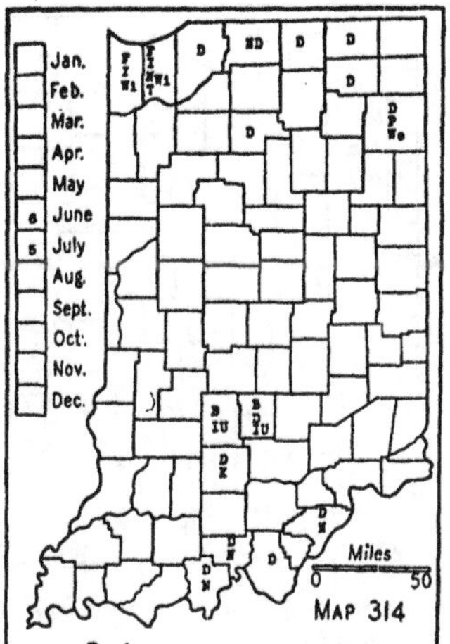

Panicum spaerocarpon Ell.

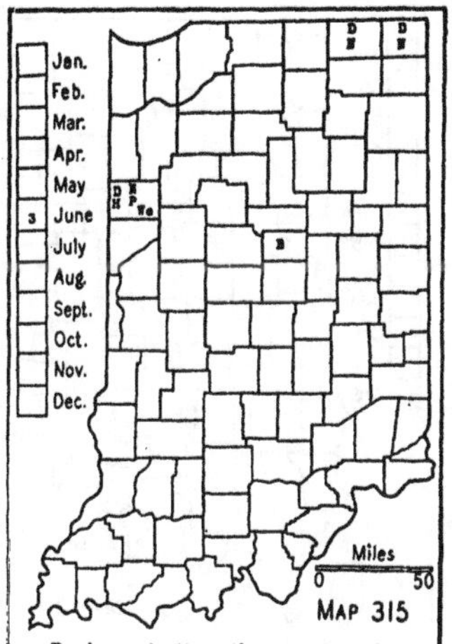

Panicum Leibergii (Vasey) Scribn.

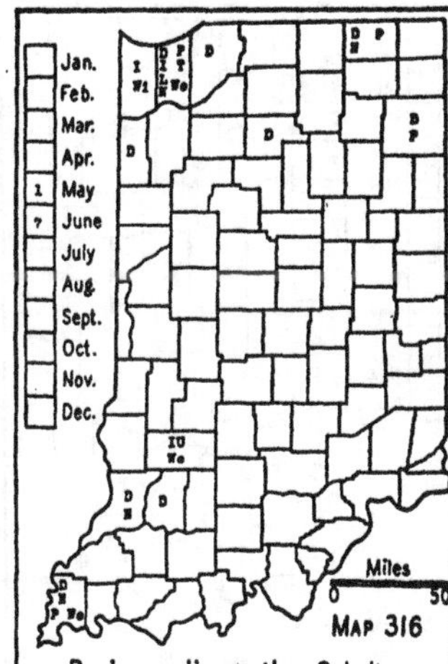

Panicum oligosanthes Schultes

culms bearded, sheaths with conspicuous white marks, and very short ligules.

Vt., Wis. to Kans., southw. to n. Fla. and Tex.; Mex. and Venezuela.

14. OLIGOSÁNTHIA

Culms rather stout; spikelets obovate, 3-4 mm long, usually papillose-hirsute, strongly 7-9-nerved; autumnal phase with the branches more or less crowded toward the summit.

Ligule less than 0.5 mm long; blades papillose-hispid above and below; spikelets papillose-hirsute..41. *P. Leibergii.*
Ligule more than 0.5 mm long; blades not papillose-hispid; spikelets not papillose.
  Culms and at least the lower sheaths with an appressed pubescence; ligules mostly 1.5 mm long with longer hairs intermixed; spikelets oblong-obovate, mostly 3.5-4 mm long and 1.7-1.9 mm wide.......................42. *P. oligosanthes.*
  Culms and sheaths with a spreading pubescence; ligules about 1 mm long; spikelets bluntly obovate, mostly 3-3.5 mm long and 2 mm wide.....43. *P. Scribnerianum.*

41. **Panicum Leibérgii** (Vasey) Scribn. Map 315. Very local in the northern part of the state, where it is found in dry, sandy or gravelly soils, usually in prairie habitats. The pH value was taken for only one specimen and it was 6.01.

N. Y. to Man. and N. Dak., southw. to Ind. and Kans.

42. **Panicum oligosánthes** Schultes. Map 316. Local in the lake area and reappearing on the low dunes of the southwestern part of the state. It grows in very sandy, dry soils on open, wooded dunes and cleared, open dunes and sand knolls. It is usually associated with *Panicum Scribnerianum* which is the more common species. These two grasses are closely related and most easily separated in the field. The leaves of this species are narrower and the upper ones are relatively longer and more spreading.

Mass. to Mo., southw. to Fla. and Tex.

43. **Panicum Scribneriànum** Nash. (*Panicum oligosanthes* var. *Scribnerianum* (Nash) Fern.) Map 317. Rather frequent in the lake area

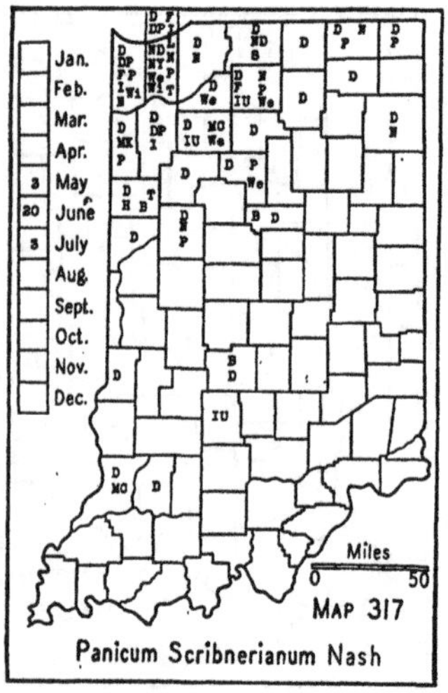

Panicum Scribnerianum Nash

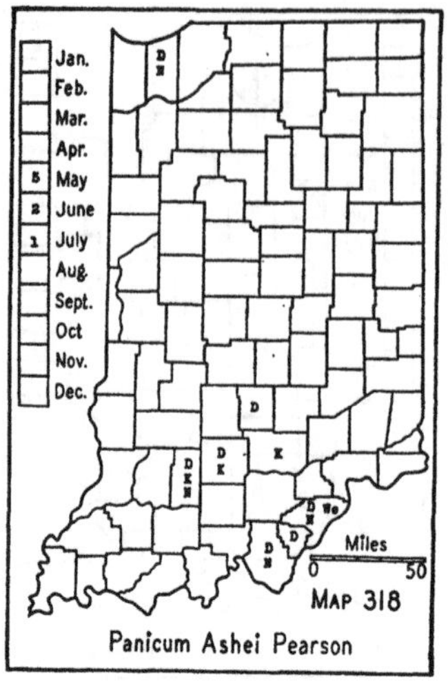

Panicum Ashei Pearson

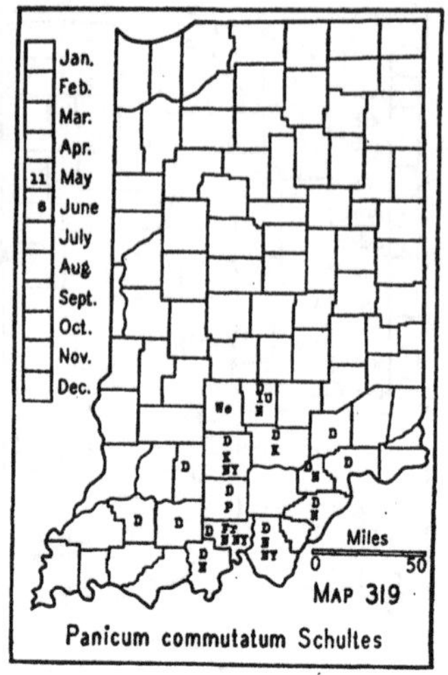

Panicum commutatum Schultes

where it is found in very sandy, dry soil on open dunes and sand hills and sometimes in rather dry, gravelly soil. Our specimens from the western part of the state are from sand dunes and sandy knolls.

Maine to B. C., southw. to Md., Tenn., Tex., and Ariz.

15. COMMUTÀTA

Culms tufted, glabrous or puberulent; ligule obsolete or nearly so; blades relatively broad, cordate at the base; spikelets pubescent.

Culms and sheaths usually densely crisp-puberulent (sometimes sparsely so); blades generally less than 12 mm wide; spikelets 2.2-2.5 (2.7) mm long...44. *P. Ashei.*

Culms and sheaths generally nearly glabrous or only sparingly puberulent (not crisp-puberulent); blades or some of them usually more than 12 mm wide; spikelets 2.5-3 mm long, generally about 2.7 mm long................45. *P. commutatum.*

44. **Panicum Àshei** Pearson. Map 318. This species, as now known in the state, is restricted to the unglaciated area, with the exception of a typical specimen which I have from Porter County found on a sandy black oak and white pine ridge about 4 miles southwest of Michigan City. In the southern part of the state it is found mostly on the crests and slopes of chestnut oak ridges.

No single character will separate Indiana specimens of this grass from those of the next. The two plants intergrade to such an extent that it is questionable whether an attempt should be made to keep them separate, even regarding one as a variety, as has been done by Fernald (Rhodora 36: 83-87. 1934). If all of our forms of this species complex are considered as one species, then the same treatment applied to borderline species in other groups would unite them. This case seems to be a decision between the "grouping" and the "splitting" of forms (species). Until an exhaustive study is made of the group, any disposition made of these plants must be mere opinion or for convenience. For these reasons I am following Hitchcock and treating our plants as two species. Such treatment leaves the problem open to future study.

Mass. to Mich. and Mo., southw. to n. Fla., Miss., and Okla.

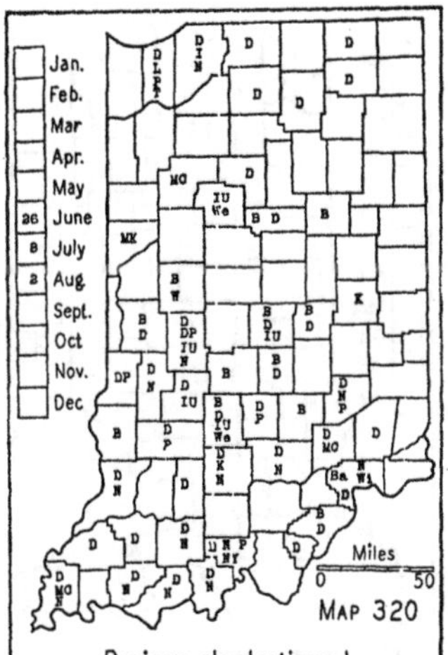

Panicum clandestinum L.

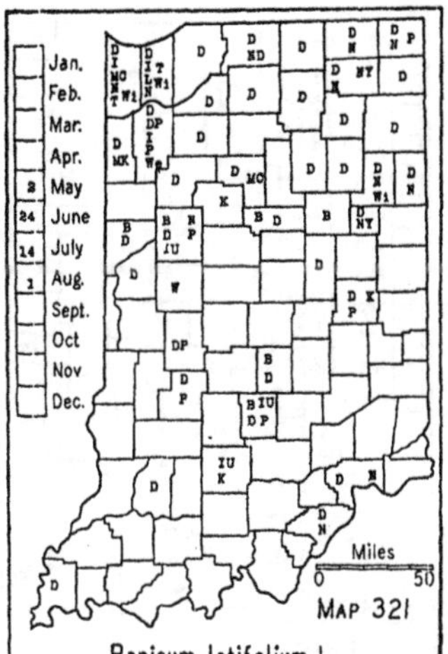

Panicum latifolium L.

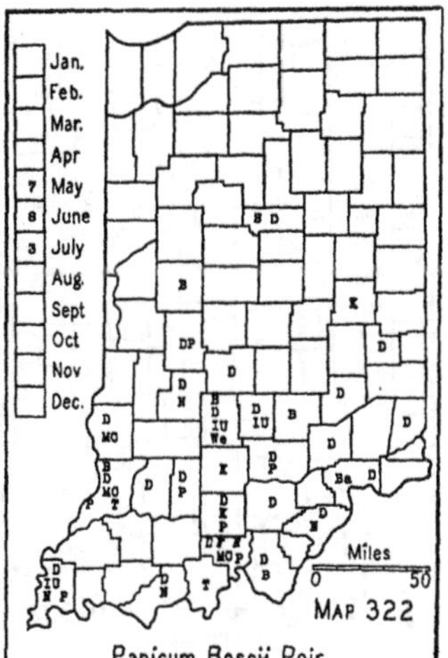

Panicum Boscii Poir.

45. **Panicum commutàtum** Schultes. Map 319. This species is restricted usually to the high hills of the unglaciated area, although it is found in Jefferson County on the bluff of the Ohio River and in Jennings County on the sandstone outcrop along the Muscatatuck River near Vernon. It is rather local except in the knobstone, where it is frequent. My no. 27633 from Clark County is exceptional in that the whole plant is soft-pubescent, including both surfaces of the leaves.

Mass. to Mich. and Mo., southw. to Fla. and Tex.

16. LATIFÒLIA

Culms stout, usually more than 50 cm high; ligules mostly less than 1 mm long; blades cordate at the base and long-acuminate at the apex, usually more than 15 mm wide; spikelets 2.7-4.5 mm long, 7-11 nerved; the autumnal phase sparingly branching at the middle nodes, becoming top-heavy, and lodging.

Sheaths, at least the lower ones and those of the branches, papillose-hispid; spikelets 2.7-3 mm long (rarely longer)............................46. *P. clandestinum.*
Sheaths glabrous or softly villous (hairs not stiff as in the preceding species).
  Nodes glabrous or nearly so; spikelets 3.2-3.7 mm long...........47. *P. latifolium.*
  Nodes retrorsely bearded; spikelets 4-4.5 mm long.
    Blades glabrous or nearly so on both surfaces...................48. *P. Boscii.*
    Blades velvety to the touch beneath...................48a. *P. Boscii* var. *molle.*

46. **Panicum clandestìnum** L. Map 320. This species is infrequent to rare in the northern part of the state; rare, local, or absent in the central counties; and frequent in most of the southern half of the state. It prefers low ground and is more abundant in areas where the soil is slightly acid. It is generally found on the moist slopes of streams and ditches. It usually forms large colonies, and often specimens with exserted panicles are absent, especially in the autumnal phase.

N. S. and Que. to Kans., southw. to n. Fla. and Tex.

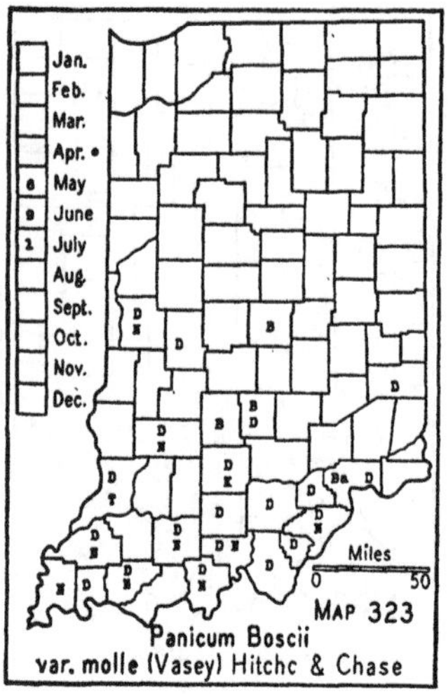

Panicum Boscii var. molle (Vasey) Hitchc & Chase

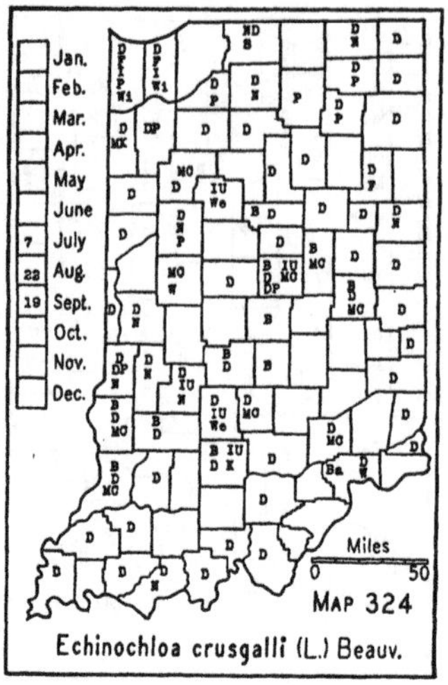

Echinochloa crusgalli (L.) Beauv.

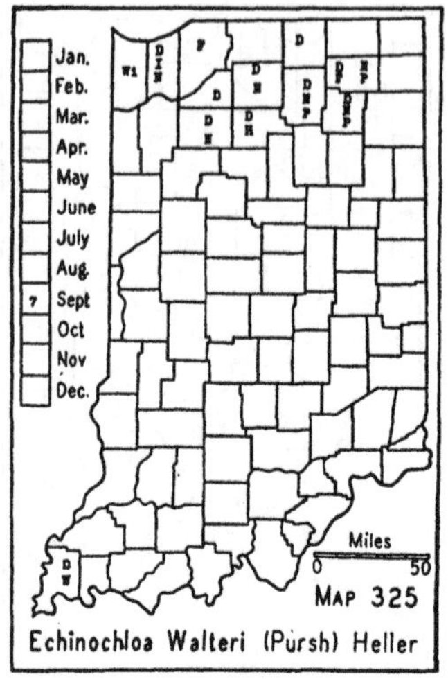

Echinochloa Walteri (Pursh) Heller

47. **Panicum latifòlium** L. Map 321. Rather frequent in dry or moist white oak and black oak woods in the lake area. Infrequent to local in the southern part of the state where it is largely replaced by the next species which is absent in our northern counties.

Maine, Que. to Minn., southw. to N. C. and Kans.

48. **Panicum Bóscii** Poir. Map 322. An infrequent species in the southern half of the state, where it is found in dry woodland, associated with black and white oak and white oak and hickory.

Mass. to Wis., and Okla., southw. to Fla. and Tex.

48a. **Panicum Boscii** var. **mólle** (Vasey) Hitchc. & Chase. Map 323. This variety has the range and habitat of the species in Indiana. It is doubtful whether it should be maintained as a variety since I have found culms from the same rootstock which would qualify for the species and the variety. The general range of the variety is nearly the same as that of the species.

133-166B. ECHINÓCHLOA Beauv.

[Hitchcock. The North American species of Echinochloa. Contr. U. S. Nation. Herb. 22: 133-153. 1920. Wiegand. The genus Echinochloa in North America. Rhodora 23: 49-65. 1921. Farwell. Notes on the Michigan flora, II. Michigan Acad. Sci. Rept. 21: 349-350. 1920.]

Sheaths glabrous (rarely the lower ones somewhat pubescent or papillose-hispid); second glume pointed, not awned; fruit ovate-elliptic, usually 1.5-2 mm wide.................1. *E. crusgalli.*

Sheaths (at least the lower ones) papillose-hispid (rarely glabrous); second glume with an awn usually 2-10 mm long (rarely shorter); fruit elliptic, generally less than 1.5 mm wide..........2. *E. Walteri.*

1. **Echinochloa crusgálli** (L.) Beauv. BARNYARD GRASS. Map 324. Frequent to common in all parts of the state. "The common name of this grass suggests that it might be a grass restricted to the vicinity of habitations,

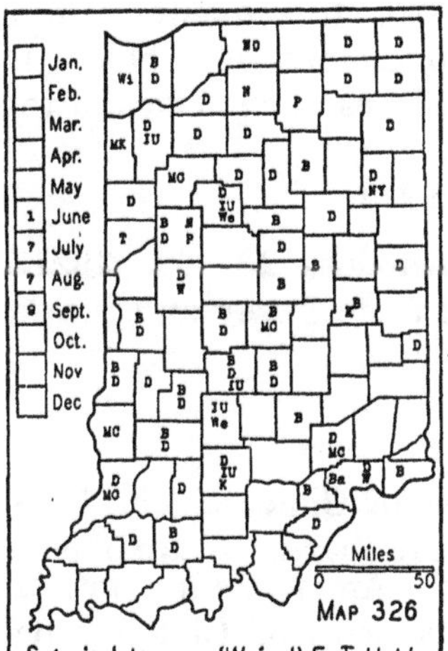

Setaria lutescens (Weigel) F. T. Hubb.

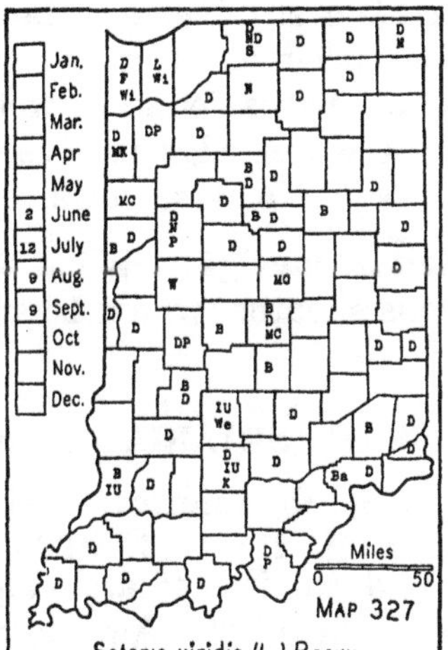

Setaria viridis (L.) Beauv.

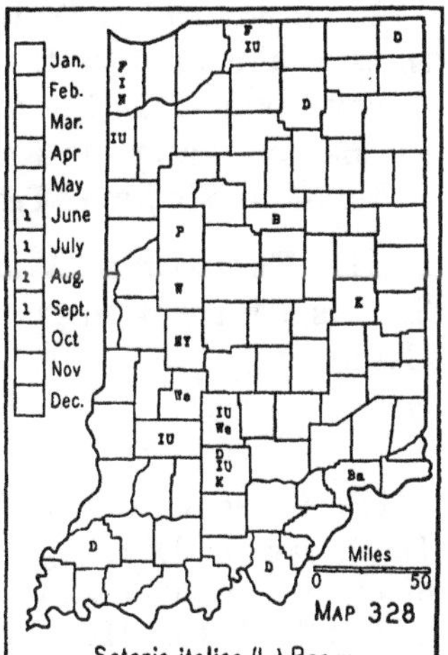

Setaria italica (L.) Beauv.

which is not true. While it is found in waste places about barns and dwellings, it is found in almost all kinds of habitats except dense shade. It prefers the sunshine. As to soil requirements, it is found from minimacid soils to the marl borders of lakes. It prefers a moist soil but will grow in wet or dry places. It is found in roadside and dredged ditches, in low places about lakes, in bayous, along streams, and in cultivated fields and pastures.

"I am regarding this species as a polymorphic one. A careful examination of more than 60 Indiana specimens shows that sheaths are usually glabrous, but sometimes the lower ones are scabrous to more or less papillose-hispid. The spikelets are usually more or less awned, the awns up to 3 cm long, but the spikelets of some panicles are all or nearly all awnless. In one specimen the primary panicle has awnless spikelets and the axillary panicle has awned spikelets. In another specimen the reverse is true. The spikelets of some panicles have scarcely any papillose hairs while those of others rarely have hairs without the papillose base. The amount and length of the pubescence vary on the same plant as well as on separate plants. The color of the spikelets varies from green to purple. In ponds and sloughs, where germination may be delayed on account of the recession of the water, I have seen mature plants only a few inches high in fruit while on the higher margin of the same pond would be plants several feet high.

"Some authors have given names to the many forms of this species. Some variations have been called species, some varieties, and some forms. The limit in assigning names seems to have been reached by Jackson who named a 'variegated purple form' of the awnless form (Guide to Nature **16**: 11. 1923). For a discussion of the so-called varieties and forms see the literature cited." (Deam, Grasses of Ind. p. 304-305, 1929.)

Hitchcock, in his manual of the grasses of the United States, also regards this species as polymorphic, but recognizes an awnless variety.

N. B. to Wash., southw. to Fla. and Calif.; Eastern Hemisphere.

2. **Echinochloa Wálteri** (Pursh) Heller. Map 325. Infrequent to local in the lake area, with one specimen from the muddy flat of a bayou in Posey County. In the lake area it is found in wet places about lakes, often in shallow water, and at the water edge in rivers.

Mass. to Fla., and Tex.; N. Y. to Wis., Iowa, and Ky.

2a. **Echinochloa Walteri f. laevigàta** Wieg. (Rhodora 23: 62. 1921.) This is a form with glabrous sheaths, which I have from Posey and Starke Counties.

135-171. SETÀRIA Beauv.

[Scribner & Merrill. The North American species of Chaetochloa. U. S. Dept. Agric. Div. Agrost. Bull. 21: 1-44. 1900. Hubbard. A taxonomic study of Setaria italica and its immediate allies. Amer. Jour. Bot. 2: 169-198. 1915. Hitchcock. The North American species of Chaetochloa. Contr. U. S. Nation. Herb. 22: 155-208. 1920. Copple & Aldous. The identification of certain native and naturalized grasses by their vegetative characters. Kansas Agric. Exper. Sta. Tech. Bull. 32: 1-73. 1932.]

Bristles below each spikelet numerous, at least more than 5, upwardly scabrous.
  Blades usually with a half twist beyond the middle; spikelets about 3 mm long, very turgid on the convex side; second glume slightly more than half as long as the spikelet. . . . . 1. *S. lutescens.*
  Blades without a twist beyond the middle; spikelets 2-2.5 mm long; second glume almost as long as the spikelet.
    Fruit disarticulating with the spikelet below the glumes, leaving a cup-shaped scar. . . . . 2. *S. viridis.*
    Fruit disarticulating above the glumes. 3. *S. italica.*
Bristles below each spikelet 1 or, by abortion of the spikelets, 2 or 3; bristles downwardly scabrous. . . . . 4. *S. verticillata.*

1. SETARIA LUTÉSCENS (Weigel) F. T. Hubb. (*Setaria glauca* and *Chaetochloa glauca* of authors.) YELLOW BRISTLEGRASS. YELLOW FOXTAIL. Map 326. A common weed throughout the state in cultivated grounds and waste places and along roads and railroads.

Nat. of Eu.; widely distributed in temperate regions.

2. SETARIA VÍRIDIS (L.) Beauv. (*Chaetochloa viridis* (L.) Scribn.) GREEN BRISTLEGRASS. GREEN FOXTAIL. Map 327. A common weed throughout the state in cultivated and waste grounds and along roads and railroads. It is not as common as the preceding species.

Nat. of Eu.; common throughout the cooler parts of the U. S., infrequent in the southern states and in the mountains; Newf. to B. C., southw. to Fla. and Calif.

3. SETARIA ITÁLICA (L.) Beauv. (*Chaetochloa italica* (L.) Scribn.) FOXTAIL MILLET. Map 328. This species has been sparingly sown as a forage crop and has escaped. For detailed information on the value of the species as a forage crop and its culture, see H. N. Vinall on Foxtail Millet (U. S. Dept. Agric. Farmers' Bull. 793).

Nat. of Eurasia; escaped in waste places and roadsides throughout the U. S.

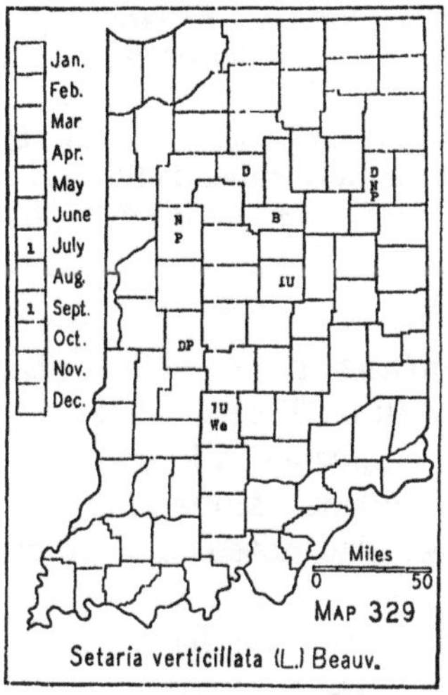

Setaria verticillata (L.) Beauv.

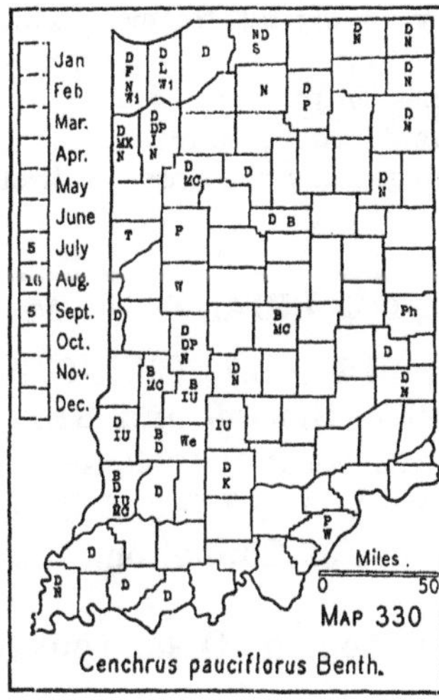

Cenchrus pauciflorus Benth.

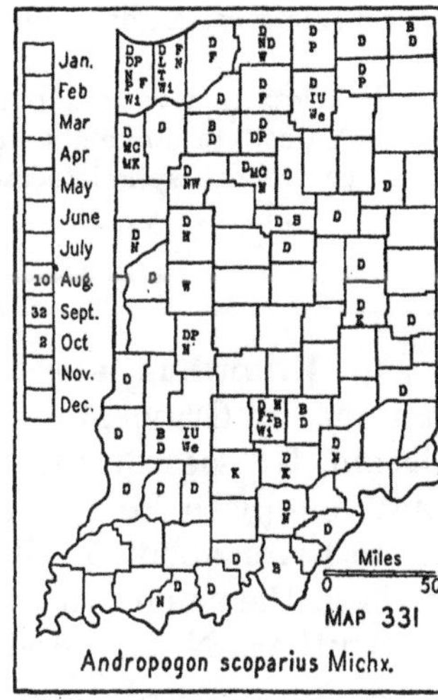

Andropogon scoparius Michx.

4. SETARIA VERTICILLÀTA (L.) Beauv. (*Chaetochloa verticillata* (L.) Scribn.) BUR BRISTLEGRASS. Map 329. This species has been reported from 7 counties. It is found sparingly (in waste places) probably throughout the state. I have known it in Wells County for 10 years. I first found it in a vacant lot in Bluffton and 10 years later I found it along the roadside outside of the city. Doubtless wherever it gets a start it will gradually spread.

Nat. of Eu.; Mass. to N. Dak., southw. to Ala., and Mo.; occasionally westw. to Calif.

137-174. CÉNCHRUS L.

[Chase. The North American species of Cenchrus. Contr. U. S. Nation. Herb. 22: 45:77. 1920.]

1. **Cenchrus pauciflòrus** Benth. (*Cenchrus carolinianus* of Gray, Man., ed. 7 in part and *Cenchrus tribuloides* of Britton and Brown, Illus. Flora, ed. 2, not L.) FIELD SANDBUR. Map 330. This sandbur prefers dry, sandy to very sandy soil and is found throughout the state where its habitat occurs. It is local where its habitat is absent and is frequent to common in the northern part of the state in the sandy areas, where it is a very obnoxious weed. It is found in cultivated grounds and waste places, in sandy railroad ballast, and along roadsides.

Maine to Oreg., southw. to Fla., Tex., and Calif.; Mexican Plateau, coastal region of tropical America, and s. S. A.

11. *ANDROPOGÒNEAE* Presl SORGHUM TRIBE

Spikelets all alike, perfect.

Rachis of the racemes not articulated; panicle fan-shaped, the panicle axis short. ....MISCANTHUS, p. 178.

Rachis articulated; panicle axis elongated................143. ERIANTHUS, p. 178.

Spikelets of two kinds, one sessile and perfect, the other pedicellate, staminate, empty, or reduced to a mere scale or pedicel.

Spikelets in slender, solitary, or digitate racemes which are terminal or lateral...145. ANDROPOGON, p. 178.
Spikelets in terminal panicles only.
Pedicellate spikelets present; culms solid..................147. SORGHUM, p. 180.
Pedicellate spikelets lacking (only the hairy pedicel present); culms hollow...148. SORGHASTRUM, p. 181.

143-112. ERIÂNTHUS Michx.

1. **Erianthus alopecuroìdes** (L.) Ell. (*Erianthus divaricatus* (L.) Hitchc. of Gray, Man., ed. 7, Britton and Brown, Illus. Flora, ed. 2, and Deam, Grasses of Ind.) SILVER PLUMEGRASS. This species is known as a native only in Perry County where I found it on a wooded slope along the Ohio River about 5 miles east of Cannelton. It was also noted in a fallow field in the same county.

Southern N. J., s. Ind., s. Mo., and Okla., southw. to Fla. and Tex.

ERIANTHUS RAVÉNNAE (L.) Beauv. RAVENNA OR PLUME GRASS. This species is a native of southern Europe and is often cultivated. There is no record of its escape. It is easily distinguished from the preceding species by having three stamens and by its scabrous sheaths.

MISCÁNTHUS SINÉNSIS Anders. EULALIA. This grass is a native of China and is often cultivated. There is no record of its escape. It is easily distinguished from *Erianthus* by the fan-shaped panicle and by the continuous rachis of the racemes.

145-134. ANDROPÒGON L.

Branches of inflorescence ending in a single raceme..................1. *A. scoparius.*
Branches of inflorescence ending in a pair or fascicle of racemes.
Racemes of each branchlet generally 3-7, 5-13 cm long; sessile spikelets 6.5-10 mm long; stamens 3..2. *A. furcatus.*
Racemes of each branchlet 2 (rarely 3 or 4), 1.5-4 cm long; sessile spikelets less than 6 mm long; stamens 1.
Awns coiled at the base; sessile spikelets generally 4-4.5 mm long, 0.7-0.8 mm wide; peduncles of the primary racemes elongated so that the racemes are borne beyond the spathes; spathes inflated, at least at maturity.....3. *A. Elliottii.*
Awns not coiled at the base; sessile spikelets 3-3.5 mm long, about 0.6 mm wide; none of the peduncles elongated so that the racemes extend beyond the spathes; spathes not inflated..............................4. *A. virginicus.*

1. **Andropogon scopàrius** Michx. (*Schizachyrium scoparium* (Michx.) Nash of Britton and Brown, Illus. Flora, ed. 2.) PRAIRIE BEARDGRASS. BROOMSEDGE. Map 331. This species occurs throughout the state in poor or impoverished soils and moist or dry, sandy soils, and is also rapidly becoming established in the better soils of the Tipton Till Plain. It is found on washed slopes and interdunal flats, in abandoned fields, and along roadsides and railroads.

The extreme variability of this species has led authors to describe many forms. My Indiana specimens show a wide range of variability, yet I hesitate to refer any of my specimens to a variety. For example, about half of my specimens are glabrous, and the other half vary from those with a few hairs on the sheaths to those with a villous pubescence. *Andropogon*

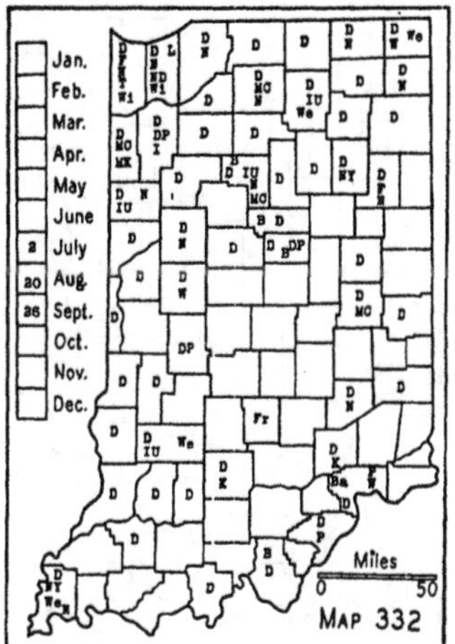

Andropogon furcatus Muhl.

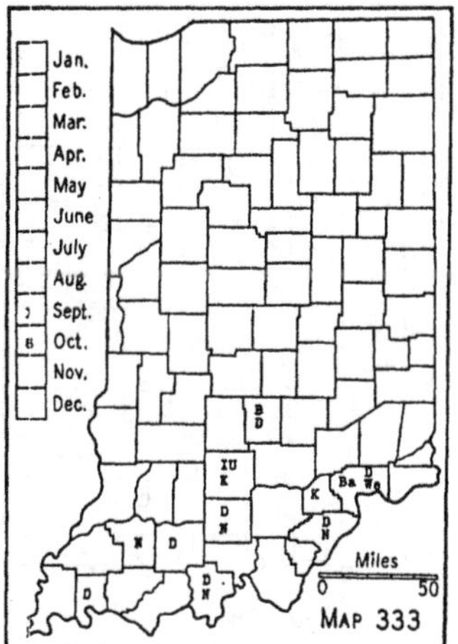

Andropogon Elliottii Chapm.

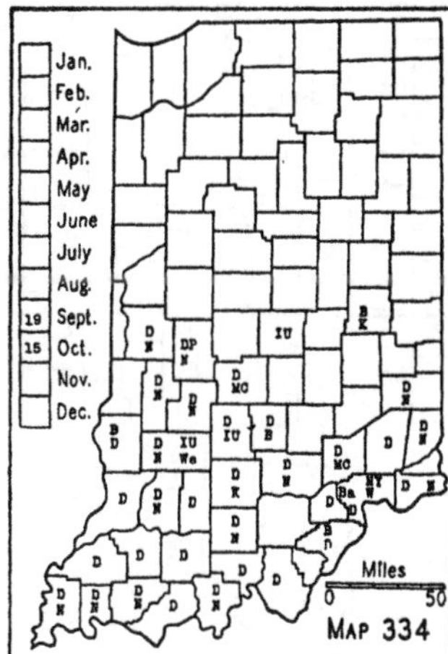

Andropogon virginicus L.

scoparius var. *frequens, Andropogon scoparius* var. *littoralis, Andropogon scoparius* var. *polycladus,* and *Andropogon scoparius* var. *villosissimus* have been reported from Indiana but Buhl (Amer. Midland Nat. **16:** 250. 1935) refers all of them to the typical form.

Plants along Lake Michigan, growing on the bases of the low dunes in West Gary, present, in the field, a striking difference because they are smaller and very glaucous. However, an examination of the floral parts shows them to be identical, or nearly so, with the typical form.

Maine, Que. to Alberta and Idaho, southw. to Fla. and Ariz.

2. **Andropogon furcàtus** Muhl. (*Andropogon provincialis* Lam. of Deam, Grasses of Ind.) BIG BLUESTEM. Map 332. Found sparingly throughout the state except in the prairie areas where it is common and where, before cultivation, it usually formed complete stands over all of the drier parts. This grass prefers a rather dry, sandy habitat but I have found it in hard, white clay soil in the Lower Wabash Bottoms and on rocky bars in streams. Outside the prairie area it is very erratic in its locations.

Maine, Que. to Sask. and Mont., southw. to Fla., Ariz., and Mex.

3. **Andropogon Ellióttii** Chapm. (*Andropogon Elliottii* var. *projectus* Fern. & Grisc.) ELLIOTT BEARDGRASS. Map 333. As now known, this species is restricted practically to the unglaciated area where it is usually found with *Andropogon virginicus.* It is most often found in dry, impoverished soil on washed slopes and in abandoned fields. A variety *projectus* has been named by Fernald & Griscom (Rhodora **37**: 139. 1935). The Indiana record is based upon my collection no. 26865. This variety is described as having the racemes on long-exserted peduncles. This is merely the early phase of the inflorescence, and late in the season the long-exserted racemes usually fall and the broad sheaths open, exposing the subsessile pairs of racemes in their axils.

Coastal Plain from N. J. to Fla. and Tex., northw. to s. Mo., Ind., and Tenn.

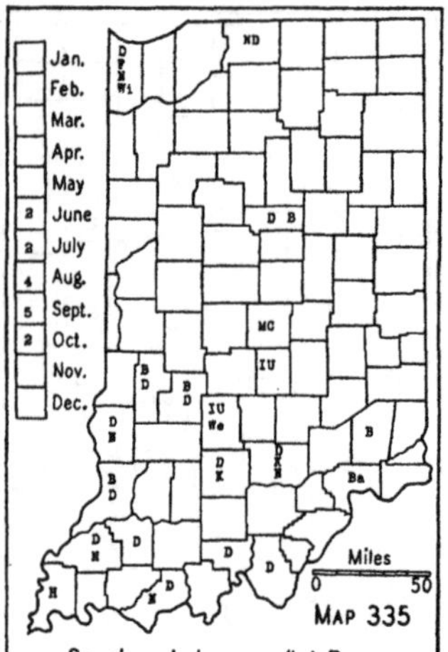

Sorghum halepense (L.) Pers.

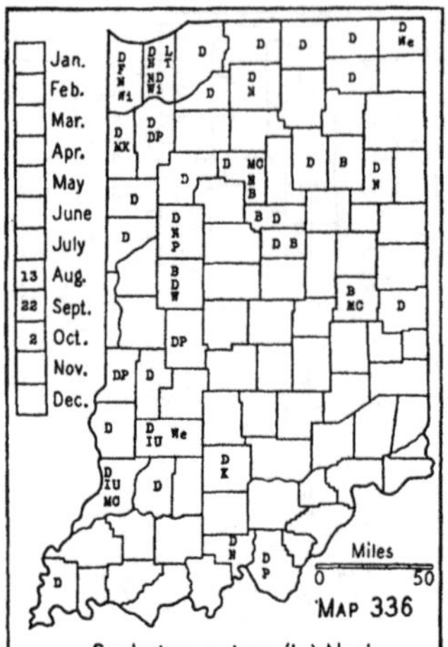

Sorghastrum nutans (L.) Nash

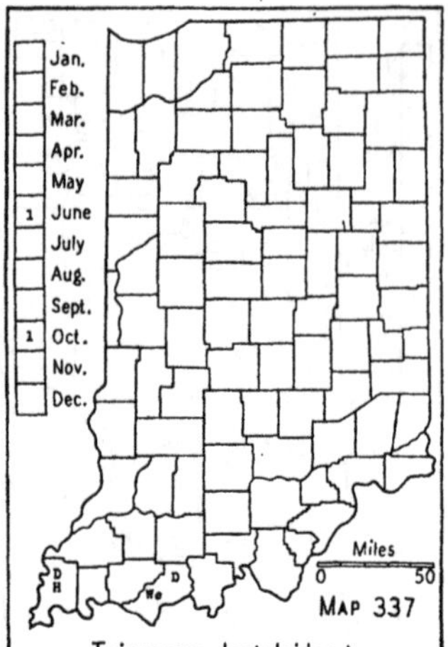

Tripsacum dactyloides L.

4. **Andropogon virgínicus** L. (Fernald. A review of Andropogon virginicus and Andropogon glomeratus. Rhodora **37**: 139-143. 1935.) BROOMSEDGE. Map 334. This species is restricted essentially to the southern half of the state where it is local to infrequent or common in slightly acid soil. It prefers moist soil but thrives also in dry situations. It is commonly found in old, worn out fields, hayfields, and pastures.

Mass., N. Y., Ind., and Kans., southw. to Fla. and Tex.; Mex.

147-134A. SÓRGHUM Pers.

Perennial, with long, creeping rootstocks; spikelets disarticulating from the pedicel at maturity. . . . . 1. *S. halepense.*
Annual; spikelets not disarticulating from the pedicel at maturity.
  Spikelets not opening and exposing the grain at maturity.
    Culms usually more than 6 mm in diameter; sheaths longer than the internodes; blades mostly more than 20 mm wide. . . . . 2. *S. vulgare* var. *Drummondii.*
    Culms usually less than 6 mm in diameter; sheaths shorter than the internodes; blades mostly less than 20 mm wide. (See no. 2.). . *S. vulgare* var. *sudanense.*
  Spikelets opening, exposing the grain at maturity. (See no. 2.) . . . . *S. vulgare.*

1. SORGHUM HALEPÉNSE (L.) Pers. JOHNSON GRASS. Map 335. Infrequent but spreading in the southwestern part of the state. It is found mostly along roadsides and railroads and sometimes in cultivated fields, these usually contiguous to streams or railroads. Several years ago I found it in large colonies in the cornfields of the Wabash Bottoms and landowners were not aware of its weedy nature. While this grass has forage crop value, it should be exterminated, because it is difficult to eradicate and carries the possibility of seeding adjacent areas where it is not desired.

Native of the Mediterranean region, and found in the tropical and warmer regions of both hemispheres. Mass. to Iowa, southw. to Fla. and Tex., and westw. to Calif.

2. SORGHUM VULGÀRE var. DRUMMÓNDII (Nees) Hitchc. CHICKEN CORN. This grass was first reported from Posey and Vanderburgh Counties in 1923. I have seen it as a common weed in the cornfields in

Point Township of Posey County where it often overtopped the corn. A pioneer in that vicinity informed me that he thought it was introduced about 1890.

Probably a native of Africa.

SORGHUM VULGARE var. SUDANÉNSE (Piper) Hitchc. SUDAN GRASS. This is an annual grass which has been recently introduced as a forage crop but there are no reports that it has escaped and become established.

Probably a native of Africa.

SORGHUM VULGARE Pers. SORGHUM. This is the cultivated sorghum, of which there are many varieties. It has been cultivated from pioneer times in this state, but there are no reports that it has perpetuated itself.

Nat. of Africa.

148-134B. SORGHÁSTRUM Nash

1. **Sorghastrum nùtans** (L.) Nash. INDIAN GRASS. Map 336. This is essentially a prairie grass and is found in "oak openings" which are remnants of prairies. It is frequent throughout the state where prairie habitats occur and is rare or absent elsewhere. It is sometimes found in marshy places and its most common associate is *Andropogon furcatus*.

Maine, Que. to Man. and N. Dak., southw. to Fla. and Ariz.; Mex.

12. *TRIPSÀCEAE* Hitchc. CORN TRIBE

157-103. TRÍPSACUM L.

1. **Tripsacum dactyloìdes** L. EASTERN GAMAGRASS. Map 337. I have found this species only twice. A few colonies were in a low, wet woods about three fourths of a mile southeast of the old Spencer School, about 10 miles southwest of Mt. Vernon, Posey County; and it was common along a ditch through a low field about 5 miles east of Lincoln City, Spencer County. I moved two colonies to Bluffton 6 years ago, and they are hardy and spreading.

Mass. to Mich., Iowa, and Nebr., southw. to Fla. and Tex.; W. I. and Mex. to Brazil.

159-102. ZÈA L.

ZEA MÀYS L. CORN. This is our cultivated corn. It appears spontaneously but does not become established. Origin probably in Central America or southeastern Mexico.

20. CYPERÀCEAE J. St. Hil. SEDGE FAMILY

Flowers all perfect, rarely some of them with stamens or pistil abortive.
  Basal empty scales of spikelets none, rarely 2, and sometimes 3 in *Eleocharis Smallii.*
    Scales of the spikelets strictly 2-ranked, conduplicate and keeled.
      Flowers without bristles; achenes beakless; inflorescence terminal.
        Spikelets few- to many-flowered, usually elongated or slender............. ....459. CYPERUS, p. 183.
        Spikelets 1-flowered (but of 3 or 4 scales), glomerate in sessile heads......... ....462. KYLLINGA, p. 190.

Flowers with bristles; achenes beaked; inflorescence axillary..458. DULICHIUM, p. 183.

Scales of the spikelets spirally imbricated.

Base of style persistent on the achene as a tubercle.

Spikelets 1; leaves reduced to sheaths; bristles usually present..469. ELEOCHARIS, p. 198.

Spikelets several or numerous; leaves blade-bearing; bristles none..471A. BULBOSTYLIS, p. 206.

Base of style not persistent as a tubercle.

Flowers without any inner scales.

Base of style enlarged; bristles none............471. FIMBRISTYLIS, p. 205.

Base of style not enlarged; bristles usually present.

Bristles 6 but each 4-6-cleft to near the base, making them appear numerous, silky, usually white, all much exserted; stamens 1-3..466. ERIOPHORUM, p. 190.

Bristles 0-8, short, not silky and only rarely whitish and long-exserted, sometimes lacking; stamens 2 or 3.............468. SCIRPUS, p. 192.

Flowers with 1 or more inner scales.

Bristles 3, barbed..................................467. FUIRENA, p. 191.

Bristles none..................................453. HEMICARPHA, p. 182.

Basal empty scales of the spikelets 3 or more.

Styles 2-cleft; enlarged base of style persistent on the achene as a tubercle.

Spikelets few-flowered; bristles usually present.....492. RHYNCHOSPORA, p. 207.

Spikelets many-flowered; bristles none................472. PSILOCARYA, p. 207.

Styles 3-cleft; enlarged base of style not persistent on the achene; bristles none...489. CLADIUM, p. 207.

Flowers all imperfect.

Pistillate flower subtended by a flat scale; achene naked, bony, and usually white...515. SCLERIA, p. 209.

Pistillate flower wholly enclosed by a sac (perigynium), the style protruding through an opening at the top...................................525. CAREX, p. 212.

453. HEMICÁRPHA Nees & Arn.

Plants growing in dense clumps, the outer culms recurved-spreading; length of an average culm (measured up to the inflorescence), 1-7 cm; height of leaves about half the average length of the culms; longest involucral bracts (those appearing as continuations of the culms) 2-4.5 cm long; average spikelets 2-4 mm long; scales of spikelets generally with short, spreading or recurved tips; achenes terete, slightly obovoid, usually about 0.6 mm long and 0.3 mm wide.....1. *H. micrantha.*

Plants growing in loose clumps, the culms erect or ascending; length of an average culm (measured up to the inflorescence), 4-9 cm; height of leaves about a third the average length of the culms; longest involucral bracts (those appearing as continuations of the culms) 1-1.5 cm long; average spikelets 4-7 mm long; scales of spikelets generally appressed; achenes terete or slightly lenticular-obovoid, usually about 0.7 mm long and 0.35 mm wide...............2. *H. Drummondii.*

1. **Hemicarpha micrántha** (Vahl) Pax. Map 338. Infrequent to rare in the area shown on the map. Found in wet, sandy places on the borders of lakes and sloughs and in ditches.

N. H., the Great Lakes area to Wash., southw. to Fla., Mex., and S. A.

2. **Hemicarpha Drummóndii** Nees. Map 339. Found only in wet sand on the borders of sloughs or in sloughs when dried up, in wet, interdunal flats in the dune area, and in a dredged ditch in Newton County.

W. Ont., Ind., Ill. to Ark., Kans., and Tex.

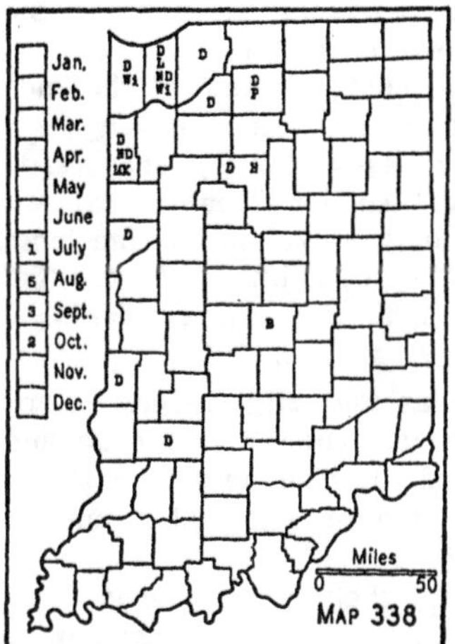

MAP 338
Hemicarpha micrantha (Vahl) Pax

MAP 339
Hemicarpha Drummondii Nees

MAP 340
Dulichium arundinaceum (L.) Britt.

458. DULÍCHIUM Pers.

1. **Dulichium arundinàceum** (L.) Britt. Map. 340. Generally found in sedge marshes or associated usually with some sedge on the low borders of lakes, sloughs, and ponds. It is rather frequent in the lake area, becoming rare south of it because its habitat is rare in southern Indiana. Newf. to Wash., southw. to Fla. and Tex.

459. CYPÈRUS [Tourn.] L.

[Geise. The Indiana species of Cyperus. Amer. Midland Nat. **15**: 241-291. 1934.]

Stigmas 2; achenes lenticular, not 3-angled; spikelets flat; scales falling from the rachis at maturity.
  Scales of spikelets stramineous, about 2 mm long, so closely imbricated as to hide the achenes even in dried specimens; achenes 0.75-1 mm long, distinctly blackish, plump, strongly compressed, strongly obovoid, transverse wrinkles distinct, superficial cells oblong.......................................1. *C. flavescens.*
  Scales of spikelets generally margined with reddish brown, 2-3 mm long; achenes lenticular, with transverse wrinkles, gray or brownish gray, mostly 1-1.4 mm long, superficial cells more or less quadrate.
    Exserted style branches many, usually exserted 2-4 mm; scales dull, thin, mostly about 2.5 mm long, rather loosely imbricated so that at least the base of the achene is visible in dried specimens........................2. *C. diandrus.*
    Exserted style branches few, usually exserted 1-1.5 mm; scales lustrous, subcoriaceous, usually 2-2.4 mm long, so closely imbricated that the achenes are hidden...3. *C. rivularis.*
Stigmas 3; achenes 3-angled.
  Scales long-acuminate at the apex, usually ending in a sharp point, the upper fourth to a third of them widely spreading or recurved; plants cespitose, mostly 3-9 cm high, fragrant when dried..............................4. *C. inflexus.*
  Scales and plants not as above.
    Scales slightly outcurved at the apex; spikelets very flat; stamens 1.
      Plants annual, 0.5-3.5 dm high; scales ovate, 3-nerved; achenes about 1 mm long and half as wide..5. *C. acuminatus.*
      Plants perennial, 4-10 dm high; scales oblong, 1-nerved; achenes oblong, about 1 mm long and 0.3 mm wide.........................6. *C. pseudovegetus.*

Scales straight on the back to the apex, sometimes a few near the apex of the spikelet with slightly curved tips in *C. dentatus;* stamens 2 or 3.

Spikelets arranged in globose heads or aggregated in short clusters at the ends of the culms or the rays, the common rachis not more than 1 cm long.

Inflorescence usually composed of 5 or 6 globose heads, usually one sessile or nearly so, the others on rays 2-5 (or more) cm long; culms leafy at the base, the leaves mostly more than 15 cm long; spikelets 4-5 mm long, 2- or 3-flowered, usually maturing a single achene; culms with cormlike bases..7. *C. ovularis.*

Inflorescence and plant not as above.

Involucral bracts recurved or widely spreading at maturity, rarely one or more erect; leaves narrowly linear, mostly less than 2 mm wide and rarely as wide as 3 mm, the lowest leaves of the culm less than 15 cm long, rarely one longer; culms below the inflorescence 0.5-1 mm in diameter.

Spikelets in a loose or close, terminal cluster, the principal ones 8-12-flowered...................................8. *C. filiculmis.*

Spikelets in compact, terminal, globose or ovoid-globose, usually solitary heads, sometimes with one or two smaller heads on short rays, in depauperate specimens the heads small and spikelets not compact; spikelets all less than 8-flowered or only a few with 8 or more flowers. ....8a. *C. filiculmis* var. *macilentus.*

Involucral bracts erect or ascending; culms usually more than 1 mm in diameter below the inflorescence; leaves linear and usually wider than those of the preceding group; spikelets usually in flat clusters.

Scales scarcely or faintly nerved, their margins reddish brown, midnerve of scale not excurrent; culms not cormlike at the base, very leafy; inflorescence umbellate; spikelets very flat; style branches exserted more than 1 mm..................................9. *C. dentatus.*

Scales strongly nerved, their margins hyaline; midnerve of scale excurrent; culms with cormlike bases; inflorescence racemose; style branches usually not exserted, or generally not more than 1 mm.

Culms, leaves, and rays smooth; leaves much shorter than the culm; spikelets 5-9-flowered; scales 2-2.5 mm long, the mucro less than 0.5 mm long; achenes 1.5-2 mm long.............10. *C. Houghtonii.*

Culms (at least below the inflorescence), margins of leaves, and rays rough; spikelets 4-16-flowered; scales mostly 3-4.5 mm long, the mucro usually 0.5-1 mm long; achenes 2.5-3 mm long............ ....11. *C. Schweinitzii.*

Spikelets arranged along an elongated rachis, the rachis usually 1-3 cm long.

Flowers remote, the successive scales not reaching the bases of the ones above on the same side of the rachilla...................12. *C. Engelmanni.*

Flowers approximate, the successive scales overlapping the bases of those above.

Scales mostly 2.75-4.5 mm long; culms with cormlike bases.

Spikelets erect or ascending, more than 2.5 mm wide; achenes ellipsoid, about 2.5 mm long and half as wide...........11. *C. Schweinitzii.*

Spikelets widely spreading or reflexed, less than 2.5 mm wide; achenes linear-oblong, mostly 1.5-2 mm long and about 0.3 mm wide except in *C. strigosus* var. *multiflorus.*

Spikelets 4-20-flowered, stramineous, very flat.........13. *C. strigosus.*

Spikelets 10-35-flowered, reddish brown, terete or nearly so; achenes 0.75 mm wide and 2 mm long....13a. *C. strigosus* var. *multiflorus.*

Scales less than 2.75 mm long; culms without cormlike bases.

Scales about 1.5 mm long, reddish brown; flowers very closely imbricated, the scales overlapping more than half their length; spikelets 10-40-

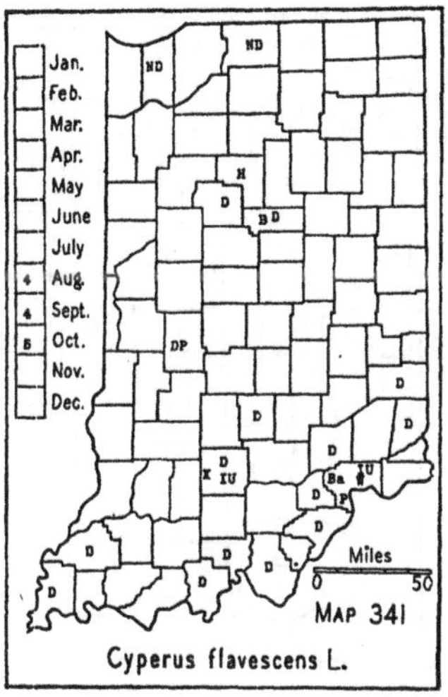

Cyperus flavescens L.

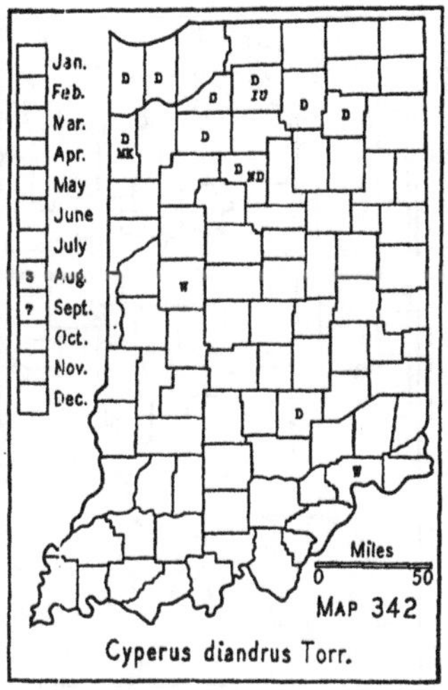

Cyperus diandrus Torr.

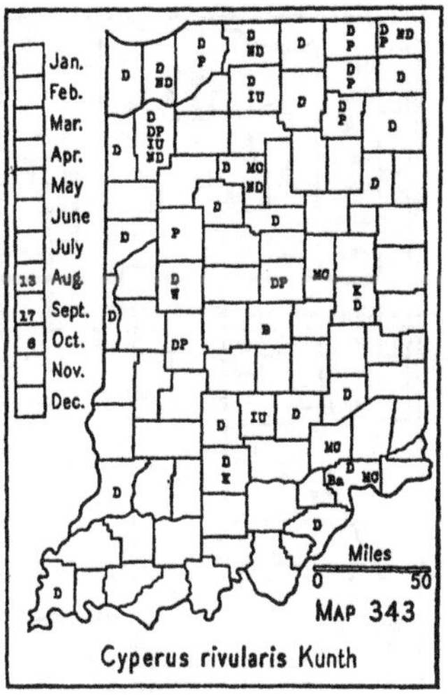

Cyperus rivularis Kunth

flowered, all of the flowers maturing achenes; achenes about 0.8 mm long and 0.5 mm wide........................14. *C. erythrorhizos.*

Scales mostly 2-2.5 mm long; flowers not very closely imbricated, the scales usually overlapping less than half their length; achenes 1-1.5 mm long.

Plants with numerous fibrous roots, annual; culms with 1-4 leaves; longest rays of umbel generally less than 5 cm long, rarely one or more of them longer; spikelets usually dense, reddish brown, at maturity easily broken into segments below the flowers.........
....15. *C. ferruginescens.*

Plants with numerous, scaly stolons that at length bear a tuber; culms very leafy; leaves usually more than 4; longest rays of umbels usually 4-13 cm long, only rarely all the rays shorter; spikelets usually stramineous, sometimes light reddish brown, at maturity not separating into segments below the flowers...16. *C. esculentus.*

1. **Cyperus flavéscens** L.* Map 341. Rare in northern Indiana and infrequent in the southern part in wet, sandy soil on bars in streams and ditches, in the outlets of springs, along ditches, and about artificial ponds.

N. Y. to Mich., southw. to Fla. and Mex.; also in Cent. Amer. and the Old World.

2. **Cyperus diándrus** Torr. Map 342. Infrequent to rare. My specimens were found in wet, sandy soil on the borders of lakes and sloughs and in mucky soil in dried-up sloughs and in like habitats along streams.

N. B. to Minn., southw. to S. C. and Kans.

3. **Cyperus rivulàris** Kunth. Map. 343. Rather frequent throughout the state in wet, sandy or gravelly soil on the borders of lakes and streams and on bars in ditches and small streams.

Maine, s. Ont. to Minn., southw. to N. C. and Ark.

× **Cyperus Nieuwlándii** Geise. (*Cyperus flavescens* × *rivularis.*) This hybrid was described by Geise (Amer. Midland Nat. 15: 245-246. 1934). She reports three specimens collected by Nieuwland in the vicinity of Chain Lakes in St. Joseph County. I have seen these specimens and their determination seems to be correct.

* Fernald (Rhodora 41: 529-530. 1939) has shown that the true species belongs to Eurasia and Africa and that the plant of eastern North America should be designated as **Cyperus flavescens** L. var. **poaeformis** (Pursh) Fern.

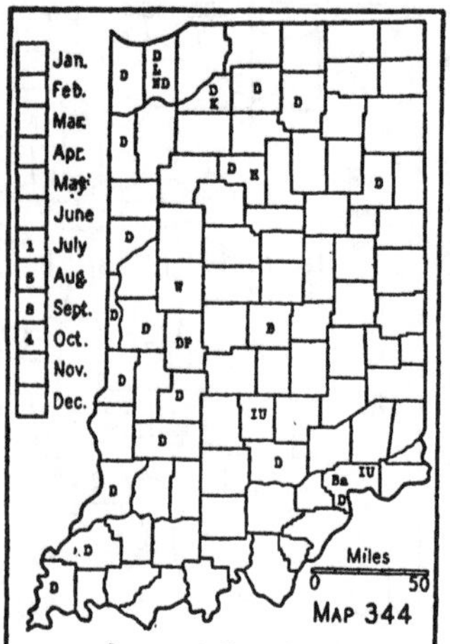

Cyperus inflexus Muhl.

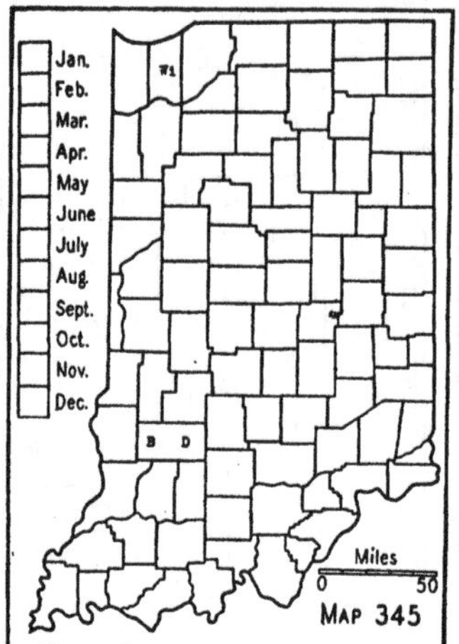

Cyperus acuminatus Torr & Hook.

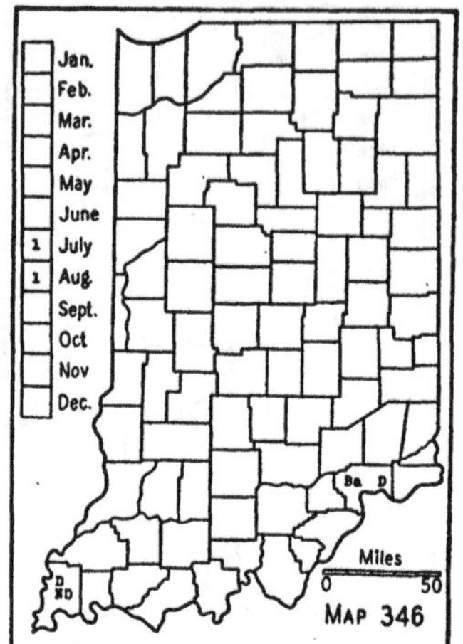

Cyperus pseudovegetus Steud.

4. **Cyperus infléxus** Muhl. (*Cyperus aristatus* Rottb.) Map 344. Infrequent in wet, sandy or muddy soil on bars in streams and ditches and on the shores of lakes and borders of sloughs. Specimens of this species when dried have a pleasing odor, similar to that of dried slippery elm leaves.

N. B. to B. C., southw. to Fla., Tex., Calif., and Mex.

5. **Cyperus acuminàtus** Torr. & Hook. Map 345. I have found this species only in Crawford and Greene Counties. I am not able to locate my Crawford County specimen now. Friesner also found it in Greene County. Geise cites a specimen from near Chesterton, Porter County, collected by E. T. Harper in 1888. This specimen is deposited in the herbarium of the University of Wisconsin. I have seen it and the determination is correct.

Ind. to N. Dak. and Wash., southw. to Ga., Tex., and Calif.

6. **Cyperus pseudovégetus** Steud. Map 346. Infrequent in ditches and swamps in Point Township of Posey County. It has been found also in Gibson, Pike, and Jefferson Counties. Where found it is usually common.

N. J. to Kans., southw. to Fla. and Tex.

7. **Cyperus ovulàris** (Michx.) Torr. Map 347. This species is found in very dry to moist, sandy habitats. It is local in the southwestern counties. It has been reported from Lake County, but Geise did not find a specimen. I believe that the Lake County report should be referred to *Cyperus filiculmis* var. *macilentus*.

N. Y. to Ill. and Kans., southw. to Fla. and Tex.

8. **Cyperus filicúlmis** Vahl. Map 348. Fernald & Griscom discuss this species and its varieties in Rhodora 37: 153-154. 1935. If I interpret their discussion correctly the distribution of this species is principally on the Atlantic slope and in the Great Plains states. My only specimen

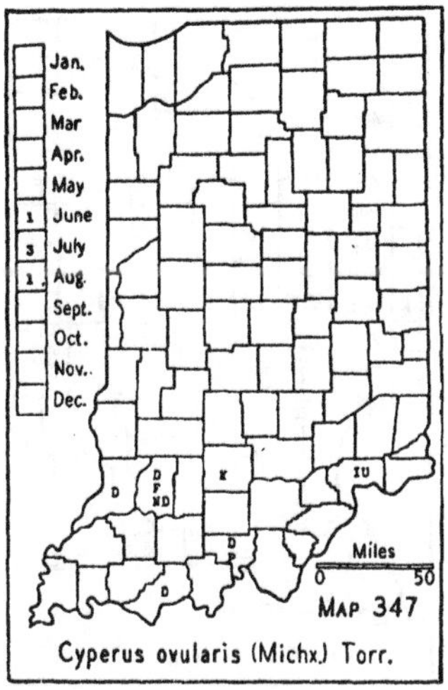

Cyperus ovularis (Michx.) Torr.

Cyperus filiculmis Vahl

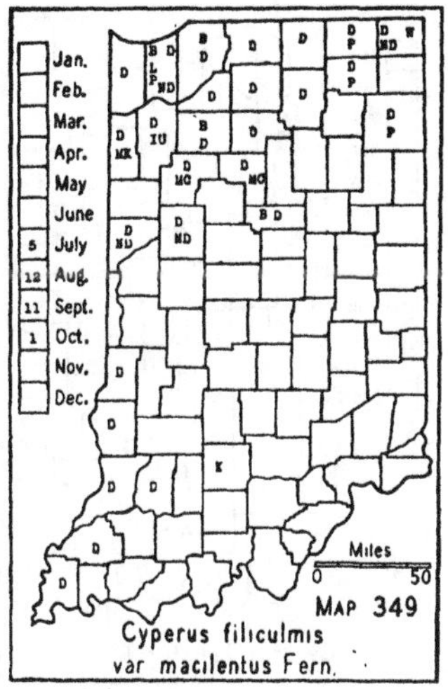

Cyperus filiculmis var. macilentus Fern.

is from a dry, sandy ridge in Gibson County. Geise (Amer. Midland Nat. 15: 254. 1934) cites specimens from Lake, La Porte, Marshall, Porter, and St. Joseph Counties, but I refer these specimens to the variety.

8a. **Cyperus filiculmis** var. **macilèntus** Fern. Map 349. This variety grows in very sandy soil and is found mostly on sand ridges and dunes, in sandy fallow fields, and in the moist intervening sandy areas between sand ridges and dunes. In its habitat it is usually frequent, elsewhere it is absent. Its distribution in the state is well represented by the map.

Cent. Maine, sw. Que. to Minn., southw. to Va., Ohio, Ind., Ill., and Mo.

9. **Cyperus dentàtus** Torr. (Including *Cyperus dentatus* var. *ctenostachys* Fern.) Map 350. This Coastal Plain species is found in only three counties. It is local but usually common where it is found. It grows in moist, sandy soil in ditches through marshes and on the wet, sandy shore of Bass Lake in Starke County. Specimens with 15-40-flowered spikelets have received a varietal name, but since both short and long spikelets are found on the same plant it is obvious that the variety is only a luxuriant form of the species.

N. S. to Ind., southw. to N. C.; principally near the coast.

10. **Cyperus Houghtònii** Torr. Map 351. This is a species of the dune area and it has been found only in Lake and Porter Counties.

Mass. to Man. and Oreg., southw. to Va., Kans., and Ariz.

11. **Cyperus Schweinítzii** Torr. Map 352. This species grows in very dry sand and has its mass distribution on the dunes near Lake Michigan. The Warren County specimen was found on the very high, gravelly bank along the Big Four Railroad about 2 miles northwest of Covington.

Western N. Y., s. Ont. to Man., southw. to Ind. and Kans.

× **Cyperus mesochòrus** Geise. (*Cyperus Houghtonii* × *Schweinitzii.*) This hybrid is described in Amer. Midland Nat. 15: 249-250. 1934. Geise

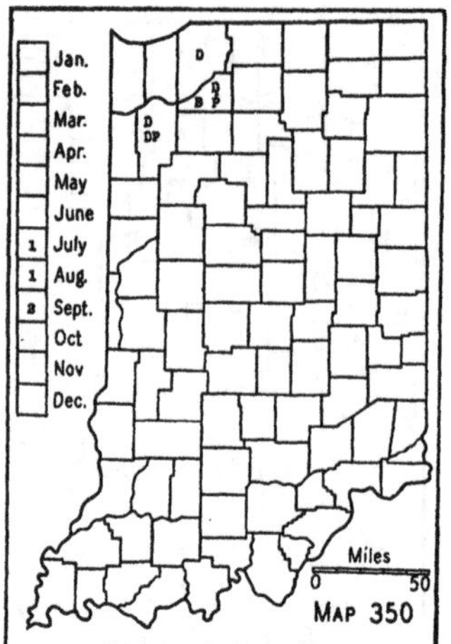

Cyperus dentatus Torr.

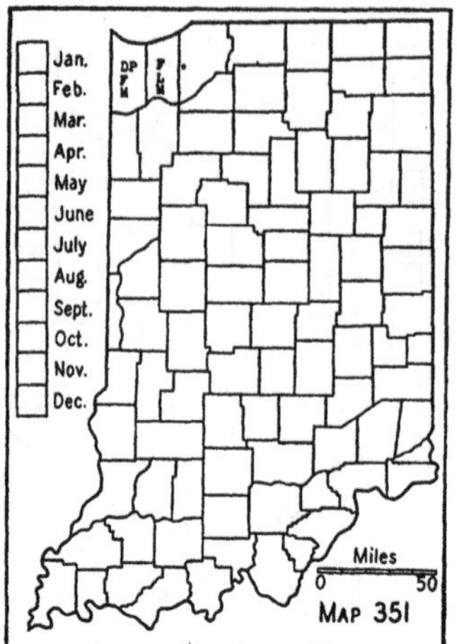

Cyperus Houghtonii Torr.

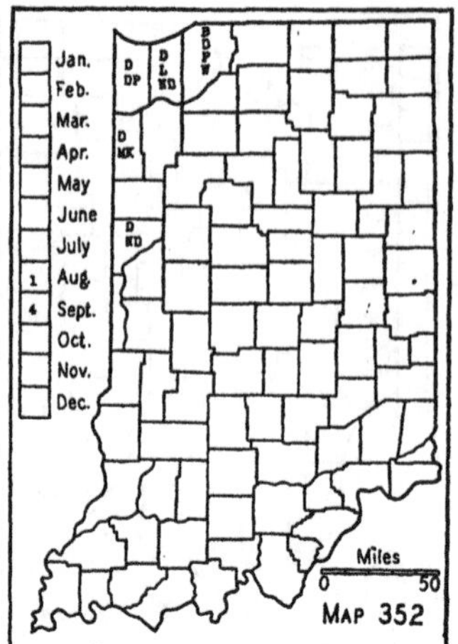

Cyperus Schweinitzii Torr.

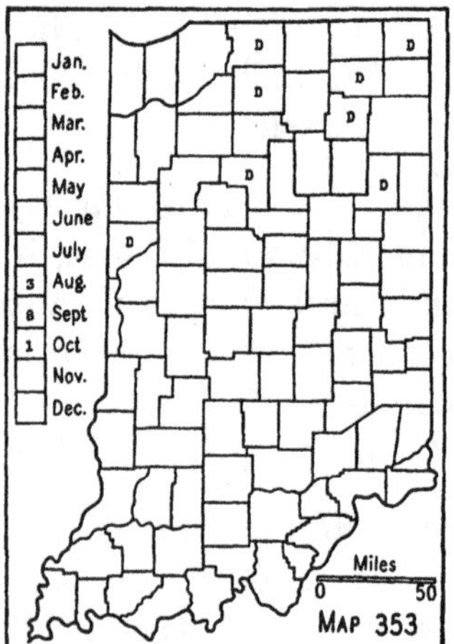

Cyperus Engelmanni Steud.

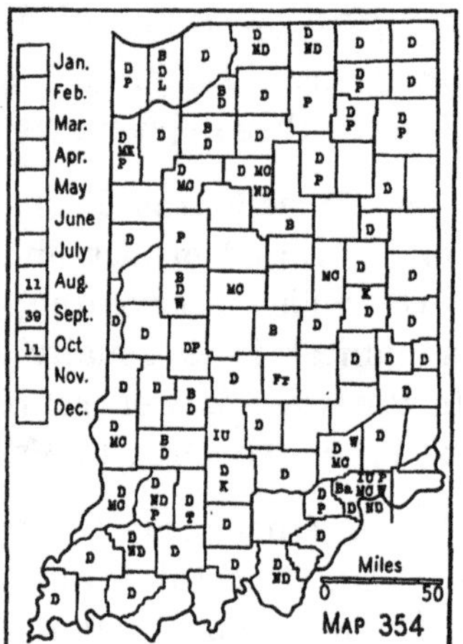

Cyperus strigosus L.

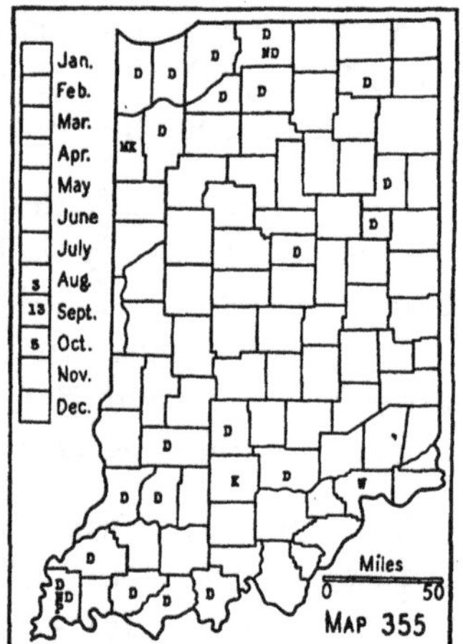

Cyperus erythrorhizos Muhl.

cites numerous specimens from Lake and Porter Counties. She also refers specimens of my collecting from La Porte, Newton, and Warren Counties to this hybrid.

12. **Cyperus Engelmánni** Steud. Map 353. Infrequent in the lake area. All of my specimens are from the wet, sandy or muck borders of lakes.

Mass. to Minn., southw. to N. J. and Mo.

13. **Cyperus strigòsus** L. (Including *Cyperus strigosus* var. *capitatus* Boeckl., *Cyperus strigosus* var. *compositus* Britt., and *Cyperus strigosus* var. *robustior* Kunth.) Map 354. This species is, without doubt, found in every county in the state. The extreme variability of this species has led authors to assign botanical names to the variations. I agree with some other authors in thinking that the forms are a matter of nutrition or of habitat and have no taxonomic value; hence I am referring all forms to the species. It is found in moist soil of almost all kinds and in all

kinds of habitats. Probably most abundant along ditches and in cornfields.

Maine, Ont. to Minn., southw. to Fla. and Tex.

13a. **Cyperus strigosus** var. **multiflòrus** Geise. This form was described by Geise in Amer. Midland Nat. **15**: 253. 1934. I collected specimens in the dried-up mucky soil on the south side of Lake Cicott, Cass County, in 1931 and 1932 which were years of severe drought. I also found a few specimens in a similar habitat on the border of an extinct lake about 2 miles north of North Liberty, St. Joseph County. The dominant associate was *Cyperus ferruginescens*. This plant is conspicuous and can be distinguished from any other *Cyperus* at a long distance. After a careful study of this form, it seems to me that it is a hybrid of *Cyperus strigosus* and *Cyperus ferruginescens*. The plants (2.5-15 cm high) are too small for *Cyperus strigosus*, and the spikelets have about twice the number of flowers that average plants of that species have. The cormlike base is a character of *Cyperus strigosus* but the terete, reddish brown spikelets belong to *Cyperus ferruginescens*.

14. **Cyperus erythrorhìzos** Muhl. Map 355. Infrequent throughout the state but usually common where it is found. It is generally found on the muddy shores of streams, in dried-up sloughs, and along ditches.

Mass. to Minn., southw. to Fla., Tex., and Calif.

15. **Cyperus ferruginéscens** Boeckl. (Rhodora 37: 148-150. 1935.) (*Cyperus speciosus* Vahl, in part, of most recent authors.) Map 356. Infrequent to frequent throughout the state. It grows in moist, wet, muddy or mucky soils of almost all kinds.

Mass. to Minn., southw. probably to Fla. and Tex.

16. **Cyperus esculéntus** L. (Including *Cyperus esculentus* var. *leptostachyus* Boeckl.) CHUFA. Map 357. Rather frequent in southern Indiana, becoming infrequent to rare in the northern part. This species prefers moist or wet, rich soil and is found along streams and in cultivated fields and truck gardens. I have seen it in dried-up sloughs where it formed complete stands. We allowed it to grow unmolested in our arboretum of about 3 acres before we knew of its weedy nature and we have been trying to exterminate it for about 10 years but still find a plant occasionally. I have noted it as a pernicious weed in truck gardens, especially along the Ohio River. The tubers are sweet and edible. They have been used as food since ancient times, having been found in Egyptian tombs dating back to 2400 years before Christ.

The species is extremely variable in the size of its spikelets. Plants with long spikelets have been named but I think they are a result of nutrition and should not receive taxonomic names. It is to be noted that plants with small inflorescences rarely mature more than a few seed while plants with large inflorescences usually mature many seed.

N. B. to Minn., Nebr., and Alaska, southw. to Fla., Tex., and Calif.; also found in the tropics; Eurasian.

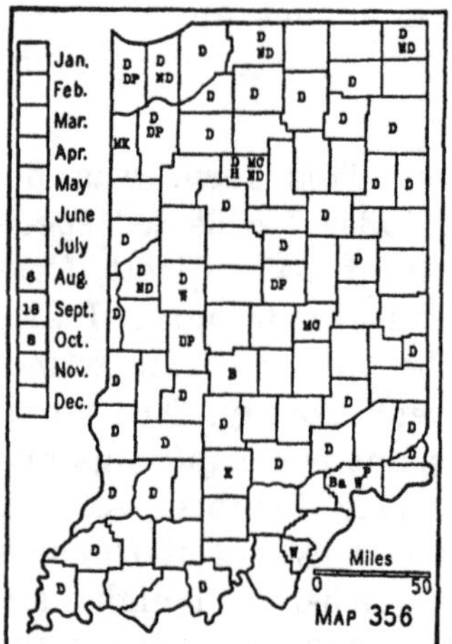

Cyperus ferruginescens Boeckl.

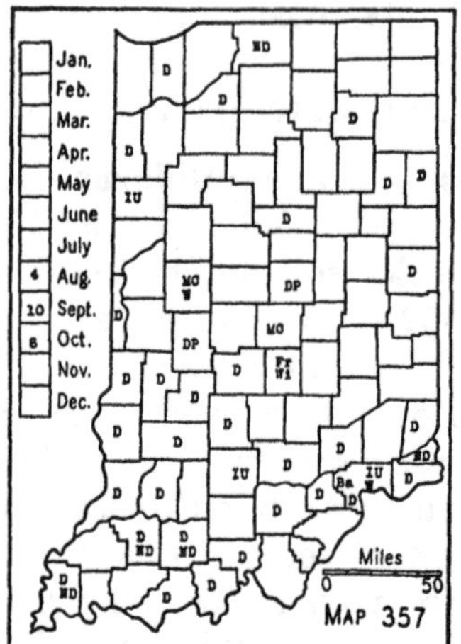

Cyperus esculentus L.

Kyllinga pumila Michx.

462. KYLLÍNGA Rottb.

1. **Kyllinga pùmila** Michx. (*Cyperus densicaespitosus* Mattf. & Kükenth. Pflanzenr. 20: 597. 1936.) Map 358. Infrequent in southern Indiana and rare or absent from many of our northern counties. It is usually found in moist or wet soil along streams, on bars in streams, along ditches, and sometimes in cornfields along streams.

Del., Ohio, Ill. to Kans., southw. to Fla. and Tex.; also W. I., Mex., and southw.

466. ERIÓPHORUM L. COTTON GRASS

Spikelets solitary; involucre none; scales lead color..................1. *E. spissum.*
Spikelets 2-several; involucre of 1-several leafy bracts.
  Leaves 1-2 mm wide, channeled their entire length; upper leaf blade shorter than its sheath; involucral bract 1; achenes ellipsoid, about 2.5 mm long..2. *E. gracile.*
  Leaves 1.5-6 mm. wide, flat at least below the middle; involucral bracts more than 1; achenes oblong-obovoid, mostly 2.5-3.5 mm long.
    Scales of spikelets with only 1 prominent rib; stamens 3; plants of May and June.
      Upper leaf sheaths dark-girdled at the summit; midrib of scales not extending to the apex, the upper part of the scale hyaline and the rib prominent below the hyaline apex..................................3. *E. angustifolium.*
      Upper leaf sheaths not dark-girdled at the summit; midrib of scales extending to the apex...4. *E. viridi-carinatum.*
    Scales of spikelets with several prominent ribs; stamen 1; bristles varying from tawny to white; plants of August and September, beginning to flower about July 15...5. *E. virginicum.*

1. **Eriophorum spíssum** Fern. (Rhodora 27: 208-209. 1925.) (*Eriophorum callitrix* of recent American authors, not Cham.) Map 359. Our specimens were found in tamarack bogs.

Baffinland and Lab. to Athabaska, southw. to Newf., N. S., N. E., mts. of Pa., n. Ind., and Wis.

2. **Eriophorum grácile** Koch. Map 360. Borders of sloughs in the dune area and elsewhere in marshes and in sphagnum in bogs.

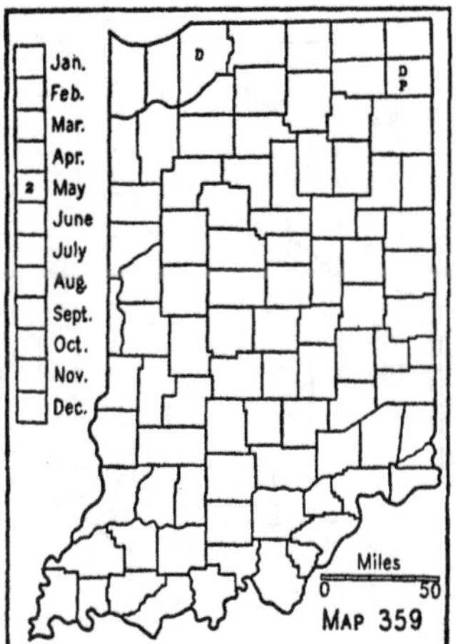

Eriophorum spissum Fern.

Eriophorum gracile Koch

Eriophorum angustifolium Roth

Newf. to B. C., southw. to Conn., Pa., Ind., Nebr., and Calif.; also in Eurasia.

3. **Eriophorum angustifòlium** Roth. Map 361. Infrequent on the borders of sloughs and in marshes and bogs.

Subarctic Amer., southw. to Maine, Ont., Ill., Iowa, and mts. of Colo. and Oreg.; also in Eurasia.

4. **Eriophorum víridi-carinàtum** (Engelm.) Fern. Map 362. Infrequent throughout our northern counties where it is usually found growing in sphagnum in open tamarack bogs and less often in sedge marshes.

Newf. to Sask. and B. C., southw. to Conn., N. Y., Ohio, Wis., Oreg., and in the mts. to Ga.

5. **Eriophorum virgínicum** L. (Including *Eriophorum virginicum* f. *album* (Gray) Wieg.) Map. 363. Since the bristles of this species vary from tawny to white with intermediate forms, I have not attempted to separate our plants on the basis of this character. Nearly all of our plants at maturity have white or whitish bristles. It is found in marshes and tamarack bogs.

Newf. to Ont. and Man., southw. to Fla. and Nebr.

467. FUIRÈNA Rottb. UMBRELLA GRASS

1. **Fuirena pùmila** Torr. (Rhodora 40: 396-398. 1938.) (*Fuirena squarrosa* of recent authors, not Michx.) Map 364. This sedge is very local, having been found in only a few places in two counties. It grows in moist sand in interdunal swamps and in wet sand on the borders of lakes. I found it to be rather frequent in wet sand on the south side of Walker Lake in Porter County.

Mass. to Mich. and Ind., southw. to Fla.

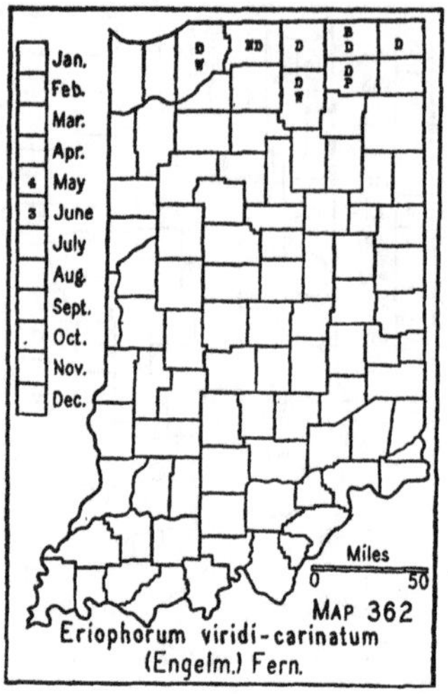

Eriophorum viridi-carinatum (Engelm.) Fern.

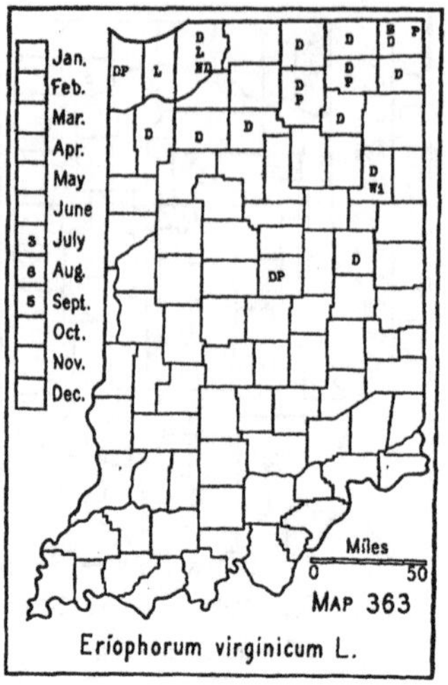

Eriophorum virginicum L.

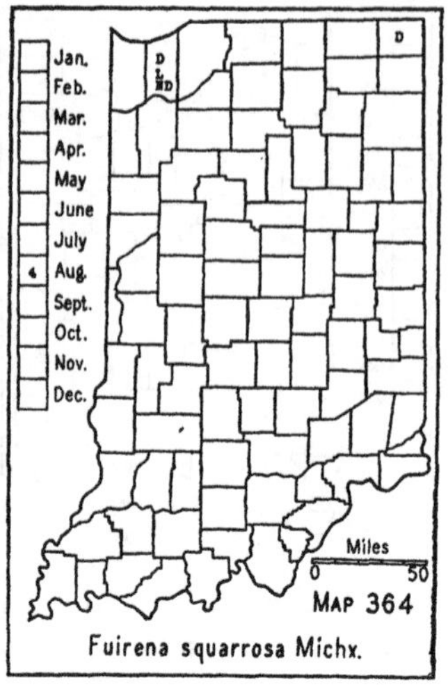

Fuirena squarrosa Michx.

468. SCÍRPUS [Tourn.] L. BULRUSH

[Sr. M. St. Leona Thornton. The Indiana species of Scirpus. Amer. Midland Nat. 15: 292-322. 1934.]

Sister Thornton's treatment of Indiana *Scirpus* seems to be comprehensive and authentic. I have seen most of the specimens she cites. I am accepting her determinations of the few I have not seen and they also are indicated on the distribution maps.

Involucral bract none. (This is Sr. Thornton's *Scirpus pauciflorus* which is now referred to *Eleocharis pauciflora* var. *Fernaldii* Svenson. (See Rhodora 36: 380. 1934.)

Involucral bract solitary (the inflorescence appearing as if on the side of the stem).

Spikelets 1, 6-13 mm long; stem cylindric, conspicuously nodulose, normally growing in shallow water but often emersed in dry weather; achenes trigonous, about 2.5 mm long and about 1.6 mm wide, brown, smooth; bristles retrorsely barbed, about equaling the achene............................1. *S. subterminalis*.

Spikelets normally more than 1.

Plants usually less than 5 dm high; annuals with tufted roots; culms terete or obtusely angled.

Culms obtusely triangular; mature involucral bract usually divaricate; achenes obovoid, unequally biconvex, about 1.7 mm long, surface black with shallow and irregular pits; bristles longer than the achene, with increasing width toward the base, mostly 0.015 mm wide near the base..........2. *S. debilis*.

Culms terete; mature involucral bract usually erect; achenes obovoid, plano-convex, 1.5-1.8 mm long, surface black without pits or with very inconspicuous ones; bristles very slender, of almost equal width, generally about 0.01 mm wide near the base.

Bristles lacking..3. *S. Smithii*.

Bristles present, usually 6, sometimes fewer, longer than the achene.........
....3a. *S. Smithii* var. *setosus*.

Plants usually more than 5 dm high; perennials with creeping rootstocks; culms triangular or terete.

Involucral bract much longer than the inflorescence.

Culms sharply triangular; involucral bract acute; achenes plano-convex, smooth; bristles shorter than the achene................4. *S. americanus*.

Culms obtusely 3-angled with concave sides; leaves nodulose; involucral bract blunt; achenes trigonous, smooth; bristles much longer than the achene. .... 5. *S. Torreyi.*

Involucral bract usually shorter than the inflorescence or merely equaling it.

Culms rather soft; inflorescence lax, usually drooping; spikelets many, ovoid, on long, drooping pedicels; achenes obovoid, 1.5-2 mm long, plano-convex; bristles usually longer than the achene.................... 6. *S. validus.*

Culms rather stiff and firm; inflorescence erect, the spikelets and pedicels erect or ascending, compact; spikelets subcylindric; achenes obovoid, in my specimens ranging from 2.3-3 mm long, unequally biconvex; bristles about equaling the achene or slightly shorter............... 7. *S. acutus.*

Involucral bracts 2 or more.

Bristles retrorsely barbed or lacking.

Spikelets large, usually 1.5-4 cm long; achenes trigonous, about 4 mm long..... .... 8. *S. fluviatilis.*

Spikelets small, generally less than 1 cm long.

Bristles scarcely longer than the achene, usually slightly shorter, rudimentary, or lacking; scales of mature spikelets with a light reddish background suffused with a lead color; achenes colorless, obovoid-oblong, trigonous, about 1 mm long.

Bristles present, about equaling the achene; lower sheaths nodulose; leaves usually 10-18 mm wide; major glomerules usually more than 7 mm in diameter.. 9. *S. atrovirens.*

Bristles lacking or rudimentary; lower sheaths not nodulose or only faintly so; leaves usually less than 10 mm wide; major glomerules usually not over 7 mm in diameter; rays of inflorescences usually longer than in the preceding; glomerules usually not so crowded...................... .... 9a. *S. atrovirens* var. *georgianus.*

Bristles twice the length of the achene; scales of spikelets rufous brown with green midribs; principal leaves usually 6-8 mm wide..... 10. *S. polyphyllus.*

Bristles smooth or with a few ascending barbs, curly.

Rays and pedicels smooth or somewhat scabrous below the involucels, not conspicuously striate, both usually drooping; scales reddish brown with strong, green midribs prolonged into sharp, short, spreading points; achenes about 1 mm long, Fawn Color (Ridgway Standard); bristles weak, about twice the length of the achene, included............................ 11. *S. lineatus.*

Rays (except the primary ones) and pedicels strongly upwardly scabrous, conspicuously striate, at least the principal rays inclined to be erect; scales of spikelets reddish, sometimes suffused with greenish black, the midrib not green, somewhat obtuse at the apex; achenes about 0.8 mm or less in length, colorless; bristles curled and much exserted beyond the scales.

Spikelets mostly sessile, in glomerules of 3-15.

Involucres and involucels reddish brown; scales reddish brown.

Spikelets ovoid, 3-6 mm long........................... 12. *S. cyperinus.*

Spikelets cylindric, 7-10 mm long.......... 12a. *S. cyperinus* f. *Andrewsii.*

Involucres and involucels drab with a blackish base.

Rays of normal length, the glomerules distinct, scales brownish, suffused with greenish black........................ 12b. *S. cyperinus* var. *pelius.*

Rays abbreviated, the glomerules crowded into dense, irregular masses. .... 12c. *S. cyperinus* var. *pelius* f. *condensatus.*

Spikelets mostly pedicellate, usually arranged in small clusters with the central one sessile and the remainder on pedicels of different lengths.

Involucels red brown or terra cotta....................... 13. *S. Eriophorum.*

Involucels dull brown, not reddish....................... 14. *S. pedicellatus.*

Involucels black. (See excluded species no. 88, p. 1031.) .... *S. atrocinctus.*

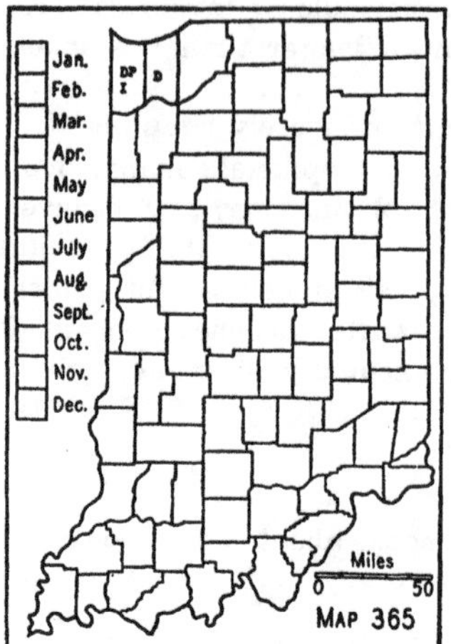

Scirpus subterminalis Torr.

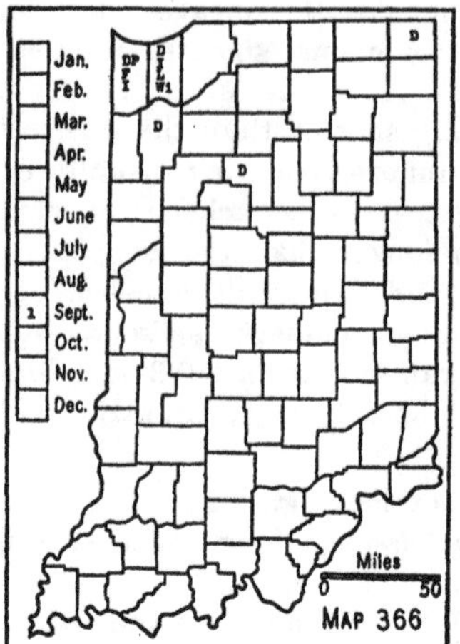

Scirpus debilis Pursh

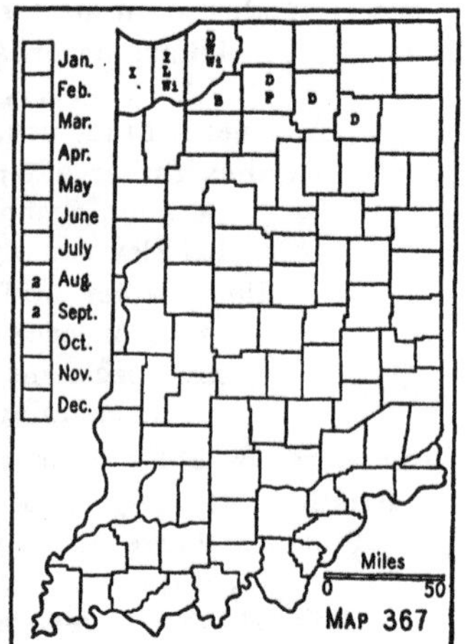

Scirpus Smithii Gray

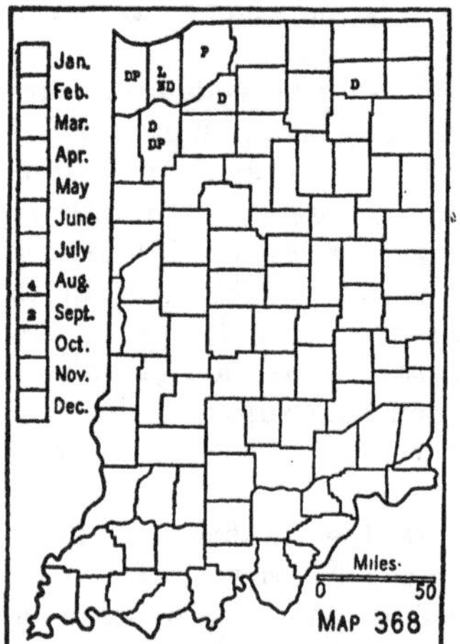

Scirpus Smithii var. setosus Fern.

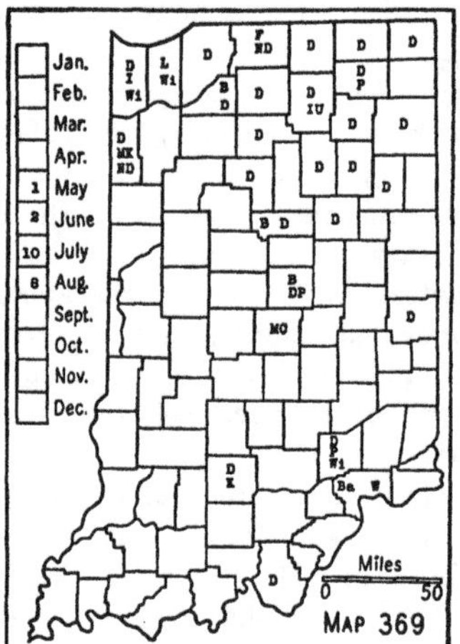

Scirpus americanus Pers.

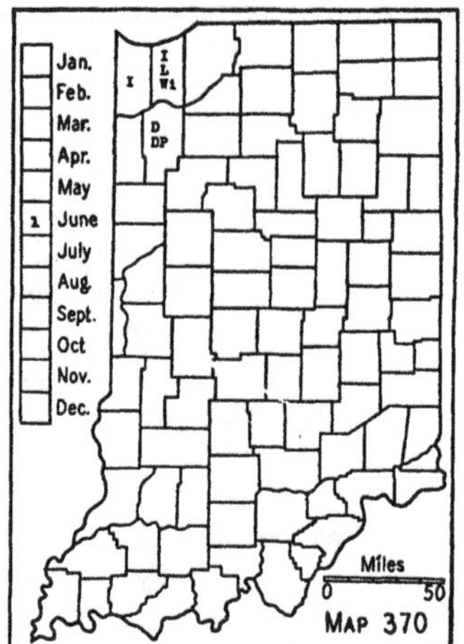

Scirpus Torreyi Olney

1. **Scirpus subterminàlis** Torr. Map. 365. My only specimens were found in a colony on the muddy border of the south side of Long Lake, Porter County, about a mile east of the Lake County line, where it was associated with *Scirpus validus*. In walking the entire length of the lake I noted only one colony. This was in very mucky soil from which the water had receded just far enough to expose the soil.

Newf. to B. C., southw. to N. J., Pa., Ind., and Idaho.

2. **Scirpus débilis** Pursh. Map 366. This species has been found in a few counties only in wet or mucky soil about sloughs in the dunes.

Maine, Ont. to Minn., southw. to Ga., Ala., and Nebr.

3. **Scirpus Smíthii** Gray. Map. 367. On the wet, sandy borders of lakes and sloughs.

Maine, Ont. to Mich., southw. to Pa., Ind., and Ill.

3a. **Scirpus Smithii** var. **setòsus** Fern. Map 368. Found in habitats

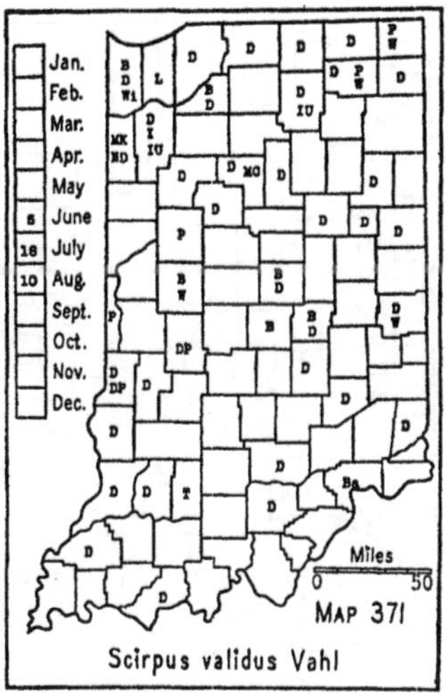

MAP 371
Scirpus validus Vahl

MAP 372
Scirpus acutus Muhl.

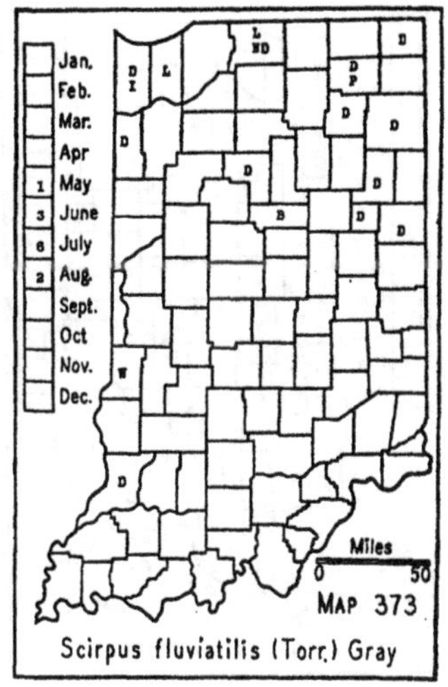

MAP 373
Scirpus fluviatilis (Torr.) Gray

similar to those in which the species is found. This variety is difficult to separate from *Scirpus debilis*. In fact, they are united in Britton and Brown, Illustrated Flora, ed. 2. The shape of the stem seems to be the only constant character. The divaricating bract of *Scirpus debilis* is very characteristic but it seems that all plants do not have a divaricating bract. The shape of the achene can not be relied upon since on the same plant one can find plano-convex as well as biconvex achenes.

Maine and Mass. to Ill.

4. **Scirpus americànus** Pers. Map 369. Frequent on the sandy shores of lakes and on gravelly bars in streams.

Throughout temperate N. A.; also found in S. A. and Eu.

5. **Scirpus Tórreyi** Olney. Map 370. Very local in a few swamps of the northwestern part of the state.

Maine to Man., southw. to R. I. and Minn.

6. **Scirpus válidus** Vahl. Map 371. This species grows in sandy or mucky soil in shallow water (usually 1-4 feet deep) in lakes and along streams. It is usually found in every lake of the state and when a lake begins to dry up it usually is the first species to occupy the area.

Throughout temperate N. A.; also found in W. I.

7. **Scirpus acùtus** Muhl. (*Scirpus occidentalis* (Wats.) Chase.) Map 372. Rather frequent in the lake area in habitats similar to those of the preceding species.

Newf. to B. C., southw. to Mass., cent. N. Y., Mo., Ariz., and Calif.

8. **Scirpus fluviátilis** (Torr.) Gray. Map 373. Infrequent in the lake area and in the Lower Wabash Valley. It is usually found in wet places about lakes, along streams, and in ditches and ponds. I have seen about five acres of it in Knox County on the west side of Swan Pond.

N. B. to the region of the Great Lakes and Minn., southw. to D. C., and Kans.

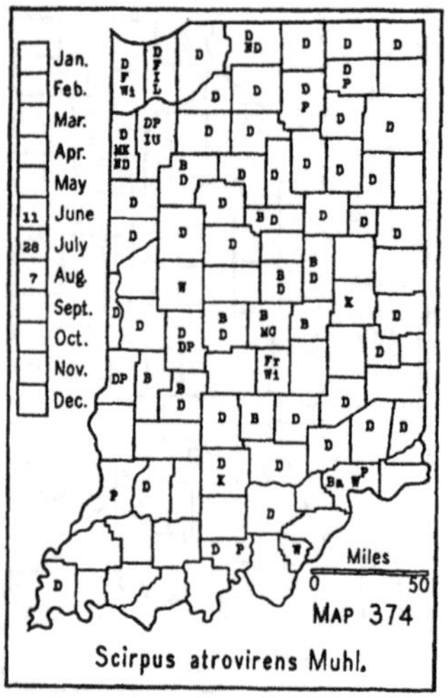

MAP 374
Scirpus atrovirens Muhl.

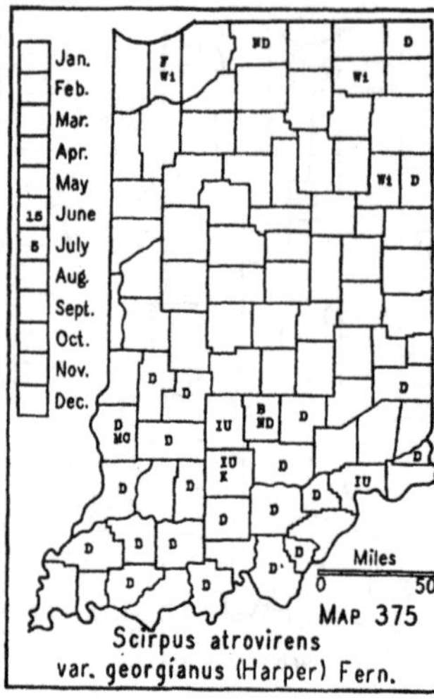

MAP 375
Scirpus atrovirens var. georgianus (Harper) Fern.

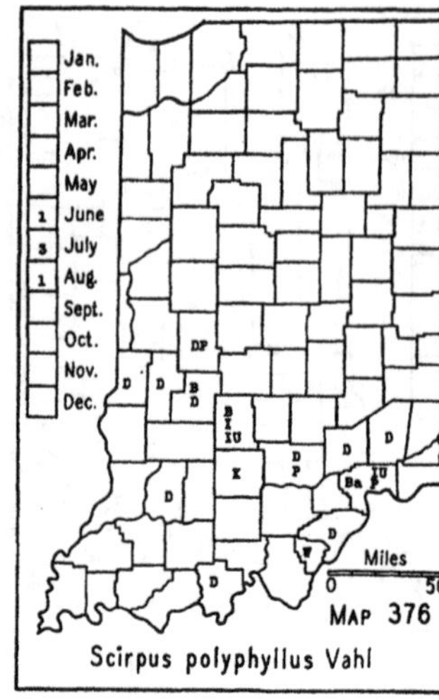

MAP 376
Scirpus polyphyllus Vahl

9. **Scirpus atróvirens** Muhl. Map 374. Frequent to common in almost all parts of the state. It is usually found in wet, mucky soil in ditches and ponds, along streams, and about lakes. One can infrequently find a specimen in which the rays of the inflorescence are short and the glomerules form a closed head. This form has received a name but I do not believe it is of taxonomic significance.

Maine to Sask., southw. to Ga. and Mo.

9a. **Scirpus atrovirens** var. **georgiànus** (Harper) Fern. (Rhodora 23: 134. 1921.) (*Scirpus georgianus* Harper.) Map 375. This variety is infrequent in the southern half of the state, becoming rare in our northern counties. The species and variety are distinct in their extremes but they so intergrade that their separation is not entirely satisfactory.

Newf. to Wis., southw. to Ga. and Ark.

9b. **Scirpus atrovirens** f. **prolíferus** Hermann. This is a viviparous form, occasionally with the species.

10. **Scirpus polyphýllus** Vahl. Map 376. Infrequent in springy places and in low beech and sweet gum woods in the southern half of the state. Its associates would indicate that it prefers a slightly acid soil. Viviparous forms are rather frequent.

Western N. E. to Minn., southw. to Ga. and Ark.

11. **Scirpus lineàtus** Michx. Map 377. This is the most common bulrush of the state. It is frequent throughout and, for the most part, is found in roadside ditches and along low roadsides. It prefers a moist or wet soil along streams, in low, open woodland and fallow fields, and about lakes and sloughs.

N. H., Ont. to Oreg., southw. to Ga. and Tex.

12. **Scirpus cyperìnus** (L.) Kunth. Map 378. Infrequent throughout the state in wet grounds of all kinds. It is more common in the lake area in wet places about lakes, in marshes, and along streams; southward it is found in roadside ditches, ponds, sloughs, sinkholes, and springy places

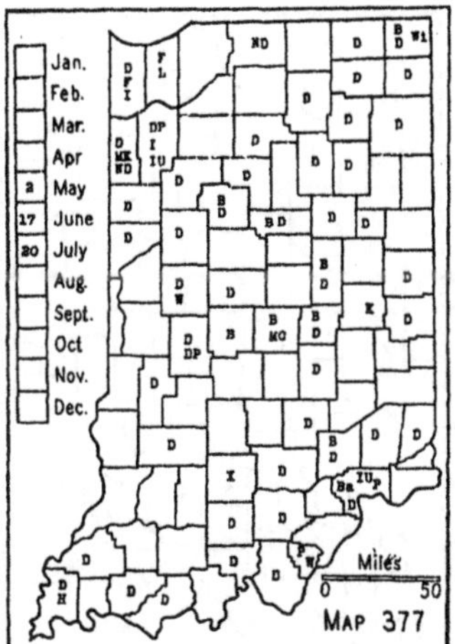

Scirpus lineatus Michx.

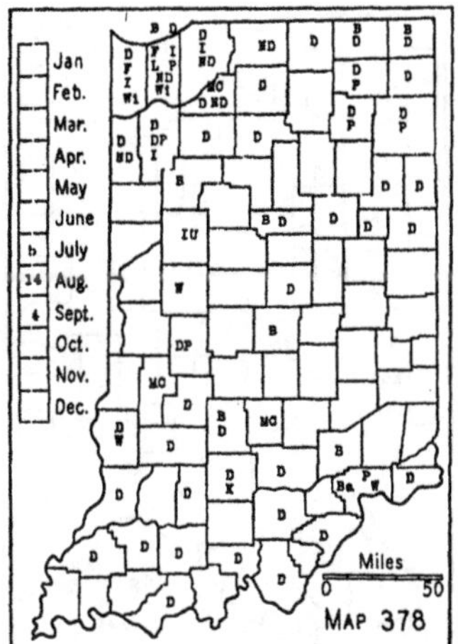

Scirpus cyperinus (L.) Kunth

Scirpus Eriophorum Michx.

and along streams. This species is extremely variable throughout its range in the grouping or segregation of the spikelets, the color of the involucre and involucels, and the color of the scales of the spikelets. Some authors do not recognize these differences while others do. I am dividing the species into the commonly recognized forms in order that those who do wish to separate these forms may have the advantage of the experience of other authors. The range of the several forms has not yet been ascertained and the range of the aggregate is given here.

Newf., Ont. to Sask., southw. to Fla. and La.

12a. **Scirpus cyperinus** f. **Andréwsii** (Fern.) Carpenter. (Dole. Flora of Vermont, p. 74. 1937.) This form has been found only in Allen County.

12b. **Scirpus cyperinus** var. **pèlius** Fern. This form is very local and is found in the habitat of the species. I have it only from Allen, Jasper, and Whitley Counties.

Newf. to Ont. and Minn., southw. to Conn., N. Y., Ind., and Wis.

12c. **Scirpus cyperinus** var. **pelius** f. **condensàtus** (Fern.) Blake. Found only in La Porte and Porter Counties. Its general range is that of the variety.

13. **Scirpus Erióphorum** Michx. Map 378-1. A botanical authority has referred to this species four sheets of my specimens of the *Scirpus cyperinus-pedicellatus* complex. It is to be noted that Britton and Brown, Illustrated Flora, ed. 2, refer this species and the next one to *Scirpus cyperinus*.

Conn. to Fla., westw. to La. and northw. in the Mississippi Valley to Ind.

14. **Scirpus pedicellàtus** Fern. Map 379. This so-called species is infrequent and is found throughout the state in habitats similar to those of *Scirpus cyperinus*.

E. Que., southw. to Conn., N. Y., Ind., and Wis.

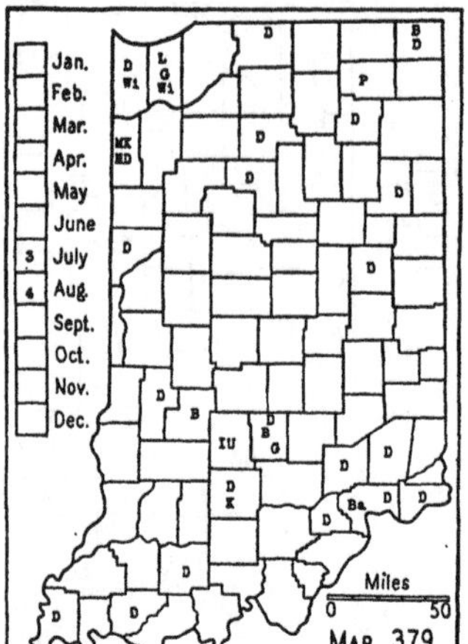

MAP 379
Scirpus pedicellatus Fern.

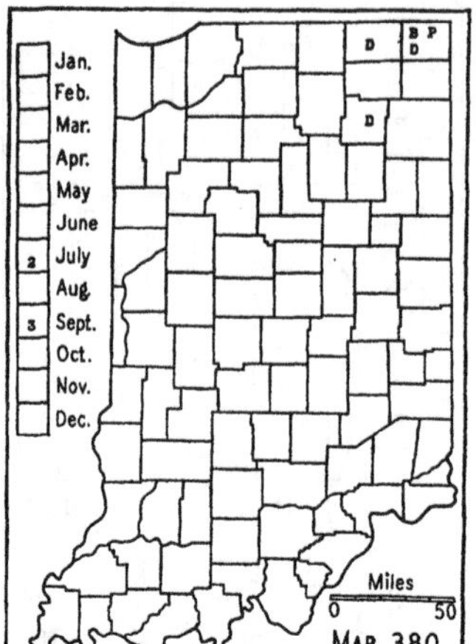

MAP 380
Eleocharis equisetoides (Ell.) Torr.

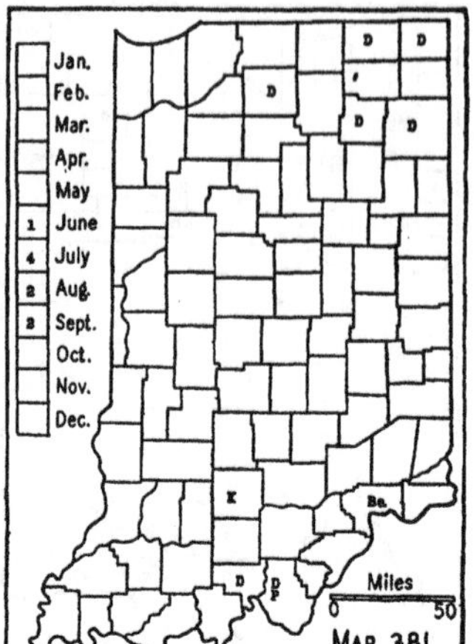

MAP 381
Eleocharis qaudrangulata (Michx.) R. & S. var. crassior Fern.

469. ELEÓCHARIS R. Br. SPIKERUSH

[Fernald and Brackett. The representatives of Eleocharis palustris in North America. Rhodora **31**: 56-77. 1929. Svenson. Monographic studies in the genus Eleocharis. Rhodora **31**: 121-135, 152-163, 167-191, 199-219, 224-242. 1929; **34**: 193-203, 215-227. 1932; **36**: 377-389. 1934; **39**: 210-231. 1937; **41**: 1-19, 43-77. 1939.]

The following key is adapted from Svenson's monographic studies of the genus. Svenson has checked the determination of all of my specimens.

Scales of mature spikelets persistent; spikelets scarcely thicker than the culms.
  Fruiting culms more than 2 mm in diameter; nerves of scales faint.
    Culms terete, with conspicuous cross-partitions................1. *E. equisetoides.*
    Culms quadrangular, without cross-partitions...2. *E. quadrangulata* var. *crassior.*
  Fruiting culms not more than 2 mm in diameter; nerves of scales distinct..3. *E. Robbinsii.*
Scales of mature spikes deciduous; spikelets thicker than the culms.
  Styles 2-cleft.
    Upper sheaths loose, with white, scarious tips...................4. *E. olivacea.*
    Upper sheaths close and firm, not scarious at the tips.
      Annual, with fibrous roots.
        Tubercle (style base) often depressed or saucer-shaped.......5. *E. geniculata.*
        Tubercle more or less conical.
          Width of tubercle less than two thirds that of the achene.
            Achenes smooth; tubercle about half as wide as the achene...6. *E. ovata.*
            Achenes pitted; tubercle about a fourth as wide as the achene..7. *E. intermedia.*
          Width of tubercle nearly or quite equal to that of the achene.
            Tubercle deltoid, a third to nearly a half as high as the body of the achene; bristles much exceeding the achene.
              Spikelets ovoid-cylindric..............................8. *E. obtusa.*
              Spikelets ellipsoid....................8a. *E. obtusa* var. *ellipsoidalis.*
            Tubercle very low, not more than a fourth as high as the body of the achene; summit of achene appearing truncate; bristles equaling the achene or rudimentary.
              Bristles about equaling the achene.................9. *E. Engelmanni.*
              Bristles absent or rudimentary.........9a. *E. Engelmanni* f. *detonsa.*

Perennial, with horizontal rootstocks.
Culms 0.5-5 mm in diameter (in dried material) at the summit of the upper sheath; basal scales of spikelet usually 2 or 3 below the thinner fertile scales; median scales acute; tubercle broadly ovate, as wide as long. ....10. *E. Smallii.*
Culms 0.5-1.5 mm in diameter at the summit of the upper sheath; basal scales of the spikelet solitary, spathiform, usually completely encircling the base of the spikelet; median scales obtuse; tubercle conical, as long as or longer than wide........11. *E. calva.*
Styles 3-cleft.
Achenes less than 2 mm long; style base not confluent with the apex of the achenes, forming a tubercle.
Surface of the achene regularly marked off by longitudinal and transverse lines.
Culms not more than 0.5 mm in diameter; achenes obscurely 3-angled; bristles equaling or longer than the achene or absent........12. *E. acicularis.*
Culms about 1 mm in diameter; achenes pyriform; bristles none........13. *E. Wolfii.*
Surface of the achene smooth or pitted, the pits arranged irregularly or in regular, longitudinal lines.
Achenes smooth.
Achenes turbinate-lenticular.
Bristles longer than the achene.
Spikelets ovoid-cylindric........8. *E. obtusa.*
Spikelets ellipsoid........8a. *E. obtusa* var. *ellipsoidalis.*
Bristles shorter than or equaling the achene........9. *E. Engelmannii.*
Achenes triangular; bristles not exceeding the achene or absent.
Mature achenes nearly black, the body not tapering toward the apex, the angles blunt; 1 mm or more long; tubercle closely capping the crown of the achene; bristles absent........14. *E. Melanocarpa.*
Mature achenes nearly black, the body not tapering toward the apex, the angles blunt; 1 mm or more long; tubercle closely capping the crown of the achene; bristles absent........14. *E. melanocarpa.*
Achenes pitted, the pits arranged irregularly or in regular, longitudinal lines.
Culms slender, erect; style bases depressed.
Culms 4-8 angled; scales of spikelets obtuse or merely acute, not conspicuously whitened at the apex.
Achenes Wax Yellow (Ridgway Standard), in age becoming golden yellow to dull orange, averaging 1-1.1 mm long (including the style base); pits of achene usually shallow; culms usually 6-8-angled........16. *E. elliptica.*
Achenes Olivaceous (Ridgway Standard); pits of achene usually deep with some of the cell-projections verrucose; culms 5-angled........17. *E. tenuis* var. *verrucosa.*
Culms flattened; scales of spikelets (except sometimes in var. *atrata*) with conspicuously whitened, often bifid, acuminate tips.
Scales chestnut brown........18. *E. compressa.*
Scales conspicuously blackened........18a. *E. compressa* var. *atrata.*
Culms capillary, diffusely spreading; scales obtuse; style base narrow-conic, about twice as long as wide; achenes finely pitted in longitudinal lines........7. *E. intermedia.*
Achenes 2-3 mm long; style base confluent with the apex of the achene, not forming a tubercle.
Culms 1-2 mm in diameter, (2) 3-10 dm long, flattened, erect, or the sterile ones reclining and often rooting at the tips; beak of achene about a third as long as the body........19. *E. rostellata.*
Culms less than 1 mm wide, 0.5-3 dm high, scarcely flattened, erect; beak of achene about a fourth as long as the body...20. *E. pauciflora* var. *Fernaldii.*

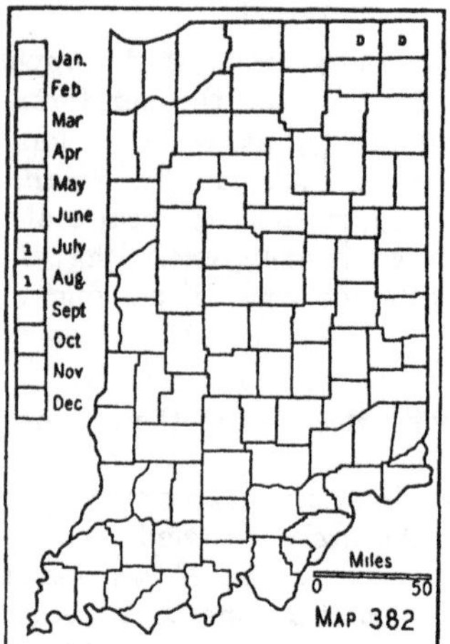

Eleocharis Robbinsii Oakes

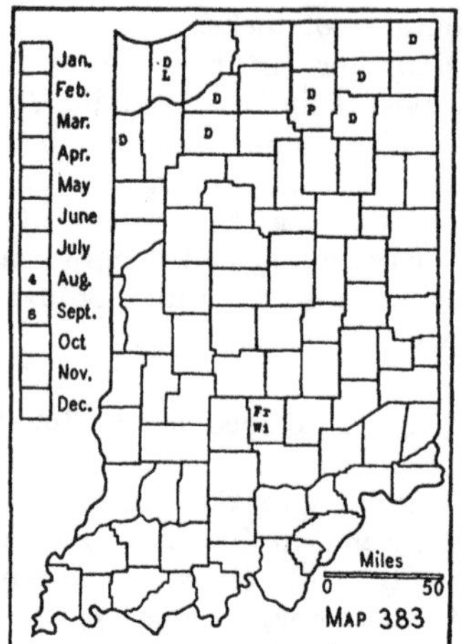

Eleocharis olivacea Torr.

Eleocharis geniculata (L.) R. & S.

1. **Eleocharis equisetoìdes** (Ell.) Torr. (*Eleocharis interstincta* of authors.) KNOTTED SPIKERUSH. Map 380. In shallow water on the sandy bottoms of some of our northern lakes.

Mass. to Fla. and Tex. and inland to Mich., Wis., and Mo.

2. **Eleocharis quadrangulàta** (Michx.) R. & S. var. **crássior** Fern. (Rhodora 37: 393. 1935.) (*Eleocharis mutata* of Britton and Brown, Illus. Flora, ed. 2, not *Scirpus mutatus* L. and *Eleocharis quadrangulata* of Indiana authors, not *Scirpus quadrangulatus* Michx.) ANGLED SPIKERUSH. Map 381. In sandy or mucky soil in shallow water or on the borders of lakes, ponds, and sinkholes.

Mass. to s. Ont., southw. to Ga., Tex., and Mexico.

3. **Eleocharis Robbínsii** Oakes. ROBBINS SPIKERUSH. Map 382. In marly soil on the borders of lakes. This species apparently does not fruit every year and it may be more frequent in Indiana than our records indicate.

N. S. and s. N. B. to Fla., chiefly along the Coastal Plain, and westw. through cent. N. Y. to Mich., Ind., and Ont.

4. **Eleocharis olivàcea** Torr. (*Eleocharis flaccida* (Reichenb.) Urban var. *olivacea* (Torr.) Fern. & Grisc. Rhodora 37: 155. 1935.) BRIGHT GREEN SPIKERUSH. Map 383. Wet, sandy or muddy, marl borders of lakes.

N. S., Ont. to Mich., southw. to Fla., Pa., Ohio, and Ind.

5. **Eleocharis geniculàta** (L.) R. & S. (Rhodora 41: 50-52. 1939.) (*Eleocharis capitata* R. Br. and *Eleocharis caribaea* (Rottb.) Blake.) Map 384. In wet, marl borders of lakes and in dried-up sloughs. In addition to the counties shown on the map, it is known in the Great Lakes area only from Washtenaw County, in southeastern Michigan and from southern Ontario.

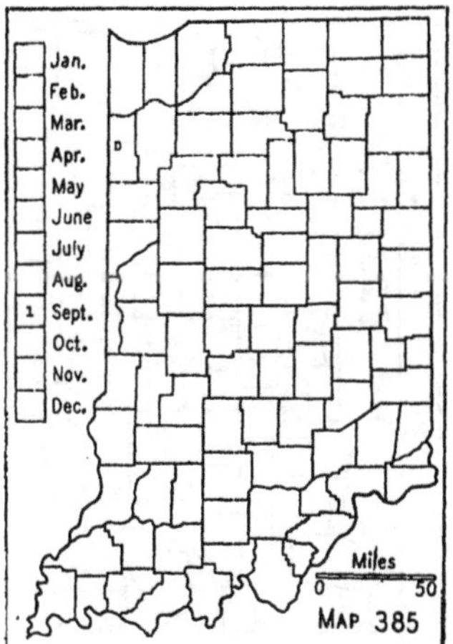

MAP 385

Eleocharis ovata (Roth) R. & S.

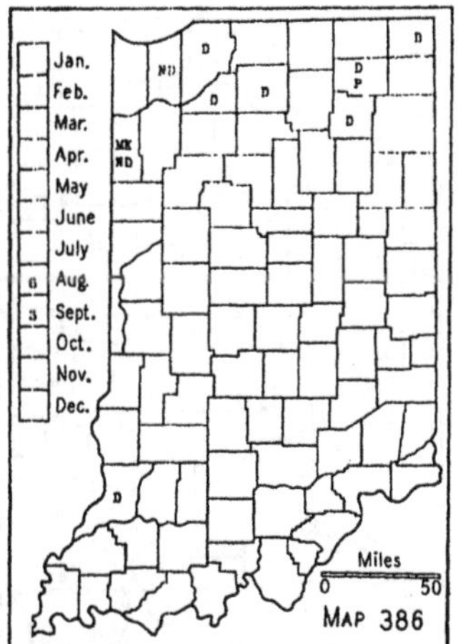

MAP 386

Eleocharis intermedia (Muhl.) Schultes

MAP 387

Eleocharis obtusa (Willd.) Schultes

6. **Eleocharis ovàta** (Roth) R. & S. OVOID SPIKERUSH. Map 385. My only specimen was collected in the bottom of a dried-up dredged ditch about 4 miles southeast of Conrad in Newton County and determined by H. K. Svenson. It has been reported from Lake and Porter Counties by Peattie but I have not seen a specimen.

Local from Newf. and e. Que. to Maine, Vt., Conn., and Mass.; also in Mich., Wis., Minn., and Wash.

7. **Eleocharis intermèdia** (Muhl.) Schultes. (Rhodora **41**: 67. 1939.) MATTED SPIKERUSH. Map 386. Muddy borders of ponds and lakes, wet, marl borders of lakes, and in the outlets of springs.

Que. to w. Ont., southw. to N. J., Pa., Ohio, and Iowa.

8. **Eleocharis obtùsa** (Willd.) Schultes. BLUNT SPIKERUSH. Map 387. Throughout the state in muddy or wet places in almost all habitats, principally in ditches, sloughs, swamps, and ponds and on the borders of streams and lakes.

The species is variable and my no. 45541 from Monroe County and no. 24288 from Posey County are here cited as exceptional plants.

Cape Breton and e. N. B. to Nebr., southw. to the Gulf of Mexico; appearing again in the northwest from B. C. to Calif.; also in the Hawaiian Islands.

8a. **Eleocharis obtusa** var. **ellipsoidàlis** Fern. (Rhodora **31**: 218. 1929.) I have a specimen from a tamarack bog in La Porte County that Svenson refers to this variety.

E. Mass. to Va. and Ind.

9. **Eleocharis Engelmánni** Steud. ENGELMANN SPIKERUSH. Map 388. In muddy places in roadside ditches and on the muddy borders of artificial and natural ponds.

S. Maine to Va., westw. through Ind., Tenn., and Mo. to Okla.

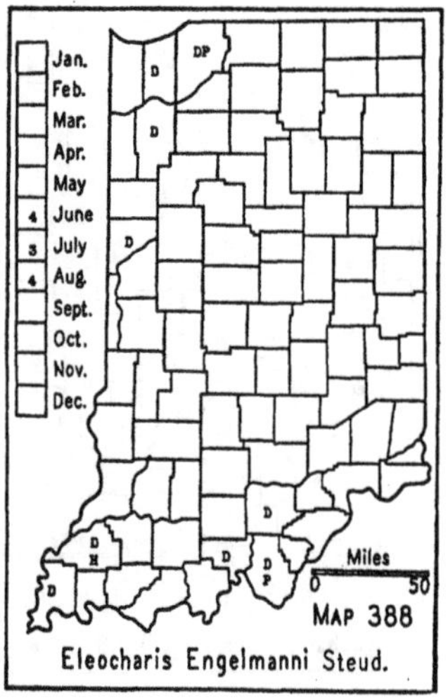

Eleocharis Engelmanni Steud.

Eleocharis Smallii Britton

Eleocharis calva Torr.

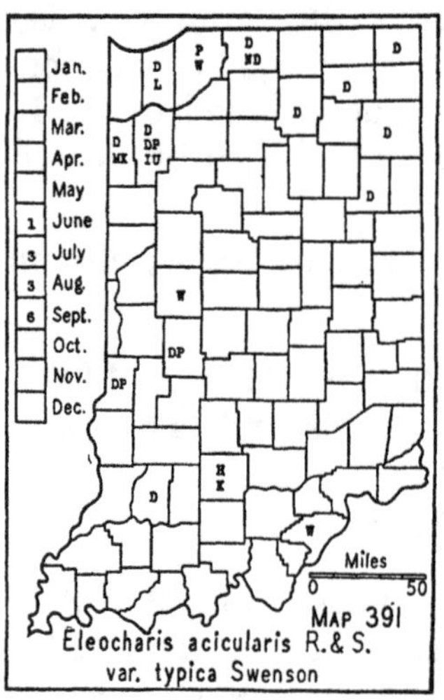

Eleocharis acicularis R.& S. var. typica Swenson

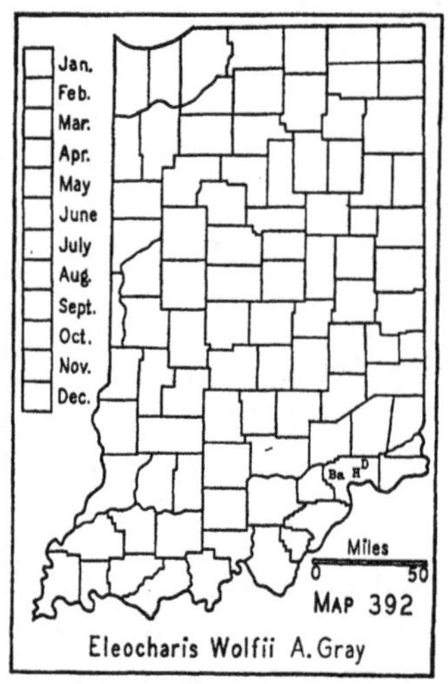

Eleocharis Wolfii A.Gray

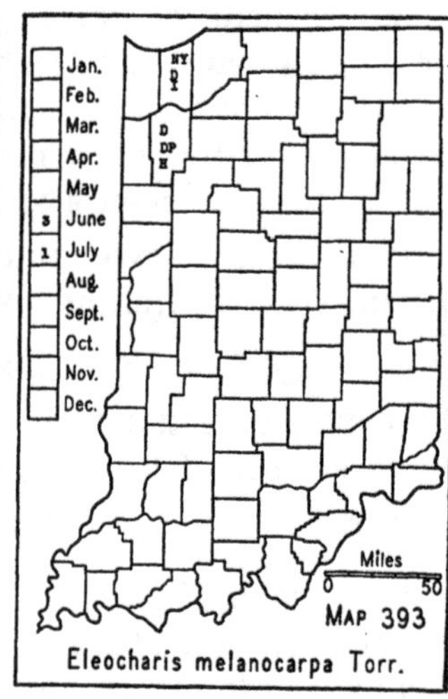

Eleocharis melanocarpa Torr.

9a. **Eleocharis Engelmanni f. detónsa** (Gray) Svenson. (*Eleocharis Engelmanni* var. *detonsa* Gray.) My specimen was collected in a field 2 miles northwest of Culver on the muddy border of a pond, where it was frequent. Also collected by E. J. Hill in La Porte County.

Mass., Pa., Mich., Ind., Ill. and Ariz.

10. **Eleocharis Smállii** Britton. (*Eleocharis palustris* in part, of Gray, Man., ed. 7 and of Indiana authors.) SMALL'S SPIKERUSH. Map 389. In muddy, peaty or wet, sandy places in ditches, sloughs, ponds, marshes, and like habitats on the borders of streams and lakes.

Sw. N. S. to Mich. and Nebr., southw. to Del., Pa., Ind., Ill., and Mo.

11. **Eleocharis cálva** Torr. (*Eleocharis palustris* var. *calva* (Torr.) Gray and *Eleocharis palustris* var. *glaucescens* of Indiana authors.) Map 390. In muddy, sandy or peaty soil in ditches, sloughs, and marshes and

on the borders of streams and lakes. In wet, stony or gravelly places along the Ohio River and in springy, marl borders of some lakes.

Que. to Alberta and Wash., southw. to Fla. and Okla., and n. Mex.; also in Hawaii and e. Asia.

12. **Eleocharis aciculàris** (L.) R. & S. var. **týpica** Svenson. NEEDLE SPIKERUSH. Map 391. In the muddy or sandy bottoms or borders of ditches, sloughs, streams, and lakes. Sometimes on the springy marl borders of lakes.

Newf. to Alaska, southw. to Fla. and Okla.

13. **Eleocharis Wólfii** Gray. WOLF'S SPIKERUSH. Map 392. My only specimens were found in Jefferson County in low, flat clearings about 3½ miles southwest of Hanover and 3 miles southeast of Hanover.

Ind. to Kans. and La.

14. **Eleocharis melanocárpa** Torr. BLACK-FRUITED SPIKERUSH. Map 393. Wet or moist, sandy borders of marshes and sloughs.

Atlantic coast from Mass. to Texas, and in nw. Ind.

15. **Eleocharis microcárpa** Torr. var. **filicúlmis** Torr. (Rhodora 39: 228-229. 1937.) (*Eleocharis Torreyana* Boeckl.) Map 394. Our only specimens were found in moist sand in the bottom of a roadside ditch about 2 miles southeast of Tefft in Jasper County.

Atlantic coast from Conn. to Fla. and Tex.; also in Cuba.

16. **Eleocharis ellíptica** Kunth. (Rhodora **41**: 65. 1939.) (*Eleocharis capitata* var. *borealis* Svenson. Rhodora **34**: 200-202. 1932.) Map 395. This sedge seems to have a wide distribution in the state. In the lake area it is found in strongly marl borders of lakes and elsewhere in moist prairie habitats.

Newf. to B. C., southw. to N. J., Tenn., Ind., and Ill.

17. **Eleocharis ténuis** (Willd.) Schultes var. **verrucòsa** (Svenson) Svenson. (Rhodora **41**: 66. 1939.) (*Eleocharis capitata* var. *verrucosa* Svenson and *Eleocharis tenuis* of authors.) Map 396. For the most part, our specimens are from wet, hard, clay soil of the borders of ponds and wet woods. Our Jasper County specimen is from a wet, interdunal flat.

Va., Ind., Ill. to Ark. and Okla., southw. to La.

18. **Eleocharis compréssa** Sulliv. (*Eleocharis acuminata* (Muhl.) Nees.) Map 397. I have only three specimens from Indiana and these are from a wide range of distance and kinds of habitats. The Ohio County specimen was found on the slope of the bank of the Ohio River, the specimen from Tipton County is from a wet, prairie habitat along the railroad just west of Goldsmith, and the specimen from Wabash County was found on the border of a small lake. It has been reported from Lake and St. Joseph Counties, but I have not seen the specimens.

W. Que. to Sask. and B. C., southw. to Ga., Okla., and the Pacific States.

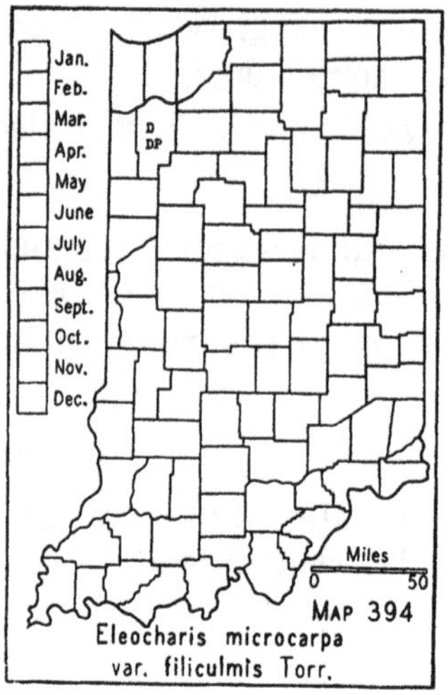

Eleocharis microcarpa var. filiculmis Torr.

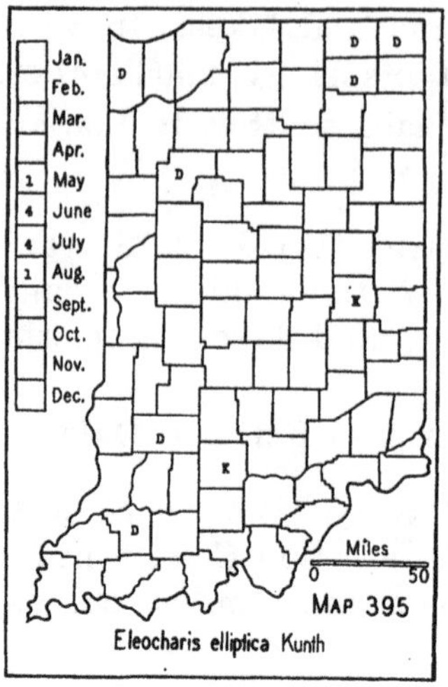

Eleocharis elliptica Kunth

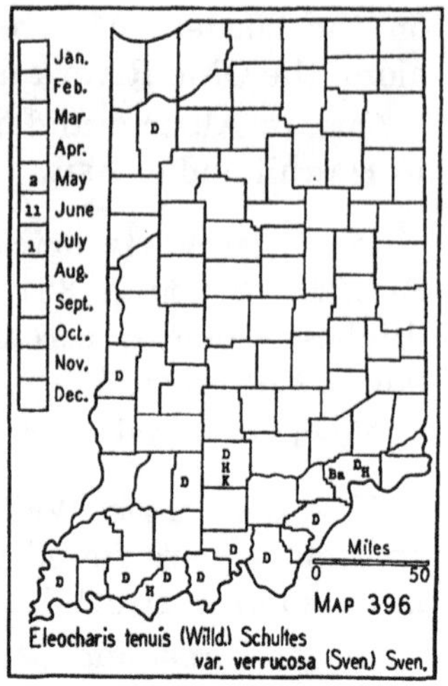

Eleocharis tenuis (Willd.) Schultes var. verrucosa (Sven.) Sven.

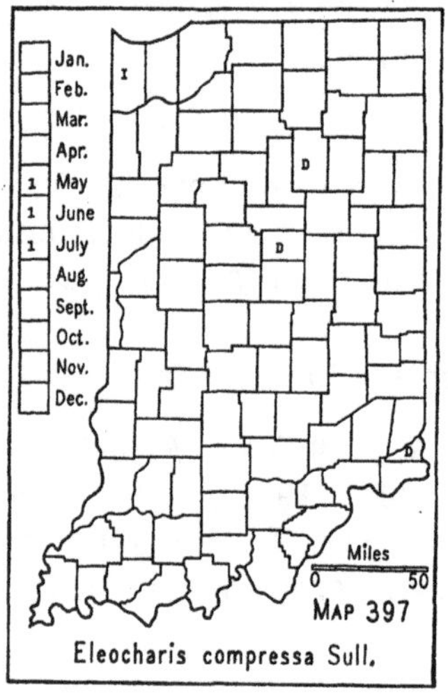

Eleocharis compressa Sull.

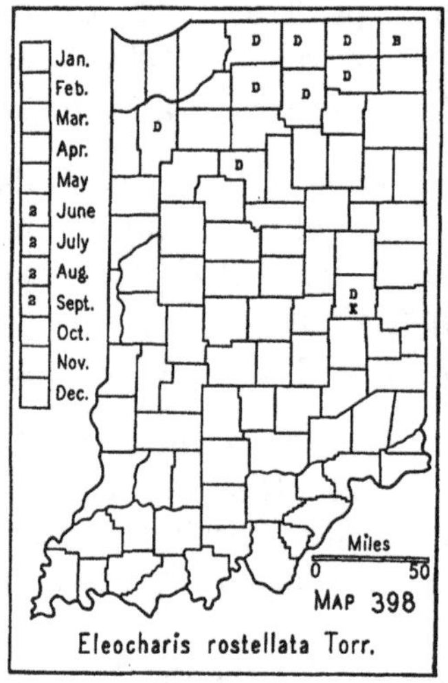

Eleocharis rostellata Torr.

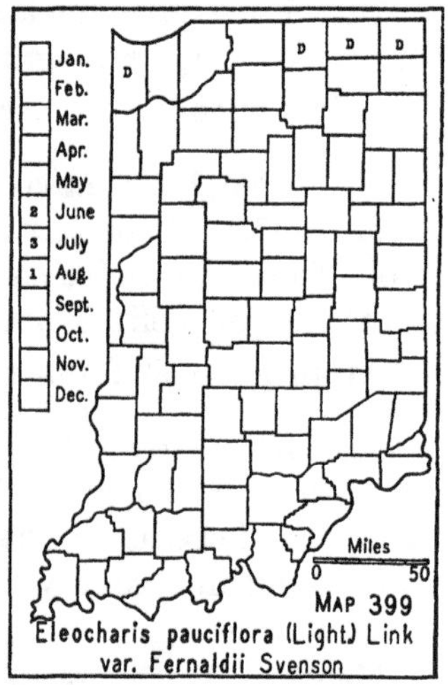

Eleocharis pauciflora (Light.) Link var. Fernaldii Svenson

18a. **Eleocharis compressa** var. **atràta** Svenson. (Rhodora 34: 218. 1932.) Under his description of this variety Svenson refers to it Bebb's specimen no. 2048 from Lake County which is in the herbarium of the University of Wisconsin. There is also a specimen in the Field Museum collected by Lansing near Indiana Harbor in 1903. It is labeled *Eleocharis acuminata* (Muhl.) Nees.

N. Mich., and Wis., southw. to N. Y., Pa., and Ind.

19. **Eleocharis rostellàta** Torr. BEAKED SPIKERUSH. Map 398. Springy marshes and wet, marl borders of lakes.

N. S. to Fla., chiefly in salt marshes along the coast; rare inland, becoming common in the alkaline regions of the West; also in Bermuda, Cuba, and Mex.

20. **Eleocharis paucifiòra** (Lightf.) Link var. **Fernáldii** Svenson. Rhodora 36: 380. 1934.) (*Scirpus pauciflorus* Lightf.) FEW-FLOWERED SPIKE-RUSH. Map 399. This sedge prefers the wet or moist, marly borders of

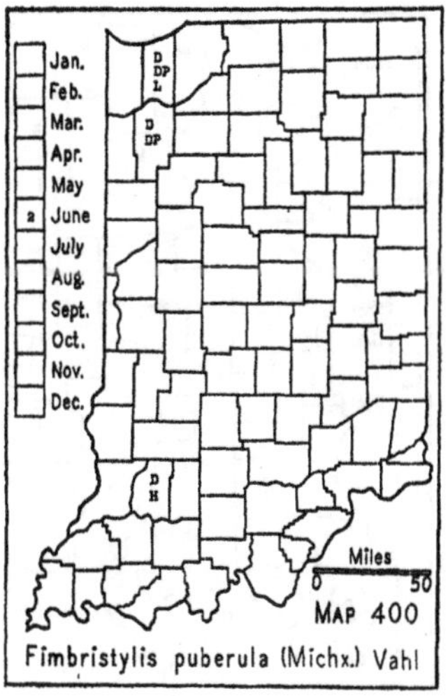

MAP 400
Fimbristylis puberula (Michx.) Vahl

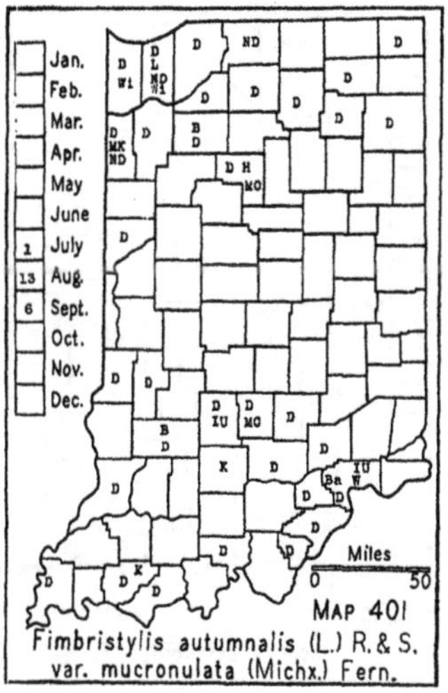

MAP 401
Fimbristylis autumnalis (L.) R. & S. var. mucronulata (Michx.) Fern.

MAP 402
Stenophyllus capillaris (L.) Britton

lakes and, where such a habitat occurs, it is often found in nearly pure stands over large areas. It is also found in a few marshes and along the borders of some of the sloughs in Lake County. It has been reported also from Newton County.

Newf. to Que., southw. to n. N. E., N. Y., Ind., and Ill.

471. FIMBRISTÝLIS Vahl

Stigmas 2; achenes lenticular.
- Scales of spikelets, at least the lower ones, puberulent or minutely pubescent; achenes slightly obovoid, truncate, about 1.5 mm long, longitudinally pitted, grayish. .... 1. *F. puberula.*
- Scales of spikelets glabrous, glossy. (See excluded species no. 94, p. 1031) .... *F. castanea.*

Stigmas 3; achenes 3-angled, colorless.
- Umbels usually simple, sometimes compound; spikelets ovoid; achenes about 0.75 mm long. (See excluded species no. 93, p. 1031) .... *F. autumnalis.*
- Umbels mostly compound; spikelets linear; achenes about 0.5 mm long. .... 2. *F. autumnalis* var. *mucronulata.*

1. **Fimbristylis pubérula** (Michx.) Vahl. Map 400. In moist, sandy soil in an interdunal flat habitat. It must be very local since I have seen it only three times.

Southern N. Y. to Fla. and La., and from Ont., Mich., Ind., and Ill. to Kans. and Tex.

2. **Fimbristylis autumnàlis** (L.) R. & S. var. **mucronulàta** (Michx.) Fern. (Rhodora **37**: 398. 1935.) (*Fimbristylis autumnalis* of some authors.) Map 401. Moist, sandy, and muddy shores of lakes, sloughs, and streams and in ditches.

This is a highly variable species both as to habit and morphological characters. When growing in the mud or in moist sand, the plant may be short and the umbels simple. When growing in its preferred habitat or among vegetation it may be 8-12 inches high. The margins of the leaves may be entire or rather closely serrate. The achenes may be plainly reticulated

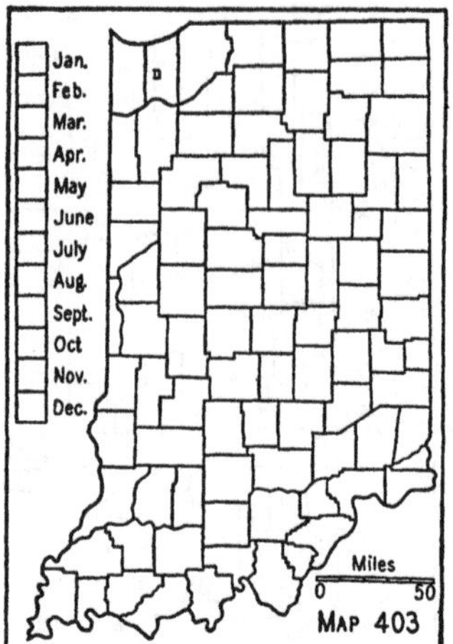

Psilocarya nitens(Vahl)Wood

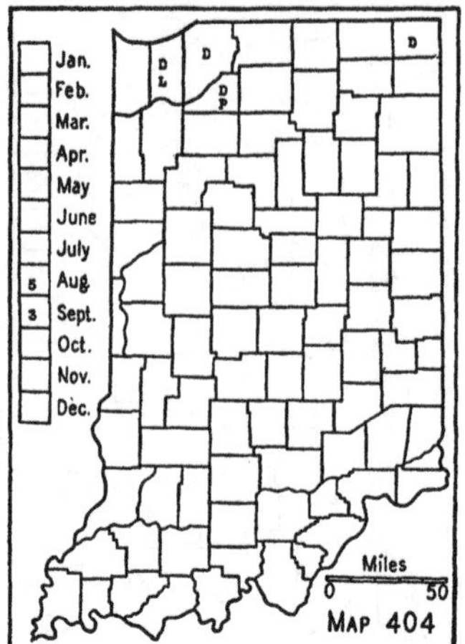

Psilocarya scirpoides Torr.

Cladium mariscoides (Muhl.) Torr.

crosswise or very faintly so, varying somewhat in length, and free of tubercles or covered with them more or less all over the surface. I have not seen a specimen of the typical form of this species and the data given in the key have been obtained from published studies. Doubtless all Indiana plants belong to this variety.

Conn. to Ill., southw. to Fla. and Tex.

471A. BULBOSTŸLIS [Kunth] C. B. Clarke

"Plant bearing crowded sessile spikelets at the bases of the leaves; spikelets of the terminal umbel rather crowded, 3-10 mm long, longer than their pedicels" (Rhodora 40: 395. 1938)..1. *B. capillaris.*

Plant not bearing sessile spikelets at the base of the leaves; lateral spikelets of umbels 2.5-6 mm long, shorter than the pedicels; pedicels unequal, 0.1-10 mm long...1a. *B. capillaris* var. *crebra.*

1. **Bulbostylis capillàris** (L) C. B. Clarke. (Rhodora 40: 395. 1938.) (*Stenophyllus capillaris* (L.) Britt.) This change of name came too late to change the name on the map. Map 402. I have the typical form of this species from Elkhart, Kosciusko, Lagrange, St. Joseph, and Starke Counties. The map shows both the typical form and the variety.

This plant is infrequent in the northern part of the state and rare in the southern part. It is found in very dry, sandy soil, usually in fallow fields and clearings, on open, sandy knolls and dunes, and the variety sometimes in residual soil on the crests of sandstone ridges and on cliffs.

Southern Maine to Minn., southw. to Va. and Mo.

1a. **Bulbostylis capillaris** var. **crèbra** Fern. (Rhodora 40: 395. 1938.) This variety has the same habitat as that of the species and ranges throughout the state. Only thorough field study will convince me that this variety is distinct in Indiana.

Md. to s. Ill., southw. to Ga., Ala., Ark., and Tex.

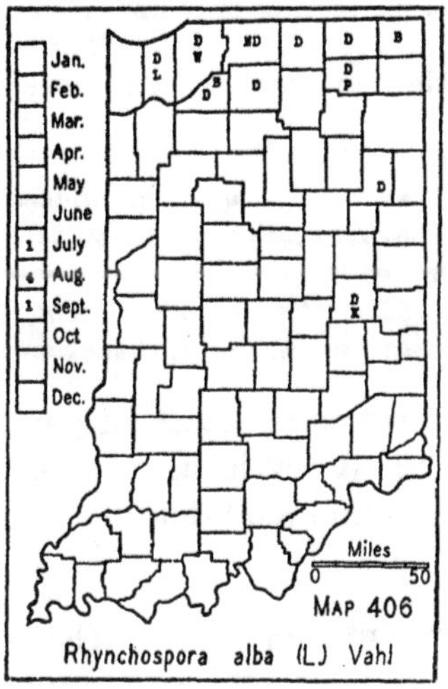

MAP 406
Rhynchospora alba (L.) Vahl

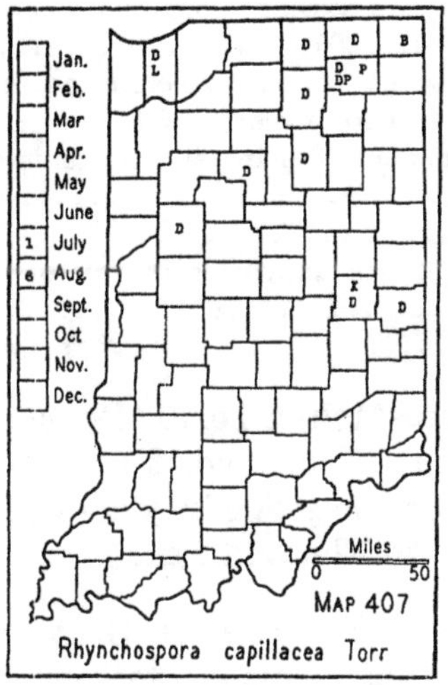

MAP 407
Rhynchospora capillacea Torr

MAP 408
Rhynchospora capillacea var. leviseta Hill

472. PSILOCÀRYA Torr.

Tubercle short, merely capping the achene, its greatest height much less than 0.5 mm; achene strongly transversely wrinkled.............................1. *P. nitens.*
Tubercle sword-shaped, more than 0.5 mm long; achene not strongly transversely wrinkled...2. *P. scirpoides.*

1. **Psilocarya nìtens** (Vahl) Wood. Map 403. In sandy soil on the borders of sloughs. My only specimen is one collected by Umbach on the border of a slough at Dune Park, Porter County, in 1899. Evidently it is very local.

Atlantic coast from Long Island, N. Y. to Fla. and along the Gulf to Tex. and along Lake Michigan in Ind.

2. **Psilocarya scirpoìdes** Torr. Map 404. This species is local but common to abundant where found. It grows in wet, sandy soil in marshes and on the borders of sloughs and lakes.

Mass. to R. I. and in n. Ind.

489. CLÀDIUM P. Br.

1. **Cladium mariscoìdes** (Muhl.) Torr. (Rhodora 25: 49. 1923.) Map 405. Rather frequent or locally common where found in the lake area in shallow water and on the wet borders of lakes and in marshes and springy places. It is usually found in very marly places.

N. S. to Ont. to Sask. and Minn., southw. to Fla., Ky., and Iowa.

492. RHYNCHÓSPORA Vahl BEAKRUSH

Mature achenes (exclusive of tubercle) 4.5-6 mm long.
Bristles longer than the achene.................................1. *R. macrostachya.*
Bristles shorter than the achene.....................2. *R. corniculata* var. *interior.*
Mature achenes (exclusive of tubercle) less than 4 mm long.
Achenes transversely wrinkled; bristles upwardly barbed.............3. *R. cymosa.*
Achenes smooth; bristles downwardly barbed or smooth.
Scales of spikelets (when fresh) white or nearly so, becoming tawny with age;

spikelets in terminal and axillary corymbose heads, perfecting only one flower; stamens 2; bristles 9-12 (20)....................................4. *R. alba.*

Scales of spikelets chestnut color; spikelets perfecting more than one flower; stamens 3; bristles 6, rarely more.

Leaves all filiform; spikelets 3-6 in terminal clusters.

Bristles barbed...5. *R. capillacea.*

Bristles not barbed..............................5a. *R. capillacea* f. *leviseta.*

Leaves wider, flat; spikelets numerous in clusters or heads.

Bristles barbed...................................6. *R. glomerata* var. *minor.*

Bristles not barbed.................6a. *R. glomerata* var. *minor* f. *discutiens.*

1. **Rhynchospora macrostàchya** Torr. (*Rynchospora corniculata* in part, of Britton and Brown, Illus. Flora, ed. 2.) Map 412. Very local in a few counties of the lake area on the sedge borders of lakes and sloughs.
Mass. to Mich., southw. to Fla. and Tex.

2. **Rhynchospora corniculàta** (Lam.) Gray var. **intèrior** Fern. (Rhodora 20: 140. 1918.) Map 409. This is a tall, coarse sedge found growing in wet woods and roadside ditches in a few of the Ohio River counties. Very local.
Ind., southw. to Ala., Ark., and Tex.

3. **Rhynchospora cymòsa** Ell. Map 410. This species has been reported from Lake and Porter Counties. In the herbarium of the University of Wisconsin there are 2 sheets from Lake County and 4 sheets from Porter County collected by Umbach. These specimens were found in wet, sandy soil along sloughs and in bogs.*
N. J., Pa. to Ill., southw. to Fla. and Tex.

4. **Rhynchospora álba** (L.) Vahl. Map 406. Mostly in the lake area. Infrequent in sedge marshes and bogs, usually on the borders of lakes.
Newf. to Alaska, southw. to Fla., Ky., and in n. Calif.

5. **Rhynchospora capillàcea** Torr. Map 407. Local in marly, springy places in the lake area, usually associated with the preceding species and with *Scleria verticillata.*
N. B., e. Que. to w. Ont., southw. to N. J., Pa., Ohio, Ind., and Mo.

5a. **Rhynchospora capillacea f. levisèta** (E. J. Hill) Fern. (Rhodora 37: 252. 1935.) Map 408. Local in a few of the northern counties. Usually found on marly borders of lakes and in interdunal flats.
Maine, Ont., Mich., and Ind.

6. **Rhynchospora glomeràta** (L.) Vahl var. **mìnor** Britt. (Rhodora 37: 401-402. 1935.) (*Rynchospora glomerata* of Gray, Man., ed. 7 and Britton and Brown, Illus. Flora, ed. 2.) Map 411. Infrequent in wet or moist sedge borders of lakes and in interdunal flats. This is our most common species of this genus and at a short distance it might be confused with *Cladium mariscoides* but the latter is much stiffer in habit.
N. B. to Ont. and Mich., southw. to Fla. and Tex.

* Collected in Newton County in 1938 by Madge McKee in swampy land about 3 miles northwest of Morocco. Specimen in her herbarium.

MAP 409 Rhynchospora corniculata var. interior Fern.

MAP 410 Rhynchospora cymosa Ell.

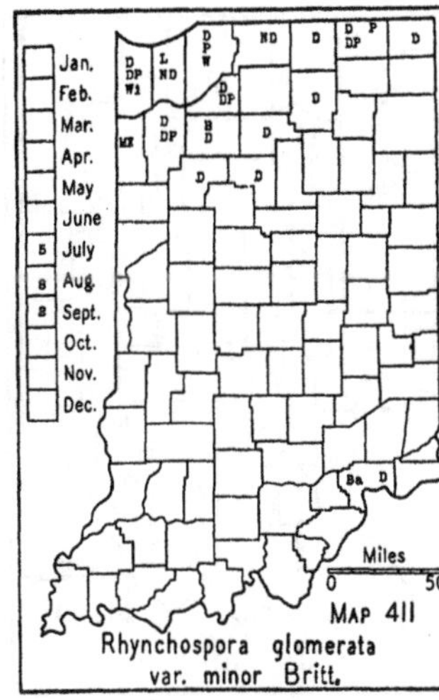
MAP 411 Rhynchospora glomerata var. minor Britt.

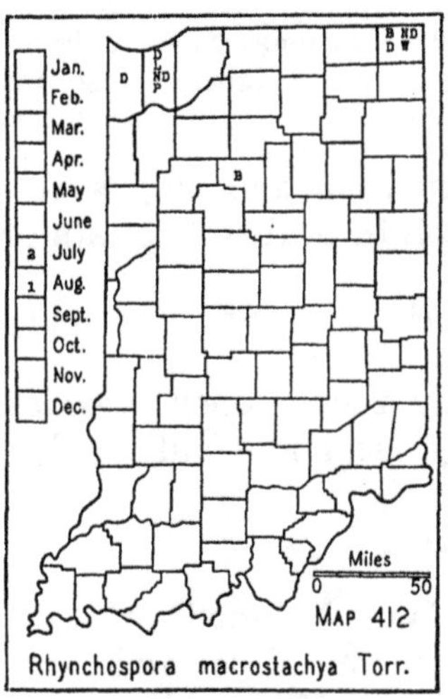
MAP 412 Rhynchospora macrostachya Torr.

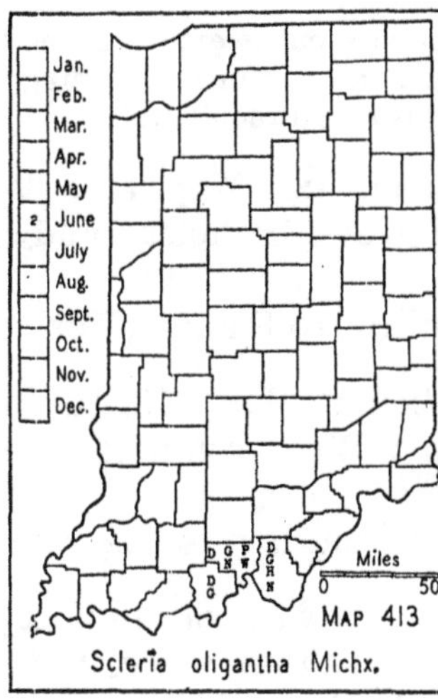
MAP 413 Scleria oligantha Michx.

MAP 414 Scleria reticularis Michx.

6a. **Rhynchospora glomerata** var. **minor** f. **discùtiens** (Clarke) Fern. (Rhodora 37: 402-403. 1935.) This form has been seen from only Lake, Porter, and Starke Counties. The habitat is that of the species.

N. J. to Ind. and southw.

515. SCLÈRIA Bergius NUTRUSH

Achenes smooth, ovoid, about 3 mm long including the basal disk (hypogynium).

Hypogynium entirely covered with a white crust.................1. *S. triglomerata.*

Hypogynium naked at the base, supporting about 9 very short tubercles just below the achene...2. *S. oligantha.*

Achenes not smooth, spheroidal, 1.5-2 mm long (2.5 mm long in *Scleria setacea*).

Achenes irregularly papillose or warty, or transversely wrinkled.

Culms, leaves, and scales densely pubescent......3. *S. pauciflora* var. *caroliniana.*

Culms, leaves, and scales essentially glabrous.................4. *S. verticillata.*

Achenes irregularly pitted.

Surface of achene glabrous; achenes 2 mm or less in length; lobes of hypogynium emarginate or cleft, somewhat obtuse; culms usually erect; peduncles sessile or short...5. *S. reticularis.*

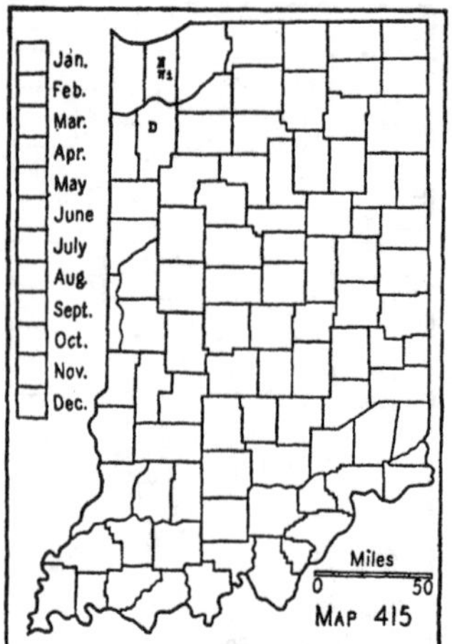

Scleria setacea Poir.

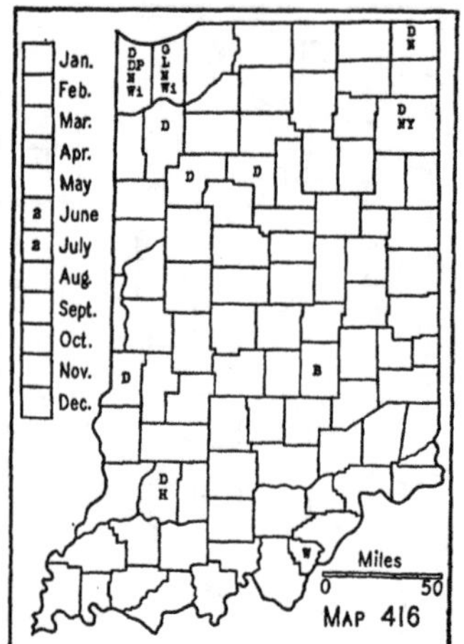

Scleria triglomerata Michx.

Scleria verticillata Muhl.

Surface of achene more or less pubescent; achenes 2-2.5 mm long; lobes of hypogynium acute or acuminate; culms spreading; axillary peduncles long and filiform, the spikelets usually drooping........................6. *S. setacea.*

1. **Scleria triglomeràta** Michx. Map 416. Very local and only a few plants found at a place. It grows in moist, sandy soil in prairie habitats or in marshes.

Vt. to Ont. and Wis., southw. to Fla. and Tex.

2. **Scleria oligántha** Michx. Map 413. On dry rocky, open, wooded slopes in three of the Ohio River counties. Very local and only a few plants found.

D. C., Va. to Ind. and Mo., southw. to Fla. and Tex.

3. **Scleria pauciflòra** Muhl. var. **caroliniàna** (Willd.) Wood. Fassett (Rhodora 35: 202. 1933) writes that two collections from Miller and three from Dune Park in the herbarium of the University of Wisconsin named *Scleria pauciflora* Muhl. should be referred to the variety. I have seen these specimens and I agree with Fassett. I have this variety also from Jasper County.

Mass. to Ga. along the coast, cent. N. Y., Ohio, and Ind. to Mo.

4. **Scleria verticillàta** Muhl. Map 417. Infrequent to frequent in marly marshes throughout the lake area. It is rather inconspicuous and is often overlooked, although where it is found it usually forms a dense stand. The report by Core (Brittonia 2: 23. 1936) for Chase from Shelby County should be referred to Shelby, Lake County. Mrs. Chase wrote me that she had never collected in Shelby County.

Mass., Ont. to Minn., southw. to Fla., Tex., Mex., and W. I.

5. **Scleria reticulàris** Michx. Map 414. In damp or wet, sandy soil in a few marshes of northwestern Indiana. Very local. This species was erroneously cited by Core (Brittonia 2: 82. 1936) as having been collected in Greene County by Nieuwland. Nieuwland's specimens of the number

cited by Core in the herbarium of the University of Notre Dame are from Porter County.

Mass. to Fla., and in n. Ind.

6. **Scleria setàcea** Poir. (*Scleria reticularis* var. *pubescens* Britt.) Map 415. There is a specimen in the herbarium of the University of Wisconsin collected by Umbach in 1908 near Dune Park, Porter County, which I am referring to this species. This specimen has the pubescent achenes and the long, filiform peduncles of the lateral spikelets but the lobes of the hypogynium are not acute or only scarcely so. Witmer Stone, in his "Flora of Southern New Jersey," on page 284 says: "The width of the leaves and pubescence of the achenes are characters which are very variable." The size of the achenes of our plants is about the same as those of *S. reticularis*. While the achenes of typical *S. reticularis* are glabrous, sometimes an achene is found which has a trace of pubescence which is a character of *S. setacea*.

This species is somewhat frequent on the moist or dry, sandy border of the west end of the second marsh from the north side of section 2 and on the east side of this section about 2½ miles southeast of Tefft, Jasper County. It is closely associated with *S. triglomerata* Michx. and *S. pauciflora* var. *caroliniana* (Willd.) Wood. This is an interdunal marsh between rather low sand hills which are covered with black oak. The marsh is covered with *Calamagrostis canadensis*. Besides the *Sclerias* already named, on the border of this marsh are found also *Hypericum adpressum* and *Panicum verrucosum*. This marsh and others nearby are noted for the number of Coastal Plain plants found in them.

Conn. to the Great Lakes and Mo., southw. to Fla. and Tex.; also in Mex., W. I., and Cent. Amer.

525. CÀREX [Dill.] L. SEDGE*

A genus of nearly 2000 species and the largest genus of vascular plants in Indiana. Few species have any economic value but the ecological role of the genus is of great importance. The species of marshes and muddy borders of lakes which form extensive colonies, and to some extent the less gregarious species, comprise an essential step in the successional stages from open water to the culmination in climax forest or prairie. Much of the fertile soils of our region today would still be barren mudflats were it not for the part played by these sedges in the conversion of the once vast boggy areas into a turf, thus enabling less hydrophytic plants to become established and add further to the fertility of the soil.

For the identification of species in this genus it is nearly always necessary to have a specimen with ripe fruit (perigynia) and as a rule the roots are also essential. In the key closely related species have been grouped for convenience into sections, roughly corresponding in size to most of the genera in other groups. After only a slight acquaintance with the sedges of an area it is generally possible to recognize at sight the group or section to which an unknown species belongs, especially since a few of the sections (*Ovales, Bracteosae, Laxiflorae, Acutae,* and *Lupulinae*) will include the great majority of the individuals found in the field.

The most recent and exhaustive treatment of the species of *Carex* in our area is K. K. Mackenzie's monograph in North American Flora 18: 1-478. 1931-35. In the following account this monograph has been freely used in the preparation of the keys and in giving distribution.

NATURAL KEY TO THE SECTIONS OF INDIANA CARICES

Stigmas two; achenes lenticular; spikes usually bisexual, the lateral sessile.......... ....Subgenus *Vignea.*

Stigmas three; achenes triangular; or if stigmas two and achenes lenticular, the lateral spikes peduncled; spikes normally unisexual.........Subgenus *Eu-Carex.*

SUBGENUS VIGNEA

Terminal or all spikes androgynous; perigynia not subterete.

Culms arising singly or few together from long-creeping rootstocks.

Heads elongate, 2-7 cm long; culms not branching; perigynia thin- or wing-margined; not plants of sphagnum bogs.

Perigynia thin- but not wing-margined, ovate-orbicular, thick-plano-convex, 3.4-5 mm long; spikes all androgynous; plants of wet habitats........... ....1. § INTERMEDIAE, p. 218.

Perigynia narrowly wing-margined, oblong-lanceolate, plano-convex, 4.75-6 mm long; lowest spikes usually pistillate, the middle staminate, and terminal androgynous; plants of dry sandy habitats.........2. § ARENARIAE, p. 218.

Heads ovoid, 0.5-1.2 cm long; culms becoming decumbent and branching; perigynia neither thin- nor wing-margined, oblong-obovate, thick-plano-convex, 2.5-3.75 mm long; plants of sphagnum bogs............3. § CHORDORRHIZAE, p. 219.

Culms cespitose, the rootstocks sometimes short-prolonged with short internodes but not long-creeping.

Perigynia abruptly contracted into the beak; culms not flaccid and not flattening in drying.

Spikes few (generally 10 or fewer), usually greenish..4. § BRACTEOSAE, p. 219.

---

* Contributed by Frederick J. Hermann, University of Michigan.

Spikes numerous, yellowish or brownish at maturity; leaf sheaths often red-dotted ventrally.

Perigynia plano-convex, thin, yellowish; bracts mostly much exceeding the spikes; leaf sheaths usually transversely rugulose ventrally............ ....5. § MULTIFLORAE, p. 224.

Perigynia thick-plano-convex or unequally biconvex, brown; bracts mostly shorter than the spikes; leaf sheaths not transversely rugulose......... ....6. § PANICULATAE, p. 225.

Perigynia tapering into the beak or, if abruptly contracted, culms flaccid and flattening in drying...............................7. § VULPINAE, p. 226.

Terminal or all spikes gynaecandrous or, if androgynous, perigynia subterete and spikes 1-3-flowered.

Perigynia without winged margins, at most thin-edged.

Perigynia 2-4 mm long.

Perigynia not thin-edged, ascending or appressed, elliptic.................... ....8. § HELEONASTES, p. 229.

Perigynia thin-edged, spreading, ovoid, usually broadest below the middle...... ....9. § STELLULATAE, p. 230.

Perigynia 4-5 mm long, narrowly lanceolate, appressed..10. § DEWEYANAE, p. 232.

Perigynia with winged margins............................11. § OVALES, p. 232.

SUBGENUS EU-CAREX

Style articulated with the achene, at length deciduous; achenes apiculate or blunt at the apex; perigynia closely enveloping the achenes or moderately inflated.

Spikes solitary, androgynous; perigynia beakless, rounded at the apex, glabrous. ....12. § POLYTRICHOIDEAE, p. 237.

Spikes one to many, when one the perigynia not as above.

Lower pistillate scales bractlike; achenes rounded at the apex, strongly constricted at the base........................13. § PHYLLOSTACHYAE, p. 238.

Lower pistillate scales not bractlike; achenes apiculate-tipped, not strongly constricted at the base.

Achenes with sides convex above, closely enveloped by the perigynia; bracts sheathless, scalelike or setaceous.................14. § MONTANAE, p. 238.

Achenes with flat or concave sides; bracts from sheathless to long-sheathing.

Achenes closely enveloped by the perigynia; bracts, when present, strongly sheathing.

Perigynia pubescent or puberulent, at least at the base of the beak.

Bracts either reduced to sheaths or absent.

Culms not dioecious; spikes more than one; bracts bladeless......... ....15. § DIGITATAE, p. 242.

Culms dioecious; spike solitary, bractless.........16. § PICTAE, p. 243.

Bracts with well-developed blades..............17. § TRIQUETRAE, p. 243.

Perigynia glabrous.................................18. § ALBAE, p. 243.

Achenes not closely enveloped by the perigynia except at the base.

Bracts long-sheathing (except in *C. prasina* of § *Gracillimae*, a species with sharply triangular perigynia which are long- and flat-beaked, nerveless except for the prominent lateral pair of nerves); achenes triangular (except in § *Bicolores*, species with pulverulent or golden yellow perigynia).

Beak of perigynium entire, emarginate, or obliquely cut and at length bidentate.

Pistillate spikes short, oblong to linear, erect or, if drooping, the perigynia acutely triangular.

Achenes lenticular; stigmas two...........19. § BICOLORES, p. 244.

Achenes triangular; stigmas three.

Perigynia with few to many strongly raised nerves.

Perigynia tapering at the base, triangular; achenes usually closely enveloped.

Rootstocks elongate, producing long horizontal stolons...20. § PANICEAE, p. 244.

Rootstocks not elongate, not producing long horizontal stolons. ....21. § LAXIFLORAE, p. 246.

Perigynia rounded at the base, suborbicular in cross section; achenes loosely enveloped........22. § GRANULARES, p. 250.

Perigynia with numerous fine impressed nerves.

Perigynia tapering at the base, constricted at the apex, obtusely triangular; achenes closely enveloped..23. § OLIGOCARPAE, p. 251.

Perigynia rounded at both ends, orbicular or orbicular-triangular in cross section......................24. § GRISEAE, p. 252.

Pistillate spikes elongate, linear to cylindric, slender-peduncled, the lower drooping.

Perigynia beakless or short-beaked; terminal spike gynaecandrous (except in *C. prasina*, and rarely in *C. gracillima*)...25. § GRACILLIMAE, p. 253.

Perigynia conspicuously beaked; terminal spike staminate (rarely with a few perigynia at the base).

Pistillate spikes narrowly linear, 3-4 mm wide; culms strongly reddish-tinged at the base, aphyllopodic...26. § SYLVATICAE, p. 254.

Pistillate spikes oblong-cylindric, 8-10 mm wide; culms not strongly reddish-tinged at the base, phyllopodic..27. § LONGIROSTRES, p. 255.

Beak of perigynium bidentate.................28. § EXTENSAE, p. 256.

Bracts sheathless or very short-sheathing (rarely the lowest long-sheathing in *C. lasiocarpa* of § *Hirtae*).

Perigynia or foliage (especially the sheaths) or both pubescent.

Beak of perigynium at most shallowly bidentate; styles very short, thickish, leaves not septate-nodulose.....29. § VIRESCENTES, p. 257.

Beak of perigynium strongly bidentate; styles long, slender; leaves septate-nodulose..........................30. § HIRTAE, p. 258.

Perigynia and foliage not pubescent.

Achenes triangular; stigmas three.

Perigynia strongly beaked, scabrous.......30a. § ANOMALAE, p. 259.

Perigynia beakless or short-beaked, not scabrous.

Perigynia transversely corrugated......31. § SHORTIANAE, p. 259.

Perigynia not transversely corrugated, papillate.

Terminal spike staminate; roots closely clothed with a yellowish felt..................................32. § LIMOSAE, p. 260.

Terminal spike gynaecandrous; roots not clothed with a yellowish felt..................................33. § ATRATAE, p. 260.

Achenes lenticular; stigmas two.

Achenes not constricted in the middle; pistillate scales obtuse to acuminate.............................34. § ACUTAE, p. 260.

Achenes constricted in the middle; pistillate scales long-awned...35. § CRYPTOCARPAE, p. 262.

Style not articulated, continuous with the achene, persistent, indurated; perigynia moderately to strongly inflated (only slightly so in some species of § *Paludosae* and § *Pseudo-Cyperi*).

Perigynia many-nerved, lanceolate, tapering into the beak.

Spikes solitary, androgynous; perigynia widely spreading or reflexed, early deciduous; leaf blades involute, 0.5 mm wide......35a. § ORTHOCERATE., p. 263.

Spikes several; perigynia ascending, not early deciduous; leaf blades flat, 5-15 mm wide......................................36. § FOLLICULATAE, p. 263.

Perigynia strongly ribbed, usually broader, generally abruptly contracted into the beak.

Perigynia finely and closely ribbed..................37. § PSEUDO-CYPERI, p. 263.
Perigynia coarsely ribbed.
Perigynia subcoriaceous; foliage or perigynia or both sometimes pubescent. ....38. § PALUDOSAE, p. 264.
Perigynia membranaceous; foliage and perigynia not pubescent (except perigynia sometimes hispidulous in *C. Grayii* of § *Lupulinae*).
Perigynia obconic or broadly obovoid, truncately contracted into a long subulate beak..................................39. § SQUARROSAE, p. 266.
Perigynia from lanceolate to ovoid or globose-ovoid, not truncately contracted.
Perigynia 7-10 mm long; achenes 2-3 mm long, 1.25-2.5 mm wide......... ....40. § VESICARIAE, p. 267.
Perigynia 10-20 mm long; achenes 2.5-6 mm long, 2-4 mm wide............ ....41. § LUPULINAE, p. 269.

ARTIFICIAL KEY TO THE SECTIONS OF INDIANA CARICES

Spike one.
Perigynia strongly inflated, sessile or nearly so, not becoming reflexed; pistillate scales persistent..................................39. § SQUARROSAE, p. 266.
Perigynia not inflated.
Pistillate scales not foliaceous; perigynia not abruptly beaked.
Pistillate scales deciduous; perigynia stipitate, at least the lower reflexed at maturity..................................35a. § ORTHOCERATES, p. 263.
Pistillate scales persistent; perigynia not reflexed.
Perigynia rounded at the apex; spike androgynous........................ ....12. § POLYTRICHOIDEAE, p. 237.
Perigynia abruptly tapering to a pointed, slightly bidenticulate apex; spike entirely staminate or pistillate..................16. § PICTAE, p. 243.
Lower pistillate scales foliaceous; perigynia abruptly beaked.................. ....13. § PHYLLOSTACHYAE, p. 238.
Spikes more than one.
Stigmas two; achenes lenticular.
Lateral spikes sessile, short; terminal spike usually androgynous or gynaecandrous. (*Vignea.*)
Culms arising singly or few together from long-creeping rootstocks; perigynia not subterete.
Heads elongate, 2-7 cm long; culms not branching; perigynia thin- or wing-margined; not plants of sphagnum bogs.
Perigynia thin- but not wing-margined, ovate orbicular, thick-plano-convex, 3-4.5 mm long; spikes all androgynous; plants of wet habitats......... ....1. § INTERMEDIAE, p. 218.
Perigynia narrowly wing-margined, oblong-lanceolate, plano-convex, 4.75-6 mm long; lowest spikes usually pistillate, the middle staminate, and terminal androgynous; plants of dry sandy habitats................. ....2. § ARENARIAE, p. 218.
Heads ovoid, 0.5-1.2 cm long; culms becoming decumbent and branching; perigynia neither thin- nor wing-margined, oblong-obovate, thick-plano-convex, 2.5-3.75 mm long; plants of sphagnum bogs.................. ....3. § CHORDORRHIZAE, p. 219.
Culms cespitose, the rootstocks occasionally somewhat prolonged with short internodes but not long-creeping (except occasionally in *C. disperma* of § *Heleonastes* which has subterete perigynia).
Spikes androgynous, many-flowered; perigynia not subterete.
Perigynia abruptly contracted into the beak; culms not flaccid and not flattening in drying.
Spikes few (generally 10 or fewer), usually greenish.................. ....4. § BRACTEOSAE, p. 219.

Spikes numerous, yellowish or brownish at maturity; leaf sheaths often red-dotted ventrally.

Perigynia plano-convex, thin, yellowish; bracts mostly much exceeding the spikes; leaf sheaths usually transversely rugulose ventrally. ....5. § MULTIFLORAE, p. 224.

Perigynia thick-plano-convex or unequally biconvex, brown; bracts mostly shorter than the spikes; leaf sheaths not transversely rugulose..............................6. § PANICULATAE, p. 225.

Perigynia tapering into the beak or, if abruptly contracted, culms flaccid and flattening in drying..........................7. § VULPINAE, p. 226.

Spikes not androgynous or, if so, perigynia subterete and spikes only 1-3-flowered.

Perigynia without winged margins, at most thin-edged.

Perigynia 2-4 mm long.

Perigynia not thin-edged, ascending or appressed, elliptic........... ....8. § HELEONASTES, p. 229.

Perigynia thin-edged, spreading, ovoid, usually broadest below the middle...............................9. § STELLULATAE, p. 230.

Perigynia 4-5 mm long, narrowly lanceolate, appressed.............. ....10. § DEWEYANAE, p. 232.

Perigynia with winged margins...................11. § OVALES, p. 232.

Lateral spikes peduncled or, if sessile, elongate; terminal spike usually staminate. (*Eu-Carex.*)

Style articulated with the achene, at length deciduous; perigynia not lustrous.

Lowest bract long-sheathing; perigynia pulverulent or golden yellow at maturity.......................................19. § BICOLORES, p. 244.

Lowest bract sheathless or rarely short-sheathing, perigynia not pulverulent or golden yellow.

Achenes not constricted in the middle; scales not long-awned, 1-nerved. ....34. § ACUTAE, p. 260.

Achenes constricted in the middle; scales long-awned, 3-nerved........... ....35. § CRYPTOCARPAE, p. 262.

Style continuous with the achene, persistent, indurated; perigynia lustrous..... ....40. § VESICARIAE, p. 267.

Stigmas three; achenes triangular.

Perigynia pubescent or scabrous.

Style articulated with the achene, at length deciduous.

Achenes closely enveloped by the perigynia; bracts sheathless or nearly so.

Perigynia obtusely triangular or orbicular-triangular in cross section; plant (except perigynia) glabrous................14. § MONTANAE, p. 238.

Perigynia sharply triangular; plant pubescent....17. § TRIQUETRAE, p. 243.

Achenes not closely enveloped by the perigynia or, if so, the bracts strongly sheathing.

Bracts sheathing, their blades absent or rudimentary; achenes closely enveloped by the perigynia....................15. § DIGITATAE, p. 242.

Bracts with well-developed blades.

Bracts sheathless or the lower short-sheathing.

Perigynia pubescent.

Beak of perigynium at most shallowly bidentate; styles very short, thickish; leaves not septate-nodulose..29. § VIRESCENTES, p. 257.

Beak of perigynium strongly bidentate; styles long, slender; leaves septate-nodulose.........................30. § HIRTAE, p. 258.

Perigynia scabrous......................30a. § ANOMALAE, p. 259.

Bracts, at least the lower ones, long-sheathing.

Beak of perigynium not strongly bidentate...21. § LAXIFLORAE, p. 246.

Beak of perigynium strongly bidentate..........30. § HIRTAE, p. 258.

Style not articulated, continuous with the achene, persistent, indurated.

Perigynia less than 1 cm long; spikes cylindric......38. § PALUDOSAE, p. 264.

Perigynia 1 cm long or longer; spikes globose .... *C. Grayii* in 41. § LUPULINAE, p. 269.

Perigynia glabrous.

Style articulated with the achene, at length deciduous.

Achenes strongly constricted at the base, rounded at the apex; lower pistillate scales bractlike .... 13. § PHYLLOSTACHYAE, p. 238.

Achenes not strongly constricted at the base, apiculate at the apex; lower pistillate scales not bractlike.

Bracts long-sheathing, at least the lower ones.

Bracts bladeless or with rudimentary blades.

Leaf blades filiform .... 18. § ALBAE, p. 243.

Leaf blades not filiform .... 21. § LAXIFLORAE, p. 246.

Bracts with well-developed blades.

Foliage, especially the sheaths, pubescent or puberulent.

Perigynia beakless or short-beaked; terminal spike gynaecandrous (rarely staminate in *C. gracillima*) ... 25. § GRACILLIMAE, p. 253.

Perigynia conspicuously beaked; terminal spike staminate .... 26. § SYLVATICAE, p. 254.

Foliage glabrous.

Beak of perigynium not bidentate, at most emarginate.

Pistillate spikes short, oblong to linear, erect or, if drooping, either on long capillary peduncles or the perigynia acutely triangular.

Perigynia with few to many strongly raised nerves.

Perigynia tapering at the base, triangular, closely enveloping the achenes.

Rootstocks elongate, often producing long horizontal stolons. .... 20. § PANICEAE, p. 244.

Rootstocks not elongate, not producing long horizontal stolons .... 21. § LAXIFLORAE, p. 246.

Perigynia rounded at the base, suborbicular in cross section, loosely enveloping the achenes .. 22. § GRANULARES, p. 250.

Perigynia with numerous fine impressed nerves.

Perigynia tapering at the base, constricted at the apex, obtusely triangular, closely enveloping the achenes .... 23. § OLIGOCARPAE, p. 251.

Perigynia rounded at both ends, orbicular to orbicular-triangular in cross section .... 24. § GRISEAE, p. 252.

Pistillate spikes elongate, linear to cylindric, on slender peduncles, the lower usually drooping; perigynia not acutely triangular.

Perigynia beakless or short-beaked; terminal spike gynaecandrous .... 25. § GRACILLIMAE, p. 253.

Perigynia conspicuously beaked; terminal spike staminate.

Pistillate spikes narrowly linear, 3-4 mm wide; culms strongly reddish-tinged at the base, aphyllopodic .... 26. § SYLVATICAE, p. 254.

Pistillate spikes oblong-cylindric, 8-10 mm wide; culms not strongly reddish-tinged at the base, phyllopodic .... 27. § LONGIROSTRES, p. 255.

Beak of perigynium bidentate.

Pistillate spikes oblong-cylindric, on slender drooping peduncles; perigynia obliquely cut, at length bidentate .... 27. § LONGIROSTRES, p. 255.

Pistillate spikes suborbicular to short-oblong, on short erect or ascending peduncles or sessile, perigynia equally bidentate. .... 28. § EXTENSAE, p. 256.

Bracts (lower) sheathless or very short-sheathing.

Terminal spike staminate (in *C. prasina* occasionally bearing a few

perigynia); perigynia appressed or ascending; leaf sheaths not septate-nodulose.
Perigynia rounded and minutely beaked at the apex; pistillate spikes oblong, 1-2.5 cm long......................32. § LIMOSAE, p. 260.
Perigynia tapering into a beak nearly the length of the body; pistillate spikes linear, 2-6 cm long. . *C. prasina* in 25. § GRACILLIMAE, p. 253.
Terminal spike gynaecandrous.
Perigynia transversely corrugated...........31. § SHORTIANAE, p. 259.
Perigynia not transversely corrugated..........33. § ATRATAE, p. 260.
Style persistent, indurated, continuous with the achene.
Perigynia subcoriaceous and firm..................38. §PALUDOSAE, p. 264.
Perigynia membranaceous.
Perigynia obconic or broadly obovoid, truncately contracted into the long, subulate beaks............................39. § SQUARROSAE, p. 266.
Perigynia from lanceolate to ovoid or globose-ovoid, not truncately contracted.
Perigynia lanceolate or ovoid-lanceolate, tapering into the beak.
Perigynia many-nerved, slightly inflated, 3 mm wide or less, yellowish green; achenes 3.5 mm long...........36. § FOLLICULATAE, p. 263.
Perigynia strongly ribbed, strongly inflated, ovoid-lanceolate, 3.5 mm wide or more, green; achenes 5 mm long. .41. § LUPULINAE, p. 269.
Perigynia broader, abruptly contracted into the beak, usually strongly ribbed.
Perigynia finely and closely ribbed........37. § PSEUDO-CYPERI, p. 263.
Perigynia coarsely ribbed.
Perigynia 7-10 mm long; achenes 2-3 mm long, 1.25-2.25 mm wide. ....40. § VESICARIAE, p. 267.
Perigynia 10-20 mm long; achenes 2.5-6 mm long, 2-4 mm wide. ....41. § LUPULINAE, p. 269.

1. § INTERMÈDIAE

Perigynia ovate-orbicular, 2.5-3 mm long, 1.5-1.75 mm wide, abruptly contracted into a beak about a fourth the length of the body.....................1. *C. Sartwellii.*
Perigynia elliptic, 4-4.5 mm long, 1-1.3 mm wide, very gradually long-beaked........ ....1a. *C. Sartwellii* var. *stenorrhyncha.*

1. **Carex Sartwéllii** Dewey. Map 418. Occasional in marshes and marly sloughs, more rarely in sandy ditches, in the northern half of the state; becoming frequent in the dune area.

Ont. and w. N. Y. to B. C., southw. to Ill., Mo., Nebr., and Colo.

1a. **Carex Sartwellii** var. **stenorrhýncha** Hermann. (Rhodora 40: 78. 1938.) Map 418a. Known only from two localities, both in Lake County: in a prairie marsh south of Sheffield St. and west of Calumet Ave., two miles north of Hammond, Deam no. 53920 (Deam Herbarium); and on a prairie east of Wolf Lake, Hermann no. 6052 (Type in Gray Herbarium).

2. § ARENÀRIAE

2. **Carex siccàta** Dewey. (*Carex foenea* Willd., according to Svenson in Rhodora 40: 325-329. 1938.) Map 419. Infrequent in the lake area in dry open sandy soil and in open black oak woods.

Maine to Wash. and Mack., southw. to N. J., Ind, Nebr., and in the mts. to Ariz.

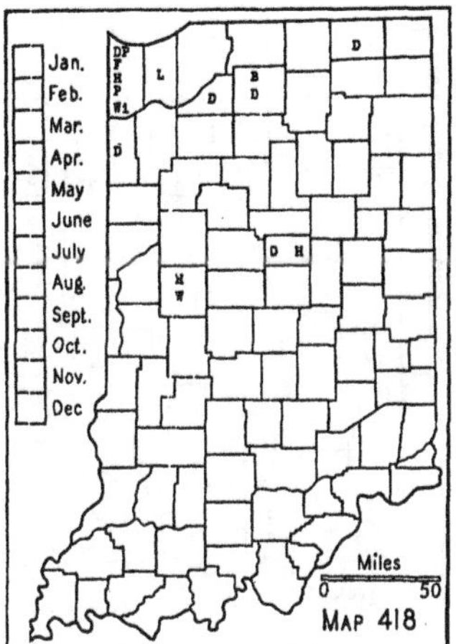

Carex Sartwellii Dewey

Carex Sartwellii var. stenorrhyncha Hermann

Carex siccata Dewey

3. § CHORDORRHÌZAE

3. **Carex chordorrhìza** L. f. Map 420. A northern species reaching the southern limit of its range in northern Indiana where it is rare. The two Indiana collections are from very wet sphagnum bogs; elsewhere in its range it is found also on peaty borders of lakes. The Indiana stations are: in a tamarack bog a mile south of Leesburg, Kosciusko County, and in an open tamarack bog west of Goose Lake, Whitley County.

Lab. and Newf. to Keewatin, southw. to N. Y., Ind., Iowa, and Sask.; also in n. Eurasia.

4. § BRACTEÒSAE

Sheaths tight, inconspicuously or not at all mottled with green and white or septate-nodulose dorsally (except sometimes in *C. Leavenworthii*); leaf blades 1-4.5 mm wide.
- Perigynia distended and spongy at the base, usually widely spreading or reflexed at maturity.
  - Beaks of perigynia smooth, scarcely exceeding the acuminate, deciduous scales. .... 4. *C. retroflexa.*
  - Beaks of perigynia minutely serrulate, much exceeding the obtuse or somewhat acute, persistent scales.
    - Stigmas long, slender, usually not twisted, light reddish; perigynium tapering into the beak; leaf blades 1-2 mm wide........................5. *C. rosea.*
    - Stigmas short, stout, strongly twisted or contorted, deep red; perigynium abruptly contracted into the beak.
      - Leaf blades 1.5-3 (averaging 2.5) mm wide; spikes with 6-20 perigynia; perigynia 3.25-4.5 mm long...........................6. *C. convoluta.*
      - Leaf blades 1-1.75 (averaging 1) mm wide; spikes with 2-6 perigynia; perigynia 2.25-3 mm long. (See excluded species no. 1, p. 271)....*C. radiata.*
- Perigynia not distended and spongy at the base, mostly ascending.
  - Inflorescence ovoid or oblong-ovoid; spikes densely capitate.
    - Scales much shorter than the bodies of the perigynia.
      - Perigynia broadest below the middle, round-tapering at the base, with raised margins ventrally; beaks long, serrulate..............7. *C. cephalophora.*
      - Perigynia broadest at the truncate-cordate base, flat ventrally; beaks short, smooth......................................8. *C. Leavenworthii.*
    - Scales from little shorter to longer than the bodies of the perigynia.

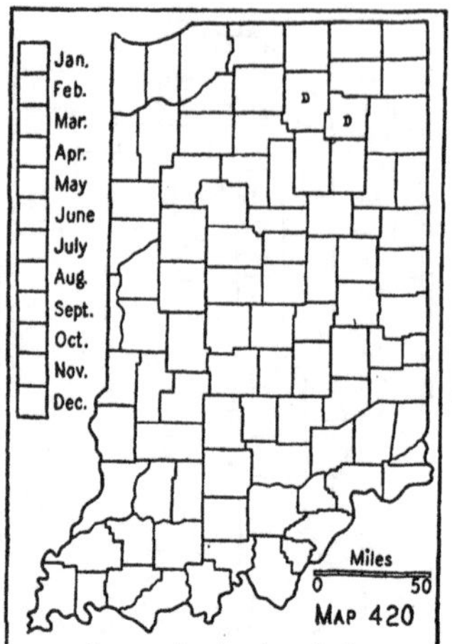

Carex chordorrhiza L.f.

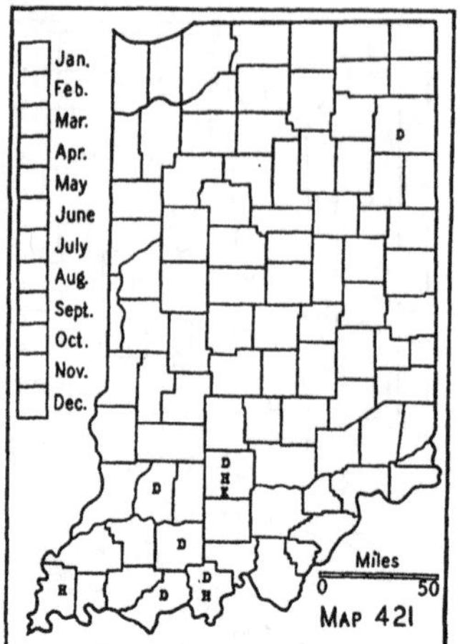

Carex retroflexa Muhl.

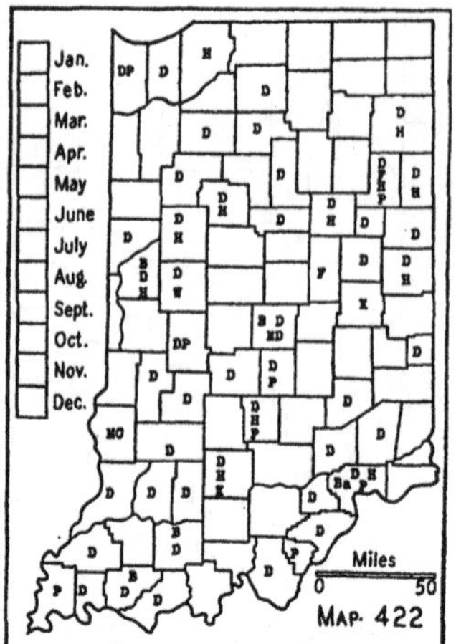

Carex rosea Schkuhr

Culms lax; perigynia faintly nerved dorsally; stigmas short. . . 9. *C. mesochorea.*

Culms stiff, erect; perigynia generally strongly nerved to ribbed dorsally; stigmas long, slender.................10a. *C. Muhlenbergii* var. *enervis.*

Inflorescence oblong or linear-oblong to elongate and interrupted; spikes not capitate.

Perigynia strongly nerved and low-convex ventrally, the margins usually slightly raised..10. *C. Muhlenbergii.*

Perigynia nerveless or nerved only at the base ventrally, the margins not raised at maturity.

Perigynia sessile, short-beaked, spreading, 3-3.5 mm long; bracts not broadly dilated at the base; scales about the length of and narrower than the bodies of the perigynia, cuspidate or short-aristate...10a. *C. Muhlenbergii* var. *enervis.*

Perigynia stipitate, long-beaked, ascending, 3.5-5 mm long; bracts broadly dilated at the base; at least the lower scales exceeding and as wide as the perigynia, long-awned. (See excluded species no. 2, p. 272)..*C. austrina.*

Sheaths loose, mottled with green and white and usually septate-nodulose dorsally; leaf blades 4.5-8 (in *C. aggregata* rarely only 3) mm wide.

Perigynia not deep green at maturity, with border raised ventrally only above the middle, the beak a fourth to a third the length of the ovate or suborbicular body or, if longer, the ventral suture deep; spikes approximate in a cylindric or ovoid head.

Perigynia ovate, tapering into the beak, typically smooth or faintly nerved dorsally, beak very sharply bidentate, the teeth about 1 mm long; pistillate scales mostly strongly awned; leaf blades rarely over 5 mm wide..11. *C. gravida.*

Perigynia broadly ovate to suborbicular, abruptly beaked, typically strongly ribbed dorsally; beak less sharply bidentate, the teeth about 0.5 mm long; pistillate scales mostly acuminate to short-awned; leaf blades often 6-8 mm wide...11a. *C. gravida* var. *Lunelliana.*

Perigynia deep green, the beak a third the length of the ovate body or more or, if rarely shorter, the ventral suture shallow.

Mature perigynia membranaceous, flat ventrally; leaf blades 3-7 mm wide; spikes approximate or the lower separate.

Scales acute or somewhat obtuse, half the length of the bodies of the perigynia; stigmas short; sheaths truncate at the mouth, the lower transversely rugulose; perigynia narrowly ovate or even elliptic with a narrow, gradually contracted beak; culms often slightly winged............12. *C. cephaloidea.*

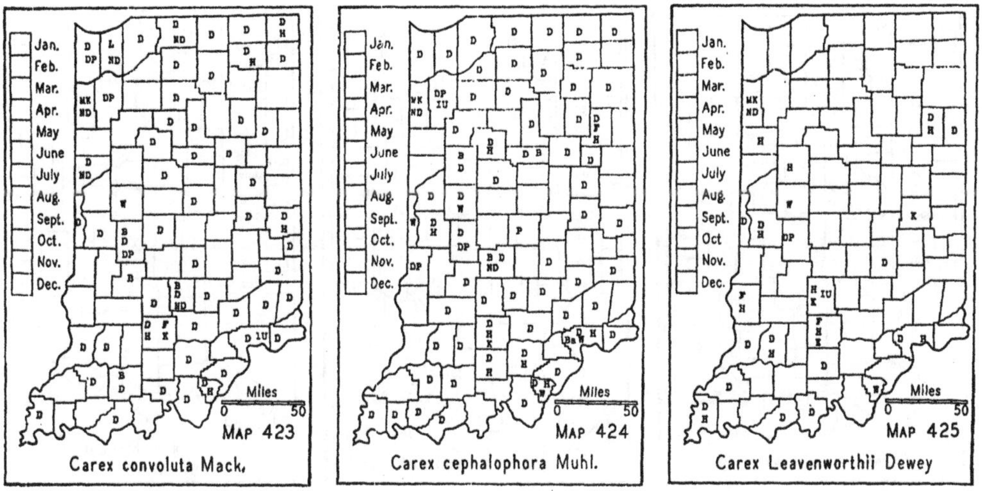

Scales acuminate or short cuspidate (rarely merely acute), about the length of the bodies of the perigynia; stigmas long and slender; sheaths concave at the mouth, usually not at all transversely rugulose; culms sulcate and white-striate up to the inflorescence, their angles usually smooth..13. *C. aggregata.*

Mature perigynia subcoriaceous, ovate, with border raised ventrally to the base, abruptly contracted into a short, stout beak, culms sulcate and white-striate only below, minutely winged or thin-margined up to the inflorescence, serrulate on the angles; leaf blades 5-10 mm wide; lower sheaths usually transversely rugulose; lower spikes usually separate; stigmas short and stout; scales short, blunt to acute..14. *C. sparganioides.*

4. **Carex retrofléxa** Muhl. Map 421. Very local in northern Indiana; frequent in the unglaciated area of the southwestern counties. A woodland species partial to dry rocky white oak woods, especially in sandstone areas.

Vt. to Mich., southw. to Fla. and Tex.

5. **Carex ròsea** Schkuhr. Map 422. Very common in both dry and moist woods. This species and *C. convoluta* are perhaps the most plentiful woodland sedges in the state as a whole.

N. S. to N. Dak., southw. to Ga. and La.

6. **Carex convolùta** Mack. (Bull. Torrey Bot. Club **43**: 428. 1916.) (*Carex rosea* of authors.) Map 423. Very common in dry and low woods of all types. Often in somewhat richer soils than *C. rosea.*

N. S. to Man., southw. to Ala., Tenn., and Ark.

7. **Carex cephalóphora** Muhl. Map 424. Very common in oak and beech-maple woods; occasional along open grassy roadsides and in thickets.

Maine to Man., southw. to Fla. and Tex.

8. **Carex Leavenwórthii** Dewey. Map 425. Frequent, except in the lake area, in open grassy, generally dry or sandy, oak woods and bordering thickets; occasionally bordering woods in clay fallow fields.

Southern N. J., sw. Ont. and Iowa to Fla. and Tex.

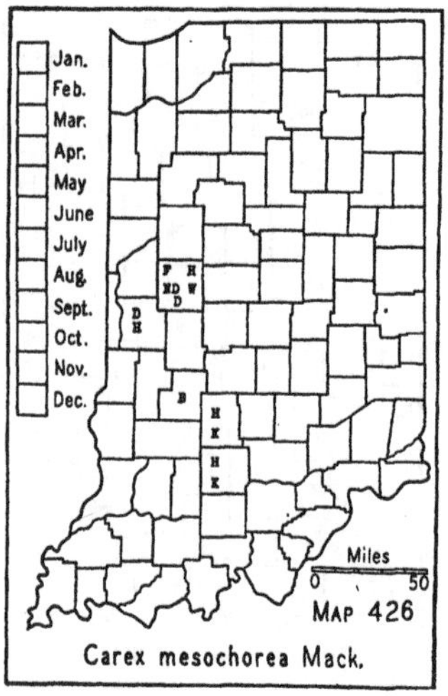

Carex mesochorea Mack.

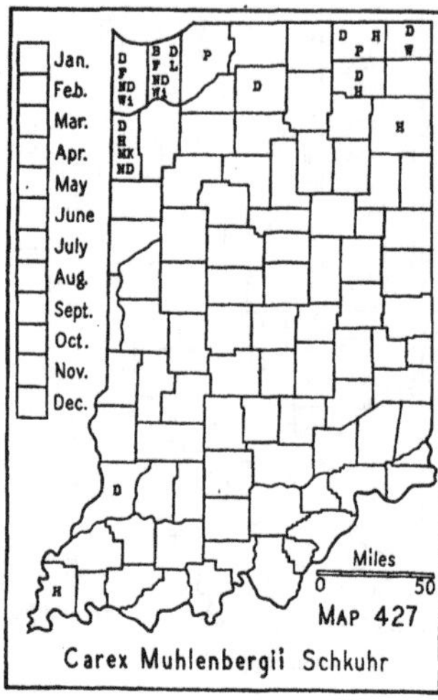

Carex Muhlenbergii Schkuhr

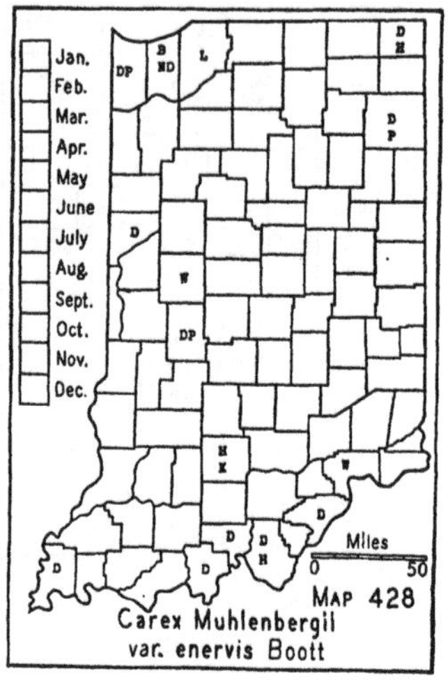

Carex Muhlenbergii var. enervis Boott

9. **Carex mesochòrea** Mack. (Bull. Torrey Bot. Club **37**: 246. 1910.) (*Carex mediterranea* Mack.) Map 426. Rare or local in pasture fields and on open, wooded, grassy slopes. At the Montgomery County station, on an open white oak ridge 5 miles west of New Market, it is associated with *Poa pratensis, Danthonia spicata, Luzula echinata* var. *mesochorea, Antennaria neglecta* and "reindeer moss."

Southern Mass. and N. Y., to D. C., Tenn., and Ind.

10. **Carex Muhlenbérgii** Schkuhr. Map 427. Frequent to common in the lake area in dry sandy fallow fields and open oak woods and on dunes; occasional in southern Indiana.

Maine to Minn., southw. to Fla. and Tex.

10a. **Carex Muhlenbergii** var. **enérvis** Boott. (*Carex plana* Mack. Bull. Torrey Bot. Club **50**: 350. 1923.) Map 428. Frequent on slopes, in sandy open woods, on wooded dunes, and in dry sandy fields. It is partial to somewhat less open habitats than the species and is less often on low or level ground, its favorite habitat being on or near the crests of wooded dunes, river bluffs, and oak ridges.

Specimens intermediate between *C. Muhlenbergii* and var. *enervis* in some or most of their characters seem to be too frequent to warrant the treatment of the latter as a species. The ventrally flat perigynium is a conspicuous character of typical var. *enervis* when fully mature or over-ripe but it is very inconstant and specimens with a pronounced raised border up to maturity are especially frequent.

Maine to Nebr., southw. to Ala. and Tex.

11. **Carex grávida** Bailey. (Including *Carex gravida* var. *laxifolia* Bailey.) Map 429. Known in Indiana only from the prairie area in the westernmost tier of counties where it is found on sandy bur oak ridges and sandy and gravelly railroad embankments.

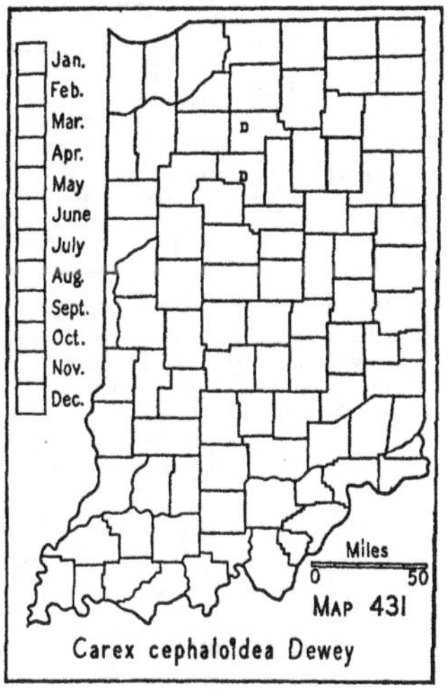

Reported from Lake County by Peattie and by Pepoon but no authentic specimens could be found. The Lake County reports were probably based upon specimens of *C. Muhlenbergii* in the Field Museum and University of Wisconsin herbaria which were collected by Umbach and distributed as *C. gravida.* The report from Fayette County by Deam was based upon a specimen of *C. aggregata* which was referred to *C. gravida* by Mackenzie.

Deam no. 43219 is intermediate between *C. gravida* and its var. *Lunelliana.*

Sw. Ont. and Ohio to N. Dak. and Wyo., southw. to Mo. and Kans.

11a. **Carex gravida** var. **Lunelliàna** (Mack.) Hermann. (Amer. Midland Nat. **17**: 855. 1936.) (*Carex Lunelliana* Mack. Bull. Torrey Bot. Club **42**: 615. 1915.) Map 430. On sandy roadsides and railroad embankments in the westernmost tier of counties where it is rare. Some of the Benton County specimens intergrade slightly with the species but the Vigo County plants from very sandy soil on a roadside knoll 5 miles north of Terre Haute are a good match with the type material of *C. Lunelliana.*

Ind. and Iowa to Tex. and N. Mex.

12. **Carex cephaloìdea** Dewey. Map 431. Rich woods. In Indiana known only from two collections by Deam: beech-sugar maple woods a mile and a half west of New Waverly, Cass County; and low woods bordering Tippecanoe River north of DeLong, Fulton County. It is probably more frequent than the few collections would indicate since it resembles the ubiquitous *C. sparganioides* so closely that it is apt to be passed by as that species.

The Tippecanoe County report by Smith is not supported by a specimen nor could any specimen be found to confirm Peattie's report from Lake County.

Specimens of *C. alopecoidea* (§ *Vulpinae*) before fully mature, and particularly when from an open habitat, often closely simulate *C. cephaloidea.* These may be most readily distinguished by their acuminate to cuspidate or aristate pistillate scales which are more than half the length

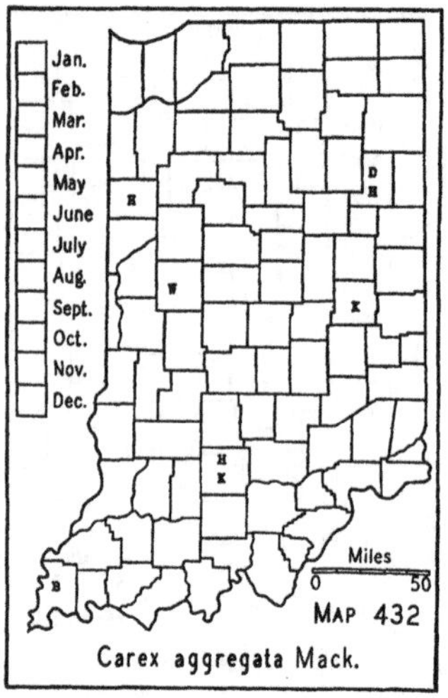

Carex aggregata Mack.

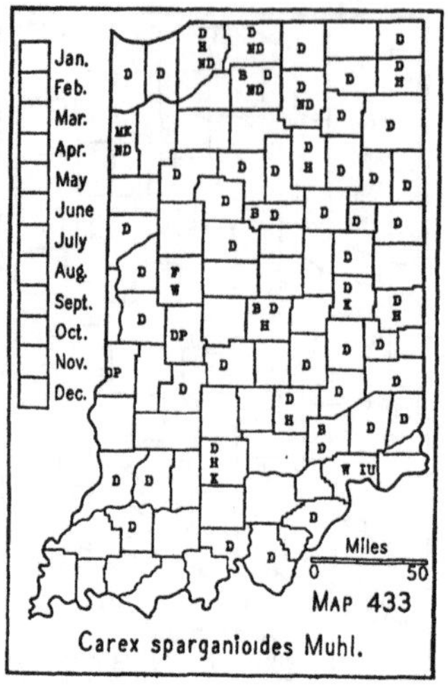

Carex sparganioides Muhl.

Carex annectens Bickn.

of the bodies of the perigynia and have a conspicuous green center. In *C. cephaloidea* the pistillate scales are obtuse or at most acute, half the length of the bodies of the perigynia or shorter, and are hyaline throughout (never becoming coppery-tinged at maturity as in *C. alopecoidea*) except for the faint green midrib.

N. B. to Minn., southw. to N. J. and Ill.

13. **Carex aggregàta** Mack. (Bull. Torrey Bot. Club 37: 246. 1910.) Map 432. Infrequent on banks of creeks, on dry grassy and partially wooded slopes, in low open woods, and as a weed in lawns. The perigynia are very susceptible to infection by a smut which often prevents their maturing.

N. J. to D. C., westw. to Kans. and Okla.

14. **Carex sparganioìdes** Muhl. Map 433. Very common in dry woods (usually sugar maple, beech or white oak), thickets, and along roadsides. One of the most abundant sedges in the state.

Que. to S. Dak., southw. to Va., Ky., and Kans.

5. § MULTIFLÒRAE

Beak of perigynium much shorter than the body; perigynium subcoriaceous; leaves usually shorter than the culms.

- Perigynia mostly broadest at the base, usually nerved dorsally, 2 mm or more wide; beak of perigynium prominent, conspicuously cleft.............15. *C. annectens.*
- Perigynia mostly broadest at or below the middle, nerveless dorsally, usually less than 2 mm wide; beak of perigynium very small, obscurely cleft; heads generally shorter and denser..................15a. *C. annectens* var. *xanthocarpa.*

Beak of perigynium about equaling the body; perigynium membranaceous; leaves normally exceeding the culms.

- Perigynia ovate, the body corky-margined to the base, contracted into the beak..16. *C. vulpinoidea.*
- Perigynia narrowly lanceolate, the body thin-edged and not at all corky-margined, tapering gradually into the beak; teeth of perigynium almost obsolete; plant low, densely cespitose; leaves narrow, rigid; inflorescence short, broad, and congested. (See excluded species no. 3, p. 272)....*C. vulpinoidea* var. *pycnocephala.*

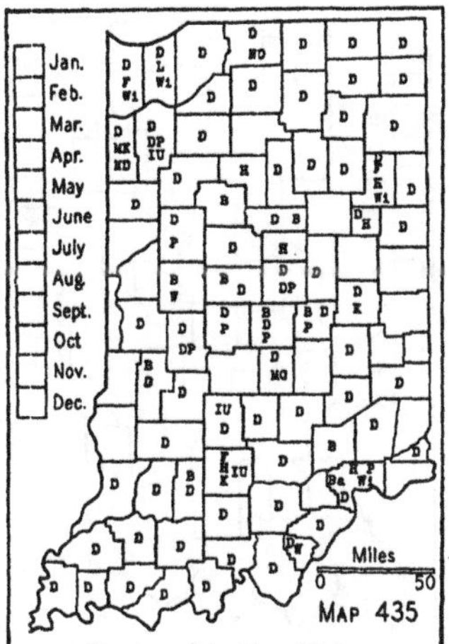

Carex vulpinoidea Michx.

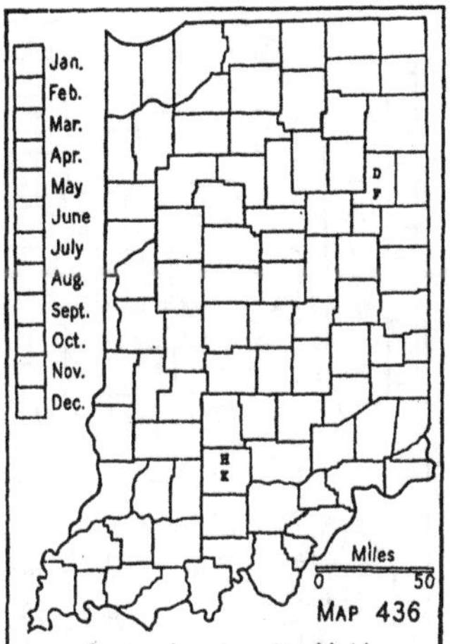

Carex decomposita Muhl.

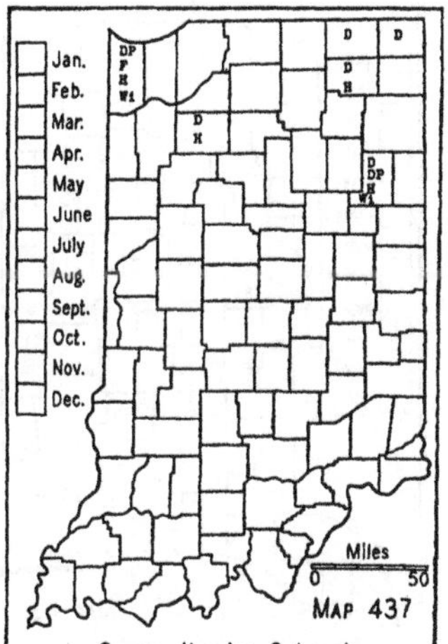

Carex diandra Schrank

15. **Carex annéctens** Bickn. (Bull. Torrey Bot. Club 35: 492. 1908.) (*C. setacea* Dewey var. *ambigua* (Barratt) Fern.) Map 434. Fairly common in the southern counties; infrequent in northern Indiana. In southern Indiana it occurs most commonly in low flat woods although it is frequently found in wet fallow clay fields; in the northern counties it is usually in marshes or pastures.

Maine to Wis., southw. to Tex. and Fla.

15a. **Carex annectens** var. **xanthocárpa** (Bickn.) Wieg. (Bull. Torrey Bot. Club 23: 22. 1896; Rhodora 24: 74. 1922.) (*Carex xanthocarpa* Bickn. and *Carex brachyglossa* Mack.) Known in Indiana from a single collection: Deam no. 42927, in a low place in an open post oak flat south of Half Moon Pond, 10 miles southwest of Mt. Vernon, Posey County.

The report from Knox County by Deam was based upon a collection determined by Mackenzie as *C. brachyglossa* but the specimen should be referred to *C. annectens*.

Maine to Iowa, southw. to Va. and Kans.

16. **Carex vulpinoìdea** Michx. Map 435. One of the commonest sedges of swampy places throughout the state. Its favorite habitat is in roadside ditches but it is found also in marshes, swamps, low open woods, and ravines, on flood plains, and banks of streams and ponds.

Newf. to B. C., southw. to Fla., Tex., Ariz., and Oreg.

6. § PANICULÀTAE

Leaf blades 3-8 mm wide; perigynia very abruptly short-beaked, tapering at the base; inflorescence usually 8-15 cm long, obviously branched.........17. *C. decomposita.*

Leaf blades 1-3 mm wide; perigynia tapering or contracted into the beak, rounded or truncate at the base; inflorescence 2.5-5 (8) cm long, obscurely branched.

Sheaths not copper-colored at the mouth; head little interrupted; perigynia 2-2.5 mm long, convex ventrally, lustrous, not concealed by the scales...18. *C. diandra.*

Sheaths copper-colored at the mouth; head interrupted; perigynia 2.5-3.5 mm long, flat or concave ventrally, dull, nearly concealed by the scales....19. *C. prairea.*

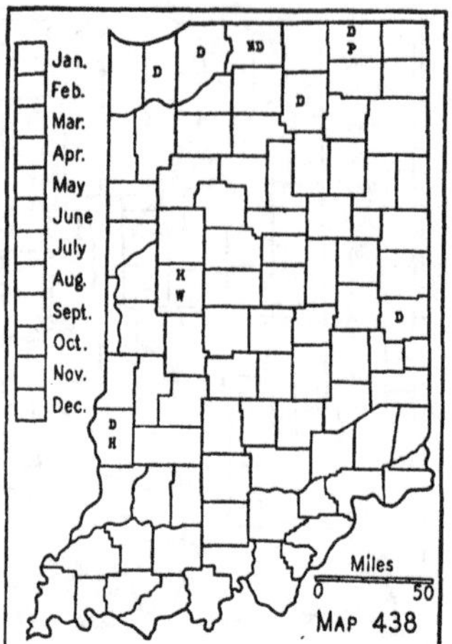

Carex prairea Dewey

Carex stipata Muhl.

Carex stipata var. maxima Chapm.

17. **Carex decompósita** Muhl. Map 436. Specimens to confirm the Marshall and Lake County reports for this very local species could not be located. From Pepoon's statement that it is an abundant species in the Chicago region it seems very likely that his report was based upon material of *C. diandra* or perhaps of *C. vulpinoidea.* Specimens of *C. vulpinoidea* collected by Umbach from the Illinois portion of the Chicago region and labeled *C. decomposita* were found in the University of Wisconsin Herbarium. *C. decomposita* is represented from Indiana by two collections: Deam, June 26, 1898, in bunches of moss on logs in a drained pond, Little's woods, Lancaster Twp., six miles northeast of Bluffton, Wells County; and Kriebel no. 2221, in a knothole at base of tree in swamp, two and a half miles northeast of Avoca, Lawrence County.

N. Y. to Mich., southw. to Fla., La., and Mo.

18. **Carex diándra** Schrank. (*Carex teretiuscula Gooden.*) Map 437. Frequent in the lake area on marly and sandy borders of lakes and in swales, marshes, or bogs. The specimen upon which Coulter's report from Daviess County was based should probably be referred to *C. prairea.* The specimen could not be located in the Indiana herbaria.

Newf. to Alaska, southw. to N. J., Ind., and Colo.; also in Eurasia.

19. **Carex pràirea** Dewey. (*Carex teretiuscula* var. *ramosa* Boott and *Carex diandra* var. *ramosa* (Boott) Fern.) Map 438. Frequent, except in southernmost counties, in marshes, tamarack bogs, marly swamps and on borders of streams or lakes.

Que. to Sask., southw. to N. J., Ind., Iowa, and Nebr.

7. § VULPÌNAE

Perigynium tapering into the beak, the body strongly nerved ventrally or perigynium very long beaked.

Perigynium 4-6 mm long, rounded at the base, strongly nerved ventrally, the beak 1-2 times the length of the body; sheaths not dotted with purple ventrally.

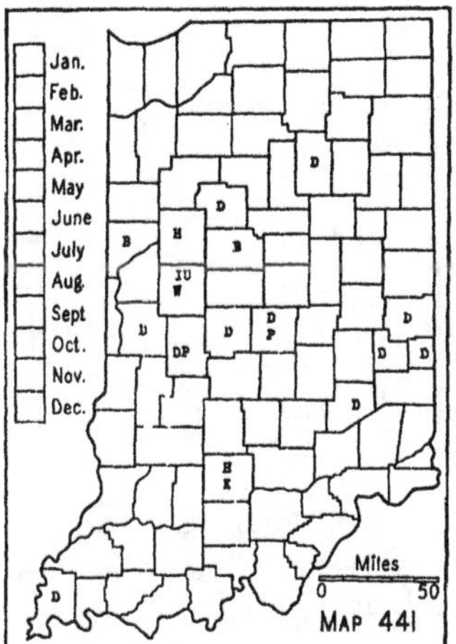

Carex laevivaginata (Kükenth.) Mack.

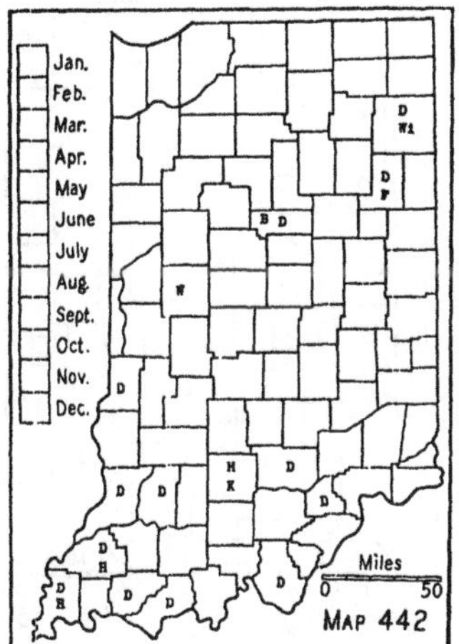

Carex Crus-corvi Shuttlw.

Carex alopecoidia Tuckerm.

Sheaths not thickened at the mouth, cross-rugulose ventrally, easily broken, prolonged upward at the mouth.

Perigynium 4-5 mm long, the beak about the length of the body; leaf blades 4-8 mm wide .... 20. *C. stipata.*

Perigynium 5-6 mm long, the beak longer than the body; leaf blades 8-15 mm wide .... 20a. *C. stipata* var. *maxima.*

Sheaths thickened (often cartilaginous) at the mouth, rarely cross-rugulose ventrally, not easily broken, concave or truncate at the mouth............... .... 21. *C. laevivaginata.*

Perigynium 6-7 mm long, abruptly enlarged below into a disklike base, obscurely nerved ventrally except at the base, the beak 2-3 times the length of the body; sheaths dotted with purple ventrally....................... 22. *C. Crus-corvi.*

Perigynium contracted into a beak not longer than the body, the body nerveless ventrally except sometimes at the base.

Sheaths not cross-rugulose ventrally; spikes yellowish or tawny at maturity; perigynium narrow, faintly nerved dorsally, the beak about the length of the body. .... 23. *C. alopecoidea.*

Sheaths cross-rugulose ventrally; spikes green; perigynium broad, usually strongly nerved dorsally, the beak generally about half the length of the body.......... .... 24. *C. conjuncta.*

20. **Carex stipàta** Muhl. Map 439. Very common in wet habitats throughout Indiana. It is usually found on borders of ponds and streams and in low woods, roadside ditches, swamps, marshes, bogs, and woodland swales.

Newf. to Alaska, southw. to N. C., Tenn., Kans., N. Mex., and Calif.

20a. **Carex stipata** var. **máxima** Chapm. (*Carex stipata* var. *uberior* Mohr and *Carex uberior* (Mohr) Mack.) Map 440. Rare; it is usually found on the borders of ponds and streams, in low woods, roadside ditches, swamps, marshes, bogs, and woodland swales.

In the western portion of its range transitional forms between this plant and *C. stipata* occur with a frequency which discourages attempts to maintain it as specifically distinct. Of the five collections known from Indiana three are typical of var. *maxima* in all their characters while two (Deam no. 36082, with leaves averaging only 7 mm wide, and Deam no. 38688,

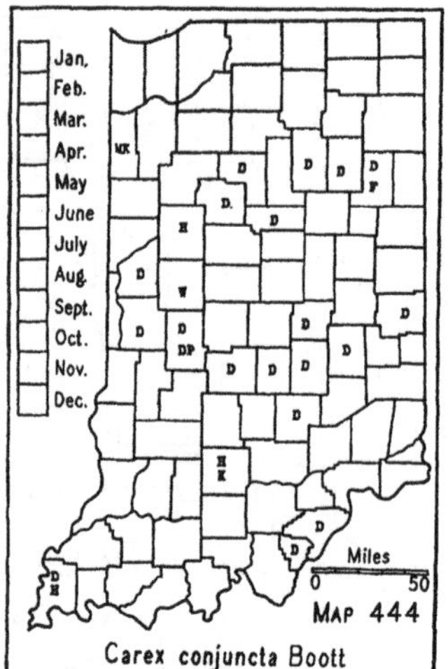

Carex conjuncta Boott

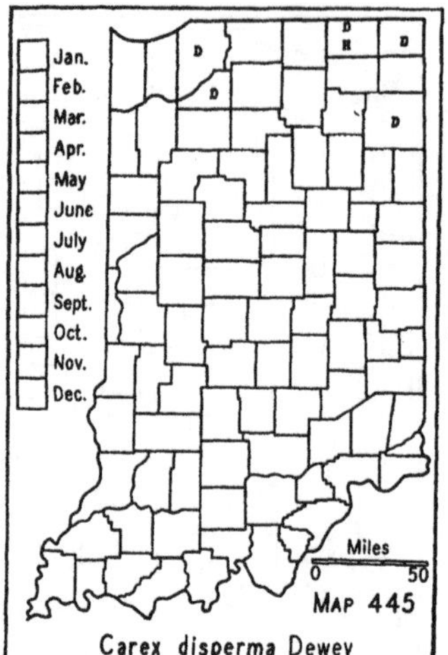

Carex disperma Dewey

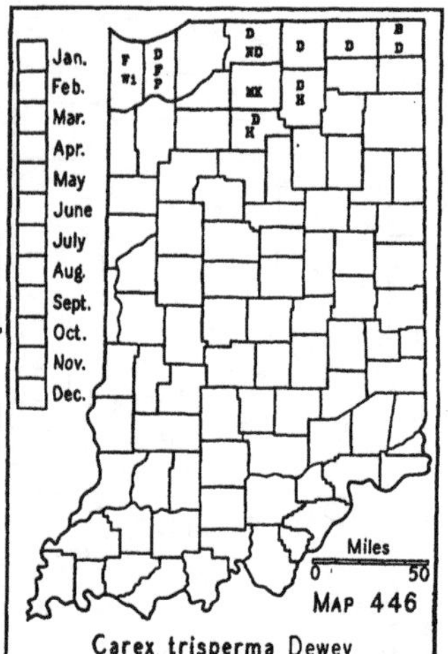

Carex trisperma Dewey

with no perigynia over 5 mm long and some less, with the beaks of the perigynia only slightly longer than the bodies but leaves averaging 10 mm wide) approach the typical form of *C. stipata.*

N. J. and Pa. southw. along the coast to Fla. and Tex., and northw. in the Mississippi Valley to Mo. and Ind.

21. **Carex laevivaginàta** (Kükenth.) Mack. (Britton and Brown, Illus. Flora, ed. 2, 1: 371. 1913. See also Fernald, Rhodora **17**: 231. 1915.) Map 441. Infrequent in wet ravines, swamps, swales in woods and on muddy banks of creeks.

Mass. to Minn., southw. to Fla. and Mo.

22. **Carex Crus-córvi** Shuttlw. Map 442. Frequent in southern Indiana in low open woods, especially flat pin oak woods; occasional in northern Indiana on borders of ponds in woods. Reported from Lake County by Peattie and by Pepoon but no specimens from the county could be located.

Tenn. southw. to Fla. and Tex.; in the Mississippi Valley from s. Mich., s. Minn., and e. Nebr. to La.

23. **Carex alopecoìdea** Tuckerm. Map 443. Known in Indiana from a single collection: Deam no. 41282, in a low place in white oak woods 3 miles south of Yorktown, Delaware County. No specimens could be found to confirm the reports by Peattie and by Pepoon from Lake County, by Pepoon from Porter County, and by Phinney from Jay, Randolph, and Wayne Counties.

Que. to Minn., southw. to N. J. and Iowa.

24. **Carex conjúncta** Boott. Map 444. Frequent to common in central Indiana; frequent elsewhere except in the lake and prairie areas. Its preferred habitat is on wooded alluvial banks of streams, but it is also found in low woods and on moist wooded slopes.

N. J. to D. C., westw. to S. Dak. and Kans.

8. § HELEONÁSTES

Spikes androgynous; perigynia unequally biconvex.................25. *C. disperma.*
Spikes gynaecandrous; perigynia plano-convex.
  Lowest bract bristlelike, many times longer than its spike; perigynia 3-3.5 mm long. ....26. *C. trisperma.*
  Lowest bract much shorter; perigynia about 2 mm long.
    Perigynia distinctly short-beaked, loosely spreading; leaves green, 1-2.5 mm wide. (See excluded species no. 5, p. 272)........................*C. brunnescens.*
    Perigynia apiculate, appressed-ascending, leaves glaucous, 2-4 mm wide.
      Spikes 6-12 mm long, remote, the lowest 2-4 cm apart; perigynia 2.3-3 mm long. ....27. *C. canescens* var. *disjuncta.*
      Spikes 4-7 mm long, subapproximate or remote; perigynia barely 2 mm long. ....27a. *C. canescens* var. *subloliacea.*

25. **Carex dispérma** Dewey. (*Carex tenella* Schkuhr.) Map 445. Frequent in the northern counties in sphagnum in tamarack bogs and on mucky borders of lakes. Reported from Putnam County by Coulter but no specimen from that county could be found.

Newf. to Yukon, southw. to N. J., Ind., N. Mex., and Calif.; also in Eurasia.

26. **Carex trispérma** Dewey. Map 446. Restricted to the tamarack bogs of the northernmost counties where it is locally plentiful in sphagnum. No specimen could be located to substantiate Coulter's report from Putnam County.

Newf. to Sask., southw. to Md., Ill., and Minn.

27. **Carex canéscens** L. var. **disjúncta** Fern. Map 447. Frequent in the counties along the northern border of Indiana in tamarack bogs or low wet woods.

Lab. to Wis., southw. to Pa. and Ind.

27a. **Carex canescens** var. **subloliàcea** Laest. Map 448. Infrequent in the northernmost counties in swampy woods and in sphagnum in tamarack bogs.

Lab. to B. C., southw. locally to Conn. and Ind.

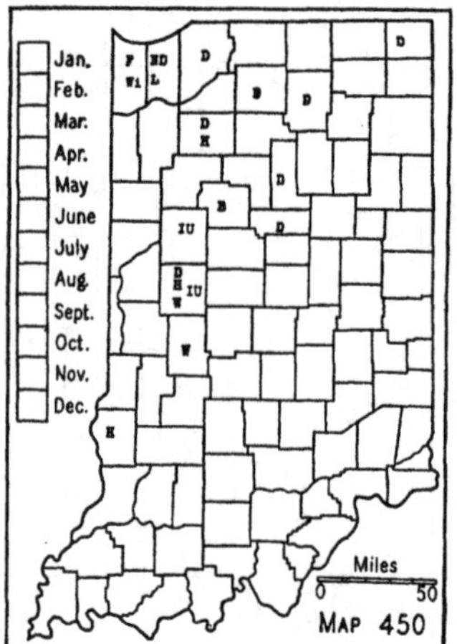

Carex interior Bailey

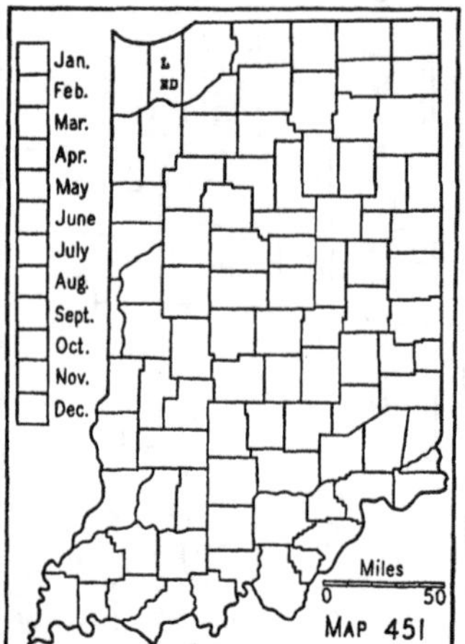

Carex Howei Mack.

Carex incomperta Bickn.

9. § STELLULÀTAE

Perigynium broadest near the middle, strongly nerved both ventrally and dorsally, with smooth beak.. 28. *C. seorsa.*

Perigynium broadest at the base, with serrulate beak.

Perigynia 2.25-3.25 mm long, the beak very shallowly bidentate.

Perigynia nerveless or few-nerved at the base ventrally, brownish or tawny, the beak with ventral false suture inconspicuous; scales obtuse; anthers 1 mm or less long; leaf blades 1-3 mm wide..........................29. *C. interior.*

Perigynia strongly nerved ventrally, deep green, the beak with ventral false suture conspicuous; scales subacute; anthers 1-1.5 mm long; leaf blades 0.25-1 mm wide..30. *C. Howei.*

Perigynia 2.75-4.75 mm long, the beak deeply bidentate.

Perigynia deep green at maturity, strongly nerved ventrally, the body suborbicular or very broadly ovate with raised margins, the beak less than half the length of the body, the teeth short, straight, rigid..........31. *C. incomperta.*

Perigynia stramineous to brown at maturity, faintly nerved ventrally, the body ovate to ovate-lanceolate (occasionally broadly ovate in *C. sterilis*).

Staminate flowers terminal, basal, or in separate spikes; margin of perigynium slightly if at all raised, setulose-serrulate toward the beak; beak of perigynium half the length of the body or more, the teeth broad, largely hyaline and soft, generally bent or twisted; scales chestnut brown, with lustrous white margins....................................32. *C. sterilis.*

Staminate flowers mostly at the base of the terminal spikes; margin of perigynium serrulate toward the beak, the teeth short, stiff; scales tinged yellowish brown, with narrow hyaline margin.

Perigynia 2.75-3.3 mm long, the beak about a third the length of the body, the teeth triangular; scales somewhat obtuse to acute.....33. *C. laricina.*

Perigynia 3.5-4 mm long, the beak about half the length of the body, the teeth subulate; scales acute to somewhat cuspidate. (See excluded species no. 9, p. 273)...................................*C. cephalantha.*

28. **Carex seòrsa** E. C. Howe. (*Carex rosaeoides* E. C. Howe.) Map 449. Rare in wet woods and tamarack bogs in the dune area. The known localities for this sedge in Indiana are: Dune Park, Keiser, and Tamarack in Porter County and Pine Station (now north Clark Street, Gary) in Lake County.

Mass. to Ga., locally westw. to Ind.

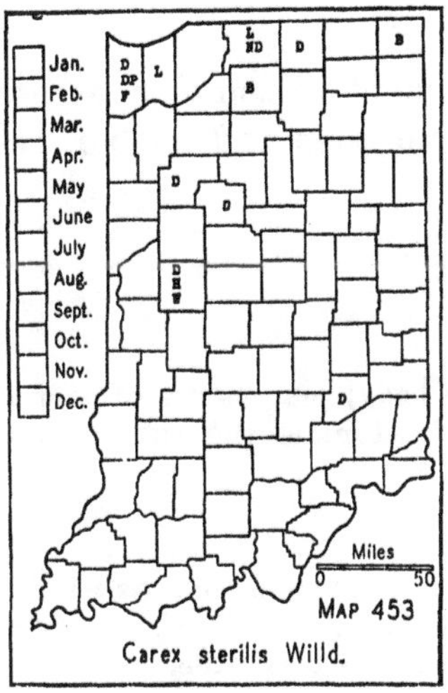

Carex sterilis Willd.

Carex laricina Mack.

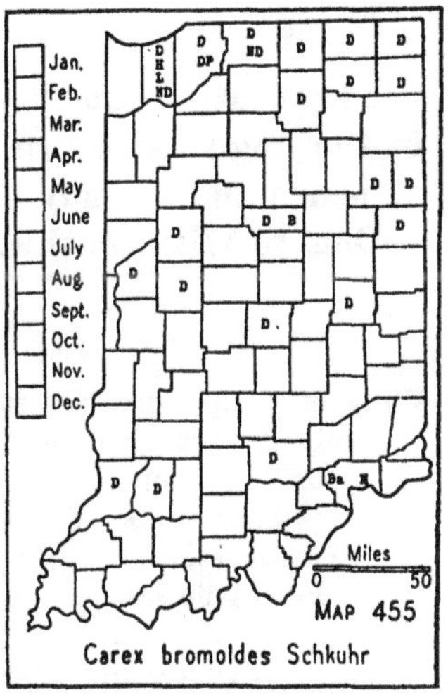

Carex bromoides Schkuhr

29. **Carex intèrior** Bailey. (*Carex scirpoides* Schkuhr, not *Carex scirpoidea* Michx.) Map 450. Frequent to common except in southern Indiana; in tamarack bogs and swamps and on springy banks.

Newf. to B. C., southw. to Pa., Ind., Kans., Calif. and Chihuahua.

30. **Carex Hówei** Mack. (Bull. Torrey Bot. Club 37: 245. 1910.) (*Carex interior* var. *capillacea* Bailey and *Carex scirpoides* var. *capillacea* (Bailey) Fern.) Map 451. Known in Indiana from a single collection by M. W. Lyon, Jr.: moist woods on dunes at Mineral Springs, Porter County, June 17, 1923.

N. S. to Fla. and La., westw., locally to Mich. and Ind.

31. **Carex incompérta** Bickn. (*Carex stellulata* var. *excelsior* Fern.) Map 452. Occasional in tamarack bogs, generally in sphagnum.

Mass. and N. Y., to Mich. and Ind., southw. to Fla. and Tex.

32. **Carex stérilis** Willd. (*Carex scirpoides* Schkuhr, in part.) Map 453. Frequent on marshy banks of streams and occasional in open swamps, bogs, and springy places in woods. Not known from the unglaciated area.

Newf. to Minn., southw. to N. J., Pa., and Ill.

33. **Carex laricìna** Mack. (N. Amer. Flora 18: 113. 1931.) Map 454. Rare, in tamarack bogs and on mucky borders of lakes in the northeastern counties.

The type collection of this species is Deam no. 10927 from a tamarack bog a mile south of Leesburg, Kosciusko County. The other two Indiana stations for it are: in a bog a mile south of Garrett, De Kalb County, and in sphagnum on the border of a small lake in Jackson Twp., Wells County.

Ont. and nw. Pa. to Wis., and southw. to Ind.

10. § DEWEYÀNAE

34. **Carex bromoides** Schkuhr. Map 455. Frequent to common except in the unglaciated area, in wet woods, swamps, and bogs and on borders of ponds and springy banks of streams.

Que. to Wis., southw. to Fla. and La.

11. § OVÀLES

Wing of perigynium not narrowed near the middle of the body; leaf blades of sterile culms erect or ascending, usually clustered toward the top; sterile culms often poorly developed.

Perigynia not obovate, widest near the middle or base.

Leaf sheaths strongly white-hyaline ventrally.

Perigynia lanceolate to narrowly ovate-lanceolate, 3 to 4 times as long as wide.. ....35. *C. scoparia.*

Perigynia ovate-lanceolate or broader, at most twice as long as wide.

Perigynia narrowly to broadly ovate, 3-4 mm long.

Leaf blades 1.5-4.5 (averaging 2.5) mm wide; sheaths not mottled with green and white dorsally.

Perigynia 3-3.5 mm long; spikes closely aggregated, not clavate at base... ....36. *C. Bebbii.*

Perigynia 3.5-4.5 mm long; spikes not aggregated, usually in a flexuous, moniliform inflorescence, clavate at base..............37. *C. tenera.*

Leaf blades 2.5-6 (averaging 4) mm wide; sheaths mottled with green and white dorsally; perigynia less abruptly beaked and beak narrower than in *C. tenera*..38. *C. normalis.*

Perigynia (2.75) 3.5-6.5 mm long, the body suborbicular.

Perigynia 3.5-5.5 mm long, thick, coriaceous or subcoriaceous, usually plano-convex.

Perigynia averaging (2.75) 3.5-4 mm long, the beak half the length of the body or more; achene 1.5 mm long, oblong-ovoid; spikes in a moniliform inflorescence..........................39. *C. festucacea.*

Perigynia 3.75-5.5 mm long, the beak less than half the length of the body; achenes 1.75-2 mm long, orbicular or suborbicular when fully mature; spikes aggregated or in a moniliform inflorescence.

Perigynia ovate, submembranaceous, few-nerved ventrally, broadest near the base, tapering into the beak, the beak broader than in *C. brevior*, especially toward the base........................40. *C. molesta.*

Perigynia broadly ovate to suborbicular, coriaceous, usually nerveless or nearly so ventrally, broadest near the middle, abruptly contracted into the beak..................................41. *C. brevior.*

Perigynia 5.6-6.5 mm long, flat and thin, nearly transparent.............. ....42. *C. Bicknellii.*

Leaf sheaths green and strongly nerved ventrally nearly to the mouth.

Scales cuspidate or even obtuse; perigynia nerveless or nearly so ventrally; spikes 2-5, aggregated into a stiff head..................43. *C. suberecta.*

Scales long-acuminate to aristate; perigynia nerved ventrally; spikes 4-8, in a very flexuous inflorescence.................................44. *C. Richii.*

Perigynia obovate, the body widest near the top.

Scales obtuse to short-acuminate; achenes sessile or substipitate; perigynia 1.5-3 mm wide.

Tips of perigynia appressed; perigynia with body rounded at apex; spikes approximate or aggregated, greenish to silvery brown.

Perigynia nerveless ventrally; spikes 5-25, densely aggregated; leaf blades of sterile culms 3.5-5 mm wide.......................45. *C. cumulata.*

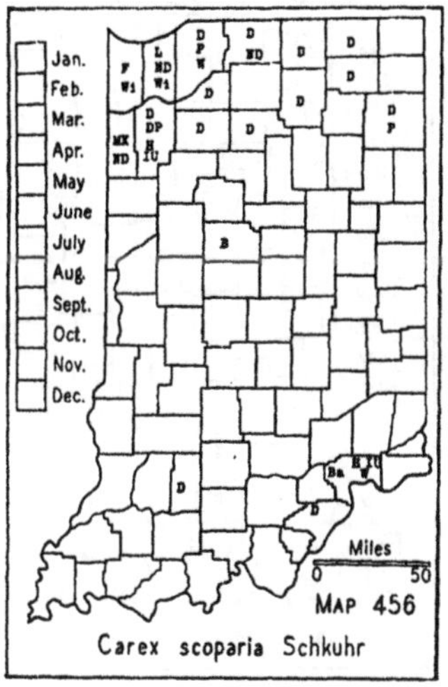

Carex scoparia Schkuhr

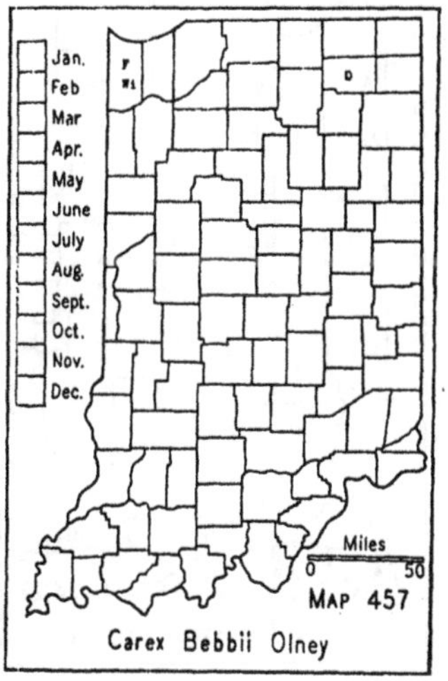

Carex Bebbii Olney

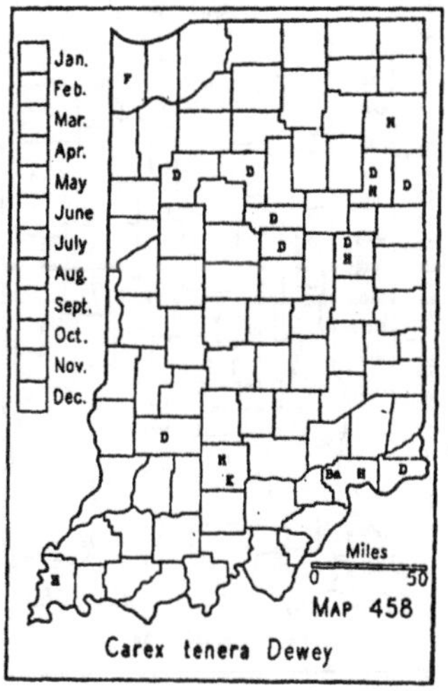

Carex tenera Dewey

Perigynia nerved ventrally; spikes 3-10, aggregated or somewhat separate; leaf blades of sterile culms 2.5-3 mm wide; scales either nearly equaling perigynia or blunt.................................46. *C. Longii.*

Tips of perigynia spreading; perigynia with body truncate-rounded at apex, very abruptly beaked; spikes not aggregated, not silvery; scales acute, conspicuously shorter than the perigynia................47. *C. albolutescens.*

Scales long-acuminate to aristate; achenes slenderly stipitate; perigynia 2.5-4 mm wide...48. *C. alata.*

Wing of perigynium rather abruptly narrowed near the middle of the body; leaf blades of sterile culms widely spreading, numerous, not clustered at the apex; sterile culms strongly developed.

Perigynia 3-7 mm long; spikes 4-15 mm long; achenes oblong-oval, 1.5 mm long; ligule much longer than wide.

Tips of perigynia appressed or ascending; perigynia thin, scarcely distended over the achenes..49. *C. tribuloides.*

Tips of perigynia recurved or widely spreading; perigynia firm, obviously distended over the achenes................................50. *C. cristatella.*

Perigynia 7-10 mm long; spikes 16-25 mm long; achenes linear-oblong, 2.5 mm long; ligule as wide as long...............................51. *C. muskingumensis.*

35. **Carex scopària** Schkuhr. (Including *Carex scoparia* var. *condensa* Fern.) Map 456. Common in marshes and open swampy places; occasional in low open woods and on sandy lake borders. This sedge is frequently the dominant plant in marshes or "sedge meadows" where it is usually associated with *Juncus effusus* var. *solutus, Juncus Dudleyi,* and *Carex vulpinoidea.*

Newf. to B. C., southw. to S. C., N. Mex., and Oreg.

36. **Carex Bébbii** Olney. Map 457. Infrequent in marshes and interdunal swales in Lake County. In Noble County a single collection was made by Deam in a ditch along a railroad a mile east of Kimmel.

Plants of *Carex Bebbii* lacking sterile culms are occasionally difficult to distinguish from *C. cristatella* especially before the perigynia are fully mature. Leaf blades of *C. Bebbii,* however, vary from 2 to 4.5 mm broad, those of *C. cristatella* from 3 to 7 mm broad. In *C. Bebbii* the pistil-

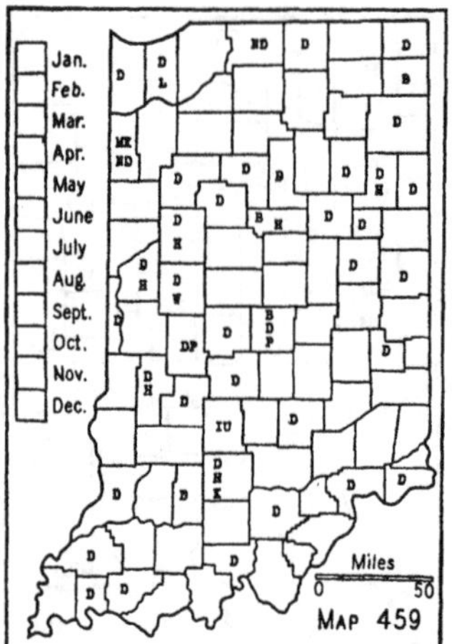

MAP 459
Carex normalis Mack.

MAP 460
Carex festucacea Schkuhr

MAP 461
Carex molesta Mack.

late scales are relatively longer, acuminate to acute or occasionally blunt; in *C. cristatella* the scales are shorter, with dilated hyaline blunt tips.

Umbach no. 3651 and Bebb nos. 541 and 874, all from Lake County, are intermediate between *C. Bebbii* and *C. cristatella* in most of their characters. Similar material from Michigan has been identified by Mackenzie as a hybrid between the two species.

Newf. to B. C., southw. to N. J., Ill., Colo., and Wash.

37. **Carex ténera** Dewey. (*Carex straminea* of recent authors, not Willd.; *Carex tenera* var. *echinodes* (Fern.) Wieg.) Map 458. Frequent in or near the lake area and in the southern counties in dry or moist, usually open, woods, on borders of ponds in woods, and along railroad ditches. Generally the heads are fewer-flowered in this sedge than in other species of § *Ovales* and this condition and the widely spreading perigynia sometimes result in a superficial resemblance to species of § *Stellulatae.*

Que. to Alberta, southw. to D. C., N. C., and Ill.

38. **Carex normàlis** Mack. (*Carex mirabilis* Dewey, not Host.) Map 459. Very common in dry or moist woods and thickets. In the eastern part of its range this species seems to be partial to dry open habitats, but in Indiana it has been most often collected in low or flat woods, shaded ravines, marshy habitats on the borders of ponds, and on the flood plains of streams.

Maine to Man., southw. to N. C. and Okla.

39. **Carex festucàcea** Schkuhr. (Bull. Torrey Bot. Club 42: 608. 1915.) Map 460. Frequent in southern Indiana in low flat woods, especially pin oak woods, and on moist wooded slopes; occasional in roadside and railroad ditches in the northern counties.

Mass. to Ind. and Iowa, southw. to Ga. and La.

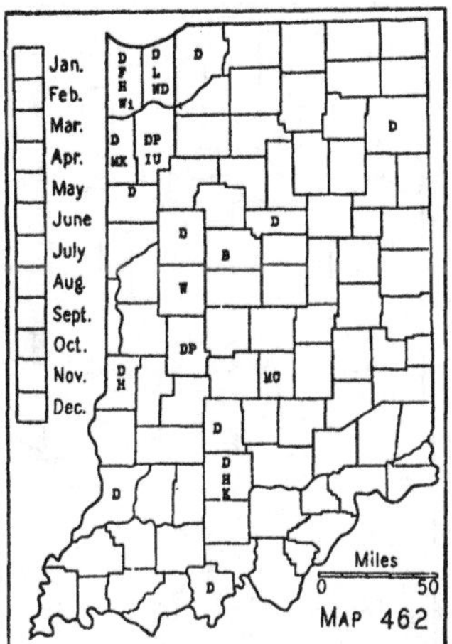

Carex brevior (Dewey) Mack.

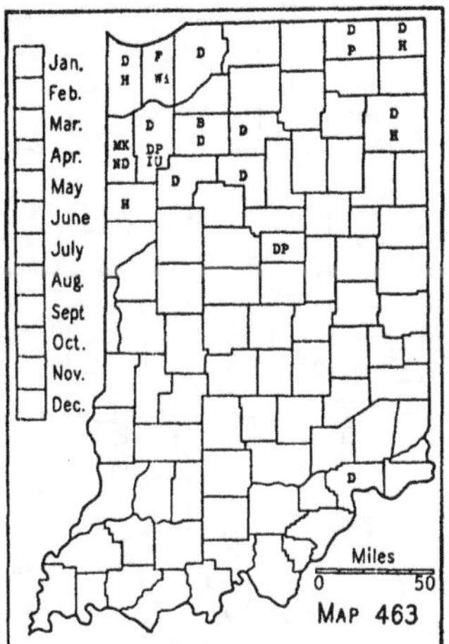

Carex Bicknellii Britton

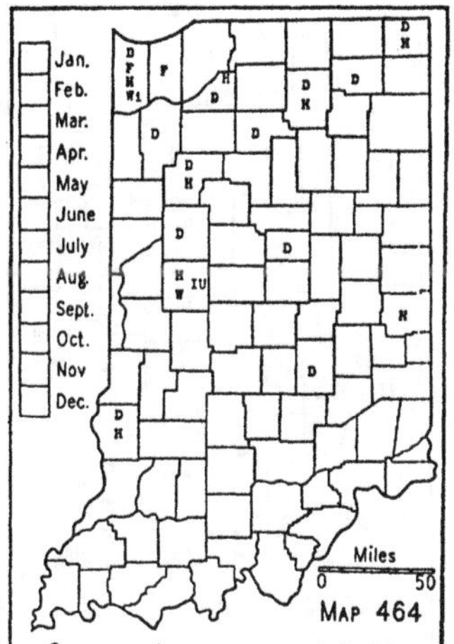

Carex suberecta (Olney) Britton

40. **Carex molésta** Mack. (N. Amer. Flora **18**: 151. 1931.) Map 461. Infrequent to rare along railroad sidings and roadsides and in ditches and dry woodlands.

N. Y. to Kans. and Nebr.

41. **Carex brévior** (Dewey) Mack. (*Carex festucacea* var. *brevior* (Dewey) Fern.) Map 462. Common in dry open woods and moist ditches and along railroads and roadsides, especially in the prairie area.

Que. to B. C., southw. to D. C., Tenn., Tex., N. Mex., and Oreg.

42. **Carex Bicknéllii** Britt. Map 463. Frequent to common along railroad sidings and grassy roadsides in northern Indiana; rare in the southern counties and not known from the unglaciated area. Occasional in low, moist sandy habitats; very rare in open woods.

Maine to Sask., southw. to Del., Ark., and Okla.

43. **Carex suberécta** (Olney) Britt. Map 464. Frequent to common, except in the southern counties, in open swamps, marshes, and moist ditches and on wet sandy borders of lakes. Not known from the unglaciated area.

Ont. to Va., Minn., and Mo.

44. **Carex Ríchii** (Fern.) Mack. (*Carex hormathodes* var. *Richii* Fern. and *Carex straminea* of Svenson, Rhodora **40**: 329-330. 1938.) Map 465. Rare and local in open swampy woods and borders of ponds in woods, less frequently in open non-calcareous marshes or swamps. The known stations are all in either the lake area or the unglaciated area.

Mass. to N. J. and D. C., westw. to Mich. and Ind.

45. **Carex cumulàta** (Bailey) Mack. (Bull. Torrey Bot. Club **49**: 366. 1922.) (*Carex albolutescens* var. *cumulata* Bailey.) Map 466. Known in Indiana only from Newton County where in 1936 a colony was found by Miss Madge McKee along a roadside ditch 3 miles northwest of Morocco. It is a local species throughout most of its range.

N. S. to N. J., westw. to Sask.

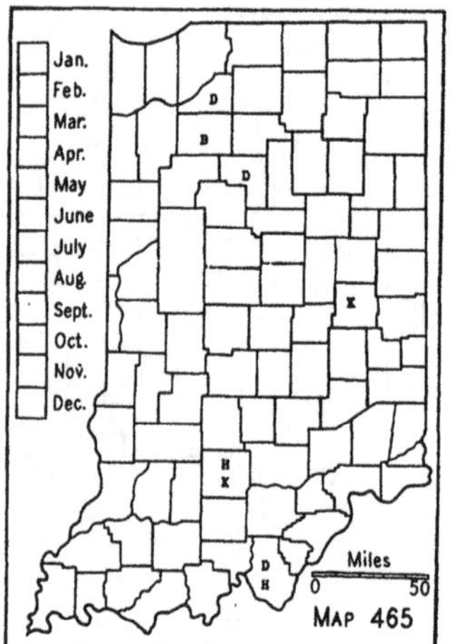

Carex Richii (Fern.) Mack.

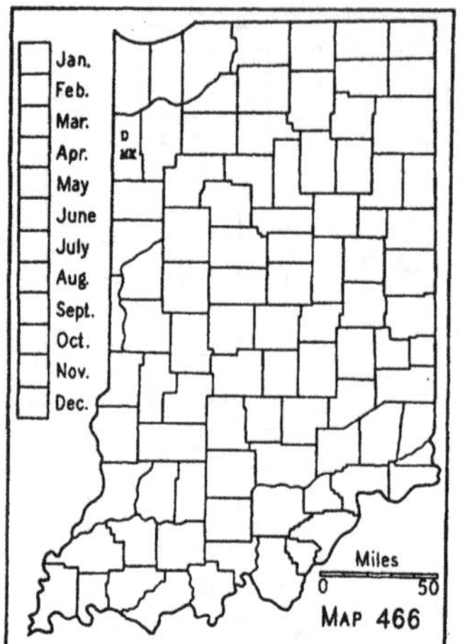

Carex cumulata (Bailey) Mack.

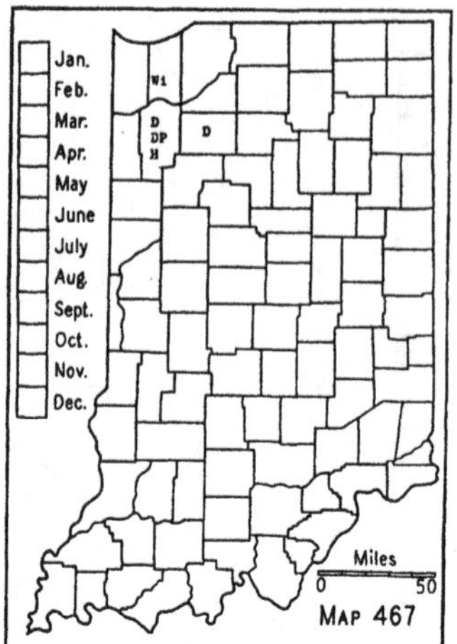

Carex Longii Mack.

46. **Carex Lóngii** Mack. (Bull. Torrey Bot. Club **49**: 372. 1922.) (*Carex albolutescens* of recent authors, not Schwein.) Map 467. Infrequent in the northwestern counties where it is found in acid swamps and sloughs, less often in sandy interdunal swales.

Mass. to Venezuela; nw. Ind. and sw. Mich.; also in Bermuda.

47. **Carex albolutéscens** Schwein. (Bull. Torrey Bot. Club **49**: 372. 1922.) (*Carex straminea* of Mack., probably not of Willd., Rhodora **40**: 329-330. 1938.) Map 468. Frequent in southern Indiana in low flat woods, associated principally with sweet gum and pin oak. It also is found rarely along the northern border of the state where it occurs in low woods, associated with beech and sugar maple, and occasionally in swamps.

N. S. southw. along the coast to Fla., westw. along the Gulf to Tex. and northw. in the Mississippi Valley to Ind. and sw. Mich.

48. **Carex alàta** Torr. Map 469. Infrequent in swamps and sandy swales in the lake area. It is seldom plentiful in any locality; frequently only one or two plants can be found at a station.

Mass. to Fla. and Tex., westw. to Mich., Ind., and Mo.

49. **Carex tribuloìdes** Wahl. (Including *Carex tribuloides* var. *sangamonensis* Clokey.) Map 470. Very common throughout the state in swamps, open marshes, low woods, and ditches and on the low borders of streams and ponds.

Que. to Minn., southw. to Fla. and La.

50. **Carex cristatélla** Britt. (*Carex cristata* Schwein., not Clairv.) Map 471. Common in low open woods, swamps, marshes, and roadside ditches and on flood plains and banks of streams. Rare in the unglaciated area.

Mass. to N. Dak., southw. to Va. and Mo.

51. **Carex muskinguménsis** Schwein. Map 472. Frequent in low wet places in woods where it often forms extensive and pure stands if not

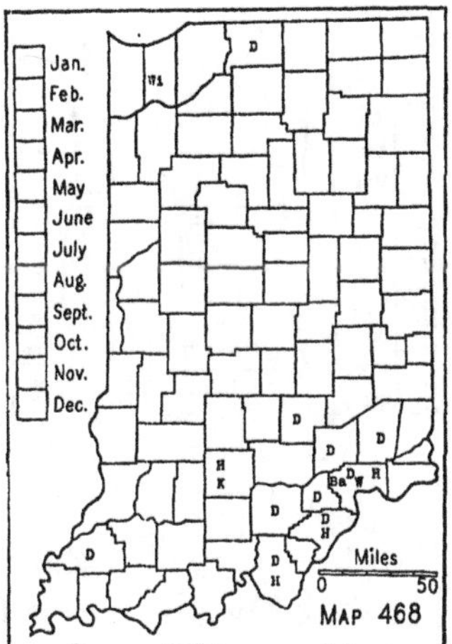

Carex albolutescens Schwein.

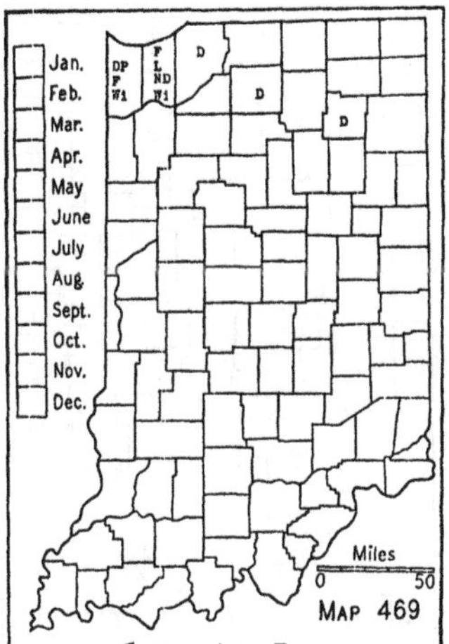

Carex alata Torr.

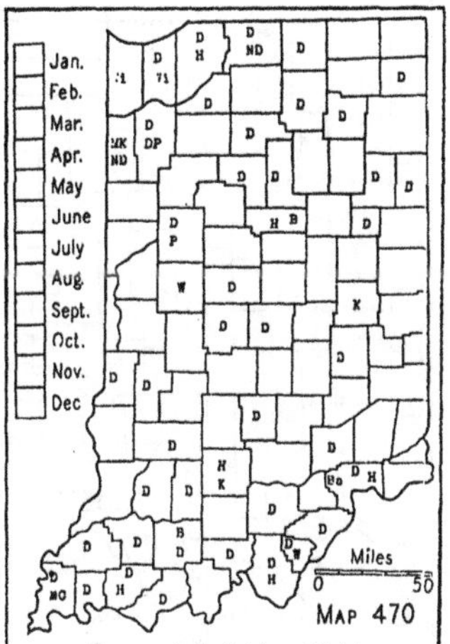

Carex tribuloides Wahl.

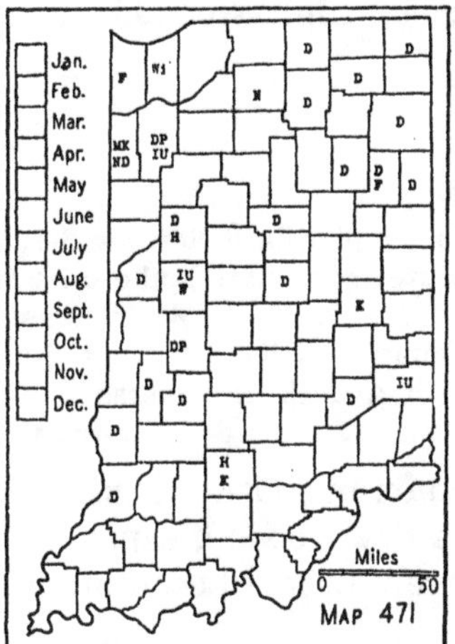

Carex cristatella Britt.

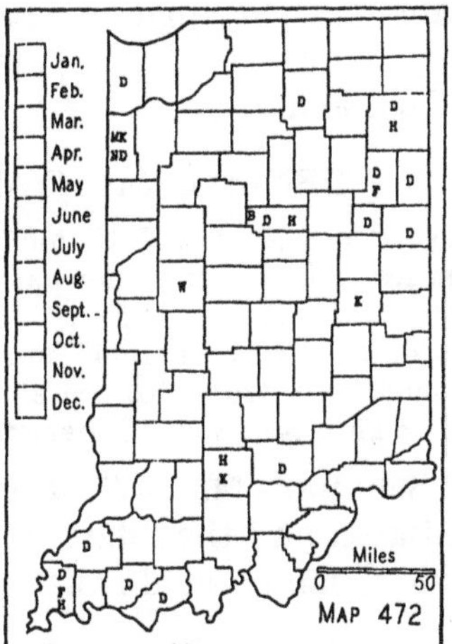

Carex muskingumensis Schwein.

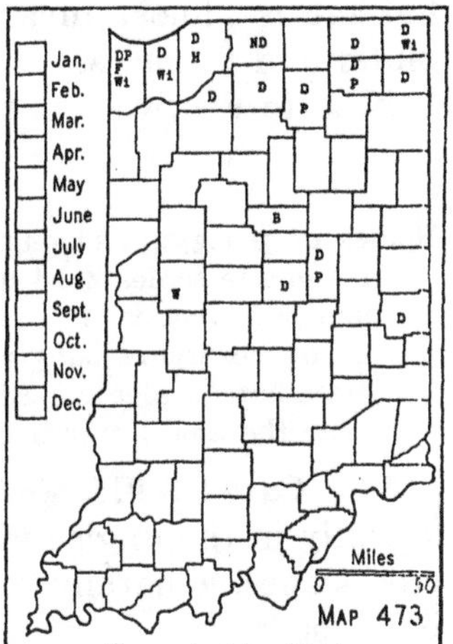

Carex leptalea Wahl.

obstructed by undergrowth. Occasional in buttonbush swamps and wet woods and on flood plains. Northward it is usually found with bur oak.

Ohio and Ky. to Man., Kans., and Ark.

12. § POLYTRICHOÌDEAE

Perigynia 2.5-3.5 mm long, slightly overlapping; achenes lustrous, obtusely angled.... ....52. *C. leptalea.*
Perigynia 4-5 mm long, strongly overlapping; achenes barely lustrous, sharply angled.. ....52a. *C. leptalea* var. *Harperi.*

52. **Carex leptàlea** Wàhl. Map 473. Common in northern Indiana in tamarack bogs and occasional in wet woods. Infrequent in central Indiana, in swamps and on banks of streams. It is generally plentiful wherever found and in tamarack bogs it is usually associated with *Carex trisperma.*

Newf. to B. C., southw. to Pa., Mo., Colo., and Calif.

52a. **Carex leptalea** var. **Hárperi** (Fern.) Stone. (*Carex Harperi* Fern.) Map 474. Rare in central and southern Indiana. In Indiana its

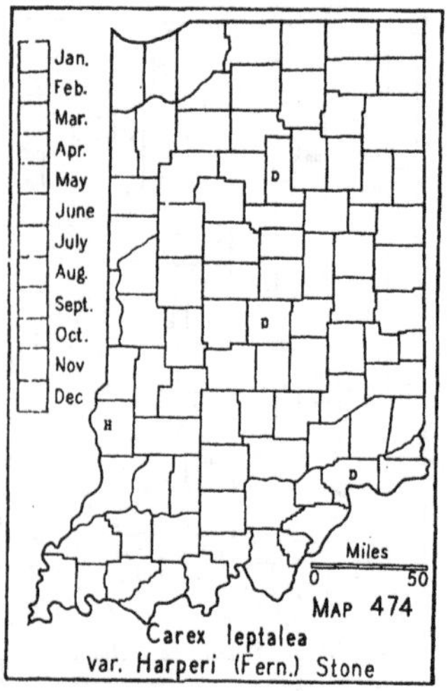

Carex leptalea var. Harperi (Fern.) Stone

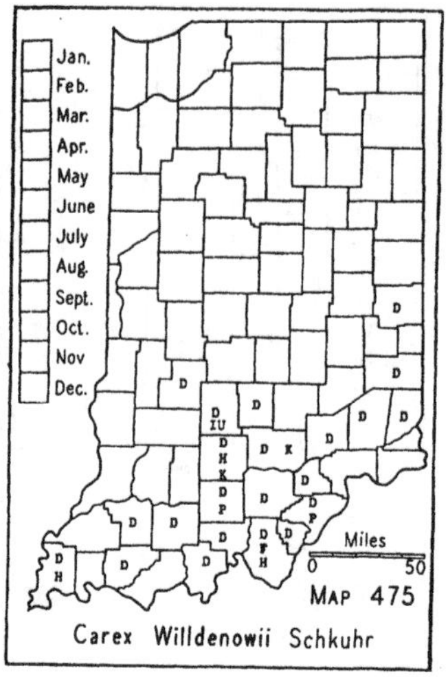

Carex Willdenowii Schkuhr

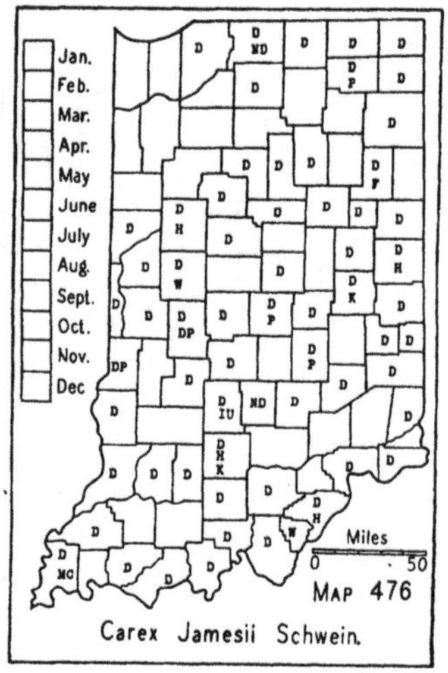

Carex Jamesii Schwein.

habitat is almost invariably at the springy bases of high wooded river bluffs and terraces.

N. J. to Fla., westw. to Ind. and Tex.

13. § PHYLLOSTÀCHYAE

Bodies of perigynia oblong-oval; lowest scale 5-15 mm long; pistillate flowers 3-10; staminate scales 2-2.4 mm long, obtuse or somewhat acute; staminate spike 0.7-0.9 mm in diameter....................................... 53. *C. Willdenowii.*

Bodies of perigynia subglobose; lowest scale 15-45 mm long; pistillate flowers 2-3; staminate scales 1.5-1.8 mm long, truncate, erose, with a dark transverse band near the apex; staminate spike 0.4-0.5 mm in diameter............ 54. *C. Jamesii.*

53. **Carex Willdenòwii** Schkuhr. Map 475. Common in southern Indiana (mostly in the unglaciated area and the "flats") on dry wooded, especially oak, slopes, generally in poor, sandy, acid soils; rarely in low beech or pin oak woods.

Vt. to Ont. and Ind., southw. to Ga. and Tex.

54. **Carex Jàmesii** Schwein. Map 476. Very common throughout Indiana except in the northwestern counties from which we have no records. It is a plant of rich woods, occurring in dry neutral soil, especially on the slopes of deep ravines. It is most frequently associated with either *Carex Hitchcockiana* or *C. oligocarpa* or both.

Ont. and N. Y. to Iowa, southw. to W. Va., Mo., and Kans.

14. § MONTÀNAE

Fertile culms all alike, elongated (7-40 cm long), bearing both staminate and pistillate spikes, basal spikes absent.

Body of perigynium elliptic to oblong-ovoid, much longer than wide; staminate spike slender.

Perigynia conspicuous in the spikes, not concealed by the scales, 2.5-3 mm long.

Staminate scales obtuse or short-acute, closely appressed, not cucullate at the tip, the midvein usually not extending to the tip; pistillate spikes usually not aggregated; culms erect............................ 55. *C. artitecta.*

Staminate scales ascending to loosely spreading, the midvein extending to the tip.

Beaks of perigynia 1.75-2 mm long; staminate spike peduncled, conspicuous,

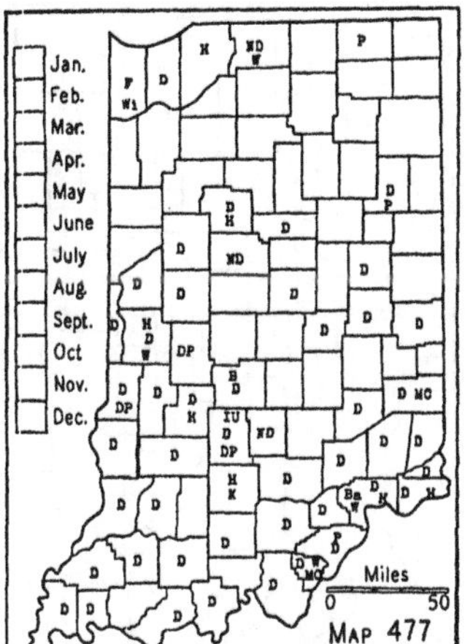

MAP 477
Carex artitecta Mack.

MAP 478
Carex artitecta var. subtilirostris Hermann

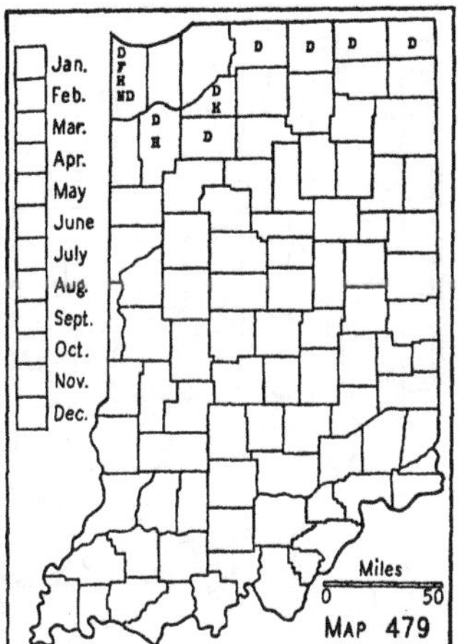

MAP 479
Carex Emmonsii Dewey

10-16 mm long, 1.6-2.3 mm wide; pistillate spikes not at all aggregated; culms erect..........................55a. *C. artitecta* var. *subtilirostris.*

Beaks of perigynia 0.5-1 mm long; staminate spike sessile, usually inconspicuous, 2-8 mm long, 1.5 mm wide; at least the upper pistillate spikes closely aggregated or congested; culms weak, more or less arcuate..56. *C. Emmonsii.*

Perigynia inconspicuous in the spikes, largely concealed by the scales, 3-4 mm long..57. *C. nigromarginata.*

Body of perigynium suborbicular to somewhat obovoid, about as long as wide.

Ligule conspicuous, longer than wide; lowest bract truncate or bifid, abruptly awned; leaf blades 2.5-4.5 mm wide; culms generally aphyllopodic, little fibrillose at the base, without long, horizontal stolons........58. *C. communis.*

Ligule short, much wider than long; lowest bract usually gradually acuminate; leaf blades 2.5 (very rarely 3) mm wide or less; culms generally phyllopodic, conspicuously fibrillose at the base, with long horizontal stolons; staminate spike stout.

Mature perigynia 1.75-2 mm wide, the body suborbicular in cross section..59. *C. heliophila.*

Mature perigynia about 1.5 mm wide, the body obtusely trigonous in cross section...60. *C. pennsylvanica.*

Fertile culms of two types, some short (1-5 cm long), partly hidden among the densely tufted bases and bearing only pistillate spikes, others elongated (5-11 cm long) and bearing staminate spikes only or both staminate and pistillate spikes.

Leaf blades rather thin, not stiff, erect or ascending, 1.5-3 mm wide; perigynia membranaceous, 2.25-4 mm long, the body short-pubescent above.

Perigynia 2.25-3.25 mm long, 1-1.25 mm wide, the beak about half the length of the body; achenes orbicular-obovoid........................61. *C. umbellata.*

Perigynia 3.25-4 mm long, the beak nearly the length of the body; achenes oblong-obovoid, minutely roughened............................62. *C. rugosperma.*

Leaf blades thick, rigid, widely spreading at maturity, 2-4.5 mm wide; perigynia subcoriaceous, 3.5-4.5 mm long, the body glabrous or very sparsely pubescent above...63. *C. tonsa.*

55. **Carex artitécta** Mack. (*Carex varia* Muhl., not Lumnitzer nor Host.) Map 477. Common in dry open woods, especially on rocky white oak slopes; occasional in thickets or low woods.

Vt. to Iowa, southw. to S. C. and Okla.

Carex nigromarginata Schwein.

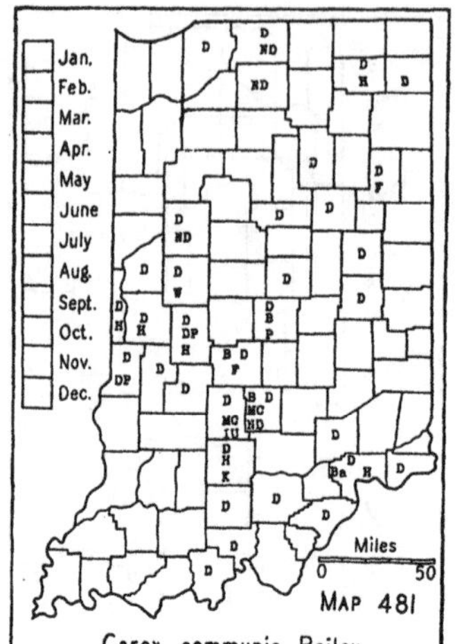

Carex communis Bailey

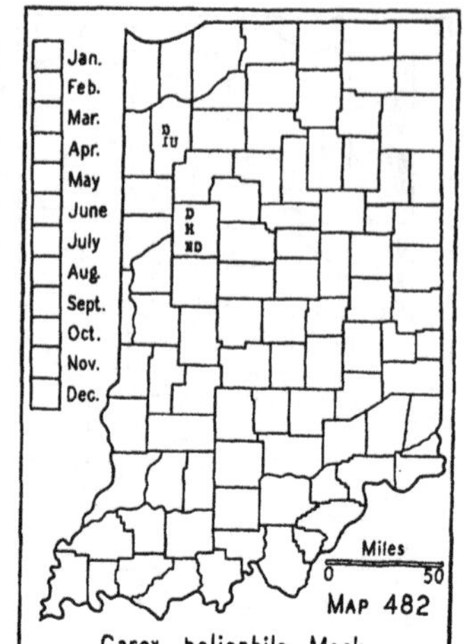

Carex heliophila Mack.

55a. **Carex artitecta** var. **subtilirόstris** Hermann. (Rhodora **40**: 79. 1938.) Map 478. Known in Indiana only from the type collection: Deam no. 54764, wooded slope along a small creek about 3 miles northwest of Clinton, Vermillion County, May 5, 1934.

Ind. and Tenn.

56. **Carex Emmόnsii** Dewey. (*Carex albicans* of authors, doubtfully of Willd., Rhodora **40**: 330-331. 1938.) Map 479. A coastal plain species found sparingly in the northern counties of the lake area. It grows in sandy open woods and on moist sandy borders of marshes or thickets in the dunes, but its preferred habitat is dry black oak ridges.

Indiana plants tend to have the culms longer and less arcuate and the pistillate spikes fewer and less congested than in the characteristic plant of the Atlantic Coastal Plain.

N. S. to Fla. mostly along the coast, and about the Great Lakes.

57. **Carex nigromarginàta** Schwein. Map 480. A southern and eastern species known in Indiana from a single collection: Deam no. 44074A, top of the wooded bluff of the Ohio River, about a quarter of a mile north of Fredonia, Crawford County, April 24, 1927.

Conn. to Tenn., Fla., and La., mostly along the coast, and northw. in the Mississippi Valley to Mo. and s. Ind.

58. **Carex commùnis** Bailey. Map 481. Common in dry woods of all types, particularly on rocky slopes. It is one of the earliest sedges to flower and fruit.

Deam no. 33881 (Gray Herbarium) is abnormal in having the leaf sheaths prolonged laterally and ventrally, forming auricles reaching almost to the summit of the ligule. Typically the leaf sheaths are deeply concave at the mouth.

N. S. to Minn., southw. to Ga., Ky., and Ark.

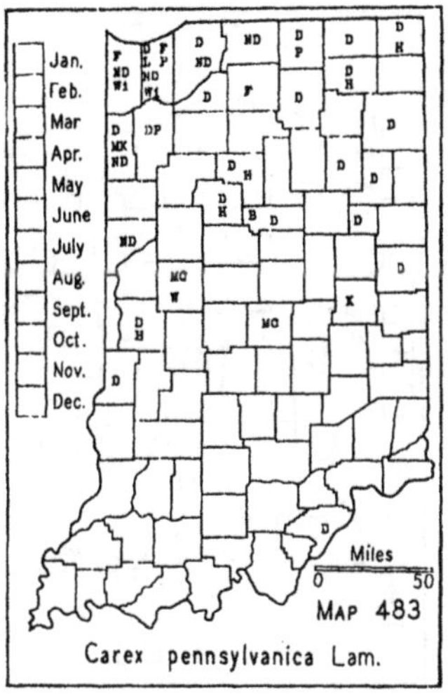

59. **Carex heliόphila** Mack. (Torreya **13**: 15. 1913.) (*Carex pennsylvanica* var. *digyna* Böck.) Map 482. A prairie species represented from Indiana by two collections by Deam: in a sandy black oak woods 2 miles southwest of Tefft, Jasper County, June 6, 1924, and on top of the high gravelly bank of Big Wea Creek terrace 4 miles southwest of Lafayette, Tippecanoe County, June 3, 1924, and May 24, 1932. At the latter station it was plentiful in an open black oak-shagbark hickory grove with such other prairie or western species as *Androsace occidentalis, Petalostemum purpureum, Arenaria patula, Opuntia Rafinesquii,* and, nearby, *Muhlenbergia cuspidata, Sporobolus clandestinus,* and *Erysimum asperum.* Other associated plants were *Festuca octoflora, Poa pratensis, Penstemon hirsutus, Houstonia longifolia,* and *Acerates viridiflora.*

Man. to Alberta, southw. to Ind., Mo., and N. Mex.

60. **Carex pennsylvánica** Lam. Map 483. Common in northern Indiana, less frequent southward, and rare or absent from the southernmost counties. Like *Carex communis* it is a species flowering in early spring, found in similar localities but preferring somewhat more open habitats and generally in more sterile soils. It usually forms rather extensive colonies, sometimes comprising the dominant floor cover in open oak woods.

N. S. to N. Dak., southw. to S. C., Tenn., and Iowa.

61. **Carex umbellàta** Schkuhr. (Bull. Torrey Bot. Club **42**: 621. 1915.) (*Carex abdita* Bickn. and *Carex umbellata* var. *brevirostris* Boott.) Map 484. Infrequent in northern Indiana in dry sandy soil, usually in open woods; frequent in southeastern Indiana on crests of rocky wooded ridges and river bluffs. This and the two following species may be more common than the records indicate because they are low, inconspicuous plants, easily overlooked by collectors.

In this species the longest peduncles are typically not over 8 cm in length and generally bear a staminate spike only. But on the dunes the prevalent form has elongated peduncles 12-20 cm in length which usually bear one or more pistillate spikes in addition to the staminate. This

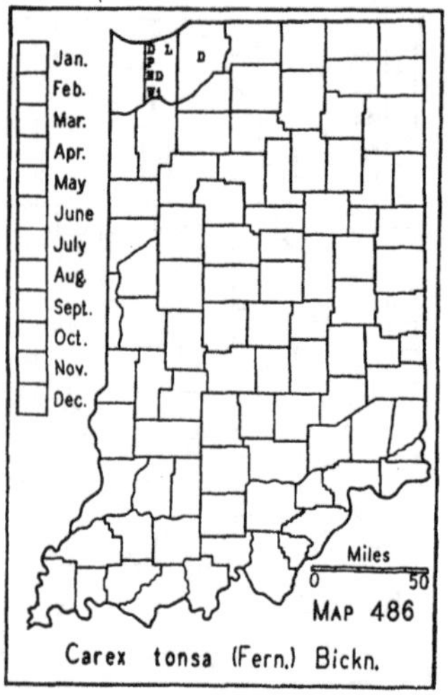

MAP 486
Carex tonsa (Fern.) Bickn.

MAP 487
Carex Richardsonii R. Br.

MAP 488
Carex picta Steud.

form is analogous to the plant which has been called *Carex umbellata* f. *vicina* (Dewey) Wieg. but the type specimen upon which that form is based is the long-beaked plant (*C. rugosperma* Mack.) so that the name cannot be applied to the Indiana plant.

Newf. to B. C., southw. to D. C. and Ill.

62. **Carex rugospérma** Mack. (Bull. Torrey Bot. Club 42: 621. 1915.) (*Carex umbellata* of many recent authors, not Schkuhr.) Map 485. Infrequent in the northern tier of counties. It is found in dry, sandy oak woods, open drained low woods, and on borders of drained marshes.

N. S. to Minn., southw. to Md. and Ind.

63. **Carex tónsa** (Fern.) Bickn. (Bull. Torrey Bot. Club 35: 492. 1908.) (*Carex umbellata* var. *tonsa* Fern.) Map 486. Frequent in the dune area on low, open dunes and in dry, open woods.

Que. to Alberta, southw. to D. C. and Ind.

15. § DIGITÀTAE

Basal spikes present; terminal spikes androgynous; pistillate scales abruptly cuspidate or short-awned. (See excluded species no. 15, p. 274.)............*C. pedunculata.*

Basal spikes absent; terminal spike staminate; pistillate scales blunt to acute........ ....64. *C. Richardsonii.*

64. **Carex Richardsònii** R. Br. Map 487. Known in Indiana only from the dunes at Pine, Lake county. Pine is now within Gary on the east side of Clark Street, an eighth of a mile south of Lake Michigan. Here on a sandy knoll at the edge of a marsh, *Carex Richardsonii* is associated with *Andropogon scoparius, Castilleja coccinea, Erigeron pulchellus, Senecio pauperculus* var. *Balsamitae, Lithospermum canescens, Potentilla fruticosa, Carex umbellata* and *C. aurea, Liatris spicata, Betula papyrifera, Pedicularis canadensis, Krigia biflora, Rhus trilobata* var. *arenaria, R. radicans, R. Vernix, Arabis lyrata, Hypoxis hirsuta,* and *Koeleria cristata.*

This is one of the rarest sedges in the eastern states where it is very local in its distribution (although its known range seems to indicate that

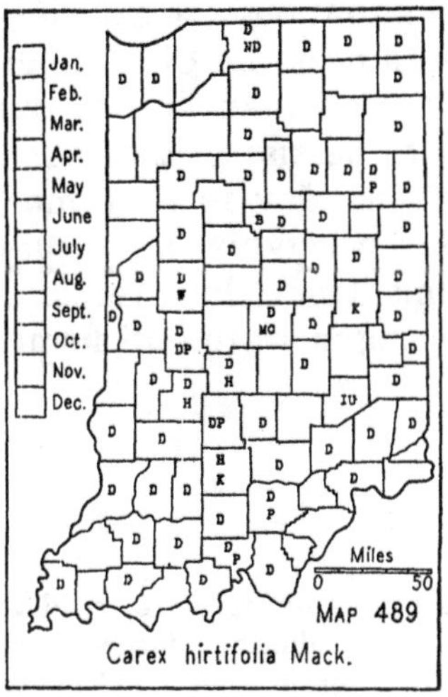

Carex hirtifolia Mack.

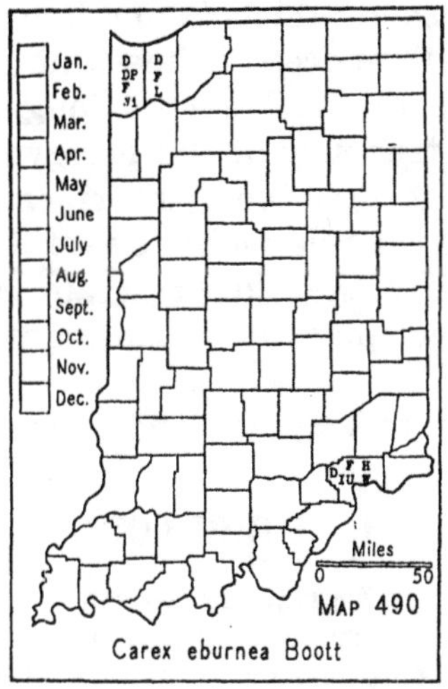

Carex eburnea Boott

Carex Garberi Fern.

it occurs generally at or near the Niagara Escarpment) and its season is very brief. After flowering it matures its fruit rapidly and then completely withers away. At Pine it is in its prime about May 30. Of the six known collections made from this station four were made on May 29 (1897; 1900; 1904, and 1935), one on May 12 (1877) and one on June 13 (1935), but at the last date the majority of the perigynia had fallen and the plants were already badly withered.

Western N. Y. and Ont. to Alberta, southw. to Ill. and S. Dak.

16. § PÍCTAE

65. **Carex pícta** Steud. Map 488. In Indiana in the unglaciated region only where it is local and largely confined to the northern half of the knob area (Chestnut Oak Upland). It is found on wooded hilltops under oak, chestnut, and beech, generally forming rather extensive colonies. Deam has noted that it "has the habit of growing in circular tufts with a hollow center" and from this characteristic the species may be readily recognized long after its flowering and fruiting season is past. It is the earliest sedge to bloom in the state, coming into flower in early April or even in late March.

Ind., Ala., and La.

17. § TRIQUÈTRAE

66. **Carex hirtifòlia** Mack. (Bull. Torrey Bot. Club **37**: 244. 1910.) (*Carex pubescens* Muhl., not Poir.) Map 489. Very common throughout the state in woodland habitats of all types, showing a slight preference for beech woods.

N. B. to Minn., southw. to D. C., Ky., and Kans.

18. § ÁLBAE

67. **Carex ebúrnea** Boott. Map 490. Apparently restricted to the northwestern and southeastern corners of the state. In the north it is known

only on the dunes in dry sandy thickets and in open situations. In southern Indiana it is found in wet crevices of limestone bluffs near the Ohio River.

Carex eburnea retains its fruit over a longer period than any of our other species due to the tendency of the perigynia to persist in the spikes long after maturity. Although the fruit ripens from May to July most of the plants have dropped relatively few of their perigynia, as a rule, by October and frequently the old prostrate culms from the preceding year will be found to have spikes in which many perigynia are still firmly attached.

Newf. to B. C., southw. to Va., Mo., and Nebr.

19. § BICÓLORES

Pistillate scales averaging three fourths the length of the perigynia or more, reddish brown, appressed; terminal spike androgynous, rarely staminate; mature perigynia white-pulverulent, elliptic-obovoid, not translucent or fleshy. . 68. *C. Garberi.*

Pistillate scales averaging half the length of the perigynia or less, generally pale yellowish brown and cuspidate, widely spreading at maturity; terminal spike staminate, rarely with a few perigynia at the base; mature perigynia golden yellow or brownish, orbicular-obovoid, translucent, fleshy. . . . . 69. *C. aurea.*

68. **Carex Gárberi** Fern. (Rhodora 37: 253. 1935.) (*Carex bicolor* of recent American authors, not All. and *Carex Hassei* of recent authors, not Bailey.) Map 491. Infrequent in the lake area (mostly in the dune region) where it grows along the wet sandy edges of swales in the dunes and on old lake beds, chiefly in calcareous soils. It is frequently associated with *Carex Crawei, C. tetanica, C. Meadii, C. viridula, and C. Haleana.* Apparently it was formerly more plentiful than at present as collections from the Indiana dunes forty and fifty years ago are much more numerous in herbaria than recent collections. At Pine, where this species is closely associated with *Carex tetanica,* plants of the latter species showing many characteristics of *C. Garberi* and plants of *C. Garberi* having characteristics of *C. tetanica* are frequent. The general aspect of such plants and the conditions under which they are found are strongly suggestive of hybridization.

Que. to Mich., Ind., and Wis.; also in Alberta and B. C.

69. **Carex aúrea** Nutt. Map 492. Frequent on the dunes in Lake County. Its habitat is often that of *Carex Garberi,* on moist sandy edges of swales and similar situations, but it is also frequently found in richer, mucky soils such as on the border of sloughs and of low wet woods.

Newf. to B. C., southw. to Conn., Ind., Nebr., N. Mex., and Calif.

20. § PANÍCEAE

Culms phyllopodic; stolons deep-seated, slender, whitish; plants of open marly or sandy habitats.

Pistillate spikes linear to linear-oblong, 3.5-4.5 mm wide; perigynia appressed or ascending, 2.5-3.5 mm long, slightly excurved and tapering to the apex, very minutely beaked or beakless; leaf blades 2-4 mm wide; culms slender. . . . . 70. *C. tetanica.*

Carex aurea Nutt.

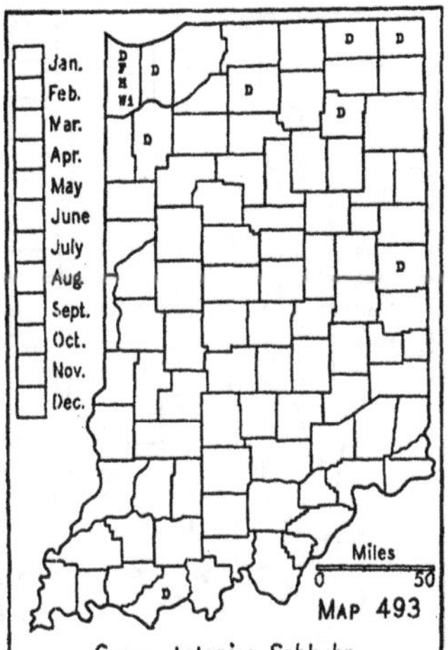

Carex tetanica Schkuhr

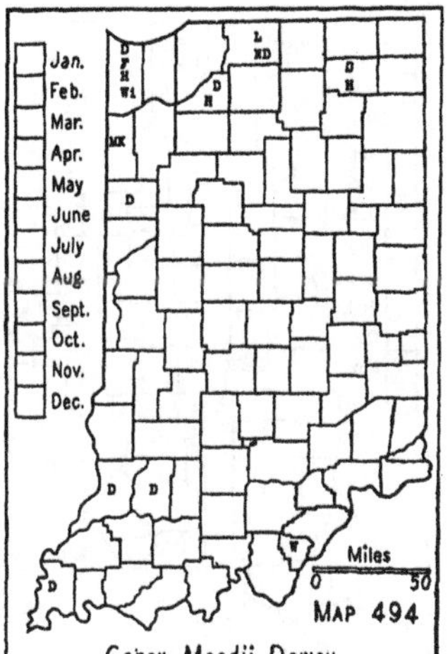

Carex Meadii Dewey

Pistillate spikes oblong or linear-oblong, 5-8 mm wide; perigynia spreading at maturity, 3-5 mm long, abruptly narrowed at the apex into a minute, more or less strongly bent beak; leaf blades 2.7-7 mm wide; culms stout....71. *C. Meadii.*

Culms strongly aphyllopodic; stolons superficial; plants of rich humus in shady woods. ....72. *C. Woodii.*

70. **Carex tetánica** Schkuhr. Map 493. Infrequent in northern Indiana in marly or sandy soils, bordering marshes and lakes; becoming frequent to locally common on the dunes where it occurs especially on low sandy interdunal flats; rare in southern Indiana, in open post oak flats.

Mass. to Alberta, southw. to Pa. and Iowa.

71. **Carex Mèadii** Dewey. (*Carex tetanica* var. *Meadii* (Dewey) Bailey.) Map 494. Infrequent in the lake area and in southeastern Indiana; frequent in the dune area. Its habitat is much that of *C. tetanica* except that *C. Meadii* also occurs in drier soils and in even more open situations. *Carex Meadii* as a rule is readily distinct from *C. tetanica* except at Pine in Lake County where the two species are closely associated and intermediate forms are frequent. The same is true of *C. tetanica* and *C. Garberi*, at this station, and, as noted under the latter species, such transitional forms may be due to hybridization.

N. J. to Sask., southw. to Ga. and Tex.

72. **Carex Wóodii** Dewey. (*Carex tetanica* var. *Woodii* (Dewey) Wood; *Carex colorata* Mack.) Map 495. Rare in the northern counties. So far this species is known in Indiana from three collections only, all by Deam: in a moist red and white oak woods 4 miles northwest of Valparaiso, Porter County, June 2, 1927, in a rich beech-maple woods 1 mile southeast of North Liberty, St. Joseph County, May 23, 1934, and June 13, 1935 and at the base of a sugar maple slope in Steuben County. At the latter locality it was associated with *Impatiens biflora, Solidago caesia, Viola canadensis, Smilacina racemosa, Polygonum virginianum, Isopyrum biternatum,* and *Caulophyllum thalictroides.*

N. Y. to Man., southw. to D. C. and Mo.

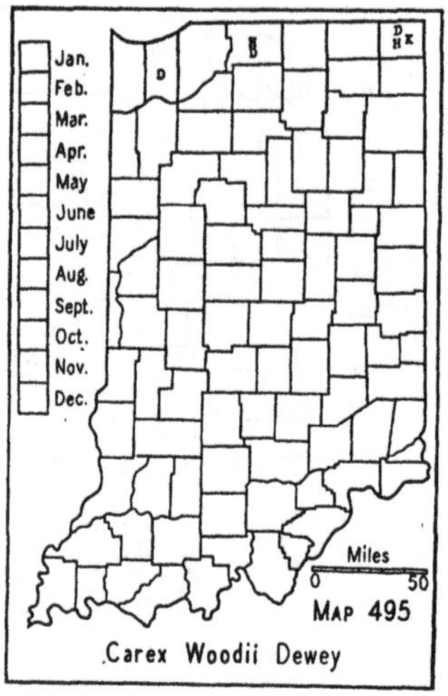

Carex Woodii Dewey

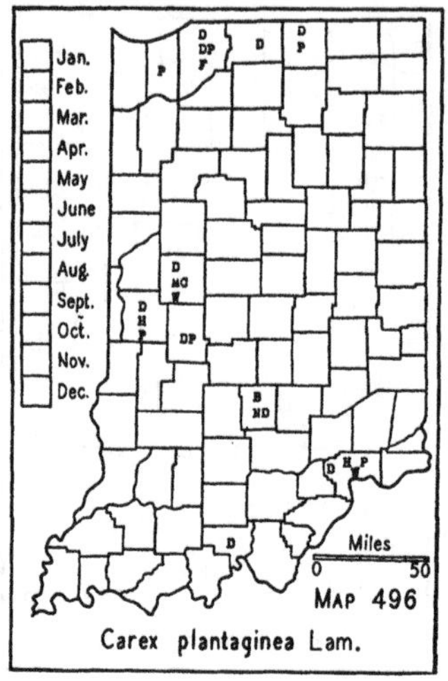

Carex plantaginea Lam.

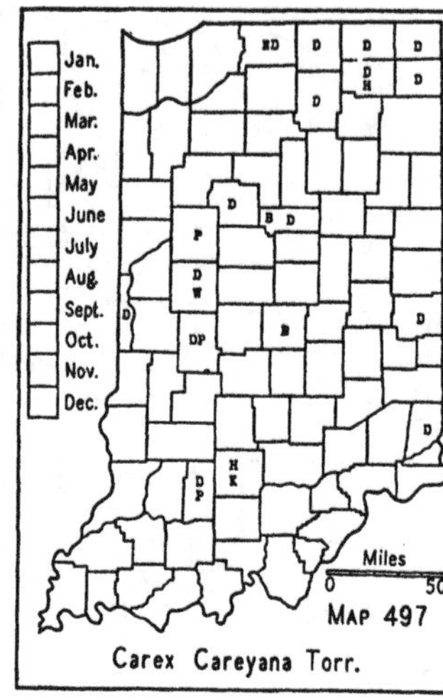

Carex Careyana Torr.

21. § LAXIFLÒRAE

Bract-sheaths, base of culms, and staminate scales strongly red-tinged.*
  Leaf blades of fertile culms rudimentary, the sheaths concave at the mouth; bracts bladeless; perigynia 4-5 mm long..........................73. *C. plantaginea.*
  Leaf blades of fertile culms well-developed, the sheaths prolonged upward at the mouth; bracts with blades well-developed; perigynia 5.3-6 mm. long..74. *C. Careyana.*
Bract-sheaths not red-tinged, base of culms rarely so; staminate scales tinged greenish white to dull reddish brown.
  Perigynia sharply triangular, short-tapering at the base, closely 35-50-nerved.
    Spikes erect, nearly sessile; leaf blades very smooth except for the margins, the larger 12-25 mm wide, those of the fertile culms much smaller than those of the sterile...75. *C. platyphylla.*
    Spikes drooping on fiiliform peduncles; leaf blades hispidulous on the veins, 2-12 mm wide, those of the fertile culms moderately smaller than those of the sterile.
      Staminate spike sessile or subsessile, inconspicuous; pistillate spikes approximate; lowest bract subspathaceous, exceeding the inflorescence..76. *C. abscondita.*
      Staminate spike peduncled, conspicuous; pistillate spikes scattered; lowest bract not at all spathaceous, not exceeding the inflorescence.
        Pistillate spikes without a staminate flower at the base; leaf blades 2-5 mm wide, erect, green....................................77. *C. digitalis.*
        Pistillate spikes with 1-2 staminate flowers at the base; leaf blades 5-12 mm wide, weak, glaucous green.
          Angles of the culms blunt, minutely serrulate only below the bracts; edges of the bract-sheaths entire; perigynia tapering at the apex, short-beaked...78. *C. laxiculmis.*
          Angles of the culms sharp, minutely serrulate; edges of the bract-sheaths minutely serrulate; perigynia rounded or round-tapering at the apex, blunt or abruptly very short-beaked....78a. *C. laxiculmis* var. *copulata.*
  Perigynia obtusely triangular (at least below), long-tapering at the base.
    Bract-sheaths smooth on the edges or shallowly serrulate; beak of perigynium straight or slightly oblique.

* This color is often called "purple" in *Carex* descriptions. It is a close match with Ridgway's "Bordeaux" which is 90% red and 10% violet.

Sterile shoots developing conspicuous culms; leaves not semi-evergreen; perigynia rather sharply angled above; pistillate spikes few-flowered, the lower on long capillary peduncles..............................79. *C. styloflexa.*

Sterile shoots reduced to tufts of leaves; leaves semi-evergreen; perigynia obtusely triangular..80. *C. laxiflora.*

Bract-sheaths strongly serrulate on the edges.

Sterile shoots reduced to tufts of leaves, not forming culms.

Pistillate scales acuminate to strongly cuspidate, more than half the length of the perigynia; beak of perigynium conspicuous, straight or oblique; leaves semi-evergreen; staminate spike peduncled, conspicuous...80a. *C. laxiflora* var. *serrulata.*

Pistillate scales broadly obovate-orbicular, half the length of the perigynia or less, strongly divergent at the base; beak of perigynium short, abruptly bent; leaves not semi-evergreen, the blades 7-30 mm wide; staminate spike sessile, very slender, inconspicuous...............81. *C. albursina.*

Sterile shoots developing conspicuous culms; leaf blades 3-12 mm wide, not semi-evergreen; pistillate scales mucronate to long-awned; beak of perigynium short, abruptly bent.

Culms not reddish-tinged at the base; lower pistillate spikes not on long capillary peduncles; staminate scales usually greenish white or slightly tinged with reddish brown; staminate spike typically sessile or very short-peduncled; perigynia obovoid, 3-4 mm long...........82. *C. blanda.*

Culms reddish-tinged at the base; lower pistillate spikes on long capillary peduncles; staminate scales strongly tinged with reddish brown; staminate spike long-peduncled; perigynia broadly obovoid, 2.5-3.2 mm long...83. *C. gracilescens.*

73. **Carex plantagínea** Lam. Map 496. Rather infrequent in the northern counties in rich woods. South of the lake area it is local and found mostly in humus on the wooded sandstone slopes of deep ravines, usually in dense shade and associated with *Hydrophyllum appendiculatum.* No specimen was found to confirm the report in Coulter's Catalogue from Tippecanoe County.

Que. to Sask., southw. to N. C. and Ky.

74. **Carex Careyàna** Torr. Map 497. Frequent but local in moist rich woods, particularly in ravines.

N. Y. to Mich., southw. to Va. and Mo.

75. **Carex platyphýlla** Carey. Map 498. All the Indiana collections of this species are from the knob area (Chestnut Oak Upland) with the exception of a single collection from Vigo County and one from Putnam County. It is found in calcareous soils on dry open woodland slopes. No specimen could be found to confirm Bradner's report from Steuben County, but the occurrence of the species in the northern counties is not improbable since it is known from southern Michigan.

Que. to Mich., southw. to N. C. and Ill.

76. **Carex abscóndita** Mack. (*Carex ptychocarpa* Steud.) Map 499. A southern and Coastal Plain species found in Indiana in the southern counties only. It is rare in dry beech woods and very rare in black-white oak woods.

Mass. to Ind., southw. to Fla. and La.

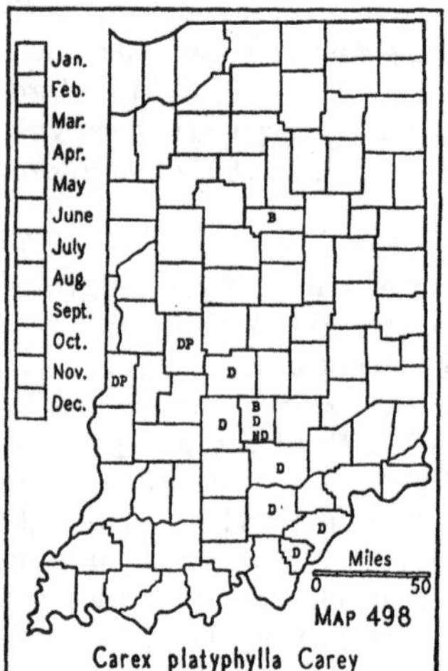

Carex platyphylla Carey

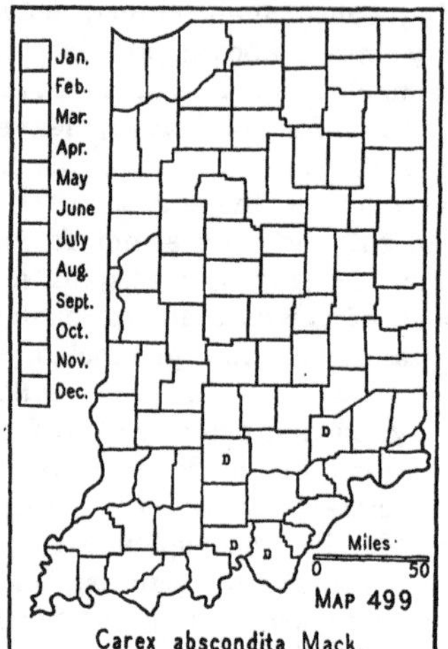

Carex abscondita Mack.

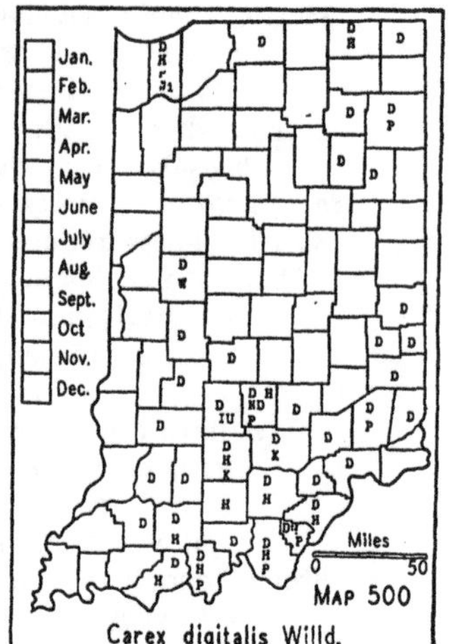

Carex digitalis Willd.

77. **Carex digitàlis** Willd. (Including *Carex digitalis* var. *macropoda* Fern. Rhodora 40: 400-401. 1938.) Map 500. Common in southern Indiana; locally frequent in the northern counties. A woodland species preferring dry beech woods but frequent also in dry or moist black or white oak woods.

The length of the peduncle of the staminate spike in this species, as in *Carex laxiculmis,* is extremely variable. An extreme form in which the staminate spike is born on a peduncle overtopping the uppermost pistillate spike and bract has been described by Professor Fernald as var. *macropoda,* and under this variety he cites Deam no. 27837 from Crawford County and no. 27119 from Perry County. In the Deam Herbarium, Deam no. 44066 from Perry County apparently represents this extreme of the species but is too immature to be placed here with certainty. Among the numerous intermediate collections Deam no. 20378 from Harrison County and no. 20592 from Washington County most nearly approximate var. *macropoda.*

78. **Carex laxicúlmis** Schwein. Map 501. Fairly common in woods and thickets. Plants intermediate between the species and the following variety are not infrequent; such are Deam nos. 844; 24750; 35708; 35924; 36407; 40669; and 51825.

Maine to Wis., southw. to N. C. and Mo.

78a. **Carex laxiculmis** var. **copulàta** (Bailey) Fern. (*Carex copulata* (Bailey) Mack.) Map 502. Frequent in eastern Indiana in dry woods, principally white oak and beech; rare in the western counties. The variety is said to be a calciphile while the species prefers neutral or only slightly calcareous soils.

N. J. to Mich. and Mo.

79. **Carex styloflèxa** Buckley. (*Carex laxiflora* var. *styloflexa* (Buckley) Boott.) Map 503. An eastern and southern species chiefly of the Coastal Plain known in Indiana from a collection by Mrs. C. C. Deam: in moist woods near Adams, Decatur County, May 13, 1911, no. 8149.

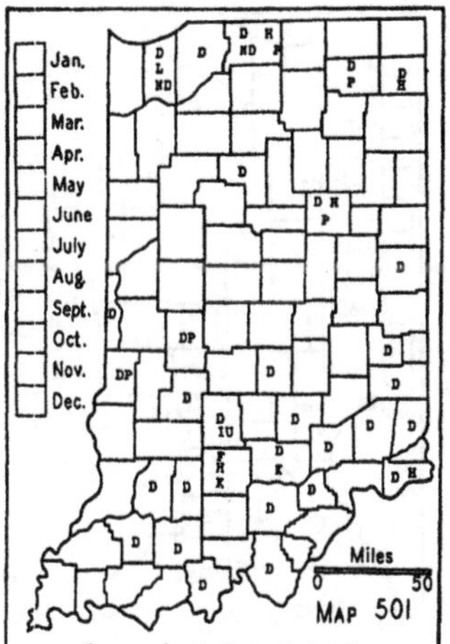

Carex laxiculmis Schwein.

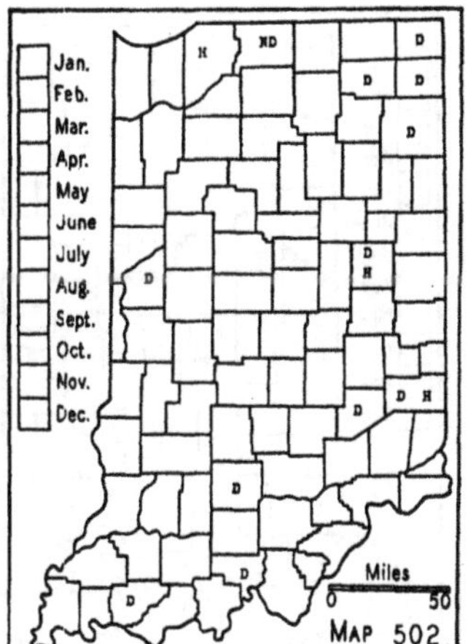

Carex laxiculmis var. copulata (Bailey) Fern.

Carex styloflexa Buckley

Conn. to Fla. and Tex., mostly along the coast, northw. in the Mississippi Valley to s. Ind.

80. **Carex laxiflòra** Lam. (*Carex heterosperma* Wahl., *Carex anceps* Muhl. and *Carex laxiflora* var. *patulifolia* (Dewey) Carey.) (Including *Carex striatula* Michx., *Carex laxiflora* var. *striatula* (Michx.) Carey, and "*Carex laxiflora*" Mack., not Lam., in Small, Manual of the Southeastern Flora.) Map 504. Common in dry woods, especially beech-sugar maple, throughout the state.

The form commonly referred to *Carex striatula* Michx. may deserve varietal recognition, at least in the southern part of its range and on the Coastal Plain where it attains the extreme of its differentiation, but in Indiana intermediates so far outnumber the extremes that all attempts to separate it even varietally have been unsuccessful.

N. S. to Mich., southw. to Fla. and Tex.

80a. **Carex laxiflora** var. **serrulàta** Hermann. (Rhodora 40: 80. 1938.) Map 505. Known from four counties all in the eastern half of the state. Its habitat is that of the species. The type collection (Deam no. 6458) came from a wooded ravine two miles northwest of Henryville, Clark County, May 25, 1910.

Pa. and Ind.

81. **Carex albursìna** Sheldon. (*Carex laxiflora* var. *latifolia* Boott.) Map 506. Common on wooded slopes, chiefly in limestone areas; rare in low, moist or alluvial woods.

Deam's collection of May 7, 1905, from Blackford County is exceptional in having the leaves semi-evergreen and rather rigid.

Que. to Minn., southw. to Va. and Ark.

82. **Carex blánda** Dewey. (*Carex laxiflora* var. *varians* of authors, not Bailey.) Map 507. Very common throughout the state, doubtless occurring in every county. In woods of all types it is the commonest species of this section of the genus. The other Indiana species of the *C. laxiflora* group

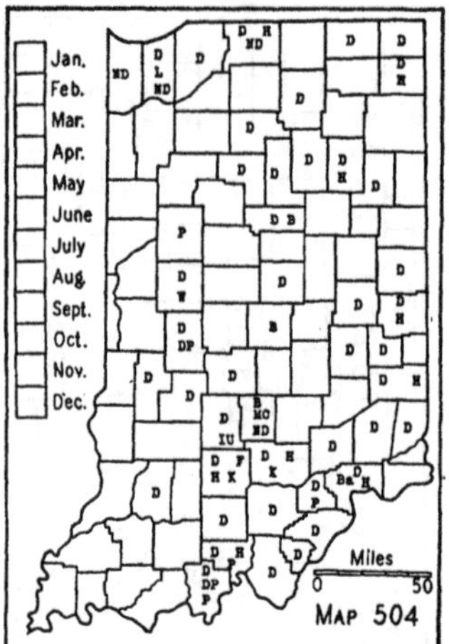

Carex laxiflora Lam.

Carex laxiflora var. serrulata Hermann

Carex albursina Sheldon

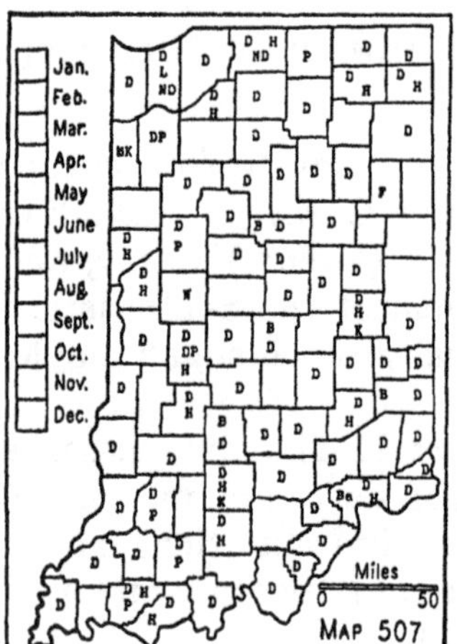

Carex blanda Dewey

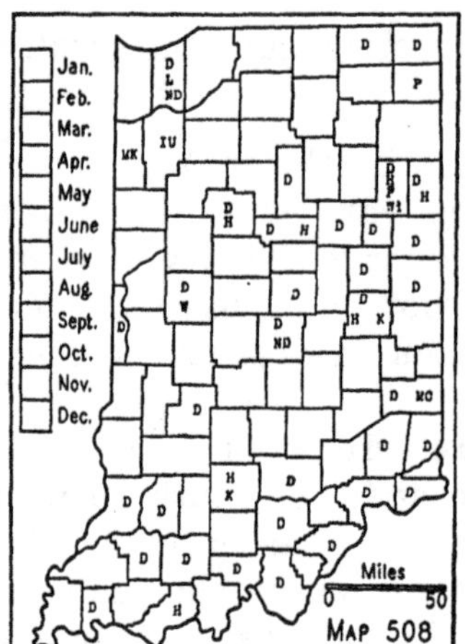

Carex gracilescens Steud.

Carex Haleana Olney

are rarely found in either very sandy or (except *C. gracilescens*) very moist woods as *C. blanda* frequently is.

Que. to N. Dak. southw. to Ala. and Tex.

83. **Carex graciléscens** Steud. ("*Carex laxiflora*" Mack., not Lam., in Britton and Brown, Illus. Flora, ed. 2 and *Carex laxiflora* var. *gracillima* of Gray, Man., ed. 7.) Map 508. Common in low woods and on wooded slopes. It is generally less plentiful where found than is *C. blanda* at its stations.

Que. to Wis., southw. to Va. and Ark.

22. § GRANULÀRES

Staminate spike short-peduncled or sessile; the two upper pistillate spikes usually contiguous; rootstocks not long-creeping.

Perigynia elliptic-obovoid to elliptic-ovoid, 2-2.5 mm long, 1-1.5 mm wide, ascending, not ventricose-squarrose, rounded at the apex, abruptly very minutely beaked... ....84. *C. Haleana*.

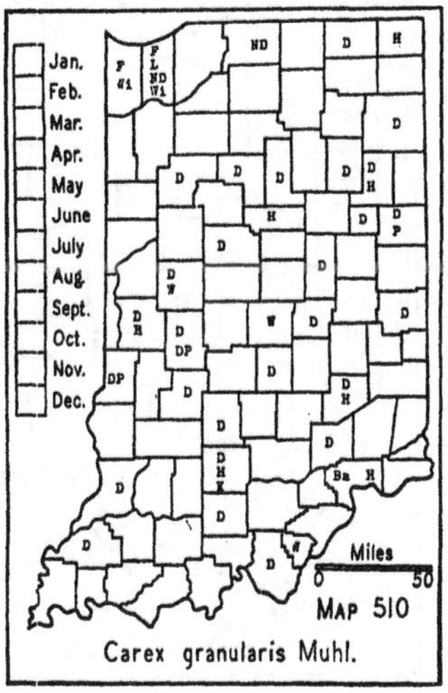

Carex granularis Muhl.

Carex Crawei Dewey

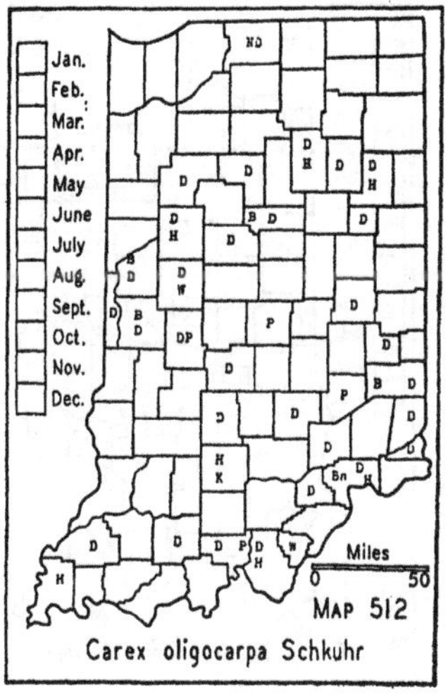

Carex oligocarpa Schkuhr

Perigynia broadly ovoid to broadly obovoid, 2.5-4 mm long, 1.5-2.5 mm wide, soon ventricose-squarrose, tapering at the apex, minutely beaked. .85. *C. granularis.*
Staminate spike long-peduncled; spikes all widely separate; rootstocks long-creeping... ....86. *C. Crawei.*

84. **Carex Haleàna** Olney. (*Carex granularis* var. *Haleana* (Olney) Porter and *Carex Shriveri* Britt.) Map 509. Infrequent in low ground, principally along creeks; occasionally on calcareous sandy shores. More frequent northward, and not known from the unglaciated area.

Que. to Sask., southw. to Va., Ind., and Kans.

85. **Carex granulàris** Muhl. Map 510. Common throughout the state in moist openings, low woods and on banks of creeks, especially in clay soils; frequent in dry open woods.

Vt. to Minn., southw. to Fla. and Ark.

86. **Carex Cráwei** Dewey. Map 511. Known in Indiana from Lake County only where it is locally plentiful on moist sandy interdunal flats. Here it is commonly associated with *Carex Garberi, C. aurea, C. Meadii,* and often with *C. viridula.*

Que. to Alberta and Wash., southw. to ne. N. J., s. Ala., Tenn., Kans., and Wyo.

23. § OLIGOCÁRPAE

Bract-sheaths glabrous, the lower 0.6-2 cm long; perigynia 4 mm or less long; leaf blades 2-4.5 mm wide; culms reddish-tinged at the base..........87. *C. oligocarpa.*
Bract-sheaths strongly hispidulous, the lower 2-6 cm long; perigynia 4.5-5 mm long; leaf blades 3-7 mm wide; culms brownish-tinged at the base. .88. *C. Hitchcockiana.*

87. **Carex oligocárpa** Schkuhr. Map 512. Common in rich woods except in the three northern tiers of counties where it is rare. It is a plant of calcareous soils and its favorite habitats are moist, wooded ravines and beech or beech-maple slopes. Occasionally it occurs on dry slopes and in open woods.

Vt. and Ont. to Iowa, southw. to Ala., Ky., and Tex.

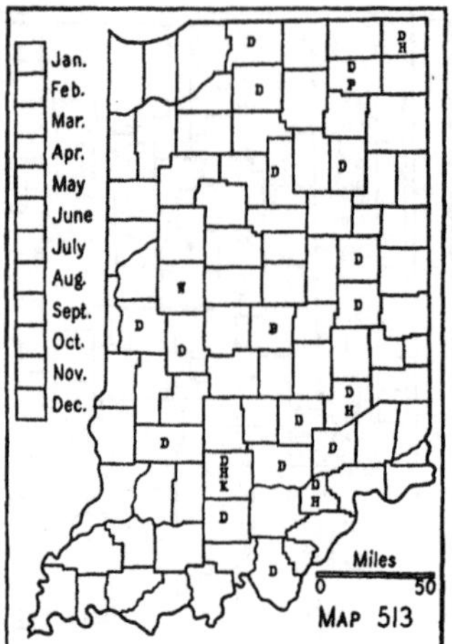

Carex Hitchcockiana Dewey

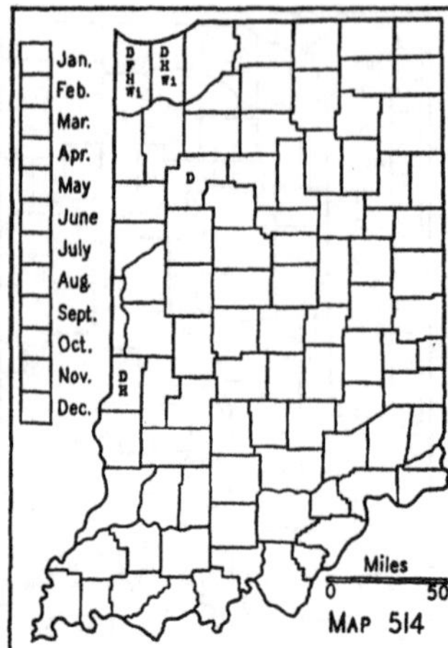

Carex conoidea Schkuhr

Carex amphiloba Steud.

88. **Carex Hitchcockiàna** Dewey. Map 513. In calcareous or neutral soils; common in rich woods and moist ravines and on river banks; rarely in dry, sandy woods. It is often associated with *Carex Jamesii, C. oligocarpa, C. blanda,* and *C. gracillima.*

Vt. and Ont. to Wis., southw. to W. Va., Ky., and Mo.

24. § GRÍSEAE

Perigynia elliptic, 1.5 mm wide; bract-sheaths minutely serrulate on the edges; peduncles of pistillate spikes rough; leaf blades 2-4 mm wide......89. *C. conoidea.*

Perigynia oblong-oval to broadly obovoid, 2- (occasionally 1.5 in *C. amphibola*) 2.5 mm wide; bract-sheaths glabrous; peduncles of pistillate spikes glabrous or nearly so; leaf blades 2-18 mm wide.

Pistillate spikes 3-12 flowered; leaves slightly if at all glaucous, thin and soft; bract-sheaths tight.

Pistillate spikes widely separated, the lower nearly basal; culms strongly reddish-tinged at the base; perigynia scarcely turgid; leaf blades 2-4 mm wide, erect; achenes slenderly stipitate..............................90. *C. amphibola.*

Lower pistillate spikes not nearly basal; culms brownish-tinged at the base; perigynia more or less turgid; leaf blades (2) 4-7 mm wide, ascending; achenes substipitate....................................91. *C. grisea.*

Pistillate spikes (12) 15-35-flowered; leaves very glaucous, thick and firm; bract-sheaths enlarged upward....................................92. *C. glaucodea.*

89. **Carex conoìdea** Schkuhr. Map 514. Infrequent in the northwestern counties in wet sandy fields and on banks of ditches. It is always a very local species and this may account for the lack of specimens or reports from northeastern Indiana where it should be found. The reports from Putnam County by Coulter, from Clark County by Baird & Taylor, and from the Lower Wabash Valley by Schneck are unsupported by specimens.

Newf. to Minn., southw. to Del., Ohio, and Iowa; also in the mts. of N. C.

90. **Carex amphíbola** Steud. (*Carex grisea* var. *angustifolia* Boott.) Map 515. Frequent in southern Indiana especially in the unglaciated area,

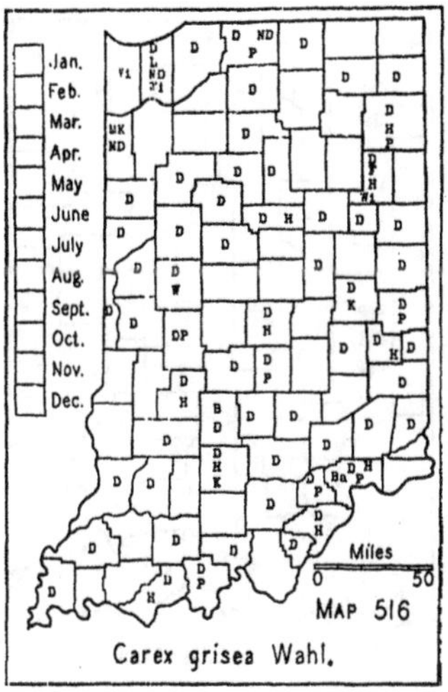

Carex grisea Wahl.

Carex glaucodea Tuckerm.

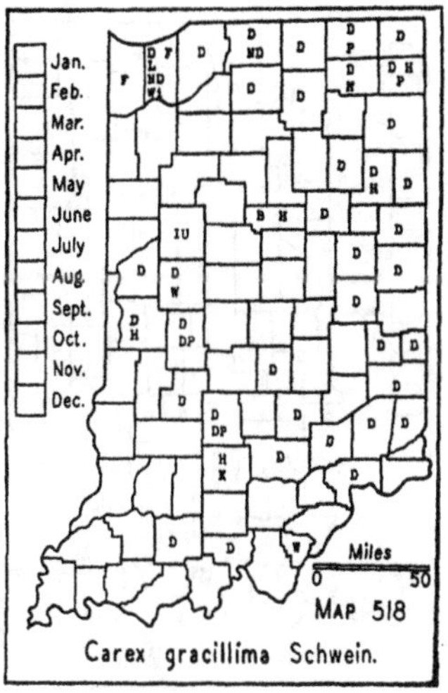

Carex gracillima Schwein.

in dry beech, beech-maple, and white oak woods. Reported from Putnam and Hamilton Counties by Wilson but no specimens could be found to authenticate these reports.

N. J. to Ind., southw. to Fla. and Tex.

91. **Carex grísea** Wahl. Map 516. Very common throughout the state in rich dry or moist woods and thickets, in ditches, on banks of streams, and along roadsides. It is extremely variable in its vegetative characteristics and in the shape and size of its perigynia.

N. B. to Ont. and Minn., southw. to Ga. and Tex.

92. **Carex glaucòdea** Tuckerm. Map 517. Frequent in southern Indiana on wooded or open hillsides in either dry or moist soils. It is partial to slopes and ridges and its most frequent habitats are abandoned roads in woods and paths on open grassy hills. No specimens could be located to authenticate the reports from Lake County by Coulter and by Peattie. All the known Indiana collections have come from within or very near the unglaciated area.

Mass. to Ont. and Ill., southw. to N. C. and Ark.

25. § GRACÍLLIMAE

Sheaths (except the lower which are dorsally somewhat hispidulous) and leaves glabrous; perigynia less than 2 mm thick.
 Bracts long-sheathing; perigynia bluntly angled, obtuse at the apex .... 93. *C. gracillima.*
 Bracts sheathless; perigynia sharply angled, tapering into a triangular, often twisted, beak nearly as long as the body .... 94. *C. prasina.*
Sheaths and often leaf blades pubescent; perigynia 2-2.5 mm thick ...... 95. *C. Davisii.*

93. **Carex gracíllima** Schwein. Map 518. Doubtless found in every county in the state. It is common in wooded ravines and in low woods of all types, although it shows a preference for open beech or beech-maple woods.

Newf. to Man., southw. to Va., Ky., and Mo.

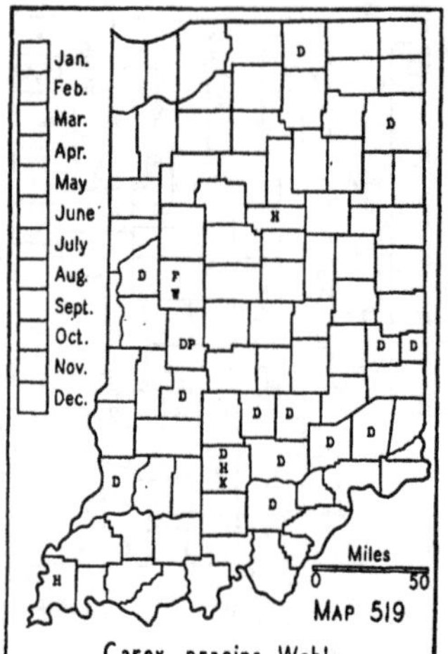

Carex prasina Wahl.

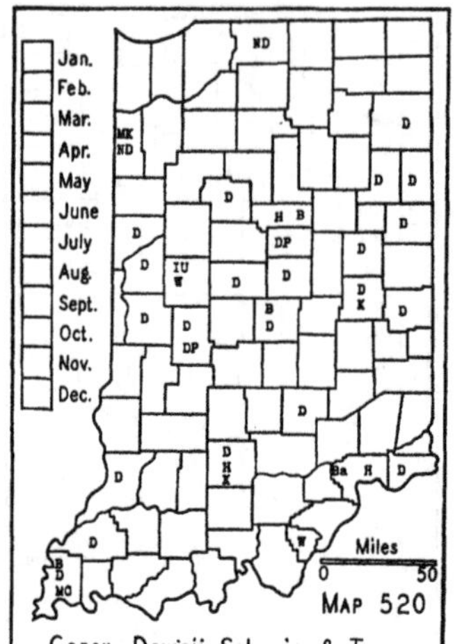

Carex Davisii Schwein. & Torr.

Carex debilis Michx.

94. **Carex prasìna** Wahl. Map 519. Infrequent, becoming frequent in the southeastern counties. A species of very wet or springy habitats in deep woods, growing generally along streamlets and frequently on bars and rocks in streams.

Que. to Mich., southw. to D. C. and Ky., and in the Alleghenies to Ga.

95. **Carex Davísii** Schwein. & Torr. Map 520. Frequent in neutral or calcareous soils in low, especially alluvial, beech and beech-maple woods and in moist roadside ditches. It sometimes superficially resembles luxuriant forms of *Carex grisea* from which it may be readily distinguished by the terminal spike which is gynaecandrous in *C. Davisii* and staminate in *C. grisea.*

Vt. to Minn., southw. to Md., Tenn., and Tex.

26. § SYLVÂTICAE

Perigynia sessile or substipitate; scales obtuse to short acuminate, usually half the length of the perigynia or less; achenes conspicuously stipitate; broadest basal leaves 2-4.5 mm wide.

Perigynia 6-10 mm long, narrowly lanceolate, broadest below the middle, very gradually tapering toward the apex or the broad portion elongate; pistillate scales mostly rounded on the back, rarely tinged with reddish brown.....96. *C. debilis.*

Perigynia 4.5-7 mm long, broadly ovate-lanceolate, broadest at the middle, abruptly tapering at both ends, the broad portion short; pistillate scales mostly keeled and tinged with reddish brown.....................96a. *C. debilis* var. *Rudgei.*

Perigynia strongly stipitate; scales strongly cuspidate or awned, usually more than half the length of the perigynia; achenes substipitate or sessile; broadest basal leaves 5-10 mm wide. (See excluded species no. 21, p. 275.)............*C. arctata.*

96. **Carex débilis** Michx. Map 521. Infrequent in southern Indiana, principally in the southeastern counties, where it is found in low wet woods, especially flat or even swampy pin oak and beech-sweet gum woods.

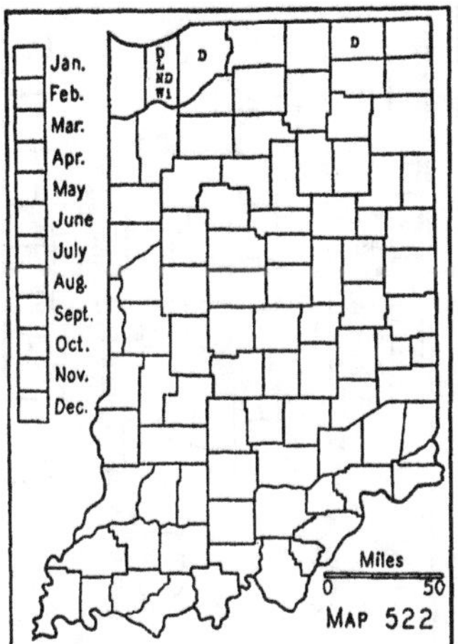

Carex debilis var. Rudgei Bailey

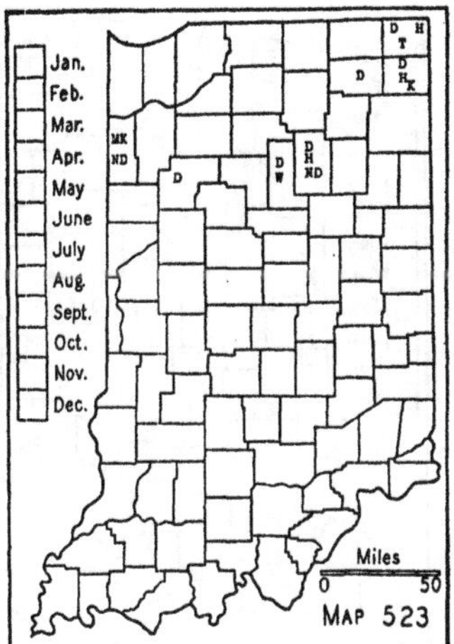

Carex Sprengelii Dewey

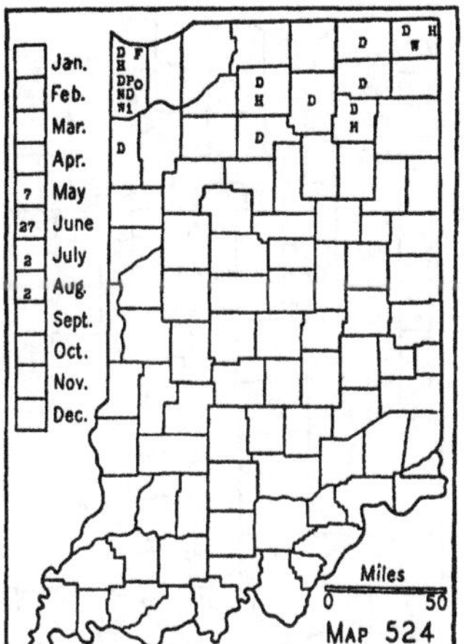

Carex Oederi var. viridula (Michx.) Kükenth.

It is not known in Indiana from the habitat ascribed to it by Mackenzie ("dry woods and copses," N. Amer. Flora **18**: 290. 1935).

Mass. and s. Ind., southw. to Fla. and Tex.

96a. **Carex debilis** var. **Rúdgei** Bailey. (*Carex flexuosa* Muhl., *Carex tenuis* Rudge, and *Carex debilis* var. *strictior* Bailey.) Map 522. Infrequent near the northern border of Indiana where it is found in low beech-maple woods. Any specimens which may have formed the basis for the report of this variety (as *C. tenuis*) from Jefferson County in Coulter's Catalogue doubtless should be referred to *C. debilis*.

Although *Carex debilis* and its variety *Rudgei* are geographically widely separated in Indiana their ranges overlap farther east.

Newf. to Wis., southw. to Va. and Mo.; also in the mts. of N. C. and Tenn.

27. § LONGIRÓSTRES

97. **Carex Sprengélii** Dewey. (*Carex longirostris* Torr.) Map 523. A local species known in the state only from the lake area. The Miami and Noble County stations are in bluegrass sod along roadsides; that of De Kalb County, on a wooded flood plain with beech and black maple; that of Steuben County, a low depression in woods; the Wabash County, the side of "Hanging Rock" on the south bank of the Wabash River, southeast of Lagro; the White County, a moist wooded bottom along the Tippecanoe River, northeast of Buffalo. Its usual habitats, outside Indiana, are rich rocky woods especially in moist depressions, and on crests of calcareous river bluffs or the tops of limestone boulders in open woods. It is often in large colonies where found.

N. B. to Alberta, southw. to Del., Pa., Nebr., and Colo.

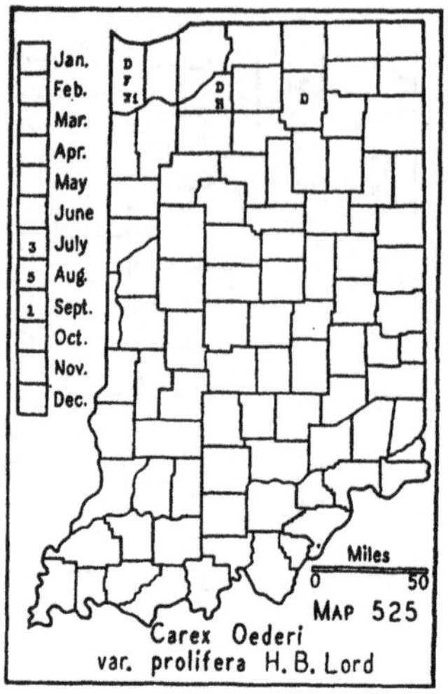

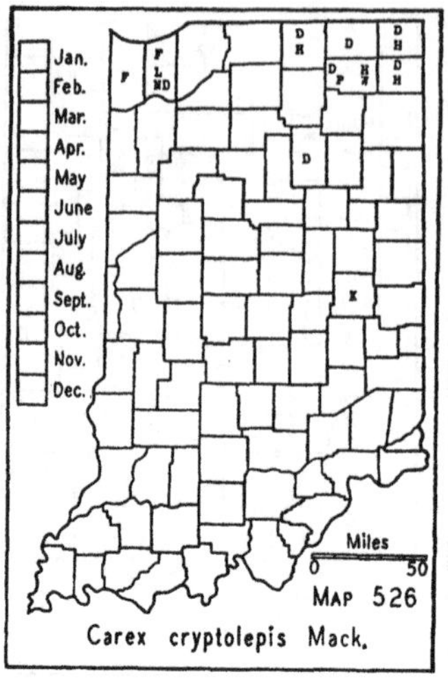

28. § EXTÉNSAE

Perigynia 2-3 mm long, little if at all deflexed, the beak much shorter than the body; spikes oblong, 4-7 mm wide.

Spikes 2-7, the lower often separate, the terminal usually staminate, conspicuous; pistillate scales usually reddish-tinged..........................98. *C. viridula.*

Spikes 4-15, mostly densely aggregated, the terminal usually androgynous with the staminate portion very small and inconspicuous; pistillate scales usually very slightly if at all reddish-tinged..............98a. *Carex viridula* f. *intermedia.*

Perigynia 3.5-6 mm long, at least the lower conspicuously deflexed, the beak equaling the body; spikes subglobose, 7-12 mm wide.

Perigynia 3.5-4.5 mm long, the beak smooth, pale at the tip; scales slightly if at all reddish-tinged, largely concealed by the perigynia; leaf blades 1-3 mm wide. ....99. *C. cryptolepis.*

Perigynia 4.5-6 mm long, the beak serrulate, reddish-tinged at the tip; scales strongly reddish-tinged, conspicuous in the spikes; leaf blades 3-5 mm wide. ....100. *C. flava.*

98. **Carex virídula** Michx. (See, Jour. Bot. **77**: 301-304. Nov. 1939.) (*Carex Oederi* var. *viridula, Carex Oederi* var. *pumila* (Coss. & Germ.) Fern., and *Carex irregularis* Schwein.) Map 524. Frequent on marly and sandy lake borders, and in swales among the dunes.

Newf. to Alaska, southw. to N. J., Ind., N. Mex., Utah, and Calif.

98a. **Carex viridula f. intermèdia** (Dudley) Hermann, comb. nov. (*Carex Oederi* f. *intermedia* Dudley, Bull. Cornell Univ. 2:117. 1886.) (*Carex chlorophila* Mack.* and *Carex Oederi* var. *prolifera* H. B. Lord.) Map 525. Infrequent in the habitats of the preceding variety.

*Of the characters employed by Mackenzie to distinguish his *C. chlorophila* from *C. viridula* very few seem to hold with any degree of constancy. A careful study of an extensive series of both plants has shown the characteristics ascribed to the leaf blades and sheaths to be wholly unreliable. The characters used in the above key to separate *C. chlorophila* from C. *viridula* (the former here considered as a form of *C. viridula*), although often well-marked, are tendencies only and they, together with a generally later flowering and fruiting date (contrary to Mackenzie's note in N. Amer. Flora 18: 303 that *C. Oederi, C. viridula,* and *C. chlorophila* "bloom and bear fruit from early summer until frost"), do not seem sufficiently constant to maintain *C. chlorophila* as a species. Umbach's collections of June 4 and 24, 1899, and Deam nos. 44412 (June 3, 1927) and 42172 (Aug. 19, 1925) are intermediate in nearly all characters, but on the basis of the early fruiting date of the first three they may be referred to *C. viridula* while the late date of the last would place it nearer f. *intermedia.*

Between *C. viridula* and f. *intermedia* there is a more or less well-defined seasonal difference in flowering and maturing of the fruit as may be seen from the collection dates with Maps 524 and 525. In Indiana *C. viridula* is in its prime in June; f. *intermedia* in August.

N. Y. to Wis., southw. to N. J. and Ind.

99. **Carex cryptólepis** Mack. (*Carex flava* var. *rectirostra* Gaudin, in part.) Map 526. Frequent in northern Indiana on marly lake borders and in marshes rich in marl; infrequent on wet sandy lake shores.

Newf. to Minn., southw. to N. J. and Ind.

100. **Carex flàva** L. Map 527. A widespread species which is common throughout most of its range but rare and very local in Indiana. The two known localities for it in the state are: marly marsh on the Wolverton Estate, 7 miles southwest of South Bend, St. Joseph County, Deam nos. 54874 and 55079; and springy wooded bank of Flat Rock River, three-fourths of a mile above St. Paul, Decatur County, Mrs. C. C. Deam nos. 10766 and 13400.

Newf. to B. C., southw. to N. J., Ind., and Mont.; also in Europe.

29. § VIRESCÉNTES

Perigynia densely pubescent; spikes about 3-4 mm thick, the lower more or less widely separated and peduncled; ligule much longer than wide.

Pistillate spikes oblong or oblong-globose, abrupt or rounded at the base, the lowest 5-20 mm long; perigynia broadly obovoid; anthers 0.8-1.6 mm long; leaves usually exceeding the culms.................................101. *C. Swanii.*

Pistillate spikes linear, attenuate at the base, the lowest 15-40 mm long; perigynia oblong-elliptic or narrowly obovoid; anthers 1.5-2.5 mm long; leaves usually shorter than the culms..................................102. *C. virescens.*

Perigynia glabrous; spikes 4-8 mm thick, contiguous or nearly so, sessile or subsessile; ligule not longer than wide.

Perigynia strongly flattened ventrally, rounded at the apex, nerved, achenes with a somewhat bent short-apiculate tip..........................103. *C. hirsutella.*

Perigynia turgid, nearly round in cross section, short-pointed at the apex, coarsely nerved or ribbed; achenes with a very abruptly bent apiculate tip or style.

Leaf blades glabrate; perigynia 2 mm long; pistillate scales not pilose, obtuse or short-cuspidate......................................104. *C. caroliniana.*

Leaf blades pubescent, especially below; perigynia 2.5-3.5 mm long; pistillate scales sparingly pilose, long-acuminate, cuspidate or awned..........105. *C. Bushii.*

101. **Carex Swánii** (Fern.) Mack. (*Carex virescens* var. *Swanii* Fern.) Map 528. Common in clearings in low woods, and in moist open oak woods; infrequent along roadsides, on flood plains, and on wooded slopes.

N. S. to Wis., southw. to N. C., Tenn., and Ark.

102. **Carex viréscens** Muhl. Map 529. Fairly common in the southern counties, especially in the knob area (Chestnut Oak Upland), on wooded bluffs, slopes, and river banks; infrequent in level woods. It is known from the lake area from a single collection and most reports from the northern third of the state were doubtless based upon material of *C. Swanii.*

Maine to Ind., southw. to Ga. and Tenn.

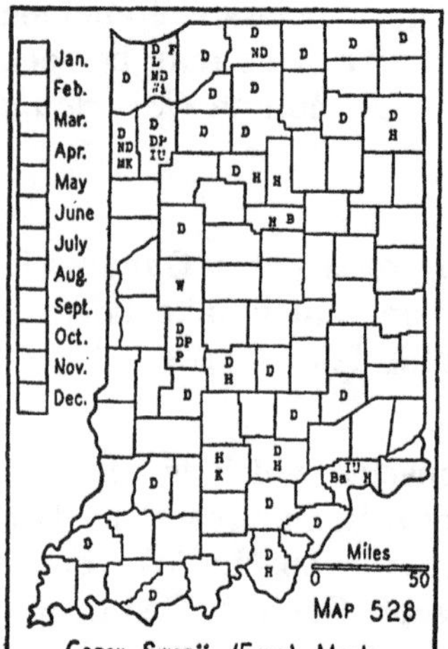

Carex Swanii (Fern.) Mack.

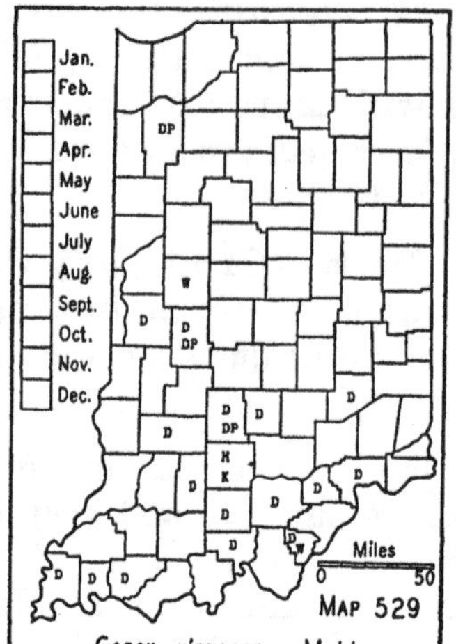

Carex virescens Muhl.

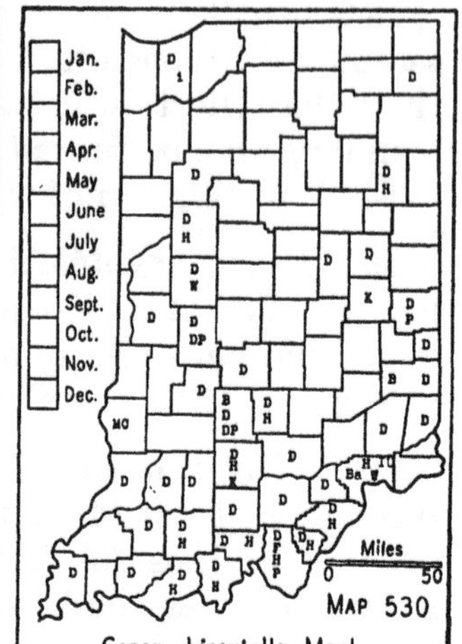

Carex hirsutella Mack.

103. **Carex hirsutélla** Mack. (*Carex triceps* var. *hirsuta* (Willd.) Bailey; *"Carex complanata"* Mack., not Torr., in Britton and Brown, Illus. Flora, ed. 2.) Map 530. Common in southern Indiana, usually in sterile soil, in dry open woods and fallow fields and along sandy roadsides; infrequent in low or flat woods; becoming less frequent northward and rare in the northernmost counties.

Maine to Mich., southw. to Ala. and Tex.

104. **Carex caroliniàna** Schwein. (*Carex triceps* var. *Smithii* Porter.) Map 531. Frequent in southern Indiana in low flat woods and in clay soil in fallow fields.

N. J. to N. C. and Tex.

105. **Carex Búshii** Mack. (Bull. Torrey Bot. Club 37: 241. 1910.) Map 532. Known in Indiana only from three stations, found by Deam, all in the unglaciated area. It is common in the Posey County locality in low, open post oak flats south of Half Moon Pond, 9 miles southwest of Mount Vernon. The two localities in Spencer County are in a low fallow field one mile north of Bloomfield (4 miles northwest of Chrisney), and in a low, open pin oak and post oak flat two miles southeast of Dale.

Mackenzie has pointed out the marked general resemblance of this species to the wholly unrelated *Carex Buxbaumii.*

Mass. to Mich., southw. to D. C., Miss., and Tex.

30. § HÍRTAE

Leaf blades flat, 2-5 mm wide; culms sharply triangular; achenes straight-apiculate. ....106. *C. lanuginosa.*

Leaf blades involute-filiform, 2 mm wide or less; culms obtusely triangular; achenes bent-apiculate..107. *C. lasiocarpa.*

106. **Carex lanuginòsa** Michx. Map 533. Common in northern Indiana in swamps, sloughs, wet ditches, open swampy woods, and on lake borders; infrequent in southern Indiana. This, like the following species, is usually plentiful where found, often being the dominant plant in a marsh or on a

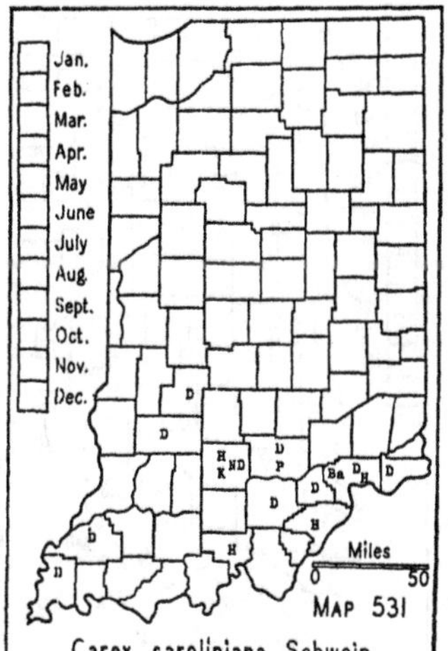

Carex caroliniana Schwein.

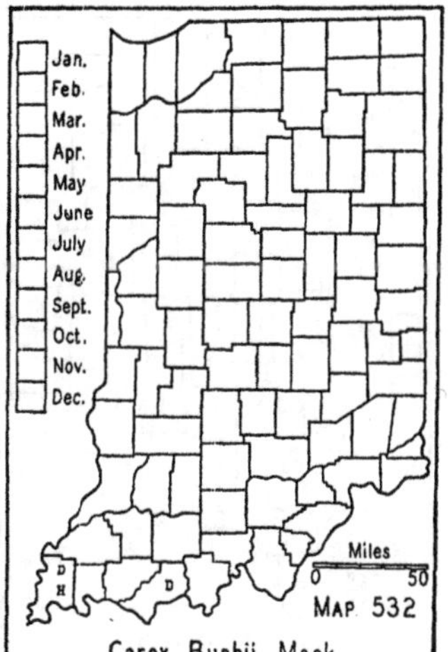

Carex Bushii Mack.

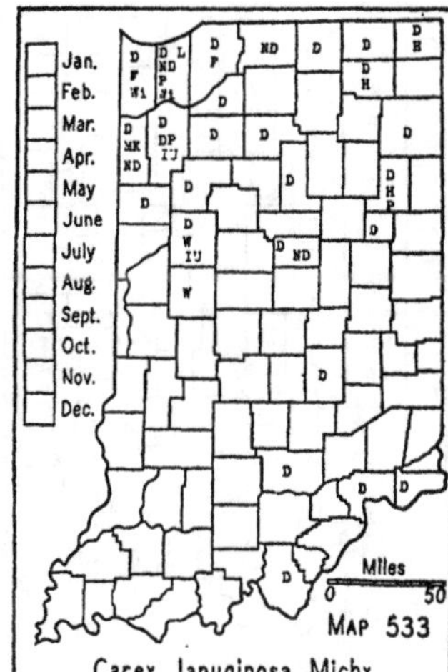

Carex lanuginosa Michx.

lake border. It is one of the most widely distributed sedges in North America.

Schneck's report from the Lower Wabash Valley is unsupported by specimens; in fact, no material has been seen from any of the southwestern counties.

N. B. to B. C., southw. to Tenn., Ark., N. Mex., and Calif.

107. **Carex lasiocárpa** Ehrh. (*Carex filiformis* of authors, not L.) Map 534. Frequent in the lake area in sloughs and sphagnum bogs and on lake borders. Like the preceding species, it often forms large colonies.

No specimen could be found to substantiate Schneck's report from the Lower Wabash Valley, an area far south of the normal range of *C. lasiocarpa.*

Newf. to B. C., southw. to N. J., Pa., Iowa, Idaho, and Wash.; also in Europe.

30A. § ANÓMALAE

Carex scabràta Schwein. (See excluded species no. 23, p. 275.)

31. § SHORTIÀNAE

108. **Carex Shortiàna** Dewey. Map 535. Common except in northern Indiana. It is found in moist open woods and roadside ditches and on banks of creeks.

Attempts to locate a specimen to support Peattie's report from Lake County have been unsuccessful and the species is not otherwise known in western Indiana north of Tippecanoe County.

Pa. to Iowa, south. to Va., Tenn., and Okla.

108a. × **Carex Dèamii** Hermann. (Rhodora **40**: 81. 1938.) A sterile hybrid between *Carex Shortiana* and *C. typhina* which was discovered by Deam in Pike County in 1926. The only known locality for it is at the edge of a low woods on the east side of the road dividing sections 17 and 18, Jefferson Twp., two miles southwest of Otwell.

Carex lasiocarpa Ehrh.

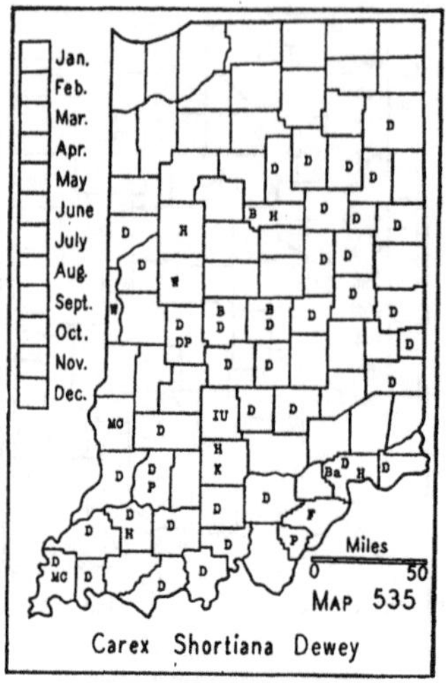

Carex Shortiana Dewey

Carex limosa L.

32. § LIMÒSAE

109. **Carex limòsa** L. Map 536. Infrequent in tamarack bogs and on mucky lake borders in northern Indiana. It is usually found in sphagnum.

Lab. and Newf. to Yukon, southw. to Del., Iowa, Mont., and Calif.; also in Eurasia.

33. § ATRÀTAE

110. **Carex Buxbaúmii** Wahl. (*Carex polygama* Schkuhr, not Gmelin.) Map 537. Rather common among the dunes; infrequent elsewhere in northern Indiana. Among the dunes it is found in swales and on interdunal flats; elsewhere in marshes and low sandy or marly openings. In southern Indiana it occurs in swampy woods.

Newf. to Alaska, southw. to Ga., Ark., Colo., Utah, and Calif.; also in Eurasia.

34. § ACÙTAE

Beak of perigynium very short, or absent, not twisted; pistillate spikes erect; culms relatively slender to the base, very rough above.

Culms strongly phyllopodic (sterile shoots sometimes aphyllopodic); fertile culms surrounded at the base by the dried-up leaves of the previous year; perigynia strongly flattened, not at all turgid, obovoid, 2.75-3.2 mm long, 1.5-2.75 mm wide . . . . 111. *C. substricta.*

Culms aphyllopodic; fertile culms all or mostly arising laterally and not surrounded at the base by the previous year's tufts of leaves; perigynia 2-2.75 mm long, 1.25-1.75 mm wide.

Perigynia inflated, strongly biconvex, brownish at maturity, broadly ovate to suborbicular, 2-2.25 mm long; pistillate scales divaricate at maturity; ligule slightly longer than wide; lower sheaths slightly if at all filamentose; stolons short, ascending; achenes suborbicular . . . . 112. *C. Haydenii.*

Perigynia not inflated, closely enveloping the achenes, unequally biconvex, green or straw colored, 2.25-2.75 mm long; stolons long, many, horizontal; achenes oblong to obovate.

Lower sheaths not filamentose ventrally, strongly septate-nodulose dorsally; ligule much wider than long; pistillate spikes (3) 4-5, the lower 2-10 cm long; perigynia ovate or obovate, 1.5-1.75 mm wide . . . . 113. *C. Emoryi.*

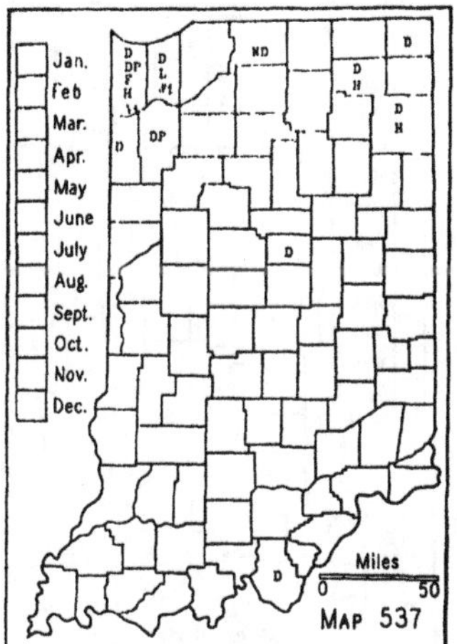

Carex Buxbaumii Wahl.

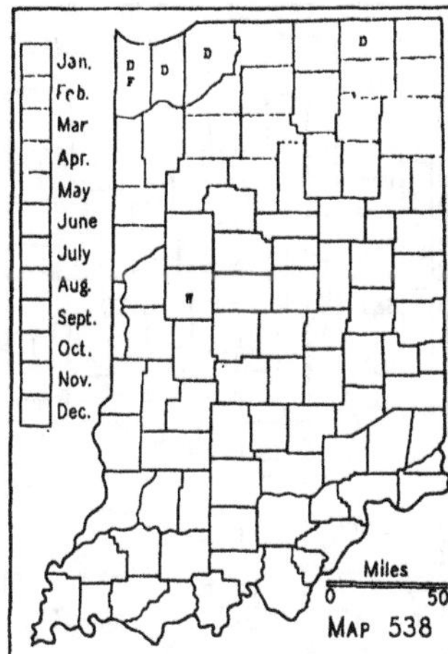

Carex substricta (Kükenth.) Mack.

Carex Haydenii Dewey

Lower sheaths filamentose ventrally; ligule much longer than wide; pistillate spikes usually 3, the lower 1-6 cm long; perigynia elliptic to narrowly or broadly ovate, 1.5 mm wide; pistillate scales appressed at maturity.
Leaf sheaths glabrous ventrally, without a minute hyaline jagged-ciliate margin at the mouth....................................114. *C. stricta.*
Leaf sheaths hispidulous ventrally (and dorsally), usually with a minute jagged-ciliate margin at the mouth; leaves usually paler green or glaucous..............................114a. *C. stricta* var. *strictior.*
Beak of perigynium prominent, twisted when dry; at least the lower pistillate spikes nodding or recurving; culms usually stout at the base, smooth above; stolons very short and ascending or none...115. *C. torta.*

111. **Carex substrícta** (Kükenth.) Mack. (In Rydb., Flora Rocky Mts. 139. 1917.) (*Carex aquatilis* var. *substricta* Kükenth.) Map 538. Infrequent but locally plentiful in the northernmost tier of counties in marshes and sloughs and on lake or river borders.

Most of the reports of *C. aquatilis* from Indiana were doubtless based upon specimens of this species.

Newf. to Wash., southw. to N. J., Ind., and Nebr.

112. **Carex Haydénii** Dewey. (*Carex stricta* var. *decora* Bailey.) Map 539. Infrequent in northwestern Indiana in wet prairies, ditches, and low clearings in open oak woods.

N. B. to Minn., southw. to N. J., Ill., and Mo.

113. **Carex Emoryi** Dewey. Map 540. Frequent along banks of creeks and in swamps, sloughs, and swales in woods. Not known from the unglaciated area.

N. J. and Va. to Man. and Colo., southw. to Tex. and N. Mex.

114. **Carex strícta** Lam. (*Carex stricta* var. *angustata* (Boott) Bailey.) Map 541. Frequent in northwestern Indiana in marshes and open swamps and on borders of creeks where it generally forms dense tussocks. Less common than the following variety except in Lake County.

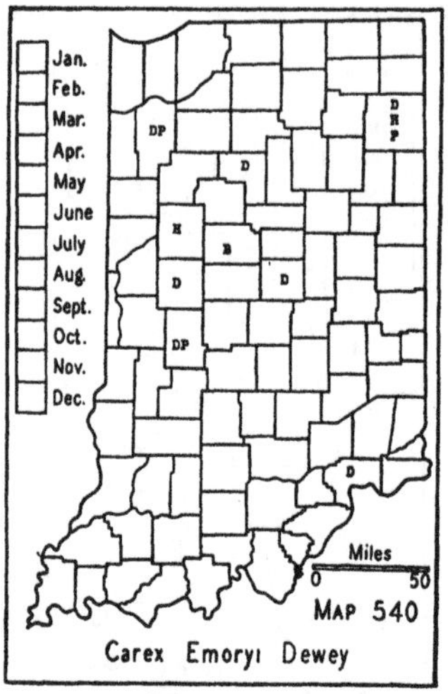

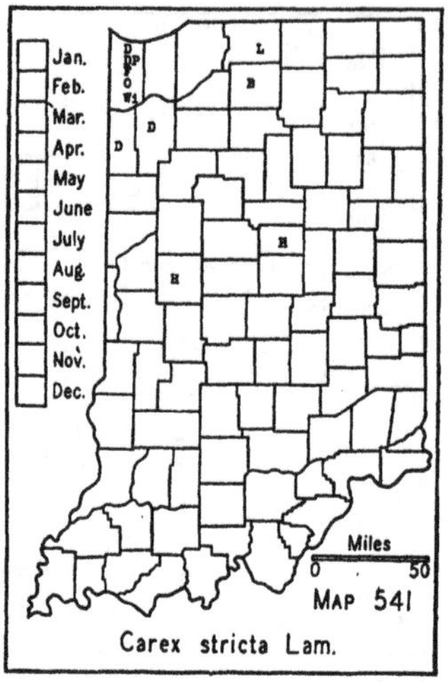

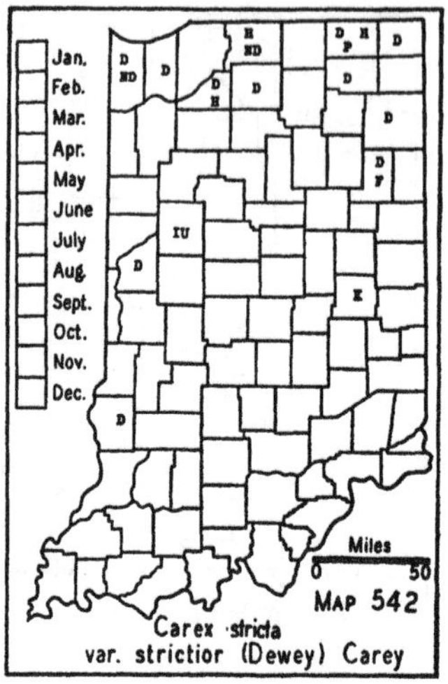

The dominant plant of "sedge meadows" is most frequently this species or var. *strictior.*

Maine to N. C., and along the Coastal Plain to Tex.; also locally in the Great Lakes region.

114a. **Carex stricta** var. **stríctior** (Dewey) Carey. (*Carex strictior* Dewey.) Map 542. Common in northern Indiana in marshes and roadside ditches, often in very marly soil; infrequent southward along the western border of the state.

This plant is reputed to grow in beds (not dense tussocks) while *C. stricta* is supposed to occur in very dense tussocks only. Field observations in Indiana, however, do not indicate that this distinction is at all reliable; *C. strictior* has often been seen to form conspicuous tussocks and *C. stricta* was frequently found in beds. The distinctions ascribed by Mackenzie to the foliage characters (leaf blades deep green, channeled and keeled toward the base in *C. stricta,* glaucous to blue-green, flat or nearly so to the base in *C. strictior*) seem to be particularly inconstant. The lowest bract is generally larger and more leaflike in var. *strictior,* but this, too, is merely a tendency. Forms which are transitional in nearly all characters are so frequent in Indiana that it seems best to regard *C. strictior* as not more than a variety.

Que. to Minn., southw. to D. C. (in the mts. to N. C. and Tenn.) and Iowa.

115. **Carex tórta** Boott. Map 543. Frequent south of the lake area on rocky beds of streams, and sand bars in creeks and on their springy banks; rare in woodland swales.

Que. to Minn., southw. to N. C., Tenn., and Ark.

35. § CRYPTOCÁRPAE

116. **Carex crinìta** Lam. Map 544. Common in swampy woods and thickets; frequent in swales, sloughs, ditches, and swamps.

Que. to Minn., southw. to N. C. and Tex.

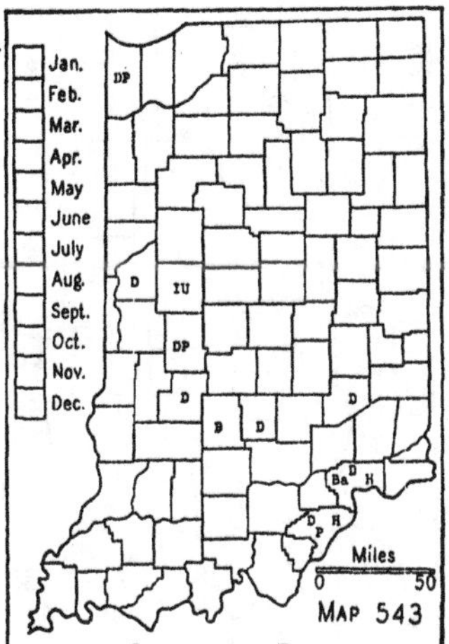

Carex torta Boott

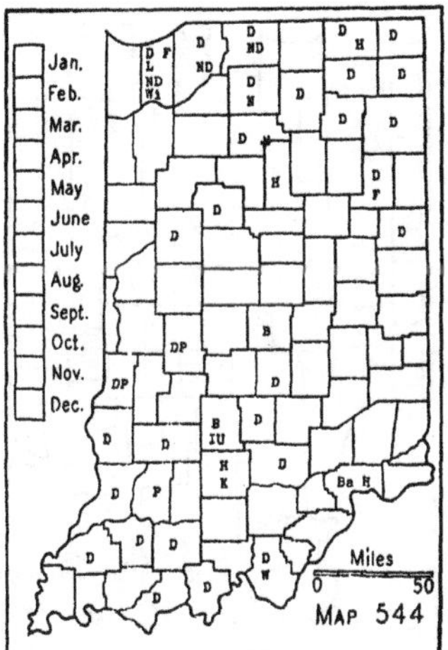

Carex crinita Lam.

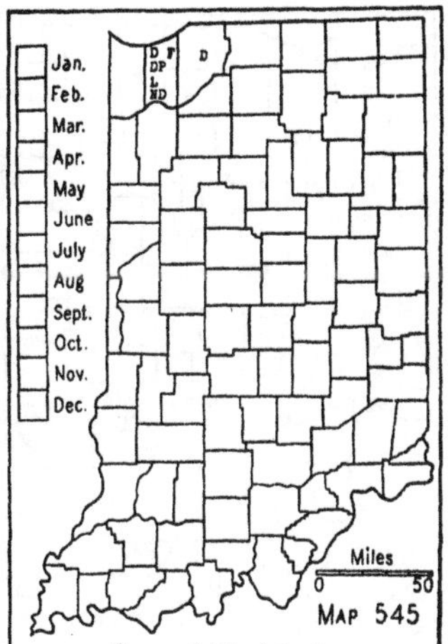

Carex folliculata L.

35A. § ORTHOCERÀTES

Carex paucifiòra Lightf. (See excluded species no. 28, p. 276.)

36. § FOLLICULÀTAE

117. **Carex folliculàta** L. Map 545. Known in Indiana only from the dune area in Porter and La Porte Counties where it is locally frequent on mucky borders of wet woods. One collection of Deam's (4 miles northeast of Michigan City) is from a sedge marsh.

Newf. to Wis., southw. to D. C. (in the mts. to N. C. and Tenn.) and Ind., but best developed on the Coastal Plain.

37. § PSEUDO-CYPÈRI

Teeth of perigynia not over 0.5 mm long; perigynia suborbicular in cross section, inflated, membranaceous, spreading; culms stoloniferous; ligule not longer than wide..118. *C. hystricina*.

Teeth of perigynia 0.5 mm or more long; perigynia flattened-triangular, scarcely inflated, coriaceous, more or less reflexed; culms not stoloniferous; ligule much longer than wide.

Teeth of perigynia recurved-spreading, 1.2-2 mm long; beaks of perigynia (exclusive of the teeth) 1.5-2.2 mm long, equaling or longer than the bodies; perigynia 6 mm long..119. *C. comosa*.

Teeth of perigynia erect or slightly spreading, 0.5-1 mm long; beaks of perigynia (exclusive of the teeth) averaging 1 mm long, shorter than the bodies; perigynia 4-5 mm long....................................120. *C. Pseudo-Cyperus*.

118. **Carex hystricìna** Muhl. (*Carex hystricina* var. *Dudleyi* Bailey and *Carex hystricina* var. *Cooleyi* Wood.) Map 546. Common in most of the glaciated area of Indiana in swamps and wet habitats of all types but usually in calcareous soils.

No corroborating specimen for Schneck's report from the Lower Wabash Valley could be found.

Que. to Wash., southw. to Va., Ky., Tex., and Calif.

119. **Carex comòsa** Boott. (*Carex Pseudo-Cyperus* var. *americana* Hochst.) Map 547. Fairly common in northern Indiana on low borders

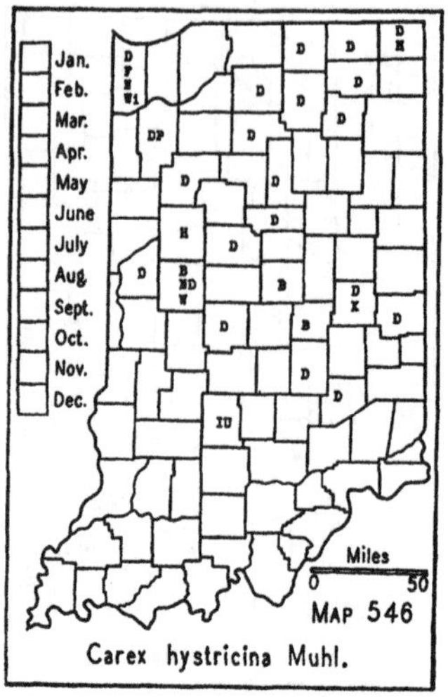

MAP 546

Carex hystricina Muhl.

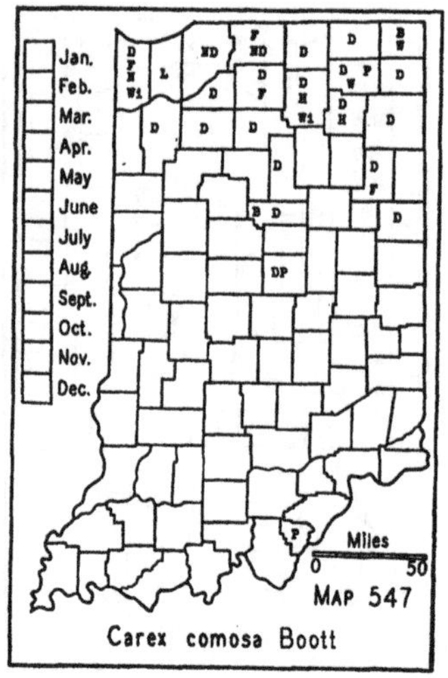

MAP 547

Carex comosa Boott

MAP 548

Carex Pseudo-Cyperus L.

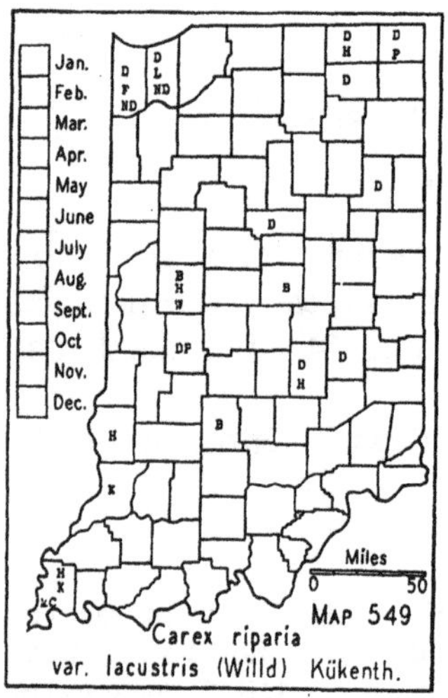

MAP 549

Carex riparia var. lacustris (Willd) Kükenth.

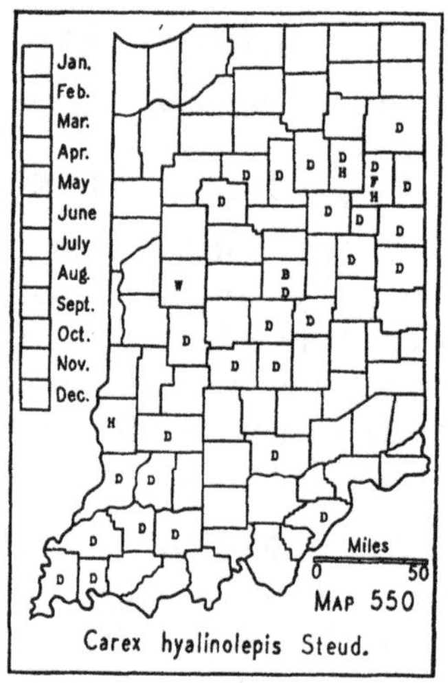

MAP 550

Carex hyalinolepis Steud.

MAP 551

Carex subimpressa Clokey

of lakes (often in shallow water) and in swamps, sloughs, and ditches. In southern Indiana it is known only from a single collection from Floyd County.

Que. to Minn., southw. to Fla. and La.; also locally in the Pacific Coast States.

120. **Carex Pseudo-Cypèrus** L. Map 548. Rare on lake borders and in sloughs and swamps in northern Indiana where it reaches the southern limit of its range. Like *C. comosa* it frequently grows in shallow water, rooted in muck.

Newf. to Sask., southw. to Conn., N. Y., Ind., and Minn.; also in Eurasia.

38. § PALUDÒSAE

Beaks of perigynia much shorter than the body, the teeth short, about 0.5 mm long, erect or nearly so; foliage glabrous.

Perigynia glabrous.

Mature perigynia strongly nerved; fertile culms aphyllopodic, strongly reddish-tinged at the base, the lower sheaths filamentose ventrally; ligule at least twice as long as wide............................121. *C. riparia* var. *lacustris.*

Mature perigynia impressed-nerved; fertile culms phyllopodic, less strongly or not at all reddish-tinged at the base, the lower sheaths not filamentose ventrally; ligule shorter or moderately (less than one and a half times) longer than wide.....................................122. *C. hyalinolepis.*

Perigynia hairy, the ribs mostly hidden by the short dense pubescence...122 a. × *C. subimpressa.*

Beaks of perigynia (including teeth) nearly as long as the body; the teeth prominent, 1-3 mm long, erect to widely spreading.

Perigynia glabrous; leaf sheaths pubescent; at least the lower leaf blades sparsely hairy below toward the base.................................123. *C. atherodes.*

Perigynia hairy; leaf sheaths and blades glabrous.............124. *C. trichocarpa.*

121. **Carex ripària** Curtis var. **lacústris** (Willd.) Kükenth. (*Carex lacustris* Willd.) Map 549. Common in calcareous soils in marshes and ditches and on borders of swamps, lakes, and streams. It often forms extensive stands in marshes. This, and to a lesser extent the following species, seem to be somewhat periodic in fruiting, at least in the Great Lakes States. Often throughout one or more seasons in a large colony, only a few plants, if any, will be found with fertile culms.

Que. to Sask., southw. to Va. and Iowa.

122. **Carex hyalinólepis** Steud. (*Carex riparia* var. *impressa* S. H. Wright and *Carex impressa* (S. H. Wright) Mack.) Map 550. Common, except in the northern three tiers of counties, in roadside ditches and wet depressions in low open woods and on flood plains and borders of ponds.

N. J. to Ont. and Nebr., southw. to Fla. and Tex.

122a. × **Carex subimpréssa** Clokey. (Rhodora **21**: 84. 1919; *Carex languinosa* × *impressa* Clokey, Torreya **16**: 199. 1916.) Map 551. Known in Indiana from collections by Deam from four counties along the northeastern border, where it is very local but usually occurs in colonies which are probably clones. It is found in ditches along roadsides or railroads and in low ground in open woods.

No verifying specimens were found for Clokey's reports from Porter and Posey Counties or for Peattie's report from Lake and Porter Counties.

Ind. and Ill.

123. **Carex atheròdes** Spreng. (*Carex trichocarpa* var. *imberbis* Gray and *Carex trichocarpa* var. *aristata* (R. Br.) Bailey.) Map 552. Rare in northern Indiana in marshes and wet prairie habitats.

Reported from Marshall County by Clark but no specimens could be located.

Ont. to Yukon, southw. to N. Y., Ind., Mo., Kans., Colo., Utah, and Oreg.; also in Eurasia.

124. **Carex trichocárpa** Muhl. Map 553. Rare in swamps, low openings, and swales in woods; chiefly in eastern-central Indiana.

Reported from Madison County by Smith but no specimens were found.

Que. and Vt. to Minn., southw. to N. J., Ind., and Iowa.

Carex atherodes Spreng.

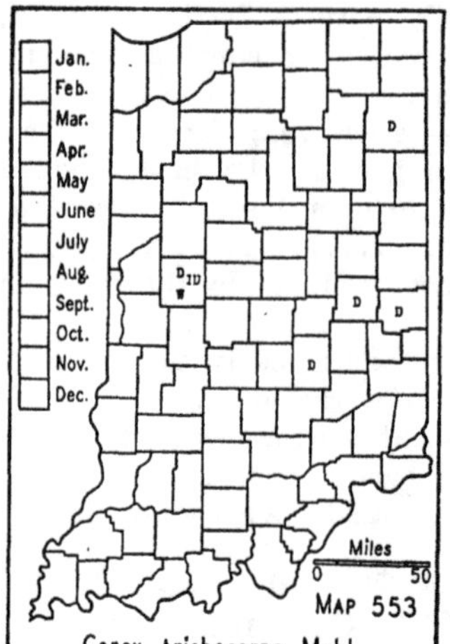

Carex trichocarpa Muhl.

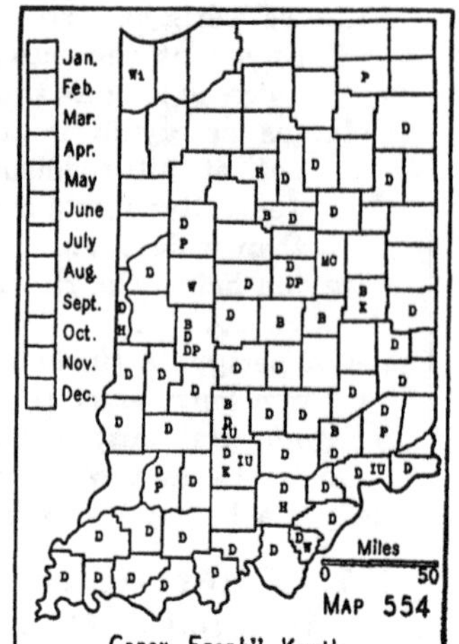

Carex Frankii Kunth

39. § SQUARRÒSAE

Perigynia shorter than the rough-awned scales, 14-20-ribbed; terminal spike usually staminate, narrowly linear, small (0.5-2.5 cm long); ligule slightly, if at all, longer than wide; achenes about 1.5 mm long..........................125. *C. Frankii.*

Perigynia much longer than the scales, several-ribbed above; terminal spike gynaecandrous; ligule much longer than wide; achenes 2.2-3 mm long.

- Beaks of perigynia mostly appressed-ascending; spikes oblong-cylindric; achenes obovoid, their sides concave; pistillate scales mostly blunt; style straight below..126. *C. typhina.*
- Beaks of perigynia widely radiating; spikes oval or oblong-oval; achenes linear-oval, their sides almost flat; pistillate scales acute to short-awned; style strongly curved or bent below....................................127. *C. squarrosa.*

125. **Carex Fránkii** Kunth. (*Carex stenolepis* Torr.) Map 554. Not known from the two northern tiers of counties. Except in the lake and prairie areas very common in ditches and low roadsides and on banks of creeks; frequent in swamps, low flat woods, ravines, marshes, and wet fallow fields.

Pa. and N. Y. to Ill. and Kans., southw. to Ga. and Tex.; also in S. A.

126. **Carex typhìna** Michx. (*Carex typhinoides* Schwein.) Map 555. Fairly common in the southern counties, infrequent in northern Indiana, and not known from the central portion of the state. Its favorite habitat is low flat woods, especially pin oak, but it is also found on borders of ponds and in marshes, swamps, and roadside ditches.

Specimens to confirm Wilson's reports from Hamilton and Tippecanoe Counties could not be found.

Que. to Wis. and Iowa, southw. to Ga. and La.

127. **Carex squarròsa** L. Map 556. Common, especially southward, in low or swampy woods and roadside ditches; frequent on wet borders of ponds and creeks.

Que. to Wis. and Nebr., southw. to N. C. and Ark.

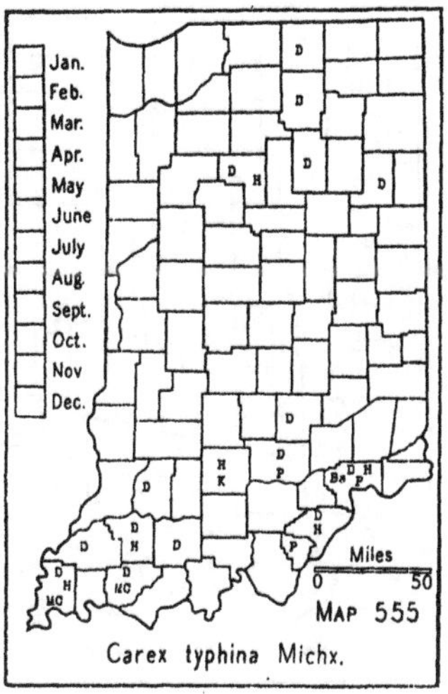

Carex typhina Michx.

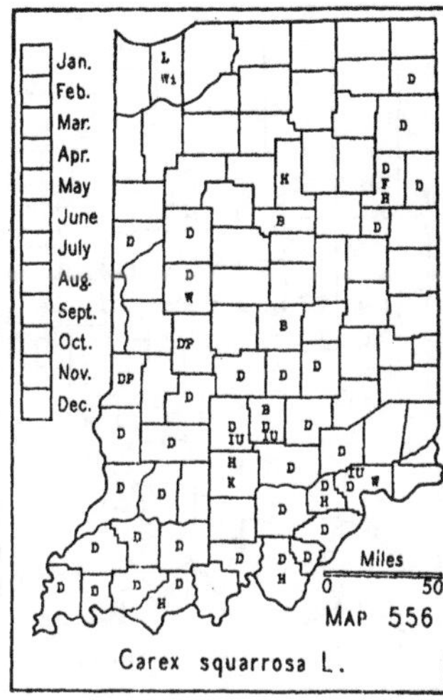

Carex squarrosa L.

Carex vesicaria L.

40. § VESICÂRIAE

Pistillate scales not rough-awned.
Pistillate spikes oblong to cylindric, 17-many-flowered; leaf blades flat or the margins somewhat revolute.
Perigynia not reflexed; bracts moderately exceeding the inflorescence.
Achenes not excavated on one side; perigynia 2.5-3.5 mm wide.
Culms sharply triangular below the spikes, rough; perigynia appressed or ascending; teeth of perigynia long or perigynia tapering gradually into the beak; lower sheaths fragile, becoming strongly filamentose; ligule much longer than wide; rootstocks short-creeping, without long horizontal stolons..128. *C. vesicaria.*
Culms bluntly triangular below the spikes, smooth; perigynia spreading to squarrose at maturity; teeth of perigynia short or perigynia abruptly contracted into the beak; lower sheaths not fragile, not becoming filamentose; ligule slightly if at all longer than wide; rootstocks with long horizontal stolons...............................129. *C. rostrata.*
Achenes deeply excavated on one side; perigynia 5-6.5 mm wide...130. *C. Tuckermani.*
Lower perigynia reflexed or widely spreading, somewhat falcate; bracts many times exceeding the inflorescence.........................131. *C. retrorsa.*
Pistillate spikes globose or short-oblong, 3-18-flowered; leaf blades involute...132. *C. oligosperma.*
Pistillate scales rough-awned...133. *C. lurida.*

128. **Carex vesicària** L. (Including *Carex vesicaria* var. *monile* (Tuckerm.) Fern.) Map 557. Infrequent in the lake area in swamps, swales, and swampy woods.

Newf. to B. C., southw. to Del., Ind., Mo., N. Mex., and Calif.; also in Eurasia.

129. **Carex rostràta** Stokes. (Including *Carex rostrata* var. *utriculata* (Boott) Bailey.) Map 558. Frequent in northern Indiana in marshes, swamps, low woods, wet roadside ditches, and swales and on borders of ponds and lakes, often in shallow water. It is a very widespread species but generally is not plentiful in any one locality.

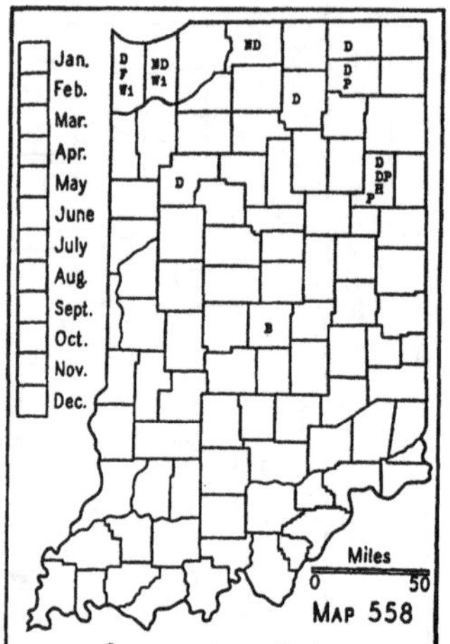

Carex rostrata Stokes

Carex Tuckermani Boott

Carex retrorsa Schwein.

Greenland to Alaska, southw. to Del., W. Va., Ind., S. Dak., N. Mex., and Calif.; also in Eurasia.

130. **Carex Tuckermáni** Boott. Map 559. Frequent in northeastern Indiana; otherwise known in the state only from the dune area and from Floyd County. It is found in swales in woods, swamps, and on borders of ponds, frequently in shallow water.

N. B. to Minn., southw. to N. J., Ind., and Iowa.

131. **Carex retrórsa** Schwein. Map 560. Known in Indiana from two collections near the northern border of the state; edge of swamp, East Chicago, Lake County, W. S. Moffatt, July 2, 1893, and, near St. Mary's Academy, Notre Dame, St. Joseph County, J. A. Nieuwland, July 9, 1913.

No corroborating specimens have been seen for the report in Coulter's Catalogue from Gibson County, Wilson's report from Hamilton County or Schneck's from the Lower Wabash Valley.

Que. to B. C., southw. to N. J., Ind., Iowa, Colo., and Oreg.

132. **Carex oligospérma** Michx. Map 561. Frequent in the dune area, otherwise quite local in northern Indiana. It prefers *Chamaedaphne* and tamarack bogs, but is found also in marshes and swales and on borders of ponds.

Newf. to Mack., southw. to Mass., Pa., and Ind.

133. **Carex lùrida** Wahl. Map 562. Very common; in southern Indiana ubiquitous in swamps, sloughs, ditches, and wet habitats of all types.

Carex lurida is frequently confused with *C. hystricina* and with *C. lupulina.* The following distinctions, in addition to those given in the key to the sections, may be useful in separating it from these. The achene of *C. lurida* is strongly rough-papillate; that of *C. lupulina* is perfectly smooth. Also the teeth of the perigynia in *C. lurida* are very short (averaging 0.5 mm long) and the stigmas all protrude from one side; in *C. lupulina* the teeth are long (0.75-2 mm) and the stigmas radiate irregularly from the orifice. In *C. hystricina* the teeth of the perigynia are longer

Carex oligosperma Michx.

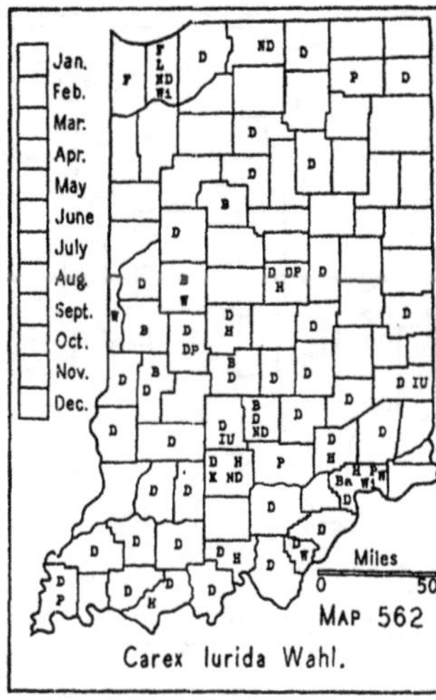

Carex lurida Wahl.

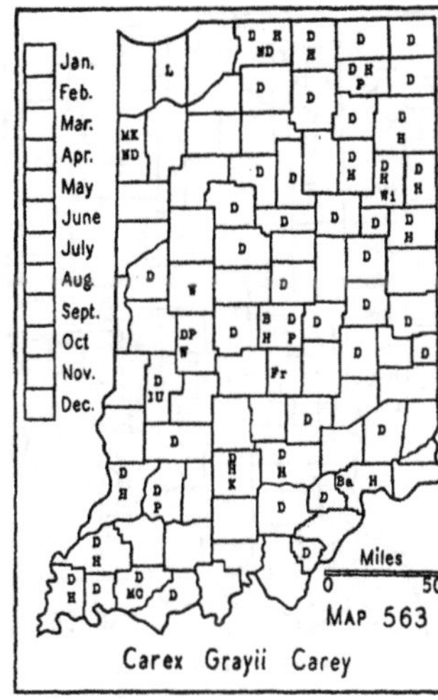

Carex Grayii Carey

and spreading or ascending; in *C. lurida* the teeth are erect or appressed, the perigynia are more abruptly beaked and the beak is longer and narrower.

N. S. to Minn., southw. to Fla., Tex., and Vera Cruz, Mex.

41. § LUPULÌNAE

Pistillate spikes globose to short-ovoid; style straight or the bend remote from the achene.
  Perigynia radiating in all directions, cuneate at the base, subcoriaceous, usually somewhat hispidulous; staminate spike usually subsessile or short-peduncled; achenes obscurely trigonous, almost suborbicular in cross section, the angles inconspicuous..134. *C. Grayii.*
  Perigynia ascending, rounded at the base, membranaceous, smooth and shining; staminate spike normally long-peduncled; achenes with blunt angles but conspicuously trigonous.
    Perigynia broadly ovoid, about half as broad (5-8 mm) as long...135. *C. intumescens.*
    Perigynia narrowly ovoid, a fourth to a third as broad (3.5-5 mm) as long...135a. *C. intumescens* var. *Fernaldii.*
Pistillate spikes oblong to cylindric; style abruptly bent immediately above the achene.
  Achenes conspicuously longer than wide, the angles not prominently knobbed, the sides shallowly concave; pistillate spikes short-oblong to oblong-cylindric.
    Culms arising one to few together from elongate rootstocks; staminate spike narrow, 2.5 mm wide, very long-peduncled; pistillate scales blunt to acute, rarely short-mucronate; leaf blades 2-6 mm wide..........136. *C. louisianica.*
    Culms cespitose; staminate spike 3-5 mm wide, sessile or short-peduncled; pistillate scales acuminate to rough-awned; leaf blades 4-15 mm wide..137. *C. lupulina.*
  Achenes not longer than wide, the angles prominently knobbed, the sides deeply concave; pistillate spikes cylindric or oblong-cylindric.
    Perigynia ascending or slightly spreading, the beak less than twice the length of the body; achenes about as wide as long..............138. *C. lupuliformis.*
    Perigynia widely spreading at maturity, the beak 2-3 times the length of the body; achenes much wider than long.............................139. *C. gigantea.*

134. **Carex Gràyii** Carey. (*Carex Grayii* var. *hispidula* Gray and

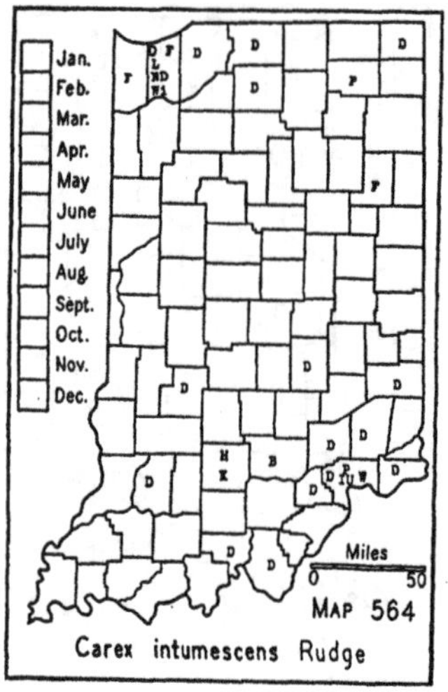

MAP 564
Carex intumescens Rudge

MAP 565
Carex intumescens var. Fernaldii Bailey

MAP 566
Carex louisianica Bailey

Carex Asa-Grayi Bailey.) Map 563. Common, but local, in low rich woods and on banks of creeks and borders of swamps. Widely distributed in the state but generally not found in abundance at any one locality. It is one of the most conspicuous of the sedges and so is apt to be collected more often than some of the inconspicuous species which may be actually more common.

The form known as var. *hispidula* shows no geographic segregation and doubtless does not merit even formal recognition. J. K. Underwood, of the University of Tennessee Agricultural Experiment Station, writes that he has observed the same plants which one year had hispidulous perigynia to be perfectly glabrous the next season.

Vt. to Wis., southw. to Ga., Tenn., and Mo.

135. **Carex intuméscens** Rudge. Map 564. Frequent to locally common in depressions in low woods (maple, beech, sweet gum or pin oak) and in flat woods.

N. H. to Wis., southw. to Fla. and Tex.

135a. **Carex intumescens** var. **Fernáldii** Bailey. Map 565. Infrequent in northern Indiana, chiefly in the lake area, in habitats similar to those of the species.

Newf. to Keewatin, southw. to Mass., N. Y., Ind., and Ill. and in the mts. to N. C.

136. **Carex louisiánica** Bailey. (*Carex Halei* Carey.) Map 566. A southern species which reaches its northwestern limit in southern Indiana where it is infrequent in low open woods, flat woods, and cypress swamps, mostly in the unglaciated area.

N. J. to Ind., southw. to Fla. and Tex.

137. **Carex lupulina** Muhl. (*Carex lupulina* var. *pedunculata* Gray.) Map 567. Very common in swamps, ditches, and low open woods and on borders of ponds and rivers.

N. S. to Minn., southw. to Fla. and Tex.

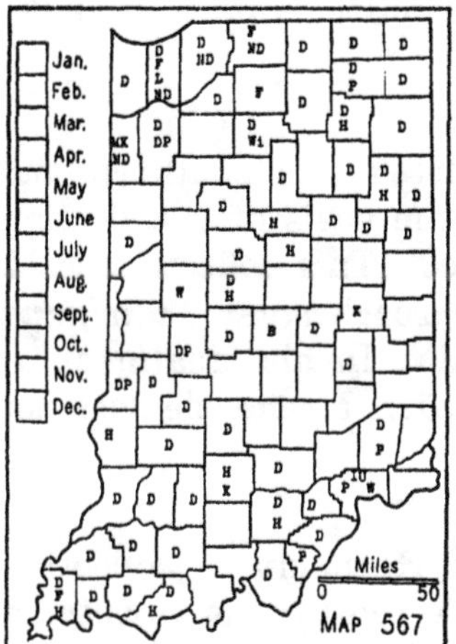

Carex lupulina Muhl.

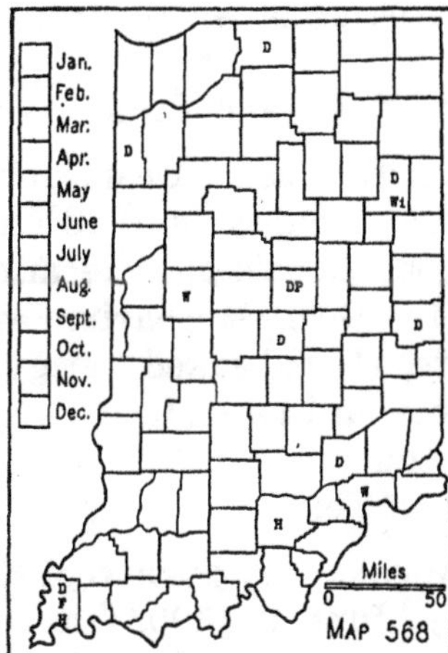

Carex lupuliformis Sartwell

Carex gigantea Rudge

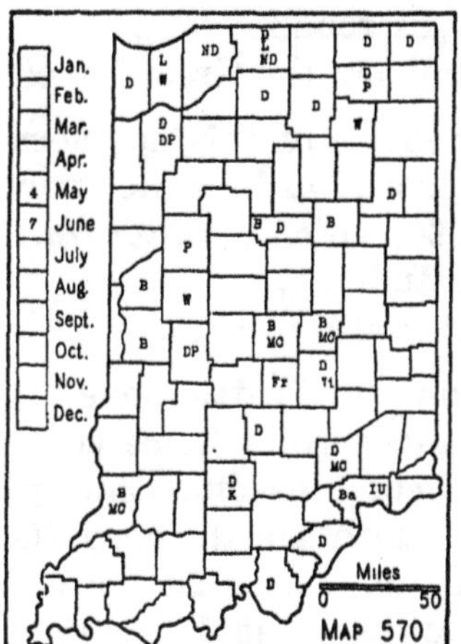

Acorus Calamus L

Symplocarpus foetidus (L) Nutt

Calla palustris L

138. **Carex lupulifórmis** Sartwell. Map 568. Infrequent and local in swampy woods, wet ditches, and buttonbush swamps, and on borders of ponds.

Vt. to Minn., southw. to Va., La., and Tex.

139. **Carex gigántea** Rudge. Map 569. Rare and local in the southern counties in cypress swamps and swampy or low open woods.

Del. to Ky., Ind., and Mo., southw. to Fla. and Tex.

EXCLUDED SPECIES

1. **Carex radiàta** (Wahl.) Dewey. Reported, as *Carex rosea* var. *radiata* Dewey, from Allen County, the Chicago region (including Lake and Porter Counties), and the Lower Wabash Valley but the reports were made before this species and *C. rosea* were clearly understood. The specimen upon which the Allen County record was based is *C. rosea* and doubtless the

specimens forming the bases of other reports should be referred to the same species.

Que. to Mich., southw. to N. C., and Tenn.

2. **Carex austrìna** (Small) Mack. Deam reported this southern and western species from Benton County in 1928 on the basis of a collection (Deam no. 43219) which Mackenzie so determined. This collection is *C. gravida.* The specimen in the Deam Herbarium approaches var. *Lunelliana* in its rather broadly ovate, short-beaked perigynia. In his treatment of the Cariceae in North American Flora (18: 57. 1931.) Mackenzie does not cite *Carex austrina* from Indiana.

Mo. and Kans. to Ark. and Tex.

3. **Carex vulpinoìdea** Michx. var. **pycnocéphala** Hermann. A collection by Deam (Steuben County, June 17, 1903) is referred to this variety in Rhodora 38: 363. 1936. Since this is the only specimen known from Indiana, however, and since it is not entirely typical it seems best to exclude it until additional and characteristic material may be found.

Mich. and Minn.; probably elsewhere on sandy shores of the Great Lakes.

4. **Carex canéscens** L. There have been many reports of this northern species from Indiana but all specimens so labeled proved to be either var. *disjuncta* or var. *subloliacea* (except E. J. Hill's no. 60, which is *C. tenera.*) Without doubt typical *C. canescens* is not found in Indiana.

Lab. to B. C., locally southw. to Conn. and n. Mich.; also in Eurasia.

5. **Carex brunnéscens** (Pers.) Poir. Both Pepoon and Peattie have reported this sedge from Lake County but no specimens from there could be located in the Indiana herbaria, nor in the Field Museum, Chicago Academy of Sciences, Gray Herbarium, National Herbarium or the herbaria of the Universities of Illinois, Wisconsin, or Michigan. Since the species is known from Ohio and from at least as far southwest as Kalamazoo County, Michigan, it is not improbable that it does occasionally reach northern Indiana.

Greenland, Lab., and Newf. to Alaska, southw. to N. J. (in the mts. to N. C.), Colo., and Wash.; also in Eurasia.

6. **Carex exìlis** Dewey. A species principally of the Coastal Plain, known in the Great Lakes region only from northern Michigan, Ontario, and Minnesota. Its occurrence in Steuben County, from which Bradner reported it, seems unlikely and it is excluded for want of a confirming specimen.

Lab. to Del.; locally inland in Vt., N. Y., Ont., Mich., and Minn.

7. **Carex stellulàta** Gooden. This and *Carex Leersii* Willd. are now considered to be synonymous with *C. muricata* L. The numerous Indiana reports of *C. stellulata* and *C. Leersii* may have been based upon almost any species of § *Stellulatae,* probably chiefly upon *C. incomperta* and *C. sterilis.*

8. **Carex muricàta** L. A boreal species known from Greenland to Newfoundland, Quebec, Alaska, and northern Eurasia. It is hardly feasible to

attempt to make any disposition of MacDougal's report from Putnam County in Coulter's Catalogue. *Carex muricata* of most American authors of that period was *C. spicata* Hudson, a European species of §*Bracteosae* which has become established locally from Nova Scotia to Virginia and Ohio.

9. **Carex cephalántha** (Bailey) Bickn. This northern and eastern species was reported by Pepoon from Lake County as *C. stellulata* var. *cephalantha* (Bailey) Fern., but no specimen could be found. Its occurrence in Indiana is improbable.

Newf. to n. Mich. and Wis., southw. to Md., also on the Pacific coast in Wash. and Vancouver Island.

10. **Carex Merritt-Fernáldii** Mack. Peattie reports this species from Dune Park (Porter County) and the Calumet District (Lake County). The only specimen which could be found bearing this name, a collection by Umbach from Lake County in the University of Wisconsin Herbarium, is *C. brevior. C. Merritt-Fernaldii* has not been found in southern Michigan and it is not likely that its range extends as far south as Indiana.

Maine to Man., southw. to Mass. and n. N. Y.

11. **Carex hormathòdes** Fern. Pepoon includes this species of the salt marshes of the Atlantic coast in his "Flora of the Chicago Region" with the statement "bogs, not common." Collections upon which this report was based could not be found but in all probability they should be referred to *C. Richii.* Deam no. 54013, from near a small creek in a field a fourth mile south of Archerville, Tippecanoe County, is more suggestive of this species than any other but the specimen is immature. No other Coastal Plain species are known from this area so that an occurrence of *C. hormathodes* here would seem to be almost certainly a chance introduction.

Along the coast, Newf. to Va., in or near salt marshes.

12. **Carex projécta** Mack. (*Carex tribuloides* var. *reducta* Bailey.) Reported from Hendricks and Marion Counties but the specimens upon which these reports were based are *C. tribuloides.*

Newf. to B. C., southw. to D. C. and Iowa.

13. **Carex foènea** Willd. (*Carex argyrantha* Tuckerm.) Reported from the Lower Wabash Valley and from Gibson and Marshall Counties. Specimens upon which Schneck's report from the Lower Wabash Valley was based were not found but they should undoubtedly be referred to some other species and the other reports also were probably based upon misidentifications.

Que. to Mich., southw. to Va. and Ohio.

14. **Carex defléxa** Hornem. A far northern species reported from Miller (Lake County) by Peattie with the statement "according to Gates." No specimen of his could be found, but in the University of Illinois Herbarium is a collection of *C. Emmonsii* which bore the label "*Carex deflexa* Hornem., sandy thicket, Miller, Ind. Agnes Chase no. 1791, May 30, 1902." The nearest known locality for *C. deflexa* is on the Keweenaw

Peninsula, Michigan, the extreme northern tip of the Upper Peninsula. The report of its occurrence in Indiana is not plausible.

Greenland to Alaska, southw. to Mass., n. Mich., and B. C.

15. **Carex pedunculàta** Muhl. Coulter says of this species, in his Catalogue, "Specimens I have examined leave no room for doubt as to its occurrence in our area," and he ascribes a record from Steuben County to Bradner and one from Noble County to Van Gorder. It is more than likely that the species occurs, or did occur, in these northern counties since it is known from Kalamazoo County, Michigan, and from Jo Daviess County, Illinois, but it must be excluded at present for lack of a confirming specimen. It should be looked for in rich beech or maple woods in the northern counties early in May as it matures early and the fruiting culms rapidly wither away.

Newf. to B. C., southw. to Va., Ill., and S. Dak.

16. **Carex lívida** (Wahl.) Willd. Reported from Clark County by Baird & Taylor and from Lake County by Pepoon. No specimens could be found and doubtless specimens forming the basis of these reports should be referred to some other species.

Sphagnum bogs, Lab. and Man. to Alaska, southw. to Conn., N. J., Mich., Idaho, and n. Calif.; also in n. Europe.

17. **Carex saltuénsis** Bailey. (*Carex vaginata* of American authors.) A boreal species reported from Lake County by Higley & Raddin and by Peattie (who ascribe the record to Hill), and by Pepoon, but no collections bearing this name could be found. A species which reaches the southern limit of its known range so much farther north is not to be expected in Indiana.

Lab. to Yukon, southw. to n. New England, n. N. Y., n. Mich., n. Minn., and B. C.

18. **Carex ormostàchya** Wiegand. (Rhodora 24: 196-197. 1922.) Deam's report in 1928 for this species from Porter County was based upon a collection (Deam no. 44381) so named by Mackenzie. This collection should be referred to *C. laxiflora*, a determination confirmed (as *C. anceps* Muhl.) by Professor Wiegand in 1935.

Que. to Minn., southw. to Mass. and Pa.

19. **Carex réctior** Mack. (N. Amer. Flora 18: 261. 1935.) (*Carex granularis* var. *recta* Dewey.) This seems questionably distinct from *C. granularis*. Mackenzie (N. Amer. Flora 18: 262. 1935.) credits it to Indiana in addition to Alabama and Louisiana but the two Indiana collections referred by him to *C. rectior* are immature. One (Deam no. 44317, Elkhart County) is so immature that it cannot be distinguished from *C. granularis* by means of his key or description; the other (Deam no. 41204, Jefferson County) is sufficiently mature to show the perigynia to be strongly ribbed and sessile, characters used by Mackenzie to distinguish *C. granularis* from *C. rectior*.

20. **Carex formòsa** Dewey. Reported from Putnam County by Grimes. The specimen upon which this report was based (Grimes no. 540, in DePauw University Herbarium) is *C. Davisii.* In Coulter's Catalogue also *C. formosa* is reported from Putnam County and the record ascribed to MacDougal. The collection which formed the basis of this report, too, should doubtless be referred to some other species.

Que. to Wis., southw. to Conn. and N. Y.; very local.

21. **Carex arctàta** Boott. Bradner reported this species from Steuben County but no specimen could be found so it must be excluded. It is known in Ohio and in southwestern Michigan (Kalamazoo County; reported also from Berrien County) so that it is quite possible that it is, or was, native in dry rich woods in northern Indiana.

Newf. to Minn., southw. to Pa. and Ohio.

22. **Carex palléscens** L. Both Pepoon and Peattie report this species from Lake County, basing the reports on a record by Hill from Berry Lake. No specimens have been seen. Smith's report from Marion County and Schneck's from the Lower Wabash Valley unquestionably must have been based upon misidentifications, and the occurrence of the species even in northernmost Indiana is very doubtful.

Newf. to Wis., southw. to N. J., Pa., and Ill.; also in Eurasia.

23. **Carex scabràta** Schwein. Reported from Lake County by Higley & Raddin and by Peattie but no Indiana specimens could be found.

N. S. to Ont. and Mich., southw. mostly in the mts. to S. C. and Tenn.

24. **Carex paupércula** Michx. A northern species reported from Pine, Lake County, by Peattie and by Pepoon (as *C. paupercula* var. *irrigua* (Wahl.) Fern.) who ascribe the record to Hill. In all probability collections upon which these reports were based, but which could not be found should be referred to *C. limosa.*

Newf. to Alaska, southw. to Pa., Minn., Colo., and Utah; also in n. Eurasia.

25. **Carex aquátilis** Wahl. A far northern and western species which has been reported from Lake, Porter, La Porte, and Marion Counties. All material forming the basis of Indiana reports should be referred to other species. *C. aquatilis* of most manuals for this area is *C. substricta* (Kükenth.) Mack.

Greenland to Alaska, southw. to Que. and in the w. mts. to N. Mex. and Calif.; also in n. Eurasia.

26. **Carex nebraskénsis** Dewey. This western sedge has been reported from Fayette, Jefferson, and Tippecanoe Counties by H. S. Jackson, apparently through the misapplication of a synonym. He lists it as the host of a rust and cites for it a correct synonym, *Carex Jamesii* Torr. But Prof. Arthur states that the rust occurs on *Carex Jamesii* Schwein., and without doubt that is the species that Jackson had.

S. Dak. and Kans. to N. Mex., Calif., and B. C.

27. **Carex crinìta** Lam. var. **gynándra** (Schwein.) Schwein. & Torr. Reported, as *Carex gynandra* Schwein., by Clark from Lake Maxinkuckee, Marshall County. Clark's specimen upon which this report was based was found in the National Herbarium and it is typical *C. crinita.*

Newf. to Wis., southw. to Fla. and La.

28. **Carex pauciflòra** Lightf. Pepoon reports this species from the Chicago region as common in bogs "southeast" (i.e. Lake or Porter Counties, Ind.), and Peattie reports it from the Calumet District (Lake County). No Indiana specimens could be found except a sheet in the herbarium of Notre Dame University bearing the label "By Mineral Springs (Porter County), Ind., J. A. Nieuwland, 1918." Since Dr. Nieuwland usually gave the exact collection date for his specimens instead of merely the year, as well as a collection number, it seems possible that this label may have been made out from memory, rather than from field notes, at a date long after the actual collection. If this were so there could be some question whether he was really certain that the specimen had been collected in Indiana. The present evidence for the occurrence of the species in the state is hardly sufficiently conclusive to admit it as a member of the Indiana flora.

Sphagnum bogs; Newf. to Alaska, southw. to Conn., Pa., and Minn., and near the Pacific coast to Wash.; also in n. Eurasia.

29. **Carex Bàileyi** Britt. (*Carex lurida* var. *gracilis* (Boott) Bailey.) Reported from Clark, Marion, and Putnam Counties. Specimens were not found but doubtless all Indiana reports were based upon incorrect determinations.

N. H. to N. Y., southw. in the mts. to Va. and Tenn.

30. **Carex comòsa** × **hystricìna** var. **Dúdleyi.** A hybrid reported from Lake County by Higley & Raddin and by Peattie. No specimens could be found.

23. ARÀCEAE Neck. ARUM FAMILY

Spadix cylindrical without an obvious spathe, borne on the side of a leaflike scape; flowers perfect, perianth present; leaves linear; rootstocks and leaves aromatic.... ....694. ACORUS, p. 277.
Spadix subtended by a spathe; leaves broader than the linear type.
Spadix globose, enveloped in a very fleshy, ovoid spathe; flowers perfect, perianth present; mature leaves large, simple, mostly 1.5-2.5 dm wide, malodorous when bruised.....................................708. SYMPLOCARPUS, p. 277.
Spadix longer than wide; flowers without a perianth; leaves, if undivided, generally less than 1.5 dm wide.
Spathes flat, divaricate, white within; spadix short-cylindric, the whole surface covered with flowers, at least the lower ones perfect.......710. CALLA, p. 277.
Spathes convolute, at least below, enveloping the spadix; spadix elongate; flowers monoecious or dioecious.
Leaves sagittate, simple; flowers covering the whole surface of the spadix...... ....747. PELTANDRA, p. 278.
Leaves not sagittate, more or less divided into 3 or more segments; upper part of spadix not flower-bearing.........................786. ARISAEMA, p. 278.

694. ÀCORUS L.

1. **Acorus Cálamus** L. SWEETFLAG. CALAMUS. Map 570. Widely distributed in the state, mostly in noncalcareous springy places along streams and about lakes. Local in the lake area and in the Tipton Till Plain and very local to rare in the unglaciated area. It is usually found in large colonies, sometimes covering acres in old stream beds. This species flowers and fruits throughout the state. In medicine, the rootstock is known as calamus.

N. S. to Ont. and Minn., southw. to Fla. and Tex.; also in Eurasia.

708. SYMPLOCÀRPUS Salisb.

1. **Symplocarpus foétidus** (L.) Nutt. (*Spathyema foetida* (L.) Raf.) SKUNKCABBAGE. Map 571. In noncalcareous springy places throughout the state although there are few records from the southwestern and unglaciated parts. While the habitat of this species is usually somewhat wetter than that of sweetflag, but otherwise similar, I have never seen them growing together. *Acorus Calamus,* however, prefers sunlight while this species prefers dense shade. The colonies vary in size, usually occupying all the available space in the habitat.

N. S. to Minn., southw. to Ga. and Iowa.

710. CÀLLA L.

1. **Calla palústris** L. WILD CALLA. Map 572. This species still occurs in La Porte County in a decadent tamarack bog about six miles west of La Porte and in Noble County in sec. 12 of Washington Township where it is found in mucky soil among *Cephalanthus* on the border of a *Chamaedaphne* bog. It was reported from two places in Noble County by Van Gorder but at both of these stations the habitat has been destroyed by drainage. It was reported in 1913 from La Porte and St. Joseph Counties

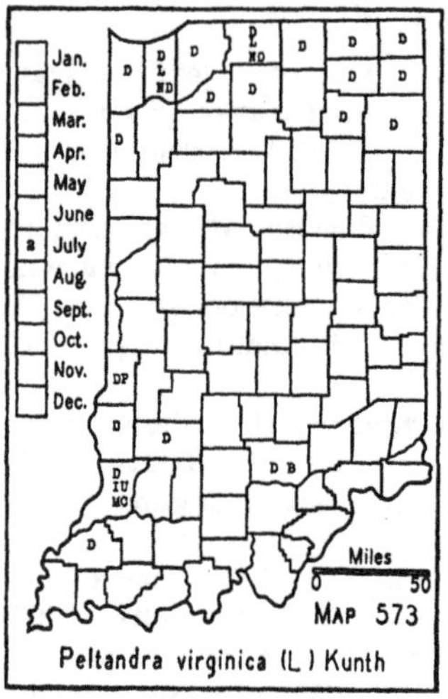

MAP 573
Peltandra virginica (L.) Kunth

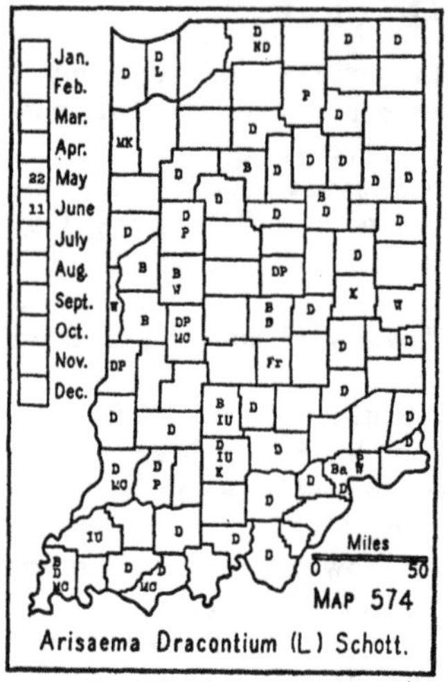

MAP 574
Arisaema Dracontium (L.) Schott.

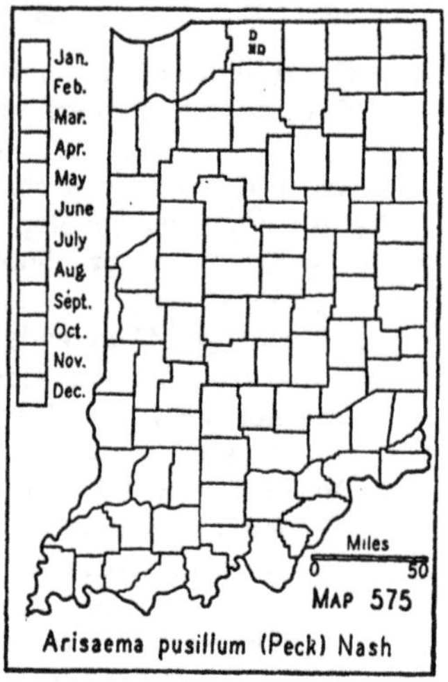

MAP 575
Arisaema pusillum (Peck) Nash

by Nieuwland, who later told me that the St. Joseph County report was an error. Peattie reported it on the authority of Nieuwland as found at Tamarack Station in Porter County, but I have not seen a specimen. There is no specimen from Porter County in the herbarium of the University of Notre Dame.

N. S. to Hudson Bay and Minn., southw. to N. J., Pa., Wis., and Iowa; also in Eurasia.

747. PELTÁNDRA Raf.

1. **Peltandra virgínica** (L.) Kunth. VIRGINIA ARROW-ARUM. Map 573. In shallow water or in wet, mucky soil on the borders of lakes and ponds and along streams. Rather frequent in the lake area but rare to absent south of this area. The leaf blades of this species are highly variable, and a wide variation can be noted between the inner and outer leaves of the same plant. Besides the typical form, Blake (Rhodora 14: 102-106. 1 pl. 1912) adds six forms, one of which has been reported from Indiana. I doubt that any of the extreme forms occur in the state although Peattie has reported f. *hastifolia* Blake from the dune region.

S. Maine to Ont. and Mich., southw. to Fla., La., and Mo.

786. ARISAÈMA Martius

Leaves pedately divided into 7-13 segments; spathe straight, narrow; spadix elongated into a caudate tip much longer than the spathe..................1. *A. Dracontium.*
Leaves 3-foliolate; lateral leaflets rarely cleft or parted; spathe hooded, rarely straight; spadix shorter than the spathe.
  Leaves green beneath; spathes purplish brown within and without; flowering about the middle of June; growing in bogs..........................2. *A. pusillum.*
  Leaves usually glaucous beneath; spathes greenish or more or less densely purplish brown within, usually green or greenish without, rarely purplish; flowering before the middle of June; not growing in bogs..............3. *A. triphyllum.*

1. **Arisaema Dracóntium** (L.) Schott. DRAGONROOT. Map 574. More or less frequent throughout the state, growing in the shade in moist, rich soil.

N. E. to Minn., southw. to Fla. and Tex.

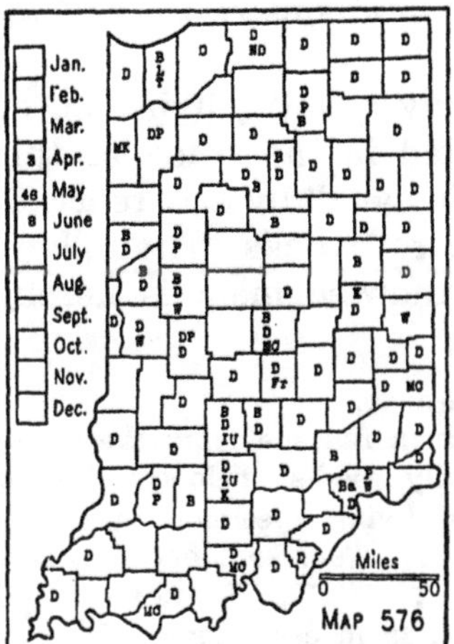

Arisaema triphyllum (L.) Schott.

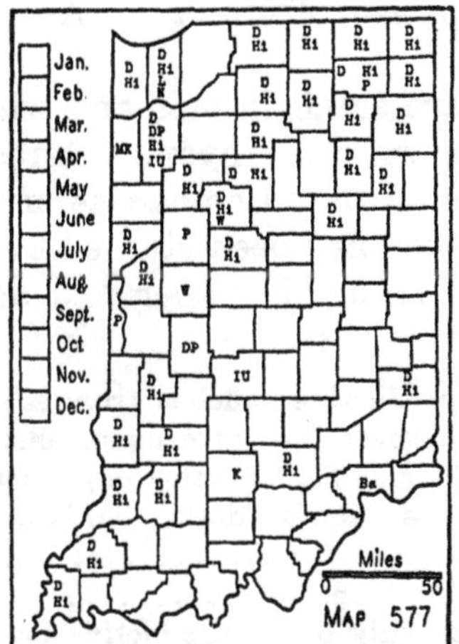

Spirodela polyrhiza (L.) Schleid.

Lemna cyclostasa (Ell.) Chevalier

2. **Arisaema pusillum** (Peck) Nash. (*Arisaema deflexum* Nieuwland & Just, Amer. Midland Nat. 12: 217-220. 1931.) Map 575. A comparison of specimens of *Arisaema deflexum* with a series of specimens of *Arisaema pusillum* from Maine, Connecticut, Pennsylvania, and New York shows no essential difference. In fact, *Arisaema pusillum* itself seems to be only an extreme form of *Arisaema triphyllum* and it is reduced to synonomy in Gray, Manual but is maintained as a species in Britton and Brown, Illustrated Flora, ed. 2. Wiegand and Eames in their flora of the Cayuga basin say: "It has not been possible to separate *A. pusillum* (Peck) Nash in this region from the species (*A. triphyllum*) by any constant characters." My opinion is that this plant as found in Indiana is only a well marked form or variety of the next species. It has been found as yet only in St. Joseph County where it grew in bogs.

Maine to N. Y. and Pa., along the coast to n. Ga. and reported in s. Mich.

3. **Arisaema triphýllum** (L.) Schott. (*Arisaema triphyllum* (L.) Torr.) JACK-IN-THE-PULPIT. Map 576. Infrequent to frequent throughout the state in moist, rich woodland. It is a shade-loving species, found from the alluvial plains to the crests of the highest ridges and seems to have no correlation with *Arisaema Dracontium* in its distribution. A study of my 69 specimens from Indiana shows that they have green and purplish spathes but very few have the hood purplish above, none flower as late as the middle of June, and none have been found in bogs. This species is extremely variable in the color of its spathe, in the shape of the blade of its hood, and in the shape of its leaflets. I have a specimen from De Kalb County with 4 leaflets and one each from Lake and Steuben Counties with the lateral leaflets parted.

N. S. to Minn., southw. to Fla., La., and Kans.

24. LEMNÀCEAE Dumort. DUCKWEED FAMILY

[Thompson, Charles Henry. A revision of the American Lemnaceae north of Mexico. Ann. Rept. Missouri Bot. Gard. 9: 1-43. 3 pl. 1898. Hicks,

Lawrence E. The Lemnaceae of Indiana. Amer. Midland Nat. **18**: 774-789. 1937.]

Plants of this family are small in size and wholly aquatic, living on or under the surface of the water. Anyone interested in the study of this family of plants should read the "Lemnaceae of Indiana" by Lawrence E. Hicks. All of my specimens have been studied by Prof. Hicks. The following text has been copied from his paper and acknowledgment is hereby made.

Plants with roots and two reproductive pouches from each node.
  Each plant of a group with several roots growing out in a fascicle from the node; plants 2.2-6.5 mm wide and 2.4-9.5 mm long, usually reddish below and with a red eye spot (the node) above; dorsal surface of living specimens a glossy green..794. SPIRODELA, p. 280.
  Each plant of a group with only one root.......... ......795. LEMNA, p. 280.
Plants without roots and with only one reproductive pouch from each node.
  Plants thick and globular..................................796. WOLFFIA, p. 281.
  Plants thin and straplike, usually submerged and attached in groups shaped like rimless wheels..796A. WOLFFIELLA, p. 282.

794. SPIRODÈLA Schleid.

1. **Spirodela polyrhìza** (L.) Schleid. GREATER DUCKWEED. Map 577. Locally abundant throughout the state in lakes, ponds, swamps, ditches, and sluggish streams. These plants are preyed upon by insects.

N. S., Ont. to B. C., southw. to Fla., Tex., and Calif.; also in Eu., Asia, and tropical Amer.

795. LÉMNA L.

Plants feather-shaped with the basal portions of the long internodes narrowed into petiolelike stems, usually submerged............................1. *L. trisulca.*
Plants oval to oblong, without petiolelike stems, connecting plants appearing sessile, usually floating.
  Shape of plants symmetrical or nearly so.
    Plants deep green, thickish, convex on both surfaces, obscurely 3-veined, cavernous throughout, appearing medium thick when pressed, sometimes reddish or purplish, especially below; margins thick........................2. *L. minor.*
    Plants usually pale green, lower surface nearly flat, obscurely 1-nerved, cavernous in the middle portions only; membranous when dried; margins thin.......... ....3. *L. minima.*
  Shape of plants unsymmetrical.
    Body of plant obliquely obovate, medium thick, usually deep green with some reddish purple, distinctly 3-veined, cavernous throughout; root sheath with lateral wing appendages.................................4. *L. perpusilla.*
    Body of plant long-oblong, thin, pale green, obscurely 1-veined; root sheath unappendaged; cavernous in the middle portions only............5. *L. cyclostasa.*

1. **Lemna trisúlca** L. SUBMERGED DUCKWEED. Map 578. Found commonly in ponds, shallow lakes, sloughs, and bogs, often growing beneath floating species, preferring cold, shaded water.

N. S., Ont. to B. C., southw. to Fla., Tex., and Calif.; also in parts of Eu., Asia, Africa, and Australia.

2. **Lemna minor** L. LESSER DUCKWEED. Map 579. Throughout the state

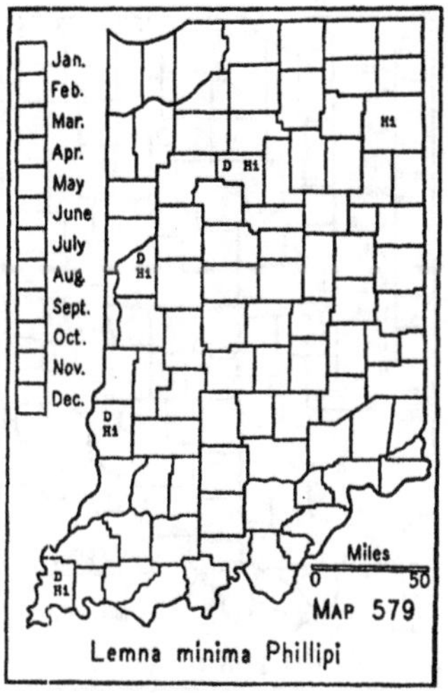

Lemna minima Phillipi

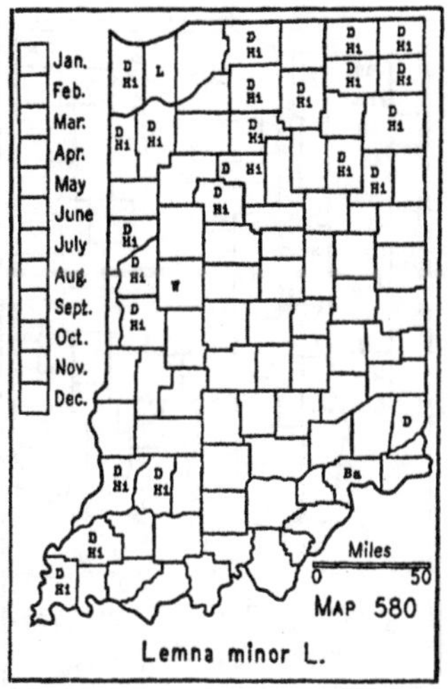

Lemna minor L.

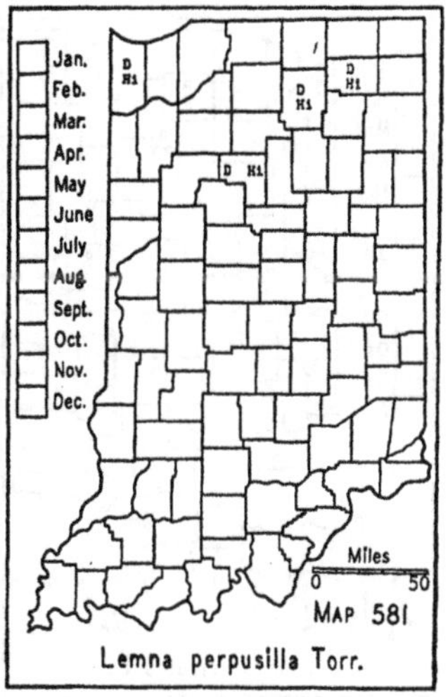

Lemna perpusilla Torr.

but more general in the lake area and in the area drained by the Wabash River.

Throughout continental America except the extreme northern part; also in Eu., Asia, Africa, and Australia.

3. **Lemna mínima** Phillipi. LEAST DUCKWEED. Map 580. The habitats are similar to those of the other species of the genus. It is known in Indiana only from Allen, Cass, and Sullivan Counties. The only Ohio record is from a pond in Paulding County within three or four miles of Allen County, Indiana.

Ohio, Ind., Minn., Wyo. to Calif., southw. to Fla., La., and Tex.; also in Mex., Cent. Amer., into S. A.

4. **Lemna perpusílla** Torr. MINUTE DUCKWEED. Map 581. Known only in the northern third of the state. The only Ohio record is from Mercer County within six miles of Randolph County, Indiana.

Mass., N. Y., Ohio, Ind., Wis., Minn. to N. Dak., southw. to Fla., Ark., and Kans.; also in S. A.

5. **Lemna cyclóstasa** (Ell.) Chevalier. PALE DUCKWEED. Map 582. This species is local in the lake area and found in organic debris in completely stagnant water in swamps and ponds.

Mass., N. Y., Ohio, Ind., Ill., Wis., Wyo. to Nev., southw. to Fla., Tex., and Calif.; also in Jamaica, Mex., Cent. Amer., and S. A.

796. WÓLFFIA Horkel

Plants globose or nearly so, not punctate, loosely cellular; upper surface convex with usually three conspicuous papules; plants not prominent above the surface of the water....................1. *W. columbiana.*

Plants more or less flattened above and gibbous beneath, brown-punctate, more compactly cellular; plants prominent on the surface of the water.

Body of plant rounded-ovate, strongly gibbous, slightly unsymmetrical; dorsal surface with a single large conical papule....................2. *W. papulifera.*

Body of plant more or less oblong with upturned acute tip (peanut-shaped), slightly gibbous, symmetrical; dorsal surface with a prominent papule near the center. ....3. *W. punctata.*

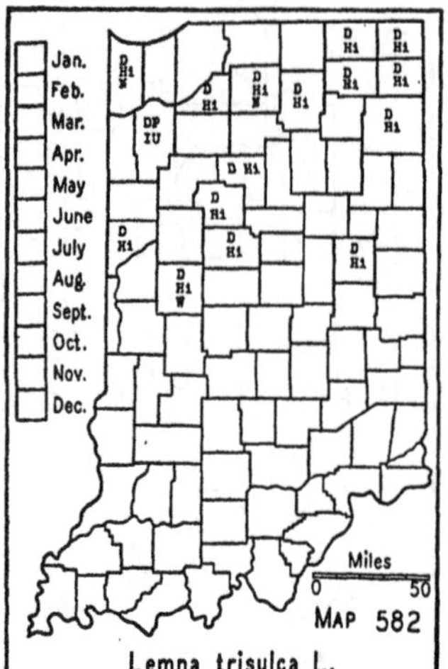

Lemna trisulca L.

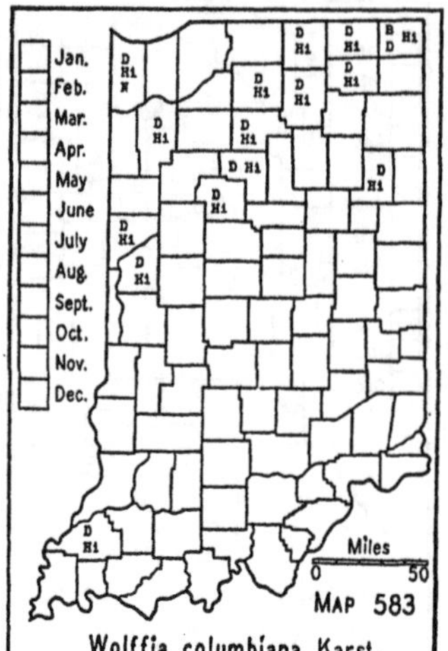

Wolffia columbiana Karst.

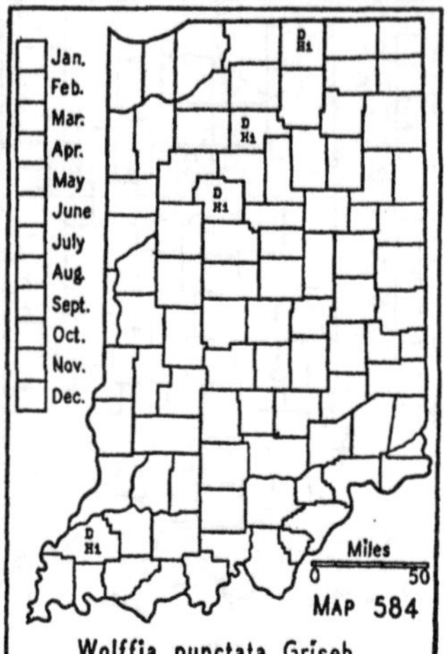

Wolffia punctata Griseb.

1. **Wolffia columbiàna** Karst. COMMON WOLFFIA. Map 583. Locally very abundant in permanently stagnant waters that abound in organic debris.

Mass., N. Y., Mich. to Minn., southw. to Fla., La., and Tex.; also in Mex., Cent. Amer., and S. A.

2. **Wolffia papulífera** Thompson. POINTED WOLFFIA. Found in isolated small colonies in permanent pools of stagnant water rich in organic matter. Known in Indiana only from Posey County. It has been found in only eight states.

Ohio, Ind., Ill., Ky., Tenn., Mo., Ark., and Kans.

3. **Wolffia punctàta** Griseb. DOTTED WOLFFIA. Map 584. Locally abundant in the habitats of the genus.

Conn., N. Y., Mich. to Minn., southw. to La. and Tex.

796A. WOLFFIÉLLA Hegelmaier

1. **Wolffiella floridàna** (J. D. Smith) Thompson. STAR WOLFFIELLA. Map 585. Restricted to wholly stagnant bodies of water and very local in the northern range of its distribution.

Ont., Mich., Wis., and Mo., southw. to Fla., La., and Tex.; also in Mex.

29. XYRIDÀCEAE Lindl. YELLOW-EYED GRASS FAMILY

826. XÝRIS [Gronov.] L. YELLOW-EYED GRASS

Base of plant bulbous; lateral sepals wingless, the keel ciliate..............1. *X. torta.*
Base of plant not bulbous; lateral sepals winged, the keel winged and erose above the middle..2. *X. caroliniana.*

1. **Xyris tórta** J. E. Smith. (*Xyris flexuosa* Muhl.) Map 586. Local in the northwestern part of the state in moist, sandy soil about lakes and in prairie habitats and fallow fields.

Maine to Minn., southw. to Ga. and Mo.

2. **Xyris caroliniàna** Walt. Map 587. In the moist, sandy borders of lakes, sloughs, and marshes. This species is very local. The fact that a few plants were found on the border of a small lake in Wells County

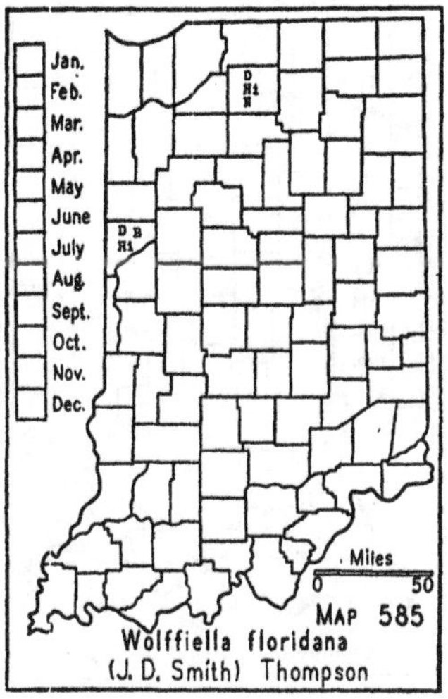

MAP 585
Wolffiella floridana (J. D. Smith) Thompson

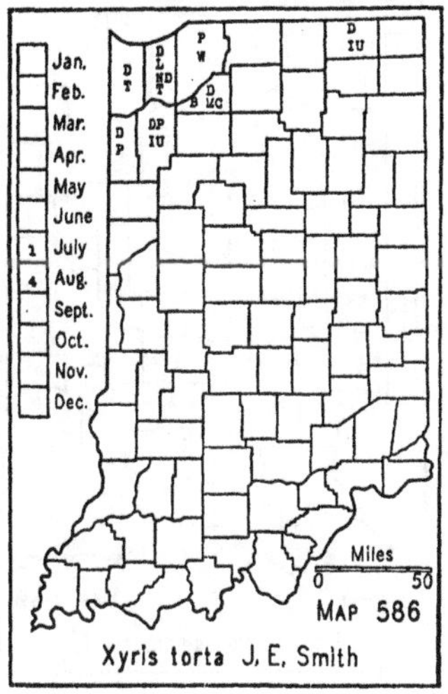

MAP 586
Xyris torta J. E. Smith

MAP 587
Xyris caroliniana Walt.

suggests that it may have been more frequent than our reports indicate because the plant is so inconspicuous.

In the Coastal Plain states from Maine to Fla. and La.; also in n. Ind. and s. Mich.

30. ERIOCAULÀCEAE Lindl. PIPEWORT FAMILY

828. ERIOCAÚLON [Gronov.] L.

1. **Eriocaulon septangulàre** With. (*Eriocaulon articulatum* (Huds.) Morong.) (Rhodora 11: 40-41. 1909.) Map 588. Local but common where found, in shallow water on the borders of lakes, usually in marly soil.

Newf. to Minn., southw. to N. J. and Ind.

33. COMMELINÀCEAE Reichenb. SPIDERWORT FAMILY

Petals unequal; perfect stamens 3; filaments naked; bracts spathelike .... 896. COMMELINA, p. 283.

Petals equal; perfect stamens 6; filaments bearded; bracts leaflike or small and scarious .... 911. TRADESCANTIA, p. 285.

896. COMMELÌNA [Plum.] L. DAYFLOWER

[Pennell. The genus Commelina [Plum.] L. in the United States. Bull. Torrey Bot. Club 43: 96-111. 1916.]

The species of this genus have not been understood, hence most of our records are of uncertain identity. I do not attempt to give the synonomy of all of our species.

Spathelike involucre open at the base, the edges not united, ciliate or minutely roughened; leaves mostly of a lanceolate or ovate-lanceolate type, 4-8 cm long.

Two posterior petals blue; anterior petal much smaller, narrow, white; capsules 2-celled; seed 2 in each cell; plants usually much branched; top of leaf sheath without long, rusty hairs .... 1. *C. communis.*

All three petals blue, the anterior one much smaller; capsules 3-celled, one cell 1-seeded and indehiscent; seed 5 .... 2. *C. diffusa.*

Spathelike involucre with the edges united at the base, the margins smooth; leaves of a linear-lanceolate or lanceolate type, or very wide (2-5 cm) and of a lanceolate or elliptic type.

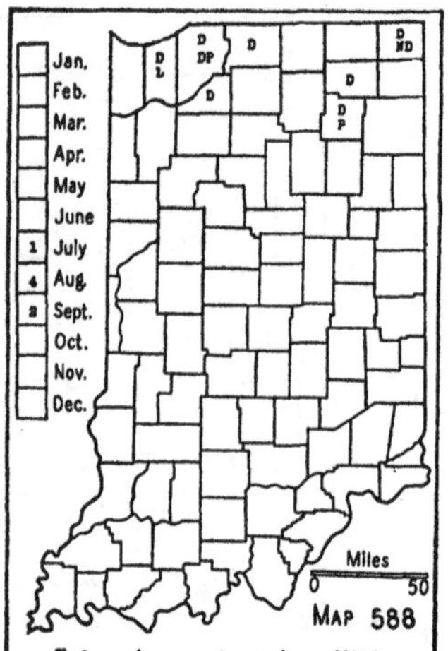

Eriocaulon septangulare With.

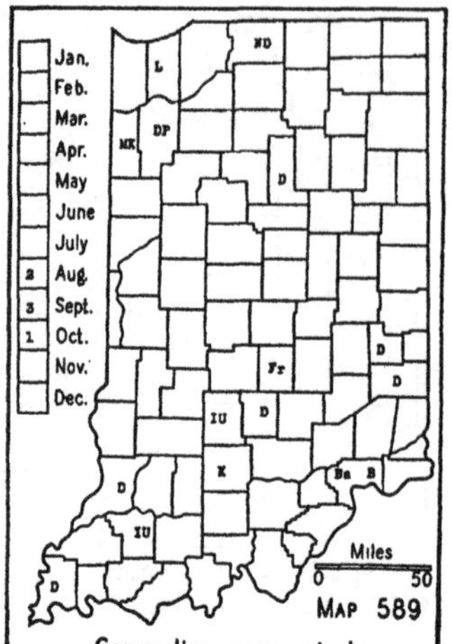

Commelina communis L.

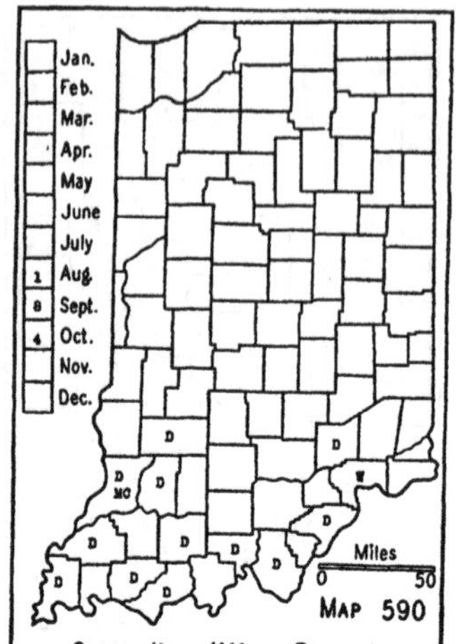

Commelina diffusa Burm. f.

Top of leaf sheath with long, rusty hairs; plants relatively robust and erect; leaves mostly lanceolate, usually 2-5 cm wide and 8-20 cm long; all petals blue and nearly equal; capsules 3-celled, one cell 1-seeded and indehiscent, the other two cells with 2 seed each, sometimes one seed aborting............3. *C. virginica.*

Top of leaf sheath without long, rusty hairs; plants rather slender, erect or ascending; leaves linear or lance-linear, smaller than those of the preceding; two petals blue, one white and much smaller; capsules 3-celled, 1 seed in each cell, 1 cell indehiscent, rotund.

Leaves linear to linear-lanceolate; posterior petals usually 10-15 (20) mm long; anterior petal about 1 mm wide; seed of an orbicular type...4. *C. angustifolia.*

Leaves lanceolate; posterior petals usually 12-25 mm long, the anterior one rudimentary, usually about 1 mm wide and 3-5 mm long..............5. *C. erecta.*

1. **Commelina commùnis** L. (Pennell. "Commelina communis" in the Eastern United States. Bartonia **19**: 19-22. 1938.) COMMON DAYFLOWER. Map 589. My specimens are mostly from moist soil along roadsides and in cultivated grounds. Three specimens were measured in the field and their measurements are as follows: blades of the blue petals 5-11 mm wide and about as long; the white one about 3 mm wide and 5 mm long.

Peattie (Amer. Midland Nat. **10**: 130. 1926. Note that in this volume there are two pages numbered 130) described and named a form of this species. He says this is a form with "the branches in whorls of 3-5 and the leaves broadly ovate; large ovate-lanceolate papery bracts accompany each verticel." Type specimen in the Field Museum, collected near Pine, Lake County, October 31, 1908, by C. W. Duesner. I have seen this specimen and I believe it is only a late autumnal form of the species.

Nat. of e. Asia; Mass. to N. C., westw. to Mo., Kans., and Tex.

2. **Commelina diffùsa** Burm. f. (Jour. Arnold Arb. **18**: 64-65. 1937.) (*Commelina longicaulis* Jacq. and *Commelina nudiflora* of Britton and Brown, Illus. Flora, ed. 2.) Map 590. In moist, wet, or muddy places along streams and in ditches and cultivated grounds in the southern part of the state. The petals are very variable in size. Five specimens were measured in the field and the measurements are as follows: the blades of the largest petals ranged from 2.25-10 mm wide and about as long, the smallest were

about 1.25-4 mm wide and nearly as long. This is a tropical species that ranges northward to the southern part of this state.

N. J. to Kans., southw. to Fla. and Tex., and in tropical Amer.

3. **Commelina virgínica** L. (*Commelina hirtella* Vahl.) VIRGINIA DAYFLOWER. Map 591. Found only in the southern part of the state in wet woods and sloughs and along streams. This is our largest species and usually forms colonies. No doubt all early reports of this species for the state should be referred to some other species. *Commelina erecta* of Gray, Man., ed. 5 is a synonym of this species and Coulter's and Young's reports for it from Jefferson County should be referred to *Commelina virginica* L.

Pa. to Kans., southw. to Fla. and Tex.

4. **Commelina angustifòlia** Michx. NARROWLEAF DAYFLOWER. Map 592. This species grows in almost pure, fine sand and is found on sand hills along roadsides, on high, sandy banks of lakes and streams, and on the open dunes about Lake Michigan. Three specimens were measured in the field and the blades of the posterior petals averaged from 17-18 mm wide and 14-17 mm long and the anterior or white petals averaged about 1 mm wide.

N. C. to Ind., southw. to Fla. and Tex.; also in Cuba.

5. **Commelina eréсta** L. (Including the reports of *Commelina crispa* Wooton from Indiana.) I found a specimen along the roadside 2 miles west of Yankeetown, Warrick County, which I am referring to this species. Pennell (Bull. Torrey Bot. Club 43: 107. 1916) reported two specimens from the dunes about Lake Michigan as *Commelina crispa* Wooton and I am including them in this species. The name of this species should not be confused with the same name applied to other species by early authors.

N. Y. to Kans., southw. to Fla. and Tex.

911. TRADESCÁNTIA [Rupp.] L. SPIDERWORT

[Anderson and Woodson. The species of Tradescantia indigenous to the United States. Contr. Arnold Arboretum 9: 1-132. 1935.]

Plants glaucous, essentially glabrous throughout, robust, mostly of a dry, sandy habitat; flowering from the first of June through the summer; sepals glabrous or with a few hairs at the apex, 8-15 mm long; pedicels 1-1.5 cm long.......... 1. *T. canaliculata*.

Plants not glaucous, more or less pubescent throughout; woodland species of a moist or dry habitat; sepals pubescent, rarely glabrous.

  Plants dwarf, generally less than 1 dm high, rarely 1.5 dm high, covered all over with long, weak hairs; sepals tinted with pink, about 1 cm long. (See excluded species no. 127, p. 1033.)........ *T. brevicaulis*.

  Plants usually more than 1 dm high, not covered all over with long, weak hairs; sepals very green.

    Stems flexuous, at least above the lowest inflorescence; leaves lanceolate, the median ones usually 2-5 cm wide; flowering from the first of June until frost; cymes both terminal and axillary; sepals mostly 6-8 mm long; pedicels 1-1.5 cm long........ 2. *T. subaspera*.

    Stems not flexuous; leaves linear or linear-lanceolate, the median ones less than 2 cm wide; sepals 8-15 mm long.

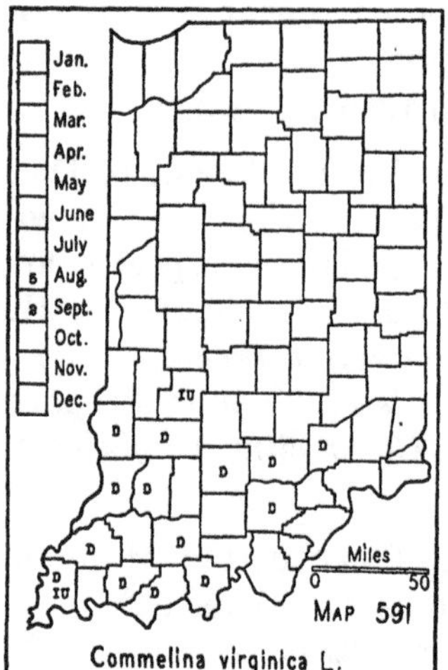

Commelina virginica L.

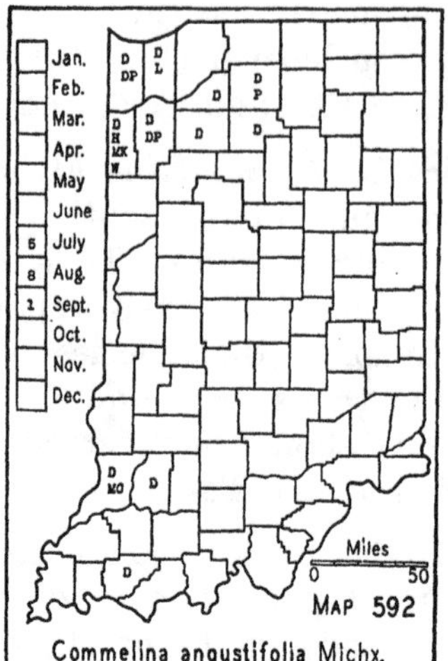

Commelina angustifolia Michx.

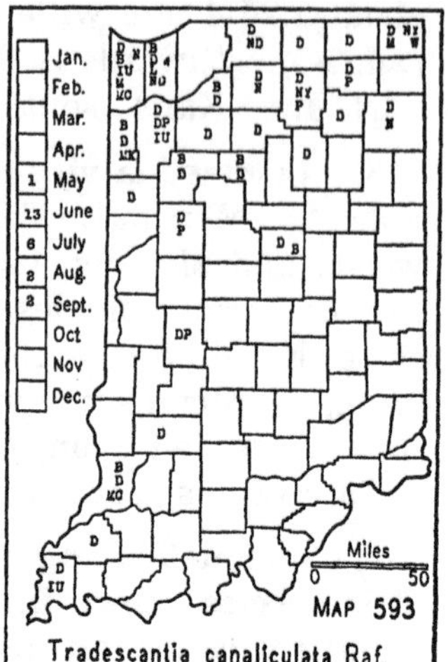

Tradescantia canaliculata Raf.

Pubescence of pedicels and sepals non-glandular; plants of dry woodland, flowering from the last of April to the first of June and usually soon dying down; cymes mostly terminal, rarely both terminal and axillary; pedicels 1.5-4 cm long...3. *T. virginiana.*

Pubescence of pedicels and sepals glandular. (See excluded species no. 126, p. 1032.)..*T. bracteata.*

1. **Tradescantia canaliculàta** Raf. (*Tradescantia reflexa* Raf. of Gray, Man., ed. 7 and Britton and Brown, Illus. Flora, ed. 2.) GLAUCOUS SPIDERWORT. Map 593. This species prefers the open and is generally found in dry, sandy or gravelly soil, along roadsides, on sand hills and high banks of lakes, and on the dunes. It is rarely found in swampy places but is frequent in moist, prairie habitats. This is a rank growing species with several color forms which have been named and which persist under cultivation.

N. C., Ohio to Minn., southw. to Fla. and Tex.

1a. **Tradescantia canaliculata** f. **albiflòra** (Slavin & Nieuwl.) comb. nov. (*Tradescantia reflexa* f. *albiflora* Slavin & Nieuwl. Amer. Midland Nat. **11**: 600. 1929.) This is a white-flowered form which is rather frequent where the species is found.

1b. **Tradescantia canaliculata** f. **Lésteri** (Standley) comb. nov. (*Tradescantia reflexa* f. *Lesteri* Standley. Rhodora 32: 32. 1930.) This is a form with "poppy-red" colored flowers which was found near Tremont, Porter County, by Lester A. Beatty.

1c. **Tradescantia canaliculata** f. **Maríae** (Standley) comb. nov. (*Tradescantia reflexa* f. *Mariae* Standley. Rhodora 32: 32. 1930.) This form with white petals, margined with lilac was found near Fowler, Benton County, by Mary Bremer.

2. **Tradescantia subáspera** Ker var. **týpica** Anderson & Woodson. (Contr. Arnold Arboretum 9: 49. 1935.) (*Tradescantia pilosa* Lehm.) ZIGZAG SPIDERWORT. Map 594. Usually infrequent but well distributed throughout the state except in the northern part from which there are no

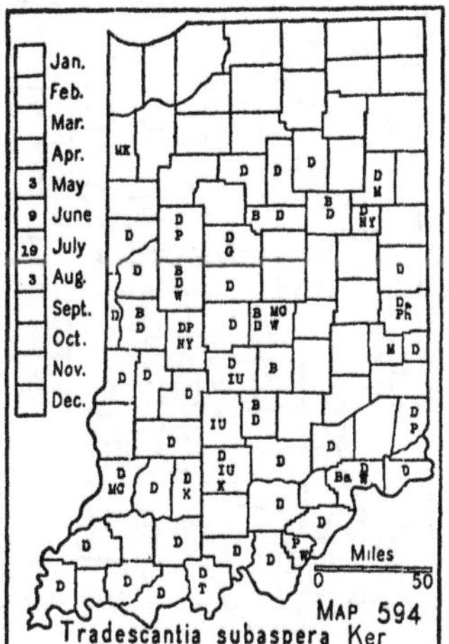

Tradescantia subaspera Ker var typica Anderson & Woodson

Tradescantia virginiana L.

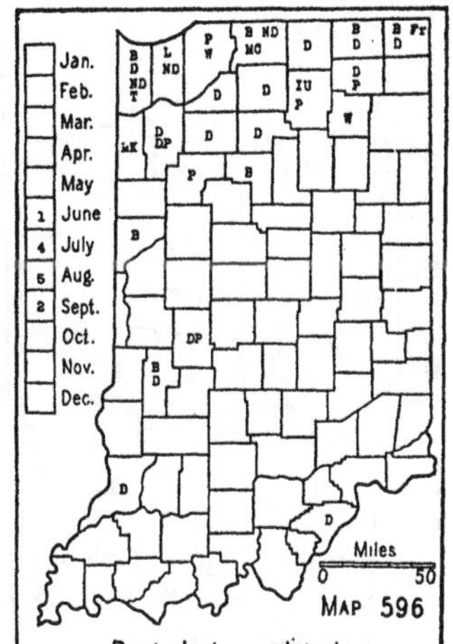

Pontederia cordata L

records or specimens. It is a woodland species and is rarely found in the open. It prefers the moist, rich, wooded terrace slopes along streams and the slopes of ravines and is less frequent in level woods.

Pa. to Kans., southw. to Fla. and La.

3. **Tradescantia virginiàna** L. VIRGINIA SPIDERWORT. Map 595. Infrequent but well distributed in the southern two thirds of the state, becoming less frequent to very rare in the northern counties. This is a woodland species and is rarely found in the open. It is usually found in dry clayey soil in white oak, white oak and black oak, and beech and sugar maple woods. White and rose colored forms are sometimes found and they persist under cultivation.

Southern N. Y. to S. Dak., southw. to Va., Ky., and Ark.

34. PONTEDERIÀCEAE Dumort. PICKERELWEED FAMILY

[Moldenke. Pontederiaceae of North America. N. Amer. Flora **19**: 51-60. 1937.]

Plants erect; leaves large, cordate to lanceolate; flowers blue, 2-lipped; stamens 6; utricle 1-seeded..922. PONTEDERIA, p. 287.

Plants floating or prostrate on mud; leaves linear, very narrow or reniform; flowers yellow, white or pale blue; perianth salver-shaped; stamens 3; capsule many-seeded. ....924. HETERANTHERA, p. 288.

922. PONTEDÈRIA L.

[Fernald (Rhodora **27**: 80. 1925) gives a key to the "Pontederias of temperate North America," which is copied here in part.]

Leaves cordate at base.

Leaves narrowly deltoid-ovate, tapering with straight sides from the base to the apex..1. *P. cordata.*

Leaves broadly ovate, gradually curved from the broad base to the blunt summit.... ....1a. *P. cordata* f. *latifolia.*

Leaves truncate to tapering at base, narrowly deltoid to linear-lanceolate............ ....1b. *P. cordata* f. *angustifolia.*

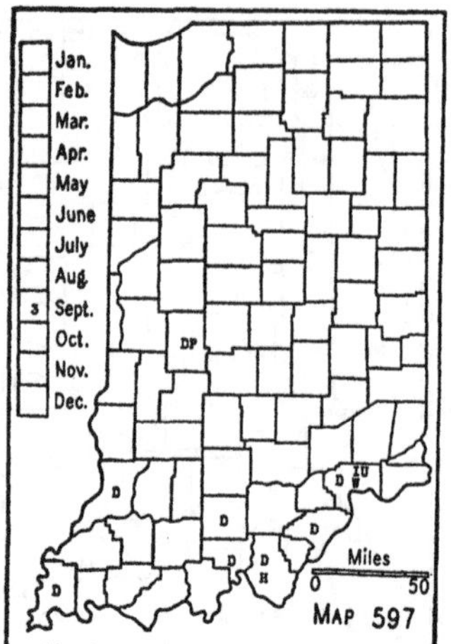

Heteranthera reniformis R. & P.

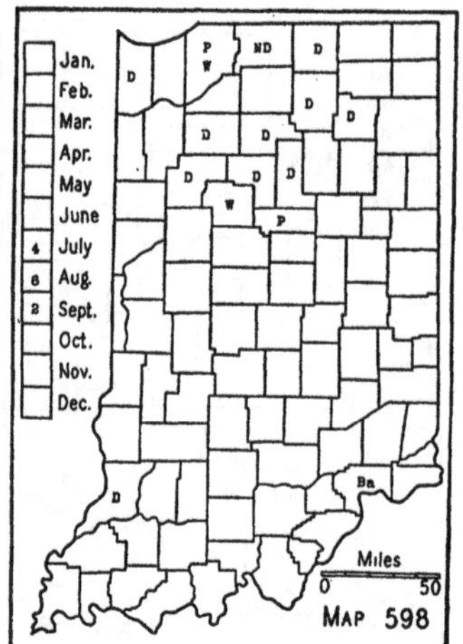

Heteranthera dubia (Jacq.) MacM.

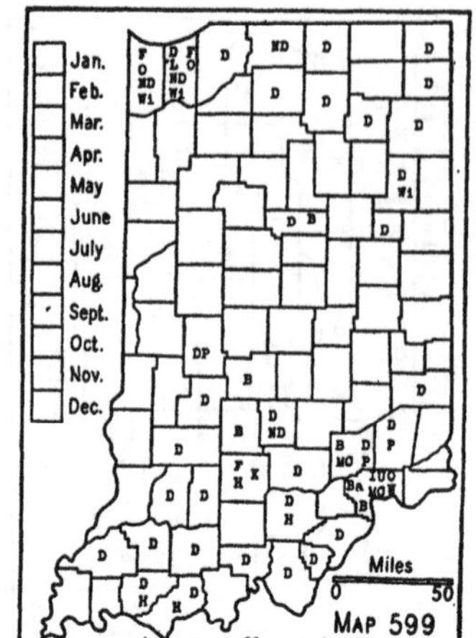

Juncus effusus L. var. solutus Fern. & Wieg.

1. **Pontederia cordàta** L. PICKERELWEED. Map 956. This species is frequent throughout most of the lake area but is absent or very local south of it. It must have its base in water most of the time, but otherwise it does not seem particular as to where it grows. It seems to grow nearly as well in almost pure marl areas as in sandy, muddy, and mucky borders. However, I find the largest specimens in mucky borders of lakes. The trimorphic flowers of this species are interesting to one who can give the time to their study.

N. S. to Minn., southw. to Va. and Tex.

1a. **Pontederia cordata** f. **latifòlia** (Farw.) House. This form, in the extreme, is well marked but our specimens seem to intergrade so much that it is a question whether the two forms should be maintained. The range is the same as that of the species.

1b. **Pontederia cordata** f. **angustifòlia** (Pursh) Solms-Laubach. This form is distinctly marked but I am not certain that another form may not be on the same rhizome. On the low, marly shore of the southeast side of Simonton Lake, Elkhart County, I made a study of this form. I found it farther out in the lake in a zone of water a little deeper than where the species grew. The species grew in a dense stand while the form was not so dense. I did not realize, until recently, the significance of this form although I have found it in several counties. Rather rare in Indiana.

P. E. I. to Wis., southw. to Ind. and southeast of the Appalachian Mts.

924. HETERANTHÈRA R. & P. MUD PLANTAIN

Leaves reniform; flowers white or pale blue........................1. *H. reniformis.*
Leaves linear; flowers yellow...2. *H. dubia.*

1. **Heteranthera renifórmis** R. & P. MUD PLANTAIN. Map 597. Very local in ponds in the southern counties. I have found it on the muddy shore of Hovey Lake, Posey County and elsewhere in natural and artificial ponds.

There is a specimen in the herbarium of DePauw University which was collected by D. T. MacDougal in Putnam County, Sept. 12, 1889.

Conn., Nebr., southw. to Ga. and Tex.; also in W. I. and to Cent. Amer.

2. **Heteranthera dùbia** (Jacq.) MacM. WATER STARGRASS. Map 598. Rather frequent in the lake area on the shallow borders of lakes, in the Tippecanoe and St. Joseph Rivers, and on sandy bars and mud flats along streams, becoming rare in the southern part of the state. It is so inconspicuous that it is usually overlooked. Sometimes it grows in great masses with such acquatic plants as *Potamogeton* and *Utricularia.* The deepest water from which I have a specimen is 4 feet but I know that it grows in even deeper water. It is most conspicuous on muddy flats in late autumn when the water of its habitat recedes.

Que. to Oreg., southw. to N. C., Ark. and Ariz.

36. JUNCÀCEAE Vent. RUSH FAMILY*

Plants glabrous; capsule usually 3-celled, with very numerous minute seeds.......... ....936. *Juncus*. p. 290.
Plants hairy or arachnoid; capsule 1-celled, with 3 large seeds....997. *Luzula*, p. 300.

936. JÚNCUS [Tourn.] L.

Inflorescence apparently growing from the side of the culm, the involucral bract terete, erect and appearing like a continuation of the culm; culm leaves reduced to bladeless sheaths. (Section GENUINI.)
Stamens 3, opposite the sepals; inflorescence greenish or stramineous; rootstocks short-creeping with inconspicuous internodes; culms densely cespitose.
Sepals not exceeding the petals, rarely exceeding the capsule, 2.5-3.5 mm long, slightly if at all spreading, about the same color as the capsule; culms not sulcate...1. *J. effusus* var. *solutus*.
Sepals exceeding both the petals and the capsule, 2.7-4 mm long, more rigid and spreading, usually lighter in color than the capsule; culms sulcate below the inflorescence.....................................1a. *J. effusus* var. *Pylaei*.
Stamens 6, opposite the sepals and petals; inflorescence dark brown at maturity; rootstocks long-creeping with conspicuous internodes; culms usually well separated, arising in a single row.
Inflorescence not diffuse, 1.5-3.5 cm long; flowers approximate or subapproximate. ....2. *J. balticus* var. *littoralis*.
Inflorescence diffuse, 4-12 cm long; flowers widely separated.................. ....2a. *J. balticus* var. *littoralis* f. *dissitiflorus*.
Inflorescence obviously terminal or, if not, the involucral bracts flat or channeled along the upper side; culm leaves with well developed blades.
Leaves flat, or in age involute, not septate (in *J. Greenei* terete but not septate).
Flowers borne singly on the branches of the inflorescence, not in heads, each with a pair of bracteoles at the base in addition to the bractlet at the base of the pedicel. (Section POIOPHYLLI.)
Inflorescence more than half the height of the plant; flowers scattered along the loose forking branches; annual.............................3. *J. bufonius*.
Inflorescence much less than half the height of the plant; perennial.
Perianth segments obtuse, appressed; leaf sheaths covering half of the stem or more..4. *J. Gerardi*.
Perianth segments acute or acuminate, usually more or less spreading; leaf sheaths covering a fourth of the stem or less.
Leaves terete; capsule much exceeding the perianth, reddish or castaneous. ....5. *J. Greenei*.
Leaves flat; capsule little if at all exceeding the perianth, green to stramineous or dull brown.
Bracts shorter than the inflorescence; flowers conspicuously secund on the usually incurved branches; capsule 3-celled; leaves usually less than a third the height of the culms.................6. *J. secundus*.
Bracts (at least the lowermost) exceeding the inflorescence; flowers not conspicuously secund; capsule 1-celled or imperfectly 3-celled; leaves usually about half the height of the culms.
Auricles at the summit of the sheaths very thin, white, and scarious, conspicuously produced beyond the point of insertion, 1-3.5 mm long; bracteoles blunt.
Flowers mostly clustered at the tips of the branches......7. *J. macer*.
Flowers scattered or somewhat secund along the branches.

* Contributed by Frederick J. Hermann, University of Michigan.

Ultimate floriferous branchlets widely spreading, 0.5-2 cm long.... .... 7a. *J. macer* f. *Williamsii*.

Ultimate floriferous branchlets elongate and ascending.

Ultimate floriferous branchlets rarely over 4 cm long; sepals and petals mostly subequal; capsule averaging three fourths the length of the acuminate sepals; plant generally stout.... ....7b. *J. macer* f. *anthelatus*.

Ultimate floriferous branchlets often 7 cm long; sepals attenuate-subulate, usually conspicuously longer than the petals; capsule averaging half the length of the sepals; plant generally slender, often lax........7c. *J. macer* f. *discretiflorus*.

Auricles at the summit of the sheaths firm, not conspicuously produced beyond the point of insertion.

Bracteoles acuminate to aristate; auricles with the very slightly produced portion membranaceous, not rigid (easily broken), stramineous, often tinged with brown or light red, occasionally somewhat cartilaginous along the sides below the summit; inflorescence generally loose; perianth from appressed to slightly spreading........8. *J. interior*.

Bracteoles blunt to acute; auricles cartilaginous, yellow, becoming brown with age, very rigid and glossy, especially the short produced portion; inflorescence generally compact; perianth widely spreading............9. *J. Dudleyi*.

Flowers in heads, not bracteolate, i.e., with only the bractlet at the base of the pedicel. (Section GRAMINIFOLII.)

Stamens not exserted in fruit; perianth exceeding the obovate, usually dull capsule; heads few (2-20), flowers many (5-10) in a head; culms cespitose, bulbous at base.......................................10. *J. marginatus*.

Stamens exserted in fruit; perianth usually shorter than the ovoid, shining capsule; heads numerous (20-100); flowers few (2-6) in a head; culms solitary or few together from an elongate, nodulose rhizome; plant taller and coarser.

Inflorescence loose; heads remote, 2-3 (rarely 6)-flowered......11. *J. biflorus*.

Inflorescence compact; heads approximate, 3-6-flowered.................... ....11a. *J. biflorus* f. *adinus*.

Leaves terete, hollow, septate. (Section SEPTATI.)

Seeds with tail-like appendages.

Heads few to many; flowers 5-50 in a head; flowers with mature fruit about 4 mm long; perianth segments subulate-tipped; capsule equaling or moderately exceeding the calyx; seed (including tails) 1-1.8 mm long, with conspicuous tails......................................12. *J. canadensis*.

Heads numerous in a diffuse panicle; flowers 3-5 in a head; flowers with mature fruit 2.5-3.5 mm long; perianth segments obtuse or nearly so, scarious-margined, less rigid; capsule usually much exceeding the calyx; seed (including tails) barely 1 mm long, the tails very short.............. ....13. *J. brachycephalus*.

Seeds without tail-like appendages.

Stamens 3, opposite the sepals.

Capsule tapering evenly to the tip or subulate-beaked, distinctly exceeding the calyx.

Heads numerous; flowers 2-7 in a head; inflorescence very large and diffuse; capsule gradually attenuate.............14. *J. diffusissimus*.

Heads few; flowers very numerous in a head; capsule subulate.......... ....15. *J. scirpoides*.

Capsule obtuse or acute at the apex, from shorter than to slightly exceeding the calyx.

Capsule half to two thirds as long as the calyx; sepals rigid, subulate, much longer than the petals; heads spherical; culms from thick, elongate rhizomes............................16. *J. brachycarpus.*

Capsule nearly equaling or exceeding the calyx; sepals and petals subequal; heads usually hemispherical; culms cespitose, not rhizomatous.

Heads 1-50; flowers several to many in a head; perianth 3-3.5 mm long; capsule acute or mucronate...................17. *J. acuminatus.*

Heads 200-500; flowers few in a head; perianth 2-2.5 mm long; capsule shorter, broader, much less rigid, blunt; nodes fewer, less conspicuous, of the same color as the culm.............18. *J. nodatus.*

Stamens 6.

Flowers solitary or in pairs, often reduced to fascicles of small leaves..19. *J. pelocarpus.*

Flowers more numerous, in heads.

Heads spherical, few, large (7-15 mm wide); capsule subulate; sepals subulate; involucral bract usually exceeding the inflorescence.

Plant low, 1-4 dm high; leaf blades erect or ascending; flowers 3-4 mm long; petals usually equaling or exceeding the sepals..20. *J. nodosus.*

Plant taller, 4-10 dm high; leaf blades divaricate; flowers 4-5 mm long; petals usually shorter than the sepals..............21. *J. Torreyi.*

Heads hemispherical, more numerous, smaller (6 mm wide or less); capsule ovoid or ellipsoid; sepals blunt or acute; involucral bract shorter than the inflorescence.

Sepals and petals acute or acuminate; capsule strongly acute; branches of the inflorescence usually widely spreading....22. *J. articulatus.*

Sepals and petals mostly obtuse, often scarious at the apex; capsule from obtuse to broadly acute or apiculate; branches of the inflorescence rarely widely spreading.

Heads loosely few-flowered, usually with one or more flowers elevated on slightly prolonged peduncles; branches of the inflorescence erect or strongly ascending...23. *J. alpinus* var. *rariflorus.*

Heads compactly and regularly several- to many-flowered; branches of the inflorescence spreading-ascending..23a. *J. alpinus* var. *fuscescens.*

1. **Juncus effùsus** L. var. **solùtus** Fern. & Wieg. (Rhodora 12: 90. 1910.) Map 599. Very common in ditches, sloughs, low fields, wet open woods, marshes, bogs and on borders of lakes. Often locally abundant.

N. S. to Wis., southw. to Fla. and Tex.

1a. **Juncus effusus** var. **Pỳlaei** (Laharpe) Fern. & Wieg. (Rhodora 12: 92. 1910.) Map 599a. Infrequent in the northern part of the lake area, except on the dunes where it is frequent. A northern variety growing in habitats similar to the preceding variety and reaching the southern limit of its range in northern Indiana.

Newf. to Wis., southw. to W. Va. and Ind.

2. **Juncus bálticus** Willd. var. **littoràlis** Engelm. Map 600. Infrequent in the dune area where it is found on the sandy borders of sloughs and lakes, in interdunal swales and marshes, and in moist depressions of the sandy beach of Lake Michigan. The elongate rootstocks of this rush, and of the following form, usually radiate in many directions from a common center and often attain a length of a yard or even several yards.

Newf. to N. Y., Pa., and the Great Lakes.

Juncus effusus L. var. Pylaei (Laharpe) Fern. & Wieg.

Juncus balticus Willd. var. littoralis Engelm.

Juncus balticus var. littoralis f. dissitiflorus Engelm.

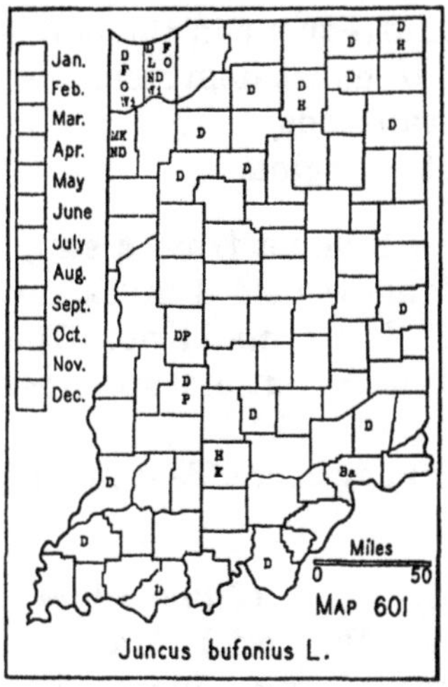

Juncus bufonius L.

Juncus Gerardi Loisel

Juncus Greenei Oakes & Tuckerm.

2a. **Juncus balticus** var. **littoralis** f. **dissitiflòrus** Engelm. (Rhodora 25: 208. 1923.) Map 600a. Confined to the dune area where it grows in the habitats of the variety but is much more common.

Range of the variety but more common inland.

3. **Juncus bufònius** L. Map 601. Common on sandy or clay roadsides and abandoned roads in open woods or marshes; frequent on low sandy lake shores, in ditches, sandy swales, and low fallow fields.

Almost throughout North America; cosmopolitan.

4. **Juncus Gerárdi** Loisel. Map 602. A Coastal Plain species which Mr. C. M. Ek found established in Howard County. He reports a colony about 5 by 10 feet (July 20, 1935) on dry open ground along the Nickle Plate Railway 4 miles east of Kokomo. It is doubtless introduced here. In the "Flora of the Indiana Dunes" by Peattie the species is reported from Lake County but no specimens could be found.

Along the coast, mostly in salt marshes, Newf. to Fla.; also on the nw. Pacific coast, in Eurasia, and n. Africa.

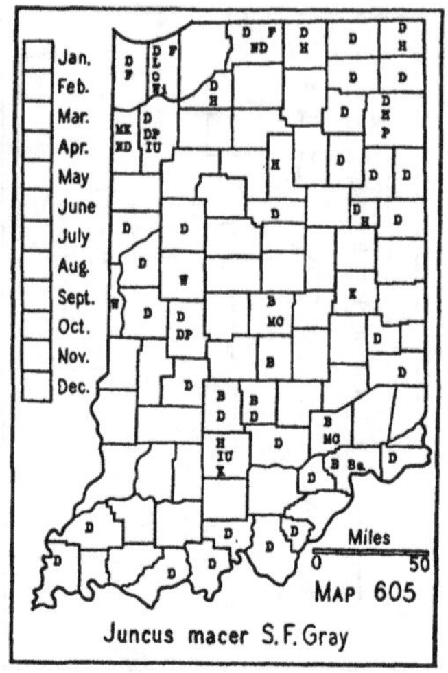

5. **Juncus Grèenei** Oakes & Tuckerm. Màp 603. Infrequent in the northwestern counties in sandy soil along low roadsides, in moist depressions on the dunes, and especially in prairie habitats along railroads.

Maine to Vt. and N. J.; locally in the Great Lakes region.

6. **Juncus secúndus** Beauv. Map 604. Known in Indiana from a single collection: wet clay border of a cattail pond in a fallow field 3 miles east of Livonia, Washington County, June 17, 1935, F. J. Hermann no. 6705. It has been reported from Putnam County by Wilson but no specimen could be found to substantiate the report.

Maine to Vt. and N. C., and in the Mississippi Valley from Tenn. to Ill. and Mo.

7. **Juncus màcer** S. F. Gray. (Jour. Bot. **68**: 367. 1930.) (*Juncus tenuis* of authors, not Willd.; including *Juncus monostichus* Bartlett.) Map 605. Very common in fields, pastures, ditches, open woods, waste places, and especially in paths and on roadsides; infrequent on banks of streams and in swampy habitats.

Juncus monostichus (originally described from Madison County) is a pathologic phase of this species in which the peculiar form of the inflorescence, the shortening of the capsules, and the tendency toward sterility are induced by a fungus infection.

Deam no. 55051 is exceptional in having the auricles scarcely prolonged, the inflorescence much congested and perianth unusually large. Intermediates between the species and its forms are frequent; thus Deam nos. 44784 and 53949, Peattie no. 2102, Lansing no. 2730, and Bechtel no. 13381 approach f. *Williamsii;* Deam no. 25456 approaches f. *anthelatus;* and Deam no. 24 approaches f. *discretiflorus.*

Almost throughout North America; adventive in Europe, South America, and Australia.

7a. **Juncus macer** f. **Williámsii** (Fern.) Hermann. (Rhodora **40**: 82. 1938.) (*Juncus tenuis* var. *Williamsii* Fern. and *Juncus macer* var.

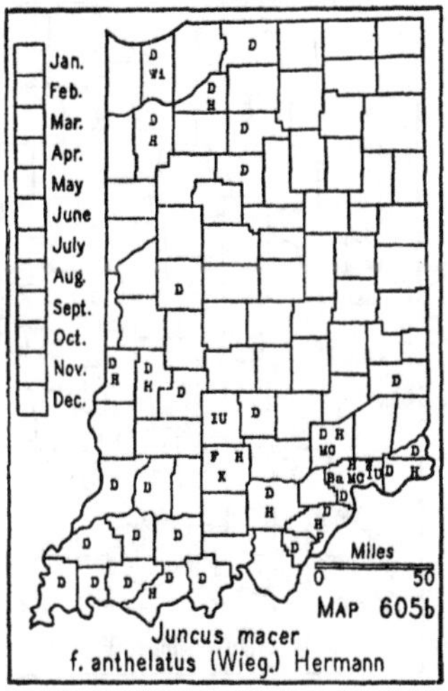

Juncus macer f. anthelatus (Wieg.) Hermann

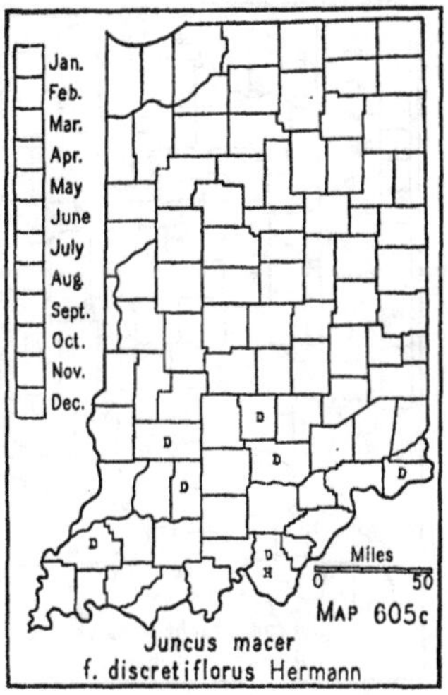

Juncus macer f. discretiflorus Hermann

Juncus interior Wieg.

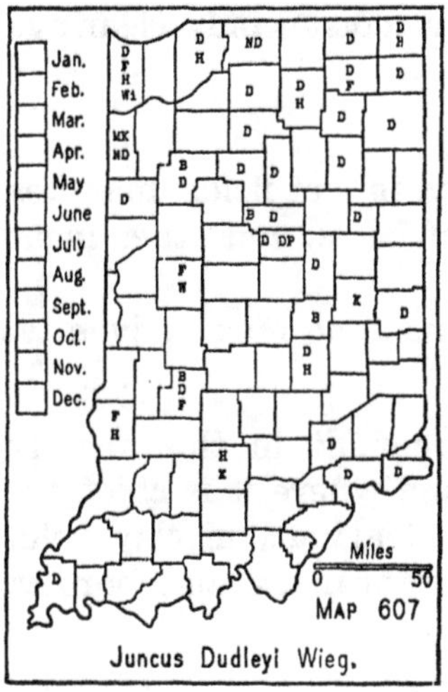

Juncus Dudleyi Wieg.

Juncus marginatus Rostk.

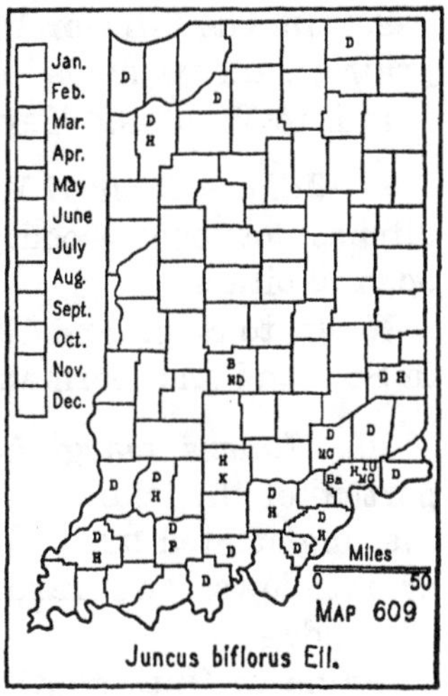

Juncus biflorus Ell.

Williamsii Fern.) Map 605a. Sporadic but infrequent in the habitats of the species.

Local but range apparently that of the species.

7b. **Juncus macer f. anthelàtus** (Wieg.) Hermann. (Rhodora **40**: 81. 1938.) (*Juncus tenuis* var. *anthelatus* Wieg. and *Juncus macer* var. *anthelatus* (Wieg.) Fern.) Map 605b. Common in most of the habitats of the species but usually in wetter soils. It is more often found in ditches and low fallow fields and on borders of swamps or ponds than is the species and much less frequently along paths or dry roadsides.

Range apparently that of the species except probably absent from arid regions.

7c. **Juncus macer f. discretiflòrus** Hermann. (Rhodora **40**: 82. 1938.) Map 605c. Frequent in southern Indiana in low woods and swamps and on wet or moist clay roadsides and banks of streams.

Southern Ind.; doubtless also in Ky., s. Ohio, and s. Ill.

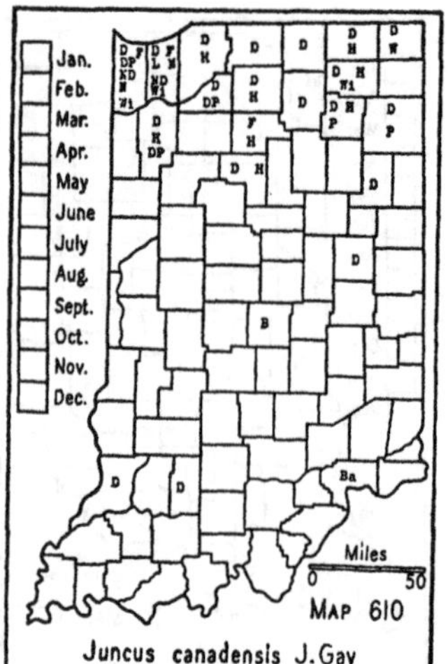

Juncus canadensis J. Gay

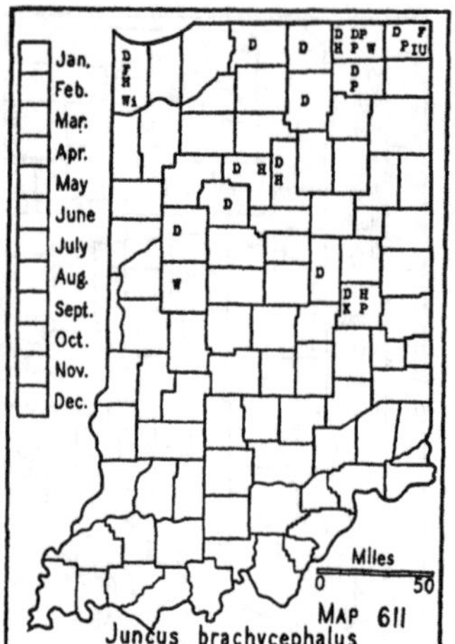

Juncus brachycephalus (Engelm.) Buchenau

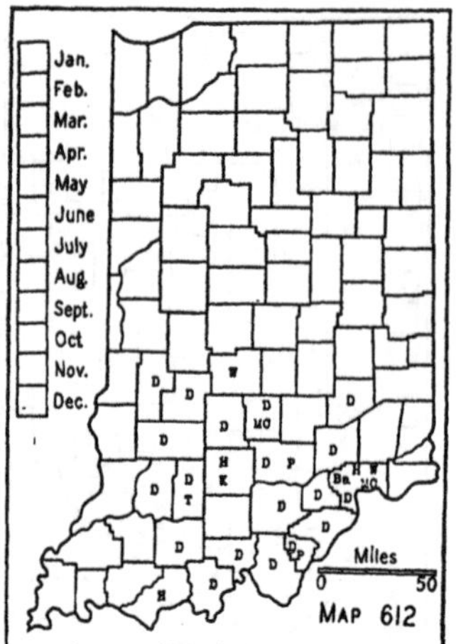

Juncus diffusissimus Buckley

8. **Juncus intèrior** Wieg. Map 606. Frequent in moist sandy clearings, prairies, fallow fields, open oak flats, and ditches.

Ind. to Wyo. and Tex.

9. **Juncus Dúdleyi** Wieg. Map 607. Very common in wet fields, marshes, ditches, low open woods, sandy or marly borders of lakes, and other moist open habitats.

Newf. to Sask. and Wash., southw. to Tenn., Kans., and Mex.; adventive in Scotland and Germany.

10. **Juncus marginàtus** Rostk. Map 608. Frequent in the western portion of the lake area and also in southern Indiana where it is chiefly in the unglaciated area. It is found in moist sandy clearings, in clay fields or meadows, and rarely in marshes and on low prairies and borders of ponds.

Maine to Ont., southw. to Fla. and Nebr.

11. **Juncus biflòrus** Ell. (Rhodora 37: 156. 1935.) (*Juncus aristulatus* of authors, not Michx. and *Juncus marginatus* var. *biflorus* (Ell.) Wood.) Map 609. Common in southern Indiana in hard white clay soils of low fallow fields and grassy meadows, in roadside ditches, and rare in open flat woods; infrequent in the lake area in moist open sandy or gravelly habitats, especially on borders of lakes.

Mass. to Mich., southw. to Fla., Tex., and Mex.

11a. **Juncus biflorus f. adìnus** Fern. & Grisc. (Rhodora 37: 156. 1935.) Deam no. 26197 from a swampy fallow field a mile and a half west of Huron, Martin County, is typical of this form.

12. **Juncus canadénsis** J. Gay. (*Juncus canadensis* var. *longicaudatus* Engelm.) Map 610. Very common in the lake area but infrequent south of it. It is found in marshes, swales, bogs, sandy or marly ditches, and on low borders or sandy shores of lakes.

Newf. to Minn., southw. to Ga. and La.

Juncus scirpoides Lam.

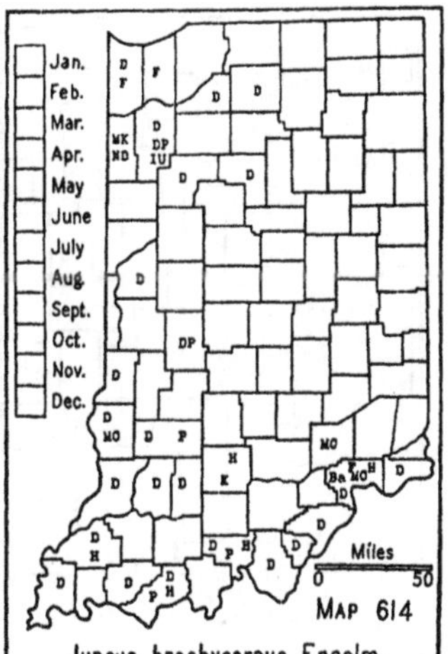

Juncus brachycarpus Engelm.

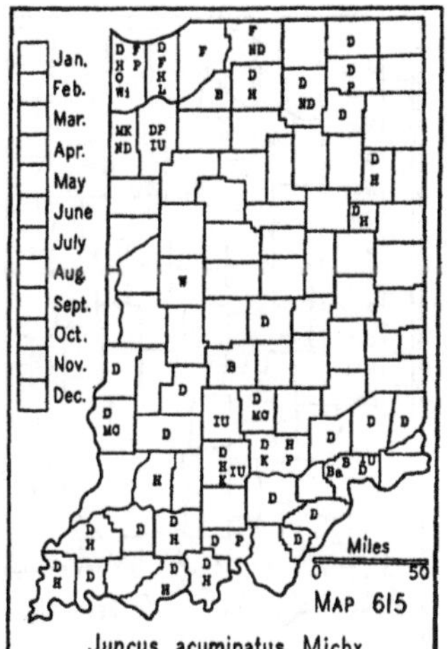

Juncus acuminatus Michx.

13. **Juncus brachycéphalus** (Engelm.) Buch. (*Juncus canadensis* var. *brachycephalus* Engelm.) Map 611. Frequent in the northern half of Indiana, becoming common in the lake area. It is often associated with other rushes, especially with *J. nodosus,* on low sandy or marly borders of lakes, in marshes and sloughs, and on springy calcareous terraces.

A form of this species having six stamens instead of the more usual three has been named *J. brachycephalus* f. *hexandrus* Martin (Rhodora 40: 460. 1938) and Deam no. 54539A in the Herbarium of the University of West Virginia is designated as the type. The six-stamened condition is frequent in *J. brachycephalus* (as in *J. canadensis* and related species) ; in fact most of the Indiana collections have at least a few of the flowers with six stamens. As a rule a single plant will have flowers predominantly either 3-stamened or 6-stamened; occasionally the number will be about equally divided between the two, but rarely, if ever, is a plant found in which all of the flowers have reverted to the 6-stamened state.

Maine to Wis., southw. to N. J., Pa., and Ill.

14. **Juncus diffusíssimus** Buckley. Map 612. Common in southern Indiana, especially in the unglaciated area, in roadside ditches, low fallow fields (mostly in hard white clay soil), swampy open woods, and along the banks of or on gravel bars in creeks.

N. Y. to Ind. and Kans., southw. to Tex. and Ga.

15. **Juncus scirpoìdes** Lam. Map 613. Known in Indiana from only the dune area where it is found in open, wet sandy habitats. Of the 22 collections seen from Lake and Porter Counties only one was made later than 1913. Previous to that date the species apparently was frequent to fairly common on the dunes.

No specimen could be found to confirm the reports of Barnes and of Coulter from Jefferson County. In all probability these reports were based upon collections of *Juncus brachycarpus*, a species common in Jefferson County and superficially resembling *J. scirpoides*. *J. brachycarpus* is the only one

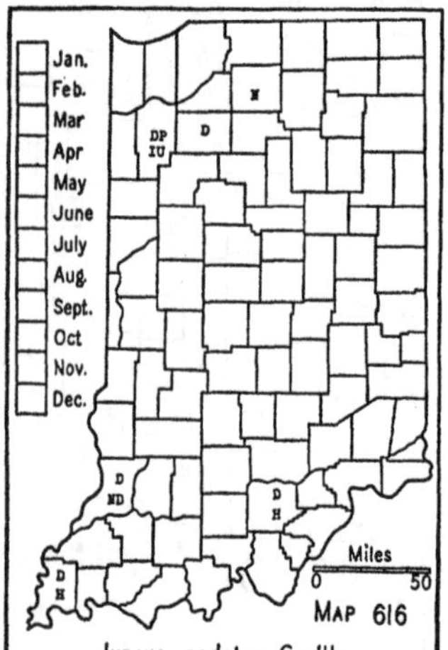

Juncus. nodatus Coville

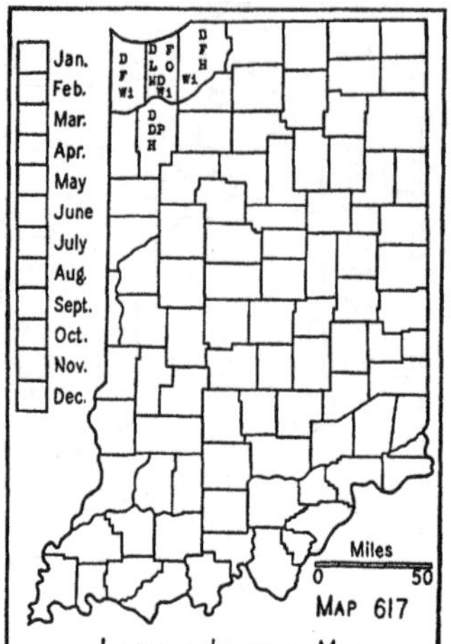

Juncus pelocarpus Mey.

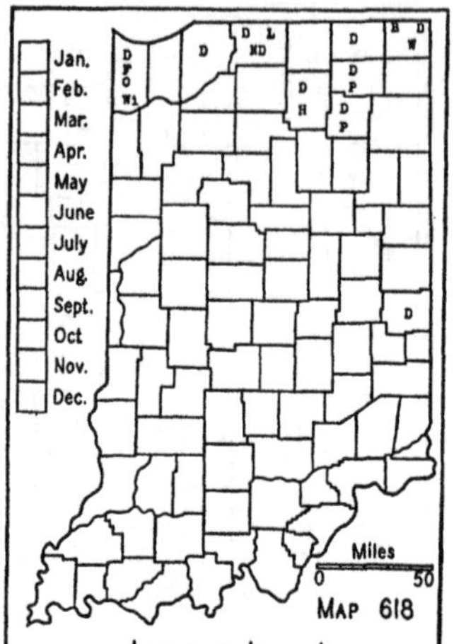

.Juncus nodosus L.

of all the species with small spherical heads to which Coulter's statement under *J. scirpoides,* "found throughout the state," is applicable.

N. Y. to Mich., southw. to Fla., Mo., and Tex.

16. **Juncus brachycárpus** Engelm. Map 614. Fairly common in southern Indiana; frequent elsewhere except in the central and eastern counties. Its favorite habitats are low fallow or grassy fields where the soil is usually a hard white clay, and sandy ditches, but it occurs also on wet roadsides and in flat woods and on the dunes in sloughs and sandy swales.

Mass. to Ont., southw. to Ga., Miss., and Tex.

17. **Juncus acuminàtus** Michx. Map 615. Very common in ditches and wet, usually more or less open, habitats of all types; frequently in shallow water in ponds or swamps. Occasionally the heads are proliferous, especially after the habitat has been flooded.

N. S. to Minn., southw. to Ga. and Tex.

18. **Juncus nodàtus** Coville. (*Juncus robustus* (Engelm.) Coville, not Wats.) Map 616. Infrequent and local. This southern species was apparently first collected in the state at its northernmost known station: along a wet railroad siding near Lake Maxinkuckee, Marshall County, J. T. Scovell and H. W. Clark no. 1468, Oct. 16, 1900. Its usual habitat in southern Indiana is on borders of ponds in low, often flooded, pin oak woods where it is associated with buttonbush and with *Carex Crus-corvi* and *C. lupuliformis.*

Northern Ind. to Kans., Okla., Tex., and La.

19. **Juncus pelocárpus** E. Mey. Map 617. Fairly common in the northwestern counties, mostly in the dune area. A species of wet open habitats, occurring on sandy or mucky borders of ponds, lakes, and swamps and in sloughs and swales. The more diffuse proliferous plants are often entirely sterile.

Newf. to N. J., Ind., and Minn.

20. **Juncus nodòsus** L. Map 618. Fairly common in the northern

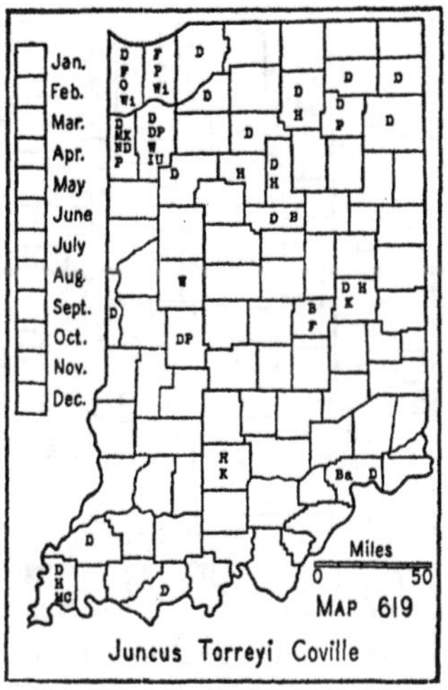

MAP 619
Juncus Torreyi Coville

MAP 620
Juncus articulatus L.

MAP 621
Juncus alpinus Vill. var. rariflorus Hartm.

counties and known from a single locality in Wayne County. It is found in a variety of wet habitats: in marshes, bogs, and swales, occasionally in ditches and sloughs, but most commonly on low sandy or marly shores.

No specimens were found to support Schneck's report from the Lower Wabash Valley.

The relatively huge grotesque heads often produced by galls in many species of § *Septati* occur with greatest frequency in this species, although they are frequent too in *J. Torreyi, J. canadensis* and *J. acuminatus.*

Newf. to B. C., southw. to Va., Ill., and Nebr.

21. **Juncus Tòrreyi** Coville. (*Juncus nodosus* var. *megacephalus* Torr.) Map 619. Common, especially in the lake area, in ditches, sloughs, and low prairies and on the borders of lakes, ponds, and creeks. It grows in both clay and sandy soils.

Mass. to Sask. and Wash., southw. to Ala., Tex., and Ariz.

22. **Juncus articulàtus** L. Map 620. Known in Indiana from a single collection: on an abandoned road through a marsh on the southeast side of Lake Wawasee, Kosciusko County, Deam no. 56408, July 19, 1935. Here it was abundant in 1935.

Specimens could not be located to corroborate the reports from Lake County made by Coulter, Deam, Peattie, and Pepoon.

Newf. to Ind. and B. C., southw. to Mass. and N. Y.; local in n. Calif.; also in Eurasia.

23. **Juncus alpìnus** Vill. var. **rariflòrus** Hartm. (Rhodora 35: 233. 1933.) (*Juncus alpinus* var. *insignis* Fries and *Juncus Richardsonianus* Schultes.) Map 621. Largely confined to the dune area in Indiana where it is often locally plentiful on wet sandy or marshy shores of lakes and ponds, on borders of sloughs, and in low sandy ditches.

Que. to B. C., southw. to Pa., Ind., Nebr., and Wash.; also in Eurasia.

23a. **Juncus alpinus** var. **fuscéscens** Fern. Map 621a. A single Indiana collection (Bebb no. 663, Clarke Junction, Lake County, Aug. 14,

MAP 621a
Juncus alpinus Vill. var. fuscescens Fern.

MAP 622
Luzula saltuensis Fern.

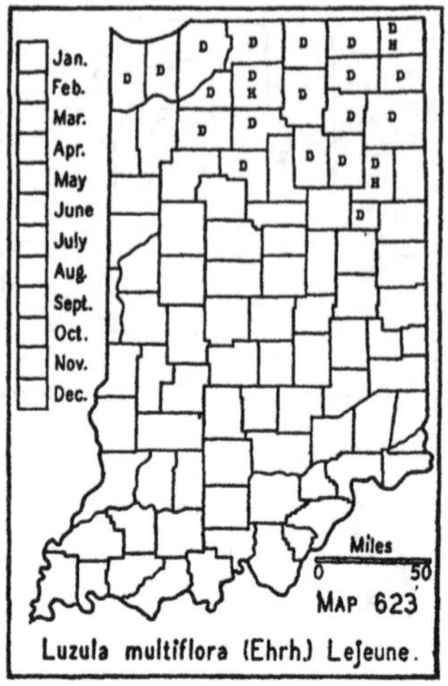

MAP 623
Luzula multiflora (Ehrh.) Lejeune.

1901) is characteristic of this variety, although transitional forms between the preceding variety and var. *fuscescens* are occasional. Its habitats are the same as those of var. *rariflorus*.

Vt. to B. C. and Mo.

937. LÙZULA DC. WOOD RUSH

[Fernald and Wiegand. The variations of Luzula campestris in North America. Rhodora 15: 38-43. 1913.]

Flowers solitary at the tips of the branches of the inflorescence...1. *L. carolinae* var. *saltuensis*.
Flowers crowded in spikelike clusters or glomerules.
Rays of umbel erect or ascending, relatively stout; heads mostly cylindric.
Cauline leaves large, (7) 9-14 cm long, 4-6 (9) mm wide; filaments equaling the anthers; perianth averaging 3 mm long, usually slightly exceeding the capsule; heads pale; base of plant rarely producing bulbs.............2. *L. multiflora*.
Cauline leaves small, 3-5.5 cm long, 2-3 mm wide; filaments shorter than the anthers; perianth averaging 2.5 mm long, shorter than the capsule; heads dark; base of plant commonly producing bulbs..2a. *L. multiflora* var. *bulbosa*.
Rays of umbel mostly strongly divergent, some elongate and filiform; heads hemispheric or short-cylindric; leaves mostly clustered at the base of the plant, the cauline small, 2-7 cm long, 1.5-3 mm wide.
Filaments half the length of the anthers or less; perianth conspicuously exceeding the capsule, generally 3 mm or more long....................3. *L. echinata*.
Filaments more than half the length of the anthers; perianth from shorter than to slightly exceeding the capsule, generally about 2.5 mm long...3a. *L. echinata* var. *mesochorea*.

1. **Luzula carolinae** S. Wats. var. **saltuénsis** (Fern.) Fern. (Rhodora 40: 404. 1938.) (*Luzula saltuensis* Fern., *Juncoides carolinae* of Britton and Brown, Illus. Flora, ed. 2, and *Juncoides pilosum* of American authors.) Map 622.* A northern plant known in Indiana from only two collections: base of a low wooded slope near pond in woods on the Douglas farm 2½ miles southeast of Hamilton, De Kalb County, Deam no. 44268, May 25, 1927, and about 4 miles north of Notre Dame, St. Joseph County, J. A. Nieuwland no. 9115, in 1909.

* The name of this plant was changed after the map was made.

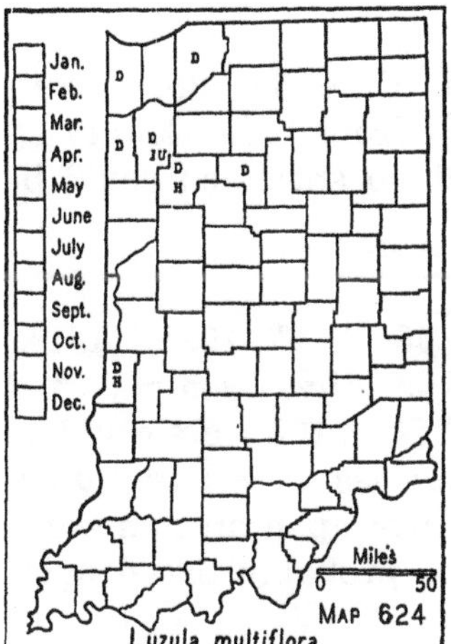

Luzula multiflora var. bulbosa (Wood) Hermann

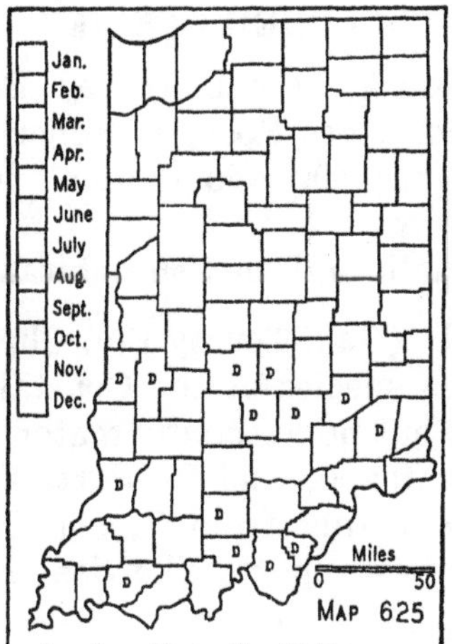

Luzula echinata (Small) Hermann

Luzula echinata var. mesochorea Hermann

Newf. to Sask., southw. to N. J. (in the mts. to Ga.), Ind., and Minn.; also in e. Asia.

2. **Luzula multiflòra** (Ehrh.) Lejeune. (Rhodora 40: 83-84. 1938.) (*Luzula campestris* var. *multiflora* (Ehrh.) Celak., *Luzula intermedia* (Thuill.) A. Nels., *Juncoides campestre* of Britton and Brown, Illus. Flora, ed. 2, in part, and *Juncoides intermedia* (Thuill.) Rydb.) Map 623. Confined to the lake area where it is very common in dry open oak woods, especially on hills or slopes, and occasionally in grassy clearings. It is frequently associated with *Carex pennsylvanica* and *C. communis*.

Newf. to Alaska, southw. to N. J., Pa., Ill., Utah, and Calif.; also in Eurasia.

2a. **Luzula multiflora** var. **bulbòsa** (Wood) Hermann. (Rhodora 40: 84. 1938.) (*Luzula campestris* var. *bulbosa* Wood and *Juncoides bulbosum* (Wood) Small.) Map 624. Known in Indiana from only the northwestern counties and apparently confined to the lake and prairie areas where its habitat, very sandy open oak woods, is common.

N. J. and Pa. to Kans., southw. to Ga. and Tex.

3. **Luzula echinàta** (Small) Hermann. (Rhodora 40: 84. 1938.) (*Luzula campestris* var. *echinata* (Small) Fern. & Wieg. and *Juncoides echinatum* Small.) Map 625. Fairly common in southern Indiana in dry oak woods, especially on wooded slopes and steep river banks.

N. J. and Pa. to Ga. and Tex., and in the Mississippi Valley at least in s. Ind.

3a. **Luzula echinata** var. **mesochòrea** Hermann. (Rhodora 40: 84. 1938.) Map 625a. The most widespread *Luzula* in Indiana; common south of the lake area and very common in the knob area. It is found in dry open woods, especially on white oak slopes, knobs or ridges, and occasionally in hard clay soil in fallow fields and clearings.

Ind., doubtless also in Ohio, Ky., and Ill.

EXCLUDED SPECIES

1. **Juncus coriàceus** Mack. (Bull. Torrey Bot. Club **56**: 28. 1929.) (*Juncus setaceus* of authors, not Rostk.) Reported from Lake County by Pepoon in the "Flora of the Chicago Region" but no specimen could be found.

Del. to Fla. and La., usually in brackish habitats.

2. **Juncus ténuis** Willd. (Bull. Torrey Bot. Club. **56**: 25-27. 1929.))*Juncus dichotomus* Ell.) This species of the Coastal Plain has been reported many times from Indiana but no authentic material from the state could be found. The reports from Jasper and La Porte Counties were based upon specimens of *J. Greenei* and that from Gibson County upon an immature specimen of *J. macer.* Other reports likewise were probably based upon errors in indentification.

Conn. to Fla. and Argentina.

3. **Juncus brevicaudàtus** (Engelm.) Fern. (*Juncus canadensis* var. *brevicaudatus* Engelm. and *Juncus canadensis* var. *coarctatus* Engelm.) Reported from Lake County by both Pepoon and Peattie but the one specimen found which had been referred to this species (a collection of Umbach's from Pine, Lake County, labeled *Juncus canadensis* var. *coarctatus,* University of Wisconsin Herbarium) is *J. alpinus* var. *rariflorus.* Indiana is considerably south of the known range of *J. brevicaudatus.*

Newf. to Minn., southw. to Conn., Pa., and W. Va.

4. **Juncus débilis** Gray. (*Juncus acuminatus* var. *debilis* (Gray) Engelm.) No specimen could be found to confirm the report from Vigo County by Blatchley of this eastern and southern species. A specimen in the Wabash College Herbarium labeled *Juncus acuminatus* var. *debilis* (Coulter no. 1918 from Hanover) was probably the basis of Barnes' report from Jefferson County. This specimen is *J. diffusissimus.*

R. I. to Fla., Miss., and Ark.

38. LILIÀCEAE Adans. LILY FAMILY

Flowers dioecious; some of the species woody vines.
  Inflorescence umbellate; fruit a 1-4-seeded berry............1151. SMILAX, p. 324.
  Inflorescence a spicate raceme; fruit a 3-celled, ellipsoid capsule, 7-10 mm long, with linear-oblong seed..........................950. CHAMAELIRIUM, p. 304.
Flowers perfect or monoecious.
  Leaves all, nearly or quite basal or lacking at flowering time.
    Flowers large, the perianth segments 6-11 cm long.
      Flowers orange................................1019. HEMEROCALLIS, p. 308.
      Flowers white....................................1103. YUCCA, p. 316.
    Flowers smaller, the perianth segments less than 6 cm long.
      Plants with solitary flowers; leaves 2, fleshy, mottled.1076. ERYTHRONIUM, p. 314.
      Plants not as above.
        Flowers deep blue, reflexed, racemose, many, divisions of perianth united; leaves narrowly linear..........................1095. MUSCARI, p. 315.
        Flowers not as above.
          Leaves 2-5, usually 2 or 3, mostly 4-10 cm wide.
            Flowers in an umbel, usually 3-6..............1117. CLINTONIA, p. 317.
            Flowers in a raceme, several, white, very fragrant; leaves 2 or 3........ ....1128. CONVALLARIA, p. 320.
          Leaves not as above.
            Stems and pedicels glandular, the glands usually blackish; leaves grass-like....................................942. TOFIELDIA, p. 304.
            Stems and pedicels not glandular.
              Plants without a bulbous base; leaves lanceolate, mostly 5-15 cm long, 7-20 mm. wide, strongly veined; flowers many, tubular, yellowish white, in a terminal, spikelike raceme; stems usually with 1 or more leaflike bracts............................1143. ALETRIS, p. 324.
              Plants not as above; leaves usually narrow-linear.
                Flowers in a long, terminal raceme, usually bluish but sometimes white; leaves long, linear, the widest usually 8-20 mm wide..... ....1087. CAMASSIA, p. 315.
                Flowers in terminal umbels or corymbose.
                  Midrib of leaves whitish; flowers corymbose..................... ....1089. ORNITHOGALUM, p. 315.
                  Midrib of leaves not whitish; flowers all in terminal umbels.
                    Bulbs globose, about 1 cm in diameter (in dried specimens), without an onionlike odor; leaves present at flowering time. ....1050. NOTHOSCORDUM, p. 311.
                    Bulbs elongate-ovoid, usually much larger than those of *Nothoscordum*, with an onionlike odor; leaves absent at flowering time, mostly 10-20 cm long and 3-6 cm wide; flowers many, white *(Allium tricoccum)*............1049. ALLIUM, p. 309.
  Leaves cauline, rarely with both basal and cauline leaves.
    Flowers large, 4-10 cm in diameter, orange or maroon purple, generally spotted within; perianth segments all similarly colored.........1072. LILIUM, p. 311.
    Flowers smaller or, if large, the calyx green.
      Leaves whorled.
        Blades of leaves parallel-veined; leaves in 2 or rarely 3 whorls; perianth segments all similar in color; rootstock white, tuberlike................ ....1135. MEDEOLA, p. 321.
        Blades of leaves net-veined; leaves 3, in a terminal whorl; sepals green; petals white, maroon or purple; rootstock dark, wrinkled..1138. TRILLIUM, p. 321.
      Leaves alternate.

Mature plants forking at the first or second leaf; leaves perfoliate or sessile; flowers solitary and from the axil of the first leaf above the fork or, if the flowers are 2, the second flower usually in the axil of the leaf above the first flower; flowers yellow or yellowish green, 15-30 mm long; capsules 3-angled or 3-winged..........................966. UVULARIA, p. 308.

Mature plants not as above.

Flowers axillary; fruit a black or red berry.

Stems simple, very rarely with axillary branches; widest leaves 1-10 cm wide; fruit a black berry...............1123. POLYGONATUM, p. 319.

Stems much branched, usually 1-2 m. high; the so-called leaves in alternate clusters, filiform, about 1 cm long; fruit a 1-seeded red berry...... ....1113. ASPARAGUS, p. 316.

Flowers in a terminal panicle or umbel.

Leaves linear, not petiolate.

Stem and inflorescence glabrous.

Plants with an onionlike odor, their bases a fleshy bulb; inflorescence a terminal umbel, consisting entirely of flowers or sometimes partly or wholly of bulblets..............1049. ALLIUM, p. 309.

Plants without the the onionlike odor; flowers in panicles.

Plants glaucous; panicle sparsely flowered; flowers mostly 8-10 mm long, shorter than their pedicels; sepals with a large, dark gland near the base..................958. ZIGADENUS, p. 306.

Plants not glaucous; panicle many-flowered; flowers mostly 5-7 mm long, longer than their pedicels; sepals lacking the black gland near the base....................957. STENANTHIUM, p. 305.

Stem and especially the inflorescence pubescent; inflorescence paniculate; fruit a 3-celled capsule................959. MELANTHIUM, p. 307.

Leaves not linear, either sessile or petiolate.

Flowers dark maroon to nearly black; panicles generally 20-50 cm long, basal stem leaves large, narrowed into long, sheathing petioles; fruit a capsule..........................960. VERATRUM, p. 307.

Flowers white; basal stem leaves lacking; fruit a globose, 1- or 2-seeded berry.

Stem leaves generally 2, rarely 3, cordate at the base, usually less than 9 cm long, the lower one generally petiolate; perianth of 4 parts...........................1119. MAIANTHEMUM, p. 318.

Stem leaves usually more than 3, generally all sessile, usually more than 9 cm long; perianth of 6 parts.....1118. SMILACINA, p. 317.

942. TOFIÈLDIA Huds.

1. **Tofieldia glutinòsa** (Michx.) Pers. Map 626. Found in wet, marly soil in a few marshes and springy places in the northern counties. Local and, where found, sometimes frequent over the entire area of its habitat.

Newf. to Minn. and Alaska, southw. to Maine, Ohio, Ind., Oreg., and in the mts. to N. C.

950. CHAMAELÍRIUM Willd.

1. **Chamaelirium lùteum** (L.) Gray. Map 627. I found a single plant in an exposed place on a limestone slope 3 miles north of Milltown, Crawford County where it was associated with *Comandra Richardsiana* and *Lithospermum croceum.* I found another specimen in a woods about 7 miles southwest of Evansville where it was closely associated with *Fagus grandifolia, Quercus alba, Cornus florida, Sassafras albidum* and *Phyto-*

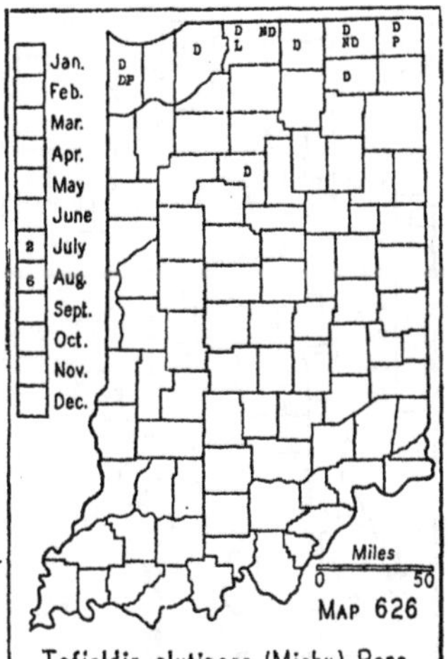

Tofieldia glutinosa (Michx.) Pers.

Chamaelirium luteum (L.) Gray

Stenanthium gramineum (Ker) Morong.

lacca americana. In both instances I found only a single specimen although I made extended search for others. Clapp reported it from the barrens near New Albany, and Barnes reported it from Jefferson County without comment. The distribution of this species is erratic and observers do not seem to understand what factors are involved. It has been reported from 15 counties in Ohio but northward it has not been reported until the Upper Peninsula of Michigan is reached.

Mass., Mich. to Nebr., southw. to Fla., Miss., and Ark.

957. STENÁNTHIUM (Gray) Kunth

Capsules reflexed at maturity, mostly 7-8 mm long; leaves mostly 6-10 mm wide; plant flowering mostly in July....................................1. *S. gramineum.*

Capsules erect at maturity, mostly 7-10 mm long; leaves mostly 10-16 mm wide; plant flowering mostly in August....................................2. *S. robustum.*

1. **Stenanthium gramíneum** (Ker) Kunth. Map 628. This species is local in Indiana and apparently so throughout its range. It is infrequent in sandy soil for half a mile in an open, black and white oak woods on the south side of the Tippecanoe River north of Rochester, Fulton County, and I found one plant in a sandy prairie habitat a mile north of Rochester. There is a small colony on a low, sandy, open black and white oak ridge between swamps in section 12 about 2½ miles southeast of Etna Green, Kosciusko County. I found it to be frequent for a short distance in sandy soil at the base of a white and black oak slope on the south side of a large swamp about 3 miles northwest of Hoover, Cass County. In the same colony I found a specimen of the next species. The remainder of my specimens were found in similar habitats. I have several times transplanted it to the open in neutral soil in our garden and it has lived for only a few years.

Va., Ind. to Mo., southw. to Fla. and Miss.

2. **Stenanthium robústum** Wats. Map 629. I have only two specimens from Indiana which I refer to this species. Data concerning this species and the preceding one are meager; some authors do not separate them and

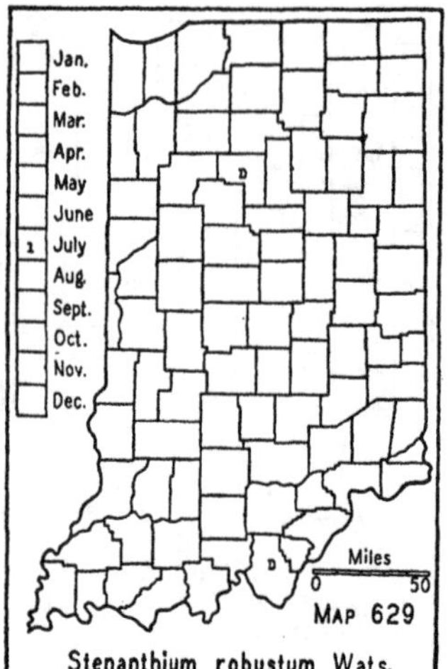

MAP 629

Stenanthium robustum Wats.

MAP 630

Zigadenus glaucus Nutt.

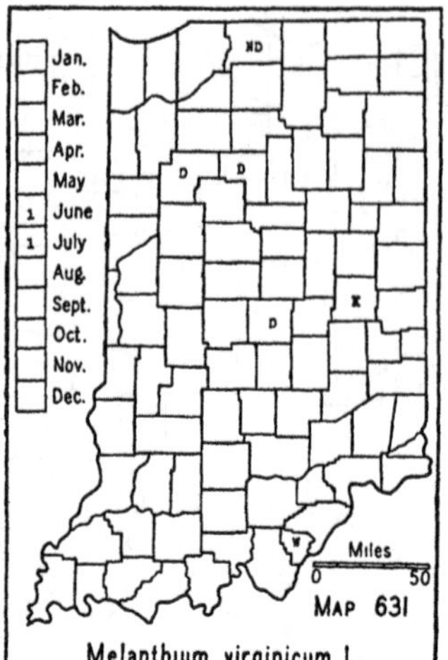

MAP 631

Melanthium virginicum L.

one has made this a mere form of the preceding. Robert Ridgway was interested in this problem and wrote me concerning it. He was firmly convinced that the two species are distinct. I quote, in part, from his letter to me dated January 13, 1925, Olney, Illinois: "I have several of the former *(Stenanthium gramineum)* transplanted from the "wilds hereabout" and one of the latter *(Stenanthium robustum)* from the Highland Nursery, North Carolina. They are planted near together, in identical soil, and all local conditions governing them are the same. The first blooms from June 21 to July 11 (average date July 1); while the last blooms from August 1-24 (average date August 18), a difference of more than six weeks."

The specific name for this species seems to be well chosen, since the whole plant is larger and more robust in all of its parts. The leaves are wider, the floral segments longer, the fruit longer, and the stigmas slightly longer. The width of the leaves and the position of the mature fruit are sufficient to distinguish the species. Since I found both species in the same colony I am not entirely satisfied that there are two species of our plants but until sufficient data are accumulated I believe it is best to separate them, placing them in the taxonomic category which the differences suggest.

Pa. and Ind., southw. to S. C., Tenn., and Mo.

958. ZIGÂDENUS Michx.

1. **Zigadenus glaúcus** Nutt. (Rhodora 37: 256-258. 1935.) (*Zigadenus chloranthus* of Gray, Man., ed. 7, not Richardson, and *Anticlea elegans* of Britton and Brown, Illus. Flora, ed. 2.) Map 630. This species is rare and local. I found a few plants in a marly place in the large swamp in a woods about 3 miles northwest of Hoover, Cass County. In Lagrange County I found a number of specimens in a marsh of about an eighth of an acre surrounded by young tamarack; the area where it was found was probably too alkaline for the tamarack, although I found it in a similar position but in a very small opening in a tamarack bog about a mile and a half southeast of Mongo. I have seen it in only two other places, and

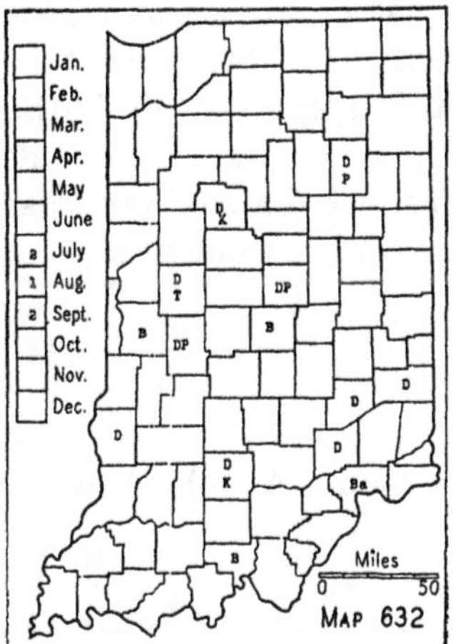

Veratrum Woodii Robbins

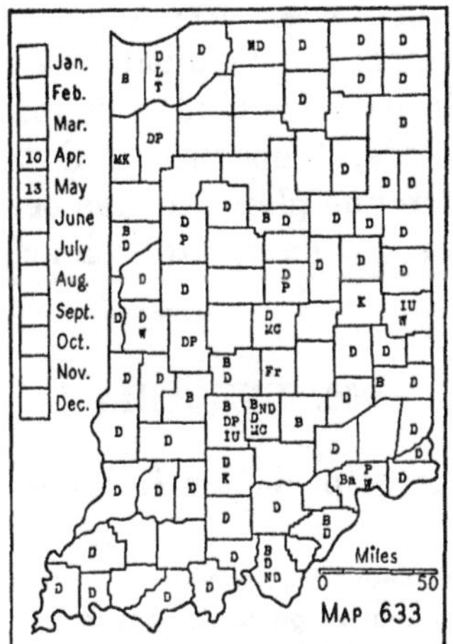

Uvularia grandiflora J. E. Smith

Uvularia sessilifolia L.

both of them were sedge marshes. Blatchley found it in a marsh near Lake James, Steuben County. Van Gorder found it in Noble County, in sec. 10 of Noble Township.

Que. to Man., southw. to N. B., n. Vt., n. N. Y., n. Ohio, and n. Ill.

959. MELÁNTHIUM L.

1. **Melanthium virgínicum** L. BUNCHFLOWER. Map 631. Rare and very local. My White County specimen was found in a wet prairie habitat a half mile east of Idaville. My Cass County specimen was found in mucky soil in a large swamp about 3 miles northwest of Hoover. I found a single specimen on a springy, gravelly slope in the open at "Crows' Nest" about 8 miles north of Indianapolis. It has also been reported from Franklin and St. Joseph Counties, from the vicinity of New Albany, and from the area of Delaware, Jay, Randolph, and Wayne Counties by Phinney, who assigns it to "wet meadows."

R. I., s. N. Y. to Minn., southw. to n. Fla. and Tex.

960. VERÀTRUM [Tourn.] L.

1. **Veratrum Woódii** Robbins. Map 632. Local throughout the area indicated on the map. Where it is found, however, it is usually frequent over a small area. It is usually found in rich soil on the south sides of deep ravines, although I found a single specimen in a crevice of a limestone cliff along the Muscatatuck River about a mile above Vernon, Jennings County. This species interested me years before I knew what it was. I had seen the large root leaves and watched to see the plant flower but could never find one. I transplanted one to our garden and it was several years before it bloomed. I have not kept a record but I estimate that it flowers about every 4 or 5 years. This plant, during the 15 or more years that it has been under observation, has increased to only two plants. The flowers are deep maroon, or some of them almost black. One specimen in particular was observed. The inflorescence was 23 inches long and three and a

half inches wide. The first flowers expanded the last of July and insects continued to visit the inflorescence until in September. Harry Dietz, a visiting entomologist, observed within five minutes 2 species of Anthomyids, 4 species of Muscids, 1 species of Ortalid, 2 species of Phorids, and 1 species of Syrphid on one plant. It has been reported also from Cass, Greene, Hamilton, Monroe, Putnam, Tippecanoe, and Vigo Counties.

Ind. to Mo.

966. UVULÀRIA L.

[Anderson and Whitaker. Speciation in Uvularia. Jour. Arnold Arb. 15: 28-42. 1934.]

Leaves perfoliate; capsules obtusely 3-angled.
Blades whitish-pubescent beneath; perianth segments smooth within or nearly so.. ....1. *U. grandiflora.*
Blades glabrous beneath; perianth segments granular-pubescent within. (See excluded species no. 132, p. 1033.) ....*U. perfoliata.*
Leaves sessile; capsules sharply 3-angled, acute at each end..........2. *U. sessilifolia.*

1. **Uvularia grandiflòra** J. E. Smith. BIG MERRYBELLS. Map 633. Infrequent to frequent in moist, rich soil throughout the state. It is never found outside of thick woodland, unless persisting after woodland has been cleared, but does well in cultivation in sun or shade. This species has been confused by some of our early authors with *Uvularia perfoliata,* the range of which is shown by recent studies to be restricted to the Allegheny Mountains and eastward to the Coast.

Que., w. N. Y. to Minn., southw. to Ga., Tenn., and Kans.

2. **Uvularia sessilifòlia** L. (*Oakesia sessilifolia* (L.) Wats.) LITTLE MERRYBELLS. Map 634. Colonies are infrequently found in the southern counties where it grows in hard, clay soil, usually associated with beech, beech and sweet gum, and lowland oaks. It propagates mostly from the rootstocks and a note on one of my labels reads: "I found, in a low woods about 3 miles southwest of Dale, Spencer County, two colonies about 10 × 20 feet and this species formed a mat over these areas."

N. B., Ont. to Minn., southw. to Ga. and Ark.

1019. HEMEROCÁLLIS L. DAYLILY

[Bailey. Hemerocallis: the day-lilies. Gentes Herbarum 2: 143-156. 1930.]

Flowers dark, tawny, fulvous or reddish orange, not fragrant, blooming in summer. ....1. *H. fulva.*
Flowers light, clear yellow or lemon color, more or less fragrant, blooming in spring and summer. (See excluded species no. 133, p. 1033.)..................*H. flava.*

1. HEMEROCALLIS FÚLVA L. TAWNY DAYLILY. Map 635. This species is ornamental and on account of its easy cultivation it has been freely planted since pioneer times. It never produces seed naturally in this country and propagates entirely by its many tuberous roots. A. B. Stout, of the New York Botanical Garden, has succeeded in producing seed by artificial pollination. He has written many articles on the species and anyone interested should read them.

This species is found infrequently in small or large colonies throughout the state along roadsides and about abandoned habitations. When it is once established, nothing can compete with it; hence it forms pure stands. The nativity of the species is not known but most authors give it as Eurasian.

N. B. to Ont., southw. to N. C. and Tenn.; escaped from cultivation.

1049. ÁLLIUM L.

Blades of leaves elliptic, usually 3-10 cm wide, 15-20 cm long, not present at flowering time..1. *A. tricoccum.*

Blades of leaves linear, terete or flat, present at flowering time.

- Leaves terete.
  - Umbels bulblet-bearing; spathe 1-valved, generally 10-30 mm long; flowers about 4 mm long; pedicels mostly 15-30 mm long; stamens slightly longer than the perianth segments; filaments petal-like, the upper half divided into 3 linear divisions, the middle division bearing an anther about 0.5 mm long; introduced species..2. *A. vineale.*
  - Umbels not bulblet-bearing; stamens included; filaments linear, entire, bearing an anther about 1 mm long.
    - Divisions of the perianth elliptic-lanceolate or oval-lanceolate, acute; pedicels mostly 8-10 mm long; plant of Eurasia. (Cultivated chive.)..*A. Schoenoprasum.*
    - Divisions of the perianth linear-lanceolate, attenuate-acuminate; perianth mostly 8-15 mm long; pedicels about 5 mm long; plant indigenous at least northward. (See excluded species no. 134, p. 1033.)..*A. Schoenoprasum* var. *sibiricum.*
- Leaves flat or keeled.
  - Umbels bulblet-bearing.
    - Spathe 1-valved, the beak more than 1 cm long, usually about 10 cm long; summit of the stem curved or coiled before flowering........3. *A. sativum.*
    - Spathe more than 1-valved, the beak short, less than 1 cm long.
      - Bulbs not multiplying; leaves flat, narrow, mostly 2-3 mm wide and keeled beneath; beak of spathe mostly 3-5 mm long; flowers few, white or pinkish, 4-6 mm long; filaments of all of the stamens entire; pedicels 10-40, usually 15-20 mm long; native species............4. *A. canadense.*
      - Bulbs producing bulblets; leaves flat, mostly 8-16 mm wide, the margins scabrous; flowers purplish; filaments of alternate stamens toothed. (See excluded species no. 135, p. 1034.)..................*A. Scorodoprasum.*
  - Umbels not bulblet-bearing.
    - Umbels nodding, the 2 bracts persistent; scapes sharply keeled; stamens conspicuously exserted....................................5. *A. cernuum.*
    - Umbels erect, the 2 bracts soon deciduous; scapes terete or nearly so; stamens about as long as the perianth segments. (See excluded species no. 136, p. 1034.)..*A. stellatum.*

1. **Allium tricóccum** Ait. WOOD LEEK. Map 636. Infrequent to rare in moist, rich soil throughout the state, although there are no records from the southwestern counties. It seems to prefer slopes and woods near streams and it is most often found associated with beech and sugar maple.

N. B. to Minn. and Iowa, southw. to Ga. and Tenn.

2. ALLIUM VINEÀLE L. CROW GARLIC. Map 637. My specimens are all from southern Indiana where it is one of the most pernicious of all weeds. A pioneer who lived in Point Township, Posey County, told me that when he was a boy (about 1860) both the garlic and wild onion were common

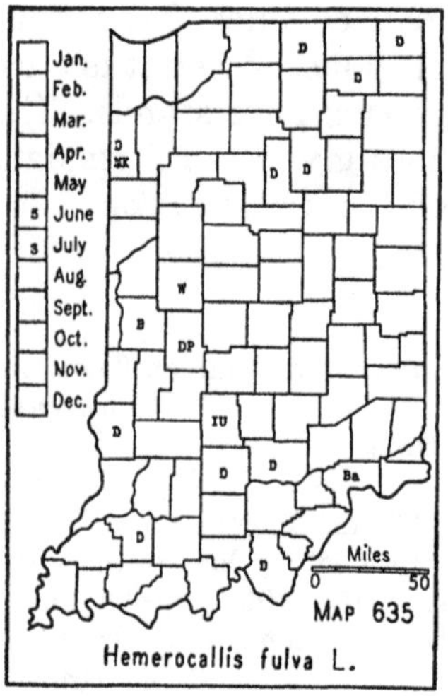

Hemerocallis fulva L.

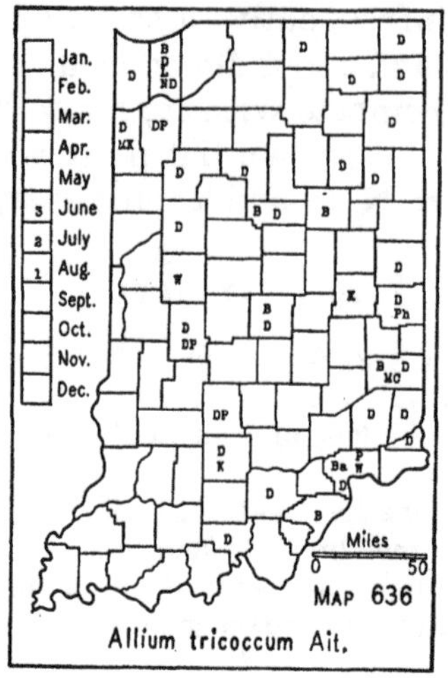

Allium tricoccum Ait.

Allium vineale L.

in the woodland. Henry Hollingsworth (Trans. Amer. Phil. Soc. **1**: 311-313. 1789, ed. 2) writes that sowing wheat stubble to oats will practically eradicate it. This indicates that it has been a weed since early times.

This species is found along roadsides and streams and in cultivated fields and pastures. It is difficult to eradicate because it propagates both by bulbs and bulblets. The principal objection to this species as well as to others of this genus is that milch cows can not be pastured where it grows because the garlic odor is transmitted to the milk. The task of ridding the soil of this and other species of this genus is a difficult one, especially if the area is subject to overflow because the bulblets are freely transported. Much literature has been published on the eradication of this species. Copies of this literature may be obtained gratis from the U. S. Department of Agriculture, Washington, D. C. and from the Purdue University Agricultural Experiment Station, West Lafayette, Indiana.

Nat. of Eu.; N. H. to Mo., southw. to Ga. and Ark.

3. ALLIUM SATÌVUM L. GARLIC. Map 638. Well established on a rocky, wooded slope in McCormick Creek State Park, Owen County. Hansen (Proc. Ind. Acad. Sci. 37: 319. 1928) writes that a number of farms, gardens, and a cemetery in Jefferson County are infested with the escaped form of the cultivated garlic (*Allium sativum* L.). It is almost certain that if a species of *Allium* becomes established it will persist unless it is destroyed by man.

Nat. of Eurasia.

4. **Allium canadénse** L. MEADOW GARLIC. Map 639. Found throughout the state. Frequent or common in the southwestern counties in moist soil in woodland and cultivated fields, where it sometimes becomes a pernicious weed.

N. B. to Minn. and Colo., southw. to Fla. and Tex.

5. **Allium cérnuum** Roth. NODDING ONION. Map 640. Infrequent throughout the state although there are no records from the southwestern counties. Where it is found, especially on the banks of streams, it is gen-

Allium sativum L.

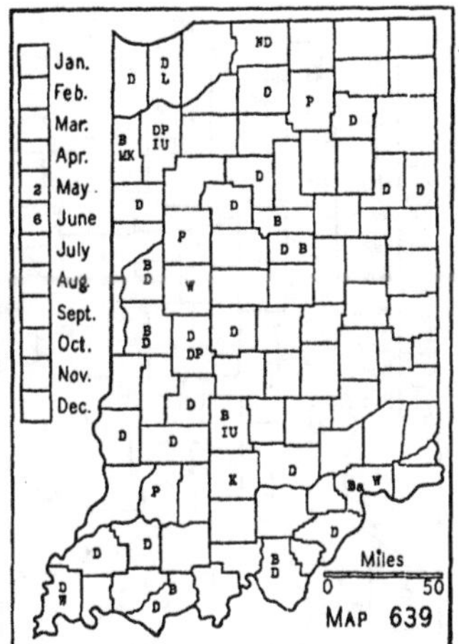

Allium canadense L.

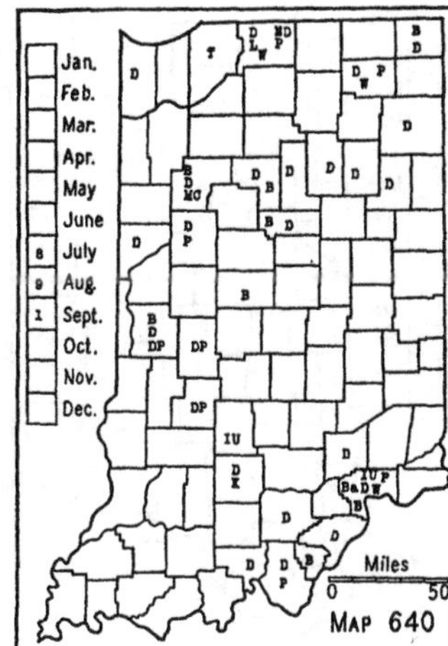

Allium cernuum Roth.

erally common except in marshes and springy places. This species has a wide range of habitat and distribution. It is usually found on the high and dry banks of streams but it is also found in low, sedge marshes, in marly springy places, and on gravelly bars in rivers. The color of the flowers ranges from white to deep pink. The white form has been named and I have it from Wabash County.

N. Y., Minn. to B. C., southw. to W. Va., Ky., N. Mex., and Calif.

1050. NOTHOSCÓRDUM Kunth FALSE GARLIC

1. **Nothoscordum biválve** (L.) Britton. FALSE GARLIC. Map 641. A few colonies of this plant have been found in the southwestern counties but I think it is much more common than our records show. Since it reproduces only by seed it may be more restricted than I think it to be. I found it to be common in alluvial bottoms about 4 miles northwest of Bloomfield, Greene County and also in low ground in the post oak flats south of Half Moon Pond in Posey County.

Va., Ohio, Ind. to Nebr., southw. to Fla. and Tex.; also in Bermuda and Jamaica.

1072. LÍLIUM L. LILY

Flowers erect; perianth segments narrowed below into claws; bulbs not rhizomatous.

Leaves lanceolate, mostly in whorls. (See excluded species no. 138, p. 1034.)....... ....*L. philadelphicum.*

Leaves linear, usually scattered on the stem except for a whorl at the summit....... ....1. *L. philadelphicum* var. *andinum.*

Flowers more or less nodding; perianth segments not clawed; bulbs rhizomatous.

Leaves all or nearly all in whorls, not bulblet-bearing in the upper axils; stems glabrous above; native species.

Blades all smooth beneath, long-attenuate at both ends; perianth segments strongly recurving from near the base, mostly 7-8 cm long; anthers mostly (17) 20-25 mm long...................................2. *L. superbum.*

Blades, at least some of them, more or less scabrous on the veins beneath; perianth segments recurving or spreading from near the middle; anthers mostly 8-12 (17) mm long, sometimes elongating after anthesis.

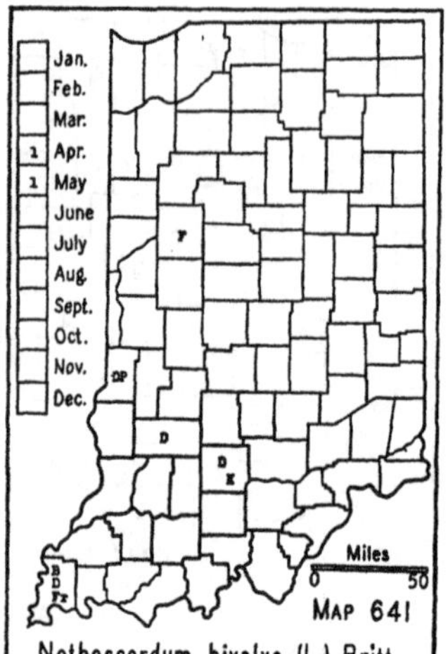

Nothoscordum bivalve (L.) Britt.

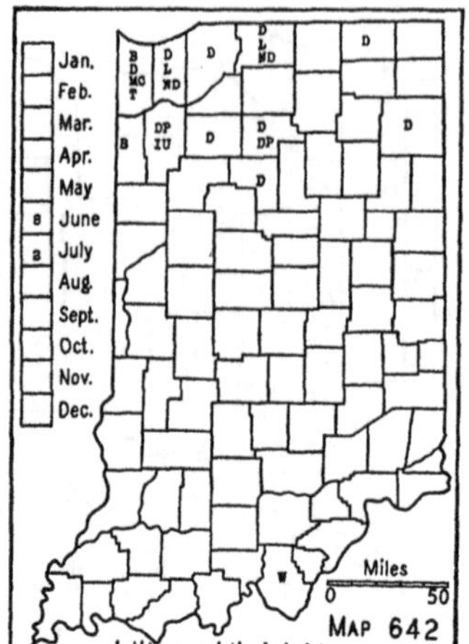

Lilium philadelphicum var. andinum (Nutt.) Ker

Lilium superbum L.

Perianth segments spreading (not recurving or scarcely so), base of perianth a reddish purple; plants of dry, wooded slopes. .3. *L. canadense* f. *rubrum.*

Perianth segments strongly recurved, orange or reddish orange on the outside; plants of a moist habitat such as prairies, marshes, and low woods........ ....4. *L. michiganense.*

Leaves all scattered, none in whorls, the upper ones usually bearing black bulblets in some of the axils; stem more or less gray-pubescent above; plants escaped from cultivation..5. *L. tigrinum.*

1. **Lilium philadélphicum** L. var. **andinum** (Nutt.) Ker. (*Lilium umbellatum* Pursh.) WESTERN LILY. Map 642. It is doubtful whether the species occurs in the state. All of the specimens I have seen belong to the variety and I think all reports of it from Indiana should be referred to the variety.

All of the reports of the variety and all of my specimens are from northern Indiana. Our reports for the species, however, extend the range to Hamilton, Vigo, Monroe, Clark, and Jefferson Counties and the Lower Wabash Valley. Prince Maximilian, June 10, 1834, reported finding *Lilium Catesbaei* in Knox County north of Hazelton. Since this species as now known is not found in Indiana and its flowering season is much later, I think this report should be referred to *L. philadelphicum* var. *andinum.* This lily is local and all the specimens I have seen were found in wet prairies or in similar habitats. Coulter, in his report from Jefferson County, says: "Common on the sand flats." There is a specimen in the herbarium of Wabash College collected in Harrison County by Clapp.

Ont. to Sask., southw. to Ohio and Ark.

2. **Lilium supérbum** L. AMERICAN TURK'S-CAP LILY. Map 643. This species has been reported from various parts of the state but I think it is very rare and that most of our reports should be referred to *Lilium michiganense.* I have found it only three times and always on wooded slopes. This species is easily confused with *Lilium michiganense* if the character of the roughness or smoothness of the under surface of the leaves is the only one used. The spreading of the perianth, which begins at the base, and the length of the anthers will easily separate them, but the

spreading of the perianth is a note often omitted, and herbarium specimens do not always clearly show this character. The map shows the location of the specimens that I have seen. Birkbeck passed through Indiana in 1817 and on page 112 of his "Notes on a journey in America from Virginia to the Illinois Territory" he says: "The road from Sholt's tavern to this place [from thirty-six miles east of Vincennes to Vincennes] is partly across barrens, that is, land of middling quality, thinly set with timber, or covered with long grass, and shrubby undergrowth; generally level and dry, and gaudy with marigolds, sunflowers, martagon lilies, and many other brilliant flowers." We have no way of determining the species of lily Birkbeck saw, but I think it was this species, although this is a mere guess.

From *Lilium michiganense* this species can be distinguished by several characters in addition to those given in the key. It is about half again as high; the leaves are more numerous in at least a few of the whorls, and all the leaves are longer and conspicuously long taper-pointed at both ends; the more numerous flowers, 3-15, are in umbels or pyramidal racemes, while the flowers in the other species are strictly in umbels, generally numbering 1-5 flowers. We have had both species moved from the wild and in cultivation for many years and the greatest number of flowers of the first has been 21 while of the second I have no record; but, as I recall, the number ranges mostly from 3 to 5 in vigorous plants, and these are always in umbels.

N. B., Ont. to Minn., southw. to Va. and Mo.

3. **Lilium canadénse** L. f. **rùbrum** Britt. (Bull. Torrey Bot. Club **17**: 125. 1890.) CANADA LILY. Map 644. This lily has been confused with the next, and possibly all, or nearly all reports for it should be referred to the next species. All of my specimens are from rocky, wooded slopes and were associated with *Vaccinium vacillans* and *Asclepias tuberosa*. This species is now considered to be Alleghanian and its distribution is not known because of its confusion with the next species. It is known to occur near Lawton in Kalamazoo County, Michigan (Nieuwland).

4. **Lilium michiganénse** Farwell. (Bull. Torrey Bot. Club **42**: 352-354. 1915.) Map 645. Infrequent throughout the northern part of the state and probably rare in the southern part or even absent from the southeastern part. It has been confused with the preceding species and our knowledge of its distribution and habitat can be now ascertained only from field studies or from existing specimens. Almost all of the reports for lilies in Indiana must be ignored on account of the recently acquired knowledge of the genus.

Lilium michiganense grows in moist prairie habitats, in mucky soil about lakes and in low woods, and in moist, black soil along roadsides and railroads. Locally it may be common over a small area. When once established it is very persistent, competing successfully with blue grass sod. I have known it to be a common plant for possibly 25 years in black, moist soil along the railroad through the old prairie north of Poneto, Wells County. It is to be noted with this species, as with the others, that the available

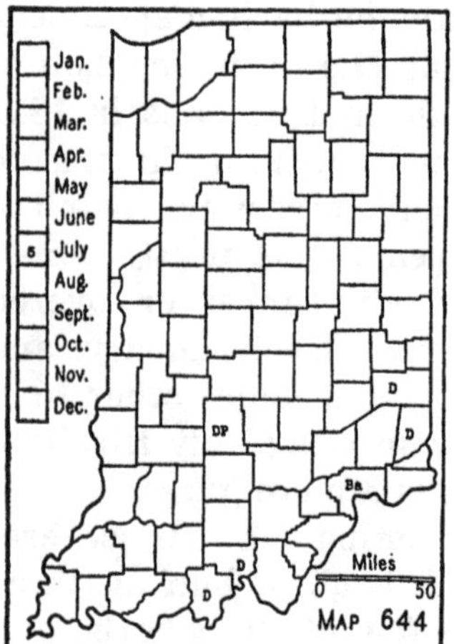

Lilium canadense L.

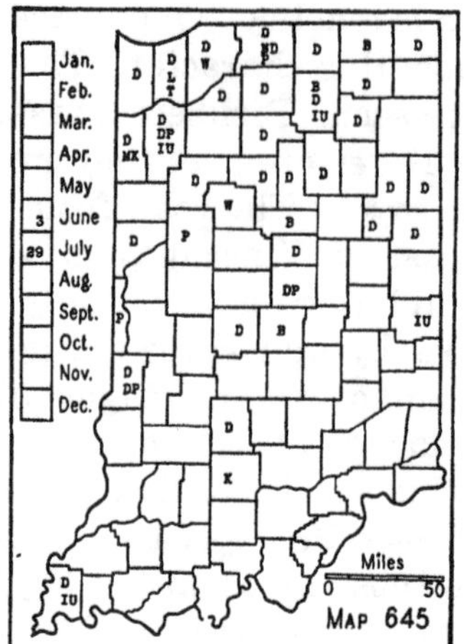

Lilium michiganense Farwell

Erythronium albidum Nutt.

amount of moisture has a marked effect upon the number of flowers on the plants. Where it is driest, most of the plants will have only one flower. The distribution is probably nearly as follows:

Ont., Mich. to Minn., southw. to Ky. and Mo.

5. LILIUM TIGRÌNUM L. TIGER LILY. Nieuwland writes (Amer. Midland Nat. 3: 106. 1913) that this species is an "escape to the woods at Notre Dame, growing perfectly wild and maintaining itself and spreading." I have paid little attention to plants of any kind that have escaped and this species may be more frequent than I know. I have a specimen which I found along a railroad about a mile south of Connersville, Fayette County.

Nat. of China and Japan.

1076. ERYTHRÒNIUM L. TROUT LILY

Flowers white or pinkish; stigmas mostly 2-3 mm long, curved outward..1. *E. albidum.*
Flowers yellow; stigmas usually shorter, erect....................2. *E. americanum.*

1. **Erythronium álbidum** Nutt. WHITE TROUT LILY. Map 646. Infrequent to frequent in moist woods throughout the state, usually more frequent and abundant in rich, alluvial flats along streams. This species and the next one have been reported from all parts of the state. It has a short flowering period, from the last of April to the first part of May, which accounts for the few specimens I have collected. This species and the next are usually called dogtooth violet in Indiana.

Ont. to Minn., southw. to Ga. and La.

2. **Erythronium americànum** Ker. COMMON TROUT LILY. Map 647. Infrequent to frequent throughout the state. Like the preceding species, where it is found it usually forms dense colonies because of its mode of reproduction. After the seed germinate, it usually takes four years' growth to produce a flower. The seedling, from the beginning of the second year, produces annual crops of runners and bulbs, each going deeper until the necessary depth, size, and vigor are reached to produce a flower, in

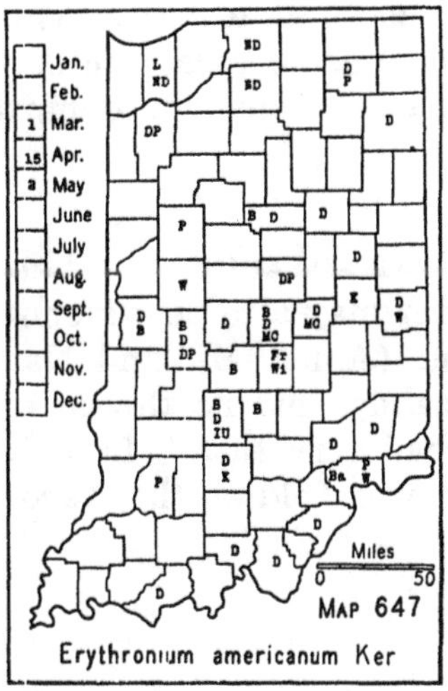

Erythronium americanum Ker

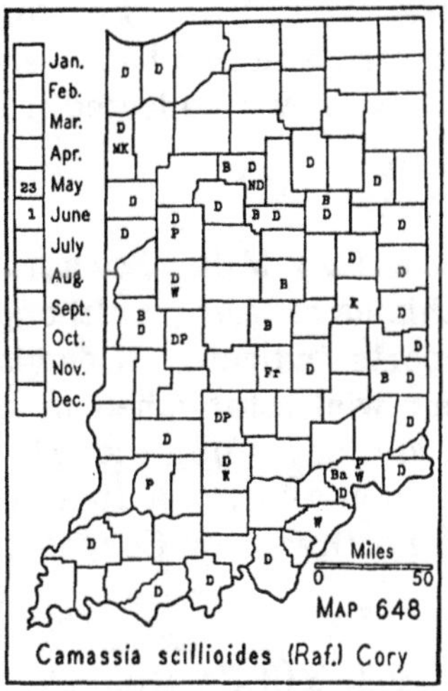

Camassia scillioides (Raf.) Cory

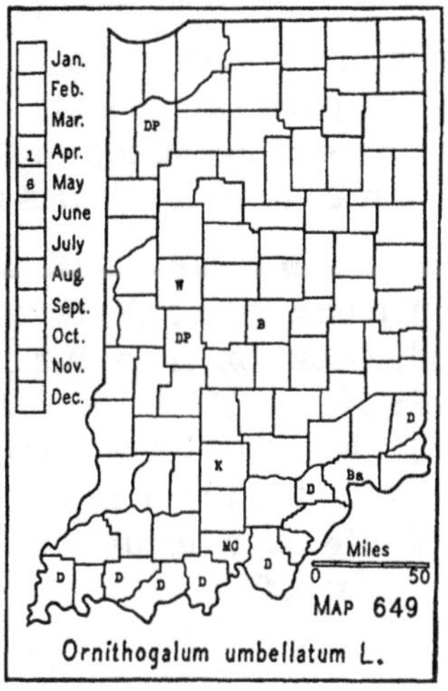

Ornithogalum umbellatum L.

addition to the leaf buds which have been produced each previous year. Considering the great number of single-leaf plants, the number of flowering ones is small. This species prefers rich, moist soil of wooded slopes in beech and sugar maple woods. It is found also in rich soil in almost all kinds of woods and is often abundant on alluvial wooded plains. In Indiana it is more frequent than the preceding species.

N. B., Ont. to Minn., southw. to Fla. and Tex.

1087. CAMÁSSIA Lindl.

1. **Camassia scillioìdes** (Raf.) Cory. (Rhodora 38: 405. 1936.) (*Camassia esculenta* (Ker) Rob. and *Quamasia hyacinthina* (Raf.) Britt.) COMMON CAMAS. Map 648. Moist, wooded slopes, usually bordering streams. It is found throughout the state, becoming rare or absent in the northern counties.

Pa. to Minn., southw. to Ga. and Tex.

1089. ORNITHÓGALUM [Tourn.] L.

1. ORNITHOGALUM UMBELLÀTUM L. COMMON STAR-OF-BETHLEHEM. Map 649. This species has been reported as an escape in many parts of the state. I have found it as an escape in considerable numbers in fallow fields and in open woodland along streams in the counties shown on the map. In some instances it covered an acre or more. The plant grows in such masses that it crowds out all other vegetation, and where it is found it should be exterminated at once.

Nat. of Eu.

1091. MUSCÀRI [Tourn.] Mill.

Flowers globose, 3-5 mm long, not fragrant, deep blue; leaves 6-13 mm wide.......... ....1. *M. botryoides.*

Flowers oblong, urn-shaped 4-5 mm long, fragrant, deep blue; leaves 2-3 mm wide. ....2. *M. racemosum.*

1. MUSCARI BOTRYOÌDES (L.) Mill. COMMON GRAPE-HYACINTH. This species is commonly cultivated and has been reported as an escape in several parts of the state. I have never collected it except in our own orchard where it has escaped.

Nat. of s. Eu. and Asia.

2. MUSCARI RACEMÒSUM (L.) Mill. STARCH GRAPE-HYACINTH. There are only two reports of this as an escape although it may also be wider in distribution than our reports indicate. Nieuwland (Amer. Midland Nat. 3: 107. 1913) says: "Very well established in a sandy field northwest of St. Mary's, Notre Dame, and spreading along a road very fast." In 1910 I found it frequent to common all over a 10-acre clover field on the Aaron Wolfe farm about 7 miles northwest of Corydon.

Nat. of Eu.

1103. YÚCCA [Rupp.] L.

1. YUCCA FILAMENTÒSA L. COMMON YUCCA. This yucca has been reported as an escape several times and remarks have been made as to its persistence and its ability to spread. It is frequently planted in cemeteries from which it has most often escaped. I recall having seen it covering a hillside near a cemetery in Crawford County near the Blue River Church. I also saw it in a woods as an escape from a cemetery in Fulton County. It is so massive that I have never collected it.

In the original Coblentz edition of "Travels in the Interior of North America" published in 1839-41, Prince Maximilian writes of his travels from Owensville, Gibson County to Vincennes, on June 10, 1834, as follows: "The region on the other side [north side of the White River, which he crossed in the vicinity of what is now known as Hazelton] changes considerably; and here appears in a now again sandy soil nearly the same plants as are found in the sandy soil and the prairies of St. Louis, with the addition of a few new ones, a fire-colored lily (Lilium catesbaei), the great-flowered lady slipper (Cypripedium spectabile), a species of Yucca, and many others." It is not known what species Maximilian saw. It may have been this one or *Yucca glauca* Nutt. both of which may have at that time extended up the Mississippi Valley into Indiana.

Nat. from N. C. along the coast to Fla. and westw. to Miss. and Tenn.; beyond this area probably escaped.

1113. ASPÁRAGUS [Tourn.] L.

1. ASPARAGUS OFFICINÀLIS L. GARDEN ASPARAGUS. Map 650. Asparagus has been reported from many counties and I have found it in several. I recall seeing only a few colonies of it, but usually single specimens here and there along roads, railroads, and streams and in fallow grounds and open woodland. We have had it in cultivation for years and I have rarely found a seedling near our cultivated plants but it is sporadic all over our field and orchard and along our fences.

Nat. of Eu.

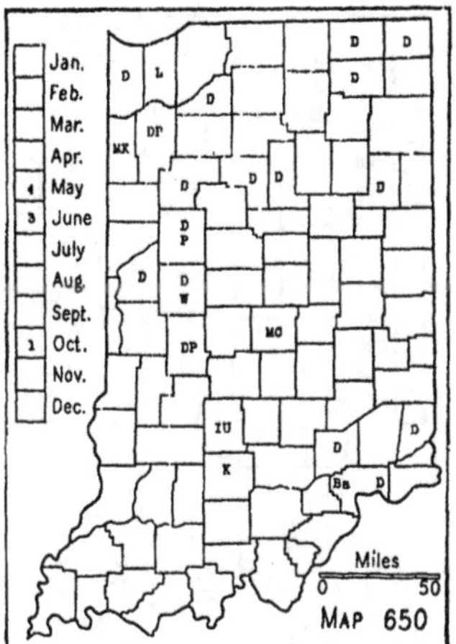

Asparagus officinalis L.

Clintonia borealis (Ait.) Raf.

Smilicina racemosa (L.) Desf.

1117. CLINTÒNIA Raf.

1. **Clintonia boreàlis** (Ait.) Raf. BLUEBEAD. Map 651. This is a very rare plant in Indiana and I have specimens from only three places. I have a specimen collected by Umbach on May 14, 1898, in full flower in a swamp near Miller, Lake County. I have another specimen discovered by M. W. Lyon, Jr., and Mrs. Lyon in a tamarack bog near Dune Park, Porter County. In 1935 I collected a specimen discovered by R. M. Tryon, Jr., in a decadent bog in the eastern part of Porter County.

This species will probably reach extinction in Indiana before long.

Lab. to Man. and Minn., southw. in the mts. to N. C.

1118. SMILACÌNA Desf.

Leaves 2-4, rarely 1; inflorescence pedunculate. (See excluded species no. 139, p. 1034.) .... *S. trifolia.*

Leaves 6-many.

  Inflorescence pedunculate, paniculate; perianth segments 1-2 mm long; leaves not glaucous.

    Panicles on a peduncle usually less than half the length of the panicle, ovoid or pyramidal, 0.7-1.7 dm long, 3-10 cm wide, three eighths to three fourths as broad as long; longest branches of panicle 2-6 cm long and with 8-24 flowers. .... 1. *S. racemosa* var. *typica.*

    Panicles on a peduncle half to one and a fourth times as long as the panicle, nearly cylindric, 4.5-8.5 (–13) cm long, 1.5-3 cm wide, a fourth to three eighths as broad as long; longest branches 1-2.5 cm long and with 6-10 flowers........ 1a. *S. racemosa* var. *cylindrata.*

  Inflorescence sessile or nearly so, racemose; perianth segments 3.4-5.5 mm long; leaves glaucous........ 2. *S. stellata.*

1. **Smilacina racemòsa** (L.) Desf. var. **týpica** Fern. FALSE SOLOMON'S-SEAL. Map 652. Infrequent to frequent throughout the state in beech and sugar maple and black and white oak woods.

This species has recently been studied by M. L. Fernald, who records his studies in Rhodora no. 478 from which I have made my key.

Que. to B. C., southw. to N. S., Va., Tenn., Ill., Mo., Ariz., and Oreg.

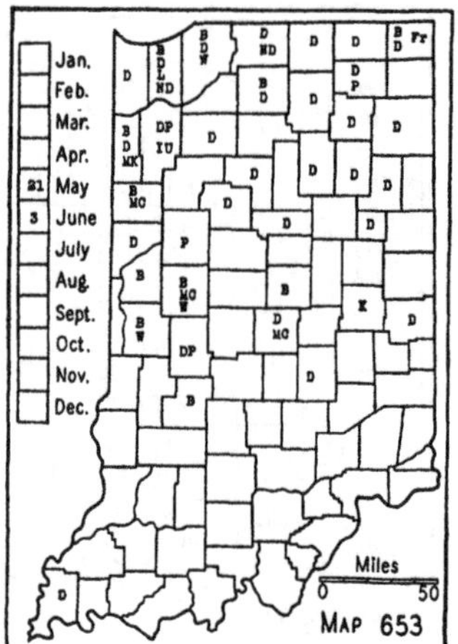

MAP 653
Smilicina stellata (L.) Desf.

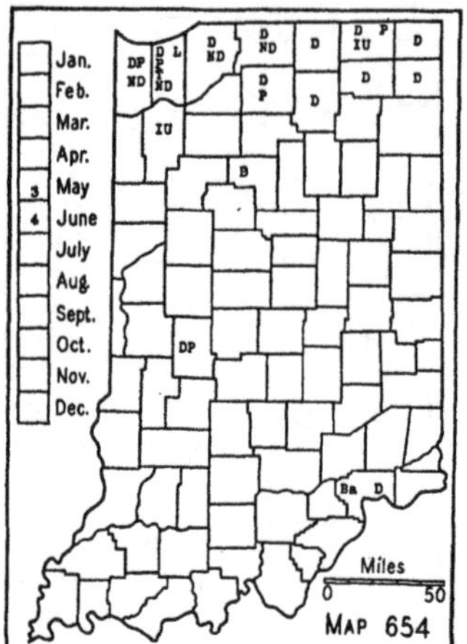

MAP 654
Maianthemum canadense Desv.

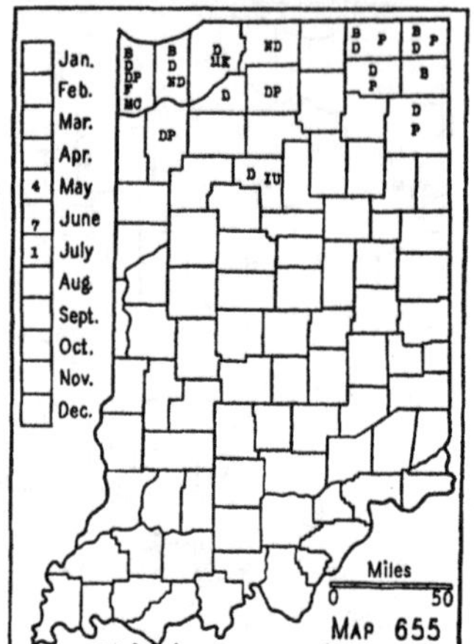

MAP 655
Maianthemum canadense var. interius Fern.

1a. **Smilacina racemosa** var. **cylindràta** Fern. (Rhodora **40**: 406. 1938.) This is the southern form of the species. Although the variety and the typical form of the species overlap with intermediate forms in Indiana, the northern or typical form of the species and the southern form are quite distinct. The two forms are found throughout the state. Both forms are given on one map because the map was made before the variety was recognized.

N. H., N. Y., s. Ont., Ohio, Ill., Kans., and Colo., southw. to Ga. and Ariz.

2. **Smilacina stellàta** (L.) Desf. STARRY FALSE SOLOMON'S-SEAL. Map 653. Infrequent to frequent in the northern half of the state and rare or absent from the southern counties. It prefers moist soil and is most often found on moist slopes and springy banks but is also found on dry banks, in black and white oak woods, and is most abundant on the sand dunes about Lake Michigan. The lower surface of the leaves is very variable, ranging from glabrous to densely short-pubescent. Two varieties of this species have been described and we have both of them, but I think, judging from the descriptions, they are only ecological forms.

Lab. to B. C., southw. to Va., Ky., Kans., and Calif.; also in Eu.

1119. MAIÁNTHEMUM [Weber in] Wiggers

[Butters. Taxonomic studies in the genus Maianthemum. Minnesota Studies in Plant Science 5: 429-444. 1927.]

Lower surface of leaves glabrous; margins of blades merely papillate or crenulate... 1. *M. canadense.*
Lower surface of leaves pubescent at least on the veins; margins of blades ciliate-pubescent.................................... 1a. *M. canadense* var. *interius.*

1. **Maianthemum canadénse** Desf. (*Unifolium canadense* (Desf.) Greene). TWO-LEAF SOLOMON'S-SEAL. Map 654. Infrequent but usually common where found in low woods and in tamarack bogs throughout the northern part of the lake area. I found it in Jefferson County in a low, flat woods in soil of a pH value of 5.6 where it was associated with

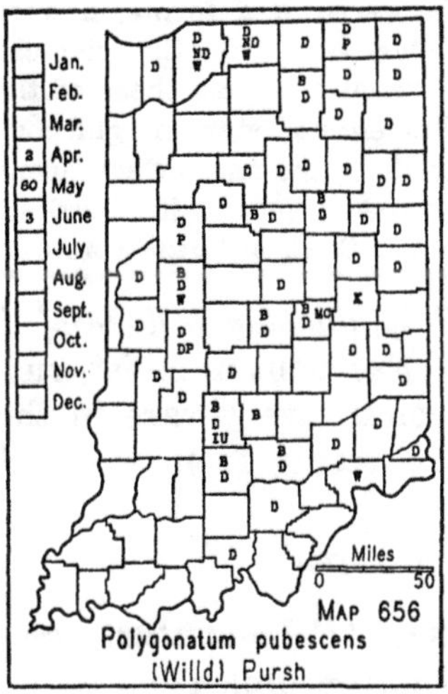

Polygonatum pubescens (Willd.) Pursh

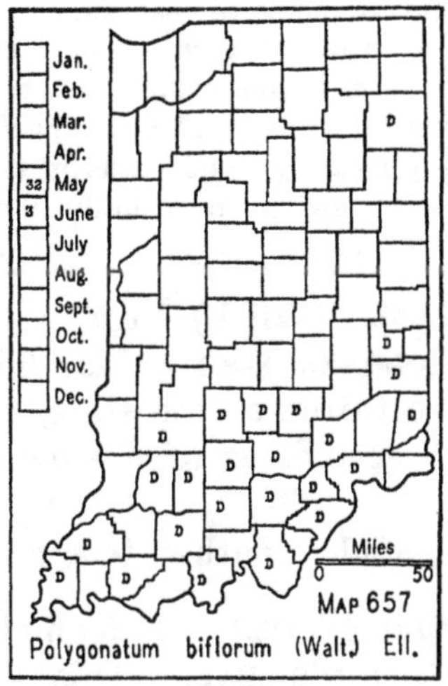

Polygonatum biflorum (Walt.) Ell.

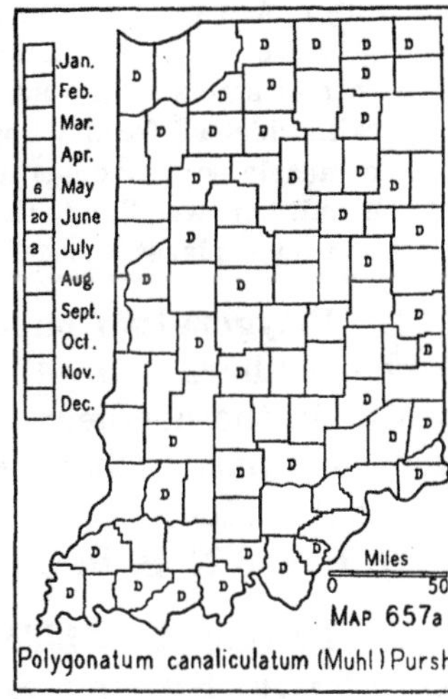

Polygonatum canaliculatum (Muhl.) Pursh

beech, sweet gum, and pin oak. Grimes reported it from Putnam County where it was associated with hemlock.

Plants of this species with 3 leaves have been named *Maianthemum canadense* f. *trifolium* (Farw.) Vict. (Contr. Lab. Bot. Univ. Montreal **14**: 17. 1929.)

Lab. to Md. and in the mts. to N. C., westw. to Minn. and northw.

1a. **Maianthemum canadense** var. **intèrius** Fern. (Rhodora **16**: 211. 1914.) Map 655. The variety has about the same range as the species in the northern part of Indiana but it is not found in the southern part of the state. The species and its variety are of nearly equal abundance but one rarely finds the two in the same colony. The mass distribution of the variety is about the Great Lakes but it has outlying posts as far east as Massachusetts and ranges westward through central Ohio to central Iowa and northward.

1123. POLYGÓNATUM [Tourn.] Hill

[Farwell. Notes on Michigan species of Polygonatum. Bull. Torrey Bot. Club **42**: 247-257. 1915. Gates. A revision of the genus Polygonatum in North America. Bull. Torrey Bot. Club **44**: 117-126. 1917. Bush. The species of Polygonatum. Amer. Midland Nat. **10**: 385-400. 1927.]

Leaves more or less puberulent beneath, at least on the veins, 6-16 in number, 2-6 cm wide and 4-13 cm long, narrow- to wide-elliptic, or slightly ovate-elliptic, generally broad at the base, rarely somewhat cuneate; peduncles usually beginning at the second leaf axil, rarely beginning at the first or at the third leaf axil, 1-4-flowered, usually 1- or 2-flowered, or a mixture of 1 and 2 flowers; flowers 7-12 mm long; filaments of stamens more or less papillose; stems usually beginning to curve above the second leaf; rhizomes near the surface; plants usually found in thick woodland. ....1. *P. pubescens.*

Leaves glabrous beneath, mostly 10-21 in number; blades of wideleaf forms 2.5-10.5 cm wide and 12-21 cm long, very broadly ovate-elliptic or broadly elliptic, those of the narrowleaf forms 1.5-3 cm wide and 6.5-14 cm long, mostly narrow-elliptic or oblong-elliptic; peduncles very variable in length, rarely more than one from an axil, the first one generally from the third to the fifth leaf axil of the wideleaf forms and from the second to the third leaf axil of the narrowleaf forms; flowers single or in twos on the narrowleaf forms and in clusters of 2-8 on the wideleaf forms; stems of the narrowleaf forms usually much more curved than those of the tall, wideleaf forms; flowers mostly 15-19 mm long; filaments of stamens glabrous or nearly so; rhizomes deep in the ground (usually 1-1.5 dm); the narrowleaf forms usually in woodland, the wideleaf forms usually along roadsides and in clearings and open places.......................................2. *P. biflorum* complex.

1. **Polygonatum pubéscens** (Willd.) Pursh. (*Polygonatum biflorum* of recent authors.) HAIRY SOLOMON'S-SEAL. Map 656. Frequent in moist, rich woods in the northern two thirds of the state, becoming very rare in the southern part. I have 86 specimens of my own collecting from which I made this study.

N. B. and N. S. to Ont., Mich., southw. to Fla. and Tex.

2. **Polygonatum biflòrum** (Walt.) Ell. (complex). (*Polygonatum commutatum.*) SMOOTH SOLOMON'S-SEAL. This species complex is found throughout the state: the tall plants with wide leaves usually along roadsides and fences and in open places in general except in cultivated fields; the small plants with narrow leaves are generally found in moist woodland. The fleshy, insipid fruit is eaten by birds and the stony seeds are widely distributed. The wideleaf form is of a somewhat weedy nature. The rhizomes are deep in the ground and if the terminal end is broken off the plant persists. For this reason it is difficult to eradicate from flower beds.

My study of this species complex was made from 155 specimens which I have collected from all parts of the state. I am not satisfied with the treatment of this species but I am not able to find differences sufficient to distinctly separate them. My specimens form a lineal series and when I have used the term wideleaf and narrowleaf forms it is in a general sense. I do not think they are all the same species and I think a character may sometime be found that will separate them satisfactorily. The genus has been monographed by three authors and my specimens have been seen by one of them but I can not accept their treatment of this complex.

Since the preceding was written a monographic study of the genus has been undertaken by Miss Ruth E. Peck who has studied all my specimens. I now learn that this complex is composed of at least *Polygonatum biflorum* (Walt.) Ell. and *Polygonatum canaliculatum* (Muhl.) Pursh. See maps 657 and 657a. I refer students of this complex to the forthcoming monograph.

A form of this species from St. Joseph County was described by McGivney (Amer. Midland Nat. 9: 662-664. 1925.) under the name of *Polygonatum commutatum* f. *ramosum* McGivney. It differs from the species by having short branches in the leaf axils and is our only report of this form.

Western N. H. to Man. and Rocky Mts., southw. to Ga., La., N. Mex., and Ariz.

1128. CONVALLÀRIA L. LILY-OF-THE-VALLEY

See excluded species no. 140, p. 1035.

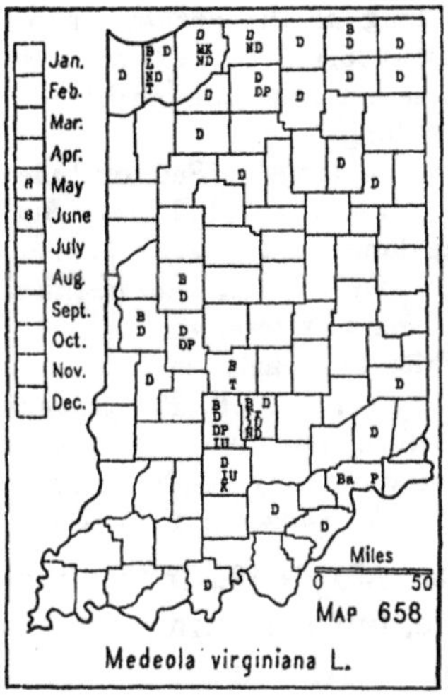

Medeola virginiana L.

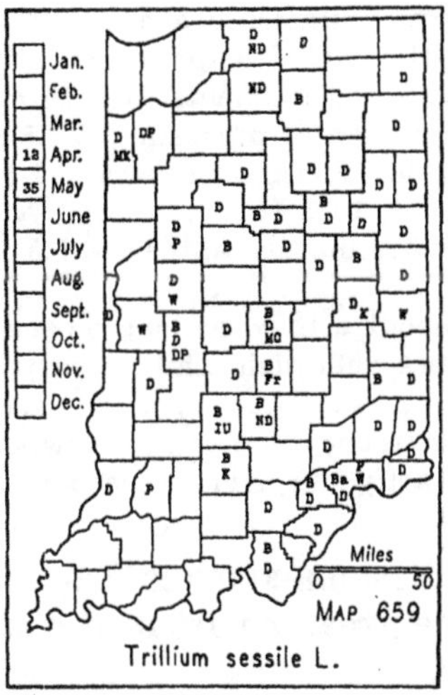

Trillium sessile L.

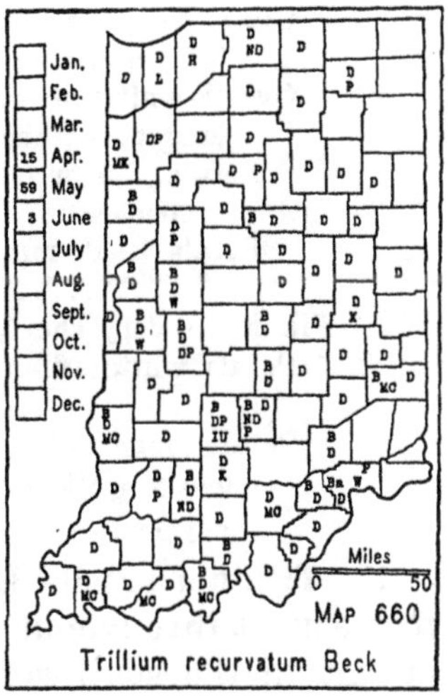

Trillium recurvatum Beck

1135. MEDÈOLA [Gronov.] L.

1. **Medeola virginiàna** L. INDIAN CUCUMBER-ROOT. Map 658. Infrequent throughout the state although there are no records from the southwestern counties. In the hilly counties it is found mostly in deep, wooded ravines and northward it is found mostly in beech woods, on rather acid, sandy flats and on the lower parts of slopes about lakes and swamps.

N. B. to Minn., southw. to Fla. and Tenn.

1138. TRÍLLIUM L.

[W. A. Anderson. Notes on the flora of Tennessee. Rhodora **36**: 119-128. 1934.] Note: Description and measurements of the floral parts in the key apply to flowers in and after anthesis.

Flowers sessile.

Leaves sessile; sepals not reflexed; petals not clawed.

Petals maroon..1. *T. sessile.*

Petals greenish yellow.......................................1a. *T. sessile* f. *luteum.*

Leaves petiolate, sometimes very shortly so; sepals reflexed; petals clawed.

Petals maroon...2. *T. recurvatum.*

Petals greenish yellow...........................2a. *T. recurvatum* f. *luteum.*

Flowers pedunculate.

Leaves petiolate, oval or ovate, obtuse; petals white; filaments as long as the ovary; among the first herbaceous plants to flower in Indiana..............3. *T nivale.*

Leaves sessile or essentially so, sometimes 1 of the 3 with a petiole a few mm long.

Stigmas slender and of uniform diameter, straight and not curved or coiled at the tip, or only slightly so, erect or spreading; petals usually very large and obovate, white, turning pink with age, their bases ascending, the upper part spreading; anthers exceeding the stigmas, mostly 10-15 mm long; ovary white, small, globose at maturity; peduncles 3-10 cm long, well above the leaves..4. *T. grandiflorum.*

Stigmas short, stout, tapering from the base to the apex, recurved or coiled at the tip, about half as long as the ovary; petals lanceolate, ovate, oblong-oval or obovate, spreading from the base; anthers usually not exceeding but only equaling the stigmas; peduncles erect, horizontal or sometimes declined beneath the leaves.

Filaments half as long as the anthers or longer.

Ovary very dark purple; filaments about half as long as the anthers, about 3.5-4 mm long. (See excluded species no. 142, p. 1035.) .... *T. erectum.*

Ovary white or nearly so; filaments two thirds as long as the anthers or longer; anthers pinkish or purplish, 2.5-6.5 mm long.

Petals 5-9 mm wide; mature anthers 2.5-4.5 mm long. (See excluded species no. 141, p. 1035.) .... *T. cernuum.*

Petals 10-17 mm wide; mature anthers 4-6.5 mm long.. 5. *T. cernuum* var. *macranthum.*

Filaments very short, about a third as long as the anthers or less.

Petals white; filaments yellowish white; ovary white or nearly so.......... .... 6. *T. Gleasoni.*

Petals purplish or maroon; filaments yellowish white, purplish or maroon; ovary white or partly purplish or maroon, rarely entirely reddish brown. .... 7. *T. Gleasoni* f. *Walpolei.*

1. **Trillium séssile** L. SESSILE-FLOWER TRILLIUM. Map 659. Infrequent to frequent throughout the greater part of the state, but becoming rare to absent in the southwestern counties. It is found mostly in rich, moist woods.

I have had plants with 4 and 5 leaves and one with greenish yellow petals under cultivation and they have come true for at least 10 years. I also have plants with 2 and 3 stems from the same rootstock. In one instance one stem has 3 leaves and the other has 4 leaves.

Pa. to Minn., southw. to Va., Tenn., Ark., and La. (Brown).

1a. **Trillium sessile f. lùteum** (Muhl.) Peattie. (Jour. Elisha Mitchell Soc. 42: 197. 1927.) This is a form with greenish yellow petals which I have found in Adams, Allen, and Wells Counties. Beyer (Torreya 27: 83. 1927) names this form f. *viridiflorum,* but since Peattie's treatment antedates Beyer's by four months, Peattie's name is used here.

2. **Trillium recurvàtum** Beck. REFLEXED-SEPAL TRILLIUM. Map 660. Infrequent throughout the state. All of my specimens are from woodland of different kinds although I recall seeing the species along the railroad south of Battle Ground, Tippecanoe County.

Ohio, Mich. to Minn., southw. to w. Tenn. and Ark.

2a. **Trillium recurvatum f. lùteum** Clute. (Amer. Bot. 28: 79. 1922.) Reported from Monroe County by Friesner, and there are specimens in the herbarium of Butler University from Lawrence and Montgomery Counties. There is a specimen from Johnson County in the herbarium of Franklin College.

3. **Trillium nivàle** Riddell. SNOW TRILLIUM. Map 661. In rocky or gravelly soil in protected places on steep, wooded slopes, usually along or near streams. It is local to very local and probably closely restricted to the area indicated on the map after the reports from Clark, Decatur, and Marion Counties are added. On account of its very early appearance it may not have been collected in many places where its occurs, and it may be more widespread than the reports indicate. Authors do not mention that the stem in cross section is hexagonal with the angles more or less winged and minutely roughened.

Western Pa. to Minn., southw. to Ky. and Iowa.

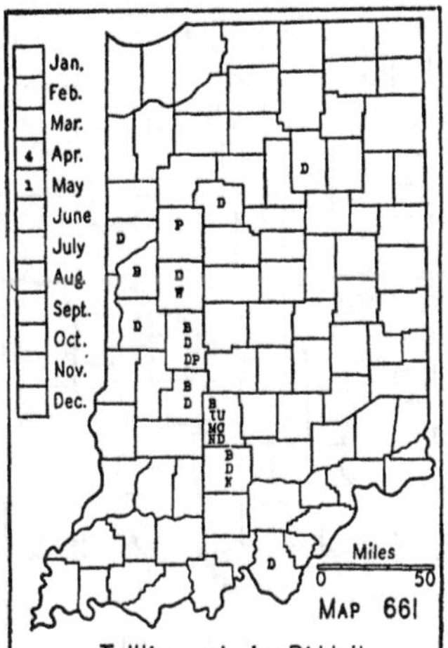

Trillium nivale Riddell

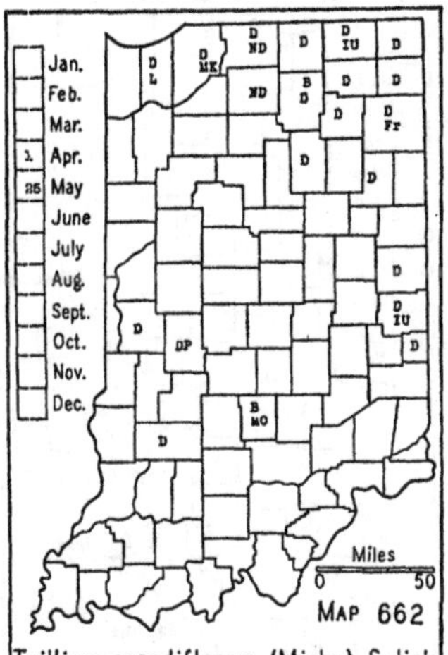

Trillium grandiflorum (Michx.) Salisb.

Trillium cernuum var. macranthum Eames & Wieg.

4. **Trillium grandiflòrum** (Michx.) Salisb. LARGE-FLOWER TRILLIUM. Map 662. This species is frequent in the northeastern part of the state, becoming infrequent to rare southward. Friesner reports it from Harrison County. It is found mostly in moist woods and is also found in springy places in woods, on moist, wooded slopes, and in marshes.

W. Vt., w. Que. to Minn., southw. to N. C. and Mo.

5. **Trillium cérnuum** L. var. **macránthum** Eames & Wieg. (Rhodora 25: 191. 1923.) PURPLE-ANTHER TRILLIUM. Map 663. I found a few specimens in a low, wet woods 8 miles east of Michigan City, La Porte County, and Peattie reports finding it in moist woods near Mineral Springs, Porter County. This variety is northern in its general distribution and in Indiana it is found only in the northern counties.

Vt., Ont. to Sask., southw. to Pa., n. Ind., n. Ill., Wis., and Minn.

6. **Trillium Glèasoni** Fern. (Rhodora 34: 21. 1932.) (*Trillium declinatum* (Gray) Gleason). Map 664. This is our common Trillium found throughout the state. The fact that the peduncles are often erect as well as horizontal or declined has led early authors to determine it also as *Trillium erectum* or *Trillium cernuum.* The early records are so confused that they should be disregarded. Friesner (Butler Univ. Bot. Stud. 1: 34-36. 1929), after an exhaustive study of this species and its forms, concluded that there were not two forms (*Trillium erectum* and *Trillium Gleasoni*) in this state, and that our plant is one highly variable species. It is now evident that the true *Trillium erectum* occurs east and south of Indiana. *Trillium cernuum,* likewise, is out of our area. Its range is chiefly along the Coastal Plain and it is represented here only by its var. *macranthum.*

Southern Mich. to s. Minn., southw. to Ohio and Mo.

6a. **Trillium Gleasoni** Fernald forma **Walpòlei** (Farw.) Deam, comb. nov. (*Trillium cernuum* var. *declinatum* Gray f. *Walpolei* Farw. Rept. Mich. Acad. Science 21: 363. 1920.) Map 664a. This is a form which is

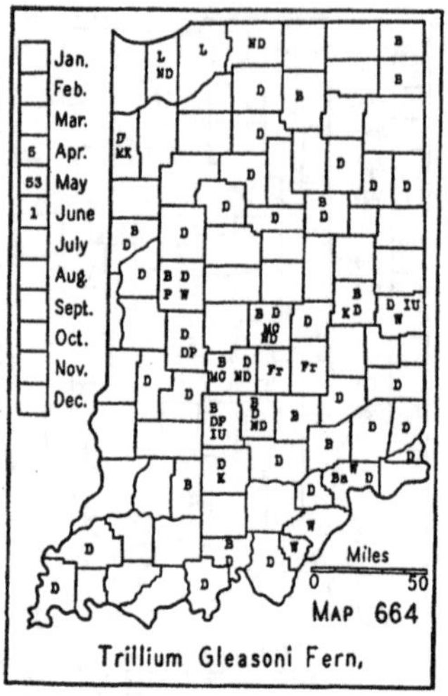

MAP 664
Trillium Gleasoni Fern.

MAP 664a
Trillium Gleasoni f. Walpolei (Farw.) Deam

MAP 665
Aletris farinosa L.

described as having the petals, and often the filaments and anthers maroon. Specimens occur in this state which also have the ovaries partly or entirely reddish brown. These should not, however, be confused with either *Trillium erectum,* which has a very dark purple ovary or with *Trillium cernuum* var. *macranthum,* which has purplish anthers. Both of the last named forms have much longer filaments than f. *Walpolei.* Doubtless this is *Trillium cernuum* var. *atrorubens* Wood. (Rept. Indiana Geol. Survey 2: 286-287. 1871.)

This form seems to occur with the species in all parts of the state.

1143. ÁLETRIS L.

1. **Aletris farinòsa** L. STARGRASS. Map 665. Infrequent throughout the northern part of the state as indicated on the map. In addition there are reports from Floyd and Vigo Counties and Schneck says it was found in prairies in the Lower Wabash Valley but is nearly extinct. It is found in moist, sandy soil in wet or moist prairies, in prairie habitats in open woods, and in open woods. I have made repeated attempts to establish this species in our garden but it fails in a few years although I have transplanted it into both neutral and sandy soils with an abundance of the original soil.

Southern Maine to Minn., southw. to Fla. and La.

1151. SMÌLAX [Tourn.] L.

[Pennell. Smilax, subgenus Nemexia (Raf.), in the eastern United States. Bull. Torrey Bot. Club 43: 409-421. 1916.]

Plants herbaceous, without prickles.

Mature leaves not glaucous beneath but sparsely pubescent with colorless hairs, ovate-oblong, very thin, yellow green, glossy above and beneath, mostly cordate at the base, sometimes subcordate or even truncate, usually long-acuminate at the apex, the margins generally erose and usually more or less ciliate with long and short, colorless hairs, blades not decurrent on the petioles or scarcely so; segments of staminate flowers mostly 4-5 mm long, lanceolate; fruit black..1. *S. pulverulenta.*

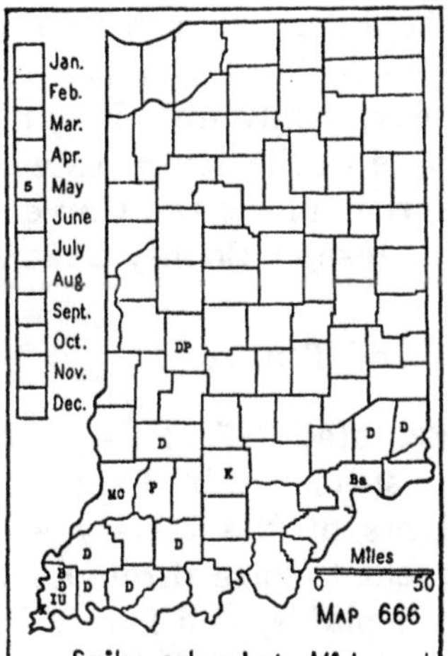

Smilax pulverulenta Michx.

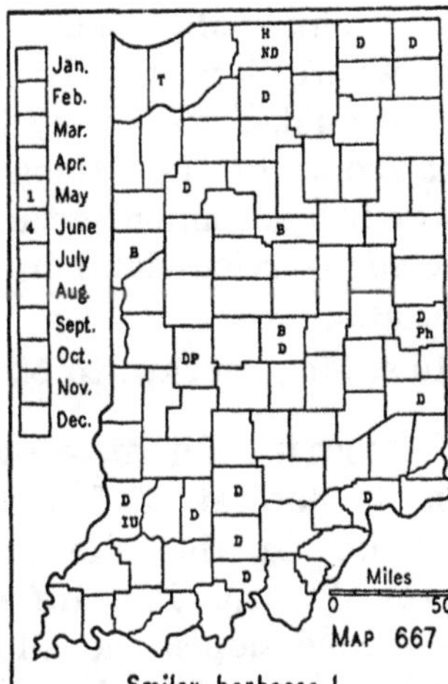

Smilax herbacea L.

Smilax herbacea var. lasioneura (Hook.) A. DC.

Mature leaves glaucous beneath, of an ovate type, bluish green, cordate, subcordate or truncate at the base, short-acuminate at the apex; margins not conspicuously erose and lacking the colorless hairs or with a few short ones; fruit glaucous.

Leaves all glabrous beneath; bracts on the stem below the leaves appressed; umbels of both pistillate and staminate plants generally with 25-80 flowers.......... ....2. *S. herbacea.*

Leaves more or less pubescent beneath with a mixture of short and long colorless hairs; bracts on the stem below the leaves loose.

Umbels of pistillate and staminate plants with more than 25 flowers, usually 30-110 flowers, sometimes fewer on branches; leaves of an ovate type, sometimes very broad and sometimes narrow, especially on branches; peduncles usually from the axils of leaves but sometimes 1-3 below the leaves, usually longer to many times longer than the petioles, rarely shorter; plants usually tall and very large ones in exposed places often bent over, generally 1-3 m long......................2a. *S. herbacea* var. *lasioneura.*

Umbels of pistillate plants generally with fewer than 20 flowers; umbels of staminate plants generally with fewer than 25 flowers (in counting the flowers add the scars on the peduncle denoting fallen flowers); mature leaves large, broadly ovate, 11-17 cm long and 9-12 cm wide, few, usually 4-9 to a plant, not tendril-bearing; peduncles 1 or 2 below the leaves or opposite the lowest leaf, shorter than the petioles; plants of low woods, 4-6 dm high...3. *S. ecirrhata.*

Plants woody, vines, with prickles.

Leaves glaucous beneath..4. *S. glauca.*

Leaves green beneath.

Stem more or less stellate-pubescent at least near the base; leaves mostly more or less contracted near the middle; denticulations and prickles of leaf margins generally colored; margin of leaf thicker than the blade......5. *S. Bona-nox.*

Stem glabrous; leaves not contracted near the middle (rarely leaves of a plant contracted); denticulations and prickles of leaf margins generally colorless; margin of leaf not thicker than the blade.

Prickles of stem all more or less flattened, the lower half green; peduncles generally shorter than the petioles; fruit more or less glaucous; seed usually 2 or 3..6. *S. rotundifolia.*

Prickles of stem round and black; peduncles longer than the petioles; fruit black, not glaucous; seed usually 1.........................7. *S. hispida.*

1. **Smilax pulverulénta** Michx. Map 666. Infrequent to rare in the southern part of the state where it is usually found in hard, dry soil on wooded slopes, associated with oaks, and rarely in dry, sandy soil. I found it also in a low post oak flat south of Half Moon Pond in the southwestern part of Posey County. There is a specimen in the herbarium of the University of Notre Dame collected by Nieuwland in St. Joseph County that I doubtfully refer to this species.

Southern N. Y., s. Ind. to s. Mo., southw. to N. C. and Tenn.

2. **Smilax herbàcea** L. SMOOTH CARRION-FLOWER. Map 667. This plant is variable in size and in its habitat. I found a specimen in Franklin County that was 15 feet long. On the whole, plants of the variety are larger than those of the typical form. It is infrequent to rare throughout the state and is found on wooded slopes and alluvial plains, and rarely in the open, usually associated with beech and white oak.

Ont. to Nebr., southw. to Ga., Ala., and Mo.

2a. **Smilax herbacea** var. **lasioneùra** (Hook.) A. DC. Map 668. The variety has the range and habitats of the species but is much more frequent, especially in sandy soil in the northern part of the state where it is somewhat frequent along roadsides.

I admit that I do not know this species and its variety and the next species. My attention was first called to them when Pennell revised this section of the genus. For several years I have been assembling the aberrant forms in our garden with the hope that a study of them would solve the problem, but I have found that the plants of this genus grow very slowly and most of the specimens I have planted are not yet old enough for study. Ordinarily each rhizome sends up one stem, although I have one plant that had 8 stems in 1936 and 9 in 1937. The number of stems from a single rhizome seems yet to be ascertained. This variety is so extremely variable that it seems that no character will hold to separate a large series of plants. I think the complex consists of several forms that might well be recognized. We have some plants that reach 4-6 feet in height that are simple and others that are usually 3-5 feet that are so much branched that they form a compact mass. Some plants will be in flower when others are just peeping through the ground. Some will have 1-3 long peduncles below all the leaves while others will have the peduncles mostly about the middle of the stem. Some plants have wide leaves, few peduncles, and only a very few tendrils, and in all the herbaria I have visited they are referred to *Smilax ecirrhata*, but they can always be separated from that species by their tendrils and, what I think to be our best character, the fewer-flowered umbels. Under this variety I have included several forms which I hope can be satisfactorily distinguished by someone in the future.

Ont. and Ohio to Wyo., southw. to Ga., Ala., and Colo.

3. **Smilax ecirrhàta** (Engelm.) Wats. Map 669. Probably infrequent to rare throughout the state, although I recall seeing it rather frequently in the Lower Wabash Valley in low woods bordering sloughs, especially in

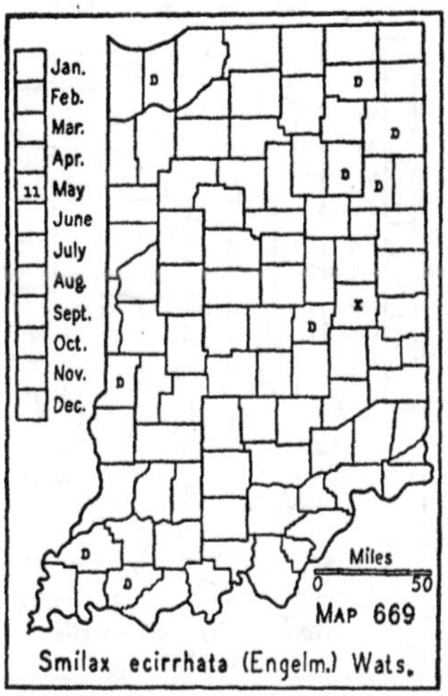

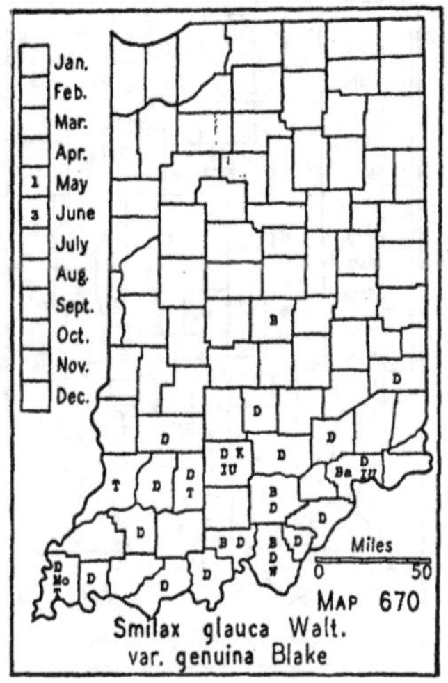

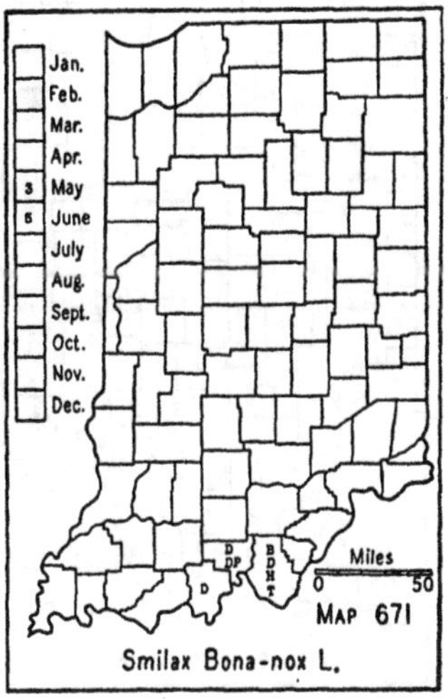

Gibson and Vigo Counties. All of my specimens are from low woods on the flood plains of streams.

Ont. to Minn., southw. to Tenn. and Mo.

4. **Smilax glaúca** Walt. var. **genuìna** Blake. SAWBRIER. Map 670. Our plant is the typical form of the species, distinguished by Blake (Rhodora **20**: 78-80. 1918) as var. *genuina*. Infrequent to common in the hilly counties of the southern part of the state and extending as far northward as Marion and Putnam Counties. It is found in open woodland and in fallow and abandoned fields. When it becomes established in cultivated ground, it is difficult to eradicate on account of its deep, tuberous rhizomes which, when broken, send up new stems.

Va. to s. Ill., southw. to Fla. and Tex.

5. **Smilax Bòna-nóx** L. FRINGED GREENBRIER. Map 671. I have seen specimens from only the counties indicated on the map and I think Andrews' report from Monroe County can safely be transferred to the next species. So far it has been found only on the high hills near the Ohio River where it is usually associated with the next species.

Va., s. Ind. to Kans., southw. to Fla. and Tex.

6. **Smilax rotundifòlia** L. ROUNDLEAF GREENBRIER. Map 672. This species is rare to infrequent in the northern counties, rare or absent in many of the counties in the Tipton Till Plain, becoming frequent to common in the southern counties. It is found in dry soil in woods, clearings, and abandoned fields where it often forms impenetrable thickets. It forms long vines, and, on account of its many prickles, it is an objectionable plant.

N. S. to Iowa, southw. to Ga. and Tex.

7. **Smilax híspida** Muhl. HISPID GREENBRIER. Map 673. An infrequent species throughout the state. It prefers a moist, rich soil.

Conn., Ont. to Minn., southw. to Va., Tenn., and Tex.

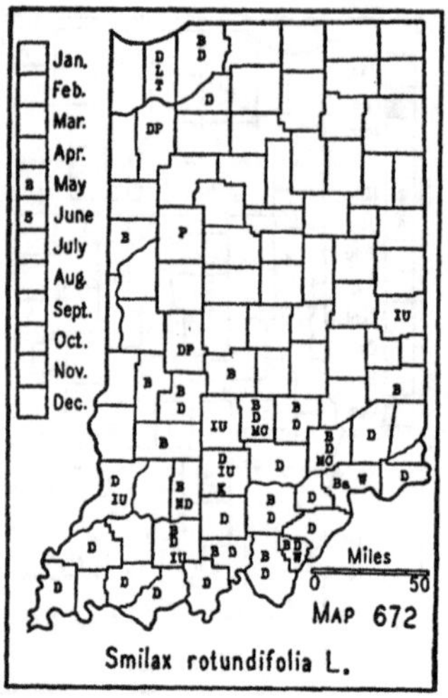

MAP 672
Smilax rotundifolia L.

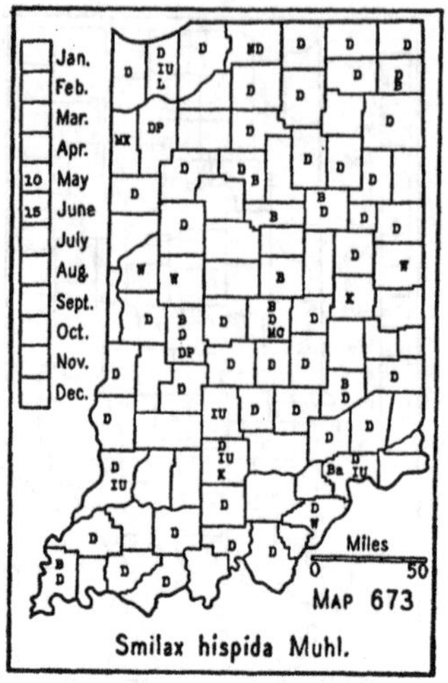

MAP 673
Smilax hispida Muhl.

MAP 674
Hymenocallis occidentalis (Le Conte) Kunth.

40. AMARYLLIDÀCEAE Lindl. AMARYLLIS FAMILY

Bulbous herbs with flowers on scapes.
 Flowers clustered; filaments united in a cup-shaped crown; anthers long-exserted. ....1194. HYMENOCALLIS, p. 328.
 Flowers solitary.
  Perianth naked in the throat......................1181. ZEPHYRANTHES, p. 328.
  Perianth with a crown in the throat....................1201. NARCISSUS, p. 329.
Bulbless herbs with rootstocks or corms.
 Tall plants with large, fleshy, basal leaves; flowers not yellow; anthers versatile..... ....1219. AGAVE, p. 329.
 Low plants with linear leaves; flowers yellow; anthers not versatile............. ....1230. HYPOXIS, p. 329.

1181. ZEPHYRÁNTHES Herb.

See excluded species no. 146, p. 1036.

1194. HYMENOCÁLLIS Salisb.

1. **Hymenocallis occidentàlis** (Le Conte) Kunth. SPIDERLILY. Map 674. The spiderlily grows in low woodland that is usually inundated each year and in soil which is comparatively free from organic matter and which becomes very hard during the summer months. The bulbs are usually 6-9 inches below the surface in a compact, blue clay. It is rather frequent in the southern part of Posey County where its habitat is frequent and local elsewhere. It is found in the Big Creek bottoms near Wadesville, Posey County, local in the bottoms along Pigeon Creek in the northern part of Warrick County and the southern part of Gibson County, and local in its habitat along Little Pigeon Creek in Spencer County. It doubtless has a range wider than the location given above, but, as I understand its habitat, it will be restricted to the peculiar low places along streams and low spots in woods of the southwestern counties. Where it is found it is comparatively abundant.

In a restricted habitat southw. from s. Ind. to Ga. and Mo.

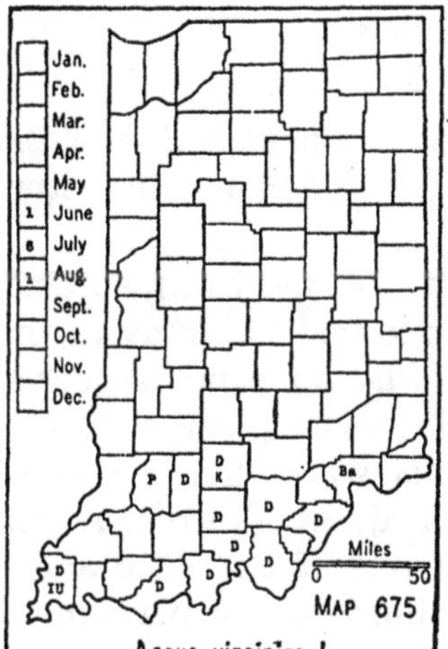

Agave virginica L.

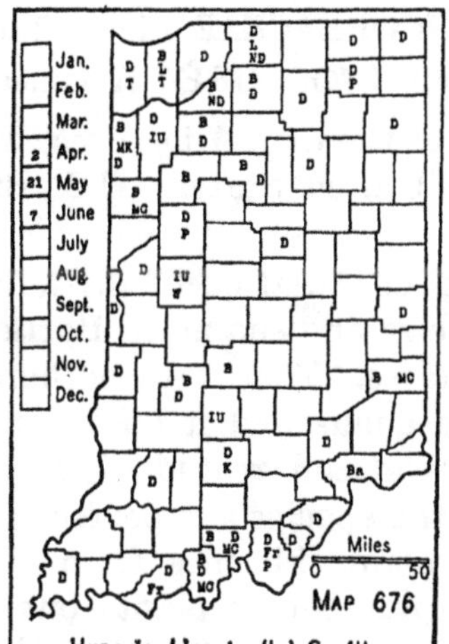

Hypoxis hirsuta (L.) Coville

Dioscorea hirticaulis Bartlett

1201. NARCÍSSUS [Tourn.] L.

Flowers white, crown small, usually much less than half as long as the perianth segments, crisped, red-edged. (See excluded species no. 147, p. 1036.) . . . *N. poeticus*.

Flowers yellow, crown equaling or exceeding the perianth segments. (See excluded species no. 148, p. 1036.) . . . . *N. Pseudo-Narcissus*.

1219. AGÀVE L.

1. **Agave virgínica L.** (*Manfreda virginica* (L.) Salisb.) FALSE ALOE. Map 675. Local but rather frequent in southern Indiana. It is generally found only in soil of low fertility in open places on the crests and spurs of post oak and black oak ridges. It is frequent also in the post oak flats of the southwestern part of Posey County. The plants are usually 3-5 feet high and not branched. It is perfectly hardy at Bluffton and does well in black loam soil. In 1932 we had one plant that was 6.4 feet high and that had a long, flowering branch at almost every node, eight branches in all. Ralph M. Kriebel found a large colony on top of a limestone bluff along White River about a mile below Tunnelton in Lawrence County, which had by actual count about 2000 individuals. Outside the range shown on the map it has been reported from Daviess, Jefferson, and Scott Counties.

Va. to s. Ohio, s. Ind. to Mo., southw. to Fla. and Tex.

1230. HYPÓXIS L.

1. **Hypoxis hirsùta** (L.) Coville. GOLDEYE-GRASS. Map 676. Infrequent throughout the state but usually common where it is found, especially in marshland in moist, prairie habitats. It seems to prefer an acid habitat but I have seen it growing in marly bogs with *Parnassia*. In southern Indiana it is found in rather sandy soil on the crests of black oak ridges, on sandstone outcrops, and in the post oak flats, while in the northern part of the state it is usually found in sandy soil at the base of black oak slopes, in mucky soil in marshes, and in moist, black sandy soil in prairies.

Sw. Maine to Sask., southw. to Fla., e. Kans., and Tex.

43. DIOSCOREÀCEAE Lindl. YAM FAMILY

1252. DIOSCORÈA [Plumier] L.

[Bartlett. The source of the drug Dioscorea, with a consideration of the Dioscoreæ found in the United States. U.S. Bureau of Plant Industry, Bull. 189: 1-29. 1910.]

The rhizomes of the species that occur in Indiana are used in medicine. Their great variation in size and shape led Bartlett to make a study of the species of the United States. There are authors who have not accepted Bartlett's division of the genus and it offers an interesting study to one with accumulated data who can restudy the genus with all the species under cultivation. All of the species are perfectly hardy at Bluffton. About 10 years ago I began to plant rhizomes from all parts of the state and I now have a considerable number of plants but failure to use permanent labels prevents me from drawing conclusions. The following key and treatment of our species should be regarded as only provisional until our species are better understood.

Lower leaves alternate, approximate, or in whorls of 3 (or 4); rhizomes linear, the older ones with lateral branches, dried ones generally 5-8 (10) mm in diameter.

Internodes, at least the lowermost, more or less spreading-pubescent with stiff, colorless or reddish brown hairs mostly 0.2-0.5 mm long; lower surface of the leaves pubescent or glabrous; staminate inflorescence axillary, in short and narrow panicles up to 6 cm long; pistillate inflorescence in axillary racemes, developing up to 10 triangular capsules; capsules usually slightly obovoid or elliptic and broader than long, up to 24 mm long; seed broadly winged, up to 16 mm long, the body of the seed mostly about 5 mm wide..................1. *D. hirticaulis.*

Internodes glabrous; staminate inflorescence axillary, in widely spreading panicles up to 12 cm long; pistillate inflorescence, capsules, and seed similar to the preceding but the racemes usually developing 5 or 6 capsules..........2. *D. villosa.*

Lower leaves in whorls of 4-9, mostly of 5-7, ovate-cordate; rhizomes more or less contorted or, if linear, with many short, knoblike branches, usually (8) 10-15 mm in diameter.

Leaves glaucous beneath, usually until maturity; leaves of lowest whorl 5-9, generally 6, the margins rarely somewhat undulate, sparsely pubescent on the principal veins beneath, rarely a plant with dense pubescence; petioles at the insertion of the blade usually more densely pubescent than the blade, glabrous nearly to the base; nodes of stem usually minutely puberulent at the base of the petioles; capsules up to 6 in a raceme, up to 30 mm long, variable in shape; sometimes broadly elliptic and obovoid ones found on the same raceme; seed up to 20 mm long, the margins colorless, body orbicular, up to 5 mm in diameter; rhizomes generally about 10 mm in diameter, contorted, extremely variable in shape, the many laterals diverging in all directions........................3. *D. glauca.*

Leaves green beneath, those of the lowest whorl generally 4-6; margins of the first whorl of leaves and often the second and third whorl conspicuously undulate; lower surface of blades glabrous (although there are specimens with the lower surface densely pubescent that are referred to this species complex); petioles of typical specimens glabrous at insertion of the blade as well as at the base; internodes generally glabrous (except the pubescent forms); capsules like the preceding but usually much larger; seed similar but larger and with a brown wing; body orbicular and about 5 mm in diameter; rhizomes mostly about 15 mm in diameter, generally of a linear type but with numerous knoblike laterals. ....4. *D. quaternata.*

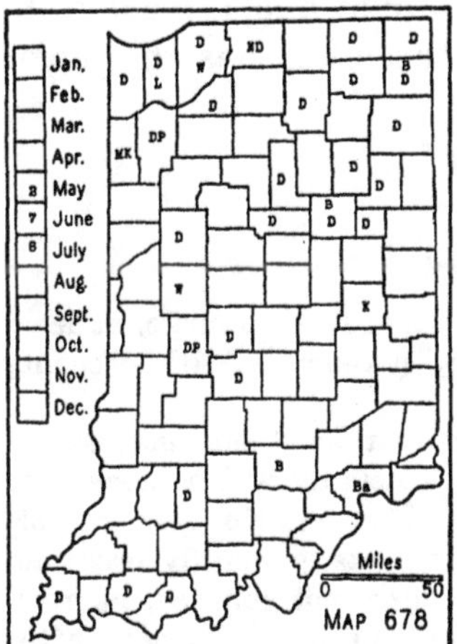

Dioscorea villosa L.

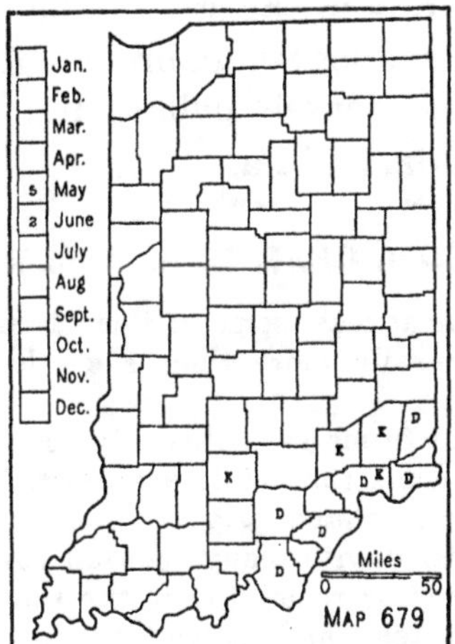

Dioscorea glauca Muhl.

Dioscorea quaternata (Walt.) Gmel.

1. **Dioscorea hirticaúlis** Bartlett. Map 677. This species is found in low woodland that usually is inundated at some time of the year, associated with pin oak, sweet gum, red maple, and black gum. Like all the other Indiana species it has both glabrous and pubescent forms and I have not seen intermediates. I think that they are distinct but a paucity of specimens does not warrant a decision in the matter.

Va., N. C. to Ga. and Ind.

2. **Dioscorea villòsa** L. WILD YAM-ROOT. Map 678. Rather frequent in the northern half of the state, becoming rare or infrequent in the southern part. It prefers moist soil of rich woodland. The lower surface of the blades of all the specimens I have seen is densely pubescent. The glabrous variety has not been found in the state. This species and the preceding can be determined definitely only when the whole plant, including the rhizome is at hand. The long, slender rhizomes with few or no laterals are conclusive in naming this species.

Mass. to Minn., southw. to Va. and Tex.

3. **Dioscorea glaúca** Muhl. (*Dioscorea quaternata* var. *glauca* (Muhl.) Fern. Rhodora **39**: 399-400. 1937.) Map 679. This species prefers slopes of deep ravines and is usually associated with beech and sugar maple. When the leaves are not glaucous beneath this species is difficult to separate from the next species. Small, in his Flora of the Southeastern States, separates them on the size of the staminate flowers. In the typical form the lower surface of the leaves is sparsely pubescent on the principal nerves; the number of leaves in the basal whorl is usually 6, their margins rarely undulate; rhizomes much branched. The wings of the seed of all of my plants are white while those of the next species are brown.

Pa. to Mo., southw. to S. C. and Ark.

4. **Dioscorea quaternàta** (Walt.) Gmel. Map 680. Infrequent in the state within the area shown on the map. It is found in wooded ravines, on the crests of chestnut oak ridges, and on the bluffs of the Ohio River.

There are two distinct forms in the state. The common and typical form has the lower surface of the leaves glabrous and the other has the lower surface of the leaves rather densely pubescent.

Va. to Mo., southw. to Fla. and La.

44. IRIDÀCEAE Lindl. IRIS FAMILY

Leaves long and narrow, all or at least some of them 1 cm wide; flowers large, at least 2 cm long; capsules generally more than 1 cm long; perennials with creeping rhizomes.
  Flowers blue to lilac (albino forms rare), usually few, more than 3 cm long; sepals recurved or spreading while the petals are erect, both usually widest above the middle; seeds irregular in shape..........................1264. IRIS, p. 332.
  Flowers orange yellow, mottled with many crimson purple spots, generally less than 3 cm long; sepals and petals remaining in the same plane, not reflexed, narrow-elliptic in shape, persistent and coiled together on top of the ovary after flowering; seed globose, black, shining..............1285. BELAMCANDA, p. 333.
Leaves long and narrow but none 1 cm wide; flowers regular, less than 2 cm long; capsules usually globose and less than 1 cm long; seeds globose, small; perennials without rhizomes..................................1286. SISYRINCHIUM, p. 334.

1264. ÌRIS [Tourn.] L. IRIS

[E. Anderson. The species problem in Iris. Ann. Missouri Bot. Gard. 23: 457-509. 1936.]

Plants less than 1.5 dm high; rootstocks slender, creeping near the surface; flowers light lavender, appearing the last of April to the last of May; perianth tube usually 4-5 cm long; sepals crested but not bearded; capsules sharply triangular...1. *I. cristata.*
Plants more than 1.5 dm high; rootstocks thickened and not very close to the surface; flowers blue to lilac, appearing the last of May to the last of June.
  Capsules hexagonal; stem decumbent or prostrate, bearing flowers from near the base .... 2. *I. brevicaulis.*
  Capsules 3-angled; stem erect, bearing flowers on the upper part.
    Petals obovate-spatulate, nearly as long as the sepals, thin in texture, readily wilting; sepals with a bright yellow, pubescent blotch at the base, the hairs longer than the thickness of the sepal; capsule usually long and narrow, inner surface dull; seed round to D-shaped, dull, with occasional broad depressions as seen under a hand lens......................3. *I. virginica* var. *Shrevei.*
    Petals lanceolate, much shorter than the sepals, firm in texture, not readily wilting; sepals usually without a conspicuous spot at the base, when present, greenish or greenish yellow, the pubescence of microscopic hairs shorter than the thickness of the sepal; capsule short and thick, inner surface shining; seed all D-shaped in outline, the surface appearing regularly pitted, the pits in definite rows under a hand lens. (Range concides very nearly with that of the northern coniferous forest, but should be sought in northern Indiana.)..*I. versicolor.*

1. **Iris cristàta** Ait. CRESTED IRIS. Map 681. Local in the knobstone area from Monroe County southward. The Randolph County record is the only one from the glacial area. I found it in a low woods with beech and white ash. In the knobstone area it is usually found on the bases of wooded slopes and where found, it generally forms large colonies.

Md., Ohio to Mo., southw. to Ga. and Tex.

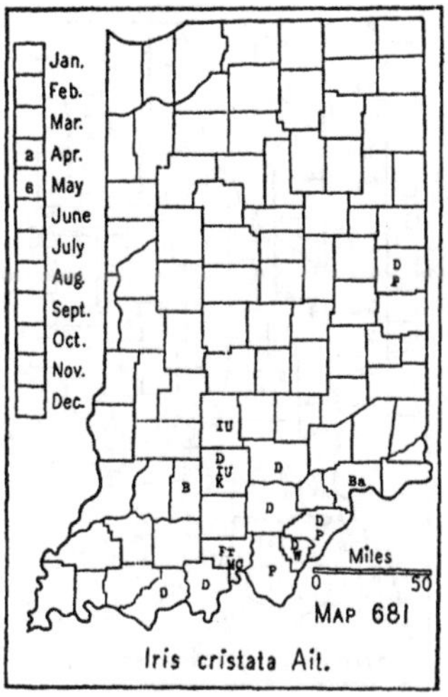

Iris cristata Ait.

Iris brevicaulis Raf.

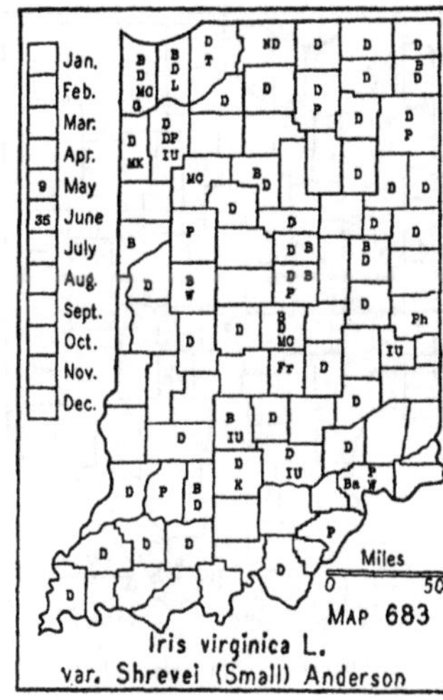

Iris virginica L. var. Shrevei (Small) Anderson

2. **Iris brevicaúlis** Raf. (Contr. Gray Herb. **114**: 41. 1936.) (*Iris hexagona* of Gray, Man., ed. 7 and *Iris foliosa* of Britton and Brown, Illus. Flora, ed. 2.) LAMANCE IRIS. Map 682. This species, as I have found it, is restricted to low, overflow land along streams and to the slopes of overflow terraces and slopes bordering streams, ponds, and sloughs. It will, no doubt, be found in suitable habitats along all of our principal streams. Where it becomes established, it usually forms large colonies.

Ohio and Ky., westw. to Ark. and Kans.; also on the Coastal Plain (Small).

3. **Iris virgínica** L. var. **Shrèvei** (Small) E. Anderson. (Ann. Missouri Bot. Gard **23**: 469. 1936.) (*Iris versicolor* in part, of Gray, Man., ed. 7 and Britton and Brown, Illus. Flora, ed. 2.) (E. Anderson. The problem of species in the the Northern Blue Flag, *Iris versicolor* L. Ann. Missouri Bot. Gard. **15**: 241-332. 1928.) VIRGINIA IRIS. Map 683. This plant has been confused, by many authors, with *Iris versicolor* L. which has a more northern range. *I. versicolor* has been found in the extreme northwestern part of Ohio and should be sought in our northern counties and it is for this reason that it appears in the key. *I. virginica* var. *Shrevei* is more or less frequent throughout the state along ditches, banks of streams, the borders of lakes, ponds, sloughs, and in low places in general.

Nieuwland (Amer. Midland Nat. 3: 115. 1913) described a variety of *I. versicolor* which he called var. *blandescens* and which, no doubt, should be referred to some form of this species.

D. C. to Minn., southw. to Fla. and Tex.

1285. BELAMCÁNDA Adans.

1. BELAMCANDA CHINENSIS (L.) DC. (*Gemmingia chinensis* (L.) Ktze.) BLACKBERRY-LILY. Map 684. This plant is an escape from cultivation and at present is restricted mostly to the southwestern part of the state where it has become well established, especially in sandy soil in the western part of Sullivan County. My specimens are mostly from the slopes of open

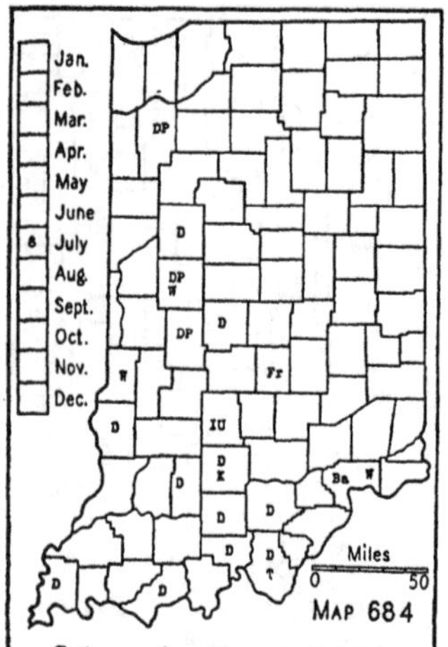

Belamcanda chinensis (L.) D.C.

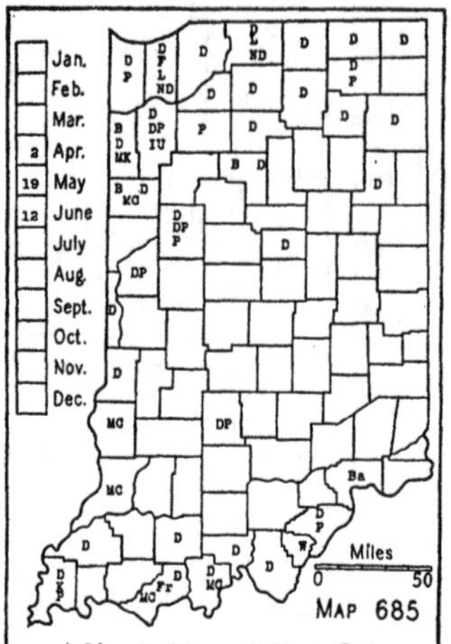

Sisyrinchium albidum Raf.

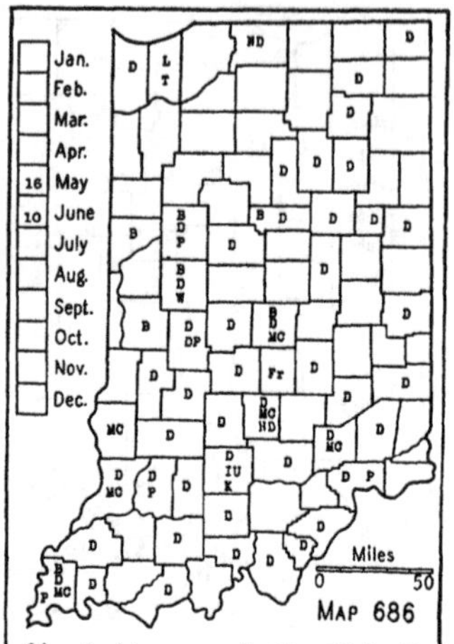

Sisyrinchium graminoides Bicknell

woodland that have a sandy soil. I found it well established over an area of about 2 acres south of Battle Ground, Tippecanoe County, where it was growing in dry, gravelly soil in open woodland.

Nat. of Asia.; Conn. to Kans., southw. to Ga. and Tex.

1286. SISYRÍNCHIUM L. BLUE-EYED-GRASS

Spathes sessile and terminal.
  Spathes 2, with a single, outer, leaflike bract........................1. *S. albidum.*
  Spathes solitary.
    Outer, elongate bract with margins free to the base; capsules pale. (See excluded species no. 153, p. 1037.)..................................*S. campestre.*
    Outer bract with the margins united above the base.
      Pedicels loosely spreading, much exceeding the inner bract; capsules 2-4 mm long. (See excluded species no. 154, p. 1037.)..............*S. mucronatum.*
      Pedicels suberect, scarcely exceeding the inner bract; capsules 4-6 mm long.... .... 2. *S. angustifolium.*
Spathes peduncled from the axil of the leaflike bract.
  Inner bract of spathe 1.5-3 cm long; stems broadly winged........3. *S. graminoides.*
  Inner bract of spathe 1-1.5 cm long; stems slender and narrowly margined; capsules beaked or beakless.......................................4. *S. atlanticum.*

1. **Sisyrinchium álbidum** Raf. Map 685. This species prefers a moist or dry, sandy soil. It is infrequent on sandy, white and black oak ridges and most frequent and abundant in moist, sandy soil of prairie habitats. It is also sometimes found in marshes. It is infrequent throughout the lake area, probably absent or rare in some of the counties of the Tipton Till Plain, and again appears sparingly in the southern counties. Most of our species seem to thrive best in full sunshine and are usually found in slightly acid soil. Most of our reports of *Sisyrinchium angustifolium* which were made before 1908 should, no doubt, be referred to this species.

Ont. to Wis., southw. to N. C., Ala., and La.

2. **Sisyrinchium angustifòlium** Mill. This species was reported from many parts of the state by early authors before our manuals recognized *Sisyrinchium albidum.* Probably most of these reports should be referred

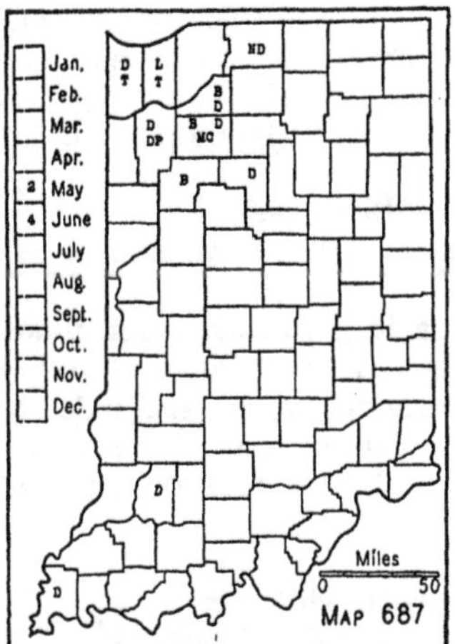

Sisyrinchium atlanticum Bicknell

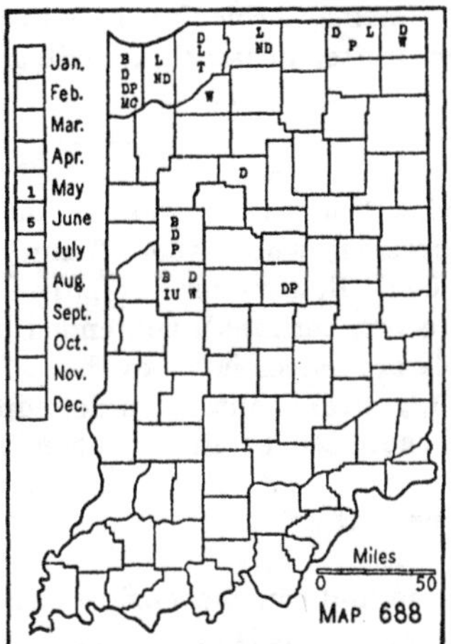

Cypripedium reginae Walt.

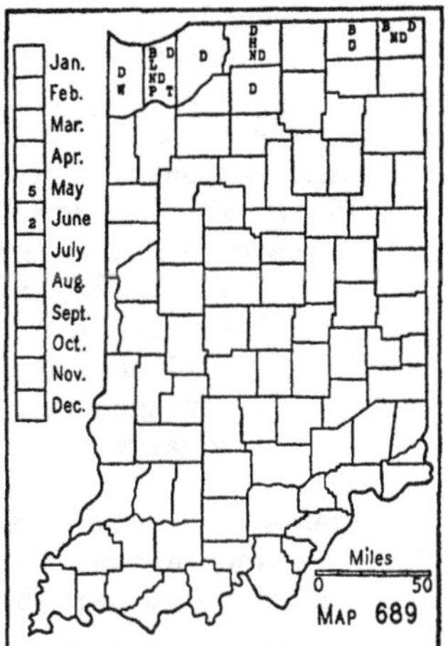

Cypripedium candidum Muhl.

to the last named species. Peattie reported this species from the Calumet Region where I, also, have found it. It is infrequent in moist soil on the low, open dunes along north Clark Street in Gary about an eighth of a mile south of Lake Michigan. Not yet known from any other county.

Newf. to B. C., southw. to Va., Pa., Mich., Minn., and in the Rocky Mts.

3. **Sisyrinchium graminoìdes** Bickn. (*Sisyrinchium gramineum* Curtis.) Map 686. Infrequent to rare in the northern part of the state, becoming frequent in the southern part. This species always has yellow roots, prefers a slightly acid soil, and is generally found in dry places in open woodland and clearings, along fence rows, and infrequently in open places with herbs and grasses of equal height.

N. S. to Minn; southw. to Fla. and Tex.

4. **Sisyrinchium atlánticum** Bickn. Map 687. This is primarily an Atlantic coast species that has possibly migrated into Indiana from the Mississippi Valley. In addition to the records on the map, it has been reported from Porter and White Counties. The species seems to be entirely distinct and is local in Indiana. It is generally found in moist, sandy soil, but my Posey County specimen was found in a moist, white clay loam on the second bottom along the Ohio River south of Caborn in a hayfield where it formed a large colony.

Maine to Fla., westw. to the Mississippi Valley and northw. to Ind. and Mich.

50. ORCHIDÀCEAE Lindl. ORCHID FAMILY

Plants with green leaves present at flowering time.
 Flower with a spur.
  Leaves all basal, 2.
   Blades usually large and rather fleshy, oblong-obovate; flowers with pinkish hoods and white or spotted lips, expanding mostly in May................ ....1396. ORCHIS, p. 339.

Blades large and usually nearly orbicular, or elliptic, not fleshy; flowers greenish yellow or greenish white, usually expanding after May..1422. HABENARIA, p. 339.
Leaves all cauline..1422. HABENARIA, p. 339.
Flower without a spur.
Plants with only one leaf. (Bracts not to be confused with leaves.)
Leaves ovate; flowers greenish white, about 3 mm long....1552. MALAXIS, p. 349.
Leaves linear or lance-oval; flowers rose or purplish.
Flowers solitary, rarely 2, terminal, subtended by a large, green bract almost as long as the flower; leaves lance-oval.............1464. POGONIA, p. 344.
Flowers generally 3-12, rarely solitary in depauperate plants, not subtended by a large, green bract; leaves linear, usually 15-30 cm long..1534. CALOPOGON, p. 348.
Plants with more than one leaf.
Flowers in racemes.
Leaves all near the base and conspicuously marked with white veins..1504. GOODYERA, p. 347.
Leaves not conspicuously veined.
Stems with bulbous bases; leaves 2, basal; flowers madder purple or yellowish green..1556. LIPARIS, p. 350.
Stems without bulbous bases.
Flowers white; leaves of a linear type, mostly less than 1 cm wide, at least the lower ones petiolate; plants mostly 2-5 dm. high..1490. SPIRANTHES, p. 345.
Flowers greenish, suffused with madder purple; median leaves of an ovate type, the largest usually 2-4 cm wide, sessile; plants usually 3-7 dm. high...................................1482. EPIPACTIS, p. 345.
Flowers not in racemes.
Plants with a whorl of 5 obovate or lanceolate leaves at the summit; flowers terminal, solitary, rarely 2, purplish...............1467. ISOTRIA, p. 344.
Plants not as above,
Leaves usually very large and long; flowers inflated, slipper-shaped, yellow, pinkish or white.........................1391. CYPRIPEDIUM, p. 336.
Leaves small, about 1 cm long, clasping, broadly ovate; flowers not inflated or slipper-shaped, usually pinkish or nearly white...1466. TRIPHORA, p. 344.
Plants without green leaves at flowering time, rarely a withered basal one persisting.
Flower solitary, terminal, rose purple.....................1474. ARETHUSA, p. 344.
Flowers not as above.
Stems bulbous at the base.
Flowers with long spurs; basal leaf purplish beneath....1560. TIPULARIA, p. 350.
Flowers without spurs; basal leaf green beneath......1642. APLECTRUM, p. 351.
Stems not bulbous at the base.
Plants with 1-several long, tuberous roots; flowers white...1490. SPIRANTHES, p. 345.
Plants with scaly or corallike rootstocks; flowers not white.
Flowers cadmium orange........................1629. HEXALECTRIS, p. 351.
Flowers more or less purplish..................1548. CORALLORRHIZA, p. 348.

1391. CYPRIPÈDIUM L. LADYSLIPPER

Plants with leafy stems; flowers 1 or 2, rarely several.
Sepals and petals shorter than the lip; tall plants of only cold, springy, and boggy places; flowers white except the variegated crimson lip..........1. *C. reginae.*
Sepals and petals equalling or longer than the lip,
Leaves 3 or 4, strongly overlapping at the base, rather narrow-elliptic or lanceolate; outside of lip white, in dried specimens generally less than

20 mm long, rarely up to 25 mm long; dried plants mostly 25-35 cm long..... .... 2. *C. candidum.*

Leaves 4 or 5, rarely only 3, not overlapping at the base or only rarely so, usually broadly oval to narrowly elliptic; lip yellow outside, generally 2-4.5 cm long in dried specimens, sometimes only 1.5 cm long in depauperate specimens; plants of dry woods or of boggy and springy places.

Lip 2-3 cm long; sepals of a madder purple color; flowers very fragrant; leaves 3 or 4...3. *C. parviflorum.*

Lip 3-5 cm long; sepals of a greenish yellow color, streaked with fine lines of madder purple; less fragrant than the preceding; leaves larger, 4 or 5, rarely 6..................................3a. *C. parviflorum* var. *pubescens.*

Plants without stems, with 2 opposite basal leaves, rarely a third near the base of the scape; leaves very variable in width and length, oval to narrowly elliptic; flowers pinkish (white in albino forms); lip usually 4-7 cm long in dried specimens; generally found only in tussocks of sphagnum in Indiana...............4. *C. acaule.*

1. **Cypripedium reginae** Walt. (*Cypripedium hirsutum* Mill.) SHOWY LADYSLIPPER. Map 688. This orchid was formerly rather frequent in its habitat in our northern counties, but now like the other species of the genus has become rare on account of drainage and grazing. It prefers a wet, cold soil and is usually found in muck in springy places or in peat in tamarack bogs, often in tussocks of sphagnum. In optimum conditions it reaches a height of 3 feet and I once measured a leaf that was nearly 7 inches wide and 1 foot long. I found it to be a common plant in a large springy area at the base of the high bank along Sugar Creek in Montgomery County. R. C. Friesner found a few plants in a marly springy area on the slope of the high bank of Flint Creek about 3 miles northwest of Westpoint, Tippecanoe County and gave me a specimen. It has been reported also from Hamilton, Kosciusko, Marshall, Noble, and St. Joseph Counties.

Newf. to Minn., southw. to Ga. and Mo.

2. **Cypripedium cándidum** Muhl. WHITE LADYSLIPPER. Map 689. This species is very local and I now know of only six places in the state where it occurs. There are no reports for it in Indiana outside of the range indicated on the map except that Schneck in 1876 reported it as occurring in the Lower Wabash Valley, saying: "Rapidly disappearing, once common here." I was informed by a reliable authority that it has been found in two places on springy banks in Tippecanoe County. It is generally found on "raised springy areas" and usually associated with *Zizia aurea.* It occurs in Porter County in a cattail mucky area.

N. Y. to s. Minn., southw. to N. J., Ky., and Mo.

3. **Cypripedium parviflòrum** Salisb. SMALL YELLOW LADYSLIPPER. Map 690. Since Indiana has been so completely drained the typical form of this species has become very rare. It is generally found in boggy places and in the dunes on the wet borders of sloughs. It is rarely found in woodland.

Newf. to Que., Man., Sask., to B. C., southw. to Ga., Ohio, Ill., Iowa, and Wash.

3a. **Cypripedium parviflorum** var. **pubéscens** (Willd.) Knight. LARGE YELLOW LADYSLIPPER. Map 691. The large-flowered variety seems to be quite distinct from the typical form for the most part but intergrading

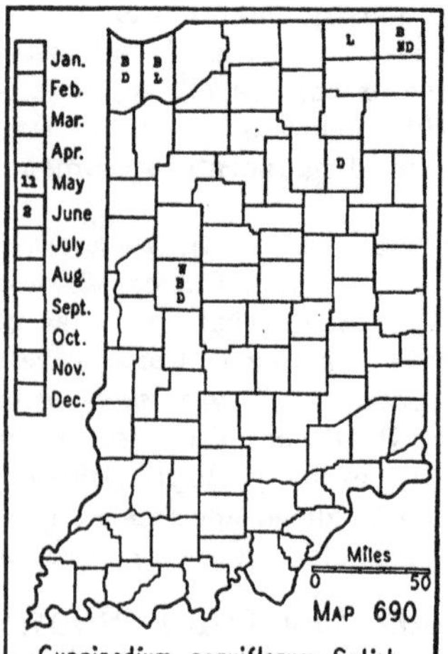

Cypripedium parviflorum Salisb.

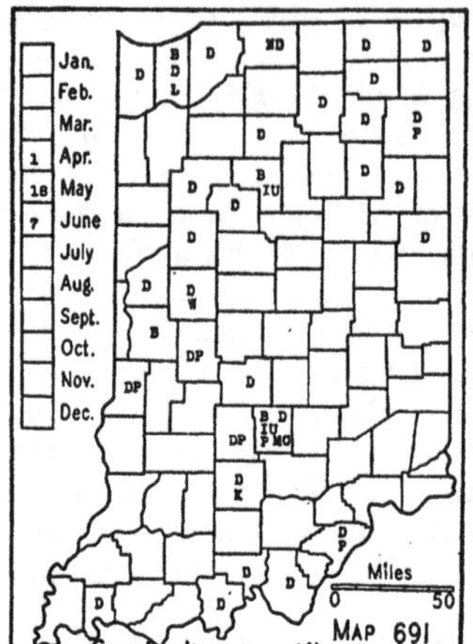

Cypripedium parviflorum var. pubescens (Willd.) Knight

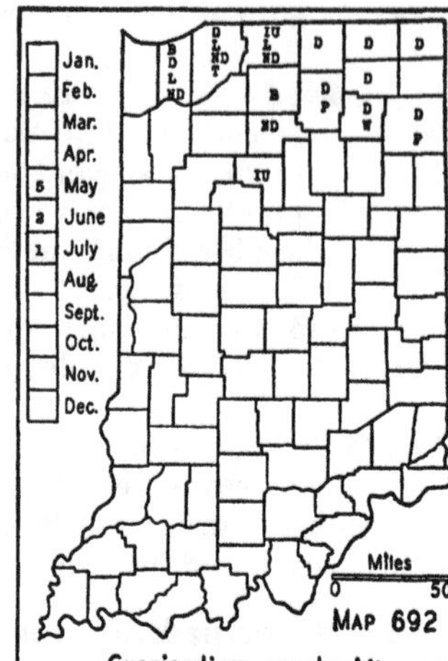

Cypripedium acaule Ait.

specimens have influenced some authors to regard it as a nutritional form. Some claim that when the large-flowered form is transplanted it will change in time to the small-flowered form. This transition is contradicted by the experience of others. Until it is proven that the one is merely a passing form of the other it is best to regard them as distinct with intergrading forms. In Indiana the habitat seems to distinguish them. The variety grows in deep leaf mold in moist or dry woods while the typical form grows for the most part in very wet or boggy places. The habitat distinction, however, does not hold even for the few specimens I have. All of my specimens of the typical form grew in boggy places except one which grew in woodland. All of my specimens of the variety grew in woodland except one that grew in a tamarack bog.

Que. to B. C., southw. to N. C., Ala., Mo., and N. Mex.

Since the preceding was written, Donovan S. Correll has published his study of the North American yellow ladyslippers in Bot. Mus. Leafl. of Harvard University 7: 1-18. 1938. He concludes that our plants are a variety of the yellow ladyslipper of Eurasia and assigns to them the name *Cypripedium Calceolus* var. *pubescens* (Willd.) Correll. He gives the range of the variety as Newf., Que. to Yukon and B. C., southw. to S. C., Ga., Miss., La., N. Mex. and Wash.

4. **Cypripedium acaúle** Ait. (*Fissipes acaulis* (Ait.) Small.) PINK LADYSLIPPER. Map 692. This species is found only in the sphagnum bogs of northern Indiana. In addition to the counties shown on the map it has been reported from Lake County. Its habitat occurs in all of these counties and also did occur in Marshall and Starke Counties, but the report from Monroe County by Andrews must be an error. It was formerly a common plant and showed great variation in the size and shape of its leaves. Since its habitat is restricted, and our sphagnum bogs are fast disappearing, it will soon become rare in our state.

Newf. to Winnipeg and Minn., southw. to N. J., Ohio, and Ind., and in the mts. to N. C. and Tenn.

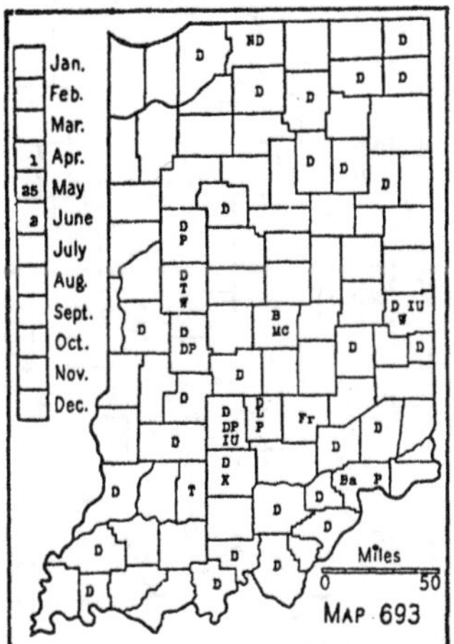

MAP 693
Orchis spectabilis L.

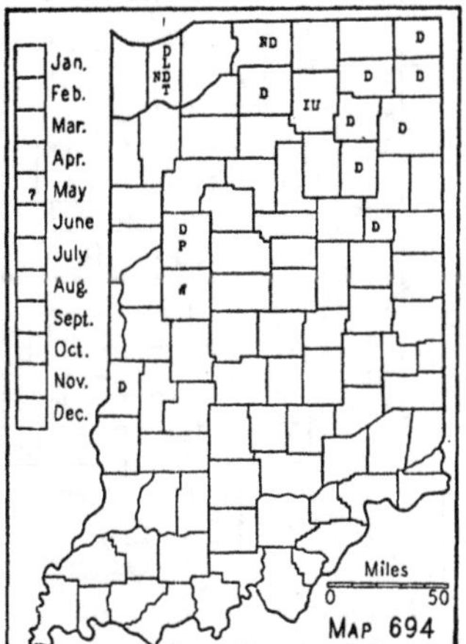

MAP 694
Habenaria viridis (L.) R. Br. var. bracteata (Muhl.) Gray

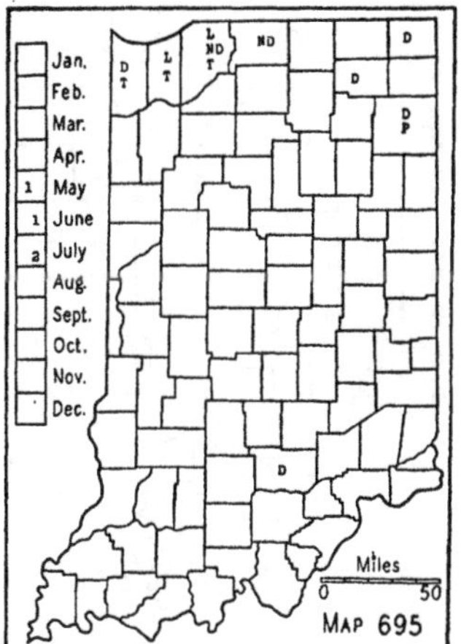

MAP 695
Habenaria flava (L.) Gray

1396. ÓRCHIS [Tourn.] L.

1. **Orchis spectábilis** L. (*Galeorchis spectabilis* (L.) Rydb.) SHOWY ORCHIS. Map 693. Infrequent to rare throughout the state except in the prairies where it is absent. It is found usually in deep leaf mold in beech and sugar maple woods and in black and white oak woods.

N. B., Que., Ont. to Minn., southw. to Ga., Tenn., and Mo.

1422. HABENÀRIA Willd.

Lip not fringed.
- Leaves cauline; stem more or less bracted above the leaves and into the inflorescence.
  - Leaves several, at least more than 2.
    - Lip lobed at the base or toothed at the apex.
      - Lip 3-toothed at the apex; spur shorter than the lip.. 1. *Habenaria viridis* var. *bracteata*.
      - Lip with a lobe on each side at the base and a median tubercle near the base; spur longer than the lip.
        - Bracts mostly longer than the flowers; lip decidedly longer than wide.. 2. *H. flava*.
        - Bracts mostly shorter than the flowers; lip about as wide as long.. 3. *H. scutellata*.
    - Lip entire, lanceolate to linear, subacute or rounded at the apex.
      - Flowers white, fragrant; lip dilated at the base................ 4. *H. dilatata*.
      - Flowers greenish, scarcely fragrant; lip not dilated at the base.. 5. *H. hyperborea*.
  - Leaves 1 or 2; lip entire at the base; bracts shorter than the flowers.. 6. *H. clavellata*.
- Leaves basal.
  - Scape bracted; flowers greenish white......................... 7. *H. orbiculata*.
  - Scape bractless; flowers greenish yellow......................... 8. *H. Hookeri*.

Lip fringed or erose-denticulate.
- Lip evenly fringed all around, not divided.
  - Flowers orange yellow; lip oblong, 1 cm long, the fringe 3-5 mm long.. 9. *H. ciliaris*.
  - Flowers white; lip narrowly ovate-lanceolate, 8-10 mm long, the fringe 0.5-1.5 mm long. (See excluded species no. 155, p. 1037.).......... *H. blephariglottis*.

Habenaria scutellata (Nutt.) F. Morris

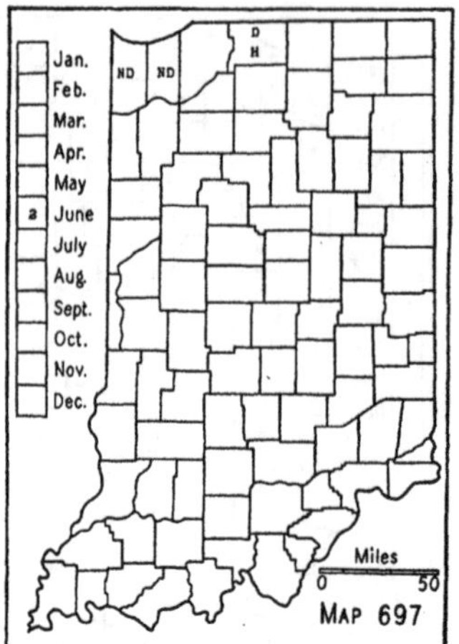

Habenaria dilatata (Pursh) Gray

Habenaria hyperborea (L.) R. Br.

Lip more or less 3-parted, the divisions fringed or erose-denticulate.
  Petals entire; flowers greenish; lip deeply parted, the divisions narrow and deeply fringed .... 10. *H. lacera.*
  Petals not entire, more or less minutely denticulate.
    Lip deeply fringed and 3-parted; fringe 2-5 mm long.
      Flowers white, not crowded on the spike .... 11. *H. leucophaea.*
      Flowers purplish, crowded on the spike .... 12. *H. psycodes.*
    Lip shallowly erose-denticulate, the teeth usually about 0.5 mm long, the terminal lobe usually bifid by an incision 2-3 mm long, 3-parted, rarely cut into 5 divisions; flowers purplish; found only in southern Indiana........ .... 13. *H. peramoena.*

1. **Habenaria víridis** (L.) R. Br. var. **bracteàta** (Muhl.) Gray. (*Habenaria bracteata* (Willd.) R. Br. and *Coeloglossum bracteatum* (Willd.) Parl.) SATYR ORCHID. Map 694. Usually not more than a single plant is found in any one locality. It occurs in moist, rich woods, in tamarack bogs or on low borders of lakes. In Noble County I found two large colonies on the low border of Crooked Lake in among *Cornus obliqua* and *Acer rubrum.* This is the only place I have seen two specimens or more in a place. It has been reported also from Lake and White Counties.

Newf. to B. C., southw. to N. C., Ohio, Ill., Mont., and Wash.; also in Japan and China.

2. **Habenaria flàva** (L.) Gray. (*Perularia flava* (L.) Farw.) TUBERCLED ORCHID. Map 695. Mostly in the lake region in tamarack bogs, marshes, and sandy, wet places. It has been reported from Marshall and Vigo Counties. It is very rare and usually a single specimen is found at a place.

N. S., Que., Ont. to Minn., southw. to Fla. and Tex.

3. **Habenaria scutellàta** (Nutt.) F. Morris. (*Perularia scutellata* (Nutt.) Small.) Map 696. On September 28, 1923, I found a large colony of this species in flower and in fruit in Posey County, growing in a bare place under a clump of buttonbush where it must have been submerged much of the year. I transferred some of it to our garden in Bluffton where

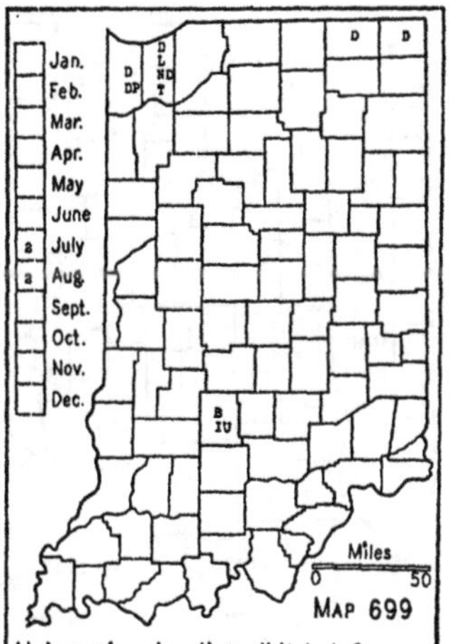

Habenaria clavellata (Michx.) Spreng.

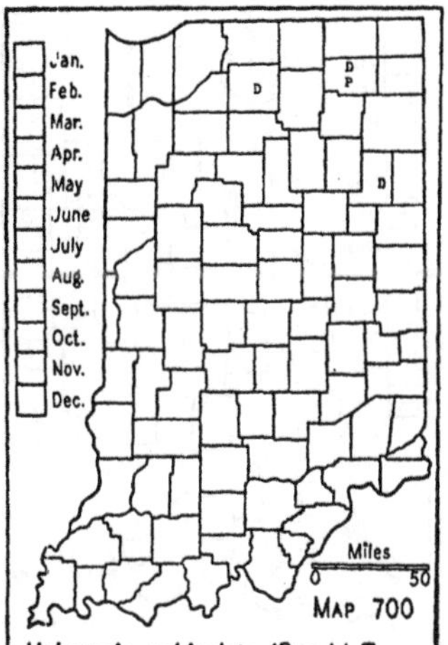

Habenaria orbiculata (Pursh) Torr.

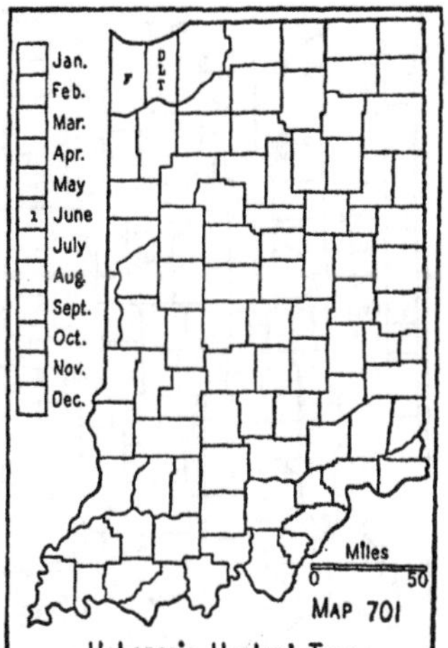

Habenaria Hookeri Torr.

it did well for several years. This is the only record I know of from Indiana.

Pa., Ind., and Ark., southw. to Fla.

4. **Habenaria dilatàta** (Pursh) Gray. (*Limnorchis dilatata* (Pursh) Rydb. of Britton and Brown, Illus. Flora, ed. 2.) WHITE BOG-ORCHID. Map 697. A few plants of this species were found in a bog on the Wolverton Estate about 7 miles southwest of South Bend, St. Joseph County. The area was heavily grazed and it will soon disappear if grazing continues. This species was reported by Nieuwland for Umbach (Amer. Midland Nat. 3: 119. 1913) but through the courtesy of N. C. Fassett the Umbach herbarium at the University of Wisconsin was searched, and no specimen was found. There are, however, specimens collected by Nieuwland in Lake and Porter Counties which are deposited in the herbarium of the University of Notre Dame. The location of this species in Indiana is the extreme southern limit of its range.

Subarctic America; Lab. to B. C. and Alaska, southw. to N. J., Ind., Minn., Mont., Idaho, Colo., and Wash.

5. **Habenaria hyperbòrea** (L.) R. Br. (*Limnorchis hyperborea* (L.) Rydb.) NORTHERN GREEN ORCHID. Map 698. Our specimens and reports are from our northern tier of counties. It must be very rare in Indiana. I have collected it only twice. Besides the counties shown on the map it has been reported from La Porte and St. Joseph Counties.

Newf. to Que., Ont. to B. C., northw. to Alaska, Iceland, and Greenland, and southw. to N. Y., Pa., Ind., Ill., Nebr., Colo., and Oreg.

6. **Habenaria clavellàta** (Michx.) Spreng. (*Gymnadeniopsis clavellata* (Michx.) Rydb.) SMALL GREEN WOOD ORCHID. Map 699. My specimens are from moist, sandy or gravelly borders of lakes and sloughs. The distribution is restricted to our northern counties with the exception of a

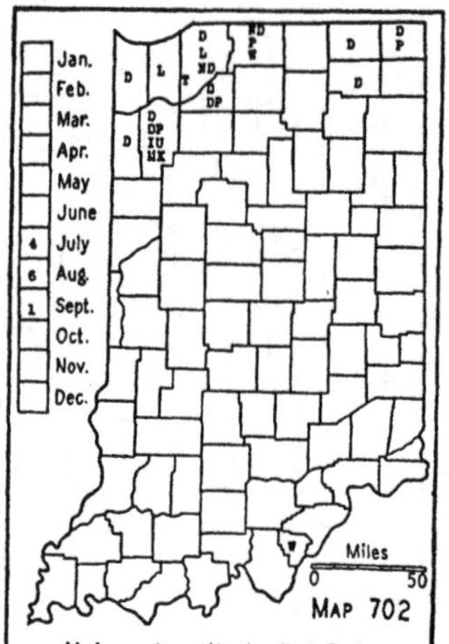

Habenaria ciliaris (L.) R. Br.

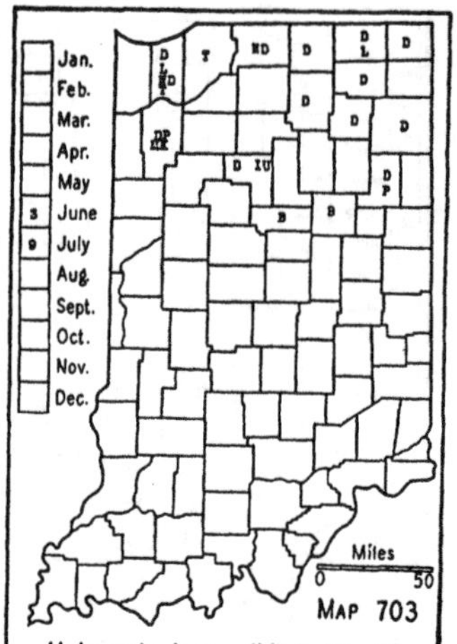

Habenaria lacera (Michx.) Lodd.

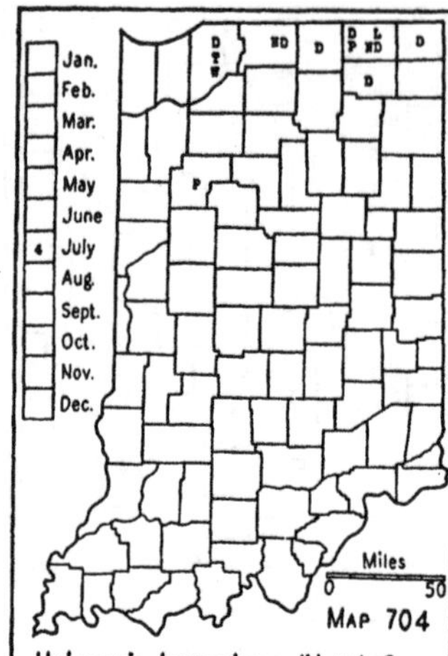

Habenaria leucophaea (Nutt.) Gray

specimen collected in a swamp in Monroe County by J. E. Potzger. There is a report from St. Joseph County which is not shown on the map.

Newf., Que., Ont. to Minn., southw. to N. Y., Fla., and La.

7. **Habenaria orbiculàta** (Pursh) Torr. (*Lysias orbiculata* (Pursh) Rydb.) LARGE ROUNDLEAF ORCHID. Map 700. This is one of our rarest orchids. I have found it twice. Van Gorder found it in Noble County and I have his specimen. It grows in very rich, sandy soil with sugar maple and beech.

Newf., Que., Ont. to B. C. and northw. to Alaska, southw. to Pa., W. Va., Md., and in the mts. to S. C., Tenn., Ohio, Ill., Mont., and Wash.

8. **Habenaria Hoókeri** Torr. (*Lysias Hookeriana* (A. Gray) Rydb.) HOOKER ORCHID. Map 701. This is also one of our rarest orchids. It has been reported from Lake, La Porte, Noble, and Porter Counties. There is a specimen in the Field Museum which was collected by Agnes Chase, June 21, 1897, east of Edgemoor (probably near what is now known as the Buffington Cement Plant or West Gary). The specimen in the Field Museum so labeled and collected by Bross in La Porte County is *Orchis spectabilis*. The report from Noble County is not supported by a specimen and I refer the report to *Habenaria orbiculata*. Dr. Lyon found a few plants in Porter County. It has been found also by R. M. Tryon, Jr., in Dunes State Park, Porter County and he has given me a specimen.

N. S., Que., Ont. to Minn., southw. to N. Y., Pa., Ohio, Ind., Wis., and Iowa.

9. **Habenaria ciliàris** (L.) R. Br. (*Blephariglottis ciliaris* (L.) Rydb.) YELLOW FRINGE-ORCHID. Map 702. In marshes, moist, sandy borders of lakes and sloughs, prairie habitats, and low, open and sandy woods. This species was formerly not rare but is now becoming scarce. Its distribution is restricted to our northern counties and besides those indicated on the map it has been reported from Marshall County.

Vt., Ont. to Mich., Ill., and Mo., southw. to Fla. and Tex.

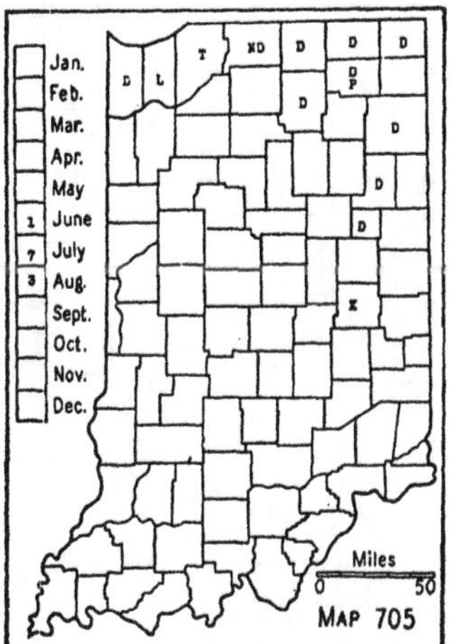

Habenaria psycodes (L.) Spreng.

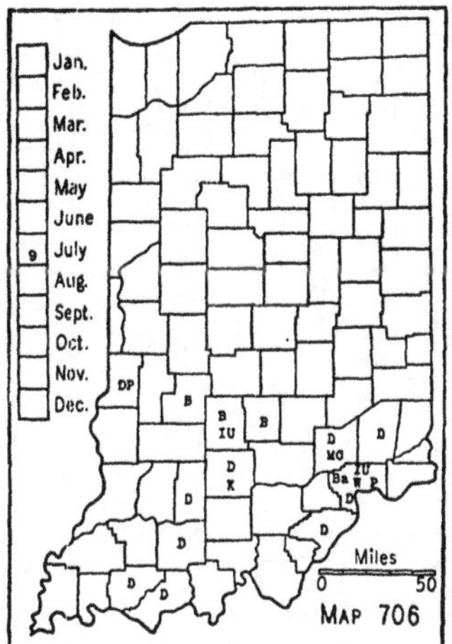

Habenaria peramoena Gray

Pogonia ophioglossoides (L.) Ker.

10. **Habenaria lácera** (Michx.) Lodd. (*Blephariglottis lacera* (Michx.) Farw.) GREEN FRINGE-ORCHID. Map 703. This species is somewhat frequent in the lake area and has a wide range of habitats. It is usually found in tamarack bogs, marshes, and marshy and springy places anywhere. I once found it under a beech tree in a woods and once in a sandy, fallow field that had been fallow for at least 25 years. In addition to the counties indicated on the map it has been reported from Fulton, Lake, and Marshall Counties.

Newf., Que., Ont. to Man., southw. to N. C., Ala., and Mo.

11. **Habenaria leucophaèa** (Nutt.) Gray. (*Blephariglottis leucophaea* (Nutt.) Farw.) PRAIRIE WHITE FRINGE-ORCHID. Map 704. Probably local in the lake area years ago but now rare. I have found it in only five places. In addition to these counties it has been reported from the following counties: Hamilton, Kosciusko, Lake, and Marshall. My specimens are from sphagnum in tamarack bogs.

N. S., Ont. to Minn., southw. to N. Y., Ohio, Ill., Mo., and La.

12. **Habenaria psycòdes** (L.) Spreng. (*Blephariglottis psycodes* (L.) Rydb.) SMALL PURPLE FRINGE-ORCHID. Map 705. Rather rare in the lake area in mucky soil about lakes and in low woods. In addition to the counties shown on the map it has been reported from Jay County. It has also been reported from Clark and Jefferson Counties but these reports doubtless should be referred to the next species.

Newf., Que., Ont. to Minn., southw. to N. C., Ohio, Ill., and Iowa.

13. **Habenaria peramoèna** Gray. (*Blephariglottis peramoena* (Gray) Rydb.) FRINGELESS PURPLE ORCHID. Map 706. This species is not infrequent in our southern counties in low, flat woods, usually associated with beech and sweet gum and pin oak. It has also been reported from Monroe County. Usually rather frequent where found.

Pa., Ohio, Ill., and Mo., southw. to N. C., Ala., and Tenn.

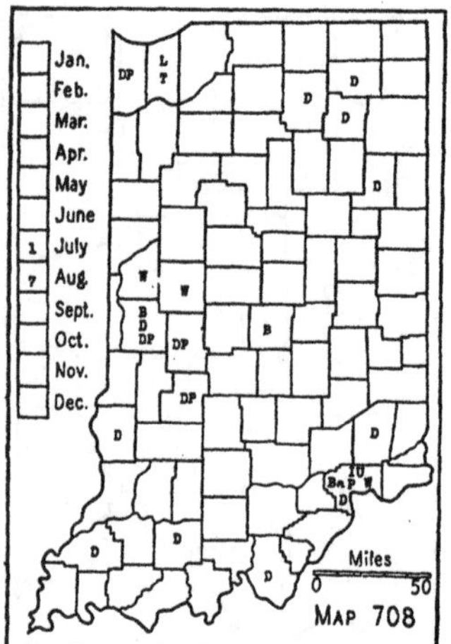

Triphora trianthophora (Sw.) Rydb.

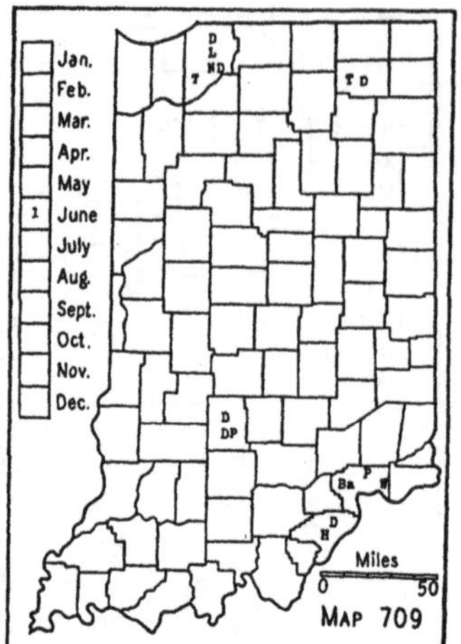

Isotria verticillata (Willd.) Raf.

Arethusa bulbosa L.

1464. POGÒNIA Juss.

1. **Pogonia ophioglossoìdes** (L.) Ker. ROSE POGONIA. Map 707. Formerly frequent in peat bogs in the lake area, now infrequent to rare on account of drainage. In addition to the counties shown on the map it has been reported from Fulton and Marshall Counties.

Newf., Que., Ont. to Minn., southw. to Fla. and Tex.

1466. TRÍPHORA Nutt.

1. **Triphora trianthóphora** (Sw.) Rydb. (*Pogonia trianthophora* (Sw.) BSP.) NODDING POGONIA. Map 708. An infrequent plant throughout the state. It is generally found in deep humus, usually in beech and sugar maple and oak woods. The flowers are mostly nodding but the fruit is erect. I once found a large colony on a bare, sandy flat in a deep, wooded ravine. Besides the counties indicated on the map it has been reported from the following counties: Fayette, Hamilton, and Steuben.

Maine to Wis., southw. to Fla., Ala., and Mo.

1467. ISÒTRIA Raf.

1. **Isotria verticillàta** (Willd.) Raf. (*Pogonia verticillata* (Willd.) Nutt.) WHORLED POGONIA. Map 709. An inconspicuous plant and apparently very rare and erratic in its distribution. In the lake area it is found in sphagnum in tamarack bogs and south of this area it has been found in white oak woods.

Maine, N. Y. to Mich., southw. to Fla. and Tex.

1474. ARETHÙSA [Gronov.] L.

1. **Arethusa bulbòsa** L. ARETHUSA. Map 710. An extremely rare plant found in sphagnum in bogs. In addition to the counties shown on the map

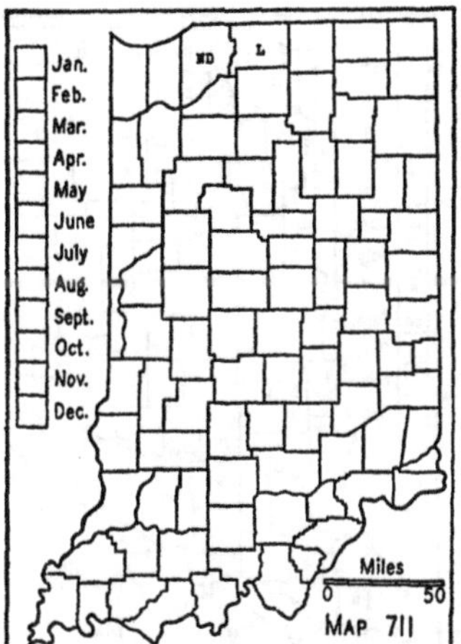

MAP 711
Epipactis latifolia (Huds.) All.

MAP 712
Spiranthes Beckii Lindl.

MAP 713
Spiranthes gracilis (Bigel.) Beck

it has been reported from Carroll and Starke Counties. It must now be very rare or almost extinct in the state.

Newf., Ont. to Minn., southw. to N. J., Pa., Ohio, Ind., and in the mts. to S. C.

1482. EPIPÁCTIS Swartz

1. EPIPACTIS LATIFÒLIA (Huds.) All. (*Serapias Helleborine* L.) BROAD-LEAF EPIPACTIS. Map 711. Reported from La Porte County by Nieuwland & Just (Amer. Midland Nat. **12**: 220. 1931). They write: "Interlaken, Laporte Co., spreading very rapidly in abundance on a dry clay hillside facing the lake, VII. 18. 1930.) (Probably introduced.)" I have seen their specimens and I am admitting it upon their statement that it is spreading rapidly. Found in 1937 by Lyon in South Bend.

Que., Ont., N. Y., and Pa.; also in Eu. Probably all, or most all of our reports are based upon escaped plants because it was formerly cultivated for its supposed medicinal qualities.

1490. SPIRÁNTHES Richard LADIES' TRESSES

Upper part of stem and lower part of the rachis of the spike glabrous; leaves basal and usually absent at flowering time; stems slender, usually less than 1 mm in diameter just below the inflorescence.

Rachis of inflorescence not twisted; stems bearing a solitary subcylindrical root.......................................1. *S. Beckii.*

Rachis of inflorescence twisted; stems bearing more than one subcylindrical root, usually 3 or more.......................................2. *S. gracilis.*

Upper part of stem and lower part of rachis of the spike more or less pubescent; leaves not all basal and some present at flowering time except in no. 3 where they may be absent; stems usually 1 mm or more in diameter just below the inflorescence.

Flowering from May through July; base of lip truncate, without callosities or with small, spreading ones.......................................3. *S. lucida.*

Flowering in Sept. and Oct.; base of lip cuneate and with two, incurved, nipplelike callosities.

Flowers about 4, not more than 5, mm long; spikes slender, less than 15 mm wide; petals lanceolate, dilated at the base.......................................4. *S. ovalis.*

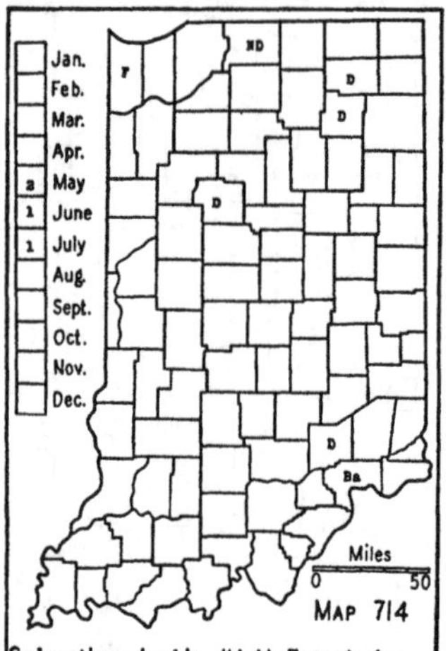

Spiranthes lucida (H. H. Eaton) Ames

Spiranthes ovalis Lindl.

Spiranthes cernua (L.) Richard

Flowers about 6, at least more than 5, mm long; spikes stout, mostly about 20 mm wide, rarely as narrow as 15 mm; petals linear, not dilated at the base.
Corolla white; spikes usually blunt; lower bracts shorter than the corolla..5. *S. cernua.*
Corolla yellowish; spikes acute; bracts longer than the corolla. (See excluded species no. 157, p. 1037.).......................*S. cernua* var. *ochroleuca.*

1. **Spiranthes Béckii** Lindl. (*Ibidium Beckii* (Lindl.) House.) BECK LADIES' TRESSES. Map 712. I have found this species in only two counties. It grew in hard, clay soil in open white and black oak woods.

Mass., Md., Ky., southw. to Fla. and Tex.

2. **Spiranthes grácilis** (Bigel.) Beck. (*Ibidium gracile* (Bigel.) House.) SLENDER LADIES' TRESSES. Map 713. This species is, no doubt, sparingly distributed throughout the state. In addition to the counties shown on the map it has been reported from Kosciusko, Noble, and Tippecanoe Counties. My specimens are from sandy or clayey soil in open, white and black oak woods and fallow fields and from sandy, black soil in a prairie habitat.

P. E. I., Que., Ont. to Man., southw. to Fla. and Tex.

3. **Spiranthes lùcida** (H. H. Eaton) Ames. (*Ibidium plantagineum* (Raf.) House.) WIDELEAF LADIES' TRESSES. Map 714. This species is local in the lake area where it is sparingly found on the springy, marl borders of lakes and in bogs elsewhere. In Jennings County in southern Indiana I found it at the base of a 75-foot cliff along the Muscatatuck River growing on narrow ledges of limestone in soil kept continually wet by seepage. In addition to the counties shown on the map it has been reported from Tippecanoe County.

Maine, Que., Ont. to Mich., southw. to Va. and Ohio.

4. **Spiranthes ovàlis** Lindl. (*Ibidium ovale* (Lindl.) House.) Map 715. This species is very rare throughout its range and I have found it in only two counties. One specimen is from the wooded bluff of the Ohio River

on the north side of Leavenworth, Crawford County. The other was found on a low, wooded promontory in the Louis B. Wilkerson woods in sec. 3 about 7 miles southwest of Rockport, Spencer County. Here it was growing under a beech tree and also under a tulip tree. Specimens from this place supplied the photograph of this species for "Our Wild Orchids" by Morris & Eames.

Ga., Ala., Miss., Tex., Okla., Tenn., Ark., Mo., and Ind.

5. **Spiranthes cérnua** (L.) Richard. (*Ibidium cernuum* (L.) House.) NODDING LADIES' TRESSES. Map 716. Well distributed in the lake area where it may be common over acres of recently drained mucky land. It prefers calcareous springy areas and marshes and is usually found in the open. In the southern part of the state it is generally found as an indidividual plant here and there growing in hard clay or sandy soil in open, black and white oak woods, on chestnut oak ridges, and less frequently in low woods with sweet gum and pin oak, and sometimes on sandy knolls in the southwestern part of the state. It is also frequent in wet prairies where such habitats occur. A variety of this species has been reported from Indiana but I am excluding it. See excluded species for a discussion of it.

N. S., Ont. to Minn., and southw. to Ga., Tex., and N. Mex.

1504. GOODYÈRA R. Br.

1. **Goodyera pubéscens** R. Br. (*Epipactis pubescens* (Willd.) A. A. Eaton and *Peramium pubescens* (Willd.) MacM.) DOWNY RATTLESNAKE-PLANTAIN. Map 717. Local in many parts of the state where its habitat exists. It generally prefers a deep humus soil that is slightly acid. I have seen it as a common plant on residual sandstone soil in Clay and Crawford Counties and only a few plants in a colony here and there in sandy soil in black oak woods. In addition to the distribution shown on the map it has been reported from Lake, Putnam, and Vigo Counties.

N. E., Que. to Minn., southw. to N. C., Ala., and Ill.

1534. CALOPÒGON R. Br.

1. **Calopogon pulchéllus** (Salisb.) R. Br. (*Limodorum tuberosum* L. in part.) GRASS-PINK ORCHID. Map 718. More or less frequent in its habitat throughout the lake area. It grows in the open in both peaty and marly springy places, in tamarack bogs, and in a moist, prairie habitat. In addition to the counties shown on the map it has been reported from Cass and White Counties.

Newf., Ont. to Minn., southw. to Fla. and Tex.

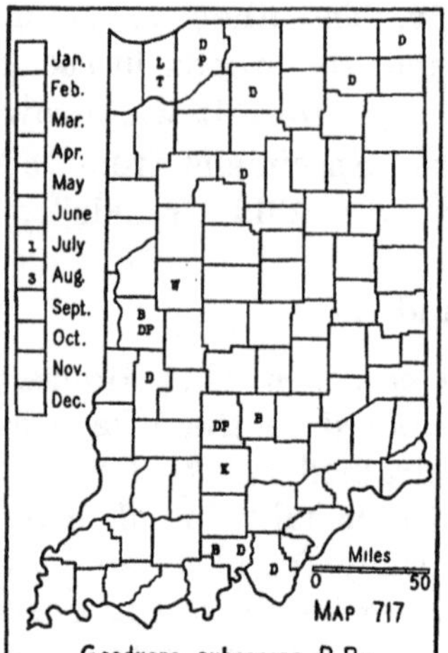

Goodyera pubescens R. Br.

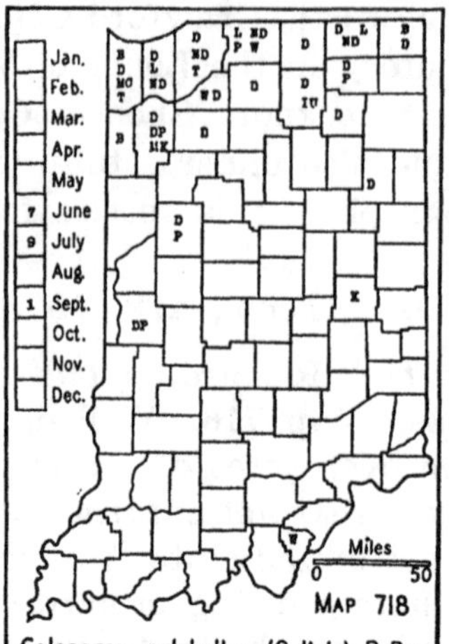

Calopogon pulchellus (Salisb.) R.Br.

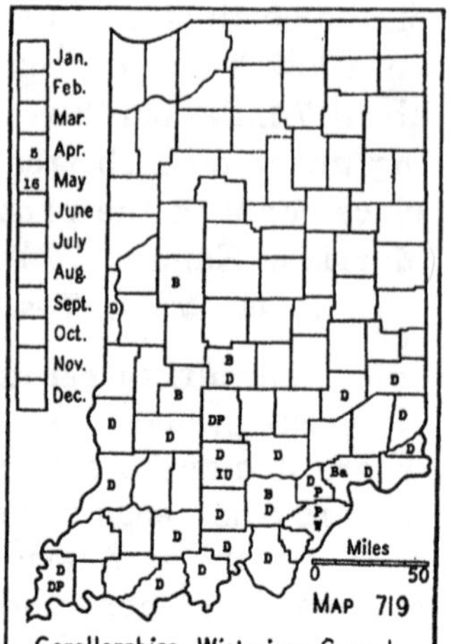

Corallorrhiza Wisteriana Conrad

1548. CORALLORRHÌZA [Haller] Chatelain

Flowering in Indiana before July 1, mostly in May and early June.
- Plants northern in their distribution, reaching only northern Indiana; flowers greenish or yellowish; perianth generally 5 mm or less in length; lip truncate at the apex. .... 1. *C. trifida.*
- Plants southern in their distribution, not yet found in northern Indiana; flowers and stem more or less purplish; perianth generally 6-8 mm long; lip notched at the apex.......... 2. *C. Wisteriana.*

Flowering in Indiana after July 1, mostly in August and September.
- Lip with a short lobe on each side at the base; mature capsule about 10 mm long.......... 3. *C. maculata.*
- Lip without lateral lobes; mature capsule about 6 mm long.......... 4. *C. odontorhiza.*

1. **Corallorrhiza trífida** Chatelain. EARLY CORALROOT. This species is admitted to our flora upon the authority of Pepoon, who says: "Frequent in the dune swale woods northeast of Dune Park" (Porter County), and upon the basis of a specimen collected by Umbach in 1892 at Miller (Lake County), now in the herbarium of the University of Wisconsin. It was also reported from Floyd County but that record should, no doubt, be referred to some other species.

Newf., Que., Sask., B. C. to Alaska, southw. to N. J., Pa., Ohio, Colo., and Oreg.; also in Eurasia.

2. **Corallorrhiza Wisteriàna** Conrad. WISTER CORALROOT. Map 719. Infrequent and rather local in the southern third of the state, rarely in small colonies, but, where found, the specimens are usually a rod or more apart. It grows in humus, generally on wooded, beech slopes, sometimes in black or black and white oak woods, and rarely in white oak woods. This is by far our most common coralroot.

Pa. to Ind., southw. to Fla. and Tex.

3. **Corallorrhiza maculàta** Raf. SPOTTED CORALROOT. Map 720. My specimens are all from the lake area except the one from Brown County. All grew in rather deep humus in black, black and white, or white oak woods. In addition to the counties shown on the map it has been reported

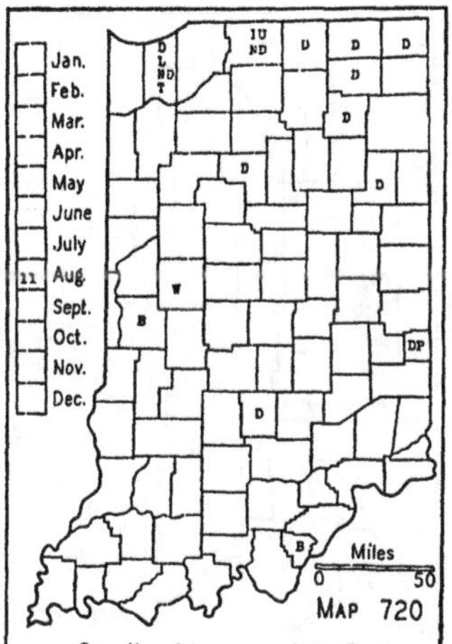

Corallorrhiza maculata Raf.

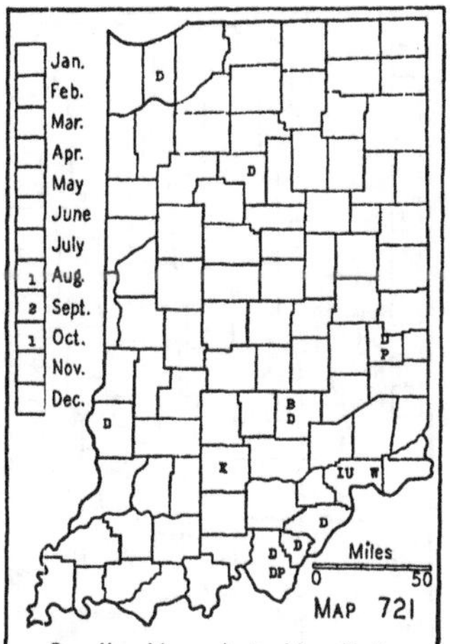

Corallorrhiza odontorhiza Nutt.

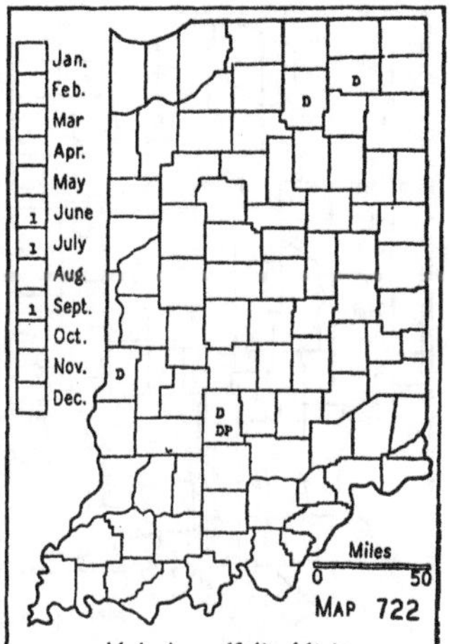

Malaxis unifolia Michx.

from Jefferson, Lake, and La Porte Counties. The report from Jefferson County may probably be wrong since the author did not report *Corallorrhiza Wisteriana* which occurs there.

Newf., Que., Sask. to B. C., southw. to Va., N. C., Ind. and Calif.

4. **Corallorrhiza odontorhìza** Nutt. LATE CORALROOT. Map 721 This species is found in slightly acid soil in bare places in fallow fields, or in rather sandy soil in deep humus in black and white oak woods. Very local in its distribution but probably found here and there throughout the state. It has been reported from other counties but wrong determinations are so frequent that to enumerate them might lead to confusion.

Southern Maine to Ont. and Mich., southw. to Fla. and Mo.

1552. MALÁXIS Sw.

Pedicels straight, mostly 3-6 mm long, longer than the ovary...........1. *M. unifolia.*
Pedicels twisted, 1-2 mm long, shorter than the ovary. (See excluded species no. 160, p. 1038.)..*M. brachypoda.*

1. **Malaxis unifòlia** Michx. (*Microstylis unifolia* (Michx.) BSP.) GREEN ADDER'S MOUTH. Map 722. I have specimens from four counties: one from Vigo County from a wooded slope, one from Monroe County from "Huckleberry Hill," one from Noble County near Pleasant Lake, and one from a clump of sphagnum in the Leesburg bog, Kosciusko County. Blatchley collected a specimen at "Huckleberry Hill" in Monroe County, June 15, 1887.

Newf. to Man., southw. to Fla., Ala., and Mo.

1556. LÍPARIS Richard TWAYBLADE

Flowers usually light madder purple; lip wedge-obovate, mostly 10 mm long; leaves elliptic or ovate; plants usually of dry ground.....................1. *L. liliifolia.*
Flowers yellowish green or light green; lip obovate or oblong, about 5 mm long; leaves elliptic-lanceolate; plants of a boggy habitat.....................2. *L. Loeselii.*

1. **Liparis liliifòlia** (L.) Richard. LILY TWAYBLADE. Map 723. Probably found in all parts of the state, being local in the northern part and

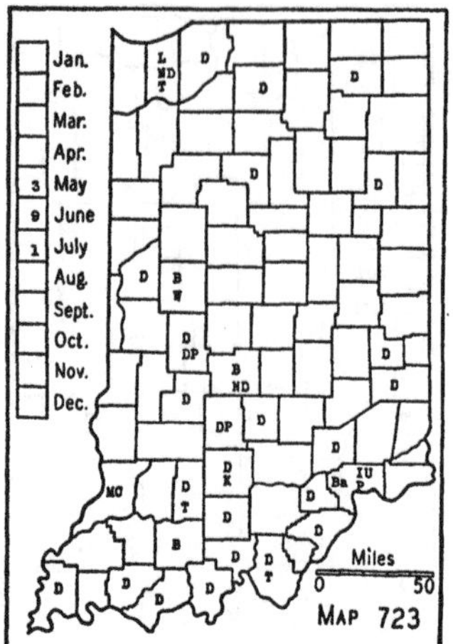

Liparis liliifolia (L.) Richard

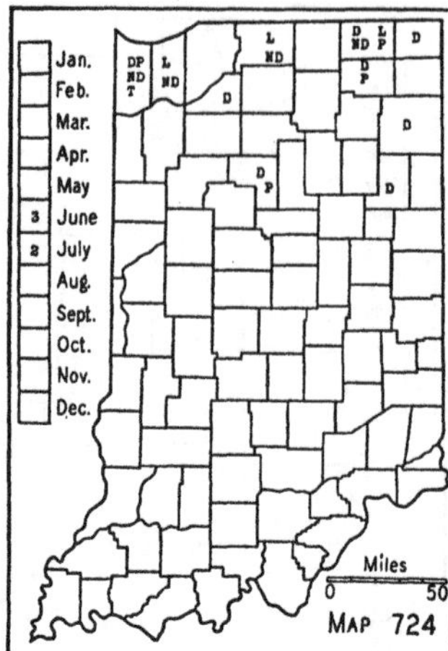

Liparis Loesellii (L.) Richard

Tipularia discolor (Pursh) Nutt.

more or less frequent in the southern part. The plant is very inconspicuous and doubtless it is more abundant than our records indicate. It evidently prefers a slightly acid soil and is generally found in deep humus in beech or white oak woods and more rarely in black and white oak woods. In Putnam County about 3 miles northwest of Greencastle, on June 3, 1910, I found it in a 19-year old Catalpa planting that had been first cultivated to strawberries and later abandoned. Here the plant was growing by the hundreds. In addition to the counties shown on the map it has been reported from these counties: Lake, Union, and Vigo.

N. H. and Mass. to Minn., southw. to Ga., Ala., and Mo.

2. **Liparis Loesèlii** (L.) Richard. Loesel Twayblade. Map 724. The majority of my specimens were found in sphagnum in tamarack bogs and in open boggy places. It is very local and is restricted to the lake area. In addition to the counties shown on the map it has been reported from Fulton, Kosciusko, La Porte, and Marshall Counties. It has already become very rare.

N. S. to Sask., southw. to N. C., Ala. and Mo.

1560. TIPULÀRIA Nutt.

1. **Tipularia díscolor** (Pursh) Nutt. (*Tipularia unifolia* (Muhl.) BSP.) Cranefly Orchid. Map 725. This species is restricted to our southern counties but it may have a wider range than the map shows. I have a memorandum that I saw it in Brown County but I did not preserve specimens. In 1938 Benjamin W. Douglass wrote me he found it near Trevlac. The leaves disappear before flowering time and look much like those of *Orchis spectabilis* or those of *Aplectrum hyemale*. *Tipularia* may easily be distinguished because the lower surface is purplish instead of green. One does not usually collect leaf specimens but in this species a leaf specimen makes a record as authentic as a flowering one. I am of the opinion that it will be found all over the unglaciated area, but very locally. Where it is found it is usually somewhat frequent but it is so inconspicuous

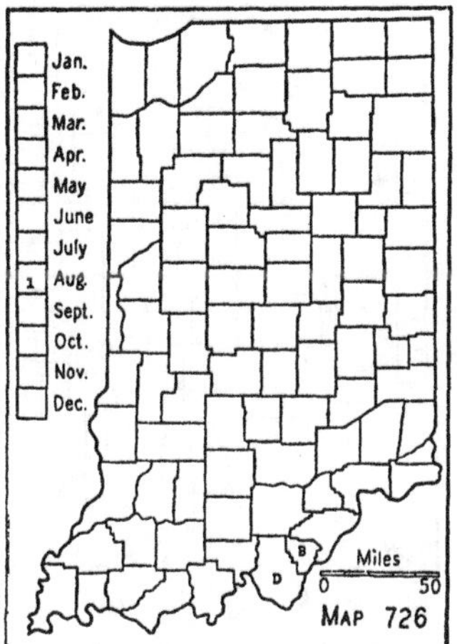

Hexalectris spicata (Walt.) Barnh.

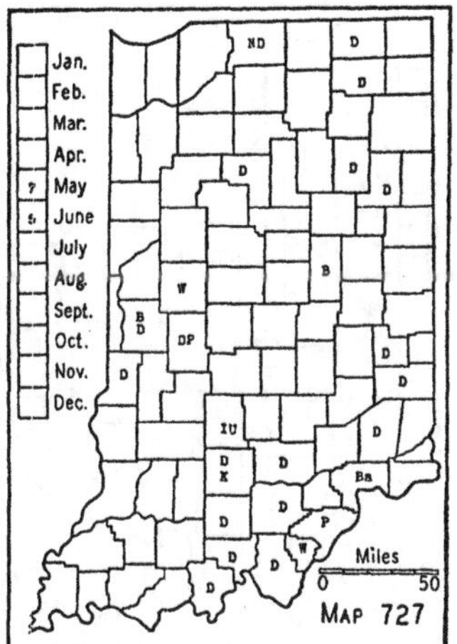

Aplectrum hyemale (Muhl.) Torr.

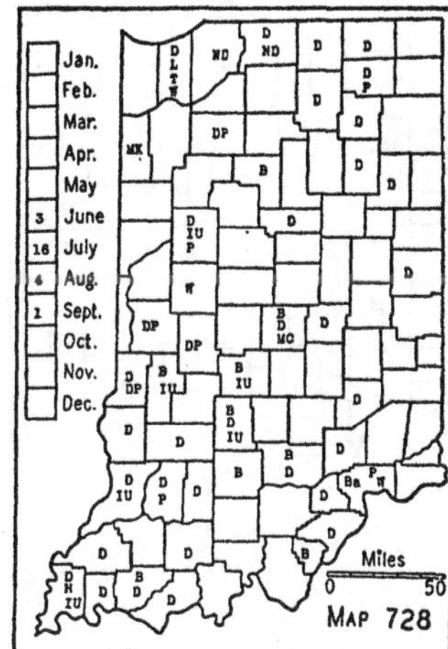

Saururus cernuus L.

that it may be overlooked. It grows in deep humus on protected slopes with beech or white oak and in black and white oak woods.

N. J., Ohio, and Ind., southw. to Fla. and Ala.

1629. HEXALÉCTRIS Raf.

1. **Hexalectris spicàta** (Walt.) Barnh. (Torreya 4: 121. 1904.) (*Hexalectris aphylla* (Nutt.) Raf.) CRESTED CORALROOT. Map 726. On August 3, 1922 I found a few scattered flowering plants on a black and white oak slope in a woods bordering the Ohio River in sec. 14 about 15½ miles southeast of Corydon. R. C. Friesner found it near Edwardsville in Floyd County August 20, 1923 and again on August 14, 1926. Blatchley reported it as growing on a high, wooded hill 2 miles south of Wyandotte Cave, Crawford County, July 25, 1896. These are our only reports.

Va. to Ind., southw. to Fla., Tex., and Ariz.; also in n. Mex.

1642. APLÉCTRUM [Nutt.] Torr.

1. **Aplectrum hyemàle** (Muhl.) Torr. PUTTYROOT. Map 727. This orchid is found sparingly throughout the state. In addition to the counties shown on the map it has been reported from the following counties: Hamilton, Lake, Marshall, Porter, Steuben, and Tippecanoe. It is found in deep humus in well protected and shaded spots in beech, black and white, and white oak woods. I recall that on the Clark County State Forest a large colony grew on a slope in a tangle of dense second growth of white oak and grape vines. During the winter the vines and excess of forest growth were removed and I never saw a plant there after that time. I have tried to grow the species at Bluffton in neutral soil in a shaded location but in a few years it disappears.

Vt. to Sask., southw. to Ga., Mo., and Kans.

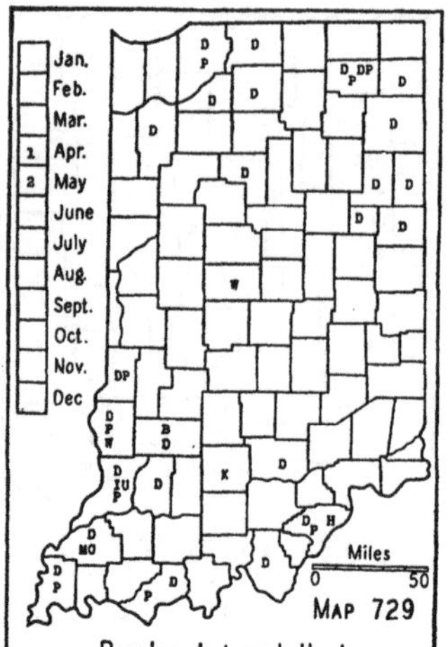

Populus heterophylla L.

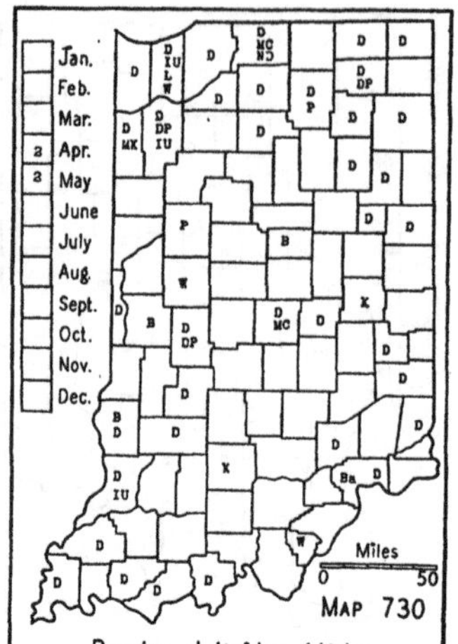

Populus deltoides Michx.

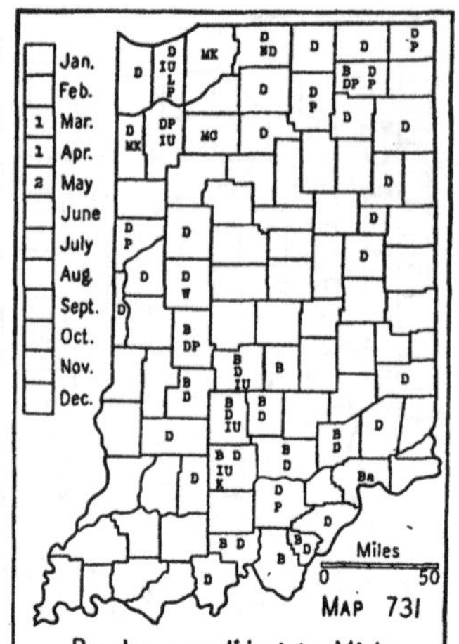

Populus grandidentata Michx.

52. SAURURÀCEAE Lindl. LIZARDTAIL FAMILY

1856. SAURÙRUS [Plum.] L.

1. **Saururus cérnuus.** L. COMMON LIZARDTAIL. Map 728. Infrequent to frequent throughout the state in wet woodland, along muddy borders of streams, and about ponds and sloughs. Where it is found it usually forms almost a complete stand over the area.

R. I. to Minn., southw. to Fla. and Tex.

56. SALICÀCEAE Lindl. WILLOW FAMILY

Buds with several scales; leaves ovate or deltoid; bracts of flowers laciniate; disk below each flower cup-shaped.........................1872. POPULUS, p. 352.

Buds with a single scale; leaves lanceolate or narrower, rarely wider; bracts of flowers entire or subentire; disk below each flower consisting of one or more glands..1873. SALIX, p. 354.

1872. PÓPULUS [Tourn.] L. POPLAR

Branchlets, outer bud scales, and lower surface of leaves white-tomentose; leaves more or less lobed (rarely a branch with unlobed leaves and these usually with about 7 coarse teeth)..1. *P. alba.*

Branchlets, outer bud scales, and lower surface of leaves not white-tomentose; leaves never lobed.

Petioles rounded and more or less channeled above.

Leaf blades 10-17 cm long, gradually narrowed toward the apex into an obtuse or merely acute point; pedicels of fruit usually 5-10 mm long..2. *P. heterophylla.*

Leaf blades 6-15 cm long; fruit nearly sessile or on pedicels up to 3 mm long.

Blades typically ovate-lanceolate, whitish, waxy, glabrous or sparsely pubescent beneath, usually acuminate at the apex; base acute, rounded or subcordate. ....3. *P. Tacamahacca.*

Blades broader, more rounded in outline, pubescent and usually with longer hairs; the apex usually acute; base more deeply cordate; teeth coarser...... ....3a. *P. Tacamahacca* var. *candicans.*

Petioles more or less flattened, especially near the blade.

Tips of the branchlets curved upward (in winter phase); mature leaves broadly deltoid and mostly more than 7 cm wide (or rhombic-ovate and cuneate at the base); stamens 20 or more; capsules 4-8 mm in diameter......4. *P. deltoides.*

Tips of the branchlets not curved upward (in winter phase); mature leaves ovate, broadly ovate to nearly orbicular, mostly less than 8 cm wide (except those of root and coppice shoots); stamens 6-12; capsules 1.5-3 mm in diameter.

Winter buds more or less pubescent, dull; young branchlets gray-tomentose at first; leaves generally with less than 12 teeth on each side (except those of root shoots).....................................5. *P. grandidentata.*

Winter buds glabrous, glossy; young branchlets glabrous or nearly so at first; leaves mostly with more than 12 teeth on each side........6. *P. tremuloides.*

1. POPULUS ÁLBA L. WHITE POPLAR. This species has been freely planted throughout the state and has escaped in all parts. It rapidly spreads from root shoots, and, when not restricted, it soon spreads in all directions, in fields and woodland in all kinds of soils except very wet ones. It is no longer planted by anyone familiar with its habit of spreading or one who knows that the branches are killed by the oyster-shell scale.

Nat. of Eurasia.

2. **Populus heterophýlla** L. SWAMP COTTONWOOD. Map 729. In Indiana it is infrequent in the lake area, local in the central part, local to frequent in the southern part, and possibly absent in the southeastern part. It becomes a tall, slender tree, 10-16 inches in diameter. It grows on the borders of ponds in woodlands which have for a subsoil a stiff blue clay, locally called "gumbo." The habitat simulates that of pin oak but I do not recall ever seeing these species growing together. It is usually associated with red maple, sweet gum, and cypress. Where there are more than a few trees it is usually found in a pure stand. It is most abundant in the sloughs of the Lower Wabash Bottoms.

Atlantic coast from Conn. to Fla., westw. to La., and northw. in the Mississippi Valley to n. Ohio, s. Mich., and Mo.

3. **Populus Tacamahácca** Mill. (*Populus balsamifera* of some recent authors.) BALSAM POPLAR. A few colonies of this poplar have been found along Lake Michigan in Lake, Porter, and La Porte Counties, and it has been found in St. Joseph County. The trees I have seen are small ones near the lake front.

Newf. and Lab. to Alaska, southw. and reaching the U. S. only on the northern border.

3a. **Populus Tacamahacca** var. **cándicans** (Ait.) Stout. (*Populus candicans* Ait.) See Jour. N. Y. Bot. Gard. *30*: 25-37. 1929. This variety is found in the eastern part of the range of the species and is found as a small tree along Lake Michigan. The clon, Balm of Gilead, originating from a specimen of this variety, has been freely planted but I do not know of any place where it is spreading.

4. **Populus deltoìdes** Michx. COTTONWOOD. Map 730. This is one of the largest trees of the state and is found throughout. It grows only in low ground about ponds, in woodland, and along streams and ditches.

N. H., w. Que. to the Rocky Mts., southw. to Fla. and Tex.

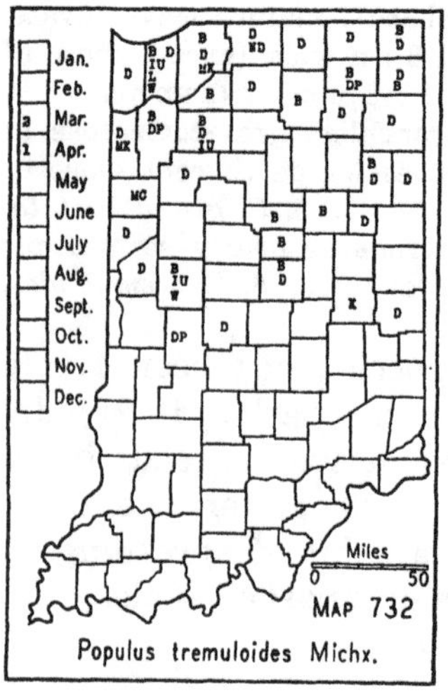

MAP 732
Populus tremuloides Michx.

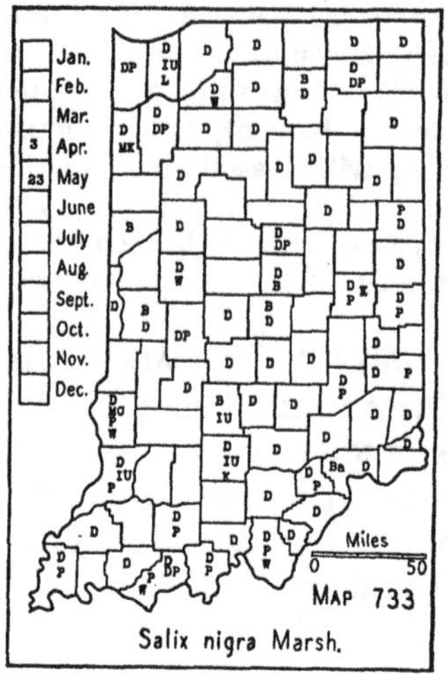

MAP 733
Salix nigra Marsh.

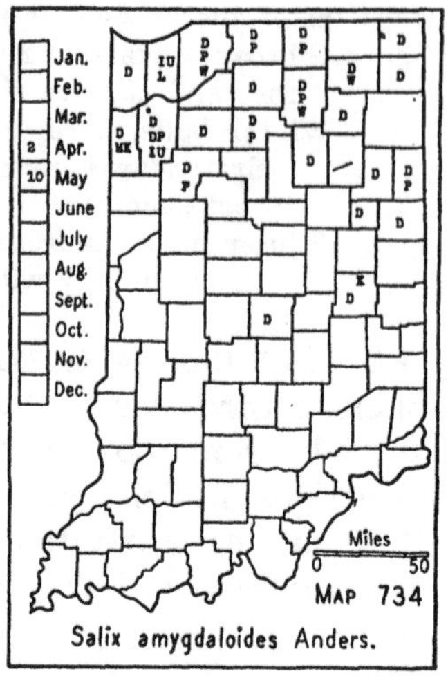

MAP 734
Salix amygdaloides Anders.

5. **Populus grandidentàta** Michx. LARGETOOTH ASPEN. Map 731. This is a tree of small or medium size found more or less frequently in the lake area and less frequently in the unglaciated area. Outside these areas it is local or absent. In the northern part of the state it is found in low ground while in the unglaciated area it is usually found on the crests of the highest ridges.

N. B. to Minn., southw. in the mts. to S. C., Ohio, Ind., and Iowa.

6. **Populus tremuloìdes** Michx. ASPEN. Map 732. This small tree is common in low ground in the lake area and I have never seen it growing on hills. It has been reported from all parts of the state but all of the specimens I have seen from the southern part of the state should be referred to the preceding species. It is doubtful whether it occurs far south of the stations shown on the map, and, if so, it will be found very locally.

Newf. and Lab. to Alaska, southw. to Tenn., Mo., Nebr., and in the mts. to Mex. and Calif.

1873. SÀLIX [Tourn.] L. WILLOW

Note: Specimens of this genus are difficult to determine because the species are dioecious, are highly variable, and freely hybridize. Hence it is advisable to collect a flowering specimen and later to collect a mature leaf specimen from the same plant to make determination easy and certain.

KEY BASED PRIMARILY ON PISTILLATE FLOWERS AND CAPSULES

Ovary glabrous; bracts of flowers usually deciduous before maturity of the capsule.
  Ovary sessile or subsessile.
    Leaves cordate at the base, silky-pubescent above............17. *S. adenophylla.*
    Leaves cuneate at the base.
      Blades more or less pubescent at flowering time; capsules 3-5 mm long.6. *S. alba.*
      Blades usually glabrous at flowering time; branches pendulous; capsules 1.5-2 mm long. (See excluded species no. 162, p. 1038.)............*S. babylonica.*

Ovary stalked, sometimes the stalk rather short.
  Stigmas sessile or subsessile (style, if any, less than 0.5 mm long); petioles without glands.
    Leaf margins entire; leaves glaucous beneath................15. *S. pedicellaris.*
    Leaf margins more or less serrate.
      Margins of leaves with widely spaced, slender, sharp teeth......8. *S. interior.*
      Margins of leaves finely and evenly serrate.
        Petioles of mature leaves 3-6 mm long.
          Blades green beneath......................................1. *S. nigra.*
          Blades whitish beneath........................5. *S. longipes* var. *Wardi.*
        Petioles of mature leaves 6-15 mm long; blades paler beneath...2. *S. amygdaloides.*
  Stigmas on distinct styles 0.5-1 mm long; petioles mostly glandular at the apex.
    Petioles not glandular; leaves glaucous beneath.
      Leaves more than 5 mm wide...........................19. S. *glaucophylla.*
      Leaves less than 5 mm wide..................16a. *S. candida* var. *denudata.*
    Petioles glandular at the apex; leaves not glaucous beneath.
      Capsules maturing after June 20, 7-9 mm long................3. *S. serissima.*
      Capsules maturing before June 20, 5-8 mm long.
        Catkins sessile....................................18. *S. cordata.*
        Catkins distinctly stalked.
          Leaves of branchlet below the catkin stalk entire...........7. *S. fragilis.*
          Leaves of branchlet below the catkin stalk serrulate.
            Catkins loosely flowered; capsules conic-subulate; pedicel twice as long as the gland. (See excluded species no. 166, p. 1038.) *S. pentandra.*
            Catkins densely flowered; capsules conic-ovoid; pedicel 2-3 times as long as the gland.
              Leaves glabrous on both surfaces.....................4. *S. lucida.*
              Leaves pubescent beneath...............4a. *S. lucida* var. *intonsa.*
Ovary pubescent; bracts of flowers persistent.
  Ovaries pedicellate.
    Stigmas sessile or subsessile.
      Catkins sessile or subsessile...............................14. *S. Bebbiana.*
      Catkins on short, leafy stalks.......................................8. *S. interior.*
    Stigmas on short styles, usually 0.25-1 mm long.
      Catkins on short, leafy stalks..............................10. *S. petiolaris.*
      Catkins sessile or subsessile, rarely with 1-3 bracts at the base, appearing on old wood mostly before the leaves.
        Bracts of flowers not darker at the apex......................16. *S. candida.*
        Bracts of flowers darker at the apex.
          Mature capsules 3-5 mm long, blunt..........................11. *S. sericea.*
          Mature capsules mostly 6-12 mm long.
            Branchlets of previous year glabrous or nearly so; catkins in flower mostly more than 2.5 cm long, 3-8 cm long in fruit....9. *S. discolor.*
            Branchlets of previous year more or less densely puberulent.
              Catkins more than 2.5 cm long; leaves 5-10 cm long and 2-3.5 cm wide, the margins somewhat toothed........9a. *S. discolor* var. *latifolia.*
              Catkins less than 2.5 cm long, usually 1-1.5 cm long; leaves smaller, more or less undulate or entire, rarely with a few teeth.
                Shrubs mostly 6-12 dm high; mature leaves erect or spreading, glabrous or glabrate above and beneath, or the midrib remaining pubescent...12. *S. humilis.*
                Shrubs mostly 4-8 dm high; mature leaves erect, pubescent above and more or less tomentose beneath; petioles about 3 mm long, for the most part shorter than those of the preceding...13. *S. tristis.*
  Ovaries sessile or subsessile; catkins appearing before the leaves.

Capsules 2-3 mm long; stigmas sessile. (See excluded species no. 167, p. 1038.) ....*S. purpurea.*

Capsules 6-8 mm long; stigmas stalked. (See excluded species no. 168, p. 1039.) ....*S. viminalis.*

KEY BASED PRIMARILY ON STAMINATE FLOWERS

Stamens 3 or more (rarely nos. 6 and 7 found here); catkins on leafy or at least on bracted stalks.

Catkins slender, mostly 8-10 mm wide at the widest diameter and 5-7 cm long; petioles not glandular.

Floral bracts generally woolly-pubescent all over the outer face, about 1.5-2 mm long; stamens usually more than twice as long as the bract........1. *S. nigra.*

Floral bracts generally woolly-pubescent only on the lower half of the outer face, usually about 2 mm long; stamens about twice as long as the bract........ ....2. *S. amygdaloides.*

Catkins stouter, mostly 11-13 mm wide at the widest diameter and 2-5 cm long; petioles glandular.

Leaves green or slightly glaucous beneath; plants of northern Indiana.

Plants flowering from the middle of May to the middle of June.

Leaves glabrous beneath..4 *S. lucida.*

Leaves somewhat rusty-pubescent beneath..........4a. *S. lucida* var. *intonsa.*

Plants flowering after the middle of June.....................3. *S. serissima.*

Leaves whitish beneath; gnarled shrubs of the rocky banks or beds of streams in southern Indiana...............................5. *S. longipes* var. *Wardi.*

Stamens 2 (sometimes 3 or 4 in nos. 6 and 7.)

Filaments more or less pubescent.

Filaments separate, not fused.

Filaments usually pubescent half their length (shrubs)...........8. *S. interior.*

Filaments usually pubescent only at the base (trees).

Young branchlets and leaves more or less silky...................6. *S. alba.*

Young branchlets and leaves glabrous or only slightly silky.....7. *S. fragilis.*

Filaments fused nearly or quite to the anthers. (See excluded species no. 167, p. 1038.)...*S. purpurea.*

Filaments glabrous, not fused at the base.

Catkins appearing with or after the leaves on leafy-bracted branchlets, these sometimes very short.

Branchlets and leaves densely pubescent, finely glandular-serrate.

Upper surface of leaves densely silky-pubescent (plants found only along Lake Michigan).................................17. *S. adenophylla.*

Upper surfaces of leaves glabrous or sparsely silky...........18. *S. cordata.*

Branchlets and leaves glabrous or glabrate.

Leaves entire; low shrubs of a bog habitat..15. *S. pedicellaris* var. *hypoglauca.*

Leaves closely glandular-serrate; shrubs also of a wet or boggy habitat, mostly in the dune area.................................19. *S. glaucophylla.*

Catkins appearing with or after the leaves, sessile or subsessile, without bracts or with 1-3 small ones at the base.

Branchlets of previous year puberulent, at least at the summit.

Anthers red.

Leaves impressed-nerved above.

Leaves woolly-pubescent above..........................16. *S. candida.*

Leaves glabrous or glabrate above........16a. *S. candida* var. *denudata.*

Leaves not impressed-nerved above.......................12. *S. humilis.*

Anthers yellow.

Bracts of flowers of a uniform, light color.................14. *S. Bebbiana.*

Bracts of flowers with darkened tips.

Anthers (dry) about 0.9 mm long..........9a. *S. discolor* var. *latifolia.*

Anthers (dry) usually less than 0.8 mm long.

Catkins 10-15 (18) mm long; young foliage somewhat tawny...10. *S. petiolaris.*

Catkins (15) 18-28 mm long; young foliage glabrous or white-pubescent.

Anthers (dry) 0.4-0.5 mm long; young foliage white-silky; hairs of bracts silky, scarcely curled or matted; twigs brittle at the base...11. *S. sericea.*

Anthers (dry) 0.6-0.8 mm long; young foliage glabrous, or, if white-pubescent, scarcely silky; hairs of the bracts curly or matted, scarcely silky; twigs tough at the base..........18. *S. cordata.*

Branchlets of previous year glabrous; anthers (dry) about 0.9 mm long...9. *S. discolor.*

KEY BASED PRIMARILY ON MATURE LEAVES AND BRANCHLETS

A. Leaves green on both sides.

Margins of leaves with unequally spaced, minute teeth.

Blades linear, acute at both ends, often somewhat falcate, less than 1 cm wide, mostly 4-6 mm wide, rarely somewhat paler beneath............8. *S. interior.*

Blades, at least some of them, linear-oblong, acute at both ends, some or many of them more than 1 cm wide, bluish green beneath...8a. *S. interior* var. *Wheeleri.*

Margins of leaves closely serrate with equally spaced teeth.

Leaves linear-lanceolate, mostly 6-12 cm long, usually 7-10 times as long as wide, rarely 2 cm wide, mostly 1-1.5 cm wide, acute or acuminate, sometimes falcate; teeth usually 6-10 per cm..................................1. *S. nigra.*

Leaves not as above.

Blades silky-pubescent on both sides, acute at the apex, subcordate at the base, ovate (plants along Lake Michigan)...................17. *S. adenophylla.*

Blades not as above.

Petioles glandular at the summit; leaves shining above, more than 2 cm wide, long-acuminate or caudate at the apex, rounded at the base.

Blades glabrous on both sides...............................4. *S. lucida.*

Blades mostly permanently more or less pubescent beneath with reddish hairs................................4a. *S. lucida* var. *intonsa.*

Petioles not glandular at the summit; leaves not shining above...18. *S. cordata.*

A. Leaves glaucous or paler beneath.

Leaves subopposite, cuneate-oblanceolate, bluish green, very smooth; stipules early deciduous. (See excluded species no. 167, p. 1038.)............*S. purpurea.*

Leaves strictly alternate.

Margins of leaves finely and distinctly serrate.

Petioles glandular at the summit (sometimes obscurely so in *S. alba*).

Leaves linear-lanceolate, 8-16 cm long, long-acuminate, glabrous, primary veins regular, ending in the border to form a rather straight line; branchlets of previous year slender, pendulous, tough. (See excluded species no. 162, p. 1038.)..............................*S. babylonica.*

Leaves not as above.

Blades ovate or ovate-oblong, 4-12 cm long, closely glandular-serrate, glabrous from the first, short-acuminate, rounded or subcordate at the base; petioles 6-10 mm long. (See excluded species no. 166, p. 1038.)..*S. pentandra.*

Blades not as above, mostly lanceolate.

Leaves glossy above, glabrous, some, or most of them, more than 2 cm wide, acute or acuminate at the apex, teeth (8) 10-20 per cm; primary veins so prominent above as to make the upper surface of dried specimens rough to the touch; native shrubs, up to 4.5 m high, flowering in late June and in July........................3. *S. serissima.*

Leaves not as above; introduced trees, flowering in May and early June (the following two species difficult to separate).

Serrations of blades rather coarse, generally 3-8 per cm and 0.4-1 mm deep; blades rather firm, glabrous or appressed-pubescent, especially beneath, acuminate or long-acuminate, the primary veins so prominent above as to make the upper surface of dried specimens rough to the touch; branchlets of previous year brittle at the base..7. *S. fragilis.*

Serrations of blades generally 6-12 per cm and 0.1-0.4 mm deep; blades thinner than the preceding, smooth to the touch above, usually more or less appressed-pubescent or glabrous, acute or acuminate at apex.

Leaves distinctly silky beneath; branchlets greenish........6. *S. alba.*

Leaves subglabrous beneath; branchlets yellowish.................. ....6a. *S. alba* var. *vitellina.*

Petioles not glandular (not to be confused with glands at the base of the blade).

Leaves glabrous on both sides. (No. 18 often so glabrous that it might be wrongly placed here.)

Blades small, broadest about the middle, usually 5-10 mm wide, rarely up to 15 mm wide, linear-lanceolate to lanceolate, teeth generally 5 or 6 per cm; mostly acuminate..........................10. *S. petiolaris.*

Blades larger, broadest above or below the middle, mostly (10) 15-30 mm wide, long-acuminate, obtuse or acute.

Blades broadest below the middle, lanceolate to broadly lanceolate or ovate-lanceolate, long-attenuate at the apex, broadly cuneate to rounded at the base, teeth mostly 7-12 per cm of margin measured midway between base and apex; stipules rarely present; petioles mostly 6-15 mm long..........................2. *S. amygdaloides.*

Blades generally broadest above the middle, or below the middle in some forms of leaves, usually oblanceolate to ovate or elliptic-lanceolate, or ovate on coppice shoots, acute at the apex, rarely acuminate, broadly cuneate or rounded at the base, cordate in the ovate type of leaves; teeth mostly 3-6 per cm; stipules usually present, especially on vegetative branchlets, subcordate to broadly reniform, 3-10 mm long, acute; petioles mostly 3-10 mm long........19. *S. glaucophylla.*

Leaves more or less pubescent, at least some of them so.

Leaves white silky-pubescent beneath, at least some of them more or less pubescent on one or both sides, especially along the midrib and toward the base on the lower surface, and pubescent at the apical end on the upper surface of terminal leaves.

Blades rather thin, small, narrow, linear-lanceolate to lanceolate, mostly 5-8 cm long and 5-10 mm wide, rarely up to 14 mm wide; pubescence, when present, usually tawny, rarely all whitish; teeth of margins usually 5 or 6 per cm; petioles 4-10 mm long; stipules none; branchlets more or less fascicled.......................10. *S. petiolaris.*

Blades mostly longer, wider, and thicker than the preceding.

Leaves silvery-pubescent beneath, the pubescence strongly or loosely upwardly appressed or sometimes glabrate or glabrous beneath, narrowly to broadly lanceolate or somewhat oblanceolate, usually about 5 times as long as wide, rounded or narrowed at the base, the lateral veins both above and beneath usually conspicuous.

Lower surface of leaves strongly upwardly appressed-pubescent, silvery (rarely somewhat glabrate in age); blades strongly tapering at the base; stipules soon deciduous...........11. *S. sericea.*

Lower surface of leaves loosely appressed-pubescent when young, usually becoming glabrous or glabrate at maturity or remaining sparsely pubescent; blades rounded at the base, stipules persistent at least on vigorous branchlets...........18. *S. cordata.*

Leaves glabrous and glaucous beneath, oblong-lanceolate or narrow-lanceolate, mostly 6-9 times as long as wide, rounded at the base; shrubs of streams near the Ohio River....5. *S. longipes* var. *Wardi.*

Margins of leaves entire, remotely dentate or serrulate, mostly revolute.

Leaves strictly glabrous, oblanceolate, rarely obovate or narrowly to broadly elliptic, 2-4 or up to 6 cm long, 1-2 cm wide, closely reticulate on both surfaces, thin, entire, often bluish beneath; apex obtuse, rounded or rarely acute (see also no. 12); small shrubs of a bog habitat..15. *S. pedicellaris* var. *hypoglauca.*

Leaves not as above.

Serrations (not undulations) of margins, if any, generally more than 0.3 mm deep; leaves large, mostly elliptic-oblanceolate, elliptic or obovate, mostly 5-10 cm long, 2-3 cm wide; petioles generally 5-20 mm long.

Branchlets of previous year and leaves entirely glabrous......9. *S. discolor.*

Branchlets of previous year and at least some of the leaves pubescent.

Blades rarely impressed-nerved above, some or most of them nearly glabrous beneath, the pubescence straight or woolly, all or at least some of the leaves with a few tawny hairs; petioles mostly 8-25 mm long................................9a. *S. discolor* var. *latifolia.*

Blades generally impressed-nerved above and strongly rugose-veined beneath, lower surface of all more or less densely woolly-pubescent; petioles mostly 4-12 mm long.......................14. *S. Bebbiana.*

Serrations (not undulations), if any, mostly less than 0.3 mm deep; leaves linear-oblanceolate or oblanceolate, rarely wider, tomentose or glabrate beneath; petioles 2-6 mm long.

Leaves generally 7-16 times as long as wide, usually densely tomentose beneath; midrib deeply impressed above; plants of a bog habitat.

Leaves pubescent above..............................16. *S. candida.*

Leaves glabrous above....................16a. *S. candida* var. *denudata.*

Leaves generally less than 7 times as long as wide; midrib not deeply impressed above; plants of a dry or prairie habitat.

Blades mostly 5-10 cm long.............................12. *S. humilis.*

Blades mostly 3-7 cm long, thicker and more tomentose beneath...13. *S. tristis.*

1. **Salix nigra** Marsh. BLACK WILLOW. Map 733. Infrequent to frequent throughout the state in low ground mostly along streams and about lakes. In southwestern Indiana along old river channels it often becomes a large tree.

The leaves of this species vary much in outline but I do not think the variations have any taxonomic value. The form with narrow and falcate leaves is known as var. *falcata* (Pursh) Torr.

N. B. to N. Dak., southw. to Fla. and Tex.

2. **Salix amygdaloides** Anders. PEACHLEAF WILLOW. Map 734. The habitat of this willow is low ground along streams and about lakes. This species is restricted to the lake area with two outposts south of that area where it was found in swamps. It is usually infrequent and only locally frequent.

Cent. N. Y. and Ont. to B. C. and the Rocky Mts., southw. to Tex. and N. Mex.

Salix serissima (Bailey) Fern.

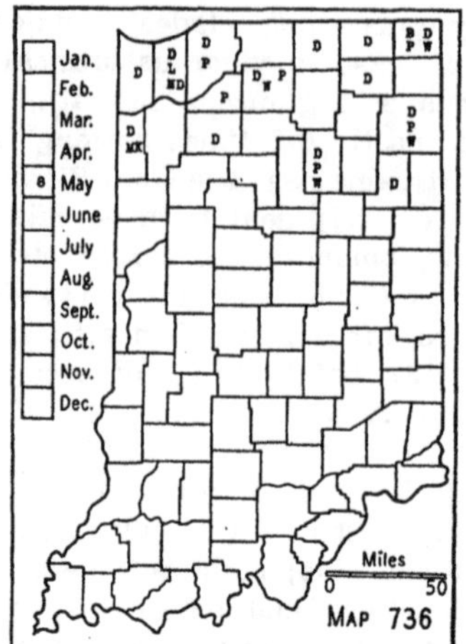

Salix lucida Muhl.

Salix lucida var. intonsa Fern.

3. **Salix seríssima** (Bailey) Fern. AUTUMN WILLOW. Map 735. This is a low ground shrub and is found along streams, about lakes, and in marshes. It is local and restricted to a few counties of the northeastern part of the state.

Newf. to Alberta, southw. to N. J., N. Y., and the Great Lakes.

4. **Salix lùcida** Muhl. SHINING WILLOW. Map 736. An infrequent willow in the lake area about lakes, along streams, and in swamps and marshes.

Lab. to Alberta, southw. to N. J., Ky., and Nebr.

4a. **Salix lucida** var. **intónsa** Fern. Map 737. A shrub 4-12 feet high; found locally in the lake area about lakes, along streams, and in swamps and marshes. Not as frequent as the species.

Newf. to Que., southw. to w. N. Y. and Ind.

5. **Salix lóngipes** Shuttl. var. **Wárdi** (Bebb) Schneid. (*Salix Wardi* Bebb.) WARD WILLOW. Map 738. This low, sprawling shrub I have found growing in the crevices of large rocks along the bank of the Ohio River about 6 miles above Cannelton, in Perry County, and in crevices of rocks in the overflow bank of Buck Creek, about 6 miles north of Laconia in Harrison County. A shrub about 2 inches in diameter and 6 feet high was found growing between layers of limestone rock, about a foot above the water from a bank about 2 feet high on the north side of Laughery Creek about a fourth mile east of Friendship, Ripley County. Good specimens are difficult to obtain because in all localities the plants are submerged during high water. The shrubs are sprawling in character because debris and ice continually keep them broken off, although they are very tough. This is a southern willow and should be sought all along the Ohio River.

Potomac River, s. Ind., s. Ill. to se. Kans., southw. to Cuba and Tex.*

* Since the manuscript was written C. R. Ball reports that this species has been found along White Water River near Brookville, Franklin County.

MAP 738
Salix longipes Shuttlw. var. Wardi (Bebb) Schneid.

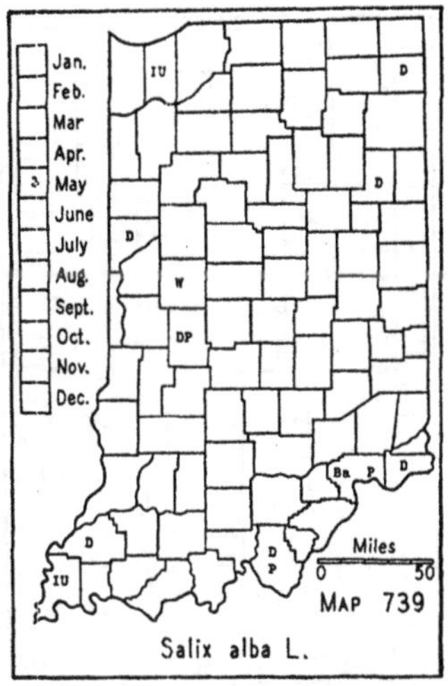

MAP 739
Salix alba L.

MAP 740
Salix fragilis L.

6. SALIX ÁLBA L. EUROPEAN WHITE WILLOW. Map 739. This is a European species that, no doubt, has been planted more or less throughout the state. I have found it as an escape only a few times although it has been reported from 12 counties besides those in which I have found it.

Nat. of Eu.

6a. SALIX ALBA var. VITELLÌNA (L.) Stokes. GOLDEN WILLOW. This willow has been reported from 10 counties, mostly by our early authors who were not careful to distinguish between escaped and planted trees. I believe it is far more common than our white willow but I have seen it only a few times where I would consider it as an escape. I doubt that it ever escapes by seed but only by means of branchlets which have been broken off and carried down streams and deposited where they are covered with mud.

Nat. of Eu.

7. SALIX FRÁGILIS L. BRITTLE WILLOW. Map 740. This European willow has been freely planted throughout the state and is found more commonly as an escape, I believe, because the branchlets are very easily broken off by wind and ice and scattered where they are covered with soil and easily propagate.

I recall the ingenious use of this species by a farmer in Wayne County who, about 1857, had planted several rows of the trees and spaced them close and in zigzag rows across a creek bottom. When I asked why he so planted them he told me that it was to catch the rails and wheat that came down the stream during floods.

Nat. of Eu.

8. **Salix intèrior** Rowlee. (*Salix longifolia* Muhl.) LONGLEAF WILLOW. SANDBAR WILLOW. Map 741. Found throughout the state along streams, especially on gravelly bars, about lakes, and along ditches. It usually forms dense colonies and often covers large areas.

Eastern Que. to Man., southw. in the interior to Va., Tenn., and Tex.; generally absent from N. E. and the Coastal Plain.

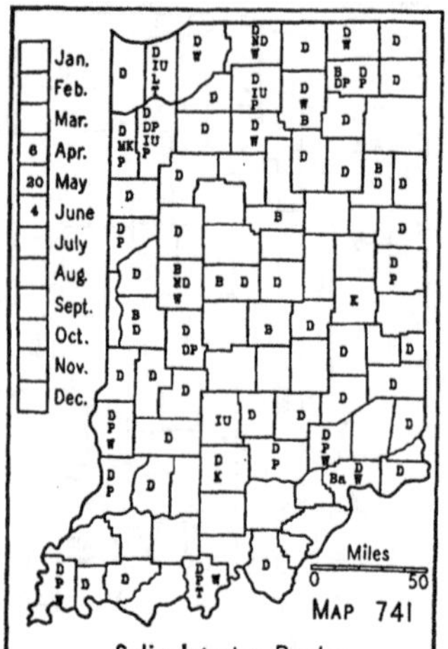

Salix interior Rowlee

Salix interior var. Wheeleri Rowlee

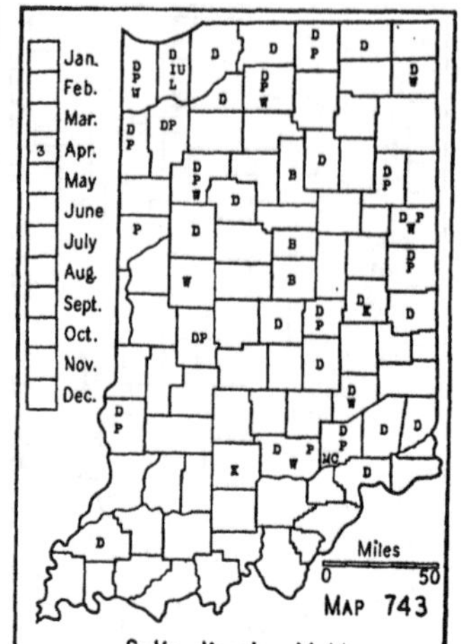

Salix discolor Muhl.

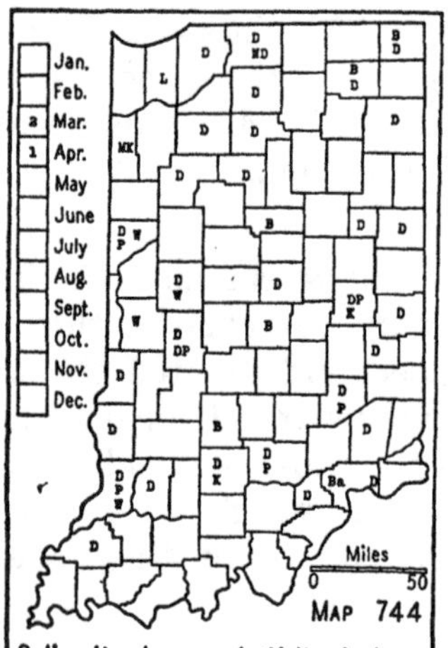

Salix discolor var. latifolia Anders.

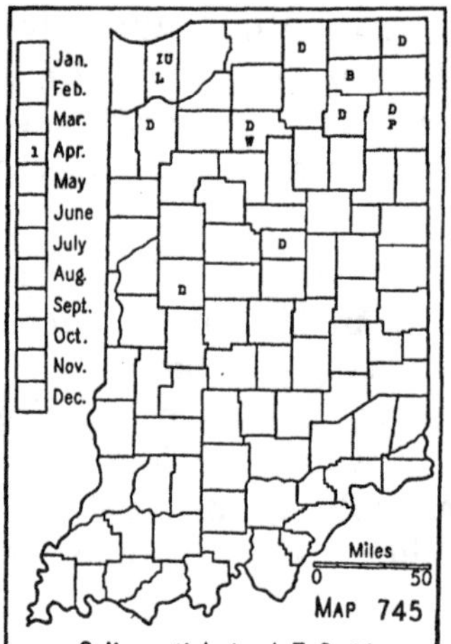

Salix petiolaris J. E. Smith

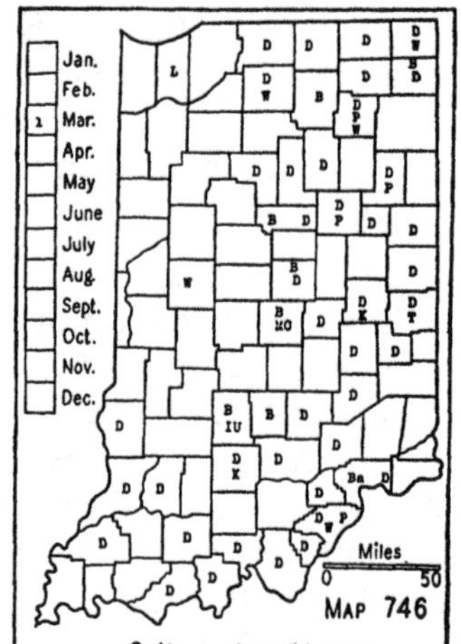

Salix sericea Marsh.

8a. **Salix interior** var. **Wheèleri** Rowlee. (*Salix longifolia* var. *Wheeleri* (Rowlee) Schneid.) WHEELER WILLOW. Map 742. This form is common along the Ohio River where it is associated with the species but may easily be distinguished at a long distance by its bluish green color.

N. B. to James Bay and e. N. Dak., southw. to Conn., Pa., and Iowa.

9. **Salix díscolor** Muhl. PUSSY WILLOW. Map 743. Found throughout the state where swampy land occurs. Frequent in the lake area and local to infrequent south of it. Usually a large shrub, it sometimes reaches a diameter of several inches a few feet above the ground.

Newf. to Man., southw. to Del. (and in the mts. to N. C.), Ill., and Mo.

9a. **Salix discolor** var. **latifòlia** Anders. (*Salix discolor* var. *eriocephala* (Michx.) Anders.) (Schneider. Jour. Arnold Arb. 2: 5. 1920.) Map 744. This variety is found throughout the state in swamps and low land in general. It is rather frequent in the lake area, becoming local to infrequent south of it. It has the same habitat as the species and both are often associated.

Probably the range of the species.

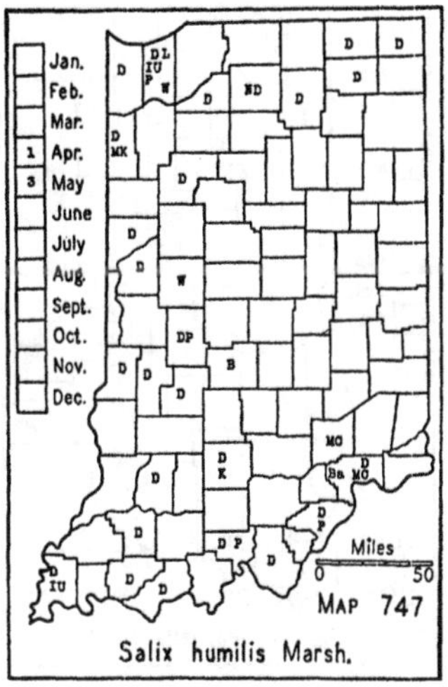

Salix humilis Marsh.

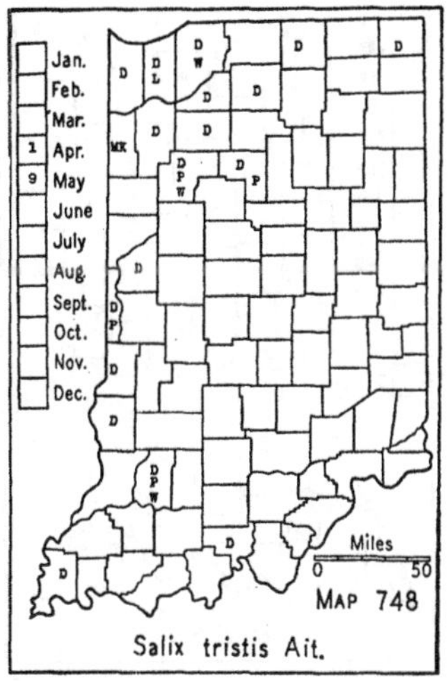

Salix tristis Ait.

Salix Bebbiana Sarg.

10. **Salix petiolàris** J. E. Smith. Map 745. This is an infrequent shrub 4-7 feet high, and found mostly in marshy and mucky land in the lake area. The species is variable and I have two named varieties from the state but I do not regard them as of taxonomic value and do not report them. I feel that of the named variations too many are ecological forms.

N. B. to N. Dak. and Man., southw. to N. J. and Tenn.

11. **Salix serícea** Marsh.* SILKY WILLOW. Map 746. Infrequent to frequent throughout the state except in the northwestern part from which there are no specimens. It is generally found in wet habitats although I have a few specimens collected from moist, sandy habitats.

N. B. and N. S. to Mich., southw. to N. C.

12. **Salix hùmilis** Marsh. PRAIRIE WILLOW. Map 747. This is a low, bushy species that grows mostly in dry, sandy habitats, usually in prairies or in similar places. It is frequent in our western prairie area, becoming local in northern and southern Indiana.

Newf. to Minn., southw. to N. C., Tenn., and Kans.

13. **Salix trístis** Ait. DWARF PUSSY WILLOW. Map 748. This is a small shrubby willow with a habit and habitat similar to the preceding species. It also has nearly the same distribution but is much less frequent and, in fact, as I understand the plant, it would be restricted to our western prairie area. Since almost all of my specimens were named by C. R. Ball, I am using his determinations to show the distribution in Indiana.

After studying my specimens carefully and noting the habitats from which they came, I have come to the conclusion that this species is merely an ecological form of the preceding species. Griggs and Schaffner both regard it as a variety of the preceding. This and the preceding species are most common in White County and I have seen them growing side by side

* After the Flora was in page proof C. R. Ball wrote me that a restudy of my specimens of willows shows that *Salix subsericea* (Anders.) Schneid. (Rhodora 11: 12. 1909) occurs in Indiana in Allen, Elkhart, Kosciusko, Lake, La Porte, and Starke Counties.

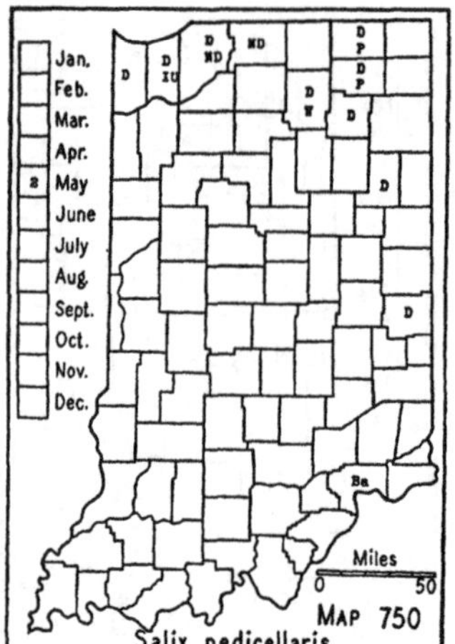

Salix pedicellaris var. hypoglauca Fern.

Salix candida Flügge

Salix adenophylla Hook.

in a strictly prairie habitat, which fact might be used to support their separation, but I am not able to find any constant structural difference.

Mass. to N. Dak., southw. to Fla., Tenn., e. Nebr. and e. S. Dak.

14. **Salix Bebbiàna** Sarg. (*Salix rostrata* Richardson.) BEBB WILLOW. Map 749. Rather frequent in the lake area and rare south of it.

Newf. to Alaska, southw. to N. J., Nebr., and Utah.

15. **Salix pedicellàris** Pursh var. **hypoglaúca** Fern. (*Salix pedicellaris* Pursh in part.) BOG WILLOW. Map 750. Infrequent in the lake area with an outlying post in the Elliott's Mill Bog in Wayne County. It is a small bog willow and usually found in sphagnum in tamarack bogs.

Newf. to B. C., southw. to N. J. (?), Pa., Ill., and Iowa.

16. **Salix cándida** Flügge. SAGE WILLOW. Map 751. This is an infrequent low willow in the bogs of the northern part of the lake area. It is generally found with sedges, cranberry, and bog-rosemary.

Newf. and Que., southw. to N. Y. and Wis.

16a. **Salix candida** var. **denudàta** Anders. This variety differs from the species in having narrower leaves, which are glabrate or glabrescent on both sides, especially above, and sometimes glaucescent beneath. I have only one specimen from a bog on the south side of Pigeon River about 2 miles east of Mongo, Lagrange County.

17. **Salix adenophýlla** Hook. (*Salix syrticola* Fern. Rhodore 9: 225-226. 1907.) (Schneider. Jour. Arnold Arb. 1: 158-160. 1920.) GLANDLEAF WILLOW. Map 752. Formerly more or less frequent along Lake Michigan at the base of the first dune on the side facing the beach. It is now nearly extinct on account of the encroachments of civilization and the attacks of the oyster-shell scale.

Lab. to James Bay, southw. to the Great Lakes, including Ohio, Ind., and Wis.

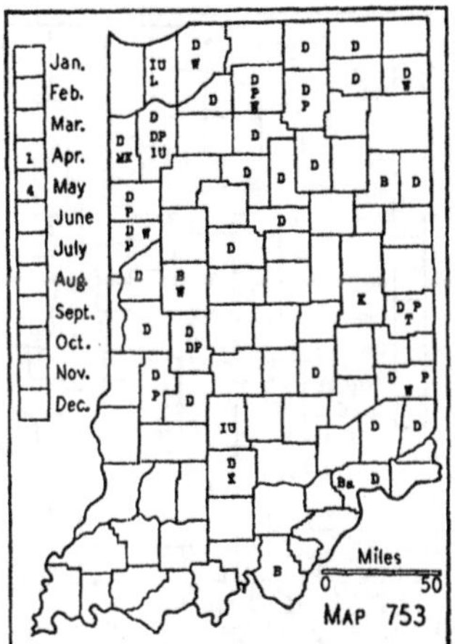

Salix cordata Muhl.

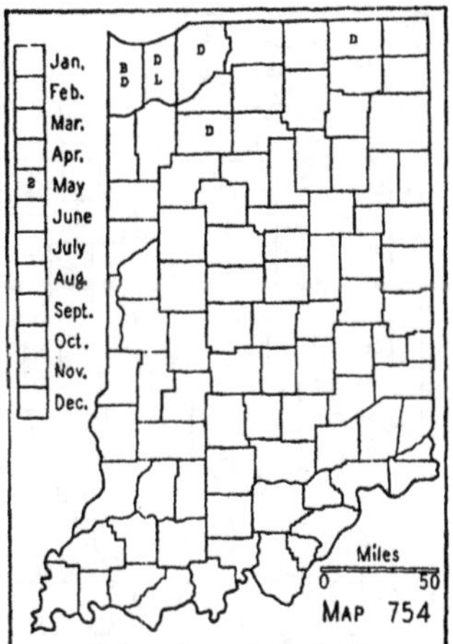

Salix glaucophylla Bebb

Comptonia peregrina (L.) Coulter

18. **Salix cordàta** Muhl. HEARTLEAF WILLOW. Map 753. This willow is infrequent throughout the lake area, becoming progressively less frequent southward and probably entirely absent from the southwestern part. It prefers a moist soil but does not demand a very wet soil such as is found in bogs and marshes. *Salix cordata* var. *angustata* Anders. is a narrowleaf form which I have from Wabash County. The species freely hybridizes and I have several specimens of each of two of its hybrids, *S. cordata* × *nigra* and *S. cordata* × *sericea*.

Newf. to B. C., southw. to Va., Mo., Colo., and Calif.

19. **Salix glaucophýlla** Bebb. BLUELEAF WILLOW. Map 754. Very local except along the sides of the dune facing Lake Michigan where it is more or less frequent. Away from the lake it is found in bogs and swamps. The variety *brevifolia* Bebb, which has been reported by Peattie and by Pepoon, is a shortleaf form which I do not regard as having any taxonomic standing.

Eastern Que. to Alberta, southw. to N. B., Maine, and the Great Lakes.

57. MYRICÀCEAE Dumort. BAYBERRY FAMILY

1874. COMPTÒNIA Banks

1. **Comptonia peregrìna** (L.) Coulter. (*Myrica asplenifolia* L.) For a discussion of the nomenclature see Rhodora 40: 408-412. 1938. SWEETFERN. Map 755. Infrequent to frequent or local in acid soils, sometimes forming large colonies. It is a shrub mostly one and a half to two and a half feet high and usually found in black, sandy soil in open places in pin oak and black oak woods.

N. B. to Sask., southw. to N. C., Tenn., and Ind.

60. JUGLANDÀCEAE Lindl. WALNUT FAMILY

Pith of twigs chambered; staminate catkins thick, sessile or short-stalked; stamens 8-40, glabrous; nuts with a network of rough projections (in ours); husk not splitting..1881. JUGLANS, p. 366.

Pith of twigs not chambered; staminate catkins slender, long-stalked; stamens 3-10, pubescent; nuts more or less angled but smooth; husk splitting..1882. CARYA, p. 367.

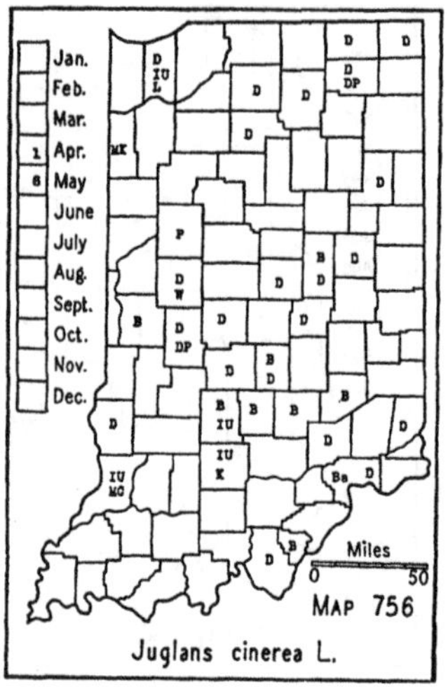

MAP 756
Juglans cinerea L.

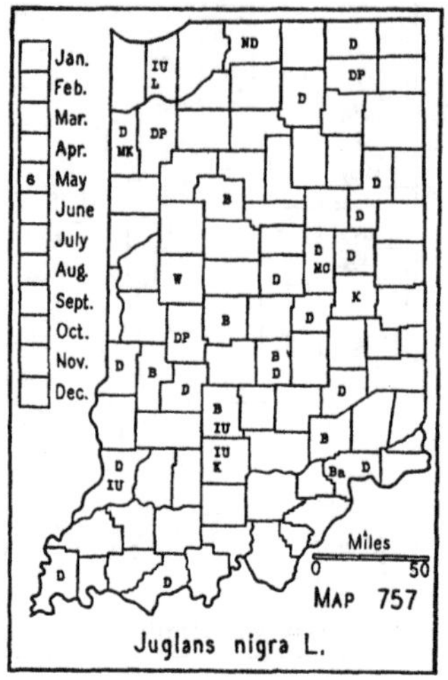

MAP 757
Juglans nigra L.

MAP 758
Carya Pecan (Marsh.) Engler & Graebner

1881. JÚGLANS L. WALNUT

Bark gray, ridges smooth; upper part of leaf-scar of last year's leaves with a mat of hairs; pith dark brown; fruit oblong, husk viscid....................1. *J. cinerea.*

Bark dark brown, ridges rough; upper part of leaf-scar of last year's leaves without a mat of hairs; pith light brown; fruit orbicular to slightly elongate, husk not viscid...2. *J. nigra.*

1. **Juglans cinèrea** L. BUTTERNUT. Map 756. An infrequent tree throughout the state and probably absent from Benton and Newton Counties. It is local in its distribution and generally only a few trees are found in a locality. I have seen it only a few times as a frequent tree and then only over small areas. Its preferred habitats are terraces and banks of streams, but it is also found in ravines and rarely in tamarack bogs. It rarely reaches a large size before the ends of the branches in the crown die. This condition may be due to civilization, since I was told by a pioneer that large trees were formerly to be found. Like the maple, the concentrated sap of this species produces sugar.

Valley of the St. Lawrence River to Nebr., southw. to the Gulf States.

2. **Juglans nìgra** L. BLACK WALNUT. Map 757. This species is probably a native of every county of the state. It is infrequent but well distributed in all parts of the state where it will grow. It will grow almost anywhere and is a native in all kinds of soils except on the hills and in the flats of the southern part and on the sand hills of the northern part. It grew to a great size. A pioneer whose veracity was unquestioned, told me that a tree 8 feet in diameter was cut near Bluffton, and 60 feet of it was used as a "dugout" in which flour and other merchandise were transported on the Wabash River from Murray to Huntington. He said he knew of another walnut tree near Montpelier that was 9 feet in diameter. It must be remembered that I have no data as to the height above the ground at which these measurements were taken.

W. Mass., Ont. to Minn., southw. to Fla. and Tex.

1882. CÀRYA Nutt. HICKORY

The specimens representing a single species of *Carya* often vary greatly in respect to the bark of both trunks and branches, size and pubescence of branchlets, number and size of the leaflets, and size and shape of the nuts. No attempt has been made to describe all of the extreme forms. Measurements refer to dried specimens.

Bud scales 4 or 6, valvate; leaflets generally curved backward (falcate).

Nuts generally elongate, nearly terete; husk thin, splitting to the base; kernel sweet; leaflets 9, 11, 13, 15 or 17, generally about 13...........1. *Carya Pecan.*

Nuts generally as broad as long, compressed, irregularly angled and reticulate; kernel bitter.

Winter buds dark reddish brown; leaflets 7, 9, 11 or 13; husk tardily splitting to about the middle. (See excluded species no. 169, p. 1039.).....*C. aquatica.*

Winter buds bright yellow, glandular; leaflets 5, 7 or 9 (11); husk usually splitting to about the middle......................................2. *C. cordiformis.*

Bud scales 6 or more, imbricated (not in pairs); leaflets not curved backward.

A. Branchlets usually stout; terminal buds large, 10-27 mm long; the year's growth usually more or less pubescent; dry husks of fruit (4) 5-10 mm thick; nuts usually strongly angled.

Prevailing number of leaflets 5 (none of the leaves with more than 5, coppice shoots might have more)..3. *C. ovata.*

Prevailing number of leaflets more than 5.

Trees of low ground; bark of young trees tight and light, that of older trees scaly, separating into long, thin plates (see exception in text); leaf stalks of leaves of the previous season usually persisting until spring (this character peculiar to this species); branchlets at first pubescent, generally becoming glabrous or nearly so at maturity, light brown; nuts usually large, compressed, generally angled, 3-6 cm long, wedge-shaped at the base; kernel sweet and not at all astringent...............4. *C. laciniosa.*

Trees usually of high ground; bark of young trees tight and dark, that of older trees tight and usually deeply furrowed, the thick ridges generally broken into short lengths which on very old trees sometimes loosen at the base; leaf stalks of the leaves of the previous season not persisting; pubescence of leaf stalks usually longer and denser than that of the preceding species, and persisting longer, often of a rusty color; branchlets more or less pubescent until maturity, reddish brown; nuts usually about half as large as the preceding and usually with a rounded base; kernel very mildly astringent...5. *C. tomentosa.*

A. Branchlets usually slender; terminal buds small, 5-12 mm long; the year's growth usually glabrous, rarely pubescent; dry husk 1-4.5 mm thick.

B. Branchlets and leaves not covered with a rusty brown pubescence when they first appear; dry husk 1-3 mm thick at thinnest point, rarely thicker.

Involucre of fruit 1-3 mm thick; winter buds glabrous or puberulous.

Prevailing number of leaflets 5, rarely 7; bark of trunk and branches tight; fruit generally smooth and usually tapering at the base to a short stem (figlike); husk not opening or splitting only above the middle; nut smooth, the shell thick, about 1.5 mm thick at the thinnest point; kernel sweet, mildly astringent...6. *C. glabra.*

Prevailing number of leaflets 7, rarely 5; bark of trunk and branches usually somewhat scaly a few feet above the ground, sometimes scarcely at all scaly; fruit granular, the sutures winged, rarely tapering at the base to a short stem (figlike); husk usually splitting to the base; nut angled or

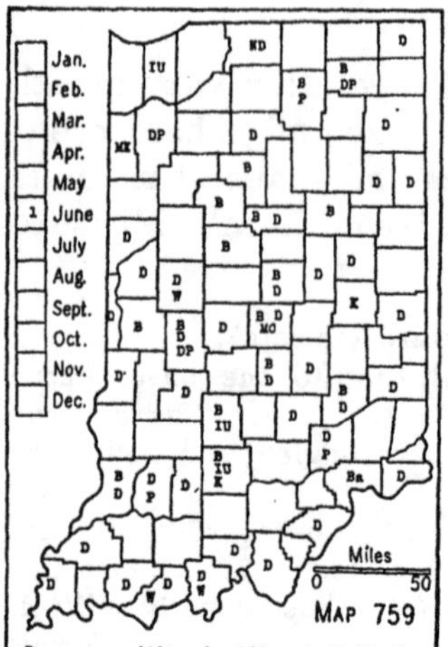

Carya cordiformis (Wang.) K. Koch

Carya ovata (Mill.) K. Koch

Carya laciniosa (Michx. f.) Loud.

smooth, the shell thin, rarely thick, thinner than that of the preceding species; kernel sweet without astringency.

C. Nuts ellipsoidal.

Inner surface of fresh husk without a resinous odor; nut rounded at base, acute at apex, broadest about the middle................7. *C. ovalis.*

Inner surface of fresh husk with a resinous odor; nuts smaller and usually more compressed than those of the preceding...7a. *C. ovalis* var. *odorata.*

C. Nuts obovoid or oblong.

D. Nuts taper-pointed or rounded at the apex, broadest above the middle.

Nut without an elongate or stipitate base....7b. *C. ovalis* var. *obovalis.*

Nut with an elongate or stipitate base..7c. *C. ovalis* var. *obovalis* f. *acuta.*

D. Nuts oblong, cordate or subcordate at the apex.

Branchlets glabrous at fruiting time......7d. *C. ovalis* var. *obcordata.*

Branchlets more or less pubescent at fruiting time..7e. *C. ovalis* var. *obcordata* f. *vestita.*

Involucres 3-4.5 mm thick; winter buds reddish brown, at least the margins of the scales pubescent.................................... 8. *C. pallida.*

B. Branchlets and leaves densely covered with a rusty brown pubescence when they first appear; dry husk 3-3.5 mm thick......9. *C. Buckleyi* var. *arkansana.*

1. **Carya Pecán** (Marsh.) Engler & Graebner. (*Carya illinoensis* (Wang.) K. Koch and *Hicoria Pecan* (Marsh.) Britt.) PECAN. Map 758. Infrequent or local in the Ohio River Bottoms as far east as Bethlehem, Clark County, up the Wabash River as far north as 4 miles south of Covington, Fountain County, up White River into Greene County, and known up the Muscatatuck River into Washington County. It was formerly a common tree in Point Township of Posey County and in the southwest part of Gibson County. Its habitat is river bottoms that are usually inundated annually.

Mississippi Valley from Ind. to Iowa, southw. to La. and Tex.

2. **Carya cordifórmis** (Wang.) K. Koch. (*Hicoria cordiformis* (Wang.) Britt.) BITTERNUT HICKORY. Generally known in Indiana as pignut hickory. Map 749. An infrequent to frequent tree throughout the state.

This species prefers a moist soil but will be found also on wooded slopes.

The species is variable in the number and size of its leaflets. The usual number of leaflets is 5 or 7, but trees with 7 or 9 leaflets are frequent. The leaflets of the greater number of trees rarely exceed 3.5 cm in width but the lateral leaflets of some trees are more than twice as wide. Sargent calls the wide-leaflet form var. *latifolia* Sarg. He says the under surface of the leaflet is usually more pubescent. This is usually true but can not be used as a character to separate the two forms. In Indiana, the forms with wide leaflets are found in the southern half of the state, especially on the wooded slopes of the hill country.

Valley of the St. Lawrence River to Nebr., southw. to the Gulf States.

3. **Carya ovàta** (Mill.) K. Koch. (*Hicoria ovata* (Mill.) Britt.) SHAGBARK HICKORY. Map 760. Infrequent to common in every county of the state. Its habitat is moist, rich woodland but it is sometimes found on slopes of hills. It is usually associated with red oak, bigleaf shagbark hickory, swamp white oak, basswood, white ash, slippery elm, sugar maple, beech, and sweet gum.

N. E., Ont. to Minn., southw. to Fla. and Tex.

3a. **Carya ovata** var. **fraxinifòlia** Sarg. (Sargent. Trees and Shrubs 2: 207. 1913.) This variety is described as "having leaflets lanceolate to slightly oblanceolate, acuminate, thick and firm in texture, lustrous above, pubescent along the midribs below, the terminal 1.4-1.5 dm long, from 4.4-5 cm wide, and raised on a slender puberulous petiolule, the lateral leaflets asymmetric at the base, sessile, those of the lowest pair 7-9 cm long, and 2.5-3 cm wide." Sargent referred specimens which I had collected from Daviess, Martin, and Wells Counties to this variety.

3b. **Carya ovata** var. **Nuttálli** Sarg. (Sargent. Trees and Shrubs 2: 207. 1913.) This variety is described as having "nut rounded, obcordate or rarely pointed at apex, rounded or abruptly pointed at the base, much compressed, prominently angled, about 1.5 cm long, and 1-1.2 cm thick; involucre 4-10 mm thick and splitting freely to the base. Except in size of the fruit there appears to be no character by which the variety can be distinguished from the common Shagbark." This variety is more or less frequent in the northeastern part of the state.

4. **Carya laciniòsa** (Michx. f.) Loud. (*Hicoria laciniosa* (Michx. f.) Sarg.) BIGLEAF SHAGBARK HICKORY. Map 761. Rare, infrequent or frequent to common throughout the state, although there are no specimens or records from the northwestern counties. I was told that it occurred in the northern part of Porter County. It may be absent from a few of these counties. This species grows in wet woodland and is usually associated with the shagbark hickory. Locally it is common and throughout the Lower Wabash Valley it is common. It is associated with many species that inhabit wet woods and in one locality in the Spencer County Bottoms southwest of Rockport I found this species and beech the dominant trees.

Exception: In the Lower Wabash Bottoms, there is a form of this hickory

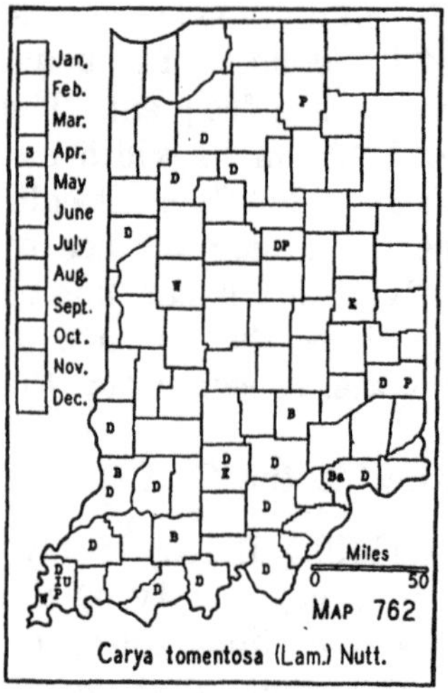

Carya tomentosa (Lam.) Nutt.

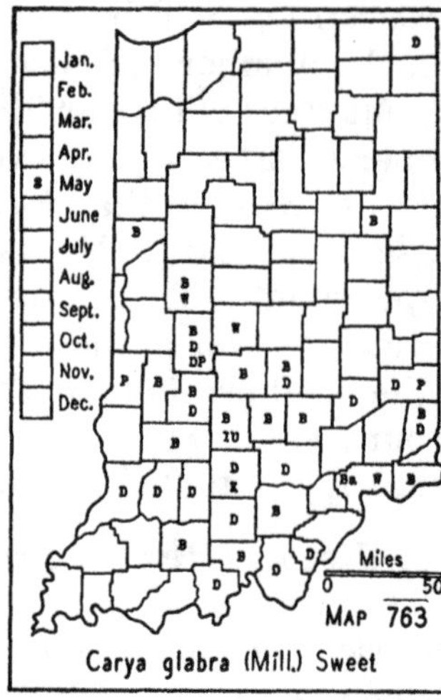

Carya glabra (Mill.) Sweet

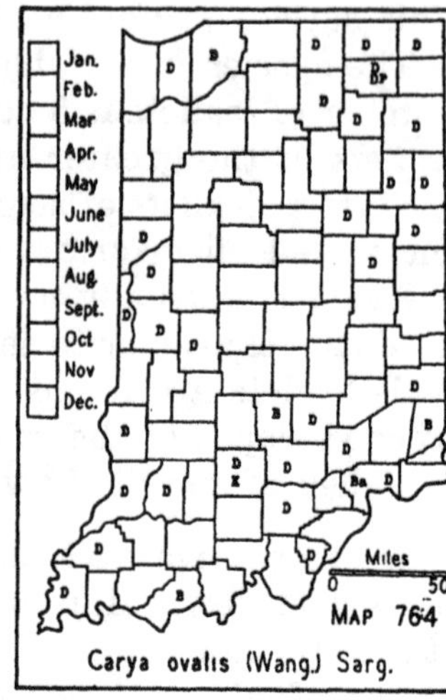

Carya ovalis (Wang.) Sarg.

that has a tight bark, like that of the mockernut hickory, otherwise it is like the species. This form has the most palatable nut of the genus. The nut is compressed, short, of more than medium size, and has the best cracking quality of all the forms. I have known the nut of this form for many years but I have not had the opportunity of working out the taxonomy of it. For many years we bought nuts from this area for table use, and I was always able to recognize this nut without mistake.

N. Y., se. Ont., to e. Iowa, and se. Nebr., southw. to W. Va., Ala., and La.

5. **Carya tomentòsa** (Lam.) Nutt. (*Carya alba* (L.) K. Koch and *Hicoria alba* (L.) Britt.) MOCKERNUT. Map 762. Very rare in the northern part of the state, becoming infrequent to frequent in the extreme southern part. It is doubtful whether all reports from the northern part of the state by other authors are authentic. It is essentially a tree of dry and usually poor soil but it is found in the lowlands of the Lower Wabash Valley where it is often associated with the preceding species. In the unglaciated area, it is generally found associated with the pignut hickory, black and white oaks, and often with the tulip tree.

E. Mass., sw. Ont., s. Mich. to se. Iowa, southw. to Fla. and Tex.

5a. **Carya tomentosa** var. **subcoriàcea** (Sarg.) Palmer & Steyermark. This variety is known from a single tree on the east bank of the cypress swamp in the southwestern part of Posey County. For several years I bought hickory nuts for table use from this area and nuts of this variety were not infrequent in the lot. It is distinguished from the species by the larger size and shape of the fruit and nut. The dried fruit is 5 cm long, oblong. The nut is oblong, 4.4 cm long, pointed at both ends, or some nuts are somewhat ovoid and more rounded at the base, little compressed, and strongly angled; shell very thick, 5 mm at the thinnest place; kernel very small and sweet.

6. **Carya glàbra** (Mill.) Sweet. (*Hicoria glabra* (Mill.) Britt.) PIGNUT HICKORY. (Generally known in Indiana as black hickory.) Map 763. This species is found principally in the southern half of the state. I think

that most of the reports of it from the northern part of the state should be referred to *Carya ovalis* or some of its many forms. One or more trees grow on the high sand bank of the north side of Lake Ann, about 5 miles northeast of Fremont, Steuben County. E. J. Palmer has verified the determination. My record from Delaware County I am now referring to *Carya ovalis* variety. This species and the next are entirely distinct, but it is impossible to name correctly herbarium specimens which are incomplete, immature, or without field data. In collecting specimens of these two species, it is desirable that a note be made whether the bark of the trunk and principal branches is tight or somewhat scaly and whether the surface of the fruit is smooth or granular. The prevailing number of leaflets also should be recorded. Fruiting specimens should not be collected until mature, usually after the first of October. Flowering specimens should always be accompanied by a fruiting specimen from the same tree.

6a. **Carya glabra** var. **megacárpa** Sarg. (Sargent. Bot. Gaz. **66**: 244. 1918.) This variety is distinguished from the type by its larger obovoid fruit, 2.5-4.5 cm long and by the husk, 2.5-3 mm thick. I have a specimen from Franklin County given this varietal name by Sargent.

Infrequent to common on hills with black and white oak. It is especially common in the knobstone area of the state.

Vt., se. Ont., s. Ind. to sw. Ill., southw. to Va., and in the mts. to Ga., n. Ala., and e. Miss.

7. **Carya ovàlis** (Wang.) Sarg. (*Carya microcarpa* Nutt. in part, and *Hicoria microcarpa* (Nutt.) Britt.) SMALL-FRUITED HICKORY. Map 764. All of the varieties are shown on the map with the species. Found throughout the state but infrequent to rare south of the lake area except on some of the sandy ridges of the southwestern part. In the lake area it is usually frequent to common on clay and sandy ridges with black and white oak.

This species is extremely variable in the character of the bark and in the shape of its fruit and nuts. The bark is generally scaly on the principal branches and on the trunk except near the base of the tree. It is usually not thick but I know of one specimen in Lagrange County that has very thick and tight bark. The nuts of this tree are almost cubical, but otherwise the tree is typical *Carya ovalis*. The nuts vary from ellipsoidal to obovoid, with the base acute or rounded, the apex acute, rounded or obcordate, little or strongly compressed, the surface from nearly smooth to strongly ridged or somewhat roughened.

Mass. to Wis., southw. to Ga., Ala., and Miss.

7a. **Carya ovalis** var. **odoràta** (Marsh.) Sarg. This variety is separated by the resinous odor of the inner surface of the fresh husk, but I have not been able to test this character. I am referring to this variety my specimens which Sargent so named. My specimens are all from the extreme northeastern part of the state, from Allen, Grant, Lagrange, Steuben, and Wells Counties.

Conn., Pa. to Mo.

Carya pallida (Ashe) Engl. & Graebn.

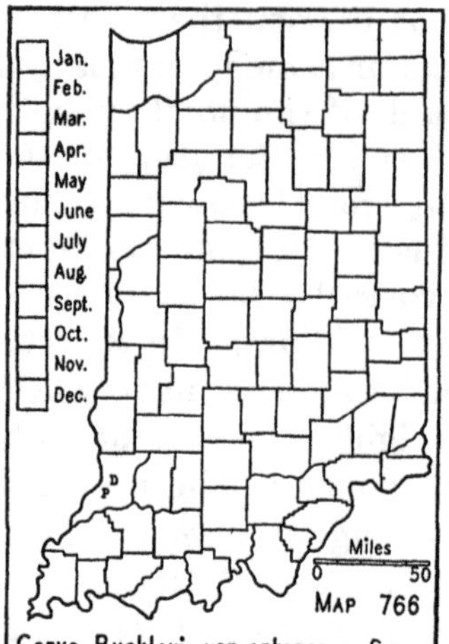

Carya Buckleyi var. arkansana Sarg.

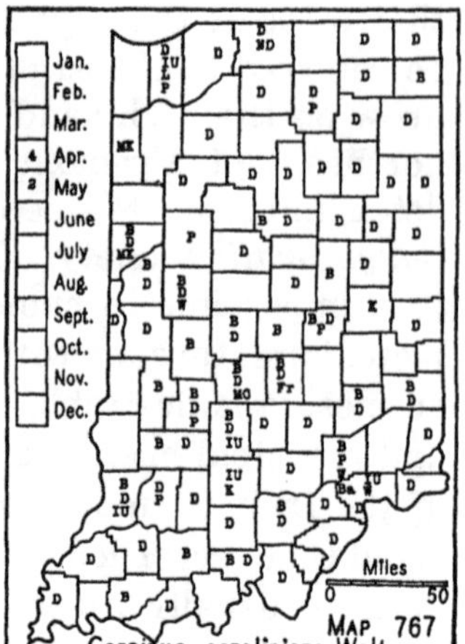

Carpinus caroliniana Walt. var. virginiana (Marsh.) Fern.

7b. **Carya ovalis** var. **obovàlis** Sarg. This form is probably found throughout the state. It is associated with the species but less frequent.
Mass. to Va. and westw. to Mo.

7c. **Carya ovalis** var. **obovalis** f. **acùta** Sarg. I have this extreme form from Steuben and Wells Counties. The Steuben County specimen is from a native tree in Pokagon State Park and is placed with this form only provisionally.

7d. **Carya ovalis** var. **obcordàta** (Muhl.) Sarg. This variety is also probably found throughout the range of the species and with it, but more rarely.
Rehder gives the distribution as Ont. to Mich.

7e. **Carya ovalis** var. **obcordata** f. **vestìta** Sarg. I collected the type from a tree in Knox County. I also have a specimen from La Porte County which I am calling this form.

8. **Carya pállida** (Ashe) Engler & Graebner. Map 765. One or more trees in the Princeton fine sand on the terrace of the Wabash River about 4 miles south of Vincennes and half a mile north of the Duncan Siding of the Chicago & Eastern Illinois Railroad. This tree is one of a few hickories and oaks on a narrow strip of land about 100 feet wide on the west of the railroad and east of the adjacent lowland. There are four hickory trees here at this station and I have made complete collections from all but I withhold their names until I can check my specimens by another collection of them.
N. J. to Ga., westw. to La. and northw. in the Mississippi Valley to Ind.

9. **Carya Búckleyi** Durand var. **arkansàna** Sarg. (Bot. Gaz. **66**: 24. 1918.) Map 766. This hickory so far has been found only in Knox County. I found one tree about 2 miles north of Decker and two trees about 4 miles south of Vincennes in a strip of woods along the railroad just north of the Duncan Siding.
Knox County, Ind., southw. in the Mississippi Valley to La. and Tex.

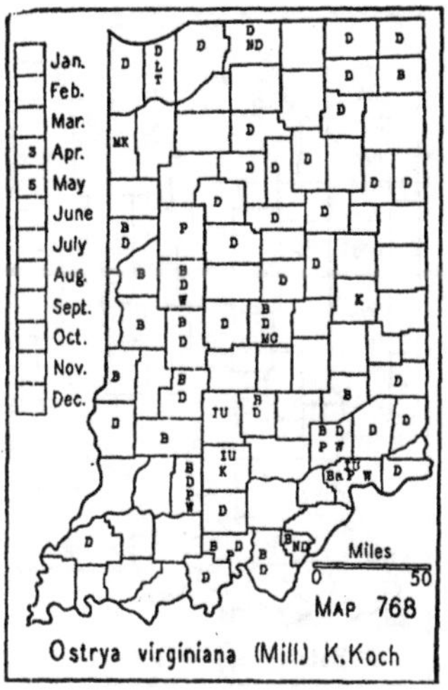

MAP 768
Ostrya virginiana (Mill.) K.Koch

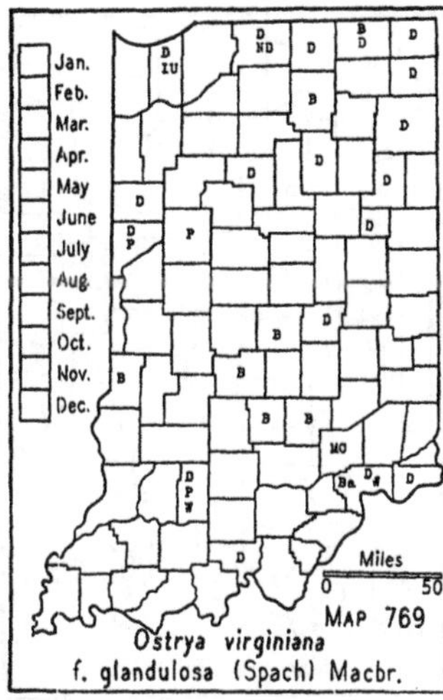

MAP 769
Ostrya virginiana f. glandulosa (Spach) Macbr.

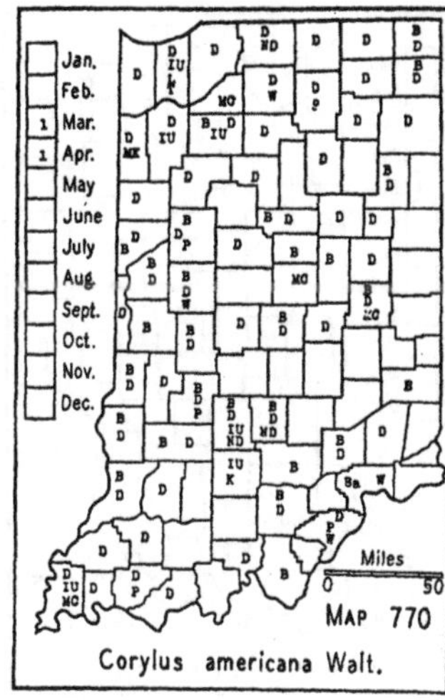

MAP 770
Corylus americana Walt.

61. BETULÀCEAE Agardh. BIRCH FAMILY

Staminate flowers solitary in the axil of each bract, without a calyx; pistillate flowers with a calyx; nut wingless.
- Small trees; leaves ovate-oblong, lower surface generally with more than 6 pairs of prominent veins; nuts 5-7 mm long.
  - Bark of tree smooth; trunk more or less grooved; lower large veins of leaves not forked; staminate aments in winter enclosed in bud scales; nut exposed, its subtending bract more or less irregularly 3-cleft......1884. CARPINUS, p. 373.
  - Bark of older trees shreddy; trunk not grooved; lower large veins of leaves generally forked; staminate aments in winter naked; nut enclosed in a bladder-like bract..1885. OSTRYA, p. 373.
- Shrubs; leaves ovate to nearly orbicular, the lower surface usually with 5 or 6 pairs of prominent veins; nuts 10-15 mm long...........1886. CORYLUS, p. 374.

Staminate flowers 3-6 in the axil of each bract, with a calyx; pistillate flowers without a calyx; nut winged.
- Winter buds sessile; stamens 2; fruiting bract deciduous at the end of the season when the nut escapes.................................1887. BETULA, p. 374.
- Winter buds stalked; stamens 4; fruiting bracts woody and persisting after the nuts escape...1888. ALNUS, p. 377.

1884. CARPÌNUS [Tourn.] L.

1. **Carpinus caroliniàna** Walt. var. **virginiàna** (Marsh.) Fern. (Rhodora 37: 425. 1935.) (*Carpinus caroliniana* of Indiana authors.) BLUE BEECH. Map 767. Often called water beech. Frequent to common throughout the state in moist woodland. It prefers a moist, rich soil but has a range of habitats in the state from the tamarack bog to the dry, black and white oak slope. It is tolerant of shade. Having no commercial value, it is regarded by foresters as a weed tree.

N. S. to Ont. and Minn., southw. to uplands of N. C. and Ark.

1885. ÓSTRYA [Micheli] Scop.

1. **Ostrya virginiàna** (Mill.) K. Koch. HOP-HORNBEAM. Map 768. In Indiana this tree is generally called ironwood. The species or its form is frequent to common in most parts of the state, although it is extremely

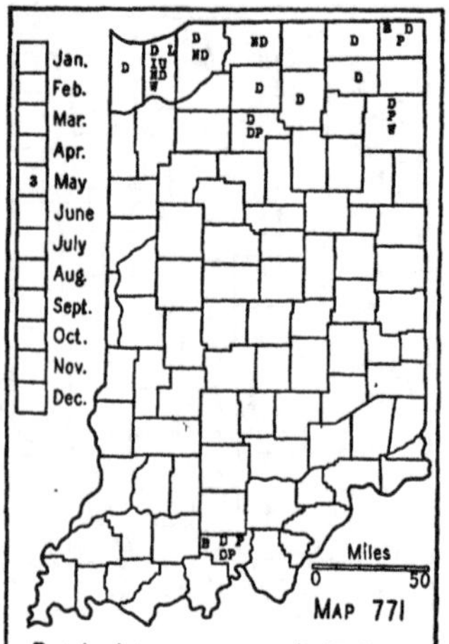

Betula lutea var. macrolepis Fern.

Betula populifolia Marsh.

Betula papyrifera Marsh.

rare in the Lower Wabash Valley. It prefers a dry soil, is of slow growth, and since it has no commercial value in Indiana, it is regarded by foresters as a weed tree.

N. S. to Man., southw. to Va., Ga., Tenn., Mo., and Okla.

1a. **Ostrya virginiana f. glandulòsa** (Spach) Macbr. (Field Mus. Nat. Hist. Publ. Bot. Ser. 4: 192. 1929.) Map 769. This form has the branchlets, petioles, peduncles, and often the midrib and veins of the lower surface of the leaves covered more or less with short, erect, reddish, glandular hairs. The form is found with the species but is not as frequent and is more northern in its distribution.

1886. CÓRYLUS [Tourn.] L.

1. **Corylus americàna** Walt. AMERICAN HAZELNUT. Map 770. Infrequent to frequent throughout the state. It adapts itself to both moist and dry soils but reaches its greatest size in the moist, black loam soils of the northern part of the state.

Maine to Sask., southw. to Fla. and Okla.

1887. BÉTULA [Tourn.] L. BIRCH

Bark of small branches usually with some wintergreen flavor; leaves with 7-15, usually 9-11 pairs of prominent veins, rounded, subcordate or narrowed at the base; mature fertile catkins generally more than 10 mm in diameter, sessile.

Outer side of scales of fruiting catkins more or less pubescent.

Scales of fruiting catkins 5-8 mm long, basal part 1-2.5 mm long.......1. *B. lutea.*

Scales of fruiting catkins 8-13 mm long, basal part 2.5-6 mm long...1a. *B. lutea* var. *macrolepis.*

Outer side of scales of fruiting catkins glabrous. (See excluded species no. 173, p. 1039.)..*B. lenta.*

Bark of small branches usually bitter, without wintergreen flavor; leaves with 4-11, usually 4-9, pairs of prominent veins, narrowed or truncate at the base; mature fertile catkins less than 10 mm in diameter (sometimes more than 10 mm in *B. nigra*), pedunculate.

Bark of trunk white, peeling in very thin strips; mature fruiting catkins drooping or spreading; wings of fruit wider than the nut.

Trunk of tree with a darkened triangular area at the base of lateral branches; leaves long-acuminate, lustrous above; staminate catkins usually solitary..... ....2. *B. populifolia.*

Trunk of tree without a darkened area at the base of lateral branches; leaves ovate, not lustrous above; staminate catkins usually 2 or 3...3. *B. papyrifera.*

Bark of trunk (tree or shrub) dark or reddish brown, not peeling off in thin strips (flaking off in thick plates in *B. nigra*); fruiting catkins erect or nearly so; wings of fruit narrower than the nut.

Bark of large specimens peeling or flaking; leaves triangular-ovate, widest below the middle, mostly with 7-9 pairs of prominent veins; bracts of mature fruiting catkins 6-10 mm long, densely pubescent; trees..................4. *B. nigra.*

Bark tight; leaves oblong-ovate, elliptic, obovate, rarely ovate, mostly with 3-6 (7) pairs of prominent veins; bracts of mature fruiting catkins 4-7 mm long, glabrous except the ciliate margins; shrubs or shrublike trees.

Blades generally with 3 or 4 pairs of distinct veins, 2-4 cm long on fruiting branchlets, obovate, rounded at the apex, rarely short-acute, cuneate at the base.

Leaves not glandular..5. *B. pumila.*

Leaves more or less glandular, glands usually plentiful on both sides........ ....5a. *B. pumila* var. *glandulifera.*

Blades generally with 6 or 7 pairs of distinct veins, mostly 5-6 cm long on fruiting branchlets, oblong-ovate, elliptic or rarely ovate, acute at the apex, mostly rounded or subcordate at the base......................6. × *B. Purpusii.*

1. **Betula lùtea** Michx. f. (*Betula alleghaniensis* Britt.) YELLOW BIRCH. After a careful study of my specimens, I believe they all belong to the variety rather than to the species. Fernald (Rhodora 24: 170. 1922) refers to two specimens of the species from Indiana.

Newf. to Man., southw. to Del., Ill., and Minn. and in the mts. of N. C. and W. Va.

1a. **Betula lutea** var. **macrolèpis** Fern. (Rhodora 24: 170. 1922.) YELLOW BIRCH. Map 771. This tree is found locally in the northern part of the state and on the sides of two deep, rocky ravines about a mile east of Taswell in Crawford County. In northern Indiana it apparently is one of the chief species in the succession after tamarack and is associated with white elm, red maple, black ash, and silver maple. All of my northern specimens have a dark bark and I believe they all belong to the dark bark form recently described by Fassett (Rhodora 34: 95. 1932) as *Betula lutea* Michx. f. forma *fallax* Fassett.

N. B. to Wis., southw. to Tenn. and Ill.

2. **Betula populifòlia** Marsh. GRAY BIRCH. Map 772. The few trees of this species found in Indiana are the remnants of a relic colony because the nearest location of this species is three to four hundred miles to the northeast. In 1911 I found a few trees in a dying condition on the border of Fish-trap Lake near La Porte in La Porte County. I have a specimen collected by Blatchley in Lake County (Ind. Geol. Rept. 22: 100. 1898). He says: "Sand ridges west of Miller's; scarce." This species has been

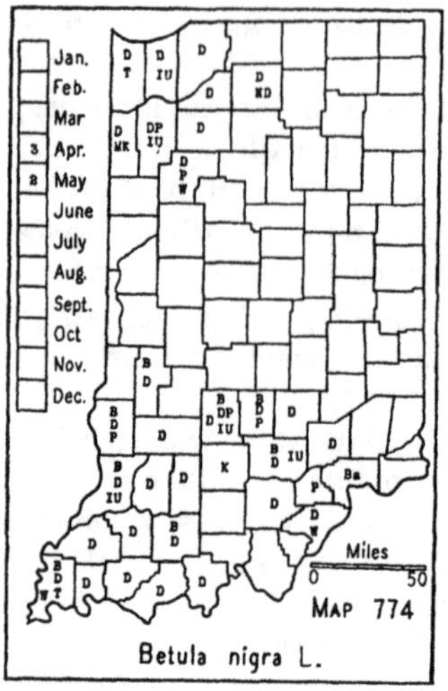

reported from St. Joseph and Tippecanoe Counties also, but these records may be based upon planted trees.

N. S. to s. Ont., southw. to Del. and Pa. and a relic colony in Ind.

3. **Betula papyrífera** Marsh. (*Betula alba* L. var. *papyrifera* (Marsh.) Spach.) PAPER BIRCH. Map 773. This is a far northern species and is found in Indiana only in the counties shown on the map. There are a few small colonies of it and it grows in rather moist, sandy soil.

Newf. to Alaska, southw. to n. Pa., cent. Mich., n. Ind., n. Wis., e. Nebr., and Wyo.

4. **Betula nìgra** L. RIVER BIRCH. Map 774. More or less frequent in all the counties bordering the Kankakee River, on the south side of Cedar Lake, Lake County, on the east shore of Lake of the Woods in Marshall County, along the Tippecanoe River in White County, and more or less frequent along some of the streams of the southwestern part of the state. It is a common tree in a few places in the "flats" of Jackson and Scott Counties.

Mass. to Minn., southw. to Fla. and Tex.

5. **Betula pùmila** L. DWARF BIRCH. Map 775. Restricted to the lake area where it is found in bogs and marshes. Infrequent to rare. It is to be noted that the under surface of the leaves of all of my specimens is glaucous.

Newf. to Wis., southw. to N. J., Ohio, and Ind.

5a. **Betula pumila** var. **glandulífera** Regel. The variety differs from the species in that the young branchlets, leaves, and bracts are covered more or less with glandular dots or resinous glands. In our area, the distinction is not always clear since in the same clump of shrubs one can often find some densely resinous specimens and others with only a minute amount of resin.

Ont. to Sask., southw. to Ind., and to se. Minn.

6. × **Betula Purpùsii** Schneider. (*Betula lutea* × *pumila* var. *glanduli-*

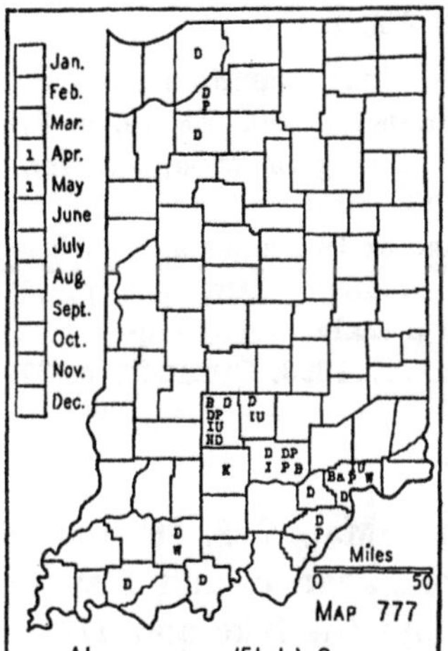

MAP 777
Alnus rugosa (Ehrh.) Spreng.

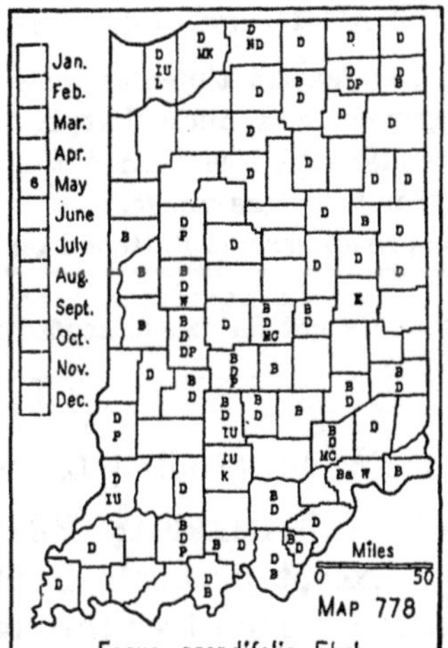

MAP 778
Fagus grandifolia Ehrh.

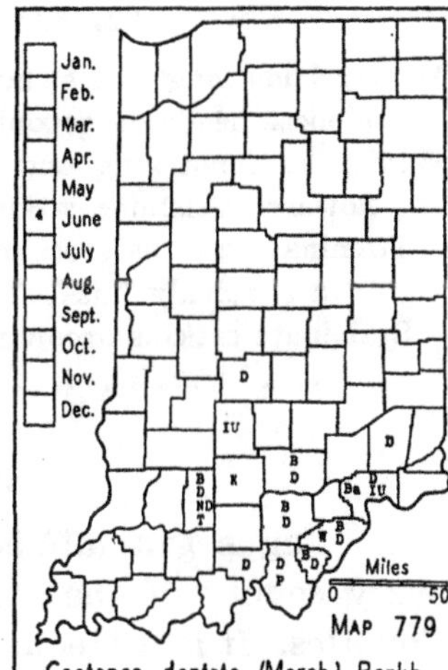

MAP 779
Castanea dentata (Marsh.) Borkh.

fera.) This is a natural hybrid. I found it in a tamarack bog about a fourth of a mile north of Mineral Springs Stop on the South Shore Electric Line, in Porter County and in a marsh about two and a half miles northwest of Porter in the same county.

The general range is unknown. Known to occur in Mich., Ind., and Minn.

1888. ÁLNUS [Tourn.] Hill. ALDER

Leaves broadly elliptic to ovate, mostly rounded at the base, acute at the apex, margins doubly serrate (that is, the 9-13 primary veins ending in the apices of large teeth which in turn are finely serrate), glaucous, glaucescent, or green beneath, deeply impressed-nerved above, not noticeably glutinous beneath; shrubs or small trees. . . . . 1. *A. incana* var. *americana.*

Leaves obovate, acute at the base, generally more or less rounded at the apex, sometimes acute, margins finely and nearly evenly serrate, green and sometimes noticeably glutinous beneath, usually not impressed-nerved above but sometimes so; shrubs . . . . 2. *A. rugosa.*

1. **Alnus incàna** (L.) Moench var. **americàna** Regel. (*Alnus incana* of Gray, Man., ed. 7 and of Britton and Brown, Illus. Flora, ed. 2.) SPECKLED ALDER. Map 776. Frequent in low ground about sloughs in the dunes near Lake Michigan and rare to very rare elsewhere in low woods or in low ground along streams. All of my specimens have the leaves more or less glaucous beneath and more or less pubescent, at least on the principal veins.

Newf. to Sask., southw. to Pa., Iowa, and Nebr.

2. **Alnus rugòsa** (Ehrh.) Spreng. (*Alnus rugosa* (DuRoi) Spreng. of Gray, Man., ed. 7, Britton and Brown, Illus. Flora, ed. 2, and Deam, Shrubs of Indiana, ed. 2.) HAZEL ALDER. Map 777. Locally in colonies but rare to infrequent in the parts of the state where it is found. It inhabits springy places in woodland or in the open. Its habitat and associates indicate that it requires a slightly acid soil.

Maine to Minn., southw. to Fla. and Tex.

62. FAGÀCEAE Drude. THE BEECH FAMILY

Winter buds long and slender, at least 4 times as long as wide; staminate flowers in globose heads on drooping peduncles; nuts sharply 3-angled..1890. FAGUS, p. 378.
Winter buds not long and slender and less than 4 times as long as wide; staminate flowers in slender catkins; nuts not as above.
Staminate catkins erect or spreading; nut flattened on one or two sides and enclosed in a prickly husk....................................1891. CASTANEA, p. 378.
Staminate catkins drooping; nuts not flattened, seated in a scaly, woody cup........ ....1893. QUERCUS, p. 379.

1890. FÀGUS [Tourn.] L. BEECH

1. **Fagus grandifòlia** Ehrh. AMERICAN BEECH. Map 778. Found in every county of the state except probably Benton, Jasper, and Newton Counties. It is a frequent to common tree throughout the lake and Tipton Till Plain areas on the ridges and hills unless these are sandy or a hard clay when they will be covered more or less with black and white oaks and hickories. In the unglaciated area it is also frequent to common but is usually found in the coves or on low hills. The higher hills with their poorer soil are usually covered with oaks and hickories. In the "flats" of the Illinoian drift it is found in low, flat woods where it is the principal species, associated with sweet gum, black gum, red maple, and oaks. Its most constant associate in the northern and central part of the state is the sugar maple.

N. S., s. Ont. to Wis., southw. to the Gulf States and Tex.

1a. **Fagus grandifolia** Ehrh. f. **pubéscens** Fern. & Rehd. This is a form with the entire under surface of the leaves more or less pubescent. It is to be noted that the leaves of none of our specimens are entirely glabrous beneath but generally have the principal veins covered with long hairs. This form is found throughout Indiana with the species.

The bark of the beech is usually smooth but sometimes a tree is found that has the bark of the lower part of the trunk broken into ridges and furrows. Usually the ridges are not continuous but in sections of a few inches in length.

1891. CASTÀNEA [Tourn.] Hill. CHESTNUT

1. **Castanea dentàta** (Marsh.) Borkh. AMERICAN CHESTNUT. Map 779. The chestnut is restricted to the part of the state indicated on the map. It is found usually on sandstone outcrops and is usually local. In 1936 it was reported from Ripley County by Dorothy Parker. On account of its excellent qualities for shingles, posts, and poles, the large trees have all been cut. It is especially valuable for its timber and nuts, but its use as a forest tree will be curtailed because the chestnut blight has already appeared in a few places in Indiana.

Maine, s. Ont., and Mich., southw. to Del., and in the mts. to Ala. and Ark.

1893. QUÉRCUS [Tourn.] L. OAK

[Dyal, Sarah C. A key to the species of oaks of Eastern North America based on foliage and twig characters. Rhodora 38: 53-63. 1936.]

Note: In collecting leaf specimens of oaks for identification, it should be kept in mind that the foliage is variable. The leaves of seedlings, coppice shoots, and vigorous shoots of old trees sometimes vary considerably in size, form, and margin. Leaves in the shade on old trees usually have the margins more nearly entire than the typical leaves. For example, on the lower and inner branches of a pin oak, leaves may be found whose lobes are not as long or longer than the undivided portion of the leaf, and this character refers them to the red oak group. In the case of *Q. bicolor* and *Q. lyrata,* while the pubescence of the under surface of the leaves is normally a white or gray tomentum, the shade leaves may be without the tomentum and may be green and merely pubescent.

Mature leaves never with bristle tips; fruit maturing the first year; inner surface of shell of nut glabrous; bark gray (except in no. 5), more or less scaly. (The White Oaks.)

Leaves glaucous and glabrous beneath at maturity (rarely a specimen retaining its pubescence until maturity)..1. *Q. alba.*

Leaves generally covered beneath with a dense, gray tomentum, often accompanied by some long, simple or fascicled hairs, rarely the tomentum lacking on the leaves of lower branches and then the surface more or less densely pubescent, rarely a specimen with leaves pubescent only on the principal veins.

Primary lateral veins of the lower surface of the leaves regularly spaced or some of the leaves with an irregular spacing; margins rather regularly sinuate-dentate or with irregular shallow lobes in no. 3.

Shrubs; leaf blades mostly 5-10 cm long; teeth of blades usually fewer than 8 to a side; fruit sessile....................................2. *Q. prinoides.*

Trees; leaf blades mostly more than 10 cm long; teeth of blades mostly more than 8 to a side (except blades from the top of some trees of *Q. Muhlenbergii*).

Lower surface of leaves mostly with 4-10 pairs of lateral veins; veins of most of the leaves not all ending in teeth of the margin; blades usually not bilaterally symmetrical; fruit on peduncles longer than the petioles; one year old branches never corky............................3. *Q. bicolor.*

Lower surface of leaves mostly with 7-12 pairs of veins; veins all ending in teeth of the margin; leaves essentially bilaterally symmetrical.

Apex of leaves of fruiting branches sharp-pointed, the sides of the apex usually forming an acute angle; fruit sessile or nearly so........... .... 4. *Q. Muhlenbergii.*

Apex of leaves of fruiting branches rounded or, if sharp-pointed, the angle formed by the sides rarely an acute angle; fruit peduncled.

Leaves dark green above and generally velvety-pubescent to the touch beneath; scales of cup free to the base; bark like that of white oak; trees of low ground................................5. *Q. Prinus.*

Leaves yellowish green and generally densely pubescent beneath but the pubescence not velvety to the touch; scales of cup free only at the tip; bark like that of the red oak; trees of high ground, usually on the crests and slopes of sandstone and knobstone ridges in Indiana...... ....6. *Q. montana.*

Primary lateral veins of the lower surface of the leaves not regularly spaced; leaves deeply lobed or pinnatifid.

Branchlets densely pubescent; leaves strongly obovate in outline; blades mostly less than 15 cm long, cut into 5 principal lobes, the two upper lateral lobes the largest and widest; the under surface usually yellow green and more or less densely pubescent with fascicled hairs, rarely with some tomentum, the upper surface often with straggling hairs; nuts mostly less than 12 mm in diameter at maturity....................................7. *Q. stellata.*

Branchlets glabrous at the end of the season or only sparsely pubescent; leaves mostly obovate in outline, rarely oblong, cut into 5-9 lobes; blades white- to gray-tomentose beneath, or those of lower branches often green and pubescent beneath; nuts more than 12 mm in diameter.

Upper scales of cup awned, forming a fringe about the cup; blades mostly 1-2.5 dm long, generally deeply lobed or pinnatifid; nuts very large, rarely nearly covered by the cup; vigorous one year old branches sometimes corky ....8. *Q. macrocarpa.*

Upper scales of the cup not awned but sometimes the upper scales forming a ragged rim about the top, which should not be mistaken for awned scales; nut usually covered or almost so by the cup; leaves generally much smaller than those of the preceding species............9. *Q. lyrata.*

Mature leaves with bristle tips; fruit maturing the second year; inner surface of shell of nut tomentose; bark dark, tight, and furrowed. (The Black Oaks.)

Leaves entire (rarely a seedling or coppice shoot with some toothed leaves)........ ....10. *Q. imbricaria.*

Leaves more or less deeply lobed, the lobes and teeth conspicuously bristle-pointed.

Mature leaves smooth beneath, except for tufts of hairs in the principal axils (rarely some of the leaves of no. 16 glabrous).

Lateral lobes of all leaves (measured along the upper side from the tip to the base of the sinus) about as long as, or slightly longer than, the undivided portion of the blade.

Cup flat on the bottom, shallow (saucer-shaped); blades not lustrous above.... ....11. *Q. borealis* var. *maxima.*

Cup rounded on the bottom.

Scales at the top of the cup closely appressed. (Should be sought in Indiana.) ....*Q. borealis.*

Scales at the top of the cup loosely imbricated, their free tips forming a fringelike border; terminal buds large, grayish-pubescent, generally somewhat 4-sided; blades lustrous above...............12. *Q. velutina.*

Lateral lobes of leaves (measured along the upper side from the tip to the base of the sinus) usually much longer than the undivided portion of the blade (lower leaves of no. 14 often not cut so deeply); blades lustrous above.

Cup flat or only slightly convex on the bottom, shallow (saucer-shaped), usually covering about a fourth of the nut.

Cup thin, usually less than 1.6 cm broad..................13. *Q. palustris.*

Cup thick, more than 1.6 cm broad (fruit resembling that of no. 11)...... ....14. *Q. Shumardii.*

Cup strongly convex on the bottom, usually covering more than a fourth to about half of the nut.

Scales at the top of the cup loosely imbricated, their free tips forming a fringelike border, generally gray-pubescent all over, never tuberculate on the back; inner bark yellow; buds large, 4-sided, gray-pubescent.... ....12. *Q. velutina.*

Scales at the top of the cup all closely appressed (in dried specimens sometimes becoming more or less loose); buds generally glabrous or nearly so, generally not so large and rarely 4-sided; lower scales usually glabrous but the upper generally pubescent.

Cup covering a fourth to a third of the nut.......................... ....14a. *Q. Shumardii* var. *Schneckii.*

Cup covering about half of the nut.

Inner bark yellowish or orange; nut generally ellipsoidal; kernel of nut yellowish or orange and very bitter............15. *Q. ellipsoidalis.*

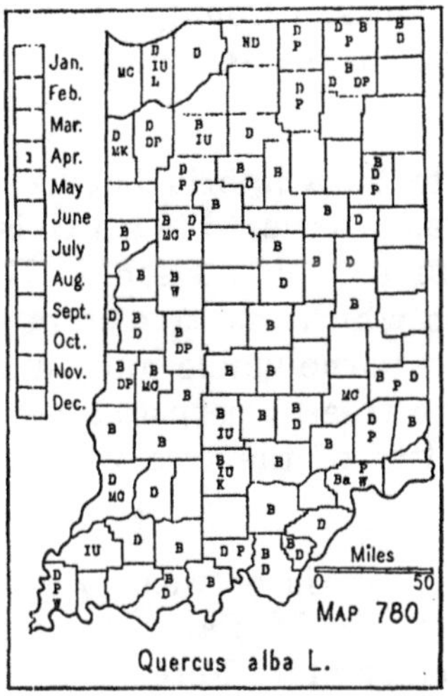

MAP 780

Quercus alba L.

MAP 781

Quercus prinoides Willd.

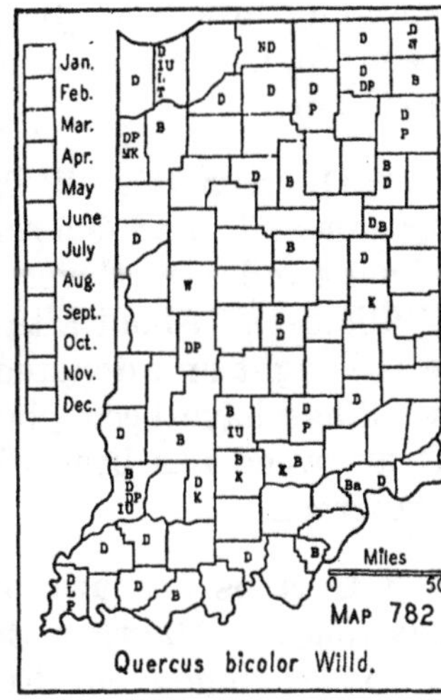

MAP 782

Quercus bicolor Willd.

Inner bark reddish or gray; nut generally ovoid; scales glossy and glabrous or nearly so; kernel white and not very bitter...16. *Q. coccinea.*

Mature leaves more or less pubescent on the whole under surface.

Leaves drooping, grayish or yellowish pubescent beneath; blades variously lobed, specimens usually having some falcate lobes; rarely specimens with 3-lobed leaves, this form more common on small trees or coppice shoots; scales of cup with a reddish brown border; nut enclosed for about a third of its length. ....17. *Q. falcata.*

Leaves brownish or rusty-pubescent beneath, sometimes appearing grayish; scales of cup without a dark border; nut enclosed for about half its length.

Blades expanded at the apex, and generally with only three lobes; mature twigs generally scurfy-pubescent......................18. *Q. marilandica.*

Blades with more than three lobes; mature twigs generally glabrous........ ....12. *Q. velutina.*

1. **Quercus álba** L. WHITE OAK. Map 780. This species is found in every county of Indiana. Knowing this fact, I have not tried to preserve specimens from every county, but have tried to secure a series of the widely varying forms. The leaves vary greatly in their lobing, especially in the depth to which the blade is cut. We have some specimens in which the width of the blade between the lobes is only 5 mm. In others, the lobes are shallow and the uncut part of the blade is 30-40 mm wide. The lower surface of the blades is glaucous and entirely glabrous at maturity. My Starke County specimen, which is pubescent over nearly the entire lower surface, is an exception. The nuts vary from 10-30 mm long.

It is found throughout the state except in low, wet grounds.

Maine, s. Ont. and Minn., southw. to Fla. and Tex.

1a. **Quercus alba** f. **latilòba** (Sarg.) Palmer & Steyermark. I am including with the species this form with the blades cut less than half way to the midrib. This form is more abundant in the northern part of the range of the species.

× **Quercus Bèadlei** Trelease. So named by William Trelease. Probably a hybrid between *Quercus alba* and *Quercus Prinus*. I found a large tree standing in a field about 3 miles east of Medora, Jackson County. It has been found in Lawrence County by Kriebel and in Knox County by Friesner.

× **Quercus Dèamii** Trelease. This is believed to be a hybrid between *Quercus alba* and *Quercus Muhlenbergii.* A tree was discovered in a woods about 4 miles northwest of Bluffton, by L. A. Williamson and his son, E. B. Williamson. About a third of an acre of ground on which this tree stands was bought and donated to the state. The tree has borne viable nuts and seedlings have been planted in the space about the tree to perpetuate it. Graft wood has been distributed so that the identity of the tree will be preserved.

× **Quercus Férnowii** Trelease. This is evidently a hybrid between *Quercus alba* and *Quercus stellata.* A tree was found by Carl M. Carpenter on a wooded ridge along Fire Lane 9 in the Brown County State Forest about 10 miles southeast of Nashville, Brown County.

× **Quercus Jackiàna** Schneider. This is evidently a hybrid between *Quercus alba* and *Quercus bicolor*. I found a specimen of this form in the woods of J. M. Hopper about 2 miles northeast of Onward, Cass County. There is another in the Deam Arboretum at Bluffton, Indiana, where it grew from Indiana seed planted there.

2. **Quercus prinoìdes** Willd. DWARF CHINQUAPIN OAK. Map 781. I found this shrub in Elkhart County while inspecting the Cooley Lake Club land in company with T. E. Shaw and Glenn B. Banks. The woods is about 6 miles northeast of Elkhart and about a quarter of a mile south of the Michigan state line. The shrub was plentiful in the north part of a cut-over woods in the southeast quarter of section 10 where it was growing in very sandy soil with black oak and white oak. I was not able to ascertain how widely it is distributed. This species has been reported from Cass County in Michigan which joins Elkhart County on the north.
Maine to Minn., southw. to N. C. and Tex.

3. **Quercus bícolor** Willd. SWAMP WHITE OAK. Map 782. This species is more or less frequent throughout the state although there are no reports from Benton, Jasper, and Newton Counties. In the northern part of the state, it is usually found on a "gumbo" hardpan soil associated most commonly with pin oak. In the southern part of the state in the "flats," it is found in hard, white clay soil with pin oak and swamp chestnut oak.
Maine, s. Ont. to Minn., southw. to Ga. and Ark.

× **Quercus Schùettei** Trelease. This is believed to be a hybrid between *Quercus bicolor* and *Quercus macrocarpa.* This hybrid is known from a specimen collected by R. M. Kriebel from a single tree in Lawrence County.

4. **Quercus Muhlenbérgii** Engelm. CHINQUAPIN OAK. Map 783. In northern Indiana this species is called sweet oak. Infrequent to rare in all parts of the state although Hill's report from Lake County is the

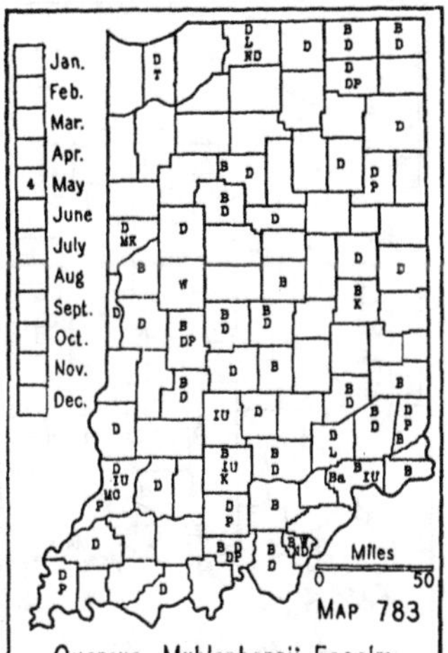

Quercus Muhlenbergii Engelm.

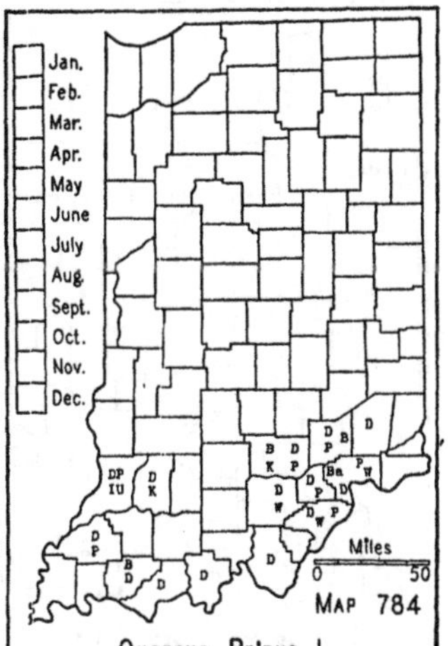

Quercus Prinus L.

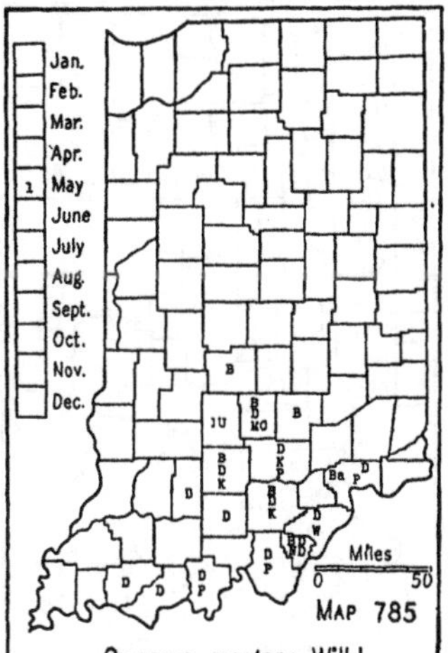

Quercus montana Willd.

only one from the northwestern part. It is generally found on the dry banks of streams, river terraces, rocky, wooded bluffs, and only rarely in level, moist woods.

Vt., s. Ont. to Wis., southw. to Fla. and Tex.

5. **Quercus Prìnus** L. (*Quercus Michauxii* Nutt.) SWAMP CHESTNUT OAK. Map 784. This species is restricted to low, flat woods of the southern part of the state. It is local in the southwestern part although it forms about 20 per cent of the stand in a few of the woods along Prairie Creek in Daviess County. It is more frequent in the "flats" of the southeastern part of the state where it is associated with sweet gum, red maple, and pin oak.

Del., s. Ind. to Mo., southw. to Fla. and Tex.

6. **Quercus montàna** Willd. (*Quercus Prinus* of Gray, Man., ed. 7 and Britton and Brown, Illus. Flora, ed. 2.) CHESTNUT OAK. Map 785. In Indiana this species is restricted to the area indicated on the map where it is found on the ridges and slopes of sandstone and of knobstone. Where it is found it is usually the dominant tree.

Maine, n. shore of Lake Erie to w. cent. Ind., southw. to Ga. and Ala.

7. **Quercus stellàta** Wang. POST OAK. Map 786. This species is, for the most part, restricted to the southwestern part of the state. In the unglaciated area it is found mostly on the crests of ridges with black oak. West of this area it is found in bottom land along the Little Pigeon Creek and in the southwestern part of Posey County on the higher bottoms. It is generally associated with white and black oak, winged elm, and mockernut hickory. In this area, it is also found sparingly on some sandy ridges.

In 1932, I found a single tree about 9 inches in diameter on the slope of the high, gravelly bank of Big Wea Creek about 4 miles southwest of Lafayette. It has been reported from Lake and Porter Counties but

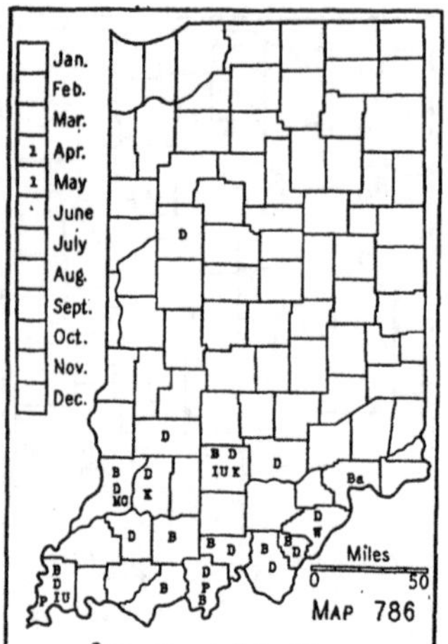

Quercus stellata Wang.

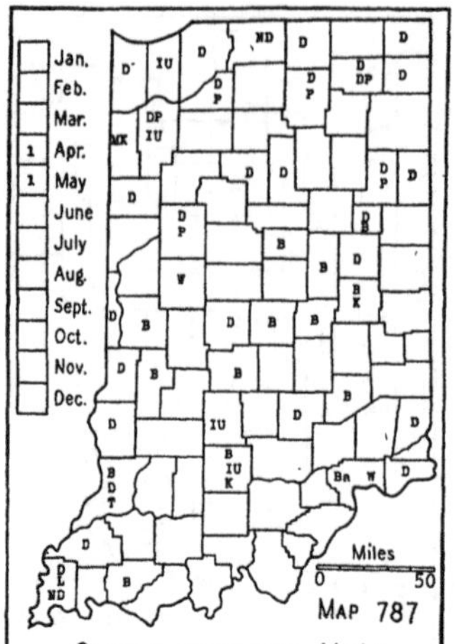

Quercus macrocarpa Michx.

Quercus lyrata Walt.

Buhl (Bull. Chicago Acad. Sci. 5: 10. 1934), in his Supplement to Pepoon, Flora of the Chicago Region, deletes these reports.

Mass. to Ind. and s. Iowa, southw. to Fla., Okla., and Tex.

8. **Quercus macrocárpa** Michx. BUR OAK. MOSSYCUP OAK. Map 787. Doubtless occurring in every county of the state, although it may be very rare in some of the hilly counties of the unglaciated area. This species is generally found in wet places in woods and along streams. It is a pioneer tree in the prairie counties where it grows both in low ground and on high ground and even on sandy ridges. In the prairie area it sometimes forms pure stands. I have noted it as a common tree in areas that undoubtedly were formerly prairies in Kosciusko, Lagrange, Noble, and Steuben Counties.

N. S. to Man., southw. to Ga., Tex., and Wyo.

8a. **Quercus macrocarpa** var. **olivaefórmis** (Michx. f.) Gray. This variety is distinguished from the typical form by its shallow cup and the long, oval nut which is often 3 cm long. The cup is semi-hemispheric and encloses the nut for about half its length. It is rare. I have specimens from Wells County, and it has been reported from Gibson and Hamilton Counties.

× **Quercus Híllii** Trelease. This is believed to be a hybrid between *Quercus macrocarpa* and *Quercus Muhlenbergii.* A single tree was found by Hill near Roby, Indiana. I report this on the authority of Sargent. I have a duplicate specimen but I believe it is only a specimen of the bur oak. I question the determination of this specimen because the last named parent of the hybrid does not occur there or, if it does, it is extremely rare.

9. **Quercus lyràta** Walt. OVERCUP OAK. Map 788. This species is very local in the southwestern counties where it grows about river sloughs and in swamps and low, wet woods. Its habitat is usually inundated each year. I have not seen it common except in a low woods along Prairie Creek about 5 miles northwest of Montgomery in Daviess County. Here it is associated

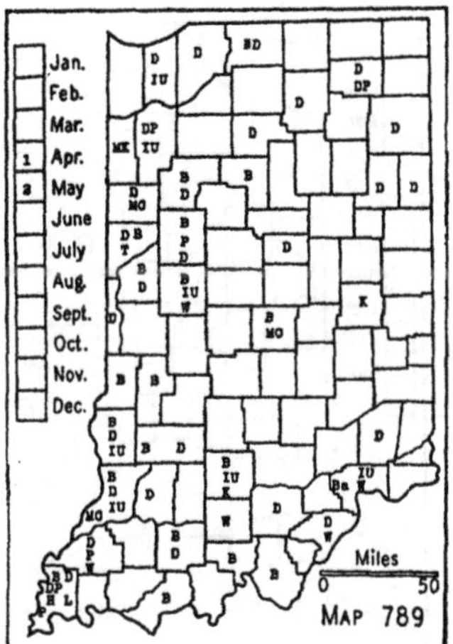

Quercus imbricarla Michx.

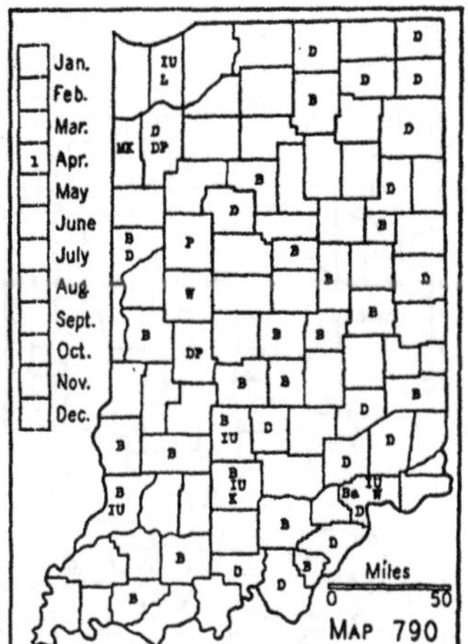

Quercus borealis var. maxima (Marsh.) Ashe.

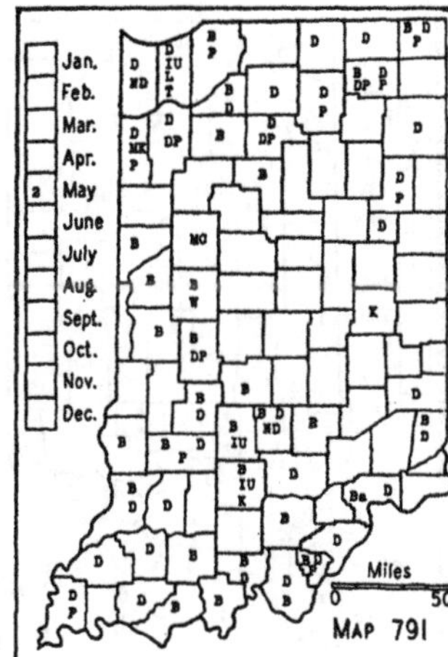

Quercus velutina Lam.

with the swamp chestnut oak. In 1931, on the bank of Slim Pond (an old river channel) in Posey County, I measured a specimen that was 56 inches in diameter at breast height, and had a clear bole of about 12 feet. Clapp writes he saw it in the vicinity of New Albany.

Md. to Iowa, southw. to Fla. and Tex.

10. **Quercus imbricària** Michx. SHINGLE OAK. Map 789. Found sparingly throughout the state. In some places it is very local and in a few areas it is frequent and locally abundant. Usually it is a tree of low ground and in some places in prairie habitats, it seems to be the pioneer tree species. In the Patoka bottoms it is usually a frequent to common tree in ground just a little higher than where the pin oak grows. On high ground it is usually closely associated with the black oak.

Pa., Mich. to Nebr., southw. to Ga. and Ark.

× **Quercus exácta** Trelease. This is believed to be a hybrid between *Quercus imbricaria* and *Quercus palustris.* I found a single tree in Posey County.

× **Quercus Leàna** Nutt. This seems to be a hybrid between *Quercus imbricaria* and *Quercus velutina.* I collected it in Lawrence County and Lake County. I also have a specimen collected by Ralph M. Kriebel from a tree in Lawrence County. Recently Kriebel has collected it in Knox County.

11. **Quercus boreàlis** Michx. var. **máxima** (Marsh.) Ashe. (*Quercus rubra* of Gray, Man., ed. 7 and Britton and Brown, Illus. Flora, ed. 2.) RED OAK. Map 790. This oak is infrequent to frequent throughout the state and even common in some parts. It may be entirely absent from Benton, Newton, and possibly Lake Counties and is rare or absent in the Lower Wabash Valley. While our map shows no specimens from the southwestern part of the state, there are reports from that part and I have seen it growing there. The paucity of specimens of this and other species of oak is due to the fact that oaks do not produce fruit every

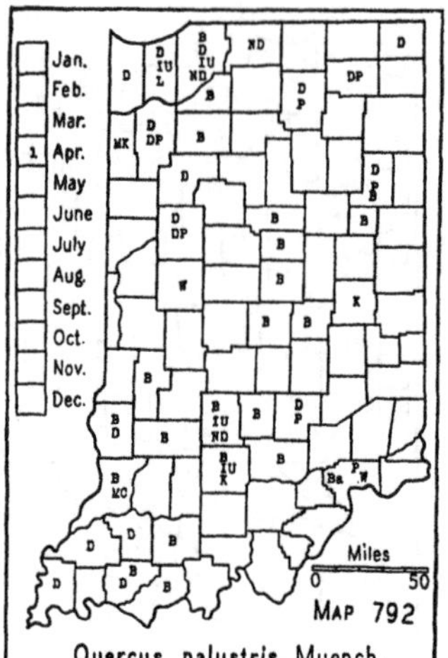

Quercus palustris Muench.

Quercus Shumardii Buckley

Quercus ellipsoidalis E.J.Hill

year. To make a good specimen it is necessary to secure a branchlet that has grown in the sun with its leaves and mature fruit. This oak, in most of its area, grows on low ground but sometimes it is found on high ground with white and black oak and on the bluffs of streams.

N. S. to Minn., southw. to Fla. and Tex.

12. **Quercus velùtina** Lam. BLACK OAK. Map 791. This species is without doubt found in every county of the state. In abundance, it ranks next to white oak, with which it is generally associated, except in very poor soil where it will be the only species or associated with post and chestnut oaks. It prefers a dry soil and is generally found on sandy and clayey ridges.

Maine, s. Ont., s. Iowa, s. Nebr., southw. to Fla. and Tex.

13. **Quercus palústris** Muench. PIN OAK. Map 792. Infrequent to common in all parts of the state. It may be absent from Benton County. It is found only in wet habitats and prefers a hard, compact, clay soil with little drainage. It is locally frequent to common in the northern part of the state and in the southern part it is abundant in the lowlands along streams and grows to great size in the low woods along the Patoka River. It is also locally common in the "flats" in the southeastern part of the state.

In Indiana there are trees with two very distinct kinds of nuts. The common form has a large nut which is depressed at the top. The other has a much smaller, ovoid nut with a conical apex. I have this form from Pike and Wells Counties.

Mass., sw. Ont., Mich., to Iowa, southw. to Va. and Okla.

14. **Quercus Shumárdii** Buckley. SHUMARD RED OAK. Map 793. Probably frequent throughout the state where its habitat occurs. Ralph M. Kriebel in 1937 studied its distribution in relation to its habitats in different soil types and found it in sixty-four counties and I am indebted to him for this information. He, however, was unsuccessful in Benton

and in several other counties in the northwestern part of the state. Since several authentic collections have been made in southern Michigan, it is believed to grow in most of our counties.

In southern Indiana it is found in well-drained bottom land along streams and on the slopes of flood plain terraces. In the general area of the Wisconsin glaciations it is not found along water courses but mostly in swampy areas on the general levels, especially in soil of the Crosby and Brookston series.

This oak, together with its variety, the Schneck oak, and the red oak, are often found growing together. They look similar and thus are often confused but can easily be separated by studying the leaves, buds, and fruit.

The leaves of red oak are dull dark green above, cut less than halfway to the midrib, 7-11-lobed, sinuses wide at the top, and the axils of the under surface have no tufts of hairs. The cup is saucer-shaped and flat on the bottom.

The leaves of the Schneck and Shumard oaks are lustrous above, cut more than halfway to the midrib, 5-7-lobed, the lobes slightly converging at the top, with tufts of hairs in the axils of the veins beneath. These two oaks differ, however, in the shape of the cups of the fruit. The cup of the Shumard oak is gray and flat on the bottom while that of the Schneck oak is rounded and deeper and the scales tinged reddish brown.

The terminal buds of the Shumard and Schneck oaks are generally grayish, somewhat compressed and acute at the apex while those of the red oak are generally blunt at the apex, smaller, shiny, and reddish brown.

The bark of the red oak has the furrows continuous, the plates wide and gray while the bark of the Schneck and Shumard oaks is dark and the furrows broken.

Atlantic States from s. Pa. to Fla., following the Gulf States to Tex. and up the Mississippi Valley to Iowa, s. Mich., and Ind.

14a. **Quercus Shumardii** var. **Schnéckii** (Britt.) Sarg. (*Quercus texana* Buckl. in part and *Quercus Schneckii* Britt.) SCHNECK RED OAK.

This variety differs from the type in its deep cup which is strongly convex on the bottom. The nuts are usually smaller than those of the type or those of the red oak. The variety in its characteristic form is easily separated from the type but there are intermediate forms in Indiana that can be called either the species or the variety. If this fact is kept in mind, controversies over determinations of this group may be avoided.

Southern Ala., La. to Tex., northw. in the Mississippi Valley to Wells County, Ind.

15. **Quercus ellipsoidàlis** E. J. Hill. JACK OAK. Map 794. The distribution of this species in Indiana is not known. It is very difficult to identify in the field unless one is familiar with it because it is easily confused with the scarlet and black oaks. I have specimens from the type tree, from a tree in Lagrange County, and from one in White County. In 1938 R. M. Kriebel made a study of its distribution in Indiana and found it through-

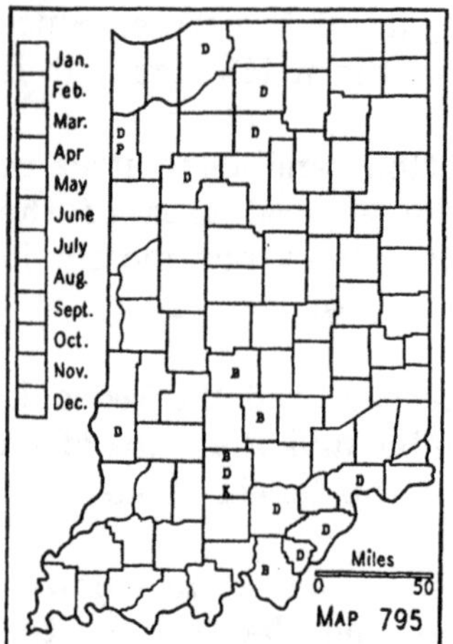

Quercus coccinea Muench.

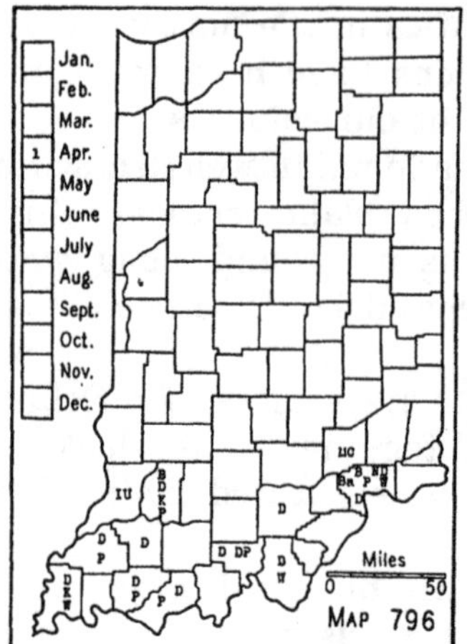

Quercus falcata Michx.

Quercus marilandica Muench.

out northwestern Indiana and in the northern tier of counties. Hill reported it as locally frequent in Lake County, especially near Liverpool. According to Hill, the tree is found on sandy and clayey uplands. Andrews' report from Monroe County may safely be ignored.

Higgins Lake, Mich. to se. Minn., southw. to nw. Ind. and nw. Mo.

16. **Quercus coccínea** Muench. SCARLET OAK. Map 795. This species is local and, no doubt, has a wider range than the map indicates. It is so often confused with the black oak that all records for it must be carefully checked. It is always intimately associated with black oak and is found in poor soil mostly on the crests of ridges. I believe it has its mass distribution in the unglaciated area, and outside of that it is a rare and local tree.

Maine, s. Ont. to s. Nebr., southw. to N. C., Ala., and Ark.

16a. **Quercus coccinea** var. **tuberculàta** Sarg. This variety differs from the typical form in that the back of the scales is prominently thickened below the middle of the turbinate cup. The upper row of scales is thin and forms a distinct marginal ring. This form has been found in Lawrence and Vanderburgh Counties.

Mass. to Ind., southw. to Tenn. and Ala.

17. **Quercus falcàta** Michx. (*Quercus rubra* of some recent authors and of Sudworth's Check List of the trees of the United States. 1927.) Map 796. All of our forms are shown on one map. The leaves of this species are extremely variable and this fact has led authors to divide it into two species and several forms. Trelease (The American oaks. Mem. Nat. Acad. Sci. 20: 201. 1924) recognized 14 forms of this species. For the benefit of those who wish to try to separate the species into groups I am giving a brief key for a few of the forms that occur in Indiana.

Leaves all obovate, usually expanded above the middle into 2 lateral lobes and 1 terminal, rather rounded lobe..........................17a. *Q. falcata* f. *triloba.*
Leaves not all obovate, most of them with more than 3 lobes, the lobes mostly acute.

Blades mostly 3-7-lobed, the lobes irregular and more or less strongly falcate.
  Pubescence of lower surface of blades whitish; blades of leaves from the lower branches not conspicuously different from those of the upper branches..17. *Q. falcata.*
  Pubescence of the lower surface of blades brownish; leaves of the lower branches conspicuously different from those of the upper ones, those of the upper branches broader at the apex than those of the next variety; leaves of the lower branches slightly obovate, and usually 7-lobed..17b. *Q. falcata* var. *leucophylla.*
Blades mostly 5-11-lobed, the lobes more regular and mostly acute, the blades mostly 12-25 cm long and 8-15 cm wide, usually very much less falcate than those of the species; pubescence of lower surface of leaves whitish to grayish..17c. *Q. falcata* var. *pagodaefolia.*

17. **Quercus falcàta** Michx. SOUTHERN RED OAK. In 1910, for four days I followed timber cutters who were making ties in Posey County. They favored me by cutting trees of this species which I had marked. This gave me the opportunity to study the leaves of the trees from the bottom to the top. This study convinced me that the species is polymorphic as to leaf form. I have found no difference in the fruit of the many forms. It is true that the three-lobed form (f. *triloba*) is the prevailing form on high ground and on sandy ridges.

This oak is restricted to the southern part of the state. In Clark and Jefferson Counties it is locally frequent in the "flats" where it is usually associated with beech, sweet gum, and black gum. In Harrison and Washington Counties I found it on high ground associated with black and post oaks. In the western part of Gibson and Posey Counties it becomes frequent and it is associated with the low ground oaks and hickories.

Along the Atlantic coast from Pa. to Fla., along the Gulf States to Tex. and up the Mississippi Valley and Ohio River Valley to s. Ind., Ohio, and W. Va.

17a. **Quercus falcata** f. **tríloba** (Michx.) Palmer & Steyermark. This form is rare and is restricted to sandy ridges and dry soil. It is to be noted that all of the coppice shoots of *Quercus falcata* I have ever seen have obovate, 3-lobed leaves.

17b. **Quercus falcata** var. **leucophýlla** (Ashe) Palmer & Steyermark. Palmer designates my no. 10339 from Posey County as belonging to this variety.

Va. to Fla., through the Gulf States to e. Tex., and northw. to Ark. and Ind.

17c. **Quercus falcata** var. **pagodaefòlia** Ell. This variety grows on low banks and in low land in close proximity to sloughs, bayous, and ponds in Gibson and Posey Counties.

Md. to n. Fla., westw. to Ark., and northw. in the Mississippi Valley to Ind.

18. **Quercus marilándica** Muench. BLACKJACK OAK. Map 797. Local and infrequent, mostly in the southwestern part of the state where it is found in poor soil on the crests of ridges or in very poor soil on sand

ridges. I found it in Point Township of Posey County on a very low ridge in a pin oak woods. It was local here; there were only a few trees and it was associated with post oak. It is usually associated with black and post oaks.

N. Y. to Nebr., southw. to Fla. and Tex.

× **Quercus Búshii** Sarg. This is a hybrid between *Quercus marilandica* and *Quercus velutina*. I found a single tree on a sandy ridge on the farm of Frank Plass about 2 miles north of Decker or just northwest of the Vollmer Siding of the Chicago & Eastern Illinois Railroad in Knox County. Seed of this tree were generously distributed in 1933 to the larger arboretums of the United States.

63. ULMÀCEAE Mirbel ELM FAMILY

Branchlets with solid pith; leaves with parallel primary veins; flowers borne on the branchlets of the preceding year............................1896. ULMUS, p. 390
Branchlets with chambered pith; leaves 3-veined at the base; flowers borne on the branchlets of the year....................................1898. CELTIS, p. 392.

1896. ÚLMUS [Tourn.] L. ELM

Inner bark mucilaginous; upper surface of the leaves very scabrous to the touch, usually densely covered with stiff, more or less erect hairs arising from large, whitish, hollow, papillose bases; branchlets densely gray-pubescent, generally becoming brownish at maturity; bud scales more or less pubescent and ciliate with rufous hairs; flowers nearly sessile; calyx densely ciliate with rufous hairs; samaras mostly suborbicular, 13-19 mm long, both sides of the body densely woolly-pubescent, the wings nearly glabrous, the margins glabrous....1. *U. fulva.*
Inner bark not mucilaginous; leaves glabrous or somewhat scabrous above; flowers on slender, jointed pedicels; samaras ciliate or pubescent all over.
One and two year old branches (at least some of them) with one or both sides covered more or less with a corky excrescence; samaras pubescent all over.
Buds ovate, not twice as long as wide, obtuse or short-pointed, dark brown; bud scales pubescent and ciliate; leaves large, 8-15 cm long, not twice as long as wide, glabrous above except along the midrib; calyx lobes 7-9, not ciliate; samaras oval, 1.5-2 cm long...............................2. *U. racemosa.*
Buds small, narrow, twice as long as wide, light brown, very sharply pointed; bud scales glabrous or merely puberulent; leaves narrow, the blades 4-8 cm long, twice as long as wide, glabrous or more or less scabrous above; calyx lobes 5, not ciliate; samaras oval, the oval part 6-10 mm long......3. *U. alata.*
One and two year old branches without corky wings; branchlets ashy gray, pubescent or glabrate, at maturity becoming light brown and glabrous or remaining pubescent; leaves more or less appressed-pubescent above (at least near the margins and the base), rarely entirely glabrous when observed under a lens, often smooth to the touch but the surface usually covered with short, appressed hairs, sometimes more or less scabrous but the hairs usually without the large, white, papillose bases, rarely a few hairs with such but not distributed over the entire surface as in no. 1; calyx not ciliate; samaras oval, about 10 mm long, both sides glabrous, the margins ciliate.......................4. *U. americana.*

1. **Ulmus fúlva** Michx. SLIPPERY ELM. Map 798. This species is found in every county of the state. It is rare to infrequent in a few of our prairie counties but frequent to common in all parts of the state outside of the oak-hickory forests and in wet woodland. Where woodland has

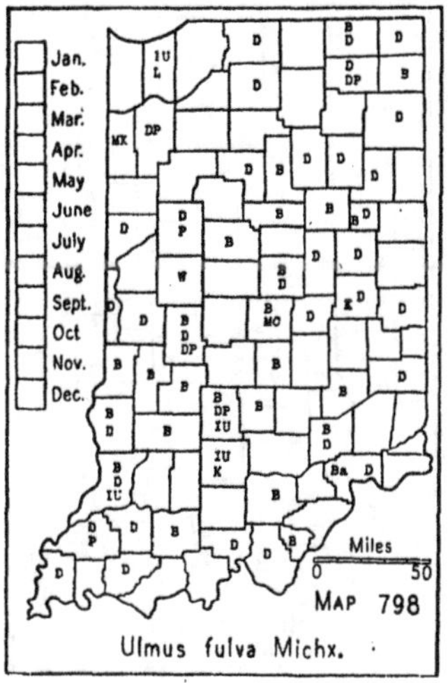

MAP 798
Ulmus fulva Michx.

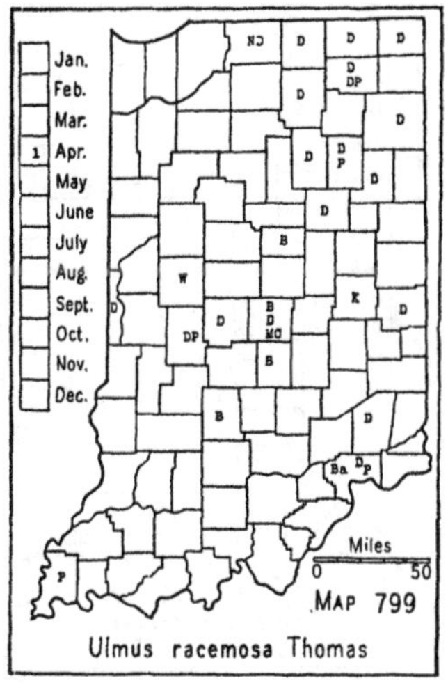

MAP 799
Ulmus racemosa Thomas

MAP 800
Ulmus alata Michx.

been heavily cut over and left for a second crop this species is usually well represented, sometimes forming the major stand. The inner bark was formerly chewed as a remedy for stomach trouble and used in medicine in powdered form for poultices.

Western Que. and w. N. E. to N. Dak., southw. to Fla. and Tex.

2. **Ulmus racemòsa** Thomas. (*Ulmus Thomasi* Sarg.) Map 799. ROCK ELM. Infrequent to frequent or rare within the area shown on the map, to which should be added Floyd, Monroe, and St. Joseph Counties. This species is found in a habitat a little drier than that of the American elm and usually in a more moist habitat than that of the slippery elm. It is almost always associated with the American elm and is difficult to distinguish from it when only the trunk and base are available as characters for separation. The American elm usually has a more buttressed base than the rock elm.

Western Que. and w. Vt. to Ont. and Minn., southw. to n. N. J., Ky., and Mo.

3. **Ulmus alàta** Michx. WINGED ELM. Map 800. Probably restricted to the area shown on the map. This species has two rather distinct habitats. In the hilly counties it is found on the sides of cliffs, on steep rocky slopes, and on the crests of high ridges. It is usually found on or near sandstone and generally associated with American chestnut and black, chestnut, and scarlet oaks. In this habitat it is usually a scrubby or small tree with the corky excrescence on the branches well developed. The other habitat is in hard, white clay flats of the southwestern counties where it is associated mostly with sweet and black gum and pin oak. In the "flats" it sometimes reaches a large size. Rarely specimens are seen which have few or no corky excrescences.

Va. to Kans., southw. to Fla. and Tex.

4. **Ulmus americàna** L. AMERICAN ELM. Map 801. Found in every county of the state. It prefers a moist or wet soil and is frequent to com-

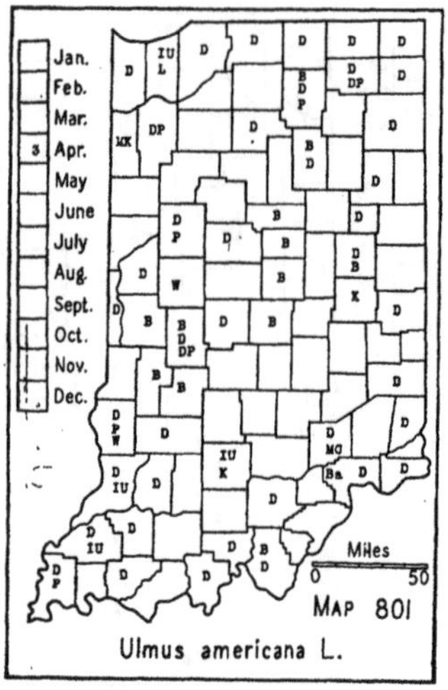

MAP 801
Ulmus americana L.

MAP 802
Celtis occidentalis var. canina (Raf.) Sarg.

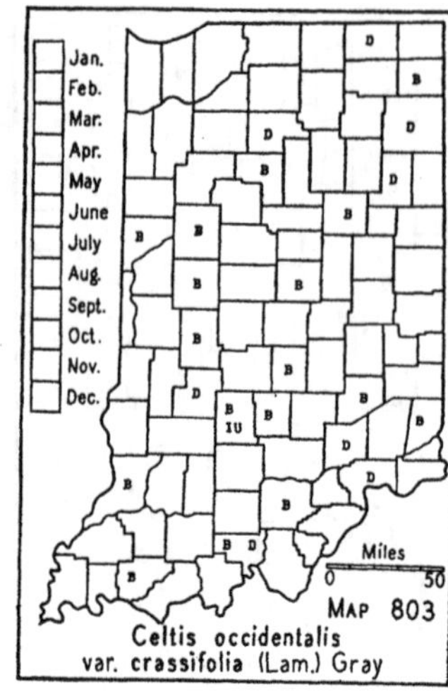

MAP 803
Celtis occidentalis var. crassifolia (Lam.) Gray

mon in such habitats throughout the state except in the dunes. This species is commonly known as white elm.

Newf. to Man., southw. to Fla. and Tex.

1898. CÉLTIS [Tourn.] L. HACKBERRY

Margins of leaves of fruiting branchlets and shoots sharply serrate all around to the base; leaf blades of an ovate to broadly ovate type, oblique at base, sometimes strongly so, those of fruiting branchlets 5-15 cm long; pedicels of fruit much longer than the petioles; nutlets 6-8 mm long; small or large trees.

Leaf blades broadly ovate, acute or short-acuminate, smooth above. (See excluded species no. 182, p. 1040.) . . . . *C. occidentalis.*

Leaf blades generally narrower than the type, apical half narrower, more curved, and long-attenuate at the apex, usually smooth above but sometimes slightly rough. . . . . 1. *C. occidentalis* var. *canina.*

Leaf blades as large as or larger than those of the preceding, more of an oblong-ovate type, very rough above. . . . . 1a. *C. occidentalis* var. *crassifolia.*

Margins of leaves of fruiting branchlets usually entire, or some with a few teeth on one side or with a few teeth on both sides but never serrate on either side to the base; margins of leaves of vegetative branchlets and shoots similar to those of fruiting branchlets, or with the margins serrate nearly all around but never serrate to the base; pedicels of fruit shorter or only slightly longer than the petioles; nutlets 5-6 mm long.

Leaves generally of an oblong-lanceolate type, generally thin, ours smooth above and medium green on both sides, not lighter or yellow green beneath; blades of fruiting branchlets mostly 4-12 cm long and 2-4.5 cm wide; mature fruit a light cherry red; medium sized trees of a wet habitat. . . . . 2. *C. laevigata.*

Leaves mostly of an ovate-lanceolate type, sometimes ovate to broadly ovate or rarely oblong-lanceolate, generally thick and yellow green beneath, generally smooth but sometimes rough above; blades extremely variable in size and shape, mostly 3-10 cm long and 2-6 cm wide, usually about half the maximum size; branchlets usually more or less pubescent; pedicels shorter or longer than the petioles; mature fruit (collected in October) a dark cherry red; trees usually 1-2.5 m high, but sometimes 4-6 m high and up to 1 dm in diameter near the base; of a dry sandy, gravelly or rocky habitat. . . . . 3. *C. pumila.*

1. **Celtis occidentàlis** L. var. **canìna** (Raf.) Sarg. (*Celtis occidentalis* in part, of Gray, Man., ed. 7 and Britton and Brown, Illus. Flora, ed. 2.)

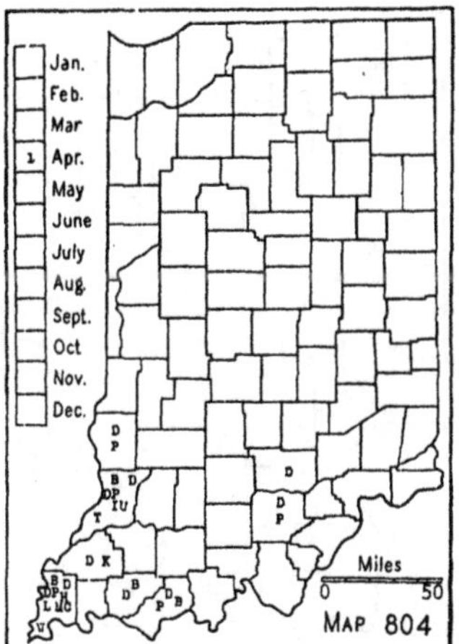

Celtis laevigata Willd.

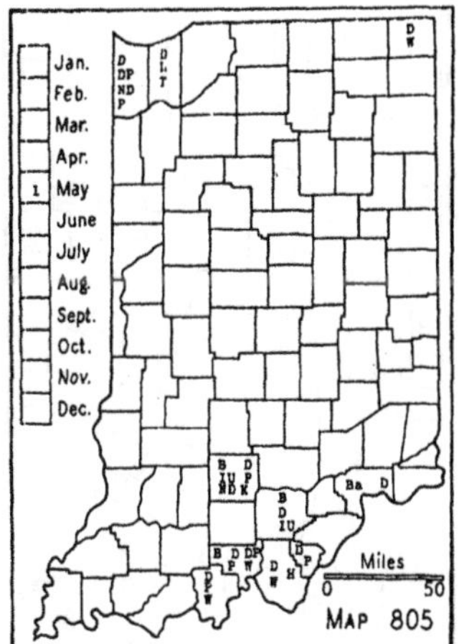

Celtis pumila (Muhl.) Pursh

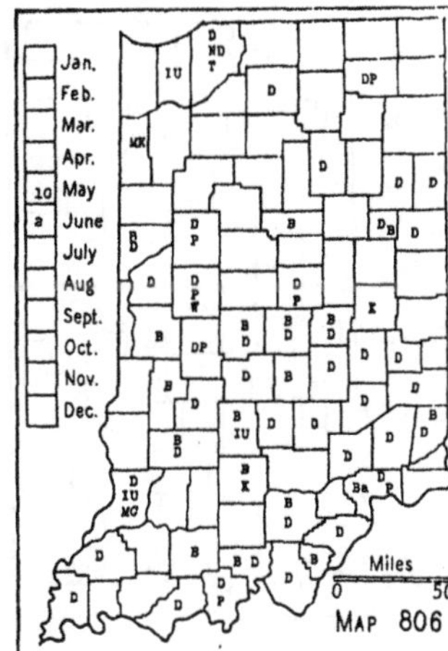

Morus rubra L.

HACKBERRY. Map 802. This tree is no doubt found in every county of the state. It prefers the moist, alluvial soil along streams but is sometimes found in sandy upland and on wooded slopes. It is rare in northern and southern Indiana but frequent along our major streams. It is infrequent to rare in low woods at a distance from a stream. This is the common hackberry in the state.

Que. to N. Dak., southw. to Mass., N. Y., Ga., and Okla.

1a. **Celtis occidentalis** var. **crassifòlia** (Lam.) Gray. (*Celtis crassifolia* Lam.) BIGLEAF HACKBERRY. Map 803. This form is found probably throughout the state with the preceding but is rare or infrequent. I am not convinced that this variety has any standing. I have found both smooth and rough leaf forms on the same tree. Undoubtedly mere roughness of leaves has little significance.

Va. and Ind. to Minn. and Wyo., southw. to N. C. and Tex.

2. **Celtis laevigàta** Willd. (*Celtis mississippiensis* Bosc of Gray, Man., ed. 7 and Britton and Brown, Illus. Flora, ed. 2.) SUGARBERRY. Map 804. Infrequent to rare or locally common in low woods in the southwestern part of the state. It is usually found in low woodlands, especially those that are more or less inundated. It is abundant in the bottoms along the Wabash River and frequent in the bottoms near the mouth of Little Pigeon Creek. It no doubt formerly followed the larger streams farther northward than our map indicates. It prefers a hard soil and is rarely found in a porous, alluvial soil. The leaves of this species are usually almost uniform but variations are found. A mile and a half northwest of Griffin, Posey County, I found a large tree that had small leaves, in size and shape like those of the next species but here and there among the leaves were typical ones. The location of the typical leaves indicated to me that the dwarfing was a matter of nutrition but in this I may be in error. The typical leaves are thin and not at all coriaceous but sometimes the leaves are more or less coriaceous. The thickening of the leaves may be due

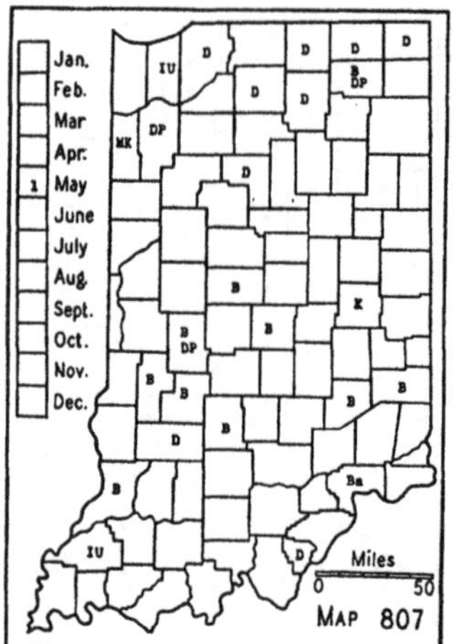

Morus alba var. tatarica (L.) Loud.

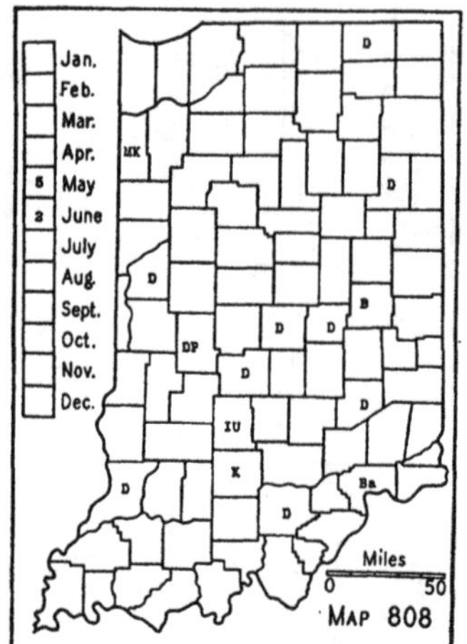

Maclura pomifera (Raf.) Schneid.

Humulus japonicus Sieb. & Zucc.

to location of the tree, because, as I now recall, trees of this sort were found in the open. In fact, most of our specimens are taken from low, round-topped trees of the open because specimens could not easily be obtained from tall, forest-grown trees. The effects of the environment of the trees must therefore have consideration. This species is usually associated with pecan, sweet gum, pumpkin ash, cane, and soft maple.

Va. to Mo. and e. Kans., southw. to the Gulf States and Tex.

3. **Celtis pùmila** (Muhl.) Pursh. (*Celtis pumila* var. *Deamii* Sarg. and *Celtis occidentalis* var. *pumila* Muhl.) DWARF HACKBERRY. Map 805. Very local and rare to infrequent where it is found. In Lake County it was rather frequent on the sand dunes about Miller. I found it in Steuben County on the high, gravelly slope east of Hogback Lake. In Lawrence County a few very small trees are found on a limestone slope in Spring Mill State Park, associated with dwarf specimens of *Zanthoxylum* and *Rhamnus lanceolata*. In Washington County a few trees were found on a wooded slope near Big Spring. In Jefferson County Miss Edna Banta found a few trees in Clifty Falls State Park near the southern end of Trail no. 1. It is found in Crawford County along Blue River near Milltown. In Harrison County it occurs on a rocky, wooded slope 3 miles east of Elizabeth. In Perry County I found it on several ridges about 7 miles east of Cannelton.

The leaves of this species are extremely variable, but no more so than its habitat.

Pa., Ind. to Mo., southw. to Fla., Ga., and Tenn.

64. MORÀCEAE Lindl. MULBERRY FAMILY

Plants woody, small trees.

Branches without spines; leaves serrate; pistillate flowers in spikes..1913. MORUS, p. 395.

Branches with spines; leaves entire; pistillate flowers in heads..1918. MACLURA, p. 395.

Plants herbaceous, tall, erect or long and twining.
  Leaves mostly 3-lobed, rarely the blade lobed deeper than to the middle or the upper ones not lobed; long, twining, perennial plants.........1972. HUMULUS, p. 396.
  Leaves 5-7-divided to near the base, the divisions narrow; tall, erect annual plants; introduced..1973. CANNABIS, p. 397.

1913. MÒRUS [Tourn.] L. MULBERRY

Leaves soft-pubescent with spreading hairs over the entire lower surface, rarely lobed except on vigorous branches or coppice growth, abruptly long-acuminate at the apex, cordate or subcordate at the base; fruit dark purple or black, mostly 2.5-3.5 cm long..1. *M. rubra.*
Leaves glabrous beneath except the midrib or midrib and principal nerves, these ciliate-pubescent with appressed hairs.
  Leaves of an ovate type, rarely lobed; fruit whitish. (See excluded species no. 183, p. 1040.)..*M. alba.*
  Leaves of an ovate type, mostly 3-5-lobed; fruit mostly 1-2 cm long, reddish or purplish.......................................2. *M. alba* var. *tatarica.*

1. **Morus rùbra** L. RED MULBERRY. Map 806. Found as scattered trees probably in every county of the state. Its distribution in the primitive forest can only be conjectured, but since it is a low, round-topped tree and very intolerant of shade, its distribution was, no doubt, very limited. It is abundantly distributed by birds and I have seen it as a frequent to common tree in second growth forests where it is soon shaded out by taller species by the time it reaches a diameter of 4-8 inches. Along fences and in fields it often reaches a diameter of 1-2 feet and usually has a clear bole of 8-10 feet. It is rarely seen in the high forest except in an opening.

Vt. to Mich. and S. Dak., southw. to Fla. and Tex.

2. MORUS ÁLBA L. var. TATÁRICA (L.) Loud. RUSSIAN MULBERRY. Map 807. This species was formerly recommended for forest planting for growing fence post timber. It is a small, crooked tree and is a failure for the purpose recommended. It is very hardy and annually produces an abundant crop of fruit which is greedily eaten by birds. Through the agency of birds this species has become widely distributed in woodland and along fences. A neighbor 3 blocks away has a large tree in his yard and each year I have the task of digging about 50-100 seedlings from our garden.

Probably introd. from Russia, hence its common name.

1918. MACLÙRA Nutt.

1. MACLURA POMÍFERA (Raf.) Schneid. (*Toxylon pomiferum* Raf. of Britton and Brown, Illus. Flora, ed. 2.) OSAGE-ORANGE. Map 808. This tree was formerly much planted for farm fences and windbreaks, especially in our prairie area. Since land has become valuable its use has been discontinued. It has sparingly escaped in all parts of the state and it is a wonder that it has not become an obnoxious weed tree. I recall that I studied two lines of large trees that were planted on each side of a deserted lane in the Ohio River bottoms in Perry County. The line of trees

was about a quarter of a mile long and the trees were mostly 10-15 inches in diameter near the base. I estimated that on the ground there were not less than 25 bushels of fruit and I assumed that the trees fruited almost annually. Yet I did not find a single seedling and I do not believe any were dug up. I made no special inquiry to ascertain the cause of the failure of reproduction.

Mo. and Kans., southw. to Tex.

1972. HÙMULUS L. Hop

[Bailey. Humulus. Manual of Cultivated Plants, 239-240. 1924.]

Petioles of principal leaves much longer than the blades; leaves 5-7-lobed; bracts of pistillate flowers greenish, usually eglandular, narrow, generally long-acuminate, their margins densely long-ciliate; parts of the staminate involucre usually very glandular, narrow, acute to acuminate; anthers eglandular........1. *H. japonicus.*

Petioles of principal leaves shorter than or only equaling the blades, rarely one or more longer; leaves usually 3-lobed; sometimes all of the leaves on the upper part of the stem unlobed; bracts of pistillate flowers glandular at least at the base, not ciliate, stramineous, not green, broad, the lower acuminate, the middle ones broadly ovate, acute or obtuse; parts of the staminate involucre glandular but the glands easily detached and often becoming eglandular, broad, obtuse; anthers more or less glandular.

Lobes of leaves short-acute at the apex or obtuse, coarsely serrate or dentate, the terminal lobe nearly as wide as long; lower surface of leaves sparsely glandular; anthers mostly with fewer than 10 glands. (See excluded species no. 187, p. 1041.) ....*H. Lupulus.*

Lobes of leaves attenuate to the apex, the teeth of the margins finer than those of the preceding species, the terminal lobe narrower at the base than at the middle, generally at least twice as long as wide; lower surface of leaves usually copiously glandular; anthers generally with more than 10 glands........2. *H. americanus.*

1. HUMULUS JAPÓNICUS Sieb. & Zucc. JAPANESE HOP. Map 809. Reported from Tippecanoe County by Wilson but probably more frequent than our knowledge of its distribution indicates. I found it along roadsides near Warsaw and Hobart. Found, also, by Chas. M. Ek in Howard County along a railroad.

Nat. of Japan; sparingly naturalized.

2. **Humulus americànus** Nutt. AMERICAN HOP. Map 810. Probably found in all or nearly all of the counties of the state. It prefers a moist and sandy soil and is found infrequently in low ground along streams, about lakes, and along roadsides. Our manuals have not separated this from the Eurasian species and all but one of our reports for the wild hop have been made under the name, *Humulus Lupulus.* I doubt that the exotic species is found in Indiana and if so, it is very rare. I have not seen a specimen of it from Indiana. Bailey says: "Lobes of leaves often 5-11," but none of the leaves of my specimens have more than 3 lobes. Since the two species have been confused by most authors I am not able to give the distribution of our native hop, but probably it is nearly the same as that given by authors for the exotic species which is as follows:

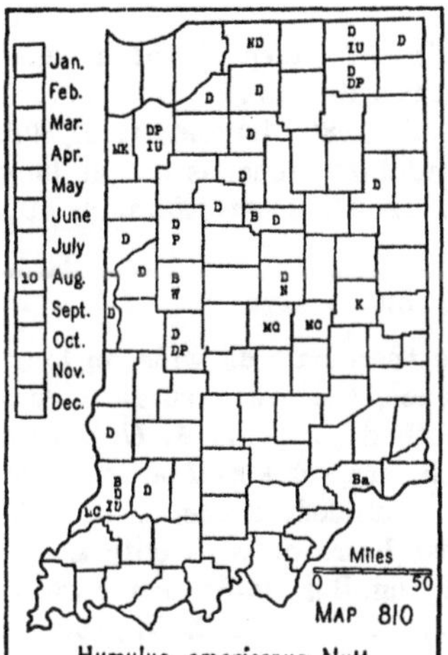

Humulus americanus Nutt.

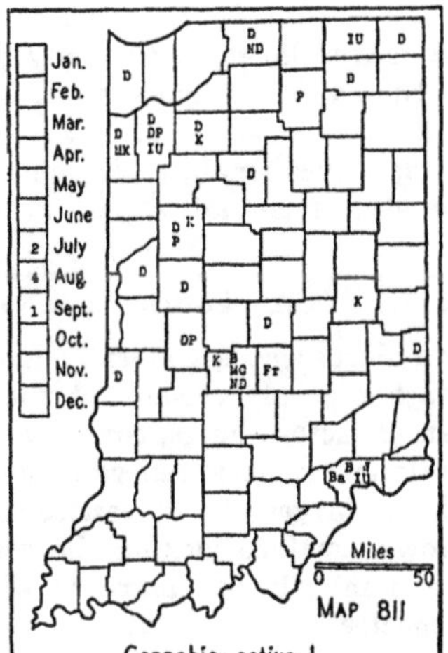

Cannabis sativa L.

Urtica procera Muhl.

N. S. to Wyo., southw. to Fla. and Ariz. The western hop is sometimes considered to be specifically distinct but is usually treated as a variety of *Humulus americanus*.

1973. CÂNNABIS [Tourn.] L.

1. CANNABIS SATÌVA L. COMMON HEMP. Map 811. This species yields a strong fibre which is extensively used for cordage. It was formerly sown in northern Indiana for its fibre. The seed of this plant are much used in commercial bird foods, and this accounts for its escape in all parts of the state. The plant grows 6-10 feet high and produces an abundance of seed; it might well be grown for winter food for birds, and people who provide feed for birds during the winter months should be interested in sowing enough hemp to produce a few sheaves of it to be used for this purpose. Hemp is also the source of the narcotic hashish or marihuana, and growing it in Indiana is now prohibited.

This species prefers a moist, rich soil but I have found it in almost all kinds of soils and locations. It is usually found in waste places, along roadsides, streams and railroads, and infrequently in fallow fields and open woods. In the Kankakee region it is frequent in low ground along fences and on ditch banks.

Nat. of Asia; naturalized from N. B. to Minn., southw. to Ga. and Kans.

65. URTICÂCEAE Reichenb. NETTLE FAMILY

Leaves opposite.

Flowers in axillary panicles.

Plants with stinging hairs, the whole plant more or less pubescent, generally 8-15 dm high; leaves generally with more than 15 pairs of teeth; achenes inclosed by the calyx..1974. URTICA, p. 398.

Plants without stinging hairs, the whole plant glabrous, generally 3-7 dm high; leaves generally with fewer than 15 pairs of teeth; achenes longer than the calyx..1984. PILEA, p. 399.

Flowers in single, axillary spikes, these with or without axillary glomerules; plants without stinging hairs, more or less pubescent throughout; achenes more or less uncinate-pubescent..................................1990. BOEHMERIA, p. 400.

Leaves alternate.

Plants with stinging hairs; leaves large, with many pairs of sharp teeth; achenes about twice as long as the calyx, oblique, the style lateral. .... 1980. LAPORTEA, p. 398.

Plants without stinging hairs; leaves small, entire and undulate; achenes not as long as the calyx, ovate, the style terminal............... 2007. PARIETARIA, p. 401.

1974. URTÌCA [Tourn.] L. NETTLE

Blades of leaves at the lower inflorescences generally more than half as wide as long, ovate to broadly ovate, rather deeply cordate at the base, the lower surface generally covered with numerous, setose hairs, coarsely toothed.......... 1. *U. dioica.*

Blades of leaves at the lower inflorescences generally much less than half as wide as long, ovate-lanceolate, oblong-lanceolate or lanceolate, rounded, truncate or subcordate at the base, the lower surface lacking the setose hairs or with only a few on the principal nerves and midrib, the margins with smaller and more regular teeth than the preceding species................................ 2. *U. procera.*

1. URTICA DIOÌCA L. There is a specimen of this species in the herbarium of the University of Notre Dame. It was collected by Nieuwland on the border of St. Joseph Lake, in St. Joseph County. He said it is established there.

Nat. of Eu.; sparingly naturalized in the e. U. S.

2. **Urtica prócera** Muhl. in Willd. (Rhodora 28: 195. 1926.) (*Urtica gracilis* of authors.) TALL NETTLE. Map 812. Infrequent to frequent in the lake area, becoming infrequent to very rare south of this area and restricted mostly to low places in the alluvial bottoms of our principal streams. It grows in rich, porous soil only in low ground and is found about lakes and ponds in low woods, in low places along unimproved roads in the lake area, in springy places throughout, and in wet places along streams.

This species is often confused with *Urtica dioica* L. which is a native of Europe and has been reported as sparingly escaped in the eastern part of the United States. It has been reported from Indiana but I am referring all of our reports except the one from St. Joseph County to this species. The two species are difficult to separate. The leaves are variable in texture, in shape of the blade and its base, in the number of setose hairs on either surface, in the number of setose hairs on the stem, petioles, and in the inflorescence, and in the size of the panicles. I have 28 specimens from Indiana and 20 of these are monoecious and 8 are pistillate. My specimens represent only the part of the plant with leaves when collected and it is probable that the lower leaves and staminate inflorescences of the pistillate specimens had fallen before the plants were collected. The density of the stand of the plants has a marked influence upon them.

N. S., Que. to N. Dak., southw. to N. C. and La.

1980. LAPÓRTEA Gaud.

1. **Laportea canadénsis** (L.) Gaud. (*Urticastrum divaricatum* (L.) Ktze.) CANADA NETTLE. Map 813. This is strictly a woodland nettle and is found more or less frequently in low, wet woods throughout the

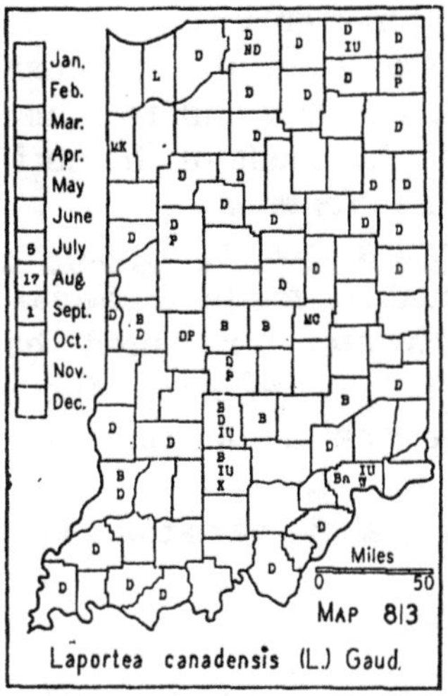

Laportea canadensis (L.) Gaud.

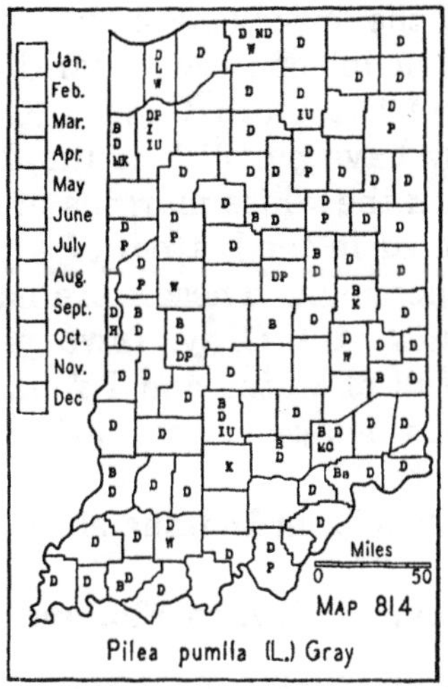

Pilea pumila (L.) Gray

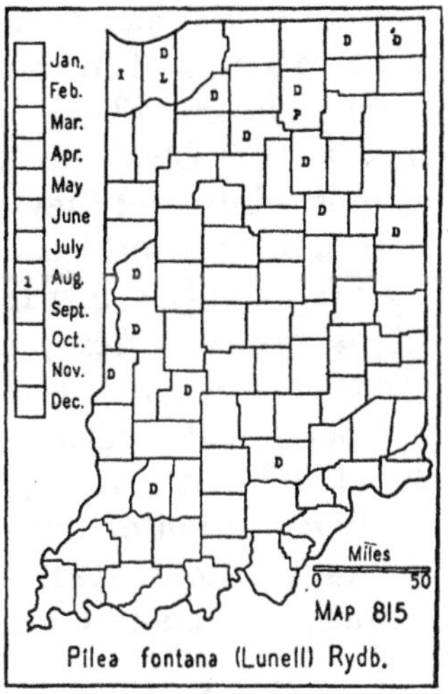

Pilea fontana (Lunell) Rydb.

state except in the hilly counties where it becomes infrequent or rare. N. B. and N. S. to Ont. and Minn., southw. to Fla. and Kans.

1984. PÌLEA Lindl. CLEARWEED

Pericarp relatively thin, the inside whitish or very light brown; fruit ovate, green (sometimes violet), the surface more or less irregularly marked with purplish brown (on immature fruit it may be dark green to brown), the total area of the markings covering about half the surface, the markings, under a 25 diameter magnification, appearing as ridges or excrescences; margins of fruit not conspicuously differing in color from the body; leaves generally cuneate at the base, rarely somewhat rounded or truncate, the number of teeth to a side of average blades 8-15; plants of moist soil and usually growing in cool, shady places....... ....1. *P. pumila.*

Pericarp relatively firm, the inside purplish; fruit ovate, blackish, dull, the surface smooth but unequally bossed all over; margins of fruit conspicuously colorless (whitish); leaves rounded, truncate or more rarely cuneate at the base, the greatest number of teeth to a blade 4-9 (10) on a side; longest petioles 0.5-6.5 cm long, varying according to the size of the plant; plants of very wet or springy habitats. ....2. *P. fontana.*

1. **Pilea pùmila** (L.) Gray. (Including *Pilea pumila* var. *Deamii* (Lunell) Fern. For a discussion of this variety see Fernald, Rhodora 38: 169. 1936.) CLEARWEED. Map 814. This plant prefers a cool, shady place in which to grow and is found in moist, rich soil throughout the state. I once found it growing on an old cypress log in a cypress swamp in Posey County. It is usually found in colonies and when a colony in rich soil is studied it will be found that the plants that are crowded are simple or with few branches at the top while those on the outside of the colony may have long branches even to the ground. Single plants in a similar habitat may be so large that they become decumbent half their length and have side branches that are nearly as long as the remainder of the main stem. The plants are variable in all their parts; the branches at the base may be short or long; the leaves are usually cuneate at the base although I have a specimen with leaves that are truncate at the base; the teeth of the margins vary from 3-17 on a side and vary from rounded to rather sharply

serrate or the margins of the lower leaves are sometimes entire; the fruits vary in size and in the amount of brown markings and are of a greenish color. My no. 48006 from Crawford County, Oct. 2, 1929, and two just like it from Clark County have purplish fruits, but the inside of the pericarp is white and they lack the white margins of *Pilea fontana*.

In a dry season I collected in the bottom of a pond a sheet of depauperate plants only a few inches high. These were named for me by a specialist as typical *Pilea pumila* (L.) Gray. Two years later I revisited the same pond when it was full of water and I found only large plants about the pond. Like all annuals delayed germination or lack of moisture produces small or dwarf plants.

My collection of 132 sheets from Indiana shows the above variations and others not mentioned.

Que., e. Canada to Minn., southw. to Fla. and Tex.

2. **Pilea fontàna** (Lunell) Rydb. (*Adicea fontana* Lunell.) Map 815. Found in favorable habitats probably throughout the state although its habitat is rarely found south of the lake area. This species grows only in very wet, springy, and boggy places, usually on the borders of lakes and streams. It commonly forms dense stands and usually is 6-12 inches high although I have a specimen more than 3 feet high. One of its favorite habitats is among dead or live cattails. This species can easily be separated from the preceding species by the white margin of the fruit and the purple color of the inside of the pericarp. It has not been recognized for a time long enough to ascertain its range.

P. E. I. to N. Dak., southw. to Fla. and Nebr.

1990. BOEHMÈRIA Jacq.

Leaves mostly broadly ovate, ascending, not folded, generally long-acuminate, smooth, smoothish or somewhat scabrous above; petioles of median leaves (20) 25-80 mm long; achenes generally yellowish green, without purplish splotches, the body usually glabrous........1. *B. cylindrica.*

Leaves narrowly ovate or ovate-lanceolate, some or all drooping and generally more or less folded, somewhat scabrous above; petioles of median leaves (3) 5-20 (25) mm long; achenes more or less splotched with purple, the wings and usually the body with uncinate hairs........1a. *B. cylindrica* var. *Drummondiana.*

1. **Boehmeria cylíndrica** (L.) Sw. FALSE NETTLE. Map 816. Infrequent to frequent throughout the state in low places in woodlands and less frequent in marshes and wet prairies.

Maine, Ont. to Minn., southw. to Fla. and Tex.

1a. **Boehmeria cylindrica** var. **Drummondiàna** Weddell. (*Boehmeria cylindrica* var. *scabra* Porter of Gray, Man., ed. 7.) DROOPINGLEAF FALSE NETTLE. Map 817. Infrequent to locally frequent in the lake area in open marshes, infrequent to local in wet places in woods and wet prairies, and rare in low places in woods or in springy places in the southern part of the state. It is to be noted that this variety intergrades into the species and in-

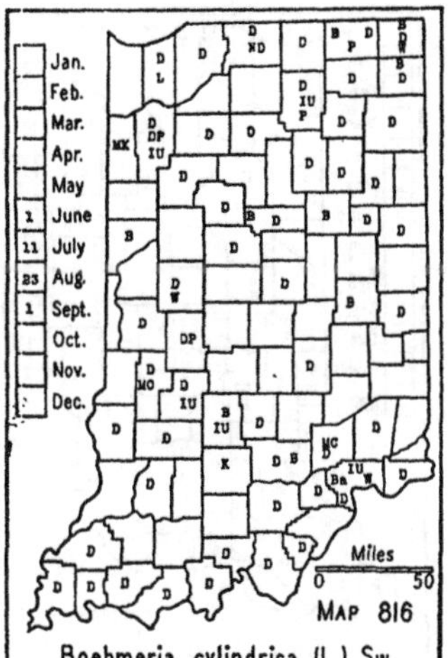

Boehmeria cylindrica (L.) Sw.

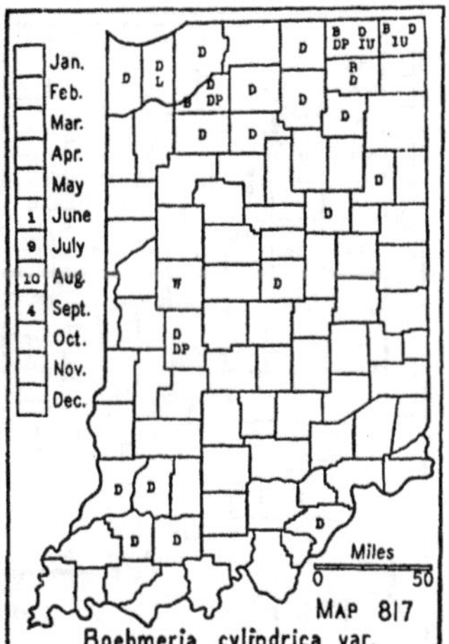

Boehmeria cylindrica var. Drummondiana Weddell

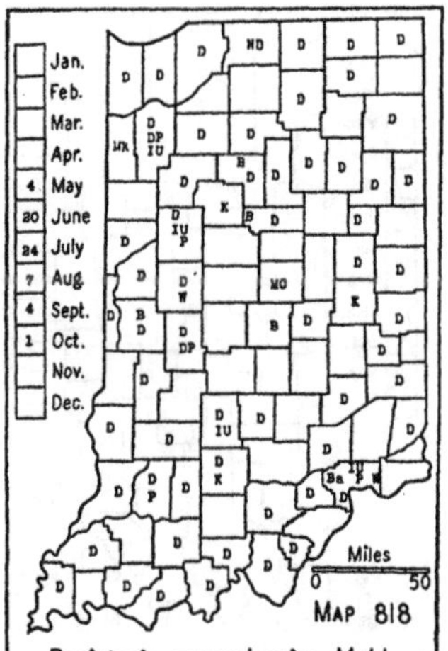

Parietaria pennsylvanica Muhl.

termediates are found that are difficult to place. The scabrous upper surface of the leaves is not a constant character and is of little value. The long-acuminate apex of the leaves generally holds for the species. The length of the petioles, drooping leaves, and purplish achenes are the most reliable characters for their separation. The wings of the achenes are variable in both the species and the variety. Sometimes they are developed more on one side than on the other; they may extend to the base on both sides or on one side only; the mass of them may be below the median line or it may be above it. On the whole, the achenes with their wings are about 1-1.25 mm wide in the species and 1.25-1.5 mm wide in the variety.

It is possible that the alkalinity of the soil has a decided influence on the plant since most of my specimens are from a more alkaline soil than are those of the species. The stem of the variety is usually much more uncinate-pubescent than the stem of the species.

Mass., N. Y., and Mich. to Kans., southw. to Fla. and Tex.

2007. PARIETÀRIA L.

1. **Parietaria pennsylvánica** Muhl. PENNSYLVANIA PELLITORY. Map 818. Infrequent to frequent throughout the state. It is usually found in colonies in dry soil in all kinds of woodland but prefers a sandy soil and is often found in large colonies in mucky or peaty areas that have been drained.

Maine, Minn. to B. C., southw. to Fla. and Mex.

67. LORANTHÀCEAE D. Don MISTLETOE FAMILY

2089. PHORADÉNDRON Nutt.

1. **Phoradendron flavéscens** (Pursh) Nutt. AMERICAN MISTLETOE. Map 819. Formerly frequent to common in the southern counties, now almost extinct. It no doubt covered the southern third of the state. There are reports from as far north as Bartholomew and Franklin Counties and Ridg-

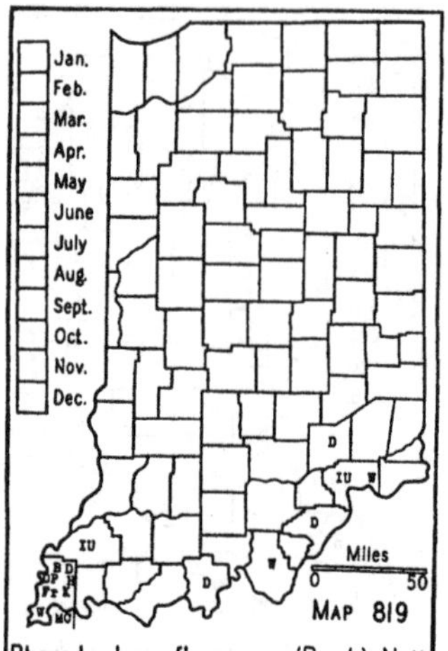

Phoradendron flavescens (Pursh) Nutt.

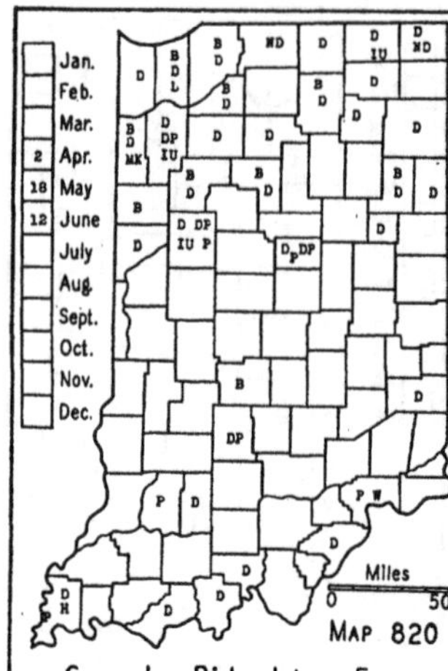

Comandra Richardsiana Fern.

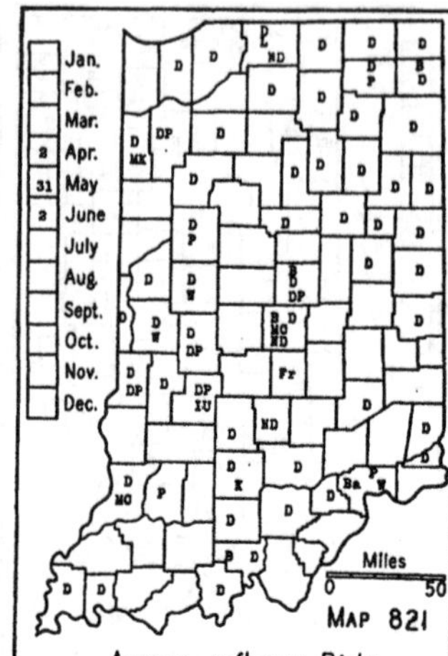

Asarum reflexum Bickn.

way says: "Fully 90 per cent of the white elm trees in the White and Wabash Rivers bottoms are affected by this parasite. I saw it on no other species except honey locust and elm." The more common hosts, however, include *Acer rubrum, Acer saccharinum, Gleditsia triacanthos, Juglans nigra, Nyssa sylvatica, Quercus palustris,* and *Ulmus americana.* I have noted walnut trees almost killed by it in both Perry and Posey Counties.

In 1934 I saw a large specimen growing on a very large native elm tree in the yard of J. F. Schmid in sec. 18 of Spencer Twp. in Jennings County. It was growing so high that I was unable to secure a specimen.

N. J., s. Ind. to Mo., southw. to Fla. and Tex.

69. SANTALÀCEAE R. Br. SANDALWOOD FAMILY

Leaves sessile; flowers in corymbiform cymes at the ends of the branches; style slender. ....2112. COMANDRA, p. 402.

Leaves on short petioles; flowers in 1-3-flowered lateral cymes; style short............ ....2112A. GEOCAULON, p. 403.

2112. COMÂNDRA Nutt.

Rootstock just beneath the surface; leaves lanceolate to ovate, thick, not paler beneath, when dried the lower surface obscurely veiny; inflorescence corymbose, 1-3 cm wide, of 1-few-flowered cymules on ascending branches....1. *C. Richardsiana.*

Rootstock underground; leaves oblong, thin, pale beneath, when dried the midrib pale beneath; inflorescence, when fully developed, an ellipsoid-oblong panicle with the cymules of smaller more numerous flowers on divergent branches. (See excluded species no. 188, p. 1041.)...*C. umbellata.*

1. **Comandra Richardsiàna** Fern. (*Comandra umbellata* in part, of Britton and Brown, Illus. Flora, ed. 2.) RICHARDS BASTARD TOADFLAX. Map 820. Infrequent in dry, sandy soil under black and white oak in northern Indiana and rare in a similar habitat in the southern counties. I have specimens from three counties which were found in black, sandy soil in prairies and a specimen from Lagrange County found in a drained tamarack bog where it was associated with tamarack and poison sumac. Most of them were seen by M. L. Fernald and he says that all of my specimens and all

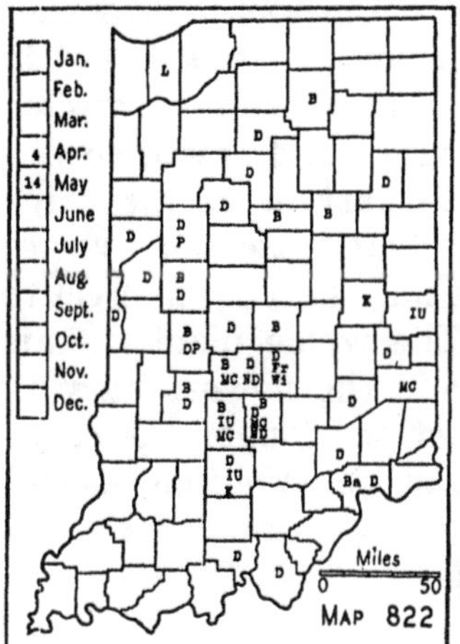

Asarum canadense L.

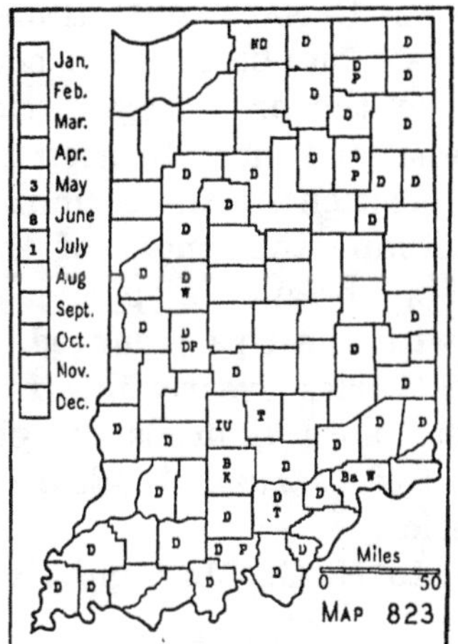

Aristolochia Serpentaria L.

Aristolochia tomentosa Sims

of those in the Gray Herbarium from west of the Allegheny Mountains belong to this species. It is doubtfully separated from *Comandra umbellata* and in Britton and Brown, Illus. Flora, ed. 2, it was regarded as a synonym. Fernald gives the range of *Comandra umbellata* as restricted to the area east of the Allegheny Mountains. Whether this species is maintained as distinct or is regarded merely as a geographical form, our specimens belong to the segregate of plants with the lower surface of the leaves not paler beneath and with a superficial rootstock.

Eastern Que. to Assina., southw. to N. Y., Ind., Mo., and Kans.

2112A. GEOCAÚLON Fern.

See excluded species no. 189, p. 1041.

74. ARISTOLOCHIÀCEAE Blume BIRTHWORT FAMILY

Acaulescent herbs; stamens 12, with more or less distinct filaments; capsule fleshy...2170. ASARUM, p. 403.

Caulescent herbs or woody vines; stamens 6, the sessile anthers adnate to the stigma; capsule dry......................................2174. ARISTOLOCHIA, p. 404.

2170. ÁSARUM [Tourn.] L.

Calyx lobes usually reflexed in anthesis, triangular, acute or short-acuminate, generally about as long or less than twice as long as the tube; internodes of the rhizomes, except the last one, generally glabrous............................1. *A. reflexum.*

Calyx lobes erect or spreading in anthesis, lanceolate, long-acuminate to caudate, much longer than the tube; internodes of the rhizomes mostly more or less pubescent...2. *A. canadense.*

1. **Asarum refléxum** Bickn. (*Asarum canadense* var. *reflexum* (Bickn.) Rob.) CURLY WILDGINGER. Map 821. Infrequent to frequent in moist, rich soils in woods throughout the state. It spreads mostly by underground stems, hence it is always found in dense colonies, usually in the lee of an old log or treetop where there is an abundance of leaf mold or in some sheltered situation on a wooded slope or in a ravine.

Conn., s. N. Y. to Mich. and Iowa, southw. to Mo. and Kans.

2. **Asarum canadénse** L. CANADA WILDGINGER. Map 822. Infrequent to rare throughout the state or absent from some areas. It is found in habitats similar to those of the preceding species but in more protected situations; hence it is restricted more to deep ravines and steep wooded slopes. The length and position of the acuminate portion of the calyx lobes are variable. In Indiana the length of the acuminate part varies from 5-20 mm and the calyx lobe and its appendage may vary from erect to spreading or spreading with the tips incurved. The whole plant in this and the preceding species varies greatly in size and the flowers vary in proportion. As a rule, the more vigorous the plant the longer the calyx lobes. All of our reports for *Asarum canadense* var. *acuminatum* Ashe I am referring to this species.

N. B. to Man., southw. to N. C., Mo., and Kans.

2174. ARISTOLÒCHIA [Tourn.] L.

Erect herbs up to 6 dm high....................................1. *A. Serpentaria.*
Long, twining, woody vines....................................2. *A. tomentosa.*

1. **Aristolochia Serpentària** L. VIRGINIA SNAKEROOT. Map 823. Infrequent to frequent in moist, rich woods throughout the state except the northwestern corner. This herb has been much used in medicine as a bitter tonic since pioneer times. The fact that the tonic was prepared by adding the roots to whiskey may have added to its popularity.

Conn. to Mich., southw. to Fla. and Tex.

2. **Aristolochia tomentòsa** Sims WOOLLY PIPE-VINE. Map 824. Local in the Lower Wabash Valley from the southwestern corner of Knox County southward. It is rather frequent along the lower course of White River in both Gibson and Knox Counties. South of Coffee Bayou in Gibson County it is rare until Point Township in Posey County is reached where it again is local. It climbs to great heights on bushes and small trees. I have seen the dead trunks of large trees shingled with it to a great height. Ridgway (Proc. Nat. Mus. **17**: 421. 1894) records the measurements of a vine found in the Lower Wabash Valley as "83 feet long and 10 inches in circumference." I measured a leaf in Posey County, the blade of which was 10 inches wide and 9 inches long. We have had it planted for years as a porch trellis and it serves this purpose well but it spreads vigorously by root suckers.

N. C., Ind., Ill., and Mo., southw. to Fla. and Okla.

77. POLYGONÀCEAE Lindl. BUCKWHEAT FAMILY

Sepals 6, the 3 inner ones much longer and enlarged in fruit (except in *Rumex Acetosella*); flowers greenish yellow, frequently tinged with red; stigmas fringed. ....2195. RUMEX, p. 405.

Sepals 5, sometimes 4, nearly equal in length; flowers purple, pink, white, greenish white, greenish pink or greenish yellow in a few species (these with linear leaves); stigmas not fringed.

Flowers in fascicles in the bracts (generally called sheaths or ocreae in this family) or solitary; if solitary, the flowers not pink and the leaves linear.

Achenes enclosed by the calyx lobes; if exserted, the leaves linear...2201. POLYGONUM, p. 407.

Achenes much exserted; plants erect, with triangular-hastate leaves.............
....2202. FAGOPYRUM, p. 418.
Flowers solitary in the bracts, rose color; stamens 8; leaves linear...............
....2203. POLYGONELLA, p. 418.

2195. RÙMEX L. DOCKS and SORRELS

[Rechinger, K. H., Jr. The North American species of Rumex. Field Mus. Nat. Hist. Publ. Bot. Ser. **17**: 1-151. 1937.]

Leaves hastate; flowers dioecious; plants generally less than 5 dm high.
Achene much exserted from the scarcely changed calyx............1. *R. Acetosella.*
Achene enclosed by the inner calyx lobes (valves). (See excluded species no. 192, p. 1042.)..*R. hastatulus.*
Leaves not hastate; flowers not dioecious; plants generally more than 5 dm high.
Inner sepals (valves) entire, crenate or denticulate.
Number of valves of fruits bearing a tubercle generally 1, these mixed more or less with fruits with 2 valves bearing a tubercle.
Leaves flat, green, tapering at the base; valves usually bearing only one tubercle.
....2. *R. altissimus.*
Leaves wavy, generally with red veins, cordate or subcordate at the base. (See excluded species no. 195, p. 1042.)........................*R. sanguineus.*
Number of valves of fruits bearing a tubercle generally 3, these mixed more or less with fruits with only 2 valves bearing a tubercle.
Leaves flat, light green.
Pedicels enlarged upward, more than twice as long as the fruit, not conspicuously enlarged at the joint; fruit maturing the last of June and first of July..3. *R. verticillatus.*
Pedicels not enlarged upward, not twice as long as the fruit............
....4. *R. triangulivalvis.*
Leaves wavy-margined or crisped, dark green.
Mature valves less than 2 mm wide. (See excluded species no. 190, p. 1041.)..
....*R. conglomeratus.*
Mature valves more than 2 mm wide.
Plants very tall, mostly 1.2-2.5 m high; median leaves generally more than 4 cm wide, narrowed at the base; pedicel longer than the fruit, the joint not conspicuously enlarged; fruit maturing in September and October..
....5. *R. Britannica.*
Plants mostly less than 1 m high; leaves rounded or cordate at the base, the median ones less than 5 cm wide; pedicel about as long as the fruit, conspicuously swollen at the joint; fruit maturing mostly in June and July...................................6. *R. crispus.*
Inner sepals or valves spinulose-dentate or pinnatifid.............7. *R. obtusifolius.*

1. RUMEX ACETOSÉLLA L. FIELD SORREL. Map 825. An abundant weed in some cultivated fields. Its presence is usually indicative of impoverished and minimacid soils. In the sandy areas of the northwestern part of the state it is an obnoxious weed, covering sometimes whole fields. It is somewhat frequent in the entire northern part of the state, rare in the central, and frequent in the southern part.

Nat. of Eu. Throughout temperate N. A.

2. **Rumex altíssimus** Wood. PALE DOCK. Map 826. Infrequent to frequent throughout the state in low ground along streams and in low ground elsewhere.

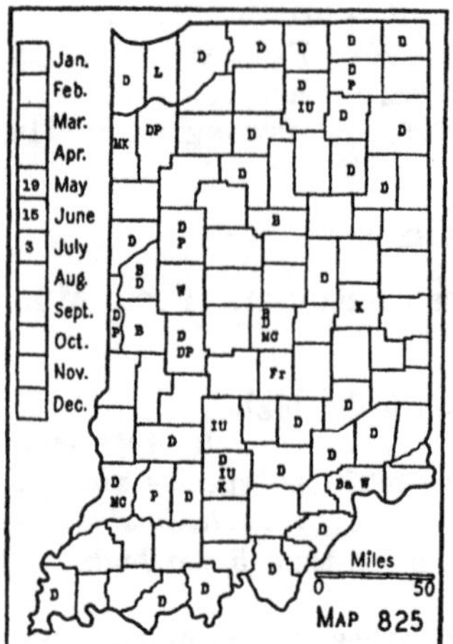

MAP 825

Rumex Acetosella L.

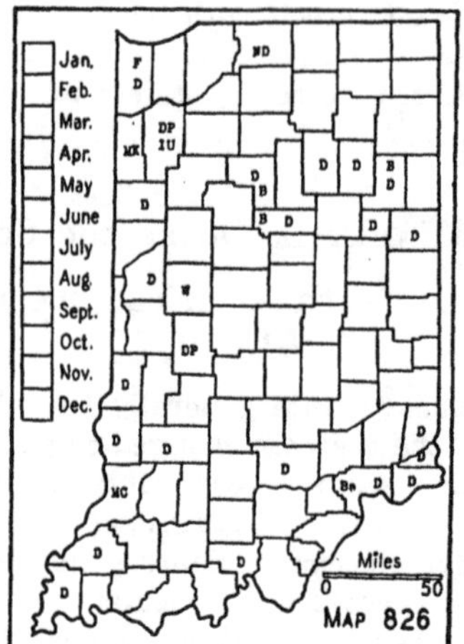

MAP 826

Rumex altissimus Wood

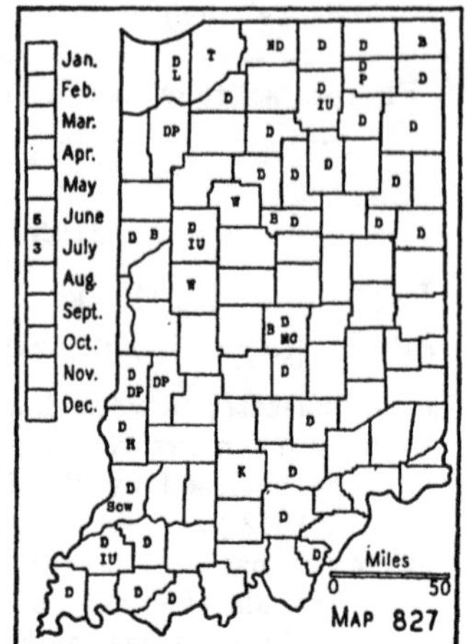

MAP 827

Rumex verticillatus L.

Peattie observed (Amer. Midland Nat. **10**: 130. 1926) that one valve of each fruit had a complete tubercle and a second valve might have an aborted tubercle. He gave this form a name, but if he had read carefully the original description of the species, he would have seen that this phenomenon was included in the description of the species. The tendency to double the number of tubercles is frequent among the fruits of this species.

Conn. to N. Dak., southw. to Md. and Tex.

3. **Rumex verticillàtus** L. SWAMP DOCK. Map 827. Found in the muddy borders of ponds, swamps, and sloughs in all parts of the state. Where it is found, it sometimes forms dense colonies.

Que. to Iowa, southw. to Fla. and Tex.

4. **Rumex trianguliválvis** (Danser) Rech. f. Map 828. (*Rumex mexicanus* of Indiana authors, not Meisn.) This species and species no. 2 are very similar and can be distinguished only by the number of valves which bear tubercles.

In addition to my records, this species has been reported only from St. Joseph County. I believe it is much more frequent, however, than our reports indicate simply because it is so easily confused with *R. altissimus* and both have the same habitat.

Newf. and Lab. to B. C., southw. to Maine, Ind., Mo., and along the Rocky Mts. to Mex.

5. **Rumex Británnica** L. GREAT WATER DOCK. Map 829. Usually in boggy or marshy places but sometimes in a habitat that is rather muddy, such as about ponds and in swamps. Infrequent. No doubt all the reports of it from southern Indiana should be transferred to some other species. In 1932, E. B. Williamson found a plant along Pigeon River in Lagrange County that had a leaf with a blade 35 inches long.

Newf., Ont., and Minn., southw. to N. J. and Kans.

6. RUMEX CRÍSPUS L. CURLY DOCK. Map 830. A common weed in low

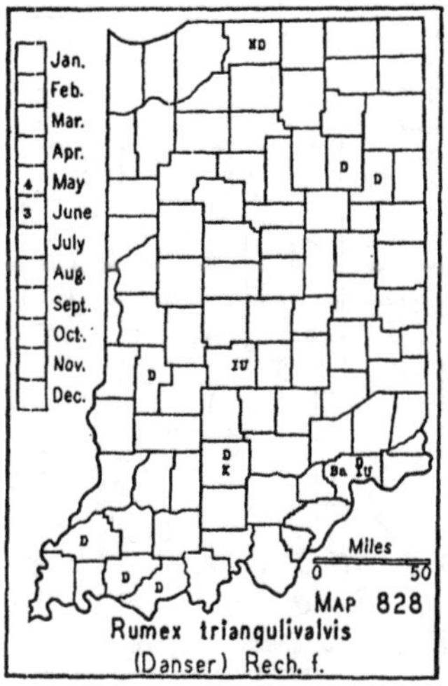

MAP 828
Rumex triangulivalvis (Danser) Rech. f.

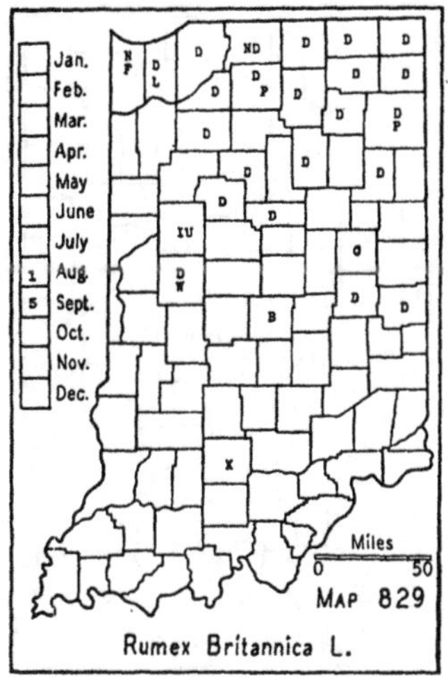

MAP 829
Rumex Britannica L.

MAP 830
Rumex crispus L.

ground in cultivated fields, along streams, and in woodland on the border of swamps, ponds, and sloughs. It is one of our most obnoxious weeds. The root was formerly official in medicine and was sold usually under the name of yellow dock. Formerly the early spring leaves were mixed with those of the dandelion and cooked for food. The mixture was called "greens." The discovery, however, that the leaves contain calcium oxalate, which is injurious, has decreased the popularity of this practice.

The farmers in Indiana usually call this plant sour dock.

The tubercles of the valves of the same plant may vary at the apex from obtuse to acute. *Rumex elongatus* Guss. is a form of this species with acute tubercles but since both acute and obtuse forms can be found on the same plant, all reports for this species should be referred to *Rumex crispus*.

Nat. of Eu. Now found throughout temperate N. A.

7. RUMEX OBTUSIFÒLIUS L. BLUNTLEAF DOCK. Map 831. Infrequent to frequent throughout the state. It is found almost everywhere in moist or rather moist soil in open woodland, fallow fields, and wasteland and along roadsides. The veins of the leaves of this species are sometimes red and I think our reports for *Rumex sanguineus* should be referred to this species.

Nat. of Eu.; Newf. to B. C. and Oreg., southw. to Fla. and Tex.

2201. POLÝGONUM [Tourn.] L. KNOTWEED, SMARTWEED

[Some recent authors divide this genus into several small genera. Since I am following Dalla Torre and Harms I am not dividing the genus.]

A. Plants not twining.
  B. Stems not armed with prickles.
    C. Flowers axillary (solitary or in clusters).
      Stems and branches terete and striate.
        Plants erect, mostly 0.4-1.5 m high, rather sparsely branched, the branches

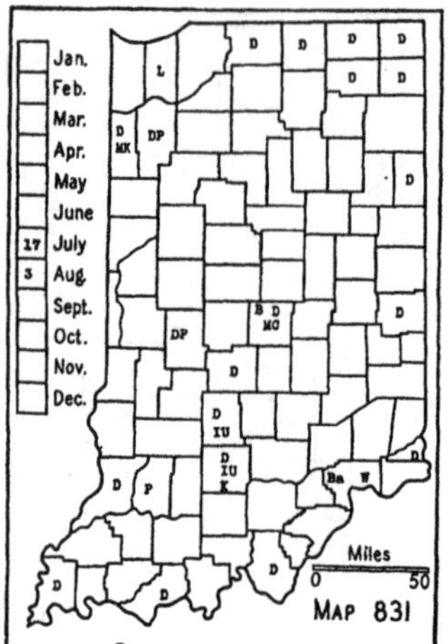

Rumex obtusifolius L.

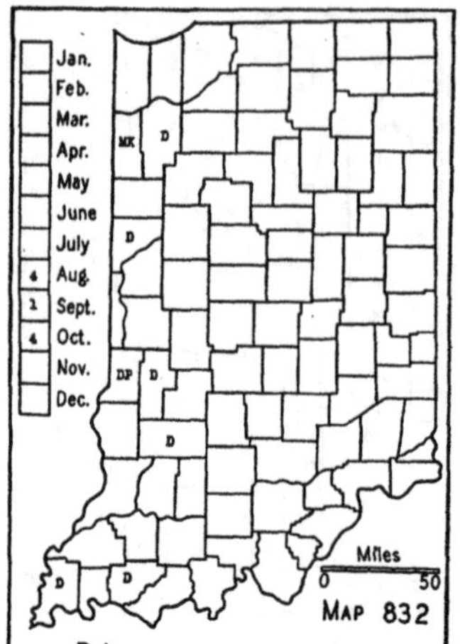

Polygonum exsertum Small

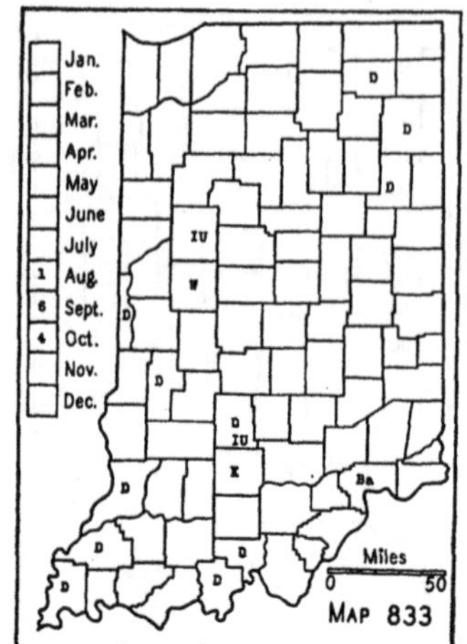

Polygonum erectum L.

stiffly ascending; leaves usually narrowly lanceolate or linear, mostly 1.5-5 cm long, usually acute or acuminate at both ends.

Sepal lobes with white or pinkish margins; normal achenes about 2.5 mm long and included in the perianth (plants of autumn often have some or all of the achenes long-exserted) .... 1. *P. exsertum.*

Sepal lobes with yellowish green margins; achenes 3-3.5 mm long (plants of autumn rarely have exserted achenes). (See excluded species no. 202, p. 1043.) .... *P. ramosissimum.*

Plants not as above.

Stems erect or ascending.

Sepal lobes with yellowish green margins; stems generally solid and erect; leaves oval, elliptic, or obovate, generally acute....... 2. *P. erectum.*

Sepal lobes with white or pinkish margins; stems generally hollow and ascending; leaves like the preceding but smaller and usually narrower .... 3. *P. monspeliense.*

Stems prostrate, or diffusely spreading.

Leaves thick, prominently veined, oblong, oval or spatulate, obtuse or rounded at the apex, usually pale; ocreae very conspicuous; faces of the achenes granular............................. 4. *P. buxiforme.*

Leaves thin, not prominently veined, lanceolate or linear, acutish, acute, or sometimes acuminate at the apex, light or dull bluish green; ocreae not conspicuous; faces of the achenes finely striate.

Perianth 2.5-3.5 mm long; achenes 2.5-3 mm long, acute; leaves 2-4 cm long, oblong-lanceolate, acute or obtusely pointed............. .... 5. *P. aviculare.*

Perianth 2-2.5 mm long; achenes 2-2.5 mm long, acuminate; leaves mostly less than 2 cm long, linear-lanceolate or linear, acute (sometimes acuminate) at the apex.................... 6. *P. neglectum.*

Stems and branches strongly angled, erect; leaves linear, sharp-pointed, minutely ciliolate.. 7. *P. tenue.*

C. Flowers in terminal spikes.

Styles short, soft, scarcely exserted, withering in fruit; leaves neither large-ovate nor acuminate.

Sheaths not ciliate, except rarely the uppermost.

Spikes 1 or 2, rarely 3; perennial, aquatic or marsh plants (sometimes persisting for years or even spreading in a terrestrial form after drainage) with long rootstocks, rooting in the mud.

Peduncles glabrous; aquatic plants with floating leaves; leaves elliptic or narrow-ovate, obtuse or subacute.................... 8. *P. natans.*

Peduncles more or less pubescent and glandular; plants semiaquatic or terrestrial; leaves ovate-oblong or ovate-lanceolate, very acute or short-acuminate.

Sheaths with herbaceous tips............8a. *P. natans* f. *Hartwrightii.*

Sheaths without herbaceous tips.....................9. *P. coccineum.*

Spikes several; annuals, preferring a rich, moist habitat.

Peduncles copiously glandular-pubescent; spikes erect; stamens 8; achenes 2.2-3.5 mm wide.

Leaves copiously strigose-pubescent beneath and often above; achenes mostly 2.2-2.8 mm wide......10. *P. pennsylvanicum* var. *genuinum.*

Leaves glabrous or glabrescent; achenes mostly 2.5-3.5 mm wide.

Glands of hairs red.........10a. *P. pennsylvanicum* var. *laevigatum.*

Glands of hairs without pigment................................ ....10b. *P. pennsylvanicum* var. *laevigatum* f. *pallescens.*

Peduncles without stalked glands, smooth or with sessile glands, rarely with a few stipitate glands; spikes drooping or erect; stamens 8; achenes 1.5-2.5 mm wide.

Lower surface of leaves glabrous or scabrous on the principal veins; peduncles glabrous or rarely covered more or less with sessile glands; spikes 3-8 cm long, drooping; achenes generally less than 2 mm wide...............................11. *P. lapathifolium.*

Lower surface of leaves (at least the lower ones) scurfy or covered with a more or less deciduous, flocculent tomentum; peduncles with sessile glands; spikes 1-3 cm long, erect; achenes more than 2 mm wide. (See excluded species no. 204, p. 1043.)..*P. tomentosum.*

Sheaths ciliate with a row of bristles.

Stem and peduncles glandular-hispid.......................12. *P. Careyi.*

Stem and peduncles not glandular-hispid.

Sepals glandular-dotted.

Achenes dull, generally triangular; spikes usually strongly arched, the flowers not far apart except toward the base of the spike, often 1 or more flowers in the axil of the next to the top leaf; flowers greenish, generally with pinkish borders; stems often reddish, the internodes short, generally 2-4 cm long; stamens 6.

Pedicels strongly exserted from the ocreolae; achenes 2-3 mm long... ....13. *P. Hydropiper* var. *projectum.*

Pedicels not strongly exserted from the ocreolae; achenes mostly 3-3.5 mm long. (See excluded species no. 200, p. 1042.)...... ....*P. Hydropiper.*

Achenes shining, generally triangular; spikes elongated, flexuous, very loosely flowered down to the first leaf but none below it; flowers greenish, rarely purplish, with white borders; stamens 3-8; stems with longer internodes than in the preceding, usually 3-8 cm long...14. *P. punctatum.*

Sepals not glandular-dotted or with only a few glands about the middle of the perianth in forms of no. 16.

Leaves lanceolate, 1-2.5 cm wide; spikes generally much less than 1 cm wide, erect or slightly flexuous; flowers 2-3 mm long.

Upper part of internodes of the stem mostly entirely glabrous; spikes erect, mostly more than 7 mm wide, the longest usually 2-4 cm long; flowers generally close together; calyx lobes rose color, usually slightly longer than the achenes; pedicels generally exserted less than 1 mm; stamens 6............... ....15. *P. Persicaria.*

Upper part of internodes of the stem generally more or less strigose below the node, usually for a third of its length; spikes generally more or less curved, mostly less than 7 mm wide, the longest generally 5-7 cm long; flowers not crowded; calyx lobes much longer than the achene, usually pink; pedicels generally exserted 1-2 mm; stamens 8 or fewer.

Achenes all triangular..................16. *P. hydropiperoides.*

Achenes both lenticular and triangular. (See excluded species no. 201, p. 1043.)....*P. hydropiperoides* var. *persicarioides.*

Leaves broadly ovate, acuminate, more than 3 cm wide; spikes stout, usually more than 1 cm wide, nodding, deep rose; flowers 3-5 mm long....................................17. *P. orientale.*

Styles long, stiff, exserted, persistent, deflexed, and hooked at the tip in fruit; spikes very long and slender, rigid, greenish; leaves large-ovate and acuminate....................................18. *P. virginianum.*

B. Stems armed with hooked prickles, reclining.

Leaves hastate; pedicels hispid and often glandular; achenes lenticular...19. *P. arifolium* var. *lentiforme.*

Leaves sagittate; pedicels generally glabrous; achenes triangular...20. *P. sagittatum.*

A. Plants twining; leaves broadly ovate, cordate at the base.

Calyx not keeled or winged in fruit; achenes dull, minutely longitudinally striate, about 3 mm long....................................21. *P. Convolvulus.*

Calyx strongly winged in mature fruit; achenes shining, surface not striate.

Mature calyx 5-8 mm long; achenes 2.5-3 mm long............22. *P. dumetorum.*

Mature calyx 7-12 mm long; achenes mostly 3.5-5 mm long......23. *P. scandens.*

1. **Polygonum exsértum** Small. Map 832. Very local. All of our specimens were found in hard, dry soil on the washed slopes of the banks of streams and sloughs. Some were very near the water and only one grew on the top of the bank. Bicknell (Bull. Torrey Bot. Club. **36**: 450. 1909.) says: "I am unable to see that *P. exsertum* is anything more than a semi-viviparous state of *Polygonum ramosissimum* Michx." My observation is that this character applies to late flowering plants of all of the species of the Section *Avicularia* which occur in Indiana. On November 14, 1932, I studied in the field several large mats of *Polygonum aviculare,* and I was able to find only exserted achenes. A study of my herbarium material showed exserted achenes on all of the plants collected late in the fall, some with a few and some with a great number of exserted achenes. Early flowering specimens of *Polygonum exsertum* show a large number of achenes of the normal form, while plants collected in September usually have few or no normal fruits.

N. B. to Minn., southw. to N. J. and Mo.

2. **Polygonum eréctum** L. Map 833. This species has been reported from all parts of the state, and no doubt is generally distributed. Since this section of the genus has been divided, however, some of the reports doubtless belong to other species. Most authors give the habitat as rich soil about dwellings and in waste places. With one exception, all of my specimens were found in moist, open woodland, usually in hard, clay soil.

Ont. to Alberta, southw. to Ga., Colo., and Tex.

3. POLYGONUM MONSPELIÉNSE Thiebaud. (?*Polygonum aviculare* var. *vegetum* of Gray, Man., ed. 7.) Map 834. My specimens are from barnyards, waste places, and roadsides. It is local, but no doubt when the

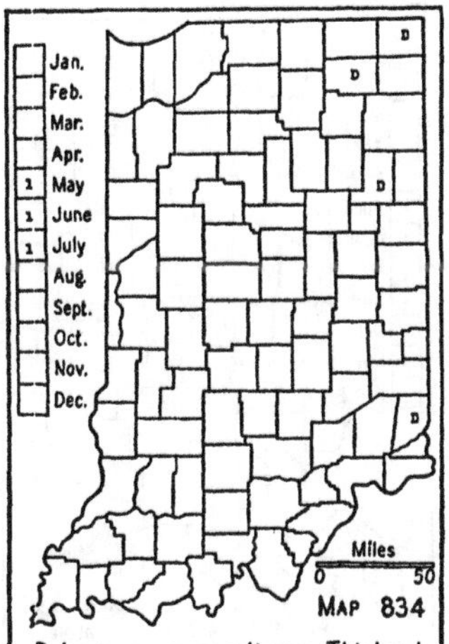

Polygonum monspeliense Thiebaud

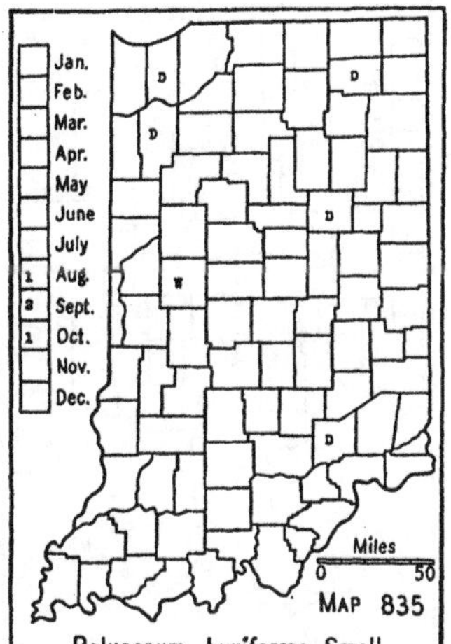

Polygonum buxiforme Small

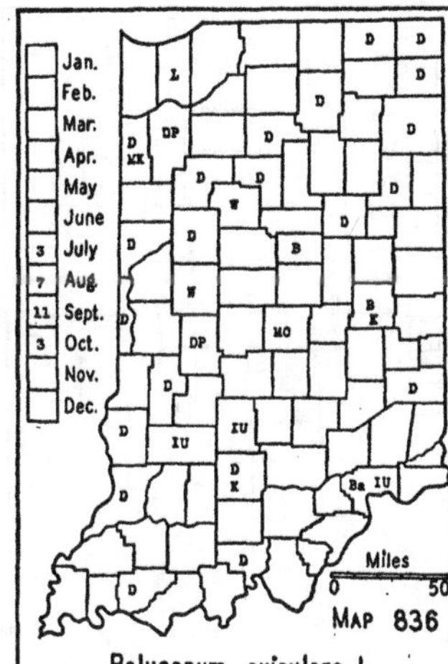

Polygonum aviculare L.

knotweeds are more thoroughly studied it will be found throughout the state.

Nat. of Eu.; becoming naturalized.

4. **Polygonum buxifórme** Small. Map 835. This species is very local but I believe when the knotweeds are studied more intensively it will be found throughout the state.

Ont. to B. C., southw. to Va. and Tex.

5. **Polygonum aviculàre** L. KNOTWEED. Map 836. An annoying weed in gardens, truck gardens, lawns, pastures, and cultivated fields. It is found, also, along logging roads in woodland, in fallow fields, and along roadsides.

Found throughout N. A. and also in Eurasia.

6. POLYGONUM NEGLÉCTUM Besser. (*Polygonum aviculare* var. *angustissimum* Meisn.) Map 837. Local. No doubt a more intensive study of the knotweeds will greatly extend its range. In sandy to very sandy soil in pastures, clearings, on black oak ridges, and along roadsides. Rydberg gives its habitat as waste places and says it is more common than *Polygonum aviculare.*

Nat. of Eu.

7. **Polygonum ténue** Michx. Map 838. This species prefers a slightly acid soil and is generally found in exposed places without ground cover and where there are very few or no other plants. In the lake area it is generally found on the crests, slopes, and bases of black and white oak ridges. South of the lake area it is generally found on sandstone bluffs, on exposed crests of chestnut oak ridges, and in sandy places similar to those in the northern part of the state.

Maine to Man., southw. to S. C., Ga., and Tex.

8. **Polygonum nàtans** A. Eaton f. **genuìnum** Stanford. (Stanford. The amphibious group of Polygonum, subgenus Persicaria. Rhodora **27**: 156-166. 1925.) Map 839. All of our reports for *Persicaria amphibia* (L.)

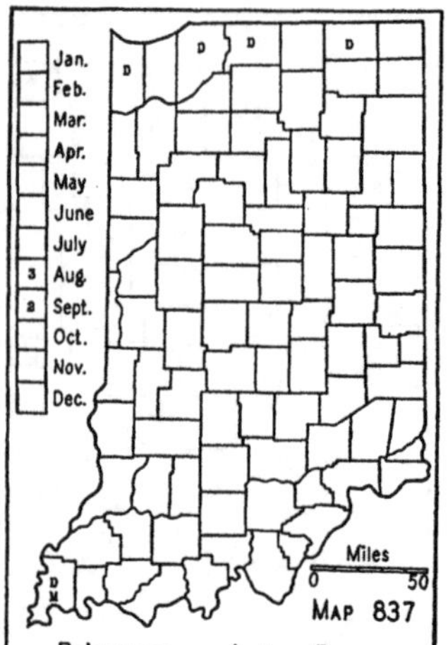

Polygonum neglectum Besser

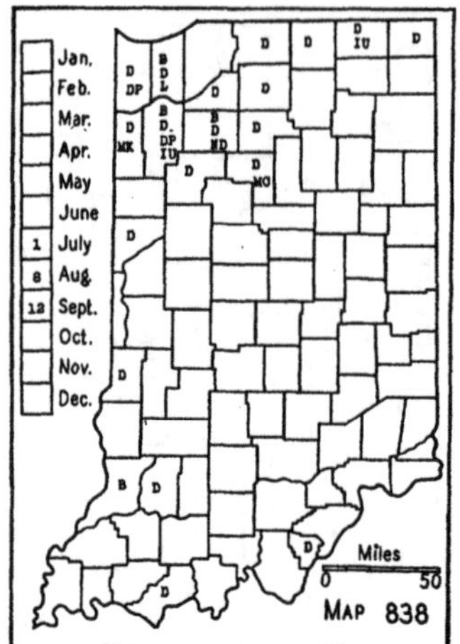

Polygonum tenue Michx.

Polygonum natans A. Eaton f. genuinum. Stanford

S. F. Gray, *Persicaria fluitans* (Eaton) Greene, *Polygonum amphibium* L., and *Polygonum amphibium* var. *aquaticum* Willd. I am referring to this species. The nomenclature of this and the next species has long been so involved that their distribution in the state can not be determined from the published records. It is, no doubt, restricted to the lake area of the state.

Newf., P. E. I., N. S., Que., southw. to Pa. and westw. across the continent to the Pacific Coast States.

8a. **Polygonum natans f. Hartwrightii** (Gray) Stanford. Map 840. I am referring to this form all of our reports for *Polygonum amphibium* var. *Hartwrightii* (Gray) Bissell, *Persicaria ammophila* Greene, *Persicaria carictorum* Nieuwl., and *Persicaria Hartwrightii* (Gray) Greene.

Mostly in sedge marshes and on the borders of lakes.

Newf. and Ont., southw. to N. Y., and westw. to the Pacific Coast States.

9. **Polygonum coccíneum** Muhl. Map 841. This species is an aggregate to which, since I am not able to separate it satisfactorily into forms and varieties, I am referring all reports from Indiana of the following: *Persicaria coccinea* (Muhl.) Greene, *Persicaria coccinea* var. *asprella* Greene, *Persicaria coccinea* var. *tanaophylla* Nieuwl., *Persicaria emersa* (Michx.) Small, *Persicaria grandifolia* Greene, *Persicaria lonchophylla* Greene, *Persicaria mesochora* var. *arenicola* Nieuwl., *Persicaria Muhlenbergii* (Wats.) Small, *Persicaria pratincola* Greene, *Persicaria tanaophylla* Nieuwl., *Polygonum coccineum* var. *pratincola* (Greene) Stanford, *Polygonum emersum* (Michx.) Britt., *Persicaria mesochora* Greene, and *Polygonum Muhlenbergii* (Meisn.) Wats.

The named variations of this species and the segregates from it are based mostly upon leaf characters, such as the general shape and base of the blades. Using these characters, I have one specimen which belongs to three species. I have a series of specimens of this species all from the same rootstock which might be referred to different species. The species is perennial. One year it may be in deep water, the next year it may be in

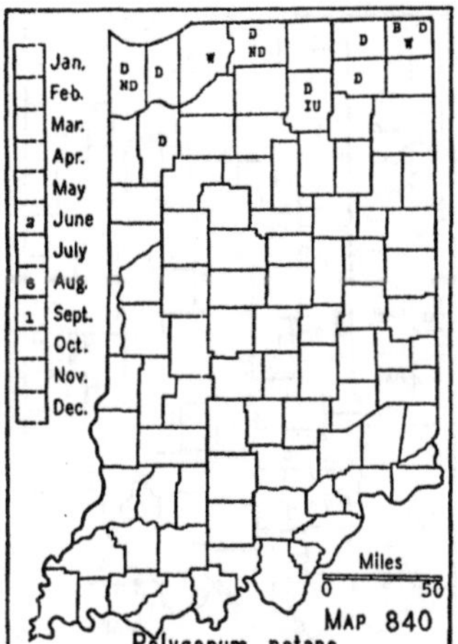

MAP 840
Polygonum natans f. Hartwrightii (Gray) Stanford

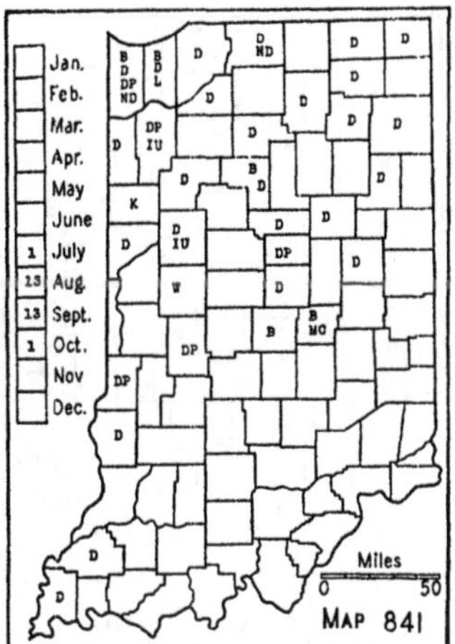

MAP 841
Polygonum coccineum Muhl.

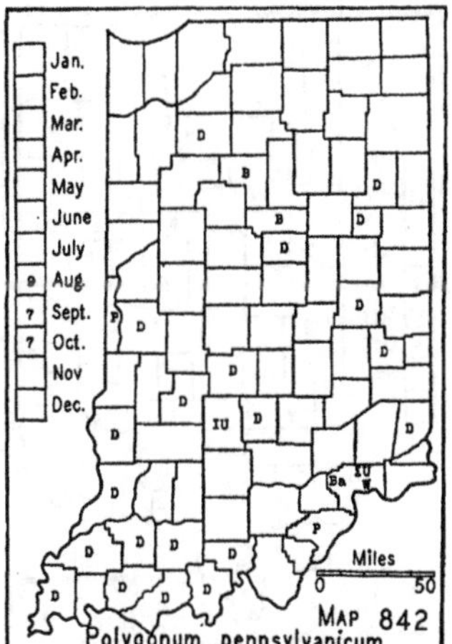

MAP 842
Polygonum pennsylvanicum var. genuinum Fern.

shallow water or for part of the year it may be on dry ground. The species has great ability to persist even when its habitat is drained, and it often advances from ditches along railroads up the banks of the fills to high ground where it seems to thrive better than in a wet habitat. The habitat and the vigor of the plants greatly change the character of the leaves. Therefore, I believe it is useless to try to name all of the many forms.

Que. and Maine, to B. C., southw. to Va., La., Calif., and Mex.

10. **Polygonum pennsylvánicum** L. var. **genuìnum** Fern. (*Persicaria pennsylvanica* (L.) Small, in part.) (Fernald. Variations of Polygonum pennsylvanicum. Rhodora **19**: 70-73. 1917, and Stanford. Polygonum pennsylvanicum and related species. Rhodora **27**: 173-184. 1925.) Map 842. Infrequent to frequent or common in low ground along streams and roadsides, in cultivated grounds, and in low grounds in general. No doubt it is found throughout the state. It has been my method to collect a single specimen of each species from each county. This species has been divided only recently and most of my collecting was done before the division was made. Since my specimens are now distributed among the three present groups, the absence of records from the northern part of the state is, I think, accidental.

This species, as well as others of the genus, varies greatly in size, depending upon habitat and date of germination of the seed. Apparently the seed do not germinate under water and when they find lodgment in areas which are submerged until summer, the delayed germination, no doubt, accounts for the smaller plants. The largest one of which I have record is my specimen no. 39887 from low ground in Gibson County which I measured in the field. The height was 86 inches above the ground and the longest branch was 82 inches long.

Coastal Plain from Mass. to Miss., northw. through the Mississippi Valley to Ont. and cent. N. Y.

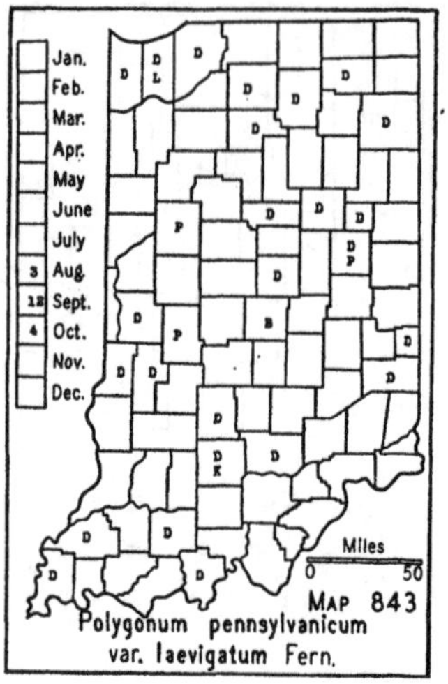

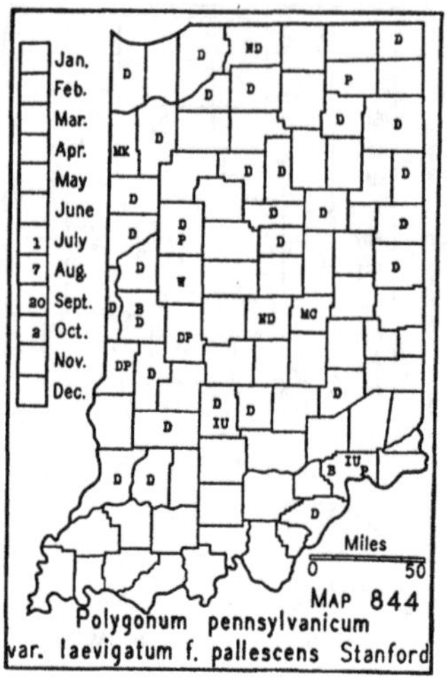

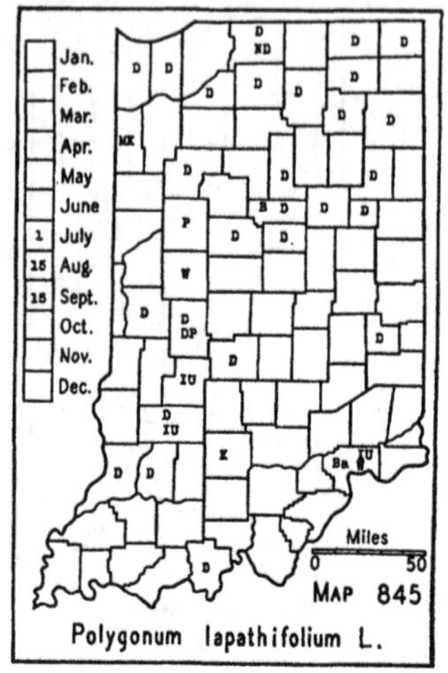

10a. **Polygonum pennsylvanicum** var. **laevigàtum** Fern. (See species references.) Map 843. Frequent throughout the state in habitats similar to those of the species.

N. B. to S. Dak. and Colo., southw. to Fla. and Tex.

10b. **Polygonum pennsylvanicum** var. **laevigatum** f. **palléscens** Stanford. (See species references.) Map 844. Frequent throughout the state in habitats similar to those of the species. It is probable that some of the specimens referred to this form belong to the preceding variety since it is difficult to distinguish this form in dried specimens.

Distribution given by Stanford is Vt. to Pa. No doubt it is frequent throughout Ind. if I understand the form.

11. **Polygonum lapathifòlium** L. (*Persicaria lapathifolia* (L.) Small.) Map 845. Frequent in low and wet grounds throughout the state, preferring the low borders of streams. It is also found in cultivated and fallow fields.

Throughout temperate N. A. and also in Eurasia.

12. **Polygonum Càreyi** Olney. (*Persicaria Careyi* (Olney) Greene.) CAREY SMARTWEED. Map 846. Very local but usually common where it is found. It prefers a black, sandy soil in pin oak and low black and white oak woods. I found it abundant in black, mucky soil in a fallow field north of Ora in Starke County. The plants are usually about a yard high with few or many branches.

This species was reported from Jefferson County by Young, but since neither Coulter nor Barnes mention it in their lists of Jefferson County plants, this report may be safely ignored. It has also been reported from Kosciusko and Noble Counties. These reports, no doubt, are correct.

Maine, Ont., and Mich., southw. to N. J., Pa., and Ohio.

13. **Polygonum Hydrópiper** L. var. **projéctum** Stanford. (*Polygonum Hydropiper* L. in part, and *Persicaria Hydropiper* (L.) Opiz.) (Stanford. Polygonum Hydropiper in Europe and North America. Rhodora 29: 77-87.

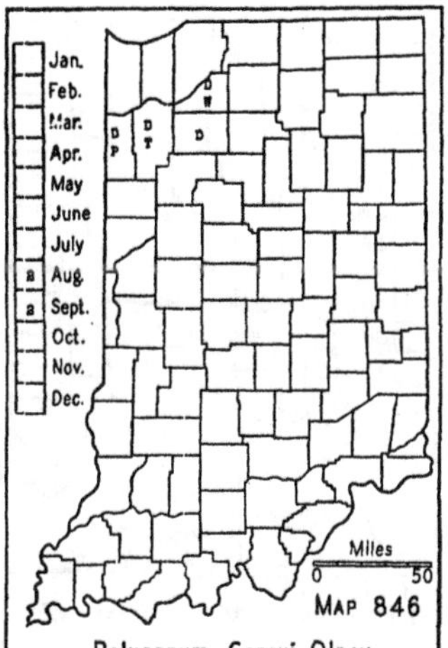

Polygonum Careyi Olney

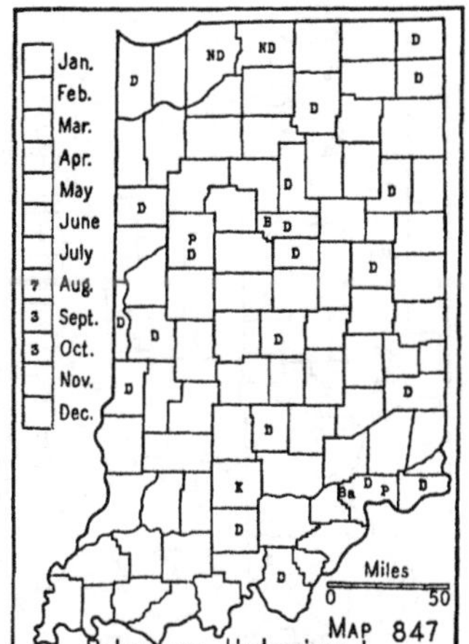

Polygonum Hydropiper L. var. projectum Stanford

Polygonum punctatum Ell.

1927.) WATER PEPPER. Map 847. Infrequent to frequent in moist soil along streams, roadsides, and ditches, about lakes, ponds, and sloughs, and in low ground in fields and woodland.

N. S. and Que. to Wis., southw. to Ga. and Okla., and westw. to Calif.

14. **Polygonum punctàtum** Ell. (*Polygonum acre* HBK. and var. *leptostachyum* Meisn. and *Persicaria punctata* (Ell.) Small.) (Stanford. Polygonum Hydropiper in Europe and North America. Rhodora 29: 77-87. 1927.) WATER SMARTWEED. Map 848. Frequent to common in all parts of the state in habitats similar to those of the preceding species.

Probably throughout N. A. except the extreme north.

15. POLYGONUM PERSICÀRIA L. (*Persicaria Persicaria* (L.) Small.) LADY'S THUMB. Map 849. Frequent throughout the state in wet ground along roadsides and streams and in woodland and fallow fields. This species begins to flower much earlier than *P. hydropiperoides.* It and others of the genus are the source of smartweed honey.

Nat. of Eu.; throughout N. A. except the extreme north.

16. **Polygonum hydropiperoìdes** Michx. (*Persicaria hydropiperoides* (Michx.) Small.) (Stanford. Polygonum hydropiperoides and P. opelousanum. Rhodora 28: 22-29. 1926.) MILD WATER PEPPER. Map 850. Frequent throughout the state in dried-up ponds and sloughs, in wet ground along streams and about lakes, and in marshes and ditches.

N. S., Que., and Minn., southw. to Fla. and Tex.

16a. **Polygonum hydropiperoides** var. **strigòsum** (Small) Stanford. This variety was reported from Indiana by Small. It is separated from the species by having a strigose-pubescent stem. The stems of the specimens at hand vary from glabrous below the nodes to densely strigose for a third of the length of the internode. One branch of a specimen may have all of the internodes glabrous and another have some of the internodes strigose below the nodes. Since a close lineal series from glabrous to

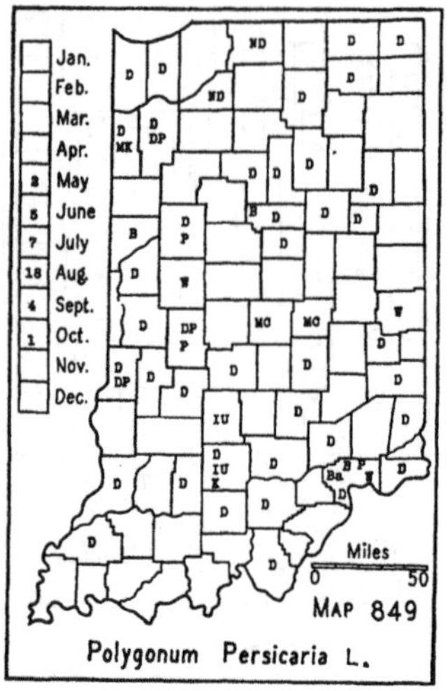

MAP 849
Polygonum Persicaria L.

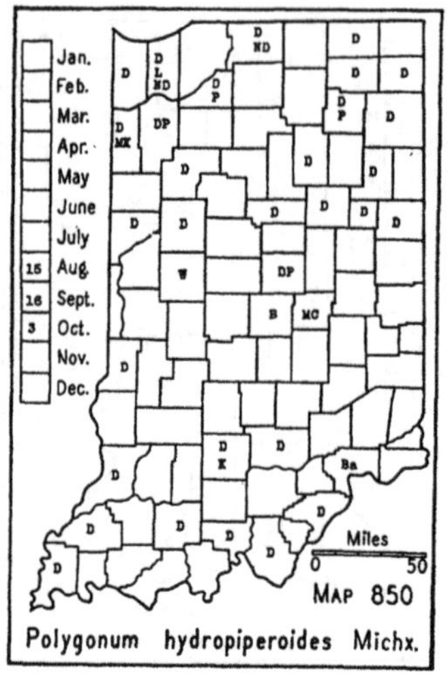

MAP 850
Polygonum hydropiperoides Michx.

MAP 851
Polygonum orientale L.

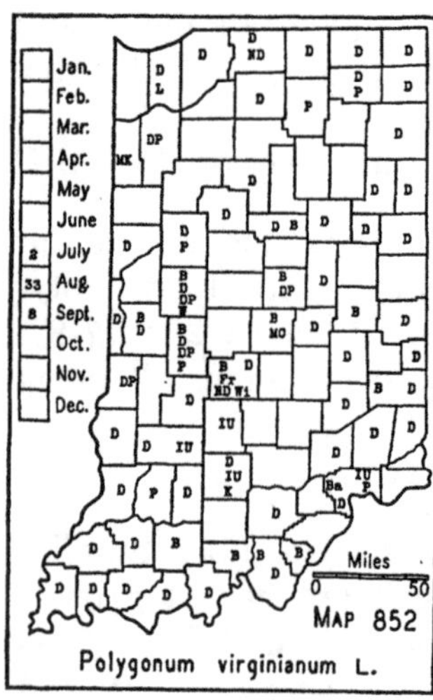

MAP 852
Polygonum virginianum L.

MAP 853
Polygonum arifolium L.
var. lentiforme Fernald & Griscom

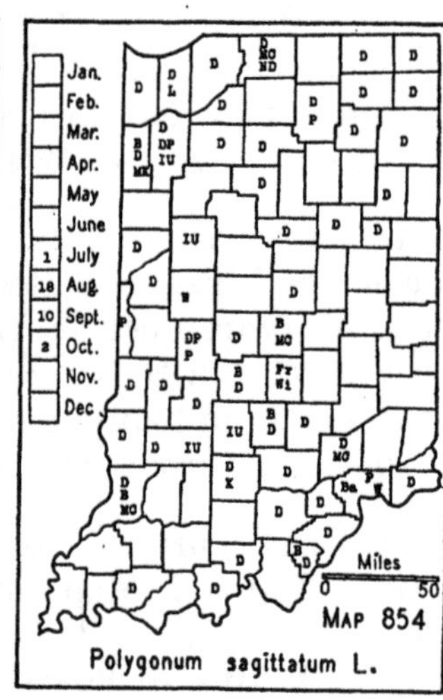

MAP 854
Polygonum sagittatum L.

densely strigose can be found, I prefer to say that the species varies from glabrous to densely strigose.

Polygonum setàceum Bald. var. **interjéctum** Fern., a closely related species, has been reported by Fernald (Rhodora **40**: 414. 1938), after the manuscript of the Flora was written, as having been found by Peattie under buttonbush at the edge of a *Chamaedaphne* bog near Rolling Prairie, La Porte County.

17. POLYGONUM ORIENTÀLE L. PRINCES-PLUME. Map 851. This species is cultivated as an ornamental and has been reported as an escape throughout the state.

Nat. of India, China, Japan; naturalized and escaped throughout eastern N. A.

18. **Polygonum virginiànum** L. (*Tovara virginiana* (L.) Raf.) VIRGINIA KNOTWEED. Map 852. This is strictly a woodland species and is

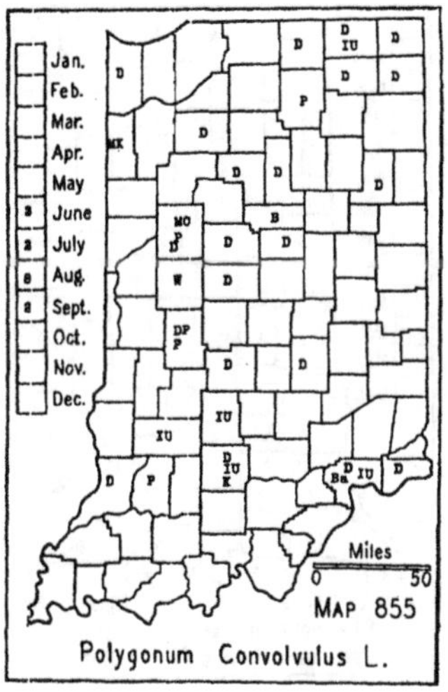

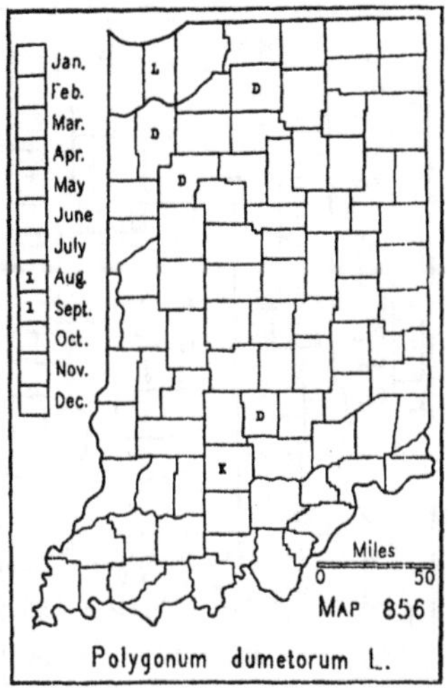

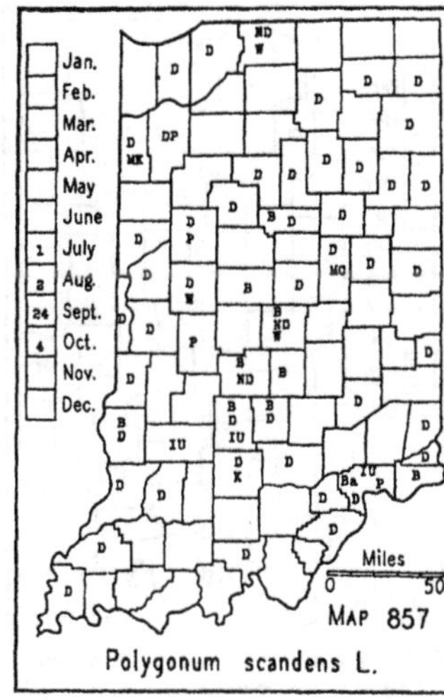

frequent throughout the state in low places in almost all types of woods.

N. S. to Minn., southw. to Fla. and Tex.

19. **Polygonum arifòlium** L. var. **lentifórme** Fern. & Grisc. (Rhodora **37**: 167. 1935.) (*Polygonum arifolium* L. in part and *Tracaulon arifolium* (L.) Raf.) HALBERDLEAF TEARTHUMB. Map 853. Infrequent to rare in springy and swampy places throughout the state. This species is much visited by honey bees.

P. E. I. to s. Ont., southw. to N. J., Pa., Ohio, Ind., and Mich.

20. **Polygonum sagittàtum** L. (*Tracaulon sagittatum* (L.) Small.) ARROWLEAF TEARTHUMB. Map 854. Frequent to infrequent throughout the state in ditches, in low ground in wooded ravines and along streams and about ponds and swamps.

Newf. to Sask., southw. to Fla. and Tex.

21. POLYGONUM CONVÓLVULUS L. (*Tiniaria Convolvulus* (L.) Webb & Moquin.) BLACK BINDWEED. Map 855. Probably infrequent in all parts of the state, although there are no reports from the southwestern part. A weed mostly of roadsides and fields, and rarely in woodland.

Nat. of Eu. Throughout temperate N. A.

22. **Polygonum dumetòrum** L. (*Tiniaria dumetorum* (L.) Opiz of Britton and Brown, Illus. Flora, ed. 2.) Map 856. This and the next species are not easily separated unless mature fruits are at hand. Some authors believe this species is a native, while others regard it as a native of Eurasia. There have been 15 reports for this species from Indiana. Some authors do not discuss it and some say that it is common. Those who say it is common have, no doubt, confused it with *P. scandens,* and I believe most of our reports should be referred to that species. One of our specimens is from a roadside and the other is from the low border of the east side of the Lake of the Woods, which is near a roadside in Mar-

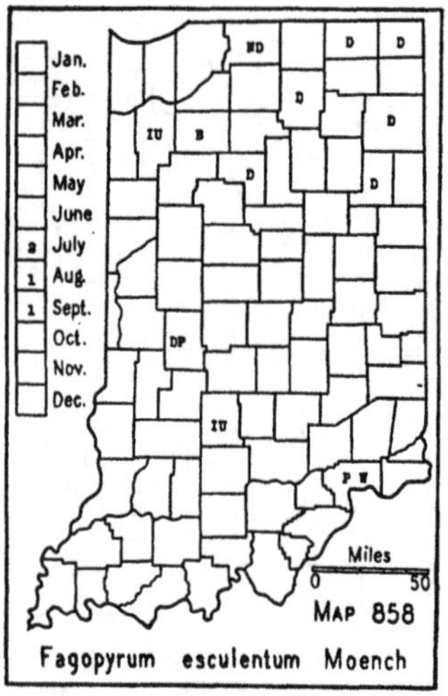

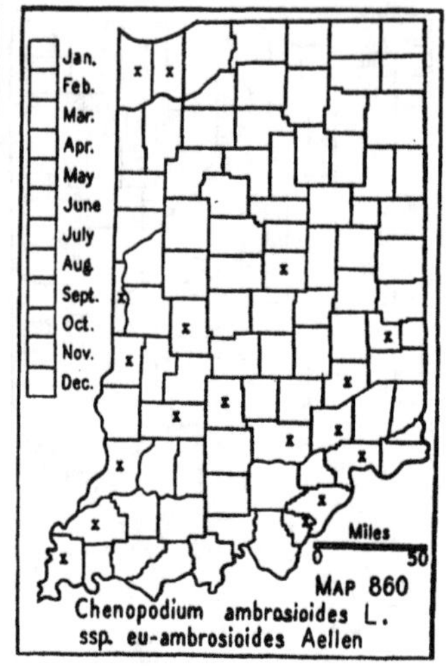

shall County. I doubt if this species is distinct from the next but I am following authors in keeping them distinct.

Temperate Eurasia and N. A.

23. **Polygonum scándens** L. (*Tiniaria scandens* (L.) Small.) CLIMBING FALSE BUCKWHEAT. Map 857. Frequent in most parts of the state in moist soil along roadsides and streams, in wooded ravines, and about lakes and ponds.

N. S. to Ont. and B. C., southw. to Fla. and Tex.

2202. FAGOPÝRUM [Tourn.] Gaertn.

1. FAGOPYRUM ESCULÉNTUM Moench. (*Fagopyrum Fagopyrum* (L.) Karst. of Britton and Brown, Illus. Flora, ed. 2.) BUCKWHEAT. Map 858. Buckwheat has been reported from 15 counties. It persists in fields where it has been cultivated or escapes to fields, roadsides, and railroads. I do not know how long it will maintain itself.

Nat. of Eu.

2203. POLYGONÉLLA Michx.

1. **Polygonella articulàta** (L.) Meisn. Map 859. Local on the dunes about Lake Michigan. It is usually found in open, exposed places.

In sands of the coast from Maine to Fla. and about the Great Lakes.

78. CHENOPODIÀCEAE Dumort. GOOSEFOOT FAMILY*

[Iljin, M. Chenopodiaceae, pp. 2-354, in Komarov, V. L. Flora URSS 6 (Centrospermae): xxxvi + 956p. 1936. Standley, P. C. Chenopodiales,

* Text contributed by Theodor Just, University of Notre Dame, Notre Dame, Indiana. The author is greatly indebted to Mr. Paul Aellen, Basel, Switzerland; to Dr. Paul C. Standley, Field Museum, Chicago, Ill.; and to Mr. C. A. Weatherby, Gray Herbarium, Cambridge, Mass., for reading his manuscript and for offering valuable criticisms.

Chenopodiaceae. North American Flora 21(1): 1-93. 1916. Ulbrich, E. Chenopodiaceae. In Engler und Prantl, Die natürlichen Pflanzenfamilien, 2. ed., **16c**: 377-584. 1934.]

Flowers perfect (or some of them pistillate); perianth mostly present.

Stem not jointed; leaves flat, not spiny; flowers without bractlets; embryo annular (or conduplicate), not spirally coiled; endosperm copious.

Flowers in clusters or panicles; calyx 3-5-toothed or -parted, obvious, persistent; fruit enclosed by or not longer than the calyx.

Fruiting calyx wingless, herbaceous, green or reddish (sometimes red and fleshy); perianth leaves free, naked; fruit free, surrounded by perianth, not hardened, indehiscent; leaves often mealy, lanceolate to ovate or deltoid or pinnately lobed to pinnate; flowers with (2) 3-5 sepals and 2-5 stamens, mostly in panicled spikes; endosperm mealy....2223. CHENOPODIUM, p. 419.

Fruiting calyx 5-cleft, horizontally winged.

Flowers paniculate; perianth keeled, developing into a broad horizontal wing at maturity; endosperm mealy; leaves ovate to lanceolate, flat, sinuate-dentate, 2-7 cm long; annuals, to 80 cm high, branches divaricate...... ....2224. CYCLOLOMA, p. 424.

Flowers spicate; each sepal with a dorsal winglike projection; endosperm absent; leaves linear or lance-linear, terete, entire..2240. KOCHIA, p. 426.

Flowers solitary in the axils of the reduced upper leaves, forming terminal narrow spikes, with 1-3 thin broad sepals; fruit oval, laterally flattened, distinctly winged (wing 0.5 mm long or more), much larger than the calyx; pericarp membranous, adherent to the vertical seed; leaves linear; caulescent annuals.. ....2245. CORISPERMUM, p. 426.

Stem jointed, fleshy, squarrosely branched; leaves fleshy, narrow, subulate, spiny-tipped; flowers 1-3 in axils of leaves, subtended by bractlets; stamens 5, free; calyx in fruit with a horizontal wing; embryo spirally and conically coiled; endosperm none; stems striate........................2269. SALSOLA, p. 427.

Flowers imperfect (unisexual, monoecious or dioecious), pistillate flower without perianth; stigmas 2 or 3, enclosed in 2 appressed triangular bractlets, these compressed, free at least above; fruiting bracts with margins often dilated and sides often muricate; utricles not winged at apex; testa coriaceous; staminate flowers in clusters, mostly spicate; calyx 3-5-parted; embryo annular; leaves lanceolate to hastate-ovate...2229. ATRIPLEX, p. 425.

2223. CHENOPÒDIUM [Tourn.] L. PIGWEED, GOOSEFOOT*

[Aellen, P. Neue adventive Chenopodien aus Schweden. Bot. Not. (Lund) **1928**: 203-210. 1928. Beitrag zur Systematik der Chenopodium-Arten Amerikas, vorwiegend auf Grund der Sammlung des United States National Museums in Washington, D. C. I. Rep. spec. nov. regn. veget. **26**: 31-64; II. loc. cit. **26**: 119-160. 1929. Die wolladventiven Chenopodien

* The satisfactory identification of the species of Chenopodium is definitely dependent upon the characteristics of the mature seeds. Consequently specimens bearing such should be collected as well as others with cauline leaves. For illustrations of the characteristics of the seeds of certain species consult especially Iljin, plate 3 facing page 56 *(C. Botrys, urbicum, hybridum, murale, album)* and Aellen, Bot. Not. **1928**: 207 *(C. missouriense)*.

The distribution of certain species is known only from a few authentic specimens whereas reports of critical groups have been discarded entirely until a more detailed study now in preparation can appear. Future collections will undoubtedly extend the range of most species and add others new to the state.

Europas. Verh. Naturf. Ges. Basel **41**: 77-104. **1930**. Nomenklatorische Bemerkungen zu einigen Chenopodien. Ostenia (Festschr. für Cornelius Osten), Montevideo, **1933**: 98-101. **1933**.]

Plants with glandular pubescence, more or less aromatic; embryo an incomplete ring.

Flowers glomerate, without pubescence; glomerules in bracteate or almost naked spikes (continuous or interrupted); perianth more or less fused; stigmas 3 or 4. Sect. *Ambrina* (Spach) Hook. f.

Spikes mostly leafy; calyx lobes slightly keeled; seed mostly horizontal, reddish brown, about 0.5 mm, with prominent wavy lines; leaves 4-18 cm long, lanceolate, coarsely toothed..........1. *C. ambrosioides* ssp. *eu-ambrosioides*.

Spikes mostly leafless, more or less elongated; calyx lobes not keeled; seed to 0.8 mm...........1a *C. ambrosioides* ssp. *eu-ambrosioides* var. *anthelminticum*.

Flowers solitary, strongly glandular pubescent, sessile in open divaricate cymes, these in loose panicles; perianth fused only in lower part; stigmas 2; seed horizontal or vertical, dark brown, 0.5-0.7 mm; leaves ovate or oblong, pinnately lobed to pinnate, lobes or leaves angled, obtuse. Sect. *Botryoides* C. A. Mey..2. *C. Botrys*.

Plants not glandular or aromatic, sometimes with a rank or heavy odor; pubescence frequently more or less mealy; embryo a complete ring.

Seeds vertical or the terminal ones occasionally horizontal.

Seeds vertical (rarely horizontal); styles filiform, one fourth to half as long as the diameter of the utricle.

Flowers in glomerules (10 mm in diam.), densely capitate; calyx very fleshy and bright red, becoming red (crimson) and berrylike in fruit; seed with slight margin, blackish brown, dull, 0.8 mm; leaves at base hastate. Sect. *Eublitum* Moq...3. *C. capitatum*.

Flowers spicate, not succulent in fruit; perianth fused to nearly the middle; stigmas 2 or 3, very long; seed almost spherical, erect (rarely horizontal), with rounded margin, blackish brown, almost smooth, 1.5 mm; leaves bright green, 5-12 cm long, triangular-hastate and acute, almost entire; perennial. Sect. *Agathophyton* (Moq.) Hook. f. (See excluded species no. 205, p. 1043.) ....*C. Bonus-Henricus*.

Seed vertical and horizontal in the same inflorescence; flowers sessile, densely glomerulate, the glomerules axillary and terminal, simple or paniculate; perianth with (3) 4 or 5 separate tips, these incompletely enclosing the fruit and not keeled; stigmas short; seed dark brown, shiny, margin slightly keeled, almost smooth, 0.6 mm; leaves on short petioles, pale, white-mealy beneath, oblong to oblong-ovate, mostly obtuse, cuneate at base, 1-4 cm long, coarsely toothed; annuals. Sect. *Pseudoblitum* Hook. f............................ ....4. *C. glaucum* ssp. *eu-glaucum*.

Seeds all horizontal; style branches short; perianth 5-tipped, herbaceous, green, fused to a varying degree, mostly keeled, sometimes winged; stigmas 2. Sect. *Chenopodia* C. A. Mey.

Seeds with characteristic alveolar depressions, black.

Seed 1 mm in diam., shiny, with surface markings less prominent towards margin; tips of perianth with a narrow, strongly winged keel; leaves small, ovate-deltoid, the strongest tooth at about the middle of the leaf, rounded at either end, upper leaves cuspidate..........5. *C. Berlandieri* ssp. *Zschackei*.

Seed larger (2 mm max.), flat, with small, narrow radial canals, often prominently developed; flowers densely glomerate, in loose foliaceous spikes; perianth tips fused to middle.

Leaves with 1 or 2 teeth in the lower half, otherwise irregularly toothed, large (7 x 4 cm max.), broadly deltoid, acuminate, mucronate..6. *C. Bushianum*.

Leaves with more teeth in the lower half, the teeth more acute............. ....6a. *C. Bushianum* f. *acutidentatum*.

Seeds with other markings.

Leaves mealy.

Leaves not entire, sinuately dentate; inflorescence dense; seed rugose-punctate, 1.5 mm in diam., black, shiny, lenticular, with acute margin; calyx lobes keeled and enclosing the fruit; plants up to 6 ft. tall..........7. *C. album.*

Leaves mostly entire.

Leaves linear or nearly so, very mealy at least beneath, with short petioles; seed black, shiny, punctate, asymmetrical; calyx lobes keeled, closely enveloping the fruit or erect; pericarp green or greenish...8. *C. pratericola.*

Leaves ovate, about as broad as long, small (1 cm long), on long petioles; seed lenticular, with rounded margin, wrinkled and finely punctate, about 1 mm in diam.; pericarp adherent; plants very fetid, densely mealy...9. *C. Vulvaria.*

Leaves green or nearly so (except *C. missouriense* var. *Bushianum*).

Seed larger than in other species, 1.5-2 mm (-3 mm max. in var.) in diam., black, lenticular, margin more or less rounded, almost smooth or with radial canals of varying depth or slightly granulate or with narrow wrinkles; perianth tips slightly keeled, incompletely enclosing the fruit; leaves with large divaricate (2-4) acute lobes, rounded or somewhat cordate at base, 4-17 cm long, to 12 cm wide, 3-5-angular-ovate, acuminate.

Inflorescence panicled, loosely branched, leafless, and terminal...10. *C. gigantospermum.*

Inflorescence contracted, spicate......10a. *C. gigantospermum* f. *Griffithsii.*

Seeds smaller.

Pericarp not firmly attached to the seed.

Perianth tips not completely enclosing the fruit, slightly keeled; pericarp rust brown, somewhat fleshy..................11. *C. Standleyanum.*

Perianth tips completely enclosing the fruit, prominently keeled; pericarp yellow.

Leaves glabrous, 5 x 3 cm; inflorescence paniculate-glomerate...12. *C. missouriense.*

Leaves mealy beneath, mostly smaller; inflorescence glomerulate-cymose. ....12a. *C. missouriense* var. *Bushianum.*

Pericarp firmly attached to the seed.

Inflorescence short, spreading, axillary, rather loose, the panicles shorter than the leaves; leaves ovate or ovate-rhombic; seed shiny (appearing dull because of firmly attached pericarp), almost black, faintly punctate, with acute margin........................13. *C. murale.*

Inflorescence suberect, moniliform, flower clusters, at least the upper ones, longer than the leaves; seed shiny, brownish black, almost smooth, finely punctate, with rounded margin.

Leaves deltoid, more or less hastate, base truncate or subcordate...14. *C. urbicum.*

Leaves longer than broad (often twice as long), base long-cuneate. ....14a. *C. urbicum* var. *intermedium.*

1. CHENOPODIUM AMBROSIOÌDES L. ssp. EU-AMBROSIOÌDES Aellen. (Rep. spec. nov. regn. veget. **26**: 34. 1929.) (*C. ambrosioides* L. s. str.) MEXICAN TEA. Map 860. A highly polymorphic species of wide tropical distribution, but adventive in the temperate zones. Found mostly in gravelly and sandy soil.

Reported from: Fayette, Floyd, Gibson, Hamilton, Monroe, Porter, Posey, and Putnam Counties.

1a. CHENOPODIUM AMBROSIOIDES ssp. EU-AMBROSIOIDES var. ANTHELMÍNTICUM (L.) Aellen. (Rep. spec. nov. regn. veget. **26**: 35. 1929.) (*C. am-*

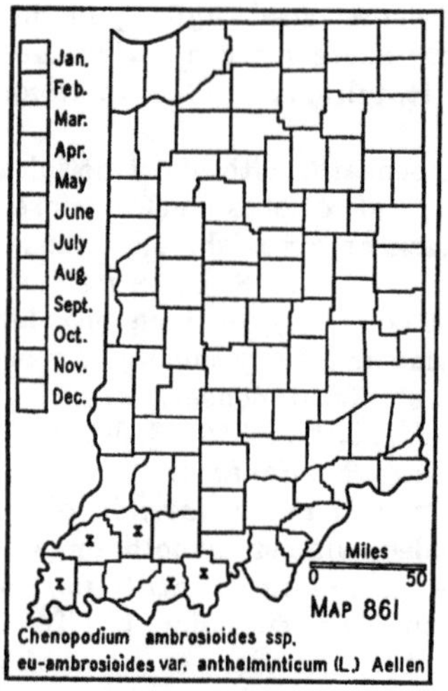

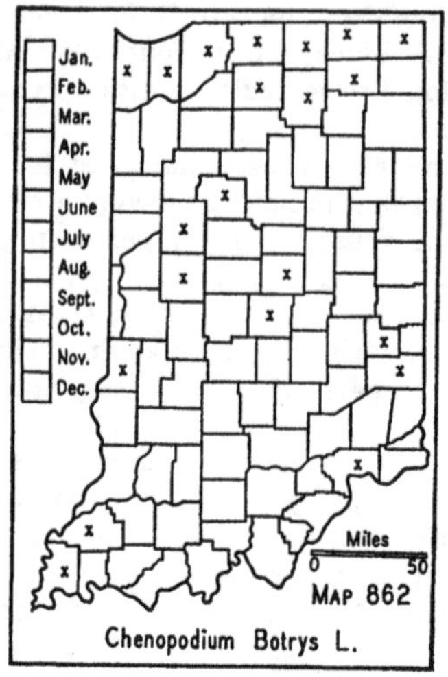

brosioides L. var. *anthelminticum* (L.) Gray.) MEXICAN TEA, STINKWEED, WORMSEED. Map 861. Specimens with fewer bracts have commonly been referred to this variety whose distribution is distinctly southern. It is probably much less common in the state than the subspecies.

2. CHENOPODIUM BÒTRYS L. FEATHER GERANIUM, JERUSALEM OAK. Map 862. Introduced in America. It grows on sandy hills, in open woods, and similar habitats.

Reported from: Fayette, Franklin, Gibson, Hamilton, Jefferson, Kosciusko, La Porte, Marion, Montgomery, Noble, Porter, Posey, St. Joseph, Tippecanoe, and Vigo Counties.

3. **Chenopodium capitàtum** (L.) Ascherson. (*Blitum capitatum* L.) STRAWBERRY BLITE, PIGWEED or SPINACH.

Reported from: Jefferson, Lake, St. Joseph, and Steuben Counties.

E. Que. to Alaska, southw. to N. J., Pa., Ill., Minn., and in the Rocky Mts. to Colo.

4. CHENOPODIUM GLAÚCUM L. ssp. EU-GLAÚCUM Aellen. (Rep. spec. nov. regn. veget. 26: 45. 1929.) OAKLEAVED GOOSEFOOT, GLAUCOUSLEAVED GOOSEFOOT. Map 863. The original occurrence of this species in America is not established as certain. Aellen, however, suggests that it is indigenous in salty places in Saskatchewan and Colorado.

Que. to Alberta, N. Mex. to Va., and Md.

Reported as *C. glaucum* L. from: Lake, Monroe, and Tipton Counties.

5. **Chenopodium Berlándieri** Moq. ssp. **Zschâckei** (Murr) Zobel. (*C. album* in part, of most American authors, not of L.) SOUTHERN WHITE PIGWEED, WOODLAND GOOSEFOOT, WOOD PIGWEED. Map 864. This species and the following are characterized by the peculiar alveolar depressions of their seeds, distinguishing them well from other speoies. The whole group however is quite polymorphic.

West of Mississippi River to Pacific coast, Canada, and Mexico, but apparently absent in the eastern U. S.

Reports discarded because of uncertainties involved.

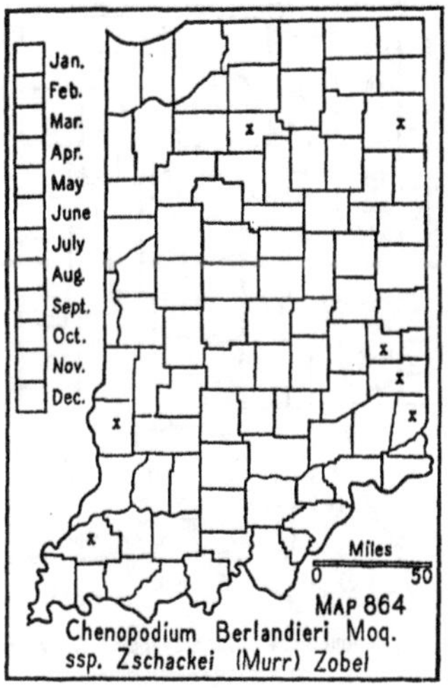

Chenopodium Berlandieri Moq. ssp. Zschackei (Murr) Zobel

Chenopodium Bushianum Aellen

Chenopodium album L.

6. **Chenopodium Bushiànum** Aellen (Rep. spec. nov. regn. veget. **26**: 63. 1929.) (*C. album* and *C. paganum* in part, of American authors, not L. or Reichenbach.) Map 865.

Allegheny region, St. Lawrence River Basin, Great Lakes, and Missouri River Basin from N. Dak. to Ark.

Reports discarded as in previous species and *C. album.*

6a. **Chenopodium Bushianum** f. **acutidentàtum** Aellen. (Rep. spec. nov. regn. veget. **26**: 119. 1929.) Aellen cites but one specimen from Indiana, Wells Co., which was collected in a truck garden.

7. **Chenopodium álbum** L. PIGWEED, LAMB'S QUARTERS, GOOSEFOOT. Map 866. Most American plants identified as *C. album* actually belong to *C. Berlandieri* ssp. *Zschackei* (Murr) Zobel. Some specifically American races, however, are found in addition to the truly European races introduced all over the world. The plants are found in sandy soils along roadsides.

8. **Chenopodium pratericola** Rydb. (*C. leptophyllum* Nutt. of most authors.) NARROWLEAF GOOSEFOOT. Map 867. Widely distributed west of the Mississippi River, eastward probably only introduced. Highly polymorphic. Found usually in sandy soil.

9. CHENOPODIUM VULVÀRIA L. STINKING GOOSEFOOT. Introduced in North America.

Reported from Monroe and Noble Counties.

10. **Chenopodium gigantospérmum** Aellen. (Rep. spec. nov. regn. veget. **26**: 144. 1929.) (*C. hybridum* of American authors, not L.) MAPLELEAVED GOOSEFOOT. Map 868. All specimens from America identified as *C. hybridum* L. should be referred to this species. Its more or less smooth seed with its relatively easily detachable perianth separates it clearly from the European plant. It occurs in sandy fallow fields, and open or moist woods.

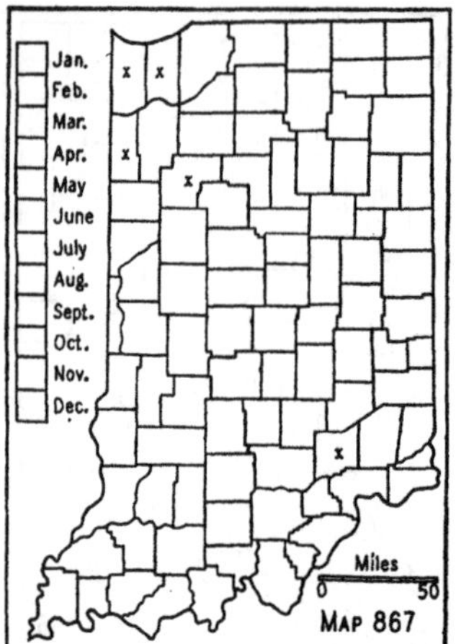

MAP 867
Chenopodium praiericola Rydb.

MAP 868
Chenopodium gigantospermum Aellen

MAP 869
Chenopodium Standleyanum Aellen

10a. **Chenopodium gigantospermum** f. **Griffíthsii** Aellen. (Rep. spec. nov. regn. veget. **26**: 147. 1929.) Aellen cites one specimen from St. Joseph County.

11. **Chenopodium Standleyànum** Aellen. (Rep. spec. nov. regn. veget. **26**: 153. 1929.) (*C. Boscianum* Moq. in part [loc.: "Texas," leg. Drummond no. 246] and of authors.) Map 869. True *C. Berlandieri* Moq. ssp. *Boscianum* (Moq.) Aellen occurs in the southern states. In sandy soil along roadsides and in open sandy woods.

Pa. to Minn., southw. to Fla. and N. Mex.

12. **Chenopodium missouriénse** Aellen. (Bot. Not., Lund, **1928**: 206. 1928.) (*C. paganum* Standley, N. Amer. Flora **21**(1): 23. 1916, in part, not Reichenbach.) Map 870. In areas formerly occupied by prairies.

12a. **Chenopodium missouriense** var. **Bushiànum** Aellen. (Rep. spec. nov. regn. veget. **26**: 156. 1929.) Aellen cites one specimen from Spencer County.

13. CHENOPODIUM MURÀLE L. NETTLELEAVED GOOSEFOOT, SOWBANE, TOWN GOOSEFOOT. Map 871. Introduced in America.

14. CHENOPODIUM ÚRBICUM L. CITY or UPRIGHT GOOSEFOOT. Map 872. Introduced in America.

Reported from Clark, Jefferson, Kosciusko, Marion, Monroe, Steuben, Tippecanoe, and Vigo Counties.

14a. CHENOPODIUM URBICUM var. INTERMÈDIUM (Mert. & Koch) Koch. Reported from Pulaski County.

2224. CYCLOLÒMA Moq. WINGED PIGWEED

1. **Cycloloma atriplicifòlium** (Spreng.) Coult. WINGED PIGWEED. Map 873. This plant occurs mostly in sand ballast along railroads and in the dunes. Characteristic in late summer.

Man. to Ind., Ark. and westw. across the plains; introd. eastw.

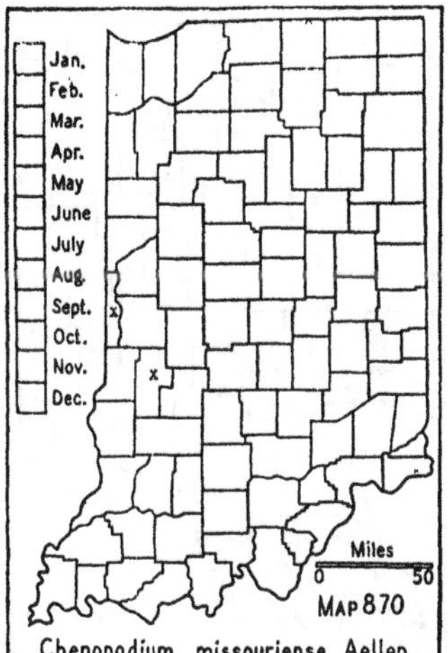

Chenopodium missouriense Aellen

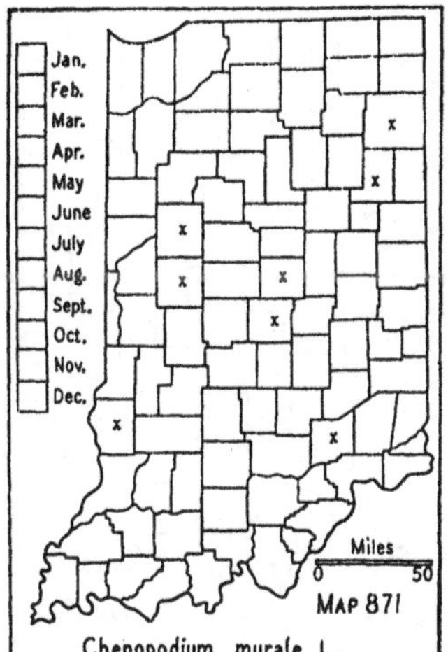

Chenopodium murale L.

Chenopodium urbicum L.

2229. ÁTRIPLEX [Tourn.] L. ORACH, SALTBUSH, SHAD-SCALES

[Collins, G. N. Seeds of Commercial Saltbushes. U. S. Dept. Agric. Div. Bot. Bull. 27. 1901. Hall, H. M. and F. S. Clements. The North American Species of Atriplex in: The Phylogenetic Method in Taxonomy. Carnegie Inst. Washington Publ. 326: 235-355. Pls. 36-58. 1923. Schreiber, Beryl O. Keys and Charts for California Species of Atriplex. California Forest and Range Exp. Sta., Techn. Note no. 8: 9p. 1938.]

Leaves green, glabrate as the rest of plant, at least the lower ones opposite, usually hastate or nearly so, only the lowest at times dentate, occasionally linear, petiolate; bracts united at the base only, with dentate foliaceous margins, the sides usually tuberculate or muricate, the teeth occasionally rather small and few; radicle inferior.

Leaves lanceolate to rounded-deltoid; plants usually decumbent; pistillate flowers all alike, bracteate.

Bracts usually smooth on the face, rhombic-oval, mostly broadly cuneate or narrowly rounded at the base; lower leaves rhombic-lanceolate, to narrowly lanceolate or oblong, not hastate, the upper ones usually entire or denticulate (occasionally slightly hastate), medium-sized, rarely with a pair of basal lobes...1. *A. patula.*

Bracts ordinarily tuberculate on the face, orbicular-deltoid or ovate-deltoid, usually truncate or broadly rounded at the base, margins mostly with a few toothlike projections; lower leaves rounded-deltoid or triangular-hastate, the upper usually more or less hastate, with basal angles or lobes, mostly large, more or less irregularly dentate; inflorescence leafless, spicate paniculate...1a. *A. patula* var. *hastata.*

Leaves linear, not hastate or but slightly so; bracts tuberculate on the face, erect...1b. *A. patula* var. *littoralis.*

Leaves gray-scurfy, all alternate, all or most dentate or hastate, never linear, ovate or rhombic-ovate, upper usually sessile, rounded or cuneate at base, cartilaginous on drying; bracts united to about the middle, ovate, acute, longer than broad, fruiting bracts rhombic or cuneate-orbicular, becoming hard and tough with dentate margins, shiny, not apiculate; radicle superior........................2. *A. rosea.*

1. **Atriplex pátula** L. ORACH, NARROWLEAF ORACH, SPEAR SCALE. Map 874. A highly variable species and linked by intermediates with its varieties.

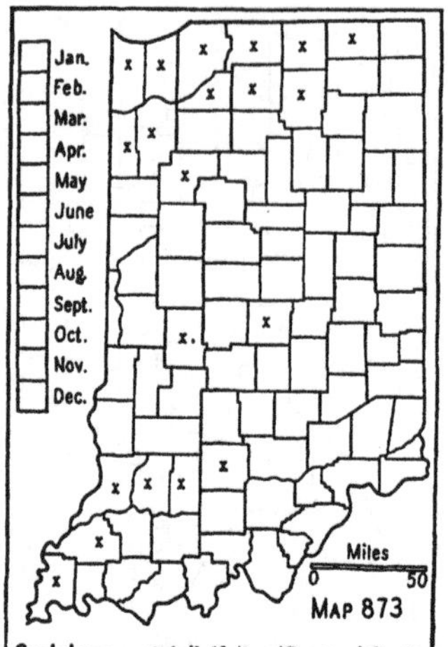

Cycloloma atriplicifolium(Spreng.) Coult.

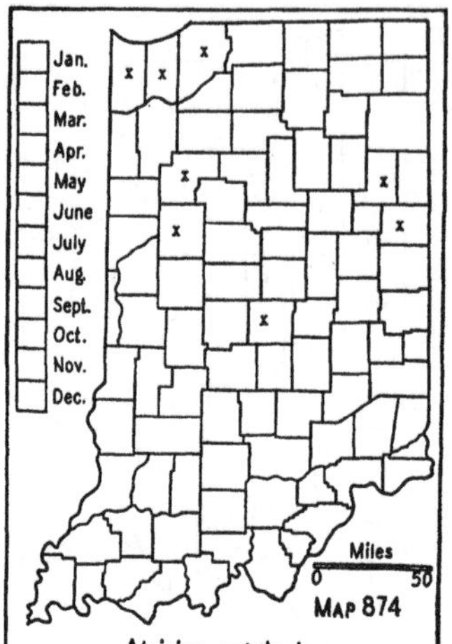

Atriplex patula L.

Atriplex patula var hastata (L.) Gray

In sandy soil and waste places.

Reported from: Lake, La Porte, Marion, and Tippecanoe Counties.

Newf. to Fla., Ala., Mo., to B. C.

1a. **Atriplex patula** var. **hastàta** (L.) Gray. (*A. hastata* L.) HALBERD-LEAVED ORACH, SPEAR ORACH. Map 874a. Same habitats as species.

Reported from: Benton, La Porte, Madison, Marshall, and Wells Counties.

Newf. to Oregon, southw. to S. C., Va., Mo., and Calif.

1b. **Atriplex patula** var. **littoràlis** (L.) Gray. (*A. littoralis* L.) Map 875. Prairie habitat, roadsides.

Reported from: La Porte, Porter, and Steuben Counties.

P. E. I. to N. J., westw. along Great Lakes.

2. ATRIPLEX RÒSEA L. RED SCALE. Reported from Porter County, probably in the state. Introduced in America.

2240. KÒCHIA Roth

1. KOCHIA SCOPÀRIA (L.) Schrad. Occasionally found on dumps, sporadic in appearance, but will never become an escape or established in the state. Introduced in America.

One specimen seen from Wells County which might be referred to the var. *trichophila* (Schinz & Thell.) Bailey. This differs from the species by its narrow, linear leaves (1-2 mm wide), by its fastigiate growth, and its bright red color in the autumn.

2245. CORISPÉRMUM [A. Juss.] L. BUG-SEED

Fruit 2-3 mm long; lower bracts much narrower than the fruit, not imbricated; spikes small, laxly flowered, slender; bracts of flowers shorter than or as long as the flowers..1. *C. nitidum.*

Fruit 3.5-4.5 (5) mm long; lower bracts equaling or longer than the flowers, imbricated; spikes broader, dense, stout; perianth parts 1-3, very rarely 5 or lacking. ....2. *C. hyssopifolium.*

MAP 875
Atriplex patula var. littoralis (L.) A. Gray

MAP 876
Corispermum nitidum Kit.

MAP 877
Corispermum hyssopifolium L.

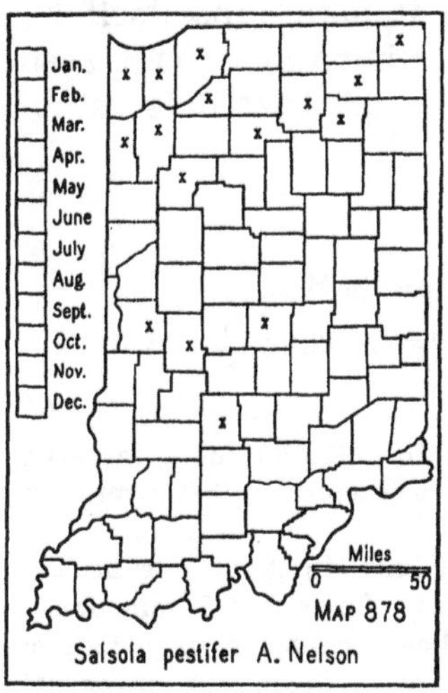

MAP 878
Salsola pestifer A. Nelson

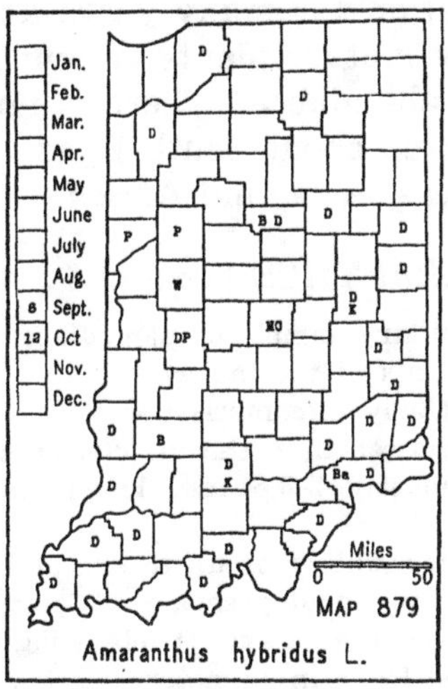

MAP 879
Amaranthus hybridus L.

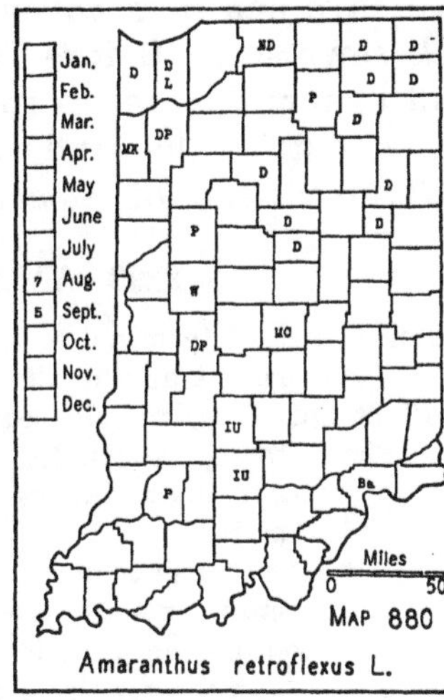

MAP 880
Amaranthus retroflexus L.

1. **Corispermum nítidum** Kit. Map 876. On sand dunes. Known from Lake and Porter Counties only.

Great Lakes, N. Dak., Idaho, southw. to Texas and Ariz.

2. **Corispermum hyssopifòlium** L. Map 877. On sand dunes; known from Lake, La Porte, and Porter Counties.

Ont. to Wash., southw. to Mo. and Mex.

2269. SÁLSOLA L. SALTWORT, RUSSIAN THISTLE

1. SALSOLA PÉSTIFER A. Nelson. (*S. Kali* L. var. *tenuifolia* G. F. W. Mey.) RUSSIAN THISTLE. Map 878. In sandy soil in waste grounds and along beaches and roadsides. Introduced in America.

79. AMARANTHÀCEAE J. St. Hil. AMARANTH FAMILY

Leaves alternate.

Ovary 3-8 seeded; filaments of stamens united into a tube....2292. CELOSIA, p. 428.

Ovary 1-seeded; filaments of stamens free.

Flowers monoecious or polygamous, all with a calyx of 5, or sometimes 3, distinct, erect sepals; sepals persistent..................2299. AMARANTHUS, p. 428.
Flowers dioecious; calyx none in the pistillate flowers; calyx of 5 sepals in the staminate flowers...................................2300. ACNIDA, p. 430.
Leaves opposite.
Flowers spicate or paniculate.
Leaves woolly beneath, sessile or nearly so, of a narrow type; flowers spicate.... ....2332. FROELICHIA, p. 431.
Leaves only sparsely pubescent beneath, with a long petiole, of an ovate type; flowers paniculate..................................2339. IRESINE, p. 432.
Flowers in dense, round heads, usually crimson or rose color..................... ....2338. GOMPHRENA, p. 432.

2292. CELÒSIA L. Cockscomb

1. CELOSIA ARGÉNTEA L. A cultivated form of this plant was reported by Nieuwland as escaped in the foreign settlement in the west side of South Bend. I have noted it from the roadside on dumps and in waste places. We have had it in cultivation for many years and it maintains itself by self sown seed. Before the mature plants are cut for burning enough seed fall to sow themselves in abundance. There is no report that it is established outside the sandy area about South Bend.

Tropical area of Americas, Asia, and Africa.

2299. AMARÁNTHUS [Tourn.] L. Amaranth

Inflorescence of terminal or axillary, simple or paniculate spikes, glomerules of flowers often present also in the axils of the leaves.
Plant spineless; utricle regularly circumscissile.
Pistillate sepals usually shorter than the utricle, or slightly longer, acuminate or acute; main bracts mostly 2-3.5 mm long; spikes generally 6-12 mm in diameter.
Bracts obtuse or acutish, equaling or up to one and a half times as long as the sepals, usually red or purple; sepals of pistillate flowers 1.5 mm long, obtuse or rounded at the apex; utricle subglobose, conspicuously longer than the sepals; seed 1 mm in diameter.....................1. *A. cruentus.*
Bracts acute or acuminate, usually twice as long as the sepals; sepals of pistillate flowers 1.5-2 mm long, acute, or the inner rarely obtuse; utricle subglobose, equaling or shorter than the sepals, very rarely exceeding them; seed 1 mm in diameter..................................2. *A. hybridus.*
Pistillate sepals obtuse or truncate, equaling or conspicuously longer than the fruit, about 3 mm long; main bracts mostly 4-6 mm long; spikes generally 8-20 mm in diameter...3. *A. retroflexus.*
Plant spiny; utricle irregularly or imperfectly dehiscent.............4. *A. spinosus.*
Inflorescence wholly of axillary glomerules.
Sepals in both staminate and pistillate flowers 4 or 5; utricle smooth; seed more than 1 mm wide, usually about 1.5 mm in diameter; plants prostrate; leaves toward the ends of the branches scarcely reduced...............5. *A. blitoides.*
Sepals 3, those of the pistillate flowers acute or acutish, those of the staminate flowers cuspidate, scarious; utricle rugose; seed about 0.8 mm in diameter; plants stout, erect, with stiff, divaricate or ascending branches; leaves toward the ends of the branches usually much reduced................6. *A. graecizans.*

1. AMARANTHUS CRUÉNTUS L. (*Amaranthus paniculatus* L.) TASSEL AMARANTH. This is a garden escape to roadsides, waste places, and dumps, which I have seen many times but never collected. I have no evi-

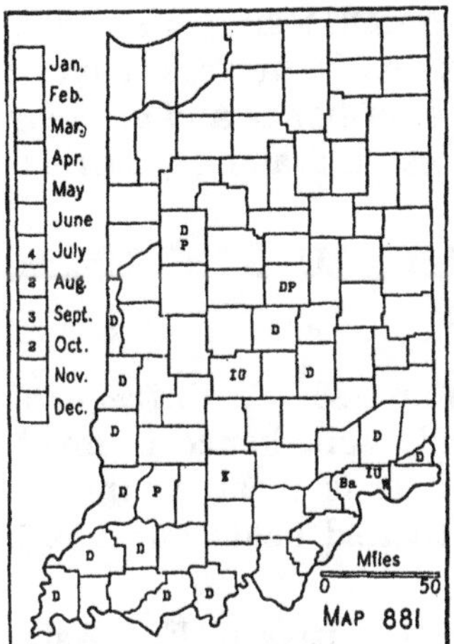

Amaranthus spinosus L.

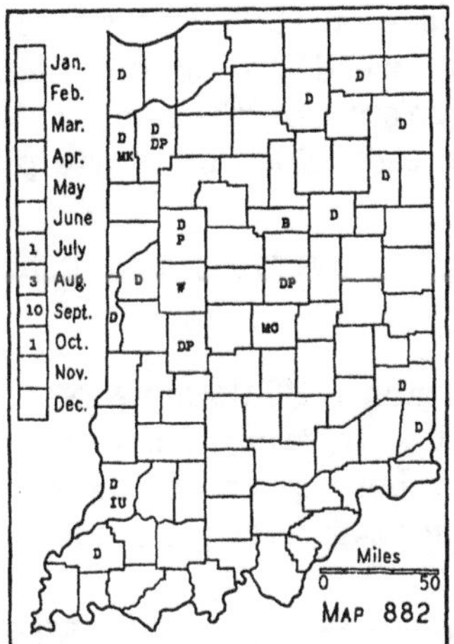

Amaranthus blitoides S. Wats.

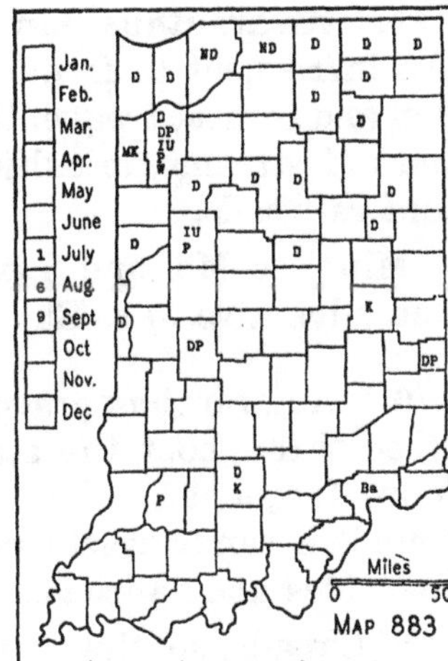

Amaranthus graecizans L.

dence that it can maintain itself in competition although it has maintained itself in our garden for several years.

Nat. of Asia; escaped or adventive in the eastern part of the U. S. as far west of N. Mex. and Ariz., southw. through the tropics to sub-tropical S. A.

2. AMARANTHUS HÝBRIDUS L. SLENDER GREEN AMARANTH. SLENDER PIGWEED. Map 879. Widely distributed throughout the state as a weed in gardens, cornfields, waste places, especially about habitations, and along roadsides and railroads. It prefers a rich, moist soil and is often, like the next species, a pernicious weed in cultivated grounds.

Found in the tropics throughout the world and naturalized throughout the U. S.

3. AMARANTHUS RETROFLÉXUS L. ROUGH GREEN AMARANTH. ROUGH GREEN PIGWEED. Map 880. Like the preceding species this one is widely distributed throughout the state as a weed in cultivated fields and waste places and along roadsides and railroads. It also prefers rich soils and is a too common weed.

Nat. of tropical America; naturalized throughout the U. S.

4. AMARANTHUS SPINÒSUS L. THORNY AMARANTH. Map 881. This is a very objectionable weed on account of its many spines. It is restricted mostly to our southern counties in barnyards and lanes where it is often very abundant. I do not understand why farmers do not try to exterminate it when first they discover it on their premises but I have never met one who was making the attempt. All who had a common name for it called it careless, a name sometimes applied to species of the pigweed family. I never could learn the origin or significance of this name and it seems to me to be very inappropriate.

Nat. of the tropics; naturalized in the U. S. from Minn. eastw.

5. **Amaranthus blitoides** Wats. PROSTRATE AMARANTH. Map 882. An infrequent to frequent weed throughout the state. It prefers a moist soil and is most frequently found on the muddy slopes of banks and gravelly bars of streams, in cultivated fields and waste places, and along roadsides and railroads.

Minn. to Mo. and Tex. and westw.; established in e. U. S., s. Canada, and adventive in s. Eu.

6. **Amaranthus graecizans** L. TUMBLEWEED. Map 883. An infrequent weed throughout the state. It prefers a dry, sandy soil, hence is much more frequent in the northern part of the state. It is most commonly found in sandy waste places, gravel pits, and cultivated fields and along roadsides and railroads.

S. Canada, southw. through the U. S. to n. Mex.; adventive in Eu., Asia, Africa, and S. A.

2300. ACNIDA L. WATER HEMP

Utricle circumscissile, verrucose all over; bracts longer than the utricle; staminate flowers mostly 3.5-4 mm long, their sepals rigid, long-acuminate, the outer conspicuously longer than the inner, their bracts 2-3 mm long; plants erect, 5-16 dm high, mostly of a moist or dry, sandy habitat.................1. *A. tamariscina.*

Utricle irregularly dehiscent or indehiscent, smooth or verrucose mostly below the middle; bracts shorter than the utricle; staminate flowers 2-2.5 mm long, their sepals thin, acute, of nearly equal length, their bracts about 1-1.5 mm long; plants erect, decumbent, or prostrate; mostly of a muddy habitat, such as muddy banks, bars in streams, and dried-up ponds and sloughs.

Plants erect, mostly 4-12 dm high; leaves of an ovate or lanceolate type, hence broadest below the middle; seed about 0.8 mm wide...........2. *A. altissima.*

Plants prostrate or decumbent; leaves obovate, spatulate or lanceolate; seed mostly 1-1.2 mm wide..3. *A. subnuda.*

1. **Acnida tamariscina** (Nutt.) Wood. Map 884. I found this species in 1919 and 1921 to be common in sandy soil about a half mile east of Lyle, Gibson County, along the roadside and in adjacent cornfields. I found it also as a common plant, 4-6 feet high, in a roadside ditch 4 miles south of Johnsonville, Warren County. This location is in the prairie area of the state. I have also a specimen collected by Umbach in ballast near Miller, Lake County. From what I can learn of the habitat of this species I think it is a native of the western part of the state. Blatchley says he found it to be common along the Wabash River in Vigo County on gravel and sandy banks which is the preferred and native habitat of the species.

Ind. to S. Dak., southw. to Tex.

2. **Acnida altíssima** Riddell. (*Acnida tuberculata* Moq.) (See North Amer. Flora 21: 122. 1917.) Map 885. Infrequent to common in all parts of the state on the muddy banks and bars of streams, on the borders of ponds and sloughs, in ditches and dried-up ponds and sloughs, and in moist, alluvial cornfields along streams. This is strictly a low ground species and is very common on the muddy slope of the bank of the Ohio River.

Ont. to Colo., southw. to Ky. and Mo.

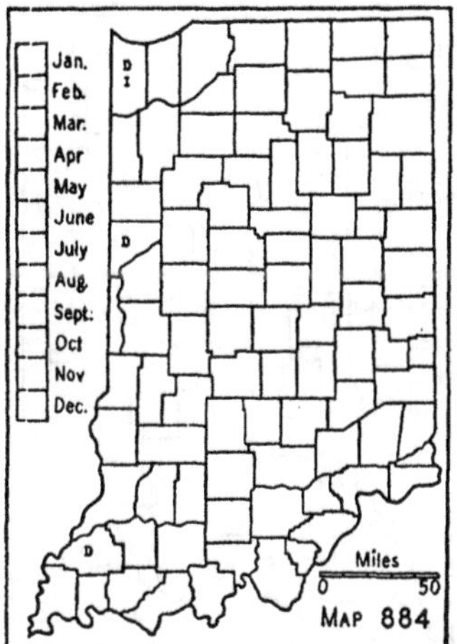

Acnida tamariscina (Nutt.) Wood

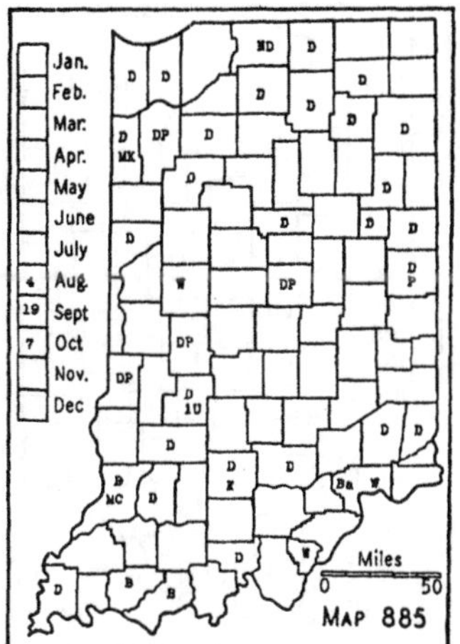

Acnida altissima Riddell

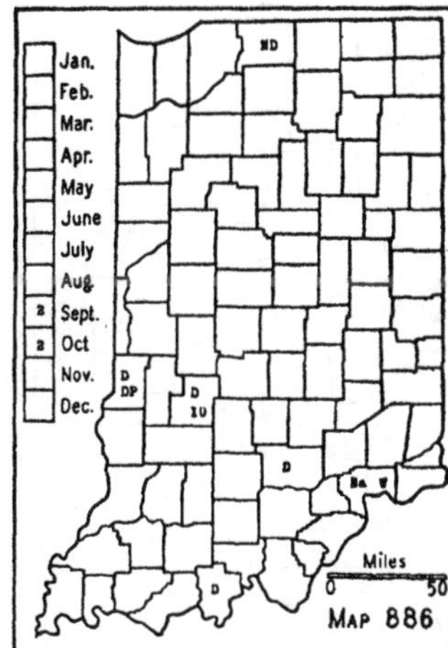

Acnida subnuda (S. Wats.) Standley

3. **Acnida subnùda** (Wats.) Standley. (*Acnida tuberculata* var. *subnuda* S. Wats.) Map 886. My specimens are from the muddy slopes and bars of our larger streams. This species is closely related to the preceding one and is separated from it primarily by its prostrate habit and by its larger seed.

Ont. to S. Dak., southw. to Tenn.

2332. FROELÍCHIA Moench

Plants much divided near the base, the erect portion usually not much longer at maturity than the long, lower, and often prostrate branches; wings of the flattened calyx tube at maturity about 1 mm wide and divided usually to the base into about 6 or 7 spinelike teeth, the tube also provided at the base with tubercles, usually one on one side and two on the opposite side..............1. *F. gracilis.*

Plants erect, tall, not divided near the base, sparingly branched; wings of the flattened calyx tube at maturity about 1 mm wide and not divided to the base but more or less deeply dentate, the tube at the base with one or both sides with a tubercle or tubercular lines.

Pubescence of the upper part of the stem of very short, brownish hairs. (See excluded species no. 207, p. 1043.)................................*F. floridana.*

Pubescence of the stems white and woolly..........................2. *F. campestris.*

1. FROELICHIA GRÁCILIS (Hook.) Moq. Map 887. In 1930 I found a few plants of this species along the Chicago & Eastern Illinois Railroad at the Duncan Switch about 4 miles south of Vincennes. The soil along the railroad here is almost a pure sand. By the fall of 1933 it had spread for a quarter of a mile and formed a complete stand at the switch and for several hundred feet to the north of it. This species will probably become a weed in the sandy area of this part of the country. In 1933 I found two small colonies in ballast along the railroad in the first mile east of Dana, Vermillion County. In 1933 Paul Weatherwax found a large colony in ballast along the railroad half a mile south of Worthington, Greene County. In 1937 Charles M. Ek found scattered plants in cinder soil in

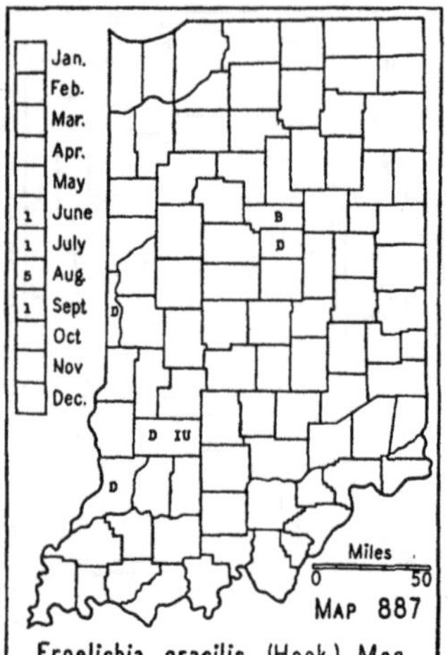

Froelichia gracilis (Hook.) Moq.

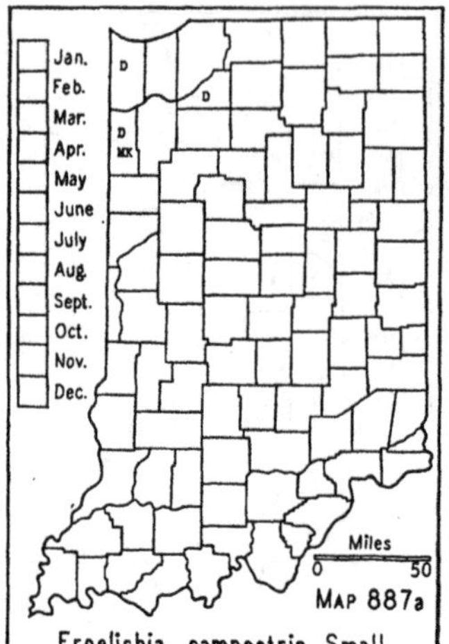

Froelichia campestris Small

Iresine rhizomatosa Standley

the railroad yards in Tipton, Tipton County. Doubtless it already has a much wider distribution in the state than our records show.

Iowa to Colo., southw. to Ark., Ariz., and Chihuahua, Mex.

2. FROELICHIA CAMPÉSTRIS Small. Map 887a. I have specimens of this species from three counties. One was collected July 7, 1900, by Umbach on railroad ballast near Aetna, Lake County. Another was collected by Miss Madge McKee along a sandy roadside about 6 miles south of Roselawn, Newton County. In 1933 I found a large area of it in very sandy soil in a fallow field in sec. 3 in Starke County about two and a half miles northeast of North Judson. This species in time will no doubt become a weed in the sandy areas of this part of the state. Doubtless introduced into the state.

Ill. and Wis. to Nebr., southw. to Mo. and Okla.

2338. GOMPHRÈNA L.

See excluded species no. 208, p. 1044.

2339. IRESÌNE P. Br.

1. **Iresine rhizomatòsa** Standley. (Proc. Washington Biol. Soc. 28: 172. 1915.) (*Iresine paniculata* of recent authors, not Kuntze.) Map 888. Very local in a few low woods and dried-up sloughs in the Lower Wabash Valley. Reported also from Clark, Floyd, and Jefferson Counties.

Md., Ind. to Kans., southw. to Ala. and cent. Tex.

80. NYCTAGINÀCEAE Lindl. FOUR-O'CLOCK FAMILY

Involucre remaining unchanged in fruit....................2347. MIRABILIS, p. 433.
Involucre enlarging and becoming membranous and reticulate in fruit................
....2348. OXYBAPHUS, p. 433.

Oxybaphus nyctagineus (Michx.) Sweet

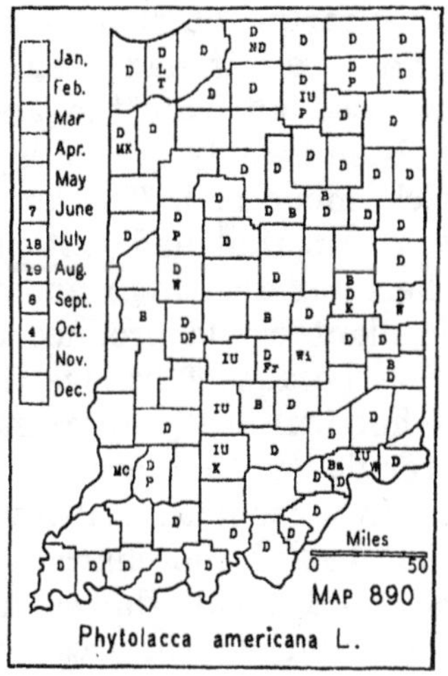

Phytolacca americana L.

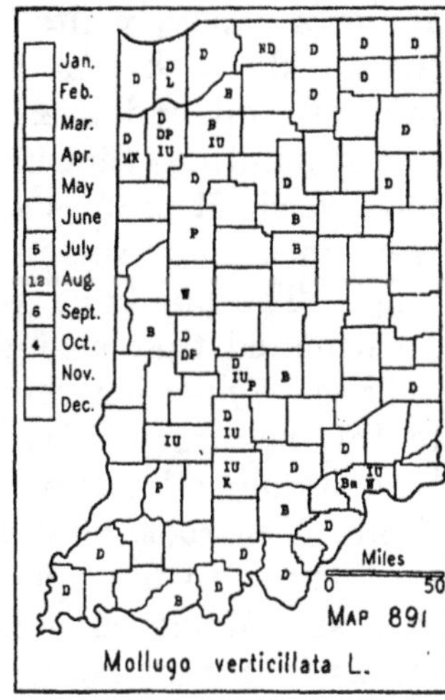

Mollugo verticillata L.

2347. MIRÁBILIS L.

See excluded species no. 209, p. 1044.

2348. OXÝBAPHUS L'Hér. UMBRELLA-WORT

Leaves petiolate and obovate or the upper bractlike and sessile......1. *O. nyctagineus.*
Leaves sessile, linear or lanceolate.
 Stem more or less hirsute as well as viscid. (See excluded species no. 211, p. 1044.) .. ..*O. hirsutus.*
 Stem glabrous below, not hirsute, viscid-puberulent above.
  Leaves narrowly linear, less than 5 mm wide, rarely wider; angles of fruit smooth. (See excluded species no. 212, p. 1044.)........................*O. linearis.*
  Leaves mostly lanceolate, sometimes linear to ovate; angles of fruit tuberculate. (See excluded species no. 210, p. 1044.).......................*O. albidus.*

1. **Oxybaphus nyctagíneus** (Michx.) Sweet. (See Field Mus. Nat. Hist. Publ. Bot. Ser. 8: 305. 1931.) (*Allionia nyctaginea* Michx.) HEARTLEAF UMBRELLA-WORT. Map 889. Infrequent to frequent in railroad ballast throughout the state. Probably found in every county. More frequent before the right-of-ways of railroads were kept clean. This species seems to require a very sandy soil. I have seen it only twice in situations other than railroad ballast. I once found it along a very sandy roadside near Roselawn in Newton County, and once in a very sandy, oat field near Kniman in Jasper County.

Ind. to Man., southw. to Tex. and Mex.; frequently adventive in e. U. S.

83. PHYTOLACCÀCEAE Lindl. POKEWEED FAMILY

2380. PHYTOLÁCCA [Tourn.] L.

1. **Phytolacca americàna** L. (*Phytolacca decandra* L.) COMMON POKEBERRY. Map 890. This plant is found throughout the state in almost all kinds of soils and habitats. Its abundance is due to birds that scatter the seed everywhere, to its ability to adapt itself to all kinds of soils, and to the fact that grazing animals do not molest it. I have seen it only a few times in a thick stand over any considerable area. I once

found a sandy, white oak clearing of about ten acres which had grown up thickly with this species after it had been grazed by hogs until the mineral soil had been exposed all over the area. In old orchards and forest plantings that have been heavily grazed by hogs, it is usually a common weed. It prefers a rich, moist soil. The largest specimens I ever saw were in a muck soil in a marsh that had just passed into the soft maple stage. The plants grew here 6-8 feet high and were wide spreading and I estimated that a single plant would produce not less than a gallon of berries. I mention this fact because I believe that in due time the fruit of this species will be of horticultural importance. Although the berries have an objectionable bitter flavor, they are not poisonous as some people think. The root, however, is poisonous. All my life I have been tasting the berries to find one that lacked the characteristic flavor, but without success. About 60 years ago I recall that a hotel keeper came to our woods to gather pokeberries and elderberries which he canned and used about half and half for making pies. If the pokeberries alone are used, some vinegar should be added. They make a very rich looking and palatable pastry. I recall eating them in pies when I was a boy. The dried berries macerated with whiskey were formerly used for rheumatism.

Southern Maine, Ont. to Minn., southw. to Fla., Ark., and Mex.

84. AIZOÀCEAE A. Br. CARPET-WEED FAMILY

2387. MOLLÙGO L.

1. MOLLUGO VERTICILLÀTA L. CARPET-WEED. Map 891. The carpet-weed is distributed throughout the state in dry or moist soils that are not covered with vegetation. It is infrequent, frequent or common where found, usually on the sandy shores of streams, in cultivated fields such as cornfields, stubble fields, and truck gardens, in ballast along railroads, along roadsides, and elsewhere in sandy soil.

Throughout temperate and tropical N. A.; also in S. A. and in the Old World.

85. PORTULACÀCEAE Reichenb. PURSLANE FAMILY

Calyx free from the ovary; capsule 3-valved.
  Leaves terete, more than 2; seed numerous.................2406. TALINUM, p. 434.
  Leaves 2, not terete, sometimes nearly so; seed not more than 6....................
  ....2412. CLAYTONIA, p. 435.
Calyx partly adnate to the ovary; capsule circumscissile.....2421. PORTULACA, p. 435.

2406. TALÌNUM Adans.

1. **Talinum rugospérmum** Holzinger. (Holzinger. Talinum rugospermum. Torreya 28: 94-95. 1928 and Fassett. Talinum teretifolium and T. rugospermum. Rhodora 30: 205-206. 1928.) PRAIRIE TALINUM. Map 892. This plant was first reported by Babcock (Lens 1: 23. 1872) as found on the sand hills at Miller and Tolleston in Lake County. On Nov. 22, 1928, Norman C. Fassett wrote me that there were five specimens in the herbarium of the University of Wisconsin collected by L. M. Umbach at Miller on the following dates: July 26, 1895; June 23, 1898; June 27, 1899; July 17,

MAP 892
Talinum rugospermum Holzinger

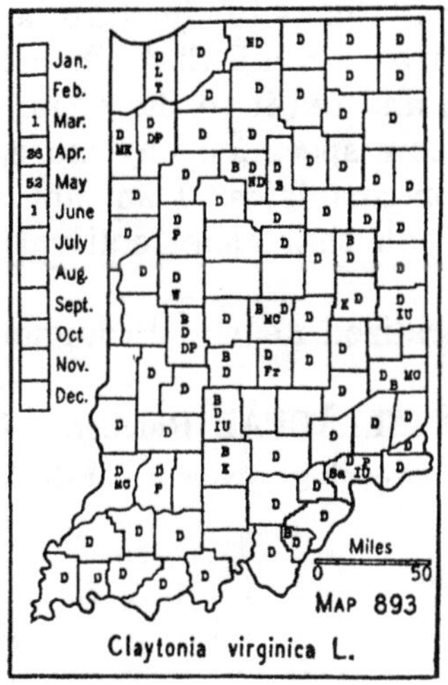

MAP 893
Claytonia virginica L.

MAP 894
Portulaca oleracea L.

1906; and August 27, 1909. I have a specimen collected by Umbach on June 27, 1899. Holzinger says the species is perennial and grows in very sandy soil.

Sandstone ledges near Duluth, Minn., southw. to cent. Ill., eastw. to Lake Michigan and Lake, Newton, and Porter Counties, Ind.

2414. CLAYTÒNIA [Gronov.] L. SPRING BEAUTY

Stem leaves linear, linear-lanceolate to lanceolate, 5-15 cm long, sessile or petiolate. .... 1. *C. virginica.*
Stem leaves oval-lanceolate to oblong-lanceolate, 2.5-5 cm long, petiolate. (See excluded species no. 214, p. 1044.) .... *C. caroliniana.*

1. **Claytonia virgínica** L. VIRGINIA SPRING BEAUTY. Map 893. Frequent to common in moist or dry woods in every county of the state. It is exteremely variable in all of its parts except the seed. It generally has only 2 stem leaves, but I have one specimen with 3 stem leaves. Stanley Coulter says: "Common with the floral parts multiplied." The dried stem leaves of my specimens vary from 2-25 mm in width and from 5-15 cm in length; some are sessile and others are petiolate. The calyx at fruiting time varies from 5-12 mm long, and the lobes from rounded to acute. I think that some of the wideleaf specimens have been reported as *Claytonia caroliniana,* which I have not found in Indiana although I have sought for it for years.

I once noted a small bird greedily eating the flowers of *Claytonia virginica.*

N. S. to Minn., southw. to Va. and Kans. (Rydberg. North Amer. Flora 21: 298. 1932).

2421. PORTULÀCA [Tourn.] L.

Leaf blades flat; axils of leaves glabrous or nearly so; petals yellow.... 1. *P. oleracea.*
Leaf blades terete or nearly so; axils of leaves pilose; petals red or purple. (See excluded species no. 216, p. 1045.) .... *P. grandiflora.*

1. PORTULACA OLERÀCEA L. COMMON PURSLANE. Map 894. This plant was formerly a common and annoying weed in gardens and cornfields.

It has been reported from all parts of the state and all authors who report it mention its weedy nature. I recall that when I was a boy 60 years ago we pulled it by the bushel and fed it to the hogs. At the present time it is rare and I very seldom see a specimen any more, although I admit that I rarely botanize gardens or cornfields. I am not able to explain its disappearance but I do not think that clean cultivation is responsible for its scarcity.

Nat. of Eu.; now naturalized nearly throughout N. A.

87. CARYOPHYLLÀCEAE Reichenb. PINK FAMILY

Sepals separate, more or less spreading; styles separate to the base; ovary sessile.
- Stipules present.
  - Leaves opposite.
    - Styles 2 .... 2475. PARONYCHIA, p. 442.
    - Styles 3 .... 2450. SPERGULARIA, p. 442.
  - Leaves whorled; styles 5 .... 2449. SPERGULA, p. 442.
- Stipules lacking.
  - Capsules opening by as many entire or at length 2-cleft valves as there are styles; petals entire or merely notched at the apex.
    - Styles as many as the sepals and alternate with them; petals of the same number or lacking .... 2433. SAGINA, p. 441.
    - Styles fewer than the sepals, rarely of the same number and then opposite them. .... 2443. ARENARIA, p. 441.
  - Capsules opening by twice as many valves or teeth as there are styles; petals deeply cleft or lacking.
    - Capsule short, ovate or oblong, opening usually by 6 valves; styles usually 3. .... 2429. STELLARIA, p. 436.
    - Capsule long, cylindric, often curved, opening at the apex usually by 10 teeth; styles usually 5 .... 2430. CERASTIUM, p. 438.

Sepals united; calyx tubular.
- Calyx naked at the base; seeds globular or reniform; embryo curved.
  - Flowers apetalous .... 2483. SCLERANTHUS, p. 444.
  - Flowers with petals.
    - Sepals with long, herbaceous tips, generally 2-3 cm long; styles 5, opposite the petals; capsules 5-toothed .... 2488. AGROSTEMMA, p. 444.
    - Sepal lobes less than 2 cm long; styles alternate with the petals.
      - Flowers bisexual or pistillate.
        - Styles 3 or 4; calyx 10-nerved; capsule several-celled at the base, 6-toothed. .... 2490. SILENE, p. 444.
        - Styles 5; calyx 10-nerved (with 10 additional fainter nerves in *Lychnis alba*); capsule 1-celled at the base, with 5 deeply bifid teeth .... 2491. LYCHNIS, p. 449.
        - Styles 2; calyx indistinctly nerved or 5-nerved; capsule 4-toothed .... 2503. SAPONARIA, p. 449.
      - Flowers unisexual, staminate, see *Lychnis alba* .... 2491. LYCHNIS, p. 449.
- Calyx subtended by 2-4 bracts; styles 2; seeds dorsally flattened; embryo nearly straight .... 2502. DIANTHUS, p. 449.

2429. STELLÀRIA L. CHICKWEEDS AND STICHWORTS

Plants glabrous.
- Median leaves broadest at the base, linear-lanceolate.
  - Plants generally 3-5 dm long, decumbent; inflorescence many-flowered, generally more than half the length of the plant, branches spreading; margins of sepals generally pubescent .... 1. *S. graminea.*

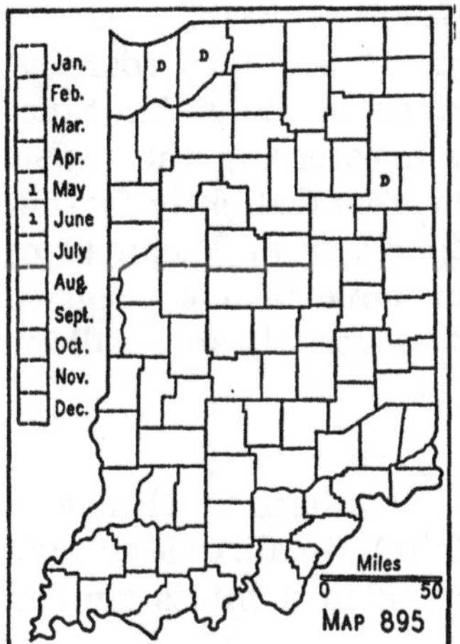

Stellaria graminea L.

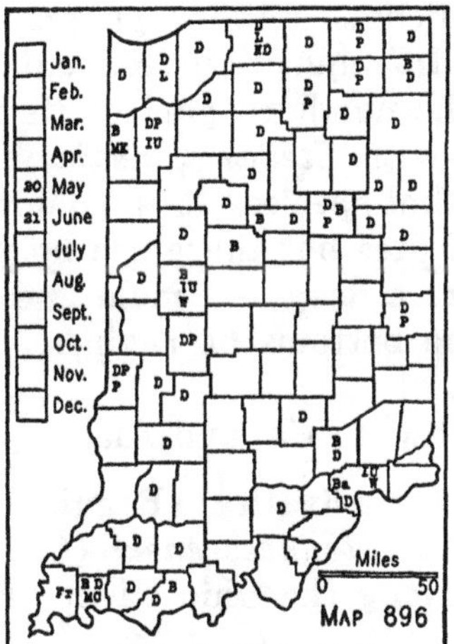

Stellaria longifolia Muhl.

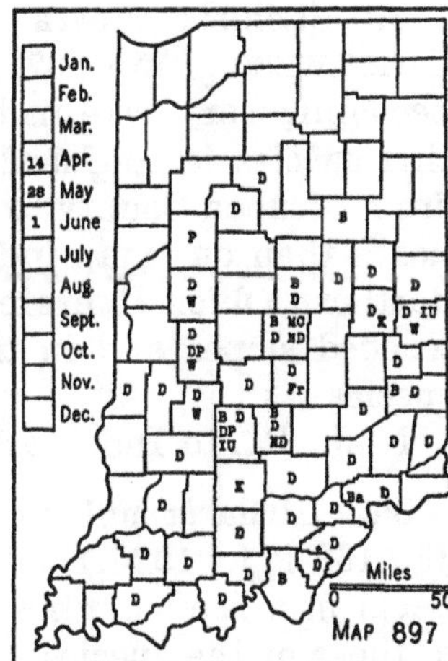

Stellaria pubera Michx.

Plants generally 1-3 dm high, erect; inflorescence few-flowered, generally much less than half the length of the plant, branches erect; margins of sepals not pubescent. (See excluded species no. 218, p. 1045.)..............*S. longipes.*

Median leaves widest above the middle, linear........................2. *S. longifolia.*

Plants pubescent in lines.

Leaves oblong, oblong-lanceolate, ovate-oblong or oval, mostly 2-10 cm long; petals longer than the sepals or equal or shorter in the variety of no. 3; stamens 10.

Median leaves of both sterile and flowering shoots sessile or subsessile; sepals 4-6 mm long, obtuse or acutish, shorter than the petals, inconspicuously, if at all, ciliate..3. *S. pubera.*

Median leaves of sterile shoots abruptly contracted into petioles 1-2 cm long; sepals 7.5-11 mm long, acute or acuminate, equaling or exceeding the petals, at least the outer ones conspicuously ciliate on the lower half.................. ....3a. *S. pubera* var. *silvatica.*

Leaves ovate, sometimes very narrowly so or oval, all but the upper ones more or less petiolate; blades 7-40 mm long; stamens 5................4. *S. media.*

1. STELLARIA GRAMÍNEA L. (*Alsine graminea* (L.) Britt.) Map 895. I found this species in La Porte County on the bank of a ditch west of the State Prison; in Porter County, I found a colony about 4 feet in diameter on the embankment of the New York Central Railroad about 3 miles west of Porter; and in Wells County I found it to be a common weed in the Six-mile Cemetery. It has been reported also from Lake County. A specimen from Jasper County so labeled in the herbarium of DePauw University proves to be *Stellaria longifolia.*

Nat. of Eurasia; Newf. to Ont. and Minn., southw. to Iowa and Md.

2. **Stellaria longifòlia** Muhl. (*Alsine longfolia* (Muhl.) Britt.) LONG-LEAF STICHWORT. Map 896. Infrequent to rare throughout the state in low or moist woodland and marshes, on the low borders of lakes, and rarely in the open along ditches.

Newf. to Alaska, southw. to Md., Ky., and La. and in the Rocky Mts.; also in n. Eu. and Asia.

3. **Stellaria pùbera** Michx. (*Alsine pubera* (Michx.) Britt.) GREAT CHICKWEED. Map 897. Infrequent to frequent in the southern counties, becoming very rare in the northern part of its range in the state. Since this species is confused with *Stellaria media,* I am referring the report from Steuben County to that species. It is, no doubt, found slightly farther north than our map indicates but there are no reports from Michigan or northern Ohio. It prefers a deep leaf mold and is found in moist soil on wooded slopes and in the bottoms of ravines. It is strictly a woodland species.
N. J., Pa. to Ind., southw. to Ga. and Ala.

3a. **Stellaria pubera** var. **silvática** (Beguinot) Weatherby. (Rhodora **26**: 169-171. 1924.) (*Alsine tennesseensis* (C. Mohr) Small.) Map 898. Local in a few counties along the Ohio River. Found in habitats similar to those of the species.
Southern Ind. to s. Tenn.

4. STELLARIA MÈDIA (L.) Cyril. (*Alsine media* L.) COMMON CHICKWEED. Map 899. Found throughout the state and reported from many counties. It is an annoying weed in lawns and in all kinds of cultivated ground, especially about habitations. My specimens, however, with a few exceptions, are from the woodland where it sometimes appears as native. It is occasional to frequent in woodland, especially in the alluvial bottoms. The species is extremely variable and our specimens might be assigned varietal names as is done by some authors, but I doubt whether they are of taxonomic value. It is to be expected that a cosmopolitan species with widely varying habitats would show conspicuous variations.
Nat. of Eurasia; throughout N. A.

2430. CERÁSTIUM L. MOUSE-EAR CHICKWEED

[Fernald & Wiegand. Studies of some boreal American Cerastiums of the section Orthodon. Rhodora **22**: 169-179. 1920.]

Bracts of the cymes with broad scarious margins and tips, rarely the lower ones wholly herbaceous; perennials.
- Petals as long as the sepals; anthers about 0.5 mm long; styles mostly 1-1.5 mm long; capsules mostly 6-10 mm long and usually slightly more than 2 mm wide, the teeth usually slightly more than 1 mm long.
  - Inflorescence hirsute with glandless hairs............1. *C. vulgatum* var. *hirsutum.*
  - Inflorescence with gland-tipped hairs...1a. *C. vulgatum* var. *hirsutum* f. *glandulosum.*
- Petals about twice as long as the sepals; anthers 0.7-1 mm long; styles 2-4 mm long.
  - Plants with the upper internode and inflorescence with gland-tipped hairs, rarely a few gland-tipped hairs below the first internode; stems generally 2.5-6 dm long; median leaves lanceolate, or oblong-lanceolate, mostly 5-10 mm wide, 2.5-5 cm long; sepals 5-7 mm long; anthers mostly 1 mm long; styles about 4 mm long; capsules mostly about 13 mm long, the teeth about 1.5 mm long; seed about 1 mm long.......................................2. *C. arvense.*
  - Plants hirsute without gland-tipped hairs, mostly 12-20 cm high; leaves all linear, rarely a few linear-lanceolate; sepals mostly 4-6 mm long; anthers about 0.7 mm long; styles about 2 mm long; capsules mostly 8-9 mm long; seed about 0.8 mm long...2a. *C.* undetermined.

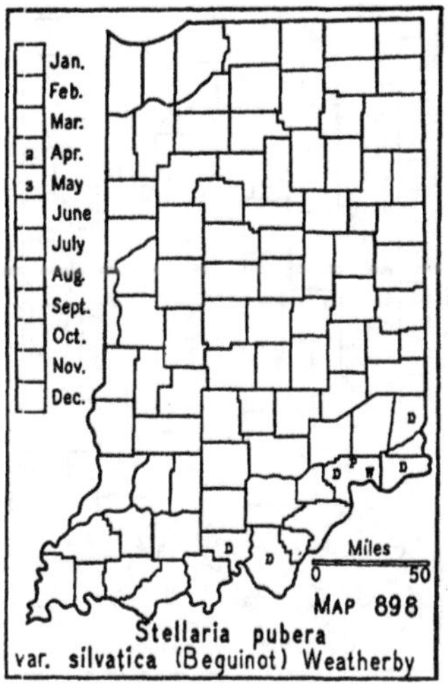

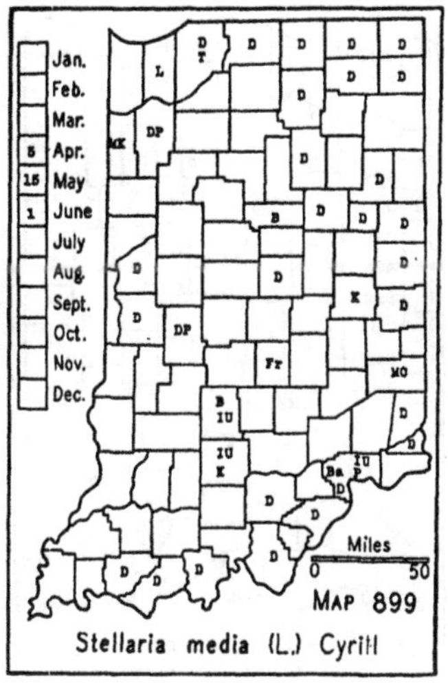

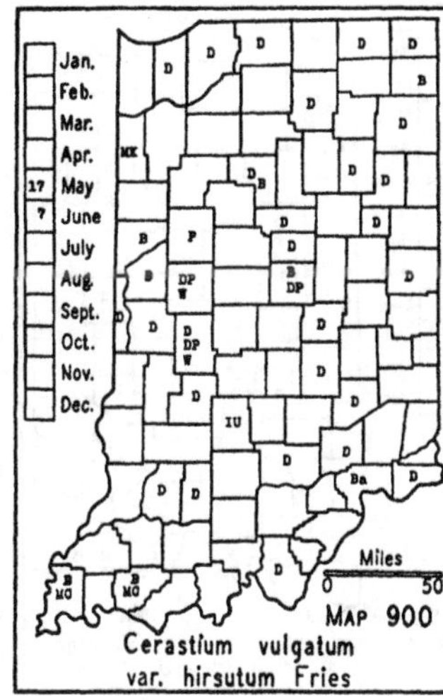

Bracts of cymes herbaceous or the upper pair sometimes slightly scarious on the margins or at the tip; petals shorter than the sepals, sometimes equaling them; anthers mostly 0.3-0.5 mm long; styles generally 0.5-1 mm long; annuals.

Median leaves ovate or obovate, usually broadly so, 4-11 mm wide and 5-25 mm long, the apex usually rounded and subacute; plants usually 1-3 dm high, lower internodes usually not glandular; sepals longer than the petals, usually all sharply acute at the apex; pedicels shorter than the calyx; capsules mostly 5-8 mm long and usually about 1.5 mm wide, the teeth generally less than 1 mm long...3. *C. viscosum.*

Median leaves lanceolate, oblong-lanceolate or oblong, 3-15 mm wide and 1-7 cm long, gradually narrowed to a subacute apex; plants generally 2-6 dm long; sepals obtuse or acute at the apex but not sharply so; petals slightly longer to twice as long as the sepals; pedicels much longer than the calyx; capsules mostly 11-13 mm long, the teeth usually 1-1.5 mm long; plants 1.5-6 dm high, the whole plant glandular-hirsute....................................4. *C. nutans.*

1. CERASTIUM VULGÀTUM L. var. HIRSÙTUM Fries. (*Cerastium vulgatum* L., in part.) COMMON MOUSE-EAR CHICKWEED. Map 900. An infrequent to common chickweed throughout the state. A weed in lawns and found in pastures, fallow fields, open woodland and along roadsides and railroads.

Fernald & Wiegand say that the U. S. forms of *Cerastium vulgatum* L. belong to this variety and its form.

Nat. of Eurasia; throughout temperate N. A.

1a. CERASTIUM VULGATUM var. HIRSUTUM f. GLANDULÒSUM (Boenn.) Druce. (*Cerastium vulgatum* L., in part.) COMMON MOUSE-EAR CHICKWEED. Map 901. Habitat and distribution the same as that of the preceding.

Nat. of Eurasia and probably local in N. A.

2. **Cerastium arvénse** L. (Pennell discusses this species and its varieties in Bartonia 12: 3-12. 1930.) FIELD CHICKWEED. Map 902. It is to be expected that this plant with a distribution throughout Europe and Asia and in North America would show a wide variation. Under this name are

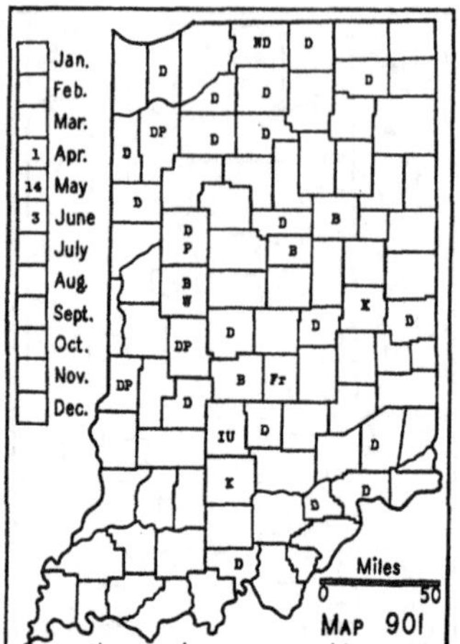

Cerastium vulgatum var. hirsutum f. glandulosum (Boenn) Druce

Cerastium arvense L.

Cerastium viscosum L.

included many forms. Some European authors have divided this species complex into several subspecies. American authors have divided it into at least 4 varieties while others do not divide it. *Cerastium arvense* var. *oblongifolium* has been reported from Indiana but the habitat ascribed to this variety precludes its appearance in Indiana. This variety is not well described so I am omitting it. Fernald & Wiegand in their article cited at the beginning of this genus said the species is a complex which they were not willing to divide. It is probable that when a larger series of specimens and more notes are at hand the forms can be delimited.

In Indiana there are two well-defined forms and it seems best to assign one to the species and separate the other from it. I have included under the species name our larger and glandular plant which has the distribution shown on the map. These plants were found in large colonies on the alluvial banks of the Mississinewa, Salamonie, and Wabash Rivers. The two northern locations belong to the glabrous form of the species.

The species and its varieties are found in the northern hemisphere around the world.

Cerastium undetermined. I have two specimens of this form which I found on high, wooded and gravelly banks of the St. Joseph River in Elkhart and St. Joseph Counties. This plant is strikingly different and has a dry soil habitat instead of a moist one. It is represented by my nos. 38515 and 38540.

3. CERASTIUM VISCÒSUM L. MOUSE-EAR CHICKWEED. Map 903. There are reports of this species from all parts of the state while the few specimens I have are from the southern part. From the number of wrongly determined specimens in our herbaria I think most authors did not understand the keys in our manuals and all reports should be evaluated on the basis of this experience.

My specimens are from fallow and pasture fields and open woodland. Nat. of Eu., N. B. to Ont., southw. to Fla., Tex., and southw.

4. CERASTIUM NÙTANS Raf. (*Cerastium longipedunculatum* Muhl.) NODDING CHICKWEED. Map 904. This species has also been reported from

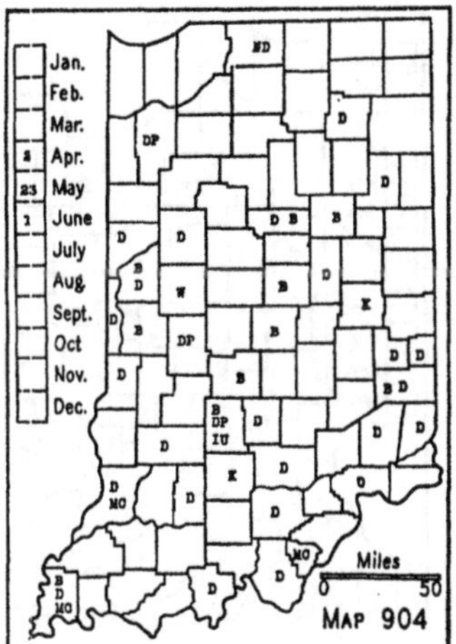

Cerastium nutans Raf.

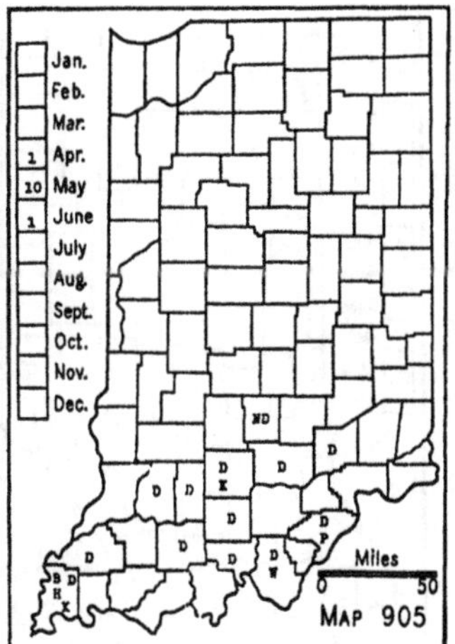

Sagina decumbens (Ell.) T. & G.

Arenaria serpyllifolia L.

nearly all parts of the state. It prefers a moist soil and is locally abundant mostly in fallow fields, pastures, and open woodland along streams.

Nat. of Eu.; throughout temperate N. A.

2433. SAGÌNA L.

1. **Sagina decúmbens** (Ell.) T. & G. (*Sagina apetala* of Amer. authors.) PEARLWORT. Map 905. Local in the southern counties in bare, sandy places in fallow fields and pastures and on the tops of river bluffs. In several instances it was intimately associated with *Plantago pusilla.* In the fields and pastures it appears as if introduced, and on the bluffs of streams far from fields it appears as if native. The plants are mostly 2-5 inches high and erect or erect from a very short, decumbent base and none are apetalous.

Mass. to Ill., and Mo., southw. to Fla. and La.

2443. ARENÀRIA L. SANDWORT

Leaves ovate, oval or oblong; capsules longer than the sepals.
- Blades less than 1 cm long, acute at the apex; seed not smooth .... 1. *A. serpyllifolia.*
- Blades mostly 1-3.5 cm long, generally obtuse at the apex; seed smooth .... 2. *A. lateriflora.*

Leaves linear or filiform; capsules shorter than the sepals.
- Plants glabrous; leaves fascicled in the axils, rigid .... 3. *A. stricta.*
- Plants glandular-pubescent, sometimes sparsely so; leaves not fascicled in the axils, soft .... 4. *A. patula.*

1. ARENARIA SERPYLLIFÒLIA L. THYMELEAF SANDWORT. Map 906. In very sandy soil along roadsides and railroads, in fallow fields, and rarely on bare spots on bluffs of streams. Naturalized in Indiana; I believe it could be found in railroad ballast in every county of the state.

Nat. of Eurasia; throughout N. A. except in the extreme north.

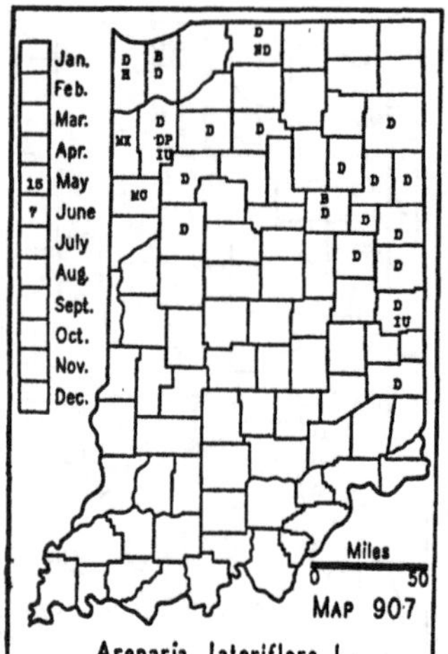

Arenaria lateriflora L.

Arenaria stricta Michx.

Arenaria patula Michx.

2. **Arenaria laterifl̀ora** L. (Woodward. On variation in Arenaria lateriflora. Rhodora **15**: 209-210. 1913. Rhodora **16**: 179-180. 1914 and St. John. Arenaria lateriflora and its varieties in North America. Rhodora **19**: 259-262. 1917.) (*Moehringia lateriflora* (L.) Fenzl.) BLUNTLEAF SANDWORT. Map 907. Local in moist woods throughout northern Indiana. It is most often found near the base of white and black oak slopes. When introduced into flower gardens, it stubbornly persists.

Arctic America southw. to N. J., Pa., Ohio, Ill. to Mo.; also in the Rocky Mts. southw. to N. Mex.; found also in Eurasia.

3. **Arenaria strícta** Michx. ROCK SANDWORT. Map 908. Local in northern Indiana where it usually grows in very sandy soil on black and white oak ridges. It is abundant on Hanging Rock along the Wabash River in Wabash County.

N. H., Ont., to Minn., southw. to Va. and Mo.

4. **Arenaria pátula** Michx. PITCHER SANDWORT. Map 909. On wooded gravelly slopes along streams and in shallow soil on sandstone bluffs. Local but very common in some of its stations.

Ind. to Minn., southw. to Ala. and Tex.

2449. SPÉRGULA L.

See excluded species no. 219, p. 1045.

2450. SPERGULÀRIA J. & C. Presl

See excluded species no. 220, p. 1045.

2475. PARONÝCHIA [Tourn.] Adans. FORKED CHICKWEED

[Fernald. Notes on Paronychia, Section Anychia. Rhodora **38**: 416-421. 1936.]

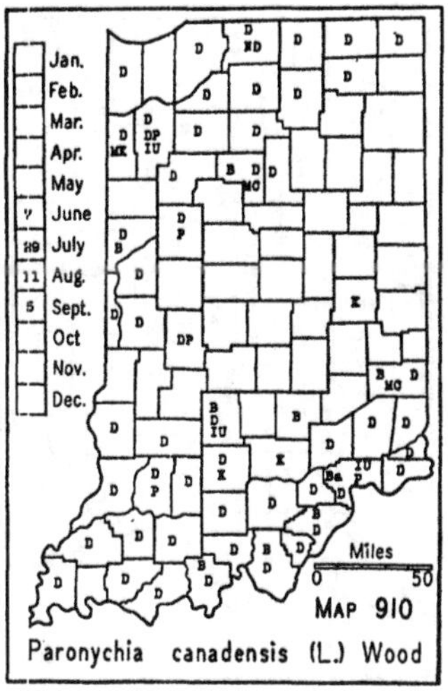

MAP 910
Paronychia canadensis (L.) Wood

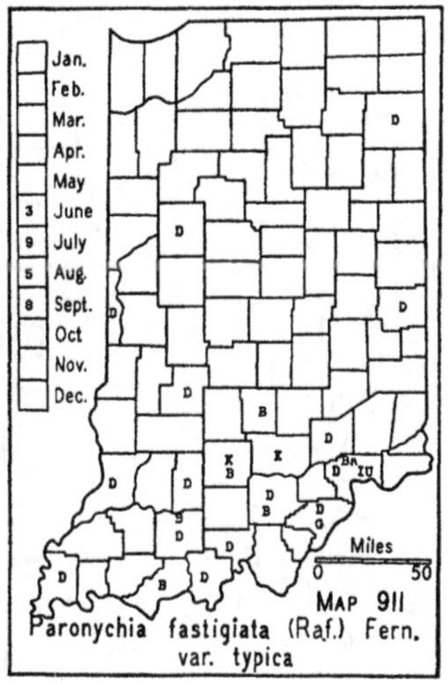

MAP 911
Paronychia fastigiata (Raf.) Fern. var. typica

MAP 912
Scleranthus annuus L.

Plants glabrous..1. *P. canadensis.*
Plants pubescent.
Stipular bracts subtending the flowers shorter than the calyx........2. *P. fastigiata.*
Stipular bracts subtending the flowers as long as or longer than the calyx.........
..2a. *P. fastigiata* var. *paleacea.*

1. **Paronychia canadénsis** (L.) Wood. (*Anychia candensis* (L.) BSP.) SMOOTH FORKED CHICKWEED. Map 910. Infrequent to rare throughout the state. This species prefers a dry, and rather sandy soil, or very sandy soil in dry places in woods, usually near the base of a large tree—which is usually a white or black oak—where the wind has kept the ground free from leaves and where the mineral soil is usually exposed. It is not absent from the central counties, as our map indicates, but it would be difficult to find it there now because woods that are not grazed are rare.

Vt., Ont. to Minn., southw. to Ga., Ark., and Kans.

2. **Paronychia fastigiàta** (Raf.) Fern. var. **týpica** Fern. (*Anychia polygonoides* Raf.) HAIRY FORKED CHICKWEED. Map 911. Infrequent northward and probably entirely absent from the northern tier of counties. It is found in dry places in sandy or gravelly soil, usually where the soil is exposed.

Mass. to Wis., southw. to Fla., Ala., and Tex.

2a. **Paronychia fastigiata** var. **paleàcea** Fern. Rhodora 38: 421. 1936.) Fernald has separated this form from the typical one because of the relative length of the bracts of the flowers. When I interpret this character as applied to my specimens I find intermediates between the two extremes. Neither this character nor others will satisfactorily separate the forms. Usually the typical form when mature is reddish in color and the flowers are much crowded on the ultimate branchlets while plants of the variety are generally greenish, usually with an erect inflorescence and the flowers are not crowded on the ultimate branchlets. It is to be noted that the variety flowers a month or more earlier than the typical form.

Del. and Pa. to Ill. and Tenn.

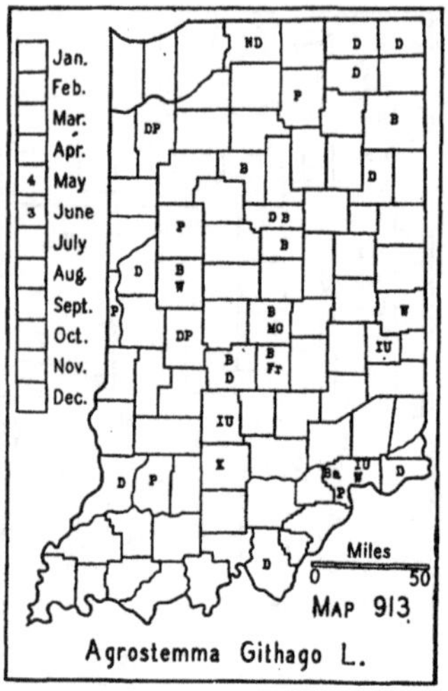

MAP 913
Agrostemma Githago L.

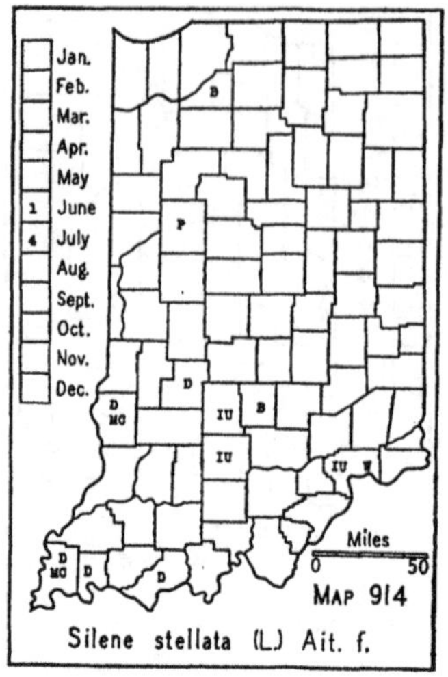

MAP 914
Silene stellata (L.) Ait. f.

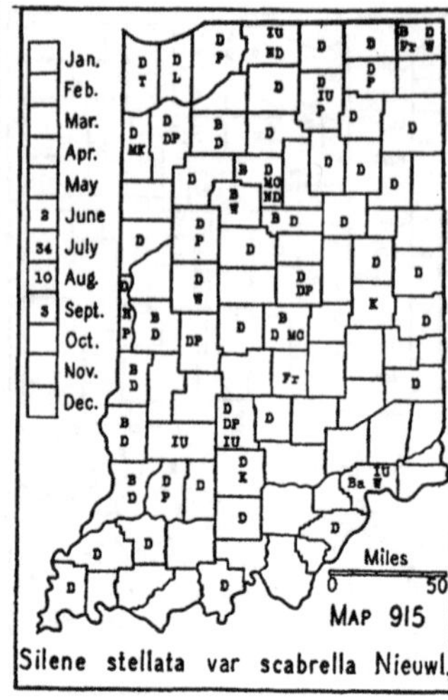

MAP 915
Silene stellata var scabrella Nieuwl.

2483. SCLERÁNTHUS L.

1. SCLERANTHUS ÁNNUUS L. KNAWEL. Map 912. This is a European weed that has been found in four places in Indiana. In 1914, Nieuwland found it as a weed at Webster Station west of Notre Dame, St. Joseph County. I have a specimen from Lagrange, which was sent to me in 1920 by the county agricultural agent who said it was a weed in an alfalfa field. I have another specimen from Lagrange County, which was sent to Purdue University from near Shipshewana. I also have a specimen sent to me in 1932 by H. C. Benke who found it near La Porte in La Porte County. No doubt this species has a wider distribution than our specimens indicate.

Nat. of Eu.; Que., Ont. to Minn., southw. to Fla.

2488. AGROSTÉMMA L.

1. AGROSTEMMA GITHÀGO L. CORN COCKLE. Map 913. A weed mostly in grain fields and fallow fields and along roadsides and railroads. It has been reported from all parts of the state and occurs in every county. It was formerly much more common than it now is because improved threshing machines separate it from the grain. When I was a boy 60 years ago it was my annual task when the cockle was in bloom to take a pair of scissors and go through the wheatfield and cut the cockle and rye. The whole plant, and especially the seed, is more or less poisonous. Wheat screenings that contained any great amount of cockle seed, when fed to poultry, have sometimes proved fatal.

Nat. of Eurasia; nearly throughout N. A.

2490. SILÈNE L. CATCHFLY

Leaves or some of them, verticillate in 4's; petals fringed.
  Leaves (except the margins and rarely a few hairs on the midrib) and stems glabrous. ....1. *S. stellata.*
  Leaves (at least the upper ones) and stems puberulent..1a. *S. stellata* var. *scabrella.*
Leaves all opposite; petals not fringed.

Calyx strongly inflated in fruit, more or less constricted at the mouth; plants glabrous throughout; leaves mostly lanceolate, generally 1-2 cm wide.
Bracts of the inflorescence leaflike; flowers few; plants not glaucous....2. *S. nivea.*
Bracts of the inflorescence much reduced; flowers numerous; plants glaucous...... ....3. *S. Cucubalus.*
Calyx not inflated or constricted at the mouth.
Plants glabrous throughout (sometimes the throat of the calyx pubescent) and usually glaucous.
Leaves less than 1 cm wide, narrowly lanceolate. (See excluded species no. 223, p. 1046.) ....*S. chlorantha.*
Leaves more than 1 cm wide.
Calyx club-shaped; capsule cylindrical. (See excluded species no. 221, p. 1046.) ....*S. Armeria.*
Calyx not club-shaped; capsule ovoid..........................4. *S. Cserei.*
Plants viscid-pubescent, pubescent or puberulent only on the lower internodes with a glutinous, colored band on the upper ones.
Flowers in racemes; pedicels mostly less than 5 mm long; annuals............ ....5. *S. dichotoma.*
Flowers not in racemes; pedicels mostly more than 5 mm long; annuals or perennials.
Whole plant not pubescent, the lowest internodes puberulent or scabrous, the upper ones glabrous with a dark, glutinous band; upper leaves linear, the lower ones lanceolate, linear-lanceolate or spatulate; corolla inconspicuous or lacking; capsules 5-8 mm long; annuals.
Leaves firm, erect or ascending; inflorescence stiffly ascending; capsules mostly 6-8 mm long.................................6. *S. antirrhina.*
Leaves thin, lax, spreading or some reflexed; inflorescence divaricate, the pedicels more filiform than in the preceding; capsules mostly 5-6 mm long...............................6a. *S. antirrhina* var. *divaricata.*
Whole plant pubescent; corolla usually very conspicuous; capsules about 1-2 cm long.
Calyx lobes mostly 5-8 mm long, linear-lanceolate; calyx in fruit ovoid or elliptical; flowers white; night-flowering annuals......7. *S. noctiflora.*
Calyx lobes mostly 2-4 mm long, ovate or triangular, acute or obtuse at the apex; calyx in fruit obovate; flowers red or pink, day-flowering; perennials.
Plants generally 1-2.5 dm high; inflorescence a terminal cyme; calyx in flower generally less than 4 mm wide at the middle (in pressed specimens). (See excluded species no. 222, p. 1046.)......*S. caroliniana.*
Plants generally more than 2.5 dm high; inflorescence cymose-paniculate or paniculate; calyx in flower generally more than 4 mm wide at the middle (in pressed specimens).
Leaves ovate, mostly clasping at the base, generally 10-20 pairs; plants erect, usually 8-12 dm high..........................8. *S. regia.*
Leaves spatulate or oblanceolate, the lower usually petiolate, the upper clasping, generally 2-4 pairs; plants ascending, generally 4-8 dm long...9. *S. virginica.*

1. **Silene stellàta** (L.) Ait. f. (*Silene stellata* in part, of Gray, Man., ed. 7 and Britton and Brown, Illus. Flora, ed. 2.) STARRY CATCHFLY. Map 914. The glabrous form of the species is the southern form and is represented in Indiana by a few specimens from the southern counties. In Sullivan County I found the species and the variety growing together. The species, like the variety, grows in dry woodland and is rarely found in

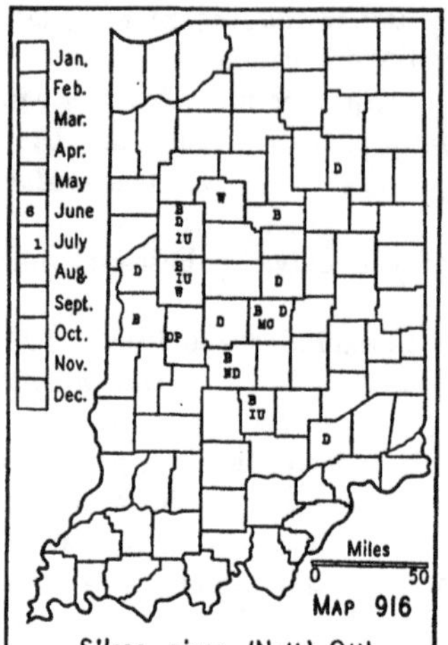

Silene nivea (Nutt.) Otth.

Silene Cucubalus Wibel

Silene Cserei Baumg.

clearings and along fences. I have no data concerning its distribution other than that it is known to occur from Pennsylvania and Indiana southward.

1a. **Silene stellata** var. **scabrélla** Nieuwland. (Amer. Midland Nat. 3: 58-59. 1913.) (*Silene stellata* in part, of Gray, Man., ed. 7 and Britton and Brown, Illus. Flora, ed. 2.) SCABROUS STARRY CATCHFLY. Map 915. Infrequent to frequent in dry woodland throughout the state. I have no data concerning its distribution. It occurs probably from Massachusetts to Minnesota and southward.

2. **Silene nívea** (Nutt.) Otth. (*Silene alba* Muhl.) SNOWY CATCHFLY. Map 916. An infrequent to rare plant probably throughout the southern two thirds of the state. There are several published records but these nearly all coincide with the distribution shown on the map. Its habitat is wooded ravines and wooded banks of streams.

Pa. to Minn., southw. to D. C. and Nebr.

3. SILENE CUCÙBALUS Wibel. (*Silene latifolia* (Mill.) Britten & Rendle.) BLADDER CATCHFLY. Map 917. My specimens were found in railroad ballast and in a pasture field. It has been reported from 7 counties.

Nat. of Eu.; N. B. to Wash., southw. to N. J. and Mo.

4. SILENE CSÈREI Baumgarten. Map 918. I have this species from Benton and Montgomery Counties. In Montgomery County, I found several large colonies in ballast, and on the right of way of the Monon Railroad about 2 miles south of New Richmond. Apparently well established here. Fassett reports a specimen from Lake County collected by Umbach which is now in the herbarium of the University of Wisconsin. I have a specimen from Lake County collected by Umbach which I am referring to this species. Charles M. Ek found it along a railroad in Howard County.

Nat. of Eu. and Asia Minor; Ohio, Ind., Wis., Iowa, Minn., and Mont.

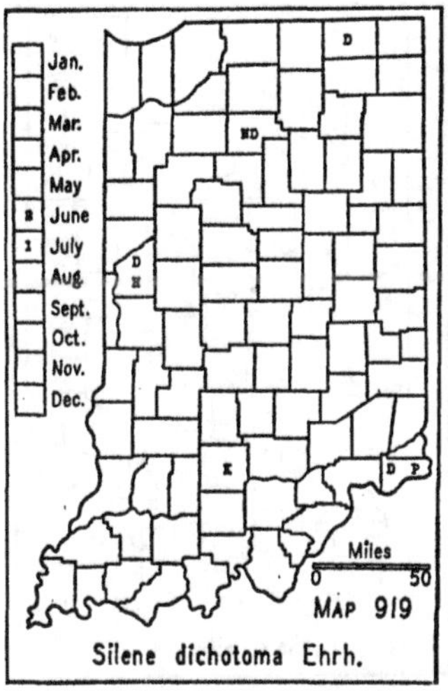

MAP 919

Silene dichotoma Ehrh.

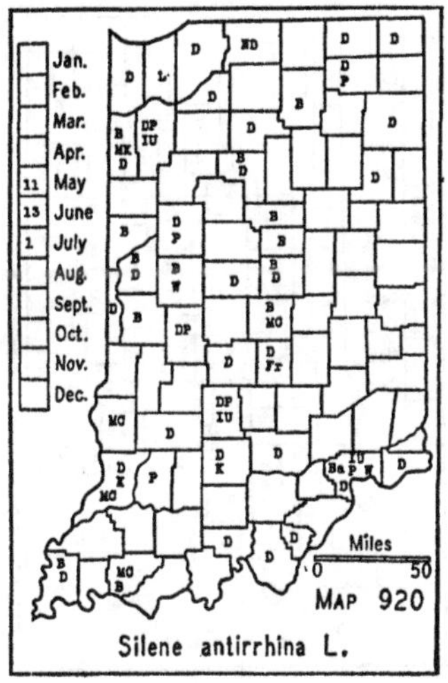

MAP 920

Silene antirrhina L.

MAP 921

Silene noctiflora L.

5. SILENE DICHÓTOMA Ehrh. FORKED CATCHFLY. Map 919. I have found this species in two places, and in each it seems to be well established. In 1915, I found it to be common along the roadside 4 miles north of Vevay in Switzerland County. In 1931 I found it to be a common weed in a very sandy alfalfa field about 4 miles southeast of Mongo in Lagrange County. I revisited the place in 1932 and I found the field in corn but the plant was frequent along the sandy roadside which bordered the field; and I found it still persisting in 1937. I believe it is well established in both places and it is entitled to membership in our flora. It has also been collected in Fountain, Fulton, and Lawrence Counties.

Nat. of Eu.; N. E. to Mo., southw. to Tex.

6. **Silene antirrhina** L. SLEEPY CATCHFLY. Map 920. This species is variable. It prefers a very sandy habitat and is frequent in railroad ballast throughout the state. It is less frequent in fallow and cultivated fields, pastures, and waste places and along roadsides. Our manuals give this species as a native of the United States. I do not believe, however, that it was a native of Indiana. Our earliest authors either do not list it or give it as a plant of waste places. M'Murtrie, who published a flora of Louisville in 1819, does not list it, nor does Riddell, who published his "Flora of the Western States" in 1835. Short, Peter & Griswold published a catalogue of the plants of Kentucky in 1833, and they do not list it. Neither do they list it in any of their four supplements, the last published in 1840. Lapham lists it from Illinois in his flora published in 1857. Dr. Clapp records that he found it in 1835 east of Corydon and in the "barrens." Young, in his catalogue of the plants of Jefferson County published in 1871, does not list it. J. M. Coulter, however, reports it in his catalogue published four years later. Schneck, who published a flora of the Lower Wabash Valley in 1876, says: "In poor grounds among cereals, common." Bradner, Phinney, and Van Gorder did not report it in their floras. Apparently it has become a frequent weed during the past 50 years. I believe it has been introduced mostly in grass and grain seed and by railroads.

A form with the internodes lacking the glutinous band is known as f.

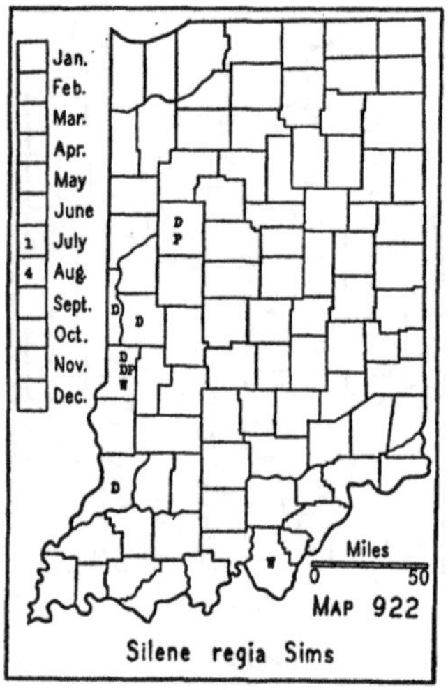

Silene regia Sims

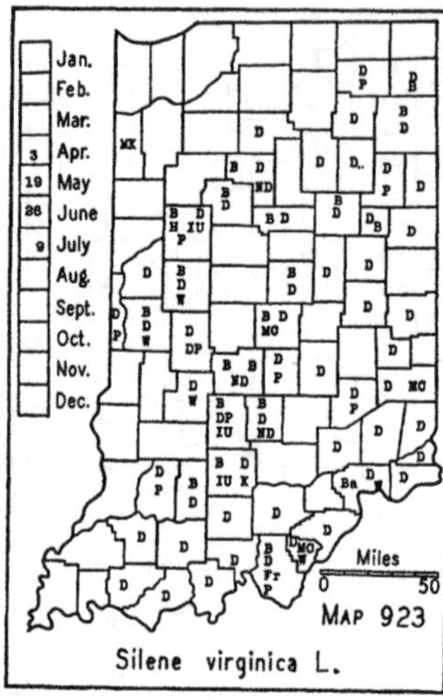

Silene virginica L.

Lychnis alba Mill.

Deaneana Fern. It occurs with the species and I found it in Posey County associated with the species and the variety.

Maine to B. C., southw. to Fla. and Mex.

6a. **Silene antirrhina** var. **divaricàta** Robinson. This variety has been reported from the dune area by Peattie. Evidently local and rare in the state. I have it from Kosciusko and Warrick Counties.

Mass. to Ill., Mo. and Kans.

7. SILENE NOCTIFLÒRA L. NIGHT-FLOWERING CATCHFLY. Map 921. This species has been reported from 8 counties as a weed of cultivated grounds and waste places. I have two specimens from open woodland.

This species much resembles *Lychnis alba* with which I think it is often confused. It can be easily separated from it by the following characters. *Silene noctiflora* has 3 styles, calyx 10-nerved, calyx lobes linear-lanceolate and mostly 4-8 mm long, and a capsule with 6 teeth. *Lychnis alba* has 5 styles, calyx with 10 strong and 10 faint nerves, calyx lobes triangular, mostly 3-5 mm long, and a capsule with 5 deeply bifid teeth. Specimens can be easily separated at any stage of growth.

Nat. of Eu.; N. B. and N. S. to Utah and Wash., southw. to Fla. and Mo.

8. **Silene règia** Sims. ROYAL CATCHFLY. Map 922. A very local plant, mostly of a dry, prairie habitat. All of my specimens are from roadsides. It has been reported from Hamilton, Vigo, and Wayne Counties and the authors say that it is scarce. It has been reported from 3 counties of Ohio.

Ohio to Mo., southw. to Tenn. and Ala.

9. **Silene virgínica** L. FIREPINK. Map 923. Frequent to common in rich woodland in all parts of the state except the northern counties of the northwestern part. Pepoon says: "Frequent in open woods from Whiting, Ind., south. Banks of the Des Plaines, abundant. (Babcock). Seems to be an error or the plant is exterminated in the Indiana territory named by Babcock." Babcock did not report this species from Indiana, so this reference

may safely be ignored. It did occur, no doubt, in the northwestern part of the state but possibly not near Lake Michigan.

N. J., N. Y., Ont. to Minn., southw. to Ga. and Mo.

2491. LÝCHNIS [Tourn.] L. CAMPION

Plants white-woolly all over; calyx teeth twisted; petals crimson. (See excluded species no. 225, p. 1046.) .... *L. Coronaria.*
Plants not white-woolly all over; calyx teeth not twisted.
  Flowers scarlet, many, in terminal clusters, opening in the morning; stem rough-pubescent, erect, simple, or little branched; capsule on a stipe about as long as the capsule. (See excluded species no. 224, p. 1046.) .... *L. chalcedonica.*
  Flowers white or pink, night-flowering, few, loosely paniculate; stem viscid-pubescent, loosely and freely branching; capsule large, ovoid, sessile in the calyx; flowers sometimes dioecious.. 1. *L. alba.*

1. LYCHNIS ÁLBA Mill. EVENING CAMPION. Map 924. A weed of fallow and cultivated fields and along roadsides. This species is frequently confused with *Silene noctiflora.* See the discussion under the latter species.

Nat. of Eu.; N. S. to Mich., southw. to N. Y. and Pa.

2502. DIÁNTHUS L. PINK

Plants more or less pubescent throughout; leaves linear; bracts 2; annuals.. 1. *D. Armeria.*
Plants glabrous, or the margins of the leaves and bracts ciliate; bracts 4; perennials.
  Leaves linear, 1-nerved; plants tufted, 2- or 3-flowered (the hardy, named garden pinks are derived from this species.) (See excluded species no. 227, p. 1046.) .... *D. plumarius.*
  Leaves lanceolate to ovate-lanceolate; stems erect, simple, or branched only above; flowers in dense, cymose clusters. (See excluded species no. 226, p. 1046.) .... *D. barbatus.*

1. DIANTHUS ARMÈRIA L. DEPTFORD PINK. Map 925. Our only report is by Nieuwland from St. Joseph County. I have found it well established in several of the southern counties along roadsides and in pastures and logging roads in woodland.

Nat. of Eu.; N. S., Ont., Mich. to Iowa, southw. to Ga.

2503. SAPONÀRIA L.

Leaves tapering at the base, sessile or short-petioled; calyx terete; flowers in rather dense, terminal and axillary clusters.. 1. *S. officinalis.*
Leaves rounded at the base, clasping; calyx strongly 5-winged; flowers in a loose corymbose cyme.. 2. *S. Vaccaria.*

1. SAPONARIA OFFICINÀLIS L. BOUNCING-BET. Map 926. This species prefers a very sandy soil in which it migrates rapidly. Found throughout the state along roadsides, railroads, and spillbanks of dredged ditches and in waste places and fallow fields. This perennial should be regarded as an obnoxious weed in the parts of the state where there is a sandy soil.

Nat. of Eu.; now throughout N. A.

2. SAPONARIA VACCÀRIA L. COW SOAPWORT. Map 927. This species has been reported from 9 counties. My specimens are from a roadside and the

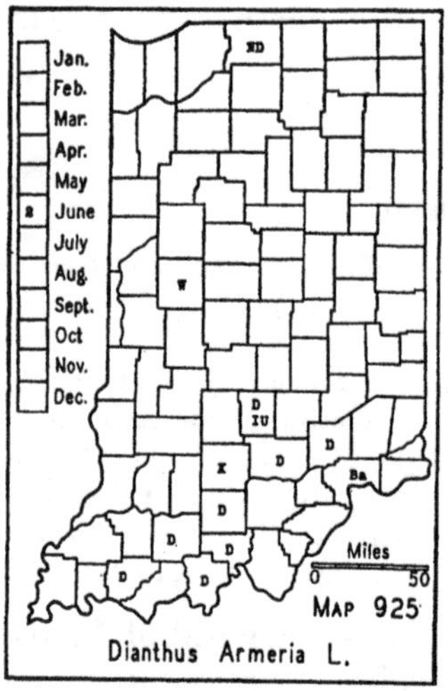

Dianthus Armeria L.

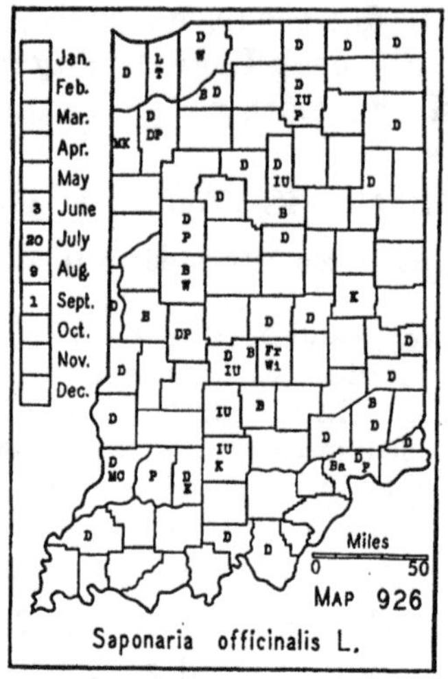

Saponaria officinalis L.

Saponaria Vaccaria L.

right of way of a railroad. While there are several reports, it is doubtful whether this species will spread a great deal or whether it will be able to maintain itself.

Nat. of Eu.; Ont. to B. C., southw. to Fla., La., and Calif.

88. NYMPHAEÀCEAE DC. Waterlily Family

Emersed leaves without a sinus, peltate.
  Leaves nearly orbicular, large, generally 3-6 dm in diameter; flowers very large, usually 1-2.5 dm wide, on erect peduncles well above the surface of the water ....2508. Nelumbo, p. 450.
  Leaves not orbicular, but oblong or oval, mostly 1-10 cm long; flowers generally less than 1.5 cm wide.
    Leaves of two kinds, the floating ones small, linear-oblong, usually 12-20 mm long, the submerged ones larger, palmately dissected; flowers white or yellowish within, generally 10-15 mm wide; stamens 3-6........2509. Cabomba, p. 452.
    Leaves of one kind, all floating, broadly oval, mostly 5-10 cm long, green above, purplish and very glutinous beneath; flowers purplish; petals mostly 12-15 mm long; stamens 12-18..........................2510. Brasenia, p. 452.
Emersed leaves with a sinus reaching to the petiole.
    Leaves orbicular; petals and stamens inserted on the sides and near the summit of the ovary, the petals showy, white, rarely a rose colored form........... ....2513. Nymphaea, p. 452.
    Leaves oval; petals and stamens hypogynous, the petals inconspicuous, yellow or with the base reddish..............................2514. Nuphar, p. 453.

2508. NELÚMBO [Tourn.] L.

1. **Nelumbo pentapétala** (Walt.) Fern. (Rhodora **36**: 23. 1934.) (*Nelumbo lutea* (Willd.) Pers.) American Lotus. Map 928. This plant is so conspicuous that botanical collectors would not overlook it. Before settlement in Indiana, it no doubt was local in nearly all parts of the state. Its habitat is the deeper ponds and shallow lakes or the borders of deeper ones. One writer suggests that its northern distribution is due to its introduction by the aborigines who used the seed and tubers for food. It is becoming extinct in Indiana. I know of five colonies and it is reported

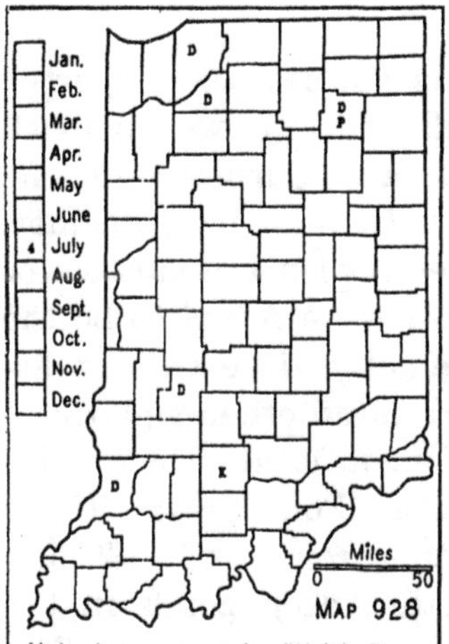

Nelumbo pentapetala (Walt.) Fern.

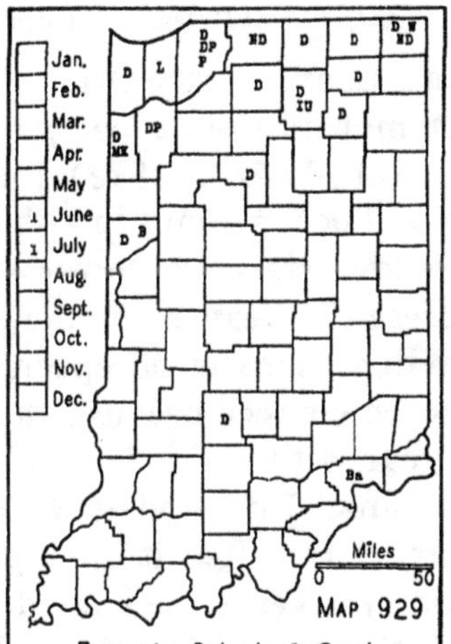

Brasenia Schreberi Gmel.

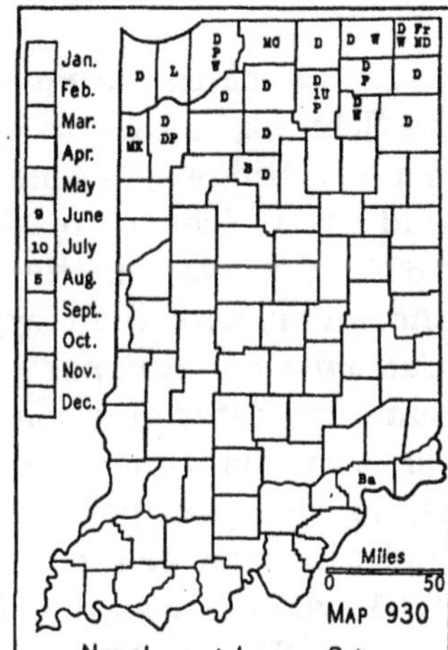

Nymphaea tuberosa Paine

still to exist in the Calumet Region. Drainage and ruthless digging of it have contributed to its disappearance. I recall that it was common in the Stodgdill Pond in Owen County and in Blue River Lake in Whitley County, but it has been gathered in these places until it is now nearly extinct. The species, however, spreads rapidly if not disturbed. In 1872 Babcock reported it as infrequent in Wolf and Calumet Lakes in Lake County. James, in a "Contribution to the Flora of Cincinnati, Ohio" writes that it was "abundant in a pond back of Jeffersonville" in Clark County in 1877. Clapp, who died in 1865, reports that it was scarce about New Albany in his time. Schneck, in 1876, reports it as "common in ponds" in the Lower Wabash Valley where it has now entirely disappeared. In 1897 Blatchley reported it as scarce in ponds in Vigo County. Ridgway mentions a trip to Foote's Pond in Gibson County in 1872, when Dr. Schneck and he measured leaves of it that were 3 feet in diameter.

There are records of the seed remaining dormant for at least 200 years and germinating (Plant Physiology 5: 225. 1930). The following quotation from Hooker's Jour. Bot. 1: 189. 1834 is instructive: "Dr. Short of Kentucky writes me . . . 'On the Ohio River, a hundred miles north of Lexington, my brother owns a considerable tract of land, a piece of which adjoining the river was subject to inundation, and in a shallow basin of 50 acres or more, the water remained throughout the year. Twenty years ago this basin was drained, sown in grass and is now a productive meadow, —the upper stratum being a tough, whitish clay. In plowing this piece of ground lately, immense quantities of the seeds of the *Cyamus* (*Nelumbo*) were turned up from among the clay in which they were embedded to a considerable depth; they are perfectly sound and hard, requiring much effort to break them open, and exhibiting, within, the cotyledons and embryo, full, plump, and apparently fresh;—none of them, however, manifest the slightest disposition to vegetate. The plant has certainly not grown there for twenty years; and the oldest resident of the neighbourhood has no recollection of having ever seen it.' "

In N. A. from Mass. to Minn., southw. to Fla. and Tex.

2509. CABÓMBA Aublet

1. **Cabomba caroliniàna** Gray. FANWORT. This species was reported by Schneck in 1876 as common in the deeper ponds of the Lower Wabash Valley. Ridgway (Amer. Nat. 6: 726. 1872) tells of a journey which he and Dr. J. Schneck made to Foote's Pond in Gibson County in September, 1871. He says: "Little, yellow, star-like flowers resting on the surface of the water, with their cypress-like leaves submerged, were found to be the *Cabomba caroliniana.*" While there is no specimen, I believe we can assume the determination to be correct because there is no other aquatic in flower at that time of the year with which it could be confused. The species is, no doubt, extinct in Indiana. I have visited Foote's Pond several times and I have never seen it and I revisited it during the drought of 1930 and 1931 when it was dry for two years. I visited other deep ponds in the Lower Wabash Valley at the same time and they were also dry. The drought probably killed many other species there.

Mich. to Mo., southw. to N. C., Fla., and Tex.

2510. BRASÈNIA Schreb.

1. **Brasenia Schrèberi** Gmel. WATERSHIELD. Map 929. Frequent, usually in 3-5 feet of water, on the borders of lakes and in a few dredged ditches in the lake area, very local south of this area. Reported in the southern part of the state by Banta from a pond in Jefferson County and by Clapp and Schneck, who say it was rare.

N. S. to Man., southw. to Fla. and Tex.; also found in Cuba, Mex., Asia, Africa, and Australia.

2513. NYMPHAÈA [Tourn.] L. WATERLILY

[Conard. The Waterlilies. 1-279. 30 pl., 81 fig. Carnegie Inst. Wash. 1905.]

Rootstocks bearing numerous, globular tubers; tubers easily detaching when mature; leaves green to purplish beneath, the veins usually more numerous and closer than in the next species; stripes on petioles conspicuous or lacking; sepals green; flowers mostly 10-23 cm in diameter, opening from 8 a. m. to 1 (2 or 3) p. m., scentless or nearly so; petals obovate to almost spatulate, generally rounded at the apex; filaments broader than the anthers; seed 2.8-4.4 mm long....1. *N. tuberosa.*

Rootstocks with few branches; branches not constricted at the base and persistent; leaves deep red to green, tinged with red beneath; stripes on petioles lacking; sepals often purplish outside; flowers mostly 7-12 cm in diameter, fragrant, opening from 6 a. m. to 12 m.; petals elliptic; inner filaments narrower than the anthers; seed 1.6-2.3 mm long. (See excluded species no. 228, p. 1046.)..*N. odorata.*

1. **Nymphaea tuberòsa** Paine. (*Castalia tuberosa* (Paine) Greene.) MAGNOLIA WATERLILY. Map 930. This species is frequent to common in the lakes and sloughs of the lake area and very local south of it because its habitat is lacking.

This species has generally been confused with *Nymphaea odorata.*

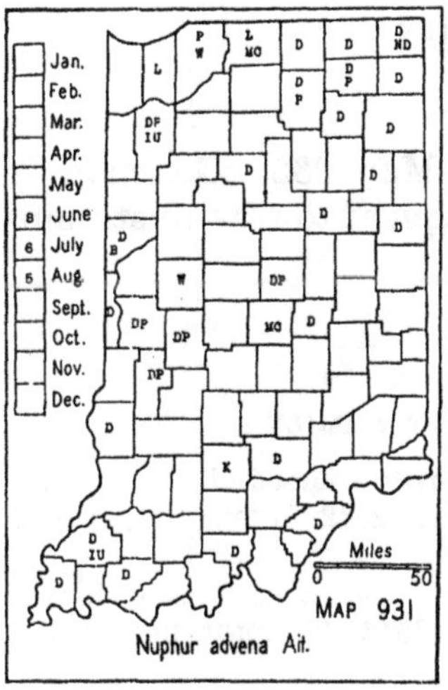

Nuphur advena Ait.

Nuphar variegata Engelm.

Ceratophyllum demersum L.

Conard wrote me in 1933 that the latter species belongs to the Coastal Plain and in the interior it does not occur as far south as Indiana.

On August 12, 1930, I found a rose colored form in the west side of Cheesborough Lake, Steuben County.

Lake Champlain to Trenton, N. J., westw. to Nebr. and Ark.

2514. NÙPHAR Smith

[Miller & Standley. The North American species of Nymphaea. Contr. U. S. Nation. Herb. **16**: i-ix. 63-108. 1912. Fernald. Nymphozanthus the correct name for the cow lilies. Rhodora **21**: 183-188. 1919. Fernald. Discusses the present nomenclature. Rhodora **39**:407-409. 1937.]

Leaves less than twice as long as wide; sinus a half to a fourth as long as the blade.
- **Petioles terete; leaf blades not floating, erect or spreading; basal sinus open, widest at the base of the lobes; sepals, petals, and fruit generally greenish yellow or yellow but with no tinge of red or only the crown reddish........1. *N. advena.***
- **Petioles conspicuously flattened; leaf blades floating; basal sinus narrow and usually nearly closed; sepals generally tinged with red inside at the base; petals usually red about half their length; fruit reddish at the base up to a half or three fourths its length..2. *N. variegata.***

Leaves more than twice as long as wide; sinus less than a fourth the length of the blade. (See excluded species no. 229, p. 1047.)....................*N. sagittifolia.*

1. **Nuphar ádvena** Ait. (*Nymphaea advena* Ait.) YELLOW SPATTERDOCK. Map 931. In shallow, running or stagnant water in lakes, streams, sloughs, and ponds throughout the state. Frequent to common in the lake area, and infrequent south of it.

N. Y. to Wis., southw. to N. C. and Tex.

2. **Nuphar variegàta** Engelm. (*Nymphaea advena* var. *variegata* (Engelm.) Fern.) VARIEGATED SPATTERDOCK. Map 932. Probably limited to the lakes of northern Indiana. I have no notes on its frequency or its associates. I have seen it in three lakes in Noble County. It is certain that it is much rarer than the preceding species.

Newf. to B. C., southw. to N. J., Pa., Ohio, and Mont.

89. CERATOPHYLLÀCEAE Gray

2516. CERATOPHÝLLUM L.

1. **Ceratophyllum demérsum** L. HORNWORT. Map 933. Common in most of the lakes of the lake region, becoming infrequent to rare in streams and ponds south of the lake region.

Throughout N. A. except the extreme north.

91. RANUNCULÀCEAE Juss. CROWFOOT FAMILY

Plants climbing; flowers white, maroon, or purple; leaves mostly compound.....................2542. CLEMATIS, p. 463.
Plants not as above.
  Leaves 3- or 4-ternate; plants dioecious; pistillate flowers white; staminate flowers greenish, whitish or purplish; fruit an achene.......2548. THALICTRUM, p. 473.
  Leaves not as above; plants not dioecious.
    Ovaries several-ovuled (1 or 2 in *Hydrastis*); fruit a follicle which sometimes resembles a berry; calyx generally petaloid.
      Flowers regular, white, scarlet or yellow.
        Leaves simple, either palmately lobed or divided, reniform or cordate.
          Flowers white, small, about 1 cm wide; leaves palmately 5-7-lobed; roots yellow; fruit red..........................2522. HYDRASTIS, p. 455.
          Flowers bright yellow or greenish yellow, large, generally 2-3 cm wide.
            Flowers bright yellow; leaves not divided; plants of springy places..........................2524. CALTHA, p. 455.
            Flowers greenish yellow; leaves divided into 7-11 lobes; plants introduced...........................2527. HELLEBORUS, p. 456.
        Leaves ternately decompound.
          Plants low, generally less than 30 cm high.
            Leaves basal and cauline, the basal ones biternate, the cauline ones ternate, alternate; flowers cauline, several axillary and terminal, the floral parts 5; roots fibrous. (Our species of this genus is often confused with *Anemonella thalictroides*. The roots of *Anemonella* are tuberous; cauline leaves in a terminal whorl; and floral parts more than 5.)..............................2532. ISOPYRUM, p. 456.
            Leaves all basal and ternate; 1-flowered, flowers on scapes..........................2534. COPTIS, p. 456.
          Plants tall, usually 0.5-2 m high.
            Flowers in terminal racemes, small, white.
              Racemes simple, generally less than 5 cm long; fruit red or white, fleshy, several-seeded, resembling a berry....2537. ACTAEA, p. 457.
              Racemes generally paniculate, usually 2-4 dm long; fruit many-seeded follicles............................2537A. CIMICIFUGA, p. 457.
            Flowers not in racemes, large, showy; petals spurred, scarlet..........................2538. AQUILEGIA, p. 458.
      Flowers irregular, blue or pinkish, except albino forms.
        Posterior sepal prolonged into a spur generally 10-15 mm long..........................2539. DELPHINIUM, p. 458.
        Posterior sepal hooded, covering the 2 petals.........2540. ACONITUM, p. 459.
    Ovaries 1-ovuled; fruit an achene.
      Leaves all radical; flowers on scapes.
        Leaves reniform, 3-lobed; scape 1-flowered..........2541B. HEPATICA, p. 462.
        Leaves linear; scape 1-flowered, the greatly elongated receptacle resembling a many-flowered spike......................2543. MYOSURUS, p. 464.

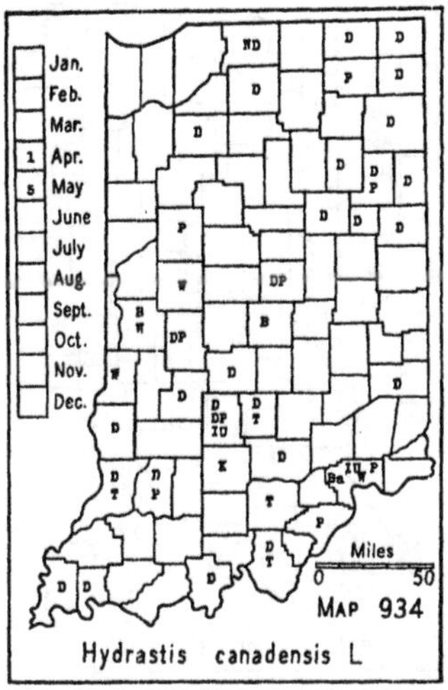

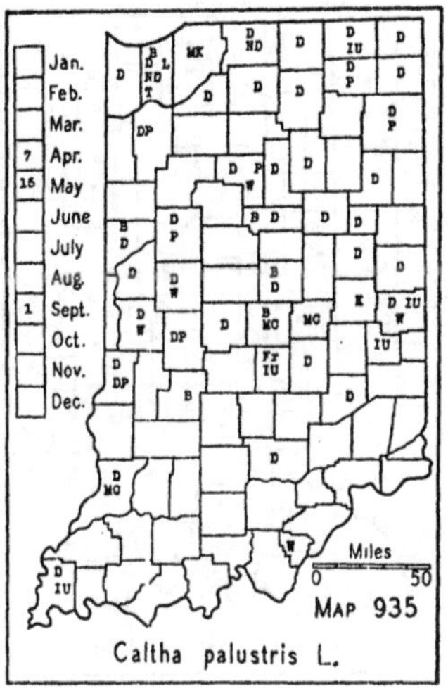

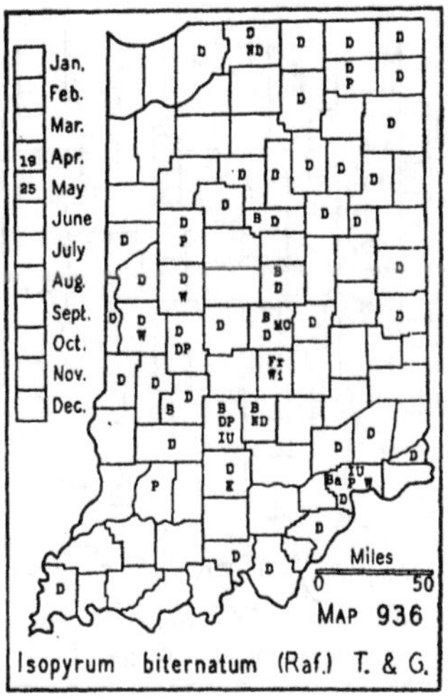

Leaves not all radical.
  Sepals and petals present; flowers yellow or white, and if white, the plants aquatic....................................2546. RANUNCULUS, p. 465.
  Sepals present, petals absent; flowers white or greenish white.
    Cauline leaves alternate, palmately lobed or ternately decompound.
      Leaves palmately lobed; flowers corymbose..2545. TRAUTVETTERIA, p. 465.
      Leaves ternately decompound with crenately lobed leaflets; flowers small, numerous, in panicles, dioecious or polygamous; achenes 5-ribbed.... ....2548. THALICTRUM, p. 473.
    Cauline leaves opposite or whorled; flowers few; sepals large, showy.
      Achenes not ribbed; leaves palmately incised, lobed, parted or divided.... ....2541. ANEMONE, p. 460.
      Achenes ribbed; leaves ternately decompound, those of the stem sessile, in a whorl near the summit..............2541A. ANEMONELLA, p. 461.

2522. HYDRÁSTIS Ellis

1. **Hydrastis canadénsis** L. GOLDENSEAL. Map 934. Infrequent to common in rich, moist woods throughout the state although there are no records or specimens from 5 of the northwestern counties. I once found it growing in a tamarack bog. From the earliest times it has been much used in medicine and now commands a high price. The root of this species and ginseng have always been valuable and for this reason they are almost extinct. I believe that goldenseal is now more rare than ginseng. Its scarcity and high price have resulted in its being cultivated.

Western N. E. to Minn., southw. to Ga., Mo., and Kans.

2524. CÁLTHA [Rupp.] L. MARSH MARIGOLD

1. **Caltha palústris** L. MARSH MARIGOLD. Map 935. Found in springy places about lakes, along streams and ditches, infrequent in swamps and ponds in woodland, and in the outlets of springs. It requires fresh water which is more or less circulating. It is frequent in the lake area, becoming less frequent to very local southward.

The leaves, both basal and cauline, vary much in the width of the basal sinus and in the margins. Some have a very wide sinus while others

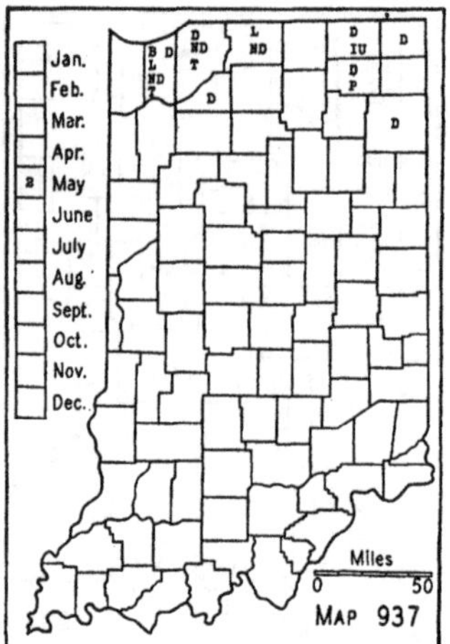

Coptis groenlandica (Oeder) Fern.

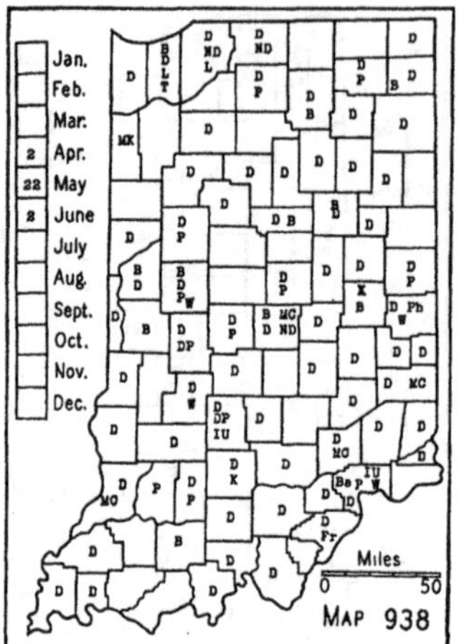

Actaea alba (L.) Mill.

Actaea rubra (Ait.) Willd.

have a narrow one. The margins vary from almost entire to acutely dentate. Nieuwland (Amer. Midland Nat. 3: 325. 1914) describes a form found in Porter County with "the upper leaves under the inflorescence somewhat lobed not unlike those of the red maple, the leaves were laciniately toothed."

Hansen reported *Caltha flabellifolia* from Kosciusko County. I am referring this report to *Caltha palustris*. See excluded species no. 230, p. 1047.

Lab. to Sask., southw. to S. C., Tenn., and Nebr.

2527. HELLÉBORUS [Tourn.] L.

See excluded species no. 231, p. 1047.

2532. ISOPỲRUM L.

1. **Isopyrum biternàtum** (Raf.) T. & G. FALSE RUE ANEMONE. Map 936. Infrequent to common in moist, rich woods throughout the state. Probably absent from a few of the northwestern counties. It prefers the beech and sugar maple type of woods.

Ont. to Minn., southw. to Fla. and Tex.

2534. CÓPTIS Salisb.

[Fernald. Coptis trifolia and its eastern American representative. Rhodora **31**: 136-142. 1929.]

1. **Coptis groenlándica** (Oeder) Fern. (*Coptis trifolia* of Gray, Man., ed. 7 and Britton and Brown, Illus. Flora, ed. 2.) GOLDTHREAD. Map 937. At one time probably infrequent to common in most of the tamarack bogs of the lake area. The tamarack bogs have nearly disappeared and those remaining have been so modified by drainage and grazing that this species has become rare and local. In Allen County its habitat has been destroyed. It was also found on sites where the tamarack bog had just passed into the

Betula lutea stage and where some of the sphagnum still remained. This plant was used in medicine.

Lab. to Alaska, southw. to Md. and Iowa, and in the mts. to N. C. and Tenn.

2537. ACTAÈA L. BANEBERRY

[Mackenzie. White-fruited Bane-berries. Torreya 28: 51-53. 1928.]

Pedicels stout, generally 3-9 mm long in flower; fruit white (sometimes purplish red but this form not yet reported from Indiana); largest seed nearly 5 mm long; leaves glabrous beneath or sometimes with a few hairs on the veins near the axils. ....1. *A. alba.*

Pedicels slender, generally 5-11 mm long in flower; fruit cherry red; largest seed nearly 4 mm long; leaves with a few scattered hairs on the midrib and lateral veins beneath........2. *A. rubra.*

1. **Actaea álba** (L.) Mill. WHITE BANEBERRY. Map 938. Frequent to very frequent throughout the state in rich woods. There is a red-fruited form of this species which has not been reported from Indiana but may have been found and reported as *Actaea rubra.* The rhizomes of this and the following species were formerly much used in medicine.

N. S., e. Que. to Minn., southw. to Ga. and Mo.

2. **Actaea rùbra** (Ait.) Willd. RED BANEBERRY. Map 939. This species is very local in a few of our northern counties. I found a single specimen in a low woods in Lagrange County and it is rather frequent on a springy, wooded terrace in the east side of Pokagon State Park. It has been reported from Noble County. A specimen from St. Joseph County is in the herbarium of the University of Notre Dame. There is a report from Tippecanoe County which could also be correct. The reports by Phinney from central-eastern Indiana and the report from Jefferson County are open to question. I believe these reports should be referred to the red-fruited form of the preceding species since they come from south of the general range of distribution and the habitat of *Actaea rubra,* although there were cold springy areas about 4 miles southeast of Richmond. It is regrettable that our early authors did not preserve specimens to validate their reports. These reports would be very interesting if they could be authoritatively interpreted.

Lab. to S. Dak., southw. to N. J., Pa., Tenn., and Nebr.

2537A. CIMICÍFUGA L.

1. **Cimicifuga racemòsa** (L.) Nutt. BLACK COHOSH. Map 940. Local to very local on wooded slopes in the southern counties. I have also found it on the wooded bank of Cedar Creek in Allen County. Outside the area indicated on the map, Coulter reported it from Kosciusko, Shelby, and Tippecanoe Counties, Higley & Raddin reported it from Pine, Lake County, and Schneck reported it from the Lower Wabash Valley and says: "Once common, now almost extinct." The plant is so conspicuous that if it was at all frequent I would have found it elsewhere in southern Indiana. The

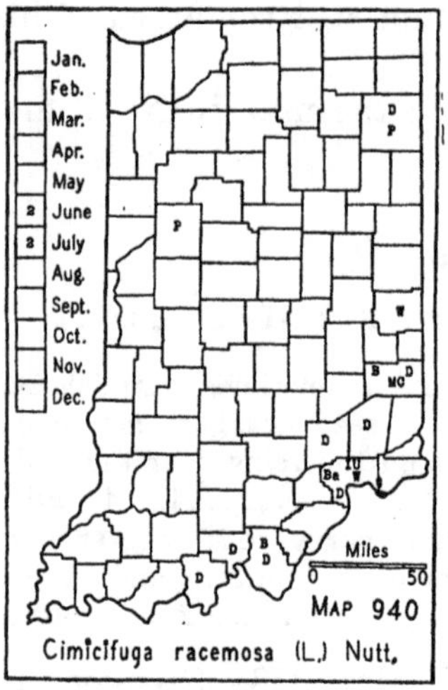

Cimicifuga racemosa (L.) Nutt.

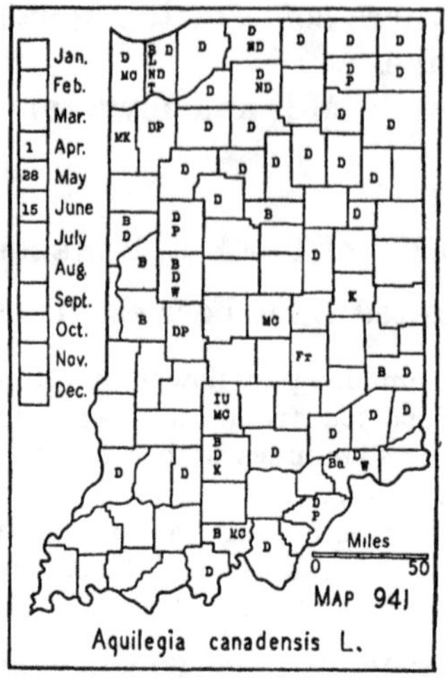

Aquilegia canadensis L.

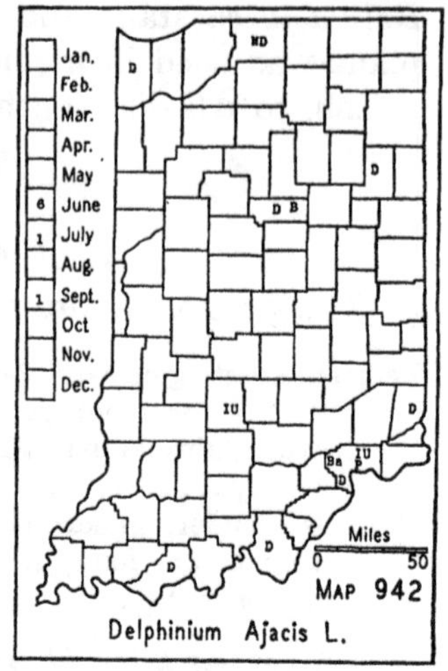

Delphinium Ajacis L.

rhizomes and roots have been much used in medicine and since the plant is so conspicuous I fear "root collectors" have almost exterminated it. It is commonly known as black cohosh and to the eclectic physician as macrotys.

Maine, Ont. to Wis., southw. to Ga. and Mo.

2538. AQUILÈGIA [Tourn.] L.

1. **Aquilegia canadénsis** L. AMERICAN COLUMBINE. Map 941. Local throughout the state on the wooded bluffs of streams, wooded slopes and banks of streams, banks and slopes of deep ravines, and rarely far removed from stream courses. I have twice found it in open tamarack bogs where it was associated with *Rhus Vernix* and *Rhamnus alnifolia.* I have also frequently found it growing in the rocky crevices of cliffs along streams. I suspected this wide difference of habitat would show some difference in the structure of the plants but I find none. The plant when taken from the wild and planted in the garden thrives and reproduces freely from seed, which fact is not entirely consistent with its restricted distribution along streams.

N. S. to Alberta, southw. to Fla. and Tex.

2539. DELPHÍNIUM L. LARKSPUR

[Wilde. Studies of the genus Delphinium. Cornell Univ. Agric. Exp. Sta. Bull. 519: 1-107. 1930.]

Pistil 1; capsules densely pubescent; leaf-segments mostly less than 2 mm wide; annual. .... 1. *D. Ajacis.*
Pistils 3; capsules glabrous or the sutures somewhat pubescent; leaf-segments more than 2 mm wide; perennial........ 2. *D. tricorne.*

1. DELPHINIUM AJÀCIS L. ROCKET LARKSPUR. Map 942. Somewhat frequent in fallow fields and open woodland near the Ohio River in the southeastern part of the state and probably very local elsewhere. Where

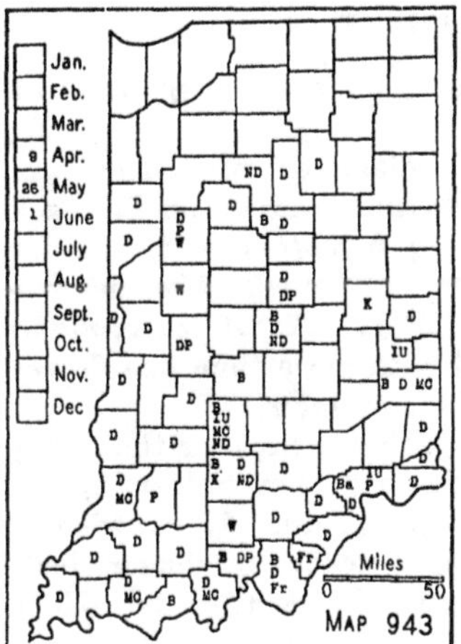

MAP 943
Delphinium tricorne Michx.

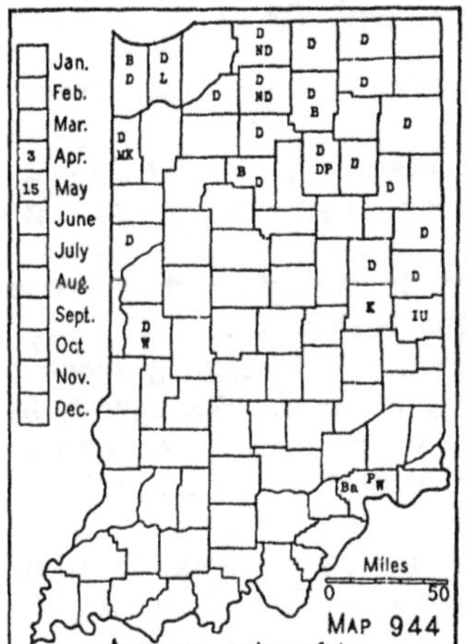

MAP 944
Anemone quinquefolia var. interior Fern.

MAP 945
Anemone caroliniana Walt.

it is common, I have seen blue, pink, and white forms of it in the same colony.

Nat. of Eu.; N. S. to Mont. and Kans., southw. to S. C.

2. **Delphinium tricórne** Michx. ROCK LARKSPUR. Map 943. Infrequent to frequent in rich soil on wooded slopes in the southern counties, becoming less frequent northward and probably very local or entirely absent from the northern two tiers of counties. It seems to have very little affinity for streams, because it is usually found near the bases of slopes of ravines as well as along streams. This wild species is easily cultivated in the garden. I have a specimen which I collected on May 1, 1910, in a woods near Wilson Creek northwest of Lawrenceburg, in Dearborn County, on which I have the following note: "In this station I estimate that there is an average of 1 plant for every square foot of space over an area of 20 acres of woodland." I have seen it in large colonies but usually only a few plants are found at a station. The plant is poisonous to stock. I met a farmer who lived a mile north of Cedar Grove in Franklin County who called the plant stagger weed and told me that he had known cattle to be killed by eating it.

Pa. to Minn., southw. to Ga. and Ark.

2540. ACONÌTUM [Tourn.] L.

1. **Aconitum uncinàtum** L. CLAMBERING MONKSHOOD. This species was reported by Short in his Fourth Supplement of the Plants of Kentucky as occurring in the "barrens" of Indiana. On January 1, 1927, I found, in the herbarium of the Philadelphia Academy of Sciences, two well preserved and ample specimens of this species collected by C. W. Short. The labels are as follows: "Barrens of Ia. near Corydon, Sept. 1840" and "Barrens of Indiana near Corydon, Oct. 1842." The identification of the specimens is correct. The species may be extinct in Indiana.

S. Pa., southw. in the mts. to Ga., westw. to Wis. and southw. to Ky.

2541. ANEMÒNE [Tourn.] L. ANEMONE

Stems generally 1-2 dm high, simple, 1-flowered, flowering in Indiana mostly in April and before May 20.

Plants from a horizontal rootstock, generally 2-3 mm in diameter; stem leaves on petioles 0.5-2 cm long; sepals generally 5, oblong or oval, glabrous on the back. ....1. *A. quinquefolia* var. *interior.*

Plants from a tuber 6-8 mm in diameter; stem leaves sessile; sepals 6-20, narrow-oblong, pubescent on the back..........................2. *A. caroliniana.*

Stems more than 2 dm high, generally branching above and with more than one flower, flowering in Indiana mostly after May 20, usually in June or later.

Stem leaves sessile or nearly so; fruiting heads orbicular; achenes broader than long, more or less pubescent with long, straight hairs, the body not hidden by the pubescence..3. *A. canadensis.*

Stem leaves on petioles more than 1 cm long; fruiting heads generally oblong; body of achene longer than wide, so densely woolly with long hairs as to be hidden by them.

Margins of the segments of the 3-parted leaves irregularly toothed or cleft above the middle only; fruiting heads more than twice as long as wide............ ....4. *A. cylindrica.*

Margins of the segments of the 3-parted leaves sharply serrate or double-serrate to below the middle; fruiting heads less than twice as long as wide........ ....5. *A. virginiana.*

1. **Anemone quinquefòlia** L. var. **intèrior** Fern. (Rhodora 37: 260. 1935.) AMERICAN WOOD ANEMONE. Map 944. Infrequent to frequent or local in rich, moist woods in the northern half of the state. I have no specimens from the southern part although there are reports of it from six southern counties.

N. Ont. to e. Man., southw. to Ohio, Ind., Ill., and Iowa.

2. **Anemone caroliniàna** Walt. CAROLINA ANEMONE. Map 945. This species was reported by Blatchley in Indiana Geol. Rept. 21: 628. 1897. He says it was reported by Miss Nora Arnold, who knew of its growing for 12-13 years in patches on a hill along Durkey's Ferry Road about 5½ miles north of Terre Haute, Vigo County. In April, 1933, I asked Prof. Fred Donaghy of the Terre Haute State Normal School to try to rediscover this species. He found it and sent me specimens. He wrote that a colony about 10 feet square was located on the slope of a bluff opposite Durkey's Ferry. The plants grew in sandy soil among grasses, had very shallow roots, and were 3-8 inches high.

Open places, Wis. to Dak., southw. to Fla. and Tex.

3. **Anemone canadénsis** L. MEADOW ANEMONE. Map 946. Found in low ground in woodland and along roadsides, mostly in alluvial soil along streams. Infrequent to frequent in the northern half of the state and in the Wabash Valley, becoming rare or absent in the hilly counties.

Cent. Maine, e. Que. to Alberta, southw. to Md., Mo., Kans., and Colo.

4. **Anemone cylíndrica** Gray. CANDLE ANEMONE. Map 947. Infrequent throughout the lake area. It prefers a very sandy or gravelly soil and is usually found in prairie habitats along railroads and roadsides and in open woodland, usually on open, black and white oak ridges. Apparently

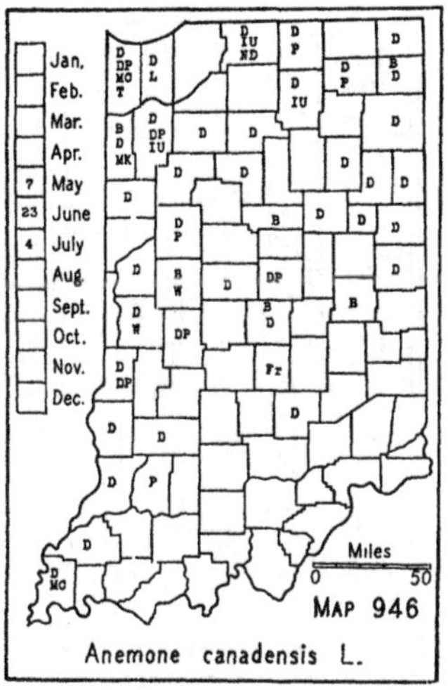

MAP 946

Anemone canadensis L.

MAP 947

Anemone cylindrica Gray

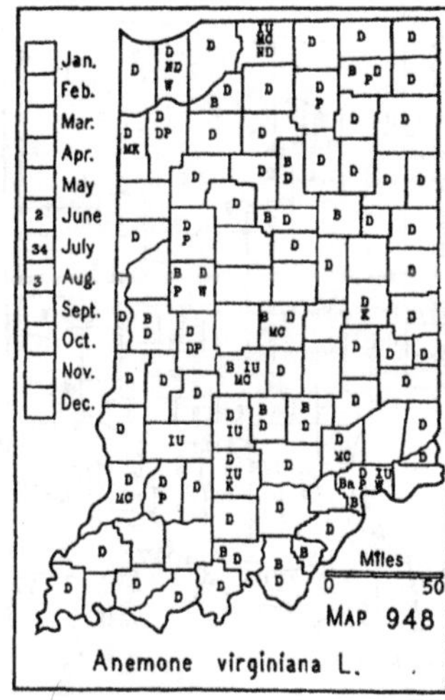

MAP 948

Anemone virginiana L.

it prefers a dry habitat, although the only place I ever saw it growing in abundance was on a gravelly bench on the north side of Wall Lake in Lagrange County, which was only a few feet above the water level. This bench was made several years ago when the water level of the lake receded. In this moist habitat the plants were common and vigorous, one of the specimens having 7 fruiting heads.

Western Maine to Sask., southw. to N. J., Pa., Ill., Mo., Kans., N. Mex., and Ariz.

5. **Anemone virginiàna** L. TALL ANEMONE. Map 948. Infrequent to frequent throughout the state. This is a woodland species and is rarely found in the open along roads and railroads. It generally grows in dry soil on wooded slopes and has a decided preference for slopes along streams.

The species varies considerably in the length of its stamens and in the size, shape, texture, and color of its sepals. Some of the variations have been given names but after a careful study of my 77 specimens from all parts of the state I have decided that the characters are too variable to be of taxonomic value.

Maine, s. Que. to Minn., southw. to Ga. and Ark.

2541A. ANEMONÉLLA Spach

1. **Anemonella thalictroìdes** (L.) Spach. (*Syndesmon thalictroides* (L.) Hoffmg.) Map 949. Infrequent to very frequent in all parts of the state except in the prairie areas and in very sandy areas. This is strictly a woodland plant which is found generally in dry soil, usually on slopes and banks along streams and ravines.

This species is variable in many ways and some of the variations have received names. Hill (Bot. Gaz. 10: 262. 1885.) wrote of finding specimens near Hobart, Lake County, with "flowers greatly doubled, of 20-30 purplish petals, alternating in whorls."

Mass., Ont. to Minn., southw. to Fla., Tenn., and Kans.

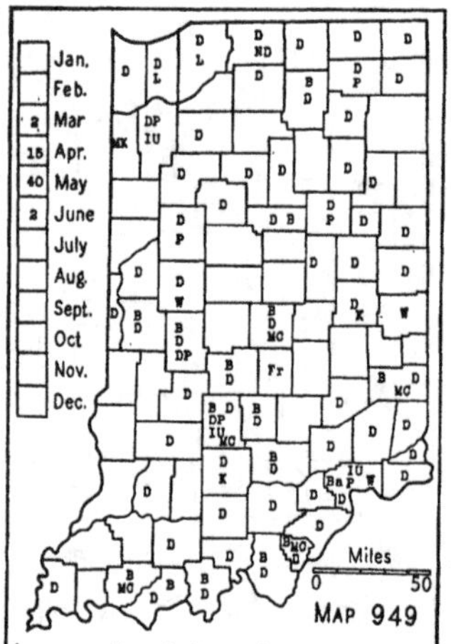

Anemonella thalictroides (L.) Spach

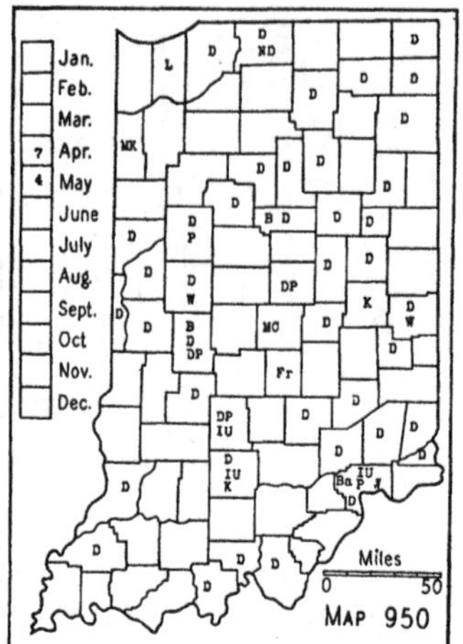

Hepatica acutiloba DC.

Hepatica americana (DC.) Ker.

2541B. HEPÁTICA [Rupp.] Hill HEPATICA

Leaf lobes acute or somewhat obtuse at the apex....................1. *H. acutiloba.*
Leaf lobes rounded at the apex....................................2. *H. americana.*

1. **Hepatica acutíloba** DC. SHARPLOBE HEPATICA. Map 950. Infrequent to frequent throughout the state except in the prairie area. The hepaticas are strictly woodland plants. This species is generally found in rich soil on wooded slopes and these most frequently near streams. The flowers vary from white to pink and purple. The leaves of both of our hepaticas vary in color from green to green mottled with maroon. Some authors regard *H. acutiloba* as only a variety of the next species, but it is entirely distinct although the characters separating it are difficult to describe. The leaves and achenes of *Hepatica acutiloba* are slightly larger than those of *Hepatica americana.* This species is a lime loving plant while the next prefers a slightly acid soil and this requirement, I believe, excludes it or makes it rare in southern Indiana.

Western N. H., w. Que. to Minn., southw. to Ga. and Mo.

2. **Hepatica americàna** (DC.) Ker. (*Hepatica triloba* of Gray, Man., ed. 7, not Chaix and *Hepatica Hepatica* (L.) Karst. of Britton and Brown, Illus. Flora, ed. 2.) (Fernald. The specific characters of Hepatica americana. Rhodora **19**: 45-46. 1917.) ROUNDLOBE HEPATICA. Map 951. Infrequent to locally frequent in the lake area, becoming local and very rare in the southern part, or possibly absent. It has been reported from several of the southern counties but there are no specimens. It is possible that the identifications are wrong since this species prefers a rather acid soil, but the habitat does occur locally in southern Indiana, and I see no reason why it should not be found. My efforts to find it have failed.

The flowers vary in color from white to pink and purple. These forms have been assigned names but Weatherby (Rhodora **27**: 131-132. 1925) found that white forms are constant but that color forms may be one color one year and another color the next year. Hence I am omitting these color form names.

N. S. to Man. and Minn., southw. to Fla. and Mo.

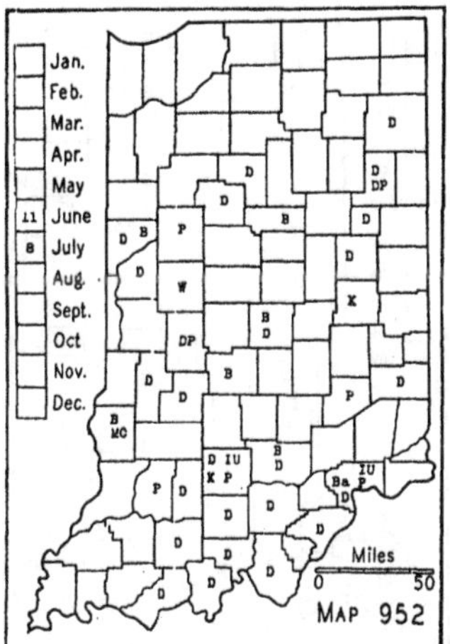

Clematis Viorna L.

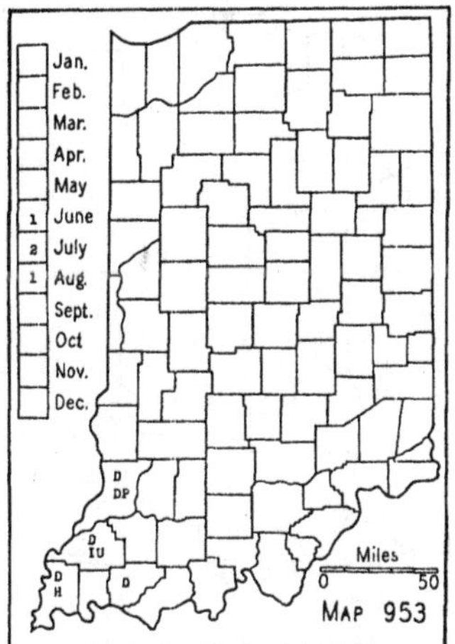

Clematis Pitcheri T. & G.

Clematis virginiana L.

2542. CLÉMATIS L.

Flowers solitary; sepals thick, purplish or pinkish purple.
Styles pubescent their entire length with long spreading hairs; hairs usually about 3 mm long at maturity, those toward the apex shorter; stigmatic lines about 0.5-1 mm long; mature stigma about 4 cm long; flowers pinkish purple.. 1. *C. Viorna.*
Styles pubescent only at the base or up to the middle with more appressed hairs; hairs usually about 2 mm long; upper part of style naked or with only a few hairs; stigmatic lines usually 1.5-2 mm long; mature stigma about 2 cm long; flowers purplish or nearly colorless above the middle............. 2. *C. Pitcheri.*
Flowers many, cymose-paniculate; sepals thin, white................. 3. *C. virginiana.*

1. **Clematis Viórna** L. (*Viorna Viorna* (L.) Small.) LEATHERFLOWER. Map 952. Infrequent to rare throughout the state except in the northern two tiers of counties where it is either absent or very rare. Clark's report from Marshall County is the only one from these counties. It is found mostly on the rocky, wooded slopes of streams.

The leaves of this species, as of the next two, are variable in the amount of the pubescence of the lower surface of the leaflets. Some are nearly glabrous while the majority are more or less densely pubescent. Plants with the apex of the leaflets long-acuminate are *Clematis Ridgwayi* Standley. I have a specimen of this form from Martin County named for me by Standley, and I have specimens from other counties which I refer to it.

Pa. to Mo., southw. to Ga. and La. (Brown. Torreya **29**: 159. 1929.)

2. **Clematis Pítcheri** T. & G. (*Viorna Pitcheri* (T. & G.) Britt.) PITCHER LEATHERFLOWER. Map 953. This species is very local in low ground in the Lower Wabash Valley and I have found it twice near the Ohio River. The reports of Blatchley from Vigo County and Clements from Daviess County are, no doubt, correct. Thompson reported it from Carroll County and this report, without doubt, should be referred to the preceding species.

Southern Ind. to Nebr., southw. to Tex.

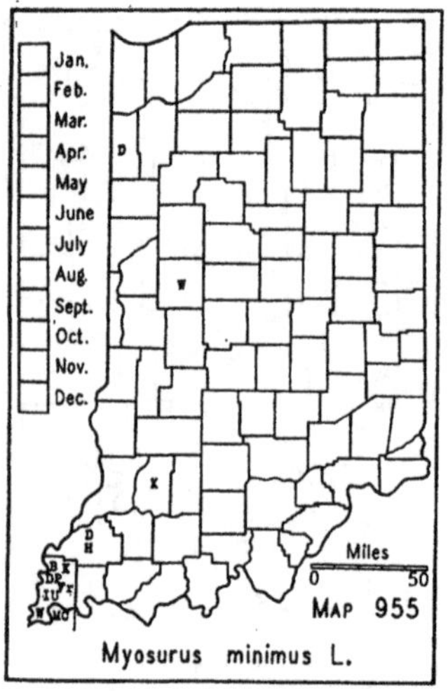

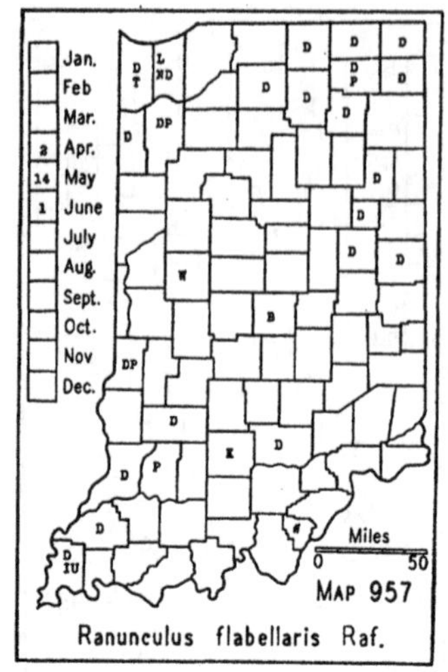

3. **Clematis virginiàna** L. VIRGINS-BOWER. Map 954. Infrequent or rarely frequent throughout the state. It is found in moist ground, generally along streams, about lakes, and along fences.

The stems of my specimens are more or less fluted and more or less densely appressed-pubescent. The upper and lower surfaces of the leaves vary from sparingly to densely pubescent, the lower surface sometimes velvety to the touch.

N. S. to Man., southw. to Ga. and La. (Brown. Torreya 29: 159. 1929.)

2543. MYOSÙRUS L.

1. MYOSURUS MÍNIMUS L. MOUSETAIL. Map 955. I have found this species in five widely separated places in Posey County. I found it abundant in very sandy soil in a forest planting of about 15 year old chestnut about 4 miles north of Mt. Vernon; in a pasture field just south of New Harmony Cemetery; in a sandy by-road along the Wabash River about 3 miles south of New Harmony; frequent in a fallow cornfield on the south side of Pitcher "Lake" about 5 miles northwest of Mt. Vernon; and common in a fallow cornfield along Black River about 2 miles east of Griffin. I found it also in Gibson County, in a fallow cornfield 6 miles west of Princeton, and recently Kriebel has collected it in Daviess County. There is a specimen from Montgomery County in the herbarium of Wabash College. The plant is very inconspicuous and since I have rarely botanized its habitat, it may be more frequent than my experience indicates. Schneck reported it from the Lower Wabash Bottoms and Benedict & Elrod reported it from Cass County, remarking that it was "scarce." In 1937 I found it in its habitat in Newton County.

I am of the opinion that if sandy, fallow cornfields are worked carefully, its distribution would be greatly extended.

Nat. of Eu. and reported from the U. S. at widely separated stations, Ont., B. C., southw. at Norfolk, Va., Fla., Tex., N. Mex., and Ill. Some of the reports may be referable to some other species. For its distribution see Greene (Amer. Midland Nat. 3: 311-316. 1914).

2545. TRAUTVETTÈRIA F. & M.

1. **Trautvetteria carolinénsis** (Walt.) Vail. FALSE BUGBANE. Map 956. This species was reported from the "knobs" on the authority of Clapp in the "Catalogue of Plants of Indiana" published in 1881. This report was repeated in Coulter's Catalogue to which was added a report for Barnes from Clark County. Dr. Clapp was an industrious and, I think, a very accurate botanist. He came to Indiana about 1817 and continued his botanical work until his death in about 1865. I was fortunate in being able to purchase his copy of Gray's Manual, first edition, and an interleaved copy of Riddell's "Flora of the Western States" in which he kept a list of the plants he collected in the vicinity of New Albany. In the Riddell's Flora he had bound 48 blank pages, on which he kept records. When he found a species in Indiana, he indicated it by a check mark in the catalogue. On one of the blank pages he summarized his work up to the end of 1840 and he recorded a total of 918 plants, which included 25 ferns. His last note was made in 1857 and whether he failed to keep records after that date I do not know. Nowhere in his books, however, does he mention collecting this plant under the name *Trautvetteria* or any of its synonyms. There is a specimen in the herbarium of Purdue University which is from the herbarium of C. R. Barnes and the label states that it was collected by A. Clapp, 1837, near New Albany, Indiana (Floyd County). There is another specimen collected in 1837 by Dr. Clapp in the herbarium of Wabash College. These specimens were, without doubt, the basis for the Floyd County report for Clapp in Coulter's Catalogue. The fact that the first specimen was in the Barnes herbarium probably led to the report of his collecting it in Clark County, where most of Barnes' collecting was done. There is not now a specimen in the Purdue herbarium which was collected by Barnes in Clark County, nor does Barnes mention this species in any of his writings. There is a specimen in the herbarium of DePauw University collected by Blatchley which was in bud June 8, 1889, and was collected in the Heckland prairie about 10 miles northeast of Terre Haute, Vigo County, and one in the Gray Herbarium bearing the following label: "Low prairies, w. Ind. E. F. Shipman, 1876."

Md., sw. Pa. to Mo., southw. to Ga.

2546. RANÚNCULUS [Tourn.] L. BUTTERCUP

The status of some of the species of this genus has been variously interpreted. The species have been divided, and the names have been changed since publication of the fifth edition of Gray's Manual and of Wood's Class-book of Botany (1885). Since these books were used by our early botanists, it is not satisfactory to accept the early reports of the species of this genus.

Plants aquatic; leaves finely dissected; achenes wrinkled.
Flowers yellow.
Leaves of submerged plants sessile or on petioles less than 1 cm long, the segments acute; leaves of emersed plants on petioles mostly 1-3 cm long; achenes

margined at the base, the beak about half the length of the body of the achene..1. *R. flabellaris.*

Leaves of submerged plants on petioles more than 1 cm long, the segments obtuse; achenes not margined, the beak about a third as long as the body of the achene. (See excluded species no. 241, p. 1049.)..................*R. Purshii.*

Flowers white.

Leaves petiolate, 2-2.5 cm long, collapsing when taken from the water...2. *R. trichophyllus.*

Leaves sessile or nearly so, usually 1-1.5 cm long, not collapsing when taken from the water...3. *R. longirostris.*

Plants terrestrial, growing in springy places and lowland, and on dry, wooded slopes; leaves not finely dissected; achenes not wrinkled.

Radical and stem leaves all entire or dentate.

Radical leaves ovate or oblong; stem leaves linear or lanceolate; blades generally less than 6 cm long and 1 cm wide; annuals.

Plants glabrous, erect at first, then trailing; petals 1-5, about 1.5 mm long; stamens 3-10; achenes brown, beakless......................4. *R. pusillus.*

Plants sparingly appressed-pubescent, erect; petals 5, 3-7 mm long; stamens usually 20-25; achenes greenish; beak about 0.2 mm long....5. *R. oblongifolius.*

Radical and stem leaves lanceolate or the lowest ones oblong, generally all denticulate; blades mostly 6-12 cm long and 5-30 mm wide; plants ascending and rooting at the lower nodes, glabrous; perennial...........6. *R. ambigens.*

Radical and stem leaves not all entire or dentate.

Achenes less than 1.7 mm wide, without a distinct margin, the style very short, not over 0.4 mm long; petals generally very small, shorter or not much longer than the sepals.

Plants glabrous (sometimes a few hairs on the stipular sheaths at the base of the leaf); annuals.

Radical leaves mostly crenate, sometimes 3-lobed or 3-parted, conspicuously cordate; calyx glabrous; petals shorter than the reflexed calyx; heads subglobose; achenes mostly 1.2-1.7 mm wide, the stigmas nearly sessile; plants of various habitats, the stems solid...............7. *R. abortivus.*

Radical leaves all lobed or parted, not conspicuously cordate; calyx pubescent; petals about equaling the calyx; heads cylindric; achenes mostly 0.8-0.9 mm wide, the stigmas nearly sessile; plants of springy or very wet places, the stems hollow...8. *R. sceleratus.*

Plants more or less pubescent.

Roots not thickened; annuals; plants of springy or very wet places; stems hollow, usually 3-9 dm high; heads of fruit cylindric; achenes many. ....8. *R. sceleratus.*

Roots thickened; perennials; plants generally of the crests of wooded ridges or of wooded slopes; stems solid, generally less than 3 dm high; heads subglobose; achenes few.............................9. *R. micranthus.*

Achenes more than 1.7 mm wide, with a distinct margin, the style more than 0.4 mm long; petals rather large except in nos. 11 and 12 (*R. recurvatus* and *R. pennsylvanicus*).

Base of stem swollen, bulblike; introduced plant................10. *R. bulbosus.*

Base of stem not swollen, not bulblike.

Flowers small; petals generally less than 3 mm long, shorter than or scarcely exceeding the sepals.

Stems solid, with a spreading pubescence; none of the radical or stem leaves divided to the base; mature heads globose; mature styles (beaks) hooked..11. *R. recurvatus.*

Stems hollow, hispid-pubescent; some or all of the radical leaves and usually some of the stem leaves divided to the base into 3 parts, the parts stalked; mature heads longer than wide; mature styles straight. ....12. *R. pennsylvanicus.*

Flowers large, 1.5-2.5 cm in diameter; petals much exceeding the sepals.
Styles in fruit less than 1 mm long, strongly curved; plant introduced, erect; radical and stem leaves 3-7-parted, usually to the base, the divisions sessile or rarely on petiolules........................13. *R. acris.*
Styles in fruit mostly 1 mm long or longer, straight or curved but not hooked except in no. 15 (*R. repens* var. *villosus*); radical and stem leaves mostly divided to the base and all of the divisions, at least the middle one, stalked.
Roots fleshy, much thickened; radical and stem leaves less than 4 cm wide, pinnately cleft or divided, the lobes or divisions narrow, linear-lanceolate; native plant of a dry, sandy habitat, generally less than 25 cm high; pubescence of the stems and petioles of our plants appressed......................................14. *R. fascicularis.*
Roots not fleshy; plants erect, trailing or creeping, mostly of a wet habitat; radical and lower stem leaves more than 4 cm wide.
Styles ending in a minute hook; introduced, creeping plants mostly of lawns and waste places.
Pubescence spreading....................15. *R. repens* var. *villosus.*
Pubescence appressed. (See excluded species no. 242, p. 1049.)...... ....*R. repens.*
Styles straight or slightly curved, not hooked at the tip.
Plants erect or ascending, never trailing and rooting at the nodes or tips, usually densely spreading-pubescent; styles 1.5-2 mm long... ....16. *R. hispidus.*
Plants erect at first, later trailing and becoming prostrate and rooting at some of the nodes or at the tips.
Stems and petioles of the later radical leaves upwardly appressed-pubescent or nearly glabrous............17. *R. septentrionalis.*
Stems and petioles of the later radical leaves spreading-pubescent.
Pubescence of the stem and of the petioles of the later radical leaves usually not dense and not retrorse; fruiting heads globose with about 20 achenes; styles 1.5-2 mm long....... ....17a. *R. septentrionalis* forma.
Pubescence of the stem and of the petioles of the later radical leaves very dense and retrorse at least on the lower parts; mature heads longer than wide (not measuring the styles), with up to 50 achenes to a head; styles 2.25-3 mm long, rarely shorter; plants robust, few-flowered, soon becoming prostrate and rooting at the nodes...................... ....17b. *R. septentrionalis* var. *caricetorum.*

1. **Ranunculus flabellàris** Raf. (*Ranunculus delphinifolius* Torr.) Map 957. Somewhat frequent in ponds, sloughs, and dredged ditches in the lake area, becoming infrequent to very local or absent from the southern part of the state. When its habitat becomes dry during the summer months, this species assumes a terrestrial form. This form has been named *Ranunculus flabellaris* f. *riparius* Fern. (Rhodora 38: 171. 1936.) (*Ranunculus delphinifolius* var. *terrestris* (Gray) Fern.) Its appearance is somewhat different from the aquatic form and I believe it has been the source of several reports for *Ranunculus Purshii.*

Central Maine, Ont., Mich., southw. to N. C. and Ark.

2. **Ranunculus trichophýllus** Chaix var. **týpicus** Drew. (Rhodora 38: 18-29. 1936.) (*Ranunculus aquatilis* var. *capillaceus* DC. and *Batrachium trichophyllum* (Chaix) F. Schultz.) Map 958. This is a species found

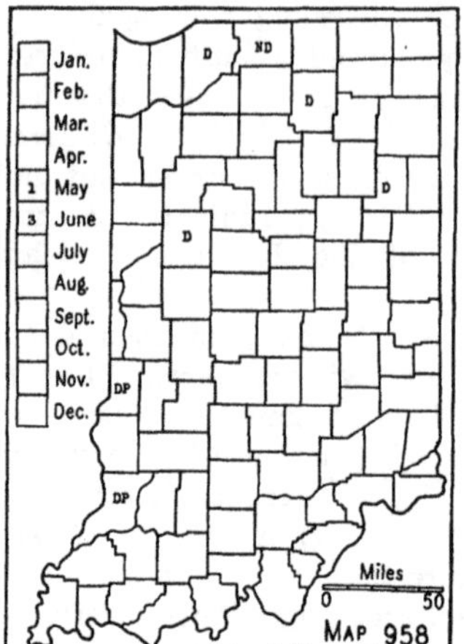

Ranunculus trichophyllus Chaix var. typicus

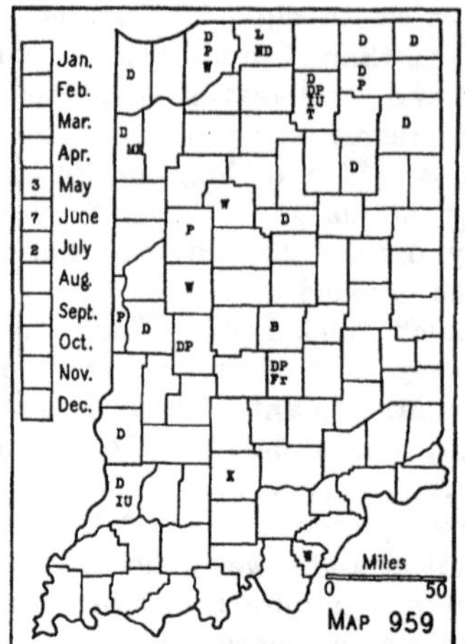

Ranunculus longirostris Godr.

Ranunculus pusillus Poir.

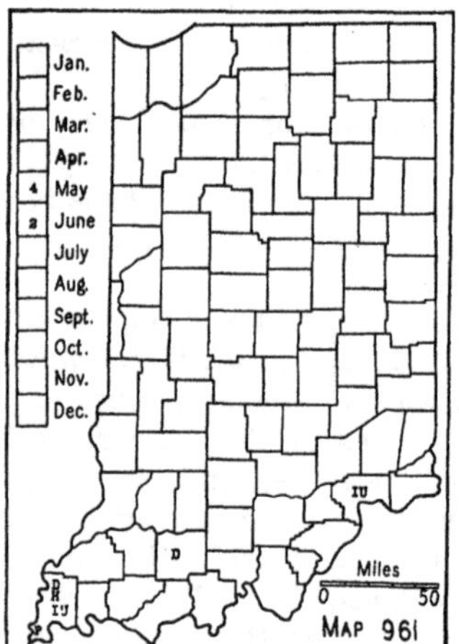

Ranunculus oblongifolius L.

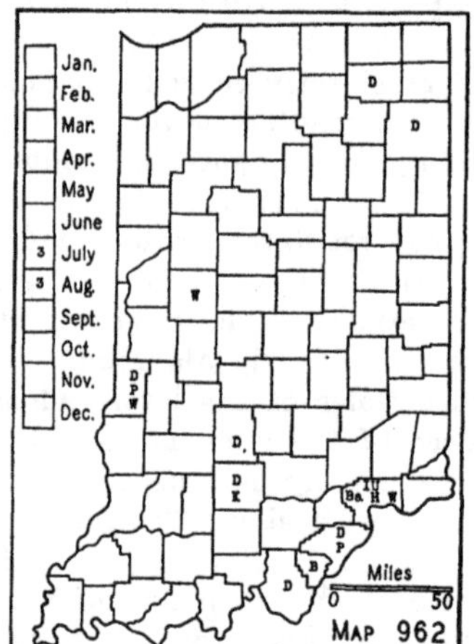

Ranunculus ambigens Wats.

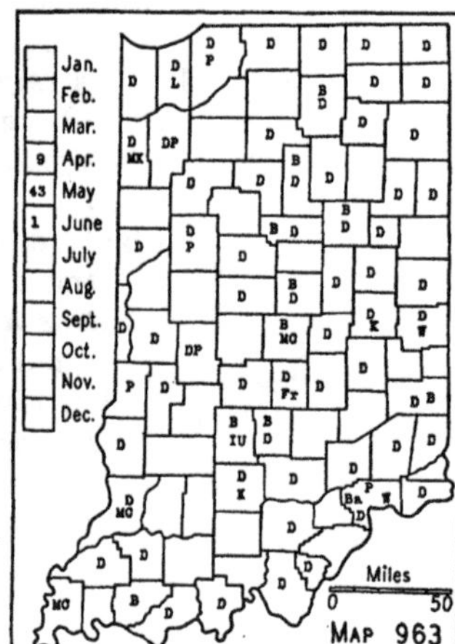

Ranunculus abortivus L.

infrequently in shallow water on the shores of lakes and bayous of streams and in ditches.

Lab., Newf., N. S., Maine, and the Coastal Plain to N. J., westw. to Minn., Alaska, and Calif.; also in temperate S. A.; Eurasia.

3. **Ranunculus longiróstris** Godr. (Rhodora 38: 42-46. 1936.) (*Ranunculus circinatus* of authors and *Batrachium circinatum* of manuals.) Map 959. This is our most common white-flowered buttercup. Its habitat is similar to that of the preceding species and it is more or less frequent in the lake area, becoming rare south of it.

Que. to Oreg., southw. to Del., Pa., Tenn., Nebr., Kans., Tex., Ariz., and N. Mex.

4. **Ranunculus pusíllus** Poir. Map 960. This species was reported from Knox County by Spillman. I found a large colony of it in a low woods about a mile east of Palmyra in Harrison County, and it is a common

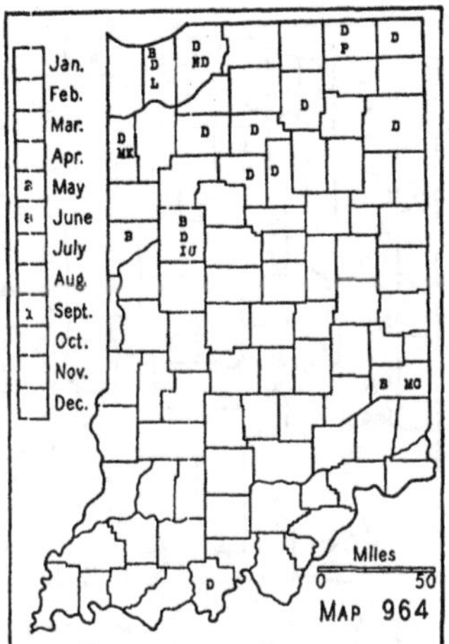

Ranunculus sceleratus L.

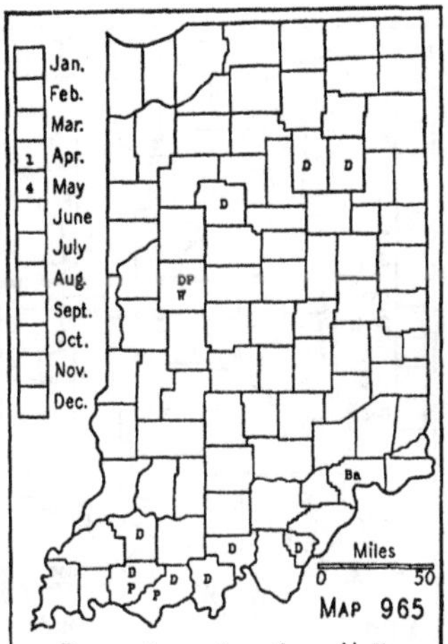

Ranunculus micranthus Nutt.

Ranunculus bulbosus L.

plant surrounding a pond of about 2 acres nearly 2 miles southeast of Palmyra. It has also been found by Edna Banta in Jefferson County.

Atlantic coast from s. N. Y. to Fla., westw. through the Gulf States to Tex., and northw. up the Mississippi Valley to Ind.

5. **Ranunculus oblongifòlius** Ell. Map 961. I found this species to be frequent in a low, open woods in the Hunley Bottoms about a mile northeast of Huntingburg in Dubois County, and in three widely separated places in Posey County where it was common in hard, white clay soil in very wet, fallow fields. Winona Welch collected it in Graebert's woods about 8 miles southwest of Mt. Vernon in Posey County, and Edna Banta found it in the "flats" in Jefferson County.

Atlantic coast from Del. to Fla., westw. to Tex., and northw. in the Mississippi Valley to Okla. and Ind.

6. **Ranunculus ámbigens** Wats. (Rhodora 38: 173-175. 1936.) (*Ranunculus laxicaulis* (T. & G.) Darby and *Ranunculus obtusiusculus* Raf.) Map 962. Very local in swamps and ponds throughout the state although there are no reports from the northwestern part. The most northern report is that from Noble County by Van Gorder. This species grows in swamps, ponds, and ditches that are full of water in the spring and become dry in summer when it flowers and fruits.

Maine to Minn., southw. to Ga. and Ark.

7. **Ranunculus abortìvus** L. SMALL-FLOWER BUTTERCUP. Map 963. This buttercup is frequent to common throughout the state and is found in all kinds of habitats except in very sandy or very wet places. Where clean cultivation is desired in lawns, orchards, and elsewhere, it is a pernicious weed.

Lab. to Man., southw. to Fla., Ark., and Colo.

8. **Ranunculus scleràtus** L. CURSED BUTTERCUP. Map 964. Local in ponds, springy places, and ditches in the lake area and absent or very

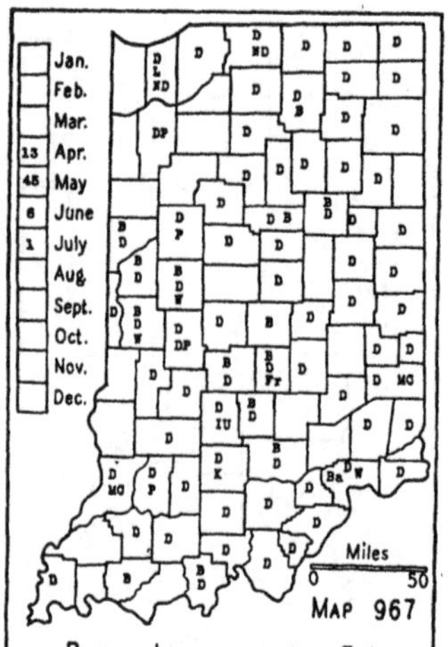

Ranunculus recurvatus Poir.

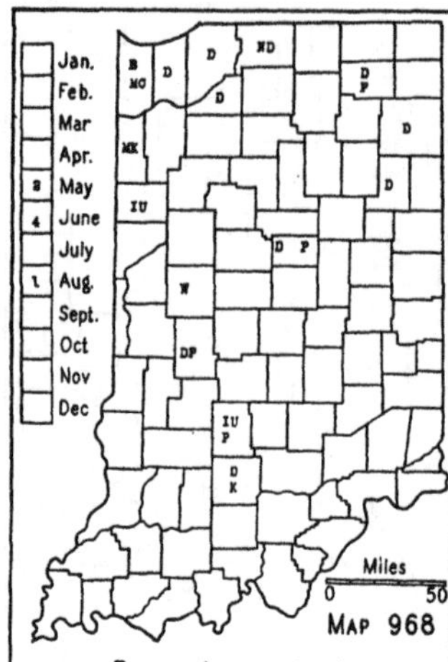

Ranunculus acris L.

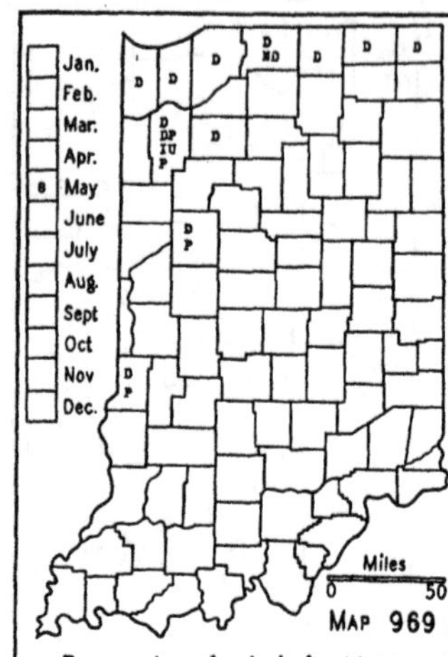

Ranunculus fascicularis Muhl.

local south of this area. Our plants vary from glabrous to pubescent all over.

Throughout N. A. except in the extreme north; also in Eurasia.

9. **Ranunculus micránthus** Nutt. (*Ranunculus cymbalistes** Greene.) Map 965. There are seven reports for this species, all from the southern half of the state. My specimens are from near the bases of slopes wooded with sugar maple and from the crests of wooded ridges. It is either very local or so inconspicuous that it is overlooked.

Maine to Minn. and Sask., southw. to Ga., Ark., and Colo.

10. RANUNCULUS BULBÒSUS L. BULB BUTTERCUP. Map 966. There are reports from seven counties. Nieuwland collected it in Lake and Porter Counties and the specimens are in the herbarium of the University of Notre Dame. Clark collected it in Marshall County and the specimen is in the National Herbarium at Washington, D. C. A specimen collected by Flora Haas in Owen County is in the herbarium of Indiana University.

Nat. of Eu.; N. E. to Ind., southw. to N. C., and La.

11. **Ranunculus recurvàtus** Poir. HOOKED BUTTERCUP. Map 967. Infrequent to frequent or even common in moist, rich woods throughout the state.

N. S. to Man., southw. to Fla., Ala., Mo., and Kans.

12. RANUNCULUS ÀCRIS L. TALL BUTTERCUP. Map 968. This species has been reported from eight counties, mostly northern. It is local along railroads and roadsides. I have seen it abundant in a few places, and in time no doubt it will become a weed in favorable habitats. My Allen County specimen, no. 47063, is exceptional in that the blades of the stem leaves are divided into three parts and these are on petiolules 0.5-3 cm long.

Nat. of Eu.; Newf. to B. C., southw. to Va. and Mo.

* Fernald makes this name a variety and calls it **Ranunculus micranthus var. cymbalistes** (Greene) Fern. (Rhodora 41: 543. 1939.)

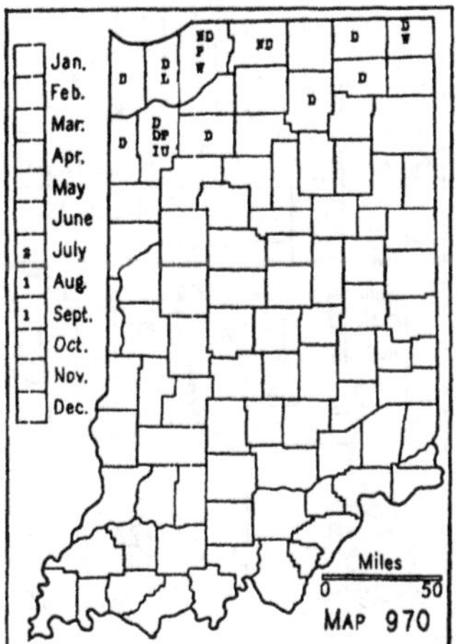

Ranunculus pennsylvanicus L. f.

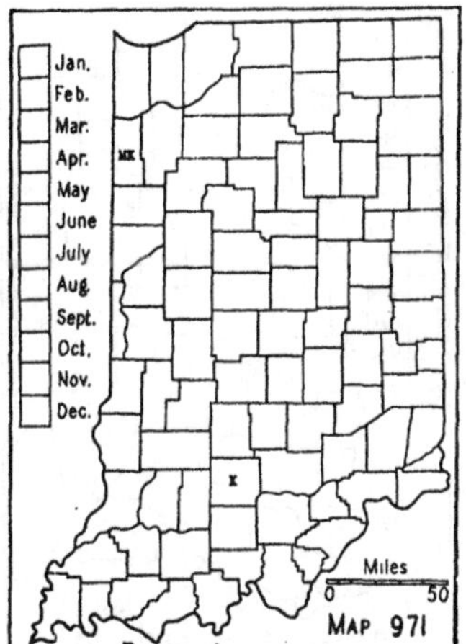

Ranunculus repens var. villosus Lamotte

Ranunculus hispidus Michx.

13. **Ranunculus fasciculàris** Muhl. TUFTED BUTTERCUP. Map 969. I have found this species in dry, sandy soil locally in only the northern counties, but it has been reported from Clark, Dearborn, Decatur, Franklin, Jefferson, Monroe, Shelby, and Vigo Counties and from the Lower Wabash Valley.

Eastern Mass., Ont., Wis., and Minn., southw. to N. C., Tex., and Kans.

14. **Ranunculus pennsylvánicus** L. f. PENNSYLVANIA BUTTERCUP. Map 970. Infrequent in wet places about lakes and in marshes and rarely in ditches in the lake area. Coulter reported it in his Catalogue for McDougal from Putnam County, but there is no specimen in the herbarium of DePauw University where MacDougal's Indiana specimens are deposited.

Newf. to B. C., southw. to Ga., Kans., and Colo.

15. RANUNCULUS RÈPENS L. var. VILLÒSUS Lamotte. (Fernald. Variations of Ranunculus repens. Rhodora **21**: 169. 1919.) Map 971. This species was reported by the majority of the earlier authors and undoubtedly all or nearly all of the reports should be transferred to other species. Coulter, in his catalogue, transferred most of them to *Ranunculus septentrionalis*. It has recently been found as a weed in a lawn at Goodland, Newton County, by Madge McKee, and as a weed in a lawn in Bedford, Lawrence County, by Ralph M. Kriebel. It doubtless is more widely distributed in our state than our reports indicate.

Nat. of Eu. and nw. N. A.; introd. in e. U. S.

16. **Ranunculus híspidus** Michx. BRISTLY BUTTERCUP. Map 972. This species, as I am treating it, seems to me to be a complex but I am not able to divide it. It has been reported from seven counties. I have 67 sheets from Indiana and these come from all parts of the state and from many habitats—from bogs to the crests of our highest and driest hills, and from the shade in woods to the brilliant sun of roadsides and prairies. The foliage varies so widely that no classification can be made on this

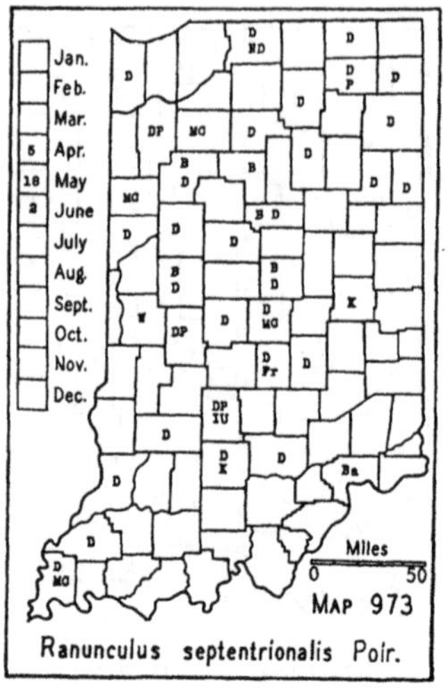

MAP 973
Ranunculus septentrionalis Poir.

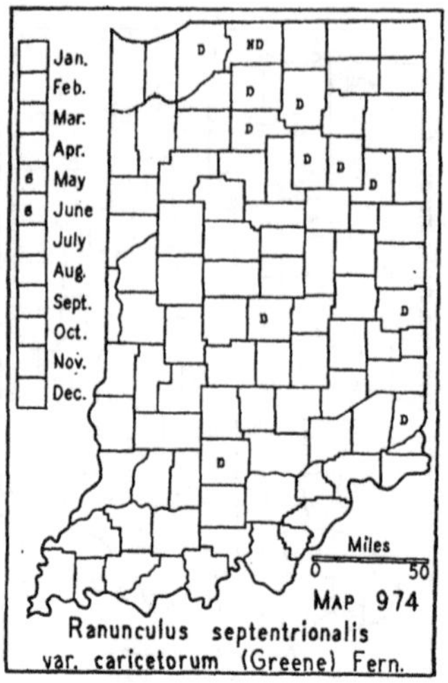

MAP 974
Ranunculus septentrionalis var. caricetorum (Greene) Fern.

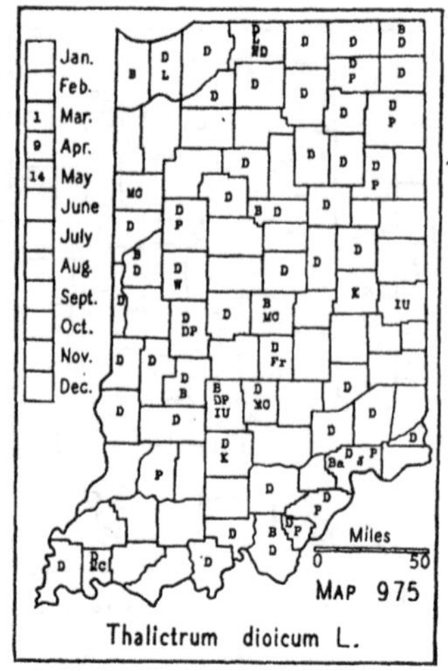

MAP 975
Thalictrum dioicum L.

character. The fruit, however, is rather constant. The petals vary from 3.5-7 mm in width and from 6-12 mm in length.

Vt., Ont., N. Dak., southw. to Ga. and Ark.

17. **Ranunculus septentrionàlis** Poir. Map 973. In separating this species from this perplexing complex I have no guide except a meager amount of literature. Poiret in his original description says the plant is 8-10 inches high and that the petioles of the radical leaves are glabrous. I infer that the description was drawn from a glabrate form and a very young specimen since he adds that the base of the stem is villous or pubescent but does not mention that later radical leaves may be quite pubescent. He says that the calyx is glabrous, furnished with sparse hairs at the summit. Of the 34 specimens which I have from Indiana, all have the entire calyx appressed-pubescent except one specimen which is glabrous throughout except for a few hairs at the summit of the peduncle. Fernald cited a specimen of *Ranunculus hispidus* var. *falsus* Fern. from Indiana and some local authors have been so naming juvenile specimens of this species. Since this species roots at the nodes it certainly does not belong to *Ranunculus hispidus* which is an erect plant.

This species is found throughout the state in wet woods and more rarely on slopes and banks. Since I have never been able to satisfactorily separate this species from the preceding one and the following variety the maps indicate only a scattered distribution.

N. B. to Man., southw. to Ga. and Kans.

17a. **Ranunculus septentrionalis** Poir. (Spreading-pubescent form.) This form is found sparsely throughout the state in habitats similar to those of the usual form of the species. I have only 6 specimens from Indiana.

17b. **Ranunculus septentrionalis** var. **caricetòrum** (Greene) Fern. (Rhodora 38: 177-178. 1936.) (*Ranunculus caricetorum* Greene.) Map 974. This variety is probably not very rare in the lake area of the state but is infrequent south of it. It is usually found in springy and rarely in drier situations.

I am following Fernald in considering this a variety of *Ranunculus septentrionalis* although it seems distinct in characters other than its pubescence. The whole plant is much more robust than its allies, has much larger fruiting heads, longer styles, and the achenes usually average 40-50 per head while the species usually has about 20. I have had the last two species and this variety under cultivation for years. While this treatment was being written during the last of June I visited my colony of this variety and I found plants 5 feet in diameter with an abundance of new plants coming from the nodes. Usually the new plants have 3-5 radical leaves 5-8 inches long and the pubescence on the petioles of all of the leaves from one node is spreading, but sometimes at other nodes some petioles have a spreading pubescence and some have a recurved pubescence toward the base. The main stem above the first node has a sparse spreading pubescence. The plant seems to have the retrorse hereditary factor of the pubescence but a quantitative statement is speculative.

Ohio and Ind. to Minn., southw. to Mo.

2548. THALÍCTRUM [Tourn.] L. MEADOWRUE

Stem leaves all petioled; flowering in April and early May, dioecious; plants generally less than 6 (8) dm high....................................1. *T. dioicum.*

Stem leaves sessile or only the lower ones petioled; generally flowering after May, dioecious or polygamous; plants generally 8-20 dm high.

- Lower surface of leaflets, especially the veins, and usually the achenes, covered more or less with stalked or sessile glands, or with both; leaflets generally thick and with revolute margins; plants generally with yellow roots......2. *T. revolutum.*
- Lower surface of leaflets without glands but more or less densely pubescent with flat, multicellular hairs, very rarely a specimen glabrous.
  - Leaflets thick or thin, the margins more or less revolute; flowering mostly in June; anthers mucronate, or at least somewhat acute, generally about 2 mm long; filaments usually more or less enlarged above; mature achenes sessile or nearly so, not reflexed.................................3. *T. dasycarpum.*
  - Leaflets thin, the margins flat; flowering mostly in July; anthers mostly 0.5-1 mm long, oblong, obtuse at the apex; filaments usually flat and much enlarged above, usually about as wide as the anther; mature achenes stipitate, reflexed. ....4. *T. perelegans.*

The treatment of the genus as given is not at all satisfactory and must be regarded as provisional. Until the many species described by E. L. Greene are understood, no satisfactory treatment is possible. I believe the "master key" to our species has not yet been discovered. The characters usually given to separate the species are not constant in a large series of specimens. What effect the habitat has upon the thickness of the leaflets I do not know but I do not think we should place thick- and thin-leaved specimens in the same category.

1. **Thalictrum dioìcum** L. EARLY MEADOWRUE. Map 975. Generally frequent throughout the state on wooded slopes. This species, like the others, shows considerable variation but I believe we do not have any of the described varieties or closely allied species.

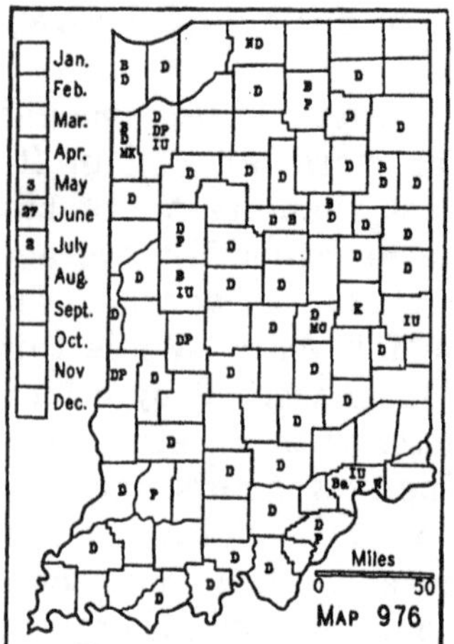

Thalictrum revolutum DC.

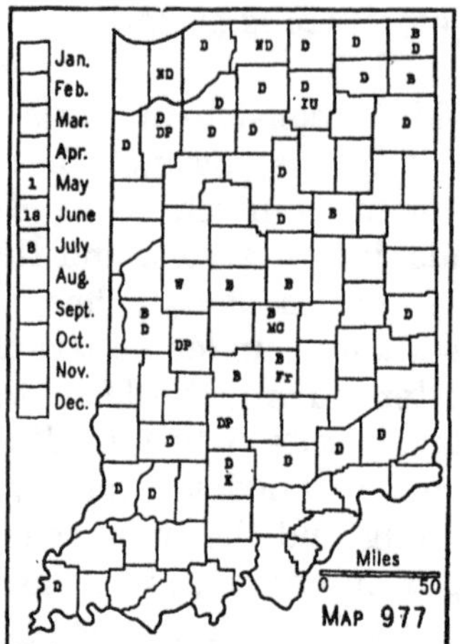

Thalictrum dasycarpum Fisch. & Lall.

Thalictrum perelegans Greene

My no. 5946, collected on the bank of Wildcat Creek west of Greentown in Howard County, I cite as unusual. My specimen arises from a node of an underground stem. The stem remaining on the specimen has nine nodes and is 16 cm long, and shows no decrease in size where it has been broken off at both ends.

Central Maine to Sask., southw. to Ala. and Mo.

2. **Thalictrum revolùtum** DC. WAXY MEADOWRUE. Map 976. Infrequent to frequent throughout the state in moist soil. A form with yellow roots (when collected) is generally found in dry soil on wooded slopes and less often in moist, sandy places. I think the form, which usually has only sessile glands, should be separated as a species or variety. It may be that it is Greene's *Thalictrum amabilis,* the type of which I have not seen. I have 70 sheets of this species from Indiana and I have failed to make a record of the color of the roots of many specimens but I think that I made a record when the color was yellow and did not when no color was evident.

Mass. to Ont., southw. to S. C., Tenn., and Mo.

3. **Thalictrum dasycárpum** Fisch. & Lall. PURPLE MEADOWRUE. Map 977. Infrequent to frequent throughout the state in moist or wet places. It is found mostly in bogs and springy places about lakes and in low places in woods and moist places along roadsides. Some of my specimens of this species had yellow roots when collected. This species shows a wide range in the texture, size, and shape of the leaflets and I think it is also a complex. The lower surface of the leaflets varies from glabrous to densely pubescent. I found a staminate specimen in Noble County that had a lavender inflorescence.

N. J., N. Dak. to Sask., southw. to Nebr. and Ariz.

4. **Thalictrum perélegans** Greene. (Greene, Leaflets of Botany 2: 59. 1910.) (*Thalictrum polygamum* of Indiana authors.) Map 978. I have seven specimens of this species and there are two from Jefferson

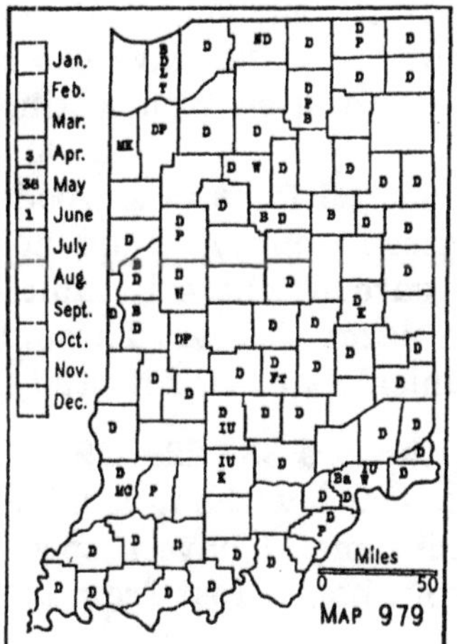

Podophyllum peltatum L.

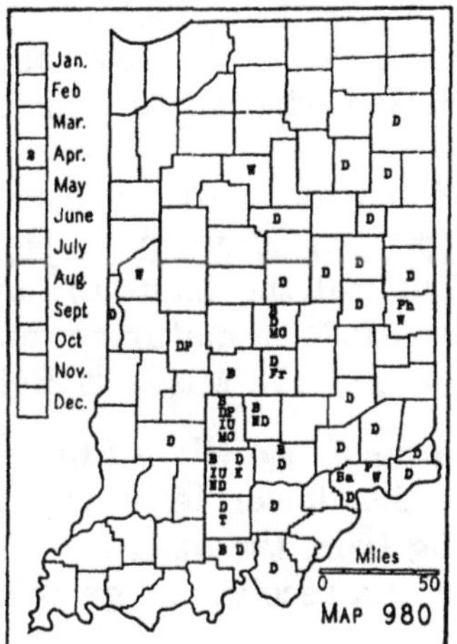

Jeffersonia diphylla (L.) Pers.

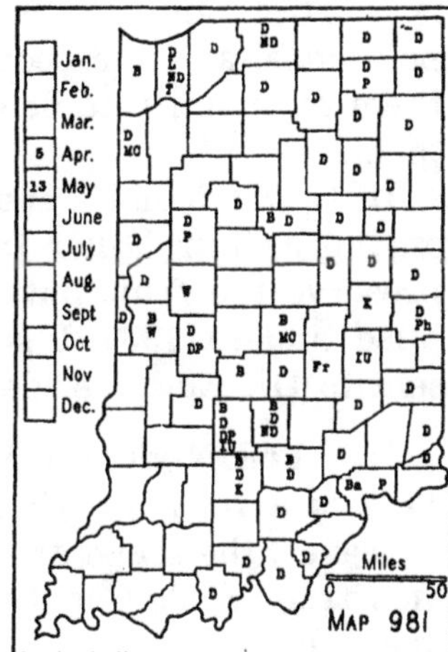

Caulophyllum thalictroides (L.) Michx.

County in the herbaria of Wabash College and Indiana University. The specimens are all from low ground in woods in the counties bordering the Ohio River. The type specimens were collected at Lemon's Gap, North Carolina, which is located at an elevation of about 3,500 feet about 13½ miles west of Hot Springs, North Carolina.

The leaflets of all of my specimens are very thin, not revolute, and more or less pubescent beneath. The short and blunt anthers on very broad filaments are very distinctive. Broad filaments alone, however, can not be used to differentiate this species because other species also have wide filaments. The achenes are usually few, mostly stipitate, and reflexed when mature. This is a tall plant of wet woods.

Ind., N. C., and Tenn.

93. BERBERIDÀCEAE T. & G. BARBERRY FAMILY

Herbs with white or greenish purple flowers.
  Petals 6-9; stamens 8-18; fruit many-seeded.
    Leaves palmately 5-9-lobed (rarely freakish forms); anthers opening longitudinally; fruit a large berry....................2558. PODOPHYLLUM, p. 475.
    Leaves all basal, the blades divided into 2 obliquely ovate leaflets; anthers opening by uplifting valves; fruit a thick-walled capsule opening by a valve at the top..2559. JEFFERSONIA, p. 476.
  Petals and stamens 6; fruit bluish black, resembling a berry, 2-seeded...2565. CAULOPHYLLUM, p. 476.
Shrubs; leaves simple; flowers yellow; fruit a 1-few-seeded, red berry..2566. BERBERIS, p. 476.

2558. PODOPHÝLLUM L. MAYAPPLE

1. **Podophyllum peltàtum** L. COMMON MAYAPPLE. Map 979. Infrequent to common throughout the state in moist woods. It often spreads and persists after woodlands are cleared because of its creeping rootstocks and the fact that no grazing animal will eat it. The rhizomes are cathartic and have long been much used in medicine. The mature fruit

is short-elliptic or suborbicular, light yellow, rarely 5.5 cm long and not poisonous, as some people think. I am very fond of them and have eaten them in quantity to determine whether any ill effect resulted from eating them.

In 1927 I found a specimen that had a maroon colored fruit in a woods on the Arthur Miller farm near Mauckport, Harrison County. I did not preserve the fruit but I moved the plant to our garden where it multiplied and in 1937 I succeeded in maturing four large fruits. These I sent to Dr. Edgar Anderson, of the Missouri Botanical Garden, for study.

1a. **Podophyllum peltatum f. aphýllum** Plitt. (Rhodora 33: 229. 1931.) This is a form in which the flower terminates a scape without a trace of leaves, the scape arising from the apex of the rhizome. I have a specimen from Wells County and a second one grew about half a foot from the one I have.

2559. JEFFERSÒNIA B. S. Barton

1. **Jeffersonia diphýlla** (L.) Pers. TWINLEAF. Map 980. Infrequent to frequent throughout the area shown on the map. Generally found in rich soil on wooded slopes and more common in the southern counties. Margins of leaflets vary from entire and undulating to 5-7-lobed.

N. Y. to Wis., southw. to Va., Tenn., and Iowa.

2565. CAULOPHÝLLUM Michx.

1. **Caulophyllum thalictroìdes** (L.) Michx. BLUE COHOSH. Map 981. Infrequent to frequent throughout the state in moist, rich woods. Since the thickened rootstocks have always been much used in medicine, it is surprising to find it as frequent as it is.

N. B. to Man., southw. to S. C., Tenn., and Mo.

2566. BÉRBERIS [Tourn.] L. BARBERRY

Leaves entire; spines generally simple; flowers in fascicles of 2-6; petals notched. .... 1. *B. Thunbergii.*
Leaves not entire, the margins more or less serrate; spines generally 3-pronged.
  Margins of leaves regularly bristly-serrate; racemes many-flowered; petals entire; two year old branchlets grayish .... 2. *B. vulgaris.*
  Margins of leaves irregularly serrate, the teeth not bristly-pointed; racemes few-flowered; petals notched; two year old branchlets reddish brown .... 3. *B. canadensis.*

1. BERBERIS THUNBÉRGII DC. JAPANESE BARBERRY. This shrub is much used as a hedge plant and for ornamental planting. Nieuwland reports it as an escape in St. Joseph County and I have found seedlings in our garden on several occasions. Since it is so commonly used, it will no doubt be found often as an escape where suitable conditions obtain.

Nat. of Japan.

2. BERBERIS VULGÀRIS L. EUROPEAN BARBERRY. This species was formerly much used as an ornamental shrub. When it was learned that

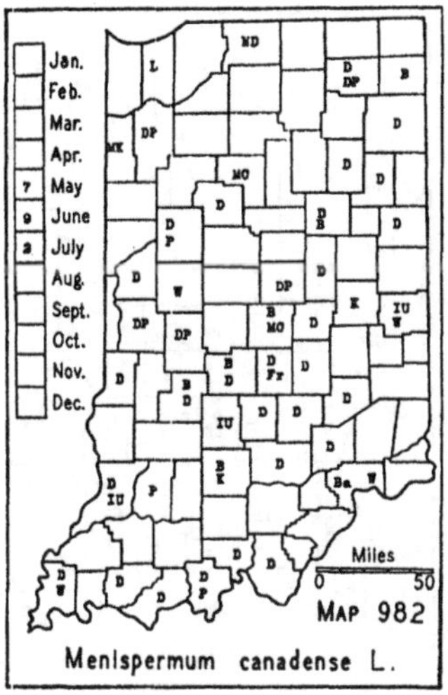

MAP 982

Menispermum canadense L.

MAP 983

Cocculus carolinus (L.) DC

MAP 984

Calycocarpum Lyoni (Pursh) Nutt.

it was responsible for the black stem rust on wheat, oats, rye, barley, and about 75 wild and cultivated grasses, which resulted in a loss of approximately fifty million dollars annually, an active campaign was started by the U. S. Department of Agriculture for its extermination in the wheat area of the U. S. See U. S. Dept. Agric. Farmer's Bull. 1544: 1-28. 1927, and Purdue Univ. Agric. Exper. Sta. Bull. 145: 1-12. 1926.
Nat. of Eu.

3. **Berberis canadénsis** Mill. ALLEGHENY BARBERRY. There is some question whether this species is a native or has been introduced. It has been found by the Barberry Eradication Agent along the Tippecanoe River in Fulton, Pulaski, White, and Tippecanoe Counties and along Wildcat Creek in Tippecanoe County. A few scattered bushes were found in Scott and Washington Counties, the habitat or exact location not being given. Nieuwland reported it from St. Joseph County, saying that a clump was found about 1894 near the bank of the St. Joseph River at the Four Mile Bridge and that it had disappeared in 1914. The discontinuous distribution indicates that it might have escaped from cultivation, although there is no evidence that this species was cultivated. Before it was eradicated, I saw long stretches of it on the bank of the Tippecanoe River and it appeared to be native. It was usually restricted to an area a few feet back from the edge of the bank and down the slope to high water mark.
Mts. of Va. to Ga., along the Alleghenies, and in Mo.

94. MENISPERMÀCEAE DC. MOONSEED FAMILY

Petals present; anthers 4-celled; seeds flat, circular with a notch.
Stamens in the staminate flowers 12-24, in the pistillate flowers represented by 6 staminodia; sepals and petals each 4-8; fruit bluish black...2567. MENISPERMUM, p. 478.
Stamens in the staminate flowers 6, in the pistillate ones reduced or lacking; sepals and petals each 6; fruit scarlet.......................2570. COCCULUS, p. 478.
Petals lacking; anthers 2-celled; stamens in the staminate flowers 12, in the pistillate ones 9, abortive; seed boat-shaped..................2590. CALYCOCARPUM, p. 478.

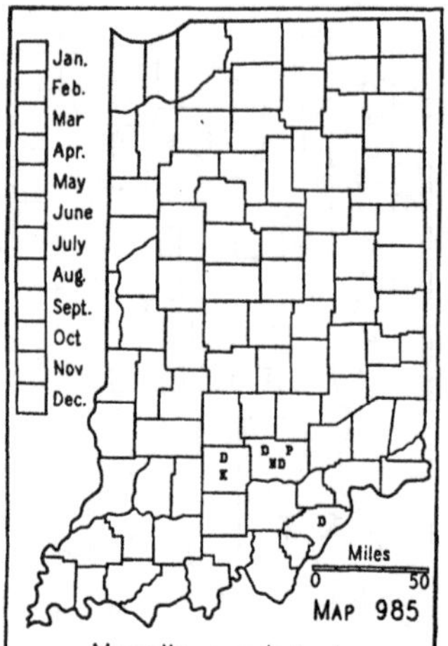

Magnolia acuminata L.

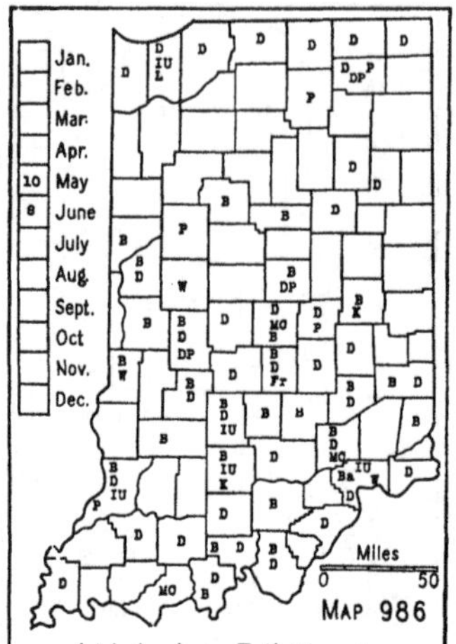

Liriodendron Tulipifera L.

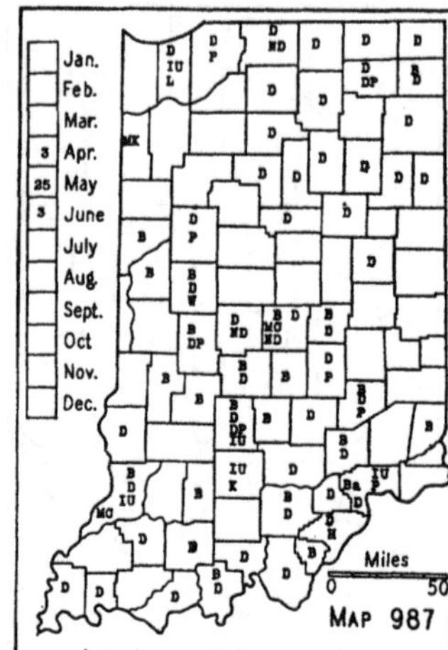

Asimina triloba (L.) Dunal

2567. MENISPÉRMUM [Tourn.] L.

1. **Menispermum canadénse** L. COMMON MOONSEED. Map 982. Infrequent to common throughout the state on the low banks of streams, in alluvial lands along streams, on fences along roadsides, and on the steep and rocky slopes of streams and ravines. Most abundant in overflow woods in the Lower Wabash Valley. This plant twines from left to right. It freezes to the ground each year throughout the state except in a few of the southwestern counties where it becomes woody. I have a specimen from Warrick County that has a stem 1 cm in diameter.

The rhizomes were formerly much used in medicine but are rarely used now. When this plant is introduced into cultivated grounds, it is almost impossible to exterminate it. Personal experience prompts this statement.

Western Que. and w. N. E. to Man., southw. to Ga., Ark., and Nebr.

2570. CÓCCULUS DC.

1. **Cocculus carolìnus** (L.) DC. (*Epibaterium carolinum* (L.) Britton.) CAROLINA SNAILSEED. Map 983. Infrequent in a few of the southwestern counties on the banks of streams, ponds, and sloughs that are usually inundated each year.

Va., Ill. to Kans., southw. to Fla. and Tex.

2590. CALYCOCÁRPUM Nutt.

1. **Calycocarpum Lỳoni** (Pursh) Nutt. CUPSEED. Map 984. Local in a few of the Ohio River counties on the low banks of streams through bottom land in thickets where it climbs to a height of 8-10 feet.

Ind., Mo., and Kans., southw. to Fla. and Tex.

95. MAGNOLIÀCEAE J. St. Hil. MAGNOLIA FAMILY

Buds silky white-pubescent; leaves entire; fruit fleshy, dehiscent.................. ....2651. MAGNOLIA, p. 479.

Buds glabrous; leaves lobed; fruit a cone of dry carpels, indehiscent until dry....... ....2654. LIRIODENDRON, p. 479.

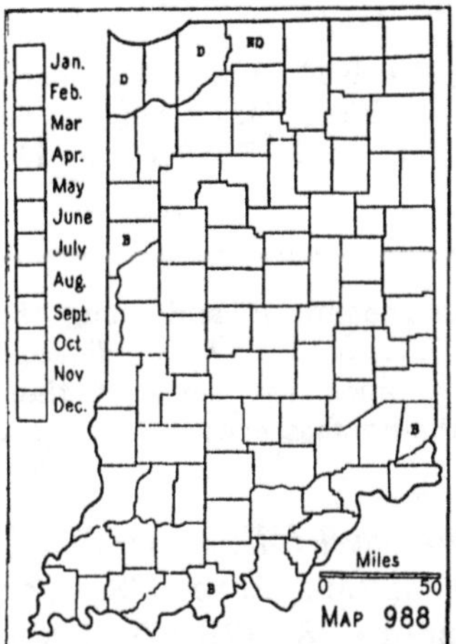

Sassafras albidum (Nutt.) Nees

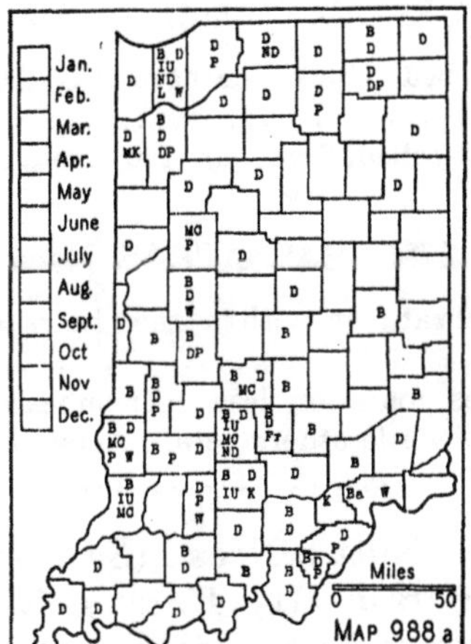

Sassafras albidum var. molle (Raf.) Fern.

Benzoin aestivale (L.) Nees

2651. MAGNÒLIA L. MAGNOLIA

1. **Magnolia acuminàta** L. CUCUMBER TREE. Map 985. This tree was very local and was probably found in all the counties south of a line joining Richmond and Vincennes. In addition to my records it has been reported from Franklin, Floyd, Jefferson, and Orange Counties. I have been told that it grew also in Crawford, Decatur, Vanderburgh, Washington, and Wayne Counties. Now known in only a few counties.

Western N. Y., s. Ont., s. Ill. to Ark., southw. to Ga. and La.

2654. LIRIODÉNDRON L. TULIP TREE

1. **Liriodendron Tulipífera** L. TULIP TREE. Map 986. This is an infrequent to frequent or common tree throughout the state although it may be absent or very local in a few of the northwestern counties. It grows in almost all kinds of soil but prefers a dry, rather sandy one where it is often a common tree in some of the southern counties. In the hilly counties it is usually found toward the bases of slopes and is almost invariably associated with beech and sugar maple, although there are exceptions where it grows with white oak, black gum, and others.

Vt., s. Ont. and s. Mich., southw. to Fla. and La.

98. ANONÀCEAE DC. CUSTARD APPLE FAMILY

2673. ASÍMINA Adans. PAPAW

1. **Asimina tríloba** (L.) Dunal. PAPAW. Map 987. The papaw is probably found in every county of the state. It is usually local in the northwestern part and in the hills of the southern part. It prefers a moist, rich soil and is usually found in colonies on account of its habit of propagating by rootshoots. The fruit is edible and is relished by most people. It is desirable for ornamental planting and is free from insect pests and diseases.

N. Y., s. Ont., s. Mich. to Nebr., southw. to Fla. and Tex.

102. LAURÀCEAE Lindl. LAUREL FAMILY

Leaves or some of them, lobed, thick; anthers 4-celled, 4-valved; fruit blue black. ....2795. SASSAFRAS, p. 480.
Leaves entire, thin; anthers 2-celled, 2-valved; fruit red........2821. BENZOIN, p. 480.

2795. SÁSSAFRAS Nees.

Lower surface of leaves glabrate or with a few hairs on the midrib and along the principal nerves...1. *S. albidum.*
Lower surface of the leaves more or less soft-pubescent, upon age some of them becoming nearly glabrous while others retain some of their pubescence........... ....1a. *S. albidum* var. *molle.*

1. **Sassafras álbidum** (Nutt.) Nees. (*Sassafras variifolium* (Salisb.) Ktze. and *Sassafras Sassafras* (L.) Karst.) SASSAFRAS. Map 988. For a discussion of this species and its variety see Rhodora 38; 178-179. 1936. My specimens pass insensibly from the glabrate to the densely pubescent form. The species and variety have no geographical range in the state.

This tree was formerly, without doubt, a native of every county of the state. It is usually found in colonies because it propagates freely by rootshoots. It is somewhat frequent in sandy soil in the northern counties, becoming rare and local south of the lake area, and frequent to common in the hilly counties of the southern part of the state. It is usually found in old, fallow and abandoned fields where it sometimes forms thickets over the whole area. The entire plant is aromatic and the bark of the root was much used by the pioneers for making sassafras tea.

Distribution probably mostly in the Mississippi Valley.

1a. **Sassafras albidum** var. **mólle** (Raf.) Fern. The pubescent form is more frequent in Indiana than the glabrate form. The distribution is probably that given for the complex by the earlier authors.

Maine, s. Ont. to Iowa, southw. to Fla. and Tex.

2821. BENZÒIN Fabricius.

1. **Benzoin aestivàle** (L.) Nees. SPICEBUSH. Map 989. Spicebush was formerly found, without doubt, in every county of the state. It is always found in wet places in woods although I found it to be abundant on the north slope of a wooded hill about 3 miles northeast of Madison in Jefferson County. In the forester's opinion it is an obnoxious shrub in the woodland.

This shrub is extremely variable as to the pubescence of the branchlets and leaves. The form with pubescent branchlets, lower surface of leaves, petioles, and pedicels has been named var. *pubescens* Palmer & Steyermark (Ann. Missouri Bot. Gard. 22: 545. 1935). Since my 73 Indiana specimens show every gradation between the glabrous and the pubescent forms I prefer to regard our specimens as belonging to a polymorphic complex.

Maine, cent. Mich. to e. Kans., southw. to Ga. and e. Tex.

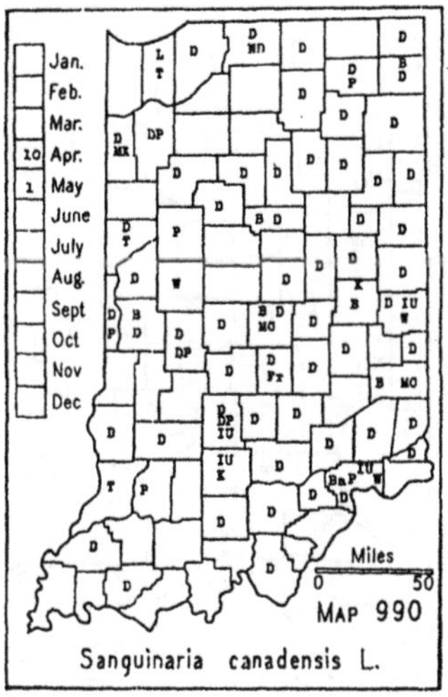

Sanguinaria canadensis L.

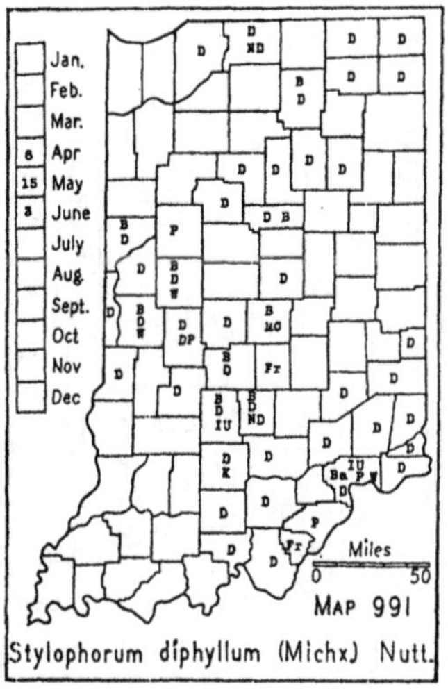

Stylophorum diphyllum (Michx.) Nutt.

Chelidonium majus L.

104. PAPAVERÀCEAE B. Juss. POPPY FAMILY

Flowers white; leaves all basal; juice of plants red........2841. SANGUINARIA, p. 481.
Flowers, leaves, and juice not as above.
  Leaves spiny-toothed....................................2852. ARGEMONE, p. 482.
  Leaves not as above.
    Flowers yellow; juice of plants yellow; pod dehiscent to the base.
      Capsule oblong, bristly; buds erect, ovoid..........2843. STYLOPHORUM, p. 481.
      Capsule linear, glabrous; buds drooping just before opening, obovoid...........
      ....2845. CHELIDONIUM, p. 481.
    Flowers not yellow; juice of plants not yellow; pods dehiscent at the top or only to the middle...................................2853. PAPAVER, p. 482.

2841. SANGUINÀRIA [Dill.] L. BLOODROOT

1. **Sanguinaria canadénsis** L. BLOODROOT. Map 990. Infrequent to frequent in all parts of the state in rich, moist woods. Bloodroot has long been much used in medicine and where it was once frequent or locally common in woodland located near where an active "root gatherer" lived, it has become extinct or rare.

N. S. to Man., southw. to Fla., Ala., Ark., and Nebr.

2843. STYLÓPHORUM Nutt.

1. **Stylophorum diphýllum** (Michx.) Nutt. CELANDINE-POPPY. Map 991. Infrequent to frequent or local in all parts of the state or possibly absent in some counties. I have not found it in the southwestern counties although I have done much collecting there.

Western Pa. to Wis., southw. to Tenn. and Mo.

2845. CHELIDÒNIUM [Tourn.] L.

1. CHELIDONIUM MÀJUS L. CELANDINE. Map 992. Three authors have reported this species as escaping from about dwellings. I found it

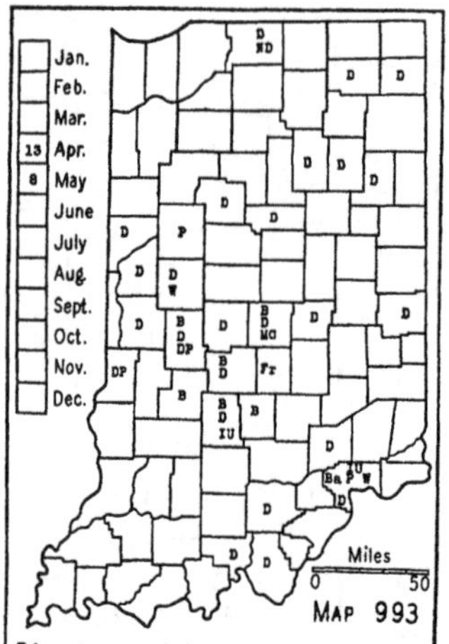

Dicentra canadensis (Goldie) Walp.

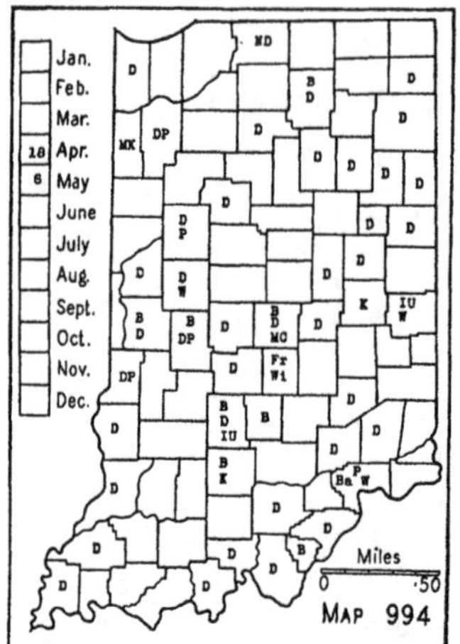

Dicentra Cucullaria (L.) Bernh.

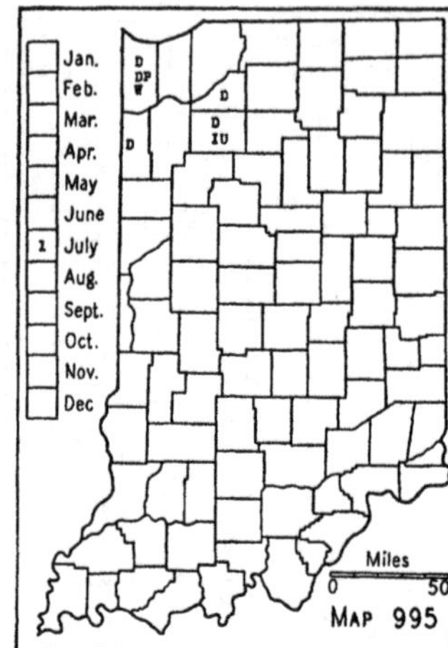

Corydalis sempervirens (L.) Pers.

in moist woods in De Kalb and Lagrange Counties where it formed a dense stand over acres. In the other counties where I found it only a few plants were found at a place. I predict that where this species becomes established in woodland, it will be the principal part of the spring flora. This plant was formerly used in medicine but is no longer official. This use is, no doubt, responsible for its cultivation and escape.

Nat. of Eu.; cent. Maine to Ont., southw. to N. C.

2852. ARGEMÒNE [Tourn.] L. PRICKLY POPPY

Corolla white; leaves not blotched. (See excluded species no. 249, p. 1050.) .... *A. intermedia.*
Corolla yellow; leaves with light blotches. (See excluded species no. 250, p. 1050.) .... *A. mexicana.*

2853. PAPÀVER [Tourn.] L. POPPY

Plant hispid, green; leaves petiolate, pinnately divided. (See excluded species no. 251, p. 1050.) .... *P. Rhoeas.*
Plant glabrous, glaucous; leaves clasping, lobed. (See excluded species no. 252, p. 1050.) .... *P. somniferum.*

104A. FUMARIÀCEAE DC. FUMITORY FAMILY

Corolla with two opposite petals spurred or saccate at the base.
  Plant scapose, low, erect; petals slightly united; corolla white or pinkish; seeds crested ....2856. DICENTRA, p. 483.
  Plant climbing; petals firmly united; corolla flesh color, scarcely saccate; seed not crested ....2857. ADLUMIA, p. 483.
Corolla with only one petal spurred or saccate at the base.
  Capsule oblong, several-seeded; seeds crested; flowers yellowish or pinkish...2858. CORYDALIS, p. 483.
  Capsule globular, 1-seeded, indehiscent; flowers deep crimson, purple, 5-7 mm long..2861. FUMARIA, p. 484.

2856. DICÉNTRA Bernh.

Corolla cordate at the base, the spurs rounded; inner petals conspicuously crested; stem from a flattened orbicular yellowish corm; leaves glaucous beneath, the segments apiculate and more or less rounded at the apex; flowers fragrant...1. *D. canadensis.*

Corolla with two widely spreading spurs; crest of inner petals minute; stem from a fleshy, loosely scaly bulb; leaves greenish beneath, sometimes more or less glaucous, the segments apiculate and acute at the apex; flowers not fragrant. ....2. *D. Cucullaria.*

1. **Dicentra canadénsis** (Goldie) Walp. (*Bicuculla canadensis* (Goldie) Millsp.) SQUIRRELCORN. Map 993. This species is found possibly throughout the state, although there are no specimens or reports from the southwestern counties. It grows in deep, rich leafmold in well drained soil, usually on wooded slopes. It is much rarer than the next species both in its distribution and in its abundance where found. This and the next species are reported to be poisonous to stock.
N. S. to Minn., southw. to N. C. and Mo.

2. **Dicentra Cucullària** (L.) Bernh. (*Bicuculla Cucullaria* (L.) Millsp.) DUTCHMAN'S-BREECHES. Map 994. Infrequent to frequent throughout the state in deep, rich leafmold and usually in well drained, moist soil.
N. S. to Minn., southw. to Ga. and Mo.

2857. ADLÙMIA Raf.

1. **Adlumia fungòsa** (Ait.) Greene. CLIMBING FUMITORY. This plant was reported from Lake County by Ball in 1884 and from Steuben County by Bradner in 1892. I have been acquainted with it in cultivation for years and I saw a specimen in 1910 in an unpastured woods about 3 miles southeast of Michigan City, La Porte County. I was not prepared to take a specimen but I returned to the same spot a few years later and found the woods heavily pastured and the specimen gone. It was, no doubt, a rare plant in northern Indiana and may yet be rediscovered.
Eastern Que. to Wis., southw. in the mts. to N. C.

2858. CORÝDALIS [Dill.] Medic.

Plants usually 4 dm or more high; flowers purplish green or rose color, tipped with yellow; mature capsules generally 3-4 cm long, usually ascending...1. *C. sempervirens.*

Plants less than 4 dm high; flowers light to bright yellow; mature capsules 1-3 cm long, becoming torulose and spreading.

Flowers 6-8 mm long; spur short; outer petals crested on the back; seed reticulate, especially near the margin..2. *C. flavula.*

Flowers 8-20 mm long; spur conspicuous; outer petals not crested on the back; seed smooth. (See excluded species no. 253, p. 1050.)....................*C. aurea.*

1. **Corydalis sempérvirens** (L.) Pers. (*Capnoides sempervirens* (L.) Borkh.) PINK CORYDALIS. Map 995. Very local in a few of the northwestern counties. Generally found in sandy soil in areas which have recently been burned. It is usually found associated with *Geranium Bicknellii.*
Newf. to Alaska, southw. to Ga., Ky., Minn., and Mont.

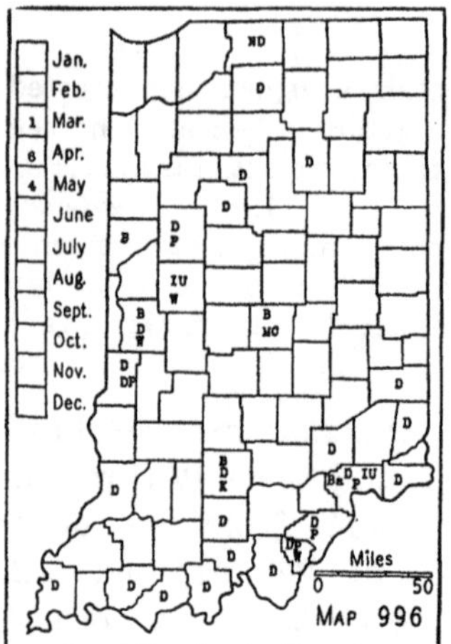

Corydalis flavula (Raf.) DC.

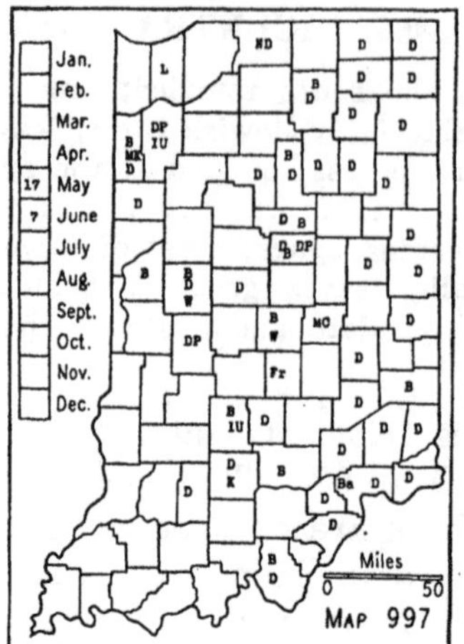

Lepidium campestre (L.) Br.

Lepidium Draba L.

2. **Corydalis flávula** (Raf.) DC. (*Capnoides flavulum* (Raf.) Kuntze.) PALE YELLOW CORYDALIS. Map 996. Local in the northern part of the state, becoming rather frequent in some of the Ohio River counties. It is found in rich, moist, sandy soil in woodland, usually on slopes and on the wooded bluffs of streams.

N. Y., sw. Ont. to Minn., southw. to Va. and La.

2861. FUMÀRIA [Tourn.] L.

See excluded species no. 254, p. 1050.

105. CRUCÍFERAE B. Juss. MUSTARD FAMILY

[Specimens of this family, in order to make determination certain, should have flowers with the color known, mature pods, and, in some species, the basal leaves. It often happens, however, that one or more of these parts are lacking. In order to compensate for the absence of one or more of these important diagnostic characters, and to employ obvious characters, omitting those so often used, such as the position of the cotyledons, the key has been expanded to its present, rather unusual form to make correct determinations possible.]

A. Flowers white, creamy white, greenish white, or purplish (flowers very small in *Lepidium densiflorum*).
  Peduncle 1-flowered, all radical....................2971. LEAVENWORTHIA, p. 501.
  Peduncles more than 1-flowered.
    Pods transversely divided into 2 cells; plants fleshy, in Indiana found only on the shore of Lake Michigan..........................2920. CAKILE, p. 490.
    Pods longitudinally divided into 2 cells.
      B. Pods short, not more than 3-3.5 times as long as wide.
        Pods flattened, if at all, parallel to a septum that is as wide as the pod.
          Pubescence of simple hairs or lacking..........2965B. ARMORACIA, p. 496.
          Pubescence not as above.
            Pods glabrous; seeds in 2 rows in each cell.........2989. DRABA, p. 502.

Pods more or less pubescent.
Beaks of pods more than 1 mm long..........3015. BERTEROA, p. 509.
Beaks of pods less than 1 mm long.
Seed 1 in each cell. (See excluded species no. 275, p. 1054.)..........
..................................3013. LOBULARIA, p. 509.
Seed more than 1 in each cell. (See excluded species no. 272, p. 1053.)
..................................2989. DRABA, p. 502.
Pods flattened at right angles to the narrow septum.
Seed 1 in each cell; pods dehiscent................2883. LEPIDIUM, p. 487.
Seed more than 1 in each cell; if only 1-seeded the pod indehiscent.
Styles very short, less than 0.5 mm long; pods very flat, thin, orbicular to obovate-orbicular, winged.
Plants stellate-pubescent; pods orbicular, not more than 4 mm wide, not strongly notched at the apex..............3006. ALYSSUM, p. 509.
Plants glabrate; leaves mostly basal, pinnatifid or incised; pods wingless, cuneate to triangular obcordate.......2986. CAPSELLA, p. 502.
Plants glabrous; leaves not mostly basal, entire or dentate; pods winged, nearly orbicular to obovate-orbicular.......2903. THLASPI, p. 488.
Styles more than 0.5 mm long; pods somewhat flattened but not thin.
Pods indehiscent, globose, 2.5 mm in diameter, surface conspicuously reticulate, generally 1-seeded; plants very pubescent. (See excluded species no. 271, p. 1053.).............2988. NESLIA, p. 502.
Pods dehiscent and not as above.
Pods globose, about 2 mm in diameter, about 4-seeded; plants finely stellate-pubescent...................2983. LESQUERELLA, p. 502.
Pods obovoid, about 4-6 mm in diameter, many-seeded; plants glabrous, at least above...............2987. CAMELINA, p. 502.

B. Pods 4-many times as long as wide.
Pods indehiscent, moniliform, up to 6-8 mm in diameter, 2- or 3-seeded, walls spongy; flowers purplish......................2950. RAPHANUS, p. 492.
Pods not as above.
Beaks of pods generally 5-10 mm long; radical leaves ternate, or palmately divided; stem leaves generally 2 or 3, mostly 2-5-parted...............
.....................................2967. DENTARIA, p. 500.
Beaks of pods less than 5 mm long; leaves not as above.
Plants stellate-pubescent, small; leaves small, simple, not clasping at the base; seeds in 2 rows in each cell...........2989. DRABA, p. 502.
Plants not as above.
Seed in 2 rows in each cell.
Plants aquatic; seeds not winged.......2965A. NASTURTIUM, p. 496.
Plants not aquatic; seeds winged in nos. 1 and 7 and excluded species no. 273, p. 1053, of......................3001. ARABIS, p. 504.
Seed in 1 row in each cell.
Seeds more than 3 mm long; leaves all simple, dentate but pinnatifid, the larger ones usually 8-13 cm long; pubescence branched; petals generally purplish, sometimes whitish, mostly 1.5-2 cm long; pods up to 12 cm long, widely spreading, contracted between the seed when mature....................3041. HESPERIS, p. 510.
Seeds less than 3 mm long; petals less than 1.5 cm long.
Plants glabrous; upper stem leaves simple, dentate; lower stem leaves more or less pinnatifid at the base, clasping; petals purplish, 6-9 mm long; pods 1.5-3 cm long, terete, widely spreading.........................2963. IODANTHUS, p. 494.
Plants not as above.
Seeds broadly or narrowly winged.
Seeds broadly winged; stems glabrous or nearly so; pods recurved-spreading or pendulous, 7-10 cm long in nos. 5 and 9 of..............................3001. ARABIS, p. 504.

Seeds narrowly winged; stems pubescent; pods spreading or erect, 2-4 cm long in nos. 2 and 3 of. .3001. ARABIS, p. 504.

Seeds wingless.

Stem leaves generally not more than 5 mm wide, entire or nearly so; basal rosette of leaves pubescent but often absent at fruiting time.

Petals 2-2.5 mm long; seed about 0.5 mm long in no. 3 of ....2917. SISYMBRIUM, p. 489.

Petals mostly 3-5 mm long; seed about 1 mm long. no. 8 of ....3001. ARABIS, p. 504.

Stem leaves more than 5 mm wide, usually more or less dentate.

Plants tall, glabrous, with leaves sagittate at the base; pods erect, 4-6 cm long in no. 6 of......3001. ARABIS, p. 504.

Plants not as above.

Plants glabrous or more or less pubescent with simple hairs.....................2966. CARDAMINE, p. 497.

Plants pubescent; hairs not simple in no. 4 of............ ....3001. ARABIS, p. 504.

A. Flowers yellow or creamy yellow.

C. Pods not more than 3 times as long as wide.

Pubescence stellate or forked; leaves entire or obscurely toothed.

Pods globose.

Pods indehiscent, 2.5 mm in diameter, surface conspicuously reticulate, generally 1-seeded..................................2988. NESLIA, p. 502.

Pods dehiscent, 2 mm in diameter, smooth, about 4-seeded.................. ....2983. LESQUERELLA, p. 502.

Pods thin, flat, orbicular, with a winged margin, 3-4 mm long, few-seeded. ....3006. ALYSSUM, p. 509.

Pubescence simple or lacking; leaves pinnatifid or toothed...2965. RORIPPA, p. 494.

C. Pods 4-many times as long as wide.

Pods indehiscent, moniliform, 2-3 cm long, much constricted between the seed when mature, up to 10-seeded, walls fleshy...............2950. RAPHANUS, p. 492.

Pods dehiscent, longitudinally 2-celled.

Seed in 2 rows in each cell.

Pods nearly beakless...........................2997. DESCURAINIA, p. 504.

Pods with beaks about 2 mm long..................2946. DIPLOTAXIS, p. 490.

Seed in 1 row in each cell.

Racemes leafy-bracted; leaves pinnatifid, with obtuse lobes................. ....2947. ERUCASTRUM, p. 490.

Racemes bractless.

D. Leaves (at least the lower stem leaves) pinnate, bipinnate, more or less pinnatifid or lobed.

Leaves oblong in outline, bipinnatifid, segments numerous, small, toothed or obtuse; pedicels mostly 1-1.5 cm long...2997. DESCURAINIA, p. 504.

Leaves not as above; seed in 1 row in each cell.

Pods flat, generally 6-15 mm long, about 1 mm wide; creeping perennials, often rooting at the lower nodes; flowers 3-4 mm long in no. 1 of.... ....2965. RORIPPA, p. 494.

Pods terete or 4-sided; annuals or perennials but not creeping.

Plants tall, widely spreading; leaves large, deeply pinnatifid, the segments very long and narrow, those of the upper leaves 1.4 mm wide and 2-5 cm long, or filiform; flowers about 6 mm wide; pods widely spreading, mostly 7-8 cm long, about 1 mm wide in no. 2 of ....2917. SISYMBRIUM, p. 489.

Plants not as above.

Plants with pods and pedicels closely appressed to the stem; pedicels

1-2 mm long; pods mostly 1.1-5 cm long, pointed, the valves with a prominent midrib; leaves runcinate-pinnatifid in no. 1 of...2917. SISYMBRIUM, p. 489.

Plants not as above.

Valves of the pods coalescing into an indehiscent, conical beak 2-12 mm long..............................2949. BRASSICA, p. 491.

Valves of the pods distinct from the beak.

Beak of pod 10-15 mm long. (See excluded species no. 260, 261, 262, p. 1052).......................2949. BRASSICA, p. 491.

Beak of pod 1-3 mm long................2961. BARBAREA, p. 493.

D. Leaves entire or dentate, not pinnatifid or lobed.

Blades conspicuously sagittate at the base, glabrous.

Plants glabrous; leaves elliptic, obtuse at the apex; pedicel and base of pod strongly curved inward..............3055. CONRINGIA, p. 510.

Plants more or less pubescent at the base; leaves oblong or ovate-elliptic, acute at the apex; pedicel and base of pod not curved inward in no. 6 of..................................3001. ARABIS, p. 504.

Blades not sagittate at the base.

Leaves linear or lanceolate, rarely more than 1.5 cm wide..3004. ERYSIMUM, p. 507.

Leaves not as above, more than 1.5 cm wide (entire-leaved forms of this genus here).......................2949. BRASSICA, p. 491.

2883. LEPÍDIUM [Tourn.] L.

[Thellung. Monograph of the genus Lepidium. Mitth. Mus. Univ. Zurich. 28: 1-340. 1906. Hitchcock. The genus Lepidium in the United States. Madroño 3: 265-320. 1936.]

Stem leaves clasping by an auriculate base.

Pods winged above and notched at the apex; styles about 0.5 mm long; annual or biennial..1. *L. campestre.*

Pods not winged above or notched at the apex; styles about 1 mm long; perennial...2. *L. Draba.*

Stem leaves petiolate or sessile, not clasping.

Pods slightly winged above, orbicular, oval, or narrowly obovate, generally 2-3 mm long; plants not glaucous, only the lower leaves ever pinnatifid; stamens usually 2.

Hairs of rachis of racemes mostly curved; petals present, spatulate-obovate to almost linear, up to 2 mm long; pods orbicular or somewhat oval, 2.5-3.1 mm wide; petals usually not more than 1.5 mm long...........3. *L. virginicum.*

Hairs of rachis of racemes stouter, erect; petals absent, or very short and narrow, rarely more than half the length of the sepals; pods slightly obovate, rarely orbicular, 1.7-2.5 mm wide..............................4. *L. densiflorum.*

Pods winged all around, generally ovate-orbicular, longer than wide, about 5 mm long; plant glaucous, leaves generally all pinnatifid; stamens 6. (See excluded species no. 255, p. 1051.)...*L. sativum.*

1. LEPIDIUM CAMPÉSTRE (L.) R. Br. FIELD PEPPERGRASS. Map 997. An infrequent to common or even abundant weed throughout the state. Our first report of it is dated in 1888. During the World War grass seed from Europe was freely sown when our pure seed law was not operative, and I think this fact accounts for its sudden appearance in such abundance. Grazing animals avoid it. I have seen clover fields of 5-10 acres that were almost pure stands of this species.

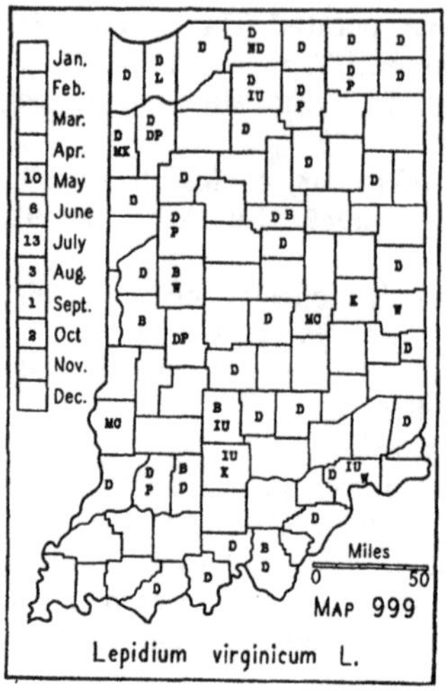

MAP 999
Lepidium virginicum L.

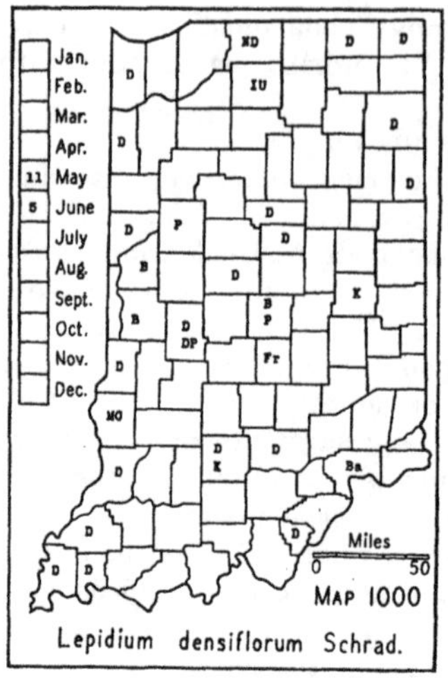

MAP 1000
Lepidium densiflorum Schrad.

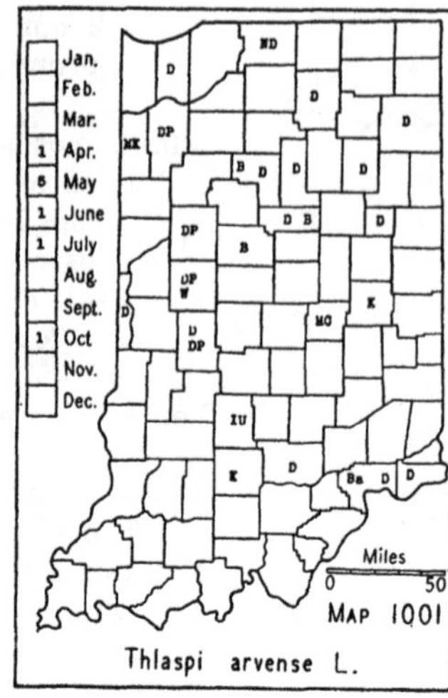

MAP 1001
Thlaspi arvense L.

Nat. of Eu.; N. B. and N. S. to Ont. and Kans., southw. to Va. and along the Pacific coast.

2. LEPIDIUM DRÀBA L. HOARY CRESS. Map 998. This species was reported by Hansen in 1927 from Wabash County and in 1925 from Rush County. In 1933 I visited both of these stations and found it to be persisting and spreading as a common weed. In 1935 it was detected along State Road 116 in Wells County by Lawrence E. Hicks of Ohio State University. A large colony was found on the south side of the road in the southwest quarter of sec. 28, west of the railroad and about 200 feet from where the road turns from a westerly direction to the northwest. The colony was about 150 feet long and was mostly on the right of way of the road with only a few plants in the border of the adjacent field. Nieuwland has collected it along a roadside north of Angola, Steuben County.

Nat. of Asia; local from N. Y. to B. C., southw. to Washington, D. C. and Calif.

3. **Lepidium virgínicum** L. var. **týpicum** C. L. Hitchcock. PEPPERGRASS. Map 999. Frequent throughout the state. Found everywhere except in dense woodland and in very wet places.

Que. to Minn. and Colo., southw. to Fla., Tex., and Mex.; also introduced as a weed into W. I. and Eu.

4. LEPIDIUM DENSIFLÒRUM Schrad. var. TÝPICUM Thellung. (*Lepidium apetalum* Willd.) Map 1000. This species is probably local or frequent throughout the state in habitats similar to those of *Lepidium virginicum* from which it can sometimes be separated only with difficulty.

Nat. of Eurasia; Maine to Ont. and B. C., southw. to Va., Tex., and Nev.

2903. THLÁSPI [Tourn.] L.

Pods elliptic, about 1 cm wide; seed rugose; lower leaves wing-petioled, the upper ones sagittate-clasping..1. *T. arvense.*
Pods obovate-orbicular, about 5 mm wide; seed smooth; leaves sessile, the upper ones subperfoliate..2. *T. perfoliatum.*

1. THLASPI ARVÉNSE L. PENNYCRESS. Map 1001. Local in many parts of the state and now possibly established in all parts, although there are no reports from the Lake Michigan area where we would most expect to find it. Most of my specimens are from railroad ballast. Hansen (Proc. Indiana Acad. Sci. **1923**: 214-215. 1924) reports it from Randolph and Switzerland Counties and calls attention to its tendency to become a weed. I have found only a few specimens at a place except along an abandoned road in Switzerland County where it was frequent.

Eu. and Russian Asia; Que. to Man., southw. to N. Y. and Kans.

2. THLASPI PERFOLIÀTUM L. PERFOLIATE PENNYCRESS. Map 1002. This species was found in 1924 by Miss Edna Banta of Brooksburg, Jefferson County, who reports it to be a frequent to common weed between Brooksburg and Madison, a distance of 8 miles, and in other places in the county. Reported also as occurring on the campus of the University of Notre Dame, St. Joseph County.

Nat. of Eu.

2917. SISÝMBRIUM [Tourn.] L.

Leaves mostly cauline, pinnate, or pinnatifid; plants generally 4-10 dm high; pedicels stout, about as large as the pod; pubescence, when present, of simple hairs; stigmas 2-lobed; seeds about 1 mm long.
- Pods 1-1.5 cm long, appressed; petals 2-3 mm long, bright yellow.
  - Pods pubescent.......1. *S. officinale.*
  - Pods glabrous.......1a. *S. officinale* var. *leiocarpum.*
- Pods more than 1.5 cm long, spreading; petals 5-8 mm long, pale yellow; leaf-segments generally longer and narrower.
  - Stems covered rather thickly all over with long hairs; cauline leaves pinnately parted, usually into 3-7 segments; lateral segments usually 0.5-1 cm wide, the terminal one large and deltoid; mature pods mostly 2-4 cm long. (See excluded species no. 256, p. 1051.).......*S. Loeselii.*
  - Stems glabrous, or with scattered long hairs, especially on the basal part; cauline leaves pinnatifid, usually cut into 9-19 segments; segments of leaves mostly 1-10 mm wide, the terminal one not large and deltoid; mature pods usually 6-8 cm long.......2. *S. altissimum.*

Leaves mostly in a basal rosette, entire or slightly toothed; plants generally not over 3 dm high; seed about 0.5 mm long.......3. *S. Thalianum.*

1. SISYMBRIUM OFFICINÀLE (L.) Scop. (*Erysimum officinale* in part, of Britton and Brown, Illus. Flora, ed. 2.) HAIRY-POD HEDGE MUSTARD. Map 1003. A weed in pastures, waste places, and open woodland and along roads and railroads. My specimens, however, are all from the northern part of the state.

Nat. of Eu.; local in the ne. U. S. and Canada.

1a. SISYMBRIUM OFFICINALE var. LEIOCÁRPUM DC. (*Erysimum officinale* in part, of Britton and Brown, Illus. Flora, ed. 2.) SMOOTH-POD HEDGE MUSTARD. Map 1004. This is a weed with habitats similar to those of the preceding species but it is much more common and is found throughout the state.

Nat. of Eurasia; widely distributed in N. A. and S. A.

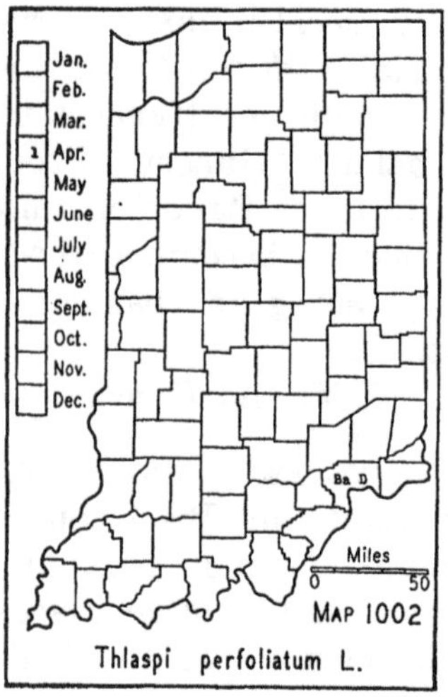

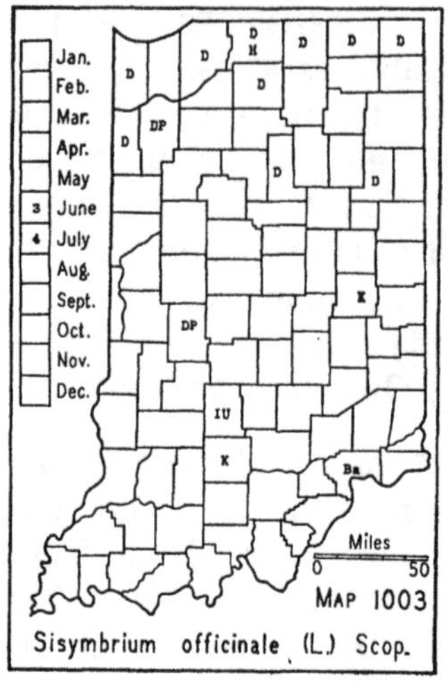

2. SISYMBRIUM ALTÍSSIMUM L. (*Norta altissima* (L.) Britt.) TUMBLE MUSTARD. Map 1005. This is a weed generally of very sandy soil and is found most often in sandy ballast along railroads. It is also found along roadsides and in waste places and fallow fields where it is sometimes abundant, especially in the sandy area of the northwestern part of the state, where it sometimes covers acres.

Nat. of Eu.; N. S. to Ont. and B. C., southw. to Va., Mo., Colo., and Oreg.

3. SISYMBRIUM THALIÀNUM (L.) J. Gay. (*Arabidopsis Thaliana* (L.) Britt.) THALE-CRESS. Map 1006. A weed of sandy soil usually found in pastures and fallow and cultivated fields. Sometimes it is common where it is found, especially in sandy, fallow cornfields. Its distribution in the state suggests that it prefers a sandy and slightly acid soil.

Nat. of Eurasia; Mass., Ont. to Minn., southw. to Ga., Mo., Ark., and Utah.

2920. CAKÌLE [Tourn.] Mill.

1. **Cakile edéntula** (Bigel.) Hook. var. **lacústris** Fern. (Rhodora 24: 23. 1922.) Map 1007. This plant is restricted to the beach area of Lake Michigan. It was formerly frequent along the beach but at present much of the beach area is within city limits or is used by children as play grounds during the summer months. The plant, consequently, has become very rare and in time will probably become extinct.

Beaches of Lakes Ontario, Erie, Huron, and Michigan.

2946. DIPLOTÁXIS DC.

See excluded species no. 257, p. 1051.

2947. ERUCÁSTRUM Presl.

See excluded species no. 258, p. 1051.

MAP 1005
Sisymbrium altissimum L.

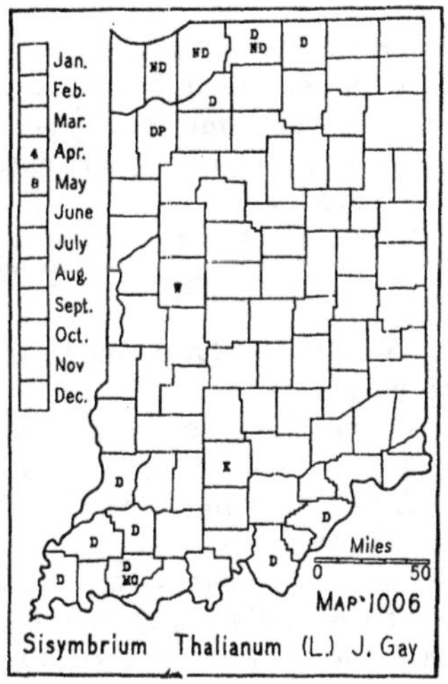

MAP 1006
Sisymbrium Thalianum (L.) J. Gay

MAP 1007
Cakile edentula var. lacustris Fern.

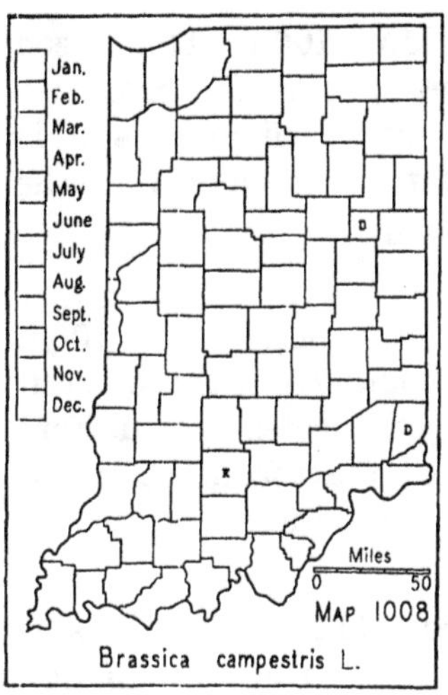

MAP 1008
Brassica campestris L.

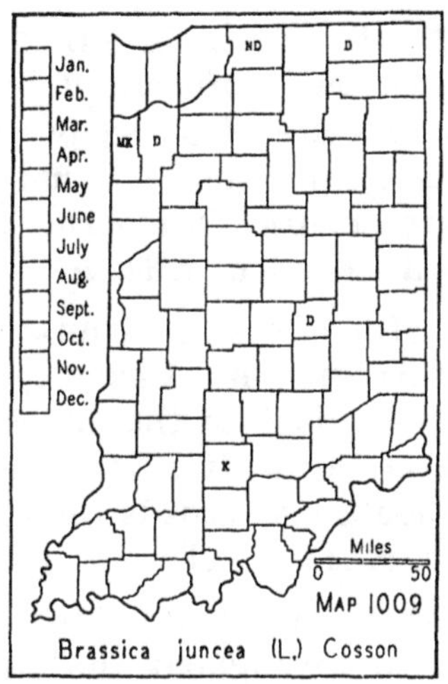

MAP 1009
Brassica juncea (L.) Cosson

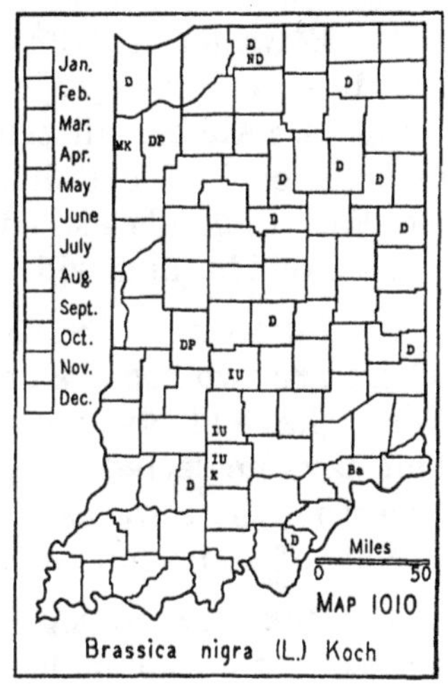

MAP 1010
Brassica nigra (L.) Koch

2949. BRÁSSICA [Tourn.] L.

[Bailey. The cultivated Brassicas. Gentes Herbarum **1**: 53-108. 1922 and **2**: 207-267. 1930.]

Upper stem leaves clasping.
  Petals (9) 10-14 mm long; pods (5) 6-10 cm long; beak 10-17 mm long, moderately stout; pedicels ascending; flowers light yellow; leaves glaucous. (See excluded species no. 260, p. 1052.) . . . . *B. Napus.*
  Petals 7-11 mm long; pods 3-8 cm long; pedicels spreading.
    Leaves glaucous, the basal ones lyrate-pinnatifid with a large-toothed lobe at the apex, glabrous or the lower with a few scattered hairs; stem leaves lanceolate, entire; root slender; sepals 5 mm long; petals 6-7 mm long; beak about 1 cm long . . . . 1. *B. campestris.*
    Leaves green or glaucous, much resembling those of the preceding species but more or less setose-hispid; root tuberous, much enlarged; petals about 7 mm long, yellow; pods 5-8 cm long, 2.5-4 mm thick, valves with 1 conspicuous nerve. (See excluded species no. 261, p. 1052.) . . . . *B. Rapa.*

Upper stem leaves not clasping.
Beak of pod terete, much narrower than the pod, a ninth to a fourth of the total length of the fruit, without a seed near the base.
Pods 3-6 cm long, 2-3.5 mm thick, somewhat 4-sided, spreading; beak 6-12 mm long; pedicels 7-10 mm long, equaling or exceeding the flowers.......... ....2. *B. juncea.*
Pods 1-2 cm long, about 1 mm thick, appressed; beak 1.5-2.5 mm long; valves with 1 conspicuous nerve; pedicels 3-6 mm long, shorter than the flowers...... ....3. *B. nigra.*
Beak of pod flat, about as wide as the body, a fourth the length of the fruit, usually containing a seed at the base.
Fruiting pedicels mostly 3-7 mm long; pods moderately slender, about 2 mm in diameter, glabrous or hispid, ascending, valves distinctly 3-nerved, the beak usually a fifth to a third the length of the fruit; lower leaves sparingly lyrate, the upper ones usually undivided..........4. *B. kaber* var. *pinnatifida.*
Fruiting pedicels about 10 mm long; pods stout, about 4 mm thick, hispid, spreading at right angles, valves indistinctly nerved, beak very broad and flat, usually more than half the length of the fruit; leaves more lyrate. (See excluded species no. 259, p. 1051.)..................................*B. hirta.*

1. BRASSICA CAMPÉSTRIS L. FIELD MUSTARD. Map 1008. This weed has been reported twice for the state and I have specimens from two counties. Almost all crucifers are of a weedy nature. Peattie says it is established in the Calumet Region and I found it to be plentiful in the old Fair Grounds at Lawrenceburg, Dearborn County.

Nat. of Eu. and widely distributed in N. A.

2. BRASSICA JÚNCEA (L.) Cosson. INDIAN MUSTARD. Map 1009. There are two reports of this weed and I have specimens from three counties. I have always very much disliked the introduced species of crucifers and have neglected to collect them. If I had appreciated the necessity of collecting these weeds, no doubt my records would be more numerous.

Nat. of Asia, but of recent introduction.

3. BRASSICA NÌGRA (L.) Koch. BLACK MUSTARD. Map 1010. This is a frequent to common weed throughout the state. It prefers a sandy soil as crucifers usually do.

Nat. of Eurasia; generally distributed throughout the U. S.

4. BRASSICA KÀBER (DC.) Wheeler var. PINNATÌFIDA (Stokes) Wheeler. (Rhodora **40**: 306-308. 1938.) (*Brassica arvensis* (L.) Rabenhorst. of Gray, Man., ed. 7 and *Sinapis arvensis* L. of Britton and Brown, Illus. Flora, ed. 2.) CHARLOCK. Map 1011. This weed has been reported from 13 counties. It is found not only in waste places and along lines of transportation but also in cultivated and fallow fields.

Nat. of Eu. and widely distributed in N. A.

2950. RÁPHANUS [Tourn.] L.

Flowers pale yellow (fading white in herbarium specimens); pods 3.5-4 mm in diameter, longitudinally grooved, 4-10-seeded, the seed-bearing part longer and more slender than that of the following species.............1. *R. Raphanistrum.*
Flowers white or purplish; pods 2- or 3-seeded, the seed-bearing part short and thick, at maturity sometimes 10 mm in diameter, not grooved. (See excluded species no. 262, p. 1052.)..*R. sativus.*

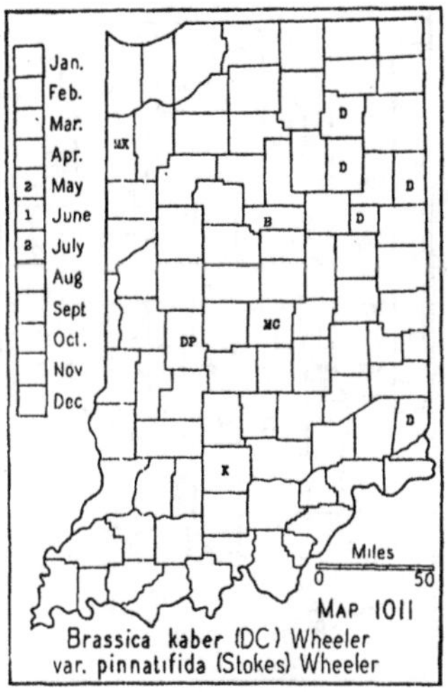

MAP 1011
Brassica kaber (DC) Wheeler var. pinnatifida (Stokes) Wheeler

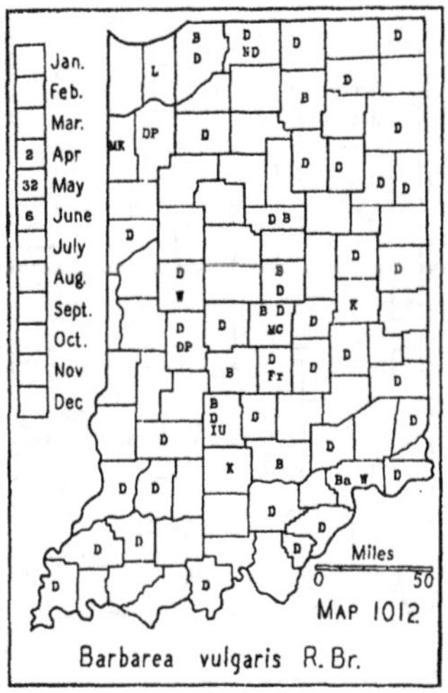

MAP 1012
Barbarea vulgaris R. Br.

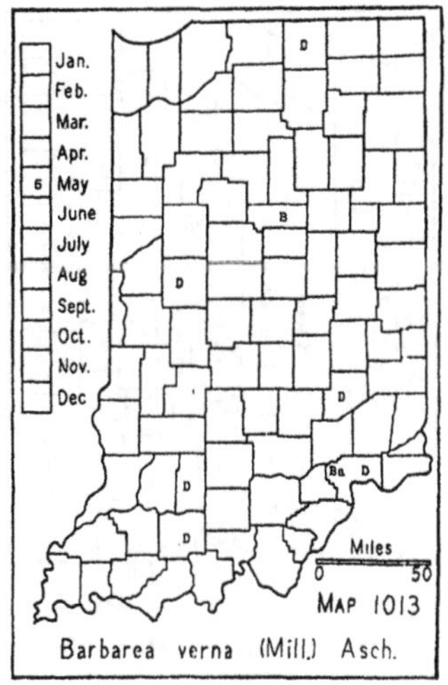

MAP 1013
Barbarea verna (Mill.) Asch.

1. RAPHANUS RAPHANÍSTRUM L. WILD RADISH. J. M. Coulter wrote of this species (Bot. Gaz. **1**: 34. 1876) that in Jefferson County it "has been found taking possession of some of our fields." Welch reports it from Jasper County. Peattie says: "A bad European weed in old fields of the Calumet District," Lake County. I have never seen it or else I did not recognize it.

Nat. of Eu. and n. Asia.

2961. BARBARÈA R.Br.

Lower leaves with 1-4 pairs of lateral leaflets, rarely entire or with 5 pairs; upper leaves generally obovate with a cuneate base, toothed, rarely pinnatifid; pedicels not as thick as the pod; flowers generally a bright yellow; mature pods 1.5-2.5 cm long, erect or spreading, obtusely angled, beak generally about 2 mm long............1. *B. vulgaris.*

Lower leaves with 5-10 pairs of leaflets, rarely one or more with as few as 4 pairs; upper leaves lyrate-pinnatifid; pedicels about as thick as the pod; flowers pale yellow; mature pods 5-7 cm long, ascending, rather sharply angled, beak generally 0.5-1 mm long............2. *B. verna.*

1. BARBAREA VULGÀRIS R. Br. BITTER WINTERCRESS. Map 1012. As treated here this species includes *Barbarea stricta* Andrz. of Gray, Man., ed. 7 and of Britton and Brown, Illus. Flora, ed. 2, not Andrz.; also *Barbarea vulgaris* var. *longisiliquosa* Carion (Rhodora **11**: 139. 1909). It is extremely variable in its leaves and in the position of the mature pods; however, the latest studies indicate that these differences are ecological (Jour. Bot. 54: 202. 1916 and 57: 304. 1919). This species is well distributed throughout the state and in some fields it forms colonies over large areas and is regarded as an obnoxious weed. It is found in fallow and cultivated fields, pastures, open woodland, and clover fields and along roadsides and railroads.

Introduced from Eu. in the Eastern and Central States but native in the north and west.

2. BARBAREA VÉRNA (Mill.) Asch. EARLY WINTERCRESS. Map 1013. Locally frequent in a few of the southern counties and probably scattered

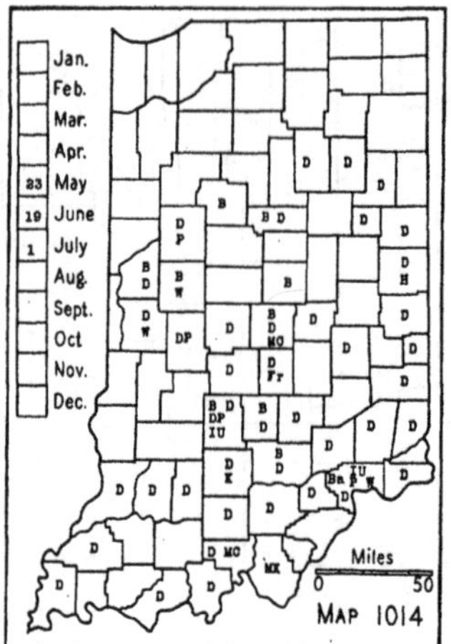

Iodanthus pinnatifidus (Michx.) Steud.

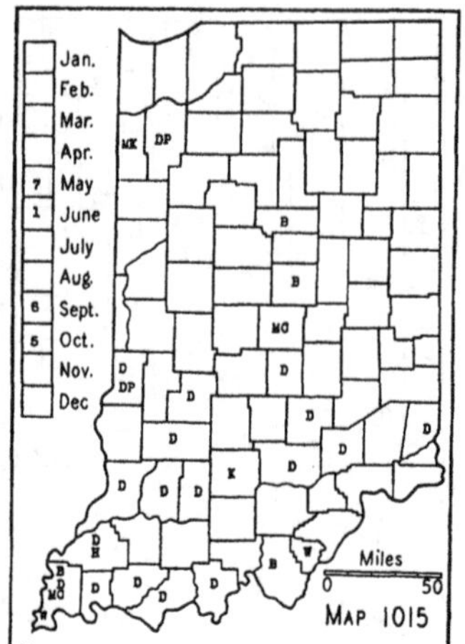

Rorippa sessiliflora (Nutt.) Hitchc.

Rorippa palustris var. glabrata (Lunell) Vict.

throughout the state. I have specimens from fallow fields, an orchard, roadsides, and railroads.

Nat. of Eu.; N. Y. to Wash., southw. to Fla. and Calif.

2963. IODÁNTHUS T. & G.

1. **Iodanthus pinnatífidus** (Michx.) Steud. PURPLE ROCKET. Map 1014. More or less frequent and locally common in moist, alluvial soil along streams and on the adjacent wooded slopes, rare elsewhere throughout the state although there are no specimens or records from the northern fourth of the state.

Western Pa. to Minn., southw. to Tenn., Mo., La., and Tex.

2965. RORÍPPA Scop.

Pedicels of mature pods not more than 3 mm long; pods oblong, somewhat flattened, mostly 6-9 mm long and about 2 mm wide.
  Styles of pods about 0.5 mm long; seed minutely pitted, about 0.5 mm wide and as long or slightly longer....................................1. *R. sessiliflora.*
  Styles of pods about 1 mm long; seed more or less pebbled. (See excluded species no. 263, p. 1052.)..*R. obtusa.*
Pedicels of mature pods mostly more than 3 mm long.
  Petioles of leaves, at least the median ones, auriculate at the base; pods widely spreading, linear, oblong-linear to ovate-oblong.
    Mature pods less than 5 mm long, the oblong type generally 3-4 mm long and about 1.5 mm wide; styles 0.5-1 mm long; seed minutely pebbled, about 0.6 mm wide and as long or longer.
      Stem and leaves glabrous, or nearly so; pods 3-4 mm long...2. *R. palustris* var. *glabrata.*
      Stem and leaves more or less pubescent; pods usually less than 3 mm long, rarely 3.5 mm long.......................2a. *R. palustris* var. *hispida.*
    Mature pods 7-15 mm long; styles 2-3 mm long. (See excluded species no. 264, p. 1052.)..*R. sinuata.*
  Petioles of leaves not auriculate at the base; perennials with creeping rhizomes; flowers bright yellow; pods usually slightly curved, mostly about 1.5 cm long and 1 mm wide; styles of mature pods 0.5-1.5 mm long........3. *R. sylvestris.*

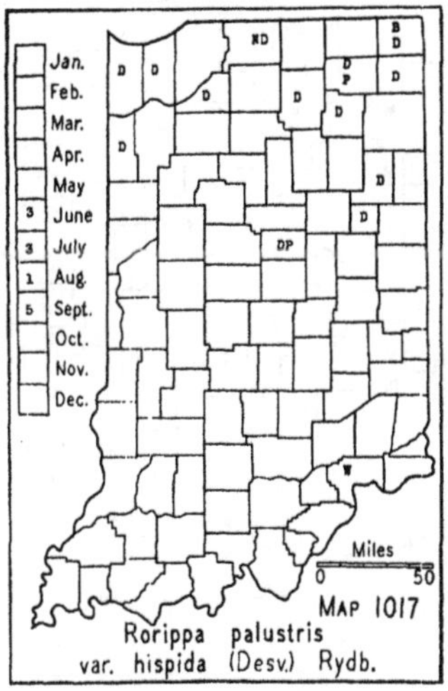

Rorippa palustris var. hispida (Desv.) Rydb.

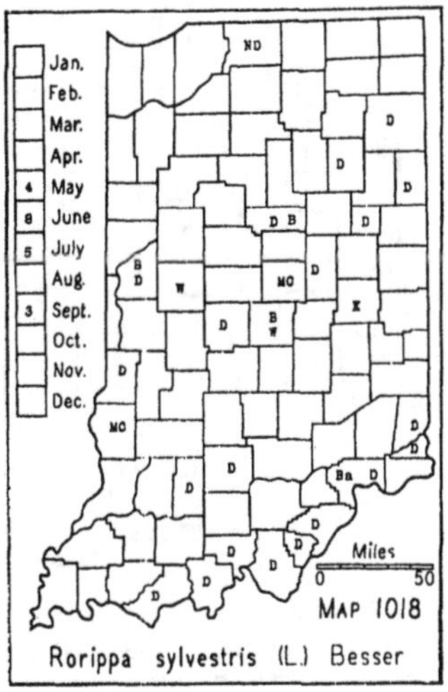

Rorippa sylvestris (L.) Besser

Nasturtium officinale R. Br.

1. **Rorippa sessiliflòra** (Nutt.) Hitchc. (*Radicula sessiliflora* (Nutt.) Greene.) SESSILE-FLOWERED CRESS. Map 1015. Infrequent but usually frequent to common where it is found. It inhabits muddy places and is often found on the borders of sloughs, ponds, and streams, usually after the water in them has been lowered by dry weather. Also found in low, flat, fallow fields. It is usually frequent on the muddy slopes of the Ohio River and, no doubt, in such a habitat it probably could be found in all of the Ohio River Counties.

Va. to Nebr., southw. to Fla. and Tex.

2. **Rorippa palústris** (L.) Bess. var. **glabràta** (Lunell) Vict.* (*Radicula palustris* (L.) Moench.) YELLOW WATERCRESS. Map 1016. Infrequent to frequent or even common in all parts of the state. It seems to have no preference for sun or shade and grows in wet places along streams, about ponds, lakes, and sloughs, and in ditches and fallow fields.

Throughout N. A. except the extreme north; also found in Eurasia.

2a. **Rorippa palustris** var. **híspida** (Desv.) Rydb.† (*Radicula palustris* var. *hispida* (Desv.) Rob. and *Radicula hispida* (Desv.) Britt.) HISPID YELLOW WATERCRESS. Map 1017. Infrequent mostly throughout the northern part of the state although it was collected by Coulter in Jefferson County. It has the habitat of the preceding species but grows in much wetter places.

Throughout temperate N. A.; also in Eurasia.

3. **Rorippa sylvéstris** (L.) Bess. (*Radicula sylvestris* (L.) Druce.) CREEPING YELLOW WATERCRESS. Map 1018. Local or infrequent on the alluvial bottoms of streams throughout the state except along the muddy slopes of the bank of the Ohio River where it is frequent to common. This is a pernicious weed and should be destroyed as soon as it is discovered. It is best exterminated by the application of some reliable weed killer.

Newf. to Ont. and Mich., southw. to Ala. and Ill.

* The latest name proposed for this plant is **Rorippa islandica var. Fernaldiana** Butters & Abbe. (Rhodora 42: 28. 1940.)

† The latest name proposed for this plant is **Rorippa islandica var. hispida** (Desv.) Butters & Abbe. (Rhodora 42: 26. 1940.)

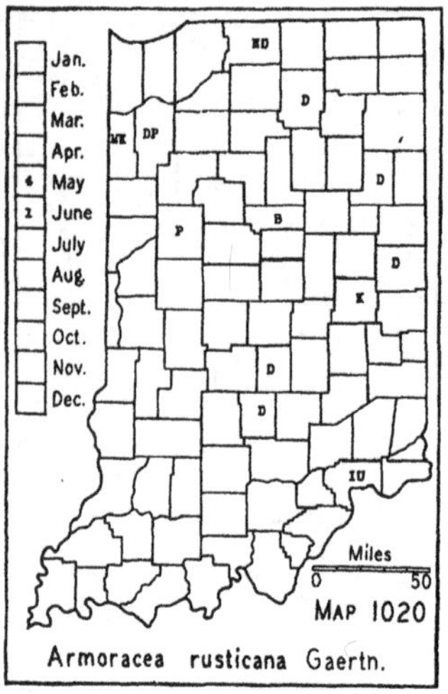

Armoracea rusticana Gaertn.

Armoracia aquatica (Eaton) Wieg.

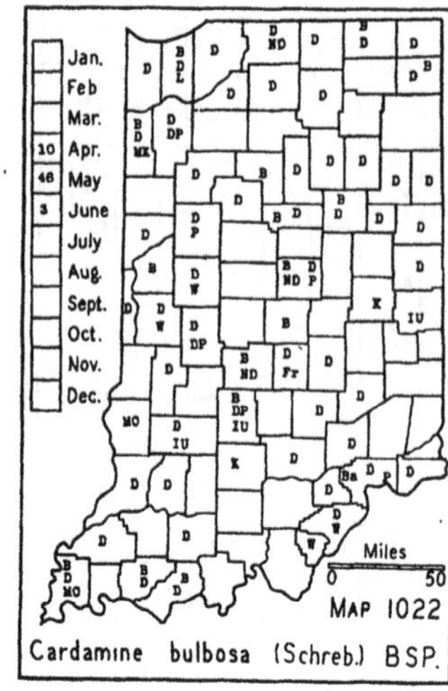

Cardamine bulbosa (Schreb.) BSP.

2965A. NASTÚRTIUM R. Br.

1. NASTURTIUM OFFICINÀLE R. Br. (*Radicula Nasturtium-aquaticum* (L.) Britten & Rendle of Gray, Man., ed. 7 and *Sisymbrium Nasturtium-aquaticum* L. of Britton and Brown, Illus. Flora, ed. 2.) WATERCRESS. Map 1019. Infrequent to rare in the glaciated area of the state, becoming rare or absent south of this area. It is found in the outlets of springs and in ditches and small streams which are fed by springs. Where it is found it is generally very abundant, often forming a complete stand over the entire surface of the water. This is the culinary watercress.

Nat. of Eurasia.

2965B. ARMORÀCIA Gaertn.

Pods short-oblong to subglobose, 2-celled; styles about 0.5 mm long; plant terrestrial, 8-12 dm high, partially aquatic when growing in ditches when the lower leaves become much dissected; basal leaves with blades 1-3 dm long; cauline leaves long and sinuate or crenate..1. *A. rusticana.*

Pods short-oblong, generally 5-7 mm long, 1-celled; styles 2-3 mm long; plant aquatic, the emersed part usually less than 6 dm long, submerged leaves capillary-dissected; cauline leaves elliptic or oblong, rarely more than 1.5 cm broad, early deciduous...2. *A. aquatica.*

1. ARMORACIA RUSTICÀNA Gaertn. (*Radicula Armoracia* (L.) Rob. and *Armoracia Armoracia* (L.) Britt.) HORSERADISH. Map 1020. Much planted and used as a condiment. It has sparingly escaped from cultivation to ditches and banks of the smaller streams throughout the state. I once found it on the bank of a pond in a clearing. I have never seen it mature seed.

Nat. of Eu. and widely spread throughout e. N. A.

2. **Armoracia aquática** (Eaton) Wieg. (Rhodora 27: 186. 1925.) (*Radicula aquatica* (Eaton) Rob. and *Neobeckia aquatica* (Eaton) Britt.) Map 1021. In stagnant water in ponds and bayous of streams and lakes. Very local but probably found in suitable habitats in many counties.

Que. and Vt. to Minn., southw. to Fla., La., and Ark.

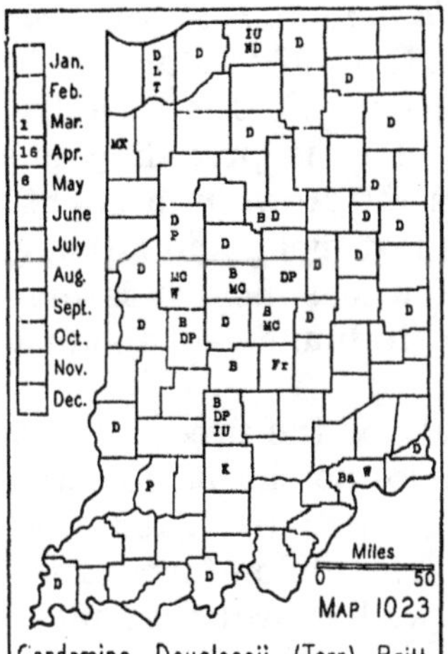

Cardamine Douglassii (Torr.) Britt.

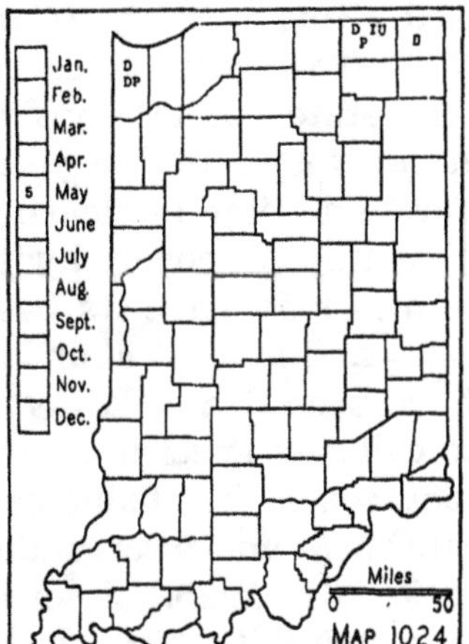

Cardamine pratensis var. palustris Wimm. & Grab.

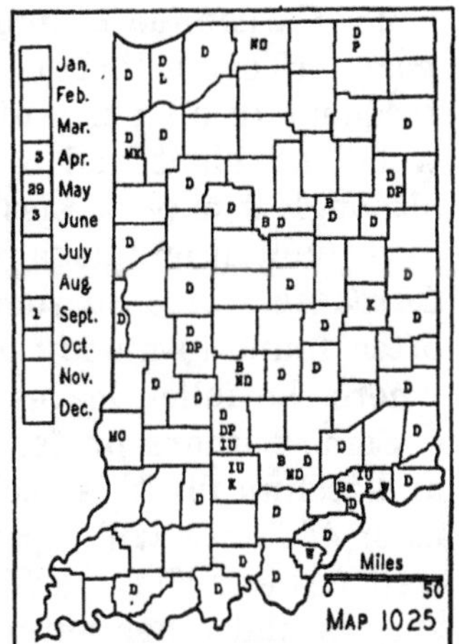

Cardamine pennsylvanica Muhl.

2966. CARDÁMINE [Tourn.] L. BITTERCRESS

Plants perennial, base tuberous; leaves not divided, sometimes those of the stem deeply toothed.
- Flowers white; stems generally 15-45 cm high, simple or much branched, pubescent at the base only, sometimes glabrous throughout or rarely more or less pubescent throughout..1. *C. bulbosa.*
- Flowers purplish (see note in text on white-flowered forms); stems generally 12-35 cm high, never branched, generally more or less pubescent all over or the upper part glabrous..2. *C. Douglassii.*

Plants without a tuberous base; leaves pinnate.
- Petals white or tinged with purple, mostly 8-13 mm long; perennials of springy places and bogs..................................3. *C. pratensis* var. *palustris.*
- Petals white, mostly 2-3 mm long; annuals or biennials.
  - Lateral leaflets of cauline leaves oblong to oval, often toothed, and usually more or less decurrent on the rachis; terminal leaflet larger, obovate, usually 3-lobed, sometimes entire or 5-7-lobed; plants always found in wet or moist soil..4. *C. pennsylvanica.*
  - Lateral leaflets of cauline leaves generally linear, or linear-oblong, entire, not decurrent on the rachis; terminal leaflet usually not larger, of the same shape as the lateral leaflets or obovate and 3-lobed; plants of dry soil or rarely in moist soil..........................5. *C. parviflora* var. *arenicola.*

1. **Cardamine bulbòsa** (Schreb.) BSP. BULB BITTERCRESS. Map 1022. Frequent to common throughout the state in low places in woodland, marshes, ditches, and springy places along streams and in wet places about ponds, sloughs, and lakes. This species seems to find its optimum in the inundated woods of the southwestern part of the state where branched specimens are more frequently found. This species, as well as the next, varies in the amount of pubescence. It is generally slightly pubescent near the base only but specimens are found which vary from entirely glabrous (with the exception of a straggling hair here and there) to pubescent up to the middle. Extremely pubescent plants, however, may be albino forms of the next species if we accept this form. Rarely a plant is found that is glabrous except for a pubescent calyx.

Eastern Mass. to Minn., southw. to Fla. and Tex.

2. **Cardamine Douglássii** (Torr.) Britt. (*Cardamine bulbosa* var. *purpurea* (Torr.) BSP.) NORTHERN BITTERCRESS. Map 1023. Rare to infrequent in all parts of the state but locally frequent. This is strictly a woodland plant and is never found in the open like the preceding one, although it may persist for some time in clearings. It grows in a slightly drier habitat and usually has its base covered with leaf mold, and although *Cardamine bulbosa* sometimes may have its base in leaf mold it is more often found in muddy places and is most abundant in inundated woodland where the fallen leaves have been floated away.

Some authors prefer to call this species a variety of the preceding one and technically this decision seems justifiable because no character except color of flower will separate them and intergrading forms are found. Farwell (Amer. Midland Nat. 9: 261. 1925.) described an albino form of this species and other authors agree with him. It is well known that plants with purplish flowers have albino forms and these are known to occur in some species of the *Cruciferae.* It is with a considerable degree of uncertainty, that albino and fruiting specimens of this species are separated from the preceding species unless a habitat description accompanies the specimen. I think, however, the species are distinct for the following reasons. (1) This species flowers 10-15 days earlier. (2) The habitat is much drier and I believe that the soil is slightly more alkaline. The preceding species is generally associated with white elm, swamp white, bur, and pin oaks, and sweet gum while this species is generally restricted to a zone slightly drier. The plant blooms in early spring when excessive rains may occur and its habitat may appear much wetter than it normally is so that wetness of soil is no criterion but the associated woody and herbaceous plants are. This species is usually found associated with beech and sugar maple, basswood, red oak, white ash, and others. (3) The plant, in a large series, is smaller in all of its parts; never (in all the specimens at hand) branched while *Cardamine bulbosa* is frequently more or less branched; cauline leaves generally 3-5 in contrast with the other species with 4-8 leaves; basal leaves are often more orbicular, smaller and thinner than in the preceding which, on the whole, has larger and more often elliptic-ovate and thicker leaves. I have studied carefully the length of the pods and the length of their beaks and they are too variable to be of taxonomic value. The seed of a long series of this species, however, are smaller.

Conn. to s. Ont. and Wis., southw. to Md. and Ky.

3. **Cardamine praténsis** L. var. **palústris** Wimm. & Grab. (Fernald in Rhodora 22: 14. 1920.) (*Cardamine pratensis* L.) CUCKOOFLOWER. Map 1024. This species inhabits tamarack bogs and marshes in a few of our northern counties. I am not certain of its abundance but I think it is rather rare and local.

Quebec to Mack., southw. to Newf., Conn., N. J., Ind., Minn., and B. C.; also in Eurasia.

4. **Cardamine pennsylvánica** Muhl. PENNSYLVANIA BITTERCRESS. Map 1025. Infrequent to frequent throughout the state in wet or moist soils.

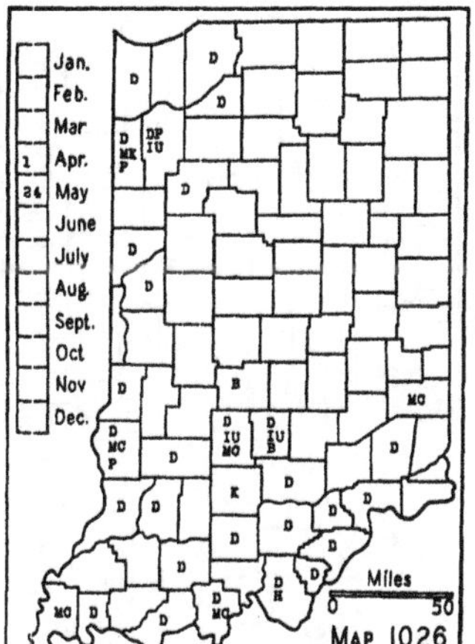

MAP 1026
Cardamine parviflora var. arenicola (Britt.) O.E. Schulz

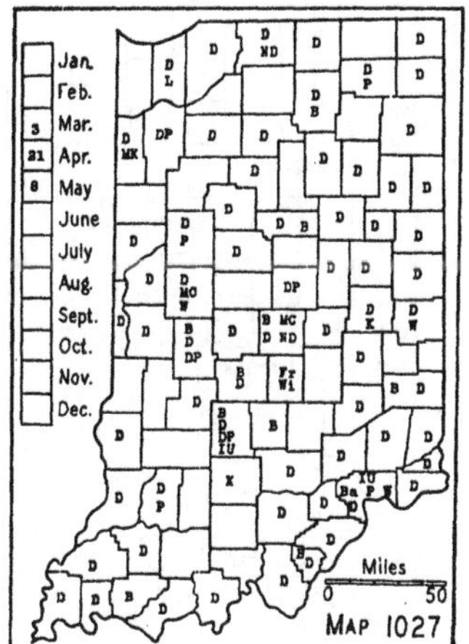

MAP 1027
Dentaria laciniata Muhl.

MAP 1028
Dentaria multifida Muhl.

This species, like the next, prefers denuded or semidenuded areas, hence it is more abundant where it is found in moist, clay soil in fallow fields. It is found in wet places in woodland, ditches, and marshes, along streams, and about lakes. Sometimes the base is immersed in water when it may be mistaken for *Nasturtium officinale* or vice versa. The species may be separated easily by the size of the flowers. The calyx of this species is about 2 mm long and petals about 3 mm long while those of *Nasturtium* are about 3 mm and 4 mm long respectively.

This species is very difficult to separate from the next. Most specimens are easily distinguishable by the habitat but we have specimens from moist, clay soil of fallow fields (the habitat of this species) that, so far as I can determine, belong to the next species. Contrary to my prejudice, I am compelled to recognize two habitats for the next species.

Lab., to Minn., and Mont. to B. C., southw. to Fla., Kans. and Calif.

5. **Cardamine parviflòra** L. var. **arenícola** (Britt.) O. E. Schulz. (Rhodora 29: 192. 1927.) (*Cardamine parviflora* L. and *Cardamine arenicola* Britt.) SMALL-FLOWER BITTERCRESS. Map 1026. This species is probably found in all parts of the state except in the rich, neutral soil of the central part. It is rare to infrequent and is usually found in dry soil in bare spots in woodland under black and white oak or in a habitat simulating this one. On these bare spots the plants may be only 2-3 inches high but on or near the border where the leaf mold and vegetation about such places begin the largest plants will be found. The plant is so delicate that it can not push its way through leaf mold or compete with much vegetation. There are, however, plants that must belong to this species that are found in moist, clay soil in fallow fields where they are usually associated with *Agrostis hyemalis, Cardamine pennsylvanica, Hordeum pusillum,* and *Arabis virginica.* I am of the opinion that both habitats have slightly acid soil.

Que. to Ga. and the Mississippi Valley to Oreg.

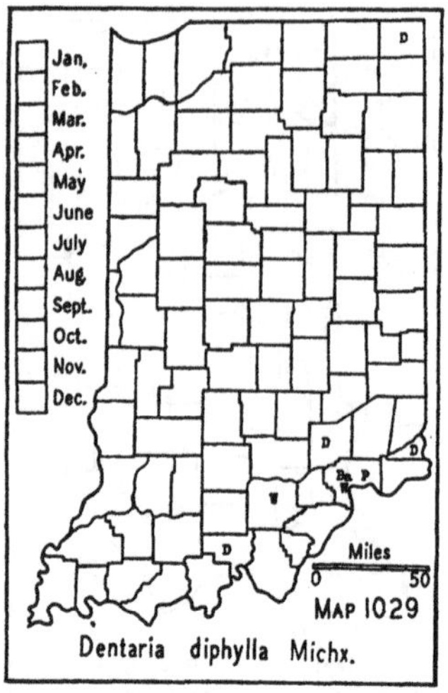

Dentaria diphylla Michx.

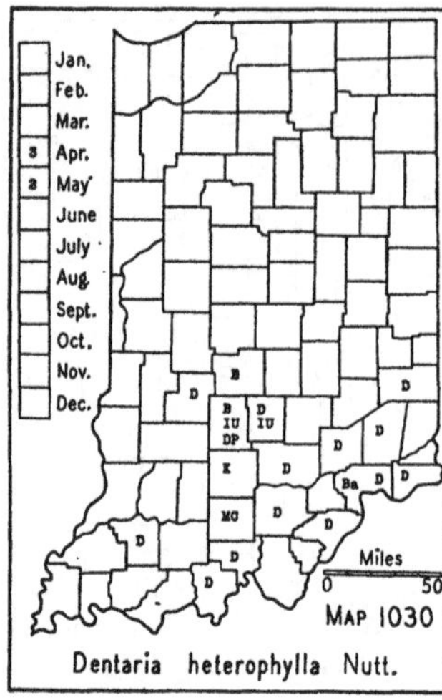

Dentaria heterophylla Nutt.

Leavenworthia uniflora (Michx.) Britt.

2967. DENTÀRIA [Tourn.] L. TOOTHWORT

Inflorescence pubescent; rootstock deeply rooted, an elongated tuber usually 1.5-3 cm long, increased annually by additional ones, forming a lineal series; basal and cauline leaves similar but the basal ones absent at flowering time; cauline leaves 3, whorled or approximate, rarely alternate, cleft nearly to or to the base into 3 nearly equal, broadly lanceolate to linear parts, the parts sometimes likewise cleft, their margins serrate, more or less incised, or entire; anthers mostly 1.5-2.3 mm long..1. *D. laciniata.*

Inflorescence glabrous.

Basal and cauline leaves similar in shape.

Rootstocks elongated tubers, deep in the soil; basal leaves biternate, all of the segments linear with smooth margins; cauline leaves 2, opposite, biternate with linear segments, margins of segments smooth, sometimes the ultimate segments cleft or divided; petals white, tinged with purple, anthers 1.5-2 mm long...2 *D. multifida.*

Rootstocks continuous, usually 5-15 cm long, not deep in the soil; basal leaves ternate, on petioles generally 7-15 cm long; leaflets rhombic-ovate, or oblong-ovate, petiolate; cauline leaves 2, rarely 3, opposite or nearly so, on petioles generally 0.5-4 cm long, ternate or deeply cleft into 3 parts; margins of all leaflets generally ciliate and coarsely crenate with bluntly mucronate teeth; anthers about 3 mm long.....................................3. *D. diphylla.*

Basal and cauline leaves not similar in shape; basal leaves much larger than the cauline ones, ternate, on petioles generally 5-15 cm long; leaflets similar to the preceding; cauline leaves 2 or 3, variously disposed, on petioles 0.5-3 cm long, ternate, deeply cleft into 2 or 3 parts, or simple, the margins of the parts generally ciliate and serrate, sparingly toothed or entire; rootstock consisting of elongated tubers near the surface of the ground; anthers generally 1.5-2.3 mm long...4. *D. heterophylla.*

1. **Dentaria laciniàta** Muhl. CUT TOOTHWORT. Map 1027. Infrequent to frequent in rich woods throughout the state. This species prefers moist soil and deep leaf mold. The variability of the plants has led authors to assign specific, varietal, and hybrid names to these variations. I can do no better than to quote J. M. Coulter (Ann. Rept. Geol. Surv. Indiana **6**: 234. 1875) who recognized these variations and said in his flora of Jefferson County: "The leaves vary from almost entire to finely dissected. Some-

times there are three leaves in a whorl; sometimes these leaves are alternate; sometimes there are four alternate leaves; often there are but two leaves either opposite or alternate. In fact there is no kind of division or position of leaves which is not represented in this species." The preceding observation applies to my specimens but I doubt that hybridization is responsible for such variations as the alternate character of the leaves, since, in our area, there are no alternate-leaved species nor entire-leaved species in the genus as now known, with which *D. laciniata* could hybridize. I believe it is best to regard the genus as a mutating one and some of the aberrant specimens as examples of reversion to ancestral forms.

Western Que. and Vt. to Minn., southw. to Fla. and La.

2. **Dentaria multífida** Muhl. Map 1028. This species was first recognized as such by Miss Edna Banta, who found it in 1935 on a rocky, wooded slope along Big Creek a mile west of Volga, Jefferson County. It was found also in Jefferson and Clark Counties by early authors who confused it with other species of the genus. They remarked about the many forms of leaves of their specimens. Schneck, in his list of plants from the Lower Wabash Valley, also calls attention to the varied leaves.

Ind. and Ohio, southw. to Ga. and Ala.

3. **Dentaria diphýlla** Michx. CRINKLEROOT. Map 1029. This species has a very restricted range and is local in the state. Where it is found, it often forms large colonies. My Steuben County plant was found in the A. E. Emerson woods about 6 miles southwest of Angola. This colony was very dense, about 3 x 6 feet, and located in very moist and sandy soil on a low, alluvial flat along a small creek. The Ohio County plants were also found in large colonies on the moist slope of Laughery Creek.

Eastern Que. to s. Ont. and Minn., southw. to S. C. and Ky.

4. **Dentaria heterophýlla** Nutt. SLENDER TOOTHWORT. Map 1030. Infrequent to frequent but locally common on moist, rich, wooded slopes in the southern part of the state. There is a report from Cass & Wabash Counties but there is no specimen. The stem leaves of this species are very variable.

N. J., Pa., and Ind., southw. to Ga. and Tenn.

2971. LEAVENWÓRTHIA Torr.

1. **Leavenworthia uniflòra** (Michx.) Britt. MICHAUX LEAVENWORTHIA. Map 1031. This species is known from only one county in the state. It was discovered by Chas. R. Barnes in 1877 at a place locally known as Denny's Lick, about a mile southeast of Charlestown, Clark County. I collected it there in mature fruit, May 1, 1918, and late in May, 1933, some members of the Indiana Academy of Science visited the place and found it abundant over several acres but it was almost past the fruiting stage. The early fruiting dates indicate that it must flower early in April. The habitat is the washed limestone slopes of a permanent pasture where it is associated with *Draba verna, Oxalis violacea,* and *Veronica arvensis.*

Ind. to Ky., Tenn., and Ark.

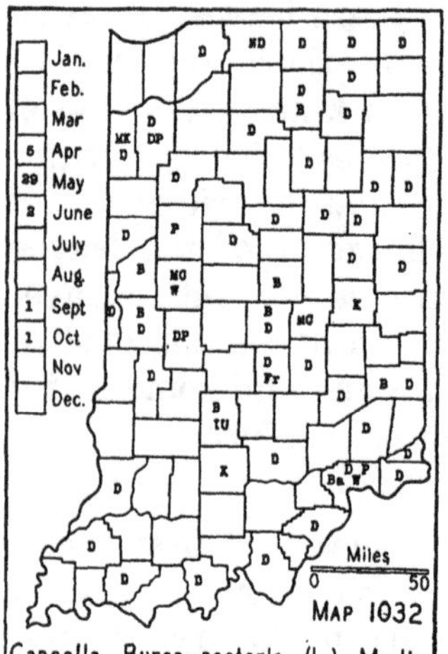

Capsella Bursa-pastoris (L.) Medic.

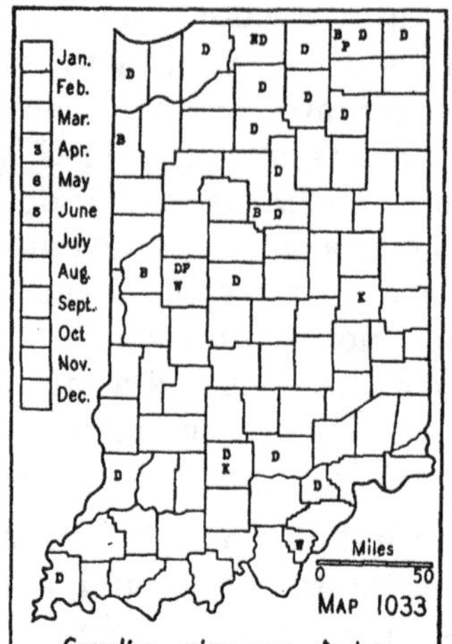

Camelina microcarpa Andrz.

Draba brachycarpa Nutt.

2983. **LESQUERÉLLA** Wats.

See excluded species no. 269, p. 1053.

2986. **CAPSÉLLA** Medic.

1. CAPSELLA BÚRSA-PASTÒRIS (L.) Medic. SHEPHERD PURSE. Map 1032. A frequent to common weed throughout the state in cultivated grounds, lawns, and pastures and along roadsides and railroads. The plant is very variable and has been the subject of much study by Almquist and Shull. Almquist, in 1920, writes as follows: "Among 370 races from different countries I was able to find at least 70 species that remained constant in culture during two or three generations. At present I have published descriptions of 200 constant forms." My specimens are variable, and, no doubt, several of the elementary species of this complex occur in Indiana.

Nat. of Eu.; widely distributed throughout the world.

2987. **CAMÉLINA** Crantz

Stems and leaves pubescent; pods 4-5 mm wide......................1.*C. microcarpa.*
Stems and leaves glabrous or with scattered hairs; pods 6-7 mm wide. (See excluded species no. 270, p. 1053.)..*C. sativa.*

1. CAMELINA MICROCÁRPA Andrz. Map 1033. This species is essentially a sandy ballast plant and is more or less frequent throughout the state. I have found it also along sandy roadsides and in sandy, fallow fields where I once found it as an abundant weed.

Nat. of Eu.; Newf. to B. C., southw. to R. I., Va., Kans., and Ariz.

2988. **NÉSLIA** Desv.

See excluded species no. 271, p. 1053.

2989. **DRÀBA** [Dill.] L.

Plants conspicuously branched and leafy to the flowers, appressed-pubescent throughout except the glabrous pods; flowers purplish; pedicels of the mature pods rarely more than 2 mm long; pods about 4 mm long, glabrous.........1. *D. brachycarpa.*

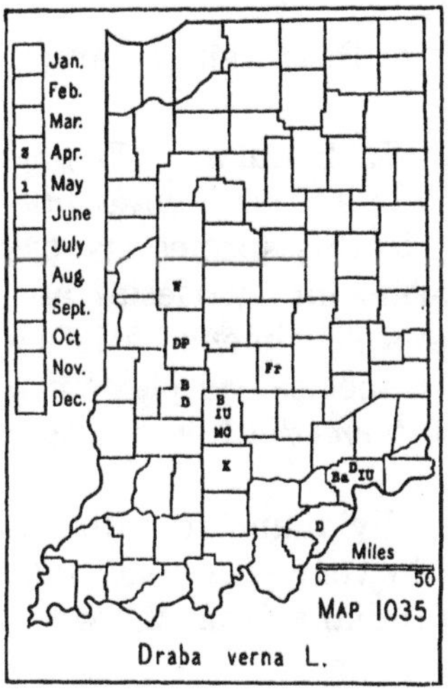

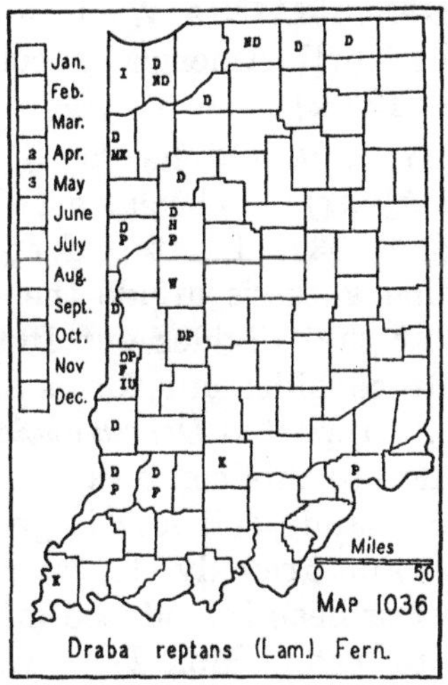

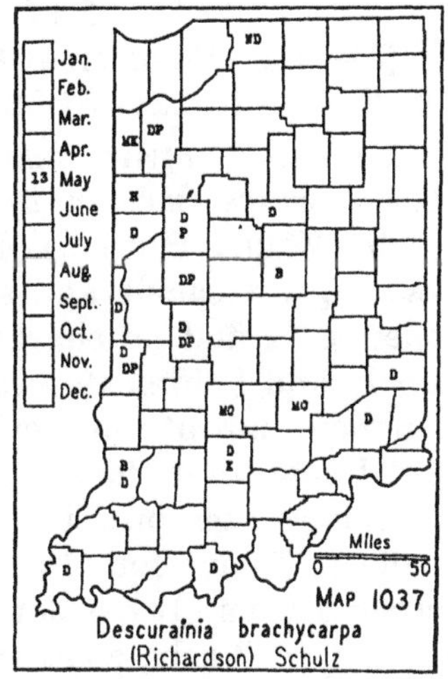

Plants not conspicuously branched, not leafy to the flowers, other parts besides the pods glabrous, pubescence not appressed; flowers white; pedicels of the mature pods more than 2 mm long; pods generally more than 4 mm long.

Pedicels of the lowest pods more than 1 cm long......................2. *D. verna.*

Pedicels of the lowest pods less than 1 cm long.

Pods glabrous..3. *D. reptans.*

Pods minutely appressed-pubescent. (See excluded species no. 272, p. 1053.)....... ....*D. reptans* var. *micrantha.*

1. **Draba brachycárpa** Nutt. SHORT-FRUITED WHITLOWGRASS. Map 1034. Found locally in dry, sandy soil in woodland pastures and on wooded slopes in a few of our southwestern counties. It has been reported by Nieuwland from Lake, La Porte, and St. Joseph Counties but I could not find specimens from these counties in the herbarium at the University of Notre Dame. Buhl (Amer. Midland Nat. **16**: 251. 1935) refers a report by Peattie from the Calumet District to *D. reptans.*

Va., Ill., Mo., and Kans., southw. to Fla. and Tex.

2. DRABA VÉRNA L. VERNAL WHITLOWGRASS. Map 1035. The common name is derived from the use of this plant in the cure of whitlow disease. Locally frequent to common as a weed in dry soil in pastures, waste places, and cultivated grounds. It has been reported from eight counties not indicated on the map. It is much more conspicuous after fruiting when the valves have fallen. The writer has seen it in several counties in this stage of growth but specimens were not collected.

Nat. of Eurasia; e. Mass. to Minn., southw. to Ga. and Tenn.

3. **Draba réptans** (Lam.) Fern. (Rhodora **36**: 368. 1934.) (*Draba caroliniana* Walt.) CAROLINA WHITLOWGRASS. Map 1036. Generally found in very dry, sandy soil in woodland pastures, fallow fields, and waste places, along roadsides, and on open, wooded dunes. The variety has been reported from the dune area. Although I have not seen a specimen, it should be sought in the state.

Eastern Mass., s. Ont., Minn. to Idaho, southw. to Ga. and Ariz.

2997. DESCURÀINIA Webb. & Barth.

[Detling. Revision of the North American species of Descurainia. Amer. Midland Nat. 22: 481-520. 1939.]

1. DESCURAINIA BRACHYCÁRPA (Richardson) O. E. Schulz.* (*Sisymbrium canescèns* var. *brachycarpon* (Richardson) Wats. and *Sophia pinnata* (Walt.) Howell.) Map 1037. Plants referred to this species are exceedingly variable and some authors divide the forms into varieties and species. I find no character that divides our Indiana specimens satisfactorily and I think it best to consider our forms as a species complex until further study of the group. I reported *Descurainia intermedia* for the state but I now refer the specimen to this complex.

This species prefers very sandy soil and is generally found in railroad ballast. I have found it also on gravelly slopes and in very sandy soil in a creek bottom. It has doubtless been introduced into Indiana from the west.

Que. to Wash., southw. to Tenn., Mo., Tex., and Calif.

3001. ÁRABIS L.

Plants with basal and stem leaves pinnatifid, much branched at the base, the branches widely spreading; plants rarely 4 dm high; pedicels rarely more than 5 mm long; pods about 2 cm long; seed narrow-winged, in 1 row.............1. *A. virginica.*

Plants not as above.

  Stem leaves more or less auricled at the base and often clasping.

    Stems and leaves more or less pubescent throughout (at least below the middle); seeds in 1 row.

      Mature pods erect or strongly ascending, 30-50 mm long; plants erect, often divided at the base; leaves pubescent mostly on the margins and midrib beneath (sometimes some of the upper ones entirely glabrous); seed oblong, about 1 mm long, narrowly winged.

        Pubescence of stem spreading or subspreading, predominantly of simple hairs. ....2. *A. pycnocarpa.*

        Pubescence of stem strictly appressed, often giving a strigose appearance, predominantly of forked hairs.......2a. *A. pycnocarpa* var. *adpressipilis.*

      Mature pods spreading; plants erect or decumbent; leaves pubescent all over, at least on the lower surface (except in *A. viridis* var. *Deamii*).

        Plants erect, not branched at the base (at least not conspicuously so), growing on dry rocky or gravelly slopes; pedicels mostly 10-18 mm long.

          Pubescence of simple hairs; basal leaves usually somewhat pinnatifid...... ....3. *A. viridis* var. *Deamii.*

          Pubescence mostly forked; basal leaves merely dentate........4. *A. patens.*

        Plants lax, conspicuously branched at the base, the branches decumbent or ascending, growing in moist, alluvial soil along streams and in moist soil at the base of slopes; pedicels mostly 1-3 mm long.........5. *A. dentata.*

    Stem leaves glabrous, or only the base of the stem and basal leaves and a few of the lowest stem leaves pubescent.

      Plants with the middle cauline leaves long and narrow, generally 5-15 mm wide and 7-13 cm long, spreading or rarely erect, entire or coarsely dentate; pedicels mostly 5-10 mm long; pods recurved, spreading, 6-11 cm long; seed in 1 row, winged, about 1.5 mm long including the wing.....6. *A. laevigata.*

      Plants not as above.

        Pods terete or 4-sided, about 1 mm wide; seed in 1 row or in some pods the seed interruptedly in 1 and 2 rows, wingless; stem glabrous or somewhat hirsute at the base; stem leaves glabrous; basal leaves more or less

* The name of this plant now becomes **Descurainia pinnata subsp. brachycarpa** (Richardson) Detling. (Amer. Midland Nat. **22**: 509. 1939.)

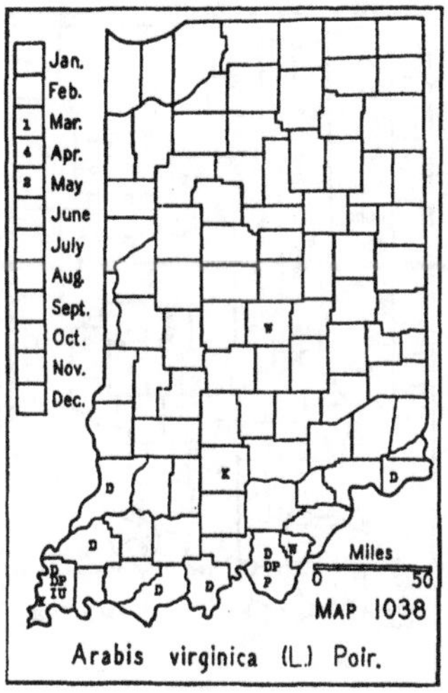

MAP 1038
Arabis virginica (L.) Poir.

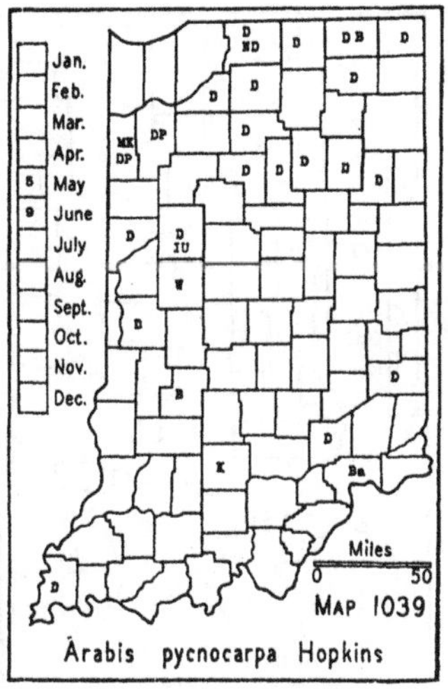

MAP 1039
Arabis pycnocarpa Hopkins

MAP 1040
Arabis pycnocarpa var. adpressipilis Hopkins

stellate-pubescent; plants simple, erect, usually 6-12 dm high.......... ....7. *A. glabra.*

Pods flattened, 1.5-2 mm wide; seeds winged; basal leaves more or less pubescent.

Mature pods erect; stem and stem leaves glabrous; seed in 2 distinct rows. ....8. *A. Drummondii.*

Mature pods spreading or loosely ascending; usually the base of the stem and a few of the lowest leaves more or less pubescent; seed, when young, in 2 distinct rows, many aborting and leaving the mature, perfect ones in 1 row almost as wide as the cell. (See excluded species no. 273, p. 1053.).....................................*A. divaricarpa.*

Stem leaves (at least the upper) without auricled bases.

Plants much branched, usually less than 3.5 dm high; basal leaves lyrate-pinnatifid; stem leaves rarely over 4 cm long, linear-oblong, obtuse, entire or with a few teeth; pedicels of fruit ascending; mature pods ascending, mostly 20-50 mm long; seed in 1 row, wingless......................9. *A. lyrata.*

Plants simple or nearly so, mostly 5-15 dm high; basal leaves dentate or lyrate-lobed; stem leaves tapered at both ends, generally much more than 4 cm long, more or less deeply toothed; pedicels of fruit recurved; mature pods pendent, mostly 50-80 mm long; seed in 1 row, broadly winged............ ....10. *A. canadensis.*

1. **Arabis virgínica** (L.) Poir. (*Arabis virginica* (L.) Trel. of Gray, Man., ed. 7 and Britton and Brown, Illus. Flora, ed. 2.) VIRGINIA ROCKCRESS. Map 1038. Locally frequent to common in fallow cornfields in the southern part of the state. It is usually in moist soil associated with *Poa Chapmaniana* and *Myosotis virginica,* indicating a slightly acid habitat which, I think, controls its distribution.

Va. to Ill., southw. to Fla. and Tex.; westw. to Calif. and Lower Calif.

2. **Arabis pycnocárpa** Hopkins. (Rhodora 39: 112. 1937.) (*Arabis hirsuta* of American authors.) HAIRY ROCKCRESS. Map 1039. Infrequent to rare in the state and probably absent from some of the central counties. It grows in sandy soil in alluvial bottoms, in crevices of rocks, and on rocky slopes and high banks of streams. Since all of my specimens are from the borders of streams and lakes, its affinity for them is apparent.

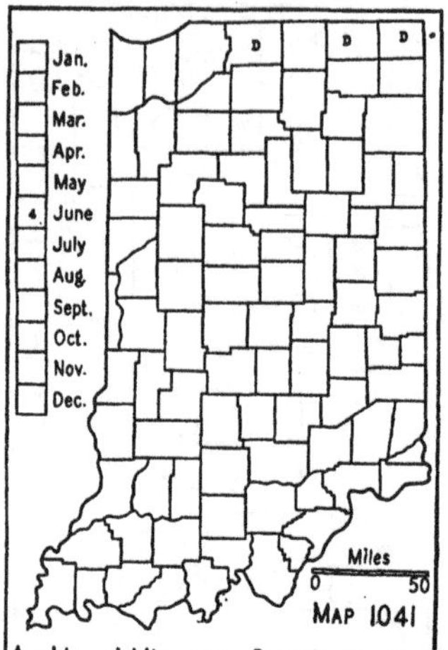

Arabis viridis var. Deamii Hopkins

Arabis patens Sulliv.

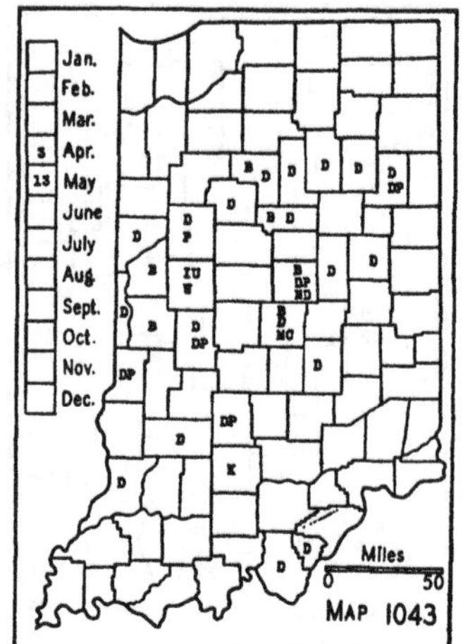

Arabis dentata T. & G.

This species is variable in its pubescence which is sometimes restricted to the lower part of the plant. Its pubescence, strict habit, and its habit of sending up several erect branches from the base usually identify it.

N. B. to Alaska, southw. to Ga., Mo., Ariz., and Calif.

2a. **Arabis pycnocarpa** var. **adpressipìlis** Hopkins. (Rhodora 39: 117-118. 1937.) Map 1040. This variety is infrequent and has the habitat of the species.

Ont. to Minn., southw. to Va., Mo. and Okla.

3. **Arabis víridis** Harger var. **Dèamii** Hopkins. (Rhodora 39: 157-158. 1937.) Map 1041. This species is probably not very rare in northern Indiana since I have it from three counties. My specimens are from dry sandy and gravelly slopes.

Ind., Wis., and Mo.

4. **Arabis pàtens** Sulliv. SPREADING ROCKCRESS. Map 1042. I have found this species in two places only. In Bartholomew County it was growing on top of a limestone rock along Clifty Creek north of Hartsville and in Harrison County there is a colony of it in the talus of the ledges of the cliff along a side road along Blue River about a half mile north of White Cloud. It is very local. It has, however, been reported from Clark, Decatur, Jefferson, and Tippecanoe Counties.

Pa. to Ind., southw. to Tenn.

5. **Arabis dentàta** T. & G. TOOTHED ROCKCRESS. Map 1043. Infrequent to rare throughout the state or absent from some counties. Locally it may be rather frequent but I have never found it so. It is usually found in moist, sandy soil of wooded, alluvial bottoms, in the talus of cliffs, and rarely in dry soil on slopes. It is restricted almost entirely to the proximity of streams and is more frequent along our major streams.

Western N. Y. to Minn., southw. to Va., Tenn., Mo., and Okla.

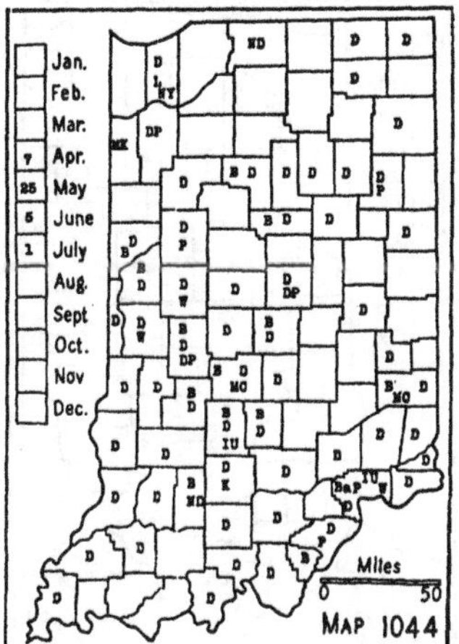

Arabis laevigata (Muhl.) Poir

Arabis glabra (L.) Bernh.

Arabis Drummondii Gray

6. **Arabis laevigàta** (Muhl.) Poir. SMOOTH ROCKCRESS. Map 1044. Frequent but never common throughout the state. It prefers a rather sandy or gravelly soil and is restricted to the wooded slopes and high banks of streams. The leaves of this species are variable and one form has been named. I am including it under the species.

Western Maine to S. Dak., southw. to Iowa and Ark.

7. **Arabis glàbra** (L.) Bernh. TOWER MUSTARD. Map 1045. Infrequent to rare in the lake region of the state and extremely rare, absent, or introduced in the southern part. My Floyd County specimen was found in a hayfield along Indian Creek near Galena. Several specimens were noted.

N. B. to B. C., southw. to n. N. J., Pa., the Great Lakes, S. Dak., Utah, and Calif.

8. **Arabis Drummóndii** Gray. DRUMMOND ROCKCRESS. Map 1046. I found two colonies of this species in gravelly soil on the slope of the north bank of the St. Joseph River, one about a half mile and one a mile and a half southwest of Bristol. Nieuwland's report for it from St. Joseph County is the only other record.

Lab. to B. C., southw to N. S., s. N. E., N. J., Ohio, Ill., Utah, and Calif.

9. **Arabis lyràta** L. LYRELEAF ROCKCRESS. Map 1047. Found in very dry, sandy soil in black oak woods and fallow fields and on open wooded dunes and sandy, roadside knolls.

Ont. to Man. and Alaska, southw. to Conn., Va., Tenn., and B. C.

10. **Arabis canadénsis** L. SICKLE-POD. Map 1048. Infrequent to rare throughout the state; probably absent from some of the central counties. It prefers a dry and rather sandy soil and is found on the crests of wooded ridges and on rocky, wooded slopes.

Eastern Mass., Vt., and Ont. to Minn., southw. to Ga., Tex., and Kans.

3004. **ERÝSIMUM** [Tourn.] L.

Petals mostly 20-30 mm long, orange yellow; pedicels and pods in our specimens ascending; pods 4-sided, 6-10 cm long..............................1. *E. asperum*.

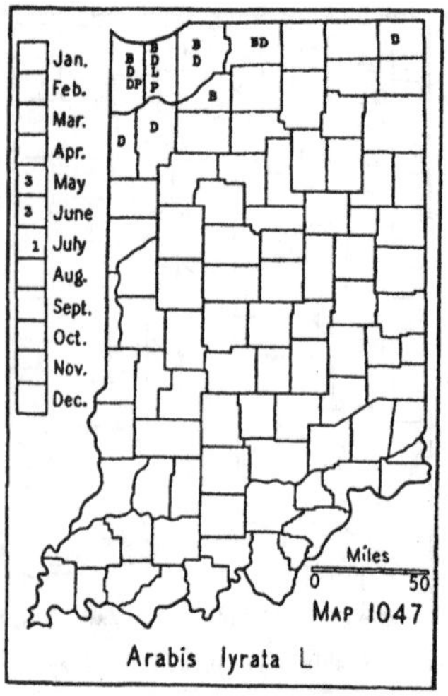

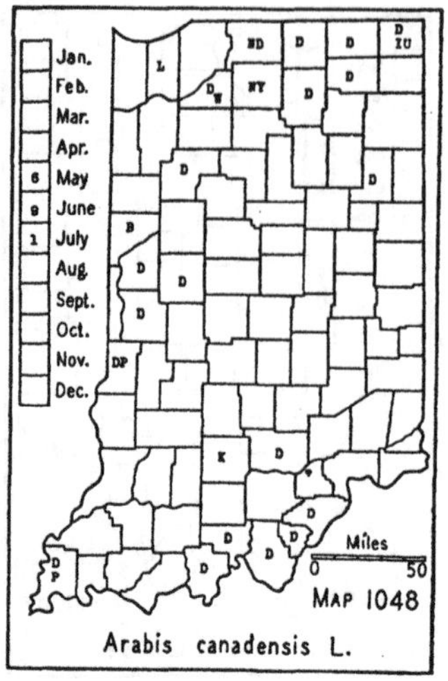

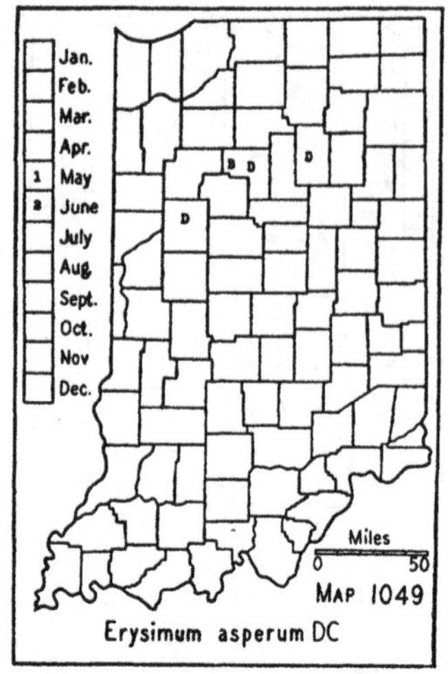

Petals less than 10 mm long, lighter yellow than the preceding.
Pedicels stout, 1 mm or more thick, almost as thick as the pod, widely spreading or ascending, 4-8 cm long; petals 6-9 mm long....................2. *E. repandum.*
Pedicels slender, ascending, mostly about 0.5 mm thick, about half as thick as the pod.
Petals 4-5 mm long; lower pedicels mostly 10-15 mm long....3. *E. cheiranthoides.*
Petals mostly 6-8 mm long; lower pedicels less than 10 mm long. (See excluded species no. 274, p. 1053.)..............................*E. parviflorum.*

1. **Erysimum ásperum** DC. (*Cheirinia aspera* (DC.) Britt. of Britton and Brown, Illus. Flora, ed. 2.) WESTERN WALLFLOWER. Map 1049. Somewhat frequent on the limestone ledges of the Wabash River west of Logansport; a few plants on a rocky, wooded slope along the Wabash River east of Wabash; and infrequent in gravelly soil of the slope and top of the high bank of Big Wea Creek about 4 miles southwest of Lafayette. Blatchley reported it as scarce on gravelly banks in Vigo County. It has also been reported from Carroll, Montgomery, and Putnam Counties. The mass distribution of the species is west of our area and it is no doubt very local in Indiana, with reports from two counties in Ohio.

Newf., Que., Sask., Colo., southw. to Ohio, Ind., Ill., and N. Mex.

2. ERYSIMUM REPÁNDUM L. (*Cheirinia repanda* (L.) Link.) TREACLE MUSTARD. Map 1050. I have found this species along a roadside, in ballast along a railroad, and in a waste place. No doubt it has a much wider distribution than the map indicates.

Nat. of Eu.; waste places about eastern seaports, and Ohio to Kans., Ariz., Utah, and Oreg.

3. **Erysimum cheiranthoides** L. (*Cheirinia cheiranthoides* (L.) Link.) WORMSEED MUSTARD. Map 1051. This species prefers a muck soil and is fast becoming established in the lake area. Where it has become well established, it forms a complete and dense stand. I found a pure stand of it about three miles north of Albion, Noble County, in muck soil in a low place in an oatfield where the oats had been drowned out. Its habit of germinating late in the season permits it to occupy low places in

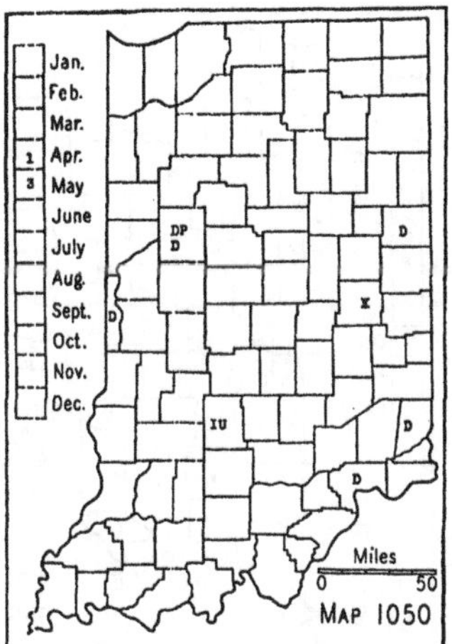

Erysimum repandum L.

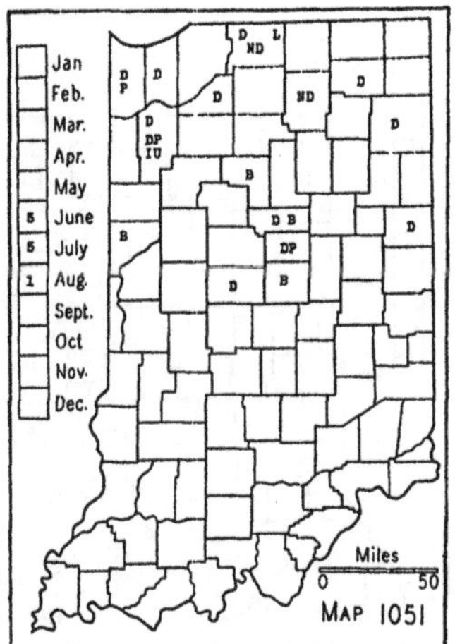

Erysimum cheiranthoides L.

Alyssum alyssoides L.

hayfields, peppermint fields, etc. It is rather frequent on the spill banks of dredged ditches and in railroad ballast. There are no reports for the state south of the area shown on the map.

Since none of the early authors reported this species, and the first report was in 1915, and because its habitat and its abundance where it is found suggest an adventive plant, I think that it has been introduced in Indiana.

Newf. to the Pacific coast, southw. to N. J., Pa., Tenn., and Mo.; found also in Eu.

3006. ALÝSSUM [Tourn.] L.

1. ALYSSUM ALYSSOÌDES L. SMALL ALYSSUM. Map 1052. A plant of sandy waste places and fallow fields. My Benton County collection is from railroad ballast where it was abundant.

Nat. of Eu.; N. H., Ont. to Iowa, southw. to Mass. and N. J.; also in the far west and about seaports.

3013. LOBULÀRIA Desv.

See excluded species no. 275, p. 1054.

3015. BERTERÒA DC.

1. BERTEROA INCÀNA (L.) DC. HOARY ALYSSUM. Map 1053. I have specimens of this weed from two places in Elkhart County and my notes say that in 1921 it was common in sandy soil along the roadside between Bristol and Elkhart. I have a specimen collected in 1920 about 2 miles northeast of Bristol and my notes say it was a common weed along the roadside and in an adjacent, fallow field. Hansen (Proc. Indiana Acad. Sci. 1923: 215. 1924) says the County Agricultural Agent reported it as a weed in a "run-down" farm in the same county. Hansen also reported a small colony in West Lafayette but it was intentionally destroyed before it

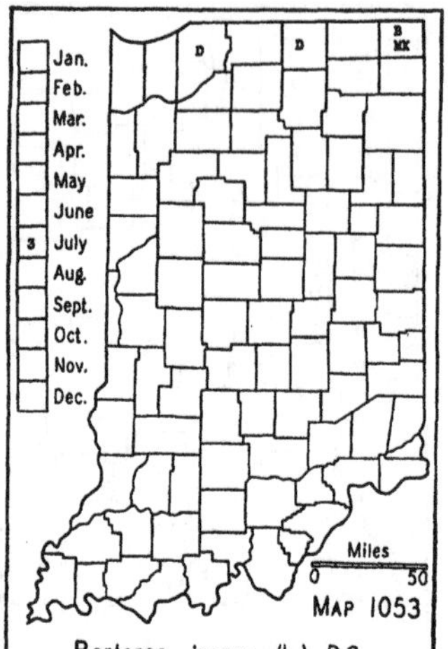

Berteroa incana (L.) DC.

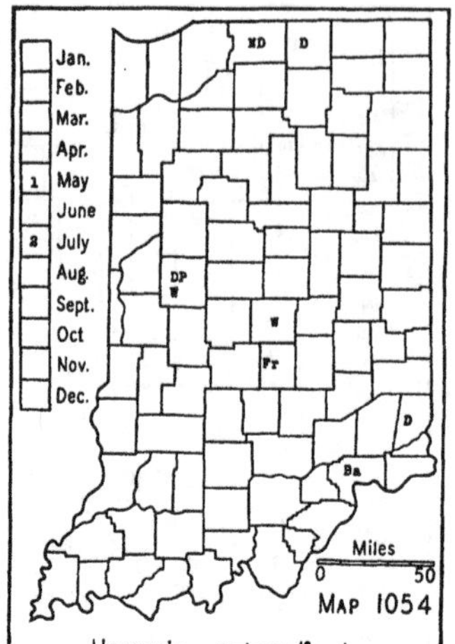

Hesperis matronalis L.

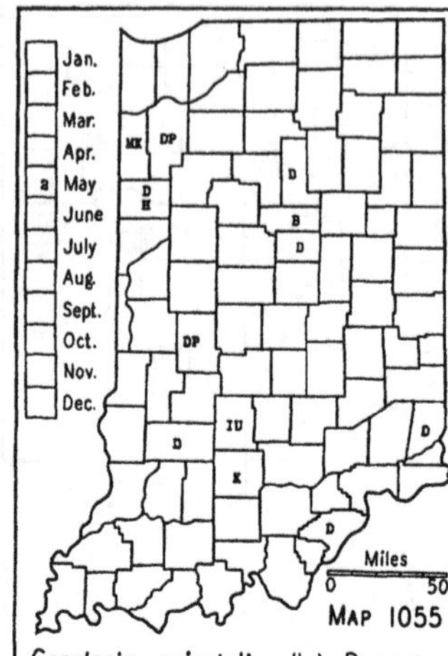

Conringia orientalis (L.) Dumort.

seeded. In 1933 I found it in Steuben County and in 1934 I found it in La Porte County. This is a pernicious weed.

Nat. of Eu.; Maine to Minn., southw. to N. J. and Mo.

3041. HÉSPERIS [Tourn.] L.

1. HESPERIS MATRONÀLIS L. DAMES ROCKET. Map 1054. This species has long been used and is still commonly planted as an ornamental plant. It was never reported, however, by our early authors. The first report is that of Grimes in 1910. Nieuwland, in 1915, reported it as escaped along the bank of the St. Joseph River in St. Joseph County. In 1921 I found it to be frequent in a wooded ravine about a half mile west of Aurora in Dearborn County. In 1933 I found it to be a common and abundant weed along the roadside and in an adjacent, fallow field just west of Aurora. Naomi Mullendore has collected it in Johnson County.

Nat. of Eu.; Maine to Iowa, southw. to N. C.

3055. CONRÍNGIA [Heist.] Adans.

1. CONRINGIA ORIENTÀLIS (L.) Dumort. HARES-EAR MUSTARD. Map 1055. All of my specimens except one are from railroad ballast. It has been reported from seven counties and all who mention its habitat except one say that it was found along railroads. Apparently this species is slow to establish itself in fields and may not become a serious pest.

Nat. of Eu.; N. B. and N. S. to Man. and Oreg., southw. to Del., Mo., and Colo.

107. CAPPARIDÀCEAE Lindl. CAPER FAMILY

Pods on long spreading stipes; stamens 4-6....................3087. CLEOME, p. 510.
Pods stipeless or on very short ascending stipes; stamens more than 6................
....3090. POLANISIA, p. 511.

3087. CLEÒME L.

Stems glabrous; leaves 3-foliolate. (See excluded species no. 276, p. 1054.)..*C. serrulata.*
Steams viscid-pubescent; leaflets 5-7. (See excluded species no. 277, p. 1054.)..*C. spinosa.*

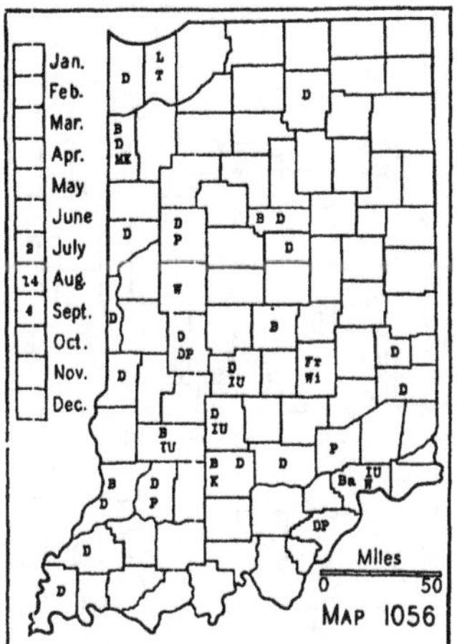

Polanisia graveolens Raf.

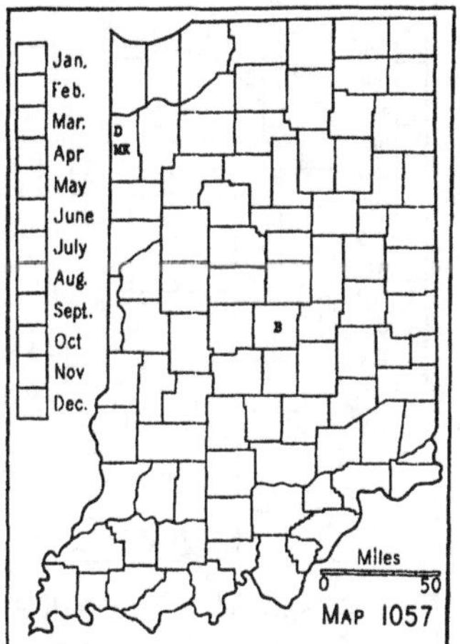

Polanisia trachysperma T. & G.

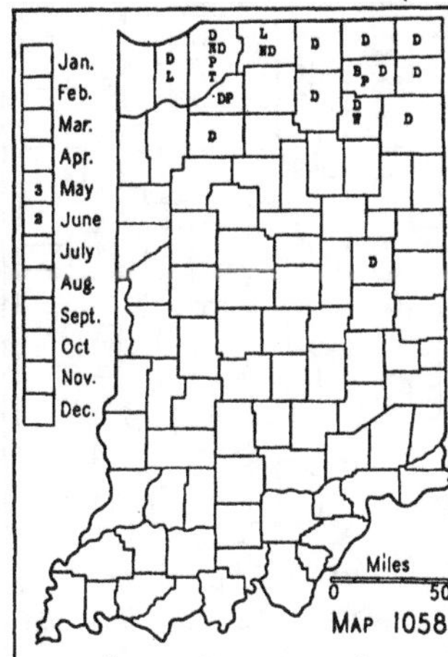

Sarracenia purpurea L.

3090. POLANÍSIA Raf.

Petals 4-5 mm long, usually nearly white; stamens mostly 5-8 mm long, usually less than 12 in number .... 1. *P. graveolens.*
Petals 8-10 mm long, light yellow; stamens mostly 9-15 mm long ... 2. *P. trachysperma.*

1. **Polanisia gravèolens** Raf. CLAMMYWEED. Map 1056. This species grows in very sandy soil and is usually found on sand and gravel bars of streams, along roadsides and railroads, and rarely in fallow or cultivated land along streams. On large sandbars it often forms extensive colonies.

Western Que. to Man., southw. to Conn., Md., Tenn., Kans., and Colo.

2. **Polanisia trachyspérma** T. & G. Map 1057. This species was found by Madge McKee in sandy soil along the roadside, 2 miles west and 1½ miles south of Lake Village, Newton County. It is a western species and there is a slight possibility of its having been introduced. The locality where it was found is within the prairie area. The nearest railroad and the nearest main highway are both two miles to the east. I think this is an eastern extension of the range of the species. Found at the same place on Sept. 4, 1938, by Indiana botanists on a field trip and specimens were collected.

This species was included in a list of plants reported from Monroe County by Andrews. No data accompanied the report and no specimen was preserved. Since the habitat does not occur in Monroe County, it must have been a waif if the plant was determined correctly.

Ind., Iowa to Mo., southw. and westw.

110. SARRACENIÀCEAE La Pyl. PITCHERPLANT FAMILY

3130. SARRACÈNIA [Tourn.] L.

1. **Sarracenia purpùrea** L. (*Sarracenia purpurea gibbosa* (Raf.) Wherry. Bartonia 15: 1-6. 1933.) COMMON PITCHERPLANT. Map 1058. This species grows in sphagnum in marshes and tamarack bogs and is restricted to the lake area. It formerly was common but is now becoming

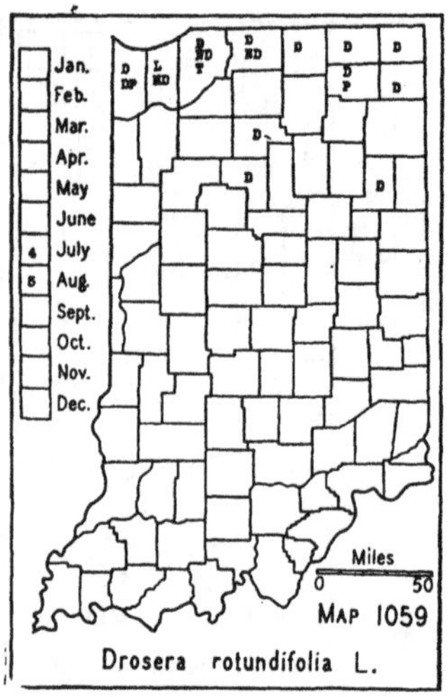

Drosera rotundifolia L.

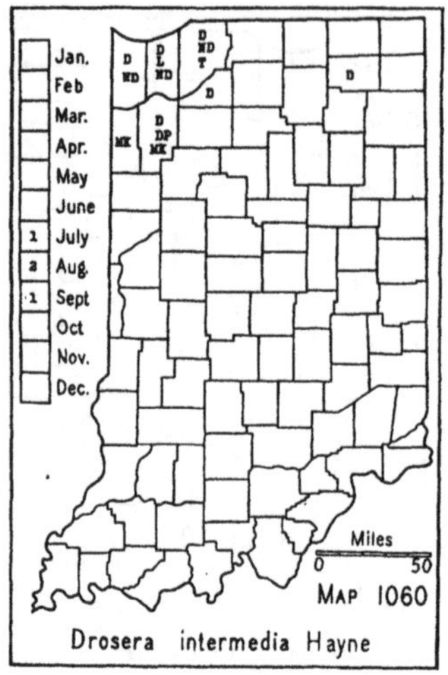

Drosera intermedia Hayne

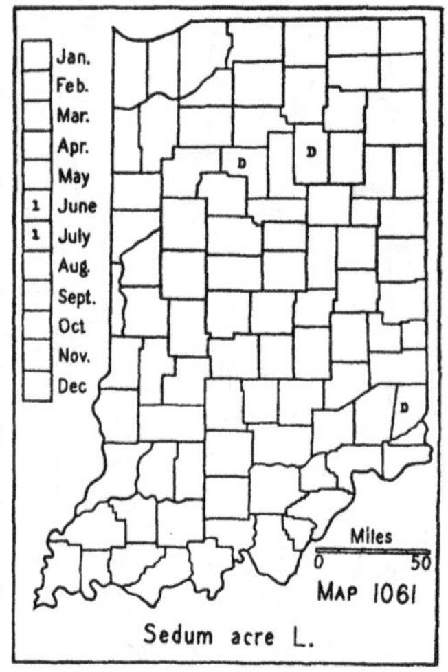

Sedum acre L.

scarce on account of drainage. My Delaware County specimen was obtained from a bog on the Emerson McCullum farm about two and a half miles southeast of Gaston.

Lab. to the Canadian Rocky Mts., southw. to Fla., Ky., the Great Lakes, and Iowa.

112. DROSERÀCEAE S. F. Gray. SUNDEW FAMILY

3136. DRÓSERA L. SUNDEW

Leaf blades suborbicular or wider than long......................1. *D. rotundifolia.*
Leaf blades 2-3 times as long as wide; stipules free; seed not appendaged, oblong..2. *D. intermedia.*

1. **Drosera rotundifòlia** L. ROUNDLEAF SUNDEW. Map 1059. Infrequent in tussocks of sphagnum moss in the open and in tamarack bogs; very rarely in moist sand with such species as *Polygala cruciata, Gaultheria procumbens,* and *Aletris farinosa.* In 1915 it was so abundant on the moist, sandy shore of Walker Lake, Porter County, that it covered acres, and at a distance, the ground looked red. This and the next species are restricted to the lake area.

Lab. to Alaska, southw. to Fla. and Calif.

2. **Drosera intermèdia** Hayne. (*Drosera longifolia* of manuals, not L.) SPATULATE-LEAF SUNDEW. Map 1060. Less frequent than the preceding species and found in the open in moist, sandy soil among sedges or in mossy places on the wet borders of lakes, and in sphagnum bogs.

Newf. to Minn., southw. to Fla. and La.

113. PODOSTEMÀCEAE Lindl. RIVERWEED FAMILY

3156. PODOSTÈMUM Michx.

See excluded species no. 278, p. 1054.

Sedum telephioides Michx.

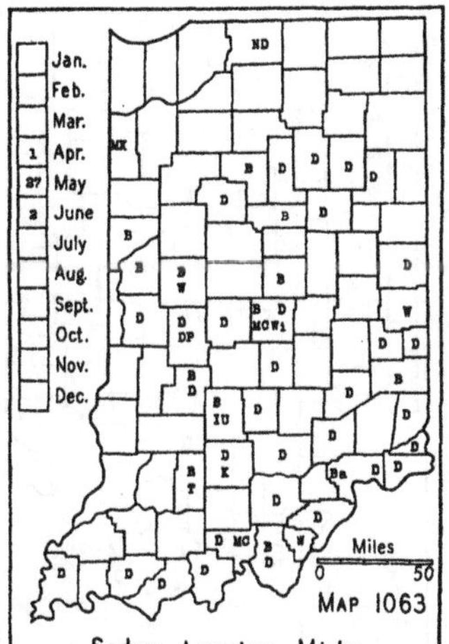

Sedum ternatum Michx.

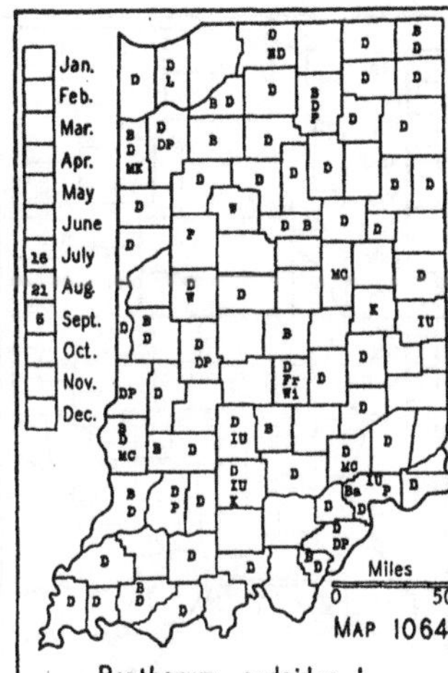

Penthorum sedoides L.

115. CRASSULÀCEAE DC. ORPINE FAMILY

Plants succulent; calyx lobes 4 or 5; petals evident; stamens 8 or 10; carpels 4 or 5, separate (in ours) ....3161. SEDUM, p. 513.
Plants not succulent; calyx lobes 5 (sometimes 6); petals usually lacking; stamens 10; carpels mostly 5 (sometimes 4 or 6), united at the base...3173. PENTHORUM, p. 514.

3161. SÈDUM [Tourn.] L. STONECROP

Flowers yellow; leaves very thick, ovate, mostly about 5 mm long; perennials..1. *S. acre.*
Flowers white, pinkish, or purplish.
  Stem leaves almost terete, linear, obtuse, slightly auriculate at the base, mostly 1.5-2 cm long; leaves of winter rosettes spatulate, about as long as the stem leaves; winter annuals. (See excluded species no. 280, p. 1054.)..*S. pulchellum.*
  Stem leaves, and those of sterile shoots, flat.
    Margins of blades more or less dentate; flowers colored.
      Stem, branches, and pedicels narrowly winged on opposite sides...2. *S. telephioides.*
      Stem, branches, and pedicels terete, not winged. (See excluded species no. 281, p. 1055.)...................................*S. Telephium* var. *purpureum.*
    Margins of leaves entire; flowers white.
      Stem leaves, at least some of them, in whorls of 3..............3. *S. ternatum.*
      Stem leaves alternate or spirally arranged. (See excluded species no. 279, p. 1054.)..*S. Nevii.*

1. SEDUM ÀCRE L. GOLDMOSS. Map 1061. Reported as an escape in 5 counties. I found it as a common plant over an acre or more in shallow soil on the bank of Pipe Creek in Cass County at Pipe Creek Falls and near the old mill at Dora in Wabash County. When it escapes from cultivation, it will persist if it finds rocky soil or a wall of stone.

Nat. of Eurasia; N. S. to Ont. and Ind., southw. to Va. and N. Y.

2. **Sedum telephioìdes** Michx. WILD LIVEFOREVER. Map 1062. We now have specimens of this species from three counties. Dr. Clapp found it "on the cliffs of the Ohio above Utica, Clark County. Sept. 22, 1837." He said he found it in only two localities. I saw his specimens in the herbaria

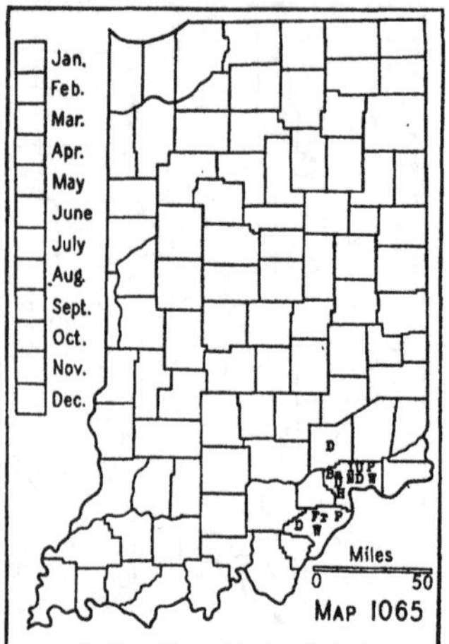

Sullivantia ohionis T. & G.

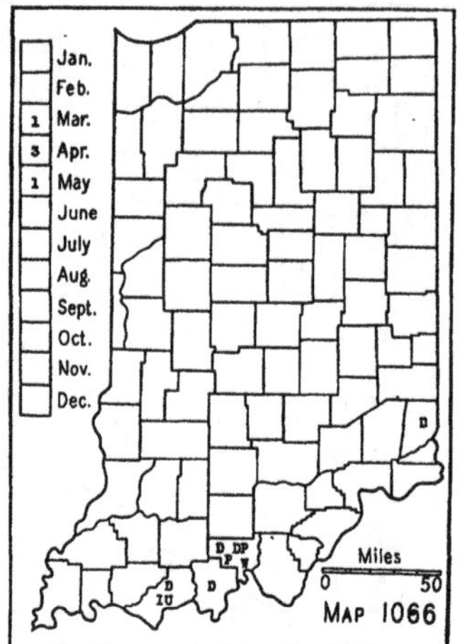

Saxifraga virginiensis Michx.

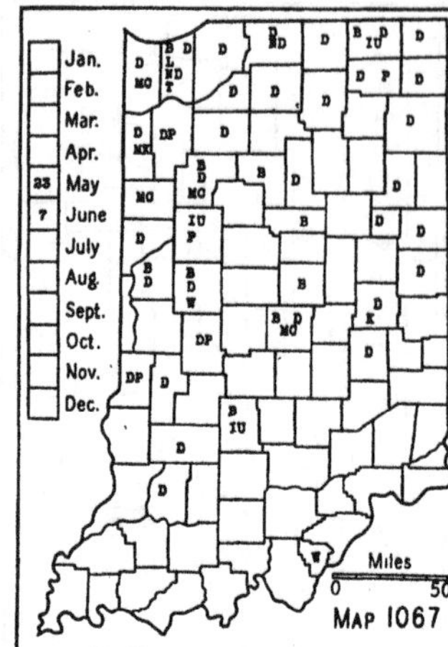

Saxifraga pennsylvanica L.

of the New York Botanical Garden, Purdue University, and Wabash College. In 1922 I found it in Harrison County on a very narrow ledge of rock near the top of the cliff along the Ohio River, in section 14 about 4 miles southeast of Laconia. The cliff at this place is about 300 feet above the river. I have had it in cultivation since that time and the flowers are nearly white to faintly pink. In 1936 R. M. Tryon, Jr., found it in Perry County on the sandstone cliffs near Magnet.

Pa., N. Y. to Ind., southw. to N. C. and Ga.

3. **Sedum ternàtum** Michx. MOUNTAIN STONECROP. Map 1063. This species is rather frequent in the southern part of the state, becoming rare or absent in the northern counties. It is found in the shade in moist soil, usually at the bases of wooded ravines and on wooded slopes and rocky, wooded cliffs.

Conn. to Mich., southw. to Ga. and Tenn.

3173. PÉNTHORUM L.

1. **Penthorum sedoìdes** L. DITCH STONECROP. Map 1064. Frequent to very frequent throughout the state in firm soil in wet places. It is found in roadside ditches, fallow fields, and low places in woodlands about ponds and sloughs.

N. B. to Minn., southw. to Fla. and Tex.

117. SAXIFRAGÀCEAE Dumort. SAXIFRAGE FAMILY

Herbs.
  Stamens 5.
    Flowers solitary on long, scapelike peduncles, white, the petals 10-18 mm long; staminodia present............................3203. PARNASSIA, p. 519.
    Flowers paniculate, small, the petals less than 10 mm long; staminodia lacking.
      Leaf blades glabrous; ovary 2-celled..............3186. SULLIVANTIA, p. 515.
      Leaf blades more or less pubescent; ovary 1-celled..3195. HEUCHERA, p. 516.

Stamens 10, rarely 8.
  Petals none; flowers sessile, axillary, usually solitary; leaves opposite........ ....3199. CHRYSOSPLENIUM, p. 519.
  Petals 5; flowers racemose or paniculate, pedicellate; leaves all basal or alternate.
    Petals entire, or subserrate.
      Capsule 2-celled, 2-beaked........................3189. SAXIFRAGA. p. 515.
      Capsule 1-celled..................................3193. TIARELLA, p. 516.
    Petals finely fringed, small; flowers racemose, white...3198. MITELLA, p. 518.
Shrubs; leaves opposite, simple.
  Stamens 20-40....................................3208. PHILADELPHUS, p. 519.
  Stamens 8-10..3217. HYDRANGEA, p. 519.

3186. SULLIVÀNTIA T. & G.

1. **Sullivantia ohiònis** T. & G. (*Sullivantia Sullivantii* (T. & G.) Britt.) OHIO SULLIVANTIA. Map 1065. This species is very local throughout its range and in Indiana it is found only in Clark, Jefferson, and Jennings Counties. It grows in the shade in the wet crevices of perpendicular, shaly cliffs in Clifty Falls State Park in Jefferson County, along Fourteen-mile Creek near its mouth in Clark County, and along the north fork of the Muscatatuck River about three fourths of a mile northeast of Vernon. It has been reported also from Carroll, Cass, and Floyd Counties.

Ohio and Ind. Reported from 7 counties in Ohio and 3 counties in Indiana.

3189. SAXÍFRAGA [Tourn.] L. SAXIFRAGE

[Johnson. Revision of North American species of the section Boraphila of Saxifraga. Univ. Minnesota Stud. Biol. Sci. 4: 1-109. 1923. Bush. Some species of Saxifraga. Amer. Midland Nat. **11**: 213-235. 1928.]

Plants of a dry habitat, less than 3 dm high; sepals reflexed.........1. *S. virginiensis*.
Plants of a wet habitat, usually 5-15 dm high; sepals ascending....2. *S. pennsylvanica*.

1. **Saxifraga virginiénsis** Michx. VIRGINIA SAXIFRAGE. Map 1066. This species is restricted to the tops and slopes of the bluffs of the Ohio River and nearby. It is local in its distribution but frequent to common where it is found. It is rather common on the top of the bluff of the Ohio River just north of Fredonia in Crawford County. There are specimens from only four counties but I think that it could be found in other counties, especially Floyd, Harrison, and Jefferson Counties and in western Vanderburgh County. Bush (Amer. Midland Nat. **11**: 215-220. 1928.) has divided my specimens into two lots. He calls one lot *Saxifraga virginiensis* and the other *Saxifraga pilosa* Haworth. I am not recognizing the latter.

N. B. to Minn., southw. to Ga. and Tenn.

2. **Saxifraga pennsylvánica** L. PENNSYLVANIA SAXIFRAGE. Map 1067. Infrequent to frequent in wet and springy places in woodland and marshes throughout the state except in the southern counties where its habitat is

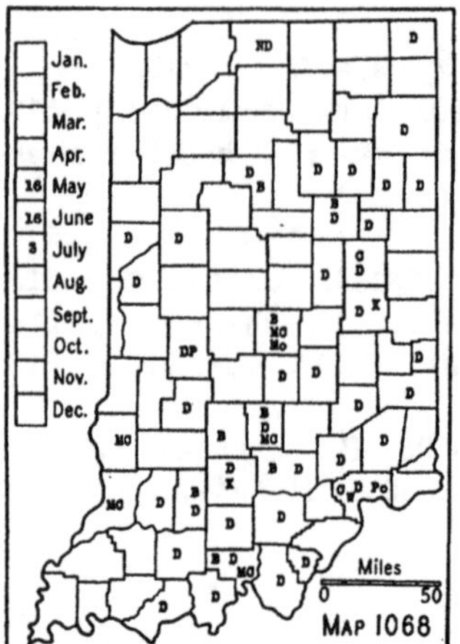

Heuchera americana var. brevipetala Rosendahl, Butters, & Lakela

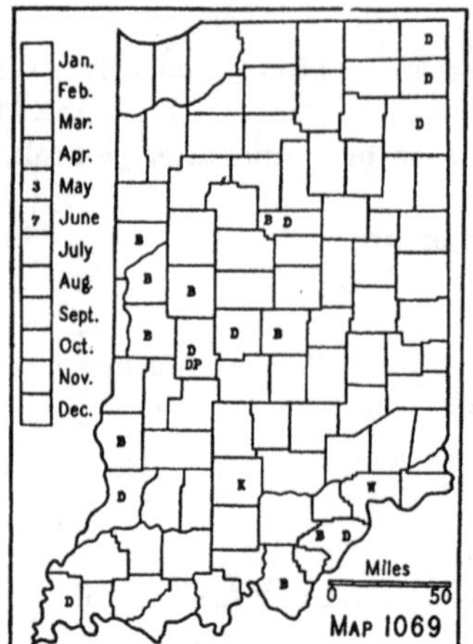

Heuchera americana var. interior Rosendahl, Butters, & Lakela

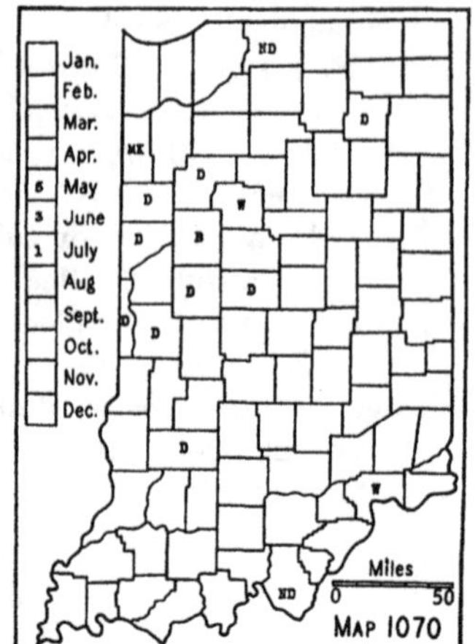

Heuchera americana var. hirsuticaulis (Wheelock) Rosendahl, Butters, & Lakela

absent or rare. No doubt it is absent from the southern counties for reasons other than the lack of habitat.

Maine, Ont. to Minn., southw. to Va. and Mo.

3193. TIARÉLLA L.

[Lakela. A monograph of the genus Tiarella L. in North America. Amer. Jour. Bot. 24: 344-351. 1937.]

(See excluded species no. 283, p. 1055.)

3195. HEÙCHERA L. ALUMROOT

[Rosendahl, Butters, and Lakela. A monograph on the genus Heuchera. Minnesota Studies in Plant Science 2: 1-180. 1936.]

All of my specimens have been named by the authors of this monograph.

The following key has been adapted from the monograph cited above.

Petioles of leaves glabrous, merely puberulent, or sparingly hirsute above .... 1. *H. americana* var. *brevipetala.*

Petioles more or less densely hirsute.

- Outside of calyx glandular-puberulent, without any long white hairs.
  - Flowers in anthesis mostly 3-4.5 mm long; hypanthium regular or somewhat oblique.
    - Flowers in anthesis 3-3.5 mm long; hypanthium regular or nearly so; petals 1-1.5 mm long, narrowly elliptic to short-spatulate, with a short, relatively broad claw .... 1a. *H. americana* var. *interior.*
    - Flowers in anthesis mostly 4-4.5 mm long; hypanthium somewhat oblique; petals 2-3 mm long, oblanceolate with narrow claws and rhombic tips .... 1b. *H. americana* var. *hirsuticaulis.*
  - Flowers in anthesis mostly 5-10 mm long; hypanthium strongly to moderately oblique.
    - Flowers mostly 5-7 mm long; hypanthium moderately oblique .... 2. *H. Richardsonii* var. *affinis.*
    - Flowers mostly 6-10 mm long; hypanthium strongly oblique .... 2a. *H. Richardsonii* var. *Grayana.*

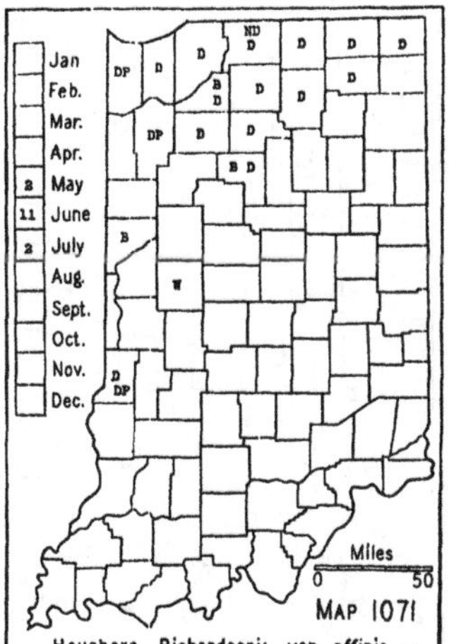

Heuchera Richardsonii var. affinis Rosendahl, Butters, & Lakela

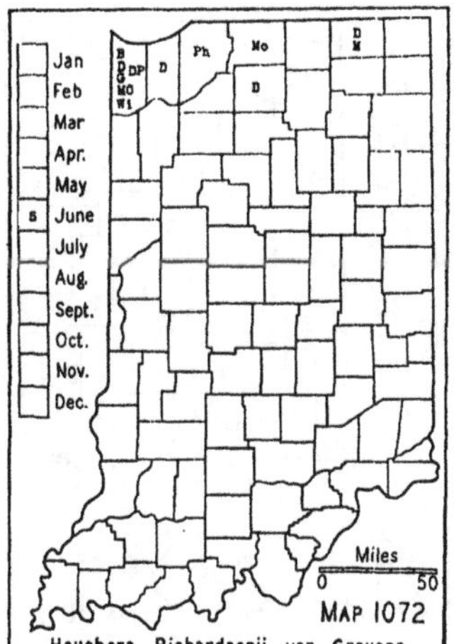

Heuchera Richardsonii var. Grayana Rosendahl, Butters, & Lakela

Heuchera villosa var. macrorhiza (Small) Rosendahl, Butters, & Lakela

Outside of calyx villous with long white hairs; plants in our area growing in the crevices of cliffs or rarely in the talus at their bases.
Lobes of leaves triangular.........................3. *H. villosa* var. *macrorhiza.*
Lobes of leaves rounded.........................4. *H. parviflora* var. *Rugelii.*

1. **Heuchera americàna** L. var. **brevipétala** Rosendahl, Butters, & Lakela. (*Heuchera americana* of many American authors.) Map 1068. This is our most common alumroot and is frequent throughout the state although there are no records from the northwestern part. It is generally found on or near the tops of wooded slopes along streams or on the slopes of ravines. According to the monographers of the genus, the typical form of this species is restricted to the Appalachian Mountains from southern Pennsylvania southward to North Carolina and Tennessee and this variety and the two following are the western allies of it.

Conn., Pa., s. Ont., and se. Mich., southw. to Md., Tenn., Mo., and Okla.

1a. **Heuchera americana** var. **intèrior** Rosendahl, Butters, & Lakela. Map 1069. The habitat of this variety is the same as that of the preceding one, but the plant is probably less frequent.

Ind. to e. Kans., southw. to w. Tenn. and n. Ark.

1b. **Heuchera americana** var. **hirsuticaúlis** (Wheelock) Rosendahl, Butters, & Lakela. (*Heuchera hirsuticaulis* Wheelock.) Map 1070. The habitat of this variety is similar to that of the preceding. This variety, in its morphology, is about midway between the species and *Heuchera Richardsonii* R. Br., which is restricted to the eastern Rocky Mountains and plains, and seems to have a range about midway between the two species with its eastern extension in west central Indiana.

Ind., Ill., and Mo.

2. **Heuchera Richardsònii** R. Br. var. **affìnis** Rosendahl, Butters, & Lakela. Map 1071. This alumroot prefers sandy soil and grows mostly on the slopes and banks of streams and lakes. Our specimens are all from

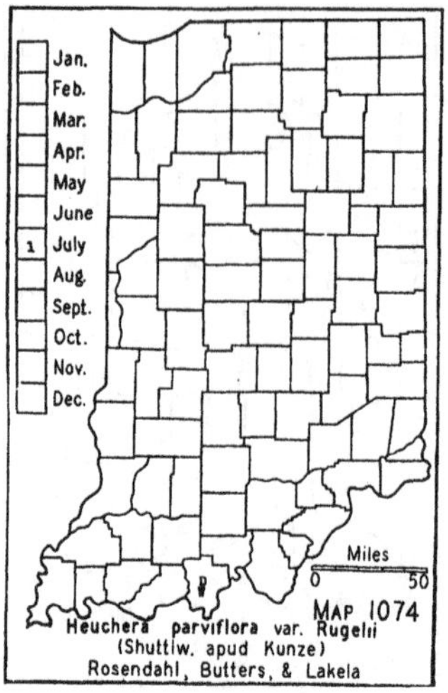

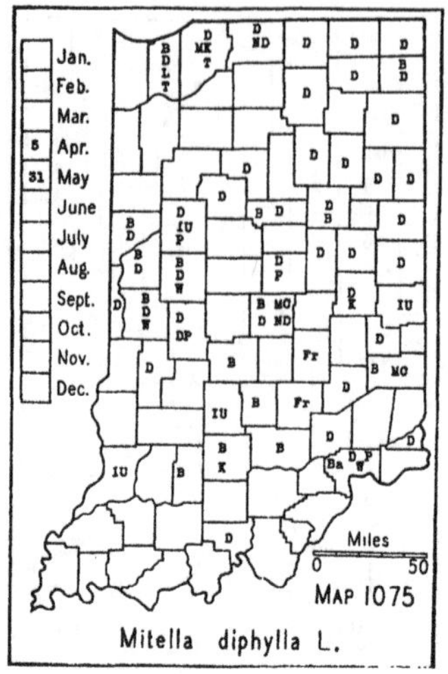

the lake area except the Vigo County specimen which was found in the Heckland Prairie.

Southern Mich. and Wis., southw. to Ill. and Mo.

2a. **Heuchera Richardsonii** var. **Grayàna** Rosendahl, Butters, & Lakela. (Rhodora 35: 117. 1933.) (*Heuchera hispida* of most authors.) Map 1072. This variety prefers a very sandy soil and is found both in moist and dry situations in the open or in woodland bordering streams and lakes.

Southern Mich. to Minn., southw. to Ind., Mo., and Kans.

3. **Heuchera villòsa** Michx. var. **macrorhìza** (Small) Rosendahl, Butters, & Lakela. (*Heuchera macrorhiza* Small.) Map 1073. Very local, in the crevices of cliffs or rarely in the talus at their bases. It is restricted to a few Ohio River Counties. Many years ago I dug some of the rhizomes from the crevices of rock and planted them in neutral garden soil at Bluffton, Wells County, and the plants are perfectly hardy and grow vigorously. On account of their large and rather compact cluster of basal leaves and large panicle of small white flowers I recommend it as a good garden plant, especially for borders.

W. Va., Ind., Ky., Tenn., and Mo., southw. to Ga. and Ala.

4. **Heuchera parviflòra** Bartl. var. **Rugélii** (Shuttlw. apud Kuntze) Rosendahl, Butters, & Lakela. Map 1074. Our only specimen of this rare alumroot is one which I found in a pocket on the perpendicular face of a sandstone cliff on a farm about 2 miles southwest of Leopold, Perry County. The leaves were quite purplish on the lower surface.

W. Va., Ind., Ill., southw. to N. C. and Ala.

3198. MITÉLLA [Tourn.] L.

Plants more than 2 dm high, generally 3-5 dm high; basal leaves ovate, middle lobe acute; stem bearing 2 opposite, sessile or nearly sessile leaves; flowers white; calyx cup-shaped...1. *M. diphylla.*

Plants generally less than 2 dm high; basal leaves reniform, the lobes rounded; stem naked, rarely with 1 or 2 leaves; flowers greenish; calyx saucer-shaped. See excluded species no. 285, p. 1055.)...................................*M. nuda.*

1. **Mitella diphýlla** L. BISHOPSCAP. Map 1075. Infrequent to frequent probably throughout the state although there are no records from the southwestern counties. It is found in moist or wet soil, usually on wooded slopes, especially on the steep slopes of deep ravines, and more rarely in flat woods.

Que. to Minn., southw. to N. C., Mo., and Iowa.

3199. CHRYSOSPLÈNIUM [Tourn.] L.

1. **Chrysosplenium americànum** Schwein. GOLDEN SAXIFRAGE. Map 1076. I found this species as a common plant in shaded woodland in cool, shallow water slowly moving through a depression in the woods, and also in adjacent pools of water in Porter County along the traction line north of Willis Stop. It has been reported from Lake, Marshall (Nieuwland reported it as found by Clark, but Clark did not report it in his list of plants in "Lake Maxinkuckee" by Evermann & Clark), Porter, and St. Joseph Counties. It is, no doubt, local in its distribution because of its peculiar habitat; however, since the plant is so inconspicuous, it may be more frequent than the reports indicate.

N. S. to Sask., southw. to Ga., Ohio, and Minn.

3203. PARNÁSSIA [Tourn.] L.

1. **Parnassia glaúca** Raf. (See Bartonia **17**:18. 1935.) (*Parnassia caroliniana* of Gray, Man., ed. 7 and Britton and Brown, Illus. Flora, ed. 2, not Michx.) Map 1077. This species is found only in wet, marly areas about lakes and in the outlets of springs. It is almost invariably associated with *Lobelia Kalmii.* Infrequent to frequent throughout the lake area, becoming rare or absent in the southern part of the state.

N. B. to Man., southw. to N. J., Pa., and Iowa.

3208. PHILADÉLPHUS L. MOCKORANGE

Flowers racemose, 5-7, fragrant; calyx lobes acuminate, 12-15 mm long. (See excluded species no. 286, p. 1055.) .... *P. coronarius.*
Flowers 1-3, usually solitary, inodorous; calyx lobes acute.
  Calyx lobes about twice as long as the tube, 8-10 mm long. (See excluded species no. 287, p. 1056.) .... *P. grandiflorus.*
  Calyx lobes about equaling the tube, about 7 mm long. (See excluded species no. 288, p. 1056.) .... *P. inodorus.*

3217. HYDRÁNGEA L. HYDRANGEA

[St. John. A critical revision of Hydrangea arborescens. Rhodora 23: 203-208. 1921.]

Lower surface of leaves more or less pubescent on the principal nerves; blades cordate, rounded, or tapering at the base.
  Blades cordate or rounded at the base.
    Corymbs without sterile flowers, or with only a few of them....1. *H. arborescens.*
    Corymbs with all the flowers sterile..............1a. *H. arborescens* var. *sterilis.*
  Blades tapering at the base......................1b. *H. arborescens* var. *oblonga.*
Lower surface of leaves pubescent over the entire surface, the pubescence more or less dense but not tomentose.........................1c. *H. arborescens* var. *Deamii.*

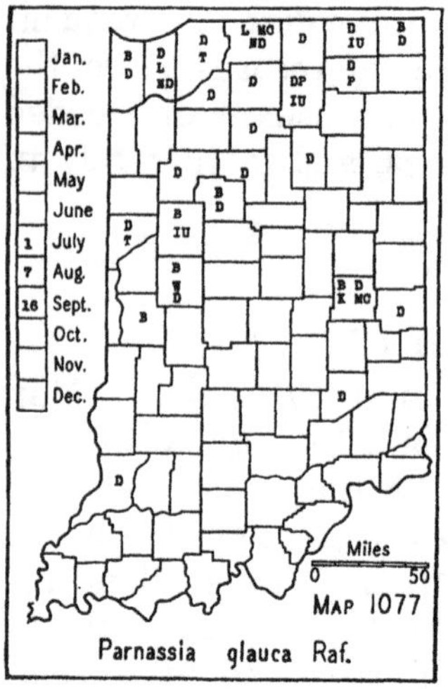

Parnassia glauca Raf.

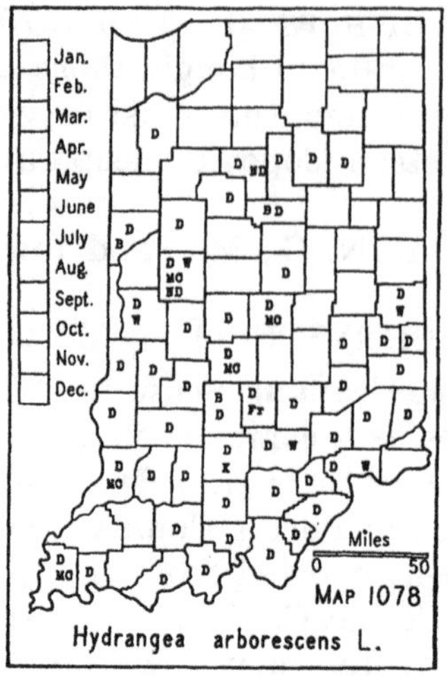

Hydrangea arborescens L.

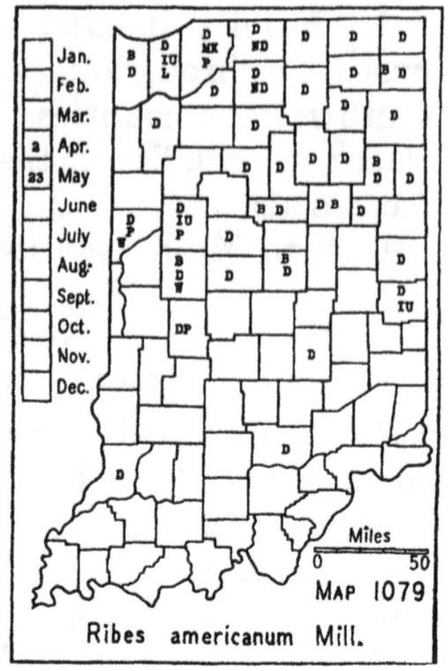

Ribes americanum Mill.

1. **Hydrangea arboréscens** L. SMOOTH HYDRANGEA. Map 1078. Rather frequent in one or more of its forms in the southern half of the state, becoming infrequent to very rare northward and possibly entirely absent from the region north of the distribution area shown on the map. This is a woodland species and is usually found in small colonies on the slopes and banks of deep ravines, cliffs, and streams.

N. Y. to Iowa, southw. to Fla. and La.

1a. **Hydrangea arborescens** var. **stérilis** T. & G. Mottier found this rare form in Monroe County and his report (Proc. Indiana Acad. Sci. **1919**: 59-62. 1921) is the only one of this variety.

1b. **Hydrangea arborescens** var. **oblónga** T. & G. At first I thought this variety was a mere environmental form. About 8 years ago I transplanted some plants into a bed in rich soil with the typical form and the plants have each year had rather small leaves with the blades tapering at the base. In this bed I have transplanted several forms and each year the plants maintain their peculiar forms. The distribution is probably the same as that of the species, and they are included on the same map.

1c. **Hydrangea arborescens** var. **Dèamii** St. John. This form is distinguished from the type only by the dense pubescence of the lower surface of the leaves. The leaves are of all shapes. Since there is no character other than pubescence to distinguish this variety, it might be better to distinguish it as a mere form of the species.

Ohio and Ind., southw. to Ga. and westw. to Okla.

117A. GROSSULARIÀCEAE Dumort. GOOSEBERRY FAMILY

[Berger. A taxonomic review of currants and gooseberries. New York State Agric. Exp. Sta. Tech. Bull. **109**: 1-118. 1924.]

Pedicels jointed beneath the ovary; flowers of racemes normally 5 or more; bractlets of flowers linear, generally 5-8 mm long; shrubs without nodal spines...3249. RIBES, p. 521.

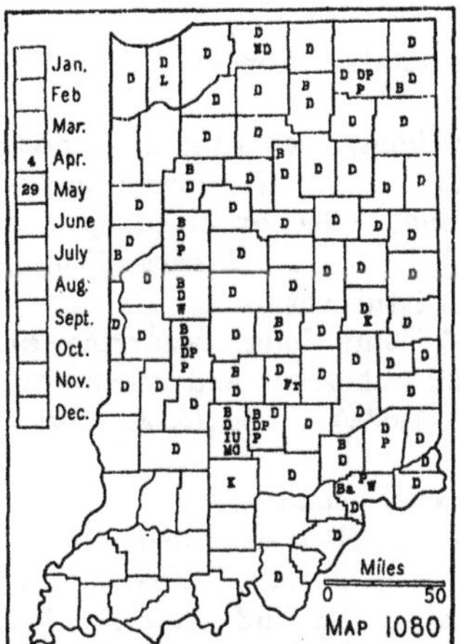

Grossularia Cynosbati (L.) Mill.

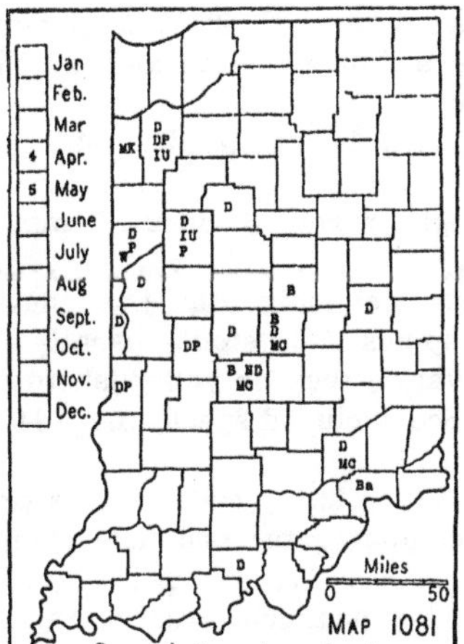

Grossularia missouriensis (Nutt.) Cov. & Britt.

Grossularia hirtella (Michx.) Spach

Pedicels not jointed beneath the ovary; flowers in clusters of 1-4, rarely 5; bractlets of flowers sheathlike, as wide as long, about 2 mm long; shrubs with nodal spines (usually lacking in *Grossularia hirtella*)............3249A. GROSSULARIA, p. 521.

3249. RÌBES L. CURRANT

Leaves with numerous resinous dots beneath......................1. *R. americanum.*
Leaves without resinous dots beneath.
  Ovary densely glandular-bristly. (See excluded species no. 289, p. 1056.).......... ....*R. glandulosum.*
  Ovary glabrous.
    Flowers purplish; weak, ascending shrubs about 0.5 m high. (See excluded species no. 292, p. 1056.)...*R. triste.*
    Flowers yellow or greenish yellow; erect shrubs more than 0.5 m high.
      Flowers greenish yellow; calyx tube obsolete; fruit red. (See excluded species no. 291, p. 1056.)..*R. sativum.*
      Flowers bright yellow; calyx tube evident; fruit black. (See excluded species no. 290, p. 1056.)..*R. odoratum.*

1. **Ribes americànum** Mill. (*Ribes floridum* L'Hér. and *Ribes americanum* f. *mesochorum* (Nieuwl.) Deam.)[1] AMERICAN BLACK CURRANT. Map 1079. Infrequent to frequent in the lake area, becoming rare southward and probably absent from most of the unglaciated area. In the central part of the state it grows generally in wet prairie habitats and springy places and in the lake area it grows in similar habitats and in mucky places and decadent tamarack bogs.

N. B. to Sask., southw. to Va., Ky., Iowa, and Nebr.

3249A. GROSSULÀRIA [Tourn.] Mill. GOOSEBERRY

Ovary with soft, glandless bristles or prickles, rarely glabrous; spreading shrubs with nodal spines; nodal spines 1-3, rarely lacking, generally about 10 mm long; our common wild gooseberry......................................1. *G. Cynosbati.*

[1] **Ribes americanum forma mesochorum** (Nieuwland) Deam, comb. nov. *Coreosma americana* var. *mesochora* Nieuwland. Amer. Midland Nat. 4: 60. 1915.

Ovary glabrous, rarely pubescent or with stalked glands.
  Stamens much exserted, more than twice as long as the petals, often exceeding the sepals; ovaries glabrous.
    Nodal spines usually present, 1-3, generally 8-20 mm long, occurring at almost every node..2. *G. missouriensis.*
    Nodal spines none or single, rarely 3, about 5 mm long. (See excluded species no. 295, p. 1057.)....................................*G. rotundifolia.*
  Stamens not exserted, at most not more than the length of the petals.
    Calyx pubescent; nodal spines 1-3, stout, usually 7-15 mm long, occurring at almost every node; ovary green or yellowish to red, pubescent and glandular-bristly or glabrous (See excluded species no. 294, p. 1057.).......*G. reclinata.*
    Calyx glabrous.
      Young branchlets generally bristly; nodal spines present, usually 3.
        Calyx tube cylindric-campanulate; peduncle elongated. (See excluded species no. 296, p. 1057.)..*G. setosa.*
        Calyx tube campanulate; peduncle scarcely exceeding the bud scales. (See excluded species no. 293, p. 1056.)..................*G. oxyacanthoides.*
      Young branchlets not bristly, rarely some bristles at the base of vigorous ones; plants rarely with nodal spines, which, if present, are subulate and 5-10 mm long..3. *G. hirtella.*

1. **Grossularia Cynósbati** (L.) Mill. (*Ribes Cynosbati* L. of Gray, Man., ed. 7.) PASTURE GOOSEBERRY. Map 1080. Infrequent to common throughout the state except the southwestern part, from which there are no records or specimens. The species prefers a moist, rich soil and is found generally in woodland. In the southern part of the state, however, it is generally found in the crevices of rocks on rocky, wooded slopes along streams.

N. B. to Man., southw. to N. C., Ala., and Mo.

2. **Grossularia missouriénsis** (Nutt.) Cov. & Britt. (*Ribes gracile* Michx. of Gray, Man., ed. 7.) MISSOURI GOOSEBERRY. Map 1081. This species has been reported from Hamilton, Kosciusko, Lake, Marion, St. Joseph, and Tippecanoe Counties. In the herbarium of DePauw University there are specimens from Jasper, Putnam, and Vigo Counties. My specimens are mostly from the central and western counties where I found them in dry soil, usually on the bluffs of streams. The Henry County specimen was found in a dry woods and the one from Tippecanoe County was found with skunkcabbage in a springy place.

Ind. to Minn., and S. Dak., southw. to Kans., Mo., and Tenn.

3. **Grossularia hirtélla** (Michx.) Spach. (*Ribes oxyacanthoides* L. in part, of Gray, Man., ed. 7.) LOW WILD GOOSEBERRY. Map 1082. This species is restricted to the northern counties and is found only in wet places, mostly in tamarack bogs. It is infrequent to very rare in the area of its distribution.

The extreme variability of this species, especially in the shape of the leaves and in the amount of pubescence on their under surface, has led to the naming of varieties of it. Writers tell us that the bases of the leaves of this species are generally more or less cuneate and that the bases of the leaves of *Grossularia oxyacanthoides* are truncate or cordate. I have two specimens with the bases of the leaves cordate but I think, because of other characters, that they belong to this species. I have one

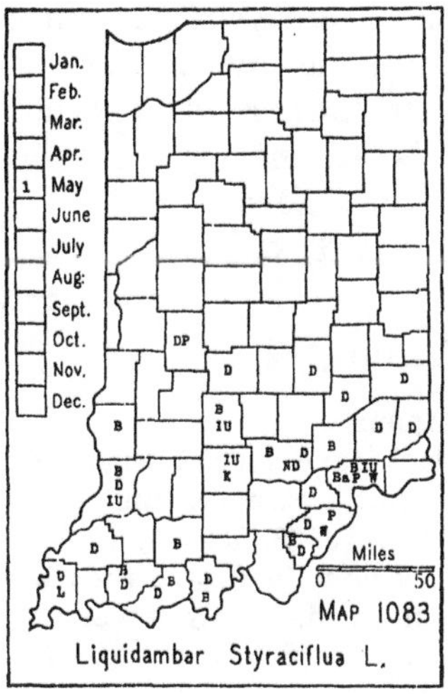

Liquidambar Styraciflua L.

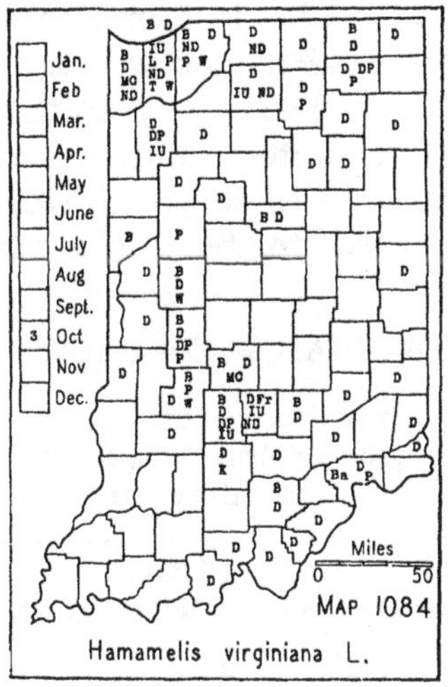

Hamamelis virginiana L.

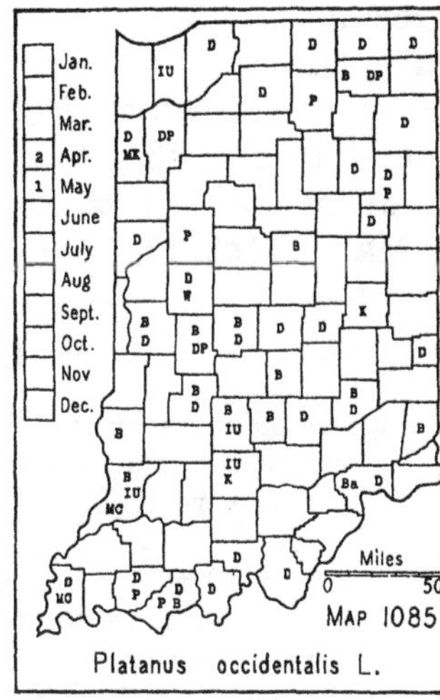

Platanus occidentalis L.

specimen that has a few stalked glands which do not properly belong to this species.

Newf. to Man., southw. to Pa., W. Va., and S. Dak.

123. HAMAMELIDÀCEAE Lindl. WITCH-HAZEL FAMILY

3309. HAMAMÈLIS L.

1. **Hamamelis virginiàna** L. (*Hamamelis virginiana* var. *angustifolia* Nieuwl. and *Hamamelis virginiana* var. *orbiculata* Nieuwl. in Amer. Midland Nat. 3: 63-64. 1913.) COMMON AMERICAN WITCH-HAZEL. Map 1084. Infrequent to locally common in the counties shown on the map. In some of the central and southwestern counties it is either very rare or absent. It generally grows in sandy soil in black and white oak woods and on the banks of streams. It prefers the shade of the forest or banks.

The branches of this shrub are the source of commercial witch-hazel but the shrub has never been commercialized in Indiana.

N. S., Ont. to Minn., southw. to Fla. and Tex.

123A. ALTINGIÀCEAE Hayne. ALTINGIA FAMILY

3298. LIQUIDÁMBAR L.

1. **Liquidambar Styracíflua** L. SWEET GUM. (Sweetgum of Standardized Plant Names.) Map 1083. Restricted to low woods in the southern half of the state. Where it is found, it is usually a frequent to common tree and associated with American beech, pin oak, American elm, river birch, and red maple. Also known as red gum by foresters and in commerce.

Conn., s. Ohio. to Mo., southw. to Fla., Tex., and in the mts. to Guatemala.

124. PLATANÀCEAE Lindl. PLANETREE FAMILY

3314. PLÁTANUS [Tourn.] L.

1. **Platanus occidentàlis** L. (*Platanus occidentalis* f. *attenuata* Sarg.) AMERICAN PLANETREE. Map 1085. In Indiana it is generally known as

sycamore. This species, no doubt, is found in every county of the state except Benton County. It grows in low woods and on the low borders of lakes and streams. While it thrives in places that are inundated, it is not found in the "flats" of southeastern Indiana. It is an infrequent to a frequent tree but rarely forms a thick stand over several acres.

Maine, Ont. to Nebr., southw. to the Gulf States and Tex.

126. ROSÀCEAE B. Juss. Rose Family

Stems armed more or less with prickles, woody; leaves compound.
  Flowers white; fruit an aggregate of drupelets, mostly black (one species with red, one with reddish purple, and one variety with amber fruit); stems biennial.... ....3353. RUBUS, p. 555.
  Flowers of all the native and most of the introduced species pink (a few introduced species with white flowers); fruit a fleshy hypanthium, red, rarely greenish; stems perennial..3389. ROSA, p. 573.
Stems not armed with prickles (thorny in *Crataegus*).
  Plants woody.
    Leaves simple.
      Shrubs or small trees generally well armed with conspicuous thorns (a few species rarely nearly thornless); flowers white, rarely pinkish, corymbose; fruit a pome, red, green streaked with red or yellowish, containing 1-5 bony carpels..3345. CRATAEGUS, p. 533.
      Shrubs or small trees without thorns.
        Flowers purplish; fruit rose purple, an aggregate of drupelets............. ....3353. *Rubus odoratus*, p. 558.
        Flowers white or pinkish; fruit not an aggregate of drupelets except in *Rubus pubescens*.
          Flowers in long or short racemes.
            Fruit a berrylike pome with 10 incomplete cells, each cell with a seed.... ....3343. AMELANCHIER, p. 531.
            Fruit a fleshy drupe with one stone...............3396. PRUNUS, p. 578.
          Flowers in umbels, umbel-like clusters, cymes or panicles.
            Flowers in terminal panicles or corymbs, usually more than 20 flowers in an inflorescence; flowers usually less than 1 cm in diameter; fruit of 1-5 follicles, each with 1-7 seed.
              Bark shreddy; spreading shrubs; leaves ovate to nearly orbicular, most of them more or less lobed; follicles inflated, 7-10 mm long, 2-3-seeded.........................3316. PHYSOCARPUS, p. 526.
              Bark not shreddy; erect shrubs; leaves narrowly oblanceolate or ovate-lanceolate; follicles not inflated, about 3 mm long, generally 2-7-seeded..................................3319. SPIRAEA, p. 526.
            Flowers in terminal or lateral clusters, fewer than 20 flowers in an inflorescence; flowers more than 1 cm in diameter; fruit a pome or drupe.
              Fruit a pome, green, yellowish green, or black.
                Fruit 2-4.5 cm wide, green or yellowish green, depressed-globose or pyriform (rarely elongate in the wild apple).
                  Petals pinkish, rarely white; styles more or less united; fruit depressed-globose, rarely elongate........3338A. MALUS, p. 528.
                  Petals white; styles free to the ovary; fruit more or less pyriform. ....3338. PYRUS, p. 528.
                Fruit about 1 cm wide, black.................3338C. ARONIA, p. 530.
              Fruit a drupe, red or purplish black............3396. PRUNUS, p. 578.
    Leaves compound.

Leaves pinnate, 1.5-3.5 cm long, silky-pubescent beneath; flowers yellow; fruit a head of many achenes........................3356. POTENTILLA, p. 565.

Leaves pinnate or 3-5-foliolate, not silky-pubescent beneath; flowers white; fruit a reddish pome or a cluster of reddish or reddish purple drupelets.

Leaves pinnate, 10-16 cm long; leaflets generally 13-17; fruit a reddish pome.. ....3338B. SORBUS, p. 529.

Leaves 3-5-foliolate; small herbaceous or semi-woody plants of a decumbent or trailing habit, of a wet or bog habitat; fruit small, usually consisting of fewer than 15 drupelets...............3353. *Rubus pubescens*, p. 558.

Plants herbaceous.

Fruit juicy, white or reddish purple; leaves 3-5-foliolate.

Fruit on naked scapes, red or white..................3354. FRAGARIA, p. 563.

Fruit terminal or in the axils of leaves.

Calyx with 5 large bracts; petals yellow; fruit red, strawberrylike........... ....3355. DUCHESNEA, p. 564.

Calyx not bracted; petals white; fruit an aggregate of drupelets (drupelets usually fewer than 15), reddish purple....3353. *Rubus pubescens*, p. 558.

Fruit dry.

Plants stemless or nearly so; leaves appearing to be all basal, trifoliate; flowers scapose, yellow; carpels few, generally 2-6, rarely up to 10, 1-ovuled...... ....3363. WALDSTEINIA, p. 568.

Plants not as above.

Flowers white in one or more long, terminal spikes or the spikes in large, terminal panicles, or greenish in dense, peduncled heads; leaves pinnately divided into 7-19 leaflets or twice or thrice pinnate; fruit a short follicle or an achene enclosed by the 4-angled calyx tube.

Leaves once pinnate; leaflets serrate or incised; flowers white in long (5-15 cm), terminal spikes or green with purple stigmas in dense, peduncled heads; fruit an achene enclosed by the calyx tube........ ....3381. SANGUISORBA, p. 573.

Leaves ternately twice or thrice pinnatifid; leaflets ovate-oblong, long-acuminate at the apex, sharply doubly serrate; flowers in spikes arranged in a large, open, terminal panicle; fruit of 2 or 3 glabrous follicles......................................3322. ARUNCUS, p. 527.

Flowers and inflorescence not as in the two preceding.

Calyx bracteolate, the bracts alternate with the calyx lobes.

Styles long, bent and jointed near the middle, at maturity the upper part deciduous, the lower part persistent and hooked at the summit; flowers white, yellow or purple.................3365. GEUM, p. 568.

Styles deciduous, not jointed or hooked; flowers yellow.................. ....3356. POTENTILLA, p. 565.

Calyx not bracteolate.

Leaves trifoliate or the upper ones 3-lobed or simple on short petioles; petals white, about 1 cm long..............3325. GILLENIA, p. 527.

Leaves pinnate; flowers many; petals yellow, pinkish or white, less than 1 cm long.

Flowers in large, cymose panicles, white or pink; fruit a 1-seeded indehiscent capsule...................3374. FILIPENDULA, p. 571.

Flowers in spikelike racemes, yellow; fruit (the enlarged calyx tube enclosing the 2 achenes) with hooked bristles..................... ....3376. AGRIMONIA, p. 571.

3316. PHYSOCÁRPUS Maxim.

Follicles glabrous or glabrate....................................1. *P. opulifolius.*

Follicles more or less densely pubescent...........1a. *P. opulifolius* var. *intermedius.*

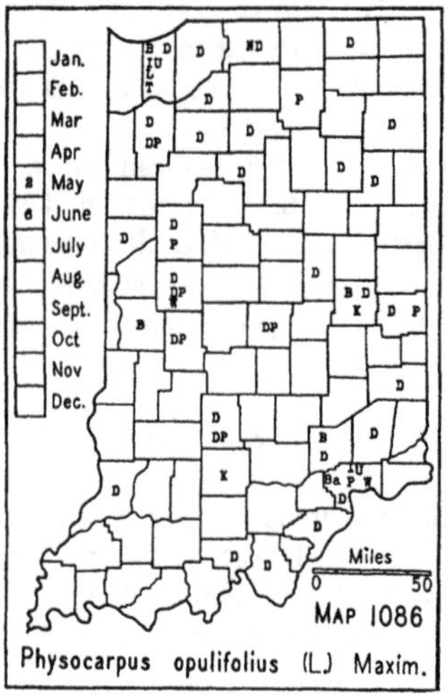

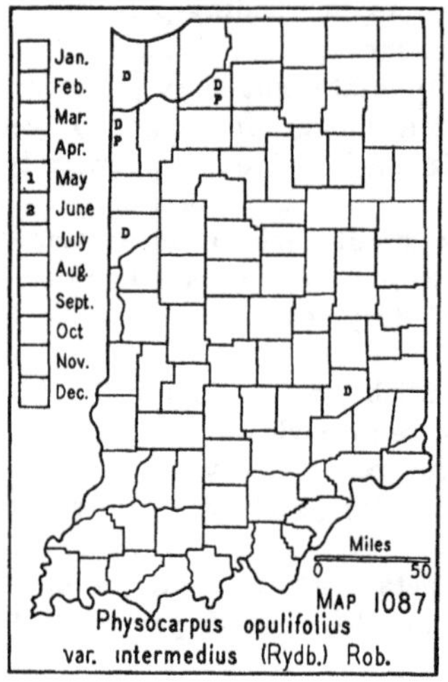

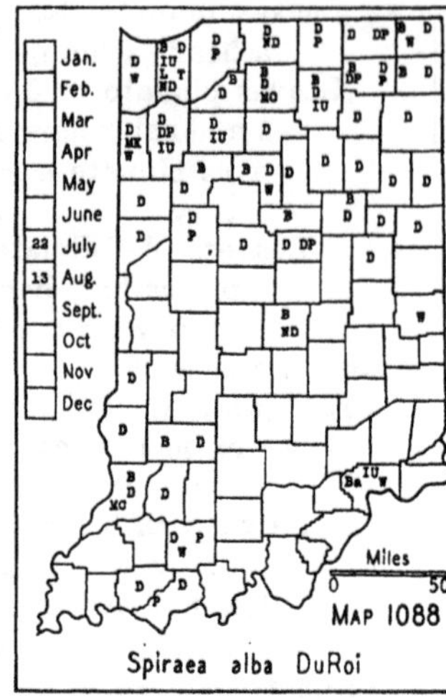

1. **Physocarpus opulifòlius** (L.) Maxim. (*Opulaster opulifolius* (L.) Kuntze of Britton and Brown, Illus. Flora, ed. 2.) COMMON NINEBARK. Map 1086. Generally local to very local in all parts of the state. It generally grows along streams a few feet above the water level, sometimes higher up on the banks, and rarely in crevices at the tops of cliffs. I once found it in a marsh along Pigeon River west of Mongo, in Lagrange County, where it was common.

Que. to ne. Minn., southw. to Ga. and Ky.

1a. **Physocarpus opulifolius** var. **intermèdius** (Rydb.) Robinson. (*Opulaster intermedius* Rydb. of Britton and Brown, Illus. Flora, ed. 2.) ILLINOIS NINEBARK. Map 1087. I am not able to separate this variety from the species by any character other than the pubescence. The number of follicles of both species and variety of all except one of my specimens is 3. The pedicels, with one exception, are densely stellate-pubescent. The habitat is similar to that of the species.

Western N. Y. to S. Dak., southw. to Ill., Mo., Ark., and Colo.

3319. SPIRAÈA [Tourn.] L. SPIREA

Leaves glabrous or sparingly pubescent beneath; calyx lobes not reflexed.
  Inflorescence puberulent.
    Inflorescence elongate, paniculate.......................................1. *S. alba.*
    Inflorescence flat-topped, corymbose. (See excluded species no. 297, p. 1057.).......
    ....*S. japonica.*
  Inflorescence glabrous or glabrate. (See excluded species no. 298, p. 1057.)..........
    ....*S. latifolia.*
Leaves tomentose beneath; calyx lobes reflexed......................2. *S. tomentosa.*

1. **Spiraea álba** DuRoi. (*Spiraea salicifolia* L. in part, of Gray, Man., ed. 7.) MEADOW SPIREA. Map 1088. This species is infrequent to frequent in the lake area and southward to the center of the state in moist, black soil about lakes, in marshes, and in roadside ditches. South of this area

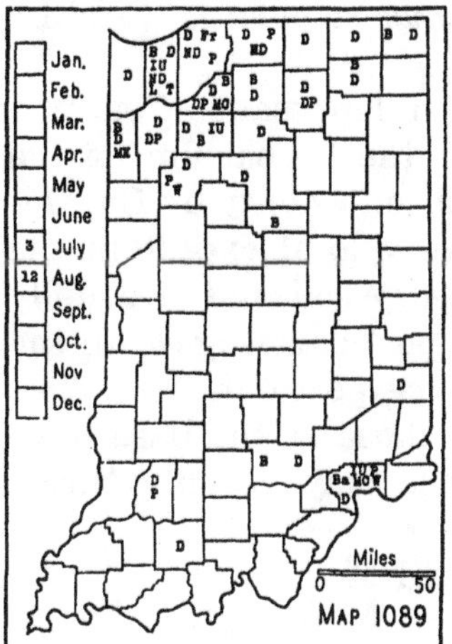

Spiraea tomentosa L.

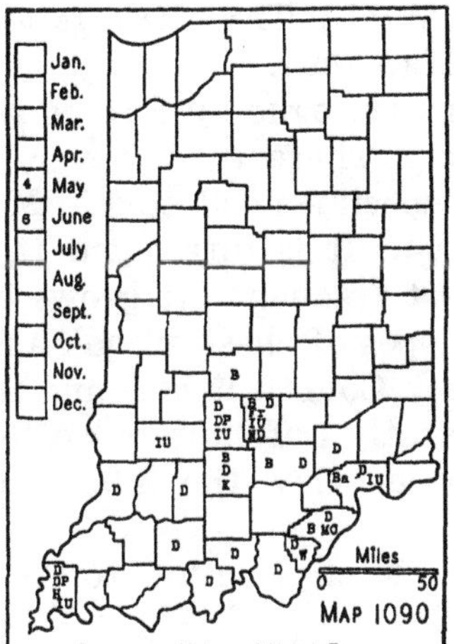

Aruncus dioicus (Walt.) Fern.

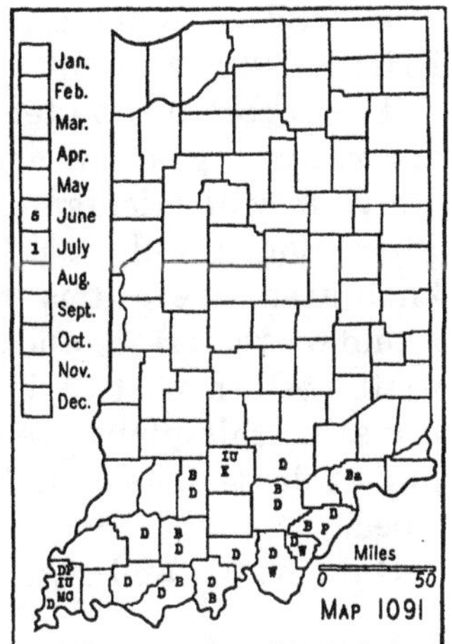

Gillenia stipulata (Muhl.) Trel.

it becomes infrequent to very rare or possibly absent in a few counties of the unglaciated region.

Ont. to Sask., southw. to N. C. and Miss.

2. **Spiraea tomentòsa** L. (*Spiraea tomentosa* var. *rosea* (Raf.) Fern.) HARDHACK. Map 1089. Hardhack requires a slightly acid soil and is frequent to common in the lake area as shown on the map. I have seen areas from an acre to not less than ten acres in extent in low flats where this species was the principal ground cover. South of this area it is absent until the pin oak and sweet gum flats of the southern counties are reached where it is sometimes found but generally in very limited numbers.

When a large colony is studied one finds that most of the specimens have elongated and narrow inflorescences but on more vigorous specimens the inflorescences are often spreading. The tomentum on the capsules varies in abundance but the capsules never become entirely glabrous.

N. S. to Man., southw. to Ga. and Kans.

3322. ARÚNCUS [L.] Adans.

1. **Aruncus dioìcus** (Walt.) Fern. (Rhodora **41**: 423. 1939.) (*Aruncus sylvester* Kost. of Indiana authors and *Aruncus Aruncus* (L.) Karst.) COMMON GOATSBEARD. Map 1090. This plant is restricted almost entirely to the unglaciated part of the state. It is local to very local and grows on the bluffs of streams and on steep, wooded slopes. It is often found clinging to the brink of the top of steep, washed slopes and often is associated with *Hydrangea*.

Ind. to Iowa, southw. to Ark. and Okla.

3325. GILLÈNIA Moench

Leaflets beneath covered more or less with sessile or stalked glands; stipules mostly foliaceous, more or less incised.................................1. *G. stipulata*.

Leaflets glandless beneath; stipules small, usually less than 5 mm wide, entire to serrate. (See excluded species no. 300, p. 1058.)....................*G. trifoliata*.

1. **Gillenia stipulàta** (Muhl.) Trel. (*Porteranthus stipulatus* (Muhl.) Britt.) INDIAN-PHYSIC. Map 1091.

The report in Coulter's Catalogue by Barnes from Tippecanoe County I am regarding as an error. There is, however, in the herbarium of the University of Michigan a specimen collected in Madison County by Charles Piper Smith, July 22, 1904. It is infrequent to very local except in the knob area where it is most frequent. It grows in dry soil and is usually found on the crests and slopes of chestnut oak and post oak ridges. In the southwestern part of Posey County it is found in the post oak flats. The leaves of this plant are mostly trifoliate but often those at the base have pinnatifid leaflets and those below the inflorescence may be only three-lobed.

Ont., N. Y., and N. J. to Mich., southw. to Ga. and Mo.

3338. PŸRUS [Tourn.] L.

See excluded species no. 301, p. 1058.

3338A. MÀLUS Mill. APPLE

Margins of leaves of sterile branchlets generally more incised than those of fruiting branchlets; teeth of leaves of both sterile and fruiting branchlets variable in size; pedicels slender, 1 mm or less in diameter at flowering time; sepals about 2 mm wide at the base.

Older leaves generally glabrous or nearly so beneath at flowering time, those at the ends of the branchlets usually more or less tomentose, all of the leaves glabrous at maturity or with some pubescence on the principal nerves.

Outer surface of calyx glabrous or nearly so at flowering time....1. *M. coronaria*.

Outer surface of calyx tomentose at flowering time..1a. *M. coronaria* var. *dasycalyx*.

Older leaves as well as those at the ends of the branchlets densely tomentose beneath, the tomentum persisting on most of the leaves until maturity..2. *M. ioensis*.

Margins of leaves of both sterile and fruiting branchlets similar in shape; teeth of leaves of both sterile and fruiting branchlets similar in size and shape; pedicels stout, 1.4-2 mm in diameter at flowering time; sepals 3-4.5 mm wide at the base. (See excluded species no. 303, p. 1058.)..............................*M. pumila*.

1. **Malus coronària** (L.) Mill. (*Malus glaucescens* Rehd. and *Malus lancifolia* Rehd. of Deam, Trees of Indiana.) WILD SWEET CRAB. Map 1092. Found throughout the state in various kinds of soil of varying amounts of moisture. It generally grows in colonies, mostly in open woodland, clearings, and wood pastures and along roadsides and fences.

The genus *Malus* of the "Trees of Indiana" was written by W. W. Eggleston. A careful restudy of my specimens convinces me that those using a local flora of this kind will be best served by regarding this species as polymorphic in many of its parts. Specimens can be found that show wide differences but these can be connected by intermediates.

The synonymy of the species is involved and is omitted unless it applies to names used in "Trees of Indiana" by Deam.

N. Y. to Mo., southw. to Ala.

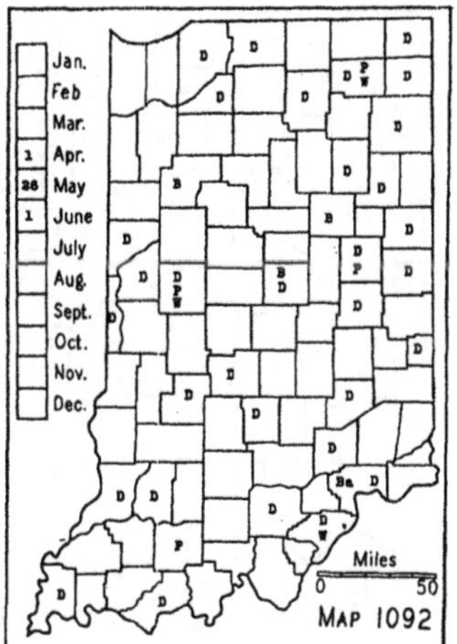

Malus coronaria (L.) Mill

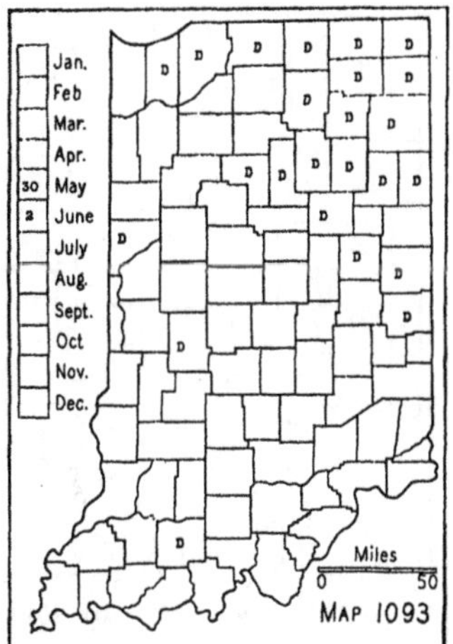

Malus coronaria var. dasycalyx Rehd.

Malus ioensis (Wood) Britt.

1a. **Malus coronaria** var. **dasycàlyx** Rehd. Map 1093. Rehder says this variety also has the leaves paler beneath than the species. It occurs throughout the state with the species.

Ont. to Ohio and Ind.

2. **Malus ioénsis** (Wood) Britt. PRAIRIE CRAB. Map 1094. This is, for the most part, a low, widely spreading tree which, according to specimens seen, is restricted mostly to the western part of the state.

Ind., Wis. to Minn., southw. to Mo.

3338B. SÓRBUS [Tourn.] L. MOUNTAIN-ASH

[Jones. A synopsis of the North American species of Sorbus. Jour. Arnold Arboretum **20**: 1-43. 1939.]

Winter buds glabrous on the back, the inner ones ciliate along the margins; branchlets at flowering time glabrate..1. *S. decora.*
Winter buds usually densely pubescent; branchlets at flowering time more or less densely pubescent. (See excluded species no. 305, p. 1059.)........*S. Aucuparia.*

1. **Sorbus decòra** (Sarg.) Schneid. SHOWY MOUNTAIN-ASH. Map 1095. The species of American mountain-ash have been poorly understood until the recent synopsis appeared. Our native species resembles *Sorbus Aucuparia,* a European species, which has sparingly escaped in northern Indiana. The European mountain-ash has been reported as only single specimens except Nieuwland and Just (Amer. Midland Nat. **12**: 221. 1930) found two colonies of about 20 trees in a woods about 6 miles southwest of South Bend and a single tree in a woods about two and a half miles northeast of Walkerton. The tree at the last station named was about five inches in diameter and approximately 35 feet high. They also report that scattered about the tree were numerous seedlings. *Sorbus decora* was first found in 1924 by Harold Orahood in a woods about a half mile northwest of Union Mills, La Porte County. The tree he found was 9 and a half inches in circumference at breast height and about 20 feet high. In 1933

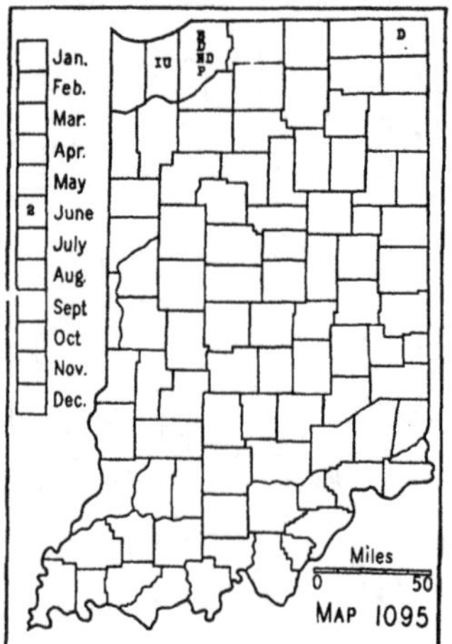

Sorbus decora (Sarg.) Schneid.

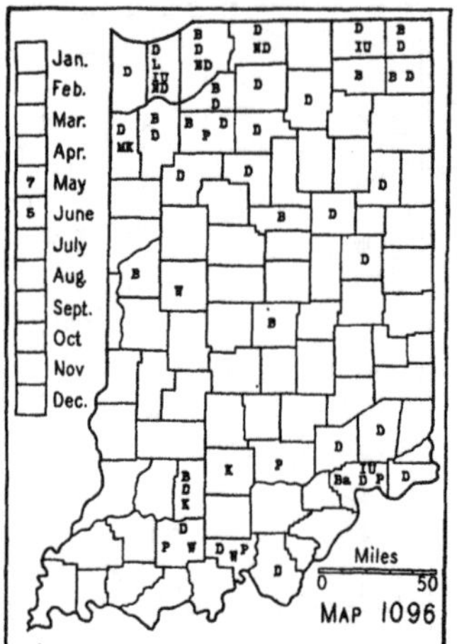

Aronia melanocarpa (Michx.) Ell.

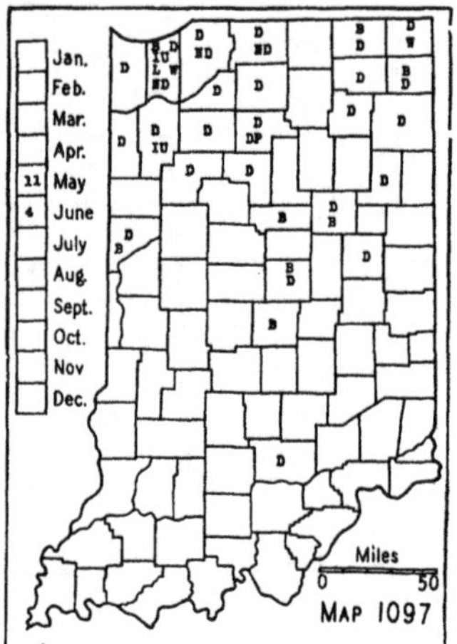

Aronia prunifolia (Marsh.) Rehd.

I found a tree on the border of a swamp in Pokagon State Park, Steuben County, that was 16½ inches in circumference and about 25 feet high.

Newf. to Minn., southw. to N. Y. and Iowa.

3338C. ARÒNIA Medic. CHOKEBERRY

Branchlets, lower surface of leaves, pedicels, and exterior of calyx glabrous or nearly so at flowering time, entirely glabrous at maturity............1. *A. melanocarpa.*

Branchlets, lower surface of leaves, pedicels, and exterior of calyx more or less densely pubescent at flowering time, remaining pubescent until maturity or some parts becoming glabrous but pubescence evident on the remaining parts, the branchlets always showing hairs on some parts..........................2. *A. prunifolia.*

1. **Aronia melanocárpa** (Michx.) Ell. (*Pyrus melanocarpa* (Michx.) Willd. and *Aronia melanocarpa* (Michx.) Britt.) BLACK CHOKEBERRY. Map 1096. This species, like the next one, grows only in slightly acid soil. In the lake area it grows in moist, sandy woods and tamarack bogs and on the borders of lakes; in the southern part of the state it grows in moist or wet, hard clay soil and on the exposed parts of sandstone cliffs. It is not frequent but common where it is found in the north, and rare to very rare in the south where there are generally only a few plants in a place or in very small colonies. It is usually 2-5 feet high.

N. S. to Mich., southw. to Fla.

2. **Aronia prunifòlia** (Marsh.) Rehder. (Jour. Arnold Arboretum **19**: 74. 1938.) (*Aronia floribunda* (Lindl.) Spach, *Pyrus arbutifolia* var. *atropurpurea* (Britt.) Rob., and *Aronia atropurpurea* Britt.) Teuscher discusses the status of this species in Torreya 33: 22-24. 1933. PURPLE CHOKEBERRY. Map 1097. Infrequent to local in the lake area and very local south of it. It is generally 3-9 feet high and grows in old tamarack bogs, swamps, and low woods. In a few places I have found it forming a dense and nearly pure stand over an acre or more and growing to a height of 6-8 feet.

Amelanchier humilis Wieg.

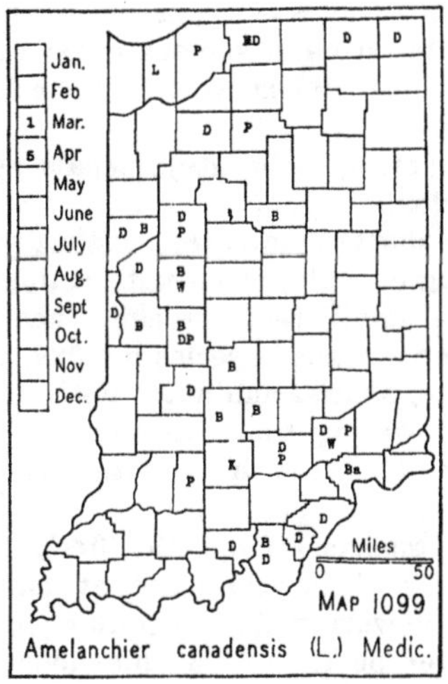

Amelanchier canadensis (L.) Medic.

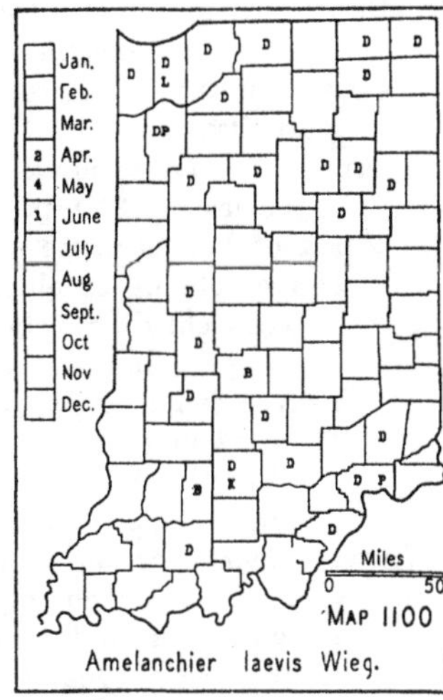

Amelanchier laevis Wieg.

The species is variable in the shape, size, and juiciness of the fruit. The common form has fruit which is nearly dry and is smaller than the juicy form.

Newf. to Mich., southw. to Fla.

3343. AMELÁNCHIER Medic. SHADBLOW

[Wiegand. Amelanchier in eastern North America. Rhodora **14**: 117-161. 2 pl. 1912. Wiegand. Additional notes on Amelanchier. Rhodora **22**: 146-151. 1920.]

The species of this genus are known to hybridize and because of this fact the determination of specimens is often difficult. K. M. Wiegand, who has made the most exhaustive study of the genus of anyone in the United States, has named nearly all of my specimens which include several hybrids. He says for accurate determination "collections should be made from the same plant at flowering time just as the petals begin to fall, at the time when the fruit is half grown, and at the maturity of the leaves. Mature ripe fruit is nearly worthless. The mature leaves are not absolutely necessary, but the other two collections are indispensable." The following key is adapted from his publications:

Teeth of leaves coarse (on average leaves 3-5 (6) per cm); veins conspicuous, usually straight, parallel and close together, short intermediate ones few or none; summit of ovary woolly; sepals revolute from the middle at the time when the petals fall; leaves rounded, obtuse or subacute at the apex.

Petals 7-10 mm long; sepals 2-3 (4) mm long; racemes erect or nearly so; leaves oval-oblong; veins usually becoming irregular just before reaching the margin; stiffly upright shrubs 0.3-1.2 m high, growing in colonies (not in clumps) from rhizomelike bases; margins of leaves serrate to below the middle. . 1. *A. humilis*.

Petals 11-20 mm long, narrow; sepals 4 mm long; racemes more or less drooping; leaves oval-orbicular; upper veins, in typical specimens, running straight to the apex of the coarse, spreading, sharp teeth; scrawny, slender, often arching shrubs, 1-2.5 m high; stems solitary or few together; margins of leaves serrate nearly to the base. (See excluded species no. 311, p. 1059.) . . . . *A. sanguinea*.

Teeth of leaves fine (on average leaves 5-12 per cm); veins irregular, unequally distant, usually with frequent, intermediate, shorter ones; summit of ovary various.
Leaves densely white-tomentose when young, becoming glabrous or nearly so at maturity.
Apex of leaves rounded (rarely subacute); sepals usually upright. (See excluded species no. 310, p. 1059.) .... *A. oblongifolia.*
Apex of leaves acute or short-acuminate; sepals usually reflexed.
Leaves oblong or oblong-obovate, often with a tinge of red; petals 6-8 mm long; tall shrub, rarely a small tree, branching near the ground or at first growing in clumps. (See excluded species no. 309, p. 1059.) .... *A. intermedia.*
Leaves ovate or obovate, short-acuminate; petals 10-14 mm long; lower pedicels 8-17 mm long at flowering time, becoming 15-25 mm long at fruiting time; petioles remaining pubescent until maturity; small trees when mature, not in clumps .... 2. *A. canadensis.*
Leaves nearly or entirely glabrous from the first, ovate, oval, or elliptical, and very acute or commonly short-acuminate at maturity, brownish purple, half grown and unfolded at flowering time; petals elongated, 10-18 mm long; summit of ovary glabrous; lower pedicels 15-33 mm long at flowering time, 30-50 mm long at fruiting time; petioles glabrous at maturity; trees or tall shrubs...... .... 3. *A. laevis.*

1. **Amelanchier hùmilis** Wieg. LOW SHADBLOW. Map 1098. Known in Indiana as low juneberry. This species grows in colonies in very sandy soil in woods and along fence rows and roadsides. I planted roots of it 10 years ago and it has grown well. A few stems have come up from each root, otherwise it has not spread. The large fruit is edible and much relished by birds as is the fruit of all the species of the genus.

Vt. to Minn. and Mack., southw. to e. and cent. N. Y., Ohio, and Nebr.

1a. **Amelanchier humilis × laèvis.** I have this hybrid from Elkhart, Fulton, Lagrange, Lake, La Porte, Porter, Starke, Steuben, and Warren Counties.

2. **Amelanchier canadénsis** (L.) Medic. DOWNY SHADBLOW. Map 1099. Known in Indiana as downy serviceberry or juneberry. This species is more or less infrequent to local throughout the state and is found generally in dry soil on the banks of streams, on wooded slopes, and rarely in level woodland. The stem is usually less than 2 inches in diameter.

N. S. and e. Maine, and from w. N. E. to Wis., southw. to Ga., La., and Mo.

2a. **Amelanchier canadensis × humilis.** I have this hybrid from Cass and De Kalb Counties.

2b. **Amelanchier canadensis × laevis.** I have this hybrid from Allen, Brown, Clark, Clay, Crawford, De Kalb, Fulton, Hendricks, Jefferson, Lagrange, La Porte, Martin, Morgan, Perry, Porter, St. Joseph, Starke, Steuben, Warren, and Whitley Counties.

3. **Amelanchier laèvis** Wieg. ALLEGHENY SHADBLOW. Map 1100. Known in Indiana as smooth serviceberry or juneberry. Frequent to infrequent in the lake area, becoming infrequent to local southward. This species, with the preceding and their hybrids, is frequent on the high dunes facing Lake Michigan. It is also found in old tamarack bogs and

interdunal flats and on the low and high banks of lakes and streams. This species is the largest of the genus in the state, sometimes reaching a diameter of 7 inches and a height of 40 feet.

Newf., N. E. to Mich., southw. to Ga., Ala., and Kans.

3345. CRATAÈGUS[1] L. HAWTHORN, THORN, RED HAW

[Britton and Brown. Illustrated Flora of the Northeastern United States, ed. 2: 294-321. 1913; Palmer. Synopsis of North American Crataegi. Jour. Arnold Arboretum **6**: 5-128. 1925; Palmer. The Crataegus problem. Jour. Arnold Arboretum **13**: 342-362. 1932; and Deam. Trees of Indiana, ed. 2: 192-228. 1932.]

Shrubs or small trees, usually found in pastures, thickets, and borders of woodland, and most abundantly in limestone regions. Many of the species are attractive on account of their flowers, foliage, and fruit, and are frequently planted in parks and private grounds. The fruit of some species is edible and is sometimes sold on the market in some sections, where it is eaten raw or used for preserves; its chief value in Indiana is for bird and game food.

Crataegus is one of the largest genera of woody plants in the number of species, and it is one of the most difficult for taxonomic treatment. Several hundred American species and varieties have been proposed, many of which are probably hybrids or only forms of polymorphic species. In many cases it is difficult to find a single constant character that can be relied upon for separating species, even though they seem to be distinct when all of the characters are considered. The dimensions of leaves, flowers, and fruit given in the descriptions are intended to cover the normal range, but it should be understood that there may be wider variations in unusually vigorous or depauperate forms.

KEY TO THE GROUPS

Nutlets not pitted on ventral surfaces; flowers (except in *Cordatae*) usually opening before the middle of May.
- Fruiting calyx persistent; fruit usually falling soon after maturity; flowers 12-25 mm in diameter.
  - Leaves of flowering branches all narrowed or acuminate at the base, mostly of an obovate, oblong, or spatulate type, broadest at or above the middle, margins merely serrate or with shallow or obscure lobes toward the apex.
    - Leaves of flowering branches usually one and a half to twice as long as wide (forms of group IV may be sought here).
      - Leaves thick and usually glossy above, unlobed except rarely on shoots; veins not conspicuously impressed; styles and nutlets 1-3 (rarely more); fruit remaining hard and dry.................................I. CRUS-GALLI.
      - Leaves thin to firm, dull above, often slightly lobed on flowering branches; styles and nutlets 2-5; fruit becoming mellow.
        - Leaves of flowering branches mostly obovate, symmetrical, with 5-7 pairs of slightly ascending, deeply impressed veins; fruit 9-16 mm in diameter (rarely larger); nutlets usually 2-4...........II. PUNCTATAE.

---

[1] The text of the genus *Crataegus* was written by Ernest J. Palmer of the Arnold Arboretum, Harvard University. The manuscript has, with his approval, been made to conform to the general style of the book.

Leaves of flowering branches mostly oblong or rhombic, often unsymmetrical, with 4 or 5 pairs of strongly ascending veins, not deeply impressed; fruit 5-10 mm in diameter; styles and nutlets usually 5..III. VIRIDES.

Leaves of flowering branches usually one to one and a half times as long as wide (except in forms of no. 10).....................IV. ROTUNDIFOLIAE.

Leaves of flowering branches abruptly narrowed, rounded, truncate, or subcordate at the base, mostly ovate, oblong-ovate or elliptic, broadest at or below the middle, usually lobed or incised.

Leaves of flowering branches usually slightly narrowed or abruptly acuminate at the base; petioles and inflorescence conspicuously glandular; usually shrubs...V. INTRICATAE.

Leaves of flowering branches usually rounded to subcordate at the base; petioles and inflorescence eglandular or slightly glandular; arborescent shrubs or trees.

Leaves thin; fruit 7-9 mm in diameter, with small, sessile calyx; flowers 12-15 mm in diameter...................................VI. TENUIFOLIAE.

Leaves firm to subcoriaceous; fruit usually 10-20 mm in diameter; flowers 16-25 mm in diameter.

Fruit with thin flesh and relatively large nutlets, remaining hard and dry; fruiting calyx large and elevated (except in no. 18)...VII. PRUINOSAE.

Fruit becoming mellow or succulent, usually edible; fruiting calyx smaller, sessile or nearly so.

Leaves barely firm; petioles and primary veins slender; fruit glabrous; styles and nutlets usually 3-4....................VIII. COCCINEAE.

Leaves firm to subcoriaceous; petioles and primary veins stout; fruit pubescent at least toward the base; styles and nutlets usually 5...IX. MOLLES.

Fruiting calyx deciduous; fruit 5-7 mm in diameter, bright red, long persistent after maturity; flowers about 10 mm in diameter, often not opening until early June...X. CORDATAE.

Nutlets pitted on ventral surfaces; fruit 6-12 mm in diameter, often long persistent after maturity; flowers usually opening after the middle of May...XI. MACRACANTHAE.

KEY TO THE SPECIES

I. CRUS-GALLI LOUD.

Leaves of flowering branches mostly of a spatulate or obovate type, broadest above the middle.

Leaves thick and glossy (except sometimes in shade), those of flowering branches 1-2.5 cm wide, usually obtuse, rounded or short-pointed at the apex.

Leaves of flowering branches mostly 1.5-2.5 cm wide.......1. *C. crus-galli*, p. 537.

Leaves of flowering branches mostly 1-1.5 cm wide.............................1a. *C. crus-galli* var. *pyracanthifolia*, p. 537.

Leaves thinner, shining above, usually pointed or acuminate at the apex.............................2. *C. pyracanthoides* var. *arborea*, p. 537.

Leaves of flowering branches broader, mostly oblong-obovate or elliptic, usually broadest about the middle.

Leaves thick, those of the flowering branches usually acute or acuminate at the apex; fruit obovoid or ellipsoid...............................3. *C. regalis*, p. 538.

Leaves thinner, those of the flowering branches usually rounded or short-pointed at the apex; fruit subglobose..........................4. *C. acutifolia*, p. 538.

II. PUNCTATAE Loud.

Foliage and inflorescence villous or pubescent; leaves dull above.
Leaves of flowering branches usually 3.5-5 cm long, 2-3 cm wide, on sterile shoots acute or acuminate and incised toward the apex; fruit 12-18 mm in diameter.........5. *C. punctata,* p. 539.
Leaves of flowering branches usually 2-3 cm long, 1.5-2 cm wide, on sterile shoots often obtuse or short-pointed and with shallow or obscure lobes; fruit 8-14 mm in diameter.........6. *C. collina,* p. 539.
Foliage and inflorescence glabrous or essentially so; leaves somewhat glossy above.
Leaves mostly rounded or abruptly pointed at the apex, slightly villous above when young; fruit subglobose, bright red at maturity.........7. *C. grandis,* p. 540.
Leaves mostly acute or acuminate at the apex, glabrous; fruit oblong or obovoid, dull red or russet at maturity.........8. *C. disperma,* p. 541.

III. VIRIDES Beadle

Leaves mostly oblong or rhombic in outline, dentate or with shallow lobes, thin, glabrous at maturity except for tufts of tomentum in the axils of the veins; fruit subglobose, 5-8 mm in diameter; nutlets usually 5......9. *C. viridis,* p. 541.

IV. ROTUNDIFOLIAE Eggl.

Leaves variable, short-obovate or suborbicular (or in the variety lance-ovate), usually incised with small, shallow lobes; fruit subglobose, 8-14 mm in diameter; nutlets 2-3.
Leaves rounded or short-pointed at the apex, from slightly longer than wide to equilateral or even wider.........10. *C. Margaretta,* p. 541.
Leaves lance-ovate or lance-elliptic, narrowed or acuminate at the apex, one to one and a half times as long as wide.........10a. *C. Margaretta* var. *angustifolia,* p. 543.

V. INTRICATAE Sarg.

Leaves and inflorescence glabrous or essentially so; fruit glabrous.
Leaves mostly ovate, rounded or abruptly contracted at the base, broadest below the middle, usually sharply lobed; fruit dull orange or bronze.........11. *C. intricata,* p. 543.
Leaves mostly oblong-lanceolate or elliptic, pointed or acuminate at both ends, usually broadest about the middle, undivided or with shallow, obscure lobes; fruit bright red at maturity.........12. *C. rubella,* p. 543.
Leaves and inflorescence villous; fruit pubescent.........13. *C. biltmoreana,* p. 544.

VI. TENUIFOLIAE Sarg.

Leaves thin, mostly ovate, more or less lobed; flowers 12-14 mm in diameter; stamens 10 or fewer; fruit 7-9 mm in diameter, becoming mellow or succulent, with a small, sessile calyx.........14. *C. macrosperma,* p. 545.

VII. PRUINOSAE Sarg.

Stamens usually about 20; fruiting calyx large and elevated; leaves glabrous (except in no. 17).
Leaves of flowering branches mostly ovate, rounded or abruptly contracted at the broad base, distinctly longer than wide, usually blue green.
Leaves glabrous; fruit subglobose to slightly pyriform, remaining hard and dry, green or dull crimson at maturity.
Leaves pointed but not conspicuously elongated at the apex; fruit usually 12-16 mm in diameter.........15. *C. pruinosa,* p. 545.

Leaves with the terminal lobe wedge-shaped and usually conspicuously elongated; fruit 10 mm or less in diameter..................16. *C. Gattingeri*, p. 546.

Leaves scabrate above when young and usually slightly villous on the veins beneath; fruit globose or depressed-globose, becoming slightly mellow, orange red at maturity..17. *C. platycarpa*, p. 547.

Leaves of flowering branches broadly ovate or deltoid-ovate, truncate or cordate at the base, often isometric or broader on shoots, usually yellow green.......... ....18. *C. rugosa*, p. 547.

Stamens 10 or fewer; fruiting calyx small and sessile; leaves scabrate above when young, glabrous beneath.................................19. *C. prona*, p. 548.

VIII. COCCINEAE Loud.

Leaves of flowering branches mostly ovate or broadly ovate, longer than wide, abruptly narrowed or rounded at the base, glabrous or nearly so at maturity; fruit obovoid or nearly globose, 10-14 mm in diameter.

Flowering corymbs and petioles villous; corymbs usually compound and many-flowered...20. *C. pedicellata*, p. 548.

Flowering corymbs and petioles glabrous; corymbs often nearly simple and few-flowered............................20a. *C. pedicellata* var. *albicans*, p. 549.

Leaves of flowering branches broadly ovate or deltoid-ovate, often isometric, truncate or subcordate at the base, permanently pubescent at least on the veins beneath; fruit subglobose, 14-20 mm in diameter................21. *C. Putnamiana*, p. 549.

IX. MOLLES Sarg.

Leaves of flowering branches ovate, oblong-ovate or oblong-elliptic, pointed or acuminate at the apex, 3-6 cm wide, scabrate or villous above, pubescent at least on the veins beneath, with stout, villous petioles; fruit 15-20 mm in diameter, pubescent.

Leaves of flowering branches mostly rounded or truncate at the broad base........ ....22. *C. mollis*, p. 550.

Leaves of flowering branches mostly oblong-elliptic, narrowed or acute at the base. ....22a. *C. mollis* f. *dumetosa*, p. 550.

Leaves of flowering branches broadly oval to suborbicular, mostly obtuse or rounded at the apex, 2.5-4 cm in width; petioles more slender; fruit glabrous or nearly so. ....23. *C. Kelloggii*, p. 551.

X. CORDATAE Beadle

Leaves glabrous, ovate or deltoid-ovate in outline, rounded to cordate at the base, usually with one or two pairs of acute, spreading lobes; flowers small, appearing after the leaves; fruit 5-7 mm in diameter, bright red, with deciduous calyx....... ....24. *C. Phaenopyrum*, p. 551.

XI. MACRACANTHAE Loud.

Leaves of flowering branches mostly ovate or elliptic, 3-4 cm wide; flowering corymbs villous or glabrate; fruit bright red, mellow or succulent at maturity.

Mature leaves firm but not subcoriaceous, veins slightly impressed above; flowering corymbs villous or tomentose; fruit obovoid to nearly globose, orange red or scarlet; thorns usually scattered and slender or branches nearly unarmed...... ....25. *C. Calpodendron*, p. 552.

Mature leaves subcoriaceous, veins conspicuously impressed above, becoming glabrous above and glabrate or finely pubescent along the veins beneath; flowering corymbs glabrate or slightly villous; fruit subglobose, dark red, becoming succulent; thorns usually numerous, long and stout.......26. *C. succulenta*, p. 552.

Leaves of flowering branches mostly obovate or elliptic, 2-3 cm wide; flowering corymbs villous or tomentose; fruit ovoid or subglobose, remaining hard, pale red or yellow green at maturity......................................27. *C. incaedua*, p. 553.

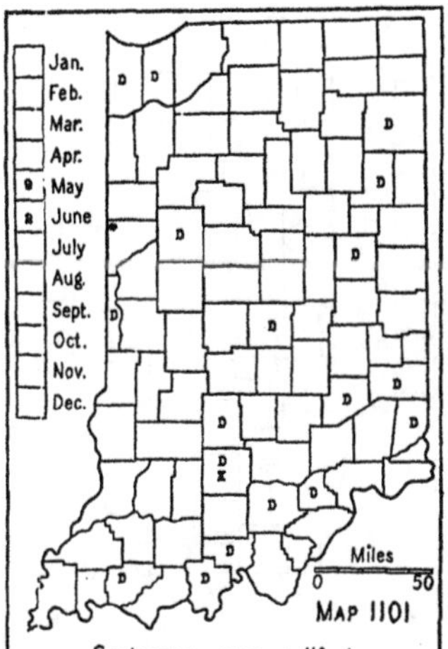

Crataegus crus-galli L.

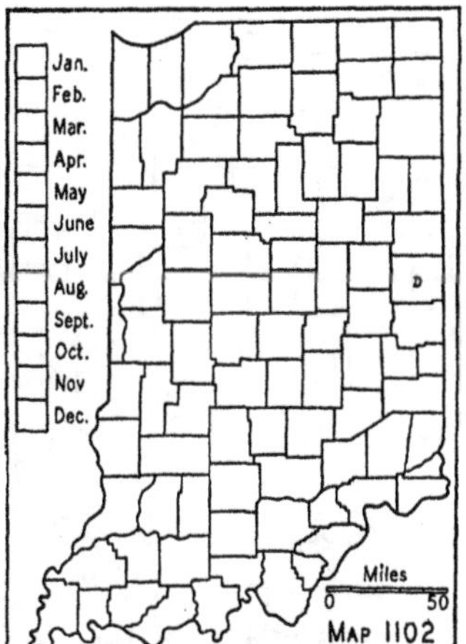

Crataegus pyracanthoides var. arborea (Beadle) Palmer

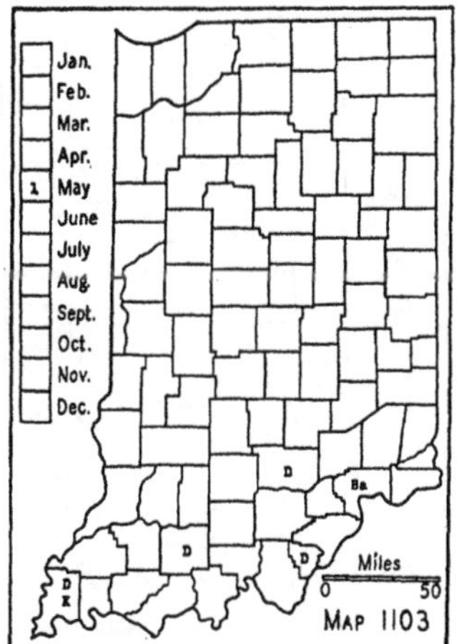

Crataegus regalis Beadle

1. **Crataegus crús-gálli** L. (*Crataegus arduennae* Sarg., *Crataegus attenuata* Ashe, and *Crataegus trahax* Ashe.) COCKSPUR THORN. Map 1101. Leaves mostly spatulate or obovate, 2-6 cm long, 1-3.5 cm wide, rounded or acute at the apex, attenuate at the base into short, slender petioles, sharply serrate to below the middle, glabrous, firm in texture, usually glossy on the upper surface; flowers 12-15 mm wide, in compound, glabrous corymbs; stamens about 10; anthers pink or creamy white; styles 1-3; fruit obovoid to subglobose, 9-12 mm in diameter, flesh thin, hard and dry, dull crimson, with dark blotches or dots; calyx sessile or nearly so; calyx lobes entire or slightly serrate near the base; nutlets 1-3, usually 2.

A small tree or rarely a stout shrub up to 6-7 m high, with slightly scaly, pale gray bark and spreading branches, forming a low, flat crown in old specimens: branchlets often flexuous and armed with numerous, long, slender thorns.

General throughout Indiana, but most common in limestone regions, growing in fertile or rocky ground in thickets and pastures, and in open woodland along small streams.

Southeastern Canada to Minn., southw. to S. C. and Ark.

1a. **Crataegus crus-galli** var. **pyracanthifòlia** Ait. Differs from the typical form only in the narrower leaves and the usually smaller fruit. This variety is known in Indiana only from Posey County, but it is likely to be found in other sections.

2. **Crataegus pyracanthoìdes** Beadle var. **arbòrea** (Beadle) Palmer. (*Crataegus arborea* Beadle and *Crataegus tenuispina* Sarg.) Map 1102. Leaves narrowly obovate or lance-obovate, 3-6 cm long, 1.5-3 cm wide, acute or short-acuminate at the apex, narrowed at the base into slender, winged petioles, serrate to below the middle with broad, shallow teeth, rather thin but firm, glabrous, glossy above; flowers 12-14 mm in diameter, in lax, mostly 6-10-flowered, glabrous corymbs; stamens 10-20; anthers usually white or cream color; styles 3-4; fruit subglobose, 7-10 mm in diameter, orange red, flesh thin; nutlets 2-4, usually 3.

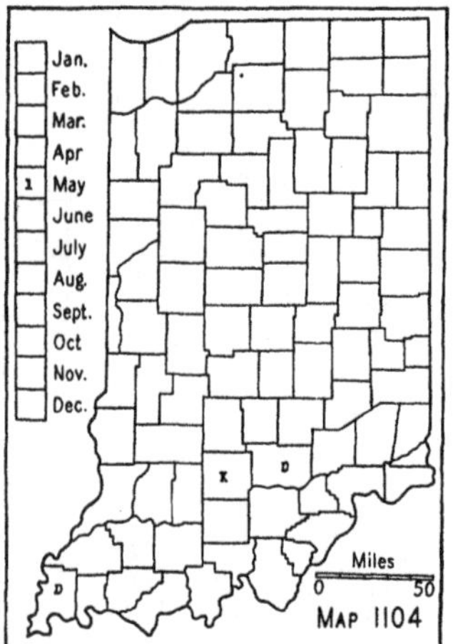

Crataegus acutifolia Sarg.

Crataegus punctata Jacq.

Crataegus collina Chapm.

Known in Indiana only from Randolph County, growing in moist, open woods.

Ala. to Mo. and Ark.

3. **Crataegus regàlis** Beadle. (*Crataegus crus-galli* of Eggleston in part, not of L. of Deam, Trees of Indiana, ed. 2, pl. 78. 1932). Map 1103. Leaves oblong-obovate or elliptic, or on shoots oval to nearly orbicular, 3-7 cm long, 2-4 cm wide, usually abruptly pointed or short-acuminate at the apex, narrowed at the base into slender petioles (8-15 mm long), sharply serrate to below the middle, firm to subcoriaceous, glabrous, shining above; flowers 14-16 mm in diameter, in lax, glabrous, many-flowered corymbs; stamens about 10; anthers white or cream color; fruit oblong or ellipsoid, 8-10 mm long, 7-8 mm thick, green or becoming dull red; calyx lobes linear, entire or nearly so, often persistent and appressed on the fruit; nutlets 2-3.

A tree sometimes 6-8 m high, with gray, slightly scaly bark and with wide-spreading branches, abundantly armed with long, spreading thorns.

Found in the southern part of Indiana growing in fertile soil along streams and in open woodland and thickets.

N. C. and Ga. to Ind., Mo. and Ark.

4. **Crataegus acutifòlia** Sarg. (*Crataegus erecta* Sarg. and *Crataegus ludoviciensis* Sarg.) Map 1104. Leaves oblong-obovate or elliptic, mostly 3.5-6 cm long, 2.5-3.5 cm wide, rounded or abruptly pointed at the apex, serrate nearly to the base with broad, shallow teeth, or on shoots sometimes obscurely lobed and with sharp, spinulose teeth, rather thin but firm, glabrous, dull or slightly glossy above; flowers 12-14 mm in diameter in lax, glabrous, many-flowered corymbs; stamens about 10-15; anthers white or pale yellow; styles 2-4; fruit subglobose or slightly elongated, 7-8 mm in diameter, dull red, firm but mellow at maturity; nutlets usually 3-4.

A tree up to 10 m high with thin, pale gray, scaly bark and with slender wide-spreading branches, usually sparingly armed with slender thorns.

This species may have originated as a hybrid between *Crataegus viridis* and *Crataegus crus-galli* or some species of the *Crus-galli* group, as is suggested by the intermediate character of the bark, foliage, and fruit, and by the fact that it is found only within the range of these species.

In Indiana it is known only from Jackson, Lawrence, and Posey Counties, growing in low, alluvial woods along the larger streams.

Southwestern Ind., s. Ill., and e. Mo.

5. **Crataegus punctàta** Jacq. (Deam. Trees of Indiana, ed. 2, pl. 80. 1932.) DOTTED HAW. Map 1105. Leaves spatulate or lance-obovate, 2.5-7 cm long, 1.5-3.5 cm wide, rounded, acute, or acuminate at the apex, attenuate at the base into winged petioles (1.5-2 cm long), sharply serrate or dentate on the upper two thirds of the blades, often incised and with shallow lobes above the middle, or on vigorous shoots deeply laciniate, firm in texture, with veins deeply impressed on the upper side, dull grayish green, scabrate above when young and pubescent along the veins beneath; flowers 16-20 mm in diameter, usually in many-flowered, compound, villous corymbs; stamens about 20; anthers red or rarely pale yellow; calyx lobes narrowly deltoid, usually entire; fruit subglobose or short-oblong and flattened at the ends, 14-20 mm in diameter, dull red with pale dots, becoming mellow; nutlets 3-4.

A tree up to 10 m high, with gray, furrowed or slightly scaly bark, and often with compound thorns on the trunk or principal branches. The branches are at first ascending but, in old trees, becoming horizontal or depressed; the branchlets villous the first season, olive brown and glabrous the second season, and finally gray, unarmed or armed with long, slender thorns.

Throughout Indiana, in thickets, pastures, and borders of woods.

Newf. and e. Canada to Minn., southw. to N. C. and Ill.

5a. **Crataegus punctata** var. **aúrea** Ait. This variety differing only in the bright yellow fruit, has been found in Jennings and Wells Counties, and should be sought in other sections.

5b. **Crataegus punctata** var. **canéscens** Britt. This variety differs in the close, copious, gray pubescence of the leaves and young branchlets. It is occasionally found throughout the range of the species, and is known from Allen, Grant, Hamilton, Howard, Marshall, Vermillion, and Wayne Counties.

6. **Crataegus collìna** Chapm. (Deam. Trees of Indiana, ed. 2, pl. 82. 1932.) (*Crataegus macropoda* Sarg. and *Crataegus sucida* Sarg.) Map 1106. Leaves narrowly obovate or oblong-obovate, 2-6 cm long, 1.5-3 cm wide, rounded or pointed at the apex, attenuate at the base, serrate on the upper two thirds of the blades, or sometimes nearly to the base, thin but firm, with veins slightly impressed above, dull green, scabrous above and villous beneath when young, at maturity glabrous above and slightly villous on the veins beneath; flowers 14-17 mm in diameter, in compound, many-flowered, villous corymbs; stamens 15-20; anthers pale yellow or rarely

MAP 1107
Crataegus grandis Ashe

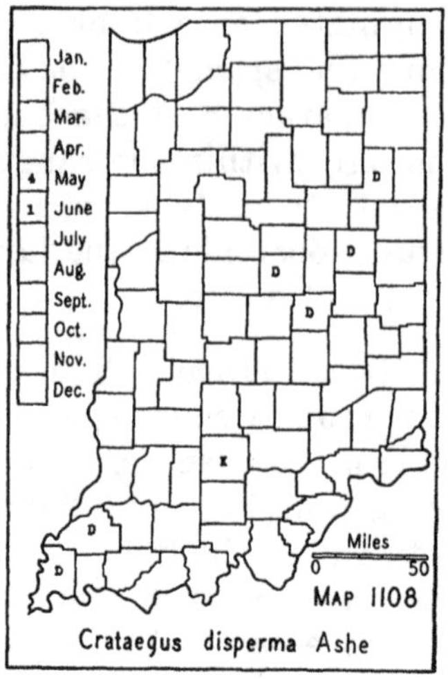

MAP 1108
Crataegus disperma Ashe

MAP 1109
Crataegus viridis L.

red; calyx lobes lanceolate, usually glandular-serrate; fruit subglobose, 8-14 mm in diameter, dull red, with thin flesh; nutlets usually 4-5.

This species has been found in Indiana only in Dearborn County, where it grew on a wooded slope along Laughery Creek, 3 miles west of Aurora.

Va. to se. Ind., southw. to N. C. and Tenn.

7. **Crataegus grándis** Ashe. (*Cratàegus cuneiformis* of Eggleston in part, not *Mespilus cuneiformis* Marsh.) Map 1107. Leaves obovate, mostly 2.5-7 cm long, 2-4 cm wide, rounded or short-pointed at the apex, cuneate and attenuate at the base into slender, winged petioles, coarsely serrate on the upper half to two thirds of the blades, otherwise entire, or sometimes obscurely lobed on shoots, glabrous or with a few scattered hairs above when young, at maturity dark green and shining above, with deeply impressed veins; flowers 14-16 mm in diameter; anthers pink or white; calyx lobes narrowly linear, entire or slightly serrate toward the base, slightly pubescent; fruit subglobose, 10-14 mm in diameter, bright crimson, flesh becoming mellow; nutlets 2-3.

A small tree 4-6 m high, or sometimes a stout shrub, with ascending, or in old specimens, horizontal, spreading branches and slender, glabrous branchlets usually armed with numerous, slender thorns.

Found in southern Indiana in thickets or borders of woods.

This species and the next one may have arisen as hybrids between some form of *Crataegus crus-galli* and *Crataegus punctata* or some related species, as suggested by Eggleston, who has grouped a number of such forms under the name *Crataegus cuneiformis* (Marsh.) Eggl. The description of *Mespilus cuneiformis* given by Marshall seems scarcely definite enough for positive identification, although it may well have applied to one of these hybrids, but since there is such a wide difference in the foliage and fruit characters between this and the next species as well as between others related to them, it seems best to distinguish them and to retain the names already published. The distribution and association of this species, as well as the shape and texture of the leaves, suggest that *Crataegus regalis* may be one of the parents.

Ohio to Ill.

8. **Crataegus dispérma** Ashe. (*Crataegus cuneiformis* of Eggleston in part, perhaps not *Mespilus cuneiformis* Marsh., *Crataegus pausiaca* Ashe, *Crataegus peoriensis* Sarg., and *Crataegus praestans* Sarg.) Map 1108. Leaves obovate or lance-obovate, mostly 2.5-6 cm long, 1.5-4 cm wide, usually pointed or acuminate at the apex, attenuate at the base into slender, winged petioles, sharply serrate on the upper part of the blades or sometimes nearly to the base, otherwise entire or with small, cuneate or spinulose lobes on shoots, rather thin but firm, glabrous or nearly so, bright green and slightly glossy above, veins moderately impressed; flowers 15-18 mm in diameter, in few-flowered or sometimes many-flowered, glabrous corymbs; stamens about 10 or sometimes 12-15; anthers pink; fruit obovoid or ellipsoid, 9-12 mm in diameter, 12-14 mm long, dark red, flesh thin, becoming mellow but dry; nutlets usually 2.

A small tree 6-8 m high, with gray, furrowed bark on the trunk and with numerous, ascending or finally spreading, horizontal branches and slender branchlets usually armed with numerous, slender thorns.

The leaves, flowers, and fruit of this species resemble somewhat more closely those of *Crataegus crus-galli* than do those of *Crataegus grandis*.

General but scattered in Indiana in open woodland, mostly along streams.

Pa. to Ill.

9. **Crataegus víridis** L. (Deam. Trees of Indiana, ed. 2, pl. 87. 1932.) (*Crataegus nitida* of Eggleston in part, not of Sarg. in Deam, Trees of Indiana. pl. 88. 1932.) Map 1109. Leaves extremely variable, elliptic, oblong-lanceolate, rhombic, or sometimes ovate on shoots, mostly 2-6 cm long, 1.5-4.5 cm wide, usually pointed or acuminate at the apex and cuneate and attenuate at the base into slender (1-2 cm) petioles, coarsely serrate on the upper two thirds or sometimes nearly to the base, undivided or sometimes with small irregular lobes, or deeply incised on shoots, thin, dark green and somewhat lustrous above, glabrous at maturity except for tufts of tomentum in the axils of the veins beneath; flowers 10-12 mm in diameter, in glabrous, many-flowered, compound corymbs; stamens about 20; anthers cream white or rarely pink; calyx lobes linear, usually entire; fruit subglobose, 5-8 mm in diameter, becoming bright red or orange red, sometimes slightly pruinose; nutlets 4-5, usually 5.

A tree sometimes 8-10 m high with a conical or depressed crown and with ascending or wide-spreading branches, pale gray bark, scaly in large, thin flakes from a cinnamon color inner layer, and slender branchlets often unarmed or sparingly armed with slender spines.

In Indiana found only in the southwestern part in alluvial bottoms.

Va. to Mo., southw. to Fla. and e. Tex.

10. **Crataegus Margarétta** Ashe. (Deam. Trees of Indiana, ed. 2. pl. 81. 1932.) (Includes *Crataegus chrysocarpa* of Eggleston, not of Ashe in Deam, Trees of Indiana, ed. 2. pl. 86. 1932, and *Crataegus Brownei* Britt.)

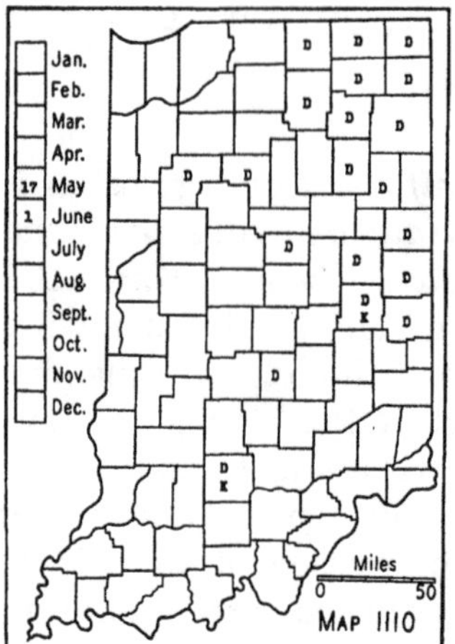

Crataegus Margaretta Ashe

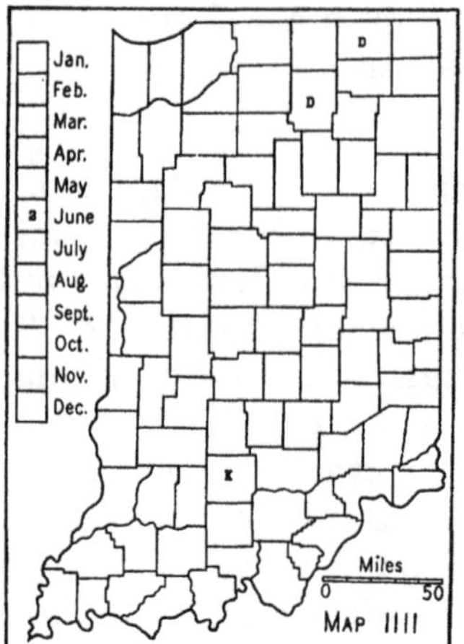

Crataegus intricata Lange

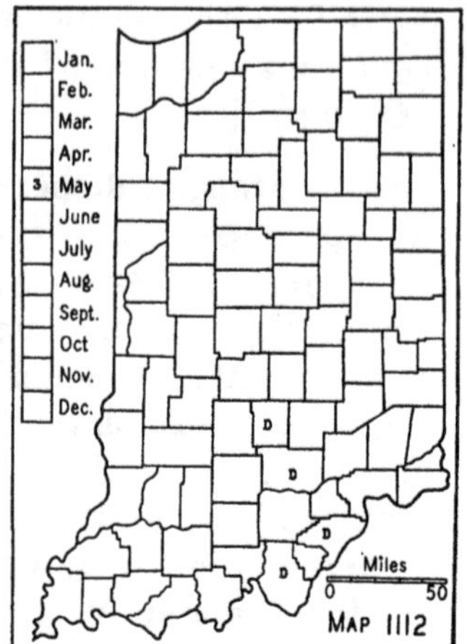

Crataegus rubella Beadle

Map 1110. Leaves variable in size and shape, short-obovate, oval, rhombic, lance-oblong, or nearly orbicular and sometimes wider than long, mostly 2-6 cm long, 1.5-5 cm wide, rounded or pointed at the apex, gradually or sometimes abruptly, contracted at the base into slender, winged petioles, coarsely serrate with broad, shallow teeth for about two thirds the length of the blades, usually incised above the middle and with shallow, rounded or triangular lobes, or sometimes undivided, slightly scabrate above when young, glabrous at maturity, firm and with veins slightly impressed above; flowers 12-15 mm in diameter, usually 6-12 in small, compact, simple or slightly branched corymbs, on glabrous or sparsely villous pedicels; stamens about 20; anthers white or cream color; calyx lobes linear-lanceolate, entire or nearly so; fruit subglobose, 7-10 mm in diameter, dull red or russet, often irregularly blotched, with thin flesh, remaining hard and dry; nutlets usually 3.

A small tree or often an arborescent shrub, up to 5-6 m high, with roughish dark gray bark and stout, ascending or spreading branches, usually sparingly armed with slender thorns, or sometimes nearly unarmed.

Crataegus Margaretta is difficult to describe because of the great variability in the shape and size of the leaves and fruit, but it is a well marked species and it is easily recognized when once known in the field. There has been considerable difference of opinion as to the relationship of this species, some botanists placing it in the *Punctatae* group, or regarding it as the type of a distinct group, but it seems most nearly related to such species as *Crataegus Dodgei, Crataegus chrysocarpa,* and *Crataegus rotundifolia,* and it is therefore retained in the *Rotundifoliae* group in this treatment.

General and frequent in Indiana, especially in the eastern and northern counties, growing in pastures, thickets, and borders of woods. In the north it is usually in dry, sandy or gravelly soil or in clay on terminal moraines and southward on rocky slopes.

Southern Ont. to Iowa, southw. to Va. and Mo.

10a. **Crataegus Margaretta** var. **angustifòlia** Palmer, var. nov.[1] Leaves oblong-lanceolate or lance-elliptic, 1-3 cm long, 0.8-2 cm wide, acute or acuminate at the apex, abruptly narrowed or acuminate at the base and decurrent on the slender petioles, which are a half to two thirds as long as the blades. Flowers and fruit like those of the typical form.

Found in northern Indiana in Elkhart and Lagrange Counties.

Specimens examined: Deam no. 38534, a quarter of a mile east of Bristol, Elkhart County (type), May 25 and September 12, 1923; Deam no. 15660, 1 mile north of Howe, Lagrange County. Type in herbarium of the Arnold Arboretum.

10b. **Crataegus Margaretta** f. **xanthocárpa** Sarg. This form differs from the typical form in having bright or pale yellow fruit. Our only specimen is from Grant County.

11. **Crataegus intricàta** Lange. (*Crataegus meticulosa* Sarg.) Map 1111. Leaves ovate or elliptic, mostly 3-6 cm long, 2.5-5 cm wide, acute at the apex, rounded or abruptly narrowed at the base, slightly decurrent on the slender (1-3 cm long), glandular petioles, coarsely serrate nearly to the base, usually incised on the upper two thirds of the blades with 2-4 pairs of shallow, steplike, lobes, thin but firm in texture, glabrous or essentially so, though sometimes with a few hairs on the upper surface when young; flowers 12-16 mm in diameter, in few-flowered, simple, corymbs, usually much exceeded by the subtending leaves; stamens about 10; anthers cream white or pink; bracts and calyx lobes glandular; fruit oblong or pyriform, or sometimes nearly globose but attenuate at the base, bronze green or becoming dull red at maturity; fruiting calyx broad and prominent; nutlets usually 3-4.

A straggling shrub 1-3 m high with dark gray, scaly bark, ascending or spreading branches, and slender branchlets usually armed with long, slender thorns.

Uncommon and scattered in northern Indiana, and known only from Kosciusko, Lagrange, and Lawrence Counties.

Vt. to Mich., southw. to Va. and Ind.

12. **Crataegus rubélla** Beadle. (*Crataegus pygmaea* Sarg. and *Crataegus meticulosa* Sarg. of Deam, Shrubs of Indiana, ed. 2. pl. 60. 1932.) Map 1112. Leaves mostly elliptic or oblong-lanceolate, 2.5-7 cm long, 1.5-4.5 cm wide, pointed or acuminate at the apex, cuneate or attenuate at the base, sharply serrate nearly to the base, obscurely lobed with 3-5 pairs of small, shallow lobes, or sometimes nearly entire, thin but firm at maturity, glabrous, yellow green; petioles slender, a fourth to half the length of the blades, glandular; flowers 18-22 mm in diameter, mostly 3-6, in compact, simple corymbs, on glabrous, glandular pedicels; bracts conspicuously glandular; stamens about 10; anthers pink or rose color; fruit oblong-obovoid or pyriform, 9-12 mm thick, 10-14 mm long, bright red or orange red at maturity; nutlets usually 3-5.

[1] A typo differt foliis oblongo-lanceolatis vel elliptico-lanceolatis, 1-3 cm longis, 0.8-2 cm latis.

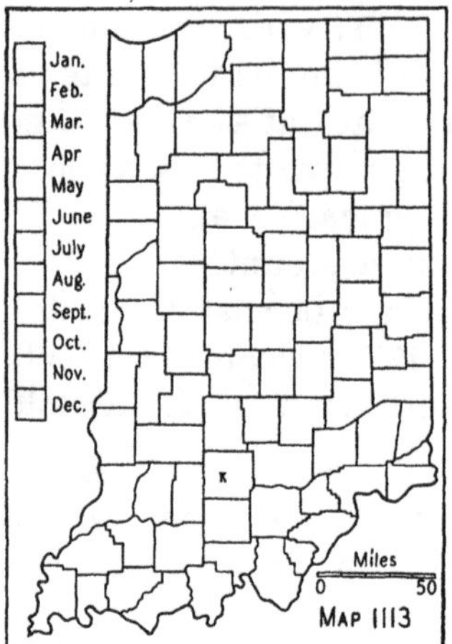

Crataegus biltmoreana Beadle

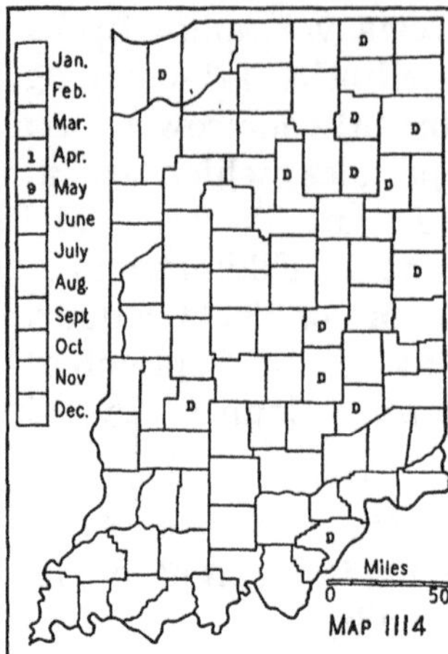

Crataegus macrosperma Ashe

Crataegus pruinosa (Wendl.) K. Koch

An irregularly branched shrub, 1-4 m high, with gray or brown gray bark, scaly on old stems; the branchlets slender, usually armed with many long, slender thorns.

Scattered and uncommon in southern Indiana, usually growing on bluffs or rocky or sandy banks of streams.

Pa. to Ind., southw. to N. C. and Ky.

13. **Crataegus biltmoreàna** Beadle. (*Crataegus intricata* of Eggleston, not Lange, in Britton and Brown, Illus. Flora, ed. 2. fig. 2366, *Crataegus modesta* Sarg., and *Crataegus villicarpa* Sarg.) Map 1113. Leaves ovate-elliptic or nearly orbicular, mostly 3-8 cm long and 2.5-6 cm wide, abruptly or acutely pointed at the apex, abruptly cuneate or rounded at the base, and usually slightly decurrent on slender (1-3 cm long), glandular petioles, coarsely serrate nearly to the base, the lower teeth glandular or gland-tipped, usually incised with 1-3 pairs of short, triangular lobes, thin, dull yellowish green, short-villous or scabrate above and pubescent at least on the veins beneath; flowers 18-22 mm in diameter, in compact, nearly simple, 3-7-flowered, villous corymbs; stamens about 10; anthers pale yellow; calyx lobes villous, conspicuously glandular-serrate or pectinate; fruit subglobose or slightly attenuate at the base, 10-15 mm in diameter, with a large, shallow calyx, pubescent, bronze green or orange red, more or less blotched with russet or brown; nutlets 3-5.

A stout shrub 1-4 m high, with brownish gray, scaly bark, ascending or spreading branches, and stout branchlets at first villous but soon becoming glabrous, olive green or brown the first season, later becoming gray and usually abundantly armed with long, slender thorns.

This has been confused with *Cratageus intricata* Lange, but examination of specimens from the type tree of that species, cultivated in the Botanic Garden at Copenhagen, Denmark, and sent us by A. Lange, shows it to be the much commoner glabrous plant described under number 11.

Rare in Indiana and known only from Lawrence and Vermillion Counties.

Vt. to Mo., southw. to N. C. and Ark.

14. **Crataegus macrospérma** Ashe. (Deam. Trees of Indiana, ed. 2. pl. 89. 1932.) (*Crataegus bella* Sarg., *Crataegus colorata* Sarg., *Crataegus ignea* Sarg., *Crataegus sextilis* Sarg., *Crataegus Egani* Ashe, *Crataegus otiosa* Ashe, *Crataegus tenera* Ashe, and *Crataegus uber* Ashe.) Map 1114. Leaves ovate, mostly 3-7 cm long, 2.5-5 cm wide, acute or acuminate at the apex, obtuse, rounded or subcordate at the base, sharply serrate nearly to the base, usually incised on the upper half or two thirds of the blades with 2-4 pairs of triangular lobes terminating in acuminate, spreading or reflexed teeth, thin, finely scabrate on the upper surface when young, otherwise glabrous; petioles slender, eglandular or with a few small glands; flowers 15-18 mm in diameter, in usually 5-10-flowered, glabrous corymbs; stamens generally 5-10; anthers pink or rose color; calyx lobes entire or slightly serrate toward the base; fruit obovoid, ellipsoid or nearly globose, 7-12 mm thick, 8-14 mm long, bright red and succulent at maturity, often slightly glaucous; calyx small and sessile; nutlets 3-5.

A small tree up to 7-8 m high, or sometimes lower and shrubby, with gray, slightly scaly bark, stiff, erect or spreading branches, and stout, often flexuous branchlets, armed with stout, curved thorns.

General but not common in Indiana, growing in pastures, thickets, and open woods, preferring well drained soils near streams.

Se. Canada to Ill., southw. to N. C. and the mts. of Ky. and Tenn.

15. **Crataegus pruinòsa** (Wendl.) K. Koch. (*Crataegus conjuncta* Sarg., *Crataegus patrum* Sarg., and *Crataegus palustris* Ashe?.) Map 1115. Leaves ovate or elliptic, mostly 4-8 cm long, 2.5-5 cm wide, pointed or short-acuminate at the apex, abruptly contracted, rounded, or subcordate at the base, sharply or coarsely serrate nearly to the base, usually incised with 2-4 pairs of shallow, triangular lobes, firm in texture, glabrous, usually bluish green; petioles slender, a third to half as long as the blades, eglandular or with a few small glands; flowers 18-22 mm in diameter, usually 6-10, in glabrous, nearly simple or somewhat branched corymbs; stamens usually about 20; anthers pink or sometimes pale yellow; calyx lobes lanceolate or narrowly deltoid from a broad base, entire or with a few shallow teeth toward the base; fruit subglobose, depressed-globose, or somewhat pyriform with an attenuate base, often 5-angled, 10-16 mm in diameter, with a broad, shallow, elevated calyx, dull or rarely bright crimson at maturity, or sometimes remaining green with dark dots and blotches, usually with a bloom; flesh thin, remaining hard and dry; nutlets usually 4-5, relatively large.

Sometimes a small tree up to 6-7 m high, or more often an arborescent shrub, with dark gray, scaly bark and intricate ascending or finally spreading branches; the branchlets slender, glabrous, usually armed with many, long, slender or stoutish thorns.

Common and general in Indiana, growing in pastures, thickets, or borders of woods, preferring dry soils along or near streams.

Que. to Man., southw. to N. C. and Ark.

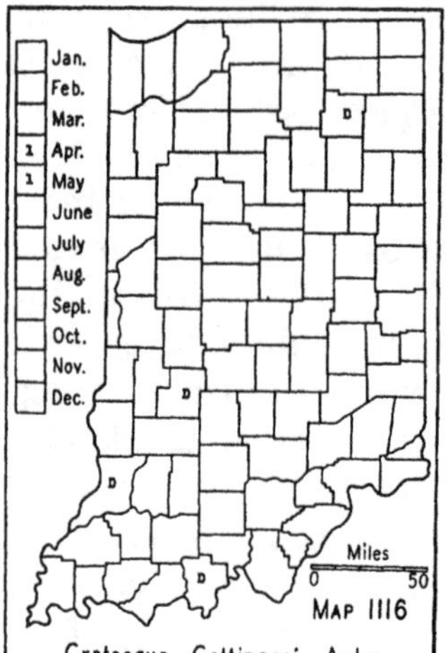

Crataegus Gattingeri Ashe

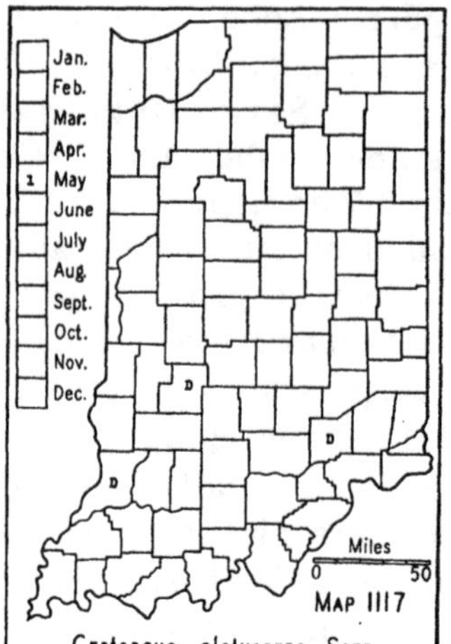

Crataegus platycarpa Sarg.

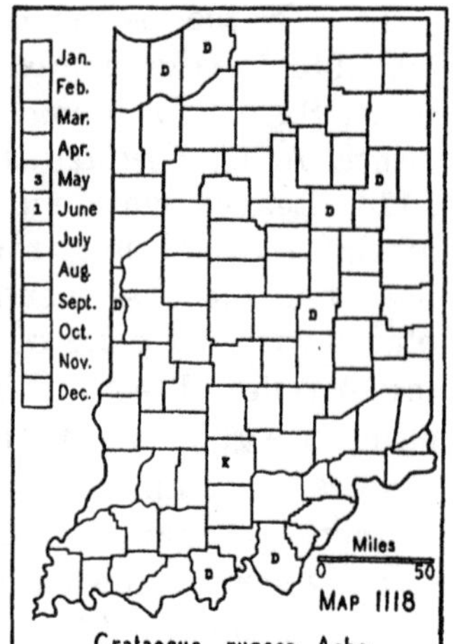

Crataegus rugosa Ashe

16. **Crataegus Gáttingeri** Ashe. (Deam. Trees of Indiana, ed. 2. pl. 93. 1932.) (*Crataegus priva* Ashe, *Crataegus vicinalis* Beadle, and *Crataegus filipes* of Eggleston, not of Ashe.) Map 1116. Leaves ovate or deltoid, variable in size, mostly 2.5-5 cm long, and 1.5-4 cm wide, acute or acuminate at the apex, abruptly narrowed, rounded or on sterile shoots, truncate or cordate at the base, sharply serrate nearly to the base, usually with 2-4 pairs of triangular lobes, the terminal one often wedge-shaped and conspicuously elongated, thin but firm, glabrous, blue green; petioles very slender, half to two thirds the length of the blades; flowers 14-16 mm in diameter, in mostly 3-7-flowered, nearly simple, glabrous corymbs; stamens about 20; anthers pink or rarely white; fruit pyriform, oblong, or nearly globose, but usually attenuate at the base, 7-10 mm thick, 8-12 mm long, with narrow, slightly elevated calyx, dull crimson, slightly pruinose, with thin flesh, remaining firm or hard; nutlets usually 4-5.

A stout shrub or sometimes a small tree up to 4-5 m high, with dark gray, slightly scaly bark, crooked, ascending or spreading branches, and slender, flexuous, glabrous branchlets, armed with numerous slender or stout thorns.

Scattered in southern Indiana, growing in thickets and on borders of woods, usually in well drained soil along streams.

W. Va. to e. Mo., southw. to Ga. and Ark.

16a. **Crataegus Gattingeri** var. **rígida** Palmer, var. nov.[1] (*Crataegus Gattingeri* of Eggleston (Deam. Trees of Indiana, ed. 2: 219-22. pl. 94. 1932), not of Ashe.) This variety differs from the type in the stouter, rigid, flexuous branchlets, and in the short, stout thorns, 1-2 cm long.

Known in Indiana only from Perry County.

Specimens examined: Indiana: Deam no. 27143, a quarter of a mile north of Cannelton, Perry County (type), April 24 and July 22, 1919. Type in herbarium of the Arnold Arboretum. Kentucky: Palmer no. 17716,

[1] A typo differt ramulis crassis rigidis spinis crassis 1-2 cm longis.

open banks and hillsides, sandy soil, Livermore, McLean County, June 2, 1920.

Sw. Ind. and w. Ky.

17. **Crataegus platycárpa** Sarg. (Rept. Missouri Bot. Gard. 19: 92. 1908.) Map 1117. Leaves mostly ovate, 2.5-6 cm long, 2-5 cm wide, acute or short-acuminate at the apex, abruptly narrowed, rounded, truncate or on shoots, sometimes cordate at the base, sharply serrate nearly to the base, usually with 2-4 pairs of shallow, lateral lobes terminating in acuminate teeth, rather thin but firm, sparingly short-villous or scabrate above when young and more or less villous on the veins beneath; petioles slender, a third to half as long as the blades, generally slightly villous and often beset with a few stalked glands; flowers 18-22 mm in diameter, in usually 3-6-flowered, nearly simple, sparsely villous corymbs; stamens about 20 or sometimes fewer; anthers red or pale yellow; fruit subglobose or depressed-globose, 12-16 mm in diameter, 10-15 mm long, bright red or orange red at maturity; calyx broad, shallow, nearly sessile or slightly elevated; flesh thick, becoming succulent; nutlets 3-5.

A tree up to 6-7 m high, with rough, gray bark and ascending or wide-spreading branches, the branchlets slender, armed with numerous long, slender thorns.

Although this species seems to agree most closely with the *Pruinosae* group, the fleshy, bright red fruit and the slight but variable pubescence of the foliage and inflorescence, so uncommon in that group, suggest that it might be a hybrid between some form of the *Pruinosae* and *Crataegus mollis*. See also note under *Crataegus* no. 23.

Indiana specimens are from both dry and moist woodland.

Southern Ind. to ne. Ark.

18. **Crataegus rugòsa** Ashe. (*Crataegus onusta* Ashe and *Crataegus superata* Sarg.) Map 1118. Leaves ovate, broadly ovate or deltoid, pointed or short-acuminate at the apex, rounded, truncate or subcordate at the base or sometimes deeply cordate on shoots, sharply serrate nearly to the base, usually with 2-4 pairs of small, lateral lobes; petioles slender, a third to two thirds the length of the blades, firm at maturity, glabrous, usually yellowish green; flowers 20-22 mm in diameter, usually in 3-6-flowered, glabrous, nearly simple corymbs; fruit subglobose or depressed-globose, 14-17 mm in diameter, with broad, shallow, slightly elevated calyx, becoming dull red, with thin flesh, remaining hard and dry; nutlets usually 4-5.

A tree up to 6-8 m high, or often a stout arborescent shrub, with dark, scaly bark and stout, ascending branches; branchlets often flexuous and armed with numerous, long, stout thorns.

Crataegus rugosa is closely related to *Crataegus pruinosa* and apparently intergrades with it, although it often looks entirely distinct in its broader leaves and somewhat larger flowers and fruit.

Generally distributed but not common in Indiana; found in thickets, pastures, and borders of woods, usually in well drained soil.

N. Y. to Iowa, southw. to N. C. and Mo.

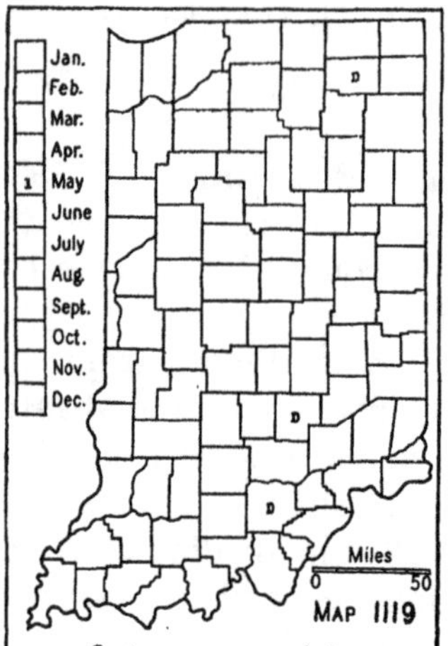

Crataegus prona Ashe

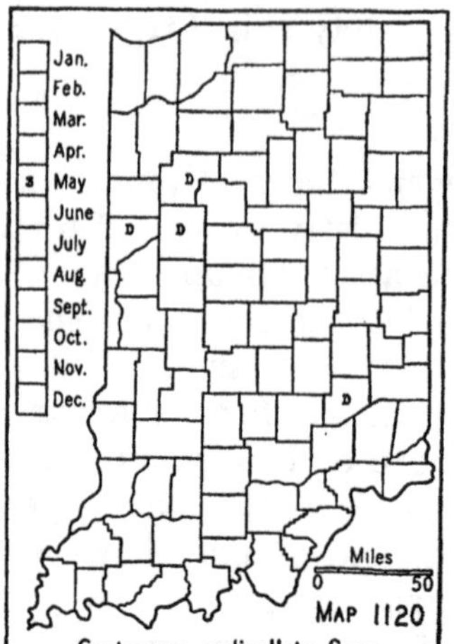

Crataegus pedicellata Sarg.

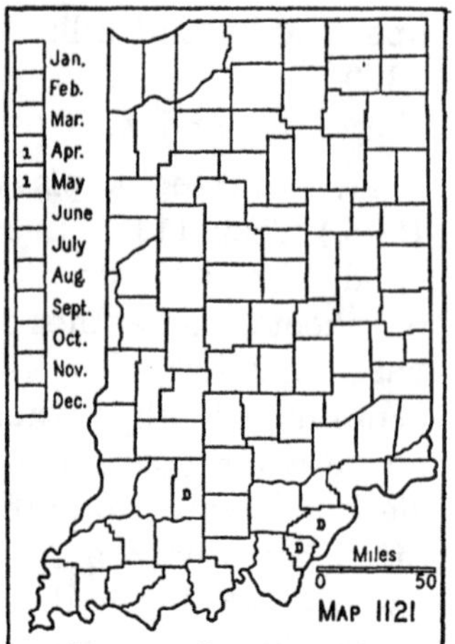

Crataegus Putnamiana Sarg.

19. **Crataegus pròna** Ashe. (*Crataegus allecta* Sarg. and *Crataegus gravis* Ashe.) Map 1119. Leaves ovate, 3-7 cm long, 2.5-6 cm wide, acute or short-acuminate at the apex, abruptly narrowed or rounded at the base, or sometimes truncate or subcordate on shoots, sharply serrate nearly to the base, usually with 2-4 pairs of obscure or shallow, triangular, lateral lobes, firm, sparsely short-villous or scabrate on the upper surface when young, glabrous at maturity; petioles slender, from a third to half the length of the blades, eglandular or with a few glands; flowers 18-20 mm in diameter, in mostly 6-10-flowered, glabrous, simple or slightly compound corymbs; stamens 10 or fewer; anthers pink or rose; fruit usually oblong or obovoid, 8-10 mm thick, 10-14 mm long, becoming crimson or orange red, with dark or russet blotches, flesh becoming mellow; calyx small and sessile or nearly so; nutlets 3-5.

A tree up to 6-7 m high, or often a stout shrub, with gray, slightly scaly bark, ascending or spreading branches, and stoutish, often flexuous, glabrous branchlets armed with numerous, long, curved thorns,

This species grows in fields and thickets in rocky or well drained soil and in dry soil on wooded slopes.

Ont. and Pa. to Mich. and Ind.

20. **Crataegus pedicellàta** Sarg. (Deam. Trees of Indiana, ed. 2. pl. 97. 1932.) (*Crataegus coccinea* of Eggleston, not of L., *Crataegus acclivis* Sarg., *Crataegus arcuata* Ashe, *Crataegus pura* Sarg., and *Crataegus sertata* Sarg.) Map 1120. Leaves ovate or broadly ovate, mostly 4-8 cm long, 3.5-7 cm wide, pointed or acuminate at the apex, rounded, truncate or subcordate at the base, sharply and rather finely serrate, usually with 3-5 pairs of small or obscure lateral lobes terminated by acuminate, spreading or reflexed teeth, scabrate or short-villous above and sometimes slightly villous on the veins beneath when young, thin and barely firm at maturity and then glabrous on both surfaces or with slight traces of pubescence beneath; petioles slender, a third to half the length of the blades, slightly villous or glabrous; flowers 16-22 mm in diameter, in compound, mostly

6-12-flowered, more or less villous corymbs; calyx lobes lanceolate, usually glandular-serrate; stamens 5-10; anthers pink or red; fruit oblong, slightly pyriform or nearly globose, 10-14 mm thick, 10-16 mm long, glabrous, bright crimson or scarlet at maturity, with soft, mellow flesh; nutlets 3-5.

A tree 6-8 m high, or often a stout arborescent shrub, with gray, slightly scaly bark, and numerous ascending or spreading branches, forming a conical or round crown; branchlets rather stout, often flexuous, and armed with numerous stout thorns.

Uncommon in Indiana and found in thickets, pastures, and borders of woods. Indiana specimens are mostly from high, wooded banks of streams.

Que. to Pa. and Ill.

20a. **Crataegus pedicellata** var. **álbicans** (Ashe) Palmer. (Dole. Flora of Vermont, 154. 1937.) (*Crataegus albicans* Ashe and *Crataegus cristata* Ashe.) Differs from the typical form in the glabrous corymbs and petioles and in the generally broader leaves.

Known in Indiana from La Porte, Steuben, and White Counties and found in habitats similar to those of the species.

N. Y. and Pa. to Ill.

21. **Crataegus Putnamiàna** Sarg. (Deam. Trees of Indiana, ed. 2. pl. 96. 1932.) (*Crataegus coccinioides* of Eggleston, not of Ashe.) Map 1121. Leaves ovate or, on shoots, deltoid in outline, acute or short-acuminate at the apex, rounded, truncate or subcordate at the base, 4-8 cm long, 3-7 cm wide, sharply and unevenly serrate with spinulose teeth nearly to the base, incised, and generally with 3-4 pairs of shallow, lateral lobes, the lowest pair sometimes enlarged and triangular on shoots, thin but firm at maturity, scabrate above when young, and permanently pubescent at least on the veins beneath; petioles slender, a third to half as long as the blades, slightly villous and usually with stalked or sessile glands; flowers 18-22 mm in diameter, in simple or rarely branched, slightly villous or glabrate corymbs; stamens about 20; anthers usually pink, sometimes white; fruit subglobose or depressed-globose, full and rounded, 12-17 mm in diameter, bright red, sometimes slightly pruinose, with thick flesh, becoming mellow but firm; calyx broad and shallow, slightly elevated; calyx lobes lanceolate, glandular-serrate, usually persistent on the fruit; nutlets 4-5, usually 5.

A tree up to 4-5 m high, or sometimes an arborescent shrub with gray, slightly scaly bark, ascending or spreading branches, and stoutish, glabrous branchlets usually sparingly armed with stout, purple thorns.

This species has been confused with *Crataegus coccinioides,* which differs from it in its glabrous, broader leaves with crisped margins, its larger flowers in glabrous corymbs, and in its larger, bright crimson, usually angular fruit with a very large calyx.

Indiana specimens are from the unglaciated area and are found in Clark and Floyd Counties in the "knobs" in open woodland, and in Martin County on a wooded slope.

Northern Ky., s. Ohio, and s. Ind.

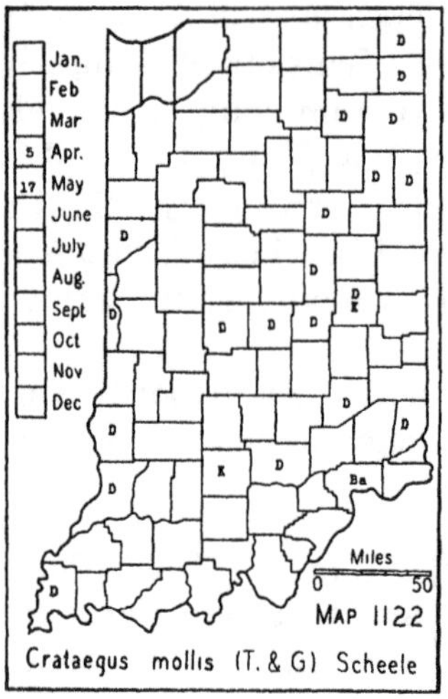

Crataegus mollis (T. & G) Scheele

Crataegus Kelloggii Sarg.

Crataegus Phaenopyrum (L. f.) Medic.

22. **Crataegus móllis** (T. & G.) Scheele. (Deam. Trees of Indiana, ed. 2. pl. 98. 1932.) (*Crataegus lanigera* Sarg., *Crataegus lasiantha* Sarg., *Crataegus umbrosa* Sarg., and *Crataegus valens* Ashe.) Map 1122. Leaves ovate, ellipsoid, or nearly orbicular, mostly 5-8 cm long, and 4-6 cm wide, acute at the apex, rounded or truncate at the base or on shoots, rarely subcordate, coarsely serrate nearly to the base, usually with 3-5 pairs of broad, shallow, lobes, firm to subcoriaceous at maturity, short-villous or scabrate above when young, permanently pubescent at least on the veins beneath; petioles stout, a third to half as long as the blades, pubescent, eglandular or rarely with a few scattered glands; flowers 20-24 mm in diameter, in compact, compound, mostly 6-16-flowered, densely tomentose corymbs; stamens about 20; anthers usually cream color, rarely pink; fruit subglobose, depressed-globose, or slightly oblong or pyriform, 15-20 mm in diameter, bright crimson or scarlet, pubescent at least toward the base; flesh thick, firm but mellow, strongly flavored and edible; calyx broad and shallow, nearly sessile; calyx lobes glandular-serrate, persistent or tardily deciduous; nutlets normally 5.

A tree up to 10-12 m high, with a trunk sometimes 3 dm in diameter; bark dark gray, rough and somewhat furrowed; branches ascending or wide-spreading, usually forming a low conical crown; branchlets villous the first season, soon glabrate, slender, nearly unarmed or sometimes armed with stout, curved thorns.

Common and generally distributed in Indiana, growing in open woods and open grounds, usually in fertile soil along streams.

Southern Ont. and Mich. to S. Dak., southw. to Tenn. and e. Okla.

22a. **Crataegus mollis** f. **dumetòsa** (Sarg.) Palmer. (*Crataegus dumetosa* Sarg.) This form differs from typical *Crataegus mollis* in the narrower, ovate or elliptic, undivided or obscurely lobed leaves, narrowed or rounded at the base and acuminate into the slightly winged petioles. It has been found in Indiana in Marion, Shelby, and Vermillion Counties, growing with the typical form.

23. **Crataegus Kélloggii** Sarg. (Sargent. Manual of Trees of North America, ed. 2: 475. fig. 432. 1922.) Map 1123. Leaves ovate, rhombic-ovate or suborbicular, mostly 2.5-6 cm long, 2-5 cm wide, rounded or abruptly pointed at the apex, abruptly narrowed, rounded or truncate at the base, sharply serrate nearly to the base, usually with 3 or 4 pairs of shallow, obscure or rounded lateral lobes, firm to subcoriaceous at maturity, scabrate above and more or less villous on the veins beneath while young, becoming glabrate or remaining slightly villous beneath; petioles slender, a third to half the length of the blades, somewhat villous or tomentose; flowers 14-17 mm in diameter, in compact, compound, villous or thinly tomentose, mostly 5-10-flowered corymbs; stamens about 20; anthers white or tinged with pink; fruit subglobose or short-ovoid, 14-20 mm in diameter, bright yellow (according to description) or red, punctate, with a slight bloom; nutlets usually 5.

A tree up to 6-7 m high, with dark, rough, deeply furrowed bark, and ascending or wide-spreading branches, forming a low, conical crown; branchlets slender, unarmed or sparingly armed with stoutish or slender purple thorns.

Known in Indiana only from Wells County where a tree was found in a clearing on the land of the Erie Stone Company about 2 miles northwest of Bluffton.

Ind. to Mo.

Crataegus Kelloggii is probably a hybrid between *Crataegus Margaretta* and *Crataegus mollis,* and may be looked for where those two species are found together. There is much variation in the characters of the fruit, foliage, and flowers, as is to be expected in hybrids. The fruit of the type tree was described as bright yellow, an unusual color in the genus and probably exceptional in this species, although it is interesting to note that a yellow-fruited form of both parent species has been found.

Crataegus mollis appears to hybridize with other species. It has been suggested that *Crataegus platycarpa* may be a hybrid between this and some species of the *Pruinosae* group (possibly *Crataegus rugosa*), and forms have been found in Lawrence County that appear to be hybrids between *Crataegus mollis* and *Crataegus pruinosa,* and possibly also between *Crataegus mollis* and *Crataegus punctata.*

24. **Crataegus Phaenopỳrum** (L. f.) Medic. (Deam. Trees of Indiana, ed. 2. pl. 99. 1932.) (*Crataegus cordata* Ait.) WASHINGTON THORN. Map 1124. Leaves ovate to deltoid in outline, sometimes appearing 3-lobed, mostly 2-6 cm long, 2-5 cm wide, acute or acuminate at the apex, rounded, truncate or cordate at the base, serrate with broad, shallow teeth, usually with 1-3 pairs of lateral lobes, the lowest pair often enlarged and with spreading, acuminate points, firm at maturity, glabrous, glossy on the upper surface; petioles very slender, a third to two thirds the length of the blades; flowers small, 10-12 mm in diameter, in glabrous, compound, mostly 10-30-flowered corymbs; stamens about 20; anthers pale yellow; fruit subglobose, 5-7 mm in diameter, long persistent, in many-fruited clusters, bright scarlet, becoming succulent at maturity; calyx relatively

large, often entirely deciduous leaving the tops of the nutlets exposed; nutlets usually 5.

A tree up to 10 m high, with brown gray, scaly bark, numerous ascending or spreading branches, forming a low, conical crown, and slender branchlets usually abundantly armed with slender thorns.

Known in Indiana only from Wayne County, where it has possibly escaped, but it should be sought as a native plant in the southern counties. This species is highly ornamental and desirable for planting on account of its abundant flowers and the brilliant color of the fruit which is produced in large, pendulous clusters, remaining on the tree until late in the season.

Va. and N. C. to Mo.

25. **Crataegus Calpodéndron** (Ehrh.) Medic. (Deam. Trees of Indiana, ed. 2. pl. 85. 1932.) (*Crataegus tomentosa* of authors but perhaps not of L. and *Crataegus structilis* Ashe.) PEAR HAW, SUGAR HAW. Map 1125. Leaves ovate, oblong-ovate or elliptic in outline, mostly 4-8 cm long, and 3-5 cm wide, pointed or short-acuminate at the apex, acute or abruptly contracted at the base and attenuate into winged petioles, sharply serrate on the upper three fourths of the blades, usually with 3-5 pairs of obscure or shallow triangular lobes, mostly above the middle, firm to subcoriaceous and with veins impressed above at maturity, scabrate above when young, and permanently pubescent at least on the veins beneath; petioles usually 1-2 cm long; flowers 12-15 mm in diameter, in loose, compound, tomentose, mostly 10-20-flowered corymbs; stamens about 20; anthers pink; fruit oblong, ovoid or nearly globose, 7-10 mm in diameter, pubescent, scarlet or orange red, flesh thin, becoming mellow; calyx relatively large, elevated; calyx lobes lanceolate, glandular-serrate, reflexed or often deciduous; nutlets 2-3, deeply pitted on the ventral surfaces.

A small tree up to 6 m high, or often an arborescent shrub, with dark, slightly scaly bark and erect or ascending branches, forming a narrow pyramidal crown; branchlets usually villous when young, soon glabrate, olive brown, becoming gray, nearly unarmed or sparingly armed with long, slender thorns.

Frequent and generally distributed in Indiana, growing in thickets or open woods, usually along streams or lakes.

Southern Ont. to Minn., southw. to N. C. and Ark.

26. **Crataegus succulénta** Schrader. (Deam. Trees of Indiana, ed. 2. pls. 83 and 84. 1932.) (*Crataegus ensifera* Sarg., *Crataegus neofluvialis* Ashe, and *Crataegus vegeta* Sarg.) Map 1126. Leaves oblong-ovate, elliptic or rhombic, mostly 5-8 cm long, and 2.5-6 cm wide, acute or short-acuminate at the apex, gradually or abruptly narrowed at the base and attenuate into short (1-2 cm), winged petioles, finely serrate except toward the base, usually with 2-5 pairs of shallow or obscure lateral lobes, coriaceous or subcoriaceous and with veins conspicuously impressed above at maturity, dark green and scabrate above when young, much paler and permanently pubescent beneath; flowers 12-15 mm in diameter, usually 15-30, in compound, villous, corymbs; stamens usually about 20; anthers

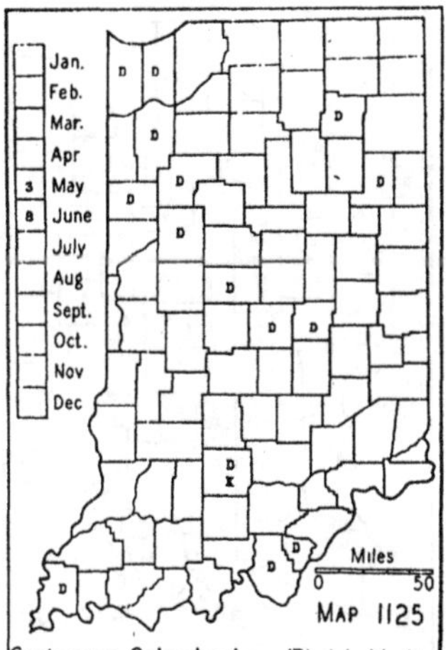

Crataegus Calpodendron (Ehrh.) Medic

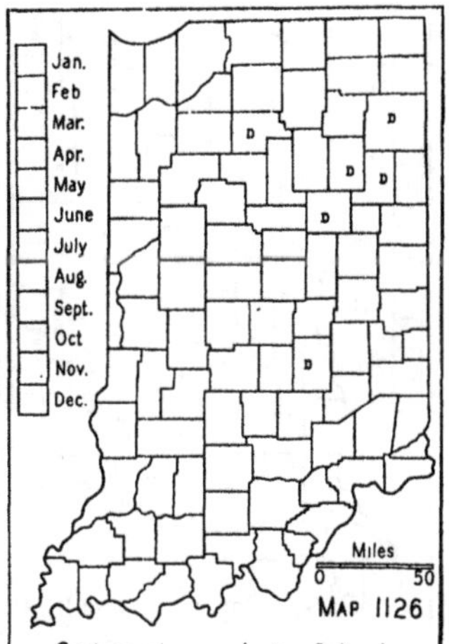

Crataegus succulenta Schrader

Crataegus incaedua Sarg.

pink or red; fruit subglobose, 9-12 mm in diameter, bright red and succulent at maturity; calyx slightly elevated; calyx lobes glandular-serrate, reflexed in fruit; nutlets 2-3, deeply pitted on the ventral surfaces.

A stout shrub or rarely a small tree up to 6-8 m high, with dark gray, scaly bark and stout ascending or slightly spreading branches; branchlets glabrous or slightly villous when young, becoming light brown or chestnut-colored at the end of the first season and finally gray, rather stout and armed with numerous long (5-9 cm), curved thorns.

Infrequent but generally distributed in Indiana, growing in thickets or on banks or bluffs of streams.

Southeastern Canada to Iowa, southw. to N. C. and Mo.

27. **Crataegus incaèdua** Sarg. (*Crataegus pudens* Sarg.) Map 1127. Leaves ovate or elliptic, mostly 3-7 cm long, and 2-5 cm wide, obtuse, acute or short-acuminate at the apex, cuneate at the base and tapering into short (0.3-1 cm) petioles, coarsely serrate except near the base, undivided except rarely on shoots, firm to subcoriaceous and with veins slightly impressed above at maturity, dark green and scabrate above when young, paler and pubescent beneath; flowers 15-18 mm in diameter, usually 8-20, in lax, compound, villous corymbs; stamens usually 10-15; anthers pale yellow; fruit subglobose or oblong, 8-12 mm in diameter, red at maturity, sometimes slightly glaucous; calyx lobes serrate or glandular-serrate, reflexed; nutlets 2-3, usually 2, sometimes with shallow pits on the ventral surfaces.

A tree up to 6-7 m high, with pale brown gray bark and ascending or spreading branches, forming a low, conical crown; branchlets villous the first season, becoming gray, usually armed with numerous, long, curved thorns.

Crataegus incaedua is probably a hybrid between *Crataegus Calpodendron* and *Crataegus crus-galli* or some species of the *Crus-galli* group.

Known in Indiana only from Harrison County, where it was found along

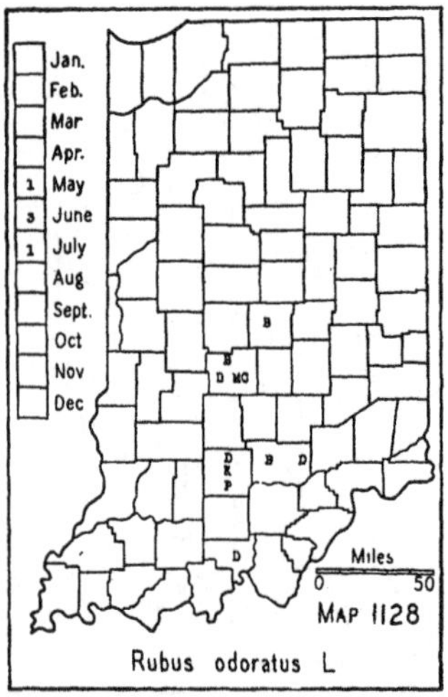

MAP 1128
Rubus odoratus L.

MAP 1129
Rubus pubescens Raf.

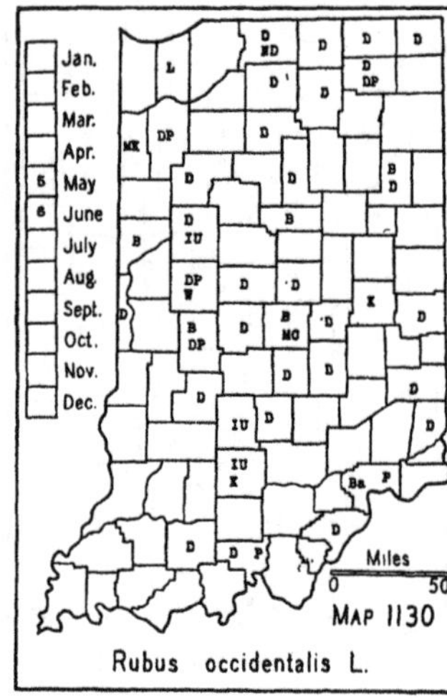

MAP 1130
Rubus occidentalis L.

a small creek at the base of a rocky, wooded slope about a mile south of Corydon Junction.

Ind. to Mo.

Excluded Species

The following species and varieties of *Crataegus,* in addition to those mentioned or disposed of as synonyms in the regular text, have been reported as having been found in Indiana or of having a range extending into the state:

1. C. **Bárrettiana** Sarg.
2. C. **berberifòlia** T. & G.
3. C. **denària** Beadle
4. C. **fecúnda** Sarg.
5. C. **tràhax** Ashe
6. C. **ovàta** Sarg.
7. C. **stramínea** Beadle
8. C. **Bòyntoni** Beadle
9. C. **Dódgei** Ashe
10. C. **gracílipes** Sarg.
11. C. **ígnea** Sarg.
12. C. **parviflòra** Sarg.
13. C. **roanénsis** Ashe
14. C. **basílica** Beadle
15. C. **grácilis** Sarg.
16. C. **beàta** Sarg.
17. C. **Jésupi** Sarg.
18. C. **Híllii** Sarg.
19. C. **sejúncta** Sarg.
20. C. **víllipes** Ashe
21. C. **Prínglei** Sarg.
22. C. **coccínea var. Ellwangeriána** (Sarg.) Eggl.
23. C. **flàva** Ait.
24. C. **spathulàta** Michx.
25. C. **Bràinerdi** Sarg.
26. C. **Déwingii** Sarg.
27. C. **macracántha** Lodd.

These species will be referred to by number so far as is practicable and where fuller discussion is unnecessary.

A re-examination of the specimens shows that numbers 1 and 5 can be referred to *C. crus-galli.* Numbers 2, 3, 7, 8, 13, 14, 17, 21, 22, 23, 24, 25, 26, and 27, are all out of range for Indiana, as these species are understood in this treatment, and the report of their occurrence is based upon erroneous determination of material. *Crataegus Engelmannii* has been considered identical with *C. berberifolia* by some botanists, although the two appear to be distinct; *C. Engelmannii* might be expected to occur in southern Indiana, although no specimens have been seen, and it was probably this species that was reported as *C. berberifolia*. *C. denaria*, so far as Indiana reports go, is probably referable to *C. acutifolia* as treated here, *C. straminea* to *C. rubella*, *C. macracantha* to *C. succulenta*, and *C. roanensis* to *C. macrosperma*. Reports of numbers 10, 11, and 12 were probably also based upon collections of *C. macrosperma*. Numbers 4, 6, 9, and 18 may ultimately be found in the state, although no authentic specimens of them have been seen; of these *C. ovata* is probably only a form or variety of *C. viridis*; *C. Dodgei* has often been confused with *C. Margaretta*, certain forms of which it closely resembles, as well as with *C. chrysocarpa*, a western species, but as it is common in parts of Michigan, it may be expected to extend into northern Indiana. Number 18 (*C. Hillii*) is found in northern Illinois and may be expected to extend into the northwestern counties, although the specimens previously identified as this species seem on re-examination to be *C. Putnamiana*, as treated here. Numbers 19 and 20 should probably be referred to *C. pedicellata*; numbers 15 and 16 are probably identical and may represent a hybrid between *C. macrosperma* and *C. pruinosa* or a related species.

3353. RÙBUS [Tourn.] L. RASPBERRIES AND BLACKBERRIES

[Bailey. Gentes Herbarum **1**: 139-200. 1923; **1**: 201-306. 1925; **2**: 269-423. 1932; **2**: 442-480. 1932; **3**: 117-148. 1933; **3**: 245-271. 1934.]

It has been my good fortune to have had all of my *Rubus* specimens pass through the hands of L. H. Bailey who has made an intensive and critical study of the species of this genus for more than forty years. He says: "Undoubtedly *Rubus* is the most baffling of the genera of North American sporophytes." Since I regard him as our foremost authority on the subject I have accepted his determinations and I am following his treatment of the species throughout. I am using his keys wherever it is possible. This study of the genus in Indiana is based upon my collection of 638 specimens.

Bailey has defined a few terms of habit of growth which I quote. "A blackberry is said to be *erect* when the general direction of the canes is perpendicular even though they may curve a little at the top. It is *ascending* when the general direction is upward but perhaps oblique or much curved. A cane is *arching* when it takes the general direction of a semicircle even though its tip or growing end may not reach the ground. It is *prostrate* when it lies prone on the ground. A prostrate cane may have *fallen* when carried to the ground by weight as of leaves, fruit, vines growing over it, or as a result of injury. Erect or ascending species may have fallen canes. A cane is *trailing* when it grows flat on the ground by habit, continuing its extension in this direction; the word is commonly erroneously employed for a cane that has merely fallen or is prostrate, and confusion results. A prostrate plant may not be a trailer. The true trailers among the blackberries usually strike root at nodes or tip." He has introduced the word *primocane* for the first year's growth and *floricane* for the fruiting or second year's growth. He suggests, also, that the direction of growth of the floricane be shown graphically on the label. In making an herbarium specimen of *Rubus* the following should be collected: the primocane, and of the floricane, at least a fifteen-inch section of the base and an equal length of the tip. It is desirable to collect a section of the longest lateral branch of the floricane if it is well developed. A note on the direction of growth of the floricanes and their range in height should be made.

Some authors believe that the species of *Rubus* freely hybridize and Brainerd, who first named my *Rubus*, named several of my specimens hybrids and I reported them as such. Bailey, whose determinations I have followed, has referred these specimens to species. These hybrids and their disposal are given in the list of excluded species.

Plants wholly unarmed (without bristles or prickles).
  Leaves simple, 3-5-lobed; petals purple............................1. *R. odoratus.*
  Leaves 3- (-5) foliolate; petals white............................2. *R. pubescens.*
Plants more or less armed with bristles or prickles or both.
  Leaves whitish beneath; ripe fruit easily separating from the receptacle as a whole. (Raspberries.)
    Floricanes arching, dark purple; primocanes and branches of the floricanes glaucous, armed with prickles but lacking long, bristlelike glandular hairs; leaves 3-foliolate or rarely pedately 5-foliolate; inflorescence corymbiform.
      Fruit black..3. *R. occidentalis.*
      Fruit amber color............................3a. *R. occidentalis* f. *pallidus.*
    Floricanes erect (sometimes old ones recurving), dark purple or reddish; primocanes and branches usually glaucous, armed with prickles and with or without long, bristlelike glandular hairs; fruit red at maturity; inflorescence a short raceme.
      Calyx lobes long-attenuate at the apex, more than 1 cm long; under surface of leaflets more or less sparsely covered with long red glandular hairs...... ....4. *R. phoenicolasius.*
      Calyx lobes acuminate at the apex, less than 1 cm long; under surface of leaflets without red glandular hairs.
        Primocanes and floricanes with prickles but lacking long, bristlelike glandular hairs; inflorescence without stipitate glands. (See excluded species no. 354, p. 1061.)..*R. idaeus.*
        Primocanes and floricanes with both prickles and long, bristlelike glandular hairs; inflorescence with stipitate glands.
          Surface of the canes, beneath the prickles and glandular hairs, more or less densely pubescent......................5. *R. idaeus* var. *canadensis.*
          Surface of the canes, beneath the prickles and glandular hairs, not pubescent..............................5a. *R. idaeus* var. *strigosus.*
  Leaves green beneath; ripe fruit not separating from the receptacle.
    Floricanes trailing and rooting more or less at the tips; flowering branches arising more or less vertically; flowers mostly with ascending pedicels; primocanes at first erect, becoming prostrate. (Dewberries.)
      Canes, branches, and petioles usually more or less densely retrorsely hispid with stiff, brown hairs, sometimes the branches and petioles glabrous or nearly so (prickles lacking).
        Petals and stamens 5.......................................6. *R. hispidus.*
        Petals and stamens 10.......................6a. *R. hispidus* f. *pleniflorus.*
      Canes, branches, and petioles more or less prickly, rarely with a few bristles.
        Pedicels glandless.
          Plants normally stout, with stiff, woody, long-trailing primocanes which are usually not conspicuously scaly-bracted at base; leaves commonly firm and coriaceous when growing in the open.............7. *R. flagellaris.*

Plants normally slender, relatively short or else comparatively weak and often with an herbaceous appearance, the bases of young primocanes bearing scalelike caducous bracts; leaves thin and soft; primocane leaflets usually 3; flowers mostly solitary, large, long-pediceled, with large tomentose calyx lobes, reflexed at full anthesis...8. *R. Enslenii.*

Pedicels with stalked glands.

Primocanes glandless, with 3-foliolate leaves, prickles small, few, and of equal size; leaflets soft-pubescent beneath; leaves of floricanes similar to those of the primocanes but less tomentose beneath, prickles small, few, sometimes the stem nearly unarmed; flowers few, 3-5, on upright pedicels; calyx lobes ascending........................9. *R. centralis.*

Primocanes usually with stalked glands (sometimes without them) and with prickles of two sizes and about 5 per cm; leaves 3-5-foliolate; floricane with two kinds of prickles and with scattered, stout glands, flowers mostly 6 or fewer on elongate pedicels which are prickly and glandular; petals elliptic; whole plant much more armed than the preceding. ....10. *R. Deamii.*

Floricanes erect, ascending or arching (rarely diffuse). (Blackberries.)

Canes erect or diffuse, glabrous, unarmed or with a few, widely scattered, weak prickles.

Canes erect. (See excluded species no. 349, p. 1061.)............*R. canadensis.*

Canes diffuse. (See excluded species no. 350, p. 1061.)...*R. canadensis* var. *Randii.*

Canes ascending or arching (at least the floricanes), well armed with stout prickles and more or less pubescent or glandular or both.

Inflorescence and petioles bearing many prominent glandular hairs; characteristic well developed flower-clusters long-racemiform with continuing axis; pedicels (except the basal ones) strongly divaricate.

Axis of well developed flower-clusters long and open, without interspersed foliage, the leaves being only at the base of the cluster; rachis prominently continuous and pedicels divaricate.........11. *R. allegheniensis.*

Axis of well developed flower-clusters leafy and shorter. Bailey says, in comparison with the preceding species, that it "lacks the narrow long-stalked leaflets, is less glandular, canes more terete and lacking the strong angles, broad-petaled flowers on very long slender pedicels (at least the lowest flowers in cluster) subtended by prominent outstanding bracts."..12. *R. impos.*

Inflorescence, petioles and other parts not bearing prominent glandular hairs, if glandular hairs are present, then few and not very large; flower-clusters various, but not long-racemiform as a rule.

Main flower-clusters standing well above the foliage and of the short-racemiform type with many flowers; floral leaves mainly at or near the base, the plant therefore representing a floriferous rather than a leafy appearance; flowers of medium size to small, with narrow petals (except in *R. pergratus*), spreading in anthesis.

Primocane leaflets of the oblong or narrow order and not cordate, sometimes small, mature leaves likely to have prominent, closely parallel side-veins; inflorescence not characteristically of the long-racemiform type.

Inflorescence interspersed with prominent simple leaves...13. *R. laudatus.*

Inflorescence without prominent simple leaves.

Leaflets of primocanes narrow, of a broad-lanceolate or ovate-lanceolate order, with curved, tapering sides, only thinly pubescent beneath; flower-clusters on unarmed pedicels or bearing only a few, weak prickles; plant not very prickly...14. *R. argutus.*

Leaflets of primocanes broad, of the ovate type, distinctly soft-pubescent beneath; flower-clusters with 8 or more flowers (in well-developed specimens), the pedicels stout unless grown in the shade, often prickly and sometimes glandular; strongly prickly plants of robust habit with thick canes..15. *R. ostryifolius.*

Primocane leaflets, at least the terminal one, broad-ovate and mostly subcordate at the base, sometimes caudate-acuminate (sterile floricane laterals should not be mistaken for primocanes); flower-clusters in characteristic forms elongated with divaricate pedicels on upper parts of plant but often in indefinite leafy clusters on lower parts. (See excluded species no. 358, p. 1062.)................*R. pergratus.*

Main flower-clusters short and somewhat hidden in the foliage, not of the long-racemiform kind or of the ascendate kind but rather corymbiform, the lower pedicels likely to be long; leaflets commonly broad and heavy; flowers prevailingly large with broad, rounded petals. (The lower flower-clusters of *R. pergratus* may seem to belong here.)

Leaflets mostly or all of a cuneate-obovate type; canes erect or nearly so, beginning to curve above the middle; prickles rather few and mostly less than 5 mm long..........................16. *R. impar.*

Leaflets not mostly cuneate-obovate, especially on the primocanes; canes arching, usually beginning to curve below the middle and the tips often touching the ground; prickles usually stouter, more curved, and often more than 5 mm long.

Floral leaflets, or leaves on flowering laterals, with broad, triangular or even obtuse serratures or teeth or, if sharp, then fine and close, not cut-toothed or jagged with narrow long teeth..17. *R. frondosus.*

Floral leaflets or leaves characteristically very strongly sharp-serrate to laciniate-dentate or jagged, the serratures or teeth narrow and deep, leaf blade usually narrow or else long-pointed; primocane leaflets on the broad order and mostly large.........18. *R. abactus.*

1. **Rubus odoràtus** L. FLOWERING RASPBERRY. Map 1128. In Indiana this species is restricted to the rocky wooded slopes of the high banks of a few streams in the counties shown on the map. I think it prefers a slightly acid soil but when transplanted to a neutral soil it becomes luxuriant.

N. S. to Mich., southw. to Ga. and Tenn.

2. **Rubus pubéscens** Raf. (Rhodora **11**: 236. 1909.) (*Rubus triflorus* Richardson of Gray, Man., ed. 7 and of Britton and Brown, Illus. Flora, ed. 2.) Map 1129. Restricted to the lake area where it is generally found in tamarack bogs and rarely in low, mucky woods.

Lab. to Alaska, southw. to n. N. J.?, Pa., Iowa, and Nebr.

3. **Rubus occidentàlis** L. COMMON BLACKCAP RASPBERRY. Map 1130. This species is a native of every county of the state, being infrequent to frequent throughout. It is found in almost all kinds of habitats but prefers moist situations.

N. B., s. Que. to Minn., southw. to Ga. and Mo.

3a. **Rubus occidentalis** f. **pállidus** (Bailey) Robinson. I have this yellow-fruited form from only Lagrange and Owen Counties; I saw a clump in the northeastern part of Steuben County but was not able to collect it.

Rubus phoenicolasius Maxim

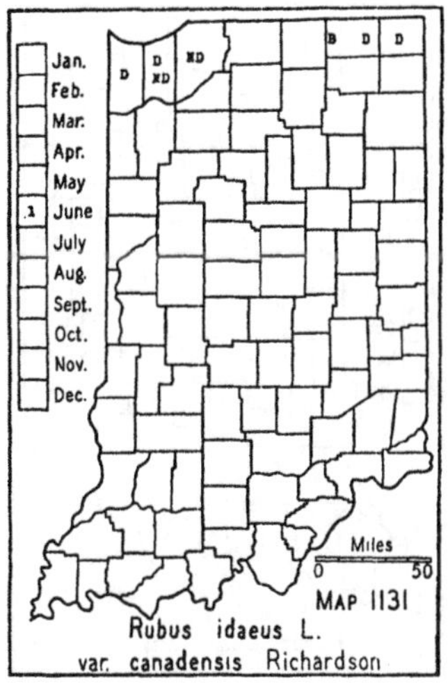

Rubus idaeus L. var. canadensis Richardson

Rubus idaeus var. strigosus (Michx.) Maxim.

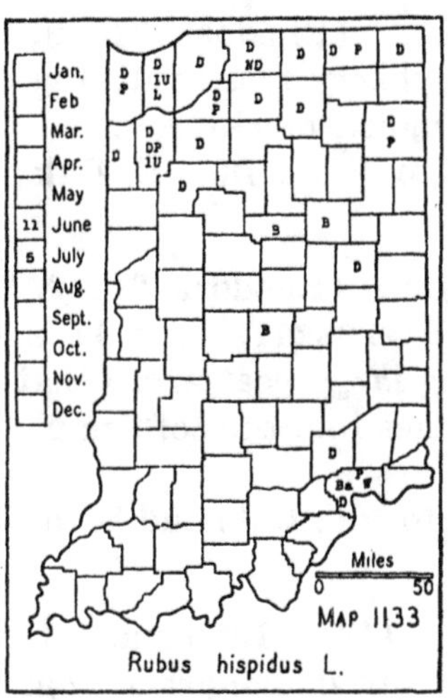

Rubus hispidus L.

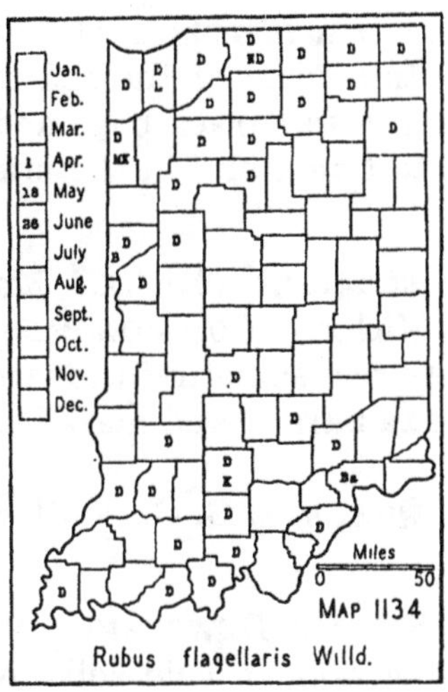

Rubus flagellaris Willd.

Rubus Enslenii Tratt

4. RUBUS PHOENICOLÀSIUS Maxim. WINEBERRY. Map 1130-1. Miss Edna Banta found this species in 1935 to be well established in Crow Hollow near Hanover, Jefferson County, and says she first observed it there in 1924. In 1932 R. C. Friesner found it established on a hillside near Marengo Cave, Crawford County. In 1938 Wm. B. Barnes sent me a specimen from the T. C. Harp farm in sec. 4 of McCameron Township, Martin County. He informs me that it is well established in deep wooded ravines near and in the vicinity of Salem Church. Since the woods about there are in the Resettlement Area, they will be protected from fire and grazing and there is little doubt that it will persist there indefinitely. It has been reported from three counties in Ohio and probably has a wider distribution in Indiana than our records show. It was introduced into the United States in 1876 and has already escaped in the eastern part of the United States.

Nat. of Korea, Japan, and n. China.

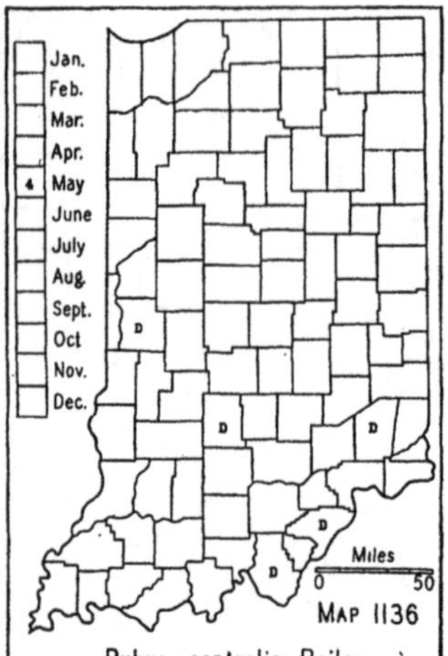

Rubus centralis Bailey

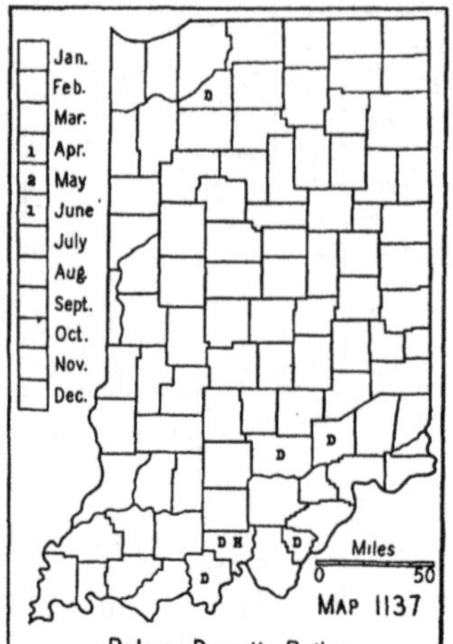

Rubus Deamii Bailey

Rubus allegheniensis Porter

5. **Rubus idaèus** L. var. **canadénsis** Richardson. (Rhodora 21: 97. 1919.) Map 1131. I have this form of the red raspberry from only four counties where it was found in tamarack bogs and moist, mucky soil.

Lab. to Alaska, southw. to Conn. and in the mts. to N. C., Ind., S. Dak., and Colo.; also in e. Asia.

5a. **Rubus idaeus** var. **strigòsus** (Michx.) Maxim. (Rhodora 21: 96. 1919.) (*Rubus strigosus* Michx.) COMMON RED RASPBERRY. Map 1132. Found throughout the lake area, sometimes covering large peat areas that have just passed out of the tamarack and marsh stages into the soft maple and white elm stages.

S. Newf. and Gaspé Co., Que., to s. B. C., southw. to Va., Ind., and Wyo.; also in e. Asia.

6. **Rubus híspidus** L. SWAMP DEWBERRY. Map 1133. Infrequent to frequent throughout the lake area in acid soils, usually in tamarack bogs or in moist, sandy soil in black oak woods, where it is usually associated with wintergreen, lowbush blueberry and black chokeberry. South of the lake area it is very local, being found principally in the hard, white, minimacid, clay soil of the Illinoian drift, especially in Jefferson and Jennings Counties.

N. S. to Minn., southw. to Ga. and Kans.

6a. **Rubus hispidus** f. **pleniflòrus** Nieuwland. (Amer. Midland Nat. 4: 69. 1915.) Known only from the type locality in St. Joseph County.

7. **Rubus flagellàris** Willd. (*Rubus villosus* Ait. and *Rubus procumbens* Muhl.) NORTHERN DEWBERRY. Map 1134. This species is found only in slightly acid soil, usually in areas where the top soil has been removed by erosion, hence mostly in fallow fields. It is more or less frequent in the lake area and frequent to common in the southern part of the state. In the lake area in the northern counties it is often found in moist, sandy, acid areas in black oak woods. In all parts of the state the foliage is variable,

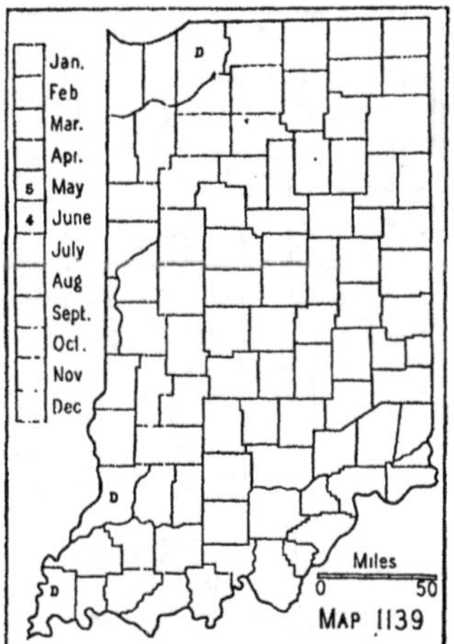

Rubus impos Bailey

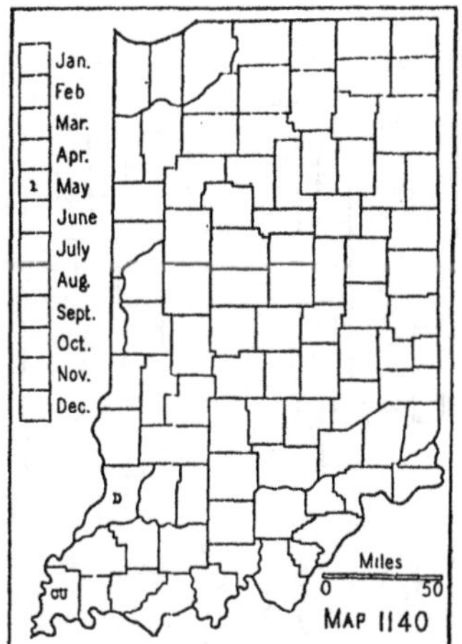

Rubus laudatus Berger

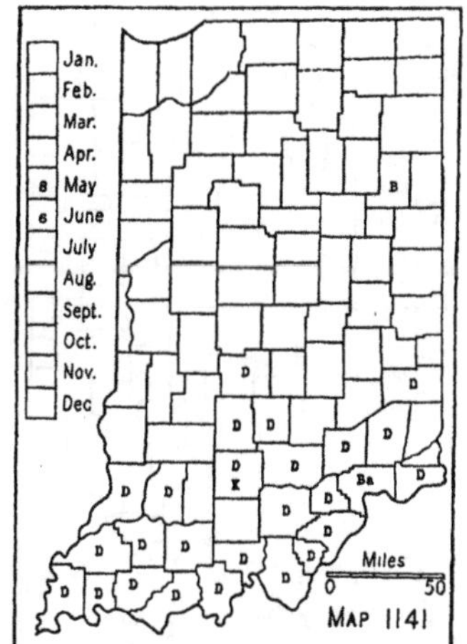

Rubus argutus Link

and this variation has led authors to segregate three forms which have been named. In the present treatment I believe it is best to regard this prostrate *Rubus* as a complex under one name.

Maine to Minn., southw. to Fla., Okla., and reported from Tex.

8. **Rubus Enslènii** Tratt. Map 1135. Our only specimens were found in very shallow soil on the cliffs in Perry County.

Eastern Mass. to Wis., southw. to Ala. and Miss.

9. **Rubus centràlis** Bailey. (Gentes Herbarum 2: 330-331. 1932.) Map 1136. The type of this species is my no. 27967 which was collected on the crest of a black and white oak ridge just east of Forest Tract 53 in the Clark County State Forest. The distribution of the species is not well known but Bailey says he has specimens from Maryland, Virginia, and Indiana.

10. **Rubus Dèamii** Bailey. (Gentes Herbarum 2: 463-464. 1932.) Map 1137. The type of this species is my no. 27799 which was collected on a washed slope in a fallow field on the north side of Little Blue River just west of the bridge across Little Blue River about a half mile south of Grantsburg in Crawford County. My no. 44636 is a topotype. My other specimens referred to this species by Bailey are shown on the map. Tennessee is the only other state from which Bailey cites specimens.

11. **Rubus allegheniénsis** Porter. ALLEGHENY BLACKBERRY. Map 1138. This is one of our most abundant highbush blackberries and is found more or less frequently probably throughout the state in almost all kinds of habitats but, like all the blackberries, it prefers open habitats.

N. S., Que. to Minn., southw. to N. C., Tenn., and Mo.

12. **Rubus impós** Bailey. (Gentes Herbarum 2: 455-456. 1932.) Map 1139. Bailey refers specimens of mine from La Porte, Knox, and Posey Counties to this species. Not known outside of Indiana.

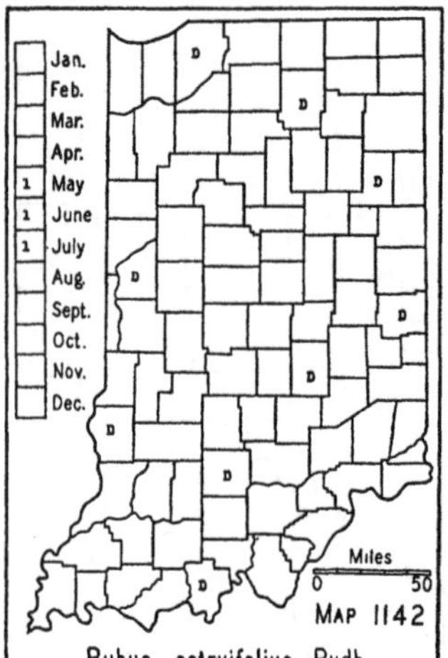

Rubus ostryifolius Rydb.

Rubus impar Bailey

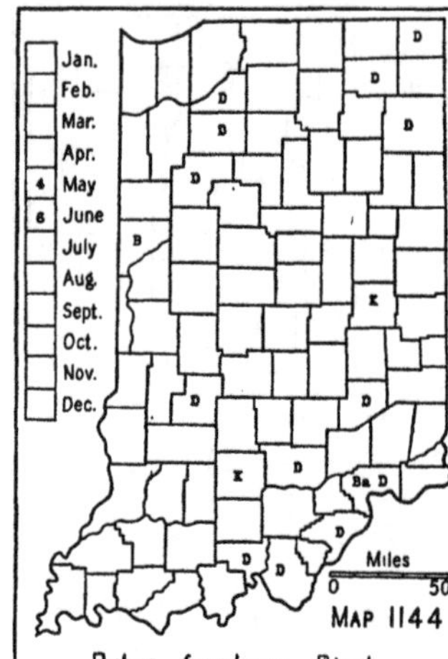

Rubus frondosus Bigel.

13. **Rubus laudàtus** Berger. (Rept. N. Y. Agric. Exp. Sta. 2: 79. 1925.) This species is fully discussed by Bailey in Gentes Herbarum 3: 265-269. 1934. Map 1140. Bailey cites two of my specimens, one from a low, flat woods in Posey County 5 miles south of Caborn and one from sandy soil on the Claypole Hill in Knox County. He gives the range as from Missouri and Kansas to eastern Illinois.

14. **Rubus argùtus** Link. HIGHBUSH BLACKBERRY. Map 1141. This species is frequent throughout the southern part of the state where it is found in white clay soil in low ground and on high ground mostly with beech and sugar maple.

Va. to s. Ind., southw. to Ga. and Tenn.

15. **Rubus ostryifòlius** Rydb. (Britton. Man. Flora North. States and Can. 497. 1901.) (*Rubus Andrewsianus* Blanchard.) Map 1142. Probably only infrequent throughout the state, mostly in black and white oak woods.

N. E. to Mich., southw. to N. C. and Kans.

16. **Rubus ímpar** Bailey. (Gentes Herbarum 3: 269. 1934.) Map 1143. This species so far is known only from the type locality which is an open, level, post oak woods just southeast of Half Moon Pond about 10 miles southwest of Mt. Vernon, Posey County. The soil of the area is a hard, white clay and is infertile and sparsely wooded mostly with post oak and an occasional black oak. The plants are slender and usually 3-5 feet high, with a few short side branches and a slightly curved summit.

17. **Rubus frondòsus** Bigel. Map 1144. Probably found more or less frequently throughout the state, growing mostly in dry soil.

N. E. to Wis., southw. to D. C. and Mo.

18. **Rubus abáctus** Bailey. (Gentes Herbarum 2: 452-455. 1932.) (Probably *Rubus recurvans* Blanchard.) Map 1145. Of our upright blackberries this species is the most arching and widest spreading, often almost as wide as long. Rather frequent in northern Indiana and in the "flats"

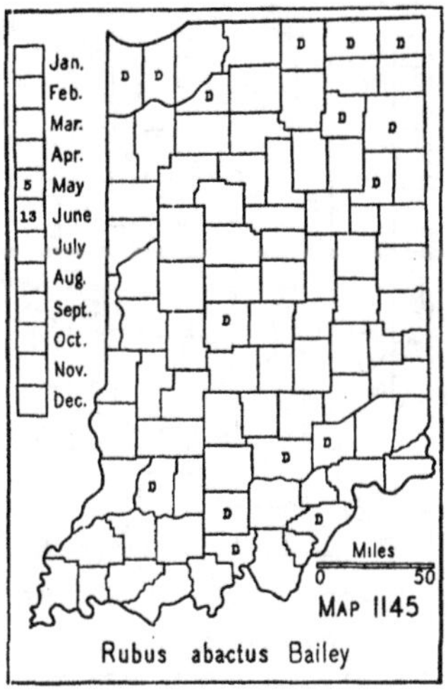

MAP 1145
Rubus abactus Bailey

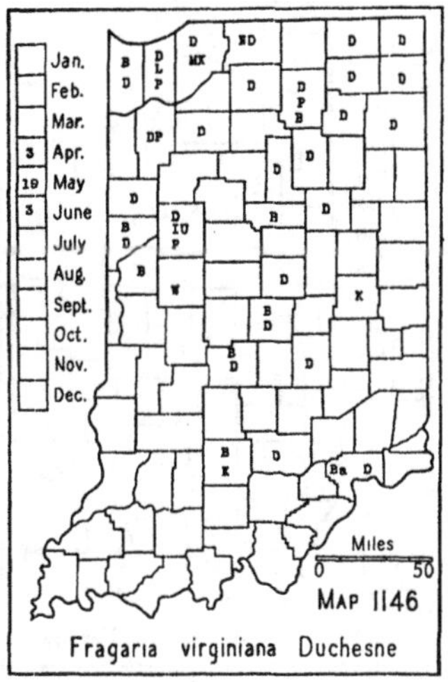

MAP 1146
Fragaria virginiana Duchesne

MAP 1147
Fragaria virginiana var. illinoensis (Prince) Gray

of the southeastern part, elsewhere it is usually infrequent. It prefers moist habitats but is also found in dry habitats. The foliage is variable. The typical and most prevalent form has leaflets with nearly regularly serrate margins, but there is also a form with sharply toothed or jagged margins. The latter form is common in Lagrange County. The range is not yet known but probably extends from New York, Ohio, Indiana, and Wisconsin to Minnesota.

3354. FRAGÀRIA [Tourn.] L. STRAWBERRY

Fruiting scape of the typical form shorter than the leaves (sometimes equaling the leaves); flowers usually 5-10, in corymbs; calyx lobes appressed or connivent on the young fruit; fruit red, subglobose; achenes in ripe fruit in pits below the surface.
  Hairs of the pedicels more or less appressed........................1. *F. virginiana.*
  Hairs of pedicels more or less widely spreading.....1a. *F. virginiana* var. *illinoensis.*
Fruiting scape of the typical form longer than the leaves (not always longer in the flowering phase); flowers few, racemelike on the scape or paniculate and the flowers more numerous; calyx lobes loosely spreading or reflexed on the young fruit; achenes on the surface of the fruit, not in pits below the surface.
  Petioles and peduncles generally copiously pubescent, the hairs of all or most of them spreading; hairs of pedicels appressed.
    Fruit red. ....2. *F. vesca.*
    Fruit white. (See excluded species no. 364, p. 1062).........2a. *F. vesca* f. *alba.*
  Petioles and peduncles generally sparsely pubescent, the hairs of all or most of them appressed; hairs of the pedicels appressed. (See excluded species no. 365, p. 1062.)...*F. vesca* var. *americana.*

1. **Fragaria virginiàna** Duchesne. VIRGINIA STRAWBERRY. Map 1146. Probably found more or less frequently throughout the state, especially in the lake area, although there are no records from the southern counties. The fact that in my early collecting I rarely collected strawberries accounts for the scarcity of my records and for the absence of records from certain parts of the state. This is true not only of this species but of the remainder of the genus. This species is found in wet, moist, and dry soils,

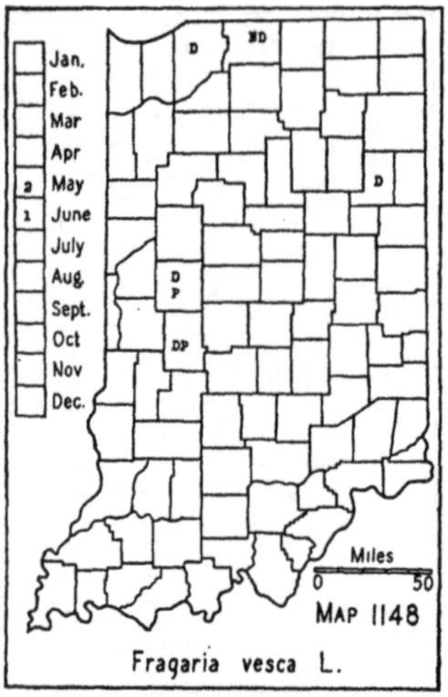

MAP 1148

Fragaria vesca L.

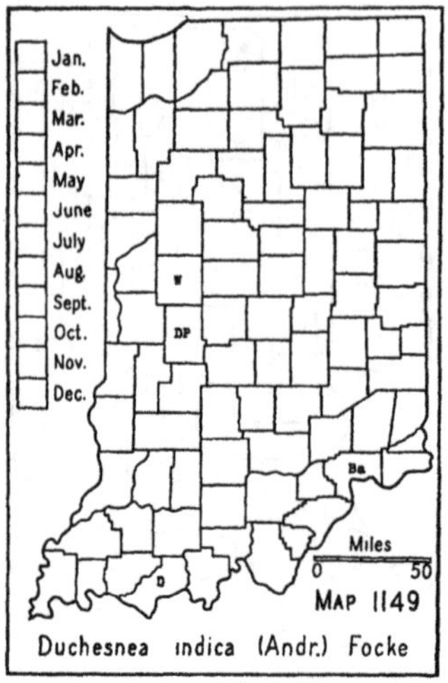

MAP 1149

Duchesnea indica (Andr.) Focke

MAP 1150

Potentilla fruticosa L.

but generally in little or no shade. I have specimens from open, wooded slopes, crevices of cliffs, roadsides, fallow fields, interdunal flats, marshes, and right of ways of railroads.

Newf. to S. Dak., southw. to Fla. and Okla.

1a. **Fragaria virginiana** var. **illinoénsis** (Prince) Gray. (*Fragaria Grayana* Vilmorin of Britton and Brown, Illus. Flora, ed. 2.) LARGE VIRGINIA STRAWBERRY. Map 1147. This variety, no doubt, is found also in all parts of the state but it is more frequent in the lake area. The habitat is similar to that of the species.

Western N. Y. to Minn., southw. to Ala., La., and Mo.

2. **Fragaria vésca** L. ALPINE STRAWBERRY. Map 1148. I have found this species only a few times although there are numerous reports of its occurrence. The strawberries are not easily differentiated and our cultivated strawberry was not separated from this species by our older authors. The cultivated strawberry often persists for a few years where it has been cultivated but does not become established.

My Montgomery County specimens were found on a wooded sandstone bluff of Sugar Creek in the "Shades." My Wells County specimen was found on an open, wooded slope. In La Porte County I found it along a roadside by a woods. This is a European species but it may be also a native of America.

Newf. to Ind., southw. to Pa. and Ky.

3355. DUCHÉSNEA J. E. Smith

1. DUCHESNEA ÍNDICA (Andr.) Focke. MOCK-STRAWBERRY. Map 1149. I found this species to be common in one place at the base of the sandstone bluff along the Ohio River in Rockport, Spencer County. I reported it from a marsh in Porter County but later discovered that my specimen was *Rubus pubescens* Raf. Peattie also reported it from the same place in Porter County, no doubt basing his report upon mine and overlooking the

fact that I had published a correction. There is, however, a specimen collected by T. G. Yuncker in the herbarium of DePauw University. It was collected along a roadside near Greencastle, Putnam County, where it was established. There is a specimen from Montgomery County in the herbarium of Wabash College. It was collected by A. R. Bechtel in Crawfordsville, where it has escaped and become established. This species is a rare introduction since there are only four records from this state and only one report from Ohio.

Nat. of Eurasia; s. N. Y. to Mo., southw. to Fla. and Ark.

3356. POTENTÍLLA L. CINQUEFOIL

Stems distinctly woody, usually 3-10 dm high; leaves pinnate, leaflets 5-7, entire. ....1. *P. fruticosa.*

Stems herbaceous.

- Leaves pinnate.
  - Mature plants erect, stout, villous-pubescent; leaflets 7-11, thick, double-serrate; terminal leaflet of lower leaves usually 3-4 cm wide; plants of a moist prairie or dry habitat........2. *P. arguta.*
  - Mature plants decumbent or in age with runners rooting at the nodes; leaflets 5-25, thin, serrate; terminal leaflet of lower leaves less than 3 cm wide; plants of a wet habitat.
    - Calyx green within; leaflets 7-25, with smaller intermediate ones, silvery-pubescent beneath but not glaucous; flowers solitary, axillary. 3. *P. Anserina.*
    - Calyx maroon within; leaflets 5-7, without smaller intermediate ones, glaucous beneath; flowers cymose........4. *P. palustris.*
- Leaves palmate.
  - Flowers cymose; peduncles short.
    - Leaflets green beneath.
      - Leaflets 5-9; plants tall, not divided at the base.
        - Petals 7-10 mm long, exceeding the calyx, pale yellow; stems green........5. *P. recta.*
        - Petals 3-7 mm long, usually shorter than the calyx or scarcely exceeding it, deep yellow; stems reddish; leaflets wider and deeper green than those of the preceding. (See excluded species no. 366, p. 1062.)........*P. recta* var. *obscura.*
      - Leaflets 3; petals shorter than the sepals, deep yellow....6. *P. monspeliensis.*
    - Leaflets silvery-pubescent beneath, 5; plants much divided at the base, the lower branches often prostrate; petals small, about 4 mm long......7. *P. argentea.*
  - Flowers solitary, axillary; stem soon becoming procumbent and usually rooting at the tips if in contact with soil.
    - First flower borne in the axil of the leaf from the first well-developed node when the stem is 1-1.5 dm high; mature stems 0.5-1 mm in diameter at the base. (See excluded species no. 367, p. 1063,)........*P. canadensis.*
    - First flower borne in the axil of the leaf from the second well-developed node when the stem is generally 1-4 dm high; mature stems 1-3 mm in diameter at the base.
      - Leaves green and more or less strigose-pubescent or slightly whitened but not copiously silvery-sericeus beneath.
        - Stems (especially when young) hirsute or villous-hirsute, with spreading or somewhat appressed hairs........8. *P. simplex* var. *typica.*
        - Stems strigose with short, appressed hairs or glabrate. (This form to be sought in Indiana.)........*P. simplex* var. *calvescens.*
      - Leaves densely silvery-sericeus beneath; pubescence of stems spreading, usually dense........8a. *P. simplex* var. *argyrisma.*

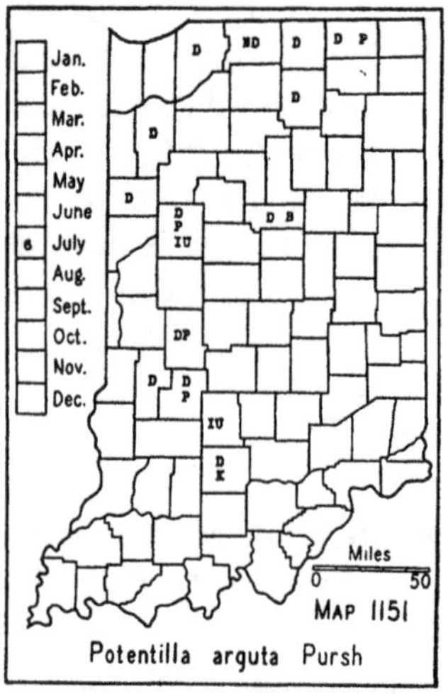

MAP 1151
Potentilla arguta Pursh

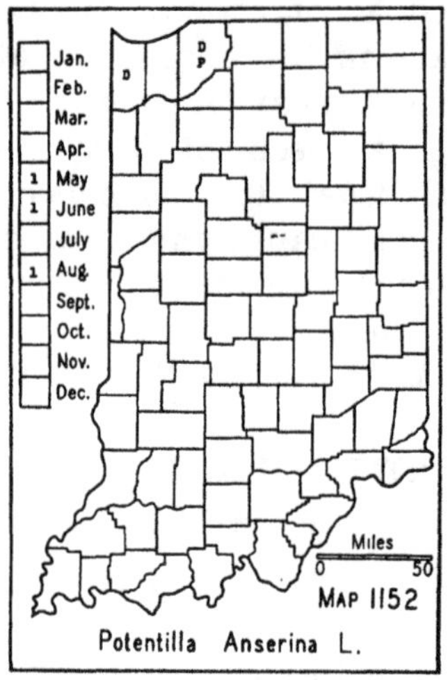

MAP 1152
Potentilla Anserina L.

MAP 1153
Potentilla palustris (L.) Scop.

1. **Potentilla fruticòsa** L. (*Dasiphora fruticosa* (L.) Rydb.) SHRUBBY CINQUEFOIL. Map 1150. This species prefers limy, springy places and marshes and is more or less frequent in the lake area with a few outlying posts south of it.

Greenland and Lab. to Alaska, southw. to n. N. J., Pa., Ill., Iowa, Ariz., and Colo.; also found in Eurasia.

2. **Potentilla argùta** Pursh. (*Drymocallis agrimonioides* (Pursh) Rydb.) Map 1151. This species is found in dry or moist sandy soil and seems to prefer a prairie habitat. It is local to very local in the area shown on the map and is most abundant in a remnant prairie in Lagrange County. My specimens are all from roadsides and fallow fields.

Eastern Que. and N. B. to Alaska, southw. to Va., Ill., Kans., and Colo.

3. **Potentilla Anserìna** L. (*Argentina Anserina* (L.) Rydb.) SILVERWEED. Map 1152. Found only in beach pools along Lake Michigan. It was formerly common just east of Michigan City but has become rare or extinct in most places. I have never found it in Porter County although there is one report. The report from St. Joseph County should possibly be referred to some other species. Grimes' specimen from Tipton County was found in the railroad yards at Tipton.

Arctic Amer., southw. to N. J., Ohio, Iowa, N. Mex., and Calif.

4. **Potentilla palústris** (L.) Scop. (*Comarum palustre* L.) (Fernald & Long. American variations of Potentilla palustris. Rhodora **16**: 5-11. 1914.) MARSH CINQUEFOIL. Map 1153. This species prefers neutral or slightly acid soils and is found mostly in marshes and swamps in the lake area although it was found also in the Bacon Bog in Marion County. It is variable in the pubescence of the leaflets. In the middle of September I studied this species on the south shore of Long Lake in Porter County where the shore is over a hundred feet wide. I was able to study the plants from near the water line back to where it was too dry for the species to

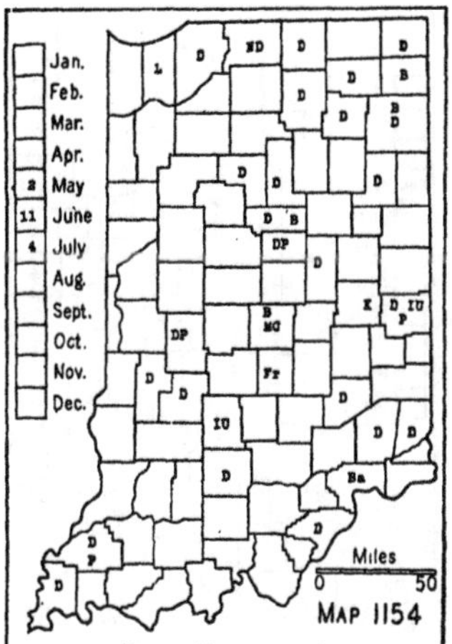

Potentilla recta L.

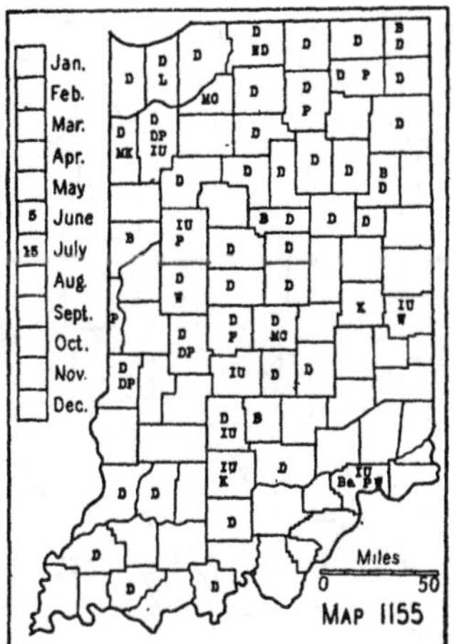

Potentilla monspeliensis L

Potentilla argentea L.

grow. I found that the leaves varied in pubescence from nearly glabrous in the wettest situations to silky-pubescent in the driest places.

Greenland and Lab. to Alaska, southw. to n. N. J., Pa., Ohio, Ind., Ill., Iowa, Wyo., and Calif.

5. POTENTILLA RÉCTA L. Map 1154. This species is rapidly becoming established in all parts of the state and has in some parts already become an annoying weed. Our first report dates back to 1905. In 1915 I made a note that I saw it only once during the year although I had driven more than 5000 miles. Now it has become more frequent along roadsides and in pasture fields and meadows.

Nat. of Eu.; Maine to Mich., southw. to Va. and Ill.

6. **Potentilla monspeliénsis** L. (Rhodora 28: 214. 1926.) (Rhodora 32: 254. 1930.) ROUGH CINQUEFOIL. Map 1155. This species is a frequent to common weed throughout the state. It is found in almost all kinds of habitats but it is not frequent in certain bottomlands. I have specimens from roadsides, cultivated and fallow fields, pastures, open woods, dry dunes, and a dense tamarack bog. It is a pernicious weed, especially in clover fields because its seed are separated from clover seed only with difficulty.

Lab. to Alaska, southw. to D. C., Mo., Kans., and N. Mex.; also in Asia.

7. **Potentilla argéntea** L. SILVER CINQUEFOIL. Map 1156. This species is restricted to dry, sandy or gravelly areas in the lake region where it is more or less frequent along roads and in pastures and open woodland.

N. S. to N. Dak., southw. to D. C., Ind., and Kans.; also found in Eu. and Asia.

8. **Potentilla símplex** Michx. var. **týpica** Fern. (*Potentilla canadensis* L. of Gray, Man., ed. 7 and Britton and Brown, Illus. Flora, ed. 2, in the major part.) (Fernald. Potentilla canadensis and P. simplex. Rhodora 33: 180-191. 1931.) COMMON CINQUEFOIL. Map 1157. Frequent to com-

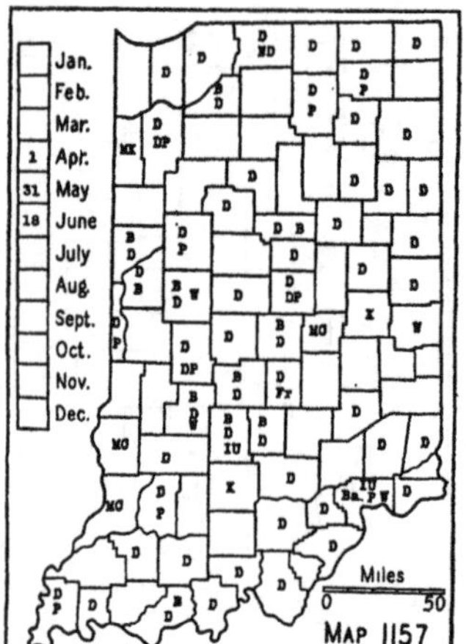
MAP 1157. Potentilla simplex Michx var. typica Fern.

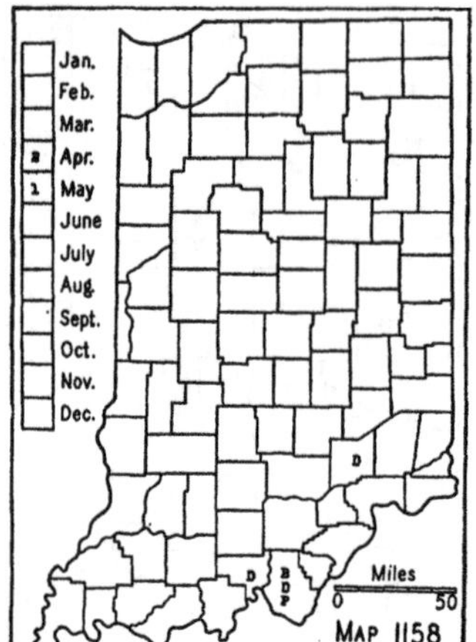
MAP 1158. Waldsteinia fragarioides (Michx.) Tratt.

MAP 1159. Geum vernum (Raf.) T & G.

mon throughout the state in dry and moist soil in almost all kinds of habitats. It becomes abundant in old fallow fields.

N. S., sw. N. B., s. Que., s. Ont. to Minn., southw. to N. C., Tenn., s. Mo., and Okla.

8a. **Potentilla simplex** var. **argyrísma** Fern. (Rhodora 33: 191. 1931.) I have this variety from Warren County on a steep, gravelly slope along the railroad west of Covington, and from Benton County in a prairie habitat.

Pa. to Ky. and Ill.

3363. WALDSTEÌNIA Willd.

1. **Waldsteinia fragarioìdes** (Michx.) Tratt. BARREN-STRAWBERRY. Map 1158. Extremely local in this state. It has been reported from only Clark and Jefferson Counties. I found it in talus at the base of a cliff along Little Blue River near the site of Carnes Mill about 2 miles south of Grantsburg in Crawford County; in talus on rocky ledges of the slope of Buck Creek where the creek parallels the road north of Dogwood in Harrison County; and in talus of the rocky slope of the North Fork of the Muscatatuck River about half a mile above Vernon, Jennings County.

N. B., Ont. to Minn., southw. to Ga., Ind., and Mo.

3365. GÈUM L. AVENS

[Fernald. Critical plants of Ontario and Michigan. Rhodora 37: 292-295. 1935.]

Styles glabrous; calyx bractless; heads conspicuously stalked in the calyx; flowering in May. .... 1. *G. vernum.*

Styles more or less pubescent (rarely one or more glabrous); calyx bracteolate; heads sessile in the calyx (short-stalked in *G. rivale*); flowering later than May.

Calyx lobes erect, purplish without; petals somewhat purplish, obovate with a long claw, 7-10 mm long; lower segment of styles densely long-pubescent.......... .... 2. *G. rivale.*

Calyx lobes reflexed, greenish without; lower segment of styles glabrous.

Lower part of stem glabrous or sparingly pubescent with spreading hairs about 1 mm long, sometimes more or less puberulent, rarely more densely pubescent; petioles of the basal leaves likewise pubescent, often more pubescent than the stem; petals white, exceeding the sepals, usually 3-6 mm long and half as wide or more.

Body of carpel sparsely appressed-pubescent as well as hispid with long hairs; upper segment of style sparsely bearded with white hairs of different lengths, usually one or more up to 0.5 mm long; simple leaves of the stem longer than wide, cuneate at the base, rarely truncate; peduncles glandless. ....3. *G. canadense.*

Body of carpel hispid above, otherwise glabrous; upper segment of style sparsely hispidulous with short hairs about half as long as the preceding; simple leaves of the stem usually wider than long, generally truncate at the base or the uppermost one cuneate; peduncles glandular..................... ....3a. *G. canadense* var. *Grimesii.*

Lower part of stem and petioles of basal leaves more or less densely pubescent with hairs about 2 mm long, spreading or somewhat retrorse; petals white or cream color and shorter than the calyx or large, bright yellow, and exceeding the calyx.

Peduncles densely puberulent, and with a few long hairs; petals cream color, 2.2-3.5 mm long, about half as wide, shorter than the sepals, usually about half as long; peduncles relatively long; heads obovate; stipules larger than in *G. canadense;* terminal leaflet usually about twice as long as the lateral ones, narrow, long-cuneate at the base, with coarse teeth...4. *G. virginianum.*

Peduncles densely puberulent, and usually densely pubescent with long, spreading hairs.

Petals bright yellow, mostly 4-8 mm long, nearly or quite as wide as long, longer than the sepals; leaves usually, as a whole, more pinnate; pubescence on stem and leaves softer, and receptacle more pubescent than in the following species....................5. *G. aleppicum* var. *strictum.*

Petals cream white, generally 2.5-5.5 mm long and about half as wide, shorter than the sepals.

Carpels glabrous...6. *G. laciniatum.*

Carpels hispid above..................6a. *G. laciniatum* var. *trichocarpum.*

1. **Geum vérnum** (Raf.) T. & G. SPRING AVENS. Map 1159. Infrequent to common in wet and moist woods throughout the state although there are no records from the northwestern counties. This species prefers moist, alluvial soil along streams, where it is often a common plant; but it grows also in less favorable habitats such as roadsides and fallow fields.

Ont. to Ill. and Kans., southw. to W. Va., Tenn., and Tex.

2. **Geum rivàle** L. PURPLE AVENS. Map 1160. This species has been found in only a few of our northern counties in tamarack bogs. Van Gorder found it in a birch marsh in Noble County and Hill reported it from La Porte County. I have had this species in cultivation in the open in garden soil for about 10 years and it thrives.

Lab. to Sask., southw. to n. N. J., Pa., Ind., and Colo.; found also in Eu.

3. **Geum canadénse** Jacq. WHITE AVENS. Map 1161. This is a woodland species well distributed throughout the state. It prefers a moist soil. Usually there are only a few plants growing at one place, but rarely it is found in dense or large colonies.

N. S. to S. Dak., southw. to Ga., La., and Kans.

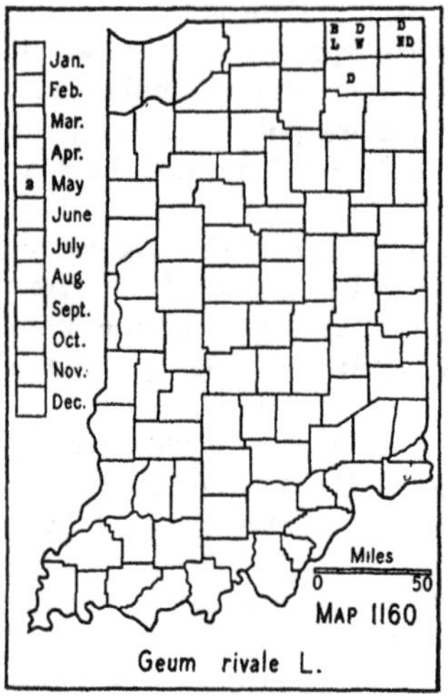

Geum rivale L.

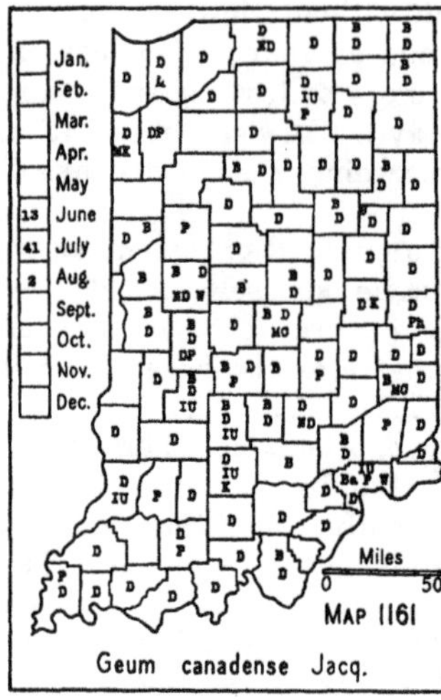

Geum canadense Jacq.

Geum canadense var. Grimesii Fern. & Weatherby

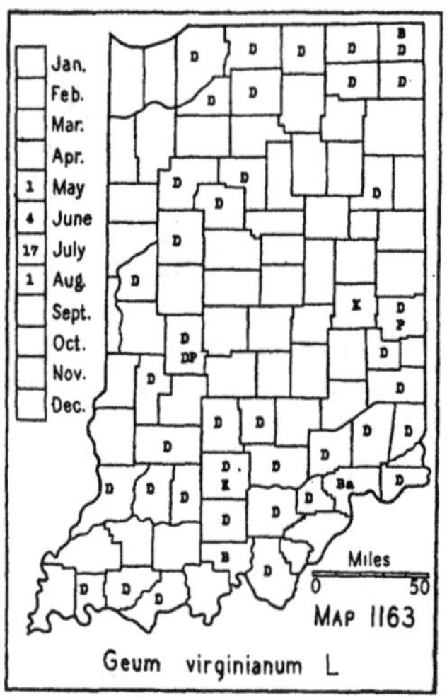

Geum virginianum L

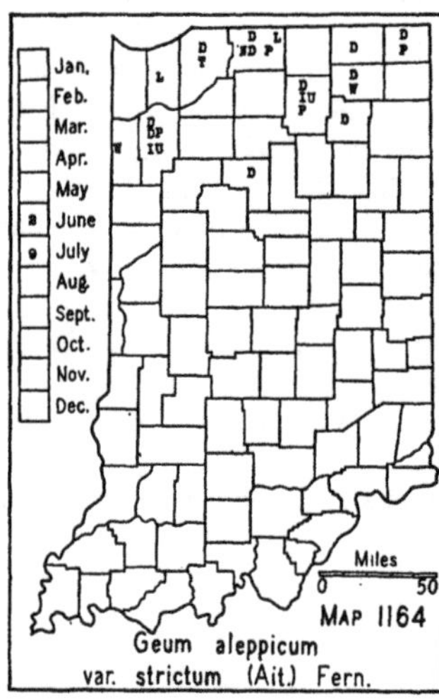

Geum aleppicum var. strictum (Ait.) Fern.

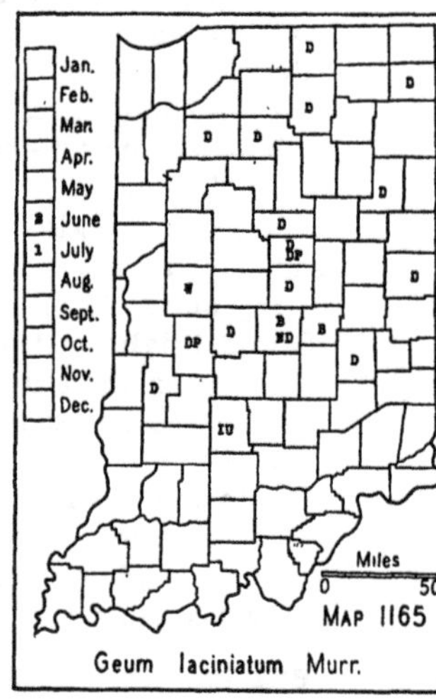

Geum laciniatum Murr.

3a. **Geum canadense** var. **Grìmesii** Fern. & Weatherby. (Rhodora 24: 49. 1922.) Map 1162. This variety is local but frequent in its habitat. It prefers a hard, white clay soil in pin oak, sweet gum, and river birch woods. It is easily distinguished in the field by its wide upper leaves.

Pa., D. C., Va., N. C., and Ind.

4. **Geum virginiànum** L. (*Geum flavum* (Porter) Bickn.) Map 1163. Infrequent in all kinds of dry woods throughout the state; rarely in wet woods or in open places.

Mass. to Ind., southw. to S. C., and Tenn.

5. **Geum aléppicum** Jacq. var. **stríctum** (Aít.) Fern. (Rhodora 37: 294. 1935.) (*Geum strictum* Ait.) YELLOW AVENS. Map 1164. All of my specimens are from the lake area where I have found it infrequently in marshes, tamarack bogs, and ditches. I have a specimen from Lagrange County with this note: "This plant had 10-12 petals to a flower and nearby plants also had more than 5 petals to a flower. Only one plant with the

normal 5 petals." Sometimes the inner row of petals is much reduced in size. This species has been reported 4 times from Clark and Jefferson Counties, the authors saying: "In meadows." Doubtless these authors meant hayfields because meadows, in the botanical sense, do not occur there. I do not believe this species occurs there but what these authors had at hand I can not determine. The manuals used by them to distinguish the species are definite as far as this species is concerned. More intensive collecting in southern Indiana may reveal the plant in a different habitat.

Newf. to B. C., southw. to N. J., Pa., Ill., Mo., and N. Mex.; also found in Asia.

6. **Geum laciniàtum** Murr. (*Geum virginianum* L. in part, of Gray, Man., ed. 7 and Britton and Brown, Illus. Flora, ed. 2.) NORTHERN ROUGH AVENS. Map 1165. This species is found more or less infrequently in low ground in woodland, on the borders of swamps, ponds, and lakes, and more abundantly in roadside ditches and along fences.

N. S. to Que., southw. to Mass. and Ind.

6a. **Geum laciniatum** var. **trichocárpum** Fern. (Rhodora 37: 292-293. 1935.) (*Geum virginianum* L. in part, of Gray, Man., ed. 7, and Britton and Brown, Illus. Flora, ed. 2.) ROUGH AVENS. Map 1166. Found in the habitats of the species but probably more frequent. The map shows the distribution of my specimens but it may also be found in the southern part of the state. It has been reported from southern Indiana by seven early authors. When these authors made their reports, however, our manuals did not separate *Geum canadense* from *Geum laciniatum* and its variety. Since *Geum canadense* is a species common to the southern counties, it is probable that all or most of these reports should be referred to *Geum canadense* or *Geum virginianum.*

N. S. to Minn., southw. to N. J. and Mo. and in the mts. to Ga.

3374. FILIPÉNDULA [Tourn.] Hill. MEADOWSWEET

1. **Filipendula rùbra** (Hill) Robinson. PRAIRIE MEADOWSWEET. Map 1167. This plant is an inhabitant of springy places and prairie swamps. The area covered by reports for the species is from Marshall County southward to the Ohio River Counties. It is frequently cultivated.

Pa. to Mich., southw. to Ga., Ky., and Iowa.

3376. AGRIMÒNIA [Tourn.] L. AGRIMONY

Leaflets (exclusive of the small, intermediate ones) generally 5-9, ovate to obovate, rhombic-ovate or elliptic-oblong; plants usually of a dry habitat.

Lower surface of leaflets subglabrous or sparsely hirsute; rachis of the inflorescence and petioles glandular-puberulent, sparsely hirsute or glabrous.

Mature fruit strongly striate, 5-6 mm in diameter, exclusive of the ring of stiff, hooked bristles; bristles in many rows, the longer ones 3.5-4 mm long and usually broadly spreading or reflexed; petals about 4 mm long, acuminate, indurated and somewhat curved inward at the tips; plants stout; root not tuberous; lower surface of leaflets plainly glandular-dotted...1. *A. gryposepala.*

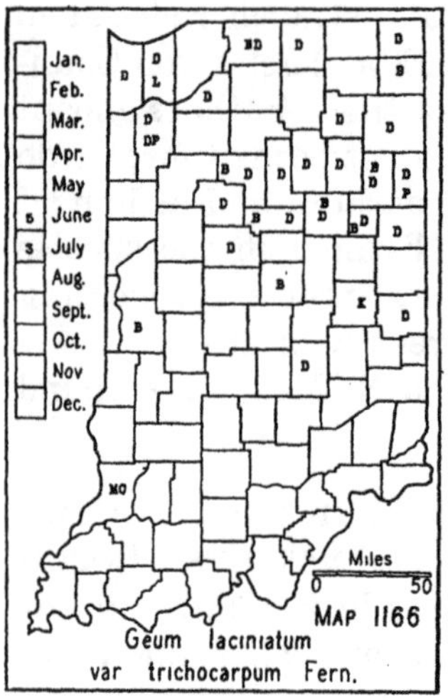

MAP 1166
Geum laciniatum var trichocarpum Fern.

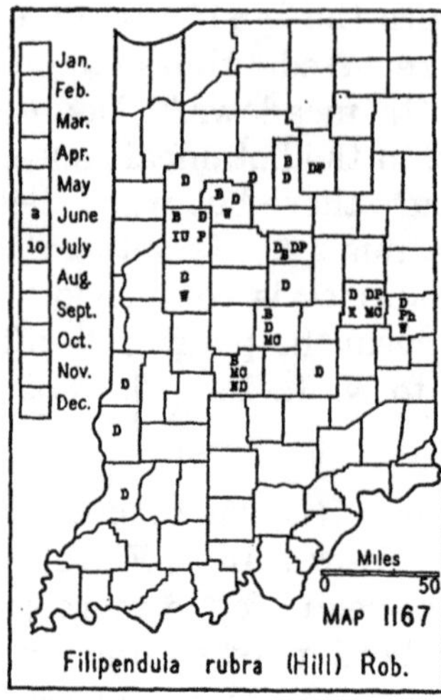

MAP 1167
Filipendula rubra (Hill) Rob.

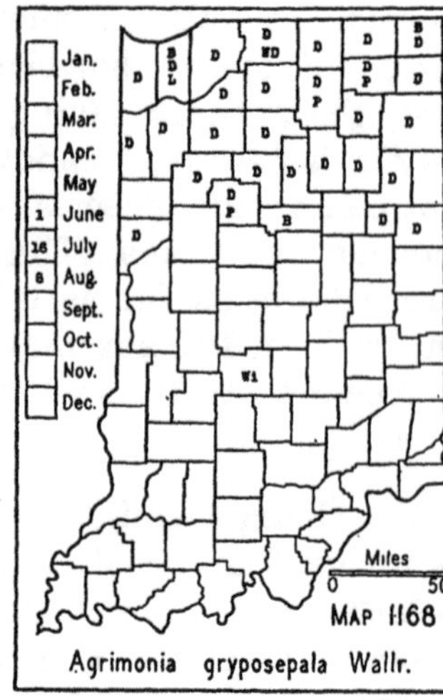

MAP 1168
Agrimonia gryposepala Wallr.

Mature fruit faintly striate, 2.5-2.8 mm in diameter, exclusive of the bristles; bristles in 3 or 4 rows, the longer ones 1.5-1.8 mm long; petals about 3 mm long, connivent, not acuminate or indurated at the tips; plants slender; root tuberous; lower surface of leaflets obscurely resinous-dotted. 2. *A. rostellata.*

Lower surface of leaflets more or less densely pubescent, especially on the veins; rachis of inflorescence and petioles closely pubescent, usually with subappressed and spreading hairs, short glandular hairs lacking.

Lower surface of leaflets velvety to the touch, the pubescence consisting of long, spreading hairs; resinous dots obscure or lacking; stipules of median leaves reniform, rather evenly but coarsely dentate; petals about 3 mm long... 3. *A. pubescens.*

Lower surface of leaflets not velvety, the pubescence consisting of long hairs but these more or less appressed; resinous dots copious and prominent; stipules of median leaves ovate, long-acuminate with a few irregular teeth below; petals about 3.5 mm long. (See excluded species no. 372, p. 1063.)..*A. striata.*

Leaflets (exclusive of the small, intermediate ones) usually 11-15, lanceolate to narrowly lance-oblong, copiously glandular beneath; fruit about 3 mm in diameter, exclusive of the ring of bristles; the lower bristles widely spreading, the upper ones the longer; stems densely hirsute; plants of wet and moist habitats.. 4. *A. parviflora.*

1. **Agrimonia gryposépala** Wallr. AGRIMONY. Map 1168. More or less frequent throughout the lake area. It is usually found in dry soil, but is also sometimes found in moist soil, especially where the soil is of a sandy nature. Since this species and the next two were formerly regarded as an aggregate, all of the reports made prior to our present manuals must be ignored because we do not know to which species they should be referred. From the specimens at hand it seems that this species is northern in its distribution, while the other two are found throughout the state.

N. B., s. N. S. and cent. Maine to Minn. and Calif., southw. to N. C., Tenn., and Mo.

2. **Agrimonia rostellàta** Wallr. Map 1169. Found throughout the state, although it is restricted to dry soil generally of thick woodland.

Conn., cent. N. Y. to Nebr., southw. to Ga., Tenn., and Mo.

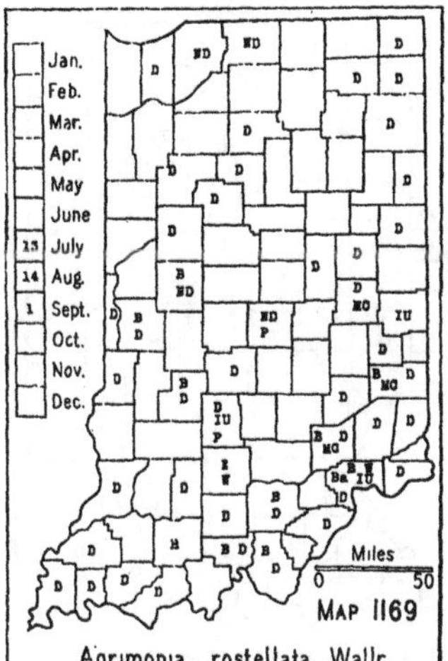

Agrimonia rostellata Wallr.

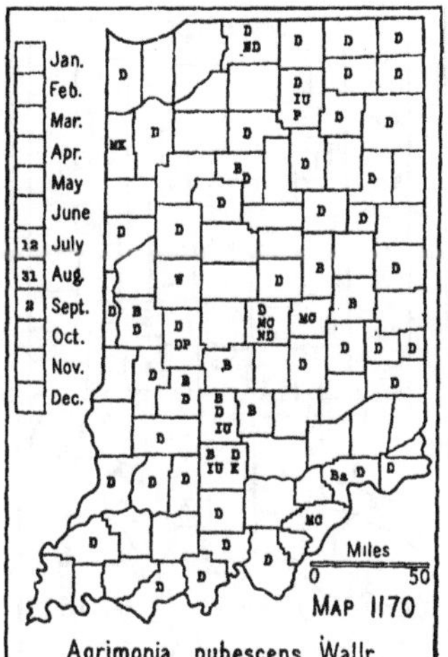

Agrimonia pubescens Wallr.

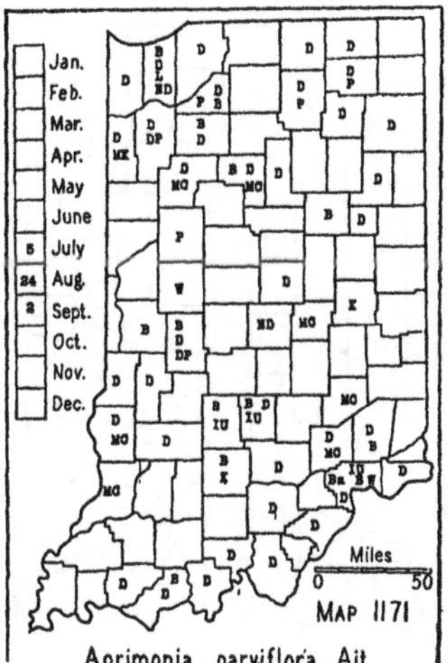

Agrimonia parviflora Ait.

3. **Agrimonia pubéscens** Wallr. (North Amer. Flora 22: 393. 1913.) (*Agrimonia mollis* (T. & G.) Britt.) Map 1170. Found throughout the state in dry soil in woodland and rarely along roadsides and in prairies.

Mass. to Mich., southw. to Ga. and Kans.

4. **Agrimonia parviflòra** Ait. SMALLFLOWER AGRIMONY. Map 1171. Throughout the state in low ground along streams, about ponds and swamps, and in marshes and roadside ditches.

Conn. to Minn., southw. to Fla., La., and Kans.

3381. SANGUISÓRBA [Rupp.] L. BURNET

1. **Sanguisorba canadénsis** L. AMERICAN BURNET. Map 1172. This species has been found in only two places in the state. Blatchley found it in Vigo County, August 1, 1892, along the Vandalia Railroad through the Heckland Prairie north of the Otter Creek Junction, about 6 miles northeast of Terre Haute, and along the same railroad near Heckland, in sec. 8, about 8½ miles northeast of Terre Haute. I found a few specimens in the last named place in 1917. In 1923 I found a few specimens in a springy place along Flint Creek about 3 miles northwest of Westpoint in Tippecanoe County.

Lab. to Man., southw. to Va. and Ind., and in the mts. to Ga.

3389. RÒSA [Tourn.] L. ROSE

Eileen Whitehead Erlanson, who has for years intensively studied the wild roses of North America, has had them under cultivation in the Botanical Garden of the University of Michigan, and has written voluminously about them, has examined all of my specimens and written the following key. It is now known that some of the species freely hybridize and when specimens of hybrids are at hand it is difficult to name them. Our native species nos. 4, 5, and 6 are extremely variable and the more conspicuous forms have been named. Some of these forms are described in

Sanguisorba canadensis L.

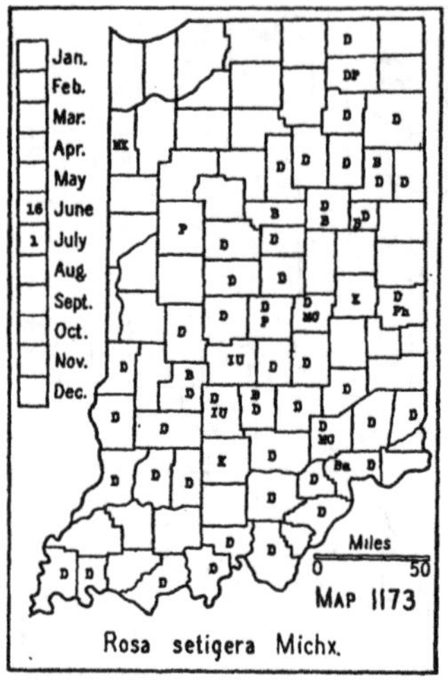

Rosa setigera Michx.

Rosa setigera var. tomentosa T. & G.

the text under the species to which they most nearly approach. It seems best to consider these variables as belonging to "species complexes" rather than to try to assign names to the many variables.

Styles united; leaves on old stems mostly with 3 leaflets, on new growth 3 or 5; stems long, recurved or semi-trailing............................1. *R. setigera.*
Styles free; leaves on old and new wood mostly with 5, 7, or 9 leaflets; stems erect, or spreading.
  Orifice (through which the styles protrude) small, 1-2 mm in diameter, surrounded by a well defined disk of fleshy tissue (introduced species).
    Orifice about 2 mm in diameter; leaflets glandular...............2. *R. rubiginosa.*
    Orifice about 1 mm in diameter.
      Leaflets large and leathery, 2-6 cm long, dark green above, pale beneath; stems low and slender; prickles short. (See excluded species no. 376, p. 1064.)..*R. gallica.*
      Leaflets small, not leathery, 1.5-4 cm long; stems tall and coarse with large prickles.
        Styles glabrous; leaflets densely glandular and pubescent beneath...3. *R. micrantha.*
        Styles pubescent; leaflets not glandular and sparsely pubescent beneath. (See excluded species no. 375, p. 1064.)..........................*R. canina.*
  Orifice in flower and fruit wide, 2-3 mm in diameter.
    Hypanthium usually glandular; calyx lobes generally deciduous from the hips.
      Shrubs of wet ground, usually 1-2 m high; branches reddish; serration of leaflets fine; flowers usually corymbose; prickles recurved....4. *R. palustris.*
      Shrubs of dry uplands, usually less than 1 m high; branches gray or greenish; serration of leaflets coarse; flowers solitary on old wood, in terminal corymbs on new canes; prickles straight..........................5. *R. carolina.*
    Hypanthium usually smooth; calyx lobes erect and persistent on the hips.
      Stems 0.5-1 m high, usually unarmed except at the base; leaflets 5 or 7, rarely 9; flowers on two year old stems only.........................6. *R. blanda.*
      Stems low, 3-5 dm high, semi-herbaceous, weak and bristly; leaflets usually 9 or 11; flowers on old wood and terminally on new shoots.........7. *R. suffulta.*

1. **Rosa setígera** Michx. PRAIRIE ROSE. Map 1173. This species is distinguished from its variety by having the lower surface of the leaflets glabrous or only the veins pubescent and the upper surface shining. It is

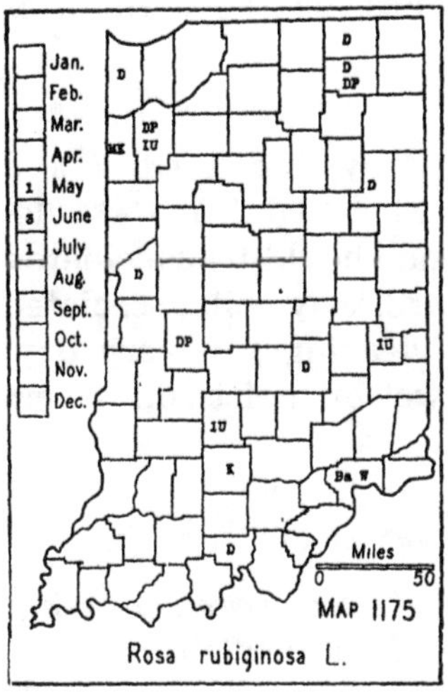

Rosa rubiginosa L.

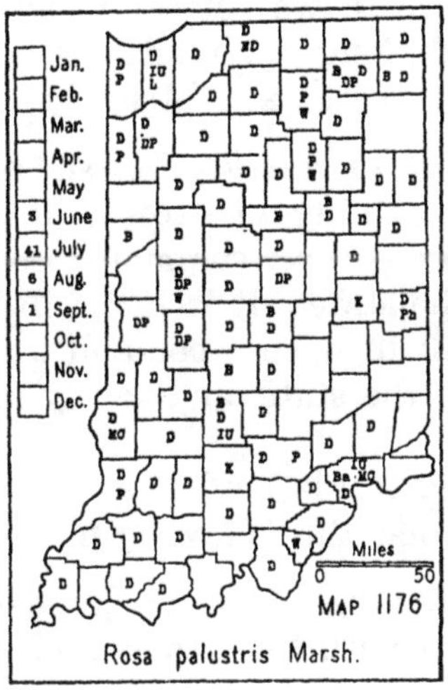

Rosa palustris Marsh.

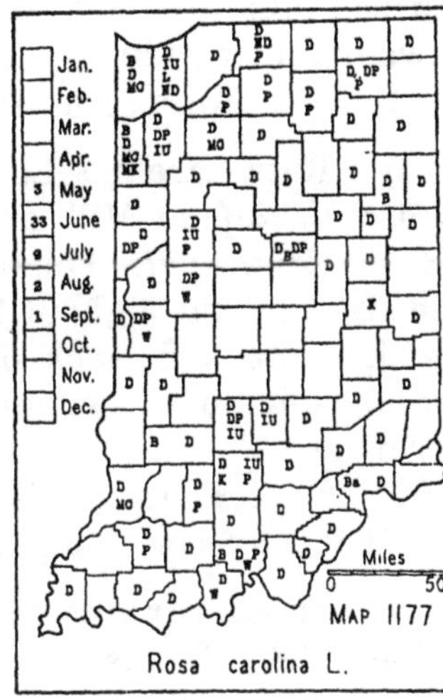

Rosa carolina L.

not as common as the variety and does not extend as far north, becoming very rare in the northern counties. I do not find where the range of this species has been determined so I am forced to give the range as found in our manuals.

N. Y. to Kans., southw. to Fla.

1a. **Rosa setigera** var. **tomentòsa** T. & G. (*Rosa rubifolia* R. Br. of North Amer. Flora 22: 491. 1918.) Map 1174. This variety is distinguished from the species by having the entire lower surface of the leaflets soft-pubescent and the upper surface dull. It is more frequent than the species and in some clearings it is often so abundant as to give them the appearance of rose gardens. The species and variety are found in open woodland, clearings, and pastures and along fences and roadsides.

Ont. to Wis., southw. to Ga. and Tex.

2. ROSA RUBIGINÒSA L. SWEETBRIER. Map 1175. This rose is much cultivated and it has escaped in all parts of the state.

Nat. of Eu.; N. S. to Ont. and Mich., southw. to Ga., Miss., and Kans.

3. ROSA MICRÁNTHA Borrer. SMALLFLOWER SWEETBRIER. I found a small colony of this rose along the roadside in Elkhart County where it had persisted, no doubt, near the site of a former habitation, although there was no evidence that a habitation ever existed here. Miss Edna Banta writes me that she has known it for ten years as an escape in Jefferson County. She says it is more or less frequent on the slope of the bluff of the Ohio River from Madison to Greasy Hollow, a few miles east of Madison.

Nat. of Eu.; sparingly escaped throughout the U. S.

4. **Rosa palústris** Marsh. (*Rosa carolina* L. of Gray, Man., ed. 7 and Britton and Brown, Illus. Flora, ed. 2.) SWAMP ROSE. Map 1176. This species is frequent to infrequent throughout the state, being more common

in the lake area. It grows in wet places about lakes, on the borders of swamps in woodland, and along low roadsides.

N. S. to Minn., southw. to Fla. and Miss.

5. **Rosa carolina** L. (*Rosa humilis* Marsh. of Gray, Man., ed. 7 and *Rosa virginiana* Mill. of Britton and Brown, Illus. Flora, ed. 2.) PASTURE ROSE. Map 1177. This is our most common rose and should be considered as frequent throughout the state. It is found in hard, clay soil, associated with white and black oak in open woodland and clearings, in very dry, sandy soils in all parts of the state, and rarely in wet or moist soil of our prairies.

Newf. to Minn., southw. to Fla. and Tex.

Of this species Indiana has the following named varieties:

5a. **Rosa carolina** var. **villòsa** (Best) Rehder. (*Rosa Lyoni* Pursh.) Map 1178. This is a form with the under surface of the leaflets more or less densely pubescent.

5b. **Rosa carolina** var. **glandulòsa** (Crep.) Farw. (*Rosa serrulata* Raf.) This form is distinguished by the double-serrate leaflets, by the secondary teeth ending in a stipitate gland, and by the entire part of the margins of the leaflets having stipitate glands. This form is not well marked in our area since specimens can be found with a few leaves with double-serrate leaflets with stipitate glands and the other leaves non-glandular or with simple serrations. I have specimens of this form from Clark, Floyd, and Franklin Counties.

5c. **Rosa carolina** var. **sabulòsa** Erlanson. This form is characterized by the trailing habit of the old stems, "by the uniformly small leaflets on old wood, and by having the hypanthium and pedicels free from hispid glands." It is represented by specimens from Lake County.

5d. **Rosa carolina** var. **Dèamii** (Erlanson) Deam, comb. nov. (*Rosa Deamii* Erlanson, Rhodora **30**: 120-121. 1928.) This rare form has been found only in Tipton County in the area of the Indian Prairie along the railroad about a half mile west of Goldsmith. It is distinguished "by the recurved stems; long prickles; thick, shining, dark green foliage with coarse serrations; large flowers and hips. All parts are coarse and well developed." I have had this form under cultivation ever since I found it.

5e. **Rosa carolina** var. **obovàta** (Raf.) Deam, comb. nov. (*Rosa obovata* Raf., Ann. Gen. Sci. Phys. 5: 217. 1820.) This form is rather frequent in Indiana, especially in the northern part. It is distinguished by its broad, oval leaflets, stout prickles, and large flowers.

6. **Rosa blánda** Ait. MEADOW ROSE. Map 1179. This species is, for the most part, restricted to the area about Lake Michigan where it is found in the low, interdunal flats and up to the very tops of the highest dunes. On account of many visitors to the high dunes it has already almost disappeared from this habitat. I have it also from Elkhart County and Grimes found it along the railroad in Putnam County where it was a migrant.

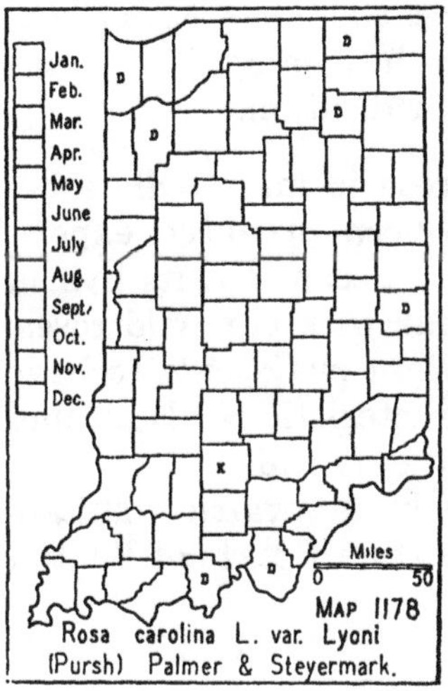

MAP 1178
Rosa carolina L. var. Lyoni (Pursh) Palmer & Steyermark.

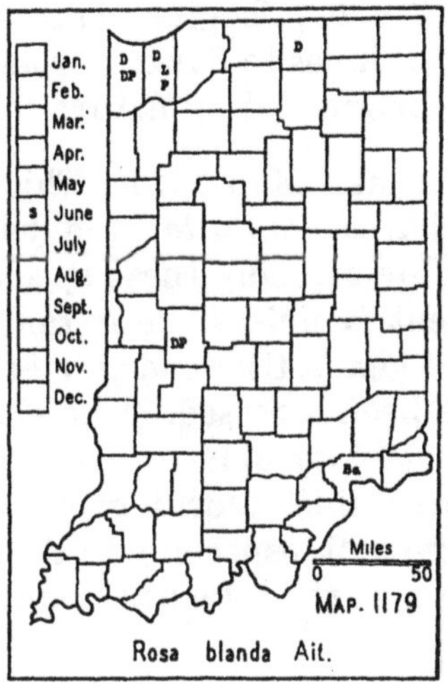

MAP. 1179
Rosa blanda Ait.

MAP 1180
Rosa suffulta Greene

It is possible that this rose may yet be found in more of our northern counties, especially in La Porte and St. Joseph Counties. For a more detailed discussion of this and other species of Indiana roses, see Deam, Shrubs of Indiana, ed. 2.

Newf. to Sask., southw. to Pa. and Ill.

Of this species complex Indiana has the following named varieties:

6a. **Rosa blanda** var. **carpohíspida** Schuette, with stipitate glands on the hypanthium. This is a rare form.

6b. **Rosa blanda** var. **glandulòsa** Schuette, with pyriform hips. I have this form from Porter County, and it has been reported from the dunes by Peattie.

6c. **Rosa blanda** var. **híspida** Farw., with densely bristly stems. Reported from the dune area of Lake Michigan by Peattie.

7. **Rosa suffúlta** Greene. (*Rosa pratincola* Greene of Gray, Man., ed. 7 and Britton and Brown, Illus. Flora, ed. 2.) (*Rosa heliophila* Greene of Deam, Shrubs of Indiana, ed. 1.) Map 1180. I have found this species only in Tipton County in Indian Prairie in moist soil along the railroad about a half mile west of Goldsmith. Chas. M. Ek has found it in several places along railroads in Howard County.

Ind. to Alberta and Man., southw. to Tex. and N. Mex.

7a. **Rosa suffulta** var. **relícta** (Erlanson) Deam, comb. nov. (*Rosa relicta* Erlanson in Rhodora 30: 116-117. 1928.) This rare form has been found in Indiana only in Tipton County in the Indian Prairie area along the railroad a short distance west of Goldsmith. "It resembles a weak *R. suffulta;* it is semi-herbaceous, the two year old wood being often semi-procumbent. It differs from *R. suffulta* in the narrow stipules, small fruit with reflexed and semi-deciduous sepals, in which characteristics it resembles *R. carolina* L." "*R. relicta* begins to flower earlier than *R. suffulta,* just after *R. blanda* and continues to flower through the summer." Like the next species, it may have originated by natural hybridization.

8. **Rosa rudiúscula** Greene. Map 1181. This rose is intermediate between *Rosa carolina* and *Rosa suffulta,* and has been produced experimentally by Dr. Erlanson by crossing these species. Because of its hybrid nature it is difficult to identify unless one is familiar with our wild roses. In former accounts of the genus the tendency of *Rosa rudiuscula* to have thick, leathery leaves has been stressed. This characteristic is also found in *Rosa carolina* and is not invariably present in the hybrid.

According to my records this hybrid grows only in prairie habitats where it is more or less frequent, especially in the northern parts of Benton County and in the southern part of Jasper County.

Ind. to Mo.

3396. PRÙNUS [Tourn.] L. CHERRIES AND PLUMS

Fruit velvety-tomentose; stone deeply sculptured and pitted; flowers subsessile, large, pinkish, appearing before the leaves. (See excluded species no. 382, p. 1064.) .... *P. Persica.*

Fruit glabrous; stone not sculptured; flowers pedicellate, smaller than the preceding, white (rarely some pinkish).

- Flowers in umbel-like clusters or somewhat corymbose, appearing before or with the leaves on branchlets of the preceding year.
  - Margins of leaves cut about 1 mm deep with sharp teeth; teeth not ending in a gland; fruit red; stone compressed.
    - Petioles glabrous beneath; branchlets glabrous; lower surface of mature leaves glabrous or rarely more or less pubescent on the principal veins; pedicels and calyx tube glabrous........ 1. *P. americana.*
    - Petioles more or less pubescent all around; branchlets puberulent, rarely becoming glabrous in autumn; lower surface of mature leaves pubescent all over; pedicels and calyx tube more or less pubescent........ 2. *P. lanata.*
  - Margins of leaves cut less than 1 mm deep with blunt or crenate teeth; teeth ending in a gland.
    - Teeth of the middle of the blades 10 or fewer per cm; calyx lobes glandular except in no. 7; fruit more than 10 mm in diameter.
      - Calyx lobes ciliate but not glandular.
        - Leaf blades generally much paler beneath; at least the basal third of the margins entire and crenate above; fruit black, about 14 mm in diameter; surface of stone marked with oblique grooves........ 3. *P. pumila.*
        - Leaf blades not paler beneath or only slightly so; the entire margins regularly and finely crenate; fruit bright red, globose, about 13 mm in diameter; surface of stone rugose........ 4. *P. angustifolia.*
      - Calyx lobes more or less glandular-ciliate.
        - Calyx lobes glabrous on both sides or with a short band of hairs at the base within; leaves dull, dark green above, abruptly pointed at the apex; flowers more than 17 mm in diameter; petioles mostly with two glands at the summit........ 5. *P. nigra.*
        - Calyx lobes pubescent both within and without; leaves lustrous above, thin, acute or acuminate at the apex; flowers less than 17 mm in diameter. .... 6. *P. hortulana.*
    - Teeth of the middle of the blades about 20 per cm; calyx lobes glandless; fruit bright red, less than 10 mm in diameter........ 7. *P. pennsylvanica.*

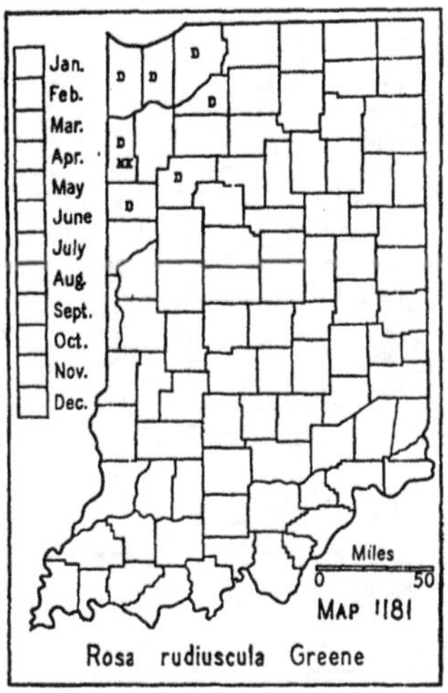

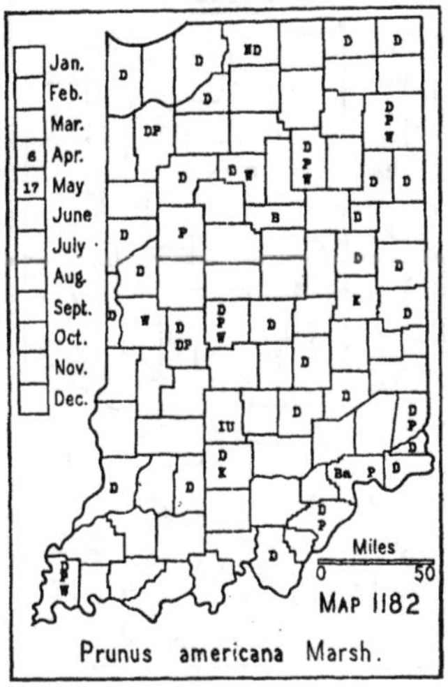

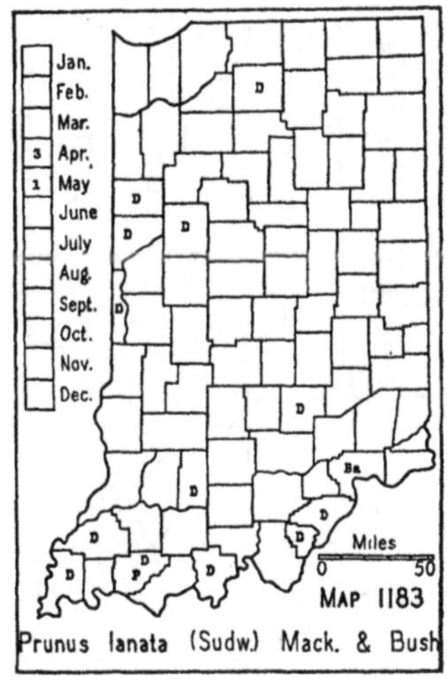

Flowers in racemes on branchlets of the present or previous year.

Flowers (15) 20-30, on long racemes, appearing on branchlets of the present season; sepals glandular; mature blades usually more than 5 cm long.

Leaf blades mostly obovate, thin, except in the variety, the margins sharply serrate; sepals wider than long, deciduous.

Rachis of racemes, pedicels of flowers, branchlets, and lower surface of leaves glabrous.. 8. *P. virginiana.*

Rachis of racemes, pedicels of flowers, branchlets, and lower surface of leaves pubescent, sometimes the branchlets nearly glabrous in autumn......... ....8a. *P. virginiana* var. *demissa.*

Leaf blades elliptic or lanceolate, thicker than those of the preceding species, the margins crenate-serrate; sepals longer than wide, not deciduous.......... ....9. *P. serotina.*

Flowers 6-12, on short racemes, appearing on branchlets of the previous year; leaf blades orbicular or orbicular-ovate, abruptly acute at the apex, generally less than 5 cm long; sepals about 2 mm long, entire, glandless..10. *P. Mahaleb.*

1. **Prunus americàna** Marsh. AMERICAN PLUM. Map 1182. More or less frequent throughout the state. It prefers moist soil and is commonly found in open woodland along streams, about ponds and lakes, and in moist, prairie habitats. In Indiana this tree rarely reaches a diameter of 8 inches and is generally 2-5 inches in diameter. Its habit of sprouting prolifically gives rise to our "plum thickets."

Conn. to Mont., southw. to Fla., Tex., and Colo.

2. **Prunus lanàta** (Sudw.) Mack. & Bush. WOOLLYLEAF PLUM. Map 1183. Probably frequent in southwestern Indiana, and infrequent to rare northward. The plums are not easy to identify, hence their collection may be neglected. It is necessary to collect the flowers and mature fruit from the same plant and this task is not as easy at it might seem. I have collected flowers from many a shrub and tree and returned at fruiting time to find that fruit had not developed.

Ind. to Okla. and southw. to the Gulf.

3. **Prunus pùmila** L. SAND CHERRY. Map 1184. Found only in the northwestern part of the state in the counties shown on the map. It is

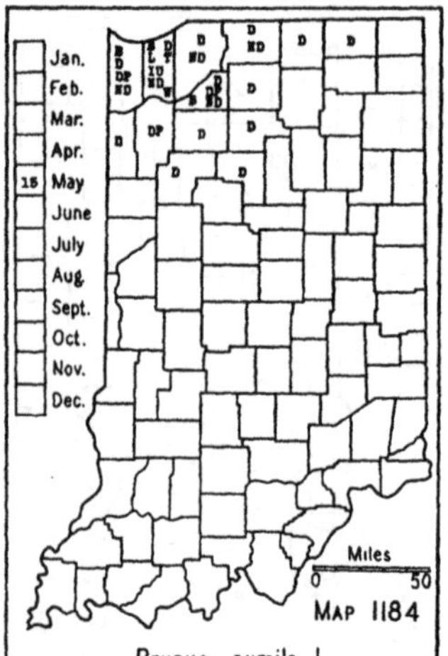

Prunus pumila L.

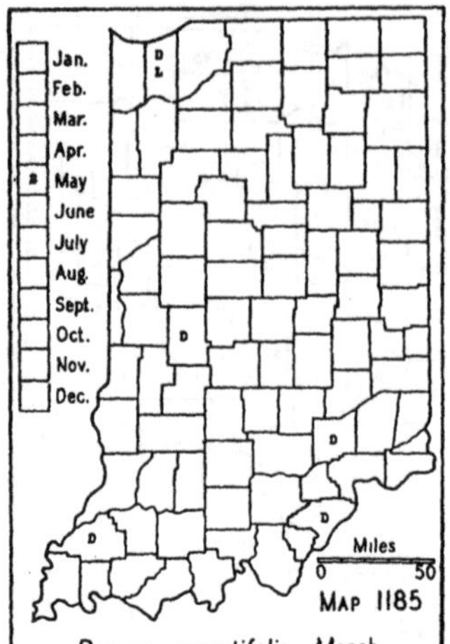

Prunus angustifolia Marsh.

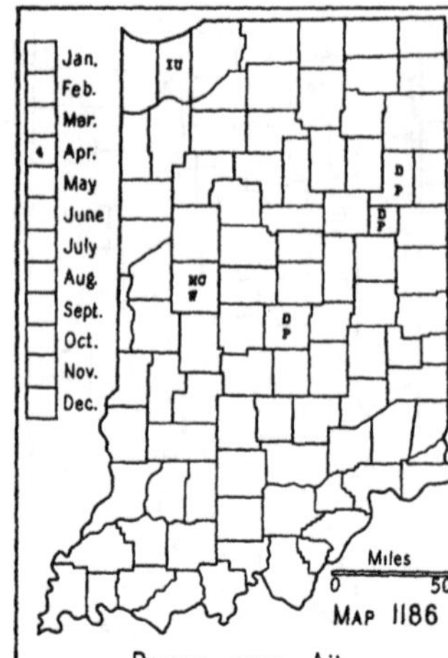

Prunus nigra Ait.

local to infrequent except on the slopes of the dunes facing Lake Michigan and on the low dunes near Lake Michigan west of Gary where it is frequent to common. In the interdunal flats a short distance from the lake large colonies may be found. Away from the lake it grows in moist, black, sandy soil and is usually about 3 feet high and erect or slightly decumbent near the base, but along the lake it is always decumbent at the base and sometimes reaches a length of 5-8 feet. I have had this species in cultivation from seed from the shore of Lake Michigan and the plants grow rapidly and are erect until they reach a height of 5-8 feet when they either become decumbent or break off near the ground.

Prunus cuneata and *Prunus susquehanae* are named forms of *Prunus pumila* which I do not regard as of taxonomic value. For a discussion of these species see excluded species nos. 381 and 383, on p. 1064.

Maine to Minn., southw. to N. Y. and Ind.

4. PRUNUS ANGUSTIFÒLIA Marsh. CHICKASAW PLUM. Map 1185. I feel positive that this species was never native to Indiana and I think that our few reports can safely be referred to naturalized plants. I have seen it persist in a fallow field in the Clark County State Forest after cultivation and spread over an area, as nearly as I can recall, of about half an acre in 30 years. I have seen it frequently in large colonies in fallow fields about former habitations. In no instance have I seen it in a place where I would regard it as native and it should be referred to the introduced species.

Sargent says: "Probably native in cent. Tex. and Okla." Now widely naturalized from Del. to Ky., southw. to the Gulf.

5. **Prunus nìgra** Ait. CANADA PLUM. Map 1186. Very local in the northern half of the state where it is found in wet woodland. All of the specimens I have seen in the wild were small, although one which I transplanted grew to a diameter of 7 inches at breast height when it was killed by borers. Its flowers are large, somewhat pinkish, and profuse; they

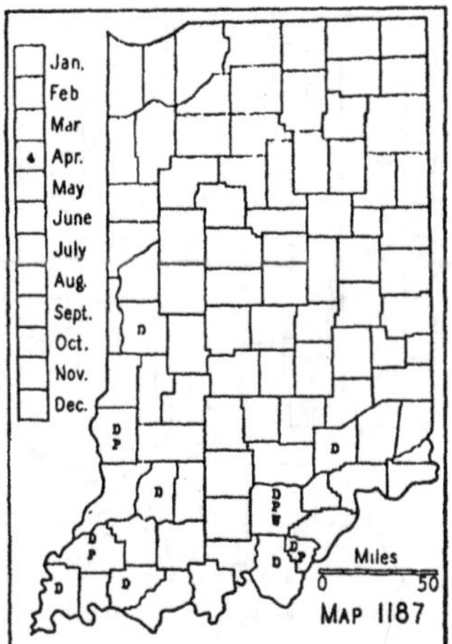

Prunus hortulana Bailey

Prunus pennsylvanica L.f.

Prunus virginiana L

appear early in April, making it the most ornamental species of the genus in this area. It suckers from the roots but not freely.

N. B. to Mass., westw. through n. Ind. to Minn.

6. **Prunus hortulàna** Bailey. HORTULAN PLUM. Map 1187. The specimens which I refer to this species are from the southwestern part of the state. My specimens were collected mostly along roadsides. In Sullivan County it is a common tree on the wooded terrace of the bank of the Wabash River. This species is said not to sucker and if this is true, I have wrongly determined a few specimens and they should be referred to *Prunus Munsoniana* Bailey which has not been reported from Indiana. I think this plum has been introduced into Indiana.

Cent. Ky. to Iowa and Kans., southw. to Tenn. and Okla.

7. **Prunus pennsylvánica** L. f. PIN CHERRY. Map 1188. This species is local and in this state is restricted to the northwestern counties where it is found in wet woodland, senescent tamarack bogs, interdunal swamps, and rarely on dry, sandy soil in the dune area. Reports for this species in Indiana outside the area shown on the map should, no doubt, be referred to some other species.

Lab. to B. C., southw. to Pa., Iowa, and in the mts., to N. C. and Colo.

8. **Prunus virginiàna** L. (*Padus nana* (DuRoi) Roem. of Britton and Brown, Illus. Flora, ed. 2.) COMMON CHOKECHERRY. Map 1189. Local to frequent in the lake area where it grows mostly in moist, alluvial soil in woodland, along streams and fences, and on the borders of interdunal swamps along Lake Michigan. Local in southern Indiana.

Newf. to S. Dak., southw. to Fla., Kans., and Tex.

8a. **Prunus virginiana** var. **demíssa** (Nutt.) Torr. The variety is frequent on the crests of the high dunes facing Lake Michigan and on the crests of the low dunes along Lake Michigan west of Gary. Here it replaces the species. I have not found it except near the lake. I found a plant about 2 inches in diameter in a tamarack bog in the Pokagon

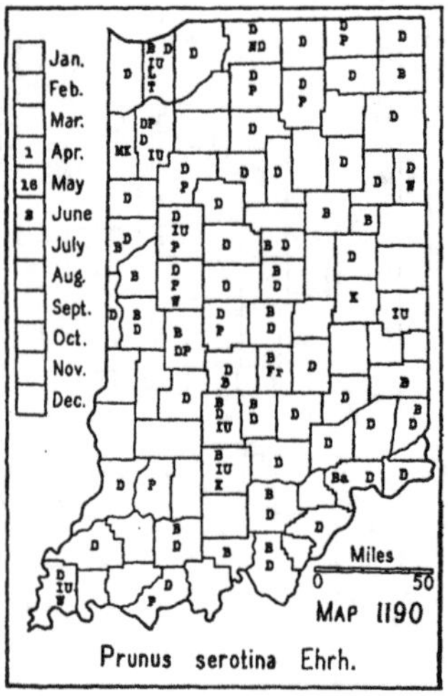

Prunus serotina Ehrh.

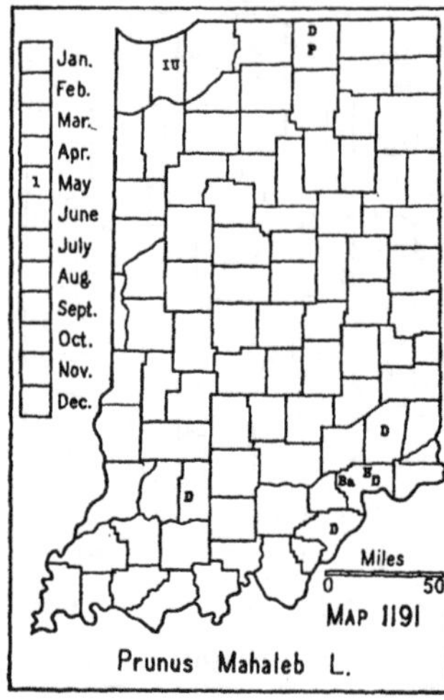

Prunus Mahaleb L.

Desmanthus illinoensis (Michx.) MacM

State Park, Steuben County. I also found a few specimens in Lagrange County in very sandy soil on the high bank of Pigeon River about 2 miles southeast of Mongo. In cultivation this variety soon forms close colonies.

Ind., Ill., n. Nebr. to B. C., southw. to Calif., N. Mex., and Tex.

9. **Prunus serótina** Ehrh. (*Padus virginiana* (L.) Mill. of Britton and Brown, Illus. Flora, ed. 2.) BLACK CHERRY. Map 1190. This species will not endure shade. It bears innumerable fruits and the seed germinate readily. The sportsmen favor this tree because its fruit is greedily eaten by birds, while the land owners condemn it because it is difficult to keep fencerows and roadsides free from it. In the primitive forest I think it was infrequent and only locally frequent in its habitat. It was found in beech and sugar maple and basswood and sugar maple habitats, usually associated with black walnut and tulip tree. It was rarely found on black and white oak ridges or in lowland woods. It is now found throughout the state in open woodland and along fences and roadsides.

N. S. to N. Dak., southw. to Fla. and Ariz.

10. PRUNUS MÀHALEB L. MAHALEB CHERRY. Map 1191. I have found this species as an escape in a few counties and it has been reported from Monroe County. I found several large trees in Clifty Creek Valley in Jefferson County. Miss Edna Banta writes me that it is a more or less frequent tree on the wooded bluffs of the Ohio River from Brooksburg to Madison in Jefferson County.

Nat. of Eu.; N. Y., Ont., and Ind., southw. to Del.

128. LEGUMINÒSAE Juss. PEA FAMILY

Leaves all simple.
  Small trees with large, cordate leaves........................3526. CERCIS, p. 585.
  Low herbs with non-cordate leaves......................3669. CROTALARIA, p. 592.
Leaves not all simple (pinnate, bipinnate or digitate).
  A. Leaves even-pinnate.
    Leaves even-pinnate or bipinnate, not ending in a tendril.
      Herbs.

Leaves bipinnate; leaflets minute, about 1 mm long; flowers small, whitish; pods about 1 cm long, strongly curved.......3450. DESMANTHUS, p. 585.

Leaves pinnate; leaflets larger, more than 5 mm long; flowers yellow; pods not strongly curved, more than 1 cm long.............3536. CASSIA, p. 586.

Trees.

Leaflets ovate, acute or acuminate; unarmed, dioecious trees; flowers in long, many-flowered racemes, about 1.5 cm long, pinkish white; pods 1-2 dm long, the valves very thick and woody......3545. GYMNOCLADUS, p. 590.

Leaflets oblong-lanceolate or oval, obtuse at the apex; trees armed with long thorns (unarmed in one rare variety); flowers polygamous, minute, in short, axillary spikes, greenish yellow; pods 3-40 cm long, 1-many-seeded, the valves not thick and woody................3544. GLEDITSIA, p. 589.

Leaves even-pinnate or bifoliolate, the rachis prolonged into a tendril, rarely the prolongation reduced to less than 1 cm long.

Styles terete, bearded only at the summit; wings and keel usually adherent..... ....3852. VICIA, p. 616.

Styles flattened, bearded along the inner face (this feature best observed in unexpanded flowers); wings of flowers usually free..3854. LATHYRUS, p. 617.

A. Leaves odd-pinnate.

Trees with 7-11 leaflets; terminal leaflets usually 6-9 cm wide; flowers in large panicles 2.5-5 dm long; pods glabrous, 4-8 cm long..3606. CLADRASTIS, p. 591.

Trees, shrubs or herbs not as above; terminal leaflets less than 6 cm wide.

B. Leaves mostly trifoliolate, or digitately 3-11-foliolate.

Stamens 10, distinct; leaflets entire; flowers large; pods inflated, generally 1-5 cm long.................................3618. BAPTISIA, p. 591.

Stamens 10, monadelphous or diadelphous (9 and 1); pods not inflated.

Leaves glandular-dotted above or beneath.

Peduncle 1-flowered; pods not wrinkled, usually 4-7-seeded............. ....3696. HOSACKIA, p. 597.

Peduncle many-flowered; pods wrinkled, 1-seeded..3703. PSORALEA, p. 597.

Leaves not glandular-dotted; pods not wrinkled.

C. Leaflets serrulate; pods 1-6-seeded, small, indehiscent or tardily dehiscent. (The clovers, melilots, alfalfa, and medic.)

Pods curved or coiled; flowers in racemes, spikes or heads, yellow or purplish; stamens free from the corolla....3688. MEDICAGO, p. 593.

Pods straight.

Inflorescence a head or spikelike; stamens adhering to the corolla. (The clovers.)......................3690. TRIFOLIUM, p. 595.

Inflorescence a raceme, white or yellow; stamens free from the corolla. (The melilots.)..............3689. MELILOTUS, p. 594.

C. Leaflets entire.

Leaflets digitately 7-11-foliolate..............3672. LUPINUS, p. 593.

Leaflets pinnately 3-foliolate.

Fruit a flat, 1-7-jointed pod, at maturity separating into as many segments as there are seed in the pod, the segments rounded above and below or the lower part angular; surface of segments densely pubescent, each hair ending in a minute hook; flowers purplish or white, never bright yellow; leaflets generally stipellate..............................3807. DESMODIUM, p. 603.

Fruit not as above.

Leaflets not stipellate.

Pods 1-seeded.

Flowers bright yellow; pods 1- or 2-jointed, ribbed lengthwise, thick, coriaceous, not symmetrical, the lower joint empty; leaflets mostly 3-8 mm wide..3802. STYLOSANTHES, p. 603

Flowers purplish or yellowish white; pods not jointed, not ribbed lengthwise, flat, symmetrical, the valves not coriaceous; leaflets mostly larger than the preceding........ ....3820. LESPEDEZA, p. 610.

Pods more than 1-seeded..............3696. HOSACKIA, p. 597.

Leaflets stipellate.

Style beardless; flowers about 12 mm long; pods less than 5 cm long; twining, herbaceous vines.

Calyx ebracteolate; leaflets of a broad, ovate type, the terminal one about as wide as long; pods densely bearded along the sutures.....................3860. AMPHICARPA, p. 620.

Calyx bibracteolate; leaflets of a narrow-ovate, oval or elliptic type, the terminal one about twice as long as wide; pods not bearded along the sutures....3882. GALACTIA, p. 621.

Style bearded lengthwise on the upper surface.

Flowers yellow, keel strongly curved but not forming a spiral; pods nearly terete, 5-seeded. (The cow peas.).......... ....3905. VIGNA, p. 623.

Flowers purplish or nearly white.

Flowers usually 1 or 2 in the axils of the leaves, mostly 4-5 cm long; calyx about 15 mm long, deciduous; pods 2-5 cm long; stipules and stipels rather conspicuous, persistent. ....3857. CLITORIA, p. 620.

Flowers less than 4 cm long.

Inflorescence of short sessile racemes in the axils of leaves; pods sessile, flat, about 1 cm wide. (The soybeans.) ....3864. GLYCINE, p. 621.

Inflorescence of racemes or umbels on long peduncles, in the axils of leaves.

Flowers in long, loose racemes, the keel spirally coiled; lower calyx lobe shorter than the tube; seed about 8 mm long, glabrous.....3901. PHASEOLUS, p. 622.

Flowers in umbel-like clusters, the keel long, strongly incurved; lower calyx lobe as long as or longer than the tube; seeds less than 7 mm long, mealy-pubescent (pubescence easily detached in one species). ....3901A. STROPHOSTYLES, p. 622.

B. Leaves with 5 or more leaflets (rarely a specimen with a few 3-foliolate leaves).

Upper or lower surface of leaflets with small, resinous dots; pods 1- or 2-seeded.

Stamens 5; leaflets (3) 5-9, 1-6 mm wide, apiculate, dotted beneath; flowers white or purplish; pods 1-seeded.........3710. PETALOSTEMUM, p. 600.

Stamens 9 or 10; leaflets 9-49.

Leaflets many, mostly less than 6 mm wide, obtuse, glabrous; pods 1-seeded....................................3709. DALEA, p. 600.

Leaflets 9-many, mostly more than 6 mm wide, more or less pubescent at least beneath.

Woody shrubs; pods not prickly.............3707. AMORPHA, p. 599.

Perennials; pods prickly..................3769. GLYCYRRHIZA, p. 602.

Upper and lower surface of leaflets without resinous dots.

Flowers in umbels on long, terminal or axillary peduncles, rose color; pods mostly 1-2 cm long, 4-angled, at maturity breaking up into 3-7 indehiscent segments........................3774. CORONILLA, p. 602.

Flowers not in umbels; pods at maturity not breaking up into indehiscent segments.

Pods mostly 3-8 mm long, 1- or 2-seeded; flowers 6-8 mm long, purplish, in dense spicate racemes, 6-12 cm long; shrubs 0.6-4 m high.........
....3707. AMORPHA, p. 599.

Pods more than 8 mm long; flowers more than 8 mm long.

Trees with spiny, woody stipules; leaflets 7-17; flowers in racemes 7-15 cm long, white, about 1.5 cm long; pods very flat, about 1 cm wide, glabrous; seed about 4 mm long............3733. ROBINIA, p. 602.

Herbs or woody vines, lacking spiny stipules; flowers not white; pods and seeds not as above.

Leaflets (3) 5-9, large, of an ovate type, generally 2-8 cm long; twining herbs or woody vines.

Twining herbs, 1-2 m long; leaflets (3) 5-7, large, the basal pair the largest, the largest blade up to 8 cm long; pods glabrous, the longest about 8 cm long, only slightly compressed; flowers maroon, many, in long axillary racemes..3874. APIOS, p. 621.

Twining, woody vines, up to 8 m long; leaflets usually 9, 3-7 cm long; flowers in rather dense racemes 15-35 cm long, lilac purple; pods 7-12 cm long............3722. WISTERIA, p. 601.

Leaflets 15-31, of a narrow type, elliptic, oval, oblong, or linear-oblong, less than 4 cm long; erect or ascending herbs, generally 3-9 dm high.

Stems, pods, and under surface of leaflets densely long-pubescent; pods 3-5 cm long..................3718. TEPHROSIA, p. 601.

Stems, pods, and under surface of leaves glabrous or glabrate, sometimes the lower surface of leaflets closely appressed-pubescent; pods glabrous, about 1.5 cm long................
.................................3766. ASTRAGALUS, p. 602.

3450. DESMÁNTHUS Willd.

1. DESMANTHUS ILLINOÉNSIS (Michx.) MacM. (*Acuan illinoensis* (Michx.) Kuntze.) ILLINOIS MIMOSA. Map 1192. I believe this species was introduced into Indiana from the west. It was first reported in 1878 from Clark County by Baird & Taylor who lived at Jeffersonville, but McMurtrie, who published a flora of the vicinity of Louisville in 1819, and Clapp, who worked intensively the area about New Albany, did not report it. It was not reported from Ohio until about 1900. Short, Peter, & Griswold did not report it from Kentucky. Riddell, who published in 1835, reports it from Kentucky on the authority of Eaton and from the area west of Indiana. Our second published record is dated in 1924. I found it, however, along a railroad in Daviess County in 1910 and along a roadside south of Charlestown in 1915. I have seen it as an abundant plant about ferries and on the rocky slopes of the bank of the Ohio River in Dearborn, Jefferson, and Perry Counties. It has been reported also from Lake, Montgomery, Putnam, and Washington Counties. Its preferred habitat seems to be rocky slopes of banks, embankments of railroads, and prairies.

Ohio to S. Dak., southw. to Ala. and Tex.

3526. CÉRCIS L.

1. **Cercis canadénsis** L. REDBUD. Map 1193. This is generally a small tree, 3-8 inches in diameter, larger ones are rare. The largest redbud I ever saw was located on the Dicksburg Hills in Knox County. It was

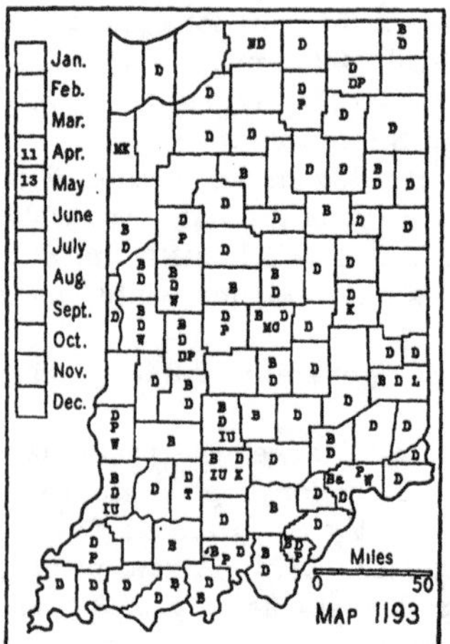

Cercis canadensis L.

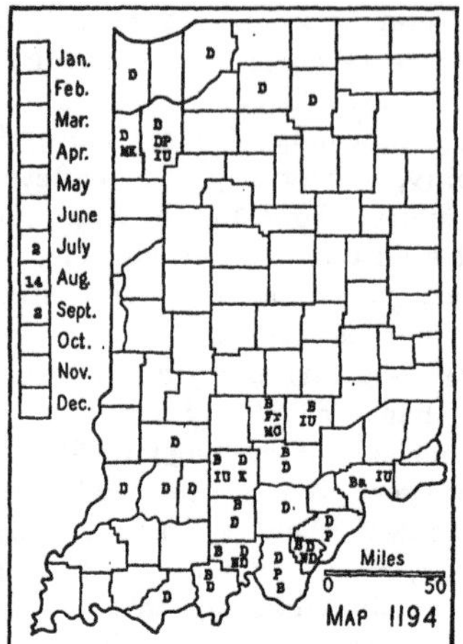

Cassia nictitans L.

Cassia nictitans var. leiocarpa Fern.

more than 2 feet in diameter at breast height. When I reported this tree to Prof. H. C. Cowles of Chicago University, he doubted the identity of the species or the measurements and made a trip to the tree and verified my measurement. It is found in woodland throughout the state, being most abundant in the southern half and infrequent to rare in the northern counties. Its preferred habitat is wooded ravines and banks of streams.

At maturity the leaves are glabrous on both surfaces with a few hairs in the axils of the veins beneath or are more or less pubescent on the lower surface. The glabrous form has been named forma *glabrifolia* Fern. (Rhodora 38: 234. 1936).

N. Y. to Iowa, southw. to Fla. and Tex.

3536. CÁSSIA [Tourn.] L.

Flowers solitary or in small clusters in the axils of the leaves; leaflets 7-20 mm long, 2-5 mm wide; pods straight, mostly less than 7 cm long, erect or ascending.
  Flowers small; pedicels 2-4 mm long; petals 3-8 mm long; stamens 5; pods 2.5-4 cm long; seed 6-9.
    Mature pods closely covered with short incurved hairs.............1. *C. nictitans.*
    Mature pods glabrous............................1a. *C. nictitans* var. *leiocarpa.*
  Flowers large; pedicels 10-20 mm long; petals 10-17 mm long; stamens 10; pods 4-7 cm long; seed 6-15.
    Stems and pedicels with short, incurved hairs; pods glabrate or with short, appressed hairs..2. *C. fasciculata.*
    Stems and pedicels with relatively long, spreading hairs; pods pubescent with relatively long, spreading hairs; whole plant larger and stouter than the preceding..............................2a. *C. fasciculata* var. *robusta.*
Flowers in axillary or terminal racemes; leaflets 2-7 cm long, 6-20 mm wide; pods more or less curved, 6-20 cm long, recurving.
  Leaflets 6 or 4, at least the terminal pair broadly obovate, 1.5-5 cm long; pods up to 2 dm long; annual. (See excluded species no. 384, p. 1065.)............*C. Tora.*
  Leaflets more than 6, oblong, elliptic or ovate to ovate-lanceolate, 2-7 cm long; pods 6-12 cm long; perennial or annual.
    Leaflets acuminate, ovate to ovate-lanceolate, mostly 4-6 pairs....3. *C. occidentalis.*
    Leaflets obtuse or acute, oblong, oblong-lanceolate or elliptic, mostly 6-11 pairs; perennial.

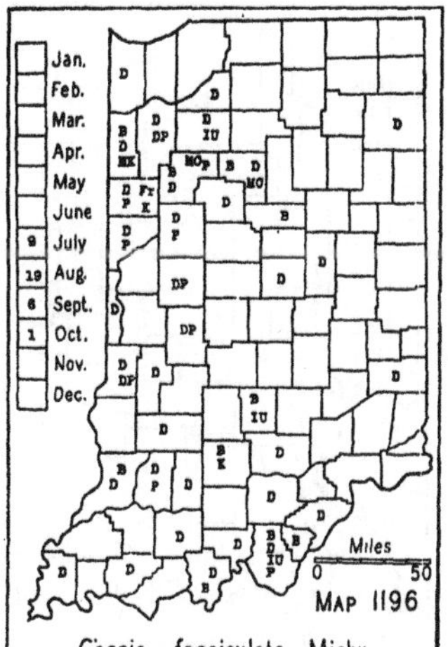

Cassia fasciculata Michx.

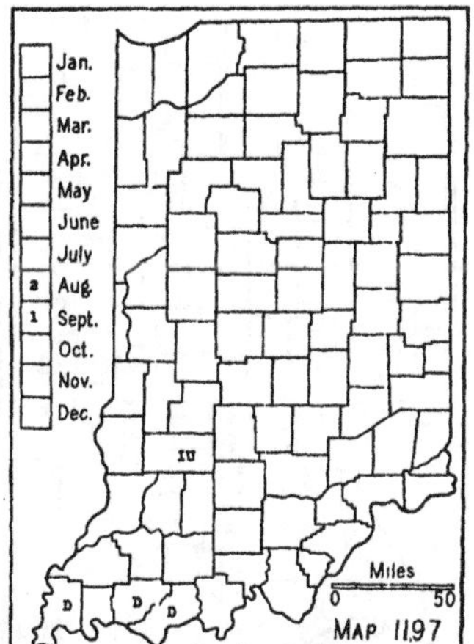

Cassia fasciculata var. robusta (Pollard) Macb.

Cassia occidentalis L.

Pubescence of stem, rachis, petioles, petiolules, and pods long and spreading; leaflets yellow green, more or less ciliate; gland of petiole light brown, on a very short pedicel; segments of pod generally as long as wide. 4. *C. hebecarpa.*

Pubescence of stem, rachis, petioles, petiolules, and pods appressed and shorter; and whole plant much more glabrate than the preceding species; leaflets dark green, the margins more or less ciliate, at least near the base or glabrous; gland of petiole dark brown, sessile; segments of pod wider than long...... .... 5. *C. marilandica.*

1. **Cassia níctitans** L. (*Chamaecrista nictitans* (L.) Moench.) SMALL-FLOWER SENSITIVE-PLANT. Map 1194. Infrequent but locally common in the unglaciated area, and northward either absent or very rare and, I think, introduced. It prefers dry, sandy or sterile soil and in the south it is usually found in open woodland on the crests or slopes of ridges, along roadsides, and in fallow fields.

Vt. to Kans., southw. to Fla. and Tex.

1a. **Cassia nictitans** var. **leiocárpa** Fern. (Rhodora 38: 423. 1936.) Map 1195. I found this variety in two places in Brown County and Friesner has also found it in Brown County. All the specimens found at the various places have both the stem and legume glabrous except one that has the stem densely pubescent as in the typical form.

Pine Mountain, Bell Co., Ky., Ind., and Ohio.

2. **Cassia fasciculàta** Michx. (*Cassia Chamaecrista* L. of manuals and *Chamaecrista fasciculata* (Michx.) Greene in part.) LARGE-FLOWER SENSITIVE-PLANT. Map 1196. This species is infrequent but locally common along roadsides and railroads in the southern and western counties, becoming rare or absent in the northeastern counties. It prefers a moist, sandy soil and, from its abundance in the prairies of our western counties, I believe it is essentially a prairie plant. Almost all of my plants are from roadsides, railroads, and fallow fields, and only a few grew along creeks and in open woodland where the seed could have come from roadsides. I believe this

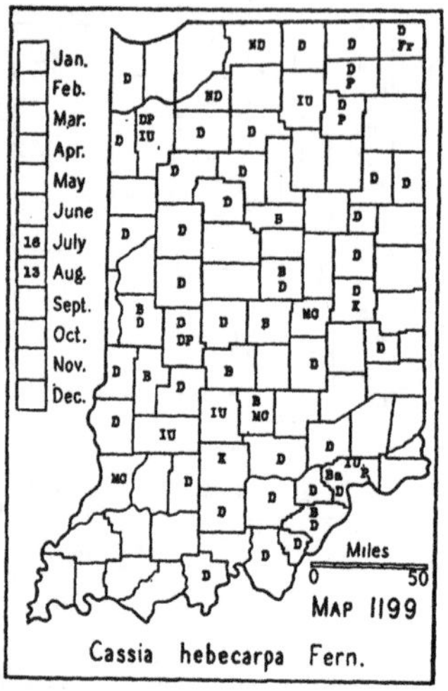

Cassia hebecarpa Fern.

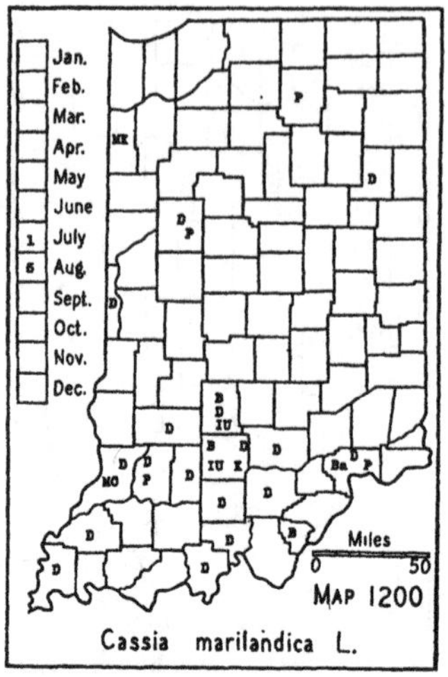

Cassia marilandica L.

Gleditsia aquatica Marsh.

plant has been introduced throughout the state except in a few of our western counties where there are prairie habitats. The preceding statement is based upon the fact that complete stands of this species may be found in suitable habitats along roadsides where the ground has been made bare recently. The dense stands show the viability of the seed and that the most important factor in reproduction is bare, sandy soil.

Mass. to Minn., southw. to Fla. and Tex.

2a. **Cassia fasciculata** var. **robústa** (Pollard) Macbride. (*Cassia Chamaecrista* var. *robusta* Pollard and *Chamaecrista fasciculata* (Michx.) Greene, in part.) STOUT LARGE-FLOWER SENSITIVE-PLANT. Map 1197. One of my specimens was found along a low roadside in Spencer County and another in a fallow field along Otter Creek in Warrick County. Probably native. Paul Weatherwax collected it in Greene County along the Illinois Central Railroad near Bloomfield.

Ohio, Ill. to Mo., southw. to Ga., Fla., and La.

3. **Cassia occidentàlis** L. COFFEE SENNA. Map 1198. Clapp, in his Medicinal Plants of the U. S., published in 1852, on page 79, says: "Two plants have been found growing on the banks of the river at this place, apparently spontaneous." There is a specimen in the herbarium of Wabash College, collected by Dr. Clapp, dated Sept. 14, 1850. There are no other reports. The seed of this species are often used as a substitute for coffee in tropical countries.

Va., Ind. to Mo., southw. to Fla. and Tex.; also southw. through the Americas to Bolivia and Paraguay and in the tropics of the Old World.

4. **Cassia hebecárpa** Fern. (Rhodora 39: 413. 1937.) (*Cassia marilandica* of authors.) WILD SENNA. Map 1199. Found infrequently throughout the state although there are no specimens from the southwestern counties. It prefers a moist soil and is found mostly along roadsides and in pastures and open woods in the alluvial bottoms along streams. In many places this species forms large colonies, especially in

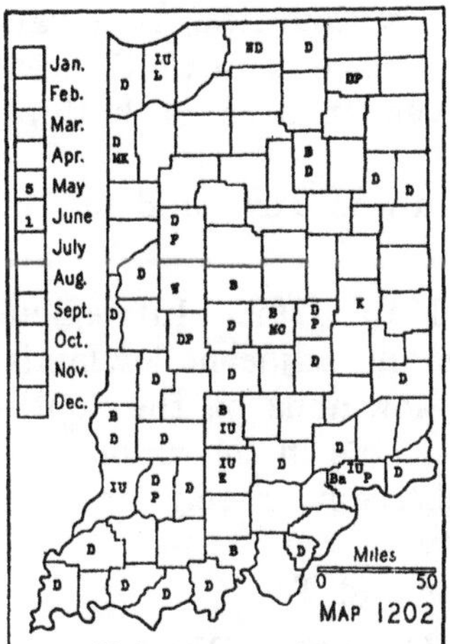

Gleditsia triacanthos L.

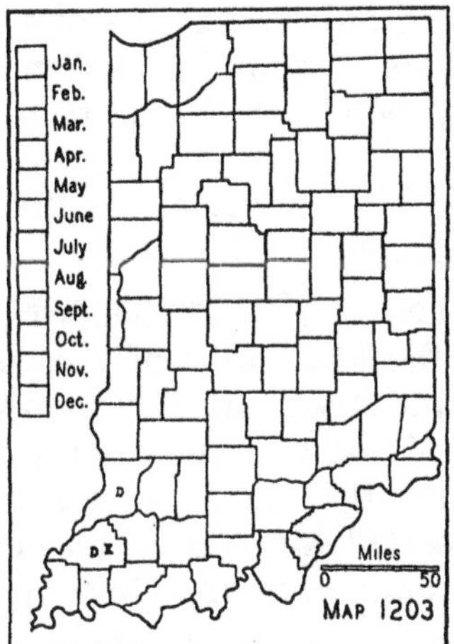

Gleditsia texana Sarg.

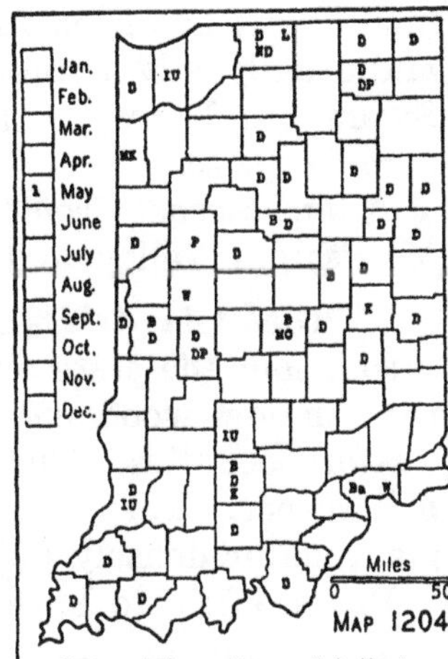

Gymnocladus dioica (L.) Koch

rather sandy soil in the alluvial bottoms of the Tippecanoe River, and elsewhere in similar habitats. It sometimes invades marshland where it is not too wet and forms complete stands. It is to be noted that grazing animals do not eat this or the next species. I have seen thick stands of this species where the blue grass was closely grazed but this plant was not eaten. The plant contains a strong purgative principle.

Mass. to Ind., southw. to N. C. and Tenn.

5. **Cassia marilándica** L. (*Cassia Medsgeri* Shafer.) Map 1200. Infrequent in the southern third of the state, becoming rare northward, and probably entirely absent from the northern counties. It is found mostly in low ground along roadsides and in low woodland and alluvial bottoms along streams. This species is often confused with the preceding one from which it is easily separated by the characters given in the key. The pubescence of *Cassia marilandica* is appressed while that of *Cassia hebecarpa* is spreading. The plant is a darker green, flowers a little later, and is not as aggressive as the preceding.

Pa. to Iowa, southw. to Ga. and Tex.

3544. GLEDÍTSIA L. HONEYLOCUST

Pods (exclusive of stipe) less than 6 cm long, 1-seeded, rarely with 2 or 3 seed; seed orbicular..1. *G. aquatica.*
Pods (exclusive of stipe) more than 6 cm long, pods of normal size with more than 3 seed; seed oval or nearly orbicular.
  Pods mostly more than 15 cm long, with pulp between the partitions or nearly wanting in the thornless form.
    Trees with thorns..2. *G. triacanthos.*
    Trees without thorns..........................2a. *G. triacanthos* f. *inermis.*
  Pods mostly 10-15 cm long, without pulp between the partitions........3. *G. texana.*

1. **Gleditsia aquática** Marsh. WATERLOCUST. Map 1201. This small tree grows on the low borders of sloughs and in swamps in a habitat so low that the base is usually more or less submerged during the winter

months. It has been found only in Gibson and Knox Counties. It is rare and usually only a single tree is found except in one place in Little Cypress Swamp in Knox County where it is common over an area of half an acre or more.

Atlantic coast from N. C. to Fla., along the Gulf to Tex., and up the Mississippi Valley to Indiana.

2. **Gleditsia triacánthos** L. HONEYLOCUST. Map 1202. Infrequent throughout the state on the low banks of streams and adjacent lowlands, rare in low woodland, and frequent in swampy lowlands of the southwestern counties. The pods of this species are variable in the amount of pubescence. They are mostly more or less pubescent, rarely entirely glabrous or densely pubescent all over at maturity.

Pa., s. Mich. to Iowa, southw. to the Gulf States and Tex.

2a. **Gleditsia triacanthos** f. **inérmis** (Pursh) Fassett. (Rhodora 38: 97. 1936.) (*Gleditsia triacanthos* var. *inermis* Pursh.) THORNLESS HONEYLOCUST. The few mature fruited specimens I have examined show that this form has straighter, shorter, and narrower pods than the species and the pods are dry within, not pulpy. The seed are elliptic-oblong, slightly compressed while the seed of the species are much larger and flatter. I have learned from nurserymen who supply western planters with the thornless form for planting that the seed of the thornless form produce about 60 per cent of seedlings without thorns. This form has been reported from Greene, Jefferson, and Lawrence Counties but I have never seen or heard of a thornless tree in northern Indiana. I saw a large tall tree near the top of a ridge in a woods in Fayette County and the remainder of the trees I have seen were in the bottoms along the Wabash River in the southwestern part of the state.

I have no data on its general distribution. Sargent writes that it is the prevailing form in Taney County, Missouri.

3. **Gleditsia texàna** Sarg. TEXAS HONEYLOCUST. Map 1203. This species was first found in Gibson and Knox Counties and was considered a hybrid of the preceding species. Later Sargent described it as a species from a colony found in Texas. In 1921 I found a fine specimen in a cornfield under which there were hundreds of 1-year seedlings which proves that if this form is a hybrid it is a fertile one. I have made no effort to determine the abundance of this species. It is, no doubt, restricted to the southwestern counties and may be very rare since I have seen only a few trees.

Ind., Ark., Miss., La., and Tex.

3545. GYMNÓCLADUS Lam.

1. **Gymnocladus dioìca** (L.) Koch. KENTUCKY COFFEETREE. Map 1204. Infrequent to very rare throughout the state. It is usually found in well drained, alluvial soil along streams and their adjacent terraces. Since the tree has the habit of sending up root suckers at a great distance from the parent tree it is often found in small colonies.

N. Y., Ont. to Minn., southw. to Tenn., Ark., and Okla.

Cladrastis lutea (Michx. f.) Koch

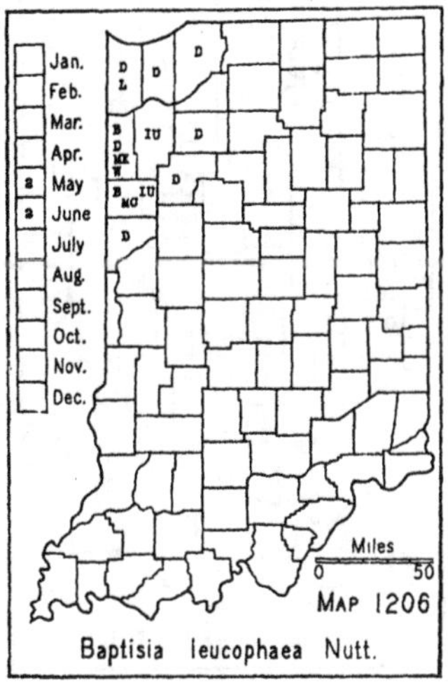

Baptisia leucophaea Nutt.

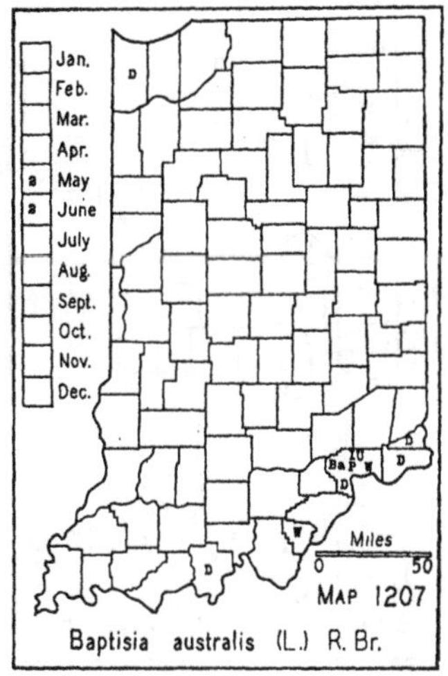

Baptisia australis (L.) R. Br.

3606. CLADRÁSTIS Raf.

1. **Cladrastis lùtea** (Michx. f.) Koch. YELLOW-WOOD. Map 1205. A single colony of this species was found in 1933 in a deep, wooded ravine in the Brown County Game Preserve. It was reported to be present also in a nearby ravine. I was informed that the former owner of the land had cut one tree and had it sawed into boards. The nearest known location of this species is 40 miles south of Evansville. The species may be exceptionally rare or may have been overlooked.

N. C., Tenn., Ky., Ind., Mo., Ala., and Ark.

3618. BAPTÍSIA Vent.

Plants densely pubescent throughout; flowers cream color............1. *B. leucophaea.*
Plants glabrous or nearly so.
Flowers lavender-violet (Ridgway Standard); calyx lobes 3-5 mm long; body of mature, dried pods mostly 4-5 cm long..........................2. *B. australis.*
Flowers white or yellow, calyx lobes less than 3 mm long; body of mature, dried pods mostly less than 3 cm long.
Leaflets 1-3 cm long; flowers yellow; calyx lobes 1-2 mm long; body of mature, dried pod usually 8-12 mm long....................3. *B. tinctoria* var. *crebra.*
Leaflets mostly 2.5-7 cm long; flowers white; calyx lobes mostly 2-3 mm long; body of mature, dried pod 2-3 cm long..........................4. *B. leucantha.*

1. **Baptisia leucophaèa** Nutt. (*Baptisia bracteata* of authors, not (Muhl.) Ell.) CREAM WILD-INDIGO. Map 1206. Infrequent to frequent in dry, sandy prairies and low, open, black oak woods throughout the northwestern part of the state, mostly as shown on the map. It was reported from Steuben County by Bradner. I have on several occasions found this species associated with *Baptisia leucantha* which flowers 1-3 weeks later.

Mich. to Minn., southw. to La. and Tex.

2. **Baptisia austràlis** (L.) R. Br. BLUE WILD-INDIGO. Map 1207. Local on the stony ledges of the slope of the bank of the Ohio River in the counties shown on the map. It is usually more or less frequent to common

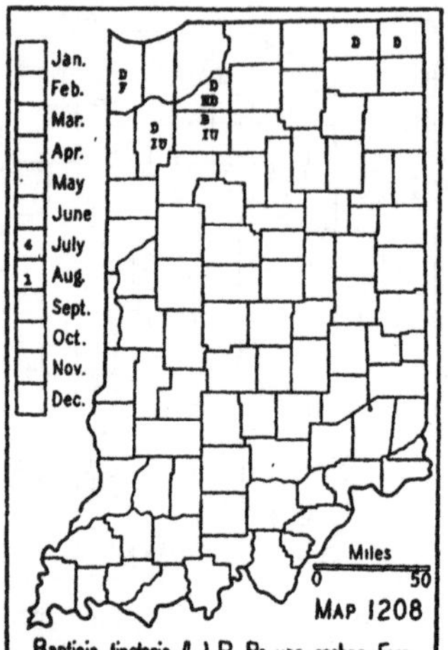

Baptisia tinctoria (L.) R. Br. var. crebra Fern.

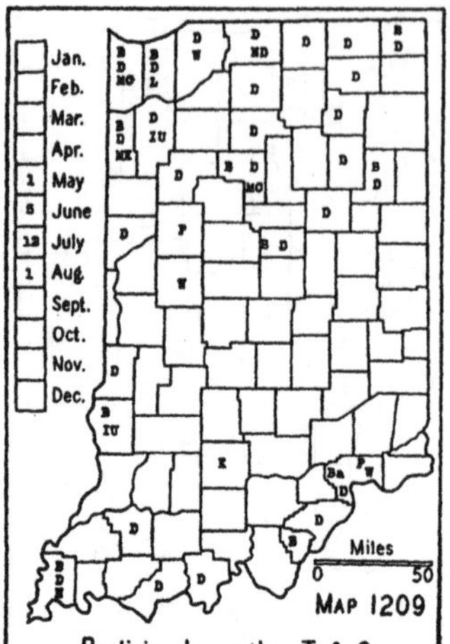

Baptisia leucantha T. & G.

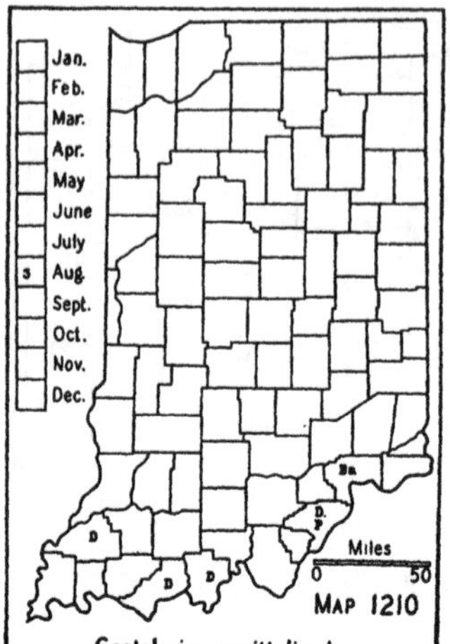

Crotalaria sagittalis L.

where its habitat occurs. About 4 miles east of Madison, Jefferson County, it forms a dense stand for about a quarter of a mile along the bank of the Ohio River where its common associate is *Desmanthus illinoensis*. In 1935 Edwin D. Hull found a colony along the New York Central Railroad tracks in Lake County. It was doubtless a railroad migrant.

Vt. to Ind., southw. to N. C. and Tenn.

3. **Baptisia tinctòria** (L.) R. B. var. **crèbra** Fern. (Rhodora 39: 414-415. 1937.) YELLOW WILD-INDIGO. Map 1208. Very local in the northwestern counties. Outside the range indicated on the map it has been reported from Kosciusko and Tippecanoe Counties. I have found it on a high, gravelly, wooded bank, in open places in woods, in prairie habitats, and most often in depressions in low, sandy black and pin oak woods where tree growth is sparse or absent. Judging from the vegetation in such depressions the soil is slightly acid.

In 1923 I found an aberrant form, probably a hybrid of this species, in Starke County with 40 flowers on the terminal raceme.

This genus is now being monographed and this plant will be given consideration.

Maine, Vt., Ont. to Minn., southw. to N. C. and La.

4. **Baptisia leucántha** T. & G. WHITE WILD-INDIGO. Map 1209. Infrequent in its habitat throughout the state and usually only a few plants found at a place. It is most commonly found in sandy soil in prairie habitats and in thin oak woods. It is also found in hard, clay soil or gravelly soil on white oak slopes, in hard, white clay of the "flats" of the southern counties, and once I found it on a bar in the Wabash River.

Ont. to Minn., southw. to N. C., Fla., and Tex.

3669. CROTALÀRIA L.

1. CROTALARIA SAGITTÀLIS L. RATTLE-BOX. Map 1210. I have collected this species four times and all of the plants were found in old, fallow

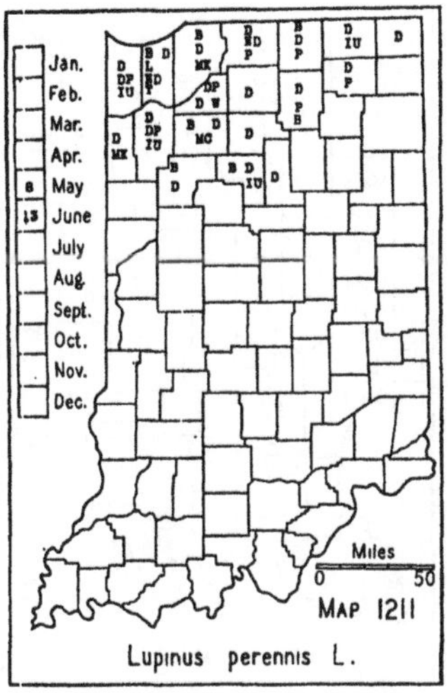

Lupinus perennis L.

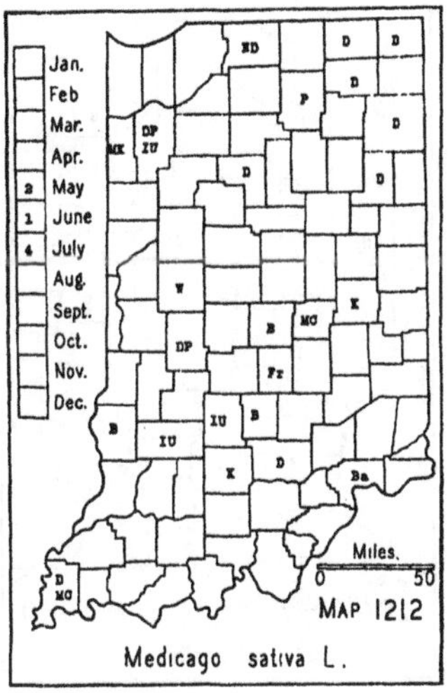

Medicago sativa L.

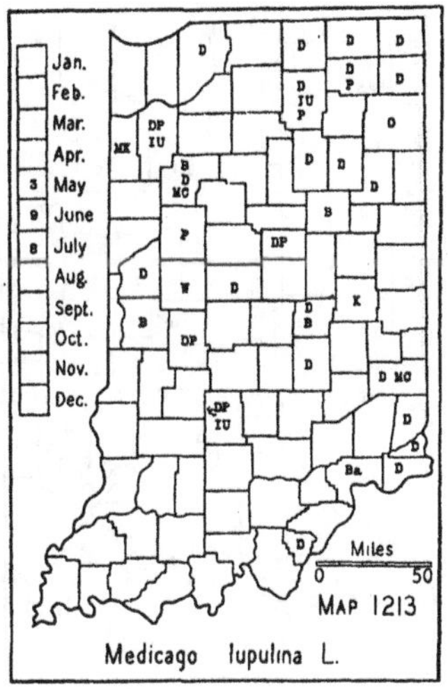

Medicago lupulina L.

fields, usually far removed from a railroad. The one in Perry County was found in dry soil in an old, fallow field about 2 miles east of Oriole where it was associated with thick stands of *Cassia fasciculata.* Pepoon and Umbach report finding two colonies along railroads in the dune area. I think this species has been introduced into Indiana, probably in grass seed or as a railroad waif.

Mass. to S. Dak., southw. to Fla. and Mex.

3672. LUPÌNUS [Tourn.] L.

1. **Lupinus perénnis** L. SUN-DIAL LUPINE. Map 1211. Infrequent to frequent in the counties indicated on the map, including Lake County, but there are no reports outside this area. This species grows only in dry and very sandy soil and is found on roadside knolls, in sandy, fallow fields, and in open, black oak woods, especially on the dunes about Lake Michigan where it formerly covered acres. The usual color of the flowers is blue but they vary from blue to bluish purple, rose, and white. In a large colony one can generally find white forms and usually individuals that are rose color. I find a note on the label of one of my specimens as follows: "Flowers pure white when collected. When taken out of the press a few hours later the flowers were rose color. After drying in a press to which heat had been applied the flowers were blue." The fact that the pubescence varies in density and in length has led to the naming of the more pubescent form which most authors now ignore. The several color forms also bear names which I am omitting.

Maine, Ont. to Minn., southw. to Fla. and La.

3688. MEDICÀGO [Tourn.] L.

Flowers blue purple; pods with 2 or 3 loose coils, 3-4 mm wide, mostly more than 1-seeded; leaflets linear-lanceolate to obovate, usually more than twice as long as wide; plants perennial, mostly erect, decumbent, or ascending, 3-14 dm high. ....1. ***M. sativa.***

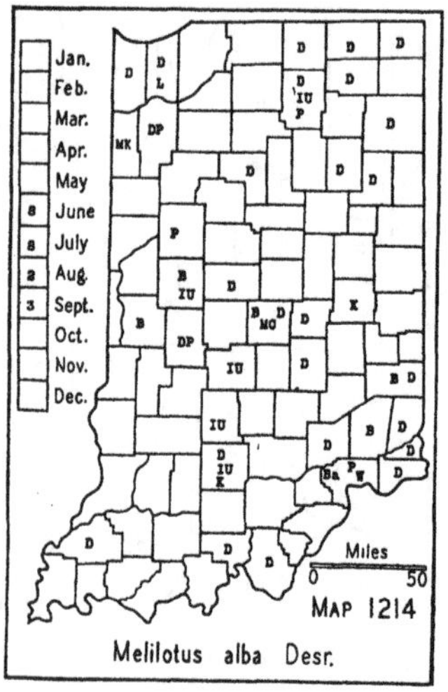

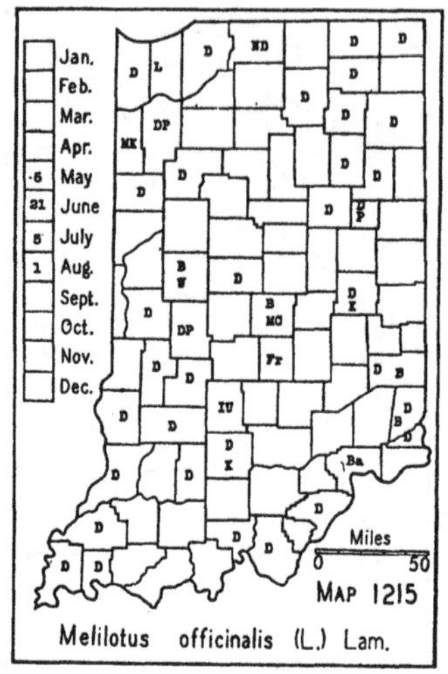

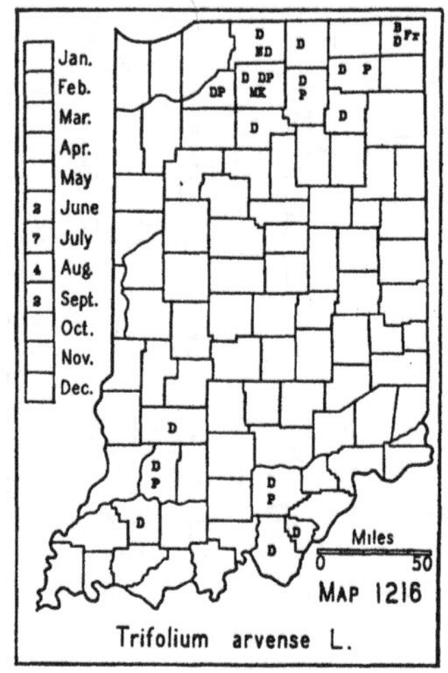

Flowers yellow; leaflets mostly broadly obovate, generally less than twice as long as wide; annual, with long, prostrate or spreading branches, mostly less than 3 dm high but the prostrate branches may be 3-7 dm long.

Pods blackish, strongly curved, prominently rugose but not armed with prickles, 1-seeded, 2-3 mm wide..2. *M. lupulina.*

Pods not blackish, with 2 or 3 coils, 5-8 mm wide; mostly more than 1-seeded. (See excluded species no. 386, p. 1065.)..............................*M. hispida.*

1. MEDICAGO SATÌVA L. ALFALFA. Map 1212. Extensively used throughout the state for grazing and fodder. It has become a frequent escape along roadsides and more rarely along railroads and in waste places and open woodland. I have rarely collected this and the next species so that the maps do not indicate the frequency with which this plant has escaped.

Nat. of Eu.: widely naturalized in the U. S. and Can.

2. MEDICAGO LUPULÌNA L. BLACK MEDIC. Map 1213. Frequent throughout the state along railroads and roadsides and in lawns, waste places, and fields. It was probably mostly introduced in clover seed and lawn grass seed.

Nat. of Eurasia; widely naturalized in N. A.

3689. MELILÒTUS [Tourn.] L.

Flowers white; wings shorter than the standard; pods glabrous, almost smooth or reticulate-alveolate; seed orbicular................................1. *M. alba.*

Flowers yellow; wings equaling the standard; pods generally with strong, transverse ridges, only slightly netted, glabrous; seed ovoid..............2. *M. officinalis.*

1. MELILOTUS ÁLBA Desr. WHITE SWEETCLOVER. Map 1214. This species has been sown for pasture and fodder and has abundantly escaped in all parts of the state to roadsides, railroads, waste places, and fields.

Nat. of Eurasia; widely naturalized throughout N. A.

2. MELILOTUS OFFICINÀLIS (L.) Lam. YELLOW SWEETCLOVER. Map 1215. This species has been sparingly sown for pasture and fodder and has escaped like the preceding species but it is much more aggressive.

Nat. of Eurasia; widely naturalized in N. A.

3690. TRIFÒLIUM [Tourn.] L. CLOVER

Flowers sessile or nearly so, crowded; corolla pink, purple or rose.
  Leaflets narrow, mostly 2-6 mm wide, linear to oblanceolate; heads mostly longer than wide; plants annual, erect .... 1. *T. arvense.*
  Leaflets mostly more than 6 mm wide, oval, ovate, obovate or cuneate-obovate; heads globose or subglobose.
    Plants pubescent, ascending, perennial; flowers purplish .... 2. *T. pratense.*
    Plants glabrous, ascending or diffuse, annual; flowers rose ...... 3. *T. resupinatum.*
Flowers on short pedicels; heads looser.
  Flowers, white, purplish or crimson.
    Heads much longer than wide; leaflets pubescent; flowers crimson. (See excluded species no. 388, p. 1065.) .... *T. incarnatum.*
    Heads globose; leaflets glabrous or nearly so; flowers white or purplish.
      Calyx lobes 2-3 mm long, about as long as the tube.
        Plants stoloniferous, creeping; heads on long peduncles, arising from prostrate stems .... 4. *T. repens.*
        Plants not stoloniferous, ascending; heads on peduncles not arising from prostrate stems .... 5. *T. hybridum.*
      Calyx lobes mostly about 4 mm long, much longer than the tube.
        Plants stoloniferous; sinuses of the calyx pubescent. (See excluded species no. 390, p. 1066.) .... *T. stoloniferum.*
        Plants not stoloniferous; sinuses of the calyx not pubescent.
          Calyx pubescent. (See excluded species no. 389, p. 1065.) .... *T. reflexum.*
          Calyx glabrous .... 6. *T. reflexum* var. *glabrum.*
  Flowers yellow.
    Terminal leaflet longer stalked than the lateral ones; stipules ovate.
      Heads 20-40-flowered; corolla conspicuously striate .... 7. *T. procumbens.*
      Heads 5-12-flowered; corolla scarcely striate .... 8. *T. dubium.*
    Terminal leaflet not longer stalked than the lateral ones; corolla conspicuously striate; stipules linear-lanceolate .... 9. *T. agrarium.*

1. TRIFOLIUM ARVÉNSE L. RABBIT-FOOT CLOVER. Map 1216. In dry sandy soil along roadsides and in pastures, open woodland, and fallow fields.

Nat. of Eurasia; Que. and Ont. to Mo., southw. to Fla. and Tenn.

2. TRIFOLIUM PRATÉNSE L. RED CLOVER. Map 1217. This species is much sown for pasture and fodder and has frequently escaped in all parts of the state to roadsides, waste places, and fallow fields.

Nat. of Eurasia; widely naturalized in N. A.

3. TRIFOLIUM RESUPINÀTUM L. STRAWBERRY CLOVER. This species was discovered north of Indianapolis by W. N. Clute, May 9, 1932. He says it occurs for a mile along the old canal and along roadsides.[1]

Greece to Persia; Mass., Pa., and Wis. southw. to Ala.

---

[1] Clute (Amer. Botanist 45:32. 1939) says: The severe winter of 1935-36 apparently killed all the plants.

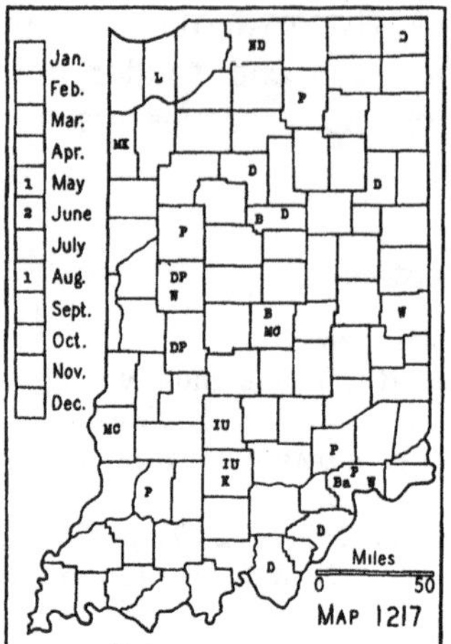

Trifolium pratense L.

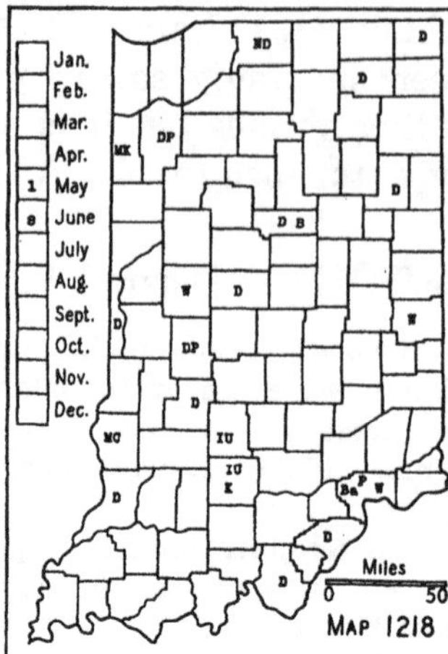

Trifolium repens L.

Trifolium reflexum var. glabrum Loj.

4. TRIFOLIUM RÈPENS L. (Erith. Monograph on White Clover. pp. 1-x, 1-150. 1924. Duckworth & Co. London.) WHITE CLOVER. Map 1218. Found throughout the state. Common in lawns, waste places and pastures and less frequent in fallow fields and open woodland and along roadsides and railroads. Erith describes several varieties and forms and, no doubt, some of them are in Indiana.

Nat. of Eurasia; widely naturalized in N. A.

5. TRIFOLIUM HÝBRIDUM L. ALSIKE CLOVER. This species has been freely sown as a pasture and fodder plant throughout the state and has escaped frequently. No effort has been made to collect this species, *Trifolium pratense* or *Trifolium repens;* consequently the maps do not indicate the frequency with which they have escaped, but no doubt all are found frequently in every county.

Nat. of Eu.; widely naturalized in N. A.

6. **Trifolium refléxum** L. var. **glàbrum** Lojacono. Map 1219. The flowers of the plants I have seen are white and odorless and the pods are about 4-seeded. I found it to be a common plant in hard, white, slightly acid, clay soil in a clearing, formerly wooded with swamp white oak and pin oak, along Little Pigeon Creek in Spencer County. I found it in great colonies in a low, flat woods 10 miles southwest of Mt. Vernon in Posey County where it was associated with post oak, agave, and *Baptisia leucantha.* Other specimens were found in dry woods, on a cliff along White River, and in dry, sandy soil in a prairie habitat in Vigo County. This is the western form of this species. The type came from Augusta, Illinois.

I have seen specimens from Va., Ohio (Wellington), Ill., Iowa, Mo., and Okla.

7. TRIFOLIUM PROCÚMBENS L. LOW HOP CLOVER. Map 1220. Probably infrequent throughout the state along roadsides and railroads and in pastures, open woodland, waste places, and fallow fields.

Nat. of Eu.; N. S. to Wash., southw. to Ga. and Miss.

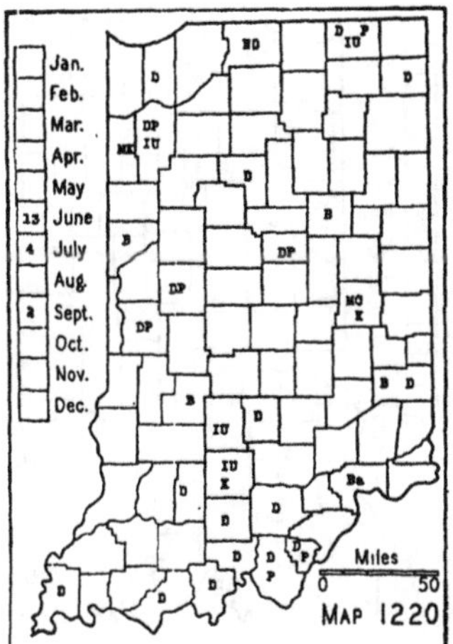

Trifolium procumbens L

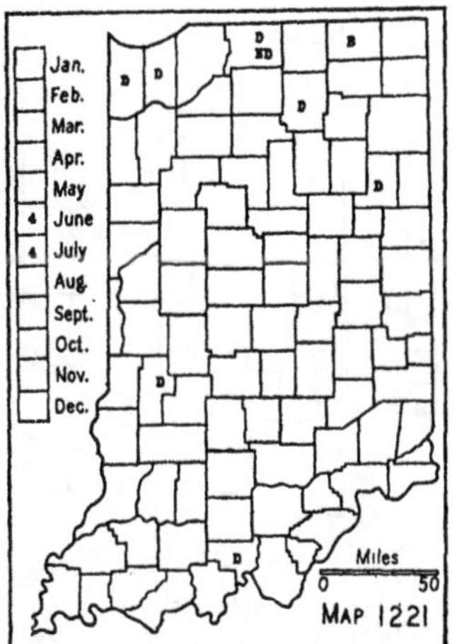

Trifolium agrarium L.

Psoralea tenuiflora Pursh

8. TRIFOLIUM DÙBIUM Sibth. LITTLE HOP CLOVER. In 1909 I found this species in gravelly soil among the cottages on the north side of Lake Wawasee. It has been collected in St. Joseph County by Nieuwland, and on May 17, 1930, Nieuwland and Just again collected it in St. Joseph County on the bank of the St. Joseph River behind St. Mary's College.

Nat. of Eu.; Mass. to Va., Tenn., and Ark., southw. to Ga. and Miss.

9. TRIFOLIUM AGRÀRIUM L. YELLOW HOP CLOVER. Map 1221. Probably introduced throughout the state. My specimens are mostly from open woods, pastures, fallow fields, and roadsides.

Nat. of Eu.; Newf. to Ont. and Iowa, southw. to Ga.

3696. HOSÁCKIA Dougl.

See excluded species no. 391, p. 1066.

3703. PSORÀLEA [B. Juss.] L.

Leaves digitately 3-5-foliolate, the petiolules of about equal length; both surfaces of the leaflets conspicuously black-dotted; pods not wrinkled.........1. *P. tenuiflora*.

Leaves 3-foliolate; terminal leaflet stalked; the lower surface not conspicuously covered with black dots; pods wrinkled.

Racemes short, 1-2 cm long, about equaling the leaves...............2. *P. stipulata*.

Racemes more than 2 cm long, mostly longer than the leaves.

Leaflets large, rhombic-ovate to rhombic-lanceolate, 5-10 cm long, 2.5-6 cm wide; pods about 1 cm long....................................3. *P. Onobrychis*.

Leaflets linear-lanceolate to elliptic, 4-7 cm long, 1-2 cm wide; pods about 4 mm long................................4. *P. psoralioides* var. *eglandulosa*.

1. PSORALEA TENUIFLÒRA Pursh. FEW-FLOWER PSORALEA. Map 1222. In 1933 I found scattered plants of this species in ballast along the railroad for a quarter of a mile about 4 miles south of Vincennes, Knox County. Nieuwland has collected it in St. Joseph County. In 1901 Stuart reported it as found along the railroad south of Lafayette. Peattie reported it from the prairies of the Calumet District near Lake Michigan.

Ill. to S. Dak. and Mont., southw. to Tex. and Sonora.

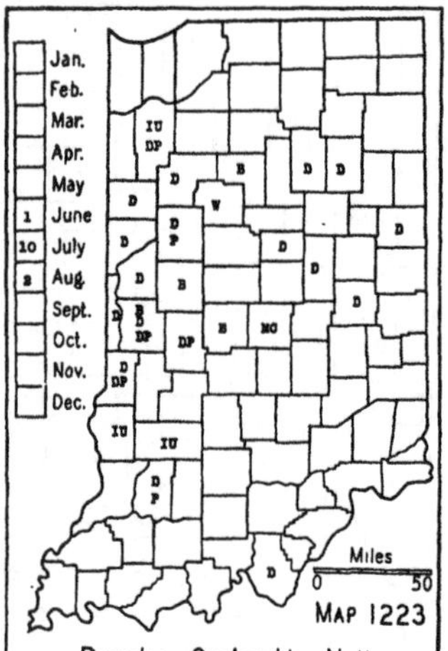

Psoralea Onobrychis Nutt.

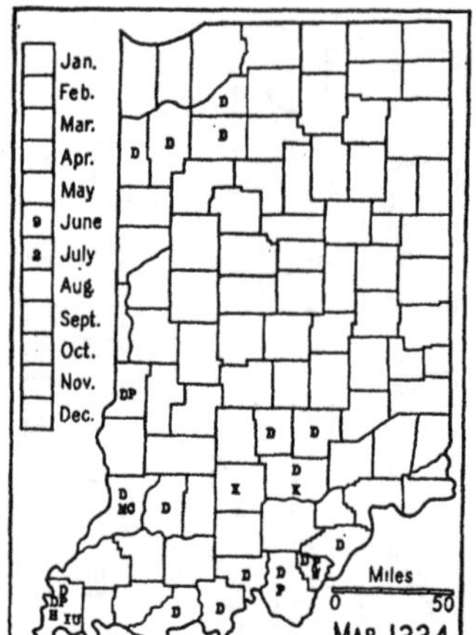

Psoralea psoralioides (Walt.) Cory var. eglandulosa (Ell.) Freeman

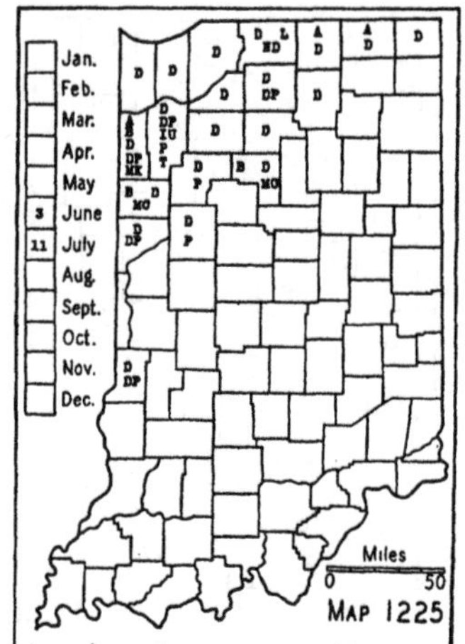

Amorpha canescens Nutt.

2. **Psoralea stipulàta** T. & G. According to Vail (Bull. Torrey Bot. Club **21**:113. 1894) the type specimen of this species was collected June 8, 1839, by Wm. Jones on Rock Island at the Falls of the Ohio River, Clark County, Ind. She also writes: "In the collections of C. W. Short, preserved in the Herb. Acad. Phila., there are notes to the effect that he never found this plant in fruit growing wild, and that he cultivated it vainly for years. His collections of *P. stipulata* in the herbaria examined, cover a period of some twenty years." J. M. Coulter wrote (Bot. Gaz. **1**: 9. 1876) that Dr. Clapp's collection contained a specimen. This specimen was collected in 1838 in the vicinity of New Albany, Floyd County, and is now in the herbarium of Wabash College. P. A. Rydberg wrote me that the specimens in the New York Botanical Garden were immature and might be some form of *Desmodium*. Thus it seems that this species, if a valid one, is extinct. Known only from the type locality.

3. **Psoralea Onóbrychis** Nutt. SAINFOIN PSORALEA. Map 1223. Probably frequent to very rare throughout the state except in the extreme northern counties. It is found mostly along roadsides and in alluvial bottoms of streams. It is essentially a prairie plant but is occasionally found in wooded areas and it is a question whether it exists in some places as a relict or is an invader.

Ohio, Ill. to Mo., southw. to N. C. and Tenn.

4. **Psoralea psoralioìdes** (Walt.) Cory var. **eglandulòsa** (Ell.) Freeman. (Rhodora **39**:426. 1937.) (*Psoralea pedunculata* Vail of Indiana authors.) Map 1224. This species has a limited distribution in Indiana but has a wide range of habitats. It is more or less frequent in the unglaciated area on the crests and upper parts of the highest ridges, usually associated with chestnut oak and black oak. In the southwestern counties it is rare and is found in dry, sandy soil or in the lowland with post oak. In the northwestern counties it is found in black and pin oak

clearings, in sandy soil near the bases of black oak ridges, and in prairie habitats.

Va., Ohio, to Kans., southw. to Fla. and Tex.

3707. AMÓRPHA L.

[Palmer, E. J., Conspectus of the genus Amorpha. Jour. Arnold Arboretum 12:157-197. 1931.]

Mr. Palmer has seen all of my specimens of *Amorpha fruticosa* and varieties and made the key to them which is used here.

Calyx lobes lanceolate, as long as or longer than the tube; shrubs of a dry, sandy or gravelly habitat, mostly less than 1 m high, densely canescent; leaflets of the upper part of the stem generally 15-20 pairs, crowded, canescent beneath, 1-2 cm long. .... 1. *A. canescens.*

Calyx lobes deltoid or half-rounded, much shorter than the tube; shrubs of moist or rocky banks, 1-4 m high, more or less pubescent; leaflets of upper part of stem mostly 7-15 pairs, rarely as many as 20 pairs, generally not crowded, 2-4 cm long.

Pubescence of petiolules and leaflets consisting of curled or matted hairs.

Leaflets mainly rounded or short-pointed at the apex........ 2. *A. fruticosa.*

Leaflets mainly truncate or emarginate at the apex........ 2a. *A. fruticosa* var. *emarginata.*

Pubescence of petiolules and leaflets consisting of short, straight, appressed hairs, or nearly absent.

Leaflets obovate or oval, not conspicuously crowded, with appressed pubescence. .... 2b. *A. fruticosa* var. *angustifolia.*

Leaflets oblong, more numerous and crowded, glabrous or nearly so. (Some excluded specimens may be this variety.)........ *A. fruticosa* var. *oblongifolia.*

1. **Amorpha canéscens** Nutt. LEADPLANT. Map 1225. This species is infrequent and is restricted to the area shown on the map. It is found in dry, sandy or gravelly soil on knolls and ridges or in a prairie habitat in the open along roadsides or in open woodland.

Mich. and Ind. to Man., southw. through the Mississippi Valley to Ark., N. Mex., and Tex.

2. **Amorpha fruticòsa** L. INDIGOBUSH. Map 1226. As represented by my specimens this species is restricted to the alluvial bottoms and banks of the Lower Wabash Valley and the moist or rocky slopes of the Ohio River. I have one specimen, however, which is from sandy soil near the Kankakee River south of Thayer, Lake County. *Amorpha fruticosa* and its varieties are locally common in the southwestern part of Posey County and in the southwestern part of Vigo County on the banks of sloughs and swamps where it is usually closely associated with buttonbush.

?Conn. to Minn., southw. to Ala. and Okla.; escaped from cultivation in the northeast.

2a. **Amorpha fruticosa** var. **emarginàta** Pursh. My only specimens of this variety are from the borders of sloughs in Gibson County.

Miss. to Ark. and Ill.

2b. **Amorpha fruticosa** var. **angustifòlia** Pursh. I have this variety

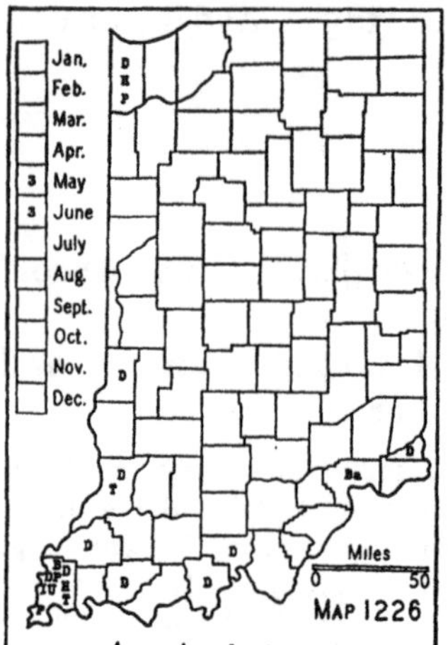

Amorpha fruticosa L.

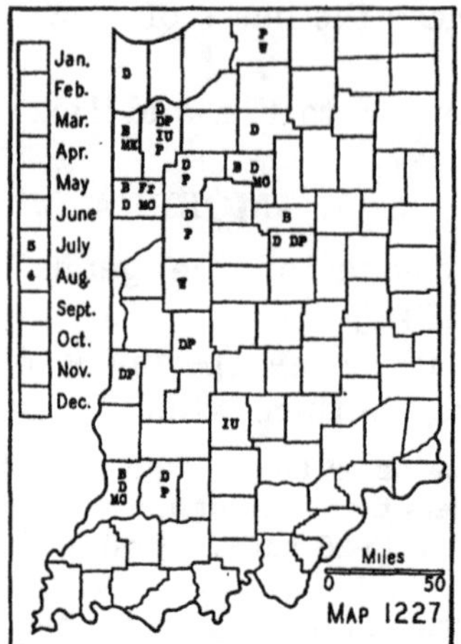

Petalostemum purpureum (Vent.) Rydb.

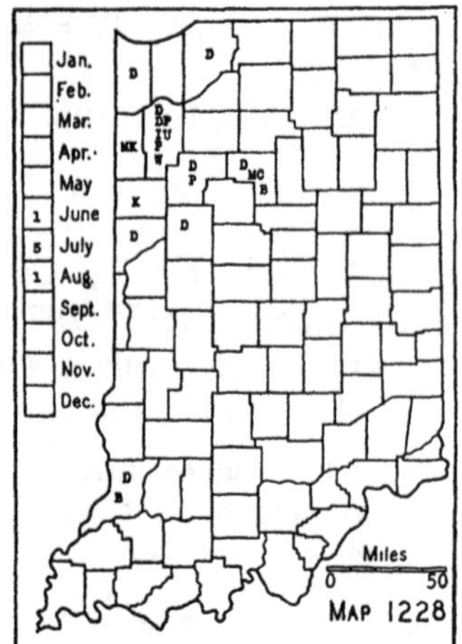

Petalostemum candidum Michx.

from Spencer, Switzerland, and Vigo Counties and Miss McKee found it in Newton County near the Kankakee River.

Wis. and Minn. to Sask., southw. to Tex., and n. Mex.

3709. DÀLEA Juss.

1. DALEA ALOPECUROÌDES Willd. On September 11, 1924, Mrs. Harry Bucklin of Brazil sent me a specimen which was collected at her summer home located in section 24 about 6 miles northeast of Brazil, Clay County. She wrote: "Frequent along the roadside and in an adjoining fallow field." I found the colony still persisting in 1934. No doubt introduced in seed of some kind since the farm is located on a little used road and not near a railroad.

Ill. to Minn., southw. to Ala., Tex., Ariz., and Mex.

3710. PETALOSTÈMUM* Michx.

Leaflets 3-5, mostly linear, involute; calyx tube densely pubescent; corolla violet or purple..1. *P. purpureum.*

Leaflets 7-9, linear, oblong, or oblanceolate, not involute; calyx tube glabrous, at least below the middle; corolla white..................................2. *P. candidum.*

1. **Petalostemum purpùreum** (Vent.) Rydb. PURPLE PRAIRIECLOVER. Map 1227. Infrequent in dry, sandy or gravelly soil in the open on sandy knolls and ridges in open, black oak woods and in prairie habitats. It is sometimes frequent on the low dunes along Lake Michigan. East of the area indicated in the map, it has been reported from Kosciusko and Marshall Counties.

Ind. to Sask., southw. to Ark., Tex., and N. Mex.

2. **Petalostemum cándidum** (Willd.) Michx. WHITE PRAIRIECLOVER. Map 1228. This species is rarer than the preceding and grows in the same habitat and usually with it.

Ind. to Sask., southw. to La. and Tex.

* Spelled Petalostemon in the International Rules of Botanical Nomenclature, p. 99.

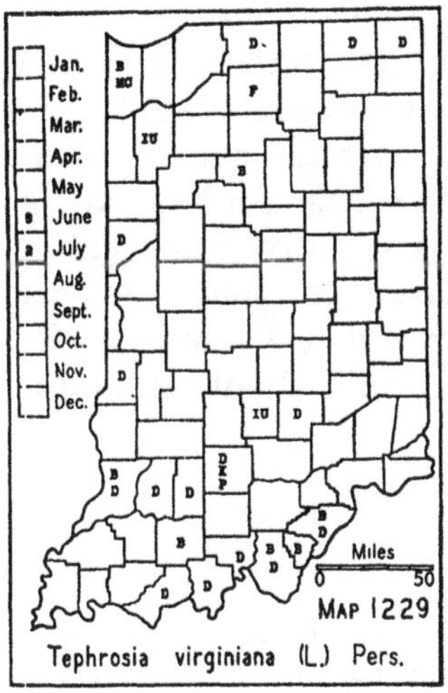

MAP 1229
Tephrosia virginiana (L.) Pers.

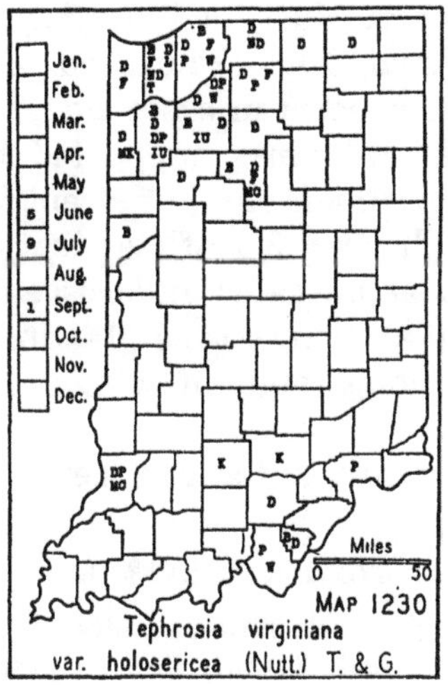

MAP 1230
Tephrosia virginiana var. holosericea (Nutt.) T. & G.

MAP 1231
Wisteria macrostachya Nutt.

3718. TEPHRÒSIA Pers.

Leaflets glabrous above...1. *T. virginiana.*
Leaflets pubescent above..........................1a. *T. virginiana* var. *holosericea.*

1. **Tephrosia virginiàna** (L.) Pers. (*Cracca virginiana* L.) SMOOTH-LEAFLET VIRGINIA GOATRUE. Map 1229. Infrequent but usually in colonies where it is found. In southern Indiana it is found on the crests of chestnut oak and black oak ridges and on sandstone outcrops. In northern Indiana it is found in very dry, sandy soil on black oak land and was formerly common on the low dunes about Lake Michigan. It is also found on sandy hills in the open or in open woodland in northwestern and southwestern Indiana. The habitats of the species and the following variety are the same and it is probable that their range is co-extensive. My attention had not been called to the variety before I undertook to write the genus. Since I usually collect only one specimen from a county, the maps do not accurately represent the distribution of the two forms. I have seen specimens of this form from the following states: Ala., Ark., D. C., Ill., Ind., Ky., Mich., Mo., N. C., N. J., Okla., Pa., S. C., Tenn., Tex., Va., and W. Va.

1a. **Tephrosia virginiana** var. **holoserícea** (Nutt.) T. & G. HAIRY-LEAFLET VIRGINIA GOATRUE. Map 1230. Habitat and distribution as discussed under the preceding species. I have seen specimens from the following states: Ark., Ill., Ind., Kans., Mass., Md., Mich., N. C., N. J., N. Y., Ohio, Okla., R. I., S. C., Va., Wis., and W. Va.

3722. WISTÈRIA Nutt.

Leaflets mostly 9, ovate or elliptic-ovate to lance-elliptic, 3-7 cm long; racemes 15-35 cm long, not dense; pedicels about 10 mm long; lower teeth of calyx at least half as long as the tube; spur of wings of corolla about as long as the claw; pods 7-12 cm long, glabrous...1. *W. macrostachya.*
Leaflets 9-15, elliptic-ovate to oblong or oblong-lanceolate, 2-5 cm long; racemes 4-10 cm long, dense; pedicels about 5 mm long; lower teeth of calyx less than half as long as the tube; spur of wings of corolla much shorter than the claw; pods 5-10 cm long, glabrous. (See excluded species no. 393, p. 1066.)..........*W. frutescens.*

1. **Wisteria macrostàchya** Nutt. (*Kraunhia macrostachys* (T. & G.) Small of Britton and Brown, Illus. Flora, ed. 2.) KENTUCKY WISTERIA. Map 1231. I collected this species in a second growth wooded ravine May 19, 1918. There were several vines supported by low trees and shrubs about 10 feet high. My specimen has pubescent branches and branchlets; 4 leaves, 15-23 cm long, all with 9 leaflets; leaflets slightly pubescent on both sides, more or less acuminate; inflorescence 21 cm long; pedicels about 10 mm long, glandular; calyx tube glandular, about 4 mm long, the longest lobes about 2 mm long; spur of wings of corolla about as long as the claw; pod glabrous.

Ind., Tenn., and Mo., southw. to La. and Tex.

3733. ROBÍNIA L.

Branchlets, petioles, and pods glabrous; flowers white............1. *R. Pseudo-Acacia.*
Branchlets and petioles bristly; pods hispid; flowers pink or purplish. (See excluded species no. 394, p. 1066.)..*R. hispida.*

1. **Robinia Pseùdo-Acàcia** L. BLACK LOCUST. Map 1232. This species has been freely planted since pioneer times and has escaped in all parts of the state. It was, no doubt, a native in the southeastern part of the state near the Ohio River.

Pa. to se. Ind. and the Ozark region of Mo., southw. to Ga., La., and Okla.

3766. ASTRÁGALUS [Tourn.] L.

1. **Astragalus canadénsis** L. (*Astragalus carolinianus* L. of Indiana authors.) CANADA MILKVETCH. Map 1233. Infrequent on the moist, clay or gravelly slopes of the high banks of our larger streams and lakes and rare in prairie habitats.

Que. to Mackenzie, southw. to N. C. and Tex.

1a. **Astragalus canadensis** var. **longilòbus** Fassett. (Rhodora 38: 94. 1936.) This variety has calyx lobes 2.5-5.5 mm long, tube 4-5 mm long. I have it from Elkhart, Gibson, Kosciusko, and Warrick Counties. All Indiana forms are on one map.

Del., Wis. to Minn., southw. to s. Ind. and Iowa.

3769. GLYCYRRHÌZA [Tourn.] L.

See excluded species no. 398, p. 1067.

3774. CORONÍLLA [Tourn.] L.

1. CORONILLA VÀRIA L. CROWNVETCH. Map 1234. This weedy perennial has been reported from Boone, Grant, Lagrange, Lake, and Marion Counties. A clump of this species was found in a waste place in Bluffton, Wells County, and, its weedy nature being unknown, was planted in our field for observation. We soon learned by its rapid spread from underground stems that it would be a pest but kept it until it flowered in order to ascertain its identity. We then destroyed it but neglected to preserve a

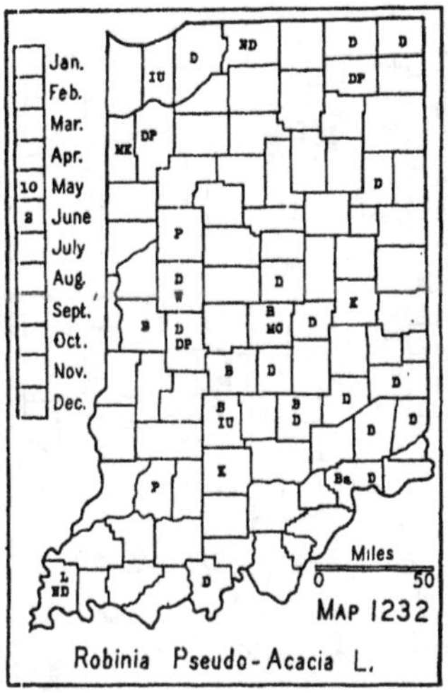

MAP 1232
Robinia Pseudo-Acacia L.

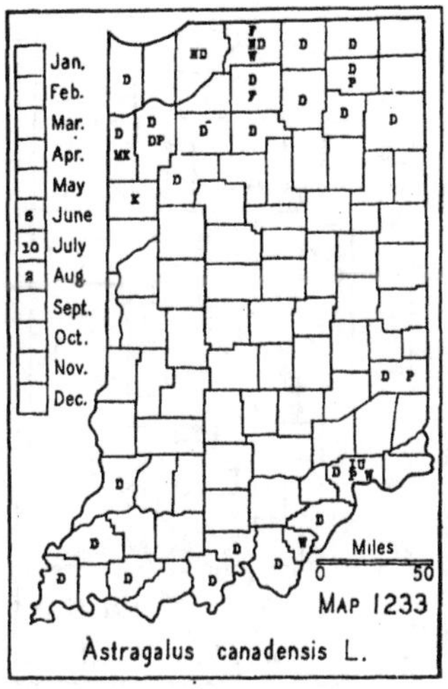

MAP 1233
Astragalus canadensis L.

MAP 1234
Corohilla varia L.

specimen. The Lagrange County specimen was collected by Nieuwland and is in the herbarium of the University of Notre Dame.

Nat. of Eu.; escaped to roadsides and waste places.

3802. STYLOSÁNTHES Sw.

Stems without long, spreading hairs or only a few toward the summit; margins of leaflets without long, hispid hairs or a few hairs on some of the leaflets near the summit of the stem and in the inflorescence........................1. *S. biflora.*

Stems more or less densely hispid with long, spreading hairs (rarely with only a few); hairs 1-2 mm long, and flat at the base; margins of leaflets hispid with similar but shorter hairs..................................1a. *S. biflora* var. *hispidissima.*

1. **Stylosanthes biflòra** (L.) BSP. PENCIL-FLOWER. Map 1235. Infrequent to rare in the southern counties on bare, open places on ridges wooded with black and white oak. Found also in a few post oak flats in the extreme southwestern counties.

N. Y. to Kans., southw. to Fla. and Tex.

1a. **Stylosanthes biflora** var. **hispidíssima** (Michx.) Pollard & Ball. Plants of the variety are larger, erect or nearly so, and with longer leaflets. My Crawford County specimen is from a dry woods near Wyandotte Cave, and the Daviess and Knox County specimens are from sand hills. I have not ascertained the range of the variety.

3807. DESMÒDIUM Desv. TICKCLOVER

Plants trailing; stem with long, spreading hairs; terminal leaflets orbicular, rounded at the apex; stipules large, ovate, attenuate, the margins ciliate with long hairs, otherwise glabrous; peduncles terminal and axillary..........1. *D. rotundifolium.*

Plants erect or decumbent at the base; terminal leaflets not orbicular or rounded at the apex.

Plants with leaves sessile or nearly so; leaflets linear or linear-oblong, reticulate beneath..2. *D. sessilifolium.*

Plants not as above.

Stylosanthes biflora (L.) BSP.

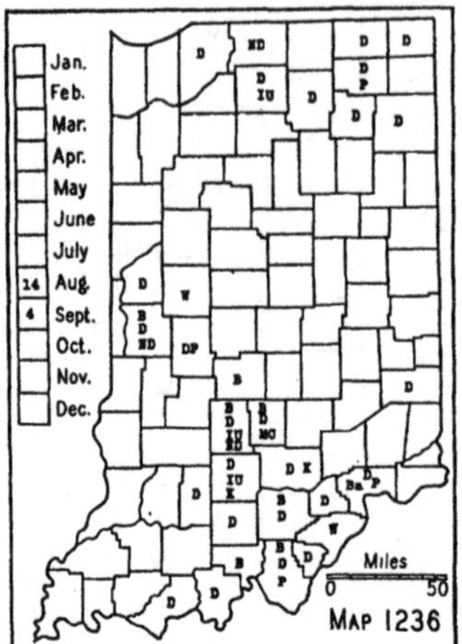

Desmodium rotundifolium (Michx.) DC.

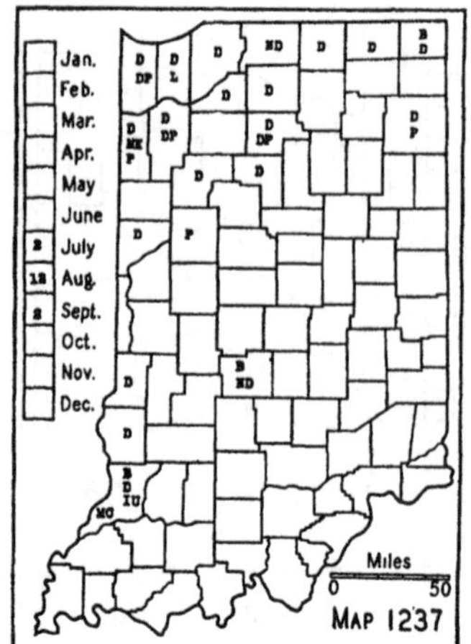

Desmodium sessilifolium (Torr.) T. & G.

Loment long-stalked, the stipe 2-3 times as long as the calyx.

Peduncles arising from the base of the stem, much longer than the leafy stem; leaflets short-acute or obtuse............................3. *D. nudiflorum.*

Peduncles terminating the leafy stem, rarely with a few axillary racemes in *Desmodium pauciflorum.*

Plants rarely more than 4.5 dm high; leaves scattered on the stem; leaflets ovate or oval, not abruptly narrowed at the apex, acute or obtuse; inflorescence usually a simple raceme, rarely paniculate; flowers few, white. ....4. *D. pauciflorum.*

Plants generally 5-10 dm high; leaves mostly clustered at the top of the stem (or base of the peduncle), terminal leaflets large, broadly ovate, abruptly contracted into a long, acuminate tip; inflorescence generally a panicle of racemes, rarely simple, usually many-flowered..........5. *D. acuminatum.*

Loment not long-stalked, the stipe less than 2-3 times as long as the calyx.

Lower surface of leaflets pubescent with hooked hairs; plants large.

Segments of loments rhomboidal, the middle mature ones mostly 8-12 mm long and 5-6 mm wide; leaflets ovate, blunt, firm but not coriaceous, the lower surface only faintly reticulated...................6. *D. canescens.*

Segments of loments oval, less than 6 mm long; leaflets ovate-oblong or ovate-lanceolate, acute, subcoriaceous, the lower surface reticulated, the primary and secondary nerves prominent........................7. *D. illinoense.*

Lower surface of leaflets glabrous or pubescent without hooked hairs; leaflets ovate, ovate-lanceolate to linear-lanceolate or oval.

Segments of mature loments mostly 8-12 mm long and 5-6 mm wide.

Leaflets glabrous beneath or essentially so, not glancous beneath, long-acuminate at the apex............................8. *D. bracteosum.*

Leaflets pubescent beneath with long, half-spreading hairs, merely acute at the apex........................8a. *D. bracteosum* var. *longifolium.*

Segments (middle) of loments less than 8 mm long.

Stipe of loment as long as the calyx, generally a half longer; segments of loment usually 2-5; stipules and bracts deciduous.

Leaflets of a lanceolate type (rarely a few ovate), oblong-lanceolate to linear-lanceolate or ovate-lanceolate, usually thin.

Segments of loments of an oval type, strongly rounded on the ventral side, usually 4-5 mm long; stems more or less pubescent at least above...9. *D. canadense.*

Segments of loments of a rhomboidal type, semi-rhomboidal on the ventral side, usually 5-8 mm long.

Stems glabrous or nearly so....................10. *D. paniculatum.*

Stems covered more or less densely with short, hooked hairs or with long, spreading hairs in addition to any short, hooked hairs that may be present................10a. *D. paniculatum* var. *pubens.*

Leaflets of an ovate type.

Leaflets glabrous and glaucous beneath; stems glabrous; inflorescence more or less puberulent; lower petioles mostly 4-8 cm long; segments of loment generally 4-6, usually about 8 mm long, of a rhomboidal type. ....11. *D. laevigatum.*

Leaflets not glaucous and more or less pubescent to velvety beneath; stems usually more or less villous, rarely somewhat glabrate.

Petioles of median leaves more than one and a half times as long as the petiolule of the terminal leaflet, generally about twice as long; leaflets more or less pubescent beneath but not velvety to the touch; stipules narrow-lanceolate, from a dilated base, long-acuminate, early deciduous; segments of loment of a rhomboidal type, very rarely of an oval type................12. *D. Dillenii.*

Petioles of median leaves less than one and a half times as long as the petiolule of the terminal leaflet, generally about as long or shorter; leaflets velvety pubescent beneath, usually conspicuously thicker and more obtuse at the apex; stipules ovate-lanceolate, acuminate, pilose and ciliate, brick red; segments of loment of an oval type, generally strongly rounded below................ ....13. *D. viridiflorum.*

Stipe of loment shorter than the calyx; segments of loment 1-5, rounded on the ventral side.

Leaflets glabrous above, glabrous beneath or with a few hairs on the principal veins, the terminal one very obtuse, ovate to narrow-ovate or oval, 15-30 mm long.......................14. *D. marilandicum.*

Leaflets more or less pubescent both above and beneath.

Terminal leaflet a little longer than wide, mostly 20-30 mm long, ovate to oval..15. *D. ciliare.*

Terminal leaflet usually twice as long as wide or longer, mostly 20-60 (75) mm long, oblong-ovate to ovate-lanceolate.

Calyx usually 4-5 mm long, the midnerve of the lobes prominently purple; segments predominantly more than 3; plants usually of moist habitats..............................9. *D. canadense.*

Calyx usually 2-3 mm long, the midnerve not prominently purple; segments fewer than 3; plants usually of dry, infertile or dry, sandy places..............................16. *D. rigidum.*

1. **Desmodium rotundifòlium** (Michx.) DC. (*Meibomia Michauxii* Vail.) PROSTRATE TICKCLOVER. Map 1236. Infrequent, but probably found in all the counties of the state in which there are sandy or clayey black oak and chestnut oak ridges. Rare in all parts except in the unglaciated region where it becomes more or less frequent. Probably absent from some of the counties of the central part whose soil is a black loam and where black oak is absent.

Eastern Mass. to Minn., southw. to Fla. and La.

2. **Desmodium sessilifòlium** (Torr.) T. & G. (*Meibomia sessilifolia* (Torr.) Ktze.) SESSILE-LEAF TICKCLOVER. Map 1237. Infrequent in the northern and western counties in very dry, sandy or gravelly soil in prairie habitats or in open woodlands that have recently been prairies.

Mass. to Ont. and Mich., southw. to Conn., Miss., and Tex.

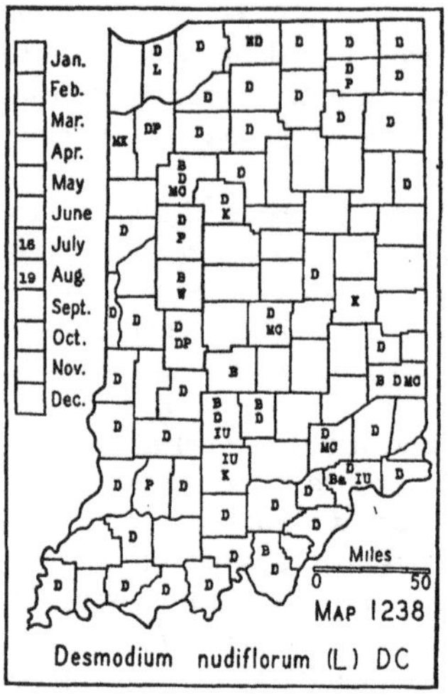

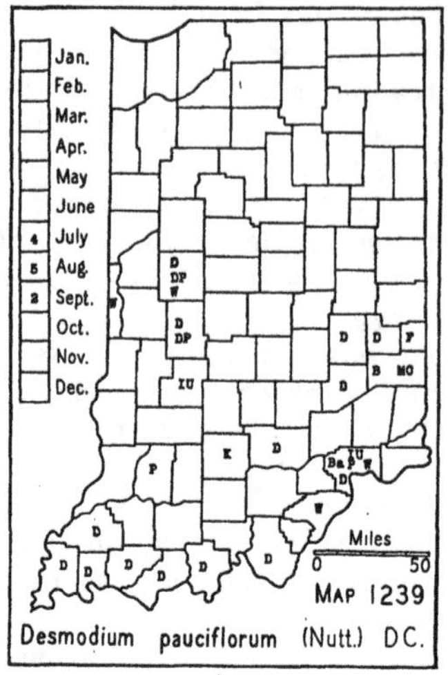

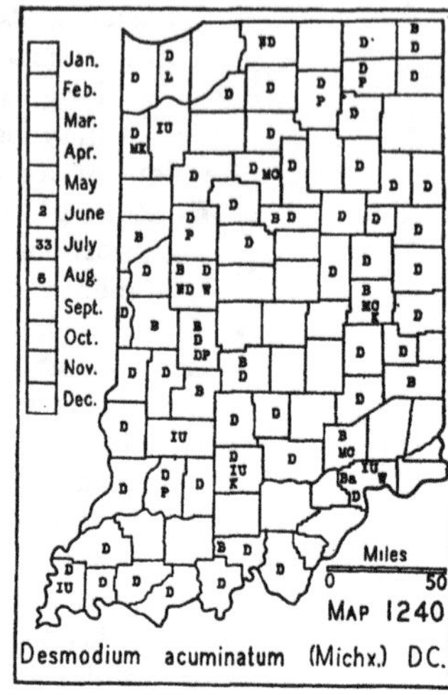

3. **Desmodium nudiflòrum** (L.) D.C. (*Meibomia nudiflora* (L.) Ktze.) NAKED-FLOWER TICKCLOVER. Map 1238. Infrequent to frequent in dry soil in black oak and black and white oak woods, and less frequent in beech woods. It is probably found in every county of the state except Benton County where there is no longer any ungrazed woodland. Rarely this species will have one or more leaflets or leaves on the flowering stem. The form with the leaves scattered on the flowering stalk is known as *Desmodium nudiflorum* f. *foliolatum* (Farw.) Fassett. When the leaves are in verticels or subverticillate the form is known as *Desmodium nudiflorum* f. *personatum* Fassett. I found a large colony of this species in a black oak woods about half a mile southeast of Sand Lake in Noble County which contained both of these forms in some abundance.

Maine to Minn., southw. to Fla., La., and Ark.

4. **Desmodium pauciflòrum** (Nutt.) DC. (*Meibomia pauciflora* (Nutt.) Ktze.) FEW-FLOWER TICKCLOVER. Map 1239. Infrequent to rare in the southern half of the state. All but two of my specimens were intimately associated with beech and were found on dry, wooded, beech slopes or in the "flats" with beech. The label on my Rush County specimens reads "common on a beech ridge two and a half miles west of Gowdy." I found a single specimen in a "post oak flat" about 10 miles southwest of Mt. Vernon, Posey County. Peattie's report from Lake County, I think, is based upon a wrong determination.

N. Y., Ont., Mich. to Kans., southw. to Fla. and Tex.

5. **Desmodium acuminàtum** (Michx.) DC. (*Meibomia grandiflora* (Walt.) Ktze.) POINTED-LEAF TICKCLOVER. Map 1240. Infrequent to frequent throughout the state (with the probable exception of Benton County) in dry, rich soil in black and white oak and beech and sugar maple woods and rarely in a moist habitat. The position of the leaflets on the stem is variable. Generally they are crowded at the summit, and more rarely there are a few smaller ones below the summit. Lunell (Amer. Midland Nat. 2: 128. 1911) described a form with "leaves not clustered at the base of

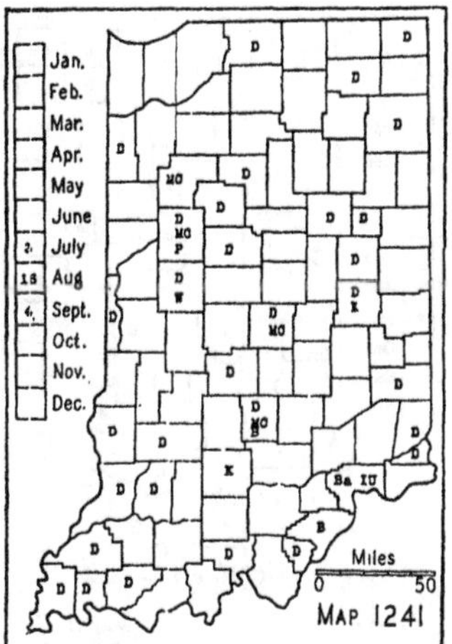

MAP 1241
Desmodium canescens (L.) DC.

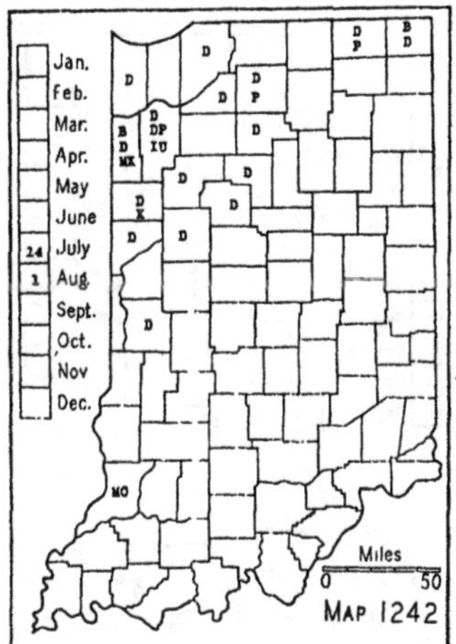

MAP 1242
Desmodium illinoense Gray

MAP 1243
Desmodium bracteosum (Michx.) DC.

the peduncle, but further down on the stem. In addition to these there is one single leaf at the base of the peduncle, and often one or sometimes two single leaves beneath this. Lastly, there are often one or two single leaves on the stem below the clustered part." He cited a specimen of mine collected in Wells County. This form is now known as *Desmodium acuminatum* f. *Chandonnetii* (Lunell) Fassett.

Maine to N. Dak., southw. to Fla., Ala., and Tex.

6. **Desmodium canéscens** (L.) DC. (*Meibomia canescens* (L.) Ktze.) HOARY TICKCLOVER. Map 1241. Frequent in dry, open habitats throughout the state, although there are no reports from the counties along Lake Michigan. This is our common, large tickclover.

Mass., Ont. to S. Dak., southw. to Fla. and Tex.

7. **Desmodium illinoénse** Gray. (*Meibomia illinoensis* (Gray) Ktze.) ILLINOIS TICKCLOVER. Map 1242. Restricted to the northern and western counties where it is infrequent. It grows on very dry, sandy or gravelly soil and is found mostly in a prairie habitat along roadsides and in open woodlands that have recently invaded prairie areas. This is closely allied to *Desmodium canadense* and may be distinguished from it by the large, persistent stipules, in contrast with the narrow, deciduous ones of *D. canadense,* and by its inflorescence. *D. illinoense* usually has a long, terminal raceme, which is much longer than the branches of the panicle, while the inflorescence of *D. canadense* is more compact and usually composed of many racemes of nearly equal length, although the main axis is sometimes much longer.

Ohio, Mich. to Nebr., southw. to Tex.

8. **Desmodium bracteòsum** (Michx.) DC. (*Meibomia bracteosa* (Michx.) Ktze.) LARGE-BRACT TICKCLOVER. Map 1243. Infrequent to frequent possibly throughout the state. Like most species of the genus it prefers the dry soil of white and black oak woods and is generally found in open places and on slopes.

N. E. to Wis., southw. to Fla. and Tex.

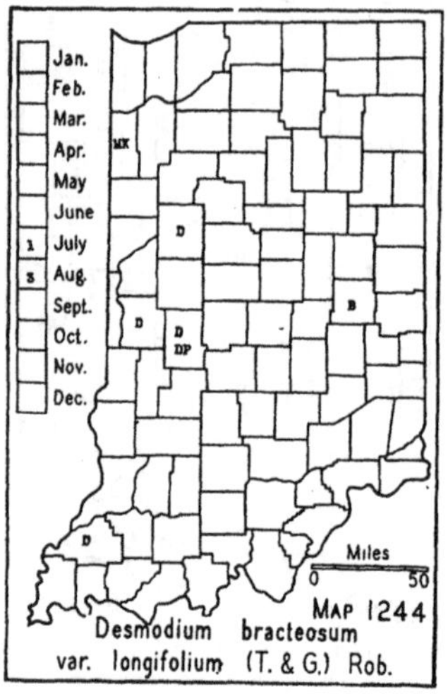

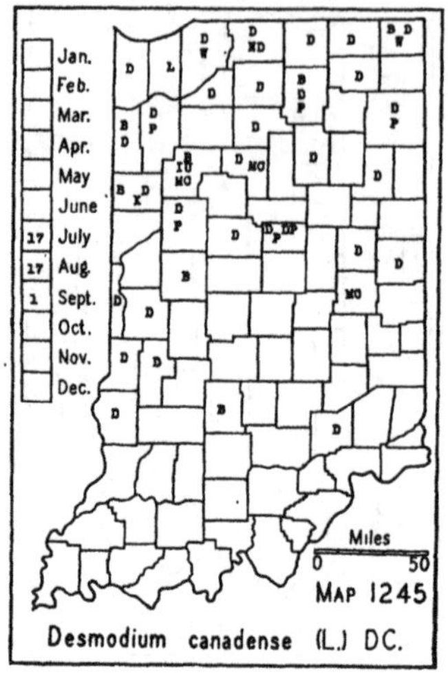

8a. **Desmodium bracteosum** var. **longifòlium** (T. & G.) Rob. Map 1244. This variety is found in the habitat of the species.

Ohio to N. Dak., southw. to Ala. and Tex.

9. **Desmodium canadénse** (L.) DC. (*Meibomia canadensis* (L.) Ktze.) CANADA TICKCLOVER. Map 1245. Rather frequent in prairie habitats throughout the state, becoming infrequent or rare in the hilly part of the unglaciated area.

N. B. to Man., southw. to N. C. and Okla.

10. **Desmodium paniculàtum** (L.) DC. (*Desmodium paniculatum* var. *angustifolium* T. & G.) PANICLED TICKCLOVER. Map 1246. This is one of our most common tickclovers and is frequent throughout the state in dry soil in woodland and clearings. It is most commonly associated with oaks in the openings on ridges, on rocky slopes, borders of woodland, and rarely in fallow fields. This species is extremely variable in the width of its leaflets and in the density of its pubescence. Plants will vary from almost glabrous to densely pubescent with both short, hooked hairs and longer ones which are not hooked. Since the forms show no geographic range in the state they are combined on one map.

10a. **Desmodium paniculatum** var. **pùbens** T. & G. This is the most vigorous and pubescent form of the species. The range and habitat are those of the species.

Maine, Ont., to Minn., and southw.

11. **Desmodium laevigàtum** (Nutt.) DC. (*Meibomia laevigata* (Nutt.) Ktze.) SMOOTH TICKCLOVER. Map 1247. This species is very local in its distribution and is probably restricted to the ridges of the unglaciated area. Potzger reported it from Monroe County.

Southern N. Y. to Mo., southw. to Fla. and Tex.

12. **Desmodium Dillènii** Darl. DILLENIUS TICKCLOVER. Map 1248. This tickclover is frequent throughout the state, preferring dry soil. It is

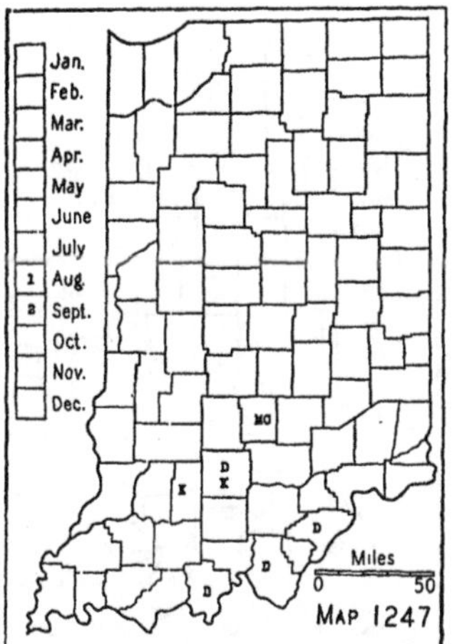

Desmodium laevigatum (Nutt.) DC.

Desmodium Dillenii Darl.

Desmodium viridiflorum (L.) Beck

usually found on high ground in open places in oak woodland and in clearings and sometimes in low oak woodland, but usually in flats.

Forms of this species with very narrow leaves so closely approach *Desmodium paniculatum* var. *pubens* that it is difficult to decide to which species they belong. In my comparatively short study of the genus I have not been able to find a single character that will definitely separate the two. Among my specimens are a few that have been assigned to this species by one authority and to the other species by another authority.

Maine to Minn., southw. to Fla. and Tex.

13. **Desmodium viridiflòrum** (L.) Beck. (*Meibomia viridiflora* (L.) Ktze.) VELVET-LEAF TICKCLOVER. Map 1249. Infrequent or rare on the slopes and crests of black and white oak ridges in a few of the southern counties. The violet purple flowers turn greenish when dried, hence the scientific name.

N. Y., Mich. to Mo., southw. to Fla. and Tex.

14. **Desmodium marilándicum** (L.) DC. (*Meibomia marilandica* (L.) Ktze.) SMOOTH SMALL-LEAF TICKCLOVER. Map 1250. Infrequent and generally on ridges in black and white oak woods in dry, sandy or gravelly soil of low fertility. No doubt absent from many of the central counties that have uniformly rich soil.

Mass. to Minn., southw. to Fla., La., and Mo.

15. **Desmodium ciliàre** DC. (*Desmodium obtusum* Muhl. and *Meibomia obtusa* (Muhl.) Vail.) HAIRY SMALL-LEAF TICKCLOVER. Map 1251. Infrequent to very local in dry, sandy or gravelly soil in open black oak woods, usually on ridges. The range in Indiana is extended by reports from Clark and Jefferson Counties.

Ont., Mich. to Nebr., southw. to Fla. and Tex.

16. **Desmodium rígidum** (Ell.) DC. (*Meibomia rigida* (Ell.) Ktze.) RIGID TICKCLOVER. Map 1252. Infrequent to very local in dry, sandy or

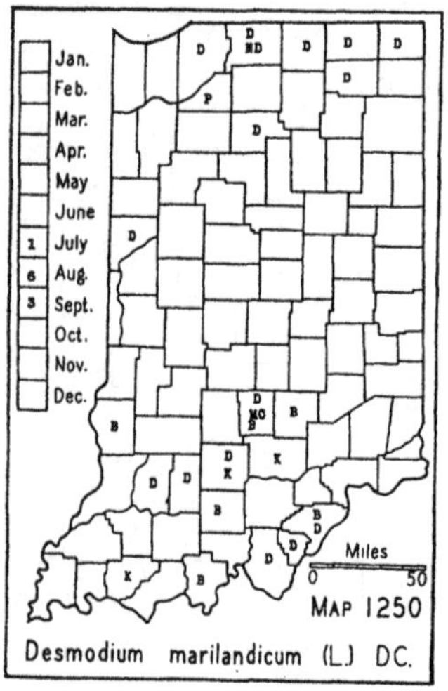

Desmodium marilandicum (L.) DC.

Desmodium ciliare DC.

Desmodium rigidum (Ell.) DC.

gravelly soil in open, black and white oak woods. This species and the preceding two have the same habitat and are often associated.

Mass., Mich. to Nebr., southw. to Fla. and Tex.

3820. LESPEDÈZA Michx. BUSHCLOVER

In a study of this genus the two kinds of pods and the relative length of the calyx and its lobes should be noted. The pods of cleistogamous flowers are usually broadly oval and short and have very short calyx lobes of nearly equal length, mostly 0.5-2 mm long, and a short, recurved style, usually less than 1 mm long. The pods of petaliferous flowers are usually not so wide and are longer, the calyx lobes more irregular in length and much longer than those of the cleistogamous flowers, and the style is much longer and not recurved.

Stipules and bracts broad, scarious, glabrous, as long as or longer than the petioles, persistent; plants annual.
- Pubescence of stems retrorsely appressed..........................1. *L. striata.*
- Pubescence of stems upwardly appressed........................2. *L. stipulacea.*

Stipules subulate-setaceous, pubescent, not scarious, generally shorter than the petioles, more or less deciduous; plants perennial.
- Calyx lobes of petaliferous flowers 4.5-8 mm long (cleistogamous flowers rare or lacking except in no. 4); pubescence of stem and petioles spreading or appressed.
  - Peduncles shorter than the leaves; leaflets narrow, elliptic-oblong to linear; flowers whitish to cream color.
    - Leaflets elliptic-oblong to almost linear, densely appressed-pubescent beneath, green and glabrous or slightly appressed-pubescent above.....3. *L. capitata.*
    - Leaflets narrowly elliptic-oblong, velvety-pubescent above and beneath...3a. *L. capitata* var. *velutina.*
    - Leaflets linear, green and glabrous or slightly appressed-pubescent above...3b. *L. capitata* var. *longifolia.*
  - Peduncles longer than the leaves; leaflets wide, orbicular to oblong.
    - Calyx mostly 4.5-5 (5.5) mm long; flowers purplish; pubescence of stem and peduncles spreading...................................4. *L. Nuttallii.*

Calyx mostly 6-9 mm long; flowers yellowish white.
Leaflets orbicular to oblong-ovate; spikes thick-cylindric, 1-1.5 cm thick...... 5. *L. hirta.*
Leaflets linear to narrowly oblong; spikes slender-cylindric, 5-8 mm thick, somewhat loosely flowered. (See excluded species no. 403, p. 1068.)...... *L. leptostachya.*

Calyx lobes of petaliferous flowers less than 4.5 mm long, those of the cleistogamous flowers 0.5-2 mm long; flowers purplish, corollas generally 6-7 mm long; pods of petaliferous flowers oval, mostly 5-7 mm long, of cleistogamous flowers suborbicular to broadly oval, mostly 4-5 mm long (sometimes 9 in *L. violacea* and *L. repens*); pubescence of stem and petioles appressed or spreading.
Pubescence of stem and petioles appressed.
Peduncles of flower clusters shorter than the petioles of their subtending leaves.
Leaflets linear to linear-oblong, appressed-pubescent above; plant virgate or more rarely virgate-branched........ 6. *L. virginica.*
Leaflets oval to elliptic-oblong, rarely suborbicular or slightly obovate, glabrous above or sometimes a few leaves with scattered hairs above; plant erect or spreading........ 7. *L. intermedia.*
Peduncles of flower clusters mostly longer than the petioles of their subtending leaves.
Plants trailing; stems usually 4-10 dm long; leaflets oval or oblong, those of the stem leaves mostly 6-14 mm wide and 10-28 mm long, usually evenly pubescent above with appressed hairs 0.2-0.4 mm long, sometimes nearly glabrate above; banner of flowers usually as long as or longer than the keel........ 8. *L. repens.*
Plants erect or somewhat spreading; stems generally 4-6 dm long, rarely up to 8 dm long; leaflets usually large, about as long as the petioles of the leaf, broadly oval to oblong, mostly 15-30 mm long, rarely up to 40 mm long, glabrous above, or sometimes glabrate, the hairs appressed and about 0.5 mm or more long; banner of flowers shorter than the keel........ 9. *L. violacea.*
Pubescence of stem and usually of petioles spreading.
Peduncles of flower clusters shorter than the leaves.
Leaflets linear to linear-oblong.
Upper surface of leaflets glabrous or strigillose with short hairs, lower surface merely appressed-pubescent; petioles of principal cauline leaves averaging 2.2 cm in length; calyx and pod commonly strigose to strigillose........ 6a. *L. virginica* f. *Deamii.*
Upper surface of leaflets tomentose-strigose with long hairs, lower surface more densely so than the upper; petioles of principal cauline leaves averaging 1.7 cm in length; calyx and pod commonly villous-canescent........ 10a. *L. Stuevei* f. *angustifolia.*
Leaflets oval to elliptic-oblong, rarely suborbicular.
Upper surface of leaflets glabrous or sparingly strigillose, lower surface strigose; petioles of principal cauline leaves nearly equaling the length of the leaflets, rarely exceeding them; peduncles of petaliferous flowers averaging 11 mm in length; calyx and pod commonly strigose or strigillose........ 7a. *L. intermedia* f. *Hahnii.*
Upper surface of leaflets tomentose-strigose, lower surface more densely so than the upper; petioles of principal cauline leaves shorter than the leaves; peduncles of petaliferous flowers short, averaging 6 mm in length; calyx and pod commonly villous-canescent....... 10. *L. Stuevei.*
Peduncles of flower clusters longer than the subtending leaves.
Plants erect; leaflets broadly oval, the largest leaflets generally 20-40 mm long; number of petaliferous flowers in a cluster generally 10-25; longest calyx lobes of petaliferous flowers or fruit usually more than 3 mm. long........ 4. *L. Nuttallii.*

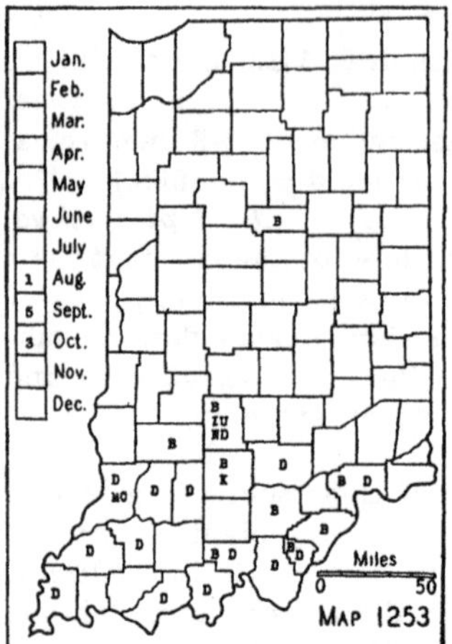

Lespedeza striata (Thunb.) H. & A.

Lespedeza stipulacea Maxim.

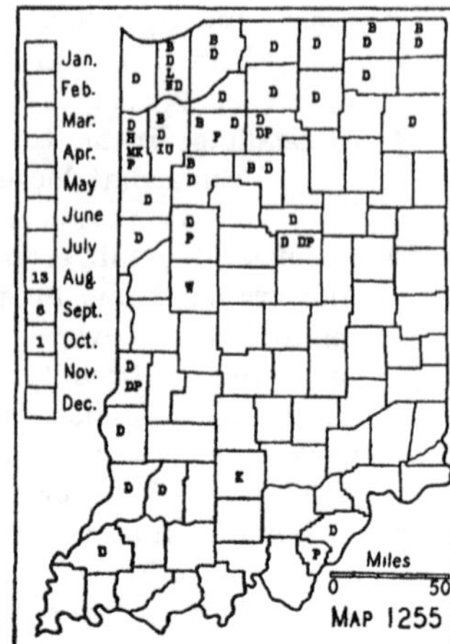

Lespedeza capitata Michx.

Plants trailing; leaflets oval to obovate-elliptic or narrowly elliptic in the variety, the largest usually 15-30 mm long; number of petaliferous flowers in a cluster usually 2-8; longest calyx lobes of petaliferous flowers less than 3 mm long.

Leaflets oval to obovate-elliptic........................11. *L. procumbens.*

Leaflets narrowly elliptic, up to 4 times as long as wide.................. ....11a. *L. procumbens* var. *elliptica.*

1. LESPEDEZA STRIÀTA (Thunb.) H. & A. JAPAN BUSHCLOVER. Map 1253. This is an introduced species that has spread with remarkable rapidity. So far it is restricted to the southern part of the state, our most northern report being from Howard County. I have been well acquainted with the Clark County State Forest of 2,000 acres since 1909. This species was never sown on the cleared land of the forest or in the neighborhood. It appeared spontaneously in the abandoned fields and soon formed dense stands over acres. The forest is so located that the seed could not have been brought in by water. I have no data as to when I first noticed it there. Its sudden and widespread appearance in Indiana is an interesting problem in distribution. Most of my specimens date from 1911-1920. It is usually found in hard, clayey soil, either moist or dry, in open woodland pastures, and fallow fields and along roadsides and railroads. It has been a boon to the grazing industry in that part of the state since it does not appear until August and September and continues until late in autumn.

Nat. of e. Asia; N. J. to Mo., southw. to Fla. and Tex.

2. LESPEDEZA STIPULÀCEA Maxim. KOREAN LESPEDEZA. Map 1254. This species was introduced into Indiana as a forage crop about 1925 and has freely escaped to roadsides and open woodland in some of the southern counties.

Nat. of Korea.

3. **Lespedeza capitàta** Michx. ROUNDHEAD BUSHCLOVER. Map 1255. Infrequent throughout the area indicated on the map in dry, sandy soil

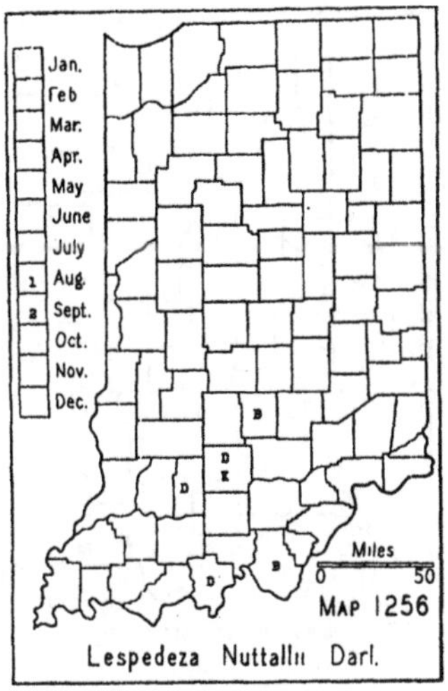

Lespedeza Nuttallii Darl.

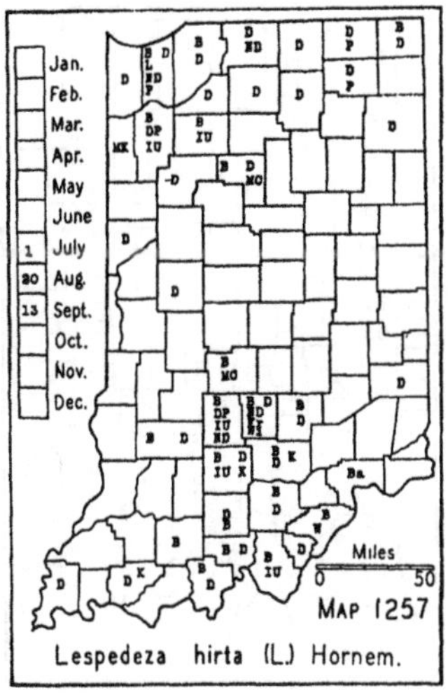

Lespedeza hirta (L.) Hornem.

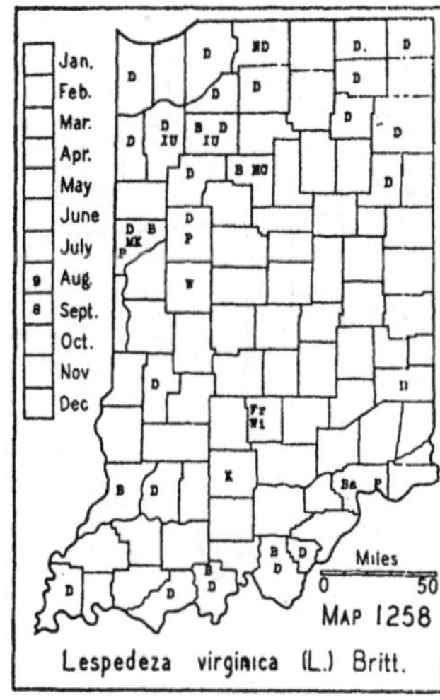

Lespedeza virginica (L.) Britt.

along roadsides and railroads and in open woodland. This is essentially a prairie plant and is found more frequently in prairie habitats.

Southern Maine to Minn., southw. to Fla. and Tex.

3a. **Lespedeza capitata** var. **velùtina** (Bickn.) Fern. This is an extreme form with velutinous leaflets and has the same habitat and range as those of the species. I have specimens from Allen and Pulaski Counties.

3b. **Lespedeza capitata** var. **longifòlia** (DC.) T. & G. This is another extreme form with long, linear leaves. I have specimens from Jasper and Lake Counties. They were found in a very dry, sandy habitat. Potzger found it in a similar habitat in Pulaski County.

Ind. to Mo., southw. to La.

4. **Lespedeza Nuttállii** Darl. NUTTALL BUSHCLOVER. Map 1256. My Martin County specimen was collected along White River about 3 miles above Shoals on the top of a thinly wooded promontory about 100 feet high, locally known as "Cedar Cliffs." My Perry County specimen was collected on a thinly wooded sandstone ridge about 7 miles east of Cannelton, locally known as the Van Buren Ridge. Kriebel has found it in ten places in Lawrence County.

N. H. to Mich. and Kans., southw. to Fla.

5. **Lespedeza hírta** (L.) Hornem. HAIRY BUSHCLOVER. Map 1257. Infrequent in dry, sandy or gravelly soil on the crests of black oak and black and white oak wooded ridges and rarely in prairie habitats. It seems not to be found in neutral or alkaline soils.

Maine, Ont. to Minn., southw. to Fla. and Tex.

6. **Lespedeza virgínica** (L.) Britt. SLENDER BUSHCLOVER. Map 1258. Infrequent in dry, clayey soil on white oak and black and white oak slopes and ridges and less frequent in post oak "flats" in southwestern Indiana. In the northwestern part of the state it is found in dry, sandy soil on black and white oak ridges and rarely on aspen flats about lakes and in the

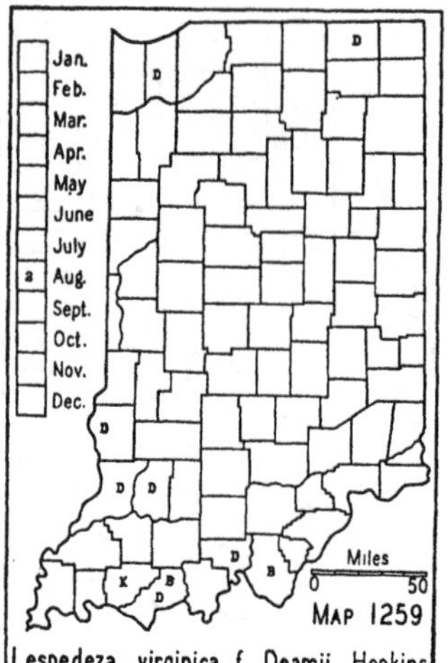

Lespedeza virginica f. Deamii Hopkins

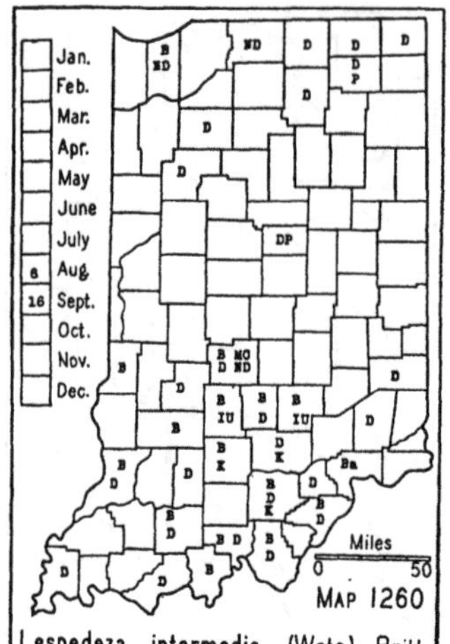

Lespedeza intermedia (Wats.) Britt.

Lespedeza intermedia f. Hahnii (Blake) Hopkins

prairie area. Its habitat suggests a slightly acid soil or one low in fertility.
N. H. to Wis., southw. to Ga. and Tex.

6a. **Lespedeza virginica** f. **Dèamii** Hopkins. (Rhodora 37: 265. 1935.) Map 1259. Found locally throughout the state, usually in sandy soil with black and white oak, in dry fallow fields, and rarely in prairie habitats.
Conn. to Ill., southw. to N. C. and Tenn.

7. **Lespedeza intermèdia** (Wats.) Britt. (Blake. Rhodora **26**: 31. 1924.) (*Lespedeza frutescens* (L.) Britt.) WANDLIKE BUSHCLOVER. Map 1260. Infrequent in dry soil in white oak and black and white oak woods throughout the state.
Maine, Ont. to Minn., southw. to Fla. and Tex.

7a. **Lespedeza intermedia** f. **Háhnii** (Blake) Hopkins. (Blake. Rhodora **26**: 32. 1924 and Hopkins. Rhodora **37**: 265. 1935.) Map 1261. Associated with the species in the southern part of the state. It was described from a specimen from Ohio County and I have specimens from Crawford, Jefferson, Lawrence, Monroe, and Sullivan Counties. This form is not well marked since the spreading pubescence of the stem may be lacking on the branches of some specimens.

8. **Lespedeza rèpens** (L.) Bart. CREEPING BUSHCLOVER. Map 1262. Infrequent on the crests and slopes of chestnut oak and post oak ridges in the southern counties. All of my specimens except the one from Gibson County are from the unglaciated region. My Posey County specimen is from the east bank of "Pitcher Lake" about 5 miles northwest of Mt. Vernon. This bank is frequently submerged and this habitat seemingly is very different from that of the hills of the counties to the east.
Conn. to Wis., southw. to Fla. and Tex.

9. **Lespedeza violàcea** (L.) Pers. BUSHCLOVER. Map 1263. Infrequent in dry, clayey soil in white oak and black and white oak woods throughout

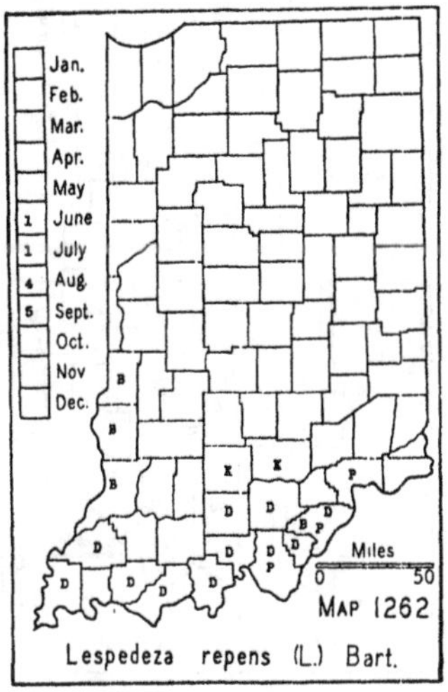

Lespedeza repens (L.) Bart.

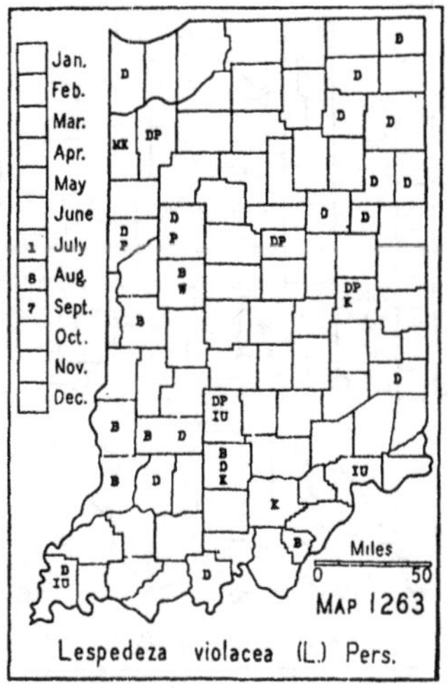

Lespedeza violacea (L.) Pers.

Lespedeza Stuevei Nutt.

the state. It has been reported frequently from the state and, no doubt, some of the reports should be referred to *L. intermedia.*

Southern N. H. to Wis., southw. to Va. and Tex.

10. **Lespedeza Stùevei** Nutt. (Blake. Rhodora **26**: 28. 1924.) STUEVE BUSHCLOVER. Map 1264. In very sandy soil in woodland in the southwestern part of Posey County. I have specimens from three woods of this area.

Vt. to Va., Ala., Tex., northw. to Ark. and Mich.

10a. **Lespedeza Stuevei** f. **angustifòlia** (Britt.) Hopkins. (Blake. Rhodora **26**: 29. 1924 and Hopkins. Rhodora **37**: 265. 1935.) My only specimen is from very sandy soil (Princeton Fine Sand) on a wooded ridge about 2 miles north of Decker, Knox County. The northeastern limit of the range of several species of the southwest occurs on this ridge. The distribution of this form is not well known.

In the original description the distribution was given as N. J. and Pa. to N. C., Mo., and Tex. Blake adds Tenn. and Ill.

11. **Lespedeza procúmbens** Michx. TRAILING BUSHCLOVER. Map 1265. Infrequent in the southern part of the state on the crests and slopes of black and white oak and chestnut oak ridges. My Warren County specimen is from the slope of the high, gravelly hill along the railroad about a mile northwest of Covington.

N. H. to Wis., southw. to Fla. and Tex. and up the Mississippi Valley.

11a. **Lespedeza procumbens** var. **ellíptica** Blake. (Blake. Rhodora **26**: 26. 1924.) My only specimen is from a sparsely wooded slope in Jefferson County at the top of the road leading down the Saluda Hill to the Ohio River about 7 miles south of Hanover.

Va., Ala., and Ind.

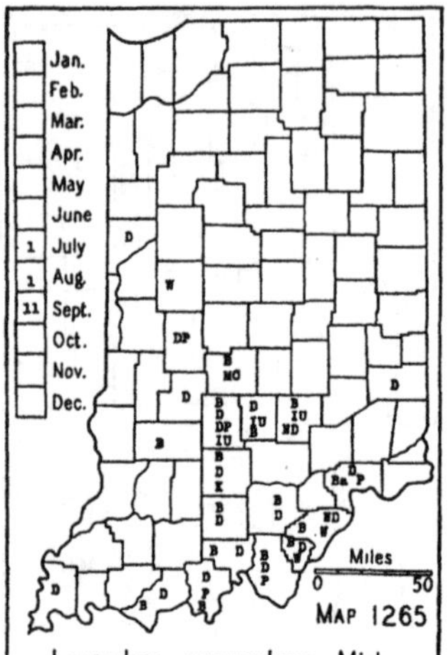

Lespedeza procumbens Michx.

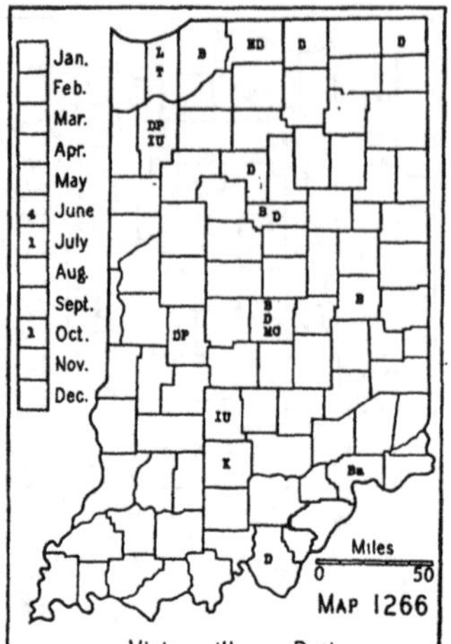

Vicia villosa Roth

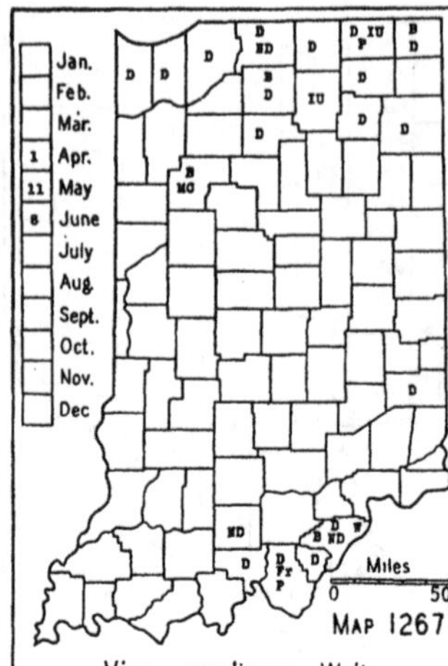

Vicia caroliniana Walt.

3852. VÍCIA [Tourn.] L. THE VETCHES

Flowers on peduncles more than 1 cm long; pods glabrous; plants perennial (except no. 1).
  Plants villous, the pubescence long and widely spreading, annual or biennial...1. *V. villosa.*
  Plants glabrous or nearly so, or puberulent with short, curved hairs.
    Mature calyx, measured to the tip of the lower lobe, less than 3 mm long; flowers numerous, less than 1 cm long, white with a blue-tipped keel; stipules linear or nearly so, rarely the lower ones with a lateral lobe........2. *V. caroliniana.*
    Mature calyx, measured to the tip of the lower lobe, 3.5-8 mm long; mature flowers generally more than 1 cm long.
      Mature calyx generally 3.5-4 mm long; stipules linear and generally with a lateral lobe near the base; flowers numerous, mostly 10-13 mm long. (See excluded species no. 405, p. 1068.).........................*V. Cracca.*
      Mature calyx mostly 6-8 mm long; peduncles mostly 4-8-flowered; flowers 15-18 mm long; stipules large, semi-sagittate, the basal section with several sharp teeth .... 3. *V. americana.*
Flowers sessile or on very short stalks, 1-3 in the axils of the upper leaves; plants annual.
  Flowers 10-18 mm long; pods 3-4 cm long, glabrous, black at maturity. (See excluded species no. 404, p. 1068.)..............................*V. angustifolia.*
  Flowers 20-30 mm long; pods 4-6 cm long, pubescent, pale tawny at maturity. (See excluded species no. 406, p. 1068.)................................*V. sativa*

1. VICIA VILLÒSA Roth. HAIRY VETCH. Map 1266. This vetch has been sparingly sown as a forage crop throughout the state and has escaped to roadsides and fallow fields. I found it to be frequent in 1930 in sandy soil along the river road about a mile west of Georgetown, Cass County. The landowner of the adjacent field told me that he had the field in hairy vetch 14 years before, therefore we have a record of its persistence for that length of time. Now frequent along the roadsides throughout northern Indiana.

Nat. of Eurasia.

2. **Vicia caroliniàna** Walt. CAROLINA VETCH. Map 1267. This species is locally frequent in the northeastern part of the state where it is usually

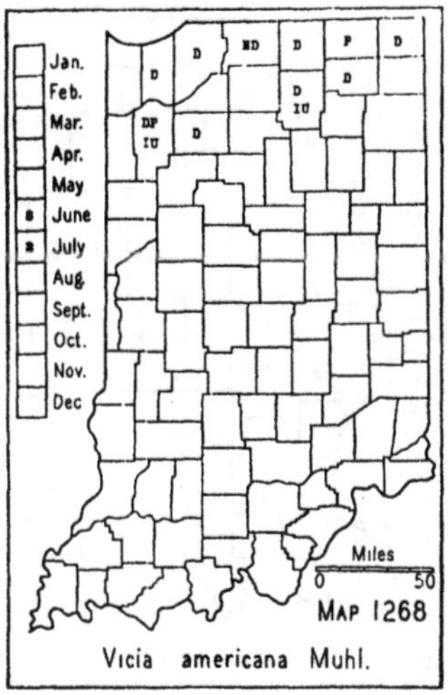

MAP 1268
Vicia americana Muhl.

MAP 1269
Lathyrus ochroleucus Hook.

MAP 1270
Lathyrus japonicus var. glaber (Ser.) Fern

found at the base or on the lower part of black and white oak slopes. In the southern part of the state it is locally frequent on wooded slopes in black and white oak woods. It can be found, no doubt, in all of the hilly counties and in more of the counties of the lake area.

Ont. to Wis., southw. to Ga. and La.

3. **Vicia americàna** Muhl. AMERICAN VETCH. Map 1268. Infrequent throughout the lake area in marshes, along moist roadsides, and on the low borders of lakes. Smith's report from Clark County should be referred to the preceding species. This species is often confused with *Lathyrus palustris*. (See that species for discussion.)

N. Y. to B. C. and the Pacific coast, southw. to Va. and N. Mex.

3854. LÁTHYRUS [Tourn.] L. PEA

Leaflets more than 2.

Whole plant more or less densely pubescent with short, weak hairs; leaflets ovate, lance-ovate or elliptic, mostly (8) 10-14, the largest on different plants varying from 18-25 mm wide and 35-50 mm long; peduncles 17-30-flowered; flowers about 15 mm long............................3a. *L. venosus* var. *intonsus*.

Whole plant glabrous except sometimes the calyx lobes ciliate and the upper surface of the petiolules of no. 5 and varieties puberulent.

Longest petioles 2-4 cm long; leaflets glaucous beneath, of an ovate type, 18-28 mm wide and 35-55 mm long; petiolules glabrous; peduncles 2-12-flowered; flowers yellow.....................................1. *L. ochroleucus*.

Longest petioles less than 2 cm long; flowers purplish.

Leaflets mostly 8-12, of a broad type, mostly 35-60 mm long and 15-28 mm wide; peduncles usually 7-25-flowered.

Leaflets green on both sides, conspicuously veiny; stipules large, the largest almost half the length of the leaflet above it; peduncles mostly 7-10-flowered.................................2. *L. japonicus* var. *glaber*.

Leaflets paler beneath, not conspicuously veiny; stipules much less than half the length of the leaflets above them; peduncles mostly 10-30-flowered. ....3. *L. venosus*.

Leaflets mostly 4-8, linear, lanceolate, elliptic, or of an ovate type; peduncles 3-9-flowered.

MAP 1271
Lathyrus venosus Muhl.

MAP 1272
Lathyrus venosus var. intonsus Butters & St. John

MAP 1273
Lathyrus palustris L.

Stems winged, generally (excluding the wings) 1.5-3 mm in diameter below the lowest peduncle; leaflets 2.5-8 cm long, linear to lanceolate; flowers 2-5 (8), 1.5-2 cm long.
Leaflets 7-23 mm wide, 2-3.5 times as long as wide...........4. *L. palustris.*
Leaflets mostly 3-9 mm wide, more than 3.5 times as long as wide...4a. *L. palustris* var. *linearifolius.*
Stems wingless, generally 0.7-1.5 mm in diameter below the lowest peduncle; flowers 3-9, 1-1.5 cm long; leaflets ovate, elliptic, or broadly lanceolate, 2-4 cm long and 6-20 mm wide...........4b. *L. palustris* var. *myrtifolius.*
Leaflets 2. (See excluded species no. 407, p. 1068.).......................*L. latifolius.*

1. **Lathyrus ochroleùcus** Hook. CREAMCOLOR PEA. Map 1269. Infrequent to rare in dry soil in black and white oak woods in the northern counties.

W. Que. to Sask., southw. to n. Pa., the Great Lakes, Mo., Wyo., and B. C.

2. **Lathyrus japónicus** Willd. var. **glàber** (Ser.) Fern. (Rhodora **34**: 181. 1932.) BEACH PEA. Map 1270. Infrequent on the beach of Lake Michigan and rather common in the Keiser Blowout in Porter County. It is fast becoming extinct on account of the building along the lake front. It was reported in 1889 from the shore of Bass Lake, Starke County, by Thompson.

Newf., Que., Minn., Man., B. C., southw. to N. J., Ohio, n. Ill. to Calif.; also in n. Eu. and Japan.

3. **Lathyrus venòsus** Muhl. SMOOTH VEINY PEA. Map 1271. There is a colony of what I think is this species in the talus of a west slope of the cliff along Blue River about half a mile north of Whitecloud, Harrison County. It has been reported from four of the northern counties but I am referring these reports to the variety.

Del., Md., and Pa. according to Butters & St. John. No doubt the range is greater than that given by these authors.

3a. **Lathyrus venosus** var. **intónsus** Butters & St. John. (Rhodora **19**: 158-159. 1917.) (*Lathyrus venosus* in part, of Gray, Man., ed. 7 and Brit-

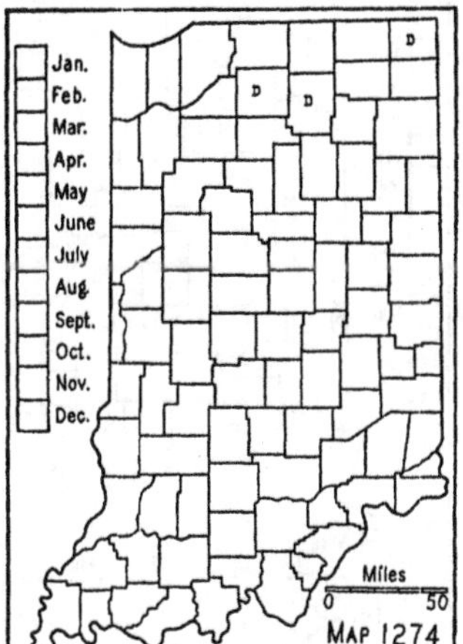

Lathyrus palustris var. linearifolius Ser.

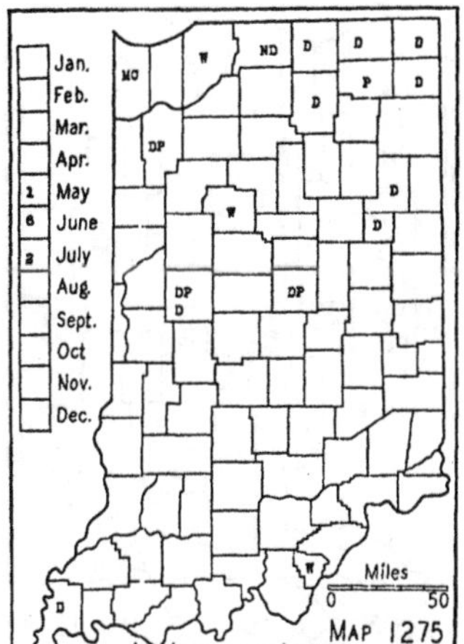

Lathyrus palustris var. myrtifolius (Muhl.) Gray

Clitoria mariana L.

ton and Brown, Illus. Flora, ed. 2.) HAIRY VEINY PEA. Map 1272. Infrequent to very rare in some of the counties of the lake area where it is generally found in dry, sandy soil in open, black oak woods, in prairie habitats, and rarely in a marshy habitat. I have had this variety under cultivation for a few years and it is spreading rapidly by underground stems. There have been four reports for the species from the northern counties but I am referring them all to the variety.

Ont. to Sask., southw. to W. Va., Tenn., and N. Dak.

4. **Lathyrus palústris** L. MARSH PEA. Map 1273. Infrequent in the lake area in swamps and marshes, on the low borders of lakes, in springy places along streams, and in a drier habitat in prairies. It has been reported from several places south of the lake area and, without doubt, it was formerly found in prairie habitats and springy places along streams and on the borders of ponds and swamps.

The species and varieties intergrade so completely that it is not possible to separate them satisfactorily.

This species and *Vicia americana* are often confused but are easily separated by the fact that *Vicia americana* usually has 8-14 leaflets while this species has 4-8. Besides the generic distinction between the styles, the basal section of the stipules of *Vicia americana* is generally sharply toothed while, if the stipules of this species are not entire or nearly so, they are never sharply toothed.

Lower St. Lawrence River to Man. and Oreg., southw. to s. Maine, Conn., cent. N. Y., Ohio, and Mo.; also in Eurasia.

4a. **Lathyrus palustris** var. **linearifòlius** Ser. Map 1274. This variety has the habitat of the species but is much less frequent.

Basin of the St. Lawrence River southw. to n. N. Y., Ind., and Minn.

4b. **Lathyrus palustris** var. **myrtifòlius** (Muhl.) Gray. MYRTLE-LEAF MARSH PEA. Map 1275. Infrequent to very rare, mostly in the lake area in habitats similar to those of the species. There are several reports

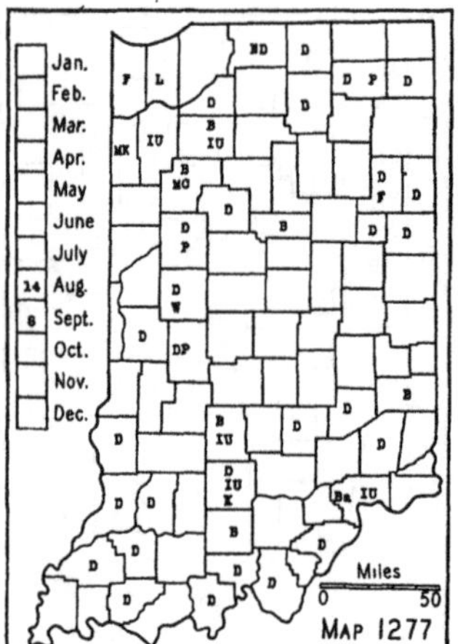

MAP 1277
Amphicarpa bracteata (L.) Fern

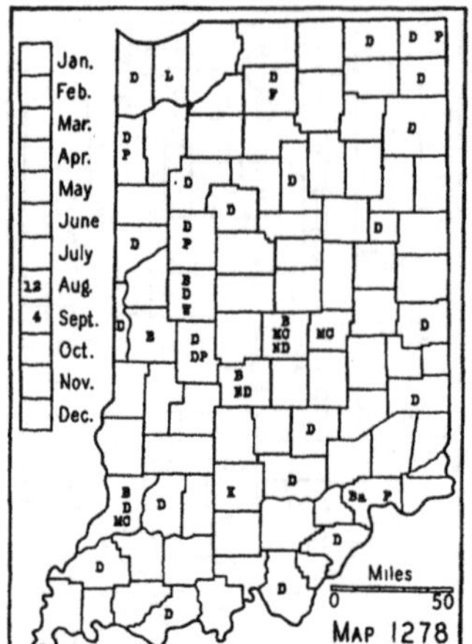

MAP 1278
Amphicarpa bracteata var. comosa (L.) Fern.

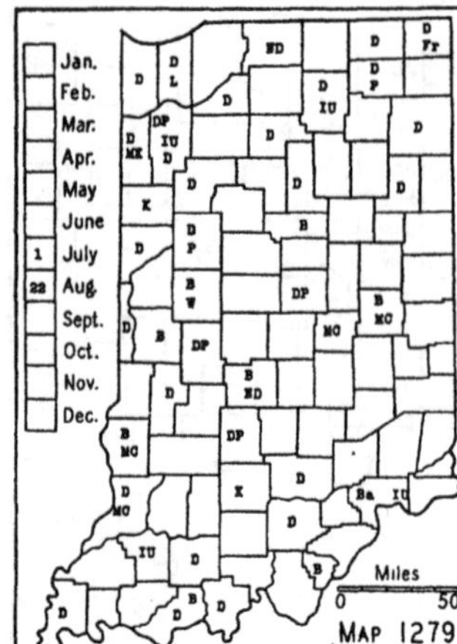

MAP 1279
Apios americana Medic.

from the dune area of Lake Michigan. My Posey County specimen is not typical and may be a southern representative of the species.

W. Que. to Wis. and Man. (?), southw. to n. N. J., Pa., N. C., and Tex.

3857. CLITÒRIA L.

1. **Clitoria mariàna** L. BUTTERFLY-PEA. Map 1276. This species is very rare on the crests of a small number of wooded, sandstone ridges in a few of our southern counties.

N. Y. to Iowa, southw. to Fla. and Tex.

3860. AMPHICÁRPA Ell. HOG PEANUT

[Fernald. Recent Discoveries in the Newfoundland Flora. Rhodora 35: 276. 1933.]

Pubescence of sutures of pods from petaliferous flowers upwardly appressed (at least on the basal half); pubescence of stem colorless and appressed, sometimes somewhat spreading in parts and slightly tawny; median stipules generally about 3 mm long; floral bracts mostly 2-2.5 mm long; calyx tube, measured to the lowest sinus, mostly about 4 mm long; mature seed about 3.5 mm long.........1. *A. bracteata.*

Pubescence of pods from petaliferous flowers downwardly appressed (at least on the basal half); pubescence of stem tawny and widely spreading; median stipules generally about 4 mm long; floral bracts usually about 3 mm long, sometimes as short as 2.5 mm; calyx tube, measured to the lowest sinus, generally 4.5-5 mm long; mature seed 3.8-5.5 mm long......................2. *A. bracteata* var. *comosa.*

1. **Amphicarpa bracteàta** (L.) Fern. (*Amphicarpa monoica* (L.) Ell. and *Falcata comosa* (L.) Ktze. of American authors.) Map 1277. Usually frequent in moist woods throughout the state. Our two species seem to intergrade and some authors regard the next one as only a race or vigorous form of this species. Schively expresses this doubt when she says the var. *comosa* is "an extremely vigorous" form of this species (Contr. Bot. Lab. Univ. Pa. 1: 356. 1897). Besides the winter pods, this species has subterranean 1-seeded pods, autumnal 1-seeded pods, and pods, which are usually 3-seeded, from petaliferous flowers.

N. B. and N. S. to Man., southw. to Fla., La., and Nebr.

Galactia volubilis var. mississippiensis Vail.

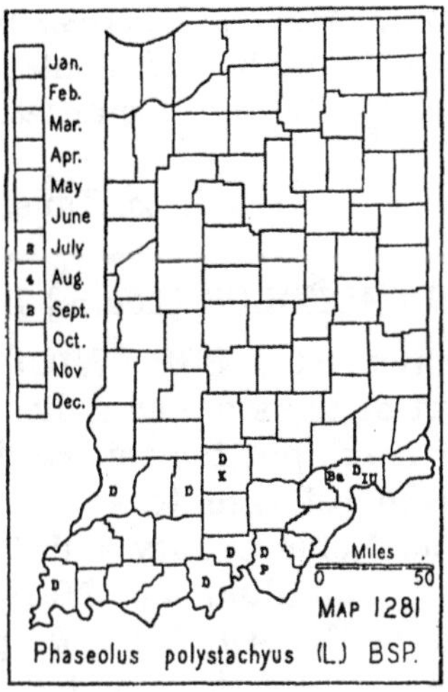

Phaseolus polystachyus (L.) BSP.

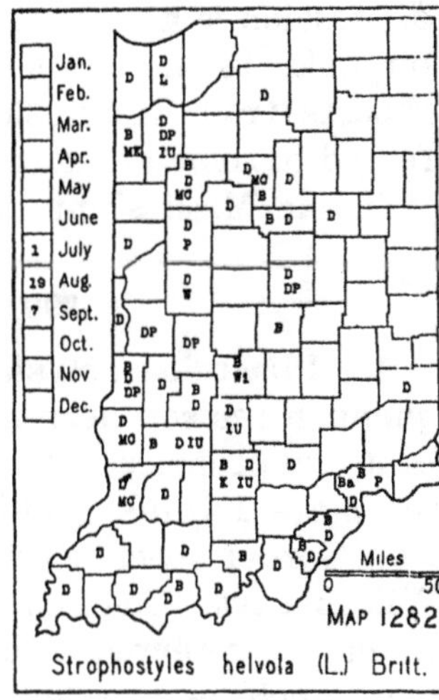

Strophostyles helvola (L.) Britt.

1a. **Amphicarpa bracteata** var. **comòsa** (L.) Fern. (Rhodora **39**: 318. 1937.) (*Amphicarpa Pitcheri* T. & G. and *Falcata Pitcheri* (T. & G.) Ktze.) Map 1278. Infrequent to frequent in moist woods throughout the state and sometimes in prairies. This species much resembles the preceding but, besides the characters given in the key to distinguish it, the plant is larger and coarser, and the leaflets especially are thicker and larger.

Maine to Mont., southw. to Ga. and Tex.

3864. GLYCÌNE L.

See excluded species no. 408, p. 1069.

3874. ÀPIOS [Boerh.] Ludwig

1. **Apios americàna** Medic. (*Apios tuberosa* Moench and *Glycine Apios* L.) POTATOBEAN. Map 1279. Infrequent to rare in all parts of the state in soil with little humus in low ground in woods about ponds, sloughs, and lakes.

N. B., N. S. to Minn., southw. to Fla. and Tex.

3882. GALÁCTIA P. Br.

Stems with a dense, spreading pubescence, decumbent, mostly twining; leaflets pubescent above. ....1. *G. volubilis* var. *mississippiensis*.
Stems glabrous or sparingly short, appressed-pubescent, prostrate; leaflets glabrous above. (See excluded species no. 409, p. 1069.) ....*G. regularis*.

1. **Galactia volùbilis** (L.) Britt. var. **mississippiénsis** Vail. DOWNY MILK PEA. Map 1280. Local on the crests of a few chestnut oak ridges of the southern part of the state. I have, however, a specimen from very sandy soil in a low depression in a very sandy woods on the Herschel Green farm about 4 miles north of Washington, Daviess County. This

depression is a small prairie of about 3 acres surrounded by red birch and pin oak. In the "flats" are a number of plants of the Coastal Plain. The reports from Kosciusko and Putnam Counties, no doubt, should be referred to some other species.

In the Mississippi Valley from Ind. to Kans., southw.

3901. PHASÈOLUS [Tourn.] L. BEAN

1. **Phaseolus polystàchyus** (L.) BSP. WILD BEAN. Map 1281. Infrequent in the southern part of the state as indicated on the map. Probably in a few counties farther north but the reports from the northern part of the state I think should be referred to some other species. I have the Van Gorder specimen from Noble County and it is *Amphicarpa bracteata.* Wilson says "Common" in Hamilton and Marion Counties. This report should be referred to one of the *Amphicarpa* species which are frequent and which he does not report. Peattie's and Pepoon's reports from Lake County may be correct. Since having seen large, entire-leaf forms of *Strophostyles helvola* labeled as this species, I suspect that some of our reports have been wrong determinations. The lower surfaces of the leaflets of *Phaseolus polystachyus* are velvety to the touch and those of *Strophostyles helvola* are not.

Conn. to Fla. and La. and northw. in the Mississippi Valley to Ind., Ill., and Mo.; also reported northw. to Minn. and Nebr.

3901A. STROPHOSTỲLES Ell. TRAILING WILD BEAN

Leaflets of an ovate type, one or both sides more or less contracted about the middle so that the leaflets appear 3-lobed or 2-lobed or one or both sides so shallowly contracted that the side or sides are only sinuate, in some specimens only a very few of the leaflets may be contracted and the remainder may be of a regular, ovate type, the upper surface nearly glabrous and the lower sparingly pubescent; calyx tube generally 2-2.5 mm long, more or less sparsely pubescent with long hairs, sometimes only the lower lobe pubescent; lower calyx lobe longer than the tube, 2.5-5 mm long; flowers mostly 7-10 mm long, the banner about 10 mm wide; pods generally 5-9 cm long, strigose-pubescent, rarely nearly glabrous; seed oblong, quadrangular, 5-6.5 mm long and about 3 mm wide.................1. *S. helvola.*

Leaflets of a narrow-ovate, lanceolate, or linear-oblong type, never with contracted sides; pods 2.5-4.5 cm long; seed mostly 3-3.5 mm long and about 2.5 mm wide.

Flowers the largest of the three species, mostly 9-12 mm long, the banner 12-16 mm wide; leaflets generally of a narrow-ovate and less often of a lanceolate type, glabrous to sparsely pubescent above and pubescent below; calyx tube 1.5-2 mm long, generally most of the surface strigose-pubescent; lower lobe of calyx tube longer than the tube, 2-2.5 mm long; pods mostly 4-4.5 cm long, strigose-pubescent or nearly glabrous; seed 3-3.5 mm long and about 2.5 mm wide..2. *S. umbellata.*

Flowers the smallest of the three species, mostly 5-6 mm long, banner 5-6 mm wide; leaflets lanceolate to linear-oblong, pubescent above and beneath; calyx tube 1-1.5 mm long, entirely covered with long, loose hairs; lower lobe of calyx longer than the tube, usually 1.5-2 mm long; pods 2.5-3.5 cm long, closely long-pubescent; seed 2.6-3 mm long and 2-2.5 mm wide............3. *S. leiosperma.*

1. **Strophostyles hélvola** (L.) Britt. Map 1282. Infrequent throughout the state as shown on the map. There are no reports from the north-

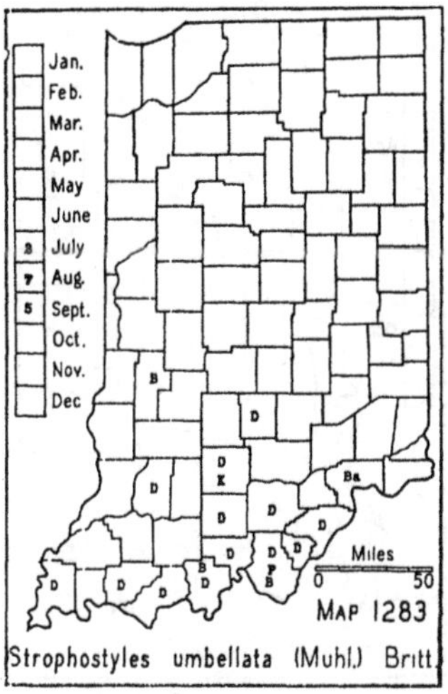
MAP 1283
Strophostyles umbellata (Muhl.) Britt.

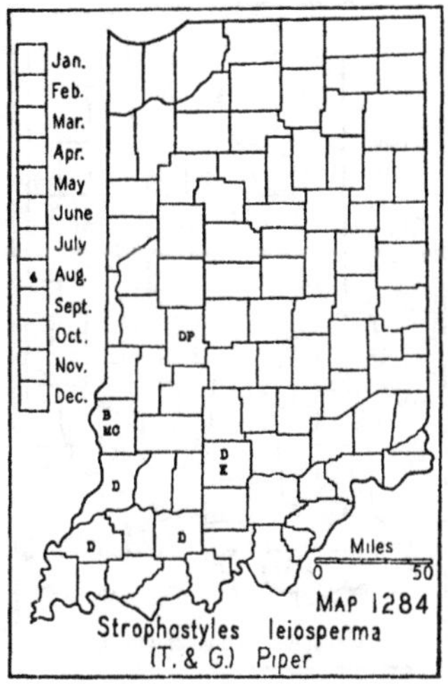
MAP 1284
Strophostyles leiosperma (T. & G.) Piper

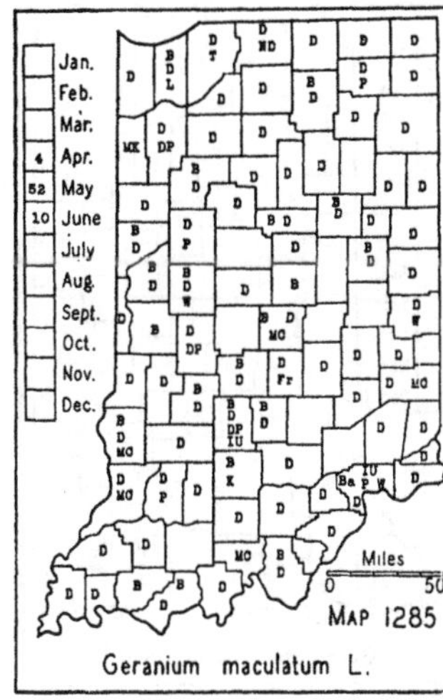
MAP 1285
Geranium maculatum L.

eastern part of the state, but it is, no doubt, more or less frequent in ballast along railroads where I rarely botanized. This species prefers a sandy or sandy, clay soil and is commonly found in ballast along railroads, along roadsides, on wooded slopes, sand bars and sandy shores of streams, on the dunes, and in fallow fields.

Que. to Minn., southw. to Fla. and Tex.

2. **Strophostyles umbellàta** (Muhl.) Britt. Map 1283. This species is rare to infrequent and has the habitat of the preceding species. It is possibly restricted to the southern counties. The specimens collected and reported from Marshall and Putnam Counties were found in ballast along railroads and may be introduced. All of my specimens are from wooded slopes and fallow fields.

Coastal Plain from L. I. to La., northw. in the Mississippi Valley to Ind. and Mo.

3. **Strophostyles leiospérma** (T. & G.) Piper. (Contr. U. S. Nation. Herb. 22: 668. 1926.) (*Strophostyles pauciflora* (Benth.) Wats.) Map 1284. This species prefers a very sandy or sandy, clay soil. I have infrequently found it in a few of the southern counties. The reports from Lake, Marshall, and Putnam Counties were of specimens collected in railroad ballast and may have been introduced. The seed of all our species are mealy-pubescent. The pubescence is easily detached in this species while in the preceding species it is persistent.

Mississippi Vally northw. to Ind. and Minn.

3905. **VÍGNA** Savi

See excluded species no. 410, p. 1069.

129. **GERANIÀCEAE** J. St. Hil. GERANIUM FAMILY

Leaves palmately cut or divided into 5 or more lobes or segments or 3-cleft; stamens 10 (5 in *Geranium pusillum*) .... 3924. GERANIUM, p. 624.
Leaves pinnate; stamens 5 .... 3928. ERODIUM, p. 626.

Geranium Robertianum L.

Geranium columbinum L.

Geranium Bicknellii Britt.

3924. GERÀNIUM [Tourn.] L. CRANESBILL

Perennial; involucral leaves mostly 7-15 cm wide....................1. *G. maculatum.*
Annual or biennial; involucral leaves mostly 3-6.5 cm wide.
  Outer mature sepals 6-10 mm long, awned.
    Leaves of 3 distinct segments which are pinnately cut or divided; carpels pale yellow, wrinkled, glabrous or nearly so..................2. *G. Robertianum.*
    Leaves palmately cut or divided into 5 or more lobes or segments; carpels black, not wrinkled, hirsute.
      Fruiting pedicels much longer than the calyx; beak of mature style column 2.5-6 mm long.
        Pedicels with minute, appressed, glandless pubescence; bodies of carpels glabrous...3. *G. columbinum.*
        Pedicels densely glandular-pilose; bodies of carpels pubescent..4. *G. Bicknellii.*
      Fruiting pedicels shorter than to slightly longer than the calyx; beak of mature style column 1-2 mm long.
        Pubescence of stem, at least the lower internodes, more or less retrorsely appressed...5. *G. carolinianum.*
        Pubescence of stem and petioles spreading...5a. *G. carolinianum* var. *confertiflorum.*
  Outer mature sepals 2.5-4 mm long, awnless.
    Stamens 5; carpels finely pubescent, not wrinkled; style column beakless..6. *G. pusillum.*
    Stamens 10; carpels glabrate, wrinkled crosswise; style column with a beak 1-2 mm long. (See excluded species no. 411, p. 1069.)..................*G. molle.*

1. **Geranium maculàtum** L. WILD GERANIUM. Map 1285. More or less frequent in moist woods. Locally abundant along railroads. The flowers vary greatly in size and color from white to deep rose pink.

Maine, Ont. to Man., southw. to Ga., Ala., and Nebr.

2. **Geranium Robertiànum** L. (*Robertiella Robertiana* (L.) Hanks.) HERB ROBERT. Map 1286. I found this species to be abundant in a very low woods of 20 acres in St. Joseph County, associated with white elm and soft maple. It was reported by Collins from Dearborn County but he left no specimen. I doubt the accuracy of the determination.

Newf. to Man., southw. to N. J. and Mo.; also in Eurasia and Africa.

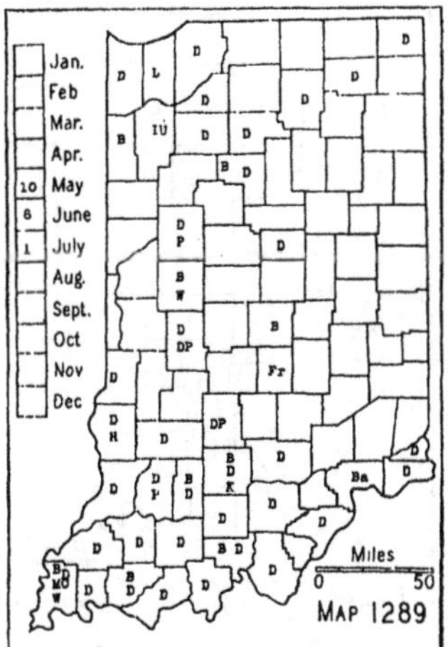

Geranium carolinianum L.

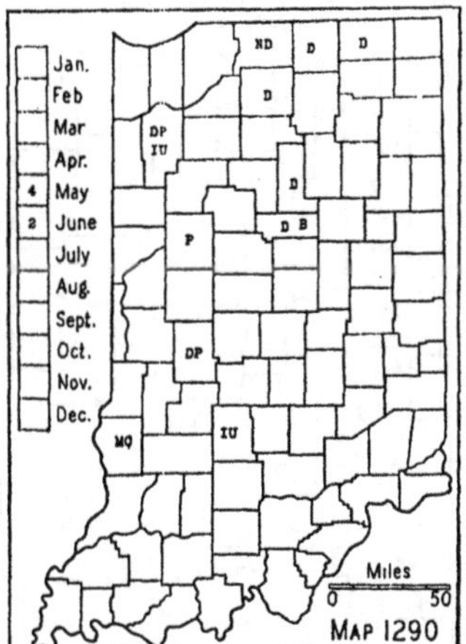

Geranium carolinianum var. confertiflorum Fern.

Geranium pusillum Burm. f.

3. GERANIUM COLUMBÌNUM L. Map 1287. This species was found in 1935 by Miss Edna Banta. It was a frequent weed in a pasture field on the Kellar farm about a mile southwest of Wirt, Jefferson County.

Nat. of Eu.; N. Y. and Ohio, southw. to Va. and W. Va.; also in S. Dak.

4. **Geranium Bicknéllii** Britt. Map 1288. In sandy soil in burned over black and pin oak woods. Frequent where found and always associated with *Corydalis sempervirens* and sometimes with *Epilobium angustifolium*. Large specimens may be three feet in diameter.

Newf. to B. C., southw. to N. E., N. Y., and Utah.

5. **Geranium caroliniànum** L. (Fernald. Geranium carolinianum and allies of northeastern North America. Rhodora **37**: 295-301. 1935.) Map 1289. This species prefers sandy to very sandy soils and is found as a weed in fallow fields, hayfields, pastures, and open, pastured woods and along roadsides and railroads. On account of its weedy nature it is debatable whether this species is a native of the state. Some of our oldest floras do not list it and others record it as found in waste places and fields and along roadsides and railroads.

Mass., Conn., s. Mich., Ill., Mo., Kan., Wyo., Idaho, and s. B. C., southw. to Fla. and s. Calif.

5a. **Geranium carolinianum** var. **confertiflòrum** Fern. (Rhodora **37**: 300. 1935.) Map 1290. In addition to this variety intermediate forms occur. This form is not very distinct in Indiana. The habitats are similar to those of the species. Fernald gives the distribution as follows:

Maine to Wis., southw. to Del., uplands of N. C., Tenn., and Mo.

6. GERANIUM PUSÍLLUM Burm. f. Map 1291. This species was found as a weed in 1902 on the grounds of Purdue University Agricultural Experiment Station, and in 1905 Wilson says: "Appears to be well established and spreading, exterminating the grass." It was found also, in 1935, well

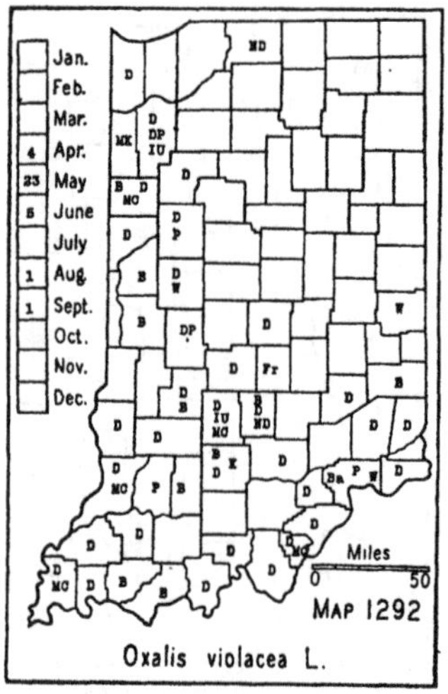

Oxalis violacea L.

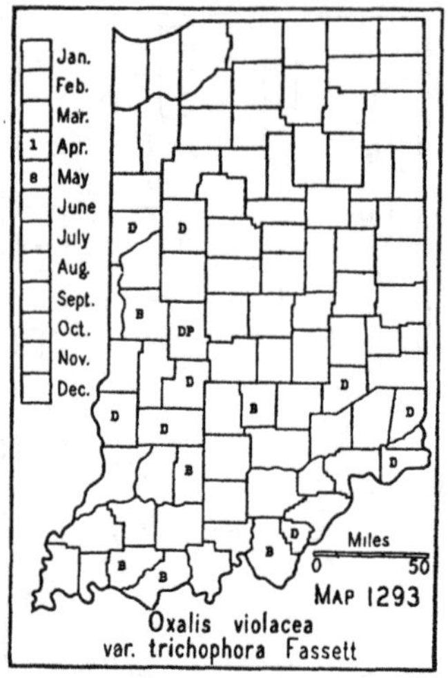

Oxalis violacea var. trichophora Fassett

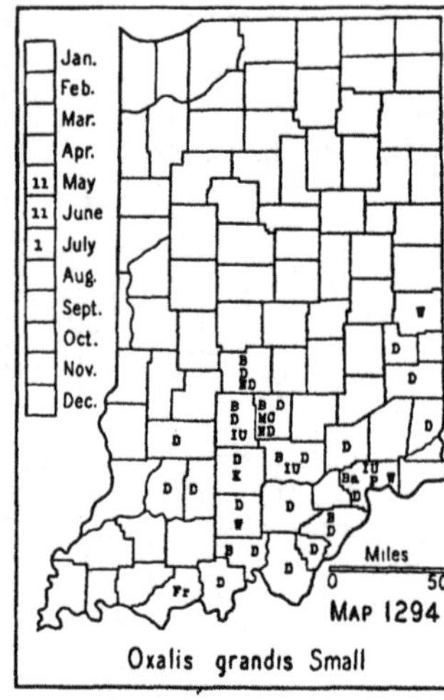

Oxalis grandis Small

established in a lawn about one and a half miles northwest of Bluffton, Wells County. Specimens from La Porte and St. Joseph Counties have been collected by Nieuwland.

Nat. of Eu.; Mass., Ont. to B. C., southw. to N. J., N. C., Nebr., and Utah.

3927. **ERODIUM** L'Hér. Storksbill

See excluded species no. 412, p. 1069.

130. **OXALIDÀCEAE** Lindl. Wood Sorrel Family

3936. **ÓXALIS** L. Wood Sorrel

[Wiegand. Oxalis corniculata and its relatives in North America. Rhodora 27: 113-130; 133-139. 1925.]

Plants with a bulblike rootstock; acaulescent; flowers violet.
- Petioles of leaves glabrous..1. *O. violacea.*
- Petioles of leaves glandular-pubescent...........1a. *O. violacea* var. *trichophora.*

Plants without a bulblike rootstock; caulescent; flowers yellow, rarely greenish yellow or green.
- Flowers 12-18 mm long; margin of leaflets usually purplish brown......2. *O. grandis.*
- Flowers 5-11 mm long; margin of leaflets not purplish brown.
  - Stems creeping on the surface of the ground; stipules usually broad and brownish purple, subscarious..3. *O. repens.*
  - Stems erect or decumbent, often with creeping rootstocks; stipules oblong, narrowly oblong, or obsolete.
    - Flowers in umbels, rarely subcymose or solitary; fruiting pedicels horizontal or deflexed but the capsules erect; capsules finely and densely pubescent, sometimes with additional villous, viscid hairs, or strigose above and glabrous below, or more rarely strigose throughout (*O. florida*).
      - Pubescence appressed or subappressed, whitish; capsules abruptly pointed, 15-25 mm long; sepals (3.5) 4-7 mm long.
        - Pubescence of capsule appressed with some longer, loose, viscid hairs..4. *O. stricta.*
        - Pubescence of capsule appressed, not viscid...4b. *O. stricta* var. *piletocarpa.*

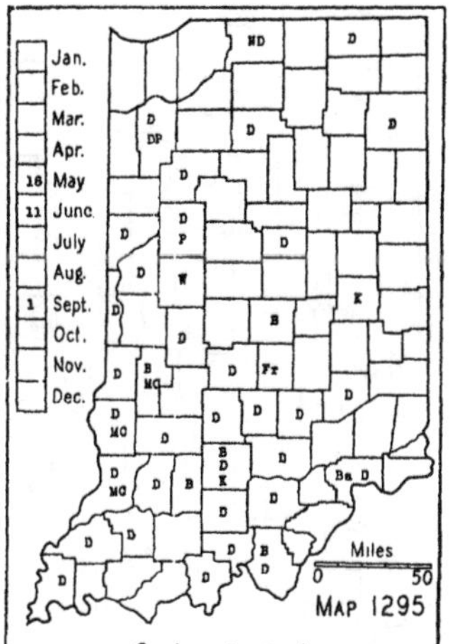

Oxalis stricta L.

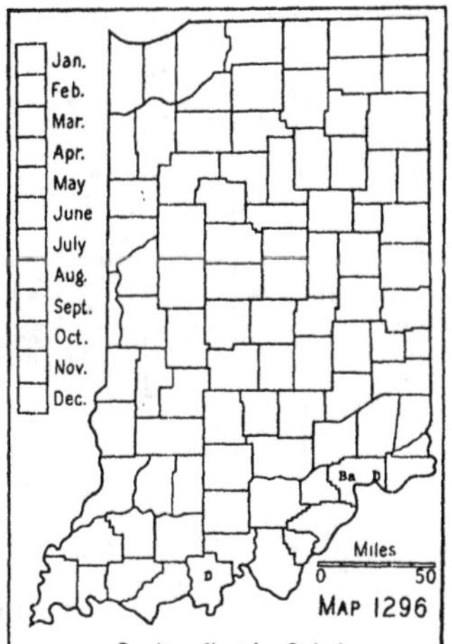

Oxalis florida Salisb.

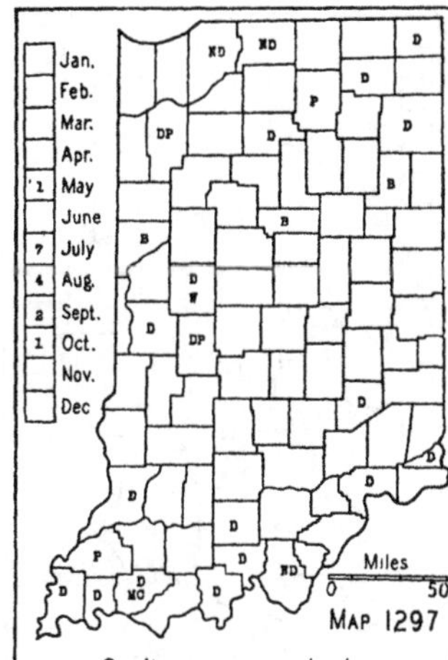

Oxalis europaea Jordan

Pubescence toward the base of the stem loosely crisp, tawny; capsules strigose toward the apex, glabrous below, rarely strigose throughout, 9-15 mm long, gradually pointed; sepals 4-4.5 mm long......5. *O. florida.*

Flowers cymose; fruiting pedicels not deflexed but sometimes widely spreading; capsules with loose spreading viscid hairs or nearly glabrous.

Upper surface of leaflets glabrous.

Pubescence of pedicels appressed.........................6. *O. europaea.*

Pubescence of pedicels spreading, usually viscid.

Stem nearly or quite glabrous................6a. *O. europaea* f. *cymosa.*

Stem villous............................6b. *O. europaea* f. *villicaulis.*

Upper surface of leaflets more or less pubescent.

Pubescence of pedicels appressed, scarcely viscid; stems with ascending pubescence or glabrate......6c. *O. europaea* var. *Bushii* f. *subglabrata.*

Pubescence of pedicels spreading, usually viscid; stems villous........... ....6d. *O. europaea* var. *Bushii* f. *vestita.*

1. **Oxalis violàcea** L. VIOLET WOOD SORREL. Map 1292. In woods in dry, sandy soils or sandy, clay soils with little competing vegetation and usually in wind swept, open places with little humus in the soil. Often on the border of woods, along roadsides, and less frequently in abandoned fields. Rarely in rich soil or in moist situations.

More or less local throughout the state although there are no records from the northeastern counties.

Eastern Mass. to Minn. and southw.

1a. **Oxalis violacea** var. **trichóphora** Fassett. (Rhodora 39: 378. 1937.) Map 1293. This is a southern form with the habitat of the species and occurring with it.

Vt., Pa., Ind., to Mo., southw. to Va. and Ark.

2. **Oxalis grándis** Small. (*Xanthoxalis grandis* Small.) GREAT YELLOW WOOD SORREL. Map 1294. Infrequent in wooded ravines and rarely in abandoned fields.

Pa. to Ill., southw. to Ga. and Ala.

3. OXALIS RÈPENS Thunb. (*Xanthoxalis corniculata* (L.) Small.) A weed introduced into greenhouses throughout the state. It has not often

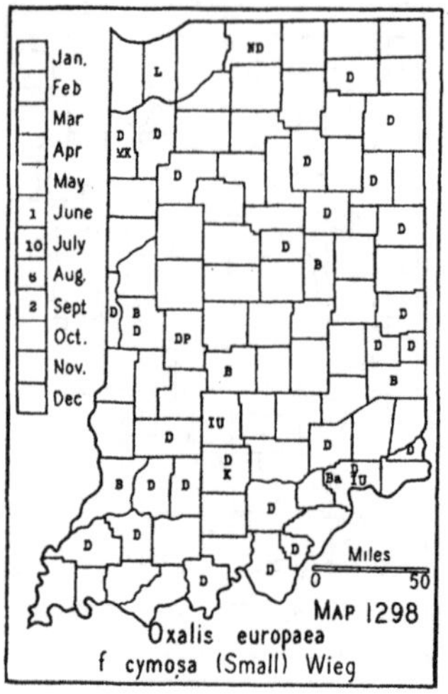

MAP 1298
Oxalis europaea f cymosa (Small) Wieg

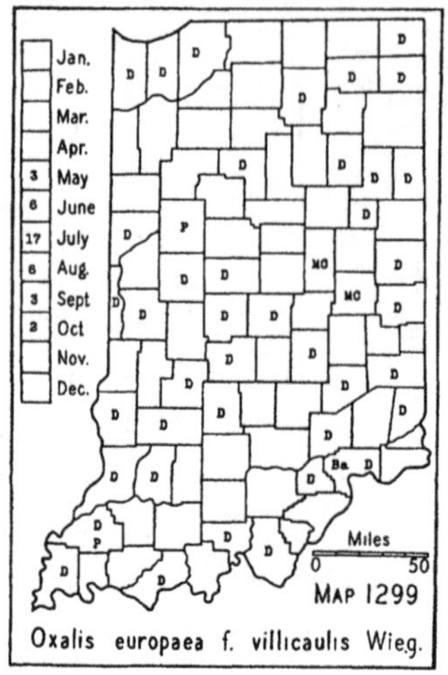

MAP 1299
Oxalis europaea f. villicaulis Wieg.

MAP 1300
Linum sulcatum Riddell

been collected, however, and the only specimen is in the herbarium of the University of Notre Dame, having been collected at Notre Dame.

Tropical regions, almost cosmopolitan.

4. **Oxalis stricta** L. (*Xanthoxalis stricta* (L.) Small.) UPRIGHT YELLOW WOOD SORREL. Map 1295. Mostly in fallow or abandoned fields, along roadsides and railroads, and less frequent in woodland; apparently preferring an impoverished soil. Throughout the state but more abundant in the southern part.

P. E. I. to B. C., southw. to Fla., Tex., and Mex.

4a. **Oxalis stricta** f. **viridiflòra** (Hus) Fern. (Rhodora 38: 425. 1936.) This is a form with green petals. It is represented in my herbarium by a specimen from Gibson County collected by Schneck.

4b. **Oxalis stricta** var. **piletocárpa** Wieg. (Rhodora 27: 123. 1925.) I have this variety from only Bartholomew and Fountain Counties. There is a specimen from Putnam County collected by Grimes, which is in the herbarium of DePauw University.

P. E. I., southw. to N. J.; also in Wyo.

5. **Oxalis flórida** Salisb. (Rhodora 27: 133. 1925.) (*Oxalis filipes* Small, *Xanthoxalis filipes* Small, and *Oxalis Brittoniae* Small.) Map 1296. Our only specimens are from a wooded bluff along the Ohio River about 6 miles above Cannelton and from a wooded bluff in Jefferson County.

I reported *Oxalis Brittoniae* Small from Steuben and Wells Counties. (Proc. Indiana Acad. Sci. 1904: 220. 1905.) I am now referring these specimens to *Oxalis europaea* Jordan.

Pepoon reported *Oxalis filipes* Small from Lake and Porter Counties for Umbach but Fassett (Rhodora 35: 200. 1933) refers the Lake County specimen to *Oxalis stricta* and says there is no specimen labeled *Oxalis filipes* from Porter County.

Maine to Fla., "mainly toward the coast but apparently not on the Coastal Plain" (Wiegand); inland in Ind.

6. **Oxalis europaèa** Jordan. (Rhodora 27: 134. 1925.) (*Oxalis corniculata* of Gray, Man., ed. 7, not L.) LADY'S SORREL. Map 1297. The flowers are generally yellow but I have one specimen with greenish and one with green petals. The color note was made when the specimens were collected.

This species is found most frequently in open beech and sugar maple woods but is also found in moister woods and in the open along roadsides.

Que. to N. Dak., southw. to Ga., Tenn., Okla., and Colo.

6a. **Oxalis europaea f. cymòsa** (Small) Wieg. (Rhodora 27: 135. 1925.) (*Xanthoxalis cymosa* Small.) Map 1298. In open or rather thick woodland, clearings, and fallow fields and along roadsides and railroads.

In a low woods on the north side of Eggwood Pond and in a low woods about two miles southeast of East Mt. Carmel in Gibson County, I collected a form with leaves that are greenish purple above and purple beneath.

Que. to Mich., southw. to N. C., Tenn., and Mo.

6b. **Oxalis europaea f. villicaúlis** Wieg. (Rhodora 27: 135. 1925.) Map 1299. This form and f. *cymosa* are our common tall woodland sorrels.

In dry or wet woodland and rarely in the open along fences and roadsides.

N. S., Mass. to Mich., southw. to Va., Tenn., and Ill.

6c. **Oxalis europaea** var. **Búshii** Small f. **subglabràta** Wieg. (Rhodora 27: 136. 1925.) My only specimen is from a low woods in Daviess County. It has also been collected in St. Joseph and Tippecanoe Counties.

Ill., Iowa, and Mo.

6d. **Oxalis europaea** var. **Bushii** f. **vestìta** Wieg. (Rhodora 27: 136. 1925.) In open woods and old fields.

Mass. and Ill.

132. LINÀCEAE Dumort. FLAX FAMILY

3945. LÌNUM [Tourn.] L. FLAX

Flowers blue; capsules mostly 7-10 mm wide. (See excluded species no. 414, p. 1069.) .... *L. usitatissimum.*

Flowers yellow; capsules mostly 2-3.5 mm wide.

- Leaves with dark glands at the base instead of stipules, 3-nerved (the outer two sometimes too close to the margin to be very distinct); margin of the outer and inner sepals more or less glandular; capsules about 3 mm wide; false septa incomplete, ciliate........ 1. *L. sulcatum.*
- Leaves without dark, stipular glands, 1-nerved; margin of the outer sepals not glandular; capsules less than 3 mm wide; false septa nearly complete, not ciliate.
  - Capsules ovoid; sepals about 2 mm long, shorter than the mature capsule........ 2. *L. intercursum.*
  - Capsules depressed at the summit.
    - Sutures of capsules slightly elevated; segments of capsules slightly concave between the sutures; cauline leaves thin, lax, acute; axis of inflorescence flexuous, the branches few, more or less flexuous; inflorescence more or less dichotomously branched, even to the ultimate divisions; outer sepals noticeably longer than the apex of the capsule, the inner ones about as long as the capsule and their margins not glandular below the middle........ 3. L. *virginianum.*

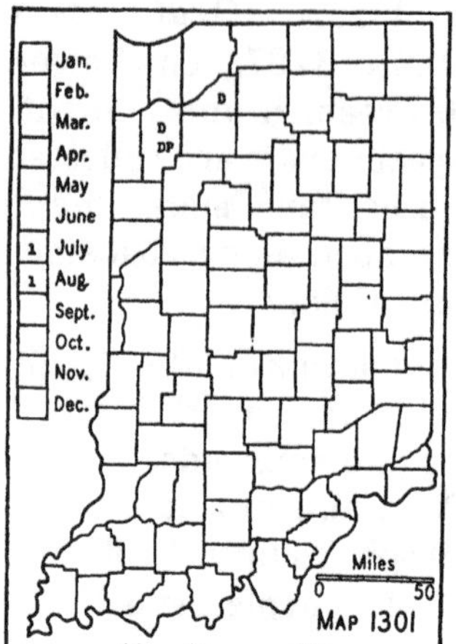

Linum intercursum Bickn.

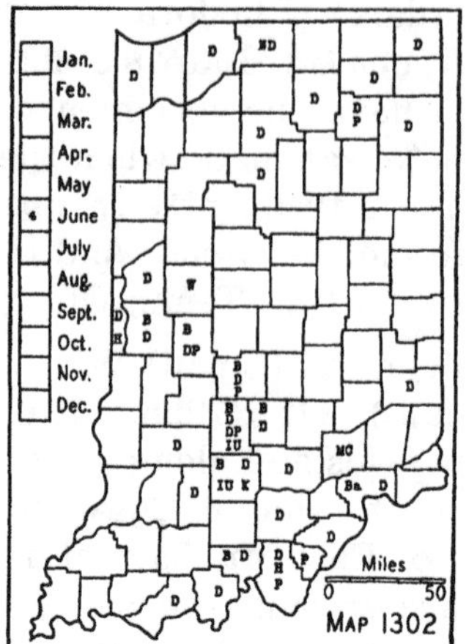

Linum virginianum L.

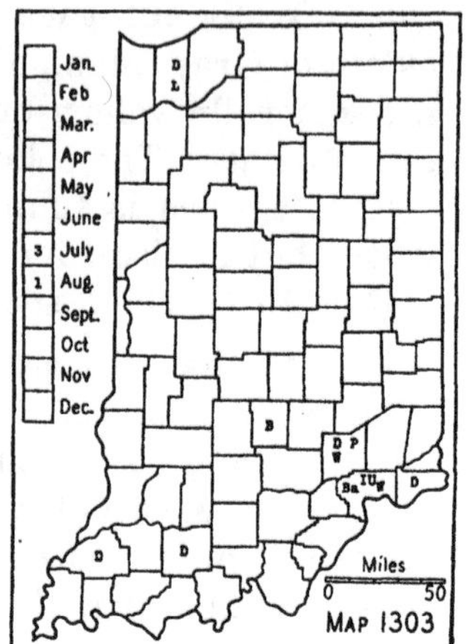

Linum striatum Walt.

Sutures of capsules not elevated; segments not concave between the sutures; inflorescence with the axis usually straight, not dichotomously branched, the branches straight and strongly ascending.

Stem leaves usually lax, thin, merely acute; leaves of branches acute or acuminate; branches ascending usually at an angle of between 40 and 60 degrees; sepals about as long as or slightly shorter than the capsule; margins of the inner sepals entire or slightly erose or glandular toward the apex, abruptly acuminate at the apex................4. *L. striatum.*

Stem leaves stiff, usually erect or strongly ascending, with rigid, acuminate tips; leaves of the branches stiff, narrow, almost appressed, apiculate at the apex; branches ascending at an angle of mostly between 60 and 75 degrees; sepals usually about a half longer than the capsule; margins of the inner sepals glandular usually to the middle or below the middle, gradually tapering at the apex into long, indurated tips..5. *L. medium* var. *texanum.*

1. **Linum sulcàtum** Riddell. Map 1300. This species was reported from Jasper and Lake Counties before our manuals recognized *Linum medium* var. *texanum* to which I believe both records should be referred. The Lake County specimen was collected by Hill, July 28, 1875, near Pine and is deposited in the herbarium of DePauw University. It is labeled *Linum sulcatum* but is *Linum medium* var. *texanum.* My specimens were collected on the low, moist, gravelly border of the north side of Wall Lake, Lagrange County; on the dry, gravelly bank of the St. Joseph River, St. Joseph County; and on the dry, gravelly, high bank of Big Wea Creek about 4 miles southwest of Lafayette, Tippecanoe County. Charles M. Ek found a colony along a railroad in Howard County. The only specimens from the southern part of the state are one collected by Clapp in Harrison County near Palmyra, July 14, 1836, which is now in the herbarium of Wabash College, and one which I collected in the same county on a post oak ridge about 5 miles southwest of Corydon.

Eastern Mass. and Vt. to Man., southw. to Ga. and Tex.

2. **Linum intercúrsum** Bickn. (Bull. Torrey Bot. Club 39: 418-420. 1912.) Map 1301. Moist, sandy soil on the borders of marshes in black oak woods. Rare.

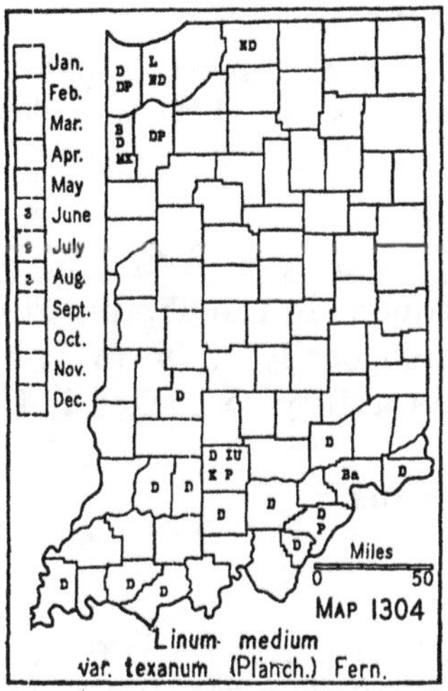

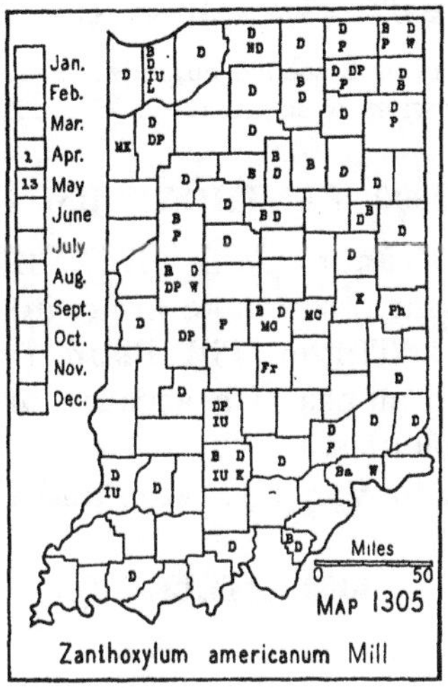

Coastal Plain from N. C. to Fla., Miss., and up the Mississippi Valley to s. Ill., and up the Kankakee Valley to nw. Ind.

3. **Linum virginiànum** L. Map 1302. Infrequent in open places on the crests and slopes of black and white oak and chestnut oak ridges in the southern part of the state; in prairie habitats in the northern part of the state; and in dry, open, white oak and white and black oak woods elsewhere.

Maine to Ont., southw. to Ga. and Ala.

4. **Linum striàtum** Walt. Map 1303. Our specimens are from mediacid, hard, white clay soil in low, beech and sweet gum woods, one from a wooded slope, and one from the border of an interdunal flat. Rare.

Mass. to Fla., Ky., Mo., and Tex.

5. **Linum mèdium** (Planch.) Trel. var. **texànum** (Planch.) Fern. (Rhodora 37: 428. 1935.) (*Linum medium* (Planch.) Britt. of Gray, Man., ed. 7 and *Cathartolinum medium* (Planch.) Small of Britton and Brown, Illus. Flora, ed. 2.) Map 1304. In southern Indiana this plant has been found on washed slopes in fallow fields, on post oak ridges, and in moist, hard clay soil on post oak flats. In the northern part of the state it is found in the prairies and lake areas in moist, sandy soil, usually associated with *Cladium mariscoides, Juncus Torreyi, Liatris spicata,* etc. Rare.

My specimens no. 41481 from Brown County may belong to the typical form. They are rather small and I hesitate to refer them to the typical form for fear they are only extremes of the variety.

Sw. Maine, s. Mich., and Ill., southw. to Fla. and Tex.

135. ZYGOPHYLLÀCEAE Lindl. CALTROP FAMILY

3978. TRÍBULUS [Tourn.] L.

See excluded species no. 415, p. 1069.

137. RUTÀCEAE Juss. RUE FAMILY

Leaves pinnate; fruit red, a 1- or 2-seeded capsule........3990. ZANTHOXYLUM, p. 632.
Leaves 3-foliolate; fruit yellowish, a samara....................4069. PTELEA, p. 632

3990. ZANTHÓXYLUM L.

1. **Zanthoxylum americànum** Mill. NORTHERN PRICKLY ASH. Map 1305. More or less frequent in the lake area; infrequent in the Tipton Till Plain; and south of the Tipton Till Plain found locally only in wet woods and on dry wooded slopes. On account of its ability to sucker it is usually found in dense colonies.

Que. to Minn., southw. to Va., Ky., Mo., and e. Kans.

4069. PTÈLEA L.

Branchlets glabrous ....1. *P. trifoliata.*
Branchlets pubescent.............................1a. *P. trifoliata* var. *Deamiana.*

1. **Ptelea trifoliàta** L. (*Ptelea mesochora* Greene.) COMMON HOPTREE. Map 1306. An infrequent shrub in all parts of the state. It is usually restricted to the alluvial banks of streams but it is found sometimes on the tops and slopes of rocky bluffs. The under surface of the leaflets is more or less pubescent on unfolding, becoming more or less glabrous at maturity. Some are glabrous with the exception of a few hairs in the axils of the veins and on the petioles; others have a straggling pubescence; and in nearly a fourth of our specimens the under surface is thickly pubescent. The pedicels of the flowers are likewise more or less densely pubescent at flowering time, and at maturity they become glabrous or remain more or less densely pubescent. There is no correlation of pubescence of the leaflets and pedicels, although the leaflets that are very pubescent at maturity also have pubescent pedicels, but nearly glabrous leaflets may have densely pubescent pedicels. Plants with leaflets remaining pubescent until maturity are *Ptelea trifoliata* f. *pubéscens* (Pursh) Fern. (Rhodora 38: 233. 1936). The pubescent form is much less frequent than the glabrate form and has no definite geographic range in Indiana, although most of our specimens are from the southern part of the state.

Conn., s. Ont. to Minn., southw. to Fla. and Kans.

1a. **Ptelea trifoliata** var. **Deamiàna** Nieuwl. This variety is found only on the dunes near Lake Michigan where, for the most part, it displaces the species although I have seen both variety and species growing together. We have both the variety and species of the same age growing at Bluffton. The variety grows more slowly and is more widely spreading than the typical form.

138. SIMARUBIÀCEAE DC. QUASSIA FAMILY

4124. AILÂNTHUS Desf.

1. AILANTHUS ALTÍSSIMA (Mill.) Swingle. (*Ailanthus glandulosa* Desf.) AILANTHUS. Map 1307. In waste places in cities and towns, in a

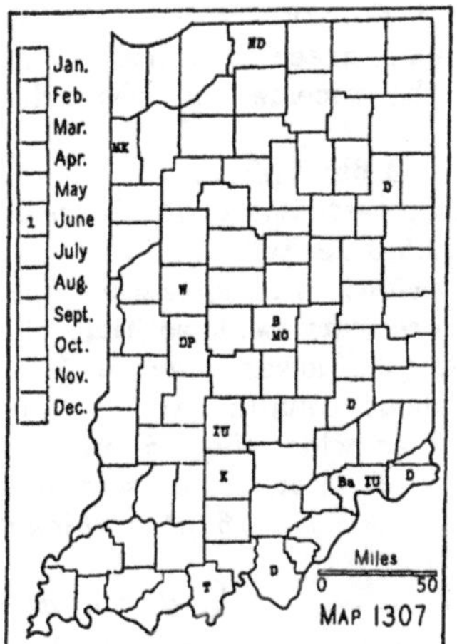

Ailanthus altissima (Mill.) Swingle

Polygala paucifolia Willd.

Polygala polygama Walt.

few places in woodland in southern Indiana, and along the wooded bluffs of the Ohio River, especially in Jefferson County.

Nat. of Asia; introduced and escaping.

145. POLYGALÀCEAE Reichenb. MILKWORT FAMILY

4273. POLÝGALA [Tourn.] L.

Lower stem leaves few, usually 2-6, scalelike, 2-8 mm long, narrower than long; leaves at the apex of the stem approximate, alternate, elliptic to oval, 10-40 mm long, 7-20 mm wide; flowers 1-3, rose purple or rarely white, mostly 10-17 mm long..1. *P. paucifolia.*

Lower stem leaves and flowers not as above.

Cleistogamous flowers present, borne on short, leafless, basal branches..2. *P. polygama.*

Cleistogamous flowers absent.

Leaves lance-ovate, lance-elliptic, narrowly lanceolate to ovate or oblong-ovate, mostly 6-30 mm wide; perennials............................3. *P. Senega.*

Leaves never ovate, rarely more than 5.5 mm wide; annuals.

Petals not united into a conspicuous cleft tube.

Racemes cylindric or conic-cylindric, acuminate or at least distinctly tapering above, mostly 2.2-6 mm broad.

Leaves all alternate; sepals suborbicular to lance-ovate; bracts of spikes persistent. (See excluded species no. 418, p. 1070.)........*P. Nuttallii.*

Leaves alternate throughout or whorled; sepals ovate; bracts of the spikes deciduous.

Raceme[1] seemingly conic, the fruits quickly falling so that the flowers and fruits present are crowded into a space 0.5-1.5 cm long; "wings" shorter than the mature capsule[2]; seed about twice as long as wide, the aril usually over half its length; leaves mostly or wholly verticillate.

Seed hirsute; capsule about 1.5 mm long, on a pedicel a third to half its length; racemes wider and looser, the sepals often purplish; plant usually 2-3 dm tall, with ascending branches and the racemes on peduncles 2-7 cm long......................4. *P. verticillata.*

[1] Pennell. Bartonia 13:9. 1932.

[2] Capsule-measurements are of the apparent size of the mature capsules, from which the ripe seed characteristically protrude.

Seed finely pubescent; capsule on a pedicel a fourth to a third its length; raceme narrow, dense, the sepals greenish white; plant with widely spreading branches and the racemes on peduncles 0.5-4.0 cm long.

Capsule about 1 mm long; plant usually 1-2 dm tall...5. *P. verticillata* var. *isocycla.*

Capsule about 1.5 mm long; plant usually 1.5-3 dm tall.......................................5a. *P. verticillata* var. *sphenostachya.*

Raceme long-cylindric, slender, the fruits more persistent so that the flowers and fruits present are scattered (the lower remote) in a slender raceme 1-5 cm long; "wings" about equaling the mature capsule; seed mostly three times as long as wide, the aril usually less than half its length; leaves mostly or all alternate or scattered on the stem and virgate branches..................6. *P. ambigua.*

Racemes capitate, ovoid, obtuse, mostly 7-17 mm broad.

Leaves whorled or the upper scattered....................7. *P. cruciata.*

Leaves all alternate...................................8. *P. sanguinea.*

Petals united into a distinct, cleft tube about 5 mm long; fruit persisting on the spike longer than in *P. sanguinea.* (See excluded species no. 417, p. 1070.)..*P. incarnata.*

1. **Polygala paucifòlia** Willd. FRINGED POLYGALA. Map 1308. The only specimens I have seen grew on the north slope of a black oak dune near Lake Michigan in Porter County. This colony was discovered by Marcus W. Lyon, Jr., who was the first to report it for Indiana. Pepoon later reported it for Lake and Porter Counties for Hill and Umbach but I have not seen these specimens. W. F. Durno, 180 N. Wacker Drive, Chicago, Ill., wrote me that on May 1, 1938, he saw the colony and estimated that there were 100 plants in bloom on that date. From his description of the location of the colony, I think it is the same colony that Dr. Lyon found in 1927. Durno also writes that there is a small colony of the white-flowered form a short distance to the southwest of this colony. In 1929 I collected a single plant for a record and at that time there were not more than 20 plants in the colony.

E. Que. to Man., southw. to Ga., Ill., and Minn.

2. **Polygala polýgama** Walt. Map 1309. Plants of this species vary greatly, from erect, from a decumbent or ascending base with only terminal spikes, or sometimes with a few lateral branches of cleistogamous flowers, to widely spreading with terminal spikes and many lateral branches with cleistogamous flowers. The latter extreme form we have from Lagrange County; it is variety *ramulosa* Farwell (Amer. Midland Nat. 11: 63. 1928).

In dry or moist sandy places in black or black and white oak woods, sandy knolls, and in moist interdunal flats. Our specimens are mostly from northwest of the Wabash River.

N. S. to Man., and southw., chiefly in the coastal region to Fla. and e. Tex.

3. **Polygala Sénega** L. SENECA SNAKEROOT. Map 1310. This plant varies greatly in size and in the width of the leaves. Plants with most of the upper blades more than 7 mm wide are referred to variety *latifolia* T. & G. Most of our plants belong to the wideleaf form. However, I am

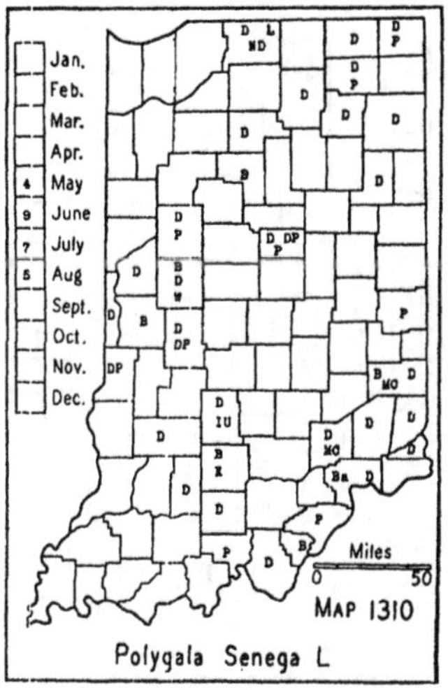

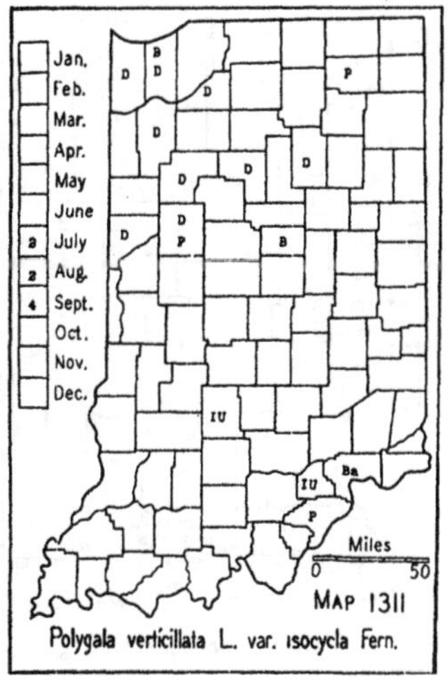

not able to separate satisfactorily the wideleaf from the narrowleaf form. Large, branched plants may have on one branch leaves of the typical form and on others leaves like those of the variety. The width of the spike is another character used to separate the two forms and it happens that in my 34 specimens the widest spike is on a plant with narrow leaves. The stems of small plants are always simple but large plants may be either simple or branched. Plants of a prairie or sandy habitat have narrower leaves than those of wooded limestone slopes.

Usually on wooded slopes along streams and about lakes. Rarely in the open in a prairie habitat.

Southern N. B. to the eastern shore of Hudson Bay, westw. to Alberta, southw. to Ga., Tenn., and Ark.

4. **Polygala verticillàta** L. (*Polygala Pretzii* Pennell.) (See Fernald's discussion of this species in Rhodora **40**: 337-338. 1938.) Map 1313. Mostly near streams and lakes in dry sandy soil in black and white oak woods; rarely in the low sedge border of lakes.

Maine to s. Mich. and Tenn.

5. **Polygala verticillata** var. **isocỳcla** Fern. (See Fernald's discussion of this species in Rhodora **40**: 334-336. 1938.) Map 1311. In poor soil in black and white oak woods and rarely in moist prairies. Rare.

Mass., Ont., and Man., southw. to Fla. and Tex.

5a. **Polygala verticillata** var. **sphenostàchya** Pennell. Map 1312. My only specimen is from a sandy roadside cut four and a half miles north and a mile and a half west of Morocco, Newton County. Other specimens have been collected in Putnam, St. Joseph, and Tipton Counties.

Ind. to Nebr. and Kans.

6. **Polygala ambígua** Nutt. Map 1314. In poor soil on open wooded slopes, in washed fallow fields, and in post oak flats.

Maine to Ala., westw. to Ind., Mo., and Okla.

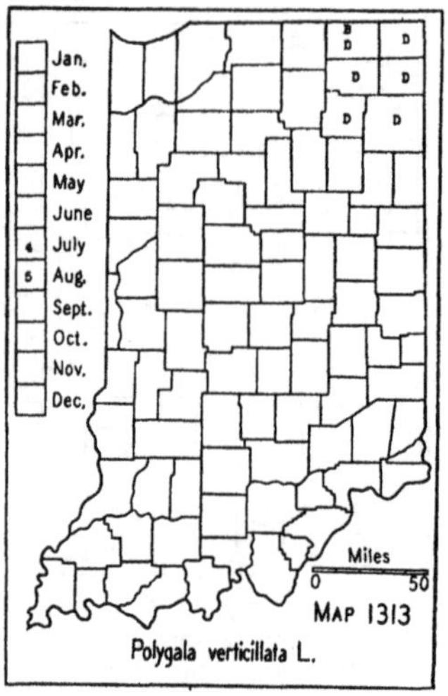

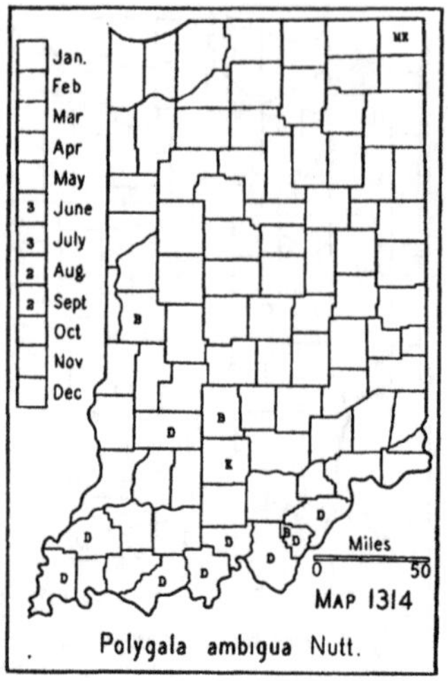

7. **Polygala cruciàta** L. Map 1315. Most often in moist sandy soil on the border between a black oak woods and a marsh and usually associated with *Gaultheria procumbens.* Sometimes in a moister location and infrequent in a moist prairie habitat. Local but usually frequent where it is found.

Maine to Minn., southw. to Fla. and La.

8. **Polygala sanguínea** L. (*Polygala viridescens* L.) Map 1316. This species has three color forms, purple, white, and intermediate. Linnaeus described the first as *Polygala sanguinea* and the last as *Polygala viridescens.* The last named plant is now regarded as a form of *Polygala sanguinea.* The white form has also been named but has not been found in Indiana. Our plants vary from almost white to purplish but most of them are more or less of a deep rose color.

This species is found in poor and slightly acid soil of old fallow fields, of open wooded slopes, of the borders of marshes, in suitable habitats along roads and railroads, and in sandy wheat stubble fields. It is usually infrequent and much scattered but I once saw it as a common plant in a moist wheat stubble field in Jasper County.

N. S., Ont. to Minn., southw. to N. C., La., and Kans.

147. EUPHORBIÀCEAE J. St. Hil. SPURGE FAMILY

Flowers not in an involucre; calyx well developed; juice not milky.
  Ovules 1 in each cell of the ovary.
    Corolla present in either the staminate or pistillate flowers or in both; pubescence stellate.
      Ovary and fruit 3- (rarely 2-4-) celled, a dehiscent capsule...4348. CROTON, p. 637.
      Ovary and fruit 1-celled, an achenelike indehiscent capsule...4350. CROTONOPOSIS, p. 638.
    Corolla none; pubescence not stellate.
      Bracts of the pistillate flowers cleft into 5-15 lobes; styles many-cleft..4407. ACALYPHA, p. 639.

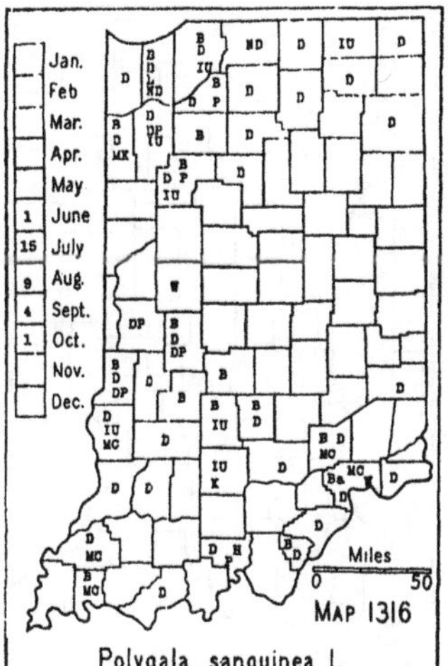

Polygala sanguinea L.

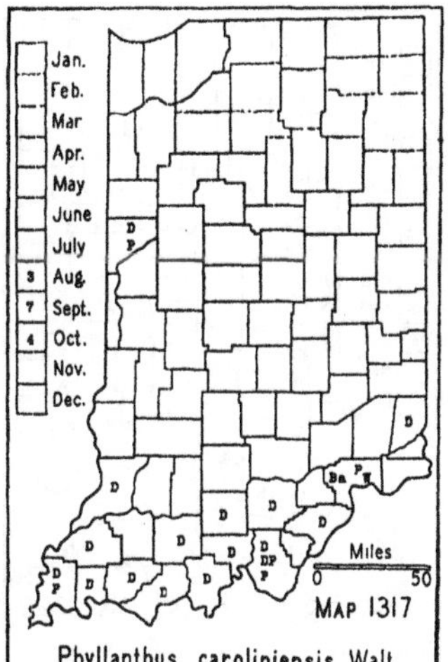

Phyllanthus caroliniensis Walt.

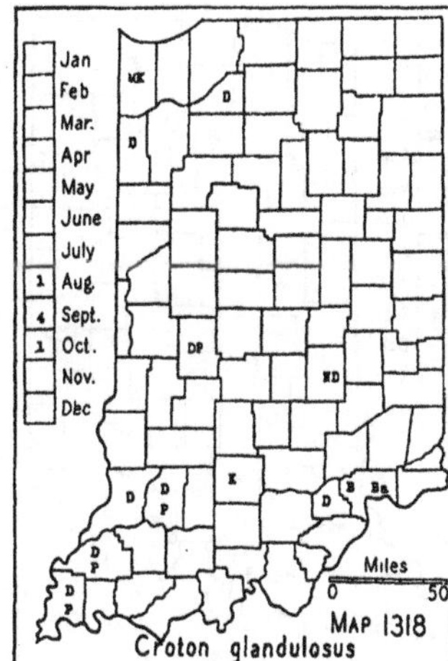

Croton glandulosus var. septentrionalis Muell.

Bracts of the pistillate flowers not cleft; styles not cleft.
Leaf blades not peltate..................................4416. TRAGIA, p. 641.
Leaf blades peltate......................................4424. RICINUS, p. 641.
Ovules 2 in each cell of the ovary......................4299. PHYLLANTHUS, p. 637.
Flowers in an involucre resembling a calyx; calyx rudimentary or lacking; juice milky...4498. EUPHORBIA, p. 641.

4299. PHYLLÁNTHUS L.

1. **Phyllanthus caroliniénsis** Walt. Map 1317. In bare spots in moist soil. Mostly in old logging roads, rarely in fallow fields, more frequent in cornfields, and sometimes on the low borders of sloughs.

Eastern Pa., cent. Ill. to se. Mo., southw. to Fla. and Cent. Amer.

4348. CRÒTON L.

[Ferguson. Crotons of the United States. Rept. Missouri Bot. Gard. **12**: 33-74. 1901.]

Leaves toothed.............................1. *C. glandulosus* var. *septentrionalis.*
Leaves entire.
Capsules clustered, erect, depressed-globose; stamens of sterile flowers 10-14...... ....2. *C. capitatus.*
Capsules mostly solitary, nodding, ovoid; stamens of sterile flowers 3-8............ ....3. *C. monanthogynus.*

1. CROTON GLANDULÒSUS L. var. SEPTENTRIONÀLIS Muell. Arg. Map 1318. This is undoubtedly a species adventive from the south. It is now found in fallow fields, roadsides, and roadside ditches. Pepoon reports it from the sand flats of Lake and Porter Counties.

Va. to Fla. and Tex.; northw. in the Mississippi Valley to Ind. and Iowa.

2. CROTON CAPITÀTUS Michx. Map 1319. Our Clark County specimen is from a fallow field and that from Martin County is from a wheat stubble field. Pepoon reports it as found in Lake County along the Wabash Railroad at Miller. Kriebel has collected it in Lawrence County and in

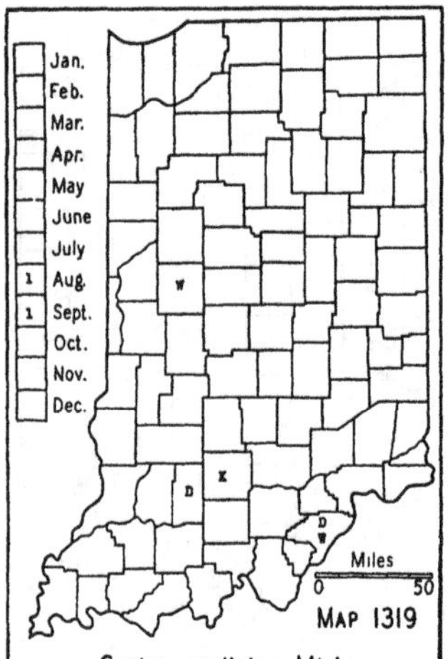

Croton capitatus Michx.

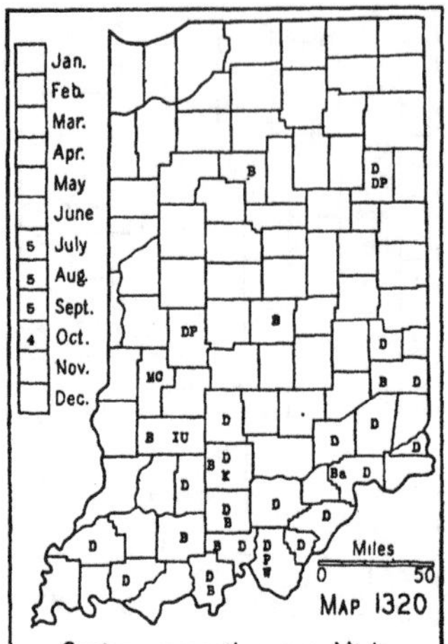

Croton monanthogynus Michx.

Crotonopsis elliptica Willd.

1929 Bechtel collected it in Montgomery County. This species also seems to be adventive from the south.

N. J. to Fla. and Tex.; northw. in the Mississippi Valley to Ill., Iowa, and Kans.

3. CROTON MONANTHÓGYNUS Michx. Map 1320. In pastures and fallow fields, along roads and railroads, and rarely in open woods. South of Laurel in Franklin County and west of Paoli in Orange County I have seen it as a pernicious weed over acres of pasture land. Stock will not eat it. On account of its weedy nature, and since it was unknown to the early botanists, I think this species is adventive in Indiana although J. M. Coulter (Bot. Gaz. 2: 146. 1877) says: "All along over the knobs, on the way to the Barrens, we encountered any quantity of *Croton monanthogynus.*" He doubtless followed an old road of travel where it may have been introduced. Dr. Clapp, who was well acquainted with the area, and who botanized the area about New Albany for 20 miles from 1832-1862, did not find this species. Riddell in his Flora of the Western States (1835) knew it only from St. Louis.

N. C., s. Ind. to Iowa and e. Kans., southw. to Fla. and Tex.

4350. CROTONÓPSIS Michx.

[Pennell. The genus Crotonopsis. Bull. Torrey Bot. Club 45: 477-480. 1918.]

Spikes short, of but 1 or 2 fruits; staminate flowers less than 1 mm broad; fruit ovoid, with an evident median vein on each side; leaves lanceolate to narrow-lanceolate, the stellate pubescence overlapping. . . . . 1. *C. elliptica.*

Spikes longer, of 3 to 6 fruits; staminate flowers more than 1 mm broad; fruit ellipsoid, without an evident vein on the side; leaves linear-lanceolate, the stellate pubescence not overlapping. (See excluded species no. 419, p. 1070.) . . . . *C. linearis.*

1. **Crotonopsis ellíptica** Willd. Map 1321. In a hard, white, minimacid, clay soil in a post and pin oak flat in Spencer County about 4 miles northwest of Chrisney. It was abundant in a 40-acre fallow field and

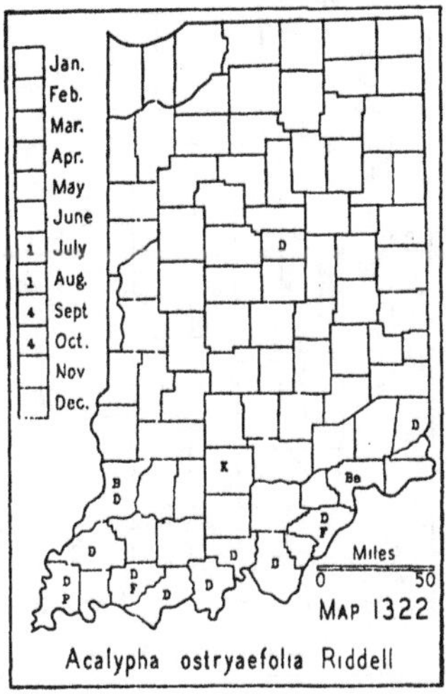

Acalypha ostryaefolia Riddell

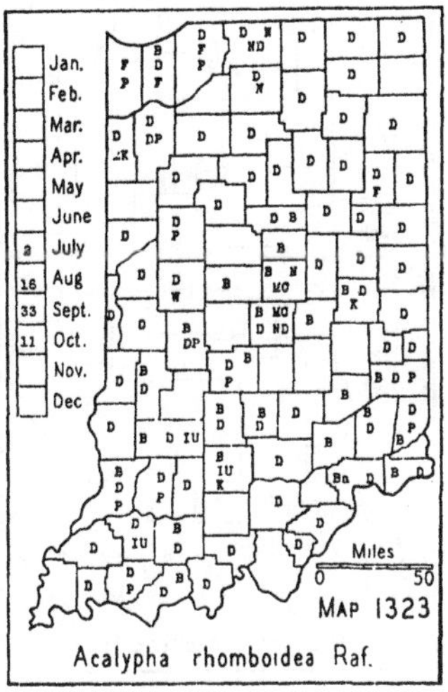

Acalypha rhomboidea Raf.

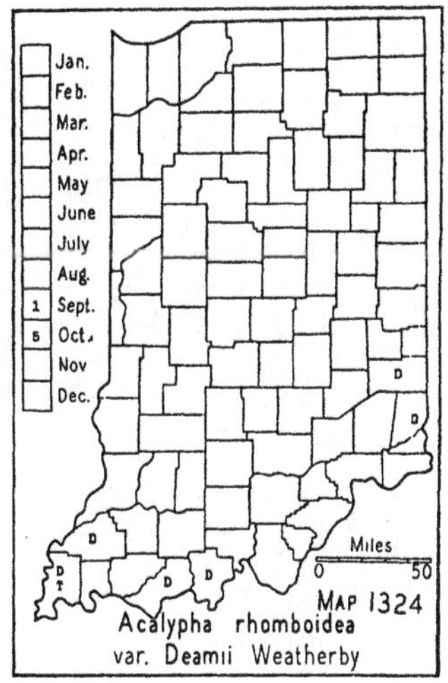

Acalypha rhomboidea var. Deamii Weatherby

scattered in an adjoining open woods, but was not found in the thick woods. Whether it was introduced into the fallow field and spread into the adjacent woods, or vice versa, I do not know.

Conn. to e. Kans., southw. to n. Fla. and cent. Tex.

4407. ACALỲPHA L. THREE-SEEDED MERCURY

[Weatherby. The group of Acalypha virginica in eastern North America. Rhodora 29: 198-200. 1927. The typification of Acalypha virginica L. Rhodora 39: 14-16. 1937.]

Staminate and pistillate flowers in separate spikes; capsules prickly. . 1. *A. ostryaefolia.*
Staminate and pistillate flowers in the same spike; capsules smooth.
  Pistillate bracts deeply cut into 5-7 (rarley 9), oblong to lanceolate acute or obtusish lobes; primary leaves mostly ovate to rhombic-ovate, glabrous except for scattered, long hairs.
    Seed 1.6-1.8 mm long. . . . . 2. *A. rhomboidea.*
    Seed 2-3 mm long. . . . . 2a. *A. rhomboidea* var. *Deamii.*
  Pistillate bracts with 9-15 lobes or teeth; primary leaves mostly ovate-lanceolate to linear, often pubescent beneath or on both surfaces.
    Pistillate bracts rather deeply cut into mostly lanceolate, very acute lobes, hispid on the nerves and margins, usually not glandular; stems with at least a few long spreading hairs in addition to the more common incurved ones; primary leaves broadly to narrowly lanceolate; petioles a third to a half as long as the blade. . . . . 3. *A. virginica.*
    Pistillate bracts shallowly cut into ovate or broadly deltoid lobes or teeth, sparsely beset with whitish stipitate or sessile red glands or both, sometimes ciliate, otherwise glabrate to coarsely pubescent; stems puberulent to pubescent with only incurved hairs; primary leaves oblong-lanceolate to linear; petioles a tenth to a fourth as long as the blades. . . . . 4. *A. gracilens.*

1. ACALYPHA OSTRYAEFÒLIA Riddell. Map 1322. Most of our specimens are from truck gardens, cornfields, and fallow fields along or near the Ohio River and near New Harmony. I have never seen it in any other habitat. In recent years it has been introduced farther north in the state. It is usually a common weed where it is found. This species seems to be adventive. It was first reported from Indiana in 1917 and none of the early

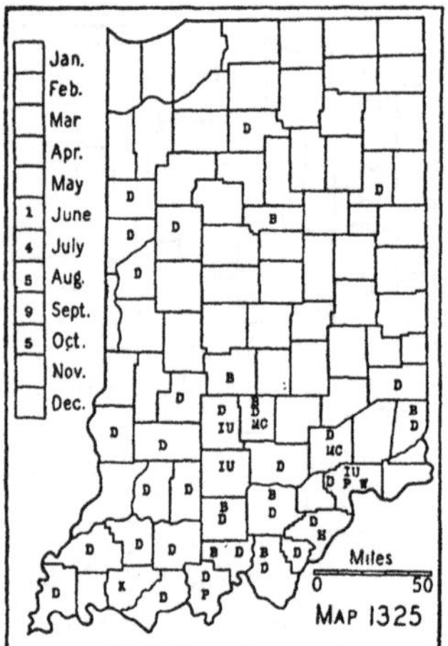

Acalypha virginica L.

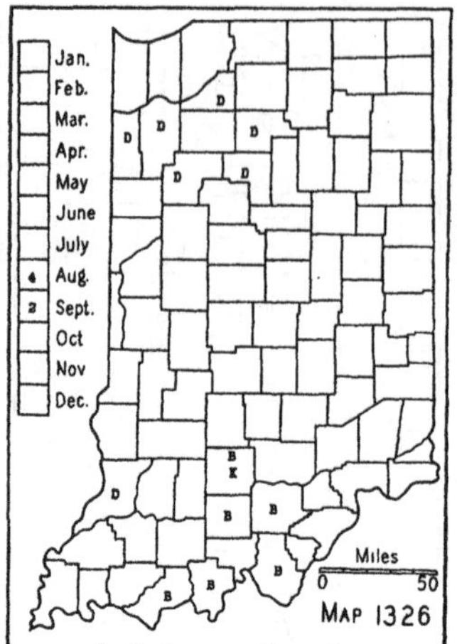

Acalypha gracilens Gray

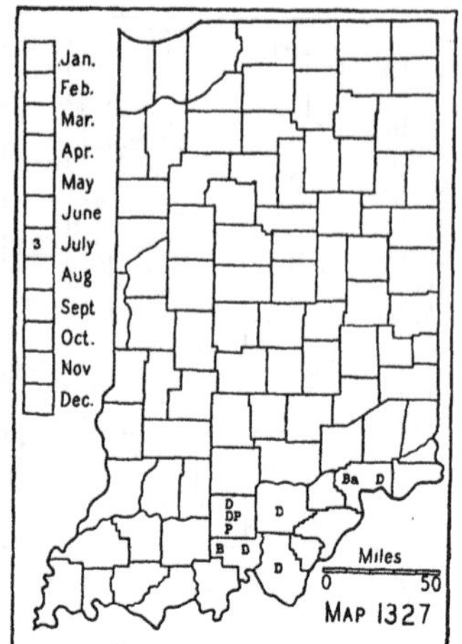

Tragia cordata Michx.

botanists had seen it. Riddell (1835) says his specimen came from a hill opposite Cincinnati. Short in his Catalogue of Kentucky Plants and his four supplements does not list it. These were published between 1833-1840. Lapham reports it from Illinois between 1836 and 1857.

N. J. to Ohio and Kans., southw. to Fla. and Mex.

2. **Acalypha rhomboìdea** Raf. (*Acalypha virginica* of recent authors, not L.) Map 1323. This species is a frequent to a common weed in all parts of the state in almost all kinds of habitats. It usually occurs in a moist black loam or sandy soil but will thrive in any kind of soil from gravel bars to dry, open, wooded slopes. It is usually found in the open and in such places it is more abundant. It is found in open, wet or dry woods, fallow or cultivated fields, and waste places and along roads and railroads.

N. S., Maine, sw. Que. to Minn., southw. to Fla., Tenn., and Kans.

2a. **Acalypha rhomboidea** var. **Dèamii** Weatherby. (Rhodora **29**: 197-198. 1927 and Rhodora **39**: 16. 1937.) Map 1324. This variety is known only from southern Indiana where it has been found in rather moist, sandy soil along Whitewater River, in Dearborn and Franklin Counties, along the Patoka River, in the talus of the sandstone cliff along the Ohio River at Rockport, on the wooded bank of the Ohio River at Derby in Perry County, and in a low place in a woods 9 miles north of Rockport. The plant is easily recognized in the field by its large drooping leaves and by the whole plant usually being at least twice as large as the typical form of the species.

3. **Acalypha virgínica** L. (Rhodora **29**: 198-200. 1927.) (*Acalypha digyneia* Raf.) Map 1325. Mostly in dry soil on open wooded slopes, associated with black and white oak, in fallow fields, and along roadsides. Rather frequent in the unglaciated area and rapidly migrating northward.

Mass., Ind. to Okla., southw. to Ga. and Tex.

4. **Acalypha grácilens** Gray. (*Acalypha gracilens* in part, of Gray, Man., ed. 7 and Britton and Brown, Illus. Flora, ed. 2.) Map 1326. In dry

or moist, sandy soil. Our specimens are from pastures and from along railroads. This species and the preceding one are united in our manuals but they are very distinct. The habitats within the known distribution are quite different. The distribution of this species in Indiana offers an interesting problem.

N. H. to Fla. and Tex., northw. in the Mississippi Valley to Ind. and Wis.

4416. TRÀGIA [Plumier] L.

1. **Tragia cordàta** Michx. (*Tragia macrocarpa* Willd.) Map 1327. On rocky, wooded bluffs of streams and black and white oak slopes. Rare.

Ind. to Mo., southw. to Fla. and Tex.

4424. RÍCINUS [Tourn.] L.

See excluded species no 420, p. 1070.

4498. EUPHÓRBIA L. SPURGE

Floral leaves with wide white margins; blades mostly 1.5-3 cm wide..1. *E. marginata.*
Floral leaves without wide white margins, narrower than the preceding.
  Glands of the involucre with a petallike appendage.
    Leaves opposite, their bases more or less oblique.
      Leaves entire, rarely slightly serrate toward the obtuse apex.
        Plants glabrous throughout.
          Leaves orbicular to orbicular-ovate........................3. *E. serpens.*
          Leaves oblong to spatulate or somewhat obovate-oblong; seeds reddish.
            Seed strongly cross-wrinkled; leaves somewhat subcordate at the base. ....4. *E. glyptosperma.*
            Seed not cross-wrinkled or only faintly so; leaves often obovate-oblong, narrowed at the base. (See excluded species no. 422, p. 1071.)....... .... *E. serpyllifolia.*
        Plants more or less pubescent, at least the capsule more or less pubescent.... ....5. *E. humistrata.*
      Leaves serrate or dentate, at least toward the obtuse apex.
        Capsules glabrous.
          Capsules 1.5 mm long or less; seeds reddish.
            Seed strongly cross-wrinkled; leaves somewhat subcordate at the base.... ....4. *E. glyptosperma.*
            Seed not cross wrinkled or only faintly so; leaves narrowed at the base. (See excluded species no. 422, p. 1071.)..............*E. serpyllifolia.*
          Capsules more than 1.5 mm long; plants ascending; seeds drab.
            Capsules mostly 2-2.25 mm long, as wide as long, not deeply 3-lobed, rounded at the summit; stems puberulent only in lines............ ....6. *E. maculata.*
            Capsules about 1.75 mm long; wider than long, retuse at the apex, deeply 3-lobed; pubescence of stems hirsute..............7. *E. vermiculata.*
        Capsules more or less pubescent.
          Leaves mostly oblong to ovate-oblong, mostly 3x7 mm to 5x13 mm, usually some or all serrate to below the middle, generally pubescent beneath; capsules usually exserted well beyond the involucre, the lobes generally rounded (sometimes acute); seed mostly 0.8-0.9 mm long, generally plainly marked after the seed is cleaned with 3 or 4 transverse rugose lines...8. *E. supina.*

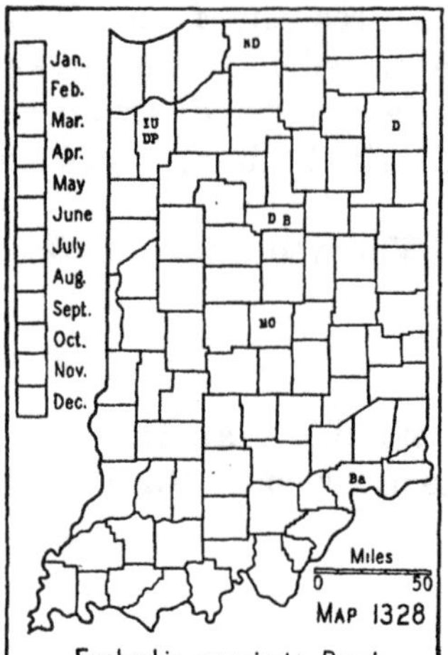

Euphorbia marginata Pursh

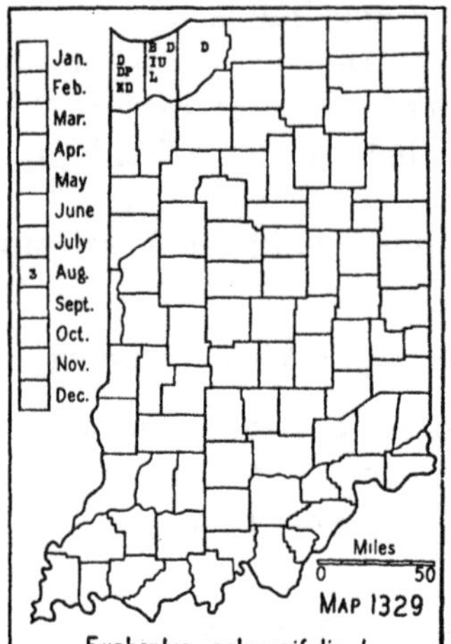

Euphorbia polygonifolia L

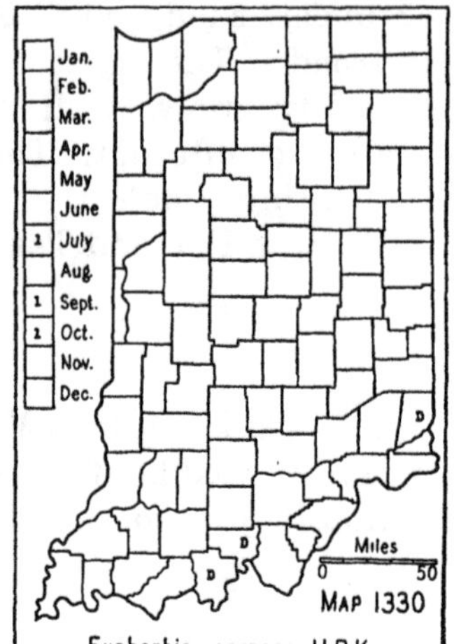

Euphorbia serpens HBK

Leaves mostly elliptic, mostly 7x12 mm to 8x16 mm, usually glabrous beneath, generally serrate only above the middle or at the apex; capsules usually not exserted beyond the involucre, the lobes mostly acute; seed 1 mm long, usually no transverse lines plainly visible after the seed is cleaned, the surface minutely roughened...........5. *E. humistrata.*

Leaves alternate or scattered on the stem, not oblique at the base. .9. *E. corollata.*

Glands of the involucre naked (with no petallike appendage).

Leaves opposite.

Leaves oblique at the base; annual, glabrous...............2. *E. polygonifolia.*

Leaves not oblique at the base; plants glabrous or pubescent.

Perennial plants with several stems from a stout rootstock, glabrous; flowering in the spring. (See excluded species no. 421, p. 1070)..*E. Ipecacuanhae.*

Annual plants; flowering in summer; stem more or less densely retrorsely puberulent in addition to long colorless multicellular hairs; leaves more or less densely short- or long-pubescent both above and beneath; glands of involucre stipitate..10. *E. dentata.*

Leaves all alternate or scattered.

Stems not topped by an umbel; stems and leaves glabrous or nearly so; glands of involucre sessile..11. *E. heterophylla.*

Stems topped by a several-rayed umbel.

Leaves serrulate..12. *E. obtusata.*

Leaves entire.

Plants perennial; seeds smooth.

Stem leaves 4-12 mm wide..............................13. *E. Esu'a.*

Stem leaves 1-3 mm wide...........................14. *E. Cyparissias.*

Plants annual or biennial; seeds pitted.

Lobes of the capsules 2-crested.........................15. *E. Peplus.*

Lobes of the capsules rounded......................16. *E. commutata.*

1. EUPHORBIA MARGINÀTA Pursh. (*Dichrophyllum marginatum* (Pursh) Kl. & Garcke.) SNOW-ON-THE-MOUNTAIN. Map 1328. Escaped from cultivation in most parts of the state but not abundantly so.

Minn. to Colo., southw. to Tex.

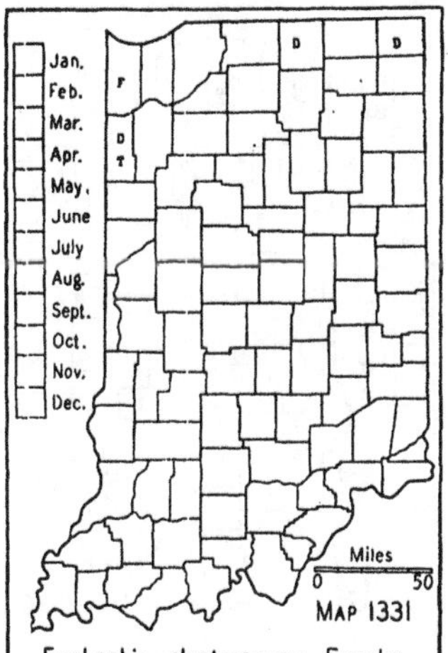

Euphorbia glyptosperma Engelm.

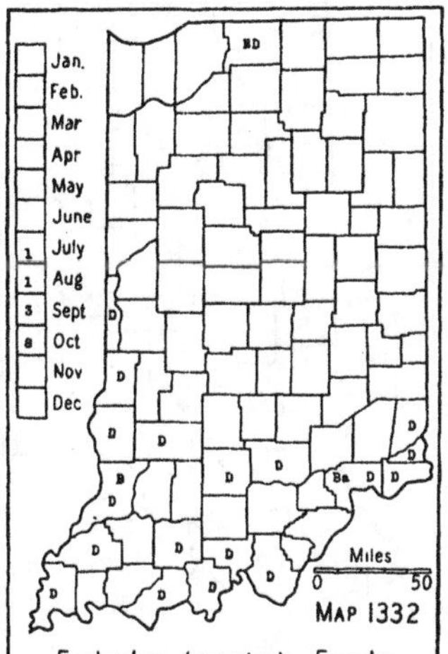

Euphorbia humistrata Engelm

Euphorbia maculata L.

2. **Euphorbia polygonifòlia** L. (*Chamaesyce polygonifolia* (L.) Small.) Map 1329. This is a small prostrate species restricted to the beach of Lake Michigan.

Atlantic coast from N. S. to Fla. and on the shores of the Great Lakes.

3. **Euphorbia sérpens** HBK. (*Chamaesyce serpens* (HBK.) Small.) Map 1330. On a rocky bar in Wilson Creek in Dearborn County and on the bank of the Ohio River and in adjoining overflow land. Infrequent.

Sw. Ont. to S. Dak., southw. to Mex.; and S. A.

4. **Euphorbia glyptospérma** Engelm. (*Chamaesyce glyptosperma* (Engelm.) Small.) Map 1331. I have found this spurge only three times although it may be rather frequent since it can easily be mistaken for other species of the genus. My specimens were found in dry sandy and gravelly soil.

Maine, Ont. to B. C., southw. to N. Y., Iowa, Tex., and Mex.

5. **Euphorbia humistràta** Engelm. (*Chamaesyce humistrata* (Engelm.) Small.) Map 1332. All of our specimens except one are from the southern half of the state. We have, however, three reports from the northern half. It is to be noted that this species is separated with difficulty from *Euphorbia supina* and this similarity may result in their confusion. In the field *Euphorbia humistrata* is notably more glaucous than *Euphorbia supina.*

Ont. to Minn., southw. to N. J., Miss., and La.

6. **Euphorbia maculàta** L. (Contr. Gray Herb. **127**: 74. 1939.) (*Euphorbia nutans* Lag., *Euphorbia Preslii* (Guss.) Arth., and *Chamaesyce Lansingii* Millsp.) NODDING SPURGE. Map 1333. A common weed in all parts of the state, usually in dry soil. It is found in fallow and cultivated grounds, along roadsides and railroads, and in open woodland and pastures.

My Randolph and Tipton County specimens were named *Chamaesyce Lansingii* Millsp. by C. F. Millspaugh and I reported them as such. I am now referring them to this species.

Mass., Ont., Wis., and Nebr., southw. to Fla. and Tex.

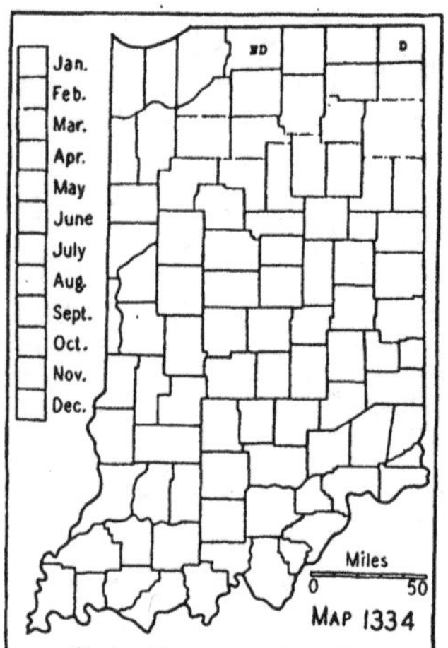

Euphorbia vermiculata Raf.

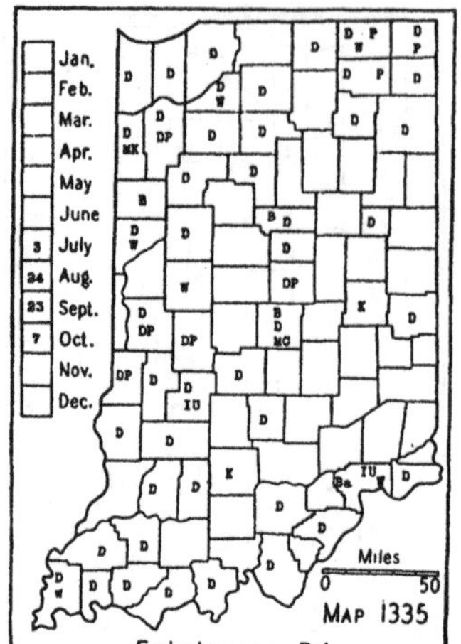

Euphorbia supina Raf.

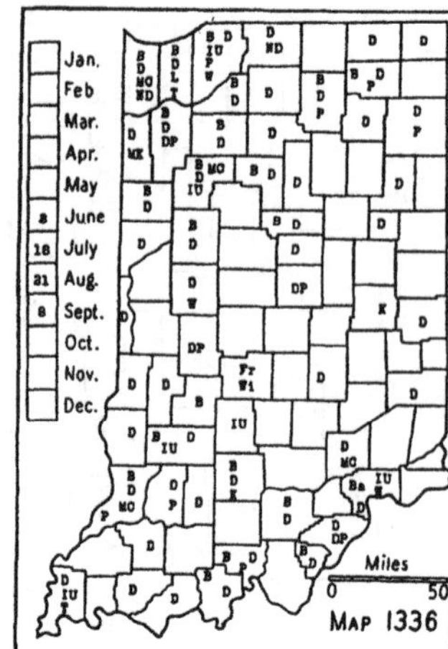

Euphorbia corollata L.

7. **Euphorbia vermiculàta** Raf. (*Euphorbia hirsuta* (Torr.) Wieg., *Euphorbia Rafinesquii* Greene, and *Chamaesyce Rafinesquii* (Greene) Small.) Map 1334. My only specimen was collected 5 miles northeast of Angola in Steuben County along a roadside just west of a crossroad where there is a small pond at the southwest intersection. Nieuwland collected it in 1910 in South Bend, St. Joseph County.

Eastern Que. to w. Ont., southw. to N. J., Ohio, and Ill.

8. **Euphorbia supìna** Raf. (Contr. Gray Herb. 127: 76. 1939.) (*Euphorbia maculata* of authors and *Chamaesyce maculata* (L.) Small.) Map 1335. A frequent weed in both moist and dry soils in all parts of the state. Mostly in fallow fields and cultivated grounds; also along roadsides and railroads, in pastures and open woodland, and on the banks of streams.

N. E., Ont. to Wyo., southw. to Fla. and Tex.; introd. west of the Rocky Mts.

9. **Euphorbia corollàta** L. (*Tithymalopsis corollata* (L.) Kl. & Garcke.) FLOWERING SPURGE. Map 1336. Infrequent but well distributed throughout the state, being more common in the lake and prairie areas. It prefers a dry sandy soil and is very rarely found in wet situations. It is found in open woodland and fallow fields and along roadsides and railroads. This species varies much as to pubescence and the width of the leaves which has induced authors to assign names to these variations. It is a perennial with a stout rootstock. It is frequently mowed off and killed above the ground by burning, especially along railroads. I have one specimen that has been repeatedly top-killed by burning; it has the crown of the rootstock much thickened and bearing many short pubescent stems. Plants that grow in very dry, exposed habitats or in very dry sand are usually more pubescent than those that grow in moister or shadier places.

Mass., Ont. to Minn., southw. to Fla. and Tex.

10. **Euphorbia dentàta** Michx. (*Poinsettia dentata* (Michx.) Small in Britton and Brown, Illus. Flora, ed. 2.) Map 1337. Along railroads and

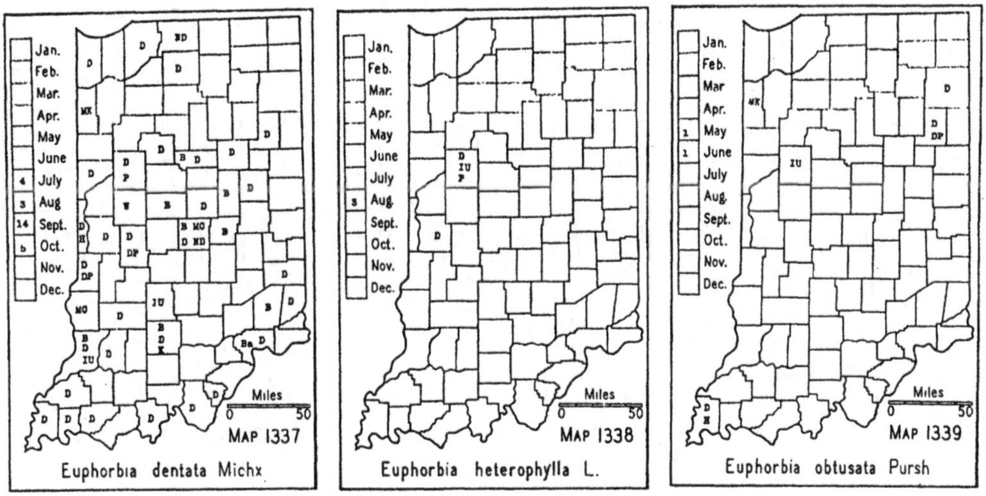

roadsides, in fallow fields, and rarely in open woodland. This species is no doubt adventive from the west. It was unknown to our earlier botanists.

The leaves vary in width and a narrowleaf form has been named.

Pa., S. Dak. to Wyo., southw. to Tenn., La., and Mex.

11. **Euphorbia heterophýlla** L. (*Poinsettia heterophylla* (L.) Kl. & Garcke.) PAINTED SPURGE. Map 1338. This species is doubtless adventive in our area. Along railroads and roadsides and in waste places.

This plant much resembles the preceding but can be separated from it by its alternate leaves, glabrous stem and leaves. The leaves of *Euphorbia heterophylla* vary from almost linear to fiddle-shaped.

Ind. to S. Dak., southw. to Fla. and Tex.

12. **Euphorbia obtusàta** Pursh. (*Tithymalus obtusatus* (Pursh) Kl. & Garcke.) Map 1339. I have this species from the dry, wooded bank of the St. Mary River south of Fort Wayne, from along the Wabash River east of Bluffton, and from the roadside on the south side of Half Moon Pond about 10 miles southwest of Mount Vernon, Posey County. Madge McKee collected it in Newton County. It was collected in Tippecanoe County by Young and the specimen is in the herbarium of Indiana University.

Pa. to Iowa, southw. to S. C. and Tex.

13. EUPHORBIA ESULA L. (*Euphorbia virgata* Wald. & Kit. in Rhodora 39: 50. 1937 and *Tithymalus Esula* (L.) Hill.) LEAFY SPURGE. Map 1340. This species has been reported as an escape, etc., in three counties. Hansen (Proc. Indiana Acad. Sci. 37: 320. 1928) says: "Specimens were collected in full flower near Winamac, where it is established along roadsides, on June 20, 1927." W. N. Clute informed me in 1936 that there is "a large colony along the canal in Indianapolis between Illinois and Meridian Streets." In 1937 I found it abundant over an area of 3 acres in a pasture about 3 miles southwest of Knox, Starke County.

Leafy Spurge is a very obnoxious weed and spreads rapidly. Its roots penetrate the soil to a depth of 5-15 feet. On account of the acrid latex

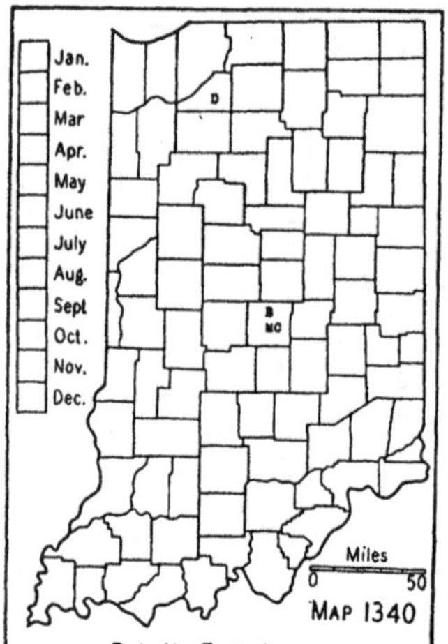

Euphorbia Esula L.

Euphorbia Cyparissias L.

Euphorbia Peplus L.

stock will not eat it. If a colony of this pest is discovered no amount of effort or expense should be spared in order to exterminate it.

Nat. of Eu.; Maine, Ont. to Mich., southw. to N. J. and Ind.

14. EUPHORBIA CYPARÍSSIAS L. (*Tithymalus Cyparissias* (L.) Hill.) CYPRESS SPURGE. Map 1341. Found in all parts of the state where it has escaped from cultivation and persisted.

It rarely produces seeds and there are no records of its doing so in Indiana. It should be closely observed and if found with mature seed the fact should be recorded.

Nat. of Eu.; Mass. to Colo., southw. to Va.

15. EUPHORBIA PÈPLUS L. (*Tithymalus Peplus* (L.) Hill.) Map 1342. I have specimens of this species only from Wells County where it is well established in north Bluffton. It was reported also by Dr. Clapp from the vicinity of New Albany, Floyd County.

Nat. of Eu.; N. B. to w. N. Y., Wis. to Iowa, southw. to N. J., W. Va., and Ala.

16. **Euphorbia commutàta** Engelm. (*Tithymalus commutatus* (Engelm.) Kl. & Garcke.) Map 1343. Infrequent on rocky or gravelly wooded slopes, mostly along streams and about lakes, sometimes in the talus of cliffs and rarely in the open, if so, usually about gravel pits.

Pa., Ont. to Minn., southw. to Fla. and Mo.

148. CALLITRICHÀCEAE L. WATER STARWORT FAMILY

4530. CALLÍTRICHE L.

Plants terrestrial; fruit on a peduncle usually a fourth to three fourths its length, wider than long, deeply notched at the apex and base, lobes with a deep margin between them; stigmas about as long as the fruit, recurving.........1. *C. Austini.*

Plants amphibious or submerged; fruit sessile, as long as or longer than wide, with a shallow notch at the apex; stigmas erect or spreading.

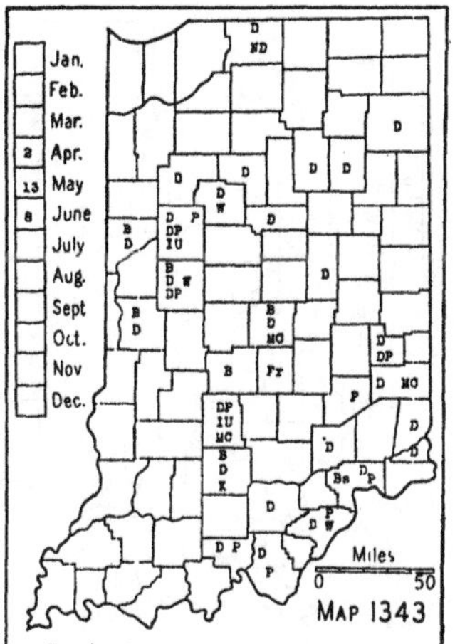

Euphorbia commutata Engelm

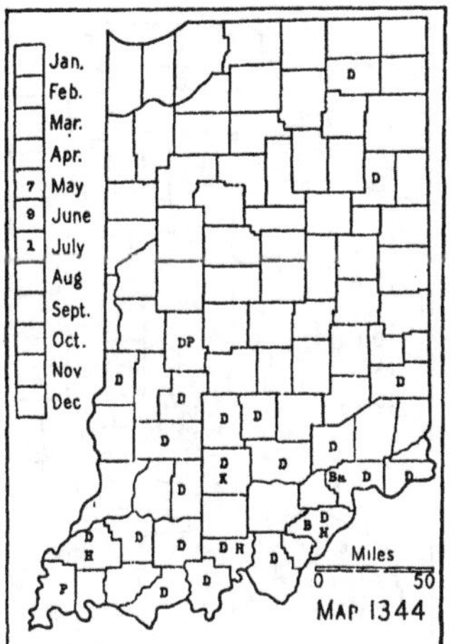

Callitriche Austini Engelm

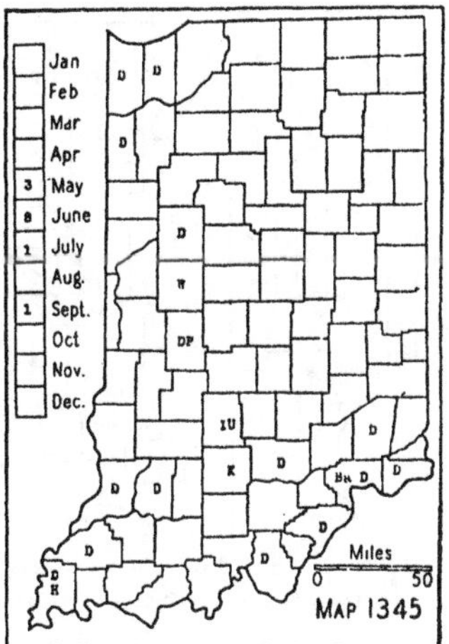

Callitriche heterophylla Pursh

Leaves usually of two forms, the submerged ones linear and 1-nerved, the floating ones obovate to broadly spatulate and 3-nerved, all more or less petioled or narrowed at the base; flowers usually between two bracts.

Fruit about as wide as long, not narrowed at the base, the lobes rounded; stigmas spreading, usually twice as long as the fruit.............2. *C. heterophylla.*

Fruit longer than broad, narrowed at the base, the mature lobes winged toward the apex; stigmas erect, about the length of the fruit. (See excluded species no. 424, p. 1071.)..*C. palustris.*

Leaves all submerged, linear, 1-nerved, not narrowed at the base, sessile; flowers without bracts; fruit with a narrow, deep notch at the apex; stigmas long, recurving, deciduous. (See excluded species no. 423, p. 1071.)..*C. hermaphroditica.*

1. **Callitriche Aústini** Engelm. (*Callitriche deflexa* var. *Austini* (Engelm.) Hegelm.) Map 1344. Frequent in southern Indiana in woodland along logging roads and in fallow cornfields. It grows only on bare spots in moist, minimacid soil, associated in logging roads usually with *Gratiola neglecta* and in fields with *Poa Chapmaniana, Alopecurus carolinianus,* and *Arabis virginica.* It is so small and inconspicuous that it is usually overlooked.

Conn. to Ind. and Mo., southw. to Del., La., and Tex.; also from Mex. to S. A.

2. **Callitriche heterophýlla** Pursh. Map 1345. Infrequent in ponds and shallow streams, which may become dry during dry seasons, and rarely in inundated woodland.

Newf. to Man., southw. to Fla., La., Mo., and Colo.

152. LIMNANTHÀCEAE Lindl. FALSE MERMAID FAMILY

4542A. FLOÉRKEA Willd.

1. **Floerkea prosperpinacoìdes** Willd. Map 1346. Locally abundant in thick woodland in rich, moist soil, usually associated with sugar maple, beech, white oak, and white elm. We have no records for the area south

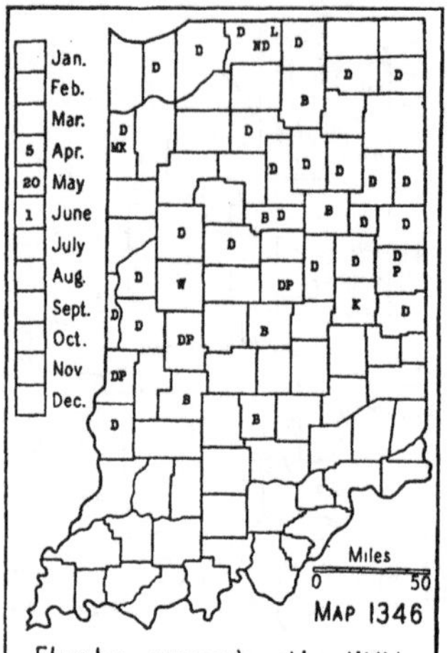

Floerkea proserpinacoides Willd.

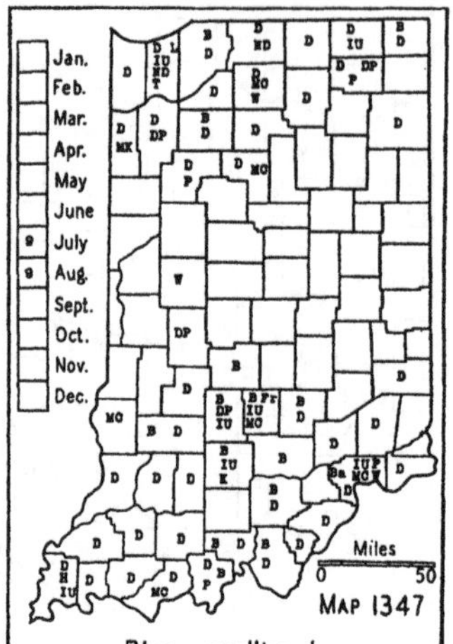

Rhus copallina L.

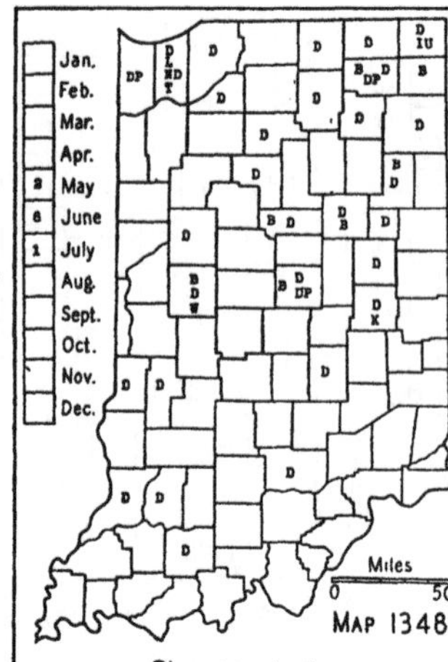

Rhus Vernix L.

of Sullivan County. This little annual when removed from the woods to a rich, shady flower garden will persist as a weed.

W. Que., Ont. and Wis., southw. to Del., Tenn., and Mo.

153. ANACARDIÀCEAE Dumort. CASHEW FAMILY

4594. RHÚS L. SUMAC

[Barkley. Monographic study of Rhus and allies in North and Central America. Annals Missouri Bot. Gard. 24: 265-496. 1937.]

Leaflets normally more than 3.
- Rachis of leaf winged; branchlets pubescent........................1. *R. copallina.*
- Rachis of leaf not winged; branchlets glabrous or pubescent.
  - Leaflets entire or nearly so; fruit from very pale green to almost colorless, glabrous........................2. *R. Vernix.*
  - Leaflets regularly serrate to the base; fruit pubescent with red hairs.
    - Branchlets below the inflorescence, lower surface of the leaflets, and rachis of leaf glabrous; branchlets more or less strongly angled; hairs of fruit mostly 0.15-0.3 mm long, obovoid, obtuse at the apex................3. *R. glabra.*
    - Branchlets below the inflorescence, lower surface of the leaflets, and rachis more or less pubescent; branchlets terete or nearly so.
      - First year branches pubescent; branchlets densely pubescent; hairs of fruit about 1 mm long, linear, and acicular at the apex.........4. *R. typhina.*
      - First year branches glabrous; branchlets varying from densely to sparingly pubescent.
        - Hairs of fruit about 1 mm long, acicular at the apex; branchlets usually densely pubescent at first........................5. × *R. pulvinata.*
        - Hairs of fruit about 0.5 mm long, acute to blunt at the apex; branchlets at first sparingly pubescent, usually soon becoming glabrous........................3a. *R. glabra* var. *borealis.*

Leaflets normally 3.
- Petiolules of leaflets generally more than 3 cm long; fruit pale green to colorless.
  - Fruit glabrous, rarely with a few hairs........................6. *R. radicans.*
  - Fruit pubescent, surface usually very papillose.....6a. *R. radicans* var. *littoralis.*

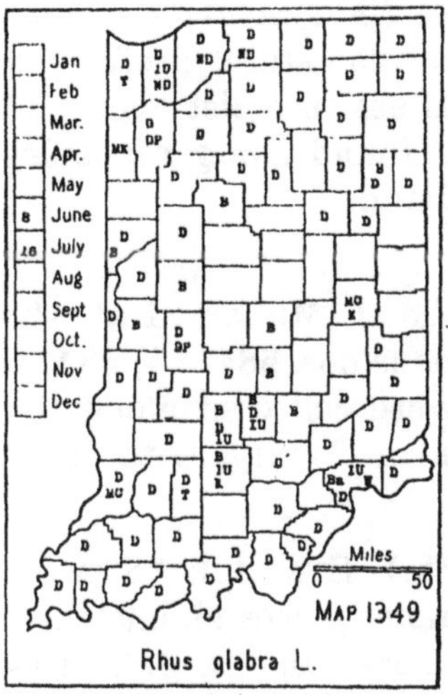

MAP 1349
Rhus glabra L.

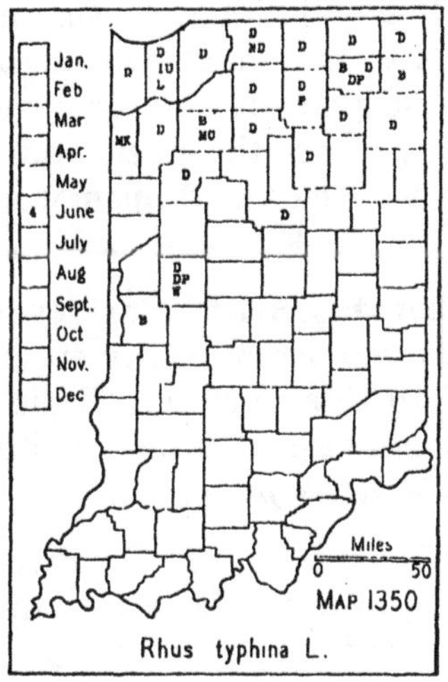

MAP 1350
Rhus typhina L.

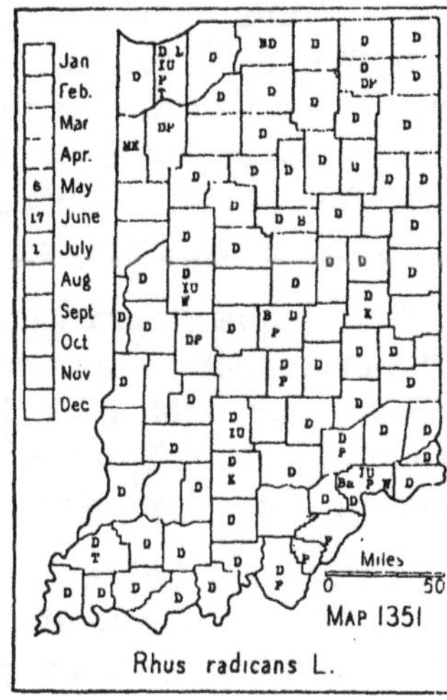

MAP 1351
Rhus radicans L.

Petiolules of leaflets less than 3 cm long; fruit red and densely pubescent.
  Terminal leaflet mostly ovate-lanceolate; flowers usually appearing before the leaves..7. *R. aromatica.*
  Terminal leaflet mostly obovate; flowers usually appearing with the leaves...8. *R. trilobata* var. *arenaria.*

1. **Rhus copallina** L. SHINING SUMAC. Map 1347. Rather frequent in some of the northwestern and southern counties, being rare or local in the remaining counties, or absent in some of the east-central counties. It prefers a rather dry sandy soil and is found mostly in the open along roadsides, fences, and railroads and in abandoned fields and open woodland.

Fernald & Griscom (Rhodora 37: 167-168. 1935) write that the typical form of this species has "lance-oblong leaflets definitely attenuate at the base" and more leaflets than our interior plant. They give the range of the typical form as along the coast from New York to Florida. They add that our form "has the comparatively few leaflets more ovate-lanceolate or short-oblong and rounded at the base," and should be known as var. *latifolia* Engler (DC. Mon. 4: 384. 1883) with a range from Maine to Michigan, southward into the upland of North Carolina and Oklahoma. Most of my specimens belong to this wide-leaved variety, and I have the typical form from Starke and from the Ohio River Counties. However, since we have forms intermediate between these two extremes, it seems best not to separate them in our area.

Maine, s. Ont. to Minn., southw. to Fla. and Tex.

2. **Rhus Vérnix** L. (*Toxicodendron Vernix* (L.) Ktze.) POISON SUMAC. POISON ELDER. Map 1348. Poison sumac is frequent in low ground about lakes and in bogs in the lake area. South of this area I have found it in springy areas as shown on the map. This species must have a springy or bog habitat in which to live. I have had the opportunity to watch the species in three bogs that were drained, and it gradually died out.

Northern N. E. to Minn., southw. to Fla. and Tex.

3. **Rhus glàbra** L. (*Rhus arbuscula* Greene and *Rhus media* Greene.) SMOOTH SUMAC. Map 1349. Infrequent throughout the state, preferring open places in dry, sandy or gravelly soil, or sometimes in moist loam and poor clay soil of hills. It prefers the open and is found along roadsides and fences and in abandoned fields and open woodland.

N. S. to N. Dak., southw. to Fla. and La.

3a. **Rhus glabra** var. **boreàlis** Britt. Barkley (Amer. Midland Nat. **19**: 598-599. 1938.) has tentatively referred my specimens nos. 58424A, 58427, 58544L, 58544M, 58544N to this variety. These were collected along the roadside about 3 miles northwest of Angola, Steuben County, with *Rhus glabra, Rhus typhina* and × *Rhus pulvinata.*

4. **Rhus typhìna** L. (*Rhus hirta* (L.) Sudworth.) STAGHORN SUMAC. Map 1350. Infrequent or local in the lake area in moist places about lakes, bogs, swamps, and low places in general, rarely on rocky slopes. In southern Indiana I have collected it on the high, wooded ridge along Sugar Creek, just east of Deer's Mill in Montgomery County, and on the bluffs of the Ohio River in Switzerland County. It has been reported also from Franklin, Knox, Posey, and Wayne Counties.

The largest specimens I ever saw were growing in dry soil in the yard of W. H. Montgomery in section 25, about 4 miles southeast of Bryant, Jay County. The larger was 30 inches and the smaller was 29 inches in circumference at breast height. These were root shoots of older trees which had formerly grown in his yard, and Mr. Montgomery estimated that these trees were about 20 years old. The bole of each was about 6 feet high. Since these trees were attractive, round-topped shade trees, this species might well be used for that purpose.

N. S., Ont. to S. Dak., southw. to Ga. and Miss.

5. × **Rhus pulvinàta** Greene. (Fedde, Rep. Spec. Nov. 5: 45. 1908.) (*Rhus glabra* × *typhina.*) Barkley discusses the status of this hybrid in Amer. Midland Nat. **19**: 589-599. 1938. He has referred my nos. 58424B, 58534A, 58534B, and 58534L to this hybrid. He refers also to it, specimens which I collected in Fulton County and some that Nieuwland collected in St. Joseph County. Doubtless this hybrid is sporadic within the range of the parent species.

6. **Rhus radìcans** L. (*Toxicodendron radicans* (L.) Kuntze, *Rhus rufescens* Greene, and *Rhus Toxicodendron* L. of Deam, Shrubs of Indiana, revis. ed. 1932.) POISON IVY. Map 1351. An infrequent to common vine throughout Indiana. It will grow anywhere except in low peaty soil. The species has two habits of growth: the one climbing and the other erect. The climbing form is the more common, being in all places where the erect form is not found. It is found mostly along fences and in open and thick woods. In the Lower Wabash Bottoms it reaches a diameter of 3 inches and climbs to the tops of the tallest trees. The erect form is usually less than 3 feet high and is found in hard, minimacid soil in some of the southern counties, where it is usually associated with sweet gum, and in the dunes along Lake Michigan. The species is extremely variable

MAP 1352
Rhus radicans var littoralis (Mearns) Deam

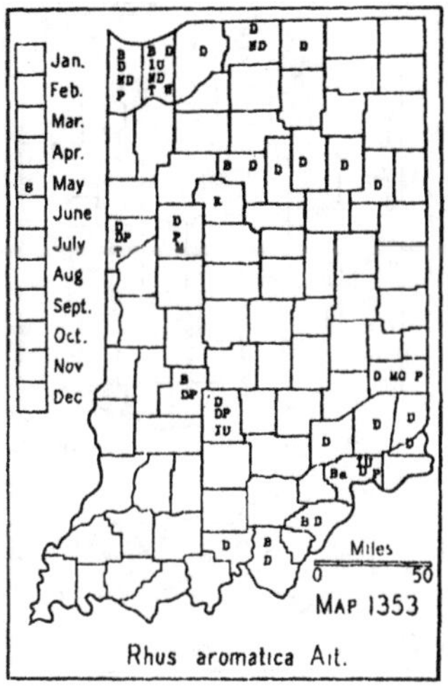

MAP 1353
Rhus aromatica Ait.

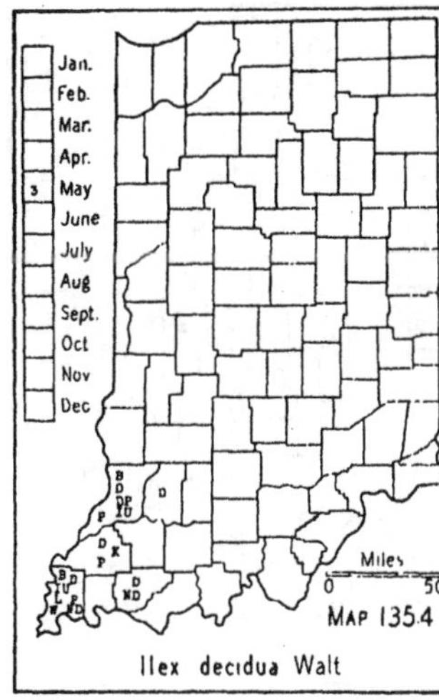

MAP 1354
Ilex decidua Walt

and some of the forms have been named. The margins of the leaflets vary from entire to serrate or somewhat lobed. The leaflets are acute but we have one specimen with a rounded apex. The fruit is subglobose but we have one specimen with elliptic fruit.

N. S. to B. C., southw. to Fla. and Mex.

6a. **Rhus radicans** L. var. **littoràlis** (Mearns) Deam, comb. nov. (*Rhus littoralis* Mearns, Proc. Biol. Soc. Wash. 15: 148. 1902.) Map 1352. This is an erect form with hairy, papillose fruit and is restricted to the Ohio River Counties.

Maine to Va., westw. to Okla.

7. **Rhus aromática** Ait. (*Rhus canadensis* Marsh.) FRAGRANT SUMAC. Map 1353. Infrequent throughout the state except on the dunes of Lake and Porter Counties where it is frequent. Found on the dunes about Lake Michigan, on the gravelly bank of the St. Joseph River, on rocky or gravelly banks and bluffs of the Wabash River and its tributaries, and in southern Indiana on bluffs and slopes of streams.

Que. to Nebr., southw. to Fla. and Tex.

8. **Rhus trilobàta** Nutt. var. **arenària** (Greene) Barkley. (Ann. Missouri Bot. Gard. 24: 408. 1937.) This shrub was formerly frequent on the low dunes near Lake Michigan in Lake County but is infrequent in Porter County. The building of Gary, Indiana Harbor, and Whiting has covered most of its original habitat.

Lake and Porter Counties in Ind. and in adjacent Ill.

157. AQUIFOLIÀCEAE Lowe. HOLLY FAMILY

Leaves never entire; petals united at the base; pedicels of fruit less than 1 cm long. ....4614. ILEX, p. 652.

Leaves entire, rarely with a few teeth; petals not united; pedicels of fruit more than 1 cm long...4615. NEMOPANTHUS, p. 653.

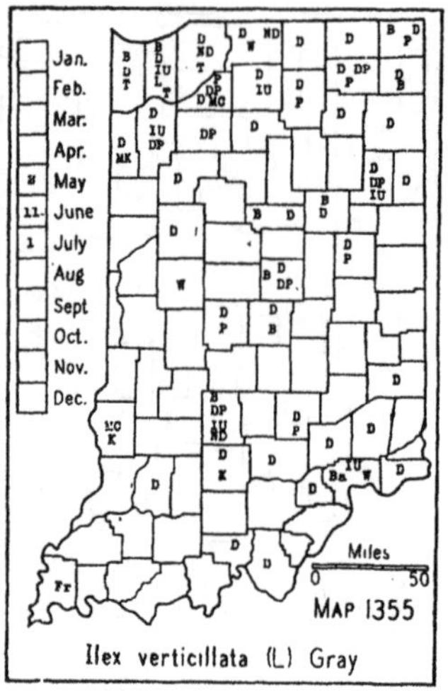

Ilex verticillata (L.) Gray

Nemopanthes mucronata (L.) Trel.

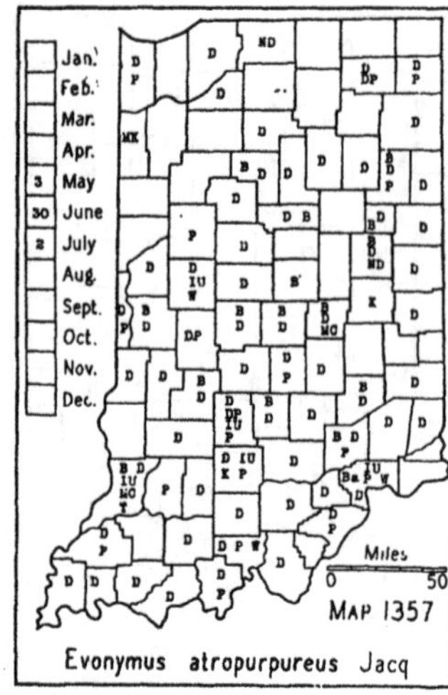

Evonymus atropurpureus Jacq.

4614. ÌLEX L. HOLLY

Calyx lobes not ciliate; nutlets ribbed....................................1. *I. decidua.*
Calyx lobes ciliate; nutlets not ribbed.
Leaves dark green, obovate or oval, mostly 2-5 cm wide, long taper-pointed at the base; blades thin, not impressed-veined above or beneath, or rarely somewhat so; pubescent beneath mostly on the veins....................2. *I. verticillata.*
Leaves yellow green, elliptic or oblanceolate, mostly 2-2.5 (4) cm wide, short taper-pointed to nearly rounded at the base; blades thick, deeply impressed-veined above and beneath, usually pubescent on the whole lower surface...2a. *I. verticillata* var. *padifolia.*

1. **Ilex decídua** Walt. POSSUMHAW. Map 1354. Infrequent in a few of the southwestern counties on the borders of sloughs, ponds, and swamps and in low woods.

Va. to s. Ill. and s. Mo., southw. to Fla. and Tex.

2. **Ilex verticillàta** (L.) Gray. (Including *Ilex verticillata* var. *tenuifolia* (Torr.) Wats. and *Ilex bronxensis* Britt.) COMMON WINTERBERRY. Map 1355. Local in the northern part of the state. It is rather frequent in some parts of the lake area in swampy and boggy places, becoming rare south of it.

This species is very variable in the shape, texture, and pubescence of the leaves, and in the color of the fruit. These variations have been named, but I am recognizing only one form. There is a yellow-fruited form which I have found once. All forms are on one map.

N. S. to Min., southw. to Fla. and Miss.

2a. **Ilex verticillata** var. **padifòlia** (Willd.) T. & G. The few specimens of this variety which I have collected were growing in wet places in hard, white, slightly acid soil in the "flats" of the southern counties. I received, however, from Mr. and Mrs. Walter Neff a specimen from Carroll County that was collected in a springy place. Specimens have also been collected in Jasper and St. Joseph Counties. This variety seems to be a complex but with a limited amount of field study I am not able to determine whether these differences are ecological or morphological.

Mass. to Minn. and southw.

4615. NEMOPÁNTHUS Raf.

1. **Nemopanthus mucronàta** (L.) Trel. MOUNTAIN HOLLY. Map 1356. In swampy and boggy places about lakes and in wet woods in the lake area. Infrequent.

Newf. to Wis., southw. to Va. and Ind.

158. CELASTRÀCEAE Lindl. STAFF-TREE FAMILY

Leaves opposite....................................4618. EVONYMUS, p. 653.
Leaves alternate....................................4625. CELASTRUS, p. 653.

4618. EVÓNYMUS [Tourn.] L.

Petioles of terminal pair of leaves over 4 mm long; capsules smooth....................................1. *E. atropurpureus.*
Petioles of terminal pair of leaves not over 4 mm long; capsules tuberculate.
Erect or ascending shrubs; terminal leaves ovate-lanceolate, thin, capsules Spinel Red (Ridgway Standard)....................................2. *E. americanus.*
Creeping shrubs with upright or ascending branches; terminal leaves obovate, firm; capsules scarlet- or orange-red (Ridgway Standard)............3. *E. obovatus.*

1. **Evonymus atropurpùreus** Jacq. WAHOO. Map 1357. An infrequent to frequent shrub on the alluvial banks of streams throughout the state. It is rarely found far from water courses or in dry situations.

N. Y. to Minn., southw. to Fla. and e. Tex.

2. **Evonymus americànus** L. BROOK EVONYMUS. Map 1358. Found in a few of the southern counties where it grows in low, flat woods with sweet gum, beech, and pin oak, and rarely in dry ground with black and white oak. There is an ascending form of this species that has leaves intermediate between this and the next species but it grows with the species and has the typical fruit. This species has been reported for some of the northern counties but I think all reports from there should be referred to *Evonymus obovatus.*

N. Y. to Ill., southw. to Fla. and Tex.

3. **Evonymus obovàtus** Nutt. RUNNING EVONYMUS. Map 1359. Infrequent in rich, moist soil, mostly in beech and sugar maple and white oak woods. It is to be noted that we have no records for the extreme southwestern counties although I have botanized this area intensively.

Ont. to Mich. and Ill., southw. to Pa. and Ky.

4625. CELÁSTRUS L.

1. **Celastrus scándens** L. AMERICAN BITTERSWEET. Map 1360. Rather frequent throughout the state in moist or dry soils. Mostly along fences and more rarely in thick woodland except in the sandy woods of the southwestern counties.

Maine to Man., southw. to N. C., Tenn., and N. Mex.

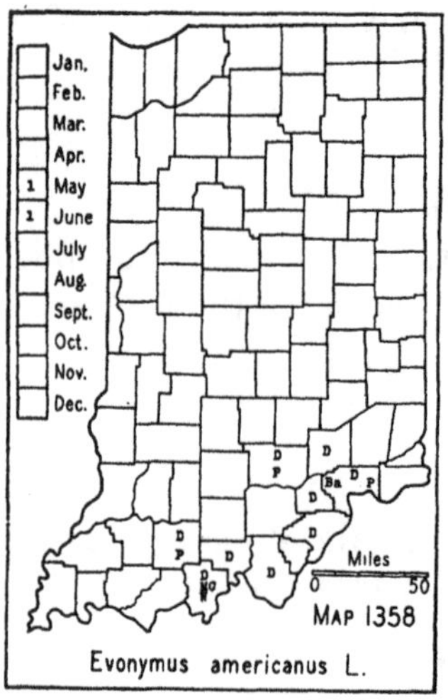

MAP 1358

Evonymus americanus L.

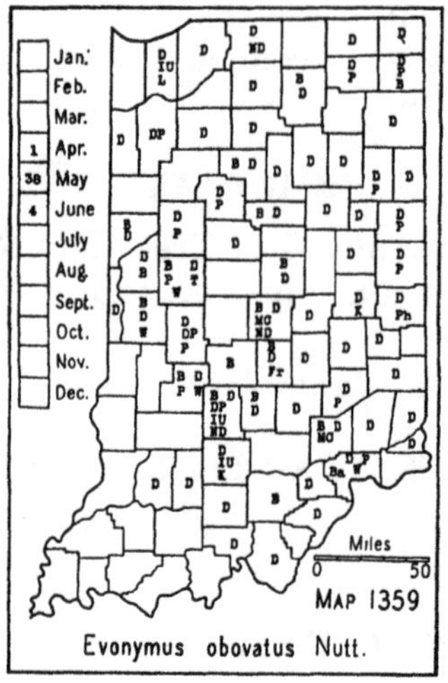

MAP 1359

Evonymus obovatus Nutt.

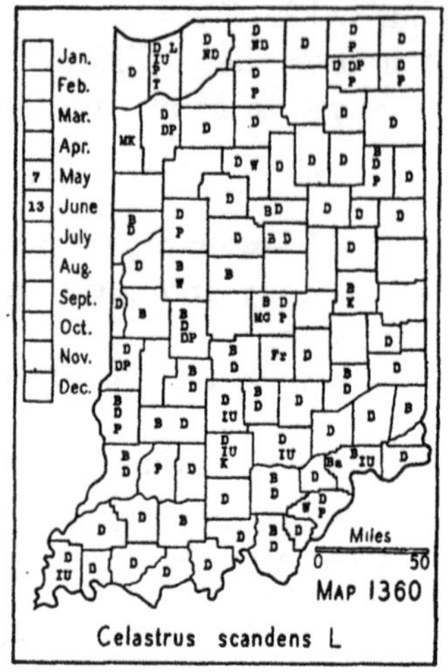

MAP 1360

Celastrus scandens L

161. **STAPHYLEÀCEAE** DC. BLADDERNUT FAMILY

4665. **STAPHYLÈA** L.

1. **Staphylea trifòlia** L. AMERICAN BLADDERNUT. Map 1361. An infrequent shrub in all parts of the state. It is restricted almost entirely to the slopes and alluvial banks of streams, and only rarely is it found in low places in woods.

Que. to Minn., southw. to N. C. and Kans.

163. **ACERÀCEAE** St. Hil. MAPLE FAMILY

4720. **ÀCER** [Tourn.] L.

Leaves 3-foliolate or pinnate.
  Branchlets green, not glaucous; fruit more or less pubescent until maturity...1. *A. Negundo.*
  Branchlets glaucous and usually more or less violet when rubbed; fruit glabrous...1a. *A. Negundo* var. *violaceum.*
Leaves simple.
  Flowers appearing before the leaves; fruit maturing in May or June.
    Petals none; fruit more or less pubescent at maturity; the terminal lobe of the leaves usually narrowest at the base....................2. *A. saccharinum.*
    Petals present; fruit glabrous at maturity; the terminal lobe of the leaves usually broadest at the base.
      Branchlets glabrous at maturity; leaves at maturity glabrous beneath except for a few hairs in the axils of the veins or rarely the entire lower surface covered more or less with a short pubescence; mature fruit generally 2-3.5 cm long..3. *A. rubrum.*
      Branchlets more or less pubescent at maturity; leaves beneath covered with a dense tomentum which remains until maturity or sometimes becoming scanty; fruit generally 4-5 cm long......3a. *A. rubrum* var. *Drummondii.*
  Flowers appearing after the leaves; fruit maturing mostly from July until September.
    Leaves yellow green beneath, the sides usually somewhat drooping; petioles of the terminal pair of leaves with expanded and more or less clasping bases (sometimes these appendages late in developing.)

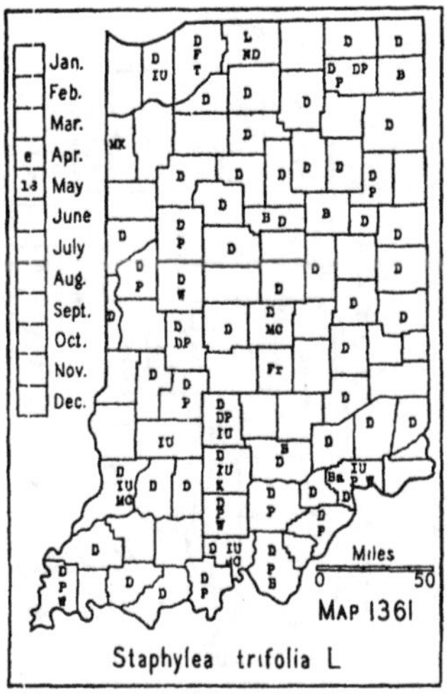

Staphylea trifolia L

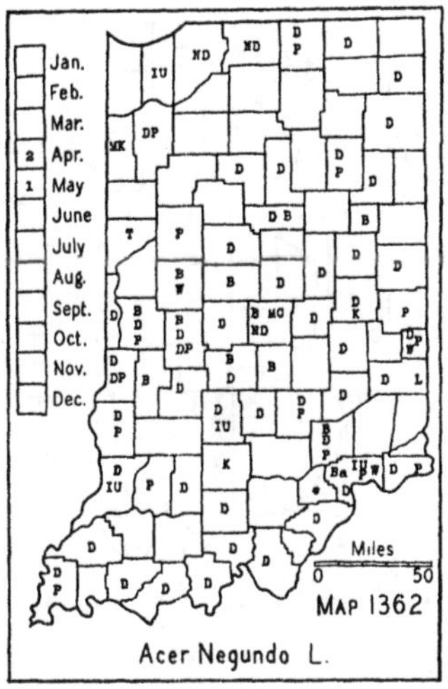

Acer Negundo L.

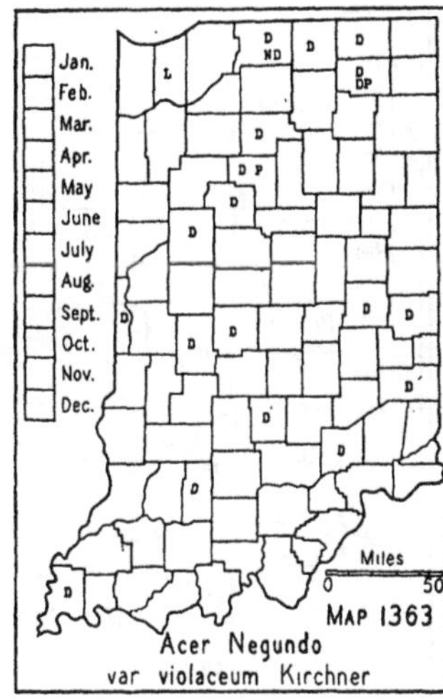

Acer Negundo var violaceum Kirchner

Leaves 5-lobed.
Petioles glabrous or with some pubescence at the base and point of insertion. ....4. *A. nigrum.*
Petioles pubescent their entire length............4a. *A. nigrum* f. *pubescens.*
Leaves 3-lobed.
Petioles glabrous..............................4b. *A. nigrum* var. *Palmeri.*
Petioles pubescent..................4c. *A. nigrum* var. *Palmeri* f. *villosum.*
Leaves not yellow green beneath, the sides not drooping; petioles of the terminal pair of leaves not expanded at the base.
Leaves 5-lobed.
Petioles glabrous...5. *A. saccharum.*
Petioles pubescent their entire length..........5a. *A. saccharum* f. *Schneckii.*
Leaves 3-lobed.................................5b. *A. saccharum* var. *Rugelii.*

1. **Acer Negúndo** L. (Nieuwland. Box-elders, real and so-called. Amer. Midland Nat. 2: 129-142. 1911.) BOXELDER. Map 1362. Usually infrequent. In low ground along streams and rarely far from them on higher ground in woods and along fences and roadsides. In some of the low woods along streams in the southwestern part of the state this species is common, and a nuisance as a forest tree because of its inferior quality.
N. E. to Minn., southw. to Fla. and e. Tex.

1a. **Acer Negundo** var. **violàceum** Kirchner. (Farwell. Botanical gleanings in Michigan. III. Amer. Midland Nat. 10: 37. 1926.) Map 1363. Found in the habitat of the species.
Mass., Ohio, n. Wis., Minn., S. Dak. to Idaho, southw. to Mo.

2. **Acer saccharìnum** L. SILVER MAPLE. Map 1364. Infrequent to frequent and locally abundant in most parts of the state. It is always found in wet places, usually in soil with little organic matter except in the lake region; along streams and about lakes and sloughs and low places in woods.
N. B. to S. Dak., southw. to Fla. and Tex.

3. **Acer rùbrum** L. RED MAPLE. Map 1365. This and the preceding species are known in commerce as soft maples in contrast to the hard or sugar maples. Infrequent, except locally, in all parts of the state. In

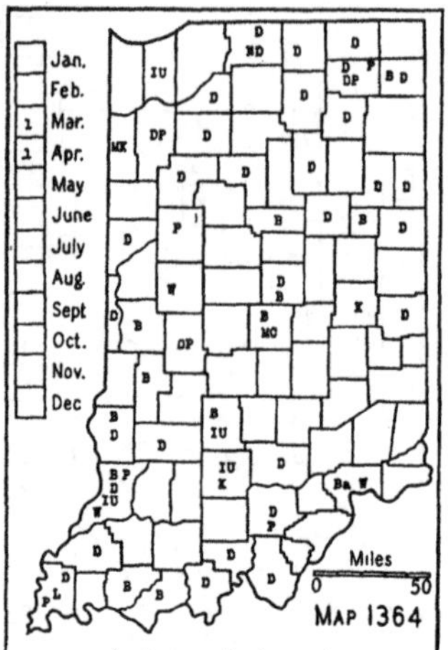

Acer saccharinum L.

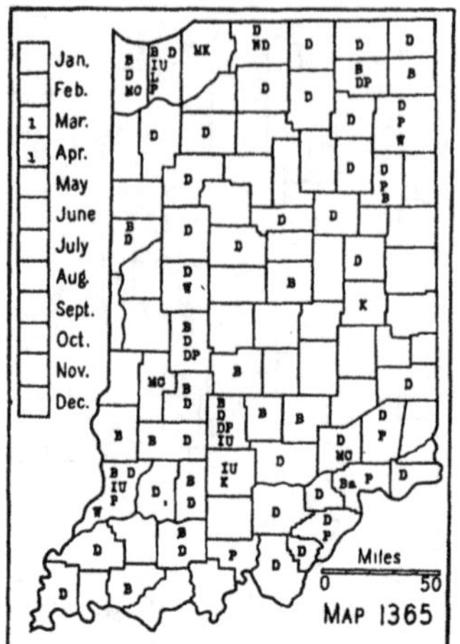

Acer rubrum L.

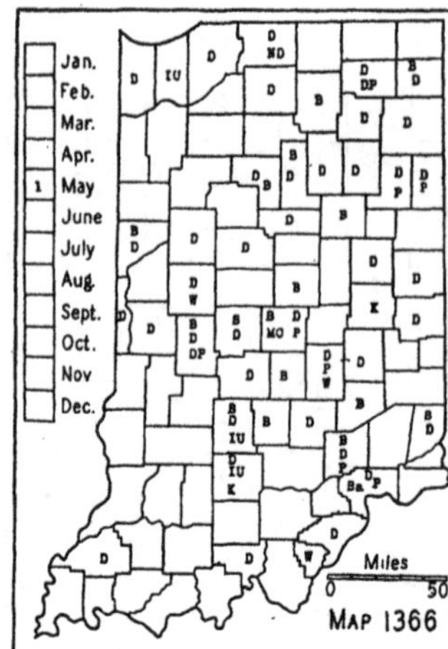

Acer nigrum Michx. f.

northern Indiana it is found both on gravelly ridges and in low ground, and rarely in bogs. In southern Indiana it is an infrequent tree on the ridges in most upland woods where it is associated with white oak, and in the "flats" in low, wet woods it may be frequent and is associated with sweet gum and beech. In the "flats" it grows to a large size and reproduces abundantly in wet, fallow fields.

Newf. to Minn., southw. to Fla. and Tex.

3a. **Acer rubrum** var. **Drummóndii** (Hooker & Arnott) T. & G. I have this variety from only the cypress swamp in Knox County and from swampy woods in the southern part of Posey County.

In deep swamps from sw. Ind., down the Mississippi Valley to La. and westw. to e. Tex.

4. **Acer nìgrum** Michx. f. BLACK MAPLE. Map 1366. Throughout the state and always assóciated with the sugar maple. Almost pure stands of the sugar maple, however, may occur with this species absent. When the two are associated, the black maple will usually be found in the moister area. Usually infrequent but locally common. I have seen specimens of this form with the glabrous petiole from the following states: Que., Maine, Vt., Mass., N. Y., Del., Va., W. Va., S. C., Ky., Tenn., Ohio, Mich., Ind., Ill., Wis., Minn., Iowa, and Mo.

4a. **Acer nigrum** Michx. f. forma **pubéscens** Deam, f. nov. A forma typica recedit petiolis pubescentibus. Petioles more or less pubescent their entire length. This form is associated with the species throughout the state but is less frequent than the glabrous form. Type in Deam Herbarium no. 58539, Randolph County, Sept. 30, 1937. I have seen specimens from the following states: Que., Ont., Vt., N. H., N. Y., W. Va., Ind., Ill., Wis., and Mo. (Atherton).

4b. **Acer nigrum** var. **Pálmeri** Sarg. (Jour. Arnold Arb. 2: 166. 1921.) This is a form with leaves 3-lobed and is far more common than the species in the northern part of the state.

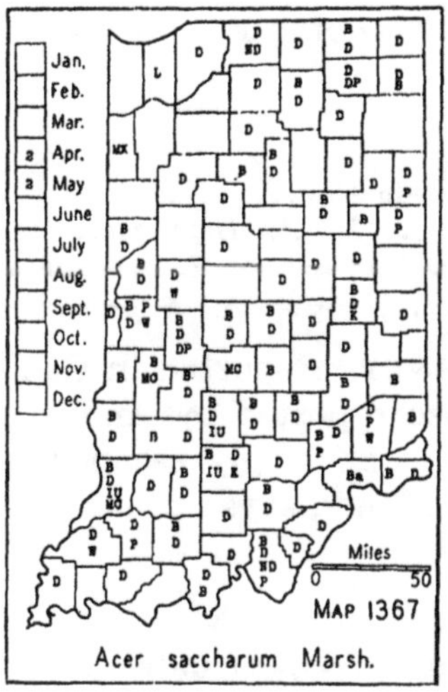

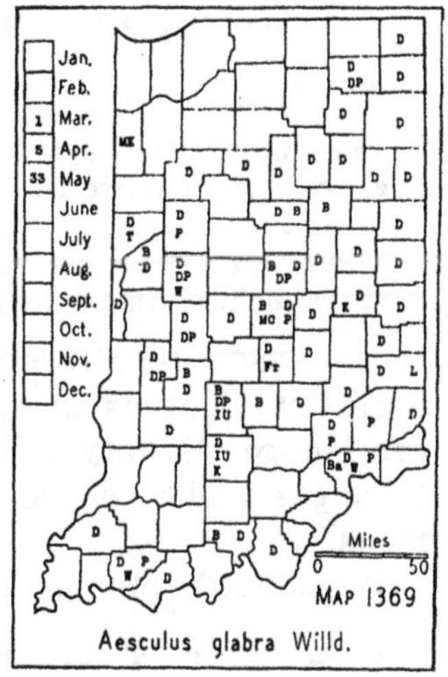

4c. **Acer nigrum** Michx. f. var. **Palmeri** Sarg. forma **villòsum** Deam, f. nov. A forma typica recedit petiolis pubescentibus. Petioles more or less pubescent their entire length. Associated with the variety but I do not know its relative frequency. Type in Deam Herbarium no. 58437, Kosciusko County, Sept. 19, 1937.

5. **Acer sáccharum** Marsh. SUGAR MAPLE. Map 1367. A frequent to common tree in all parts of the state. It is absent in the "flats" and on the crests of the ridges in the unglaciated area. It is usually associated with beech or in some of our northern woods the beech is replaced by basswood, red oak, and white ash. The species is very variable in leaf outline and in the pubescence of the petiole and the lower surface of the leaves. Several forms based upon these characters have been named. The sugar maple in Indiana has the lower surface of the leaves glaucous while in the northern range of its distribution it has the lower surface of the leaves green. To distinguish the two forms, Sargent (Bot. Gaz. **67**: 233. 1919) named the glaucous form var. *glaucum*.

Newf. to Man., southw. to Ga. and Tex.

5a. **Acer saccharum** Marsh. forma **Schnéckii** (Rehder) Deam, comb. nov. (*Acer saccharum* var. *Schneckii* Rehder and Sargent, Trees and Shrubs 2: 256. 1913.) Map 1368. This is a form found with the species and is restricted to a few of the southwestern counties. The few trees I have seen usually have a decidedly whiter bark and the trunk and branches have a more gnarled and twisted appearance. I have had it under cultivation since 1919 and seed from this tree shows that the seedlings do not come true (all with pubescent petioles).

Wis., Ind., Ill., Mo., w. Ky., and w. Tenn.

5b. **Acer saccharum** var. **Rugélii** (Pax) Rehder. This is a form with 3-lobed leaves that is infrequent throughout our area.

164. HIPPOCASTANÀCEAE T. & G. HORSE-CHESTNUT FAMILY

4721. AÉSCULUS L.

Anthers protruding from the flowers; fruit warty.
Flowers white, blotched with red, yellow, or purple; introduced..1. *A. Hippocastanum.*
Flowers yellow or greenish yellow; native............................2. *A. glabra.*
Anthers included in the flowers; fruit smooth........................3. *A. octandra.*

1. AESCULUS HIPPOCÁSTANUM L. HORSE-CHESTNUT. Reported in Coulter's Catalogue as escaping from cultivation. Also reported by Andrews for Monroe County without data. It is reported as sparingly escaped in Michigan and Schaffner, in his Catalogue of Ohio Plants, says: "No specimens."

Introduced from Asia through Eu.

2. **Aesculus glàbra** Willd. OHIO BUCKEYE. Map 1369. Usually a frequent tree in all parts of Indiana although it appears to be absent from a few of the northwestern counties. Because it is poisonous to stock, land owners from the earliest times have tried to exterminate it. It is usually associated with beech, sugar maple, and American linden.

The pubescence on the under surface of the leaflets is quite variable as to abundance and duration. Trees with the leaflets permanently pubescent beneath are known as *Aesculus glabra* f. *pallida* (Willd.) Fern.

West of the Allegheny Mts. from Pa. to Iowa, southw. to Ala. and Okla.

3. **Aesculus octándra** Marsh. YELLOW BUCKEYE. Map 1370. Infrequent on wooded slopes near the Ohio River from Dearborn to Crawford Counties. Phinney's report for Delaware and Jay Counties was an error.

The flowers vary in color from deep yellow to purple or reddish purple. This variation, added to the fact that the species begins flowering when it is shrublike and ultimately grows to be a very large forest tree, has led authors to describe several species and forms. A purple flowered form was reported by Young[1] for Jefferson County under the name of *Aesculus flava* var. *purpurascens.* This color form is now known as *Aesculus octandra* f. *virginica* (Sarg.) Fern. (Rhodora 39: 318. 1937.)

Pa. to Iowa, southw. to Ga., Okla., and Tex.

165. SAPINDÀCEAE R. Br. SOAPBERRY FAMILY

Herbaceous plants. (See excluded species no. 432, p. 1072)...4726. CARDIOSPERMUM, p. 658.
Woody plants..4824. KOELREUTERIA, p. 658.

4726. CARDIOSPÉRMUM L.

See excluded species no. 432, p. 1072.

4824. KOELREUTÈRIA Laxm.

1. KOELREUTERIA PANICULÀTA Laxm. GOLDENRAIN-TREE. Escaped from cultivation in New Harmony, in Posey County.

[1] Botany of Jefferson County. Indiana Geol. Surv. Rept. 2: 255. 1871.

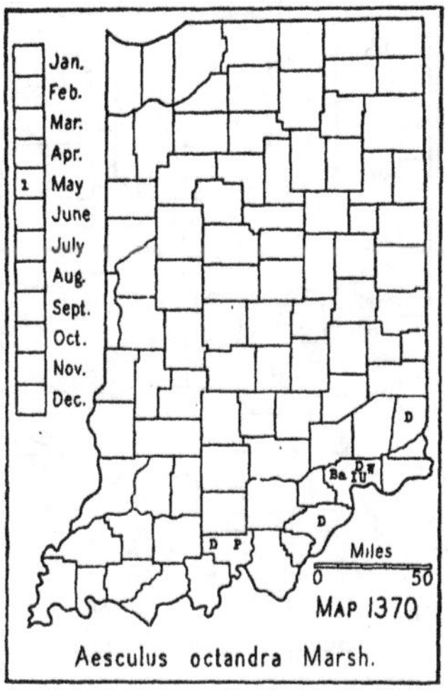

Aesculus octandra Marsh.

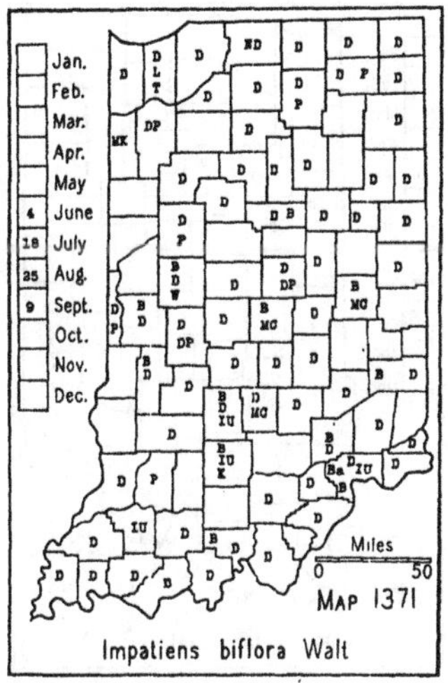

Impatiens biflora Walt

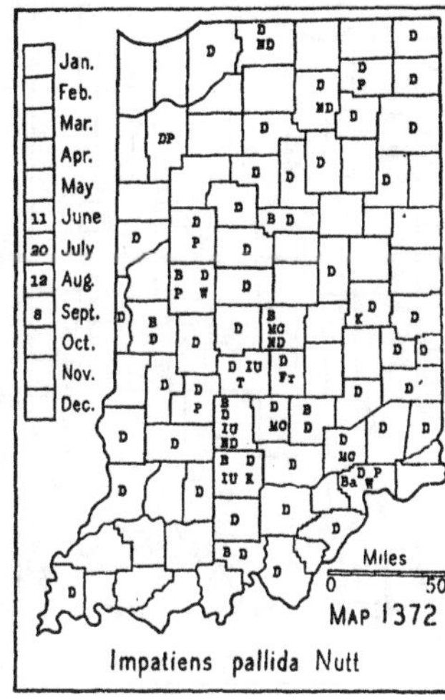

Impatiens pallida Nutt

Nat. of Asia. Introduced by McClure in the early settlement of New Harmony and planted in his yard near his gate. Since it had no common name it was referred to as the gate tree, the name by which it is still popularly known in New Harmony.

168. BALSAMINÀCEAE Lindl. TOUCH-ME-NOT FAMILY

4856. IMPÀTIENS [Rivin.] L.

Flowers orange yellow, thickly dotted with reddish brown; sac longer than broad..... ....1. *I. biflora.*
Flowers pale yellow, sparingly dotted with reddish brown; sac broader than long..... ....2. *I. pallida.*

1. **Impatiens biflòra** Walt. SPOTTED TOUCH-ME-NOT. Map 1371. Usually in large colonies or covering large areas, in wet or moist woodland and along streams.

This species has several named color forms but as yet none have been reported for this state.

Newf. to Sask., southw. to Fla. and Nebr.

2. **Impatiens pállida** Nutt. PALE TOUCH-ME-NOT. Map 1372. This plant is usually larger than the preceding and grows in drier situations. Usually in large colonies in moist places in beech woods or with other species in damp, shady woods.

N. Maine and w. N. E. to Sask., southw. to Ga. and Kans.

169. RHAMNÀCEAE Dumort. BUCKTHORN FAMILY

Leaves pinnately-veined; flowers greenish yellow; fruit a drupe.......... ....4875. RHAMNUS, p. 659.
Leaves triple-veined; flowers white; fruit a dry capsule......4877. CEANOTHUS, p. 661.

4875. RHÀMNUS [Tourn.] L. BUCKTHORN

Pedicels pubescent; flowers mostly in peduncled cymes..............1. *R. caroliniana.*
Pedicels glabrous; flowers not in peduncled cymes.

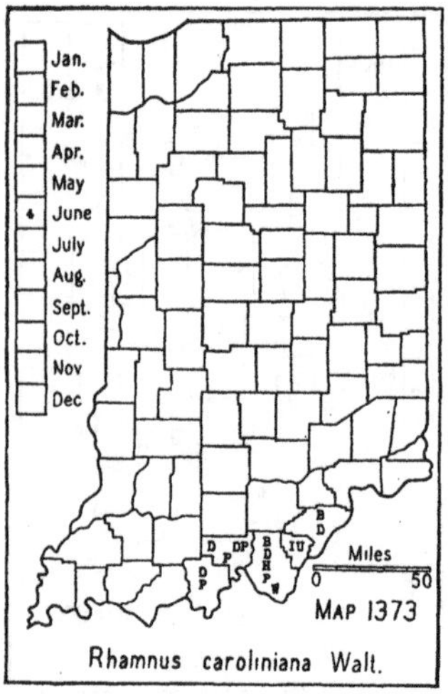

Rhamnus caroliniana Walt.

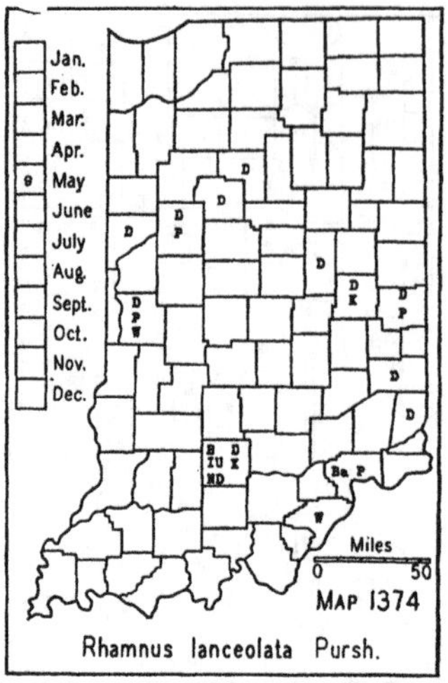

Rhamnus lanceolata Pursh.

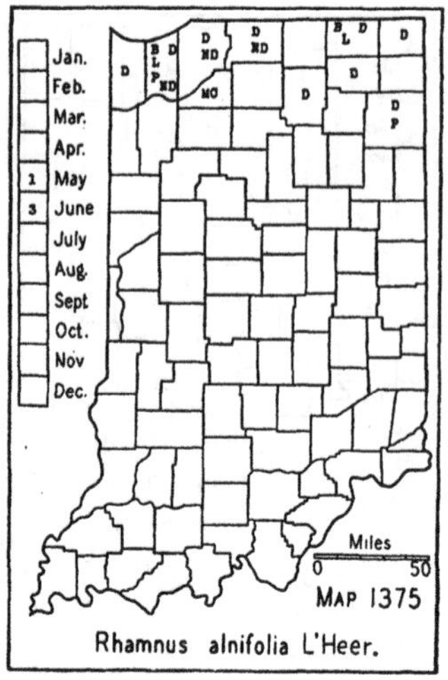

Rhamnus alnifolia L'Heer.

Leaves mostly with 3 pairs of veins; flowers 4-parted; nutlets 3 or 4. (See excluded species no. 433, p. 1072.) .... *R. cathartica.*
Leaves mostly with 4-10 pairs of veins; flowers 4- or 5-parted; nutlets 2 or 3.
  Flowers 4-parted; nutlets 2; leaves pubescent beneath........... 2. *R. lanceolata.*
  Flowers 5-parted; nutlets 3.
    Leaves crenate-serrulate; petals none; nutlets grooved on the back; leaves glabrous or glabrate beneath................................ 3. *R. alnifolia.*
    Leaves entire or nearly so, glabrous beneath; petals 5; nutlets smooth on the back.. 4. *R. Frangula.*

1. **Rhamnus caroliniàna** Walt. CAROLINA BUCKTHORN. Map 1373. On rocky wooded slopes near the Ohio River and rarely inland on gentle wooded slopes.
   Va. to Nebr., southw. to Fla. and Tex.

1a. **Rhamnus caroliniana** var. **móllis** Fern. (Rhodora 12: 79. 1910.) This is a form with the under surface of the leaves more or less densely pubescent. Specimens in our area show all grades of pubescence.

2. **Rhamnus lanceolàta** Pursh. LANCE-LEAF BUCKTHORN. Map 1374. This species is very local and has two distinct habitats. One is on limestone cliffs and rocky or gravelly wooded slopes and the other is in springy places with skunk cabbage, and in alluvial soil along streams.
   Pa., Ind., and Nebr., southw. to Ala. and Tex.

3. **Rhamnus alnifòlia** L'Hér. ALDER BUCKTHORN. Map 1375. A small shrub in or on the borders of tamarack bogs and at the moist bases of dunes in the dune area. Rare.
   Newf. to B. C., southw. to N. J., W. Va., Ill., and Calif.

4. RHAMNUS FRÁNGULA L. GLOSSY BUCKTHORN. Freely escaping about Interlaken in La Porte County. First reported by J. A. Nieuwland. In 1938 S. W. Witmer of Goshen College found a colony about 15 feet high in low, mucky soil in sec. 34 about 2 miles north of Goshen, Elkhart County.
   Nat. of Eu.

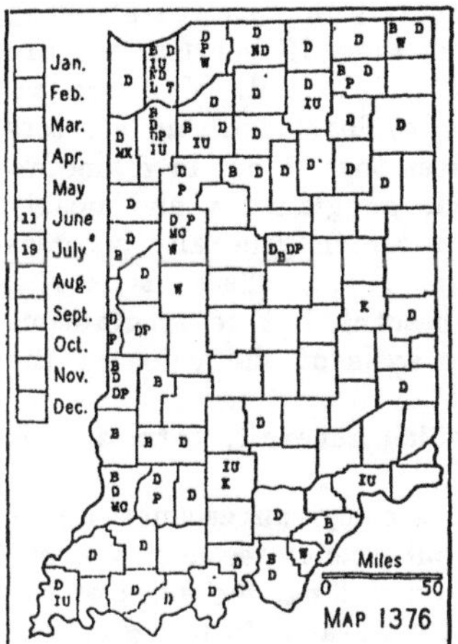

Ceanothus americanus L.

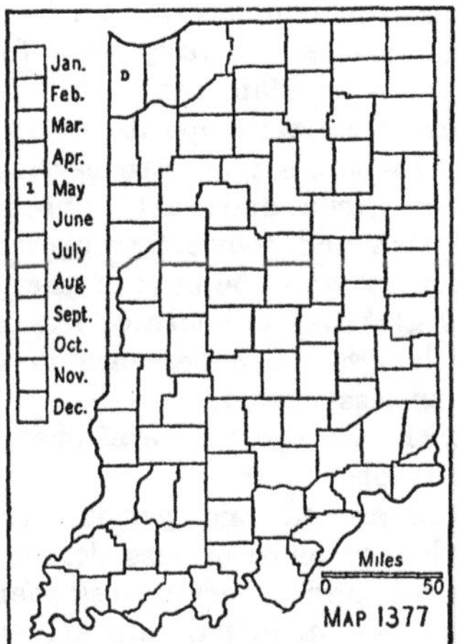

Ceanothus ovatus Desf.

Vitis Labrusca L.

4877. CEANÒTHUS L.

Leaves ovate to ovate-oblong, generally pubescent all over the lower surface; peduncle pubescent, longer than the subtending leaf; capsules generally crested or roughened on the angles; seed more than 2 mm long, smooth...........1. *C. americanus.*
Leaves elliptic-lanceolate, glabrous or nearly so beneath; peduncle glabrous, generally shorter than the subtending leaf; capsules not crested or roughened on the angles; seed 2 mm or less in length, surface pitted........................2. *C. ovatus.*

1. **Ceanothus americànus** L. NEW JERSEY TEA. Map 1376. Infrequent throughout the state in dry situations, being more frequent in the sandy soil of the lake area, especially in the dune area. South of the lake area it prefers the slopes and crests of black and white oak ridges and is found also on limestone and sandstone bluffs.

Maine to Man., southw. to Fla. and Tex.

2. **Ceanothus ovàtus** Desf. INLAND NEW JERSEY TEA. Map 1377. Our only specimens come from the low dunes along Lake Michigan between Pine and Miller in Lake County.

Vt. to Man., southw. to D. C., Ill., and Tex.

170. VITÀCEAE Lindl. GRAPE FAMILY

Leaves simple.
Pith of branches interrupted at the nodes; petals cohering at the summit and falling off as a cap without separating; berry spheroidal in shape (except in *Vitis Labrusca*), black or bluish black, edible.................4909. VITIS, p. 661.
Pith of branches not interrupted at the nodes; petals distinct and expanding before falling off; berry oblate-spheroidal in shape, bluish, not edible.................
....4916. AMPELOPSIS, p. 665.
Leaves palmately compound.........................4915. PARTHENOCISSUS, p. 664.

4909. VÌTIS [Tourn.] L. GRAPE

[Bailey. The species of grapes peculiar to North America. Gentes Herbarum 3: 149-244. 1934.]

Vines having a tendril or flower cluster opposite each leaf (rarely the tendrils aborting and falling off); fruiting clusters generally with fewer than 15 berries; mature berries generally 1 cm or more in diameter......................1. *V. Labrusca.*

Vines lacking a tendril or flower cluster opposite every third leaf; fruiting clusters generally with more than 15 berries; mature berries less than 1 cm in diameter.

Under surface of the leaves green and rusty pubescent or glaucous and nearly glabrous, or with a whitish and cobwebby pubescence which generally becomes rusty where the veins converge at the base of the leaf..........2. *V. aestivalis.*

Under surface of the leaves glabrous or more or less pubescent, but never glaucous or rusty pubescent (although the pubescence in the axils of the veins of the leaf may become more or less rusty).

Leaves without lobes or with two short, lateral ones, which generally form a wide sinus with the terminal lobe.

Branchlets more or less angled and permanently densely pubescent; under surface of mature leaves more or less densely pubescent; teeth of margin of blades short, convex, and generally less than 3 mm long, the sides of the teeth which end the two principal lateral veins generally forming an obtuse angle.. 3. *V. cinerea.*

Branchlets not conspicuously angled; under surface of the leaves glabrous, or pubescent, generally only along the veins; teeth of margin of blades sharp, generally more than 3 mm long, the sides of the two ending the two principal lateral veins, generally forming an acute angle..........4. *V. vulpina.*

Leaves mostly with two lateral lobes, generally acute, and usually forming an acute sinus with the terminal lobe.

Branchlets bright red (shade forms sometimes greenish); fruit without a bloom; in Indiana found only in the Lower Wabash Valley...5. *V. palmata.*

Branchlets not red; fruit with a bloom; found throughout the state...6. *V. riparia.*

1. **Vitis Labrúsca** L. FOX GRAPE. Map 1378. This species is apparently restricted to the northwestern and southeastern parts of the state. In the northwestern part it is found in low ground in woods, usually associated with pin and black oaks or in the dunes mostly on the mucky borders of streams and ditches. In the southeastern part it is generally found in the "flats" in woodland or along roadsides. At least in this section of the state it prefers the mediacid soils and is associated with sweet gum, pin oak, and beech.

The color of the fruit of this species is usually purplish black. The late Wm. Henderson, a grower and collector of medicinal plants, who lived about 11 miles northeast of Greensburg, found in Franklin County a wild plant of this species that bore yellow green fruit.* He sent me seed and a part of the original plant which I now have growing. Seed were planted and seedlings were widely distributed to botanical gardens. The seedlings do not all have green fruit.

N. E. to Ind., southw. to Ga. and Tenn.

2. **Vitis aestivàlis** Michx. SUMMER GRAPE. Map 1379. Throughout the state. Rare to infrequent in the northern two thirds of the state, becoming frequent to common in the southern part. It is usually found in dry situations in woodland or in the open.

N. H. to Kans., southw. to Fla. and Tex.

2a. **Vitis aestivalis** var. **argentifòlia** (Munson) Fern. (Rhodora 38: 428. 1936.) (*Vitis bicolor* Le Conte.) This variety is generally regarded as

* Vitis Labrusca f. alba (Prince) Fern. (Rhodora 41: 431. 1939.)

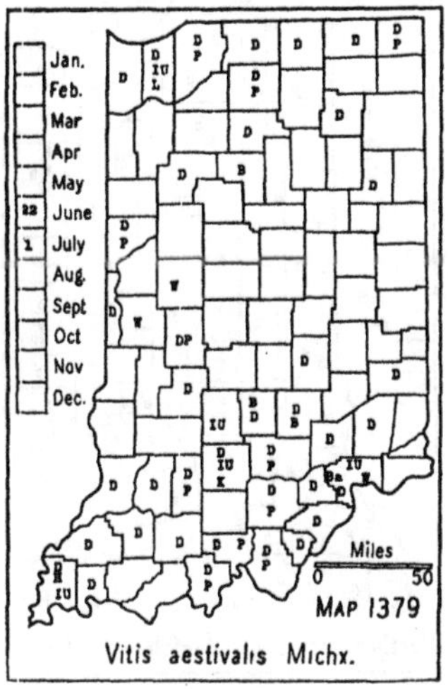

Vitis aestivalis Michx.

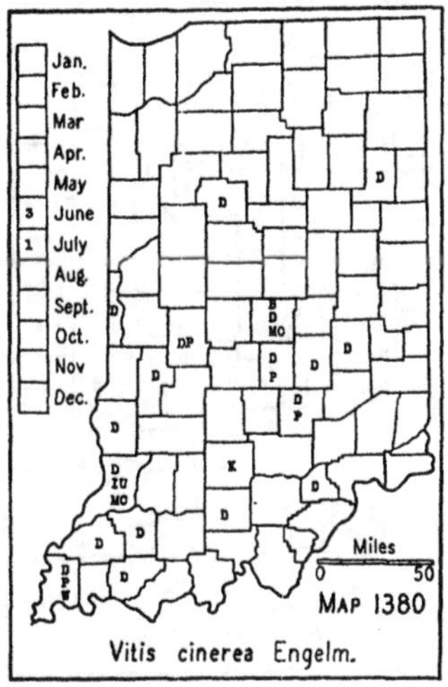

Vitis cinerea Engelm.

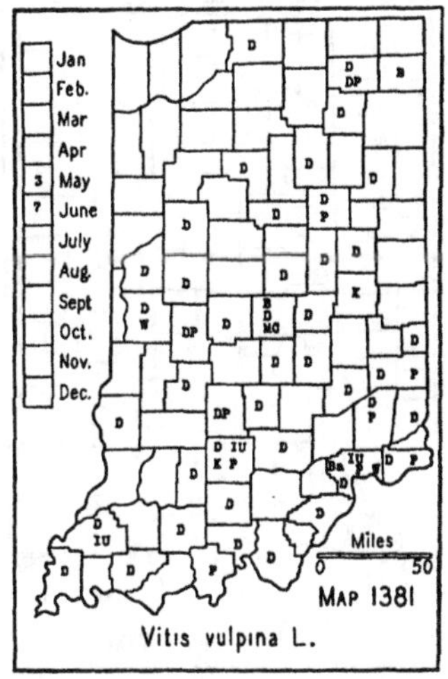

Vitis vulpina L.

a northern form of *Vitis aestivalis* and is separated from it by the less dense pubescence and glaucous color of the under surface of the leaves. Since my specimens show all degrees of intermediates between the two extremes I think this variety is merely a form of the species and I have grouped them on one map. We have this form from the Ohio River to Lake Michigan but it becomes progressively more frequent toward the northern part of the state.

3. **Vitis cinèrea** Engelm. SWEET WINTER GRAPE. Map 1380. This species is usually associated with *Vitis aestivalis* from which it was not formerly separated. It is more common in the southwestern counties and in rather sandy soil. Pepoon's report from Lake County should be referred to *Vitis Labrusca.* (Rhodora 35: 302. 1933.)

Va., sw. Ohio to Nebr., southw. to Fla., La., and Tex.

4. **Vitis vulpìna** L. Rhodora 41: 431-434. 1939.) (*Vitis cordifòlia* Michx.) FROST GRAPE. Map 1381. Throughout the state although we have no specimens from the extreme northwestern part. This species prefers the open and is commonly found in dry soil along fences.

I measured a specimen in Perry County, that was 9 inches in diameter at breast height.

N. Y. to Nebr., southw. to Fla. and Tex.

5. **Vitis palmàta** Vahl. CATBIRD GRAPE. Map 1382. This species is local and has been found in only Knox, Gibson, Posey, and Montgomery Counties on the low borders of sloughs and ponds. It is usually associated with buttonbush.

In the Mississippi Valley from Ind. to La. and Tex.

6. **Vitis ripària** Michx. (*Vitis vulpina* of Gray, Man., ed. 7 and Britton and Brown, Illus. Flora, ed. 2.) (See Rhodora 41: 431-434. 1939.) RIVERBANK GRAPE. Map 1383. Throughout the state. Although we have few specimens from the unglaciated area, it occurs there at least along the Ohio River. It prefers alluvial soil.

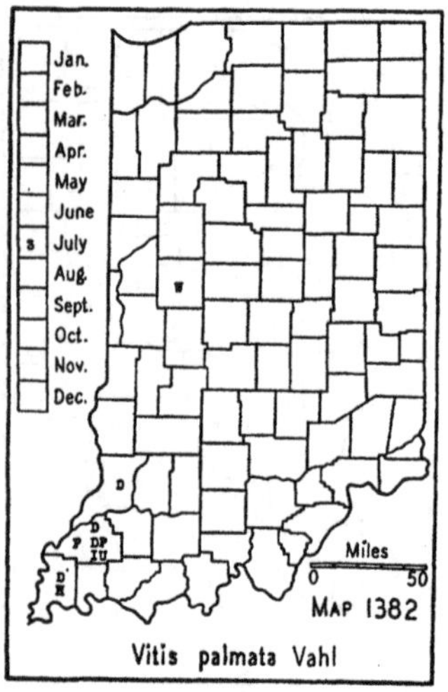

MAP 1382
Vitis palmata Vahl

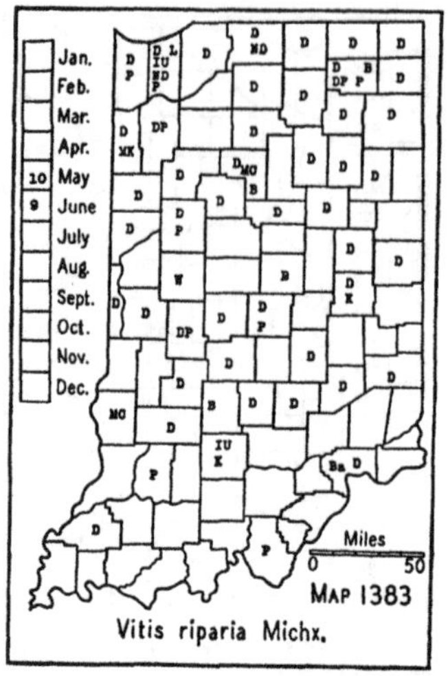

MAP 1383
Vitis riparia Michx.

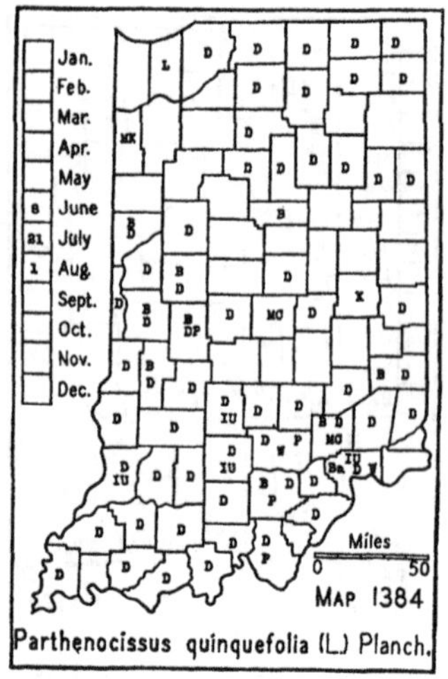

MAP 1384
Parthenocissus quinquefolia (L.) Planch.

along streams but it is found also in moist soil along fences and woodland.

N. B. to Man., southw. to Va., Ark., and Tex.

6a. **Vitis riparia** var. **syrtícola** (Fern. & Wieg.) Fern. (Rhodora **41**: 431. 1939.) This is merely a more pubescent form occurring on the dunes along Lake Michigan and found once along the Kankakee River in Lake County.

4915. PARTHENOCÍSSUS Planch.

Leaves dull above; tendrils adhesive; inflorescence not dichotomously branched; fruit 5-7 mm in diameter; seed 1-3.............................1. *P. quinquefolia.*
Leaves glossy above; tendrils not adhesive; inflorescence dichotomously branched; fruit mostly 8-10 mm in diameter; seed 3 or 4.....................2. *P. inserta.*

1. **Parthenocissus quinquefòlia** (L.) Planch. (*Psedera quinquefolia* (L.) Greene of Gray, Man., ed. 7 and *Parthenocissus quinquefolia* (L.) Planch. of Britton and Brown, Illus. Flora, ed. 2 as to name but not figure.) VIRGINIA CREEPER. FIVE-LEAF IVY. Map 1384. More or less frequent throughout the state along fences and in clearings and woodland. It is more common in beech and sugar maple and sandy black and white oak woodland. It also is usually frequent in the "flats," if not too wet, and in the Lower Wabash Valley where it reaches its largest size. The reports from the dune area should probably be referred to *P. inserta.* (See Buhl, Amer. Midland Nat. **16**: 251. 1935.)

N. E. to Wis. and Mo., southw. to Fla. and Mex.

1a. **Parthenocissus quinquefolia** f. **hirsùta** (Donn) Fern. (Rhodora 41: 664. 1939.) This is a form of the species with the branchlets, tendrils, petioles, and leaves pubescent, at least while young. My specimens range from glabrous to pubescent and show all degrees of pubescence.

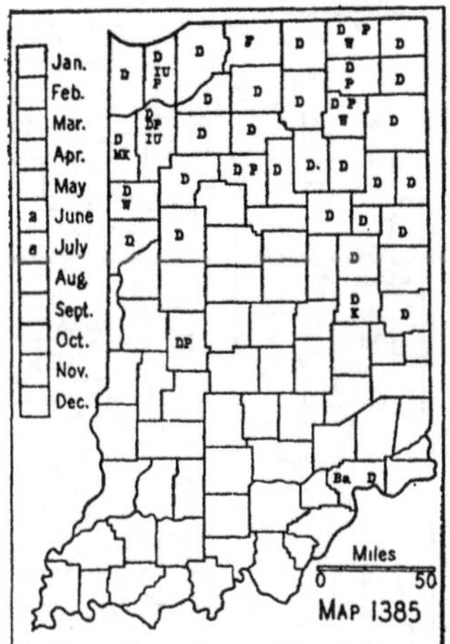

Parthenocissus vitacea (Knerr) Hitchc.

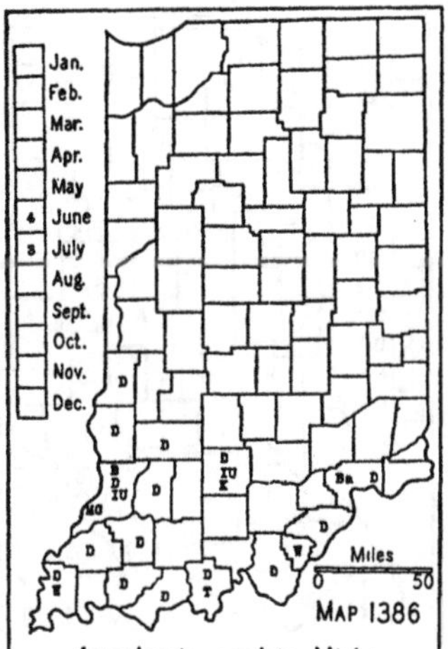

Ampelopsis cordata Michx.

Tilia americana L.

Since the distribution of the forms shows no geographic or habitat range, all forms are shown on the map as belonging to the species.

2. **Parthénocissus insérta** (Kerner) Fritsch. (Jour. Arnold Arb. 20: 419. 1939.) (*Parthenocissus vitacea* (Knerr) Hitchc. and *Psedera vitacea* (Knerr) Greene.) Map 1385. Rather frequent in the open throughout the lake area, usually on fences.

Canada and N. E. to Man. and Wyo., southw. to N. Y., Kans., and Tex.

4916. AMPELÓPSIS Michx.

1. **Ampelopsis cordàta** Michx. (*Cissus Ampelopsis* Pers.) HEARTLEAF AMPELOPSIS. Map 1386. Restricted to the flood plains of the streams of the southern and southwestern part of the state. In cultivation it has proven hardy as far north as Bluffton. In the Lower Wabash Valley it climbs to the tops of the tallest trees and reaches a diameter of 3 inches.

Va. to Nebr., southw. to Fla. and Tex.

174. TILIÀCEAE Juss. LINDEN FAMILY

4964. TÍLIA [Tourn.] L.

Leaves glabrous beneath or with tufts of hairs in the axils of the veins..1. *T. americana.*
Leaves densely white- or gray-pubescent beneath................2. *T. heterophylla.*

1. **Tilia americàna** L. (*Tilia glabra* Vent. of Deam, Trees of Indiana.) AMERICAN LINDEN. BASSWOOD. Map 1387. More or less frequent to common in all parts of the state. It prefers a moist habitat and in the lake area it often formed 10-15 per cent of the stand of the original forests. In the hills of the southern part of the state it is often found on the rocky bluffs of streams.

N. B. to Man., southw. to Ga. and w. Tex.

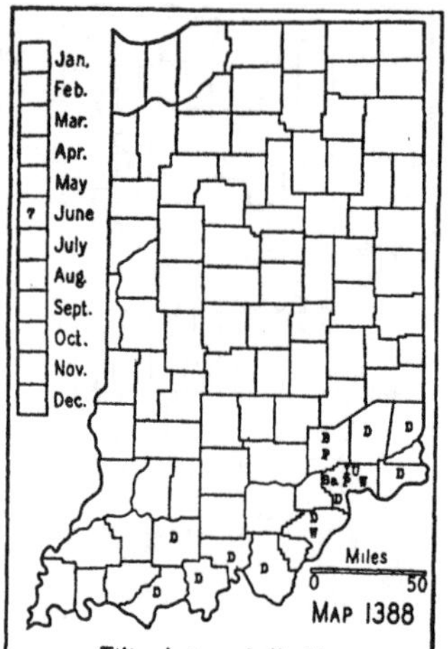

Tilia heterophylla Vent.

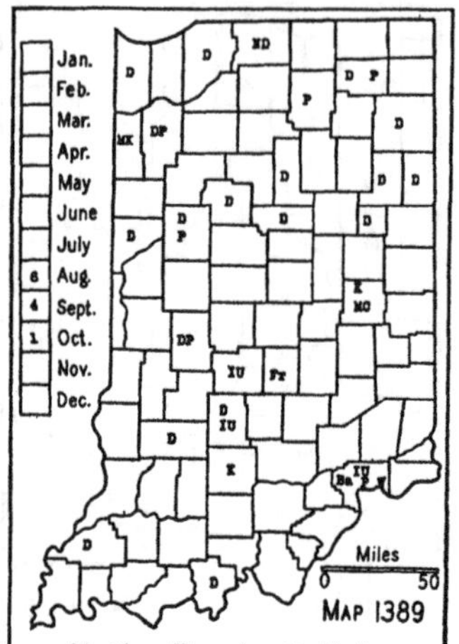

Abutilon Theophrasti Medic

Malva sylvestris var. mauretiana (L.) Boiss.

2. **Tilia heterophýlla** Vent. WHITE BASSWOOD. Map 1388. An infrequent tree on the bluffs and slopes of ravines and streams in a few of the southern counties.

W. Va. to Ind., southw. to Fla. and Ala.

175. MALVÀCEAE Neck. MALLOW FAMILY

Calyx without involucral bracts.
  Leaves mostly 6-15 cm long, cordate at the base.
    Leaves not lobed; flowers yellow........................4983. ABUTILON, p. 666.
    Leaves lobed; flowers white or pinkish..................4994. NAPAEA, p. 668.
  Leaves less than 6 cm long, from tapering to cordate at the base; flowers yellow..4998. SIDA, p. 668.
Calyx with involucral bracts.
  Involucral bracts of calyx 2 or 3.
    Leaves oblong-lanceolate or linear; flowers yellow......4995. MALVASTRUM, p. 668.
    Leaves not oblong-lanceolate or linear; flowers not yellow.
      Blades orbicular in outline; petals obovate; ovaries not beaked..4992. MALVA, p. 667.
      Blades triangular-hastate; petals truncate; ovaries beaked..4992A. CALLIRHOË, p. 668.
  Involucral bracts of calyx 6 or more.
    Stamen column anther-bearing at the summit; fruit composed of 15-20 carpels..4991. ALTHAEA, p. 667.
    Stamen column anther-bearing below the summit; fruit a 5-celled capsule..5013. HIBISCUS, p. 669.

4983. ABÙTILON [Tourn.] Mill.

1. ABUTILON THEOPHRÁSTI Medic. (*Abutilon Abutilon* (L.) Rusby of Britton and Brown, Illus. Flora, ed. 2.) VELVET-LEAF. INDIAN MALLOW. Map 1389. A frequent to common weed found throughout the state. It is found in cultivated grounds, stubble fields, open woodland, and waste places, and along roadsides and railroads.

Nat. of India; in all but the colder parts of N. A.

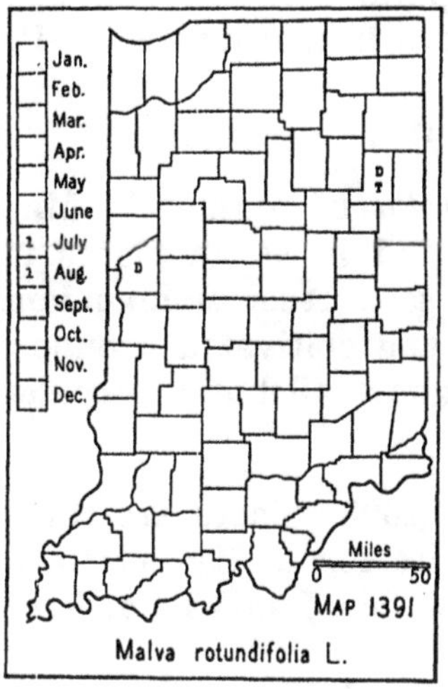

Map 1391
Malva rotundifolia L.

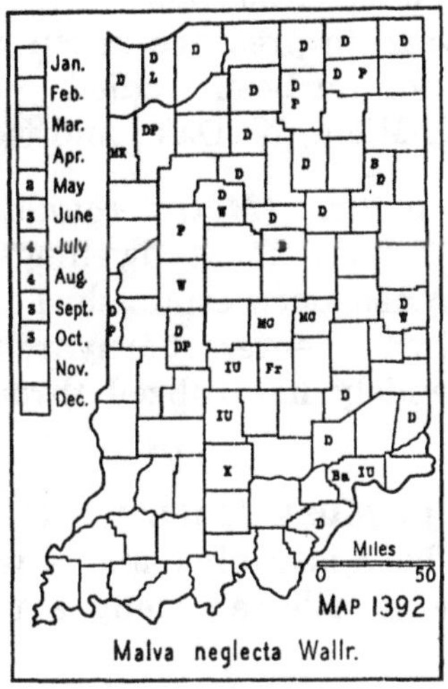

Map 1392
Malva neglecta Wallr.

Map 1393
Malva moschata L.

4991. ALTHAÈA L.

See excluded species no. 437, p. 1073.

4992. MÁLVA [Tourn.] L. MALLOW

Leaves crisped or puckered on the margins; flowers and fruit sessile. (See excluded species no. 439, p. 1073) .... ***M. crispa.***
Leaves not crisped or puckered on the margins; flowers and fruit not sessile.
  Leaves with 5-9 shallow lobes.
    Plants erect, 1.5-9 dm high.
      Petals purplish or rose, about 20 mm long.
        Leaves with triangular lobes. (See excluded species no. 440, p. 1073) .... ***M. sylvestris.***
        Leaves with rounded lobes............... **1. *M. sylvestris* var. *mauretiana.***
      Petals pinkish white, about 5 mm long; carpels 8-11 (usually 10), more or less pubescent and rugose on the back; suture between carpels more or less crested.. **2. *M. rotundifolia.***
    Plants procumbent; petals about 10 mm long, pale blue; carpels 12-15, more or less pubescent but smooth or faintly reticulate on the back; suture between carpels not raised, usually depressed...................... **3. *M. neglecta.***
  Leaves deeply 5-7-lobed.
    Stem leaves 5-parted, the lobes 1- or 2-parted; carpels pubescent..**4. *M. moschata.***
    Stem leaves 5-parted, the lobes incised or toothed but not parted; carpels glabrous. (See excluded species no. 438, p. 1073) .... ***M. Alcea.***

1. MALVA SYLVÉSTRIS L. var. MAURETIÀNA (L.) Boiss. (Rhodora **12**: 140. 1910.) HIGH MALLOW. Map 1390. A rare garden escape. It is said that this variety is the form of the species that has escaped in the eastern United States and probably all of our reports should be referred to it. All specimens seen belong to the variety.

Nat. of Eurasia; widely but sparingly escaped in e. N. A.

2. MALVA ROTUNDIFÒLIA L. (Rhodora 39: 98-99. 1937.) (*Malva pusilla* Smith.) (Bergman. Comments on Malva rotundifolia L. and its allies. Minnesota Bot. Stud. 4: 437-441. 1916.) ROUNDLEAF MALLOW. Map 1391. This species has the same habitat and is closely allied to the

following species from which it has not been separated in our manuals. No doubt search will reveal its presence in our area in many places. Its general distribution has not yet been studied.

Nat. of Eu.; N. J., Pa., Mich., N. Dak., and westw.

3. MALVA NEGLÉCTA Wallr. (*Malva rotundifolia* of authors.) Map 1392. A frequent weed mostly about dwellings and in waste places in all parts of the state. When once established it soon becomes abundant and a pest, especially when it enters barnyards and cultivated grounds.

Nat. of Eurasia and widely naturalized throughout all but the colder parts of N. A.

4. MALVA MOSCHÀTA L. MUSK MALLOW. Map 1393. There are three reports from Indiana. There is a pink and a white form. My specimen is the white form and was collected along a roadside 5 miles north of Goshen in Elkhart County.

Nat. of Eu.; Newf. to B. C., southw. to N. J., Va., and Wis.

4992A. CALLÍRHOË Nutt.

1. **Callirhoë triangulàta** (Leavenw.) Gray. CLUSTERED POPPY-MALLOW. Map 1394. Indiana is included in the range of this species in Gray, Man., ed. 7 and in Britton and Brown, Illus. Flora, ed. 2. There were, however, no specimens in the Gray Herbarium or in the herbarium of the New York Botanical Garden until I sent them specimens in 1933. There are no other reports. In July, 1933, Scott McCoy found it plentiful in sandy soil along the C. E. & I. Railroad about a mile north of Oaktown, Knox County. I visited this place in August, 1933, and found the species well established at three places. An attempt to dig some plants convinced me that it has been there for some time and is doubtless established.

Ill. to Minn., southw. to N. C. and Tex.

4994. NAPAÈA [Clayt.] L.

1. **Napaea dioìca** L. GLADE MALLOW. Map 1395. Local and very rare. Alluvial banks of the Wabash River and moist roadsides. I have had the staminate form in cultivation since 1918 and the pistillate form since 1931.

Pa. to Minn., southw. to Va., Tenn., and Iowa.

4995. MALVÁSTRUM Gray

See excluded species no. 441, p. 1073.

4998. SÌDA L.

Leaves ovate-lanceolate or ovate-oblong................................1. *S. spinosa.*
Leaves palmately 3-7-lobed. (See excluded species no. 442, p. 1073)...*S. hermaphrodita.*

1. **Sida spinòsa** L. PRICKLY SIDA. Map 1396. An infrequent weed in cultivated fields, truck gardens, waste places, open woodland, and pas-

Callirhöe triangulata (Leavenw.) A. Gray

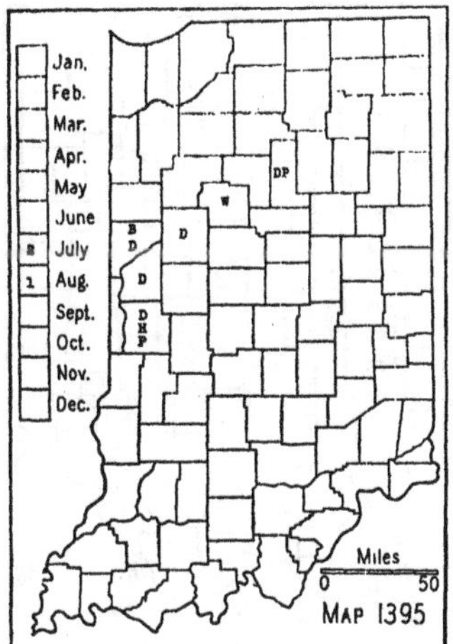

Napaea dioica L.

Sida spinosa L.

tures, and along roadsides and railroads throughout the state except in the northern counties where it may be rare or absent. Some authors believe this species to be adventive from the south. Our earliest authors list it and Dr. Clapp in 1852 says: "Very common in the vicinity of New Albany." I am considering it a native, at least in the southern part of the state.

Mass. to Mich. and Kans., and southw. to Fla. and Tex.; tropical America.

5013. HIBÍSCUS L.

Tall, perennial herbs, 1-2 m high; calyx not inflated about the capsules.
- Stems and leaves glabrous; seed pubescent........................1. *H. militaris.*
- Stems and lower surface of leaves pubescent; seeds not pubescent.
  - Bractlets densely short stellate-pubescent on the outside, the margins not ciliate with longer, simple hairs; flowers pink..................2. *H. Moscheutos.*
  - Bractlets densely short stellate-pubescent on the outside, the margins more or less ciliate with longer, simple hairs; flowers white, red within at the base.
    - Leaves glabrous or nearly so above; capsules glabrous or nearly so...3. *H. palustris.*
    - Leaves velvety-pubescent above; capsules densely stellate-pubescent...4. *H. lasiocarpos.*

Low, hairy annuals, mostly 1-5 dm high; calyx inflated and enveloping the capsule. ....5. *H. Trionum.*

1. **Hibiscus militàris** Cav. SOLDIER ROSEMALLOW. Map 1397. Frequent to common on the muddy shores of sloughs, ponds, and our larger streams. It is to be noted that this species is rapidly migrating. I have known well the shores of the Wabash River near Bluffton for a distance of five miles since 1880. The first colony of this species was noted in 1897 and it is now common all along the muddy shores and on the muddy bars in the river. In the early history of the state our streams were clear and when the forests were removed the streams became muddy and sediment was deposited on the shores and on the gravelly and rocky bars which made a suitable habitat for this species. This same thing is true of several

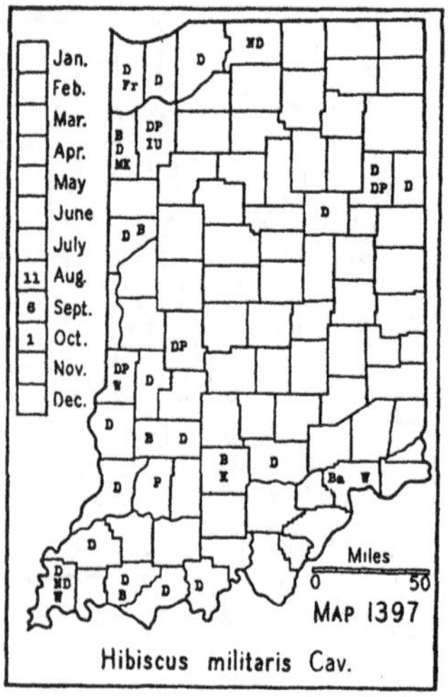

Hibiscus militaris Cav.

Hibiscus Moscheutos L.

Hibiscus palustris L.

other plants, especially *Dianthera americana,* which is now clogging some of our smaller streams.

Pa. to Minn., southw. to Fla. and La.

2. **Hibiscus Moscheùtos** L. COMMON ROSEMALLOW. Map 1398. Local in the lake area of the state on the borders of streams and in roadside ditches, millraces, and drained ponds. This species prefers a habitat with more organic matter than the preceding species. Usually the colonies are small but once I saw in a drained pond in Wabash County three acres of a complete stand of this species.

It has been reported also from Clark, Jefferson, Knox, and Posey Counties. All of these reports, however, were made before our manuals separated *Hibiscus palustris* from this species and doubtless all of these reports should be referred to *Hibiscus palustris.*

Along the Atlantic coast from Mass. to Fla.; inland from Ont. and Mich. to Mo.

3. **Hibiscus palústris** L. (*Hibiscus oculiroseus* Britt. and *Hibiscus palustris* L. f. *oculiroseus* (Britt.) Fern.) Map 1399. In ditches and ponds in the southwestern part of the state. It is local but common where it is found. The largest colony I ever saw was in hard, clay soil in a buttonbush pond of about 3 acres in a low, flat pin oak woods about 10 miles southwest of Mt. Vernon. I think all of the reports for *Hibiscus Moscheutos* from southern Indiana should be referred to this species. I saw it from a train window in a ditch in Jennings County.

Along the Atlantic coast and up the Mississippi Valley to s. Ind.

4. **Hibiscus lasiocárpos** Cav. HAIRY-FRUITED ROSEMALLOW. Map 1400. I have found this species only on the muddy borders of ponds in the southwestern part of Knox County, and in the southern part of Gibson County. It has been reported from Daviess, Hamilton, and Vigo Counties. It is probable that these reports should all be referred to the preceding

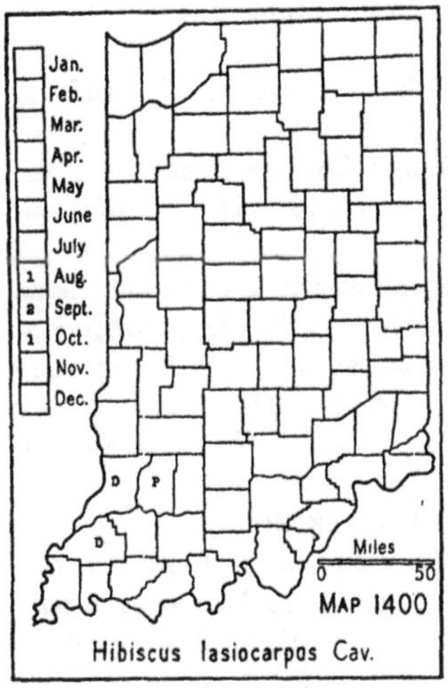

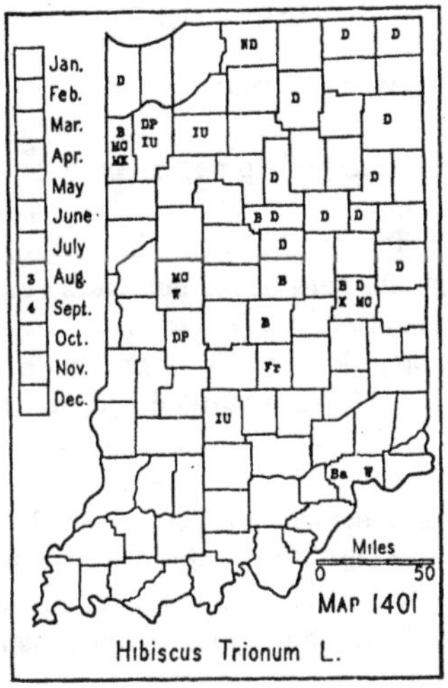

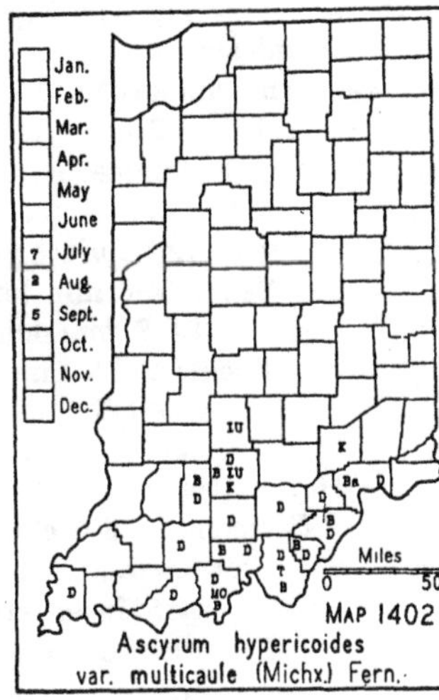

species. A specimen reported from Parke County by Grimes is in the herbarium of DePauw University and is *Hibiscus palustris.*

Ind. to Mo., southw. to Fla. and Tex.

5. HIBISCUS TRIÒNUM L. FLOWER-OF-AN-HOUR. Map 1401. An obnoxious weed in rather sandy soil in cultivated grounds in many parts of the state, especially in the glaciated area. I can recall the time when I rarely saw it but now in certain areas it forms a complete stand in cornfields. Although it is an annual, when once established, it is difficult to exterminate on account of its numerous seeds and their unusual viability.

Nat. of Eu.; N. S. to S. Dak., southw. to Fla. and Kans.

187. HYPERICÀCEAE Lindl. ST. JOHNSWORT FAMILY

Sepals 4, the outer two very much larger and bractlike; petals 4...5167. ASCYRUM, p. 671.
Sepals 5; petals 5....................................5168. HYPERICUM, p. 671.

5167. ÁSCYRUM L.

1. **Ascyrum hypericoìdes** L. var. **multicaúle** (Michx.) Fern. (Rhodora 38: 433. 1936.) (*Ascyrum hypericoides* of authors.) ST. ANDREW'S CROSS. Map 1402. Usually in poor soil on open black and white oak ridges and rarely elsewhere. I have one specimen collected in an old fallow field. Infrequent.

Mass. to Nebr.; southw. to Fla. and Tex.

5168. HYPÉRICUM [Tourn.] L. ST. JOHNSWORT

Petals yellow, convolute in the bud.
 Styles 5; capsules 5-celled.
  Tall perennials; leaves mostly 4-8 cm long; flowers generally 4-5 cm wide; capsules 1.7-2.5 cm long..........................1. *H. Ascyron.*
  Shrubs, usually 4-7 dm high; leaves mostly 2.5-4.5 cm long; flowers 2-2.5 cm wide; capsules 6-9 mm long..........................2. *H. Kalmianum.*

Styles 3 (rarely 4); capsules 3- (rarely 4-) celled, or incompletely 3-celled in *H. frondosum.*
  Stamens numerous, 15-40; flowers mostly 7-25 mm wide (except in *H. majus).*
    Shrubs 0.5-2 m high.
      Sepals foliaceous, very unequal, more than 6 mm long; flowers mostly 3-5 cm wide..3. *H. frondosum.*
      Sepals not foliaceous; flowers 1-2 cm wide.
        Sepals 4-6 mm long; flowers 1.5-2 cm wide; capsules about 1 cm long; seed 1.5-2 mm long....................................4. *H. prolificum.*
        Sepals 2-2.5 mm long; flowers 1-1.5 cm wide; capsules 4-6 mm long; seed about 1 mm long. (See excluded species no. 444, p. 1073)............ .... *H. densiflorum.*
    Herbs (sometimes woody toward the base).
      Stamens in 3-5 clusters; styles separate; petals with black dots.
        Flowers and leaves many, the upper leaves usually not over 6 mm wide; introduced species...............................5. *H. perforatum.*
        Flowers and leaves few, the upper leaves usually more than 6 mm wide; native species.
          Sepals 3-4 mm long; petals with the black dots in lines..6. *H. punctatum.*
          Sepals 5-7 mm long; petals with the black dots on the margins. (See excluded species no. 446, p. 1074)....................*H. graveolens.*
      Stamens not in clusters; styles distinct, more or less united; petals without black dots.
        Lateral nerves of the under surface of the leaves obscure or wanting; leaves linear or nearly so..............................7. *H. dolabriforme.*
        Lateral nerves of the under surface of the leaves visible; leaves wider than linear.
          Leaves and sepals more or less revolute................8. *H. adpressum.*
          Leaves and sepals not revolute.
            Sepals inclosing or nearly inclosing the capsule.....9. *H. denticulatum.*
            Sepals much shorter than the capsule.
              Stems somewhat woody at the base; leaves oblong to linear-oblong, thick; seed about 2 mm long................10. *H. cistifolium.*
              Stems herbaceous; leaves elliptic, oval, lanceolate to narrow-lanceolate, thin; seeds about 0.5 mm long.
                Leaves spreading, elliptic-oblong, obtuse, broadest near the middle, pinnately veined. (See excluded species no. 445, p. 1074)..... ....*H. ellipticum.*
                Leaves ascending, lanceolate to narrow-lanceolate, broadest below the middle, with 5-7 strong lateral veins, at least toward the base.......................................11. *H. majus.*
  Stamens few (5-12, rarely more than 12); flowers not over 7 mm broad.
    Bracts of the inflorescence foliaceous, appearing as a continuation of the stem leaves...12. *H. boreale.*
    Bracts of the ultimate branchlets of the inflorescence subulate.
      Leaves scalelike or linear-subulate, strongly ascending.
        Leaves scalelike; capsules much exceeding the calyx....13. *H. gentianoides.*
        Leaves linear-subulate, 6-20 mm long; capsules slightly exceeding the calyx. ....14. *H. Drummondii.*
      Leaves not scalelike or linear-subulate.
        Leaves linear, 3-nerved...............................15. *H. canadense.*
        Leaves not as above.
          Leaves ovate-oblong or short-elliptic, obtuse; stems usually diffusely branched; only the ultimate branchlets of the inflorescence subulate-bracted; capsules 2.8-4 mm long...................16. *H. mutilum.*

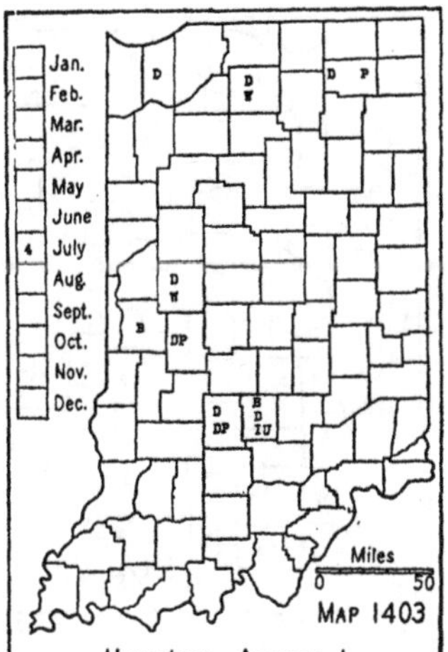

Hypericum Ascyron L.

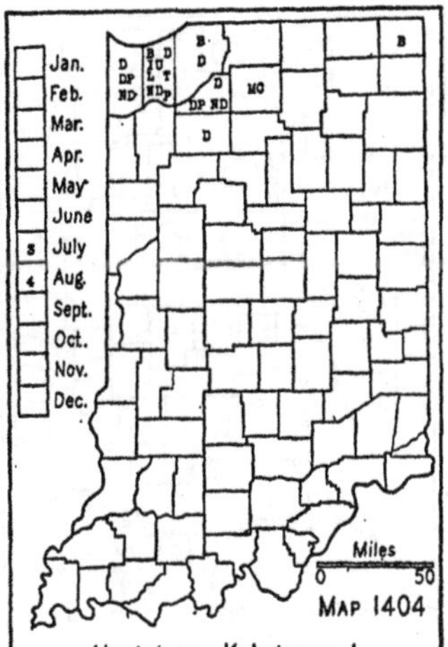

Hypericum Kalmianum L.

Hypericum frondosum Michx.

Leaves ovate and acute or the lower oval and obtuse; stems generally simple, if branched, the branches strict; inflorescence subulate-bracted; capsules 4-5 mm long. (See excluded species no. 447, p. 1074.) .... *H. gymnanthemum.*

Petals pinkish or reddish purple, imbricate in the bud.

Leaves sessile or partly clasping, not conspicuously narrowed at the base.

Leaf blades broadest at the base or below the middle, mostly 2-5 cm long, copiously glandular beneath.

Sepals acute at the apex; styles 2-3 mm long................17. *H. virginicum.*

Sepals obtuse or rounded at the apex; styles 0.5-1 (-2) mm long.............. ....17a. *H. virginicum* var. *Fraseri.*

Leaf blades broadest above the middle, mostly 5-15 cm long, not glandular below or only with a few glands; mature sepals about 5 mm long................ ....18. *H. tubulosum.*

Leaves petiolate, copiously glandular beneath, conspicuously narrowed at the base... ....18a. *H. tubulosum* var. *Walteri.*

1. **Hypericum Ascyron** L. GIANT ST. JOHNSWORT. Map 1403. Moist alluvial banks of streams. All of our specimens were growing in dense shade. Rare.

W. Que. to Man., southw. through Vt. and N. Y. to Pa., Ill., Mo., and Kans.

2. **Hypericum Kalmiànum** L. KALM HYPERICUM. Map 1404. In the open in moist, sandy soil in a few of the northern counties. Local. Does well in cultivation in a black loam soil for a short time.

Que. and along the Great Lakes to Wis., southw. to N. Y. and Ill.

3. **Hypericum frondòsum** Michx. (Jour. Arnold Arb. **19**: 149. 1938.) (*Hypericum aureum* Bartr.) GOLDEN ST. JOHNSWORT. Map 1405. A few plants of this southern species were found in 1935 by Miss Edna Banta along "Brough's Trail" in Clifty Falls State Park, Jefferson County. This species is doubtless a native here because the location is more than a mile from any habitation of consequence and it is not known to be in cultivation anywhere in the vicinity.

S. C., Ind. to Tenn., southw. to Ga. and Tex.

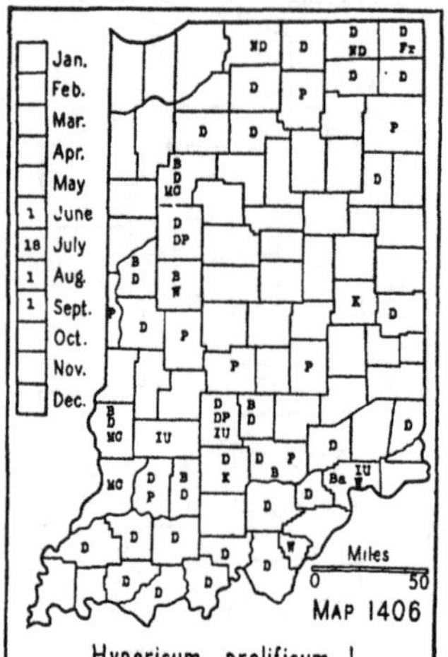

Hypericum prolificum L.

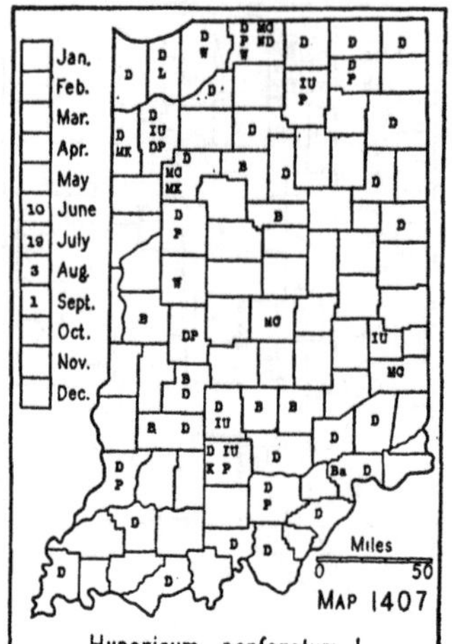

Hypericum perforatum L.

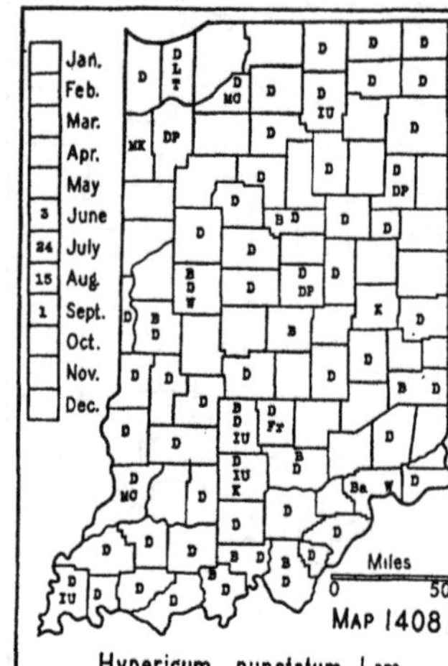

Hypericum punctatum Lam.

4. **Hypericum prolíficum** L. SHRUBBY ST. JOHNSWORT. Map 1406. Throughout the state although it seems to be rare or absent about Lake Michigan. In moist soil, usually in woods, along streams, and about swamps. More rarely in dry or moist soil along roadsides and on wooded slopes. It was noted in Crawford and Martin Counties where it had abundantly invaded abandoned fields. I also saw it in crevices of sandstone of a high cliff in Crawford County.

S. Ont. to Minn., southw. to Ga. and Miss.

5. HYPERICUM PERFORÀTUM L. COMMON ST. JOHNSWORT. Map 1407. An infrequent to common weed throughout the state, being more abundant in the northwestern part. It prefers sandy and poor or worn out soils. Chiefly along roadsides and in pastures, open woods, and fallow fields.

In California this species is known as Klamath weed and it has become a very obnoxious weed in many parts of that state. It is poisonous to stock but probably not fatal.

Nat. of Eu.; widely naturalized in N. A.

6. **Hypericum punctàtum** Lam. Map 1408. Widely distributed but never occurring in any abundance in any one place. In moist or dry soils in level woodland and on open wooded slopes, along roadsides, and in abandoned fields.

E. Que. to Minn., southw. to Fla. and Tex.

7. **Hypericum dolabrifórme** Vent. Map 1409. I have a specimen collected on July 13, 1899, by W. S. Blatchley in the vicinity of Wyandotte Cave.

Dry, barren hills of Ind., Ky., and Tenn. to Ga.

8. **Hypericum adpréssum** Bart. Map 1410. In moist, sandy soil on the borders of marshes and in ditches in the northern part of Jasper County. Reported by Schneck from the Lower Wabash Valley.

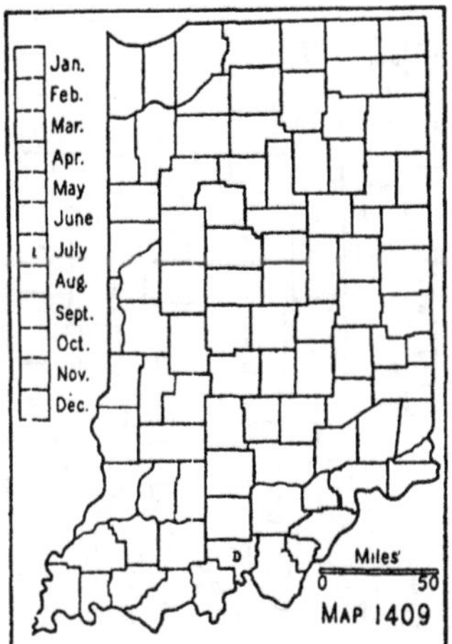

Hypericum dolabriforme Vent.

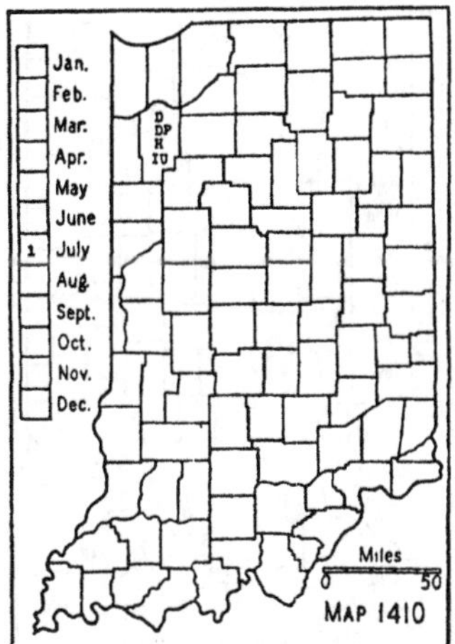

Hypericum adpressum Bart.

Hypericum denticulatum Walt.

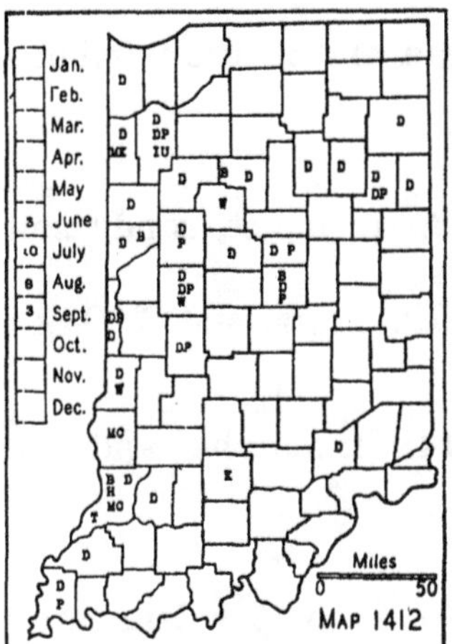

Hypericum cistifolium Lam.

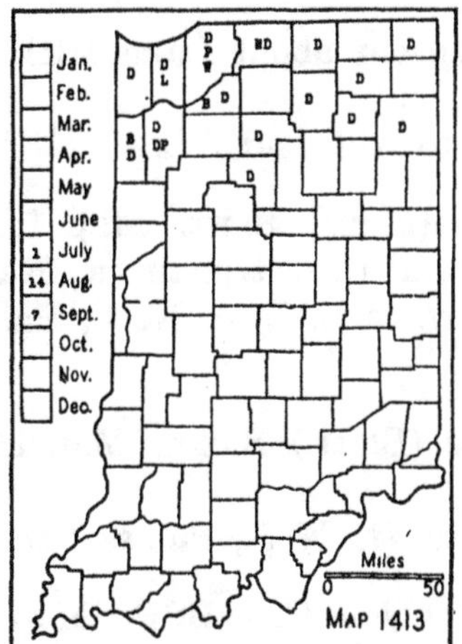

Hypericum majus (Gray) Britt.

Hypericum boreale (Britt.) Bickn.

When this perennial is compelled by weather conditions to grow with the base of the plant submerged for the greater part of the season, the submerged part becomes spongy. This form is known as *Hypericum adpressum* var. *spongiosum* Robinson. It occurs with the species but in wetter situations.

Atlantic coast from Mass. to Ga. and up the Mississippi Valley from La. to Ind.

9. **Hypericum denticulàtum** Walt. (*Hypericum virgatum* Lam.) Map 1411. In hard, white, minimacid clay soil in low, flat woods, associated with pin and post oaks. The report by Clark from Marshall County should be referred to some other species. Add Posey County to the map.

N. J., Pa., and Ill., southw. to Fla. and Tenn.

10. **Hypericum cistifòlium** Lam.* Map 1412. Alluvial and rocky, wooded banks of streams and in sandy soil along roadsides and in prairies. Infre-

* Svenson (Rhodora **42**: 17-18. 1940) has shown that this name should be applied to the Atlantic Coastal Plain plants while our plants should be called **Hypericum sphaerocarpum Michx.**

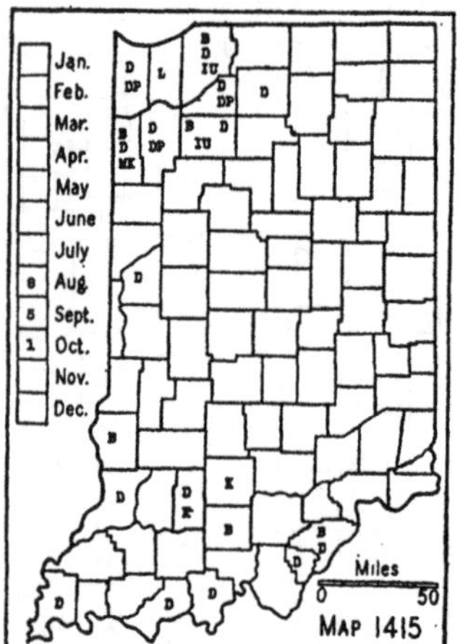

Hypericum gentianoides (L.) B S P.

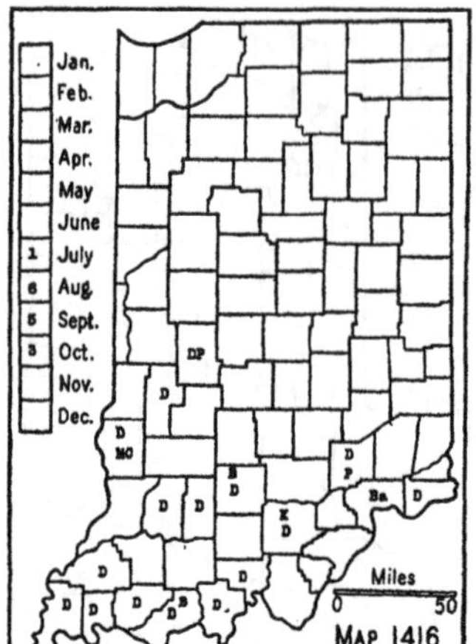

Hypericum Drummondii (Grev. & Hook.) T.& G.

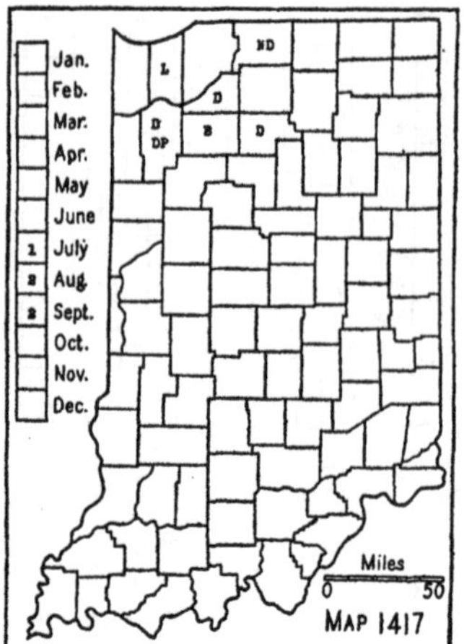

Hypericum canadense L.

quent. It is apparently absent about Lake Michigan and in the northern counties.

Ohio to Iowa, southw. to Ala. and Ark.

11. **Hypericum màjus** (Gray) Britt. Map 1413. In moist, sandy soil about lakes and swamps and in wet prairies. Rare.

E. Que. to Man., southw. to L. I., n. N. J., Pa., Ill., Iowa, and S. Dak.; also in e. Wash.

12. **Hypericum boreàle** (Britt.) Bickn. Map 1414. In marshes and wet sandy places about lakes.

Newf. to Ont., southw. to Vt., N. J., Pa., and Ind.

13. **Hypericum gentianoìdes** (L.) BSP. Map 1415. In the lake area this species is found in moist, sandy soil in low places in black and white oak woods and in prairies. In the southern part of the state it is found in minimacid soil on the crests of chestnut oak and black oak ridges and in low ground in the pin and post oak flats.

Maine, sw. Ont. to Ill., southw. to Fla. and La.

14. **Hypericum Drummóndii** (Grev. & Hook.) T. & G. Map 1416. Usually in hard, white, slightly acid, clay soil in low, flat, wheat stubble, hayfields, and fallow fields and in poor soil in exposed places on the crests of wooded ridges.

Va., Ill., and Iowa, southw. to Ga. and Tex.

15. **Hypericum canadénse** L. Map 1417. Sandy soil on the low borders of swamps and lakes. Very rare. Often confused with *Hypericum majus.*

Newf. to Man., southw. to Ga., Ky., and Wis.

16. **Hypericum mùtilum** L. Map 1418. In northern Indiana in moist, sandy soil about lakes and swamps, in low woods, cornfields, and wet prairies. In the southern part of the state it is more frequent in minimacid

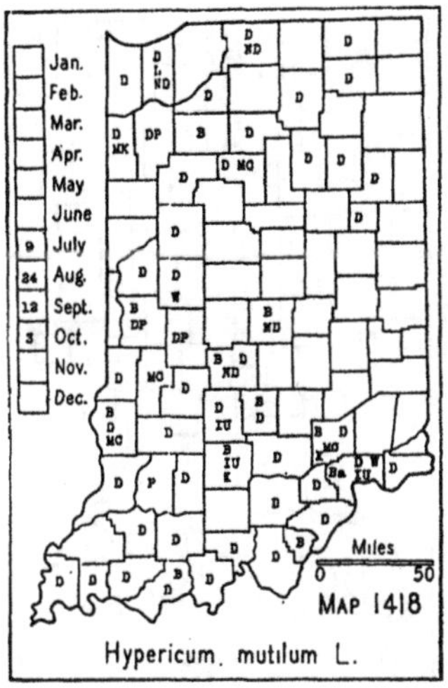

Hypericum mutilum L.

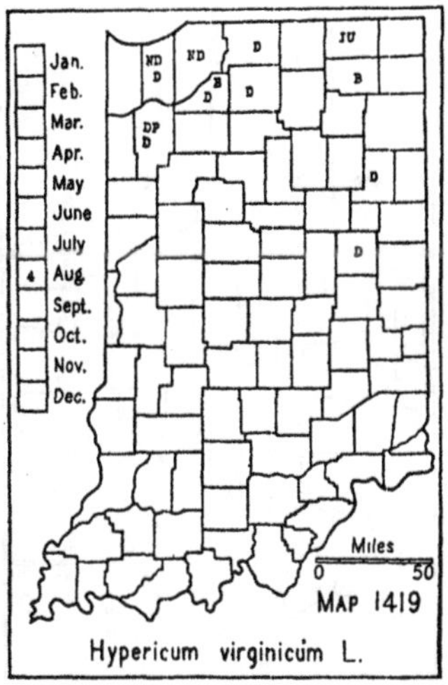

Hypericum virginicum L.

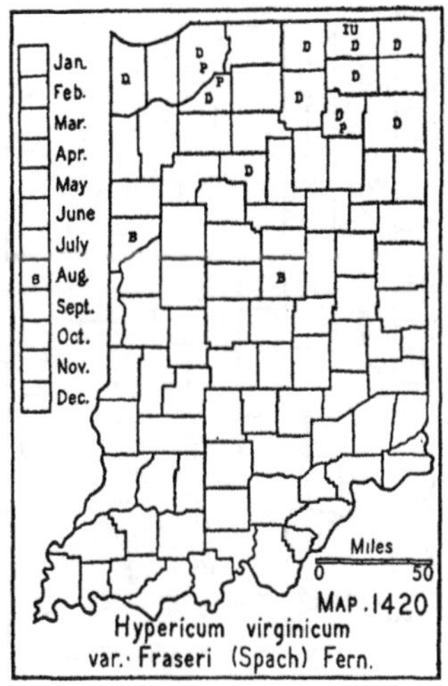

Hypericum virginicum var. Fraseri (Spach) Fern.

soil in cultivated or fallow fields, roadside ditches, pin oak and post oak flats, and more rarely on wooded slopes.

N. S. to Man., southw. to Fla. and Tex.

17. **Hypericum virgínicum** L. Map 1419. Frequent to infrequent in the interdunal flats about Lake Michigan and on the marshy borders of lakes and in swamps throughout the lake area.

N. S. to Ind., southw. to Fla.

17a. **Hypericum virginicum** var. **Fràseri** (Spach) Fern. (Rhodora 38: 434. 1936.) Map 1420. The variety has the habitat of the species and almost the same range in Indiana.

Newf. and Lab. to Man., southw. to Mass., Conn., Pa., Ind., Ill., Iowa and Nebr.

18. **Hypericum tubulòsum** Walt. (Rhodora 38: 435-436. 1936 and Jour. Arnold Arb. **19**: 279. 1938.) Map 1421. In swampy woods or on the borders of swamps. I have seen this species growing on an old cypress log in a cypress slough. Rare.

N. J. and Md. to Ind. and Mo., southw. to Fla. and La.

18a. **Hypericum tubulosum** var. **Wálteri** (Gmel.) Lott. (Jour. Arnold Arb. **19**: 279. 1938.) Map 1422. In low places in low, flat woods and in swamps. Rare. The lower surface of the leaves of my specimens are very glaucous and not glandular or only sparingly so near the margins. The axillary flowers in the species are mostly in 1's and 2's while those of the variety are mostly in 3's. The sepals of the species are about 3 mm long while those of the variety are about 5 mm long.

Ky. and Ind. to Mo., southw. to Fla. and Ala.

189. ELATINÀCEAE Lindl. WATERWORT FAMILY

5231. ELATÌNE L.

See excluded species no. 448, p. 1074.

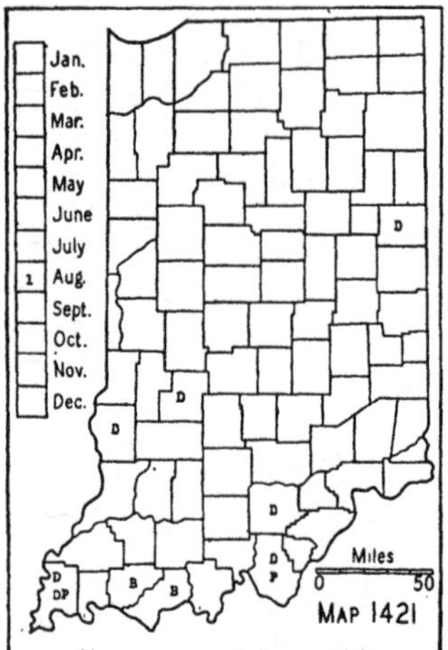

Hypericum tubulosum Walt.

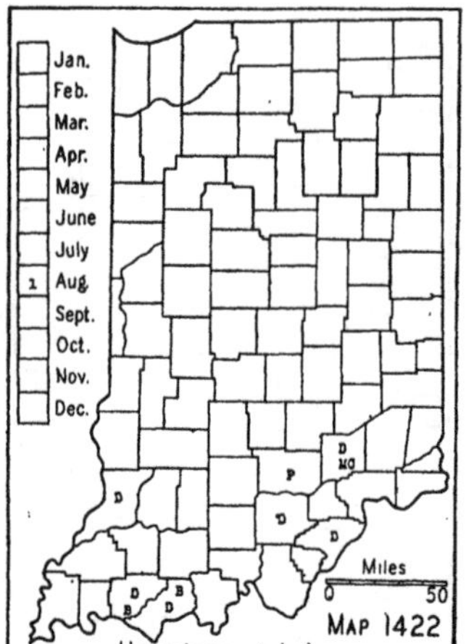

Hypericum tubulosum var. Walteri (Gmel.) Lott

Helianthemum canadense (L.) Michx

193. CISTÀCEAE Lindl. ROCKROSE FAMILY

Petals 5, yellow, fugacious, or lacking.
  Leaves lanceolate, oblong, or oblanceolate; styles short.......................... ....5245. HELIANTHEMUM, p. 678.
  Leaves subulate or scalelike; styles long..................5247. HUDSONIA, p. 679.
Petals 3, not yellow, persistent; styles none..................5248. LECHEA, p. 679.

5245. HELIÁNTHEMUM [Tourn.] Mill. ROCKROSE

Petaliferous flowers bright yellow, solitary or rarely 2, 2-4 cm wide, their capsules 6-9 mm in diameter, flowering usually in early June and soon much overtopped by the branches and becoming lateral, the outer 2 sepals much shorter than the inner, or wanting; capsules of the apetalous flowers of two sizes, the terminal ones 3-4 mm in diameter and the lateral ones much smaller; seed papillose..1. *H. canadense.*

Petaliferous flowers pale yellow, 5-12 in a short terminal cymose raceme, 1.5-2.5 cm wide, flowering the last of June or in July, the outer 2 sepals as long as or nearly as long as the inner ones, their capsules 3-5 mm in diameter, these flowers rarely overtopped by the later branches; capsules of the apetalous flowers not of two kinds, about 2 mm in diameter; seed reticulate.................2. *H. Bicknellii.*

1. **Helianthemum canadénse** (L.) Michx. (*Crocanthemum canadense* (L.) Britt.) Map 1423. Generally in open black and white or black oak woods and usually in very dry sandy soil or rarely in dry gravelly soil. Infrequent.

Maine to Wis., southw. to N. C. and Miss.

2. **Helianthemum Bicknéllii** Fern. (Rhodora 21: 36-37. 1919.) (*Helianthemum majus* BSP., *Helianthemum Walkerae* (Evans) Lyon, and *Crocanthemum majus* (L.) Britt.) Map 1424. In habitats and soils similar to those of the preceding species but much less frequent.

N. S. to Minn., southw. to S. C., Tex., and Colo.

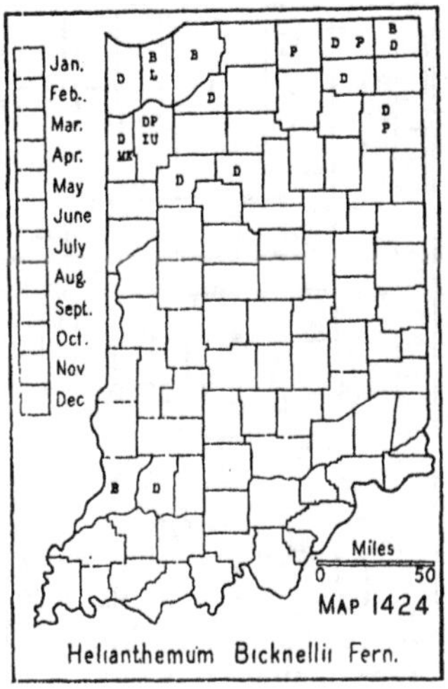

Helianthemum Bicknellii Fern.

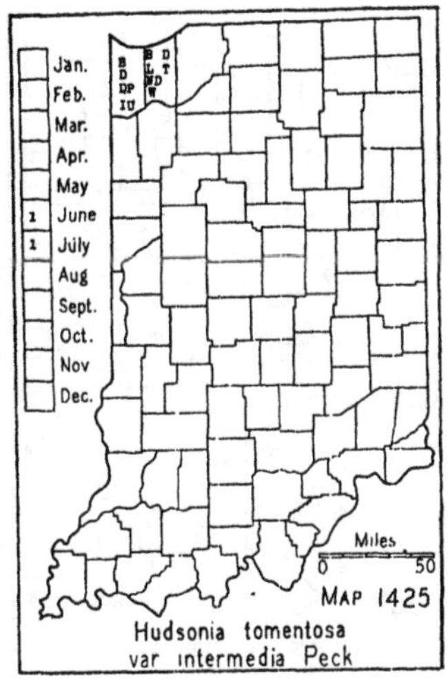

Hudsonia tomentosa var intermedia Peck

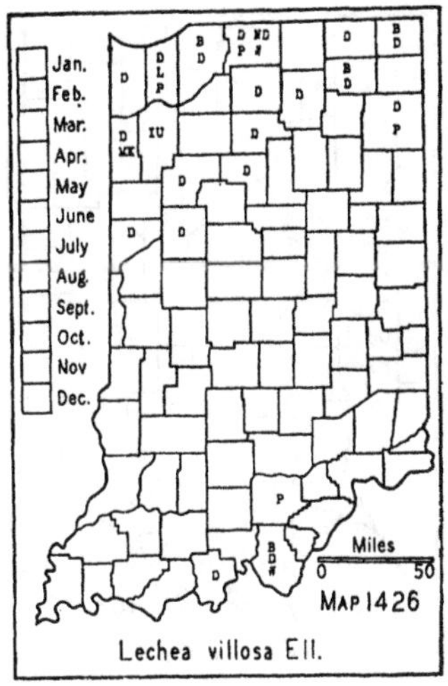

Lechea villosa Ell.

5247 HUDSÒNIA L.

1. **Hudsonia tomentòsa** Nutt. var. **intermèdia** Peck. WOOLLY HUDSONIA. Map 1425. In the open in almost pure sand and restricted to the dune area of Lake and Porter Counties.

N. B. to Man., southw. to N. C., the Great Lakes, and N. Dak.

5248. LÈCHEA [Kalm] L. PINWEED

[Hodgdon, A. R. A taxonomic study of Lechea. Rhodora **40**: 29-69, 87-131. 1938.]

All of my specimens were named by A. R. Hodgdon.

Pubescence of stems spreading..1. *L. villosa.*

Pubescence of stems appressed.

Calyx with the outer (narrow) sepals longer than the inner ones.

Blades of both cauline and basal leaves of an oval or elliptic type.....2. *L. minor.*

Blades of both cauline and basal leaves of a linear or subulate type...3. *L. tenuifolia.*

Calyx with the outer (narrow) sepals shorter than the inner ones.

Inner sepals 1-nerved; capsules longer than the sepals; basal leaves oval...4. *L. racemulosa.*

Inner sepals 3-nerved; capsules not conspicuously longer than the sepals.

Plants canescent.

Panicle strict, fastigiate, nerves of inner sepals faint on both surfaces; basal leaves linear-oblong, sparsely pubescent beneath on midrib and margin; pedicels mostly 2-3 mm long...........................5. *L. stricta.*

Panicle with primary branches spreading, nerves of inner sepals plainly visible on both surfaces; basal leaves oblong or ovate-oblong, entire under surface of blades pubescent; pedicels mostly 1-2 mm long. (See excluded species no. 450, p. 1074.)........................*L. maritima.*

Plants green or reddish green, strongly pubescent but not canescent.

Branches of panicle nearly erect or slightly spreading, usually up to 30 degrees and rarely spreading as widely as 45 degrees; capsules globose or depressed-globose, about 2 mm in diameter. (See excluded species no. 449, p. 1074.)...*L. intermedia.*

Branches of panicle usually spreading, generally to about 45 degrees; capsules short-elliptic to obovoid, about 1-1.5 mm in diameter...6. *L. Leggettii* var. *moniliformis.*

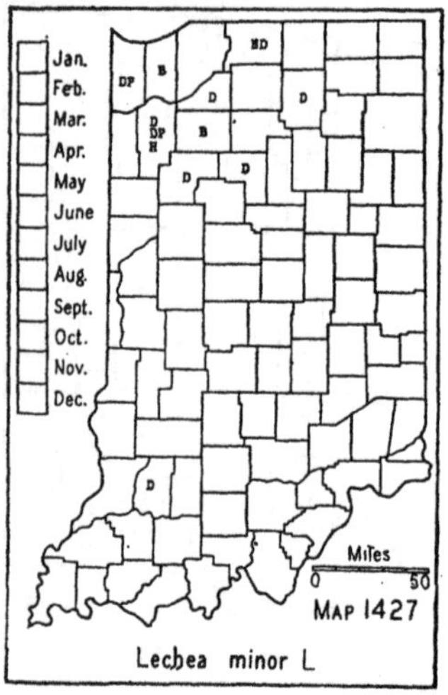

Lechea minor L.

Lechea racemulosa Lam.

Lechea tenuifolia Michx.

1. **Lechea villòsa** Ell. LARGE PINWEED. Map 1426. In dry or moist, sandy soil, generally on open, wooded slopes or crests of black oak and chestnut oak ridges and on low dunes or in interdunal flats. Infrequent in the lake area, rare in the knob area, and probably absent from most of the central counties of the state. It was no doubt present in some of the prairies of the central counties but these are now all under cultivation.

Vt. and Mass. to s. Ont. and Nebr., southw. to Fla., Tex., and n. Mex.

2. **Lechea minor** L. Map 1427. In very dry, sandy soil on wooded slopes or at their bases in moist, sandy soil. Rare.

N. H. and Vt. to Fla. and Miss.; also in Ont., Mich., and Ind.

3. **Lechea racemulòsa** Lam. Map 1428. In clay soil in black and white oak woods in the knobstone area and in dry, sandy soil in the lake area. (See Rhodora 40: 100. 1938.) Very local.

N. Y. to Ind., southw. to Fla. and Tenn.

4. **Lechea tenuifòlia** Michx. Map 1429. In poor soil on the crests and slopes of black and white and chestnut oak ridges in the southern part of the state and in dry or moist, sandy soil in pin oak woods and on the dunes in the northern part.

N. H. to Wis. and Nebr., southw. to Fla. and Tex.

5. **Lechea stricta** Leggett. Map 1430. Reported by Peattie from Lake County. A. R. Hodgdon, who monographed the genus, found specimens in the Field Museum from Lake and Starke Counties collected by E. J. Hill.

Western N. Y. to Ill. and Minn.

6. **Lechea Leggéttii** Britt. & Holl. var. **monilifórmis** (Bickn.) Hodgdon. (Rhodora **40**: 118-119. 1938.) (*Lechea Leggettii* Britt. & Holl. of Indiana authors). Map 1431. In moist sandy soil in depressions in black oak woods and in prairie habitats. Restricted to the sandy areas of the lake area.

Mass. to Ind. and Mich., southw. to N. C.

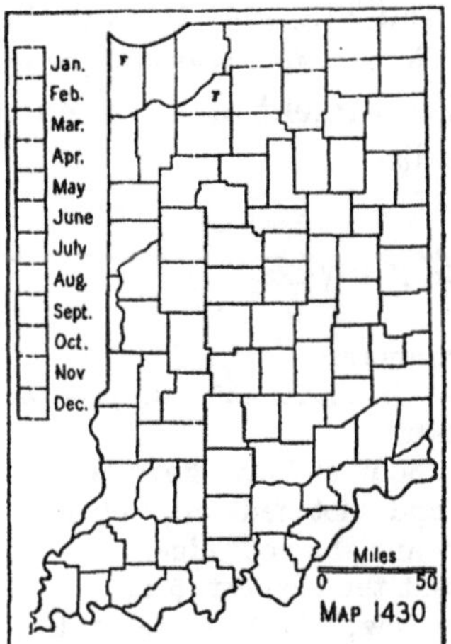

Lechea stricta Leggett

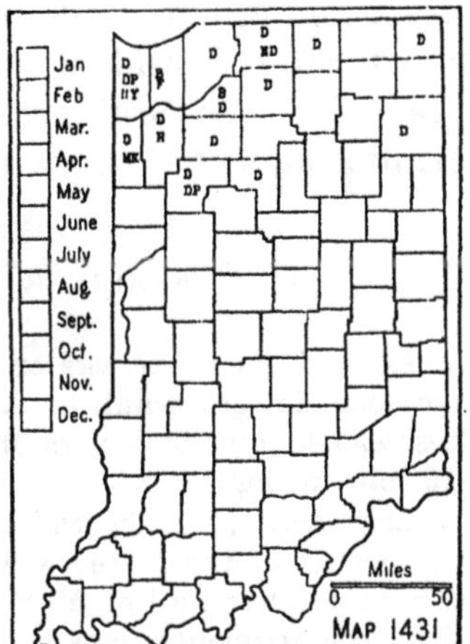

Lechea Leggettii Britton & Hollick var. moniliformis (Bicknell) Hodgdon

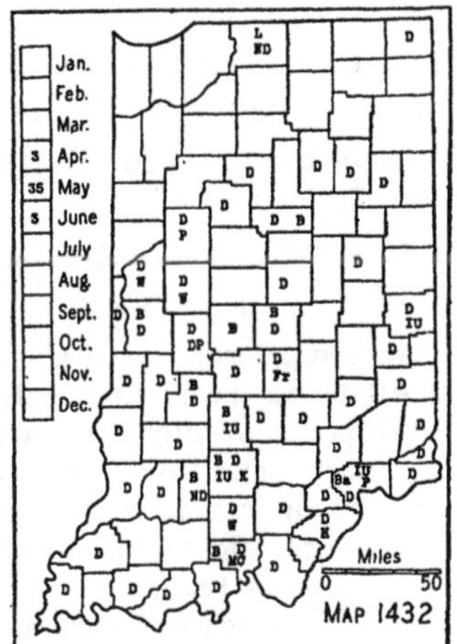

Hybanthus concolor (Forst.) Spreng.

198. VIOLÀCEAE DC. VIOLET FAMILY

Sepals not auricled; petals fleshy, green, lower one not spurred; stamens united into a sheath..5271. HYBANTHUS, p. 681.

Sepals auricled; petals not fleshy, colored or white, the lower one spurred (except in *V. eriocarpa*); stamens distinct..........................5274. VIOLA, p. 681.

5271. HYBÀNTHUS Jacq.

1. **Hybanthus cóncolor** (Forst.) Spreng. (*Cubelium concolor* (Forst.) Raf.) GREEN VIOLET. Map 1432. Generally associated with beech in beech and sugar maple or beech and oak woods. It prefers the rich soil of wooded slopes, especially their bases. It is usually not frequent or common where it is found. I saw it once in a rather open woods, however, where it formed almost a closed stand over nearly half an acre. This was in an open beech and white and black oak woods in Steuben County.

The stem of this species is normally pubescent all over or nearly so. Within the range of the species occurs a form with the "stems glabrous throughout, to slightly pubescent in narrow lines, or very sparsely hispidulous at the top." This form should be sought in Indiana and is known as f. *subglabratus* Eames. (Rhodora 32: 140. 1930.)

Conn. to s. Ont. and Mich., southw. to N. C. and Kans.

5274. VÌOLA [Tourn.] L. VIOLET

The violet group is one of the most admired groups of native plants and is at the same time one of the most difficult of determination. It is well known that some of the species freely hybridize, thus making positive identification of some forms difficult or impossible. Ezra Brainerd, who was our foremost student of violets and wrote the keys for *Viola* in our manuals, determined most of my violets until his death. Since my specimens have been determined by him, I have followed his keys and descriptions as closely as possible. Some recent authors believe that some of the species recognized by Brainerd are not tenable. The species most difficult to separate are our numbers 2 and 4, 8 and 10, 16 and 17a, and 18 and 19. Dr. E. L. Greene was also a profound student of violets and named some of my specimens. Since I have followed Brainerd's treatment of the genus, I have made Greene's determinations conform with it.

KEY TO SPECIES BASED ON PETALIFEROUS FLOWERS

Plants stemless; leaves and scapes all from rootstocks or runners.
All petals beardless; cleistogamous flowers wanting.................1. *V. pedata.*
All petals usually not beardless; cleistogamous flowers present.
Rootstocks stout, (2.5) 3-10 mm in diameter, short, without stolons; flowers blue or violet, sometimes nearly white, rarely white and blotched with blue or purple, or in one species a part of the flower white and the rest blue.
Leaves truncate or cuneate at the base, all or all except the earliest, 5-11 (-15)-lobed or -parted, all of the segments usually narrow; plants more or less pubescent.
Spurred petal glabrous; leaves 5-11-lobed or -parted; flowers on peduncles shorter than the leaves. (See excluded species no. 456, p. 1075.)..........
....*V. palmata.*
Spurred petal bearded; leaves 3-parted or -divided, each segment again usually 3-cleft or -parted and these generally further divided into 2-4 linear segments; flowers on peduncles generally longer than the leaves.
....2. *V. pedatifida.*
Leaves cordate at the base.
Blades or at least some of them, lobed or parted.
Leaves or some of them, 3-lobed or -parted, the segments large and usually more or less lobed or deeply toothed or the middle one entire; plants pubescent; spurred petal glabrous or bearded..........3. *V. triloba.*
Leaves 3-7-lobed or -parted, middle segment broad, acute, serrate, the basal ones sinuately serrate; plants glabrous; spurred petal glabrous. (See excluded species no. 460, p. 1075.).......................*V. viarum.*
Blades not lobed or cut, most of the vernal ones more than 2.5 cm wide.
Beard of lateral petals generally of strongly clavate hairs; spurred petal glabrous, shorter than the lateral petals; flowers on peduncles much exceeding the leaves, with a dark eye; cleistogamous flowers on long slender, erect peduncles; plants glabrous, of cold springy places and bogs...4. *V. cucullata.*
Beard of lateral petals of cylindrical or only slightly clavate hairs; spurred petal glabrous or bearded, equaling or longer than the lateral petals; flowers on peduncles shorter than or only slightly exceeding the leaves, with a white or yellow eye; cleistogamous flowers on prostrate, spreading peduncles; plants of wet or moist places.
Plants essentially glabrous, sometimes some of the petioles and some of the leaves more or less pubescent (the pubescence a sign of hybridization).
Spurred petal glabrous or nearly so.
Vernal leaves ovate-deltoid; flowers on peduncles generally as long as or longer than the leaves, pale violet to nearly white, with a darker band above the pale eye; sepals with a white margin; cleistogamous flowers on short, prostrate peduncles, their capsules dotted with brown; plants of wet places................
.....................................5. *V. missouriensis.*
Vernal leaves reniform to ovate; flowers on peduncles usually shorter than or as long as the leaves and in some early specimens slightly longer, deep purple, with a white or yellow eye; margins

of sepals not so white as those of the preceding species; cleistogamous flowers on short spreading peduncles, their capsules green or dark purple; plants usually of moist habitats........ ....6. *V. papilionacea.*

Spurred petal villous; vernal leaves ovate, blunt or attenuate at the apex; flowers on peduncles shorter than the leaves, sometimes as long as or longer than the leaves, violet, with a white eye; cleistogamous flowers on ascending peduncles, their capsules purplish; plants of moist or wet habitats......................7. *V. affinis.*

Plants more or less pubescent.

Leaves pubescent above and beneath, not purplish beneath, not appressed to the ground; petioles pubescent, at least on the upper part...8. *V. sororia.*

Leaves pubescent above, otherwise glabrous; leaves mostly appressed to the ground, purplish beneath.................9. *V. hirsutula.*

Leaves cordate or truncate at the base, at least some of them more or less dentate or cut at the base, the vernal ones generally less than 2.5 cm broad except in *V. viarum.*

Spurred petal glabrous; vernal leaves broadly deltoid, mostly more than 2.5 cm broad; plant glabrous. (See excluded species no. 460, p. 1075.)........ ....*V. viarum.*

Spurred petal bearded.

Leaves lanceolate, glabrous or nearly so or sometimes pubescent, the basal lobes generally prominently toothed or incised; blades usually shorter than their petioles...............................10. *V. sagittata.*

Leaves ovate-oblong, pubescent, the basal lobes entire or only slightly but sharply toothed; blades shorter than or as long as their petioles...... ....10a. *V. sagittata* var. *ovata.*

Rootstocks slender, rarely wanting, 2-4 mm in diameter near the summit, 1-1.5 mm farther back; flowers white with purplish veins.

Leaves lanceolate or linear-lanceolate, tapering at the base into the margined petioles; plants glabrous.............................11. *V. lanceolata.*

Leaves ovate to oblong, slightly cordate, rounded, or tapering at the base; plants glabrous or pubescent.......................12. *V. primulifolia.*

Leaves deeply cordate at the base.

Leaves glabrous above and beneath; upper petals broadly ovate; peduncles usually much longer than the leaves; beard of lateral petals absent or rudimentary...13. *V. pallens.*

Leaves more or less pubescent on one or both surfaces; peduncles usually shorter than the leaves or a few longer.

Leaves slightly fleshy, spreading, the base markedly heart-shaped with short lobes closely approximate, the margins scalloped evenly, surface not rugose, but with scattered hairs above; petioles and peduncles purplish; upper petals narrow; capsules usually roundish-ovate, purplish, mottled or blotched........................14. *V. blanda.*

Leaves not spreading, thin, with a scattered pubescence on both surfaces, and on petioles and peduncles. (See excluded species no. 454, p. 1075.) ....*V. incognita.*

Leaves not spreading, thin, with a scattered pubescence above, often only a few hairs on the lobes, glabrous beneath; the peduncles and petioles glabrous; lateral petals bearded, upper pair obovate; seed smooth, brown, 2 mm long; capsule elliptic......15. *V. incognita* var. *Forbesii.*

Plants with leafy stems.

Styles enlarged at the summit; spur short (2-4 mm long) or none.

Styles bearded at the summit; stipules entire or nearly so; flowers yellow or white within with a yellowish base and pale violet without.

Flowers yellow; stipules ovate to lanceolate, usually not scarious; capsules 9-14 mm long.
Plants branched at the base, nearly glabrous; root leaves 1-3; margins of stem leaves usually with 25-30 teeth; stipules narrowly ovate.
Capsules more or less woolly........................16. *V. eriocarpa.*
Capsules glabrous........................16a. *V. eriocarpa* f. *leiocarpa.*
Plants usually not branched at the base, densely pubescent; root leaves usually wanting; margins of stem leaves usually with 30-45 teeth; stipules broadly ovate.
Capsules woolly.............................. 17. *V. pubescens.*
Capsules glabrous........................17a. *V. pubescens* var. *Peckii.*
Flowers white within with a yellowish base, pale violet without; stipules sharply linear-lanceolate, scarious; capsules 4-6 mm long...18. *V. canadensis.*
Styles globose at the summit, hollow; stipules large and leaflike, laciniate at the base.
Upper leaves and middle lobe of stipules entire or nearly so..19. *V. Kitaibeliana* var. *Rafinesquii.*
Upper leaves and middle lobe of stipules plainly crenate.
Petals large, 2 or 3 times as long as the sepals; petals yellow on unfolding, at first the upper ones, then the lateral ones, and finally the spur becoming blue or purple violet as far as the yellow throat............20. *V. tricolor.*
Petals usually shorter than the sepals; petals roundish and usually entirely yellow...21. *V. arvensis.*
Styles not enlarged at the summit; spur long (4-12 mm long); stipules bristle-toothed; flowers violet to white.
Spurs 4-8 mm long; lateral petals bearded; style bent and bearded at the tip; auricles of sepals about 2 mm long.
Flowers white or cream colored; sepals more or less fimbriate (rarely entire); stipules 1.5-2.5 cm long, toothed throughout................22. *V. striata.*
Flowers violet or purple; sepals entire; stipules 1-3 cm long, toothed toward the base...23. *V. conspersa.*
Spurs 8-13 mm long; lateral petals beardless; styles straight and smooth; flowers purplish, spotted with a darker violet......................24. *V. rostrata.*

Key to the Species Based on Cleistogamous Flowers and Fruit

Plant stemless; leaves and scapes from rootstocks or runners.
Cleistogamous flowers wanting.................................1. *V. pedata.*
Cleistogamous flowers present.
Rootstock stout, (2.5) 3-10 mm in diameter, short; without stolons or runners.
Cleistogamous flowers on prostrate peduncles, their capsules mostly purplish, sometimes green; leaves cordate, the margins crenate, lobed, or cut.
Leaves truncate at the base, all, or all except the earliest, 5-11-lobed or -parted, the segments variously toothed or cleft, the middle one usually the widest; plants pubescent; seed brown, about 2 mm long. (See excluded species no. 456, p. 1075.) ....*V. palmata.*
Leaves cordate at the base, at least some of them 3-lobed or -parted, the segments large and usually more or less lobed or deeply toothed, or the middle one entire; plants pubescent; seed buff or brown, about 2 mm long..3. *V. triloba.*
Leaves cordate at the base, none cut.
Capsules 6-8 mm long; seed light brown, about 2 mm long; leaves appressed to the ground, purplish beneath, pubescent above, otherwise glabrous..9. *V. hirsutula.*
Capsules 10-16 mm long; seed light buff or dark brown, about 2 mm long; leaves not appressed to the ground, not purplish beneath, glabrous to pubescent.

Leaves ovate-deltoid, attenuate to apex; seed bright buff; plants glabrous. ....5. *V. missouriensis.*

Leaves broader, usually acute or abruptly pointed; seed dark brown.

Plant glabrous or some part more or less pubescent..6. *V. papilionacea.*

Plant pubescent....................................7. *V. sororia.*

Cleistogamous flowers on ascending peduncles, rarely nearly erect, their capsules 4-7 mm long, purplish; sepals much shorter than the capsules, their auricles 0.5-2 mm long; seed about 1.7 mm long, light buff; leaves cordate, uncut..7. *V. affinis.*

Cleistogamous flowers on erect peduncles, their capsules green or yellow.

Leaves ovate to reniform, cordate, glabrous, uniformly and inconspicuously crenate, acute; cleistogamous capsules oblong, 10-15 mm long, their sepals nearly as long as the capsules, often ciliolate at the apex, auricles 2-4 mm long; seed black, about 1.4 mm long............4. *V. cucullata.*

Leaves lobed (at least some of them) or the margins sharply incised or toothed toward the subcordate or truncate base.

Blades of mature leaves lanceolate to ovate-oblong, glabrous or finely pubescent.

Leaves lanceolate, glabrous or nearly so, the basal lobes generally prominently toothed or incised; blades usually shorter than their petioles. ....10. *V. sagittata.*

Leaves ovate-oblong, finely pubescent, the basal lobes entire or slightly but sharply toothed; blades usually as long as their petioles...... ....10a. *V. sagittata* var. *ovata.*

Blades of mature leaves 3-parted or -divided, each segment divided into linear segments; leaves of late summer not so deeply divided and the segments not so deeply cut; cleistogamous flowers yellowish, their peduncles commonly shorter than the petioles; seed about 2 mm long. ....2. *V. pedatifida.*

Rootstocks slender, rarely wanting, 2-4 mm in diameter at the summit, 1-1.5 mm in diameter farther back, often rather long and creeping.

Leaves lanceolate or linear-lanceolate, tapering at the base into the margined petioles; plants glabrous; seed dark brown...............11. *V. lanceolata.*

Leaves ovate to oblong, slightly cordate, rounded, or tapering at the base; plants glabrous or pubescent; cleistogamous capsules green; seed reddish brown, about 1.5 mm long...........................12. *V. primulifolia.*

Leaves deeply cordate at the base.

Blades of leaves glabrous above and beneath; cleistogamous capsules ellipsoid-cylindric; seed almost black, about 1 mm long............13. *V. pallens.*

Blades more or less pubescent on one or both surfaces.

Leaves slightly fleshy, spreading, with scattered hairs above; petioles and peduncles purplish; seed black, short-ovate, minutely rugose, acute at the base, 1.2-1.6 mm long..........................14. *V. blanda.*

Leaves not spreading, thin, with a scattered pubescence above and beneath and on the petioles and peduncles; seed long-elliptic, smooth, blunt at the base, 1.6-1.9 mm long. (See excluded species no. 454, p. 1075.)... .... *V. incognita.*

Leaves not spreading, thin, with a scattered pubescence above, often only a few hairs on the lobes, glabrous beneath; peduncles and petioles glabrous; seed as in the typical species...15. *V. incognita* var. *Forbesii.*

Plants with leafy stems.

Stipules not leaflike, either entire or nearly so or bristle-toothed.

Stipules scarious, entire or ciliolate; capsules generally puberulent, 4-6 mm long; seed about 2 mm long, brown..........................18. *V. canadensis.*

Stipules green, sometimes the margin slightly scarious, entire, more or less ciliolate or with a few crenate teeth but never bristle-toothed.

Plants branched at the base, nearly glabrous; root leaves 1-3; margins of the stem leaves usually with 25-30 teeth; stipules narrow-ovate.

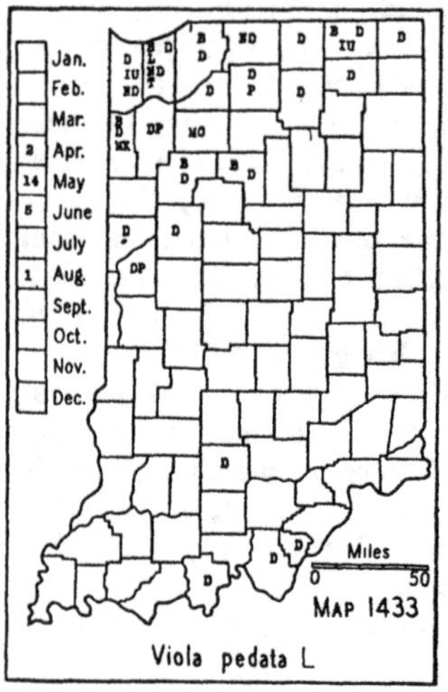

Viola pedata L.

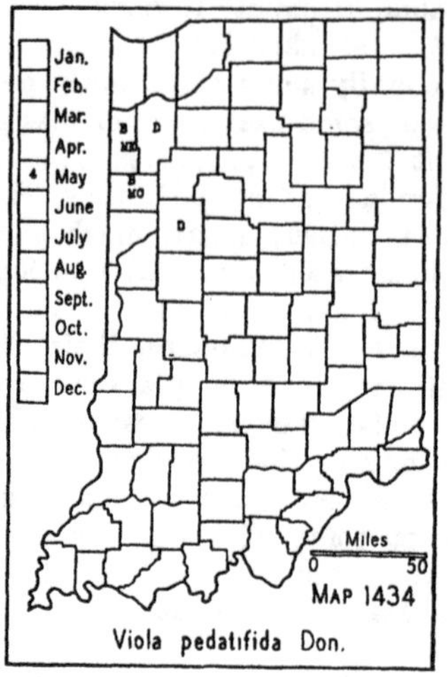

Viola pedatifida Don.

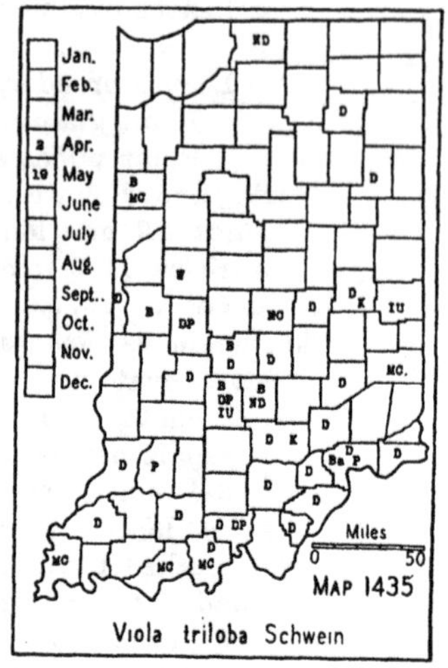

Viola triloba Schwein

Capsules more or less woolly ....16. *V. eriocarpa.*
Capsules glabrous............................16a. *V. eriocarpa* f. *leiocarpa.*

Plants usually not branched at the base, softly pubescent; root leaves generally absent; margins of stem leaves usually with 30-45 teeth; stipules broad-ovate.

Capsules woolly... 17. *V. pubescens.*
Capsules glabrous..........................17a. *V. pubescens* var. *Peckii.*

Stipules green, bristle-toothed; capsules glabrous.

Leaves with round-crenate teeth; stipules (1) 1.5-2.5 cm long, toothed throughout; auricles of sepals about 2 mm long; seed about 2 mm long...22. *V. striata.*

Leaves with flattened crenate teeth; stipules 0.5-1.3 cm long or rarely longer, toothed mostly toward the base; auricles of sepals about 1 mm long; seeds 1.5-1.8 mm long.

Capsules 4-5 mm long, light brown, splotched with a darker brown; seed bone color, splotched with brown, about 1.5 mm long....23. *V. conspersa.*

Capsules mostly 5.5-7 mm long; seed 2 mm or more long.....24. *V. rostrata.*

Stipules large, leaflike, laciniate at the base.

Upper leaves and middle lobes of stipules entire or nearly so; seed light brown, about 1 mm long......................19. *V. Kitaibeliana* var. *Rafinesquii.*

Upper leaves and middle lobes of stipules plainly crenate; introduced from Europe.

Leaves cordate at the base...............................20. *V. tricolor.*
Leaves cuneate at the base...............................21. *V. arvensis.*

1. **Viola pedàta** L. (*Viola pedata* var. *concolor* Holm.) BIRDFOOT VIOLET. Map 1433. Locally frequent in the lake area in very sandy or gravelly soil in the dunes and open woodland and along roadsides. Probably absent or very rare south of the lake area until the southern part of the state is reached where it has been found in a few counties in rather sandy soil on the crests of ridges. Here it is usually associated with chestnut oak, post oak, black oak, and Virginia pine.

Mass. to Minn., southw. to Fla. and La.

1a. **Viola pedata** var. **linearíloba** DC. A form with all of the leaf-segments linear. Our manuals tell us that this leaf-form is correlated with

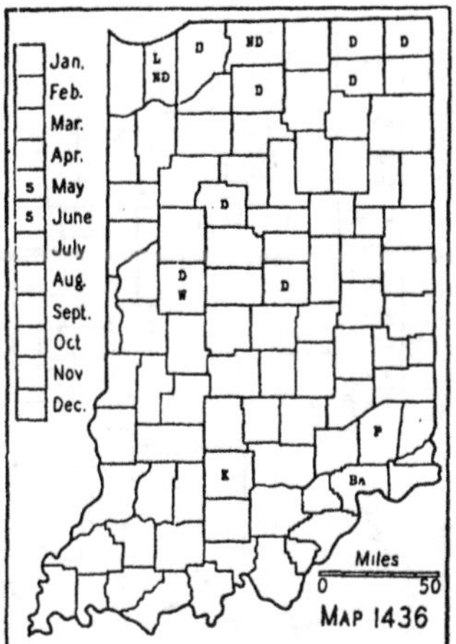

Viola cucullata Ait.

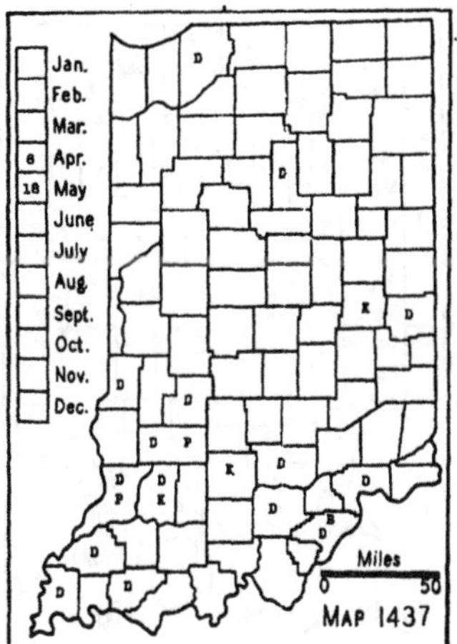

Viola missouriensis Greene

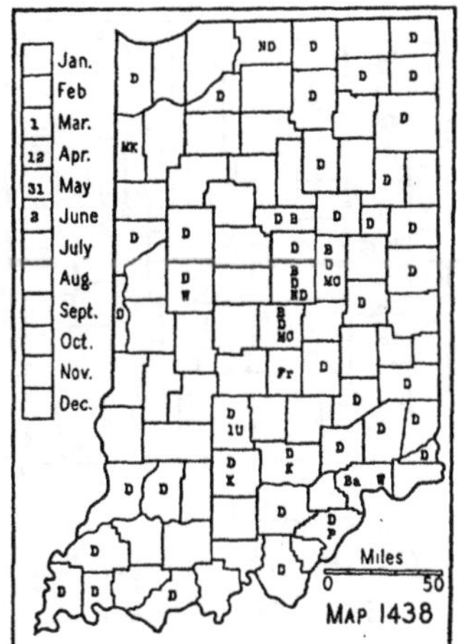

Viola papilionacea Pursh

flowers, having all of the petals of the same color. This form is rather rare in Indiana.

2. **Viola pedatífida** Don. PRAIRIE VIOLET. Map 1434. In dry prairies and open woodland. Very rare.

Prairies from Ohio to Sask., southwestw. to N. Mex. and Ariz.

2a. **Viola pedatifida** × **sorória** Brainerd. From Warren and White Counties.

3. **Viola tríloba** Schwein. THREE-LOBED VIOLET. Map 1435. Rather frequent in the southern part of the state, becoming infrequent or absent in the northern part. No doubt many of the reports for *Viola palmata* should be referred to this species. In rich, dry woods, usually found in beech and sugar maple, beech and oak, and black oak and white oak woodland.

Vt. and N. Y. to Ind., southw. along the mts. to Ga. and Ala.

3a. **Viola triloba** var. **dilatàta** (Ell.) Brainerd. A form in which the pedately cut leaves have more numerous and deeper incisions. I have it from Brown, Harrison, Lawrence, Monroe, and Sullivan Counties.

Mo., eastw. to Ind. and the coast and southw. to Fla. and La.

4. **Viola cucullàta** Ait. MARSH BLUE VIOLET. Map 1436. In tamarack bogs and marshy places in the lake area and in springy places and on wet, rocky ledges in southern Indiana. Rather rare. Most of the many reports for this species should be referred to other species.

In cold bogs and springs from Que. and Ont., southw. to Ga.

4a. × **Viola festàta** House. (*Viola cucullata* × *sagittata* Brainerd.) I have this hybrid from Lagrange County.

5. **Viola missouriénsis** Greene. Map 1437. In wet, hard, white, and slightly acid clay soil, either in woodland with sweet gum or in the open on sweet gum land; less frequent in wet woodland and springy places. In-

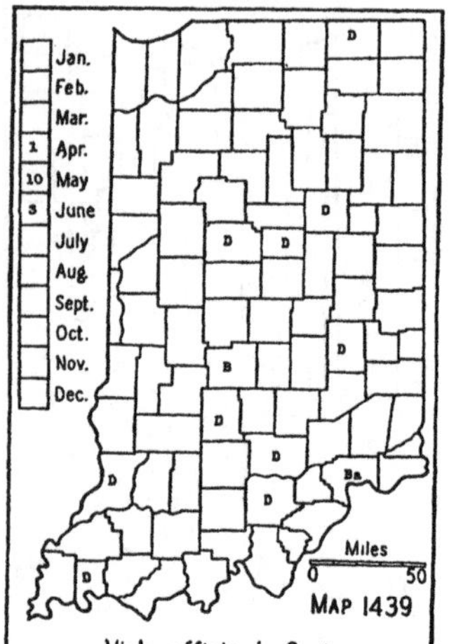

MAP 1439
Viola affinis LeConte

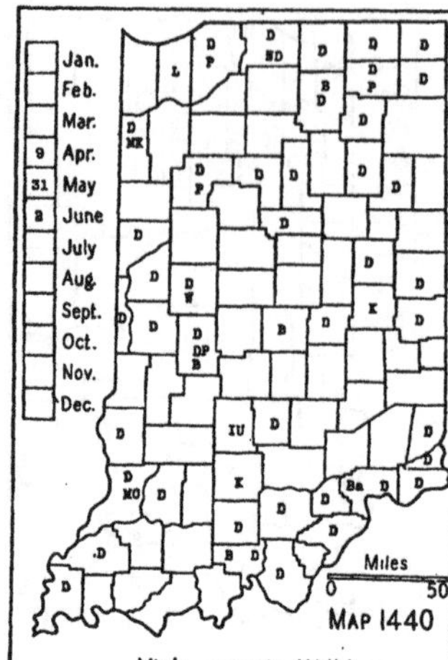

MAP 1440
Viola sororia Willd.

MAP 1441
Viola hirsutula Brainerd

frequent but often locally abundant in the southern part of the state and rare in the northern part.

Mississippi Valley from Ind. to se. Kans., southw. to La. and Tex.

5a. **Viola missouriensis** × **sororia** Brainerd. I have this hybrid from Knox, Sullivan, Tippecanoe, and Vanderburgh Counties.

5b. **Viola missouriensis** × **triloba.** I have this hybrid from Daviess, Greene, Knox, and Lawrence Counties. Brainerd named my specimens but did not describe the hybrid.

6. **Viola papilionàcea** Pursh. BUTTERFLY VIOLET. Map 1438. One of our most common blue violets. Frequent in all parts of the state except the area about Lake Michigan for which there are no records. In moist woodland and clearings and along roadsides. This species is probably best considered as a glabrate form of *Viola sororia.*

Maine to Minn., southw. to Ga. and Okla.

6a. × **Viola nàpae** House. (*Viola papilionacea* × *sororia* Brainerd.) I have this hybrid from Tippecanoe and Vermillion Counties.

6b. × **Viola variábilis** Greene. (*Viola papilionacea* × *triloba* Brainerd.) I have this hybrid from Clay, Grant, and Randolph Counties.

6c. **Viola papilionacea** × **pedatifida** Brainerd. I have this hybrid from Jasper County.

7. **Viola affìnis** LeConte. Map 1439. In moist, alluvial soil, usually in woodland but also in the open and in prairies. Infrequent.

N. E. to Wis., southw. to Ga. and Ala.

7a. **Viola affinis** × **triloba** Brainerd. I have this hybrid from Clark County.

8. **Viola soròria** Willd. DOWNY BLUE VIOLET. Map 1440. This is our most common blue violet. Throughout the state in rich, moist or dry woodland and in the open in various habitats.

Que. and N. E. to Minn., southw. to N. C. and Okla.

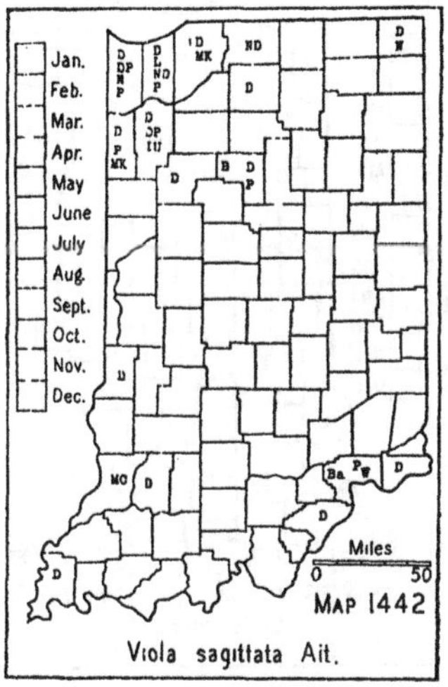

Viola sagittata Ait.

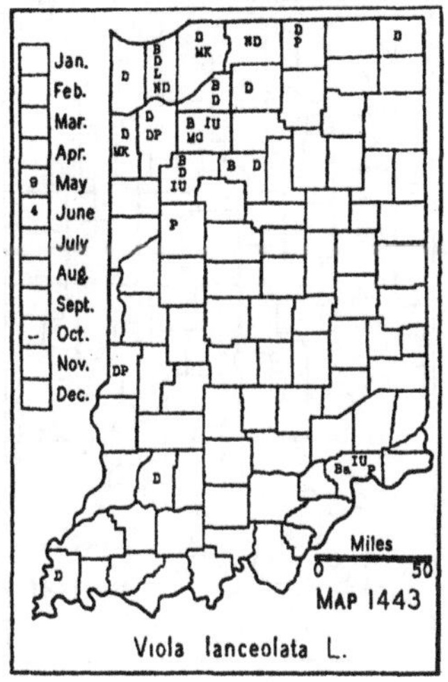

Viola lanceolata L.

Viola primulifolia L.

9. **Viola hirsùtula** Brainerd. Map 1441. My only specimen was collected on a black oak and Virginia pine slope about 2 miles northwest of Bennettsville in Clark County. On this same slope I collected the following hybrids, all of which were named by Dr. Brainerd.

9a. × **Viola cordifòlia** (Nutt.) Schwein. (*Viola hirsutula* × *papilionacea* Brainerd.)

9b. × **Viola díssita** House. (*Viola hirsutula* × *triloba* Brainerd.)

9c. **Viola hirsutula** × **missouriensis** (never described).

10. **Viola sagittàta** Ait. ARROWLEAF VIOLET. Map 1442. In northern Indiana this species is generally found in black, moist, sandy soil in the open or in open woods. Rather local. Probably absent in many of the central counties, appearing again in southeastern Indiana in slightly acid soil in the sweet gum flats; in the knobstone area on or near the crests of Virginia pine and chestnut oak ridges; and in southwestern Indiana in the post oak flats. Rare in southern Indiana.

Mass. to Minn., southw. to Ga. and La.

10a. **Viola sagittata** var. **ovàta** (Nutt.) T. & G. (*Viola fimbriatula* Smith of Gray, Man., ed. 7 and Britton and Brown, Illus. Flora, ed. 2.) This variety is a form with more ovate and shorter leaf blades and is more or less densely pubescent. It insensibly grades into the typical form.

In Indiana mostly near Lake Michigan and in Posey County.

10b. **Viola sagittata** × **sororia** Brainerd. Lake County.

11. **Viola lanceolàta** L. LANCELEAF VIOLET. Map 1443. Rather local but usually frequent to abundant where it is found in the lake area. Usually in a sandy black loam soil in the open in marshes, on the borders of swamps, and in bogs. Probably absent in most of the counties immediately south of the lake area but common on the slightly acid soil of the flats of southern Indiana. Here it is locally abundant in old fallow, wet

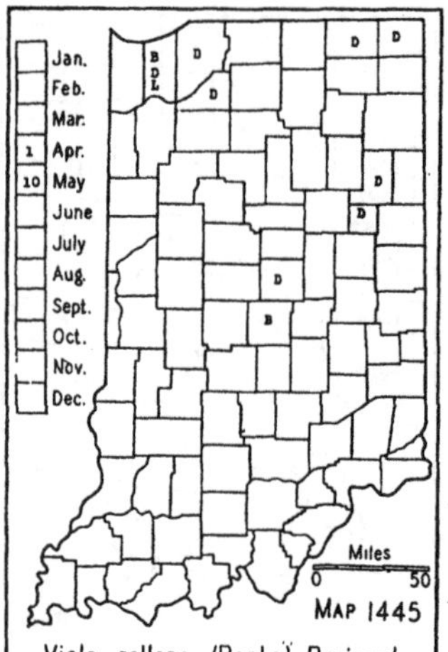

Viola pallens (Banks) Brainerd

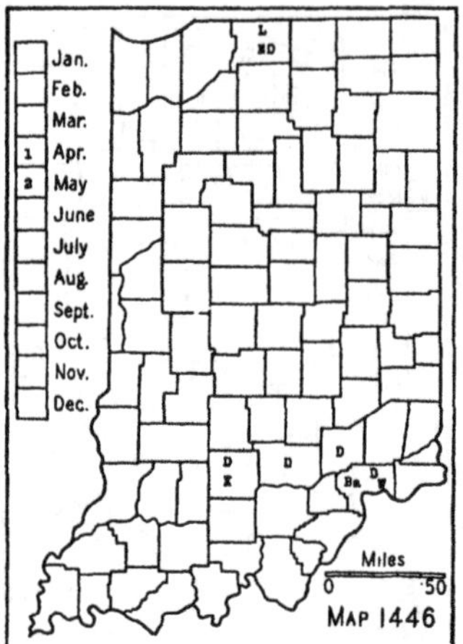

Viola blanda Willd

Viola incognita var. Forbesii Brainerd

fields and in low, open, sweet gum, red maple, and beech woods. Also in the southwestern part of the state in pin oak and swamp white oak flats where it is very rare.

N. S. to Minn., southw. to the Piedmont Plateau.

12. **Viola primulifòlia** L. PRIMROSELEAF VIOLET. Map 1444. In moist, black sandy soil on the margins of swamps. Very rare.

N. B. to Fla. and Tex., also in Ind.

13. **Viola pállens** (Banks) Brainerd. SMOOTH WHITE VIOLET. Map 1445. Local in the lake area in bogs and springy or wet places, usually associated with tamarack or aspen. It is often associated with *Viola incognita* var. *Forbesii* which it very much resembles.

Lab. to Alberta, southw. to S. C., Tenn., and in the mts. to Colo.

14. **Viola blánda** Willd. REDSTEM WHITE VIOLET. Map 1446. In slightly acid soil in sweet gum, red maple, and beech woodland; more rarely in dry ground with beech and oaks. Nieuwland found it growing in moss in a tamarack bog in St. Joseph County.

This species has been reported frequently from Indiana but no doubt many of the reports should be referred to other species. In my exchanges and in the herbaria I have examined I find many specimens labeled this species that should be referred to *Viola incognita* and its variety.

W. Que. and w. N. E. to Minn., southw. to the mts. of Ga.

15. **Viola incógnita** Brainerd var. **Fórbesii** Brainerd. HAIRY WHITE VIOLET. Map 1447. Generally in mucky soil on the shady borders of lakes and in bogs and marshes. Very local but in colonies. For the typical species, see list of excluded species.

Que. to Wis., southw. to Mass. and Tenn.

16. **Viola eriocárpa** Schwein. (*Viola scabriuscula* Schwein.) STEMMED YELLOW VIOLET. Map 1448. Rather frequent in moist, rich woods through-

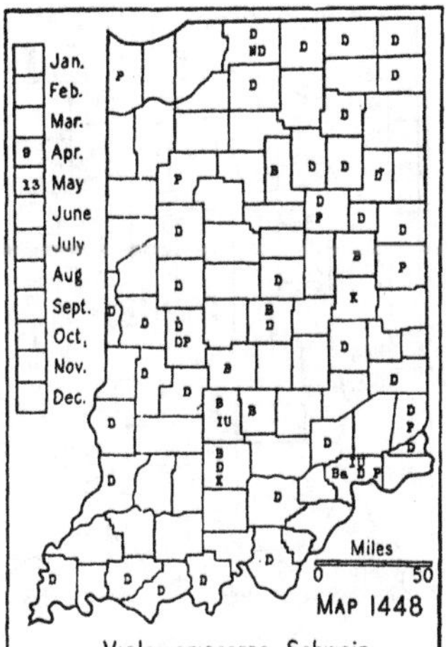

Viola eriocarpa Schwein.

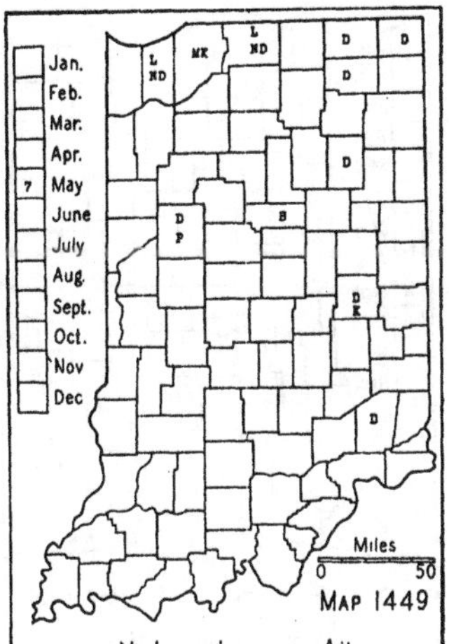

Viola pubescens Ait.

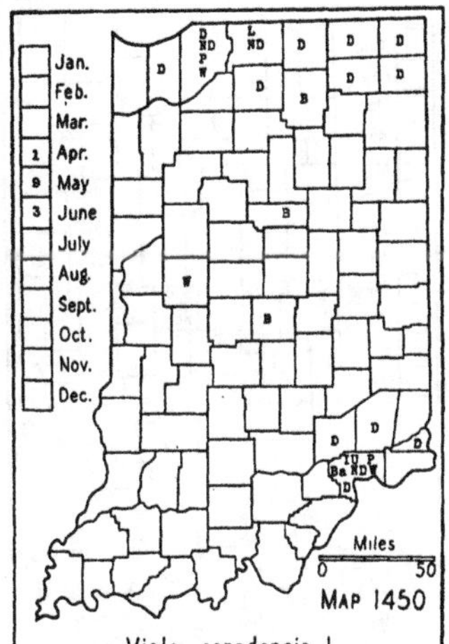

Viola canadensis L.

out the state. It is more frequent and abundant in beech and sugar maple and white oak woods.

Most of our specimens are more pubescent than the typical form, in fact many so closely approach *Viola pubescens* in pubescence that it seems wrong to place them with this species.

Conn., s. Ont. to Minn., southw. to Md. and Okla.

16a. **Viola eriocarpa** Schwein. forma **leiocárpa** (Fern. & Wieg.) Deam, comb. nov. (*Viola eriocarpa* var. *leiocarpa* Fern. & Wieg. in Rhodora 23: 275. 1921.) This is a form with glabrous capsules. In my Indiana fruiting specimens I have 19 sheets with woolly capsules and 28 sheets with glabrous capsules. The forms have no specific geographical range in Indiana.

17. **Viola pubéscens** Ait. STEMMED DOWNY YELLOW VIOLET. Map 1449. In rich, moist woods. Rare.

N. S. to N. Dak., southw. especially in the mts. to Va. and Mo.

The separation of this species from the preceding is not at all satisfactory. The characters used in their separation are not constant and it appears from my specimens that all characters fail about equally, so that a preponderant character is absent. If it is true that this species has no long root leaves and never branches at the base, then I have only 3 specimens of it from Indiana. But we have specimens that are much branched that are as pubescent as any we have. It is likewise with other characters, such as width of the stipules and the number of the teeth of the leaf margin.

17a. **Viola pubescens** var. **Péckii** House. (N. Y. State Mus. Bull. 243-244: 50. 1923.) The form with glabrous capsules. I have it from Steuben County.

18. **Viola canadénsis** L. CANADA VIOLET. Map 1450. Almost invariably found in beech and sugar maple woods and rarely in white ash and red

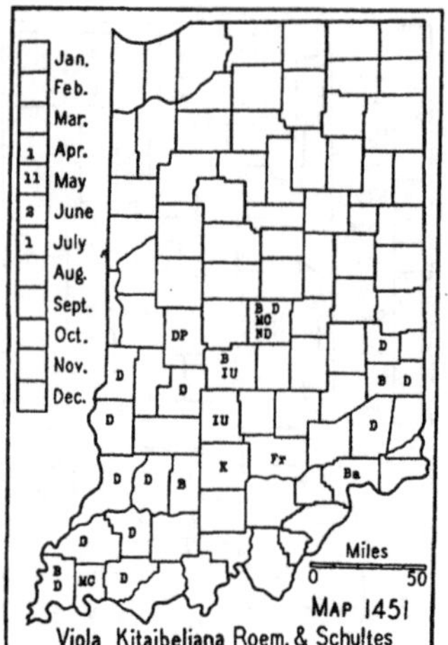

Viola Kitaibeliana Roem. & Schultes var. Rafinesquii (Greene) Fern.

Viola arvensis Murr.

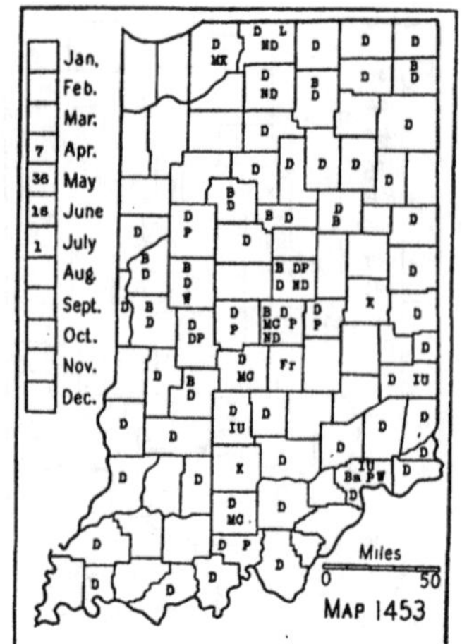

Viola striata Ait.

oak or in black oak and white oak woods. Usually in large colonies and rather frequent in the northern counties, becoming rare or absent until the southern counties are reached. Here it is found in similar habitats and is as abundant as in the northern part. This species does well in cultivation and flowers from May until freezing weather.

N. B. to Sask., southw. to S. C., Ala., Nebr., and in the Rocky Mts. to Ariz. and N. Mex.

19. VIOLA KITAIBELIÀNA Roem. & Schultes var. RAFINÉSQUII (Greene) Fern. (Rhodora 40: 443-446. 1938.) (*Viola Rafinesquii* Greene.) FIELD PANSY. Map 1451. Infrequent but fast becoming more widely scattered. Most abundant in sandy soil or in sandy clay soil in woodland, fallow, and cultivated fields and along roadsides. There are no reports for it north of Tippecanoe County, although it occurs in Michigan.

Nat. of Eu.; Conn. to Nebr., southw. to the Gulf States.

20. VIOLA TRÍCOLOR L. GARDEN PANSY. This species has been reported from the following counties: Clark (Baird & Taylor); Jefferson (Barnes, Coulter); Knox (Spillman); Shelby (Ballard); St. Joseph (Nieuwland); and for the Lower Wabash Valley (Schneck). Nieuwland says it maintains itself in St. Joseph County.

Nat. of Eu.

21. VIOLA ARVÉNSIS Murr. FIELD PANSY. Map 1452. This species was reported by Nieuwland as maintaining itself in St. Joseph County. In 1916 I found about a ten-acre field of it in Lagrange County. At first I thought the owner was growing this species for its seed but I learned that he had sown clover seed and that this species was so abundant that it had practically crowded out the clover. This place was revisited in 1930 and I found that it had disappeared.

Nat. of Eu.

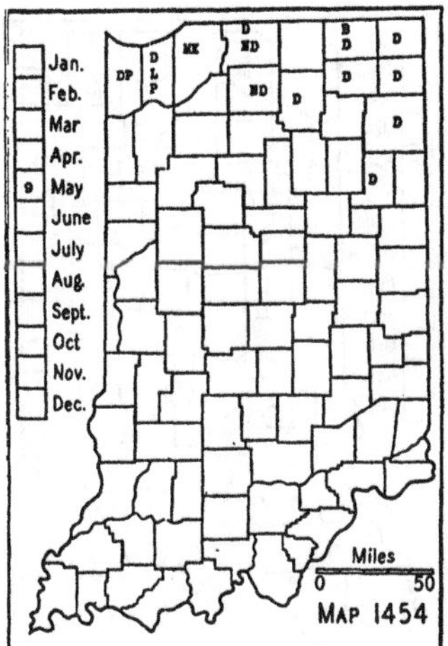

Viola conspersa Reich.

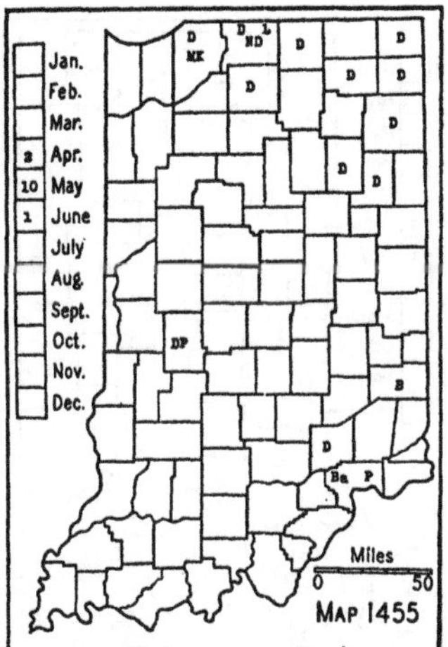

Viola rostrata Pursh

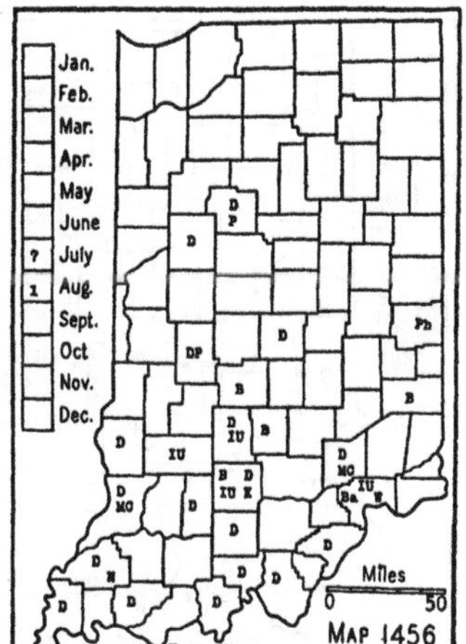

Passiflora lutea var. glabrifolia Fern.

22. **Viola striàta** Ait. CREAM VIOLET. Map 1453. Our most common stemmed violet. It grows both in the open and in the shade, preferring the moist alluvial soil of woodland along streams and elsewhere.

N. Y. to Minn., southw. to Ga. and Mo.

23. **Viola conspérsa** Reichenb. DOG VIOLET. Map 1454. Rather frequent in the northeastern part of the state in wet or moist woodland. Rare or absent in the southern part of the state.

E. Que. to Minn., southw. to Ga.

24. **Viola rostràta** Pursh. LONGSPUR VIOLET. Map 1455. Usually in moist rich soil at the bases of slopes in thick woodland. Rather frequent in the northeastern part of the state, becoming rare or absent in many of the southern counties.

W. Que. to Minn., southw. to Ga.

203. PASSIFLORÀCEAE Dumort. PASSIONFLOWER FAMILY

5372. PASSIFLÒRA L.

[Killip. The American species of Passifloraceae. Field Museum Nat. Hist. Bot. Ser. 19: 1-613. 1938.]

Leaves obtusely 3-lobed above the middle, the lobes entire; flowers yellowish; fruit about 1.2 cm in diameter............................1. *P. lutea* var. *glabriflora*.

Leaves deeply 3-5-lobed, the lobes serrate; petals whitish; corona purplish; fruit much inflated, about 3.5 cm in diameter............................2. *P. incarnata*.

1. **Passiflora lùtea** L. var. **glabriflòra** Fern. (Rhodora 41: 436. 1939.) YELLOW PASSIONFLOWER. Map 1456. Moist or dry woods, usually on wooded slopes bordering streams. Infrequent to rare as far north as Carroll County.

Pa. to Ill., southw. to Fla. and Tex.

2. **Passiflora incarnàta** L. MAYPOP. Map 1457. Very rare. It is locally common, however, on the rocky open slope of the Ohio River about midway between Cannelton and Tell City. I saw it in Crawford County near

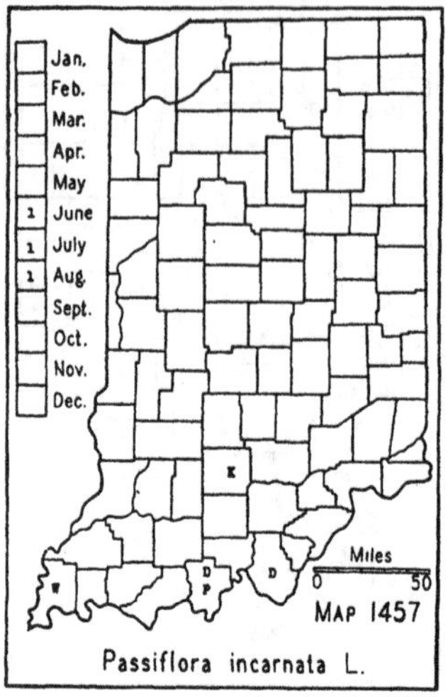

Passiflora incarnata L.

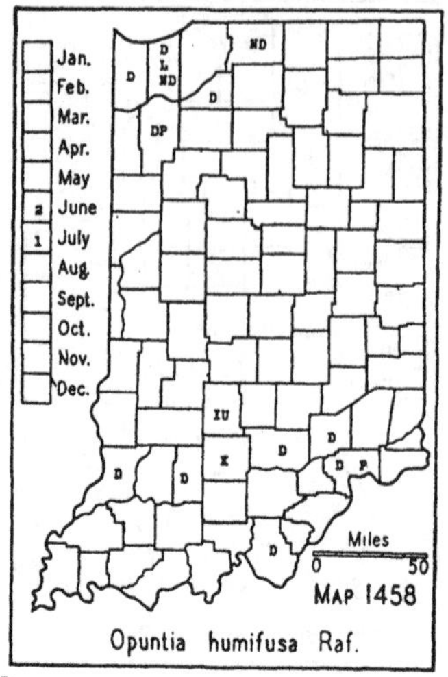

Opuntia humifusa Raf.

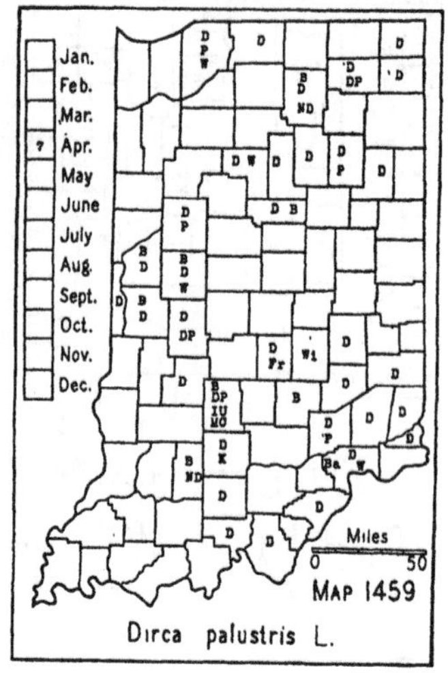

Dirca palustris L.

Wyandotte Cave but I was not prepared to preserve a specimen. I have it from two places along the Ohio River above Cannelton in Perry County and from one place along the Ohio River about 3 miles above Mauckport in Harrison County. Charles M. Ek found a large colony in hard, clay soil along a railroad embankment a quarter of a mile north of Galveston, Cass County.

Pa. to Mo., southw. to Fla. and Tex.

210. CACTÀCEAE Lindl. CACTUS FAMILY

5417. OPÚNTIA [Tourn.] Mill.

1. **Opuntia humifùsa** Raf. PRICKLY PEAR. Map 1458. Usually in very sandy soil but in Harrison, Jefferson, and Jennings Counties it is found in a friable clay soil. It forms large colonies and becomes an obnoxious weed, especially on the bluff of the Ohio River north of Madison.

Mass., s. Ont., n. Ill. to Mo., southw. to Va. and Tenn., and in the mts. to Ga. and Ala.

The Indiana species of *Opuntia* are not well understood. The most recent revision of the genus would refer all of our reports of *Opuntia vulgaris* Mill. to this species (under the synonymous name, *Opuntia Opuntia* (L.) Karst). (See Britton and Rose. The Cactaceae 1: 127-129. 1919.)

214. THYMELAEÀCEAE Reich. MEZERUM FAMILY

5448. DÍRCA L.

1. **Dirca palústris** L. LEATHERWOOD. Map 1459. Infrequent throughout the state except in the northwestern and southwestern parts from which there are no records. In the northern part of the state it is usually found in rich soil, in beech and sugar maple woods, generally carpeted with a deep leaf mold, more rarely in wet woods, and in a tamarack bog in Steuben County. In the southern part, it usually occurs on the lower part of wooded slopes along streams. An exceptional habitat is its occur-

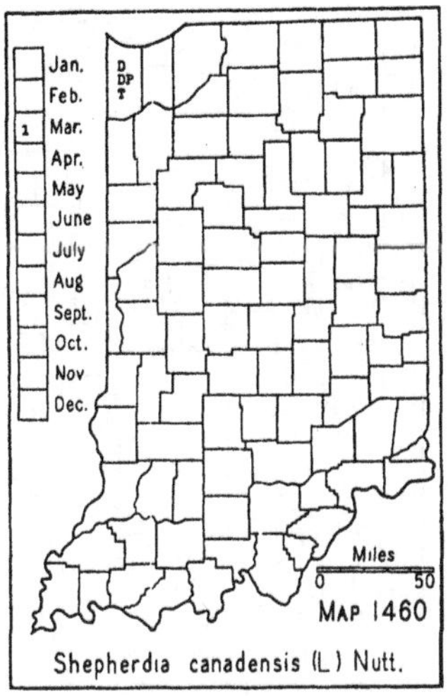

MAP 1460
Shepherdia canadensis (L.) Nutt.

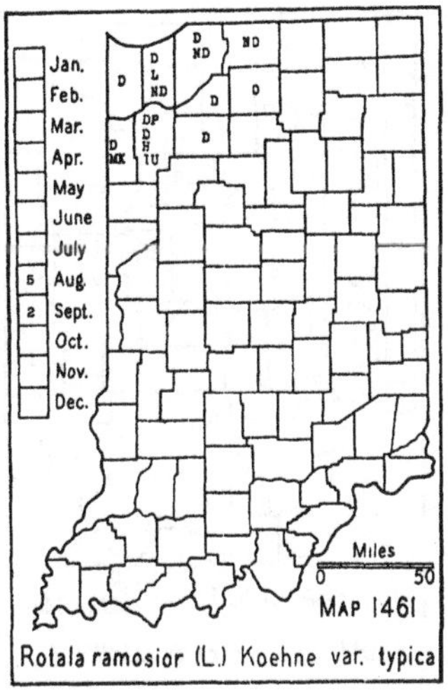

MAP 1461
Rotala ramosior (L.) Koehne var. typica

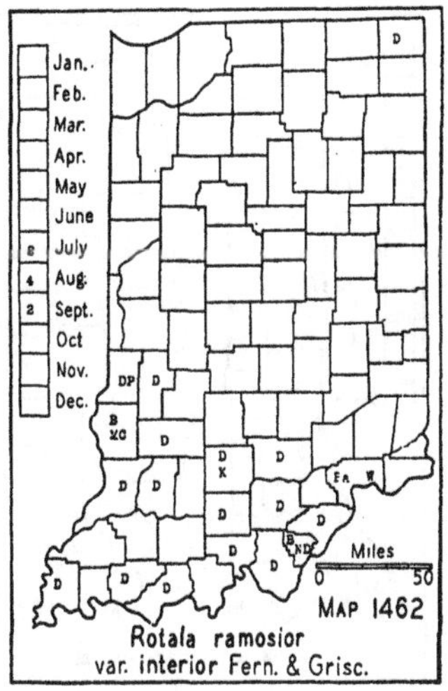

MAP 1462
Rotala ramosior var. interior Fern. & Grisc.

rence under hemlock trees on a low sandstone cliff along the Muscatatuck River between Vernon and North Vernon, Jennings County, where it was growing with its roots in the crevices of the sandstone cliff. It is most frequent in Parke County where a creek bears its name.

N. B. to Ont. and Minn., southw. to Fla., Tenn., and Mo.

215. ELAEAGNÀCEAE Lindl. OLEASTER FAMILY

5471. SHEPHÉRDIA Nutt.

1. **Shepherdia canadénsis** (L.) Nutt. (*Lepargyrea canadensis* (L.) Greene.) RUSSET BUFFALOBERRY. Map 1460. Near the bases of low dunes near Pine in Lake County where it is infrequent. In 1906 I found it about 2 miles east of Indiana Harbor. City development is fast encroaching upon its native area and it will soon become extinct in Indiana. It is the first shrub of Indiana to bloom and it is soon followed by leatherwood and certain species of willow.

Newf. to Alaska and B. C., southw. to N. S., Maine, Vt., n. and w. N. Y., Ind., and in the Rocky Mts. to N. Mex.

216. LYTHRÀCEAE Lindl. LOOSESTRIFE FAMILY

Flowers regular; petals equal; plants not glandular-pubescent.
 Flowers small, axillary, solitary or few; low or erect herbs.
  Calyx tube short, campanulate or hemispheric, not striate, about 3 mm or less long in flower; petals 0-4; plants commonly less than 5.5 dm high.
   Plants of wet habitats, not collapsing when uprooted; petals 4; calyx tube with appendages in the sinuses.
    Flowers solitary and sessile in the axils of the leaves; capsules 4-celled, septicidal..5473. ROTALA, p. 696.
    Flowers in 1 or 2, three-flowered clusters, the clusters sessile or nearly so (flowers sometimes solitary on the upper part of the stem or on the branches); capsules 2-4-celled, bursting irregularly..5474. AMMANNIA, p. 697.

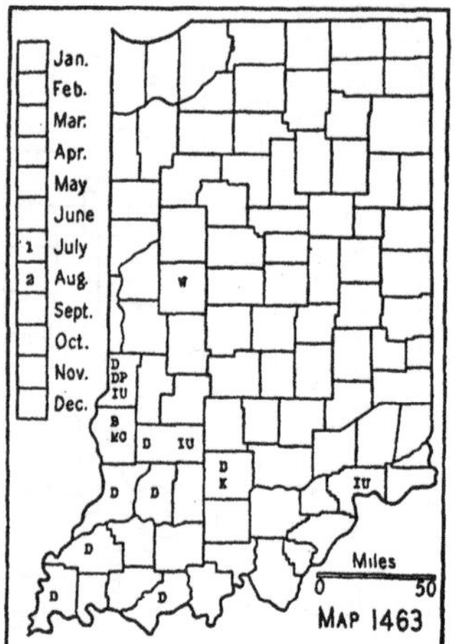

Ammania coccinea Rottb.

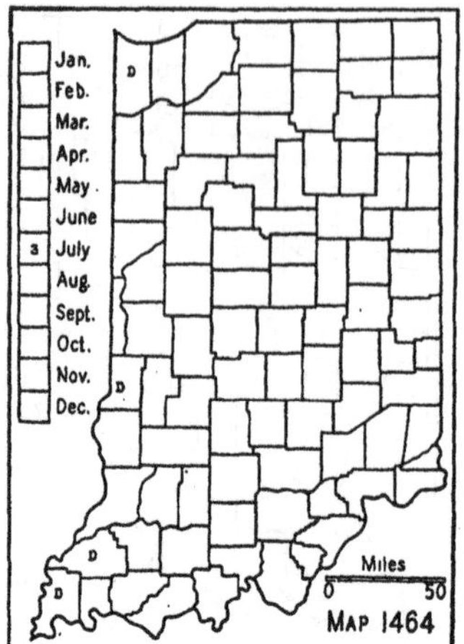

Didiplis diandra (Nutt.) Wood

Lythrum alatum Pursh.

Plants aquatic, collapsing when taken from the water; petals none; calyx tube without appendages; capsules indehiscent..............5475. DIDIPLIS, p. 697.

Calyx tube cylindric, striate, generally 4-5 mm long in flower; petals 5 or more; plants commonly more than 6 dm high...............5476. LYTHRUM, p. 697.

Flowers large, in axillary cymes; long, curving, semi-shrubby plants..5488. DECODON, p. 698.

Flowers irregular and unsymmetrical; petals unequal; plants glandular-pubescent. ....5478. CUPHEA, p. 698.

5473. ROTÀLA L.

[Fernald & Griscom. The variations of Rotala ramosior. Rhodora 37: 168-169. 1 pl. 1935.]

Plants low, simple or branched, rarely more than 3 dm high; leaves usually ascending, the larger ones 1.5-4 (5) mm wide; capsules 2-3.3 mm wide, 2-4 mm long; bractlets subulate, 0.5-1.4 mm long......................1. *R. ramosior* var. *typica*.

Plants robust, up to 4.5 dm high, simple or branched; lower leaves usually widely spreading or reflexed, the larger ones 5-10 mm wide; capsules (3.2) 3.8-4.4 mm wide, 3.5-5 mm long; bractlets linear-lanceolate, 1.6-2.4 (4) mm long...1a. *R. ramosior* var. *interior*.

1. **Rotala ramòsior** (L.) Koehne var. **týpica** Fern. & Grisc. Map 1461. All the specimens I have seen are from the wet sandy areas of the northwestern part of the state. It is local and is found in ditches and on the borders of sloughs.

Coastal Plain from Mass. to Fla. and Tex.; sands of s. Mich., n. Ind., Ill., and Minn.; also in Wash. and Oreg.

1a. **Rotala ramosior** var. **intèrior** Fern. & Grisc. Map 1462. My specimens are all from southern Indiana except a typical one from Steuben County. Infrequent to local and found in mud in ditches, on borders of sloughs, on muddy shores of streams and artificial ponds, and in wet woods and fallow fields.

N. Y. to Iowa, southw. to Fla., La., and Okla.

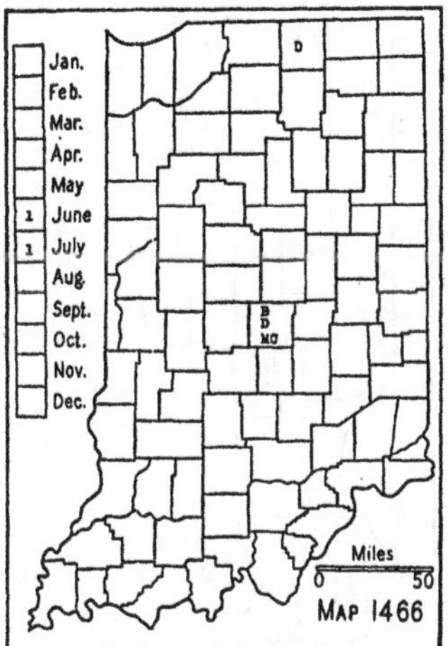

Lythrum Salicaria L.

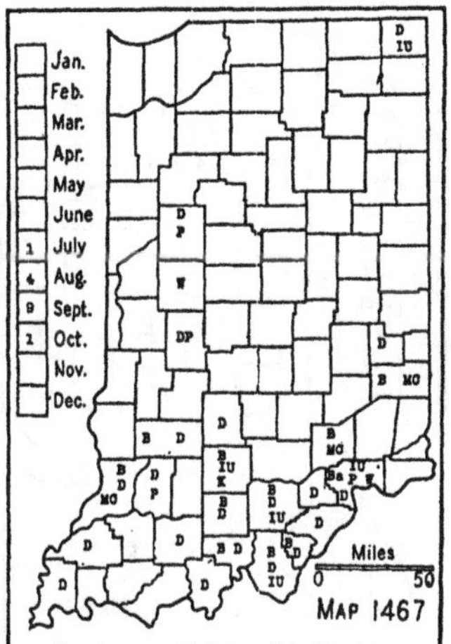

Cuphea petiolata (L.) Koehne

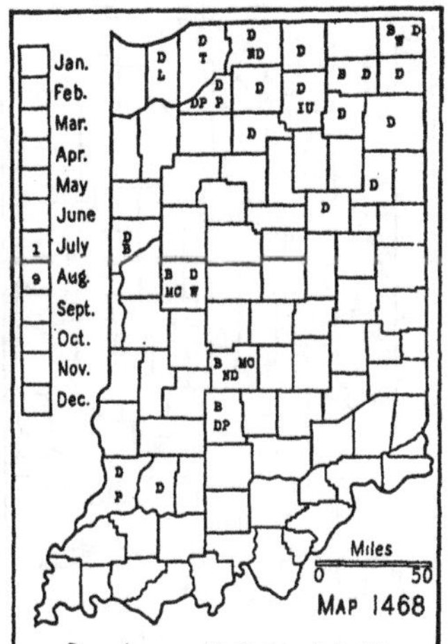

Decodon verticillatus (L.) Ell.

5474. AMMÁNNIA [Houston] L.

1. **Ammannia coccínea** Rottb. Map 1463. Muddy borders of sloughs, ponds, bayous, reservoirs, and streams and in dredged ditches. Local but often abundant where found.

N. J., Ohio to S. Dak., southw. to Fla. and Tex.; also in Mex. and Brazil.

5475. DÍDIPLIS Raf.

1. **Didiplis diándra** (Nutt.) Wood. WATER PURSLANE. Map 1464. Very rare. In stagnant water in ponds.

N. C. to Fla., westw. to Tex., and up the Mississippi Valley to Minn. and Wis.

5476. LÝTHRUM L.

Flowers solitary in the axils of the upper leaves; petals and stamens 5-7. . 1. *L. alatum.*
Flowers in a terminal, spicate panicle, crowded, whorled; petals 6; stamens 12, rarely 8-10. . . . . 2. *L. Salicaria.*

1. **Lythrum alàtum** Pursh. WINGED LYTHRUM. Map 1465. Essentially a plant of the open. Mostly in sandy soil in prairies, marshes, and low borders of lakes and in roadside ditches. Frequent in the lake and prairie areas, becoming infrequent to rare in the southern counties where its habitat is rare.

Southeastern N. E., Ont. to Minn., southw. to Ga., La., and Colo.

2. LYTHRUM SALICÀRIA L. PURPLE LOOSESTRIFE. Map 1466. Reported in 1925 by R. C. Friesner as well established along a small stream about a mile southeast of Irvington in Marion County. Also reported for Lake County by Pepoon. Common in a springy bayou of the Little Elkhart River just north of Middlebury, Elkhart County.

Nat. of Eu.; N. S. to Ont., southw. to N. Y., Del., and D. C.

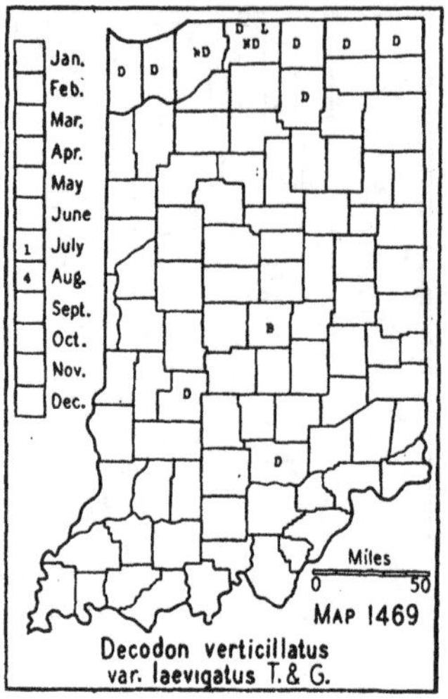

MAP 1469
Decodon verticillatus var. laevigatus T. & G.

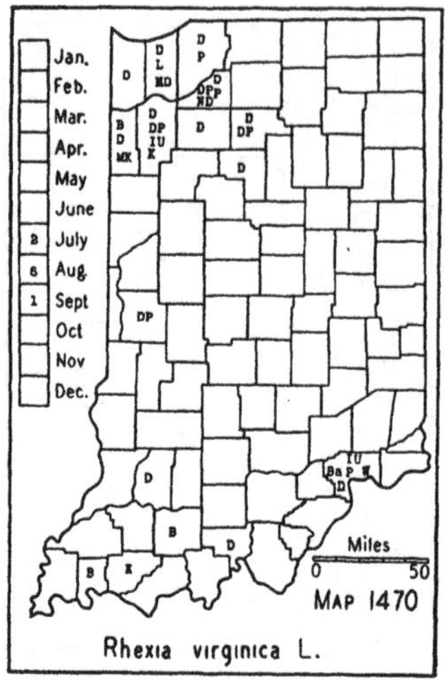

MAP 1470
Rhexia virginica L.

MAP 1471
Rhexia mariana L. var. leiosperma Fern. & Grisc.

5478. CÙPHEA P. Br.

1. **Cuphea petiolàta** (L.) Koehne. (*Parsonia petiolata* (L.) Rusby.) CUPHEA. Map 1467. Prefers sandy soil in dry situations but adapts itself to moist conditions. Rather frequent in the southern half of the state in open woodland, pastures, and fallow fields and along roadsides. Our Steuben County specimen was collected in 1928 by Anna May Weatherwax on the border of a cornfield along Little Crooked Lake northwest of Angola. No doubt this specimen was adventive. The species has a weedy nature. I had it under cultivation several years ago and it reproduced so abundantly that I feared it might become a weed so I exterminated it.

N. H. to Kans., southw. to Ga. and La.

5488. DÉCODON J. F. Gmel.

Stem and lower surface of leaves pubescent......................1. *D. verticillatus.*
Stem and lower surface of leaves glabrous.........1a. *D. verticillatus* var. *laevigatus.*

1. **Decodon verticillàtus** (L.) Ell. HAIRY SWAMP LOOSESTRIFE. Map 1468. In the mucky or peaty borders of lakes, bogs, and swamps. Infrequent in the lake area, and very rare southward.

Coastal Plain from Maine to Fla., inland in sw. Ont., Ind. and Ill.

1a. **Decodon verticillatus** var. **laevigàtus** T. & G. (*Decodon verticillatus* in part, of Gray, Man., ed. 7 and Britton and Brown, Illus. Flora, ed. 2.) SMOOTH SWAMP LOOSESTRIFE. Map 1469. On the mucky or peaty borders of lakes, bogs, and swamps. Rather rare and not so frequent as the pubescent form.

N. E. to Wis., southw. to Va. and Tenn.

223. MELASTOMÀCEAE R. Br. MELASTOMA FAMILY

5664. RHÉXIA L.

[Fernald & Griscom. Rhexia in northeastern America. Rhodora 37: 169-173. 1 pl. 1935.]

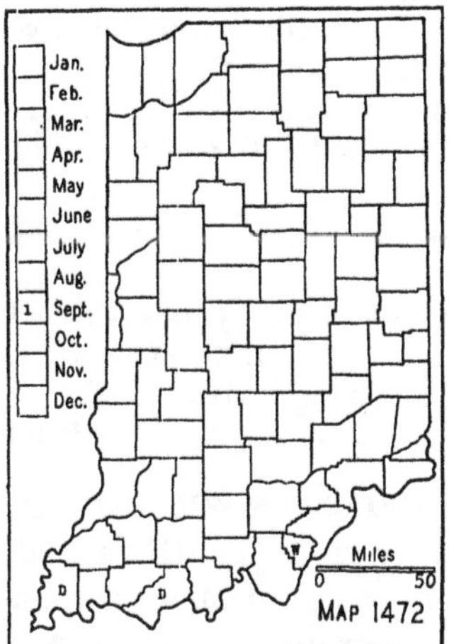

Jussiaea decurrens (Walt.) DC.

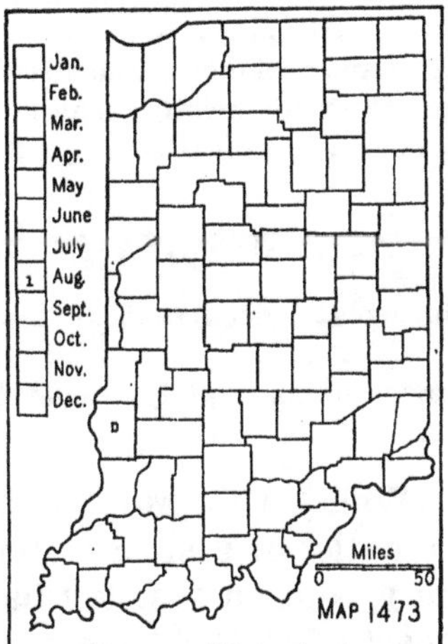

Jussiaea diffusa Forsk.

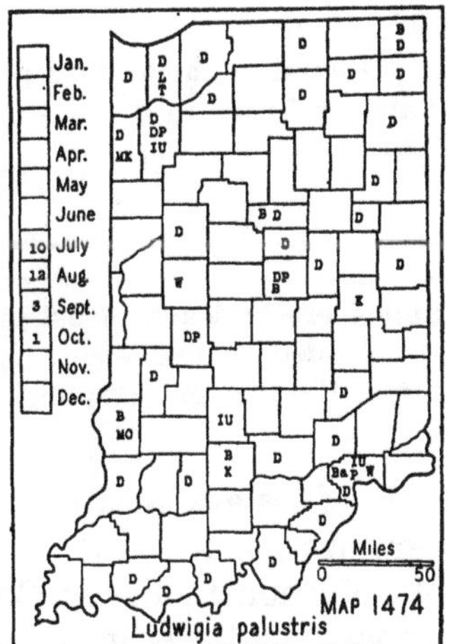

Ludwigia palustris var. americana (DC.) Fern & Grisc.

Neck of capsule shorter than the body; stem and branches usually only sparsely pubescent; seed muricate, 0.65-0.8 mm long. . . . . .1. *R. virginica.*
Neck of capsule as long as or longer than the body; stem and branches usually copiously spreading-pubescent; seed 0.5-0.6 mm long, the papillae depressed. . . . . .2. *R. mariana* var. *leiosperma.*

1. **Rhexia virgínica** L. COMMON MEADOWBEAUTY. Map 1470. In northern Indiana found in moist, slightly acid, black, sandy loam soil in treeless areas in black and white oak woods, prairies, chokeberry thickets, and borders of marshes and lakes. It is local, but where it is found it may be very abundant over acres. In the southern part of the state it is very local and found in the "flats" in fallow fields which were wooded with beech and sweet gum or in low, open, flat, sweet gum, red maple, and pin oak woods.

Along the coast from N. S. to Fla.; inland from sw. Ont. to se. Iowa, southw. to La. and Mo.

2. **Rhexia marìana** L. var. **leiospérma** Fern. & Grisc. (Rhodora **37**: 171-172. 1935.) MARYLAND MEADOWBEAUTY. Map 1471. Restricted to a few southern counties and found in moist and usually rather sandy soil in roadside ditches and hayfields and along railroads.

Ind., Ill., Ky., Tenn., Mo., Ark., and Tex.

224. ONAGRÀCEAE Dumort. EVENING-PRIMROSE FAMILY

Parts of the flower in fours or more numerous.
 Calyx tube not prolonged beyond the ovary.
  Calyx persistent on the fruit; seeds without a tuft of hairs at the summit.
   Capsules mostly 10-20 mm long; stamens twice as many as the petals. . . . . .5791. JUSSIAEA, p. 700.
   Capsules less than 10 mm long; stamens 4. . . . . .5793. LUDWIGIA, p. 700.
  Calyx deciduous; seeds with a tuft of hairs at the summit. . . . . .5795. EPILOBIUM, p. 702.
 Calyx tube prolonged beyond the ovary.

Flowers yellow (pink or white in *Oenothera speciosa*); fruit not deciduous, dehiscent..5804. OENOTHERA, p. 703.
Flowers light to dark pink; fruit deciduous, indehiscent.......5819. GAURA, p. 707.
Parts of flower in twos; stamens 2; fruit bristly; leaves opposite..5828. CIRCAEA, p. 709.

5791. JUSSIAÈA L.

Stems erect; leaves mostly lanceolate, decurrent at the sessile base; petals 4; pod 4-sided, club-shaped..................................1. *J. decurrens.*
Stems floating or creeping; leaves of an oval type; petals 5; pod cylindric..2. *J. diffusa.*

1. **Jussiaea decúrrens** (Walt.) DC. Map 1472. PRIMROSE-WILLOW. In very wet, sandy soil in the outlet of a spring about 10 miles southwest of Mt. Vernon in Posey County, and on a sandy bar in a small stream in a woods about 4 miles southeast of Hatfield, Spencer County.

Md. to Fla., westw. to Tex. and up the Mississippi Valley to Ill. and Ind.

2. **Jussiaea diffùsa** Forskal. FLOATING PRIMROSE-WILLOW. Map 1473. In 1935 I found this species to be common in the artificial lake in Shakamak State Park, Sullivan County. I did not investigate how extensively it was established but the border was well stocked with it as far as I traversed it. This lake is artificial and was made only a few years ago. No one seems to know when or how it got its start in the lake. I was informed by a workman that aquatic plants had been collected from the sloughs along Eel River and put into the lake. The interesting thing is that this species has so well adapted itself to its new habitat that it now grows abundantly on the shore of a lake which a few years ago was a poor, fallow field. Since no effort will be made to exterminate it, it is established in this place.

Ind., Ill. to Kans., southw. to Fla. and Tex.

5793. LUDWÍGIA L.

Leaves all opposite...................................1. *L. palustris* var. *americana.*
Leaves alternate.
Flowers showy; petals bright yellow; capsules on short pedicels, quadrangular, the angles slightly winged and greenish, the mature ones mostly 4-5 mm wide between the angles; plants more or less pubescent with short, incurved hairs. ....2. *L. alternifolia.*
Flowers inconspicuous; petals none or small, yellowish or green; capsules sessile, cylindric, subglobose or obpyramidal, mostly 2-4 mm wide, the angles, if any, rounded, and the sides with a shallow groove.
Capsules cylindric, about 2 mm in diameter, about twice as long as wide..3. *L. glandulosa.*
Capsules subglobose or obpyramidal, not twice as long as wide.
Plants glabrous or nearly so; bractlets of the capsules usually as long as or longer than the capsules; sepals about half as long as the capsules...4. *L. polycarpa.*
Plants pubescent; bractlets of the capsules usually about half as long as the capsules, more rarely minute or up to two thirds as long as the capsules. ....5. *L. sphaerocarpa* var. *Deamii.*

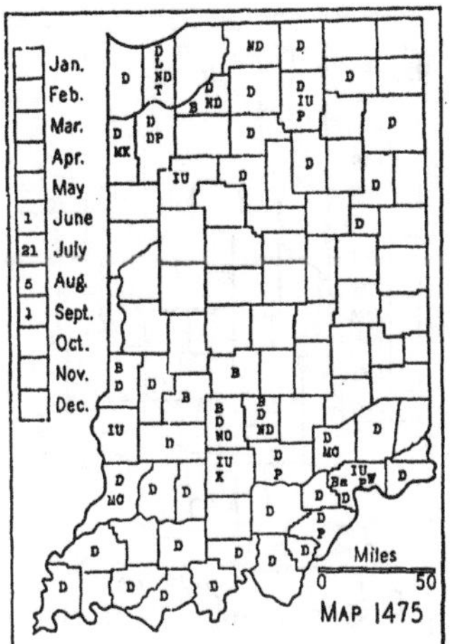

Ludwigia alternifolia L.

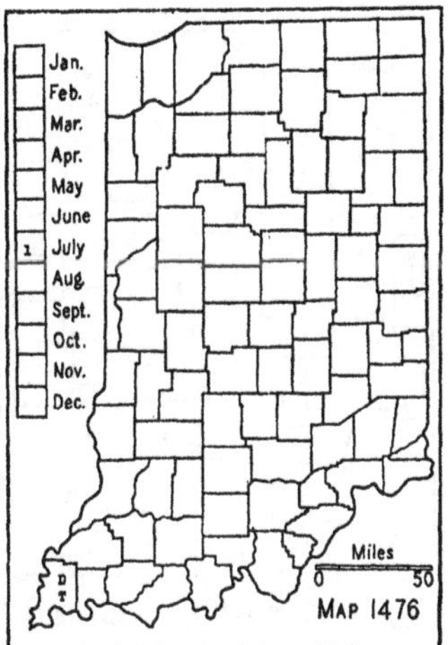

Ludwigia glandulosa Walt.

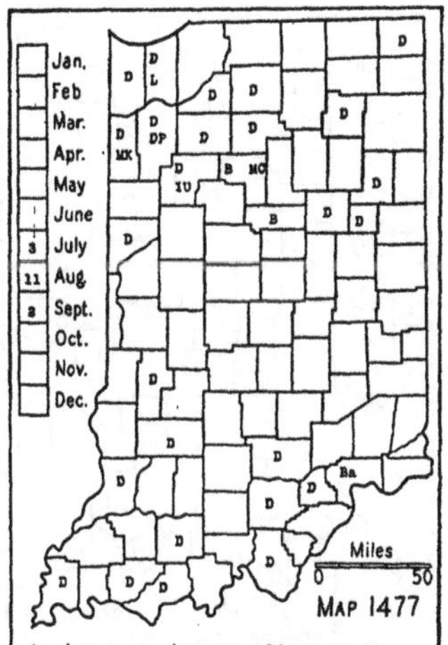

Ludwigia polycarpa Short & Peter

1. **Ludwigia palústris** (L.) Ell. var. **americàna** (DC.) Fern. & Grisc. (Rhodora 37: 176-177. 1 pl. 1935.) (*Ludwigia palustris* in part, of Gray, Man., ed. 7 and *Isnardia palustris* in part, of Britton and Brown, Illus. Flora, ed. 2.) MARSH PURSLANE. Map 1474. All Indiana reports for *Ludwigia palustris* should be referred to this variety. Frequent to common throughout the state, mostly in beds of ditches and small streams and on the muddy borders of ponds, swamps, sloughs, lakes, and streams. Sometimes in mucky soil in marshes.

N. S. to s. Que., Minn., and Oreg., southw. to Ga., La., Tex., e. Wash., e. Oreg., ne. Calif to Mex. and Guatemala; also in Bermuda.

2. **Ludwigia alternifòlia** L. Map 1475. In wet places along streams, about lakes, ponds, sloughs, and in wet woodland, fallow fields, and roadside ditches. Throughout the state but usually only a specimen or two at a place.

Mass. to Fla.; and in the interior from sw. Ont. to Kans. and Tex.

3. **Ludwigia glandulòsa** Walt. Map 1476. In swampy woods and dried-up sloughs. Known only from Posey County.

Gulf Coast from Fla. to Tex.; up the Mississippi Valley to Ill. and Ind.

4. **Ludwigia polycárpa** Short & Peter. Map 1477. In the muddy borders of ponds, sloughs, swamps, streams, lakes, and in dredged and roadside ditches. No doubt to be found in every county of the state but infrequent and rarely many specimens at a place.

E. Mass., sw. Ont. to Minn., southw. to Tenn. and Kans.

5. **Ludwigia sphaerocárpa** Ell. var. **Dèamii** Fern. & Grisc. (Rhodora 37: 174-175. 1935.) (*Ludwigia sphaerocarpa* in part, of Gray, Man., ed. 7 and of Britton and Brown, Illus. Flora, ed. 2.) Map 1478. Mucky or muddy borders of marshes, streams, and interdunal flats. Rare.

Plants growing in water often develop corky bases.

Nw. Ind.

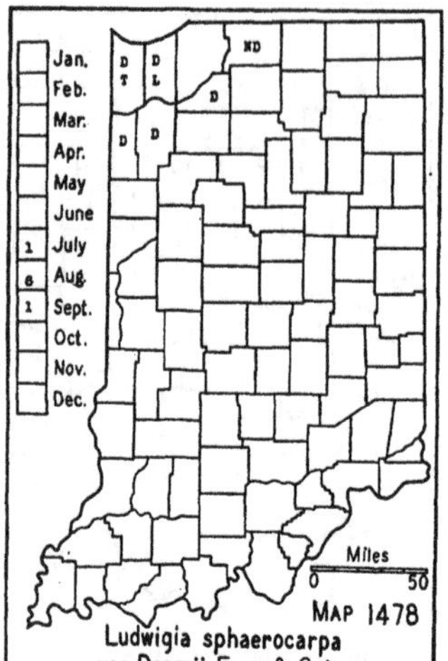

Ludwigia sphaerocarpa var. Deamii Fern. & Grisc.

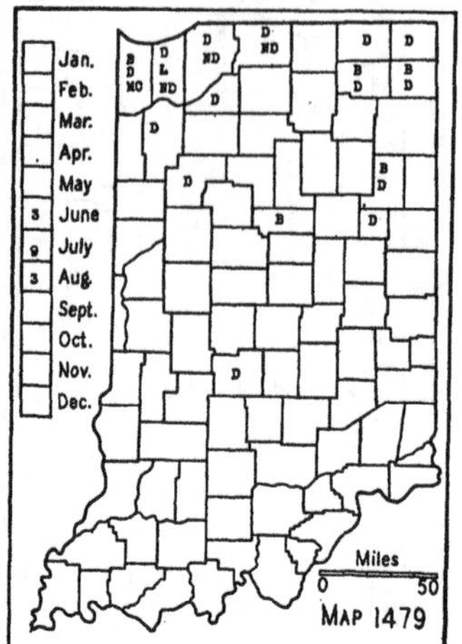

Epilobium angustifolium L.

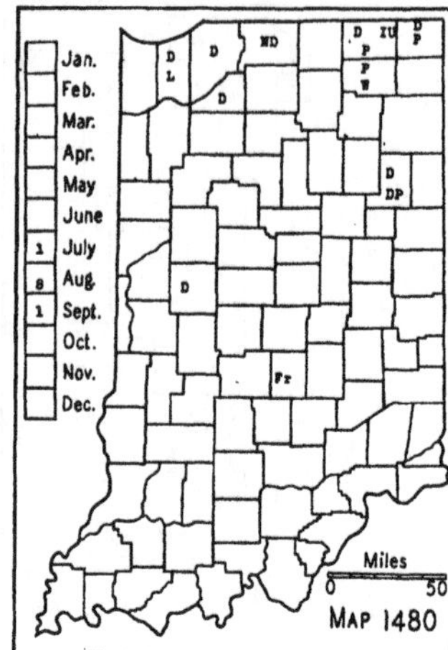

Epilobium strictum Muhl.

5795. EPILÒBIUM L.

Flowers large; petals entire, 10-20 mm long; stigmas 4-lobed.....1. *E. angustifolium.*
Flowers small; petals notched at the summit, less than 10 mm long; stigmas entire.
Stems terete, without decurrent lines from the leaf bases; leaves linear or lanceolate, entire or nearly so, the margins revolute.
Capsules and stems with a spreading pubescence....................2. *E. molle.*
Capsules and stems with a short, curved pubescence................3. *E. densum.*
Stems with decurrent lines from the leaf bases; leaves lanceolate or ovate-lanceolate, serrate, the margins flat.
Seed obconic-fusiform, 1.3-1.5 mm long, beakless, not striate but papillate in longitudinal lines, coma tawny in dried specimens but often whitish in immature or green specimens; flowers pinkish; leaves narrowed at the base. ....4. *E. coloratum.*
Seed obovoid, about 1 mm long, with a very short beak; dried coma white; flowers bluish pink or bluish purple; leaves rounded at the base, more remotely serrate and thinner than no. 4.............5. *E. glandulosum* var. *adenocaulon.*

1. **Epilobium angustifòlium** L. (*Chamaenerion angustifolium* (L.) Scop.) GREAT WILLOWHERB. Map 1479. Infrequent to rare in the open in newly made clearings and in wet soil about lakes, marshes, and interdunal flats. Rarely in dry sandy soil in woodland.

Greenland to Alaska, southw. to N. C., Ind., Kans., Ariz., and Calif.; also found in Eu. and Asia.

2. **Epilobium stríctum** Muhl. (*Epilobium molle* Torr.) Map 1480. In sedge marshes and bogs. Rare. It has been reported from Gibson, Jefferson, and Monroe Counties.

E. Que. to Alberta, southw. to Va., Ill., and Minn.

3. **Epilobium dénsum** Raf. (*Epilobium lineare* Muhl.) Map 1481. In bogs and sedge marshes. Infrequent.

E. Que. to Alberta, southw. to Del., W. Va., Kans., and Colo.

4. **Epilobium coloràtum** Muhl. WILLOWHERB. Map 1482. Frequent throughout the state except in the southern counties. In wet soil in road-

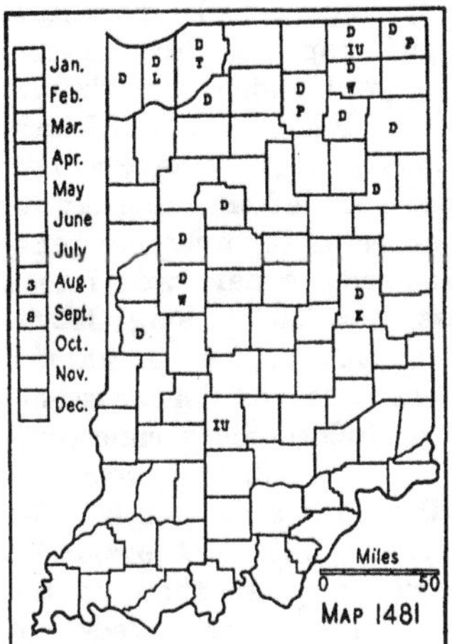

Epilobium densum Raf.

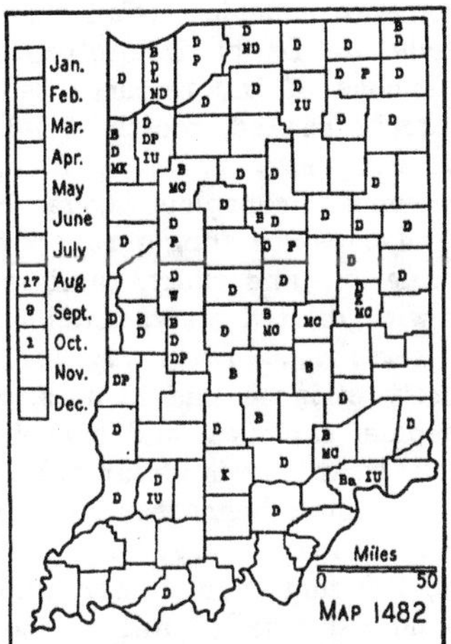

Epilobium coloratum Muhl.

Epilobium glandulosum Lehm. var. adenocaulon (Haussk.) Fern.

side and dredged ditches and wet woods, and on the borders of lakes, ponds, and streams.

N. S. to Wis., southw. to S. C., Tenn., Kans., and Nebr.

5. **Epilobium glandulòsum** Lehm. var. **adenocaúlon** (Haussk.) Fern. (Rhodora **20**: 34. 1918.) (*Epilobium adenocaulon* Haussk.) Map 1483. This variety was reported from Kosciusko County by Chipman (Proc. Indiana Acad. Sci. **1896**: 155. 1897). He says that he found two specimens and that these were sent to William Trelease, who had recently monographed the genus, and that Trelease reported that they were this species. It was also reported from Kosciusko County by Clark, and from Lake County by Pepoon and by Peattie. I found this variety to be abundant in a springy place at the base of the south bank of the southeast side of Lake Pleasant about 4 miles northeast of Orland, Steuben County. Some of the mature plants were up to 3 feet high.

Newf. to B. C., southw. to Del., W. Va., the Great Lakes, Nebr., Colo., and Calif.

5804. OENOTHÈRA L.[1]

Plants with stems.
 Flowers yellow; flower buds erect.
  Stamens of equal length; seeds in two rows in each cell; capsules subquadrangular, the angles broadly rounded.
   Capsules, when mature, 4-6 mm in diameter, tapering upward from a thickish base; seeds with sharp angles and not strongly pitted.
    Sepal-tips[2] terminal, hence connivent in the bud; seeds mostly 1.2-1.6 mm long.
     Stems not conspicuously angled, mostly reddish, somewhat strigulose and also hirsute, or almost glabrous; leaves green; hypanthium, sepals, capsules, and branches of inflorescence more or less hirsute and strigulose to glabrous, but with quite evident gland-tipped hairs among the others.

[1] I wish to thank P. A. Munz for his great assistance in constructing my key and in naming my specimens.

[2] Species under this lead probably formerly included in *O. biennis* of authors.

Bracts of inflorescence conspicuous in the bud, tending to be persistent and foliaceous in fruit; inflorescence and capsules hirsute and strigose; branches of inflorescence tending to be long and simple; capsules scarcely beaked, mostly 2.5-3.5 mm long; leaves thickish. ....1. *O. pycnocarpa.*

Bracts of inflorescence inconspicuous, deciduous soon after anthesis; inflorescence and capsules subglabrous; branches of inflorescence tending to be numerous and short, fastigiate; capsules narrowed into a beaklike tip with dilated apex, usually less than 2.5 cm long; leaves thin..2. *O. nutans.*

Stems angled, densely strigose and hirsute, grayish; leaves subcanescent; hypanthium, sepals, capsules, and branches of inflorescence canescent-strigose, scarcely if at all glandular-pubescent........3. *O. canovirens.*

Sepal-tips[1] not quite terminal, hence separate in the bud; seed mostly 1.8-2 mm long..4. *O. cymatilis.*

Capsules, when mature, 2-3 mm in diameter, more slender than those of the preceding group, and essentially uniform their entire length; seeds not angled.

Leaves denticulate or subentire; flowers in terminal spikes; seed indistinctly and shallowly pitted....................................5. *O. rhombipetala.*

Leaves sinuate-toothed or pinnatifid; flowers in the axils of foliage leaves; seed deeply and distinctly pitted........................6. *O. laciniata.*

Stamens of two lengths; seeds not in two distinct rows, clustered, not crested; capsules conspicuously angled or ridged.

Capsule on a stipe equaling or somewhat exceeding the permanently pubescent body; hairs on the capsule appressed or curved inwardly. (See excluded species no. 468, p. 1077)....................................*O. fruticosa.*

Capsule sessile or the stipe shorter than the body, pubescent to glabrate, the hairs straight.

Hairs of stems below the inflorescence spreading, usually 1-2.5 mm long (stems rarely glabrate); leaves mostly 10-25 (35) mm wide; internode of stem and branches below the lowest flower usually longer than the one below it; flowers usually subtended by foliaceous bracts; pubescence, if any, of capsule and calyx tube straight and glandless; calyx tube 12-18 mm long; petals of earlier flowers 20-25 mm long; capsules sessile or essentially so. ....7. *O. pilosella.*

Hairs of stems below the inflorescence curved or appressed, usually less than 1 mm long; leaves mostly 5-15 mm wide; internode of stem and branches below the lowest flower usually much elongated and much longer than the one below it; flowers usually subtended by linear bracts; pubescence of capsule and calyx straight, with glands; calyx tube 5-10 mm long; petals of earlier flowers 5-18 mm long; capsules manifestly on stipes.

Petals of earlier flowers mostly 12-18 mm long; buds and tip of inflorescence erect or nearly so; inflorescence when in fruit much less than half the height of the plant..................8. *O. tetragona* var. *longistipata.*

Petals of earlier flowers mostly 5-9 mm long; buds and tip of inflorescence nodding; inflorescence in fruit usually more than half the height of the plant..9. *O. perennis.*

Flowers pink or white..10. *O. speciosa.*

Plants without stems...11. *O. triloba.*

1. **Oenothera pycnocárpa** Atkinson & Bartlett. (Rhodora **15**: 83. 1913.) Map 1484. EVENING-PRIMROSE. This is the common form of the *Oenothera biennis* complex in Indiana. It is found throughout the state and in all kinds of habitats. It and the next three species are regarded as obnoxious

[1] This species probably *O. muricata* of authors.

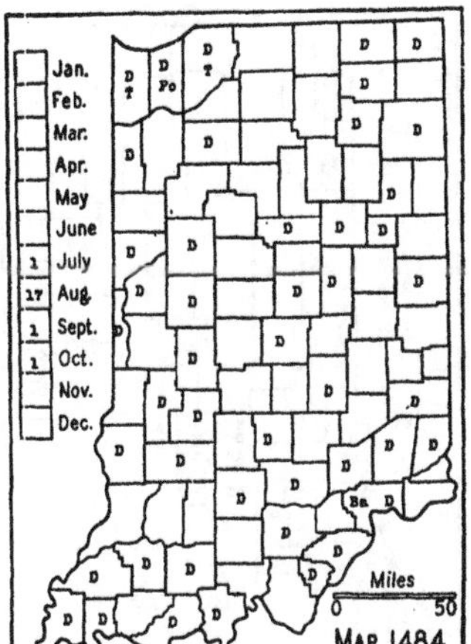

MAP 1484
Oenothera pycnocarpa Atkinson & Bartlett

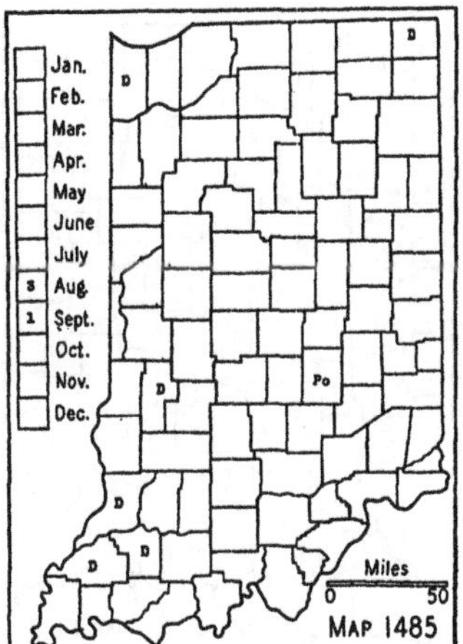

MAP 1485
Oenothera nutans Atkinson & Bartlett

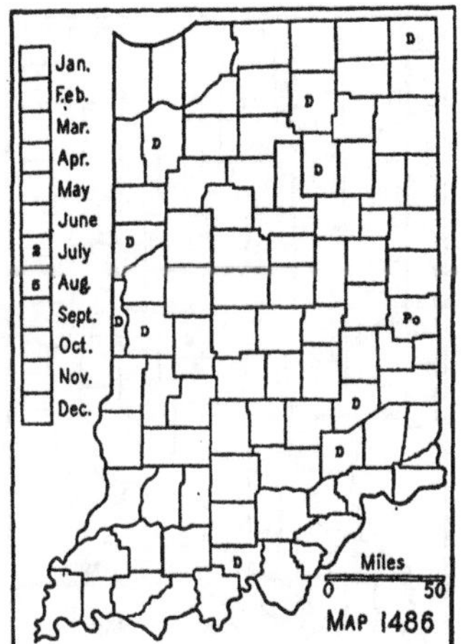

MAP 1486
Oenothera canovirens Steele.

weeds. Each plant bears a great number of seed and self-sown seedlings will appear many years afterward. The status of this and the next three species is not yet definitely determined. Some authors regard them simply as varieties of *Oenothera biennis* but I am regarding them as species as did the authors who described them. The plants are exceedingly variable and only an expert can name them with any degree of certainty. I have a large number of specimens which I am not including in this treatment because I can not satisfactorily name them.

N. E. to Minn. and southw.

2. **Oenothera nùtans** Atkinson & Bartlett. (Rhodora **15**: 83. 1913.) Map 1485. This form is less common than the preceding one as is shown by the map. The plants have the same habitats as those of the preceding.

Widely distributed in e. N. A.

3. **Oenothera canovìrens** Steele. (Contr. U. S. Nation. Herb. **13**: 365. 1911.) Map 1486. This is the form that has been regarded as *Oenothera strìgosa* (Rydb.) Mack. & Bush. It is more common than the preceding species and has the same habitats. Probably widely distributed in eastern N. A.

4. **Oenothera cymátilis** Bartlett. (Cybele Columbiana, p. 51, 1914.) Our only specimen of this species was collected by the late Carl Buhl in an old tamarack bog about 7 miles west of La Porte, La Porte County. Since it is reported from both Illinois and Michigan, it is doubtless more or less frequent in the dune area.

5. **Oenothera rhombipétala** Nutt. (*Raimannia rhombipetala* (Nutt.) Rose.) Map 1487. In very sandy soil along roadsides and in fallow fields and in the dune area in open woodland, along roadsides, and in waste places, becoming plentiful where conditions permit it to spread.

Ind. to Minn., southw. to Tex.

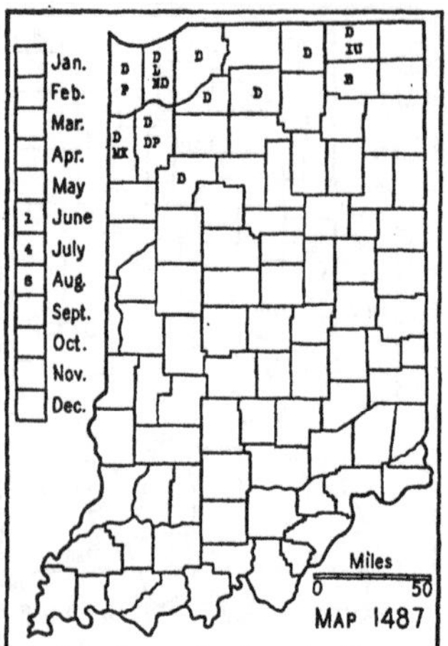

Oenothera rhombipetala Nutt.

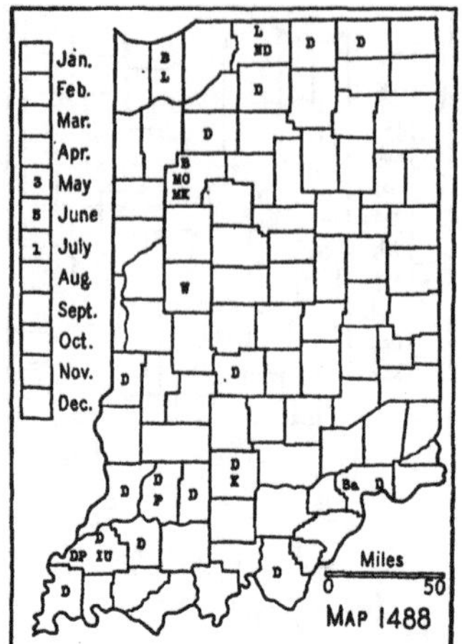

Oenothera laciniata Hill

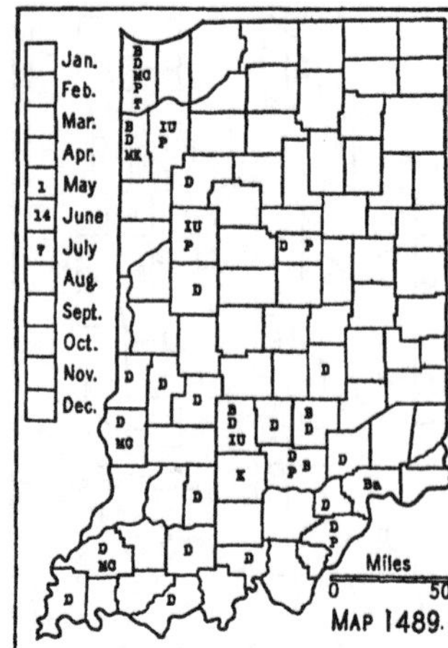

Oenothera pilosella Raf.

6. **Oenothera laciniàta** Hill. (*Raimannia laciniata* (Hill) Rose.) Map 1488. In sandy to very sandy soil along roadsides and in fallow fields. We have one specimen from an open woods and one from hard, white clay soil in a fallow field. It has already become a weed in some kinds of soils and in time it will doubtless become a weed throughout the state.

Maine to S. Dak., southw. to Fla., and Tex.

7. **Oenothera pilosélla** Raf. (*Oenothera pratensis* (Small) Rob. and *Kneiffia pratensis* Small.) SUNDROPS. Map 1489. Usually common where found. In low ground in open woods, open marshy places, and fallow fields of slightly acid soil and along roadsides.

Ohio to Iowa, southw. to Mo. and Ark. Recorded from eastern N. E.

8. **Oenothera tetrágona** Roth. var. **longistipàta** (Pennell) Munz. (Probably in part, *Oenothera fruticosa* of Gray, Man., ed. 7 and *Kneiffia fruticosa* of Britton and Brown, Illus. Flora, ed. 2.) Map 1490. Usually in moist, black sandy soil in prairie habitats. In marshes and along roadsides.

The species varies from dwarf and simple-stemmed plants to those with long spreading branches, the last form sometimes decumbent. I am citing my no. 48906 as exceptional. It seems to be a hybrid in that it has the pubescence of *O. pilosella,* flowers intermediate, otherwise as this variety.

N. E. to Minn., southw. to Ga. and Ind. When this species is studied further, the range may be changed.

9. **Oenothera perénnis** L. var. **týpica** Munz. (*Oenothera pumila* L. and *Kneiffia pumila* (L.) Spach.) Map 1491. Our specimens from southern Indiana were found in hard, white clay soil and those from the northern part were in a wet, prairie habitat.

N. S. to Man., southw. to Ga. and Kans.

10. OENOTHERA SPECIÒSA Nutt. (*Hartmannia speciosa* (Nutt.) Small.) Map 1492. I have collected this species from the side of a railroad, a

MAP 1490
Oenothera tetragona var. longistipata (Pennell) Munz

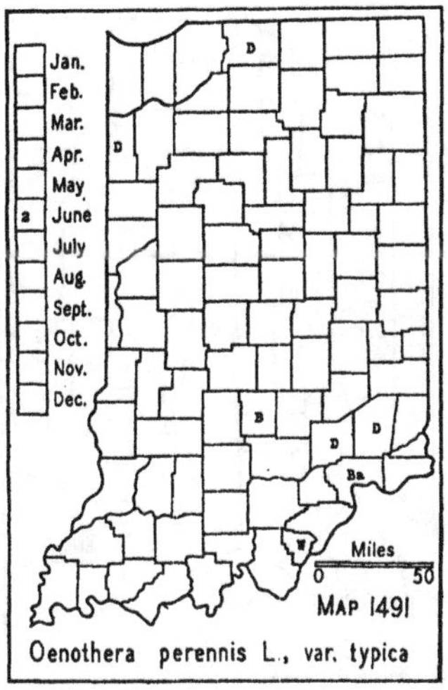

MAP 1491
Oenothera perennis L., var. typica

MAP 1492
Oenothera speciosa Nutt.

roadside to which it had escaped from a cemetery, and a roadside where it was common, and also in an adjoining alfalfa field. I planted some of it and when, in two years, it had spread by underground rootstocks over a large area, it became necessary to destroy it and the process required three years of careful work. Since I never permitted it to seed I do not know its ability to propagate from seed but most species of this genus should be regarded with suspicion. This species, no doubt, in time will become an obnoxious weed.

Mo. and Kans., southw. to Tex.; introd. eastw. to Ind. and reported from Ohio.

11. **Oenothera tríloba** Nutt. (*Lavauxia triloba* (Nutt.) Spach). Map 1493. Dry, rocky, wooded bluff of the Ohio River near Madison, Jefferson County. I have a specimen, too, collected by Wm. H. Rudder near Salem, Washington County. No doubt a native of some of the hills of southern Indiana. Biennial in Indiana; flowering from spring sown seed.

Ind., Ky. to Kans., southw. and westw. to Miss., Tex., and Mex.

5819. GAÚRA L.

[Munz. A revision of the genus Gaura. Bull. Torrey Bot. Club **65**: 105-112 and 211-228. 1938.]

Fruit sessile or nearly so; leaves pubescent.
  Flowers 3-4 mm wide; fruit glabrous, somewhat terete, not with four symmetrical sides; leaves ovate-lanceolate; plants not branching at the base. .1. *G. parviflora.*
  Flowers 8-10 mm wide; fruit pubescent, with four symmetrical sides; leaves narrower than those of the preceding species.
    Plants not branching at the base, mostly 60-120 cm high; flowers white, turning pink or pinkish; fruit without a necklike base. . . . . .2. *G. biennis.*
    Plants branching at the base, mostly 15-50 cm high; flowers rose colored, turning scarlet; fruit with a necklike base nearly as long as the body. .3. *G. coccinea.*
Fruit on short pedicels; leaves glabrous or nearly so. . . . . .4. *G. filipes.*

1. GAURA PARVIFLÒRA Dougl. Map 1494. In 1910 I found this plant well established in Indianapolis along White River near the Vandalia

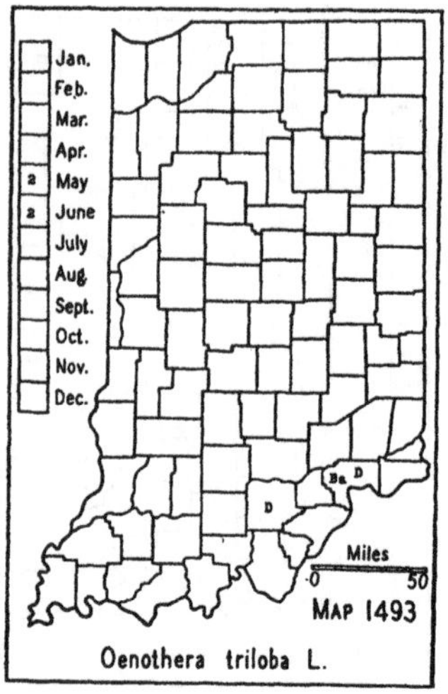

Oenothera triloba L.

Gaura parviflora Dougl.

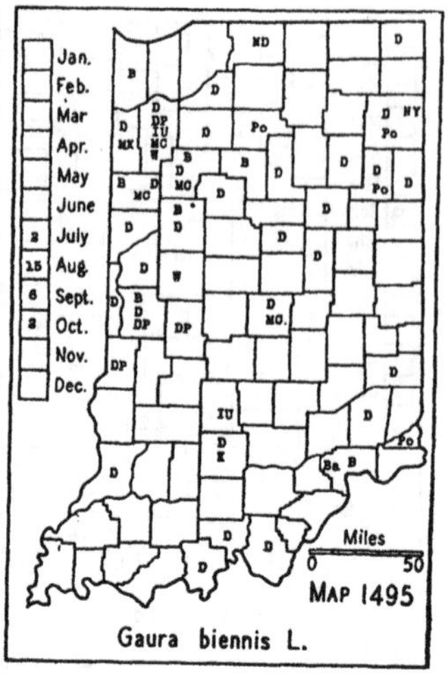

Gaura biennis L.

Railroad. Smith found it before this date well established in another section of Indianapolis. Standley, in 1930, found it to be plentiful in vacant lots in East Chicago, Lake County. He found it also in Porter County along a railroad. Chas. M. Ek found it well established east of Peru in Miami County.

Ill., S. Dak. to Wash., southw. to Utah, Ariz., Tex., and Sonora, Mex.

2. **Gaura biénnis** L. Map 1495. Usually in the open in alluvial soil along or near streams in open woods, in prairies, and more rarely in fallow fields or on washed slopes.

Conn., Que., Ont. to Minn., southw. to Ga., Ark., and Nebr.

3. GAURA COCCÍNEA Nutt. ex Pursh. Map 1496. This species was reported from Fayette County by Hessler, who found it along a railroad, and said that it soon died out. Hill found it in Porter County along a railroad near Crisman. I found it in 1930 in White County in ballast along the railroad about a mile east of Idaville. This colony was first discovered in 1929 by Mr. and Mrs. Walter Neff. Nieuwland found it to be well established along a railroad near Lydick, St. Joseph County.

Man. to Minn., Nebr., Mo., and Mex., westw. to Mont. and Ariz. and into Mex.

4. **Gaura filipes** Spach. Map 1497. I found this species in 1929 in an open place on a sand hill in the edge of a black oak woods about a mile southwest of Plainville, Daviess County. It was reported by Wilson as being common in Hamilton and Marion Counties. It was also reported by Phinney from the area of Delaware, Jay, Randolph, and Wayne Counties. He says: "Fields and woods. Common." He also reports *Gaura biennis* and says: "Fields and woods. Rare." Phinney used Gray's Manual, edition 5, for his determinations and in it the distinction between the two species is not very clear and he may have confused them. I think both Phinney and Wilson made wrong determinations but I am not able to account for their errors. Munz (Bull. Torrey Bot. Club **65**: 217. 1938) determined my

Gaura coccinea Pursh

Gaura filipes Spach

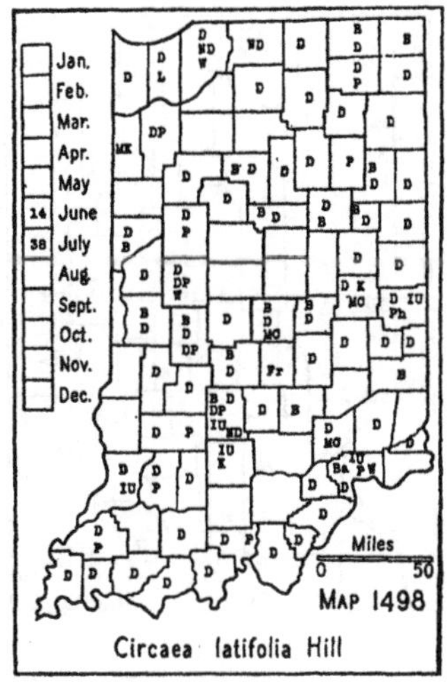

Circaea latifolia Hill

specimen, collected near Plainville, Daviess County as *Gaura filipes* var. *major* T. & G. In 1938 Kriebel collected plants from the exact place where I collected my specimen and the sepals of all the specimens are less than 7 mm long, so I am referring all my specimens to the species.

Ind. to S. C., southw. to Fla. and Miss.

5828. CIRCAÈA [Tourn.] L.

Leaves dark green above, rather firm, rounded or subcordate at the base; mature pedicels strongly reflexed; calyx lobes mostly 1.4-2 mm wide; disk cuplike, prolonged about 0.5 mm beyond the perianth; anthers 0.7-1 mm long; stigma shallowly 2-lobed; mature fruit compressed-pyriform, with 3-5 longitudinal corrugations on each face, 3.5-5 mm wide (including the hairs) .... 1. *C. latifolia.*

Leaves pale green above, flaccid, cordate or subcordate at the base; mature pedicels spreading or only slightly reflexed; calyx lobes 0.8-1.7 mm wide; disk inconspicuous; anthers 0.2-0.8 mm long; stigma deeply cleft; fruit clavate, not corrugated, 1-3 mm long (including the hairs).

Rootstocks slender; calyx lobes 1.2-1.7 mm wide; petals 2.3-3.5 mm long; anthers 0.5-0.8 mm long; fruit unequally 2-celled, 1.5-3 mm wide (including the hairs). (See excluded species no. 472, p. 1077) .... *C. canadensis.*

Rootstocks tuberous-thickened; calyx lobes 0.8-1.2 mm wide; petals 2-2.5 mm long; anthers 0.2-0.3 mm long; fruit 1-celled, 1-1.5 mm wide (including the hairs) .... 2. *C. alpina.*

1. **Circaea latifòlia** Hill.* (*Circaea lutetiana* of authors, not L.) (See Rhodora **17**: 222. 1915 and **19**: 87. 1917.) ENCHANTER'S NIGHTSHADE. Map 1498. In woodland of almost all kinds, preferring wooded ravines and beech and sugar maple woods. Infrequent to frequent throughout the state except on the crests of black oak and chestnut oak ridges, on the dunes, and in prairies. The sepals of this plant are usually green, but sometimes are rose purple.

N. B., N. S., and Maine to Minn., southw. to N. C., Tenn. and Okla.

---

* The name now proposed for this plant is **Circaea quadrisulcata** (Maxim.) Franch. & Sav. var. **canadensis** (L.) Hara. (Rhodora **41**: 386-387. 1939.)

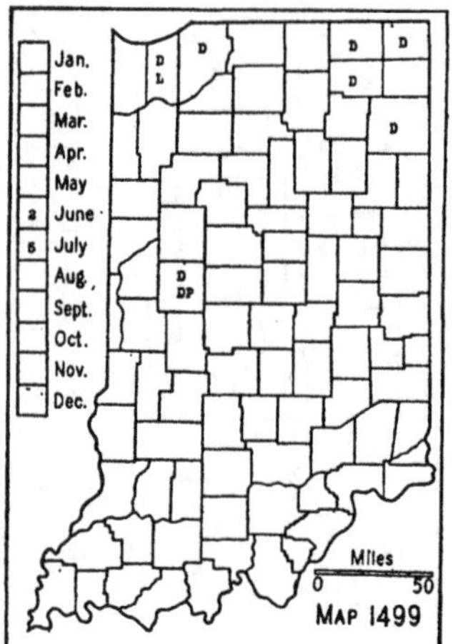

Circaea alpina L.

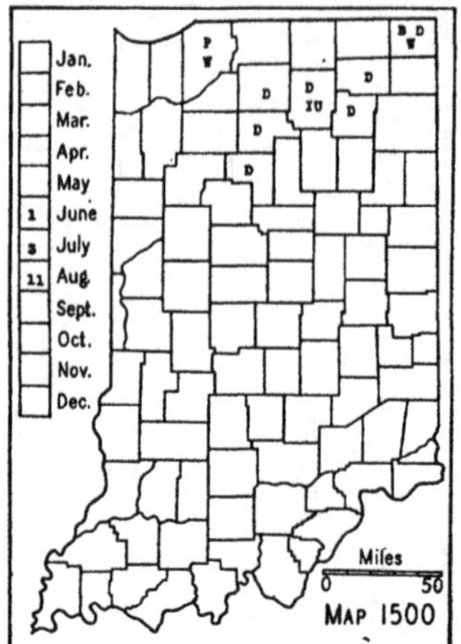

Myriophyllum exalbescens Fern.

Myriophyllum heterophyllum Michx.

2. **Circaea alpìna** L. Map 1499. Very local but often common where found. In bogs and on old logs in swamps and in very damp places such as deep ravines.

S. Lab. to James Bay and Alaska, southw. to N. E., Ga., the Great Lakes and S. Dak.; found also in Eurasia.

225. HALORAGIDÀCEAE Klotsch & Garcke. WATER-MILFOIL FAMILY

Leaves in whorls (sometimes scattered in *Myriophyllum scabratum*).
  Plants with immersed leaves dissected; emersed leaves not entire; stamens 4 or 8; fruit 4-celled. . . . . .5834. MYRIOPHYLLUM, p. 710.
  Plants with all the leaves entire; stamen 1; fruit 1-celled. . . .5837. HIPPURIS, p. 712.
Leaves alternate; fruit 3-angled. . . . . .5835. PROSERPINACA, p. 711.

5834. MYRIOPHÝLLUM [Vaill.] L. WATER-MILFOIL

Bracts or floral leaves entire, sparingly dentate or serrate.
  Bracts shorter than or rarely as long as the flowers or fruit, ovate to very broadly ovate, entire with a narrow, brown, chartaceous margin or sparingly dentate; stamens 8; carpels 2-3 mm long, smooth; rachis of leaf about the width of the divisions. . . . . .1. *M. exalbescens.*
  Bracts about twice as long as the flowers or fruit or even longer, linear-oblong, finely serrate; stamens 4; carpels 1-1.5 mm long, papillose, 2-ridged on the back; stigmas prominent; rachis of leaf slightly broader than the divisions. . . . . .2. *M. heterophyllum.*
Bracts pectinate.
  Bracts about 5 times as long as the flowers or fruit; stamens 4; fruit 1-1.5 mm long, with 2 ridges on the dorsal side, the lateral faces slightly roughened; divisions of the leaf mostly wider than the rachis. . . . . .3. *M. scabratum.*
  Bracts as long as or up to 2.5 times as long as the flowers or fruit; stamens 8; carpels 2.5-3 mm long, smooth; stigmas prominently recurved; rachis of leaf slightly broader than the divisions. . . . . . . .4. *M. verticillatum* var. *pectinatum.*

1. **Myriophyllum exalbéscens** Fern. (Rhodora **21**: 120. 1919.) (*Myriophyllum spicatum* L. of Gray, Man., ed. 7 and Britton and Brown, Illus. Flora, ed. 2.) Map 1500. This is our most common species of the genus

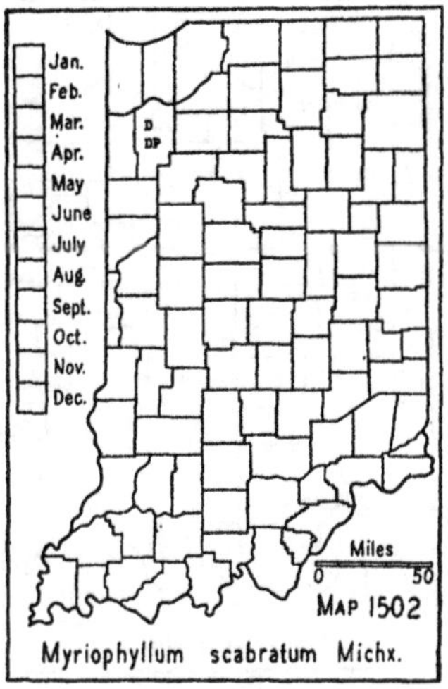

and is doubtless infrequent to frequent in most of our lakes although I have botanized several lakes without finding it. In addition to the counties in which I have found it, it has been reported from Lake County. I have found it only in lakes.

Lab. to Alaska, southw. to Conn., N. Y., Ind., Kans., N. Mex., Ariz., and Calif.

2. **Myriophyllum heterophýllum** Michx. Map 1501. This species is infrequent or locally frequent in the lake area. It is found in lakes and more frequently in dredged ditches.

N. J. to Fla., near the coast; also from cent. N. Y., Ont. to Minn., southw. to Mo. and Tex.

3. **Myriophyllum scabràtum** Michx. (*Myriophyllum pinnatum* of Britton and Brown, Illus. Flora, ed. 2.) Map 1502. Our only recent report for this species is from Jasper County where I found it in the old channel of the Kankakee River half a mile west of the Tefft Bridge. It is doubtless very rare in Indiana.

Coastal Plain from Mass. to Fla. and Tex., and northw. to Ind. and Iowa.

4. **Myriophyllum verticillàtum** L. var. **pectinàtum** Wallr. Map 1503. Our only specimens are from Crooked Lake, Steuben County, but doubtless it is more widely distributed. The reports, made by earlier authors, of *Myriophyllum pectinatum* from Fulton, Kosciusko, Marshall, Steuben, and Vigo Counties and from the dune area should probably be referred to this variety.

Newf. to Ont. and Wash., southw. to Md., Ill., and Utah.

5835. PROSERPINÀCA L.

[Fernald & Griscom. Proserpinaca palustris and its varieties. Rhodora 37: 177-178. 1935.]

Angles of fruit rounded............................1. *P. palustris* var. *amblyogona.*
Angles of fruit sharp..................................1a. *P. palustris* var. *crebra.*

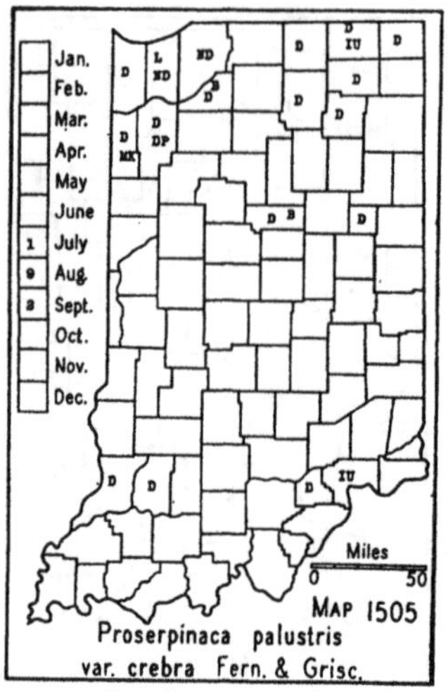

MAP 1505
Proserpinaca palustris var. crebra Fern. & Grisc.

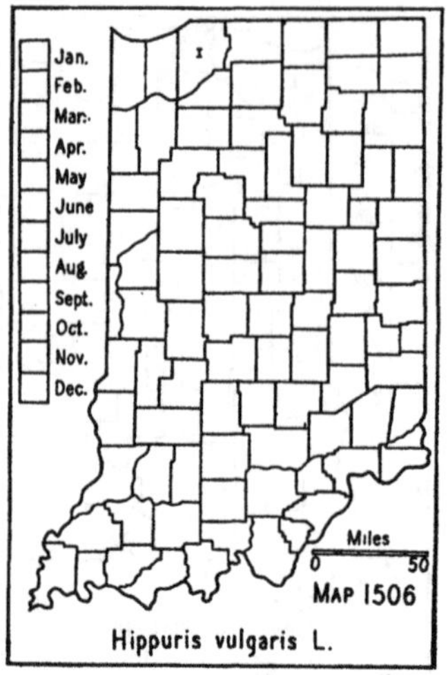

MAP 1506
Hippuris vulgaris L.

MAP 1507
Aralia nudicaulis L.

1. **Proserpinaca palústris** L. var. **amblyógona** Fern. (Rhodora **11**: 120. 1909.) Map 1504. This is a form with the angles of the fruit rounded. My Kosciusko County specimen might be referred to this variety although it is not well marked. Fernald cited O. E. Lansing's no. 2509 from a ditch, Roby, Lake County, Indiana as the type.

Lake Huron, Ont., Ind., and Mo.

1a. **Proserpinaca palustris** var. **crèbra** Fern. & Grisc. (Rhodora **37**: 177-178. 1935.) (*Proserpinaca palustris* in part, of Gray, Man., ed. 7 and in part, of Britton and Brown, Illus. Flora, ed. 2.) MERMAID WEED. Map 1505. All reports for *Proserpinaca palustris* L. should be referred to this variety or the preceding one. In the low sedge borders of lakes, in swamps, dried-up ponds and sloughs, and cypress swamps. Infrequent to rare.

N. S. to Wis., southw. to Ga. and Okla.

5837. HIPPÙRIS L.

1. **Hippuris vulgàris** L. MARE'S TAIL. Map 1506. Our only specimen is one collected by E. J. Hill, July 5, 1880 in a millpond near Otis, La Porte County. This specimen is in the herbarium of the University of Illinois. The species has been reported from Kosciusko and Lake Counties and from the Lower Wabash Valley by Schneck, who says it is "rare in ponds and streams."

Lab. to Alaska, southw. to N. S., Maine, Vt., N. Y., Ind., Ill., Nebr., N. Mex. and Calif.; also in Eurasia and S. A.

227. ARALIÀCEAE Vent. GINSENG FAMILY

Leaves alternate, compound, the ultimate divisions pinnate; inflorescence compound; carpels 5; fruit black.............................5881. ARALIA, p. 713.

Leaves 3 (rarely 4) in a whorl, palmately 3-7-foliolate; inflorescence simple; carpels 2 or 3; fruit red or yellowish...........................5883. PANAX, p. 714

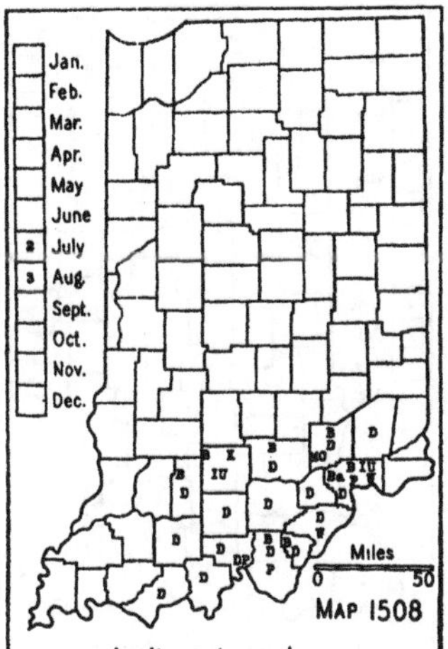

Aralia spinosa L.

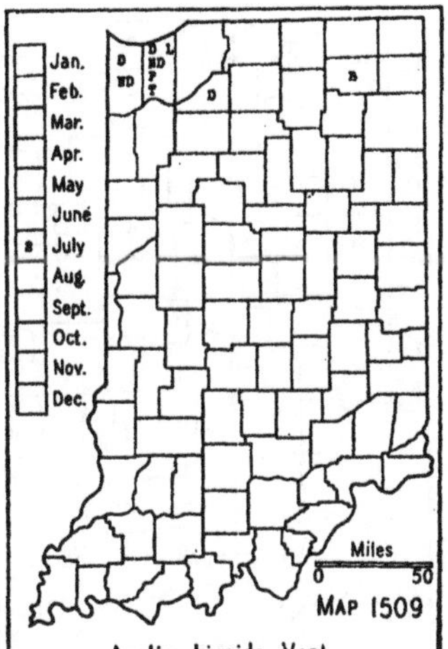

Aralia hispida Vent.

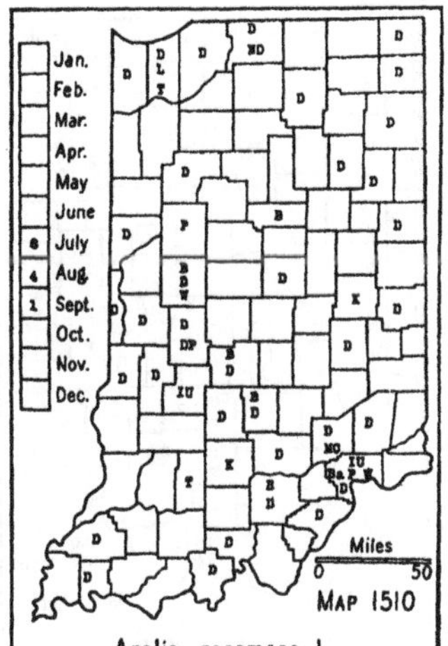

Aralia racemosa L.

5881. ARÀLIA [Tourn.] L.

Flowers in 2-7 umbels on a naked scape..........................1. *A. nudicaulis.*
Flowers in umbels but not on a naked scape.
  Plants with prickles or bristles (at least near the base).
    Shrubs or small trees with numerous stout prickles................2. *A. spinosa.*
    Herbs with bristles (at least near the base).......................3. *A. hispida.*
  Plants without prickles or bristles..............................4. *A. racemosa.*

1. **Aralia nudicaúlis** L. WILD-SARSAPARILLA. Map 1507. Infrequent to rare in the northern counties in moist soil on the borders of marshes, bogs, and lakes; south of the lake area it is very rare, being restricted to a few rocky, wooded bluffs.

The rays of the umbels and the principal veins of the under surface of the leaves of Indiana specimens are pubescent.

Newf. to Man., southw. to Ga., Mo., Colo., and Idaho.

2. **Aralia spinòsa** L. DEVILS-WALKINGSTICK. Map 1508. Somewhat frequent in most of the counties indicated on the map and probably rare or absent in most of the remaining southern counties. Found on the crests and slopes of black and white oak ridges and in low ground in hard, white clay soil with sweet gum and beech.

Southern N. Y., Ind. to Mo., southw. to Fla. and Tex.

3. **Aralia híspida** Vent. BRISTLY ARALIA. Map 1509. In fine, sandy soil at the bases of wooded dunes and in sandy, burned-over areas. Rare.

Newf. to Hudson Bay, southw. to N. C., W. Va., Ind., and Minn.

4. **Aralia racemòsa** L. AMERICAN SPIKENARD. Map 1510. In rich, level woodland and on wooded slopes throughout the state. Formerly frequent to common but soon becoming extinct in woods where hogs are admitted because they are very fond of the roots. The roots were formerly much used in medicine for man and beast for respiratory ailments.

N. B. and N. S. to Minn., southw. to Ga., Mo., and S. Dak.

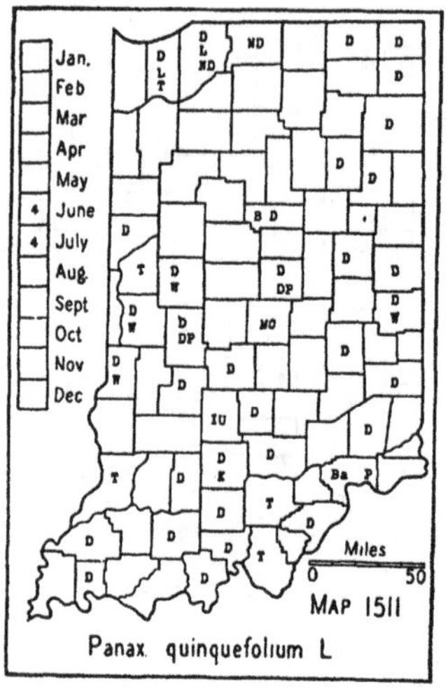

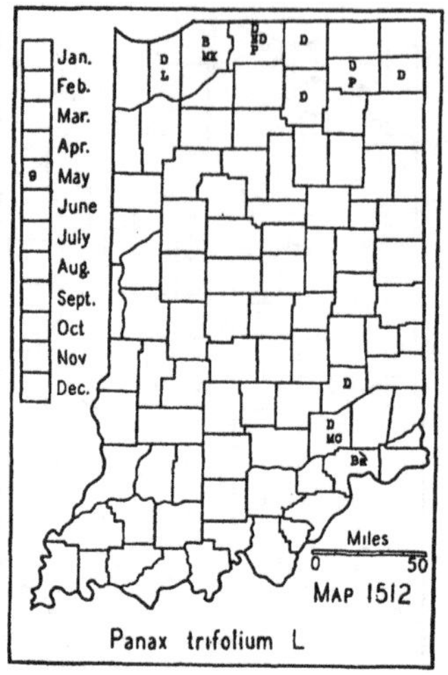

5883. PÀNAX L.

Leaflets 5, stalked, obovate or ovate, abruptly acuminate, the larger ones 8-14 cm long; fruit red..1. *P. quinquefolium.*

Leaflets 3-5, sessile or nearly so, obtuse or acute, narrow-oval, the larger ones 1.5-6 cm long; fruit yellowish..2. *P. trifolium.*

1. **Panax quinquefòlium** L. AMERICAN GINSENG. Map 1511. Formerly frequent to common in rich woods throughout the state. From the earliest times it was dug for its large roots which were shipped mostly to China for use as a medicine. The earliest pioneers received 25 cents a pound for the dried roots. The fact that the price has steadily advanced, until it now sells for about 16 dollars a pound, has resulted nearly in its extinction.

Que. and Ont. to Minn., southw. to Pa., Mo., and in the mts. to Ga.

2. **Panax trifòlium** L. DWARF GINSENG. Map 1512. Leaves usually 3, sometimes 4. Mostly in moist, rich beech and sugar maple woods and rarely in wet places in woods. It is rather local in a few of the northern counties and reappears in slightly acid soil in Decatur and Jennings Counties. The plant is rather inconspicuous and may be more common than the reports indicate.

N. S. to Minn., southw. to Del., Md., Ill., Iowa, and along the mts. to Ga.

228. UMBELLÍFERAE B. JUSS. PARSLEY FAMILY*

Leaves all simple.

Inflorescence capitate..5923. ERYNGIUM, p. 718.

Inflorescence a distinct umbel, not capitate.

Leaves orbicular-peltate or reniform; umbels simple...5893. HYDROCOTYLE, p. 716.

Leaves perfoliate; umbels compound....................5994. BUPLEURUM, p. 721.

Leaves not all simple.

Ovary and fruit variously armed with bristles or with hooked or barbed prickles.

Ovary and fruit armed with bristles; fruit several times longer than wide..5941. OSMORHIZA, p. 719.

* The manuscript of this family was read, and the key to the genera written by Mildred E. Mathias, Research Associate, University of California.

Ovary and fruit armed with hooked or barbed prickles; fruit not several times longer than wide.
  Plants glabrous; leaves palmately 3-7-foliolate; flowers perfect or staminate, ...5918. SANICULA, p. 717.
  Plants pubescent; leaves pinnately decompound; flowers all perfect.
    Ovary and fruit flattened laterally, covered with hooked prickles; rays of umbels short..................................5945. TORILIS, p. 720.
    Ovary and fruit flattened dorsally, covered with barbed bristles; rays of umbels long......................................6142. DAUCUS, p. 728.
Ovary and fruit not armed.
  Fruit several times longer than wide; flowers white.
    Leaves trifoliolate with ovate leaflets; plants perennial, 3-9 dm high; involucels absent....................................6015. CRYPTOTAENIA, p. 723.
    Leaves ternately compound with pinnatifid leaflets; plants annual, usually 2-5 dm high; involucels present..............5935. CHAEROPHYLLUM, p. 718.
  Fruit less than twice as long as wide; flowers white or yellow.
    Leaves palmately or ternately divided, or the lower simple and the upper ternate, or the lower palmate and the upper not cut or ternate, then pinnate.
      Leaves densely tomentose; outer petals of the umbel larger and 2-cleft..... ....6122. HERACLEUM, p. 728.
      Leaves usually glabrous; outer petals of the umbel not larger nor 2-cleft.
        Central flower and fruit of umbellule sessile.........6008. ZIZIA, p. 721.
        Central flower and fruit of umbellule not sessile.
          Plants small, from a bulblike tuber; involucre leafy................. ....5960. ERIGENIA, p. 720.
          Plants taller, from elongated roots; involucre usually absent, never leafy.
            Leaflets entire; plants glaucous and glabrous; involucel bracts usually absent..............................6031. TAENIDIA, p. 723.
            Leaflets not entire; plants glabrous or pubescent, never glaucous; involucel bracts present.
              Flowers yellow; calyx teeth prominent....6076. THASPIUM, p. 725.
              Flowers white; calyx teeth small or obsolete.
                Plants annual; fruit about 3 mm long....6048. AETHUSA, p. 724.
                Plants perennial; fruit 4-6 mm long.
                  Leaves finely divided; involucral bracts linear or absent; plants of bogs and springy places...6081. CONIOSELINUM, p. 726.
                  Leaves not finely divided; leaflets serrate or sometimes incised; involucral bracts absent; plants of moist to dry habitats.
                    Rays of umbel glabrous; fruit flattened laterally.......... ....6070. LIGUSTICUM, p. 724.
                    Rays of umbel densely scabrous or densely short-pubescent; fruit flattened dorsally.........6082. ANGELICA, p. 726.
    Leaves pinnately divided.
      Involucre present.
        Stems abundantly specked with purple............5970. CONIUM, p. 720.
        Stems never specked with purple.
          Leaflets filiform; roots tuberous; garden escape..6020. CARUM, p. 723.
          Leaflets linear to lanceolate; roots not tuberous or a fascicle of tubers.
            Leaf margins regularly and sharply serrate to the base, usually with 2-6 teeth to the cm; fruit about 3 mm long...6038. SIUM, p. 724.
            Leaf margins remotely and irregularly dentate, usually only above the middle or entire, usually 1 or 2 teeth to the cm; fruit usually 5-6 mm long..........................6107. OXYPOLIS, p. 727.
      Involucre usually absent.

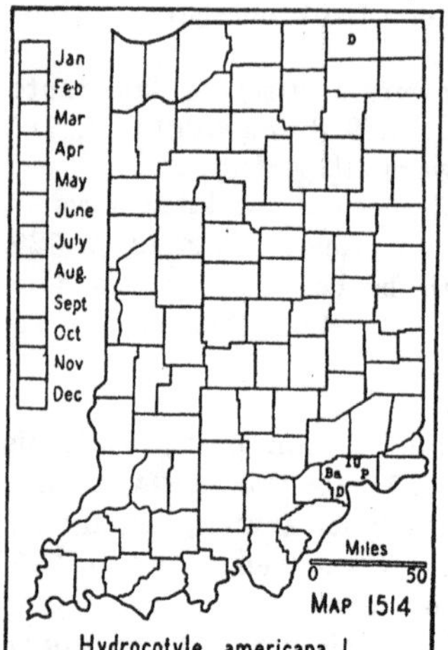

Map 1514 Hydrocotyle americana L

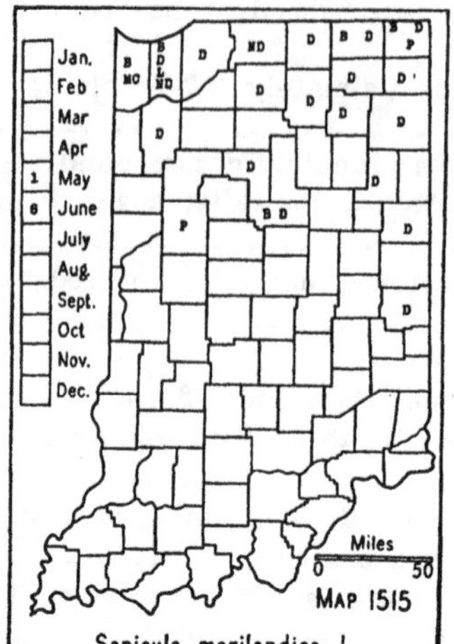

Map 1515 Sanicula marilandica L

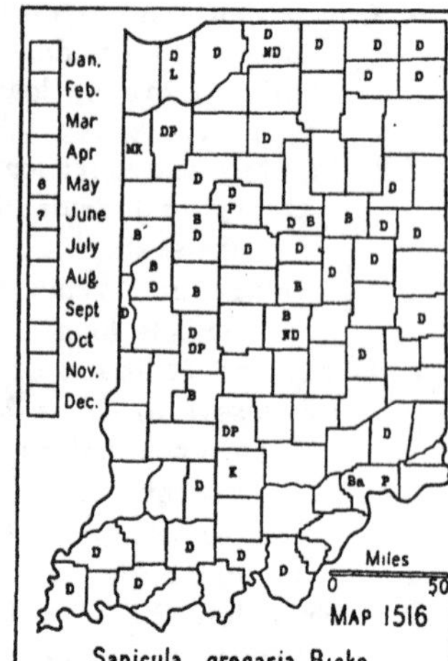

Map 1516 Sanicula gregaria Bickn

Flowers white; fruit flattened laterally.
  Leaflets toothed; stems streaked with purple or with bulblets in the axils of upper leaves........................6011. CICUTA, p. 722.
  Leaflets entire; stems not streaked with purple, bulblets absent...6036. PERIDERIDIA, p. 723.
Flowers yellow; fruit flattened dorsally.
  Leaves once pinnate; fruit broadly ovate, usually 5-7 mm long...6120. PASTINACA, p. 727.
  Leaves 2 or 3 times pinnate; fruit slightly obovate, 6-10 mm long...6102. POLYTAENIA, p. 727.

5893. HYDROCÓTYLE [Tourn.] L. WATER PENNYWORT

Leaves peltate..1. *H. umbellata.*
Leaves not peltate.
  Umbels pedunculate; fruit sessile. (See excluded species no. 477, p. 1078)...*H. rotundifolia.*
  Umbels sessile or nearly so; fruit pedicellate....................2. *H. americana.*

1. **Hydrocotyle umbellàta** L. UMBELLATE PENNYWORT. Map 1513. Common on sandy beaches and in the outlets of a few lakes in the northern counties.

N. S. to Fla., westw. to Ark. and Tex.; also in Oreg. and Calif.

2. **Hydrocotyle americàna** L. AMERICAN PENNYWORT. Map 1514. In 1933 I found a small colony of this species in the tamarack border of the east side of Cogg Lake, Lagrange County. It grew in the shade in sphagnum around the base of a small tamarack with *Menyanthes, Sarracenia,* and *Vaccinium macrocarpon.* It was at maximum anthesis on July 15.

This species was reported in 1878 from Jefferson County by Barnes, but no data accompanied the report. There is a specimen in the herbarium of Indiana University which was collected by Young in Jefferson County, September, 1875. In 1935 it was again found in Jefferson County by Miss Edna Banta. It was growing in shallow soil on a high ledge of rock

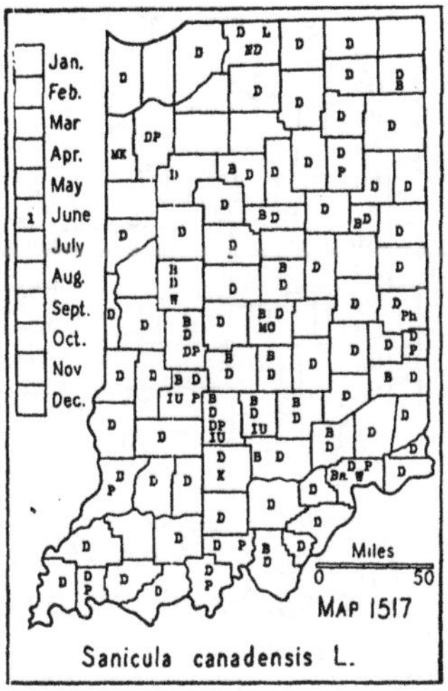

MAP 1517
Sanicula canadensis L.

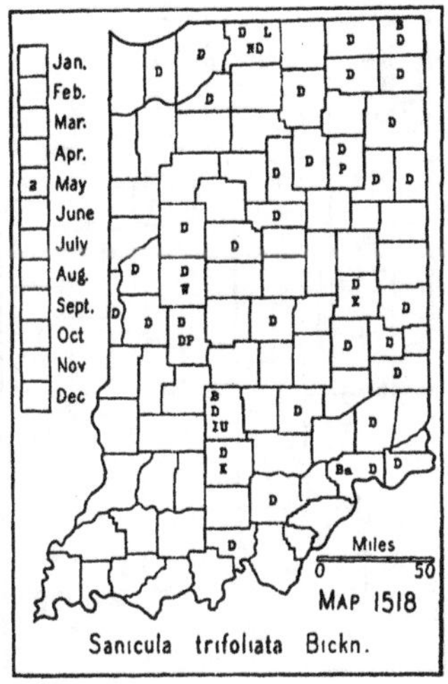

MAP 1518
Sanicula trifoliata Bickn.

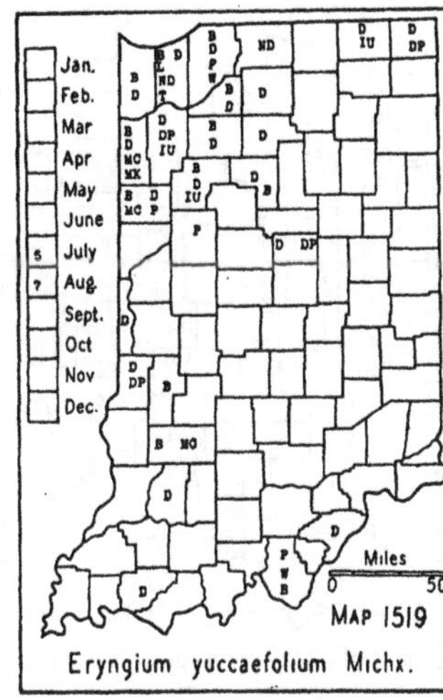

MAP 1519
Eryngium yuccaefolium Michx.

on the border of a seeping spring on the north slope of Big Creek, a mile south of Lancaster.

Newf. to Wis., southw. to N. J., Pa., Ohio, Ind., and in the mts. to N. C.

5918. SANÍCULA L. Sanicle

Styles much exceeding the bristles of the fruit, recurved.
- Calyx teeth of staminate flowers lanceolate, 1-1.6 mm long, indurated at the apex, sharp-pointed; flowers whitish green; fruit sessile..........1. *S. marilandica.*
- Calyx teeth of staminate flowers triangular-ovate, 0.5-0.7 mm long, not indurated at the apex, obtuse or acute; flowers yellowish green; fruit pedicellate...2. *S. gregaria.*

Styles shorter than the bristles of the fruit.
- Pedicels of staminate flowers 2-3 mm long; fruit subglobose; leaf segments 3-5, rather narrow; calyx inconspicuous in mature fruit...........3. *S. canadensis.*
- Pedicels of staminate flowers about 4 mm long; fruit elliptic; leaf segments generally 3, rather broad; calyx forming a conspicuous beak on mature fruit...4. *S. trifoliata.*

1. **Sanicula marilándica** L. Map 1515. Infrequent in moist or dry woods, usually associated with white oak. I have seen no specimens from the southern part of the state.

Newf. to B. C., southw. to Ga., Colo., and N. Mex.

2. **Sanicula gregària** Bickn. Map 1516. Infrequent throughout the state, usually associated with some species of oak.

N. B. and N. S. to S. Dak., southw. to Ga. and La.

3. **Sanicula canadénsis** L. Map 1517. Our most common sanicle. Frequent in all parts of the state except in the northern counties where it is infrequent to rare. Moist or dry woods, usually associated with some species of oak but sometimes in beech and sugar maple woods.

N. H. to Minn. and S. Dak., southw. to Fla., Nebr., and Tex.

4. **Sanicula trifoliàta** Bickn. Map 1518. Infrequent to probably rare in all parts of the state. In moist or dry woods, more often in beech and sugar maple woods than in white and black oak woods.

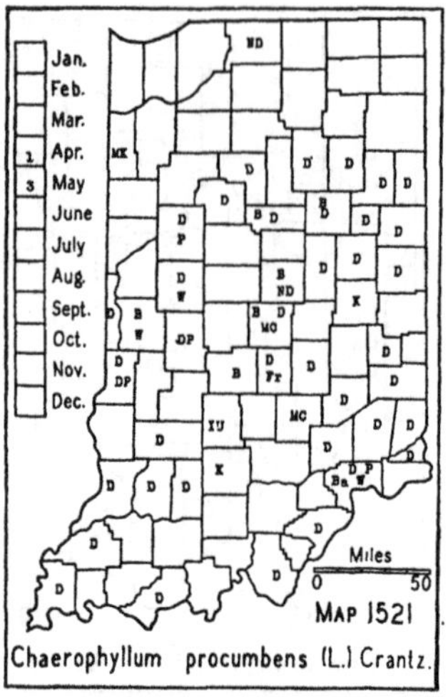

Chaerophyllum procumbens (L.) Crantz.

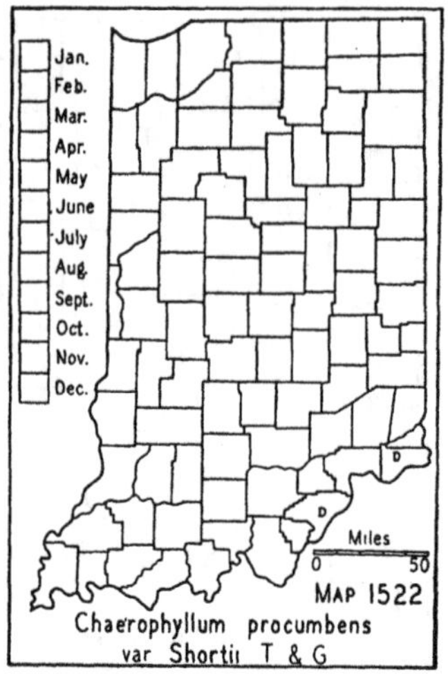

Chaerophyllum procumbens var. Shortii T. & G.

Chaerophyllum Tainturieri Hook.

Cent. Maine and from w. N. E. to Ont. and Minn., southw. to Tenn. and in the mts. to N. C.

5923. ERÝNGIUM [Tourn.] L.

1. **Eryngium yuccaefòlium** Michx. BUTTON-SNAKEROOT. Map 1519. Infrequent to frequent in moist sandy soil in prairie habitats in northwestern and western Indiana and in the southern counties in dry oak woods which, for the most part, were formerly known as the barrens.

Conn. to Minn., southw. to Fla. and Tex.

5935. CHAEROPHÝLLUM [Tourn.] L.

Stems glabrous or nearly so; under surface of leaves glabrous or nearly so, the segments of the leaflets oblong, bluntish or rounded at the apex; umbels on long peduncles; pedicels not clavate.

Fruit 7-8 mm long, glabrous .... 1. *C. procumbens.*

Fruit about 6 mm long, densely pubescent with short, spreading hairs .... 1a. *C. procumbens* var. *Shortii.*

Stems distinctly villous; under surface of leaves villous, the segments of the leaflets lanceolate or oblanceolate, acutish; umbels sessile; pedicels clavate .... 2. *C. Tainturieri.*

1. **Chaerophyllum procúmbens** (L.) Crantz. CHERVIL. Map 1521. Frequent on the alluvial flood plains of streams throughout the state except in the northern counties where it is rare. Where it is found it is usually abundant and sometimes an annoying weed.

N. Y., Mich. to Iowa, southw. to N. C., La., and Ark.

1a. **Chaerophyllum procumbens** var. **Shórtii** T. & G. Map 1522. Abundant on the alluvial bank of the Ohio River above the mouth of Fourteenmile Creek in Clark County. The variety flowers about 10 days earlier than the species which grew in abundance about a fourth mile from where the variety was found. The two were not mixed. The species was badly

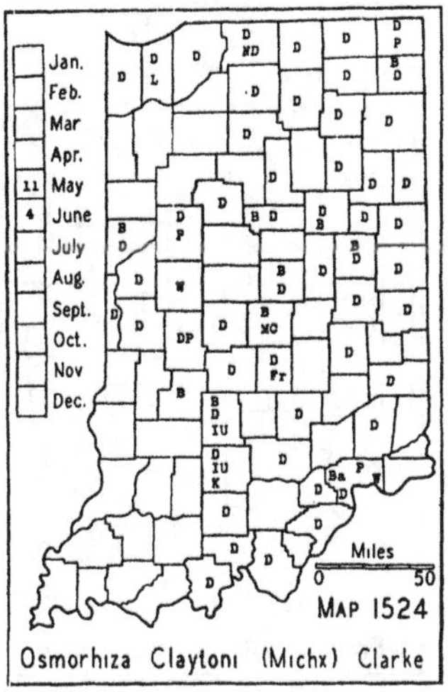

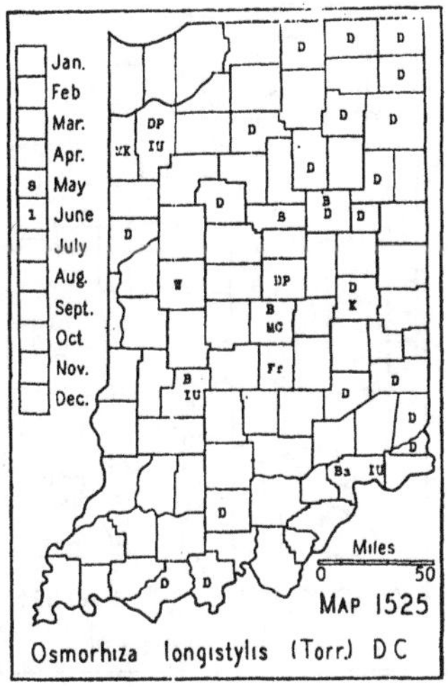

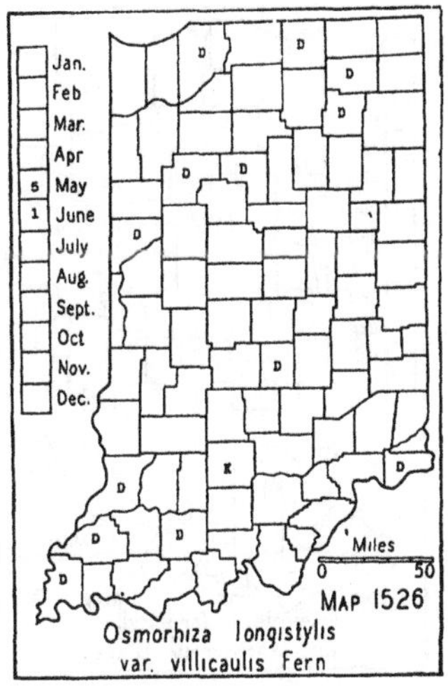

attacked by a rust but the variety was free from it. Another location is in Switzerland County on the wooded flood plain of a small creek about a mile and a half northwest of Vevay.

Pa. to Va., westw. to Ind. and Ky.

2. **Chaerophyllum Tainturièri** Hook. Map 1523. Abundant along the L. & N. Railroad about 6 miles west of Solitude in Posey County. At this place the railroad runs along the base of Brewer Hill in the alluvial flood plain of the Wabash River.

Va. to Mo., southw. to the Gulf.

5941. OSMORHÌZA Raf.

Stylopodium and style 0.8-1.5 mm long, not longer than the expanded petals, recurved in flower, becoming straight and erect when mature..............1. *O. Claytoni.*
Stylopodium and style 3-4 mm long, longer than the expanded petals, usually divergent when mature.
- **Stems glabrous or the upper part with some pubescence............2. *O. longistylis.***
- **Stems villous with hairs about 1-2 mm long.......2a. *O. longistylis* var. *villicaulis.***
- **Stems puberulent with hairs mostly about 0.5 mm long..2b. *O. longistylis* var. *brachycoma.***

1. **Osmorhiza Clàytoni** (Michx.) Clarke. (*Washingtonia Claytoni* (Michx.) Britt.) SWEET CICELY. Map 1524. Frequent throughout the state except in the southwestern counties. In moist or dry woods of all kinds except in very sandy places.

The stems are usually more or less villous but sometimes nearly glabrous.

E. Que. and N. S. to S. Dak., southw. to N. C., Ala., to Mo., and Kans.

2. **Osmorhiza longistỳlis** (Torr.) DC. (*Washingtonia longistylis* (Torr.) Britt.) SWEET ANISE. Map 1525. Infrequent in moist or dry woods and probably found throughout the state.

E. Que. to Assin., southw. to N. C., Ala., Kans., and Colo.

MAP 1527
Osmorhiza longistylis var. brachycoma Blake

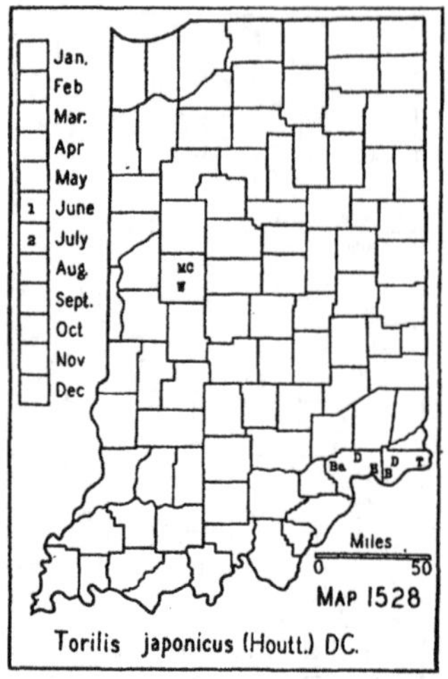

MAP 1528
Torilis japonicus (Houtt.) DC.

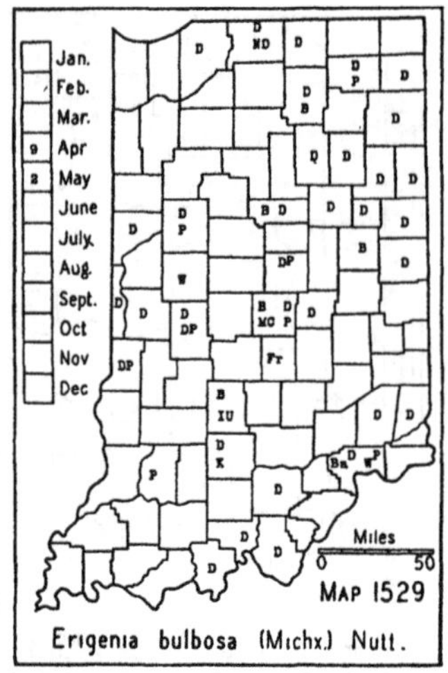

MAP 1529
Erigenia bulbosa (Michx.) Nutt.

2a. **Osmorhiza longistylis** var. **villicaúlis** Fern. Map 1526. Infrequent, probably throughout the state in rich or dry woods.

Del. to N. Dak., and Nebr., southw. to Va., Mo., and Okla.

2b. **Osmorhiza longistylis** var. **brachycòma** Blake. (Rhodora 25: 110. 1923.) Map 1527. Our only specimen is from a woods about 12 miles northeast of Indianapolis now known as "Woollen's Garden of Birds and Botany."

Ont., N. Y., D. C., Pa., Maine, Ohio, and Ind.

5945. TÓRILIS Adans.

1. TORILIS JAPÓNICUS (Houtt.) DC. (Rhodora 40: 291-292. 1938.) (*Torilis Anthriscus* Gmel.) ERECT HEDGE PARSLEY. Map 1528. This species was first collected in Montgomery County by A. R. Bechtel. It was common on a wooded bank of Sugar Creek about 3 miles north of Crawfordsville in 1926, and in 1927 it was collected in Jonathan Winters' woods about 2 miles northwest of Darlington. In 1935 it was discovered by Miss Edna Banta to be a frequent to common weed along the road and adjacent areas along the river bluff between Brooksburg and Madison in Jefferson County.

Nat. of Eu.; N. Y., N. J., Okla., Tex., and Oreg.

5960. ERIGENÌA Nutt.

1. **Erigenia bulbòsa** (Michx.) Nutt. HARBINGER-OF-SPRING. Map 1529. Frequent to rare in rich woods throughout the state, usually found in beech and sugar maple woods and sugar maple and basswood woods.

Western N. Y., s. Ont. to Minn., southw. to Kans., Ark., and Ala.

5970. CONÌUM L.

1. CONIUM MACULÀTUM L. POISON HEMLOCK. Map 1530. I have seen this plant in cultivation twice but the owners were not aware of its poisonous character. Local along roadsides and alluvial banks of streams and

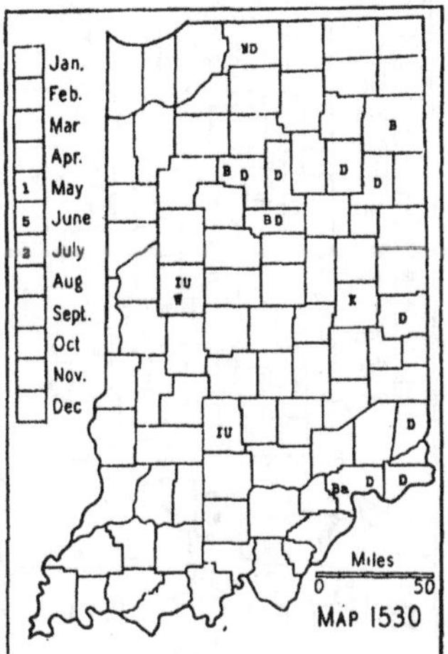

Conium maculatum L.

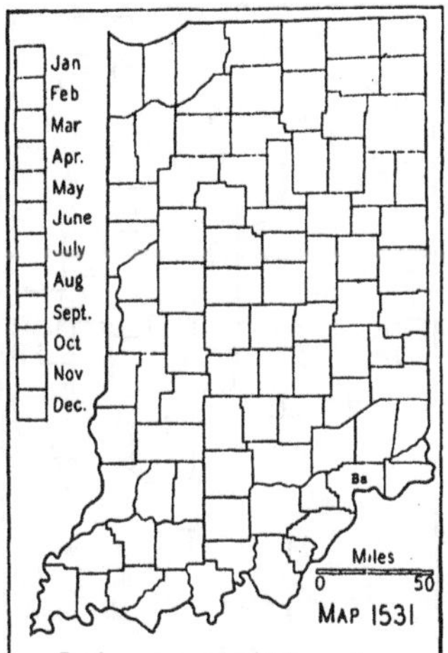

Bupleurum rotundifolium L.

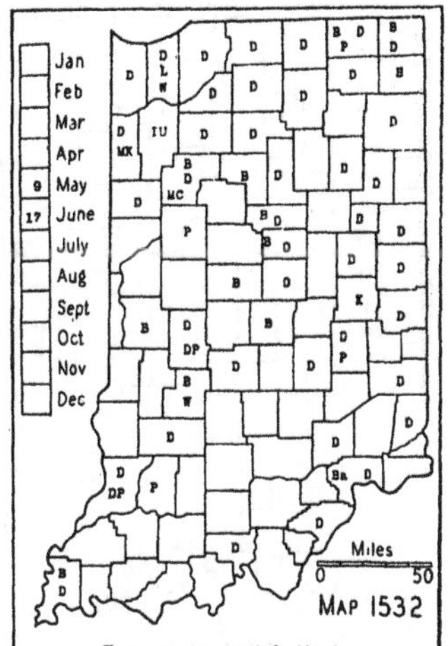

Zizia aurea (L.) Koch

locally abundant along the old canal in Huntington, Wabash, and Miami Counties and found, no doubt, farther down the canal. Noted along the Ohio River in Dearborn County and as a weed in fields between Madison and Hanover in Jefferson County.

Nat. of Eu.; N. S. to Ont. and Mo., southw. to Del., Pa., and Tex.; also in Calif., Mex., and S. A.

5994. BUPLEÙRUM [Tourn.] L.

1. BUPLEURUM ROTUNDIFÒLIUM L. HARE'S EAR. Map 1531. This species was discovered in 1933 by Miss Edna Banta who says it is frequent along Lost Fork Creek about 3 miles east of Brooksburg, Jefferson County. She adds that the weed was known on her father's farm for at least five years.

Nat. of cent. Eu. to n. Africa and w. Asia; N. Y. to N. C., westw. to Ind., Ky., Tenn., Mo., Ark., and S. Dak.

6004. SPERMÓLEPIS Raf.

See excluded species no. 478, p. 1078.

6008. ZÍZIA Koch

Basal and cauline leaves 2-3-ternate; leaflets very acute; central flower of umbellules sessile; fruit oval, about 4 mm long................................1. *Z. aurea.*
Basal leaves simple, suborbicular or broadly ovate; cauline leaves with 3-5 rather bluntish leaflets; central flower of umbellules sessile; fruit ovate, about 3 mm long..2. *Z. aptera.*

1. **Zizia aúrea (L.) Koch.** GOLDEN ALEXANDERS. Map 1532. Rather frequent in moist soil in most parts of the state. Moist woodland and very often in moist places along roadsides. Usually in small colonies. Gray, Man., ed. 5 and Wood, Classbook of 1865 did not make the distinction between this species and *Thaspium trifoliatum* var. *flavum* very clear and both were known to authors as *Thaspium aureum* Nutt. Consequently both

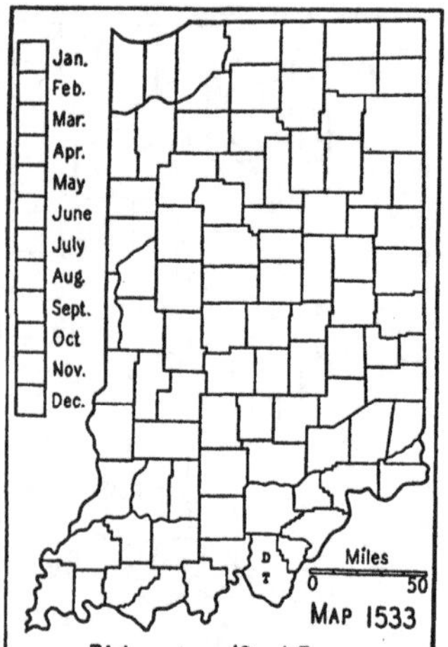

Zizia aptera (Gray) Fern.

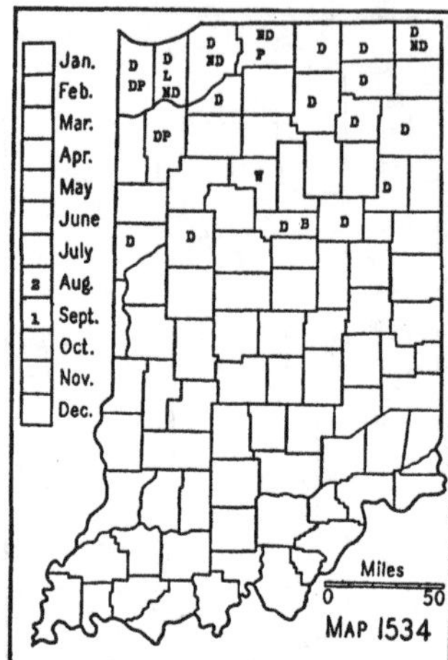

Cicuta bulbifera L.

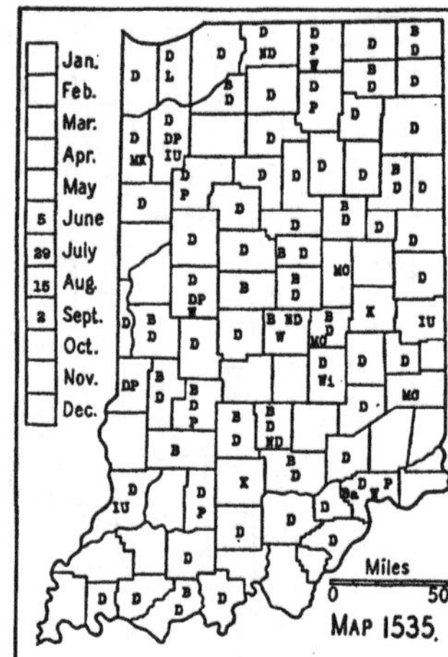

Cicuta maculata L.

species were reported under the last name by authors before our present manuals were in use so that it is impossible to know which species the author had in hand.

E. Que. to Alberta, southw. to Fla., Ark., and Tex.

2. **Zizia áptera** (Gray) Fern. (Rhodora 41: 441-444. 1939.) (*Zizia cordata* (Walt.) DC.) HEARTLEAF ALEXANDERS. Map 1533. Our only specimens are from an open, wooded slope in Harrison County about 3 miles east of Elizabeth. It was reported from Steuben County by Bradner, but no doubt this report should be referred to *Thaspium trifoliatum* var. *flavum* which is found there and which he did not report. *Zizia aptera* is often confused with *Thaspium trifoliatum* var. *flavum* which also has thickened, white, and glabrous margins of the leaflets, but from which it can easily be distinguished because *Zizia aptera* has a sessile central flower in each umbellule.

Conn. to Alberta, southw. to Ga., Mo., Colo., and Oreg.

6011. CICÙTA L.

Leaflets narrowly linear; fruit orbicular, 1.5-2 mm long; axils of upper leaves bulblet-bearing..1. *C. bulbifera.*

Leaflets lanceolate, rarely some narrowly ovate; fruit oblong, 2.8-3.9 mm long; axils of leaves not bulblet-bearing..2. *C. maculata.*

1. **Cicuta bulbífera** L. Map 1534. Restricted to the lake area of the northern part of the state. Miry, mucky, or sandy borders of lakes, ponds, and swamps. Infrequent. Grows in wetter situations than does the next species.

Newf. to B. C. southw. to Md., Ind., Nebr., and Oreg.

2. **Cicuta maculàta** L. WATER HEMLOCK. Map 1535. This plant is poisonous and each year in this state there are reports of the death of stock due to eating it. A man in Wells County, mistaking the tuberous roots for sweet anise, ate them and died. Frequent throughout the state in low ground about lakes and ponds, in low woods, and in and along ditches.

Que. to Man., southw. to Fla. and Tex.

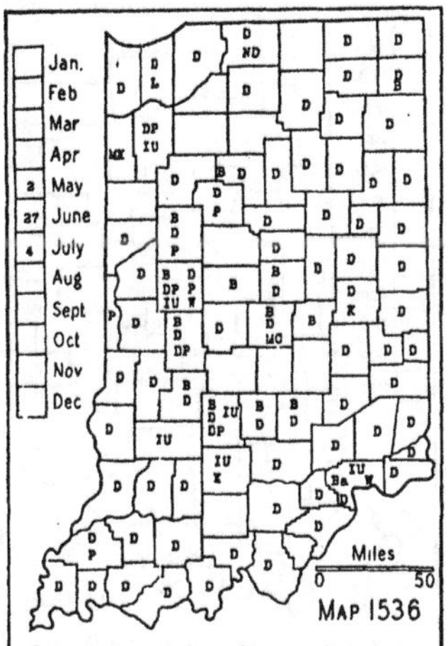

Cryptotaenia canadensis (L.) DC.

Carum Carvi L.

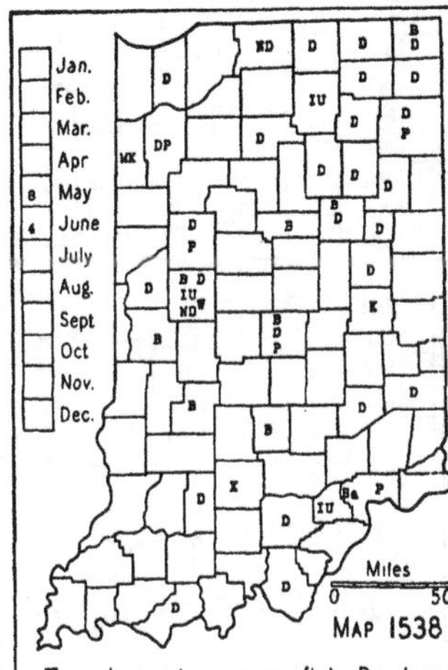

Taenidia integerrima (L.) Drude

6015. CRYPTOTAÈNIA DC.

1. **Cryptotaenia canadénsis** (L.) DC. (*Deringa canadensis* (L.) Ktze.) HONEWORT. Map 1536. Frequent to common in moist, rich woods throughout the state. Frequent almost everywhere in woods except on very dry slopes, in very sandy soil, and in very wet woodland. Experience has shown that this species and *Sanicula* should not be introduced into wild flower gardens because both soon become weeds.

Que. to La., westw. to S. Dak., Kans., and Ark.

6020. CÀRUM L.

1. CARUM CÁRVI L. CARAWAY. Map 1537. Cultivated for its aromatic seed which are used as a condiment. Sparingly escaped.

Nat. of Eu.; Newf. to B. C., southw. to Pa., Mont., and Oreg.

6031. TAENÍDIA Drude

1. **Taenidia integérrima** (L.) Drude. Map 1538. Infrequent to rare in all parts of the state. Generally on the upper part and crests of wooded slopes bordering streams and usually in clay or gravelly soil, associated with white oak or with white and black oaks.

W. Que. and w. N. E. to Minn., southw. to Ga. and Miss.

6036. PERIDERÍDIA Reichenb.

1. **Perideridia americàna** (Nutt.) Reichenb. (*Eulophus americanus* Nutt.) Map 1539. A rare plant in a few of our western counties. Our only reports are from Jasper and Vigo Counties. I have collected it in Benton, Newton and Spencer Counties. It is a typical prairie plant and is usually found in such a habitat. My Spencer County specimen, however, was found among large post oaks in a low, flat, post oak, pin oak,

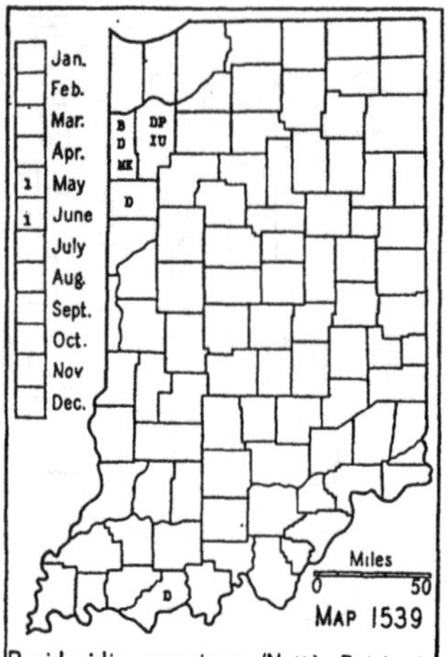

Perideridia americana (Nutt.) Reichenb.

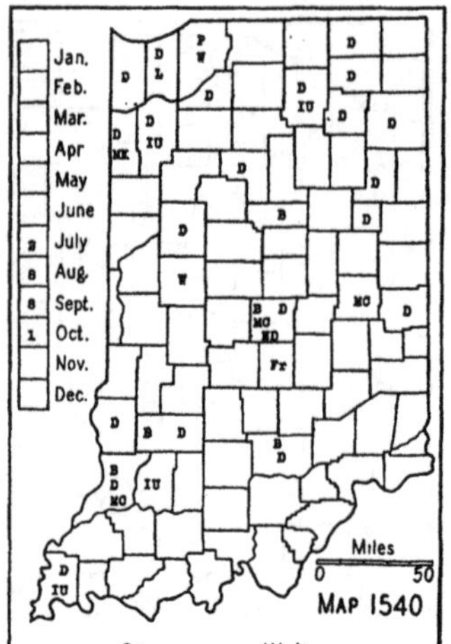

Sium suave Walt.

Ligusticum canadense (L.) Britt.

and swamp white oak flat about 4 miles northwest of Chrisney. This area is a low, flat woods about a half mile wide on the east side of Little Pigeon Creek. The soil is a hard, white, slightly acid clay in which are found several plants distinctly southwestern in their distribution.

W. Ind. to Mo., southw. to Tenn. and Ark.

6038. SÌUM [Tourn.] L.

1. **Sium suàve** Walt. (Rhodora **17**: 131. 1915.) (*Sium cicutaefolium* Schrank.) WATER PARSNIP. Map 1540. In the shallow water of the borders of ponds and sloughs and in springy places along streams, swampy woodland, and ditches.

Newf. to B. C., southw. to Fla., La., and Calif.

6048. AETHÙSA L.

See excluded species no. 479, p. 1078.

6070. LIGÚSTICUM L.

1. **Ligusticum canadénse** (L.) Britt. Map 1541. Our only specimen was found in a moist place near the crest of a high wooded slope along the Ohio River about 3 miles east of Elizabeth in Harrison County, and near the road leading up the bluff from Stuart's Landing. This species was reported in Coulter's Catalogue for Wilson from Hamilton County. Since this species much resembles *Thaspium barbinode* which, no doubt, is more or less frequent in Hamilton County and which was not reported by Wilson, and since the habitat of *Ligusticum canadense* does not occur in Hamilton County, there is little doubt but Wilson made an error in determination. I have not been able to locate Wilson's specimens. I never received answers to the letters I sent him.

S. Pa. and s. Ind. to Mo., southw. to Ga. and Ala.

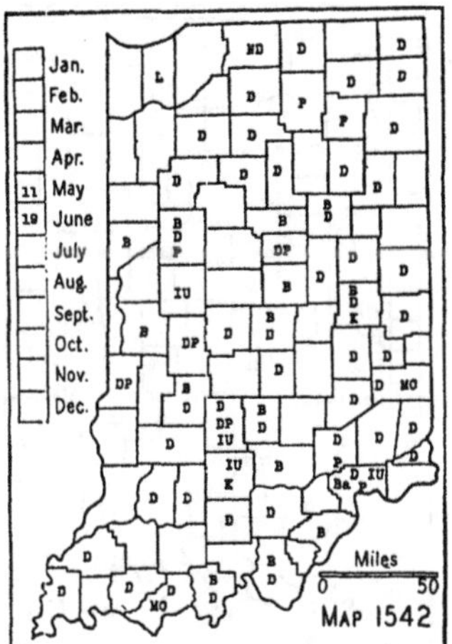

Thaspium trifoliatum var. flavum Blake

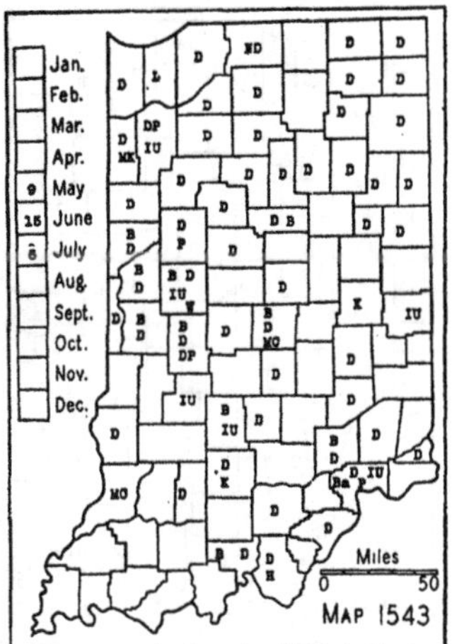

Thaspium barbinode (Michx.) Nutt.

Conioselinum chinense (L.) B S P.

6076. THÂSPIUM Nutt.

Margins of leaves and leaflets white and smooth; plants glabrous throughout; root-leaves mostly cordate; basal leaves simple and cordate or ternate; leaflets thickish, crenate; flowers deep yellow; fruit globose-ovoid, about 4 mm long..1. *T. trifoliatum* var. *flavum.*

Margins of leaves and leaflets green and more or less ciliate; plants puberulent at least at the nodes; root leaves ternate; basal leaves mostly biternate; segments of leaflets ovate to lanceolate, with a cuneate base, thin, incised, coarsely toothed or ternately parted; flowers light yellow; fruit 4-6 mm long.........2. *T. barbinode.*

1. **Thaspium trifoliàtum** (L.) Britt. var. **flàvum** Blake. (Rhodora **20**: 53. 1918.) (*Thaspium aureum* Nutt. and *Thaspium trifoliatum* (L.) Britt.) Map 1542. Frequent in woodland throughout the state. Usually restricted to wooded slopes along or near streams. It is to be noted that all Indiana specimens have yellow flowers.

N. Ohio and Md. to Wyo., southw. to Ga. and Ark.

2. **Thaspium barbinòde** (Michx.) Nutt. (*Thaspium barbinode* var. *angustifolium* Coult. & Rose.) Map 1543. Frequent throughout the state in rich woods and infrequent in prairie habitats, preferring alluvial soil along streams and wooded slopes.

This species is extremely variable in all of its parts and the form with narrow leaf-segments has been named. Plants that grow in rich soil in shady places usually have the leaf-segments large and ovate while plants that grow in poor soil and prairie habitats usually have the leaf-segments narrow. The nodes, peduncles, umbels, and furrows of the fruit are generally more or less pubescent with short, stout, colorless hairs (sometimes only granulose). The nodes are always pubescent and rarely can a plant be found that has the inflorescence nearly glabrous. Sometimes the pubescence is conspicuous in the inflorescence and on the veins of the lower surface of the leaflets. The flowers are sometimes cream-colored, and the fruit varies in size and pubescence. I am not able to correlate the pubes-

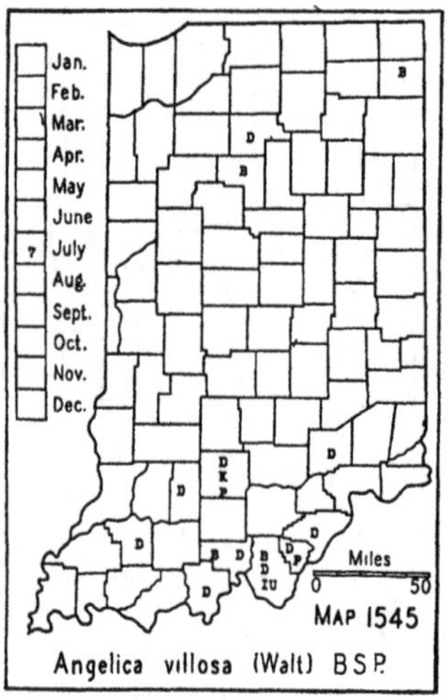

Angelica villosa (Walt) BSP.

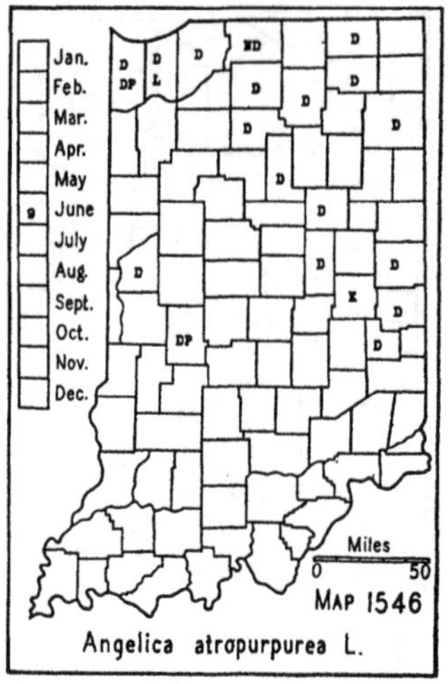

Angelica atropurpurea L.

Polytaenia Nuttallii DC.

cence with any other character and have concluded that we have a polymorphic species whose variations are due to soil and exposure.

N. Y. to Minn., southw. to Fla. and Ark.

6081. CONIOSELÌNUM Hoffm.

1. **Conioselinum chinénse** (L.) BSP. Map 1544. Very rare. Two of my specimens were collected in dense shade on the south banks of streams in springy places in sandy-gravelly soil made miry by large springs, and I found it in a tamarack bog in La Porte County. Grimes collected a specimen in damp soil on a rocky slope along Sugar Creek in Montgomery County. It has also been reported from Carroll and Noble Counties and from the area of Delaware, Jay, Randolph, and Wayne Counties.

Newf. to Minn., southw. to Pa. and Ind., and in the mts. to N. C.

6082. ANGÉLICA L.

Stems pubescent above; leaflets of median leaves mostly less than 2.5 cm wide; umbels densely pubescent; fruit pubescent............................1. *A. villosa.*

Stems glabrous above; leaflets of median leaves mostly more than 2.5 cm wide; umbels densely scabrous; fruit glabrous............................2. *A. atropurpurea.*

1. **Angelica villòsa** (Walt.) BSP. HAIRY ANGELICA. Map 1545. Infrequent in the unglaciated region on barren wooded slopes and appearing again in the northern counties in a dry, sandy, prairie habitat.

W. Mass. to Minn., southw. to Fla., Tenn., and Mo.

2. **Angelica atropurpùrea** L. PURPLESTEM ANGELICA. Map 1546. Infrequent in the northern two thirds of the state, being more frequent in the northern counties. In marshes, in mucky soil about lakes and ponds, and in alluvial bottoms along streams.

Newf. to Minn., southw. to Del., Ill., and Iowa.

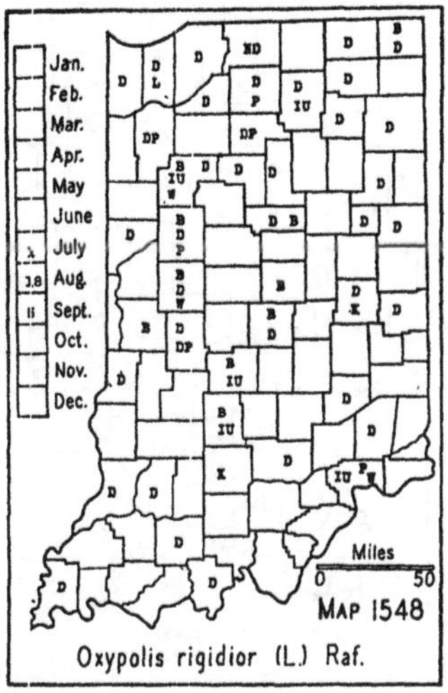

Oxypolis rigidior (L.) Raf.

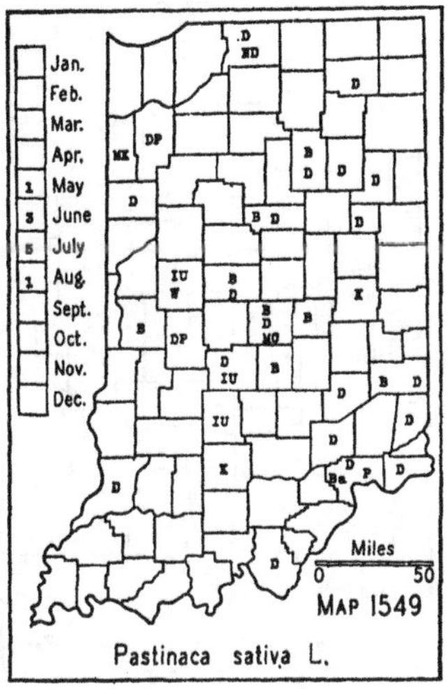

Pastinaca sativa L.

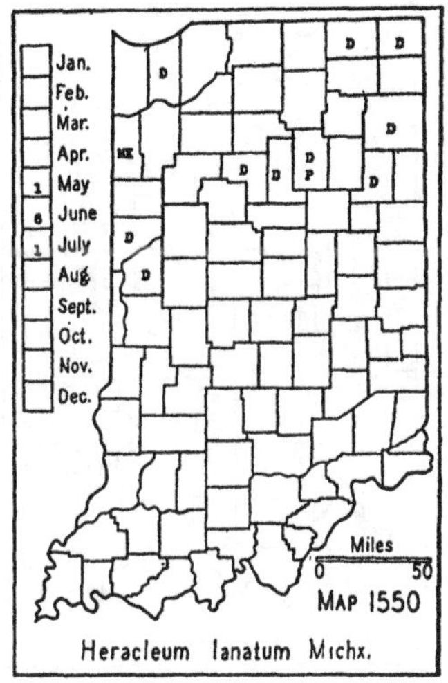

Heracleum lanatum Michx.

6102. POLYTAÈNIA DC.

1. **Polytaenia Nuttállii** DC. (*Pleiotaenia Nuttallii* (DC.) Coult. & Rose.) Map 1547. In a prairie habitat in a few of the northwestern counties. Also found by Dr. Clapp in the "barrens west of the lake" (Harrison County, southwest of Palmyra). His specimen is in the herbarium of Wabash College. Rare.

Mich. (?) to Wis., southw. to Ala. and Tex.

6107. OXÝPOLIS Raf.

1. **Oxypolis rigídior** (L.) Raf. COWBANE. Map 1548. Infrequent throughout the state in marshes, wet borders of lakes, wet woods, swamps and wet interdunal flats. In some large marshes it is common and in these a study of its variation may be made. I have done so and found plants of varying size with all of the leaflets toothed, usually above the middle, some with nearly all of the leaflets entire or some with all of the leaflets entire. Usually the larger plants have larger leaves and the leaflets are more toothed, and the leaves of the smaller plants have fewer teeth or are entire. The form with entire leaflets has been given a varietal name but I believe this is only a form of the species without taxonomic significance. The plants with entire leaflets conform to the type in other characters. The width of the widest leaflets of the plants with entire leaflets is as follows: 2.5 mm; 5.5 mm; 6 mm; 7 mm; 8 mm; and 15 mm.

Wilson reports this species as "common" in Hamilton and Marion Counties. No doubt this report should be referred to *Cicuta maculata* which is a common plant in those counties and which he does not report.

N. Y. to Minn., southw. to the Gulf.

6120. PASTINÀCA L.

1. PASTINACA SATÌVA L. PARSNIP. Map 1549. A weed throughout the state. Rare in only a few areas and infrequent to common along roadsides

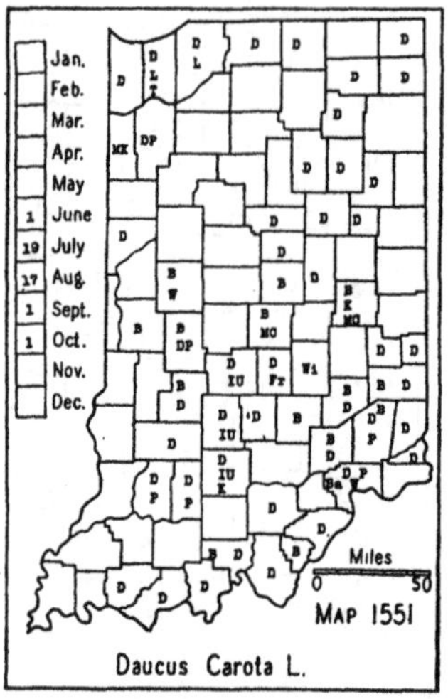

Daucus Carota L.

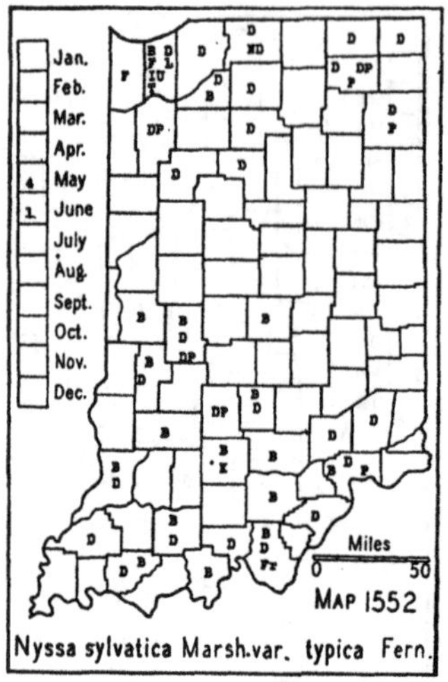

Nyssa sylvatica Marsh. var. typica Fern.

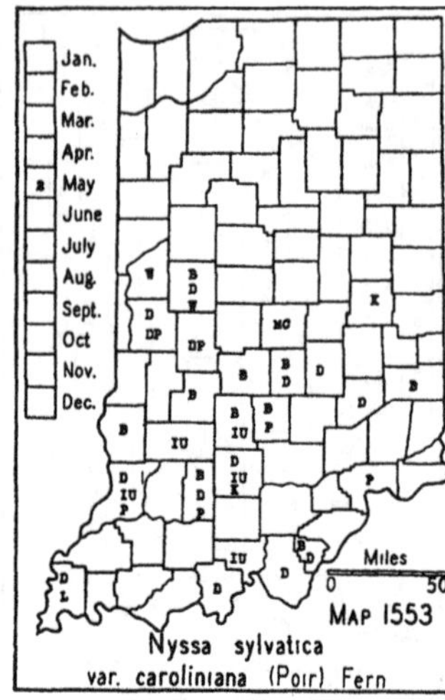

Nyssa sylvatica var. caroliniana (Poir) Fern

and railroads, in bottom land along streams, in hayfields, pastures, and waste places. The parsnip is common in cultivation.

The juice of this plant is said to be poisonous to the skin (Rhodora 4: 188. 1902.)

Nat. of Eu.; in all parts of N. A.

6122. HERACLÈUM L.

1. **Heracleum lanàtum** Michx. COW PARSNIP. Map 1550. An infrequent or rare plant in moist, rich soil along streams, about lakes, and along roadsides.

Lab. to Alaska, southw. in the mts. to Ga., Nev., Kans., Utah, and Calif.

6142. DAÚCUS [Tourn.] L.

1. DAUCUS CARÒTA L. COMMON CARROT. Map 1551. Infrequent to common in all parts of the state but rarely found in a prairie habitat. The flowers vary in color from white to yellow; 11 of my 33 specimens have one or more purple flowers in the inflorescence; rarely there is an inflorescence with rose colored flowers. Millspaugh has named the rose colored form, forma *rosea.* Farwell calls the form without purple flowers, forma *epurpurata.* (See Grier. Variation in the flower of the wild carrot. Torreya 22: 64-66. 1922.) Often called Queen Anne's-lace.

Nat. of Eurasia; throughout N. A.

229. CORNÀCEAE Link. DOGWOOD FAMILY

Flowers 5-merous, polygamo-dioecious; leaves alternate; stone of fruit oblong, about 7 mm long....................6151. NYSSA, p. 729.

Flowers 4-merous, perfect; leaves opposite except in *Cornus alternifolia* which has a fruit with a suborbicular stone about 5 mm long............6159. CORNUS, p. 729.

6151. NÝSSA L.

[Fernald. The varieties of Nyssa sylvatica. Rhodora 37: 433-437. 1935.]

Lower surface of leaves smooth, not papillate or rarely so, glabrous, glabrate, or rarely densely pubescent on young specimens; leaves firm or subcoriaceous when mature, short-acute or blunt at the apex, lustrous above; green branchlets usually bending when flexed to a right angle; wood difficult to split.....1. *N. sylvatica* var. *typica.*

Lower surface of leaves papillose, glabrous, glabrate or more or less pubescent, especially on the veins; leaves not firm or subcoriaceous when mature, usually acuminate at the apex or some blunt; green branchlets usually breaking when flexed to a right angle; wood easy to split...........1a. *N. sylvatica* var. *caroliniana.*

1. **Nyssa sylvática** Marsh. var. **týpica** Fern. BLACK GUM. Map 1552. Infrequent to rare in the northern two thirds of the state and frequent to common in the southern part. It is found in both dry and wet soils, apparently preferring slightly acid soils. It is erratic in its distribution and is found in several tree associations. The leaves of coppice shoots and sometimes those of seedlings are often more or less lobed.

West-cent. Maine, s. Ont., s. Mich., se. Wis. to n. Mo., southw. to Fla. and ne. Tex.

1a. **Nyssa sylvatica** var. **caroliniàna** (Poir.) Fern. Map 1553. This variety is infrequent in the southern part of the state. It is one of the cove type and prefers a richer soil than does the typical form. Pioneers have always insisted that there were two kinds of black gum. They distinguish them by their splitting qualities. The form very difficult to split was known as the black gum, and the form that split "like poplar" was known as yellow gum. The bark of the variety much resembles that of the tulip tree, and the branches are usually ascending.

Chester County Pa. to Essex County, Ont., southw. to N. C., Miss., and e. Tex.

6159. CÓRNUS [Tourn.] L.

Inflorescence capitate, surrounded by a large 4-leaved, white, petaloid involucre; fruit red.
 Plants subherbaceous, low, mostly less than 2 dm high.............1. *C. canadensis.*
 Plants arborescent, small trees..................................2. *C. florida.*
Inflorescence cymose, without an involucre; fruit not red.
 Pith of branchlets and of one and two year old branches white.
  Leaves alternate...3. *C. alternifolia.*
  Leaves opposite.
   Leaves broadly ovate, generally having 7-9 pairs of veins, usually woolly-pubescent beneath at maturity; branchlets yellow green and usually more or less blotched with longitudinal, purplish spots; fruit bluish..4. *C. rugosa.*
   Leaves ovate, oblong-ovate or lanceolate-ovate, generally having 3-6 pairs of veins, woolly or appressed-pubescent beneath at maturity; branchlets reddish or grayish.
    Under surface of leaves mostly woolly at maturity.
     Leaves not rough above; pith of last year's branchlet wide, more than a third the diameter of the branchlet; branches bright red............ ....5a. *C. stolonifera* var. *Baileyi.*
     Leaves rough above; pith of last year's branchlet small, usually less than a third the diameter of the branchlet; branches gray (this species rarely with a white pith)...............................6. *C. asperifolia.*

Cornus canadensis L.

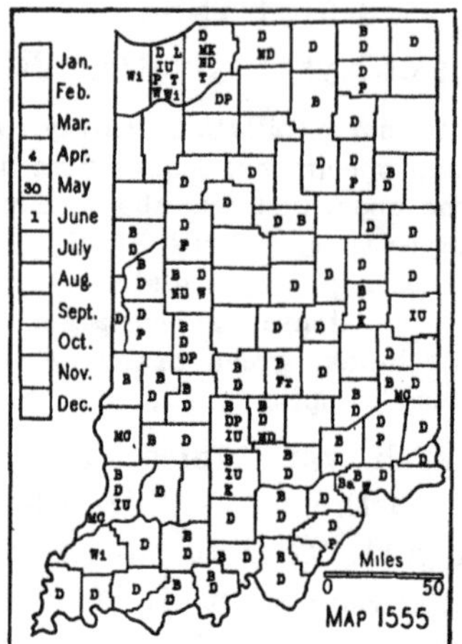

Cornus florida L.

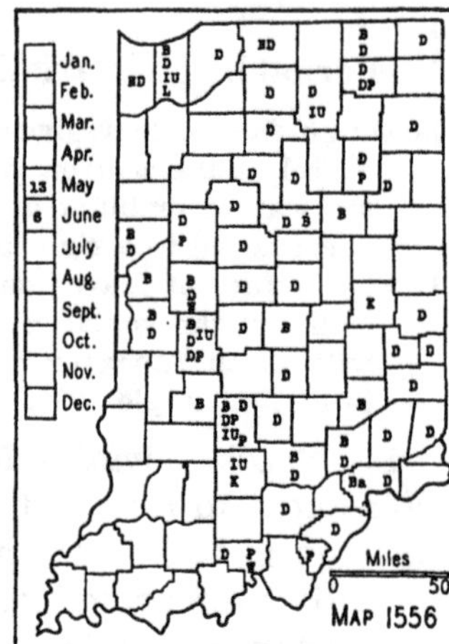

Cornus alternifolia L.f.

Under surface of leaves appressed-pubescent or glabrous at maturity.
Year old branches bright red; pith wide, usually more than a third the diameter of the branch; under surface of leaves thickly appressed-pubescent (sometimes old leaves becoming somewhat glabrous; fruit white..5. *C. stolonifera.*
Year old branches gray or dull, reddish brown; pith narrow, usually less than a third the width of the branch; under surface of leaves sparsely clothed with a short pubescence or almost glabrous.
Fruit white; inflorescence more or less appressed-pubescent with colorless hairs; peduncles 2-2.5 cm long; under surface of leaves more or less farinose (this species generally with a brownish pith, especially of two-year-old branchlets)................7. *C. racemosa.*
Fruit blue; inflorescence glabrous or appressed-pubescent; peduncles 2.5-7 cm long; under surface of leaves green; pith of branchlets white..8. *C. stricta.*
Pith of branchlets and of one and two year old branches tawny, sometimes white in the branchlets and in one year old branches of nos. 6 and 8.
Leaves rough above, woolly-pubescent beneath; fruit white......6. *C. asperifolia.*
Leaves not rough above, appressed-pubescent beneath; fruit white or bluish.
Pubescence of the under surface of the leaves consisting of colorless hairs, only those of the midrib sometimes reddish.
Branches reddish brown; branchlets densely pubescent; calyx lobes 0.75-1 mm long; fruit bluish..9. *C. obliqua.*
Branches gray; branchlets glabrous or glabrate; calyx lobes less than 0.75 mm long, usually minute or up to about 0.5 mm long; fruit white........
....7. *C. racemosa.*
Pubescence of under surface of the leaves consisting of reddish hairs; fruit bluish..10. *C. Amomum.*

1. **Cornus canadénsis** L. (*Chamaepericlymenum canadense* (L.) Asch. & Graebn.) BUNCHBERRY. Map 1554. Found only in Lake and Porter Counties near Lake Michigan. Very rare. I have it only from the Mineral Springs bog in Porter County where it was formerly common. McCaslin's report from Jay County and Scott's report from Kosciusko County no doubt should be referred to some other species, probably to *Medeola virginiana.*

Lab. to Alaska, southw. to N. J., W. Va., Ind., Minn., Colo., and Calif.

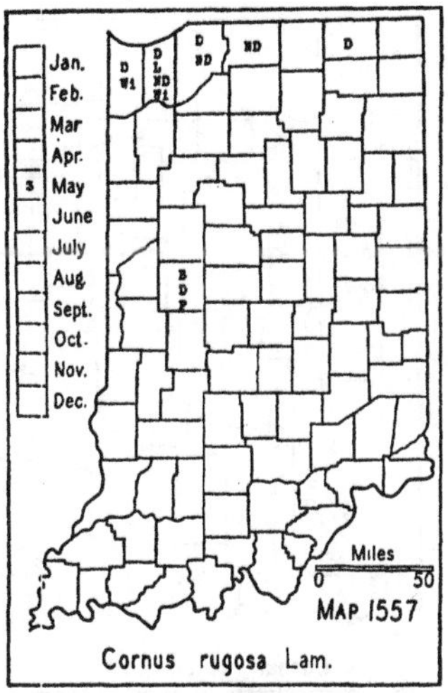

Cornus rugosa Lam.

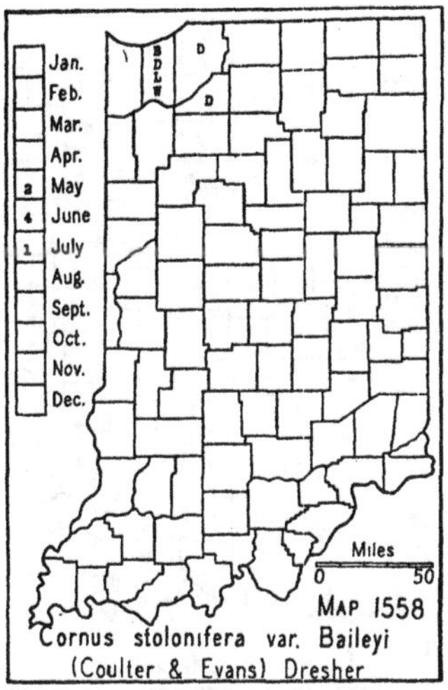

Cornus stolonifera var. Baileyi (Coulter & Evans) Dresher

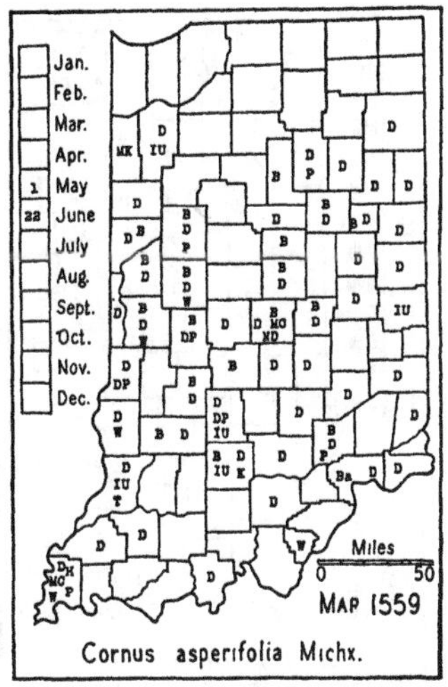

Cornus asperifolia Michx.

2. **Cornus flórida** L. (*Cynoxylon floridum* (L.) Raf.) FLOWERING DOGWOOD. Map 1555. Frequent to common in dry woods throughout the state except in the northwestern part where it is absent from the sandy black oak woods. The largest tree I have seen was in Warrick County, which had a clear bole of 10 feet and measured 40 inches in circumference at four and a half feet above the ground.

S. Maine and Ont. to Minn., southw. to Fla. and Tex.

3. **Cornus alternifòlia** L. f. PAGODA DOGWOOD. Map 1556. Infrequent to rare in the greater part of the state. We have only one record for the southwestern part of the state and none for the prairie counties. It usually grows in moist rich soil at the base of usually rocky, wooded slopes along or near streams where it may be locally frequent. The largest specimen seen was in Warren County which was 4 inches in diameter at breast height, and had a clear bole of 6 feet.

Newf. to Minn., southw. to Ga., Ala., and Mo.

4. **Cornus rugòsa** Lam. (*Cornus circinata* L'Hér.) ROUNDLEAF DOGWOOD. Map 1557. Found in the counties indicated on the map. The reports for other counties are, no doubt, errors in determination. It is infrequent on the moist shady slopes in the dunes near Lake Michigan, on the high sandy bank of Pigeon River west of Mongo in Lagrange County and in a low sandy woods north of Pigeon River 3 miles east of Mongo, and on the crest of a wooded ridge along Sugar Creek about a mile east of the Shades in Montgomery County. The Montgomery county plant was found in a relict area with *Pinus Strobus, Gaultheria procumbens,* and *Rhus typhina.*

E. Que. to Man., southw. to Va., Ind., Ill., Iowa, and N. Dak.

5. **Cornus stolonífera** Michx. RED-OSIER DOGWOOD. Map 1560. Infrequent to rare in swamps and wet places, mostly in the lake area. Nos. 2, 3, 5, and 7 flower about 2 weeks earlier than the other species.

Lab. to Mackenzie, southw. to Va., Ky., Iowa, Nebr., N. Mex., Ariz., and Calif.

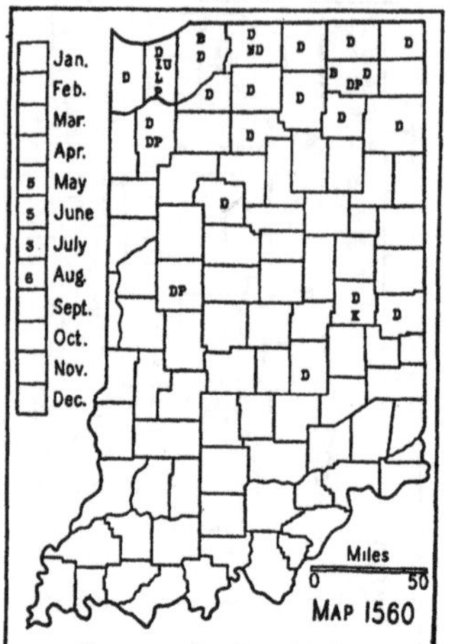

Cornus stolonifera Michx.

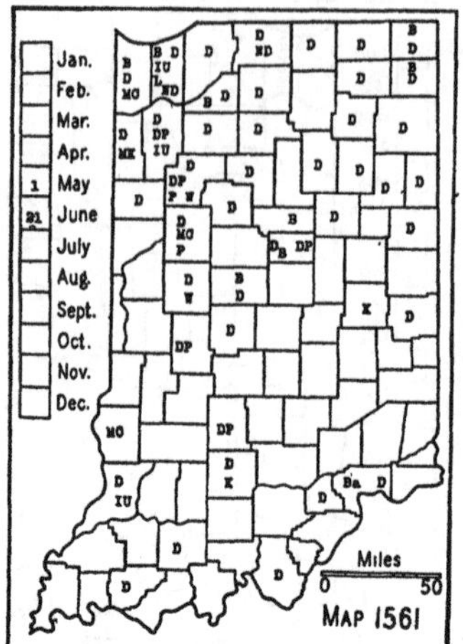

Cornus racemosa Lam.

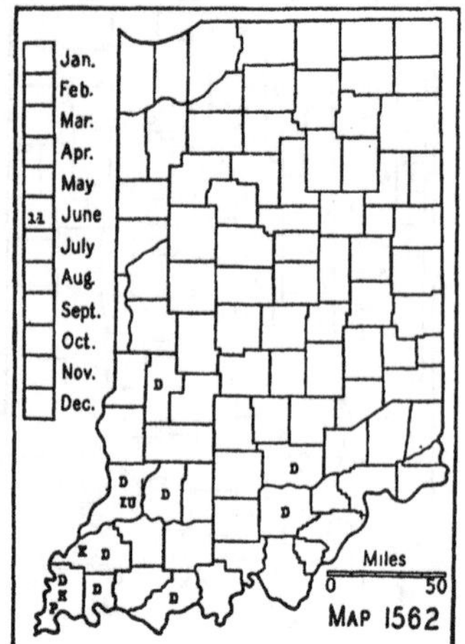

Cornus stricta Lam.

5a. **Cornus stolonifera var. Bàileyi** (Coulter & Evans) Drescher. (Trans. Wisconsin Acad. Sciences 28: 190. 1933.) (*Cornus Baileyi* Coult. & Evans.) BAILEY DOGWOOD. Map 1558. I reported this variety from Lagrange County but I am now referring that specimen to *Cornus stolonifera.* All of my specimens are from the dune area bordering Lake Michigan except one from Starke County which was collected in low ground along the Kankakee River.

Great Lakes Region from Ont., westw. to S. Dak.

6. **Cornus asperifòlia** Michx. ROUGHLEAF DOGWOOD. Map 1559. Infrequent throughout the state except in the northern tier of counties where it may be absent or rare. Banks of streams, borders of ponds and lakes, in wet woods, and along moist roadsides.

Ont. to S. Dak., southw. to Fla. and Tex.

7. **Cornus racemòsa** Lam. (*Cornus paniculata* L'Hér. of Gray, Man., ed. 7 and *Cornus femina* Mill. of Britton and Brown, Illus. Flora, ed. 2.) GRAY DOGWOOD. Map 1561. More or less frequent in the lake area, becoming rare or absent in the southern counties. It grows in both dry and wet places, preferring drained marshes. It is often found in moist or dry sandy or gravelly soil along roadsides and fences, in clearings, and in low ground about lakes and streams.

Cent. Maine to Ont. and Minn., southw. to N. C., Tenn., and Nebr.

8. **Cornus strícta** Lam. STIFF DOGWOOD. Map 1562. Local but usually frequent where it is found. In low woods, usually with pin oak, sweet gum, and cypress. The map shows all reports of this species. The specimen from Porter County appears to be this species and it no doubt will be found along the Kankakee River.

Va. to Fla., westw. to Mo.

9. **Cornus obliqua** Raf. (*Cornus Amomum* of most authors.) PALE DOGWOOD. Map 1563. In the lake area frequent to common in low places

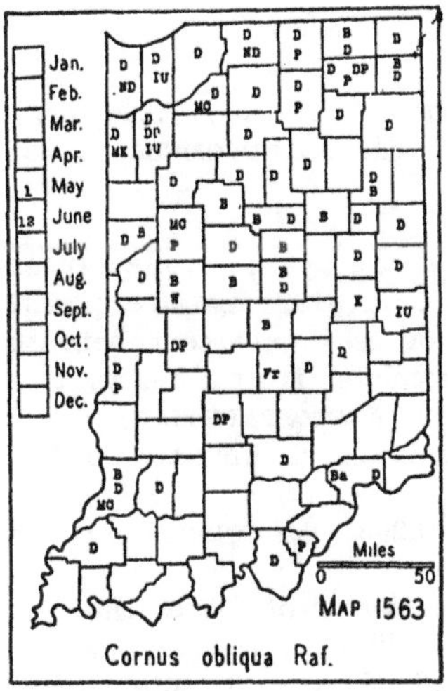

Cornus obliqua Raf.

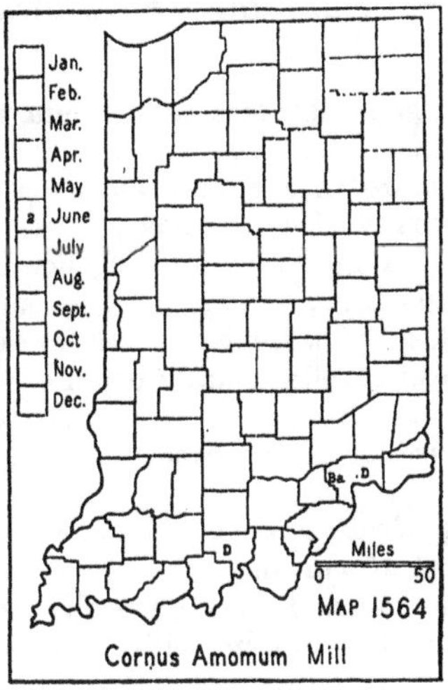

Cornus Amomum Mill

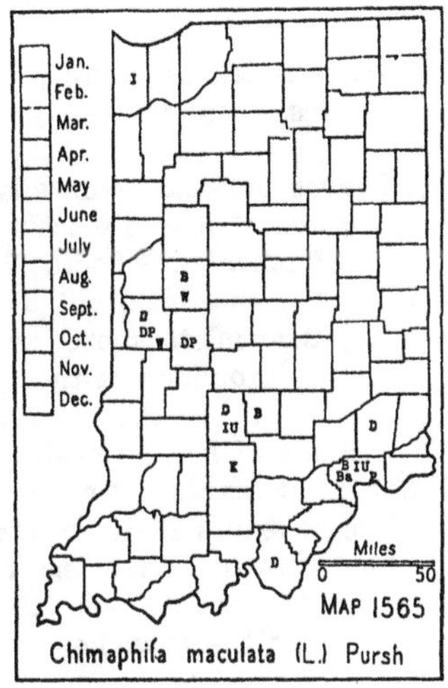

Chimaphila maculata (L.) Pursh

about swamps, ponds, and lakes and along streams. South of this area it becomes infrequent to rare, especially in the unglaciated region.

Que. to Alberta, southw. to Pa. and Mo.

10. **Cornus Amòmum** Mill. SILKY DOGWOOD. Map 1564. Our only specimens are from the bank of the Ohio River in Crawford and Jefferson Counties.

Newf. to Fla., westw. to Ky.

233. ERICÀCEAE DC. HEATH FAMILY

Ovary superior.

Plants saprophytic, without green color; pollen grains simple; anthers horizontal, opening by 2 transverse slits; fruit a capsule.......................................1. Subfamily MONOTROPOIDEAE, p. 733.

Plants with green foliage; pollen grains compound.

Corolla polypetalous; anthers inverted, dehiscing by basal (apparently apical) pores; fruit a capsule....................2. Subfamily PYROLOIDEAE, p. 733.

Corolla gamopetalous; anthers erect, dehiscing by apical pores (except *Oxydendrum* whose anthers open by chinks); fruit a berry or capsule.......................................3. Subfamily ERICOIDEAE, p. 734.

Ovary inferior; pollen grains compound; corolla gamopetalous; fruit a berry..4. Subfamily VACCINOIDEAE, p. 734.

1. Subfamily **Monotropoìdeae**

Plants white or somewhat tinged with pink; corolla polypetalous...6169. MONOTROPA, p. 737.

2. Subfamily **Pyroloìdeae**

Leaves scattered, lanceolate or oblanceolate; flowers in corymbs or umbels; styles very short; valves of capsule with smooth edges............6166. CHIMAPHILA, p. 734.

Leaves basal, suborbicular or elliptic; styles long; valves of capsules with cobwebby margins..6167. PYROLA, p. 735.

3. Subfamily **Ericoideae**

Margin of leaves entire.
  Leaves sessile or nearly so, generally less than 8 mm wide..6199. ANDROMEDA, p. 738.
  Leaves petioled, generally more than 8 mm wide.
    Blade of leaves mostly narrowed at the base.
      Branchlets smooth; leaves glabrous beneath, acute at the apex; corolla saucer-shaped; fruit a capsule..........................6192. KALMIA, p. 737.
      Branchlets more or less pubescent; leaves more or less pubescent beneath, rounded at the apex; corolla usually ovoid or urceolate; fruit a berry...6212. ARCTOSTAPHYLOS, p. 739.
    Blade of leaves mostly cordate at the base................6205. EPIGAEA, p. 739.
Margin of leaves not entire.
  Trees; leaves 10-15 cm long; fruit a capsule.............6203. OXYDENDRUM, p. 738.
  Low shrubs; leaves less than 10 cm long.
    Shrubs rarely over 1.5 dm high; leaves generally in a cluster of 3-5 at the ends of the branches, more than 1.5 cm wide; fruit berrylike...6206. GAULTHERIA, p. 739.
    Shrubs usually 4-9 dm high; leaves scattered along the branches, generally less than 1.5 cm wide; fruit a capsule...............6200. CHAMAEDAPHNE, p. 738.

4. Subfamily **Vaccinoideae**

Under surface of leaves and calyx tube with resinous scales; ovary 10-celled...6215. GAYLUSSACIA, p. 740.
Under surface of leaves and calyx tube without resinous scales; ovary 4-5-celled...6216. VACCINIUM, p. 740.

6166. CHIMÂPHILA Pursh

Leaves mostly cuneate-oblanceolate, the midrib above bordered with whitish green, generally acute at the apex, the taper beginning about the middle of the blade, the margins usually with not more than 8 teeth to a side; dilated part of filaments merely ciliate...1. *C. maculata.*
Leaves lanceolate or ovate-lanceolate, the midrib above not bordered with whitish green, rounded or acute at the apex, the taper beginning well above the middle of the blade, the margins with more than 8 teeth to a side; dilated part of filaments villous.......................................2. *C. umbellata* var. *cisatlantica.*

1. **Chimaphila maculàta** (L.) Pursh. STRIPED PIPSISSEWA. Map 1565. An infrequent to rare plant of high ground, associated with either black oak or beech. There is a specimen in the herbarium of the University of Illinois collected in 1883 by E. J. Hill in a pine woods near Edgemoor (now in west Gary).

Maine(?) and Mass. to Ont. and Minn., southw. to Ga. and Miss.

2. **Chimaphila umbellàta** (L.) Bart. var. **cisatlántica** Blake. (Rhodora 19: 241. 1917.) (*Chimaphila umbellata* (L.) Nutt. of manuals in part.) COMMON PIPSISSEWA. Map 1566. An infrequent to rare plant of our northern counties. I have seen no specimen from south of White County. All of our specimens are from moist or dry, sandy black oak or black and white oak woods.

N. S. to Ga., westw. to the Pacific coast.

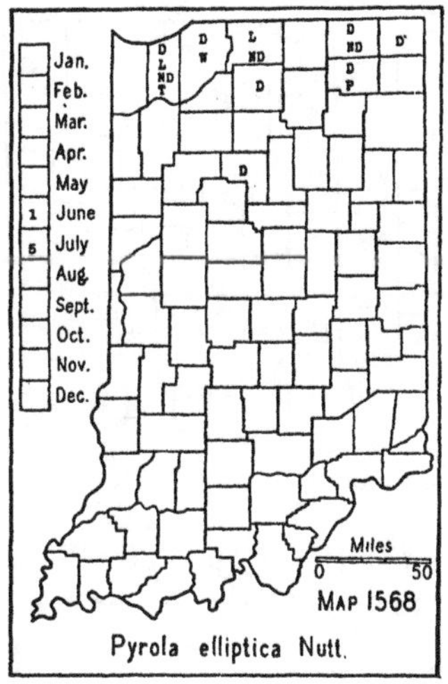

6167. PÝROLA [Tourn.] L.

Styles straight; petals connivent; racemes secund....................1. *P. secunda.*
Styles strongly declined; petals spreading; racemes not secund.
Cauline bracts none, or 1-3, narrowly lanceolate, long-acuminate, not sheathing at the base; calyx lobes ovate-triangular, little or not at all longer than broad.
Blades of leaves oval, 3-8 cm long, longer than the petioles; anthers blunt, orange. ....2. *P. elliptica.*
Blades of leaves suborbicular, 1-3 cm long, shorter than the petioles; anthers with a neck or point...3. *P. chlorantha.*
Cauline bracts 1-5 (rarely none), ovate-lanceolate, their bases somewhat sheathing the stem; leaf blades usually shorter than the petioles; sepals at least a half longer than wide.
Sepals ovate-lanceolate, blunt or acute, twice as long as wide; petals white, rarely pinkish-tinged, 6.5-10.5 mm long; bracts of flowers about as long as the pedicels..............................4. *P. rotundifolia* var. *americana.*
Sepals triangular, acute or acuminate, about 1.5 times as long as wide; petals pink, about 5 mm long.
Leaf blades cordate at the base. (See excluded species no. 483, p. 1079.)........ ....*P. asarifolia.*
Leaf blades subtruncate, rounded or tapering at the base.................. ....5. *P. asarifolia* var. *incarnata.*

1. **Pyrola secúnda** L. Map 1567. This species has been reported from Lake, Porter, and Steuben Counties. There are specimens from Lake and Porter Counties, collected by Nieuwland, now deposited in the herbarium of the University of Notre Dame. The Lake County specimen was collected at Miller, June 24, 1916; the Porter County specimen was collected at Mineral Springs June 14, 1911. The leaves of these specimens are narrowed at the apex instead of rounded; the secund racemes contain more than 10 flowers; the styles are straight; the basal cauline bracts are involute and lanceolate-acuminate.

This species reaches the southern limit of its range in northern Indiana. The Steuben County report may have been correct, but the report from

Pyrola chlorantha Swartz

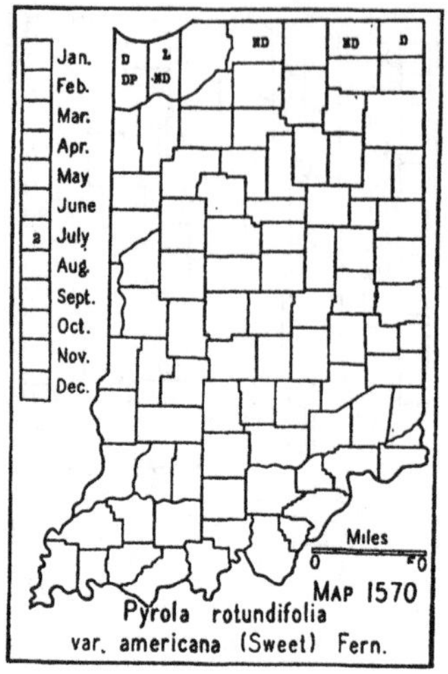

Pyrola rotundifolia var. americana (Sweet) Fern.

Pyrola asarifolia var. incarnata (Fisch.) Fern.

Monroe County by Dudley may safely be disregarded. See explanation under excluded species no. 484, p. 1079.

Lab. to Alaska, southw. to Md., Mich., n. Ind., Nebr., and Calif.

2. **Pyrola ellíptica** Nutt. SHINLEAF. Map 1568. An infrequent to rare plant in some of the counties of the lake region. It is usually found in cool, shady places in sandy soil at the base or on the lower part of black and white oak slopes where these border a lake, swamp, or pond. Where it is found it is usually frequent to common. This is by far our most common species of the genus.

Newf, to B. C., southw. to D. C., Ill., Iowa., and N. Mex.

3. **Pyrola chlorántha** Swartz. Map 1569. A specimen of this species was collected by Hill, May 25, 1878, in sandy woods near Whiting, Lake County. It is in the herbarium of DePauw University. It has more recently been collected by Nieuwland & Just in a tamarack swamp on the north side of Bass Lake, St. Joseph County, June 18, 1930. The specimens are very young but seem to be sufficiently distinct. Both specimens belong to the typical form.

Lab. to B. C., southw. to D. C., Ill., Nebr., and in the mts. of Ariz.

4. **Pyrola rotundifòlia** L. var. **americàna** (Sweet) Fern. (Rhodora 22: 122. 1920.) (*Pyrola americana* Sweet.) ROUNDLEAF PYROLA. Map 1570. A rare plant of a few of our northern counties. In shady places in moist, sandy soil, usually at the bases of wooded dunes or wooded slopes.

P. E. I. to S. Dak., southw. to Ga. and Ohio.

5. **Pyrola asarifòlia** Michx. var. **incarnàta** (Fisch.) Fern. Map 1571. Our only specimen was found in a tamarack bog on the southwest side of Tamarack Lake in Steuben County.

Newf. to Alaska, southw. to Vt., cent. N. Y., Wis., Colo., and Calif.

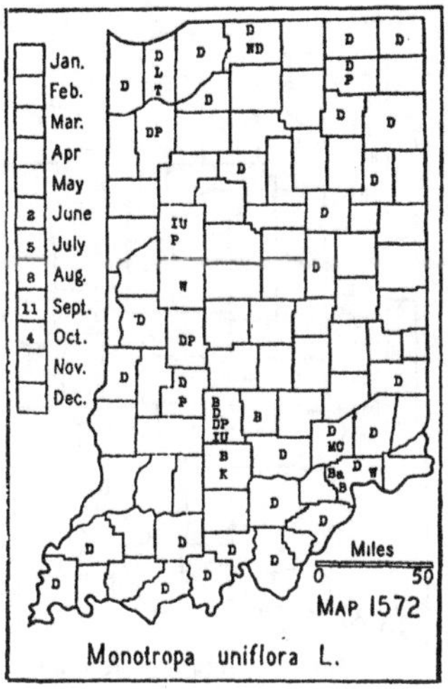

Monotropa uniflora L.

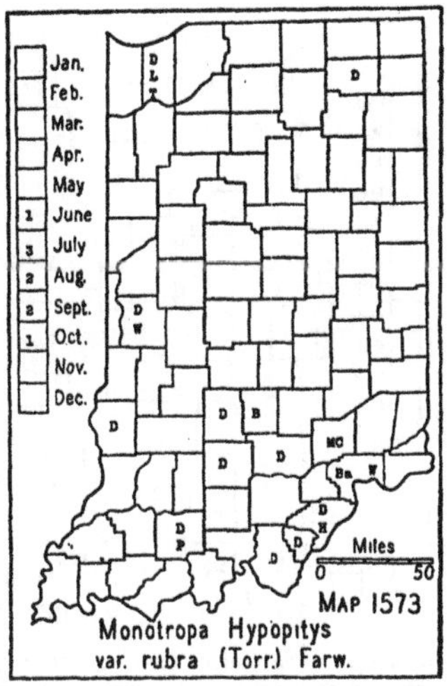

Monotropa Hypopitys var. rubra (Torr.) Farw.

Kalmia latifolia L

6169. MONÓTROPA L.

Flowers solitary; style shorter than the ovary, glabrous; stigmas naked............. ....1. *M. uniflora.*
Flowers racemose; style longer than the ovary, pubescent; stigmas more or less retrorsely bearded..................................2. *M. Hypopitys* var. *rubra.*

1. **Monotropa unifl̀ora L.** INDIAN PIPE. Map 1572. A saprophyte on humus in several types of habitat but usually in black and white oak woods. I once found a large clump of large plants growing in sphagnum in a tamarack bog. The species is well distributed in the state but ordinarily infrequent. In the low woods on the north side of the Kankakee River south of Schneider in Lake County, however, it was so common that it reminded one of a woods in winter when the snow was on the ground. Acres of this woods were carpeted with it. I revisited this woods several years at the same time of the year but I was able to find only a plant here and there.

Newf. to B. C., southw. to Fla. and Mex.; also in eastern Asia.

2. **Monotropa Hypópitys** L. var. **rùbra** (Torr.) Farw. (Amer. Midland Nat. **10**: 39. 1926.) (*Monotropa Hypopitys* L. of Gray, Man., ed. 7 and *Hypopitys lanuginosa* (Michx.) Nutt. of Britton and Brown, Illus. Flora, ed. 2.) PINE-SAP. Map 1573. Saprophytic on slightly acid humus in dry or moist woods. Infrequent to rare, possibly in all parts of the state. In addition to the counties indicated on the map it has been reported from Cass, Franklin, Hamilton, Marion, Vigo, and White Counties. Usually found sparingly in black and white oak woods. I found it in Clark County, however, as an abundant plant in a low, flat, beech and sweet gum woods where the soil is a hard, white, slightly acid clay. The stigmas of all of our Indiana plants are pubescent.

Que. to B. C., southw. to Fla., La., and Mex.

6192. KÂLMIA L.

1. **Kalmia latifòlia** L. MOUNTAIN-LAUREL. Map 1574. A few colonies have been found in Clark, Crawford, and Perry Counties. It was reported

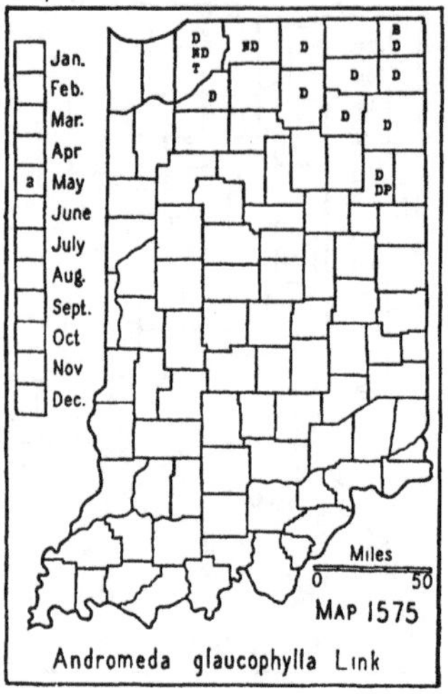

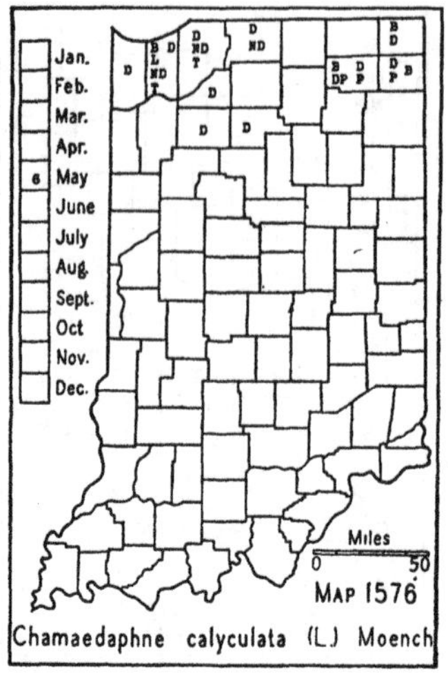

by Clapp as found "near Lafollette's in the vicinity of New Albany," and by the Editors of the Botanical Gazette in a Flora of Indiana (p. 17, 1881.) for Dudley. The last record can safely be ignored since it is known that Dudley confused his records.

6199. ANDRÓMEDA L.

1. **Andromeda glaucophýlla** Link. (*Andromeda Polifolia* of Britton and Brown, Illus. Flora, ed. 2.) DOWNY BOG-ROSEMARY. Map 1575. A rare plant of bogs in a few of the northern counties. It is now extinct in Wells County because of draining and, no doubt, has or will soon become extinct in several other counties for the same reason.

Lab. to Man., southw. to N. J., Pa., and Minn.

6200. CHAMAEDÁPHNE Moench

1. **Chamaedaphne calyculàta** (L.) Moench. LEATHERLEAF. Map 1576. In bogs and blueberry marshes in our northern counties. It is local but where found it may cover acres to the exclusion of almost all other kinds of vegetation.

Lab. to Alaska, southw. to Ga., Ill., Minn., and B. C.; Eurasia.

6203. OXYDÉNDRUM DC.

1. **Oxydendrum arbòreum** (L.) DC. SOURWOOD. Map 1577. There are a few trees of this species in two localities about five miles apart about seven miles northeast of Cannelton in Perry County. The largest trees were located on the Walter Hafele farm in Township 6, Range 2 W., section 21. They were associated with beech near the base of a sandstone slope and measured 41½ inches in circumference at breast height, had a clear bole of about 25 feet, and were about 55 feet high. There is a specimen in the herbarium of Wabash College collected in the vicinity of New Albany, Floyd County, June 14, 1836, by Dr. A. Clapp. The report by

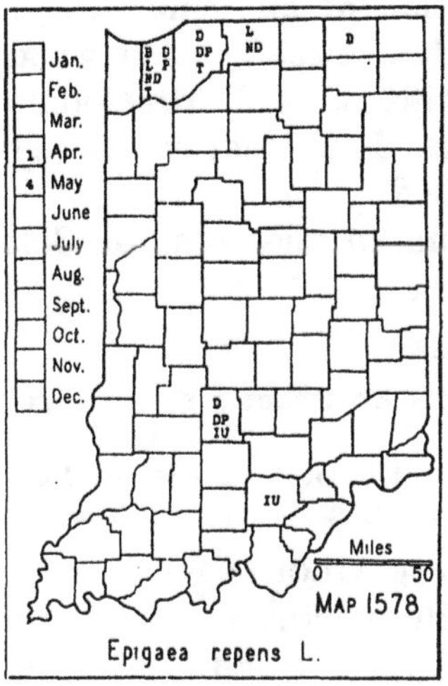

Epigaea repens L.

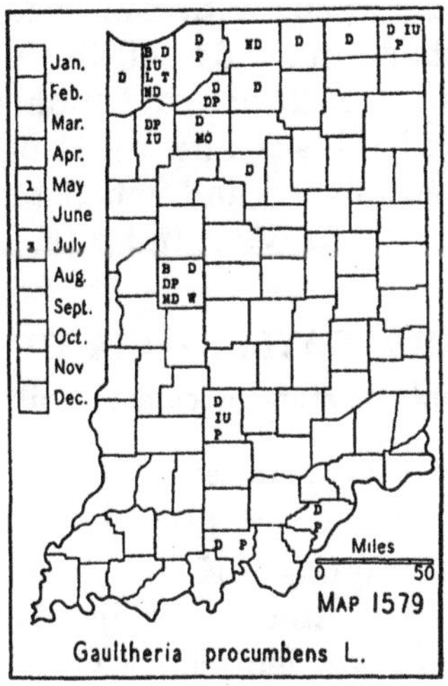

Gaultheria procumbens L.

Arctostaphylos Uva-ursi var. coactilis Fern. & Macb.

Dudley for Monroe County may be safely ignored. (See explanation under excluded species no. 484, p. 1079.)

Pa. to Ind., southw. to Fla. and Ala.

6205. EPIGAÈA L.

1. **Epigaea rèpens** L. TRAILING-ARBUTUS. Map 1578. Very local and limited in quantity at each station where I have seen it. In addition to the counties indicated on the map it has been reported from Lake, Marshall, and Montgomery Counties. In northern Indiana it grows in moist and very sandy soil in protected places in woodland. In the southern part of the state it grows in slightly acid soil on shady slopes on or close to the sandstone outcrops, usually associated with black and white oaks.

Newf. to Sask., southw. to Fla. and Ky.

6206. GAULTHÈRIA [Kalm] L.

1. **Gaultheria procúmbens** L. WINTERGREEN. Map 1579. Rare to frequent in some of the northern counties. Southward it has been found in only a few places in a few counties as relicts on sandstone outcrops. Its preferred habitat in Indiana is rather moist and very sandy black oak flats. It is also found on dry, sandy black and white oak slopes. The usual form of the leaf is obovate to oval but plants with nearly orbicular and narrow-elliptic leaves are found. The extremes in leaf form have been given botanical names but I do not consider our plants as coming within the range of the named forms.

Newf. to Man., southw. to Ga. and Ala.

6212. ARCTOSTÁPHYLOS Adans.

1. **Arctostaphylos Ŭva-úrsi** (L.) Spreng. var. **coáctilis** Fern. & Macb. (Rhodora **16**: 212. 1914.) (*Arctostaphylos Uva-ursi* (L.) Spreng. and *Uva-Ursi Uva-Ursi* (L.) Britt.) BEARBERRY. Map 1580. Restricted to the

dune area about Lake Michigan, with the exception of one small colony which I found in dense shade in sandy soil in the Margaret Trasker woods about two and a half miles southeast of Union Mills in La Porte County, where it was associated with black and white oaks. It is local but usually forms large mats when established and not disturbed.

Newf. to Yukon, southw. to Va., Ind., Ill. and in the mts. to Colo. and Calif.

6215. GAYLUSSÀCIA HBK.

Leaves green beneath; filaments of stamens ciliate; fruit black without a bloom, rarely with a slight bloom..1. *G. baccata.*

Leaves glaucous beneath; filaments of stamens glabrous; fruit dark blue. (See excluded species no. 486, p. 1079.)..*G. frondosa.*

1. **Gaylussacia baccàta** (Wang.) K. Koch. BLACK HUCKLEBERRY. Map 1581. Found only in silicious and acid soils. In the northern part of the state it is usually found on wooded slopes with black oak or in black and pin oak woods, and rarely in tamarack bogs. In the "knobs" it is generally associated with chestnut oak and dryland blueberry; and in the "flats" it is found with sweet gum and pin oak.

Newf. to Man., southw. to Ga., Ill., and Wis.

1a. **Gaylussacia baccata f. leucocárpa** (Porter) Fern. This is a form with white to pinkish fruit. I found a single colony of it on a rocky wooded slope of Bear Creek near Fountain in Fountain County. The fruit was light rose color and about a half larger than that of the typical form.

6216. VACCÍNIUM L. BLUEBERRY AND CRANBERRY

Stems not creeping or trailing; leaves not evergreen (in Indiana); corolla 5-toothed or lobed; fruit not red.

- Corolla open-campanulate, 5-lobed; anthers with long tubes, 2-awned on the back; fruit not edible.
  - Anthers exserted at anthesis; leaves not glossy above, usually glaucous beneath; fruit green, greenish yellow or light purple, usually with a bloom.
    - Branchlets and under surface of leaves more or less densely pubescent..1. *V. stamineum.*
    - Branchlets and under surface of leaves glabrous..1a. *V. stamineum* var. *neglectum.*
  - Anthers not exserted at anthesis; leaves glossy above, never glaucous beneath; fruit black, without a bloom..2. *V. arboreum.*
- Corolla cylindric, cylindric-ovoid, or urceolate, 5-toothed; anthers awnless, included; fruit edible.
  - Shrubs of a boggy or wet habitat (except 3b), up to 4 m high, usually associated with chokeberry, winterberry, and buttonbush.
    - Branchlets glabrous or pubescent in lines; under surface of leaves glabrous or more or less pubescent along the principal veins until maturity.
      - Margin of leaves entire or slightly erose, rarely a few leaves somewhat glandular ciliate-serrulate or glandular bristly-ciliate..3. *V. corymbosum.*
      - Margin of leaves glandular ciliate-serrulate or glandular bristly-ciliate.
        - Blades green on both surfaces.........3a. *V. corymbosum* var. *amoenum.*
        - Blades green above and glaucous beneath..3b. *V. corymbosum* var. *pallidum.*
    - Branchlets and under surface of leaves densely pubescent..3c. *V. corymbosum* var. *atrococcum*

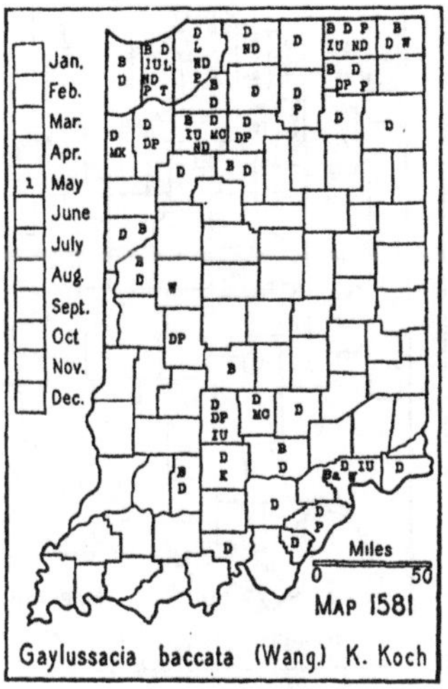

Gaylussacia baccata (Wang.) K. Koch

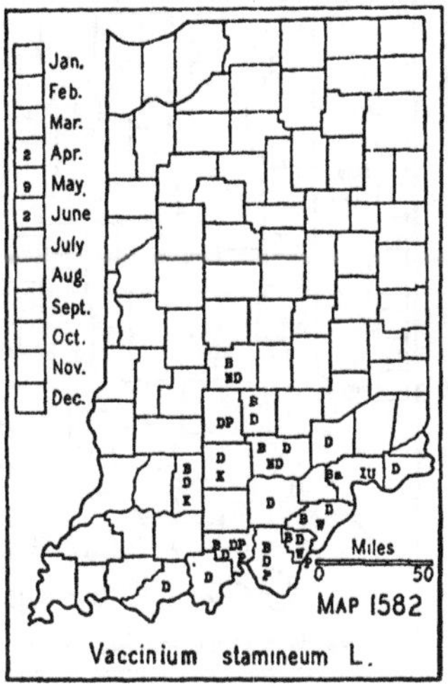

Vaccinium stamineum L.

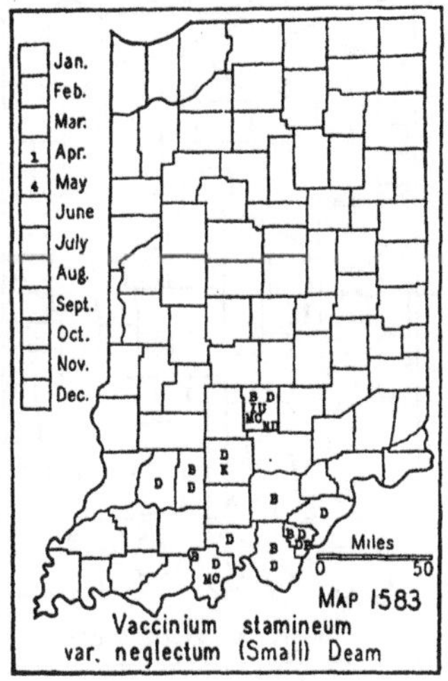

Vaccinium stamineum var. neglectum (Small) Deam

Shrubs of dry soils or moist sandy soil, mostly less than 5 dm high except in 5a.
Leaves glabrous or pubescent on the midribs beneath at fruiting time.
Fresh leaves green beneath, serrulate; fruit with a bloom...4. *V. angustifolium.*
Fresh leaves glaucous beneath, entire or serrulate; fruit with or without a bloom.
Leaves oblong-lanceolate to oblong-elliptic, serrulate; fruit without a bloom. ....4a. *V. angustifolium* var. *nigrum.*
Leaves obovate to oval or broadly oblong, entire or serrulate; fruit with a bloom...5. *V. vacillans.*
Leaves more or less pubescent all over the under surface at fruiting time.
Blades obovate to oval, ovate or broadly oblong, usually about 30 mm long and 15 mm wide, never all of them entire; shrubs up to 1.5 m high...5a. *V. vacillans* var. *crinitum.*
Blades oval to narrowly elliptic or lanceolate, usually about 20 mm long and 10 mm wide, margins always entire; low shrubs, generally less than 0.5 m high...6. *V. canadense.*
Stems trailing and creeping; leaves evergreen; corolla 4-parted; fruit reddish.
Bracts of the pedicel above the middle of the pedicel, generally green, flat, 3 mm long or longer, and more than 1 mm wide; margin of leaves only slightly inrolled, elliptic in outline..............................7. *V. macrocarpon.*
Bracts of the pedicel generally about the middle of the pedicel or lower, mostly somewhat colored, generally very thin, involute; margins of leaves strongly inrolled, becoming triangular in outline....................8. *V. Oxycoccos.*

1. **Vaccinium stamíneum** L. (Ashe. Polycodium. Jour. Elisha Mitchell Scien. Soc. **46**: 196-213. 1931.) (*Polycodium stamineum* (L.) Greene.) DEERBERRY. Map 1582. A shrub mostly of wooded slopes in the unglaciated region where it is generally associated with black and chestnut oaks and sometimes with Virginia pine. I have it also from a woods in the "flats" of Switzerland County about 2 miles southeast of Fairview, where it was associated with white oak, and from a low woods in an old lacustral bed in Crawford County about 3 miles northwest of Leavenworth where it was associated with pin oak, sweet gum, and red maple.

Mass. to Ont., southw. to n. Ga., and westw. to Ind., Ky., and Tenn.

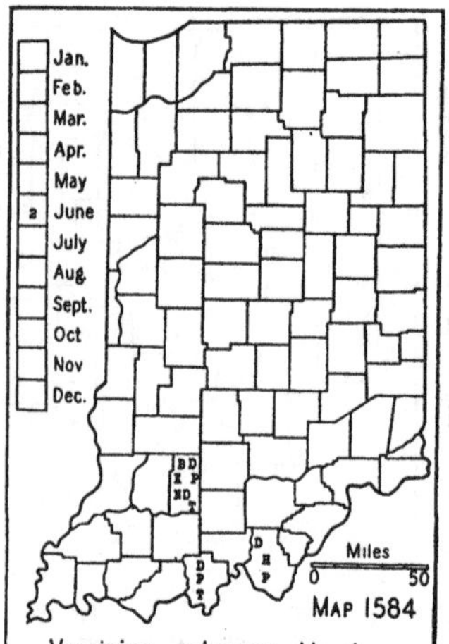

Vaccinium arboreum Marsh.

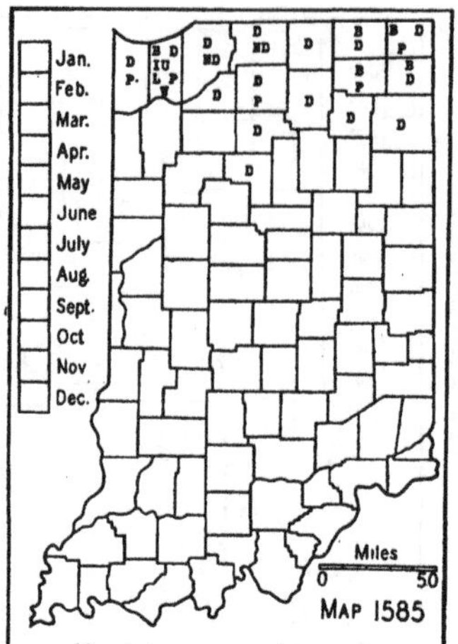

Vaccinium corymbosum L.

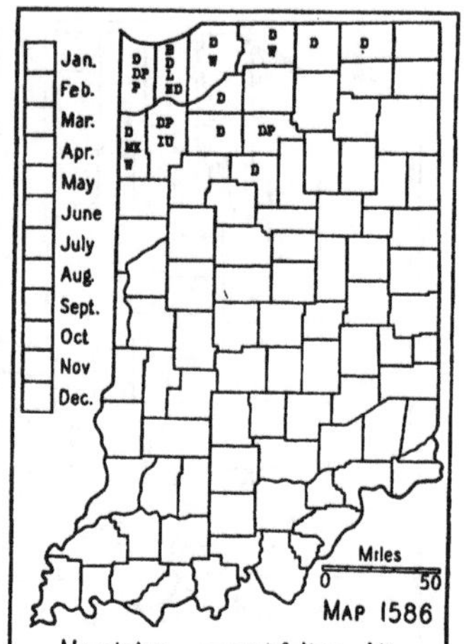

Vaccinium angustifolium Ait.

1a. **Vaccinium stamineum** var. **negléctum** (Small) Deam. Map 1583. This variety seems to be merely a glabrous form of the species and my specimens show that its range in Indiana is much the same as that of the species.

Pa. to middle Ga., westw. to Ind., Ky., and Tenn.

2. **Vaccinium arbòreum** Marsh. (*Batodendron arboreum* (Marsh.) Nutt.) FARKLEBERRY. Map 1584. This is a straggling shrub up to 9 feet high, usually found in shallow soil on sandstone ridges and bluffs where it is associated with post and black oaks.

Va. to Ind., southw. to Fla. and Tex.

3. **Vaccinium corymbòsum** L. HIGHBUSH BLUEBERRY. Map 1585. This species is restricted to the lake area where it was formerly frequent to common over large areas. It grows in boggy and swampy places in tamarack bogs, marshes, and interdunal sloughs. Before the lake area was drained it covered hundreds of acres of swamp land but there now remain only a few small blueberry marshes.

Maine to Minn. and southw. to Fla. and La.

The following varieties are of questionable value:

3a. **Vaccinium corymbosum** var. **amoènum** (Ait.) Gray. This form has been reported from Lake County by Hill and from Montgomery County by Grimes. I have seen the Grimes specimen, which is now in the herbarium of DePauw University, and it should be referred to *Vaccinium corymbosum* var. *pallidum.*

The range of the variety is given as the same as that of the species. I have it from De Kalb, Lagrange, La Porte, and Porter Counties.

3b. **Vaccinium corymbosum** var. **pállidum** (Ait.) Gray. This form has been reported from Indiana but since its range is given as Virginia to South Carolina, it is doubtful whether it occurs in Indiana. There is, however, a shrub mostly 3-5 feet high in the "knobs" of the unglaciated area that I place here for want of better determination. I do not believe that these plants belong to *Vaccinium corymbosum* but are, of themselves, a unit, which may belong to the *Vaccinium pallidum* of Small's "Flora of the Southeastern United States." Of the Indiana species, they seem to be nearest related to *Vaccinium vacillans*. There are probably two species or varieties in this complex. The leaves are mostly oval or obovate to elliptic, acute, acuminate, or somewhat obtuse, glabrous or more or less pubescent all over, the margins ciliolate-serrulate; fruit usually black, subglobose, sweet, and one form with glaucous fruit, about 8 mm wide and 9 mm long. It is usually associated with chestnut oak. The specimens in the DePauw University herbarium collected by Grimes on the "Devil's Backbone" in Montgomery County belong here. They were reported as *Vaccinium corymbosum* var. *amoenum*.

3c. **Vaccinium corymbosum var. atrocóccum** Gray. I have specimens from Lagrange, La Porte, Kosciusko, Starke, and Steuben Counties which I refer to this variety.

N. B. to Ont., southw. to N. J., Pa., and Ala.

4. **Vaccinium angustifòlium** Ait. (*Vaccinium pennsylvanicum* Lam.) LOWBUSH BLUEBERRY. Map 1586. Erect or ascending shrubs, usually 10-20 inches high. Its preferred habitat is sandy white oak, black and white oak, and pin and black oak woods. It is usually associated with dryland blueberry.

Newf. to Sask., southw. to Va., Ill., and Wis.

4a. **Vaccinium angustifolium var. nìgrum** (Wood) Dole. I studied for two successive years, both in flower and in fruit, a large colony of this variety in Starke County, growing in an acre or more of shrubs of the typical species. The following differences were noted. The leaves were more or less glaucous, both on unfolding and at fruiting time; the corolla was about 0.5 mm wider; and the fruit longer than wide, and black with little or no bloom. I collected this variety also in La Porte County.

5. **Vaccinium vacíllans** Kalm ex Torrey. DRYLAND BLUEBERRY. Map 1587. Erect, branching shrubs up to 4 feet high, usually 12-20 inches high. This species is restricted to the lake region and to the sandstone and knobstone area of the southern part of the state. Its preferred habitat is a dry sandy soil and it is rarely found in a moist soil unless it is that of a sandy black and pin oak woods in the lake region. In the southern part of the state it is generally associated with white, black, scarlet, and chestnut oaks, and Virginia pine.

N. S. to Mich., southw. to Ga., Tenn., and Kans.

5a. **Vaccinium vacillans var. crinìtum** Fern. (Rhodora 13: 236. 1911.) In this variety the branchlets and under surface of the leaves are generally more or less permanently pubescent. It is much taller and the leaves are larger. I have it from Clark, Floyd, Jackson, Pulaski, St. Joseph, and Washington Counties. I do not know its general range.

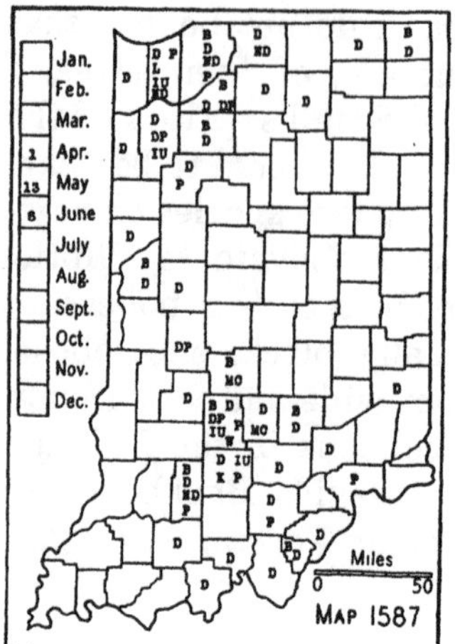

MAP 1587

Vaccinium vacillans Kalm ex Torrey

MAP 1588

Vaccinium canadense Kalm

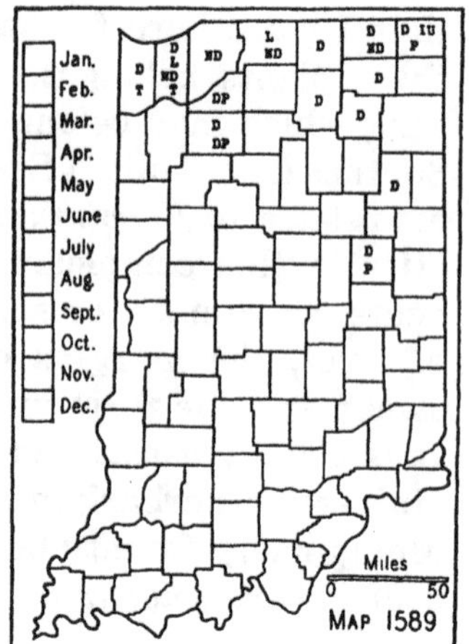

MAP 1589

Vaccinium macrocarpon Ait.

6. **Vaccinium canadénse** Kalm. CANADA BLUEBERRY. Map 1588. This species is distinguished by its dwarf size, densely pubescent branchlets, and narrow leaves which are entire and densely pubescent beneath. Our only authentic record for Indiana is that of a colony on the north slope of a wooded headland along Bear Creek near Fountain, Fountain County. The area where it is located is used as a summer resort and since the plant is exposed it will doubtless soon disappear. Associated with this species at this place was a form of it about 1 dm taller, with leaves all of a narrow form, and with fruit usually oblong, black, and without a bloom. I find in literature no reference to this form.

Lab. to Man., southw. in the mts. to Va. and Ill.

7. **Vaccinium macrocárpon** Ait. (*Oxycoccus macrocarpos* (Ait.) Pursh.) CRANBERRY. Map 1589. In boggy and marshy places, usually associated with sphagnum. Formerly there were large areas of "cranberry marshes" in Indiana but now the species has become rare.

Newf. to Wis., southw. to N. J., W. Va., and Ark.

8. **Vaccinium Oxycóccos** L. (*Oxycoccus Oxycoccus* (L.) MacM.) Map 1590. Found only in boggy places associated with sphagnum. It is exceedingly rare and, no doubt, will soon become extinct except possibly in the La Porte County station.

Arctic regions, southw. to Pa., Ind., and Wis.

237. PRIMULÀCEAE Vent. PRIMROSE FAMILY

Corolla and calyx with erect or spreading segments.
- Plants stemless....................................6321. ANDROSACE, p. 745.
- Plants with leafy stems.
  - Plants aquatic; immersed leaves pectinate...............6327. HOTTONIA, p. 745.
  - Plants terrestrial or of marshes; leaves entire.
    - Leaves opposite or in whorls (rarely a few of the lower ones alternate).
      - Leaves 5-10 in a whorl at the top of the stem; stem erect, usually 8-18 cm long; flowers white..............................6333. TRIENTALIS, p. 750.

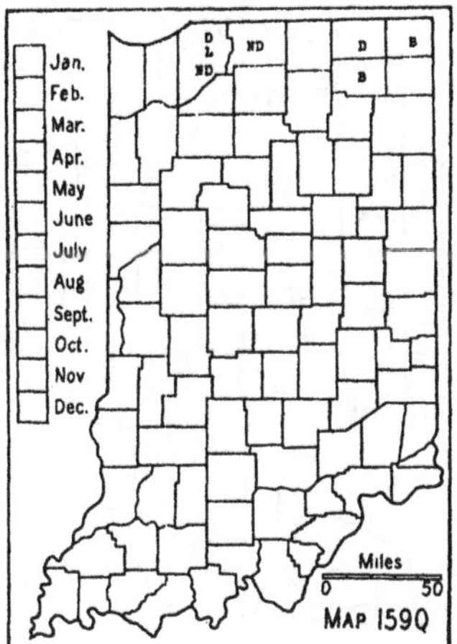

Vaccinium Oxycoccos· L

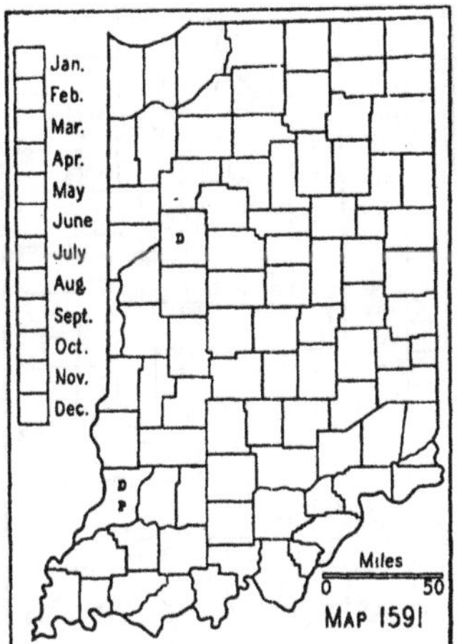

Androsace occidentalis Pursh

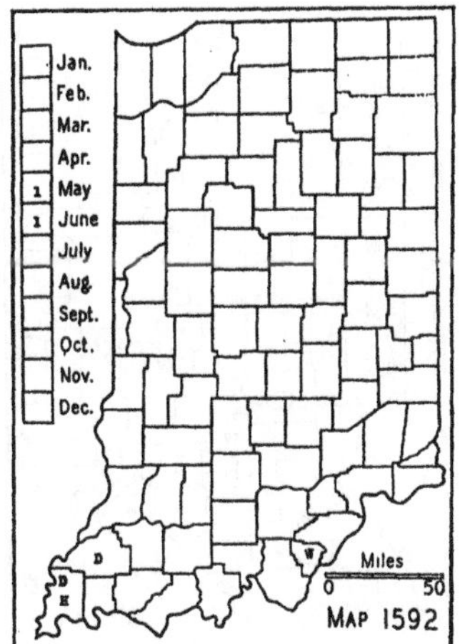

Hottonia inflata Ell.

Leaves opposite or in whorls along leafy stems; plants generally more than 18 cm long, if shorter the plants trailing; flowers yellow or scarlet (rarely white in *Anagallis*).

Flowers scarlet, rarely white; plants annual; leaves less than 2 cm long; capsules circumscissile.......................6338. ANAGALLIS, p. 750.

Flowers yellow; plants perennial; leaves more than 2 cm long (except in *Lysimachia Nummularia*, a creeping plant); capsules opening by valves. ....6330. LYSIMACHIA, p. 746.

Leaves alternate.

Flowers in axillary racemes, on long pedicels, corolla white; leaves mostly 2-8 cm long; capsule opening by 5 apical valves......6328. SAMOLUS, p. 746.

Flowers solitary, axillary, sessile, corolla pinkish; leaves 4-8 mm long; capsule circumscissile.............................6339. CENTUNCULUS, p. 750.

Corolla segments reflexed; stamens exserted, connivent, forming a cone; plants scapose......................................6341. DODECATHEON, p. 751.

6321. ANDRÓSACE [Tourn.] L.

[St. John. Revision of certain North American species of Androsace. Dept. of Mines, Canada, Memoir **126**: no. 4, Biol. Ser. pp. 45-55. 1922.]

1. **Androsace occidentàlis** Pursh. Map 1591. Reported for the state by Dorner (Proc. Indiana Acad. Sci. **1903**: 119. 1904), who says "somewhat abundant in lowland near Wea Creek." I have one of his specimens, and one from Knox County, without data, collected by W. S. Blatchley. It is to be noted that the area along Big Wea Creek has many western species such as *Muhlenbergia cuspidata, Arenaria patula, Lithospermum incisum,* and others.

W. Ont. to s. Sask. and B. C., southw. to Ind., Ill., Mo., Okla., and Ariz.

6327. HOTTÒNIA [Boerh.] L.

1. **Hottonia inflàta** Ell. WATER VIOLET. Map 1592. This species has been collected only three times in Indiana. A specimen was found in flower on May 17, 1901, by Dr. Schneck in a shallow pond in Gibson County near Lyle Station. I now have this specimen. I collected several specimens

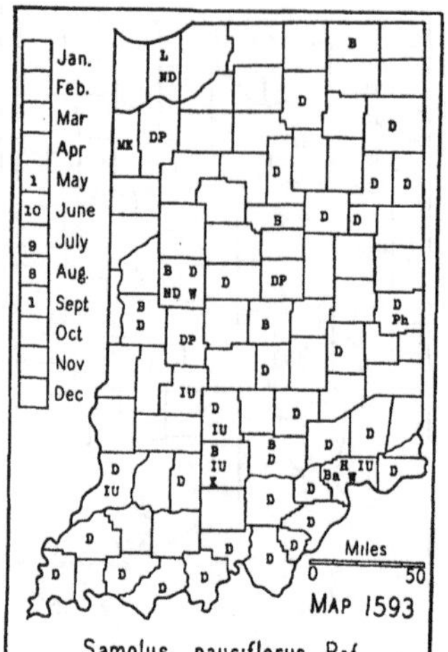

Samolus pauciflorus Raf.

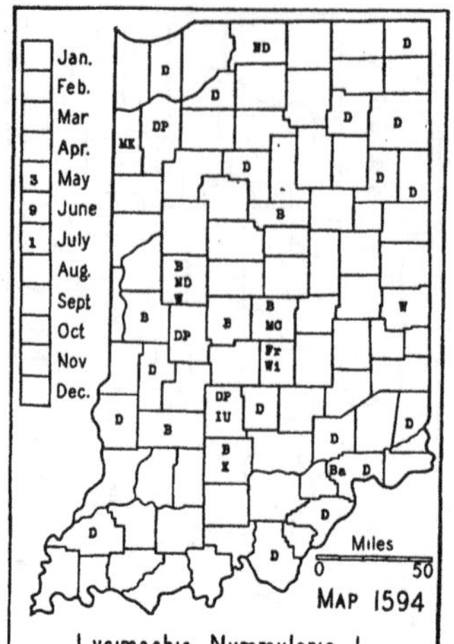

Lysimachia Nummularia L.

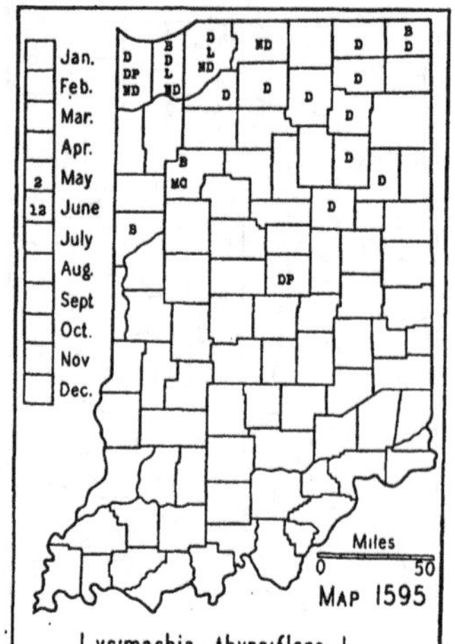

Lysimachia thyrsiflora L.

in flower on June 15, 1935 in Posey County where it was common in a low area in the pin oak woods belonging to Mrs. Nola Erwin, in sec. 5 of Point Township. There is a specimen in the herbarium of Wabash College collected by Dr. Clapp near New Albany, Floyd County, in 1838.

Maine and N. H. to cent. N. Y. and Fla., westw. to Mo. and La.

6328. SÅMOLUS [Tourn.] L.

1. **Samolus paucifl̀orus** Raf. (*Samolus floribundus* HBK.) WATER PIMPERNEL. Map 1593. In wet places throughout the state. While I have no specimen from the northwestern part of the state, there are several records for that section. It is usually found on muddy and sandy bars and banks of streams, in ditches, low places in woods, and cultivated fields. Although it produces an abundance of seed, it is never abundant and is usually only an occasional or infrequent plant.

N. B. to Fla., westw. to B. C., Calif., and Tex.; also in Mex., West Indies, and S. A.

6330. LYSIMÀCHIA [Tourn.] L.

[Fernald. The identity of Lysimachia lanceolata. Rhodora 39: 438-442. 1937.]

Leaves dotted above; staminodia none or very rudimentary.
  Leaves orbicular-ovate to orbicular, not twice as long as wide in our specimens, all cordate or subcordate at the base, rounded at the apex; stems creeping and sometimes rooting at the nodes........................1. *L. Nummularia.*
  Leaves and stems not as above.
    Flowers in dense, axillary, spikelike clusters from the axils of leaves of the middle of the stem................................2. *L. thyrsiflora.*
    Flowers axillary or in terminal racemes.
      Leaves mostly whorled; flowers axillary on long pedicels......3. *L. quadrifolia.*
      Leaves opposite (rarely a few of the lower ones alternate); flowers in loose terminal racemes................................4. *L. terrestris.*
Leaves not dotted above; five slender staminodia between the fertile stamens.
  Blades of median leaves more than 7 mm wide, plainly pinnately veined.

Blades of median leaves ovate to ovate-lanceolate, 2-6 cm wide, rounded or subcordate at the base, the margins rigidly short-ciliate; petioles of median leaves mostly about 2 cm long, rarely as short as 1 cm, the margins thickly and strongly long-ciliate their entire length........................5. *L. ciliata.*

Blades of median leaves lanceolate, sometimes broadly so, 0.5-3 cm wide, long or short taper-pointed at the base, the margins scabrous, rarely somewhat short-ciliate; petioles (if any) of the median leaves generally less than 2 cm long, the margins not so closely or strongly ciliate as those of the preceding species, often only the basal part ciliate; calyx lobes 5-9 mm long.

Median and two lateral veins of calyx lobes very obscure when seen from the outside under an 8-diameter magnification; plants stoloniferous, from a long decumbent rootstock, 15-65 cm high, generally 20-30 cm high, usually of dry woods and prairies, rarely in wet or moist situations, simple or with short branches usually above the fifth to the seventh node, the branches generally shorter than the subtending leaf, rarely the main stem dividing into several branches below the fifth node; blades below the second and up to the sixth internodes generally much shorter and of different shape from those above, varying from nearly orbicular to broad-oblong, elliptic or lanceolate, all petiolate; blades above the fifth internode (rarely down to the second or up to the seventh) usually much longer than the lower blades, up to 15 cm long, lanceolate to narrow-elliptic, long taper-pointed at the base, sessile, subsessile or rarely petiolate.................6. *L. lanceolata.*

Median and two lateral veins of calyx lobes plainly visible and usually distinct when seen from the outside under an 8-diameter magnification; plants 40-120 cm high, with long branches throughout; usually the largest and longest leaves at the base of the plant (not so if they are stipular leaves), all petiolate or sometimes a few at the top subsessile; lower and upper blades of the same shape, mostly short taper-pointed at the base or sometimes the upper ones with a long taper-pointed base..................7. *L. hybrida.*

Blades of median leaves linear, mostly 2-7 mm wide, 1-nerved or very obscurely pinnately veined, more or less involute, the margins smooth, sessile or sometimes the lower ones petiolate; branches usually longer than their subtending leaves; calyx lobes plainly 1-nerved, 4-6.5 mm long............8. *L. longifolia.*

1. LYSIMACHIA NUMMULÀRIA L. MONEYWORT. Map 1594. Frequent in low ground along streams, ditches, roadsides, and elsewhere. I have seen it form a carpet in low, open woods along streams, crowding out all other herbaceous vegetation. When it becomes established in a pasture field, it chokes out the native grass and is very difficult to exterminate. Since the plant is not palatable to stock, my advice to land owners is to exterminate it at any cost.

Nat. of Eu.; Newf. to Wis., southw. to N. J., Va., and Ill.

2. **Lysimachia thyrsiflòra** L. (*Naumbergia thyrsiflora* (L.) Duby.) WATER LOOSESTRIFE. Map 1595. In mucky or peaty soil in bogs and marshy places and less frequent in low, sandy borders of lakes. Usually found in shallow water.

This species has been placed in another genus by some authors, assuming the presence of staminodia, but this character is not constant. (Rhodora 22: 193. 1920.)

No doubt Andrews' report of this species from Monroe County should be referred to some species which occurs in that county, and which he has

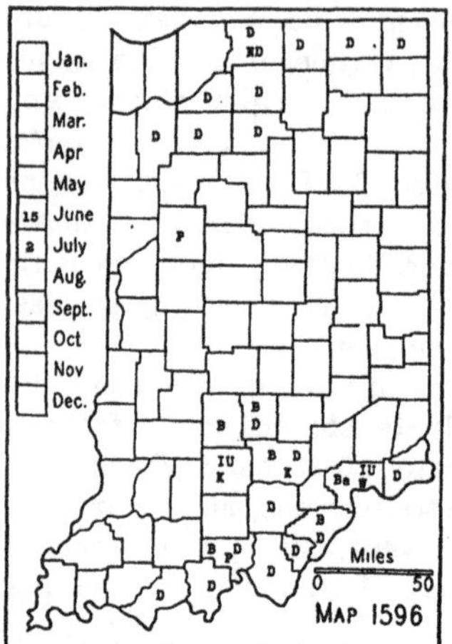

Lysimachia quadrifolia L.

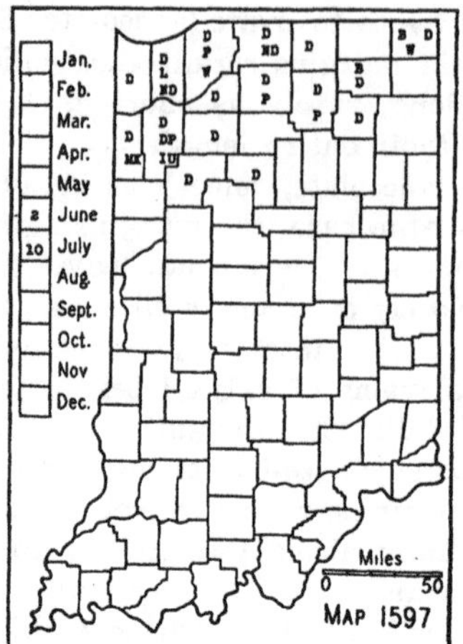

Lysimachia terrestris (L.) BSP.

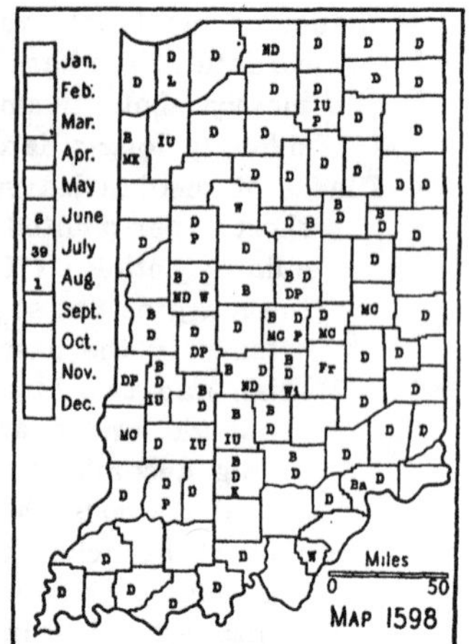

Lysimachia ciliata L.

failed to report. This species is possibly restricted to the lake region of the state.

Que. to Sask. and Alaska, southw. to Pa., Mo., and Calif.; also in n. Eu. and n. Asia.

3. **Lysimachia quadrifòlia** L. WHORLED LOOSESTRIFE. Map 1596. Found generally in dry, sandy soil, associated mostly with black oak or with black and white oaks, and once I found it in a sedge marsh. In the southern part of the state it is found in dry soil on black and white oak ridges and sometimes in old worn out fields. It is only an infrequent plant where found and never forms close stands. It is interesting to note that there are no records for the area about Lake Michigan, although we should expect it there. It is absent throughout the central part of the state because the soil is not sufficiently acid. Add Wells County to the map.

Our Indiana specimens are all more or less pubescent.

N. B. to Minn., southw. to Ga. and Mo.

4. **Lysimachia terréstris** (L.) BSP. SWAMPCANDLE. Map 1597. An infrequent plant on mucky borders of lakes, marshes, and sloughs, and more rarely on wet, sandy borders of lakes. We have one specimen from the very wet marly border of a lake. Instead of flowering, this species sometimes develops bulblets in the axils of the leaves; also sometimes the lower leaves are alternate when normally they would be opposite.

A form of this species occurs in which the flowers are in the axils of foliaceous bracts. Our specimen from Pulaski County is of this form.

There is no evidence or specimen to support the Monroe County record.

Newf. to Man., southw. to Ga. and Ark.

5. **Lysimachia ciliàta** L. (*Steironema ciliatum* (L.) Raf.) FRINGED LOOSESTRIFE. Map 1598. Frequent to abundant in swampy woodland, wet prairies, wet borders of streams, and wet roadsides.

N. S. to B. C., southw. to Fla., Ala., Kans., N. Mex., and Ariz.

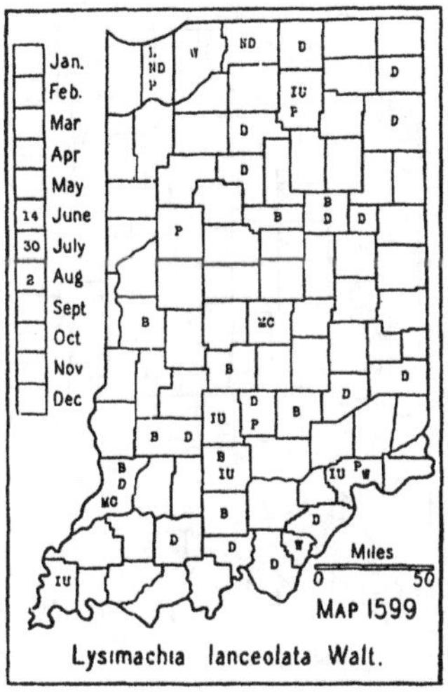

Lysimachia lanceolata Walt.

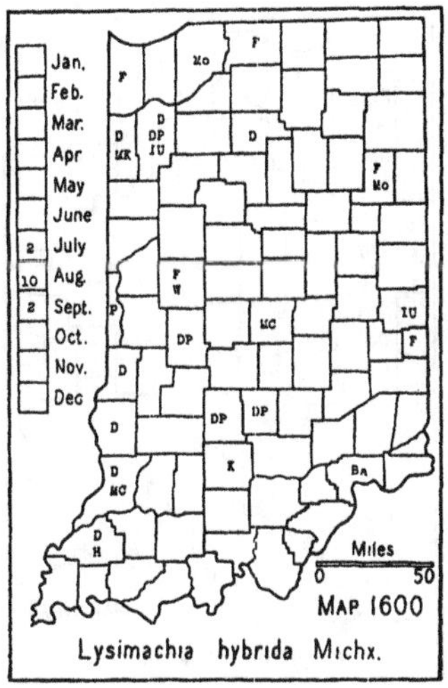

Lysimachia hybrida Michx.

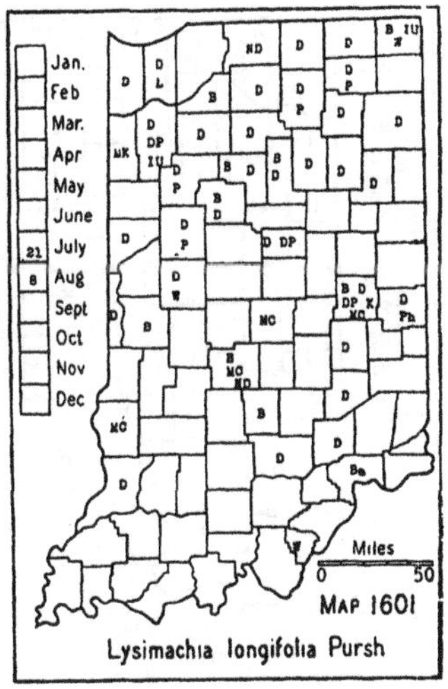

Lysimachia longifolia Pursh

6. **Lysimachia lanceolàta** Walt. (*Steironema heterophyllum* Michx. and *Steironema lanceolatum* (Walt.) Gray.) (Fernald. The identity of Lysimachia lanceolata. Rhodora 39: 438-442. 1937.) Map 1599. Rather frequent in small colonies in dry soil on the crests and slopes of black and white oak ridges, in dry prairies, and rarely in moist soil and then usually in a slightly acid soil and usually associated with black chokeberry or sweet gum. In the woods it is most commonly associated with black and white oak. Small plants usually have their leaf blades more or less folded inward.

The fact that this species is difficult to separate from the next one led me to place it under cultivation. I have found it very responsive to light and moisture. I think these two factors and temperature greatly change the appearance of the mature plants. The plants send up one or two sets of basal leaves late in autumn or early winter. These leaves are usually short and obtuse and have long petioles. In mild winters when the crown of the plant is protected these basal leaves persist, sometimes until maturity. In some plants these early leaves are killed and no leaves will be seen from the first few short internodes. Crowding, too, has much the same effect in killing off the early leaves. The length of the internodes is easily accounted for when the habitat and moisture are known. The next species usually grows in very wet places, usually inundated more or less until late spring. Many plants begin their growth under water and the submerged leaves die off and are later replaced by stipular leaves that are smaller and usually much narrower. In 1937 it was very wet and one bed of my plants set their principal cauline leaves at the third node, but usually these leaves begin about the fifth node.

Pa., Ohio, s. Mich., southw. to Fla. and Tex.

7. **Lysimachia hỳbrida** Michx. Map 1600. Infrequent to rare in the bottoms of ditches, in ponds and swamps, and on the muddy borders of sloughs and streams.

Que. to w. Ont. and N. Dak., southw. to Fla. and Tex.

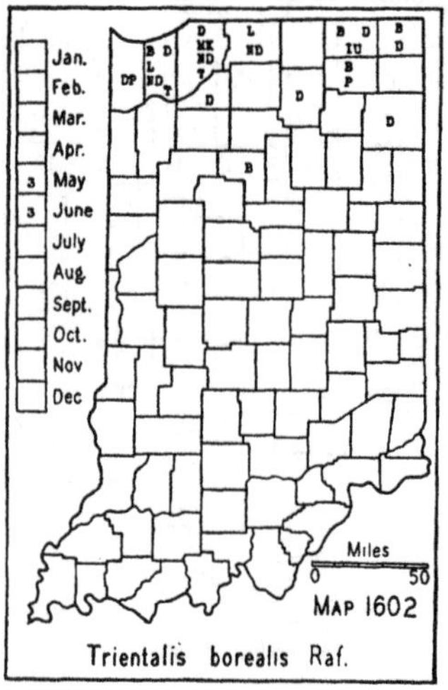

Trientalis borealis Raf.

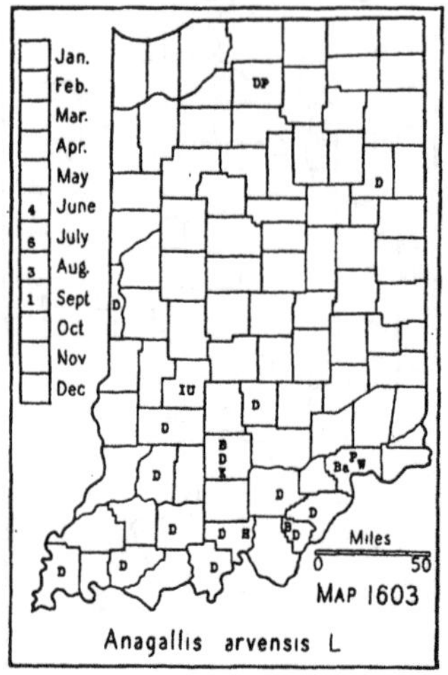

Anagallis arvensis L.

Centunculus minimus L.

8. **Lysimachia longifòlia** Pursh. (See Pflanzenfam. IV, **237**: 279. 1905.) (*Steironema quadriflorum* (Sims) Hitchc.) Map 1601. Rather frequent in the lake area in marshes and springy areas about lakes and along streams. Farther south it is infrequent to local in springy places. It is sometimes found in wet prairies.

N. Y. to Man., southw. to Va. and Mo.

6333. TRIENTÀLIS [Rupp.] L.

1. **Trientalis boreàlis** Raf. (Rhodora **11**: 236. 1909.) (*Trientalis americana* (Pers.) Pursh.) STAR FLOWER. Map 1602. In deep humus, usually in tamarack and birch bogs, under white pine in swamps, or in low woods which border the preceding habitats. Local but frequent to common where found.

Lab. to Man., southw. to Va. and Ill.

6338. ANAGÁLLIS [Tourn.] L.

1. ANAGALLIS ARVÉNSIS L. SCARLET PIMPERNEL. Map 1603. Our specimens are from clover, wheat, and abandoned fields, waste places about habitations, and rarely in open places in nearby woods.

Nat. of Eurasia; Newf. to Fla., westw. to the Pacific coast.

6339. CENTÚNCULUS [Dill.] L.

1. **Centunculus mínimus** L. CHAFFWEED. Map 1604. This plant is usually one and a half to four inches high and so minute as to be easily overlooked. It has been reported from Floyd and Jefferson Counties. Where I have found it, it is always a common to abundant plant. I believe it to be local, however, because of its habitat, for it apparently prefers a minimacid soil. It is found in bare places in open woods, usually associated with black oak, in bare places in pastured woods, along paths in woods, and in abandoned fields.

Ill. and Minn. to B. C., southw. to Fla., Tex., and Mex.; also in Eu. and S. A.

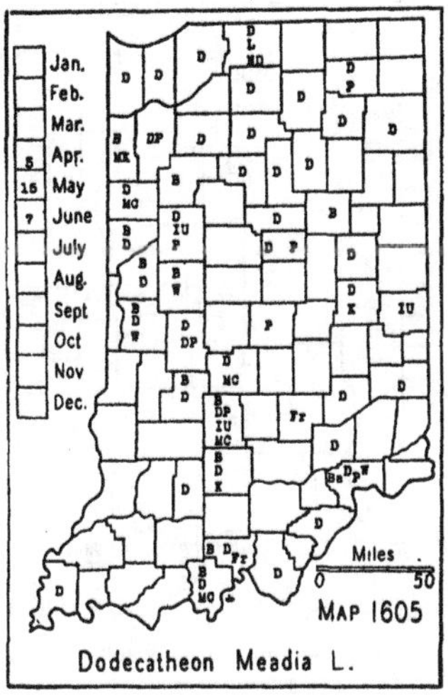

Dodecatheon Meadia L.

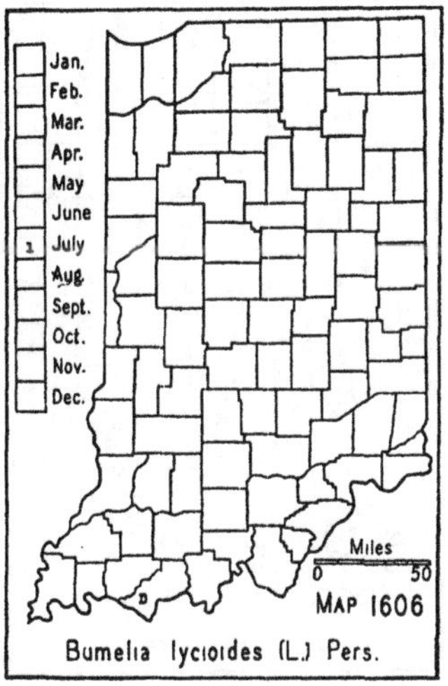

Bumelia lycioides (L.) Pers.

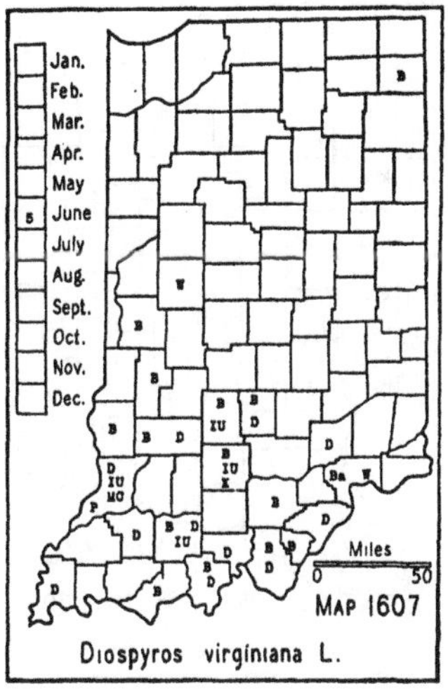

Diospyros virginiana L.

6341. DODECÀTHEON L.

1. **Dodecatheon Mèadia** L. COMMON SHOOTINGSTAR. Map 1605. Mostly on high, wooded banks and bluffs of streams and in prairies, more rarely on wooded slopes, and very rarely in marshes.

The flowers vary in color from white to deep pink. Plants with white flowers are known as f. *alba* Macbride (Field Museum Nat. Hist. Publ. Bot. Ser. 8: 129. 1930.)

Pa. to Man., southw. to Ga. and Tex.

239. SAPOTÀCEAE Reichenb. SAPODILLA FAMILY

6374. BUMÈLIA Swartz

1. **Bumelia lycioìdes** (L.) Pers. BUCKTHORN BUMELIA. Map 1606. Our only station for this shrub or small tree is the talus slope of the sandstone cliffs of the Ohio River about 3 miles above Cannelton, Perry County. When I found it in 1912 there was one specimen about 10 feet high and several other specimens of lesser height. The area has been pastured and in 1929, there were only two small specimens surviving.

Va. to s. Ill., southw. to Fla. and Tex.

240. EBENÀCEAE Vent. EBONY FAMILY

6406. DIOSPỲROS L.

1. **Diospyros virginiàna** L. COMMON PERSIMMON. Map 1607. This tree was doubtless a native of southern Indiana from Franklin County to Parke County and southward. Probably introduced northward. For the most part it is a scattered tree throughout this area, occurring more frequently and in greater abundance in the unglaciated area. It is found most frequently in dry ground but in the southwestern counties it is found in low ground where it reaches its greatest size. In old abandoned fields it forms thickets, due to its ability to spread from root shoots.

Conn. to s. Iowa, southw. to Fla. and Tex.

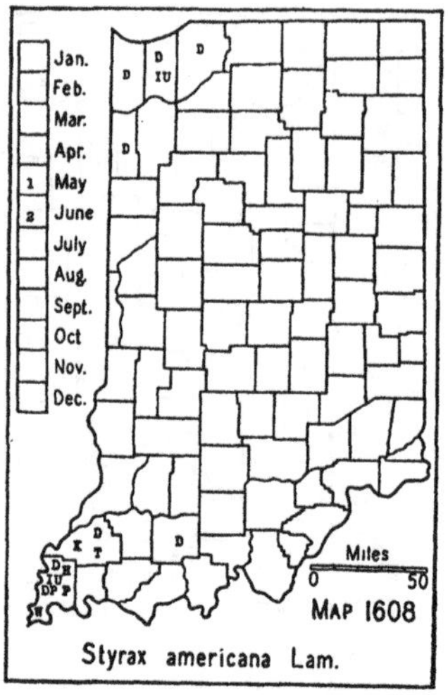

Styrax americana Lam.

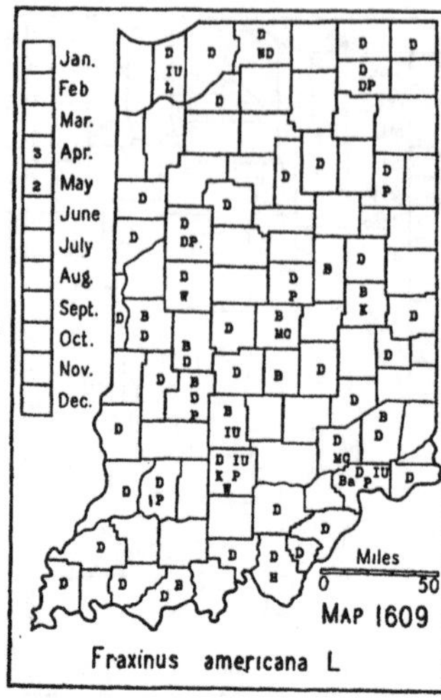

Fraxinus americana L

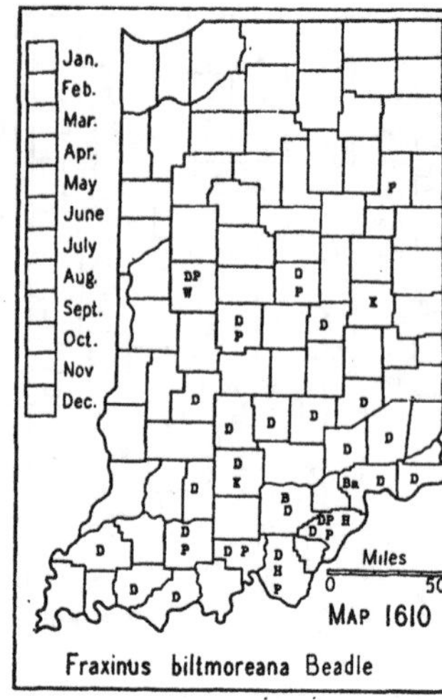

Fraxinus biltmoreana Beadle

241. STYRACÀCEAE A.DC. STORAX FAMILY

6411. STÝRAX [Tourn.] L.

1. **Styrax americàna** Lam. AMERICAN SNOWBELL. Map 1608. Swampy woods and in woodland along streams that usually overflow annually. Local. Its distribution in Indiana offers an interesting problem.

Va. to Fla. and La., and northw. in the Mississippi Valley to the Kankakee River Valley in Ind.

243. OLEÀCEAE Lindl. OLIVE FAMILY

Leaves compound; fruit dry, a samara........................6420. FRAXINUS, p. 752.
Leaves simple; fruit fleshy, a drupe.......................6427. FORESTIERA, p. 754.

6420. FRÁXINUS [Tourn.] L. ASH

Bark of mature trees furrowed; fruit not winged to the base.
  Body of fruit robust, round and rather abruptly passing into the wing; terminal buds deltoid.
    Branchlets glabrous; axis of the leaves glabrous, rarely somewhat pubescent...1. *F. americana.*
    Branchlets pubescent; axis of leaves pubescent, at least until nearly mature..2. *F. biltmoreana.*
  Body of fruit flattened and gradually passing into the wing, gradually tapering from the wings to the base; terminal buds longer than wide.
    Branchlets glabrous or nearly so and usually smaller than those of *F. americana.*...3. *F. lanceolata.*
    Branchlets velvety-pubescent, at least when young.
      Calyx of fruit less than 3 mm long; body of samara just below the wing less than 3 mm wide, rarely 4 mm wide, usually 1.4-2.5 mm wide; samaras 3-4.5 cm long...................................4. *F. pennsylvanica.*
      Calyx of fruit more than 3 mm long, generally 4-5 mm long; body of samara just below the wing more than 3 mm wide, usually 4-5 mm wide; samaras generally 4-6 cm long..................................5. *F. tomentosa.*
Bark of mature trees scaly or flaky; fruit winged to the base.
  Branchlets and small branches usually 4-angled; leaflets on very short stalks..6. *F. quadrangulata.*
  Branchlets and branches round; leaflets sessile.........................7. *F. nigra.*

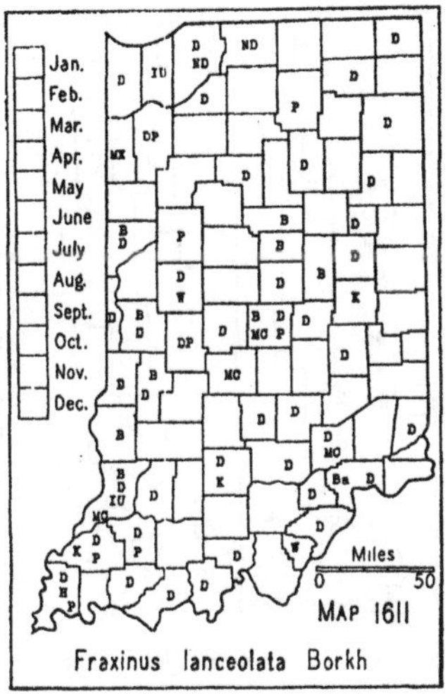

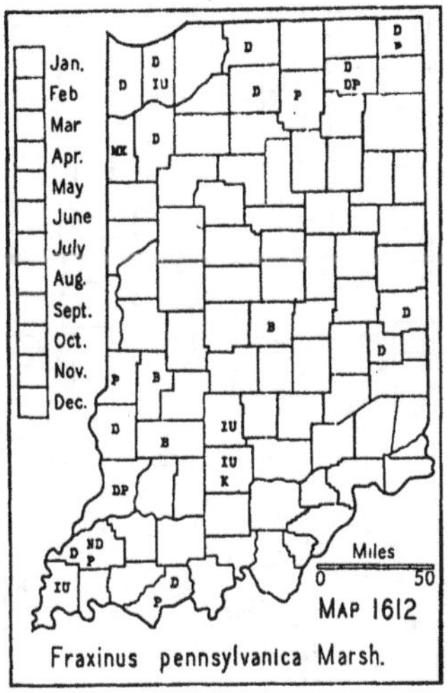

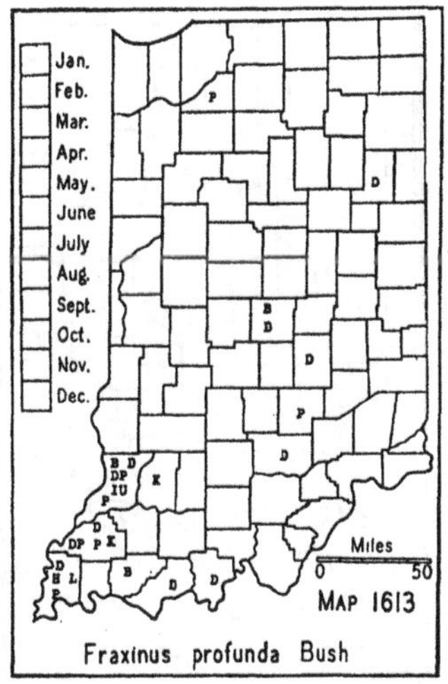

1. **Fraxinus americàna** L. WHITE ASH. Map 1609. Frequent to common on uplands in the beech and sugar maple type of forest and rarely in the black oak and hickory type except in the coves. In the northern part of the state in the level woods it is always a frequent tree in the beech and sugar maple type and in the white oak, red oak, basswood type of woodland.

A form of this species with reddish purple fruit is known as f. *iodocarpa* Fern. It is found throughout the eastern part of the state.

N. S. to Ont. and Minn., southw. to Fla. and Tex.

2. **Fraxinus biltmoreàna** Beadle. BILTMORE ASH. Map 1610. This species is not as frequent as the white ash but in certain habitats it is a common tree. I am certain that I have seen this species as far north as Yellow River in Marshall County but I was not able to collect a specimen.

Pa., Ind., and Mo., southw. to Ga., and Ala.

3. **Fraxinus lanceolàta** Borkh. GREEN ASH. Map 1611. Frequent and locally common in low ground along streams, in swamps, and in low woods. It often forms a complete stand. Its most constant associates are white elm and soft maples.

Maine, Que., to Sask., southw. to Fla, and Tex.

4. **Fraxinus pennsylvánica** Marsh. Fernald (Rhodora 40: 452-454. 1938) discusses this species and its varieties. RED ASH. Map 1612. In dry or moist soil, usually on or near the banks of streams and lakes.

N. S. to Man., southw. to Ga., Miss., and Okla.

5. **Fraxinus tomentòsa** Michx. f. (Rhodora 40: 450-452. 1938.) (*Fraxinus profunda* of authors and *Fraxinus profunda* var. *Ashei* E. J. Palmer.) PUMPKIN ASH. Map 1613. In swamps, ponds, sloughs, and overflow land along streams. Its most constant associates are pecan, red maple, white elm, shellbark hickory, green ash, and cypress. Infrequent to common in its habitat.

S. Ind. to s. Ill., southeastern Mo., southw. to Fla. and La.

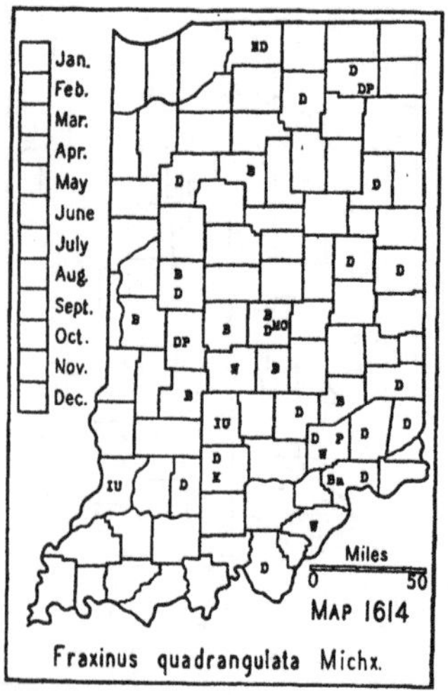

Fraxinus quadrangulata Michx.

Fraxinus nigra L.

Forestiera acuminata (Michx.) Poir.

6. **Fraxinus quadrangulàta** Michx. BLUE ASH. Map 1614. Found sparingly throughout the state, although we have no records from the northwestern counties. It is generally found on high ground and where its distribution is limited, it is usually restricted to the high banks of streams.

Ont. to Iowa, southw. to Tenn. and Ark.

7. **Fraxinus nìgra** L. BLACK ASH. Map 1615. Rather local but usually of considerable abundance in its preferred habitat. Found in wet and swampy woods throughout the lake area; southward it becomes an infrequent tree of swampy places. There are no records for the unglaciated area except in the White River Valley. The species is more frequent in northern Indiana than our map indicates.

Newf. to Lake Winnipeg, southw. to W. Va., Ind., and northwestern Ark.

6427. FORESTIÈRA Poir.

1. **Forestiera acuminàta** (Michx.) Poir. (*Adelia acuminata* Michx.) TEXAS ADELIA. Map 1616. Low borders of sloughs, swamps, and river banks. It is usually associated with buttonbush. Very local.

Sw. Ind. to Mo., southw. to Tex.

245. LOGANIÀCEAE Dumort. LOGANIA FAMILY

6453. SPIGÈLIA L.

1. **Spigelia marilándica** L. PINKROOT. Map 1617. Our only known station for this plant is a post oak flat just south of Half Moon Pond about 10 miles southwest of Mt. Vernon, Posey County. It is frequent here over several acres. The report for Marion County is doubtless an error.

Ohio, s. Ind. to Mo., southw. to Fla. and Tex.

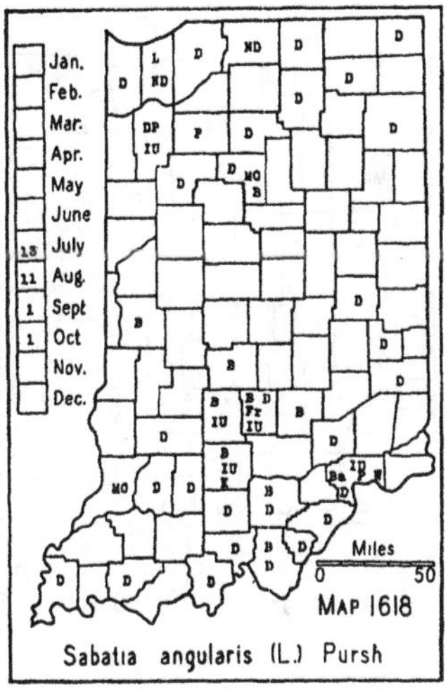

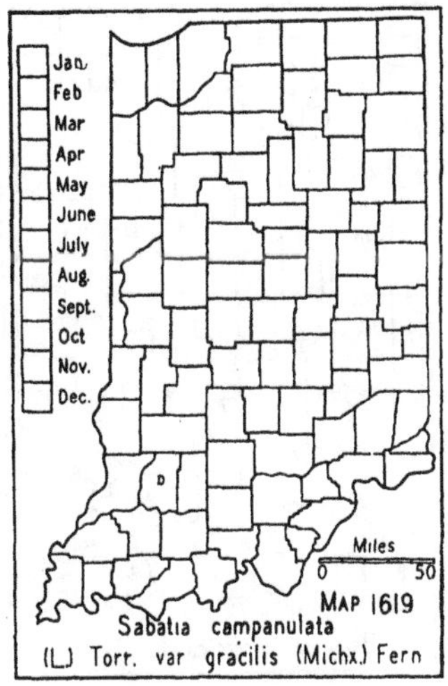

246. GENTIANÀCEAE Dumort. GENTIAN FAMILY

Leaves simple, sessile, opposite or whorled or if scalelike sometimes some of them alternate.
  Plants not filiform; leaves not scalelike.
    Plants less than a meter tall; leaves not in whorls.
      Lobes of corolla much more than half as long as the tube. .6494. SABATIA, p. 755.
      Lobes of corolla not more than half as long as the tube.
        Flowers pink to rose purple; styles usually deciduous; anthers twisted...... ....6496. CENTAURIUM, p. 756.
        Flowers blue, greenish white or yellowish; styles persistent; anthers not twisted.
          Plants not more than 16 cm tall; basal leaves reduced to scales; corolla lobes imbricated in bud........................6502. OBOLARIA, p. 756.
          Plants more than 16 cm tall; basal leaves not reduced to scales; corolla lobes convolute in bud......................6509. GENTIANA, p. 757.
    Plants more than a meter tall; leaves mostly in whorls of 4. .6512. FRASERA, p. 760.
  Plants filiform; leaves scalelike...........................6501. BARTONIA, p. 756.
Leaves 3-foliolate; petioles alternate....................6543. MENYANTHES, p. 760.

6494. SABÀTIA Adans. ROSE GENTIAN

Branches opposite.
  Leaves cordate-clasping; stem strongly 4-angled..................1. *S. angularis.*
  Leaves linear and linear-oblong, sessile; stem slightly 4-angled. (See excluded species no. 494, p. 1080)...*S. brachiata.*
Branches alternate, rarely one opposite; plant diffusely branched at maturity; leaves linear or the lowest lance-linear; calyx lobes setaceous, as long as the corolla lobes. ....2. *S. campanulata* var. *gracilis.*

1. **Sabatia angulàris** (L.) Pursh. Map 1618. In Indiana this plant has two distinct habitats. In the lake area, including our Henry County specimen, all of our specimens with one exception were found on the moist sandy or peaty borders of lakes and swamps. In 1938 I found it to be a common plant on an open, pastured black and white oak ridge a half mile northwest of Disko, Fulton County. In the southern part of the state this species is frequent to common in hard, dry, clay soil in old fallow fields,

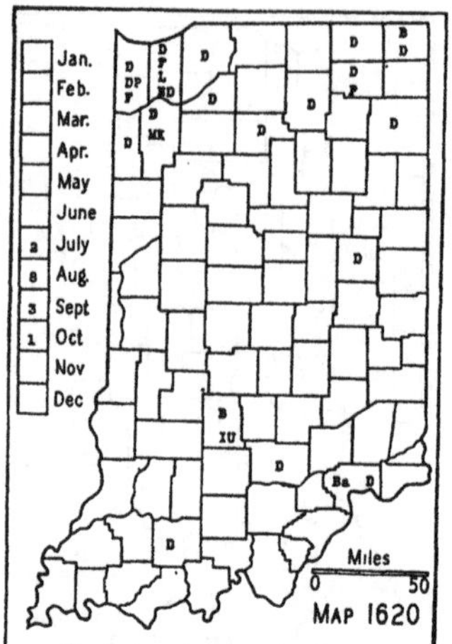

Map 1620
Bartonia virginica (L.) BSP.

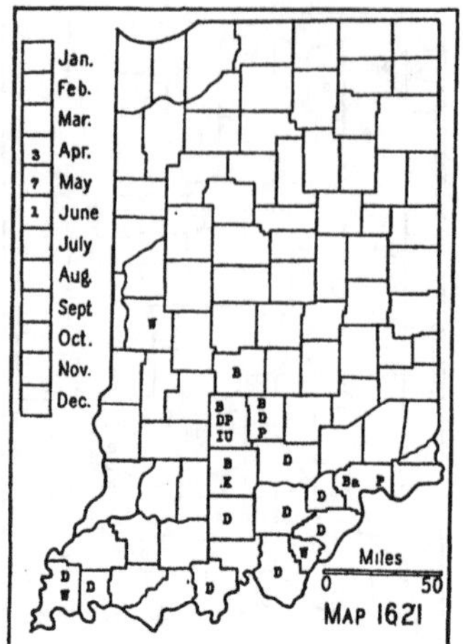

Map 1621
Obolaria virginica L.

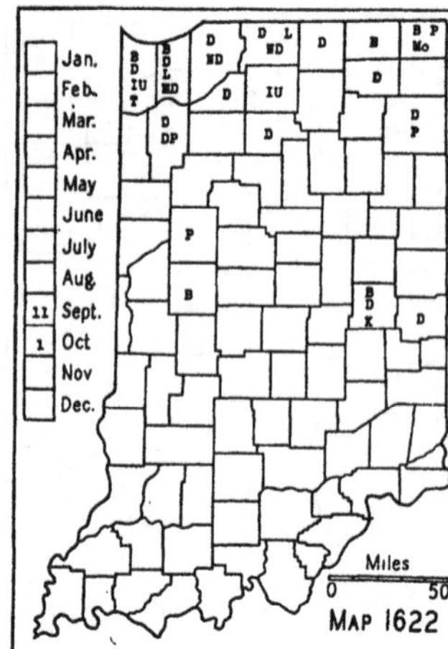

Map 1622
Gentiana crinita Froel.

and in this habitat it reaches its greatest size. It is also infrequently found on exposed open places on the crests or slopes of wooded ridges.

N. Y. to Ont. and Mich., southw. to Fla. and La.

2. **Sabatia campanulàta** (L.) Torr. var. **grácilis** (Michx.) Fern. (Rhodora 39: 444. 1937.) (*Sabatia gracilis* (Michx.) Salisb.) Map 1619. Our only specimens were found in a small colony in rather acid soil in a low, sandy flat in a woods 4 miles north of Washington, Daviess County. Here it was closely associated with *Betula nigra, Spiraea tomentosa, Viola lanceolata, Viola sagittata, Rhexia mariana* var. *leiosperma, Rhexia virginica, Linum medium* var. *texanum,* and *Hieracium Gronovii.*

Nantucket, Mass. to Fla. and west to La. in salt marshes and brackish swamps, rarely inland in fresh-water swamps; also on the summits of the southern Alleghenies; Bahamas; Cuba.

6496. CENTAÚRIUM Hill

See excluded species nos. 495 and 496, p. 1080.

6501. BARTÒNIA Muhl.

1. **Bartonia virgínica** (L.) BSP. Map 1620. Infrequent in the lake area and very rare south of it. Usually found in clumps of sphagnum in bogs and more rarely in moist habitats in very sandy, minimacid soil in open places in black and white oak woods, growing in moss with *Polygala cruciata, Gaultheria procumbens,* and *Aronia melanocarpa.*

The petals vary in color from greenish yellow to yellow or rose purple.

N. S. to Minn., southw. to Fla. and La.

6502. OBOLÀRIA L.

1. **Obolaria virgínica** L. PENNYWORT. Map 1621. This very inconspicuous plant is usually rare and only a few specimens are found in a colony. It is sometimes frequent, however, and on April 26, 1927, I found it to be a common plant in a small field on a wooded slope in Harrison

County. This field had not been cultivated for more than 20 years and had reforested mostly to tulip trees 4-6 inches in diameter. It prefers rather sandy soil of exposed places, although it is often found in places with a thick cover of leaves but in such situations it is never abundant. It has been reported as far north as Parke and Putnam Counties.

N. J. to Ill., southw. to Ga. and Tex.

6509. GENTIÀNA [Tourn.] L. GENTIAN

Corolla without plaits, lobes or teeth in the sinuses.
  Peduncles more than 2 cm long; corolla lobes fringed.
    Leaves ovate to ovate-lanceolate; corolla lobes deeply fringed around the summit; ovary lanceolate ....1. *G. crinita.*
    Leaves linear to lance-linear; corolla lobes fringed at the sides, the summit sparingly and shortly fringed or merely dentate; ovary elliptic.....2. *G. procera.*
  Peduncles less than 2 cm long; corolla lobes not fringed.
    Calyx lobes mostly 2-3.5 mm long. (See excluded species no. 499, p. 1081)........ .... *G. quinquefolia.*
    Calyx lobes mostly 4-8 mm long..............3. *G. quinquefolia* var. *occidentalis.*
Corolla with plaits in the sinuses.
  Margins of leaves and calyx lobes scabrous or ciliate; flowers blue; seeds winged.
    Corolla nearly truncate at the summit, the narrow lobes almost obsolete or at least 2.5-3 mm shorter than the plaits; margins of the wide, whitish, wedge-shaped plaits minutely fimbriate-dentate.........................4. *G. Andrewsii.*
    Corolla with distinct, broad, rounded or acute lobes 1.5-10 mm long; margins of the less conspicuous 2-cleft plaits dentate.
      Stamens cohering more or less in a ring about the style; calyx lobes oblanceolate, widest about the middle, usually 1.5-3 mm wide and 7-12 mm long, strongly ciliate on the margins, rather abruptly acuminate; stems usually glabrous or some internodes more or less puberulent in lines; corolla 3-5 cm long, the erect, mostly acute lobes 2-3 mm longer than the whitish plaits. ....5. *G. Saponaria.*
      Stamens free (sometimes adhering in dried specimens because of pressure applied in drying); calyx lobes linear, usually 6-8 mm long and about 1 mm wide, the margins scabrous or somewhat minutely ciliate, long taper-pointed from about the middle; entire stem usually puberulent in lines; corolla usually 2.5-3.5 cm long, the conspicuous, acute, slightly spreading lobes 5-7 mm longer than the bluish plaits.........................6. *G. puberula.*
  Margins of leaves and calyx lobes smooth; flowers white or yellowish; seed winged or wingless.
    Base of leaves cordate and closely clasping; calyx lobes ovate or narrow-ovate; corolla lobes broad, about twice the length of the broad, toothed appendages; seed winged ....7. *G. flavida.*
    Base of leaves narrowed; calyx lobes linear, unequal; corolla lobes ovate, much exceeding the small, sparingly toothed appendages; seed wingless......... ....8. *G. villosa.*

1. **Gentiana crinita** Froel. FRINGED GENTIAN. Map 1622. An infrequent to common plant in open, springy places, marshes, interdunal flats and on the sandy borders of sloughs in the dune area. This is a much admired plant and attempts to naturalize it usually fail because it is so exacting in its habitat. It was formerly common in certain marshes but since these have been grazed it has disappeared or only a few plants have been able to persist.

Cent. Maine to N. Dak., southw. to Ga., Ohio, and Iowa.

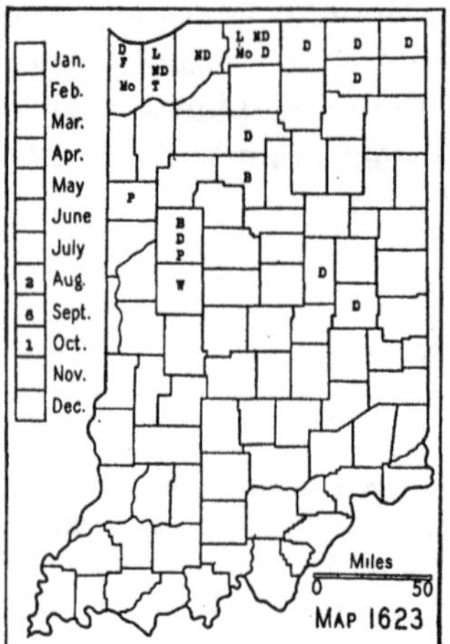

Gentiana procera Holm

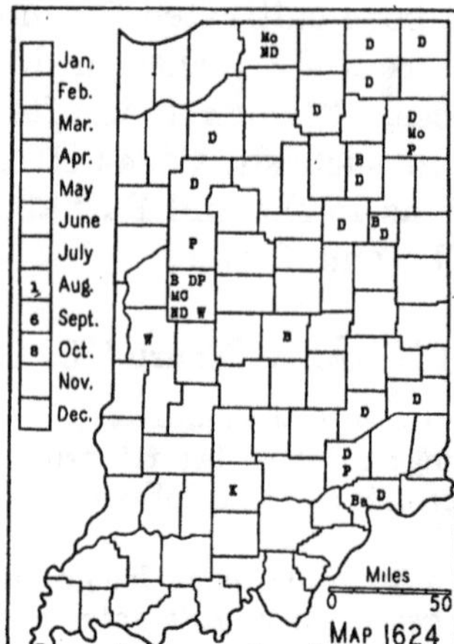

Gentiana quinquefolia var. occidentalis (Gray) Hitchc.

Gentiana Andrewsii Griseb.

2. **Gentiana procèra** Holm. LESSER FRINGED GENTIAN. Map 1623. A rare plant of habitats similar to those of the preceding. Our southernmost plant was collected in a springy place along White River near Anderson by Ray Dawson. When the two species (*Gentiana crinita* and this species) grow together this species is found in wetter situations.

N. Y. and Ont. to Man., Minn., and S. Dak.

2a. **Gentiana procera** f. **laevicàlyx** Fern. (Rhodora 32: 221. 1930.) This is a form with a smooth calyx. Our only specimen is from the sedge border of the north side of Bruce Lake in Fulton County.

3. **Gentiana quinquefòlia** L. var. **occidentàlis** (Gray) Hitchc. (*Aloitis occidentalis* Greene and *Aloitis mesochora* Greene.) Map 1624. *Gentiana quinquefolia* is a highly variable species which has led authors to name variants. The calyx lobes vary from 4-8 mm long, acute to acuminate, linear to narrow-ovate, leaving a wide open sinus or overlapping. The form with wide and overlapping calyx lobes is rare and more western in the state. No doubt this species, as well as other species, is more common than our records indicate because the plants are in flower usually after the season for botanical collecting is over. It is local but common where it is found.

Maine, Ont., and Mich., southw. to Fla. and Mo.

4. **Gentiana Andréwsii** Griseb. (*Dasystephana Andrewsii* (Griseb.) Small.) CLOSED GENTIAN. Map 1625. Infrequent in all parts of the state except in the knobstone area, in low woods, roadside ditches, low ground about lakes, and interdunal flats. Flowers vary in color from light to dark blue although I have one specimen from Steuben County that is maroon.

Mass., Que., and Nebr., southw. to Md. and Mo.

5. **Gentiana Saponària** L. (*Dasystephana Saponaria* (L.) Small.) SOAPWORT GENTIAN. Map 1626. In the southern part of the state it is found in hard, white clay soil in low, flat woods, usually associated with pin oak,

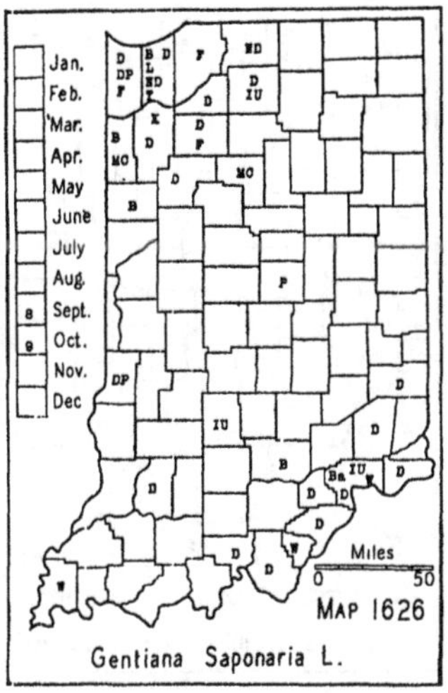

Gentiana Saponaria L.

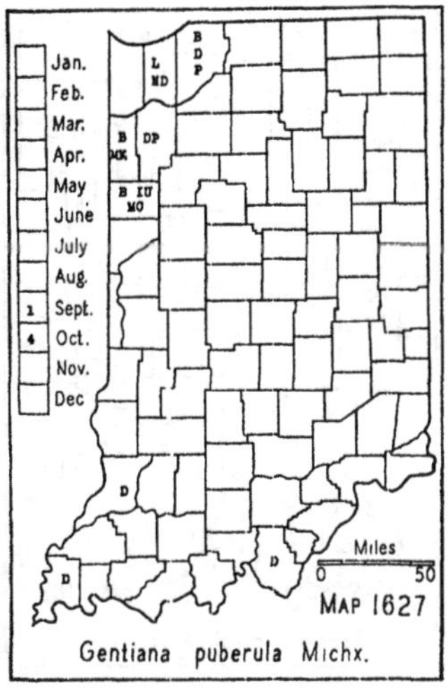

Gentiana puberula Michx.

Gentiana flavida Gray

sweet gum, beech, and red maple, and in the northwestern part of the state it is found in moist black sand in interdunal flats about Lake Michigan, in sandy flats in black and white oak woods, and in moist prairie habitats. Infrequent, but where found several specimens may be found here and there, growing singly.

Conn., Ont. to Minn., southw. to Fla. and La.

6. **Gentiana pubérula** Michx. (*Dasystephana puberula* (Michx.) Small.) DOWNY GENTIAN. Map 1627. This is supposed to be a prairie plant. Infrequent to very rare in moist, black, sandy soil in the open, usually along roadsides and railroads and in fallow fields, low open woods, and marshes. In two different years I made a collection of this species on an open, rocky, black and post oak slope of the high hill at Stewart's Landing about 3 miles east of Elizabeth in Harrison County. This seems to be an unusual habitat but it was associated with other prairie plants such as *Liatris, Viola pedata,* and *Andropogon furcatus.* Very rare here.

Md. to Minn., southw. to Ga. and Kans.

7. **Gentiana flávida** Gray. (*Dasystephana flavida* (Gray) Britt.) YELLOWISH GENTIAN. Map 1628. This species has been reported from Cass, Lake, Marshall, Monroe, Noble, Porter, St. Joseph, Steuben, Tippecanoe, and Vigo Counties. I have it from the border of a very sandy black and white oak woods in Lagrange County, from the roadside of a little used road along a woods on a ridge in Perry County, and from moist soil near Mineral Springs in Porter County.

Ont. to Minn., southw. to Va., Ky., and Mo.

8. **Gentiana villòsa** L. (*Dasystephana villosa* (L.) Small.) Map 1629. I have two collections of this species from Harrison County. It was first discovered by Mrs. Chas. C. Deam. Both collections were made in the southeastern part of the county in black and white oak woods. Only a few specimens were found.

N. J., Pa., and Ind., southw. to Fla., and La.

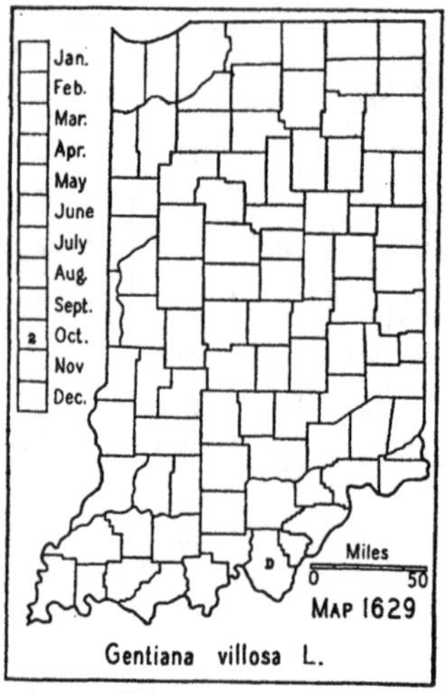

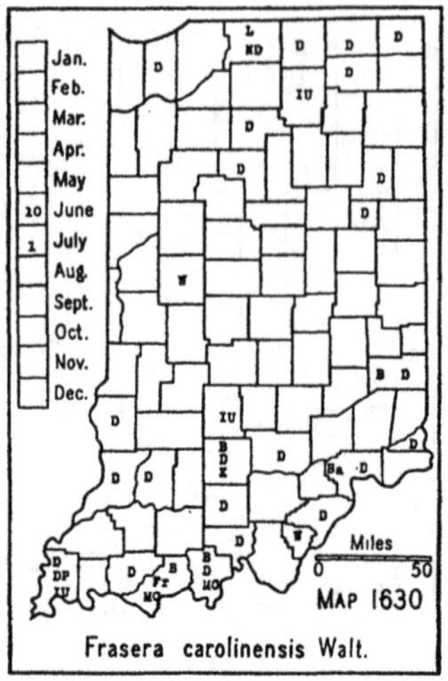

6512. FRÀSERA Walt.

1. **Frasera carolinénsis** Walt. AMERICAN COLUMBO. Map 1630. Infrequent to rare in all parts of the state. It is usually 4-8 feet high. Generally in dry, clay soil, associated with white and black oaks.

N. Y., Ont., and Wis., southw. to Ga. and Tenn.

6543. MENYÁNTHES [Tourn.] L.

[Fernald. Menyanthes trifoliata var. minor. Rhodora **31**: 195-198. 1929.]

1. **Menyanthes trifoliàta** L. var. **mìnor** Raf. (*Menyanthes trifoliata* L. of American authors.) BUCKBEAN. Map 1631. Frequent in tamarack bogs and marshes in the lake area before these were drained; now becoming rare.

Lab. to the Rocky Mts., southw. to Va., Nebr., and Mo.

247. APOCYNÀCEAE Lindl. DOGBANE FAMILY

Leaves alternate..6591. AMSONIA, p. 760.
Leaves opposite.
Plants small, evergreen, creeping or trailing, the erect stems usually 1-2.5 dm high; leaf blades mostly 1.5-4 cm long; flowers axillary, solitary, blue.......................6598. VINCA, p. 761.
Plants not evergreen, long climbing vines or large erect or diffuse perennials; leaf blades more than 5 cm long; flowers terminal, not solitary, white, cream or pink.
Climbing vines; calyx glandular within; corolla funnel-shaped; filaments slender. ....6667. TRACHELOSPERMUM, p. 761.
Erect or rarely diffuse perennials; calyx not glandular within; corolla bell-shaped or cylindric; filaments short, broad, and flat.........6684. APOCYNUM, p. 762.

6591. AMSÒNIA Walt.

[Woodson. A monograph of the genus Amsonia. Ann. Missouri Bot. Gard. **15**: 379-434. 1928.]

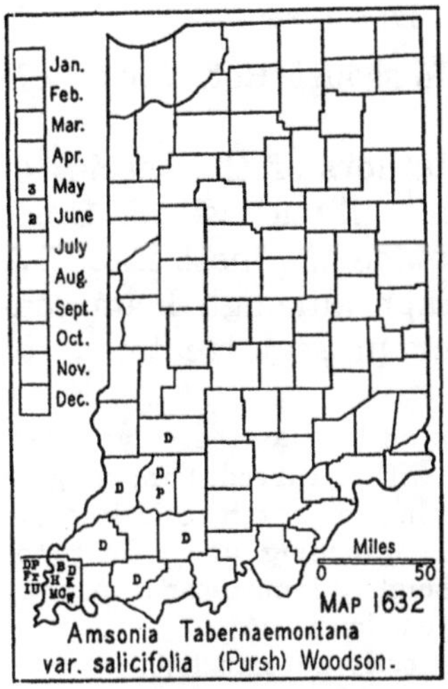

Amsonia Tabernaemontana var. salicifolia (Pursh) Woodson.

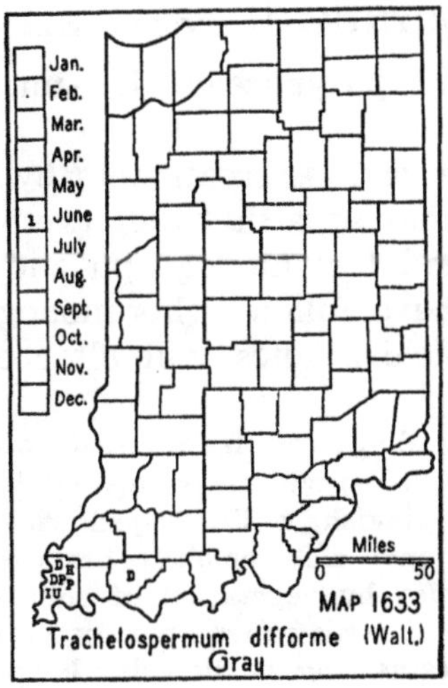

Trachelospermum difforme (Walt.) Gray

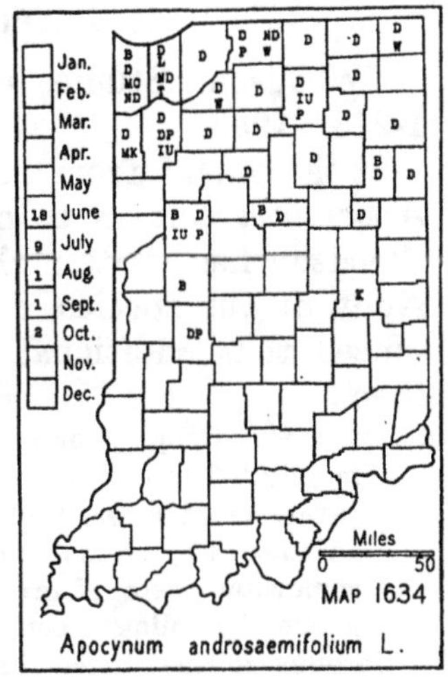

Apocynum androsaemifolium L.

1. **Amsonia Tabernaemontàna** Walt. var. **salicifòlia** (Pursh) Woodson. WILLOW AMSONIA. Map 1632. Frequent to infrequent in our southwestern counties in compact soils in low open woodland and along ditches. It is perfectly hardy in cultivation in northern Indiana.

Va., N. C., S. C., Ga., Ala., La., Ky., Tenn., Ind., Ill., Mo., Ark., and Tex.

6598. VÍNCA L.

1. VINCA MÌNOR L. COMMON PERIWINKLE. A trailing, woody perennial which spreads rapidly in some localities. I have no evidence that it spreads by seed. The largest colony known to me is in Spring Mill State Park. Here it covers acres of native forest land and forms so dense a stand that it not only makes difficult the reproduction of the forest trees but in many places excludes almost all kinds of vegetation. For this reason it should not be permitted to escape to woodland.

Most commonly used in cemeteries from which it often escapes. On account of this use it is commonly called graveyard myrtle or myrtle. Naturalized in all parts of Indiana.

Nat. of Eu.

6667. TRACHELOSPÉRMUM Lemaire

1. **Trachelospermum diffórme** (Walt.) Gray. Map 1633. A climbing vine, herbaceous in Indiana, growing in compact, clay soil in low, open woods in southwestern Indiana. The flowers are yellowish and very fragrant. The first specimen I ever found was detected by its odor. I noted a peculiar, pungent fragrance and in searching for the source I found this vine, at least a rod distant. Very rare in our area.

Del. to Fla., westw. to Ind., Mo., and Tex.

6684. APÓCYNUM L. INDIAN HEMP. DOGBANE

[Woodson. Studies in Apocynaceae. I. Ann. Missouri Bot. Gard. **17**: 1-213. 1930.]

It is evident from the various treatments by authors of the species of *Apocynum* that occur in our area that they are not well understood. Since Woodson has written the latest monograph and made the most intensive study of our species, I have followed his monograph although I am convinced that additional field studies will alter our present treatment of them.

Stem leaves drooping or spreading; corollas at least twice as long as the calyx lobes, mostly 3-6 mm long in dried specimens, campanulate, with recurved, spreading or rarely erect lobes, pink, pinkish, white striped with pink, or rarely colorless (in 2a and 2b); stems unevenly dichotomously branched, spreading at maturity or somewhat erect if crowded by vegetation of equal height; inflorescence usually above the foliage; coma of seeds tawny, mostly 1.5-2 cm long.

Corollas at least three times as long as the calyx lobes, mostly 4-6 mm long in dried specimens, pink, pinkish or white striped with pink, the lobes recurved, strongly fragrant; stems conspicuously dichotomously branched; stem leaves and sometimes those of the branches drooping; follicles 6-15 cm long, straight or nearly so; seed about 2 mm long..........................1. *A. androsaemifolium.*

Corollas about twice as long as the calyx lobes, mostly 3-4 mm long in dried specimens, faintly pink or colorless in 2a and 2b, the lobes generally spreading; stems more or less dichotomously branched; stem leaves spreading, rarely ascending; follicles 7-15 cm long, straight, divergent, or somewhat falcate; seed about 4 mm long.

Plants not glabrous throughout.

Calyx glabrous without, the lobes pinkish.......................2. *A. medium.*

Calyx hirtellous without, the lobes colorless......2a. *A. medium* var. *sarniense.*

Plants glabrous throughout; calyx lobes colorless..2b. *A. medium* var. *leuconeuron.*

Stem leaves ascending; corollas less than twice as long as the calyx lobes, usually 2.5-4 mm long in dried specimens, white or greenish white, tubular, sometimes the tube very short, the lobes erect; the cymose inflorescences conspicuously overtopped by sterile branches.

Stem leaves evidently petiolate, narrowed to the base, or the very lowest sessile and obtuse at the base; follicles relatively long and usually falcate, 12-20 cm long; coma of seed 20-30 mm long.

Plant pubescent, at least the lower surface of the leaves.

Plant glabrous except the lower surface of the leaves........3. *A. cannabinum.*

Plant pubescent on both surfaces of the leaves and in the inflorescence. ....3a. *A. cannabinum* var. *pubescens.*

Plant entirely glabrous...................3b. *A. cannabinum* var. *glaberrimum.*

Stem leaves sessile or subsessile, cordate or subcordate at the base and often clasping, oblong to lanceolate; follicles relatively short and straight, 4-10 cm long; coma of seed white, 8-20 mm long.

Leaves glabrous on both surfaces............................4. *A. sibiricum.*

Leaves pubescent on the lower surface, at least along the veins................ ....4a. *A. sibiricum* var. *Farwellii.*

1. **Apocynum androsaemifòlium** L. SPREADING DOGBANE. Map 1634. More or less infrequent in rather sandy or gravelly soil throughout the northern counties along roadsides and in open woodland, where it is associated with black and white oak. In southern Indiana it has a similar habitat but becomes rare to very rare.

Throughout temperate N. A.

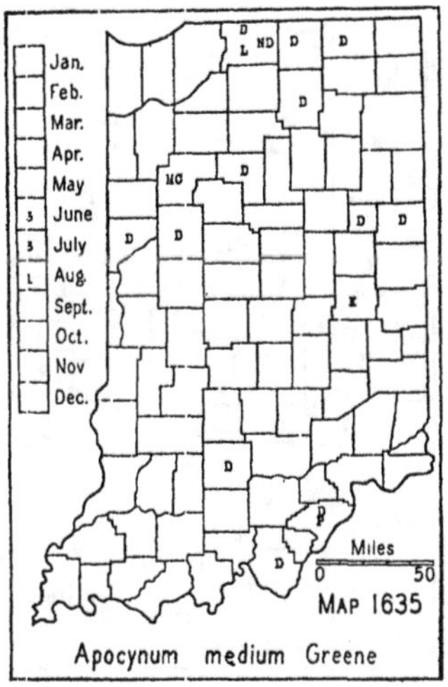

Apocynum medium Greene

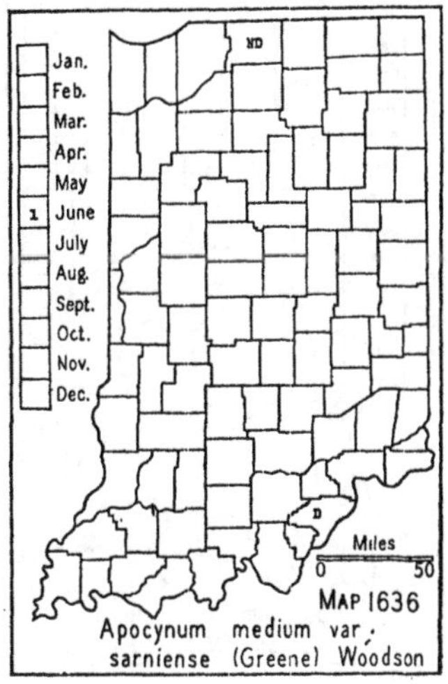

Apocynum medium var. sarniense (Greene) Woodson

Apocynum medium var leuconeuron (Greene) Woodson

2. **Apocynum mèdium** Greene.[1] Map 1635. Moist places in open woodland, along roadsides, and in prairie habitats.

N. B. to Fla., westw. to Que., Iowa, Nebr., and Tex.

2a. **Apocynum medium** var. **sarniénse** (Greene) Woodson. Map 1636. Habitat that of the species.

Woodson says: "Apparently a spontaneous variety, collected in sw. Ont., se. Mich., n. Ind., and s. B. C."

2b. **Apocynum medium** var. **leuconeùron** (Greene) Woodson. Map 1637. Habitat similar to that of the species.

Woodson says: "Upper Miss. Valley, eastw. to s. Mich. and n. Ind."

3. **Apocynum cannabìnum** L. HEMP DOGBANE. Map 1638. In moist or dry situations in almost all kinds of soils in all parts of the state. It is usually found in small colonies along roadsides and fences, in hayfields, fallow fields, and open woodland. I have found specimens of this and other species of the genus growing in adverse ecological conditions, that have much reduced leaves and a decumbent habit. I refer to such habitats as ballast of railroads, gravel pits, and gravel bars of streams.

Woodson says: "Generally throughout the eastern half of the U. S."

3a. **Apocynum cannabinum** var. **pubéscens** (Mitchell) A. DC. Map 1639. Found in habitats and places similar to those of the species. Infrequent.

Woodson says: "Generally throughout the southeastern and central U. S., in north-central Calif.; also in s. Ont."

[1] After the text of this genus was written, Anderson writes that this species is a fertile hybrid betwten *Apocynum androsaemifolium* and *Apocynum cannabinum*. See Anderson. An experimental study of hybridization in the genus Apocynum. Ann. Missouri Bot. Gard. 23: 159-168. 1936.

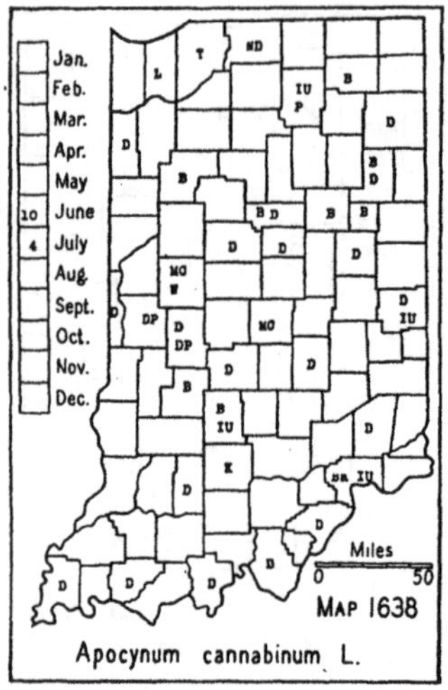

MAP 1638
Apocynum cannabinum L.

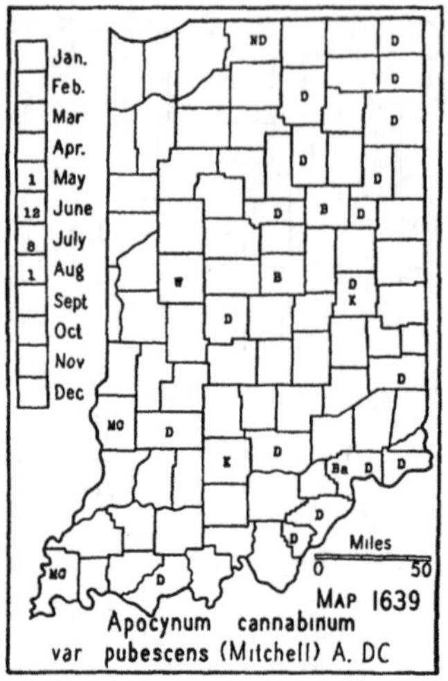

MAP 1639
Apocynum cannabinum var pubescens (Mitchell) A. DC

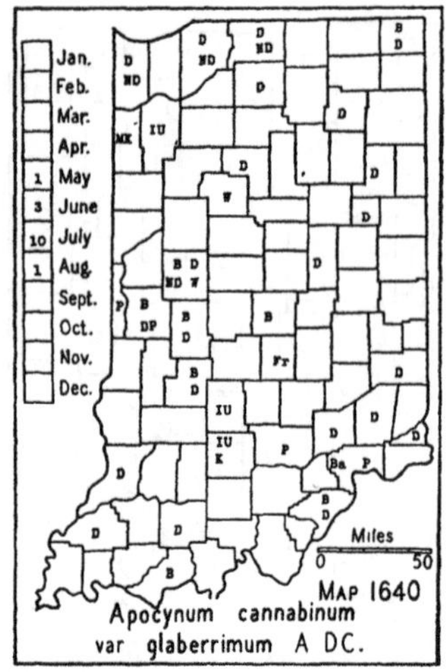

MAP 1640
Apocynum cannabinum var glaberrimum A DC.

3b. **Apocynum cannabinum** var. **glabérrimum** A. DC. Map 1640. Habitats and distribution similar to those of the species. This variety is more frequent than either the species or the pubescent variety.

Woodson says: "Common in every state in the U. S. and sparingly in Can."

4. **Apocynum sibíricum** Jacq. Map 1641. Habitat and distribution similar to those of the other species. Infrequent.

Newf. and s. Canada, westw. to Wyo., and southw. to Va. and Tex.

4a. **Apocynum sibiricum** var. **Farwéllii** (Greene) Fern. Map 1642. A study of our specimens shows several of them to be pubescent on the lower surface of the leaves. Our specimens vary from densely velvety-pubescent to pubescent mostly along the veins with scattered hairs between. Habitat similar to that of the other species.

Woodson says: "Cent. N. Y., e. Mich., and n. Ind.; apparently collected in Tex."

248. ASCLEPIADÀCEAE Lindl. MILKWEED FAMILY

Erect or decumbent herbs.
- Corona hoods without an incurved horn within.............6787. ACERATES, p. 764.
- Corona hoods each with an incurved horn within...........6791. ASCLEPIAS, p. 765.

Twining vines.
- Corolla lobes erect..................................6812. AMPELAMUS, p. 769.
- Corolla lobes rotate.................................6943. GONOLOBUS, p. 770.

6787. ACERÀTES Ell. GREEN MILKWEED

Leaves many, scattered alternately; sepals about 2 mm long; crown of column 0.5-1 mm long; hoods 1.5-2.5 mm long, mostly about 2 mm long...........1. *A. hirtella.*

Leaves opposite; sepals about 3 mm long, lanceolate, acute (narrower and more acute than those of the preceding species); crown sessile; hoods about 4 mm long. ....2. *A. viridiflora.*

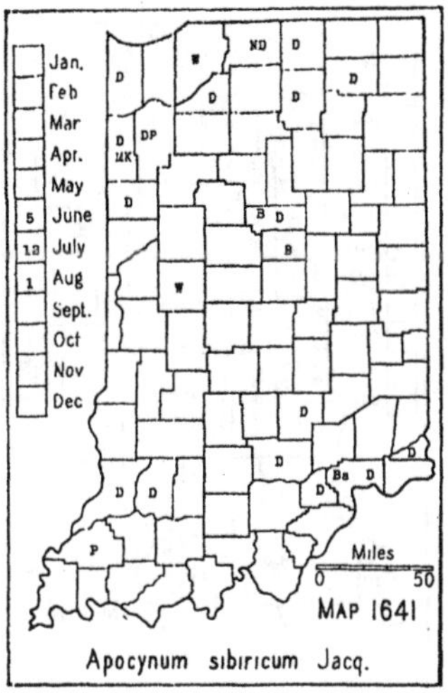

MAP 1641
Apocynum sibiricum Jacq.

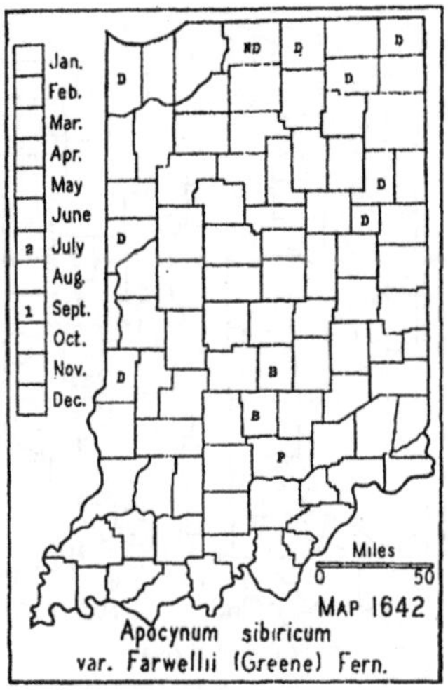

MAP 1642
Apocynum sibiricum var. Farwellii (Greene) Fern.

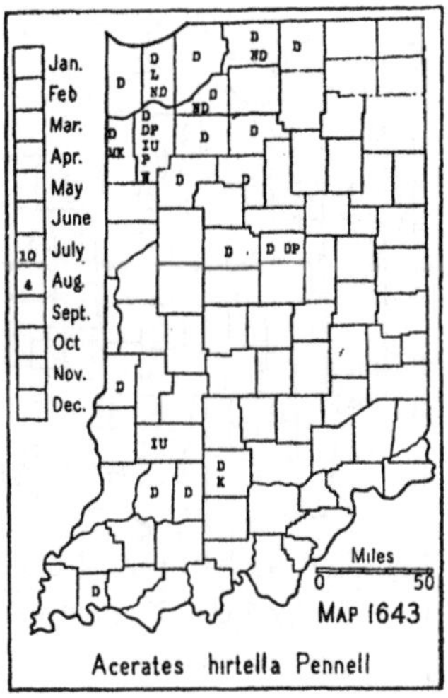

MAP 1643
Acerates hirtella Pennell

1. **Acerates hirtélla** Pennell. (Bull. Torrey Bot. Club **46**: 184-185. 1919.) (*Acerates floridana* (Lam.) Hitchc., in part, of Gray, Man., ed. 7 and Britton and Brown, Illus. Flora, ed. 2.) Map 1643. In sandy soil along roadsides and railroads and in fallow fields. Infrequent in the lake area and in the western part of the state. Mostly in prairie habitats.

Mississippi Valley, range not determined; probably from Mich. southw. and westw. to Okla.

2. **Acerates viridiflòra** (Raf.) Eaton. Map 1644. In sandy soil along roadsides and railroads and in fallow fields. It prefers sandy soil and is apparently a prairie plant but it is sporadic in the southern part of the state.

Acerates viridiflora var. *lanceolata* (Ives) Gray is a variety which has been reported from Indiana but which I am referring to the species, as some authors do, because I am not able to separate the two. I have one specimen with some of the leaves 5 cm wide and one specimen with leaves 8 mm or less in width, and other specimens with leaves that connect these extremes. Since the width of the leaves is the only distinguishing character, it is best to consider the species as one with variable foliage. A very narrowleaf form occurs on the low dunes in west Gary, Lake County.

Ohio and Ont. to Minn., southw. to Fla. and Tex.

6791. ASCLÈPIAS L. MILKWEED

Leaves filiform-linear, mostly in whorls of 3-6; flowers white........1. *A. verticillata.*
Leaves not as above.
  Leaves alternate or a few opposite; flowers orange, sometimes very pale orange or rose color...2. *A. tuberosa.*
  Leaves opposite or sometimes in nos. 3 and 7 with 1 or 2 whorls of 3 or 4 leaves.
    Leaves sessile or clasping, broad and cordate at the base.
      Blades rather small, the median and upper pairs rarely more than 7 cm long, ovate to lanceolate, gradually tapering from below the middle to an acute apex, flat, 3 or 4 pairs, rarely a whorl, the margins scabrous; umbels termi-

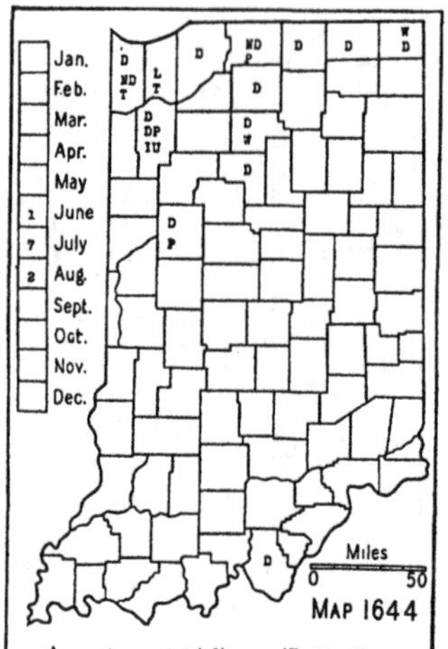

Acerates viridiflora (Raf) Eaton

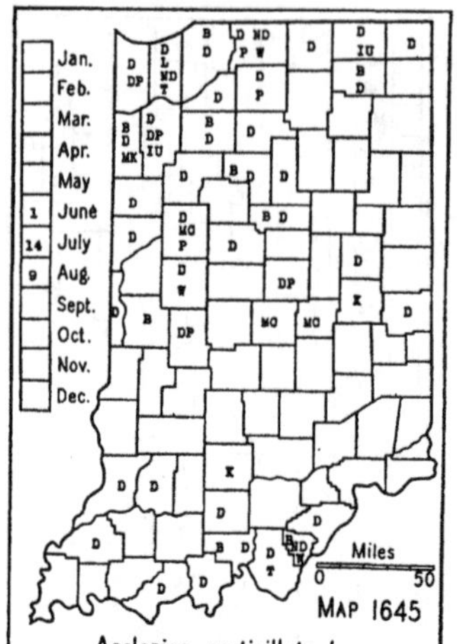

Asclepias verticillata L.

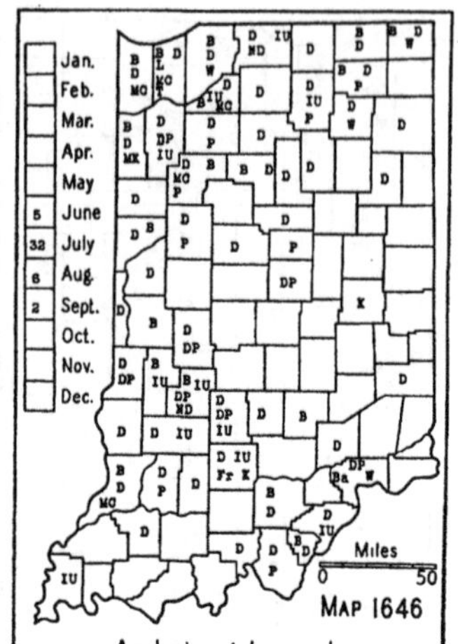

Asclepias tuberosa L.

nal, solitary; peduncles generally less than twice as long as the length of the upper pairs of leaves; pedicels pubescent, usually less than 1.5 cm long; corollas greenish white............................3. *A. Meadii.*

Blades large, the median and upper pairs usually much more than 7 cm long, oblong or oblong-ovate, usually abruptly rounded near the apex to a rounded or short-acute apex; corollas generally purplish.

Leaves flat, the margins smooth; umbels terminal and solitary or a few axillary; peduncles generally shorter than the subtending upper pair of leaves; pedicels glabrous, mostly 2-3 cm long; pods glabrous. ....4. *A. Sullivantii.*

Leaves generally with a ruffled margin; umbels terminal and solitary or a few axillary; peduncles generally several times longer than the upper pair of leaves; pedicels pubescent, mostly 2-5 cm long; pods puberulent....... ....5. *A. amplexicaulis.*

Leaves more or less petioled, generally narrowed at the base, sometimes rounded or truncate at the base but never cordate.

Reflexed petals 3-5 mm long.

Flowers rose purple (very rarely greenish white); plants of low ground. ....6. *A. incarnata.*

Flowers white or pinkish.

Plants mostly with 3-5 pairs or whorls of leaves, usually 1 or 2 pairs in whorls of 3 or 4 leaves............................7. *A. quadrifolia.*

Plants with more than 6 pairs of leaves, none of them in whorls; plants of very low ground...................................8. *A. perennis.*

Reflexed petals more than 5 mm long.

Flowers white or tinged with pink.

Umbels dense; pedicels straight, generally less than 15 mm long; leaves with a short tip at the strongly rounded or blunt apex..9. *A. variegata.*

Umbels loose; pedicels more or less drooping, more than 15 mm long; leaves mostly long-acuminate at the apex..............10. *A. phytolaccoides.*

Flowers deep or light purple or lavender.

Hoods usually exceeding the stigmas about 1.5 mm, with a tooth on each side about the middle..................................11. *A. syriaca.*

Hoods usually exceeding the stigmas about 2-3 mm, without a tooth on each side about the middle..........................12. *A. purpurascens.*

1. **Asclepias verticillàta** L. HORSETAIL MILKWEED. Map 1645. Infrequent in dry, sandy soil or in moist, prairie habitats in the lake area,

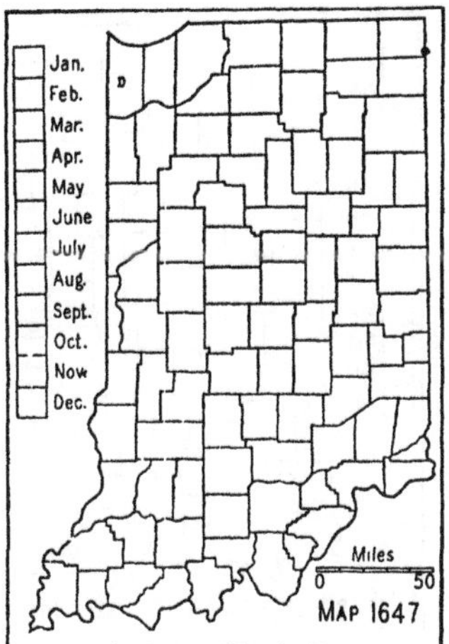

Asclepias Meadii Torr.

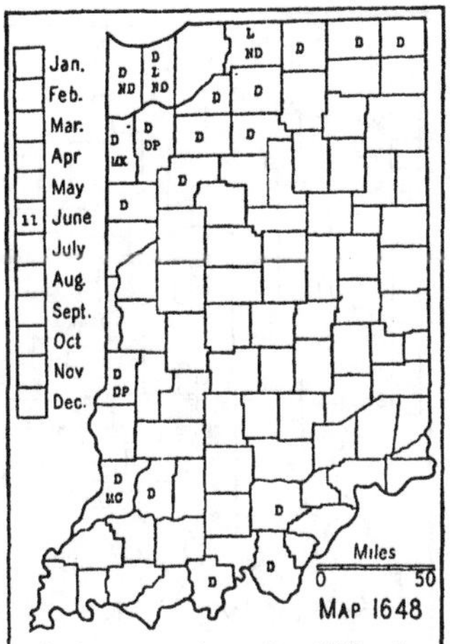

Asclepias amplexicaulis J E.Smith

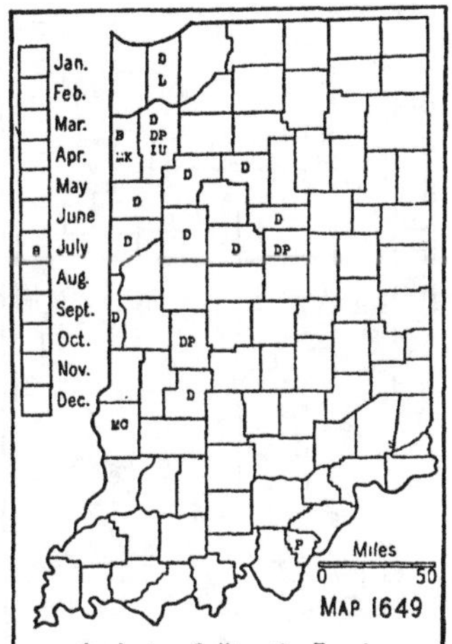

Asclepias Sullivantii Engelm.

mostly along roadsides and railroads, becoming rarer southward. In the southern part of the state it is found on washed slopes, sandy, wooded ridges, along roadsides in clay or on sand hills, and in the hard, clay flats. In 1935 I noted this species to be abundant in the old Beaver Lake bottom in Newton County.

Maine, Ont. to Sask., southw. to Fla. and Mex.

2. **Asclepias tuberòsa** L. BUTTERFLYWEED. PLEURISY ROOT. Map 1646. Infrequent throughout the state except in the sandy areas of the lake region where it becomes frequent. It is usually found on dry, sandy, roadside knolls, or in dry, sandy, open woodland and in moist or dry, sandy prairies.

This species is variable in habit, sometimes almost erect, usually somewhat ascending, or rarely almost decumbent. It is also variable in the number, position, and shape of the leaves on the stem. The stem divides at the top, usually into 2-5 parts, these sometimes much elongated and spreading, and each bearing 1-5 umbels of flowers.

N. H. to Minn., southw. to Fla., Tex., and Ariz.

2a. **Asclepias tuberosa f. bícolor** Standley. (Rhodora 32: 33. 1930.) This is a color form from Porter County recently described by Standley. In it the corolla is bright yellow and the remainder of the flower is generally orange.

3. **Asclepias Mèadii** Torr. MEAD MILKWEED. Map 1647. The only record from Indiana is that of a specimen collected July 3, 1888, in dry ground near Crown Point, Lake County, by Dr. M. A. Brannon. This specimen came into the hands of S. C. Wadmond of Delavan, Wisconsin, who was generous enough to donate it to me. This species is either very rare or not recognized by collectors.

Ind. to Iowa and Wis.

4. **Asclepias amplexicaúlis** J. E. Smith. Map 1648. In the lake area this milkweed is found in very sandy soil on roadside knolls, in very sandy,

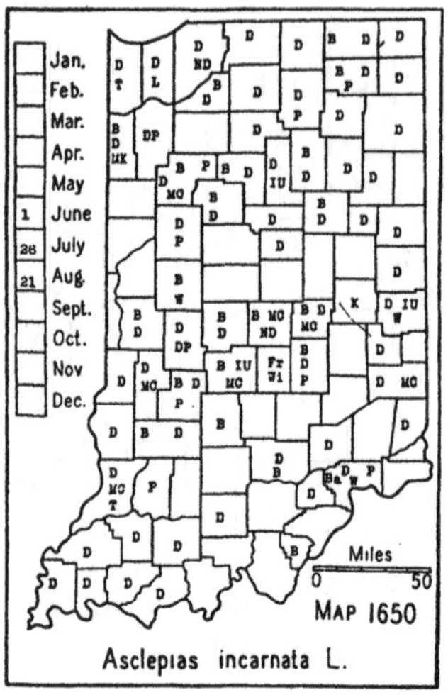

Asclepias incarnata L.

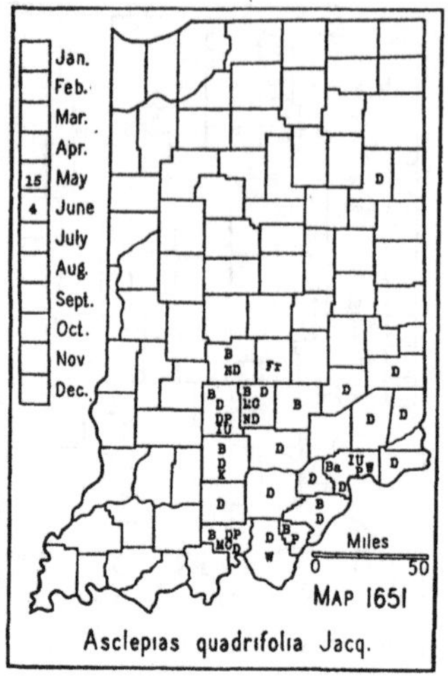

Asclepias quadrifolia Jacq.

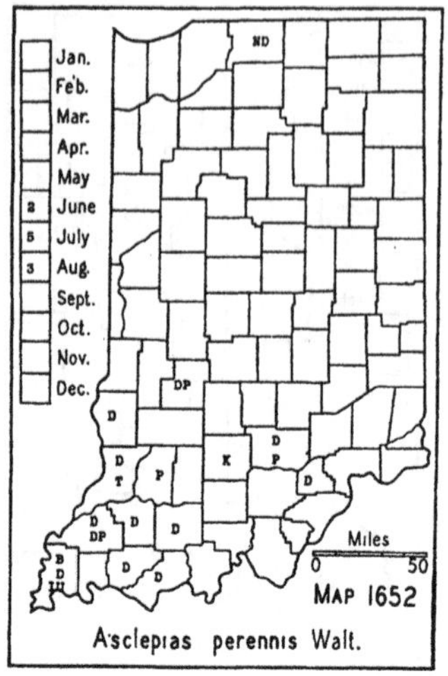

Asclepias perennis Walt.

fallow fields, and in prairie habitats. In the southern part of the state it is found in similar habitats and on the crests of sandstone ridges in open woodland, and very rarely in hard, clay soil. Probably entirely absent from many counties of the Tipton Till Plain.

N. H. to Fla., westw. to Minn., Nebr., and Tex.

5. **Asclepias Sullivántii** Engelm. SMOOTH MILKWEED. Map 1649. Locally frequent in prairie habitats along roadsides and railroads in a few of the western counties. Rarely in other than a prairie habitat.

S. Ont. to Ohio, westw. to Minn., Nebr., and Kans.

6. **Asclepias incarnàta** L. SWAMP MILKWEED. Map 1650. Infrequent to common throughout the state in roadside ditches, along streams and railroads, on the borders of lakes, ponds, and swamps, and in low, open woodland and sometimes in fallow fields.

N. B. to Sask., southw. to Ga. and Kans.

7. **Asclepias quadrifòlia** Jacq. Map 1651. Infrequent in the southern part of the state in dry woodland. The reports from northern Indiana are dubious and if it occurs there it is very rare. There are no reports or specimens from the southwestern part of the state.

N. H. to Ont. and Minn., southw. to N. C. and Ark.

8. **Asclepias perénnis** Walt. Map 1652. Infrequent in swampy woods, and about sloughs and ponds, mostly in the southwestern counties.

Ind. to Fla., westw. to Mo. and Tex.

9. **Asclepias variegàta** L. Map 1653. A rare plant of dry, open woodland in the southern counties. Usually in sandy to very sandy soil and rarely more than a single specimen in a place. The report by Van Gorder from Noble County may be correct, although I bought his herbarium and found no specimen.

L. I. to Fla., westw. to Ind. and La.

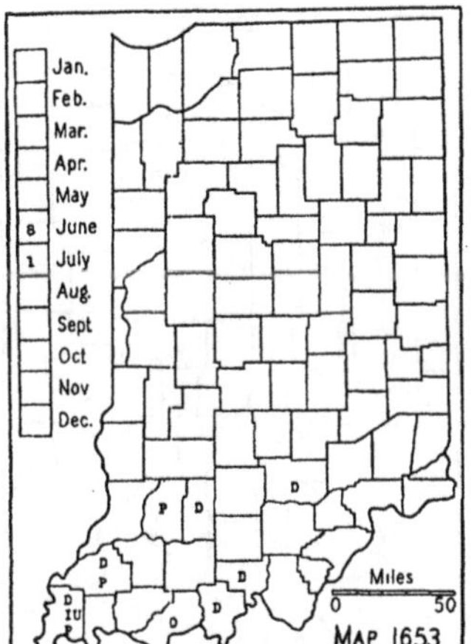

MAP 1653

Asclepias variegata L.

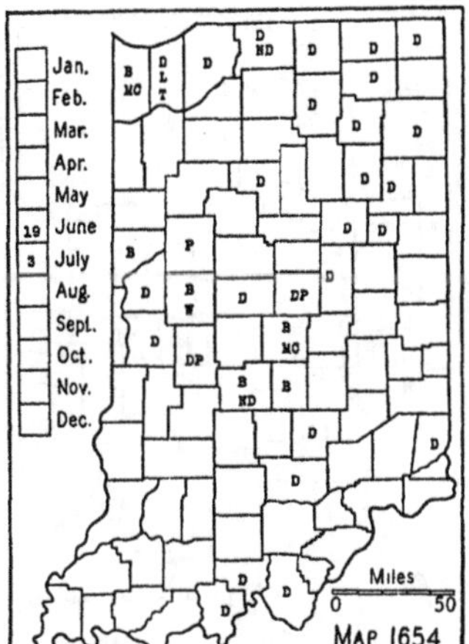

MAP 1654

Asclepias phytolaccoides Pursh

MAP 1655

Asclepias syriaca L.

10. **Asclepias phytolaccoìdes** Pursh. (*Asclepias exaltata* (L.) Muhl. of Britton and Brown, Illus. Flora, ed. 2.) POKE MILKWEED. Map 1654. This is strictly a woodland species and is more or less infrequent, and associated for the most part with white oak. Ordinarily only a single plant or two are found at a place.

Maine to Minn., southw. to Ga. and Ark.

11. **Asclepias syrìaca** L. COMMON MILKWEED. Map 1655. Frequent to common in all parts of the state. Less frequent in the less calcareous soils. Usually in moist soil along roadsides and railroads, often common in cultivated fields, especially oatfields, and in fallow fields and open woodland.

This species is variable as to width and shape of the leaves and the density and length of the tubercles on the follicles.

N. B. to Sask., southw. to Fla., Tex., and Ariz.

12. **Asclepias purpuráscens** L. PURPLE MILKWEED. Map 1656. Infrequent throughout the state. Usually only one or a few plants are found together. It has various habitats. The most common one is a rather dry, and usually somewhat sandy soil in open woodland and along roadsides. Also found in damp, open woodland about swamps and lakes and even in tamarack bogs.

N. H. to N. C., westw. to Minn. and Ark.

6812. AMPÉLAMUS Raf.

1. **Ampelamus álbidus** (Nutt.) Britt. (Bull. Torrey Bot. Club **21**: 314. 1894.) (*Gonolobus laevis* Michx.) BLUEVINE. Map 1657. Mostly on the banks and alluvial plains of streams and in cultivated fields in southern Indiana. It is an obnoxious weed in corn and cultivated fields in the "bottoms." In 1938 County Agent Mervin F. Smith found it well established in a cornfield a mile south of Uniondale, Wells County. As a weed it is as difficult to eradicate as our common bindweed. The beekeepers widely

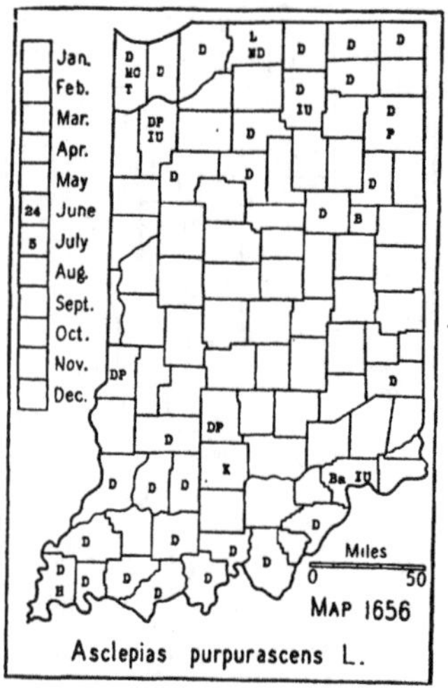

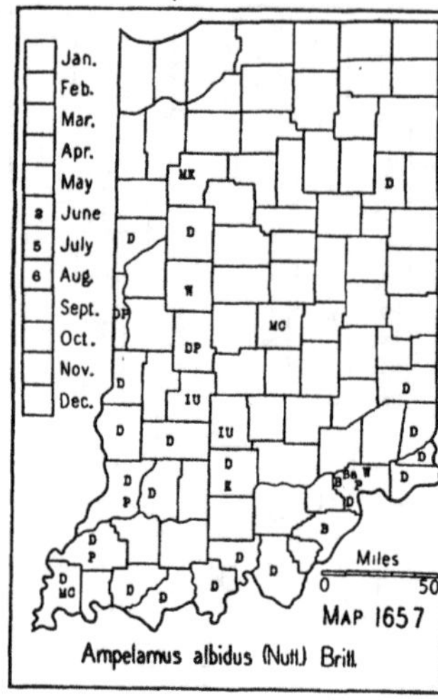

publicized this plant as an excellent honey plant under the name of bluevine. We introduced it for this purpose at Bluffton and some seed escaped and we have been trying to exterminate it now for eight years without success. If the Indiana beekeepers responded to the appeal to plant this plant, it is, no doubt, now well established in all parts of the state.

Pa. to Ill. and Kans., southw. to Fla. and Tex.

6943. GONÓLOBUS Michx.

[Perry, Lily M. Gonolobus within the Gray's Manual range. Rhodora 40: 281-287. 1938.]

Pedicels and fruit glabrous; flowers greenish yellow; fruit angular but not warty. .... 1. *G. gonocarpos.*

Pedicels and fruit pubescent, the pubescence consisting mostly of minute stalked glands; flowers crimson purple; fruit both angular and warty......2. *G. obliquus.*

1. **Gonolobus gonocárpos** (Walt.) Perry. (*Vincetoxicum gonocarpos* Walt.) Map 1658. Climbing vines in low woodland and in cultivated fields.

Va. to Ind., southw. to S. C., Ala., La., and Tex.

2. **Gonolobus oblìquus** (Jacq.) Schultes. (*Vincetoxicum obliquum* (Jacq.) Britt.) Map 1659. All of my specimens are from rocky wooded slopes except one which was found in a low woods in Posey County, associated with the preceding species.

Pa., Ohio, Ind., southw. to Ga., Tenn., and Mo.

249. CONVOLVULÀCEAE[1] Vent. MORNING-GLORY FAMILY

Plants leafless, parasitic, twining; stems filiform, yellowish or flesh colored; corolla whitish, very small, less than 5 mm long................6968. CUSCUTA, p. 771.

Plants with green leaves, not parasitic; corolla usually colored, more than 5 mm long.

Stigmas 2, linear or oblong; calyx with two large subtending bracts (small in *Convolvulus arvensis* and far down on the peduncle)....6993. CONVOLVULUS, p. 774.

[1] T. G. Yuncker has critically reviewed the species of this family, occurring in Indiana, and has cited specimens which are not in my herbarium. I have indicated these on the maps with a "Y".

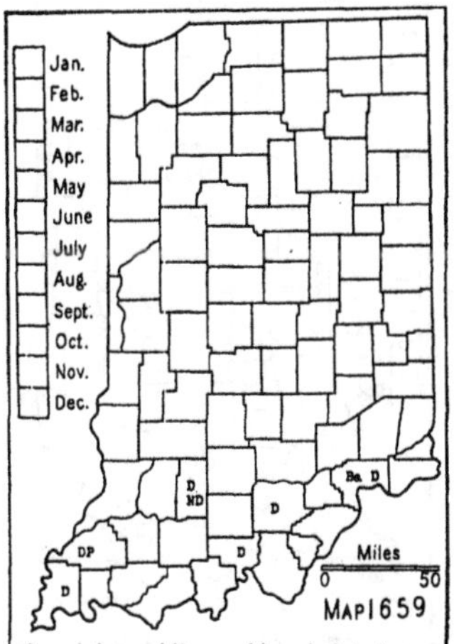

Gonolobus obliquus (Jacq.) Schultes

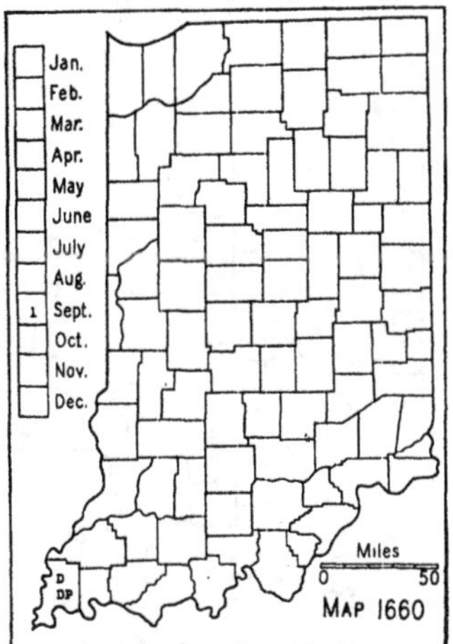

Cuscuta cuspidata Engelm.

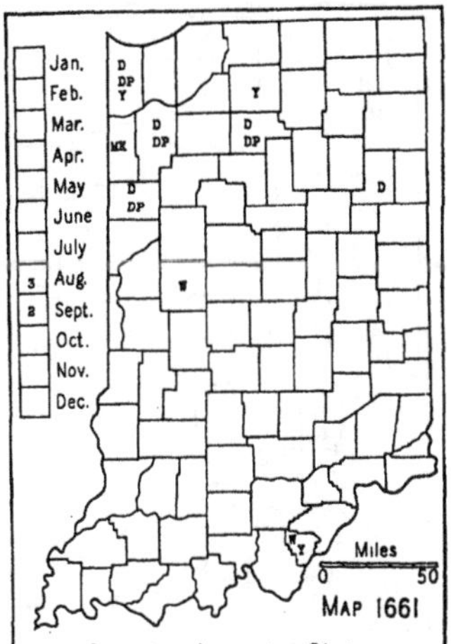

Cuscuta glomerata Choisy

Stigmas capitate; calyx without or with small subtending bracts.
  Calyx lobes not broadly rounded and without a subtending bract; corolla mostly funnel-shaped; stamens and styles not exserted.......7003. IPOMOEA, p. 776.
  Calyx lobes broadly rounded and with a subtending bract about 3 mm long; corolla tubular, not expanded at the base, limb salver-shaped; stamens and style exserted.......................................7005. QUAMOCLIT, p. 777.

6968. CUSCÙTA [Tourn.] L. DODDER

[Yuncker. The genus Cuscuta. Mem. Torrey Bot. Club **18**: 113-331. 1932. Yuncker. Notes on our Indiana Dodders. Proc. Indiana Acad. Sci. **1919**: 157-163. 1921.]

Stigmas linear; capsules circumscissile; plants usually found parasitic on legumes (usually clover or alfalfa); although reported for Ind. we have no specimens. (See excluded species no. 510, p. 1082).........................*C. Epithymum.*
Stigmas capitate; capsules not circumscissile.
  Sepals generally 5, distinct; flowers subtended by one or more bracts.
    Flowers on bracteolate pedicels, in loose panicles................1. *C. cuspidata.*
    Flowers closely sessile, in densely compact clusters.
      Inflorescence dense, ropelike; bracts oblong, scarious, acute, with recurved tips. ....2. *C. glomerata.*
      Inflorescence less dense and not particularly ropelike; bracts orbicular or oval (shape of the sepals), not scarious, obtuse, closely appressed...3. *C. compacta.*
  Sepals united, at least at the base; flowers 4- or 5-parted; flowers not subtended by bracts.
    Flowers commonly 5-parted.
      Corolla lobes reflexed, acute, with inflexed tips; capsules globose or depressed-globose (not pointed).
        Calyx lobes broadly overlapping at the sinuses to form angles...4. *C. pentagona.*
        Calyx lobes not overlapping to form angles at the sinuses; flowers mostly 2-3 mm long...5. *C. campestris.*
      Corolla lobes acute, erect to spreading; capsules ovoid or globose, not depressed but usually pointed...6. *C. Gronovii.*

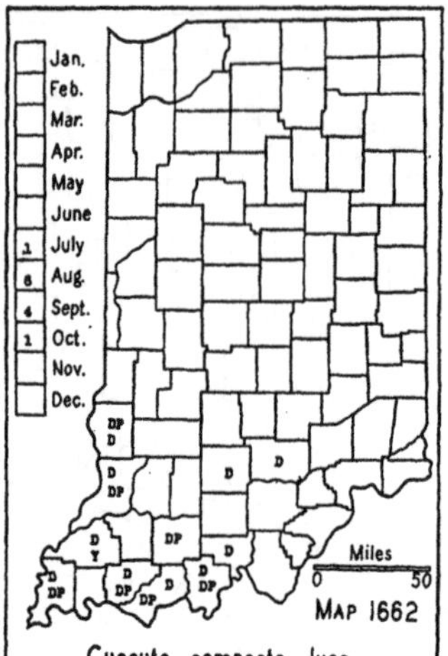

Cuscuta compacta Juss.

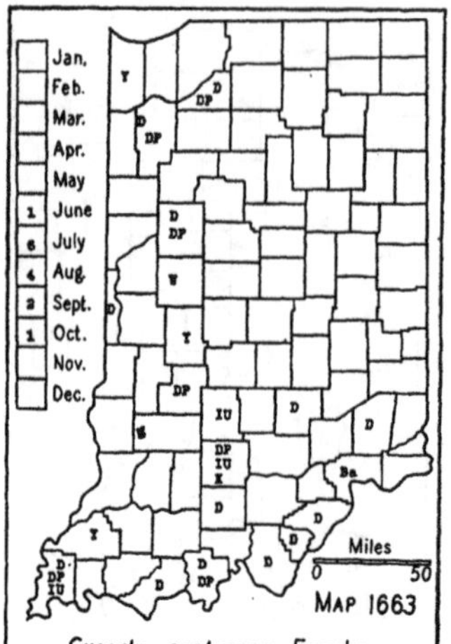

Cuscuta pentagona Engelm.

Cuscuta campestris Yuncker

Flowers commonly 4-parted (or 3-parted).
- Withered corolla remaining more or less persistent as a cap at the apex of the capsule; infrastaminal scales well developed............7. *C. Cephalanthi.*
- Withered corolla remaining at the base of the capsule, around it or early deciduous; infrastaminal scales rudimentary.
  - Flowers fleshy, papillate, about 2 mm long; on pedicels longer or shorter than the flowers; calyx lobes acute; corolla lobes with inflexed tips...8. *C. Coryli.*
  - Flowers glabrous, about 2-2.5 mm long, subsessile; calyx lobes obtuse; corolla lobes without inflexed tips........................9. *C. Polygonorum.*

1. **Cuscuta cuspidàta** Engelm. CUSPIDATE DODDER. Map 1660. My only specimens are from Posey County. Yuncker reported it from Vigo County (Proc. Indiana Acad. Sci. **1920**: 229. 1921.) This is a southwestern species.

Hosts: Prefers species of *Compositae;* specimens reported are on *Ambrosia.*

Ind., Colo., Utah, southw. to La. and Tex.

2. **Cuscuta glomeràta** Choisy. GLOMERATE DODDER. Map 1661. On hosts of low ground, mostly in marshes.

Hosts of my specimens are: 1 on *Apios,* 1 on *Asclepias syriaca,* 3 on *Aster,* 4 on *Helianthus,* and 1 on *Solidago.*

Mich. and Ind. to S. Dak. and Nebr., southw. to Miss. and Tex.

3. **Cuscuta compácta** Juss. COMPACT DODDER. Map 1662. On hosts mostly of low ground. Hosts of my species are as follows; 3 on *Campsis radicans,* 5 on *Cephalanthus occidentalis,* 2 on *Rhus radicans;* 1 on *Salix,* 2 on *Sassafras albidum,* and 1 on *Vitis cinerea.*

N. H. to Okla., southw. to Fla., and Tex.

4. **Cuscuta pentágona** Engelm. (*Cuscuta arvensis* Beyrich.) FIELD DODDER. Map 1663. This is our common field dodder and must be regarded as an obnoxious weed. It is commonly found on clover.

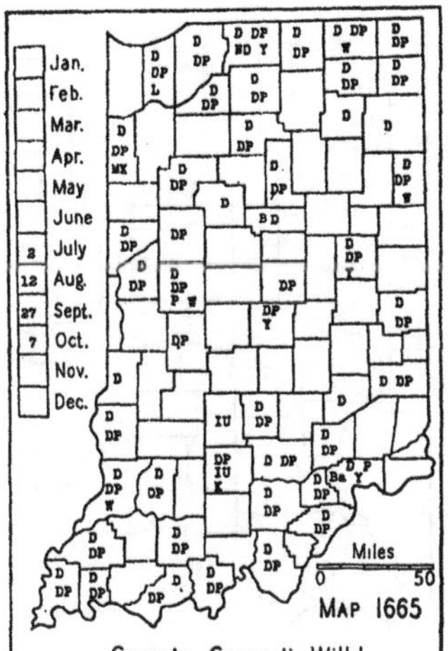

Cuscuta Gronovii Willd.

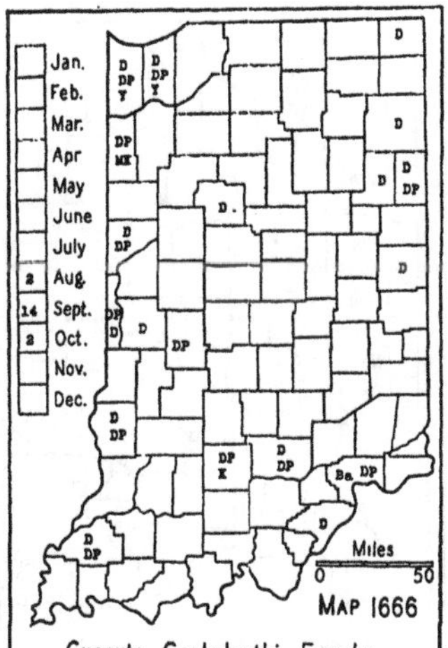

Cuscuta Cephalanthi Engelm.

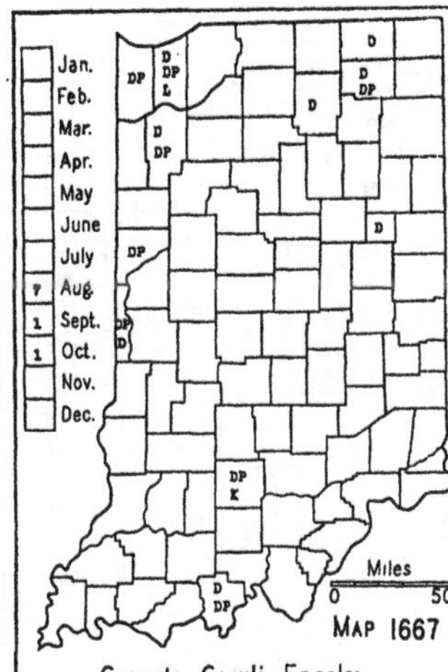

Cuscuta Coryli Engelm.

Hosts of my specimens are as follows: 1 on *Daucus Carota,* 3 on *Euphorbia corollata,* 1 on *Lespedeza,* 1 on *Lespedeza striata,* 8 on *Trifolium pratense.*

Mass. to Fla., westw. to Calif.

5. **Cuscuta campéstris** Yuncker. (*Cuscuta pentagona* var. *calycina* Engelm.) Map 1664. Yuncker referred my specimens named *Cuscuta pentagona* var. *calycina* to this species.

U. S., W. I., and S. A.

6. **Cuscuta Gronòvii** Willd. (*Cuscuta Gronovii* var. *vulvivaga* Engelm.) GRONOVIUS DODDER. Map 1665. This species is parasitic mostly on plants of a wet habitat. The following is a list of the hosts of my specimens; 3 on *Aster,* 4 on *Boehmeria,* 2 on *Campsis,* 3 on *Cephalanthus,* 1 on *Dianthera,* 1 on *Decodon,* 1 on *Eupatorium,* 1 on *Helianthus,* 1 on *Hypericum,* 19 on *Impatiens,* 1 on *Lactuca,* 1 on *Laportea,* 2 on *Mentha,* 1 on *Phytolacca,* 1 on *Polygonum,* 1 on *Rubus,* 2 on *Saururus,* 2 on *Salix,* 3 on *Solidago,* and 1 on *Vernonia.*

This species is variable in the size and proportion of its flower parts. Two specimens of var. *vulvivaga* from Steuben County were cited by Yuncker (Univ. of Illinois Biol. Monographs **6**: 66. 1921) but in 1932 he refers this variety to the species. Accordingly, a report from Porter County by Lyon is referred to *C. Gronovii* by Buhl (Amer. Midland Nat. **16**: 252. 1935).

N. S. to Man., southw. to Fla., Tex., and Ariz.

6a. **Cuscuta Gronovii** var. **calyptràta** Engelm. "This variety differs from the typical form in the usually longer corolla lobes less than half as long as the tube. The calyx lobes are oval-oblong and commonly serrated. The withered corolla caps the capsule" (Mem. Torrey Bot. Club **18**: 175. 1931). I found it in Clark County on *Solidago* and in Sullivan County on *Saururus.*

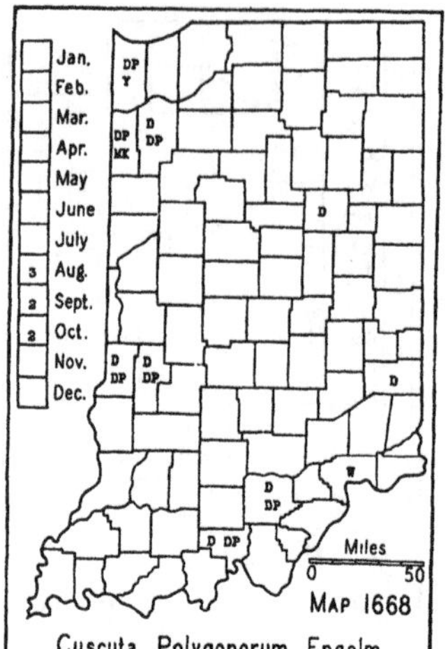

Cuscuta Polygonorum Engelm.

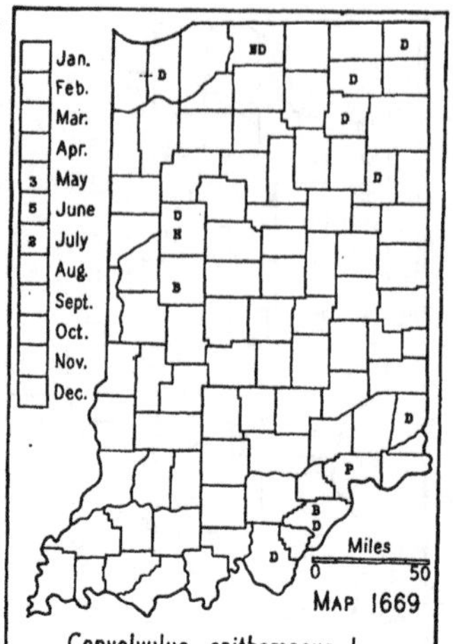

Convolvulus spithamaeus L.

Convolvulus japonicus Thunb.

7. **Cuscuta Cephalánthi** Engelm. BUTTONBUSH DODDER. Map 1666. This species, also, prefers plants of low ground for hosts. The host plants of my specimens are as follows: 7 on *Aster,* 1 on *Cephalanthus,* 1 on *Dianthera,* 1 on *Physostegia,* 1 on *Rhus,* 3 on *Salix,* and 1 on *Teucrium.*

Maine to Wash. and Oreg., southw. to Va., Tenn., and Tex.

8. **Cuscuta Córyli** Engelm. HAZEL DODDER. Map 1667. On plants about lakes and in low woods. The hosts of my specimens are as follows: 1 on *Aster,* 1 on *Campsis,* 1 on *Corylus,* 1 on *Prunella,* 1 on *Sanicula,* 2 on *Solidago,* and 2 on *Stachys hyssopifolia.*

R. I. to Man., southw. to Va., Tex., and Ariz.

9. **Cuscuta Polygonòrum** Engelm. (*Cuscuta obtusiflora* of Gray, Man., ed. 7.) SMARTWEED DODDER. Map 1668. On plants of low ground about ponds and lakes and in low woods. The hosts of my specimens are as follows: 1 on *Aster,* 4 on *Bidens,* 1 on *Polygonum,* and 1 on *Xanthium.*

Md. to Minn. and Nebr., southw. to Tenn., and possibly Tex.

6993. CONVÓLVULUS [Tourn.] L. BINDWEED

Bracts large, surrounding and inclosing the calyx; stigmas oval or oblong.
- **Plants short, erect or ascending; petioles mostly less than a fourth as long as the blade** .... **1. *C. spithamaeus.***
- **Plants long, trailing or twining; petioles mostly more than a fourth as long as the blade.**
  - **Flowers double; plants escaped from cultivation** .... **2. *C. japonicus.***
  - **Flowers single; plants native.**
    - **Peduncles mostly much longer than the petioles; flowers commonly only 1 in an axil.**
      - **Leaves glabrous, sometimes somewhat pubescent, rarely densely pubescent; basal lobes mostly hastate (turned outward); peduncles not wing-angled.** .... **3. *C. sepium.***
      - **Leaves densely pubescent; basal lobes rounded or sagittate (lobes not turned outward)** .... **4. *C. repens.***
    - **Peduncles mostly shorter than the petioles; flowers commonly 2 in each axil.** .... **3a. *C. sepium* var. *fraterniflorus.***

Bracts small, much smaller than the calyx and at some distance below the flower; stigmas filiform .... **5. *C. arvensis.***

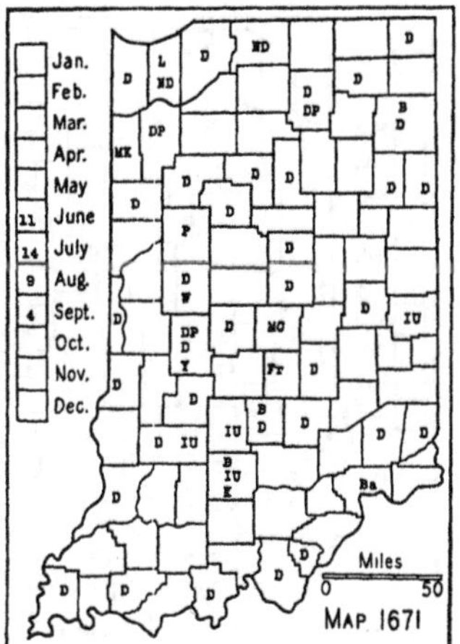

Convolvulus sepium L

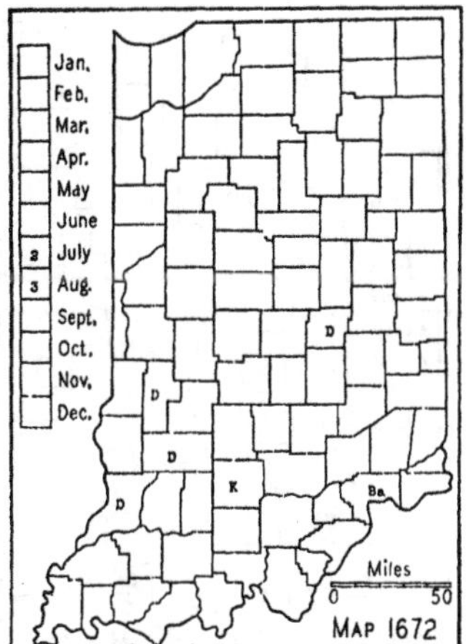

Convolvulus sepium var. fraterniflorus Mack. & Bush.

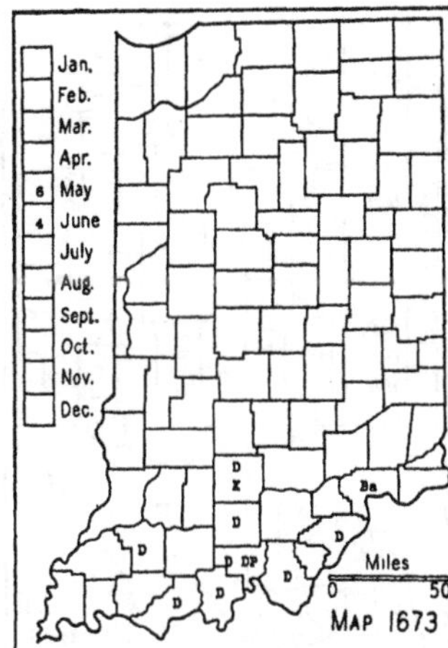

Convolvulus repens L.

1. **Convolvulus spithamaèus** L. Map 1669. This species is variable in the density of its pubescence, the shape of the leaves, and the length of the stem. Generally in poor clay soil in bare places on open wooded slopes and rarely in sandy soil in prairie habitats.

N. S. to Man., southw. to Fla. and Ky.

2. CONVOLVULUS JAPÓNICUS Thunb. ROSE CONVOLVULUS. Map 1670. In moist waste places. Escaped from cultivation in Marion, Monroe, Putnam, and Tipton Counties.

Nat. of Asia.

3. **Convolvulus sèpium** L. HEDGE BINDWEED. Map 1671. I was told by a farmer who had lived in several places in Carroll County that this species is known there to the farmers as gopher weed. It is a pernicious weed. It prefers a moist alluvial soil. Frequent to common in cultivated fields, along roadsides and railroads, and in waste places, fallow fields, and open woodland along streams.

N. S. to B. C., southw. to N. C., Kans., and N. Mex.

3a. **Convolvulus sepium** var. **fraterniflòrus** Mack. & Bush. (*Convolvulus fraterniflorus* Mack. & Bush.) Map 1672. This variety is rare in Indiana and has the habitat of the species.

Ind. to Mont., southw. to Ark. and N. Mex.

4. **Convolvulus rèpens** L. (*Convolvulus sepium* var. *pubescens* (Gray) Fern.) Map 1673. In poor clay soil in fallow fields and on bare places on open wooded slopes. Reported by Peattie for the Calumet District. Rare.

E. Que. to Fla., westw. at least to Ind.

5. CONVOLVULUS ARVÉNSIS L. FIELD BINDWEED. Map 1674. This species is extremely variable in the shape of its leaves. We have specimens with leaves varying from 5 to 35 mm in width. It is an obnoxious weed wherever it is found. It is a plant mostly of waste places and along roadsides, railroads, streets, and alleys.

Nat. of Eu.

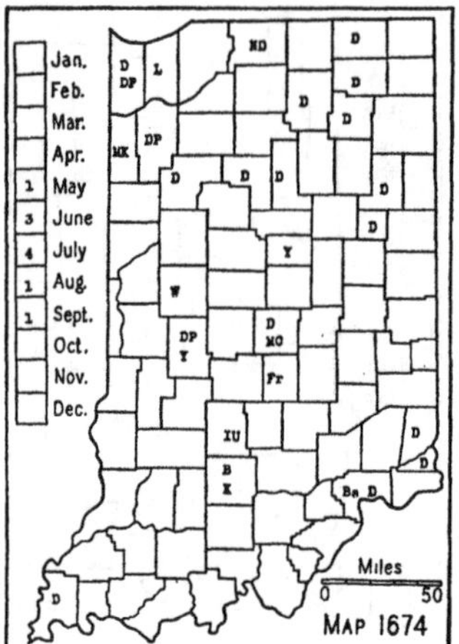

MAP 1674
Convolvulus arvensis L.

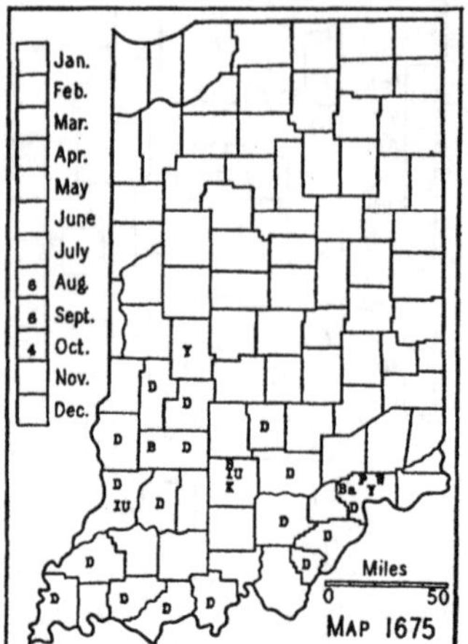

MAP 1675
Ipomoea lacunosa L.

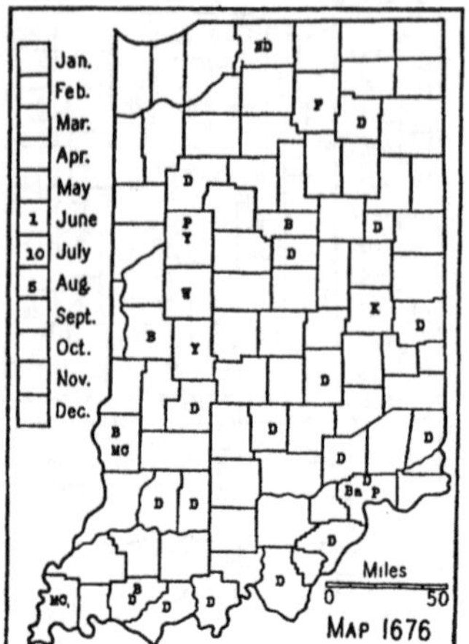

MAP 1676
Ipomoea pandurata (L.) G. F. W Mey.

7003. IPOMOÈA L. MORNING-GLORY

Flowers small, 1-2 cm long, usually about 1.5 cm long, varying in color from white to pink and purplish; capsules pubescent; seed glabrous..............1. *I. lacunosa.*
Flowers large, more than 3 cm long, varying in color from white to purple; capsules glabrous; seeds more or less pubescent.
  Stems, petioles, and peduncles glabrous.
    Under surface of most of the leaves sparsely pubescent at least on the principal nerves ....2. *I. pandurata.*
    Under surface of leaves glabrous...............2a. *I. pandurata* var. *rubescens.*
  Stems, petioles, and peduncles with long, retrorse hairs.
    Flowers mostly less than 4 cm long; base of calyx tube with more or less rufous hairs; calyx lobes long, attenuate-caudate, with spreading tips; leaves always 3-lobed...3. *I. hederacea.*
    Flowers mostly more than 4 cm long; base of calyx with colorless hairs; calyx lobes ovate-oblong to lanceolate, the tips not spreading; leaves never lobed except in the form.
      Leaves entire...4. *I. purpurea.*
      Leaves 3-lobed..................................4a. *I. purpurea* forma.

1. **Ipomoea lacunòsa** L. Map 1675. Infrequent but common enough where it is found. Usually in cultivated fields and in the wild in moist or wet places along streams. The leaves vary from entire to angled or 3-lobed and the flowers from white to pink, mostly white in our area.

Pa., Ill. to Kans., southw. to S. C. and Tex.

2. **Ipomoea panduràta** (L.) G. F. W. Mey. WILD POTATO VINE. Map 1676. More or less frequent throughout the state although there are no records for the area north of the Kankakee River. It prefers moist soil along streams but is also found on washed slopes and infrequently in dry woodland.

This species develops a tuber of enormous size at a great depth in the ground. The largest I have seen was about 24 inches long and about 5 inches in circumference in the middle. It was elliptic in shape. I was told of one that was of incredible size and I forbear to give dimensions for fear an error was made in identification or measurements. Vines 9 to 15

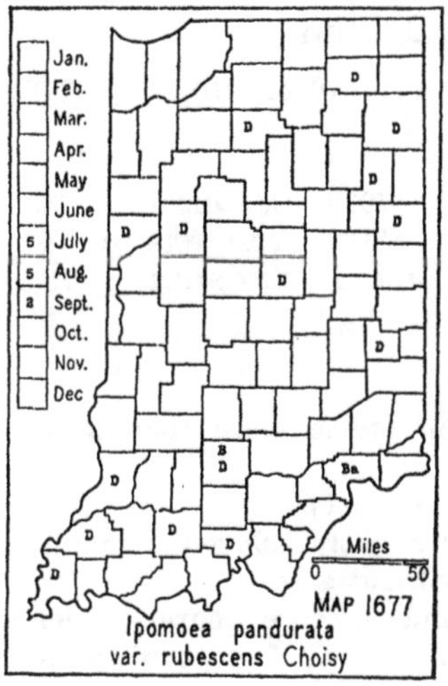

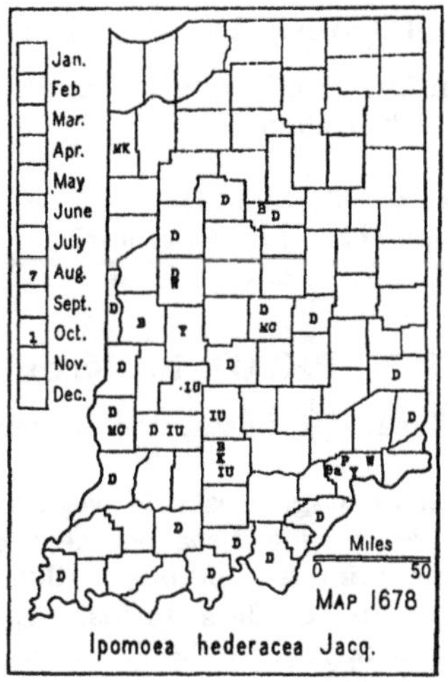

feet long are not uncommon and these crawl upon bushes 3 to 9 feet high.

The variety and species have not been separated long enough to ascertain the range of either. In Indiana the ranges of the two are practically coextensive. The variety is distinct in our area and we have no intergrading specimens. The range of the species is taken from our manuals.

Conn., Ont., Mich., and Kans., southw. to Fla. and Tex.

2a. **Ipomoea pandurata** var. **rubéscens** Choisy. (Rhodora 20: 65. 1918.) Map 1677. The habitat is that of the species.

3. IPOMOEA HEDERÀCEA Jacq. IVYLEAF MORNING-GLORY. Map 1678. A local, infrequent or frequent vine of cultivated and fallow fields, along roadsides, and rarely in open woodland.

Nat. of tropical America; now established from Maine to Nebr., southw. to Fla. and Mex.

4. IPOMOEA PURPÙREA (L.) Roth. COMMON MORNING-GLORY. Map 1679. Reported as an escape from all parts of the state. I have seen it as a pernicious weed in cornfields in several counties. I have not collected it as often as I saw it; so our map does not indicate its frequency in the state. The leaves of this species are sometimes 3-lobed.

Nat. of tropical America; now found from N. S. to Nebr., southw. to Fla. and Tex.

4a. IPOMOEA PURPUREA forma. . . . This is a form with 3-lobed leaves which I have found in Kosciusko and Wells Counties.

7005. QUÁMOCLIT [Tourn.] Moench

1. QUAMOCLIT COCCÍNEA (L.) Moench. (*Ipomoea coccinea* L.) SCARLET STARGLORY. Map 1680. A rare escape in cultivated fields and along roadsides.

Nat. of tropical America; now established from R. I. to Mo., southw. to Fla. and Tex.

250. POLEMONIÀCEAE[1] DC. PHLOX FAMILY

Leaves opposite, simple and entire; corolla salver-shaped.........7014. PHLOX, p. 778.
Leaves alternate; flowers not salver-shaped.
Leaves simple.
Leaves entire; flowers lilac purple to white..............7015. COLLOMIA, p. 783.
Leaves cut into filiform segments; flowers reddish........7016. IPOMOPSIS, p. 783.
Leaves compound, the leaflets entire; flowers blue........7017. POLEMONIUM, p. 784.

7014. PHLÓX L. PHLOX

Leaves ovate, lanceolate or linear (if linear, the flowering stems more than 2 dm high).
Plants at flowering time without long, prostrate, vegetative shoots.
Leaves mostly more than 2 cm wide, broadest near or slightly below the middle, lateral veins widely spreading and plainly visible without a lens.
Corolla tube generally more or less pubescent; inflorescence more or less pubescent, the hairs rarely glandular....................1. *P. paniculata.*
Corolla tube glabrous; inflorescence densely glandular-pubescent, the hairs glandular...2. *P. amplifolia.*
Leaves mostly less than 2 cm wide, generally broadest about a fourth of their length above the base, sometimes broadest near the middle, lateral veins strongly ascending, rarely visible without a lens.
Plants glabrous or nearly so.
Stems green, rarely with purple spots; inflorescence (measured from the tips of the calyx lobes) as wide as long or not more than twice as long as wide.
Stems at flowering time with 3 or 4 pairs of leaves............3. *P. ovata.*
Stems at flowering time with more than 4 pairs of leaves.
Upper leaves lanceolate to ovate; calyx 6-11 mm long................ ....4. *P. carolina* var. *triflora.*
Upper leaves linear to lanceolate; calyx 6-8 mm long...5. *P. glaberrima.*
Stems generally purple spotted; inflorescence (measured from the tips of the calyx lobes) more than twice as long as wide........6. *P. maculata.*
Plants more or less pubescent.
Leaves linear to lanceolate or some nearly ovate, acuminate; bracts spreading, scattered through the moderately compact cyme; hairs of inflorescence fine or exceptionally coarse, sometimes gland-tipped, rarely lacking; calyx awns often long.
Hairs of inflorescence glandular (eglandular in occasional colonies); calyx lobes narrow and long-awned or occasionally broader and shorter awned; plant of eastern U. S. generally..................7. *P. pilosa.*
Hairs of inflorescence consistently eglandular; calyx lobes broad (narrow in occasional colonies), short-awned or long-awned in occasional colonies.
Hairs fine and lustrous; plant of northwest....7a. *P. pilosa* var. *fulgida.*
Hairs coarse (lacking in occasional colonies); plant of southwest...... ....7b. *P. pilosa* var. *amplexicaulis.*
Leaves oblong-elliptic or sometimes lanceolate, obtusish to acuminate; bracts ascending, closely surrounding the decidedly compact cyme; hairs of inflorescence coarse, never gland-tipped; calyx awns short. (See excluded species no. 511, p. 1082).............................*P. amoena.*
Plants at flowering time with long, prostrate vegetative shoots.
Leaves of vegetative shoots on long petioles; blades lanceolate to ovate, long taper-pointed at both ends; stems at flowering time usually with about 3

[1] All of my *Polemoniaceae* have been seen and named by E. T. Wherry, of the University of Pennsylvania.

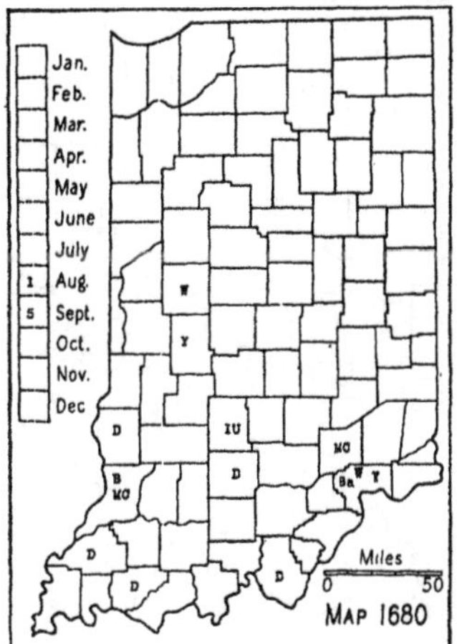

Quamoclit coccinea (L.) Moench

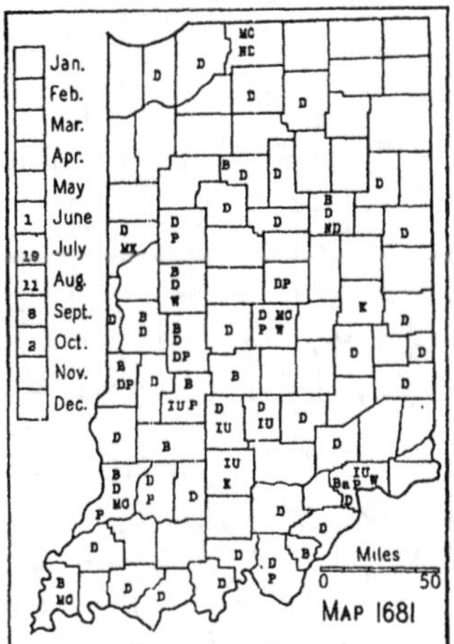

Phlox paniculata L.

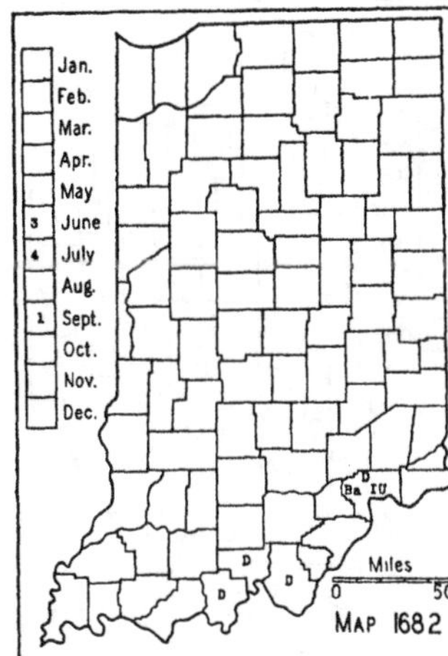

Phlox amplifolia Britt.

pairs of leaves, the lower ones narrowed at the base and usually more or less petiolate..3. *P. ovata.*

Leaves of vegetative shoots sessile or the lower on short petioles; blades lanceolate to ovate-lanceolate, elliptic or obovate, narrowed at the base but not long taper-pointed, subacute or rounded at the apex.

Apex of leaves of vegetative shoots acute or subacute; corolla tube glabrous, petals usually notched; plant found throughout Indiana......8. *P. divaricata.*

Apex of leaves of vegetative shoots rounded; corolla tube pubescent, petals never notched; plant found to the southeast of Indiana. (See excluded species no. 512, p. 1082)....................................*P. stolonifera.*

Leaves linear, acute or subulate, more or less fascicled; low, diffuse plants, creeping or decumbent, generally less than 2 dm high.

Nodes few; lower leaves 20-60 mm long, sharply acute; sinuses of corolla lobes averaging 3 mm deep...9. *P. bifida.*

Nodes numerous, crowded; lower leaves 8-20 mm long; sinuses of corolla lobes averaging 1 mm deep...10. *P. subulata.*

1. **Phlox paniculàta** L. GARDEN PHLOX. Map 1681. Infrequent throughout the state except in the northern counties; in wooded flood plains of streams and rarely in wet woods away from streams or elsewhere. I have a narrowleaf form from a wooded cliff. The northern tier of counties in Indiana has been botanized more thoroughly than any other area of the state, yet the species has been found in only a few of these counties because it reaches its northern limit in northern Indiana.

Pa. to Ill. and Kans., southw. to Fla. and La.

2. **Phlox amplifòlia** Britt. Map 1682. We have had this species under cultivation for eleven years and it is very thrifty, has a long blooming period, and is one of the best phloxes for cultivation.

Found locally on open, wooded slopes and on wooded flood plains of streams in a few of the Ohio River Counties.

Ind. to Mo., southw. to Tenn.

3. **Phlox ovàta** L. (Wherry. Bartonia 13: 25-29. 1932.) MOUNTAIN PHLOX. Map[1] 1683. A few colonies on slopes in white and black oak

[1] The maps in *Polemoniaceae* include the specimens not seen by me but seen by E. T. Wherry and his records are indicated by "W".

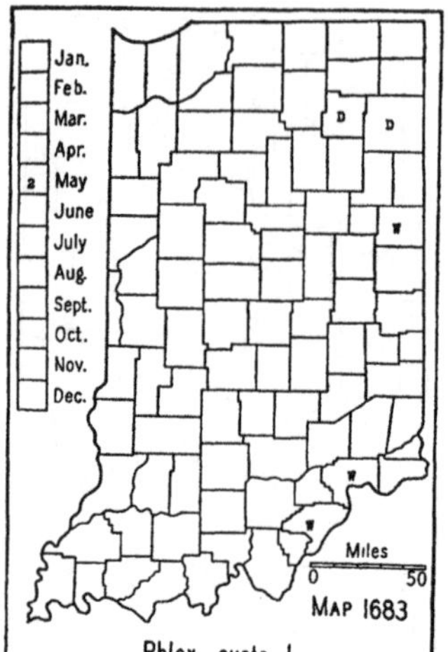

Phlox ovata L.

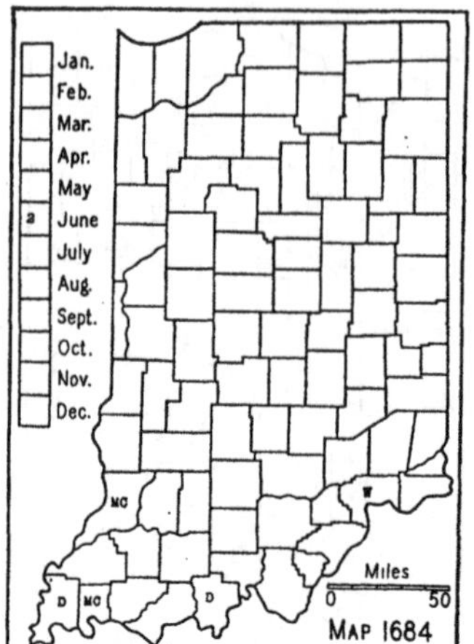

Phlox carolina var. triflora (Michx.) Wherry

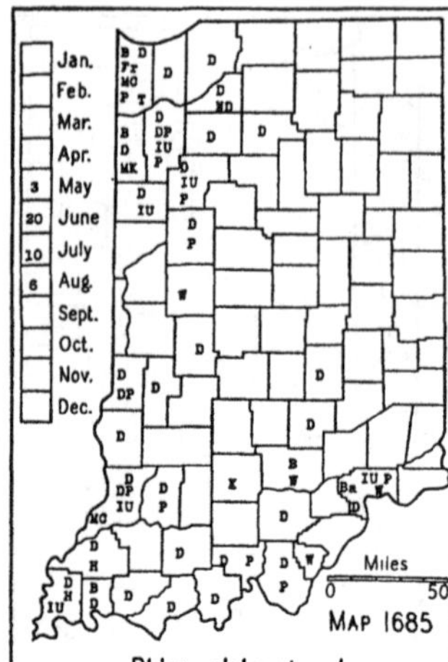

Phlox glaberrima L.

woods in a few of the eastern counties. Phinney's report for Jay County can not be verified.

This is an Applachian Mountain species, extending from e. Pa. to nw. Ohio and n. Ind., southw. to Ga. and Tenn.

4. **Phlox carolìna** L. var. **triflòra** (Michx.) Wherry. (Wherry. Bartonia 13: 30-37. 1932.) Map 1684. Low woods and moist, wooded ravines. Very rare in Indiana.

This variety of the species ranges from Md. to Ind., southw. to N. C.

5. **Phlox glabérrima** L. (Wherry. Bartonia 14: 14-19. 1932.) SMOOTH PHLOX. Map 1685. Infrequent in prairie habitats in the northwestern part of the state and in the Illinoian area, especially in the southwestern part of the state, in hard, clay soil in low woods. Usually frequent to even common where it is found. Generally in low, wet woods and along roadsides in southern Indiana, and mostly along roadsides and railroads in the northwestern part. I collected an albino form of this species which I planted and it has done well in cultivation for nearly four years. It seems to prefer a slightly acid soil.

Wherry divides this species into two varieties, a northern and a southern one, as follows:

Sepals 5.5-7.5 mm long, united to about two thirds their length; calyx lobes thus 1.5-3 mm long.................................... *P. glaberrima* var. *interior* Wherry.

Sepals 6.5-8.5 mm long, united a half to two thirds their length; calyx lobes thus 2.5-4 mm long. ....*P. glaberrima* var. *melampyrifolia* (Salisbury) Wherry.

The first variety is the northern form of the species and extends as far south as Kentucky, hence all Indiana plants belong to this variety. The second variety is the southern representative of this species and has not yet been found as far north as Indiana.

Se. Va. to se. Wis., southw. to n. Fla. and e. Tex.

6. **Phlox maculàta** L. (Wherry. Bartonia 14: 20-26. 1932.) SWEET WILLIAM PHLOX. Map 1686. An infrequent plant but usually frequent to common where it is found. It generally occurs in open, springy places,

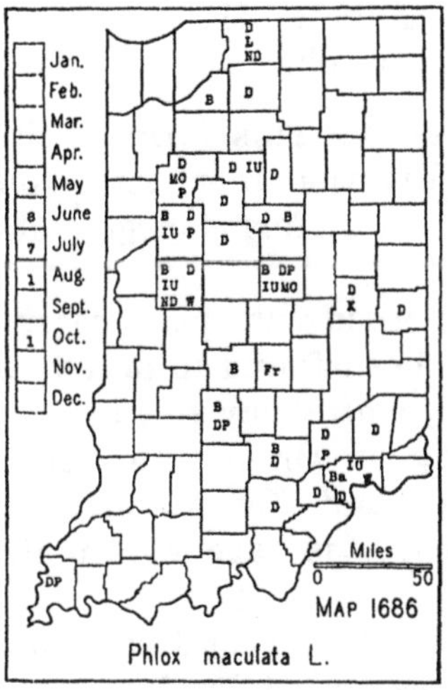

MAP 1686

Phlox maculata L.

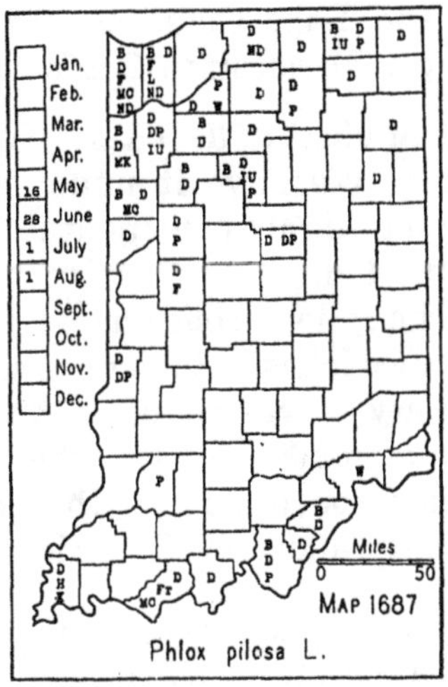

MAP 1687

Phlox pilosa L.

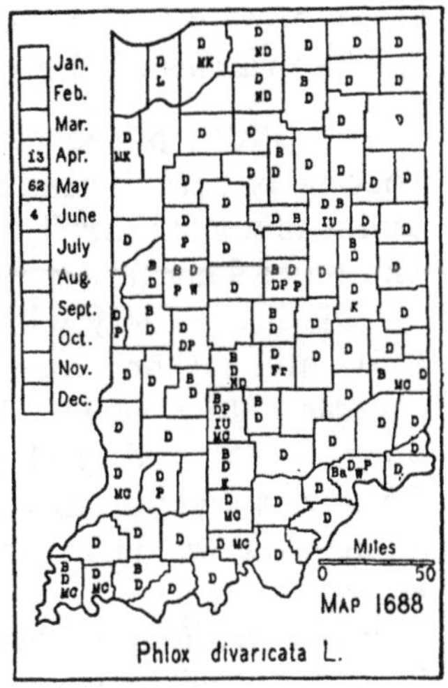

MAP 1688

Phlox divaricata L.

although in some of the southern counties in the Illinoian area it is found in hard, white clay soil in low, flat, beech and sweet gum woods, where it is usually common.

Wherry divides the species into two varieties as follows:

Nodes few (about 7-15) and often remote; upper leaves tending to become broadly lanceolate and cordate; blooming chiefly in early summer...*P. maculata* var. *odorata* (Sweet) Wherry.

Nodes numerous (about 15-30), and often crowded; upper leaves tending to become linear- or oblong-lanceolate; blooming chiefly in late summer..*P. maculata* var. *pyramidalis* (Smith) Wherry.

The first variety is the northern form and extends southward in Indiana to Jennings County. The second variety is the southern form of the species and extends northward in Indiana to Jackson and Wayne Counties.

Que. to Minn., southw. to N. C. and Mo.

7. **Phlox pilòsa** L. (Wherry. Bartonia 12: 36-47. 1931.) DOWNY PHLOX. Map 1687. A frequent plant in the lake area and where it is found it often forms complete stands, notably along railroads. In the Tipton Till Plain, it is infrequent to rare, becoming infrequent to frequent on the crests and open slopes of wooded ridges along the Ohio River. It is also sometimes found in wet places in the south. In the northern part of the state it is most commonly found in rather dry, open, sandy woods, along roadsides and railroads, and less frequently in moist prairies, marshes, and bogs.

Wherry writes me that "the common *Phlox pilosa* in Indiana is what I call variety *virens,* which is the same thing as variety *'typica'* in the sense that it is the Linnean plant on which the species was founded."

The species and its varieties are found from Conn. to N. Dak., southw. to Fla. and Tex.

7a. **Phlox pilosa** var. **fúlgida** Wherry. This is a variety with the inflorescence densely clothed with fine nonglandular hairs; calyx lobes broadish. I have a single specimen referable to this variety. It is from a roadside about 10 miles northwest of Fort Wayne. It was also found by

Scott McCoy in Benton and Lake Counties. The range of the variety is in the prairies of the Upper Mississippi Basin to Manitoba.

7b. **Phlox pilosa** var. **amplexicaúlis** (Raf.) Wherry. This form of the species is rare in Indiana. We have it under cultivation in neutral soil and it is a very thrifty and a highly ornamental plant. It forms large mats and has a long flowering period.

This is a variety with the inflorescence densely clothed with long, coarse hairs and rather broad calyx lobes. So far, it has been found in only two counties. In Spencer County it was found in hard, white clay soil in low, open, white and post oak woods along Little Pigeon Creek. In Perry County it was on an open wooded slope with beech, and it was abundant along this woods in an old fallow field. This variety is rare within its range.

Phlox argillacea Clute & Ferriss. The authors of this species say that it is distinguished by its "lighter green leaves, greater height, less compact flower clusters, restricted habitat, and, above all, pale flowers and later and longer season of bloom." Wherry refers this to a form of the variable species, *Phlox pilosa,* and says "the only way they can be distinguished is by the greater vigor of the former, a difference of horticultural but not of taxonomic significance."

S. Ind., Tenn., and La. to Tex.

8. **Phlox divaricàta** L. (Wherry. Bartonia **12**: 25-35. 1931.) BLUE PHLOX. Map 1688. The flowers of this species vary greatly in intensity of color, length of corolla tube, and the form of the corolla lobes. The apical end of the lobes is usually marked with a sinus 1-3 mm deep but there are plants with the lobes rounded or rounded and mucronate. This round-lobed form, var. *Laphami* Wood, is restricted mostly to our western counties. The largest specimens of this species belong to this form and are found in low, wet woods and wet, alluvial soil which is usually slightly acid.

Albino forms are not infrequent. We have had an albino form in cultivation for about 15 years and it remains true.

Frequent in most moist woods throughout the state. It occurs in every county of the state although it may not be native in Benton County. It is a plant of the woodland and is rarely found in the open. It prefers a neutral soil, shuns sandy habitats, and is rarely found in swampy places.

Vt., Que. to Minn., southw. to Fla, and Tex.

9. **Phlox bífida** Beck. (Wherry. Bartonia **11**: 29-35. 1929.) CLEFT PHLOX. Map 1689. Within the range of the species, the number of gland-tipped hairs varies greatly. In the northern part of its range the young growth, at least, has an abundance of gland-tipped hairs. This form has been named *P. bifida* var. *glandifera* Wherry, and has been reported from St. Joseph County by Sr. Elizabeth McDonald. In the center of its range the glands become fewer and may be present only on the pedicels, and in the southeastern part of its range the plants may be glandless. A thinly pubescent to glabrous extreme is found among Harrison, Montgomery,

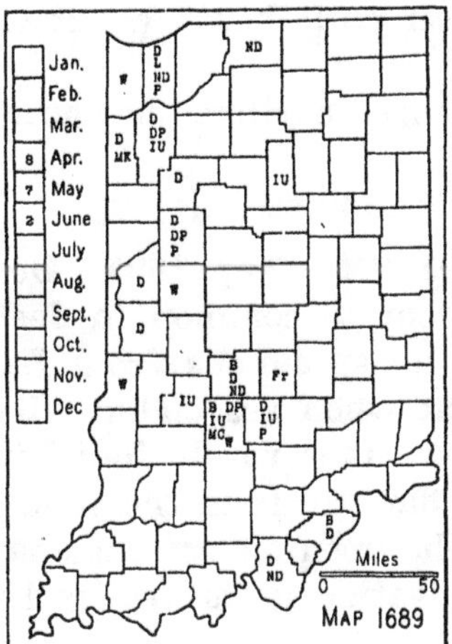

Phlox bifida Beck

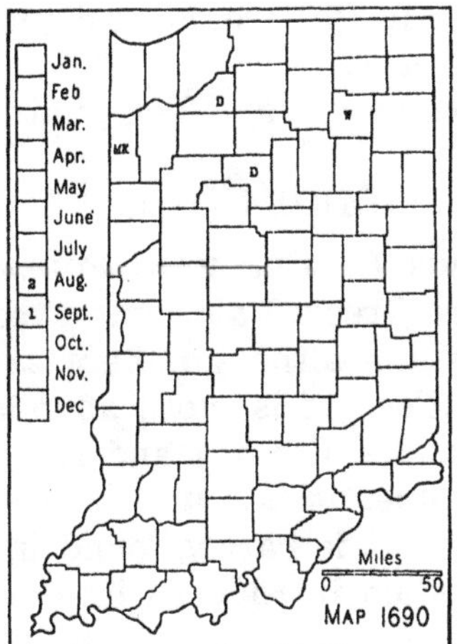

Ipomopsis rubra (L.) Wherry

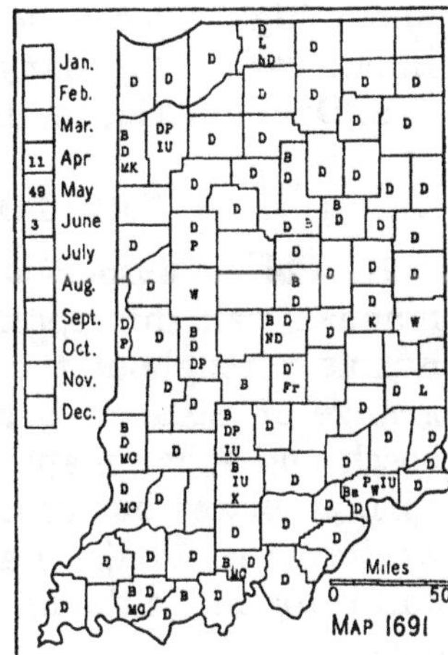

Polemonium reptans L.

and St. Joseph County specimens and is known as *Phlox bifida* var. *stellaria* (Gray) Wherry.

S. Mich. to Iowa, southw. to Tenn. and Ark.

10. **Phlox subulàta** L. (Wherry. Bartonia **11**: 18-28. 1929.) MOSS PINK. This species has been reported from six counties and probably all of them should be regarded as escapes since the natural distribution is mostly to the east of Indiana, and those making the reports do not give the habitat. The St. Joseph County specimen, however, grew along the St. Joseph River and may be native. It has been much used for planting on graves in cemeteries. Since it has proved very hardy and prolific, its escape is to be expected. I found it on a wooded slope along a creek and I traced it back to a cemetery on the bank a short distance away.

Cent. N. Y. to s. Mich., southw. to w. N. C.

10a. **Phlox subulata** var. **ciliàta** (Brand) Wherry. This is a form with the hairs of the inflorescence normally glandless; corolla purple, averaging 11 mm long with lobes 8 mm long and 5.5 mm wide. It has been reported from St. Joseph County by Sr. McDonald for Nieuwland. The specimen was collected 6 miles north of Notre Dame, near the Michigan boundary.

7015. COLLÒMIA Nutt.

(See excluded species no. 513, p. 1082.)

7016. IPOMÓPSIS Michx.

1. IPOMOPSIS RÙBRA (L.) Wherry. (Bartonia 18: 56. 1936.) (*Gilia rubra* (L.) Heller.) STANDING CYPRESS. Map 1690. In my herbarium there is a specimen from Cass County and there are specimens from two places in Starke County. Four of the specimens I have seen are from sandy roadside knolls and one I collected was on a cleared sand hill in a large black oak woods about a mile south of Koontz Lake, Starke County. It has escaped in the vicinity of Morocco, Newton County. This plant is biennial

and I highly recommend it for ornamental planting. It has sown itself in our garden for many years.

S. Dak. to Ark., southw. to Fla. and Tex.; naturalized northw. and eastw.

7017. POLEMÒNIUM [Tourn.] L. POLEMONIUM

1. **Polemonium réptans** L. CREEPING POLEMONIUM. Map 1691. Our manuals call this species Greek Valerian. Frequent to common in deep humus throughout the state, although there are no specimens from Lagrange or Steuben Counties. It is more abundant when associated with beech and sugar maple and white oak and red oak. It is rarely found in springy places but sometimes it is found in low, flat woods in the southwestern part of the state. It is rarely found in the open along roadsides and railroads, although when introduced into cultivation it thrives in the open.

Two old pioneers told me that the root was a diuretic and a specific for kidney disorders.

N. Y. to Minn., southw. to Ga. and Kans.

251. HYDROPHYLLÀCEAE Lindl. WATERLEAF FAMILY

Flowers in scorpioid cymes or in loose racemes.
- Corolla lobes convolute in the bud; blades of median stem leaves generally more than 8 cm long; plants not glandular; ovary-placentae dilated..7021. HYDROPHYLLUM, p. 784.
- Corolla lobes imbricated in the bud; blades of median stem leaves generally less than 8 cm long; ovary-placentae not dilated............7025. PHACELIA, p. 786.

Flowers solitary, opposite the leaves...........................7023. ELLISIA, p. 785.

7021. HYDROPHÝLLUM [Tourn.] L. WATERLEAF

Stem leaves mostly orbicular in outline, more or less deeply 5-7-lobed (sometimes the terminal leaf 3-lobed or the lower one with a pair of pinnae below the main body of the leaf).
- Peduncles longer than the petioles of the stem leaves; stem leaves usually more than 3; stems usually densely pubescent; flowers purplish; lower pedicels mostly 5-15 mm long; calyx lobes densely pubescent, with long reflexed appendages in the sinuses..1. *H. appendiculatum.*
- Peduncles shorter than the petioles of the stem leaves; stem leaves usually 2 or 3; flowers white or nearly so; stems nearly glabrous; lower pedicels mostly 5-10 mm long, glabrous or nearly so; calyx lobes without reflexed lobes in the sinuses, or with very short ones....................................2. *H. canadense.*

Stem leaves oblong, longer than broad, pinnately lobed or pinnately divided into 5-7 lobes.
- Plants nearly glabrous, the pubescence scant, short, and appressed; hairs of the stem usually 0.5-1 mm long; lateral lobes of leaves more or less ovate, strongly narrowed at the base; flowers more or less colored, sometimes white...3. *H. virginianum.*
- Plants densely pubescent, the hairs of the stem usually about 2 mm long; lateral lobes of leaves oblong, scarcely narrowed at the base; flowers white..4. *H. macrophyllum.*

1. **Hydrophyllum appendiculàtum** Michx. APPENDAGED WATERLEAF. Map 1692. Infrequent to common in all parts of the state, although we have no reports for the area near Lake Michigan. It prefers deep leaf mold

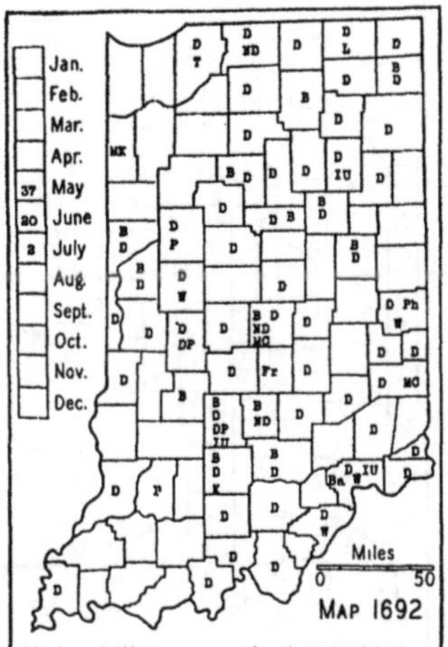

Hydrophyllum appendiculatum Michx

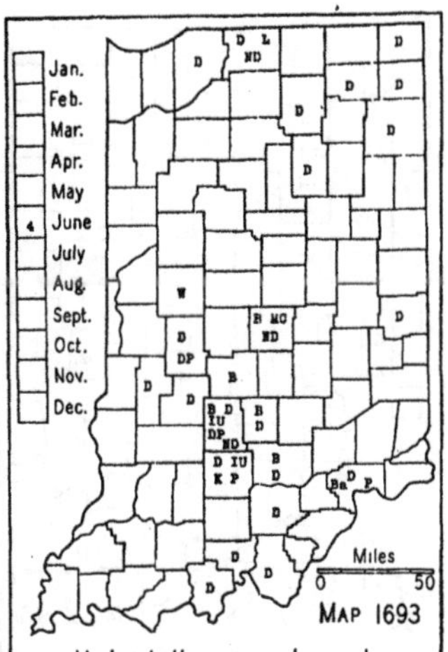

Hydrophyllum canadense L.

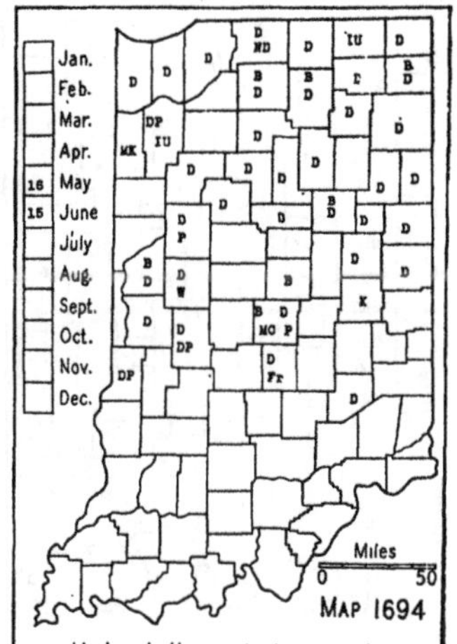

Hydrophyllum virginianum L.

and is most abundant in beech and sugar maple woods. On rich, wooded slopes of ravines, alluvial plains, and rarely in exposed places on open wooded slopes. Not found on poor black oak slopes.

All of the waterleafs do well in cultivation.

N. Y., Ont. to Minn., southw. to N. C. and Kans.

2. **Hydrophyllum canadénse** L. BROADLEAF WATERLEAF. Map 1693. Infrequent to rare in deep humus in moist soil, usually toward the bases of deep wooded ravines. Generally associated with beech and usually forming large colonies. In cultivation where it is relieved of competition it spreads rapidly.

Sw. Vt. to Ont. and Ill., southw. to N. C. and Ky.

3. **Hydrophyllum virginiànum** L. VIRGINIA WATERLEAF. Map 1694. Infrequent throughout Indiana except the southwestern part. It prefers moist soil in woodland, although it is sometimes found along roadsides and railroads. Usually found in alluvial flood plains and in moist woods of nearly any composition, although it is more frequent in beech and sugar maple and white oak woods.

Que. to S. Dak., southw. to S. C. and Kans.

4. **Hydrophyllum macrophýllum** Nutt. LARGELEAF WATERLEAF. Map 1695. Infrequent in most parts of the state, although there are no records from the northwestern part. It prefers deep humus and is usually found on the slopes of deep ravines, generally associated with beech.

My Dearborn County specimen no. 5697 has purplish flowers, the corolla glabrous without, the calyx lobes scarcely dilated at the base and short pubescence on these lacking or nearly so.

Va., Ohio, and Ill., southw. to Ala. and Tenn.

7023. ELLÍSIA L.

1. **Ellisia Nyctèlea** L. (*Nyctelea Nyctelea* (L.) Britt. of Britton and Brown, Illus. Flora, ed. 2.) NYCTELEA. Map 1696. Very local but common

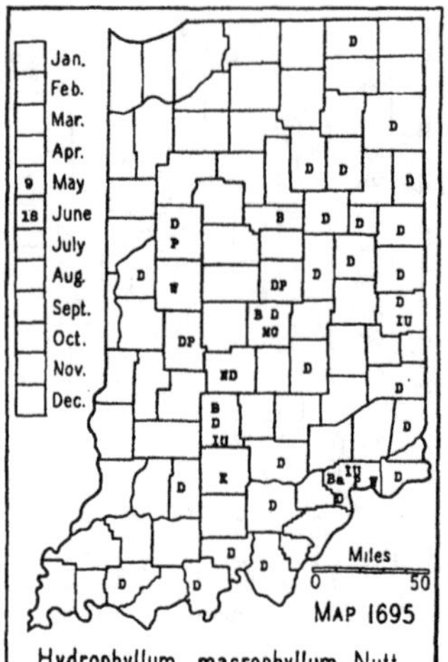

Hydrophyllum macrophyllum Nutt.

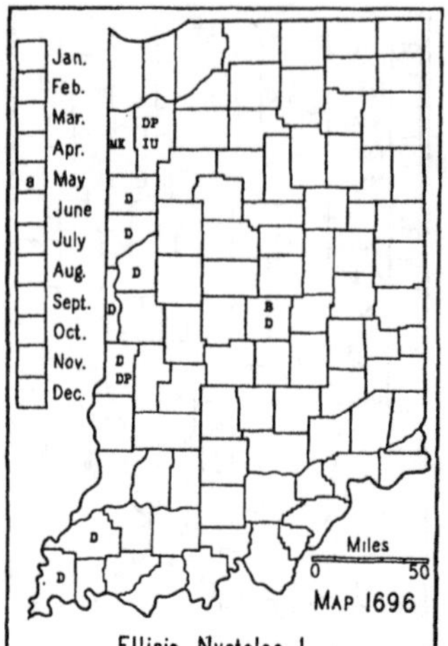

Ellisia Nyctelea L.

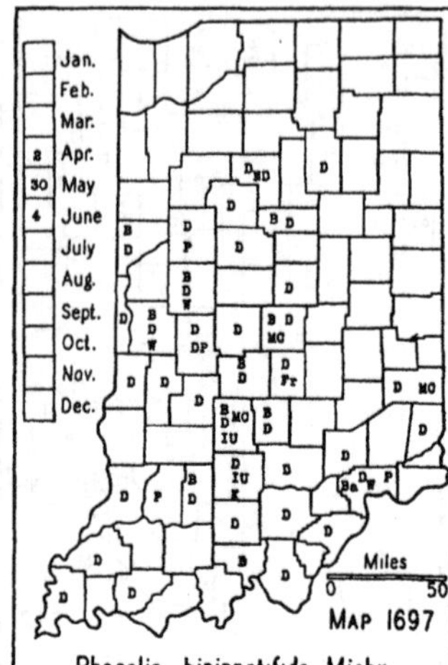

Phacelia bipinnatifida Michx.

where found. Most of my specimens are from wooded flood plains and terrace banks of the Wabash River although I found it in a woods in Benton County. Welch reported it for Fountain Park in Jasper County and it has been reported for the Calumet Region.

N. J., Minn. to Sask., southw. to Va., Nebr., and Kans.

7025. PHACÈLIA Juss. PHACELIA

Inflorescence with a copious pubescence of spreading, glandular, hairs; corolla lobes entire..1. *P. bipinnatifida.*
Inflorescence with a rather sparse, appressed pubescence.
  Calyx lobes pubescent over the entire outer surface.
    Filaments of stamens pubescent; upper leaves sessile, only the lower ones on long petioles; calyx lobes of an elliptic type, about 4 mm long, obtuse, much shorter than the corolla. (See excluded species no. 514, p. 1082).....*P. dubia.*
    Filaments of stamens glabrous; upper leaves long-petiolate; calyx lobes of a linear type, about 5 mm long, almost as long as the corolla......2. *P. Covillei.*
  Calyx lobes glabrous on the back, the margins ciliate with long, spreading hairs; corolla lobes fringed..3. *P. Purshii.*

1. **Phàcelia bipinnatífida** Michx. Map 1697. Infrequent throughout the area shown on the map which covers all of our reports. Probably absent from the area east and north of the stations indicated. The only report from Ohio is from Hamilton County, near Cincinnati. It prefers a moist, rich soil, usually that of wooded slopes along streams. The bruised plant is ill-scented.

Ohio, Ill. to Mo., southw. to Ga., Ala., and Tenn.

2. **Phacelia Covíllei** Wats.* COVILLE PHACELIA. Map 1698. An extremely rare and local plant. I have specimens from the low woods about Little Cypress Swamp in Knox County as follows: It was first found by Schneck on May 16, 1896, when it was in fruit. He again found it in flower on April 20, 1903. Blatchley found it in flower April 23, 1903. I found it in fruit May 23, 1926, and in flower on April 19, 1927. I made a study of it on the ground and made the following notes: Length of corolla 4 mm,

* The name of this plant now becomes **Phacelia ranunculacea** (Nutt.) Constance. (Rhodora **42**: 39. 1940.)

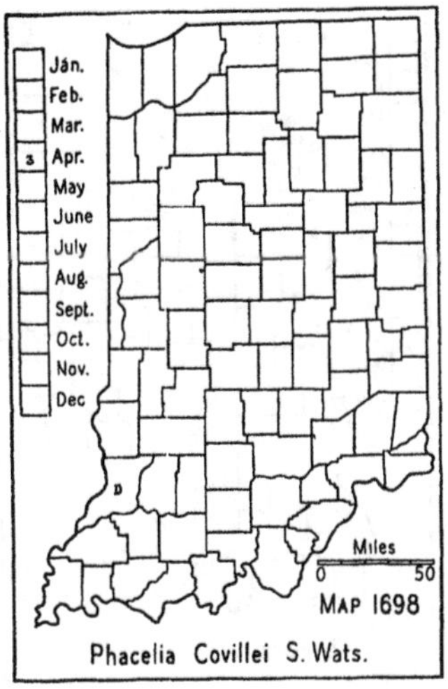

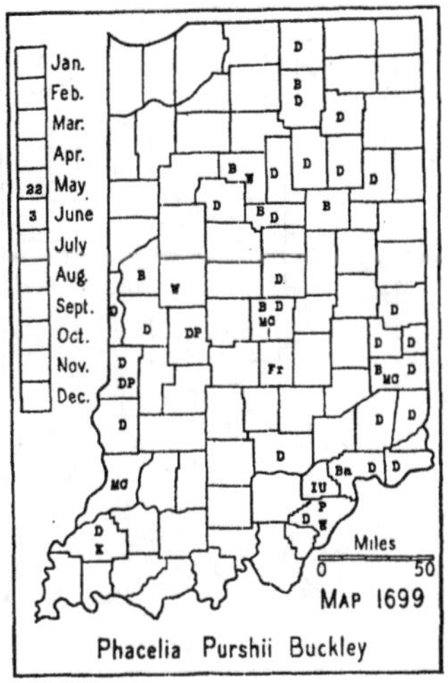

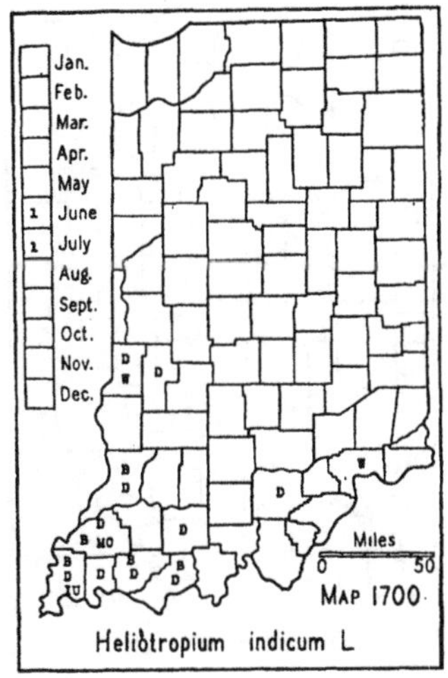

expanse of corolla 4 mm, light Wisteria Violet (Ridgway) ; filaments glabrous; capsules about 4 mm wide and 3 mm long, 2- or 3-seeded; surface of seed not reticulated in lines.

The area where the plant grows is usually inundated each year for weeks at a time. It is associated with *Carya laciniosa, Liquidambar,* and *Quercus Prinus.*

Known only from Knox County, Ind., along the Potomac River above Washington, D. C., and Arlington County, Va.

3. **Phacelia Púrshii** Buckley. PURSH PHACELIA. Map 1699. Our records indicate that this species is restricted to the alluvial flood plains, banks, and slopes of the terraces of streams. Found in sandy soil in the locations indicated, along roadsides, and in clover fields. It is the most abundant in the White Water River Valley. I have seen it by the acre along this river in Franklin and Union Counties. It has become so abundant in some places that farmers have reported it as an obnoxious weed. It can not stand competition but when once established it will persist if bare soil exists. We have had it in our meadow along the Wabash River for 25 years. Wood's Classbook of all editions except the first gives Miami Mist for its common name. Fisher[1] says it was so called in western Ohio.

Pa. to Minn., southw. to N. C., Ala., and Miss.

252. BORAGINÀCEAE Lindl. BORAGE FAMILY

[Johnston. A synopsis of the American native and immigrant borages of the subfamily Boraginoideae. Contr. Gray Herb. Harvard Univ. **70**: 1-55. 1924.]

Flowers white or yellow.
  Nutlets armed with prickles.
    Calyx lobes in anthesis about 1 mm long; flowers usually 2 mm long; seeds mostly 3-5 mm long.................................7073A. HACKELIA, p. 790.
    Calyx lobes in anthesis about 3 mm long; flowers more than 2 mm long; seed mostly 5-6 mm long. White-flowered form of *Cynoglossum officinale*....p. 789.

[1] Torreya **23**: 106. 1923.

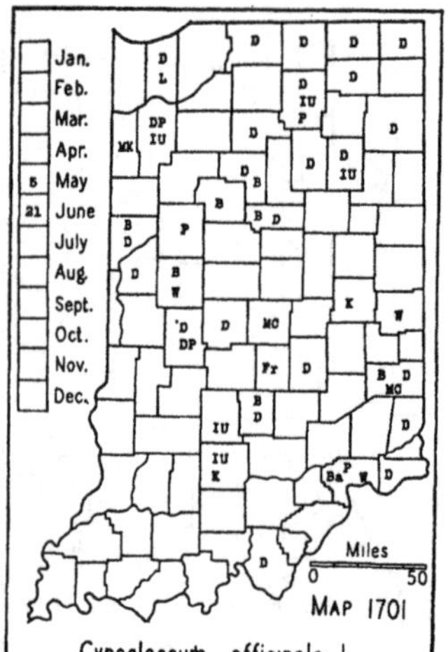

Cynoglossum officinale L.

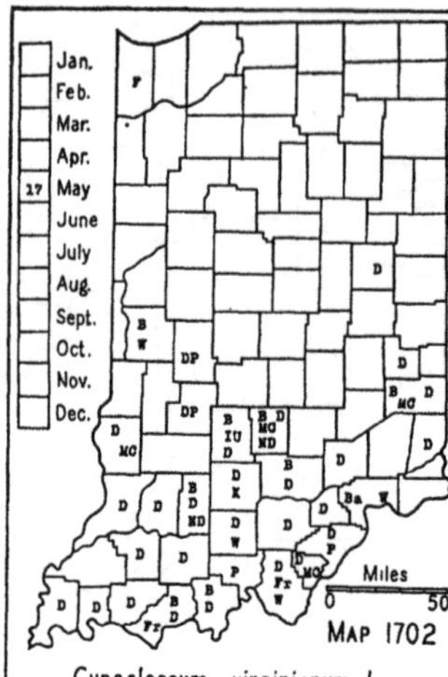

Cynoglossum virginianum L.

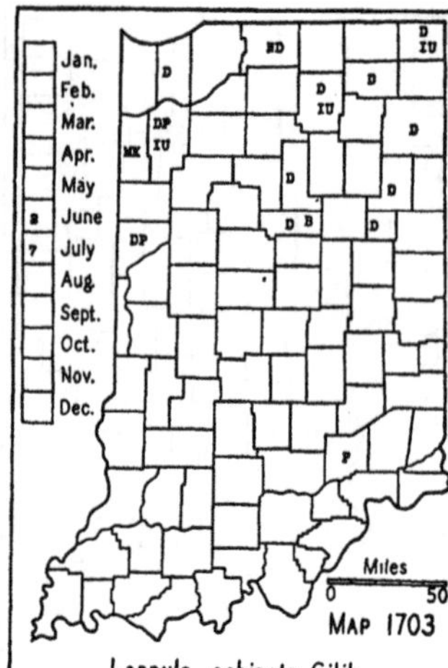

Lappula echinata Gilib.

Nutlets not armed with prickles.
Racemes without bracts.
Plants large and coarse with long, acuminate leaves..7090. SYMPHYTUM, p. 790.
Plants not large and coarse; leaves mostly obtuse...7100. *Myosotis virginica*, p. 791.
Racemes with bracts.
Lobes of corolla spreading, rounded..............7109. LITHOSPERMUM, p. 792.
Lobes of corolla erect, acute......................7113. ONOSMODIUM, p. 794.
Flowers blue to purplish.
Nutlets armed with prickles.
Leaves more than 1 cm wide; nutlets (exclusive of prickles) 4-7 mm long..7064. CYNOGLOSSUM, p. 788.
Leaves mostly less than 1 cm wide; nutlets (exclusive of prickles) less than 4 mm long..7073. LAPPULA, p. 789.
Nutlets not armed with prickles.
Flowers regular.
Corolla generally 2-2.5 cm long......................7102. MERTENSIA, p. 792.
Corolla less than 2 cm long.
Stem leaves on long petioles....................7052. HELIOTROPIUM, p. 788.
Stem leaves sessile or on short petioles..............7100. MYOSOTIS, p. 790.
Flowers irregular.
Corolla bent at about the middle; stamens included......7094. LYCOPSIS, p. 790.
Corolla not bent; stamens exserted......................7118. ECHIUM, p. 794.

7052. HELIOTRÒPIUM [Tourn.] L. HELIOTROPE

1. HELIOTROPIUM ÍNDICUM L. INDIA HELIOTROPE. Map 1700. Infrequent in moist, sandy, open, generally alluvial woods of the southern part of the state; more rarely in waste places and fallow fields.

Nat. of India; naturalized in U. S. from N. C., Ky., to Mo., southw. to Fla. and Tex.

7064. CYNOGLÓSSUM [Tourn.] L. HOUNDSTONGUE

Flowers maroon, purple or rarely white; stems leafy to the top; nutlets flat...1. *C. officinale.*
Flowers blue; stems leafless above; nutlets convex..................2. *C. virginianum.*

1. CYNOGLOSSUM OFFICINÀLE L. COMMON HOUNDSTONGUE. Map 1701. More or less frequent in dry soil in pasture fields and woods pastures, on open wooded slopes, and along roadsides and railroads. This is a species that one usually, by preference, neglects to collect, and this fact, no doubt, accounts for the lack of specimens from the southwestern part of the state. There are specimens with white flowers from Kosciusko and Noble Counties.

Nat. of Eurasia; now naturalized in N. A. from Que., Ont., Man., and Oreg., southw. to N. C., Ala., and N. Mex.

2. **Cynoglossum virginiànum** L. WILD COMFREY. Map 1702. Frequent in the southern half of the state and rare in the northern part. My Lagrange County specimen is not shown on the map. It is strictly a woodland species found mostly on wooded slopes of white oak, black and white oak, and beech. Careful measurements of our specimens in anthesis show the following results. In 3 specimens the calyx was from 2-2.5 mm in length, the corolla from 11-14 mm in width, the lobes orbicular, and the sinuses closed; in 7 specimens the calyx was 3 mm long, the corolla from 11.5-16 mm wide, the lobes orbicular, and the sinuses closed; and in 4 specimens the calyx was 3.5-4 mm long, corolla 14-16 mm wide, the lobes orbicular, and the sinuses closed. Not included in the preceding measurements I have a specimen from Franklin County, no. 34008, with a calyx 2.5 mm long, corolla 9 mm wide, the lobes oblong, and the sinuses open; and one specimen from Jennings County with a calyx 2 mm long, corolla 9 mm wide, the lobes oblong, and the sinuses open.

Cynoglossum boreale Fern., a northern species, is described as having a calyx 2-2.5 mm long; corolla 6-8 mm wide, the lobes ovate-oblong, and the sinuses open. Our Franklin and Jennings County specimens belong, no doubt, to this species. The preceding measurements convince me that our specimens belong to one variable species. Johnston (Contr. Gray Herb. Harvard Univ. **70**: 34. 1924), in his synopsis of the genus, says: "All the vegetative characters of this species [*Cynoglossum boreale*] can be matched, after a short search, among undubitable material of *C. virginianum.*"

Peattie reported *Cynoglossum boreale* from the dune area but I have not seen his specimen if he preserved one. Buhl (Amer. Midland Nat. **16**: 262. 1935) says this report lacks confirming specimens.

S. Conn. to Mo., southw. to Fla. and La.

7073. LÁPPULA [Rivin.] Moench

1. LAPPULA ECHINÀTA Gilib. (*Lappula Lappula* (L.) Karst.) Map 1703. This species has been reported from all parts of the state although my specimens are all from the northeastern part. It prefers a sandy soil and is generally found in ballast along railroads and roadsides, in waste places and fallow fields, and rarely in pastures or open woods.

Nat. of Eurasia; naturalized in N. A. from N. S. to B. C., southw. to N. J., Kans., and Calif.

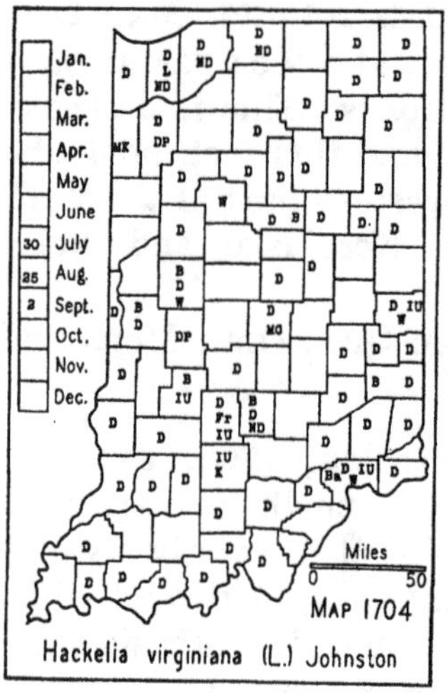

Hackelia virginiana (L.) Johnston

Myosotis scorpioides L.

Myosotis laxa Lehm.

7073A. HACKÈLIA Opiz. STICKSEED

1. **Hackelia virginiàna** (L.) I. M. Johnston. (*Lappula virginiana* (L.) Greene.) Map 1704. This is strictly a woodland species and is found throughout our area in dry woods of all kinds, although it is most abundant in beech and sugar maple and white oak woods. It is rarely found in very wet or springy places.

Maine, w. Que. to Minn., southw. to Ga., La., and Kans.

7090. SÝMPHYTUM [Tourn.] L.

See excluded species no. 516, p. 1082.

7094. LYCÓPSIS L.

See excluded species no. 517, p. 1083.

7100. MYOSÒTIS [Rupp.] L. FORGET-ME-NOT

Hairs of calyx few, short, straight, closely appressed, nonglandular; annuals.
- Calyx lobes shorter than the tube; corolla large, 5-9 mm wide; style usually much exceeding the nutlets .... 1. *M. scorpioides.*
- Calyx lobes about equaling the tube; corolla rather small, 3-6 mm wide; style clearly exceeding the nutlets .... 2. *M. laxa.*

Hairs of calyx, at least the lower ones, with hooked tips; annuals or perennials.
- Fruiting pedicels longer than the calyx. (See excluded species no. 518, p. 1083) .... *M. arvensis.*
- Fruiting pedicels not longer than the calyx.
  - Calyx very unequally cleft; flowers white; nutlets 1.5-2.5 mm long.
    - Fruiting calyx usually less than 5 mm long; nutlets about 1.5 mm long .... 3. *M. virginica.*
    - Fruiting calyx generally 5-7 mm long; nutlets 2-2.5 mm long .... 3a. *M. virginica* var. *macrosperma.*
  - Calyx nearly regular; flowers blue; nutlets about 1 mm long .... 4. *M. micrantha.*

1. MYOSOTIS SCORPIOÌDES L. TRUE FORGET-ME-NOT. Map 1705. Reported by Pepoon as common on the shores of the Calumet River at Clarke, in Lake County. Also reported by Blatchley as an escape in Vigo County,

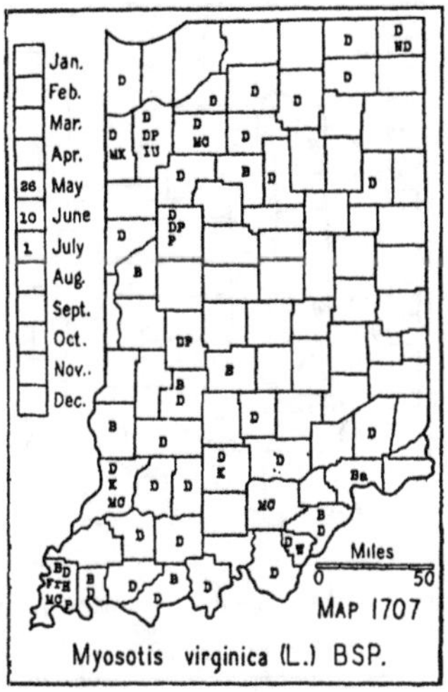

MAP 1707
Myosotis virginica (L.) BSP.

MAP 1708
Myosotis virginica var. macrosperma (Engelm.) Fern.

MAP 1709
Myosotis micrantha Pall.

and by Young for Jefferson County. I found a large colony along the St. Joseph River just west of the Elkhart County line.

Nat. of Eu.; now naturalized in N. A. from Newf. to Que., southw. to Ga. and La.; also in Calif. and B. C.

2. **Myosotis láxa** Lehm. Map 1706. Frequent in the Mineral Springs Bog in Porter County and also reported from Lake County by Pepoon as occurring on the banks of a cold brook near Miller and on the margin of the Little Calumet River. Probably restricted to these two counties.

Newf., Ont., and Ind., southw. to Ga. and Tenn.; also on the Pacific coast from Calif. to B. C.; and in Chile.

3. **Myosotis virgínica** (L.) BSP. Map 1707. An infrequent plant in the northern and southern parts of the state. There are no records from the area about Lake Michigan or from the central part of the Tipton Till Plain. It is generally found in open places in noncalcareous soils. In the northern part it is found in bare spots on the crests and slopes of black oak ridges; on lower ground, it is usually found in depressions in sandy soil in open places in black and white oak woods, generally associated with *Gaylussacia baccata, Vaccinium vacillans, Houstonia longifolia,* etc., and more rarely found along railroads and roadsides. In the southern part of the state it is most commonly found in white clay soil in fallow fields, where it is sometimes abundant, associated with *Alopecurus caroliniana, Callitriche Austini, Arabis virginica,* and *Poa Chapmaniana.* It is also found in bare places in low, flat, post oak woods and in bare places on the crests of black and white oak ridges.

Maine, Ont. to Minn., southw. to Fla. and Tex.

3a. **Myosotis virginica** var. **macrospérma** (Engelm.) Fern.* Map 1708. This is a much larger plant than the species with larger calyx and seed and is found in wet woods, associated with white elm, ash, and river birch; in drier woods with black and white oak; also on wooded slopes.

Va. to Ind., southw.

* Fernald has recently restored this plant to specific rank, **Myosotis macrosperma** Engelm. (Rhodora **41**: 558. 1939.)

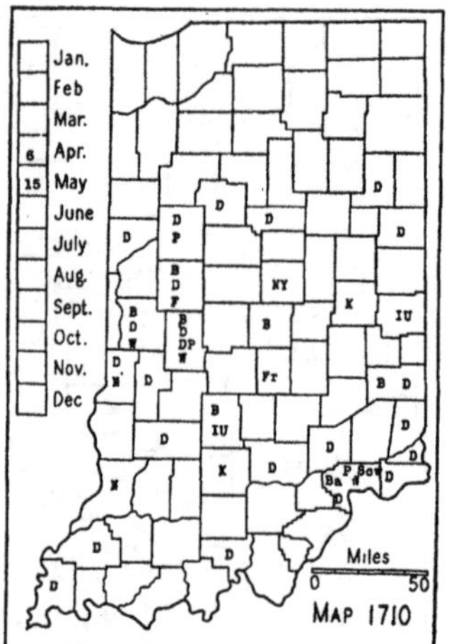

Mertensia virginica (L.) Link

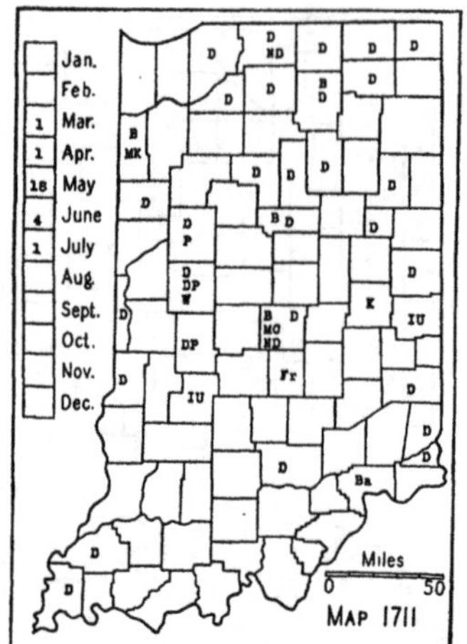

Lithospermum arvense L.

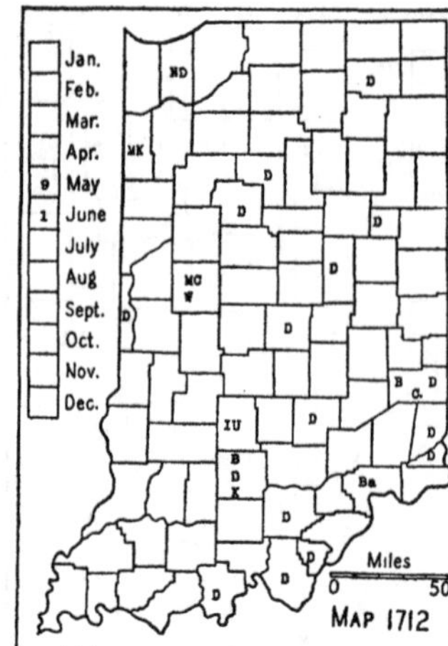

Lithospermum latifolium Michx.

4. MYOSOTIS MICRÁNTHA Pall. Map 1709. Found in a white oak woods pasture in Lagrange County, common in blue grass along Eel River at Cataract Falls in Owen County, and a common weed in Turkey Run State Park about the Administration Building.

Nat. of Eu.; now established in N. H., Mass., N. Y., N. J., Ohio, Mich., Ont., and Ind.

7102. MERTÉNSIA Roth BLUEBELL

1. **Mertensia virgínica** (L.) Link. (*Mertensia virginica* (L.) DC.) (Williams. A monograph of the genus Mertensia in North America. Ann. Missouri Bot. Gard. 24: 17-159. 16 fig. 1937.) VIRGINIA BLUEBELL. Map 1710. Local throughout the area indicated on our map but usually frequent to abundant where it is found. All but three of our specimens are from wooded flood plains and wooded terraces of streams. These three are from rather sandy woods. I have seen it also as a common plant on a rocky, wooded hillside about three miles southeast of Dillsboro, Dearborn County. This species does well in cultivation but prefers partial shade.

N. Y., Ont. and Minn., southw. to Ala. and Kans.

7109. LITHOSPÉRMUM [Tourn.] L. GROMWELL

Flowers less than 10 mm long.

Corolla white or nearly so, without crests in the throat, lobes glabrous without; nutlets gray, dull, wrinkled, and roughened........................1. *L. arvense.*

Corolla greenish white or yellow, with crests in the throat, lobes pubescent without; nutlets ivory white, glossy, smooth, more or less impressed-punctate.

Leaves lanceolate, firm, acute, veins obscure beneath; corolla greenish white; nutlets about 3 mm long. (See excluded species no. 519, p. 1083)..*L. officinale.*

Leaves ovate to ovate-lanceolate, thin, acuminate, veins prominent beneath; corolla yellow; nutlets about 4 mm long............................2. *L. latifolium.*

Flowers more than 10 mm long.

Leaves narrowly linear, acute, appressed canescent-pubescent; corolla light yellow, tube 15-33 mm long, lobes more or less fimbriate or erose, with well developed crests in the throat of the corolla.............................3. *L. incisum.*

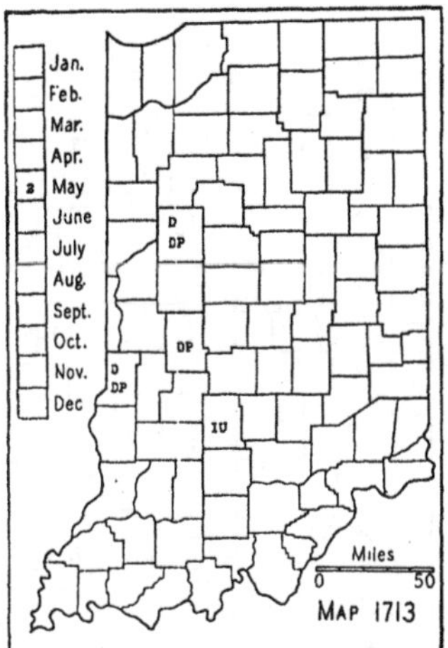

Lithospermum incisum Lehm.

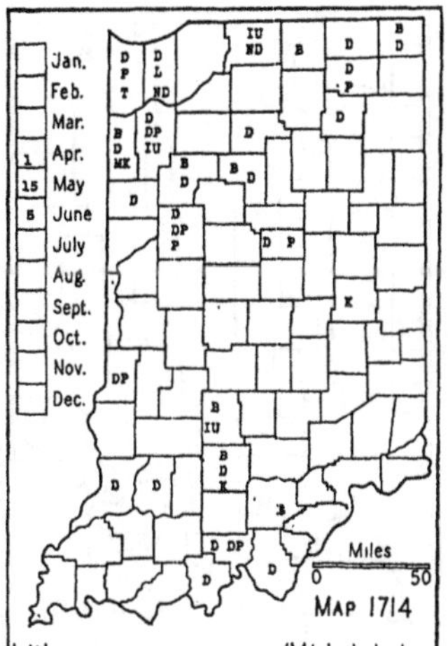

Lithospermum canescens (Michx.) Lehm.

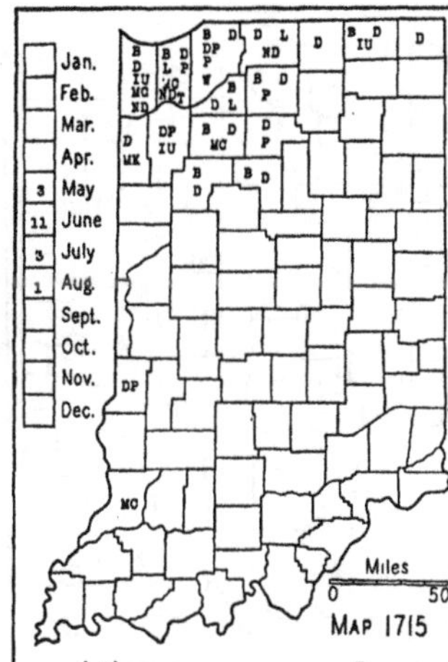

Lithospermum croceum Fern.

Leaves mostly oblong, lanceolate or narrow-lanceolate, rarely linear, mostly obtuse; corolla light to deep orange yellow, tube less than 15 mm long, generally without well developed crests in the throat.

Corolla orange yellow, the ring of glands at the base within not hairy; leaves closely appressed canescent-pubescent above, the hairs about 0.6 mm long and not with a conspicuous papillose base; calyx lobes in anthesis 5-6 mm long; nutlets mostly 2.5-3 mm long............................4. *L. canescens.*

Corolla generally light yellow, the ring of glands at the base within hairy; leaves loosely appressed-pubescent above; hairs fewer than in the preceding species, about 1 mm long, arising from a conspicuous, transparent, papillose base; calyx lobes at anthesis more than 6 mm long; nutlets 3.5-4 mm long...5. *L. croceum.*

1. LITHOSPERMUM ARVÉNSE L. CORN GROMWELL. Map 1711. Frequent to common in all parts of the state, mostly in sandy soil along roadsides and railroads and in waste places, fallow, and cultivated fields.

Nat. of Eu. and adjacent Asia and Africa; Maine to Mont., southw. to Fla. and La.; also in B. C., Calif. and Utah.

2. **Lithospermum latifòlium** Michx. Map 1712. An infrequent to rare plant throughout the state. Generally on wooded slopes adjacent to streams and rarely in comparatively level woods. It has no particular tree associates, but is more often associated with the oaks.

Western N. Y. to s. Minn., southw. to e. Tenn. and Kans.

3. **Lithospermum incìsum** Lehm. (Kew Bull. **1934**: 59. 1934.) (*Lithospermum angustifolium* Michx. and *Lithospermum linearifolium* Goldie.) Map 1713. This rare species has been reported from only Putnam, Steuben, Tippecanoe, and Vigo Counties. Grimes' specimens from Putnam and Tippecanoe Counties are in the herbarium of DePauw University. Found in sandy and gravelly open places. Infrequent along the roadside southwest of Lafayette, on the gravelly terrace of Big Wea Creek west of the Wabash Railroad.

Ont. to Man. and B. C., southw. to Tex. and Chihuahua, Mex.

4. **Lithospermum canéscens** (Michx.) Lehm. PUCCOON. Map 1714. Infrequent in dry, sandy prairie habitats, on dry, sandy knolls along road-

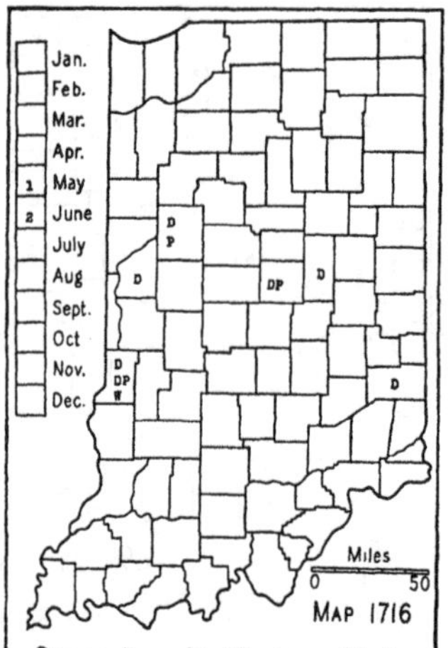

Onosmodium hispidissimum Mack.

Echium vulgare L.

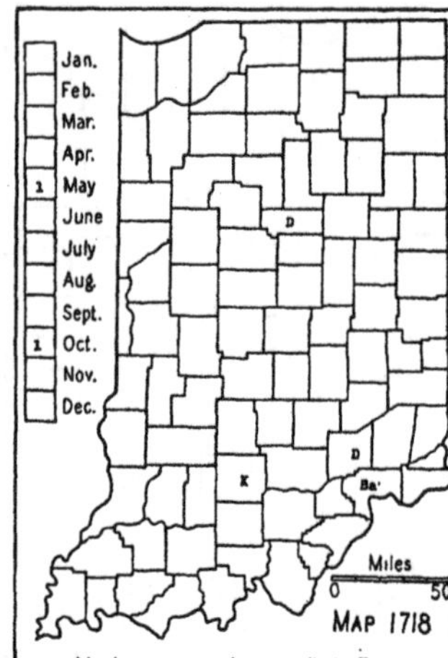

Verbena canadensis (L.) Britt.

sides and railroads, on the crests and slopes of open, wooded ridges in the area of the sandstone outcrops, and rarely in moist prairie habitats. Probably absent from many of the counties of the area of the Tipton Till Plain.

S. Pa. to Sask., southw. to Ala., and Tex.

5. **Lithospermum cròceum** Fern. (Rhodora 37: 329. 1935.) (*Lithospermum Gmelini* of Gray, Man., ed. 7 and *Lithospermum carolinense* of Britton and Brown, Illus. Flora, ed. 2.) Map 1715. Infrequent throughout the sandy area of the lake region. There are reports from southern Indiana but this and the preceding species have always been confused and I believe that most of the southern reports should be referred to *Lithospermum canescens*. Usually in very sandy soil in open black oak woods, along roadsides and railroads, and in sandy prairie habitats.

N. Y. to Man., southw. to Fla., Tex., and Mex.

7113. ONOSMÒDIUM Michx.

1. **Onosmodium hispidíssimum** Mack. FALSE GROMWELL. Map 1716. Until 1905 our manuals did not properly distinguish our species of this genus. Previous to 1905 three species were reported from six counties in Indiana and no doubt all of these should be referred to this species. For a discussion of this subject see excluded species. All the specimens that I have seen were collected in dry, gravelly soil in open woods or along roadsides. Rare, and only a few specimens at a location.

Cent. N. Y. to Minn., southw. to Ga. and Tex.

7118. ÈCHIUM [Tourn.] L.

1. ECHIUM VULGÀRE L. BLUEWEED. BLUE THISTLE. Map 1717. This species has been reported from five counties besides those shown on the map. It prefers sandy soil and is found mostly in fallow fields, along roadsides and railroads, and sometimes in open woods and woods pastures. In 1920 I noted a five-acre field of it in Lagrange County where it was so abundant that on June 21, when it was in flower, the whole field presented

a sky blue appearance. I observed this field in later years and the owner had been able to exterminate it entirely. It has become well established in Lagrange County and is found in many places, especially in the vicinity of Mongo and Brushy Prairie. I cultivated this plant one year and the largest one stood 28 inches high, and had 22 branches, the longest of which was 42 inches long. Needless to say I did not permit it to mature seed.

Nat. of Eu.; naturalized in N. A. from N. B., Ont. to Nebr., southw. to Ga. and Tex.

253. VERBENÀCEAE J. St. Hil. VERVAIN FAMILY

[Perry. A revision of the North American species of Verbena. Ann. Missouri Bot. Gard. 20: 239-362. 1933.]*

Calyx tubular; limb of corolla 5-lobed, regular or nearly so; fruit in long or short spikes and not very dense; fruit splitting into 4 nutlets....7138. VERBENA, p. 795.

Calyx short, 2-cleft; limb of corolla 4-lobed, 2-lipped; fruit in short or long, dense heads; fruit splitting into 2 nutlets........................7145. PHYLA, p. 798.

7138. VERBÈNA [Tourn.] L. VERVAIN

Flowers 15-25 mm long; anthers of the longer stamens gland-tipped; calyx 8-10 mm long.

- Bracts mostly shorter than the calyx; limb of corolla mostly 15-25 mm wide........ ....1. *V. canadensis.*
- Bracts mostly longer than the calyx; limb of corolla mostly 8-15 mm wide. (See excluded species no. 523, p. 1083)..........................*V. bipinnatifida.*

Flowers 4-10 mm long; anthers of the longer stamens not gland-tipped; calyx mostly less than 5 mm long.

- Bracts shorter than the flowers; spikes filiform or slender; plants erect or diffuse in *Verbena officinalis.*
  - Spikes filiform; fruiting calyx about 2 mm long or less; fruit scattered.
    - Plants diffuse annuals; leaves incised or pinnatifid, sessile; fruiting calyx less than 2 mm long; flowers purplish. (See excluded species no. 524, p. 1084) ....*V. officinalis.*
    - Plants erect perennials; leaves serrate (rarely incised), petiolate; fruiting calyx about 2 mm long; flowers white (rarely pink or purplish).
      - Leaves strigose-hirsute beneath or glabrate, the colorless hairs 1-1.3 mm long, appressed or subappressed; mature inflorescence with stiff, strigose ascending branches; mature calyx 2-2.3 mm long; mature nutlets definitely ribbed on the back.................................2. *V. urticaefolia.*
      - Leaves densely pubescent beneath with short hairs about 0.3 mm long; mature inflorescence with loosely ascending or spreading, puberulent branches; mature calyx usually less than 2 mm long; mature nutlets about 1.5 mm long and quite smooth on the back....2a. *V. urticaefolia* var. *leiocarpa.*
  - Spikes slender; fruiting calyx more than 2 mm long; flowers blue (rarely albino forms); fruit densely imbricated.
    - Stems glabrous or sparingly rough-pubescent, the hairs mostly less than 0.5 mm long; leaves lanceolate or narrower.
      - Leaves on petioles mostly 1.5-2 cm long, acuminate; fruiting calyx generally 1.5-3 mm long; seed 1.5-1.7 mm long.....................3. *V. hastata.*
      - Leaves sessile, mostly obtuse; fruiting calyx 3.5-4.5 mm long, usually about 4 mm long; seed about 2.5 mm long......................4. *V. simplex.*

---

* H. L. Moldenke examined all my specimens of this family.

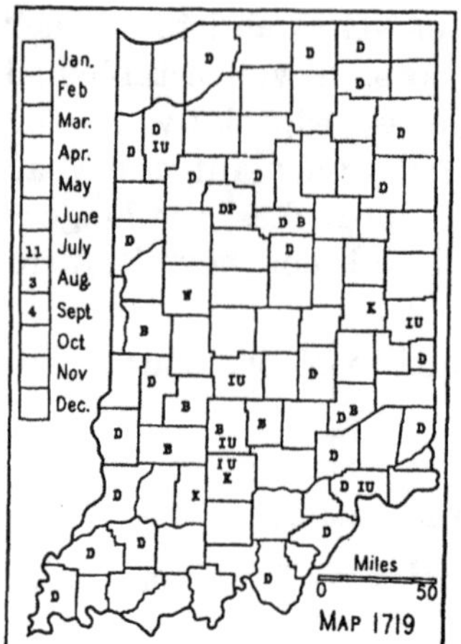

Verbena urticaefolia L.

Verbena urticaefolia var. leiocarpa Perry & Fern.

Verbena hastata L.

Stems densely soft-pubescent, the hairs mostly 0.75-1 mm long; fruiting calyx generally 4-5 mm long; seed about 2.5 mm long; leaves thick, rigidly ascending, mostly ovate, densely soft-pubescent................5. *V. stricta.*

Bracts longer than the flowers; spikes thick, dense; plants usually spreading, never erect...6. *V. bracteata.*

1. VERBENA CANADÉNSIS (L.) Britt. ROSE VERBENA. Map 1718. This species has been reported from 6 counties. It has long been cultivated and much used for planting on graves in cemeteries whence it doubtless frequently escapes. I have found this species growing on the slope of a creek bank below an old cemetery in which I found it to be common. It has abundantly escaped from cemeteries in Jefferson County and it was found by Chas. M. Ek as an escape from a cemetery in Howard County. I believe it is an escape in Indiana.

Va. to Ill. and Kans., southw. to Fla. and Tex.

2. **Verbena urticaefòlia** L. WHITE VERVAIN. Map 1719. This species doubtless occurs in every county. It is found in almost all kinds of soil except in very wet places; it is generally found in open woods, along logging roads in thick woodland, in fallow fields and waste places, and along roadsides and railroads. All of the species of Indiana vervains are extremely variable, especially in the leaf margins and color of flowers. Evidence of hybridization is frequent. I have a specimen with pink flowers from Wells County.

N. B. to Nebr., southw. to Fla. and Tex.

2a. **Verbena urticaefolia** var. **leiocárpa** Perry & Fern. (Rhodora 38: 441-443. 1936.) Map 1720. This variety has the habitat of the species and is about as widely distributed in Indiana.

Fernald gives the range as from Conn. to S. C.

3. **Verbena hastàta** L. BLUE VERVAIN. Map 1721. This species is found throughout the state. It prefers a moist soil in the open. It is frequent to common in the lake area in moist places about lakes, in marshes, moist, sandy prairie habitats, interdunal flats, low, open woods, roadside ditches,

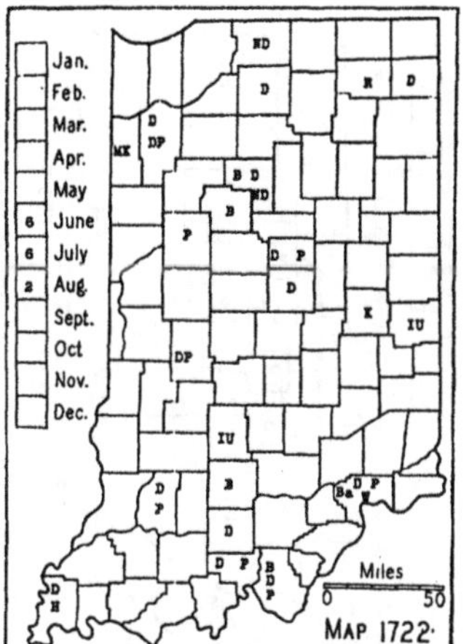

MAP 1722
Verbena simplex Lehm

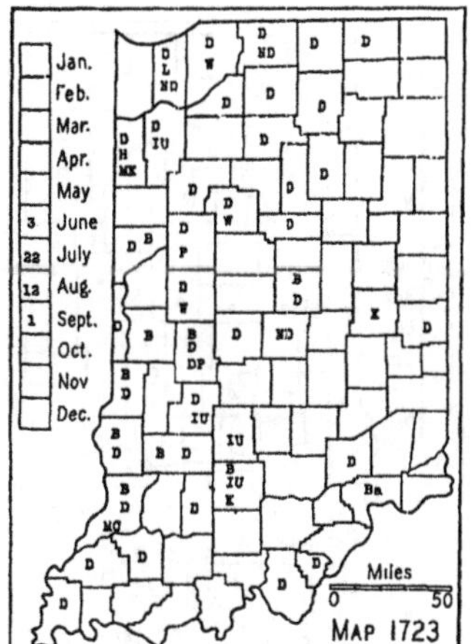

MAP 1723
Verbena stricta Vent

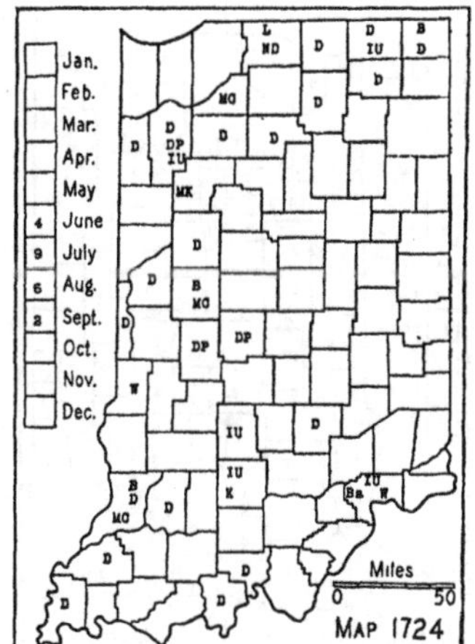

MAP 1724
Verbena bracteata Lag. & Rodr.

and even in the moist, marl border of a lake. It is less frequent south of the lake area and is found in moist places along roadsides, in clearings, fallow fields and low open woods.

I have specimens with white flowers from La Porte and Warrick Counties.

N. S. to B. C., southw. to Fla., Nebr., and Ariz.

3a. × **Verbena Engelmánnii** Moldenke. (*Verbena hastata* × *urticaefolia.*) I collected this hybrid in a prairie habitat along the roadside 2 miles south of Circleville, Clinton County and in Warrick County.

4. **Verbena símplex** Lehm. (*Verbena angustifolia* Michx.) NARROW-LEAF VERVAIN. Map 1722. This species has been reported from 9 counties and doubtless is found infrequently throughout the state. It prefers dry and rather sandy soil in the open, although I have one specimen from a dried-up slough. It is generally found along roadsides and railroads, in fallow fields, and on open, washed, wooded slopes. It is usually found associated with *Verbena stricta.*

Vt., Ont., and Nebr., southw. to Fla. and Okla.

4a. × **Verbena moechìna** Moldenke. This is the name recently proposed for the commonly occurring natural hybrid between *Verbena simplex* and *Verbena stricta.* I have it from Daviess, Harrison, Marion, Orange, and Washington Counties.

5. **Verbena stricta** Vent. HOARY VERVAIN. Map 1723. Found throughout the state although there are no records or specimens from some of the central counties. It is almost exclusively found in very sandy soil along roadsides, rarely along railroads, in sandy pastures, waste places, and fallow fields. I believe it has migrated into northern Indiana, and were it not for the fact that Michaux,[1] who spent August 18, 1795 botanizing along the Wabash River in the vicinity of Vincennes, reported finding

[1] Michaux. Travels west of the Alleghenies. Thwaite's ed. p. 67. 1904.

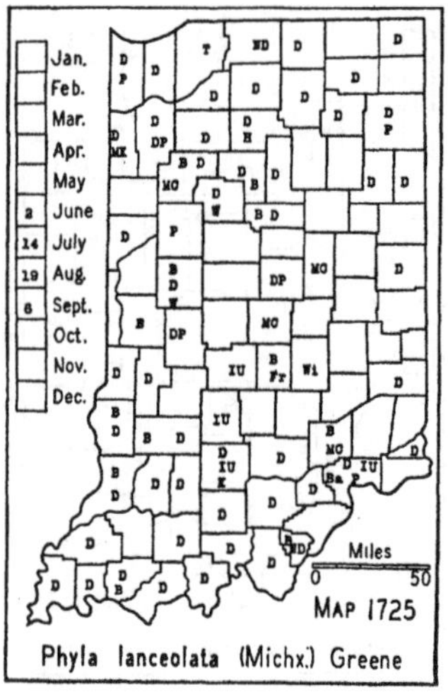

Phyla lanceolata (Michx.) Greene

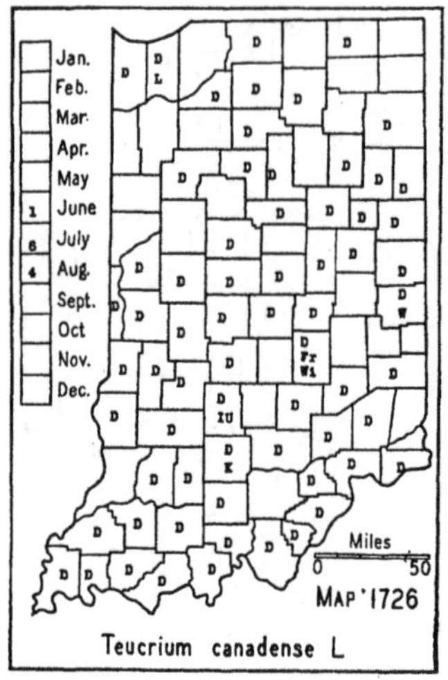

Teucrium canadense L.

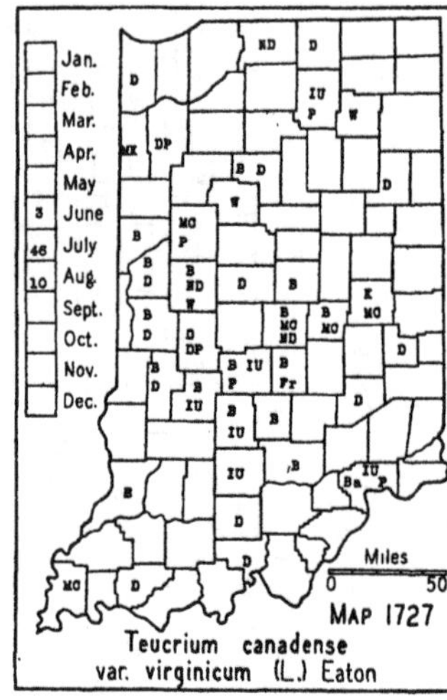

Teucrium canadense var. virginicum (L.) Eaton

Verbena bracteata, Verbena hastata, Verbena stricta, and *Verbena urticaefolia,* I should believe it had invaded the whole state in recent years.
Mass. to Mont., southw. to Okla. and N. Mex.

6. **Verbena bracteàta** Lag. & Rodr. (*Verbena bracteosa* Michx.) LONG-BRACT VERVAIN. Map 1724. This species is an infrequent plant throughout the state in sandy places, mostly along roadsides and in waste places. Sometimes it is found in sandy pastures, in ballast along railroads, and on the slopes of the banks of the Ohio River, especially at boat landings.
Maine to Alberta, southw. to Fla. and westw. to Calif.

6a. × **Verbena Perriàna** Moldenke. (*Verbena bracteata* × *urticaefolia.*) I have specimens of this hybrid from Fulton, Kosciusko, Lagrange, and Lawrence Counties.

7145. PHỲLA Lour.

1. **Phyla lanceolàta** (Michx.) Greene. (Pittonia **4**: 47. 1899.) (*Lippia lanceolata* Michx. and *Lippia lanceolata* Michx. var. *recognita* Fern. & Grisc. Rhodora **37**: 178. 1935.) Map 1725. Found in various habitats throughout the state. Infrequent on the muddy borders of streams, lakes, ponds, and bayous, and in ditches, usually growing with grasses and sedges. When it has competition it does not root at the nodes but when it grows on the muddy borders of banks and on sandbars it becomes a creeping plant up to a yard long, rooting at the nodes.
E. Pa., s. Ont. to Iowa and Nebr., southw. to Fla., Tex., and adjacent Mex.

254. LABIÀTAE B. Juss. MINT FAMILY

A. Ovary of 4 united nutlets; nutlets laterally attached; styles not basal.
Flowers in dense terminal spikes; leaves mostly 4-14 cm long, regularly serrate..7212. TEUCRIUM, p. 800.
Flowers axillary, 1-3 in an axil; leaves 3-8 cm long, entire or with a few irregular teeth.

Calyx nearly regular; corolla about 5 mm long; stamens scarcely exceeding the corolla; seed slightly pubescent at the summit........7217. ISANTHUS, p. 801.

Calyx strongly 2-lipped, the two lower teeth not reaching the base of the 3 upper teeth; corolla more than 5 mm long; stamens exserted; seed glabrous at the summit......................................7218. TRICHOSTEMA, p. 801.

A. Ovary of 4 distinct or nearly distinct nutlets; nutlets basally attached; styles basal.

Calyx 2-lipped, gibbous at the base, both lips entire, not hairy in the throat; stamens 4......................................7234. SCUTELLARIA, p. 802.

Calyx either 2-lipped or regular, 4- or 5-toothed, if gibbous at the base, then hairy in the throat *(Hedeoma)*.

Stamens and style included in the corolla tube; calyx teeth aristate and recurved; flowers in dense axillary clusters...................7238. MARRUBIUM, p. 806.

Stamens and style not included in the corolla tube.

B. Upper lip of corolla concave.

Anther-bearing stamens 4.

Stamens strongly exserted beyond the corolla; tall, coarse herbs; inflorescence of long and usually dense, terminal spikes...7241. AGASTACHE, p. 806.

Stamens not exserted beyond the corolla.

Stems generally 0.5-1.5 m tall; leaves usually 7-15 cm long and 1.5-3 cm wide, sharply and coarsely serrate; inflorescence of closely flowered, long, terminal spikes; flowers purplish, mostly 15-25 mm long...7257. PHYSOSTEGIA, p. 808.

Stems, leaves, and inflorescence not as above.

Plants low and diffuse, stoloniferous, or creeping and rooting at the nodes; at least the lower leaves petioled and cordate.

Flowers axillary, generally 1-3 in an axil; leaves reniform; petioles of about equal length...................7249. GLECOMA, p. 807.

Flowers, leaves, and petioles not as above.

Calyx 15-nerved; flowers large, usually 2-3.5 cm long...7243. MEEHANIA, p. 807.

Calyx about 5-nerved; flowers usually less than 2 cm long...7271. LAMIUM, p. 810.

Plants not as above.

Corolla generally 3-4 cm long, colorless or greenish yellow; leaves on long petioles, cordate at the base.......7259. SYNANDRA, p. 809.

Corolla less than 2.5 cm long, usually pinkish or purplish; leaves rarely cordate at the base.

Plants canescent-pubescent; under surface of leaves velvety to the touch...............................7247. NEPETA, p. 807.

Plants not as above.

Calyx teeth rigid and spine-tipped, spreading at maturity.

Leaves pinnately crenate...........7270. GALEOPSIS, p. 809.

Leaves incised or lobed............7273. LEONURUS, p. 810.

Calyx teeth not rigid and spine-tipped; erect at maturity.

Calyx closed in fruit; bracts large, ovate-orbicular...7254. PRUNELLA, p. 808.

Calyx not closed in fruit; bracts not as above...7281. STACHYS, p. 810.

Anther-bearing stamens 2.

Calyx regular, 15-ribbed, generally hairy in the throat...7296. MONARDA, p. 814.

Calyx 2-lipped, 12-13-ribbed, not hairy within the throat.

Teeth of the calyx of our species not equal; bracts not ciliate with long hairs...................................7290. SALVIA, p. 813.

Teeth of the calyx equal; bracts ciliate with long hairs...7297. BLEPHILIA, p. 816.

B. Upper lip of corolla flat, or the corolla regular.
C. Flowers in axillary whorls or clusters, or these forming terminal spikes. Corolla more or less 2-lipped; upper lip erect or spreading, the lower lip also spreading.
Anther-bearing stamens 2.
Calyx teeth equal; stamens long-exserted........7323. CUNILA, p. 821.
Calyx teeth not equal; stamens not exserted....7302. HEDEOMA, p. 817.
Anther-bearing stamens 4.
Calyx 15-nerved.............................7313. HYSSOPUS, p. 819.
Calyx 10-13 nerved.
Stamens curving more or less, ascending under the upper lip.
Corolla tube curved upward; calyx 13-nerved, not hairy in the throat. ....7304. MELISSA, p. 818.
Corolla tube straight; calyx 10-13-nerved, usually hairy in the throat.............................7305. SATUREJA, p. 818.
Stamens straight.
Plants tall, erect; calyx nearly regular...7317. PYCNANTHEMUM, p. 819.
Plants low, creeping at the base; calyx 2-lipped..7319. THYMUS, p. 820.
Corolla nearly regular, 4- or 5-toothed.
Anther-bearing stamens 2......................7326. LYCOPUS, p. 821.
Anther-bearing stamens 4......................7328. MENTHA, p. 823.
C. Flowers in terminal panicled racemes or spikes; corolla 2-lipped.
Anther-bearing stamens 2; corolla yellow, lower lip not fimbriate; native.. ....7331. COLLINSONIA, p. 826.
Anther-bearing stamens 4; corolla usually purplish, the lower lip fimbriate; introduced.................................7332. PERILLA, p. 826.

7212. TEÙCRIUM [Tourn.] L. GERMANDER

Pubescence of the upper part of stem and of the inflorescence sparse or dense, consisting mostly of recurved hairs about 0.5-0.75 mm long, glandless.
Bracts exceeding the calyx; leaves narrowed at the base..........1. *T. canadense.*
Bracts not exceeding the calyx, equaling it or shorter; leaves usually more or less rounded at the base.......................1a. *T. canadense* var. *virginicum.*
Pubescence of the upper part of stem and of the inflorescence dense, consisting usually of spreading or slightly recurved hairs about 1 mm long or longer.
Hairs of the inflorescence mostly gland-tipped..................2. *T. occidentale.*
Hairs of the inflorescence mostly without glands.......2a. *T. occidentale* var. *boreale.*

1. **Teucrium canadénse** L. (Rhodora 35: 295. 1933.) (*Teucrium littorale* Bickn. and *Teucrium canadense* var. *littorale* (Bickn.) Fern.) AMERICAN GERMANDER. Map 1726. Infrequent throughout the state in moist soil along roadsides, in low, open woods, especially along streams, about lakes, and in fallow and cultivated fields. The amount of pubescence of the stem and calyx varies greatly.

N. B. to Ind., southw. to Fla.

1a. **Tcucrium canadense var. virgínicum** (L.) Eaton. (*Teucrium canadense* of Gray, Man., ed. 7, not L.) Map 1727. Rather frequent throughout the state in habitats similar to those of the species. I admit that the distinction between the species and the variety is not very constant since the bracts of the flowers become progressively shorter toward the end of the raceme. The lowest bracts may be conspicuously longer than the calyx while the remainder may be shorter. Other characters that have been

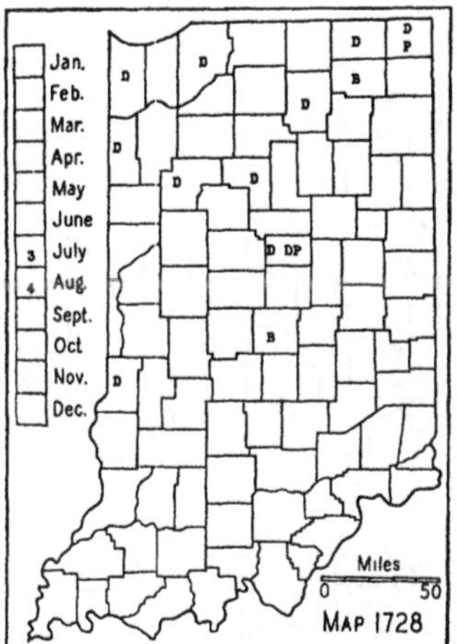

Teucrium occidentale Gray

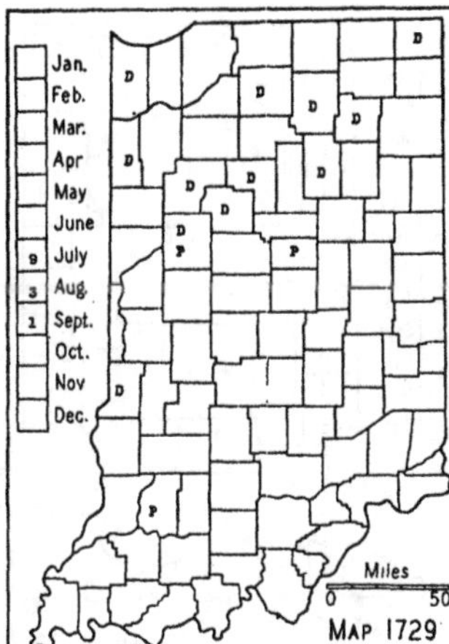

Teucrium occidentale var. boreale (Bickn.) Fern.

Isanthus brachiatus (L.) BSP.

given to separate them are not constant enough to be of much assistance. Probably only a form of the species in our area. The range is probably N. E. to Minn., southw. to Tex.

2. **Teucrium occidentàle** Gray. Map 1728. Infrequent and found mostly in low ground about lakes and in prairie habitats.

Maine to B. C., southw. to Pa., Ohio, Mo., N. Mex., and Calif.

2a. **Teucrium occidentale** var. **boreàle** (Bickn.) Fern. (Rhodora **10**: 85. 1908.) Map 1729. Infrequent to rare in the habitats of the species.

Northern N. H. to Wash., southw. to w. N. Y., Ill., and Tex.

7217. ISÂNTHUS Michx.

1. **Isanthus brachiàtus** (L.) BSP. FALSE PENNYROYAL. Map 1730. My specimens, supplemented by reports from 6 other counties made by other authors, show that this species is an infrequent to rare plant throughout the state. It may, however, be more frequent than our records indicate because it closely resembles the common pennyroyal and may not be distinguished easily. It is generally found in bare, gravelly or sandy places about gravel pits, in old lake beds, along roadsides and railroads, washed places in fallow fields, and on open, wooded slopes. Usually common in large colonies where it is found.

Vt., Que. to Minn., southw. to Ga. and Tex.

7218. TRICHOSTÈMA L.

1. **Trichostema dichótomum** L. BLUECURLS. Map 1731. I have found this rare species on the dry, sandy spill bank of the Kankakee River west of the Tefft Bridge in Jasper County, on a wooded ridge in Harrison County, in the Princeton fine sand in an open black and white oak woods 4 miles south of Vincennes, and on a slight rise in a post and pin oak woods in the "flats" about 4 miles northwest of Chrisney in Spencer County. It

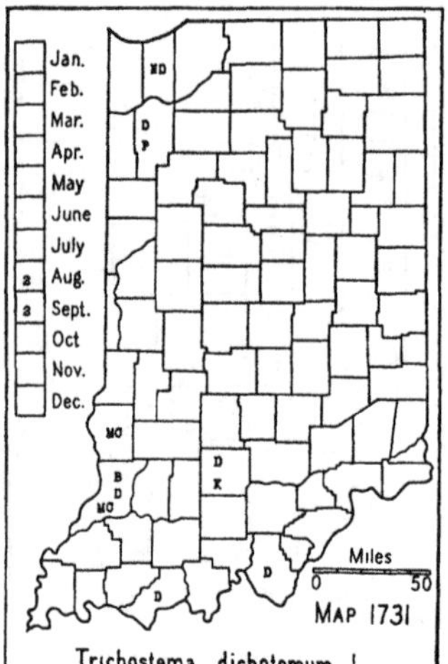

Trichostema dichotomum L.

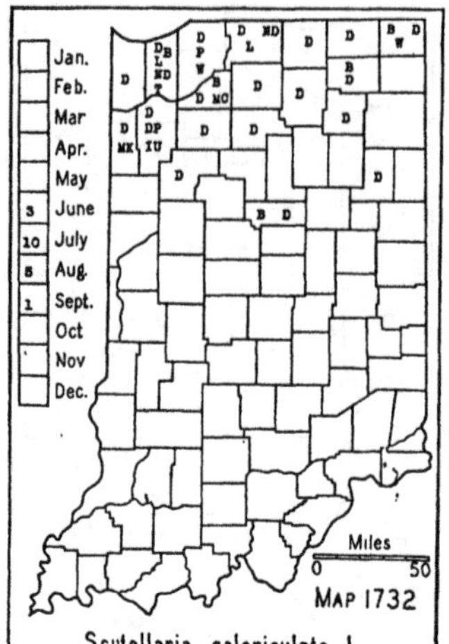

Scutellaria galericulata L.

Scutellaria nervosa Pursh.

has also been collected in Lawrence County by Kriebel and in Porter County by Nieuwland.

Maine to n. Ind., southw. to Fla. and Tex.

7234. SCUTELLÀRIA [Rivin.] L. SKULLCAP

[Penland. Notes on North American Scutellarias. Contr. Gray Herb. Harvard Univ. **71**: 61-79. 1924. Leonard. The North American species of Scutellaria. Contr. U. S. Nation. Herb. **22**: 703-748. 1927.]

Median stem leaves sessile or nearly so or on petioles up to 3 mm long, sometimes the lowest leaves on longer petioles.
  Stems with a retrorse pubescence; plants of a marsh habitat; corollas 17-22 mm long. ....1. *S. galericulata.*
  Stems glabrous or, if pubescent, not as above; plants not of a marsh habitat; corollas less than 15 mm long.
    Median and lower leaves coarsely crenate..........................2. *S. nervosa.*
    Median and lower leaves entire or mostly so.
      Pubescence of stems and pedicels spreading.
        Lower surface of leaves more or less dotted with sessile glands.............. ....3. *S. parvula.*
        Lower surface of leaves not dotted with sessile glands..........4. *S. australis.*
      Pubescence of stem and pedicels upwardly appressed....................... ....5. *S. Leonardi.*
Median stem leaves on petioles more than 5 mm long.
  Internodes of stems and lower surface of leaves glabrous or with only a straggling pubescence on the angles of the stem and on the veins of the blades.
    Inflorescence a terminal loose raceme. (See excluded species no. 525, p. 1084).... ....*S. serrata.*
    Inflorescence not as above.
      Median stem leaves rounded or subcordate at the base, acute to acuminate at the apex; inflorescence in loose axillary racemes; flowers 5-8 mm long; plants of wet places.................................6. *S. lateriflora.*
      Median stem leaves cordate at the base, obtuse to very obtuse at the apex; inflorescence of single flowers in the axils of the leaves; flowers mostly 15-18 mm long; plants not of wet places...................7. *S. saxatilis.*

Scutellaria parvula Michx.

Scutellaria australis (Fassett) Epling

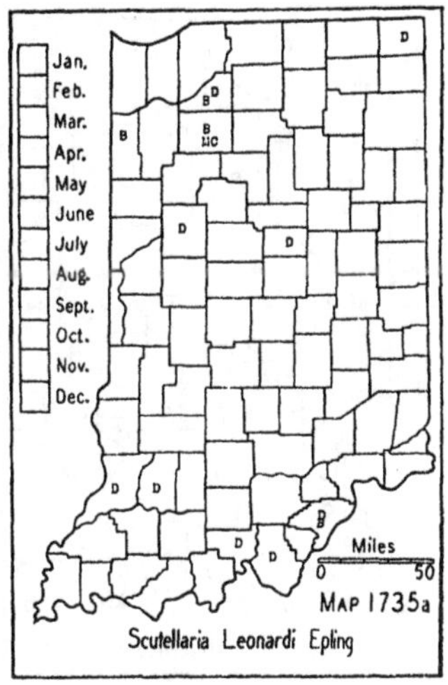

Scutellaria Leonardi Epling

Internodes of stems and under surface of leaves more or less densely pubescent.
Bracts of racemes lanceolate or linear; hairs of inflorescence mostly recurved or appressed; calyx canescent, rarely with a few gland-tipped hairs...8. *S. incana.*
Bracts of racemes ovate or oblong-spatulate; hairs of inflorescence spreading and mostly glandular; calyx densely pubescent with glandular, spreading hairs.
Leaves narrowed or rarely truncate at the base, sometimes the lowest leaves cordate; blades beneath conspicuously and densely covered with oil glands; middle internodes of the stem conspicuously longer; bracts oblong-spatulate. ....9. *S. ovalifolia.*
Leaves deeply cordate at the base; blades beneath not conspicuously and densely covered with oil glands; middle internodes of the stem not noticeably longer; bracts ovate..10. *S. ovata.*

1. **Scutellaria galericulàta** L. (*Scutellaria epilobiifolia* Hamilton; Fernald in Rhodora 23: 85-86. 1921.) Map 1732. Found in marshes about lakes, between dunes, in tamarack bogs, about swamps in woods, and in low borders of dredged ditches. The known specimens of this species restrict its distribution to the lake area of the state. No doubt the report from Clark County should be referred to some other species.

Newf. to B. C., southw. to N. J., Ohio, Nebr., and Ariz.

2. **Scutellaria nervòsa** Pursh. Map 1733. Rather frequent in the southern third of the state in moist soil in woodland, becoming rare northward, and probably absent from our northern counties. It is found in moist woods of all kinds. I have a white-flowered specimen from Spencer County. I have had it in cultivation for many years, and it is so prolific that each year many seedlings must be weeded out.

Pa., Ill. to Mo., southw. to Va., Ala., and La.

3. **Scutellaria párvula** Michx. Map 1734. My specimens are from dry soil on the top of high wooded banks of streams.

Ont. to Iowa, southw. to Ala., La., and Tex.

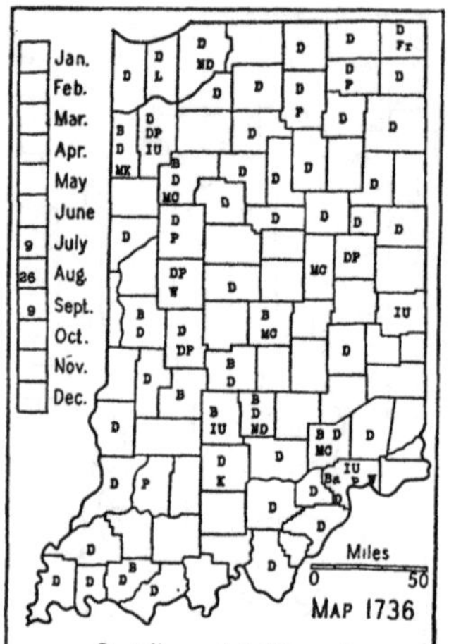

Scutellaria lateriflora L.

Scutellaria saxatilis Riddell

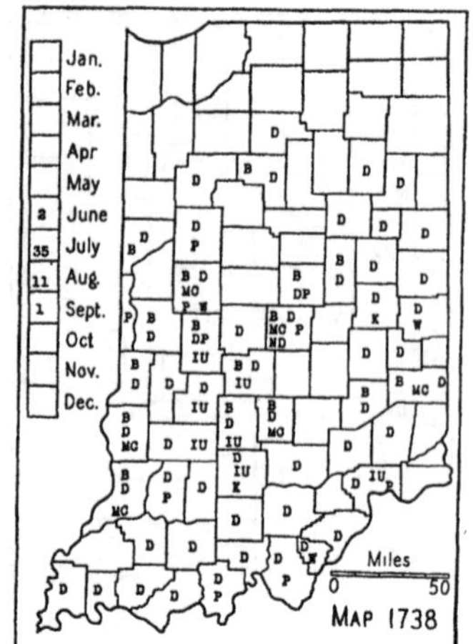

Scutellaria incana Spreng

4. **Scutellaria austràlis** (Fassett) Epling. (Amer. Jour. Bot. **26**: 21-22. 1939.) (Rhodora 39: 378-379. 1937.) Map 1735. On the crests of wooded ridges and in the post oak flats of the southwestern counties.

Ind. to Kans. and Okla., southw. to Tenn., Ala., and e. Tex.

5. **Scutellaria Léonardi** Epling. (Amer. Jour. Bot. **26**: 20-21. 1939.) (*Scutellaria parvula* var. *ambigua* (Nutt.) Fern., Rhodora 3: 198-201. 1901.) Map 1735a. In dry clayey soil on the crests of wooded ridges and in dry, black, sandy soil of prairies.

Maine to N. Dak., southw. to Tenn., Mo., and Kans.

6. **Scutellaria laterifl̀ora** L. SKULLCAP. Map 1736. Frequent throughout the state on the low, wet borders of lakes, ponds, and swamps, in low wet woods, and dredged ditches, and sometimes in roadside ditches. It is commonly found on the inner zone of vegetation of swamps and ponds which become dry in summer.

This species is the one used in medicine, The whole plant is used and 15 grains of the powdered plant is an average dose. It is used as a nervine and tonic.

Newf. to B. C., southw. to Fla. and N. Mex.

7. **Scutellaria saxátilis** Riddell. Map 1737. A rare plant throughout its general range. I have found it only twice. Once I found it in the detritus of a sandstone cliff along Little Blue River at the site of the old Carnes Mill, about 2 miles southeast of Grantsburg, in Crawford County. There are two vigorous colonies here, growing in the shade of the cliff and surrounding trees. I found it again in the detritus at the base of a cliff of the Ohio River, about 2 miles south of Fredonia, in Crawford County. This station is just south of a picnic ground 2 miles south of Fredonia. Here is a small colony, growing in the dense shade of the cliff and woods.

It has been found in the following states: Del., Md., D. C., Va., W. Va., N. C., Ohio, Ind., Ky., Tenn., and Ark.

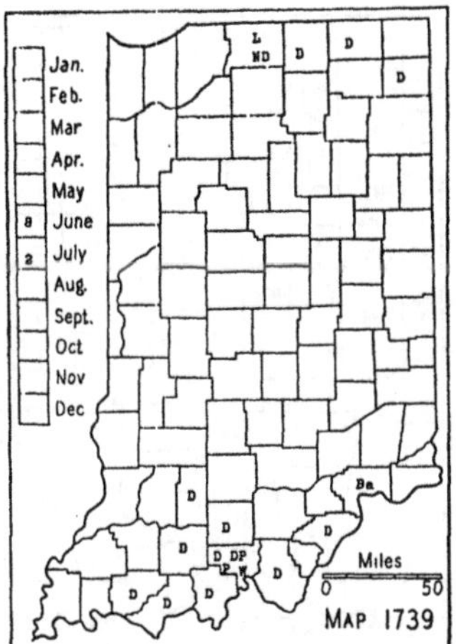

Scutellaria ovalifolia Pers

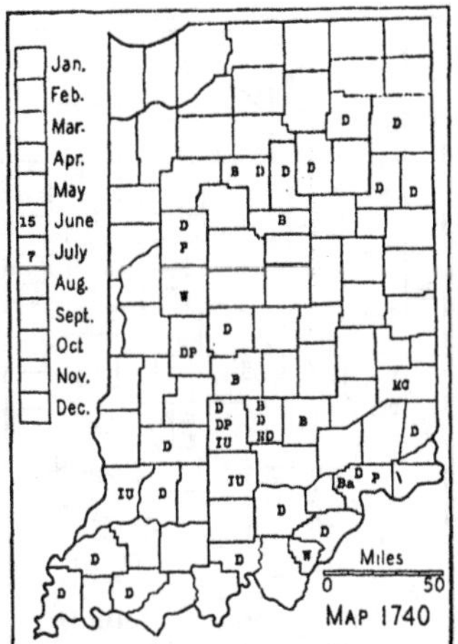

Scutellaria ovata Hill

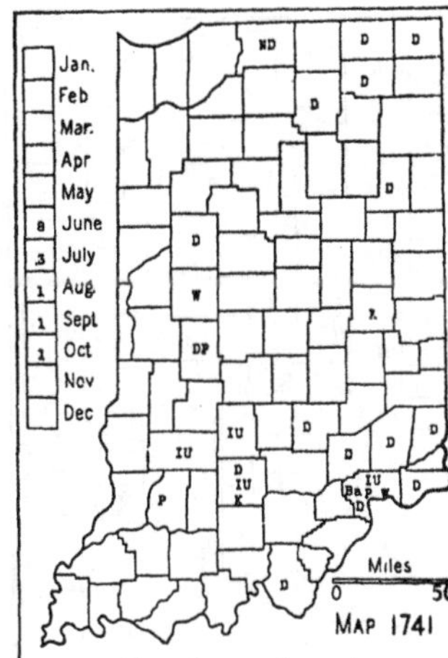

Marrubium vulgare L

8. **Scutellaria incàna** Spreng. (*Scutellaria canescens* Nutt. and *Scutellaria incana* Muhl.) Map 1738. Infrequent throughout the state except the northern part, from which we have no records. Usually found in dry soil in black and white oak and in beech and sugar maple woods. Rarely along roadsides and in wet situations. About half of my specimens from the southern part of the state have stems more or less glandular-pubescent.

Pa., Ont., and Wis., southw. to Fla., and Kans.

9. **Scutellaria ovalifòlia** Pers. (*Scutellaria pilosa* Michx. and *Scutellaria pilosa* var. *hirsuta* (Short) Gray of Coulter's Cat. 1900.) Map 1739. Infrequent to rare in the area of the southern part of the state shown by the map. On black and white oak and beech and sugar maple wooded slopes. Probably entirely absent from most of the Tipton Till Plain and the lake area. I have specimens from dry, white oak woods in De Kalb and Lagrange Counties. Nieuwland collected it in St. Joseph County. It has been reported from 3 other counties, Dearborn, Floyd, and Putnam. The great variation in the length of the pubescence led to the naming of the extreme pubescent form but the species is now regarded as variable enough to include this form.

Southern N. Y. to Mich., southw. to Fla. and Tex.

10. **Scutellaria ovàta** Hill. (*Scutellaria versicolor* Nutt. and *Scutellaria cordifolia* Muhl.) Map 1740. Infrequent in beech and sugar maple and white oak and beech woods throughout the state, although we have no specimens from the northern counties. We have Van Gorder's record from Noble County which is the only one north of the range shown on the map. It is to be noted that this species is a preferred food for insects and it is often very difficult to secure an herbarium specimen which is not badly eaten by them. The bracts of the flowers are variable in size. We have one specimen from Tippecanoe County which has large, broadly cordate bracts about 8 mm. long.

Pa. to Minn., southw. to Fla. and Kans.

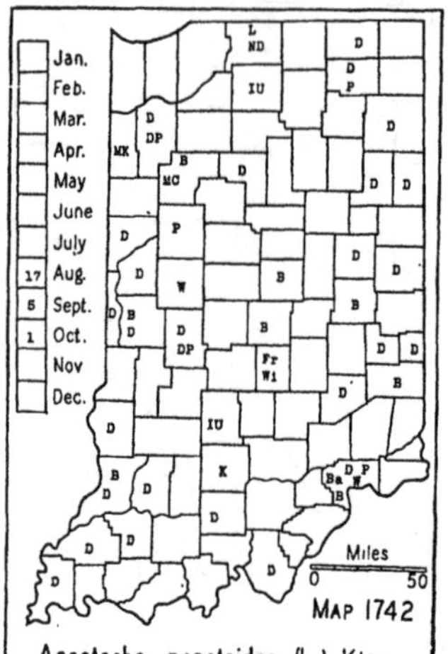

Agastache nepetoides (L.) Ktze

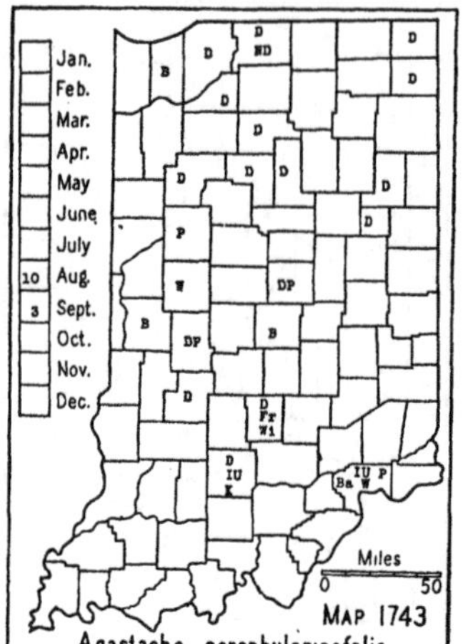

Agastache scrophulariaefolia (Willd.) Ktze.

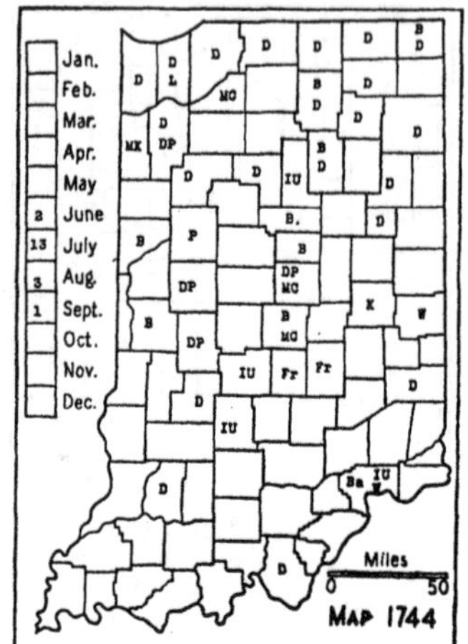

Nepeta Cataria L.

7238. MARRÙBIUM [Tourn.] L.

1. MARRUBIUM VULGÀRE L. COMMON HOREHOUND. Map 1741. This plant has been cultivated for its medicinal properties since pioneer times. It has escaped from gardens to barnyards, roadsides, open woods, and woods pastures.

I have found it to be common in hogyards where the hogs had killed all other vegetation without disturbing this plant.

Nat. of Eu.; naturalized in N. A. from Maine, Ont., B. C., southw. to N. C., Ala., N. Mex., and Calif.

7241. AGÁSTACHE Clayton

Leaves glaucous white and minutely tomentose beneath; flowers blue. (See excluded species no. 526, p. 1084) .... *A. Foeniculum.*

Leaves green beneath and glabrous or short-pubescent; flowers greenish yellow, purplish or pinkish.

Blades more or less densely short-pubescent beneath; bracts broadly ovate, green, gradually acuminate; flowers greenish yellow; calyx 4-5 mm long, the teeth 1-1.5 mm long, green, obtuse or acute........ 1. *A. nepetoides.*

Blades glabrous beneath or with a few hairs along the veins; bracts nearly orbicular, abruptly acuminate, somewhat colorless; flowers purplish or pinkish; calyx 6-7 mm long, the teeth 2-3 mm long, somewhat colorless, long-acuminate........ 2. *A. scrophulariaefolia.*

1. **Agastache nepetoìdes** (L.) Ktze. GIANT HYSSOP. Map 1742. Infrequent to rare and generally found in rather open woods, in moist and usually sandy soil along streams; sometimes in moist open woods and fallow fields, and along roadsides.

E. Mass., w. Que., and S. Dak., southw. to Ga., Tenn., Kans., and Ark.

2. **Agastache scrophulariaefòlia** (Willd.) Ktze. FIGWORT GIANT HYSSOP. Map 1743. A rare plant and found more often in the northern part of the state. It is generally found in somewhat moist and sandy soil in dry, open woods and along roadsides. The flowers vary greatly in color.

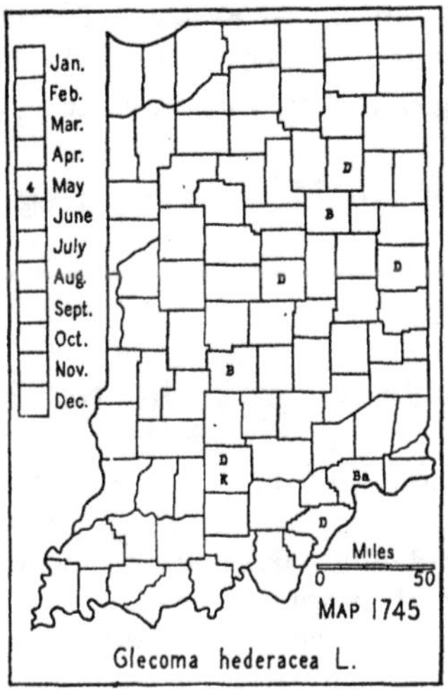

Glecoma hederacea L.

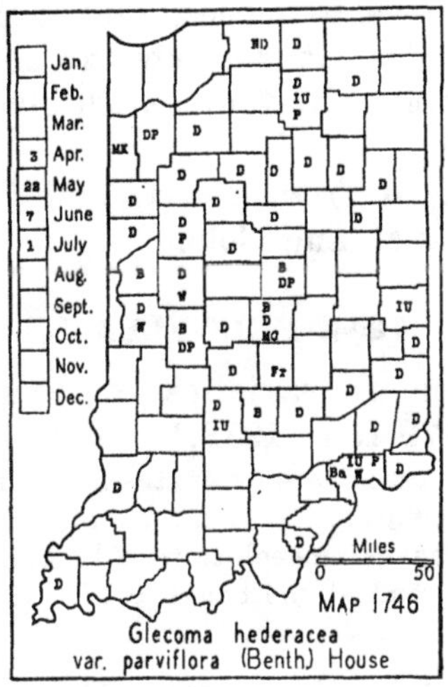

Glecoma hederacea var. parviflora (Benth.) House

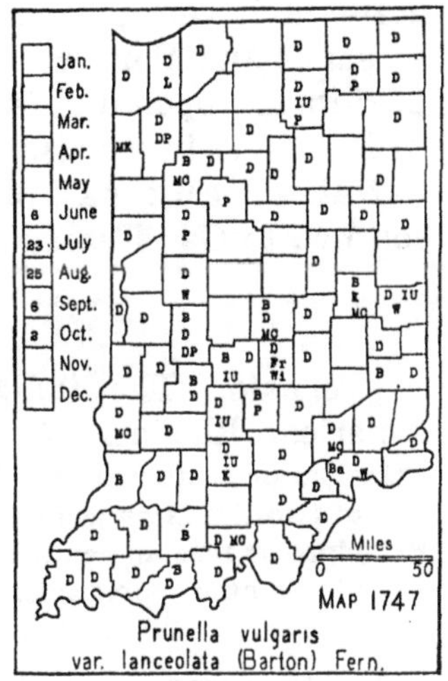

Prunella vulgaris var. lanceolata (Barton) Fern.

They are usually purplish but sometimes nearly white with a tinge of purple or pink.

N. H., Ont., and Wis., southw. to N. C., Ky., and Mo.

7243. MEEHÀNIA Britt.

See excluded species no. 527, p. 1084.

7247. NÉPETA [Rivin.] L.

1. NEPETA CATÀRIA L. CATNIP. Map 1744. Prefers a moist or dry sandy soil and is frequent to common in all parts of the state. Too frequent in waste places about dwellings, truck gardens, pastures, and open woodland. At our home we regard it as an obnoxious weed, and even by persistent efforts can not entirely eradicate it.

Nat. of Eurasia; naturalized from Newf., Que., Oreg., southw. to Ga., Kans., and Utah.

7249. GLECÒMA L.

Corollas mostly 1.6-2.2 cm long.................................1. *G. hederacea.*
Corollas mostly 1-1.5 cm long.......................1a. *G. hederacea* var. *parviflora.*

1. GLECOMA HEDERÀCEA L. (*Nepeta hederacea* (L.) Trev.) LARGE-FLOWER GROUND-IVY. Map 1745. Fernald (Rhodora 23: 289. 1921) separated the large-flowered form of this species from the small-flowered one. Most of our reports for the species were made before the separation was made. These reports show the species to be all over the state. However, my specimens and observations of recent years show that the large-flowered form is rare in the state. Habitat the same as that of the variety.

Nat. of Eu.; naturalized from P. E. I. to Conn. and N. Y. I have not been able to investigate further its general range.

1a. GLECOMA HEDERACEA var. PARVIFLÒRA (Benth.) House. SMALL-FLOWER GROUND-IVY. Map 1746. More or less frequent throughout the state in lawns, gardens, waste places, and moist, open woodland along streams, and along roadsides. It is an obnoxious weed wherever found. It prefers the open and is generally found with bluegrass and herbs, and not in leaf mold in woods.

Nat. of Eu.; naturalized from Newf. and Ont. to Minn. and Oreg., southw. to Ga., Tenn., Kans., and Colo.

7254. PRUNÉLLA L.

[Fernald. The indigenous varieties of Prunella vulgaris in North America. Rhodora 15: 179-186. 1913.]

Principal or median cauline leaves ovate or ovate-oblong, rounded at the base, two fifths to two thirds as wide as long....................................1. *P. vulgaris.*

Principal or median cauline leaves lanceolate to oblong, gradually narrowed or cuneate at the base (sometimes broad at the base), a fifth to half as wide as long........ ....1a. *P. vulgaris* var. *lanceolata.*

1. PRUNELLA VULGÀRIS L. SELFHEAL. This is the European plant and is described by Bentham as having the "stem procumbent or creeping, and rooting at the base, with ascending flowering branches, sometimes 2 or 3 inches, rarely near a foot high." Smith described a variety *minor* as having "stems a span high, erect or ascending, etc." Clute (Amer. Bot. 3: 10. 1902) described *Prunella vulgaris* var. *nana* as a plant of lawns, creeping and rooting at the nodes. I have found this plant an obnoxious weed in a lawn in Bluffton, Indiana. It is also well established in a lawn at 206 Wakewa St., South Bend, St. Joseph County and probably in other places throughout the state. No doubt many of our reports for this species, however, should be referred to the native variety.

Nat. of Eu.; naturalized from Newf. and Que. to Minn., southw. to N. C. and Mex.

1a. **Prunella vulgaris** var. **lanceolàta** (Bart.) Fern. (Rhodora 15: 179-186. 1913.) AMERICAN SELFHEAL. Map 1747. Frequent throughout the state in moist or rather dry open woods, fallow fields, waste places, hayfields, and along roadsides and railroads. It prefers rather sandy and moist soil near streams and in ravines, and is usually found in grassy places. It adapts itself to almost all kinds of soils and situations.

Newf. to B. C., southw. to Fla., La., and Ariz.

7257. PHYSOSTÈGIA Benth. FALSE-DRAGONHEAD

Upper leaves of stem not conspicuously reduced in size; calyx densely puberulent with stiff, straight hairs; flowers 8-20 mm long.

Flowers usually 15-20 mm long; leaves sessile......................1. *P. speciosa.*

Flowers usually 8-15 mm long; leaves petiolate. (See excluded species no. 528, p. 1084)..*P. parviflora.*

Upper leaves of stem greatly reduced in size; calyx densely puberulent with stiff, straight hairs and covered more or less with stipitate glands of about the same length as the hairs (sometimes the glands few); flowers generally 20-30 mm long. ....2. *P. virginiana.*

1. **Physostegia speciòsa** (Sweet) Sweet. (*Physostegia virginiana* in part, of Gray, Man., ed. 7 and *Dracocephalum virginianum* in part, of Brit-

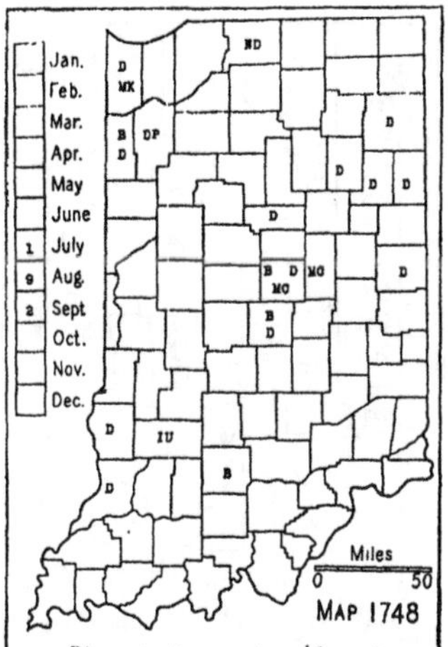

Physostegia speciosa (Sweet)

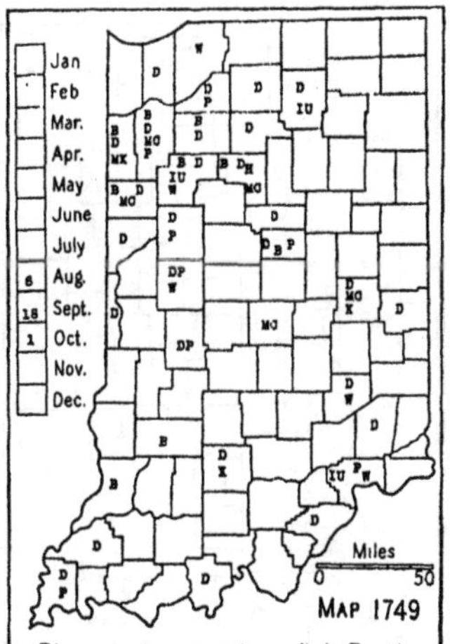

Physostegia virginiana (L.) Benth.

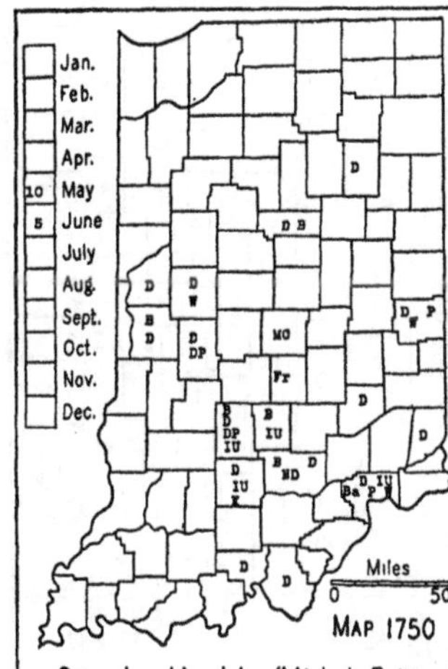

Synandra hispidula (Michx.) Britt.

ton and Brown, Illus. Flora, ed. 2.) TALL CLUSTER FALSE-DRAGONHEAD. Map 1748. Infrequent in moist soil mostly along streams. The distribution of this and the next species is not known because I did not separate the two species before I made my study of the genus. The two species have always been confused and their range is not known. This species flowers about 10 days earlier than the next and is strongly stoloniferous. The general range of the two species is given as follows: Vt., Que. to Minn., southw. to Fla. and Tex.

2. **Physostegia virginiàna** (L.) Benth. (*Physostegia virginiana* in part, of Gray, Man., ed. 7 and *Dracocephalum virginianum* in part, of Britton and Brown, Illus. Flora, ed. 2.) VIRGINIA FALSE-DRAGONHEAD. Map 1749. Infrequent throughout the state in moist, sandy soil in prairie habitats, in moist soil on wooded banks of streams, in moist borders of lakes, and more rarely on rocky, open, wooded slopes.

Both species do well in cultivation in good, black loam soil. They are easily propagated from seed.

7259. SYNÁNDRA Nutt.

1. **Synandra hispídula** (Michx.) Britt. Map 1750. This mint is local and rare in the state but where it is found there are usually several plants in a colony or it is found growing for some distance in its restricted habitat. It grows in deep leaf mold in cool, moist places, usually toward the bases of deep, wooded ravines. Rarely found in level woods where it grows in dense shade.

Ohio to Ill., southw. to Va. and Tenn.

7270. GALEÓPSIS L.

See excluded species no. 530, p. 1084.

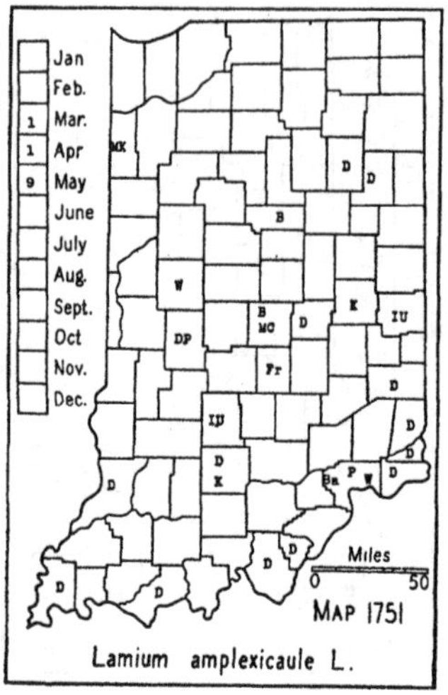

Lamium amplexicaule L.

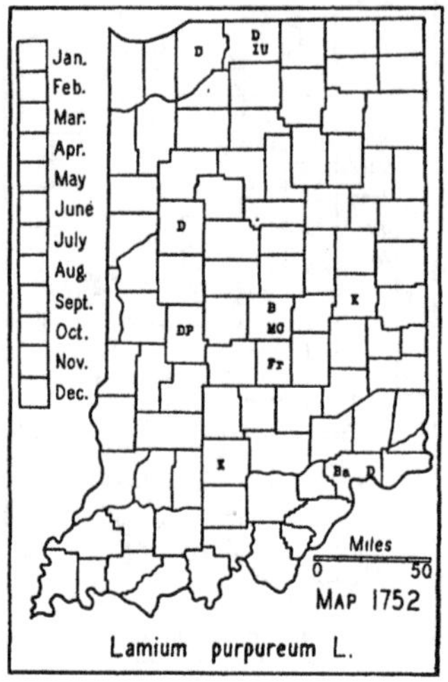

Lamium purpureum L.

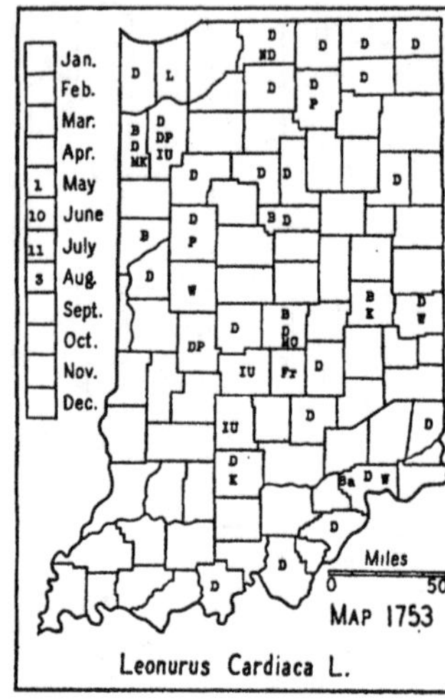

Leonurus Cardiaca L.

7271. LÀMIUM [Tourn.] L. DEADNETTLE

Leaves sessile or clasping on the upper part of the stem..........1. *L. amplexicaule.*
Leaves all petiolate.
  Flowers red or purplish..........................,............2. *L. purpurem.*
  Flowers white. (See excluded species no. 531, p. 1084)..................*L. album.*

1. LAMIUM AMPLEXICÁULE L. HENBIT. Map 1751. Frequent to common, at least in the southern part of the state, in sandy soil. It grows in waste places, gardens, truck gardens, fallow fields, cornfields, pastures and open woods, and along roadsides and railroads.

Nat. of Eurasia; N. B. to B. C., southw. to Fla., Ark., and Calif.

2. LAMIUM PURPÙREUM L. PURPLE DEADNETTLE. Map 1752. We have a single report for this species. I have specimens from four counties: from Jefferson County, frequent along the roadside along a creek northeast of Brooksburg; from St. Joseph County, in a cemetery along the St. Joseph River near South Bend; and from Tippecanoe County from nursery grounds in West Lafayette.

Nat. of Eurasia; Newf., N. C., Pa., and Mo.

7273. LEONÙRUS L. MOTHERWORT

1. LEONURUS CARDÌACA L. COMMON MOTHERWORT. Map 1753. Frequent throughout the state in waste places, fallow fields, and open woods and along roadsides and fences.

Nat. of Eu.; N. S. to N. Dak. and Utah, southw. to N. C. and Kans.

7281. STÀCHYS [Tourn.] L. HEDGE NETTLE[1]

Lower and upper leaves sessile, the median subsessile, rarely on petioles as long as 8 mm.
  Leaves entirely glabrous or thinly hispid with stiff hairs only, neither pubescent nor glandular.

[1] This key and the synonymy were contributed by Carl Epling of the University of California at Los Angeles. He has seen and named all of my specimens.

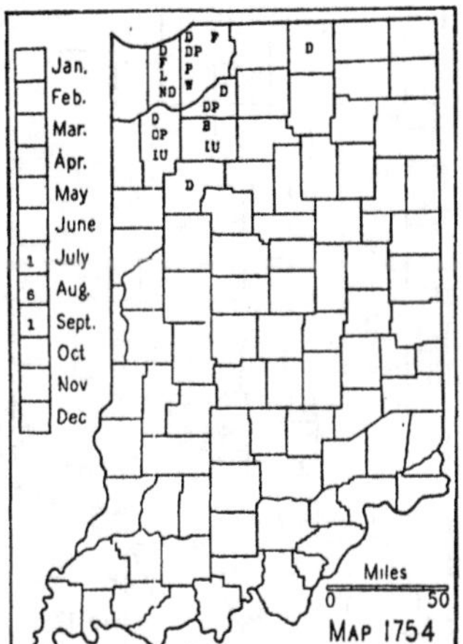

MAP 1754
Stachys hyssopifolia Michx.

MAP 1755
Stachys aspera Michx.

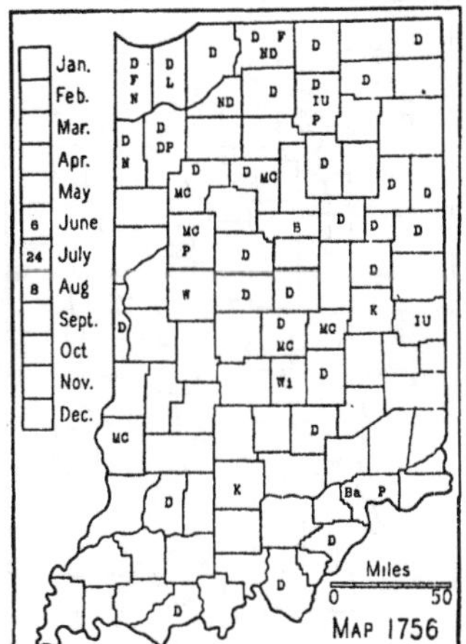

MAP 1756
Stachys hispida Pursh

Calyx essentially glabrous, sometimes bearing a few bristles, especially toward the base; leaves entire or serrulate, essentially glabrous.

Leaves rarely more than 6 mm wide, mostly entire............1. *S. hyssopifolia.*

Leaves mostly 8-12 mm wide, usually serrulate....................2. *S. aspera.*

Calyx definitely hispid; leaves hispid, rarely glabrate, mostly 1.5-4 cm wide...3. *S. hispida.*

Leaves definitely pubescent, even velvety.

Leaves tending to be elliptical and narrowed below the middle, mostly 2-4 cm wide, sometimes wider........................4. *S. palustris* var. *homotricha.*

Leaves tending to be oblong, particularly below the middle, usually less than 2.5 cm wide. (See excluded species, no. 532, p. 1085)................*S. palustris.*

Lower leaves on petioles usually 1-2 cm long, the median ones on petioles 1-7 cm long, the uppermost rarely sessile.

Plants essentially glabrous, sometimes thinly hispid on the margins of the stems. ....5. *S. tenuifolia.*

Plants definitely pubescent.

Stems clothed with both long and short hairs; leaves prevailingly ovate or obovate; calyx teeth deltoid and shorter than the tube..................6. *S. Riddellii.*

Stems clothed on the angles with stiffish hairs, otherwise glabrous; leaves prevailingly oblong or tending to be oblong rather than ovate; calyx teeth in shape similar to those of the preceding species.............7. *S. Clingmanii.*

1. **Stachys hyssopifòlia** Michx. Map 1754. A local plant in the northwestern counties, usually common where it is found. It is usually found in moist, open, sandy places about lakes, and was found once along a sandy roadside. Its habitat is doubtless minimacid because the plants most often associated with it are *Rhexia virginica, Aletris, Hypericum gentianoides, Polygala cruciata, Gaultheria,* and *Vaccinium angustifolium.*

Mass. to Mich., southw. to Fla.

2. **Stachys áspera** Michx. (*Stachys hyssopifolia* var. *ambigua* Gray and *Stachys ambigua* Britt., not Smith.) Map 1755. Our only specimen is from railroad ballast about a quarter of a mile east of Dana in Vermillion County. Common here in one place but not noted again between Dana and Hillsdale, a distance of 6 miles.

Mass. to Wis., southw. to Ga. and Ky.

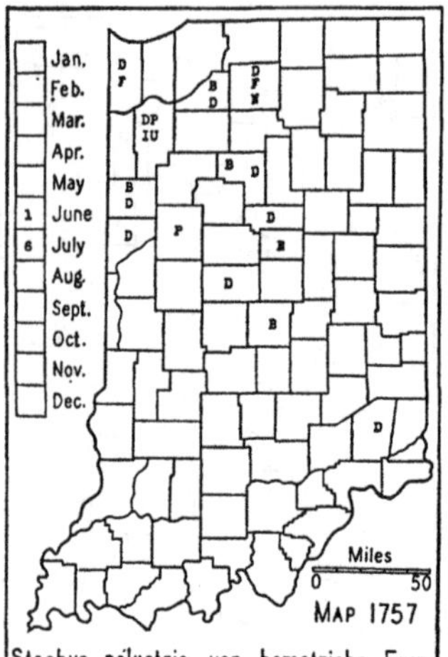

Stachys palustris var. homotricha Fern.

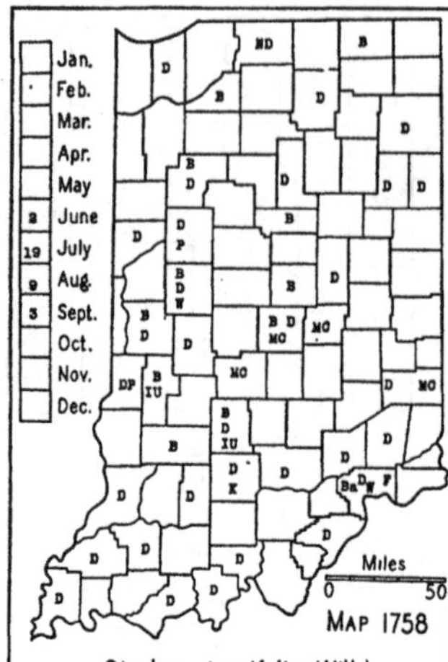

Stachys tenuifolia Willd.

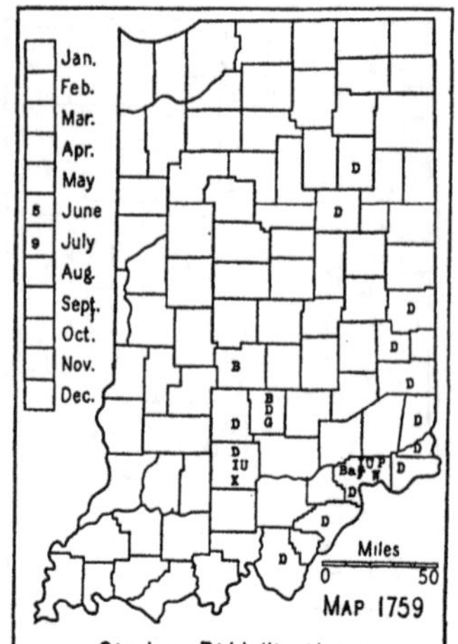

Stachys Riddellii House

3. **Stachys híspida** Pursh. (*Stachys tenuifolia* var. *aspera* (Michx.) Fern. and *Stachys aspera* of authors, not Michx.) Map 1756. Frequent throughout the state in low places in woods, wet borders of lakes, ditches along roadsides and railroads, and sometimes in fallow fields.

E. Mass., Vt., and Ont., southw. to Fla. and La.

4. **Stachys palústris** L. var. **homótricha** Fern. (Rhodora **10**: 85. 1908.) Map 1757. An infrequent and local plant. It prefers a moist or wet sandy habitat, especially a prairie. Found in open places about lakes and along roadsides and railroads. My Ripley County specimen was found along the B. & O. Railroad one mile west of Osgood.

Newf. to Mackenzie, southw. to N. C., Ohio, Ill., and Colo.

5. **Stachys tenuifòlia** Willd. Map 1758. Infrequent throughout our area. Generally found in low woods, moist ravines, rarely in open places and on banks of streams. When it grows in very dense shade, the plant usually becomes weak and decumbent and is more branched than when it grows in open places or in the sun.

N. Y. to Iowa and Kans., southw. to N. C. and La.

6. **Stachys Riddéllii** House. (*Stachys cordata* of Gray, Man., ed. 7 and Britton and Brown, Illus. Flora, ed. 2.) Map 1759. A rare plant found usually on moist or dry wooded slopes.

Ohio, Ind., and Ill., southw. to N. C. and Tenn.

7. **Stachys Clíngmanii** Small. (Small, Flora Southeastern United States, p. 1032. 1903.) Map 1760. This is a local species of various habitats. I have specimens from dry oak slopes, moist sugar maple and beech woods, and from hard white clay soil in a sweet gum "flat" in Clark County where I found specimens 5 feet high.

W. Va., Ind., and Ill., southw. to N. C. and Tenn.

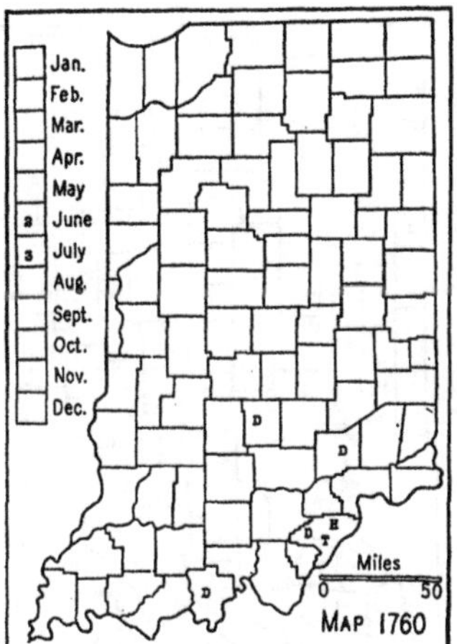

Stachys Clingmanii Small

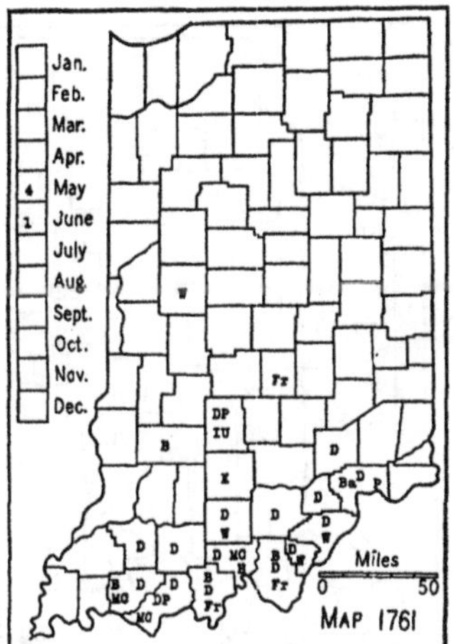

Salvia lyrata L.

Salvia reflexa Hornem

7290. SÁLVIA [Tourn.] L. SAGE

Stems with 1-3 pairs of leaves, the leaves mostly basal....................1. *S. lyrata.*
Stems leafy, bearing more than 3 pairs of leaves.
 Leaves narrowly oblong or lanceolate, less than 2 cm wide.
  Annual; pubescence of pedicels retrorsely appressed................2. *S. reflexa.*
  Perennial; pubescence of pedicels spreading. (See excluded species no. 533, p. 1085) ....*S. Pitcheri.*
 Leaves oblong-lanceolate, ovate or oval, more than 2 cm wide.
  Stems glandular-pubescent above. (See excluded species no. 535, p. 1085)........ ....*S. urticaefolia.*
  Stems not glandular-pubescent.
   Corollas about 1 cm long.
    Pedicels shorter than the calyx..............................3. *S. sylvestris.*
    Pedicels about as long as the calyx. (See excluded species no. 536, p. 1085)... ....*S. verticillata.*
   Corollas about 2 cm long. (See excluded species no. 534, p. 1085)............. ....*S. pratensis.*

1. **Salvia lyràta** L. LYRELEAF SAGE. Map 1761. Infrequent in a few of the south central counties. It is usually found in dry soil but sometimes in moist soil in the "flats." It is generally found in open, bare places on slopes in woods associated with black and white oak, or with beech. It was noted as a weed in a woods pasture about a mile east of Charlestown in Clark County. Here the soil was shallow and underlaid by limestone.

Conn. to Ill., southw. to Fla. and Tex.

2. SALVIA REFLÉXA Hornem. (Fedde Rept. Spec. Nov. **110**: 102. 1938.) (*Salvia lanceaefolia* Poir.) Map 1762. This western annual has been found established in a few counties. I found it to be well established in sandy soil in a barnyard about half a mile northeast of Leiter's Ford in Fulton County, and in a sandy truck garden along the Wabash River in Gibson County.

Ohio to Mont., southw. to Ariz. and Mex.

3. SALVIA SYLVÉSTRIS L. Map 1763. I found this sage in a pasture field about a half mile north of Culver, Marshall County in 1920 and in 1937

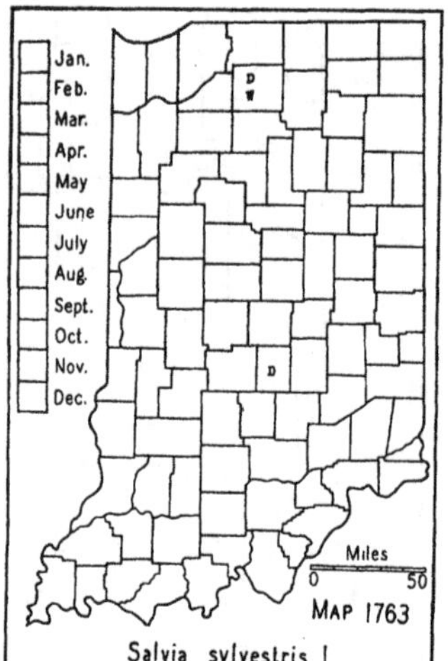

Salvia sylvestris L.

Monarda Bradburiana Beck

Monarda clinopodia L.

it still persisted there. In 1936 Ralph M. Kriebel found it as a common weed along State Road 37 near Waverly, Morgan County. He informed me it was well established in an adjacent field also. J. E. Potzger told me that he had observed it as a common weed at the same place for several years.

Native of Eurasia.

7296. MONÁRDA L. Beebalm

Calyx lobes generally 2.5-3 mm long; leaves sessile or nearly so....1. *M. Bradburiana.*
Calyx lobes generally less than 2.5 mm long.
- Flower clusters on long, terminal peduncles (rarely some clusters from the upper axils, these on long peduncles); anthers exserted; leaves petiolate.
  - Calyx without hairs in the throat; corolla bright red, showy. (See excluded species no. 537, p. 1085)..*M. didyma.*
  - Calyx with hairs in the throat; corolla white, yellowish white or purplish.
    - Corolla white or yellowish white; stems generally simple, not branched; flowering mostly in June....................................2. *M. clinopodia.*
    - Corolla purplish; stems branched, rarely simple; flowering mostly in July or later.
      - Stem crisp-pubescent and more or less villous-hirsute..........3. *M. fistulosa.*
      - Stem crisp-puberulent only......................3a. *M. fistulosa* var. *mollis.*
- Flower clusters all axillary; anthers not exserted.....4. *M. punctata* var. *villicaulis.*

1. **Monarda Bradburiàna** Beck. Map 1764. A very rare plant in this state and probably confused with other species. Our older manuals did not recognize *Monarda clinopodia* and when the flowers lose their color, the species are separated with difficulty. It has been reported from Clark, Franklin, Jefferson, and Putnam Counties, for the area of Delaware, Jay, Randolph, and Wayne Counties, and for the area of the Lower Wabash Valley. All of these reports were made before 1886. Schneck, in 1872, says for the Lower Wabash Valley: Open woods and fence-rows, common. I have it from clayey roadsides and dry wooded slopes from Knox and Posey Counties.

Ind. to Kans., southw. to Ala., Tenn., and Ark.

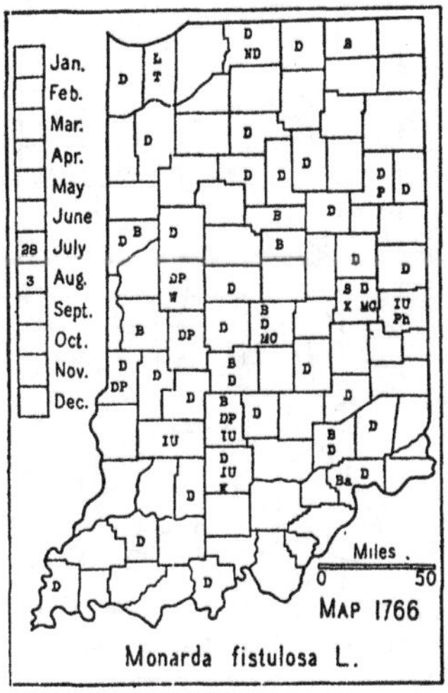

MAP 1766
Monarda fistulosa L.

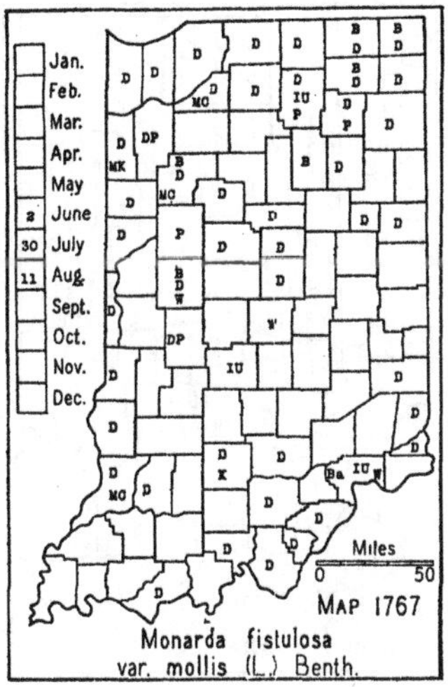

MAP 1767
Monarda fistulosa var. mollis (L.) Benth.

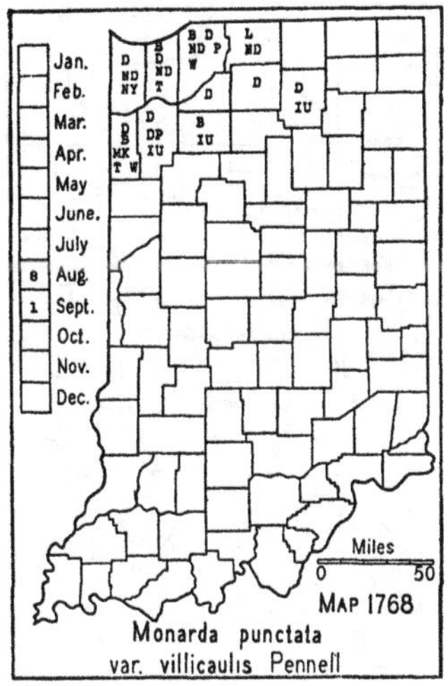

MAP 1768
Monarda punctata var. villicaulis Pennell

2. **Monarda clinopòdia** L. Map 1765. Infrequent in a few of the southern counties. It is usually found in dry, white and black oak woods and less frequently in beech woods. The flowers are generally white or yellowish white.

Ont., N. Y. to Ill., southw. to Ga. and Ky.

3. **Monarda fistulòsa** L. WILD BERGAMOT. Map 1766. Infrequent to frequent throughout the state. In moist sandy soil in prairie habitats or in extinct lake bottoms, it is often abundant over several acres. It prefers a moist, sandy soil but is found also in dry, gravelly soil and on rocky wooded slopes. It is generally found in moist places along streams, usually in open woods; in deep wooded ravines, in fallow fields, on open rocky wooded slopes, and along roadsides and railroads.

A careful study of my 75 specimens shows that I have specimens of the typical form of the species and its variety. Nearly all are intermediate in the kind and quantity of pubescence; plants typical of the variety will have some villous hairs about the nodes or on the petioles. The calyx tube varies from 5 to 9 mm long, its teeth from 1 to 2 mm long, and the surface is more or less densely glandular-puberulent. The leaves vary from ovate with rounded bases to those that are ovate-lanceolate to lanceolate with rounded, truncate or cuneate bases. Ordinarily the deep woods forms have broad leaves while those of dry habitats have narrower leaves. The color of the flowers also varies from a light to a deep purple.

Maine to Minn., southw. to Fla. and Tex.

3a. **Monarda fistulosa** var. **móllis** (L.) Benth. (*Monarda mollis* L.) HAIRY WILD BERGAMOT. Map 1767. The habitat and distribution are the same as that of the species. The under surface of the leaves of the typical form is velvety to the touch. The pubescence of the branches of the stem and under surface of the leaves in the variety is densely canescent and the hairs on the stem are not at all spreading.

4. **Monardo punctàta** L. var. **villicáulis** Pennell. (Bull. Torrey Bot. Club 46: 186. 1919.) (*Monarda punctata* L. of Gray, Man., ed. 7 and Britton

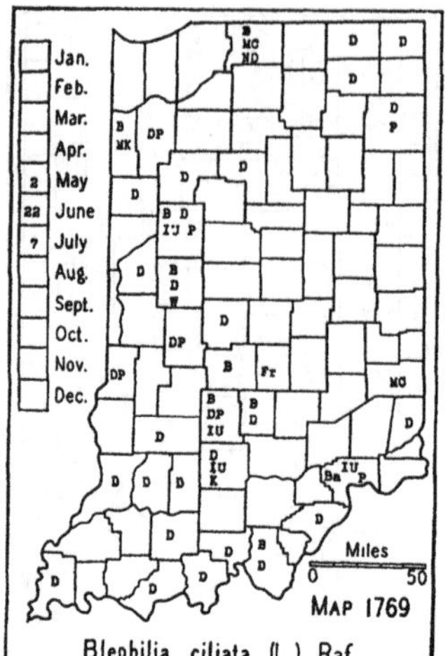

Blephilia ciliata (L.) Raf.

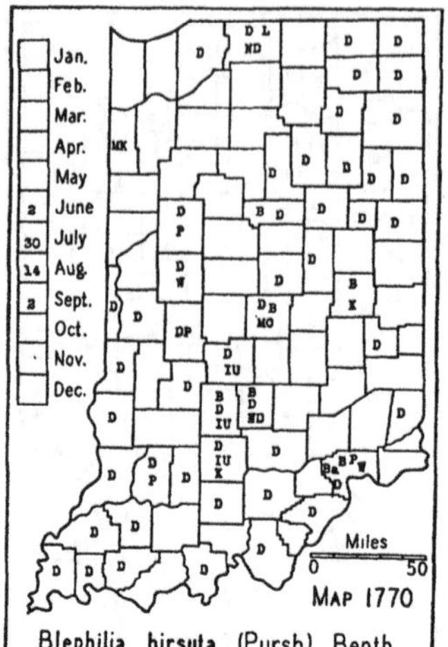

Blephilia hirsuta (Pursh) Benth.

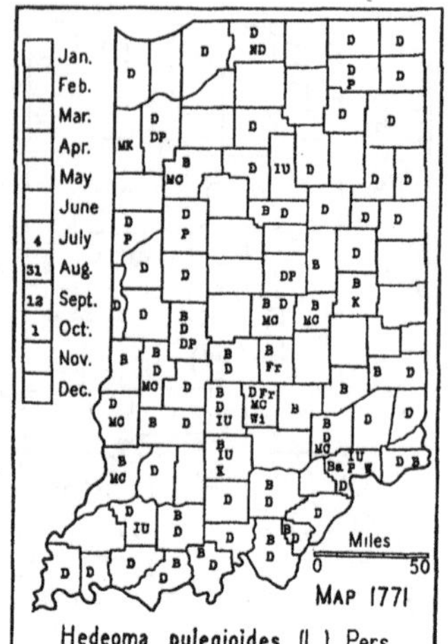

Hedeoma pulegioides (L.) Pers.

and Brown, Illus. Flora, ed. 2.) HORSEMINT. Map 1768. Found in the open on sandy knolls and dunes. Where it is found, it is usually abundant. Found also by Chas. M. Ek in Kokomo, Howard County, on the siding of the old plate glass works. Introduced here from glass sand.

The herb and oil have long been used in medicine.

N. Y. to Minn., southw. to Fla. and Tex.

7297. BLEPHÍLIA Raf. BLEPHILIA

Stems generally simple; leaves usually only slightly serrate; upper leaves generally lanceolate or oblong, sessile or on petioles up to 5 mm long; outer bracts ovate, acute or acuminate, colored, as long as the calyx; corolla generally hairy............ 1. *B. ciliata.*

Stems generally with 2 or more branches; leaves generally sharply serrate; upper leaves ovate to ovate-lanceolate, on petioles generally 10-15 mm long; outer bracts mostly linear, with aristate tips usually shorter than the calyx; corolla generally not hairy............ 2. *B. hirsuta.*

1. **Blephìlia ciliàta** (L.) Raf. Map 1769. Found throughout the state although we have no reports for the counties bordering Lake Michigan. This is a species generally of open dry places but sometimes it is found in moist places in dense shade such as the base of wooded ravines where it develops long, stoloniferous branches which root at each node. These creeping branches have leaves which vary greatly in shape, some truncate and even cordate at the base. Blatchley had such a specimen from Monroe County, which I now have, which he reported to be *Meehania cordata.* The specimen is the creeping form of this species which had not yet developed a flowering head. Also when it grows in dense shade it sometimes develops a pubescence much like that of the next species. This species rarely develops branches. I have one specimen with axillary heads on peduncles up to 5 cm long. I have an albino specimen from Noble County. I recommend this species highly for cultivation both for its beauty and for its long flowering period.

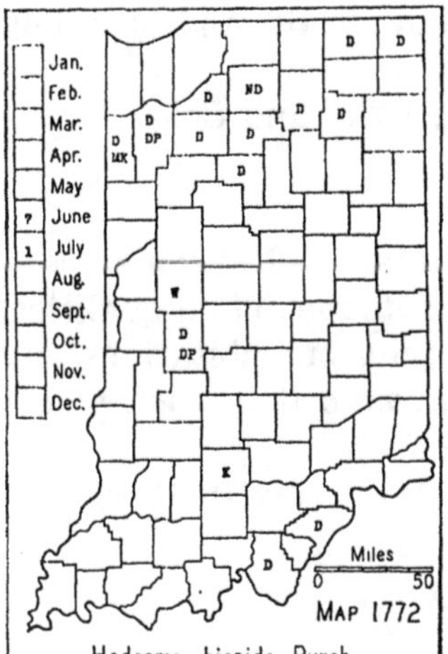

Hedeoma hispida Pursh

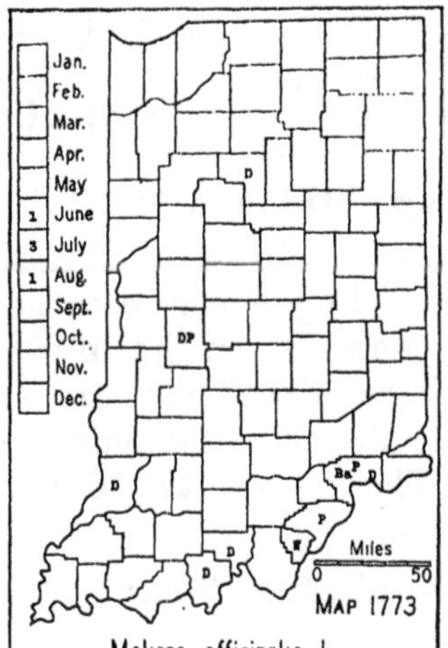

Melissa officinalis L.

Satureja vulgaris (L.) Fritsch

It is generally found in dry open woods, clearings, fallow fields, and along roadsides.

Vt. to Minn., southw. to Ga., Ala., and Mo.

2. **Blephilia hirsùta** (Pursh) Benth. Map 1770. This is strictly a woodland species and is very rarely found elsewhere. Found throughout the state although we have no records for the northwestern part of the state west of La Porte County. It is generally found in rich woods in deep leaf mold. It is notable that the foliage is eaten by insects to such an extent that whole specimens are usually difficult to obtain.

Vt., Que., and Minn., southw. to Ga. and Tex.

7302. HEDEÒMA Pers.

Leaves petiolate, ovate to linear-lanceolate, mostly 3-12 mm wide, sometimes up to 20 mm wide, at least the wider ones serrate; bracts mostly shorter than the pedicels, obtuse; teeth of the upper lip generally triangular, acute; seed smooth, broadly ovate, generally slightly less than 1 mm long...........1. *H. pulegioides.*

Leaves sessile or the lowest petiolate, linear, 1.5-3 mm wide, rarely wider, entire; bracts linear-subulate, generally longer than the pedicels; upper teeth of calyx generally subulate; seed oblong, generally 1-1.3 mm long, surface reticulate under a magnification of 20 diameters...................................2. *H. hispida.*

1. **Hedeoma pulegioìdes** (L.) Pers. AMERICAN PENNYROYAL. Map 1771. Frequent to common in all parts of the state and probably found in every county. This species prefers dry soil and is found in dry, open woods of all kinds, sometimes in low woods, fallow fields and along roadsides and railroads. It usually flowers about two weeks or more later than the next species.

N. S., Que., and N. Dak., southw. to Fla., Ala. and Ark.

2. **Hedeoma híspida** Pursh. ROUGH PENNYROYAL. Map 1772. Infrequent in Indiana. The plant is inconspicuous and no doubt is more common than our map indicates. It is found only in dry, usually very sandy soil, in rather acid habitats in open black oak woods, open wooded crests of ridges,

barren places in fallow, clayey fields, in sandy, fallow fields, and on sandy spill-banks of dredged ditches.

Conn., N. Y., Ill. to Sask., southw. to Tenn., La., Ark., and Colo.

7304. MELÍSSA [Tourn.] L.

1. MELISSA OFFICINÀLIS L. COMMON BALM. Map 1773. This plant was frequently cultivated by the pioneers because of its medicinal qualities. It has, in some instances, persisted on the sites of pioneer habitations and sometimes escaped, especially to roadsides. I have found it in such places and also on open, wooded hills near the Ohio River.

Nat. of Eu.; Maine to Fla., westw. to Mo. and Ark.; also in Oreg. and Calif.

7305. SATURÈJA [Tourn.] L. SAVORY

Plants puberulent or pubescent.
- Leaves linear or linear-oblong; bracts shorter than the pedicels; annuals.......... ....1. *S. hortensis.*
- Leaves ovate; bracts shorter or longer than the pedicels; perennials.
  - Bracts very small, linear, shorter than the pedicels, introduced plant. (See excluded species no. 539, p. 1086)..................................*S. Nepeta.*
  - Bracts setaceous, longer than the pedicels; native plant............2. *S. vulgaris.*

Plants glabrous.
- Leaves linear or the lowest spatulate, those of the runners broader and longer petioled; stem leaves generally entire; corolla less than 10 mm long.......... ....3. *S. glabra.*
- Leaves oblong or elliptic, sparingly serrate, short-petioled; corolla more than 10 mm long. (See excluded species no. 538, p. 1086)..................*S. glabella.*

1. SATUREJA HORTÉNSIS L. SUMMER SAVORY. This species has been rather extensively cultivated as a kitchen herb and for its medicinal properties and no doubt it sometimes escapes. It was reported for Clark County in 1878 and for Porter County in 1930 by Lyon. I have seen his specimen. Evidently it is only an occasional escape or the reports would be more numerous.

Nat. of Eu.; N. B. to Mich. and Ky.

2. **Satureja vulgàris** (L.) Fritsch. BASIL. Map 1774. In wet woods and moist roadsides. Rare. Probably introduced.

Newf. to Man., southw. to Mass., Va., and Ind.; also in Eurasia.

3. **Satureja glàbra** (Nutt.) Fern.* Map 1775. Common in moist sandy soil on the dune just south of Pine, in Lake County; local in the crevices and in the talus at the base of the limestone escarpment of the Wabash River below Logansport and Georgetown in Cass County; and in the Elliott's Mill Bog about 4 miles southeast of Richmond. It has also been reported for Porter County and for Clark and Jefferson Counties. This species is easily cultivated and because of its stoloniferous habit, it soon spreads and covers the ground or rocks on which it is planted. We have had it in cultivation for several years and it is perfectly hardy and is admired by visitors.

Ont. and w. N. Y. to Minn., southw. to s. Ind., Mo., Ark., and Tex.

---

* The latest name proposed for this plant is **Satureja glabella** var. **angustifolia** (Torr.) Svenson. (Rhodora **42**: 7. 1940.)

MAP 1775
Satureja glabra (Nutt.) Fern.

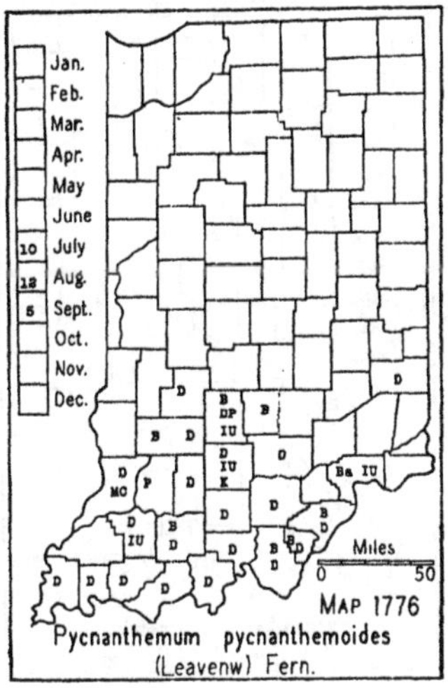

MAP 1776
Pycnanthemum pycnanthemoides (Leavenw.) Fern.

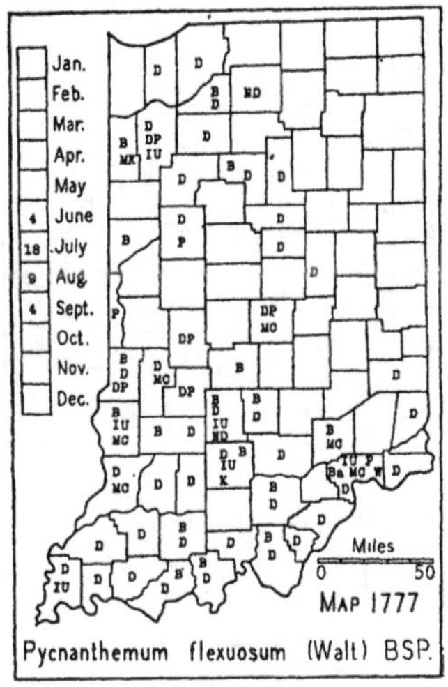

MAP 1777
Pycnanthemum flexuosum (Walt.) BSP.

7313. HYSSÒPUS [Tourn.] L. HYSSOP

See excluded species no. 540, p. 1086.

7317. PYCNÁNTHEMUM Michx. MOUNTAIN-MINT

Leaves ovate to ovate-lanceolate, the upper floral ones whitened beneath and mostly 1-2 cm wide; fruiting cymose clusters generally loose and 1-2 cm wide or up to 4 cm wide.
- Lower pair of calyx teeth 1.5-2.5 mm long, upper pair a third to half as long........ ....1. *P. pycnanthemoides.*
- Lower pair of calyx teeth 1-1.5 mm long, upper pair a half to two thirds as long. (See excluded species no. 542, p. 1086)..........................*P. incanum.*

Leaves linear to lanceolate, the floral ones not whitened and less than 1 cm wide; fruiting heads less than 1 cm wide; calyx teeth nearly glabrous to densely pubescent but never with long hairs (1-2 mm long).
- Stems, upper branches, and leaves glabrous, rarely the upper part of the stem and branches minutely puberulent or scabrous, sometimes the margins of the leaves scabrous; blades mostly 1-3.5 mm wide; calyx teeth triangular-lanceolate, long-acuminate, nearly glabrous to puberulent or pubescent, not long and densely pubescent to the apex; corolla about as long as the calyx tube....2. *P. flexuosum.*
- Stems, upper branches, and leaves pubescent, generally densely so, and very rarely the pubescence scant; bracts and calyx teeth generally woolly- or matted-pubescent.
  - Stems pubescent on the angles only; upper leaves canescent or densely puberulent above; median leaves generally densely pubescent beneath, the blades mostly 5-15 mm wide; calyx teeth generally triangular-lanceolate, acuminate, generally densely pubescent to the apex................................3. *P. pilosum.*
  - Stems pilose with spreading hairs; upper leaves green and glabrous or nearly so above; median leaves more or less pubescent on the principal veins beneath, rarely nearly glabrous, the blades mostly 2-10 mm wide; calyx teeth generally short, triangular-ovate or somewhat narrower, not usually densely pubescent to the apex..4. *P. virginianum.*

1. **Pycnanthemum pycnanthemoìdes** (Leavenw.) Fern. (*Koellia pycnanthemoides* (Leavenw.) Kuntze.) Map 1776. In dry soil in open woods and fallow fields and along roadsides. This is a very conspicuous plant, usually about a meter high. It is restricted to the southern part of the state and has been reported as *Pycnanthemum incanum.*

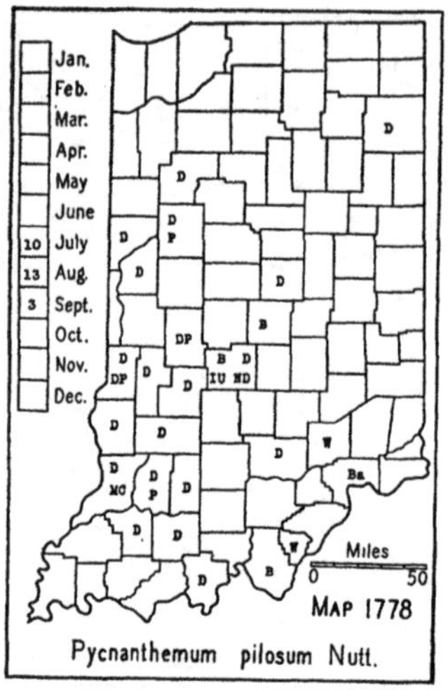

Pycnanthemum pilosum Nutt.

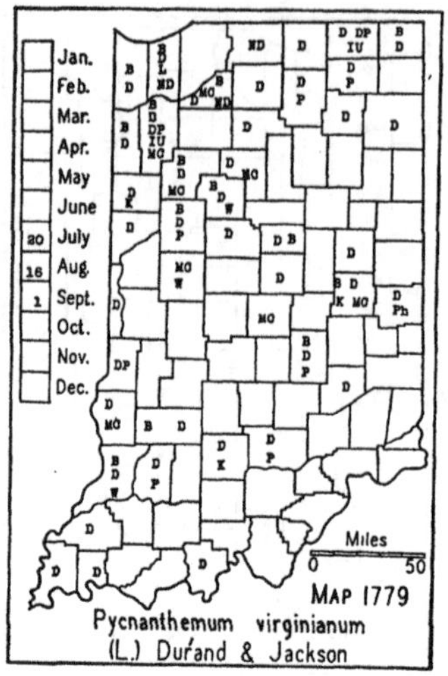

Pycnanthemum virginianum (L.) Durand & Jackson

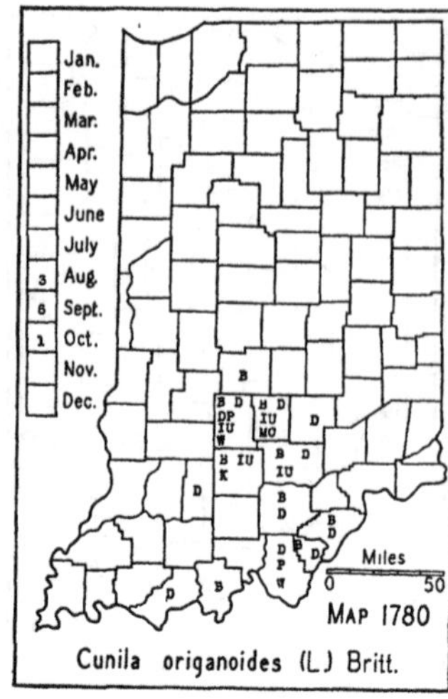

Cunila origanoides (L.) Britt.

I was informed by a pioneer of Perry County that he was cured of incontinence of urine by a tea made of this plant.

Va. to Ind., southw. to Ga. and Tenn.

2. **Pycnanthemum flexuòsum** (Walt.) BSP. (*Koellia flexuosa* (Walt.) MacM.) SLENDER MOUNTAIN-MINT. Map 1777. This species has a wide range of habitats. It is generally found in dry soil on the crests and slopes of black and white oak ridges, in dry fallow fields, and in dry soil along roadsides and railroads. It is also found in moister situations but usually in sandy or hard, sandy, clay soil along roadsides and in prairie habitats.

This species is likely to be confused with *Pycanthemum virginianum,* from which it may be distinguished easily by being glabrous or nearly so, and by having long, glabrous calyx teeth. *Pycnanthemum virginianum* is generally pubescent, and its calyx teeth are merely acute and generally long white-pubescent to the tip.

Cent. Maine to Minn., southw. to Fla. and Tex.

3. **Pycnanthemum pilòsum** Nutt. (*Koellia pilosa* (Nutt.) Britton of Britton and Brown, Illus. Flora, ed. 2.) HAIRY MOUNTAIN-MINT. Map 1778. Generally in dry sandy soil, along roadsides, in open woods, and rarely in moist places.

This species is a great favorite of the honey bee.

Pa., Ont., and Iowa, southw. to Ga., Ark., and Kans.

4. **Pycnanthemum virginiànum** (L.) Durand & Jackson. (*Koellia virginiana* (L.) MacM. of Britton and Brown, Illus. Flora, ed. 2.) VIRGIANA MOUNTAIN-MINT. Map 1779. This species is generally found in low ground about lakes and ponds, in marshes, low open woods, roadside ditches, and frequently in moist, sandy prairie habitats.

Cent. Maine to N. Dak., southw. to Ga., Ala., and Kans.

7319. THỲMUS[Tourn.] L. THYME

See excluded species no. 544, p. 1086.

7323. CUNILA L. STONEMINT

1. **Cunila origanoides** (L.) Britt. STONEMINT. Map 1780. An infrequent plant of the unglaciated area on the crests and slopes of chestnut oak and black and white oak ridges.

I recommend this plant for rock gardens.

N. Y. to Mo., southw. to Fla. and Tex.

7326. LYCOPUS [Tourn.] L. BUGLEWEED

Leaves all more or less serrate, not even the lowest incised; tops of nutlets scarcely oblique, more or less tuberculate, at least the outer margin with one or more tubercles (not very conspicuous in *Lycopus asper*).

Calyx lobes lanceolate or deltoid, blunt, shorter than the mature nutlets, rarely equaling them.

Root of year old plants ending in a subterranean tuber, usually simple; roots of older plants not ending in a tuber, stoloniferous and usually many of the stolons bearing tubers; older plants much and widely branched, usually 20-50 cm high except sometimes taller when growing in a crowded environment; leaves usually sessile, sometimes on petioles up to 7 mm long, the widest ones ranging from 8-25 mm in width, the greatest number of teeth to a side generally 5-7; nutlets usually 1-1.5 mm long.................... 1. *L. uniflorus.*

Root of year old plants not ending in a subterranean tuber; older plants strongly stoloniferous, rarely with subterranean tubers; older plants usually simple or sparingly branched, 30-90 cm high; leaves usually petiolate (some plants with all the leaves sessile), petioles up to 20 mm long, widest leaves ranging from 18-55 mm in width, the greatest number of teeth to a side 7-14; nutlets usually 1.6-2 mm long.. 2. *L. virginicus.*

Calyx lobes narrow, very acute to acuminate, longer than the mature nutlets.

Leaves sessile.

Stems glandular-puberulent, most of the internodes more than 2 cm long; leaves ovate to lanceolate, the margins coarsely and irregularly serrate, generally with 3-5 teeth to the side; outer bracts minute, filiform, much shorter than the calyx... 3. *L. sessilifolius.*

Stems pubescent with flat, multicellular hairs, most or all of the internodes less than 2 cm long; leaves oblong-lanceolate, the margins regularly and sharply serrate, generally with 6-8 teeth to a side; outer bracts conspicuous, usually as long as or longer than the calyx. (See excluded species no. 545, p. 1087) .... *L. asper.*

Leaves petiolate... 4. *L. rubellus.*

Leaves (at least the lowest) generally more or less incised or pinnatifid; nutlets mostly 1-1.5 mm long, top very oblique, not tuberculate but the outer margin raised and entire (rarely slightly undulate), the inner angle not raised.

Stems glabrous or with a few short and long hairs................. 5. *L. americanus.*

Stems more or less densely pubescent with long, spreading, multicellular hairs.... .... 5a. *L. americanus* var. *Longii.*

1. **Lycopus uniflorus** Michx. Map 1781. All of my specimens except one are from the lake area. The Parke County specimen was found east of Rosedale, in "Nigger Legs" prairie, which is now drained. It is frequent throughout the lake area and usually common where found. It grows mostly on the borders of lakes in the moist, sandy or marly shores. I have seen this species common in the litter on the shore of a lake, while in a zone back of the litter, *Lycopus virginicus* was found; but the two species were restricted to two separate zones. It is also found in marshes, sphag-

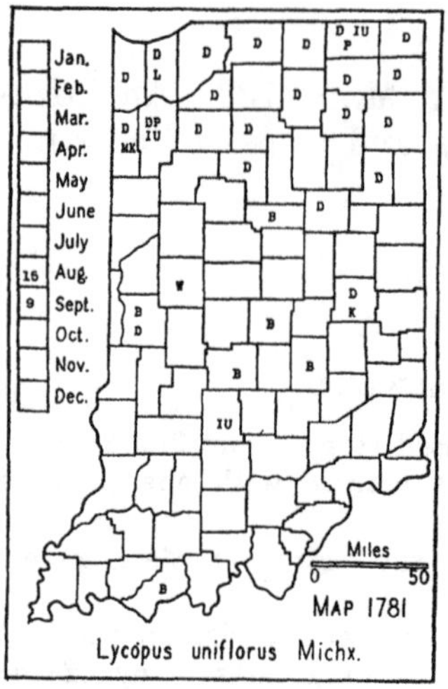

Lycopus uniflorus Michx.

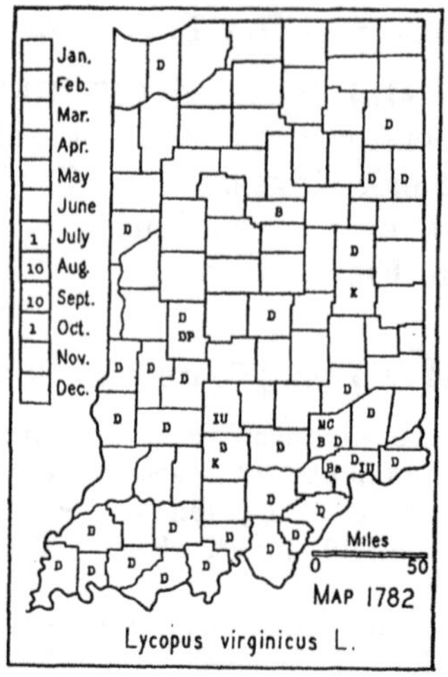

Lycopus virginicus L.

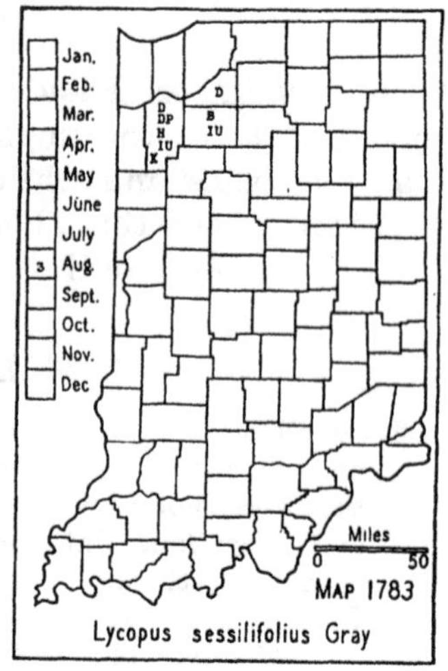

Lycopus sessilifolius Gray

num bogs, and mucky places. The tuber on this species has always interested me, and a few years ago I planted one year old seedlings with a tuber and the second year I found that the tuber had decayed and, in clay soil, the plant was, as usual, stoloniferous with many subterranean tubers. Two year old plants were planted in clay, and they were more proliferous and grew an incredible number of tubers. The limited number of tubers in their native habitat is doubtless due to lack of nutrients. This species is not satisfactorily separated from the next one and more study is needed on all parts, especially on the flowers.

According to Gray's Manual, the distribution is as follows:

Newf. and Lab. to B. C., southw. in the mts. to Va., Mich., Minn., Nebr., Wyo., and Oreg.

2. **Lycopus virgínicus** L. Map 1782. This species is frequent in the southern part of the state and infrequent to local in the northern part. It grows in wet places in woodland, in ditches, and on the muddy borders of sloughs and streams.

N. H. to Nebr., southw. to Fla., Miss., and Mo.

3. **Lycopus sessilifòlius** Gray. Map 1783. This species is local. It has been found in Jasper County in marshes about two and a half miles southeast of Tefft, and in Starke County in a marsh near Bass Lake and in a roadside ditch south of San Pierre.

Costal Plain from Mass. to Fla. and Miss. and in n. Ind.

4. **Lycopus rubéllus** Moench.* Map 1784. Found sparingly throughout the state. It is usually found in dried-up swamps and ponds in woods, often on old logs, and in the moss on the bases of trees that are growing in or on the border of ponds.

Vt. to Minn., southw. to Fla., La., and Ark.

5. **Lycopus americànus** Muhl. AMERICAN BUGLEWEED. Map 1785. Infrequent to frequent throughout the state. It is found in wet and moist ground in all kinds of habitats. This species was reported by some of our

* Variety **arkansanus** (Fres.) Benner (Bartonia 15: 50. 1935.) occurs in Posey County.

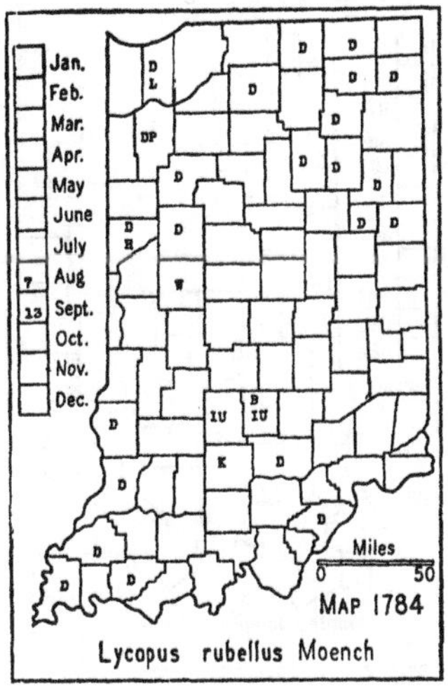

Lycopus rubellus Moench

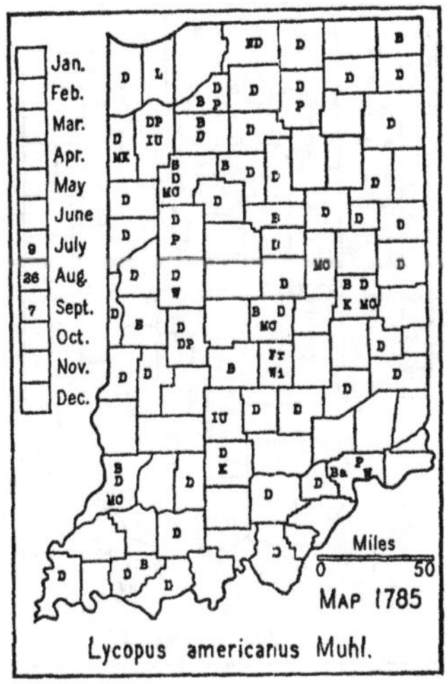

Lycopus americanus Muhl.

Lycopus americanus var. Longii Benner

early authors as *Lycopus europaeus* L. when some of our manuals did not separate this species from the European plant. All of our Indiana reports for the European species should be referred to *Lycopus americanus.*

Newf. to B. C., southw. to Fla., Tex., and Calif.

5a. **Lycopus americanus** var. **Lóngii** Benner. (Bartonia **16**: 46-47. 1935.) Map 1786. This variety is not well marked on account of a lineal series of intermediates, but is amply distinct in the extremes. In Indiana it is restricted mostly to our northern counties where it grows in habitats similar to those of the species but usually in slightly wetter situations.

Long Island, N. Y., Pa., Del.; and in nw. Ohio, s. Mich., and n. Ind.

7328. MÉNTHA [Tourn.] L. MINT

Whorls of flowers in terminal spikes, or some in the axils of the upper leaves.
Plants glabrous or nearly so.
Leaves sessile or nearly so; calyx generally about 1.5 mm long......1. *M. spicata.*
Leaves petiolate; calyx mostly 2.5-3.5 mm long....................2. *M. piperita.*
Plants canescent, woolly-pubescent or pubescent.
Leaves petiolate; petioles about 1 cm long. (See excluded species no. 546, p. 1087) ....*M. aquatica.*
Leaves sessile or nearly so.
Spikes canescent.
Leaves acute, the margins sharply serrate.....3. *M. longifolia* var. *mollissima.*
Leaves rounded at the apex, the margins incised. (See excluded species no. 548, p. 1087)...............................*M. longifolia* var. *undulata.*
Spikes not canescent, leaves obtuse.........................4. *M. rotundifolia.*
Whorls of flowers axillary.
Upper leaves much smaller than the lower ones. (See excluded species no. 547, p. 1087.)...*M. Cardiaca.*
Upper leaves not conspicuously reduced.
Stems more or less pubescent; calyx more or less pubescent, especially the teeth; corolla usually 4-5 mm long.
Pedicels glabrous; calyx lobes usually less than 1 mm long......5. *M. arvensis.*
Pedicels retrorsely pubescent; calyx lobes usually more than 1 mm long....... ....5a. *M. arvensis* var. *sativa.*
Stems glabrous; calyx glabrous or nearly so; corolla about 2 mm long.......... ....6. *M. gentilis.*

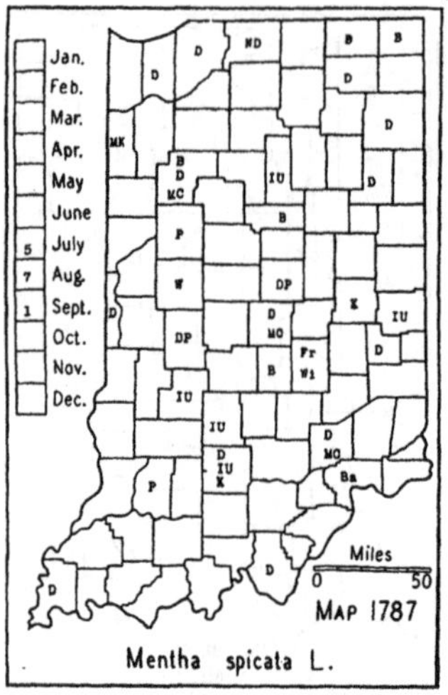

MAP 1787
Mentha spicata L.

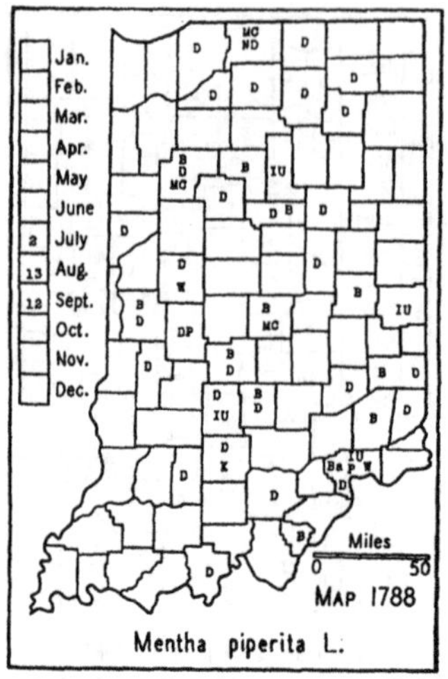

MAP 1788
Mentha piperita L.

MAP 1789
Mentha longifolia Huds. var. mollissima Borkh.

1. MENTHA SPICÀTA L. SPEARMINT. Map 1787. This plant was cultivated by the pioneers for its medicinal properties and has escaped in many parts of the state. It is generally found in wet places along roadsides and streams, and about lakes.

Na. of Eurasia; N. S. to Wash., southw. to Fla. and Calif.

2. MENTHA PIPERÌTA L. PEPPERMINT. Map 1788. This species has been commonly cultivated for its medicinal properties and as a kitchen herb; of recent years it has been cultivated on a commercial scale for its volatile oil which is now extensively used as a flavoring agent. This species is regarded as of hybrid origin. It does not produce viable seed and is propagated by stolons. For this reason its escape is limited although it has widely escaped, especially in the northern part of the state, where it has been cultivated. It prefers moist situations and is found along fences, roadsides, and streams, and about lakes.

Nat. of Eu.; N. S. to Minn., southw. to Fla. and Ark., also in Calif., Bermuda, and Jamaica.

3. MENTHA LONGIFÒLIA (L.) Huds. var. MOLLÍSSIMA Borkh. Map 1789. I found this mint in 1923 in moist soil along a recently graded roadside 3 miles southwest of Packerton, in Kosciusko County, where it formed a complete stand on both sides of the road for about 200 feet. I first found it in 1916 in sandy soil along the roadside north of Wolf Lake in Noble County. In 1922 and 1938, I revisited this place and the plant still persisted although the owner of the land had tried to exterminate it. In 1922, I found a colony about 2 rods long along the roadside in section 13, about 4 miles north of Modoc, in Randolph County.

Nat. of Eu.; I do not know the distribution of the variety in the U. S.

4. MENTHA ROTUNDIFÒLIA L. APPLE MINT. Map 1790. In 1923, I found a small colony of this plant in dry, sandy soil near Clarke, in Lake County. Peattie reports finding this same colony a few years later and also says that it is established at Gibson, in Lake County. Clark reported it

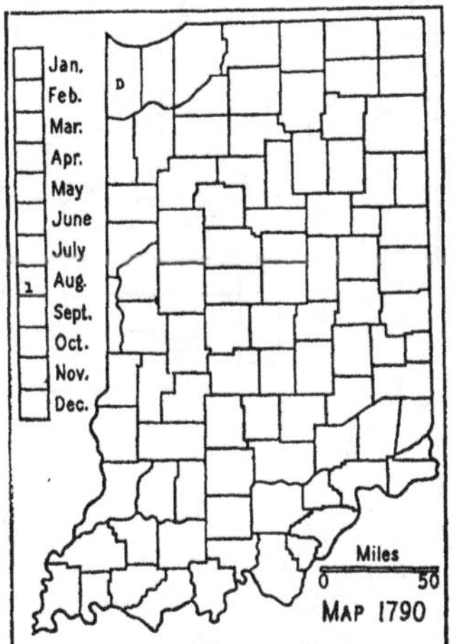

Mentha rotundifolia L.

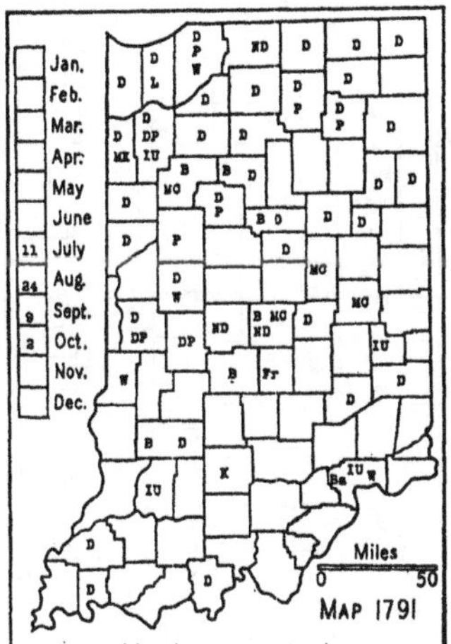

Mentha arvensis L.

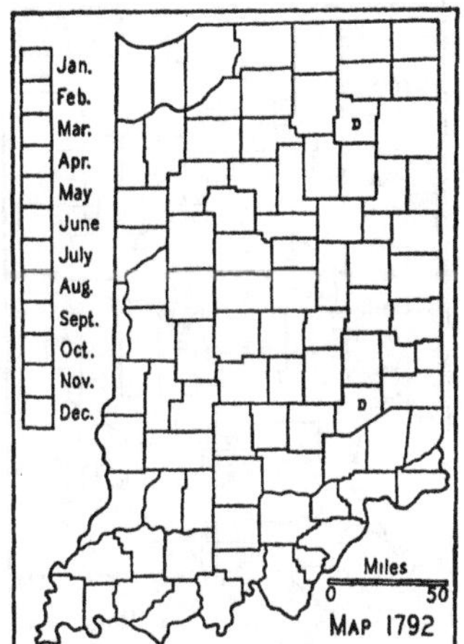

Mentha arvensis var. sativa Benth.

from Kosciusko County. No doubt it will persist in Lake County and can safely be admitted to our flora.

Nat. of Eu.; Maine to Ark., southw. to Fla. and Tex.

5. **Mentha arvénsis** L. FIELD MINT. , Map 1791. This species is found throughout the state although it is more frequent in the northern part. It is very variable and I am regarding it as a species complex. I am referring to it all of my specimens which I had formerly named *Mentha arvensis*, *Mentha arvensis* var. *canadensis*, and *Mentha arvensis* var. *glabrata*. I have only 72 sheets from which to make a study and I find that this number of specimens together with the meager amount of literature available are not sufficient to justify a satisfactory conclusion. I find that Victorin in his "Flore Laurentienne" treats the plants of his area as one species and calls them *Mentha canadensis*. The plants, as a whole, differ widely from each other, and large and widely spreading plants differ greatly in the parts of the same plant. Bentham, in his monograph of *Labiatae*, described seven varieties, one of which I am recognizing because it has one character which seems to be constant. Doubtless the plants of America are different from those of Europe and Asia and it may have been wiser to have followed other authors in calling our plant *Mentha arvensis* var. *canadensis* or *Mentha canadensis* as Victorin did. Without convincing proof I prefer to be conservative and use the old name and await the report of a monographic study of the genus. All Indiana forms are shown on one map.

Newf. to B. C., southw. to Pa., N. Mex., and Calif.; Eurasia.

5a. **Mentha arvensis** var. **sativa** Benth. Map 1792. I have only two sheets of this variety. One is from Decatur County and one is from Whitley County. The upper parts of the stems and branches of these specimens are pubescent all over; the leaves are of an ovate type, strongly rounded or subcordate at the base; the calyx lobes are very sharp and 1-2 mm long; and the corollas pubescent without.

Distribution not known.

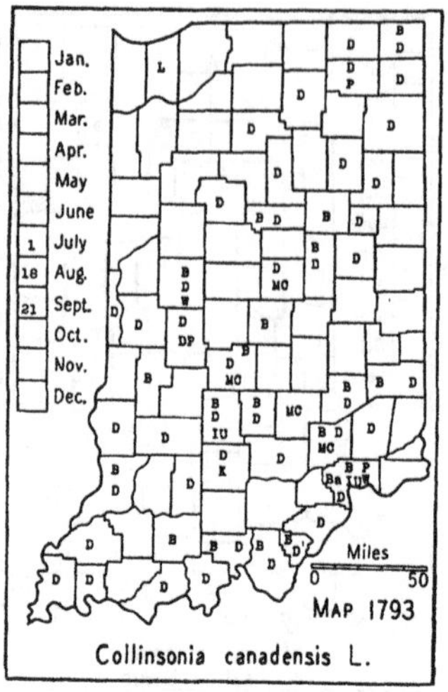

MAP 1793
Collinsonia canadensis L.

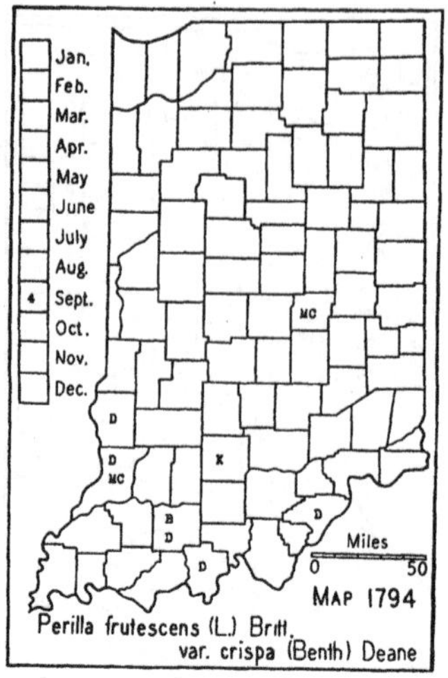

MAP 1794
Perilla frutescens (L.) Britt. var. crispa (Benth) Deane

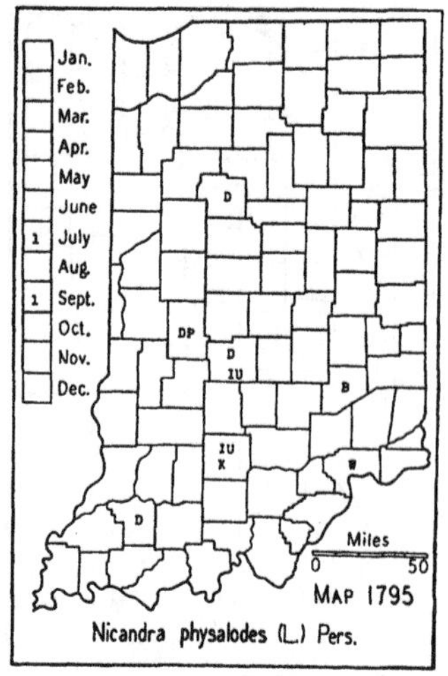

MAP 1795
Nicandra physalodes (L.) Pers.

6. MENTHA GENTÌLIS L. This species has been reported from four counties and since it is a frequent escape in other states, it is given a place in our flora. It has been reported from Clark, Jefferson, and White Counties. I have it from Decatur County.

Nat. of Eu.; N. S. to Iowa, southw. to N. C. and Tenn.

7331. COLLINSÒNIA L. HORSEBALM

1. **Collinsonia canadénsis** L. CITRONELLA HORSEBALM. STONEROOT. Map 1793. Throughout the state in dry, rich woods although we have but one record from the northwestern counties. The thickened, hard rootstock is much used in medicine for kidney and urinary disorders.

W. Que. to Wis., southw. to Fla., Mo., and Kans.

7332. PERÌLLA L.

1. PERILLA FRUTÉSCENS (L.) Britt. var. CRÍSPA (Benth.) Deane. (Rhodora 25: 40. 1923.) PURPLE PERILLA. Map 1794. Sparingly escaped from gardens. We have had only four county records. I noted it covering at least half an acre in an open woods in Perry County. In a sandy woods about 3 miles northwest of Bicknell, in Knox County, it was a common weed over the greater part of a 20 acre woods. This woods had been heavily grazed by hogs. It is to be noted that even hogs do not disturb it and if it once gets started it may become a permanent plant. It is an annual which is frequently grown as a border plant. I have grown it for years and have allowed a few plants to seed themselves annually but have never noted seedlings except in the flower beds.

Nat. of the Himalayas, Burma, China, and Japan; Conn. to Mo., southw. to Fla. and Tex.

256. SOLANÀCEAE Pers. NIGHTSHADE FAMILY

Flowers less than 3.5 cm long; fruit a fleshy or dry berry.

Plants woody; trailing or climbing shrubs, with or without thorns; fruit a dry berry...7379. LYCIUM, p. 827.

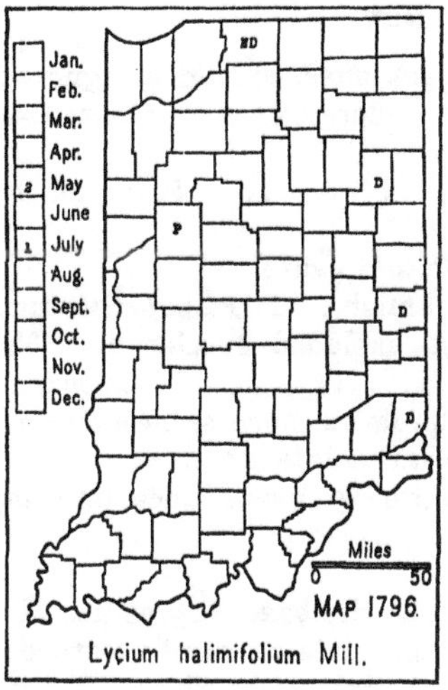

MAP 1796
Lycium halimifolium Mill.

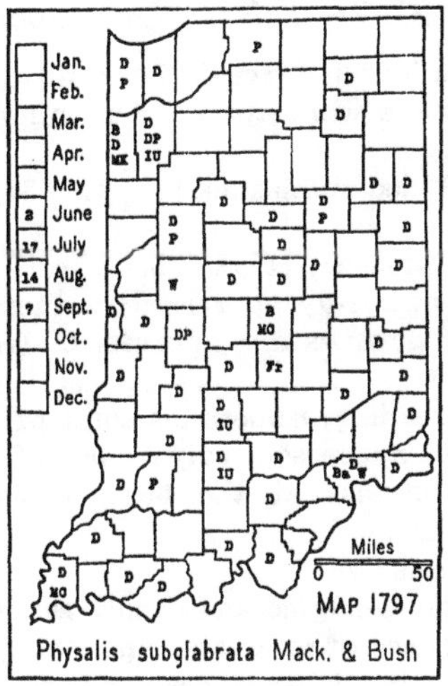

MAP 1797
Physalis subglabrata Mack. & Bush

MAP 1798
Physalis virginiana Mill.

Plants herbaceous (sometimes partly woody and climbing in *Solanum Dulcamara*); fruit a fleshy berry.
Flowers large, generally 2.5 cm long, purplish; fruit enclosed in the inflated calyx; calyx with 5 large wings, the wings sagittate at the base..7377. NICANDRA, p. 827.
Flowers smaller, generally much less than 2 cm long; fruit naked or enclosed in an inflated calyx.
Anthers not opening by pores; flowers yellow; fruit enclosed in an inflated calyx; calyx without sagittate wings but often somewhat 5-angled..7401. PHYSALIS, p. 828.
Anthers opening by pores; flowers purplish, white or yellow; fruit a berry, not enclosed by the calyx.........................7407. SOLANUM, p. 829.
Flowers 3.5 cm long or longer; fruit a capsule.
Plants glabrous; capsules prickly..........................7415. DATURA, p. 831.
Plants viscid-pubescent; capsules not prickly...............7436. PETUNIA, p. 832.

7377. NICÀNDRA Adans. APPLE-OF-PERU

1. NICANDRA PHYSALÒDES (L.) Pers. (*Physalodes physalodes* (L.) Britt.) APPLE-OF-PERU. Map 1795. This species has been reported from 10 other counties than those in which I have found it. Two authors report on its habitat and say: "In waste grounds." I have found it three times and each time in a cornfield where it was frequent to abundant. I have not been able to revisit any of these stations to learn whether it has persisted. Since it is rarely or no longer cultivated, I believe its spread will be limited.

Nat. of Peru; N. S. to Ont., southw.

7379. LÝCIUM L. MATRIMONY-VINE

1. LYCIUM HALIMIFÒLIUM Mill. COMMON MATRIMONY-VINE. Map 1796. This species has been reported from 7 counties, and all of the reports but one are nearly 40 years old. I do not believe this vine is any longer cultivated and doubtless its spread will be slow. I have seen it only once in abundance or far from a habitation.

Nat. of Eu.; Ont. to Minn., southw. to Va. and Kans.

7401. PHÝSALIS L. GROUNDCHERRY

All Indiana records for *Physalis*, except the more recent ones, should be ignored because most of them were made while the species as now understood were treated as aggregates by the older manuals.

Peduncles glabrous. (See excluded species no. 551, p. 1087) . . . . *P. angulata.*
Peduncles pubescent.
  Pubescence of peduncles appressed; calyx lobes short, deltoid-ovate.
    Pubescence of peduncles upwardly appressed; anthers bluish. . . .1. *P. subglabrata.*
    Pubescence of peduncles downwardly appressed. (See excluded species no. 552, p. 1087) . . . . *P. ixocarpa.*
  Pubescence of peduncles spreading, sometimes some of the hairs more or less curved toward the apex but not appressed; calyx lobes of a lanceolate type.
    Leaves narrowed more or less at the base, lanceolate to ovate-lanceolate or oblanceolate; perennials.
      Blades usually 2-6 cm long.
        Fruiting calyx obtusely 5-angled and deeply sunken at the base; leaves usually lanceolate, sometimes of an ovate type, entire, undulate or rarely toothed; margins of the calyx lobes densely ciliate with hairs less than 0.5 mm long without intervening long hairs. . . . . .2. *P. virginiana.*
        Fruiting calyx scarcely angled and scarcely sunken at the base; leaves oblanceolate or spatulate, subentire or rarely undulate. (See excluded species no. 553, p. 1087) . . . . *P. lanceolata.*
      Blades usually 6-12 cm long; margins of calyx lobes densely ciliate with short hairs with the addition of a few hairs about 1 mm long. . . . .3. *P. nyctaginea.*
    Leaves rounded, truncate or subcordate at the base, broadly ovate; annuals or perennials.
      Annuals; anthers bluish, 1-2 mm long; filaments not dilated.
        Stem sharply angled; leaf blades short-acuminate at the apex, rounded at the base, margin entire, undulate or with a few blunt teeth, the margin never toothed to the base of the blade; corolla usually 10-12 mm in diameter, the limb recurving; anthers blue, generally 1.5-2 mm long. . . . . .4. *P. pubescens.*
        Stem obtusely angled; blades thicker, acute or obtuse at the apex, cordate at the base, the margins of most of the blades more or less strongly sinuate-toothed to the base; anthers blue, generally 1-1.5 mm long. .5. *P. pruinosa.*
      Perennials; anthers greenish white or greenish yellow, often turning bluish after anthesis or rarely blue, about 3 mm long; filaments blue, dilated above.
        Upper part of plant densely pubescent with short, terete, glandular hairs about 0.5 mm long, with a few much longer, flat, jointed, often glandular hairs, the long hairs generally more numerous on the lower part of the stem; anthers yellow. . . . . .6. *P. heterophylla.*
        Upper part of plant densely pubescent with long, flat, jointed, sometimes glandular hairs, lacking the short, terete, glandular hairs; anthers yellowish white, turning purplish after anthesis. . . . . .7. *P. ambigua.*

1. **Physalis subglabràta** Mack. & Bush. SMOOTH GROUNDCHERRY. Map 1797. A frequent to common weed in cultivated ground, fallow ground, clover fields, waste places, open woods, and pastures and along roadsides and railroads. By far the most common groundcherry of the state.

R. I., Ont. and Minn., southw. to Ga. and Colo.

2. **Physalis virginiàna** Mill. VIRGINIA GROUNDCHERRY. Map 1798. Plants of dry, usually very sandy soil. Found mostly in railroad ballast, fallow fields, open wooded slopes, and along roadsides.

N. Y., Ont. to Man., southw. to Fla. and Tex.

Physalis nyctaginea Dunal

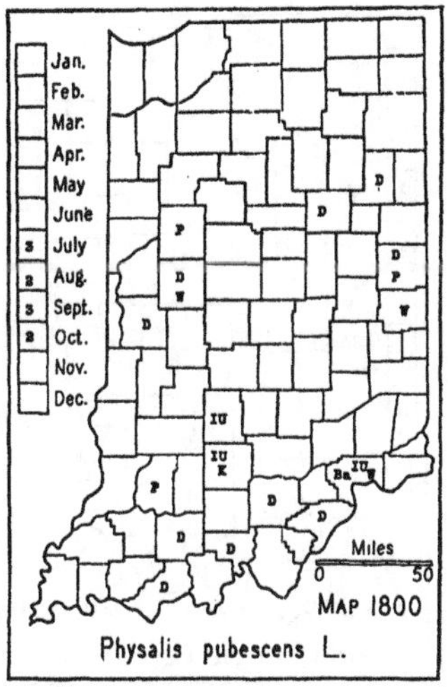

Physalis pubescens L.

Physalis pruinosa L.

3. **Physalis nyctagínea** Dunal. Map 1799. I have only one specimen and it was collected in a shady, black and white oak woods about 4 miles east of Bloomington, Monroe County.

R. I. to Iowa, southw. to La.

4. **Physalis pubéscens** L. COMMON GROUNDCHERRY. Map 1800. Usually in cultivated ground such as cornfields and less frequently on open wooded slopes and in alluvial bottoms.

Pa. to Calif., southw. to Fla. and Mex.

5. **Physalis pruinòsa** L. Map 1801. In moist soil in clearings, alluvial bottoms, pastures, and fallow fields.

Mass. to Iowa, southw. to Fla. and Tenn.

6. **Physalis heterophýlla** Nees. Map 1802. This species prefers dry, sandy soil and is found on wooded slopes and along roadsides.

N. B. to Sask., southw. to Fla. and Tex.

7. **Physalis ambígua** (Gray) Rydb. Map 1803. This species also prefers a dry, sandy soil and is found in cultivated and fallow fields, along roadsides and railroads, and in open woods. It usually has not been separated from the preceding species.

Vt. to Iowa, southw. to Tenn.

7407. SOLÀNUM [Tourn.] L. NIGHTSHADE

Plants more or less prickly; pubescence stellate.
  Perennial; corolla violet or white; fruit naked.......................1. *S. carolinense.*
  Annual; corolla yellow; fruit partly covered by the spiny calyx......2. *S. rostratum.*
Plants not prickly; pubescence not stellate.
  Annual; a low, much branched, and often spreading plant; flowers white; berries black. ....3. *S. nigrum.*
  Perennial; a tall, climbing, semi-woody plant; flowers purplish or white; berries scarlet. ....4. *S. Dulcamara.*

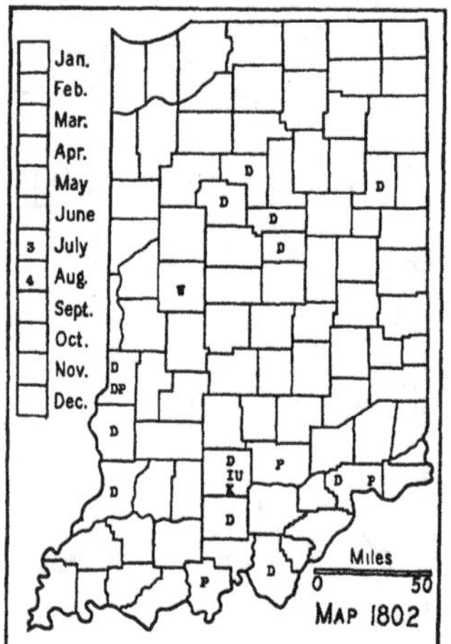

Physalis heterophylla Nees

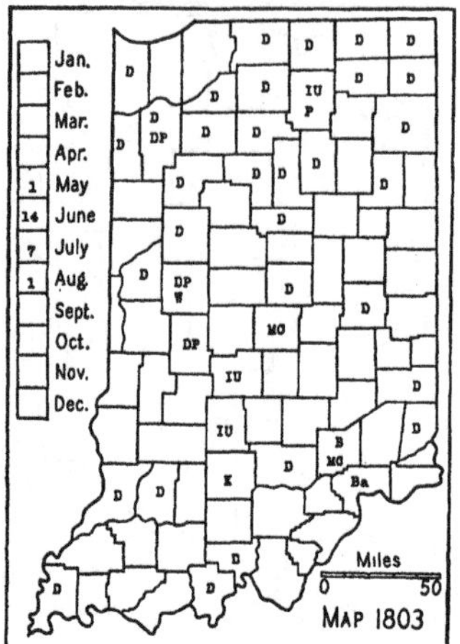

Physalis ambigua (A. Gray) Rydb.

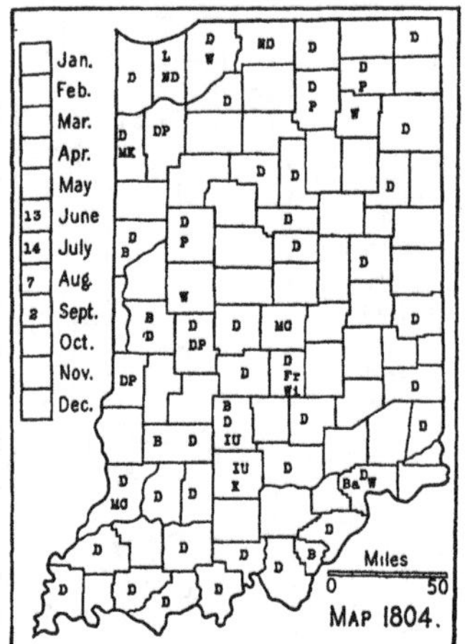

Solanum carolinense L.

1. **Solanum carolinénse** L. HORSE NETTLE. Map 1804. The root and fruit of this nightshade are used in medicine. An obnoxious weed, more or less frequent to abundant throughout the state. It prefers a sandy soil. Found mostly in cultivated and fallow fields, waste places, and sometimes in open woods. There is little doubt that this species is native to Indiana because it was reported in 1834 by Clapp from the vicinity of New Albany, and in 1819 by Thomas from the vicinity of Vincennes. The early botanists of the southern part of the state reported it as common in that area, but the botanists of northern Indiana reported it as rare. It has, no doubt, been introduced in later years at least in the northern part of the state.

Mass. to Nebr., southw. to Fla. and Tex.

2. SOLANUM ROSTRÀTUM Dunal. BUFFALO BUR. Map 1805. This species has been reported from 11 counties but none of the authors state its abundance or whether it persisted. I have found it in 6 counties, and in 3 counties I found only a single plant; in two counties it covered large barnyards; in St. Joseph County, however, it was an abundant and common weed in sandy soil over 2-3 acres in a large barnyard and an adjoining truck garden. The owner despaired of ever being able to eradicate it. It has been collected by Bechtel in Montgomery County where it is established. Miss Edna Banta informs me that it is a weed on a farm near Brooksburg, Jefferson County. No doubt this species can safely be regarded as a permanent introduction.

S. Dak. to Tex. and Mex.; adventive eastw. to N. H., southw. to Fla.

3. **Solanum nìgrum** L. COMMON NIGHTSHADE. Map 1806. Infrequent throughout the state. Sometimes frequent to common in woods pastures. Ordinarily the plant is not grazed but when it is eaten in sufficient quantity, it proves fatal. Sheep are frequently killed by it. It is found in open woods, pastures, fallow and cultivated fields and along roadsides and railroads. The berries are poisonous and there are records where death of children resulted from the eating of the fruit.

N. S. to Alberta, southw. to Fla. and Tex.

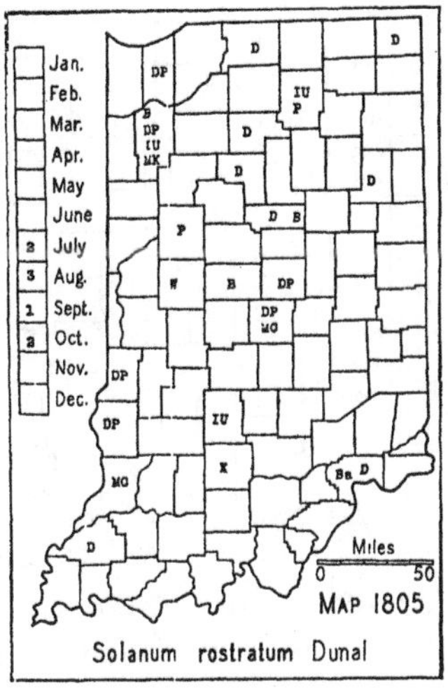

Solanum rostratum Dunal

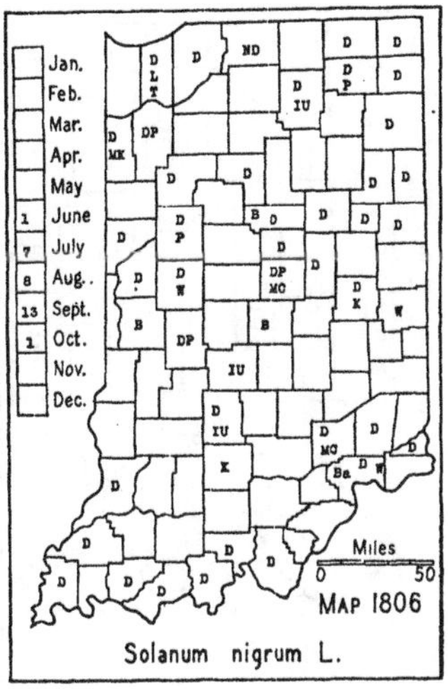

Solanum nigrum L.

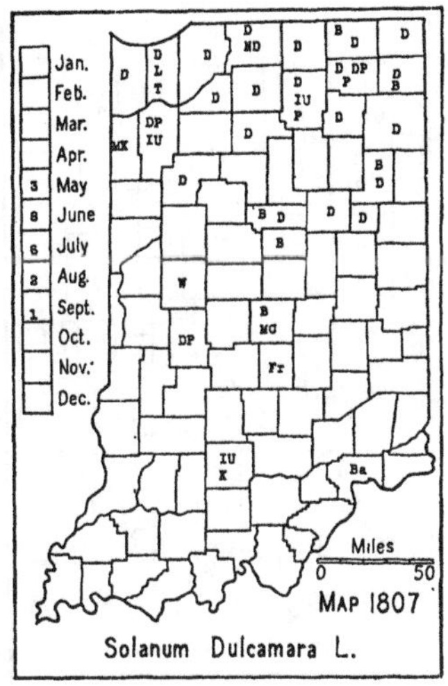

Solanum Dulcamara L.

4. **Solanum Dulcamàra** L. BITTER NIGHTSHADE. BITTERSWEET. Map 1807. This is the true bittersweet of medicine, and should not be confused with *Celastrus scandens* which is also called bittersweet.

This species is more or less frequent in the lake area and is practically confined to it although it is reported from 6 of the southern counties. It is found in swamps, bogs, and low woods and along low roadsides. Authors say it is adventive from Europe but all of our early authors found it and its habitat suggests that it is native. It is, no doubt, native in Indiana.

This species varies greatly in the amount of pubescence of the branchlets, varying from almost glabrous to rather densely pubescent but the pubescence not quite dense and long enough to make our specimens belong to the pubescent variety. The young branchlets are used in medicine. White-flowered forms are found occasionally.

N. S. to Minn. and Wash., southw. to Pa., Ga., and Kans.

7415. DATÙRA L.

1. DATURA STRAMÒNIUM L. (*Datura Tatula* L.) DATURA. JIMSONWEED. Map 1808. The whole plant is very poisonous when taken internally, yet it is much used in medicine externally. Plants are found with white flowers and green stems and with purple flowers and purplish stems; some plants have capsules with all of the prickles of the same length while other plants have capsules with the lower prickles of the capsule shorter. Until recently the two plants have been regarded as separate species. In recent years much genetic study has been given these two forms and the result of this study shows that the two characters used to separate the species do not correlate, so it seems best to regard the two forms as a polymorphic species.

This species has been reported from all parts of the state. It was formerly much more abundant than at present. Its decrease may be due partly to the changed habitat, but probably more to the fact that farmers recognize its poisonous character and destroy it. When I was a boy it

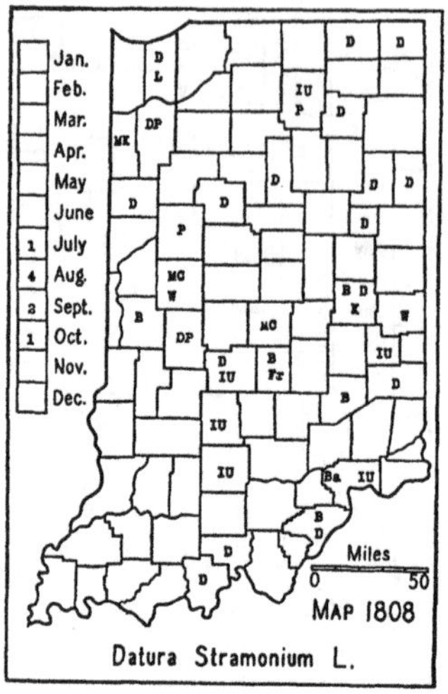

MAP 1808
Datura Stramonium L.

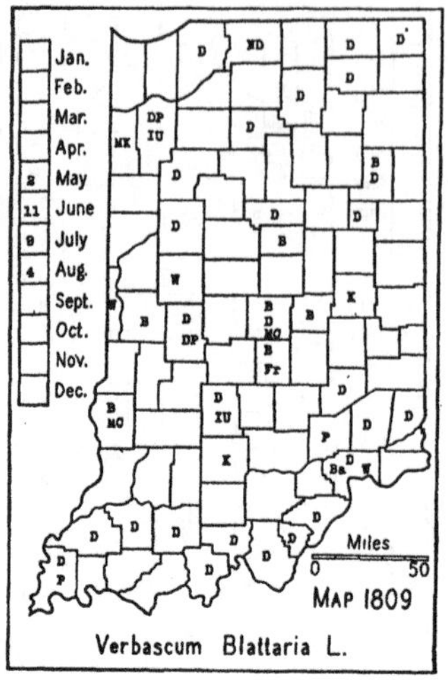

MAP 1809
Verbascum Blattaria L.

MAP 1810
Verbascum phlomoides L.

was a common weed of barnyards and hogyards but now it is rarely seen; when it is found there it is usually a common weed, because all stock avoid it. It is found sometimes in waste places and cultivated fields.

Nat. of tropical regions; N. S. to Minn., southw. to Fla. and Tex.

7436. PETÙNIA Juss.

See excluded species no. 560, p. 1088, for a discussion of the species.

257. SCROPHULARIÀCEAE Lindl. FIGWORT FAMILY[1]

[Pennell. The Scrophulariaceae of Eastern Temperate North America. Monograph no. 1. Acad. Nat. Sci. Philadelphia i-xiv. 650p. 155 maps. 1935.]

Anther-bearing stamens 5; corolla rotate; leaves alternate....7460. VERBASCUM, p. 834.
Anther-bearing stamens 2 or 4; leaves opposite, verticillate or alternate.
  Corolla spurred, saccate or gibbous on the lower side at the base.
    Leaves ovate with widely spreading auricles at the base; stems densely spreading-pubescent....................................7479. KICKXIA, p. 835.
    Leaves of a linear type, entire.
      Flowers solitary in the axils of the leaves; stems glandular-pubescent throughout; pedicels about 10 mm long............7484. CHAENORRHINUM, p. 836.
      Flowers in terminal racemes; stems glabrous; pedicels about 5 mm or less long...7480. LINARIA, p. 835.
  Corolla not spurred, saccate or gibbous at the base.
    Fertile stamens 2.
      Leaves opposite or in whorls.
        Leaves mostly in whorls of 3-6, rarely opposite.......................... ....7579A. VERONICASTRUM, p. 849.
        Leaves opposite.
          Capsules flattened and more or less notched at the apex................. ....7579. VERONICA, p. 845.
          Capsules not flattened or notched at the apex.

[1] Dr. F. W. Pennell, who has made a lifelong study of this family, has identified nearly all of my specimens. His profound study of the family leads me to accept his nomenclature throughout. I have used his keys to genera that occur in Indiana in a condensed form, and I hereby wish to express grateful acknowledgement.

Calyx 2-bracteolate, the bracts longer than the calyx lobes; sterile filaments stout, short or almost lacking......7542. GRATIOLA, p. 843.
Calyx not bracteolate; sterile filaments slender, 2-lobed..7562. LINDERNIA, p. 844.

Leaves alternate.

Basal leaves present at flowering time, large, on long petioles, the blades usually cordate at the base and 5-15 cm long; stem leaves sessile and much reduced, usually 1-2.5 cm long.............7583A. BESSEYA, p. 850.

Basal and stem leaves not as above.................7579. VERONICA, p. 845.

Fertile stamens 4.

Trees (introduced)...............................7513. PAULOWNIA, p. 842.

Herbs.

Leaves alternate.

Leaves sessile, 3-5-lobed or cleft, 2-6 cm long.....7631. CASTILLEJA, p. 856.

Leaves, all but the uppermost, petiolate, pinnately parted, 6-10 cm long. ....7648. PEDICULARIS, p. 857.

Leaves mostly opposite.

Flowers all axillary.

Leaves obovate or orbicular, entire; plants aquatic or of muddy shores. ....7548. HYDRANTHELIUM, p. 844.

Leaves not entire; plants not aquatic.

Leaves serrate, 4-10 cm long.................7524. MIMULUS, p. 842.

Leaves pinnately parted into 3-7 linear segments, 1-2 cm long...... ....7545. LEUCOSPORA, p. 844.

Flowers not all axillary, at least some or all in terminal spikes, racemes or panicles.

Leaves and bracts entire, linear, sessile..........7604. GERARDIA, p. 850.

Leaves not as in the preceding.

Plants mostly 1.5-5 dm high; lower leaves petiolate, smaller than the upper cauline ones; flowers not more than 15 mm long, whitish, half blue or purplish, never yellow.

Lower leaves petiolate, upper ones much larger, sessile or clasping; flowers in the upper axils and in a terminal cluster, the lower lip blue.............................7503. COLLINSIA, p. 836.

Lower and upper leaves short-petiolate; flowers in terminal spikes, subtended by large, foliaceous bracts.......................... ....7635. MELAMPYRUM, p. 857.

Plants usually more than 5 dm high, if shorter, the lower leaves sessile and similar to the upper cauline ones.

Blades of leaves regularly serrate (rarely laciniate toward the base), or nearly entire with a few short teeth or with 1 or 2 long auricles at the base; flowers never yellow.

Flowers sessile in terminal and axillary spikes.

Tall, glabrous or partly pubescent plants; flowers usually more than 20 mm long..................7507. CHELONE, p. 838.

Low, hispid or scabrous plants; flowers mostly less than 10 mm long.

Flowers leafy-bracted, about 2.5 cm long; capsules about 10 mm long..................7604A. TOMANTHERA, p. 853.

Flowers minutely bracted, about 1.5 cm long; capsules about 8 mm long.....................7622. BUCHNERA, p. 856.

Flowers pedicellate in terminal panicles, thyrses, cymes or racemes.

Cauline leaves on petioles 1-8 cm long, of an ovate type; flowers maroon color................7505. SCROPHULARIA, p. 837.

Cauline leaves sessile or clasping, of a lanceolate type; flowers mostly purplish or white.........7508. PENSTEMON, p. 839.

Blades of leaves 1 or 2 times parted, pinnatifid, pinnately lobed or at least some on the stem with 1 or 2 auricles at the base; flowers yellow.
Flowers sessile.
Plants large, coarse, of a dry habitat; leaves large and irregularly cut; flowers less than 20 mm long...7602. DASISTOMA, p. 850.
Plants slender, of a wet habitat; leaves of a lanceolate type, rather regularly pinnately lobed, the lobes short...7648. PEDICULARIS, p. 857.
Flowers on short pedicels and more than 20 mm long...7604B. AUREOLARIA, p. 854.

7460. VERBÁSCUM [Bauhin] L. MULLEIN

[Murbeck. Monographie der Gattung Verbascum. 630p. 31 pl. Lund (Sweden), 1933-34.]

Stem, pedicels, and calyx with simple, glandular hairs, otherwise glabrous..1. *V. Blattaria.*
Stem, pedicels, calyx, and leaves more or less pubescent with stellately branched, nonglandular hairs.
Inflorescence interrupted, at least the lower clusters somewhat remote; pedicels several to an axil, up to 10 mm long; leaves slightly decurrent on the stem; corollas 25-40 mm wide....................................2. *V. phlomoides.*
Inflorescence densely crowded; pedicels usually 1 to an axil, very short or lacking; leaves long-decurrent on the stem; corollas 15-22 mm wide......3. *V. Thapsus.*

1. VERBASCUM BLATTÀRIA L. MOTH MULLEIN. Map 1809. Infrequent to frequent or locally common throughout the state. It is spreading every year. It is found mostly in pastures, fallow ground, and hayfields and along roadsides. There are two forms, a yellow-flowered one, the typical form, and a white-flowered one (f. *albiflora* (G. Don) House). Since my labels do not always give the color of the flower, unfortunately, I am not able to give their ratio of abundance. My recollection is, however, that the yellow form is much more common. Through neglect we permitted the white form to become established in our three-acre arboretum about 10 years ago. Since then I have endeavored to exterminate it by digging every plant as soon as discovered, and not a single plant has been permitted to seed. The viability of the seed is shown, however, by the fact that a few plants were found last year. It might be added that I have never seen a yellow-flowered plant in the tract. I have seen large areas of this species and I do not recall that I ever saw the two forms growing together, although this is quite possible.

Nat. of Eu.; naturalized throughout the U. S.

2. VERBASCUM PHLOMOÌDES L. Map 1810. About 1925 Mr. Walter Neff and Mrs. Ivy Neff discovered this species as a common weed in the Cedarville Cemetery and nearby pastures and roadside in Carroll County, about two or two and a half miles southwest of Burnettsville. Mrs. Neff has written of the discovery and described the plant (Amer. Bot. 36: 85-87. 1930). At that time the name was still in controversy. I visited this colony in 1929 and found that it formed an almost complete stand in a pasture of two to three acres and that it was scattered in pasture fields for a distance

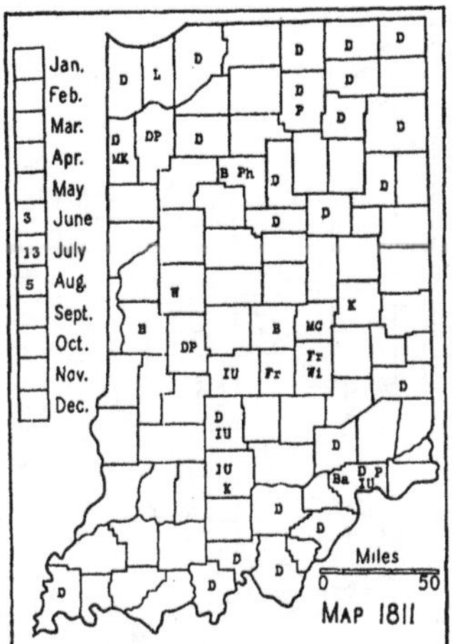

Verbascum Thapsus L.

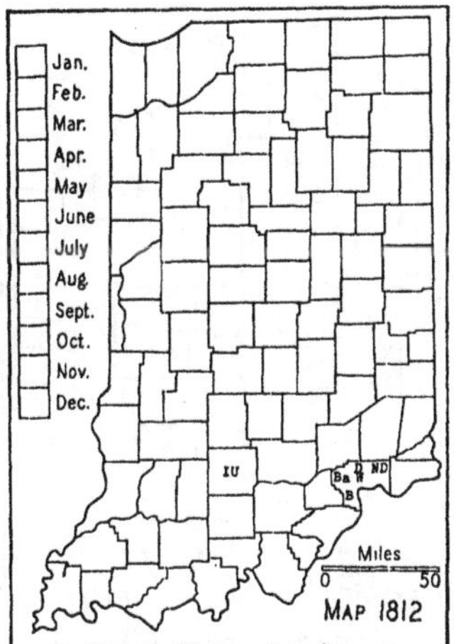

Kickxia Elatine (L.) Dumort.

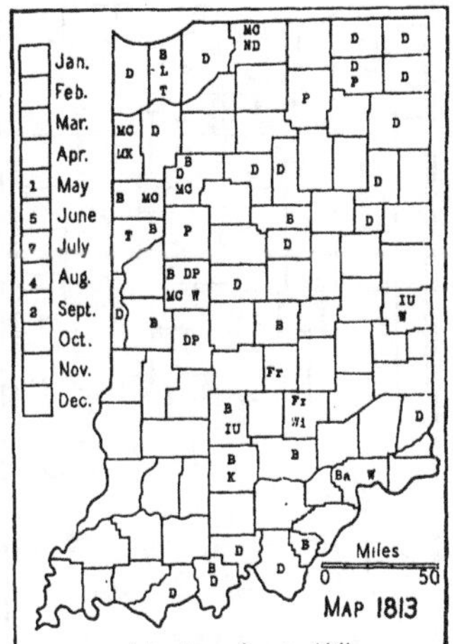

Linaria vulgaris Hill

of about three miles. I sent specimens collected from this colony to Pennell who sent them to Murbeck for determination. Murbeck, in 1936, identified them as *Verbascum phlomoides* L. In 1937 I found this mullein common along an east and west road two miles north of Rochester, Fulton County and in several places along the Tippecanoe River south of Talma.

Nat. of Eu.

3. VERBASCUM THÁPSUS L. MULLEIN. Map 1811. Frequent to common throughout the state. This mullein prefers a dry, sandy or gravelly soil and is found principally in pastures, idle fields, and waste places along roadsides. It is a common weed of pastures because stock do not eat it.

Nat. of Eu.; naturalized nearly throughout temperate N. A.

7479. **KÍCKXIA** Dumort.

1. KICKXIA ELATÌNE (L.) Dumort. (*Linaria Elatine* (L.) Mill.) Map 1812. This species was found in 1925 by R. C. Friesner in Clifty Falls State Park, Jefferson County. It was well established about half a mile north of Tunnel Falls. It has also been reported from Ohio and Vanderburg Counties by Hansen (Proc. Indiana Acad. Sci. 34: 257. 1925.) There is a specimen in the herbarium of Indiana University collected by Wible in Lawrence County.

Nat. of Eu.; naturalized from Mass. to La., mostly near the coast and inland to n. N. Y., Ind., and Mo.

7480. **LINÀRIA** [Bauhin] Mill. TOADFLAX

Flowers yellow, 2-3 cm long; capsules mostly 6-8 mm long .... 1. *L. vulgaris*.
Flowers blue or rarely white, about 1 cm long; capsules about 2-5 mm long........ .... 2. *L. canadensis*.

1. LINARIA VULGÀRIS Hill. COMMON TOADFLAX. Map 1813. This species prefers dry, sandy soil and has escaped from cultivation to roadsides and pastures throughout the state. It has become a weed in some of the eastern

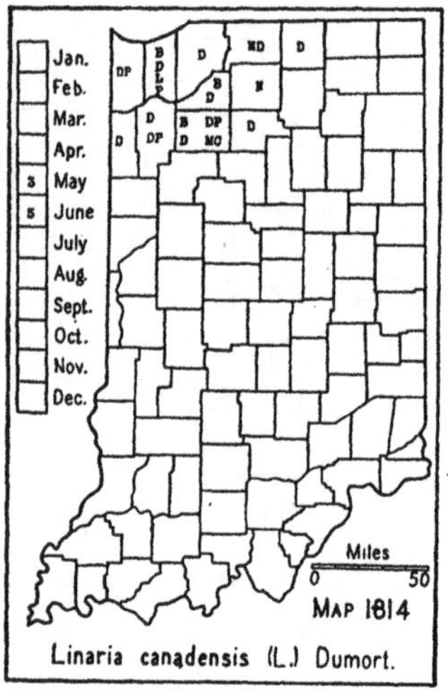

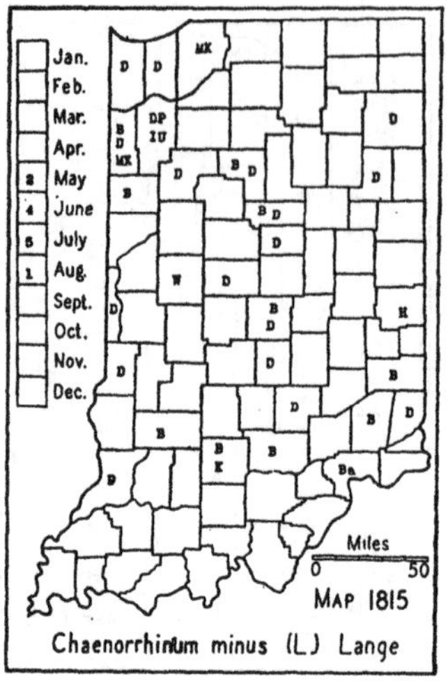

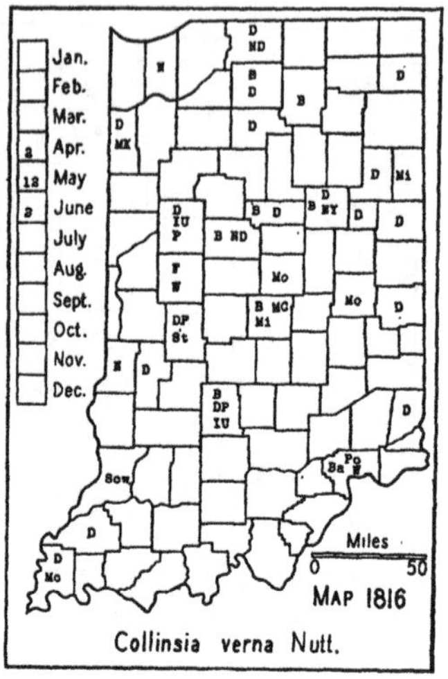

states, and I have seen large colonies of it in Indiana in sandy soil in pastures. It is difficult to eradicate and, for this reason, should be exterminated as soon as it is detected. It is commonly called butter and eggs.

Nat. of Eu.; naturalized from Newf., Que. to B. C., southw. to Fla., Tex., and Calif.; most common in the ne. U. S.

2. **Linaria canadénsis** (L.) Dumort. Map 1814. Usually in dry and almost pure sand in fallow fields and on open dunes. Sometimes in fallow fields it grows in such abundance that the landscape is blue. This species should still be sought in Kosciusko, Lagrange, and Steuben Counties.

N. S. to S. Dak., southw. to Fla. and Tex.; along the Pacific coast from B. C. to Calif.

7484. CHAENORRHÌNUM Reich.

1. CHAENORRHINUM MÌNUS (L.) Lange. (*Linaria minor* (L.) Desf.) Map 1815. This species is reported to have been introduced in 1874 at Camden, New Jersey. Since that time it has spread extensively and is always found in cinder or sand ballast along railroads. I first found it in Vigo County in 1918. I have seen it spread from a few plants along the traction line in Wells County until the railroad bed for miles in flowering season is blue with it.

Nat. of the Mediterranean region; now naturalized from N. S. and Que. to Wis. and Iowa, southw. to N. J., Ohio, and Ill.

7503. COLLÍNSIA Nutt.

[Newsom. A revision of the genus Collinsia. Bot. Gaz. 87: 260-301. 1929.]

1. **Collinsia vérna** Nutt. BLUE-EYED-MARY. Map 1816. This species is well distributed throughout the state but rather local and usually abundant where it is found. Its preferred habitat is moist, rich soil and it is most

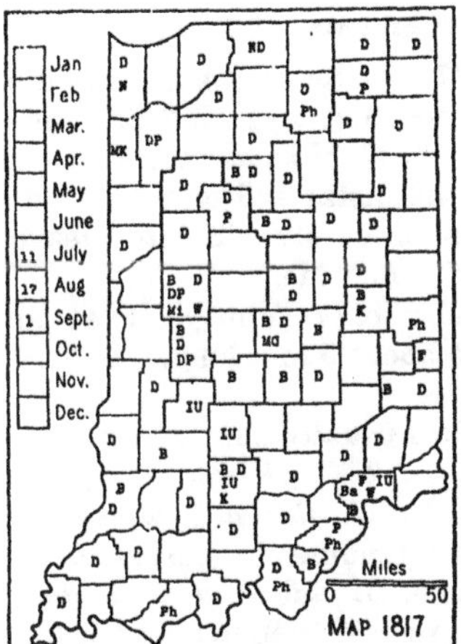

MAP 1817
Scrophularia marilandica L.

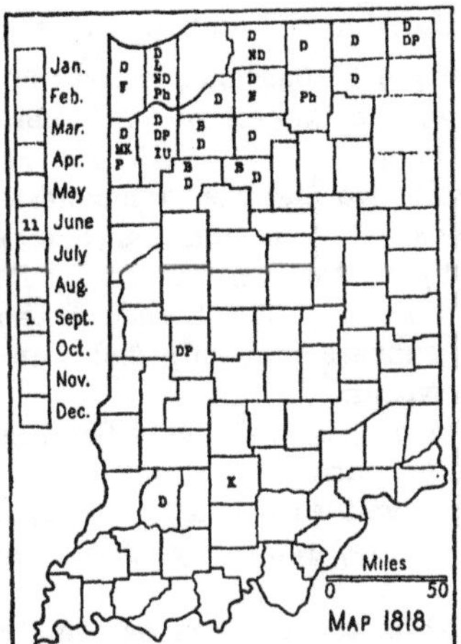

MAP 1818
Scrophularia lanceolata Pursh

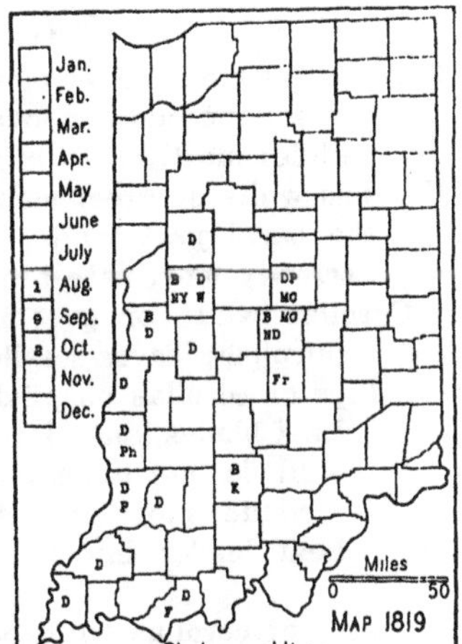

MAP 1819
Chelone obliqua var. speciosa Pennell & Wherry

often found in alluvial bottoms and on stream terraces, most often associated with sugar maple and white oak. It is much less frequently found in woodland not adjacent to streams.

N. Y., Ont., Mich. to Wis., southw. to Va., Ky., Mo., Ark., and Kans.

7505. SCROPHULÀRIA [Bauhin] L.

Sterile stamen brown or purplish; corolla 5-8 mm long, dull; panicle usually broad, 5-18 cm wide, its branches usually spreading; plant flowering mostly after July 15; capsules 4-7 mm long, usually glossy; stem with each side deeply grooved; leaves slender-petioled, the petioles mostly 3-8 cm long and scarcely margined. .... 1. *S. marilandica.*

Sterile stamen greenish yellow; corolla 7-11 mm long; panicle narrowly elongate, 4-8 cm wide, its branches relatively stout and ascending; plant flowering mostly in June; capsules dull, 6-9 mm long; stem with the sides flat or with a very shallow groove; leaves on stout, margined petioles, the petioles 1-3 cm long......... .... 2. *S. lanceolata.*

1. **Scrophularia marilándica** L. Map 1817. Frequent throughout the state. It is usually found in open woodland in moist or dry soils of varying fertility. Sometimes it is found along roadsides and in fallow fields.

The leaves of this species vary from essentially glabrous to densely pubescent. For the convenience of those who wish to recognize the extremely pubescent form by a name, Pennell has called it f. *neglecta* (Rydb.) Pennell. I have this form from Clark, Gibson, and Knox Counties.

Maine to Minn., southw. to S. C., La., and Okla.

2. **Scrophularia lanceolàta** Pursh. (*Scrophularia leporella* Bickn.) Map 1818. Except for three widely separated locations, all of our specimens are from the area north and west of the Wabash River. It is infrequent to rare and usually found in moist or dry and very sandy soil. It is generally found in open, black oak woods or on wooded slopes, and less frequently along roadsides and on the right of way of railroads.

Cape Breton Island to B. C., southw. to N. C., Okla., N. Mex. and Calif.

7507. CHELÒNE [Tourn.] L. TURTLEHEAD

Corollas purple or reddish purple throughout, mostly 30-37 mm long; sepals ciliolate; leaves lanceolate to ovate, the largest on each plant varying from 3-7 cm wide; petioles mostly 5-15 mm long........................1. *C. obliqua* var. *speciosa.*
Corollas white or greenish white throughout or purple at the distal end or rarely only the basal part white, mostly 20-25 mm long; sepals obscurely ciliolate; leaves narrowly lanceolate to elliptic, the largest usually 8-25 mm wide.
  Corollas white or greenish at the distal end, externally only faintly, if at all, purplish; leaves sessile, subsessile or on petioles up to about 5 mm long.
    Lips of corollas purplish within.
      Leaf blades lanceolate or elliptic, relatively firm; spikes usually short.
        Blades not tomentose beneath.........................2. *C. glabra* var. *typica.*
        Blades more or less tomentose beneath...........2a. *C. glabra* f. *tomentosa.*
      Leaf blades linear-lanceolate, relatively thin; spikes tending to elongate..... ....2b. *C. glabra* var. *elongata.*
    Lips of corollas white within, the corollas externally greenish yellow; leaf blades linear to narrowly lanceolate, mostly 1-2 cm wide.
      Blades glabrous or pubescent only on the veins beneath. ....2c. *C. glabra* var. *linifolia.*
      Blades densely pubescent over the entire lower surface...................... ....2d. *C. glabra* var. *linifolia* f. *velutina.*
  Corollas purple at the distal end; petioles 5-20 mm long; leaf blades lanceolate to elliptic-oval, the largest 2-6 cm wide..............2e. *C. glabra* var. *elatior.*

1. **Chelone obliqua** L. var. **speciòsa** Pennell & Wherry. (Bartonia **10**: 19. 1929.) (*Chelone obliqua* of Gray, Man., ed. 7 and Britton and Brown, Illus. Flora, ed. 2.) ROSE TURTLEHEAD. Map 1819. Usually found in low woods and less frequently in springy places in woodland.
Ind. to Iowa and Ark.

2. **Chelone glàbra** L. var. **týpica** Pennell. WHITE TURTLEHEAD. Map 1820. Pennell has divided *Chelone glabra* into several varieties and forms, five of which he cites from Indiana. For the benefit of those who wish to study this species intensively I have listed these forms and given their distribution. The species and its forms grow in wet woods, springy places about lakes, along streams, and in marshes.
Newf., n. Ont. to Minn., southw. to Ga. and Ala.

2a. **Chelone glabra** f. **tomentòsa** (Raf.) Pennell. I have this form from Porter and Spencer Counties.

2b. **Chelone glabra** var. **elongàta** Pennell & Wherry. (Bartonia **10**: 22. 1929.) I have this variety from Dubois, Jennings, and Spencer Counties and Kriebel has collected it in Lawrence County.
Ohio to Ill. and Tenn.

2c. **Chelone glabra** var. **linifòlia** Coleman. (Cat. Fl. Pl. S. Mich. 27. 1874.) Map 1821. This is the common form of the species in our area.
S. Ont. to Man., southw. to Ohio, Ind., and Ill.

2d. **Chelone glabra** var. **linifolia** f. **velùtina** Pennell & Wherry. I have this form from Carroll, Elkhart, Lake, La Porte, and Miami Counties.

2e. **Chelone glabra** var. **elàtior** Raf. (Raf. Med. Fl. **2**: 118. 1830.) I have this form from only Clark County.
Pa., Ind., and Ala.

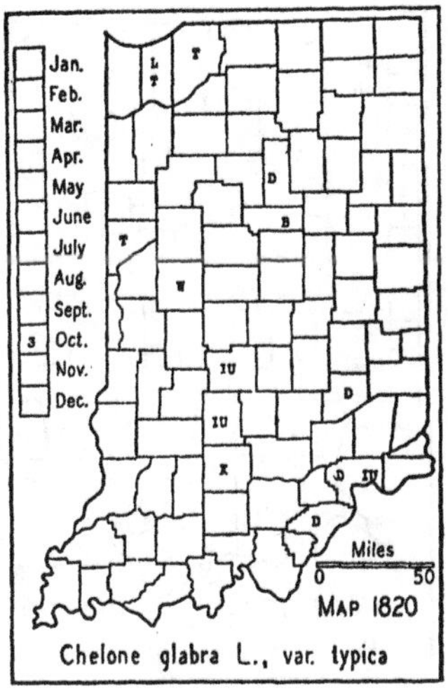

Chelone glabra L., var. typica

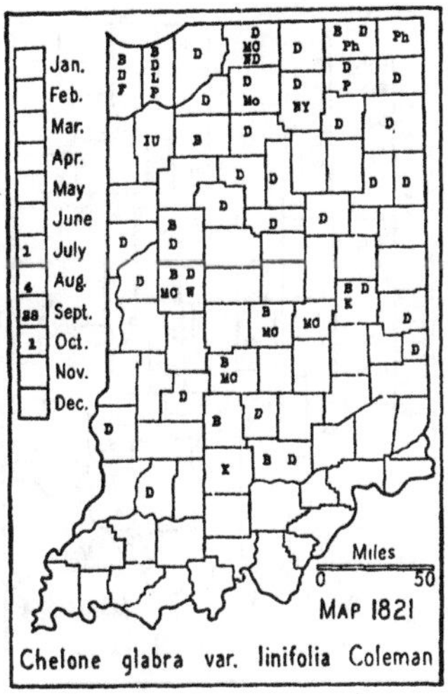

Chelone glabra var. linifolia Coleman

Penstemon tubaeflorus Nutt.

7508. PENSTÈMON Mitchell PENSTEMON

[Key adapted from Pennell's Monograph, loc. cit.]

Corolla glandular-puberulent within on all sides, the throat slightly inflated, obscurely or not at all ridged within, white throughout, 20-25 mm long, the lobes strongly spreading; sepals 3-4 mm long, triangular-ovate, acuminate......1. *P. tubaeflorus.*

Corolla pubescent with glandless hairs within over the bases of the anterior lobes, the throat more inflated and plainly ridged within.

Throat of corolla much inflated and only slightly ridged within anteriorly, the anterior lobes of the corolla little exceeding the posterior ones; sterile filament slightly to moderately bearded.

Corollas purplish (rarely only faintly so), (15) 20-35 mm long, the throat moderately inflated; sepals linear-attenuate, 5-12 mm long, usually conspicuously widely spreading in flower, generally the margins not plainly scarious; anthers never bearded; cauline leaves widely lanceolate to ovate; inflorescence sparingly glandular..........................2. *P. calycosus.*

Corollas white or faintly tinged with purple, 13-30 mm long, the throat amply inflated; sepals ovate or ovate with acuminate tips, 3-9 mm long (conspicuously scarious and caudate-tipped in *Penstemon Digitalis*); anthers usually bearded with a few, stiff, white hairs on the dorsal part (hairs not to be confused with the teeth of the sutures).

Corollas 20-30 mm long; inflorescence decidedly glandular; sepals in anthesis 5-8 mm long, caudate-acuminate, plainly scarious-margined; stem somewhat shining, glabrous, and slightly glaucous; usually found along roadsides, in pastures, and fallow fields, and rarely in woodland...........3. *P. Digitalis.*

Corollas mostly 13-23 mm long; inflorescence in anthesis glabrous or slightly glandular; stems dull, finely pubescent or glabrous; sepals in anthesis 2-5 mm long.

Sepals becoming 5-9 mm long at maturity, more than half the length of the capsule, oval with caudate tips, not at all or only scarcely scarious-margined; corollas 17-23 mm long; lower blades lanceolate, acuminate, rather sharply serrate, the basal ones usually few at anthesis...... ....4. *P. alluviorum.*

Sepals only 2-4 mm long at maturity, less than half the length of the capsule, ovate or somewhat acuminate, plainly scarious-margined; corollas 15-20 (22) mm long; lower blades oblong or oval, rounded, entire or slightly denticulate, the basal ones usually many at anthesis........5. *P. Deamii.*

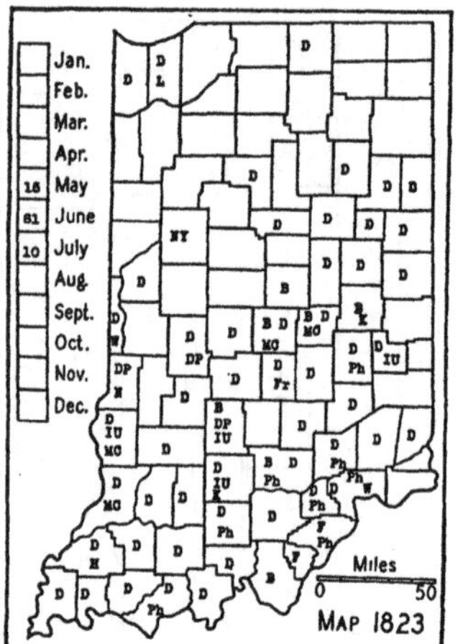

MAP 1823
Penstemon calycosus Small

MAP 1824
Penstemon Digitalis Nutt.

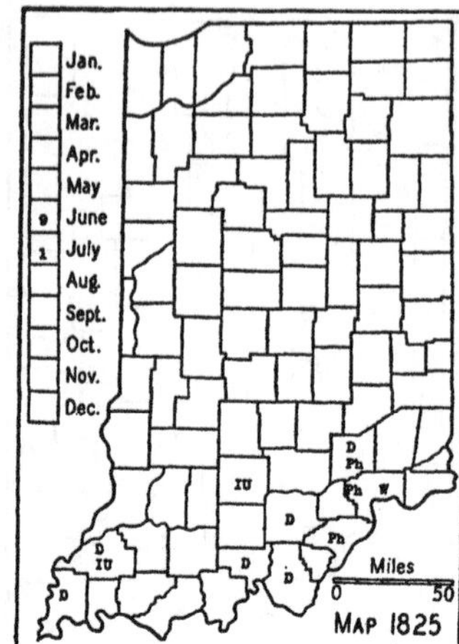

MAP 1825
Penstemon alluviorum Pennell

Throat of corolla narrow, flattened and strongly ridged within, the anterior lobes of the corolla projecting considerably beyond the posterior ones; sterile filament more densely bearded.

Orifice to the throat of the corolla open; cells of anthers longer than wide; lower surface of leaves more or less short-pubescent or pubescent on the midrib and along the principal veins.

Throat of corolla moderately inflated; corolla 20-32 mm long; lower surface of leaves sparsely pubescent all over with long hairs or only on the midrib and larger veins............................6. *P. canescens* var. *typicus*.

Throat of corolla slightly inflated; corolla 17-22 mm long; lower surface of leaves densely pubescent all over with short hairs............7. *P. pallidus*.

Orifice to the throat of corolla closed by the uparching lower lip; corolla 23-28 mm long; cells of anthers about as wide as long; lower surface of leaves usually soon glabrous or only the midrib with long hairs........8. *P. hirsutus*.

1. **Penstemon tubaeflòrus** Nutt. TUBE PENSTEMON. Map 1822. I have this species from only three places in two counties where I found it in dry and very sandy soil on an open, oak, sand ridge and along the railroad about 5 miles south of Vincennes. It was also collected in Vigo County by Evermann.

Ind. to Kans. and Tex.; probably introduced in the Atlantic States from Maine to Pa.

2. **Penstemon calycòsus** Small. Map 1823. This is our most common species and could possibly be found in every county of the state, although it is less frequent in our northern counties. While it prefers moist, alluvial soil along streams and bases of slopes, it is found also on dry, wooded slopes, along roadsides and railroads, and in fallow fields.

Maine, Mich., to Ill., and along the coast to Pa., southw. to n. Ala., Miss., Mo., and Ark.

3. **Penstemon Digitàlis** Nutt. (*Penstemon laevigatus* var. *Digitalis* (Sweet) Gray of Gray, Man., ed. 7 and *Penstemon Digitalis* (Sweet) Nutt. of Britton and Brown, Illus. Flora, ed. 2.) FOXGLOVE PENSTEMON. Map 1824. This species is somewhat frequent throughout the state in both moist and dry soils in various habitats. It often forms large colonies, espe-

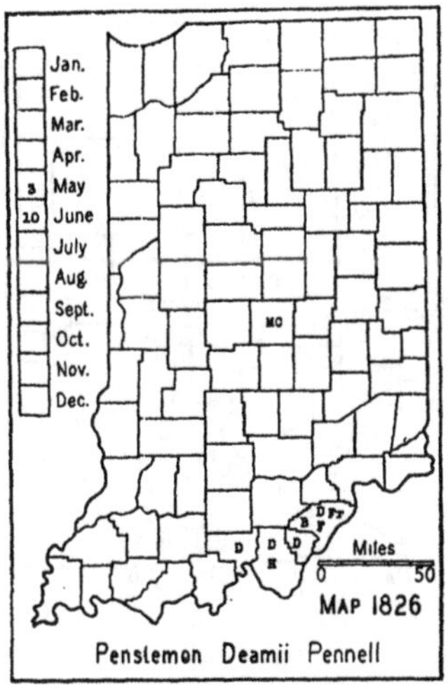

Penstemon Deamii Pennell

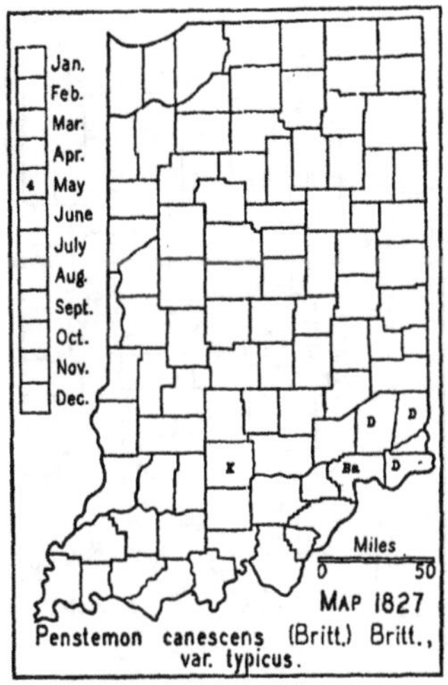

Penstemon canescens (Britt.) Britt., var. typicus.

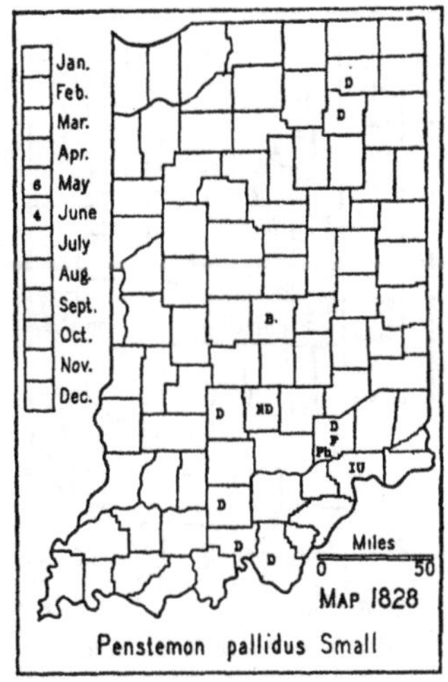

Penstemon pallidus Small

cially in fallow fields in the Illinoian drift area where it is most frequent.

Maine to S. Dak., southw. to n. Ala. and Tex.; absent from the Coastal Plain from Va. southw. and westw. to Ala.

4. **Penstemon alluviòrum** Pennell. Map 1825. This and the next species are white-flowered, rarely partly purple tinged, and are restricted to our southern counties. The species is local and is found in both moist and dry soil, mostly on open slopes in our area.

Along the Ohio and Mississippi Rivers and s. Ohio, s. Ind., and Ky., southw. to Ark., Mo., and Tenn.

5. **Penstemon Dèamii** Pennell. Map 1826. This rare penstemon is local or infrequent in the "knobs" of a few counties along the Ohio River. It is found in poor soil in fallow fields and in open places on oak ridges. In 1932 it was found by Scott McCoy in rich soil in a field in the north part of Indianapolis, Marion County. It is difficult to reconcile these two widely different kinds of habitat for this plant.

Ind. and Ill. (Pennell).

6. **Penstemon canéscens** (Britt.) Britt. var. **týpicus** Pennell. Map 1827. Found locally on white oak slopes in our southeastern counties.

S. Pa. to se. Ind. and n. Ala., thence eastw. to the Coastal Plain line.

7. **Penstemon pállidus** Small. Map 1828. Infrequent to local on dry, wooded or washed slopes. My specimens are mostly from the southern part of the state.

Maine, Mich., n. Ill. to Iowa, southw. to Ga., Tenn., and Ark.

8. **Penstemon hirsùtus** (L.) Willd. EASTERN PENSTEMON. Map 1829. Restricted almost entirely to sandy, gravelly or rocky soils on the dry banks or rocky bluffs along streams and about lakes. Where it is found it is usually frequent to common and may be found in suitable habitats along streams for miles and may be absent in intervening habitats.

Maine to Wis., southw. to Va. and Tenn.

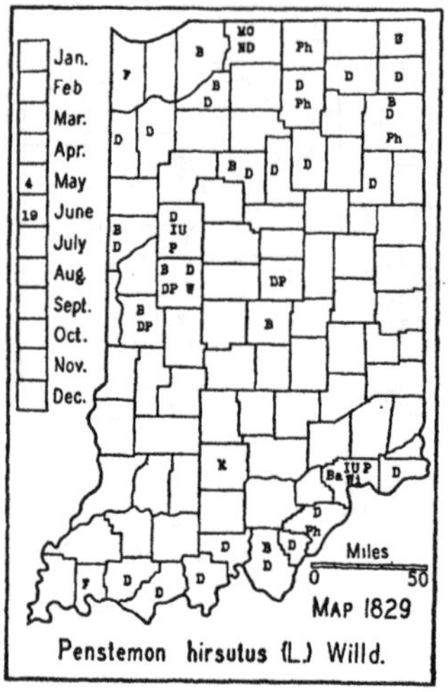

Penstemon hirsutus (L.) Willd.

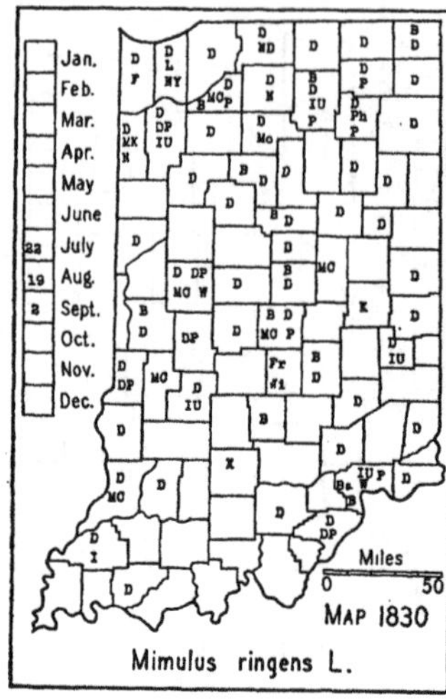

Mimulus ringens L.

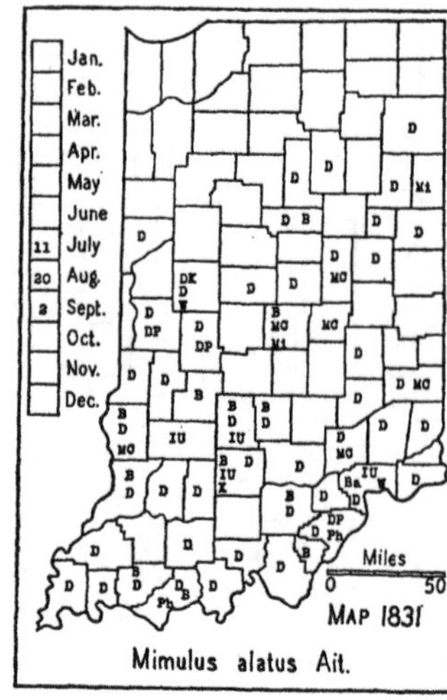

Mimulus alatus Ait.

7513. PAULÒWNIA Sieb. & Zucc.

1. PAULOWNIA TOMENTÒSA (Thunb.) Steud. ROYAL PAULOWNIA. This species has been introduced in a few Ohio River towns and is apparently hardy. I know of a planted tree that is hardy on the "knobs" about 3 miles northwest of Henryville, Clark County. In 1925 I found a tree in a woods just east of no. 10 school about 5 miles southeast of Laconia, Harrison County. It was 10 inches in diameter with a clear bole of about 30 feet. This tree was surrounded by several rootshoots, one of the largest of which I cut off. It measured slightly more than 2 inches in diameter near the base and was 5 years old. In 1935 I again visited this place and found that the large tree had been cut but that there were many seedlings along the roadside about 125 feet distant where the mineral soil was exposed. In 1935 some specimens were sent to me from a "black jack" ridge about 3 miles south of Livonia, Washington County. The letter accompanying the specimens said that there were a few small trees about 15 feet high. Ralph M. Kriebel writes that there are a few trees planted in Bedford, Lawrence County, and that in 1935 he found it as an escape in four abandoned stone quarries in the vicinity of Bedford. It was found growing in the "grout" (small chips of limestone) of these quarries. This habitat observation is very significant and worthy of further investigation.

Some recent authors place this genus in *Bignoniaceae*. (Campbell. The relationships of Paulownia. Bull. Torr. Bot. Club 57: 47-50. 1930.)

Nat. of Japan.

7524. MÍMULUS L. MONKEYFLOWER

[Grant. A monograph of the genus Mimulus. Ann. Missouri Bot. Gard. 11: 99-399. 1924.]

Stems erect; leaves lanceolate, oval or ovate; flowers Hortense Violet (Ridgway).
  Leaves clasping; angles of stem wingless.........................1. *M. ringens.*
  Leaves petiolate; angles of stem more or less winged.................2. *M. alatus.*
Stems diffuse; leaves nearly orbicular; flowers lemon yellow. (See excluded species no. 565, p. 1089)...............................*M. glabratus* var. *Fremontii.*

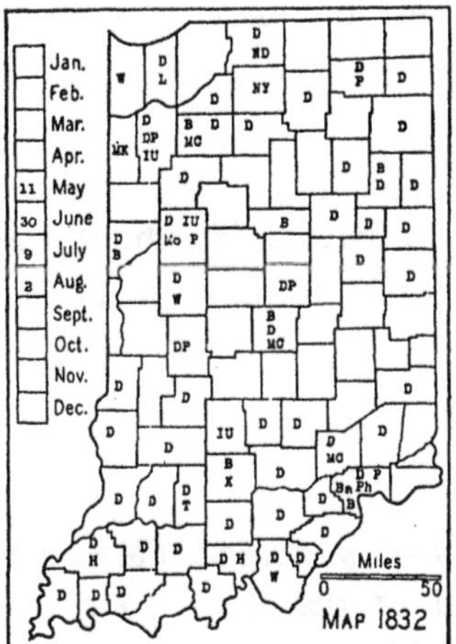

Map 1832

Gratiola neglecta Torr.

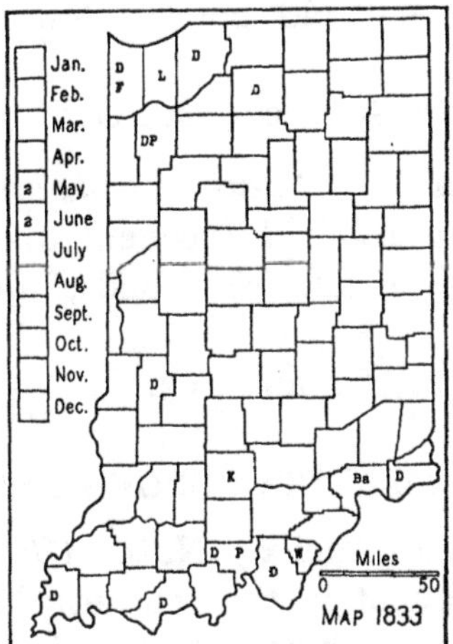

Map 1833

Gratiola virginiana L.

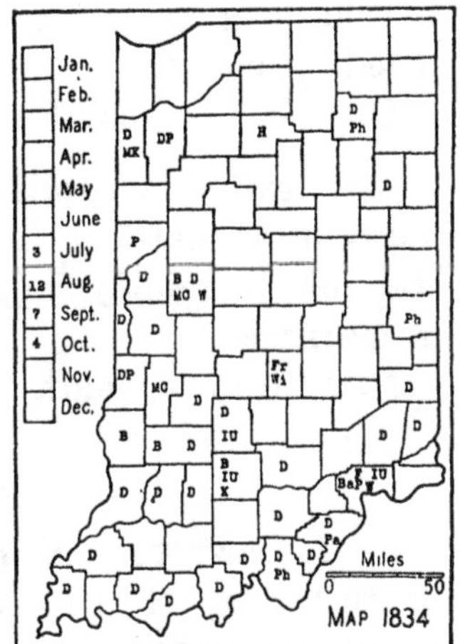

Map 1834

Leucospora multifida (Michx.) Nutt.

1. **Mimulus ríngens** L. Map 1830. Frequent to almost common in the northern two thirds of the state, becoming infrequent to rare in the southern counties. It is found in wet soils both in the open and in the woodland. Usually found in moist or wet soil along streams, in ditches, and about lakes and ponds.

Cape Breton Island, James Bay, and Man., southw. to Ala., La., Okla., and Colo.

2. **Mimulus alàtus** Ait. Map 1831. This species is more or less frequent in all of the southern half of the state, becoming less frequent northward until the Wabash River is reached. North of the Wabash River there are specimens from only Allen and Warren Counties.

Conn. to e. Nebr., southw. to n. Fla. and e. Tex.

7542. GRATÌOLA [Bauhin] L.

Pedicels slender, 10-25 mm long in fruit; stems relatively slender, glandular-puberulent or sometimes nearly glabrous.................................1. *G. neglecta.*

Pedicels stout, mostly less than 5 mm long, rarely up to 12 mm long at maturity; stems relatively fleshy, glabrous or rarely somewhat glandular above...2. *G. virginiana.*

1. **Gratiola neglécta** Torr. (*Gratiola virginiana* of Gray, Man., ed. 7 and Britton and Brown, Illus. Flora, ed. 2, not L.) Map 1832. Infrequent to frequent throughout the state except in the northern tier of counties. It grows in moist or wet soil in bare places, and, where it is found, it often forms large colonies. My notes say "abundant over an acre or more in acid soil in a one year old, fallow cornfield in Warren County," and "abundant over more than an acre in moist, sandy soil in a fallow field about 2 miles northwest of Monticello, White County." It is most often found

in old logging roads, about old hog wallows, and on the borders of dried-up ponds.

Que. to B. C., southw. to Ga., Tex., and Calif.

2. **Gratiola virginiàna** L. (*Gratiola sphaerocarpa* Ell. and *Gratiola mesochora* Peattie.) Map 1833. Infrequent to very rare throughout the state on the muddy shores of artificial ponds, in ditches, and in wet places in marshes.

N. J. to Iowa, southw. to Fla. and Tex.

7545A. LEUCÓSPORA Nutt.

1. **Leucospora multífida** (Michx.) Nutt. (*Conobea multifida* (Michx.) Benth.) Map 1834. Frequent in the southern half of the state, soon becoming infrequent to rare northward and probably absent or very rare in our northern counties. While well distributed, it is rarely found in colonies but more or less as scattered plants, except on stretches of the slope of the bank of the Ohio River, where it may be present for considerable distances. It prefers a moist, sandy soil and is almost entirely restricted to bare places on sand bars and muddy shores of streams and rarely in cultivated fields and open woodland.

Ohio to Iowa and Kans., southw. to Ga. and s. Tex.

7548. HYDRANTHÈLIUM HBK.

1. **Hydranthelium rotundifòlium** (Michx.) Pennell. (Pennell. Monograph Scrophulariaceae of eastern North America, p. 629. 1935.) (*Bacopa rotundifolia* (Michx.) Wettst. and *Bramia rotundifolia* (Michx.) Britt.) Map 1835. Infrequent in sink holes in Lawrence, Orange, and Washington Counties, but not seen in sink holes in other counties where the same habitat occurs. The specimen found in Warrick County was found in the old canal near Millersburg. It was very common in the ponds where it was found, although stock had injured it.

Ind. and Tenn. to Mont., southw. to Colo. and Tex.

7562. LINDÉRNIA All.

Mature seed pale yellow, averaging 0.4 mm long and mostly twice or three times as long as wide; leaf blades 1-3 cm long, the lower ones generally narrowed at the base; pedicels shorter or longer than the leaves; later corollas falling unopened, the flowers cleistogamous.

Pedicels as long as or longer than the leaves, 10-20 mm long, usually divaricately spreading..1. *L. dubia* var. *typica.*

Pedicels mostly shorter than the leaves, usually less than 10 mm long, and mostly ascending..1a. *L. dubia* var. *major.*

Mature seed a brownish yellow, averaging 0.3 mm long and mostly one and a half to two times as long as wide; leaf blades 0.5-1.5 cm long, nearly all rounded and widest near the base; pedicels much longer than the leaves; corollas all opening, not cleistogamous..2. *L. anagallidea.*

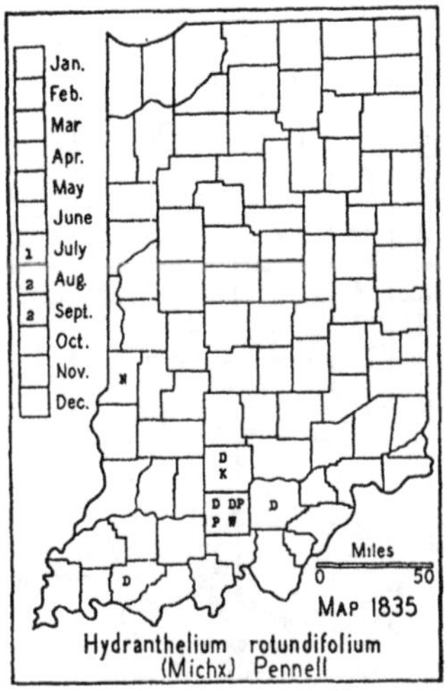

MAP 1835 Hydranthelium rotundifolium (Michx.) Pennell

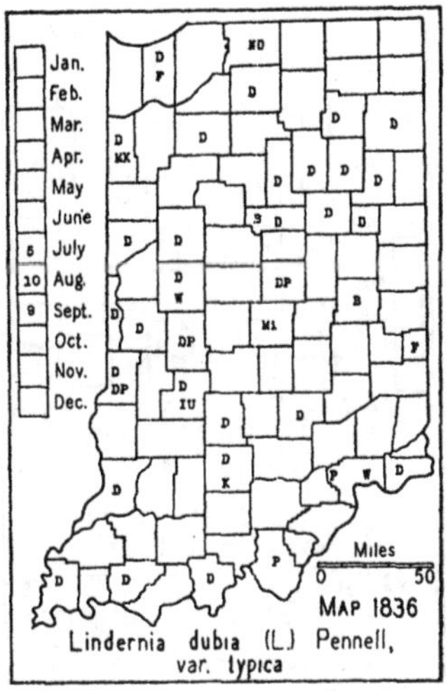

MAP 1836 Lindernia dubia (L.) Pennell, var. typica

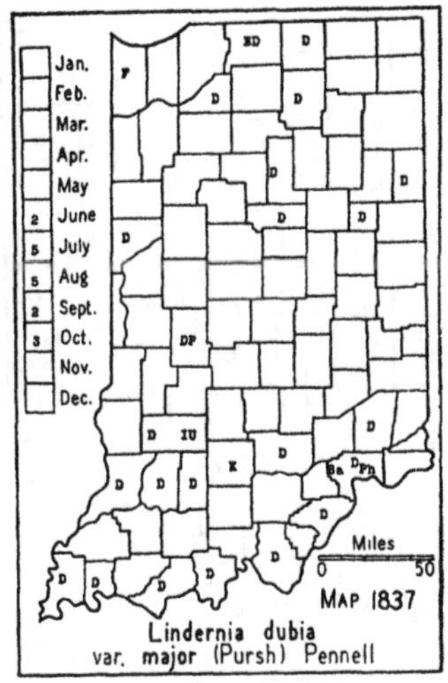

MAP 1837 Lindernia dubia var. major (Pursh) Pennell

1. **Lindernia dùbia** (L.) Pennell var. **týpica** Pennell. (*Ilysanthes dubia* (L.) Barnh.) Map 1836. Probably infrequent to frequent throughout the state. It grows in moist or muddy bare places about ponds, on bars and on the banks of streams and ditches, and in logging roads in woodland.

W. Vt. to e. N. Dak., southw. to Fla. and e. Tex.

1a. **Lindernia dubia** var. **màjor** (Pursh) Pennell. Map 1837. This form is probably nearly as common as the preceding one and as well distributed. The two forms of this species are not well marked and it is sometimes difficult to say to which form a specimen belongs. The habitats are the same as those of the preceding variety.

N. S. to Minn., southw. to Fla. and La.

2. **Lindernia anagallídea** (Michx.) Pennell. (*Ilysanthes anagallidea* (Michx.) Rob.) Map 1838. This species is infrequent and all of my specimens are from the western and southern parts of the state. The habitats are the same as those of the preceding varieties, but it prefers a more sandy soil.

N. H. to N. Dak., southw. to Fla. and Tex.

7579. VERÓNICA [Bauhin] L. SPEEDWELL

Leaves of the stem and of the branches below the flowers opposite, those subtending the flowers alternate (rarely a few flowers in the axils of opposite leaves in *Veronica persica*); flowers solitary in the axils of the upper leaves.
- Styles hidden between the lobes of the capsules, appearing obsolete, less than 0.5 mm long.
  - Stems glabrous; capsules glabrous....................1. *V. peregrina* var. *typica*.
  - Stems glandular-pubescent; capsules of Indiana specimens glabrous, those west of our area more or less pubescent...........1a. *V. peregrina* var. *xalapensis*.
- Styles mostly 0.5-2 mm long.
  - Pubescence of the stem curved upward, the hairs short, mostly 0.2-0.4 mm long and not conspicuously multicellular; perennial from a subterranean rhizome.......2. *V. serpyllifolia*.
  - Pubescence of stem spreading or partly upwardly curved, the spreading hairs not more than 0.5 mm long and conspicuously multicellular; the curved hairs, if any, like those of the preceding species; annual.

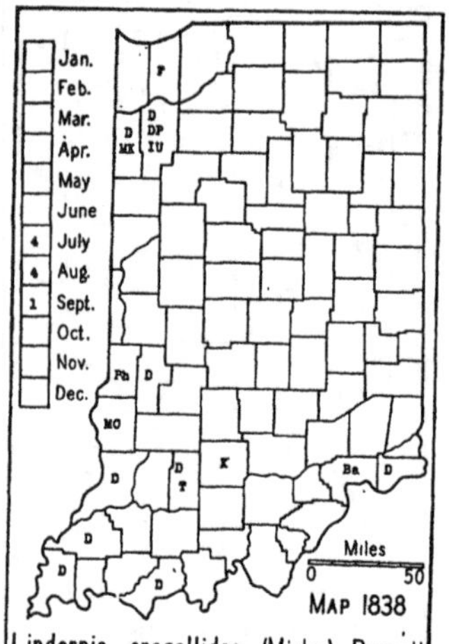

MAP 1838
Lindernia anagallidea (Michx.) Pennell

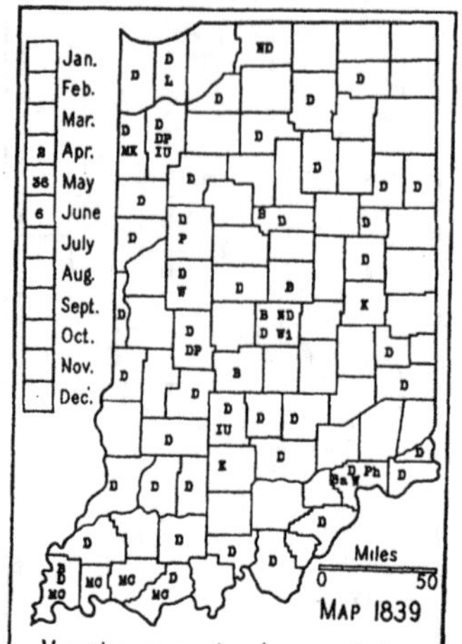

MAP 1839
Veronica peregrina L., var. typica

MAP 1840
Veronica peregrina var. xalapensis (HBK.) Pennell

Pedicels shorter than the calyx; leaves sessile except the lowermost ones....... ....3. *V. arvensis.*

Pedicels longer than the calyx.

Lobes of capsule somewhat acutish, strongly divergent, the most distal point of each near the lateral margin; styles as long as or exceeding the capsules, usually 1.5-2 mm long; pedicels mostly 12-25 mm long..4. *V. persica.*

Lobes of capsule rounded, not strongly divergent, the most distal point about midway between the style and the lateral margin; styles usually not exceeding the capsules, about 1 mm long. (See excluded species no. 567, p. 1089)..*V. agrestis.*

Leaves all opposite; flowers in axillary racemes—these sometimes reduced to a single flower or sometimes appearing terminal.

Capsules pubescent; stem, leaves, pedicels, and sepals pubescent; plants of dry soil.

Leaves oval or obovate, petiolate; pedicels shorter than the calyx.............. ....5. *V. officinalis.*

Leaves ovate, nearly or quite sessile; pedicels longer than the calyx............ ....6. *V. Chamaedrys.*

Capsules glabrous or with a few minute gland-tipped hairs; stems, leaves, pedicels, and sepals glabrous or sparsely glandular-puberulent (or rarely pubescent in *Veronica scutellata*); plants more or less semi-aquatic.

Capsule conspicuously wider than long, strongly two-lobed, mostly 4-5 mm wide; pedicels filiform and reflexed in fruit.......................7. *V. scutellata.*

Capsule not conspicuously wider than long, and scarcely or not at all two-lobed.

Leaves all petiolate; plants strictly glabrous; pedicels 5-13 mm long.......... ....8. *V. americana.*

Leaves not all petiolate, at least the upper ones sessile and clasping; pedicels 3-6 mm long; plants more or less glandular-pubescent, at least in the inflorescence.

Sepals acute or acuminate; capsule scarcely wider than long, not at all or only slightly notched at the apex; leaf blades lanceolate to ovate-lanceolate, serrate with close teeth (four or more to 1 cm), those of young autumnal shoots petiolate.

Pedicels, rachis, and upper part of stem glabrous or nearly so; sepals acute to slightly acuminate; styles 1.5-3 mm long; leaf blades oblong-ovate, mostly widest about the middle, slightly serrate to nearly entire. (See excluded species no. 568, p. 1089)..............*V. Anagallis-aquatica.*

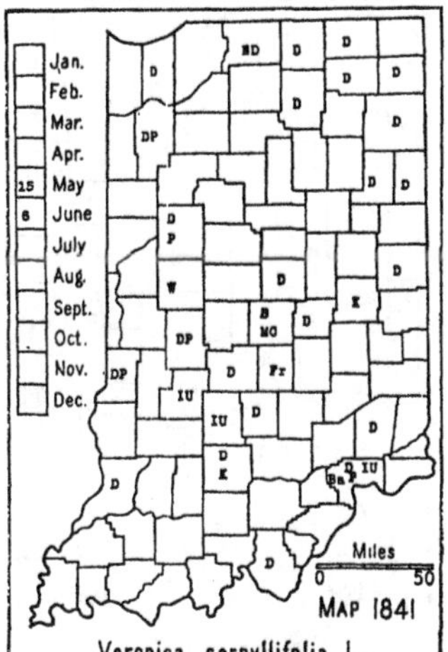

Veronica serpyllifolia L.

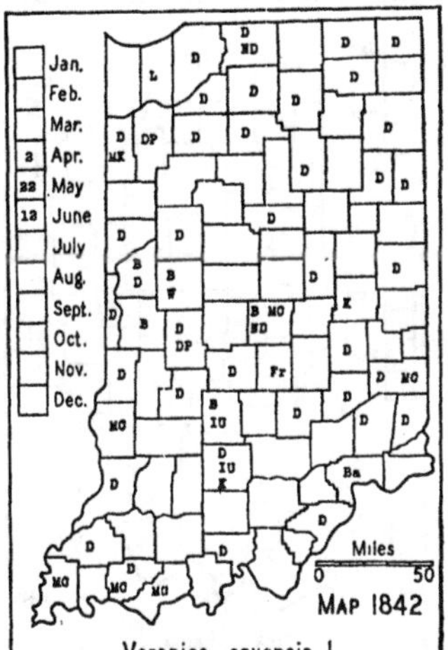

Veronica arvensis L

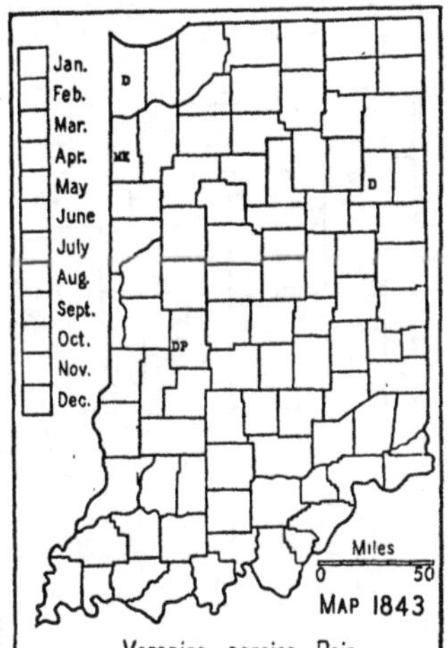

Veronica persica Poir

Pedicels, rachis, and upper part of stem pubescent with gland-tipped hairs; sepals strongly acuminate; styles 1-1.5 mm long; leaf blades lanceolate or broadly lanceolate, widest near the base, usually more strongly serrate..9. *V. glandifera.*

Sepals obtuse to acutish; capsule obviously wider than long, plainly notched at the apex; leaf blades linear-lanceolate, crenate-serrate with remote teeth, all clasping..10. *V. connata.*

1. **Veronica peregrìna** L. var. **týpica** Pennell. PURSLANE SPEEDWELL. Map 1839. This is a common and obnoxious weed in cultivated grounds throughout the state except the northern counties where it is rare. When once it becomes well established in gardens, it is difficult to eradicate or to keep in control.

Maine to Minn., southw. to Fla. and Tex.; also introduced on the Pacific coast from B. C. to Oreg.; Bermuda Islands, W. I., and w. Eu.

1a. **Veronica peregrina** var. **xalapénsis** (HBK.) Pennell. Map 1840. This is a variety rare to infrequent in all parts of the state except in the sandy areas where it is more frequent but I have never seen it entirely displace the species.

Que. to Yukon, southw. to Ga., Calif., and Guatemala; also in S. A.

2. VERONICA SERPYLLIFÒLIA L. THYMELEAF SPEEDWELL. Map 1841. Infrequent to frequent throughout the state in pastures, open woodland, fallow fields, and lawns.

Nat. of Eu. and Asia; naturalized in N. A. from Newf., Que. to Minn., southw. to Ga. and Mo., and on the Pacific coast from B. C. to n. Calif.

3. VERONICA ARVÉNSIS L. CORN SPEEDWELL. Map 1842. Frequent to common in all parts of the state in pastures, open woodland, fallow fields, lawns, and waste places.

Nat. of Eurasia; Newf. to Minn., southw. to Ala. and Tex., also in Colo., Utah, and on the Pacific coast from B. C. to Calif.

4. VERONICA PÉRSICA Poir. Map 1843. This species has been found in only four counties in the state. I first found it as a lawn weed in Bluffton in 1917 and later in two other parts of Bluffton a half mile

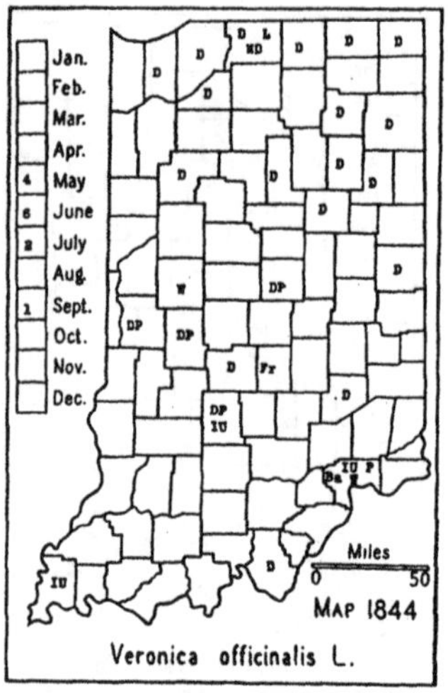

Veronica officinalis L.

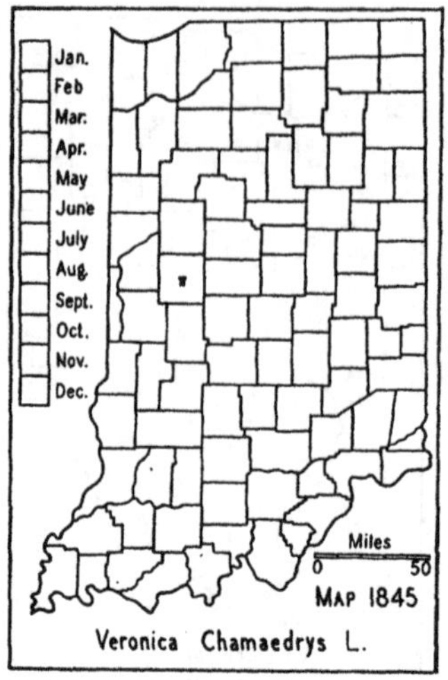

Veronica Chamaedrys L.

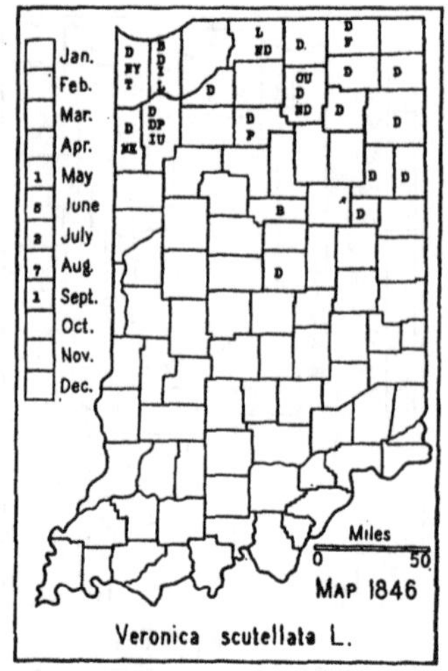

Veronica scutellata L.

distant. The fact that it was still persisting in 1936 shows it to be well established in this locality. It has been found in Goodland, Newton County, by Madge McKee. Grimes found it in waste ground in Russellville, Putnam County. It doubtless could be found in many other places.

Nat. of Eurasia; Newf. to Man., s. Alaska, southw. to Fla., Tex., and Calif.

5. VERONICA OFFICINÀLIS L. COMMON SPEEDWELL. Map 1844. This species prefers a dry and rather sandy soil. It is infrequent in the lake area and progressively less frequent to rare southward. It is most frequent in pastures and open woodland.

Nat. of Eurasia; Newf. to Wis., southw. to N. C. and Tenn.; also in e. S. Dak., and near the Pacific coast in Wash. and Oreg.

6 VERONICA CHAMAÈDRYS L. GERMANDER SPEEDWELL. Map 1845. This species was first collected in 1924 by A. R. Bechtel in the Crawford woods near Crawfordsville, Montgomery County. He made a second collection in 1935 and he says it is well established there.

Nat. of Eu.; Newf. to Wis., southw. to N. C. and Tenn.; also in e. S. Dak. and near the coast in Wash. and Oreg.

7. **Veronica scutellàta** L. SKULLCAP SPEEDWELL. Map 1846. Infrequent to frequent in the lake area with two stations south of it. It prefers the dried-up borders of ponds that are well covered with old leaves. While it sometimes grows in marshes and in muck it prefers to root in decaying vegetation.

Newf. to Mackenzie, southw. to Va., Ill., Colo., and Calif.

8. **Veronica americàna** (Raf.) Schwein. May 1847. This is a water loving plant which is found in swampy places. Our only specimen was collected by Nieuwland in the Mineral Springs Bog, Porter County.

Newf. to Alaska, southw. to N. C., Mo., Calif., and Mex.

9. **Veronica glandifera** Pennell.* (Torreya 19: 170. 1919.) Map 1848.

* For this plant Fernald proposes the name **Veronica Anagallis-aquatica f. anagalliformis** (Boreau) G. Beck. Rhodora 41: 564. 1939.

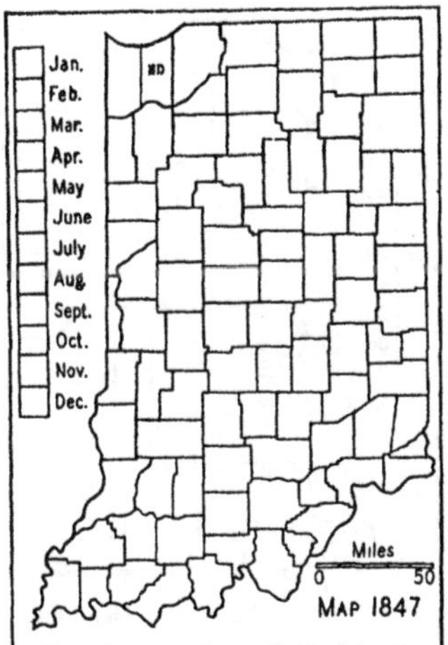

Veronica americana (Raf.) Schwein.

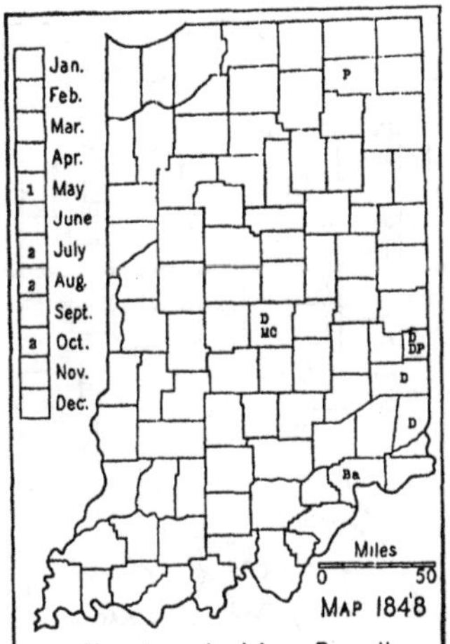

Veronica glandifera Pennell

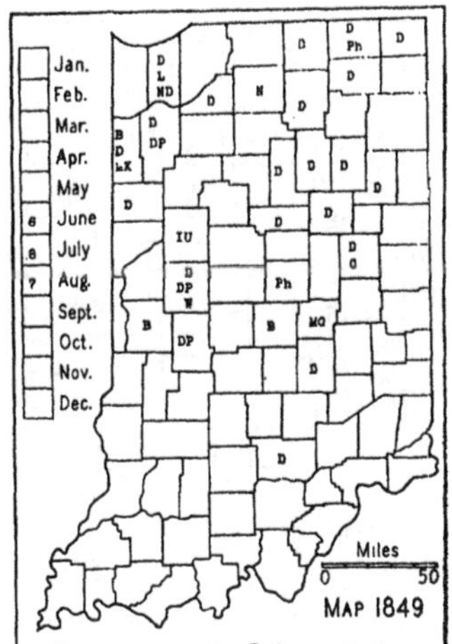

Veronica connata Raf., var. typica

I have this species from three of our southeastern counties where I found it on the borders of small streams and in the outlets of springs. Scott McCoy also found it along Crooked Creek in Marion County about three miles northwest of Indianapolis.

Pa. to Ind., southw. to N. C.

10. **Veronica connàta** Raf. var. **týpica** Pennell.† Map 1849. This species has been sparingly found in the lake area and in a few places south of it. It is found on the muddy shores of dredged ditches, streams, and lakes.

Mass. and Ont. to N. Dak., southw. to Pa., Tenn., and Okla.

7579A. VERONICÁSTRUM [Heist.] FABRICIUS

1. **Veronicastrum virgínicum** (L.) Farw. (*Veronica virginica* L. and *Leptandra virginica* (L.) Nutt.) CULVER'S-PHYSIC. Map 1850. Somewhat frequent in the lake and prairie areas of the state and infrequent to rare or absent elsewhere. It is found in small colonies or as scattered plants and possibly originally occurred in small prairie openings in every county of the state. Its moisture requirements vary from those of a marsh to a dry, wooded slope. The plant is peculiar in that one can rarely predict where it may be found and that it has no special plant associate. On the whole, it seems to prefer sandy soil and prairie habitats.

The plants of this species vary greatly and some of the extremes have been named, but Pennell regards the species as polymorphic. The leaves vary from 3-6 in a whorl; the pubescence of the stem and lower surface of the leaves from glabrous to velutinous; and the flowers from white to purplish. The plant has long been used in medicine and is known to the eclectic physician as Leptandra, one of its generic synonyms.

Mass. to Man., southw. to Fla. and Tex.

† For this plant Fernald proposes the name **Veronica salina** Schur. Rhodora 41: 568. 1939.

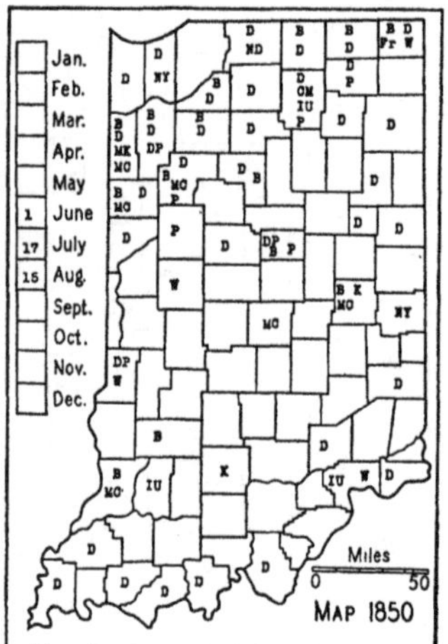

Veronicastrum virginicum (L.) Farw.

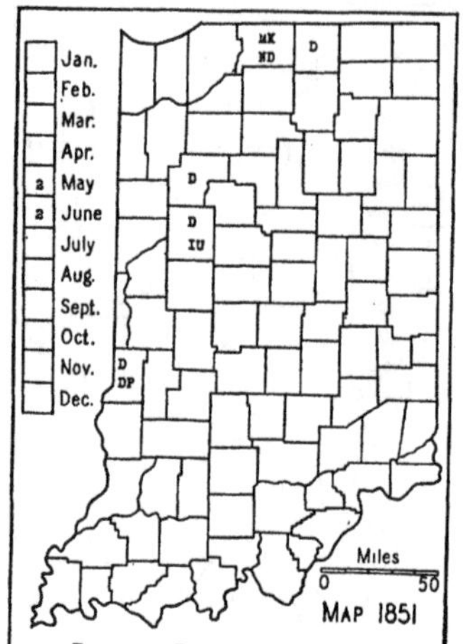

Besseya Bullii (Eaton) Rydb.

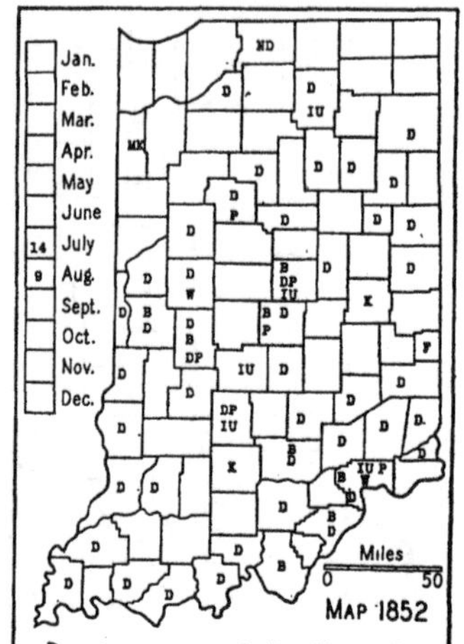

Dasistoma macrophylla (Nutt.) Raf.

7583A. BÉSSEYA Rydb.

1. **Besseya Búllii** (Eaton) Rydb. (*Synthyris Bullii* (Eaton) Heller.) Map 1851. Very local. Usually only one or a few plants are found at a place. Seemingly it prefers a slightly acid and gravelly soil and is found on or near the brink of high, gravelly banks of streams. In White County I found it on the east bank of the Tippecanoe River about a mile northeast of Buffalo, where scattered plants were found for about 50 feet and associated with *Berberis canadensis* and *Pedicularis canadensis*. Both this species and the *Berberis* were restricted to the edge of the bank.

S. Mich. to s. Minn., southw. to s. Ohio, Ill., and Iowa.

7602. DASÍSTOMA Raf.

1. **Dasistoma macrophýlla** (Nutt.) Raf. (*Seymeria macrophylla* Nutt., and *Afzelia macrophylla* (Nutt.) Ktze.) Map 1852. Infrequent to frequent except in the northern counties, where it is rare or absent. This is a woodland species found principally in dry soil along streams. It is a coarse perennial usually turning black on drying and soon beginning to disintegrate so that specimens more than twenty years old become very brittle. The largest plant I have seen was 7 feet high, and another large plant was 6 feet high with a lower side branch 4 feet long.

Ohio to Nebr., southw. to n. Ala. and ne. Tex.

7604. GERÁRDIA L. GERARDIA

Pedicels, in flower, mostly shorter or 1-2 times as long as the calyx.

Calyx lobes short and acuminate, 0.5-2 mm long, usually about 1 mm long; sinuses between the calyx lobes usually very broad and rounded; corollas 20-35 mm long; capsules globose or globose-ovoid; branches widely spreading....1. *G. purpurea.*

Calyx lobes longer, 1.5-3.5 mm long, mostly about 2 mm long, acute; sinuses between the calyx lobes usually acute or rounded and much narrower than those of the preceding species; corollas 15-25 mm long; capsules globose or cylindric.

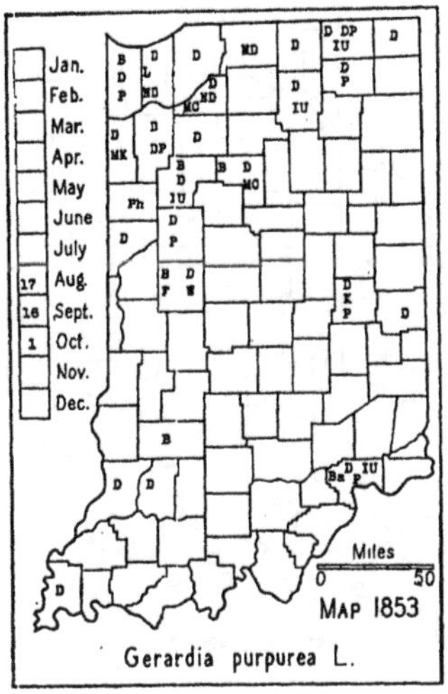

MAP 1853
Gerardia purpurea L.

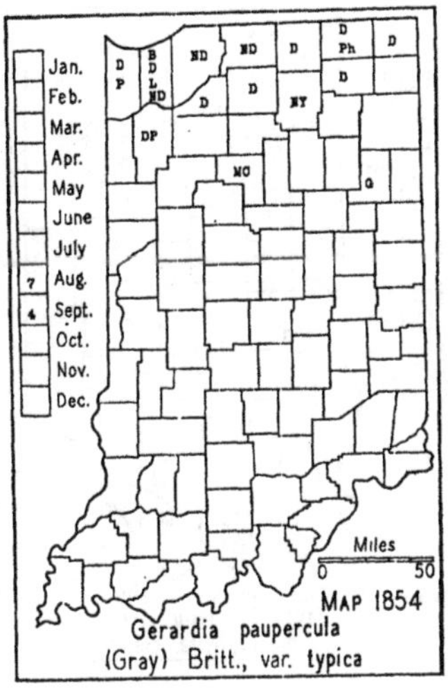

MAP 1854
Gerardia paupercula (Gray) Britt., var. typica

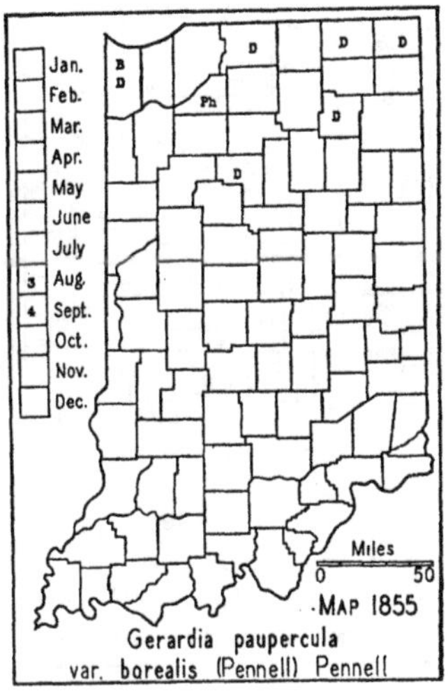

MAP 1855
Gerardia paupercula var. borealis (Pennell) Pennell

Capsules cylindric, decidedly longer than wide; corollas 18-25 mm long, the upper lobes only slightly spreading; pedicels and branches strongly ascending; leaves very scabrous above. (See excluded species no. 569, p. 1090)..*G. aspera*.

Capsules globose or nearly so; corollas 15-20 (23) mm long, the upper lobes reflexed-spreading; pedicels and branches ascending or somewhat spreading; leaves densely scabrous to nearly glabrous above.

Anthers rather densely white-villous; corolla 15-20 (23) mm long, campanulate, its throat relatively broad, the lobes spreading; in older buds the corolla broadly rounded, its profile being widely oblong to obovate-oblong; styles 7-10 mm long..............................2. *G. paupercula* var. *typica*.

Anthers sparingly villous-pubescent with pale brownish or white hairs; corolla 10-17 mm long, tubular-campanulate, its throat narrower, the lobes projecting or slightly spreading; in older buds the corolla narrowly rounded, its profile being oblong to oblanceolate; styles 6-8 mm long..2a. *G. paupercula* var. *borealis*.

Pedicels, in flower, 2-6 times as long as the calyx.

Seed dark brown or blackish; plants relatively dark green, tending to blacken in drying; corolla purplish; leaves mostly 1-6 mm wide.

Calyx lobes 0.2-1 mm long; capsules usually 3-4 mm long; anthers villous; axillary fascicles not developed..........................3. *G. tenuifolia* var. *typica*.

Calyx lobes mostly 1-2 mm long; capsules usually 5-7 mm long.

Axillary fascicles not at all or only slightly developed; anthers densely villous; leaves and branches spreading..........3a. *G. tenuifolia* var. *macrophylla*.

Axillary fascicles usually conspicuously developed; anthers sparingly villous; leaves and branches ascending, the former usually more decidedly scabrous on the upper surface.....................3b. *G. tenuifolia* var. *parviflora*.

Seeds yellow or yellowish brown; plants yellowish green, not tending to blacken in drying; corollas pinkish.

Stems conspicuously striate-angled, the angles scabrellous; corolla lobes truncate; stigmas 1.5-2 mm long; capsules 4-5 mm long..............4. *G. Skinneriana*.

Stems less conspicuously striate-angled, the angles glabrous or nearly so; corolla lobes more or less emarginate; stigmas 1-1.5 mm long; capsules 3-4 mm long.. ....5. *G. Gattingeri*.

1. **Gerardia purpùrea** L. (*Agalinis purpurea* (L.) Britt.) PURPLE GERARDIA. Map 1853. This species is infrequent to frequent in the northwestern part of the state, where its habitat is frequent, and local or absent in other parts of the state where its habitat is absent. Its preferred

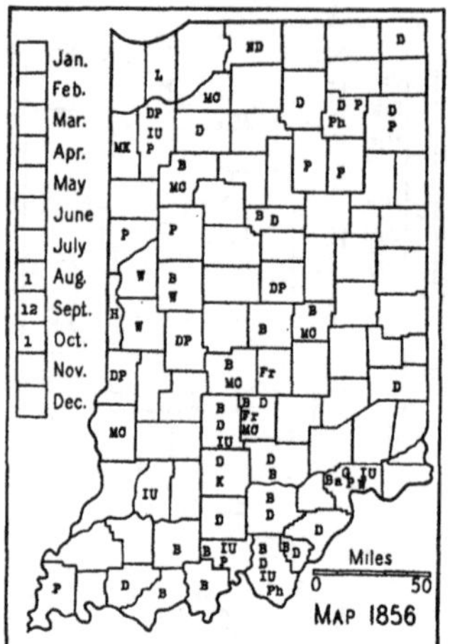

Gerardia tenuifolia Vahl, var. typica

Gerardia tenuifolia var. macrophylla Benth.

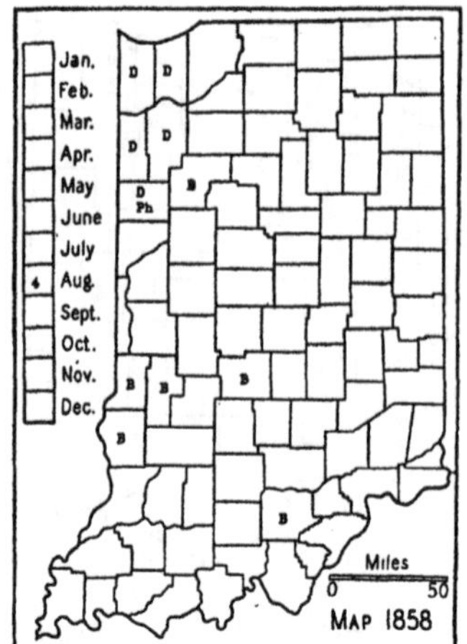

Gerardia tenuifolia var. parviflora Nutt.

habitats are moist, sandy soil on interdunal flats, in marshes, and springy places, wet prairies, and, in the southern part of the state, in hard, white clay soil in wet, open sweet gum woods and fallow fields.

Mass. to Minn., southw. to Fla. and Tex.

2. **Gerardia paupércula** (Gray) Britt. var. **týpica** Pennell. (*Agalinis paupercula* (Gray) Britt.) Map 1854. Infrequent in the lake area where it is found on the low borders of lakes or on interdunal flats. It grows in moist, sandy or gravelly places and on marshy shores.

N. B. to Pa. and Minn.

2a. **Gerardia paupercula** var. **boreàlis** (Pennell) Pennell. (Proc. Acad. Nat. Sci. Philadelphia **81**: 159. 1929.) Map 1855. The variety is less frequent than the typical form and is found in similar habitats.

Que. to Minn., chiefly in the St. Lawrence Valley and in the Upper Mississippi Valley.

3. **Gerardia tenuifòlia** Vahl var. **týpica** Pennell. (*Agalinis tenuifolia* (Vahl) Raf.) Map 1856. This species is probably found throughout the state although it may not be present in the dune area. Infrequent in the northern counties and frequent in the southern counties. With the exception of an intermediate form all of my specimens were found on white and black and white oak slopes and on chestnut oak ridges.

The extreme variability of this species has led authors to divide it into species and varieties. The well known botanist, E. L. Greene, found a very wideleaf form near Ridgeville, Indiana, which he described as a new species. The forms seem to intergrade and are so perplexing that I have copied the section of Pennell's key to this species and its varieties and I have indicated my specimens on the maps as he has named them.

Maine to Mich. and Mo., southw. to Ga. and La.

3a. **Gerardia tenuifolia** var. **macrophýlla** Benth. (*Agalinis Besseyana* Britt.) Map 1857. This variety is a very common form of the species and is somewhat frequent throughout the state. It is found in both dry

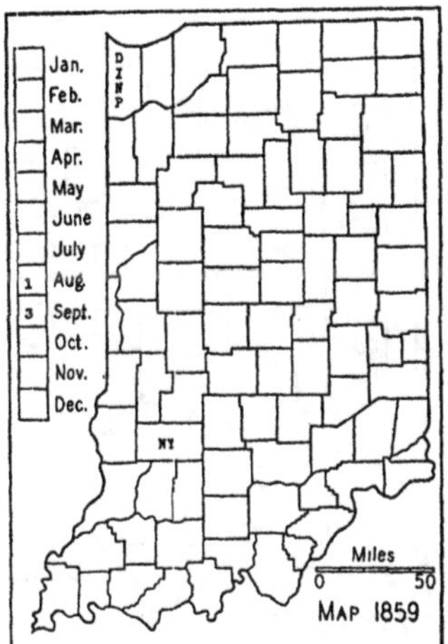

Gerardia Skinneriana Wood

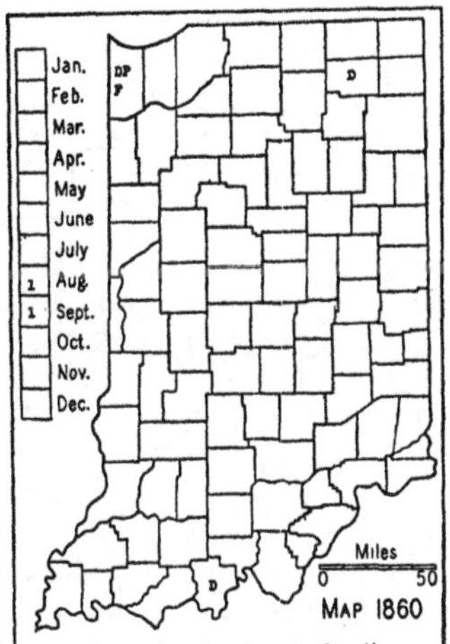

Gerardia Gattingeri Small

Tomanthera auriculata (Michx.) Raf.

and moist habitats but is more common in moist places about lakes and on alluvial areas and banks of streams. It is also found on moderate slopes in woodland.

Pa. to se. Minn., southw. to Miss. and Okla.

3b. **Gerardia tenuifolia** var. **parviflòra** Nutt. Map 1858. This variety is found in habitats similar to those of the preceding variety.

Que. to Man., southw. to Wyo. and Okla.

4. **Gerardia Skinnerìana** Wood. (*Agalinis Skinneriana* (Wood) Britt.) SKINNER GERARDIA. Map 1859. This rare species is known only from the type locality in Greene County and in moist sands of the northwestern part of Lake County.

Ont. to Wis., southw. to se. Kans.

5. **Gerardia Gáttingeri** Small. (*Agalinis Gattingeri* Small.) GATTINGER GERARDIA. Map 1860. Known only from three widely separated counties. In sterile soil at the bases of wooded slopes or on the crests of ridges.

Ont. to Minn., southw. to Ala. and Tex.

7604A. TOMANTHÈRA Raf.

1. **Tomanthera auriculàta** (Michx.) Raf. (*Gerardia auriculata* Michx. and *Otophylla auriculata* (Michx.) Small.) Map 1861. I found this species in a wet prairie in Benton County about a mile southwest of Fowler. It was first found by Elmore Barce who told me where to look for it. It was found in Lake County by H. C. Benke. Pennell reports a specimen of this species in the herbarium of the Academy of Natural Sciences of Philadelphia, which was collected by Dr. Short in the "barrens" of Indiana. There is a specimen in the herbarium of Wabash College collected by A. Clapp in the vicinity of New Albany, Floyd County, in 1837.

N. J. to Minn., southw. to Ala. and Tex.

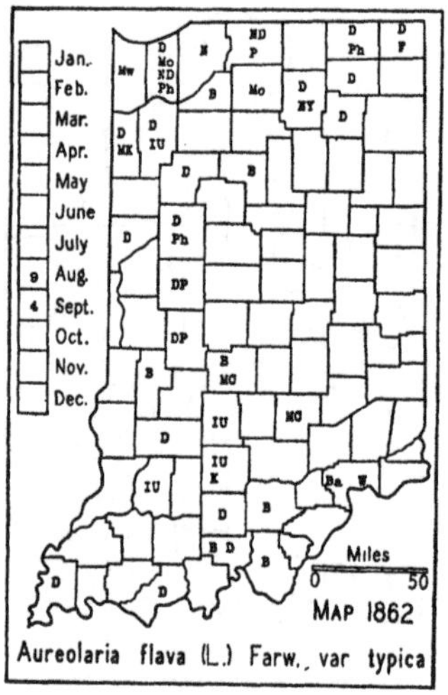

MAP 1862

Aureolaria flava (L.) Farw., var typica

MAP 1863

Aureolaria flava var. macrantha Pennell

MAP 1864

Aureolaria virginica (L.) Farw.

7604B. AUREOLÀRIA Raf.

Stems glabrous and more or less glaucous; capsules glabrous.
- Calyx lobes 2-5 mm long; corollas 35-40 mm long............1. *A. flava* var. *typica.*
- Calyx lobes 5-14 mm long; corollas 35-60 mm long......1a. *A. flava* var. *macrantha.*

Stems more or less puberulent or pubescent, at least above the base, some with glandular hairs also; capsules more or less pubescent.
- Pubescence glandless; perennials.
  - Capsule pubescent at maturity, 12-15 mm long; pedicels 1.5-3 mm long; corollas 30-35 mm long; flowering in July..........................2. *A. virginica.*
  - Capsule glabrous, 15-23 mm long; pedicels 3-25 mm long; corollas mostly 45-55 long; flowering in August......................3. *A. grandiflora* var. *pulchra.*
- Pubescence more or less glandular.
  - Upper part of stems closely pubescent, not at all or only slightly glandular; leaves puberulent, scarcely or not glandular; capsules narrowly ellipsoid, usually 9-11 mm long.........................4. *A. pedicularia* var. *typica.*
  - Upper part of stems glandular-pubescent to hirsute; leaves glandular-puberulent to pubescent; capsules ellipsoid to broadly ellipsoid, usually 11-15 mm long.
    - Glands scattered through the pubescence of the upper portions of the stems; capsules mostly 11-12 mm long..........4a. *A. pedicularia* var. *intercedens.*
    - Glands crowded in the pubescence of the upper portions of the stems; capsules mostly 11-15 mm long..................4b. *A. pedicularia* var. *ambigens.*

1. **Aureolaria flàva** (L.) Farw. var. **týpica** Pennell. (*Gerardia virginica* in part, of Gray, Man., ed. 7 and *Dasystoma virginica* in part, of Britton and Brown, Illus. Flora, ed. 2.) SMOOTH FALSE FOXGLOVE. Map 1862. Infrequent to frequent in the lake area, less frequent in the unglaciated area, and probably local or absent in the intervening area. It prefers very sandy soil but is found also in clayey soil. This plant, as well as the next two, are supposed to be parasitic on the roots of species of the white oak group of oaks and are found on slopes and ridges wooded with these oaks.
Maine to Wis., southw. to Ala.

1a. **Aureolaria flava** var. **macrántha** Pennell. (*Gerardia virginica* in part, of Gray, Man., ed. 7 and *Dasystoma virginica* in part, of Britton and Brown, Illus. Flora, ed. 2.) SMOOTH FALSE FOXGLOVE. Map 1863. This variety is rare in the northern part of the state, becoming somewhat fre-

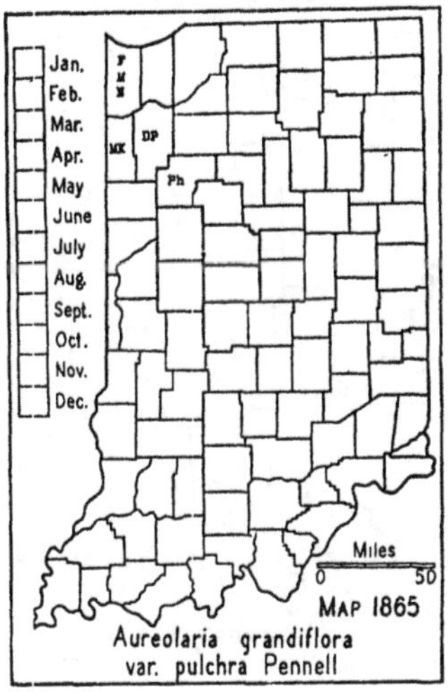

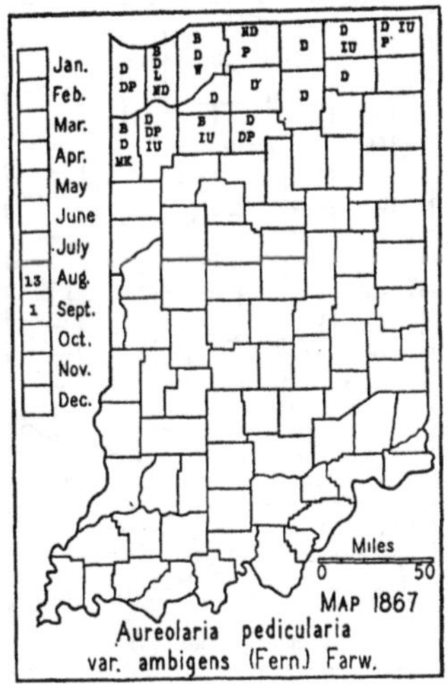

quent in the southern part in the unglaciated area. This variety, like the species, grows on white oak and chestnut oak slopes and ridges.

Ont. to Mo.; southw. to Ala. and La.

2. **Aureolaria virgínica** (L.) Farw. (*Gerardia flava* of Gray, Man., ed. 7 and *Dasytoma flava* of Britton and Brown, Illus. Flora, ed. 2.) DOWNY FALSE FOXGLOVE. Map 1864. This species seems to be localized in a few of the northern counties, where it is infrequent, and there are a few specimens from some of the Ohio River Counties. Probably it is rare or absent elsewhere. It is found in slightly acid soils on white oak slopes.

N. H. to Mich., southw. to Fla. and La.

3. **Aureolaria grandiflora** (Benth.) Pennell var. **púlchra** Pennell. (*Gerardia grandiflora* in part, of Gray, Man., ed. 7 and *Dasystoma grandiflora* in part, of Britton and Brown, Illus. Flora, ed. 2.) Map 1865. This variety is known from specimens from only four counties. It is a western plant which barely enters our state.

Wis. to Ind. and Mo.

4. **Aureolaria pediculària** (L.) Raf. var. **týpica** Pennell. (*Gerardia pedicularia* in part, of Gray, Man., ed. 7 and *Dasystoma pedicularia* in part, of Britton and Brown, Illus. Flora, ed. 2.) Map 1866. Known in Indiana only from specimens collected in very sandy soil in a black and white oak clearing in the northern part of Newton County. This form is found mostly along the Atlantic coast.

Maine to Minn., southw. along the coast to N. C.

4a. **Aureolaria pedicularia** var. **intercèdens** Pennell. (*Gerardia pedicularia* in part, of Gray, Man., ed. 7 and *Dasystoma pedicularia* in part, of Britton and Brown, Illus. Flora, ed. 2.) This form of the species is known from only two of our northwestern counties—Lake and Newton Counties. The habitat is the same as that of the next variety.

Mass. to Minn., southw. to N. C.

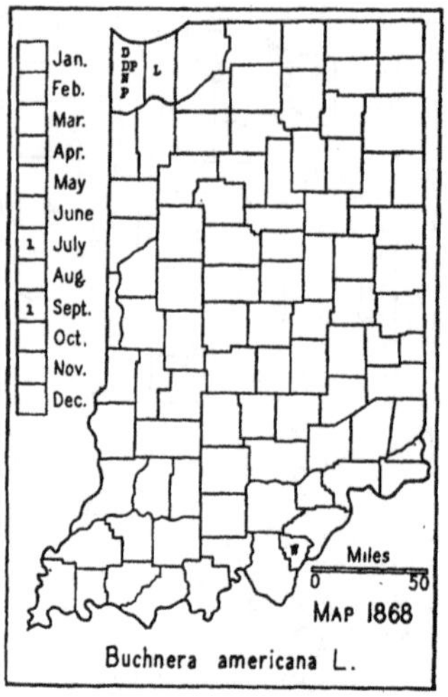

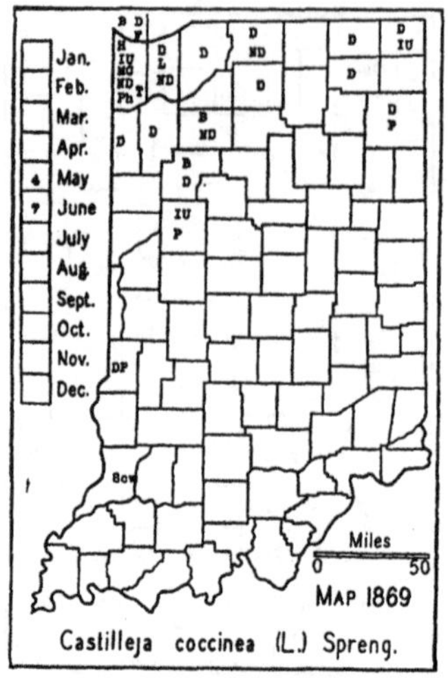

4b. **Aureolaria pedicularia** var. **ámbigens** (Fern.) Farw. (*Gerardia pedicularia* var. *ambigens* Fern. and *Dasystoma pedicularia* in part, of Britton and Brown, Illus. Flora, ed. 2.) Map 1867. This variety is infrequent in northern Indiana and its range probably can be extended only to a few counties adjacent to those shown on the map. There are reports from Tippecanoe and White Counties. It is found only in very sandy, slightly acid soil. It usually grows in oak woods in low areas surrounded by *Vaccinium angustifolium* or in somewhat moist, rarely dry, sandy places near the bases of oak slopes.

Nw. Ohio to se. Minn.

7622. BÚCHNERA L.

1. **Buchnera americàna** L. Map 1868. This plant is extremely rare in Indiana. The only recent specimens are from the low dunes near Lake Michigan in Lake and Porter Counties. It formerly was frequent on the low dunes at Pine, now the north end of Clark Street in Gary, but in 1935 search was made for it and only a few plants were seen. It will soon be extinct at this station, and only a few plants have been seen in Porter County. In the Wabash College herbarium are two sheets collected by Dr. A. Clapp Aug. 6, 1835, in the "barrens" (in Floyd or Harrison Counties).

N. Y., Ont., and Ill., southw. to Fla., Kans., and Tex.

7631. CASTILLÈJA Mutis

1. **Castilleja coccínea** (L.) Sprèng. INDIAN PAINTBRUSH. Map 1869. This is an infrequent to rare species in the lake area with a few reports from the southwestern border of the state. It grows in moist, sandy, slightly acid soils, usually in marshes or wet prairie habitats. It is generally found in small colonies, but I saw acres of it in a drained marsh that had been closely pastured for a few years. It is still somewhat frequent on the borders of sloughs between the low dunes near Lake Michigan

Map 1871 Melampyrum lineare var. pectinatum Pennell

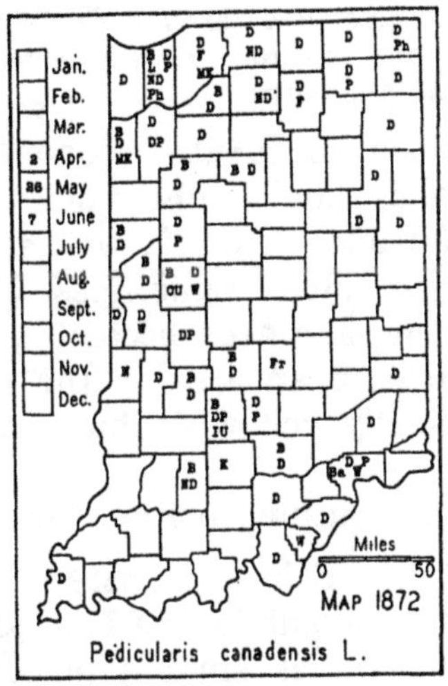

Map 1872 Pedicularis canadensis L.

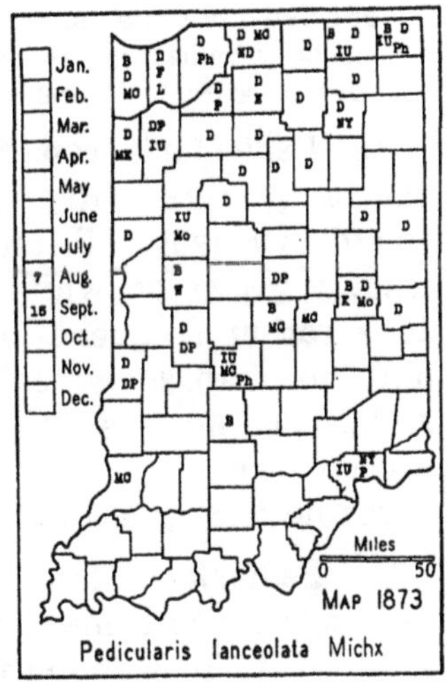

Map 1873 Pedicularis lanceolata Michx

in Lake County; elsewhere it is rare or extinct. Red is the common color form, and yellow is infrequent.

N. H. to Man., southw. to Fla. and La.

7635. MELAMPỲRUM [Bauhin] L.

[Key from Pennell's Monograph.]

Upper bracts slightly or moderately fimbriate-dentate near base, the teeth shorter than the width of the blade; capsule acute to slightly attenuate, only slightly decurved; seed 3-4 mm long, black to blackish.

Main stem leaves linear-lanceolate to nearly ovate; corolla 9-13 mm long; bracts very variable in amount of fimbriation............1. *M. lineare* var. *latifolium.*

Main stem leaves linear or lanceolate-linear; corolla 6-9 mm long; bracts only slightly fimbriate. (See excluded species no. 571, p. 1090)......*M. lineare* var. *typicum.*

Upper bracts usually conspicuously fimbriate-dentate near base, with teeth frequently as long as the width of the blade; capsule acuminate-attenuate, more strongly decurved; seed 2-3 mm long, brown or dark brown, more rarely blackish...1a. *M. lineare* var. *pectinatum.*

1. **Melampyrum lineàre** Lam. var. **latifòlium** (Muhl.) Beauv. Map 1870. The two varieties of *Melampyrum* are very rare in Indiana. This one is the more frequent and is found in moist, slightly acid soil, usually at the bases of slopes and often associated with *Polygala cruciata, Aureolaria pedicularia* var. *ambigens,* and *Aletris farinosa.*

N. S. to Minn., southw. to Ga.

1a. **Melampyrum lineare** var. **pectinàtum** Pennell. Map 1871. This variety has been found only in moist, acid soil about Lake Michigan.

Mass. to Va. and nw. Ind.

7648. PEDICULÀRIS [Bauhin] L.

Plants flowering mostly in May, of a dry habitat, rarely found elsewhere; stems, at least the upper part, hirsute or pubescent; leaves pinnately lobed; capsules lanceolate, three times as long as the calyx.......................1. *P. canadensis.*

Plants flowering the last of August and through September, of a wet habitat; stems usually pubescent below and glabrous above; leaves pinnately parted; capsules ovate, scarcely longer than the calyx........................2. *P. lanceolata.*

1. **Pedicularis canadénsis** L. EARLY WOODBETONY. Map 1872. This species prefers a dry, sandy, and slightly acid soil, although it is often found in clayey soil, and I once found it well established in a marsh. It usually grows on white oak slopes, sometimes with beech, along streams. It is rather frequent in the lake area becoming less frequent southward and our map shows a large, open area in the southwestern part of the state. The flowers are usually yellowish, but plants with reddish flowers are not rare.

Maine and Que. to Man., southw. to Fla., Tex., and Chihuahua.

2. **Pedicularis lanceolàta** Michx. SWAMP WOODBETONY. Map 1873. This species is somewhat frequent in the lake area, becoming very local south of it. It is found in marshes, springy places in general, and ditches.

Mass. to Man., southw. to N. C., Mo., and Nebr.

258. BIGNONIÀCEAE Pers. TRUMPET-CREEPER FAMILY

Leaves compound; anther-bearing stamens 4; our species vines.
 Leaves with a tendril; leaflets 2, margins entire; flowers about 5 cm long; pods flat..................7705. BIGNONIA, p. 858.
 Leaves without a tendril; leaflets 7-13, margins serrate; pods terete..................7714. CAMPSIS, p. 858.
Leaves simple; anther-bearing stamens 2; trees..............7727. CATALPA, p. 859.

7705. BIGNÒNIA [Tourn.] L.

1. **Bignonia capreolàta** L. (*Anisostichus capreolata* (L.) Bureau.) CROSSVINE. Map 1874. This vine climbs trees to a height of 60 feet, and prefers full sunlight. It is recommended for ornamental planting because of its large and highly colored, though ill-scented, flowers which appear the last of May. It grows on wooded slopes and alluvial bottoms along streams. Without doubt Thompson's record from Carroll County either should be referred to the next species or considered a cultivated specimen.

Miss Edna Banta informs me that this species grows along the Ohio River in Jefferson County, 2 miles east of Madison. We have had it in cultivation for 9 years and in that time it has climbed a walnut tree to a height of 35 feet.

Va. to s. Ill., southw. to Fla. and La.

7714. CÁMPSIS Lour. TRUMPET-CREEPER

1. **Campsis radìcans** (L.) Seemann. (*Tecoma radicans* (L.) Juss. and *Bignonia radicans* L.) TRUMPET-CREEPER. Map 1875. A vine trailing or climbing to a length of 40 feet. It is infrequent in woodland except in a few of the Lower Wabash Valley counties where it may be more or less frequent. It is rare to infrequent in all of northern Indiana. This species, however, produces an abundance of seed which grow easily when they come in contact with exposed soil, and it has become one of the most despised plants in the Lower Wabash Bottoms where it is known as shoe-

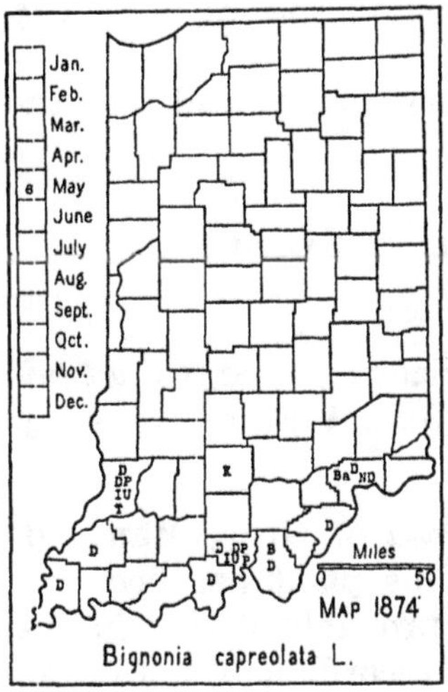

Bignonia capreolata L.

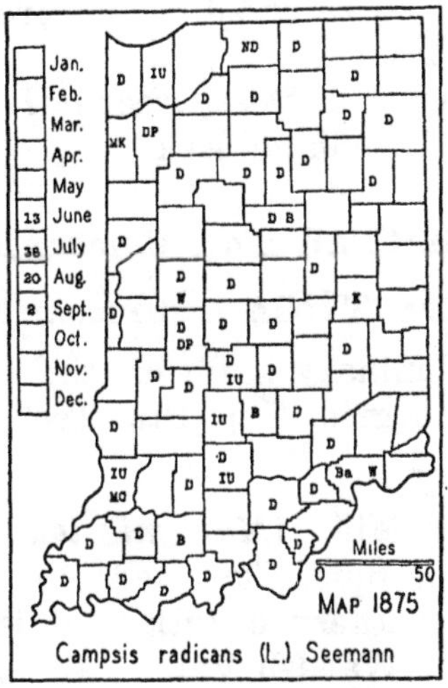

Campsis radicans (L.) Seemann

Catalpa bignonioides Walt.

strings and hell vine. It grows so rapidly that in one or two years it is difficult to cultivate ground in which it becomes established. It prefers alluvial bottoms and wherever this vine is noted in such a habitat it should be destroyed or the capsules gathered and burned before the seed escape. It is ornamental and has been widely planted which accounts for its distribution. I doubt that it was a native of more than the Ohio River Counties and the Lower Wabash Valley. It is still being planted and recommended for ornamental planting but only by persons who are ignorant of its potential weedy nature. My advice is to exterminate it wherever found and never permit the vine to mature seed.

Pa. to Iowa, southw. to Fla. and Tex.

7727. CATÁLPA Scop. CATALPA

Bark of old trees thin and scaly; odor of bruised leaves fetid; lower lobe of corolla entire...1. *C. bignonioides.*
Bark of old trees fissured and ridgy; odor of bruised leaves not fetid; lower lobe of corolla notched at the apex....................................2. *C. speciosa.*

1. CATALPA BIGNONIOÌDES Walt. (*Catalpa Catalpa* (L.) Karst.) COMMON CATALPA. Map 1876. This species has been freely planted as an ornamental and, no doubt, does escape. I have seen it freely escaping along a roadside in Johnson County and abundantly so in a few sandy, fallow fields in northwestern Elkhart County. It is not recommended for ornamental planting. If a species of catalpa is desired it is best to use the next species.

Ga. to Fla. and westw. to Miss.; introduced northward.

2. **Catalpa speciòsa** Warder. HARDY CATALPA. Map 1877. This is a forest tree and was a native of the Lower Wabash Valley. I think I was reliably informed by a pioneer of Perry County who told me that it was a native in the lower valley of Deer Creek. The tree is not readily distinguished from the preceding species and reports for this species from counties not indicated on the map should be regarded with suspicion.

Doubtless they should all be referred to the preceding species or to cultivated trees of this species.

Ohio Valley from the mouth of Deer Creek in Perry County, Ind., and the Mississippi Valley to se. Mo. and ne. Ark.

260. MARTYNIÀCEAE Link. UNICORN PLANT FAMILY

7784. MARTÝNIA L. UNICORN PLANT

[Van Eseltine. A preliminary study of the unicorn plants (*Martyniaceae*). New York State Agric. Exp. Sta. Tech. Bull. **149**: 1-41. 15 fig. 1929.]

1. **Martynia louisiánica** Mill. (*Martynia louisiana* of Gray, Man., ed. 7 and Britton and Brown, Illus. Flora, ed. 2.) This plant has been reported from five of the southern counties. It is probably native in the Ohio Valley and the Lower Wabash Valley. Clapp records it in his list in 1834. A specimen in the herbarium of Indiana University was collected by Young in Jefferson County in 1877. Authors report it as being found in sandy habitats, especially along the Ohio and Wabash Rivers. It has been cultivated and one finds it as an occasional escape. I have found it three times in Wells County but in each instance it was introduced by some means or other. Once it was introduced with some strawberry plants.

Del. to Nebr., southw. to Fla. and Mex.

261. OROBANCHÀCEAE Lindl. BROOMRAPE FAMILY

Flowers all perfect and complete.
  Plants glabrous; flowers in a thick, scaly spike; calyx deeply cleft in front; stamens exserted..7790. CONOPHOLIS, p. 860.
  Plants glandular-pubescent; calyx 5-cleft; stamens included..7791. OROBANCHE, p. 860.
Flowers of two sorts, the lower cleistogamous and fertile, the upper complete but usually sterile..7792. EPIFAGUS, p. 862.

7790. CONÓPHOLIS Wallr.

1. **Conopholis americàna** (L. f.) Wallr. CANCER-ROOT. Map 1878. Infrequent to frequent or local in all parts of the state. It is inconspicuous, and for this reason it often may be overlooked, and, no doubt, it is more frequent than the reports indicate. In Indiana it is parasitic on species of oak, probably most commonly on the white oak. It is generally found in woods with a deep cover of leaves, in such places as ravines and on protected slopes.

S. Maine to Mich., southw. to Fla. and Tenn.

7791. OROBÁNCHE [Tourn.] L. BROOMRAPE

[Achey. A Revision of the section Gymnocaulis of the genus Orobanche. Bull. Torrey Bot. Club **60**: 441-451. 1933.]

Flowers racemose or the upper sessile and the lower on pedicels up to 1 cm long; each flower with 1 or 2 long bracts below the calyx..............1. *O. ludoviciana.*

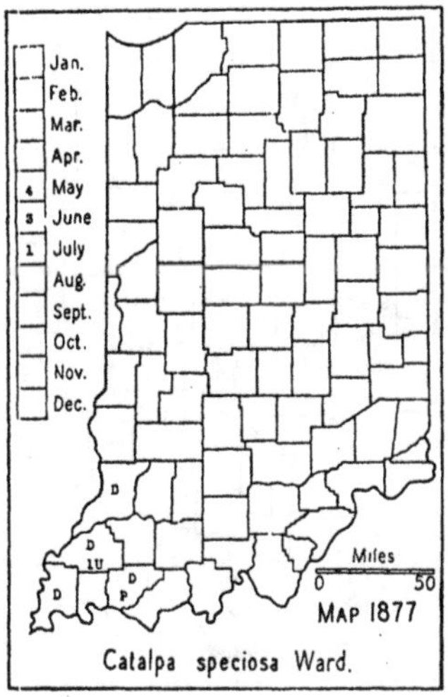

Catalpa speciosa Ward.

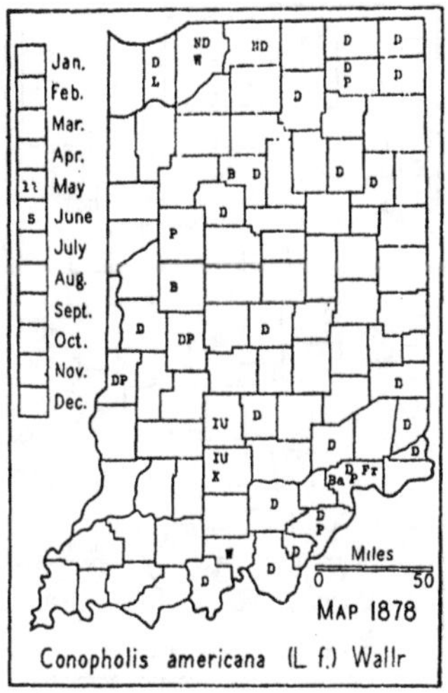

Conopholis americana (L. f.) Wallr.

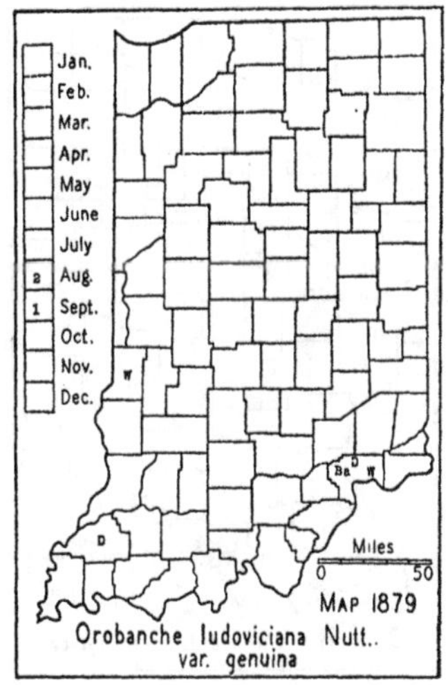

Orobanche ludoviciana Nutt. var. genuina

Flowers solitary on long, naked peduncles, without bracts.

Cauline scales glabrous; stems very short; peduncles 1-4, erect; calyx lobes lanceolate, acuminate..2. *O. uniflora.*

Cauline scales pubescent; stems usually 4-10 cm long; peduncles several; calyx lobes broad, acute..3. *O. fasciculata.*

1. **Orobanche ludoviciàna** Nutt. var. **genuìna** G. Beck. (Munz. The North American species of Orobanche, section Myzorrhiza. Bull. Torrey Bot. Club 57: 620. 1930.) Map 1879. A very local plant in Indiana. It has been reported from Jefferson and Vigo Counties and from the Lower Wabash Valley. Usually a parasite on the roots of *Ambrosia trifida.* I found it as a common plant on this host on the east bank of Goose Pond in Gibson County. Miss Edna Banta found it on the roots of tobacco plants in Jefferson County. In 1938 I found it in Knox County.

Sask. southw. to Ind. and Tex.

2. **Orobanche uniflòra** L. var. **týpica** Achey. (*Thalesia uniflora* (L.) Britt.) ONE-FLOWER BROOMRAPE. Map 1880. This is a rare plant but probably found throughout the state. It has been reported from ten widely separated counties. The plants I have found were parasitic on white and chestnut oak. Bechtel found it parasitic on the roots of a *Solidago* in Montgomery County.

Newf. to Ont. and possibly B. C., southw. to Ga., Tex., and Calif.

3. **Orobanche fasciculàta** Nutt. var. **týpica** Achey. (*Thalesia fasciculata* (Nutt.) Britt.) CLUSTERED BROOMRAPE. Map 1881. This species has been found only in Lake and Porter Counties in the dunes bordering Lake Michigan. Most of the collections have been made on the low dunes just south of Pine, in Lake County. Locally it is common. I have seen it only on the low dune south of Pine where it was common on the roots of *Artemisia caudata.*

Ind. to Minn. and B. C., southw. to Nebr. and Calif.

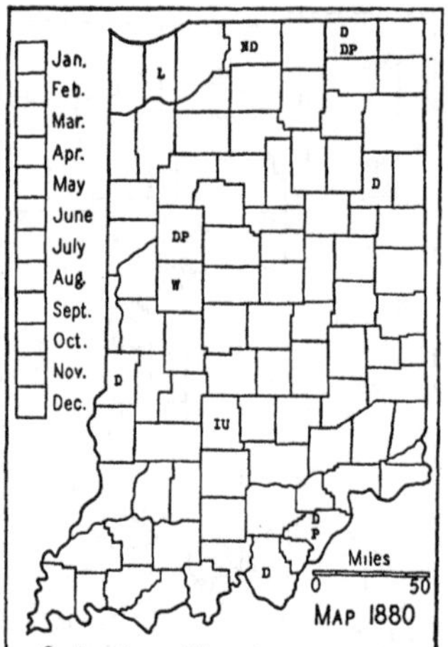

Orobanche uniflora L., var. typica

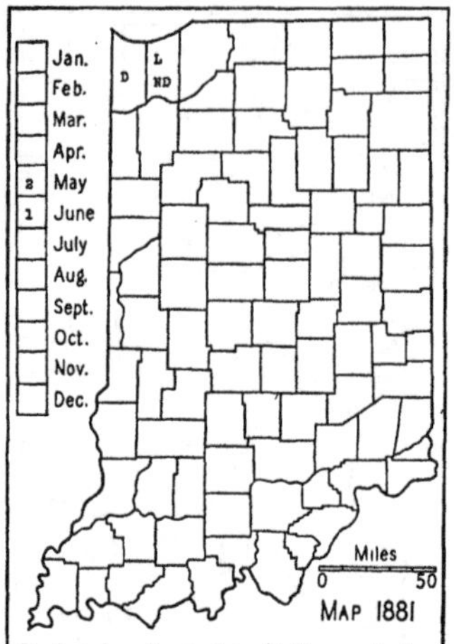

Orobanche fasciculata Nutt., var. typica

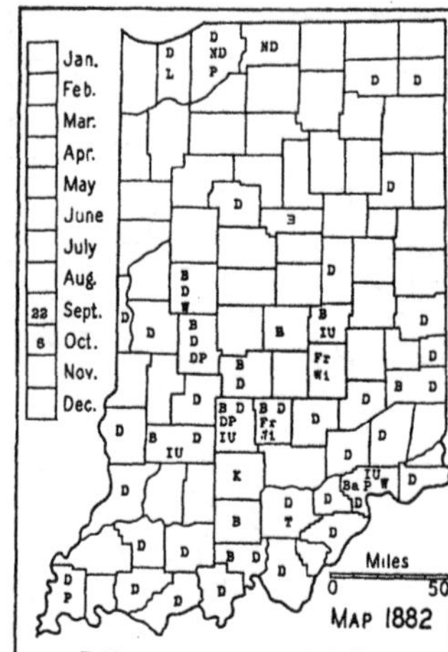

Epifagus virginiana (L.) Bart.

7792. EPIFÀGUS Nutt.

1. **Epifagus virginiàna** (L.) Bart. (*Leptamnium virginianum* (L.) Raf.) BEECHDROPS. Map 1882. Frequent to common in all parts of the state where the beech tree grows. It is parasitic on the roots of the beech and is not found elsewhere.

N. B. and Ont. to Wis., southw. to Fla. and La.

264. LENTIBULARIÀCEAE Lindl. BLADDERWORT FAMILY

7901. UTRICULÀRIA L.[1]

Scapes naked (except some small, scaly bracts).
- Flowers all purplish.
  - Stems 3-9 dm long, free-floating with copious whorled leaves; scapes 2-4-flowered. .... 1. *U. purpurea.*
  - Stems 0.5-3 dm long, rooting in marly mud or sand and not free-floating; scapes with a solitary flower and appearing as a single plant with a few, very small leaves at the base, these rarely bladder-bearing.......... 2. *U. resupinata.*
- Flowers yellowish.
  - Bracts at the base of the pedicel accompanied by a pair of bractlets; calyx closing in fruit.......... 3. *U. cornuta.*
  - Bracts at the base of the pedicel not accompanied by bractlets; calyx not closing in fruit.
    - Stem creeping on the bottom in shallow water; corolla 4-12 mm long.
      - Pedicels ascending in fruit; spur and palate of the corolla conspicuous.
        - Segments of the leaves capillary; upper lip of the corolla equaling the lower one which is about 6 mm long.......... 4. *U. gibba.*
        - Segments of the leaves linear, flat, the margins bristle-toothed; bladders on separate branches; upper lip of the corolla about half as long as the lower one which is 10-15 mm long.......... 5. *U. intermedia.*
      - Pedicels recurved in fruit; spur a mere sac; palate obsolete; corolla 4-8 mm long.......... 6. *U. minor.*
    - Stem submerged or free-floating; corolla 14-20 mm long.......... 7. *U. macrorhiza.*

Scapes with a whorl of elongated, floating bladders formed of inflated petioles; flowers yellow.......... 8. *U. radiata.*

[1] Dr. J. H. Barnhart, of the New York Botanical Garden, has named nearly all of my specimens.

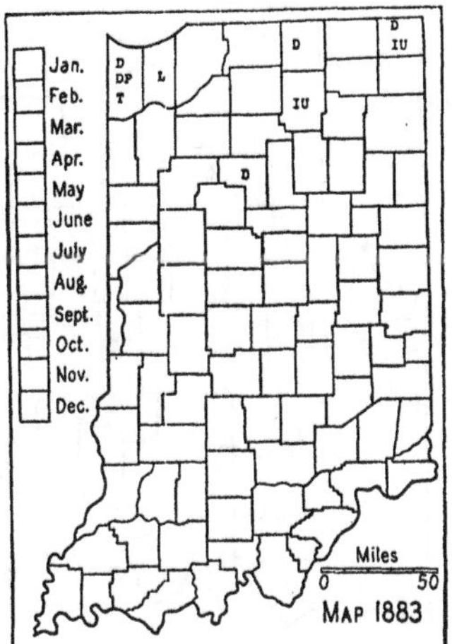

Utricularia purpurea Walt.

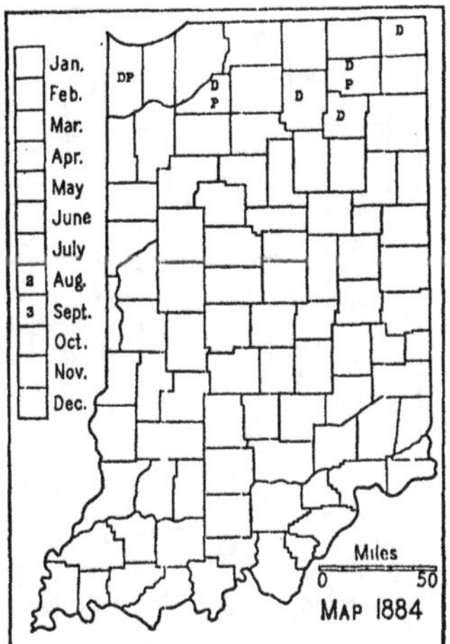

Utricularia resupinata B. D. Greene

Utricularia cornuta Michx.

1. **Utricularia purpùrea** Walt. (*Vesiculina purpurea* (Walt.) Raf.) PURPLE BLADDERWORT. Map 1883. Local and sometimes common. In 1930, in a bayou of about 5-10 acres in Lake Cicott, Cass County, it was common in about 3-5 feet of water, associated with *Utricularia macrorhiza, Nuphar advena,* and *Nymphaea tuberosa.* It has been reported also from Marshall County.

Maine to Fla. and La., near the coast; also in Mich., Ind. to Minn.

2. **Utricularia resupinàta** B. D. Greene. (*Lecticula resupinata* (B. D. Greene) Barnhart.) Map 1884. Reported from Lake, Marshall, Noble, and Whitley Counties. It is local and grows on wet, sandy or marl borders of lakes or in shallow water up to 10 inches deep.

N. B. to w. Ont. and Pa., southw. to S. C. and Fla.

3. **Utricularia cornùta** Michx. (*Stomoisia cornuta* (Michx.) Raf.) HORNED BLADDERWORT. Map 1885. This is one of our rarest species. It has been found only in St. Joseph, Elkhart, and Lake Counties. It was formerly common on the wet, sandy borders of sloughs near Lake Michigan in Lake County.

Newf. to Minn., southw. to Fla. and Tex.

4. **Utricularia gíbba** L. HUMPED BLADDERWORT. Map 1886. This is a small species usually found on the wet sandy or mucky borders of lakes and sloughs. A few years after the highway was built around Bass Lake, Starke County, I found this species and *Utricularia resupinata* by the thousands in the bottom of the moist, sandy roadside ditch. It was no doubt frequent throughout the lake area before it was drained, and rare elsewhere.

Maine to Mich., southw. to Fla. and Tex.

5. **Utricularia intermèdia** Hayne. Map 1888. This species has been reported from Kosciusko and Marshall Counties by Clark, and from Lake County by Peattie and by Pepoon. In a letter from J. H. Barnhart of the New York Botanical Garden, dated June 11, 1932, he says that there are

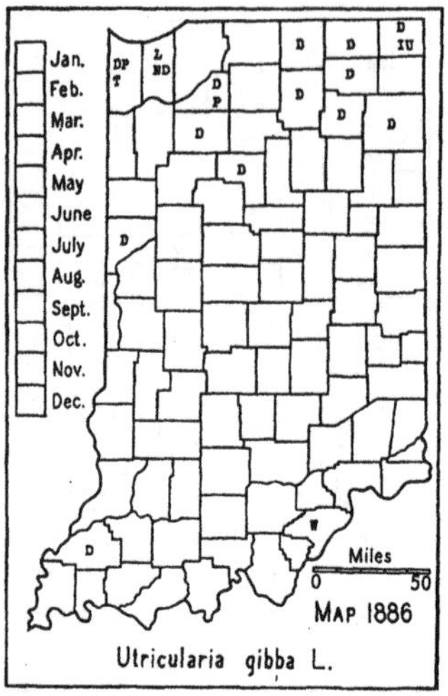

MAP 1886
Utricularia gibba L.

MAP 1888
Utricularia intermedia Hayne

MAP 1889
Utricularia minor L.

two specimens from Indiana in that herbarium. One is from Marshall County collected by Scovell & Clark near Lake Maxinkuckee, August 13, 1900. The other is one from Lagrange County, which I collected June 1, 1916. I also have it from Elkhart, Lake, La Porte, and Steuben Counties.

Newf. to B. C., southw. to N. J. and Calif.; also in Eu.

6. **Utricularia minor** L. LESSER BLADDERWORT. Map 1889. This species has been reported from Lake, Marshall, Noble, and Porter Counties. My only specimens were collected on the border of Kellogg Lake, in the northeastern corner of Steuben County and on the low marsh border of a lake in Elkhart County.

Circumpolar and southw. in America to Conn., N. Y., Pa., Ohio, Ind., Colo., and Calif.

7. **Utricularia macrorhìza** LeConte. (*Utricularia vulgaris* var. *americana* Gray.) GREATER BLADDERWORT. Map 1890. This species was, no doubt, frequent to common in shallow water throughout the lake region before it was drained, and local elsewhere in shallow water in suitable habitats.

Newf. to Yukon, southw. to Md., Mo., Okla., Ariz., and Lower Calif.

8. **Utricularia radiàta** Small. Map 1891. This species was first found August 10, 1924, by Winona Welch in Walker Township of Jasper County, in a roadside ditch through the old basin of Clear Lake about 3 miles south of Tefft. I found it at the same place in 1937. This is another of the Coastal Plain species found in this immediate vicinity and the record is added evidence of the migration of plants through the Mississippi and Kankakee River Valleys.

Maine to Tex., mostly near the coast.

266. ACANTHÀCEAE J. St. Hil. ACANTHUS FAMILY

Plants not growing in water or on muddy shores and bars in streams; leaves wider than linear-lanceolate.

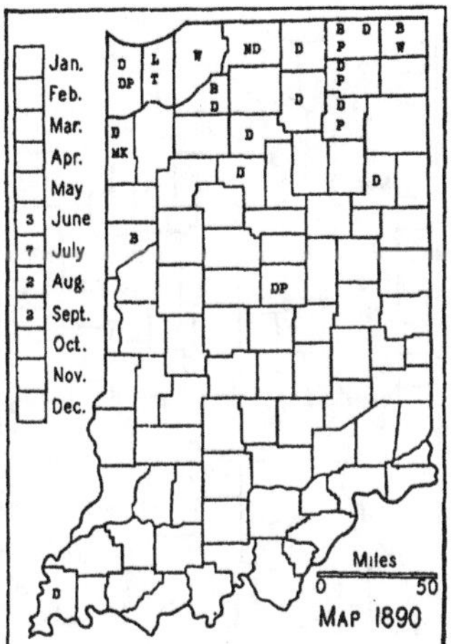

Utricularia macrorhiza Le Conte

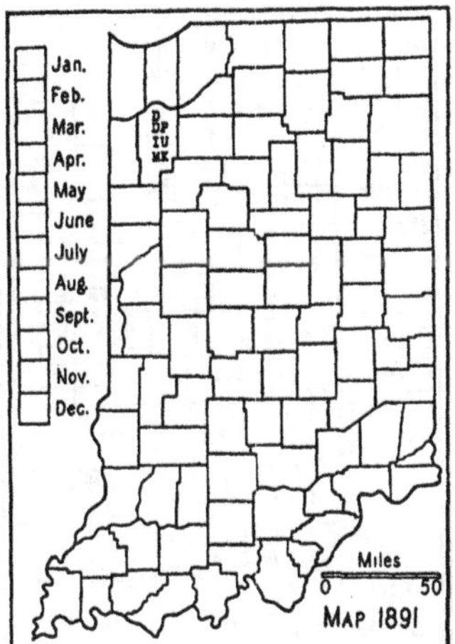

Utricularia radiata Small

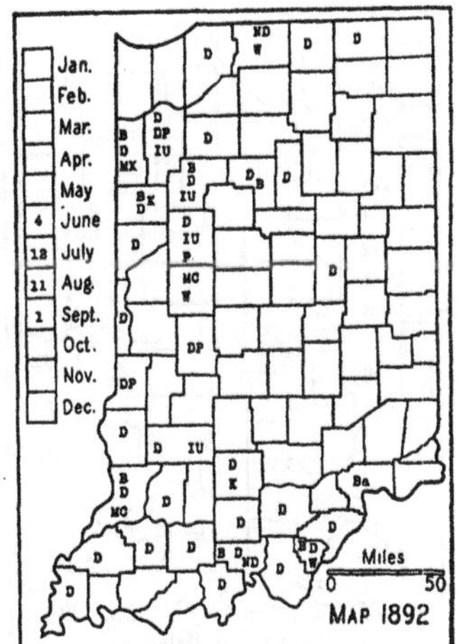

Ruellia caroliniensis (Walt.) Steud.

Corolla convolute in the bud, nearly regular; stamens 4; seeds densely short-pubescent..7965. RUELLIA, p. 865.
Corolla imbricate in the bud, bilabiate; stamens 2; seeds prickly..8031. DIAPEDIUM, p. 866.
Plants growing in the water or on muddy shores and bars in streams; leaves linear-lanceolate, elongated, gradually acute or acuminate......8094. DIANTHERA, p. 866.

7965. RUÉLLIA [Plumier] L. RUELLIA

Calyx lobes linear-filiform, much exceeding the capsules, generally 2-2.5 cm long and rarely 1 mm wide; plants villous and sometimes with shorter, recurving hairs.
Corolla tube much longer than the calyx lobes..................1. *R. caroliniensis.*
Corolla tube shorter or nearly as long as the calyx lobes; flowers smaller than those of the species............................1a. *R. caroliniensis* var. *parviflora.*
Calyx lobes linear-lanceolate, about equaling the capsules, generally 1.5-2 cm long; plants glabrate or sparingly pubescent.
Flowers solitary in the axils of the leaves; none of the flowers cleistogamous...2. *R. strepens.*
Flowers clustered in the axils of the leaves: some or all of the flowers cleistogamous; calyx lobes usually much more glandular..........2a. *R. strepens* f. *cleistantha.*

1. **Ruellia caroliniénsis** (Walt.) Steud. (Blake. Neglected names in Walter's Flora. Rhodora **17**: 137. 1915.) (*Ruellia ciliosa* Pursh.) HAIRY RUELLIA. Map 1892. This species prefers a dry, sandy soil and is found mostly in open places along roadsides and railroads, on dry, open wooded slopes, and in prairie habitats.

N. J. to Mich. and Kans., southw. to Fla. and La.

1a. **Ruellia caroliniensis** var. **parviflòra** (Nees) Blake. Map 1893. Has the habitat of the species.

Md. to Ind., southw. to Fla. and Tex.

2. **Ruellia strèpens** L. SMOOTH RUELLIA. Map 1894. This species prefers moist, alluvial soil in open woodland along streams, but is also found in moist, open woodland and rarely in moist, open places. It is sometimes found in dry soil on the slopes of high, wooded banks and ridges.

Pa. to Wis. and Kans., southw. to Fla. and Tex.

Map 1893
Ruellia caroliniensis var. parviflora (Nees) Blake

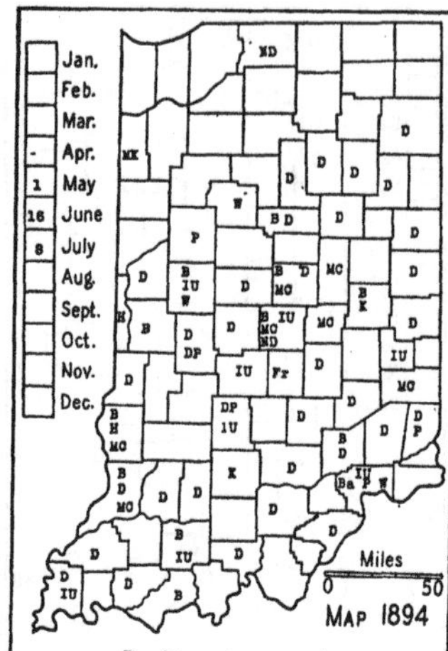

Map 1894
Ruellia strepens L.

Map 1895
Diapedium brachiatum (Pursh) Kuntze

2a. **Ruellia strepens** f. **cleistántha** (Gray) McCoy. There are specimens of this form from Adams, Carroll, Dubois, Gibson, Harrison, Madison, Marion, Putnam, Perry, Spencer, Sullivan, and Warrick Counties. McCoy has shown that this form is only the autumnal phase of the species. (Amer. Botanist 43: 22-24. 1937.)

8031. DIAPÈDIUM Konig

1. **Diapedium brachiàtum** (Pursh) Ktze.* Map 1895. Common at the base of a wooded slope and the adjoining alluvial bottoms along Little Pigeon Creek on the farm of Nathan Taylor about a half mile west of Yankeetown in Warrick County. This is the only known station in Indiana.

N. C. to Fla., Mo., Kans., and Tex.

8094. DIANTHÈRA [Gronov.] L. WATER WILLOW

1. **Dianthera americàna** L. DENSE-FLOWERED WATER WILLOW. Map 1896. This species grows in dense colonies on the muddy shores of streams and on muddy bars in streams and rarely on the low border of lakes. It is local to frequent in the counties indicated on the map. It appears to be absent or rare in the extreme northern part and rare in the southwestern part of the state.

Que., Ont. to Wis., southw. to Ga. and Tex.

268. PHRYMÀCEAE Schauer LOPSEED FAMILY

8115. PHRỲMA L. LOPSEED

1. **Phryma Leptostàchya** L. LOPSEED. Map 1897. This plant prefers beech and sugar maple woods and is less frequent in black and white oak woods. Infrequent to frequent in woodland with deep leaf mold. Found throughout the state although there are no records for the area near Lake Michigan.

N. B. to Man., southw. to Fla. and Kans.; found also in e. Asia.

* The name of this plant now becomes **Dicliptera brachiata** (Pursh) Spreng.

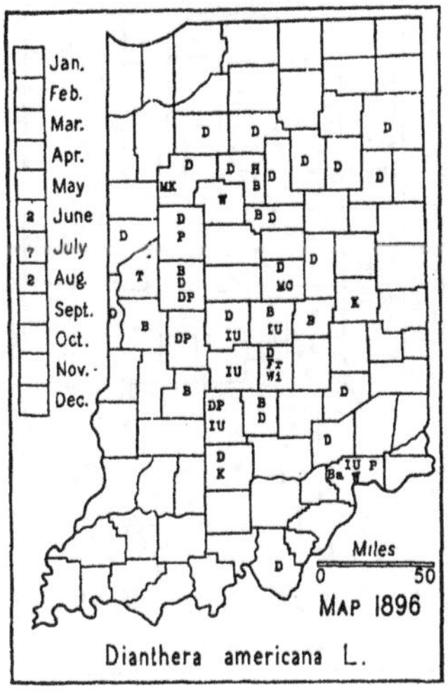

MAP 1896

Dianthera americana L.

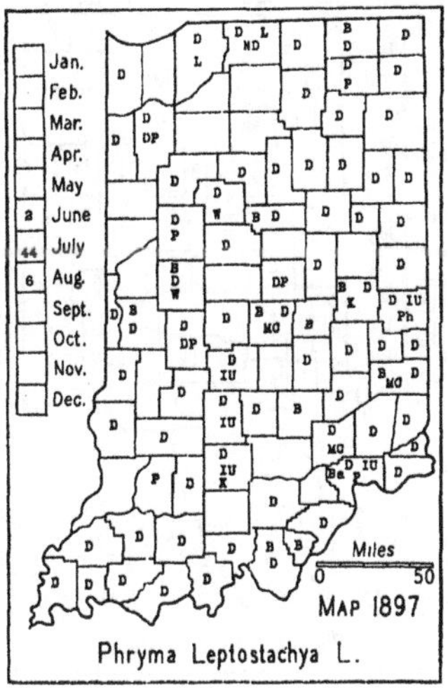

MAP 1897

Phryma Leptostachya L.

MAP 1898

Plantago cordata Lam.

269. PLANTAGINÀCEAE Lindl. PLANTAIN FAMILY

[Pilger, Robert. Plantaginaceae. Das Pflanzenreich, **IV**. 269. 1937.]

8116. PLANTÀGO [Tourn.] L. PLANTAIN

Plants acaulescent; flowers spicate or capitate at the ends of scapes.
  Corolla lobes spreading or reflexed in fruit, not closed over the top of the capsule.
    Leaves broad, abruptly contracted, generally into long petioles; petioles much longer than half the length of the blades; spikes long and slender; seeds not hollowed on the inner face.
      Ribs of the broad leaves arising from the midrib; capsule 2-4-seeded.......... ....1. *P. cordata.*
      Ribs of the leaves free to the contracted base; capsule generally more than 4-seeded.
        Bracts and sepals broad and rounded, obtuse; capsule ovate, about 2.5 mm long, circumscissile near the middle; seed mostly 7-15, usually about 0.5 mm wide and 1 mm long..............................2. *P. major.*
        Bracts and sepals narrow, subacute; capsule elliptic-oblong, about 3.5 mm long, circumscissile much below the middle; seed 4-9, about 0.5-0.8 mm wide and 1-1.5 mm long.
          Scapes and leaves entirely glabrous.......................3. ***P. Rugelii.***
          Scapes and usually the lower surface of the leaves more or less pubescent. ....3a. ***P. Rugelii* var. *asperula.***
    Leaves lanceolate or linear; capsule 2-seeded; seeds concave on the inner face.
      Lower bracts at least twice as long as the flowers...............4. *P. aristata.*
      Lower bracts less than twice as long as the flowers.
        Leaves lanceolate, linear-lanceolate to linear-elliptic, more than 5 mm wide; mature spikes generally about 8 mm wide near the base; bracts not silky-pubescent.
          Spike at beginning of anthesis narrowly ovoid-conic, tapering to the apex, in fruit cylindric and obtuse, 1.5-8 cm long; leaf blades 0.5-2.3 dm long, 0.6-4 cm wide, glabrous or sparsely pubescent above.......... ....5. *P. lanceolata.*
          Spike at beginning of anthesis subglobose, rounded to the apex, in fruit subglobose to cylindric and obtuse, 0.5-2.3 cm long; leaf blades 0.2-1.2 dm long, 0.3-2 cm wide; upper surface gray with abundant, long hairs. ....5a. *P. lanceolata* var. *sphaerostachya* f. *eriophora.*

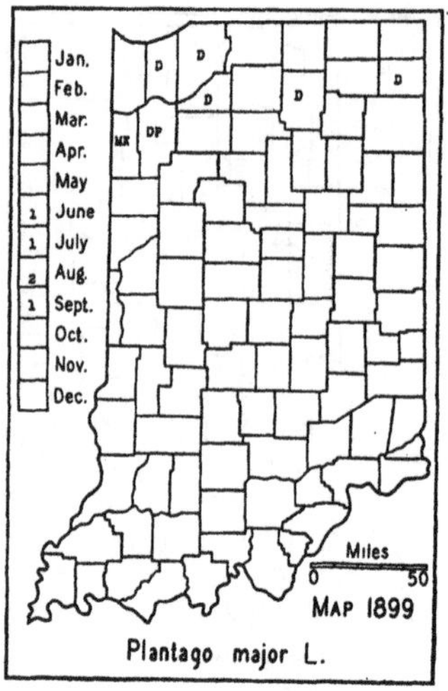

MAP 1899
Plantago major L.

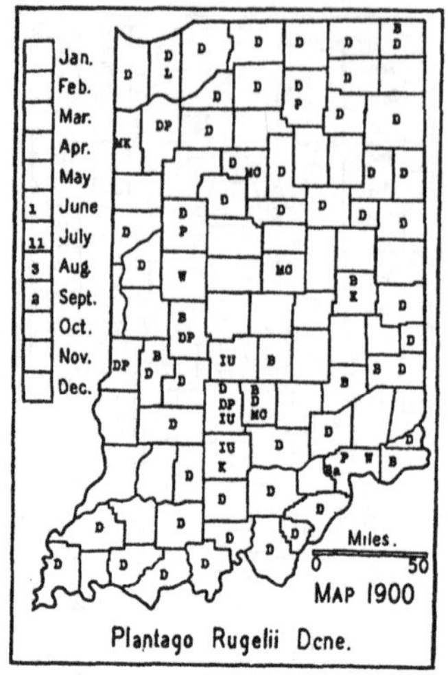

MAP 1900
Plantago Rugelii Dcne.

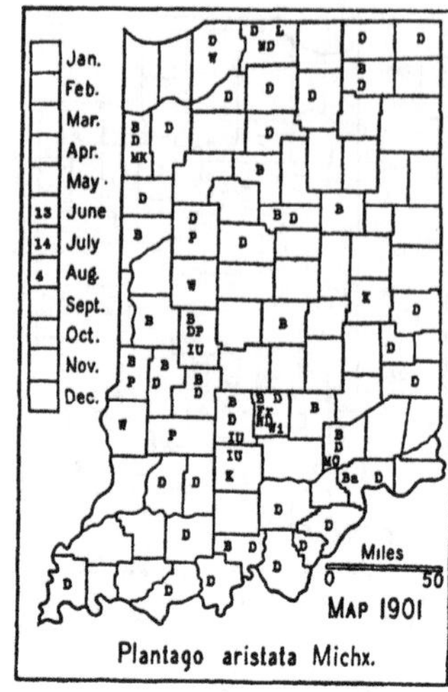

MAP 1901
Plantago aristata Michx.

Leaves linear, generally much less than 5 mm wide; mature spikes usually 5-6 mm wide near the base; bracts silky-pubescent..........6. *P. Purshii.*

Corolla lobes erect and closed over the top of the capsule.

Leaves spatulate to obovate; stamens 4; seed concave on the inner face, the surface not plainly punctate....................................7. *P. virginica.*

Leaves linear-filiform; stamens 2; seed umbilicate on the inner face, the surface plainly pitted, about 1.5 mm long............................8. *P. pusilla.*

Plants with erect, leafy stems; flowers capitate at the ends of axillary peduncles. (See excluded species no. 576, p. 1091)..............................*P. indica.*

1. **Plantago cordàta** Lam. HEARTLEAF PLANTAIN. Map 1898. Reported from Kosciusko County and from the Lower Wabash Valley. It is extremely rare. I have found it in a small open ditch in Wells County, in a low woods in Knox County which was inundated much of the time, and in an open ditch in a woods in the southeast corner of Whitley County.

N. Y., Ont., and Minn., southw. to Ala., La., and Mo.

2. **Plantago màjor** L. COMMON PLANTAIN. Map 1899. In moist or dry waste places. Rare in Indiana. It has been reported by many authors but I believe that most of the reports should be transferred to *Plantago Rugelii,* which without doubt occurs in every county of the state, although our early authors failed to report it.

I collected an abnormal specimen of *Plantago major* on a knoll on the south side of the Pennsylvania Railroad about a quarter of a mile east of Winona Lake Station in Kosciusko County. It has five scapes and three of them have a whorl of leaves just below the spike. One spike has three leaves, one has five, and the third has six, the largest 5 cm long, but these are not in a whorl. All of my specimens of this species are more or less pubescent.

Newf. to B. C., southw. to Fla. and Calif.

3. **Plantago Rugélii** Dcne. RUGEL PLANTAIN. Map 1900. Frequent to common throughout the state in lawns, waste places, fields, and open woodland and along roadsides and railroads.

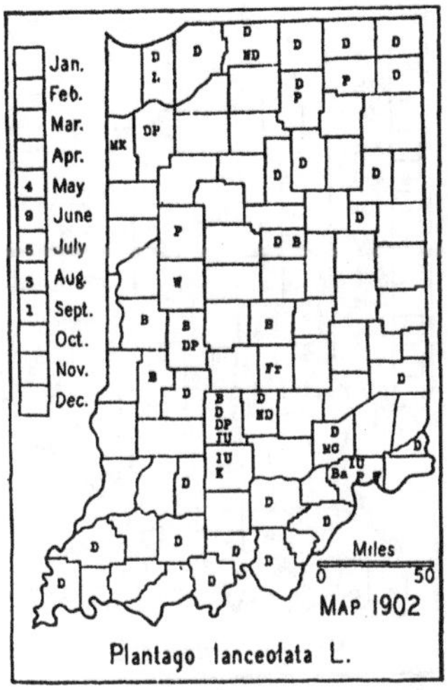

MAP 1902
Plantago lanceolata L.

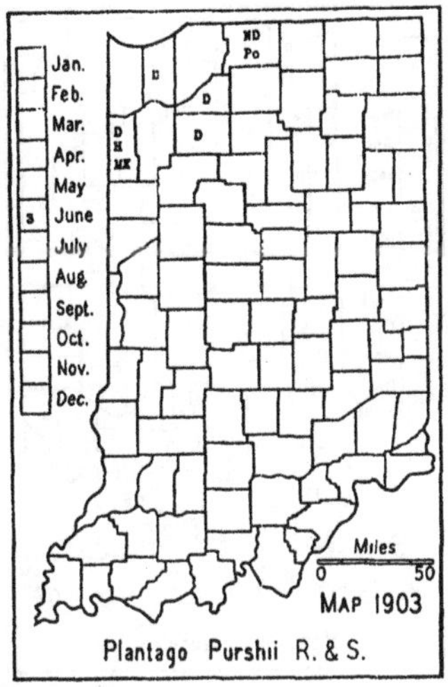

MAP 1903
Plantago Purshii R. & S.

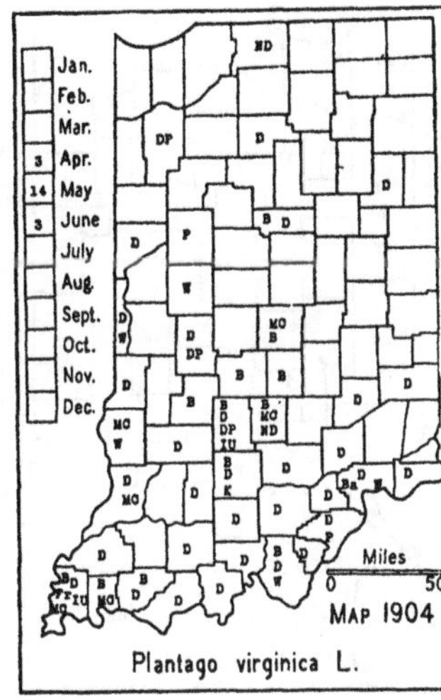

MAP 1904
Plantago virginica L.

The inflorescence of this species sometimes branches and I have noted plants with as many as five branches.

N. B. to N. Dak., southw. to Fla. and Tex.

3a. **Plantago Rugelii** var. **aspérula** Farw. (Papers Michigan Acad. Sci. 1: 99. 1923.) This variety is not well marked on account of the many intergrading forms. Most of my specimens are entirely glabrous but some of them are rather densely pubescent on the scapes and lower surface of the blades, while some of them are pubescent on the scapes only.

The variety is not geographically separated in Indiana and is included in the map of the species.

4. **Plantago aristàta** Michx. BRACTED PLANTAIN. Map 1901. This species prefers slightly acid soil and is a good indicator of soil of this kind. It is generally a common plant where it is found and is regarded as a weed. It is found in fallow fields, on washed slopes, and sometimes on the crests of ridges in open woodland.

Maine to B. C., southw. to Fla. and N. Mex.

5. PLANTAGO LANCEOLÀTA L. BUCKHORN. ENGLISH PLANTAIN. Map 1902. A pernicious weed found throughout the state. It is especially troublesome in clover fields because the seed are separated with difficulty from the clover seed. A farmer in Perry County, however, told me that the young plants are much relished by stock, and he said that he always sowed the seed in his pastures to increase the forage.

It is found in cleared grounds almost everywhere except in very sandy or very wet soils.

Nat. of Eu.; Newf. to B. C., southw. to Fla. and Kans.

5a. PLANTAGO LANCEOLATA var. SPHAEROSTÀCHYA Mert. & Koch f. ERIÓPHORA (Hoffmansegg & Link) Beck. (Rhodora 24: 204. 1922.) I found this form as a frequent plant in a pasture field on the east side of Pleasant

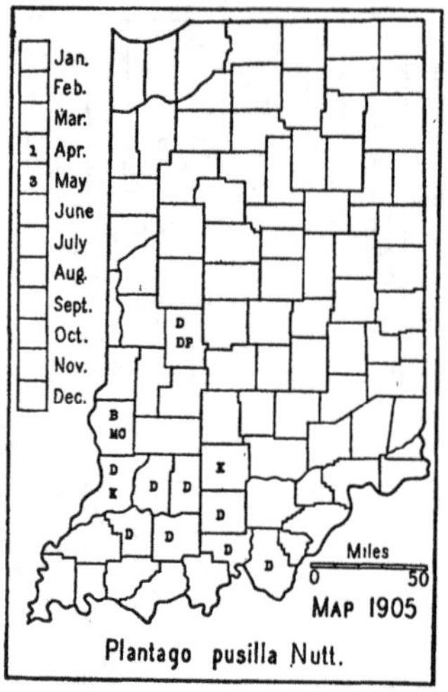

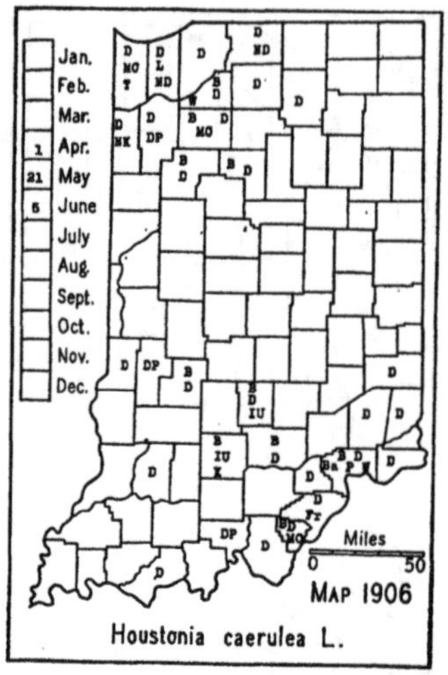

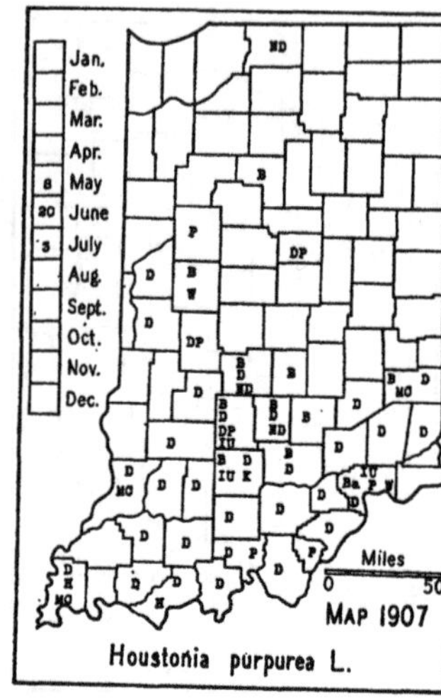

Lake, Noble County, and a few specimens on the sandy bank of the north side of Simonton Lake, Elkhart County.

N. S., s. N. E., and Oreg.

6. **Plantago Púrshii** R. & S. PURSH PLANTAIN. Map 1903. This species grows in very sandy soil in pastures and fallow fields. I found acres of it in a fallow, sandy field in Starke County.

Ont. to B. C., southw. to Tex. and Ariz.

7. **Plantago virgínica** L. Map 1904. Rare or absent from most of our northern counties, becoming frequent to common in the southern counties. It is found in both dry and moist habitats, but prefers slightly acid, moist soil and is usually found in fallow fields and pastures.

Conn. to Mich., southw. to Fla., Kans., and Ariz.

8. **Plantago pusílla** Nutt. (Pilger. Plantaginaceae, **IV.** 269. Page 74. 1937.) Map 1905. This species prefers slightly acid and dry soil and is generally found in pastures where it is often frequent to common. Scott's report for the Leesburg swamp in Kosciusko County should be referred to some other species.

Man. to Alberta, southw. to Ind. and Okla.

270. RUBIÀCEAE B. Juss. MADDER FAMILY

Leaves opposite or sometimes in verticels of 3 in *Cephalanthus*.
  Flowers terminal, pedicellate, solitary, cymose or in heads.
    Herbs; leaves less than 3 cm wide.
      Plants evergreen, creeping; leaves broadly ovate to orbicular, glabrous, cordate at the base; fruit composed of 2 united drupes, red..8451. MITCHELLA, p. 873.
      Plants not evergreen, erect or ascending; leaves not as above; fruit a dry capsule with the calyx tube adnate at least half the length of the capsule...8141. HOUSTONIA, p. 871.
    Shrubs; leaves more than 3 cm wide; flowers white, in heads...8230. CEPHALANTHUS, p. 872.

Flowers axillary, sessile or nearly so.
Plant and capsule pubescent..............................8471. DIODIA, p. 873.
Plant and capsule glabrous..............................8475. SPERMACOCE, p. 874.
Leaves in whorls of 4 to 8..............................8486. GALIUM, p. 874.

8141. HOUSTÒNIA L. HOUSTONIA

Flowers solitary on filiform peduncles usually 2-5 cm long..............1. *H. caerulea.*
Flowers cymose or in small clusters, peduncles shorter than those of the preceding.
Fruit (when mature) conspicuously broader than long; sepals mostly 3-6 mm long; leaf blades ovate, lanceolate or narrow-lanceolate..............2. *H. purpurea.*
Fruit (when mature) not conspicuously broader than long; leaf blades of the stems linear to narrow-oblong.
Calyx lobes about 1 mm long, usually shorter than the mature capsule; leaves linear; capsule free only at the apex; stems tufted, from a hard or woody root. ....3. *H. angustifolia.*
Calyx lobes usually more than 2 mm long, longer than the mature capsule; leaves linear or narrow-oblong; stems not from a woody root.
Stem leaves linear to narrow-oblong, more or less glabrous, sometimes the blades rough-pubescent all over, in lines or only the margins roughened, regularly ciliate, but the radical leaves narrowly oval or oblong, the margins not regularly ciliate..................................4. *H. longifolia.*
Stem leaves few and distant (the lower internode sometimes 5 cm long), rather thick, obscurely 1-nerved, oblanceolate or linear-oblong, the radical ones oval or oblong-spatulate, tapering into a petiole, ciliate, sometimes all of the leaves ciliate on the margins (rough-pubescent margins of leaves must not be confused with ciliate margins). (See excluded species no. 579, p. 1091)..*H. canadensis.*

1. **Houstonia caerùlea** L. BLUETS. Map 1906. Mostly in the northwestern and southeastern parts of the state. This species prefers a slightly acid soil and is usually found in black, sandy soil in woodland or pastures in the northwestern part of the state, and in open woodland and fallow fields in the southeastern part of the state. Where it is found it is usually common, sometimes covering acres.

N. S. to Ont. and Wis., southw. to Ga., Ala., and Mo.

2. **Houstonia purpùrea** L. MOUNTAIN HOUSTONIA. Map 1907. My specimens are mostly from the southern half of the state where it is more or less frequent and occurs mostly on slopes in white and black oak woods. In the northern part of the state it has been reported from Cass, Kosciusko, Lake, and Tippecanoe Counties.

This species, like *Houstonia longifolia,* is extremely variable. I have specimens with the leaves varying from narrowly lanceolate to broadly ovate. The calyx lobes vary from 3-6 mm in length. The plant that has the widest leaves has the longest calyx lobes. The plants are more or less pubescent and an extreme form which is densely pubescent all over is f. *pubescens* (Britt.) Fern. (Rhodora 38: 444. 1936.) I have the form from Perry and Pike Counties.

Md. to Iowa, southw. to Ala., Ga., and Ark.

3. **Houstonia angustifòlia** Michx. NARROWLEAF HOUSTONIA. Map 1908. I found this species in very shallow soil on top of the high cliff of the Ohio

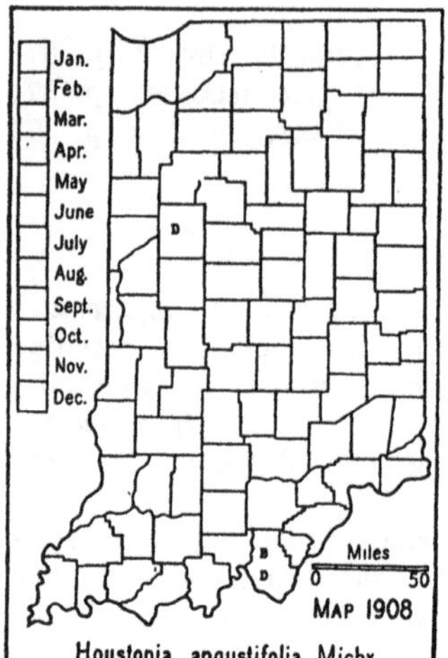

Houstonia angustifolia Michx.

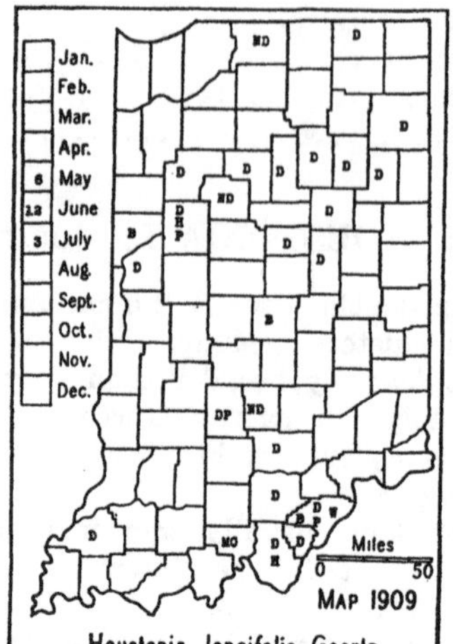

Houstonia longifolia Gaertn.

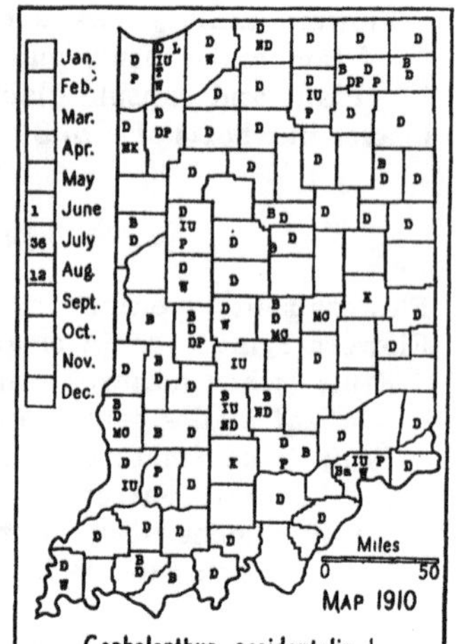

Cephalanthus occidentalis L.

River in a woods in Harrison County in sec. 14 about 4 miles southeast of Laconia. It occurs also on the slope of the high gravelly terrace of Big Wea Creek about 4 miles southwest of Lafayette, Tippecanoe County.

Ind. to Kans., southw. to Fla. and Tex.

4. **Houstonia longifòlia** Gaertn. LONGLEAF BLUETS. Map 1909. Probably infrequent throughout the state on bare places on the crests of wooded ridges, on washed wooded slopes, and more rarely in moist, low woodland. It is commonly found on bare, gravelly places in woodland of all kinds but most commonly on the terraces of streams.

This species is variable in the form of the leaves and in the amount of pubescence. Some leaves are nearly glabrous but the leaves of many specimens are more or less harsh-pubescent above; some are more or less pubescent all over above; some are pubescent only on the nerves and margin, and there are a few with only the margin rough-pubescent. I have not seen a specimen with the basal leaves strictly ciliate. I have specimens with the leaf margins rough-pubescent but the pubescence is not in lines and I do not class them as ciliate. *Houstonia canadensis* has been reported 13 times from Indiana and I had all my specimens so labeled, but after a careful restudy of the material, I changed all of them to *Houstonia longifolia*. I do not believe we have typical *Houstonia canadensis* in Indiana. Reported by many Indiana authors as *Houstonia ciliolata* Torr.

Maine to Man., southw. to Ga. and Mo.

8230. CEPHALÁNTHUS L. BUTTONBUSH

Branchlets, peduncles, petioles, and under surface of the leaves glabrous..........
..1. *C. occidentalis.*
Branchlets, peduncles, petioles, and under surface of the leaves more or less pubescent.
...1a. *C. occidentalis* var. *pubescens.*

1. **Cephalanthus occidentàlis** L. COMMON BUTTONBUSH. Map 1910. Throughout the state in ponds and swamps and on the borders of lakes and streams. Found also in Tippecanoe County on a high gravelly slope about 4 miles southwest of Lafayette.

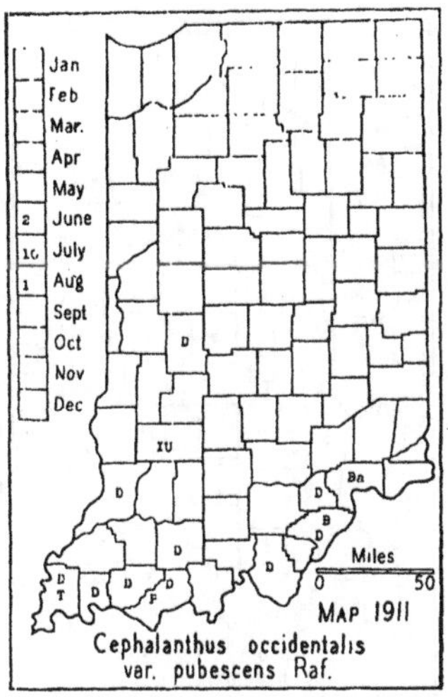

Cephalanthus occidentalis var. pubescens Raf.

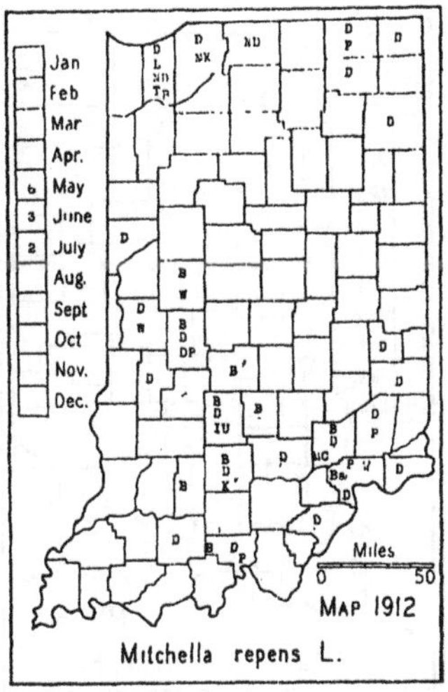

Mitchella repens L.

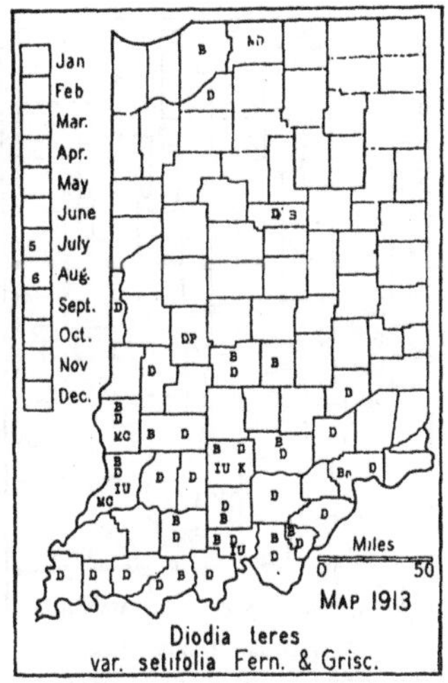

Diodia teres var. setifolia Fern. & Grisc.

Unusual common names are Pond Dogwood (Lower Wabash Valley), Flowering Ash (Shelby County), and Swamp Sycamore (Jay County) because the fruit resembles that of the sycamore.

N. B., Ont. to Calif., southw. to Fla. and Tex.; also in e. Asia.

1a. **Cephalanthus occidentalis** var. **pubéscens** Raf. HAIRY BUTTONBUSH. Map 1911. This variety is found principally in the southern half of the state. Our only report from northern Indiana is that of Peattie from Lake County. I have looked closely for this in Lake County without success. I could not find Peattie's specimen.

The habitat is the same as that of the species but it is rarely found with it.

Ind., southw. to Ga., La., and Tex.

8451. MITCHÉLLA L. PARTRIDGEBERRY

1. **Mitchella rèpens** L. PARTRIDGEBERRY. Map 1912. In all parts of the state where there are areas of slightly acid soil. I have never seen it associated with lime loving plants. The map covers the area of all of the reports except one in Lake County where its habitat occurs. It is usually found in low, flat sweet gum and beech woods, on the crests and slopes of sandstone ridges, and in the black sand of black and pin oak woods of the northern part of the state.

N. S., Ont., and Minn., southw. to Fla., Ark., and Tex.

8471. DIÒDIA [Gronov.] L. BUTTONWEED

[Fernald & Griscom. Notes on Diodia. Rhodora 39: 306-308. 1937.]

1. **Diodia tères** Walt. var. **setifòlia** Fern. & Grisc. (Rhodora 39: 307. 1937.) (*Diodia teres* Walt. of authors.) ROUGH BUTTONWEED. Map 1913. Generally in hard, poor clay soil in pastures, fallow and wheat stubble fields, and clearings, on the crests of open woods, and along roadsides and

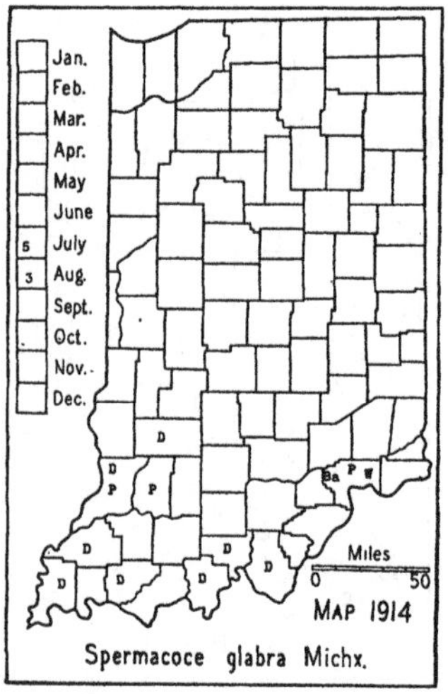

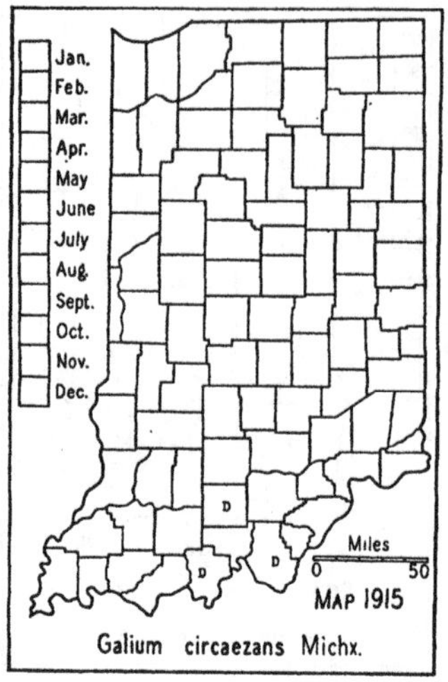

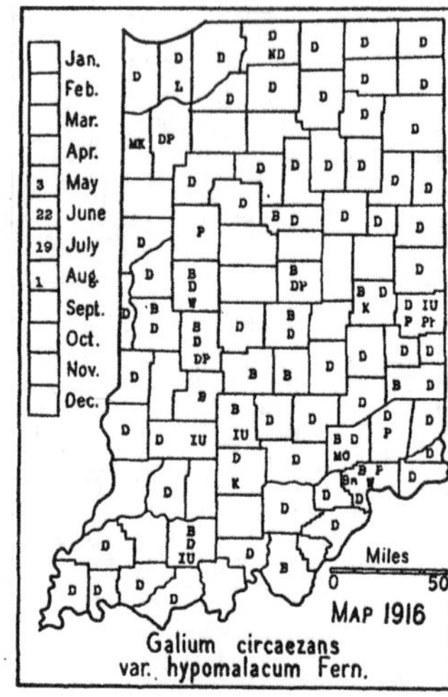

railroads. It is restricted mostly to the southern half of the state although we have a few reports from the northern part of the state along railroads.

The pubescence of the stem of all my specimens is dense, spreading, and less than 0.5 mm long. A few specimens have in addition a few long spreading hairs about 1 mm long.

Southern Mich. to Texas; represented east of Indiana and south to Fla. and west to Tex. by the typical form of the species.

8475. SPERMACÒCE [Dill.] L.

1. **Spermacoce glàbra** Michx. SMOOTH BUTTONWEED. Map 1914. Infrequent to frequent on the muddy slopes of rivers, ponds, and sloughs and in very low, open woods. Reported, also, from Clark and Jefferson Counties.

Southern Ohio, Ill., and Ark., southw. to Fla. and Tex.

8486. GÀLIUM L. BEDSTRAW

Fruit uncinate or more or less hispid.

Leaves 3-nerved, at least at the base (obscurely so in a form of no. 4), not cuspidate.

Flowers along the primary branches of the inflorescence, sessile or nearly so, greenish yellow or purple.

Plants more or less pubescent; upper leaves generally ovate to ovate-lanceolate, obtuse or rarely somewhat acuminate; corolla greenish yellow, 2-2.5 (3) mm wide, generally pubescent.

"Larger leaves 1.5-2.5 cm long and 0.7-1.4 cm broad, the nerves beneath sparingly short-hispid to glabrous."* ....1. *G. circaezans* var. *typica*.

"Larger leaves 2-5 cm long and 1-2.5 cm broad, their nerves conspicuously long-hirsute beneath."* ....1a. *G. circaezans* var. *hypomalacum*.

Plants glabrous or nearly so; upper leaves generally lanceolate to narrow-lanceolate, generally long-acuminate or acuminate; flowers deep purple, rarely greenish yellow, 3-6 mm wide, glabrous...........2. *G. lanceolatum*.

Flowers pedicellate, paniculate.

Leaves oval, more or less pubescent, the median ones usually 6-12 mm wide; corolla greenish purple.................................3. *G. pilosum*.

*** Free translation from Rhodora 39: 450. 1937.**

Leaves narrowly lanceolate, mostly less than 5 mm wide; corolla white.
Fruit covered with long, straight hairs......... .4. *G. boreale* var. *typicum.*
Fruit covered with short, appressed or incurved hairs..4a. *G. boreale* var. *intermedium.*
Fruit glabrous or glabrate.................4b. *G. boreale* var. *hyssopifolium.*
Leaves 1-nerved.
Blades cuspidate at the apex.
Leaves about 8 in a whorl, narrowly oblanceolate; flowering mostly before the middle of June; annuals....................................5. *G. Aparine.*
Leaves 6 in a whorl, elliptic-lanceolate; flowering after the middle of June; perennials..6. *G. triflorum.*
Blades rounded at the apex....................................7. *G. obtusum.*
Fruit glabrous.
Flowers yellow; introduced species (See excluded species no. 586, p. 1092)..*G. verum.*
Flowers purple. (See excluded species no. 583, p. 1091).............*G. latifolium.*
Flowers white.
Flowers in terminal panicles; plants erect.......4b. *G. boreale* var. *hyssopifolium.*
Flowers solitary, in 2's or 3's; plants usually weak, reclining or ascending on other plants, or diffuse.
Leaves cuspidate or sharply acute at the apex.
Stems glabrous; leaves in whorls of 8 on the stems and in whorls of 6 on the branches; perennial. (See excluded species no. 584, p. 1091)..*G. Mollugo.*
Stems more or less scabrous on the angles.
Margins of leaves upwardly roughened.
Annual; leaves mostly ascending; flowers about 1 mm wide; fruit about 1 mm wide; introduced species.......................8. *G. parisiense.*
Perennial; stems weak, in clusters or dense mats, less than 5 dm high; leaves mostly spreading, linear, mostly less than 2 mm wide; flowers about 2 mm wide; fruit about 2 mm wide; native species of rich woodland......................................9. *G. concinnum.*
Margins of leaves retrorsely hispid; stems long, climbing on other plants sometimes to a height of 1.5 m; leaves oval or slightly oblanceolate, usually 2.5-5 mm wide; plants of a wet habitat........10. *G. asprellum.*
Leaves blunt at the apex.
Corolla 4-lobed, the lobes acute, 1 mm or more long; peduncles and pedicels glabrous.
Leaves ascending or spreading, 1.5-2.5 cm long; stem strictly glabrous; fruit 2-celled, mostly 2.5-3.5 mm wide (one cell often not developing)...7. *G. obtusum.*
Leaves usually all strongly reflexed, mostly less than 11 mm long, rarely longer; fruit 2-celled, usually about 1.5 mm wide; pedicels short...11. *G. labradoricum.*
Corolla 3-lobed, rarely 4-lobed, the lobes obtuse, usually 0.5-1 mm long; fruit 2-celled, 1.5-2.75 mm wide.
Flowers mostly in 2's and 3's; pedicels usually straight, glabrous, mostly 2-5 (6) mm long...............................12. *G. tinctorium.*
Flowers solitary; mature pedicels generally arcuate, scabrous, usually 5-10 mm long. ....13. *G. trifidum.*

1. **Galium circaèzans** Michx. var. **týpicum** Fern. WILD LICORICE. Map 1915. Fernald (Rhodora **39**: 449-450. 1937) has divided this species into a northern and southern form. He designates the southern form as the typical form of the species. Out of my 73 Indiana specimens I am referring all but three to the variety.

R. I., Conn., N. Y., s. Mich., southw. to Fla. and Tex.

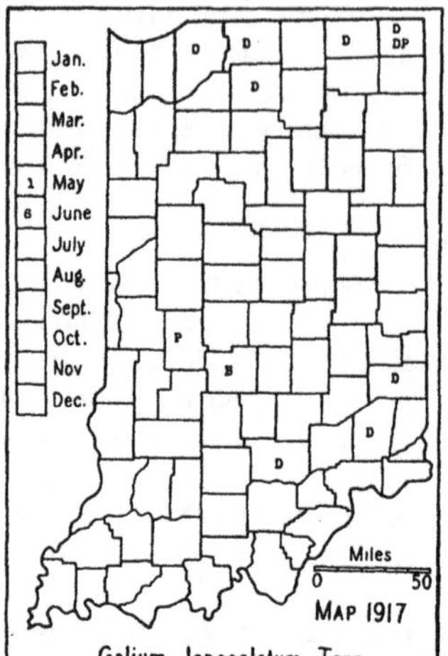

Galium lanceolatum Torr.

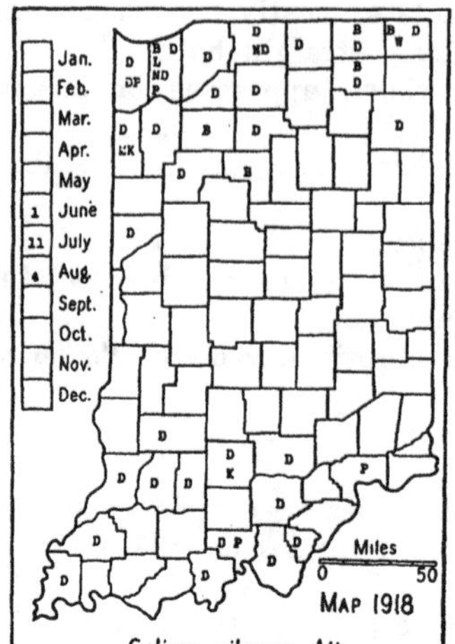

Galium pilosum Ait.

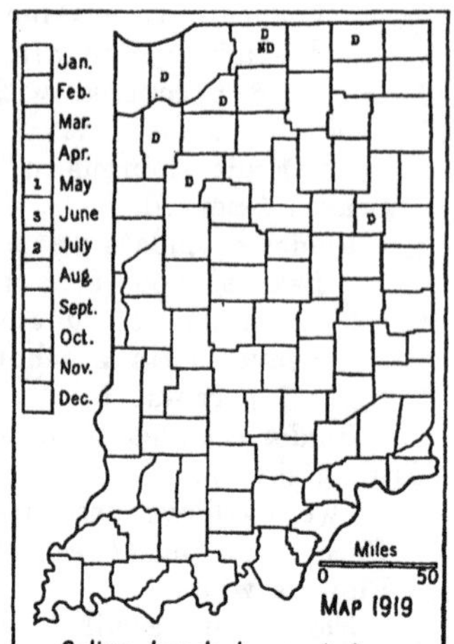

Galium boreale L., var. typicum

1a. **Galium circaezans** var. **hypomálacum** Fern. (Rhodora 39: 450. 1937.) WILD LICORICE. Map 1916. The variety is frequent in moist, rich woods throughout the state.

Maine, sw. Que. to Minn. and Nebr., southw. to N. C., Ky., Mo. and Okla.

2. **Galium lanceolàtum** Torr. WILD LICORICE. Map 1917. Very local; in moist or dry woods, usually associated with beech and sugar maple.

Maine to Minn., southw. to N. C. and Ky.

3. **Galium pilòsum** Ait. HAIRY BEDSTRAW. Map 1918. Infrequent throughout the lake area in dry, sandy soil, usually associated with black and white oak; rarer in the southwestern part of the state, where it is generally found in rather sandy soil on the crests and slopes of black oak ridges; apparently absent from the Tipton Till Plain.

N. H., Ont., Mich., Ill., and Kans., southw. to Fla. and Tex.

4. **Galium boreàle** L. var. **týpicum** Beck von Man. (Fernald. The varieties of Galium boreale. Rhodora 30: 106-10. 1928.) NORTHERN BEDSTRAW. Map 1919. Restricted to the lake area. Our specimens are from moist, sandy soil along railroads and roadsides and one is from a tamarack bog.

N. H. to B. C., and Alaska, southw. to N. Y., N. Mex., and Oreg.

4a. **Galium boreale** var. **intermèdium** DC. Map 1920. Local in the lake area in moist, sandy soil in black oak woods, on borders of lakes, and along roadsides.

N. E. to Ont., southw. to Del. and Ind.

4b. **Galium boreale** var. **hyssopifòlium** (Hoffm.) DC. Map 1921. Restricted to the lake area and found in dry, sandy soil along railroads and roadsides, and less frequently in bogs and marshes.

Gaspé Peninsula, s. Que., N. Dak. to Vancouver Island, southw. to n. N. J., Ohio, Mo., and Oreg.

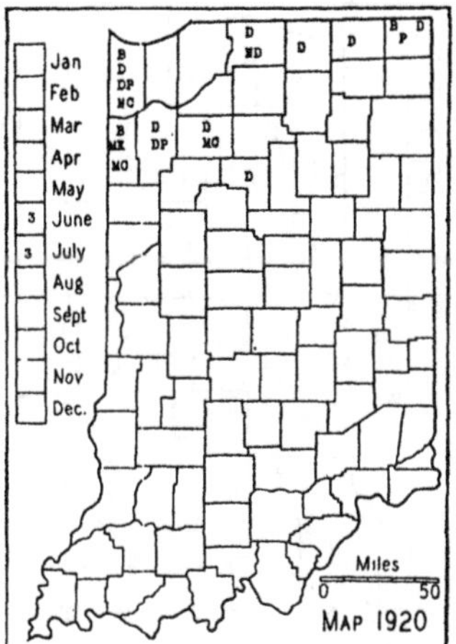

MAP 1920
Galium boreale var. intermedium DC

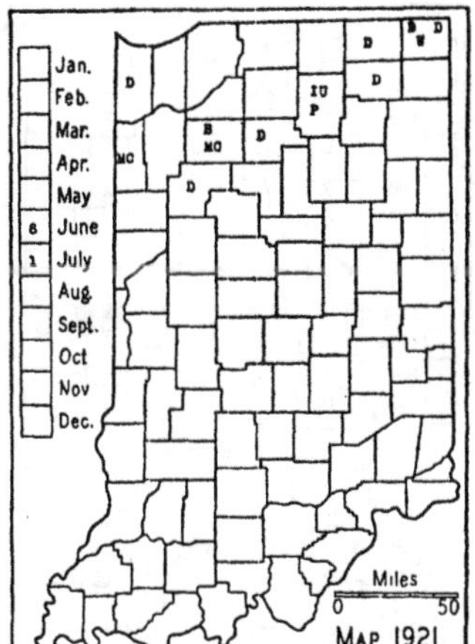

MAP 1921
Galium boreale var. hyssopifolium (Hoffm.) DC.

MAP 1922
Galium Aparine L

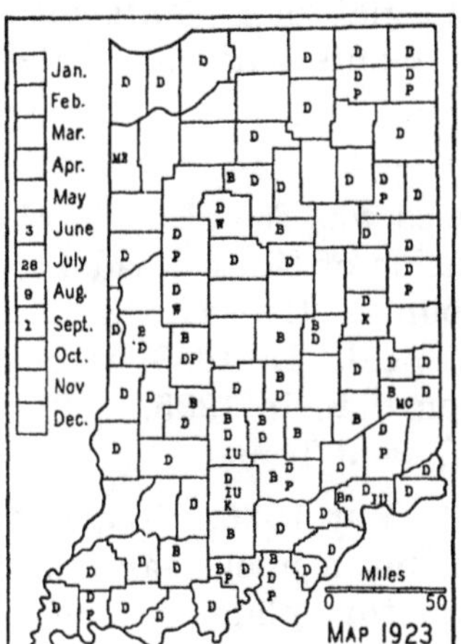

MAP 1923
Galium triflorum Michx

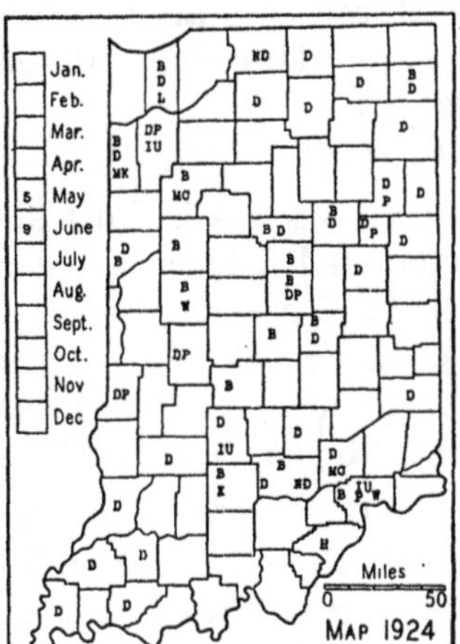

MAP 1924
Galium obtusum Bigel

MAP 1925
Galium parisiense L

5. **Galium Aparìne** L. CLEAVERS. Map 1922. Throughout the state in moist woods. It is more common in moist woods and in some places it will form dense stands, especially in alluvial flats; also found along roadsides and sometimes it is an annoying weed in gardens. It is undoubtedly a native of Indiana.

N. B. to B. C., southw. to Fla., Tex., and Calif.; also in Eurasia.

6. **Galium triflòrum** Michx. SWEET-SCENTED BEDSTRAW. Map 1923. Frequent in moist woods throughout the state.

Greenland and Newf. to B. C., southw. to Fla., La., Colo., and Calif.

7. **Galium obtùsum** Bigel. (Rhodora 37: 443-445. 1935.) (*Galium tinctorium* of Gray, Man., ed. 7 and Britton and Brown, Illus. Flora, ed. 2.) Map 1924. Frequent to infrequent in wet woods throughout the state. There is a form common in a low, wet woods in section 17 of Point Township, Posey County, that has the fruit more or less hispid. I studied this plant where it was common over several acres and found the fruit to be

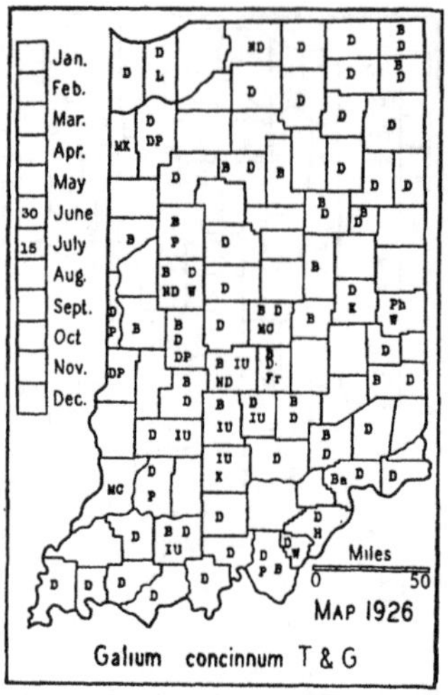

MAP 1926
Galium concinnum T & G

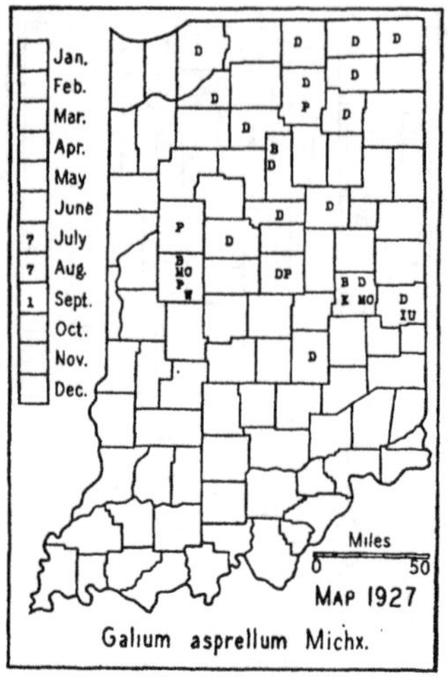

MAP 1927
Galium asprellum Michx.

MAP 1928
Galium labradoricum Wiegand

very variable. There were plants with all of the fruit glabrous, plants with some of the fruit more or less glabrous, and some plants with all of the fruit rather densely pubescent.

N. S. to Mich. and Nebr., southw. to N. C. and Ariz.

8. GALIUM PARISIÉNSE L. Map 1925. This species was found by R. C. Friesner in 1935 in an abandoned field on the east side of State Road 135 about 0.3 mile north of Bean Blossom, Brown County. He says that it was common over at least three acres of a large field. In 1939 he revisited this place and reported it even more common than in 1935.

Nat. of Eu.; sparingly introduced, Va., N. C., Ind., and Tenn.

9. **Galium concínnum** T. & G. PRETTY BEDSTRAW. Map 1926. Frequent in rich, dry woods throughout the state. In 1916 I met a man near Lake George in Steuben County who makes a tea from this plant and who says that it is an infallible remedy for kidney disorders and dropsy.

N. J. to Minn., southw. to Va. and Ark.

10. **Galium asprĕllum** Michx. ROUGH BEDSTRAW. Map 1927. A plant mostly of springy places and swamps where it usually climbs upon vegetation to a height of 3 to 5 feet. It is very local and is found mostly in the northern half of the state.

Newf. to Minn., southw. to N. C., Ill., Mo., and Nebr.

11. **Galium labradòricum** Wieg. Map 1928. Local. Generally in sphagnum in tamarack bogs, marshes, and sedge borders of lakes.

Lab. to Wis., southw. to n. Conn., N. Y., and Ind.

12. **Galium tinctòrium** L.* (*Galium Claytoni* Michx.) Map 1929. Infrequent in all parts of the state in swampy woods, about ponds, and along ditches.

Newf., N. Y., and Mich. to Nebr., southw. to Fla. and Tex.

13. **Galium trífidum** L. Map 1930. Very local; in marshes and bogs. There are 18 reports of it from south of the lake area. I believe all of

*For a discussion of this species see Rhodora 41: 388. 1939.

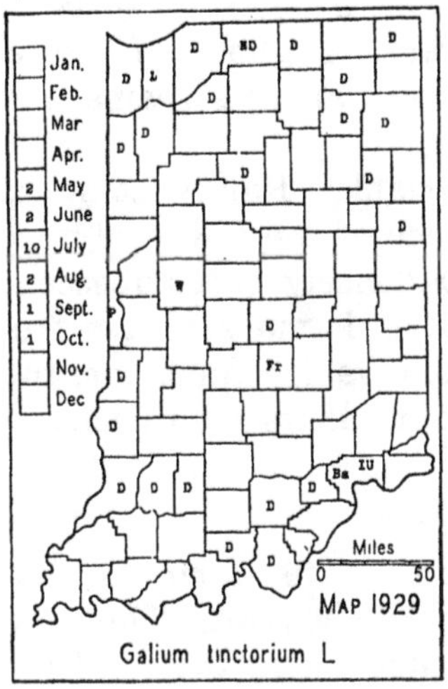

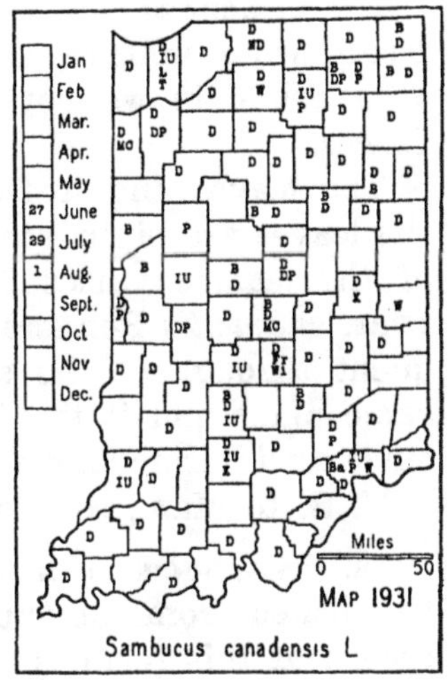

these reports should be referred to some other species although we now have two specimens from south of the lake area. One from Jefferson County and another which was collected by Wilson in Hamilton County in 1899 now in the herbarium of DePauw University. The books used by the early botanists did not enable them to easily separate this species from those closely allied to it, and that fact may have been responsible for some of the early reports.

Newf. and Lab. to B. C., southw. to n. and w. N. E., N. Y., Ohio, Ind., Colo., and Calif.

271. CAPRIFOLIÀCEAE Vent. HONEYSUCKLE FAMILY

Leaves compound...8515. SAMBUCUS, p. 879.
Leaves simple.
  Flowers in compound cymes.................................8516. VIBURNUM, p. 880.
  Flowers not as above.
    Plants trailing, evergreen, semi-herbaceous; flowers on long peduncles; leaves rounded-oval to nearly orbicular, generally less than 1 cm wide, sparsely crenate...8520. LINNAEA, p. 887.
    Plants not as above.
      Stems herbaceous, erect; flowers sessile, axillary, 1-3 in a cluster, pale yellow or reddish purple, 8-18 mm long; fruit fleshy, 9-14 mm in diameter...8517. TRIOSTEUM, p. 884.
      Stems woody.
        Leaves serrate; fruit a capsule.....................8524. DIERVILLA, p. 890.
        Leaves not serrate; fruit a berry.
          Flowers regular, mostly 4-6 mm long, sessile or on short pedicels, in clusters, short spikes or racemes, terminal or from the axils of the upper leaves; erect or spreading shrubs..............8518. SYMPHORICARPOS, p. 887.
          Flowers irregular, over 1 cm long, capitate, spicate or in pairs; mostly twining vines, one species erect...............8523. LONICERA, p. 888.

8515. SAMBÙCUS [Tourn.] L. ELDER

Pith of plant white; mature fruit black or greenish yellow............1. *S. canadensis.*
Pith of plant brown; mature fruit bright red..........................2. *S. pubens.*

1. **Sambucus canadénsis** L. AMERICAN ELDER. ELDERBERRY. Map 1931. Fruit, when mature, a purple black. In moist soil throughout the state. It is found in wet, open woodland, about lakes, and along streams and fences.

The leaves and leaflets of this species are variable. Rarely some of the leaves are bipinnate at the base. The pubescence of the lower surface of the leaflets varies from slightly pubescent to densely soft-pubescent (var. *submollis* Rehder). The densely pubescent form is more or less frequent throughout the state. The pubescence often varies much on the same plant and it is of no advantage to divide our plants on this basis since all intermediate forms can be found.

N. S. to Man., southw. to Fla., Kans., and Ariz.

1a. **Sambucus canadensis** f. **chlorocárpa** Rehd. This form is distinguished from the typical form by its greenish yellow fruit. The only record of this form is that of a colony which I found along the roadside about a half mile northwest of Helmer, Steuben County. I found it in a colony of the typical form. I have had it under cultivation since 1923, and new plants from its seed have the characteristic greenish yellow fruit.

2. **Sambucus pùbens** Michx. (*Sambucus racemosa* L. of Gray, Man., ed. 7 and of Britton and Brown, Illus. Flora, ed. 2.) SCARLET ELDER. Map 1932. This species is restricted to the lake area where it is generally found in moist woods, in swamps where it is frequently associated with black ash, and rarely on dry ground where I found it associated with beech and sugar maple.

Newf. to B. C., southw. to Pa., Iowa, Colo., and Calif.; also in the mts. to Ga.

2a. **Sambucus pubens** f. **cálva** Fern. (Rhodora 35: 310. 1933.) This is a form with glabrous leaves and branchlets. I have a specimen from Noble County.

"Occasional through the range of the species."

2b. **Sambucus pubens** f. **xanthocárpa** (Cockerell) Fern. (*Sambucus pubens* var. *xanthocarpa* Nieuwl.) This is a yellow fruited form found west of South Bend, St. Joseph County, by Nieuwland.

8516. VIBÚRNUM [Tourn.] L. VIBURNUM

Leaves 3-lobed, rarely one or more pairs not lobed.
  Branchlets glabrous; petioles glabrous, with 2 glands near the base of the leaf blade; fruit red..1. *V. trilobum.*
  Branchlets pubescent; petioles pubescent, glandless; fruit black....2. *V. acerifolium.*
Leaves not lobed.
  Leaves without stipules, finely serrate or more or less crenulate-dentate, glabrous above and beneath (no. 6 usually having some rusty tomentum on the veins, midrib, and petiole); petioles flat and more or less margined; stones flat, without grooves on the sides (except in no. 3).
    Blades more or less crenulate-dentate, some more or less entire or only the upper half crenulate; cymes on peduncles 1-2 cm long............3. *V. cassinoides.*
    Blades finely and sharply serrate; cymes sessile or on short peduncles, these rarely 2 cm long.

Sambucus pubens Michx.

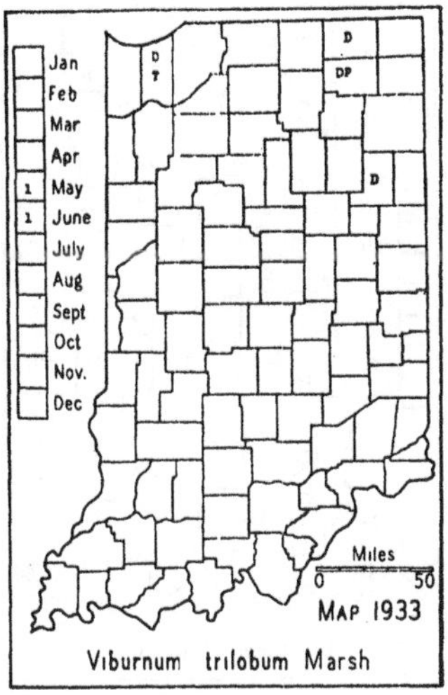

Viburnum trilobum Marsh

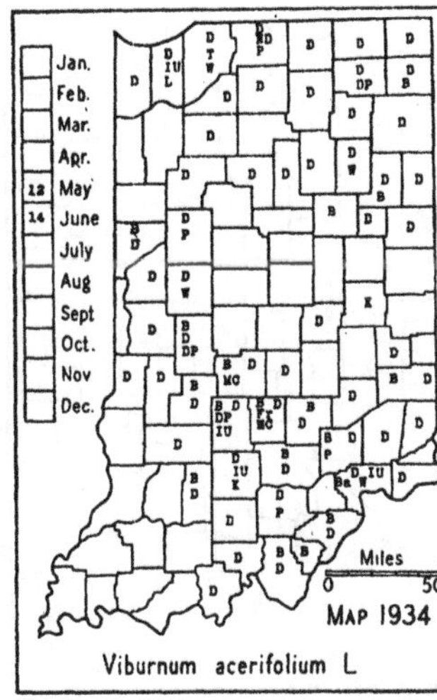

Viburnum acerifolium L

Buds not scurfy-punctate during the summer phase, the surface not porous, usually somewhat glossy; leaves thin, acute or acuminate; subapical margins of principal blades concave; under surface of blades and petioles glabrous or some of the petioles under flowering cymes with rusty tomentum.

Leaves caudate-acuminate, rarely some or all short-pointed at the apex; margins of petiole wide and wavy; stamens usually exserted half their length; shrubs of springy or wet places....................4. *V. Lentago.*

Leaves merely acute or obtuse at the apex; margins of petiole narrow, not so wide or wavy as the preceding; stamens usually exserted about a fourth their length; shrubs usually of moist places and generally near streams. ....5. *V. prunifolium.*

Buds scurfy-punctate or rusty-pubescent, the surface porous, not at all glossy; leaves very thick, usually rounded at the apex or short-acute; subapical margins of principal blades convex or straight; under surface of petioles and often the midrib, veins, or the whole under surface of the blade covered more or less with a rusty tomentum......................6. *V. rufidulum.*

Leaves with or without stipules, more or less pubescent both above and beneath but the pubescence never of a rusty color; petioles rounded; sides of stones of fruit more or less grooved.

Teeth of leaves rarely more than 9 to a side; petioles less than 1 cm long; peduncles of cymes mostly 0.5-2 cm long; cymes mostly 3-5 cm broad.

Under surface of leaves pubescent only on the midrib and on the principal veins or very sparingly between the veins....................7. *V. affine.*

Under surface of leaves densely pubescent over the entire surface.......... ....7a. *V. affine* var. *hypomalacum.*

Teeth of leaves mostly more than 9 to a side; petioles longer than 1 cm; peduncles of cymes mostly 2-9 cm long; cymes mostly 2-7 cm broad.

Bark exfoliating; leaves cordate at the base......................8. *V. molle.*

Bark not exfoliating; leaves not cordate at the base.

Under surface of leaves pubescent all over........9. *V. pubescens* var. *Deamii.*

Under surface of leaves pubescent only along and on the principal veins, rarely sparsely pubescent between the veins...................... ....9a. *V. pubescens* var. *indianense.*

1. **Viburnum trilobum** Marsh. (*Viburnum Opulus* var. *americanum* (Mill.) Ait. and *Viburnum Opulus* of Britton and Brown, Illus. Flora, ed. 2.) AMERICAN CRANBERRYBUSH. CRAMP BARK. Map 1933. Restricted to the lake area where it is found in low woods and on the borders of

Viburnum cassinoides L

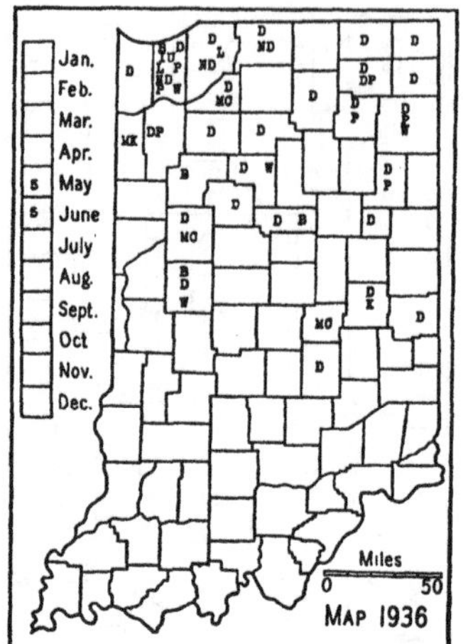

Viburnum Lentago L

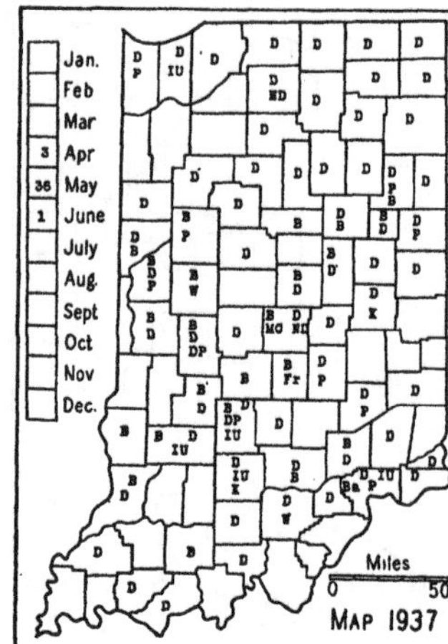

Viburnum prunifolium L.

lakes and streams. It has been reported from Kosciusko, La Porte, Steuben, and Tippecanoe Counties. The Tippecanoe County report is probably based upon a cultivated plant. I found no specimen.

N. B. to B. C. and southw. to N. Y., Ind., S. D., and Oreg.

2. **Viburnum acerifòlium** L. MAPLELEAF VIBURNUM. Map 1934. In dry woods in all parts of the state although there are neither records nor specimens from 10 of the southwestern counties. It is usually a shrub from 2-5 feet high but I measured a specimen in St. Joseph County that was 7 feet high.

N. B. to Minn., southw. to Ga. and Ala.

2a. **Viburnum acerifolium** f. **ovàtum** Rehder. (Jour. Arnold Arboretum 5: 241. 1924.) This is a form with leaves ovate, remotely dentate, and subcordate. I found it in a white oak woods about 2 miles east of Grayford in Jennings County.

3. **Viburnum cassinoìdes** L. WITHE-ROD. Map 1935. So far as it is known this species is restricted to the northern tier of counties. Found in low, sandy, black and white oak woods. Very local. There is a specimen from La Porte County in the herbarium of the University of Notre Dame. Buhl (Amer. Midland Nat. **16**: 252. 1935) refers the report of Pepoon to *V. affine* or its variety.

Newf. to Man. and Minn., southw. to N. J., Fla., and Ala.

4. **Viburnum Lentàgo** L. NANNYBERRY. Map 1936. Rather frequent in the lake area and occasional in boggy places south of it. It has been reported from the southern part of the state but I believe all or at least most of the reports should be referred to *Viburnum prunifolium.* This species is difficult for some to distinguish from *Viburnum prunifolium.* Most of the southern reports say that the specimens were found in dry woods which is not the habitat of this species. I have seen it only in wet woods and springy places.

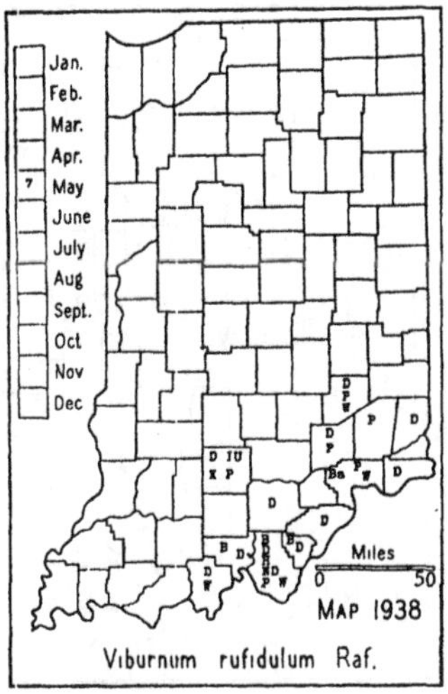

Viburnum rufidulum Raf.

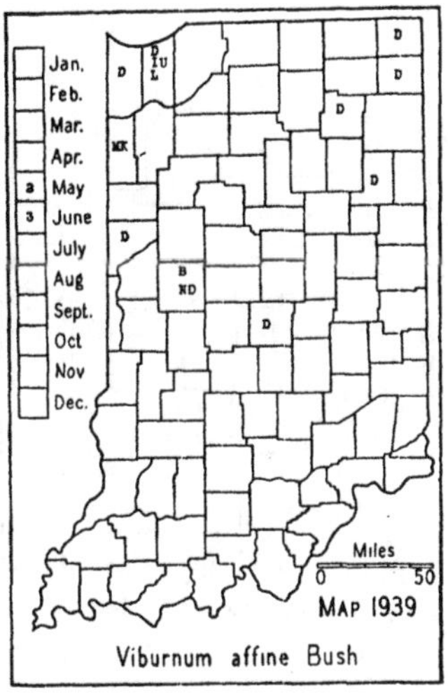

Viburnum affine Bush

Viburnum affine var. hypomalacum Blake

In 1923, I measured a specimen in the Clarence Tumm woods 7 miles east of Michigan City that was 16.5 inches in circumference at 40 inches above the ground, and was 20 feet high.

Que. to Man., southw. to N. J., Ind., Kans., and Colo.; in the mts. to Ga.

5. **Viburnum prunifòlium** L. BLACKHAW. Map 1937. This species, no doubt, was native in every county of the state. More or less frequent in moist woods throughout the state, except in the hilly counties where it becomes more or less rare, and its place is taken by *Viburnum rufidulum.*

Conn. to Iowa, southw. to Ga. and Tex.

6. **Viburnum rufídulum** Raf. SOUTHERN BLACKHAW. Map 1938. Restricted to the southern part of the state and possibly not extending far beyond the area indicated on the map. All of our specimens are from the slopes of rocky, wooded ravines.

N. J., Ind., Mo., and Kans., southw. to Fla. and Tex.

7. **Viburnum affìne** Bush. (*Viburnum pubescens* (Ait.) Pursh, in part, of Gray, Man., ed. 7 and Britton and Brown, Illus. Flora, ed. 2.) MISSOURI VIBURNUM. Map 1939. In clay soil on white oak slopes and their bases, in sandy soil on the crests and slopes of wooded ridges, and in moist places at their bases. All of our specimens and reports are from Marion County and northward.

Ind. to Minn. and Mo.

7a. **Virburnum affine** var. **hypomálacum** Blake. (Rhodora 20: 14. 1918.) (*Viburnum pubescens* (Ait.) Pursh, in part, of Gray, Man., ed. 7 and Britton and Brown, Illus. Flora, ed. 2.) Map 1940. Usually in dry clay or sandy soil on wooded slopes and restricted to the lake area.

Que. to Man., southw. to Ga. and Ill.

8. **Viburnum mólle** Michx. KENTUCKY VIBURNUM. May 1941. Very local. On gravelly or rocky, wooded slopes, usually bordering streams.

Ind. to Iowa, southw. to Ky. and Mo.

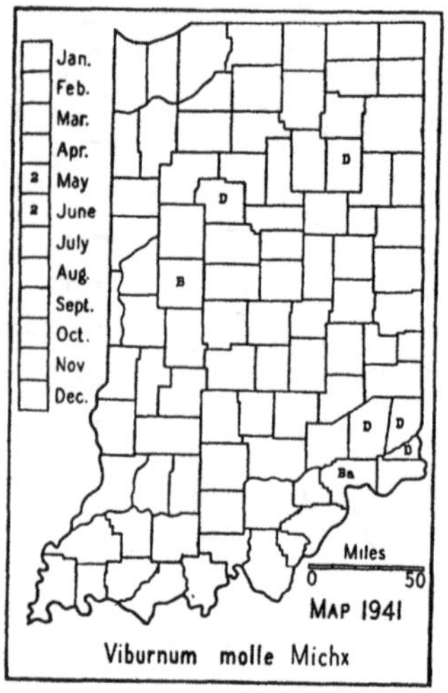

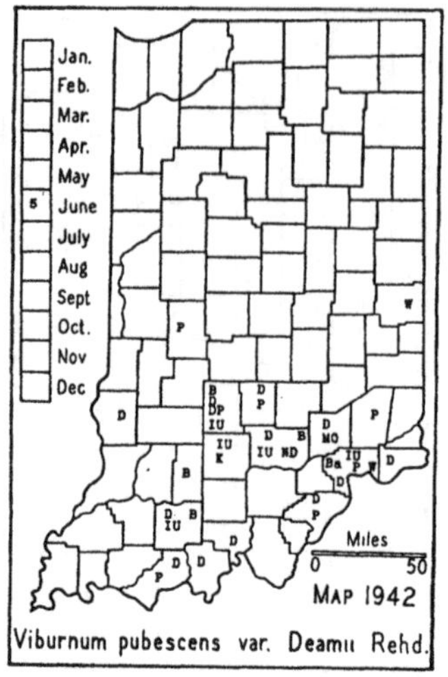

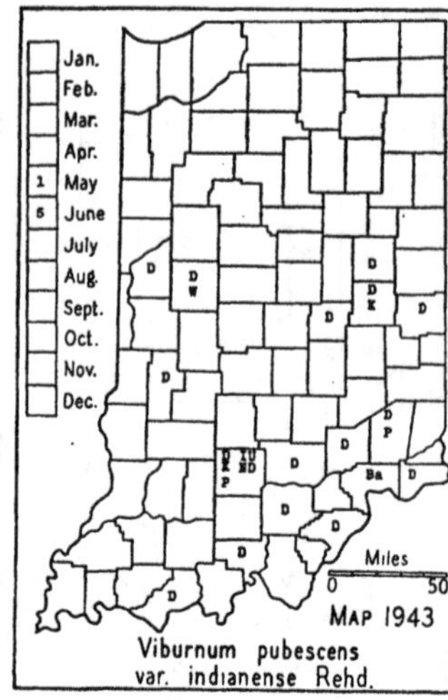

9. **Viburnum pubéscens** (Ait.) Pursh var. **Dèamii** Rehd. (Jour. Arnold Arboretum 5: 58-59. 1924.) Map 1942. This shrub is found in the southern half of the state in hard, clay soil, associated with sweet gum, black gum, pin oak, and beech. It is also found in the knobstone area toward the bases of wooded slopes where it is associated with oaks, or with beech and maple.

So far as known it is restricted to Ind., Ky. and ne. Mo.

9a. **Viburnum pubescens** var. **indianénse** Rehd. (Jour. Arnold Arboretum 5: 59. 1924.) Map 1943. Usually found in low woods, associated with beech, red maple, and sweet gum; with beech, white ash, shagbark hickory, and sugar maple; and with white elm, ash, and red oak. Rarely is it found on dry, rocky, wooded slopes and once it was found in a springy terrace along Sugar Creek in Montgomery County.

Known in Indiana from Montgomery County southward, and in southwestern Ohio (Braun).

8517. TRIÓSTEUM L. HORSEGENTIAN

[Wiegand. Triosteum perfoliatum and related species. Rhodora 25: 199-203. 1923.]

Sepals finely and, for the most part, evenly pubescent; stipules of leaves rarely reaching the tips of the sepals; flowers 2-6 at each node; corolla pale to deep purple, 8-15 (17) mm long, densely puberulent, more or less glandular; stem densely pubescent with short and more or less glandular hairs and with a sparser and longer pubescence or villous with few or no short hairs; leaves narrowly to broadly ovate-oblong, finely strigose to subglabrate above, sometimes with hairs 1 mm long or less.

Leaves velvety beneath.

Principal leaves usually connate-clasping; stem densely short-pubescent with a mixture of longer and shorter hairs, the shorter ones often nearly all glandular; sepals usually narrow (in flower 0.9-2 mm wide), generally very acute; corolla purplish, often greenish on the lower part, firm, the mouth 5-6 mm wide, usually not flaring; fruit usually 6 at some of the nodes, especially the lower

ones, short-pubescent, glandular, the hairs mostly not more than 0.5 mm long, the longest hairs glandular, surface dull; fruit maturing in late October, 12-14 mm long, and 11-12 mm wide, subglobose, Ochraceous-Orange (Ridgway). ....1. *T. perfoliatum.*

Principal and other leaves usually not connate, narrowed below into winged, sessile, hardly clasping bases, generally less velvety; stem usually spreading-villous, with few or no short hairs; sepals generally broader (in flower 1.5-2.8 mm wide), obtuse or acute, sometimes purple tinged; corolla purplish red, often lighter, the mouth 7-9 mm wide, usually more flaring; fruit usually maturing less than 6 at a node, sometimes only 2 or 2 or 4 perfect ones and the others aborted, villous with spreading hairs 0.75-2 mm long, the longest hairs not glandular, interspersed with shorter, glandular ones, surface rather glossy; fruit maturing from early August to early October, 11-18 mm long and 8.5-15 wide, ellipsoid-ovoid, Brazil Red (Ridgway).

Cauline hairs up to 1.5 mm long; longest hairs of fruit 0.75-1 mm long..... ....2. *T. aurantiacum.*

Cauline hairs 1.5-2 mm long or longer; pubescence of fruit 1-1.5 (2) mm long; hairs on upper surface of leaves up to 1 mm long................... ....2a. *T. aurantiacum* var. *illinoense.*

Leaves glabrous or nearly so beneath, pubescence of sepals and corolla usually less than that of *T. aurantiacum;* sepals usually acute........................ ....2b. *T. aurantiacum* var. *glaucescens.*

Sepals hispid-ciliate, otherwise glabrous or sparingly short-hispid; stipules of leaves usually extending beyond the sepals; flowers usually 2 at each node; corolla yellow, 14-18 mm long, loosely villous, slightly glandular, the lobes large and broad; stem setose-hispid, the hairs nearly all long (longest hairs 1.5-2.8 mm long); leaves not perfoliate, lanceolate to oblong-lanceolate, distinctly hispid-strigose above with hairs 0.8-1.8 mm long; fruit maturing in September, about 8.5 mm long and 6.5 mm wide, ellipsoid, Mars Orange (Ridgway)...................3. *T. angustifolium.*

1. **Triosteum perfoliàtum** L. COMMON HORSEGENTIAN. Map 1944. There are reports of this species from all parts of the state but since those from the southern part were made before the species was divided, the only way to consider the distribution is from the specimens at hand. It is generally found in dry clay or sandy soil on white oak slopes and rarely in moist ground.

Wiegand, in his article cited above, considers the following species, *T. aurantiacum,* as a variety of *T. perfoliatum.* However, they seem sufficiently distinct in Indiana to warrant treatment as species. Colonies of both species growing in the Deam garden were observed for four years, and notes were kept concerning them.

T. perfoliatum was found to bloom from two to three weeks later than *T. aurantiacum,* and to mature its fruit correspondingly later. It is a light orange yellow when beginning to mature, gradually darkening until it is Ochraceous-Orange (Ridgway). The fruit persists well into November, resisting a temperature as low as 26° without harming the plant or causing the fruit to shrivel or drop. By the latter part of October, the fruit of *T. aurantiacum* began to dry up and to fall.

Mass., Wis., Nebr., south to D. C. and in the mts. to N. C., Kans., and Mo.

2. **Triosteum aurantìacum** Bickn. (Torreya 1: 26. 1901.) Map 1945. Infrequent possibly throughout the state. It is generally found in rather open, dry woods and more rarely in moist, sandy places in a prairie habitat.

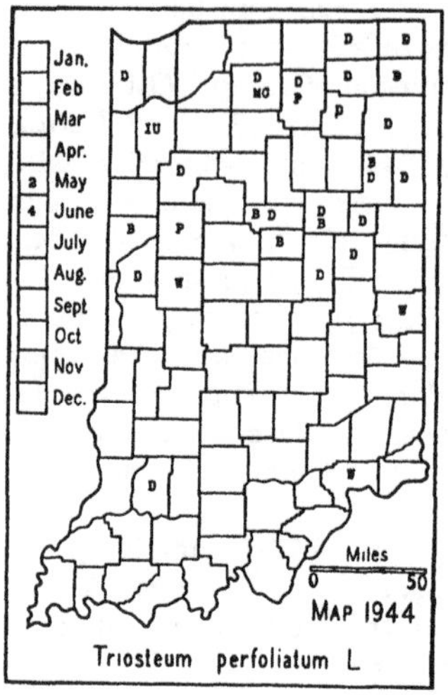

Wiegand places this species as a variety of the preceding one on the basis that there are intergrading plants and some plants of the non-perfoliate group are sometimes perfoliate and sometimes the pubescence characters are not constant. It has been observed that, as is shown in Bicknell's description, the leaves of *T. aurantiacum* are sometimes perfoliate but it is true only of the upper ones, not the middle or principal leaves, while in *T. perfoliatum,* it is the middle leaves which are perfoliate and if there is any difference in the leaves, it is the upper ones which are narrowed. The fruit characters, however, their color, shape, and pubescence, the time of flowering and maturing of fruit, and various other combinations of characters are sufficient to separate the two without depending upon the types of leaves although they are helpful when understood.

Que. to Conn., and in the mts. to Va., and from N. Y. to Ill. and Wis.

2a. **Triosteum aurantiacum** var. **illinoénse** (Wieg.) Palmer & Steyermark. (Rhodora **40**: 133. 1938.) Map 1946. This variety occurs throughout the state. Nearly all of our specimens are from rocky, wooded slopes bordering streams. A few are from dry woods. The pubescence of the fruit, stem, and leaves is longer than that in the species.

Ohio to Mo.

2b. **Triosteum aurantiacum** var. **glaucéscens** Wieg. (Rhodora **20**: 116. 1918.) Map 1947. Our only specimen is from the wooded bank of Raccoon Creek south of Russellville, Putnam County. Both surfaces of the leaves are entirely glabrous; the pubescence of the stem is as in *T. aurantiacum.*

Cent. N. Y. to Pa. and Ind.

3. **Triosteum angustifòlium** L. YELLOW-FLOWER HORSEGENTIAN. Map 1948. This species is undoubtedly restricted to the southern half of the state. I have found it on dry, wooded slopes only, and usually associated with black and white oak.

Conn. to Md., and in the uplands to Ala. and Tenn., westw. to Ill. and Mo.

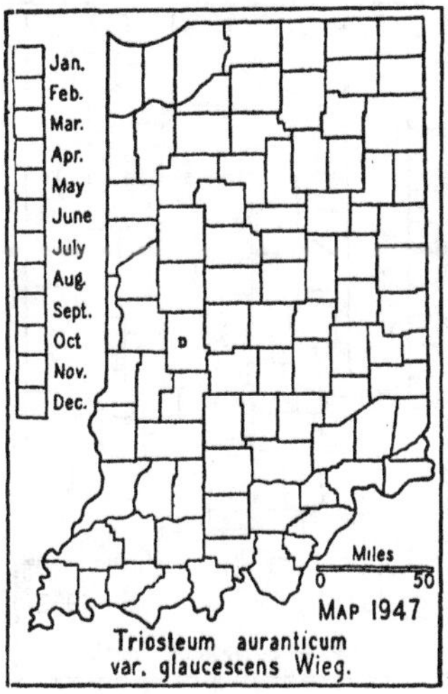

MAP 1947
Triosteum aurantiacum var. glaucescens Wieg.

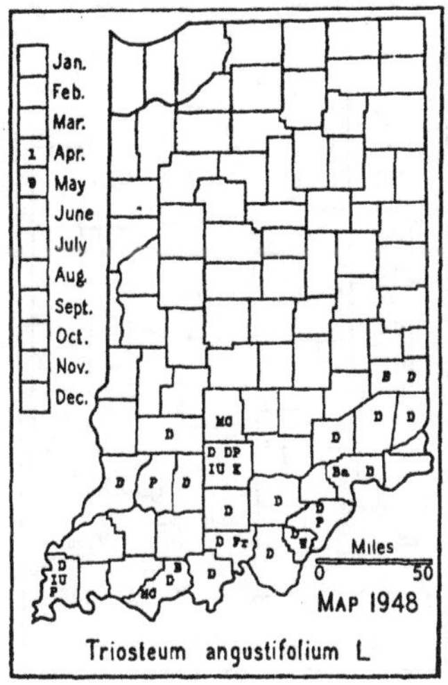

MAP 1948
Triosteum angustifolium L.

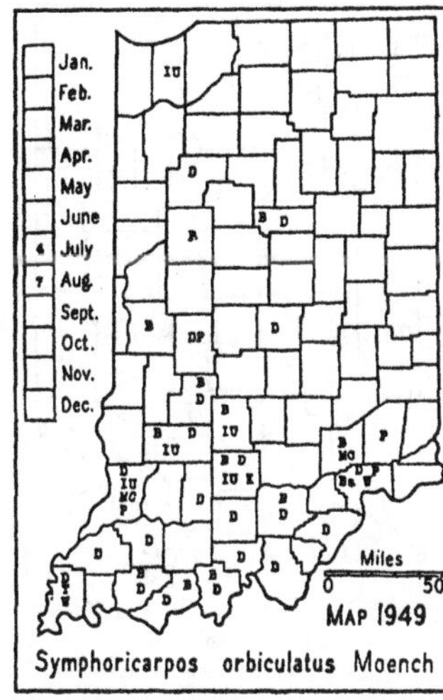

MAP 1949
Symphoricarpos orbiculatus Moench

8518. SYMPHORICÁRPOS [Dill.] Ludwig SNOWBERRY

Style bearded; fruit coral red....................................1. *S. orbiculatus.*
Style glabrous; fruit white......................................2. *S. rivularis.*

1. **Symphoricarpos orbiculàtus** Moench. (*Symphoricarpos Symphoricarpos* (L.) MacM.) CORALBERRY. Map 1949. In Indiana generally called buckbush. It is native, probably only in the southern half of the state although it is now found as an escape in the northern part. Since it is freely planted and produces an abundance of fruit, it is strange that it does not escape more often than it has.

N. J. to S. D., southw. to Ga. and Tex.

1a. **Symphoricarpos orbiculatus f. leucocarpus** (D.M. Andrews) Rehd. (Jour. Arnold Arb. 21: 277. 1940.) This form was reported to me in 1937 by Harold W. Reynolds as scattered on the border of the grounds about the old Reid schoolhouse, on the Thomas Brewer farm, eleven and a half miles north of Salem, Washington County. Mr. Reynolds reports he has known the colony about 25 years and that his sister had known of it for about 40 years.

2. SYMPHORICARPOS RIVULARIS Suksdorf. (*Symphoricarpos racemosus* var. *laevigatus* Fern. GARDEN SNOWBERRY. This species is commonly planted as an ornamental shrub and has been reported as an escape in Franklin, Jefferson, and Steuben Counties and in the Lower Wabash Valley. I have found it twice as an escape.

Alaska to Calif., eastw. to Mont. Cultivated and escaping.

8520. LINNAÈA [Gronov.] L.

1. **Linnaea boreàlis** L. var. **americàna** (Forbes) Rehd. (*Linnaea americana* Forbes.) TWINFLOWER. Map 1950. This plant grows in bogs and rarely on dry land. I found a single small colony in very sandy soil, growing in a patch of *Gaylussacia baccata* on a north slope, wooded with black

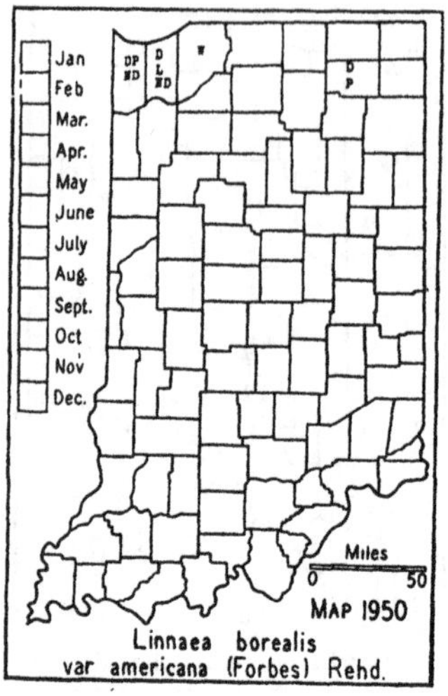

MAP 1950
Linnaea borealis var americana (Forbes) Rehd.

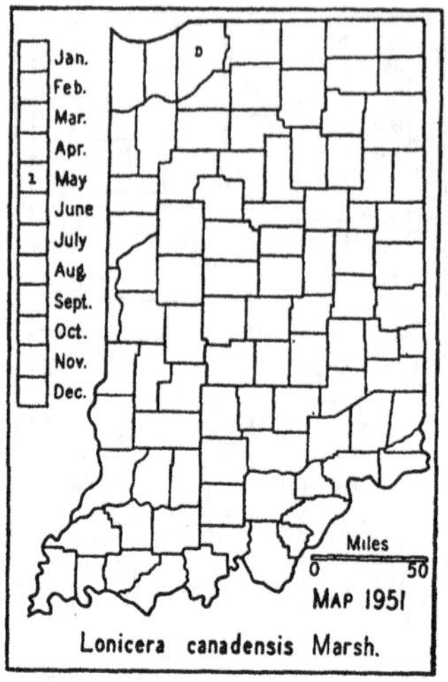

MAP 1951
Lonicera canadensis Marsh.

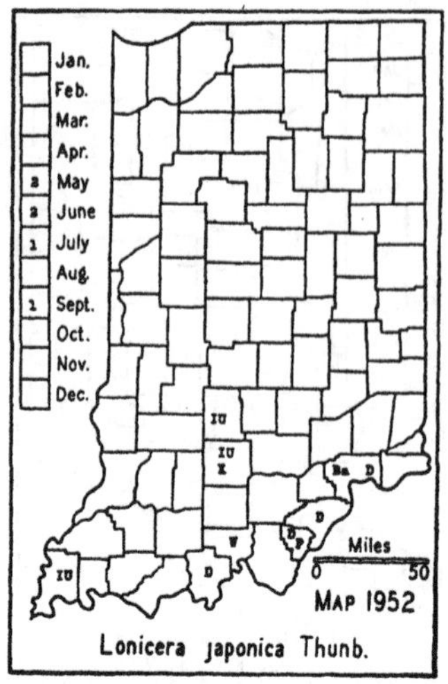

MAP 1952
Lonicera japonica Thunb.

and white oak, about a mile south of Alcinda in Noble County. This slope bordered a land-locked blueberry swamp and I believe this plant may have been a frequent plant about the swamp before the water level was lowered. The plant grew about 8 feet above the water level. I found this colony in 1916, but when I revisited the place in 1929, although the colony still persisted, it was almost extinct. I searched the border of this swamp for other colonies but failed to find any.

So far as is known, this plant will soon become extinct in Indiana except in Porter County.

Lab. to B. C. and Alaska, southw. to N. J., Md., Pa., Ind., and Minn.

8523. LONÍCERA L. HONEYSUCKLE

The honeysuckles are much used in ornamental planting for covering trellis work. Foreign species, however, are mostly used. They can be propagated by sowing the seed in the fall, and by cuttings.

Flowers in pairs on axillary peduncles; none of the leaves connate-perfoliate; upright shrubs or twining vines.
  Bracts linear or wanting; upright shrubs.
    Leaves with the margins more or less ciliate, the base rounded....1. *L. canadensis*.
    Leaves with margins not ciliate, more or less narrowed at base. (See excluded species no. 594, p. 1093)..*L. oblongifolia*.
  Bracts leaflike; twining vines..2. *L. japonica*.
Flowers terminal, in dense clusters or interrupted spikes; upper leaves of flowering branchlets connate-perfoliate; twining vines.
  Leaves pubescent both above and beneath, the margins green and ciliate. (See excluded species no. 593, p. 1093)..*L. hirsuta*.
  Leaves not pubescent both above and beneath, the margins cartilaginous and not ciliate.
    Flowers more than 3 cm long, the limb nearly equally 5-lobed; anthers scarcely exserted; cultivated and possibly escaped. (See excluded species no. 595, p. 1093)..*L. sempervirens*.
    Flowers less than 3 cm long; the tube 2-lipped and spreading; anthers conspicuously exserted; native vines.

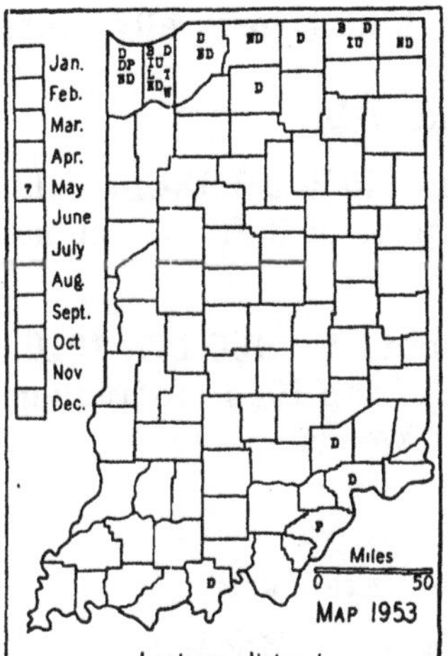

Lonicera dioica L.

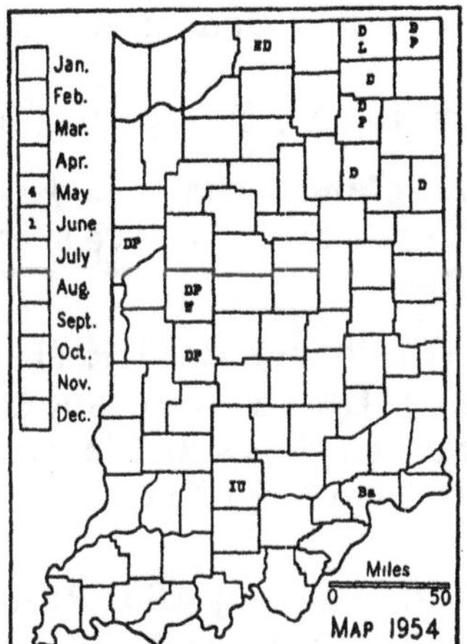

Lonicera dioica var. glaucescens (Rydb.) Butters

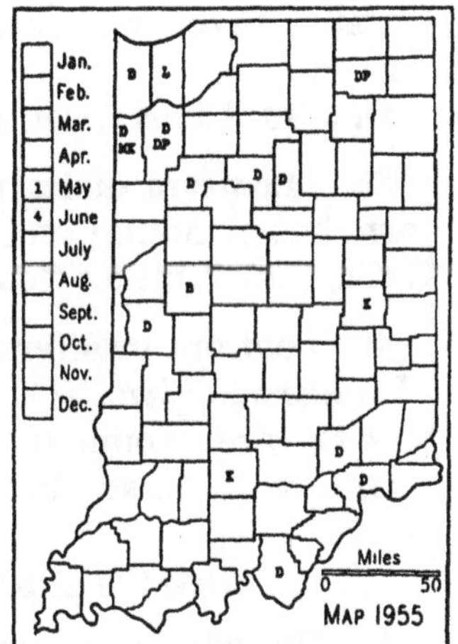

Lonicera prolifera (Kirchner) Rehd.

Upper surface of disk not glaucous; disk usually somewhat rhombic or elliptic, usually tapering at the ends or sometimes rounded; leaves usually of an oblong type; flowers generally more or less purplish and usually glabrous or nearly so on the outside, or entirely pale yellow and more or less pubescent; whorls of flowers approximate.

Leaves nearly or quite glabrous beneath; outside of corolla glabrous or nearly so; corolla more or less purplish.........................3. *L. dioica.*

Leaves pubescent beneath; outside of corolla villous; corolla pale yellow, rarely purplish..............................3a. *L. dioica* var. *glaucescens.*

Upper surface of disk glaucous; disk usually of an orbicular type with rounded ends; leaves more of an oval type; flowers always pale yellow, glabrous without; whorls of flowers usually separated and forming a short spike.... ....4. *L. prolifera.*

1. **Lonicera canadénsis** Marsh. AMERICAN FLY HONEYSUCKLE. Map 1951. This species has been reported from Pine Station, Lake County, and I found a single specimen in La Porte County. This is one of our rarest shrubs and may soon be extinct.

E. Que. to Sask., southw. to Pa., Ind., Wis., and Minn.

2. LONICERA JAPÓNICA Thunb. JAPANESE HONEYSUCKLE. Map 1952. This species has been cultivated for a long time and is still common in cultivation. Where it is planted it persists under the most adverse circumstances and usually spreads rapidly by rootshoots. I have seen it only twice where I felt sure that it was an escape from seed. No doubt it is permanently established in Indiana because the task of destroying it is too great.

Nat. of e. Asia; escaped in Conn. to Ind., southw. to Fla.

3. **Lonicera dioìca** L. LIMBER HONEYSUCKLE. Map 1953. This species is restricted mostly to the lake area where it is infrequent mostly in swampy and springy places and is absent or very rare south of the lake area.

Maine to Man., southw. to Ga. and Nebr.

3a. **Lonicera dioica** var. **glaucéscens** (Rydb.) Butters. (*Lonicera glaucescens* Rydb.) Map 1954. Infrequent in the northeastern part of the state in most soil, usually about swamps and even in bogs. South of this area it becomes rare and local and is found on wooded bluffs, generally along streams.

Ont. to Alberta, southw. to Pa., N. C., Ohio, and Nebr.

3b. **Lonicera dioica var. glaucescens f. dasýgyna** (Rehder) Deam. This form has glandular and hirsute fruit. I have specimens of it from Steuben, Wells, and Whitley Counties.

4. **Lonicera prolífera** (Kirchner) Rehder. (Rhodora **12**: 166-167. 1910.) (*Lonicera Sullivantii* Gray.) GRAPE HONEYSUCKLE. Map 1955. Infrequent to very rare. Found on wooded slopes and in sandy woods.

Ont. to Man., southw. to Tenn. and Iowa.

8524. DIERVÍLLA [Tourn.] Mill. BUSH-HONEYSUCKLE

1. **Diervilla Lonícera** Mill. (*Diervilla Diervilla* (L.) MacM.) BUSH-HONEYSUCKLE. Map 1956. This shrub grows in very sandy soil about Lake Michigan where it is more or less frequent. South of the lake it becomes very rare. In Fountain and Montgomery Counties a few specimens were found on the crests of wooded sandstone ridges. In slightly acid soil this species does well in cultivation.

Newf. to Man., southw. to s. N. E. and Wis., and in the mts. to Ga.

273. VALERIANÀCEAE Batsch VALERIAN FAMILY

Sepals minute or lacking; fruit 3-celled, but only one cell seed-bearing; leaves not pinnatifid..8529. VALERIANELLA, p. 890.

Sepals inrolled in flower, in fruit forming a crown or pappuslike fringe; ovary 1-celled, 1-seeded; some leaves pinnatifid........................8532. VALERIANA, p. 891.

8529. VALERIANÉLLA [Tourn.] Mill. CORNSALAD

[Dyal, Sarah C. Valerianella in North America. Rhodora **40**: 185-212. 1938.]

Corolla blue; bracts ciliate and obtuse; fruit wider than long............1. *V. olitoria.*

Corolla white; bracts not ciliate, acute; fruit longer than wide.

  Fruit triangular-pyramidal, mostly 2.5-4 mm long............2. *V. chenopodifolia.*

  Fruit oblong-tetragonal, mostly 1.5-2.1 mm long.

    Corolla 3-5 mm long, conspicuous above the bracts...............3. *V. intermedia.*

    Corolla 1.5-2 mm long, inconspicuous above the bracts. (See excluded species no. 598, p. 1093)..*V. radiata.*

1. VALERIANELLA OLITÒRIA (L.) Poll. CORNSALAD. This species was reported more than 60 years ago as a garden escape in Jefferson County by Barnes and by Young. There are two specimens in the herbarium of Wabash College which were collected by J. M. Coulter in 1877. In 1936 it was discovered by Miss Edna Banta in the Big Creek Bottoms about a mile west of Volga, Jefferson County. The plant is said to be cultivated for salad, although I have never seen it in cultivation.

Nat. of Eu.

2. **Valerianella chenopodifòlia** (Pursh) DC. Map 1957. Our only specimens were collected by Nieuwland in Studebaker's woods, St. Joseph

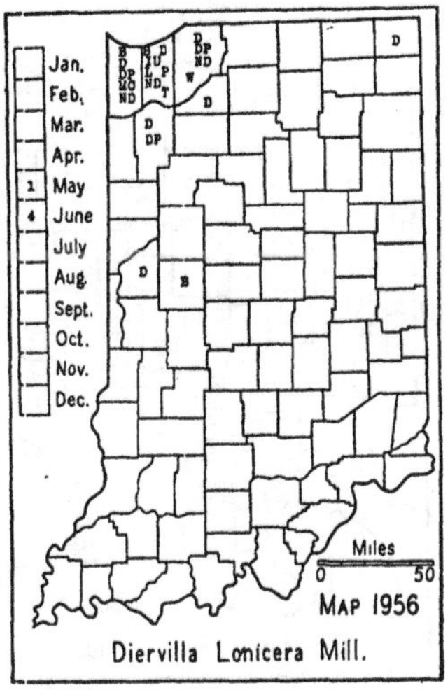

Diervilla Lonicera Mill.

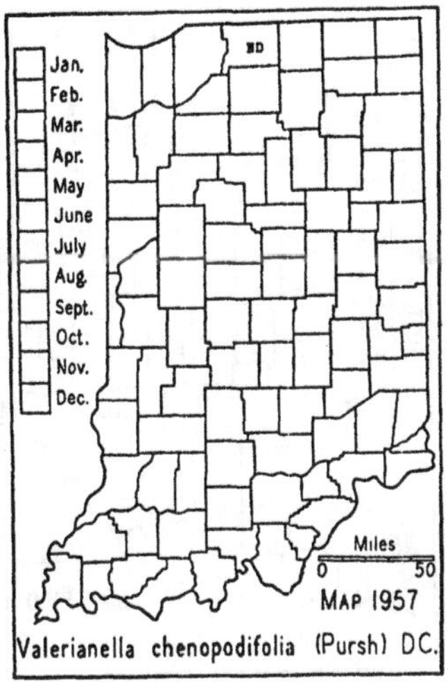

Valerianella chenopodifolia (Pursh) DC.

Valerianella intermedia Dyal

County. One specimen was collected June 4, 1912, and another was collected July 17, 1919. This cornsalad was reported in 1895 from Hamilton and Marion Counties by Wilson but I have not seen specimens to confirm this report.

N. Y., Pa., Ont., Ohio, and Ind.

3. **Valerianella intermèdia** Dyal. (Rhodora 40: 202-204. 1938.) (*Valerianella radiata* of most Indiana authors.) Map 1958. In low ground in creek bottoms, fields, and open woods. Infrequent but usually abundant where it is found. The fruit is either glabrous or pubescent. In Indiana the glabrous form is more frequent.

Mass. and Conn. to Ill., southw. to N. C. and Ky.

8532. VALERIÀNA [Tourn.] L. VALERIAN

Corolla tube 10-20 mm long, slender; basal leaves and those of the runners cordate at the base..1. *V. pauciflora.*

Corolla tube less than 10 mm long; basal leaves not cordate at the base.

- Roots fusiform; leaves thickish, the basal ones mostly entire, parallel-veined, stem leaves more or less parted; fruit crowned with the plumose calyx teeth; native plants of a springy and marly habitat..........................2. *V. edulis.*
- Roots fibrous; leaves thin, all more or less divided into 7-25 segments; fruit without plumose calyx teeth; introduced plants, escaped from gardens. (See excluded species no. 599, p. 1094)..............................*V. officinalis.*

1. **Valeriana paucifl̀ora** Michx. LARGE-FLOWER VALERIAN. Map 1959. Infrequent but generally abundant where it is found; in moist, shaded, wooded ravines and wooded, alluvial plains.

Pa. to Mo., southw. to W. Va. and Tenn.

2. **Valeriana édulis** Nutt. EDIBLE VALERIAN. Map 1960. In Indiana this species grows in marly springy places, and I have found it in only three places. It has been destroyed in one or two of these places by drainage

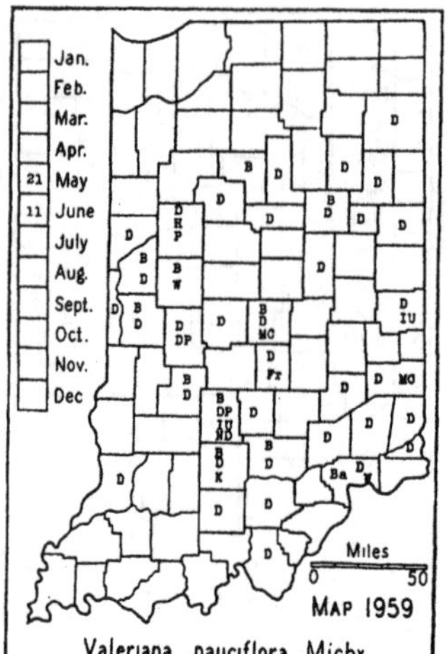

Valeriana pauciflora Michx.

Valeriana edulis Nutt.

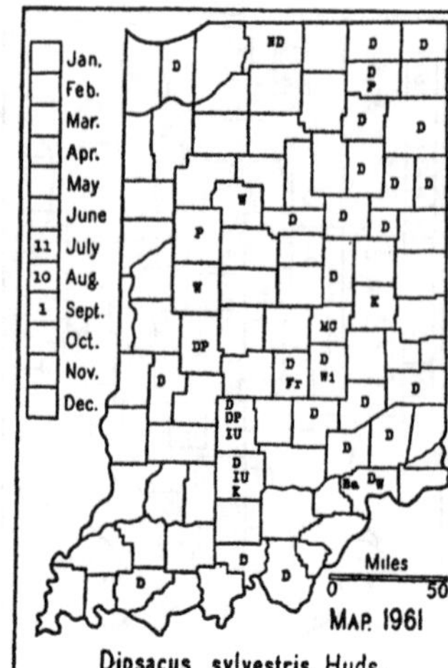

Dipsacus sylvestris Huds.

and grazing, but still persisted in 1935 in a marly springy place in eastern Cass County.

Ont. to B. C., southw. to Ohio and Iowa, and in the Rocky Mts. to Ariz. and N. Mex.

274. DIPSACÀCEAE Lindl. TEASEL FAMILY

8540. DÍPSACUS [Tourn.] L.

1. DIPSACUS SYLVÉSTRIS Huds. COMMON TEASEL. Map 1961. An obnoxious weed along roadsides, on the banks of streams, and in waste places, fields, and open woods.

Nat. of Eu. and Asia; Maine to Mich., southw. to N. C. and Ind.

275. CUCURBITÀCEAE B. Juss. GOURD FAMILY

Corolla large, yellow, generally 6-15 cm long; stem trailing...8622. CUCURBITA, p. 892.
Corolla small, less than 6 cm long, white or greenish; stem high-climbing by tendrils.
  Fruit glabrous; tendrils simple........................8562. MELOTHRIA, p. 892.
  Fruit prickly or spiny; tendrils divided.
    Stem and leaves glabrous; fruit an inflated pod, dehiscent at the apex or bursting irregularly; usually 4-seeded..................8629. ECHINOCYSTIS, p. 893.
    Stem and leaves more or less pubescent; fruits 3-10 together, indehiscent, 1-seeded. ....8637. SICYOS, p. 893.

8562. MELÒTHRIA L.

1. **Melothria péndula** L. Map 1962. Wooded bluffs of the Ohio River. Rare. It has been reported from Clark and Jefferson Counties.

Pa. to Mo., southw. to Fla. and La.

8622. CUCÚRBITA [Tourn.] L. GOURDS, SQUASHES, PUMPKINS

Leaves deeply 3-5-lobed. (See excluded species no. 602, p. 1094)....*C. Pepo* var. *ovifera.*
Leaves merely angulate or slightly 3-5-lobed. (See excluded species no. 601, p. 1094) ....*C. foetidissima.*

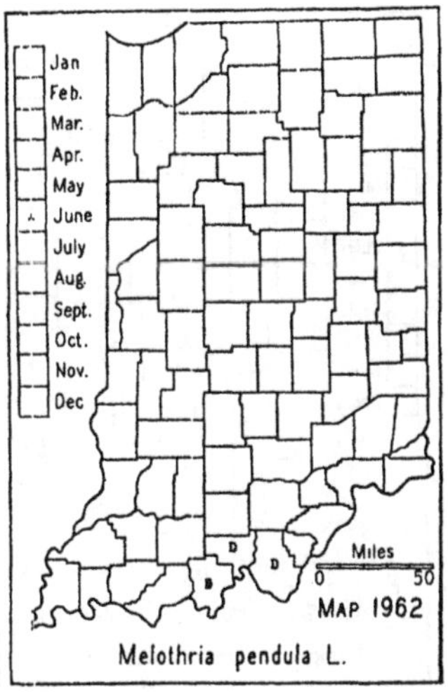

Melothria pendula L.

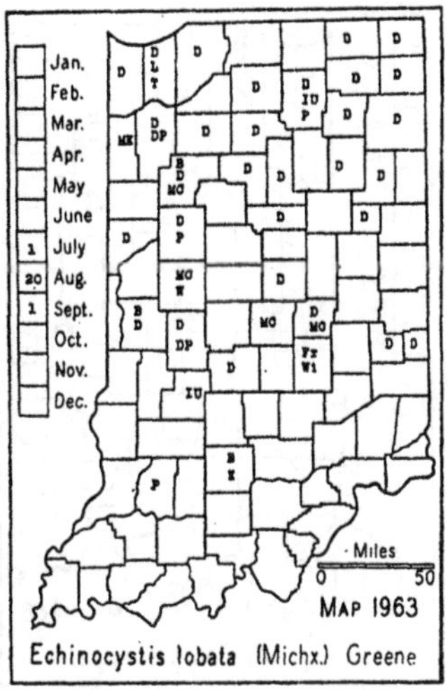

Echinocystis lobata (Michx.) Greene

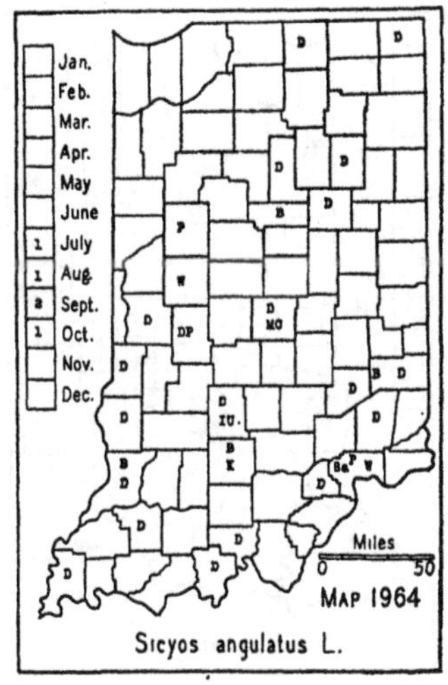

Sicyos angulatus L.

8629. ECHINOCÝSTIS T. & G.

1. **Echinocystis lobàta** (Michx.) T. & G. (*Micrampelis lobata* (Michx.) Greene.) WILD BALSAM-APPLE. Map 1963. Infrequent in low ground along streams and about lakes and ponds throughout the state although there are no published records from the southwestern part. Sometimes cultivated as an ornamental vine.

N. B. to Man. and Mont., southw. to Pa., Ga., Ky., Kans., and Tex.

8637. SÍCYOS L. ONE-SEEDED BUR CUCUMBER

1. **Sicyos angulàtus** L. Map 1964. Probably found throughout the state. It prefers moist soil along streams in open woodland and in cultivated fields. It is rare in the northern part of the state, becoming abundant in cornfields in the Lower Wabash Bottoms where it is regarded as one of the most objectional of all weeds because the spines of the fruit stick through clothing, and in husking corn the hands of workmen are injured.

S. Maine and w. Que. to S. Dak., southw. to Fla., e. Kans., and Tex.

276. CAMPANULÀCEAE Juss. BELLFLOWER FAMILY

Leaves petiolate or, if sessile, very narrow................8644. CAMPANULA, p. 893.
Leaves sessile, usually clasping, with wide, cordate bases....8649. SPECULARIA, p. 896.

8644. CAMPÁNULA [Tourn.] L. BELLFLOWER

Flowers nearly sessile, in spikes or racemes.
  Corolla rotate; style declined; capsule with apical pores............1. *C. americana.*
  Corolla campanulate; style straight; capsule with pores at the base...2. *C. rapunculoides.*
Flowers long-peduncled, solitary or in loose panicles.
  Plants of marshes, bogs or wet ground; stems weak, usually supported by adjacent vegetation; corolla 5-12 mm long.
    Leaves linear, 2.5-6 cm long, long-acuminate at the apex; calyx lobes in flower usually 1.5-3 mm long; corolla mostly 10-12 mm long, blue...3. *C. uliginosa.*

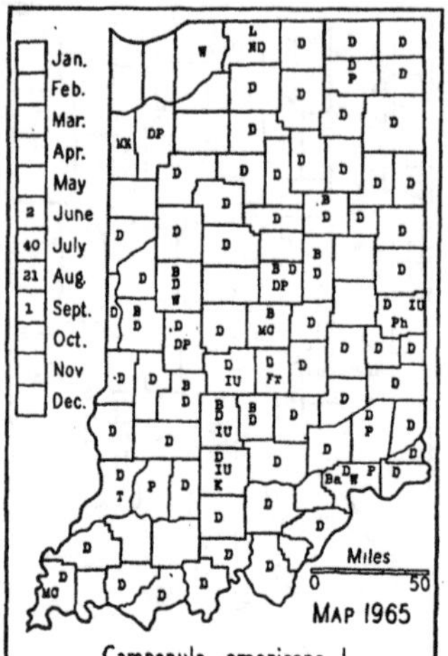

Campanula americana L

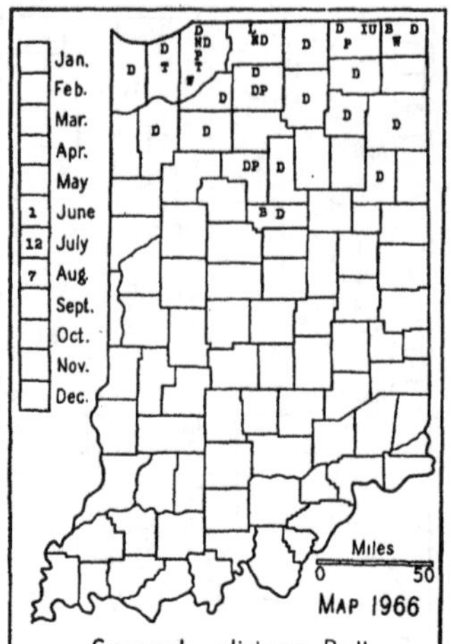

Campanula uliginosa Rydb

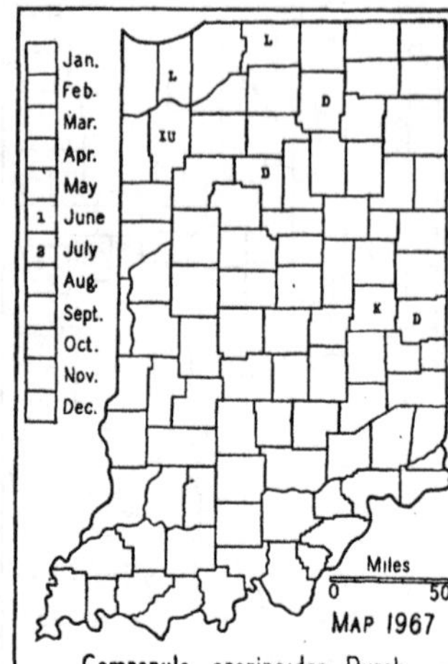

Campanula aparinoides Pursh

Leaves lanceolate or linear-lanceolate, 2-5 cm long, subacute at the apex; calyx lobes in flower generally 1-2 mm long; corolla 5-8 mm long, white........ ....4. *C. aparinoides.*

Plants of dry, gravelly or sandy places or on rocky ledges; stems decumbent only at the base; corolla more than 15 mm long.

Base of stem pubescent all over for at least 10 cm..........5. *C. rotundifolia.*

Base of stem glabrous or pubescent only in lines.............................. ....5a. *C. rotundifolia* var. *intercedens.*

1. **Campanula americàna** L. TALL BELLFLOWER. Map 1965. In moist, rich woods throughout the state. It is easy to grow in cultivation and, if the ground is bare when it ripens its seed, it sows itself abundantly. In the spring hundreds of plants will grow and these can be thinned out and the surplus be used for "greens."

Ont. and N. Y. to Nebr., southw. to Fla. and Ark.

2. CAMPANULA RAPUNCULOÌDES L. FALSE RAMPION. In 1937 I found a colony about 200 feet long and from one to three feet wide on the east side of a north and south road near the south side of sec. 11 about five and a half miles southeast of Angola, Steuben County. No house was near and no evidence of a former habitation near. I found it also as an escape along the railroad near Bluffton, Wells County.

2a. CAMPANULA RAPUNCULOIDES var. UCRÁNICA (Bess.) K. Koch is a glabrous variety which I found as an escape along a roadside in Lake County. I have not been able to revisit this place to learn if it has persisted or not.

Nat. of Eurasia.

3. **Campanula uliginòsa** Rydb. BLUE MARSH BELLFLOWER. Map 1966. In the low borders of lakes and in marshes in the lake area. Rather frequent where its habitat exists.

Que. to Man., southw. to Conn., N. Y., and Ind.

4. **Campanula aparinoìdes** Pursh. WHITE MARSH BELLFLOWER. Map 1967. This species was confused with the preceding one and they were not

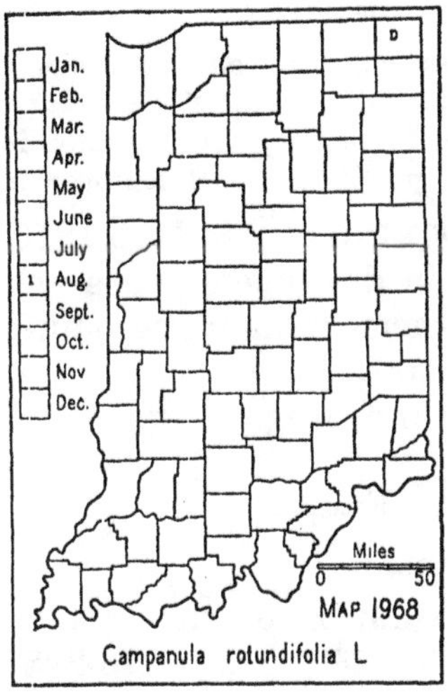

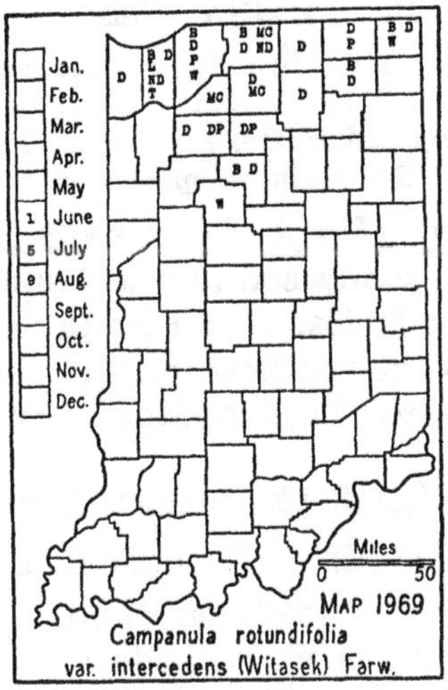

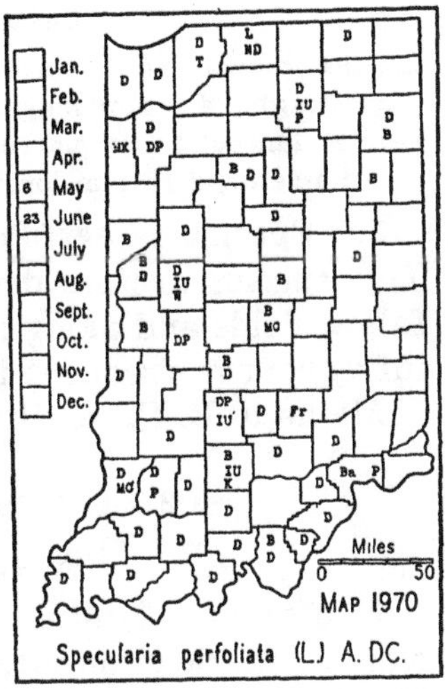

separated until 1901. *Campanula aparinoides* is more southern in its distribution. Since most of our reports were made before the preceding species was described it is impossible to say to which one, the reports should be referred. I believe all, or almost all, reports from the lake area should be referred to *Campanula uliginosa*. *Campanula aparinoides* is rare in northern Indiana.

N. B. to Colo., southw. to Ga. and Ky.

5. **Campanula rotundifòlia** L. (Malte. Critical notes on plants of Arctic America. Rhodora **36**: 188-190. 1934.) HAREBELL. Map 1968. In sandy or gravelly soil usually on the slopes and ledges of banks of streams and lakes. This is the true species and has the stems densely pubescent at the base and is found in Europe and western America and rarely inland or in eastern America. I found it on the gravelly bank of the south side of Lake Gage in Steuben County, and on the slope of the high, wooded bank of the south side of North Twin Lake in Lagrange County where it was growing side by side with the glabrous form which was common, while the species was rare.

Boreal regions southw. to N. J., Great Lakes Region, and Tex. (Ann. Missouri Bot. Gard. **20**: 797. 1933.)

5a. **Campanula rotundifolia** var. **intercèdens** (Witasek) Farw. Map 1969. This is the glabrous form of the species; it is found in the interior and eastern North America. The habitat is the same as that of the species. I found it common, however, on the north side of the Wabash River below Georgetown in Cass County, in the upper crevices of the 10 foot rock cliff which forms the bank of the river.

This variety is more or less frequent throughout most of the lake area where its habitat is well represented.

The var. *arctica* Lange has been reported by Peattie as occurring in the dunes. This is merely a dwarf form of the preceding variety and I regard it as an ecological variation without taxonomic significance.

8649. SPECULÀRIA [Heist.] Fabricius

Leaves rounded or ovate, clasping by the cordate base; capsules ellipsoid, short, straight, 4-6 mm long..1. *S. perfoliata.*
Leaves lanceolate or narrowly lanceolate, sessile; capsules cylindric, about 1 mm in diameter, 8-15 mm long. (See excluded species no. 604, p. 1094).....*S. leptocarpa.*

1. **Specularia perfoliàta** (L.) A. DC. VENUS LOOKING-GLASS. Map 1970. This species prefers dry, sandy soil and is found in dry, open woods and fallow fields and along roadsides. It is rare in northern Indiana, becoming infrequent to frequent in the southern part of the state.

Maine to B. C., southw. to Fla., La., Mex., Ariz., and Oreg.

276A. LOBELIÀCEAE Dumort. LOBELIA FAMILY

8694. LOBÈLIA [Plumier] L. LOBELIA

[McVaugh. Studies in the taxonomy and distribution of the eastern North American species of Lobelia. Rhodora 38: 241-263. 1 pl.; 276-298; 305-329; 346-362. 1936.]

Corolla tube more than 6 mm long.
  Corolla 3-4 cm long; flowers red, rarely white; calyx lobes linear, straight, not auricled at the base..1. *L. Cardinalis.*
  Corolla less than 3 cm long; flowers blue, rarely white.
    Calyx lobes broadly linear, more or less folded together, making them crooked, with a broad, recurving auricle on each side at the base; stem glabrous or slightly pubescent on the lower half; leaves long-tapered at the base, glabrous or sparingly pubescent above and beneath; flowers usually about 2 cm long. ....2. *L. siphilitica.*
    Calyx lobes linear, flat, straight, without auricles at the base; stem densely pubescent all over; leaves very short-tapered at the base, densely pubescent above and beneath; flowers mostly 1-1.5 cm long..............3. *L. puberula.*
Corolla tube less than 6 mm long.
  Leaves linear, mostly 1-2 mm wide; plants of a wet, marly habitat......4. *L. Kalmii.*
  Leaves more than 3 mm wide; plants of a dry habitat, sometimes in a moist habitat or in dried-up wet places.
    Stem densely long-pubescent all over, usually branched; median leaves mostly ovate-lanceolate, repand-dentate or denticulate; pods inflated, usually about 5 mm wide, without ridges..5. *L. inflata.*
    Stem glabrous, pubescent in lines or short-pubescent below; leaves mostly entire or with a few denticulations, rarely the whole blade coarsely denticulate; pods not inflated or rarely so, mostly less than 3 mm wide, more or less ribbed. ....6. *L. spicata.*

1. **Lobelia Cardinàlis** L. CARDINAL-FLOWER. Map 1971. Infrequent throughout the state in low ground in woods, along ditches, and on the borders of lakes and ponds.

Southern N. B. to Ont. and Kans., southw. to Fla. and Tex.

1a. **Lobelia Cardinalis** f. **álba** (A. A. Eaton) St. John. This is a white-flowered form of the species which has been reported from the dune area by Peattie.

2. **Lobèlia siphilítica** L. LARGE BLUE LOBELIA. Map 1972. Rather frequent in low ground throughout the state in woodland, along roadsides and ditches, and about lakes.

Maine, Ont. to S. Dak. and Mo., southw. to N. C. and Ala.

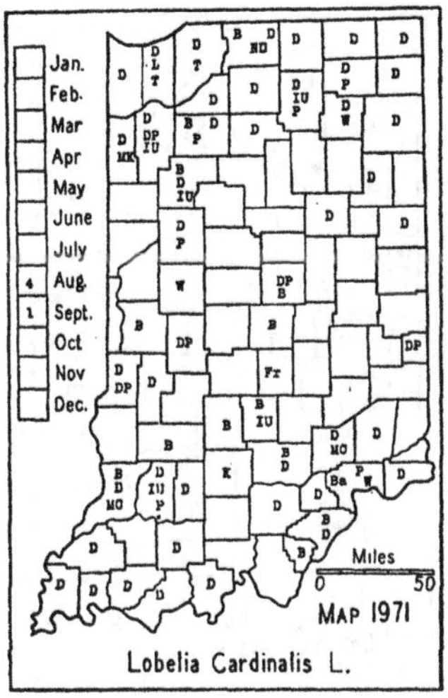

Lobelia Cardinalis L.

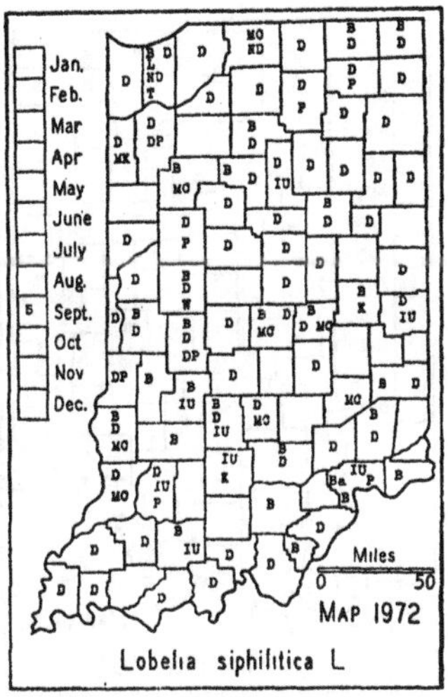

Lobelia siphilitica L.

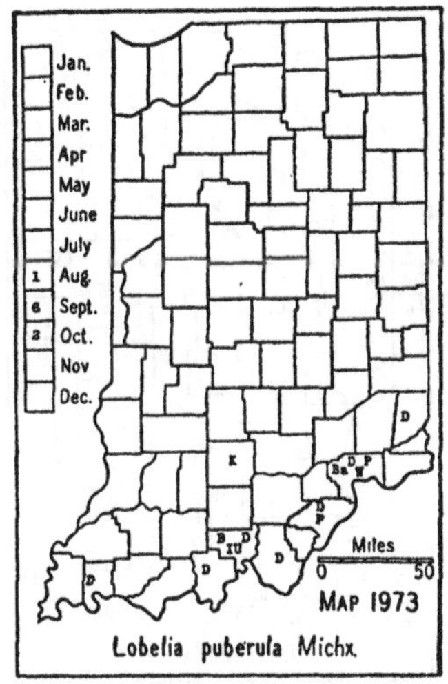

Lobelia puberula Michx.

2a. **Lobelia siphilitica** f. **albiflòra** (Britt.) House. This a form with white flowers. Bradner reported it from Steuben County and Peattie reported it from the dune area. I have found this form twice in Wells County.

3. **Lobelia pubérula** Michx. Map 1973. Infrequent to rare in its range in Indiana. It is generally found in dry woods but I have found it also in "flat woods" associated with beech and sweet gum. I believe this species is restricted to the southern part of the state. It has been reported from the northern counties by some authors, but I believe all of these reports should be transferred to *Lobelia spicata* var. *leptostachys.* It was reported from Tippecanoe County upon the authority of Hussey. I have seen his specimen, which is in the herbarium of Purdue University, and it belongs to *Lobelia spicata* var. *leptostachys.* Pepoon reported it from Porter County upon the authority of Umbach and I refer this report also to *Lobelia spicata* var. *leptostachys* which he does not report and which occurs there. Buhl (Amer. Midland Nat. **16**: 252. 1935) says there are no confirming specimens for this latter report.

N. J. to Ill. and Mo., southw. to Fla. and Tex.

4. **Lobelia Kálmii** L. KALM LOBELIA. Map 1974. Locally frequent to common in calcareous habitats such as wet, marly borders of lakes and in marshes. It is usually associated with *Parnassia glauca.*

Newf. to Man. and Minn., southw. to N. J., Ohio, Ind., and Iowa.

5. **Lobelia inflàta** L. LOBELIA. Map 1975. The dried plant and seed of this species have been used in medicine for about one hundred and fifty years. The plant is known to the medical profession and to the laity as lobelia, and for this reason no other name should be used. Some authors call it Indian tobacco but, since several other plants have the same name, it seems inadvisable to supplant a well known name by one which is misleading. Infrequent to frequent throughout the state in various situations. It is usually found in poor soils in open woods and fallow fields, and some-

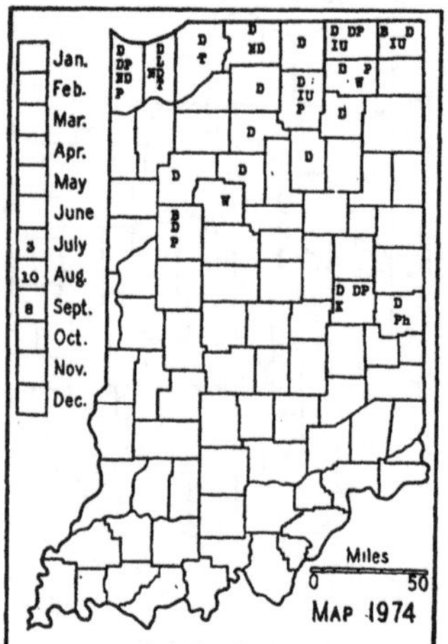

MAP 1974
Lobelia Kalmii L.

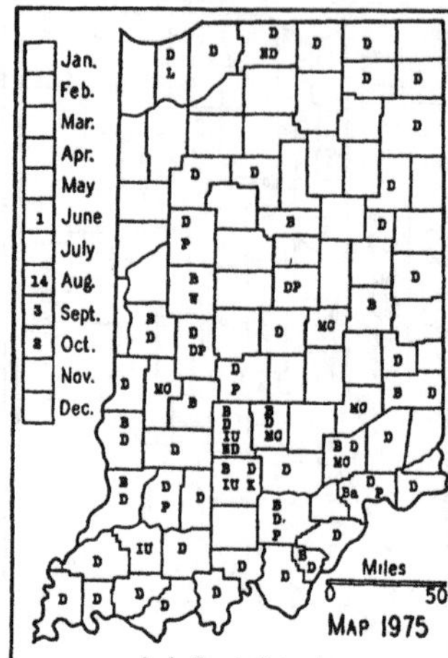

MAP 1975
Lobelia inflata L.

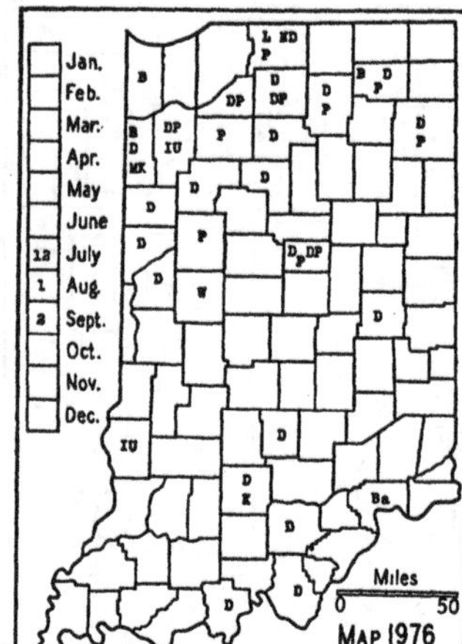

MAP 1976
Lobelia spicata Lam. var. leptostachys (A.DC.) Mack. & Bush

times as a weed in cultivated fields. It is also found in rich soil and in dried-up sloughs.

Lab. to Sask., southw. to Ga., Kans., and La.

6. **Lobelia spicàta** Lam. McVaugh says: "This is a species with at least five well defined phases," four of which occur in Indiana. A key is given to separate these variations.

Sinus between the calyx lobes with a conspicuous deflexed appendage...6a. *L. spicata* var. *leptostachys.*
Sinus between the calyx lobes without a deflexed appendage, rarely a sinus with a short one.
  Lobes of the calyx more or less ciliate with short stiff hairs...6b. *L. spicata* var. *hirtella.*
  Lobes of calyx not ciliate.
    Anthers blue; calyx flattish in anthesis; racemes dense, many-flowered; capsules short-hemispheric.............................6c. *L. spicata* var. *originalis.*
    Anthers white; calyx roundish in anthesis; racemes few-(10-30) flowered; capsules globose, often somewhat inflated............6d. *L. spicata* var. *campanulata.*

6a. **Lobelia spicata** var. **leptóstachys** (A. DC.) Mack. & Bush. Map 1976. Infrequent to rare in all parts of the state in dry soil in rather open woods, clearings, prairies, and fallow fields and along roadsides.

W. Va. to Wis., southw. to Ga., and Ark.; rare or absent on the eastern Coastal Plain.

6b. **Lobelia spicata** var. **hirtélla** Gray. Map 1977. Local in prairie habitats or in marly, springy places in northern Indiana.

Gaspé Peninsula, southw. to N. Y. and westw. to Ind., and thence northwestw. to Alberta and southw. to e. Kans.

6c. **Lobelia spicata** var. **originàlis** McVaugh. This form of the species is very local. My specimens are from dry, gravelly oak slopes in Lagrange, Steuben, and Warren Counties, and from an interdunal flat in Lake County.

N. B. to Pa., Mich., and Minn., southw. to Ga. and Mo.

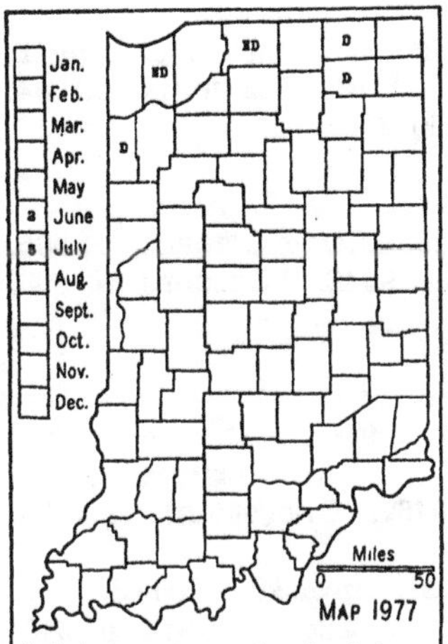

MAP 1977
Lobelia spicata var. hirtella A.Gray

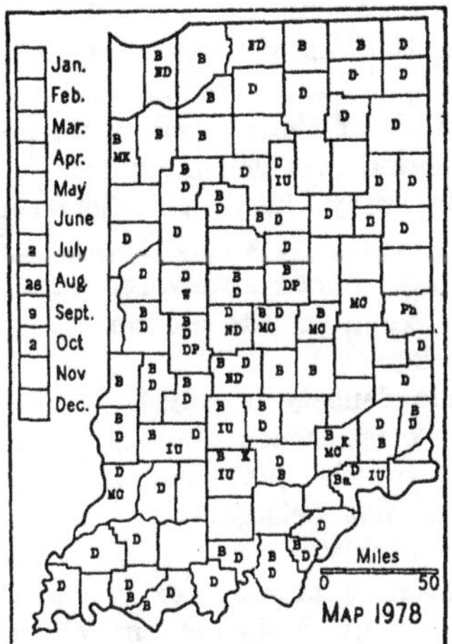

MAP 1978
Vernonia altissima Nutt.

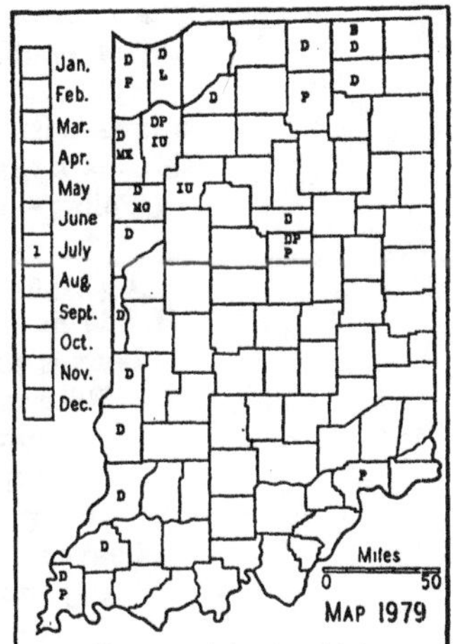

MAP 1979
Vernonia fasciculata Michx

6d. **Lobelia spicata** var. **campanulàta** McVaugh. (Rhodora 38: 316. 1936.) This extremely local and widely distributed plant is found in dry, open woods in Clark, Lagrange, and Lake Counties. Our Clark County station is the southern limit of its range in the United States.

Maine, n. Mich., n. Wis., and Minn., southw. to Pa., N. J., and Ind.

280. COMPÓSITAE Adans. COMPOSITE FAMILY

[The following key is adapted from the one found in Gray, Manual, ed. 7.]

Corolla tubular in all of the perfect flowers, regularly 5 (rarely 3 or 4)-parted; flowers ligulate only in the marginal or ray-flowers, which, when present, are either pistillate only, or neutral (with neither stamens nor pistil); sap of plants not milky..SERIES I. TUBULIFLORAE.

Corolla ligulate in all of the flowers of the head; flowers all perfect; sap of plants milky; leaves alternate.............................SERIES II. LIGULIFLORAE.

Series I. Tubuliflorae DC.

A. Staminate and pistillate flowers in separate heads (rarely so in *Cirsium*).
  Pappus capillary.
    Leaves not prickly, entire.
      Basal leaves much larger than the cauline and differing from them in shape; plants usually less than 4 dm high...8978. ANTENNARIA, p. 950.
      Basal leaves lacking at flowering time or, if present, similar to the cauline ones in shape and size......................8983. ANAPHALIS, p. 953.
    Leaves prickly, not entire; heads large...............9462. CIRSIUM, p. 999.
  Pappus none.
    Pistillate heads 1-flowered, developing an obovoid achene armed with 4-8 tubercles or straight spines....................9146. AMBROSIA, p. 960.
    Pistillate heads forming an oblong or oval bur covered with hooked or straight prickles....................................9148. XANTHIUM, p. 962.

A. Staminate and pistillate flowers not in separate heads.
  B. Heads discoid, rays or ligulate flowers none or very inconspicuous; corollas all tubular.

C. Pappus composed of bristles.
  Pappus double, the outer bristles very short, the inner ones longer. ....8751. VERNONIA, p. 904.
  Pappus-bristles all of the same length, not in 2 series.
    Heads aggregated into dense clusters.
      Foliage not spiny; heads few-flowered...8775. ELEPHANTOPUS, p. 905.
      Foliage spiny; heads 1-flowered..........9442. ECHINOPS, p. 998.
    Heads not aggregated into clusters.
      Leaves prickly.
        Receptacle densely bristly.
          Pappus-bristles not plumose...........9461. CARDUUS, p. 999.
          Pappus-bristles plumose...............9462. CIRSIUM, p. 999.
        Receptacle not bristly..............9467. ONOPORDUM, p. 1003.
      Leaves not prickly.
        Pappus-bristles plumose or conspicuously upwardly barbed.
          Corollas whitish; heads corymbose......8825. KUHNIA, p. 910.
          Corollas rose colored, purple, or white; heads racemose or spicate..........................8826. LIATRIS, p. 911.
        Pappus-bristles not plumose or with inconspicuous barbs.
          Stems twining; leaves opposite, triangular-hastate; flowers flesh to pale purplish............8818. MIKANIA, p. 910.
          Stems not twining.
            Involucral bracts scarious throughout; plants more or less white-woolly.
              Plants dioecious; pistillate heads with a few perfect flowers in the center; perennial....8983. ANAPHALIS, p. 953.
              Plants not dioecious; all of the flowers fertile, the central ones perfect, surrounded by pistillate ones; annual or perennial...............8992. GNAPHALIUM, p. 954.
            Involucral bracts not scarious throughout or if so, the plants not white-woolly.
              Involucral bracts hooked..........9452. ARCTIUM, p. 998.
              Involucral bracts not hooked.
                Bracts of involucre in 1 series.
                  Length of bracts about 15 mm...9389. ERECHTITES, p. 994.
                  Length of bracts about 10 mm or less.
                    Cauline leaves pinnatifid....9411. SENECIO, p. 996.
                    Cauline leaves not pinnatifid..9409. CACALIA, p. 994.
                Bracts of involucre in more than 1 series.
                  Central disk flowers sterile; anthers tailed at the base; bruised foliage malodorous..8941. PLUCHEA, p. 949.
                  Central disk flowers fertile; anthers not tailed at the base; bruised foliage not malodorous.
                    Leaves large, triangular, the lower ones hastate at the base ....9409. CACALIA, p. 994.
                    Leaves not as above.
                      Leaves opposite or whorled; plants usually with resinous dots; stigmatic lines only at the base of the minutely and uniformly pubescent style branches.........8816. EUPATORIUM, p. 905.
                      Leaves alternate; plants not resinous; stigmatic lines extending to the tips of the style branches or their appendages.

Corollas of disk flowers 5-lobed; achenes terete or angled.........8849. **SOLIDAGO**, p. 914.
Corollas of disk flowers 4-parted; achenes flat. ....8901. **ERIGERON**, p. 947.

C. Pappus not composed of bristles, entirely lacking or a mere crown, or composed of scales.
Achenes crowned with 2-4 stiff awns...........9237. **BIDENS**, p. 981.
Achenes not crowned with stiff awns.
Receptacle bristly or chaffy.
Heads many in long, terminal, bracted spikes....9141. **IVA**, p. 959.
Heads not in long, terminal, bracted spikes.
Leaves mostly opposite, large, thin, lobed, malodorous when bruised.........................9122. **POLYMNIA**, p. 955.
Leaves alternate, thick, not lobed, nor malodorous when bruised. ....9138. **PARTHENIUM**, p. 959.
Receptacle naked.
Involucral bracts in 1 series; leaves finely dissected.
Achenes conspicuously obovoid, about 4 mm long, densely pubescent; receptacle nearly flat.......................... ....9292. **HYMENOPAPPUS**, p. 986.
Achenes slightly obovoid, about 1 mm long, glabrous; receptacle conic-oblong.....................9339. **MATRICARIA**, p. 990.
Involucral bracts in more than 1 series.
Heads chiefly nodding, in spikes, racemes, or panicles........... ....9358. **ARTEMISIA**, p. 992.
Heads erect, corymbose.
Leaves bipinnatifid.............9341A. **TANACETUM**, p. 991.
Leaves not bipinnatifid, crenate-dentate..................... ....9341. **CHRYSANTHEMUM**, p. 990.

B. Heads with rays, i.e., the marginal flowers or some of them, with ligulate corollas.
D. Leaves opposite or whorled, at least the lower ones (sometimes somewhat variously disposed in *Silphium*).
Achenes crowned with 2-6 slender awns; rays yellow.
Plants terrestrial, although often in very wet places............. ....9237. **BIDENS**, p. 981.
Plants aquatic, floating...............9237A. **MEGALODONTA**, p. 985.
Achenes not crowned with awns.
Leaves finely dissected...................9312. **DYSSODIA**, p. 988.
Leaves not dissected.
Involucres small, less than 6 mm high; rays white.
Peduncles with a spreading, glandular pubescence............ ....9246. **GALINSOGA**, p. 986.
Peduncles with an upwardly appressed, nonglandular pubescence. ....9166. **ECLIPTA**, p. 964.
Involucres large, generally more than 6 mm high, if less, the leaves divided (*Coreopsis*); rays yellow (whitish in *Polymnia canadensis*).
Achenes flat, winged, notched at the apex.
Disk flowers sterile..................9131. **SILPHIUM**, p. 956.
Disk flowers fertile.................9227. **COREOPSIS**, p. 979.
Achenes compressed or quadrangular, not winged or notched.
Leaves large, thin, lobed; achenes blackish, about 7 mm long, elliptic, strongly compressed, and longitudinally striate or brownish, about 4 mm long, strongly compressed, and 3-ribbed......................9122. **POLYMNIA**, p. 955.

Leaves not thin or lobed; achenes somewhat quadrangular-obovoid, 3-4-sided, truncate at the summit.

Ray flowers pistillate, fertile, the rays persisting.................................... 9157. HELIOPSIS, p. 963.

Ray flowers neutral, sterile, the rays deciduous.................................... 9200. HELIANTHUS, p. 970.

D. Leaves alternate.

E. Pappus of terete awns or bristles.

Pappus of terete awns.

Involucres very glutinous; bracts glabrous and hooked; awns smooth and deciduous; receptacle not chaffy.................................... 8833. GRINDELIA, p. 913.

Involucres not glutinous; bracts not hooked, not deciduous; receptacle chaffy.................... 9237. BIDENS, p. 981.

Pappus of capillary bristles.

Involucres with conspicuous, oblong glands; leaves dissected. .... 9312. DYSSODIA, p. 988.

Involucres lacking conspicuous glands.

Rays many, 40-200, arranged in more than 1 row.................................... 8901. ERIGERON, p. 947.

Rays fewer than 40, arranged in one row.

Bracts of involucre in 1 series, sometimes with a few bractlets at the base of the involucre; rays yellow.................................... 9411. SENECIO, p. 996.

Bracts of the involucre in more than 1 series.

Rays yellow.

Involucres 2.5-10 cm in diameter ..9061. INULA, p. 955.

Involucres 0.5-2.3 mm in diameter.

Pappus double, the outer row of very small, chaffy bristles, much shorter than the inner row of numerous capillary bristles.................................... 8844. CHRYSOPSIS, p. 914.

Pappus simple, of numerous slender and equal bristles. .... 8849. SOLIDAGO, p. 914.

Rays violet, purple, blue, or white.

Bracts few, 3-6, firm and thick, the inner ones blunt or rounded and green at the apex, the green part more or less inconspicuously glandular-punctate; heads corymbose; flowers white.................................... 8904. SERICOCARPUS, p. 949.

Bracts not as above; flowers mostly colored.................................... 8900. ASTER, p. 928.

E. Pappus none or a cup or crown, or of thin chaffy scales.

Plants with the stem leaves decurrent; pappus of 5-8 thin, 1-nerved chaffy scales, the nerve usually ending in a bristle or point. .... 9305. HELENIUM, p. 987.

Plants not as above.

Receptacle naked.

Rays yellow........................ 9227. COREOPSIS, p. 979.

Rays white.

Leaves entire or mostly so; involucres less than 8 mm wide. .... 8892. BOLTONIA, p. 928.

Leaves coarsely dentate to pinnatifid; involucres more than 8 mm wide........... 9341. CHRYSANTHEMUM, p. 990.

Receptacle chaffy, at least at the summit.

Rays sterile, neutral or rarely pistillate; disk flowers perfect and fertile.

Receptacle flat or nearly so; rays yellow..9227. COREOPSIS, p. 979.
Receptacle convex to columnar.
Involucral bracts with thin, scarious margins..9330. ANTHEMIS, p. 988.
Involucral bracts distinctly herbaceous.
Pappus consisting of 2 flat awns.
Awns of pappus decidedly deciduous....................................9200. HELIANTHUS, p. 970.
Awns of pappus persistent...9215. ACTINOMERIS, p. 978.
Pappus none or merely a crown of short teeth.
Rays pistillate, rose colored (rarely yellow)............................9178B. BRAUNERIA, p. 968.
Rays neutral, yellow to brownish red or orange.
Achenes 4-sided, marginless; leaves not pinnately divided..............9178. RUDBECKIA, p. 964.
Achenes flattened and margined; leaves pinnately parted...............9178C. RATIBIDA, p. 969.
Rays fertile, pistillate.
Disk flowers also fertile, their achenes maturing.
Leaves simple; rays yellow.
Leaves serrate...............9218. VERBESINA, p. 979.
Leaves entire...................9253. MADIA, p. 986.
Leaves dissected or bipinnately parted; flowers white, rarely pinkish.
Heads (with rays expanded) more than 1 cm wide; achenes tuberculate, terete; annual.......................................9330. ANTHEMIS, p. 988.
Heads (with rays expanded) less than 1 cm wide; achenes smooth, flat; perennial.....9332. ACHILLEA, p. 989.
Disk flowers not fertile; mature achenes flat.
Rays 5, obcordate, scarcely exceeding the disk, whitish. ....9138. PARTHENIUM, p. 959.
Rays more than 5, yellow, much longer than the disk; achenes with wide margins...9131. SILPHIUM, p. 956.

Series II. Liguliflorae DC.

Pappus none..9556. SERINIA, p. 1004.
Pappus composed of scales or of both scales and bristles.
Pappus of scales only; flowers blue, rose colored or white...9553. CICHORIUM, p. 1004.
Pappus composed of scales and bristles; flowers yellow........9560. KRIGIA, p. 1004.
Pappus composed of either bristles or hairs.
Bristles plumose (seen best when mature and dry).
Plants scapose...............................9572. HYPOCHAERIS, p. 1006.
Plants not scapose..............................9579. TRAGOPOGON, p. 1006.
Bristles simple, at most scabrous.
Achenes spinulose at the summit.....................9592. TARAXACUM, p. 1006.
Achenes not spinulose at the summit.
Achenes flat or flattish.
Achenes beaked; flowers light yellow................9596. LACTUCA, p. 1008.
Achenes narrowed at the apex or truncate.
Flowers blue (rarely cream color in *Lactuca spicata*)..9596. LACTUCA, p. 1008.
Flowers yellow............................9595. SONCHUS, p. 1007.

Achenes columnar, often slender.
 Flowers cream color, whitish, or pale purplish; heads pendulous...9606. PRENANTHES, p. 1014.
 Flowers yellow or reddish; heads erect.
  Achenes beaked.............................9604. PYRRHOPAPPUS, p. 1013.
  Achenes not beaked.
   Pappus white.................................9605. CREPIS, p. 1013.
   Pappus tawny ....9607. HIERACIUM, p. 1016.

8751. VERNÒNIA Schreb. IRONWEED

Note: The Indiana ironweeds are difficult to separate into species because there are so many intergrading forms, which are due, possibly, to hybridization. My study was made with a lens of 28 diameter magnification and was restricted to my 123 specimens from Indiana. Duplicates of most of my specimens have been seen by H. A. Gleason, who revised the genus (North Amer. Flora 33: 32-95. 1922) and he writes that Indiana has only the three species.

Under surface of leaves (except the midrib and principal veins which are usually more or less pubescent) subglabrous to minutely pubescent with one-celled, conical hairs, the hairs more or less appressed.
 Inflorescence paniculate, the branches widely spreading; under surface of leaves not punctate, rarely specimens more or less punctate................1. *V. altissima.*
 Inflorescence fastigiate (the heads in close clusters); under surface of leaves conspicuously punctate....................................2. *V. fasciculata.*
Under surface of leaves (except the midrib and principal veins which are pubescent) pubescent with the one-celled, appressed hairs and with few to many multicellular hairs, all over the lower surface; also the lower surface of the blades more or less densely punctate...3. *V. missurica.*

1. **Vernonia altíssima** Nutt. TALL IRONWEED. Map 1978. Probably found in every county of the state although there are no authentic reports from the northwestern counties. Usually frequent to common or abundant in the eastern part of the lake area and in the Tipton Till Plain, becoming rare in the unglaciated area. It has a wide range of habitat and is found in dry, open or moist woodland, fallow fields and prairies, and rarely in dried-up sloughs and swamps.

Pepoon's reports for Hill and for Umbach from Porter County are referred by Fassett (Rhodora 35: 202. 1933) to *V. missurica* Raf.

N. Y., Ohio, and Mo., southw. to S. C., Ga., and La.

1a. **Vernonia altissima** var. **lilacìna** Clute. (Amer. Bot. 36: 225. 1930.) This is a form with "pinkish-lavender" flowers which was found on the campus of Butler University at Indianapolis.

The flowers of this species are variable, ranging from purplish, the normal color, to colorless (white). In the white form the bracts are usually green with no trace of purple. I have seen this form several times. In a pasture of about five acres in Montgomery County it was noted repeatedly. I have collected a rose colored form in Lagrange County. I have had the white and rose colored forms in cultivation for several years and as far as I have been able to determine, they continue the same color forms.

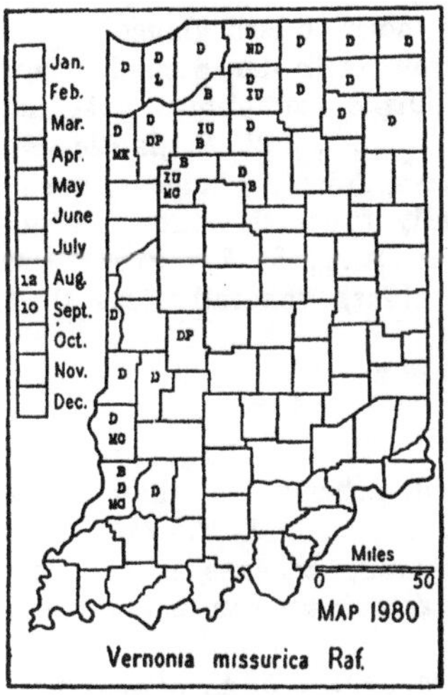

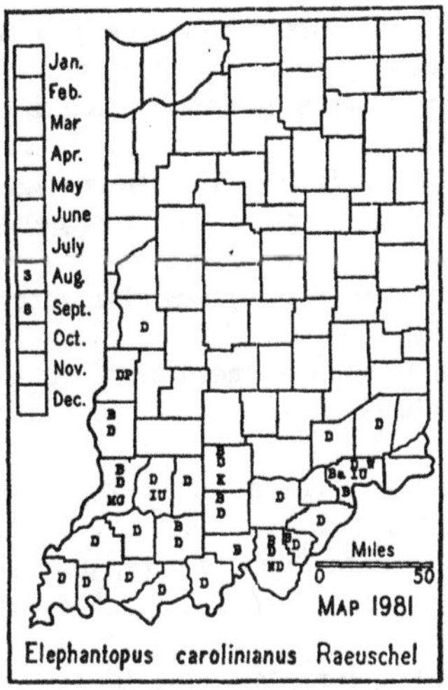

2. **Vernonia fasciculàta** Michx. Map 1979. This species prefers a wet or prairie habitat and is found in sloughs in the Lower Wabash Bottoms, in wet marshes, and moist prairie habitats. It is infrequent and probably entirely absent from the southeastern part of the state although there are reports for it from that area.

Ohio to Minn., southw. to Okla.

3. **Vernonia missùrica** Raf. (*Vernonia illinoensis* Gleason and *Vernonia altissima* var. *taeniotricha* Blake.) Map 1980. Infrequent to frequent, usually in dry places and rarely in wet places. It is generally found along roadsides and railroads, in pasture fields, and less frequently in open woods.

Ont. to Iowa, southw. to Ala., Miss., and N. Mex.

3a. **Vernonia missurica** f. **cárnea** Standley (Rhodora 32: 33. 1930.) is a form with "rose or flesh colored" flowers which was reported by Standley as found in Porter County.

8775. ELEPHÂNTOPUS [Vaill.] L. ELEPHANT'S-FOOT

1. **Elephantopus caroliniànus** Raeuschel. (*Elephantopus carolinianus* Willd.) ELEPHANT'S-FOOT. Map 1981. Frequent to infrequent or rare in dry and usually more or less sandy soil, mostly in black and white oak woods but also in beech woods, and along roadsides. It was once found in a hogyard where the hogs had destroyed all the vegetation except this species, which they had not molested.

N. J. to Ill., and Kans., southw. to Fla. and Tex.

8816. EUPATÒRIUM [Tourn.] L.

Leaves verticillate in 3's-6's, or the upper opposite, petiolate; tall plants generally 1-3 m high.

Florets 9-15, rarely 8 or more than 15; inflorescence flat-topped; flowers generally pinkish purple; florets scarcely exserted at anthesis; stems generally solid, rarely hollow, not glaucous; leaves mostly in 4's or 5's, rarely in 3's or 6's. ....1. *E. maculatum.*

Florets 5-7, rarely 3, 4, or 8; inflorescence convex; stems more or less glaucous.
Stems hollow, plainly glaucous, purple (except when growing in dense shade), not darker at the nodes; flowers generally pinkish purple, sometimes greenish yellow..2. *E. fistulosum.*
Stems solid (with a pith), rarely hollow toward the base, faintly glaucous, green with the nodes generally purplish; flowers generally greenish yellow, rarely pinkish purple..3. *E. purpureum.*
Leaves opposite (rarely ternate or the upper alternate); plants generally less than 1.5 m high.
Involucral bracts imbricated in 2 or more series, the outer shorter.
Involucral bracts oblong, obtuse.
Heads 12-15-flowered.
Leaves, at least the lower, slender-petioled, 3-nerved; achenes less than 2 mm long..4. *E. serotinum.*
Leaves sessile, the cuneate base entire, many-nerved; achenes more than 2 mm long..........................4a. *E. serotinum* var. *polyneuron.*
Heads 5-flowered.
Leaves strongly 3-nerved, long-tapering at the base.........5. *E. altissimum.*
Leaves pinnately veined, rounded and sessile at the base...6. *E. sessilifolium.*
Involucral bracts lanceolate, acute.
Flowers white.
Leaves connate-perfoliate..............................7. *E. perfoliatum.*
Leaves not connate-perfoliate.
Leaves, at least the upper, truncate or rounded at the base...7a. *E. perfoliatum* f. *truncatum.*
Leaves cuneate at the base, sessile, smaller than in the species or other forms..........................7b. *E. perfoliatum* var. *cuneatum.*
Flowers purplish..........................7. *E. perfoliatum* f. *purpureum.*
Involucral bracts in 1 or 2 series, all equal or nearly so.
Flowers white; leaves ovate, large, thin, generally abruptly narrowed at the base, or rarely truncate or subcordate; bracts generally obtuse.......8. *E. rugosum.*
Flowers pink or violet purple; leaves ovate or deltoid-ovate, truncate, cordate, subcordate, or shortly narrowed at the base.
Receptacle flat; leaves deltoid-ovate; outer bracts mostly less than half as long as the inner, acute..................................9. *E. incarnatum.*
Receptacle conical; leaves ovate; outer bracts nearly as long as the inner ones, long-acuminate..................................10. *E. coelestinum.*

1. **Eupatorium maculàtum** L. (Wiegand. Eupatorium purpureum and its allies. Rhodora 22: 57-70. 1920 and Wiegand & Weatherby. The nomenclature of the Verticillate Eupatoria. Rhodora 39: 297-306. 1937.) SPOTTED-STEM JOE-PYE-WEED. Map 1982. In wet ground or springy places along streams and ditches, about lakes, and in marshes and wet woods throughout the lake area of the state. South of this area it is rare or absent.

The Joe-pye-weeds have been misunderstood, and all or most all of the reports should be ignored, because, as far as I know, none of our authors knew of the existence in our area of three species of this group or had keys which would separate them.

Newf., Que., Mich. to B. C., southw. to Pa., Ill., and N. Mex.

2. **Eupatorium fistulòsum** Barratt. PURPLE-STEM JOE-PYE-WEED. Map 1983. Infrequent to frequent in springy and wet places in woods and along streams in the southern half of the state. Formerly I did not recognize the

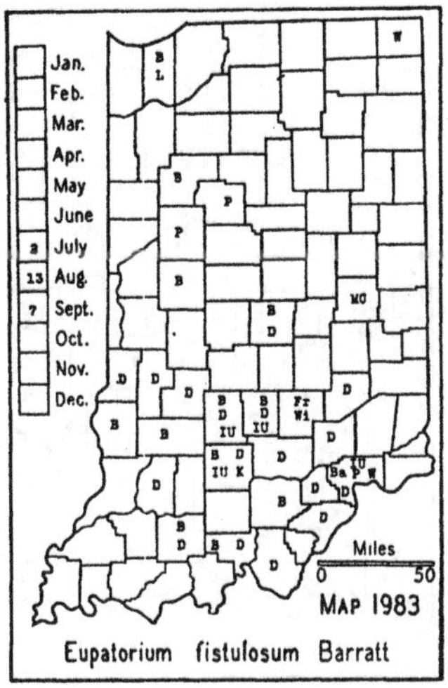

Eupatorium fistulosum Barratt

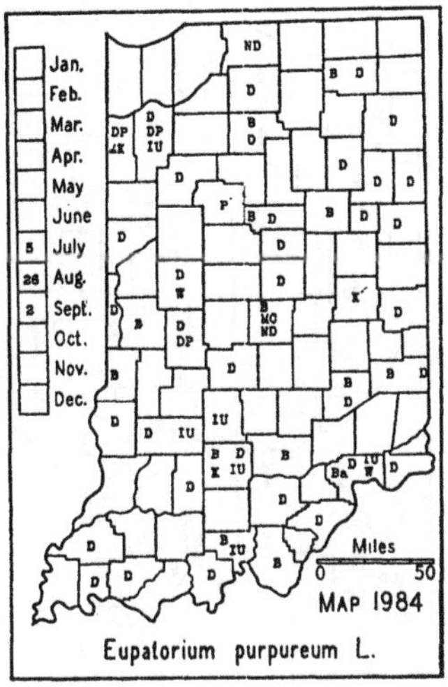

Eupatorium purpureum L.

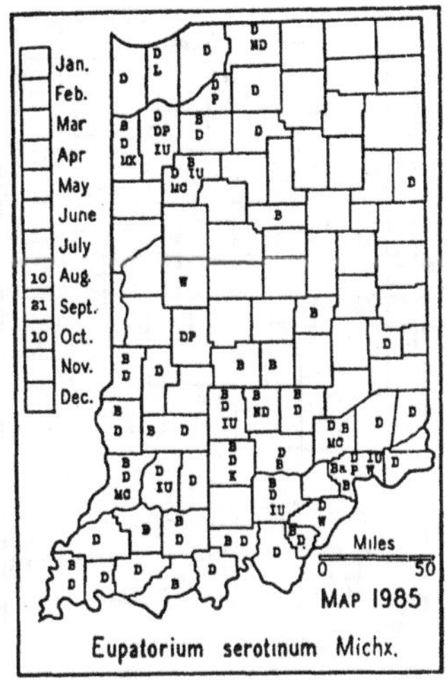

Eupatorium serotinum Michx.

three species of the Joe-pye-weeds and, if I had, I might have been able to extend the range of this species farther north in the state.

S. Maine, R. I., w. Pa., and Ohio, southw. to Fla. and Tex.

3. **Eupatorium purpùreum** L. (Probably *Eupatorium purpureum* var. *amoenum* (Pursh) Gray of Gray, Man., ed. 7 and *Eupatorium falcatum* Michx.) GREEN-STEM JOE-PYE-WEED. Map 1984. Infrequent, probably throughout the state in moist or dry soil, usually in wooded ravines, open woodland, and clearings; also near the bases of slopes bordering wet grounds.

Mass., Ont., Wis., and Nebr., southw. to Ga. and Okla.

4. **Eupatorium serótinum** Michx. LATE EUPATORIUM. Map 1985. This species prefers a slightly acid soil and is more or less frequent to common in such habitats. It prefers a moist, white clay or moist, black, sandy soil. In the "flats" in the southern part of the state, it sometimes covers acres of fallow or pasture lands. Stock do not eat this species nor any other species of *Eupatorium* unless they are forced to do so by scarcity of food. It is, no doubt, rare or absent from the area where it is not represented on the map.

Del. to Minn., southw. to Fla. and Tex.

4a. **Eupatorium serotinum** var. **polyneùron** F. J. Hermann (Rhodora 40: 86. 1938.) This form was found by Edna Banta in Jefferson County in 1933. It was found in hard, white, moist, clay soil in a flat beech woods on the Schumann farm about 3 mi. northeast of Hanover. This is the only known station for it.

5. **Eupatorium altíssimum** L. Map 1986. Very local but not rare where it is found. Most of my specimens are from high, wooded banks of streams; frequent in Henry County in one place at the base of a high slope that borders a marsh, and frequent in a prairie habitat in Benton County. Its habitats and locations in Indiana suggest that it is a prairie plant.

Pa. to Minn., southw. to N. C. and Tex.

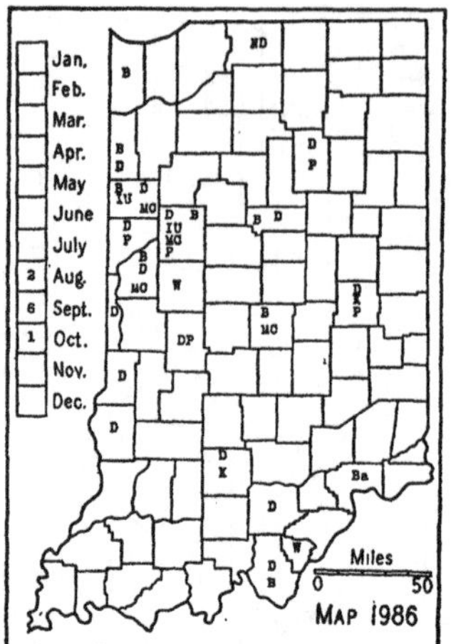

Eupatorium altissimum L.

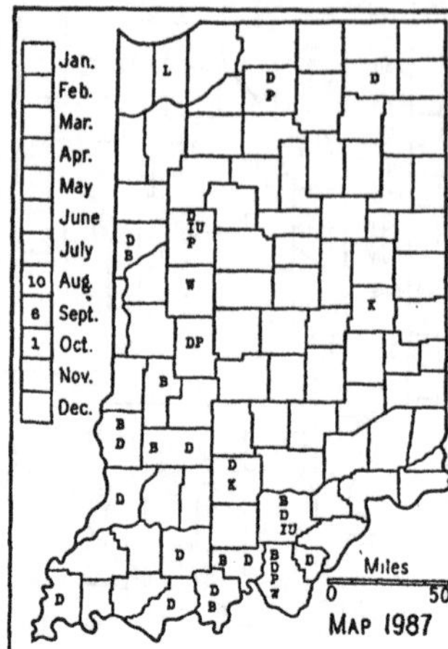

Eupatorium sessilifolium L.

Eupatorium perfoliatum L.

6. **Eupatorium sessilifòlium** L. UPLAND BONESET. Map 1987. Infrequent in many parts of the state. It is generally found on high, wooded ridges and in dry, sandy woods. It is usually common where it is found.

Vt. and Mass. to Ill., southw. to Ga., Ala., and Mo.

7. **Eupatorium perfoliàtum** L. BONESET. Map 1988. Frequent to common in all parts of the state in low ground in woodland, pastures, and fallow fields and along ditches and roadsides. The tops and leaves were formerly an official drug and the pioneers freely used a warm infusion of it as a diaphoretic.

The leaves and flowers are variable and these variations have been named. I have included all forms in the one map.

7a. **Eupatorium perfoliatum** f. **truncàtum** (Muhl.) Fassett. This is a form with at least the upper leaves separate, truncate or rounded at the base. A few of my specimens belong to this form.

7b. **Eupatorium perfoliatum** var. **cuneàtum** Engelm. This is a form with leaves smaller, narrowed at the base, and separate; heads fewer-flowered. I found it along a moist, sandy roadside in Newton County about 2 miles north of Lake Village.

7c. **Eupatorium perfoliatum** f. **purpùreum** Britt. A form with purplish flowers. It is not as common as the typical form which has white flowers.

N. S. and N. B. to Man., southw. to Fla. and Tex.

8. **Eupatorium rugòsum** Houtt. (Rhodora 40: 293. 1938.) (*Eupatorium urticaefolium* Reich.) WHITE SNAKEROOT. Map 1989. Frequent to common in most of the dry and moist woods of the state. It is more common in beech and sugar maple and black and white oak woods. This plant is poisonous to grazing animals and if it is eaten in a sufficient quantity it proves fatal. A symptom of having eaten too much of this weed is a trembling of the animal and because of this characteristic, the disease has

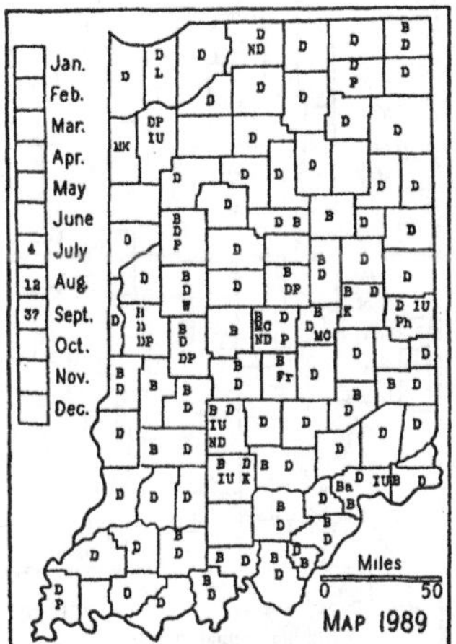

Eupatorium rugosum Houtt.

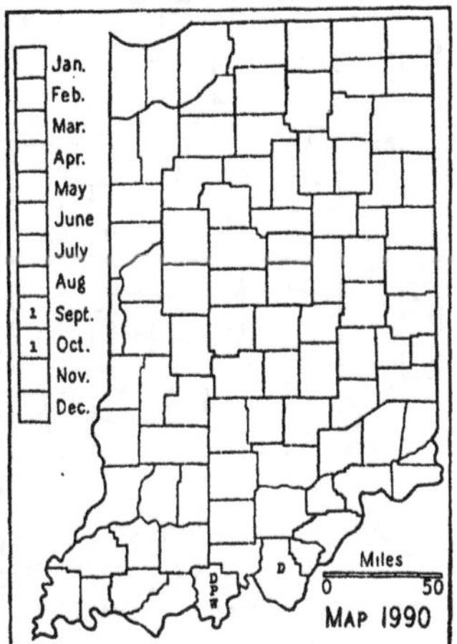

Eupatorium incarnatum Walt.

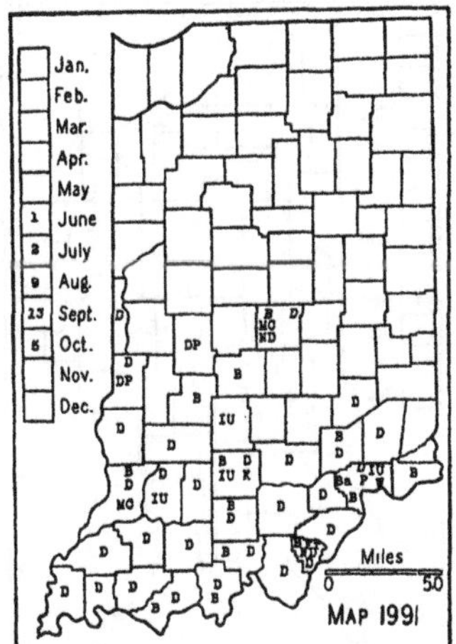

Eupatorium coelestinum L.

been called "trembles." The plant is frequently eaten by sheep and by cattle when the pasturage becomes scarce, and many of those animals are killed in Indiana each year by this weed. When it is eaten by milch cows, the poisonous principle (a barium salt) is communicated to the milk; such milk, when consumed by people, has the same effect as the plant has upon stock. The pioneers called it "milk sickness," and many of them died from drinking too much of the affected milk. A pioneer informed me that a family of four in my own county died from this cause.

Indiana specimens show some variation in leaf form. All of my specimens are generally densely short-pubescent in the inflorescence and on the upper half of the stem, and in a few plants the stem is villous. (See Rhodora **10**: 87. 1908.) The leaves of all of my specimens are abruptly cuneate at the petiole except in my Lake and Warren County specimens in which they are slightly cordate at the base.

N. B. to Nebr., southw. to Fla. and La.

9. **Eupatorium incarnàtum** Walt. Map 1990. I have found this species in only three places in Indiana; in Harrison County, in the rather moist talus of a cliff along Blue River about half a mile north of White Cloud; and in Perry County, more or less frequent for a mile in moist places in the roadside ditch at the base of the high, wooded bluff along the Ohio River about 5 miles above Cannelton, and along the moist roadside of an abandoned road on the crest of the "German Ridge" about 6 miles east of Cannelton.

Va., s. Ind. to Mo., southw. to Fla. and Mex.

10. **Eupatorium coelestìnum** L. MISTFLOWER. Map 1991. Rather infrequent in the southern half of the state. My only specimen from northern Indiana was one which I found on the moist, north bank of Tippecanoe Lake, in Kosciusko County. It was in a habitat which suggested it was native although it might have been seeded there from one of the cottages on the lake, the nearest one being about 150 feet to the east. It prefers a moist, hard, clay soil and is found in roadside ditches and moist places

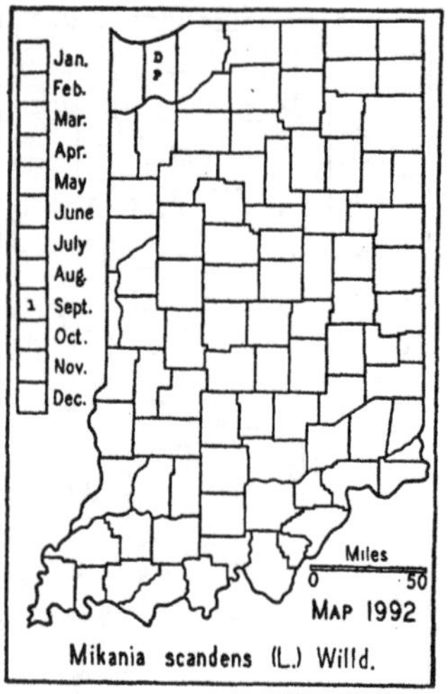

MAP 1992
Mikania scandens (L.) Willd.

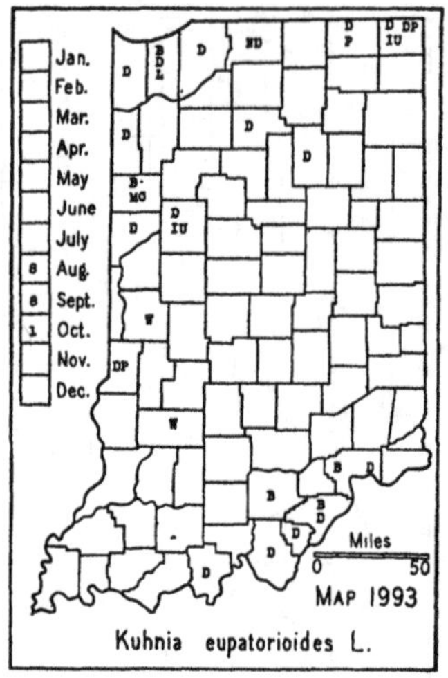

MAP 1993
Kuhnia eupatorioides L.

MAP 1994
Kuhnia eupatorioides var corymbulosa T & G

along streams. This species is frequently cultivated and it is possible that some of our roadside plants are escapes.

N. J., s. Ohio, Ind. to Kans., southw. to Fla. and Tex.

8818. MIKÀNIA Willd.

1. **Mikania scándens** (L.) Willd. May 1992. Common in low ground along the Kankakee River at the Baum Bridge, south of Kouts, in Porter County. Here in 1915 it was common especially along the old channel of the river where it was found climbing usually on buttonbush or on tall weeds. Blatchley reported that it was abundant about 50 feet south of the bridge over Sandy Hook Creek about 5 miles east of Hebron, in Porter County. It was reported by Schneck from the Lower Wabash bottoms, and Coulter's Catalogue reports it from Putnam and Tippecanoe Counties on the authority of MacDougal and Wright, respectively. It is, no doubt, very local.

Maine, southw. near the coast to Fla. and through the Gulf States to Tex., northw. in the interior to Okla., n. Ind., s. Mich., and in N. Y.

8825. KÙHNIA L.

Leaves puberulent, sometimes nearly glabrous, sparingly dentate or entire, the lower ones often on short petioles; heads mostly 8-10 mm long, rather loosely clustered... ....1. *K. eupatorioides.*

Leaves pubescent or tomentulose, distinctly dentate (those of the upper branches sometimes entire), veiny, sessile; heads mostly 12-16 mm long, densely clustered... ....1a. *K. eupatorioides* var. *corymbulosa.*

1. **Kuhnia eupatorioides** L. FALSE BONESET Map 1993. Locally infrequent to common in very sandy soil on open, wooded dunes and along roadsides in the extreme northern part of the state and in a few counties to the south of this area. In the southern part, and in a few of the central counties, it is found on high, wooded river bluffs, and on the crests and slopes of open, wooded ridges.

N. J. to Minn., southw. to Ga. and Tex.

1a. **Kuhnia eupatorioides** var. **corymbulòsa** T. & G. Map 1994. Usually found in sandy to very sandy soil in dry prairie habitats and infrequently on high, gravelly banks of streams.

Prairies and plains from Ind. westw. and southw.

8826. LIÀTRIS Schreb. GAYFEATHER

Pappus very plumose; bracts acute or acuminate; corolla lobes pubescent within.
 Stems and peduncles pubescent; bracts lanceolate, thick, stiff, long sharp-pointed, squarrose....................1. *L. squarrosa.*
 Stems and peduncles glabrous; bracts broadly oval, thin, all but the outer abruptly short-pointed, appressed....................2. *L. cylindracea.*
Pappus barbellate (not obviously plumose to the naked eye); bracts acute, obtuse or rounded; corolla lobes not pubescent within.
 Heads oblong, mostly about 10 mm long, in dense spikes, 3-15-flowered.
  Rachis of spike pubescent.
   Involucral bracts merely acute, their broad tips spreading or recurving....................3. *L. Bebbiana.*
   Involucral bracts ending in long-acuminate tips. (See excluded species no. 608, p. 1095)....................*L. pycnostachya.*
  Rachis of spike glabrous, rarely puberulent; involucral bracts obtuse, appressed....................4. *L. spicata.*
 Heads hemispheric or campanulate, mostly 15-20 mm long, 15-45-flowered, generally loosely racemose, subcorymbose or sometimes the heads subsessile; bracts obtuse or rounded; rachis of inflorescence pubescent.....5. *L. scariosa* (complex).

1. **Liatris squarròsa** Willd. (*Lacinaria squarrosa* (L.) Hill.) Map 1995. Local in southern Indiana where it is found in poor soil on black oak ridges or in almost pure sand on a black oak ridge in Daviess County; in the northern part of the state it has been reported as occurring on the dunes near Lake Michigan. My specimens are variable in the pubescence of the leaves and of the bracts, and the length of the peduncle. Some of the heads are sessile and some are on peduncles up to 5 cm long. I have a specimen from Perry County with the stem almost glabrous, the leaves glabrous, and the bracts glabrous except the ciliate margins. It also has very narrow leaves and closely approaches the glabrate form of this species, known as *Liatris glabrata* Rydb.

Pa. to S. Dak., southw. to Fla. and Tex.

2. **Liatris cylindràcea** Michx. (*Lacinaria cylindracea* (Michx.) Ktze.) Map 1996. Local in northern Indiana on open sand hills and dunes and near Lake Michigan on dry interdunal flats. It was reported by Schneck as rare in prairies in the Lower Wabash Valley but since the Indiana side of the Wabash River has no real prairies in the territory where Dr. Schneck collected, I believe that his report should go to the Illinois side of the river.

W. Ont. to Minn., southw. to Ohio and Mo.

3. **Liatris Bebbiàna** Rydb. (Brittonia 1: 99. 1931.) (*Liatris pycnostachya* Michx. of Indiana authors and Gray, Man., ed. 7 and *Lacinaria pycnostachya* (Michx.) Ktze. of Britton and Brown, Illus. Flora, ed. 2.) CATTAIL GAYFEATHER. Map 1997. A very rare species of prairies. *Liatris pycnostachya,* which does not occur in Indiana, has been reported from

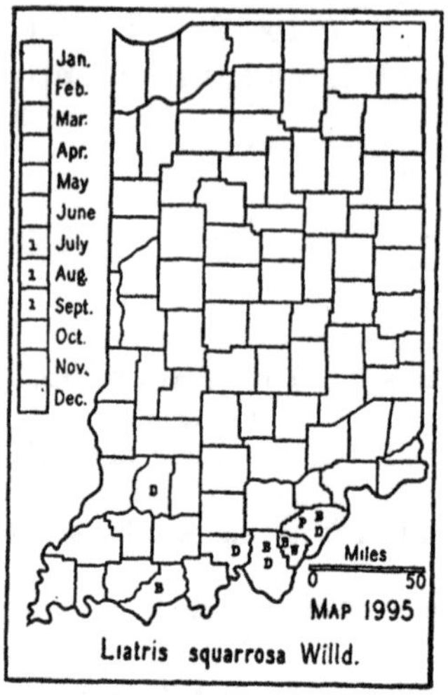

Liatris squarrosa Willd.

Liatris cylindracea Michx.

Liatris Bebbiana Rydb.

Jasper, Marshall, St. Joseph, and Vigo Counties and Schneck reported it from the Lower Wabash Valley. Doubtless all of these reports should go to this species. A specimen in the herbarium of Purdue University collected in St. Joseph County by Barnes and labeled *L. pycnostachya* is *L. Bebbiana.* I have seen the Jasper County specimen and it belongs here. Blatchley reported it from Vigo County. He collected his specimen in the Haeckland Prairie in 1889 and it is now in the herbarium of Butler University. I collected my specimen in the same place in 1917. Since the Lower Wabash Valley has no prairies on the Indiana side of the river where Dr. Schneck collected, it is best to refer his specimen to Illinois.

Prairies of Ind., westw. to Nebr. and Mo.

4. **Liatris spicàta** (L.) Willd. (*Lacinaria spicata* (L.) Ktze.) SPIKE GAYFEATHER. Map 1998. This species and others of this genus are often called blazing star. In the northern half of the state this species grows generally in marshy places and in moist prairies. In the prairies it is often so abundant that it gives a rose purple color to the landscape. In the southern part of the state it is local and is found in the "flats" and in sandy soil on open, wooded slopes.

The rachis of all of my plants is quite glabrous. Kriebel's no. 3958 from Greene County has the rachis closely puberulent.

Mass. to Minn., southw. to Fla. and La.

5. **Liatris scariòsa** (L.) Willd. (*Lacinaria scariosa* (L.) Hill., *Lacinaria Deamii* Lunell, *Lacinaria scariosa intermedia* Lunell, *Lacinaria scariosa* var. *Nieuwlandii* Lunell, *Lacinaria scariosa* var. *petiolata* Lunell, *Lacinaria scariosa* var. *praesignis* Lunell, and *Lacinaria scariosa* var. *strictissima* Lunell.) Map 1999. Lunell in his revision of the genus described new species and new varieties and cited Indiana specimens in the Deam Herbarium. I am regarding this polymorphic species as a complex. E. S. Steele had my specimens and after working on this genus for several

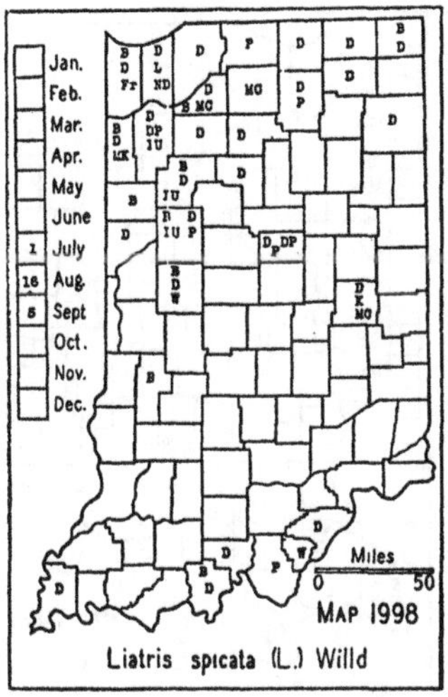

Liatris spicata (L.) Willd

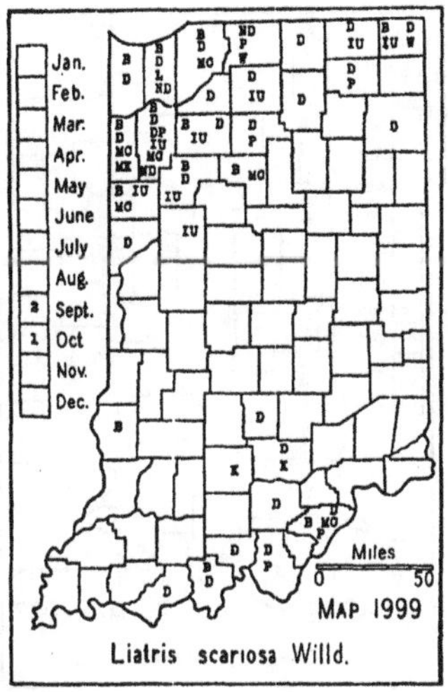

Liatris scariosa Willd.

Grindelia squarrosa (Pursh) Dunal

years, he wrote a manuscript of about 500 pages in which he described many species of this complex. In my collection of about 400 sheets I have many type specimens and varieties which he proposed to publish. I was told by a geneticist that he estimated this species contains at least 100 elemental species. It at once becomes evident that a detailed account of this group would be out of place in a work of this kind.

Infrequent to frequent or common in prairie habitats in northern Indiana, in moist or dry, sandy soil in fallow fields, in open woods, in prairie habitats, and along roads and railroads. In the southern part of the state it is local and is found in dry, sandy clay soil on ridges or on open, wooded slopes.

This and the preceding species are easily cultivated and their inflorescences are commonly seen on the market. They prefer a sandy, well drained soil. In the event that the corms are forced to the surface by freezing during the winter they should be replanted in the spring, the depth depending upon the soil.

Maine to Man., southw. to Fla. and Tex.

5a. **Liatris scariosa** f. **Bénkei** Macbride. This is a white flowered form reported from Lake County. I have specimens from Fulton and White Counties.

8833. GRINDÈLIA Willd.

[Steyermark. Studies in Grindelia I. Ann. Missouri Bot. Gard. 21: 227-30. 1934. Studies in Grindelia II. Ann. Missouri Bot. Gard. 21: 433-608. 1934.]

Cauline leaves mostly 2-4 times longer than wide, ovate to broadly oblong.......... ....1. *G. squarrosa.*

Cauline leaves (4-4.5) 5-7 times longer than wide, linear-oblong, or oblong to lanceolate..........1a. *G. squarrosa* var. *serrulata.*

1. GRINDELIA SQUARRÒSA (Pursh) Dunal. BROADLEAF GUMPLANT. Map 2000. This is a western species that is becoming established in Indiana as

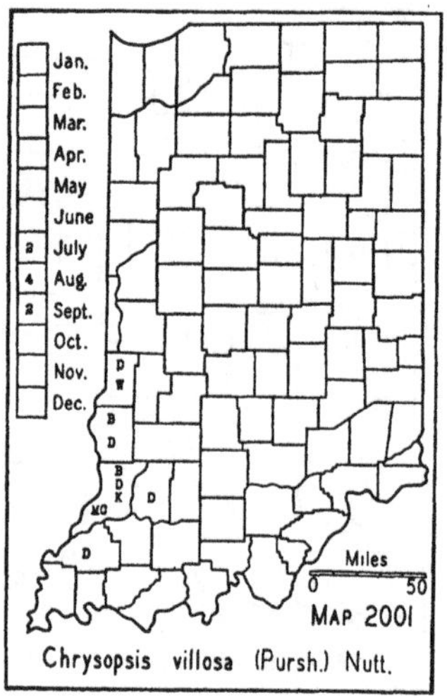

Chrysopsis villosa (Pursh.) Nutt.

Solidago squarrosa Muhl.

Solidago Buckleyi T. & G.

a weed in sandy fallow fields and waste places. Paul Standley writes that it is frequent in vacant lands in the vicinity of Whiting in Lake County.

Ne. Mich. to N. Dak. and Idaho, southw. to Tex. and Mex.

1a. GRINDELIA SQUARROSA var. SERRULÀTA (Rydb.) Steyermark. The variety, like the species, is becoming established in fallow fields, alfalfa fields, and waste places in northern Indiana. I have specimens from St. Joseph, Steuben, and Wells Counties.

My specimens were determined by J. A. Steyermark who studied the genus.

Both the species and variety are western plants that have been introduced and have established themselves.

8844. CHRYSÓPSIS Nutt.

1. **Chrysopsis villòsa** (Pursh) Nutt. HAIRY GOLDEN-ASTER. Map 2001. Locally common in very sandy soil along roadsides in the counties indicated on the map. This species is easily cultivated and is worthy of cultivation, but, no doubt, it would soon spread if a congenial habitat were nearby unless the seed were collected before they became mature.

Ill. to Minn. and Man., southw. to Ala. and N. Mex.

8849. SOLIDÀGO L. GOLDENROD

[Friesner. The genus Solidago in northeastern North America. Butler Univ. Bot. Stud. 3: 1-64. 1933.]

The following key has been adapted from Friesner's study and grateful acknowledgment is given.

A. **Heads all distinctly pedicellate.**

Bracts of the involucre, at least the outer ones, squarrose.

Basal rosettes conspicuous; blades of rosettes and lower stem leaves 5-15 cm long and 3-7 cm wide, on margined petioles of nearly equal length; upper stem leaves sessile or subsessile; pedicels about 5 mm long; pubescence on back of bracts appressed and eglandular or with a few glands........ ....1. *S. squarrosa.*

Basal rosettes usually lacking; blades of the lower stem leaves generally much smaller than the median ones, usually 7-12 cm long and 1-2 cm wide, all sessile or subsessile.

Pedicels usually about 2 mm long; pubescence of bracts eglandular. (See excluded species no. 617, p. 1096)........................*S. petiolaris.*

Pedicels mostly 5-10 mm long; pubescence of bracts rather dense, short, spreading, and glandular..............................2. *S. Buckleyi.*

Bracts of the involucre erect, not squarrose.

B. Inflorescence axillary, i. e., in clusters or short racemes from the axils of ordinary leaves.

Achenes glabrous or sparsely pubescent at maturity.

Stems densely pubescent from the base through the inflorescence; pubescence multicellular, spreading on the lower part of the stem, and on the upper part usually upwardly subappressed; leaves not thick, densely pubescent below and pubescence usually equally as dense above but the hairs shorter; involucres mostly 4.5-5 mm long, glabrous or pubescent; mature achenes 1.5-2.2 mm long, those of the ray flowers generally longer.

Rays white...3. *S. bicolor.*

Rays yellow..4. *S. hispida.*

Stems glabrous below the inflorescence but the axis of the inflorescence pubescent; leaves thick, glabrous or nearly so both above and below. ....5. *S. erecta.*

Achenes densely pubescent.

Involucres 3.5-5 mm high.

Stems glabrous, glaucous, terete, usually more or less branched; leaves of a lanceolate type...............................6. *S. caesia.*

Stems glabrous or somewhat pubescent above but not glaucous, more or less angular and zigzag, rarely branched; leaves usually of a broadly ovate type............................7. *S. latifolia.*

Involucres 6-9 mm high; plants found in Indiana only along Lake Michigan.

Blades of basal rosette obovate to narrow-obovate, rounded at the apex or some of them acute, on margined petioles about half the length of the blades; inflorescence narrow and compact, spikelike in appearance, less than 5 cm wide; heads mostly in small clusters..... ....8. *S. Deamii.*

Blades of the basal rosette oblanceolate, acute, on margined petioles almost as long as the blades; inflorescence usually of a paniculate type, mostly 5-15 cm wide (small specimens may be narrower); heads comparatively few, usually racemosely disposed on the branchlets, rarely a few heads in a cluster..................... ....9. *S. racemosa* var. *Gillmani.*

B. Inflorescence not axillary, either racemose, paniculate or corymbose.

C. Inflorescence racemose or paniculate.

D. Heads secund, i. e., racemes one-sided.

E. Leaves triple-nerved, i. e., one pair of lateral veins decidedly more prominent than the others.

Involucres 2-2.8 mm high.

Stems glabrous throughout. (See excluded species no. 621, p. 1096) .... *S. rupestris.*

Stems more or less densely pubescent.

Stems glabrous below the inflorescence; leaves pubescent on the nerves beneath .... 10. *S. canadensis.*

Stems densely pubescent throughout; leaves densely cinerous-puberulent beneath ..... 10a. *S. canadensis* var. *gilvocanescens.*

Involucres 3-8 mm high.

Stems entirely glabrous, including the inflorescence.

Plants flowering mostly from the last of August through September, of a prairie habitat; plants strongly stoloniferous, the stolons forming sterile branches, usually with a terminal rosette of leaves; leaves of stolons linear-lanceolate, mostly 7-15 mm wide, the veins conspicuous beneath, the margins usually strongly and evenly serrate above the middle...... .... 11. *S. glaberrima.*

Plants flowering mostly from the first of July to the latter part of August; not strongly stoloniferous but with root stocks at flowering time, these with elliptic, serrate leaves, the blades sometimes 15 cm long on petioles of equal length.... .... 11a. *S. juncea.*

Stems and inflorescence not entirely glabrous.

Stems usually glabrous to the inflorescence or if pubescent not rough.

Rootshoot leaves always present at flowering time; plants flowering mostly from the first of July through August, usually of a dry habitat.................. 11a. *S. juncea.*

Rootshoot leaves absent at flowering time; plants flowering mostly from the last of August through September, usually of moist or wet places.

Leaves glabrous or somewhat scabrous above, pubescent at least on the midrib beneath.............. 12. *S. gigantea.*

Leaves glabrous above and beneath....................... .... 12a. *S. gigantea* var. *leiophylla.*

Stems at least partly rough or scabrous.

Stems minutely rough-pubescent above; leaves rigid, glabrous, oblong-lanceolate, the lower ones mostly serrate toward the apex. (See excluded species no. 622, p. 1096) .... *S. Shortii.*

Stems distinctly scabrous or pubescent their entire length.

Cauline leaves of a lanceolate type.......... 13. *S. altissima.*

Cauline leaves of an oblong or oblanceolate type.

Stems scabrous, green; rays 3-7. (See excluded species no. 619, p. 1096) .... *S. radula.*

Stems grayish pubescent or canescent; rays 5-9.

Lower stem leaves 3-6.25 times as long as wide; involucres 3-4.5 mm high............ 14. *S. nemoralis.*

Lower stem leaves 7-10 times as long as wide; involucres 4.5-5.5 mm high.... 14. *S. nemoralis* var. *decemflora.*

E. Leaves not triple-nerved, more or less pinnately veined, although some of the leaves may have short and faint lateral veins.

a. Stems glabrous or more or less pubescent above the middle; pappus of bristles longer than the achenes.

Stems strongly angled, at least below the middle, glabrous; leaves usually very thick and the upper surface strongly scabrous, the hairs (under a 28 diameter magnification) appearing stout, conical, and arising from a callous base; plants of a springy or wet habitat........................15. *S. patula.*

Stems terete (strongly striate in *Solidago rugosa* var. *celtidifolia*).

Plants flowering mostly from the first of July to the last of August, with large rootshoot leaves at flowering time, these often with blades 15 cm long on petioles of equal length; upper cauline leaves linear or narrowly elliptic and entire or with only a few teeth.....................11a. *S. juncea.*

Plants flowering mostly from the last of August through September, usually lacking the rootshoots at flowering time; upper cauline leaves not of a linear type, usually short-elliptic or short-lanceolate and serrate on both margins.

Leaves pubescent beneath, sharply serrate to the inflorescence; achenes pubescent.

Blades of leaves usually firm to thick, antrorsely scabrous above with stout projections much less than 0.5 mm long, these arising from a callous base making the upper surface very rough to the touch, both surfaces of the blade more or less strongly pitted, that is, with the areas between the veinlets more or less sunken, giving the appearance of deeply hammered metal; plants of wet places, stoloniferous.

Leaves tapering at the base................17. *S. rugosa.*

Leaves rounded at the base.

Blades more than two and a half times as long as wide; pedicels mostly with 1-5 bracts.................
......................17a. *S. rugosa* var. *aspera.*

Blades less than two and a half times as long as wide; pedicels mostly with 3-9 bracts.................
..................17b. *S. rugosa* var. *celtidifolia.*

Blades of leaves usually thin, pubescent above with scattered, curved hairs 0.5-1 mm long, rarely glabrous; leaves mostly long-acuminate or short-acuminate at the apex, tapering to a sessile or subsessile base, neither the upper nor the lower surface having the areas between the veinlets sunken; plants of dry soil, not stoloniferous........................16. *S. ulmifolia.*

Leaves glabrous beneath; basal leaves lanceolate, sometimes broadly so, the cauline ones similar in shape but narrower and quickly diminishing in size upward, the margins entire or some more or less serrate; fresh leaves usually with a greasy texture; achenes glabrous or strigose.

Axis of the inflorescence pubescent........18. *S. uniligulata.*

Axis of the inflorescence glabrous......................
......................18a. *S. uniligulata* var. *levipes.*

a. Stems pubescent, scabrous-puberulent or somewhat hispid. rarely glabrous above the base.

Pappus of long bristles, much longer than the achenes.

Lower stem leaves and usually those of rootshoots oblanceolate.

Lower stem leaves 3-6.25 times as long as wide; involucres 3-4.5 mm high........14. *S. nemoralis.*

Lower stem leaves 7-10 times as long as wide; involucres 4.5-5.5 mm high.................... ....14a. *S. nemoralis* var. *decemflora.*

Lower stem leaves ovate or oblong-lanceolate to elliptic.

Leaves tapering at the base..........17. *S. rugosa.*

Leaves rounded at the base.

Blades more than two and a half times as long as wide; pedicels mostly with 1-5 bracts........ ....17a. *S. rugosa* var. *aspera.*

Blades less than two and a half times as long as wide; pedicels mostly with 3-9 bracts........ ....17b. *S. rugosa* var. *celtidifolia*

Pappus as long as or shorter than the achenes.

Pappus (about 0.5 mm long) about half as long as the achenes; leaves of rootshoots broadly ovate, cordate at the base................19. *S. sphacelata.*

Pappus (about 1 mm long) as long as the achenes.... ....20. *S. ovata.*

D. Heads not secund; plants of a dry sandy soil, mostly of northern Indiana or of crests of the higher ridges of southern Indiana; stems glabrous to the inflorescence; leaves glabrous except the ciliate margins; achenes glabrous when mature.

Lowest stem leaves oval, never widest above the middle; corolla lobes mostly 0.8-1.1 mm long; mature achenes mostly 1-2.3 (2.5) mm long............................21. *S. speciosa.*

Lowest stem leaves usually slightly broadest above the middle; corolla lobes mostly 1.4-1.8 mm long; mature achenes mostly (2) 2.5-3.5 mm long...........................5. *S. erecta.*

C Inflorescence corymbose, not at all racemose.

Leaves of an ovate, oval or oblong type; pubescent above and below; plants of a dry habitat...........................22. *S. rigida.*

Leaves of a linear or lanceolate type, glabrous both above and below; leaves of the rootshoots half the length of the plant or longer; plants of a boggy or marshy habitat.

Plants entirely glabrous except for the margins of the leaves; leaves of the rootshoots obtuse at the apex; cauline leaves not clasping at the base, always flat..........23. *S. ohioensis.*

Plants usually pubescent in the inflorescence; leaves of the rootshoots acute at the apex; cauline leaves sheathing at the base, some or all somewhat folded, at least the lower ones usually recurving................................24. *S. Riddellii.*

A. Heads sessile or subsessile; inflorescence a corymb.

Largest leaves 5-nerved, i.e., with 3 prominent and 2 less distinct nerves.

Stem, branches, pedicels, and leaves minutely and usually densely short-pubescent............................25. *S. graminifolia* var. *Nuttallii.*

Stem, branches, and leaves except the margins glabrous. (See excluded species no. 613, p. 1095)..................................*S. graminifolia.*

Largest leaves 3-nerved, i.e., with a prominent midrib and 2 faint lateral nerves.

Heads glomerate, in clusters of 3-7; lower branches floriferous...26. *S. media.*

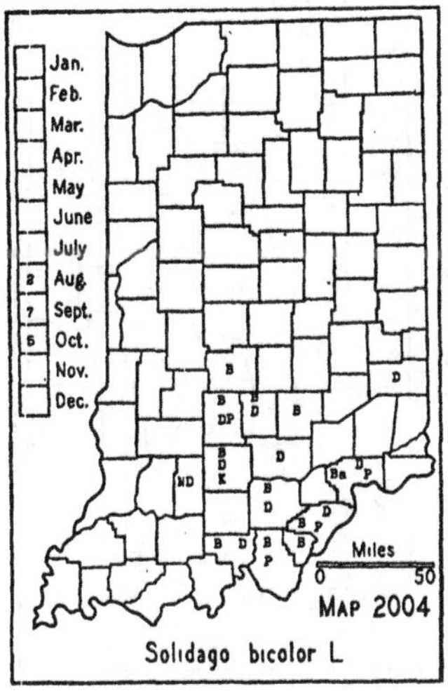

Solidago bicolor L

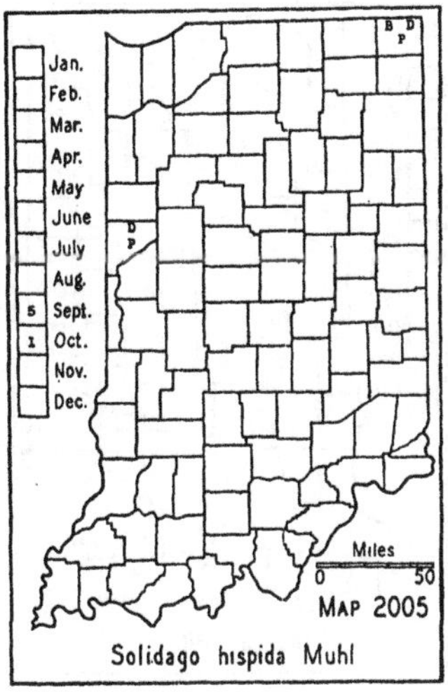

Solidago hispida Muhl

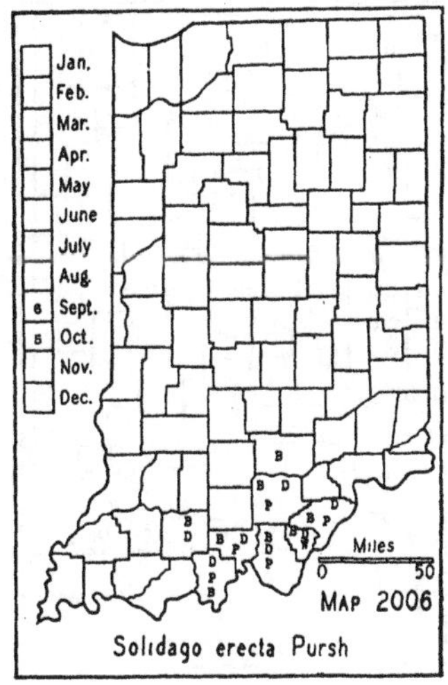

Solidago erecta Pursh

Heads not glomerate, nearly all on separate pedicels; lower branches sterile or only sparsely floriferous.............................27. *S. remota.*

1. **Solidago squarròsa** Muhl. Map 2002. In Indiana this goldenrod is known from only Clark and Floyd Counties. It is frequent in the Clark County State Forest about 3 miles northwest of Henryville on the south side of a deep hollow just north of the fire tower. This wooded hollow is probably 150 feet deep and the goldenrod is found here and there from the top to the bottom of the slope facing north. At the top of the slope it is associated with *Pinus virginiana, Quercus montana, Quercus velutina,* and *Vaccinium.* There is a specimen in the herbarium of Purdue University collected by A. Clapp in 183? in the "barrens" (probably in Floyd County).

N. B. to Ont., N. Y., Ohio, s. Ind., southw. to N. C.

2. **Solidago Búckleyi** T. & G. (Flora of North America 2: 198. 1841-1843.) Map 2003. In 1935 I found a colony of this species about 3 feet in diameter in hard, white clay soil in a low, flat oak woods about three fourths of a mile southeast of the Spencer school house or about 8 miles southwest of Mt. Vernon, Posey County. I transplanted some of it at Bluffton where it has proved to be hardy and grows vigorously, flowering in October.

W. Va., s. Ind., s. Ill., and Mo., southw. to Ala.

3. **Solidago bícolor** L. WHITE GOLDENROD. Map 2004. I am following other authors in maintaining this goldenrod and the next as species although I do not believe they are of specific rank. I believe this species is only an albino form of *Solidago hispida.* I prefer to regard it as a fertile strain of *S. hispida* that has lost its power to produce yellow rays. I think this assumption is supported by the fact that there is a general reduction of the number of rays in the colorless forms. I have tried to separate this plant from the next one and I find that all characters used by other authors fail.

Outside of Jefferson County *S. bicolor* is restricted chiefly to the unglaciated area of the state and is only rarely found a few miles outside

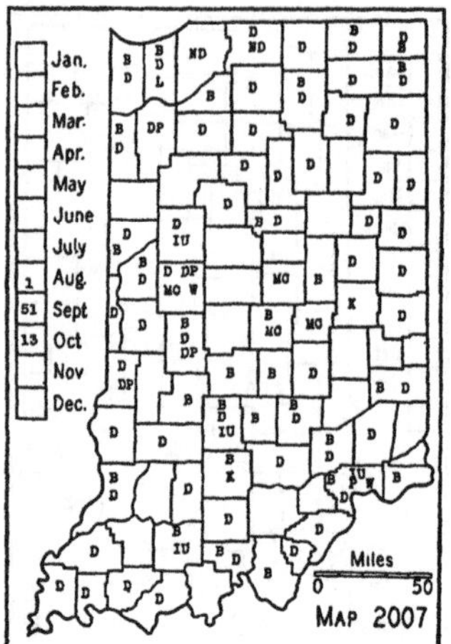

Solidago caesia L

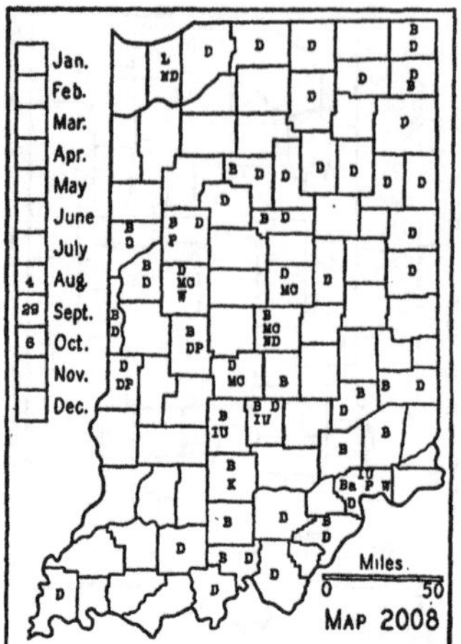

Solidago latifolia L.

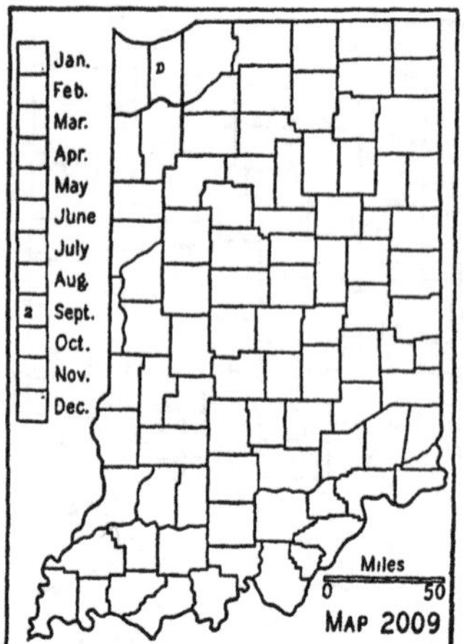

Solidago Deamii Fern.

of it. It is rather local and is found only on the crests and slopes of oak ridges or rarely in fallow fields.

P. E. I. to Mich. and Minn., southw. to Ga. and Mo.

4. **Solidago híspida** Muhl. Map 2005. This goldenrod is very rare in Indiana. I have it from dry sandy and gravelly wooded banks of lakes in Steuben County and from a sandstone outcrop in Warren County.

Newf. to Man., southw. to Ga. and Ark.

5. **Solidago erécta** Pursh. Map 2006. Restricted to the unglaciated area and found on the crests of chestnut oak ridges underlaid with sandstone or in soil of weathered sandstone. It is often associated with *Solidago bicolor*.

N. J., Pa., and Ind., southw. to Ga. and Ala.

6. **Solidago caèsia** L. WREATH GOLDENROD. Map 2007. Frequent throughout the state in both dry and moist woods. Sometimes it forms large colonies. The plants vary from simple to widely branched forms which are often found in the same colony. These forms have been given names but I do not believe they are of taxonomic value. Since they occur throughout the state, all forms are shown on one map.

N. S., Ont. to Minn., southw. to Fla. and Tex.

7. **Solidago latifòlia** L. (*Solidago flexicaulis* L.) BROADLEAF GOLDENROD. Map 2008. This goldenrod occurs frequently throughout the state in both dry and moist woods. It is found in colonies because it propagates mostly by stolons.

Newf. to N. Dak., southw. to Ga. and Kans.

8. **Solidago Dèamii** Fern. (Rhodora 38: 204-205. 1936.) DEAM GOLDENROD. Map 2009. Known only from dunes near Lake Michigan in Lake and Porter Counties.

9. **Solidago racemòsa** Greene var. **Gillmani** (Gray) Fern. (*Solidago Fisheri* Steele, *Solidago racemosa* Greene of Indiana authors, and *Solidago*

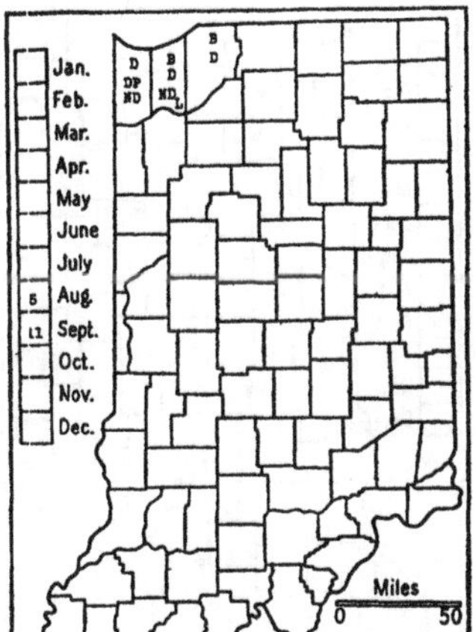

MAP 2010
Solidago racemosa Greene var. Gillmani (Gray) Fern.

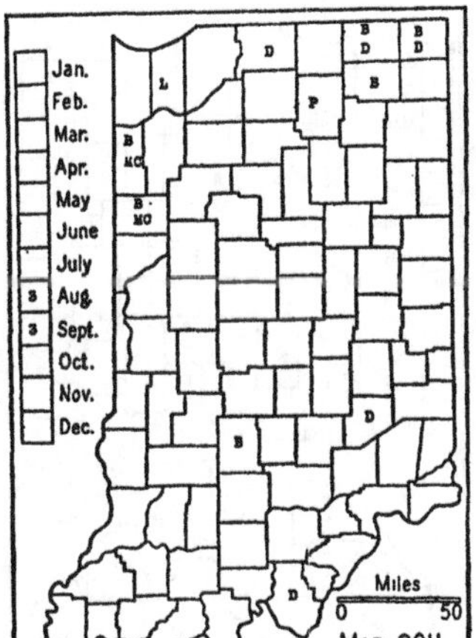

MAP 2011
Solidago canadensis L.

MAP 2012
Solidago canadensis var. gilvocanescens Rydb.

Gillmani (Gray) Steele of Indiana authors.) GILLMAN GOLDENROD. Map 2010. An infrequent goldenrod on open dunes bordering Lake Michigan. At present it is common on the dunes just east of the Dunes State Park, Porter County.

Authors write that the stems are glabrous to the inflorescence. My specimens are all more or less or have been, densely appressed-pubescent. The abrasive force of moving sand, however, has detached the hairs from the stems of some of my specimens so that to a casual observer they appear glabrous. Close inspection, however, will usually show many hairs on protected parts of the stem and the many hair scars prove that the plants were pubescent. The glabrate specimens match those grown in places protected from shifting sand. This species is highly variable in all parts and it is possible that the preceding species should be included in it. In 1937 I made a special effort to collect this species in large series. In so doing I found the roots of a few plants infested with aphids. In my collection of former years, I have several sheets with small heads and with many undeveloped flowers. Might it not be that at least some of the variation in these plants is nutritional and due to badly infested roots? The plants found this year that were infested were normal but the aerial effect of aphids on cultivated asters is well known.

Dunes and rocks on the borders of the Great Lakes.

10. **Solidago canadénsis** L. CANADA GOLDENROD. Map 2011. My Steuben County specimen is the only one I have that I regard as typical. The remainder are atypical forms that are nearer the typical form than the variety. My Steuben County specimen is from the mucky border of a lake and the remainder are from dry slopes.

Newf. to N. Dak., southw. to Va. and Ky.

10a. **Solidago canadensis** var. **gilvocanéscens** Rydb. Map 2012. This form is found in various habitats ranging from alluvial banks, open woods,

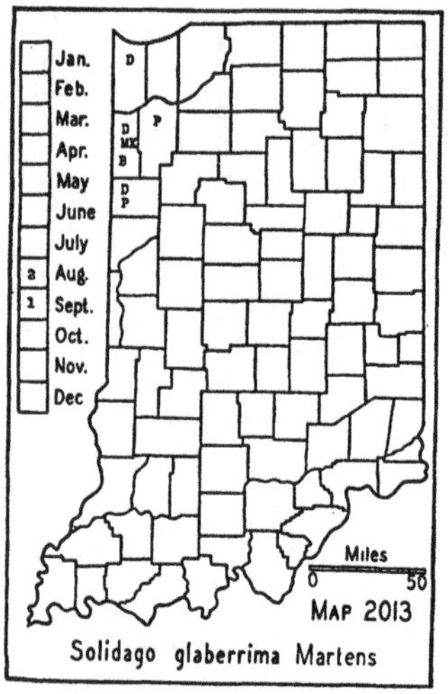

MAP 2013
Solidago glaberrima Martens

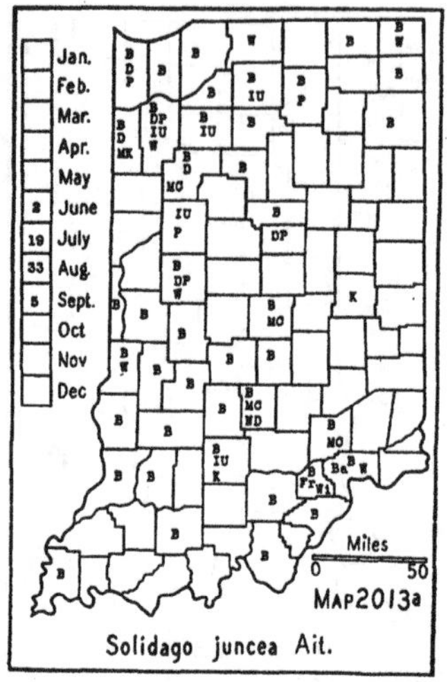

MAP 2013a
Solidago juncea Ait.

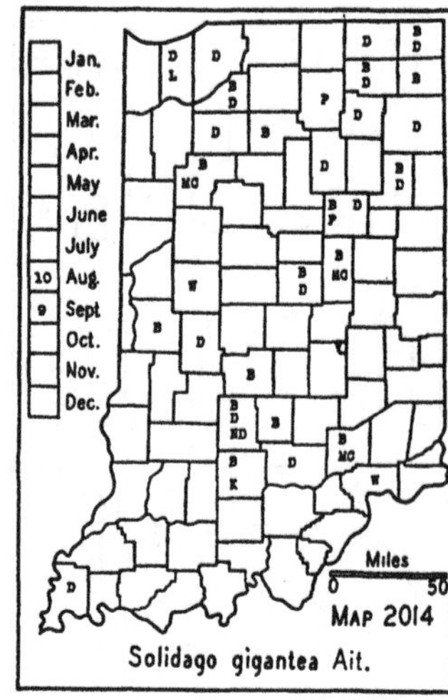

MAP 2014
Solidago gigantea Ait.

and crests of hills to roadsides. It is difficult to separate from forms of *Solidago altissima.*

Mass., N. Y. to Mich., southw. to Va., Md., and Ind.

11. **Solidago glabérrima** Martens. Map 2013. Local and restricted to the prairie area of the state. It is generally found in colonies because it suckers freely from the roots as does *Solidago juncea.*

This is a perplexing species to name because it closely resembles four other species. The original description calls for smooth plants with 3-nerved, serrulate leaves that are shining-punctate below. This species is much like small specimens of *Solidago juncea* but differs in being glabrous throughout, in having leaf margins sharply serrate; and in having the upper leaves more crowded and elongate. It differs from *Solidago missouriensis* Nutt. in having the branches of the inflorescence spreading or recurving instead of being erect. *Solidago moritura* Steele differs in that the leaves are not triple-nerved, at least the lateral nerves, if present, are short and faint.

Mich. to Man., southw. to Mo., Tex., and Ariz.

11a. **Solidago júncea** Ait. EARLY GOLDENROD. Map 2013a. This is our early goldenrod and is more or less frequent throughout the state. It is generally found in small colonies in dry soils along roadsides, railroads, and fences and on dry gravelly or clayey knolls in open woodland.

N. B. to Hudson Bay and s. Sask., southw. to S. C. and Mo.

12. **Solidago gigantèa** Ait. (Rhodora **41**: 457. 1939.) (*Solidago serotina* var. *gigantea* (Ait.) Gray.) Map 2014. Frequent to infrequent in the lake area and infrequent to local south of it. The habitat is the same as that of the variety although it is usually found in wetter places.

Newf. to Que. and Wis., southw. to S. C. and Tex.

12a. **Solidago gigantea** var. **leiophýlla** Fern. (Rhodora **41**: 457. 1939.) (*Solidago serotina* Ait.) Map 2015. Frequent in the glaciated area but less frequent south of it. It prefers a moist rich soil and is usually found in low

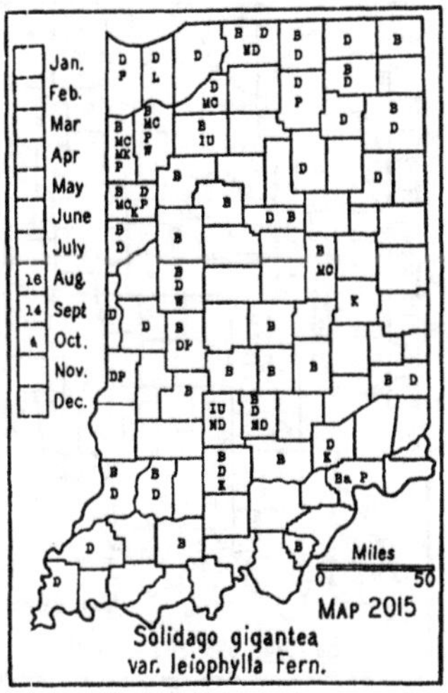

Solidago gigantea var. leiophylla Fern.

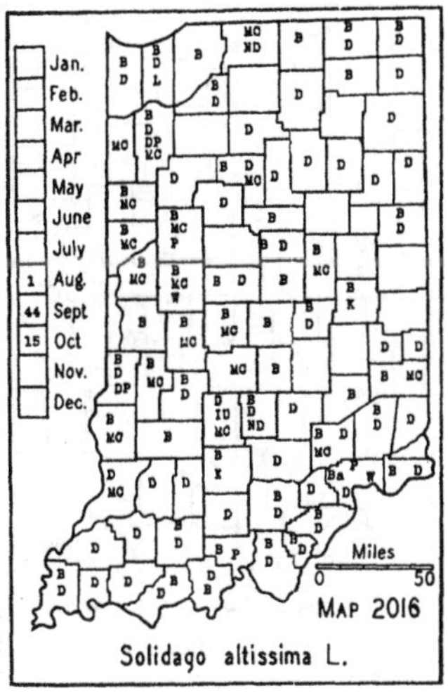

Solidago altissima L.

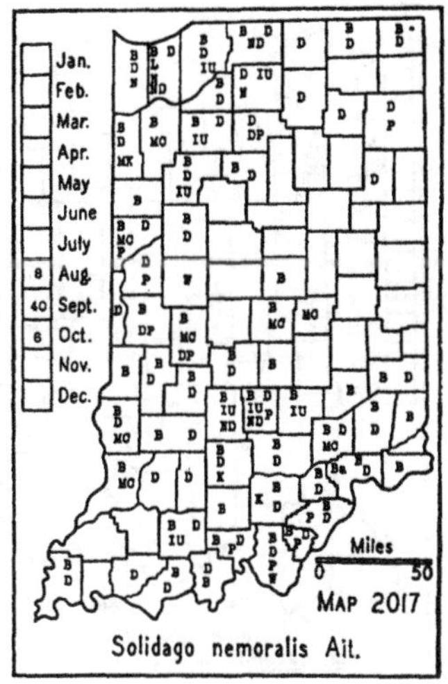

Solidago nemoralis Ait.

places about lakes and along streams. It is sometimes found in marshes and rarely in dry woods.

Newf. to B. C., southw. to Ga., Tex., and Oreg.

13. **Solidago altíssima** L. TALL GOLDENROD. Map 2016. This goldenrod is frequent to common in every county of the state. It prefers a moist rich soil but adapts itself to all kinds of soils and habitats.

Newf. to Alberta, southw. to Fla. and Tex.

14. **Solidago nemoràlis** Ait. OLD-FIELD GOLDENROD. Map 2017. Frequent to common in every county of the state. It prefers a poor, dry, clay or sandy soil and is a common weed in fallow fields. It is frequent in open woodland and along roadsides.

Newf. to Sask., southw. to Fla. and Ariz.

14a. **Solidago nemoralis** Ait. var. **decemflòra** (DC.) Fern. (Rhodora 38: 226. 1936.) (*Solidago longipetiolata* Mack. & Bush.) Map 2018. Frequent on the dunes bordering Lake Michigan and local elsewhere in the lake area in dry sandy or gravelly soil.

W. Ont. to n. Alberta, southw. to Ky., Ark., Tex., and Ariz.

15. **Solidago pátula** Muhl. ROUGHLEAF GOLDENROD. Map 2019. Infrequent in the lake area and local south of it. It is found in springy places, bogs, and marshes and rarely about ponds, hence it becomes local in southern Indiana because its preferred habitat is lacking.

Maine to Ont. and Minn., southw. to Ga., Ala., and Tex.

16. **Solidago ulmifòlia** Muhl. ELMLEAF GOLDENROD. Map 2020. Frequent in every county of the state, although the map shows no specimens from a few central counties that have not been botanized. This is a woodland species and is found in dry soil on the crests of ridges, on wooded slopes, and on the high banks of streams.

I am citing my no. 54623 as exceptional. In 1933 I found this plant in a sandy black and white oak woods on the northeastern side of Simonton

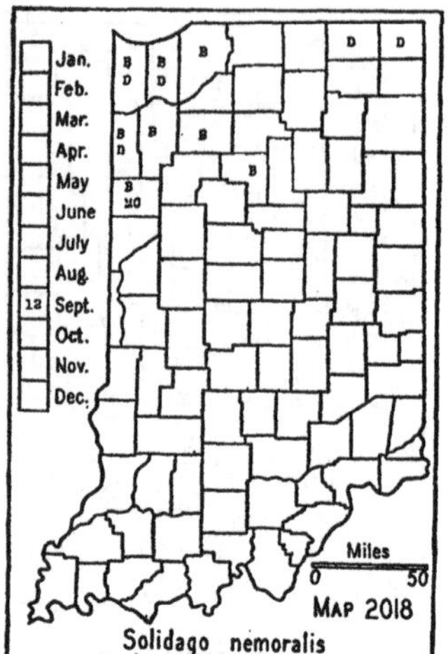

MAP 2018
Solidago nemoralis var. decemflora (DC.) Fern.

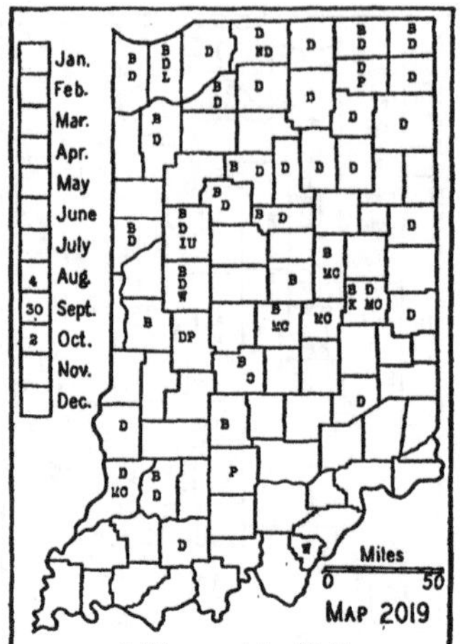

MAP 2019
Solidago patula Muhl.

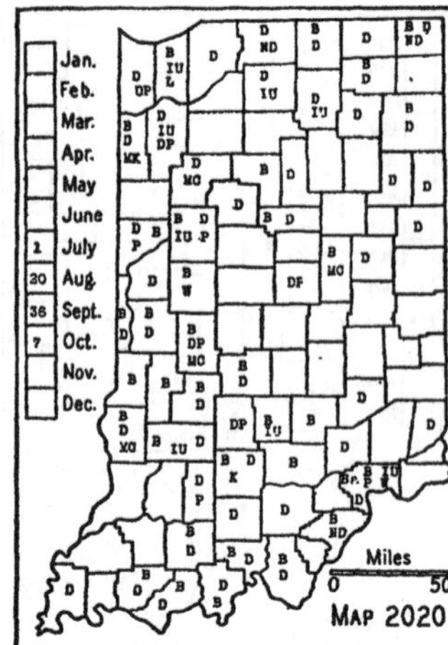

MAP 2020
Solidago ulmifolia Muhl.

Lake, Elkhart County. In 1935 I again collected it under my no. 56864. This form covered an area about 50 feet wide and 125 feet long. It was associated with a thick stand of *Solidago caesia* which covered an acre or more. The leaves of this form are on distinct short petioles, the base rounded, the teeth of the margin fewer and wide apart, the blades distinctly much longer than in the typical form. It has been suggested to me by a student of the genus as a possible hybrid of *Solidago caesia* and *Solidago ulmifolia.*

N. S. to Minn., southw. to Ga. and Tex.

17. **Solidago rugòsa** Mill. Map 2021. Infrequent in the lake area and local south of it. In the northern part of the state it is found mostly on the wet or moist borders of lakes, bogs, and marshes. In the southern part it grows in wet woodland.

It is to be noted that Indiana plants differ from that shown in plate 426 of Rhodora, 1938. The leaves of our plants are not oblanceolate but are of a lanceolate, ovate, or elliptic type and the surface is more or less rugose both above and beneath. The pubescence of the upper surface of the leaves is sparse and consists of simple, short, stout, colorless, conical hairs, arising from a papillose base and is usually more or less appressed. The pubescence of plants I have seen from New England consists of multicellular, flattened trichomes similar to those of *Solidago ulmifolia* and the surface of the leaves is not conspicuously rugose. The trichomes of the New England plants arise mostly from veinlets while ours arise mostly from the spaces enclosed by the veinlets. The blades of Indiana plants are usually thick while those of New England plants are thin.

Newf. to Ont., southw. to Va. and La.

17a. **Solidago rugosa** var. **áspera** (Ait.) Fern. (Rhodora **17**: 7. 1915.) Map 2022. This variety has a limited distribution in the state and has much the same habitats as the species but grows in slightly drier soil.

Most authors define the specific name of this species as "wrinkled." As I understand this definition, the axis of the wrinkle would be longer than

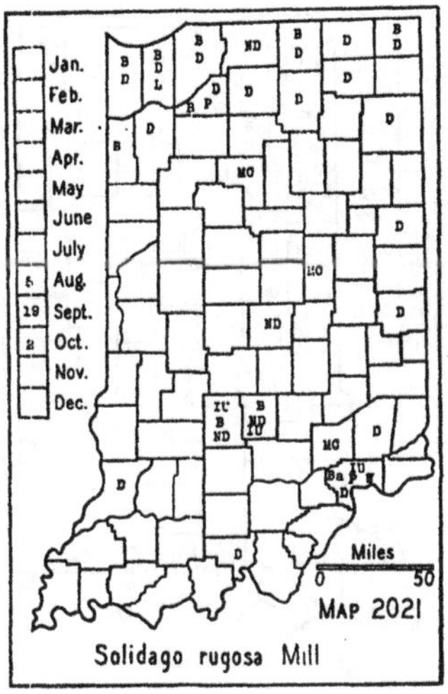

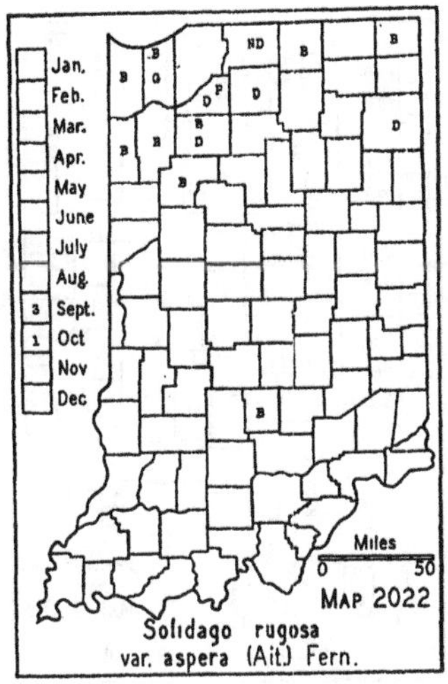

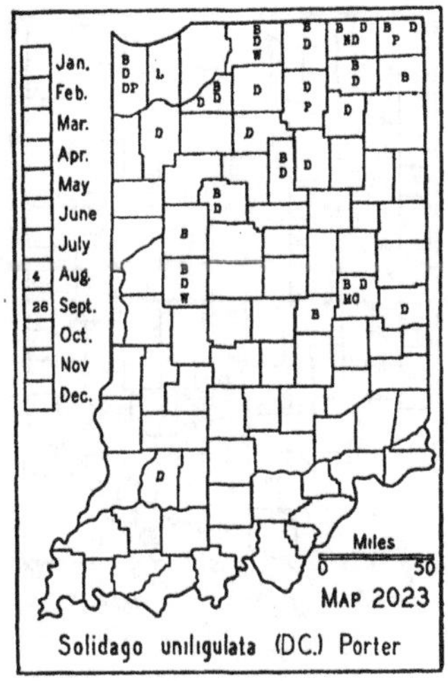

wide which does not agree with the facts. The character described is the sunken area between the veinlets and has nothing to do with the prominent lateral veins. The surfaces of the blades appear as "hammered metal" without a design.

Maine, Ohio, to Mo., southw. to Fla. and Tex.

17b. **Solidago rugosa** var. **celtidifòlia** (Small) Fern. (Rhodora 38: 223-224. 1936.) The range of this variety is given by Fernald to include Indiana. This range is based upon a specimen labeled "Valparaiso, Indiana, Sept. 17, 1927. Benke, 5096" in the Gray Herbarium. I have not seen this specimen.

Va., Ind. to Ark., southw. to Ga. and Tex.

18. **Solidago uniligulàta** (DC.) Porter. (*Solidago uniligulata* var. *neglecta* (T. & G.) Fern., *Solidago uliginosa* Nutt. of Indiana authors, and *Solidago stricta* Ait. of early Indiana authors.) Map 2023. Infrequent throughout the lake area and in a few springy places south of it. This goldenrod is strictly a bog and marsh plant. It is conspicuously variable in size, in branching of the inflorescence, and in the number of rays to a head. I have had an opportunity to study it in several places where it grew in abundance. One place was a decadent marsh on the south side of Little Long Lake, Noble County. This marsh covered about an acre and in places large colonies of *Cornus* and *Salix* were established on the border. In the center of the marsh, which was the wettest part, grew very slender plants of this species, while in the drier part on the border of the shrub zone, grew larger and branched plants. Between these two extreme habitats intermediate plants were found. I collected a large series for future study which has convinced me that the difference in the plants was a result of environment. The reason for the difference, I do not know. I have found this species in both marl and peaty habitats.

Newf. to Minn., southw. to N. C., Ohio, Ind., and Ill.

18a. **Solidago uniligulata** var. **lévipes** Fern. (Rhodora 17: 7. 1915.) I think this is merely a glabrous form of the species and is found with it.

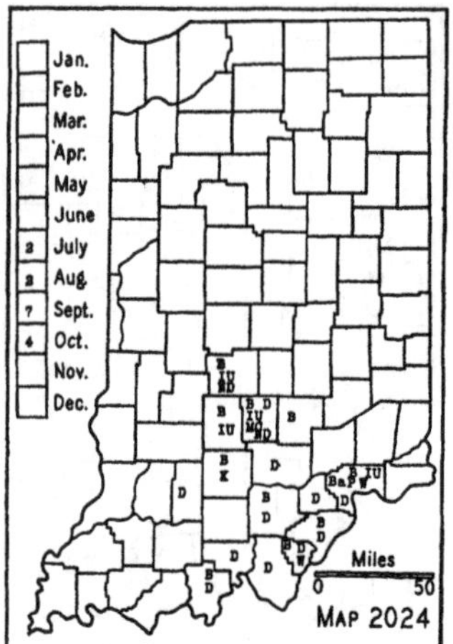

Solidago sphacelata Raf.

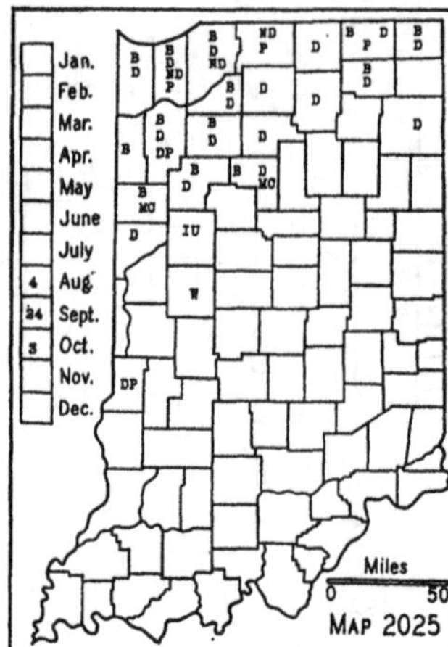

Solidago speciosa Nutt.

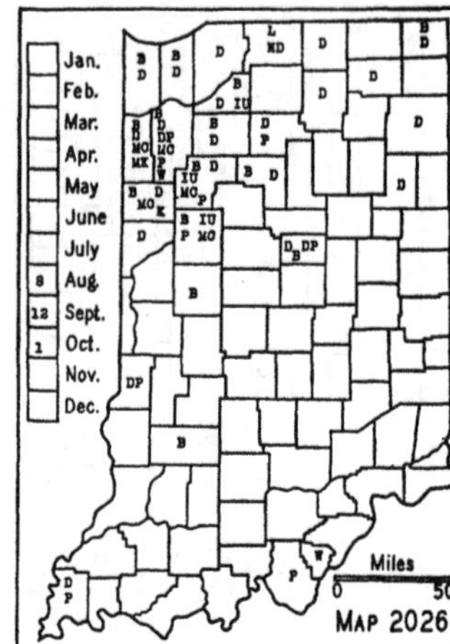

Solidago rigida L.

19. **Solidago sphacelàta** Raf. (*Brachychaeta sphacelata* (Raf.) Britt.) Map 2024. Restricted mostly to the unglaciated region where it is usually found in poor clayey soil on the crests and slopes of ridges and on the tops of high banks along streams.

I have had this goldenrod in cultivation many years and I regard it as the most beautiful of the genus in our area. In good clay loam it grows to a height of about three feet with many long spreading or recurving branches. It begins to flower about the middle of September and continues until killing frost. It self sows in exposed soil but I have never found it as an escape although no effort has been made to prevent it.

I am keeping this species in the genus *Solidago* because it is known to hybridize with *Solidago ulmifolia* and I do not like bigeneric hybrids.

Va. to s. Ind., southw. to e. Ga. and Ala.

20. **Solidago ovàta** Friesner. Our only specimens of this species were collected by Ray C. Friesner on a clayey wooded ridge west of Centerton, Morgan County and in a similar habitat in Brown County. Friesner has had this form under cultivation in the Butler University botanical garden and his study of the plant convinces him that it is a hybrid of *Solidago sphacelata* Raf. and *Solidago ulmifolia* Muhl. I quite agree with him on the status of the plant.

Known only from the type locality in Morgan County and from one collection in Brown County.

21. **Solidago speciòsa** Nutt. (*Solidago rigidiuscula* and *Solidago speciosa* var. *rigidiuscula* of Indiana authors.) Map 2025. Infrequent to frequent in the lake area and absent or local south of it. It grows only in sandy or gravelly soil and is found in open wooded dunes, open black and white oak woods, and in sandy prairies.

N. S. to Minn., southw. to N. C., Ark., and Kans.

22. **Solidago rígida** L. (*Solidago rigida* f. *magna* Clute.) STIFF GOLDENROD. Map 2026. Infrequent in prairie and decadent prairie habitats

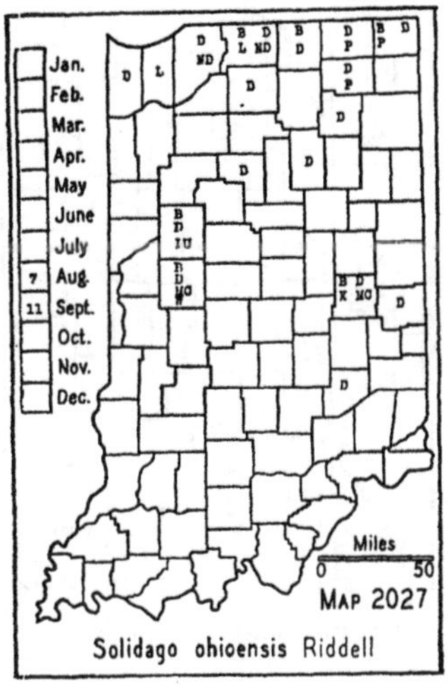

MAP 2027
Solidago ohioensis Riddell

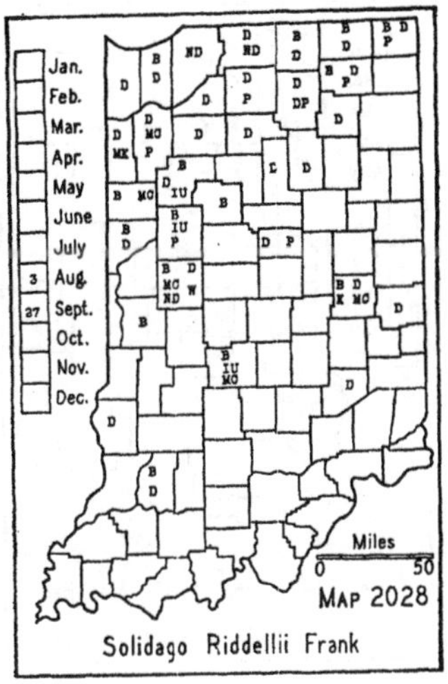

MAP 2028
Solidago Riddellii Frank

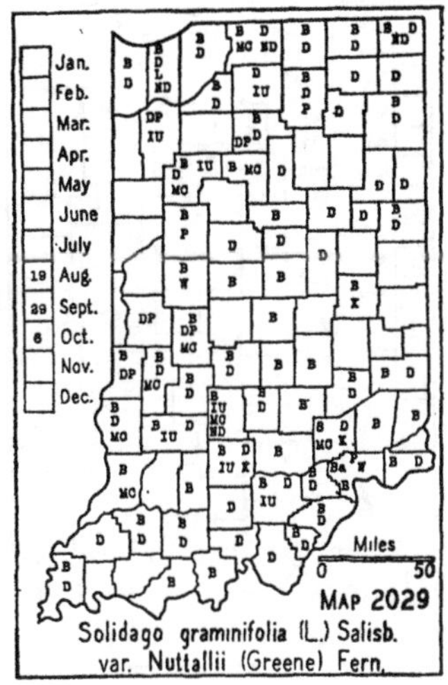

MAP 2029
Solidago graminifolia (L.) Salisb. var. Nuttallii (Greene) Fern.

in northern Indiana and very local in southern Indiana in similar habitats. Now found mostly along roads and railroads.

Mass. to Sask., southw. to Ga. and Tex.

23. **Solidago ohioénsis** Riddell. Map 2027. Infrequent in marly marshes in the lake area and very local in springy places south of it. It is usually common where it occurs. I once saw a colony of about five acres on the wide marl border of the south side of Lake Pleasant which is located just south of the Michigan State line in Steuben County. Only the most tolerant calciphiles were associated with it such as *Triglochin maritima, Eleocharis pauciflora* var. *Fernaldii, Juncus brachycephalus,* and *Lobelia Kalmii.* This species is always indicative of a limy soil and if the soil is not too alkaline *Lobelia Kalmii* and *Parnassia glauca* will be found with it.

N. Y., Ont. to Wis., southw. to Ohio and Ill.

24. **Solidago Riddéllii** Frank. RIDDELL GOLDENROD. Map 2028. Infrequent in the lake area and local south of it where its habitat occurs. It is found in springy and marshy places that are somewhat alkaline. It is often found closely associated with *Solidago ohioensis,* which flowers about 10 days earlier, but in a wetter habitat. This zonal distribution is often quite conspicuous. It is to be noted that where this species grows in numbers the plants vary greatly in size, doubtless due to some habitat factor.

Ont. to Minn., southw. to Ohio, Ill., and Mo.

25. **Solidago graminifòlia** (L.) Salisb. var. **Nuttállii** (Greene) Fern. (*Solidago graminifolia* of early Indiana authors, *Solidago hirtella* (Greene) Bush, and *Euthamia hirtella* Greene.) Map 2029. Infrequent to frequent in every county of the state. It prefers a moist rich soil but adapts itself to almost all kinds of soils and habitats. It is usually found in large colonies where its spread is not limited. Frequent along roadsides and railroads, in open places in alluvial soil along streams, in open woodland, and in fallow fields.

Mass., Ont. to Minn., southw. to N. J. and Tenn.

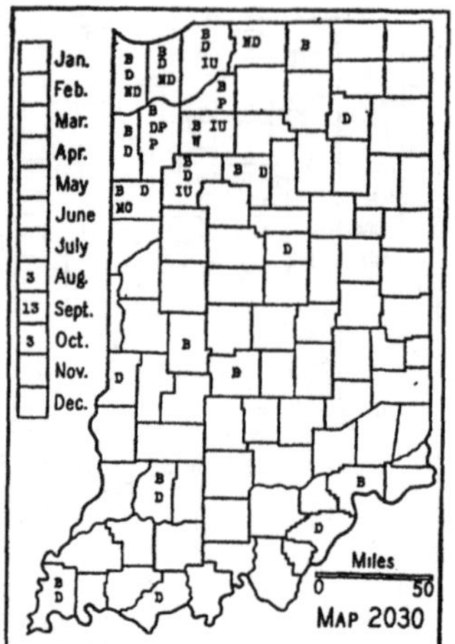

Solidago media (Greene) Bush

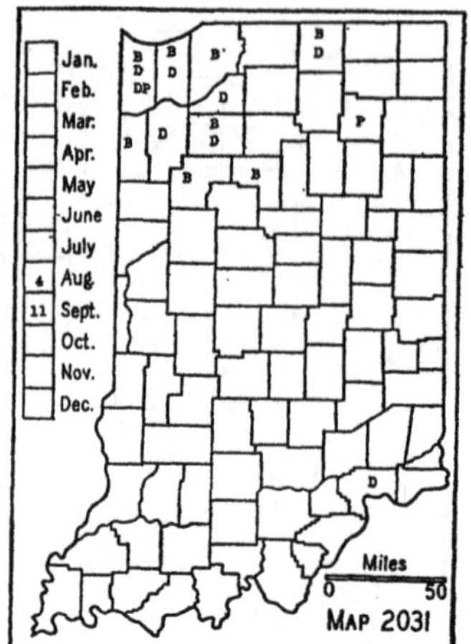

Solidago remota (Greene) Friesner

Boltonia asteroides (L.) L'Her.

26. **Solidago mèdia** (Greene) Bush. Map 2030. This species prefers the moist soil of prairie habitats and is found also about lakes and in the southern part of the state in flat woods in a slightly acid soil.

Ind. to Minn., southw. to Mo.

27. **Solidago remòta** (Greene) Friesner. Map 2031. This goldenrod also prefers the moist soil of prairie habitats but is found also in dry sand and in wet woods. Restricted mostly to northwestern Indiana.

This and the preceding species are closely allied and in the extremes are difficult to separate and both species may be considered only as varieties of *Solidago graminifolia.* When the literature is considered it is apparent that authors are far from unanimous concerning the status of the species of the section *Euthamia* of the genus *Solidago.* I have made no field study of the group and my conclusions have been drawn from the literature and from my specimens.

Ind. and Wis.

8892. BOLTÒNIA L'Hér.

1. **Boltonia asteroìdes** (L.) L'Hér. WHITE BOLTONIA. **Map 2032.** Infrequent in moist soil in prairie habitats along roadsides and streams and about lakes, ponds, and sloughs. Rare or absent from the eastern part of the state.

Conn. to S. Dak., southw. to Fla. and La.

8900. ÁSTER [Tourn.] L. ASTER

[Burgess. Species and variations of Biotian Asters. Mem. Torrey Bot. Club **13**: 1-419. 1906. Wiegand. Aster lateriflorus and some of its relatives. Rhodora **30**: 161-179. 1928 and Aster paniculatus and some of its relatives. Rhodora **35**: 16-38. 1933.]

A. Basal and lower leaves or some of them, cordate or subcordate and slender-petioled, mostly of an ovate-cordate type and long-petioled; upper cauline blades essentially similar, but with shorter petioles or even sessile.

Upper stem leaves not cordate-clasping; bracts glabrous or with ciliate margins except nos. 1 and 2.

Peduncles and branches of the inflorescence more or less glandular........ ....1. *A. macrophyllus.*

Peduncles and branches of the inflorescence not glandular.

Rays white.

Bracts linear, all acute or acuminate, mostly less than 0.6 mm wide; peduncles usually short and many-bracted.................... ....6a. *A. sagittifolius* var. *urophyllus.*

Bracts various (the lower mostly ovate, obtuse or acute, the middle ones oblong, obtuse, and the inner ones linear and obtuse or acute), more than 0.6 mm wide; peduncles generally naked.

Leaves slightly scabrous above, glabrate beneath; involucres mostly 6-7 mm long; lower bracts ovate, obtuse or subacute, the middle ones mostly narrow-oblong, generally 1-1.3 mm wide, obtuse, the inner ones linear, obtuse or acute. (See excluded species no. 627, p. 1097)................................A. *divaricatus.*

Leaves very scabrous above, rather densely hispidulous beneath; involucres mostly 7-8 mm long; bracts generally obtuse and wider than the preceding, the middle ones generally 1.4-1.8 mm wide.... ....2. *A. furcatus.*

Rays blue or violet.

Plants usually not leafy below, internodes of the stem long; upper leaves linear and subulate-pointed; leaves of branches linear and closely appressed, subulate-pointed; lower cauline leaves subcordate or narrowed at the base, thick, their margins entire or shallow-serrate, scabrous above and beneath; inflorescence paniculate, the branches widely spreading; heads usually secund and terminating long scaly peduncles; bracts with short-acute, green tips.......... ....3. *A. azureus.*

Plants not agreeing with the preceding description.

Leaves entire (sometimes some leaves with a part of the margin serrate), thick, firm, ovate-lanceolate or lanceolate, glabrous or slightly scabrous above (rarely very rough), more or less short-pubescent beneath, sometimes glabrous; petioles not margined at the base; bracts linear, short-acute................4. *A. Shortii.*

Leaves more or less sharply serrate, thin or firm.

Bracts short-acute, mostly with colored tips; leaves thin; lowest stem leaves of an ovate to broadly ovate type, generally strongly cordate and sharply serrate, the petioles usually not margined..................................5. *A. cordifolius.*

Bracts long taper-pointed, rarely some of them with colored tips; leaves firm; lower stem leaves usually not deeply cordate, mostly of an ovate-oblong type, usually less serrate and teeth not so long, generally all or at least the upper ones with margined petioles.

Stems of plants essentially glabrous or pubescent in lines........ ....6. *A. sagittifolius.*

Stems of plants usually densely short-pubescent............... ....7. *A. Drummondii.*

Upper stem leaves and those of the branches more or less sessile and cordate-clasping; lower stem leaves usually with margined, clasping petioles; bracts pubescent, rarely glabrous......................8. *A. undulatus.*

A. Basal leaves not both cordate and petiolate; lower stem leaves relatively narrow, cordate and sessile, more or less narrowed at the base and sessile or narrowed at the base and more or less petiolate in nos. 12 and 27 and rarely a few leaves petiolate in other species.

B. Stem leaves with their bases more or less cordate and clasping.

C. Involucral bracts and sometimes the peduncles glandular.

Bracts narrowly linear, long-attenuate at the apex, mostly 6-9 mm long, more or less suffused with purple; achenes about 1.5 mm long; plants of moist soil........................9. *A. novae-angliae.*

Bracts oblong-linear, merely acute at the apex or some of the inner ones with acuminate tips, without purple color, mostly 4-6 mm long; achenes about 2 mm long; plants of very dry habitats.

Stem leaves mostly 15-25 mm wide, the basal lobes developed so that the leaves appear perfoliate.....................10. *A. patens.*

Stem leaves mostly 5-10 mm wide, their bases merely clasping.

Pubescence of stem and branches dense, widely spreading...11. *A. oblongifolius.*

Pubescence of stem and branches not dense, upwardly appressed ....11a. *A. oblongifolius* var. *rigidulus.*

C. Involucral bracts and peduncles not glandular.

D. Stems entirely glabrous or sometimes pubescent in lines on the upper part or in the inflorescence.

E. Leaves more or less serrate.

Blades abruptly narrowed below the middle so as to form a broad-margined, entire petiole......12. *A. prenanthoides.*

Blades not as above.

Leaves gradually narrowed to a narrow base, rarely clasping; heads mostly 15-20 mm wide; involucre 4.5-5.5 mm long. ....18a. *A. paniculatus* var. *simplex.*

Leaves gradually narrowed to a wide clasping base; heads large, mostly more than 20 mm wide; involucre 6 mm long or more........................15. *A. lucidulus.*

E. Leaves entire or some with a few short teeth near the middle.

Tips of the bracts squarrose or recurved-spreading. (See excluded species no. 631, p. 1098)............*A. novi-belgii.*

Tips of the bracts not squarrose or spreading.

Plants glaucous, glabrous or with a few lines of hairs on the upper parts; leaves thick, very smooth, entirely glabrous except the scabrous margins; the middle bracts short-acute, with indurated tips, the green area short-rhomboidal................................13. *A. laevis.*

Plants not as above.

Bracts of nearly equal length, mostly 6-8 mm long.

Leaves thin, usually less than 18 mm wide; branches generally longer than the subtending leaves; heads few or solitary at the ends of the branches; bracts in 1 or rarely 2 rows; lower part of stem generally 2-3 mm in diameter..............14. *A. longifolius.*

Leaves thick, generally 15-30 mm wide; branches mostly shorter than the subtending leaves; heads usually several and somewhat in clusters towards the ends of the branches; bracts in 2 rows; lower part of stem more than 4 mm in diameter.......15. *A. lucidulus.*

Bracts imbricated in 3-5 rows, of several lengths.

Veinlets of under surface of blades conspicuous, the areas enclosed by them about as long as wide; tall plants with reddish brown stems; branches and branchlets very leafy; leaves a yellowish green, thick, glossy, slightly revolute with strongly involute, indurated tips, covered above more or less with short, stout, forward-pointing hairs, especially near the margins and at the apical end; flowers in dense, racemose clusters toward the ends of the branches; rays light lavender.

Leaves of the branches linear or linear-lanceolate or narrowly elliptic-lanceolate, very acute.

Leaves of stem and branches lanceolate to narrowly elliptic-lanceolate, those of the primary branches 6-10 times as long as broad....16. *A. praealtus.*

Leaves of the stem and branches linear or nearly so, those of the primary branches about 11 times as long as broad..16a. *A. praealtus* var. *angustior.*

Leaves of the branches, at least the ultimate, broadly elliptic-lanceolate or oval, often obtuse......... ....16b. *A. praealtus* var. *subasper.*

Veinlets of the under surface of the blades not conspicuous, the areas enclosed by them longer than broad; stem and branches not very leafy; leaves dark green, thinner than those of the preceding, usually flat and not involute at the tip, not glossy, more or less pubescent above but the hairs not as stout as those in the preceding species; heads usually not clustered, mostly white, rarely colored.

Involucre 5-7 mm high, hemispheric; inflorescence subcorymbose, not crowded; heads large, spread of rays 15-25 mm, rays commonly 30 or more; lobes of disk flowers short, about 25% of the total length of the limb; leaves always linear; plants usually of a marsh habitat........17. *A. junceus.*

Involucre 3-5.5 mm high, turbinate; inflorescence paniculate, heads numerous, of medium size or smaller, spread of rays 10-20 mm; rays usually less than 30; lobes of disk flowers moderately deep, 40%-50% of the total length of the limb; leaves linear to lanceolate; plants of moist or dry habitats.

Heads of medium size, spread of rays 12-20 mm; involucre (4) 4.5-5.5 mm high; rays 6-11 mm long.

Leaves linear, 12 times as long as broad or longer. ....18. *A. paniculatus.*

Leaves lanceolate, less than 12 times as long as broad......18a. *A. paniculatus* var. *simplex.*

Heads smaller, spread of rays 10-16 mm; involucre 3-4 (4.5) mm high; rays 4.4-8.5 mm long...... ....19. *A. interior.*

D. Stems pubescent more or less over the entire surface, not in lines.

Leaves glabrous above and beneath, margins scabrous, narrow, mostly 2-3.5 mm wide and 2-4 cm long; flowers usually few; involucre about 7 mm high; bracts thick, mostly obtuse or merely acute; rays violet..................20. ***A. linariifolius.***

Leaves pubescent above and beneath (sometimes some of the leaves more or less glabrous in no. 23), otherwise not as above.

Involucral bracts about 5 mm long, imbricated in about 3 or 4 rows, linear, with long-acuminate points, more or less tinged with purple. (See excluded species no. 625, p. 1097) ....A. *amethystinus.*

Involucral bracts not as above:

Involucres mostly 6-9 mm long, their bracts essentially of the same length, in 2 loosely imbricated rows; plants of wet places with stems 5 mm or more in diameter near the base.

Branches usually much exceeding the subtending leaves.......................................21. *A. puniceus.*

Branches shorter than the subtending leaves.

Leaves elongate-lanceolate, hispid on the midrib beneath. ....21a. *A. puniceus* var. *demissus.*

Leaves subrhomboidal, smooth or sparingly hispidulous beneath..............21b. *A. puniceus* var. *compactus.*

Involucres not as above; plants of a dry habitat; stems less 5 mm in diameter, near the base.

Rays violet purple; median stem leaves more than 12 mm wide, their basal lobes usually developed so that the leaves appear perfoliate; heads large, 20 mm wide or more, solitary or a few together at the ends of long branches, rarely racemose...............10. *A. patens.*

Rays white; median stem leaves mostly less than 5 mm wide; heads small, 6-8 mm wide.

Pubescence of stem dense and spreading..22. *A. exiguus.*

Pubescence of stem not dense, upwardly appressed...23. *A. ericoides.*

B. Stem leaves sessile or sometimes the lower on very short petioles (petiolate in no. 27), not at all clasping.

a. Leaves more or less pubescent over the entire under surface.

Blades silky-pubescent above and beneath................24. *A. sericeus.*

Blades not silky-pubescent above and beneath.

b. Involucral bracts (at least the outer ones) and leaves of the branchlets with mucronate tips.

Rays blue or violet.

Plants glabrous or nearly so; bracts with recurving tips. (See excluded species no. 631, p. 1098)..............*A. novi-belgii.*

Plants pubescent.

Bracts glandular.

Pubescence of stem and branches spreading...11. *A. oblongifolius.*

Pubescence of stem and branches not dense, upwardly appressed. ....11a. *A oblongifolius* var. *rigidulus.*

Bracts not glandular, linear and long-acuminate, more or less tinged with purple. (See excluded species no. 625, p. 1097. ....*A. amethystinus.*

Rays white.

Stems more or less densely pubescent.

Bracts (at least the lower ones) with recurved tips, stout, hispid and hispid-ciliate or only hispid-ciliate; heads small, densely clustered.

Pubescence of stems dense and spreading; bracts, at least the outer ones, hispid on the back..............22. *A. exiguus.*

Pubescence of stem not dense, upwardly appressed; bracts generally glabrous on the back...............23. *A. ericoides.*

Bracts appressed, not stout; heads larger than in the preceding, usually not in clusters.

Bracts generally more than 7 mm long. (See excluded species no. 633, p. 1098)........................*A. polyphyllus.*

Bracts mostly less than 7 mm long.
Leaves linear to linear-lanceolate; large plants with long, wide-spreading branches................25. *A. pilosus.*
Leaves lanceolate to oblong-lanceolate; inflorescence not as large and spreading....25a. *A. pilosus* var. *platyphyllus.*
Stems glabrous or pubescent in lines.
Plants large and bushy............25b. *A. pilosus* var. *demotus.*
Plants simple, small, usually about 3-6 dm high, branches short, heads few. (A northern form, see excluded species no. 632, p. 1098)..........................*A. pilosus* var. *Pringlei.*
b. Involucral bracts without mucronate tips.
Bracts with green tips and midribs; leaves lanceolate to ovate-lanceolate, serrate in the middle; inflorescence paniculate, heads mostly racemose; corolla tube campanulate.
Inner bracts of the involucre 3.2-4.6 mm long; lobes of the disk-corollas (0.8) 1-1.2 mm long; heads racemose on long, spreading branches..............................26. *A. missouriensis.*
Inner bracts 2.8-3 mm long; lobes of the disk-corollas 0.7-1 mm long; plant more strict, with a more abundant small, ascending rameal leaves and smaller, more densely racemose heads...... ....26a. *A. missouriensis* var. *thyrsoides.*
Bracts without green tips or the midrib only somewhat green; leaves lanceolate to elliptic, much larger, entire; inflorescence composed mostly of compound corymbs, generally flat-topped; corolla tube funnel-shaped..............................27. *A. umbellatus.*
a. Leaves glabrous beneath or pubescent only on the midrib.
Plants with a flat-topped inflorescence, composed generally of compound corymbs; leaves mostly 1-3 cm wide, 5-12 cm long, margins entire, ultimate areolae conspicuous, very small, usually less than 0.5 mm in diameter; pappus-bristles in 2 series, the outer very short........ ....27. *A. umbellatus.*
Plants not as above.
Plants branched at the top, with stiff, linear leaves, mostly 2-3.5 mm wide, 2-4 cm long; flowers few; involucres about 7 mm high, in several series, thick, mostly obtuse or merely acute; rays violet. ....20. *A. linariifolius.*
Plants not as above.
Involucral bracts subequal, mostly 8-10 mm long, in 1 or 2 rows.
Annual; pappus much longer than the disk flowers. (See excluded species no. 626, p. 1097).......................*A. angustus.*
Perennial; pappus about as long as the disk flowers; plant glabrous or nearly so; leaves lanceolate to linear-lanceolate............ ....14. *A. longifolius.*
Involucral bracts not subequal, less than 8 mm long, usually in 3 or 4 series.
Plants with a white, flat-topped inflorescence; involucre less than 5 mm high, the bracts fleshy and closely appressed; leaves linear-lanceolate or linear, with 3 longitudinal veins usually visible; plants of the dunes about Lake Michigan...28. *A. ptarmicoides.*
Plants not as above.
Heads in more or less 1-sided racemes.
Plants with (9) 11-12 (14) rays; leaves lanceolate to elliptic-lanceolate or oval-lanceolate; heads mostly 7-10 mm wide, usually on short branchlets 1-10 mm long, the branchlets mostly shorter than the subtending leaves; involucre 4-5.5 mm long; corolla of disk flowers goblet-shaped, its lobes 1-1.6 mm long; lobes 50%-75% of the total length of the limb.

Leaves lanceolate to broadly lanceolate, less than 8.3 times as long as wide....................29. *A. lateriflorus.*

Leaves linear to linear-lanceolate, more than 8.3 times as long as wide.......29a. *A. lateriflorus* var. *angustifolius.*

Plants not as above.

Involucres 4-5.5 mm long; heads on long, ascending branchlets, 12-15 mm wide (including the rays); branchlets longer than the subtending leaves, usually 1-2 cm long or up to 4 cm long or longer; lobes of disk flowers 0.4-0.8 mm long, 21%-36% of the total length of the limb; limb funnel-shaped; leaves of branchlets abruptly smaller than the cauline, very small, linear, generally mucronate-pointed.

Leaves of the branches and branchlets mostly spreading or reflexed; cauline leaves 4-7 mm wide, 3-5 cm long; rays 19-26.........................30. *A. dumosus.*

Leaves of branches and branchlets mostly ascending; cauline leaves linear, 5-7.6 mm wide, 7-11 cm long; inflorescence rather small and terminal, the branches ascending; rays 13-16 (20)........................ ....30a. *A. dumosus* var. *strictior.*

Involucres 3-3.6 mm long; heads numerous, 6-10 mm wide, mostly on very short branchlets, the branchlets longer or shorter than the subtending leaves; leaves of the branchlets abruptly smaller than the cauline, linear, with indurated tips; lobes of disk flowers 0.6-0.8 mm long, 38%-41% of the total length of the limb; limb funnel-shaped; rays 15-22 (25)................31. *A. vimineus.*

Heads not in 1-sided racemes; inflorescence paniculate, heads scattered or somewhat clustered at the ends of the branches in no. 16.

Veinlets of under surface of blades conspicuous, the areas enclosed by them about as long as wide; tall plants with reddish brown stems; branches and branchlets very leafy; leaves a yellowish green, thick, glossy, slightly revolute with strongly involute, indurated tips, covered above more or less with short, stout, forward-pointing hairs, especially near the margins and at the apical end; flowers in dense racemose clusters toward the ends of the branches; rays light lavender.

Leaves of the branches linear or linear-lanceolate or narrowly elliptic-lanceolate, very acute.

Leaves of stem and branches lanceolate to narrowly elliptic-lanceolate, those of the primary branches 6-10 times as long as broad......................16. *A. praealtus.*

Leaves of the stem and branches linear or nearly so, those of the primary branches about 11 times as long as broad..............16a. *A. praealtus* var. *angustior.*

Leaves of the branches, at least the ultimate ones, broadly elliptic-lanceolate or oval, often obtuse............... ....16b. *A. praealtus* var. *subasper.*

Veinlets of the under surface of the blades not conspicuous, the areas enclosed by them longer than broad; stem and branches not as leafy as the preceding; leaves dark green, thinner than the preceding, usually flat, and not involute at the tip, not glossy, more or less pubescent above, but the

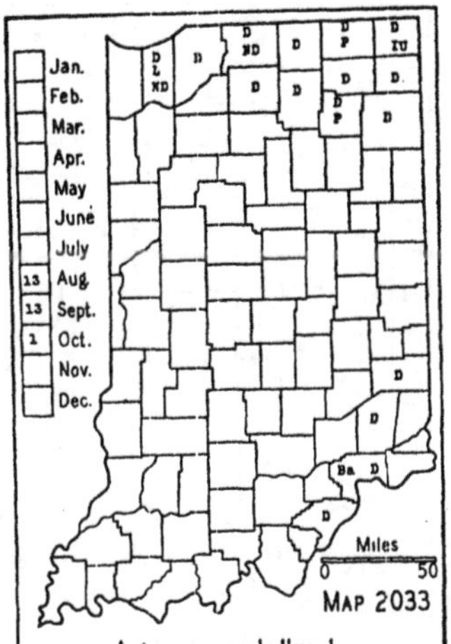

Aster macrophyllus L.

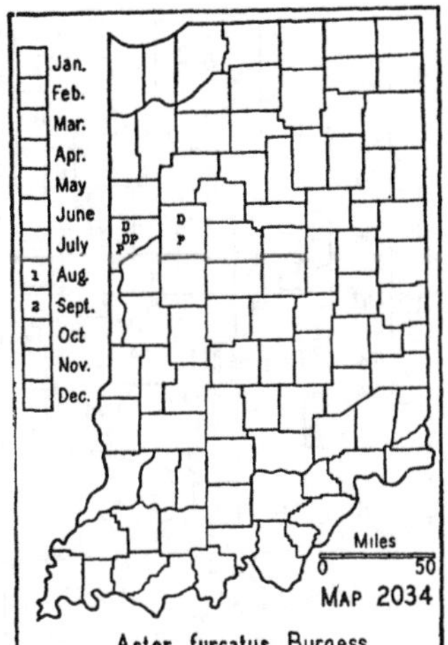

Aster furcatus Burgess

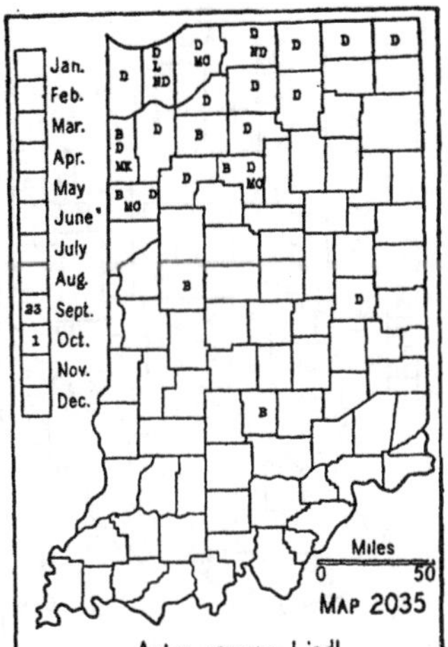

Aster azureus Lindl.

hairs not as stout as those in the preceding species; heads usually not clustered, mostly white, rarely colored.

Involucre 5-7 mm high, hemispheric; inflorescence subcorymbose, not crowded; heads large, spread of rays 15-25 mm, rays commonly 30 or more; lobes of disk flowers short, about 25% of the total length of the limb; leaves always linear; plants usually of a marsh habitat. .17. *A. junceus.*

Involucre 3-5.5 mm high, turbinate; inflorescence paniculate, heads numerous, of medium size or smaller, spread of rays 10-20 mm; rays usually fewer than 30; lobes of disk flowers moderately deep, 40%-50% of the total length of the limb; leaves linear to lanceolate; plants of moist or dry habitats.

Heads of medium size, spread of rays 12-20 mm; involucre (4) 4.5-5.5 mm high; rays 6-11 mm long.

Leaves linear, 12 times as long as broad or longer...... ....18. *A. paniculatus.*

Leaves lanceolate, less than 12 times as long as broad. ....18a. *A. paniculatus* var. *simplex.*

Heads smaller, spread of rays 10-16 mm; involucre 3-4 (4.5) mm high; rays 4.4-8.5 mm long................ ....19. *A interior.*

1. **Aster macrophýllus** L. BIGLEAF ASTER. Map 2033. In our northern counties, colonies of this aster are infrequently found in sandy or gravelly soil on black and white oak slopes and in the dune area, at the base of such slopes. In the southern part of the state, I have found large colonies in three counties on black and white oak slopes. It is difficult to explain why it has not been found in other of our southern counties since its habitat apparently exists in many of them.

The extreme variability of this species has given rise to the publication of several varieties, three of which have been reported from Indiana. I have studied my specimens rather carefully and have had the species under cultivation for years. I prefer to regard it as a polymorphic species. N. B. to Minn. and N. C.

The following three varieties have been reported from Indiana:

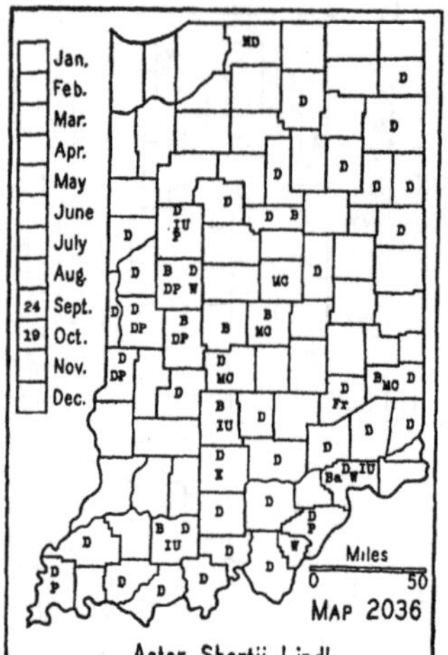

Aster Shortii Lindl.

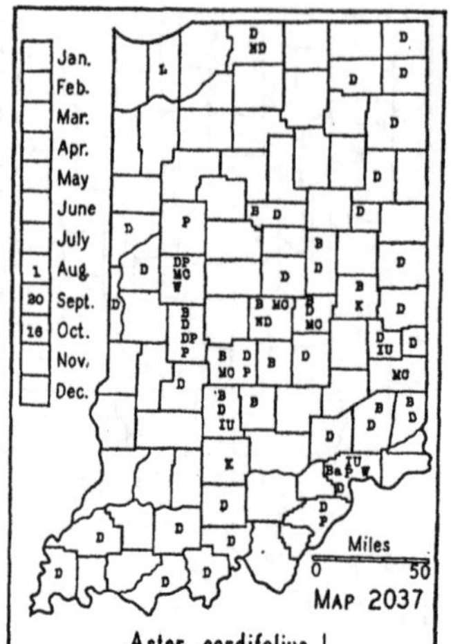

Aster cordifolius L.

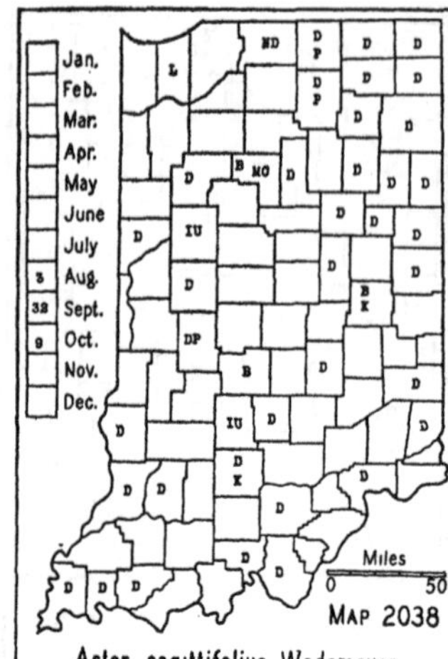

Aster sagittifolius Wedemeyer

1à. **Aster macrophyllus** var. **iánthinus** (Burgess) Fern. This variety is described as having thin leaves and minute glands, these rarely stipitate. I reported it from Clark County.

Maine to Ont., southw. to W. Va. and Ind.

1b. **Aster macrophyllus** var. **pinguifòlius** Burgess. This variety is described as having many of the basal leaves very smooth (almost greasy). This form was reported from the dune area. I have a few specimens that have this character.

Maine to N. Y. and westw.

1c. **Aster macrophyllus** var. **velùtinus** Burgess. This variety is described as having villous-pubescent stems and leaves pilose beneath, all but the lowest truncate or tapering at the base. This form was reported from the dune area.

Throughout the range.

2. **Aster furcàtus** Burgess. FORKING ASTER. Map 2034. My specimens are from a bluff along Pine Creek in Warren County and from a moist place near Wildcat Creek in Tippecanoe County. Lyon writes that his specimen from Porter County is deposited in the S. F. Blake herbarium. This species is evidently very rare in this state.

Ind. to Mo.

3. **Aster azùreus** Lindl. AZURE ASTER. Map 2035. Infrequent to frequent in sandy soil in open, black and white oak woods and in the dunes. Found rarely in moist soil and once a specimen was found in a marly marsh in Henry County. This species is easily distinguished from closely related asters by the long, linear stem leaves just below the inflorescence and the appressed, linear leaves of the branches. I believe reports for this species from southern Indiana should be referred to some other species.

Western N. Y. and Ont. to Minn., southw. to Ga. and Tex.

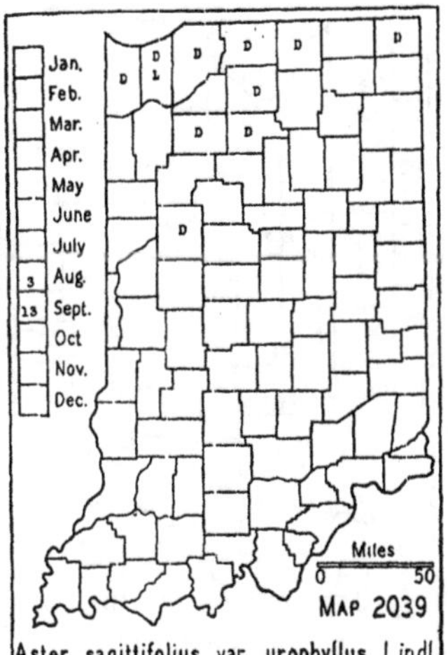

Aster sagittifolius var urophyllus Lindl.

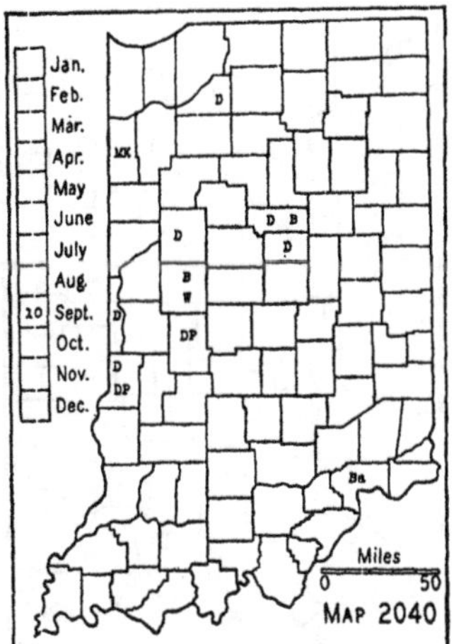

Aster Drummondii Lindl.

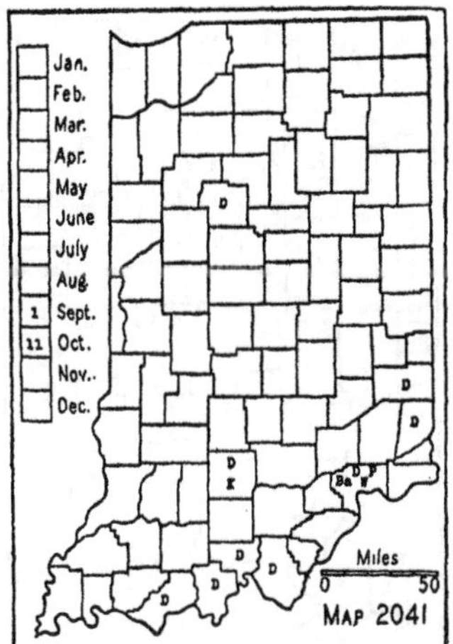

Aster undulatus L.

4. **Aster Shórtii** Lindl. (*Aster Shortii* Hook.) SHORT'S ASTER. Map 2036. Infrequent to frequent in dry woods throughout the state, although there are no specimens or records from the northern tier of counties. It is more common toward the bases of wooded slopes. Very variable in the width of the leaves and the pubescence of the under surface of the blades, which varies from a dense, short, harsh pubescence to only a few hairs on the midrib. The bracts are usually more or less densely pubescent, at least ciliate, and generally the rhomboidal, green tip is also pubescent, usually short-acute, rarely acuminate or some of the lower ones subulate. My no. 19155 from Franklin County, collected about 2 miles west of Metamora, is cited as exceptional. This plant is glabrous to the inflorescence; above that it is only slightly pubescent and then only in lines. The leaves are narrow-lanceolate and long-acuminate, entirely glabrous both above and beneath, the margins ciliate and most of them more or less shallow-serrate to about the middle; bracts very narrow, the widest 0.5-0.6 mm wide, long-acuminate, some of the lower subulate-pointed, glabrous or minutely and finely ciliate toward the apex. It seems to agree with the description of *Aster camptosorus* Small.

Pa. to Wis. and Iowa, southw. to Ga. and Tenn.

5. **Aster cordifòlius** L. BLUE WOOD ASTER. Map 2037. Infrequent to frequent throughout the state in dry woods. This is also a highly variable species and several varieties have been described. None of them have been reported, and I hesitate to report the variation in my specimens under varietal names.

N. S. and N. B. to Ont. and Minn., southw. to Ga. and Mo.

6. **Aster sagittifòlius** Wedemeyer ex Willd. ARROW ASTER. Map 2038. Infrequent to frequent in some places throughout the state except the northwestern part, where the variety takes its place. It is found mostly in dry, white oak and black and white oak woods.

N. B. to Ont., N. Dak., southw. to N. J., Ga., and Mo.

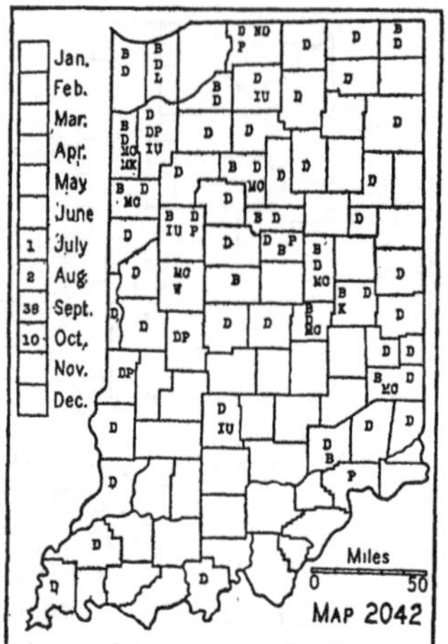

Aster novae-angliae L.

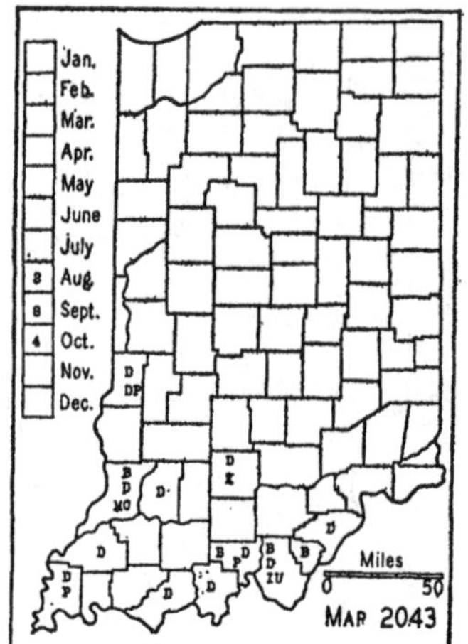

Aster patens Ait.

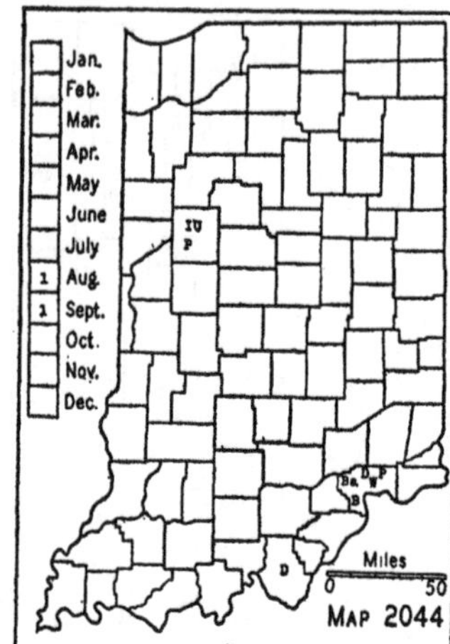

Aster oblongifolius Nutt.

6a. **Aster sagittifolius** var. **urophýllus** Lindl. WHITE ARROW ASTER. Map 2039. This variety is infrequent to frequent in very sandy soil, usually in open woodland and in the dunes. It is distinguished from the species by its white rays, closer inflorescence, and the under surface of the leaves, which is more glabrous than in the typical form. In fact, the whole plant has a more glabrous aspect.

N. Y. to Minn.

7. **Aster Drummóndii** Lindl. DRUMMOND ASTER. Map 2040. Very local in dry, open woods. Reported from the Calumet District by Peattie. This species seems to be merely a pubescent form of the preceding one but its range does not coincide with that of *A. sagittifolius.*

Ohio to Minn., southw. to Ky. and Tex.

8. **Aster undulàtus** L. WAVYLEAF ASTER. Map 2041. A rare aster found on the crests of high, open ridges and on high, wooded banks. It has been reported from five other counties besides those indicated on the map, but I have seen no specimens to verify these reports. My experience indicates that it is very local.

N. B. to Ont. and Minn., southw. to Fla. and La.

9. **Aster nòvae-ángliae** L. NEW ENGLAND ASTER. Map 2042. This is a species of moist, rich soil and is found throughout the state. It is frequent to rather common in the northern part of the state, becoming infrequent or rare in the hill area of the southern part. It is more generally found in marshy places, along moist roadsides, and in prairie habitats.

Maine to Sask., southw. to S. C., Ala., and Kans.

9a. **Aster novae-angliae** f. **ròseus** (Desf.) Britt. This is a form with rose colored rays. I have found it a few times, and I have also found a white rayed form.

10. **Aster pàtens** Ait. SPREADING ASTER. Map 2043. Local in the southwestern part of the state on the crests of open, wooded ridges, usually with black and white oak or in very sandy soil on wooded, sandy knolls,

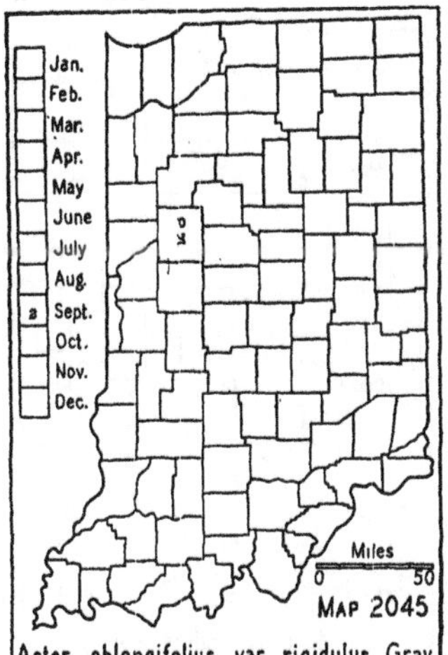

Aster oblongifolius var. rigidulus Gray

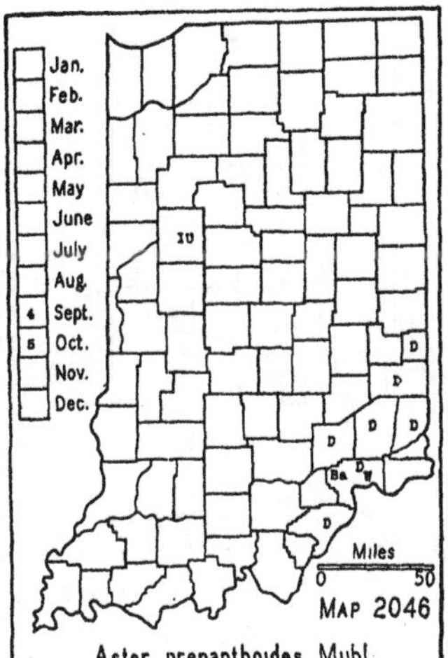

Aster prenanthoides Muhl.

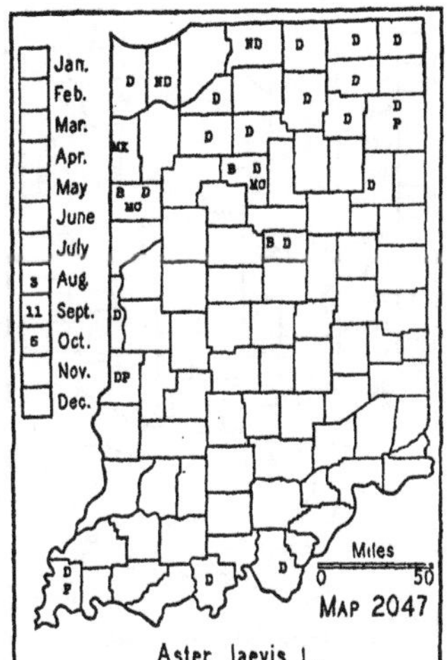

Aster laevis L.

and terraces. None of my specimens have the pedicels, small branches, or small leaves of the branches glandular. The inflorescences vary from those with the branches terminating in a single head to those with 20-25 heads. The leaves are also variable. In one specimen the leaves are narrowed at the base into a margined, clasping petiole.

Maine to Minn., southw. to Fla. and Tex.

11. **Aster oblongifòlius** Nutt. OBLONG-LEAF ASTER. Map 2044. Found on high, wooded bluffs of the Ohio River. A report from Clark County is, no doubt, correct. It has also been reported from Noble and Wayne Counties but these reports doubtless are based upon wrong determinations. The report from Tippecanoe County should be referred to the variety.

Bluffs and prairies from Pa. to Minn., N. Dak., and Colo., southw. to Va. and Tex.

11a. **Aster oblongifolius** var. **rigídulus** Gray. Map 2045. I found this variety on the high, gravelly slope of the flood plain of Big Wea Creek about 4 miles southwest of Lafayette. Associated with it were other western plants such as *Muhlenbergia cuspidata, Linum sulcatum, Lithospermum incisum,* and *Houstonia angustifolia.* No doubt the Tippecanoe County report for the species was made from a specimen collected in this vicinity and should be referred to this variety.

Ind., Wis., S. Dak. to Colo., southw. to Tex.

12. **Aster prenanthoìdes** Muhl. CROOKED-STEM ASTER. Map 2046. Infrequent on wooded flood plains and in roadside ditches in a few counties of the southeastern part of the state. It has been reported from a few of the central counties and no doubt its range will be extended in Indiana, although I believe it is a rare species in the state.

Mass. to Minn., southw. to Va., Ky., and Iowa.

13. **Aster laèvis** L. SMOOTH ASTER. Map 2047. Infrequent to rare in all parts of the state. It is generally found on white and black oak ridges and on bluffs of streams, in clayey soil or more often in very sandy soil.

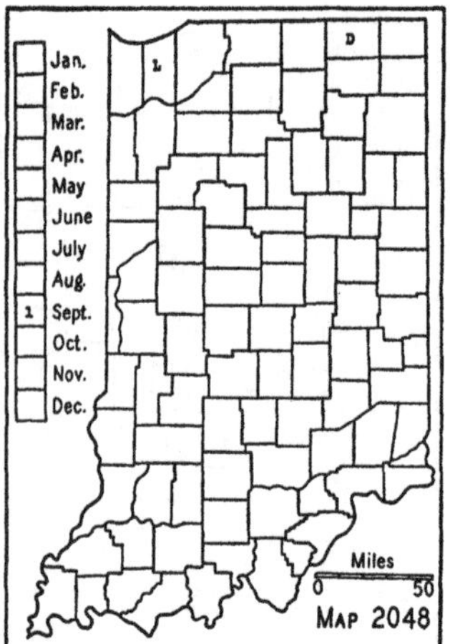

Aster longifolius Lam.

Aster lucidulus (Gray) Wieg.

Aster praealtus Poir

It is also found in prairie habitats and in Posey County I found it on the bank of a pond that usually overflows each year. The great variation of this species in the shape and width of the leaves (1-4 cm wide), and in the form of the inflorescence has resulted in the description of 9 varieties. The involucres of my specimens are usually 6-7 mm long. The upper bracts are mostly 1-1.4 mm wide, abruptly acute (rarely acuminate) and usually with slightly spreading tips. An exception is my no. 11970, collected August 4, 1912, along the railroad about a mile east of Dana in Vermillion County. In this the inflorescence is fastigiate and very leafy; the involucres are 8-10 mm long and the bracts are narrower than those of the typical form and are long-acuminate.

Maine to Sask., southw. to Va., Ala., La., Mo., and Colo.

13a. **Aster laevis** var. **falcàtus** Farw. (Rept. Michigan Acad. Sci. 21: 370. 1920.) Farwell describes this variety as follows: "Panicle usually shorter and ovate; median stem leaves usually broadest at the auriculate base, linear or oblong-lanceolate, under three fourths inch wide and often 6 inches long, some of them falcate; small subulate leaves as in the preceding variety" (var. *laevigatus*). I have this variety from the wooded sand hills about 3 miles south of Ft. Wayne, Allen County, and from an upland woods about 2 miles south of Oriole, Perry County. The specimens are deposited in the Gray Herbarium.

14. **Aster longifòlius** Lam. LONGLEAF ASTER. Map 2048. This species has been reported from Cass, Noble, and Porter Counties. The specimen collected Sept. 13, 1926, by Dr. Lyon in a subdunal marsh at Tamarack, in Porter County, and one which I collected on the low border of Cogg Lake, about 4 miles south of Lagrange, Lagrange County, are the only specimens which I have seen.

Lab. to Sask., southw. to n. N. E., Ont., Great Lake Region, and Mont.

15. **Aster lucídulus** (Gray) Wieg. (Rhodora 26: 4. 1924.) (*Aster puniceus* var. *lucidulus* Gray of Gray, Man., ed. 7.) GLOSSYLEAF ASTER. Map

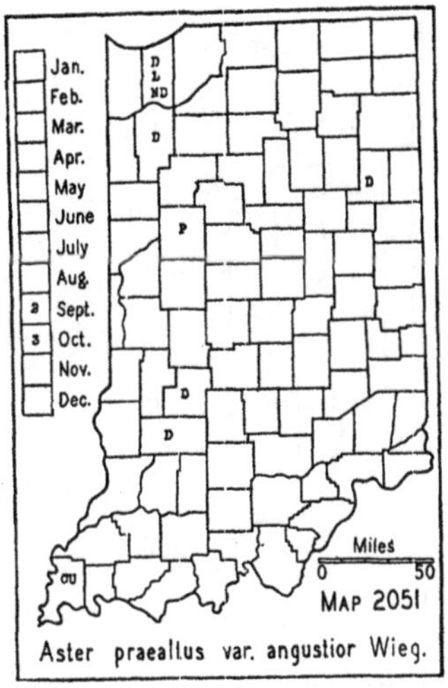

Aster praealtus var. angustior Wieg.

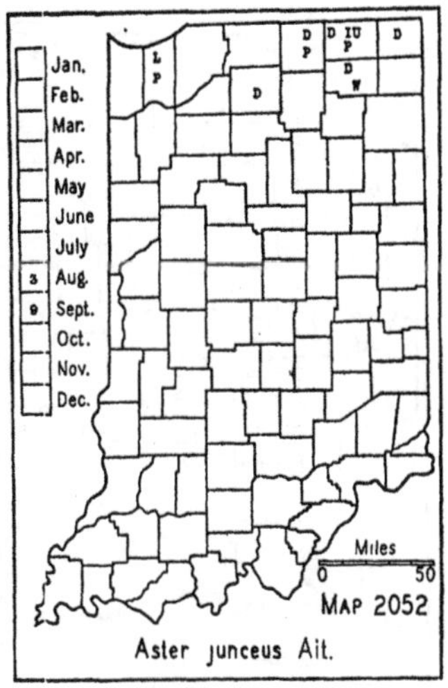

Aster junceus Ait.

Aster paniculatus Lam.

2049. This species seems to be restricted to springy and marshy places in the northern part of the state.

N. E. to Wis. and Ill.

15a. **Aster lucidulus** (Gray) Wieg. f. **fírmus** (Nees) comb. nov. (*Aster firmus* Nees, Gen. et Sp. Asterearum: 66. 1832.) This form of the preceding species has sharply serrate leaves. I have specimens from Allen, Marshall, and Steuben Counties which I refer to this form.

16. **Aster praeáltus** Poir. (Rhodora 35: 21-24. 1933.) (*Aster salicifolius* Ait. of recent authors.) Map 2050. This species is essentially an inhabitant of moist, prairie habitats. Infrequent in moist prairie habitats and less often in moist, black loam about lakes and in marshes and in low, open woods.

Ohio to Wis. (?) and Kans., southw. to Ky., Tex., and n. Mex.

16a. **Aster praealtus** var. **angústior** Wieg. (Rhodora 35: 24. 1933.) Map 2051. This variety is distinguished from the species by having narrower leaves and is found in similar habitats. It is apparently local in its distribution.

Mass., Ind., and Ill.

16b. **Aster praealtus** var. **subásper** (Lindl.) Wieg. This variety was reported by Wiegand (Rhodora 35: 25. 1933) from Indiana, as collected by Dr. Clapp, who did his collecting in the vicinity of New Albany.

Ind. and Ill., southw. to Tex.

17. **Aster júnceus** Ait. RUSH ASTER. Map 2052. Infrequent to frequent in marshes in northern Indiana. With one exception, all of my specimens have white flowers. This species is variable in the size of the heads and branching of the stem.

N. S. to B. C., southw. to N. J., Ohio, and Colo.

18. **Aster paniculàtus** Lam. (Rhodora 35: 28-32. 1933.) PANICLED ASTER. Map 2053. This aster has been reported from all parts of the state but my specimens are mostly from the northern half of the state. All

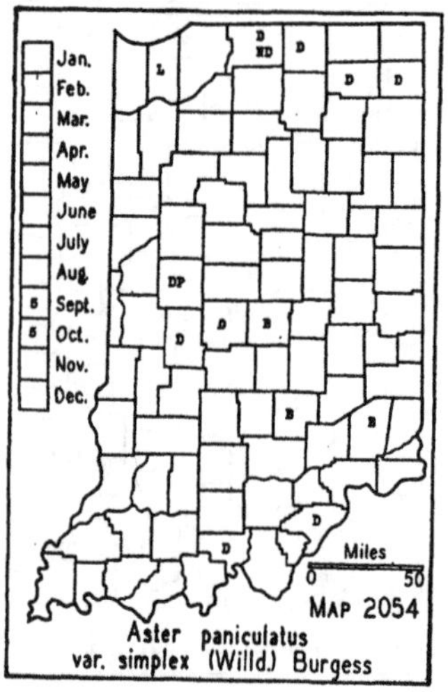

MAP 2054
Aster paniculatus var. simplex (Willd.) Burgess

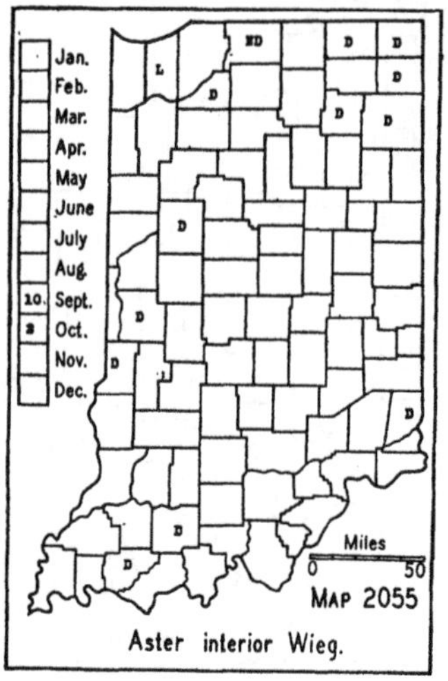

MAP 2055
Aster interior Wieg.

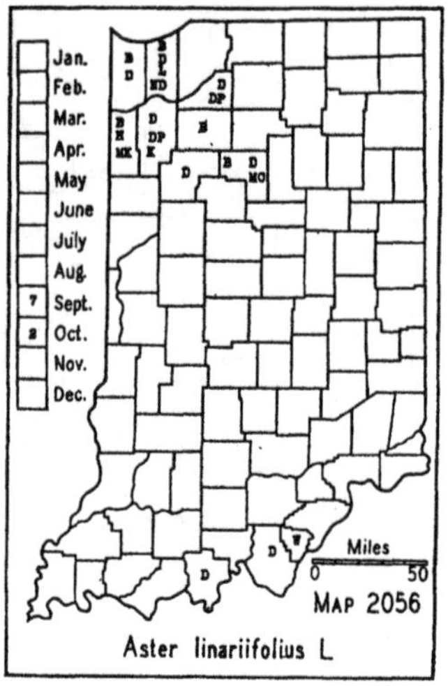

MAP 2056
Aster linariifolius L.

of my specimens have white flowers, with the exception of one from Henry County. This species prefers a moist habitat and is found in a wide range of situations, but rarely, if ever, in woodland unless it is open. Most of my specimens are from roadside ditches and marshes.

N. B., N. S., cent. Que. to Wis., southw. to N. J., e. Pa., n. Ohio, n. Ill. to Mo.

18a. **Aster paniculatus** var. **símplex** (Willd.) Burgess. (Rhodora 35: 32-34. 1933.) Map 2054. This is a more southern and western form of the species. All of my specimens are from moist places in woodland.

N. B. and Que., S. Dak., and Nebr., southw. to Va., W. Va., and Mo.

19. **Aster intèrior** Wieg. (Rhodora 35: 35-36. 1933.) (*Aster Tradescanti,* in part, of Gray, Man., ed. 7.) Map 2055. This is a species of moist woodland. It is found throughout the state, and is, no doubt, frequent to common in most parts.

N. Y. to Wis. (?) and Ill., southw. to Mo. and La.

20. **Aster linariifòlius** L. (*Ionactis linariifolius* (L.) Greene.) STIFF-LEAF ASTER. Map 2056. Infrequent in a few of the northwestern counties associated with black and white oak, on dunes, sandy ridges, and knolls. Very rare on the crests of ridges in a few of our southern counties. It has been reported also from Floyd, La Porte, Marshall, Putnam, and Vigo Counties.

Maine to Minn., southw. to Fla. and Tex.

21. **Aster puníceus** L. PURPLE-STEM ASTER. Map 2057. Frequent in the northern part of the state, becoming infrequent to very rare in the southern part. It is an inhabitant of springy places along streams and about lakes and swamps. It rarely forms large colonies and sometimes grows to great height. In Noble County, I measured a specimen that was 9 feet high.

Newf., Ont. to Man., southw. to Ga. and Tenn.

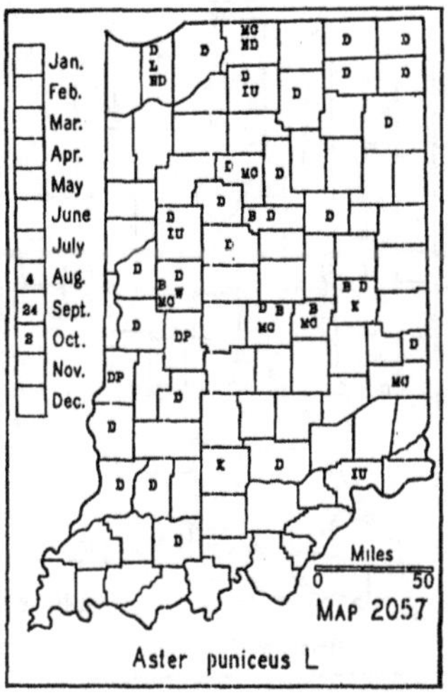

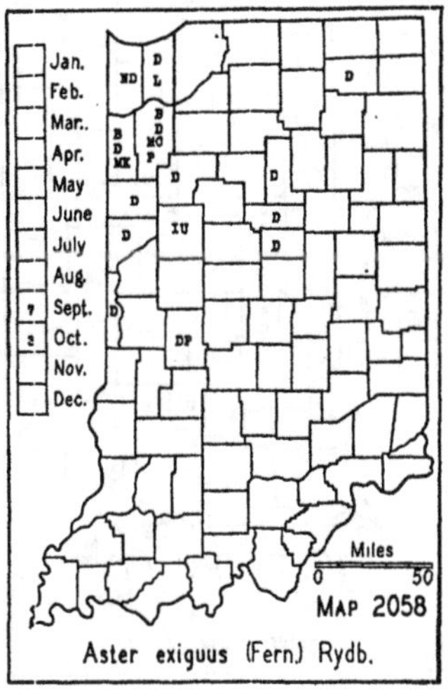

21a. **Aster puniceus** var. **demíssus** Lindl. This variety has elongate-lanceolate leaves that are usually as long as or longer than the branches. Peattie reported it from La Porte County and I have it from Grant, Lagrange, and Owen Counties. Buhl (Bull. Chicago Acad. Sci. 5: 9. 1934) was in error in reporting Peattie's collection as from Porter County. Peattie's report was from Trail Creek, Michigan City, which is in La Porte County.

21b. **Aster puniceus** var. **compáctus** Fern. This is a form with subrhomboidal leaves that are usually as long as or longer than the branches. I have it from only Parke County where I found it in the remnant of Nigger Legs Prairie about a mile east of Rosedale.

22. **Aster exíguus** (Fern.) Rydb. (Rydberg. Bull. Torrey Bot. Club 28: 505. 1901.) (*Aster multiflorus* var. *exiguus* Fern. and *Aster multiflorus* Ait., in part.) Map 2058. Infrequent along roadsides in prairie habitats. I have one specimen from a high, gravelly, wooded bank on the north side of Diamond Lake in Noble County.

Vt. to Wash., southw. to Pa., Ariz., and Mex.

23. **Aster ericoides** L. (Blake. Rhodora 32: 138. 1930.) (*Aster multiflorus* Ait., of Gray, Man., ed. 7 and *Aster multiflorus*, in part, of Britton and Brown, Illus. Flora, ed. 2.) WREATH ASTER. Map 2059. An infrequent plant in dry soil, mostly in prairie habitats, in the western part of the state.

Maine to Mont., southw. to Ga. and Mex.

24. **Aster seríceus** Vent. SILKY ASTER. Map 2060. All of our specimens and reports come from the six counties shown on the map. Found in very sandy soil on wooded slopes or low dunes. Restricted mostly to the dunes near Lake Michigan.

Ind. to Minn. and Man., southw. to Tenn. and Tex.

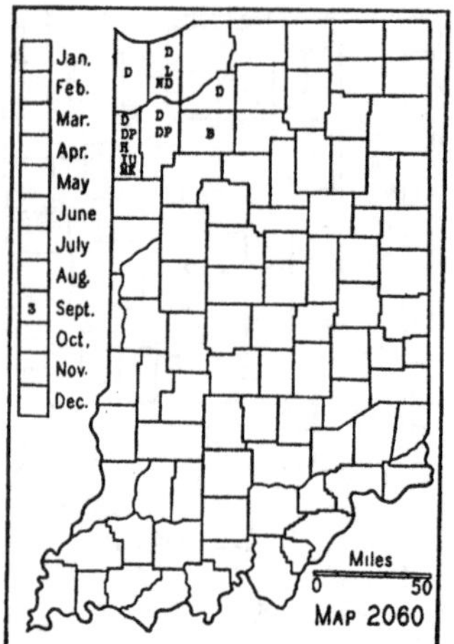

Aster sericeus Vent.

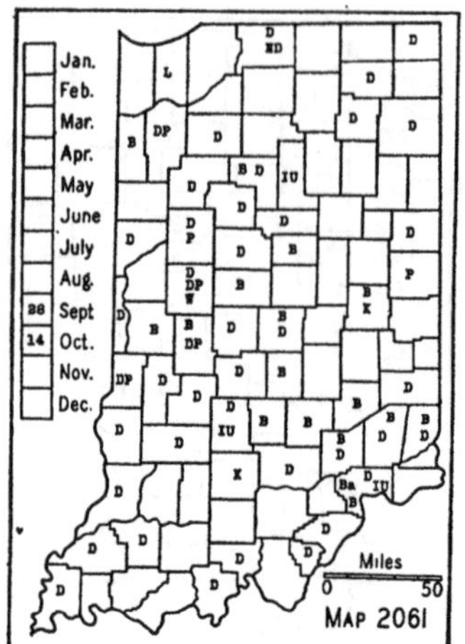

Aster pilosus Willd.

Aster missouriensis Britt.

25. **Aster pilòsus** Willd. (*Aster ericoides* var. *villosus* T. & G. and *Aster ericoides* of authors, not L.) HEATH ASTER. Map 2061. In southwestern Indiana this species is called goodbye meadow, which is a very appropriate name for it there because it soon forms dense stands in fallow fields and in meadows (hayfields). Frequent to abundant in all parts of the state in dry soil, in fallow fields, meadows, and open woodland and along roadsides. It should be regarded as an obnoxious weed because of its ability to crowd out other vegetation and because of its success in spreading widely by means of its wind-borne seed. It has a wide range of habitats, but is most at home in a clay soil. In good soil it reaches a height of over 3-4 feet, while a depauperate specimen growing in hard soil along the roadside may not be over a foot high. It is, also, rather variable. The involucres of my 40 specimens vary from 3.5-6.5 mm long. The bracts vary from 3-5 series.

Maine to Minn., southw. to Fla.

25a. **Aster pilosus** var. **platyphýllus** (T. & G.) Blake. (*Aster ericoides* var. *platyphyllus* T. & G.) This variety was described by Torrey and Gray in the Flora of North America 2: 124. 1841, and they cite a specimen from Indiana collected by Dr. Clapp, who did his collecting in the vicinity of New Albany. It was also reported by Lyon from Porter County, and Peattie duplicated Dr. Lyon's report. I have seen this specimen and it is the common form of the species. This variety is described, in part, as follows: "Cauline leaves pubescent-hirsute, lanceolate; the lower ones oblong-spatulate", and with larger heads. I have specimens from Clark and Kosciusko Counties which I refer to this variety. These have leaves which are 20-30 mm wide. A specimen from Owen County approaches this variety, and my Jennings County specimen has a leaf 18 mm wide, while those of ordinary specimens are mostly less than 8 mm wide.

Ohio to Mich. and Ill., and southw.

25b. **Aster pilosus** var. **demòtus** Blake. (*Aster ericoides* in part of Gray, Man., ed. 7.) This variety is glabrous or nearly so; otherwise it is like

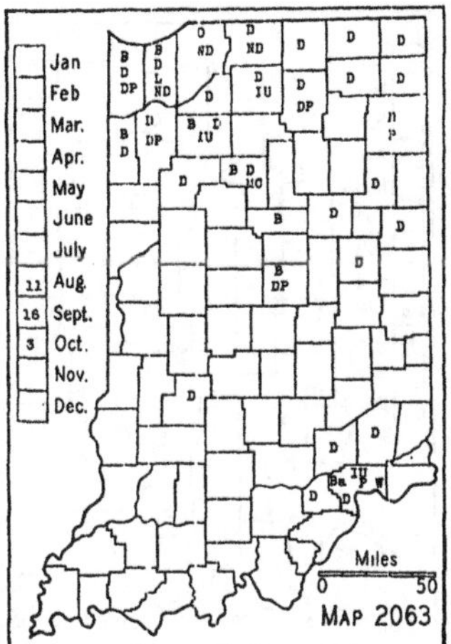

Aster umbellatus Mill

Aster ptarmicoides (Nees) T & G.

Aster lateriflorus (L.) Britt.

the species. My specimens are all sparsely pubescent in lines. I have specimens from Greene, Harrison, Jay, Posey, Spencer, and Wells Counties which I refer to this variety.

Maine to Ont., southw. to N. C. and Mo.

25c. **Aster pilosus** f. **pulchéllus** Benke. (Rhodora 34: 11. 1932.) This is a form with rose red rays which Benke reported from Porter County.

26. **Aster missouriénsis** Britton. (Rhodora 30: 177. 1928.) Map 2062. Probably infrequent in Indiana. My specimens are from low woodland bordering streams.

Mich. to S. Dak., southw. to Tenn. and Mo.

26a. **Aster missouriensis** var. **thyrsoìdes** (Gray) Wieg. (Rhodora 30: 177. 1928.) My specimen no. 26479 from the Kankakee River in Porter County is provisionally referred to this variety.

Ind. and Ill. to Tenn.

27. **Aster umbellàtus** Mill. (*Doellingeria umbellata* (Mill.) Nees.) FLAT-TOP ASTER. Map 2063. Infrequent to somewhat frequent in marshes and low places in woodland and in moist, prairie habitats throughout the lake area. It is also found in a few of our southern counties in low, flat woods.

Our specimens vary considerably in the size of the heads and in the pubescence of the under surface of the leaves. A few plants are glabrous but the greater number are more or less pubescent beneath with straggling, coarse hairs. The leaves of my La Porte County specimens are almost hirsute but none of our plants have leaves that are puberulent beneath.

Newf. to Sask., southw. to Ga. and Iowa.

28. **Aster ptarmicoìdes** (Nees) T. & G. WHITE UPLAND ASTER. Map 2064. This species grows in almost pure sand on the low dunes in Lake and Porter Counties. It was formerly frequent in the shifting sands near

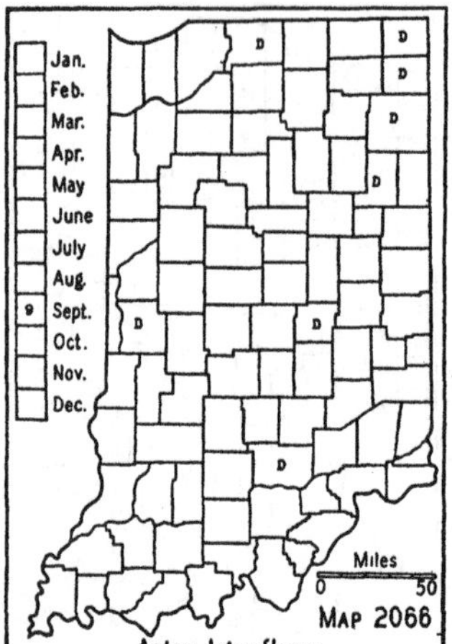

Aster lateriflorus var angustifolius Wieg.

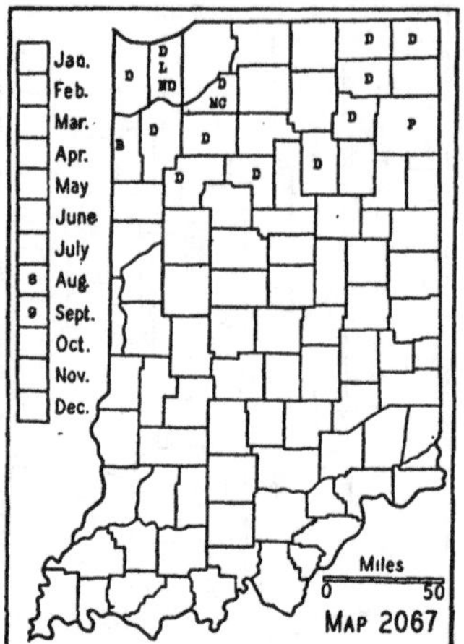

Aster dumosus L.

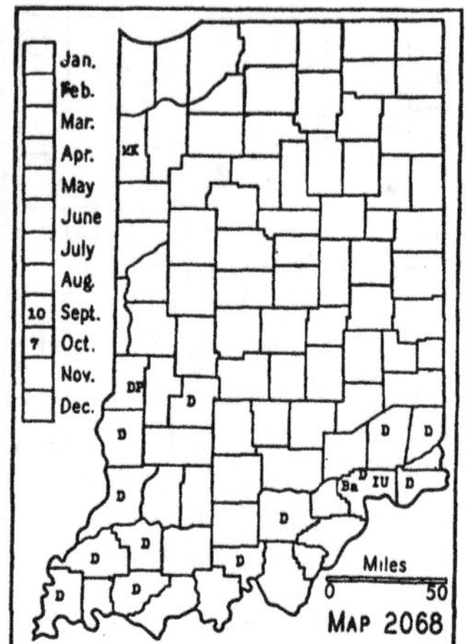

Aster vimineus Lam.

Indiana Harbor, becoming rare in Porter County. It was reported in Coulter's Catalogue on the authority of Conner & Laben, as occurring in Happy Hollow near Lafayette, in Tippecanoe County. I doubt this determination and, in the absence of a specimen, it is best to restrict its distribution in Indiana to the dune area.

Mass., Ont. to Sask., southw. to Ind., Mo., and Colo.

29. **Aster laterifl̀orus** (L.) Britt. (Rhodora 30: 172-173. 1928.) WHITE WOODLAND ASTER. Map 2065. This species is our common woodland aster. It is found in both dry and moist places, usually preferring white oak woodland.

P. E. I., N. S., Que., southw. to Conn., Pa., and Ind., and in the mts. to N. C.

29a. **Aster lateriflorus** var. **angustifòlius** Wieg. (Rhodora 30: 174. 1928.) Map 2066. This is a narrowleaf form of the species and, like it, prefers the woodland, although both of them are sometimes found in the open, mostly along roadsides.

Western N. E., Ont. to Wis., southw. to N. Y. and Ind.

30. **Aster dumòsus** L. (Rhodora 30: 165. 1928.) Map 2067. Infrequent or probably rare in the lake area of the state, where it occurs in moist, very sandy soil.

Maine and along the coast to N. J. and westw. to Ind. and probably in the mts. to N. C.

30a. **Aster dumosus** var. **stríctior** T. & G. I have this variety from Allen and Jasper Counties and it has been reported from Porter County on the authority of Umbach. My four specimens are from moist, very sandy soil.

Western N. Y. and w. Ont. to Mich. and Ill.

31. **Aster vimíneus** Lam. (Rhodora 30: 168. 1928.) SMALL WHITE ASTER. Map 2068. All of my specimens are from the southern part of the state, where they are usually found in a hard, white, moist, slightly acid,

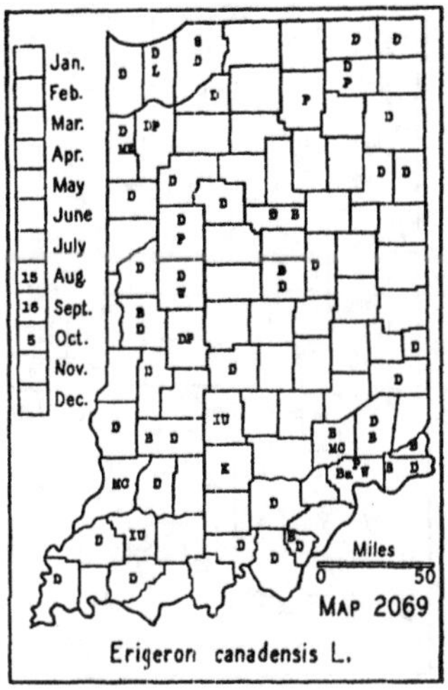

Erigeron canadensis L.

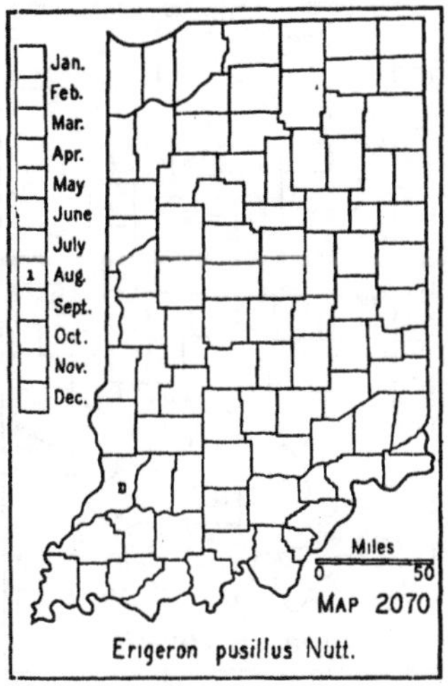

Erigeron pusillus Nutt.

Erigeron divaricatus Michx.

clay soil on the borders of ponds, in low woodland and fallow fields, and along roadsides. Most of my specimens are from low places in beech and sweet gum woods in the southeastern part and from low, post oak, shingle, and pin oak woods in the southwestern part. It has been reported from several counties of northern Indiana, but I believe many or all of the reports should be referred to some other species. I have seen the Porter County specimen and it is *Aster dumosus*.

Maine to Ind., southw. to Va.

31a. **Aster vimineus** var. **subdumòsus** Wieg. In this variety, the heads are solitary on the ends of long, slender, more uniformly bracted peduncles. The leaves of the branches and branchlets are linear and acute. The rays are somewhat more numerous, 17-30. A specimen of this variety was found in an open, low, flat woods in Daviess County, where it was associated with other southern species.

Ind., Ill., Mo., and southw. to Ala.

8901. ERÍGERON L. FLEABANE

Heads small, generally about 5 mm wide; rays scarcely exceeding the disk.
  Stems erect, mostly 3-20 dm high.
    Involucral bracts with attenuate, whitish tips; stems usually more or less densely pubescent.......................................1. *E. canadensis.*
    Involucral bracts with minute, purple tips; stems glabrous or nearly so...2. *E. pusillus.*
  Stems diffuse, mostly 1-3 dm high..............................3. *E. divaricatus.*
Heads large, generally 1.5-3.5 cm broad; rays much exceeding the disk.
  Leaves sessile and clasping; rays colored, bluish or pinkish.
    Stems simple; heads generally 2-7 (9), rarely solitary; bracts more or less glandular; rays about 50, bluish purple, mostly about 1 mm wide; corollas 4.5-5.5 mm long, about 1 mm wide..............................4. *E. pulchellus.*
    Stems branched above; heads generally 6-35, rarely more or fewer; bracts not glandular, rarely with a trace of glands; rays generally 100-150, rose purple, sometimes lighter, generally about 0.5 mm wide; corollas mostly 2.5-3 mm long, about 0.5 mm wide..............................5. *E. philadelphicus.*
  Leaves sessile, not at all clasping; rays white, rarely tinged with purple.

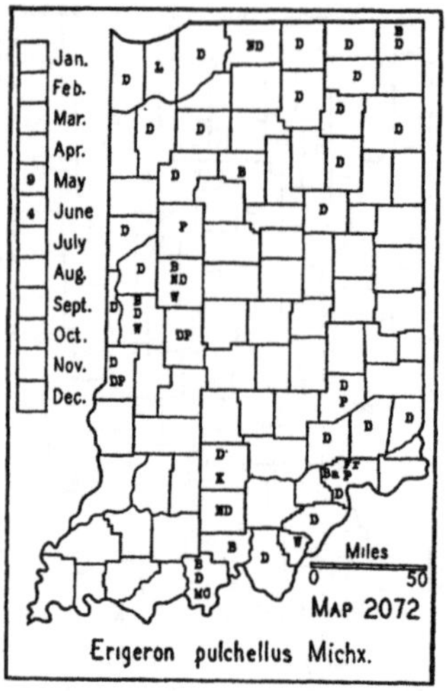

MAP 2072
Erigeron pulchellus Michx.

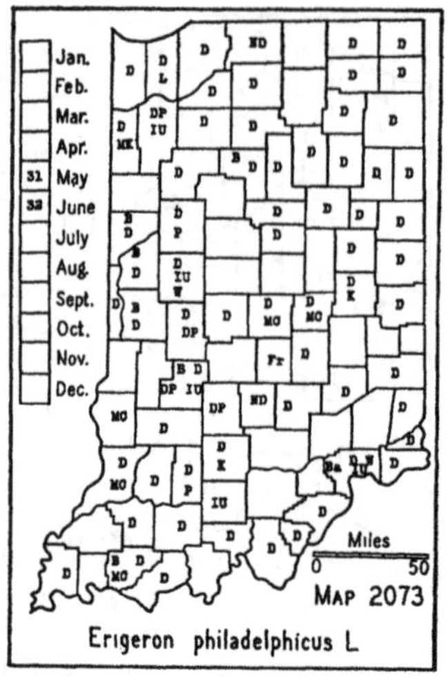

MAP 2073
Erigeron philadelphicus L.

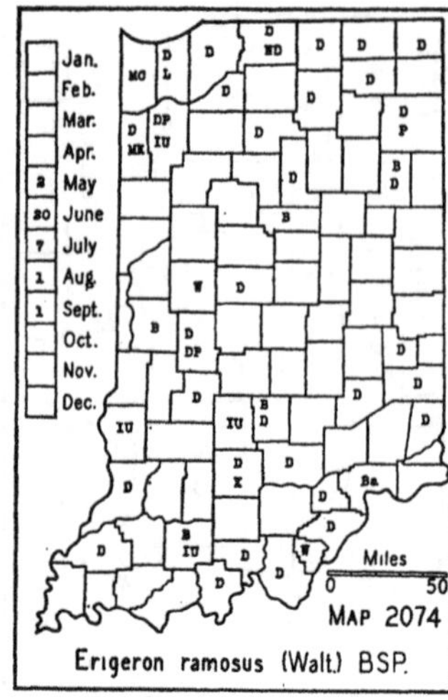

MAP 2074
Erigeron ramosus (Walt.) BSP.

Median and upper leaves entire, rarely with a few teeth, linear or narrow-lanceolate, margins not conspicuously ciliate; pubescence of the middle part of stem usually appressed, the hairs mostly about 0.5 mm long..........6. *E. ramosus.*

Median and upper leaves toothed in the middle, ovate-lanceolate to narrow-lanceolate, the margins conspicuously, coarsely hispid-ciliate; pubescence of the middle part of the stem usually sparse, the hairs usually spreading, mostly 1-2 mm long, about twice as long as those of the preceding species...........
....7. *E. annuus.*

1. **Erigeron canadénsis** L. (*Leptilon canadense* (L.) Britt.) CANADA FLEABANE. Map 2069. This plant bears several other common names, not one of which is applicable to it. It is a frequent to a common weed in cultivated grounds throughout the state. It is also infrequent to frequent in clearings and open woodland.

Throughout N. A. except in the extreme North; spread also to other countries.

2. **Erigeron pusíllus** Nutt. Map 2070. My only specimen of this fleabane was found on a dune of Princeton fine sand along the railroad about 4 miles south of Vincennes. It is to be noted that on this same dune I have found *Stylosanthes biflora* var. *hispidissima, Carya Buckleyi* var. *arkansana,* and two other hickories that I have not been able to name, which apparently belong to the south or southeast.

Mass. to Fla., and southw. to S. A., westw. to Ky., Tex., and Calif.

3. **Erigeron divaricàtus** Michx. (*Leptilon divaricatum* (Michx.) Raf.) SPREADING FLEABANE. Map 2071. All of my specimens of this species but two were found in dry soil in pasture fields. It is avoided by grazing animals and for this reason may be detected in a field at a long distance. It prefers a sandy or prairie habitat. In Indiana there are no reports east of the counties shown on the map. My opinion is that this plant has been introduced into northern Indiana within the past 25 years.

Ohio to Minn., southw. to La. and Tex.

4. **Erigeron pulchéllus** Michx. ROBIN'S PLANTAIN. Map 2072. The common name is very inappropriate since it is not a plantain. Infrequent to rare in all parts of the state in open places in woodland or in clearings,

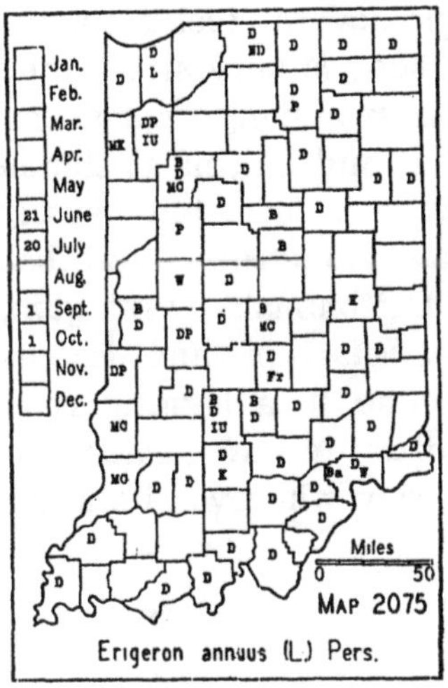

MAP 2075
Erigeron annuus (L.) Pers.

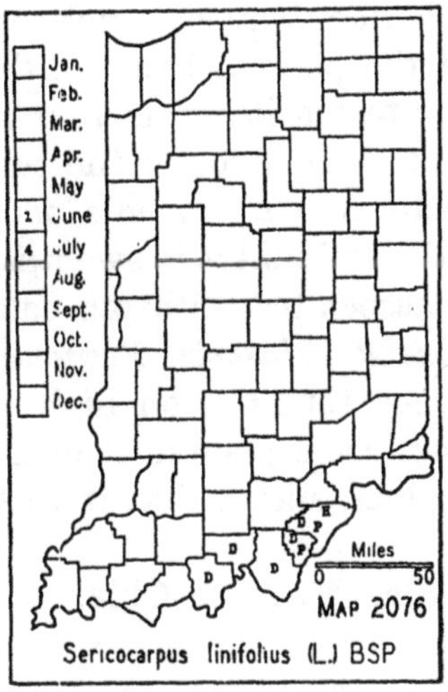

MAP 2076
Sericocarpus linifolius (L.) BSP

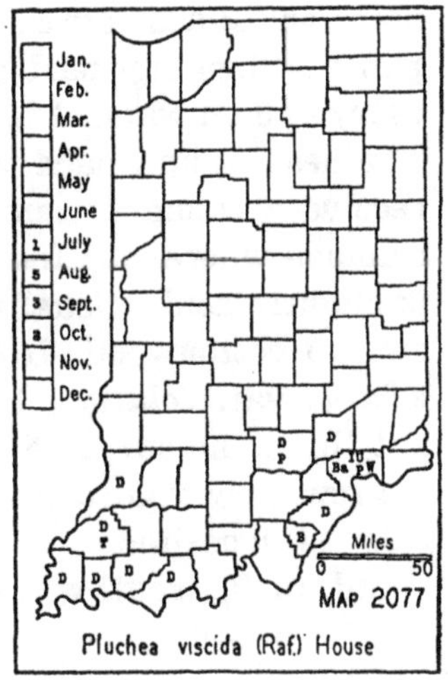

MAP 2077
Pluchea viscida (Raf.) House

more often on or near the banks of streams. It is perennial by stolons and these are often conspicuous in late summer.

S. Maine to Minn., southw. to Fla. and La.

5. **Erigeron philadélphicus** L. PHILADELPHIA FLEABANE. Map 2073. Frequent to common throughout the state in moist grounds in open woods, in creek bottoms, in open woodland pastures, in moist meadows, marshes, fallow fields, and prairies. It is rarely found in dry soil.

Lab. to B. C., southw. to Fla. and Calif.

6. **Erigeron ramòsus** (Walt.) BSP. NARROWLEAF WHITETOP. Map 2074. Frequent to infrequent or even rare in dry, open woodland, sandy and gravelly fallow fields, and dry prairies; rare to infrequent in hayfields and along roadsides; rare in moist habitats.

Newf. to B. C., southw. to Fla., Tex., and Calif.

7. **Erigeron ánnuus (L.)** Pers. WHITETOP. Map 2075. Infrequent to frequent in open woods and clearings throughout the state. Also a frequent to common weed in hayfields and waste cleared grounds and along roadsides. In some hayfields it is an obnoxious weed.

N. S. to Man., southw. to Ga., Ky., and Mo.

8904. SERICOCÁRPUS Nees

1. **Sericocarpus linifòlius** (L.) BSP. NARROWLEAF WHITE-TOP ASTER. Map 2076. Rare in barren, upland woods in a few counties shown on the map. The records from Kosciusko and Vigo Counties, no doubt, should be referred to some other species.

Maine to s. Ind., southw. to Ga. and La.

8941. PLÙCHEA Cass

1. **Pluchea víscida** (Raf.) House.* (Amer. Midland Nat. 7: 129. 1921.) (*Pluchea petiolata* Cass.) INLAND MARSH FLEABANE. Map 2077. This

* Fernald (Rhodora 41: 459-461. 1939) shows that the proper name for this plant is **Pluchea camphorata** (L.) DC. *Pluchea camphorata* of authors is **P. marilandica** (Michx.) Cassini.

plant emits a disagreeable odor which is noticeable several feet from the plant. When any part of the plant is bruised, the odor is very strong and every one on whom I have tested it agrees that it is extremely unpleasant. The nearest approach to it is the odor of the skunk, and I think it should receive a common name to suggest its vile odor. It is local but usually common where it is found. Its habitat is swamps and sloughs in a soil that is slightly acid. Usually associated with pin oak, buttonbush, sweet gum, swamp cottonwood, *Hibiscus palustris, Panicum stipitatum,* and *Juncus effusus* var. *solutus*. I once found it on high ground in a logging road but this is no surprise, because I planted it in Bluffton in our garden and it grew very vigorously which shows that it will grow wherever its seeds may be deposited.

Md. to Ill., southw. to Fla., Mo., and Okla.

8978. ANTENNÀRIA Gaertn.[1]

Rosette leaves (those of the previous year) comparatively small; blades 0.5-1.4 cm wide and 2-4.5 cm long, lower surface with only the midrib prominent; exserted portion of styles 0.5-3 (3.5) mm long.

Middle and upper stem leaves terminated by a flat or merely involute scarious appendage; rosette leaves gradually tapering to the sessile base, oblanceolate to spatulate-oblanceolate, or narrowly obovate, subacute, rarely rounded, 1-nerved beneath.

Stolons decumbent, the leaves much reduced except the apical ones which become enlarged at maturity; upper surface of rosette leaves dull, glabrous or somewhat tomentose; upper surface of leaves of stolons and stems tomentose; pistillate heads 1-8, usually crowded into a cluster, later becoming racemose; staminate inflorescence a terminal cluster of heads; pistillate corollas 4.5-5.5 mm long; achenes mostly 1.2-1.5 mm long; exserted portion of styles 1-3 (3.5) mm long; style branches mostly 0.5-0.8 mm long..............1. *A. neglecta.*

Stolons ascending, leafy throughout; upper surface of rosette leaves, stolons, and stem a bright green and generally glabrous from the first or loosely tomentose (the tomentum tardily deciduous); pistillate inflorescence of 3-18 heads in a corymb; pistillate corollas 4.5-5.5 mm long; exserted portion of style mostly 1-1.5 mm long. (To be sought in northern Indiana.)............*A. canadensis.*

Middle and upper stem leaves subulate-tipped or mucronate, without a scarious appendage (except sometimes on the bracteal leaves of the inflorescence); stolons at once ascending, leafy throughout but the terminal leaves the largest; leaves of the stolons abruptly contracted below the middle into a petiolelike base, tomentose above as are those of the stem; inflorescence generally of 5-8 heads in a terminal corymb; corollas of pistillate flowers 3.7-5 mm long; achenes 1.1-1.5 mm long; exserted portion of style usually 0.5-1.5 mm long..........2. *A. neodioica.*

Rosette leaves comparatively large; blades 1.4-2.5 cm wide and mostly 3-5 cm long, with 3-5 somewhat prominent ribs beneath (leaves of young stolons much smaller).

Inflorescence of the pistillate and staminate plants each consisting of a single terminal head; lobes of the corolla of pistillate flowers conspicuously glandular under a sixteen diameter magnification, lobes of staminate flowers generally 0.7-1 mm long (longer and larger than those of any other Indiana species)..3. *A. solitaria.*

[1]Adapted mostly from the key in the "Flora of the Cayuga Basin" by Wiegand and Eames. The measurements are those of specimens in the Deam herbarium and, no doubt, the range of measurements would be changed if a larger series had been measured.

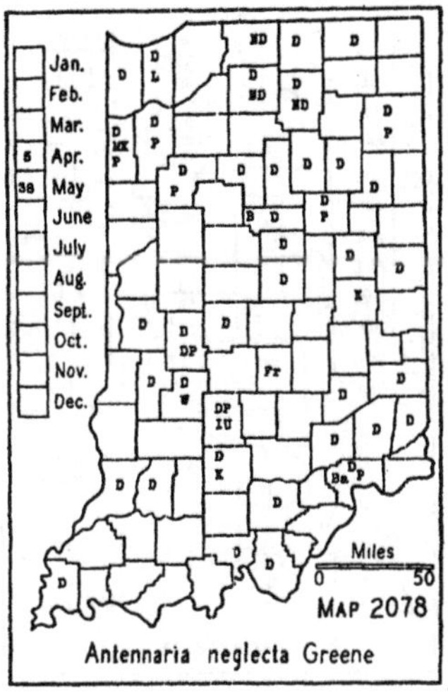

Antennaria neglecta Greene

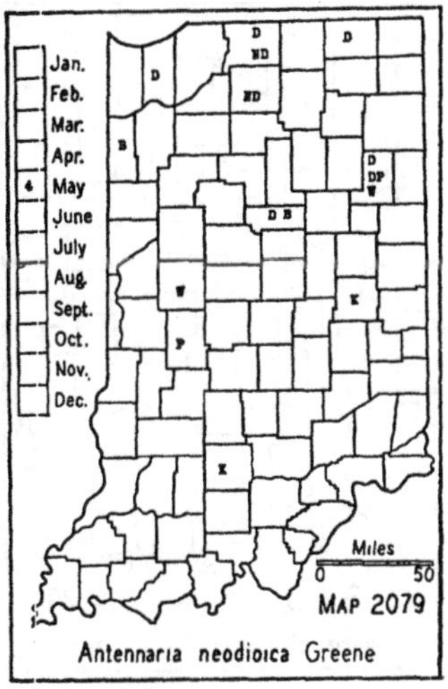

Antennaria neodioica Greene

Antennaria solitaria Rydb.

Inflorescence and corollas not as above.

Upper surface of leaves of stolons and stems bright green, glabrous or nearly so from the beginning or loosely tomentose (the tomentum tardily deciduous); stems and stem leaves and usually also the stolons (especially toward the base) generally covered more or less with purplish glands, sometimes the glands very few, restricted to a few on the stem and on only the margins of some of the stem leaves; leaves mostly rhombic-obovate, rounded or subacute at the apex; involucres of pistillate plants mostly 7-8 mm long; corollas of pistillate heads 4.5-5.8 mm long, those of staminate heads 3.5-4 mm long; achenes mostly 1.5-1.7 mm long............................4. *A. Parlinii.*

Upper surface of leaves of stolons, stem, and rosette dull, dark green, tomentose or those of the rosette weathered glabrous.

Pistillate heads small; involucres 5-7 mm long; stems slender, stem leaves distant, basal leaves obovate to roundish oblong-ovate or more rarely oblanceolate, mature corollas of pistillate heads generally 3-4.3 mm long; achenes mostly 1.2-1.5 mm long...........................5. *A. plantaginifolia.*

Pistillate heads larger, their involucres 6.5-10 mm long; mature pistillate corollas 4.5-6 mm long (-7 mm in *A. munda*); achenes mostly 1.2-1.8 mm long.

Principal rosette leaves of an ovate or elliptic type; blades broadest at or below the middle, mostly 1.5-3.5 cm wide except in *A. munda.*

Blades of the rosette leaves tapering to a subacute apex.......6. *A. fallax.*

Blades of principal rosette leaves rounded at the apex...6a. *A. fallax* var. *calophylla.*

Principal rosette leaves mostly broadest above the middle, "spatulate or narrowly spatulate-obovate, rounded at the apex; blades 2-6 cm long, 1.3-5 cm wide, 3-5-nerved, thinly canescent-tomentulose above; involucres 8-10 mm long; flower bracts in 3 or 4 series, brownish or purplish at the base, the outer obtuse, the inner acute; corollas 5.5-7 mm long; achenes 1.5-1.8 mm long; longest pappus 8-9 mm long."[1]...................7. *A. munda.*

1. **Antennaria neglécta** Greene. PUSSYTOES. Map 2078. In dry soil on open, wooded slopes, on dry knolls along roadsides, in dry pastures where it is most common, in the sandy soil of drained lake basins, and in sandy, dry prairies. Staminate plants are nearly as frequent as the pistillate ones.

[1] A free translation of the original description. (Fernald. Rhodora 38: 229-230. 1936.)

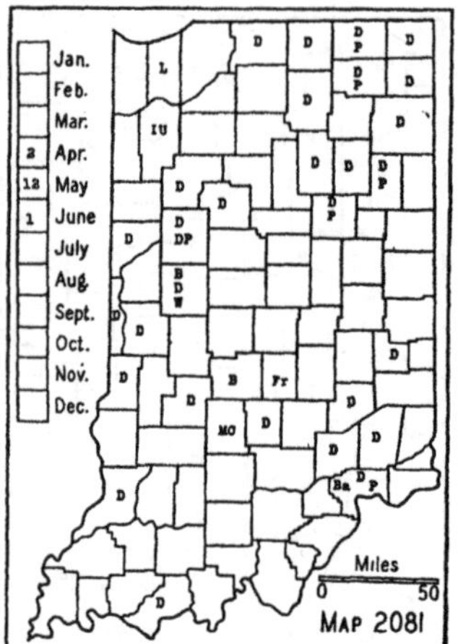

Map 2081
Antennaria Parlinii Fern.

Map 2082
Antennaria plantaginifolia (L.) Richards

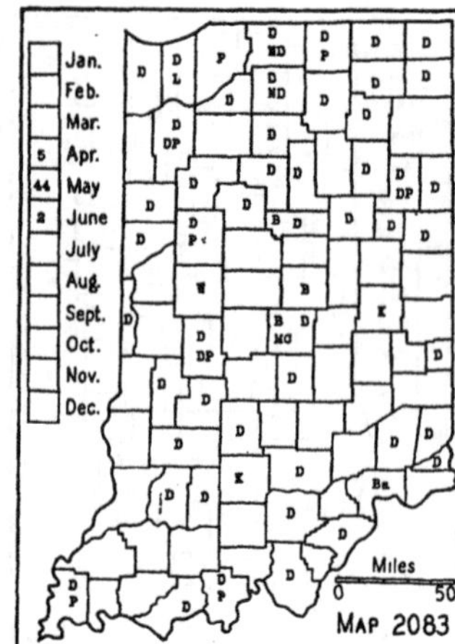

Map 2083
Antennaria fallax Greene

All of the species usually grow in poor ground where there is little competition with other plants and when once the plant becomes established, it soon forms a complete mat because of its stoloniferous habit of growth and because it is not eaten by grazing animals. It is sometimes called everlasting which is a very appropriate name for it because when it becomes established it is everlasting, and also because the leaves of the rosette remain green during the winter.

Maine to Minn., southw. to Va., Ind., Mo., and Kans.

2. **Antennaria neodioica** Greene. PUSSYTOES. Map 2079. In dry clay or sandy soil in open places in woodland and pastures and along roadsides. This species is more northern than the preceding one and is restricted mostly to the northern part of the state. I have not seen any staminate plants.

Newf. and N. S. to Wis., southw. to Va., Ont., and n. Ind.

3. **Antennaria solitària** Rybd. SINGLE-HEAD PUSSYTOES. Map 2080. Infrequent to rare on the crests or slopes of chestnut oak ridges of a few of the southern counties. Staminate plants rather rare.

Pa. to s. Ind., southw. to Ga. and La.

4. **Antennaria Párlinii** Fern. PARLIN PUSSYTOES. Map 2081. Rather frequent throughout the state in dry, clay or sandy soil in open places in woodland, on the tops of high, wooded banks of streams, on roadside knolls, and in pastures. Staminate plants as frequent as the pistillate ones.

Antennaria Parlinii var. *arnoglossa* (Greene) Fern. is a more glabrous form of the species which I reported from Vermillion County. I now include this form with the species.

N. S. to Ont. and Iowa, southw. to Va., Ohio, Ill., and in the mts. to Ga.

5. **Antennaria plantaginifòlia** (L.) Richards. PLANTAIN-LEAF PUSSYTOES. Map 2082. In dry soil on wooded slopes. I believe this species is rare in Indiana and that it is often confused with the next species.

Maine to Minn., southw. to Va., Tenn., Mo., and in the mts. to Ga.

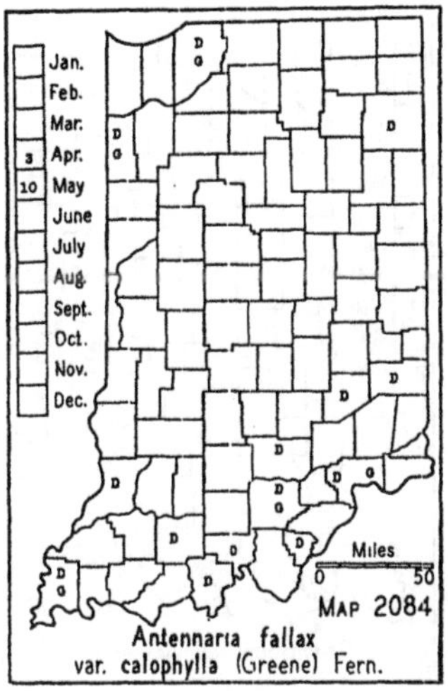

Antennaria fallax var. calophylla (Greene) Fern.

Antennaria munda Fern.

Anaphalis margaritacea (L.) Gray var revoluta Suksd. f arachnoidea Fern

6. **Antennaria fállax** Greene. PUSSYTOES. Map 2083. This species closely resembles the preceding one and was not separated from it in Britton and Brown, Illustrated Flora, edition 2. This is the most common species of the genus in the state. Frequent in all parts of the state in dry clay or sandy soil in open woodland and pastures and on roadside knolls.

Que. to Minn., southw. to Va., Ind., Miss., and Tex.

6a. **Antennaria fallax** var. **calophýlla** (Greene) Fern. (Rhodora 38: 230. 1936.) (*Antennaria calophylla* Greene. Pittonia 3: 347. 1898.) Map 2084. This variety is more frequent in the southern counties and according to Fernald "ranges from Georgia to Texas, coming north to Virginia, Indiana, Illinois, and Missouri, in the northern states passing insensibly into *A. fallax.*"

7. **Antennaria múnda** Fern. (*Antennaria occidentalis* of authors, not Greene.) (Rhodora 38: 229-230. 1936.) Map 2085. The only specimen of this species from Indiana which I have seen is the one I collected in Porter County. Fernald writes me that a duplicate of this number belongs to this species. I have not been able to make an intensive study of this and the preceding species. I have not seen a key that will definitely separate them. In this complex I am also including our reports of *Antennaria occidentalis* Greene.

Cent. Maine to Que., westw. to Thunder Bay District., Ont., southw. to Mass., Conn., Va., ne. Pa., cent. and w. N. Y., n. Ind., and Minn.

8983. ANÁPHALIS DC. EVERLASTING

1. **Anaphalis margaritàcea** (L.) Gray var. **revolùta** Suksd. f. **arachnoìdea** Fern. *(Rhodora 40: 219. 1938.) PEARLY EVERLASTING. Map 2086. This is a northern species which has been reported several times from Indiana. In nearly every instance the author has failed to report *Gnaphalium obtusifolium* which doubtless occurs in every county of the state. Without doubt all or most of our authors have confused the two plants.

* The latest name for this plant is **Anaphalis margaritacea** var. **intercedens** Hara. (Rhodora 41: 391. 1939.)

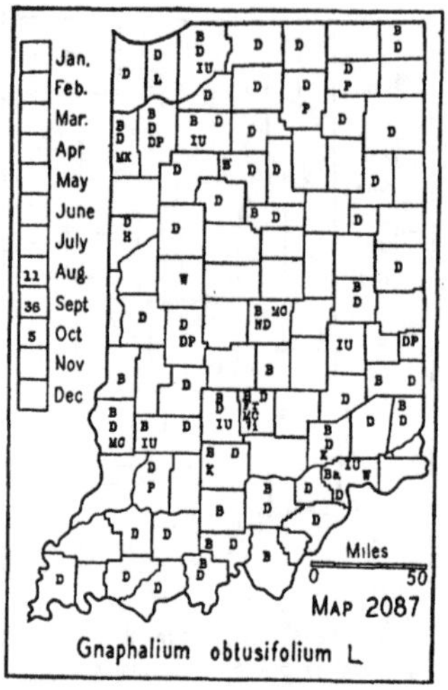

Gnaphalium obtusifolium L.

Gnaphalium Macounii Greene

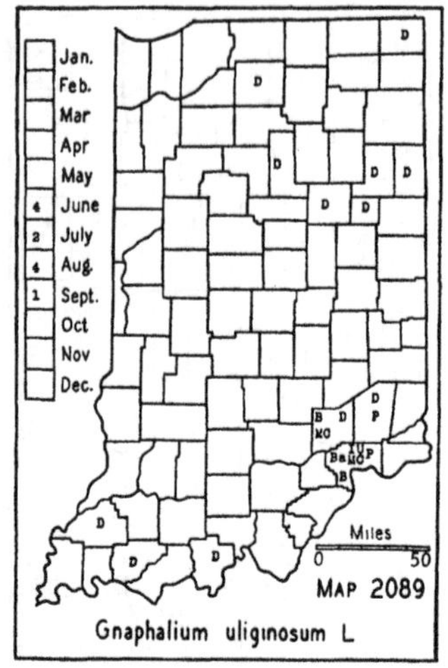

Gnaphalium uliginosum L.

When a study is made of the two, it is easy to understand how they could be confused. This species is perennial, stoloniferous, has papery, white, finely striate, spreading involucral bracts while in *Gnaphalium* the bracts are yellowish white or brownish, not striate, and subappressed, and this species lacks the balsamic odor which is characteristic of *Gnaphalium obtusifolium.* I have *Anaphalis margaritacea* from Elkhart County and Potzger has collected it in Morgan County. These are the only specimens I have seen.

Newf. to Alaska, southw. to Va., Kans., and Oreg.

8992. GNAPHALIUM L. CUDWEED

Bristles of the pappus distinct; bracts white or light brown.
- Plants tall, erect, simple below, with a large, more or less paniculate corymb; achenes smooth; bracts pearly white.
  - Stems white-tomentose; leaves not decurrent on the stem; outer bracts obtuse. .... 1. *G. obtusifolium.*
  - Stems white-tomentose only in the inflorescence, the main stem green, glandular-viscid; leaves decurrent on the stem; outer bracts mostly with short-acute tips. .... 2. *G. Macounii.*
- Plants low, generally less than 2.3 dm high, diffusely branched above the base; achenes scabrous; bracts light brown; plants of dried-up muddy places.......... 3. *G. uliginosum.*

Bristles of the pappus united at the base; bracts more or less purplish; inflorescence a terminal and usually an interrupted spike.......... 4. *G. purpureum.*

1. **Gnaphalium obtusifolium** L. (*Gnaphalium polycephalum* Michx. Gray, Man., ed. 7.) OLD-FIELD BALSAM. Map 2087. Throughout the state in dry soil, mostly in pasture fields, fallow fields, and open woodland. The plant has several common names but I believe old-field balsam is the most appropriate because it is the only species of the genus in Indiana that has a balsamic odor by which it is easily distinguished.

I knew of a case where a person who was suffering with flux and had been given up by the attending physician was cured by drinking copious draughts of milk in which this herb had been boiled.

N. S. to Man., southw. to Fla., Kans., and Tex.

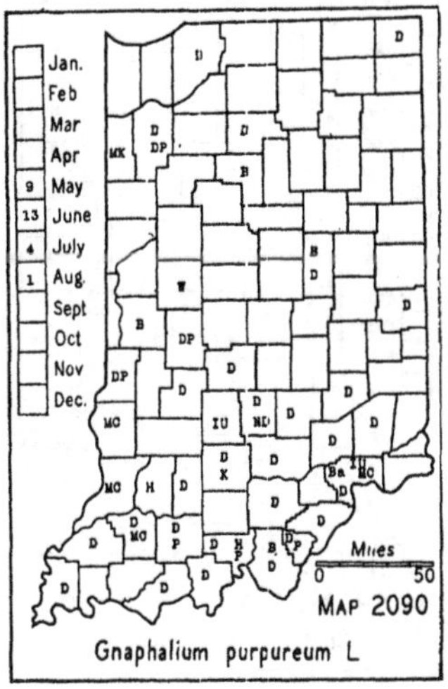

Gnaphalium purpureum L

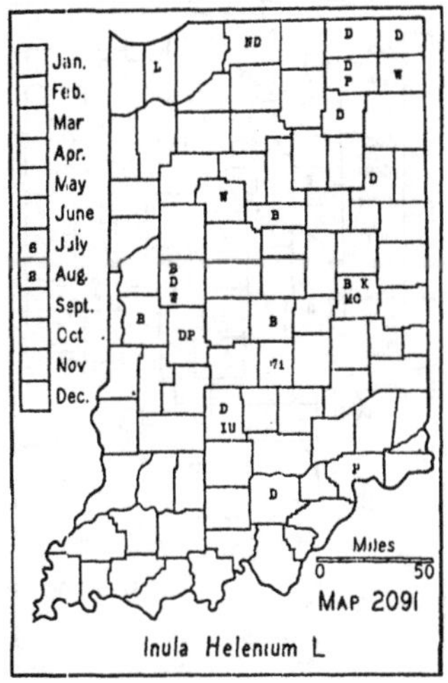

Inula Helenium L

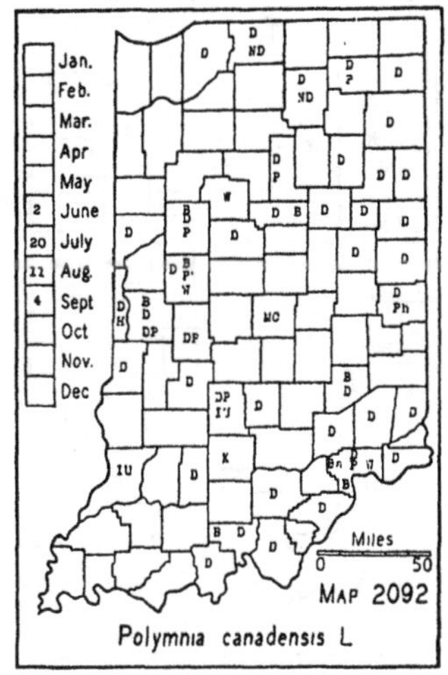

Polymnia canadensis L

2. **Gnaphalium Macoùnii** Greene. (*Gnaphalium decurrens* Ives.) WINGED CUDWEED. Map 2088. This is a northern species. May specimens are from open, sandy woods and I found a very sandy, fallow field of about five acres that was covered with old-field balsam and this species. This species was rare and found in the moister situations. The plants were much taller and, in most instances, with several branches from near the base that were almost as large as the central stem.

E. Que. to B. C., southw. to Pa., Ohio, Ind., and Minn.

3. **Gnaphalium uliginòsum** L. LOW CUDWEED. Map 2089. A rare or infrequent plant throughout the state. It is usually found in dried-out muddy places, such as hog wallows in lanes, in open woods, and along river banks.

I believe this species and *G. obtusifolium* and *G. purpureum* are rapidly spreading since their habitat is becoming more frequent.

Newf. to Sask. and B. C., southw. to Va., Ind., and Colo.

4. **Gnaphalium purpùreum** L. PURPLISH CUDWEED. Map 2090. This species prefers a dry, sandy soil and is more or less frequent in fallow fields and open woodland in the southern half of the state, becoming infrequent to very rare in the northern half of the state but being rapidly introduced.

Maine to Minn., southw. to Fla. and Tex.

9061. ÍNULA L.

1. INULA HELÈNIUM L. ELECAMPANE. Map 2091. This plant has medicinal qualities and was commonly cultivated by the pioneers. It has escaped in all parts of the state to roadsides, pastures, and open woodland.

Nat. of Eu.; N. S. to Minn., southw. to N. C. and Mo.

9122. POLÝMNIA L. LEAFCUP

Rays whitish, usually small and shorter than the involucre, sometimes all well developed in f. *radiata* and about 10 mm long; plants generally 7-12 dm high, glandular,

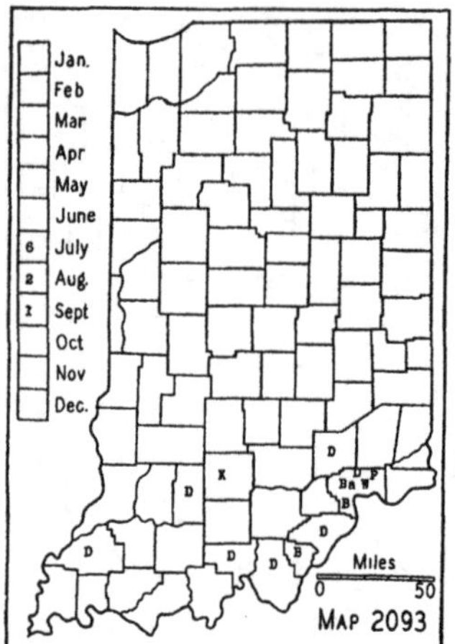

MAP 2093
Polymnia Uvedalia L

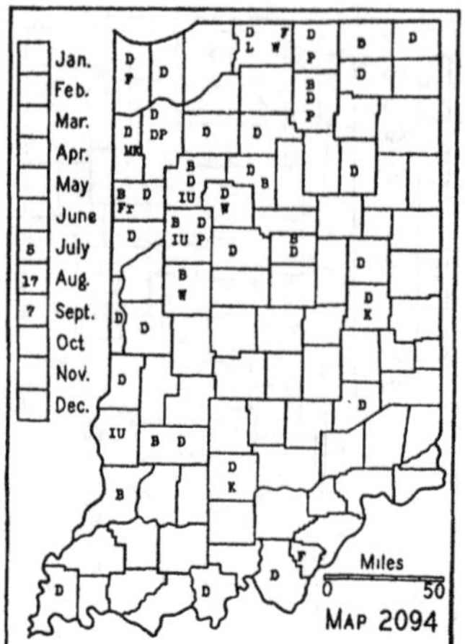

MAP 2094
Silphium terebinthinaceum Jacq.

MAP 2095
Silphium perfoliatum L

strongly scented, odor characteristic and unpleasant; leaves thin, pinnately lobed; achenes mostly 3-4 mm long, 3-ribbed and finely striated or the striae obscure.... ....1. *P. canadensis*.

Rays yellow, 12-22 mm long; plants generally 1.5-2.5 m high; glandular only in the inflorescence; leaves firm, palmately lobed; achenes mostly 6-8 mm long, coarsely striated...2. *P. Uvedalia*.

1. **Polymnia canadénsis** L. WHITE-FLOWER LEAFCUP. Map 2092. This species is found, no doubt, in every county except possibly in a few of the prairie counties. It is strictly a woodland species and prefers a moist soil covered with leaf mold in thick woodland. It is rarely found on steep slopes without leaf mold or in open woodland, but is often found in overflow land along streams.

W. Vt. to Minn., southw. to N. C., Tenn., and Ark.

1a. **Polymnia canadensis f. radiàta** (Gray) Fassett. This is a form in which the ligules of the heads are fully developed, usually being about 1 cm long. Found with the species but rare.

2. **Polymnia Uvedàlia** L. YELLOW-FLOWER LEAFCUP. Map 2093. Restricted to the southern part of the state where it is found on wooded slopes in places exposed to the sun, usually toward the base of a slope but not always so. It is infrequent and grows in colonies. In 1931 in Harrison County, I found it as a common weed in an orchard of Wm. W. Jacobs about a mile west of Glidas. The orchard was on the south side of a woods where the species was common and from which it had escaped into the orchard. The owner was making strenuous efforts to eradicate it.

N. Y. to Ind., southw. to Fla. and Tex.

9131. SÍLPHIUM L. ROSINWEED

[Perry. Notes on Silphium. Rhodora **39**: 281-297. 1937.]

Stem leafless or nearly so, scaly above; very tall plants with large basal leaves.
Leaves cordate, dentate....................................1. *S. terebinthinaceum*.
Leaves pinnatifid or lobed. (See excluded species no. 646, p. 1099)................ ....*S. terebinthinaceum* var. *pinnatifidum*.

Stem leafy throughout; large basal leaves wanting.
Stems quadrangular; large plants with the upper leaves large and connate-perfoliate. ....2. *S. perfoliatum.*
Stems more or less terete, the upper leaves not very large or connate.
Leaves pinnately parted, large, all alternate; involucres generally 2-3 cm broad.
Involucre and peduncle glandless.............................3. *S. laciniatum.*
Involucre and peduncle glandular.............3a. *S. laciniatum* var. *Robinsonii.*
Leaves not parted or pinnatifid, entire, dentate or serrate, generally opposite or whorled or sometimes some of them alternate; involucres mostly 1-1.5 cm broad; involucral bracts ciliate.
Outer involucral bracts glabrous on both faces; stems terete, glabrous and usually very glaucous; leaves lanceolate, oblong-lanceolate or lanceolate-ovate, opposite or usually the middle ones in 3's or 4's, tapering at the base into a distinct petiole except the upper ones which are sessile; petioles mostly 0.3-3 cm long.
Plant with at least the upper surface of the leaves pubescent; leaves chiefly verticillate though often opposite or alternate...........4. *S. trifoliatum.*
Plant glabrous; leaves usually opposite......4a. *S. trifoliatum* var. *latifolium.*
Outer involucral bracts more or less pubescent on one or both faces, sometimes the whole involucre also glandular; stems more or less compressed to 4-sided, usually somewhat scabrous or pubescent, rarely glabrous, but scabrous in the inflorescence; leaves opposite, rarely a few ternate or alternate, ovate to lanceolate-ovate, all sessile or the lower ones on very short petioles, narrowed, rounded or cordate-clasping at the base.
Involucral bracts pubescent but not glandular..............5. *S. integrifolium.*
Involucral bracts glandular-pubescent........5a. *S. integrifolium* var. *Deamii.*

1. **Silphium terebinthinàceum** Jacq. DOCK ROSINWEED. Map 2094. Found here and there in all parts of the state but frequent to common in the prairie area. It has a very wide range of habitat—from the crest of a wooded sandstone ridge to a marsh. It is generally found in a prairie habitat along roadsides and railroads and rarely on open, wooded or gravelly slopes.

Ont. to Minn., southw. to Ga. and La.

2. **Silphium perfoliàtum** L. CUP ROSINWEED. Map 2095. Found throughout the state although it may be absent from a few of our northern counties. It is generally found on the alluvial banks of streams and on the low borders of lakes. It is found both in woodland and in open places and sometimes in low ground along roadsides.

Mass. to Minn., southw. to Ga., Miss., and Okla.

3. **Silphium laciniàtum** L. COMPASSPLANT. Map 2096. This species is a true prairie plant and in its distribution in Indiana it was restricted to the prairie areas. All of our prairie areas are under cultivation and it is found now only in moist prairie habitats along railroads and roadsides. The published records extend its range somewhat beyond that shown on the map. Beyond the area indicated by the map, it has been reported from the area of Delaware, Jay, Randolph, and Wayne Counties, and from Elkhart, Knox, and Noble Counties.

Ohio and Ind. to Minn., southw. to Okla. and Tex.

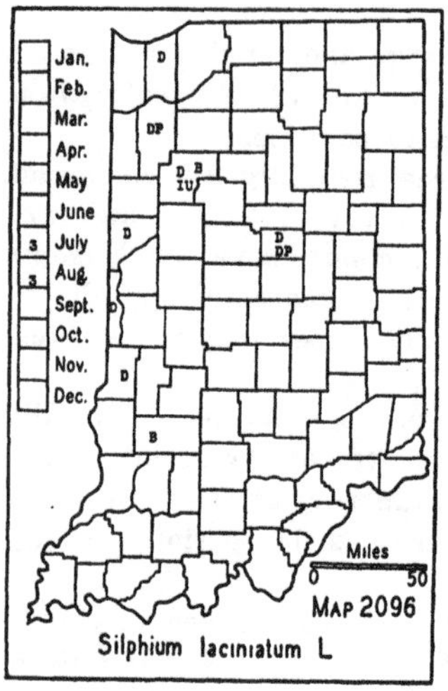

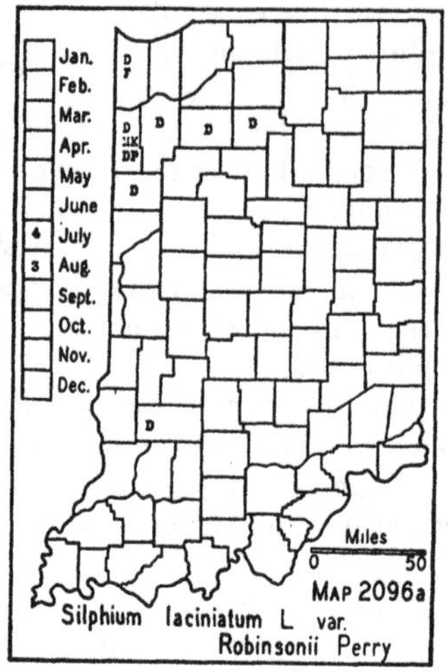

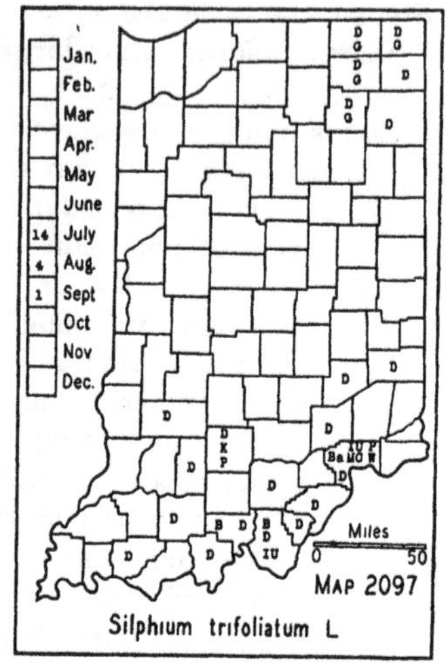

3a. **Silphium laciniatum** var. **Róbinsonii** Perry. (Rhodora 39: 297. 1937.) Map 2096a. In Indiana this variety has the same habitat as the species.

Ind., southw. to La.

4. **Silphium trifoliàtum** L. WHORLED ROSINWEED. Map 2097. Infrequent to rare in dry soil on open, wooded slopes in two widely separated parts of the state. Beyond the area shown on the map, it has been reported from Carroll, Cass, Knox, and La Porte Counties. This species, like the next, is variable and in certain forms it is separated from it with difficulty. In the typical form, the leaves are lanceolate, dark green, some whorled, and generally with nearly entire margins. Non-typical plants may have only opposite leaves or some alternate ones, and narrow-ovate blades. The inner face of the achenes is glabrous or pubescent, usually the latter.

Pa., Ont., Ind., southw. to N. C. and Tenn.

4a. **Silphium trifoliatum** var. **latifòlium** Gray. This is a glabrate form of the species and is not well marked. Some specimens will have most of the leaves glabrous above and one or more scabrous above. I have this form from Crawford, Decatur, and Washington Counties.

Ohio and Ind., southw. to S. C. and Ala.

5. **Silphium integrifòlium** Michx. ENTIRE-LEAF ROSINWOOD. Map 2098. Infrequent to rare, rarely frequent, throughout Indiana although there are no records from the northeastern part. It is generally found in dry soil on open wooded slopes, frequent in prairie habitats, and on the low wooded dunes along Lake Michigan. The plants are variable in the width and margins of the leaves and in the pubescence of the stem, leaves, and involucre. Some plants have stems with a few ternate or alternate leaves. The inner face of the achenes is either glabrous or pubescent, mostly more or less pubescent. I have had all forms of this species under cultivation for many years to study them.

Ohio to Minn., southw. to Miss. and Tex.

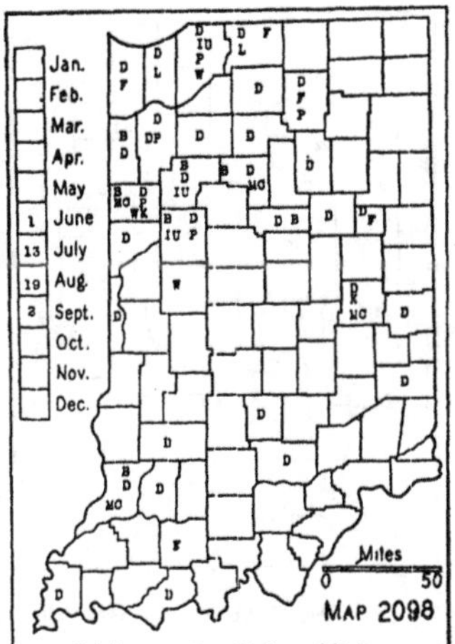

Silphium integrifolium Michx

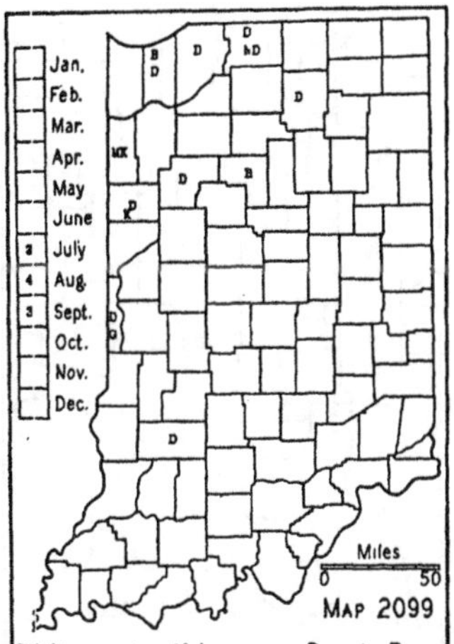

Silphium integrifolium var. Deamii Perry

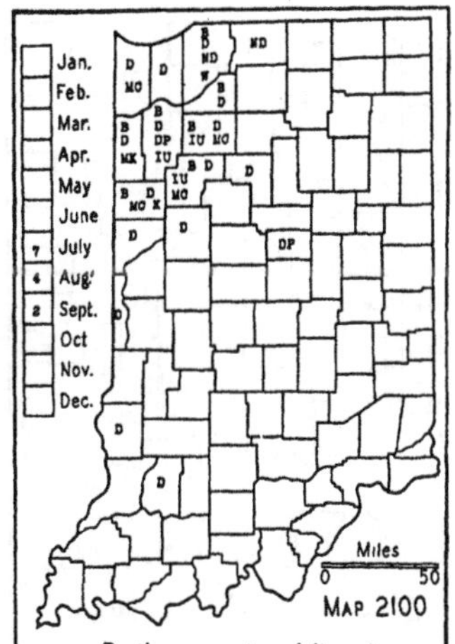

Parthenium integrifolium L.

5a. **Silphium integrifolium** var. **Dèamii** Perry. (Rhodora 39: 287. 1937.) Map 2099. Found throughout the range of the species but rarely found closely associated with it in Indiana.

Ind. and Wis., southw. to Ala., Miss., and Ark.

9138. PARTHÈNIUM L.

1. **Parthenium integrifòlium** L. AMERICAN FEVERFEW. Map 2100. This is one of our typical prairie plants. Since all of our original prairies are under cultivation, this plant is found now only in prairie habitats along roadsides and railroads. I have a few specimens collected in "oak openings," which means that the plants are relicts. The range in Indiana is extended by published records from Clark, Floyd, Jefferson, and Marshall Counties.

Md. to Minn., southw. to Ga. and Ark.

9141. ÌVA L.

Heads subtended by leafy bracts; fertile flowers with evident corollas......1. *I. ciliata.*
Heads not subtended by leafy bracts; corollas of fertile flowers rudimentary or lacking. ....2. *I. xanthifolia.*

1. **Iva ciliàta** Willd. Map 2101. Known only from Gibson and Posey Counties. I found it to be frequent to common in hard, clay soil in a field on the border of Pitcher's "Lake," along the roadside on the south side of Half Moon Pond, and along the roadside for a mile or more along the Wabash River in the vicinity of Bone Bank, Posey County.

Ind. to Nebr., southw. to La. and N. Mex.

2. **Iva xanthifòlia** Nutt. Reported by Hansen as found along a ditch in Tippecanoe County, and by Peattie as found in the Calumet District. Although I have not seen a specimen, I am admitting this species because there is little possibility of a wrong determination.

Ont. and Mich. to Sask., southw. to Tex. and Utah; introduced in the East, from Maine to Del.

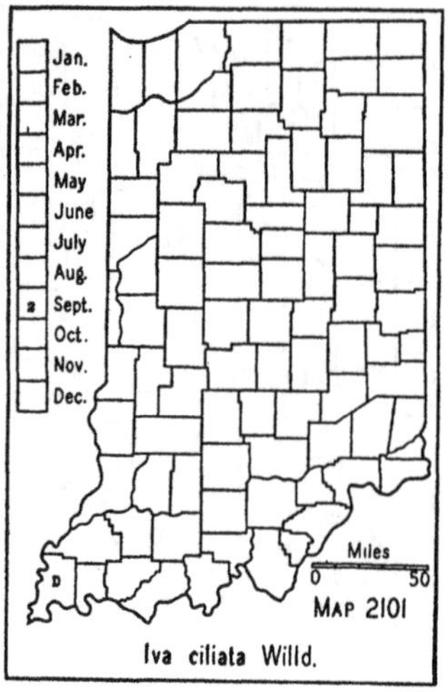

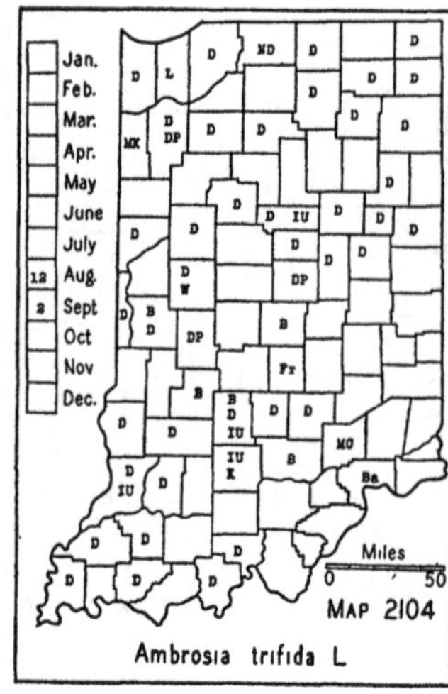

9146. AMBRÒSIA [Tourn.] L. RAGWEED

Staminate heads sessile; leaves not divided, mostly less than 1 cm wide; fruit 4-angled, each angle ending in a short prickle........................1. *A. bidentata.*

Staminate heads on short pedicels; leaves 1-2-pinnatifid, 3-5-lobed or, if undivided, more than 1 cm wide.

Leaves 3-5-lobed or undivided..2. *A. trifida.*

Leaves pinnatifid or bipinnatifid.

Stem leaves petiolate; fruit with 5-7 short prickles about 0.5 mm long, the beak generally 1-1.5 mm long; annual, branches widely spreading...3. *A. elatior.*

Stem leaves sessile or with short petioles; fruit without prickles or with 1-4 very short ones less than 0.5 mm long, the beak generally about 0.5 mm long; perennial, branches ascending, more compact..............4. *A. coronopifolia.*

1. AMBROSIA BIDENTÀTA Michx. LANCELEAF RAGWEED. Map 2103. Generally found in hard, white clay soil in low land in fallow fields, in open woodland, and along roadsides. Pioneers have told me that they did not note this species until the past ten years. Blatchley reports that it was first noted in 1895. Schneck in 1876 reports it as "common in prairies." This western species is slowly migrating eastward. Where it is found, it usually forms dense stands. I was told by a farmer that stock will not eat it, although they will eat other species of ragweed. It is restricted to the southwestern part of the state. There are records of its occurrence in Clay and Vigo Counties.

Ind. to Nebr., southw. to La. and Tex.

2. **Ambrosia trífida** L. GREAT RAGWEED. Map 2104. This ragweed, without doubt, occurs in every county of the state. It is usually abundant in the alluvial bottoms of streams. Found in low, open places in cultivated and fallow fields and woodland. It grows to giant size and J. M. Coulter records measuring a specimen 18 feet high. A form with entire leaves is often found growing with the species. I believe that it is a depauperate form of the species and it has no taxonomic significance.

Que. to Man., southw. to Fla. and N. Mex.

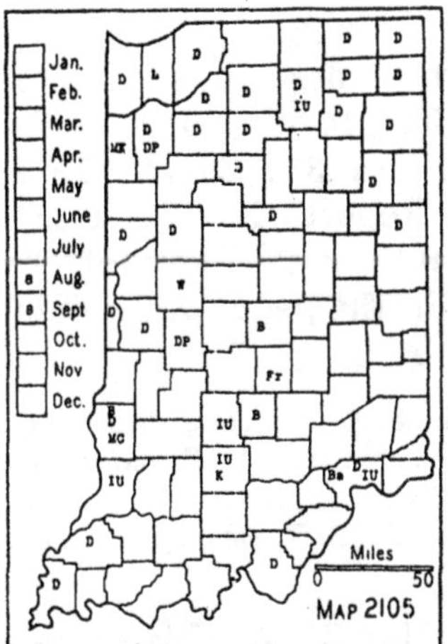

Ambrosia elatior L.

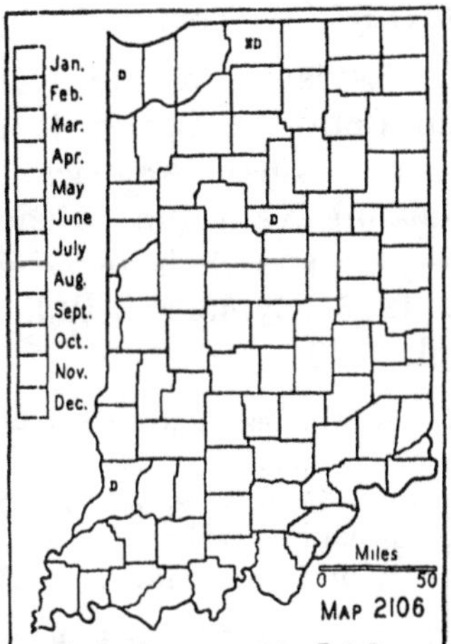

Ambrosia coronopifolia T. & G.

Xanthium spinosum L.

3. **Ambrosia elàtior** L. (*Ambrosia artemisiifolia* L. and *Ambrosia elatior* var. *artemisiifolia* (L.) House.) (For a discussion of this species see Jones. Studies on Ambrosia. Amer. Midland Nat. 17: 673-700. 1936 and Fernald & Griscom. Ambrosia artemisiaefolia and its variations in temperate North America. Rhodora 37: 184-185. 1935.) COMMON RAGWEED. Map 2105. I am referring all of our reports under whatever name reported to this species. As Jones has pointed out, it is a highly variable plant, producing pistillate plants and also plants bearing both stamens and pistils.

An abundant weed everywhere in cultivated and fallow fields, waste places, roadsides, and almost any place where the ground is not covered with a sod of grass. Milch cows are usually kept out of pastures and stubble fields where it is abundant because when they eat this plant the milk has a nauseating taste.

The ragweeds are a few of the species whose pollen causes autumnal hay fever. On account of the abundance of these plants they have the credit of being the chief cause of this disease.

N. S. to B. C., southw. to Va., Colo., and Wash.

4. AMBROSIA CORONOPIFÒLIA T. & G. (*Ambrosia psilostachya* of Gray, Man., ed. 7 and Britton and Brown, Illus. Flora, ed. 2.) WESTERN RAGWEED. Map 2106. This is a western species that has been reported several times in Indiana, probably mostly as a railroad migrant. It has been reported from the dunes by Peattie and I have a specimen from near Hammond. It has also been reported from Jefferson and Marion Counties, and from the Lower Wabash Valley. In 1933 I found it as a common weed in dry, sandy soil along the C. E. & I. Railroad just south of Emison, Knox County. It is common in sandy soil in a woods about 1 mile southeast of Notre Dame, St. Joseph County (Nieuwland). Ek has found it in two places along railroads in Howard County.

Mich. to Sask., southw. to Idaho and n. Mexico; introduced into Conn.

9148. XÁNTHIUM [Tourn.] L. COCKLEBUR

[Millspaugh & Sherff. Revision of the North American species of Xanthium. Field Museum Botanical Series 4: 9-51. 14 pl. 1919. Millspaugh & Sherff. Xanthium. North American Flora 33: 37-44. 1922. Widder: Die Arten der Gattung Xanthium. Fedde Repertorium 20: 1-221. Tafel 4, Karte 4. 1923. Wiegand & Eames. Xanthium.[1] Flora of the Cayuga Lake Basin: 414. 1926. Symons. Studies in the genus Xanthium. Bot. Gaz. 81: 121-147. 3 pl. 1926.]

Leaves attenuate at both ends, mostly less than 2 cm wide, with a strong, 3-pronged prickle at the base of each....................................1. *X. spinosum.*
Leaves cordate or ovate, more than 3 cm wide, without prickles at the base.
  Body of fruit and prickles glabrous or nearly so..............2. *X. pennsylvanicum.*
  Body of fruit and prickles more or less hispid......................3. *X. italicum.*

1. XANTHIUM SPINÒSUM L. SPINY COCKLEBUR. Map 2107. This species has been reported from Clark, Franklin, Jefferson, and Putnam Counties. Young, in 1875, said that it was spreading and not uncommon ten years before in Jefferson County, mostly along roadsides. I have traveled all of the principal roads of Jefferson County and have done considerable botanical work there and I have never seen it. This would indicate that for some reason it is not spreading. It may be that landowners have recognized the plant as a weed and eradicated it.

I found it in a hogyard and along the roadside near Mauckport in Harrison County and in a hogyard and an adjacent pasture and roadside east of Cannelton in Perry County.

Nat. of Eu.; Maine to Ont. and Mo., southw. to Fla. and Tex.

2. **Xanthium pennsylvánicum** Wallr. SMOOTH-BODY COCKLEBUR. Map 2108. Frequent to common throughout the state in moist places along streams, about lakes, in dried-up ponds, cornfields, and cultivated grounds in general. This species and the next are very annoying weeds in the cornfields of the Lower Wabash Bottoms.

Mass. to Wash., southw. to Fla., Tex., and Calif.

3. **Xanthium itálicum** Moretti. HAIRY-BODY COCKLEBUR. Map 2109. The distribution, frequency, and habitat in Indiana are the same as those of the preceding species.

Que., Sask., and Wash., southw. to W. Va. and Calif., southw. into Mex. to Oaxaca, and in Eu.

[1] Wiegand says: "Several years ago I undertook a revision of the American Xanthiums, making use of the material in the Gray Herbarium. After a prolonged but unsuccessful effort to prepare a satisfactory treatment, the problem was laid aside. I am now greatly in doubt as to the existence of more than one real species in the group represented by *X. chinense* Mill., *X. pennsylvanicum* Wallr., *X. italicum* Mor., and other related forms." He discusses all of the Indiana forms (as I understand him) except *Xanthium spinosum* and he refers them all to one species which he calls *Xanthium orientale* L. I have made a limited study of our forms in the field and I have decided to treat all of our native forms under two species. Since Symons' studies show that the species will hybridize, it seems best to regard our species as complexes until study defines the species. The synonymy is so badly involved that it is useless to repeat it.

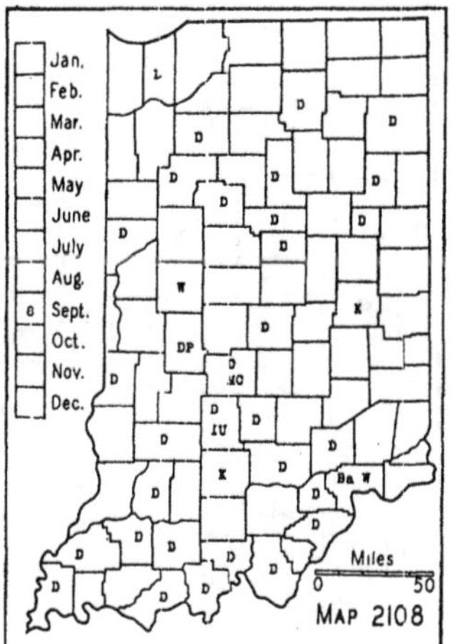

Xanthium pennsylvanicum Wallr.

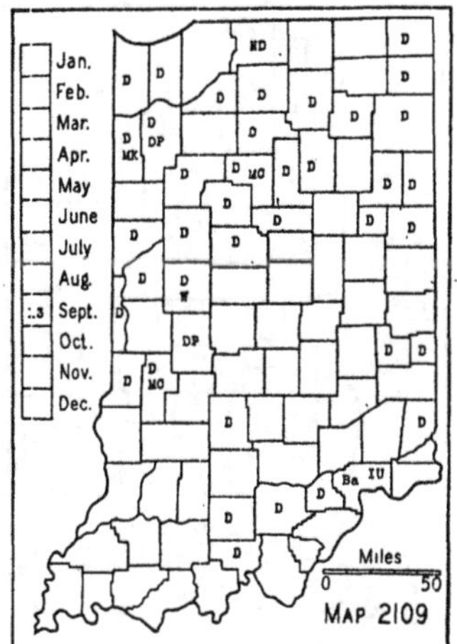

Xanthium italicum Moretti.

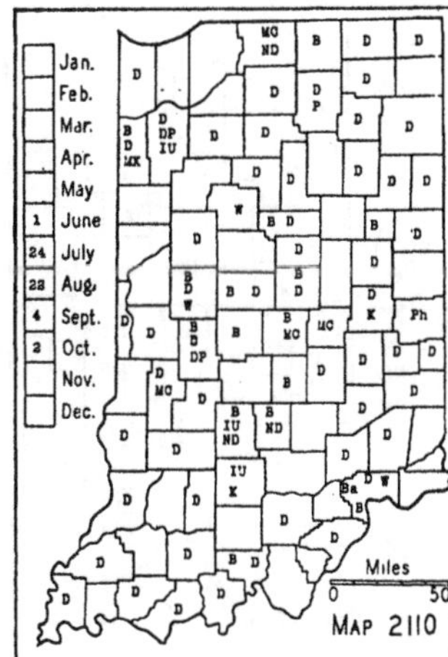

Heliopsis helianthoides (L.) Sweet

9157. HELIÓPSIS Pers.

1. **Heliopsis helianthoides** (L.) Sweet. SUNFLOWER HELIOPSIS. Map 2110. Found throughout the state, preferring open areas and moist soils. It is most frequently found in open woodland in the alluvial bottoms along streams and along roadsides. I have it from 47 counties, and I find that my specimens are highly variable, and I am not able to divide them on the characters given in the manuals. The leaves are not smooth on any of my specimens. All of them are more or less scabrous to the touch either above or beneath. Sometimes the upper surface is more scabrous than the lower, and in other specimens the reverse is true. The width of the largest median leaf varies from 3 to 10 cm. The apex of the leaves varies from acuminate to short-acute. The margins are variously cut, and the number of teeth is exceedingly variable. The petioles of the median leaves are from 1 to 4 cm long. The stems of all of my specimens are comparatively smooth, only rarely with a somewhat scabrous internode. The heads vary from less than 1 cm to 2 cm wide and are very variable on the same plant. The number of heads varies from one to many. The crown of the achene varies in height and smoothness.

Heliopsis scabra Dunal has been often reported from the state, and I have two specimens, one from Hamilton County and one from Tipton County, that some authors might refer to this species. While this species in its extremes seems to be distinct, I believe that all Indiana material should be referred to one highly variable complex. The descriptions of the two species by different authors show that the differences between them are slight and that there are exceptions to these differences.

My plants with largest leaves, longest petioles, and smoothest leaves are from deep woodland, and those with the smallest and most scabrous leaves are from prairie habitats, although larger intermediates are also found in prairie habitats.

Nieuwland, however, on June 24, 1909, found a specimen of what I would call the typical form of *Heliopsis scabra* Dunal along the Grand Trunk

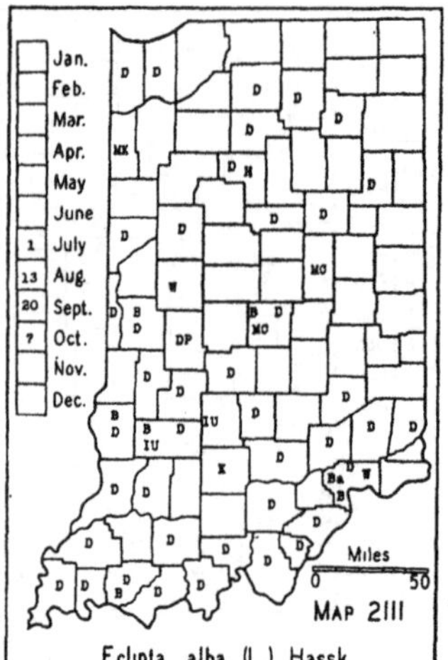

Eclipta alba (L.) Hassk.

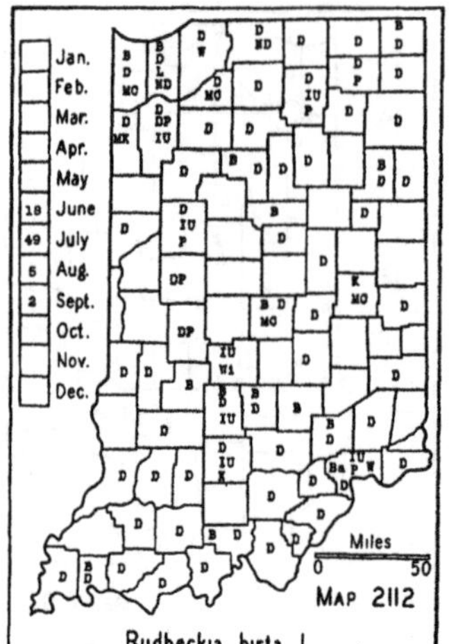

Rudbeckia hirta L.

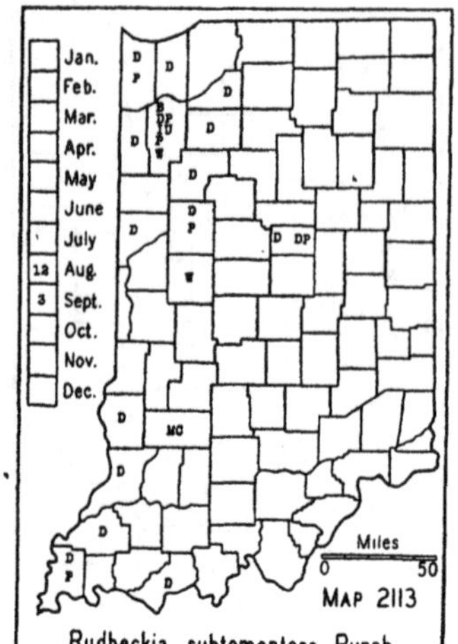

Rudbeckia subtomentosa Pursh

Railroad, St. Joseph County. I am regarding this specimen as a migrant and we have no record that other plants were left and perpetuated themselves.

N. Y., Ont., and N. Dak., southw. to Fla., Tenn., and Mo.

9166. ECLÍPTA L.

1. **Eclipta álba** (L.) Hassk. (*Verbesina alba* L.) YERBA DE TAJO. Map 2111. Local throughout the state, but frequent to common along the bank of the Ohio River and in the Lower Wabash Bottoms. This is a southern species which is migrating northward and it may be absent as yet from the northern tier of counties. It prefers the muddy shores of streams, ponds, and sloughs but is found also in low places in cultivated fields.

Mass. to Nebr., southw. to Fla., Tex., and Mex., and southw.

9178. RUDBÉCKIA L. CONEFLOWER

Corolla lobes recurved after anthesis, about 0.5 mm long, usually 0.3-0.4 mm long.
- Chaff of disk acute, hispid-ciliate on the margins and on the back at the summit; plants flowering mostly from the middle of June to the middle of August; style branches long and subulate at anthesis; leaves not divided........1. *R. hirta.*
- Chaff of disk acute, not strictly ciliate but the margins at the summit and the back for half its length densely covered with short-stalked glands, and with a few colorless, hispid hairs; plants flowering mostly from the middle of August to the first of October; style branches short and obtuse at anthesis; at least the lower leaves 3-lobed, rarely none of the leaves lobed....................2. *R. subtomentosa.*

Corolla lobes erect or some of the outer ones spreading after anthesis, more than 0.5 mm long except in *Rudbeckia palustris.*
- Heads greenish; chaff truncate and densely glandular-pubescent at the summit; plants generally 1-2 m high; at least the lowest leaves 3-7-parted, the segments variously toothed...3. *R. laciniata.*
- Heads purplish; chaff not truncate at the summit; plants generally less than 1 m high; leaves entire, toothed, or 3-lobed.
  - Chaff long-acuminate or aristate at the apex, glabrous; some or all of the lowest leaves petiolate, some or all more or less 3-lobed................4. *R. triloba.*
  - Chaff merely acute at the apex; no 3-lobed leaves.

Summit of chaff (except the outer rows) more or less ciliate on the margins or pubescent within.

Chaff more or less pubescent within and usually more or less pubescent without; involucral bracts hispid above; corollas about 4 mm long, the lobes about 1 mm long; achenes about 2 mm long; plants mostly on the crests of wooded ridges; blades of stolons narrow, lanceolate or elliptic-lanceolate, generally less than 3 cm wide or in a wet season under cultivation somewhat ovate and up to 5 cm wide............................5. *R. fulgida.*

Chaff not pubescent within, rarely with a few hairs, and glabrous without; involucral bracts glabrous above; plants of moist places.

Largest leaves of stolons mostly 3-8 cm wide, of an ovate or oval type, their petioles mostly 7-17 cm long; median and upper cauline leaves sessile or on short, margined petioles; achenes 3 mm long.

Ligules of rays 1.5-2 cm long; disk usually less than 14 mm wide; chaff mostly with narrow-triangular tips...................6. *R. umbrosa.*

Ligules of rays 2-4 cm long; disk 12-18 mm wide; chaff mostly with wide-triangular tips...................................7. *R. Sullivantii.*

Largest leaves of stolons less than 3 cm wide, lanceolate, or elliptic-lanceolate, narrowed to a very long cuneate base, remotely shallow-crenate, their petioles mostly 2-12 cm long; median and upper cauline leaves usually narrowed to long, margined petioles; achenes 2-2.5 mm long. ....8. *R. palustris.*

Summit of chaff with erose or denticulate margins, sometimes one or more of the outer ones with a few ciliate hairs.

Stems more or less densely retrorsely hirsute, appressed-pubescent above; involucral bracts hispid above; ligules of rays less than 25 mm long; leaves of stolons ovate or oval, narrowed or rounded at the base...9. *R. Deamii.*

Stems glabrate, sparingly hispid, or rarely spreading or upwardly pubescent; involucral bracts glabrous above; ligules of rays 20-40 mm long, usually more than 25 mm long; leaves of stolons large, of an ovate type, cordate or subcordate at the base............................7. *R. Sullivantii.*

1. **Rudbeckia hírta** L. BLACK-EYED SUSAN. Map 2112. This species I am regarding as a species complex. The plants in our area are variable. Some are annual and flower mostly in June and July; these probably belong to the typical form. They are found in all parts of the state and are more or less frequent in both the glaciated and unglaciated areas. They are usually found in fallow fields, prairie habitats, and open black oak woods and along roadsides and railroads. I have found them in acid marshes and once in great numbers on the marl border of a lake. It is to be noted that the border of the lake was more than 100 feet wide and the plants covered about an acre. Those that grew in the moist part of the border were simple, usually bearing but one head while those on the beach margin where it was dry were branched and had many heads.

These early flowering plants also vary in the size and shape of the leaves and in the color of the rays. Miss Edna Banta found a specimen in Jefferson County which I determined as *Rudbeckia bicolor,* overlooking the fact that this species sometimes has flowers with the base of the rays a maroon color. I am now referring her specimen to *Rudbeckia hirta.* The bracts of this form are mostly 10-12 mm long, rarely one up to 20 mm long. The rays are mostly 20-35 mm long. The heads are on long peduncles and well developed ones are 15-22 mm wide.

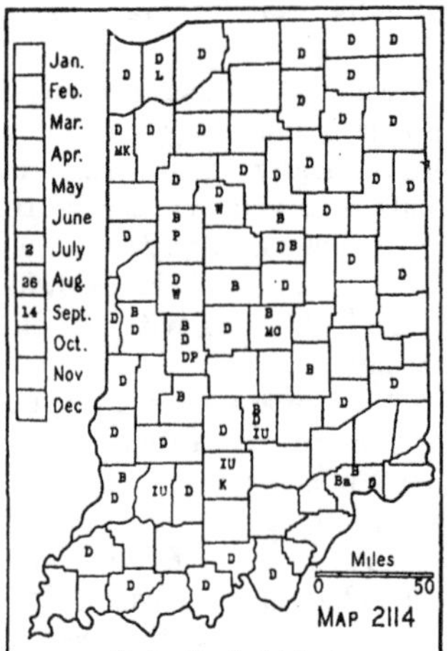

Rudbeckia laciniata L.

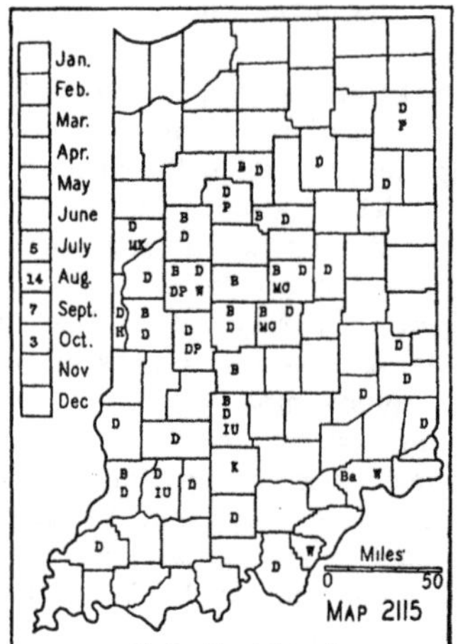

Rudbeckia triloba L.

Rudbeckia fulgida Ait.

On September 26, 1932, I was collecting along the roadside near Blocher in Jefferson County and my attention was directed to large flowering specimens of this species. The date of flowering and the mammoth size of the specimens attracted my attention. I measured the longest ray of the specimen I collected and it was 57 mm long. I dug several plants, brought them home, and planted them in our garden where they have been ever since. I find that they are perennial and in cultivation they are very prolific. In 1936 I made 12 full specimens from one plant. The plants have long root leaves, the blades tapering at both ends, 3-5 cm wide and about 15 cm long, on petioles 10-20 cm long. The heads are mostly 15-18 mm wide, with involucral bracts 10-20 mm long and are on long petioles except one bushy plant which has the many heads on short petioles. Here in our garden this plant begins to bloom about the middle of August and continues until killing frost.

I have tried to find the correct name for my plants but have failed to satisfy myself. Fernald (Rhodora 29: 458. 1937) published a key to *Rudbeckia hirta* L. which I am not able to fit to our plants. He regards the typical form of the species as having the "pubescence of the lower leaf-surface variously spreading, with broad open glabrous areas between the conspicuous green bulbous bases of the trichomes." He refers to *Rudbeckia hirta* var. *sericea* plants of this complex that have the "pubescence of both leaf-surfaces closely appressed (or chiefly so), the crowded hairs chiefly parallel with the midrib, with minute or obscure bulbous bases." He does not give the range of this variety but I can not make it apply to our plants because Moore's original description calls for plants with subulate involucral bracts three fourths of an inch (20 mm) long, while the bracts of our plants are not subulate and are mostly 10-12 mm long.

I find no description to fit my Jefferson County plants and I regard them unique, requiring further study to place them.

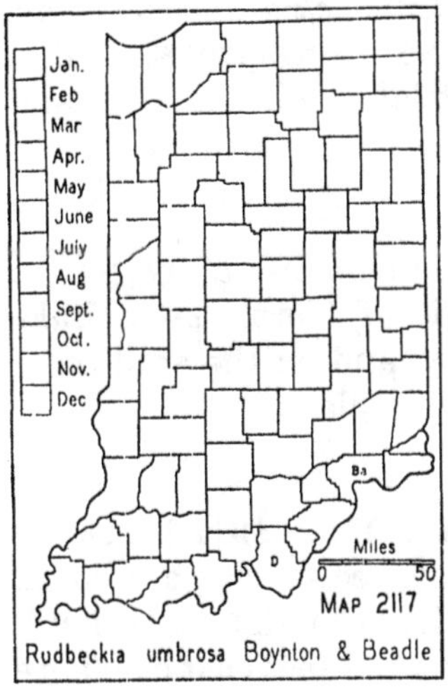

Rudbeckia umbrosa Boynton & Beadle

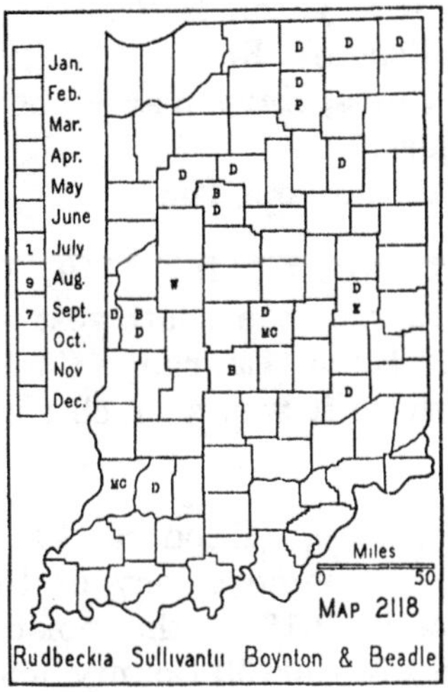

Rudbeckia Sullivantii Boynton & Beadle

Rudbeckia palustris Eggert

I have had all our species of *Rudbeckia* under cultivation for several years and this autumnal form of this species baffles me. I hope to continue and to increase my observation of it.

N. E. to Man., southw. to Fla., Colo., and Tex.

2. **Rudbeckia subtomentòsa** Pursh. SWEET CONEFLOWER. Map 2113. Infrequent in rather wet prairie habitats in the northwestern part of the state, mostly along roadsides; and in the southwestern part of the state in low, open woods, where it is usually associated with prairie plants.

Ind., Wis. to Kans., southw. to La. and Tex.

3. **Rudbeckia laciniàta** L. CUTLEAF CONEFLOWER. Map 2114. Goldenglow is a cultivated form of this species. In our area, this species varies in the pubescence of the lower surface of the leaves from glabrous to densely short-pubescent. The rays of our plants are spreading. Infrequent, but usually in large colonies, on the moist, alluvial bottoms of streams in the open or in woods, and rarer in low woodland and about lakes.

W. Maine to Man. and Idaho, southw. to Fla., Colo., and Ariz.

4. **Rudbeckia tríloba** L. BROWN-EYED SUSAN. Map 2115. Infrequent, but usually in large colonies in the open or wooded, moist banks of streams and in moist wooded ravines. Throughout the state although there are no reports or specimens from the northern counties.

N. J. to Minn., southw. to Ga., La., and Kans.

5. **Rudbeckia fúlgida** Ait. ORANGE CONEFLOWER. Map 2116. This is a rare species found in dry, open woodland. It is slender, usually 4-8 dm high, and grows in colonies.

N. Y., Pa. to Mo., southw. to Fla. and Tex.

6. **Rudbeckia umbròsa** Boynton & Beadle. Map 2117. My specimens are from the low, moist border of a small creek about a mile southeast of Corydon Junction (New Salisbury) in Harrison County. This species, no doubt, has a wider distribution in Indiana.

W. Va. to Ky., southw. to N. C. and Tenn.

7. **Rudbeckia Sullivántii** Boynton & Beadle. (*Rudbeckia speciosa* var. *Sullivantii* (Boynton & Beadle) Rob.) SULLIVANT CONEFLOWER. Map 2118. Local in moist, wet, or springy places about lakes and marshes and along streams and roadsides.

Ohio to Mich., southw. to Ala. and Tenn.

8. **Rudbeckia palústris** Eggert. Map 2119. Common in sandy soil in the wet, sandy, sedge border of the southwest side of North Twin Lake about 2 miles northwest of Howe, Lagrange County, and on the spill bank of the inlet of this lake where it was a much smaller plant.

Ind. to Tenn. and Mo.

9. **Rudbeckia Dèamii** Blake. (Rhodora **19**: 113-115. 1917.) DEAM CONEFLOWER. Map 2120. A single colony of this species was found in 1914 and in the same place in 1916 on the moist slopes of the north bank of Wildcat Creek in section 1 in Carroll County, about 150 feet east of where the creek is crossed by the Delphi and Frankfort pike, about 9 miles southeast of Delphi. The type locality was visited in later years and the species had disappeared. I have searched up and down the creek from this place and I have never been able to find additional specimens. In September, 1932, I found a large colony of it in a roadside ditch about a mile and a half southwest of Williamsport in Warren County.

Known only from Ind.

9178A. BRAUNÈRIA Necker

Leaves of an ovate type, the lower ovate, the upper ovate-lanceolate, abruptly narrowed at the base, 5-nerved, the margins, at least some of them, more or less serrate; stems in our specimens smooth or nearly so; awn of chaff about as long as its body. .... 1. *B. purpurea.*

Leaves of a lanceolate type, attenuate at the base, entire or somewhat denticulate; stems scabrous or rough-pubescent; awn or chaff shorter than its body.

Rays drooping, mostly 3-6 cm long; tips of awns of chaff indurated and some of them more or less hooked.. 2 *B. pallida.*

Rays spreading or slightly reflexed, mostly 1.5-4 cm long; tips of awns of chaff indurated, straight or nearly so.............................. 3. *B. angustifolia.*

1. **Brauneria purpùrea** (L.) Britt. (*Echinacea purpurea* (L.) Moench of Britton and Brown, Illus. Flora, ed. 2.) PURPLE CONEFLOWER. Map 2121. In prairie habitats and woodland. Very local. The prairie seems to be its preferred habitat. The published records are from the area of Delaware, Jay, Randolph, and Wayne Counties, the Lower Wabash Valley, the "barrens" of Floyd and Harrison Counties, and from Carroll, Cass, Franklin, Marshall, Tippecanoe, and Vigo Counties. In the woodland I have seen only isolated specimens.

Pa., Mich. to Iowa, southw. to Ga., Ala., and Ark.

2. **Brauneria pállida** (Nutt.) Britt. (*Echinacea pallida* (Nutt.) Britt.) PALE-PURPLE CONEFLOWER. Map 2122. All of our reports say that this species was found along railroads, and it is probably a railroad migrant in this state. I found it along the railroad east of Dune Park in Porter

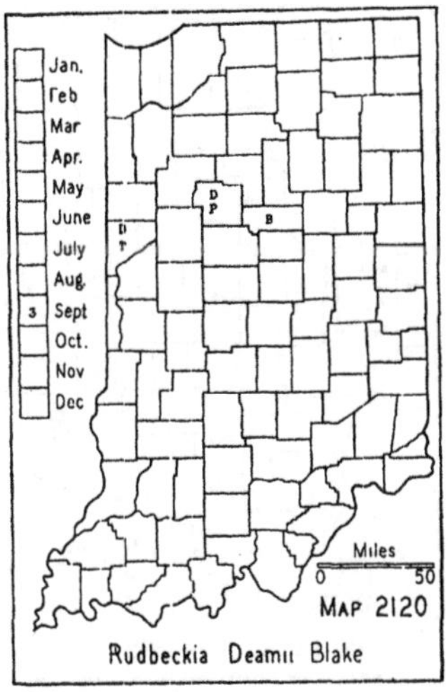

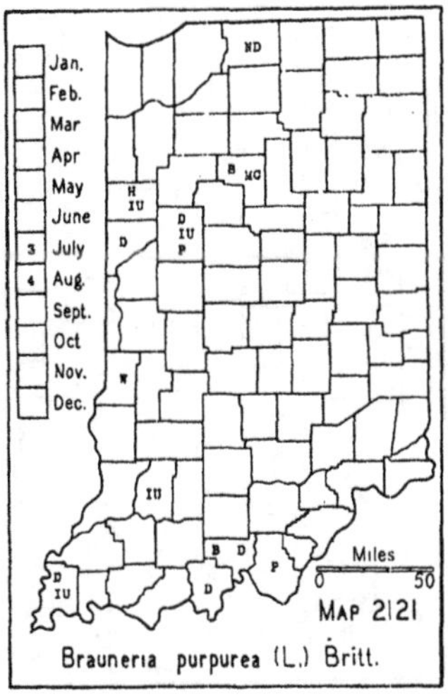

County. Peattie reports it from Lake and Porter Counties and Nieuwland reports it from La Porte and St. Joseph Counties.

Mich. to Iowa, southw. to Ala. and Tex.

3. **Brauneria angustifòlia** (DC.) Heller. (*Echinacea angustifolia* DC.) NARROWLEAF PURPLE CONEFLOWER. This species was reported by Hill as having been found in a prairie near Durham, La Porte County, July 4, 1892. His specimen is in the herbarium of DePauw University. In 1936 R. M. Tryon, Jr. tried to rediscover it but failed. He reports the area now all under cultivation. Peattie reported it from Lake County but I have not seen his specimen.

Minn. to Sask., southw. to Tex.; essentially a prairie plant and probably introduced eastward.

9178B. RATÍBIDA Raf.

Stigmas lanceolate, acute; heads broadly elliptic, about 1.5 times as long as wide at maturity; rays light to bright yellow, generally 2.5-5 cm long; leaf segments mostly lanceolate; angles of achene not ciliate........................1. *R. pinnata.*

Stigmas obtuse; heads cylindric, generally more than twice as long as wide at maturity; rays yellow, generally 1.5-2.5 cm long, or partly or wholly brownish purple in f. *pulcherrima;* leaf segments mostly linear; angles of achene usually more or less ciliate..2. *R. columnifera.*

1. **Ratibida pinnàta** (Vent.) Barnh. (*Lepachys pinnata* (Vent.) T. & G.) (Ann. Missouri Bot. Gard. 22: 75. 1935.) GRAY-HEAD CONEFLOWER. Map 2123. Infrequent to frequent in all parts of the state, although it may be rare in the Lower Wabash Bottoms. It is generally found in dry or gravelly soil along streams and roadsides and in prairie habitats, where it is rarely absent. It usually forms large colonies and sometimes becomes a weed.

N. Y. to Man., southw. to Fla. and Tex.; adventive eastw. to Mass.

2. RATIBIDA COLUMNÍFERA (Nutt.) Woot. & Standl. (*Lepachys columnaris* (Sims) T. & G. and *Ratibida columnaris* (Sims) D. Don.) (Rhodora 40: 353-356. 1938.) LONG-HEAD CONEFLOWER. Map 2124. I have a specimen

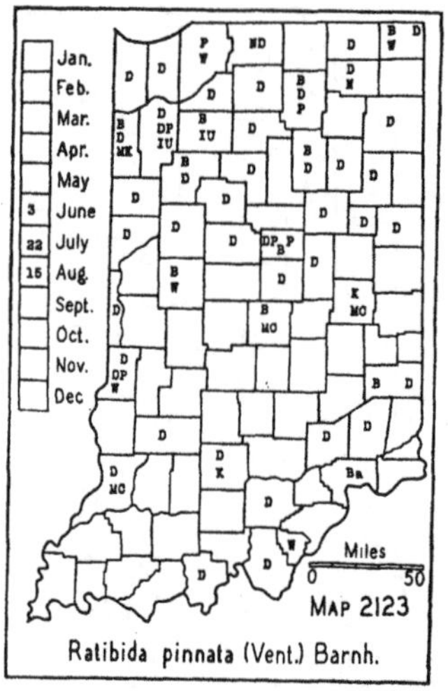

Ratibida pinnata (Vent.) Barnh.

Ratibida columnifera (Nutt.) Woot & Standl

Helianthus angustifolius L.

of this species collected in 1929 by Robert Hessler along the B. &. O. Railroad about a mile and a half east of Irvington in Marion County. The rays are entirely yellow. Hessler found only two specimens. Peattie reports it as naturalized in the Calumet District of Lake County but he does not tell us how abundant it is there. This species may be only a railroad migrant.

Dry prairies, Minn. to Sask. and B. C., southw. to Tenn., Tex., and Ariz.

9200. HELIÁNTHUS L. SUNFLOWER

[Watson. Contributions to a monograph of the genus Helianthus. Papers Michigan Acad. Sci. 9: 305-475. 1929. Johns. Heliantheae of Iowa, III. University of Iowa studies in natural history, New Ser. 295: 337-416. 1935.]

I have given this genus considerable study not only in the herbarium and in the field but I have had most of our species under cultivation for several years for observation. Prof. Elba E. Watson named all of my specimens up to 1936 and I had for study the large collection of Ralph M. Kriebel which was named by Watson. I at first attempted to construct a key to our species using Watson's determinations. This I was not able to do. Prof. Watson in his monograph says: "Related species have a most perplexing tendency to fade into one another and in such a way that, while the typical extremes are readily enough recognized, there will always be a large number of plants that will not fully satisfy the definition of either of two species, and that can be as logically placed with one as with the other. This is flagrantly true of three groups" which ne discusses in detail. I agree with the preceding statement.

The sunflowers are extremely responsive to soil, light, moisture, and crowded conditions. Some authors credit hybridization for many departures from the normal species. I have not seen a specimen which I believe to be a hybrid. I have had 12 species under cultivation for a number of years and to prevent them from spreading I restricted them to their beds about three feet in diameter by bands of galvanized iron placed below the surface of the ground. In a few years the space in the bed became occupied and the plants began to crowd. In a bed of *Helianthus grosseserratus* I measured one plant 9 feet high with an inflorescence 2 feet long while several other plants in the same bed were but 3 feet high with a single head. In a bed of *Helianthus divaricatus* which usually has only a few heads I found one with more than 50 heads and some with a single head. I have observed unusual plants in the field. Once I found a whole colony of a species which normally has a simple stem that had branches at every node. In my beds I have cut back plants at different dates to learn what the response would be and have found it had no perceptible effect upon the degree of pubescence or length of the hairs and little or no effect upon the size, shape, and serration of the leaves.

Quantitative characters such as pubescence may be quite variable within one species and in another may be quite constant. The shape and length of the involucral bracts vary so greatly in most species that they can not be safely relied upon as characters, although in *Helianthus rigidus* the involucre is constant enough to characterize the species. Ordinarily the color of the plant is very significant although we do have both bluish green and grayish green plants of the same species. The leaves are mostly opposite, in some species more or less alternate, and rarely a specimen with ternate leaves. A study of herbarium material often reveals aberrant specimens which can not be named satisfactorily because the growth environment is not known. I have excluded 15 species that have been reported for the state. For a discussion of these see excluded species.

Cauline leaves linear, less than 1 cm wide and more than 10 times as long as wide; disk flowers red..1. *H. angustifolius.*

Cauline leaves ovate, lanceolate or linear-lanceolate, less than 10 times as long as wide.

  Receptacle flat; annuals; leaves alternate; disk flowers red.

    Heads 1-2.5 cm wide; chaff toward the center of the head conspicuously white-bearded; achenes appressed-pubescent all over at maturity...2. *H. petiolaris.*

    Heads more than 2.5 cm wide; chaff not white-bearded; achenes glabrous or only slightly pubescent at maturity..................................3. *H. annuus.*

  Receptacle convex; perennials; leaves mostly opposite or mostly alternate.

    Plants generally with fewer than 7 internodes below the inflorescence; leaves usually large, long-tapering at the base, on petioles mostly 3-10 cm long; inflorescence on vigorous plants paniculate with heads on long peduncles; depauperate plants usually with 1-3 heads and often on short peduncles...4. *H. occidentalis.*

    Plants not as above, internodes more than 7.

      Corolla lobes of disk flowers reddish, never yellow.

        Bracts of involucre oblong, glabrous on the back; cauline leaves deep green, rounded and blunt at the apex, rarely acute. (See excluded species no. 659, p. 1100)..*H. atrorubens.*

        Bracts of involucre of an ovate type, usually glabrous on the back, sometimes scabrous, shorter than the disk and appressed; cauline leaves gray green, long taper-pointed at the apex..............................5. *H. rigidus.*

      Corolla lobes of disk flowers yellow, never dark colored.

        Heads small, the disk rarely more than 7 mm wide; leaves ovate-lanceolate, thin, the lower surface conspicuously resin-dotted; petioles 1-3 cm long; rays 5-7, 1 cm long; usually flowering in August and in early September. ....6. *H. microcephalus.*

Heads not conspicuously small, more than 8 mm wide.

Leaves sessile or subsessile, rarely a few on petioles up to 3 mm long.

Lateral nerves converging with the midrib at the base of the blade, sometimes above the base; blades broadly rounded, truncate or subcordate at the base, long attenuate at the apex; stems glabrous and often glaucous, somewhat scabrous above; disk less than 1.5 cm wide; outer involucral bracts usually attenuate; peduncles more or less angled; pubescence of peduncles more or less spreading, shaggy in appearance (due to the various lengths of the hairs and the tendency of some to be appressed upward or downward)............7. *H. divaricatus.*

Lateral nerves converging with the midrib about a fourth the length of the blade above the base (rarely at the base or obscurely so—Welch no. 881); stems more or less villous with spreading hairs.

Leaves rounded at the base, usually slightly clasping, generally all opposite up to the inflorescence, soft gray-canescent on both surfaces; stems generally densely villous; rays usually 18-26...8. *H. mollis.*

Leaves narrowed at the base to a sessile or subsessile base, not at all clasping, at least a few of the upper leaves below the inflorescence alternate..9. *H. doronicoides.*

Leaves all petiolate, the petioles very short in some species and others with long, margined petioles.

Internodes of stem generally more than 20, rarely as few as 15 in depauperate plants; leaves mostly alternate, lanceolate or oblong-lanceolate.

Stems glabrous; leaves not scabrous above or only slightly so; lower and median cauline leaves with long tapering bases with petioles usually 1-2 cm long, the upper leaves with short-petioled to sessile blades; blades above the widest portion gradually tapering to the apex in straight lines, except for the short acuminate tip; lower surface of blades densely pubescent with soft, short, more or less appressed hairs..........................10. *H. grosseserratus.*

Stems scabrous or hairy at least above; leaves scabrous above.

Blades, or at least some of them, conduplicate and often falcate, decurrent on the petiole to the base, gray-canescent on both sides; pubescence dense, scabrous-setose; inflorescence racemose; flowers on short peduncles, one terminal and one from each of the upper axils; usually flowering late........ ....11. *H. Maximiliani.*

Blades neither conduplicate nor falcate; lower and median leaves with short tapering bases with short petioles, the longest not more than 1 cm long, the upper ones sessile; blades gradually tapering to an acuminate point above the widest portion, the margin of one or both sides toward the apex forming a shallow arc; lower surface of the blades sparsely pubescent, usually with long, nearly erect, stiff hairs, sometimes the hairs shorter and subappressed but always stiff; inflorescence not racemose; bracts rather evenly ciliate with multicellular hairs 1 or more mm long (no other Indiana species has bracts with such long hairs)..............
....12. *H. giganteus.*

Internodes of stem fewer than 20; leaves mostly opposite; blades ovate, ovate-lanceolate, rarely lanceolate.

Bracts closely appressed, ovate, merely acute, generally glabrous on the back, shorter than the disk.............5a. *H. rigidus* f. *flavus.*

Bracts not as above.

Plants with the blades of median leaves broadest at the base or a short distance above it, subcordate, truncate, rounded or slightly decurrent at the base, usually thick, strumose-hispid above, hispid beneath or slightly soft-pubescent to the touch, lateral nerves converging with the midrib less than 1 cm above the base; petioles generally less than 1 cm long; pubescence of peduncles and midribs generally spreading; peduncles generally short, robust, conspicuously clavate; bracts linear-lanceolate, lanceolate or narrowly ovate, very loose, sometimes recurving, about as long as the disk or a little longer; plants generally 7-10 dm high; stems scabrous-hispid, scabrous-hirsute or sometimes nearly smooth, the pubescence on the stems spreading or even retrorse on the lower internodes........................13. *H. hirsutus.*

Plants not as above; leaves on short or long petioles, usually long-decurrent at the base; peduncles rarely with a spreading pubescence.

Margins of leaves generally coarsely and regularly dentate-serrate; median and lower leaves large, usually ovate, sometimes narrower, long taper-pointed at the base, the lateral nerves converging with the midrib within the decurrent base; petioles usually 2.5-8 cm long, measured from the convergence of the lateral veins to the base of the petiole; upper leaves smaller, sessile or on short, decurrent petioles.

Stems smooth at least below, usually green; leaf blades thin, generally subglabrous beneath, the hairs restricted mostly to the principal veins and closely appressed, sometimes the lower surface rather closely pubescent with short, ascending hairs; yellow glands on the lower surface of blades usually lacking; heads small, the disk usually less than 1.5 cm wide; bracts loose and many recurving, generally as long as or longer than the disk; plants usually of dry, open woodland. ....14. *H. decapetalus.*

Stems scabrous-hispid, sometimes glabrescent except the inflorescence, usually reddish, especially in the inflorescence or greenish yellow throughout; leaf blades firm, the lower surface generally densely covered with short, erect or semi-erect hairs and yellow glands; heads usually rather large, the disk 1-1.5 cm wide; bracts exceedingly variable, usually linear-lanceolate, loosely appressed but some widely spread- or recurving, as long as or much longer than the disk, sometimes wider and shorter and much resembling those of *Helianthus rigidus*, at least the inner ones dark colored, sometimes almost black; inflorescence varying greatly in size but usually large, the leaves always alternate, the internodes usually more or less zigzag, and the branches usually more or less compressed; roots often bearing tubers; plants usually of moist, open, sunny places........15. *H. tuberosus.*

Margins of leaves more or less irregularly and shallowly serrate, or nearly entire; median and lower leaves usually ovate-lanceolate, rarely ovate or lanceolate, usually large, thick, firm, opposite, with a short or long tapering base, the lateral veins converging with the midrib within the decurrent base; lower surface of blades more or less densely covered with short hairs, these sometimes strumose in part, rarely sparsely pubescent, upper leaves sessile or subsessile; stems usually glabrous and and often glaucous, rarely scabrous at least above, generally green; heads of medium size, the disk usually about 1 cm wide or up to 1.5 cm wide; bracts usually broadly lanceolate, acuminate or rarely merely acute, usually shorter than the disk or some of them as long as or longer, loose and usually a few more or less spreading; rays 9-20, usually conspicuous; roots never thickened or tuberous..........16. *H. strumosus.*

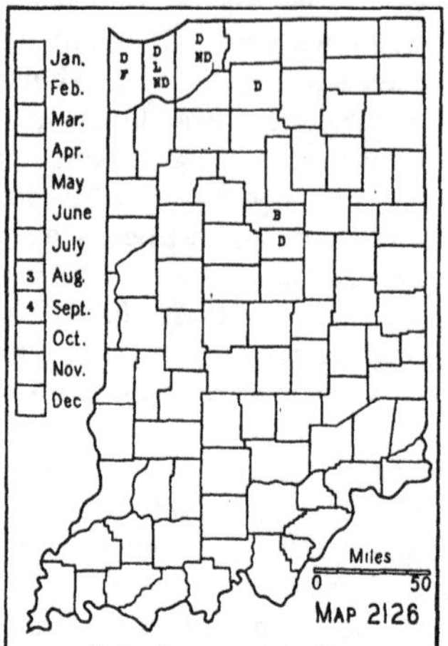

Helianthus petiolaris Nutt.

Helianthus annuus L.

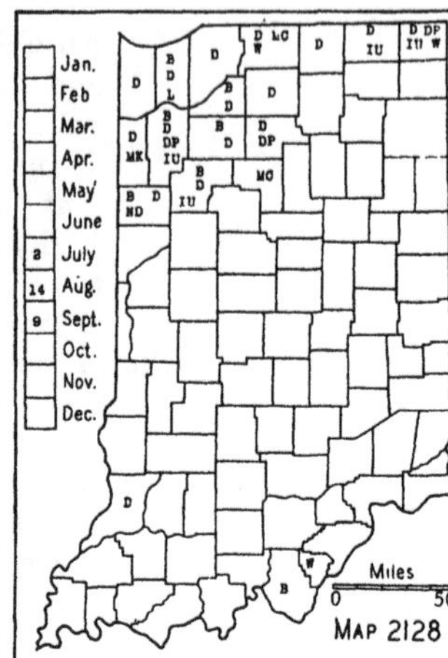

Helianthus occidentalis Riddell

1. HELIANTHUS ANGUSTIFÒLIUS L. Map 2125. In 1931 in Pike County I found this species rather common over an area of about three acres in a large creek-bottom pasture field. I noted that the cattle did not eat it. I found it also in a low pasture field in Jefferson County. Doubtless it has been introduced although the field in Pike County is far removed from a railroad and it is the second field back from a little used road. No doubt it came in through grass seed. I introduced it in our garden which has neutral soil. It gradually died out in four years. Doubtless it requires a slightly acid soil, the kind in which I found it.

N. J. to Mo., southw. to Fla. and Tex.

2. HELIANTHUS PETIOLÀRIS Nutt. Map 2126. This species probably has just begun to invade the state. It was first reported in 1900. I began to botanize the dune area in 1905 but I did not find it until 1925. It grows in very sandy soil and within the area of its distribution in the state where the sand has been disturbed it has become an abundant weed in cities and along roads and railroads.

Man. to Tex., westw. to Calif.; introduced eastw.

3. HELIANTHUS ÁNNUUS L. COMMON SUNFLOWER. Map 2127. This species has been cultivated more or less for many years in all parts of the state and there are reports of its escape from all parts of the state. I doubt whether it is a native of the state although in 1922 I found it to be a common weed along a sandy roadside and in an adjoining sandy, fallow field about 2 miles northeast of Jacksonville, Vermillion County. The plants were comparatively small, mostly from four to six feet high. Phinney in 1883 reported it as common in the prairies in Delaware County but most authors report it as an escape.

Minn. to Tex. and westw.; becoming introduced eastw.

4. **Helianthus occidentàlis** Riddell. (*Helianthus illinoensis* Gleason.) Map 2128. This species is local but not rare in the lake area. It is always found in very sandy soil and usually in moist places such as low depres-

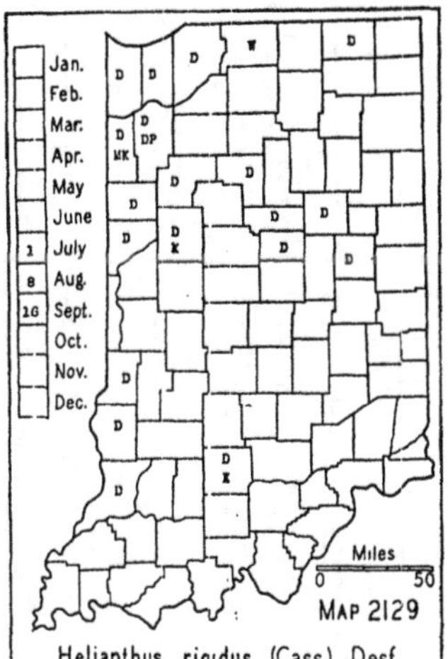

Helianthus rigidus (Cass.) Desf.

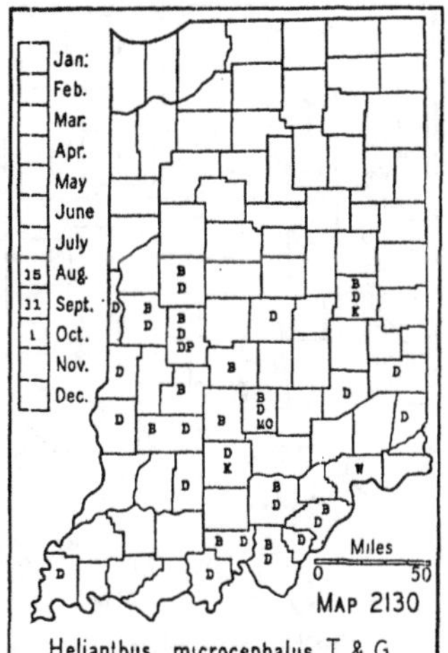

Helianthus microcephalus T. & G.

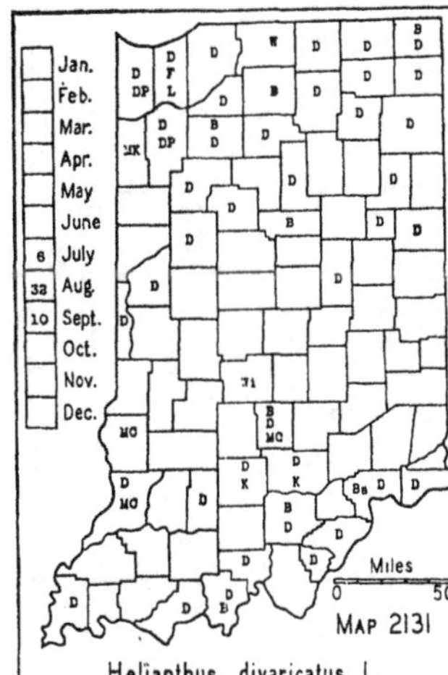

Helianthus divaricatus L.

sions in black oak woods, at the bases of the slopes of black oak woods, and sometimes on sandy knolls and ridges. In addition to the area shown on the map it has been reported from Vigo County where it doubtless formerly occurred.

Ohio to Minn., southw. to Ga. and Ark.; introduced into N. E. and N. J.

5. **Helianthus ŕigidus** (Cass.) Desf. (*Helianthus scaberrimus* of Indiana authors.) PRAIRIE SUNFLOWER. Map 2129. This is a typical prairie species and is frequent in the "western prairie" area of the state. It is also local in other parts of the state in relict prairie areas.

According to Watson it is common in plains and moist prairies from western Canada to Texas; introduced and becoming common east of the Mississippi River but rare in the eastern United States.

5a. **Helianthus rigidus** (Cass.) Desf. forma **flàvus**, f. nov., disco flavo. (*Helianthus laetiflorus* Pers. of Indiana authors.) Disk flowers yellow. The type, Deam, no. 57312, was found in a prairie habitat in Newton County, Indiana, and is deposited in the Deam Herbarium. So far as I can determine the species and form are exactly alike except in the color of the disk flowers. I suspected this and in 1936 I cruised the western part of the state for over a thousand miles to learn if both the red and yellow forms could be found in the same colony. At last I found a small red colony with a single yellow flower in it.

I am interpreting this form as an "albino" of the species. I believe the yellow flowered form is simply a strain of the species that is no longer able to develop the anthocyanin of the species. For an exhaustive treatment of the subject see Onslow's "The Anthocyanin Pigment of Plants."

Probably throughout the range of the species.

6. **Helianthus microcéphalus** T. & G. SMALL WOOD SUNFLOWER. Map 2130. This is strictly a woodland sunflower and is well but sparsely distributed in the southern part of the state. It does not form colonies like most of our sunflowers and usually only a single specimen or a few are

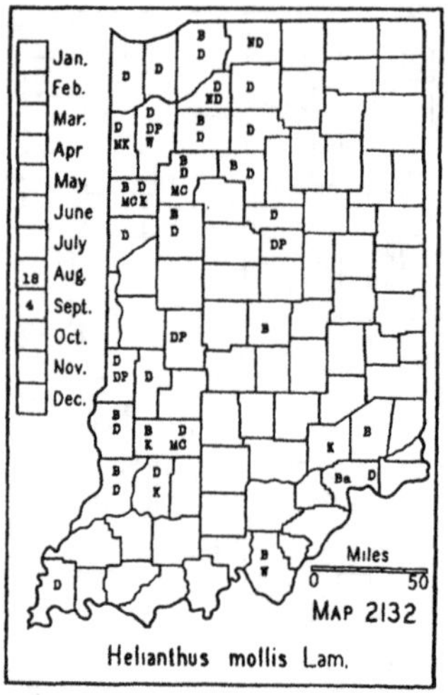

Helianthus mollis Lam.

Helianthus doronicoides Lam.

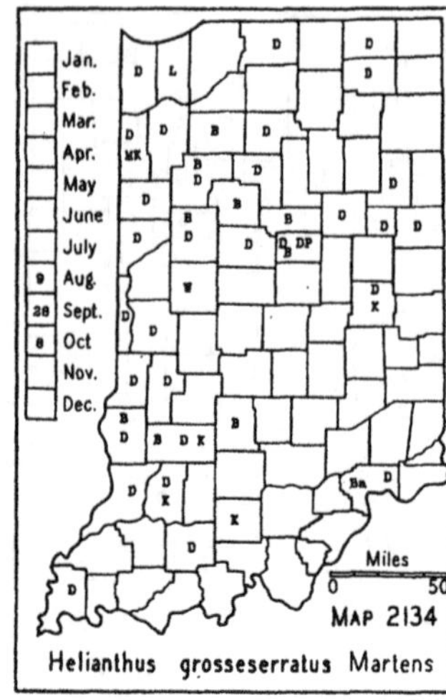

Helianthus grosseserratus Martens

found together. Its habitat is a dry or sandy wooded slope, and it is usually associated with black and white oak. It was reported from Porter County by Peattie but I have not seen a verifying specimen.

Pa., Ind. to Mo., southw. to Fla. and Miss.

7. **Heliànthus divaricàtus** L. Map 2131. This is one of the most common of our sunflowers. It is usually rather frequent on the crests and slopes of white oak and white and black oak ridges and in the sun along roadsides and fences. It is rarely found in moist rich soil except in the prairies.

Maine to Lake Winnipeg, southw. to Fla. and La.

8. **Helianthus móllis** Lam. ASHY SUNFLOWER. Map 2132. This sunflower is generally found in black sandy soil in prairie habitats but is sometimes found in moist, hard, white clay soil in the Illinoian drift area. It is rather frequent in the northwestern counties and along the western part of the state, becoming very rare or absent in the central counties and local in the southern counties. Doubtless it prefers a slightly acid soil because it soon dies out when transferred to a neutral soil.

Mass. to Iowa, southw. to Ga. and Tex.

9. **Helianthus doronicoìdes** Lam. Map 2133. Doubtless this is one of our rarest sunflowers. In 1930 I found a colony in an old woods pasture in Marshall County and Miss Madge McKee found it in a prairie habitat along the railroad near Goodland, Newton County. There are a few reports of it from the state but the species of sunflowers were not well understood by our earlier authors so that it is best not to rely upon them. I transplanted the specimen I found and it is still growing vigorously. Watson, who visited me, said it was the first wild specimen he had ever seen and writes that it is rare.

N. J. to Mo.

10. **Helianthus grosseserràtus** Martens. SAWTOOTH SUNFLOWER. Map 2134. This sunflower prefers the moist, black, sandy soil of prairie habitats and is frequent in them in the western part of the state. It grows also in

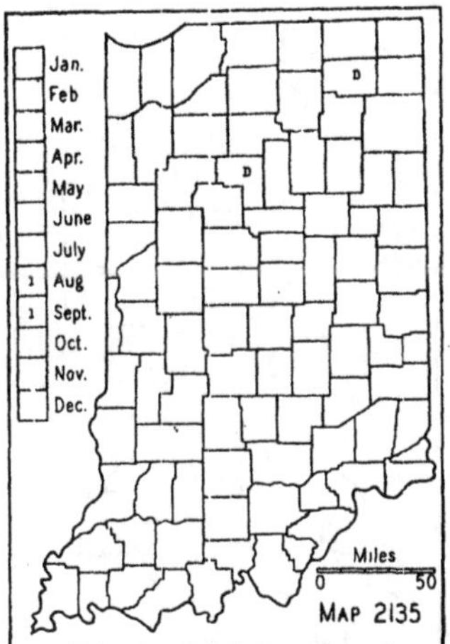

Helianthus Maximiliani Schrad.

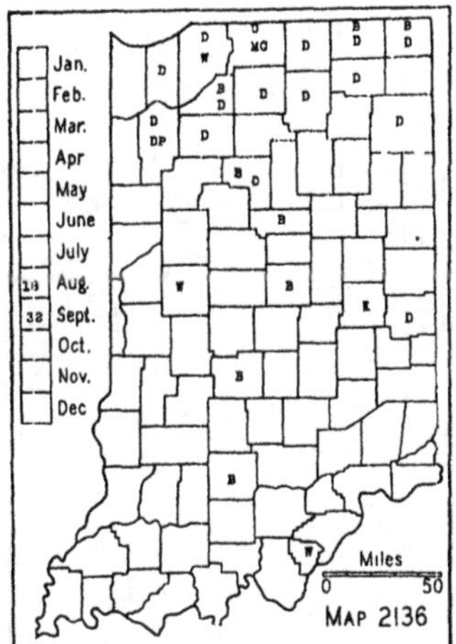

Helianthus giganteus L.

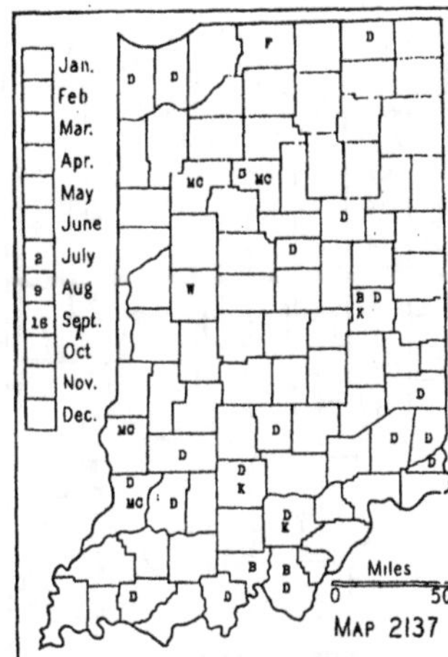

Helianthus hirsutus Raf.

moist, hard, white clay soil in the western part of the Illinoian drift with other typical prairie plants. It is now found mostly along dredged ditches, roadside ditches, and streams and in low woods.

Maine to N. Dak., southw. to Va. and Okla. (Watson).

11. HELIANTHUS MAXIMILIÀNI Schrad. MAXIMILIAN SUNFLOWER. Map 2135. This sunflower has probably been introduced into Indiana. I found a few plants on the sandy shore of the east side of Diamond Lake in Noble County with no habitation within half a mile. A large colony was found by Charles M. Ek along the Pennsylvania Railroad in Cass County about 7 miles northwest of Kokomo. It has been reported from Lake County by Peattie but I have not seen his specimen. It was also reported from St. Joseph County by McDonald. This report was based upon my specimen so named by Watson which I am now referring to *Helianthus giganteus.*

Minn. and Sask., southw. to Tex.; adventive eastw.

12. **Helianthus gigantèus** L. GIANT SUNFLOWER. Map 2136. This sunflower grows in moist or wet mucky soils and is generally found in places such as decadent tamarack bogs, marshes, low borders of lakes, and in wet prairie habitats. It is frequent in our northern counties, becoming local south of the lake area. Doubtless it occurs in southwestern Indiana although I have not seen a specimen.

Maine to Sask., southw. to N. C. and Colo.

13. **Helianthus hirsùtus** Raf. Map 2137. In the southern part of the state this sunflower generally grows in dry clay soil on the crests and slopes of open black and white oak woods and is found also in like soil conditions along roadsides and fences. In the northern part of the state it is generally found in dry sandy soil on slopes in open black and white oak woodland and in like soil habitats along roadsides.

Pa. to Wis. and Kans., southw. to Ga. and Tex.

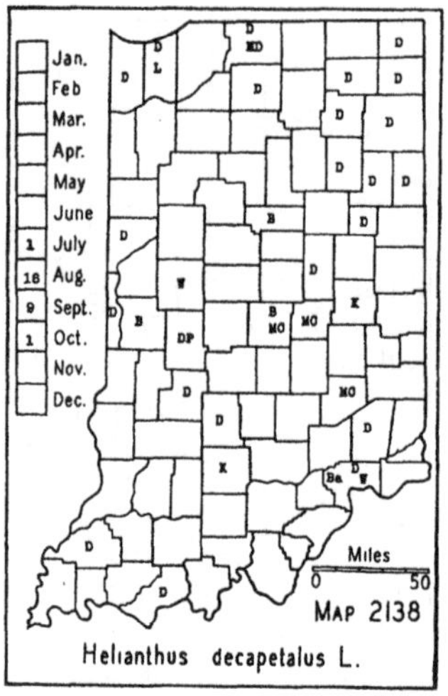

Helianthus decapetalus L.

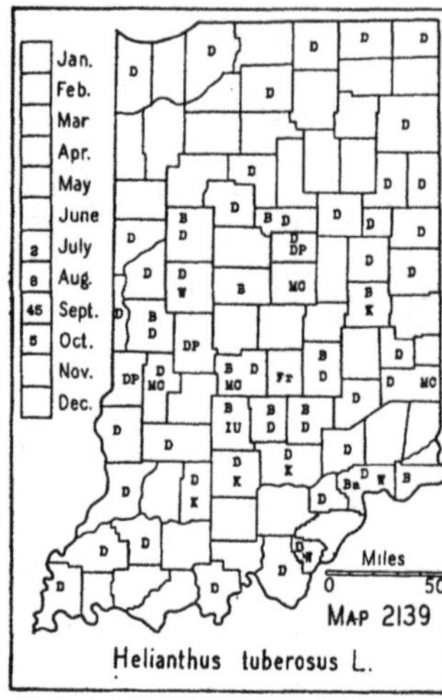

Helianthus tuberosus L.

Helianthus strumosus L.

14. **Helianthus decapétalus** L. THINLEAF SUNFLOWER. Map 2138. This sunflower is usually found in dry woods with oaks and less frequently with sugar maple. It is rarely found in the open or in moist locations. Frequent to infrequent throughout the state.

Cent. Maine, w. Que. to Minn., southw. to Ga., Tenn., and Mo.

15. **Helianthus tuberòsus** L. JERUSALEM ARTICHOKE. Map 2139. This is one of our most common sunflowers and is frequent throughout the state. It grows in the open in moist soil along streams, ditches, and roadsides.

This species is quite variable and gives more trouble in naming than any other species. Some authors rely upon the tuberous roots for identification. In September, 1936, which was a dry year, I dug in our garden 25 specimens and then went to an old fence row and dug many more and I failed to find a single tuber. I regard the dark color of the bracts and the pubescence on the lower surface of the leaves as the most reliable characters for the identification of this species. The shape and length of the bracts are too variable to consider although in some instances they are quite characteristic and are confirmatory characters.

N. A. east of the Rocky Mts., according to Watson.

16. **Helianthus strumòsus** L. Map 2140. This is a frequent sunflower in the lake area in dry woods and in dry sandy soil along roadsides and elsewhere in a similar soil. South of the lake area it becomes rare, local or absent.

Specimens of this species with short petioles and the blades of the leaves nearly round at the base closely approach *Helianthus divaricatus*. In separating the two species I have relied upon the convergence of the lateral veins of the leaves. In this species they always converge slightly above the base while in *Helianthus divaricatus* they converge at the base of the blade.

Maine, Ont. to Minn., southw. to Ga. and Ark.

9215. ACTINÓMERIS Nutt.

1. **Actinomeris alternifòlia** (L.) DC. (*Verbesina alternifolia* (L.) Britt.) YELLOW IRONWEED. Map 2141. Infrequent to frequent or locally

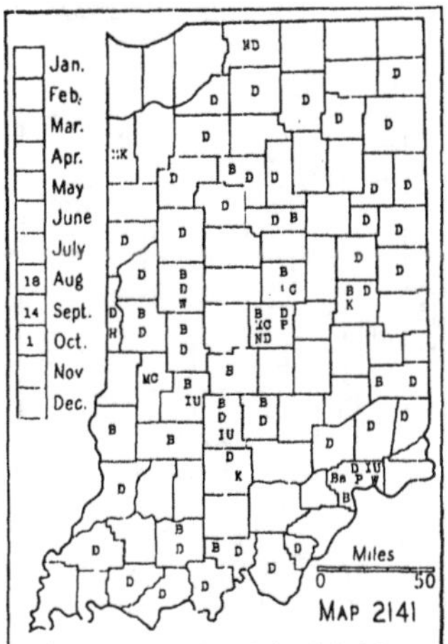

Actinomeris alternifolia (L.) DC

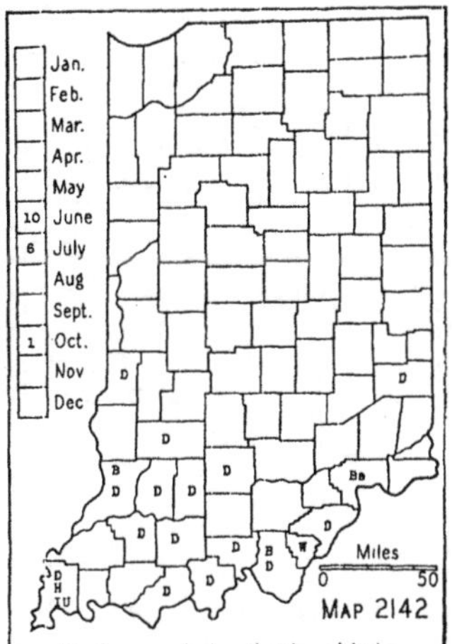

Verbesina helianthoides Michx

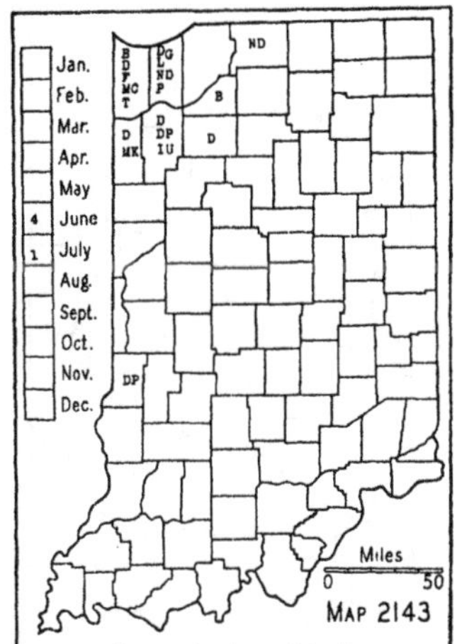

Coreopsis lanceolata L.

common throughout the state, although there are no reports from the northwestern counties. It is a coarse weed preferring moist situations, and found usually in alluvial soil along streams in open woodland and pastures.

N. Y. and Ont. to Iowa, southw. to Fla. and La.

9218. VERBESÌNA L. CROWNBEARD

1. **Verbesina helianthoìdes** Michx. (*Phaethusa helianthoides* (Michx.) Britt.) Map 2142. Infrequent in the southern part of the state. It is generally found on open black and white oak slopes and less frequently on level ground in sandy soil in open woodland and along roadsides.

Ohio to Iowa, southw. to Ga. and Tex.

9227. COREÓPSIS L.

[Sherff. Revision of the genus Coreopsis. Field Mus. Nat. Hist. Publ. Bot. Ser. **11**: 277-475. 1936.]

Leaves entire.
  Plants essentially glabrous..1. *C. lanceolata.*
  Plants pubescent, at least the leaves and base of the stem..........................
  ....1a. *C. lanceolata* var. *villosa.*
Leaves not entire.
  Leaves sessile, deeply 3-cleft, but not to the base....................2. *C. palmata.*
  Leaves petiolate (at least the lower ones).
    Style tips truncate or nearly so; outer involucre shorter than the inner; rays yellow with a more or less crimson brown base; achenes linear-elliptic, about 2 mm long, wingless; leaves bipinnately parted, the segments linear-lanceolate to linear; annual. (See excluded species no. 672, p. 1101)........*C. tinctoria.*
    Style tips cuspidate; outer and inner involucres nearly equal; rays yellow the entire length; achenes winged; leaves 3-5-parted; perennial.
      Rays palmately lobed; achenes orbicular, about 2-5 mm long; lower leaves simple.
      ....3. *C. grandiflora.*
      Rays entire, blunt; achenes elliptic, mostly 5-6 mm long; none of the leaves simple.

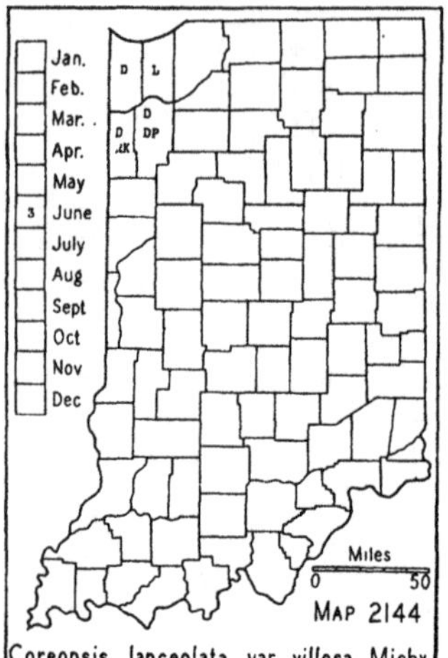

Coreopsis lanceolata var. villosa Michx.

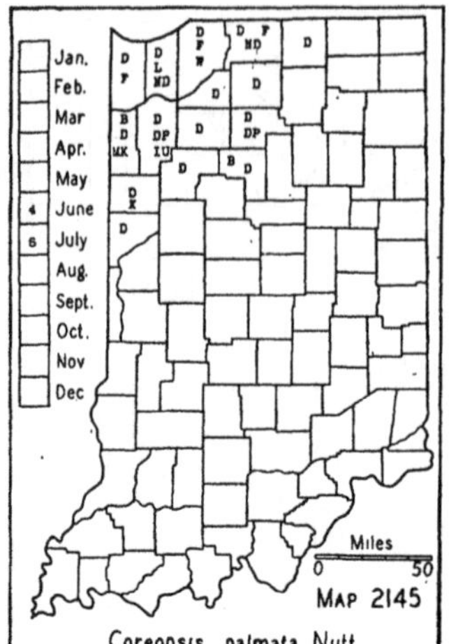

Coreopsis palmata Nutt.

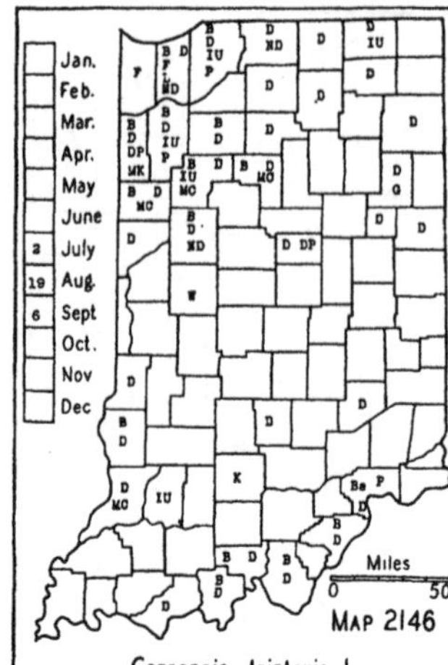

Coreopsis tripteris L.

Leaves and involucres glabrous..4. *C. tripteris.*
Leaves and involucres not glabrous.
Lower surface of blades and involucres more or less pubescent..4a. *C. tripteris* var. *Deamii.*
Lower surface of blades pubescent and involucres glabrous..4b. *C. tripteris* var. *intercedens.*

1. **Coreopsis lanceolàta** L. LANCE COREOPSIS. Map 2143. Infrequent in very sandy, dry soil on open dunes and knolls in the northwestern counties. The species and variety are sometimes closely associated. The species is much cultivated in gardens and doubtless our Marion County report should be considered a garden escape. It has been reported also from St. Joseph, Steuben, and Vigo Counties, where no doubt, it is native. The Vigo County specimen was collected by Blatchley at Five-mile Pond. I have, however, very thoroughly botanized Steuben County without finding it; it may be a garden escape in this county.

Mich. and Lake Superior, southw. to Fla., Ala., La., se. Tex., and n. N. Mex.

1a. **Coreopsis lanceolata** var. **villòsa** Michx. (*Coreopsis crassifolia* Ait.) Map 2144. My specimens and our reports of this variety are all from the few northwestern counties shown on the map. The habitat is the same as that of the species, but the variety is less frequent.

Va., S. C., Ill., Mo., and w. Ark., southw. to Fla., Ala., and La.

2. **Coreopsis palmàta** Nutt. FINGER COREOPSIS. Map 2145. All of my specimens and all of the reports are from the northwestern part of the state. It is infrequent and is found in dry, sandy soil in open woodland and in sandy, prairie habitats.

Ind., Minn. to Man., southw. to Okla.

3. COREOPSIS GRANDIFLÒRA Hogg. BIG COREOPSIS. According to Nieuwland, this species is well established along the Lincoln Highway near South Bend and in a few other places in St. Joseph County. Doubtless it has been introduced from the west. The seed may have been scattered

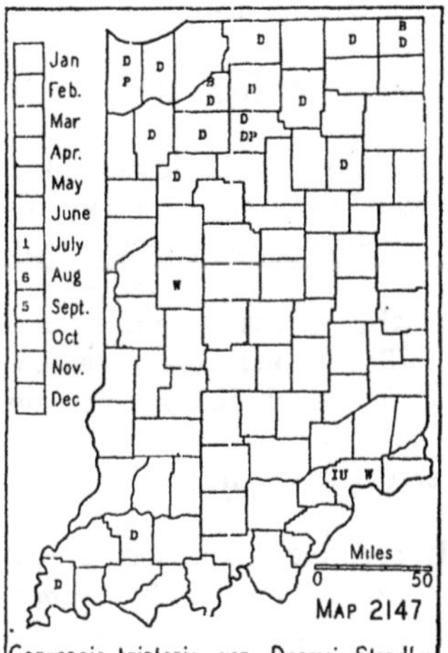

Coreopsis tripteris var. Deamii Standley

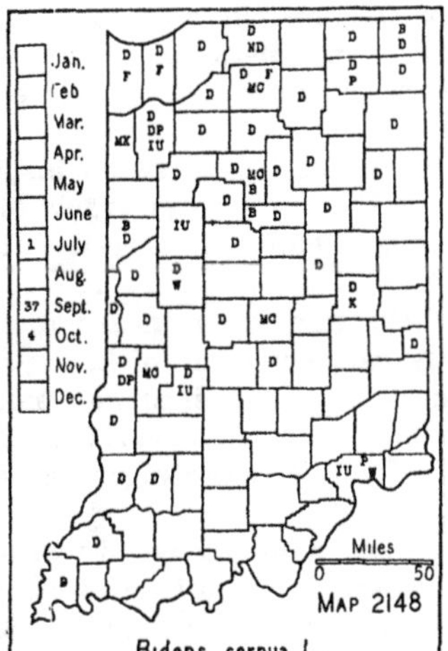

Bidens cernua L.

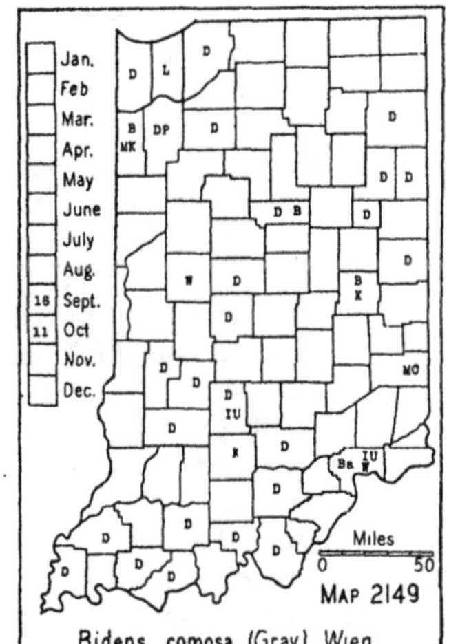

Bidens comosa (Gray) Wieg.

here along the highway by some sentimental, trans-continental tourists who acted upon the ill advice published in a magazine a few years ago. It was recommended that tourists should scatter seeds of conspicuous flowers along the roadsides from coast to coast and from the Gulf of Mexico northward. This produced a storm of indignation from botanists who knew that such a procedure would destroy the natural range of species.

The species has also been found about 3 miles south of Fort Wayne in an open woods which has been used for years as a dump.

Md. to Mo. and e. Kans., southw. to Ga. and Tex.

4. **Coreopsis trípteris** L. TALL COREOPSIS. Map 2146. Frequent in the lake area, where it is usually found in very sandy soil in open woodland and fallow fields, in prairie habitats, and along roadsides. It is rare to local in the southern part of the state, where it is found in small prairie areas or in open woodland in the knobstone area.

Mass., s. Ont., and Wis., southw. to Ga., Miss., w. La. and e. Kans.

4a. **Coreopsis tripteris** var. **Dèamii** Standl. (Rhodora 32: 33. 1930.) Map 2147. This variety has nearly the same distribution as the species, but it is much less frequent. I have had both it and the species under cultivation for more than ten years. In addition to the characters given in the key, the variety may be separated at a distance by its darker green color and earlier flowering period.

Pa., Mich., Ill., and Mo., southw. to N. C., Ga., and Ark.

4b. **Coreopsis tripteris** var. **intercèdens** Standl. (Rhodora 32: 33. 1930.) This form has been reported from the dune area, and I have a specimen from Whitley County. Doubtless it is rare. I believe this is only a glabrate form of the preceding variety.

Ill., Ind., Md., and N. C.

9237. BÌDENS L.

[Sherff. The Genus Bidens. Field Mus. Nat. Hist. Publ. Bot. Ser. **16**: 1-709. 1937.]

Leaves all simple, sometimes the median and basal ones cleft or 3-parted but, if parted, the terminal segment not petiolate.

Rays large and showy, longer than the disk; heads somewhat nodding at anthesis; stamens exserted.

Stem erect, rarely decumbent, usually somewhat hispid at least on the lower internodes; leaves connate at the base; outer bracts of heads unequal; rays wanting or less than 1.7 cm long; chaff with a yellow tip; margins of achenes pale. ....1. *B. cernua.*

Stem decumbent at the base and often floating, smooth; leaves sessile; outer bracts subequal, rarely exceeding the disk; rays 1.5-3 cm long; chaff reddish brown at the tip; margins of achenes not pale. (See excluded species no. 673, p. 1102)............*B. laevis.*

Rays scarcely exceeding the disk or wanting; heads erect; stamens included or exserted.

Stamens included; disk-flowers pale yellow, 4-toothed; corollas yellowish green; inner mature achenes mostly 7-8 mm long, usually 3-awned.......2. *B. comosa.*

Stamens exserted; disk-flowers orange yellow, 5-toothed; inner mature achenes mostly about 6 mm long, the outer 3-awned, the inner 4-awned.

Primary cauline leaves 3-lobed, the remaining ones usually not lobed, subsessile or with short, margined petioles............3. *B. connata.*

Primary cauline leaves simple, as are all the other leaves, rarely a cauline leaf lobed; petioles usually long and scarcely margined............3a. *B. connata* var. *petiolata.*

Leaves pinnate.

Inner mature achenes 12-18 mm long, 4-angled, somewhat quadrangular, narrowed at the top; awns 4, with retrorse barbs............4. *B. bipinnata.*

Inner mature achenes not as above.

Achenes with upwardly barbed or hispid awns or bidentate with the sides of the achenes upwardly pubescent.

Outer involucral bracts 3-5, generally 4, glabrous or nearly so; heads small; rays wanting; inner mature achenes 1-1.5 mm wide; awns mostly 1-2 mm long............5. *B. discoidea.*

Outer involucral bracts generally 8 or more, ciliate or hispid; rays showy, usually about twice as long as the head.

Inner mature achenes 1-2 mm wide, usually about 1.5 mm wide; awns generally 0.5-3 mm long............6. *B. coronata.*

Inner mature achenes generally 2.5-3.5 mm wide.

Awns generally 2.5-4 mm long............7. *B. aristosa.*

Awns mere teeth, usually about 0.5 mm long.....7b. *B. aristosa* var. *mutica.*

Achenes with downwardly barbed awns.

Rays small, inconspicuous; outer bracts mostly spatulate, the larger ones generally 1.5-3 mm wide, ciliate but not hispid.

Outer involucral bracts 6-8; inner mature achenes 6-9.5 mm long, 2-2.75 mm wide; awns 3-5.5 mm long, generally 3-4 mm long.........8. *B. frondosa.*

Outer involucral bracts 10-16, foliaceous, longer and wider than in the preceding; inner mature achenes mostly 8-12 mm long; awns generally 4-5.5 mm long............9. *B. vulgata.*

Rays conspicuous, generally twice as long as the disk; outer involucral bracts linear, or very narrow and widest below the middle, all generally less than 1.5 mm wide, densely hispid.

Outer involucral bracts 8-10, ciliate, shorter than the inner bracts............7a. *B. aristosa* var. *Fritcheyi.*

Outer involucral bracts 12-20, coarsely hispid, mostly longer than the inner bracts............10. *B. polylepis* var. *retrorsa.*

1. **Bidens cérnua** L. NODDING BUR-MARIGOLD. Map 2148. More or less frequent in the northern part of the state and becoming infrequent to rare

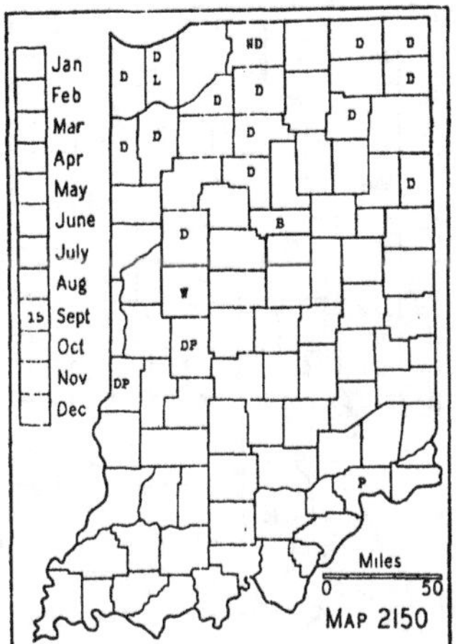

MAP 2150

Bidens connata Muhl

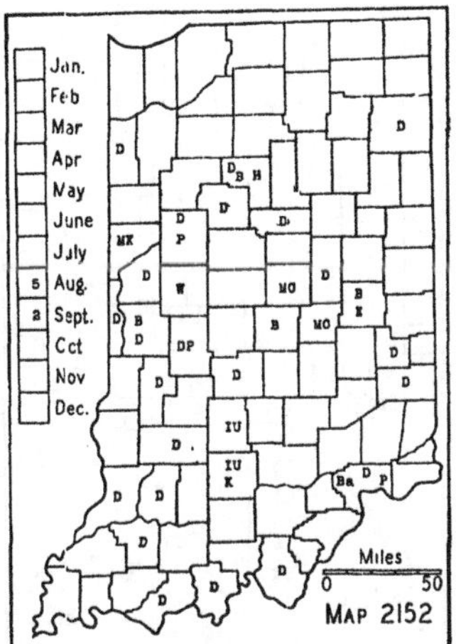

MAP 2152

Bidens bipinnata L.

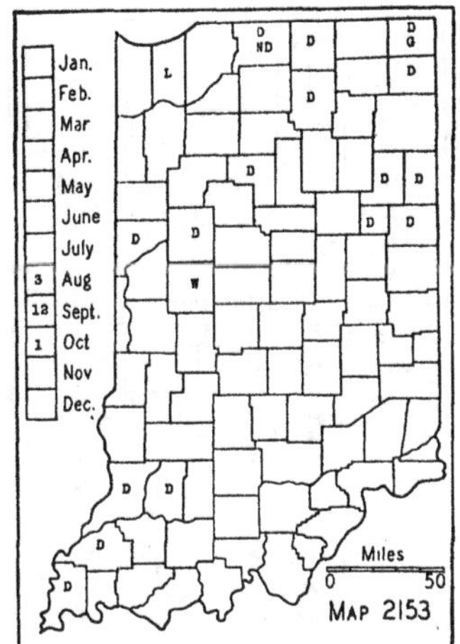

MAP 2153

Bidens discoidea (T. & G.) Britt

south of the lake area. It is found in wet places in marshes, in bogs and ditches, and on the borders of lakes, rivers, creeks, ponds, and swamps. On the whole, this species prefers a wetter and more springy habitat than the next two species. The leaves are variable in their shape and in the number and length of the teeth of their margins. Some authors have segregated these variations as varieties but I have not thought them worthy of naming.

P. E. I. to Hudson Bay and B. C., southw. to N. C., Mo., and Calif.

2. **Bidens comòsa** (Gray) Wieg. Map 2149. Probably more or less infrequent to frequent throughout the state in moist or wet habitats about lakes, along streams and ditches, and on borders of ponds and swamps.

N. S. to Minn., southw. to N. J., Ky., and Colo.

3. **Bidens connàta** Muhl. Map 2150. The typical form of this species is apparently rare in Indiana. I have specimens from only Lagrange and Starke Counties. My specimens are from the moist, sandy shores of lakes.

Que. to Mich., southw. probably to Ga., Mo., and Nebr.

3a. **Bidens connata** var. **petiolàta** (Nutt.) Farw. (See Sherff, Monograph genus Bidens, p. 257. 1937.) My plants are all from the lake area, although there are reports for it throughout the state. It is found mostly in wet places in woods, dried-up swamps and ponds, and less frequently on the borders of lakes and streams. It must be kept in mind that since the species of *Bidens* were not well separated by our older manuals, there were many wrong determinations of the species as now understood.

Range the same as that of the species but more frequent.

4. **Bidens bipinnàta** L. SPANISH NEEDLES. Map 2152. Infrequent throughout the state, although there are no records from the dune area or the extreme northern counties. It is found in both moist and very dry, sandy places. I have never seen it abundant, only once common over a small area, and only once in a cultivated field. All of my specimens are from open woodland and along railroads and canals.

R. I. to Nebr., southw. to Fla., Kans., and Ariz.

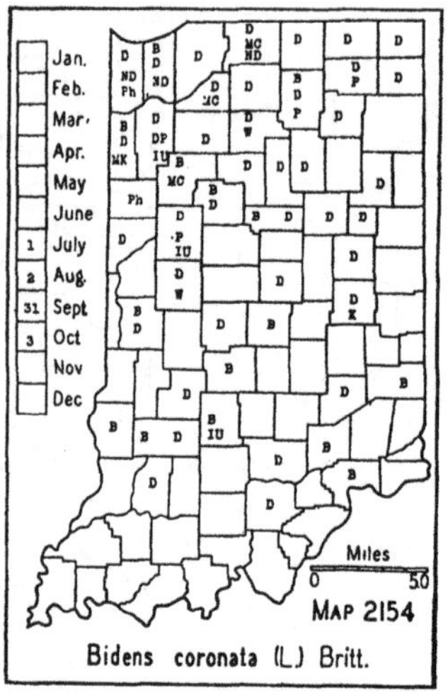

MAP 2154
Bidens coronata (L.) Britt.

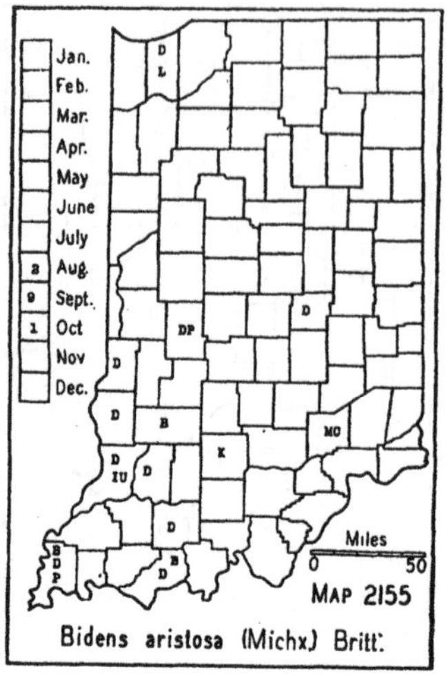

MAP 2155
Bidens aristosa (Michx.) Britt.

MAP 2156
Bidens aristosa var. Fritcheyi Fern.

5. **Bidens discoìdea** (T. &. G.) Britt. Map 2153. Infrequent throughout the lake area and probably local in the remainder of the state. Most of my specimens grew on old logs in dried-up swamps. The species is generally found on the borders of dried-up swamps, in wet woods, and on the borders of lakes.

N. S., s. Que. to Minn., southw. to Va., Ohio, La., and Tex.

6. **Bidens coronàta** (L.) Britt. (*Bidens trichosperma* (Michx.) Britt.) Map 2154. Fernald (Rhodora **40**: 348-351. 1938) has divided this species into four varieties. His publication came too late for me to study our Indiana specimens, although I find we have both the typical form and var. *tenuiloba* (Gray) Sherff. Frequent in the lake area and local southward. In the lake area before drainage it sometimes covered acres of marsh land and was the source of "Spanish Needle" honey. Where it is found, it usually forms dense colonies. Its habitat is in marsh land, tamarack bogs, springy places, and low places along streams and ditches. Very narrow-leaved forms are regarded by some authors as belonging to a variety, but I have not recognized this vegetative fluctuation.

Mass. to Minn., southw. to Ga., and Ky.

7. **Bidens aristòsa** (Michx.) Britt. Map 2155. This is a western species that has invaded the western part of the state. My Hancock County specimen was found along a railroad. Our specimens are from low roadsides and are mostly from low, fallow fields which have a hard, white, clay soil.

Maine to Minn., southw. to Va., Mo., and s. Tex.

7a. **Bidens aristosa** var. **Frítcheyi** Fern. (Rhodora **15**: 78. 1913.) Map 2156. Our specimens of this species are all from moist roadsides.

Ind. and Ky., westw. to Ill. and Mo.

7b. **Bidens aristosa** var. **mùtica** Gray ex Gattinger. (Rhodora **15**: 78. 1913.) Map 2157. In wet prairie habitats and along the Kankakee River in

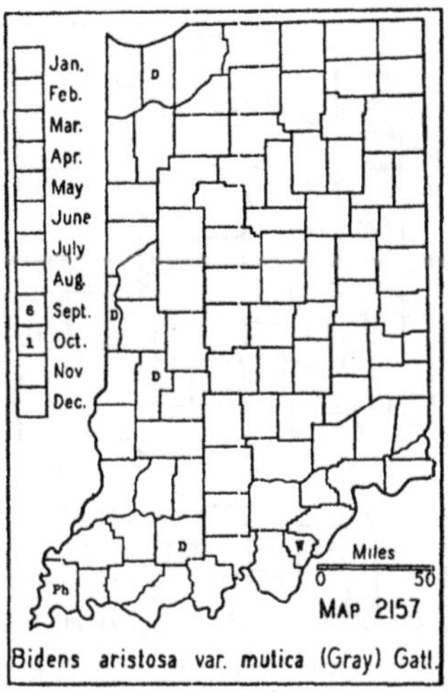

Bidens aristosa var. mutica (Gray) Gatt.

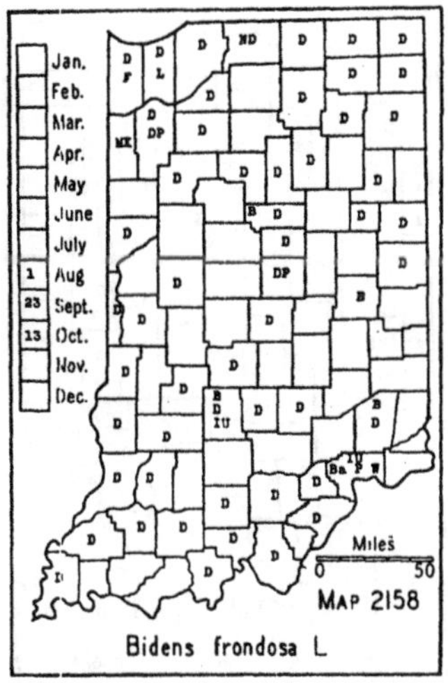

Bidens frondosa L.

Bidens vulgata Greene

Porter County, in a prairie habitat in Vermillion County, and in wet, hard clay soil in fallow fields in other places.

Mass. and Va., westw. to Ill. and Mo.

8. **Bidens frondòsa** L. Map 2158. Frequent to common or abundant throughout the state in moist places in stubble and fallow fields, woodland, and ditches and along roadsides.

Newf. to B. C., southw. to Fla., Tex., and Colo.

9. **Bidens vulgàta** Greene. Map 2159. Frequent to common in all parts of the state although there are no records from the dune area. It is found usually in a moist habitat in woodland, stubble and fallow fields, and waste places and along roadsides. This species varies greatly in the density and harshness of its pubescence. The var. *puberula* (Wieg.) Greene has been reported from Indiana but I am now referring these reports to the species.

Que. to B. C., southw. to N. C., Colo., and Calif.

10. **Bidens polýlepis** Blake var. **retrórsa** Sherff. (Sherff. The genus Bidens, p. 220. 1937.) In 1921, I collected this form in Owen County along the roadside about a half mile north of Coal City, and in 1932, I found large colonies of it along the roadside just south of Coal City. It grew in hard, white, clay soil and, no doubt, it has a wider distribution than our collections indicate.

Ohio and Ind. to Mo.

9237A. MEGALODÓNTA Greene

1. **Megalodonta Béckii** (Torr.) Greene. (*Bidens Beckii* Torr. of Gray, Man., ed. 7.) WATER MARIGOLD. Map 2160. Floating in still, shallow water of bayous of lakes and rivers. This species has been reported from Fulton, Kosciusko, Lake, Marshall, Starke, Steuben, and Whitley Counties. Doubtless it was formerly found throughout the lake area but the settlement of all lake fronts has destroyed it. Another reason why it is not commonly

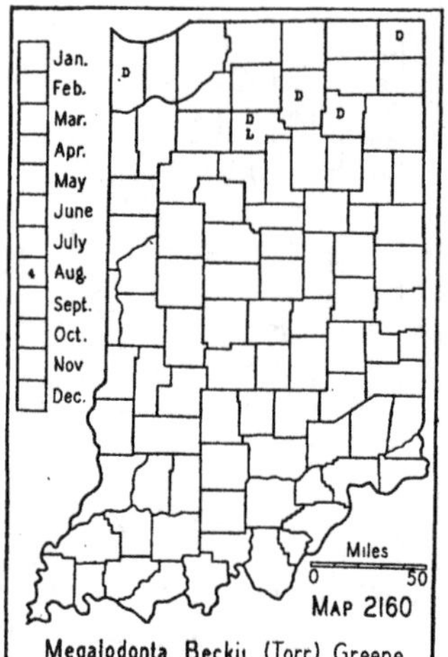

Megalodonta Beckii (Torr.) Greene

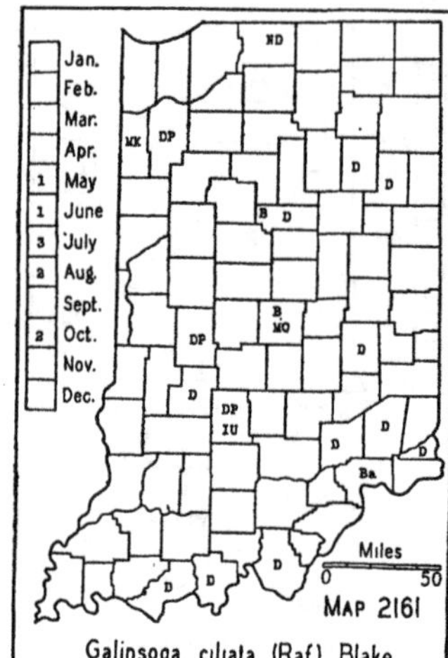

Galinsoga ciliata (Raf) Blake

Hymenopappus carolinensis (Lam.) Porter

reported is because it is inconspicuous except at its flowering time, which is of short duration.

Que. to Man., southw. to N. J. and Mo.

9246. GALINSÒGA R. & P.

1. GALINSOGA CILIÀTA (Raf.) Blake. (Rhodora 24: 35. 1922.) (*Galinsoga parviflora* Cav. var. *hispida* DC.) QUICKWEED. Map 2161. This pernicious weed was first reported in 1911 from Putnam and Ripley Counties. Since that time it has been discovered in several other counties. It is probably found in cultivated fields in every county along the Ohio River. I found it to be a common weed in the park and adjacent lots in Rushville, Rush County, in 1925. This weed will, no doubt, eventually become a pest in all parts of the state.

Nat. of tropical America; throughout the U. S. and s. Canada, southw. to S. A.

9253. MÀDIA Molina

1. MADIA CAPITÀTA Nutt. This species was found July 21, 1929, by Paul C. Standley on an open bank in Dune Forest at Tremont, Porter County. He says: "About a dozen plants." It is undoubtedly a migrant, but on account of its weedy nature it may become established.

Weed in waste places from Oreg. to Calif.

9292. HYMENOPÁPPUS L'Hér.

1. HYMENOPAPPUS CAROLINÉNSIS (Lam.) Porter. Map 2162. This species was found first by Blatchley in 1890 in Vigo County on a sandy hillside northeast of the Seventh Street Bridge across Lost Creek. I found it in 1930 in three sandy, fallow fields in Starke County, three miles north and one and a half miles east of North Judson. I also found a colony in this vicinity in an open, sandy woods. It is probably established here. It

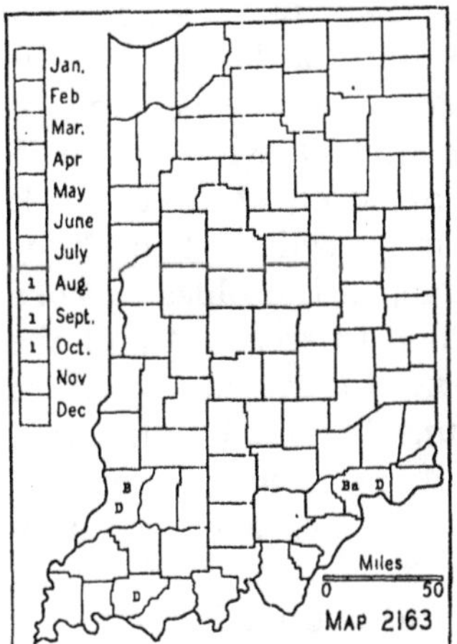

Helenium tenuifolium Nutt.

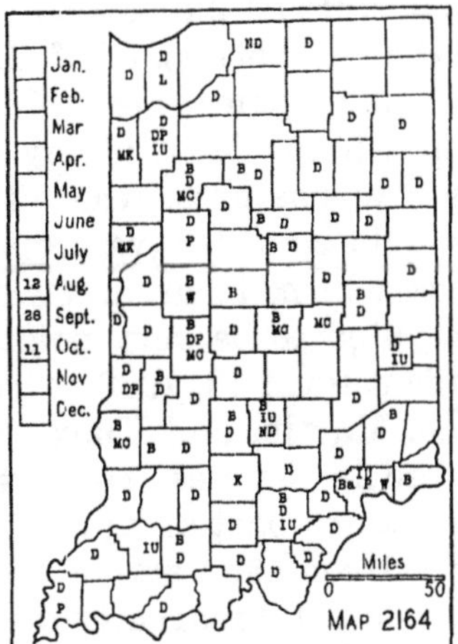

Helenium autumnale L

Helenium nudiflorum Nutt

was found in 1925 in Vigo County in a pasture north of Terre Haute by A. R. Bechtel.

S. C., n. Ind. to Kans., southw. to Fla. and Tex.

9305. HELÈNIUM L.

Disk of heads yellow; rays fertile.
 Leaves all linear-filiform, entire.................................1. *H. tenuifolium.*
 Leaves lanceolate to ovate-lanceolate, sometimes broadest above the middle.........
 ....2. *H. autumnale.*
Disk of heads brownish purple; rays sterile, yellow or partly or wholly purple..........
 ....3. *H. nudiflorum.*

1. HELENIUM TENUIFÒLIUM Nutt. BITTERWEED. Map 2163. This species has only recently appeared in this state, and, no doubt, in time will become an obnoxious weed at least in the southern part of the state. I first found it in 1931 in a 3-acre hogyard and it covered at least a fourth of the area. Hogs in the yard did not feed upon it. It is a native of the southern states and is introduced northward.

Mass. to s. Ind. and Mo., southw. to Fla. and Tex.

2. **Helenium autumnàle** L. COMMON SNEEZEWEED. Map 2164. I have included all of the forms of this complex species under this name. The plants show a wide variation in the shape and size of the leaves and in the number of heads on each plant, their size, and the length of the rays. The heads of some plants are about 8 mm wide and others are about 16 mm; the rays of some plants are about 6 mm long while others will have rays about 20 mm long. The pappus of the achenes is extremely variable, as is also the color of the hairs on the bodies of the achenes, these varying from white to reddish brown. This species is frequent throughout the state but is never found in very large colonies and never becomes dominant as do the other two species. It is said to be poisonous to stock. It grows in moist soil, usually in the open, along ditches and streams and about lakes and ponds.

W. Mass., w. Que., Man. to Oreg., southw. to Fla. and Nev.

3. **Helenium nudiflòrum** Nutt. PURPLEHEAD SNEEZEWEED. Map 2165. Apparently restricted to the southern part of the state, although Peattie cites specimens found in Lake County near Miller. It is a weed and is likely to appear almost anywhere. Most of my specimens were found in moist, hard, white clay soil in pastures, where it often covered acres. It seems to prefer a slightly acid soil. I have a specimen collected in Posey County in 1878 by Schneck which, to my knowledge, is the oldest record of it in the state. It is reputed to be very poisonous to stock. I add the following note which I made August 19, 1933: "Today I traveled over U. S. Road 50 through Lawrence, Martin, and Knox Counties, and I found this species to be a common weed in the western part of Lawrence County, in Martin County, and in the eastern part of Knox County. I noted it in many fields where it formed almost complete stands over 3-5 acres. I saw hogs and cattle in some of the fields but apparently they did not eat it."

Conn., Mich. to Mo., southw. to Fla. and Tex.

9312. DYSSÒDIA Cav.

1. DYSSODIA PAPPÒSA (Vent.) Hitchc. FETID MARIGOLD. Map 2166. This species has been reported from all parts of the state. A few authors remark about its relative abundance. J. M. Coulter (Bot. Gaz. 2: 146. 1877) in a report covering a trip through Floyd and Harrison Counties says: "Hardly absent from the roadside for a 30-mile trip." Schneck, in his report of the plants of the Lower Wabash Valley, says: "Along the roadsides in considerable numbers. This appears to be a new-comer in our locality." Blatchley, in his flora of Vigo County, published in 1897, says: "Roadsides and railways: common." I do not recall that I have ever found more than a few plants at a place, and I have found it only once during the past 20 years, although I have been most active in collecting. I am of the opinion that the plant is disappearing from our area, probably on account of the present method of taking care of our highways. Most of my plants are from highways, two are from pastures, and one is from a wooded bank. It is evidently adventive in the state, and its future behavior with us is a subject well worth recording. It is worthy of note that the achenes of all of my specimens are densely upwardly appressed-pubescent except those of my Perry County specimen, which are glabrous.

Ill. to Minn. and Mont., southw. to La. and Ariz.

9330. ÁNTHEMIS [Micheli] L.

Chaff awl-shaped; achenes glandular-tuberculate; fresh plants with a fetid odor....... ....1. *A. Cotula.*

Chaff not awl-shaped or lacking; achenes not glandular-tuberculate; fresh plants not fetid.

Achenes ribbed all around, usually with 10 ribs; achenes 1.5-2 mm long; chaff linear-lanceolate, abruptly cuspidate, slightly shorter than the tubular flowers....... ....2. *A. arvensis.*

Achenes ribbed on the inner surface only, the ribs 3, indistinct; achenes 1-1.5 mm long; chaff, if present, oblong, obtuse, sometimes lacerate at the summit....... ....3. *A. nobilis.*

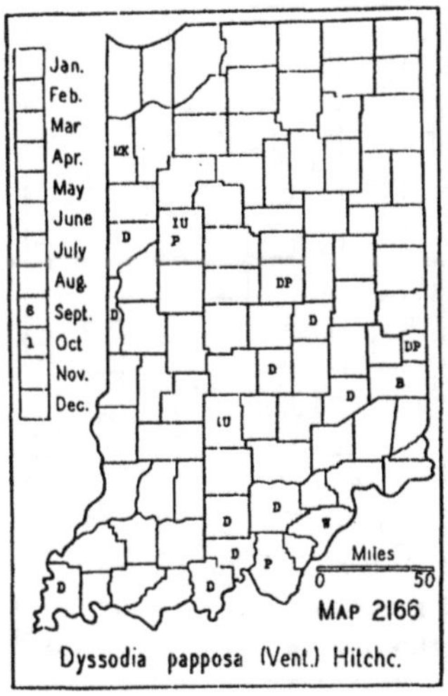

MAP 2166
Dyssodia papposa (Vent.) Hitchc.

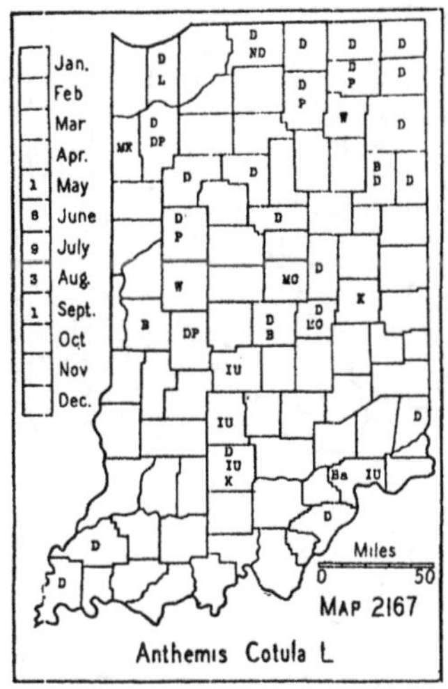

MAP 2167
Anthemis Cotula L.

MAP 2168
Anthemis arvensis L.

1. ANTHEMIS CÓTULA L. DOGFENNEL. Map 2167. This species is doubtless found in every county of the state. It is usually found in waste grounds about habitations. It is also found along roadsides and in fallow fields and waste places in general. I can remember that, when I was a boy, every barnyard was white with dogfennel during its season of flowering, as were most roadsides which, at that time, were new, rich earth. In recent years one rarely sees this species. I have no scientific data concerning its distribution, but I believe it is fluctuating in its abundance. It had almost disappeared until a few years ago when it began to reappear, and now it seems to be becoming abundant. I have discussed this subject with other observers and they agree in the preceding observation.

Nat. of Eu., Africa, and the Orient; throughout the U. S. and s. Canada.

2. ANTHEMIS ARVÉNSIS L. FIELD CAMOMILE. Map 2168. This species has been reported from Clark, Monroe, and St. Joseph Counties.

Nat. of Eu. and the Orient; Maine, Que., B. C., southw. to Fla. and Oreg.

3. ANTHEMIS NÒBILIS L. COMMON CAMOMILE. Map 2169. I found this species in La Porte County, where it covered an acre in very sandy soil in a yard and adjacent nursery. Peattie reported it as escaped in the Calumet District. The plant is used in medicine and cultivated in gardens, especially by people who still grow their own medicinal herbs.

Nat. of Eu.; R. I., southw. to N. C. and Tenn.

9332. ACHILLÈA [Vaill.] L.

1. **Achillea Millefòlium** L. COMMON YARROW. Map 2170. This is a polymorphic species. Our species vary greatly in the pubescence of the stem, leaves, and involucre, in the shape and size of the heads, in the color of the margins of the bracts, and in the shape of the inflorescence. Plants with pinkish rays are not infrequent. These variations have led authors to describe several forms of this species. It is an obnoxious weed, especially in pastures, although some faddists recommend it for lawns. It spreads

MAP 2169
Anthemis nobilis L.

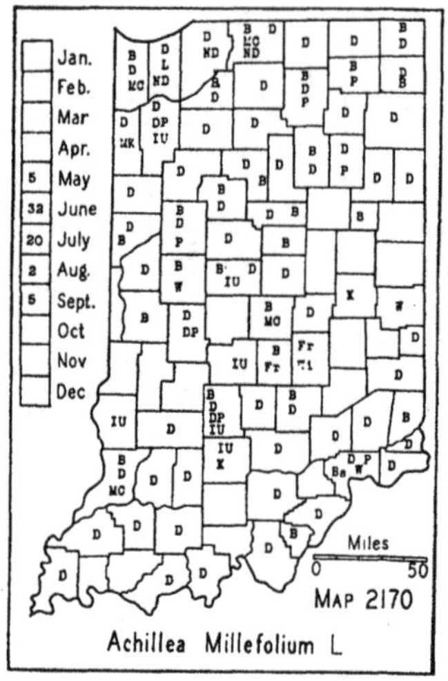

MAP 2170
Achillea Millefolium L.

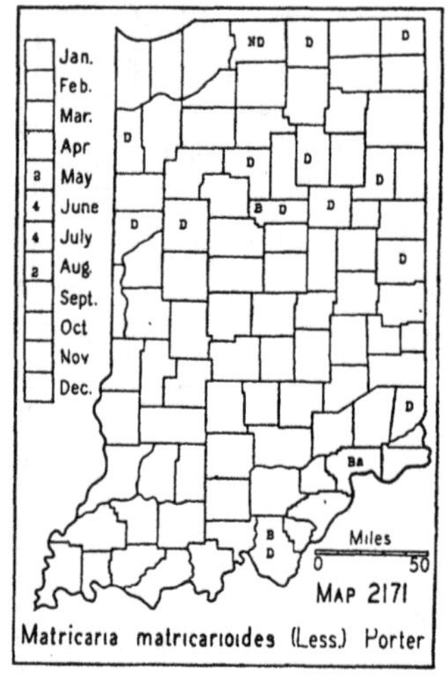

MAP 2171
Matricaria matricarioides (Less.) Porter

by creeping rootstocks and is difficult to exterminate. It is found everywhere in dry soil except in deep woodland and cultivated fields.

Eurasian, and by most authors regarded also as a native. Now found throughout the U. S.

9339. MATRICÀRIA [Tourn.] L.

1. MATRICARIA MATRICARIOÌDES (Less.) Porter. (*Matricaria suaveolens* (Pursh) Buchenau.) RAYLESS CAMOMILE. Map 2171. This species has been reported from only three counties yet I believe it may be found throughout the state. The decumbent habit of the plant and its rayless heads have, I believe, led collectors to pass it by, thinking that such specimens were trampled down or non-flowering specimens of *Anthemis Cotula*. I know that I so regarded the species for many years until I discovered my error. All of my specimens are from barnyards except one which is from a roadside.

Adventive from the Pacific coast.

9341. CHRYSÁNTHEMUM [Tourn.] L.

Heads few or solitary, mostly 3-5 cm wide; rays white, spreading; leaves glabrous, pinnately incised.
- Basal leaves spatulate-obovate, on long slender petioles, the blades crenate-dentate; middle and upper stem leaves oblong or oblanceolate, coarsely and regularly crenate or dentate above, with larger, spreading teeth at base. (See excluded species no. 675, p. 1102) .... *C. Leucanthemum.*
- Basal leaves pinnatifid, subpinnatifid or coarsely and irregularly toothed; middle and upper stem leaves narrowly oblong or oblanceolate, conspicuously subpinnatifid at base.............................1. *C. Leucanthemum* var. *pinnatifidum.*

Heads numerous, less than 1.5 cm wide; leaves puberulent, crenate-serrate or pinnately parted.
- Leaves oblong, crenate-serrate; heads rayless, 6-8 mm wide; inner bracts with colorless, scarious, erose tips.....................2. *C. Balsamita* var. *tanacetoides.*
- Leaves pinnately parted; heads mostly 12-20 mm wide, inner bracts more or less scarious-tipped and brownish. (See excluded species no. 676, p. 1102)........ ....*C. Parthenium.*

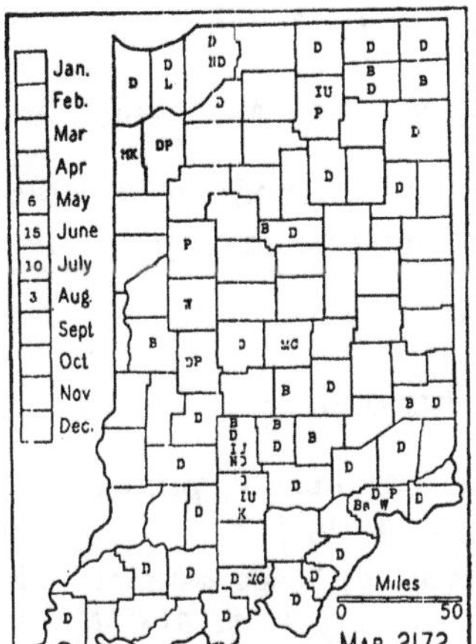

MAP 2172
Chrysanthemum Leucanthemum var pinnatifidum Lecoq & Lamotte

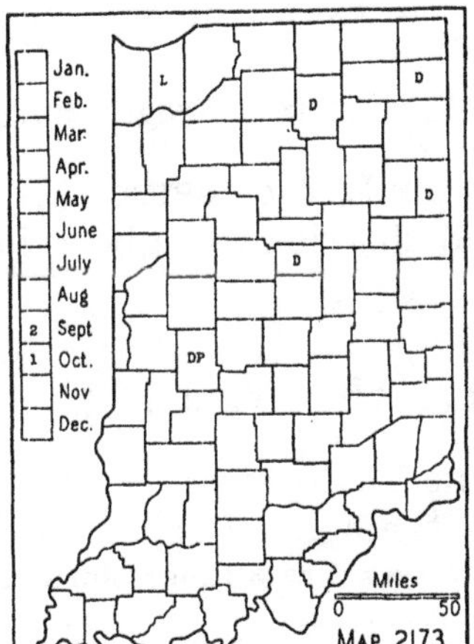

MAP 2173
Chrysanthemum Balsamita var. tanacetoides Boiss

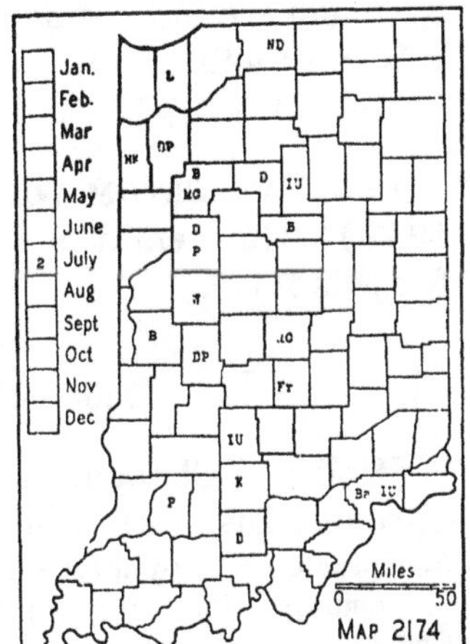

MAP 2174
Tanacetum vulgare L.

1. CHRYSANTHEMUM LEUCÁNTHEMUM L. var. PINNATÍFIDUM Lecoq & Lamotte. (For a discussion of species and variety see Rhodora 5: 177-181. 1903.) OXEYE DAISY. Map 2172. This plant is now found throughout the state. I can remember when it was very rare or absent in northern Indiana, but it has now become well established in all parts, especially on washed slopes in pastures. It is a common weed in the southern part of the state, especially in the worn-out fields and pastures of the limestone area. Not common in the southwestern counties. Clapp, in 1852, writes: "Rare in the vicinity of New Albany." J. M. Coulter, in 1875, writes: "Is becoming more abundant (in Jefferson County) every year and almost takes possession of certain old pastures." On account of its showy flowers it has been much cultivated and I believe its spread can be, for the greater part, attributed to this cause. I have never seen the typical form of the species.

Nat. of Eu.; Newf. and Que., southw. to N. J., and doubtless more widely distributed.

2. CHRYSANTHEMUM BALSÁMITA L. var. TANACETOÌDES Boiss. COSTMARY. Map 2173. This species possesses medicinal qualities and for this reason was formerly much cultivated in gardens, from which it has occasionally escaped. There are five reports for the state. When once established, it is able to maintain itself.

Nat. of the Old World: N. S. to Mich., southw. to N. Y. and Ind.

9341A. TANACÈTUM [Tourn.] L. TANSY

Ultimate divisions of the leaves sparingly incised-serrate................1. *T. vulgare.*
Ultimate divisions of the leaves finely and closely denticulate, many of the teeth incurved..1a. *T. vulgare* f. *crispum.*

1. TANACETUM VULGÀRE L. COMMON TANSY. Map 2174. This is a medicinal plant which has been cultivated in gardens since pioneer times. It has escaped in all parts of the state. Apparently it propagates entirely

by underground stems since it is found so sparingly and about the site of a former habitation.

Nat. of Eu.; N. S. to Minn., and Oreg., southw. to Ga., Mo., and Nev.

1a. TANACETUM VULGARE L. f. CRÍSPUM (L.) Fern. (Rhodora 38: 235. 1936.) The remarks and distribution given for the species apply also for this form.

9358. ARTEMÍSIA [Tourn.] L. WORMWOOD

[Hall and Clements. The Phylogenetic Method in Taxonomy, pp. 31-156. Carnegie Institution of Washington Publication 326. 1923.]

Receptacle villous-pubescent; leaves 2 or 3 times pinnately parted; heads 2-3 mm high, usually wider than high....................................1. *A. Absinthium.*
Receptacle glabrous.
  Leaves glabrous or nearly so on both surfaces, 1-3 times pinnatifid or dissected.
    Leaves once pinnatifid, the segments 1-4 mm wide; flowers on short, leafy axillary spikes or in glomerules, the clusters much shorter than the leaves; plants biennial..2. *A. biennis.*
    Leaves more than once pinnatifid or pinnately parted; the segments mostly less than 1 mm wide.
      Heads about 4 mm wide; involucre pubescent; plants perennial. (See excluded species no. 677, p. 1102)................................*A. Abrotanum.*
      Heads 2-3 mm wide; involucre glabrous; plants annual or biennial.
        Ultimate segments of leaves mostly about 1 mm long or 2-3 mm long; central flowers perfect; annual..................................3. *A. annua.*
        Ultimate segments of leaves linear, 5-20 mm long; central flowers sterile; biennial...4. *A. caudata.*
  Leaves densely woolly on one or both surfaces.
    Blades pinnately parted into 5-7 narrow, entire segments. (See excluded species no. 679, p. 1102)..*A. Carruthii.*
    Blades lanceolate or linear, serrate or entire, not pinnatifid (sometimes the basal leaves pinnatifid).
      Leaves green and glabrate above, at least in age, lanceolate, entire or the lower ones pinnatifid with lanceolate, linear-lanceolate, often falcate divisions; involucre about 3 mm high; achenes about 1 mm long. (See excluded species no. 681, p. 1102)...*A. ludoviciana.*
      Leaves white-tomentose on both surfaces.
        Involucres 3-4 mm high, 2-3.5 mm wide; upper leaves linear and entire, the lower ones oblanceolate, usually serrate toward the apex, 5-10 cm long; achenes about 1 mm long...........................5. *A. gnaphalodes.*
        Involucres 4-5 mm high, 4-7 mm wide; leaves linear or linear-lanceolate, entire, 6-15 cm long. (See excluded species no. 680, p. 1102)......*A. longifolia.*

1. ARTEMISIA ABSÍNTHIUM L. COMMON WORMWOOD. Map 2175. There are five reports of this species having escaped to roadsides, and I have seen it a few times and collected it once. I believe it may be considered established, especially in the sandy areas of northern Indiana.

Nat. of Eu.; Newf. to Hudson Bay and Mont., southw. to N. C., Ohio, and N. Dak.

2. **Artemisia biénnis** Willd. BIENNIAL WORMWOOD. Map 2176. Reported from 14 localities within the state and three authors report it as

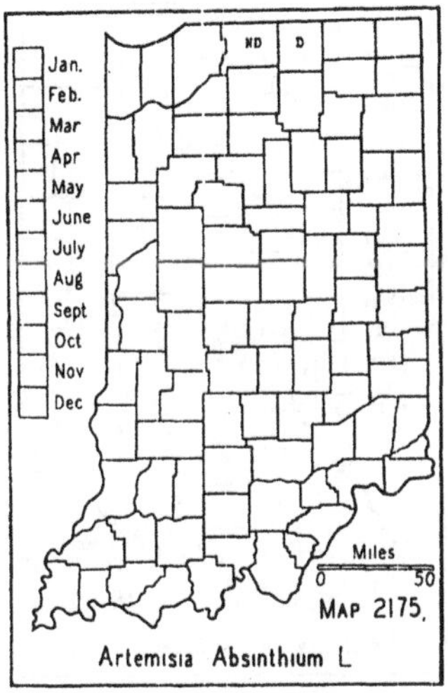

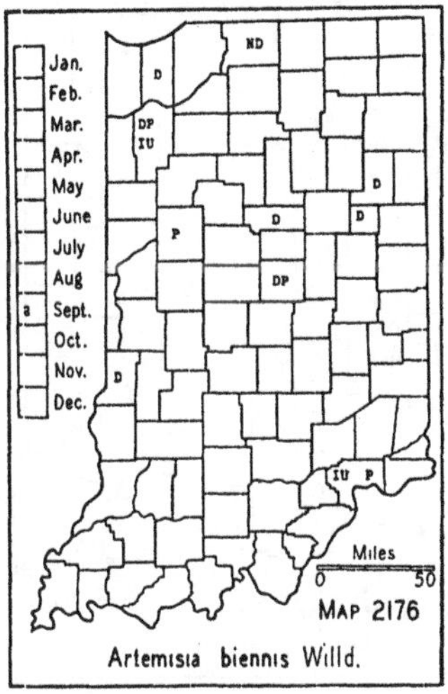

common in waste places. I have found it only five times, and then only a specimen or two at a place.

N. S. to B. C., southw. to N. J. and Calif.

3. ARTEMISIA ÁNNUA L. SWEET WORMWOOD. Map 2177. Local throughout southern Indiana and there are reports of it from Lake and Montgomery Counties. A very common weed half way up the slope of the bank of the Ohio River along almost the entire length of the river in this state. I believe that it will become an obnoxious weed in many places in the southern part of the state. It prefers moist, muddy banks and dry, sandy soils. Frequent about habitations.

Nat. of Asia; N. B., Ont. to Calif., southw. to Va. and Tenn.

4. **Artemisia caudàta** Michx. Map 2178. This species requires a dry, sandy soil and is usually found in the open on sandy knolls, on open sand dunes, on slopes bordering lakes and streams, and rarely in a prairie habitat. It is restricted to the lake area and is very local except in the dune area where it is frequent. This species is one of the hosts of the parasitic plant, *Orobanche fasciculata* which I have found only at Pine, Lake County.

Que., Ont. to Man., southw. to Fla. and Tex.

5. ARTEMISIA GNAPHALÒDES Nutt. (*Artemisia ludoviciana* of authors in part, not Nutt.) Map 2179. Reported by Peattie as rare in the Calumet region. In 1923 I found a colony about 4 feet square along the railroad about 2 miles north of Rochester, Fulton County. In 1930 I found it scattered over a large area in a fallow field in Newton County about 6 miles southwest of Fair Oaks. I have not been able to check its persistence at either of these locations, but I believe it is established at the Newton County location. In 1935 I found a small colony near the top of the 160 foot bluff of the Wabash River at Merom, Sullivan County.

Ont. to Alberta, southw. to Tex. and Mex.; introduced eastw. to N. H. and Del.

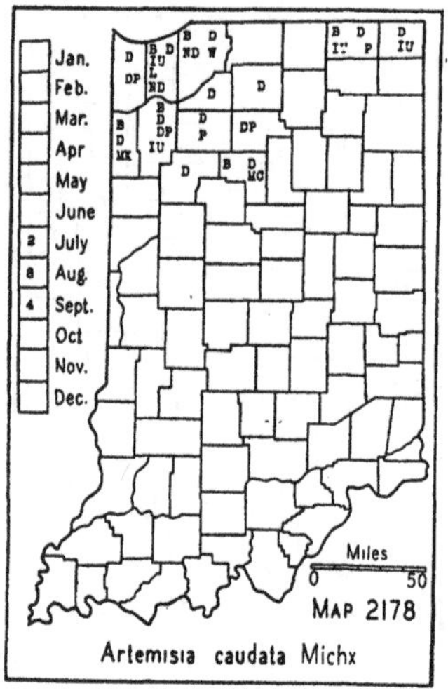

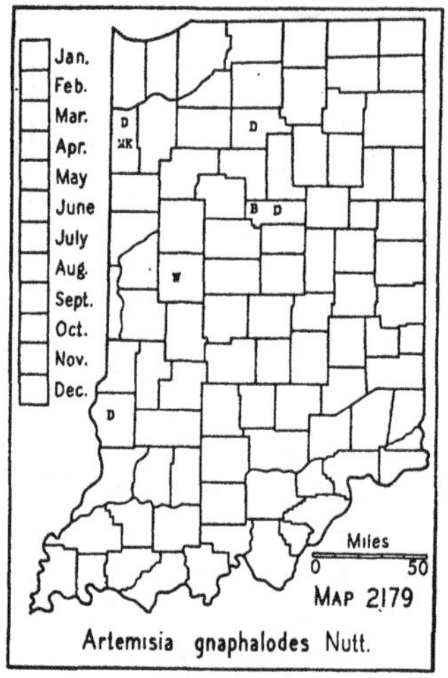

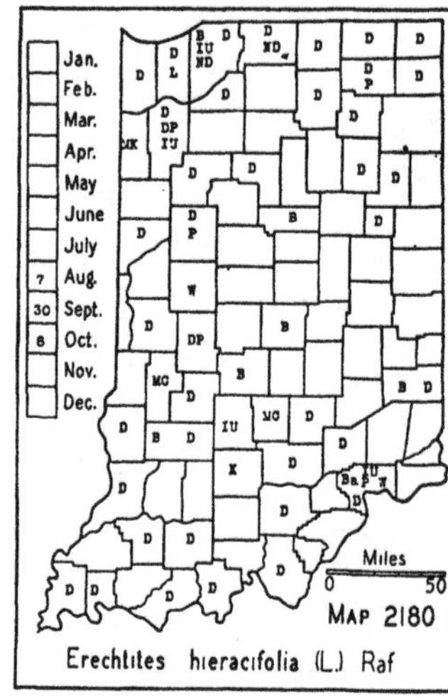

9389. ERECHTÌTES Raf.

[Fernald. The genus Erechtites in temperate North America. Rhodora 19: 24-27. 1917.]

1. **Erechtites hieracifòlia** (L.) Raf. FIREWEED. Map 2180. Infrequent to frequent throughout the state. Found in many habitats and in dry and moist soils. It is often found in burned-over areas in woodland and in marsh land, where it frequently forms dense stands. It is in such an area that the variation of the species can be best studied. Varieties have been described, but my studies convince me that ours is a polymorphic species. I have seen the form with reduced upper leaves growing close beside a specimen which had long leaves up to the inflorescence. In the same colony leaves may be found with bases clasping or not clasping. Individuals with the upper leaves reduced is the common form, and those with the upper leaves not reduced is less frequent.

P. E. I. to Ont., southw. to Fla. and Tex.

9409. CACÀLIA L. INDIAN PLANTAIN

Lower leaves hastate, the upper ones triangular-lanceolate; involucral bracts 12-15; heads 20-30-flowered; receptacle flat............................1. *C. suaveolens.*

Lower leaves not hastate, the upper ones not triangular-lanceolate; involucral bracts 5; heads 5-flowered; receptacle appendaged in the center.

Leaves, at least the lower ones, cordate or reniform at the base, palmately veined.

Plants more or less glaucous, of a dry habitat; stems terete or inconspicuously furrowed; sinuses between the teeth of the margins of the leaves not ciliate. ....2. *C. atriplicifolia.*

Plants not glaucous, of a dry or wet habitat; stems conspicuously furrowed; sinuses between the teeth of the margins of the leaves ciliate.....3. *C. Muhlenbergii.*

Leaves of an oval type, green on both sides, thick, strongly 5-7-nerved, the nerves of a parallel type, margins entire or with short teeth...............4. *C. tuberosa.*

1. **Cacalia suavèolens** L. Map 2181. Local near the dunes about Lake Michigan, and then very local until the southern part of the state is reached, where it is very local to infrequent. In addition to my records,

MAP 2181

Cacalia suaveolens L.

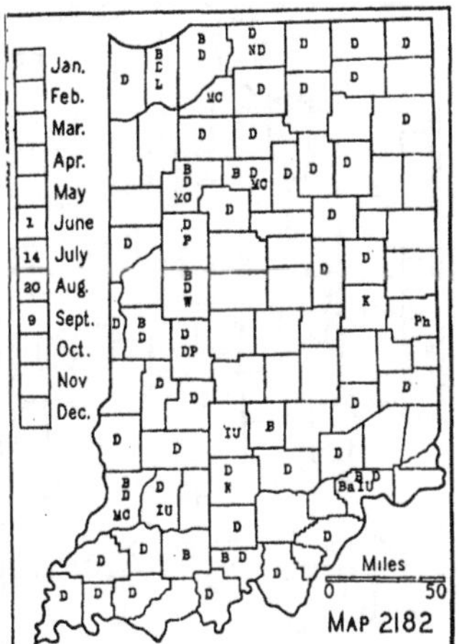

MAP 2182

Cacalia atriplicifolia L.

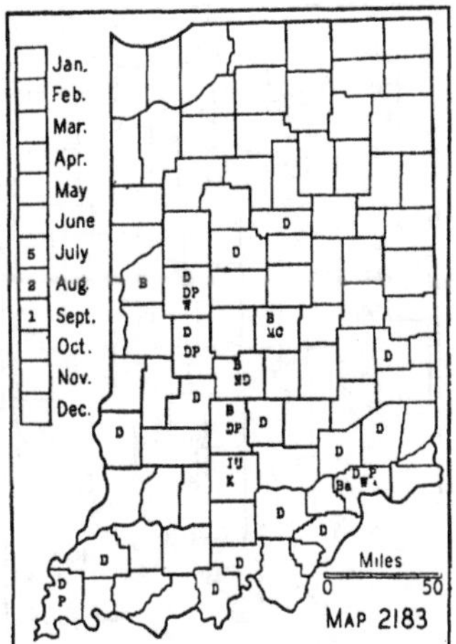

MAP 2183

Cacalia Muhlenbergii (Sch. Bip.) Fern.

it has been reported from Hamilton, Lake, La Porte, and Tippecanoe Counties. It is always found in moist or wet grounds, usually near a stream, and it spreads rapidly by underground stems. I planted this and the next two species in alluvial soil in bottomland, and the other two lived only a few years, while *Cacalia suaveolens* has spread about a foot each year through an adjacent bluegrass sod.

Mass. to Minn., southw. to Fla. and Tenn.

2. **Cacalia atriplicifòlia** L. Map 2182. Infrequent to frequent throughout the state. It prefers dry, open woodland, especially clayey oak slopes. It has a varied habitat, however, ranging from the woodland to the moist prairie habitat.

N. J. to Minn., southw. to Fla. and Kans.

3. **Cacalia Muhlenbérgii** (Sch. Bip.) Fern. (Rhodora **40**: 356-357. 1938.) (*Cacalia reniformis* Muhl. and *Mesadenia reniformis* (Muhl.) Raf.) Map 2183. Infrequent to local in the southern two thirds of the state. It prefers the moist, rich soil of beech slopes but it is found also in other types of moist soil, even in springy places. Ordinarily only a few plants are found at a place, but in 1921 I was asked by a land owner, who lived about five miles southeast of Greensburg, Decatur County, to identify an obnoxious weed which he had in his woods on an open beech ridge and which proved to be a vigorous growth of this species in almost a pure stand over an acre or more.

N. J. to Minn., southw. to Ga. and Ala.

4. **Cacalia tuberòsa** Nutt. Map 2184. This is a local species found only in marly springy places. Where it is found, it is usually a common plant. Its absence in the northeastern part of the state is of interest. If it occurs there, it is rare, because I have collected intensively in these counties without finding it.

Ohio, Ont. to Minn., southw. to Ala., La., and Tex.

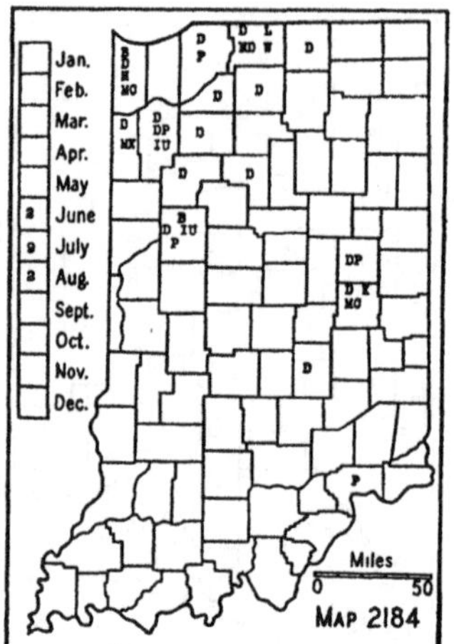

Cacalia tuberosa Nutt.

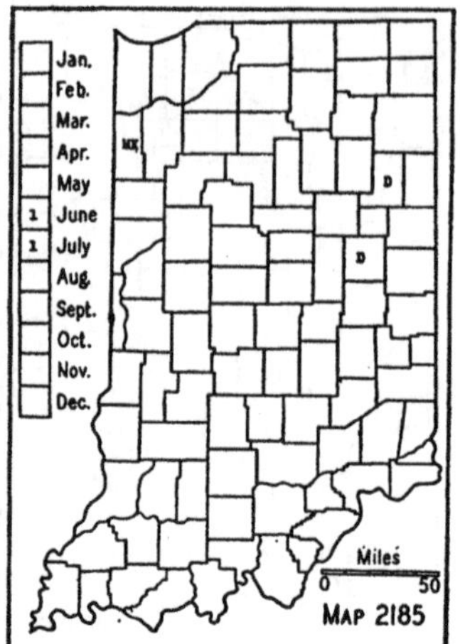

Senecio vulgaris L.

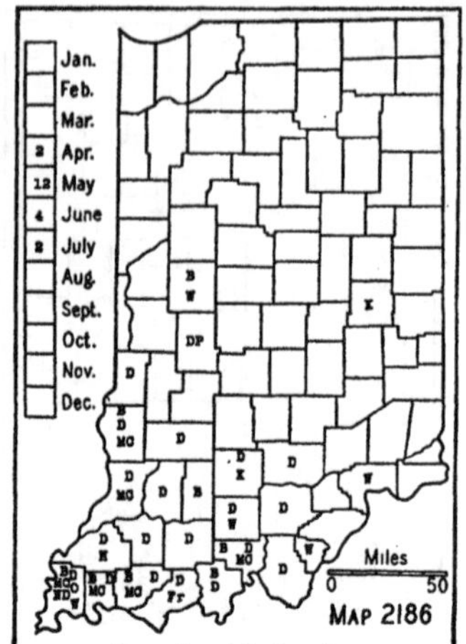

Senecio glabellus Poir.

9411. SENÈCIO [Tourn.] L.

[Greenman. Monograph of the North and Central American species of the genus Senecio-Part II. Ann. Missouri Bot. Gard. 2: 573-626. 1915; 3: 85-194. 1916.]

Plants leafy to the top, the leaves gradually diminishing upward; leaf blades mostly pinnately parted; annuals.
  Heads discoid, the numerous bracteoles of the calyx black-tipped, the principal involucral bracts about 7 mm long, sometimes black-tipped..........1. *S. vulgaris.*
  Heads radiate; involucres with few or no bracteoles, not black-tipped, the principal involucral bracts about 5 mm long..............................2. *S. glabellus.*
Plants usually with many large basal leaves, the cauline few and much smaller; perennials.
  Leaves and stems more or less persistently tomentose; basal leaves petiolate, elliptic to oblong-lanceolate, mostly 15-40 mm wide, crenate-dentate; cauline leaves deeply and irregularly pinnatifid..............................3. *S. plattensis.*
  Leaves and stems glabrous or essentially so at maturity.
    Basal leaves obovate, sometimes one or rarely all subrotund to oblong-elliptic, usually glabrous, narrowed at the base, rarely some subcordate; plants usually of dry soil on slopes and banks..............................4. *S. obovatus.*
    Basal leaves rotund-ovate, oblong-ovate to oblong-lanceolate, cordate to narrowed at the base, glabrous or glabrate; plants of a wet or moist habitat, rarely of a dry, sandy soil.
      Lower leaves and those of rosettes usually large, round-ovate, the principal ones deeply cordate at the base; plants usually of a wet habitat...5. *S. aureus.*
      Lower leaves ovate to oblong-lanceolate, shallowly cordate, subcordate or long-narrowed at the base.
        Basal leaves and those of rosettes usually subrotund or of an ovate type, subcordate or truncate at the base.............5a. *S. aureus* var. *gracilis.*
        Basal leaves cuneate at the base..........6. *S. pauperculus* var. *Balsamitae.*

1. SENECIO VULGÀRIS L. COMMON GROUNDSEL. Map 2185. I first found this species in 1919 on private grounds in the northern part of Muncie, Delaware County. In 1929 I found it to be well established in the west part of Bluffton, Wells County. I revisited the area in 1935 and found that it is spreading. Madge McKee found it in Goodland, Newton County.

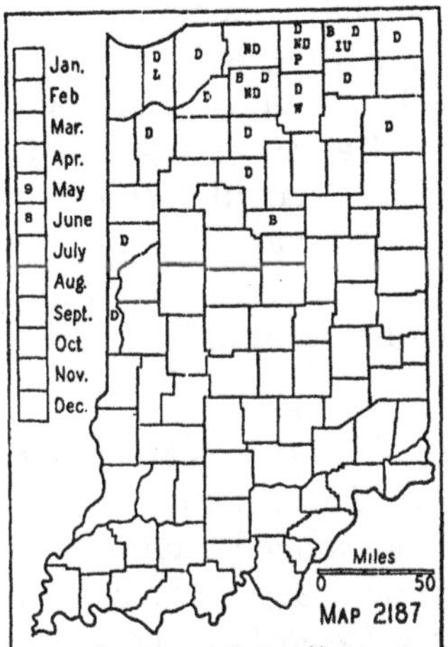

Senecio plattensis Nutt.

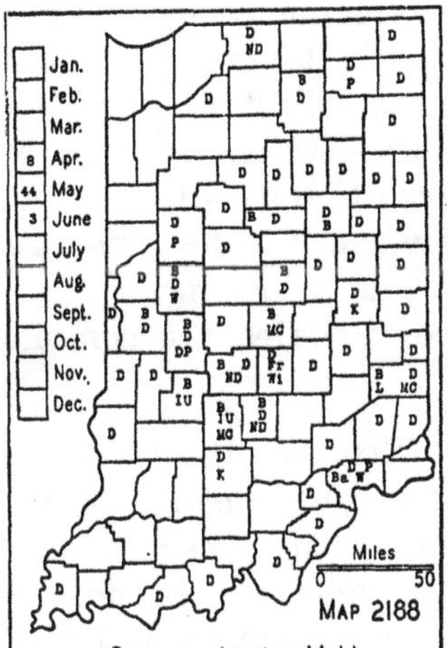

Senecio obovatus Muhl

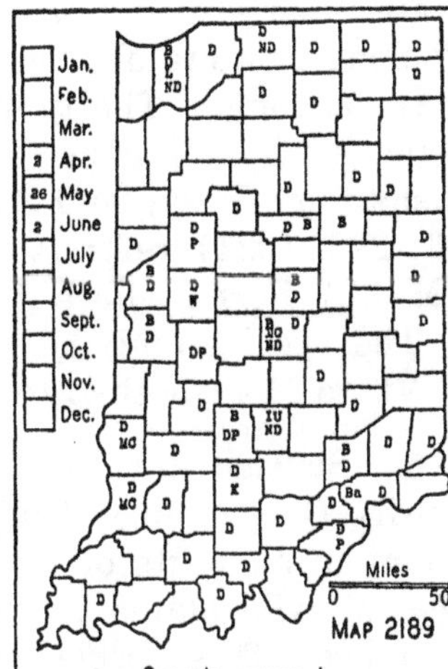

Senecio aureus L.

Thus year after year we add European weeds to our flora and make it necessary to increase our efforts to grow wanted vegetables and ornamental plants.

Nat. of Eu.; Newf. to Hudson Bay, Minn., and B. C., southw. to N. C. and Ind.

2. **Senecio glabéllus** Poir. BUTTERWEED. Map 2186. This plant was not known to Schneck in 1876 in the Lower Wabash Valley, and in 1897 Blatchley reported it as scarce in Vigo County, but it was collected by Clapp in 1838 at New Albany. Evidently it is migrating into the state at a rapid pace. It is now a common to abundant weed in most of the area shown on the map. It prefers cultivated and fallow ground. In the springtime, fields not plowed, oatfields, and wheatfields are sometimes yellow with it.

N. C., Ind. to Mo. southw. to Fla. and Tex.

3. **Senecio platténsis** Nutt. Map 2187. This species is infrequent to rare in the area shown on the map. It is found in dry, sandy soil on open, black and white oak ridges, in moist soil between ridges, and in sandy prairie habitats.

Sw. Ont. to Sask., southw. to La. and Tex.

4. **Senecio obovàtus** Muhl. (*Senecio obovatus* var. *rotundus* Britt. and *Senecio obovatus* var. *umbratilis* Greenman.) ROUNDLEAF GROUNDSEL. Map 2188. Infrequent probably throughout the state, although there are no records from the northwestern counties. Where it is found, it is usually common to abundant over small areas. Its preferred habitat is clayey or rocky slopes and dry clayey banks along streams. It is also found in moist soil in various habitats.

The variety *rotundus* Britt. is a form with subrotund basal leaves. A study of large colonies of this species convinces one of the futility of trying to keep this variety separate from the typical form since both forms may be found in the same colony. The colonies, however, are usually of one form and the two forms are about equally distributed throughout the state. Variety *umbratilis* Greenman usually has basal leaves which are "oblong-ovate to oblong-elliptic, 2-8 cm long, 1.5-5.5 cm wide, with petioles 2-12 cm long." The type was collected near New Albany and it has been reported from Porter County. Greenman has referred some of my specimens from Posey and Starke Counties to this variety. Fernald (Rhodora 23: 299. 1921) refers this variety to *Senecio pauperculus* var. *Balsamitae* (Muhl.) Fern., where it seems to belong.

The species and the two varieties are combined on one map.

Vt. to Mo., southw. to Fla. and Tex.

5. **Senecio aúreus** L. (*Senecio aureus* var. *semicordatus* (Mack. & Bush) Greenman.) GOLDEN GROUNDSEL. Map 2189. Infrequent to frequent in all parts of the state. It is usually found in wooded ravines on wet, alluvial plains along streams, wet borders of ponds, bogs, lakes, and marshes. This species is also variable and var. *semicordatus* has been segregated. This variety is described by Greenman as having the "lower leaves rotund-ovate to oblong-ovate, 1-8 cm long, 1-4 cm broad, usually rounded at the apex, shallowly cordate." Greenman has referred some of my specimens from Lagrange and Wells Counties to this variety. Since I am not convinced that this variety has taxonomic value, I am including it in the species.

Lab., Ont. to N. Dak., southw. to Fla. and Tex.

5. **Senecio aúreus** L. (*Senecio aureus* var. *semicordatus* (Mack. &
is infrequent to rare, and I am not able to say with certainty whether it is found with the species or not. My recollection is that it is not. It is always found in very wet places such as bogs and marshes and never common where it is found. I have made no study of it in the field to ascertain how closely it is related to the species. Greenman gives the distribution as: "Occurring with the species."

6. **Senecio paupérculus** Michx. var. **Balsámitae** (Muhl.) Fern. (Rhodora 23: 299. 1921.) (*Senecio Balsamitae* Muhl. and *Senecio pauperculus* Michx. in part.) Map 2191. This plant prefers moist, mucky or sandy soil and is usually found in fallow fields where it sometimes covers acres (near Griffith, Lake County). Less frequent in prairie habitats along roadsides and in open flats in woods. Local in its distribution. Reported also from St. Joseph and Wabash Counties.

N. S. to Minn., southw. to Va. and Mo.

9442. ÉCHINOPS L.

See excluded species, no. 685, p. 1103.

9452. ÁRCTIUM L. BURDOCK

[Fernald and Wiegand. A synopsis of the species of Arctium in North America. Rhodora 12: 43-47. 1910.]

Involucre less than 2 cm wide; outer bracts about 10 mm long, inner bracts not exceeding the flowers; heads racemose; petioles usually hollow.........1. *A. minus.*

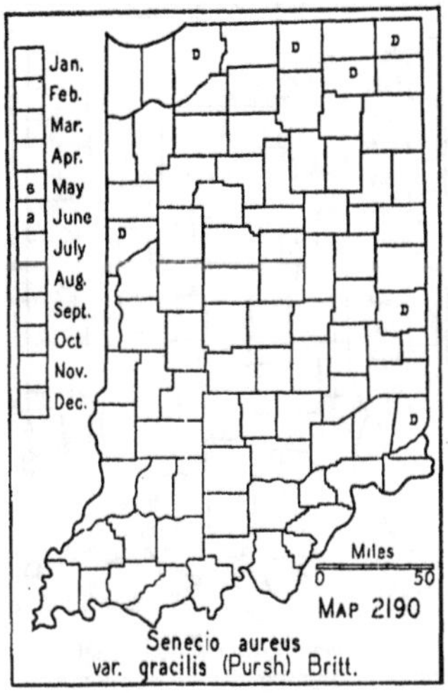

MAP 2190
Senecio aureus var. gracilis (Pursh) Britt.

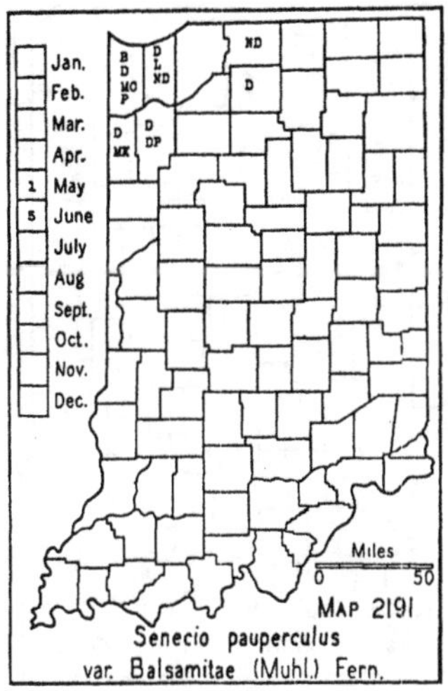

MAP 2191
Senecio pauperculus var. Balsamitae (Muhl.) Fern.

MAP 2192
Arctium minus (Hill) Bernh.

Involucre about 2.5 cm wide; outer bracts 3-5 mm long, inner bracts at least equaling the flowers; heads corymbose; petioles usually solid. (See excluded species no. 686, p. 1103) .... *A. Lappa.*

1. ARCTIUM MÌNUS (Hill) Bernh. COMMON BURDOCK. Map 2192. Infrequent to frequent throughout the state. Commonly found in rich soil about habitations, but also found along roadsides, in waste grounds, and open woodland. Since it is avoided by stock, it has little to prevent its spreading. The roots are used in medicine.

Nat. of Eu.; throughout the U. S. and s. Canada.

9461. CÁRDUUS [Tourn.] L.

1. CARDUUS NÙTANS L. MUSK THISTLE. Map 2193. Hansen (Proc. Indiana Acad. Sci. 34: 257. 1925) reports that this species was found established in a few fields east of Elkhart. Miss Edna Banta, in 1934, found it in a pasture field along Lost Fork Creek near Brooksburg, Jefferson County. She writes that it has been known in this locality for about 17 years and it is spreading, since no determined effort has been made to exterminate it. In 1935 Kriebel found it in Posey County in a pasture between Hovey Lake and Half Moon Pond.

Nat. of Eu.; N. B. and Que. to Pa.

9462. CÍRSIUM [Tourn.] Mill. THISTLE

Upper surface of leaves copiously covered with stiff, upwardly appressed, acicular bristles of unequal lengths, otherwise glabrous; involucral bracts all ending in prickles .... 1. *C. vulgare.*

Upper surface of leaves glabrous or with weak, multicellular hairs (woolly in *C. Pitcheri*).

Leaves woolly on both surfaces, pinnately parted, usually into 12-20 linear or linear-oblong, entire segments ending in a prickle and usually with a prickle near the base; segments of leaves revolute; flowers cream color; plants found only on the dunes near Lake Michigan .... 2. *C. Pitcheri.*

Carduus nutans L.

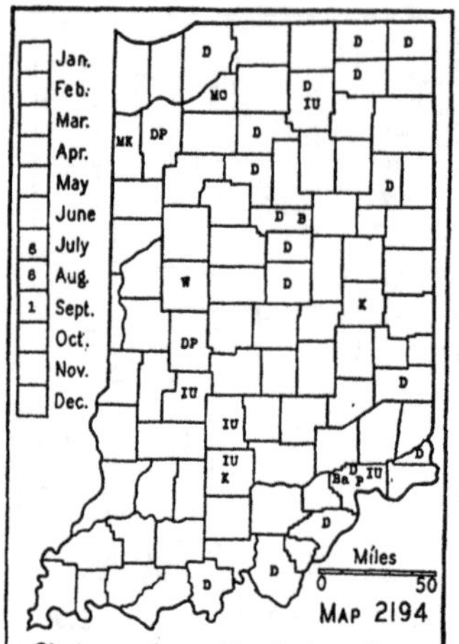

Cirsium vulgare (Savi) Airy-Shaw

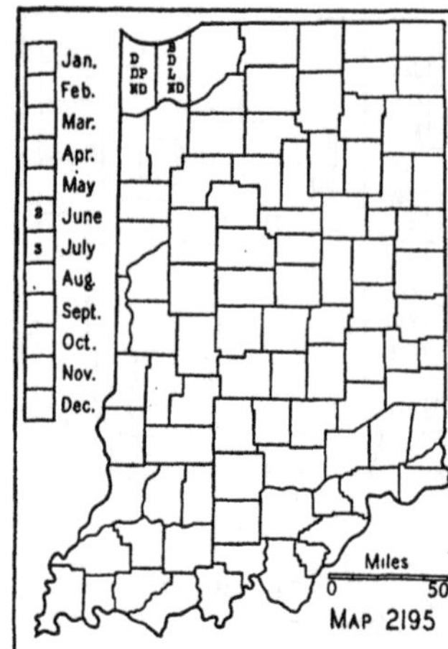

Cirsium Pitcheri (Torr.) T. & G.

Leaves and flowers not as above.
Largest involucres usually not more than 13 mm wide; leaves glabrous above and beneath or woolly beneath and tardily glabrous; perennial with deep, creeping rootstocks.
Leaves more or less deeply pinnatifid..............................3. *C. arvense.*
Leaves of stem sinuate-pinnatifid, somewhat ruffled; leaves of branches subentire or dentate, minutely spinose................3a. *C. arvense* var. *mite.*
Leaves all glabrous, linear-oblong, lanceolate or obovate, the margins entire or setose-spinulose..........................3b. *C. arvense* var. *integrifolium.*
Leaves denticulate, lower surface white-tomentose; peduncles lanate.........
....3c. *C. arvense* var. *vestitum.*
Largest involucres usually more than 13 mm wide; lower surface of the leaves tomentose; plants without deep, creeping rootstocks.
Plants low and very stout, mostly 3-6 dm high; heads usually 1-3, rarely 5, very large; involucres 4-6 cm wide and about 4 cm high; tips of inner involucral bracts crisped, dilated, pubescent, with laciniate-ciliate margins; plants flowering mostly from June 12 to July 28, flowers mostly Purplish Lilac (Ridgway Standard)..................................4. *C. Hillii.*
Plants generally taller; heads smaller and more numerous.
Tips of outer involucral bracts not prickly, sometimes with a mucro about 0.5 mm long..5. *C. muticum.*
Tips of outer involucral bracts prickly, the tips usually 2-8 mm long.
Heads on long, bracted peduncles; peduncles mostly 1-4 dm long; involucral bracts more or less scabrous......................6. *C. virginianum.*
Heads terminating leafy branches; peduncles short, less than 1 dm long.
Leaves deeply pinnatifid, with linear-lanceolate lobes; leaf-margins revolute..7. *C. discolor.*
Leaves entire or sparingly sinuate-lobed, or the basal ones sometimes deeply pinnatifid; leaf-margins generally flat.......8. *C. altissimum.*

1. CIRSIUM VULGÀRE (Savi) Airy-Shaw. (Fedde Rept. Spec. Nov. 43: 302-315. Apr. 15, 1938.) (*Cirsium lanceolatum* (L.) Hill of Indiana authors.) BULL THISTLE. Map 2194. This species is biennial. It no doubt has become established in every county of the state. It formerly was common in pastures and clearings, and frequent along roadsides and in fields, open woodland, and waste places. As nearly as I can remember, about 25 years ago it began to disappear, and in a few years it had prac-

tically disappeared. Its disappearance was due to the butterflies *Vanessa cardui* and *Pyrameis cardui* whose eggs are laid in the flowering heads, the larvae eating the seed. This thistle is now infrequent to rare in the state and I believe will be held in check by its natural enemy. In 1938 I have noted more specimens than for many years.

Nat. of Eurasia; Newf. to Oreg., southw. to Fla., Nebr., and Calif.

2. **Cirsium Pítcheri** (Torr.) T. & G. PITCHER THISTLE. **Map 2195.** This species grows in almost pure, dry sand and is restricted to the dunes near Lake Michigan in Lake and Porter Counties. It probably did occur in La Porte County but I have not seen a specimen from this county. It is commonest in the blow-outs. I planted seed in our garden, and when the plants were three years old I measured the largest one. This plant had a spread of 65 inches, and had 14 decumbent and radiating branches and 1 upright stem that was 28 inches high. The lateral branches had 110 heads, each branch with 5-10 heads. The upright stem had 14 heads. This specimen had about ten times the number of heads that an average specimen has and was several times larger.

Shores of Lakes Michigan, Huron, and Superior.

3. CIRSIUM ARVÉNSE (L.) Scop. CANADA THISTLE. **Map 2196.** This species is infrequent to frequent in the lake area and is more or less local south of this area. Since it is a very obnoxious weed, farmers have been made acquainted with it, and they usually exterminate it as soon as possible. There is a state law against harboring it, but the law is not enforced, and only occasional arrests are made. This species propagates by underground stems and spreads rapidly. It is variable, and several varieties have been described. I have not collected all of them, but three varieties have been reported as established, and it is safe to assume that they will persist until destroyed by force. Several bulletins have been published describing the species and its varieties and give methods for its eradication. One to be recommended is Bulletin 414 of the Ohio Agricultural Experiment Station, by Freda Detmers, published in 1927.

Nat. of Eu.; Newf. and B. C., southw. to Va., Nebr., and Utah.

3a. CIRSIUM ARVENSE var. MÌTE Wimm. & Grab. This variety was reported by Hansen (Proc. Indiana Acad. Sci. 34: 256. 1925) as established in Blackford, Grant, and Henry Counties.

3b. CIRSIUM ARVENSE var. INTEGRIFÒLIUM Wimm. & Grab. This variety was reported by Hansen (Proc. Indiana Acad. Sci. 35: 199. 1926) as established in Grant and Hancock Counties. It is regarded by some authorities as a species, and, from its appearance and behavior, I believe it is of specific rank. I know of a large colony in Wells County that was treated with chemicals for two years and still it persisted. I have not visited the colony recently.

3c. CIRSIUM ARVENSE var. VESTÌTUM Wimm. & Grab. This variety was reported from Grant and Hancock Counties by Hansen.

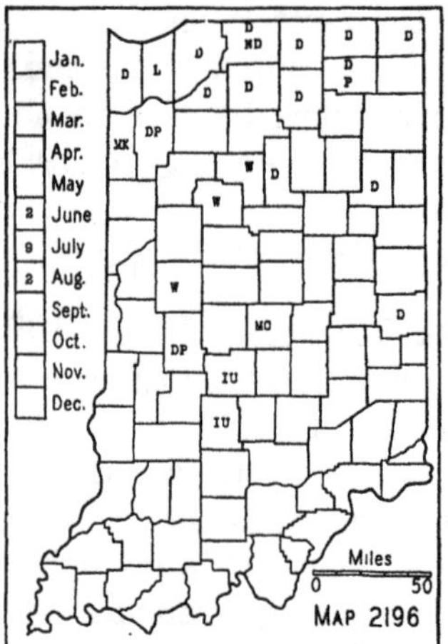

Cirsium arvense (L.) Scop.

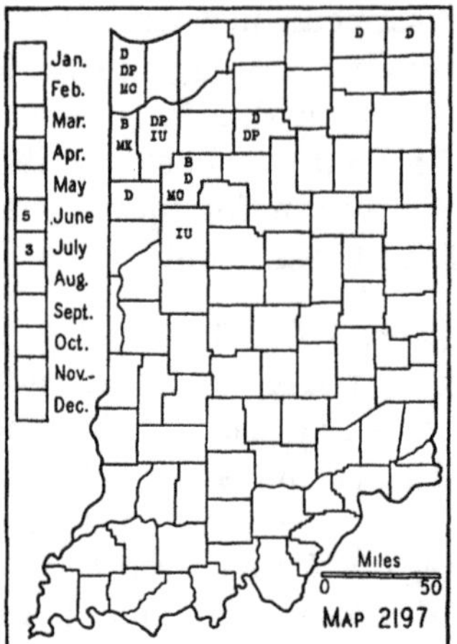

Cirsium Hillii (Canby) Fern.

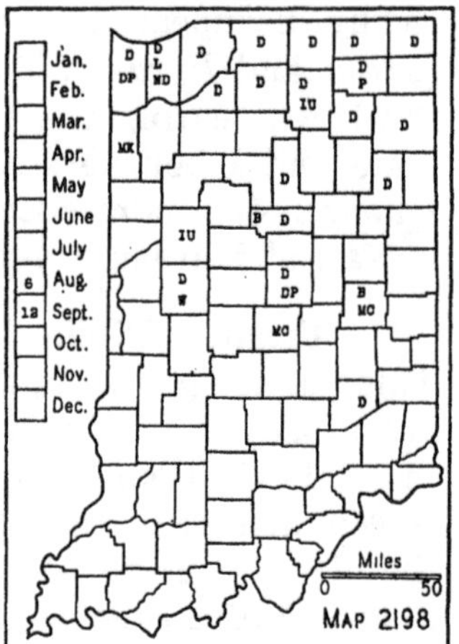

Cirsium muticum Michx.

4. **Cirsium Híllii** (Canby) Fern. (See Hill. Rhodora **12**: 211-214. 1910.) Map 2197. This thistle prefers dry, sandy or gravelly soil and all of our records are from within the area shown on the map. My specimens from Benton, Fulton, and White Counties are from the right-of-way of railroads. The others are from open dunes, open woodland, and the high bank of a stream.

Ont. to Man., southw. to Pa. and Iowa.

5. **Cirsium mùticum** Michx. SWAMP THISTLE. Map 2198. This species is found in boggy places, marshes, and swamps, often in marly soils. It is infrequent in the lake area, becoming rare or absent southward.

Newf. to Sask., southw. to Fla. and Tex.

6. **Cirsium virginiànum** (L.) Michx. VIRGINIA THISTLE. Map 2199. I have found this species only on wooded slopes. It is rare, and Phinney's report for it from the area of Delaware, Jay, Randolph, and Wayne Counties I refer to some other species. This is a southern plant, and Phinney did not report all of the species that are common in his area.

Va., Ohio, and Ind., southw. to Fla. and Tex.

7. **Cirsium díscolor** (Muhl.) Spreng. FIELD THISTLE. Map 2200. Infrequent to frequent throughout the lake area, becoming rare southward and very local, if found at all, in the hill country. It prefers a moist soil rich in humus and is often a common plant in mucky soil that has recently been drained. It is found in its habitat along roadsides and streams and in marshes and swamps.

N. B. to S. Dak., southw. to Ga., Mo., and Nebr.

8. **Cirsium altíssimum** (L.) Spreng. TALL THISTLE. Map 2201. This species is infrequent to rare throughout the state. It is more frequent in southern Indiana and is most common on wooded slopes along streams. This is really a woodland and dry soil species, but it is also found in the open and even in springy places.

Mass. to Minn., southw. to Fla. and Tex.

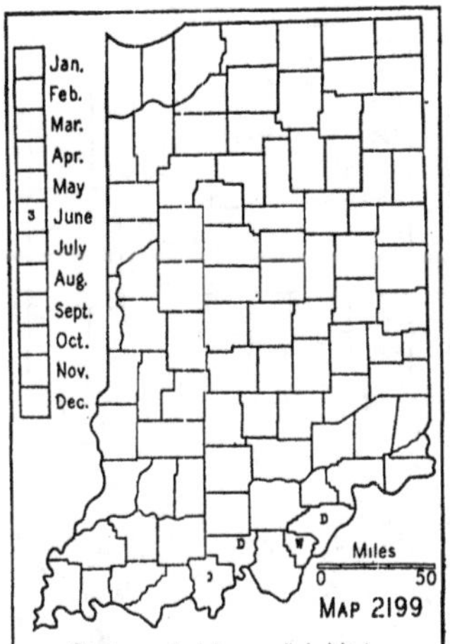

Cirsium virginianum (L.) Michx.

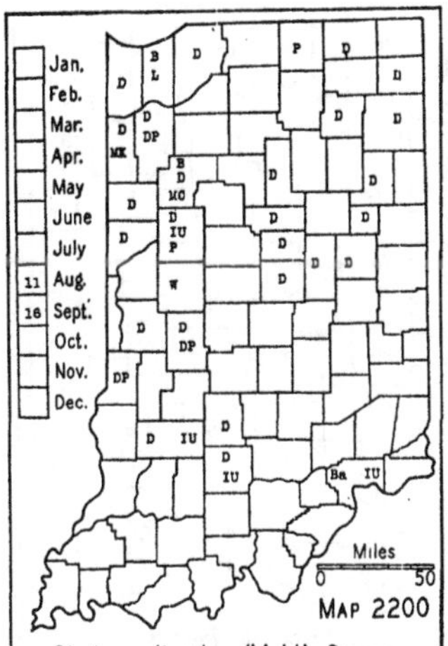

Cirsium discolor (Muhl.) Spreng.

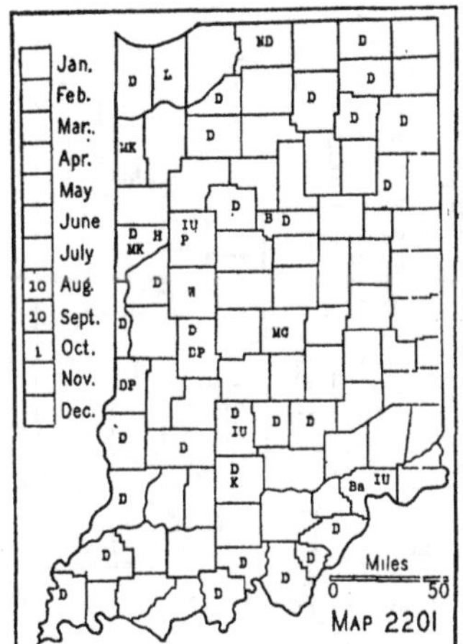

Cirsium altissimum (L.) Spreng.

9467. ONOPÓRDUM [Vaill.] L.

1. ONOPORDUM ACÁNTHIUM L. COTTONTHISTLE. Map 2202. In 1910 I found this species to be a common plant along the roadside about a quarter of a mile north of Rosedale, Parke County. I passed along this road in 1918, and it was still plentiful. In 1929 I found a large colony on the bluff of the Ohio River near the roadside on the south side of Glendale Cemetery, Dearborn County. There is a specimen from Jefferson County collected by Stanley Coulter in the herbarium of Wabash College. It has been reported also from Clark and Marion Counties.

Nat. of Eurasia; N. B. and N. S. to Ont. and Mich., southw. to N. J.

9476. CENTAÚREA L.

Some of the species of this genus have long been cultivated in gardens, and seeds from them have found their way to roadsides and dumps. Specimens from these have been collected and reported, but unfortunately, little or no information accompanies the reports. No data are given as to how long the species has been found in the same place or as to the size of the colony. An annual species has been reported twice as found in alfalfa fields. Since alfalfa is usually mowed from two to three times a year, an annual would have little chance to perpetuate itself. It seems best to regard all of our reports as garden escapes or chance introductions. I prefer to be too conservative rather than to be too hasty in accepting exotic species as a part of our flora.

Bracts of the involucre (at least the lower ones) tipped with stout prickles; flowers yellow; annual. (See excluded species no. 695, p. 1104) . . . . *C. solstitialis.*

Bracts of the involucre lacerate or fimbriate, not prickly.

Leaves pinnatifid into linear segments; bracts ribbed, pectinate at the black tip only; annual or biennial. (See excluded species no. 693, p. 1104) . . . . *C. maculosa.*

Leaves entire, denticulate or some of the lower ones lobed.

Tips of the involucral bracts not dilated; leaves linear or linear-lanceolate, entire or remotely denticulate; annual. (See excluded species no. 691, p. 1104) . . . . *C. Cyanus.*

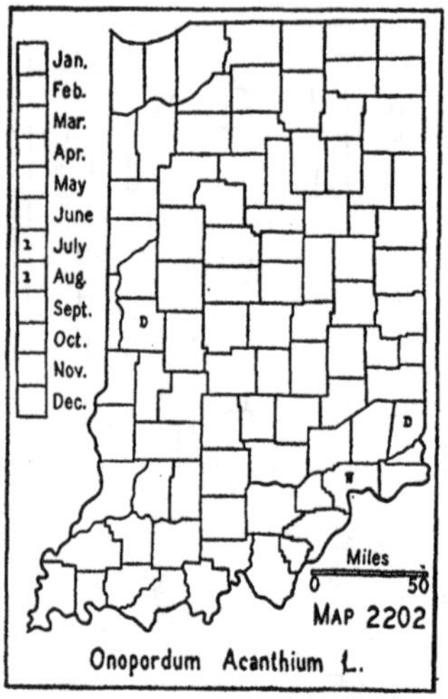

Tips of the involucral bracts much dilated.
Bracts irregularly denticulate or lacerate for half their length or more. (See excluded species no. 692, p. 1104) .... *C. Jacea.*
Bracts pectinate or ciliate at the summit only. (See excluded species no. 696, p. 1105) .... *C. vochinensis.*

9553. CICHÒRIUM [Tourn.] L.

1. CICHORIUM INTYBUS L. CHICORY. Map 2203. This species is now found throughout the state and in many parts has become an obnoxious weed. When once established, I have found from personal experience that it is very difficult to eradicate. Our first reports for it say: "an escape from gardens." In recent years it doubtless has been introduced in grass and other seeds. The dried roots are used as a substitute for coffee, and it has been cultivated for that purpose. My bitter experience with it compels me to advise against its use in the flower garden and to exterminate it wherever it is found. All of my specimens are from hard, dry clay or dry, sandy soils. Plants with white flowers, forma *alba* Farwell, are sometimes found. In a colony extending for nearly a half-mile in hard, clay soil along an unimproved road in Allen County I estimated that 40 per cent of the plants were white-flowered.

Nat. of Eu.; N. S. to Wash., southw. to Fla., Tex., and Calif.

9556. SERÍNIA Raf.

1. **Serinia oppositifòlia** (Raf.) Ktze. Map 2204. This species was found in flower on May 12, 1935, by Scott McCoy. It was growing in wet soil on the border of a woods along State Road 62 a few miles east of Boonville, Warrick County.

Va., Ill., Mo. to Kans., southw. to Fla. and Tex.

9560. KRÍGIA Schreb.

Plants stemless or nearly so; flowers on scapes.
Plants bearing tubers; tubers usually one to a plant, globose, about 1 cm in diameter; basal leaves mostly 4-20 cm long; involucres 10-14 mm long; pappus of 10-15 narrow, oblong, white scales and 15-20 longer bristles.......... 1. *K. Dandelion.*

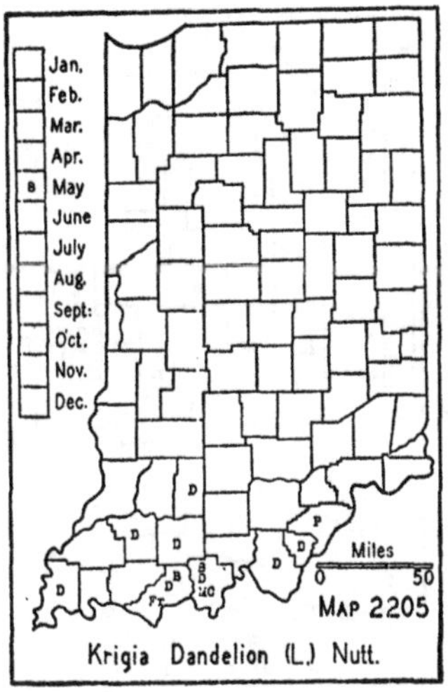

MAP 2205
Krigia Dandelion (L.) Nutt.

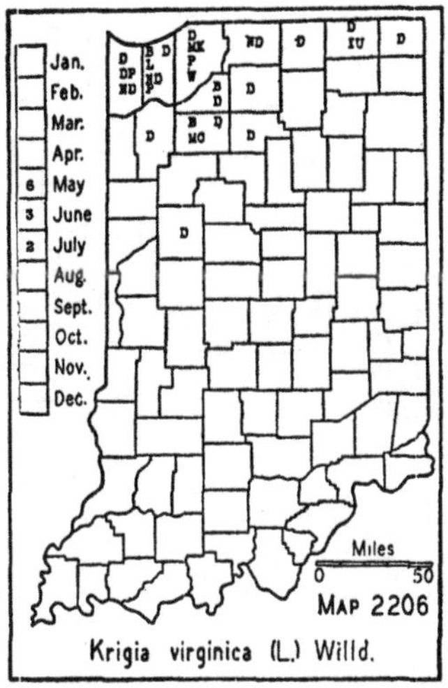

MAP 2206
Krigia virginica (L.) Willd.

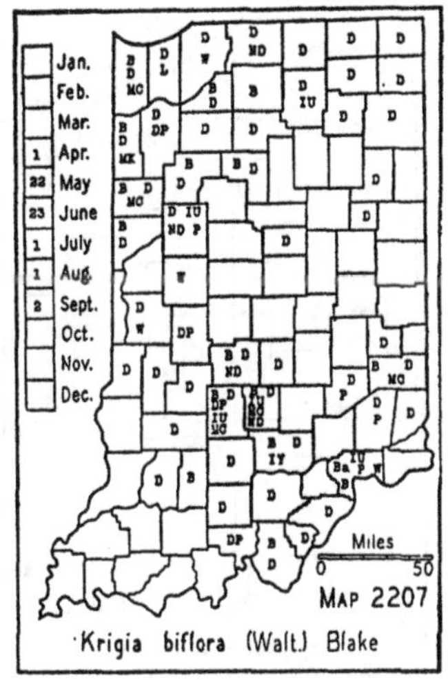

MAP 2207
Krigia biflora (Walt.) Blake

Plants not bearing tubers; basal leaves mostly 3-6 cm long; involucres usually 5-7 mm long; pappus of 5-7 short, rounded, white scales and an equal number of longer bristles.......................................2. *K. virginica.*

Plants with stems, usually 1-6 dm high, with 1-3 conspicuous, clasping leaves; flowers on peduncles arising from the axils of cauline leaves; involucre usually less than 1 cm long; pappus similar to that of *Krigia Dandelion*.............3. *K. biflora.*

1. **Krigia Dandèlion** (L.) Nutt. (*Cynthia Dandelion* (L.) DC.) Map 2205. In sandy clay soil, usually in open woodland on the crests of black oak and chestnut oak ridges. It is found also in low ground in the post oak flats. It is restricted to the southern counties, and Wilson's report for it in Hamilton County I regard as an error in determination, since he does not report *Krigia biflora* which should be found there. It has been reported also from Clark and Jefferson Counties.

N. J., Md., Ill., Mo. to Kans., southw. to Fla. and Tex.

2. **Krigia virgínica** (L.) Willd. DWARF-DANDELION. Map 2206. This species grows only in dry, sandy soil and is generally found on open wooded dunes, ridges, sand hills and in sandy, fallow fields. It is infrequent to frequent in its habitat.

Maine, Ont. to Minn., southw. to Fla. and Tex.

3. **Krigia biflòra** (Walt.) Blake. (Rhodora **17**: 137. 1915.) (*Krigia amplexicaulis* Nutt. and *Cynthia virginica* (L.) D. Don.) Map 2207. Frequent to common throughout the state. It has a wide range of habitats, growing in soils ranging from clay to sand, and from marshes to the crests of black and white oak ridges. It prefers open black and white oak wooded slopes, open wooded dunes, interdunal flats, and moist, sandy prairies. I have 52 specimens from Indiana and all of them have the peduncles and involucres perfectly glabrous except 6 specimens which are more or less densely glandular-pubescent on the peduncles below the flower and at the base of the involucre. This glandular form is one of the varieties which was named by Farwell (Amer. Midland Nat. **12**: 76. 1930) and which apparently has no taxonomic significance.

Mass., Ont. to Man., southw. to Ga. and Kans.

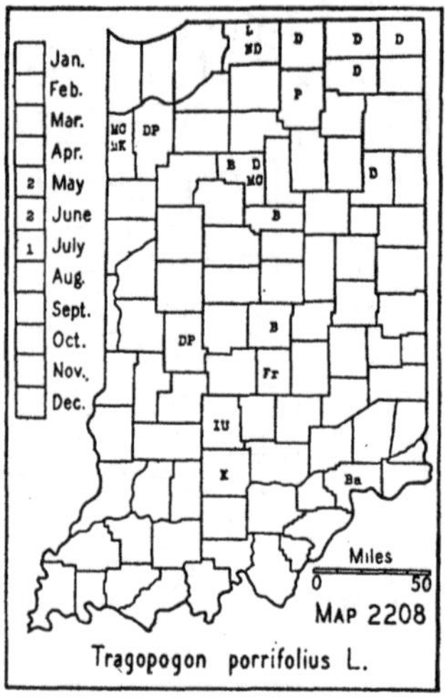

Tragopogon porrifolius L.

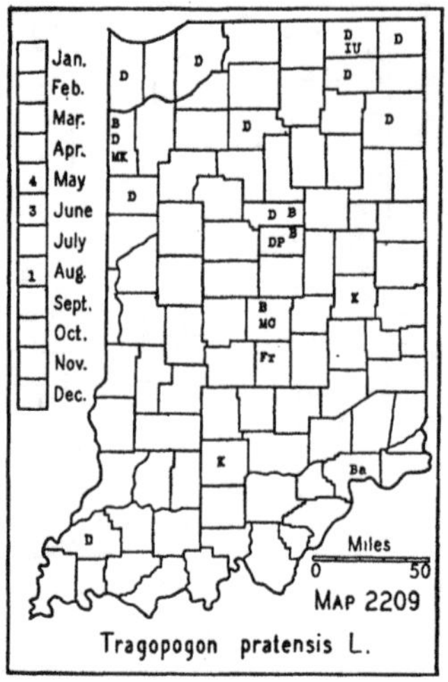

Tragopogon pratensis L.

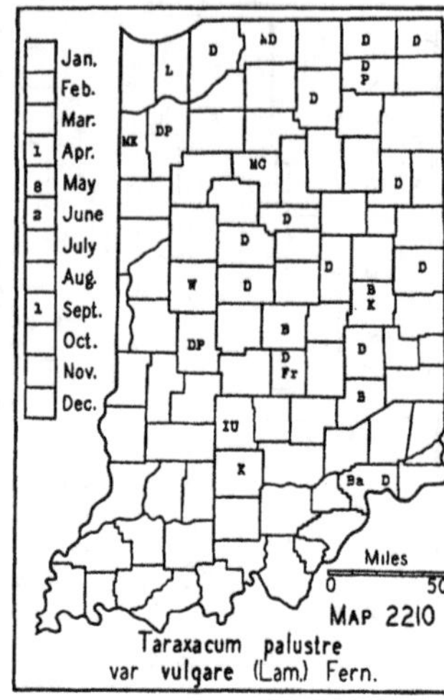

Taraxacum palustre var vulgare (Lam.) Fern.

9572. HYPOCHAÈRIS [Vaill.] L. CAT'S-EAR

See excluded species no. 697, p. 1005.

9579. TRAGOPÒGON [Tourn.] L.

Flowers purple; peduncle thickened at the summit; involucral bracts usually 40-60 mm long, rarely 25-40 mm long, much longer than the rays..........1. *T. porrifolius.*
Flowers yellow; peduncles usually not conspicuously thickened at the summit; bracts usually 25-35 mm long, as long as or shorter than the rays........2. *T. pratensis.*

1. TRAGOPOGON PORRIFÒLIUS L. VEGETABLE-OYSTER. Map 2208. This species is infrequent to rare as yet in the state but is becoming well established in the northeastern part. It is found along roadsides and railroads and in waste places and fallow fields. I found it in a waste place in Bluffton in 1897, and it still persists and has spread over a much larger area.

Nat. of Eu.; Ont. to Minn. and B. C., southw. to Ga. and Calif.

2. TRAGOPOGON PRATÉNSIS L. GOATSBEARD. Map 2209. Becoming established in our northern counties along roadsides and railroads and in waste places and fallow fields.

Nat. of Eu.; N. B. and N. S to Man., southw. to N. J., Ohio, and Colo.

9592. TARÁXACUM [Haller] Ludwig DANDELION

[Fernald. Taraxacum in Eastern America. Rhodora 35: 369-386. 1933.]

Seed greenish buff, the space between the muricate body and the base of the beak about 0.5 mm long; few or none of the outer series of bracts with a callosity on the back near the summit; leaves usually shallowly divided, their basal parts usually not reddish................................1. *T. palustre* var. *vulgare.*
Seed reddish, the space between the muricate body and the base of the beak about 1 mm long; all or nearly all of the outer series of bracts with a callosity on the back near the summit; leaves deeply divided usually to the midrib, their basal parts reddish..2. *T. laevigatum.*

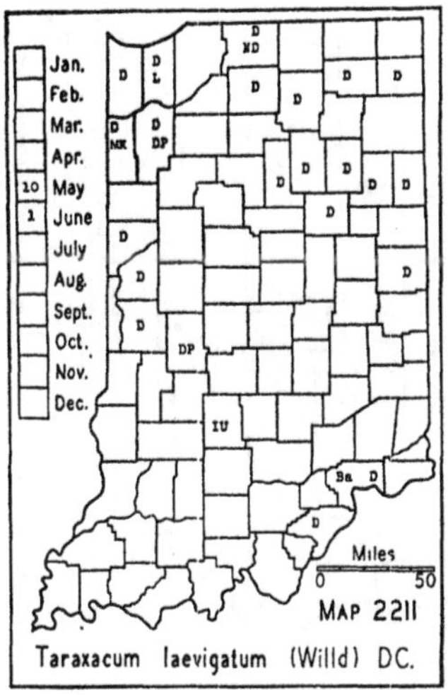

Taraxacum laevigatum (Willd) DC.

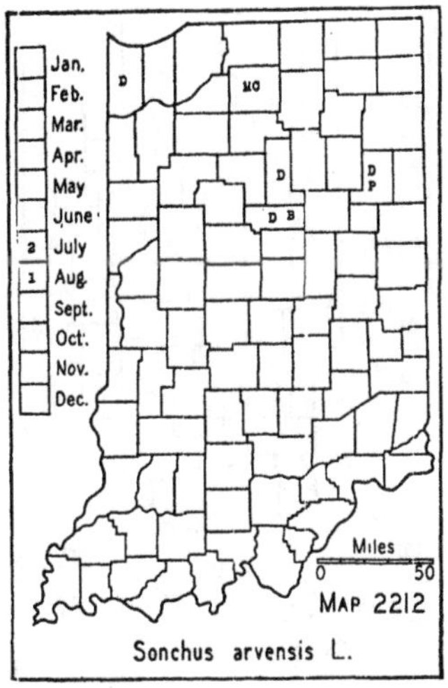

Sonchus arvensis L.

Sonchus arvensis var. glabrescens. Guenth. Grab. & Wimm.

1. TARAXACUM PALÚSTRE (Lyons) Lam. & DC. var. VULGÀRE (Lam.) Fern. (*Taraxacum officinale* Weber and *Leontodon Taraxacum* L.) DANDELION. Map 2210. A frequent to common obnoxious weed found throughout the state except in the southwestern part, where it is less frequent. It is a common weed in lawns, orchards, and fields and along roadsides. The outer series of involucral bracts of the variety are recurved even in the bud while those of the species are appressed until maturity. Although the species has not been reported west of Pennsylvania, it may be found westward and in Indiana.

Nat. of Eu. and Asia; throughout s. Canada and the U. S.

2. TARAXACUM LAEVIGÀTUM (Willd.) DC. (*Taraxacum erythrospermum* Andrz. and *Leontodon erythrospermum* (Andrz.) Britt.) RED-SEED DANDELION. Map 2211. This species is frequent in northern Indiana and is possibly well distributed in sandy soils throughout the state. It prefers a more sandy soil than the preceding, although it is adaptive as to habitat. Found in habitats similar to those of *Taraxacum palustre* var. *vulgare*.

Nat. of Eu.; Maine to Ont. and Alberta, southw. to N. C., Tenn., and Wyo.

9595. SÓNCHUS [Tourn.] L. SOW THISTLE

Flowering heads about 4 cm in diameter, orange to lemon yellow; achenes 2-3 mm long; perennials with creeping rootstocks.
- Involucre and peduncles more or less glandular-setose................1. *S. arvensis.*
- Involucre and peduncles glabrous...................1a. *S. arvensis* var. *glabrescens.*

Flowering heads 1.2-2.5 cm in diameter, pale yellow; achenes 1-1.5 mm long; annuals.
- Auricles of leaves acute; achenes transversely wrinkled and with more than 3 longitudinal lines to a side................ ....2. *S. oleraceus.*
- Auricles of leaves rounded; achenes not transversely wrinkled and with 3 longitudinal lines to a side...3. *S. asper.*

1. SONCHUS ARVÉNSIS L. FIELD SOW THISTLE. Map 2212. Frequent along U. S. Highway 12 south of Buffington, about 2 miles southeast of

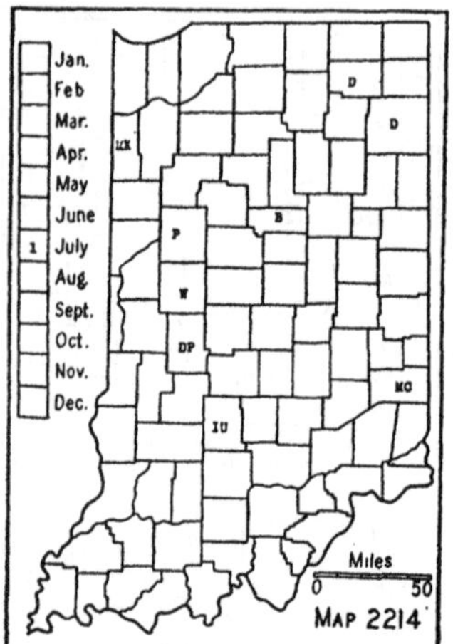

Sonchus oleraceus L.

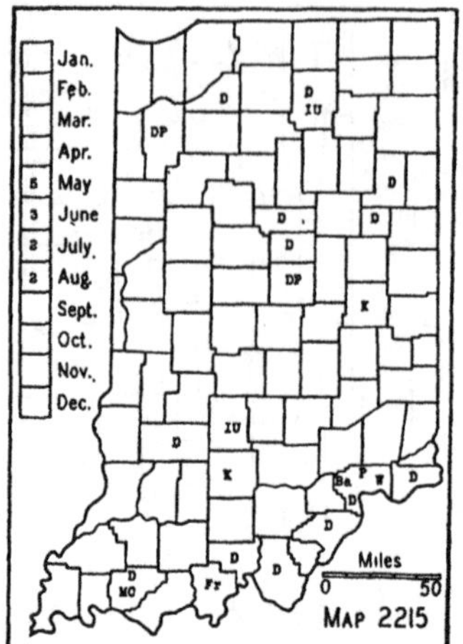

Sonchus asper (L.) Hill

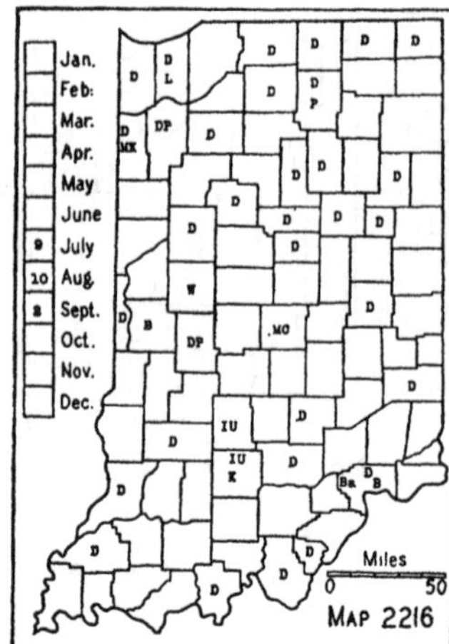

Lactuca Scariola L.

Indiana Harbor, Lake County. There is also a large colony along the Lake Erie Railroad just south of Bluffton, Wells County. If left undisturbed this species and its variety spread rapidly.

Nat. of Eu.; Newf. to Minn. and B. C., southw. to N. J., Colo., and Utah.

1a. SONCHUS ARVENSIS var. GLABRÉSCENS Guenther, Grab. & Wimm. (Rhodora 30: 19. 1928.) SMOOTH FIELD SOW THISTLE. Map 2213. As shown by the map this form of the sow thistle is becoming well established in the state. Most of my specimens are from highways and usually the colony is near a dwelling. This variety is ornamental when in flower and while I do not know that it has been used as a garden plant, I strongly suspect it has, because of the proximity of most of the colonies to habitations. Obviously its principal mode of propagation is by the multiplication of rhizomes since the colonies noted are closed and few separate plants observed.

Special stress should be placed upon the eradication of the few colonies we now have, or in due time this weed will be ubiquitous in the state. In nearly every instance where I have found it I have informed the owner of the land of the dangerous character of the plant, and I have also notified the county agricultural agent of its existence. I have no data concerning the general distribution of the variety.

Nat. of Eu.

2. SONCHUS OLERÀCEUS L. COMMON SOW THISTLE. Map 2214. There are reports of this species being found throughout the state. In my early botanical work I did not collect what I considered common garden weeds, and in most instances this accounts for the comparative paucity of specimens of these common plants. This species is a weed and prefers rich soil. It is found mostly in gardens, truck gardens, waste places, and fallow fields and along railroads and roadsides.

Nat. of Eu.; now a weed throughout the world except in the extreme north.

3. SONCHUS ÁSPER (L.) Hill. SPINYLEAF SOW THISTLE. Map 2215. This is more common than the preceding species and found in similar habitats. Nat. of Eu.; now a weed in all cultivated parts of the world.

9596. LACTÙCA [Tourn.] L. LETTUCE

Achenes with filiform beaks, flat, with thin margins.
 Margins of leaves, and usually their midribs prickly.
  Achenes light brown, not shining.
   Sides of achenes with 3-6 longitudinal, scabrous ribs, hispid at the summit; beak about as long as the body of the achene; rays yellow.
    Leaves runcinate-pinnatifid, the segments mostly 5 or 7........1. *L. Scariola.*
    Leaves spinulose-denticulate, not pinnatifid...1a. *L. Scariola* var. *integrata.*
   Sides of achenes with one rib, otherwise similar to the preceding; rays pinkish purple..2. *L. campestris.*
  Achenes black, shining. (See excluded species no. 702, p. 1105)........*L. virosa.*
 Margins of leaves not prickly but the leaves sometimes with sharp teeth.
  Achenes light brown, body 2.5-3 mm long, their beaks twice as long as their bodies, usually with 6-9 longitudinal, scabrous ribs, not hispid at the apex...3. *L. saligna.*
  Achenes dark brown with a mottling of black, minutely and closely marked with transverse ridges, with one prominent longitudinal ridge on each face.
   Involucres 10-14 mm long; mature achenes 5-6 mm long; pappus 5-7 mm long.
    Leaves all, or at least the lower ones, more or less lobed.
     Leaves with linear-falcate, usually entire lobes; upper unlobed leaves, if any, linear or linear-lanceolate; base of leaf sagittate or auriculate..4. *L. canadensis* var. *typica.*
     Leaves with broadly falcate, or obovate and obliquely truncate, entire or toothed lobes; upper leaves similar or unlobed and lanceolate, rarely oblanceolate or obovate, entire or toothed, sagittate, and clasping at the base............................4a. *L. canadensis* var. *latifolia.*
    Leaves all unlobed, lanceolate, oblong, oblanceolate or obovate, entire or denticulate, the lowest sometimes with shallow lobes.
     Cauline leaves lanceolate to ovate-lanceolate, entire or toothed.
      Base of leaf sagittate, clasping.........4b. *L. canadensis* var. *integrifolia.*
      Base of leaf tapering, not sagittate...4c. *L. canadensis* var. *integrifolia* f. *angustata.*
     Cauline leaves oblanceolate or obovate, usually toothed, sagittate, and and clasping at the base..............4d. *L. canadensis* var. *obovata.*
   Involucres 16-22 mm long; mature achenes 7-9 mm, including the beak; pappus 9-12 mm long. (See excluded species no. 700, p. 1105)..........*L. hirsuta.*
Achenes without filiform beaks, beakless or essentially so; rays bluish, sometimes cream color in *Lactuca spicata.*
 Pappus white.
  Leaves coarsely and unevenly dentate, acuminate at the apex, long taper-pointed at the base, sessile, not sagittate, more or less sparingly pubescent above and beneath, the pubescence usually restricted to the principal veins, sometimes nearly glabrous, and rarely the pubescence conspicuous.......5. *L. villosa.*
  Leaves lyrately cut into 3-6 segments, the terminal segment usually the largest and triangular with about equal sides, sometimes one or more of the upper leaves not lobed, glabrous or nearly so.......................6. *L. floridana.*
 Pappus tawny; very large plants with bluish or cream color flowers; latex white or tawny.

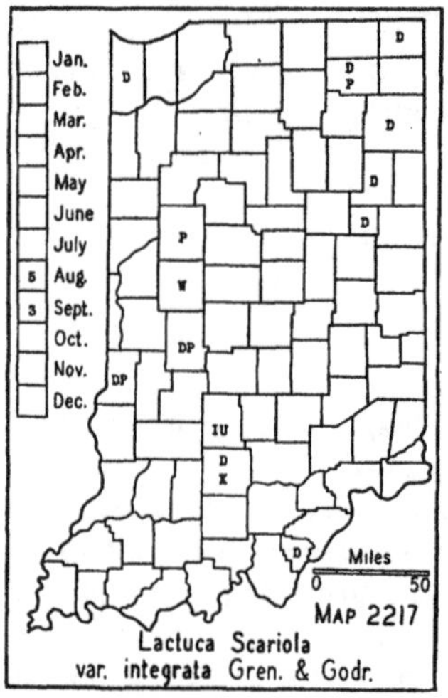

MAP 2217
Lactuca Scariola var. integrata Gren. & Godr.

MAP 2218
Lactuca campestris Greene

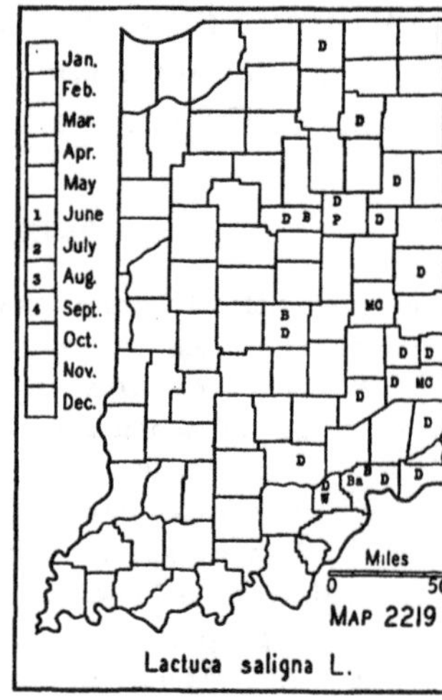

MAP 2219
Lactuca saligna L.

Leaves irregularly pinnatifid, segments 3-12........................7. *L. spicata.*
Leaves not divided or the lower ones sinuate.....7a. *L. spicata* var. *integrifolia.*

1. LACTUCA SCARÌOLA L. PRICKLY LETTUCE. Map 2216. Frequent to abundant throughout the state. It seems to be periodic in its abundance. Some years it is rarely seen and other years it is a common weed. It is found along roadsides and railroads, in waste places in general, and in fallow and cultivated fields. It is usually found in greatest abundance in wheatfields and oatfields. It seems that it appeared in Indiana about 1890, and in a few years it had become an obnoxious weed throughout the state.

Nat. of Eu.; N. E. to Ga. and Tenn., westw. to Calif.

1a. LACTUCA SCARIOLA var. INTEGRÀTA Gren. & Godr. Map 2217. I can not separate this form of the prickly lettuce from the preceding one in reports, so I must rely upon the data of the specimens at hand. It has the same habitat as the species and I believe that it is rather infrequent in the state.

Nat. of Eu.; probably local throughout the greater part of the U. S. I have a specimen from N. Mex., and Jepson reports it from Calif.

2. **Lactuca campéstris** Greene. (Pittonia 4: 37-38. 1899.) Map 2218. I found two plants of this species in the old lake basin of Beaver Lake about 200 feet west of the bridge on U. S. Highway 41 over the Beaver Lake ditch about 3 miles south of Lake Village. It was growing in dry, sandy soil near the ditch. Evidently it is scarce in this area because a companion and I searched for several hours for more of it. The two plants were about 100 feet apart on opposite sides of the ditch. This species is conspicuously different from all others of the genus because of its leafy and short stem, and its broad rather naked inflorescence, which is well above the leaves.

It was described from specimens collected in southwestern Minnesota and is a western plains and prairie plant. Its distribution is not known because most authors do not separate it from *Lactuca ludoviciana* which has yellow flowers.

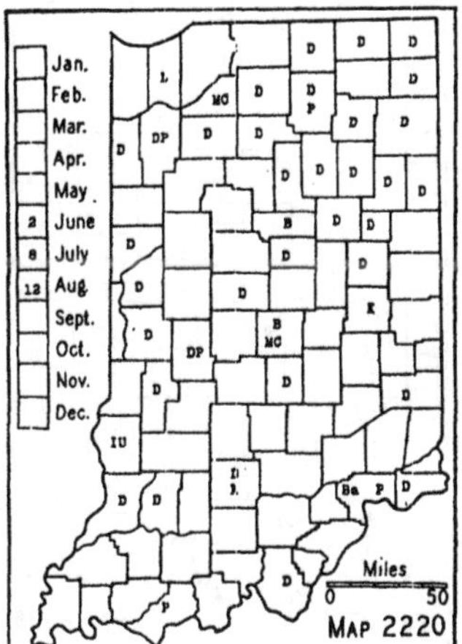

Lactuca canadensis L., var. typica

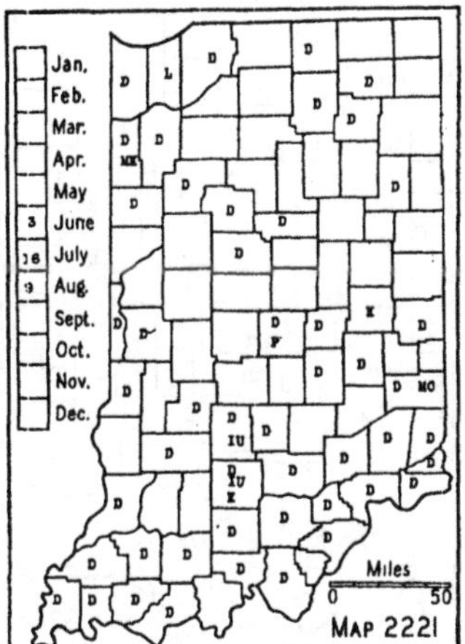

Lactuca canadensis var. latifolia O. Kuntze

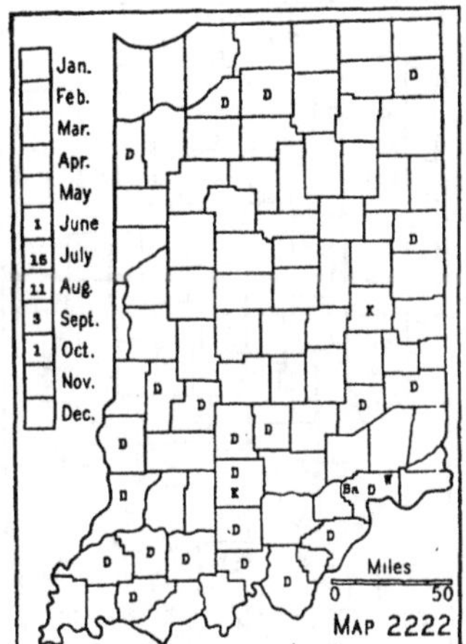

Lactuca canadensis var integrifolia (Bigel) Gray

3. LACTUCA SALÍGNA L. Map 2219. Local but abundant where it is found. No doubt it has a wider distribution than our map indicates. It prefers a sandy soil and in such a habitat it is a common weed in ballast for miles along railroads. Found along railroads and roadsides and waste places and pastures.

Nat. of Eu.; I have not been able to ascertain its distribution in the U. S. although it is definitely reported from Ohio, Mich., Mo., and Calif.

4. **Lactuca canadénsis** L. var. **týpica** Wieg. (Rhodora 22: 10. 1920.) (*Lactuca canadensis* L. in part.) Map 2220. Infrequent to frequent in dry, open woodland throughout the state; also along railroads and rarely in fallow fields. Wiegand describes also f. *angustipes* of this variety which I have not seen in Indiana.

N. S. to B. C., southw. to Ga., Ala., La., Ark., and Colo.

4a. **Lactuca canadensis** var. **latifòlia** O. Ktze. (Rhodora 22: 10. 1920.) (*Lactuca canadensis* L. in part.) Map 2221. Infrequent to frequent throughout the state although I do not have a specimen from the dune area. It is found chiefly in moist or dry, open woodland and occasionally along roadsides and railroads.

Wiegand describes also f. *exauriculata* of this variety with leaves which are not sagittate at the base. I have not seen it in Indiana.

P. E. I. to Wis., southw. to Fla. and Okla.

4b. **Lactuca canadensis** var. **integrifòlia** (Bigel.) Gray. (Rhodora 22: 10. 1920.) (*Lactuca sagittifolia* Ell.) Map 2222. Infrequent to frequent in moist or dry, open woodland in the southern part of the state. I have only a few specimens from the northern part. It is also found along roadsides and fences.

P. E. I. to Wis., southw. to Ga., Ill., Okla., and Nebr.

4c. **Lactuca canadensis** var. **integrifolia** f. **angustàta** Wieg. (Rhodora

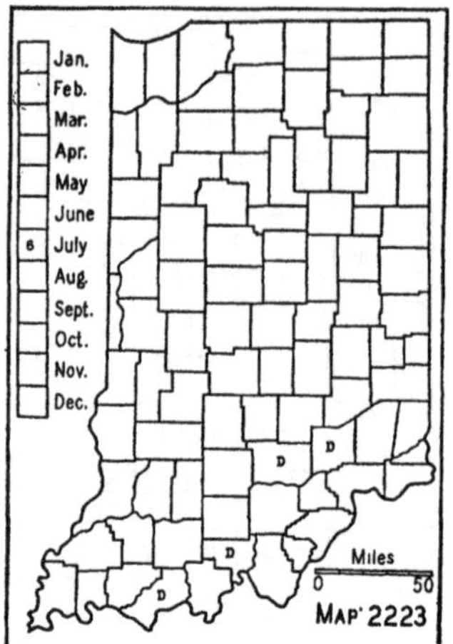

Lactuca canadensis. var obovata Wieg.

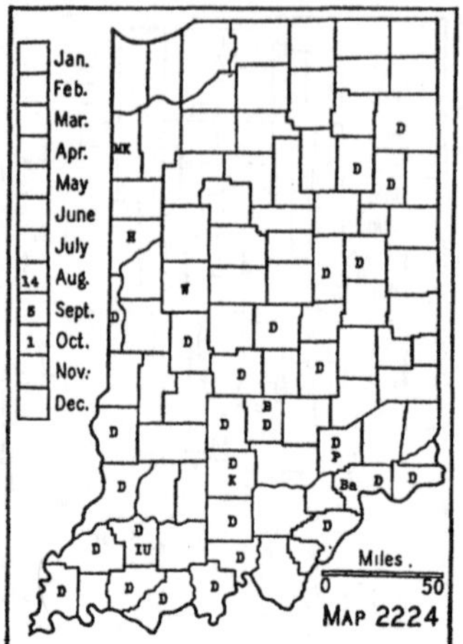

Lactuca villosa Jacq.

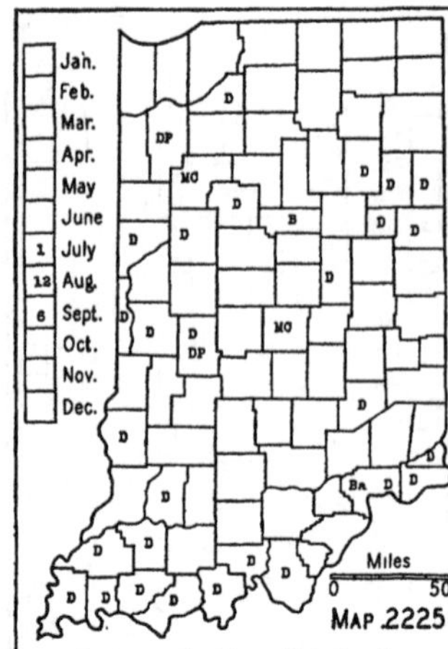

Lactuca floridana (L.) Gaertn.

22: 10. 1920.) My only specimen was found along a trail in Clifty Falls State Park.

Mass., Conn. to Del., N. Y., and Ill.

4d. **Lactuca canadensis** var. **obovàta** Wieg. (Rhodora 22: 10. 1920.) (*Lactuca integrifolia* of Gray, Man., ed. 7, not Bigel.) Map 2223. Infrequent in open woodland and along roadsides in the southern part of the state. Wiegand also describes f. *stenopoda* of this species with the leaves not sagittate. I have not found this form in Indiana.

Maine to Ind. and Nebr., southw. to N. J. and Okla.

5. **Lactuca villòsa** Jacq. Map 2224. This species is infrequent to frequent in the southern part of the state, becoming infrequent, local, or absent in the northern counties.

N. Y. to Nebr., southw. to Fla. and La.

6. **Lactuca floridàna** (L.) Gaertn. Map 2225. I have what I believe to be this species from the greater part of the state. It is frequent at least in the southern part and is usually found in woodland. It prefers shady woods along streams. Our manuals describe it as having the achene with a short, narrow beak. All of my specimens are beakless, at least none with a beak longer than 0.3 mm.

N. Y. to Minn., southw. to Fla. and Tex.

7. **Lactuca spicàta** (Lam.) Hitchc. Map 2226. This is our largest species and normal size specimens range from 6-10 feet high. It is a woodland species, preferring rather moist, rich soil. It is rather frequent in the northern part of the state where its flowers are usually cream color. In the southern part of the state it becomes infrequent. Throughout its range it is also found along roadsides.

Newf. to Man., southw. to N. C., Tenn., Iowa, and Colo.

7a. **Lactuca spicata** var. **integrifòlia** (Gray) Britt. Map 2227. The variety is much smaller in stature and very local. Its habitat is the same as that of the species.

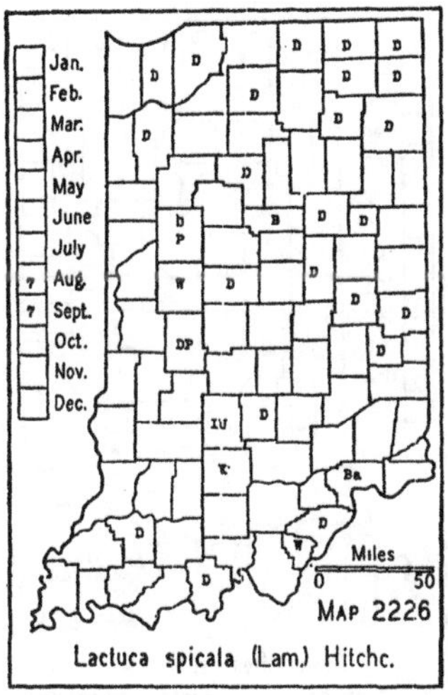

Lactuca spicata (Lam.) Hitchc.

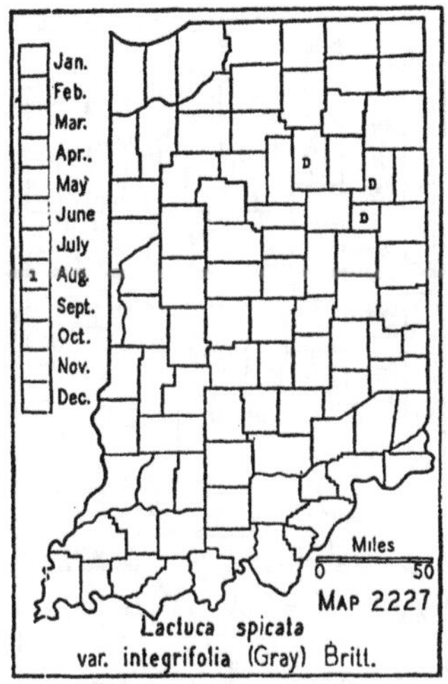

Lactuca spicata var. integrifolia (Gray) Britt.

Pyrrhopappus carolinianus (Walt.) DC.

9604. PYRRHOPÀPPUS DC.

1. **Pyrrhopappus caroliniànus** (Walt.) DC. FALSE DANDELION. Map 2228. My only specimens are from a low, flat fallow field along Big Creek about one and three fourths miles south of Wadesville, Posey County. In this field are small areas where there is no vegetation, called by the land owner salt spots. In this hard, white clay soil several specimens were collected.

This species was reported from White County by Heimlich. Since White County has no habitats similar to the one in which I found my specimen, I question the identification. White County is far north of the known range of the species. If Heimlich found it, I believe it must have been introduced. Andrews reported it from Monroe County, but since he preserved no specimen, and since the habitat is lacking and the place is north of the range of the plant, the report is disregarded.

Del., Ky. to Mo., southw. to Fla. and Tex.

9605. CRÈPIS L. HAWKBEARD

Involucre pubescent; seed about 2.5 mm long.........................1. *C. capillaris.*
Involucre glabrous; seed about 5 mm long.............................2. *C. pulchra.*

1. CREPIS CAPILLÀRIS (L.) Wallr. Map 2229. I found this species well established in Fairview Cemetery at Bluffton in 1923. In 1935 it had spread, which shows that when it is established it will persist unless diligent efforts are made to eradicate it. It would seem that constant mowing in a cemetery would kill it, but it thrives nevertheless. It is plentiful in the lawn of St. Mary's College, St. Joseph County.

Nat. of Eu.; Conn. to N. Y., southw. to N. J. and Ohio.

2. CREPIS PÚLCHRA L. Map 2230. This species was found in Jefferson County in 1934 by Miss Edna Banta. She reports that it is a common roadside weed from Eagle Hollow east of Madison eastward along the River Road to Morris Chapel, a mile east of Brooksburg, a distance of about 8 miles.

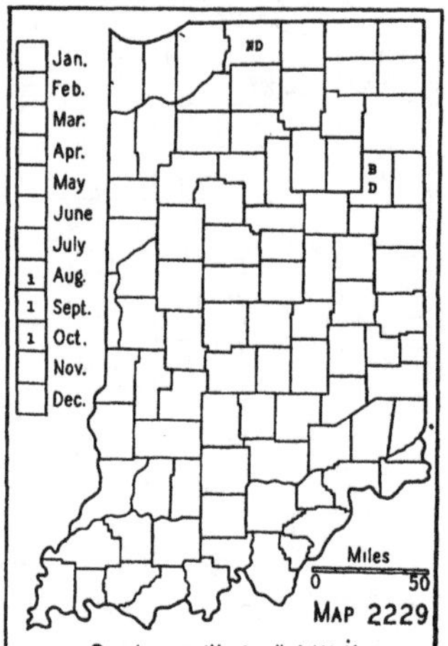

Crepis capillaris (L.) Wallr.

Crepis pulchra L.

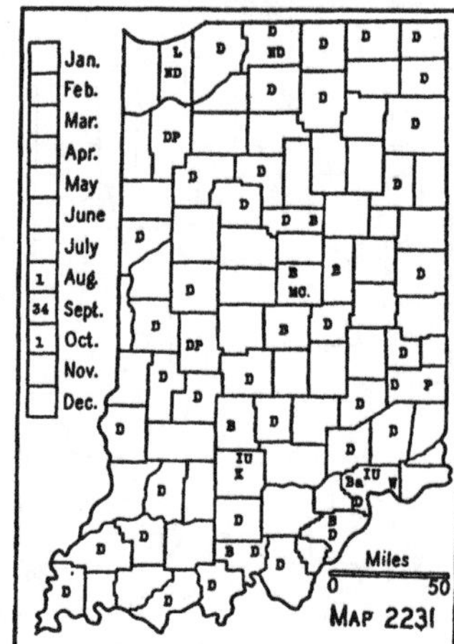

Prenanthes altissima L.

The size and vigor of the plants sent me, being much branched and over three feet high, suggest that it is well established and will be able to compete with any native vegetation and forever be a weed in our state. I collected it in the same area in 1937.

Nat. of Eu.; no recent distribution given for the U. S.

9606. PRENÁNTHES [Vaill.] L.

Involucre glabrous.
- Inner bracts 5; flowers 5 or 6; pappus of mature achenes light brown (Antimony Yellow to Ochraceous-Buff of Ridgway, Standard.) .... 1. *P. altissima.*
- Inner bracts 6-8; flowers 8-12.
  - Involucre greenish white .... 2. *P trifoliolata.*
  - Involucre purple tinged; plant glaucous .... 3. *P. alba.*

Involucre pubescent.
- Heads 8-16-flowered.
  - Stems and lower surface of leaves glabrous; flowers purplish....... 4. *P. racemosa.*
  - Stems and lower surface of leaves rough-pubescent; flowers cream color............ 5. *P. aspera.*
- Heads 25-35-flowered; flowers cream color............ 6. *P. crepidinea.*

1. **Prenanthes altíssima** L. Map 2231. Doubtless formerly found in every county of the state. It is a woodland plant, preferring dry soil, and found principally in oak woods and less frequently in beech and maple woods. The great variation in the shape of the leaves of this species and the two following has led to the naming of several varieties, none of which I consider worthy of a name. I have found the most diverse forms in the same colony. The blades may be undivided, merely dentate, parted with the divisions not stalked, divided into three parts with the two lateral ones stalked and often deeply parted or deeply lobed, cordate or cuneate at the base, and there are many intermediate forms.

The pappus of the species varies somewhat in color but no plants have been found with a white or even of a sordid color. It is to be noted that the pappus does not acquire its characteristic color until it is mature. Fernald described a variety of this species with cinnamon brown pappus

Prenanthes trifoliolata (Cass.) Fern.

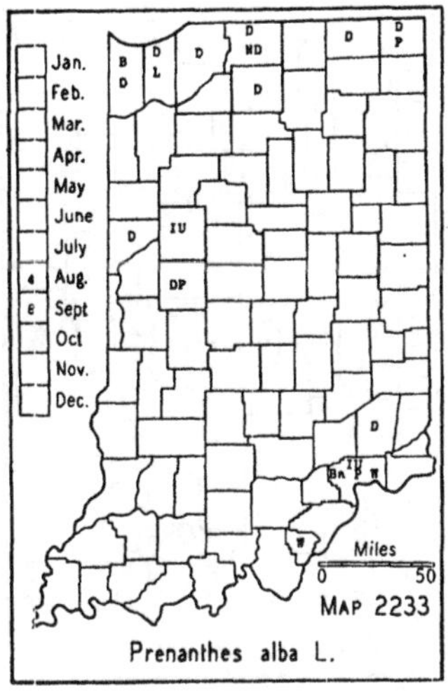

Prenanthes alba L.

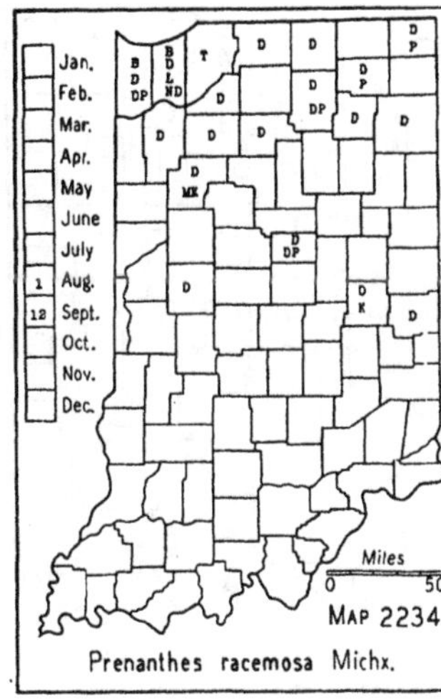

Prenanthes racemosa Michx.

and cited Indiana within its range. In none of our specimens is the pappus as dark as Cinnamon-Brown of Ridgway's Standard.

Newf. to Man., southw. to Ga. and Tenn.

2. **Prenanthes trifoliolàta** (Cass.) Fern. Map 2232. My only specimen is from a woods in Porter County about 5 miles southwest of Michigan City. Peattie reported it from La Porte County. Clark reported it from Marshall County, but, since he did not report *Prenanthes altissima,* this report should no doubt be referred to the latter. It has been reported also from White County. Doubtless it is rare in Indiana.

Newf. and Que. to N. Y. and Mo., southw. to N. C. and in the mts. to Tenn.

3. **Prenanthes álba** L. Map 2233. This species has its mass distribution in the lake area, becoming local southward. It is rather infrequent in the lake area and is found mostly in moist, sandy soil in woodland and sometimes in marshes. This plant has a common name assigned to it which rightfully belongs to another plant. Since the other plant is a medicinal plant it claims the common name, and, since it is confusing to have two plants with the same common name, I do not mention it here.

S. Maine to Sask., southw. to Ga., Tenn., and Ill.

4. **Prenanthes racemòsa** Michx. Map 2234. All of my specimens of this species are from the northern part of the state where it is infrequent and found in marshes and moist prairie habitats.

N. B., Que. to Man., southw, to N. J., Mo., and Colo.

5. **Prenanthes áspera** Michx. Map 2235. Very local in the area shown on the map. It has been reported from Clark, Jefferson, and Steuben Counties by early authors. My specimens were found along roadsides in dry, sandy soil in prairie habitats. Late in 1938 Kriebel and I found it in hard, white clay soil in Spencer County.

Ohio to S. Dak., southw. to Tenn. and La.

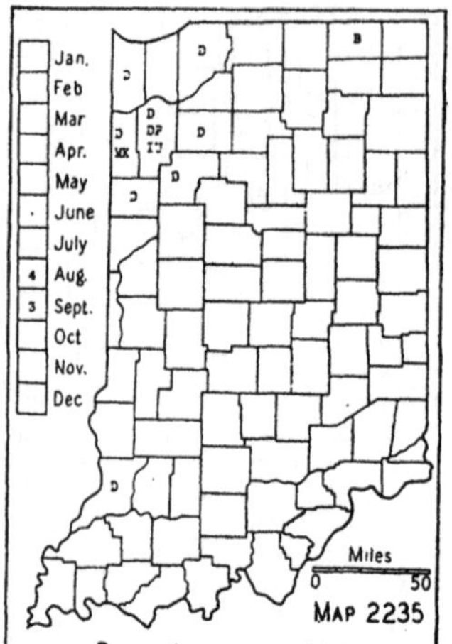

Prenanthes aspera Michx.

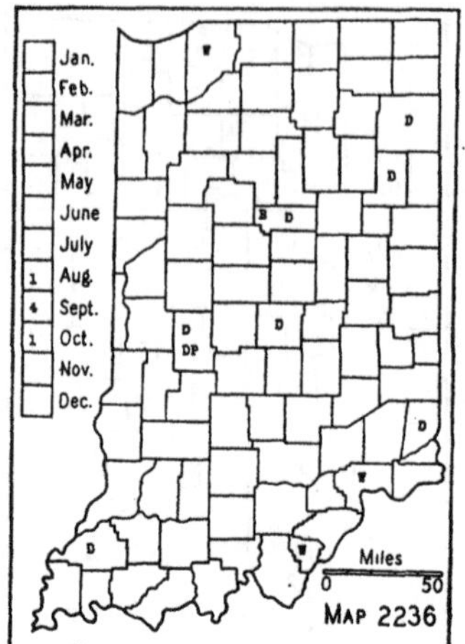

Prenanthes crepidinea Michx.

Hieracium aurantiacum L.

6. **Prenanthes crepidínea** Michx. Map 2236. Very local throughout the state. In addition to the counties shown on the map it has been reported from Carroll, Clark, Fayette, Franklin, Tippecanoe and Wabash Counties. All of my specimens with one exception are from the alluvial banks of streams. I have never seen more than one specimen at a place.

Western N. Y. to Minn., southw. to Ky. and Kans.

9607. HIERÀCIUM [Tourn.] L. HAWKWEED

Leaves all basal; flowers on a scape, orange........................1. *H. aurantiacum.*
Leaves mostly cauline.
Cauline leaves more than 20, usually 25-45, all sessile, the margins coarsely dentate, the teeth usually 1-5 mm long; basal leaves without long, margined petioles; mature heads large, usually more than 15 mm wide; plants restricted to the northern part of the state.................................2. *H. canadense.*
Cauline leaves fewer than 20, their margins entire or merely denticulate; mature heads mostly less than 15 mm wide; plants not restricted to the northern part of the state.
Mature achenes narrowed at the summit.
Pubescence of the lower part of the stem and lower leaves spreading, the hairs generally much less than 5 mm long; inflorescence racemosely paniculate; achenes mostly 3-3.75 mm long..........................3. *H. Gronovii.*
Pubescence of the lower part of the stem and lower leaves appressed, the hairs mostly 5-15 mm long; inflorescence corymbose-paniculate; achenes mostly 3.5-4 mm long...4. *H. longipilum.*
Mature achenes not narrowed at the summit.
Inflorescence glabrous or with a few glandular hairs at the summit of the pedicel and on the involucre; cauline leaves narrow-oblong, acute; flowers 12-20; pedicels slender.
Cauline leaves 1-3 (plants probably scapose with large cauline bracts); basal leaves generally purple-veined and purplish; inflorescence a loose, divided corymb..5. *H. venosum.*
Cauline leaves usually more than 10, basal leaves not purplish; inflorescence an elongated, lax panicle............................6. *H. paniculatum.*
Inflorescence densely glandular-pubescent; flowers mostly 40-50; pedicels stout; achenes 2-2.5 mm long; cauline leaves elliptic, obtuse........7. *H. scabrum.*

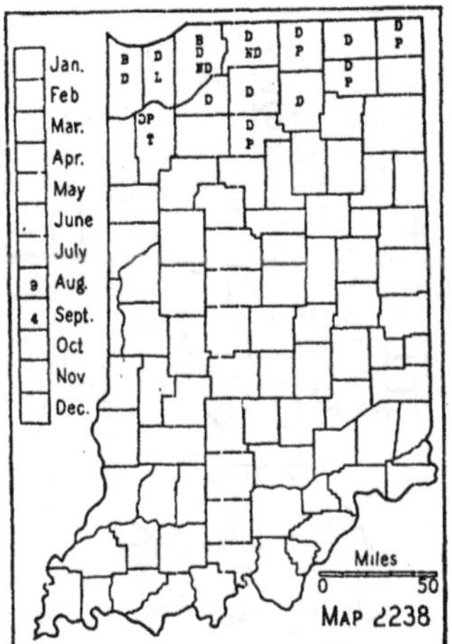

Hieracium canadense Michx.

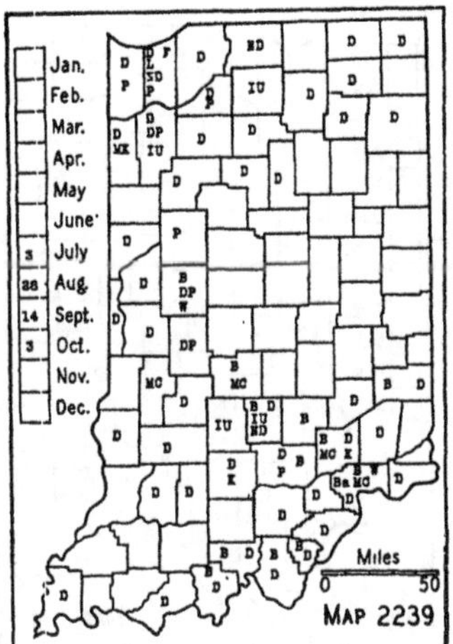

Hieracium Gronovii L.

Hieracium longipilum Torr.

1. HIERACIUM AURANTÌACUM L. ORANGE HAWKWEED. DEVILS-PAINT-BRUSH. Map 2237. In 1934 I found this hawkweed in the sandy commons on the south side of Simonton Lake in Elkhart County. In 1935 I found it in a sandy, waste field and in an adjoining open woodland on the north side of Weber Lake in Steuben County. It is an obnoxious weed in the eastern states and, unfortunately, it is now cultivated as an ornamental plant in Indiana. It will doubtless soon escape in many parts of the state if it has not already done so. Usually called Devil's-paint-brush.

Nat. of Eu.; Newf. to Mich., southw. to N. J. and Pa.; also in Colo. and on the Pacific coast.

2. **Hieracium canadénse** Michx. CANADA HAWKWEED. Map 2238. Infrequent in the lake area in very dry, sandy soil, on slopes wooded with black and white oak and on open dunes and rarely in a prairie habitat. Young's report from Jefferson County should no doubt be referred to some other species.

Lab. to B. C., southw. to N. J., Pa., Ind., S. Dak., and Oreg.

3. **Hieracium Gronòvii** L. GRONOVIUS HAWKWEED. Map 2239. Infrequent to frequent in the lake area, infrequent in the southern part of the state, and local, rare or absent in many of the counties in the Tipton Till Plain where the soil is too alkaline for it. It seems to prefer a slightly acid soil and this fact accounts for its being found on washed wooded slopes, interdunal flats, and the hard, white, sandy, clay loam of the Illinoian drift.

Fernald & Griscom (Rhodora 37: 185-186. 1935) report a variety of this species, var. *foliosum* Michx., as being found in southern Indiana. The variety is described as having more cauline leaves which extend nearly to the inflorescence. We have a few plants answering this description, but I do not regard them as worthy of a varietal name. The cauline leaves on our plants vary from few to many, the variation apparently due, for the most part, to nutrition. I have seen very vigorous and large plants that exceeded the size of average plants, growing in moist, clay soil in

MAP 2241
Hieracium venosum L.

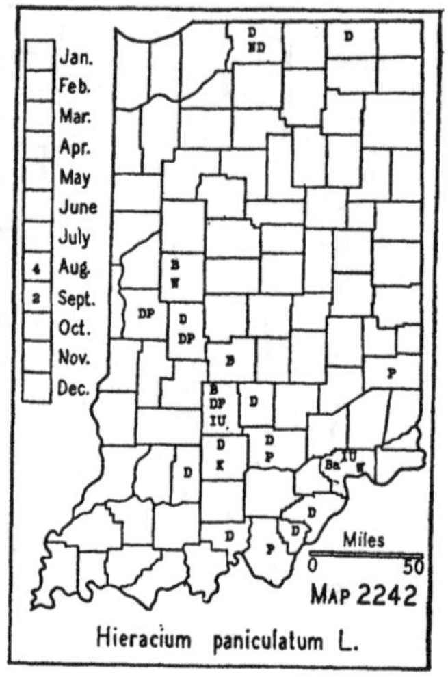

MAP 2242
Hieracium paniculatum L.

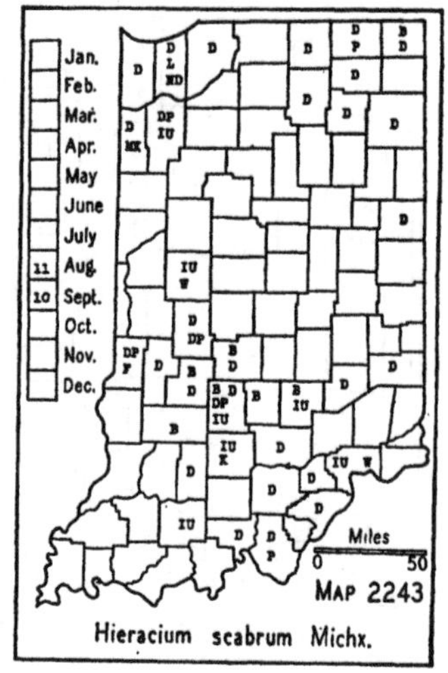

MAP 2243
Hieracium scabrum Michx.

fallow fields. The plant is highly variable even as to the number of achenes. Since I am not recognizing this variety, its distribution in the state is not given, although I have specimens from the northern to the southern border.

Mass., Ont. to Kans., southw. to Fla. and Tex.

4. **Hieracium longípilum** Torr. LONG-BEARD HAWKWEED. Map 2240. This species has been reported from the dune area, Lower Wabash Valley, and from Parke, St. Joseph, Steuben, and Vigo Counties. I have seen the Parke County specimen and it should be referred to *Hieracium Gronovii.* This hawkweed is a weed in sandy fallow fields in the vicinity of Heaton Lake, Elkhart County, and in several fields in northeastern St. Joseph County. It is probably more widely distributed. I noted it as frequent in the old Beaver Lake Basin about 3 miles south of Lake Village, Newton County.

Ont. to Minn., southw. to Ind. and Tex.

5. **Hieracium venòsum** L. RATTLESNAKE-WEED. **Map 2241.** I have found this species in three places in Clark County and in no other place. A few plants were found on the crests of ridges with chestnut oak and Virginia pine. It has also been reported from Floyd, Jefferson, Monroe, and White Counties. I have not been able to check any of these reports. I have searched the herbaria of the Field Museum and of the University of Illinois for the Lake County specimen, but I did not find it. N. C. Fassett says there is no specimen in the Umbach herbarium at the University of Wisconsin. Specimens of *Hieracium Gronovii* might be mistaken for this species.

S. Maine to Man., southw. to Ga., Ky., and Nebr.

6. **Hieracium paniculàtum** L. Map 2242. Infrequent to very local in the counties shown on the map. It has been reported from Jefferson, Johnson, Lake, and Monroe Counties. N. C. Fassett says that the Umbach specimen from Lake County is *H. canadense.* It prefers slightly acid and

sandy soil. Usually found on black and white oak slopes, although I found several specimens in a moist, level, sandy woods along Pigeon River in Lagrange County. A glandular variety of this species has been described, but all of our specimens are glandless.

N. S. and cent. Maine to Mich., southw. to Ga. and Ala.

7. **Hieracium scàbrum** Michx. (For described varieties of this species see Rhodora **16**: 182-183. 1914.) Map 2243. Infrequent throughout the area indicated on the map. This species also prefers slightly acid and sandy soil. It is usually found in dry soil on slopes with black and white oak and rarely on sandy flats with the same associates. Besides the counties shown on the map, it has been reported from Fayette, Parke, Tippecanoe, and White Counties.

N. S. to Minn., southw. to Ga., Iowa, Nebr., and Kans.

EXCLUDED SPECIES.

This list contains native species reported from Indiana of which no specimen could be found to confirm the report. Reports of foreign trees, shrubs, and plants found in yards, cemeteries, parks, and gardens have been excluded. This list contains many foreign species that have been reported without data to confirm the fact that they have become established and have become a part of our flora. Among these are many ornamental plants that have escaped to or have been thrown into streets and alleys and have not become a part of our flora. It contains, also, many incorrect determinations which are explained in the text.

1. LYGÒDIUM PALMÀTUM (Bernh.) Sw. This fern was reported by Collett in Rept. Indiana Geol. Surv. 5: 256. 1874. This report was changed by the same author to *Camptosorus rhizophyllus* (L.) Link in Rept. Indiana Geol. Surv. 7: 400. 1876.

2. **Dryópteris Linnaeàna** C. Chr. (*Phegopteris Dryopteris* (L.) Fée and *Dryopteris Dryopteris* (L.) Britt.) OAKFERN. This species was reported from Allen County upon the authority of J. A. Sanford by the editors of the Botanical Gazette in a supplement to the Flora of Indiana, published in April, 1882. Since this is our only report and there is no verifying specimen, the species is excluded.

Newf. to Alaska, southw. to Va., Kans., Colo., and Oreg.

3. DRYOPTERIS PHEGÓPTERIS (L.) C. Chr. (*Phegopteris polypodioides* Fée.) NARROW BEECHFERN. This fern was first reported from La Porte County in 1873 by Babcock. It was next reported from Putnam County by Coulter on the authority of Underwood. Pepoon says in his "Flora of the Chicago Region" that it is abundant in Porter County north of Port Chester. Behrens reported it as frequent throughout Turkey Run State Park, Parke County. I searched the Babcock herbarium for it in 1929 but I could not find a specimen. I am excluding it because I believe that depauperate forms of *Dryopteris hexagonoptera* have been mistaken for it.

Newf. to Alaska, southw. to N. Y., Va., Wis., Iowa, and Wash.; also in Eurasia.

4. DRYOPTERIS SIMULÀTA Davenport. (*Aspidium simulatum* Davenport.) Reported from the dune area by Peattie and by Pepoon. Fassett (Rhodora **35**: 200. 1933) says that the specimen upon which these records are based is *Dryopteris Thelypteris* var. *pubescens*. This species is not found west of the Allegheny Mountains.

Maine, Vt. to Md.

5. DRYOPTERIS FÌLIX-MÁS (L.) Schott. (*Aspidium Filix-mas* (L.) Sw.) MALE FERN. There are three reports of this fern from Jefferson County, and I reported it from Wells County. I now refer the Wells County specimen to *Dryopteris Goldiana* and, no doubt, the Jefferson County reports should be referred to some other species, since the range of *Dryopteris Filix-mas* does not extend to our area.

Newf., N. S., n. Vt., Lake Huron, Lake Superior, N. Dak. to Ariz., and northw.

6. **Dryopteris Clintoniàna × spinulòsa** Benedict. Reported by Nieuwland (Amer. Midland Nat. **2**: 277. 1912) from La Porte County, based upon Deam's specimen no. 8751. I am now referring this specimen to *Dryopteris spinulosa* var. *fructuosa* (Gilbert) Trudell.

7. DRYOPTERIS SPINULOSA var. AMERICÀNA (Fisch.) Fern. (*Aspidium spinulosum* var. *dilatatum* f. *anadenium* Rob. and *Dryopteris dilatata* (Hoffm.) Gray.) This form was reported in 1912 from Porter County by Hill and from Parke County in 1928 by Behrens. Since the range is to the north of Indiana and I have not seen a specimen, I am excluding it.

Lab. to B. C., southw. to the uplands of N. E., Pa., Mich., Idaho, and Oreg. and in the mts. to N. C. and Tenn.

8. DRYOPTERIS THELÝPTERIS (L.) Gray. This species is restricted to Europe, western Siberia, eastern Asia, southward to the Himalayas and southern China. Fernald has recently shown that the American marsh shield fern is distinct from the typical species. The Indiana reports for the marsh shield fern have been under several names but they should now all be referred to *Dryopteris Thelypteris* var. *pubescens* (Lawson) A. R. Prince.

9. ATHÝRIUM FÌLIX-FÉMINA (L.) Roth. This species is now regarded as belonging to Europe and probably western North America. All of the many reports of it from Indiana should be referred to *Athyrium asplenioides* and *Athyrium angustum* and its forms. For a discussion of the Lady Ferns see Butters' treatment (Rhodora **19**: 170-207. 1917).

Specimens have been reported bearing the following names: *Asplenium Filix-femina* var. *angustum* Moore, *Asplenium Filix-femina fissidens* Doell, *Asplenium Filix-femina* var. *Michauxii* Mett., and *Asplenium Filix-femina ovatum* Roth. The authors have not cited specimens, and, since they cannot be traced in order to be identified, these reports are valueless and should be dropped.

10. CHEILÁNTHES TOMENTÒSA Link. WOOLLY LIPFERN. Reported from Martin County by Craw (Butler Univ. Bot. Studies **2**: 160. 1932). I have

seen the specimen and it should be referred to *Cheilanthes lanosa* (Michx.) Watt. *Cheilanthes tomentosa* has its range south of our area.

Va. to Ga., westw. to Tex., Ariz., and Mex.

11. ADIÁNTUM PEDÀTUM var. ALEÙTICUM Rupr. This variety of the maidenhair fern was reported by Behrens (Proc. Indiana Acad. Sci. **37**: 377. 1928) from Turkey Run State Park, Parke County. He writes: "Common maidenhair fern. The Maidenhair is distributed throughout the park in the uplands. It is one of the most common of the ferns found in Turkey Run." When the range of this fern is considered this report becomes absurd.

Alaska southw. in the high Sierra Mts. to Nev., Que., and n. N. E.

12. MARSÍLEA QUADRIFÒLIA L. PEPPERWORT. Grimes, in 1911, reported this species from an old mill pond on the south side of the Vandalia Railroad in Greencastle, Putnam County. He remarks that it was fast disappearing due to drainage and subsequent encroaching of vegetation. He adds that it was first detected in 1904 by Dr. Banker. A specimen collected by Banker, dated October 11, 1905, is in the herbarium of DePauw University and bears the following information: "Transplanted from Connecticut to Ithaca, N. Y. by W. R. Dudley; from there to Columbus, Ohio, by W. A. Kellerman; and to here in 1903 by Mel T. Cook." In 1937 I searched for it but failed to find it. It has probably disappeared.

Nat. of Eu. and Asia.

13. EQUISÈTUM SYLVÁTICUM L. WOOD HORSETAIL. Reported from Dearborn County by Collins. In Coulter's Catalogue it was placed in a list for which there were no verifying specimens. Since this species is more northern in its distribution and there is no confirming specimen, it is excluded.

Schaffner says it is found in the northern part of the North Temperate Zone.

14. **Equisetum praténse** L. MEADOW HORSETAIL. This species was reported from St. Joseph County by McDonald. I have seen the specimen and I refer it to *Equisetum arvense* L.

N. S., Que. to Alaska., southw. to N. J., Iowa, and Colo.; also in Eurasia.

15. LYCOPÒDIUM CLAVÀTUM L. Reported from Lake County by Ball, and Pepoon includes it in his "Flora of the Chicago Region" and says: "Found near Miller, Ind., by Higley." It is not included in the Higley & Raddin Flora. Peattie, in his "Flora of the Indiana Dunes," says: "Perhaps an error." Buhl (Amer. Midland Nat. **16**: 250. 1935) adds: "Northern in range, reports in error." Since confirming specimens are lacking, the species is dropped from our flora.

Lab. to Alaska, southw. to N. C., Mich., and Wash.; also in Eurasia and tropical Amer.

16. LYCOPODIUM COMPLANÀTUM L. GROUNDCEDAR. This species has been reported from the area about Lake Michigan and from Monroe and Putnam Counties. Since the distribution of this species, as now understood,

is north of Indiana, I am referring all of our records to *Lycopodium flabelliforme* (Fern.) Blanchard.

Newf. to Alaska, southw. to n. Ont., n. Mich., n. Wis., and n. Wash.

17. **Lycopodium lucídulum** var. **occidentàle** (Clute) Wilson. (Rhodora 34: 170. 1932.) (*Lycopodium porophilum* Lloyd & Underwood (Bull. Torrey Bot. Club 27: 150. 1900) in part but not as to type specimen and *Lycopodium lucidulum* var. *porophilum* (Lloyd & Underwood) Clute in part.) Buhl refers Peattie's report from the Indiana Dunes to the species. It grows in moist pockets in sandstone cliffs and on ledges.

Mich., Minn. and Wis.

18. LYCOPODIUM TRISTÀCHYUM Pursh. This clubmoss was reported from the Indiana Dunes by Peattie. Since there are no confirming specimens, and the range of the species is to the north of Indiana, I am excluding it.

Newf. to Lake Superior, southw. to Del. and in the mts. to Ga.; also in Eu.

19. ISÒËTES BRAÚNII Durieu, In Crawford County, in a small pond in the corner of a field on the Nathan Bowman farm, about 5 miles south of Marengo, and a half mile east of the Pilot Knob School, I collected a quillwort that was placed in this species by a specialist. Later my specimens were named *Isoëtes Engelmanni* A. Br. by Norma Pfeiffer in her monograph. I reported this specimen as *I. Braunii* (Proc. Indiana Acad. Sci. **1916**: 315. 1917) but now refer it to *I. Engelmanni.*

North America.

19a. ISOËTES FOVEOLÀTA A. A. Eaton. A specimen from an old stream bed in a low woods in Harrison County, 1 mile east and 4 miles south of Palmyra, was named *Isoëtes foveolata* by an authority. Later the same specimen was referred by Norma Pfeiffer to *Isoëtes Engelmanni* A. Br. In the meantime I had reported it as *I. foveolata* (Proc. Indiana Acad. Sci. **1916**: 315. 1917). I am now referring these specimens to *I. Engelmanni.*

N. H.

20. **Pìnus echinàta** Mill. SHORTLEAF PINE. It is doubtful whether this species occurs in Indiana and until it is definitely established that it is native, all references to it should be referred to *Pinus virginiana.* It was planted about 1917 on the old Schlamm farm in the Clark County State Forest among *Pinus virginiana.* This statement is made to prevent its being reported in the future as a native plant. An old timber buyer told me that in the vicinity of Borden, Clark County, there were two kinds of "scrub" pines but in a search for them he failed to find the two species for me.

Long Island, N. Y., W. Va., sw. Ill., s. Mo., se. Okla., southw. to Fla. and Tex.

21. PINUS RESINÒSA Ait. NORWAY PINE. This pine has been reported as an escape in Wabash County, but there is no verifying evidence.

Mass., Mich. to Minn. and northw.

22. PINUS RÍGIDA Mill. PITCH PINE. Reported from Clark County, but there is no verifying evidence that it was a native of the state.

Maine to Ont., southw. to Va. and e. Ohio and in the mts. to Ga. and Tenn.

23. ABIES BALSÀMEA (L.) Mill. BALSAM FIR. Reported from Porter County (Proc. Indiana Acad. Sci. **1900**: 141. 1901). The report was an error in quoting from Cowle's paper on the dunes of Lake Michigan. Not found south of Michigan.

24. CHAMAECÝPARIS THYOÌDES (L.) BSP. SOUTHERN WHITE CEDAR. See Deam's "Trees of Indiana," ed. 2, p. 306. 1932, for details of reports. The range of the species is east of the Allegheny Mountains and no doubt it never was a native of Indiana.

25. JUNÍPERUS COMMÙNIS L. A small tree of pyramidal habit up to 35 feet high occurs in certain parts of North America and Eurasia. The upright form of the juniper does not occur in Indiana and all reports for it should be referred to *Juniperus communis* var. *depressa* Pursh.

26. SPARGÀNIUM ANGUSTIFÒLIUM Michx. (*Sparganium simplex* Fern. & Eames as to plant, not Huds.) Reported by Buhl (Amer. Midland Nat. **16**: 248. 1935) as found in the Indiana Dunes. He cites a specimen in the herbarium of the University of Illinois and one in the herbarium of the University of Notre Dame. I have seen both specimens and I refer them to other species. Fassett writes me that there are three of Umbach's specimens in the herbarium of the University of Wisconsin labeled *Sparganium simplex* Huds. and they should be referred to *Sparganium americanum* and *Sparganium chlorocarpum* var. *acaule*.

Newf., Que. to Alaska, southw. to Conn., N. Y., uplands of n. N. J. and Pa., Mich., n. Wis., Colo., and Calif.

27. SPARGANIUM MÍNIMUM Fries. This species was reported fifty years ago from Lake County by Babcock and by Hill. The recent reports, I believe, are based upon these old reports. Nieuwland reported it from St. Joseph County, but the specimen proves to be *Sparganium chlorocarpum* var. *acaule*. The range of this species seems to be north of Indiana.

Newf., Que., Man., and Alaska, southw. to Conn., cent. and w. N. Y., uplands of n. N. J. and Pa., Mich., n. Wis., Utah, and Oreg.; Eurasia.

28. **Potamogèton dimórphus** Raf. This species was reported from Vigo County by Blatchley. Since I have not seen the Blatchley specimen, it can not be properly referred, and it is best to omit this report.

29. POTAMOGETON FILIFÓRMIS Pers. This species has been reported from Lake and Marshall Counties. Since this is a northern species, and there are no specimens from Indiana, it is excluded.

Newf. to Alaska, southw. to Maine, Pa., and Colo.

30. POTAMOGETON PERFOLIÀTUS L. There are early reports from Kosciusko, Lake, and Marshall Counties. As now understood, this species

occurs far north of our area, and doubtless our reports should be referred to segregates of this species.

31. **Potamogeton pusíllus** L. There are old reports of this species from the Lower Wabash Valley and from Lake, Kosciusko, Marshall, and Starke Counties. The species, as now understood, is divided into several varieties, one of which is reported from Indiana. This species so closely resembles some others that reports can not be accepted without verifying specimens.

32. **Potamogeton Vàseyi** Robbins. Reported from the Lower Wabash Valley and from Porter County. This species, as now understood, occurs north of our area, but comes so close to us that it should be sought in Indiana. Since there is no verifying specimen, the reports are dropped.
N. B. to e. Minn., southw. to Maine, Conn., Pa., n. Ohio, s. Mich., and n. Ill.

33. SAGITTÀRIA ENGELMANNIÀNA J. G. Smith. This species has been reported from 8 counties. These reports should be referred, doubtless, to a narrow-leaf form of *Sagittaria latifolia.* Since Indiana is outside the range of the species, and there are no confirming specimens, the species is excluded. It is said to grow in shallow water, and its distribution as now understood is restricted to the Coastal Plain from Massachusetts to Virginia.

34. SAGITTARIA LONGIRÓSTRA (M. Micheli) J. G. Smith. I reported this species (Proc. Indiana Acad. Sci. **1916**: 316. 1917) as found in Vigo County. Since I am not now able to locate the specimen, I cannot tell to what species it should be referred. At the Gray Herbarium this species of Small's "Flora of the Southeastern United States" is referred to *Sagittaria australis,* and the true *Sagittaria longirostra* of J. G. Smith is considered a broadleaf *Sagittaria Engelmanniana.* Hence this name is dropped from our flora.

35. SAGITTARIA PUBÉSCENS Muhl. Reported from Hamilton and Marion Counties by Wilson who says: "Common." I am not able to account for such a report since the range of this species is said to be south of Indiana. Since there is no verifying specimen, it is excluded.
N. J., Pa., Tenn., southw. to Fla. and Ala.

36. ELODÈA NUTTÁLLII (Planch.) St. John. This species was reported by Peattie from Lake County. It is now regarded as a synonym of *Anacharis occidentalis* (Pursh) Vict. to which I am referring this report.

37. ELODEA PLANCHÒNII Caspary. I reported this species from Knox County. It is now regarded as the pistillate form of *Anacharis canadensis* (Michx.) Planch. to which I now refer it.

38. **Limnòbium Spóngia** (Bosc) L. C. Richard. Reported from Lake County by T. H. Ball in his "History of Lake County" (1884, p. 170). This species is within our area, but since no specimens were preserved, it is necessary to exclude it.

Lake Ontario to Ill., southw. to Fla. and Tex.

39. ARUNDINÀRIA TÉCTA (Walt.) Muhl. SMALL CANE. There have been a few reports for this species for the state. Since the species of cane were not understood until recently, and it is now known that this species is restricted to the Atlantic Coastal Plain, it is excluded.

Coastal Plain from Md. to Fla. and La.

40. BRÒMUS ARVÉNSIS L. A specimen from Jefferson County was so named for me and I reported it as such. I am now referring this specimen to *Bromus japonicus* Thunb.

41. BROMUS ÁSPER Murr. This species was reported by McDonald from St. Joseph County (Amer. Midland Nat. 15: 208. 1934). Hitchcock now refers this species to *Bromus ciliatus* L.

42. BROMUS ERÉCTUS Huds. This species was reported from Tippecanoe County by Wilson (Proc. Indiana Acad. Sci. 1905: 166. 1906). Wilson says the determination was made by the Bureau of Plant Industry, Washington, D. C. They are not now able to find this specimen at Washington, and, no doubt, it has been referred to some other species. There is no other record and, in the absence of a verifying specimen, the species is dropped from our flora.

43. BROMUS RACEMÒSUS L. This species has been reported from Clark, Gibson, Jasper, Jefferson, Noble, and Vigo Counties, from the area of Delaware, Jay, Randolph, and Wayne Counties, and from the Lower Wabash Valley. I have seen the Jasper County specimen, which is now deposited in the herbarium of DePauw University, and I am referring it to *Bromus secalinus*. I have the Vigo County specimen and it is an immature specimen of *Bromus secalinus*. All of the records except the Jasper County one were made years ago when Gray's Manual, ed. 5 was used, and when the species were not divided as they now are. This species is very rare in the United States, and since we have no specimens, I believe we can safely exclude it for the present.

44. GLYCÈRIA MELICÀRIA (Michx.) Hubbard. (*Glyceria Torreyana* (Spreng.) Hitchc. in Gray, Man., ed. 7 and *Panicularia Torreyana* (Spreng.) Merrill of Britton and Brown, Illus. Flora, ed. 2.) This species has been reported from Clark and Noble Counties and from the area of Delaware, Jay, Randolph, and Wayne Counties. It has not been found west of the western slopes of the Appalachian Mountains and, in the absence of a verifying specimen, it is excluded.

N. B. to ne. Ohio, southw. to the mts. of N. C.

45. GLYCERIA OBTÙSA (Muhl.) Trin. Troop reported this species in his "Grasses of Indiana" as "found in wet places in the southern counties". In a letter dated January 4, 1917, Troop writes that the record was based upon a Ripley County collection. Since this is an Atlantic coast species, the identification must have been wrong, or else the plant was a waif. It is excluded since there is no confirming specimen.

Near the coast from N. S. to N. C.

46. ERAGRÓSTIS HIRSÙTA (Michx.) Nees. A robust specimen of *Eragrostis capillaris* from Posey County was named *Eragrostis hirsuta* for me and I reported it as such. I now refer it to *Eragrostis capillaris*.

Md. to Mo., southw. to Fla. and e. Tex.

47. ERAGROSTIS MEXICÀNA (Hornem.) Link. MEXICAN LOVEGRASS. A specimen of this species was found by Umbach many years ago in the vicinity of Clarke, Lake County. It is his no. 3837, and I am considering it as a waif since we have had no additional reports.

Tex. to Ariz.; introduced into Del. and Iowa.

48. ERAGROSTIS PILÒSA (L.) Beauv. INDIA LOVEGRASS. This is a European species which has escaped to all parts of the eastern United States and which has been confused with our native species, *Eragrostis pectinacea*. I have not seen a specimen from Indiana, and I believe all of our reports should be referred to *Eragrostis pectinacea*.

Nat. of Eu.; Mass. to Colo., southw. to Fla. and Tex.; southw. through Mex. and W. I. to Argentina.

49. ERAGROSTIS POAEOÌDES Beauv. (*Eragrostis minor* Host of Gray, Man., ed. 7 and *Eragrostis Eragrostis* (L.) Karst. of Britton and Brown, Illus. Flora, ed. 2.) This species has been reported several times by the early authors when it was not separated from *Eragrostis cilianensis*. I have seen no specimen, and I believe all reports should be referred to *Eragrostis cilianensis* from which it is usually not distinguished.

Nat. of Eu.; Vt. to Iowa, southw. to Ga., Tex., Ariz., and Calif.

50. **Eragrostis trichòdes** (Nutt.) Nash. This is a western species that has been reported, but I have seen no specimen. It will doubtless be found on the sand dunes of the southwestern part of the state.

Ill. to Nebr., southw. to Tex.

51. CATABRÒSA AQUÁTICA (L.) Beauv. J. C. Arthur, in his "Manual of Rusts of the United States," page 150, reported this species as occurring in Indiana. I have not investigated this report but I assume that there has been an error in determination since the range of the plant is far from Indiana.

Newf., Lab. to Alberta, southw. to N. Dak., e. Oreg., and n. Ariz.

52. CYNOSÙRUS CRISTÀTUS L. In 1933 Madge McKee found this species in a lawn at 656 North Eighth Street, Lafayette, Tippecanoe County; it had no doubt been introduced in some lawn seed.

Nat. of Eu.; Newf. to Mich., southw. to Va., Wash., and Oreg.

53. AGROPỲRON CANÌNUM (L.) Beauv. This is a European species which I believe our authors have confused with our native species, *Agropyron subsecundum*. The glumes of the foreign species have 3 nerves instead of 4-7 nerves.

Nat. of Eu.; found in ballast near Portland, Oreg.

54. AÈGILOPS CYLÍNDRICA Host. JOINTED GOATGRASS. A colony of this grass was found in June 1938 by Chas. M. Ek along the Nickel Plate Rail-

road in Kokomo, Howard County. The colony was one to one and one half feet wide and about 80 feet long, located between the main track and a siding about midway between Lock and Ohio Streets. Undoubtedly a railroad migrant.

Introduced from Europe; Mo., Kans., Okla., Colo., and N. Mex.

55. HÓRDEUM VULGÀRE L. BARLEY. Barley was formerly a staple crop in Indiana, but it is now rarely grown. It is an annual and sometimes appears spontaneously in fields and waste places, but it will not persist. There are no reports for the state, and it is given in the key in order to prevent errors in determination.

The origin of our cultivated barleys is lost in antiquity.

56. LÒLIUM TEMULÉNTUM L. DARNEL. Wilson reported this species as found on the streets of Lafayette. There is no specimen.

Nat. of Eu.; common on the Pacific coast and occasional throughout the eastern U. S.

57. TRISÈTUM PENNSYLVÁNICUM (L.) Beauv. This species was reported from Clark County by Baird & Taylor. Since there is no specimen, the report must be ignored.

Mass. to Ohio, southw. on the Coastal Plain to Fla. and westw. to Tenn.

58. DESCHÁMPSIA FLEXUÒSA (L.) Trin. Reported from Clark County by Baird & Taylor. Since there is no confirming specimen, this species is dropped from our flora.

Greenland to Alaska, southw. to N. C., Mich., Wis., and Okla.; also in Eurasia.

59. AÌRA PRAÈCOX L. This grass was reported from "southern Indiana" by Lapham (Trans. Wisconsin Agric. Soc. 3: 469. 1854). There is no subsequent record and I have not seen a specimen.

Coastal species from N. J. to Va. and Vancouver to Calif.

60. AVÈNA FÁTUA L. WILD OAT. This grass was reported by C. P. Smith (Proc. Indiana Acad. Sci. **1905**: 301. 1906). He writes: "About a half dozen plants were found along the Monon Railroad at the State Fair Grounds." I searched this area for two different years and I could not find it. I am assuming that it was a migrant and that it has not established itself.

Nat. of Eu.; Maine to Pa., Mo., and westw.; common on the Pacific coast.

61. AVENA SATÌVA L. OAT. This is our cultivated oat which is an annual and is often found as a volunteer but it does not maintain itself.

62. DANTHÒNIA COMPRÉSSA Austin. This species was included in Troop's "Grasses of Indiana." Troop wrote me that the specimen came from Lagrange County. Since there is no specimen, it is excluded.

N. S. to Que., southw. to mts. of N. C.

63. CALAMAGRÓSTIS CINNOÌDES (Muhl.) Bart. My specimen no. 9014 was named this species for me by an authority and I reported it as new to

Indiana. I am now referring this specimen to *Calamagrostis inexpansa.* McDonald has reported it from St. Joseph County, but since this species, as now known, is an eastern and southern species and does not occur in the Great Lakes region, I believe it is safe to refer this report to a form of *Calamagrostis inexpansa.*

Maine to N. Y. and southw. to Ala.

64. AGRÓSTIS CANÌNA L. VELVET BENT. Reported from St. Joseph County. There are no data concerning this species except that it was found at Notre Dame.

Nat. of Eu.; Que., southw. to Del. and Mich.

65. AGROSTIS PERÉNNANS var. ELÀTA (Pursh) Hitchc. Reported from Tippecanoe County, but, since this variety belongs to the Atlantic coast from New York to Mississippi, it is excluded.

66. AGROSTIS SPÌCA-VÉNTI L. A specimen of *Agrostis Elliottiana* from Orange County was erroneously referred to this species and so reported.

Nat. of Eu.; Maine to Md., Ohio, and Oreg.

67. AGROSTIS STOLONÍFERA L. I reported this species but I am now referring my specimens to other species.

Newf. to Alaska, southw. to N. Y. in the East, and to Oreg. in the West; apparently native in n. N. A.

68. **Cínna latifòlia** (Trev.) Griseb. DROOPING WOODREED. This species was reported from Steuben County by Bradner who did not report the common *Cinna arundinacea.* Doubtless he confused the two species. Peattie says: "Reported from Clarke", Lake County. Since I have not seen a specimen, I am excluding it, although Indiana is within the possible range of the species.

Lab., Newf. to Alaska, southw. to Conn. (in the mts. to N. C.), Mich., Ill., S. Dak., and the Rocky Mts. to n. Mex., Utah, and cent. Calif.

69. SPORÓBOLUS VIRGÍNICUS (L.) Kunth. This species was reported as *Agrostis virginicus* L. by Riddell in his "Supplement of Ohio Plants," on page 28, in 1836. He says: "Culms procumbent, 1 ft. high, New Albany, Clapp". There is no specimen.

Along the coast from Va. to Fla. and Tex.; W. I. to Brazil.

70. HELEÓCHLOA SCHOENOÌDES (L.) Host. Umbach found a colony of this species along the railroad near Clarke, Lake County, many years ago. I have not been able to investigate this colony to learn whether it has persisted or not. Since there are no data concerning its persistence, and it is so sparingly introduced, I believe it is best to regard these specimens as waifs.

Nat. of Eu.; Mass. to Del., Mich., and Ill.

71. DACTYLOCTÈNIUM AEGÝPTIUM (L.) Richt. There are a few reports of this species, but I believe authors have confused it with *Eleusine indica,* to which I am referring our reports. There are no verifying specimens.

It is reported to be found in waste places similar to those in which *Eleusine indica* grows.

Nat. of the Old World; Maine, N. J., Coastal Plain from N. C. to Fla., Ill., and Ariz.

72. CHLÒRIS VERTICILLÀTA Nutt. WINDMILL GRASS. A few clumps of this grass were found by Clark in Marshall County in the depot grounds at Culver. Clark's specimen is in the National Herbarium. I have searched for this grass several times at the place cited but I have never been able to find it. Probably extinct. I am regarding it as a waif.

Mo. to Colo., southw. to La. and N. Mex.; introduced in Md., Ill., Ind., and Calif.

73. PHÁLARIS CANARIÉNSIS L. CANARY GRASS. There are a few reports of this species having been found in the state but there are no data to show that it is able to maintain itself. The seed of this grass are used in commercial birdseed. It is usually found on dumps and waste places where it has found lodgement from bird cages.

Nat. of the Mediterranean region; N. S. to Alaska, southw. to Va., Kans., Wyo., and Calif.

74. ZIZANIÓPSIS MILIÀCEA (Michx.) Doell & Aschers. This species was reported from Steuben County but the report should be referred to *Zizania*.

Md. to Ky. and Okla., southw. to Fla. and Tex.

75. PÁSPALUM CILIATIFÒLIUM Michx. This species was reported from Vigo County by Blatchley before *Paspalum pubescens* was recognized. I have the Blatchley specimen and it is *Paspalum pubescens*. There is a specimen of Blatchley's labeled *Paspalum ciliatifolium* in the herbarium of DePauw University and it also is *Paspalum pubescens*.

N. J. to Fla., Tenn., Ark., and Tex.

76. PASPALUM LAÈVE Michx. This species has been reported, but doubtless all reports should be referred to *Paspalum circulare* which was not reported and which occurs in the area from which the reports were made. The range of *Paspalum laeve* does not include Indiana.

N. J. and Pa. to Fla., Ark., and e. Tex.

77. PASPALUM SETÀCEUM Michx. There are several reports for this species, but doubtless all should be referred to some other species. Blatchley reported it from Vigo County. I have his specimen and it is *Paspalum pubescens*.

Coastal Plain from Long Island to Fla. and Tex.; also in Mex.

78. PASPALUM SUPÌNUM Bosc. I had specimens so named for me from Greene, Monroe, Orange, and Perry Counties. I am now referring these specimens to *Paspalum pubescens*.

N. C. to Fla. and westw. to La.

79. PÁNICUM AMÀRUM Ell. This species was reported from Vigo County by Coulter upon the authority of Blatchley. This is an Atlantic coast species. There is no specimen.

Atlantic coast from Conn. to Ga., s. Miss., and Tex.

80. PANICUM MILIÀCEUM L. BROOMCORN MILLET. This species has been reported from Indiana but there is no evidence that it has become established anywhere.

Nat. of the Old World; escaped in the northeastern states and occasionally in other parts.

81. **Panicum scopàrium** Lam. I refer our reports of this species to *Panicum Scribnerianum.* For a discussion of this subject see Deam's "Grasses of Indiana," p. 335.

Mass. to Ky., Mo., and Okla., southw. to Fla. and Tex.; Cuba.

82. **Panicum Tuckermáni** Fern. (Rhodora 21: 112-114. 1919.) This species is reported from Indiana in Hitchcock's Manual. His report is probably based upon two of my specimens which he has so named. I have studied the descriptions of this species as given by Hitchcock, Fernald, Wiegand, and Victorin and, as I understand them, they do not agree. The duplicate specimens of the numbers which I sent to Hitchcock seem to me to be only forms of *Panicum Gattingeri,* hence I am excluding it from our flora. This may be a valid species but I do not believe the specimens at hand belong to it as it is described.

Maine and Que. to Conn. and N. Y.; Ind. and Wis.

83. CYPÈRUS COMPRÉSSUS L. This species was reported from Jasper County by Welch but the specimen is now referred to *Cyperus dentatus* Torr.

Coastal Plain from Pa. to Fla. and Tex.

84. CYPERUS FÈRAX Richard. (Rhodora 37: 148-150. 1935.) (*Cyperus ferax* Richard, in part, of Gray, Man., ed. 7.) Fernald & Griscom, in a study of this species, show that it is restricted to the brackish and saline shores from northern Massachusetts, southward to tropical America, and on the Pacific coast from California southward, and that our interior plants which formerly have been referred to this species should be referred to *Cyperus ferruginescens* Boeckl.

85. CYPERUS FLAVÍCOMUS Michx. This species was reported from Jefferson County by Barnes and by Coulter. The range of this species does not include Indiana and the report should be referred to some other species. Our early authors should not be censured for making a few errors in determination. It is surprising that they did not make more when it is known that they had no authentic specimens for comparison and that the manuals of their time gave short descriptions and these often applied to aggregates.

Va. to Fla.

86. CYPERUS HYSTRICÌNUS Fern. Reported from Jasper County by Welch. No specimen so labeled can be found in the herbarium of DePauw University, where a complete collection of Welch's Jasper County specimens is deposited or elsewhere.

N. J. to Ga.

87. CYPERUS MICRODÓNTUS Torr. This species was reported from Carroll County by Thompson. This report, without doubt, should be referred to some other species.

Coastal Plain from N. J. to Fla. and Tex.

88. SCÍRPUS ATROCÍNCTUS Fern. This species was reported from Porter County by Pepoon in his "Flora of the Chicago Region." He based his report upon a specimen collected by Hill at Dune Park. The report is excluded for want of a confirming specimen.

Newf. to Hudson Bay and Sask., southw. to Conn., Pa., Mich., and Iowa.

89. SCIRPUS MICROCÁRPUS Presl. This species was reported by Scott from the Leesburg Swamp in Kosciusko County. Since its range is north and west of our area and there is no confirming specimen, the report should be referred to some other species.

90. SCIRPUS ROBÚSTUS Pursh. This species was reported by Sr. McDonald in her list of St. Joseph County plants (Amer. Midland Nat. 15: 209. 1934), but Sr. Thornton, who made an intensive study of Indiana *Scirpus,* did not find a specimen. Since this is a plant of the salt marshes of the Atlantic coast, this report no doubt should be referred to *Scirpus fluviatilis.*

Mass. to Fla. and Tex.

91. ELEÓCHARIS PALÚSTRIS (L.) R. & S. This species as known by the early authors is now interpreted to be a complex of which *Eleocharis Smallii* Britt. and *Eleocharis calva* Torr. are the representatives found in Indiana. All of the reports from Indiana for this species should be referred to one of the two last named species. *Eleocharis palustris* is a species found north of our area.

Eurasia; Newf. to B. C., southw. to N. E., n. Mich., N. Dak., and Oreg.

92. ELEOCHARIS TÉNUIS (Willd.) Schultes. This species as now understood has a range along the Atlantic coast from Nova Scotia to South Carolina and westward into West Virginia. It was reported from Delaware, Jay, Randolph, and Wayne Counties by Phinney, from Jefferson County by J. M. Coulter, from Kosciusko County by Clark, from Lake County in Peattie's Flora, from Noble County by Van Gorder, and from Vigo County by Blatchley. All of these reports should be referred to the varieties of this species or to other species.

93. FIMBRISTÝLIS AUTUMNÀLIS (L.) R. & S. (*Fimbristylis Frankii* Steud. of Gray, Man., ed. 7 and *Fimbristylis geminata* (Nees) Kunth of Britton and Brown, Illus. Flora, ed. 2.) This species has been reported from 7 counties but I believe all of these reports should be referred to *Fimbristylis autumnalis* var. *mucronulata.* I have not seen a specimen.

Maine to Ont., southw. to Ga., Tenn., and La.

94. FIMBRISTYLIS CASTÀNEA (Michx.) Vahl. Reported from Lake and Porter Counties by four authors, but all the specimens I have seen are

Fimbristylis puberula. I am excluding it for want of a verifying specimen. According to Britton and Brown, Illus. Flora, ed. 2, this is an Atlantic Coastal Plain species.

N. Y. to Fla.

95. RHYNCHÓSPORA CORNICULÀTA (Lam.) Gray. Reported by Coulter for Clapp. This report should no doubt be referred to the var. *interior* Fern. The species was also reported by Pepoon from Porter County. Fassett (Rhodora 35: 202. 1933) writes that he examined three of Umbach's specimens so labeled, which were collected in the vicinity of Dune Park, and found them to be *Rhynchospora macrostachya.* No doubt, Pepoon's data are founded upon these specimens. Buhl (Chicago Acad. Sci. Bull. 5: 10. 1934) refers Pepoon's report to *Rhynchospora macrostachya.*

96. CÁREX RADIÀTA (Wahl.) Dewey. See no. 1, page 271.
97. CAREX AUSTRÌNA (Small) Mack. See no. 2, page 272.
98. **Carex vulpinoìdea** var. **pycnocéphala** Hermann. See no. 3, page 272.
99. CAREX CANÉSCENS L. See no. 4, page 272.
100. **Carex brunnéscens** (Pers.) Poir. See no. 5, page 272.
101. CAREX EXÌLIS Dewey. See no. 6, page 272.
102. CAREX STELLULÀTA Gooden. See no. 7, page 272.
103. CAREX MURICÀTA L. See no. 8, page 272.
104. CAREX CEPHALÁNTHA (Bailey) Bickn. See no. 9, page 273.
105. CAREX MERRITT-FÉRNÁLDII Mack. See no. 10, page 273.
106. CAREX HORMATHÒDES Fern. See no. 11, page 273.
107. CAREX PROJÉCTA Mack. See no. 12, page 273.
108. CAREX FOÈNEA Willd. See no. 13, page 273.
109. CAREX DEFLÉXA Hornem. See no. 14, page 273.
110. **Carex pedunculàta** Muhl. See no. 15, page 274.
111. CAREX LÍVIDA (Wahl.) Willd. See no. 16, page 274.
112. CAREX SALTUÉNSIS Bailey. See no. 17, page 274.
113. CAREX ORMOSTÀCHYA Wieg. See no. 18, page 274.
114. CAREX RÉCTIOR Mack. See no. 19, page 274.
115. CAREX FORMÒSA Dewey. See no. 20, page 275.
116. **Carex arctàta** Boott. See no. 21, page 275.
117. CAREX PALLÉSCENS L. See no. 22, page 275.
118. CAREX SCABRÀTA Schwein. See no. 23, page 275.
119. CAREX PAUPÉRCULA Michx. See no. 24, page 275.
120. CAREX AQUÁTILIS Wahl. See no. 25, page 275.
121. CAREX NEBRASKÉNSIS Dewey. See no. 26, page 275.
122. CAREX CRINÌTA var. GYNÁNDRA (Schwein.) Schwein. & Torr. See no. 27, page 276.
123. CAREX PAUCIFLÒRA Lightf. See no. 28, page 276.
124. CAREX BÀILEYI Britt. See no. 29, page 276.
125. CAREX COMÒSA × HYSTRICÌNA var. DÚDLEYI. See no. 30, page 276.

126. TRADESCÁNTIA BRACTEÀTA Small. There are two reports of this species for the state made by authors who were not botanists, and I believe both should be referred to *Tradescantia virginiana.* I have a speci-

men which I found in my strawberry patch and it was, no doubt, introduced with some plants which I received from the west.

Anderson and Woodson (Contr. Arnold Arb. **9**: 86. 1935) cite a specimen from Indiana collected by Mason, April, 1877, near French Lick Springs, Orange County, and now deposited in the herbarium of the Field Museum. I have seen this specimen and it is, beyond a doubt, *Tradescantia virginiana* as originally labeled. The lower surface of the bracts is densely short-pubescent all over, which is a character of *Tradescantia virginiana.* Someone with the signature of G. D. has written above the label, "*Tradescantia bracteata* Small." This specimen does not bear the verification label of Anderson & Woodson as do all the specimens which passed through their hands. Nor does the specimen show any mark that such a label was ever attached.

The Indiana record for this species based upon this specimen is distinctly an error.

Minn. to S. Dak., southw. to Mo., Kans., and Tex.

127. TRADESCANTIA BREVICAÚLIS Raf. There are four reports for this species and doubtless all of them should be referred to *Tradescantia virginiana.* The characters used in our manuals are not usually sufficient to separate this species from *Tradescantia virginiana,* hence the error of authors. A specimen collected in Tippecanoe County by Grimes and labeled *Tradescantia brevicaulis* is in the herbarium of DePauw University and proves to be *Tradescantia virginiana.*

Ill. to Kans., southw. to Tenn. and Tex.

128. JÚNCUS CORIÀCEUS Mack. See no. 1, page 302.

129. JUNCUS TÉNUIS Willd. See no. 2, page 302.

130. JUNCUS BREVICAUDÀTUS (Engelm.) Fern. See no. 3, page 302.

131. JUNCUS DÉBILIS Gray. See no. 4, page 302.

132. UVULÀRIA PERFOLIÀTA L. WOOD MERRYBELLS. This species has been reported by 15 authors, most of whom also reported *Uvularia grandiflora.* It is now known that this species does not occur west of the Allegheny Mountains. Hence all of our reports should be transferred to *Uvularia grandiflora.*

Coastal Plain and Allegheny Mountains from Mass. to Fla.

133. HEMEROCÁLLIS FLÀVA L. (Stout. The Lemon Daylily (Hemerocallis flava L.): its origin and status. Jour. New York Bot. Gard. **36**: 61-68. 1935.) LEMON DAYLILY. This plant has been reported only from the Lower Wabash Valley by Schneck. He writes: "Sparingly escaped from gardens." It so rarely escapes that our manuals give it no range in the United States.

Nat. of Eurasia.

134. ALLIUM SCHOENÓPRASUM var. SIBÍRICUM (L.) Hartm. (*Allium sibiricum* L. of Britton and Brown, Illus. Flora, ed. 2.) This variety is native to the area north of Indiana and was reported from Porter County

by Lyon who says: "Not far from an abandoned house." Peattie reported this plant as *A. sibiricum* saying that it occurred as a weed in Lake County. I have seen the Lyon specimen and it belongs to *A. Schoenoprasum,* the common garden chives. I believe the Peattie report should also be referred to the species. The variety has not been reported south of the Upper Peninsula of Michigan. Since the garden chives multiplies so rapidly that it must be divided and some of it discarded, it seems odd that it has not been reported more often than it has been.

Newf. to Alaska, southw. to N. E. and the Great Lakes region.

135. ALLIUM SCORODÓPRASUM L. This species was reported by Welch from Jasper County. Her specimen was determined by J. M. Greenman. I have seen the specimen and the determination seems to be correct. Since this is the only report, I am excluding it until there is another which makes it definite that it is established.

Nat. of Eu.

136. ALLIUM STELLÀTUM Ker. Standley (Rhodora 34: 174. 1932) found a large colony of this species on a railroad embankment between McCool and Porter, Porter County. Since it is established in a place where it is likely to be destroyed, I believe we should wait until there is a report of it where it has a chance to persist and become permanent. Standley says: "Doubtless an introduction".

Ill. and Minn., southw. to Mo. and Kans.

137. LÍLIUM CÀTESBAEI Walt. Prince Maximilian writes in the original Coblentz edition of his travels, published in 1839-41, on his trip from Owensville, Gibson County to Vincennes on June 10, 1834, as follows: "The region on the other side [north side of White River, which he crossed in the vicinity of what is now known as Hazelton] changes considerable; and here appears in a now again sandy soil nearly the same plants as are found in the sandy soil and the prairies of St. Louis, with the addition of a few new ones, a fire-colored lily (Lilium catesbaei), the great-flowered lady slipper (Cypripedium spectabile), a species of Yucca, and many others".

This species flowers much later than the date given above and doubtless this report should be referred to *Lilium philadelphicum* or its variety *andinum.* Small gives the distribution of this species as follows:

Pinelands and acid swamps, Coastal Plain from N. C. to Fla. and La.

138. **Lilium philadélphicum** L. ORANGECUP LILY. This species has been reported from various parts of Indiana but all the specimens I have seen belong to the variety. A recent study of its distribution has not been made.

139. **Smilacìna trifòlia** (L.) Desf. Pepoon reported this species from Lake County for Moffatt, who collected it at Clarke and Pine, and for Babcock, who collected it at Berry Lake, Gibson, and Pine. It possibly did occur about Lake Michigan and may now be extinct. Butters, however, in his studies of *Maianthemum* (Minn. Studies in Plant Science 5:

437. 1927) found 3-leaf forms of *Maianthemum* labeled *Smilacina trifolia.* This discovery suggests that plants reported as *Smilacina trifolia* may have been wrongly determined. Buhl (Amer. Midland Nat. **16**: 251. 1935) says Pepoon's reports lack confirming specimens.

Lab. to B. C., southw. to n. N. J., Pa., Ohio, and Mich.

140. CONVALLÀRIA MAJÀLIS L. LILY-OF-THE-VALLEY. This species was reported as an escape in Lake County by Hill and as a doubtful escape in St. Joseph County by Nieuwland. Andrew's report I am ignoring because no data accompany the report. Since this species will persist for years where it was planted about habitations although the buildings are removed, authors must be careful to ascertain that the site was not that of a former habitation. There is no proof that it has established itself anywhere by propagation.

Nat. of Eu., also from Va. to S. C.; common in cultivation and probably escaped.

141. TRÍLLIUM CÉRNUUM L. Eames and Wiegand (Variations in Trillium cernuum. Rhodora 25: 189-191. 1923) have shown that this species belongs to the area east of the Allegheny Mountains and that our form of the species is represented by *Trillium cernuum* var. *macranthum* which is a northern form and is known so far in Indiana only from La Porte and Porter Counties.

142. TRILLIUM ERÉCTUM L. This species has been reported many times. Specimens of *Trillium Gleasoni* with erect peduncles have, no doubt, been confused with this species. The true *Trillium erectum* has much longer filaments and a small, very dark purple ovary which should not be confused with the larger, lighter red brown ovary of *Trillium Gleasoni* f. *Walpolei*. *Trillium erectum* occurs south and east of our area.

Ne. U. S., southw. in the Appalachians to Tenn.

143. SMÌLAX LANCEOLÀTA L. LANCELEAF GREENBRIER. This species was reported to have been found along the railroad in Decatur County. This may have been a wrong determination; if not, the plant was probably a migrant because the range of the species is south of Indiana.

Va. to Ark., southw. to Fla. and Tex.

144. SMILAX PSEÙDO-CHÌNA L. LONGSTALK GREENBRIER. Pennell (Bull. Torrey Bot. Club **43**: 410. 1916) has shown that this species should be regarded as a synonym of *Smilax herbacea* L. The few reports for it in Indiana should be transferred to the last named species.

145. SMILAX WÁLTERI Pursh. CORAL GREENBRIER. I reported this species for Indiana on the authority of Schneck. Later I had an opportunity to study the specimen and I found it to be *Smilax rotundifolia* L. It was also reported from Posey County by Ridgway. Ridgway wrote me that he was certain that he saw it on a sandy ridge near the Wabash River and north of Coffee Bayou in Gibson County. I have searched this area for it but most of the area has been cleared and if it was present at one time, it has been exterminated.

N. J. to Fla., and westw. to La.

146. **Zephyránthes Atamásco** (L.) Herbert. (*Atamasco Atamasco* (L.) Greene of Britton and Brown, Illus. Flora, ed. 2.) ATAMASCO-LILY. M'Murtrie, in his "Sketches of Louisville", in a list of the plants of the vicinity of Louisville, published in 1819, lists this species as being found in Indiana. Riddell, in his "Flora of the Western States", published in 1835, on page 87, repeats this record. While the known range of this species is south of Indiana, there is a large area where its habitat occurs north of Jeffersonville in Indiana and I have no doubt that it was found there when primitive conditions existed. No one has thoroughly botanized this area in recent years and it is barely possible that it may yet be found.
Va. to Fla. and westw. to Miss.

147. NARCÍSSUS POÉTICUS L. POETS NARCISSUS. Reported from Monroe County by Andrews without any data. This species has been freely planted about habitations and will persist for a long time but there is no evidence that it has become established.
Nat. of Eu.

148. NARCISSUS PSEÙDO-NARCÍSSUS L. COMMON DAFFODIL. This species was also reported from Monroe County by Andrews without any data. Like the preceding species it has been freely planted and may be seen persisting about old habitations that have been abandoned but there is no evidence that it has become established.
Nat. of Eu.

149. IRIS HEXÁGONA Walt. This species was reported from the vicinity of New Albany by J. M. Coulter for Clapp (Bot. Gaz. 1: 9. 1876). As now known, this is a Coastal Plain species and no doubt this record should be referred to *Iris brevicaulis.*

150. IRIS PSEUDÁCORUS L. This iris was reported from Jasper County by Welch. It has been used for ornamental planting since pioneer times and it may be seen persisting in gardens and elsewhere about old habitations. Miss Welch reports that it has escaped to a wet place along Carpenter Creek in Fountain Park which was established in 1895. Since this is our only report and its spread is not likely, I believe it is best to await additional reports before we give it a place in our flora.
Nat. of Eu.

151. **Sisyrínchium apiculàtum** Bickn. Reported by Nieuwland from St. Joseph County. This species is not regarded as distinct by most authors and I agree with them and refer this report to *Sisyrinchium atlanticum.* It seems to be a smaller plant with an apiculate capsule. The species of this genus vary greatly in size and in various parts, especially in the shape and pubescence of the capsule. The habitat and precipitation, no doubt, have a very marked influence especially on the vegetative parts.
Muskegon County, Mich.

152. **Sisyrinchium apiculatum** var. **mesochòrum** Nieuwl. (Amer. Midland Nat. 3: 116. 1913.) This form I am also referring to *Sisyrinchium atlanticum* Bicknell.

Known only from the type locality—Webster's Crossing near Notre Dame, Ind.

153. SISYRINCHIUM CAMPÉSTRE Bickn. This species has been reported by Nieuwland from Porter and St. Joseph Counties, but I refer his specimens to *Sisyrinchium albidum.* Buhl (Amer. Midland Nat. **16**: 251. 1935) says there are no confirming specimens from the dune area.

Prairies, Wis., N. Dak., southw. to Mo. and N. Mex.

154. **Sisyrinchium mucronàtum** Bickn. This species was reported from Tippecanoe County by Grimes. I have the specimen or a duplicate of it and it should be referred to *Sisyrinchium albidum* Raf. There is a specimen in the herbarium of DePauw University collected by Grimes in Tippecanoe County which proves to be *Sisyrinchium albidum.*

Mass. to Mich., southw. to Va. and Pa.

155. **Habenària blephariglóttis** (Willd.) Torr. (*Blephariglottis blephariglottis* (Willd.) Rydb. of Britton and Brown, Illus. Flora, ed. 2.) WHITE FRINGE-ORCHID. Reported from Marshall County by Nieuwland for Clark (Amer. Midland Nat. 3:120. 1913). Clark did not report it in his list of the plants of Lake Maxinkuckee, which was published in 1920. It is evident that the information is not consistent and it is advisable to entirely disregard this report.

Newf. to Ont., Mich., and Ohio, southw. to N. C. and Miss.

156. HABENARIA FIMBRIÀTA (Dryander) R. Br. in Aiton. (*Blephariglottis grandiflora* (Bigel.) Rydb. of Britton and Brown, Illus. Flora, ed. 2.) LARGE PURPLE FRINGE-ORCHID. Reported from Clark County by Baird & Taylor. Since this species belongs to the Atlantic coast these authors, doubtless, mistook a large specimen of *Habenaria peramoena* Gray, which is frequent in the lowlands of that county, for this species.

Newf., Que., N. E., N. Y., southw. to N. J., W. Va., N. C., and Tenn.

157. SPIRÁNTHES CÉRNUA var. OCHROLEÙCA (Rydb.) Ames. This variety was reported by Ames in 1905 from Jefferson, Lake, and Steuben Counties. In 1933 it was reported by Price & Welch from Monroe County. Ames (Rhodora 23:78. 1921) in a critical study of the species and variety says, "There is only one sure guide that I have found satisfactory, namely, polyembryonic seeds for the species and normal seeds for the variety". Later, Ames, in his "Orchids of the United States and Canada," published in 1924, gives the distribution of the variety as N. S., N. H., and Mass., southw. to N. C.; Mo. (Palmer). Since Ames is now restricting the variety to the Coastal Plain, and none of my specimens come within the variety according to Rydberg's original key, I am excluding it.

158. SPIRANTHES PRAÈCOX (Walt.) Wats. This species was reported by Baird & Taylor from Clark County. Since they did not report *Spiranthes cernua* which is within the area, and since the range of the species which they do report is outside our area, doubtless their report should be referred to *Spiranthes cernua.* Bradner's report from Steuben County should also be referred to the same species.

N. J. to Fla., and westw. to Tex.

159. **Goodyèra rèpens** R. Br. This species was reported from Steuben County by Bradner. Since he did not report *Goodyera pubescens* which occurs in the county and since the species he reported belongs to Europe and to the Rocky Mountains of the far northwest, Bradner doubtless erred in his determination.

160. **Maláxis brachypòda** (Gray) Fern. (*Malaxis monophyllos* (L.) Lindl. of Gray, Man., ed. 7 and of Britton and Brown, Illus. Flora, ed. 2.) Reported from Porter County by Pepoon, who says: "A few plants in a cold tamarack swamp near Dune Park in the vicinity of the Lake Shore and Michigan Southern Railroad. (Clarke)". It was reported from Carroll County by Thompson but there is no specimen. Coulter reported it from Floyd County for Clapp. I have a list of plants collected by Clapp but this species is not in it. In the absence of verifying specimens I believe I am justified in excluding this northern orchid, although our area is included in the range of the species by Ames.

Newf., Que. to Man., southw. to Pa., Mich., Minn., Tex., and Calif.; also in Eu. and Asia.

161. SÀLIX ÁLBA var. COERÙLEA (Smith) Koch. CRICKETBAT WILLOW. This form of the white willow has been reported from Jefferson and Putnam Counties. These are old reports and specimens may have been taken from cultivated trees. At least there is not sufficient evidence to warrant admission to our flora.

162. SALIX BABYLÓNICA L. BABYLON WEEPING WILLOW. This species has been reported from 4 counties. Most of these reports are early reports and may have been from cultivated trees. The evidence is not sufficient to admit it to our flora.

Nat. of China and introd. into cultivation about 1730.

163. SALIX EXÍGUA Nutt. This species was reported from St. Joseph County but, no doubt, this report should be referred to a narrowleaf form of *Salix interior*. *Salix exigua* is much like a narrowleaf *Salix longifolia* and replaces it in the west.

W. Mont. to B. C., southw. to Colo., N. Mex., and Calif.

164. **Salix longifòlia** var. **argyrophýlla** Anders. This was reported from White County. The variety is no longer recognized and should be dropped.

165. SALIX MISSOURIÉNSIS Bebb. Reported from White County by Heimlich, but C. R. Ball disposes of this report by referring it to some other species. See Deam's "Shrubs of Indiana," p. 357. 1932.

Ky. to Mo., Iowa, and Nebr.

166. SALIX PENTÁNDRA L. LAUREL WILLOW. Reported from Porter and St. Joseph Counties as a possible escape.

Nat. of Eu. and escaped in e. U. S.

167. SALIX PURPÙREA L. PURPLE WILLOW. This species has been reported from some counties by early authors. No doubt all reports were based upon plants in cultivation because those authors did not distinguish between escaped plants and plants persisting after cultivation.

168. SALIX VIMINÀLIS L. COMMON OSIER. This willow has been reported from 4 counties. All are early reports and most of them may have been shrubs under cultivation.

Nat of Eurasia.

169. **Càrya aquática** Nutt. WATER HICKORY. There are three reports from the state, but those from Fountain and Parke Counties may be safely ignored. Prince Maximilian reported finding it during his sojourn in the vicinity of New Harmony and I believe it did occur in the cypress swamp in Point Township of Posey County. The habitat is there and it has been found just west in Gallatin County, Illinois. I found an old nut which I believe belongs to this species, but I am not sure of its identity, although it was found in the cypress swamp which is its likely habitat. At present, the swamp has been heavily cut over, but in due time I believe this species will be found in Indiana.

Va. to Ill., southw. to Fla. and Tex.

170. CARYA BÚCKLEYI var. VILLÒSA Sarg. Reported by Pepoon under the name of *Carya glabra* var. *villosa* (Sarg.) Rob. as occurring in La Porte County, and Peattie referred to this record. I believe this report should be referred to a form of *Carya ovalis* with pubescent branchlets.

S. Ill., southw. to Ark. and Okla.

171. CARYA MYRISTICAEFÓRMIS Nutt. NUTMEG HICKORY. This species also was reported by Prince Maximilian from the same area as *Carya aquatica*. The nut of this species is so easily identified that it is unlikely that an error would be made in its identification. No specimen, however, has yet been found.

Ark. and Mex.

172. CÓRYLUS CORNÙTA Marsh. (*Corylus rostrata* Ait.) BEAKED HAZELNUT. This species was reported by David Thomas in a list of plants found in the vicinity of Vincennes in 1818. There are no other reports and I believe that a mistake was made in the identification.

Que. to Sask., southw. to Mo. and Ga.

173. BÉTULA LÉNTA L. SWEET BIRCH. This species has been reported from Fulton, Gibson, Lake, Miami, Noble, Porter, Posey, St. Joseph, and Steuben Counties. All of these reports should be referred to some other species. Buhl (Amer. Midland Nat. **16**: 251. 1935) refers the Lake and Porter County specimens to *B. lutea*.

The range of this species as now understood is s. Maine, nw. Vt., e. Ohio, e. Ky., and Tenn., n. Del., and in the mts. to Ga. and Ala.

174. × **Betula Sandbérgii** Britt. of Deam's "Shrubs of Indiana," ed. 1. I now refer the report of this hybrid to × *Betula Purpusii* Schneider.

175. ALNUS GLUTINÒSA Gaertn. (*Alnus vulgaris* Hill of Gray, Man., ed. 7 and *Alnus Alnus* (L.) Britt. of Britton and Brown, Illus. Flora, ed. 2.) This was reported by Nieuwland from St. Joseph County. He says: "Cultivated. Probably spreading from the roots of a cultivated specimen."

Nat. of Eu.

176. CASTÀNEA PÙMILA (L.) Miller. CHINQUAPIN. This species was given a place in our flora in Coulter's Catalogue upon the authority of Sargent, Ridgway, and Schneck. Ridgway, in giving an additional list of the trees of the Lower Wabash Valley, says, "There is some doubt as to No. 16, Castanea pumila, which is given on Prof. Sargent's authority; but there is a possibility of an error having been made from the circumstances that the name "chinquapin" is in that region almost universally applied to the fruit of Quercus Muhlenbergii." The Posey County record was based upon a specimen in Dr. Schneck's herbarium, which proves to have been taken from a cultivated tree near Poseyville.

177. QUÉRCUS CÀTESBAEI Michx. TURKEY OAK. Riddell in his "Supplement to Plants of Ohio" on p. 25, reports this species on the authority of Clapp as growing on the "knobs" near New Albany. This report, no doubt, should be referred to some other species.

N. C. to Fla. and La.

178. QUERCUS ILICIFÒLIA Wang. BEAR OAK. Reported from the Leesburg Swamp by Scott. This, without doubt, should be referred to some other species.

Maine to Va., westw. to Ohio and Ky.

179. QUERCUS NÌGRA L. WATER OAK. This species has been reported by a few authors but the reports should be referred to some other species.

Del. to Fla., westw. to Ky. and Tex.

180. QUERCUS PHÉLLOS L. WILLOW OAK. There have been six reports for this species from Indiana. They should, no doubt, be referred to *Quercus imbricaria.*

N. Y. to Fla., westw. to Mo. and Tex.

181. QUERCUS TEXÀNA Buckley. TEXAS RED OAK. There have been six reports of this oak from Indiana and all of them should be referred to *Quercus Shumardii* var. *Schneckii* Sarg. or to some other species.

Cent. and w. Tex.

182. **Céltis occidentàlis** L. Reported from all parts of Indiana, but the species as understood by Sargent is not our tree. Indiana is within the range of the species. Sargent has named all of my specimens as belonging to the var. *canina,* and since he has been recognized as our leading authority on trees I have followed him although I do not believe the varieties are valid.

N. E. to N. Dak., southw. to Va., Mo., and Kans.

183. MÒRUS ÁLBA L. (Nakai. Morus alba and its allies in the herbaria of Linnaeus, Thunberg, and others. Jour. Arnold Arboretum 8: 234-238. 1927.) WHITE MULBERRY. There are several reports for this species but I believe most of them should be referred to the Russian mulberry which is a rather common escape. This species is no longer planted and I have found it in only Jasper and Jefferson Counties.

Nat. of Eu.

184. MORUS NÌGRA L. BLACK MULBERRY. There are three reports of this species as a native tree and of course all are wrong determinations. I have planted the species twice at Bluffton and it is only semi-hardy.

Nat. of w. Asia.

185. PAPỲRIUS PAPYRÍFERA (L.) Kuntze. (*Broussonetia papyrifera* (L.) Vent. of Gray, Man., ed. 7.) PAPER-MULBERRY. Reported as an escape in Gibson County. I had a specimen, purporting to be this species sent to me from Vanderburgh County and it proved to be the Russian mulberry. This is a small round-headed tree often planted in lawns and along streets in Evansville, Mt. Vernon, and New Harmony where it has proven to be hardy but we have no evidence that it has escaped.

China and Japan.

186. FÌCUS CARÌCA L. COMMON FIG. I found several shoots, 3-6 feet high, of this species in 1918 along the Southern Railroad in Gibson County. The plant was visited a few years later and it still persisted. It was again visited and it had disappeared. In 1932 Chas. O. McBride, of Bedford, sent me a specimen which he said was from a clump growing on the top of a pile of stones. He adds: "It does not seem to be hardy in this location." The determination of this specimen was made by Alfred Rehder of the Arnold Arboretum. It is reported to be hardy as far north as Tennessee.

Nat. of Asia.

187. HÙMULUS LÙPULUS L. COMMON HOP. This species has been reported from 23 counties. I believe all of these reports should be referred to our native species and I am excluding the introduced species from our flora because I have not seen a specimen. Our native species is not recognized by our manuals although the differences seem to me to be convincing.

Eurasia.

188. COMÁNDRA UMBELLÀTA (L.) Nutt. BASTARD TOADFLAX. There have been 30 reports of this species from 20 counties. According to Fernald, its range is east of the Allegheny Mountains which means that all of our reports should be referred to *Comandra Richardsiana.*

Cent. Maine, southw. to Ga.

189. GEOCÁULON LÍVIDUM (Richardson) Fern. (Rhodora 30: 23-24. 1928.) (*Comandra livida* Richardson of Gray, Man., ed. 7 and Britton and Brown, Illus. Flora, ed. 2.) This plant was reported from Marshall County by Clark. Since the range of this species is north of Indiana I assume that Clark made an error in determination.

Lab. to Alaska, southw. to s. N. B., e. Maine, mts. of N. E., n. Mich., Sask., Alberta, and B. C.

190. RÙMEX CONGLOMERÀTUS Murray. This species was reported from Jefferson County by Barnes. Since we have no specimen and the species does not belong to our area, I exclude it. A specimen in the herbarium of Wabash College collected by J. M. Coulter in Jefferson County is *Rumex obtusifolius.*

Va. to S. C.; also in Calif. and Wash.

191. RUMEX ELONGÀTUS Guss. This species was reported by Andrews for Monroe County. Since Andrews did not preserve a specimen it is impossible to correctly refer this report. Doubtless an error.

192. RUMEX HASTÁTULUS Baldwin. This dock was reported from Monroe County by Andrews. He preserved no specimen and since the range of the species is along the Atlantic coast from Mass. to Fla. and Tex., and up the Mississippi Basin to Ill., I exclude it.

193. RUMEX OCCIDENTÀLIS S. Wats. Reported from Clark and Jefferson Counties under the name of *Rumex longifolius* DC. Since Indiana is not within the range of the species, these reports are referred to other species.

Lab. to Alaska, southw. to Maine and Ont., and in the Rocky Mountains to Tex. and Calif.

194. RUMEX PERSICARIOÌDES L. Coulter reported this species from Marion County without any data. Since Indiana does not have the habitat of the species, the specimen must be regarded as a waif.

Seashore, Que. to S. C.; and in brackish and saline places; Ont. to Ill. and westw.

195. RUMEX SANGUÍNEUS L. Reported from the Lower Wabash Valley by Schneck and from Hamilton and Marion Counties by Wilson, who says, "Common." Both of these reports no doubt should be referred to the red-veined form of *Rumex obtusifolius*.

Native of Eu. and sparingly found in the U. S. It has been reported from Mass., N. Y., and from Va. to La.

196. RHÈUM RHAPÓNTICUM L. GARDEN RHUBARB. Reported by Peattie as escaped in the Calumet District. Buhl (Amer. Midland Nat. **16**: 251. 1935) says it is "only a nonpersistent garden escape". After cultivation, rhubarb will persist for years in a suitable habitat. Since this is our only report and authors rarely report it as an escape, it is best to exclude it until there is additional evidence that it escapes freely enough to be considered a part of our flora.

Nat. of Eu.

197. POLÝGONUM ARIFÒLIUM L. According to Fernald & Griscom (Rhodora **37**: 167. 1935) the typical form is restricted to the southeastern part of the United States south of the District of Columbia. Hence our plants should all be referred to the var. *lentiforme* Fern. & Grisc.

198. **Polygonum atlánticum** (Rob.) Bickn. This species was reported from White County by Heimlich. I am referring this report to ***Polygonum exsertum.***

199. **Polygonum aviculàre** var. **arenástrum** (Bor.) Rouy. Reported by Peattie from "sandy roadsides, Pine and perhaps elsewhere" in Lake County. Since I have seen no specimen, I am excluding the variety.

200. POLYGONUM HYDRÓPIPER L. (*Persicaria Hydropiper* (L.) Opiz of Britton and Brown, Illus. Flora, ed. 2.) (Stanford. Polygonum Hydropiper

in Europe and North America. Rhodora 29: 77-87. 1927.) In the absence of a verifying specimen the typical form of this species is excluded. All reports from Indiana are referred to *Polygonum Hydropiper* var. *projectum* Stanford.

Nat. of Eu.; introduced in Newf., N. S., Mass.; also in Oreg., Wash., and Idaho.

201. POLYGONUM HYDROPIPEROÌDES var. PERSICARIOÌDES (HBK.) Stanford. (*Persicaria persicarioides* (HBK.) Small of Britton and Brown, Illus. Flora, ed. 2.) (Rhodora 28: 27. 1926.) This plant has been reported from Marshall and St. Joseph Counties. I have seen no specimen. Since the range of the variety as now understood is far to the west of our area, I am excluding it.

Plains of Nebr. to Tex. and N. Mex.; also in Mex.

202. POLYGONUM RAMOSÍSSIMUM Michx. This species has been reported from Jefferson, Lake, Porter, and Vigo Counties. It is believed that these reports were confused with *P. exsertum,* so, lacking a specimen, it is excluded. Rydberg, in his "Flora of the Prairies and Plains," gives the range as Man. to B. C., southw. to Ill., N. Mex., and Nev. It may be found occasionally, however, as a waif eastward.

203. POLYGONUM SETÀCEUM Baldwin. Reported from Jefferson County by Young in 1871 before the present concept of the species. This report should be referred, no doubt, to some other species.

S. C. to Mo., southw. to Fla. and Tex.; also in Asia.

204. POLYGONUM TOMENTÒSUM Schrank. This species was reported by Nieuwland from St. Joseph County under the name of *Persicaria tomentosa* (Schrank) Bicknell. A small, sterile specimen was found along the I. I. & I. Railroad near South Bend, bearing the data 2733, June 1. I believe the specimen is correctly identified but since it was found along the railroad I am regarding it as a waif.

Newf. to N. Y., Colo., and B. C.; also in Eu.

205. CHENOPÒDIUM BÒNUS-HENRÌCUS L. Reported for Monroe County by Andrews. Since there is no confirming specimen, it is excluded.

Nat. of Europe.

206. AMARÁNTHUS LÍVIDUS L. Reported from Jefferson County by Young. This is the only non-spiny amaranth he reported and John M. Coulter in his catalogue of the plants of Jefferson County reports only *Amaranthus retroflexus.* Since Coulter had access to Young's plants and does not mention the species Young reported, I assume that Coulter discovered the mistake and made the correction without comment. This is a tropical species and has been found as an escape a few times about our eastern seaports.

Tropical S. A., e. Asia, and n. Africa.

207. FROELÍCHIA FLORIDÀNA (Nutt.) Moq. A native of the southern states. I reported this species before it was separated from *Froelichia*

campestris, on the basis of a specimen collected by Umbach in ballast near Aetna, Lake County. Peattie and Pepoon also reported it and probably both reports are based upon material from this locality. I am now referring my Umbach specimen to *Froelichia campestris*.

Coastal Plain, Ga. to Fla. and westw. to Miss.; adventive in Del.

208. GOMPHRÈNA GLOBÒSA L. GLOBE-AMARANTH. This species was reported without any data from Monroe County by Andrews. It is grown frequently in gardens as an "everlasting" flower and this report is no doubt based upon a chance escape. Reported as an escape in the Gulf States and southward.

Nat. of s. Asia.

209. MIRÁBILIS JALÀPA L. COMMON FOUR-O'CLOCK. Reported in 1914 by Nieuwland as found near old dump piles south of South Bend, St. Joseph County. In 1922, I found several fine plants on a dump along Big Vermilion River about a half mile northwest of Eugene, Vermillion County. I regard these reports as of waifs and wait until it is reported to be self-sustaining before I consider it part of our flora.

W. Tex., Mex., Cent. Amer., southw. through tropical S. A. Widely cultivated and probably naturalized in some of our southern states.

210. OXÝBAPHUS ÁLBIDUS (Walt.) Sweet. (*Allionia albida* Walt. of Britton and Brown, Illus. Flora, ed. 2.) Peattie reports this species as found "along railroad tracks, Michigan City". I have not seen his specimen and I do not deem it wise to report a single railroad migrant as a part of our flora.

S. C. to Tenn. and Kans., southw. to Fla. and Tex.

211. OXYBAPHUS HIRSÙTUS (Pursh) Sweet. (*Allionia hirsuta* Pursh of Britton and Brown, Illus. Flora, ed. 2.) This species was reported by Coulter for Jenkins as found in Wabash County. This was, no doubt, a migrant and since there are no additional reports, we should regard the species as not yet established in this state.

Wis. to Man. and Wyo., southw. to Mo., Tex., and N. Mex.

212. OXYBAPHUS LINEÀRIS (Pursh) Robinson. (*Allionia linearis* Pursh of Britton and Brown, Illus. Flora, ed. 2.) Reported in 1902 by Dorner as established along the Wabash Railroad near Lafayette. I have not been able to visit this place to determine whether it still persists. I have one of Dorner's specimens, which is correctly named.

S. Dak. to Mont., southw. to w. Mo., Tex., Ariz., and Mex.; rarely adventive eastw.

213. TALÌNUM TERETIFÒLIUM *Pursh.* This species was reported from Lake County by Babcock (Lens 1: 23. 1872) as found on sand hills at Miller and Tolleston before *Talinum rugospermum* was described. It is now known that the Indiana *Talinum* belongs to the last named species.

Pa., southw. to Ga., Ala., and Tenn.

214. **Claytònia caroliniàna** Michx. CAROLINA SPRING BEAUTY. This species has been reported from Clark, Franklin, Jefferson, Lake, and

Steuben Counties. J. M. Coulter, in discussing the report from Jefferson County, says: "I very much doubt the genuineness of the specimens placed under this species. I strongly suspect it to be nothing more than an extreme form of *Claytonia virginica*". I agree with him and believe that all of our Indiana reports should be transferred.

Woods, especially in the mountains, from N. S. to Minn., southw. to N. C. and Ga.

215. **Claytonia robústa** (Somes) Rydb. Reported from Indiana by Rydberg (Flora of North Amer. **21**: 298. 1932). I have not seen his specimen or investigated this report because I believe this species is only a form of *Claytonia virginica.*

Ind. to Iowa, southw. to Mo.

216. PORTULÀCA GRANDIFLÒRA Hook. COMMON PORTULACA. This species has been reported from Monroe and St. Joseph Counties. Schneck wrote that it escaped from gardens to the streets in the Lower Wabash Valley. The two reports were, no doubt, of garden escapes, and the species has not yet become established.

Nat. of S. A.; naturalized in e. and w. N. A.

217. STELLÀRIA AQUÁTICA (L.) Scop. This species was found on June 30, 1924, by Madeline Gullion in a roadside ditch north of Ellettsville just north of the crossing of the Gosport Road and the Chicago, Indianapolis, and Louisville Railway. Since this is our only report and it is evidently a railroad migrant, the species is excluded until further reports are made.

Nat. of Eu.; Que. and Ont., southw. to Pa. and Mich.; also in B. C.

218. STELLARIA LÓNGIPES Goldie. (*Alsine longipes* (Goldie) Coville of Britton and Brown, Illus. Flora, ed. 2.) Reported from Noble, St. Joseph, Tippecanoe, and Vigo Counties. Since there are no specimens and since the range of the species is far to the north of Indiana, I am referring these reports to some other species.

Lab., N. S. to Que. and Minn. to Alaska, southw. in the Rocky Mts. to Colo. and Calif.

219. SPÉRGULA ARVÉNSIS L. SPURREY. On July 4, 1912, I found a dense colony of this species about 150 feet long along the road between Cannelton and Derby about 6 miles from Derby in Perry County. In 1934 it was found in Crawfordsville by A. R. Bechtel. It was reported from Clark County by Baird & Taylor. There has been only one report from Ohio. Because the reports of its distribution do not show that it is of an aggressive, weedy nature, I prefer to leave it with the excluded species until there are additional reports.

Nat. of Eu.; e. Canada to Calif. and southw. to S. C.

220. SPERGULÀRIA RÙBRA (L.) J. & C. Presl. (*Tissa rubra* (L.) Britt. of Britton and Brown, Illus. Flora, ed. 2.) SAND SPURREY. In 1914, this species was reported by Nieuwland as found on the road from Notre Dame to Lost Lake. This is a European plant which in due time will possibly be-

come a part of our flora. Following the rule that usually a single report for a plant in the state does not make it a part of our flora, I am excluding it for the present.

Newf. to B. C., southw. to Calif. and Va. It has been found only locally in this extensive area.

221. SILÈNE ARMÉRIA L. SWEET WILLIAM CATCHFLY. In 1876, Schneck, in his "Flora of the Lower Wabash Valley," says: "Escaped from gardens". In 1914, Nieuwland reported it from St. Joseph County. Since I have no evidence that it is established, I am excluding it.

Nat. of Eu.; spontaneous in waste places and gardens, N. B. and Ont. to Mich., southw. to N. J. and Ohio.

222. SILENE CAROLINIÀNA Walt. This species was reported from Jefferson County by Young in 1871. I am of the opinion that this report is correct but in the absence of a specimen I must exclude it. Reported also from Tippecanoe County, but probably a garden escape.

Maine, cent. N. Y., Pa., Ohio, Ky., southw. to Ga.

223. SILENE CHLORÁNTHA (Willd.) Ehrh. Deam & Weatherwax found this species in ballast along the Monon Railroad about 2 miles north of Harrodsburg in Monroe County. I also found it in Knox County in railroad ballast about 4 miles south of Vincennes. I regard this species as a waif.

Nat. of Eu.

224. LÝCHNIS CHALCEDÓNICA L. MALTESE CROSS. Reported by Peattie as an escape in the Calumet District. A single report of an escape without any data should not admit a species to our flora, therefore I am excluding it. Buhl (Amer. Midland Nat. **16**: 251. 1935) says it is only a "non-persistent garden escape," and deletes the report.

Nat. of Japan; escaped from cultivation.

225. LYCHNIS CORONÀRIA (L.) Desr. ROSE CAMPION. MULLEIN PINK. In 1914, this species was reported by Nieuwland as persisting after cultivation in St. Joseph County. In 1921, I found a colony about 50 feet long and 10 feet wide on the slope of the wooded bank of the St. Joseph River one and a half miles northwest of Bristol. The colony was far removed from any habitation.

Nat. of Eu.; locally established from Maine to Mich.; also in Oreg. and Wash.

226. DIÁNTHUS BARBÀTUS L. SWEET WILLIAM PINK. Reported by Nieuwland as escaped locally about Notre Dame.

Nat. of Eu.; as yet regarded by authors as a garden escape.

227. DIANTHUS PLUMÀRIUS L. Nieuwland reported this species as an escape near Notre Dame.

Nat. of Austria to Siberia; not mentioned in our manuals.

228. NÝMPHAEA ODORÀTA Ait. (*Castalia odorata* (Ait.) Woodville & Wood of Gray, Man., ed. 7 and *Castalia odorata* (Dryand) Woodville &

Wood of Britton and Brown, Illus. Flora, ed. 2.) This species has been reported many times from Indiana but it is now believed that it belongs to the Coastal Plain and to the area north of Indiana.

Newf. to Man., southw. to Fla. and La., and in the interior westw. to Mich.

229. NÙPHAR SAGITTIFÒLIA (Walt.) Pursh. (*Nymphaea sagittifolia* Walt. of Gray, Man., ed. 7 and Britton and Brown, Illus. Flora, ed. 2.) This species was reported from the deeper ponds of the Lower Wabash Valley upon the authority of Schneck. Miller & Standley say it is probable that the range of the species is restricted to North Carolina and South Carolina, and that all reports of it from outside this area should be referred to some other species. Dr. Schneck was a very careful botanist and I believe he found it. The description of the vegetation of the Lower Wabash Valley left by Robert Ridgway tells us that it contained southern forms that have become extinct. This same area had birds and animals which belong to a zone much farther south and which disappeared at an early date. The plant population of the former cypress swamps and deeper ponds of the Lower Wabash Valley will never be known.

230. CÁLTHA FLABELLIFÒLIA Pursh. This species was reported by Hansen as found on the farm of John E. Steffe near Warsaw in Kosciusko County. I have revisited the place and I refer the plants I found there to *Caltha palustris*. They are only an extreme form of that species.

In mountain springs, s. Pa., n. N. J., and Md.

231. HELLÉBORUS VÍRIDIS L. GREEN HELLEBORE. This species was reported from the Lower Wabash Valley (we do not know whether it was on the Indiana or Illinois side) by Schneck, who lived at Mt. Carmel, Illinois. He says: "Found in a fence-row . . .; evidently escaped from a garden nearby". Andrews reported it without any data from Monroe County. Doubtfully established in Indiana.

Adv. from Eu.

232. NIGÉLLA DAMASCÈNA L. LOVE-IN-A-MIST. Reported as a garden escape in Jefferson County and in the Lower Wabash Valley. I have no evidence that it is established.

Introd. from Eurasia.

233. AQUILÈGIA VULGÀRIS L. EUROPEAN COLUMBINE. There are four reports of this species as a garden escape. The most recent one was made nearly fifty years ago when it was a custom to throw garden rubbish over the fence into an unimproved street where it might remain or at least be raked into a pile nearer the center of the street and be burned, leaving the seeds behind or scattered. I believe all of our reports of garden escapes should be so regarded and not included in our flora.

Adv. or naturalized from Eu.; in the e. U. S.

234. DELPHÍNIUM CAROLINIÀNUM Walt. (*Delphinium azureum* Michx.) (Bull. Torrey Bot. Club 65: 28. 1938.) This species was reported by Collins from Dearborn County, by Phinney from Wayne County, by Young from Jefferson County, and by J. M. Coulter in his catalogue of plants of Jeffer-

son County upon the authority of Young. These reports were made before 1890 when *Delphinium Ajacis* was not in our manuals. Since *Delphinium carolinianum* has a range south of Indiana while *Delphinium Ajacis* is known to be a common escape in southeastern Indiana, it is safe to refer these records to *Delphinium Ajacis.* Benedict & Elrod, geologists, reported this species from Cass & Wabash Counties. Since they did not report *Delphinium tricorne,* which occurs in this area, it seems safe to refer these reports to that species.

Va. to Mo., southw. to Fla. and Tex.

235. DELPHINIUM EXALTÀTUM Ait. TALL LARKSPUR. This species has been reported from Dearborn County by Collins, from Wayne County by Phinney, and from Cass & Wabash Counties by Benedict & Elrod. All of these reports are over 40 years old. Benedict and Elrod listed only 92 species in their partial list of the plants of those two counties. They were geologists and nearly all of their list consisted of the commonest trees and herbs; among these, there is known to have been at least one error in determination, and it is quite probable that *Delphinium exaltatum* was also wrongly determined. According to Wilde, who has made the most recent study of the genus covering our species, it does not occur in our area. I am excluding it for this reason and because there is no specimen.

Atlantic coast of America, Pa., Ohio, and Va.

236. DELPHINIUM CONSÓLIDA L. (Long. Delphinium Consolida in America, with a consideration of the status of Delphinium Ajacis. Rhodora 18: 169-177. 1916.) (*"Delphinium Consolida* L., a European species which has a glabrous style and capsule, is widely recorded as naturalized in the eastern United States, and was admitted to our first edition; but all specimens examined prove to be *Delphinium Ajacis".* Britton and Brown, Illus. Flora, ed. 2, vol. 2: 93. 1913.) FIELD LARKSPUR. Reported from about ten counties and all reports except one are about fifty years old. They should all be referred, no doubt, to some other species, and most likely to *Delphinium Ajacis* L.

Nat. of Eu.

237. ANEMÒNE PARVIFLÒRA Michx. This species was reported from Steuben County by Bradner. Since he did not report *Anemone cylindrica* Gray, I believe that he mistook a depauperate specimen of this species for *Anemone parviflora* which has a range far to the north of Indiana.

Lab. to Alaska, southw. to n. Mich., Wis., Minn,. and in the mts. to Colo.

238. **Ranúnculus cymbalístes** Greene. Described by Greene (Amer. Midland Nat. 3: 333. 1914). This species is undoubtedly the same as *Ranunculus micranthus* Nutt. and I am referring the name to the synonomy of that species.

239. RANUNCULUS FLÁMMULA L. This species was reported from the vicinity of New Albany by Clapp in 1852 and from Jefferson County by Young in 1871. It is a European plant which has been reported in North

America only from Newfoundland. These reports from Indiana are undoubtedly due to an error but I am not able to determine what species these authors had.

240. RANUNCULUS MACOÚNII Britt. This species was reported in Coulter's Catalogue for Blatchley from Monroe County. Coulter says the specimen is in the herbarium of DePauw University, but in an examination of that herbarium in 1935 I was not able to find a specimen so labeled.

Ont. to Iowa and B. C., southw. to N. Mex. and Utah.

241. RANUNCULUS PÚRSHII Richards. PURSH BUTTERCUP. This species has been reported from five counties but I believe all of the reports should be referred to the terrestrial form of *R. flabellaris*. It was reported from Marshall County by Clark, and his specimen is in the National Herbarium. I asked S. F. Blake to examine it in 1933 and he reports that it belongs to *R. flabellaris* Raf.

N. S., Ont. to Colo., Oreg., and Alaska.

242. RANUNCULUS RÈPENS L. This species was reported in 1878 by Baird & Taylor from Clark County, in 1871 by Young from Jefferson County, in 1875 by Coulter from Jefferson County, and in 1878 by Barnes from Jefferson County. Since the range of the species is from Arctic America southward to New Jersey, Pennsylvania, and Michigan, and in the Rocky Mountains to Colorado, and since these reports were made before our manuals recognized the variety *villosa,* they doubtless should be referred to the variety.

243. THALÍCTRUM POLÝGAMUM Muhl. This species has been reported from all parts of the state. The reports should be referred to other species since this species, as now understood, has a range east of Indiana.

244. ADÒNIS AUTUMNÀLIS L. PHEASANTEYE. This species was reported in 1876 from the Lower Wabash Valley by Schneck. He says: "Escaping from gardens to fields and roadsides". Since there are no additional reports, it should not yet be recognized as a member of our flora.

Fugitive from Eu.

245. CALYCÁNTHUS FÉRTILIS Walt. SMOOTH SWEETSHRUB. This shrub was reported without any data in 1878 by Baird & Taylor in a list of plants of Clark County. This may have been a garden escape and since there are no other records and Indiana is outside the range of the species, it is excluded.

Pa. to N. C., e. Tenn., Ga., and Ala.

246. CALYCANTHUS FLÓRIDUS L. COMMON SWEETSHRUB. Reported in 1878 from Clark and Jefferson Counties but excluded for the same reasons as is the preceding species.

Va. and N. C. to Fla., Ala., and Miss.

247. BENZÒIN MELISSIFÒLIUM (Walt.) Nees. This shrub was reported from the Lower Wabash Valley by Ridgway. He expresses doubt as to

the correctness of the determination. Since there are no other reports and the range of the species is outside the state, it is excluded.

N. C. to s. Ill. and Mo., southw. to Fla. and Ala.

248. MACLÈYA CORDÀTA (Willd.) R. Br. (*Bocconia cordata* Willd.) PLUMEPOPPY. A colony of this species was discovered July 14, 1933, by Charles M. Ek of Kokomo on the high, dry bank of Wildcat Creek about 4 miles west of Kokomo. The colony was far from a habitation and growing as if wild.

China and Japan.

249. ARGEMÒNE INTERMÈDIA Sweet. This species was reported by Pepoon as found along railroads near Miller. He says: "Evidently a railroad 'stray' ".

Plains of S. Dak. to Wyo., southw. to Tex., and in n. Mex.

250. ARGEMONE MEXICÀNA L. MEXICAN POPPY. Reported by Nieuwland in 1914 as found in a clover field near Notre Dame in St. Joseph County. Also reported by Schneck in 1876 as "escaped from flower garden". I found a single specimen in sandy soil along a roadside far removed from a residence in Sullivan County. I do not believe that there is sufficient evidence that this species has become established so I am excluding it.

Nat. of tropical Amer.; Mass. to Pa., southw. to Fla. and Tex.; also introduced into Africa, East Indies, and Australia.

251. PAPÀVER RHOÈAS L. CORN POPPY. Reported in 1914 by Nieuwland as escaped from gardens.

Nat. of Eu.; Maine to N. Dak., southw. to Va. and Nebr.

252. PAPAVER SOMNÍFERUM L. OPIUM POPPY. This species has been reported seven times and the authors who comment upon it say that it was spontaneous near dwellings. I do not believe it has become established anywhere and while it may be found as a garden escape, it should not be regarded as a part of our flora.

Nat. of Mediterranean region.

253. CORÝDALIS AÚREA Willd. (*Capnoides aureum* (Willd.) Kuntze of Britton and Brown, Illus. Flora, ed. 2.) GOLDEN CORYDALIS. Reported from Floyd, Jefferson, and Cass & Wabash Counties. All but the first were referred to *Corydalis flavula* in Coulter's Catalogue. I have the books in which Clapp kept a record of the plants that he found and since his records were made before *Corydalis flavula* was recognized, of course he was forced to name his plant *Corydalis aurea.* This is a northern species while *Corydalis flavula* is more southern and is frequent in the southern counties. All of our reports, no doubt, should be referred to *Corydalis flavula.*

E. Que. to Alaska, southw. to Vt., Pa., Wis. and Mo. and in the Rocky Mts. to Ariz.

254. FUMÀRIA OFFICINÀLIS L. COMMON FUMITORY. This species was reported from Franklin, Putnam, and Wayne Counties about 40 years

ago. It was a garden escape, no doubt, which should be excluded. It was found on a sand hill along a roadside in Porter County in 1934 by W. B. Welch. The specimen is in the herbarium of Wabash College.

Nat. of Eu.; reported from Newf. to Fla. and the Gulf States.

255. LEPÍDIUM SATÌVUM L. GARDEN CRESS. This species was reported by Phinney as an escape in Wayne County. Since there are no additional reports, it may not yet be established anywhere. It is the cultivated cress of gardens and is likely to become an established escape.

Nat. of Eu.; local from Que., N. Y. to B. C.

256. SISÝMBRIUM LOESÈLII L. This plant was found in flower on July 6, 1930, in St. Joseph County by Marcus Lyon, Jr., and J. A. Nieuwland. We have no data that it is established.

Nat. of Eu.

257. DIPLOTÁXIS TENUIFÒLIA (L.) DC. This European weed was first reported by Erlanson for Grimes (Proc. Indiana Acad. Sci. **1923**: 139. 1924) as found along the Monon Railroad 2 miles north of Roachdale in Montgomery County. The specimen can not be found at DePauw University, but there are two specimens collected by Grimes in Warren County along the roadside 2 miles east of Pine Village, August 29, 1914. The second record was made by Nieuwland & Just (Amer. Midland Nat. **12**: 220. 1931) who say that it was found July 8, 1930, near the Kankakee River in sec. 19 of Greene Township of St. Joseph County. The specimen is in the herbarium of the University of Notre Dame, and I refer it to *Erysimum cheiranthoides* L. These are the only records. There is no evidence that it is established in the state.

Nat. of Eu.

258. ERUCÁSTRUM GÁLLICUM (Willd.) O. E. Schulz. (*Erucastrum Pollichii* Schimp. & Spenn. Rhodora **13**: 11. 1911.) On September 1, 1930, Paul C. Standley found two plants on a railroad embankment near the west boundary of Porter County at the intersection of State Road 53. He gave one specimen to me and the other is deposited in the herbarium of the Field Museum. In 1937 Chas. M. Ek found a few plants in the railroad yards in Tipton and also a few plants in dry soil along the railroad 5 miles west of Tipton. This species is probably already established as a ballast migrant.

Nat. of Eu.

259. BRÁSSICA HÍRTA Moench. (Rhodora **40**: 306. 1938.) (*Brassica alba* (L.) Rabenh., *Brassica alba* (L.) Boiss. of Gray, Man., ed. 7, and *Sinapis alba* L. of Britton and Brown, Illus. Flora, ed. 2). WHITE MUSTARD. This species has been reported from various parts of the state, mostly by early authors whose reports are now 50 years old. I have never seen it and there is no evidence that it is able to maintain itself. The seed are a household commodity for use in pickling and this fact accounts for its wide distribution.

Nat. of Eu. but not yet considered established.

260. BRASSICA NÀPUS L. RAPE. This species has been reported from two counties. I have found it a few times. It has been rather extensively sown for green feed for hogs and if such fields are permitted to remain fallow the year following the sowing, sometimes in sandy places many plants may appear, but there is no evidence that it has become established anywhere. Found also along roadsides.

Nat. of Eu.

261. BRASSICA RÀPA L. TURNIP. There are reports of this species from two counties. I have found it several times in fields where turnips had been grown the previous year but there is no record of its being established.

Nat. of Eu.

262. RÁPHANUS SATÌVUS L. GARDEN RADISH. This is an annual and may persist for a year or two but there is no evidence that it has become established anywhere. In 1916 I found it to be plentiful in an oatfield 1 mile south of Alexandria in Madison County.

Nat. of Asia.

263. **Rorìppa obtùsa** (Nutt.) Britt. (*Radicula obtusa* (Nutt.) Greene of Gray, Man., ed. 7 and Britton and Brown, Illus. Flora, ed. 2.) This species has been reported from Clark, Jefferson, and Tippecanoe Counties. These reports are all more than 50 years old and since our manuals of that period did not very distinctly separate this species, I believe all of these reports should be referred to some other species. There is no Indiana specimen.

Mich. to Mont., southw. to Tex. and Calif.

264. **Rorippa sinuàta** (Nutt.) Hitchc. (*Radicula sinuata* (Nutt.) Greene of Gray, Man., ed. 7 and Britton and Brown, Illus. Flora, ed. 2.) This species was reported by Wilson (Proc. Indiana Acad. Sci. **1905**; 170. 1906) who says a colony had persisted for fifteen years along the Big Four Railroad west of Greencastle. Doubtless this report should be referred to *Rorippa sylvestris*. In the absence of a verifying specimen I am omitting this species from our flora.

Ill. and Minn. to Wash., southw. to Tex. and Ariz.

265. CARDÁMINE HIRSÙTA L. Reported by 20 authors and all of the reports except one are 40 or 50 years old and were made when our manuals did not recognize *Cardamine pennsylvanica* to which I refer all of these reports. As now understood, this name is restricted to an introduced plant which is of rare occurrence.

Nat. of Eurasia.

266. **Cardamine flexuòsa** With. Now regarded as a semi-aquatic form of *Cardamine pennsylvanica* Muhl. to which I am referring it. It was reported by Smith from Marion County (Proc. Indiana Acad. Sci. **1905**: 157. 1906).

267. CARDAMINE PRATÉNSIS L. This is a plant of Eurasia and all of our reports for it were made before our American plant was recognized as a variety of it. Hence all of our reports are referred to the variety.

268. DENTÀRIA MÁXIMA Nutt. This species was reported from Jefferson County by Barnes. J. M. Coulter, who no doubt saw the specimen upon which this report was made, includes it under his *D. laciniata*.
S. Maine to Mich. and Pa.

269. LESQUERÉLLA GLOBÒSA (Desv.) Wats. Reported without comment from Monroe County by Andrews. Since he did not preserve a specimen or publish any data, the species is not included in our flora.
Ky. and Tenn. to e. Mo.

270. CAMÉLINA SATÌVA (L.) Crantz. Reported from 7 counties but all reports but one were made when the manuals did not separate *Camelina microcarpa* Andrz. from this species. I have seen the Hussey specimen reported from Tippecanoe County and it is *Camelina microcarpa*. Possibly all of the reports of this species should be referred to *Camelina microcarpa*. In the absence of an Indiana specimen, I am excluding it from our flora.
Nat. of Eu.

271. NÉSLIA PANICULÀTA (L.) Desv. BALL MUSTARD. This species was reported from Lake County by Pepoon who says Moffatt found it in 1894 along the Pennsylvania Railway near Clarke, and adds that it is not now at the original station and is "seemingly not a permanent introduction." Peattie also reports it from the dune area without data. I have a specimen collected by Moffatt in Lake County but he gives no specific location.
Nat. of Eu.; Que. to Man. and B. C., southw. in the east to Pa.

272. **Dràba réptans** var. **micrántha** (Nutt.) Fern. (*Draba caroliniana* var. *micrantha* (Nutt.) Gray of Gray Man., ed. 7.) This variety has been reported from the dune area by Peattie but I have not seen a specimen. Buhl (Amer. Midland Nat. **16**: 251. 1935) refers this report to the typical form. I believe, however, that it may occur in Indiana and a verifying specimen should be sought.
Ill. to Minn., Mont. to Wash. and southw. to La. and Calif.

273. ARABIS DIVARICÁRPA Nelson. (*Arabis brachycarpa* (T. & G.) Britton.) PURPLE ROCKCRESS. This species has been reported under the name of *Arabis brachycarpa* from five counties but I have not seen a specimen. This species and its allies are so closely related that they are difficult to separate unless good specimens are at hand. The known range of the species is to the north of Indiana.
Que. to Man. and Assin., southw. to Vt., w. N. Y., Ill., Minn., and in the Rocky Mts. to Colo.

274. ERÝSIMUM PARVIFLÒRUM Nutt. (*Cheirinia inconspicua* (Wats.) Britt. of Britton and Brown, Illus. Flora, ed. 2.) Reported by Erlanson for Grimes (Proc. Indiana Acad. Sci. **1923**: 139. 1924) as having been found on the track of the Monon Railroad a short distance north of Greencastle,

Putnam County, June 4, 1911. I have a specimen which E. L. Greene collected in 1917 in ballast along the railroad in Plymouth, Marshall County. These are the only records, and the species is obviously a railroad migrant which does not maintain itself.

Ont., Man., B. C., and Alaska, southw. to Kans., Colo., and Nev.; adventive farther east.

275. LOBULÀRIA MARÍTIMA (L.) Desv. (*Koniga maritima* (L.) R. Br. of Britton and Brown, Illus. Flora, ed. 2.) SWEET ALYSSUM. Reported as an escape in a few counties but it does not persist.

Nat. of Eu.; Vt. to Pa. and on the Pacific coast.

276. CLEÒME SERRULÀTA Pursh. PINK CLEOME. Schneck found this species in the Lower Wabash Bottoms but he left no data. I found a single specimen on the fill of the bridge across the wet prairie about 4 miles southwest of Bluffton, Wells County. It was far removed from a habitation but I believe it was introduced in dumpings along the fill.

Ill., Minn. to Sask., southw. to Mo., N. Mex., and Ariz.

277. CLEOME SPINÒSA L. SPIDERFLOWER. This species has been reported three times as a garden escape but not since 1904. In 1933 I found three specimens about a hundred feet apart in a wet pasture field on the east side of the C. E. & I. Railroad about 4 miles north of Decker, Knox County. This area was formerly a part of a great cypress swamp which has been drained and is now farmed and is more than a quarter of a mile from the nearest habitation. Since there is no conclusive evidence that it has become established, it is best to continue to regard it as an escape.

Nat. of tropical America; in waste places from s. N. Y. to Fla., westw. to Ind., Ark., and La.

278. **Podostèmum ceratophýllum** Michx. RIVERWEED. This plant has been reported by Peattie as rare in the Grand Calumet and Little Calumet Rivers. I have asked two careful collectors to search for it in these streams but they did not find it. I have looked for it in the Tippecanoe and Eel Rivers but I did not find it. If Peattie collected specimens I do not know where they are located. There is no specimen in the Field Museum, and it seems best to exclude the species.

N. B. to Ont. and Minn., southw. to Ga. and Ala.

279. SÈDUM NÈVII Gray. In discussing the distribution of this species, Howe (Torreya 5: 115. 1905) says: "Collected originally in southwestern Virginia, but since found to extend to Indiana." This is the only reference I have found which ascribes it to Indiana. Neither of our manuals include Indiana in its range, so it may be safely omitted from our flora.

Va. to Ill. and Mo., southw. to Ala.

280. SEDUM PULCHÉLLUM Michx. TEXAS STONECROP. Ball reported this species from Lake County in 1884. Since this is far north of the natural range of the species, it must have been a garden escape. I have had it in cultivation for many years and it abundantly reseeds itself each year and

will persist in flower beds. Indiana is given in the natural range of the species and although I have searched carefully for it in its habitat along the Ohio River, I have failed to find it. I have not been able to find an herbarium specimen from Indiana, so I am excluding it.

Va., Ky., Mo., and Kans., southw. to Ga. and Tex.

281. SEDUM TELÈPHIUM var. PURPÙREM L. LIVEFOREVER. This species has been reported as an escape from several parts of the state. It has escaped from dwellings and cemeteries.

Nat. of Eu. and w. Asia; Que. to Ont. and Mich., southw. to Md. and Ind.

282. ASTÍLBE BITERNÀTA (Vent.) Britton. ASTILBE. This species was reported by Young in a flora of Jefferson County but he did not report *Aruncus dioicus*, which very much resembles this species. J. M. Coulter and C. R. Barnes later published floras of the same county and reported *Aruncus dioicus* but did not report *Astilbe biternata.* Since Astilbe is found in the area to the southeast of Indiana and *Aruncus dioicus* is frequent in Jefferson County, where Young did the most of his collecting, it is almost certain that he confused the two plants.

Mts. of Va. to N. C., Ga., and Tenn.

283. TIARÉLLA CORDIFÒLIA L. ALLEGHENY FOAMFLOWER. Reported from the area of Delaware, Jay, Randolph, and Wayne Counties by Phinney, who says: "Common in rich woods." Also reported from St. Joseph County, upon the authority of Rothert, by Nieuwland who adds: "I have nowhere found it within the region." While Indiana is within the possible range of the species, I believe that it has been confused with some form of *Heuchera.*

N. E., Ont. to Minn., southw., especially in the mts., to Ga. and Ark.

284. HEÙCHERA VILLÒSA Michx. This species was reported from Clark County by Baird & Taylor and was also reported in Coulter's Catalogue upon the authority of Barnes. Since the typical form of the species as now understood occurs in the southern Appalachian Mountains, these records must be referred to some other species.

285. MITÉLLA NÙDA L. Higley & Raddin report this species from Lake County as growing in "moist rich woods east of Berry Lake in 1884 and at Miller in 1886." Pepoon reported it as common in Mineral Springs bog in Porter County. Buhl, in his Supplement to Pepoon's "Flora of the Chicago Region" (Bull. Chicago Acad. Sci. 5: 10. 1934) refers this report to *Mitella diphylla.*

Lab. to Alaska, southw. to Conn., Pa., Mich., Minn., and Mont.

286. PHILADÉLPHUS CORONÀRIUS L. SWEET MOCKORANGE. Reported by Nieuwland as an escape at Notre Dame. Since this species is universally planted throughout the state and this is the only report, its escape at Notre Dame may, for the present, be regarded as exceptional.

Nat. of cent. Eu.; escaped from gardens in Va. and Ohio, and sparingly in the Middle and Eastern States.

287. PHILADELPHUS GRANDIFLÒRUS Willd. BIG SCENTLESS MOCKORANGE. Reported from Clark and Jefferson Counties. The specimens were, no doubt, from cultivated plants or possibly garden escapes.

Pa. to Va., Tenn., and Fla.

288. PHILADELPHUS INODÒRUS L. SCENTLESS MOCKORANGE. Reported from Clark and Lawrence Counties. Both reports, no doubt, are from planted or escaped specimens. In 1915 I collected a specimen in Lawrence County from the yard of the George Donaldson home near Mitchell. It must have been planted before 1883. The home burned many years ago and the yard and orchard have now all grown up to large forest trees but the *Philadelphus* still persists.

Va. to Ky., southw. to Ga. and Miss., principally in the mts. Escaped from cultivation in Pa.

289. RÌBES GLANDULÒSUM Grauer. (*Ribes prostratum* L'Hér. of Gray, Man., ed. 7.) SKUNK CURRANT. This species was reported from Jefferson County by Young as *Ribes prostratum*. Since the range of this species is far to the north of this county, it is best to refer this report to some other species.

Lab. and Newf. to Athabasca, southw. to n. N. E., Mich., Minn., and along the mts. to N. C.

290. RIBES ODORÀTUM Wendl. (*Ribes aureum* Pursh of Gray, Man., ed. 7.) GOLDEN CURRANT. This species has been reported as an escape in two counties and I have seen it in two counties. It has been rather common in cultivation for a long time and since it has not, by this time, escaped to any great extent, I doubt if it will become a part of our flora.

Minn., S. Dak., Mo., and Tex., westw. to the Rocky Mts.

291. RIBES SATÌVUM (Reichenb.) Syme. (*Ribes vulgare* Lam. of Gray, Man., ed. 7 and Britton and Brown, Illus. Flora, ed. 2.) COMMON RED CURRANT. This species has been reported as an escape by six authors. It has been under cultivation since pioneer times and if it is to become a part of our flora it would have done so long ago. I have seen it as an escape only once.

Nat. of Eu.; escaped from cultivation, Mass. to Ont., southw. to Va. and Wis., and in Oreg. and B. C.

292. RIBES TRÍSTE Pallas. SWAMP RED CURRANT. This species was reported from Clark and Jefferson Counties by Stanley Coulter, who says: "No herbarium specimens have been examined." Since the range of this species is far north of these counties, this report should be referred to some other species.

Newf. to Alaska, southw. to N. J., Mich., S. Dak., and Oreg.; also in n. Asia.

293. GRÓSSULARIA OXYACANTHOÌDES (L.) Mill. This species was reported before *Grossularia hirtella* was recognized by our manuals. Since

the range of this species is north of our area, I am referring all reports of it to *Grossularia hirtella.*

Newf., Hudson Bay to B. C., southw. to n. Mich. and N. C.

294. GROSSULARIA RECLINÀTA (L.) Mill. (*Ribes Grossularia* L. of Gray, Man., ed. 7.) EUROPEAN OR GARDEN GOOSEBERRY. I found a large colony of this species in a wooded ravine in the "knobs" near Brownstown, Jackson County. This is the only record so it can not be regarded as established.

Nat. of Eu.; along roadsides in e. N. J. and se. N. Y.

295. GROSSULARIA ROTUNDIFÒLIA (Michx.) Cov. & Britt. (*Ribes rotundifolium* Michx. of Gray, Man., ed. 7.) ROUNDLEAF GOOSEBERRY. Reported from Clark and Jefferson Counties. Coulter, in his Catalogue of the Plants of Indiana, says: "All of the specimens labeled Ribes rotundifolium that have come to my notice are to be referred to Ribes gracile" which is now known as *Grossularia missouriensis.* A specimen labeled *Grossularia rotundifolia* collected by A. H. Young in Tippecanoe County is in the herbarium of Indiana University and proves to be *Grossularia missouriensis.* Since the range of the species reported is far from our area, it is best to exclude it.

Rocky woods, mostly in the mts. from Mass. to N. C.

296. GROSSULARIA SETÒSA (Lindl.) Cov. & Britt. (*Ribes setosum* Lindl.) BRISTLY GOOSEBERRY. This species was reported by Wolcott & Montgomery as found in the Mineral Springs bog in Porter County. I have not seen the specimen and since the range of the species is west of our area, without doubt the determination is incorrect, and it is excluded.

Berger gives the distribution as "Cent. Western North America."

297. SPIRAÈA JAPÓNICA L. f. JAPANESE SPIREA. I found this species in 1919 as a frequent shrub on the wooded bluff of the Ohio River about 6 miles east of Cannelton, Perry County. In 1923 I found several colonies in a deep, wooded ravine near Dodd Post Office which is about a mile farther up the river. It is well established in this vicinity where it seemed perfectly hardy. I transplanted some of it to our home in Bluffton where it has been growing vigorously ever since. Since the species is not extensively planted, it may not become a common escape.

Nat. of Japan; Conn. to Pa.

298. SPIRAEA LATIFÒLIA (Ait.) Borkh. PINK MEADOW SPIREA. This species was reported from White County by Heimlich. He questioned the identification and believed it belonged to *Spiraea alba,* to which it no doubt did belong. Other reports should be referred to the same species.

Newf. to Sask., southw. to Va. and w. Pa.; the common spirea of N. E. and the Coastal Plain.

299. **Spiraea tomentòsa** var. **ròsea** (Raf.) Fern. This variety, instead of the species, was reported from the Dune Region by Peattie. I have seen his specimen,which is in the herbarium of the Field Museum, and it is the common form of the species in Indiana.

300. **Gillènia trifoliàta** (L.) Moench. (*Porteranthus trifoliatus* (L.) Britt. of Britton and Brown, Illus. Flora, ed. 2.) BOWMANSROOT. Reported from Clark County by Baird & Taylor and from the Lower Wabash Valley by Schneck. The fact that both authors also reported *Gillenia stipulata* supports their reports. Schaffner in his latest list of the plants of Ohio carries the species but says: "No specimens." The record from Michigan is based upon a report by N. W. Winchell who says this specimen was deposited in the herbarium of the University of Michigan. J. H. Ehlers, curator, writes me that Winchell's specimen is not there. In the absence of specimens, the species is excluded.

N. Y. to Mich., southw. to Ga. and Mo.

301. PỲRUS COMMÙNIS L. COMMON PEAR. Nieuwland and Wilson reported this species as escaping. I have seen a few small trees in woodland and along fence rows but I do not believe those of the woodland in Indiana are able to maintain themselves. The pear has had ample time to escape in the state and if it has done so and is maintaining itself someone would make mention of the fact.

Nat. of Eu. and w. Asia; often escaped and naturalized.

302. MÀLUS ANGUSTIFÒLIA (Ait.) Michx. SOUTHERN CRAB. Reported from Indiana but since this is a southern species our records must belong to some other species.

Va. to Fla. and Miss.

303. MALUS PÙMILA L. COMMON APPLE. This species has been reported from several counties but I am excluding it because no author says that it is maintaining itself. I have seen fruiting specimens along fences and in woodland but search failed to show any offspring. In St. Joseph County about 6 miles southwest of South Bend I saw possibly 25 trees in moist, sandy soil in the Rupel woods where it joined a marl, treeless swamp. Since the trees were approximately the same size the indications were that they had not reproduced and that they might have all been planted there. In Elkhart County on the north side of Simonton Lake there is a small colony of trees but there is no evidence that any of them were self sown. Since during the past years millions of apple cores have been cast aside along roadsides and fences and in public grounds and woodland, it is surprising that we do not find many more "wild apples" than we do. Sufficient time has elapsed for someone to have found it where it is maintaining itself but I find no such record. I believe it is best to treat as occasional escapes species of this kind where millions of seed are scattered throughout the state on all kinds of soils and no reproduction follows.

Nat. of Eu. and w. Asia; cultivated since ancient times.

304. SÓRBUS AMERICÀNA Marsh. The first specimen of this genus which I found was collected in La Porte County and I named it this species. Nieuwland, upon my authority, reported it in a list of "Local plants." I now refer the specimen to *Sorbus Aucuparia* L.

Newf. to Man., southw. to N. C. and Mich.

305. SORBUS AUCUPÀRIA L. My specimens of this species were named by a recognized authority and I reported them as such. I found a small tree in La Porte County in an open woods which has been cleared. I found a small tree in St. Joseph County and it was later killed by the San Jose scale. Lyon, Nieuwland, and Just found a colony in a woods in St. Joseph County that was apparently established. It shoud be placed upon probation before admitting it as established.

Eu. to w. Asia and Siberia.

306. SORBUS SCOPULÌNA Greene. Reported by Nieuwland & Just from St. Joseph County. I am referring reports of this species from Indiana to *Sorbus Aucuparia* L. *S. scopulina* is a far western species and probably does not reach Indiana.

S. D., N. Mex., Ariz., and Oreg.

307. SORBUS SUBVESTÌTA Greene. I am referring all reports of this species from Indiana to *Sorbus Aucuparia* L. and *Sorbus decora* (Sarg.) Schneid.

Type locality in Minn.

308. ARÒNIA ARBUTIFÒLIA (L.) Ell. Reported by some of the early authors but no doubt all reports should be referred to one of the species given in the text. This species has red fruit and is very much like *Aronia floribunda* which, when better known, may be regarded as only a variety of this species. I believe, however, that the red-fruited form does not occur in Indiana.

Mass. to w. Minn., southw. to Fla. and Tex.

309. AMELÁNCHIER INTERMÈDIA Spach. This shadbush was reported from Lake and St. Joseph Counties by Nieuwland. These reports should no doubt be transferred to some other species.

N. S., n. Mich., to Minn., southw. to N. C.

310. AMELANCHIER OBLONGIFÒLIA (T. & G.) Roem. There are 4 reports for this species or for forms whose names are now referred to this species. Since its distribution, according to Wiegand, is restricted to the Atlantic Coastal Plain no doubt the reports should be referred to some other species.

Coastal Plain from s. Maine to S. C. and possibly to Ga.

311. AMELANCHIER SANGUÍNEA (Pursh) DC. There are 4 reports for this species under this name and names now referred to this species. Since this species as understood by Wiegand does not occur in our area, these reports undoubtedly should be referred to some other species.

Maine, Que., Ont., and Minn., southw. through N. Y. along the mts. to n. Ala.

312-338. Crataegus species. Twenty seven species are listed on page 554.

339. RUBUS ALLEGHENIÉNSIS × ARGÙTUS. This hybrid was reported by me in Proc. Indiana Acad. Sci. **1915**: 136. 1916. It was based upon my no. 15883 which Bailey refers to *Rubus argutus* Link.

340. RUBUS ALLEGHENIENSIS × RECÚRVANS. This hybrid was reported by me in Proc. Indiana Acad. Sci. **1916**: 320. 1917. My Allen County specimen no. 19871, Bailey refers to *Rubus abactus* Bailey. My other specimens, no. 20235 from De Kalb County, no. 19948 from Elkhart County, no. 21283 from Franklin County, and no. 20088 from Lake County, Bailey refers to *Rubus allegheniensis* Porter.

341. RUBUS ALÚMNUS Bailey. The reports of this species in Deam's "Shrubs of Indiana," ed. 2, p. 117. 1924, from Knox and La Porte Counties are now referred to *Rubus impos* Bailey.

342. RUBUS ARGUTUS × INVÌSUS. I reported this hybrid in Proc. Indiana Acad. Sci. **1915**: 136. 1916. It was based upon my no. 10825 which Bailey refers to *Rubus allegheniensis.*

343. RUBUS ARGUTUS × PROCÚMBENS. This hybrid was reported by me in Proc. Indiana Acad. Sci. **1915**: 136. 1916. It was based upon my no. 9210 from Decatur County which Bailey now refers to *Rubus frondosus.*

344. RUBUS ARGUTUS × RECURVANS. This hybrid was reported by me in Proc. Indiana Acad. Sci. **1916**: 320. 1917. The Porter County record was based upon my no. 20032 which is now referred to *Rubus abactus* Bailey. The Wayne County record was based upon my no. 20249 and is now referred to *Rubus ostryifolius* Rydb.

345. RUBUS ANDREWSIÀNUS Blanchard. Reported from St. Joseph County by Nieuwland in Amer. Midland Nat. 4:70. 1915. Bailey is now referring this species to *Rubus ostryifolius* Rydb.

346. RUBUS BAILEYÀNUS Britt. I reported this species in Proc. Indiana Acad. Sci. **1916**: 319. 1917 from Allen, Bartholomew, Clark, Crawford, Elkhart, Harrison, Lagrange, Marshall, Starke, and Steuben Counties. These are now referred to *Rubus flagellaris.* It was reported by McDonald from St. Joseph County in Amer. Midland Nat. **15**:223. 1934. I think all of our reports for this species should now be referred to *Rubus flagellaris* because Bailey says: "The name has been applied generally in the North to forms of *R. flagellaris.*" Bailey (Gentes Herbarum 2:325. 1932) now regards this species as belonging to the area to the southeast of us.

347. RUBUS BAILEYANUS × ENSLÈNII. This hybrid was reported by me in Proc. Indiana Acad. Sci. **1918**: 147. 1919. It was based upon my no. 22894 from Vanderburgh County which is now referred to *Rubus flagellaris.*

348. RUBUS BETULIFÒLIUS Small. I reported this species in Proc. Indiana Acad. Sci. **1916**:319. 1917. My Pike County record was based upon my no. 16967, and the Posey County record was based upon my no. 16850, both of which are now referred to *Rubus argutus.*

349. RUBUS CANADÉNSIS L. This species was reported by J. M. Coulter in Indiana Geol. Rept. **6**:242. 1875. I have seen his specimen in the herbarium of Wabash College and it is *Rubus hispidus.* The report from Monroe County by Andrews can not be checked because he preserved no specimen, but doubtless the report should be referred to some other species. I have not been able to verify the report from the dunes (Flora of the Indiana Dunes, p. 227, 1930) by Peattie, who says: "Frequent throughout." If it is frequent throughout it is rather surprising that no one else has collected it. Since the species is a northern one, this report may be authentic, but I am excluding it for lack of a verifying specimen.

350. RUBUS CANADENSIS var. RÁNDII Bailey. (Gentes Herbarum **3**:261. 1934.) Peattie (Flora of the Indiana Dunes, p. 227, 1930) reported *Rubus Randii* (Bailey) Rydb. as found in "shaded and somewhat sandy ground, Dune Park, and perhaps elsewhere." Bailey has made a recent study of this form and restricts its distribution to the type locality in Maine. Thus it is evident that what Peattie had at hand is some related form and I exclude it for lack of a verifying specimen.

351. RUBUS ENSLENII × FRONDÒSUS. This hybrid was reported from Grant County (Fairmount) by Brainerd & Peitersen in Vermont Agric. Exp. Sta. Bull. **217**:82. 1920. I have no duplicate, therefore I can not determine to which species I would now refer it.

352. RUBUS FLÒRIDUS Tratt. I reported this species in Proc. Indiana Acad. Sci. **1916**:319. 1917 from Harrison County under no. 20518. Bailey refers this number to *Rubus argutus* Link.

353. RUBUS FLORÍCOMUS Blanchard. This species was reported by Peattie (Flora of the Indiana Dunes, p. 226, 1930). Bailey now refers this species to *Rubus ostryifolius* Rydb. which is recognized in the text.

354. RUBUS IDAÈUS L. Reported by Nieuwland (Amer. Midland Nat. **4**:70. 1915) as an escape near Hudson Lake, La Porte County. This is our common cultivated red raspberry.

355. RUBUS IDAEUS var. ANÓMALUS Arrh. This variety was reported from the dune area by both Peattie and Pepoon. Greene described *Batidea heterodoxa* from a collection made May 29, 1897 by Umbach in a woods near Clarke, Lake County. Peattie paraphrases a part of the description and cites the name as a synonym. Pepoon also cites the Umbach collection. Fernald (Rhodora **21**:96. 1919) refers Greene's species and this variety as used by Peattie and Pepoon to *Rubus idaeus* var. *strigosus* (Michx.) Maxim.

356. RUBUS INVÌSUS Bailey. This species was reported by me (Proc. Indiana Acad. Sci. **1915**: 137. 1916) from Brown and Clark Counties. Both specimens are now referred to *Rubus flagellaris* Willd.

357. RUBUS LACINIÀTUS Willd. This species was found August 4, 1935 by Scott McCoy in a sterile, fallow field near the Bird Sanctuary at Lake Maxinkuckee, Marshall County.

Origin unknown but it was known before 1770. It is sometimes cultivated and often escapes.

358. RUBUS PERGRÀTUS Blanchard. This species was reported from Putnam County for Grimes in Proc. Indiana Acad. Sci. **1923**:142. 1924, and from White County by Heimlich in Proc. Indiana Acad. Sci. **1922**:286. 1923. I have not seen these specimens and since Bailey in his discussion of the species (Gentes Herbarum 2: 399. 1932 and 2: 44. 1932) is not convinced that this species is entirely distinct from its related species or that it occurs in our area, I prefer to omit it until its status is more definite.

359. RUBUS PROCÚMBENS Muhl. This species has been reported from several counties by several authors. Bailey refers this species to *Rubus flagellaris* Willd.

360. RUBUS RECÚRVANS Blanchard. I reported this species (Proc. Indiana Acad. Sci. **1915**: 137. 1916) from Elkhart, Lagrange, and Whitley Counties. Bailey now refers these reports to *Rubus abactus* Bailey.

361. RUBUS TRIVIÀLIS Michx. Reported from Parke County by Esten (Butler Univ. Bot. Stud. 2: 192. 1932). Since the range of this species is far to the south of Indiana I refer this report to some other species.

362. RUBUS VILLÒSUS Ait. Reported by many of the early authors and some of the later authors. This species has had several interpretations placed upon it in the history of the study of the genus so it is useless to try to identify these reports from published records.

363. RUBUS VILLOSUS var. HUMIFÙSUS T. & G. This variety has been reported mostly by our very early authors when the status of the variety was a complex and it is now impossible to identify the reports from published records without the specimens.

364. FRAGÀRIA VÉSCA f. ÁLBA (Ehrh.) Rydb. Found by Mrs. H. E. Bucklin on the Bucklin farm about 6 miles east of Brazil in Clay County. This was an abandoned farm when purchased and Mrs. Bucklin has no positive record of the plant. She sent me a few plants in 1926 and they have multiplied freely until the present time, long after cultivation was withdrawn. Fruit elongate-ovoid with a neck.

365. FRAGARIA VESCA var. AMERICÀNA Porter. (*Fragaria americana* (Porter) Britton of Britton and Brown, Illus. Flora, ed. 2.) AMERICAN STRAWBERRY. This form has been reported from Putnam County by both Grimes and Wilson. I have seen the Grimes specimen, which is in the herbarium of DePauw University and it is typical *Fragaria vesca.* The report from Wells County is now referred to the species.

366. POTENTÍLLA RÉCTA var. OBSCÙRA Koch. Wilson reported a specimen of *Potentilla sulphurea* Lam. from Putnam County which may be this

variety but I have not seen his specimen. This is a European variety which is slowly becoming established in the United States.

367. POTENTILLA CANADÉNSIS L. This species has often been reported for the state but as the species is now understood it does not occur in our area and undoubtedly all of our reports should be referred to *Potentilla simplex* var. *typica.*

Maine to S. C., east of the Appalachian Mts. and inland across N. Y. to sw. Ont. and n. Ohio.

368. GÈUM MACROPHÝLLUM Willd. This is a northern species which Blatchley reported from Vigo County and Grimes reported from Tipton County. Blatchley reported *Geum canadense, Geum vernum,* and *Geum macrophyllum.* Since he did not report all the species that might occur there, it is best to refer his report to one of the species which he did not report. I have seen the Grimes specimen from Tipton County and it is *Geum laciniatum.*

Newf. to Alaska, southw. to N. Y., n. Mich., Colo., Mo., and Calif.

369. GEUM PÉCKII Pursh. Wilson reported this species from Hamilton and Marion Counties saying it was common. Since Wilson reported only *Geum canadense, Geum vernum,* and *Geum Peckii* and did not report *Geum virginianum* of our manuals, which occurs in that area, it is safe to refer his record to *Geum laciniatum* or its variety.

White Mts. of N. H. and Mt. Kineo, Maine.

370. FILIPÉNDULA ULMÀRIA (L.) Maxim. EUROPEAN MEADOWSWEET. In 1923 I found a large clump of this species on the fill to the approach of a small bridge about one and three fourths miles southeast of Mongo in Lagrange County. The nearest habitation was about 40 rods away and this species was not growing there. This is the only time I have found it as an escape.

Nat. of Eurasia; Que., southw. to Mass., N. Y., and Ohio.

371. AGRIMÒNIA MICROCÁRPA Wallr. This species was reported from Clark and Marion Counties. The range of the species is sufficient reason to regard these reports as wrong determinations.

Pa. to Fla., westw. to Tex.

372. AGRIMONIA STRIÀTA Michx. This was reported by Andrews from Monroe County and by Wilson from Hamilton and Marion Counties. There is no specimen and these reports are doubtless based on wrong determinations.

Newf. to Sask., southw. to W. Va., Ill., Nebr., S. Dak., Wyo., and N. Mex.

373. SANGUISÓRBA MÌNOR Scop. (*Poterium Sanguisorba* L. of Britton and Brown, Illus. Flora, ed. 2.) SMALL BURNET. This species was found in a field in the southeast corner of Lawrence County in the summer of 1914 and reported by M. L. Fisher, who said it was introduced in grass seed.

Nat. of Eurasia; Maine to w. N. Y. and Md.

374. RÒSA ACICULÀRIS Lindl. PRICKLY ROSE. This species has been reported from Lake County by Cowles, Hill, and Pepoon and from Porter County by Nieuwland for Cowles. Mrs. Erlanson writes me that this species does not occur in Indiana and that all reports should be referred to other species or more probably to some natural hybrid.

375. ROSA CANÌNA L. DOGBRIER. I have found this species as an escape in Harrison, Lagrange, and St. Joseph Counties. Also reported from St. Joseph County by Nieuwland.

Nat. of Eurasia.

376. ROSA GÁLLICA L. FRENCH ROSE. I have found this rose as an escape in La Porte and Tipton Counties.

Nat. of Eu.; N. E. to Ind.

377. ROSA MULTIFLÒRA Thunb. JAPANESE ROSE. In 1933 I found a specimen of this species far removed from a dwelling on the wooded border of a small stream about 4 miles southwest of Canaan, Jefferson County.

Nat. of Japan and China; Md., Ala., and Costa Rica.

378. ROSA PIMPINELLIFÒLIA L. In 1932 I found a colony of this species along the Monon Railroad about one and an eighth miles south of Ladoga in Montgomery County. It seems to be spontaneous here.

Nat. of Eurasia; sparingly naturalized, N. H. to Ont. and Ill.

379. PRÙNUS ANGUSTIFÒLIA var. WÁTSONI (Sarg.) Waugh. I now refer to the species the specimens formerly called this variety. See Deam's "Shrubs of Indiana," ed. 2.

380. PRUNUS CÉRASUS L. SOUR CHERRY. This species is no doubt sometimes spontaneous in Indiana but there are only two or three reports which I do not believe are sufficient to regard it as established.

Nat. of Eu.

381. **Prunus cuneàta** Raf. In the Proc. Indiana Acad. Sci. 1920:227. 1921 I referred the broadleaf forms of my *Prunus pumila* to this species. I now regard these specimens as broadleaf forms of *Prunus pumila.* Fernald regards this species as a synonym of *Prunus susquehanae* Willd. (Rhodora 25:73. 1923).

382. PRUNUS PÉRSICA (L.) Stokes. (*Amygdalus persica* L. of Britton and Brown, Illus. Flora, ed. 2.) PEACH. Reported as spontaneous from three counties. Since there are annually thousands of peach seed cast aside along roadsides, fences and in fields, waste places, and woodland, the surprising thing is that this species is not common or at least frequent. It is, no doubt, more common than reports indicate but I do not believe it should be included in our flora.

Nat. of Asia.

383. **Prunus susquehánae** Willd. Reported from the dunes area by Peattie. Fernald (Rhodora 25:74. 1923) cites Hill's specimen no. 117 from a dune near Indiana Harbor as belonging to this species. I believe

all of the sand cherries of Indiana belong to the same species and I have placed them under the name of *Prunus pumila.* Peattie, in his key, says the fruit of *Prunus pumila* is "purple or black" and that of *Prunus susquehanae* is "claret red." I have noted the difference in color in that the fruit of all the plants are claret red at first and at maturity are black or purplish, turning from a red to black just as do species of *Aronia, Amelanchier,* and *Rhamnus.* To find one plant with black fruit and one with claret red fruit is not evidence of two species unless supported by other differences.

384. **Cássia Tòra** L. This species has been reported from Clark, Floyd, and Jefferson Counties. It does not appear in a list of plants collected by Dr. Clapp in the vicinity of New Albany. Although Indiana is included in its range in Gray's Manual, there is no specimen in the Gray Herbarium. Since I have not been able to find a specimen it is excluded.

Pa. to Ind. and Mo., southw. to Fla. and Tex.; also from Mex. to Bolivia and in the tropics of the Old World.

385. BAPTÍSIA ÁLBA (L.) R. Br. Reported from Floyd County on the authority of Clapp. Before the publication of Gray's Manual in 1840 this species was not separated from *Baptisia leucantha.* Clapp, in his *Medicinal Plants of the U. S.,* published in 1852, had dropped this species and reported *Baptisia leucantha.*

Atlantic Coastal Plain from N. C. to Fla.

386. MEDICÀGO HÍSPIDA Gaertn. This species was reported without data from Monroe County by Andrews. Since no specimen was preserved and there is no evidence that it is established, it is excluded.

Nat. of Eurasia; sparingly found in the Atlantic Coast and Pacific Coast States and rarely found in the interior; more common in the Gulf States and southw.

387. MEDICAGO HỲBRIDA (Pourr.) Traut. This species was reported by Hansen (Proc. Indiana Acad. Sci. **1923**: 216. 1924) as found along fence rows on the Purdue University Agricultural Experiment Station farm. Search for it in 1934 showed that it had entirely disappeared.

Nat. of Eu.

388. TRIFÒLIUM INCARNÀTUM L. CRIMSON CLOVER. There are only two reports of this species and it is doubtful whether either specimen was an escape. At least there is no evidence that it has become established.

Nat. of Eu.; naturalized along the Atlantic coast from Maine to Va.

389. TRIFOLIUM REFLÉXUM L. This species was reported by Blatchley, Coulter, and Schneck. I have the Blatchley specimen and it should be referred to the variety. It is doubtful whether the species occurs in Indiana and I believe all of our reports belong to the variety. Since our manuals do not separate the glabrous form from the pubescent one, the range of the species in them includes the range of the variety. I have seen specimens from North Carolina, Missouri, Florida, Georgia, and Texas.

390. **Trifolium stoloníferum** Muhl. Higley & Raddin reported this species as found along the railroad near Indiana Harbor. Coulter reports it from Marion County on the authority of Copeland but no data are given. In the absence of a verifying specimen the species is excluded from Indiana.

Ohio to Iowa, southw. to Tenn., Mo., and Kans.

391. HOSÁCKIA AMERICÀNA (Nutt.) Piper. This species was found by Fred Donaghy "in an old fallow field bordering the Pennsylvania Railroad a mile or so east of Brazil on August 22, 1934." This is our first report and further observation is required to ascertain whether it has become a member of our flora. In 1935 I found a few specimens in a sand pit along a railroad in Porter County.

Dry soil, Minn. to N. Dak., Idaho, Mo., Ark., Tex., N. Mex., and Sonora, Mex.

392. AMÓRPHA NÀNA Nutt. Through some error this species was reported from a gravelly slope on the east side of Winona Lake in Kosciusko County. I have collected and studied specimens from this slope and this colony belongs to *Amorpha canescens* Nutt.

Iowa to Sask., southw. to Kans. and N. Mex.

393. WISTÈRIA FRUTÉSCENS (L.) Poir. (*Krauhnia frutescens* (L.) Small of Britton and Brown, Illus. Flora, ed. 2.) This species was reported from the Lower Wabash Valley by Schneck; Jay County by Phinney who says: "Scarce"; Kosciusko County by Coulter; and White County by Heimlich. The reports from northern Indiana may be escapes while those of the Lower Wabash Valley may be native. Heimlich wrote me that it was abundant in White County near Norway and Buffalo along the Tippecanoe River and also east of Monon. I have searched the Tippecanoe River for it at the places named and I also went along Monon Creek east of Monon for more than a mile but I failed to find it.

Unless a specimen is found, this species will be regarded as extinct, or the report assumed to have been based upon an escape or a wrongly determined plant. It is to be noted that the manuals of the time when the reports were made, except that of Heimlich, did not recognize *Wisteria macrostachya,* to which species, no doubt, the Lower Wabash Valley reports should be referred.

Coastal Plain from Va. to Fla. and Ala.

394. ROBÍNIA HÍSPIDA L. ROSE ACACIA. This species was reported (Amer. Bot. 40: 81. 1934) as persisting on the site of an abandoned habitation east of Gary, Lake County. I have been told, also, that it was growing in Scott County, northeast of Scottsburg in the yard of an abandoned home, where it was well established. There is a large, dense colony near the base of a wooded dune a short distance north of the Baltimore & Ohio Railroad about 8 miles west of Chesterton, Porter County. Madge McKee reports a rank thicket of it along the roadside in sec. 32 of McClellan Township, Newton County. In 1937 I noted a colony a hundred

feet long on the border of an old orchard near Culver. This species is widely cultivated and may escape.

395. ASTRÁGALUS GLYCYPHÝLLOS L. This species was reported in 1926 from Fulton County by Hansen. A large colony was found on the farm of Louis Murray in sec. 23, Newcastle Twp. I visited the place in 1934 and found it common over an area of several square rods. I interviewed Mr. Murray who said it had persisted for more than 75 years and had always been known there as "Fits Root."

Cent. Eu. and w. Asia.

396. ASTRAGALUS PLATTÉNSIS Nutt. Reported in Coulter's Catalogue as a migrant found near Lafayette Junction, Tippecanoe County.

Minn. to Colo. and Tex.

397. ASTRAGALUS TENNESSEÉNSIS Gray. (*Geoprumnon tennesseense* (Gray) Rydb. of Britton and Brown, Illus. Flora, ed. 2.) This species was reported from Tippecanoe County by Stuart (Proc. Indiana Acad. Sci. **1901**: 283. 1902.) He writes: "This plant was collected in sandy bottom land along Wea Creek, about four miles south of Lafayette, and some two hundred yards down stream from the Wabash Railroad bridge. Not very abundant . . . in fruit the latter part of May." I have not been able to find it here.

Ill., Tenn., and Ala.

398. GLYCYRRHÌZA LEPIDÒTA (Nutt.) Pursh. WILD LICORICE. A specimen of this species was collected by Edwin D. Hull on July 17, 1934 along the Wabash Railroad near the eastern limit of Lake County.

Hudson Bay and Minn. to Mo., N. Mex., and westw.; also as a migrant eastw.

399. AESCHYNÓMENE VIRGÍNICA (L.) BSP. Reported from Lake County without comment by T. H. Ball in a "History of Lake County," p. 167, 1884. Since the range of the species is outside Indiana, I regard this report as an error in determination.

Coastal Plain from N. J. to Fla. and Tex.

400. DESMÒDIUM GLABÉLLUM (Michx.) DC. (*Meibomia glabella* (Michx.) Ktze. of Britton and Brown, Illus. Flora, ed. 2.) Reported from Vigo County by Blatchley as frequent along the canal near Five-mile Pond and along the roadside near Heckland. I have seen no specimens and I have not been able to revisit the stations mentioned.

Coastal Plain from Mass. to Ala.

401. LESPEDÈZA ANGUSTIFÒLIA (Pursh) Ell. This species was first reported from Lake County by Hill who discusses it at length (Bot. Gaz. **9**: 47. 1884). It was also reported from Vigo County by Blatchley. I have the Blatchley specimen and it is not this species. The Tippecanoe County record is based on a specimen too immature for correct determination. I have not seen the Cass County specimen. The species, as now understood, is an Atlantic coast species and does not occur in our area; so, no doubt, all of our reports should be referred to some other species, probably to some narrow leaflet form of *Lespedeza capitata*.

402. **Lespedeza capitàta** var. **stenophýlla** Bissel & Fern. Reported from White County by Heimlich. He says the determination was made at the Gray Herbarium. I have tried to rediscover this form where Heimlich said he found it but all I could find are narrowleaf forms of the species. *Lespedeza capitata* is so variable in the form of its leaflets and the amount and length of its pubescence that it is a question whether it is advisable to assign names to extreme forms.

403. LESPEDEZA LEPTOSTÀCHYA Engelm. This species was reported by Peattie as found "in dry open soil, prairies of the Calumet District." I have seen no specimen from Indiana, and Buhl (Amer. Midland Nat. **16**: 251. 1935) says the Peattie report lacks confirming specimens. I have searched for it several times in the remnant prairie north of Hammond but failed to find it.

Prairies of Ill. to Minn. and Iowa.

404. VÍCIA ANGUSTIFÒLIA Reichard. Reported from Cass and Marion Counties but there are no specimens to support these reports. It has, however, been found by Chas. M. Ek in both Cass and Howard Counties and I have specimens.

Nat. of Eu., w. Asia, and n. Africa; naturalized throughout the eastern states.

405. VICIA CRÁCCA L. This species has been reported from Monroe, Steuben, and Tippecanoe Counties. No doubt all of these reports should be referred to some other species. There is a specimen in the herbarium of DePauw University which was collected by Grimes along the New York, Chicago & St. Louis Railroad (Nickel Plate Road), 2 miles north of Tipton, Tipton County. Since this is the only specimen, I am regarding this species as a railroad migrant and not as an established plant of our flora.

Nat. of Eurasia; probably native in the north, Newf. to Minn. and B. C., southw. to N. J., Ky., and Iowa.

406. VICIA SATÌVA L. Reported by Grimes as a weed in Russellville, Putnam County. I have not been able to learn whether the species has persisted or not. Charles M. Ek collected it along a railroad in Howard County. This species is very variable and 19 varieties are recognized in U. S. Dept. Agric. Bull. 1289: 1-20. 1925. Anyone interested in these varieties or in its cultivation should consult this bulletin.

Nat. of Eu., w. Asia, and n. Africa; becoming naturalized especially in the southern states and the Pacific coast.

407. LÁTHYRUS LATIFÒLIUS L. PERENNIAL PEA. In 1918 I found a colony of this species in the dense woods east of the old Donaldson home which is now included in Spring Mill State Park, Lawrence County. This species had escaped from the Donaldson garden into the woods and had persisted there for more than 30 years. In 1937 I found a colony along state road 152 in Tippecanoe County, doubtless started from a root dragged from a colony near a house nearby.

Nat. of Eu.; escaped in Conn., D. C., and Wis.

408. GLYCÌNE SÒJA Sieb. & Zucc. SOY BEAN. This plant has been reported from Jasper County. It has been extensively sown throughout the state and is found spontaneous here and there but there is no evidence that it is established anywhere.

Nat. of China and Japan.

409. **Galáctia regulàris** (L.) BSP. Reported by Phinney from the area of Delaware, Jay, Randolph, and Wayne Counties. This report should no doubt be referred to some other species.

N. Y. to Kans., southw. to Fla., Miss., and Okla.

410. VÍGNA SINÉNSIS (L.) Endl. COMMON COWPEA. This species was reported by Schneck to have escaped in the Lower Wabash Valley. It has been commonly sown throughout the state and since there have been no additional reports I am concluding that Schneck's report was of a casual escape.

Nat. of Asia.

411. GERÀNIUM MÓLLE L. This species was reported by Hansen (Proc. Indiana Acad. Sci. **36**: 251. 1927), who says it was established along the roadside near Battle Ground in Tippecanoe County. There is no specimen. It was, however, collected on the campus of Indiana University and a specimen is in the herbarium of that University.

Nat. of Eu.

412. ERÒDIUM CICUTÀRIUM (L.) L'Hér. STORKSBILL. This species was reported by Schneck from the Lower Wabash Valley. He says: "Escaped from gardens, very rare." There is a specimen collected in St. Joseph County in 1917 by Nieuwland in the herbarium of the University of Notre Dame. Probably a chance escape.

Nat. of Eu.

413. OXALIS MONTÀNA Raf. (Rhodora **22**: 143-144. 1920.) (*Oxalis Acetosella* L. of Gray, Man., ed. 7 and Britton and Brown, Illus. Flora, ed. 2.) This species was reported from La Porte County by the Editors of the Botanical Gazette in 1881 in a catalogue of the plants of Indiana. It was also reported by Collins from Dearborn County. There are no specimens.

Deep woods in N. S. and e. Que. to Sask., southw. to N. E., N. Y., and in the mts. to N. C.

414. LÌNUM USITATÍSSIMUM L. FLAX. Before the advent of ready made clothing, flax was universally grown by the pioneers for its tough fiber which was woven into cloth. It is now grown mostly for its seed from which an oil is obtained which is used most largely in the manufacture of paints. The crop is reported to quickly exhaust the soil, and is no longer grown in the state. When it was widely cultivated it was a common escape principally along roadsides and railroads, but persisted only for the year.

Nat. of Eu.

415. TRÍBULUS TERRÉSTRIS L. A colony of this plant was found along the Nickel Plate Railroad just south of Bluffton, Wells County, in 1927. The place was revisited in 1930 and the colony had disappeared. A colony

was discovered in 1929 by E. D. Hull, along the south end of Henry Street, Gary, in Lake County. I visited this place in 1934 and the plant was found not only persisting but spreading. Since this is the only colony now known, however, it seems best to exclude the species until other colonies are found.

Nat. of Eu.; occasional in Atlantic Coast States, also Ill. to Kans. and Nebr.

416. ZANTHÓXYLUM CLÀVA-HÉRCULIS L. HERCULES-CLUB. Reported from Fountain County by Brown, a geologist, under the name of *Zanthoxylum carolinianum*. There is no doubt that this record should be transferred to *Zanthoxylum americanum*.

Va. to Fla., westw. to Tex. and Ark.

416a. PTÈLEA TRIOLIÀTA var. MÓLLIS T. & G. of authors is referred to *Ptelea trifoliata* var. *Deamiana* Nieuwl. See Amer. Midland Nat. 2: 178-180. 1912.

417. POLÝGALA INCARNÀTA L. Reported from Eggleston, Indiana by Higley and Raddin. Since Eggleston is in Illinois, the authors evidently made a mistake. Since, however, this species has been reported three times as coming from the area about Chicago, and since it was a native of the original prairie, it should be sought in Indiana. There is no specimen in the Gray Herbarium.

N. J., s. Ont., Wis., and Nebr., southw. to Fla., Ark, and Mex.

418. POLYGALA NUTTÁLLII T. & G. Reported from Jefferson County by J. M. Coulter and C. R. Barnes. Probably a wrong determination was made since the known range of this species is south of our area. There is no specimen.

S. Mass. to Ga., westw. to Ala. and Ark.

419. CROTONÓPSIS LINEÀRIS Michx. Reported by Meyncke as "common" in Franklin County. The two species of *Crotonopsis* were not separated in the manuals of Meyncke's time, and since we have no specimen, we have no way of knowing what he had at hand.

Coastal Plain, S. C. to Fla. and e. Tex.; inland near the Mississippi River to southeastern Mo. and Ill.

420. RÍCINUS COMMÙNIS L. COMMON CASTOR-BEAN. Our only report is that of Young from Jefferson County who says: "Commencing to escape into roads and streets." I have never seen it as an escape and since we have no additional records it seems best to regard it, for the present, as a chance escape.

Introd. from the Tropics.

421. EUPHÓRBIA IPECACUÁNHAE L. (*Tithymalopsis Ipecacuanhae* (L.) Small of Britton and Brown, Illus. Flora, ed. 2.) This species is listed in both Gray's Man., ed. 7 and Britton and Brown's Illus. Flora, ed. 2 as occurring in southern Indiana. In a catalogue of the plants of Indiana published in 1881, it was given as found on "the knobs," and in Marion County, without the names of the collectors. It was reported in 1819 by Dr. Mc-

Murtrie in a flora of Louisville but Dr. Clapp, who knew more than anyone else about the flora of the "barrens" of Indiana, does not report it. There are no specimens in the Gray Herbarium nor in the herbarium of the New York Botanical Garden to validate its inclusion in the manuals. Not being able to find a specimen anywhere I am excluding it.

Conn. to Fla.; also barrens of s. Ind. (Gray, Man., ed. 7). Gattinger reported it for Tenn.

422. EUPHÓRBIA SERPYLLIFÒLIA Pers. (*Chamaesyce serpyllifolia* (Pers.) Small of Britton and Brown, Illus. Flora, ed. 2.) Reported from Clark County by Baird & Taylor and from Monroe County by Andrews. Indiana is outside the range of the species and it is probable that the reports are based upon wrong determinations.

N. Mich., Wis., S. Dak., southw. to Mo., Tex., and Mex.

423. CALLÍTRICHE HERMAPHRODÍTICA L. (*Callitriche autumnalis* L. of Gray, Man., ed. 7 and Britton and Brown, Illus. Flora, ed. 2.) Reported by Deam in Proc. Indiana Acad. Sci. **1920**: 227. 1921. The plant was wrongly named and the correction was made in the Proc. Indiana Acad. Sci. **1923**: 221. 1924.

424. CALLITRICHE PALÚSTRIS L. Reported from Lake County by Peattie and by Pepoon, but I have not been able to find their specimens.

Practically throughout the U. S. and Can.; almost cosmopolitan.

425. **Rhús arbúscula** Greene. This species was described by E. L. Greene (Washington Acad. Sci. 8: 184. 1906) from specimens collected on the east shore of Lost Lake, south of Culver, Marshall County. I have inspected this colony and I refer it to *Rhus glabra* L.

426. RHUS ASHEI (Small) Greene. Nieuwland collected some specimens from the wooded bank of the St. Joseph River, St. Joseph County, which Barkley refers to this species. I have seen these specimens and I think they belong to *Rhus glabra*. This species is known only from specimens collected in North Carolina by Ashe.

427. **Rhus gymnóclada** Greene. This was published by Greene (Fedde, Rep. Spec. Nov. 5: 45. 1908). He cites a specimen collected by Clark by the shore of Lake Maxinkuckee, Marshall County. Barkley (Ann. Missouri Bot. Gard. **24**: 330. 1937) refers this species to × *Rhus pulvinata*.

428. ILEX GLÀBRA (L.) Gray. INKBERRY. Reported by Babcock (Lens **1**: 144. 1872) as common near Miller in Lake County. Higley and Raddin in their flora say there is no specimen from Indiana in the Babcock herbarium. I also examined the Babcock herbarium and found no specimen.

Near the coast from N. S. to Fla.

429. ILEX MONTÀNA (T. & G.) Gray. (*Ilex monticola* Gray and *Ilex mollis* Gray.) Reported by Young as "rather rare in Jefferson County." He also reports *Ilex verticillata*. J. M. Coulter published a list of the plants of the same county a few years later, and says he had access to Young's collection, but he does not list this species. Barnes published a list of the

plants of the same county subsequent to that of Coulter. He was in close communication with the preceding authors, but he does not list this species. It thus appears that Coulter and Barnes did not agree with Young in his determination of this form of *Ilex*. Also reported by Scott from a tamarack bog near Leesburg in Kosciusko County. This is an Appalachian Mountain species and since Scott did not report *Ilex verticillata* which I have collected in this bog, it is fairly safe to presume that this report should be transferred to *Ilex verticillata*.

Mountain woods from N. Y. and Pa. to Ga. and Ala.

430. ILEX OPÀCA Ait. Reported in Coulter's Catalogue for the Lower Wabash Valley on the authority of Robert Ridgway. Ridgway told me that he had never seen it in Indiana as a wild plant, so there must have been some confusion in the records.

Atlantic coast from Mass. to Fla. and the Mississippi Valley from Ill. to the Gulf and west to Tex.

431. ACER PENNSYLVÁNICUM L. This tree was reported from the environs of New Harmony by Prince Maximilian under the name of *Acer striatum*. Since there is no preserved specimen, it is excluded. Robert Ridgway told me that it occurred just across the Wabash River in Illinois.

N. S. to Lake Superior, southw. in the Great Lakes region and in the mts. to Ga.

432. CARDIOSPÉRMUM HALICÁCABUM L. BALLOONVINE. Reported from Clark County (Baird & Taylor); Jefferson County (Barnes, J. M. Coulter, and Young); and from the Lower Wabash Valley (Schneck). I have it from Wells County. This is an occasional garden escape and probably not yet established. All reports and specimens date back more than 30 years.

Introd. from the Tropics and escaped from gardens.

433. RHÁMNUS CATHÁRTICA L. COMMON BUCKTHORN. Reported from Wayne County by Phinney and from Monroe County by Andrews. Neither author makes any comment so we are at a loss to know whether it is an escape and, if so, how successfully it has maintained itself. A. R. Bechtel found it as an escape in Montgomery County. We planted it in our aboretum and before we observed it hundreds of seedlings came up. We at once destroyed these and the parent pistillate trees.

Introd. from Europe and will escape in Indiana if cultivated.

434. VÌTIS ROTUNDIFÒLIA. Michx. MUSCADINE GRAPE. A specimen of *Cissus Ampelopsis* was referred to this species by Deam in Proc. Indiana Acad. Sci. **1911**: 372. 1912. This was an error and was corrected in Proc. Indiana Acad. Sci. **1912**: 83. 1913.

435. VITIS RUPÉSTRIS Scheele. SAND GRAPE. A form of *Vitis vulpina* found in the dunes bordering Lake Michigan was referred to this species in Coulter's Catalogue upon the authority of L. H. Bailey. This was a wrong determination of *Vitis riparia* var. *syrticola* (Fern. & Wieg.) Fern.

S. Pa. to Mo. and southw.

436. TÌLIA EUROPÈA L. Reported by Phinney (Indiana Geol. Rept. **11**: 148. 1882) as one of the "more common and important trees observed" in Delaware County. He also lists *Tilia americana.* The report of the European Linden as a common tree in Delaware County serves as an example to warn against the acceptance of any report until it is carefully considered.

437. ALTHAÈA RÒSEA (L.) Cav. HOLLYHOCK. Reported by Wilson as escaped from gardens in Hamilton and Marion Counties. It was collected in Benton County by W. S. Rhoades. We have no evidence that the species is able to maintain itself, so it is excluded.

Introd. from China.

438. MÁLVA ALCEA L. HOLLYHOCK MALLOW. Reported by Clark as found in Marshall County. He says: "A few plants which have escaped from seed of some old garden near Culver."

Introd. from Eu.

439. MALVA CRÍSPA L. CURLY MALLOW. Reported in Coulter's Cataloge from Putnam County by Underwood. Probably a garden escape.

Nat. of Eu.

440. MALVA SYLVÉSTRIS L. This species has been reported from Indiana several times but recent studies show that probably all of the reports for this species should be transferred to var. *mauretiana* in the text.

441. MALVÁSTRUM ANGÚSTUM Gray. Reported from the area of Delaware, Jay, Randolph, and Wayne Counties by Phinney. He says: "Rare. August. Dry grounds. Distinguished from *Sida* by its notched petals." Since the range of this species is to the south of Indiana and this is our only record, we are regarding this report as of a waif. There is no specimen.

Tenn. to Iowa and Kans.

442. SÌDA HERMAPHRODÌTA (L.) Rusby. Reported by Bradner from Steuben County without any comments. This species was at one time cultivated and he may have found an escape. There is no specimen.

Glades and river banks, Pa. to Tenn.; rare.

443. HIBÍSCUS SYRÌACUS L. SHRUB-ALTHAEA. In 1911 I found a specimen 2 inches in diameter and 10 feet high in a wooded ravine southwest of Hanover in Jefferson County. This plant was certainly an escape and there is no other record.

Nat. of Asia.

444. HYPÉRICUM DENSIFLÒRUM Pursh. Reported in 1898 by Blatchley as occurring on the banks of the Wabash River below Fort Harrison, Vigo County. I have Blatchley's specimen which was collected Oct. 12, 1896; it has neither flower nor fruit and is badly broken. It is a small specimen and apparently of this species but it may be a narrowleaf form of *Hypericum cistifolium.* It was also reported in Coulter's Catalogue for Cunningham from Tippecanoe County. In the absence of specimens from

which a positive determination may be made, it is best to exclude it for the present.

Pine barrens of N. J. to glades of Ky., Ark., and southw.

445. HYPERICUM ELLÍPTICUM Hook. Reported by Wilson from Hamilton and Marion Counties. He says: "Common." He also reports *Hypericum mutilum* and says: "Common." *Hypericum mutilum* and *Hypericum punctatum,* which he does not report, are the only two natives species that would be common in those counties. The range of *Hypericum ellipticum* is to the north of Indiana and all evidence suggests a wrong determination.

N. B. to Man., southw. to Pa., Mich., Wis., and Minn.

446. HYPERICUM GRAVÈOLENS Buckl. Reported for Monroe County by Andrews. The range of this species is the mountains of southwestern Virginia, Tennessee, and North Carolina. Andrews preserved no specimen, so the species is excluded from our list.

447. **Hypericum gymnánthemum** Engelm. & Gray. Reported from Lake County by Higley & Raddin and also by Pepoon. Probably confused with *Hypericum majus.* I have not seen a specimen.

N. J. and e. Pa. to Fla. and Tex.; northw. in the Mississippi Valley to Ohio.

448. **Elatìne americàna** (Pursh) Arn. (Fernald. The genus Elatine in eastern North America. Rhodora **19**: 10-15. 1917.) This plant was reported by Peattie as rare about ponds in the Calumet district, and also reported by Pepoon in his flora of the Chicago region. Bradner reported it from Steuben County and Schneck reported it from the Lower Wabash Valley. I am excluding it from our flora for lack of a confirming specimen. I have searched diligently for it for several years without success.

449. **Lèchea intermèdia** Leggett. Reported from Cass (Hessler), Lake (Blatchley and Coulter's Catalogue for Hill), Marshall (Clark), Steuben (Bradner), and Vigo Counties (Blatchley). These are all old reports and I have not seen a specimen. A. R. Hodgdon, who recently monographed the genus, wrote me that it occurred near Chicago in Illinois and in Lucas County, Ohio. It doubtless occurs in northern Indiana but I have not seen a specimen.

N. B. and N. S. to Wis., southw. to Pa.

450. LECHEA MARÍTIMA Leggett. Reported by Higley and Raddin from Lake County. Hodgdon, who recently monographed the genus, wrote me that the species is restricted to the Atlantic seaboard, which excludes it from Indiana.

451. **Viola crássula** Greene. I reported this species from Steuben County in Proc. Indiana Acad. Sci. **1905**: 186. 1906. I am now referring this specimen to *Viola cucullata.*

452. VIOLA EMARGINÀTA LeConte. I have a small specimen collected May 22, 1910, along a ditch just west of the State Prison in La Porte County, and which was named for me by Ezra Brainerd. I feel doubtful as

to the determination of this small, young specimen and I prefer to exclude the species until a more authentic specimen is available.

Southern N. Y., southw. to n. Ga. and westw. to Okla.

453. VIOLA HASTÀTA Michx. This species was reported from Clark County by Baird & Taylor. Stanley Coulter discussed this report in Proc. Indiana Acad. Sci. **1899**: 107. 1900 and said that the report doubtless was based on a wrong determination. There is no specimen.

Mts. of Pa. to Ohio, southw. to Fla.

454. VIOLA INCÓGNITA Brainerd. This species was reported from Porter and St. Joseph Counties by Nieuwland & Kaczmarek and from Porter County by Pepoon. I have not seen the species in the state and I am referring these reports to the variety.

Lab. to N. Dak., southw. along the mts. to Tenn.

455. VIOLA NEPHROPHÝLLA Greene. I reported this species from Grant and Noble Counties in Proc. Indiana Acad. Sci. **1915**: 139. 1916. I do not have these specimens and none are to be found in the Brainerd herbarium. Doubtless Brainerd changed the names and failed to report the change. In the absence of positive evidence I am excluding it from our flora.

Newf. and the Great Lakes to Wash., southw. to Conn. and in the mts. to Colo. and Calif.; also in Ariz. and N. Mex.

456. **Viola palmàta** L. PALMATE VIOLET. *Viola pedatifida* and *Viola triloba* and their hybrids may be easily mistaken by amateurs for this species. I have several specimens named *Viola palmata* but Prof. A. Gershoy says they are hybrids of other species. This species has been variously reported for the state but as I understand the species, all or most of our reports should doubtless be referred to other species or their hybrids.

457. **Viola pedatífida** × **soròria** Brainerd. I reported this hybrid from Wells County in Proc. Indiana Acad. Sci. **1915**: 139. 1916. I do not have this specimen and it cannot be found in the Brainerd herbarium. Doubtless Brainerd changed the name and did not report the change. In the absence of positive evidence, I am excluding it.

458. **Viola pratíncola** Greene. I reported this species from Lake and Porter Counties in Proc. Indiana Acad. Sci. **1916**: 320. 1917. Ezra Brainerd says: "Appears not to be specifically distinct from *Viola papilionacea* to which I am referring my specimens."

459. VIOLA ROTUNDIFÒLIA Michx. Reported from Dearborn (Collins) and Jefferson (Young) Counties. Coulter discusses these reports (Proc. Indiana Acad. Sci. **1899**: 108. 1900) and says the reports are based upon wrong determinations. There are no specimens.

Maine to Lake Huron, southw. along the Alleghenies to Ga.

460. VIOLA VIÀRUM Poll. Our only specimen was collected in very sandy soil along the railroad about 4 miles south of Vincennes. Dr. Brainerd named my specimen and says: "Much like the type collection, St. Louis

along the R. R., July 15, 1899." I found it closely associated with *Viola affinis* × *sororia.* I sent the specimen to A. Gershoy who says that he does not know the species. I think it best to relegate this report to the excluded species until the identity of the species is established beyond a doubt.

461. RHÉXIA MARIÀNA L. According to Fernald & Griscom (Rhodora 37: 169-173. 1935), the typical form of the species is a Coastal Plain plant with a range from Massachusetts to Florida.

462. LUDWÍGIA HIRTÉLLA Raf. This species was reported from Jefferson County by Young and from White County by Heimlich. It belongs to the pine barrens of the Atlantic coast. No doubt these authors confused some hairy form of our native species with this species which is far from our area. There is a specimen from Young's herbarium which is now at Indiana University; it was collected by a Dr. Fretz in New Jersey, and evidently was received in exchange. It is to be noted that in Indiana *Ludwigia alternifolia* is always more or less pubescent and sometimes rather densely so, although our manuals call it glabrous or nearly so. Prof. Heimlich may have had at hand *Ludwigia sphaerocarpa* var. *Deamii* which is densely pubescent and which at that time had not been reported for the state.

Pine barrens, N. J. to Fla. and Tex.

463. LUDWIGIA PALÚSTRIS (L.) Ell. According to Fernald & Griscom (Rhodora 37: 176. 1935), the typical form of this species is found in Europe and adjacent Asia and Africa and our form belongs to var. *americana* (DC.) Fern. & Grisc.

464. LUDWIGIA SPHAEROCÁRPA Ell. According to Fernald & Griscom (Rhodora 37: 173-174. 1935) the typical form of this species is found on the Coastal Plain from Rhode Island to Florida and Louisiana, and the Indiana plant belongs to var. *Deamii* Fern. & Grisc.

465. LUDWIGIA VIRGÀTA Michx. Reported from Jefferson County by Young. There is no specimen.

Dry pine lands, N. C. to Fla. and La.

466. EPILÒBIUM PALÚSTRE L. I reported this species from Steuben and Wells Counties (Proc. Indiana Acad. Sci. **1904**: 220. 1905). I now refer the specimens on which this report was made to *Epilobium molle* Torr. This species was also reported from Hamilton County by Wilson, who says: "Common." He does not report *Epilobium coloratum* which is our common species and to which no doubt this report should be referred. It was reported also from Steuben County by Bradner before our manuals made the separation of our species definite.

Newf. to Alaska, southw. to Mass., Ont., Lake Superior, Colo., and Wash.

467. OENOTHÈRA GRANDIFLÒRA Ait. Reported from Putnam County by Wilson, who found it along the Big Four Railroad; from Kosciusko County by Clark, who says: "A patch, probably of recent introduction, was found in moist soil near Warsaw." I have seen no specimen but no doubt this

species will sooner or later become established in the state if it has not already done so. Britton and Brown in Illus. Flora, ed. 2, say: "Large-flowered races of the preceding species *(Oenothera biennis)* have been mistaken for it." Therefore our reports may not be authentic.

468. OENOTHERA FRUTICÒSA L. (*Oenothera linearis* Michx. of Gray, Man., ed. 7 and *Kneiffia linearis* (Michx.) Spach of Britton and Brown, Illus. Flora, ed. 2.) Reports of this species for Indiana no doubt should be referred to some other species and I am referring the reports from Jasper County by Welch and from Lake County by Pepoon and by Peattie to *Oenothera tetragona* var. *longistipata.*

469. OENOTHERA OAKESIÀNA Robbins. Reported from Putnam County by Cook, who found it in a quarry in Greencastle. In this habitat it should be regarded as a waif.

Sandy fields, e. Mass. to Long Island.

470. OENOTHERA ALBICAÚLIS Pursh. Reported from Hamilton County by Wilson, who found it along a railroad. He reported it also from Tippecanoe County where he collected it in a meadow (probably a hayfield) east of Lafayette. There are no specimens nor any evidence that it has become established.

Dry plains and prairies, Sask. to w. Minn., westw. and southw.

471. **Oenothera tríloba** var. **parviflòra** Wats. This variety was reported in Coulter's Catalogue for Blatchley as found in Monroe County.

Munz (American Jour. Bot. **17**: 360. 1930) now refers this variety to the species which see on page 707.

472. CIRCAÉA CANADÉNSIS Hill. (Rhodora **19**: 87. 1917.) (*Circaea intermedia* Ehrh. of Gray, Man., ed. 7 and Britton and Brown, Illus. Flora, ed. 2.) Reported from Lake and Porter Counties by Pepoon. Since he does not report *Circaea alpina* L. which is known to occur there, and since the range of this species is far to the east of Indiana, I believe that Pepoon confused the species. Buhl (Amer. Midland Nat. **16**: 252. 1935) says there are no confirming specimens.

E. Que. and N. S. to w. Mass and N. Y.; found also in Eu.

473. **Myriophýllum hùmile** (Raf.) Morong. This species was reported from the area of Delaware, Jay, Randolph, and Wayne Counties by Phinney. There is no specimen.

Coastal Plain from Maine to Md.; also reported in the interior from Ind., Ill. to Tenn.

474. **Myriophyllum humile** var. **capillàceum** (Torr.) Fern. Reported from the Lower Wabash Valley by Schneck who says "in ponds, not rare." There are no specimens.

475. MYRIOPHYLLUM VERTICILLÀTUM L. All of our reports for this species were made before the last editions of our manuals were published. Our latest manuals regard this species as Eurasian and refer the American plants to var. *pectinatum* Wallr.

476. PROSERPINÀCA PALÚSTRIS L. According to Fernald and Griscom (Rhodora 37: 177. 1935) all reports for this species in Indiana should be referred to the varieties which will be found in the regular text. The species belongs to the Coastal Plain of the southeastern United States.

477. HYDROCÓTYLE ROTUNDIFÒLIA Roxb. In the herbarium accessions of the New York Botanical Garden published in the Journal of the New York Botanical Garden **23**: 184. 1922, there is the following: "1 specimen of *Hydrocotyle rotundifolia* from Indiana (Given by A. A. Hansen)." There are no other data. This is an Asiatic species and has been reported as established in lawns in Evansville.

478. SPERMÓLEPIS PÀTENS (Nutt.) Robinson. This species was found June 24, 1898, by L. M. Umbach along the railroad near Miller and first reported for him by Coulter & Rose in Contr. U. S. Nation. Herb. **7**: 72. 1900. It was later reported by Pepoon who says: "B. & O. Ry. at crossing of the Little Calumet River, northeast of Miller; common locally. (Umbach, Pepoon.)" I visited this place about July 1, 1930, and I was not able to find it. This species should no doubt be regarded as a migrant since it is found along a railroad out of its range and has not been able to spread or probably to maintain itself.

479. AETHÙSA CYNÀPIUM L. FOOL'S PARSLEY. This species was reported by Erlanson for Grimes (Proc. Indiana Acad. Sci. **1923**: 149. 1924) as having been found in a ravine in Putnam County. The specimen which was reported has been located in the herbarium of DePauw University and it is *Osmorhiza Claytoni*, so the species must be excluded. *Aethusa Cynapium* is a poisonous plant naturalized from Europe.

N. S. to Pa., Minn., and Ont.

480. THÁSPIUM PINNATÍFIDUM (Buckley) Gray. This species was reported from Marshall County by Clark. This determination, no doubt, should be referred to the narrow leaflet form of *Thaspium barbinode*.

Ky. to N. C. and Ala.

481. NÝSSA AQUÁTICA Marsh. This species has been reported by several authors and all of the reports should be referred to *Nyssa sylvatica* except those of Ridgway and Schneck which may be correct, but there is no specimen. The cypress swamps in Knox County furnish the proper habitat for the species and it may have occurred there.

Many years ago I questioned Michael Catt, 83 years old at that time, who had lived for about 75 years on the border of the cypress swamp in Knox County and he told me that he was positive that the tupelo gum was an occasional tree in the cypress swamp west of Decker. In my botanical experience I have met several people who were positive that this species existed, but upon investigation, I found all reports to be erroneous. In 1931 I found an old timber buyer who was positive that it occurred in Goose Pond in Gibson County and when he showed me the tree, it proved to be *Populus heterophylla*. With conflicting reports, it is best to exclude it.

Along the Atlantic coast from Va. to Fla., west through the Gulf States to Tex., and northw. in the Mississippi Valley to Ill.

482. NYSSA BIFLÒRA Walt. This was reported by Pepoon for Umbach from Dune Park in Porter County. N. C. Fassett has examined the specimen, which is in the herbarium of the University of Wisconsin, and he refers it to *Nyssa sylvatica.* (Rhodora 35: 200. 1933.)

Pine barrens of the Coastal Plain from N. C. to Fla. and westw.

483. PÝROLA ASARIFÒLIA Michx. Reported from Lake and Porter Counties. These reports may be correct but probably they should be referred to the variety. I have not seen a specimen.

E. Que. to Yukon, southw. to N. S., n. N. E., n. N. Y., n. Mich., and Colo.

484. RHODODÉNDRON NUDIFLÒRUM (L.) Torr. Reported for W. R. Dudley from Monroe County by the Editors of the Botanical Gazette in a Flora of Indiana on page 17, 1881. Evidently Dudley, who collected also in the vicinity of Ithaca, New York, confused his records, because the species does not occur in Indiana.

N. H. to N. Y., southw. along the mts. to Fla. and La.

485. KÁLMIA ANGUSTIFÒLIA L. This species was reported on the same authority as the preceding and is excluded for the same reason.

Lab. to Hudson Bay, southw. to Ga. and Mich.

486. GAYLUSSÀCIA FRONDÒSA (L.) T. & G. DANGLEBERRY. Reported from Clark County by Baird & Taylor and by C. P. Smith, and from Monroe County by F. M. Andrews. The range of the species does not extend west of the Allegheny Mountains and all reports of it should be referred to some other species.

487. SÁMOLUS VALERIÁNDI L. This species has been found in the United States only in ballast at Philadelphia. It has been reported from Clark County by Baird & Taylor; from Fayette County in Coulter's Catalogue for Hessler; and from Monroe County by Andrews. No doubt all of these reports should be referred to our native species, especially in such cases where the author did not report our native species.

Europe and Asia.

488. HALÈSIA CAROLÌNA L. GREAT SILVERBELL. Reported in Coulter's Catalogue from Vanderburgh County without quoting authority. The only other reference I can find to this species is that of Brendel who says: "*Halesia tetraptera (Halesia carolina)* has been found near Evansville on the Ohio, and might be sought in the south part of the state of Illinois." (Trans. Illinois Agric. Soc. 3: 600. 1859.)

489. FRÁXINUS CAROLINIÀNA Mill. WATER ASH. This ash was included in Coulter's Catalogue upon the authority of Dr. Schneck. I have seen no specimen and since its range is south of Indiana, the species is omitted.

490. SYRÍNGA VULGÀRIS L. COMMON LILAC. The lilac has been reported from Jasper County by Welch and from White County by Heimlich. I have never seen it escape from cultivation. I have seen it, however, persist on the site of deserted habitations until the area was reforested. I believe

that the presence of this species, when investigated, will show that it was planted.

Introd. from Europe.

491. CHIONÁNTHUS VIRGÍNICA L. WHITE FRINGETREE. This species was reported from Clark County by Smith. Investigation convinces me that the plant reported had persisted about an old squatter habitation.

492. LIGÚSTRUM VULGÀRE L. EUROPEAN PRIVET. This species was reported from Monroe County by Andrews without any data; from Montgomery County by Grimes, who says: "Roadsides and waste places"; and from the Lower Wabash Valley by Schneck, who says "found occasionally in woods and fields." Specimens have been found in St. Joseph County. I have never seen this species as an escape and I believe when a rigid investigation is made it will be found as a relict from some habitation, as in the case of the two preceding species.

Introd. from Europe.

493. POLYPRÈMUM PROCÚMBENS L. This species was reported for the vicinity of Lawrenceburg by Collins (Indiana Geol. Rept. **16**: 382. 1889). No data were given. Since the range is south of our area and there is no specimen the species is excluded.

Md. to Fla., Tex., and Mo.; also adv. in N. J. and Pa.

494. SABÀTIA BRACHIÀTA Ell. (*Sabbatia concinna* Wood of Wood, Class-book of Botany, ed. 2: 451. 1847.) This species was described by Wood in his second edition of his Class-book of Botany as occurring in "dry, grassy prairies, Ia., abundant." The Editors of the Botanical Gazette in a catalogue of the Plants of Indiana repeat this report. Coulter, in his Catalogue, reports it from Jefferson County for J. M. Coulter but J. M. Coulter does not give it in his list of the plants of Jefferson County. Both of our late Manuals cite it as found in Indiana. I have written to the curators of both the Gray Herbarium and the New York Botanical Garden and they have no specimen from Indiana. Since there is no existing specimen, the species is excluded.

N. C. to Fla. and La.

495. CENTAÚRIUM PULCHÉLLUM (Sw.) Druce. Reported from the Calumet District of the dunes, without any definite locality, by Peattie (Flora of the Indiana Dunes, p. 303, 1930.) I have not seen his specimen. I have a specimen of this species collected by Agnes Chase in 1901 in a pastured prairie in South Chicago, Illinois. Since this is an introduced species it is best to wait to see whether it establishes itself.

496. CENTAURIUM UMBELLÀTUM Gilib. This species was reported by Babcock as local along a roadside south of Michigan City (Lens **1**: 148. 1872). There are three small specimens of this collection in the Gray Herbarium. Since it has not been seen since that time it is best to exclude it. I have watched the roadsides about Michigan City for years with the hope that I might find it but I have failed.

Nat. of Eu.

497. BARTÒNIA IODÁNDRA Rob. This species is endemic in Newfoundland. I reported it from Steuben County but I am now referring the specimen to *Bartonia virginica.*

498. GENTIÀNA LINEÀRIS Froel. (*Dasystephana linearis* (Froel.) Britt. of Britton and Brown, Illus. Flora, ed. 2.) This species was reported by Clapp in his Catalogue of Medicinal Plants of the United States, p. 160. 1852. He refers to finding his specimen on the barrens. At that time the distinction between this species and *Gentiana puberula* was not clear, and I believe he had the latter because I collected it there and because Clapp did not report *Gentiana puberula.* It was also reported by Ball for Lake County in 1884. Since Ball did not report *Gentiana puberula,* which occurs there, I refer this report also to that species.

499. GENTIANA QUINQUEFÒLIA L. I am referring all reports of this species to its variety *occidentalis* (Gray.) Hitchc. I have not seen the typical form in the state.

S. Maine to Ont.; southw. to Fla.

500. APÓCYNUM ANDROSAEMIFÒLIUM var. INCÀNUM A. DC. All reports of this variety should be referred to the species.

501. **Apocynum cinèreum** Nieuwland. (Amer. Midland Nat. 3: 56-57. 1913.) This species was reported for Lake County. Woodson (North Amer. Flora 29: 191. 1938) refers this species to *Apocynum cannabinum* var. *pubescens* (R. Br.) A. DC.

502. **Apocynum isophýllum** Greene. (Greene. Leaflets of Bot. 2: 166. 1912.) Reported for St. Joseph County. Woodson (North Amer. Flora **29**: 192. 1938) refers it to *Apocynum cannabinum* var. *glaberrimum* A. DC.

503. **Apocynum platyphýllum** Greene. (Greene. Leaflets of Bot. 2: 167. 1912.) Reported for Wells County. Woodson now refers this to *Apocynum cannabinum* L. (North Amer. Flora **29**: 191. 1938.)

504. **Apocynum tomentéllum** Nieuwl. (Amer. Midland Nat. 3: 55-56. 1913. On page 166 of same publication he changes name to *Apocynum tomentulosum.*) Woodson (North Amer. Flora **29**: 191. 1938) refers this species to *Apocynum cannabinum* var. *pubescens* (R. Br.) A. DC.

505. **Asclèpias decúmbens** L. This species was reported from Knox County by Thomas and from Marshall County by Clark. It is doubtfully distinct from *Asclepias tuberosa* and I have not seen any specimens answering the description of *Asclepias decumbens.*

506. ASCLEPIAS INCARNÀTA var. PÚLCHRA (Ehrh.) Pers. There are a few reports of this variety for Indiana but I am referring all of them to pubescent forms of *Asclepias incarnata* L. The variety *pulchra* as now known is confined to the Atlantic Coastal Plain.

507. ASCLEPIAS SPECIÒSA Torr. Reported by Andrews from Monroe County. This is a western milkweed and since Andrews did not report *Asclepias purpurascens,* which is known to occur in Monroe County, I am referring this report to that species.

508. GONOLOBUS CAROLINENSIS (Jacq.) Schultes. This species was reported from the environs of New Harmony by Prince Maximilian under the name of *Gonolobium hirsutum.* Doubtless this report should be referred to some other species.

Md. and Va., southw. to Fla. and Tenn.

509. GONOLOBUS SHORTII Gray. I reported this species but Miss Perry refers my specimens to *Gonolobus obliquus* (Jacq.) Schultes.

Pa. to Ky., southw. to Ga.

510. CUSCÙTA EPÍTHYMUM Murr. FLAX DODDER. Reported by me from the Lower Wabash Valley upon the authority of Schneck. There are no specimens. Reported from Putnam and Ripley Counties by Blatchley in his weed book. The Ripley County report was made upon my authority and I find that the determination was wrong. No doubt the Putnam County report should also be referred to some other species.

511. PHLÓX AMOÈNA Sims. This species has been reported from Jefferson County and from the Lower Wabash Valley. Since the distribution of the species is south of our area, it is excluded from our flora.

Va. to e. Ky., southw. to Fla. and Miss.

512. PHLOX STOLONÍFERA Sims. Reported from Indiana in Wood's Class-book of Botany, ed. 2, and carried in all succeeding editions. It was reported on the authority of Plummer. Since the distribution of this species is to the southeast of Indiana, it is dropped from our flora.

Pa., southw. to the mts. to Ga. and Tenn.

513. COLLÒMIA LINEÀRIS Nutt. In 1935 Charles M. Ek found a colony approximately a hundred feet long on both sides of the Pennsylvania Railroad about 2 miles northwest of Kokomo, Howard County. Doubtless introduced here.

N. B., Wis. to B. C., southw. in the Rocky Mts. to Calif., but introduced east of the Rocky Mts.

514. **Phacèlia dùbia** (L.) Small. Our only report is one from Monroe County made by Andrews. It is excluded for lack of a confirming specimen.

N. Y. and Pa. to Mo. and Kans., southw. to Ga. and Tex.

515. LÁPPULA REDÓWSKII (Hornem.) Greene var. OCCIDENTÀLIS (Wats.) Rydb. I reported this variety from Porter County but I am now referring the specimen to *Lappula echinata* Gilib.

Nat. of Asia and Amer.; nat. of the U. S. west of our area, from Sask., N. Dak. to Okla. and N. Mex., but introd. into several states east of its range.

516. SÝMPHYTUM OFFICINÀLE L. COMMON COMFREY. This is a medicinal herb which was cultivated in gardens by the pioneers, but it is now very rarely or never cultivated. It was reported as an escape by some of the early botanists but I have seen it only once and that was along the roadside near a house. The colony may have persisted from an old garden. It was reported as being common in woods in Jefferson County. I doubt

that it has been able to establish itself. In 1932 Scott McCoy found a specimen in a wooded ravine near Indianapolis.

Nat. of temperate Eu.; naturalized from Newf. to Que. and Mont., southw. to N. C. and La.

517. LYCÓPSIS ARVÉNSIS L. Andrews reported this species from Monroe County without any data and Young did likewise for Jefferson County in 1871. Since there is only one record from Ohio, it is certain that this species rarely escapes, and since it is not planted, there is reason to believe that it will never become so well established in Indiana that it can be called a member of our flora.

Nat. of temperate Eu.; naturalized from N. B. to Minn., southw. to Va. and Colo.; also in Calif.

518. MYOSÒTIS ARVÉNSIS (L.) Hill. This plant was reported in 1892 by Benedict & Elrod as "found growing sparsely in Cass County, near Lake Cicott, and in Bethlehem Township. It seems to prefer the sandy ridges and sandy fields, and was not seen elsewhere." As it has not been reported since or elsewhere it is best to place it with the excluded species.

Newf. to Minn., south to W. Va.; and in Eu.

519. LITHOSPÉRMUM OFFICINÀLE L. This species has been reported by several authors but I believe that they have confused it with *Lithospermum arvense.* Meyncke reported it from Franklin County but did not report *Lithospermum arvense.* Riddell reported it from the vicinity of New Albany on the authority of Clapp in his "Supplement of Ohio Plants," on page 27, 1836. Since there are no specimens, and since there are no specimens from Ohio, although reported there, I believe it best to exclude this species.

Nat. of Eu.; e. Que. to Minn., south to N. J.

520. ONOSMÒDIUM MÓLLE Michx. Reported from Clark County by Baird & Taylor in 1878. Since this species is known only from the cedar barrens of Kentucky and Tennessee, the report should doubtless be referred to *Onosmodium hispidissimum.*

521. ONOSMODIUM OCCIDENTÀLE Mack. This species was reported from Hamilton County by Grimes. His specimen is in the herbarium of DePauw University and it should be referred to *Onosmodium hispidissimum* Mack.

Ill. to N. Dak., Alberta, southw. to Kans., Tex., and N. Mex.

522. ONOSMODIUM VIRGINIÀNUM (L.) A. DC. Reported by Phinney from the area of Delaware, Jay, Randolph, and Wayne Counties in 1883. This species is entirely out of our range, and I refer this report to *Onosmodium hispidissimum.*

Conn. to Fla., westw. along the Gulf to La.

523. VERBÈNA BIPINNATÍFIDA Nutt. DAKOTA VERBENA. This species was reported from the Calumet District, on railroad embankments. Since the range of the species is far to the west of Indiana, and since it has been found only along railroads, I regard it as a migrant.

S. Dak. to Mo. and Mex.

524. VERBENA OFFICINÀLIS L. EUROPEAN VERVAIN. This species was reported from Fayette and Jefferson Counties about 40 years ago. There are no later reports and in the absence of verifying specimens, and since it has not been reported from Ohio, it is best to regard it as a waif.

Nat. of Eu.; naturalized from Maine to Fla., Tenn., and Tex. Also on the Pacific coast.

525. **Scutellària serràta** Andr. Reported from Fayette County by Hessler and from the Lower Wabash Valley by Schneck. This species much resembles *Scutellaria incana* and these authors may have confused these two. There are no specimens.

Southern N. Y. and Pa. to Ill., southw. to S. C. and Tenn.

526. AGÁSTACHE FOENÍCULUM (Pursh) Ktze. This species was reported from Clark County by Baird & Taylor and reported without comment from Jay County by Phinney. It is a native far to the west of our area and these references should be regarded as of escapes from cultivation or as of migrants. It has been reported as an escape in Essex County, N. Y.

Ill. to Man., Alberta, southw. to Colo.

527. **Meehània cordàta** (Nutt.) Britt. This species was reported by Blatchley from Monroe County in a manuscript which is on deposit at Indiana University. He says: "June, shaded banks of streams." I have the specimen upon which this record was made and it is a creeping form of *Blephilia ciliata* which is not yet in flower. Andrews also reported this species from Monroe County but since Andrews only listed the species without any data the report is valueless.

Pa. to Ill., southw. to N. C. and Tenn.

528. PHYSOSTÈGIA PARVIFLÒRA Nutt. This species was reported from Marion County by Douglass and from Putnam County by Grimes. I have seen the specimens upon which this report was made and the Douglass specimen is *Physostegia speciosa* and the Grimes specimen is *Physostegia virginiana.*

529. LEÙCAS MARTINICÉNSIS R. Br. This introduced species was reported by Collins (Ann. Rept. Indiana Dept. Geol. and Nat. Hist. **16**: 379. 1889) from Dearborn County. Our present manuals do not list this species, so it must be a rare escape.

530. GALEÓPSIS TETRÀHIT L. There are reports of this species from Franklin and Jefferson Counties, but they were made more than 50 years ago. Evidently the plant was a waif and has not become established. There is no specimen. There is only one record from Ohio.

Nat. of Eu.; in waste places and on ballast from Newf. to B. C. and Alaska, southw. to N. C., W. Va., and Mich.

531. LÀMIUM ÁLBUM L. WHITE DEADNETTLE. This species was reported by Andrews from Monroe County but since there are neither specimens nor data the species is excluded.

Nat. of Eu.; waste places in Ont. to Mass. and Va.

532. STÀCHYS PALÚSTRIS L. There are many reports for this species but according to Fernald (Rhodora 23: 289. 1921) this is a European plant which has been introduced from Newfoundland to Quebec and Ottawa, southward, chiefly near the coast, to New Jersey. Probably most of the reports from Indiana should be referred to *Stachys palustris* var. *homotricha* Fern.

533. SÁLVIA PÍTCHERI Torr. I have the private copy of Dr. J. Schneck's list of the plants of the Lower Wabash Valley in which he recorded additions to his list. In 1912, I published the additions and this species was among them. He says he found it in Gibson County on the Martin Meyer farm which is located two and a half miles south of the bridge of the Southern Railroad over the Wabash River. This is a plant of the dry plains, and in the absence of more data or of a specimen, we exclude it from our flora. The area where the plant was found was originally heavily wooded but the plant might have been introduced.

Mo., Kans., Colo., southw. to Tex.

534. SALVIA PRATÉNSIS L. Clute reported this species as found in a pasture about 12 miles south of Indianapolis. This is our only report and we have no evidence that it has become established. It is a cultivated species and its escape may be expected.

Nat. of Eu.

535. SALVIA URTICAEFÒLIA L. This species was reported by Riddell for Clapp in the "Supplement to Ohio Plants" on page 27, 1836. "Found among the knobs, New Albany, rare." I have Clapp's catalogue of the plants that he found and in it he records "found on the knobs near Mr. Jones." I do not question this record but, following the rule that I include only reports supported by a specimen, I am compelled to exclude it.

Pa. to Ky., southw. to Ga. and La.

536. SALVIA VERTICILLÀTA L. In 1935 Charles M. Ek found a large colony of this species along the Pennsylvania Railroad about a quarter of a mile northwest of Galveston, Cass County. Doubtless introduced.

Nat. of Eu.

537. MONÁRDA DÍDYMA L. OSWEGO BEEBALM. This species has long been under cultivation as an ornamental plant and kitchen herb and easily escapes. Schneck reported it as a garden escape for the Lower Wabash Valley in 1874. Higley and Raddin reported it on the authority of Brennan for Lake County. Peattie also reported it for the Calumet District but this report may be based upon the Higley and Raddin report. There is no Indiana specimen in the Field Museum. Phinney reported it in 1883 for the area of Delaware, Jay, Randolph, and Wayne Counties. He says: "July. Moist places, rare." This is the only reference in which it might be considered a native plant, but because it is known that Phinney often did not distinguish between native and cultivated plants, this reference is doubtful. From the ease with which this plant might escape, and because it is so conspicuous that it certainly would be noticed by any amateur

botanist, I believe we can dispose of this species as a rare escape in the state and not as a native.

Western Que., Ont., and Mich., southw. to Ga., Ala., and Tenn.

538. **Saturèja glabélla** (Michx.) Briquet. This species was reported for Clark County and for the area of Delaware, Jay, Randolph, and Wayne Counties. It was also included in the manuals but there are no Indiana specimens in the Gray Herbarium nor in the New York Botanical Garden. The range is given in our manuals as Ind., Ky., and Ark. No doubt it will be found in southern Indiana but in the absence of a specimen I exclude it.

539. SATUREJA NÉPETA (L.) Scheele. This species was reported for Franklin County by Meyncke in 1885. Since we have no subsequent reports and no specimen, it is excluded.

Nat. of Eu.; locally naturalized from Md. to Ark.

540. HYSSÒPUS OFFICINÀLIS L. HYSSOP. Reported in 1878 for Clark County by Baird & Taylor. We have no subsequent report. This plant was cultivated by pioneers for its medicinal qualities and may have at that time sometimes escaped, but since it is no longer cultivated, there is little probability of it becoming established.

Nat. of Eu.; locally established from Ont. and Maine, to N. C. and on the Pacific coast.

541. PYCNÁNTHEMUM CLINOPODIOÌDES T. & G. (*Koellia clinopodioides* (T. & G.) Ktze. of Britton and Brown, Illus. Flora, ed. 2.) Reported for Clark County by Baird & Taylor but, since the range of the species is outside of Indiana, and they did not report *Pycnanthemum pilosum* which occurs here, I refer this report to the latter species.

Conn. to Pa., Va., and Tenn.

542. PYCNANTHEMUM INCÀNUM (L.) Michx. (*Koellia incana* (L.) Ktze. of Britton and Brown, Illus. Flora, ed. 2.) This species was reported for six southern counties, but the authors did not report *Pycnanthemum pycnanthemoides*. I believe that these authors confused the two, and for the lack of a specimen, I exclude the species.

Maine to Ont., southw. to Fla., Ala., and Mo.

543. PYCNANTHEMUM MÙTICUM (Michx.) Pers. (*Koellia mutica* (Michx.) Britt. of Britton and Brown, Illus. Flora, ed. 2.) Andrews reported this species for Monroe County without any data. Schneck reported it also for the Lower Wabash Valley. Since its range is outside our area, and we have no specimen, I exclude it.

Maine to Pa., Va., and Fla. and Mo.

544. THÝMUS SERPÝLLUM L. THYME. This species has not yet been reported for Indiana but I have had a large colony in Fairview Cemetery at Bluffton under observation for ten years. It was a large colony when I first found it and it has been gradually spreading since that time. My advice is never to let it escape because it will be almost as difficult to

exterminate as ground-ivy. It is established in the grounds of the University of Notre Dame, St. Joseph County.

545. LÝCOPUS ÁSPER Greene. This species was reported from St. Joseph County by McDonald for Nieuwland. I have seen this specimen and it should be referred to *Lycopus americanus* Muhl.

Mich., Man., and B. C., southw. to Kans., Ariz., and Calif.

546. MÉNTHA AQUÁTICA L. There are two specimens of this European species in the herbarium of Indiana University. They were collected by A. H. Young in July, 1881, but they were never reported. Since these are our only specimens and no data were given, it seems best to exclude the species.

Nat. of Eu.; N. S. to Pa. and Ga.

547. MENTHA CARDÌACA Gerarde. I found this species in 1922 in a pasture field in Spencer County. It was reported from Porter County by Peattie on the authority of Churchill. Since there are no data concerning the ability of this species to maintain itself it is best to regard it as a migrant.

Nat. of Eu.

548. MENTHA LONGIFÒLIA var. UNDULÀTA (Willd.) Fiori & Paoletti. In 1923 I found this mint as a common plant in sandy soil along the roadside on the site of a former habitation about 2 miles northwest of Monticello, White County. A few years later I again noted it at the same place. Since there are no other records, I regard this as only a chance introduction.

Nat. of Eu.

549. HÝPTIS RADIÀTA Willd. This species was reported from Jefferson County by Young (Rept. Indiana Geol. Surv. **2**: 273. 1871). If the identification was correct, no doubt it was a migrant.

N. C. toward the coast to Tex.

550. HYOSCÝAMUS NÌGER L. BLACK HENBANE. This medicinal plant was found by F. J. Hermann, June 20, 1935. It was an escape along the road about 2 miles west of Angola, Steuben County.

Nat. of Eu.; N. S. to Ont., southw to N. Y. and Mich.

551. PHÝSALIS ANGULÀTA L. This species has been reported from the dune area by Peattie and by Pepoon. Since it is a western species and it was found in ballast, I regard these plants as migrants.

Va. to Iowa, southw. to Fla. and Tex.

552. PHYSALIS IXOCÁRPA Brotero. TOMATILLO. E. D. Hull reported this species in American Botanist **41**: 27. 1935. A few plants were adventive in Gary, Lake County where they had escaped from cultivation.

Introd. from the southwest.

553. PHYSALIS LANCEOLÀTA Michx. There are several reports for this species. The plant very much resembles *Physalis virginiana* and may have

been confused with it. Since it is a western species, it is best to wait until we secure authentic specimens before it is admitted to our flora.

S. C. to Ill., S. Dak., southw. to Kans. and N. Mex.

554. PHYSALIS PERUVIÀNA L. PERUVIAN GROUNDCHERRY. I found this species in Crawford County near a barn and was told by the owner that they had cultivated it for its fruit. It could easily escape but it should not be considered a part of our flora until it has established itself.

Nat. of S. A.

Several other species of *Physalis* have been reported from Indiana but they may all be referred to the synonomy of some one of the species treated in the text.

555. SOLÀNUM HETERODÓXUM Dunal. Reported from Monroe County for Hessler in Coulter's Catalogue. This species is not a native of the United States and if the plant was not wrongly identified, it must have been a migrant.

556. SOLANUM TÓRREYI Gray. Reported from Hancock County by Douglass. It is a western species and since it closely resembles *Solanum carolinense,* I omit it from our flora until a check upon the identification can be made.

Ark. to Kans., southw. to Tex.

557. SOLANUM VIRGINIÀNUM L. Riddell reported this species for Clapp in 1836 in his "Supplement to Ohio Plants" on page 27. I have the book in which Dr. Clapp kept his records and he records this species for 1834, but later he scratched it out. This is a dubious species and authors agree that Linnaeus described a plant foreign to Virginia, and its identity has not been satisfactorily established.

558. DATÙRA MÈTEL L. I reported this species from the Lower Wabash Valley for Schneck who says: "Occasionally spontaneous." This is a native of the tropics and there is no evidence that it is a part of our flora.

559. PETÙNIA AXILLÀRIS (Lam.) BSP. PETUNIA. This species was reported by Schneck as an escape from gardens but he does not say whether it maintained itself. It is an annual and only a chance escape.

Nat. of Brazil.

560. PETUNIA VIOLÀCEA L. PETUNIA. Reported both by Peattie and by Pepoon on the authority of Hill as persisting on the site of an old camp in the dunes. I noted it once persisting about an abandoned dwelling in the dunes in Porter County, but in this instance it had no competition in its sandy habitat and I doubt that it would persist long. This is an annual and without doubt would fail to maintain itself; hence it is excluded.

561. ANTIRRHÌNUM MÀJUS L. SNAPDRAGON. This species has been reported as a garden escape, but there are no data concerning its persistence.

Nat. of Eu.

562. CHELÒNE LŸONI Pursh. This species was reported from the Lower Wabash Bottoms by Schneck. The report was made when the species was not understood, and it should, no doubt, be referred to *Chelone obliqua* var. *speciosa.*

Eastern Appalachian Mts., w. N. C., and adjacent S. C. to Tenn.

563. CHELONE OBLÌQUA L. Reported by several authors before the status of the species was understood. All reports should be referred to *Chelone obliqua* var. *speciosa.*

Coastal Plain from Md. to Ala.

564. PENSTÈMON LAEVIGÀTUS Soland. Reported from all parts of the state before the present division of the genus. This species, as now understood, is restricted to the Blue Ridge province of the eastern Appalachians.

Pa. to Fla. and e. Miss.

565. **Mímulus glabràtus** var. **Fremóntii** (Benth.) Grant. This variety was reported by Higley & Raddin as being found in July, 1885, near Miller, Lake County. Since Grant and Pennell, who searched every herbarium in which a specimen might be deposited, did not find one, and R. M. Tryon, Jr., searched the herbarium of the Chicago Academy of Science, it is excluded for lack of verifying evidence. I have no doubt that the species did formerly occur in Indiana.

Ont. to Man., southw. into Mex.

566. MIMULUS VISCÍDULA var. TÝPICA Pennell. (*Gratiola viscosa* Schwein.) This species was reported from Jefferson County by Young, who says: "Rather plentiful. All the specimens I have found have the peduncle a little longer than the leaves." Young's specimens so named are in the herbarium of Indiana University, and they should be referred to *Gratiola neglecta* Torr.

Del. to n. Ga. and e. Tenn.

567. VERÓNICA AGRÉSTIS L. This species was reported from Putnam County in Coulter's Catalogue, upon the authority of MacDougal. There is no specimen of MacDougal's in the herbarium of DePauw University. There is, however, a specimen so labeled collected by Lewis & Bridges May 2, 1888, and it proves to be *Veronica arvensis.* Since I have not seen an Indiana specimen, the species is excluded. It has been reported from Lake County by Standley, but I have seen his specimens and I am referring them to *Veronica persica.*

Nat. of Eurasia; Newf. to Mich. and Pa.

568. VERONICA ANAGÁLLIS-AQUÁTICA L. Reported several times, but all of the specimens so named which I have seen I am referring to one or the other of our aquatic species. The specimen collected by Grimes in Putnam County is in the herbarium of DePauw University and is *Veronica connata.*

Nat. of Eurasia; Maine to Wash., southw. to N. C., Tex., and Ariz.

569. GERÁRDIA ÁSPERA Dougl. (*Agalinis aspera* (Douglass) Britt. of Britton and Brown, Illus. Flora, ed. 2.) This species was reported by Schneck as found in wet prairies in the Lower Wabash Valley, and by Peattie as found in the Indiana dunes "in sandy soil of the Post-Tolleston beaches." Pennell, in his studies of this species, has examined all the specimens in all of the leading herbaria of the United States and has not found a specimen from Indiana. Hence our reports are referred to other species.

Man. to Ill. and Okla.

570. AUREOLÀRIA LAEVIGÀTA (Raf.) Raf. (Probably *Gerardia laevigata* Raf. of Gray, Man., ed. 7 and *Dasystoma laevigata* Raf. of Britton and Brown, Illus. Flora, ed. 2.) This species was reported from Porter County by Pepoon, and the specimen has been examined by Fassett, who refers it to *Aureolaria flava*. It was reported from Marshall County by Clark, and doubtless this report also should be referred to *Aureolaria flava.*

Pa. and Ohio to Ga. and Tenn.

571. MELAMPỲRUM LINEÀRE Desr. var. TÝPICUM Pennell. All reports of *Melampyrum* made before the publication of Peattie's "Flora of the Indiana Dunes" were made before authors recognized the variety. Since Pennell gives the range of the typical form of the species as being outside our area, doubtless all reports should be transferred to one of the varieties.

Newf. to N. Y., Minn., and B. C.

572. UTRICULÀRIA BIFLÒRA Lam. This species was reported by Scovell (Proc. Indiana Acad. Sci. **1899**: 130. 1900) as occurring in Little Lake near Lake Maxinkuckee. As now understood, this species occurs along the coast from Massachusetts to Florida and Louisiana, and, no doubt, this report should be referred to some other species. Our manuals of that date did not make very clear the distinction between the species.

573. UTRICULARIA CLEISTÓGAMA (Gray) Britt. This species was reported by me for Dr. J. Schneck. Dr. Barnhart has examined the specimen and writes that it is a depauperate specimen of *Utricularia gibba.*

574. UTRICULARIA INFLÀTA Walt. This species was reported from Jasper County by Welch (Proc. Indiana Acad. Sci. **36**: 219. 1927). I have seen the specimens reported and they should be referred to *Utricularia radiata* Small.

575. RUÉLLIA PEDUNCULÀTA Torr. Clute (Amer. Bot. **36**: 169. 1930) reported this species from Marion County under the name of *Ruellia longipedunculata.* The specimen is in the herbarium of Butler University, and seems to be a variation of *Ruellia strepens* mentioned in Gray, Synoptical Flora, 1886, which has short peduncles or peduncles of varying lengths, but they are not as long as those of *R. pedunculata* which equal the leaves. I have one specimen which has peduncles of three lengths.

Ill. and Mo. to Ark. and La.

576. PLANTÀGO ÍNDICA L. Charles M. Ek found this species in 1935 in dry, cindery ballast in the yards of the Pennsylvania Railroad at Kokomo, Howard County, and also in Cass County, 4 miles northwest of Galveston in ballast of the Pennsylvania Railroad. I am regarding this species as a railroad migrant.

Nat. of cent. and s. Eu.; Pa., Ohio, Ind., Mich., and Iowa.

577. PLANTAGO ELONGÀTA Pursh. Reported four times for Indiana but doubtless all reports should be referred to *Plantago pusilla.*

N. Dak. to Utah, southw. to Nebr. and Okla.

578. PLANTAGO SPARSIFLÒRA Michx. This species was reported from Clark and Jefferson Counties by the earlier authors before our manuals made clear the distinction between the species, *Plantago elongata* and *Plantago pusilla.* These reports should be referred, no doubt, to some other species. A specimen collected by J. M. Coulter in Jefferson County is in the herbarium of Wabash College and proves to be *Plantago Rugelii.*

Coastal Plain, N. C. to Fla.

579. **Houstònia canadénsis** Willd. (*Houstonia ciliolata* Torr. of Gray, Man., ed. 7 and Britton and Brown, Illus. Flora, ed. 2.) This species has been reported from Indiana thirteen times, but I believe all of these reports should be referred to *Houstonia longifolia.* I have not seen a typical specimen of this species but our manuals say that at least the basal leaves of this species are ciliate and Gray, Manual, ed. 7 says: "hirsute-ciliate." I have seen no specimens which have this character or which look like the plant shown in the colored plate in Torrey's Flora of New York.

Maine to N. D., southw. to W. Va. and Ark.

580. **Houstonia lanceolàta** (Poir.) Britton. This species was reported by Daubenmire from Parke County, but he later referred the specimen to *Houstonia purpurea.*

Maine to Ill., southw. to Okla. and Ala.

581. HOUSTONIA TENUIFÒLIA Nutt. Reported from Marion County by Douglass. I have seen the specimen upon which this record was made and it is a glabrate form of *Houstonia longifolia* Gaertn.

E. Ohio to Va., southw. to N. C. and Tenn.

582. GÀLIUM APARÌNE var. VAILLÁNTII (DC.) Koch. This variety was reported from Marshall County by Clark, who says that he found it in a marsh. Since this species rarely grows in marshes, and since the habitat is that of *Galium labradoricum,* I believe that he has confused the two.

Ont. to B. C., southw. to Mo., Ariz., and Calif.

583. GALIUM LATIFÒLIUM Michx. Andrews reported this species from Monroe County but he did not report *Galium lanceolatum,* which should be found in that county. Since the range of this species is in the Appalachian Mountains, I believe that Andrews confused it with *Galium lanceolatum.*

Mts. of Pa. to Ga. and Tenn.

584. GALIUM MOLLÙGO L. I reported this species from Marshall County where I found several large colonies in a pasture field about a half mile

north of Culver. Since this report I found a large colony in Jefferson County along a creek near where it parallels a road about 3 miles east of Canaan. Data concerning the time when these colonies were introduced and how long they will persist will determine whether the species is established in the state.

Nat. of Eu.; naturalized from Newf. to Vt., Pa., Ohio, N. J., and Va.

585. GALIUM UNIFLÒRUM Michx. Reported from Jefferson County by Young in 1871. This report doubtless should be referred to some other species.

S. C. to Fla., and Tex.

586. GALIUM VÈRUM L. I reported this species from Noble County. A large colony was found along a roadside a mile west of Kendallville. Since this is our only record, it is not included in our flora.

Nat. of Eu.; naturalized from Maine to Ont., southw. to N. J.

587. SAMBÙCUS NÌGRA L. EUROPEAN ELDERBERRY. A colony of this elderberry was found October 2, 1937 by Ray C. Friesner along the Nickel Plate Railroad 0.8 mile west of Goldsmith, Tipton County. Prof. Friesner has kindly permitted me to make this first report of this species for the state. This is our first record for the state and will be regarded as a garden escape.

Eu., northern Africa and western Asia.

588. VIBÚRNUM CÁNBYI Rehd. I reported this shrub from Brown, Clark, Jennings, and Ripley Counties (Proc. Indiana Acad. Sci. 1912: 84. 1913). The specimens were named for me by Rehder, who now refers them to *Viburnum pubescens* var. *indianense.*

589. VIBURNUM DENTÀTUM L. ARROWWOOD. Reported from Indiana several times by our early authors but the range of this species as now understood is east of Indiana and doubtless all of our reports should be referred to some other species. Buhl (Amer. Midland Nat. **16**; 252. 1935) refers Peattie's and Pepoon's reports from the dune area to *Viburnum affine* or its variety.

590. VIBURNUM NÙDUM L. SMOOTH WITHE-ROD. Reported from Steuben County by Bradner in 1892. Since its range is southeast of Indiana, this report should no doubt be referred to some other species.

591. VIBURNUM PUBÉSCENS (Ait.) Pursh. DOWNY VIBURNUM. The range of this species as understood by recent authors is east of Indiana, and all of our reports should be referred to some other species.

592. SYMPHORICÁRPOS OCCIDENTÀLIS Hook. WESTERN SNOWBERRY. This species was reported from Jefferson County by Coulter and by Young. Since its range is to the west and north of Indiana, these plants must have been escapes from gardens.

Mich. to B. C., southw. to Ill., Colo., and Kans.

593. LONICÈRA HIRSÙTA Eat. HAIRY HONEYSUCKLE. Reported from Steuben County by Bradner. Since he did not report the other two species which I have seen in this county, and since there is no verifying specimen, I believe it best to refer this record to some other species. It was also reported from Kosciusko County by Clark. He says: "Found, but not in flower, in the tamarack northeast of the lake." Clark's specimens are supposed to have been preserved in the National Herbarium, but a letter from the Curator, dated March 21, 1924, says that his specimen cannot be found there.

Vt. to Man., southw. to Pa., Ohio, and Minn.

594. LONICERA OBLONGIFÒLIA (Goldie) Hook. SWAMP FLY HONEYSUCKLE. This species was reported from Marshall County by Clark. He says: "Rather rare; one plant found on the south shore of the lake. Throughout the tamarack swamps of northern Indiana one comes frequently across a honeysuckle which is probably this species." Clark's specimens are supposed to have been preserved in the National Herbarium, but a letter from the Curator, dated March 21, 1924, says that the Clark specimen cannot be found there.

595. LONICERA SEMPÉRVIRENS L. TRUMPET HONEYSUCKLE. This species has been reported from Clark, Franklin, Jefferson, St. Joseph, Tippecanoe, and Wayne Counties. J. M. Coulter, in his Flora of Jefferson County, says: "Sparingly spontaneous." I feel certain that it is not a native of Indiana, and do not believe it has escaped to the extent that it will become a permanent part of our flora. Some of the above reports were made by authors who did not distinguish between cultivated and native plants, so we have no way of knowing to just what extent it has escaped. I have never seen it as an escape, but noted it in Jefferson County along a fence where there was formerly a dwelling.

596. LONICERA TATÁRICA L. TARTARIAN HONEYSUCKLE. R. M. Kriebel informed me that he found a bush of this species at the edge of a marsh three fourths of a mile east of Mt. Summit, Henry County. It was in flower on May 15, 1937 and in fruit on June 20, 1937. I am surprised to learn that this is our only record of this plant escaping.

S. Russia to Altai and Turkestan.

597. LONICERA XYLÓSTEUM L. Found as an escape in 1937 by R. C. Friesner in a decadent tamarack bog about a mile south of Garrett, De Kalb County. It has been reported by McDonald as found by J. A. Nieuwland and by P. E. Hebert on the bank of the St. Joseph River in St. Joseph County.

Eu. to Altai.

598. VALERIANÉLLA RADIÀTA (L.) Dufr. This species has been reported for all parts of the state. A recent revision of the genus shows that our plant is *Valerianella intermedia* Dyal and that our reports should be referred to this species.

Pa. to Kans., southw. to Fla. and Tex.

599. VALERIÀNA OFFICINÀLIS L. COMMON VALERIAN. GARDEN HELIOTROPE. Cultivated in gardens for ornament. This species was reported without any comment, from Monroe County by Andrews. Since there are no other reports, it is excluded from our flora.

Nat. of Eu.; escaped to roadsides in N. Y., N. J., and Ohio.

600. VALERIANA SEPTENTRIONÀLIS Rydb. This species was reported under the name of *Valeriana sylvatica* Banks by four authors about 60 years ago. Since these authors did not report *Valeriana intermedia* which is a native of the state, they no doubt confused the names of the plant, and I accordingly exclude *Valeriana septentrionalis* from the Indiana flora.

B. C., southw. in the Rocky Mts. to N. Mex.

601. CUCÚRBITA FOETIDÍSSIMA HBK. MISSOURI GOURD. Pepoon reports that this plant was found along the Wabash Railroad near Miller, Lake County, and persisted for eight years, when cold destroyed it. Peattie reported it for the Calumet District, but doubtless he had reference to this report without giving credit. I regard this species as a railroad migrant.

S. Dak. to Tex.; westw. to Calif.

602. CUCURBITA PÈPO var. OVÍFERA Alefeld. PEAR GOURD. Peattie reported this variety as "nat. on pure sand around Tremont" (Porter County). I regard this as a migrant or as an escape from some garden. Buhl (Amer. Midland Nat. **16**: 252. 1935) says this report is of a non-persistent garden escape.

603. CAMPÁNULA DIVARICÀTA Michx. This species was reported from the Lower Wabash Valley as occurring "in hilly woods; rare" by Schneck. It was also reported from Monroe County by Andrews. I bought the Indiana specimens of the Schneck herbarium and the collection did not contain a specimen of this species. Andrews preserved no specimen. Since there is no verifying specimen, the species is excluded.

Va. to Ky. and southw.

604. SPECULÀRIA LEPTOCÁRPA (Nutt.) Gray. This species was found by Charles M. Ek, July 10, 1935, on an embankment of the New York, Chicago & St. Louis Railroad (Nickel Plate Road), about 2 miles southeast of Sharpsville, Tipton County. Without doubt it was a railroad migrant.

Mont., Colo., Mo., and Kans. to Tex.

605. VERNÒNIA NOVEBORACÉNSIS Willd. Reported mostly by our early authors who did not understand the species. Its range is restricted, in general, to the Atlantic coast. Pepoon's report of Umbach's specimen from Lake County should be referred to *V. missurica* Raf. (Rhodora **35**: 210. 1933.)

Mass. to Ohio, southw. along the Atlantic coast to Miss. and the Gulf.

606. EUPATÒRIUM HYSSOPIFÒLIUM L. Reported in a "Catalogue of the Plants of Indiana," published by the editors of the Botanical Gazette and C. R. Barnes in 1881. They say: "From specimens in the herbarium of Lafayette High School. Locality not certain but probably Tippecanoe Co."

This report was repeated in Coulter's Catalogue. It seems that the information is very uncertain and, if found in Tippecanoe County, it must have been a waif.

Mass. to Va., e. Ky., southw. to Fla. and Tex.

607. BRICKÉLLIA GRANDIFLÒRA (Hook.) Nutt. This western composite was reported from Hamilton County by Wilson. He later said that this record should be referred to *Cacalia suaveolens* L.

608. LIÀTRIS PYCNOSTÀCHYA (Michx.) Ktze. This species as now understood does not occur in Indiana and reports for it are referred to *Liatris Bebbiana* Rydb.

609. AMPHIÁCHYRIS DRACUNCULOÌDES (DC.) Nutt. On Sept. 21, 1930 H. C. Benke found a few plants of this species on the outskirts of La Porte, La Porte County. Since this is a western species, I am regarding it as a waif until there are additional reports.

Mo. and Kans., southw. to Tex. and N. Mex. Adventive at Easton, Pa.

610. SOLIDÀGO ARGÙTA Ait. Reported by several of our early Indiana authors instead of *Solidago juncea* Ait. which was not recognized at that time. In 1927 it was reported by Pepoon from Lake and Porter Counties. Buhl and Fassett write that the Pepoon report should be transferred to *Solidago patula.* Doubtless all reports should be transferred to some other species. General distribution not definitely known.

Maine to Ont., southw. to Ohio, N. C., Tenn., and Ala.

611. SOLIDAGO FISTULÒSA Ait. This species was reported by Young from Jefferson County as *Solidago pilosa* Walt. Since there is no confirming specimen I refer this report to some other species.

N. J. to Fla. and La.

612. SOLIDAGO GLOMERÀTA Michx. Reported from the "knobs near New Albany" by Riddell for Clapp in "Supplement to Ohio Plants," page 28, 1836. I have Dr. Clapp's copy of Riddell's "Flora of the Western States" in which he records that he found it September 17, 1834. His specimen is in the herbarium of Wabash College and I refer it to *Solidago erecta* Pursh.

Cliffs and rocky woods, Blue Ridge, N. C. and Tenn.

613. **Solidago graminifòlia** (L.) Salisb. This species was reported frequently by the earlier authors but as now understood it belongs to the area east and north of Indiana. Some Indiana specimens, however, may belong to this species.

N. S. to N. Y., and westw. to Mich.

614. SOLIDAGO MISSOURIÉNSIS Nutt. Reported from Indiana but I am referring all reports to *Solidago glaberrima* Martens.

S. Dak., Colo., Oreg., and Wash. (Rydberg), although Nuttall gives Ark.

615. SOLIDAGO ODÒRA Ait. (*Solidago suaveolens* Schoepf.) Reported from Indiana by Blatchley, Editors of the Botanical Gazette, and Schneck. Since this species, as now understood, does not occur in Indiana, reports for it should be referred to other species.

N. S., N. Y. to Mo., southw. to Fla., Tex., and Okla.

616. **Solidago perglàbra** Friesner. This is a species that Friesner segregated from the section *Euthamia*. I believe this section has been divided too much. I think the specimens I have seen can safely be referred to *Solidago media* or some may be the true *Solidago graminifolia*. More definite data must be at hand before the *Euthamia* species can be separated with certainty.

W. Va., Mich., Ind. and Ill.

617. SOLIDAGO PETIOLÀRIS Ait. Reported from Clark County by Baird & Taylor and from Jefferson County by Barnes, J. M. Coulter, and Young. There are no verifying specimens. Since this species has a range to the south and southwest of Indiana, it is probable it was confused with one of the squarrose-bracted species.

N. C., s. Ill., Mo., and Kans., southw. to Fla. and Tex.

618. SOLIDAGO PUBÉRULA Nutt. Reported by Young from Jefferson County. The report should be referred to some other species.

Que., southw. to Pa. and Fla., chiefly near the coast.

619. SOLIDAGO RÁDULA Nutt. This species was reported from Jefferson County by Barnes and by Coulter for Young. It was reported from Marshall County by Hessler in 1896 and by Clark in 1920, who writes: "A few plants on the east side of Lake Maxinkuckee." I found specimens so labeled from Jasper County but all the specimens I saw in herbaria are *Solidago rigida*. Since this species has a range far to the west of Indiana I think it can be safely excluded.

Sw. Ill., Kans., southw. to La. and Tex.

620. SOLIDAGO RÁNDII (Porter) Britt. This species was reported by McDonald from St. Joseph County. If this report is based upon Nieuwland's no. 2260, labeled *Solidago Randii*, it should be referred to *Solidago rugosa* var. *aspera* (Ait.) Fern.

Maine, and in the mts. to Va., and westw. to Mich.

621. **Solidago rupéstris** Raf. Reported from Clark and Floyd Counties. Riddell in his "Supplement to Ohio Plants" published in 1836, on page 36 says: "A plant 16-18 inches high, flowering in September, found in rocky situations on the north bank of the falls of the Ohio." I have a book in which Dr. Clapp recorded that he also found it on the north shore of the falls.

Britton and Brown (Illus. Flora, ed. 2) refer this species to *Solidago canadensis* L. I have not seen a specimen.

W. Va., Ky., and Ind.

622. SOLIDAGO SHÓRTII T. & G. This species was described from specimens collected by C. W. Short in 1840 on Rock Island, one of the islands of the falls of the Ohio, which is located in about the middle of the Ohio River south of Clarksville, Clark County, Indiana. Since the southern boundary of Indiana is low water mark of the north side of the Ohio River, Rock Island is technically in Kentucky and this species must be excluded because it has never been found in Indiana. There are three specimens in

the Gray Herbarium and two specimens in the herbarium of the New York Botanical Garden and possibly others elsewhere. So far as I can learn this species is known only from Rock Island.

623. SOLIDAGO TENUIFÒLIA Pursh. Since this species, as now understood, has a range to the east of Indiana, reports of it should be referred either to *Solidago media* or to *Solidago remota.*
N. S. to Fla.

624. SOLIDAGO ULIGINÒSA Nutt. Reported mostly by our early authors but I am referring all reports to *Solidago uniligulata.* As I understand this species it does not come as far south as Indiana in our longitude.
Newf. to the mts. of N. C., westw. to Mich. and Minn.

625. **Aster amethỳstinus** Nutt. AMETHYST ASTER. This is an ambiguous aster with a wide range but of only local occurrence. It has been found to be a hybrid of *Aster novae-angliae* and *Aster multiflorus* (of our manuals). See Rhodora **41**: 190-192. 1939. I have a specimen from Massachusetts which is undoubtedly this species. I also have in my exchange from other states, specimens so labeled which evidently belong to this species. It was reported many years ago from Steuben County, by Bradner, but I believe this record can safely be ignored. It was reported from Lake County in 1930 by Peattie, who says "It has been collected rarely near Clarke." I have not been able to see a specimen, and following the rule of excluding all species unless I know of an authentic specimen, I exclude it.
Mass., N. Y., Pa., Ill., Iowa, and Nebr.

626. ASTER ANGÚSTUS (Lindl.) T. & G. This is a western and northern aster which is spreading eastward. According to Gray's Manual, ed. 7, it has reached Chicago. Peattie reports it from the dune area but cites no specimen. It seems that Peattie reported species for which he had no verifying specimen and for that reason, I exclude it until I see or learn of an authentic specimen that was found in Indiana.

627. **Aster divaricàtus** L. This species has been reported from Clark, Monroe, Noble, and Porter Counties. It may occur in Indiana, but I have not seen a specimen. Buhl (Amer. Midland Nat. **16**: 252. 1935) refers Peattie's report from Porter County to *A. furcatus.*
Que. to Man., southw. to Ga. and Tenn.

628. ASTER IMPERIÀLIS M'Murtrie. (M'Murtrie. Sketches of Louisville including a Florula Louisvillensis, p. 213. 1819.) This species was described from a single specimen found by its author on the bank of Blue River (probably on the boundary between Harrison and Crawford Counties). Apparently the description applies to some species of *Erigeron.* Since the species is in doubt, and in the absence of a specimen, it is dropped.

629. **Aster laterifloòrus** var. **glomerèllus** (T. & G.) Burgess. This I regard as an ecological form of the species and place it in the synonomy of the species. Reported from Porter County by Peattie.

630. **Aster Lowrieànus** Porter. This species was reported from Monroe County by Andrews but he preserved no specimen. It occurs in Ohio and probably in Indiana.

Conn. to Ont., southw. to N. C. and Ky.

631. ASTER NÒVI-BÉLGII L. NEW YORK ASTER. This aster has been reported from three counties. Since this species belongs to the Atlantic Coastal Plain, it is evident that these reports should be transferred to some other species.

Newf., Maine to Ga., mainly near the coast.

632. **Aster pilòsus** var. **Prínglei** (Gray) Blake. This variety was reported from Lake County, but I have not seen a specimen. I believe a depauperate specimen of the species has been mistaken for the variety.

633. **Aster polyphýllus** Willd. This species was reported to have been found by Hill near Whiting. I have seen his specimen, which is in the herbarium of DePauw University, and it is not this species.

Maine, Ont., and Wis., southw. to Pa. and N. C.

634. ASTER TENUIFÒLIUS L. Reported from a few counties by our early botanists. Since the species is restricted in its distribution to the Atlantic coast doubtless all reports should be referred to some other species.

Salt marshes from Mass. to Fla.

635. ASTER TRADESCÁNTI L. K. M. Wiegand, who has made an intensive study of the group of asters of which this species was considered a part, writes that this species was so indefinitely defined that its description can not be applied without doubt. Consequently, he proposes to drop the name and refer the plate at least in part to *Aster lateriflorus*. It has been reported from all parts of the state.

636. ASTER TRADESCANTI var. FOLIÒSUS (Ait.) Gray. This variety was reported from Porter County by Peattie. Since the application of the name is in doubt, and I have not seen a specimen, the report is ignored. Buhl (Amer. Midland Nat. **16**: 252. 1935) refers this report to *A. Tradescanti,* which name is dropped in this treatment.

637. ASTER TURBINÉLLUS Lindl. Reported from Monroe County by Andrews, but there is no specimen. The species, as understood by Burgess and Gray, has a range west of Indiana.

Prairies of Ill. to Kans., southw. to La.

638. **Aster vimíneus var. foliòsus** (Ait.) Gray. This variety was reported from Franklin County by Meyncke, and from Porter County by Pepoon. The status of the variety is questioned and Wiegand says "the standing of the variety is not entirely clear." I think it best to drop it for the present.

639. ERÍGERON ÀCRIS L. Reported from Monroe County by Andrews. No doubt this should be referred to some other species.

Lab. to Alaska, southw. to Maine, Ont., and in the Rocky Mts. to Colo. and Utah.

640. ERIGERON VÉRNUS (L.) T. & G. Reported from Monroe County by Andrews. This is a marsh plant of the southeastern part of the United States, and no doubt the report should be referred to some other species.

Va. to Fla. and La.

641. PLÙCHEA CAMPHORÀTA (L.) DC. Reports for this species should be referred to *Pluchea petiolata* Cass.

Salt marshes along the Atlantic coast from Mass. to Fla. and along the Gulf to Tex. and Mex.

642. PLUCHEA FOÉTIDA (L.) DC. Reports for this species also should be referred to *Pluchea petiolata* Cass.

Swamps along the Atlantic coast from N. J. to Fla., and along the Gulf to Tex. Mainly near the coast.

643. ANTENNÀRIA OCCIDENTÀLIS Greene. I reported this species from Cass County but I am now referring my specimen to another species. Greene referred to this species a specimen collected in Lake County by Moffatt which is now in the National Herbarium. I have not seen it. Lyon and Peattie both report it but I have not seen their specimens. Fernald (Rhodora 38: 229. 1936) gives the range of this species to the west of Indiana.

644. **Antennaria Wílsonii** Greene. This species was described by Greene from a specimen collected by Wilson near Cold Creek in Hamilton County in June, 1911. I have not seen the specimen. From the description, I believe it to be a form of *Antennaria neglecta* Greene.

645. SÍLPHIUM ASTERÍSCUS var. LAEVICAÚLE DC. This form was reported from Montgomery County by Grimes. I have seen the specimen and I am referring it to the alternate-leaf form of *Silphium integrifolium*. The range of this species as now known is to the south of our area.

646. **Silphium terebinthinàceum** var. **pinnatífidum** (Ell.) Gray. This variety has been reported from Hamilton County and by Peattie from the dune area. Doubtless these reports should be referred to *Silphium laciniatum*. This seems to be a drastic disposal of these reports but once I investigated the report of a botanist of this variety and found it was the last named species.

Ohio, Tenn., and Ala.

647. **Xánthium americànum** Walt. The synonomy of the species of *Xanthium* is so involved that the application of names to forms is extremely uncertain. In the absence of specimens of this species and the following excluded ones, it is impossible to say to which of our two native species they belong.

648. **Xanthium canadénse** Mill. This species has been reported from 10 counties and is referred to one of our native species.

649. **Xanthium commùne** Britt. This species is referred as in the preceding.

650. XANTHIUM ECHINÀTUM Murr. This species has been reported from 4 counties and is referred to one of our native species.

651. **Xanthium púngens** Wallr. This species I refer to *Xanthium pennsylvanicum* Wallr.

652. **Xanthium strumàrium** L. This species has been reported from 15 counties and I refer it, also, to one of our native species.

653. **Heliópsis scàbra** Dunal. Reported from all parts of the state. I believe these specimens have been so named on account of the roughness of the leaves. This character alone, however, is not sufficient to separate them from *Heliopsis helianthoides* to which I think all of our Indiana reports should be referred. This species is western and our native species insensibly grades into it. The range of *Heliopsis scabra* is given as follows.
Maine, Man. to B. C., southw. to N. J., Tenn., and Ark.

654. RUDBÉCKIA SPECIÒSA Wenderoth. SHOWY CONEFLOWER. This species is interpreted differently by authors, and in consequence, the range of the species differs. Boynton & Beadle (Small's "Flora of the Southeastern States," 1903) give the range as "Pa. to Va. and N. C." In Gray's Manual, ed. 7, it is given as "N. J. and Pa. to Ga. and Mo." In Britton and Brown's Illus. Flora, ed. 2, it is given as "N. J. to Mich., south to Ala. and Ark." All of the reports for this species, no doubt, should be referred to some other species.

655. **Heliánthus altíssimus** L. E. E. Watson so named specimens I collected in Lagrange County which I am now referring to *Helianthus giganteus* L.

656. **Helianthus ambíguus** (T. & G.) Britt. I reported this species upon the basis of specimens so named for me by E. E. Watson. I now refer three of them to *Helianthus hirsutus* Raf. and two of them to *Helianthus strumosus* L.

657. **Helianthus ámbulans** Watson. I am referring my specimens so named by Watson to *Helianthus strumosus* L.

658. **Helianthus arenícola** Watson. Several of my specimens, so named by Watson, I am now referring to other species.

659. HELIANTHUS ATRÓRUBENS L. This species was reported from Lake County by Peattie. Buhl writes that a confirming specimen is lacking. Since the range of this species is far to the southeast of our area, I believe this report can safely be ignored.

660. **Helianthus boreàlis** Watson. I reported this species from Lagrange and Steuben Counties upon the authority of Watson. I now refer these specimens to other species.

661. **Helianthus exasperàtus** Watson. I reported this species from La Porte and Warren Counties. I now refer these specimens to other species.

662. **Helianthus gigantèus** var. **microcéphalus** Peattie. Peattie reported this variety from Lake County. I have not seen his specimen but doubtless it is only a depauperate specimen of this variable species.

663. **Helianthus glaúcus** Small. Watson referred all of my specimens which I had named *Helianthus microcephalus* T. & G. to this species. He writes that the small-flowered sunflower of the interior should be known as *Helianthus glaucus* and that the Coastal Plain form is the real *Helianthus microcephalus*. I believe that until the genus is better understood it is best not to recognize this species, at least as a species.

664. **Helianthus instábilis** Watson. Watson named many of my specimens this species. I now refer them all to *Helianthus grosseserratus* Martens.

665. HELIANTHUS LAETIFLÒRUS Pers. Watson referred several of my specimens to this species. I am excluding the species from our flora and referring specimens so named to the yellow flowered form of *Helianthus rigidus* (Cass.) Desf.

666. **Helianthus leptocaúlis** (Wats.) Blake. I reported this species from two counties upon the authority of E. E. Watson. I now refer these specimens to other species.

667. HELIANTHUS TOMENTÒSUS Michx. This species has been reported by five of our early authors. Its range as now understood is to the southeast of Indiana.

668. **Helianthus tracheliifòlius** Mill. This species has been reported by six of our early authors. Since I have not been able to find a confirming specimen, it is excluded.

669. **Helianthus vírilis** Watson. The specimens so named for me by Watson I am now referring to *Helianthus hirsutus* Raf.

670. COREÓPSIS AURICULÀTA L. This species has been reported from Clark and Steuben Counties. The known distribution of this species is south of Indiana, but it may be found in the southern part of the state. Since there is no verifying specimen, it is excluded.

Va., Ky. to Ill., southw. to Fla. and La.

671. COREOPSIS MÀJOR Walt. Reported from Lake County without any data concerning its distribution. If the determination was correct, the plants were doubtless migrants, since the range is south of our area.

Va. to Ky., southw. to Fla. and Miss.

672. COREOPSIS TINCTÒRIA Nutt. Reported from St. Joseph County by McDonald as an escape along roadsides. Since it is not stated that it has become established, I believe it best to consider it as a temporary escape. Nieuwland says that it probably will not maintain itself. There is a specimen in the herbarium of Purdue University collected by Dorner in Tippecanoe County, which is labeled, "Escape."

Minn. to Alberta, southw. to Nebr., Ariz., La., and Tex.

673. BÌDENS LAÈVIS (L.) BSP. This is an Atlantic coast species but there are reports for it from some of the Mississippi Valley States. It has been reported from Indiana several times but I have seen no specimens. I have examined the Barnes specimen from Jefferson County, which is in the herbarium of Purdue University, and it is *Bidens cernua.*

Mass. to Ga., cent. N. Y. and Calif.

674. BIDENS MÌTIS (Michx.) Sherff. This species was reported from Jefferson County by Young under the name of *Coreopsis arguta* Pursh. It is restricted in its distribution to the southeastern part of the United States, and since there is no specimen, it is excluded.

675. CHRYSÁNTHEMUM LEUCÁNTHEMUM L. Reported from all parts of the state. Doubtless all these reports should be referred to the variety which is found throughout the state.

Nat. of Eu.; Newf. and e. Que. to N. J.; rare southw.

676. CHRYSANTHEMUM PARTHÈNIUM (L.) Bernh. FEVERFEW. This plant was formerly cultivated in gardens on account of its medicinal qualities. It sparingly escaped, but, since there are no data to prove that it has maintained itself, it is excluded.

Nat. of Eu.; N. B. and Ont., southw. to N. J. and Ohio; also in Calif.

677. ARTEMÍSIA ABRÓTANUM L. SOUTHERNWOOD. This species has been reported from Hamilton, Jefferson, and Monroe Counties. Since there is no evidence that this species has escaped and has established itself, it is excluded and regarded only as a garden escape. I have seen the Hamilton County specimen and it is *Artemisia biennis.*

Nat. of Eu.; Mass. to s. Ont., N. Y., and Nebr.

678. ARTEMISIA CANADÉNSIS Michx. Reported from the dune area of Lake and Porter Counties. As now understood, it is a form of *Artemisia borealis* and belongs to the Hudson Bay Region. Reports for it from Indiana should be referred to *Artemisia caudata.*

679. ARTEMISIA CARRÙTHII Wood. (*Artemisia kansana* Britt.) Pepoon reports that in 1899 Umbach found a colony of this plant along the B. & O. Railroad near Miller, Lake County. Peattie reports it, doubtless referring to the same report. Since there are no additional data I regard it as a railroad migrant. In 1935 N. C. Fassett reported that there was no specimen in the Umbach herbarium at the University of Wisconsin.

Mo., Colo., Utah, and Tex.

680. ARTEMISIA LONGIFÒLIA Nutt. Pepoon reports that Umbach collected this species along the Pennsylvania Railroad at "Clarke Junction," Lake County (now near the intersection of Clark and Fourth Streets, Gary). Since there is no evidence of its persistence, it is excluded. In 1935 N. C. Fassett reported that there was no specimen in the Umbach herbarium.

Man., Idaho, Colo., and Wash.

681. ARTEMISIA LUDOVICIÀNA Nutt. Reported from Lake County, and in 1921 I found a few colonies along the railroad about a mile and a half

southwest of Plymouth, Marshall County. I visited this colony a few years later, and it was spreading but still on the right of way. It is well established here and will persist unless it is destroyed. I exclude it until there are more records of its occurrence.

Minn., Utah, southw. to Tex. and Ariz.

682. **Senècio aúreus** var. **semicordàtus** (Mack. & Bush) Greenman. This variety was reported by Buhl (Bull. Chicago Acad. Science 5: 9. 1934) from Lake County, Indiana, upon the authority of Greenman. Buhl was in error, since Greenman cited a Lake County, Illinois, specimen (Ann. Missouri Bot. Gard. 3: 130. 1916).

683. **Senecio obovàtus** var. **umbratìlis** Greenman. The type specimen of this variety was collected by Clapp in the vicinity of New Albany and is deposited in the Gray Herbarium. Fernald (Rhodora 23: 29. 1921) refers this variety to *Senecio pauperculus* var. *Balsamitae* (Muhl.) Fern.

684. SENECIO PALÚSTRIS (L.) Hook. This species was reported from Clay County by Coulter (Proc. Indiana Acad. Sci. 1896: 166. 1897). As now understood, it is northern in its distribution and reaches the United States only in the northwest.

Lab. to Alaska, southw. to n. Wis., N. Dak., and Iowa.

685. ECHINOPS SPHAEROCÉPHALUS L. COMMON GLOBETHISTLE. This plant was reported by McDonald as being found at Chain Lakes, St. Joseph County. Since this is our first report, and no data are given concerning its establishment, I am regarding it as a casual garden escape. Like many other garden plants, however, it may escape and become a permanent part of our flora. Paul C. Standley informs me that it is well established in Kankakee County, Illinois.

Nat. of Eu.; our manuals give no data concerning its distribution in the U. S.

686. ARCTIUM LÁPPA L. GREAT BURDOCK. This species has been reported from many counties, but I believe all reports should be referred to *Arctium minus*. In the absence of confirming specimens, it is excluded from our flora.

Nat. of Eu.; N. E. and east central states, possibly farther westward.

687. CÍRSIUM PÙMILUM (Nutt.) Spreng. (*Cirsium odoratum* (Muhl.) Britt. of Britton and Brown, Illus. Flora, ed. 2.) This thistle has been reported from the dune area, but I refer these reports to *Cirsium Hillii*. Collins reported it from Dearborn County and Coulter reported it from Marion County on the authority of Wilson. Both of these reports should go to some other species. This is regarded as a trans-Alleghenian species.

N. E. to Pa., Del., and N. C.

688. CIRSIUM SPINOSÍSSIMUM (Walt.) Scop. (*Cirsium horridulum* Michx. of Britton and Brown, Illus. Flora, ed. 2.) YELLOW THISTLE. Reported from Putnam County on the authority of MacDougal by Coulter, who later said the specimen should be referred to *Cirsium vulgare*. I re-

ported it from the Lower Wabash Valley for Schneck. This report was taken from Dr. Schneck's notes and a specimen was not seen, hence this report should be dropped.

Coastal Plain from Maine to Fla., and along the Gulf to Tex.

689. CIRSIUM UNDULÀTUM (Nutt.) Spreng. This species was reported from Lake County by Hill and by Higley & Raddin before *Cirsium Hillii* was known to these authors. The reports should be transferred to *Cirsium Hillii.*

Lake Huron to Assina. and Alberta, southw. to Kans., N. Mex., and Ariz.

690. SÍLYBUM MARIÀNUM (L.) Gaertn. (*Mariana mariana* (L.) Hill of Britton and Brown, Illus. Flora, ed. 2.) MILK THISTLE. In 1905 I found a single plant of this species in a truck garden on a farm 2 miles east of Bluffton, Wells County. It has been reported also from Jasper County by Welch, who says it was introduced in radish seed. Since this species is regarded as an occasional escape, it seems best so to regard our two reports.

Nat. of s. Eu.; an occasional escape from Ont. to Ala. and on the Pacific coast where it is naturalized in Calif.

691. CENTAÚREA CYÀNUS L. CORNFLOWER. We have this species reported from seven counties as a garden escape or without data. In 1937 I found it to be frequent in the railroad yards in east Goshen, Elkhart County.

Nat. of Eu. and the Orient.; Que. to Nebr., southw. to Va., and Calif.

692. CENTAUREA JÀCEA L. BROWN KNAPWEED. Reported as a garden escape in the Calumet District of Lake County. There is no definite information concerning it. In 1921 I collected a specimen in a pasture just east of Bluffton and in 1935, and after the area had been severely grazed for 14 years, I found it still plentiful.

Nat. of Eu. and Siberia; naturalized in various parts of N. A.

693. CENTAUREA MACULÒSA Lam. SPOTTED KNAPWEED. Hansen (Proc. Indiana Acad. Sci. **36**: 251. 1927) reported this species as a weed near Atlanta, Hamilton County. There is no information concerning how long it has been known in this area. It is a species we can expect to become established soon if it has not already done so.

Nat. of Eu.; Mass., Vt., Mich., Wis. to Minn., southw. to N. J. and Pa.

694. CENTAUREA MOSCHÀTA L. SWEET-SULTAN. This species was reported as occurring about Mineral Springs but the report lacked specific data.

Nat. of Asia; not yet reported in our manuals of botany.

695. CENTAUREA SOLSTITIÀLIS L. YELLOW STAR THISTLE. Reported as found in alfalfa fields in Dearborn County (Proc. Indiana Acad. Sci. **1905**: 175. 1906) and Gibson County (Proc. Indiana Acad. Sci. **33**: 215. 1924). No data are given as to how long it was found in the locality.

Nat. of Eu. and Algeria; Mass., Ont., Iowa, to Calif.; southw. to Fla.

696. CENTAUREA VOCHINÉNSIS Bernh. TYROL KNAPWEED. This species was found by Kriebel in Lawrence County in 1936. It was common for a distance of about a hundred yards along U. S. Highway 50, and an adjacent worn out field about 4 miles southwest of Bedford. Also reported from Notre Dame, St. Joseph County.

Nat. of Eu.; Mass. to Ont. and s. N. Y.

697. HYPOCHAÈRIS RADICÀTA L. Reported from St. Joseph County by McDonald as found at Notre Dame, where Nieuwland says that it is well established. In 1897 I found a specimen in a yard in Bluffton, but I have not seen a specimen since that time. Blatchley reported it from Monroe County where he found it on the campus of Indiana University in 1887.

Nat. of Eu.; Newf. to Ohio, southw. to N. J. and Pa., also in Colo. and on the Pacific coast.

698. APÁRGIA AUTUMNÀLE (L.) Hoffm. (*Leontodon autumnale* L. of Gray, Man., ed. 7.) Coulter reported this species for me from Wells County, but a reexamination of the specimen shows it to be *Hypochaeris radicata* L.

Nat. of Eu.; Newf. to Mich., southw. to Pa. and Ohio.

699. SÓNCHUS ULIGINÒSUS Bieb. I reported this species from Noble County (Proc. Indiana Acad. Sci. **1922**: 264. 1923). I am now referring the specimen to *Sonchus arvensis* var. *glabrescens* Guenth., Grab. & Wimm. (Rhodora **30**: 19. 1928.)

700. LACTÙCA HIRSÙTA Muhl. This species has not been correctly treated in our manuals. It has been confused with *Lactuca canadensis* from which it has been separated principally on the pubescence of stem and leaves. Fernald and Wiegand made a study of the two species (Rhodora **12**: 145-146. 1910) and found the length of the involucre, achenes, and pappus were the true characters to separate them. Too, the range of this species does not include Indiana.

Que. to Ala. and Tex., chiefly east of the Allegheny Mts., especially along the Coastal Plain.

701. LACTUCA SATÌVA L. Hansen (Proc. Indiana Acad. Sci. **36**: 251. 1927) writes: "Near Anderson there is an infestation of a plant that appears to be a wild form of the common garden lettuce, *Lactuca sativa* L. On one farm where the plant infests about five acres of land and is very thick in places, the farmer considers it a bad weed." There is no other record of our garden lettuce becoming a weed and I believe this report should be referred to some other species.

702. LACTUCA VIRÒSA L. This species has been reported several times, and I believe authors who reported it have followed Britton and Brown's Illus. Flora, ed. 2, whose *Lactuca virosa* is our *Lactuca Scariola*. As I understand this species it has black, shining achenes and has not been found in Indiana but has been found in several places in the United States. See Dewey's discussion of this species and *Lactuca Scariola* and its variety in Rhodora **7**: 12. 1905.

703. PRENÁNTHES SERPENTÀRIA Pursh. Reported from Clark, Jefferson, and Steuben Counties by early authors. It probably does not come into our area. In the absence of verifying specimens our records are referred to other species.

Mass. to Fla. and west. to Ky. and Miss.

704. HIERÀCIUM MARIÀNUM Willd. Reported from the dunes of Lake and Porter Counties by Pepoon, upon the basis of Umbach's specimens. Fassett (Rhodora 35: 201. 1933) says the Umbach specimens should be referred to *Hieracium Gronovii* L. and *Hieracium scabrum* Michx.

N. H. to Ohio, southw. to Miss.

705. HIERACIUM SCRÍBNERI Small. Small, in his "Flora of the Southeastern United States," includes Indiana in the range of this species, but I have no other data concerning its occurrence in Indiana.

Blue Ridge Mts. to Ind., southw. to Ga. and Ala.

225a. GYPSÓPHILA MURÀLIS L. A single large specimen of this garden species was reported by Hull (Amer. Botanist 44: 162. 1938) as found along the Hobart Road north of East Gary, Lake County. This is a garden escape.

Nat. of Eu.

581a. **Richárdia scàbra** L. MEXICAN-CLOVER. I am indebted to H. A. Gleason for calling my attention to this species. He found that it was included in Small's "Flora of the Southeastern States" as found in Indiana. Upon investigation I found that A. A. Hansen received a fragmentary specimen from Henry County and that he wrote N. L. Britton about it on Sept. 12, 1922. A fragmentary specimen is now deposited in the Gray Herbarium, bearing the data, "Henry County, Sept. 15, 1922."

Adv. from the tropics; N. C. to Ark., southw. to W. I., Mex., and Argentina.

SUMMARY OF SPECIES, VARIETIES, FORMS, AND HYBRIDS

In this tabulation, plants represented in Indiana by only a variety are listed as species.

Families	Genera		Species		Varieties		Forms		
	Native	Introduced	Native	Introduced	Native	Introduced	Native	Introduced	Hybrids
PTERIDOPHYTA									
Ophioglossaceae	2	......	6	......	3	......	1	......	......
Osmundaceae	1	......	3	......	......	......	3	......	......
Polypodiaceae	16	......	31	......	5	......	7	......	4
Salviniaceae	1	......	1	......	......	......	......	......	......
Equisetaceae	1	......	8	......	......	......	......	......	......
Lycopodiaceae	1	......	5	......	......	......	......	......	......
Selaginellaceae	1	......	2	......	......	......	......	......	......
Isoëtaceae	1	......	1	......	......	......	......	......	......
SPERMATOPHYTA									
GYMNOSPERMAE									
5. Taxaceae	1	......	1	......	......	......	......	......	......
6. Pinaceae	6	......	9	......	......	......	......	......	......
ANGIOSPERMAE									
Monocotyledoneae									
8. Typhaceae	1	......	2	......	......	......	......	......	......
10. Sparganiaceae	1	......	4	......	1	......	......	......	......
11. Potamogetonaceae	2	......	22	1	3	......	1	......	......
12. Najadaceae	1	......	3	......	......	......	......	......	......
14. Juncaginaceae	2	......	3	......	......	......	......	......	......
15. Alismaceae	4	......	11	......	1	......	3	......	......
17. Hydrocharitaceae	2	......	3	......	......	......	......	......	......
19. Gramineae	52	10	171	40	16	1	3	......	......
20. Cyperaceae	15	......	215	......	23	......	6	......	4
23. Araceae	5	......	7	......	......	......	......	......	......
24. Lemnaceae	4	......	10	......	......	......	......	......	......
29. Xyridaceae	1	......	2	......	......	......	......	......	......
30. Eriocaulaceae	1	......	1	......	......	......	......	......	......
33. Commelinaceae	2	......	8	......	......	......	3	......	......
34. Pontederiaceae	2	......	3	......	......	......	2	......	......
36. Juncaceae	2	......	26	......	4	......	5	......	......
38. Liliaceae	20	5	42	9	3	......	3	......	......
40. Amaryllidaceae	3	......	3	......	......	......	......	......	......
43. Dioscoreaceae	1	......	4	......	......	......	......	......	......
44. Iridaceae	2	1	7	1	......	......	......	......	......
50. Orchidaceae	16	1	39	1	1	......	......	......	......
Dicotyledoneae									
52. Saururaceae	1	......	1	......	......	......	......	......	......
56. Salicaceae	2	......	23	3	5	1	......	......	......
57. Myricaceae	1	......	1	......	......	......	......	......	......
60. Juglandaceae	2	......	11	......	7	......	2	......	......
61. Betulaceae	5	......	10	......	2	......	2	......	1
62. Fagaceae	3	......	20	......	5	......	3	......	9

Families	Genera		Species		Varieties		Forms		
	Native	Intro-duced	Native	Intro-duced	Native	Intro-duced	Native	Intro-duced	Hybrids
63. Ulmaceae	2		7		1				
64. Moraceae	2	2	2	4					
65. Urticaceae	5		6	1	1				
67. Loranthaceae	1		1						
69. Santalaceae	1		1						
74. Aristolochiaceae	2		4						
77. Polygonaceae	3	1	24	9	2		2		
78. Chenopodiaceae	4	2	12	9	3	2	2		
79. Amaranthaceae	3	2	6	7					
80. Nyctaginaceae	1		1						
83. Phytolaccaceae	1		1						
84. Aizoaceae		1		1					
85. Portulacaceae	2	1	2	1					
87. Caryophyllaceae	6	5	14	16	4		1	1	
88. Nymphaeaceae	5		6						
89. Ceratophyllaceae	1		1						
91. Ranunculaceae	16	1	40	6	1				
93. Berberidaceae	4		4	2			1		
94. Menispermaceae	3		3						
95. Magnoliaceae	2		2						
98. Annonaceae	1		1						
102. Lauraceae	2		2		1				
104. Papaveraceae	2	1	2	1					
104A. Fumariaceae	3		5						
105. Cruciferae	11	12	31	25	2	1			
107. Capparidaceae	1		2						
110. Sarraceniaceae	1		1						
112. Droseraceae	1		2						
115. Crassulaceae	2		3	1					
117. Saxifragaceae	7		11		6				
117A. Grossulariaceae	2		4						
123. Altingiaceae	1		1						
123A. Hamamelidaceae	1		1						
124. Platanaceae	1		1						
126. Rosaceae	19	1	90	7	23		5		3
128. Leguminosae	26	6	68	20	16		4		
129. Geraniaceae	1		4	2	1				
130. Oxalidaceae	1		5	1	3		4		
132. Linaceae	1		5						
137. Rutaceae	2		2		1				
138. Simarubiaceae		1		1					
145. Polygalaceae	1		8						
147. Euphorbiaceae	5	1	18	8	1				
148. Callitrichaceae	1		2						
152. Limnanthaceae	1		1						
153. Anacardiaceae	1		6		2				1
157. Aquifoliaceae	2		3		1				
158. Celastraceae	2		4						
161. Staphyleaceae	1		1						
163. Aceraceae	1		5		4		3		

	Families	Genera		Species		Varieties		Forms		
		Native	Intro-duced	Native	Intro-duced	Native	Intro-duced	Native	Intro-duced	Hybrids
164.	Hippocastanaceae	1		2	1			1		
165.	Sapindaceae		1		1					
168.	Balsaminaceae	1		2						
169.	Rhamnaceae	2		5	1	1				
170.	Vitaceae	3		9		2		1		
174.	Tiliaceae	1		2						
175.	Malvaceae	4	2	7	6					
187.	Hypericaceae	2		18	1	2				
193.	Cistaceae	3		9						
198.	Violaceae	2		21	3	4		1		12
203.	Passifloraceae	1		2						
210.	Cactaceae	1		1						
214.	Thymelaeaceae	1		1						
215.	Elaeagnaceae	1		1						
216.	Lythraceae	6		6	1	2				
223.	Melastomaceae	1		2						
224.	Onagraceae	6		26	3					
225.	Haloragidaceae	3		6		1				
227.	Araliaceae	2		6						
228.	Umbelliferae	19	6	29	6	3				
229.	Cornaceae	2		11		2				
233.	Ericaceae	12		25		6		1		
237.	Primulaceae	7	1	13	2					
239.	Sapotaceae	1		1						
240.	Ebenaceae	1		1						
241.	Styracaceae	1		1						
243.	Oleaceae	2		8						
245.	Loganiaceae	1		1						
246.	Gentianaceae	6		14				1		
247.	Apocynaceae	3	1	6	1	5				
248.	Asclepiadaceae	4		17				1		
249.	Convolvulaceae	3	1	14	5	3		1		
250.	Polemoniaceae	2	1	10	1	8				
251.	Hydrophyllaceae	3		8						
252.	Boraginaceae	6	3	11	7					
253.	Verbenaceae	2		6	1	1				3
254.	Labiatae	19	7	49	18	7	1			
256.	Solanaceae	2	3	10	4					
257.	Scrophulariaceae	21	4	45	12	12		2		
258.	Bignoniaceae	3		3	1					
260.	Martyniaceae	1		1						
261.	Orobanchaceae	3		5						
264.	Lentibulariaceae	1		8						
266.	Acanthaceae	3		4		1		1		
268.	Phrymaceae	1		1						
269.	Plantaginaceae	1		7	1	1			1	
270.	Rubiaceae	6		19	1	4		1		
271.	Caprifoliaceae	7		19	2	5		6		
273.	Valerianaceae	2		4	1					
274.	Dipsacaceae		1		1					

Families	Genera		Species		Varieties		Forms		
	Native	Intro-duced	Native	Intro-duced	Native	Intro-duced	Native	Intro-duced	Hybrids
275. Cucurbitaceae	3	......	3	......	......	......	......	......	......
276. Campanulaceae	2	......	5	1	1	1	......	......	......
276A. Lobeliaceae	1	......	6	......	4	......	2	......	......
280. Compositae	46	18	202	44	52	6	10	1	1
	587	103	1,838	302	279	13	95	3	38

The result of the study of the Indiana flora is shown in the following table.

Flora of 1881 lists 1,194 native species; 140 introduced.
Flora of 1900 lists 1,400 native species; 177 introduced; 188 excluded.
Flora of 1940 lists 1,838 native species; 302 introduced; 707 excluded.

To the last flora should be added 292 varieties and 98 forms.

What a census of our flora will show 25 years hence is mere conjecture. I believe our native flora will never exceed 1,900 native species. Some of the present species may be reduced in rank to varieties while some varieties may be elevated to species but the number of new native species discovered will be few. There are a number of species found in Michigan just north of our border which may be found in Indiana. Doubtless there are some southern species as yet undiscovered in the unglaciated region. The introduction of foreign plants will steadily increase, and western species will become established because of the interstate highway traffic. Our pure seed law will lessen introduction in grass and grain seed but I believe we already have many species established that came to us in cheap imported seed during the World War that have not been discovered. A rich field to botanize for foreign and western plants will be railroads, highways, land about factories, cemeteries, and tourist camps.

The study for this flora is based upon Indiana specimens or duplicates seen in the following private and public herbaria.

Banta, Edna	1,018
Butler University	9,347
DePauw University	3,736
Field Museum of Natural History	704
Franklin College	326
Hermann, Frederick J.	804
Illinois (University of)	231
Indiana University	3,710
Kriebel, Ralph M.	1,719
Lyon, Marcus W., Jr.	972
McCoy, Scott	1,986

McKee, Madge	840
National Museum	1,534
New York Botanical Garden	86
Northwestern University	25
Notre Dame (University of)	2,231
Oberlin College	27
Purdue University	3,961
Tryon, Rolla M., Jr.	27
Wabash College	2,677
Weatherwax, Paul	332
Wisconsin (University of)	643

This list may be divided into two groups, private herbaria and public herbaria, to which must be added the number of Indiana specimens now in the Deam herbarium in order to complete the total of specimens examined in the preparation of this flora. These totals are as follows:

Private herbaria	7,698
Public herbaria	29,238
Deam herbarium	47,648
Total number of specimens examined	84,584

Note for the 2020 Printing

The following sections are not printed here. They are available on BlackburnPress.com under Flora of Indiana

- **Lists of new forms and new combinations**
- **Names of collecting places not in current use**
- **List of Indiana collectors whose specimens have been seen or referred to in the Flora**
- **Bibliography**

GLOSSARY[1]

Glossary of words used in the botanical keys with a definition of the meaning assigned to them in the Flora.

Acaulèscent. Stemless or apparently so, or having the stem below the surface of the ground.

Accúmbent (cotyledon). Having the edges against the radicle.

Achène. A small, dry, hard, 1-celled, 1-seeded, indehiscent fruit.

Acícular. Slenderly needle-shaped.

Acùminate. Gradually tapering to a point.

Acùte. Ending in a point.

Ádnate. Having one organ attached wholly or in part to another.

Advèntive. Imperfectly naturalized.

Álternate. Not opposite.

Álveolate. Closely pitted.

Áment. A catkin or scaly spike; refers to inflorescences.

Amplèxicaul. Clasping the stem.

Anástomosing. Veins crossing so as to form a network.

Andrógynous (inflorescence). Composed of both staminate and pistillate flowers, with the staminate flowers above the pistillate.

Ánnual. Of only one year's duration.

Annual (winter). A plant from autumn-sown seed which blooms and fruits the next year.

Aphyllópodic (*Carex*). Lower leaves bladeless or rudimentary.

Ánther. The part of a stamen containing the pollen.

Anthèsis. The time at which the flower expands and frees the pollen.

Apètalous. Without petals.

Apículate. Ending in a short, pointed tip.

Apprèssed. Lying flat against another organ.

Aráchnoid. Like a cobweb.

Árcuate. Moderately curved.

Arèolate. Having a network of small spaces, usually marked out by veinlets.

Arèola. One of the small spaces in an areolate surface

Arístate. Tipped with an awn.

Artículated. Jointed.

Ascènding. Growing obliquely upward, or upcurved.

Attènuate. Slenderly tapering.

Aúricle. An ear-shaped appendage.

Auriculate. With a basal lobe.

Awn. A bristle-shaped appendage.

Áxil. The angle formed by a leaf or branch with the stem.

Beaked. Ending in a prolonged tip.

Berry. A fruit with the entire pericarp fleshy.

Bidèntate. Two-toothed.

Biènnial. Of two years' duration.

Bipínnate. Twice pinnate.

Blade. The flat, expanded part of a leaf.

Bract. A diminutive leaflike structure subtending a flower or flower cluster or attached to divisions of an inflorescence.

Brácteole. A diminutive bract or a secondary bract.

Branch. As applied to woody plants, any division or subdivision from the stem except the growth of the season.

Branchlet. As applied to woody plants, the growth of the season.

Cálciphile. A plant reaching its optimum in a soil more alkaline than neutral.

Callósity. A small, hard protuberance.

Càlyx. The outer of two series of floral leaves, sometimes the inner series lacking.

Canalículate. Longitudinally channeled.

Canèscent. Hoary with a gray pubescence.

Cápillary. Hairlike.

Cápsule. A dry, dehiscent fruit of more than one carpel.

Cárpel. A simple pistil or one member of a compound pistil.

Castàneous. Chestnut color.

Caúdate. With a slender tail-like appendage.

Caúline. Belonging to the stem.

Cell. Any structure containing a cavity, as an anther, ovary, etc.

Cèspitose. Growing in tufts.

Chaff. A thin, dry scale.

Chartàceous. Papery in texture.

Cíliate. With marginal hairs.

Cíliolate. Minutely ciliate.

Circumscíssile. Dehiscing by a regular transverse circular line of division.

Clàvate. Having a terete, longitudinal body larger at one end than at the other; club-shaped.

Cleistógamous. Fertilized in the bud by its own anthers.

Cleft. Cut about halfway to the midvein.

[1]For those who wish a more complete and illustrated glossary, I recommend Lindley's "Illustrated Dictionary of Botanical Terms", 1848. Republished by Alice Eastwood, price 50 cents. Address California Academy of Science, San Francisco, Calif.

Clòne. A plant propagated vegetatively by cutting, budding, layering, or grafting.

Cólorless. Without distinct color, opaque or translucent.

Condùplicate. Folded together lengthwise.

Cónnate. Similar organs more or less united.

Cónvolute. Rolled up longitudinally.

Córdate. Heart-shaped.

Coriàceous. Leathery in texture.

Corólla. The inner of the two series of floral leaves.

Córymb. A convex or flat-topped flower cluster with the pedicels or rays arising from different points on the axis, with the marginal flowers blooming first.

Cotylèdon. A rudimentary leaf of the embryo.

Crènate. Scalloped with rounded teeth.

Cùcullate. Hooded, or resembling a hood.

Cúlm. The stem of grasses and sedges.

Cùneate. Wedge-shaped.

Cúspidate. Tipped with a sharp, rigid point.

Cỳme. A convex or flat-topped flower cluster with the central flowers unfolding first.

Decíduous. Falling away at the close of the growing season.

Decúmbent. Having the stem or branches on an incline with their growing ends erect.

Dehíscence. The opening of an ovary or anther sac to discharge its contents.

Dèltoid. Broadly triangular.

Dèntate. Toothed, with the teeth projecting outward.

Dentículate. Dentate but the teeth very small.

Depaúperate. Starved or smaller than normal size.

Dichótomous. Forking regularly into two nearly equal branches.

Diffùse. Loosely spreading.

Dioècious. Unisexual, with the two kinds of flowers on separate plants.

Disk. The enlargement of the receptacle at or around the base of the pistil; in *Compositae* the tubular flowers of the head as distinct from the ray flowers.

Dissècted. Cut or divided into numerous segments.

Distínct. Not united; separate; evident.

Divàricate. Separated by a wide angle.

Dórsal. Upon or relating to the back or outer surface of an organ.

Drupe. A simple fruit, usually indehiscent, with fleshy pericarp and the seed portion hard or bony.

Drùpelet. A diminutive drupe.

Ecológical. Concerning the relation of plants to their environment.

Ellíptic. Oval; in the form of an ellipse.

Ellípsoid. A solid body, elliptic in longitudinal section.

Emárginate. Notched at the apex.

Endógenous. Forming new tissue within.

Epígynous. Adnate to or borne on the summit of the ovary.

Epiphýtic. Growing on other plants, but not parasitic.

Eròse. As if gnawed.

Exógenous. Forming new tissue in layers outside the older tissue.

Exsèrted. Prolonged beyond surrounding organs.

Fálcate. Scythe-shaped.

Fárinose. Covered with meal-like powder.

Fáscicle. A dense cluster.

Fastígiate. Erect and close together.

Fíbrillose. Abounding with fine fibers.

Fílament. The stalk of a stamen which supports the anther.

Fíliform. Threadlike.

Filamèntose. Composed of threads.

Fímbriate. Fringed.

Fláccid. Lax; weak.

Flèxuous. Zigzag; bending alternately in opposite directions.

Foliàceous. Similar to leaves.

Fóllicle. A fruit with a single carpel dehiscing along one suture.

Flòret. A small flower such as one of a grass or sedge; one of a dense cluster.

Frond. The leaf of a fern.

Fruit. The seed-bearing product of a plant.

Gamopètalous. With the petals more or less united.

Gíbbous. Enlarged or swollen on one side.

Glàbrate. Almost without hairs.

Glàbrous. Devoid of hairs.

Glánd. A secreting cell, or group of cells; any protuberance having the appearance of such an organ.

Glándular. With glands or gland-like.

Glaúcous. Covered with a fine bluish or white bloom.

Glómerule. A dense capitate cyme.

Glúme. Bract at the base of the spikelet in grasses and sedges.

Gynaecándrous. In *Carex*, having a spike with the upper flowers pistillate and the lower ones staminate.

Hábit. General appearance of a plant.

Hábitat. A plant's natural place of growth.

Hástate. Like an arrowhead but with the basal lobes diverging.

Haustòria. The specialized roots of parasites.

Head. A dense cluster of sessile or nearly sessile flowers.

Herb. A plant with no persistent woody stem above the ground.

Hírsute. Pubescent with rather coarse or stiff hairs.

Híspid. Beset with rigid or bristly hairs.

Hỳaline. Thin and translucent, rarely transparent.

Hypánthium. An enlargement of the torus under the calyx.

Hypógynous. Borne at the base of the ovary.

Ìmbricate. Overlapping.

Impèrfect. Flowers with either stamens or pistils, but not with both.

Incìsed. Cut sharply and irregularly, more or less deeply.

Inclùded. Not projecting beyond the surrounding parts.

Incùmbent (cotyledon). Lying with the back of one against the radicle.

Indehíscent. Not opening.

Ìndurated. Hardened.

Indùsium. The membrane covering a sorus in ferns.

Infloréscence. The flowering part of a plant.

Innovàtion. An offshoot from a stem (grasses).

Ìnternode. The portion of a plant between two nodes.

Intrórse. Facing inward.

Ìntrodùced. Brought intentionally from another region, as for cultivation.

Ìnvolucre. A circle or collection of bracts surrounding a flower cluster or head, or a single flower.

Ìnvolute. Rolled inward.

Irrègular (flower). Having one or more of the organs of the same series unlike or unequal.

Keel. A central dorsal ridge; the two anterior united petals of a papilionaceous flower.

Làbiate. Provided with a liplike organ.

Lacíniate. Cut into narrow lobes or segments.

Lacùna. Small depression or pit on a surface.

Lánceolate. Several times longer than wide, tapering at both ends, widest about a third above the base.

Làtex. The milky sap of certain plants.

Lèaflet. One of the divisions of a compound leaf.

Lègume. The fruit of the *Leguminosae*, formed of a simple pistil.

Lèmma. The lower of the two bracts inclosing the flower in grasses.

Lígule. A thin appendage between the base of a leafblade and its sheath (grasses); the oblong appendage of the ray flowers in *Compositae*.

Línear. Elongated like a blade of grass, with nearly parallel sides and at least six times as long as wide.

Lòbed. Divided to about the middle.

Lòment. A jointed legume, the constrictions usually between the seed.

Membranàceous. Thin and semi-transparent.

Mídrib (midvein). The central rib or vein of a leaf or other organ.

Moníliform. Like a string of beads.

Monoècious. With stamens and pistils in separate flowers on the same plant.

Mùcronate. With a short, sharp, abrupt tip.

Nàtive. Indigenous to the area where it is found.

Náturalized. Not indigenous to the region where found, but so well established as to have become a part of the flora.

Node. The juncture of two internodes.

Nódulose. Knotty.

Nut. An indehiscent 1-seeded fruit with a hard or bony pericarp.

Nútlet. A diminutive nut.

Obcórdate. Inversely heart-shaped.

Oblánceolate. Inverse of lanceolate.

Óblong. Longer than broad and with nearly parallel sides.

Obòvoid. Inversely ovoid.

Obtùse. Blunt or rounded at the end.

Òcrea (*Polygonum*). The sheathing, united stipules.

Ochrèola (*Polygonum*). The ocreae subtending flowers.

Òvary. The part of a pistil containing the ovules.

Òvoid. Shaped like a hen's egg.

Òvule. The body which after fertilization becomes the seed.

Pàlea. The upper bract which, with the lemma, incloses the flower in grasses.

Pálmate. Diverging radiately like the fingers.

Panículate. A loose, irregular, compound inflorescence with pedicellate flowers.

Papilionàceous (*corolla*). Having a standard, wings, and keel, as in the peculiar corolla of many *Leguminosae*.

Pápillose. With minute, blunt projections.

Páppus. The bristles, awns, teeth, etc. surmounting the achene in *Compositae*.

Parasitic. Growing upon and deriving nourishment from another plant.

Pèctinate. Pinnatifid with narrow closely set segments; comblike.

Pèdicel. The support of a single flower.

Pedúncle. A primary flower stalk, supporting either a cluster or a single flower.

Pèltate. Shield-shaped; a flat organ with a stalk on its lower surface.

Perènnial. Lasting year after year.

Pèrfect (flower). Having both stamens and pistil.

Pèrianth. The sepals and petals considered collectively.

Pèricarp. The ripened wall of an ovary.

Perigýnium. The structure inclosing the achene in the genus *Carex.*

Pètal. One of the divisions of the corolla.

Pètiole. The support of a leaf.

Phaenógamous. Having flowers with stamens and pistils and producing seed.

Phyllópodic. With lower leaves of fertile culms normally blade-bearing, in the genus *Carex.*

Pìlose. With long, soft hairs.

Pínna. A primary division of a pinnately compound leaf.

Pínnate. Having leaves divided into leaflets or segments along a common axis.

Pinnátifid. Pinnately cleft to the middle or beyond.

Pístil. The seed-bearing organ of a flower, consisting of the ovary, stigma, and style when present.

Pístillate. With pistils, usually used to mean without stamens.

Plàno-cónvex. Flat on one side and curved on the other.

Plùmose. Resembling a plume or feather.

Póllen. The fecundating grains contained in the anther.

Polypètalous. Having separate petals.

Pome. A fleshy fruit of the apple type.

Procùmbent. Trailing or lying on the ground, but without rooting at the nodes.

Príckle. A spiny outgrowth from the bark or rind of a plant.

Próstrate. Lying flat on the ground.

Pubèrulent. Minutely pubescent.

Pubèscent. Provided with hairs.

Pulvèrulent. Powdered; appearing as if covered with minute grains of dust.

Púnctate. Dotted with depressions or with translucent glands or colored dots.

Racème. A simple inflorescence of pediceled flowers upon a common more or less elongated axis.

Rácemose. In racemes or resembling a raceme.

Rachílla. The axis of the spikelet in grasses.

Ràchis. The axis of a compound leaf, spike, or raceme.

Ray. One of the branches of an umbel; the flat marginal flowers in *Compositae.*

Recèptacle. The termination of the flower stalk, bearing the floral organs.

Recúrved. Curved downward or backward.

Reflèxed. Bent backward abruptly.

Règular. Having the members of each part alike in size and shape.

Rèniform. Kidney-shaped.

Retículate. In the form of a network.

Retrórse. Turned backward or downward.

Retùse. With a shallow notch at a rounded end.

Rèvolute. Rolled backward.

Rhìzome. A prostrate or subterranean stem, usually rooting at the nodes and becoming erect at the apex.

Root. The underground part of a plant which supplies it with nourishment.

Rootstock. Same as rhizome.

Rùgose. Wrinkled.

Rúgulose. Somewhat wrinkled.

Ságittate. Shaped like an arrowhead with the basal lobes directed downward.

Sámara. A simple, indehiscent, winged fruit.

Sáprophyte. A plant that grows on dead organic matter.

Scàbrous. Rough to the touch.

Scale. A minute, rudimentary or vestigial leaf.

Scape. A peduncle arising from the ground, naked or without proper foliage.

Scàrious. Thin, dry, and translucent, not green.

Scórpioid (inflorescence). Coiled up in the bud, unrolling in growth.

Sècund. Borne along one side of an axis.

Seed. The ripened ovule (non-technical definition). Used in the plural sense for any number of ripened ovules of the same species.

Seeds. The plural form refers to a collection of seed of more than one species.

Sèpal. One of the divisions of a calyx.

Sèptate. Provided with partitions.

Sèrrate. Having sharp teeth pointing forward.

Sèrrulate. Finely serrate.

Sèssile. Without a stalk.

Setàceous. Bristlelike.

Sètose. Bristly.

Sheath. A tubular envelope, as the lower part of a leaf in grasses.

Shrub. A woody perennial, usually with several stems.

Silíque. The name of certain fruits of the *Cruciferae.*

Sínuate. With strongly wavy margins.

Sìnus. The space between two lobes.

Sòrus (pl. *sori*). A heap or cluster, applied to the fruit dots of ferns.

Spàdix. A spike with a fleshy axis.

Spàthe. A bract, usually more or less concave, subtending a spadix.

Spìcate. Arranged in a spike or resembling a spike.

Spike. A simple inflorescence with the flowers sessile or nearly so upon an elongated common axis.

Spìkelet. A small or secondary spike; the characteristic unit of the inflorescence of a grass.

Spine. A sharp, woody or rigid outgrowth from the stem.

Sporángium. A spore case.

Spreading. Diverging nearly at right angles.

Spur. A hollow projection of a floral organ.

Squárrose. With the parts spreading or their tips recurved.

Stàmen. The organ of a flower which bears the pollen.

Stándard. The upper, usually broad, petal of a papilionaceous corolla.

Stèllate. Starlike.

Stem. The main ascending axis of a plant.

Stèrile. Without spores or without seed.

Stígma. The summit of a pistil to which pollen grains become attached.

Stìpe. The stalklike support of a pistil; the leaf stalk of a fern.

Stípitate. Provided with a stipe.

Stípule. An appendage at the base of a petiole, often adnate to it.

Stòlon. A basal branch rooting at the nodes.

Stoloníferous. Producing or bearing stolons.

Stramíneous. Straw colored.

Strìgose. With appressed or ascending, sharp, stiff hairs.

Style. The narrowed top of the pistil which connects the ovary to the stigma.

Stylopòdium. A disklike expansion at the base of the style, as in *Umbelliferae.*

Súbulate. Awl-shaped.

Súcculent. Soft and juicy.

Súlcate. Grooved longitudinally.

Supèrior (*ovary*). Free from the calyx.

Sùture. A line of dehiscence.

Téndril. A slender coiling organ by which climbing plants are attached to a supporting body.

Terète. Circular in cross section.

Tèrnate. Divided into three segments, or arranged in threes.

Throat. The orifice of a gamopetalous corolla or calyx.

Thýrsus. A compact panicle.

Toméntose. Densely pubescent with the hairs matted.

Tòrus. The receptacle of a flower.

Trígonous. Three-angled.

Trúncate. Ending abruptly as if cut off transversely.

Tùber. A thick, short, underground branch or part of a branch, having eyes or buds.

Tùbercle. The persistent base of a style in some *Cyperaceae.*

Tùberous. Having the character of a tuber; tuberlike in appearance.

Túrgid. Swollen or tightly drawn.

Ŭmbel. An inflorescence in which the peduncles or pedicels of the cluster arise from the same point.

Ŭmbellate. In or like an umbel.

Ŭmbellule. A secondary umbel.

Ŭmbonate. Bearing a stout projection in the center; bossed.

Ŭncinate. Hooked or in the form of a hook.

Ŭndulate. With wavy margins.

Ŭrceolate. Urn-shaped.

Ŭtricle. A bladderlike organ; a 1-seeded fruit with a loose pericarp.

Válvate. Meeting by the margins in the bud, not overlapping; dehiscent by valves.

Vein. A thread of fibro-vascular tissue in a leaf or other organ.

Velùtinous. Velvety; with a dense, fine pubescence.

Véntral. Belonging to the anterior or inner face of an organ; the opposite of dorsal.

Véntricose. Swelling unequally, or inflated on one side.

Vérrucose. Covered with wartlike elevations.

Verticillate. With three or more leaves or branches at a node; whorled.

Víllous. Bearing long, soft hairs.

Vírgate. Wand-shaped; slender, straight, and erect.

Víscid. Glutinous; sticky.

Woolly. With long and tortuous or matted hairs.

SOME HABITAT AND DISTRIBUTION TERMS USED

These terms are defined and discussed in the sense they are used in the flora. Many terms have been omitted from this list because their meaning seems obvious.

Abandoned fields. See fields.

Alluvial banks. See streams.

Ballast. See railroads.

Banks of streams. See streams.

Barrens. It is my opinion that this term was used by early authors and pioneers to designate remnant prairies. I have heard it applied to a relict prairie in Noble County, and to the black and scarlet oak areas of Floyd and Harrison Counties.

The subject seems to be of sufficient importance to warrant a few notations. The old buffalo trail passed through the last named barrens, crossed the Ohio River at the Ohio Falls and continued southward into Kentucky to the salt springs and big barrens near Bowling Green.

The Kentucky barrens were described in 1802 by Michaux. J. M. Coulter (Bot. Gaz. 2: 145-146. 1877) wrote of the "barrens" of southern Indiana and located them in the corners of Clark, Floyd, Harrison, and Washington Counties. In the same article he lists 22 species of plants which he collected in this area, and these are all essentially prairie plants. About 1920 I studied the region more critically. Three very old men who were born and had spent all of their lives in the neighborhood of the barrens said that when they were boys the barrens were covered with "scrub oak" and a few hickory trees and that thickets of hazel and wild plum fringed the sinkholes. They said the "scrub oak" were not tall enough to hide a man on horseback. They also directed me to a small area along the roadside about five miles southeast of Corydon which, in their unanimous opinion, had never been plowed. Here I found *Andropogon furcatus* and *Sorghastrum nutans,* typical prairie grasses. Prof. Fred Breeze accompanied me on a trip over this area and the big barrens near Bowling Green, Kentucky. He was of the opinion that the geological formations in both areas were similar.

Bayou lake. See slough.

Bluffs. See streams.

Bog. The terms bog and boggy places have been so widely and loosely applied that their meaning is not specific. I am restricting the term bog to areas where the surface soil is organic (peat) and the soil water is acid in contrast to habitats that have a mineral soil. Its application in the flora can be best understood when its formation and growth are given. A bog has its beginning when such vegetation as sedges and riparian aquatics begin to invade any body of water and form a mat over it. This is the first stage of the quaking bog. The mat stage of a bog is soon followed by the establishment of sphagnum, perennials, shrubs such as cranberries, swamp loosestrife, willows, dogwoods, highbush blueberry, alders, poison sumac, and lastly of tree species such as tamarack which is the principal tree in Indiana bogs. In Indiana, arborvitae, white pine, and *Chamaedaphne* are rarely the dominant woody species. In due time the surface of a bog builds up and becomes dry when the tamarack species begins to wane and low ground broadleaf species such as soft maples, yellow birch, white elm, and others take possession. These soon build up on top of the peat soil a muck soil which displaces the bog botanical area. Bogs are usually designated by the dominant species growing in them, such as sphagnum, tamarack, arborvitae, white pine, and *Chamaedaphne* bogs. Indiana bogs are mostly about lakes and along streams and have their water table at or near the surface. This is usually the level of the water in the adjacent lake or stream.

Branch. See streams.

Clearings. See woods.

Cliffs. See streams.

Common. See distribution.

Creeks. See streams.

Cultivated fields. See fields.

Dense woods. See woods.

Distribution of plants. Volumes have been written on this subject. Plants within their area of distribution are distributed as their habitat is distributed. The number of individuals at a station depends much upon a season maturing a great amount of viable seed, which is followed by a season with the optimum conditions for germination and development. Poorly developed seed and adverse conditions for germination and growth result in a paucity of individuals. Probably only once in a life time will one find some species abundant. To confirm this statement I will cite two personal experiences. In 1937 I saw *Polygala verticillata* so thick in an open blue grass sod in an open wooded pasture that the whole surface was white over an area of at least two acres. In Indiana this plant is rarely found in numbers exceeding 25 specimens at a place. On another occasion I saw *Monotropa uniflora* so abundant that the ground was white with it over several acres. I revisited the same place at the same date on two successive years and one year found no plants and one year found a few plants. The seasonal variation should be kept in mind. Some plants seem to have cycles of abundance, probably the result of fortuitous and co-ordinating optimum conditions for growth. Annuals fluctuate most of all the types of vegetation.

The following terms, which are also used by other authors, are here defined in the sense I use them:

Abundant. Occurring in large numbers in various places throughout the range of the species.

Common. Plentiful in all parts of its range.

Frequent. Evenly distributed throughout its range, but not plentiful.

Infrequent. Only occasional throughout its range.

Local. Species whose habitat is restricted or infrequent in the state, but the number of individuals at a station may vary from a few to many.

Rare. Plants apparently not restricted to a particular habitat yet extremely rare in Indiana, such as *Anemone caroliniana, Chamaelirium luteum,* and *Trautvetteria carolinensis.*

Dunes. Ridges or hills of wind-blown sand. They vary greatly in extent and in height, from a few feet high to 192 feet (Mt. Tom in Dunes State Park). Dunes are located mostly along Lake Michigan and in the Kankakee River Valley, and when mentioned elsewhere the locality is given. In the dunes and the areas between them grow some species not found elsewhere. The dunes in Lake County are, for the most part, low and those near the lake were wooded mostly with jack pine, birch, and oak. The high dunes in Porter County near Lake Michigan were wooded mostly with white and black oaks, jack and white pines, and basswood. The dunes in the Kankakee River Basin are wooded mostly with oaks.

Fallow fields. See fields.

Farm pasture. See pastures.

Fields. This term is applied to areas larger than truck gardens that have been or are cultivated (exclusive of pasture fields). An *abandoned field* is one that is no longer being cultivated because it has become too rough by erosion or too sterile by sheet washing. A *fallow field* is one that lies idle because of non-cultivation or has lost part of its fertility which will be restored if left idle for a few years. A *cultivated field* is one that is being cultivated during the current season.

Flats. The flats are level, poorly drained areas in the undissected part of the Illinoian drift section of southeastern Indiana and along Little and Big Pigeon Creeks and the Patoka River in the southwestern part of the state. The soil is a very finely divided white clay with high water-holding capacity.

Fork. See streams.

Frequent. See distribution of plants.

Gardens. Gardens are small areas about habitations where vegetables and sometimes some flowers are grown. *Truck gardens* are larger tracts where vegetables are grown both for home consumption and for the market.

Gravel pits. See ponds.

Hayfields. Hayfields are fields devoted to growing of forage. In Indiana they are called *meadows.* Since the latter term is popularly not used in a botanical sense, to avoid confusion it has not been used in the flora.

Infrequent. See distribution of plants.

Interdunal flats. The flat area connecting the bases of two dunes is known as an interdunal flat, which is wet during the rainy season, becoming dry in summer.

Knobs. This is a local name for the dissected topography of the unglaciated region.

Lake. A natural lake is a depression on the surface of the earth partially filled with water and which never becomes dry (at least in Indiana.) It usually has both an inlet and an outlet stream. The shore is usually sandy or gravelly in places and mucky with spatterdock or waterlilies on the border in other places. In some part it must have water too deep for the white waterlily to grow, which is usually 6 to 8 feet. According to origin, lakes may be divided into two types, natural and artificial. Our natural lakes are all located in the lake area. In the lake area are several lakes made for water power purposes, such as Koontz Lake in Starke County, Sylvan Lake in Noble County, and Shafer Lake in White County. In recent years several large artificial lakes have been made in southern Indiana for recreational purposes. See definition of ponds and sloughs.

Local. See distribution of plants.

Marsh. A marsh is a wet, level, treeless area covered mostly with sedges and grasses and generally fringed more or less with willows, pale dogwood, or other shrubs of a like habitat. This habitat is what some authors call a meadow. Marshes have a mineral or mucky soil.

Meadows. See hayfields.

Oak openings. See prairies.

Old river channels. See slough.

Pastures. A *farm pasture* is a field of any kind devoted to grazing, permanent or temporary. A *woodland pasture* is a woods of any kind that is being grazed.

Pond. A pond is a body of water in a natural or artificial depression of the earth, except a lake or slough, that is not as deep as a lake, but which rarely, if ever, becomes dry. There are several kinds of ponds and each usually has a specific name. A *typical pond* is the nucleus part of a swamp that rarely or never goes dry. I do not recall ever seeing any vegetation in them other than spatterdock, but they usually have some buttonbush on their borders. There are many artificial ponds in the Illinoian drift area. These are made to retain water for stock and are commonly called water holes, although some are made to supply water for boilers. The vegetation in these is usually abundant, if not disturbed, usually consisting of *Eleocharis, Lophotocarpus calycinus,* and *Sagittaria. Gravel pits* are cavities left on the surface after some of the gravel of the substratum has been removed. They vary greatly in size and depth. One in Wells County of about five acres is a true lake. I have not been able to study their vegetation. Most of them are used as swimming holes and are kept free of vegetation for that purpose. I know of one small pit about 30 years old that is full of cattails. The water in them seems to be fresh and I see no reason why lake species would not come into them. *Millponds* are made by damming a stream for power purposes. They are usually full of lake species of vegetation, including spatterdock, waterlilies, pickerel weed, *Potamogeton, Ceratophyllum,* and *Myriophyllum.*

Prairies. Prairies are naturally treeless areas, either wet or dry. *Dry prairies* are always dry and covered mostly with big bluestem grass. Ours are mostly eastern extensions into Indiana of the Great Western Prairie. *Wet prairies* have a black, sandy, muck soil and, during the winter months, are usually covered with water which disappears by late spring. They are covered mostly with little bluestem grass and prairie cordgrass which are the source of marsh hay. This type of prairie covers much of the Kankakee region, and parts of Jasper, Newton, Benton, Tippecanoe, and Warren Counties. *Oak openings* are remnants of dry prairies in northern Indiana where bur oak was the invading tree species.

Railroads. Term applied to the right-of-way of all kinds of railroads. *Ballast* refers to the filled-in or built-up part upon which the rails are laid.

Rare. See distribution of plants.

Reservoirs. Storage basins of water used mostly for city water supply. These I have not studied.

Rivers. See streams.

Rivulet. See streams.

Roadsides. This term refers to the right-of-way of all kinds of public thoroughfares used by vehicles.

Shore. The margin adjacent to lakes, ponds, sloughs, and streams.

Sinkhole. A funnel-shaped cavity in the earth's surface made by the dissolving of the underlying limestone. The cavity varies in size from less than an eighth to five acres. Sinkholes are frequent to common in most of the limestone area of the unglaciated region. The water in them varies in depth up to several feet, depending upon the amount of rainfall and the seepage. I have never seen any vegetation in them when they are located in deep woodland, doubtless because the decay of many leaves prevent it. Those in fields are disturbed more or less by stock which destroys much of their vegetation. Over a period of many years I have made a list of plants I have seen or collected in them and it is a large and variable one. All are usually surrounded more or less with *Eleocharis* and often *Sagittarias, Junci,* and *Lophotocarpus calycinus.* In the deep parts *Potamogeton* is often found. Constant filling in from the washing of the surrounding slopes frequently contaminates the water so that vegetation will not thrive.

Slough. A slough is an elongated basin filled with water, rarely exceeding a depth of six feet. Sloughs are remnants of old river channels that have not been filled by sedimentation and are usually short, although a few are about a mile long. They may be divided into high and low sloughs. Those that frequently overflow are usually devoid of vegetation other than spatterdock and have little or no vegetation on their banks. High sloughs are those that do not annually overflow such as Half Moon Pond which has both spatterdock and waterlilies and vegetation on its banks. *Bayou lakes* have the same origin, but are more circular in shape. Our most notable example is Hovey Lake in Posey County. For want of a better name I am calling also the water channels between the dunes near Lake Michigan sloughs.

Soil. The soft outer surface of the earth composed of minute particles of various rocks, organic matter, and solutes. This soft envelope or any part of it is called soil (not ground) when a relation between it and plants is expressed.

Spring. The discharge of water from a subterranean stream forms a spring. The volume discharged may be of sufficient size to form a small stream like Donaldson Cave or the volume may be reduced to form a pool and a small rivulet. Some springs never reach the surface, but discharge below the surface, their water escaping as seepage. These hidden springs are most frequent about lakes and along rivers, forming springy areas which are erroneously called boggy places if they discharge in mineral soil. The water of a springy place differs from the stagnant water of a swamp.

Springy areas. See spring.

Stone quarry. A cavity made in the surface of the earth for the purpose of getting stone for highways and building purposes. If they fill with water they usually have no soil on their borders except at their outlet so they can not support vegetation. I have seen *Potamogeton* in several, but do not recall any other species except cattails.

Streams. A stream is the generic term used to denote a volume of water moving from one point to another, usually by the force of gravity. The volume of moving water may vary greatly. The largest surface volumes are known as rivers, and these may have many tributaries which in turn are usually known as rivers. Streams of a size less than that of a river are known as *creeks, branches, forks,* and *rivulets.* The volume of water a stream carries varies from flood to low water stages. The channels of most streams in an early geologic time were much wider than now as shown by wide intervening valleys between the old terrace banks. *Banks of streams* include the area at the top of the channel as far back as quantity of light and moisture differentiates the vegetation, and all the slope of the channel from the top to the bed of the stream. The common meaning of a *bank* is the top and slope of the channel from high water mark to the bed of the stream. The term is often subdivided as top of the bank, upper or lower slope, and muddy base above or below water. The slope may be interrupted by narrow level areas (benches) or by wide level areas (alluvial flood plains or overflow banks). The top and slope of the old channel above flood stage of the present channel I call *terrace bank.*

Bluff. Sometimes the meander of a stream encountered high land and by undercutting the slope becomes very steep. If the slope is clay it will usually be devoid of vegetation. If it is mostly rock, the slope will usually consist of one or a series of cliffs of varying height. I have arbitrarily called the steep bank of a stream a bluff when the top of the bank is more than 15 feet above high water mark. Rocky bluffs often have a peculiar flora.

Cliff. A cliff is a perpendicular exposure of rock. I have arbitrarily placed the minimum height at 10 feet and the maximum height in Indiana may not exceed 150 feet. The banks of the Ohio River are up to 250 feet high or higher and may consist of one slope or of a series of cliffs and slopes. In the crevices and solution holes on the face of the cliffs are found plants peculiar to them, such as *Asplenium pinnatifidum, Lycopodium Selago* var. *patens,* and *Sedum telephioides.*

Strip mine pit. Depression made in the process of surface mining of coal and a failure to leave the surface in a level condition. These depressions are usually a series of longitudinal v-shaped channels. The area at any one place will vary from a few acres to 40 or more. These I have not studied, but I recall that many of the older ones were full of cattails.

Swale. I am not able to distinguish between swale and marsh. L. M. Umbach, who did much collecting among the dunes, refers to the troughs between the dunes that were wet and filled mostly with sedges and grasses as swales. I have accepted his name for this type of habitat and restricted its use to the dune area.

Swamp. A depression in the surface of the earth where water accumulates and is retained for weeks or months is a swamp. These are usually in woodland because those occurring in cultivated lands generally have been drained, but if not extinct they will have a different flora in and about them. The character of the soil and subsoil determines in part the length of time the water will be retained. Swamps usually become dry in late spring or early summer because of evaporation. The part of a swamp that does not become dry is a pond. A pond and the deeper parts of a swamp are devoid of tree growth, but are fringed by buttonbush or willows. The vegetation on the wet border of a swamp is usually distinctive, and that occurring in the zone formerly covered by water differs from that on the border. Swamps in cultivated land are little more than mud basins and when they become dry the weed seeds blown and washed into the basin germinate and in due time weeds form a cover.

Terrace bank. See streams.

Thick woods. See woods.

Truck garden. See gardens.

Waste place. Term applied to non-cultivated areas, usually about habitations or within a city or town limit, such as town lots and unimproved streets.

Woods. A woods is a term used to designate all areas covered with tree growth. The species and growth vary greatly according to the soils, topography, and cutting of the trees. A *dense woods* is one with a dense and high canopy, usually made by large trees. A *thick woods* is one thickly set with medium to small size trees. A *thicket* is an area devoid of large trees and densely set with very small trees or shrubs. It is also applied to clearings that have grown up with a thick stand of *Crataegus* and dense stands of blackberries or roses. A *clearing* is a woods that has had all or nearly all the trees removed, making the area ready for farming or grazing. If the final steps are delayed, as they often are, the area will grow up and become a thicket. Woods are often designated by the dominant species, as sugar maple woods, beech woods, and oak-hickory woods.

Woodland pasture. See pastures.

Average date of first killing frost in autumn based on data from cooperative and other stations to 1930.

(Courtesy U. S. Weather Bureau Office, Indianapolis, Indiana.)

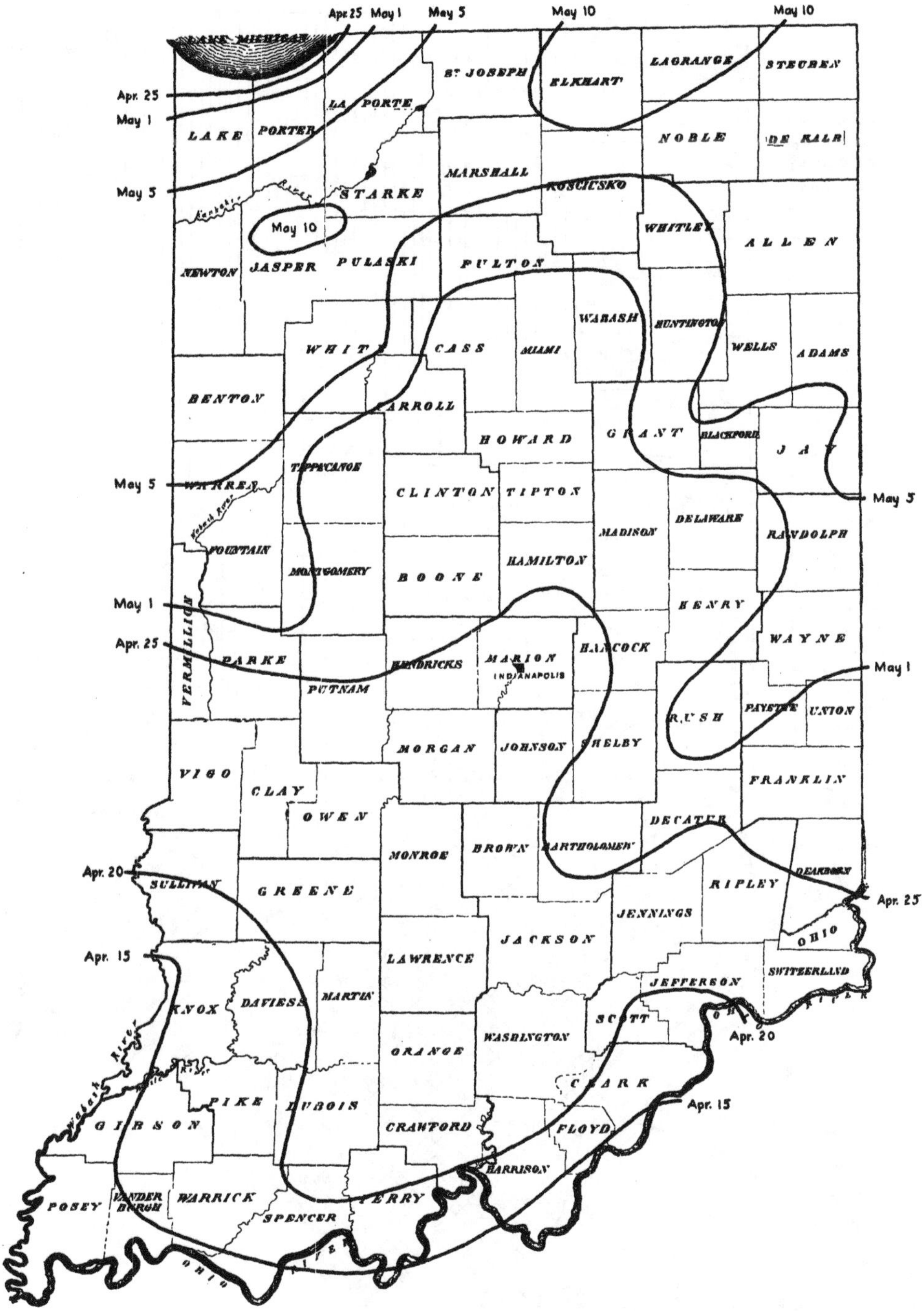

Average date of last killing frost in spring based on data from cooperative and other stations to 1930.

(Courtesy U. S. Weather Bureau Office, Indianapolis, Indiana.)

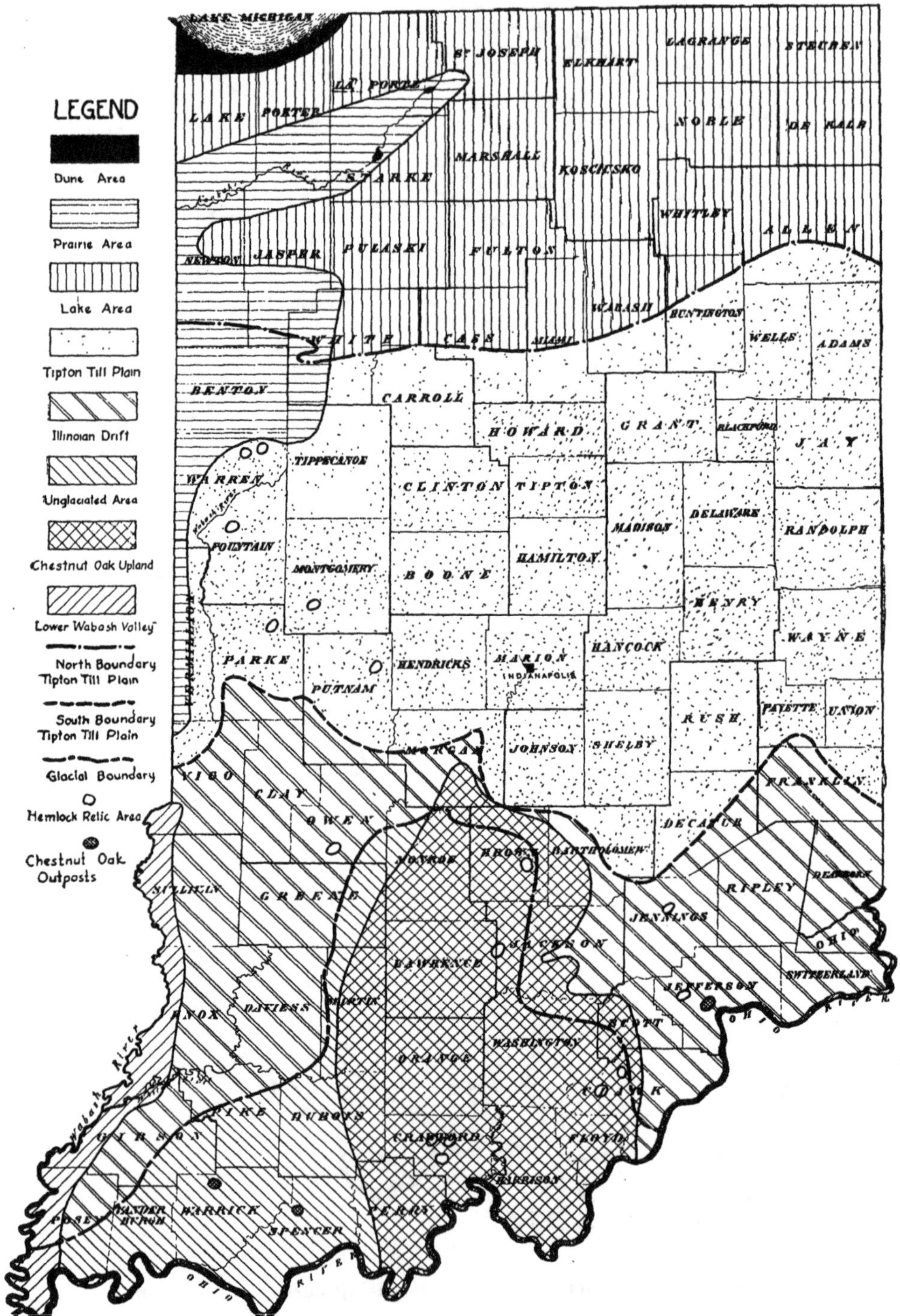

Floral Areas of Indiana

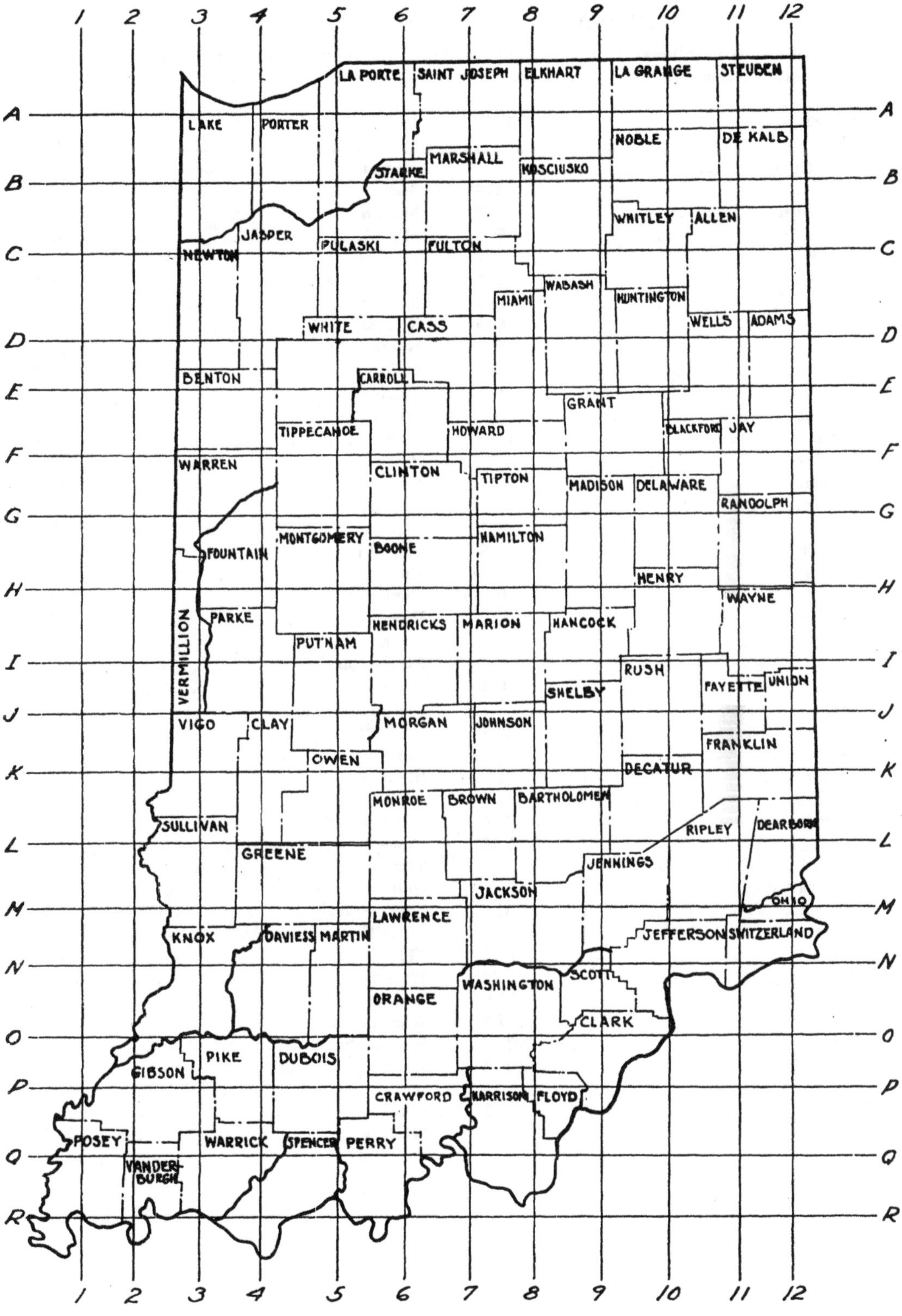

Finding County Map of Indiana

ALPHABETICAL LIST OF COUNTIES OF INDIANA

Each county is followed by a figure and a letter which represent horizontal and vertical lines whose intersection falls within the county.

Adams	D 12	Lawrence	M 6
Allen	C 11	Madison	G 9
Bartholomew	L 8	Marion	I 7
Benton	E 4	Marshall	B 7
Blackford	F 10	Martin	N 5
Boone	H 6	Miami	D 8
Brown	L 7	Monroe	L 6
Carroll	E 6	Montgomery	H 5
Cass	D 6	Morgan	K 6
Clark	O 9	Newton	C 3
Clay	K 4	Noble	B 10
Clinton	G 6	Ohio	M 12
Crawford	P 6	Orange	O 6
Daviess	N 4	Owen	K 5
Dearborn	L 12	Parke	I 4
Decatur	K 10	Perry	Q 6
De Kalb	B 11	Pike	P 4
Delaware	G 10	Porter	A 4
Dubois	P 5	Posey	Q 1
Elkhart	A 8	Pulaski	C 5
Fayette	J 11	Putnam	I 5
Floyd	P 8	Randolph	G 11
Fountain	H 4	Ripley	L 11
Franklin	K 11	Rush	J 10
Fulton	C 7	St. Joseph	A 7
Gibson	P 2	Scott	N 9
Grant	F 9	Shelby	J 9
Greene	M 4	Spencer	R 4
Hamilton	H 8	Starke	B 6
Hancock	I 9	Steuben	A 11
Harrison	Q 8	Sullivan	L 3
Hendricks	I 6	Switzerland	N 11
Henry	H 10	Tippecanoe	F 5
Howard	F 7	Tipton	G 8
Huntington	D 10	Union	J 12
Jackson	M 8	Vanderburgh	Q 2
Jasper	C 4	Vermillion	I 3
Jay	F 11	Vigo	K 3
Jefferson	N 10	Wabash	D 9
Jennings	M 9	Warren	F 3
Johnson	J 8	Warrick	Q 3
Knox	N 3	Washington	O 7
Kosciusko	B 8	Wayne	I 11
La Grange	A 10	Wells	D 11
Lake	A 3	White	D 5
La Porte	A 5	Whitley	C 10

Index

The scientific names of the species, genera, families, tribes of grasses, sections of *Carex, Juncus,* and *Crataegus* admitted to the Indiana flora are printed in bold face. The page on which the botanical description of a species, genus or tribe is given is numbered in bold face. Synonyms are printed in italics. Latin binomials in the text are also in italics. Common names and names of excluded species are in roman. Varieties are abbreviated to var. and forms to f.

C

G

L

M

N

O

Q

S

W

www.ingramcontent.com/pod-product-compliance
Lightning Source LLC
LaVergne TN
LVHW062201200926
848484LV00006B/1294